D1403241

The Plant-book is internationally accepted as an essential reference text for anyone studying, growing or writing about plants. In over 20 000 entries this comprehensive dictionary provides information on every family and genus of seed-bearing plant (including gymnosperms) plus ferns and other pteridophytes, combining taxonomic details with invaluable information on English names and uses. In this new edition each entry has been updated to take into consideration the most recent literature and almost 2500 new entries have been added, ensuring that *The Plant-book* continues to rank among the most practical and authoritative botanical texts available.

THE PLANT-BOOK

To Anne Sing

D.J. MABBERLEY

Wadham College, University of Oxford; Rijksherbarium, University of Leiden; Royal Botanic Gardens Sydney

THE PLANT-BOOK

A portable dictionary of the vascular plants

utilizing Kubitzki's *The families and genera of vascular plants* (1990–), Cronquist's *An integrated system of classification of flowering plants* (1981) and current botanical literature arranged largely on the principles of editions 1–6 (1896/97–1931) of Willis's *A dictionary of the flowering plants and ferns*

SECOND EDITION

completely revised, with almost 2500 additional new entries

CAMBRIDGE UNIVERSITY PRESS

PUBLISHED BY THE PRESS SYNDICATE OF THE UNIVERSITY OF CAMBRIDGE
The Pitt Building, Trumpington Street, Cambridge, United Kingdom

CAMBRIDGE UNIVERSITY PRESS
The Edinburgh Building, Cambridge CB2 2RU, UK
40 West 20th Street, New York, NY 10011–4211, USA
477 Williamstown Road, Port Melbourne, VIC 3207, Australia
Ruiz de Alarcón 13, 28014 Madrid, Spain
Dock House, The Waterfront, Cape Town 8001, South Africa

http://www.cambridge.org

First published 1987
Reprinted with corrections 1998, 2000, 2002

Printed in the United Kingdom at the University Press, Cambridge

Typeset in 7/9 pt Palatino [SE]

A catalogue record for this book is available from the British Library

Library of Congress Cataloguing in Publication data
Mabberley, D. J.
The plant-book: a portable dictionary of the vascular plants
utilizing Kubitzki's The families and genera of vascular plants
(1990–), Cronquist's An integrated system of classification of
flowering plants (1981), and current botanical literature arranged
largely on the principles of editions 1–6 (1896/97–1931) of Willis's
A dictionary of the flowering plants and ferns / D.J. Mabberley.–
2nd ed., completely rev.
p. cm.
ISBN 0 521 41421 0 (hardback)
1. Botany–Nomenclature. 2. Plant names, Popular—Dictionaries.
3. English language—Dictionaries—Latin, Medieval and modern.
I. Title.
QK11.M29 1997
580'.1'4–dc21 96-30091 CIP

ISBN 0 521 41421 0 hardback

Contents

Preface

This dictionary has been written as an attempt at filling the gap felt by botanists both professional and amateur, horticulturists, ecologists and all those journalists and other writers who need a replacement for the early editions of J.C. Willis's *Dictionary of the flowering plants and ferns*. Those early editions have been called 'the most remarkable botanical works of reference ever written – true vade-mecums for every botanist's pocket' (the late Professor P.W. Richards in *Journal of Ecology* 63 (1975)368), but the last one in the style of the original was published in 1931. It is scarce but still useful. As a schoolboy, buying the 1914 reprint of edition 3 'neatly rebacked 3/6' at Heynes's Bookshop (now the Dhaka Brasserie – *O Tempora! O Mores!*) in Cheltenham, I became gripped by tropical botany. As Professor Richards continued, 'Today it is probably impossible to compile such a useful single volume'. My book must therefore be seen merely as a grateful attempt to provide a handy text covering the vascular (the pejorative 'higher' has been avoided in this edition) plants, their botany and relationships, their uses and their common names.

J.C. Willis

It seems worthwhile to insert a brief appreciation of the man who wrote the book which inspired the present one.

John Christopher Willis was the son of a Liverpudlian, who emigrated to America, and was the cousin of N.P. Willis, the poet; his mother was Scottish. Willis was born at Birkenhead in 1868 and was educated there and then at University College, Liverpool, and at Gonville and Caius College, Cambridge. For a time he was assistant in the Department of Botany at Glasgow and, in 1896, he was appointed Director of the Royal Botanic Garden, Peradeniya, Sri Lanka, a post he held for fifteen years. From 1912 to 1915, he was Director of the Botanic Garden at Rio de Janeiro and afterwards he worked in retirement at Cambridge, eventually going to live in Montreux, where he died, ninety years of age, in 1958. Further details of his life are to be found in the works cited by R. Desmond, *British and Irish botanists and horticulturists*, p. 664 (1977).

At Peradeniya, Willis was a vigorous director, taking a broad interest in plant science, particularly its application in tropical agriculture on which he was an authority. His first researches were on temperate plants, however, notably floral biology and dispersal, areas suggested to him by Sir Francis Darwin at Cambridge. At Peradeniya, he took up the study of the remarkable tropical aquatic family of flowering plants, Podostemaceae. These highly specialized hydrophytes resemble liverworts but their vegetative morphology is so complicated and diverse that it led Willis to question a complete reliance on Darwinian Natural Selection as an explanation for such diversity under what seemed to him to be uniform conditions of fast-flowing water in warm climates. His ideas are getting a welcome re-airing today: see Rutishauser on 'fuzzy morphology' in *Canadian J. Botany* 73(1995)1305–17. Willis went further and argued that many characters of plants seemed 'unadapted' and that, as there were no intermediate stages between certain character states (e.g. three-, four- and five-petalled flowers), large mutations rather than accumulations of small ones in the Darwinian way seemed a simpler explanation. He felt that small populations could lead to rapid evolution and in this he has some followers today. With the constant questioning of evolutionary processes, particularly the pin-pointing of the differences between mechanisms in plants and animals, his work bears rereading.

Combining his loss of faith in Natural Selection with his flair for statistics, he produced a shaky theory of 'Age and Area', which held that taxa with the greatest distribution were the oldest. Although this caused considerable debate between the two World Wars, it is now largely, and probably best, forgotten. Willis combined his interest in statistics with his favourite pastime, travelling, in his non-botanical publications – *Ceylon, a handbook for the resident and traveller* (1907) and *The tube–bus guide to London* (1928), a remarkable compilation which went through three editions. Indeed, this meticulous, methodical approach, which led to the *Dictionary*, had been long manifest for, as an undergraduate, he drew up abstracts and condensations of textbooks. These were to be the basis of the book which he completed on his voyage out to Sri Lanka. Willis is now largely remembered for this, his *Dictionary of the flowering plants and ferns*, as the sixth edition was called, though it first appeared in two slim volumes, one issued before Christmas 1896, the other after it, as *A manual and dictionary of the flowering plants and ferns*, one in the series of Cambridge Botanical Handbooks. Five more editions were published in Willis's lifetime, the last of them in 1931 and reprinted several times: curiously, Willis married in the year of the first edition and his wife died in the year of publication of the last. The contents of the first volume, a treatise on many aspects of botany, were later reduced and more-or-less incorporated in the text of the *Dictionary*. In 1966, Mr Airy Shaw of Kew published a revised edition in a larger format (and a similar eighth edition in 1973), dealing with all names ever used at the generic, family and some higher levels. The production of such a nomenclator meant that a good deal of the general information had to be excluded and many of the articles pruned. Some of the omitted information was gathered together and partially updated as *A dictionary of useful and everyday plants* (1974), compiled by F.N. Howes of Kew. Mr Howes died before the book went to press and, at the suggestion of Peter Sell, I saw it through to publication.

The Plant-book

With increased travel in both temperate and tropical countries, an explosion in plant material available for gardens and as houseplants with a (fortunate) interest in trying more exotic foods, aromatherapy and herbalism amongst the general public as well as the rise of a generation of young botanists, agriculturists and foresters, keen for the whole plant yet enmeshed in the complexities of DNA and other biochemistry at school, the need for an updated book of the 'Willis' type in the spirit of the early editions seems greater than ever.

Here, then, is an attempt to present all currently accepted generic and family names of flowering plants (including gymnosperms) and ferns and other Pteridophyta, as well as a wide range of English names encountered in the literature. Willis largely based his taxonomic entries on the works of the Engler school and, in abstracting from those, he was able to indicate the tribal and subfamilial position of each genus. This was indicated by numbers referring to the numbered subfamilies and tribes under the family accounts and therefore allowed an immediate broadening of the significance of the generic entry as well as a reference to the Englerian work. Often, there was a further reference to the position in the work of Bentham and Hooker. Both systems are used in herbaria. But the latter is well over a century old and the former is in German, a language regrettably little known among modern young botanists and others, as well as being largely out of date. Furthermore, undergraduate texts are increasingly based on the systems of Takhtajan or Cronquist, and as this book is aimed primarily not at the herbarium taxonomist, who has access to other books with which to compare the traditional system by which his or her herbarium is arranged, I have decided to follow the system of Cronquist, *An integrated system of classification of flowering plants* (1981) as modified by Kubitzki (see below), for this is in English and has modern circumscriptions of families and fresh descriptions with valuable bibliographies. Indeed it represents a landmark in angiosperm systematics and a synopsis of it, as revised by Kubitzki, is set out at the end of the dictionary. I have relied very heavily upon it. As there are a number of other modern systems of arrangement of the angiosperms, which make up the bulk of the entries in this book, it has not been a simple decision to plump for one, and where there is widespread dissent from Cronquist's view, I have tried to indicate this.

Although the other seed-plants are less controversial in the way in which they are arranged, the pteridophytes, particularly the ferns, have no generally agreed classification. I have therefore followed the system set out in K. Kubitzki's *The families and genera of vascular plants. I. Pteridophytes and gymnosperms* (1990), which I have also used for the gymnosperms; moreover the second volume of this invaluable work has been heavily relied on in updating the angiosperm entries of taxa covered in that book. It is to be hoped that the remaining volumes of this monumental endeavour, a new 'Genera Plantarum', will be published with all speed.

The inclusion of vernacular names has been problematic, too, particularly as many 'book' names for common wild plants have become established in the English language. Moreover, a number of Latin names are used in the vernacular in a sense different from their current technical one. Some are taxonomically or nomenclaturally outmoded, some are misidentifications, some are pre-Linnaean and some are spelling errors or even specific epithets. Such include: acidanthera, afrormosia, alyssum, amaryllis, arum, aster, azalea, aubretia, bartonia, calla, chrysanthemum, cineraria, croton, coleus, dimorphotheca, epiphyllum, eulalia, fremontia, funkia, geranium, gloxinia, godetia, goldfussia, hortensia, ilex, jacobaea, kentia, kochia, laurustinus, leylandii, macrocarpa, mesembryanthemum, mespilus, mimosa, moluccana, montana, montbretia, nasturtium, poinciana, poinsettia, pyrethrum, retinospora, rochea, smilax, spinifex, statice, stephanotis, stevia, syringa, tuberose, utile, verbena and wellingtonia. Many are in use in horticulture or are timbers etc.: the words are part of the English language and no amount of insistence by academic botanists will remove them. They are therefore in this book. Not that the professionals are immune from whimsy worthy of Edward Lear's *Manypeeplia* and *Smalltoothcombia* (*Nonsense botany*, 1870), though fortunately most of *Damrongia*, *Johnson-sea-linkia*, *Microweedia* (Toots), *Monosporascus* including *M. cannonballus*, *Muchmoria*, *Otherodendron*, *Perplexia*, *Toolongia* and *Wigwamma* are either referring to cryptogams or have been long lost in synonymy.

I have written the entries with the help of modern Floras, handbooks, monographs and periodicals, particularly dwelling on the literature published since 1970: indeed, as comprehensive a scan of the pertinent botanical literature of this period as could be made by one man has been attempted. It would be impossible to cite all sources of information but the major ones are listed under 'Acknowledgement of sources' on p. 783. As I have attempted to cover only commonly encountered synonyms as well as currently accepted names, I have been able to dispense with an enormous number of entries of the type which make Mr Airy Shaw's book so invaluable as a nomenclator. Orchid generic hybrid names are also excluded. Generic names encountered in the older literature and not found in this book, should be sought first in *Index nominum genericorum* (1979) and in Mr Airy Shaw's book, where their identity may be indicated.

A major departure from the pattern in the entries in Willis's book goes a little way, I hope, to compensate for the impossibility of including modern tribal and subfamilial positions of all (but see below) individual genera in this handbook. Where a revision of a genus has been published in recent years, or an older one is still widely quoted, I have indicated the place of publication in a very abbreviated form. It has always seemed to me that this would give the reader with a need to follow up the literature a valuable start and that this device was an omission from Willis's book. Since the publication of the first edition of *The Plant-book*, it has been possible to examine and now cite very many hundreds more such. The abbreviations are explained, as are others used in the entries, on p. 799. I have also developed a cross-reference file from genera to families so that the estimated size of families in terms of numbers of genera and species more exactly reflects the information actually set out in the text under the constituent generic headings. Moreover, it has been possible to indicate the subfamilial and tribal positions of many thousands of genera on modern lines thanks not only to Kubitzki's work but also recent major reviews of many families (including four of the biggest five) such as Anacardiaceae, Araceae, Asclepiadaceae, Commelinaceae, Compositae (the biggest of all), Flacourtiaceae, Gramineae, Iridaceae, Labiatae, Orchidaceae (the biggest family of monocotyledons), Palmae, Rubiaceae and Sapotaceae. It has also been necessary to scan a further 180 journals on top of those examined for the first edition. The result is that some 625 new author names appear in Abbreviations 2. and the text is over 20% longer with almost 2500 additional new entries and almost all of the original entries have had to be updated in the light of newly published work, so that this edition is a considerable advance on the first.

Readers who compare this edition with the first will see that there has been an enormous taxonomic effort (notably insightful being the rush of studies on DNA, particularly that of the chloroplasts), over the last decade, but that this has sometimes had the apparently detrimental effect of the changing of names of plants well known in commerce: foodplants, timbers, drugs, fibres and ornamentals. Because of the new provisions (see under *Pseudolarix* in *The Plant-book* ed. 1) made in the *International Code of Botanical Nomenclature* at the Tokyo Congress in 1993, names can be protected from upset due to the unearthing of earlier valid ones so that the laymen's cavil against the scholastic principles of the *Code* are now redundant. Nonetheless, recent taxonomic opinion, largely as a result of a putative phylogenetic methodology (cladistics) combined with the sometimes revolutionary findings of chemotaxonomy (see above) often reinforcing the findings from classical work, notably that of the intricate structure of seeds brought together in the late E.J.H. Corner's monumental *Seeds of dicotyledons* (1976), has led to the changing of the Latin names of couch grass, epiphyllum, greenheart, gum tragacanth, ipecacuanha, iroko, senna pods, stephanotis, Sturt's desert pea, verbena, wallflower and watercress for example. It is tempting to deprecate name-changes, but when these are based on a sound advance in scientific understanding, it is perhaps unreasonable to argue for the outmoded. One example will suffice: until recently, the genus *Castanospermum* (Leguminosae) was thought to comprise a single species, *C. australe*, the Moreton Bay Chestnut, from eastern Australia. The tree was found to contain an important pharmaceutical (castanospermine used in the fight against AIDS) but not in great quantities, so it was important to screen its closest allies, which taxonomists ascertained were American species; these were found to have the drug in more parts of the trees and are now placed in *Castanospermum*, but up until then they had been put in a separate genus, *Alexa*, which had to be 'sunk' and names of the American trees thereby changed, reflecting scientific advance. For a cogent discussions of these issues, see C.J. Humphries (1991) 'The implication of pragmatism for systematics', *Regnum vegetabile* 123 [*Improving the stability of names: needs and options*] 313–22 and A. Minelli (1995) 'The changing paradigms of biological systematics...', *Bull. Zool. Nomenclature* 52: 303–9. Nonetheless, such changes are taken up reluctantly as can be seen from the eloquent *tour-de-force* by Christopher Lloyd in *Country Life* 6 Oct. 1994, p. 72.

Reunification and realignment at the family level are likely to lead to improved understanding but with less public upset. As was pointed out in the first edition of this book, some well-known families (e.g. Umbelliferae), largely temperate and therefore first to be described (see below), are really parts of more diversified largely tropical families (in this case Araliaceae), which in cladistic parlance are 'paraphyletic'. Monophyletic groups are established when the temperate and tropical are (re-)combined, though because the older name is that applied to the temperate element, the whole will therefore bear that based on European plants (an odd quirk of the colonial legacy) – Umbelliferae including Araliaceae, Cruciferae including Capparidaceae, Urticaceae including Moraceae, Cannabaceae and Cecropiaceae, etc. (see W.S. Judd *et al.* (1994), 'Angiosperm family pairs: preliminary phylogenetic analyses', *Harvard Pap. Bot.* 5: 1–51). Again there has been some reluctance to take up these ideas so that, for example, Urticaceae are treated in the narrow sense and Fumariaceae are kept distinct from Papaveraceae in Kubitzki's book, which text I have followed so that the current literature can be made more easily accessible to readers.

Indeed, for this reason, I myself have tried not to innovate and have been conservative in the splitting of families and genera. This has generally been the philosophy at the family level in both Kubitzki's and Cronquist's work and is convincingly argued by the late Professor C.G.G.J. van Steenis in his 'Doubtful virtue of splitting families' in *Bothalia* 12(1978)425–7; nevertheless the current fashion is to fragment Liliaceae (which I have somewhat reluctantly followed because the last word clearly has not been said, though Kubitzki's forthcoming volume on monocotyledons will have this philosophy and it is increasingly accepted in popular works) and Loganiaceae and possibly Euphorbiaceae. Nevertheless, I have so arranged the text that those wishing to keep a broad view of such families can do so with the help of this book. Although attempting to mirror current opinion at the generic level, I have, when faced with conflicting views, taken the conservative line in maintaining larger genera. From a field-worker's point of view this is more satisfactory in any case – the *Gestalt* of a fig is usually unmistakable but to split the genus *Ficus* into several on the basis of characters only revealed by lenses seems academic self-indulgence. I entirely agree with P.H. Davis and V.H. Heywood (1963) *Principles of angiosperm taxonomy*, p. 106, 'When in doubt whether to accord generic rank to a group, there is much to be said for the *subgenus* as a suitable category; it draws attention to the group in the classification and at the same time allows people to continue to use the old binomial'. There are exceptions where current use has confirmed the splitting as in the case of *Senna* segregated from *Cassia*, *Astracantha* from *Astragalus* and the shattering of *Aechmea*, *Chrysanthemum*, *Eupatorium* (though now, Humpty-Dumpty-like, being put back together again), *Helichrysum* & *Vernonia*, but perhaps the more common trend, particularly from cladistic and other modern work, is indeed to reunite small genera with larger ones, such as *Gladiolus*, *Justicia*, *Silene*, *Strobilanthes*, *Tabernaemontana* or, less spectacularly, recent floristic treatments of genera in Cruciferae, Labiatae and Umbelliferae, for example. Nonetheless, there has been a worry that the attempt to recognize monophyletic groups could lead to a collapsing of the whole system, for example in discussing the uniting of *Eucalyptus*, *Angophora* and *Corymbia* (Myrtaceae) which they keep distinct, K.D. Hill and L.A.S. Johnson (*Telopea* 6(1995)188) comment, 'it could be argued on such a basis that the whole of the Myrtaceae, or the Anthophyta [angiosperms], or even all life, should be in one genus'.

By comparison with many wholly tropical groups, however, many families commonly represented in Europe comprise large numbers of small genera: the Cruciferae and Umbelliferae mentioned above are notable. This splitting is largely due to pre-Linnaean folk-taxonomy and is explained in S.M. Walters's masterly 'The shaping of angiosperm taxonomy', *New Phytologist* 60(1961)74–84. Starting afresh today, many might be swept into a small number of large genera, but the fashion has lately been to split even further – some of the generic names raised for splits, of, say *Thlaspi* (Cruciferae), are not included here, however. On the other hand, a number of other genera (and families) have grown and grown by 'chaining' and are still heterogeneous. Even so, there are few genera with large numbers of (over 500) species: *Acacia, Allium, Anthurium, Asplenium, Astragalus, Berberis, Bulbophyllum, Carex* (equal biggest), *Cassia, Centaurea* (s.l.), *Cousinia, Crotalaria, Croton, Cyathea, Cyclosorus, Dendrobium, Epidendrum, Eria, Erica, Eucalyptus, Eugenia, Euphorbia* (equal biggest), *Ficus, Habenaria, Helichrysum, Impatiens, Indigofera, Justicia, Lepanthes, Miconia, Oncidium, Oxalis, Peperomia, Phyllanthus, Piper, Pleurothallis, Polygala, Potentilla, Psychotria, Ranunculus, Rhododendron, Salvia, Schefflera, Sedum, Selaginella, Senecio* (shorn but still one of the biggest at 1250), *Silene, Solanum, Stelis, Syzygium* and *Vernonia*. It is such genera as these, often with a wide range of morphological form, that can give us a sound basis for hypotheses on evolutionary processes. Many of them have no modern monographs. It is impossible to grapple with a group of this size in a Ph.D. thesis or short-term research periods, or in the writing of regional Floras, important as this may be. Who has the time to use his or her intellect to unravel the stories of these genera? The taxonomy of the vascular plants still has a long way to go.

The first edition of *The Plant-book* was typed with the author's right index finger on a Brother electric typewriter: this edition was prepared on a word-processor working from a text gained from optically scanning the first and I am grateful to Denis Filer and Cheryl Howes for computing help here. In the preparation of this new edition, I am indebted to many other people too, notably the Librarians in the Department of Plant Sciences, University of Oxford, Rijksherbarium, Leiden, Linnean Society of London, Royal Botanic Gardens Kew, Natural History Museum London, Royal Botanic Gardens Sydney and to many people who suggested amendments, additions or improvements, especially Annette Aiello, the late S. Balasubramaniam, Kathy Bilton, Mary Burke, B.L. Burtt, The Bodyshop, A. Borhidi, L. Boulos, Ian Clarke, Mrs J.H. Creech, Quentin Cronk, Rhys Gardner, Tom Gilard, the late Jan Gillett, S. Hammer, Alfred Hansen, Jeffrey Harborne, Alistair Hay, Murray Henwood, D. Herbst, the late Ru Hoogland, Hsuan Keng, Colin Hughes, Camilla Huxley-Lambrick, Bob Johns, Douglas Kent, Rogier de Kok, Elsie Mabberley, Laura Mabberley, Marcus Mabberley, Victoria McMichael, Rogers McVaugh, Margaret Matthews, David Middleton, Jeremy Montagu, Josef Mullins, Robert Peden, C.J. Perraton, David Philcox, Peter Placito, John Potter, Hugh Prendergast, Christian Puff, Vernon Reynolds, E. Sanoja, Karen Sidwell, John Silba, M.G. Simpson, Rosemary Smith, B.E. Smythies, Clive Stace, George Staples, Tony Symons, Tim Synnott, Malcolm Walker, F.R. Whatley, John Whitehead, Emma Wilson and Rosemary Wise, but above all to Anne Sing, who, often in trying circumstances, has once again sorted and filed a myriad pieces of information in preparation for my revision, and to whom this second edition is dedicated. My mother, my daughter, my son and Andrew Drummond gave me the motivation to complete this revision; to them and to Professor Kubitzki and the late Professor Cronquist I am grateful for their support throughout the production of this book. I would also like to put on record my appreciation of the work of Philip Glass, Malcolm McLaren, Franz Schubert (1797–1828) and Carl Maria von Weber (1786–1826), which has kept me sane during some of the more tedious episodes inevitable in revising a dictionary such as this.

In conclusion, I would once again offer the lexicographer's lament, following Dr Johnson, who pointed out that readers only remark on a dictionary when what they seek cannot be found, but would add that any suggestions for additions, improvements or emendations, backed by reference to published materials, where appropriate, will be gratefully received in the hope that future editions might more nearly approach the ideals set by Willis in his *Dictionary*.

Oxford, July 1996

How to use this book

This book attempts to present all currently accepted generic and family names (found for example in Airy Shaw's book, Brummitt's *Vascular plant families and genera* (1992), Greuter *et al.*'s *Names in current use for extant plant genera* (1993)), and commonly used English names of flowering plants (including gymnosperms) and ferns (as well as other pteridophytes), excluding wholly fossil groups.

Each generic entry includes the family to which the genus is assigned, the number of species within the genus, its distribution and other details of botanical, horticultural, agricultural, medicinal or other economic importance as well as English names applied to species within the genus. Where there has been a recent or recently cited monograph of the genus, the place of its publication is added in abbreviated form. Even with this reference, however, it is always worth referring to the family entry, where additional botanical information will be found.

Each family entry includes a statistic of the number of genera followed by an oblique line and the number of species. Other information given includes the classification of the family and its sub-division, the principal genera, and the distribution, botanical details and main uses of plants within the family.

English names are merely cross-referenced to the generic entries, which should be followed up as any further information on that species and its relations will be found only there.

General abbreviations and abbreviations for authors' names are given on p. 799 and p. 809, respectively. However, some additional explanation may be useful. 'R' refers to recent revisions, reviews, synopses or keys of the genus or family concerned, while N, E, S, W refer to compass points and not political divisions, so that, for example, 'S Afr.' = southern Africa. 'Warm' is taken to mean subtropical and/or warm temperate, while 'SE As.' is mainland SE Asia (Indochina), 'Mal.' (Malesia) is the area covered by *Flora malesiana* (Malay Peninsula to Bismarck Archipelago), 'Papuasia' is New Guinea to and including the Solomon Islands, 'Macaronesia' comprises the groups of islands off the African coast in the north Atlantic and 'Eur.' (Europe) is used in the sense of *Flora europaea*. Of unusual signs, ~ before a generic name indicates that the name used in the entry is sometimes included in, has recently been included in or is very close to the generic name and single quotes round a name mean that the name has been used (widely) in the wrong sense. Brackets around a number in a floral formula indicate that the parts are united (usually by some intercalated tube and not 'fused', a term which like 'anomalous' for rare or unusual, has been avoided).

Technical terms have been kept as few as possible as it has not proved feasible to provide a glossary. A comprehensive modern manual of terms is not available but most will be found in B.D. Jackson, *A glossary of botanic terms* (1900) which has been reprinted recently.

Sample entries

For users unfamiliar with the condensed style of dictionaries such as this, the following generic, family and English name entries are set out *in extenso* as examples to aid comprehension.

Anisodus Link ex Sprengel (~ *Scopolia*). Solanaceae (1). 6 temp. E As. *A. stramoniifolius* (Wallich) G. Don f. – yak-fodder in Himal.

Anisodus described by Curt Polycarp Joachim Sprengel (1766–1833), who validated the name first suggested by Johann Heinrich Friedrich Link (1767–1851). Closely allied to and sometimes included in genus *Scopolia* [see that entry]. Family Solanaceae [see that entry for further details], subfamily Solanoideae. Six species indigenous in temperate east Asia. One of these is *Anisodus stramoniifolius*, first described in another genus by Nathaniel Wallich (or

Nathan Wolf, 1786–1854) and first referred to *Anisodus* by George Don [the Younger] (1798–1856), used as fodder for yaks in the Himalaya

Erythroxylaceae Kunth. Dicots – Rosidae – Linales. 4/240 trop. esp. Am. Glabrous trees & herbs often with alks incl. cocaine. Lvs spirally arr. (opp. in *Aneulophus*), simple, entire; stipules intrapetiolar. Fls small, reg., usu. bisexual, 5-merous, often heterostylous, solit. or in axillary fascicles; K a tube with imbricate or valvate lobes, C imbricate & usu. with adaxial ± basal appendages, disk 0, A 10 usu. forming a tube, anthers with longit. slits, $\underline{G}$ ((2 or) 3) with as many locs & styles (± connate), ovule 1(2) in 1 fert. loc., axile, pendulous, anatropous to hemitropous, bitegmic. Fr. a 1-seeded drupe; seed with straight embryo in copious (rarely 0) starchy endosperm. x = 12
Genera: *Aneulophus, Erythroxylum, Nectaropetalum, Pinacopodium*
Close to Linaceae
Erythroxylum a source of narcotics etc.

Erythroxylaceae described by Karl Sigismund Kunth (1788–1850). Dicotyledonae, subclass Rosidae, order Linales. Four genera with 240 species indigenous in the tropics, especially tropical America. Glabrous trees and herbs often with alkaloids including cocaine. Leaves spirally arranged (but opposite in species of *Aneulophus*), simple, entire; stipules intrapetiolar. Flowers small, regular, usually bisexual with parts in fives, often heterostylous, borne singly or in axillary fascicles; calyx a tube with imbricate or valvate lobes, petals imbricate and usually with adaxial, more-or-less basal appendages, disk absent, stamens ten usually forming a tube, their anthers with longitudinal slits; ovary superior, with rarely two and usually three united carpels with as many locules and more-or-less connate styles, ovules usually one (rarely two) axile, pendulous in the only fertile locule, anatropous to hemitropous, bitegmic. Fruit a one-seeded drupe; seed with a straight embryo in copious (rarely absent) starchy endosperm. Basal chromosome number 12
Genera: *Aneulophus, Erythroxylum, Nectaropetalum, Pinacopodium* [see those entries]
Closely related to Linaceae [see that entry]
Species of *Erythroxylum* a source of narcotics etc.

money plant *Epipremnum pinnatum* 'Aureum'; **m.wort** *Lysimachia nummularia*; **Cornish m.w.** *Sibthorpia europaea*

money plant *Epipremnum pinnatum* cultivar Aureum; **moneywort** *Lysimachia nummularia*; **Cornish moneywort** *Sibthorpia europaea*

The entries to which these refer should then be examined. For example, under **Epipremnum**:

Epipremnum Schott. Araceae (III 2). c. 15 SE As. to W Pacific (alleged fossils in Oligocene of N Egypt). Lianes, some medic. & cult. orn. esp. *E. pinnatum* (L.) Engl. (Indomal. to W Pacific) – like *Monstera deliciosa* with perforated lvs, many cvs esp. 'Aureum' (? orig. Solomon Is., 'money plant' – rarely flowering so owners of fl. pls considered 'in the money'), irreg. variegated, widely planted in trop.

Epipremnum described by Heinrich Wilhelm Schott (1794–1865). Family Araceae [see that entry], subfamily Monsteroideae. Approximately 15 species indigenous in south-east Asia to the western Pacific (alleged fossils known from the Oligocene of northern Egypt). Lianes, some of medicinal value and some culitvated as ornamentals especially *Epipremnum pinnatum* which was first described in another genus by Carl Linnaeus (von Linné, 1707–1778) and first transferred to *Epipremnum* by Heinrich Gustav Adolf Engler (1844–1930). It is indigenous in Indomalesia to the western Pacific, resembling *Monstera deliciosa* [see entry for *Monstera*], with perforated leaves; there are many cultivars grown especially 'Aureum', possibly first found in the Solomon Islands and known as 'money plant' because it rarely flowers and owners of flowering plants are considered to be 'in the money'; it has irregularly variegated leaves and is widely planted in the tropics

Bibliographic note on the first edition of *The Plant-book*

The first printing (pp. 706, 1987) was on thick paper ('*The Plant-brick*') without the rounded page-corners called for by the cover; it was reprinted in 1989 (pp. 707) with many corrections and on the planned fine paper with rounded corners. The second reprint (1990) included a small number of new corrections but returned to the thick paper and was issued with a new wrapper with a cover photograph by Heather Angel: the two reprints of 1993 and the fifth of 1996 (on finer paper) were similar. All these states bear the same ISBN number.

Note on the reprint of the second edition

I am indebted to the following for pointing out errors or improvements incorporated in this reprint: Alistair Hay, Philippe Morat, Robert Ornduff, Paul van Rijckevorsel, Rudi Schmid, Leo Schofield, Diethard Storch, Malcolm Walker and H.P. Wilkinson. The opportunity has also been taken to update the entries relating to the commercially important genus *Citrus*.

Sydney, June 1998

Note on the second reprint of the second edition

I am grateful to the following for pointing out further amendments or improvements here incorporated: Wayne Cherry, Aljos Farjon, Alistair Hay, Peter Heenan, Dale Johnson, Marcus Mabberley and Jiří Paclt. In the light of recent research, the opportunity has also been taken to update entries relating to the commercially significant genera *Cinchona* (quinine) and *Malus* (apples).

Sydney, April 2000

A

Aa Reichb.f. (~ *Altensteinia*). Orchidaceae (III 3). 25 Andes

Aaron's rod *Verbascum thapsus*

Aaronsohnia Warb. & Eig. Compositae (Anth.-Mat.). 2 N Afr.

abacá *Musa textilis*

abarco (wood) *Cariniana* spp. esp. *C. pyriformis*

Abarema Pittier (~ *Archidendron*). Leguminosae (II 5). 20 OW trop. (incl. *Klugiodendron*). Fls with 1 or 2 ovaries in same infl.

Abasaloa Llave & Lex. = ? *Eclipta*

Abassian boxwood *Buxus sempervirens*

Abatia Ruíz & Pavón. Flacourtiaceae. 9 trop. Am. mts; in Andes above timber line. *A. rugosa* Ruíz & Pavón lvs source of black dye (Peru)

Abbo rubber *Ficus lutea*

Abdominea J.J. Sm. Orchidaceae (V 16). 1 Malay Pen., Java: *A. minimiflora* (Hook.f.) J.J. Sm. R: OB 95(1988)51

Abdulmajidia Whitm. Lecythidaceae. 2 Malay Pen.

abé *Canarium schweinfurthii*

Abebaia Baehni = *Manilkara*

abel *Canarium schweinfurthii*

abele *Populus alba*

Abelia R. Br. Caprifoliaceae. 30 Himal. to E As., Mex. R: A. Rehder (1911) Plantae Wilsonianae 1(1911)122. Cult. orn. esp. *A.* × *grandiflora* (André) Rehder with variable K (*A. chinensis* R. Br. (K5) × *A. uniflora* R. Br. (K usu. 2), both China) & domatia

Abeliophyllum Nakai. Oleaceae. 1 Korea: *A. distichum* Nakai cult. orn., fewer than 20 left in wild. Like *Forsythia* but fls white, fr. winged, 1 of v. few Korean endemic genera (cf. *Megaleranthis, Pentactina*)

Abelmoschus Medik. (~ *Hibiscus*). Malvaceae. 15 OW trop. *A. esculentus* (L.) Moench (bandakai, gobbo, gombo, gumbo, okra or lady's-fingers), ed. young fr.: allopolyploid (2n = 130), one genome poss. from diploid *A. tuberculatus* Pal & Singh (2n = 58); *A. manihot* (L.) Medik. (aibika, Mal., ? domesticated in New Guinea) – cordage like jute; *A. moschatus* Medik. – grown for musky seeds (ambrette)

abem *Berlinia* spp.

Aberemoa Aublet = *Guatteria*

Abies Miller. Pinaceae. 49 N temp. (Eur. 5) to Vietnam, C Am. Firs. Lvs borne on branches without short shoots, those on laterals twist into horiz. plane. Cones mature in 1 yr. R: NRBGE 46(1989)59. *A. alba* Miller (silver fir, whitewood, S. Eur. mts) – tallest Eur. tree (to 350 yrs old) form. much grown for constr. work & telegraph poles & favoured by Greeks & Romans for building fast warships, esp. for oars of triremes (loses lower branches early), but since 1900 much attacked by aphids and now replaced by *A. grandis* (D. Don) Lindley (white or giant fir, NW Am.), reaching 100 m, introd. GB 1834 & >62 m tall by 1989, *A. alba* source of Alsatian or Strasburg turpentine (Vosges), oil used in bath preps and med. esp. resp. when inhaled, the principal Christmas tree of the Cont.; *A. balsamea* (L.) Miller (balsam fir, N Am.) pulp contains juvabione, homologue of insect juvenile hormone, & used in Am. (not Brit.) paper prods, also source of Canada balsam used in microscop. preps. Many other imp. sources of timber, resins; *A. cephalonica* Loudon (Greek fir, Greece) used for ships in Anc. Greece; *A. firma* Siebold & Zucc. (Japanese fir); *A. fraseri* (Pursh) Poiret (Fraser fir, she balsam, Alleghanies); *A.lasiocarpa* (Hook.) Nutt. (alpine fir, W US); *A. magnifica* A. Murr. (red fir, NW Am.); *A. nebrodensis* (Lojac.) Mattei (*A. alba* ssp. *n.*) reduced to 20 trees in N Sicily but being replanted there; *A. nordmanniana* (Steven) Spach (Caucasian fir, Nordmann Christmas tree, E Medit.) – needles held longer than those of *Picea abies*; *A. pindrow* Royle (Himal. fir); *A. procera* Rehder (*A. nobilis*, noble fir, NW Am.) often sold as 'larch'; *A. sachalinensis* (Schmidt) Masters (N Japan) – needles mixed with those of *Picea jezoensis* to yield Jap. pine-needle oil

Abildgaardia Vahl = *Fimbristylis*

abir scented powder used in Hindu cerem., largely ground rhiz. *Hedychium spicatum*

abiu *Pouteria cainito* & other *P.* spp.

Abobra Naudin. Cucurbitaceae. 1 temp. S Am.: *A. tenuifolia* (Gillies) Cogn. Cult. orn. dioec. cli.

Abolboda Humb. Xyridaceae (Abolbodaceae). 17 trop. S Am., marshy savanna

Abolbodaceae Nakai. See Xyridaceae

Aboriella Bennet (*Smithiella*). Urticaceae. 1 E Himal. (Abor Hills)

Abortopetalum Degener = *Abutilon*

aboudikro *Entandrophragma cylindricum*

Abrodictyum C. Presl = *Cephalomanes*

Abroma Jacq. (*Ambroma*). Sterculiaceae. 2 trop. As. to Aus. *A. augusta* (L.) L.f. (devil's cotton, Mal.) – bark a source of jute-like fibre

Abromeitia Mez = *Fittingia*

Abromeitiella Mez. Bromeliaceae (1). 3 trop. S Am. Cult. orn. saxic. R: Plantsman 12(1990)180

Abronia Juss. Nyctaginaceae (III). 33 SW US to N Mex. Ground r. of some spp. form. eaten by N Am. Indians. Cult. orn.

Abrophyllum Hook.f. ex Benth. Grossulariaceae (Escalloniaceae). 1-2 E Aus.

Abrotanella Cass. Compositae (Sen.-Blen.). 20 New Guinea, Aus., NZ, temp. S Am.

Abrus Adans. Leguminosae (III 5). 17 pantrop. *A. precatorius* L. (coral pea, crab's eyes, Indian liquorice, jequerity seeds, lucky or Paternoster beans) cont. alks, used as basis of contraceptive & abortifacient in India; r. yields a poor quality liquorice subs.; seeds used as beads & weights (rati, India cf. *Adenanthera*) but v. poisonous esp. in contact with wounds or eyes due to abrin, a toxic glycoprotein, inhibiting protein synthesis (esp. in cancer cells), 0.5 g fatal in humans but detoxified above 65°C

absinthe *Artemisia absinthium*

Absolmsia Kuntze. Asclepiadaceae (III 4). 1 SW China; 1 Borneo

abura *Mitragyna stipulosa*

Abuta Aublet. Menispermaceae (II). 32 trop. S Am. R: MNYBG 22,2(1971)30. Some ed. fr., med. (dangerous, some with ecdysteroids), arrow poisons. *A. imene* (Mart.) Eichler bark used in kind of curare (Colombia); *A. rufescens* Aublet (white Pareira root, Guianas) – med. (urinogenital tract)

Abutilon Miller. Malvaceae. 100+ trop. & warm (1 Eur – *A. theophrasti* Medik., imp. fibre pl. in China (chingma, Chinese jute or hemp, Indian mallow, Manchurian jute, velvetleaf) but major agric. weed in Am. introd. from China pre-1750 & causing losses of >$343m per annum by 1980s). No epicalyx, some bird-poll. Many cult. orn. esp. *A. hybridum* Siebert & Voss, cultigen, others medic. & fibre pls. Fls of *A. esculentum* A. St-Hil. (Braz.) ed.

Abutilothamnus Ulbr. Malvaceae. 1 trop. S Am.

Acacallis Lindl. = *Aganisia*

Achmena H.P. Fuchs = *Erysimum*

acacia *Acacia* spp., (Aus.) *Albizia* spp., weedy *Senna* spp.; **cedar a.** *Paraserianthes toona*; **false a.** *Robinia pseudoacacia*; **rose a.** *R. hispida*; **sweet a.** *Acacia farnesiana*

Acacia Miller. Leguminosae (II 4). 1200 trop. & warm esp. Aus. (900, wattles). New, T.R. (1984) *Biology of acacias*; Mem. Bot. Survey S Afr. 44(1979)1 (African spp.); Simmons, M.H. (1981) *Acacias of Aus.* Classification into 3 subg. (s.t. segregate genera, *Racosperma* & *Senegalia*, recog.) based on morph., seed chem., gums & chromosomes:

subg. *Acacia*. Lvs bipinnate, spines stipular, s.t. swollen & inhabited by ants. Distr. widely

subg. *Aculiferum* Vassal. Lvs bipinnate, unarmed or with prickles. Distr. widely

subg. *Heterophyllum* Vassal. Lvs bipinnate or phyllodic, without prickles, s.t. with stipular spines. Mainly Aus.

A few twiners & hook-climbers, most are trees & shrubs of dry regions, where they can dominate e.g. *A. xanthophloea* Benth. (fever trees of Kenya Rift Valley), *A. nilotica* (L.) Del. (Indian plains where almost only tree) & *A. gerrardii* Benth. (trop. Afr. to Iraq) the only native tree sp. in Kuwait. Phyllodic species with bipinnate lvs when seedlings: phyllode develops through intercalary growth in midrib region, s.t. decurrent on stem as in *A. alata* R. Br. (W Aus.); phyllotaxis of phyllodes decussate, spiral, fasciculate or whorled as in *A. verticillata* (L'Hérit.) Willd. (SE Aus.) or 'chaotic' as in *A. conferta* A. Cunn. ex Benth. (NE Aus.). White powdery gum from phyllodes eaten by Aus. aborigines. Myrmecophilous spp. like *A. sphaerocephala* Cham. & Schldl. (C Am.) with extrafl. nectaries on petioles &

yellow sausage-shaped food-bodies on leaflet tips, attractive to ants which discourage climbing plants & attack would-be grazers but are not exclusive to any particular *Acacia* sp. Many imp. prods – timber, fuel, forage (but some Am. & Aus. (c. 5%) spp. cyanogenic & *A. berlandieri* Benth. (Mex.) has toxic amines), tanbark (some tannins molluscicidal, while those of *A. nigrescens* Oliv. (SE Afr.) imp. killer of browsing kudu, dying of indigestion, the levels of tannins allegedly rising when trees attacked), dyestuffs, gums, scents & cult. orn. though some pestilential weeds, esp. Aus. spp. in SW Cape (S Afr.) where poll. occurs & there are no seed-pests. *A. acuminata* Benth. (raspberry jam (tree), SW Aus.) – fragrant durable timber; *A. aneura* F. Muell. ex Benth. (mulga, Aus.) – dark heartwood (boomerangs etc.) & seeds (ground & eaten) used by aborigines; *A. auriculiformis* A. Cunn. ex Benth. (? = *A. spirorbis* Lab., trop. Aus., New Guinea) – fuel crop tree; *A. baileyana* F. Muell. (Cootamundra wattle, E Aus.); *A. catechu* (L.f.) Willd. (catechu, cutch, India to China) – heartwood for tanning, dyeing, for treating fishing nets & sails, med. & as masticatory with betel (true khaki cloth is dyed & shrunk with it), fuelwood (khayer); *A. cornigera* (L.) Willd. (bull-horn thorn, myrmecophyte, C Am., natur. US); *A. dealbata* Link (blue or silver wattle, Aus.) – florists' 'mimosa' imported to London from S France in winter; *A. dodonaeifolia* (Pers.) Balbis (S Aus.) – twigs with resinous ridges; *A. drepanalobium* Harms (whistling thorn, myrmecophyte (whistling due to wind blowing over ant-holes), E Afr.) – source of gum arabic subs.; *A. eburnea* (L.f.) Willd. (cockspur thorn, Arabia to India) – cult. orn., long spines; *A. farnesiana* (L.) Willd. (cassie, opopanax, popinac, warm Am. now pantrop., introd. Aus. before Eur. settlement) – source of ess. oil (cassie ancienne) for perfumery (used in 'violet' scent), esp. cult. in S France; *A. greggii* A. Gray (cat's-claw, N Am.) – common desert plant; *A. homalophylla* A. Cunn. ex Benth. (*A. omalophylla*, yarran, E Aus.) – durable timber; *A. howittii* F. Muell. (sticky wattle, E Victoria) – viscid cult. orn.; *A. implexa* Benth. – E Aus. hickory, lightwood; *A. koa* A. Gray (Hawaii) – with 2 other Hawaiian endemics bearing extrafl. nectaries attractive to ants, though Hawaii has no native ant spp. (? phylogenetic inertia), *A. heterophylla* Willd. (Réunion) 18 000 km away perhaps conspecific, traditional timber for surfboards but also furniture etc.; *A. longifolia* (Andr.) Willd. (Sydney golden wattle, SE Aus.) – cult. orn., lime-tolerant stock for other spp., pest in S Afr. fynbos controlled by gall-forming wasp; *A. mangium* Willd. (mangium, Mal.) – fuelwood; *A. mearnsii* De Wild. (black wattle, SE Aus.) – imp. tanbark, widely cult. (tannin 60–65% of extract); *A. melanoxylon* R. Br. (E Aus. blackwood, Aus.) – sup. timber; *A. modesta* Wallich (N India, Afghanistan, Pakistan) – Amritsar gum; *A. nilotica* (L.) Del. (babul, trop. Afr., natur. India) – imp. tanbark & gum arabic subs., timber & forage, used by Anc. Egyptians, fr. (Gambia or sant pods) ed. & tanning (30% tannin, molluscicidal – used on field scale & algicidal); *A. paradoxa* DC (*A. armata*, kangaroo thorn, temp. Aus.) – hedging; *A. pendula* A. Cunn. ex G. Don f. (myall, E Aus.) – fine violet-scented timber; *A. prominens* A. Cunn. ex G. Don f. (golden rain wattle, SE Aus.); *A. pubescens* (Vent.) R. Br. (hairy wattle, NSW); *A. pycnantha* Benth. (golden wattle, SE Aus.) – extrafl. nectaries attractive to poll. birds, tree source of tanbark & gum; *A. retinodes* Schldl. (wirilda, S & SE Aus.) – cult. orn.; *A. salicina* Lindl. (cooba, wirra, Aus.); *A. saligna* (Labill.) H.L. Wendl. (*A. cyanophylla*, Port Jackson (willow), W Aus., natur. elsewhere in Aus.) – coastal sandhills rehabilitation in Libya, pest in S Afr. fynbos where controlled by gall-forming fungus; *A. senegal* (L.) Willd. (arid trop. Afr.) – tapped trees source of true gum arabic for lozenges, gum sweets, adhesives, inks, watercolours & med.; *A. seyal* Del. (shittim, trop. Afr. to Egypt) – ed. gum, timber used for Ark of the Covenant & coffins of pharaohs; *A. spectabilis* A. Cunn. ex Benth. (Mudgee wattle, E Aus.); *A. terminalis* (Salisb.) Macbr. (*A. elata*, cedar wattle, SE Aus.); *A. victoriae* Benth. (gundabluey, Aus.) – emergency fodder

Acaena Mutis ex L. Rosaceae. c. 100 S hemisph. N to Calif. (1) & Hawaii (1), many natur. elsewhere, e.g. *A. novae-zelandiae* Kirk ('*A. anserinifolia*', SE Aus., NZ) in GB, introd. early 1900s in fleeces from Aus. or NZ. R: BB 74(1910–11). Bidgee-widgee (Aus.), bidi-bidi (NZ); *A. eupatoria* Bitter (Urug.) used to control fertility

açai *Euterpe edulis*

acajou *Anacardium occidentale*; *Guarea guidonia* or other 'mahoganies'

Acalypha L. Euphorbiaceae. c. 430 trop. & warm (Afr. 50). Mostly monoec. or dioec. shrubs, stigmas much branched. R: Pflanzenr. IV.147.XVI(1924)12. *A. hispida* Burm.f. (Mal.) – cult. orn., only female known, infls coloured by branched red stigmas; *A. indica* L. (S As.) – roots irresistible to cats; *A. wilkesiana* Muell. Arg. (= *A. tricolor* Veitch, beefsteak plant, copperleaf, Jacob's coat, Pacific) – widely cult. trop. orn. with mottled lvs

Acalyphopsis Pax & K. Hoffm. = *Acalypha*

Acampe Lindley (~ *Saccolabium*). Orchidaceae (V 16). 5 Indomal., 1 (*A. rigida* (Sm.) P. Hunt) ext. to Afr.

Acamptoclados Nash = *Eragrostis*

Acamptopappus (A. Gray) A. Gray (~ *Haplopappus*). Compositae (Ast.-Sol.). 2 SW N Am. deserts. R: Madrono 35(1988)247

Acanthaceae Juss. Dicots – Asteridae – Scrophulariales. 229/3450 largely trop., inc. open country & deserts, extending to Medit., US, Aus. Herbs, inc. some aquatics (even mangrove – *Acanthus*) & twiners, shrubs, few trees, s.t. with unusual sec. thickening. Many typical of ground flora of trop. forests & some with gregarious flowering (*Mimulopsis*, *Strobilanthes*). Lvs usu. opp. (spiral in *Nelsonioideae*) with cystoliths, seen as streaks in lvs (not in *Nelsonioideae*, *Mendoncioideae* nor *Thunbergioideae* nor tribe *Acantheae* of Acanthoideae), s.t. spiny. Bracts & bracteoles usu. present, oft. coloured & ± encl. fl. Fls ⚥, usu. zygom. or 2-lobed (upper lip s.t. absent e.g. *Acanthus*), usu. with nectariferous disk below ovary. K (4–6), usu. 5), to 16-lobed; C (4 or 5); A 2, 4 or 5 (*Pentastemonacanthus*), epipetalous; 1 or more staminodes oft. present; 1 anther lobe often smaller than other, connective oft. long, pollen with diverse architecture, imp. in generic delimitation. $\underline{G}$ (2, 2-loc. with axile placentation, each with 2–∞ (esp. *Nelsonioideae*) usu. anatr. ovules in 2 rows; style often with 2 stigmas. Fr. usu. 2-loc., explosive loculicidal capsule, usu. with seeds on hook-like funicular jaculators (not *Mendoncioideae*); seeds non-endospermous (exc. *Nelsonioideae*, where oily & ruminate), with large embryos, the testa s.t. (e.g. *Blepharis*, *Crossandra*) with hairs or scales becoming sticky on wetting (cf. *Linum*). x = 7–21

Classification & large genera:

I. **Nelsonioideae** (ovules ∞, jaculators 0): *Staurogyne*

II. **Thunbergioideae** (ovules 4, jaculators 0): *Thunbergia*

III. **Acanthoideae** (ovules 2–∞, jaculators): *Aphelandra*, *Dicliptera*, *Justicia*, *Ruellia*, *Strobilanthes*

[IV. **Mendoncioideae** (cystoliths 0, jaculators 0) = Mendonciaceae]

Closely allied to Scrophulariaceae with no meaningful distinction between them: *Nelsonioideae* are perfectly intermediate & are here by tradition, though poss. polyphyletic with *Nelsonia* referable to Scrophulariaceae. Subfam. *Mendoncioideae* treated as separate family, Mendonciaceae here, though molecular work supports inclusion (AJB 82(1995)266)

Many cult. orn. esp. *Aphelandra*, *Crossandra*, *Eranthemum*, *Fittonia*, *Hypoestes*, *Hygrophila*, *Justicia* & *Thunbergia*

Acanthambrosia Rydb. = *Ambrosia*

Acanthanthus Y. Ito = *Echinopsis*

Acanthella Hook.f. Melastomataceae. 2 trop. S Am.

Acanthephippium Blume (*Acanthophippium*). Orchidaceae (V 11). 15 trop. As. to New Caled. (1) & Fiji

Acanthocalycium Backeb. = *Echinopsis*

Acanthocalyx (DC) Tieghem (~ *Morina*). Morinaceae. 3 Sino-Himal. R: BBMNHB 12(1984)13

Acanthocardamum Thell. Cruciferae. 1 S Iran

Acanthocarpus Lehm. Xanthorrhoeaceae. 1 SW Aus.: *A. preissii* Lehm.

Acanthocephalus Karelin & Kir. Compositae (Lact.-Crep.). 2 C As.

Acanthocereus (A. Berger) Britton & Rose. Cactaceae (III 1). 6+ SE US to NE Braz. Heterogeneous? *A. pentagonus* (L.) Britton & Rose ed. fr.

Acanthochiton Torrey = *Amaranthus*

Acanthochlamydaceae (S.L. Chen) P.C. Kao = Anthericaceae

Acanthochlamys P.C. Kao. Anthericaceae (Acanthochlamydaceae; Velloziaceae). 1 SW China: *A. bracteata* P.C. Kao

Acanthocladium F. Muell. (~ *Helichrysum*). Compositae (Gnap.-Cass.). 1 SE Aus.: *A. dockeri* F. Muell. R: OB 104(1991)81

Acanthococos Barb. Rodr. = *Acrocomia*

Acanthodesmos C. Adams & du Quesnay. Compositae (Vern.-Vern.). 1 Jamaica: *A. distichus* C. Adams & du Quesnay

Acanthogilia A. Day & Moran. Polemoniaceae. 1 Baja Calif.: *A. gloriosa* (Brandegee) A. Day & Moran – desert shrub with persistent woody-spinose lvs & winged seeds. R: PCAS 44(1986)111

Acanthogonum Torrey = *Chorizanthe*

Acantholepis Less. Compositae (Card.-Ech.). 1 S W & C As. Annual

Acantholimon Boiss. Plumbaginaceae (II). 165 E Med. to C As., gravelly deserts & mts (Eur. 1). R: S. Mobayen (1964) *Revision taxon. du genre A.* Cult. rock-pls

Acantholippia Griseb. (~ *Lippia*). Verbenaceae. 6 arid S Am. R: Darw. 22(1980)511, Dominguezia 3(1982)1

Acanthomintha (A. Gray) A. Gray. Labiatae (VIII 2). 3 Calif.

Acanthonema Hook.f. Gesneriaceae. 2 W Afr. Epiphyllous infls

Acanthopale C.B. Clarke (~ *Strobilanthes*). Acanthaceae. 15 palaeotrop.

Acanthopanax (Decne. & Planchon) Miq. = *Eleutherococcus*

Acanthophippium Blume. See *Acanthephippium*

Acanthophoenix H. Wendl. Palmae (V 4a). 1 Masc.: *A. rubra* (Bory) H. Wendl. – cult. orn., form. felled for palm hearts & now rare in wild

Acanthophora Merr. = *Aralia*

Acanthophyllum C. Meyer. Caryophyllaceae (III 1). 56 SW & C As., Siberia. Many xeroph. with prickly lvs, some cult. orn.

Acanthopsis Harvey. Acanthaceae. 7 S Afr.

Acanthopteron Britton = *Mimosa*

Acanthorhipsalis (Schumann) Britton & Rose = *Lepismium*

Acanthorrhinum Rothm. Scrophulariaceae (Antirrh.). 2 NW Afr.

Acanthorrhiza H. Wendl. = *Cryosophila*

Acanthoscyphus Small = *Oxytheca*

Acanthosicyos Welw. ex Hook.f. Cucurbitaceae. 2 S trop. Afr. *A. horridus* Welw. ex Hook.f. (nara, narras) of sand-dunes with roots to 12 m long & ed. fr. & seeds; *A. naudinianus* (Sonder) C. Jeffrey imp. food & water source, Kalahari

Acanthospermum Schrank. Compositae (Helia.-Mel.). 6 trop. Am., intr. OW esp. *A. hispidum* DC. R: CUSNH 20(1921)383. Some med., esp. *A. australe* (Loefl.) Kuntze (Uruguay), contraceptive

Acanthosphaera Warb. = *Naucleopsis*

Acanthostachys Klotzsch. Bromeliaceae (3). 1 S trop. Am.

Acanthostelma Bidgood & Brummitt (~ *Neuracanthus*). Acanthaceae. 1 Somalia

Acanthostyles R. King & H. Robinson (~ *Eupatorium*). Compositae (Eup.-Dis.). 2 E S Am. R: MSBMBG 22(1987)74

Acanthosyris (Eichler) Griseb. Santalaceae. 3 temp. S Am. *A. falcata* Griseb. ed. fr.

Acanthothamnus Brandegee. Celastraceae (Canotiaceae). 1 Mex.

Acanthotreculia Engl. = *Treculia*

Acanthoxanthium (DC) Fourr. = *Xanthium*

Acanthura Lindau. Acanthaceae. 1 Brazil

Acanthus L. Acanthaceae. 30 trop. & warm OW, S Eur. (3). Mostly thorny xeroph.; fl. with lower lip only; extrafl. nectaries. *A. ilicifolius* L. OW mangrove, bird-poll. in Aus.; *A. mollis* L. (bear's breech, Eur.) cult. orn. bee-poll.; *A. spinosus* L.(oyster pl. (Aus.), E Medit.) – supposed origin of leaf motif in Corinthian capitals

Acareosperma Gagnepain. Vitaceae. 1 SE As.

acaroid resin See *Xanthorrhoea*

Acarpha Griseb. Calyceraceae. 8 temp. S Am.

Acaulimalva Krapov. Malvaceae. 19 Andes. R: Darw. 19(1974)9

Acca O. Berg (~ *Psidium*). Myrtaceae (Myrt.). 6 S Am. *A. sellowiana* (O. Berg) Burret (*Feijoa s.*, pineapple guava, S Brazil, Paraguay, Uruguay, N Arg.) – ed. fr. for preserves

Accara Landrum (~ *Psidium*). Myrtaceae (Myrt.). 1 Brazil: *A. elegans* (DC) Landrum. R: SB 15(1990)221

Aceituna *Quassia simarouba*

Acelica Rizz. = *Justicia*

Acentra Philippi (~ *Hybanthus*). Violaceae. 1 Arg.

Acer L. Aceraceae. 111 N temp. & trop. mts (Eur. 14, many E As., 1 W Mal.). R: D.M. van Gelderen *et al.* (1994) *Maples of the World*. Trees & shrubs (maples) with entire or palmately lobed lvs, occ. pinnate (sect. *Negundo*), s.t. with domatia, & fr. a pair of long-winged samaras. Monopodial & sympodial spp., the latter able to occupy shady habitats. Some imp. timbers, sources of syrup & many cult. orn. inc. street trees, often with bright autumn colours. Wood hard, usu. white, used for flooring, esp. squash courts & dancefloors, furniture, musical instruments, gunstocks, shoelasts etc.; figured wood known as papapsco, fiddle-back, & bird's eye maple (burrs of *A. saccharum*), used in veneers (a Victorian fad); bark ground for bread by E Canad. Indians. *A. campestre* L. (field or hedge maple, Eur., inc. Br.) – can form coppicing clumps >4 m diam. in English woods, timber form. for bowls, spoons, cutlery handles, Saxon harps & tobacco pipes (Ulmer pipes); *A. negundo* L. (box elder, N Am. to Guatemala) – pinnate lvs, cult. orn., wood soft; *A. palmatum* Thunb. (Japanese maple, E As.) – cult. orn. inc. many dwarf cvs; *A. pensylvanicum* L. (moosewood, N Am.) – cult. orn.; *A. platanoides* L. (Norway maple, Eur., W As.) – timber, cult. orn.; *A. pseudoplatanus* L. (sycamore (app. first confused with sycomore (*Ficus sycomorus*) by Shakespeare), great or Scottish maple, Eur., W As.) – timber for turnery, joinery, polo mallets, violin-backs, when stained grey = harewood (as in Hepplewhite furniture), early nectar source for bees (resultant honey green), early plantation tree in Scotland & now agressive weed of Br. woodland; *A. rubrum* L. (red maple, N Am.) – populations of constant males with some labile to females, fewer constant females & even fewer labile to males though some fluctuate widely in their sexuality, timber; *A. saccharum* Marshall (rock or sugar or striped maple, N Am.) – timber & maple syrup & sugar, tapped in early spring (Massachusetts used to produce 200 000 kg per annum), wind- & insect-poll., one of most spectacular trees of the fall (gold to crimson); *A. tataricum* L. ssp. *ginnala* (Maxim.) Wesm. (*A. ginnala*, E As.) – poss. source of polyphenols

Aceranthus Morren & Decne = *Epimedium*

Aceraceae Juss. Dicots – Rosidae – Sapindales. 2/113 N temp. & trop. mts. Trees & shrubs. Lvs opp., pinnate (*Dipteronia, Acer* sect. *Negundo*), palmate or simple, then freq. palmately-lobed or at least -veined, epidermis often mucilaginous, petiolate, stipules usu. 0. Infls corymbiform, umbelliform, s.t. racemose. Fls reg., some or all functionally or truly unisexual, entomophilous, rarely anemophilous, hypogynous or staminate fls s.t. perigynous; K 5, rarely 4, rarely connate; P 5, rarely 4 or 6 or 0, oft. like K; A 8, less oft. 4, 5, 10 or c. 12, within or without a nectary-disk, which is rarely 0; $\underline{G}$ (2), 2-loc., styles 2; ovules usu. 2 in each loc., anatr. to almost orthotr., bitegmic. Fr. a winged schizocarp, often a double samara; seeds usu. solit., endosperm 0

Genera: *Acer, Dipteronia*

Oligocene pollens referred here; lvs common in Miocene. Close to (& poss. best included in) Sapindaceae, esp. tribe *Harpullieae*

Aceras R. Br. Orchidaceae (IV 2). 1 Eur., N Afr.: *A. anthropophorum* (L.) Aiton f. (man orchid, GB)

Acerates Elliott = *Asclepias*

Aceratium DC. Elaeocarpaceae. 20 E Mal. to Solomon Is., N Aus., Vanuatu (1). *A. oppositifolium* DC (Indomal.) – ed. fr.

Aceratorchis Schltr.(= ?). Orchidaceae (IV 2). 2 Tibet, China

acerola *Malpighia glabra*

Acetosa Miller = *Rumex*

Acetosella (Meissner) Fourr. = *Rumex*

acha *Digitaria exilis*

Achaenipodium Brandegee = *Verbesina*

Achaetogeron A. Gray = *Erigeron*

Acharia Thunb. Achariaceae. 1 S Afr.

Achariaceae Harms. Dicots – Dilleniidae – Violales. 3/3 S Afr. Subshrubby, stemless or climbing herbs. Lvs spirally arranged, usu. palmately-lobed. Fls solitary or in racemes, hypogynous, regular, unisexual (plants monoecious). K 3–5, C (3–5), A same no. as C, alternating with lobes, attached to tube, connective broad; $\underline{G}$ (3–5), 1-loc. with 2–several ovules on each parietal placenta, style ± deeply lobed. Fr. septicidal-capsular; seeds arillate with straight embryo & copious endosperm

Genera: *Acharia, Ceratiosicyos, Guthriea*. Close to Passifloraceae

Acharitea Benth. = *Nesogenes*

Achasma Griffith = *Etlingera*

Achatocarpaceae Heimerl (~ Phytolaccaceae). Dicots – Caryophyllidae – Caryophyllales. 2/6 warm Am. Dioec. trees or shrubs with regular sec. thickening, s.t. spiny. Lvs spirally arranged, simple, entire, exstip. Fls in axillary racemes or panicles or ramifl.; K 4 or 5, persistent in fr., P 0, A 10–20, filaments s.t. connate at base, pollen diff. from other Caryophyllales, $\underline{G}$ (2), l-loc. with 1 basal, campylotropous ovule & 2 distinct styles. Fr. a berry; seed exarillate, with curved embryo surrounding perisperm, without true endosperm

Genera: *Achatocarpus, Phaulothamnus*

Unilocular ovary would be anomalous in Phytolaccaceae

Achatocarpus Triana. Achatocarpaceae. 5 Mex. to Arg.

Achetaria Cham. & Schldl. Scrophulariaceae. 5 trop. Am.

Achillea L. Compositae (Anth.-Ach.). 115 Euras. (Eur. 52), Medit., few natur. N Am. Alks. Many med. & cult. orn. (yarrow), e.g. *A. erba-rota* All. ssp. *moschata* (Wulfen) Richardson (*A. moschata*, C Alps) – med., source of Iva liqueur & bitters; *A. filipendulina* Lam. (W to C As., yellow fls). *A. millefolium* L. (milfoil, Eur., W As., widely natur. temp.) – med., tobacco subs. (Sweden), ed. (Germany), herbal treatment of arthritis; *A. ptarmica* L. (sneezeweed, sneezewort, Euras.) – form. med. & salad pl., 'double' forms = bachelor's buttons; *A. santolina* L. (NE Afr. to Iran) – Bedouin med. pl., poss. source ess. oil

Achimenes Pers. Gesneriaceae. 22 trop. Am. Diff. spp. poll. by birds, moths, bees & butterflies. Cult. orn. e.g. *A. erecta* (Lam.) H.P. Fuchs (*A. coccinea, A. pulchella*), parent of many hybrids, & *A. grandiflora* (Schiede) DC (hot water plant)

achira *Canna indica* (*C. edulis*)

Achlaena Griseb. = *Arthropogon*

Achlyphila Maguire & Wurd. Xyridaceae. 1 Venez.

Achlys DC. Berberidaceae (II 2; Podophyllaceae). 3 Japan, W N Am. R: BMT 29(1913)169. Perianth aborts early; unstable arr. of organ primordia leads to irreg. floral phyllotaxis & unstable A nos. *A. triphylla* (Sm.) DC (deerfoot, vanilla leaf, W N Am.) cult. orn.

Achnatherum P. Beauv. = *Stipa*

Achnophora F. Muell. Compositae (Ast.-Ast.). 1 Aus.: *A. tatei* F. Muell.

Achnopogon Maguire, Steyerm. & Wurd. Compositae (Mut.-Mut.). 2 Venez., Guyana. Shrubs

Achoriphragma Soják = *Parrya*

Achradelpha Cook = *Pouteria*

Achradotypus Baillon = *Pycnandra*

Achras L. = *Manilkara*

Achrouteria Eyma = *Chrysophyllum*

Achuaria Gereau. Rutaceae. 1 Peru: *A. hirsuta* Gereau

Achudemia Blume = *Pilea*

Achyrachaena Schauer. Compositae (Hele.-Mad.). 1 NW US. Pappus scales silvery – 'everlastings'

Achyranthes L. Amaranthaceae (I 2). 6–8 OW trop. & subtrop. (Eur. 1). *A. arborescens* R. Br. (Norfolk Is.) – 4-merous fls like *Nototrichium* with *A. mangarevica* Suesseng. (Gambier Is., S Pacific) – poss. extinct, only tree-like spp. in genus; *A. aspera* L. – lvs ed. (Java), ash source of salt, branches used as toothbrushes (Arabia)

Achyrobaccharis Schultz-Bip. ex Walp. = *Baccharis*

Achyrocalyx Benoist. Acanthaceae. 4 Madag.

Achyrocline (Less.) DC. Compositae (Gnap.-Gnap.). 32 trop. Afr., Madag., S Am. R: OB 104(1991)148. Some contraceptives (Uruguay)

Achyronychia Torrey & A. Gray. Caryophyllaceae (I 2). 2 SW US, Mex.

Achyropappus Kunth (~ *Bahia*). Compositae (Hele.-Cha.). 1 Mex.

Achyropsis (Moq.) Hook.f. Amaranthaceae (I 2). 6 trop. & S Afr. R: NS 16(1960)100

Achyroseris Schultz-Bip. = *Scorzonera*

Achyrospermum Blume. Labiatae (VI). c. 10 OW trop.

Achyrothalamus O. Hoffm. Compositae (Mut.-Mut.). 1 E trop. Afr. Herb

Aciachne Benth. Gramineae (16). 3 high Andes (N Arg. to Costa Rica). R: NJB 7(1987)667

Acianthera Scheidw. = *Pleurothallis*

Acianthus R. Br. Orchidaceae (IV 1). 27 New Guinea, Aus. (4, 1 ext. to NZ), New Caled. (17). Mainly fly-poll., cult. orn.

Acicarpha Juss. Calyceraceae. 5 trop. Am.

Acidanthera Hochst. = *Gladiolus*

Acidocroton Griseb. Euphorbiaceae. 10 Carib.

Acidonia L. Johnson & B. Briggs. Proteaceae. 13 SW Aus.

Acidosasa Chu & C.S. Chao ex Keng f. (~ *Sasa*). Gramineae (1a). 6 S China, Vietnam (1). R: APS 29(1991)517. Bamboos

Acidoton Sw. Euphorbiaceae. 6 trop. Am.

Acilepidopsis H. Robinson = *Vernonia*

Acineta Lindley. Orchidaceae (V 10). 20 trop. Am. Cult. orn. *A. chrysantha* (Morren) Lindley & Paxton (C Am.) – fragrance 45% cineole, attracting male euglossine bee pollinators

Acinos Miller = *Clinopodium*

Acioa Aublet. Chrysobalanaceae (3). 4 N S Am. with ed. oily seeds. R: PTRSB 320(1988)112. *A. edulis* Prance, long known as oil seed coll. in flooded Amazon forests not described until 1975. For Afr. spp. see *Dactyladenia*

Aciotis D. Don. Melastomataceae. 30 trop. Am.

Aciphylla Forster & G. Forster. Umbelliferae (III 8). 39 NZ (37), Aus. (2). Spiny rock-pls (cult.), dioec., spines poss. deterred moas from grazing. Many hybrids, some with spp. of *Anisotome* & *Gingidia*. R: TRSNZ 84(1955)1. *A. squarrosa* Forster & G. Forster (NZ) – bayonet plant, speargrass

Acisanthera P. Browne. Melastomataceae. 17 trop. Am.

Ackama A. Cunn. = *Caldcluvia*

ackee = akee

Acleisanthes A. Gray. Nyctaginaceae (III). 7 SW N Am. R: Wrightia 5(1976)261

Aclisia E. Meyer ex C. Presl = *Pollia*

Acmadenia Bartling & H.L. Wendl. Rutaceae. 33 SW to E Cape. R: JSAB 48(1982)169

Acmanthera (A. Juss.) Griseb. Malpighiaceae. 6 S Am. R: CUMH 11,2(1975)41, 17(1990)39

Acmella Rich. ex Pers. (~ *Spilanthes*). Compositae (Helia.-Zinn.). 30 trop. with some (now) pantrop. weeds esp. *A. uliginosa* (Sw.) Cass. R: SBM 8(1985). *A. oleracea* (L.) R.K. Jansen (*Spilanthes o.*, Braz. or Pará cress) – salad veg. though spilanthol produces tingling of tongue, local medic., cultigen perhaps derived from *A. alba* (L'Hérit.) R.K. Jansen (C Peru); *A. oppositifolia* (Lam.) R.K. Jansen ed. Chaco Indians, cult. orn.

Acmena DC (~ *Syzygium*). Myrtaceae. 15 Mal., Aus. (7). Some cult. orn. esp. *A. smithii* (Poir.) Merr. & Perry (lilly pilly, lilli-pilli, E Aus.) – fr. ed.

Acmenosperma Kausel = *Syzygium*

Acmispon Raf. = *Lotus*

Acmopyle Pilger. Podocarpaceae. 2 New Caled., Fiji. R: Phytol.M 7(1984)10

Acnistus Schott. Solanaceae (1). 1 trop. Am.: *A. arborescens* (L.) Schldl.

Acoelorrhaphe H.L. Wendl. Palmae (I 1b). 1 C Am.: *A. wrightii* (Griseb. & H.A. Wendl.) Becc. – cult. orn.

Acokanthera G. Don f. (~ *Carissa*). Apocynaceae. 5 Arabia, trop. E & S Afr. R: KB 37(1982)41. Arrow poisons & drugs – cardiac glycosides, incl. ouabain with effect of digitalin. Wood extract mixed with *Euphorbia* latex & *Acacia* gum and applied to arrow. Ouabain can be absorbed through skin; death can follow 20 mins after injection into bloodstream, effected by murderers coating prickly fr. of *Tribulus terrestris* L. with extract & leaving these in path of barefoot victims. *A. oppositifolia* (Lam.) Codd (*A. venenata*) – ordeal-poison (wood up to 1.1% ouabain)

acom *Dioscorea bulbifera*

Acomastylis E. Greene = *Geum*

Acomis F. Muell. Compositae (Gnap.-Ang.). 4 Aus. R: OB 104(1991)123

aconite *Aconitum* spp.; **winter a.** *Eranthis hyemalis*

Aconitella Spach = *Delphinium*

Aconitum L. Ranunculaceae (I 4). c. 300 N temp. (Eur. 7). Alks, poisonous (though *A. bisma* (Buch.-Ham.) Rapaics (Himal.) not, tubers med.) & drug source in med. inc. comm. cough mixtures. Cult. orn. (R: GH 6(1945)463; aconites, monkshood). *A. ferox* Wallich ex Ser. (N India, med.); *A. lycoctonum* L. (badger's bane, wolfsbane, Eur., N Afr.) – cult. orn., fls purple-lilac; *A. napellus* L. (W & C Eur.) – source of aconitine for heart treatment, form. administered to criminals, fly control since before 1240; *A. volubile* Pallas ex Koelle (E As.) – twiner (in cult. '*A. volubile*' other spp. e.g. *A. hemsleyanum* Pritzel (C & W China))

Aconogonon (Meissner) Reichb. = *Persicaria*

Aconogonum Reichb. = *Persicaria*

Acopanea Steyerm. = *Bonnetia*

Acoraceae Martinov (~ Araceae). Monocots – Arecidae – Arales. 1/2 OW & N Am. Marsh or emergent aquatic rhizomatous herbs, aromatic; raphides 0. Lvs linear, distichous; venation parallel. Infl. a cylindric spadix-like spike subtended by erect leaf-like bract & borne on leaf-like peduncle. Fls herm.; P 6 free; A 6 with free flattened filaments & horseshoe-shaped anthers; G (3), each loc. with several apical ovules

Genus: *Acorus*. *Gymnostachys* may be referable here

Form. incl. in Araceae but habit & other morphological as well as molecular features exclude it; it has been argued that the fam. is 'basal' to the Monocots in being intermediate between several orders, having similarities with Araceae & Typhaceae but the presence of oil idioblasts & absence of raphides reminiscent of Zingiberaceae though it is outstanding in its well-developed perisperm devoid of starch

Acoridium Nees & Meyen = *Dendrochilum*

acorn single-seeded fr. of *Quercus* spp.

Acoroides Sol. ex Kite = *Xanthorrhoea*

Acorus L. Acoraceae. 2 OW & N Am. wetlands. Rhiz. sympodial, scented. Lvs iris- or grass-like ('gladiolus' of Med. GB). *A. calamus* L. (calamus, sweet flag, N temp. & India to New Guinea) – rhiz. used med. (esp. modern Chinese herbalism) since Hippocrates (460–377 BC; also in Tutankhamun's tomb), form. for toothache, tonic & in dysentery, also spread on hall & church floors ('rushes') & in the 'oil of holy ointment' for anointing altars & sacred vessels in Old Testament ('sweet calamus' of Exodus XXX), hung up at night in Sumatra to keep evil spirits from children, effective insecticide (di- & tri-terpenes) & still imp. flavouring in Continental eaux-de-vie. Pls in Eur. (& temp. India) all triploid (diploids in As., both in Canada, tetraploids in trop. S & temp. E As.), clonal intr. (Russia C11, Poland C13, Hungary 1517 & distrib. Clusius (Holland) 1574) reaching GB from Lyon with Gerard & natur. 1660

Acosmium Schott. Leguminosae (III 2). 17 trop. Am. R: NRBGE 29(1969)349. Alks; construction timber

Acosta Adans. = *Centaurea*

Acostaea Schltr. Orchidaceae (V 13). 4 C Am. R: MSB 15(1986)15, 24(1987)1

Acostia Swallen. Gramineae (34b). 1 Ecuador: *A. gracilis* Swallen

Acourtia D. Don. Compositae (Mut.-Nass.). 65 S US to C Am., WI. R: Phytol. 27(1973)228, 38(1978)456. Herbs. *A. microcephala* DC (Calif.) cult. orn.

Acrachne Wight & Arn. ex Chiov. Gramineae (31d). 3 OW trop.

Acradenia Kipp. Rutaceae. 2 E Aus. R: JAA 58(1977)171. Cult. orn.

Acrandra O. Berg = *Campomanesia*

Acranthera Arn. ex Meissner. Rubiaceae (I 8). 35 Indomal., esp. Borneo. R: JAA 28(1947)261. Some rheophytes

Acreugenia Kausel = *Myrcianthes*

Acridocarpus Guillemin & Perrottet. Malpighiaceae. 30 trop. Afr., Madag. (1), New Caled. (1!)

Acriopsis Blume. Orchidaceae (V 9). 6 SE As. to Solomon Is. R: OM 1(1986)1. *A. javanica* Reinw. ex Blume (Sumatra to New Guinea) – infusion used for fever, cult. orn.

Acrisione R. Nordenstam (~ *Senecio*). Compositae (Sen.-Tuss.). 2 C Chile. R: BJ 107(1985)581. Trees & shrubs

Acritochaete Pilger. Gramineae (34d). 1 trop. Afr. mts

Acritopappus R. King & H. Robinson (~ *Eupatorium*). Compositae (Eup.-Ager.). 13 E Brazil. R: MSBMBG 22(1987)138

Acrobotrys Schumann & K. Krause. Rubiaceae (I 5). 1 Colombia

Acrocarpus Wight ex Arn. Leguminosae (I 1). 1 Indomal.: *A. fraxinifolius* Arn. (mt. rain forests India, Burma to Java) – imp. timber (hard, brown) used esp. for tea-chests (Darjeeling), promising plantation sp. (C Am.). R: Blumea 38(1994)314

Acrocephalus Benth. Labiatae (VIII 4b). 130 OW trop. *A. robertii* Robyns used as copper indicator in Katanga

Acroceras Stapf (~ *Panicum*). Gramineae (34b). 19 OW trop. (12 Madag. endemics). *A. macrum* Stapf (Nile grass, SE Afr.) – cult. pasture grass, highveld, Transvaal

Acrochaene Lindley = *Monomeria*

Acrochaete Peter = *Setaria*

Acroclinium A. Gray = *Rhodanthe*

Acrocoelium Baillon = *Leptaulus*

Acrocomia C. Martius. Palmae (V 5e). c. 20 trop. Am. Monoec. prickly, fast-growing cult. orn., seeds sometimes disp. by cattle. *A. mexicana* Karw. ex C. Martius (coyoli palm, C Am.) ed. fr. (cooked), oil source, palm-sap wine; *A. totai* C. Martius (gru-gru, mbocarya, Paraguay palm, NE Arg., Paraguay) imp. source of palm kernel oil, locally used for soap

Acrodiclidium Nees & C. Martius = *Licaria*

Acrodon N.E. Br. Aizoaceae (V). 4 SW Cape. R: Bothalia 16(1986)212. Cult. orn.

Acroglochin Schrader ex Schultes. Chenopodiaceae (I 1) 2 C & E As., Himal. Fr. mass prickly with axes not terminating in fls

Acrolophia Pfitzer. Orchidaceae (V 9). 9 SW Cape to Zululand

Acronema Falc. ex Edgew. Umbelliferae (III 8). 25 Sino-himal.

Acronychia Forster & G. Forster. Rutaceae. 43 Indomal., W Pac. R: JAA 55(1974)469. Alks. Some medic. pls SE As., used for caulking boats. Cult. orn.

Acropelta Nakai = *Polystichum*

Acrophorus C. Presl. Dryopteridaceae (I 2). 2 SE As. to Fiji

Acrophyllum Benth. Cunoniaceae. 1 E Aus.: *A. australe* (A. Cunn.) Hoogl. – cult. orn.

Acropogon Schltr. (~ *Sterculia*). Sterculiaceae. 29 New Caled. R: BMNHN 4, 8(1986)357, 10(1988)93. Often remarkable unbranched pachycauls

Acroptilon Cass. (~ *Centaurea*). Compositae (Card.-Cent.). 1 SW Eur. to C As.: *A. repens* (L.) DC – noxious allelopathic weed introd. N Amer., Aus.

Acrorchis Dressler. Orchidaceae (V 13). 1 C Am.: *A. roseola* Dressler

Acrorumohra (H. Ito) H. Ito = *Arachniodes*

Acrosanthes Ecklon & Zeyher. Aizoaceae (I). 4 SW Cape. R: JSAB 25(1959)23

Acrosorus Copel. Grammitidaceae. 7 Mal. to Samoa

Acrostemon Klotzsch. Ericaceae. 8 SW & S Cape

Acrostichaceae Mett. ex Frank = Pteridaceae

Acrostichum L. Pteridaceae (VI). 3 pantrop. mangrove. Marsh ferns. Upper pinnae reduced and bearing sporangia ('acrostichoid' condition achieved in unrelated fern groups). *A. aureum* L. (pantrop.) – young lvs ed. (Malesia), old ones used for thatch (Vietnam)

Acrosynanthus Urban = *Remijia*

Acrotome Benth. ex Endl. Labiatae (VI). 8 trop. & S Afr.

Acrotrema Jack. Dilleniaceae. 10 Indomal. esp. Sri Lanka. Herbs s.t. with finely divided lvs

Acrotriche R. Br. Epacridaceae. 14 temp. Aus. R: Telopea 1(1980)421. Cult. orn.

Acrymia Prain. Labiatae (IV). 1 Malay Pen.

Acsmithia Hoogl. = *Spiraeanthemum*

Actaea L. Ranunculaceae (I 2). 8 N temp. (Eur. 2). Beetle-poll. Carpel 1. Fr. a berry. *A. spicata* L. (baneberry, Herb Christopher, black cohosh, Euras.) – v. poisonous, medic., Radix Christopherianae form. used against skin disease, asthma, rheumatism & esp. St Vitus Dance

Actephila Blume. Euphorbiaceae. 35 Indomal., China, N Aus. Seeds, anomalous in *Phyllanthoideae*, more nearly approach Guttiferae or Geraniales

Actinanthella Balle. Loranthaceae. 1 S trop. Afr.

Actinanthus Ehrenb. = *Oenanthe*

Actinea A. Juss. = *Helenium*

Actinidia Lindley. Actinidiaceae. 40 Indomal., E As. Climbers. R: JAA 33(1952)1. Monoterpenoid alk., actinidine, physiologically active in cats attracted to pls, beta-phenylethyl alc. inducing them to salivate being a widely used ingredient in scents long known to be attractive to them. *A. arguta* (Siebold & Zucc.) Miq. & *A. deliciosa* (A. Chev.) Liang & A.R. Ferg. ('*A. chinensis*', Chinese gooseberry, kiwi-fruit, yangtao) of E As. widely cult. for fr., the latter esp. in NZ, notably large-fruited 'Hayward', also used in liqueurs & 'champagne' (form. for wine in China), all commercial orchards derived from NZ stocks orig. planted before 1910. Apparently bisexual, pl. actually dioec. though some bisexual clones in cult. Cult. orn. incl. *A. kolomikta* (Rupr. & Maxim.) Maxim. with naturally varieg. lvs in males and *A. polygama* (Siebold & Zucc.) Maxim., both attractive to cats

Actinidiaceae Gilg & Werderm. Dicots – Dilleniidae – Ericales. 3/340 trop. & warm & As. mts. Trees, shrubs & lianes with raphides in parenchyma; indumentum of scales & hairs. Lvs simple, spirally arr. Infl. usu. cymose, axillary or on old wood, s.t. fls solitary. Fls bisexual or unisexual, hypogynous; K 5, rarely 4, 6 or 7, imbricate, persistent in fr., C 5, rarely 4, 6 or 7, imbricate, A ∞ or as few as 10, often in 5 clusters opp. petals, anthers s.t. dehiscing by pores, $\underline{G}$ (3–30 or more) with as many locules as carpels with ∞ axile unitegmic anatr. ovules. Fr. a berry or rarely a loculicidal capsule; seeds with copious oily or proteinaceous endosperm

Genera: *Actinidia, Clematoclethra, Saurauia*

Despite recent placement in Ericales, seed studies suggest closest affinity with Theaceae

Actiniopteridaceae Pichi-Serm. = Pteridaceae

Actiniopteris Link. Pteridaceae (III). 5 Afr. (most) to Sri Lanka. In xeric conditions, esp. *A. radiata* (Sw.) Link

Actinobole Endl. Compositae (Gnap.-Ang.). 4 Aus. R: OD 104(1991)127

Actinocarya Benth. Boraginaceae. 2 NW Ind., Tibet, Pakistan

Actinocheita F. Barkley. Anacardiaceae (IV). 1 Mex.

Actinocladum McClure ex Söderstrom. Gramineae (1b). 1 Brazil savanna: *A. verticillatum* (Nees) Söderstrom, regenerating from 2 ground-level nodes after fire

Actinodaphne Nees. Lauraceae (II). 100 Indomal., E As. *A. hookeri* Meissner (trop. As.) seeds with 75% oil (90% lauric acid)

Actinodium Schauer. Myrtaceae (Lept.). 1–2 W Aus. Pollen like Santalaceae

Actinokentia Dammer. Palmae (V 4h). 2 New Caled. R: Allertonia 3(1984)320

Actinolema Fenzl. Umbelliferae (II 1). 2 E Medit.

Actinomeris Nutt. = *Verbesina*

Actinophloeus (Becc.) Becc. = *Ptychosperma*

Actinorhytis H. Wendl. & Drude. Palmae (V 4m). 2 Papuasia. *A. calapparia* (Blume) R. Scheffer widely cult. in SE As. & W Mal. for magic, seeds chewed like *betel* in Java

Actinoschoenus Benth. (~ *Fimbristylis*). Cyperaceae. 4 Madag., Sri Lanka, China

Actinoscirpus (Ohwi) R. Haines & Lye. Cyperaceae. 1 Indomal.: *A. grossus* (L.f.) Goetghebeur & D. Simpson (*Scirpus g.*) – used for sleeping-mats, bags etc., locally medic.

Actinoseris (Endl.) Cabrera. Compositae (Mut.-Mut.). 6 SE Brazil, Uruguay, NE Arg.

Actinospermum Elliott = *Balduina*

Actinostachys Wallich = *Schizaea*

Actinostemma Griffith. Cucurbitaceae. 1 India to Japan. Cooking oil

Actinostemon C. Martius ex Klotzsch. Euphorbiaceae. 40 trop. Am.

Actinostrobus Miq. ex Lehm. Cupressaceae. 3 SW Aus. Cult. orn. R: Phytol.M 7(1984)11

Actinotinus Oliver = *Viburnum*. The species *A. sinensis* Oliver was based on a trick, the infl. of a *Viburnum* having been inserted in the term. bud of an *Aesculus* (cf. *Papilionopsis*, *Stalagmitis*)

Actinotus Labill. Umbelliferae (I 1b). 15 Aus. (15, 14 endemic), NZ. Heads congested & surrounded by bracts suggesting a Composita. *A. helianthi* Labill. (flannelflower, Aus.) – cult. orn., protected in Aus.

Actites Lander (~ *Sonchus*). Compositae (Lact.-Son.). 1 Aus.: *A. megalocarpus* (Hook.f.) Lander. R: Telopea 1(1976)129

Acunaeanthus Borh., J.-Komlódi & Moncada (~ *Neomazaea*). Rubiaceae (I 5). 1 Cuba: *A. tinifolius* (Griseb.) Borh. R: ABASH 26(1981)282

Acyntha Medikus = *Sansevieria*

Acystopteris Nakai (~ *Cystopteris*). Dryopteridaceae (II). 2 Indomal. to Japan

Ada Lindley. Orchidaceae (V 10). 15 Colomb. Andes. *A. aurantiaca* Lindley easily cult.

Adam and Eve *Aplectrum hyemale*; **Adam's hood** (W Afr.) *Ficus* spp.; **A.'s flannel** *Verbascum thapsus*; **A.'s needle** *Yucca* spp.

Adansonia L. Bombacaceae. 8 Afr.(1), Madag.(6), NW Aus.(1). Trees with charac. swollen trunks; fls open at dusk, some in 30 secs. *A. digitata* L. (baobab, Judas's bag, monkey bread, trop. Afr.) lives up to 2000 yrs, the fls of equal nos of right & left-twisting convolute C, though right on average have more stamens, poll. by bats but also visited by insects & bushbabies; fr. distrib. by mammals esp. baboons & elephants, seeds with enhanced germ. thereafter; bark used for cloth, inner bark for rope, fruit & seeds for fuel, dried fr. provides drink rich in citric & tartaric acids, fresh fr. eaten by baboons, many med. & other uses, even as rooms (prison-cells, lavatories etc. – see KB 37(1982)173). *A. gregorii* F. Muell. (*A. gibbosa*, gourd-tree, Aus.) – ed. seeds, trunk a watersource (up to 300 litres per trunk) for birds & aborigines. Madag. spp. seeds water-disp.(?), 'Longitubae' largely hawkmoth-poll., but also visited by birds & lemurs, these the main pollinators of *A. grandidieri* Baillon ('Brevitubae'); *A. gregorii* interm. in fl. structure & poll. unknown but visited by honeyeaters & (?) marsupials

Adarianta Knoche = *Pimpinella*

adder's meat *Stellaria holostea*; **a.'s mouth** *Malaxis*; **a.'s tongue** *Ophioglossum* spp.

Addisonia Rusby = *Helogyne*

Adelaster Lindley ex Veitch = *Fittonia*

Adelia L. Euphorbiaceae. 15 trop. & warm Am.

Adeliopsis Benth. = *Hypserpa*

Adelobotrys DC. Melastomataceae. 25 trop. Am.

Adelocaryum Brand = *Cynoglossum* + *Lindelofia*

Adelodypsis Becc. = *Dypsis*

Adelonema Schott = *Homalomena*

Adelonenga Hook.f. = *Hydriastele*

Adelopetalum Fitzg. = *Bulbophyllum*

Adelosa Blume. Labiatae (?I). 1 Madag.

Adelostemma Hook.f. Asclepiadaceae (III 1). 3 Burma, China

Adelostigma Steetz. Compositae (Pluch.). 2 trop. Afr.

Adenacanthus Nees = *Strobilanthes*

Adenandra Willd. Rutaceae. 18 SW Cape. R: OB 32(1972). *A. fragrans* (Sims) Roemer & Schultes cult. orn., aromatic lvs used as tea in Cape

Adenanthe Maguire, Steyerm. & Wurd. = *Tyleria*

Adenanthellum R. Nordenstam. Compositae (Anth.-Tham.). 1 S Afr.

Adenanthemum R. Nordenstam = *praec.*

Adenanthera L. Leguminosae (II 3). 12 trop. As., Pacif. R: NJB 12(1992)85. Seeds hard, bright red or red & black. Timber good. *A. pavonina* L. (red sandalwood, SE As. to Aus.) – street tree, seeds used as beads (Circassian seeds) & weights by goldsmiths (A. = goldsmith in Arabic). NB Ganda system of India based on *Abrus precatorius* (q.v.), but derived from a system with double that seed weight & *Adenanthera* seeds are twice as heavy as *Abrus* seeds

Adenanthos Labill. Proteaceae. 32 S & W Aus. R: Brunonia 1(1978)303. Cult. orn. 'woolly bush'

Adenarake Maguire & Wurd. Ochnaceae. 1 Venez. R: BJ 113(1991)183

Adenaria Kunth. Lythraceae. 1 Mex. to Arg.

Adenia Forssk. Passifloraceae. 94 OW trop., esp. Afr. R: MLW 71-18(1971). Many Afr. spp. with thorns or swollen stems, some med. *A. ellenbeckii* Engl. (E Afr.) – fr. ed.; *A. palmata* Lam. has an inner cambium producing medullary steles whilst outer cambium continues to divide

Adenium Roemer & Schultes. Apocynaceae. 5 trop. & subtrop. Afr., Arabia. R: MLW 80-12(1980). Thick-stemmed xerophytes with cardiac glycosides effective as fish & other poisons and in ordeals. *A. multiflorum* Klotzsch (impala lily, trop. Afr.) – fish stupifier, Transvaal; *A. obesum* (Forssk.) Roemer & Schultes (desert rose, E Afr. to S Arabia) – arrow & ordeal poison, cult. orn. 'specially protected' in Zimbabwe

Adenoa Arbo. Turneraceae. 1 E Cuba: *A. cubensis* (Britton & P. Wilson) Arbo, on serpentine

Adenocalymna C. Martius ex Meissner. Bignoniaceae. 60 trop. Am. Big lianes with 4-ribbed stems due to 'anomalous' vascular structure

Adenocarpus DC. Leguminosae (III 31). 15 trop. Afr. mts, Canaries, Med. (Eur. 4). R: BSB II 41(1967)67, OBPC 5(1989)69. Alks. Cult. orn.

Adenocaulon Hook. Compositae (Mut.-Nass.). 5+ S As., Guatemala, Arg.

Adenochilus Hook.f. Orchidaceae (IV 1). 2 Aus., NZ

Adenochlaena Boivin ex Baillon = *Cephalocroton*

Adenocline Turcz. Euphorbiaceae. 8 S Afr.

Adenocritonia R. King & H. Robinson (~ *Eupatorium*). Compositae (Eup.-Crit.). 1 Jamaica: *A. adamsii* R. King & H. Robinson. R: MSBMBG 22(1987)302

Adenodaphne S. Moore = *Litsea*

Adenoderris J. Sm. Dryopteridaceae (?I). 2 Cuba & Jamaica

Adenodolichos Harms. Leguminosae (III 10). 15 trop. Afr.

Adenoglossa R. Nordenstam. Compositae (Anth.-Mat.). 1 NW Cape

Adenogramma Reichb. Molluginaceae. 10 Afr. R: JSAB 21(1995)84

Adenogrammaceae (Fenzl) Nakai = Molluginaceae

Adenolisianthus (Progel) Gilg = *Irlbachia*

Adenolobus (Benth.) Torrey & Hillc. Leguminosae (I 3). 2 SW Afr.

Adenoncos Blume. Orchidaceae (V 16). 15 Mal., Vietnam

Adenoon Dalz. Compositae (Vern.-Vern.). 1 Indomal.

Adenopappus Benth. Compositae (Hele.-Pect.). 1 Mex.

Adenopeltis Bertero ex A. Juss. Euphorbiaceae. 2 temp. S Am.

Adenophaedra (Muell. Arg.) Muell. Arg. Euphorbiaceae. 4 trop. S Am.

Adenophora Fischer. Campanulaceae. 40 temp. Euras. (Eur. 2). Differs from *Campanula* only in disk at style-base. Some ed. roots & cult. orn. *A. liliifolia* (L.) Besser (*A. communis*, C Eur. to Manchuria) – cult. root-crop (Japan)

Adenophorus Gaudich. = *Grammitis*

Adenophyllum Pers. (~ *Dyssodia*). Compositae (Hele.-Pect.). 9 SE US, Mexico

Adenoplea Radlk. = *Buddleja*

Adenoplusia Radlk. = *Buddleja*

Adenopodia C. Presl (*Pseudoentada*, ~ *Entada*). Leguminosae (II 3). 10 trop. Afr. & Am. R: KB 41(1986)73

Adenoporces Small = *Tetrapterys*

Adenorachis (DC) Nieuwl. = *Photinia*

Adenosciadium H. Wolff. Umbelliferae (III 8). 1 SW Arabia

Adenosma R. Br. Scrophulariaceae. 15 China, Indomal., Aus.

Adenostachya Bremek. = *Strobilanthes*

Adenostegia Benth. = *Cordylanthus*

Adenostemma Forster & G. Forster. Compositae (Eup.-Aden.). 24 trop. Am., Afr., with 1 pantrop. weed, *A. lavenia* (L.) Kuntze (*A. viscosum*), with sticky pappus suited to animal dispersal, also source of a blue dye. R: MSBMBG 22(1987)58

Adenostoma Hook. & Arn. Rosaceae. 2 Calif. & nearby Mex. incl. *A. fasciculatum* Hook. & Arn. shrub char. of chaparral

Adenostyles Cass. Compositae (Senec.-Tuss.). 4 C & S Eur. R: Phyton 23(1983)141 ('*Cacalia*')

Adenothamnus Keck. Compositae (Hele.-Mad.). 1 Baja Calif.

Adesmia DC. Leguminosae (III 14). 230 S Am. R: Darw. 13(1964)9, 14(1967)463

Adhatoda Miller = *Justicia*

Adiantaceae Newman = Pteridaceae

Adiantopsis Fée (~ *Cheilanthes*). Pteridaceae (IV). 7 trop. Am.

Adiantum L. Pteridaceae (V). 150 cosmop. (Eur. 1) esp. trop. Am. (maidenhair ferns), some med. & cult. orn. (C.J. Goudey (1985) *M. fs in cultivation*), many NW spp. weedy in OW oil palm & tea plantations, all Aus. spp. protected. *A. aethiopicum* L. (warm OW) – med. & abortifacient; *A. capillus-veneris* L. (cosmop.) – med. & flavouring, in hair-tonics & syrups esp. 'Sirop de Capillaire', a cure-all, cult. orn. (many cvs); *A. pedatum* L. (N Am., E As.) – black petioles used in Am. Indian basket-weaving; *A. raddianum* C. Presl (*A. cuneatum*, Am., Afr.) – fertility control in Uruguay, cult. orn., many cvs

Adina Salisb. Rubiaceae (I 2). 3 E As. R: Blumea 24(1978)357

Adinandra Jack. Theaceae. 70 Indomal. R: JAA 28(1947)1

Adinauclea Ridsd. Rubiaceae (I 2). 1 Sulawesi, Moluccas: *A. fagifolia* (Havil.) Ridsd.

Adinobotrys Dunn = *Callerya*

Adipe Raf. = *Bifrenaria*

Adipera Raf. = *Senna*

Adiscanthus Ducke. Rutaceae. 1 Amaz. Braz.: *A. fusciflorus* Ducke

adlay ed. forms of *Coix lacryma-jobi*

Adlumia Raf. ex DC. Fumariaceae. 1 E N Am. (*A. fungosa* (Aiton) Britton, Sterns & Pogg., biennial climber, cult. orn., alks), 1 Korea & Manchuria. R: OB 88(1986)19

Adolphia Meissner. Rhamnaceae. 2 SW N Am. Chaparral

Adonis L. Ranunculaceae (II 1). 26 temp. Euras. (Eur. 10). R: ANMW 66(1963)51, Webbia 25(1971)299. Cardiac glycosides. *A. annua* L. (*A. autumnalis*, S Eur., SW As., pheasant's eye) – cult. orn. with scarlet petals, each with dark basal spot, form. coll. in S Eng. for Covent Garden market, London & sold as 'red morocco'; *A. vernalis* L. & other spp. (Eur.) med. with same effect as digitalin

Adoxa L. Adoxaceae. 4 N temp. (Eur. 1: *A. moschatellina* L., moschatel, townhall clock), S to W Himal., Colorado & Illinois, seeds disp. by snails). R: BBR 7,4(1987)96

Adoxaceae E. Meyer. Dicots – Asteridae – Dipsacales. 2/5 N temp. Small perennial herbs. Lvs 3-foliolate. Fls mostly 5, herm., greenish; K 2 (lateral fls) or 3 (term. fl.), C (5 & ± irreg., lat. fls, or 4 and reg., term. fl.), the lobes imbricate, each with a nectary at base. Stamens attached to corolla tube, divided almost to base, pollen rather like that of *Sambucus*, $\overline{G}$ (35 rarely (2), usu. 4 in term. fl., 5 in lat., with 1 pend. anatrop. ovule with single integument. Fr. a dry drupe with distinct stones; seeds with copious oily endosperm. x = 9. R: BBR 7,4(1987)93

Genera: *Adoxa, Sinadoxa*

App. closest to Caprifoliaceae & that fam. prob. best dismembered with *Sambucus* & *Viburnum* united with A.

Adrastaea DC (~ *Hibbertia*). Dilleniaceae. 1 NE Aus.: *A. salicifolia* DC

Adriana Gaudich. Euphorbiaceae. 3–5 Aus. (bitterbushes)

Adromischus Lemaire. Crassulaceae. 26 S Afr. R: Bothalia 12(1978)382, 633. Succ. herbs & subshrubs allied to *Cotyledon* but infls racemose & corolla-tube slender. Cult. orn. often with spotted lvs

Adrorhizon Hook.f. Orchidaceae (V 13). 1 Sri Lanka

adrue (Carib.) *Cyperus articulatus*

adzuki bean *Vigna angularis*

Aechmanthera Nees. Acanthaceae. 3 Himal. *A. gossypium* (Nees) Nees flowers gregariously every 12 yrs in Mussoorie Hills

Aechmea Ruíz & Pavón. Bromeliaceae (3). (Excl. *Chevalieria, Lamprococcus, Macrochordion, Ortgiesia, Platyaechmea, Podaechmea*, q.v.; incl. *Streptocalyx*) c. 85 trop. Am. R: FN 14(1979)1766. Epiphytes, scape usu. well-developed. Many cult. orn. esp. room plants

Aedesia O. Hoffm. Compositae (Vern.-Vern.). 3 trop. Afr.

Aegialitidaceae Lincz. = Plumbaginaceae

Aegialitis R. Br. Plumbaginaceae (II). 2 Indomal. to Aus., mangroves

Aegiceras Gaertner. Myrsinaceae. 2 OW trop. Bark & seeds sources of effective fish-poison saponin. *A. corniculatum* (L.) Blanco in mangrove, viviparous

Aegicerataceae Blume = Myrsinaceae

Aegilemma Löve = *Aegilops*

Aegilonearum Löve = *Aegilops*

Aegilopodes Löve = *seq.*

Aegilops L. (~ *Triticum*). Gramineae (23). 21 Med. (Eur. 10) to C As. & Pakistan. R: FR 91(1980)225. Close to *Triticum* and 2 spp. prob. contributed genomes to breadwheat

Aeginetia L. Orobanchaceae. c. 3 Indomal., E As. Parasitic on roots of monocots esp. Gramineae. *A. indica* L. (Indomal. to Japan) – pest of sugar-cane etc., converting sucrose to reducing sugars

Aegiphila Jacq. Labiatae (II). c. 150 trop. Am. R: Brittonia 1(1934)245. Some dioec.

Aegle Corr. Serr. Rutaceae. 3 Indomal. Thorny decid. trees, alks. *A. marmelos* (L.) Corr. Serr.

(bael, beli, Bengal quince, golden apple, India & Burma) – Hindu sacred tree, good timber, fr. pulp for drinks & in treatment of dysentery & as soap, flowers for scenting water, bark for gum, fruit rind for ess. oil (mycotoxic), unripe fr. a yellow dye

Aeglopsis Swingle. Rutaceae. 5 trop. Afr. Cult. orn., poss. *Citrus* stock

Aegokeras Raf. (*Olymposciadium*, ~ *Seseli*). Umbelliferae (III 8). 1 Turkey

Aegopodium L. Umbelliferae (III 8). 7 temp. Euras. (Eur. 1). *A. podagraria* L. (ground elder, goutweed, bishop weed, Eur., natur. N Am.) – intr. GB by Romans as potherb (boiled like spinach) and med. for gout, now a pernicious weed of gardens; 'Variegatum' (inexplicably) cult. widely US

Aegopogon Humb. & Bonpl. ex Willd. Gramineae (33c). 3 S US to Arg. & Peru & New Guinea (Mt Michael). R: Univ. Wyoming Publ. 13(1948)17

Aegopordon Boiss. Compositae (Card.-Card.). 1 SW As.

Aellenia Ulbr. = *Halothamnus*

Aeluropus Trin. Gramineae (31c). 3-4 Med. (Eur. 2) to N China & India. Halophytes

Aenictophyton A.T. Lee. Leguminosae (III 26). 1 NW Aus.

Aenigmopteris Holttum (~ *Tectaria*). Dryopteridaceae (I 3). 5 Mal. R: Blumea 30(1984)3

Aeolanthus C. Martius = *seq.*

Aeollanthus C. Martius ex Sprengel. Labiatae (VIII 4c). (Inc. *Icomum*) 43 trop. & warm Afr. Some with spirally-arr. lvs. Explosive poll. mechanism. Some ed. & cult. *A.myrianthus* Bak. (*A. gamwelliae*, ninde, trop. Afr.) distilled for oil rich in geraniol, subs. for palmarosa; *A. suaveolens* C. Martius ex Sprengel (trop. Afr.) sold in Brazil as local medicine (macassá) for skin & eye problems

Aeoniopsis Rech.f. = *Bukiniczia*

Aeonium Webb & Berth. Crassulaceae. 31 Macaronesia (most), Med. to Tanzania & Arabia. R: Ho-Yih Liu (1989) *Systematics of A.* Close to *Sempervivum* but caulescent usu.; many hybrids in wild. Cult. orn. succulents esp. *A. arboreum* (L.) Webb & Berth. (Morocco) – natur. in Medit. (& NZ where an aggressive coast colonist), sap used to harden fishermen's lines

Aequatorium R. Nordenstam (~ *Gynoxys*). Compositae (Sen.-Tuss.). c. 30 Colombia to Arg. Trees & shrubs. R: OB 44(1978)59

Aerangis Reichb.f. (~ *Angraecum*). Orchidaceae (V 16). c. 60 trop. & S Afr. (26, 1 ext. to Sri Lanka, R: KB 34(1979)239), Madag. Cult. orn., some with long spurs e.g. *A. ellisii* (B.S. Williams) Schltr. (Madag.)

Aeranthes Lindley. Orchidaceae (V 16). 30 Masc. Cult. orn.

Aerides Lour. Orchidaceae (V 16). 20 trop. & E As., W Mal. Cult. orn.

Aeridostachya (Hook.f.) Brieger = *Eria*

Aerisilvaea Radcl.-Sm. Euphorbiaceae. 1 Tanzania: *A. sylvestris* Radcl.-Sm. R: KB 45(1990)149

Aerva Forssk. Amaranthaceae (I 2). 10 warm & trop. OW. Some ed. *A. javanica* (Burm.f.) Schultes (trop. As.) – fluffy infls used to stuff pillows in Middle E; *A. lanata* Juss. (trop. As.) – exported from Sri Lanka for use in kidney disease

Aesandra Pierre = *Diploknema*

Aeschrion Vell. Conc. = *Picrasma*

Aeschynanthus Jack. Gesneriaceae. c. 140 Indomal. (China 34). Epiphytic subshrubs or lianes with fleshy lvs. Seeds with long hairs or emergences at chalazal end. Cult. orn.

Aeschynomene L. Leguminosae (III 13). 150 trop. & warm. Shrubs with sensitive lvs & lightweight pith, that of *A. aspera* L. (emergent aquatic with bulk of aerenchyma being sec. xylem, trop. As.) & *A. indica* L. (trop. & warm OW – seeds fatal to humans & pigs) being used in sunhats (shola or sola), floats etc. and charcoal of latter in fireworks & gunpowder. *A. elaphroxylon* (Guillemin & Perrottet) Taubert (trop. Afr.) component of sudd, wood (ambatch) used for rafts & floats; *A. hispida* Willd. (S N Am.) – lightest wood known (specific gravity 0.044)

Aesculus L. Hippocastanaceae. SE Eur. (1), India & E As. (5), N Am. (7). R: Plantsman 6 (1985)228. Buckeyes, horse-chestnuts, cult. orns, some fish-poisons (US) perhaps due to aesculin, a coumarin. *A. californica* (Spach) Nutt. (Californian buckeye, Calif.) – seeds form. much eaten by Calif. Indians; *A. flava* Sol. (*A. octandra*, yellow b., E N Am.) – valuable timber; *A. hippocastanum* L. (horse-chestnut, Balkans to Himal., widely cult.) – propolis from buds, nectar guides in fls change from yellow to red as fls age, some med. uses, wood limited use, seeds the conkers (i.e. conquerors) of children's games, parent (with *A. pavia* L.) of sexual hybrid, *A.* × *carnea* Zeyher ('Plantierensis' a sterile backcross with *A. hippocastanum*), and with *A. flava* Sol. of graft hybrid, *A.* + *dallimorei* Sealy, form. fed to ailing horses by Turks, bark yields black dye for silk & cotton; *A. indica* (Cambess.)

Hook. (Indian horse-chestnut, Himal.) – horse med., molluscicidal saponins; *A. pavia* L. (red horse-chestnut, N Am.) – widely cult.

Aetanthus (Eichler) Engl. Loranthaceae. 10 N Andes

Aetheocephalus Gagnepain = *Athroisma*

Aetheolaena Cass. (~ *Lasiocephalus*). Compositae (Sen.-Sen.). c.15 S Am.

Aetheolirion Forman. Commelinaceae. 1 Thailand. Climber with elongate capsules of winged seeds

Aetheopappus Cass. = *Centaurea*

Aetheorhiza Cass. Compositae (Lact.-Son.). 1 Medit. inc. Eur.

Aethephyllum N.E. Br. Aizoaceae (V). 1 SW Cape. R: BBP 61(1986)436

Aethiocarpa Vollesen. Sterculiaceae. 1 Somalia: *A. lepidota* Vollesen

Aethionema R. Br. Cruciferae. 70 Med. (Eur. 9). *A. heterocarpum* Gay has 2 kinds of fr.: dehiscent, many-seeded and indehiscent, 1-seeded. Many cult. orn. esp. *A.* × *warleyense* Bergmans

Aethusa L. Umbelliferae (III 8). 1 Eur., W As., N Afr.: *A. cynapium* L. (fool's parsley) – poisonous weed (coniine), used med. with caution

Aetoxylon (Airy Shaw) Airy Shaw. Thymelaeaceae. 1 Borneo

Aextoxicaceae Engl. & Gilg. Dicots – Rosidae – Celastrales. 1/1 Chile. Tree, dioec. with peltate indumentum. Lvs subopp. to spirally arr., simple. Fls in axillary racemes, each enclosed in bracteole in bud; K 5, rarely 4 or 6, strongly imbr., C 5, rarely 4 or 6, imbr., A 5, rarely 4 or 6 reniform nectary glands, G̲ (2) with bifid style & 2 anatr. bitegmic, pendulous ovules in 1 loc. only, each with a massively beaked nucellus protruding beyond integuments. Fr. a dry drupe with single stone & seed with ruminate endosperm. n = 16

Genus: *Aextoxicon*

Variously assigned by authors but morphologically like Celastrales, though nucellar beak like some Euphorbiaceae

Aextoxicon Ruíz & Pavón. Aextoxicaceae. 1 Chile: *A. punctatum* Ruíz & Pavón – commercial timber (olivillo)

afara or **white a.** *Terminalia superba*; **black a.** *T. ivorensis*

Affonsea A. St-Hil. = *Inga*

Afgekia Craib. Leguminosae (III 6). 3 S China, Burma, Thailand. Named after the collector Arthur Francis George Kerr!

afo *Poga oleosa*

Afrachneria Sprague = *Pentaschistis*

Afraegle (Swingle) Engl. Rutaceae. 4 W Afr. Spiny trees & shrubs, source of oil for food

Aframmi Norman. Umbelliferae (III 8). 2 Angola

Aframomum Schumann. Zingiberaceae. 50 trop. Afr. Fls last 1 day or less; nectaries at ovary apex ('stylodia'). Seeds spices: *A. angustifolium* (Sonn.) Schumann (Madagascar cardamom); *A. melegueta* Schumann (Guinea grains, grains of paradise, alligator, malagueta or melagueta pepper, W Afr.) the best

Afrardisia Mez = *Ardisia*

Afraurantium A. Chev. Rutaceae. 1 W Afr.

Africa or **African blackwood** *Dalbergia melanoxylon*; **A. bowstring hemp** *Sansevieria* spp.; **A. breadfruit** *Treculia africana*; **A. canarium** *Canarium schweinfurthii*; **A. corn lily** *Ixia* spp.; **A. ebony** *Diospyros* spp.; **A. elemi** *Boswellia frereana*; **A. golden walnut** *Lovoa trichilioides*; **A. hemp** *Sparrmannia africana*; **A. kino** *Pterocarpus erinaceus*; **A. lily** *Agapanthus africanus*; **A. lotus** *Ziziphus lotus*; **A. love-grass** *Eragrostis curvula*; **A. mahogany** *Khaya* spp. esp. *K. antotheca*, *K. senegalensis* etc.; **A. marigold** *Tagetes erecta*; **A. millet** *Pennisetum glaucum*; **A. padouk** *Pterocarpus soyauxii*; **A. peach** *Sarcocephalus latifolius*; **A. pencil cedar** *Juniperus procera*; **A. rosewood** *Guibourtia demeusei*, *Pterocarpus erinaceus*; **A. rubber** *Landolphia* spp.; **A. satinwood** *Zanthoxylum macrophyllum*; **A. teak** *Milicia excelsa*, *Oldfieldia africana*; **A. tulip-tree** *Spathodea campanulata*; **A. violet** *Saintpaulia ionantha*; **A. walnut** *Lovoa trichilioides*; **A. whitewood** *Annickia* spp. esp. *A. chlorantha*; **A. yellow-wood** *Podocarpus* spp.

afrikander (S Afr.) *Gladiolus* spp.

Afrobrunnichia Hutch. & Dalziel = *Brunnichia*

Afrocarpus (Buchholz & N. Gray) Page (~ *Podocarpus*). Podocarpaceae. 3 trop. & S Afr. R: NRBGE 45(1988)383

Afrocalathea Schumann. Marantaceae. 1 W Afr.

Afrocarum Rauschert (*Baumiella*). Umbelliferae (III 8). 1 S trop. Afr.: *A. imbricatum* (Schinz) Rauschert

Afrocrania (Harms) Hutch. = *Cornus*

Afrodaphne Stapf = *Beilschmiedia*

Afrofittonia Lindau. Acanthaceae. 1 trop. W Afr.
Afroguatteria Boutique. Annonaceae. 1 trop. Afr.: *A. bequaertii* (de Wild.) Boutique
Afroknoxia Verdc. = *Knoxia*
Afrolicania Mildbr. = *Licania*
Afroligusticum Norman. Umbelliferae (III 8). 1 trop. Afr.
Afrolimon Lincz. (~ *Limonium*). Plumbaginaceae (II). 7 S Afr.
Afropteris Alston. Pteridaceae (III). 2 W Afr., Seychelles
Afrorhaphidophora Engl. = *Rhaphidophora*
afrormosia *Pericopsis elata*
Afrormosia Harms = *Pericopsis*
Afrosersalisia A. Chev. = *Synsepalum*
Afrosison H. Wolff. Umbelliferae (III 8). 3 trop. Afr.
Afrostyrax Perkins & Gilg. Huaceae. 2 trop. Afr. Bark & seeds imp. local garlicky flavouring
Afrothismia (Engl.) Schltr. Burmanniaceae. 4 trop. Afr.
Afrotrewia Pax & K. Hoffm. Euphorbiaceae. 1 W trop. Afr.
Afrotrichloris Chiov. Gramineae (33b). 2 Somalia
Afrotrilepis (Gilly) Raynal. Cyperaceae. 2 trop. W Afr. *A. pilosa* (Boeck) Raynal 'arborescent'
Afrotysonia Rauschert (*Tysonia*). Boraginaceae. 3 S & E Afr. R: NRBGE 43(1986)467
Afrovivella A. Berger = *Rosularia*
Afzelia Sm. Leguminosae (I 4). 12 OW trop. (Mal. 2). Comm. timbers (aligna, apa) e.g. *A. rhomboidea* (Blanco) S. Vidal (Malacca teak, Mal.), *A. quanzensis* Welw. (chamfuta, trop. Afr.) – seeds (red & black) used as beads (lucky beans)
Afzeliella Gilg = *Guyonia*
Agalinis Raf. (~ *Sopubia*). Scrophulariaceae. 40 trop. & warm Am., ± parasitic on roots of other pls. *A. purpurea* (L.) Pennell (US) – cult. orn.
Agallis Philippi. Cruciferae. 1 Chile
Agalmyla Blume. Gesneriaceae. c. 6. Mal. Epiphytes. Cult. orn.
Aganippea Moçiño & Sessé ex DC = *Jaegeria*
Aganisia Lindley. Orchidaceae (V 10). 3 trop. Am. Diminutive cult. orn.
Aganonerion Pierre ex Spire. Apocynaceae. 1 SE As.
Aganope Miq. = *Ostryocarpus*
Aganosma (Blume) G. Don f. Apocynaceae. 8 SE As. to C Mal. Coffee & tea subs.
Agapanthus L'Hérit. Alliaceae (Amaryllidaceae; Liliaceae s.l.). 9 S Afr. R: JSAB suppl. 4 (1965)1. Popular tub-pls, esp. *A. africanus* (L.) Hoffsgg. (Africa lily), though this less often grown than *A. praecox* Willd. (*A. orientalis*) with many named cvs. Thick rhiz., winged seeds
Agapetes D. Don ex G. Don f. (~ *Vaccinium*). Ericaceae. 95 trop. As. to W Pacif. Close to *Vaccinium*. Everg. cult. orn. shrubs, some with lvs used in a tea (India)
Agarista D. Don ex G. Don f. (~ *Leucothoe*). Ericaceae. Afr. to Masc. 1 (*Agauria*), Am. 29. R: JAA 65(1984)255
agarwood *Aquilaria malaccensis*
Agastache Clayton ex Gronov. Labiatae (VIII 2). 22 C & E As., N Am. to Mex. R: SBM 15(1987). Bold cult. orn., some used as flavouring (N Am.) – *A. foeniculum* (Pursh) Kuntze (*A. anethiodora*, anise or giant hyssop), basis of a drink & local med.
Agastachys R. Br. Proteaceae. 1 Tasmania
Agasyllis Sprengel. Umbelliferae (III 9). 1 Caucasus
Agatea A. Gray. Violaceae. 1 (variable) W Pacif.
Agathelpis Choisy. Globulariaceae. 2 SW Cape. R: SAJB 56(1990)471
Agathis Salisb. Araucariaceae. 20 Sumatra to NZ (1) & Fiji (New Caled. 5); kauri & dammar. R: PSE 135(1980)41. Imp. trop. timber trees, producing copals: *A. australis* (D. Don) Steud. (kauri, NZ); *A. dammara* (Lamb.) Rich. (bendang, bindang, damar minyak, E India or Manila copal, W & C Mal.) – bark burned to deter mosquitoes; *A. macrophylla* (Lindley) Masters (*A. vitiensis*, dakua, Fijian kauri, Solomon Is. to Fiji); *A. microstachya* J. Bailey & C. White (Queensland) – to 1000 yrs old. Saplings easily grown pot-pls, though diff. spp. scarcely distinguishable then. Lvs broad, entire, leathery; cones take 2 yrs to mature; trees usu. monoec.
Agathisanthemum Klotzsch (~ *Hedyotis*). Rubiaceae (IV 1). 5–6 trop. Afr., Comoro Is.
Agathophora (Fenzl) Bunge = *Halogeton*
Agathosma Willd. Rutaceae. 135 S Afr. esp. SW Cape. R: JSAB 16(1950)55. Heath-like shrubs with ess. oils, for which some cult. to give buchu oil (diuretic). Cult. orn. *A. betulina* (Bergius) Pill. principal source of buchu, also used in artificial blackcurrant flavourings, urinary antiseptic (volatile oil: diosphenol) also used in herbal treatment of arthritis

Agati Adans. = *Sesbania*

Agauria (DC) Hook.f. = *Agarista*

Agavaceae Dumort. Monocots – Liliidae – Asparagales. 13/210 trop. & warm esp. arid Am. Sparsely branched trees to coarse short-stemmed ± herb. pls with bulb-like corms with polyfructosans (no starch) & vessels in roots & s.t. lvs, usu. with sec. thickening. Lvs spirally arr., usu. in rosettes, often narrow, s.t. prickly, often leathery-succulent. Fls usu. bisexual & ± reg., 3-merous in spike-like racemes to thyrses; P usu. with basal tube; A 3 + 3, anthers introrse with longit. slits, pollen occ. released in tetrads; $\underline{G}$ to $\overline{G}$ with several to ∞ ovules. Fr. capsule or berry; seeds many

Genera: *Agave, Beschorneria, Furcraea, Hesperaloe, Hesperocallis, Hosta, Leucocrinum, Manfreda, Polianthes, Prochnyanthes, Pseudobravoa, Samuela, Yucca*

The woody genera (all Am.) have been referred to subfam. Yuccoideae ($\underline{G}$) and subfam. Agavoideae ($\overline{G}$). With the inclusion of Hostaceae (Funkiaceae; $\underline{G}$), form. excluded on habit, A. is more natural. *Cordyline, Dracaena, Nolina* and allies (locules with 1–6 ovules) now referred to Dracaenaceae

Many imp. fibre-pls (esp. *Agave, Furcraea*) & cult. orn. esp. *Agave, Hosta, Polianthes, Yucca*

Agave L. Agavaceae. 100+ S US to trop. S Am. Short-stemmed pachycauls with spectacular tall, fast-growing infls. H.S. Gentry (1982) *As of continental N. Am.* Fibre-pls (tampico fibre etc.), cult. orn. incl. hybrids e.g. *A.* × *taylori* B.S. Williams (*A. geminiflora* (Tagl.) Ker × *A. filifera* Salm-Dyck, Mex.), sources of pulque (fermented sap) & mescal or tequila (distilled), form. imp. in all aspects of SW US life: bud (heart) ed., fls cooked in tortillas, soap source and for stunning fish; cloudy pulque or tequila imp. vitamin source; lvs poss. use in paper-making. Flowering shoots hapaxanthic, pl. reprod. by suckers or bulbils from infl. as well as seed. *A. americana* L. (century plant – through erroneous belief it flowers when 100 yrs old (actually 10–20), maguey, American aloe, E Mex.) – widely cult. (even S Br.) orn. (in Eur. (Padua) since 1561), hedgepl., with varieg. cvs; *A. cantala* Roxb. ex Salm-Dyck (cantala, Manila maguey, Bombay aloe, ? Mex.) – widely cult. OW for fibre softer & finer but weaker than sisal used in hard fibre twines; *A. fourcroydes* Lemaire (henequen, Yucatan) – fibre for binder twine etc., waste for paper pulp; *A. funkiana* K. Koch & Bouché (Jaumave ixtle or fibre, Mex.) – fibre for brushes; *A. inaequidens* K. Koch (Mex.) – fasciated infls to 1.4 m wide recorded; *A. lecheguilla* Torrey (lecheguilla, tula ixtle, Texas & N Mex.) – fibre for brushes; *A. letonae* F.W. Taylor (Salvador henequen, C Am.) – fibre finer than *A. fourcroydes*, used in sacking; *A. sisalana* Perrine (sisal, Bahama hemp, E Mex.) – widely cult. for fibre esp. Tanzania (industry orig. derived from 62 plantlets introd. 1893), wax a carnauba subs. for car- & shoe-polishes in India, molluscicidal saponins; *A. vera-cruz* Miller (Mex.) – comm. source of fructose, cordage

agba *Gossweilerodendron balsamiferum*

agboin *Piptadeniastrum africanum*

Agdestidaceae (Heimerl) Nakai = Phytolaccaceae

Agdestis Moçiño & Sessé ex DC. Phytolaccaceae (IV). 1 Mex. to Honduras: *A. clematidea* Moçiño & Sessé ex DC, climber, cult. orn. R: JAA 66(1985)35

Agelaea Sol. ex Planchon. Connaraceae. 7 trop. Afr. (5), As. (2). Lianes or scramblers, some med. R: AUWP 89-6(1989)136, BJBB 61(1991)72

Agelanthus Tieghem = *Tapinanthus*

Agenium Nees. Gramineae (40g). 4 Brazil to Arg. R: FR 43(1938)80

Ageomoron Raf. = *Caucalis*

Ageratella A. Gray ex S. Watson. Compositae (Eup.-Alom.). 2 Mex. R: MSBMBG 22(1987)233

Ageratina Spach. Compositae (Eup.-Oxy.). c. 290 E US (3), C & W S Am., 2 natur. OW. R: MBSMBG 22(1987)428. x = 17. *A. adenophora* (Sprengel) R. King & H. Robinson (Crofton weed (S Afr.), Mexican devil, Mex.) – aggressive weed in S Afr., Aus. & NZ (gall fly introd. to control it), causing acute lung disease in grazing horses & skin allergies in humans; *A. riparia* (Regel) R. King & H. Robinson (*Eupatorium a.*, C Am.) – bad weed trop. As.

Ageratinastrum Mattf. Compositae (Vern.-Vern.). 2 trop. Afr.

Ageratum L. Compositae (Eup.-Ager.). 44 trop. Am. R: AMBG 58(1971)6, 62(1975)901. x = 10. *A. conyzoides* L. (billy-goat weed (Aus.), now pantrop. weed) – folk med. (e.g. diarrhoea in Burundi), fls ed. Mafia Is., Tanzania; *A. houstonianum* Miller (*A. mexicanum*, C Am., WI) widely cult. orn. edging pl. with blue (cvs with white or pink) fls., contains precocenes 1 & 2 (based on 2,2-dimethylcromene), which interfere with juvenile hormone activity & cause precocious metamorphosis of insects, & an oil v. toxic against *Fusarium* wilts of *Cajanus cajan*

Agiabampoa Rose ex O. Hoffm. = *Alvordia*

Agianthus E. Greene (~ *Streptanthus*). Cruciferae. 1 Calif.

Aglaia Lour. Meliaceae (I 5). 105 Indomal., W Pacif. Some unbranched pachycauls & rheophytes. Some timbers & ed. fr. (i.e. seed arils) locally eaten e.g. *A. edulis* (Roxb.) Wall. (Indomal.) & *A. exstipulata* (Griff.) Theob. (SE As., W. Mal.). R: KBAS 16(1992). *A. cucullata* (Roxb.) Pellegrin (Indomal.) – timber (tasua, Thailand); *A. korthalsii* Miq. (Himal. to Sulawesi) – cult. fr. tree Malay Pen.; *A. odorata* Lour. (SE As.) – cult. shrub inc. topiary work, male fls used to scent tea & linen; *A. sexipetala* Griff. (*A. aspera*, SE As. to New Guinea) – ed. aril, local timber

Aglaodorum Schott. Araceae (VI 3). 1 W Mal.: *A. griffithii* (Schott) Schott, tidal mud-flats, viviparous

Aglaomorpha Schott. Polypodiaceae (II 1). (Incl. *Drynariopsis*) 32 Indomal. Large epiphytes, cult. orn., some humus-collecting, esp. *A. cornucopiae* (Copel.) Roos

Aglaonema Schott. Araceae (VI 3). 21 Indomal. R: SCB 1(1969)1. Cult. orn. esp. *A. commutatum* Schott (C Mal.), many cvs & *A. pictum* (Roxb.) Kunth 'Tricolor' (Sumatra); *A. modestum* Engler (S China, SE As.) – long cult. in E for good luck

Aglossorrhyncha Schltr. Orchidaceae (V 13). 6 E Mal., Papuasia

Agnirictus Schwantes = *Stomatium*

Agnesia Zulonga & Judziewicz (~ *Olyra*). Gramineae (4). 1 Amaz.: *A. lancifolia* (Mez) Zulonga & Judziewicz. R: Novon 3(1993)306

Agonandra Miers ex Benth. Opiliaceae. 10 C & trop. S Am. *A. brasiliensis* Miers ex Benth. (Brazil) – seeds an oil source, partially hydrogenated yielding a rubber subs.

Agonis (DC) Sweet. Myrtaceae (Lept.). 12 W Aus. Cult. orn. (willow myrtles)

Agoseris Raf. Compositae (Lact.-Mic.). c. 17 W N Am. (9), temp. S Am. (2). Cult. orn. *A. aurantiaca* (Hook.) E. Greene form. eaten by N Am. Indians

Agrianthus C. Martius ex DC. Compositae (Eup.-Gyp.). 6 Brazil. R: MSBMBG 22(1987)118

Agrimonia L. Rosaceae. 15 N temp. (Eur. 3). to C & S Afr. (1). Agrimony. *A. eupatoria* L. (common a., OW) – folk med. in Eur. (liver disease), source of yellow dye; *A. procera* L. ('*A. repens*', '*A. odorata*', As. Minor) – cult. orn. ('scented a.')

agrimony *Agrimonia* spp.; **common a.** *A. eupatoria*; **hemp a.** *Eupatorium cannabinum*; **scented a.** *A. procera*

Agriophyllum M. Bieb. Chenopodiaceae (I 6). 6 Eur. (1) to C As. *A. gobicum* Bunge (Mongolia, Siberia) – seeds imp. food in Mongolia, also forage crop; *A. latifolium* Fischer & Meyer (C As.), a tumbleweed

Agriphyllum Juss. = *Berkheya*

Agrocharis Hochst. Umbelliferae (III 3). 4 NE & trop. to S Afr. Poss. toxic to cattle

Agropyron Gaertner. Gramineae (23). c. 15 temp. OW (Eur. 6). (*A. repens* etc., see *Elymus*)

Agropyropsis (Battand. & Trabut) A. Camus. Gramineae (18). 1 Algeria

Agrostemma L. Caryophyllaceae (III 3). 2 Med. Corn cockle. *A. brachylobum* (Fenzl) Hammer, diploid, & *A. githago* L., tetraploid, form. troublesome weed but now scarce in GB where introd. Seeds poss. poisonous due to saponins. Cult. orn.

Agrostis L. Gramineae (21d). 220 temp. (Eur. 25), trop. mts. R: SBU 17,1(1960). Bent grasses, imp. in lawns & pastures. *A. canina* L. (velvet bent, brown b., Eur.) – for lawns esp. putting greens; *A. capillaris* L. (*A. tenuis*, common b., brown top, colonial b., Rhode Is. b., Eur.) – pasture & lawns esp. bowling greens, selected metal-tolerant strains in spoil-tip reclamation; *A. gigantea* Roth (red top, black b., Euras.) – pastures & lawns; *A. stolonifera* L. (creeping b., fiorin, N temp.) – lawns, found in coffin of child mummified not later than 21st Dynasty. See *Lachnagrostis* & *Calamagrostis* (wetter sites) with which *A.* intergrades

Agrostistachys Dalz. Euphorbiaceae. 8–9 India to W Mal. Ants nest around leaf-bases. *A. longifolia* Benth. ex Hook.f. (S As.) – lvs to 60 cm long used for thatch in Sri Lanka

Agrostocrinum F. Muell. Phormiaceae (Liliaceae). 1 SW Aus.: *A. scabrum* (R. Br.) Baillon

Agrostophyllum Blume. Orchidaceae (V 13). 85 Seychelles to W Pacif.

aguacate *Persea americana*

aguassú *Orbignya* spp.

ague root *Aletris farinosa*; **a. weed** *Gentianella quinquefolia*, *Eupatorium cannabinum*

Aguiaria Ducke. Bombacaceae. 1 Brazil

Ahernia Merr. Flacourtiaceae (2). 1 Hainan, Philippines

Ahouai Miller = *Thevetia*

ahun *Alstonia congensis*

Ahzolia Standley & Steyerm. = *Sechium*

ai *Blumea balsamifera*

aibika *Abelmoschus manihot*

Aichryson Webb & Berth. Crassulaceae. 15 Macaronesia to Morocco. Succ. pls, some esp. *A.* ×

domesticum Praeger (? *A. tortuosum* (Aiton) Praeger × *A. punctatum* (C. Sm.) Webb & Berth.) & 'Variegatum' (form. common cottage window-sill pl.) cult. orn.

Aidia Lour. Rubiaceae (II 1). 18 trop. As. & Afr. (8, 3 restricted to S. Tomé. R: BMNHN 4,8(1986)259

Aidiopsis Tirv. (~ *Randia*). Rubiaceae (II 1). 1 Mal. R: BMNHN 4,8(1986)287

Aidomene Stopp. Asclepiadaceae (III 1). 1 Angola: *A. parvula* Stopp

aielé *Canarium schweinfurthii*

Ailanthus Desf. Simaroubaceae. 5 As. to Aus. *A. altissima* (Miller) Swingle (tree of heaven, China, natur. in N Am., C & S Eur.) – cult. orn. introd. 1751, used as street tree & in soil conservation, prob. fastest-growing deciduous hardy tree in GB with lvs to 1.3 m long on saplings, weedy in many countries (e.g. E Aus., where causing dermatitis), allelopathic, usu. dioec., males with foetid lvs & fls, form. a cv. used as host (*'A. vilmoriniana'*) for Chinese silkmoth, *Samia cynthia* (Drury) in production of Shantung silk (now from *Morus*, though tree still in old Chinese cemeteries in Calif.), honey with cat smell but delicious on standing; *A. excelsa* Roxb. (India) – wood for matches; *A. malabarica* DC (mattipaul, India) – resin used as incense in Hindu temples, lvs yield a black dye; *A. moluccana* DC (Mal.) – 'ailanto', 'reaching for the sky', hence Engl. vern. name

Ainea Ravenna. Iridaceae (III 4). 2 Mex.

Ainsliaea DC. Compositae (Mut.-Mut.). 40 E As. to W Mal.

Ainsworthia Boiss. (~ *Tordylium*). Umbelliferae (III 11). 2 E Med. R: NRBGE 31(1971)109

Aiouea Aublet. Lauraceae (I). 21 trop. Am. R: FN 31(1982)89, AMBG 75(1988)402. Polyphyletic?

Aiphanes Willd. Palmae (V 5e). 38 trop. Am. Prickly monoec. cult. orn. e.g. *A. erosa* (Linden) Burret (macaw palm, Barbados)

Aipyanthus Steven = *Arnebia*

air fern *Sertularia argentea* L., a marine zoophyte (animal) sold in US; **a. plant** *Kalanchoe pinnata, Tillandsia ionantha*; **a. potato** *Dioscorea bulbifera*

Aira L. Gramineae (21b). 10 Eur. (8) & Med. to Iran but widespread as weeds. Some cult. – 'hair grasses'

Airopsis Desv. Gramineae (21b). 1 S Eur., NW Afr.

Airosperma Lauterb. & Schumann. Rubiaceae (?III 5). 6 New Guinea (4), Fiji (2)

Airyantha Brummitt. Leguminosae (III 2). 1 Afr. (Guineo- Congolian), 1 NW Borneo to SW Philippines

Aisandra Airy Shaw = *Diploknema*

Aistocaulon Poelln. ex H.J. Jacobsen = *Nananthus*

Aistopetalum Schltr. Cunoniaceae. 1 New Guinea

Aitchisonia Hemsley ex Aitch. Rubiaceae (IV 12). 1 S to C Iran, Afghanistan: *A. rosea* Hemsl. ex Aitch.

Aitonia Thunb. = *Nymania*

Aitoniaceae (Harvey) Harvey = Meliaceae

Aizoaceae Martinov. Dicots – Caryophyllidae – Caryophyllales. (Excl. Molluginaceae) 128/1850 trop. & subtrop., mostly in S Afr. &, to a lesser extent, Aus. Succ. herbs, more rarely shrubs or subshrubs, rarely spiny (*Eberlanzia*) with betalains & not anthocyanins, often with C_4 photosynthesis & crassulacean acid metabolism (C_3 in *Tetragonia*). Lvs simple, opp. to spiral, usu. ± succ. with centric rather than bifacial structure, rarely stipulate (e.g. *Trianthema*). Fls solit. or in small cymes, bisexual, rarely pl. monoec. P (3–)5(–8), often basally united with A forming a tube, A (4)5–∞, when ∞ outer ones petaloid 1–6 whorls, the flowers resembling Compositae, nectaries commonly in a ring at inner base of A; $\underline{G}$ or $\overline{G}$(2–5–∞) with as many locules, rarely 1-loc., ovules 1 – (usu.) ± ∞ in each, campylotr. to almost anatr., bitegmic. Fr. usu. loculicidal capsule, often enclosed in persistent calyx; seeds with large embryo curved round copious starchy or ± oily & proteinaceous perisperm, without true endosperm, sometimes arillate. x = 8, 9

Classification & principal genera

I. **Aizooideae** (petaloid A 0, capsule loculicidal or septicidal, aril 0): *Aizoon*

II. **Sesuvioideae** (petaloid A 0, capsule circumscissile, seeds arillate): *Sesuvium, Trianthema*

III. **Tetragonioideae** (Tetragoniaceae; petaloid A 0, nut): *Tetragonia*

IV. **Mesembryanthemoideae** (petaloid A usu. united by a tube, placentation central, nectaries shell-shaped to tubular): *Aptenia, Mesembryanthemum* (here treated in the narrow sense: form. incl. spp. now referred to over 110 segregates in IV. & V., many recognizable by aspect; R: SAJB 57(1991)95)

V. **Ruschioideae** (petaloid A usu. free, placentation basal or parietal, nectaries crest-shaped): *Argyroderma, Carpobrotus, Cephalophyllum, Conophytum, Delosperma, Dinteranthus, Drosanthemum, Faucaria, Lampranthus, Lithops*

Compared with related Cactaceae, most A. are leaf-succulents. *Tetragonia* a potherb, many cult. orn. esp. spp. of *Carpobrotus, Conophytum, Dorotheanthus, Gibbaeum, Lampranthus, Lithops, Ruschia*

Aizoanthemum Dinter ex Friedrich. Aizoaceae (I). 4 N Namibia to S Angola

Aizoon L. Aizoaceae (I). 11 S Spain (1), As. Minor, N, E & S (15, R: JSAB 25(1959)31) Afr., introd. elsewhere. *A. canariense* L. (N & S Afr. to S As.) used as food by Tuareg

Aizopsis Grulich = *Sedum*

Ajania Polj. Compositae (Anth.-Art.). 34 C & E As. R: BZ 68(1983)207. Heterogeneous

Ajaniopsis C. Shih (~ *Artemisia*). Compositae (Anth.-Art.). 1 Tibet, China

ajowan *Trachyspermum ammi*

Ajuga L. Labiatae (IV). 50 OW esp. temp. (Eur. 10) but also lowland Mal. Bugle. C with no upper lip; ecdysteroids. *A. chamaepitys* (L.) Schreber (ground pine, yellow b., Euras.) – med.; *A. iva* (L.) Schreber (Medit.) – poss. antimalarial; *A. reptans* L. (common b., Eur.) – cult. orn. (ground cover), many cvs

Ajugoides Makino (~ *Ajuga*). Labiatae (IV). 1 Japan

Akania Hook.f. Akaniaceae. 1 NE Aus. (fossil from Palaeocene of Arg.): *A. bidwillii* (Hogg) Mabb. (turnipwood)

Akaniaceae Stapf. Dicots – Dilleniidae – Capparidales. 1/1 NE Aus. Small tree with alkaloids. Lvs spirally arr., pinnate, leaflets toothed. Fls bisexual in panicles. K 5, imbr.; C 5, convolute; A usu. 8, 5 outer opp. K; no nectary-disk; G 3-loc. with style & 3-lobed stigma & 2 superposed ± anatr., pendulous, bitegmic ovules per locule. Fr. a loculicidal capsule; seeds with copious endosperm, smelling of bitter almonds

Only sp.: *Akania bidwillii*

Form. thought allied to Sapindaceae (Rosidae) but rbcL sequences & presence of glucosinolates show it close to Bretschneideraceae in Capparidales (AusSB 5(1992)717)

akeake *Dodonaea viscosa; Olearia avicenniifolia*

Akebia Decne. Lardizabalaceae. 5 temp. E As., monoec. twiners. Cult. orn. esp. *A. quinata* (Houtt.) Decne. (fls vanilla-scented) & *A. trifoliata* (Thunb.) Koidz. (fls scentless). Female fls (usu. lower) larger; follicles fleshy, ed. (insipid)

akee *Blighia sapida*

Akersia Buin. = *Cleistocactus*

ako *Antiaris toxicaria*

akom *Terminalia superba*

akomu *Pycnanthus angolensis*

Akrosida Fryx. & Fuertes (~ *Bastardia*). Malvaceae. 1 Brazil: *A. macrophylla* (Ulbr.) Fryx. & Fuertes

akund fibre *Calotropis procera*

al *Morinda citrifolia*

Aladenia Pichon = *Farquharia*

Alafia Thouars. Apocynaceae. 26 trop. Afr., Madag. Latex of *A. perrieri* Jum. used as soap in Madag.

Alajja Ikonn. Labiatae (VI). 1 S C As. R: NS 8(1971)274

Alamania Lex. Orchidaceae (V 13). 1 Mex.

alan *Shorea albida*

alang-alang *Imperata cylindrica*

Alangiaceae DC. Dicots – Rosidae – Cornales. 1/21 OW trop. (fossils in Eocene of Eur.). Trees, less often shrubs or lianoid, with laticifers, sometimes thorny & with alks. Lvs spirally arr. simple, pinnately or palmately veined or lobed. Fls in axillary cymes, bisexual usu.: K (4–10) or obs.; C 4–10, sometimes connate at base; A 4–10 alt. with C or 2–4 times as many as C round prominent epigynous disk; $\overline{G}$ (2); ovules solit., pendulous in each locule, or 1 locule empty, anatr., unitegmic. Fr. a drupe with 1-seeded stone; seeds with straight embryo & copious endosperm (oily). x = 8, 11

Only genus: *Alangium*

DNA sequencing suggests close affinity with Cornaceae s.s. Distinctive in Cornales with laticifers and alks

Alangium Lam. Alangiaceae. 21 trop. Afr., China to E Aus. & New Caled. (1). R: BJB Buit. III, 16(1939)139. Cult. orn., timber and med. esp. *A. salviifolium* (L.f.) Wangerin (trop. As.) – ipecacuanha subs. in India; *A. villosum* (Blume) Wangerin (muskwood, NE Aus.) – local scented timber

Alania Endl. Anthericaceae (Liliaceae s.l.). 1 Brisbane Water & Blue Mts, SE Aus.: *A.*

endlicheri Kunth. R: FA 45(1987)279

Alantsilodendron Villiers (~ *Dichrostachys*). Leguminosae (II 3). 8 Madag.

Alaska pine *Tsuga heterophylla*; **A. yellow cedar** *Chamaecyparis nootkatensis*

Alatoseta Compton. Compositae (Gnap.). 1 S Afr.: *A. tenuis* Compton. R: OB 104(1991)46

albarco *Cariniana* spp.

albardine *Lygeum spartum*

Alberta E. Meyer. Rubiaceae (III 5). 6 S Afr. (1), Madag. (5). *A. magna* E. Meyer (SE Afr.) – cult. orn. with reddish fls, seeds need 1 yr 'after-ripening' on tree

Albertinia Sprengel. Compositae (Vern.-Vern.). 1 Brazil

Albertisia Becc. Menispermaceae (I). 19 trop. & warm Afr. (13), SE As. (6)

Albertisiella Pierre ex Aubrév. = ? *Pouteria*

Albidella Pichon = *Echinodorus*

Albizia Durazz. Leguminosae (II 5). 118 warm As. (35), Afr. (48) & Am. (35). See also *Paraserianthes*. Trees, shrubs or lianes, unarmed, arils 0. Timber ('acacia' in Aus.) and shade trees for tea, coffee etc., gums. *A. anthelmintica* Brongn. (trop & S Afr.) – molluscicidal saponins; *A. chinensis* (Osbeck) Merr. (India to Thailand & Indonesia) – tea-shade; *A. grandibracteata* Taubert (nongo, trop. E Afr.) – timber; *A. julibrissin* Durazz. (silk tree, Iran to Japan) – cult. orn., long-selected (cf. *Melia azedarach*), some cvs hardy in GB, seeds 147 yrs old germ. on herbarium sheets at Natural Hist. Mus. (London) after bomb damage in World War II; *A. lebbeck* (L.) Benth. (trop. As., natur. Carib., Afr., East Indian walnut, kokko, siris) – extrafl. nectaries on lvs attractive to (? defending) ants; *A. saman* (Jacq.) F. Muell. (*Samanea saman*, rain tree, trop. Am.) – poss. form. disp. by extinct Pleistocene mammals, associated with *Azteca* ants in Costa Rica, much planted street-tree with shallow grooved bark colonized by epiphytes & 'rain' excreted by Homoptera, distinctive leaflet movements circadian oscillations & light-change-induced through massive fluxes of K^+, Cl^- & H^+ ions through pulvinar motor-cell membranes

Albizzia Benth. = *Albizia*

Albovia Schischkin = *Pimpinella*

Albraunia Speta (~ *Antirrhinum*). Scrophulariaceae. 3 SW As. R: D.A. Sutton, *Rev. Antirrhineae* (1988)131. Pls distrib. as small tumbleweeds with seeds retained in capsules

Albuca L. Hyacinthaceae (Liliaceae s.l). c. 50 Afr., Arabia. Some cult. orn., e.g. *A. canadensis* (L.) Leighton (S Afr.!)

Alcantara Glaz. ex G. Barroso. Compositae (Vern.-Vern). 1 Brazil. *A.* an illeg. name

Alcantarea (Mez) Harms = *Vriesea*

Alcea L. Malvaceae. 50 Med. to C As. (Eur. 5). Hollyhocks, cult. orn. (to 7.39 m, UK record), esp. *A. rosea* L., poss. hybrid of *A. setosa* (Boiss.) Alef. (Crete & Turkey) and *A. pallida* (Willd.) Waldst. & Kit. (C & SE Eur.), or an As. sp., some med. uses

Alchemilla L. Rosaceae. 250 N temp., trop. mts (Eur. c. 118). Many (inc. all UK spp.) apomicts & high polyploids; fls inconsp., green, apetalous, with epicalyx. Some cult. orn. esp. *A. mollis* (Buser) Rothm. (E Carpathians to Caucasus) with large lvs & consp. guttation. *A. vulgaris* L. (*A. acutiloba* Opiz, lady's mantle, Eur.) – medic.

Alchornea Sw. Euphorbiaceae. 70 trop. Some fish-disp. in Amaz. *A. cordifolia* (Schum. & Thonn.) Muell. Arg. (Christmas bush, trop. Afr.) – thicket-forming, foodpl. of sitatunga, source of black dye; *A. laxiflora* (Benth.) Pax & Hoffm. (trop. & S Afr.) – stem a chewing-stick in Nigeria; *A. triplinervis* (Sprengel) Muell. Arg. (Amaz.) – insect antifeedants; *A. villosa* Muell. Arg. (Mal.) – bark fibre used for string

Alchorneopsis Muell. Arg. Euphorbiaceae. 3 trop. Am.

Alcimandra Dandy = *Magnolia*

Alciope DC ex Lindley. Compositae (Sen.-Tuss.). 3 SW Cape

alcohol fermentation of sugars & starches gives ethanol, further purified by distillation. Wide range of plant sources for industrial alc. or potable, e.g. *Saccharum* for car fuel (Brazil), grapes, cereals, incl. maize, rice, barley, rye, potatoes, sugar-beet, bananas, *Agave* & many palms incl. *Arenga, Borassus, Caryota, Cocos, Elaeis, Nypa, Phoenix*

Aldama Llave. Compositae (Helia.-Helia.). 2 C & trop. S Am. R: SB 14(1989)580

alder *Alnus* spp.; **black, common** or **Eur. a.**, *A. glutinosa*; **a. buckthorn** *Frangula alnus*; **green a.** (Am.) *A. crispa*, (Eur.) *A. viridis*; **grey a.** *A. incana*; **hazel a.** *A. rugosa*; **Italian a.** *A. cordata*; **Japanese a.** *A. japonica*; **Oregon a.** *A. oregona*; **red a.** *A. oregona*; **white a.** *A. incana*

Aldina Endl. Leguminosae (III 1). 15 trop. S Am. R: MNYBG 8(1953)103

Aldrovanda L. Droseraceae. 1 C Eur., As. to NE Aus.: *A. vesiculosa* L. – rootless aquatic with whorls of lvs, each with a trap like *Dionaea* for insect-capture, comprising pair of concave lobes connected by midrib each with 6–10 trigger hairs; 2 (1 in young lvs) hairs struck leads to Ca^{2+} ion influx in midrib cells perhaps activating membrane ATPase responsible for K^+ transport leading to turgor loss & lobes coming together

Aldrovandaceae Nakai = Droseraceae

alecost *Tanacetum balsamita*

Alectorurus Makino = *Comospermum*

Alectra Thunb. Scrophulariaceae. 40 trop. Afr., As. R: NBGB 15(1941)423. Some dyes

Alectryon Gaertner. Sapindaceae. 34 (incl. *Heterodendrum*) E Malesia (R: Blumea 33(1988)313, Aus.(13, 12 endemic), NZ, New Caled. to Hawaii. Seeds black with red arils; some timbers & cult. orn. incl. *A. oleifolius* (Desf.) S. Reynolds (*Heterodendrum o.*, Aus. rosewood, boonaree, boonery, Aus.) – fr. eaten fresh by aboriginals, imp. fodder tree in inland Aus., & *A. excelsus* Gaertner (titoki, NZ) – tough wood for tool-handles etc.; *A. myrmecophilus* Leenh. (NE New Guinea) – twigs myrmecophilous

alehoof *Glechoma hederacea*

Aleisanthia Ridley. Rubiaceae (I 5). 2 Malay Pen.

Alepidea Delaroche. Umbelliferae (II 1). 20 trop. & esp. S Afr. R: BN4(1949)257. Some medic.

Alepidocalyx Piper = *Phaseolus*

Alepidocline S.F. Blake. Compositae (Helia.-Gal.). 4 trop. Am. R: Phytol. 69(1990)387

Alepis Tieghem (~ *Elytranthe*). Loranthaceae. 1 NZ

Aleppo galls *Quercus pubescens* etc.; **A. pine** *Pinus halepensis*

alerce *Fitzroya cupressoides, Austrocedrus chilensis* (Chile), *Tetraclinis articulata*

Aletes J. Coulter & Rose. Umbelliferae (III 8). 15–20 W N Am.

Aletris L. Melanthiaceae (Liliaceae s.l.). 10 E As., Mal., N Am. Stargrass. Some cult. orn. *A. farinosa* L. (ague root, colic root, unicorn root, N Am.) – medic. esp. diuretic

Aleurites Forster & G. Forster. Euphorbiaceae. (Incl. *Vernicia*) 5 Indomal., W Pacif. Grown for shade & drying oil (known to Confucius) from the seeds. *A. fordii* Hemsley & *A. montana* (Lour.) E. Wilson (poorer quality), C As. to Burma, sources of tung oil used in paints & quick-drying varnishes, widely cult.; *A. moluccana* (L.) Willd. (*A. triloba*, candle-nut oil tree, candleberry, balucanat, kukui nut, Otaheite walnut; SE As., widely grown (prob. from 'Hoabinhian' times (6000–1000 BC) on), even on otherwise useless land, & natur. in trop.) – source of China wood oil, candle-nut oil, lumbang oil with similar uses, also in curry & comm. shampoos, & seeds used like conkers (*Aesculus hippocastanum*) in Malay Pen., soot from burnt ones form. used in tattooing in Tonga, State tree of Hawaii

Aleuritopteris Fée = *Cheilanthes*

Alexa Moq. = *Castanospermum*

alexanders *Smyrnium olusatrum*

Alexandra Bunge. Chenopodiaceae (III 2). 1 C As.: *A. lehmannii* Bunge

Alexandrian senna *Senna alexandrina*

Alexeya Pakhamova = *Paraquilegia*

Alexfloydia B. Simon (~ *Panicum*). Gramineae (34b). 1 Coffs Harbour, NSW: *A. repens* B. Simon. R: Austrobaileya 3(1992)670

Alexgeorgea Carlq. Restionaceae. 3 SW Aus. R: AusSB 3(1990)752. Geocarpic

Alexitoxicon St-Lager = *Vincetoxicum*

alfa = halfa

alfalfa *Medicago sativa*

Alfaroa Standley. Juglandaceae. 7 Mex. to Colombia. R: AMBG 65(1978)1078

Alfredia Cass. Compositae (Card.-Card.). 5 C–E As.

algaroba *Ceratonia siliqua* (Eur.); *Prosopis* spp. esp. *P. chilensis, P. glandulosa* (Am.)

algarobilla *Balsamocarpon brevifolium*

Algerian fibre *Chamaerops humilis*; **A. grass** *Stipa tenacissima*

Algernonia Baillon. Euphorbiaceae. 3 Brazil

Alhagi Gagnebin. Leguminosae (III 15). 1–3 Med. (Eur. 1–2) to Nepal. Xerophytes. Withered pls blow around in dry season; branches exude sap, hardening into brownish lumps (manna), some effective laxatives

Alibertia A. Rich. ex DC. Rubiaceae (II 1). 35 trop. Am. Some ed. fr.

Aliciella Brand = *Gilia*

Aliella Qaiser & Lack (~ *Phagnalon*). Compositae (Gnap.) 3 Morocco mts. R: BJ 106(1986)488

Alifana Raf. An older name for *Brachyotum* (q.v.)

Aligera Suksd. Valerianaceae. 15 W N Am.

aligna *Afzelia* spp.

Aliniella Raynal = *Alinula*

Alinula Raynal. Cyperaceae. 4 C & E Afr. R: BJBB 58(1988)457

Aliopsis Omer & Qaiser. Gentianaceae. 1 C As.: *A. pygmaea* Omer & Qaiser, 1–5cm tall. R: Willd. 21(1991)190

Alisma L. Alismataceae. 9 N temp. (Eur. 4), Aus. (2). R: AMN 58(1957)470. Water plantains, esp. *A. plantago-aquatica* L. (N temp.) sometimes cult. orn., medic. & ed., though base believed poisonous

Alismataceae Vent. Monocots – Alismatidae – Alismatales. (Incl. Limnocharitaceae) 14/100 cosmop. esp. N temp. Perennial (annual in seasonal water), usu. glabrous aquatics (some free-floating) & marsh pls with rhizomes, vessels only in roots, laticifers present. Lvs spirally arr., often dimorphic, juvenile linear submerged & mature linear to ovate or sagittate emergent, usually with petiole, sheathing at base. Infls usu. thyrsoid with whorls of branches or umbel-like. Fls hermaph. to unisexual (dioec. in *Burnatia*), solit. or in cymose umbels. K 3; C 3, usu. white, ephemeral, occ. 0 in female fls of *Burnatia* & *Wiesneria*; A 3 or 6–∞, when 6 in pairs next to petals, when ∞ in dense whorls, developing centripetally (centrifugally in *Butomopsis*, *Hydrocleys* & *Limnocharis* s.t. separated as Limnocharitaceae), anthers extrorse, pollen 9–29-porate, rarely 2(3)-porate (*Caldesia*); $\underline{G}$ 3 or 6–∞, each carpel (sometimes distally unclosed) with 1 (rarely more) anatropous to campylotropous bitegmic ovules. Fr. a head of achenes (or follicles), usu. dehiscent; seeds without endosperm, embryo horseshoe-shaped. x = 5–13

Chief genera: *Alisma*, *Echinodorus*, *Sagittaria*. Generic distinctions unsatisfactory esp. *Baldellia*, *Caldesia*, *Ranalisma* & *Echinodorus*

Family form. considered allied to Ranunculaceae but similarities are superficial

Some cult. orn. & ed. lvs

Alistilus N.E. Br. Leguminosae (III 10). 2 S trop. Afr., Madag.

alizarin *Rubia tinctorum*

alkali grass (Am.) *Distichlis*, *Puccinellia* or *Zigadenus* spp.

alkaloids small organic compounds, many derived from amino acids, & containing heterocyclic nitrogen. Some are deterrent to animals & several are of importance in medicine: they are also involved in moving nitrogen around the plant. Generally named after the genus of the species where originally found, e.g. digitalin (*Digitalis*), nicotine (*Nicotiana*), strychnine (*Strychnos*). Colchicine (from *Colchicum*) is used in plant breeding to double chromosome numbers somatically: it inhibits spindle functioning in mitosis so that chromosomes do not separate to form 2 separate nuclei, but remain together, forming 1

alkanet *Alkanna lehmannii*, *Pentaglottis sempervirens*; **bastard a.** *Lithospermum arvense*

Alkanna Tausch. Boraginaceae. 25–30 S Eur. (17), Med. to Iran. *A. lehmannii* Tineo (*A. tinctoria*, *A. tuberculata*, Med., alkanet) – roots source of red dye, form. used to tint inferior port and colour thermometer fluids, chemists' shop bottles and to detect fats (now replaced by Sudan IV etc.)

Allaeophania Thwaites = *Hedyotis*

Allagopappus Cass. Compositae (Inul.). 2 Canary Is.

Allagoptera Nees. Palmae (V 5b). 5 Brazil, Parag. Some cult. orn.

Allamanda L. Apocynaceae. 12 trop. Am. R: RBB 9(1980)125. Seeds hairy. *A. cathartica* L. (S Am.) – widely cult. orn. liane

Allanblackia Oliver ex Benth. Guttiferae (III). 10 trop. Afr. Seeds a source of oils, used as butter subs. during World War II. *A. floribunda* Oliver (W Afr., kisidwe), *A. oleifera* Oliver (C Afr., kagné butter), *A. stuhlmannii* (Engl.) Engl. (E Afr., mkani fat), bark a source of red dye

Allantoma Miers. Lecythidaceae. 1 NE S Am.

Allantospermum Forman. Ixonanthaceae. 1 Mal., 1 Madag. Sometimes placed in Simaroubaceae

Allardia Decne (~ *Waldheimia*). Compositae (Anth.-Can.). 8 C & E As.

Alleghany blackberry *Rubus allegheniensis*

Alleizettea Dubard & Dop = *Danais*

Alleizettella Pitard. Rubiaceae (II 4). 1 SE As.

Allemanda L. = *Allamanda*

Allenanthus Standley. Rubiaceae (III 4). 2 C Am.

Allendea Llave & Lex. = *Liabum*

Allenrolfea Kuntze. Chenopodiaceae (II 2). 3 SW US

Allexis Pierre. Violaceae. 3 trop. W Afr. *A. cauliflora* (Oliver) Pierre has explosive 1-seeded fr.

allgood *Chenopodium bonus-henricus*

allheal *Stachys palustris*, *Valeriana officinalis*

Alliaceae Batsch ex Borkh. (Liliaceae s.l.). Monocots – Liliidae – Asparagales. 30/850 subcosmop. excl. Aus. Perennial herbs with bulb or corm, rarely rhiz. Lvs spirally arr. or distichous (e.g. *Agapanthus*), usu. lanceolate to linear, s.t. fistulose. Raphides or allylic sulphides (onion smell) present. Infl. with terete scape with umbel of condensed helicoid cymes (raceme in *Milula*) subtended by involucre of (1)2+ bracts s.t. forming a calyptra in bud. Fls 3-merous, usu. bisexual, reg. (irreg. in many Gillesioideae); P 3 + 3; A 3 + 3 (s.t. all but 2 or 3 staminodal) borne at P bases, anthers with longit. dehiscence; G(3) with 1 (trilobed) style, each locule with 2–∞ campylotropous to anatropous ovules. Fr. a loculicidal capsule with few to ∞ often angled seeds

Classification & principal genera

Agapanthoideae (rhizomes, S Afr.): *Agapanthus, Tulbaghia*

Allioideae (bulbs or corms with membranous or fibrous coats, fls reg., subcosmop. esp. Am.)

Brodiaeeae (infl. subtended by 3+ bracts, A 3 + 3 (3 s.t. staminodes), Am. esp. N): *Brodiaea, Leucocoryne, Triteleia*

Allieae (infl. subtended by (1)2 bracts, A 3 + 3): *Allium, Nothoscordum, Tristagma*

Gillesioideae (bulbs, infl. subtended by 2 bracts, P ±unequal, up to 4 staminodes, filaments usu. with tube or cup, S Am. esp. Chile, mostly with 1 or 2 spp.): *Gilliesia*

Distinct from Amaryllidaceae in possession of saponins and perhaps close to Hyacinthaceae though there are many similarities with certain genera of Anthericaceae, demonstrating that the fragmenting of Liliaceae s.l. has raised new problems

Many cult. orn. and foodpls (*Allium*)

Alliaria Heister ex Fabr. Cruciferae. 2 Eur. (1), temp. As. R: JAA 69(1988)218. *A. petiolata* (M. Bieb.) Cavara & Grande (Eur., Jack-by-the-hedge, garlic mustard, hedge garlic) – form. used as flavouring (higher in vitamin A than spinach & in C than orange juice)

alligator apple *Annona* spp.; **a. pear** *Persea americana*; **a. pepper** *Aframomum melegueta*; **a. weed** *Alternanthera philoxeroides*; **a. wood** *Guarea guidonia*

Allionia L. Nyctaginaceae (III). 2 C & W USA to Chile & Arg. Fr. glandular like *Pisonia*

Allioniella Rydb. = *Mirabilis*

Allium L. Alliaceae (Liliaceae s.l.) c. 690 Euras.(Eur. 115 excl. *Nectaroscordum*) esp. C As., Am., to Afr., Sri Lanka & Mex. (R (sect. *A.*): B. Mathew (1996) *A review of A. sect. A.*). Mostly perenn. bulbous or rhiz. herbs, with strong smell (largely aliphatic disulphides) when bruised, delaying growth of salmonella in meat noted by Pasteur (1858); hybrids almost unknown. Many cult. ed. (e.g. *A. altaicum* Pallas (Siberia, Mongolia), *A. ramosum* L. (C As.) & *A. victorialis* (Euras.) in Mongolian markets) & orn. (D. Davies (1992) *The ornamental onions*). *A. ampeloprasum* L. (Euras., N Afr.) – 3 Groups: Ampeloprasum = Levant garlic, large bulbs used as seasoning, Porrum = leek, narrow bulbs & leaf-bases eaten (15 000 t prod. per annum in UK alone), known to Sumerians, infusion used to wash windows deters flies (saponins effective against leek moth larvae), that of seeds as a kind of Anglo-Saxon toothpaste, outer layers of leeks giving a hair-bleach, national emblem of Wales, commemorating victory of Cadwallader over Saxons (AD 640), when worn as distinguishing markers, Kurrat = kurrat, lvs ed.; *A. ascalonicum* – name misapplied to shallot; *A. canadense* L. (Canada garlic) – bulbs form. consumed by Am. Indians; *A. cepa* L. (cultigen, onion, poss. derived from *A. oschaninii* B. Fedtsch. but known from Dead Sea Chalcolithic, (c. 3500–3000 BC) – 3 groups: Cepa = onion, single bulbs to over 5 kg without bulbils in infl. (6 types grown at time of Pliny) – world prod. 12m tons per annum (esp. Med., Japan, US), much dehydrated or powdered as well as marketed fresh, damaged tissue releasing sulphurous volatiles (hence weeping – avoided by breathing through nose only), 10 compounds interacting to inhibit platelet aggregation & thereby blood-clotting & thrombosis, Aggregatum = shallot ('*A. ascalonicum*' the original 'scallion', a name now referred to 'spring onions' (young Cepa) or (US) leek), multiplier or potato o., produces lateral bulbs & none in infl. (often sterile), widely used as pickled o., Proliferum (? hybrid origin involving *A. fistulosum*) = Egyptian or tree o., prop. by bulbils in infl., all groups with some med. uses (extracts antibacterial); *A. cernuum* Roth (lady's leek, N Am.) – ed. (strong); *A. chinense* G. Don f. (chiao t'ou (China), rakkyō (Japan), As.) – pickled; *A. fistulosum* L. (Welsh (i.e. German *welsche*, foreign) o., Japanese bunching o., cultigen natur. on turf roofs in Norway) – ed. lvs for salads; *A. moly* L. (moly, S Eur.) – cult. orn. yellow fls; *A. oleraceum* L. (field garlic, Eur.) – occ. gathered for food; *A. sativum* L. (garlic, cultigen, poss. derived from *A. longicuspis* Regel (C As.) but known from Dead Sea Chalcolithic, c. 3500–3000 BC and 2/3 types cult. at time of Pliny) – strong flavour assoc. with French & Iberian (100 000 t prod. per annum in Spain alone) cooking also in chewing-gum and garlic salt for cooking, bactericidal & much used med. (e.g. toothache in Sumatra, insect bites in S Eur.) since time before Galen, against infections but allicin also effective in circulatory & respiratory disorders & (?) tumours, and in hort. (since time of Pliny) where effective against foot-rot of runner beans, also in superstition worn round neck to ward off Trolls etc. (var. *ophioscorodon* (Link) Doell is true rocambole); *A. schoeno-*

prasum L. (chives, Euras.) – lvs used in cooking; *A. scorodoprasum* L. (sandleek, Euras.) – bulbs ed.; *A. sphaerocephalon* L. (round-headed garlic or leek, Eur. & Med.) – bulbs ed.; *A. tuberosum* Rottler ex Sprengel (garlic chives or Chinese c., SE As.) – ed.; *A. ursinum* L. (ramsons, wild or wood garlic, Euras.) – medic., lvs cross over in development so that 'lower' surfaces are on top; *A. vineale* L. (crow or false garlic, Eur.) – bad weed, molluscicidal steroid saponins

Allmania R. Br. ex Wight. Amaranthaceae (I 2). 1–2 trop. As.

Allmaniopsis Süsseng. Amaranthaceae (I 2). 1 E Kenya

Alloburkillia Whitm. = *Burkilliodendron*

Allocalyx Cordemoy = *Monocardia*

Allocarya E. Greene = *Plagiobothrys*

Allocaryastrum Brand = *Plagiobothrys*

Allocassine N. Robson. Celastraceae. 1 SE trop. & S Afr.: *A. laurifolium* (Harvey) N. Robson

Allocasuarina L. Johnson (~ *Casuarina*). Casuarinaceae. 58 Aus. (esp. S). R: JABG 6(1982)73. Dioec. or monoec. Sheoaks (some); *A. luehmannii* (R. Baker) L. Johnson (E Aus.) – bull oak

Allocheilos W.T. Wang. Gesneriaceae. 1 SW China: *A. cortusiflorum* W.T. Wang

Allochilus Gagnepain = *Goodyera*

Allochrusa Bunge ex Boiss.(~ *Acanthophyllum*). Caryophyllaceae (III 1). 7 W & S As.

Alloeochaete C.E. Hubb. Gramineae (25). 6 S trop. Afr. R: KB 30 (1975)570. *A. oreogena* Launert (Mt Mlanje) with tree-trunk-like tussocks to 1.5 m

Alloispermum Willd. (~ *Calea*). Compositae (Helia.-Gal.). 7 Am. R: Phytol. 38(1978)411, 68(1990)134

Allolepis Söderstrom & Decker. Gramineae (31c). 1 S US & Mex.

Allomaieta Gleason. Melastomataceae. 1 Colombia

Allomarkgrafia Woodson. Apocynaceae. 6 trop. Am. R: AMBG 76(1989)924

Allomorphia Blume (~ *Oxyspora*). Melastomataceae. 25 SE As.

Alloneuron Pilger (~ *Loreya*). Melastomataceae. 7 S Am.

Allophyllum (Nutt.) A.D. & V. Grant. Polemoniaceae. 5 W US. R: Aliso 3(1955)93

Allophylus L. Sapindaceae. 1 very polymorphic sp., *A. cobbe* (L.) Räusch., trop. & warm, which at a local level can be subdivided into c. 175 apparently distinct biological spp. R: Blumea 15(1967)301. Cult. orn., useful wood, somewhat medic., fr. used to stun fish in New Ireland

Alloplectus C. Martius. Gesneriaceae. 65 trop. Am. Cult. orn. close to *Columnea*

Allopterigeron Dunlop. Compositae (Pluch.). 1 trop. Aus.

Allosanthus Radlk. Sapindaceae. 1 Peru

Alloschemone Schott (~ *Scindapsus*). Araceae (III 2). 1 Brazil

Alloschmidia H. Moore. Palmae (V 4m). 1 NE New Caled.: *A. glabrata* (Becc.) H. Moore. Allied to *Clinostigma*

Allosidastrum (Hochr.) Krapov., Fryx. & D. Bates (~ *Pseudabutilon*). Malvaceae. 4 trop. Am. R: BSBM 48(1988)23

Allostigma W.T. Wang. Gesneriaceae. 1 S China: *A. guangxiensis* W.T. Wang

Allosyncarpia S.T. Blake. Myrtaceae (Lept.). 1 N Aus.

Alloteropsis J. Presl. Gramineae (34b). 6 OW trop. R: SB 13(1988)587. *A. semialata* (R. Br.) Hitchc. – diploid (2n = 18) C_3, hexaploid (2n = 64) C_4!

Allotropa Torrey & A. Gray. Ericaceae (Monotropaceae). 1 W US

Allowissadula Bates. Malvaceae. 9 Texas & Mex. R: GH 11(1978)329

Allowoodsonia Markgraf. Apocynaceae. 1 Solomon Is.

Alloxylon P. Weston & Crisp (~ *Oreocallis*). Proteaceae. 4 New Guinea & E Aus. (3 endemic – satin oak)

allseed *Radiola linoides*; **four-leaved a.** *Polycarpon tetraphyllum*

allspice *Pimenta dioica, Chimonanthus praecox*; **Calif. a.** *Calycanthus occidentalis*; **Carolina a.** *C. floridus*, **a. jasmine** *Gelsemium* spp.; **wild a.** *Lindera* spp.

allthorn *Koeberlinia spinosa*

Alluaudia (Drake) Drake. Didiereaceae. 6 S & SW Madag.

Alluaudiopsis Humbert & Choux. Didiereaceae. 2 S & SW Madag.

Almaleea Crisp & P. Weston (~ *Pultenaea*). Leguminosae (III 24). 5 SE Aus. R: Telopea 4(1991)307

Almeidea A. St-Hil. Rutaceae. 6 NE S Am.

almeidina see *Euphorbia tirucalli, Hevea brasiliensis*

almon *Shorea almon*

almond *Prunus dulcis*; **Barbados a.** *Terminalia catappa*; **bitter a.** *P. dulcis* var. *amara*; **country a.** (Ind.) *Terminalia catappa*; **Cuddapah a.** *Buchanania lanzan*; **dog a.** *Andira inermis*; **earth a.**

Cyperus esculentus; **Ind. a.** *Terminalia catappa*; **Java a.** *Canarium luzonicum*; **red a.** *Alphitonia* spp.; **Russian dwarf a.** *P. tenella*; **wild a.** (S Afr.) *Brabejum stellatifolium*

almondette *Buchanania lanzan*

Almutaster Löve & D. Löve = *Aster*

Alniphyllum Matsum. Styracaceae. 3 SW China, Taiwan, SE As. Cult. orn.

Alnus Miller. Betulaceae (I). 25 N temp. (Eur. 4), S to Assam, SE As. & Andes. R: JAA 71(1990)18. Alders: useful timber, roots with nitrogen-fixing actinomycete symbionts, twigs accumulate gold, pollen significant in hay-fever, cult. orn. Male catkins, females short, becoming woody cones with 5-lobed scales & minute winged nutlets. Fls before lvs in many spp. but several, notably trop. spp., flower in autumn when lvs still present (cf. *Quercus*). Planted spp. include: *A. acuminata* Kunth (C Am.) – fuelwood; *A. cordata* Desf. (Italian a., Corsica & S Italy); *A. crispa* (Aiton) Pursh (green a. (Am.), E N Am.); *A. glutinosa* (L.) Gaertner (black, common or Eur. a., Euras., Med.) – wood for carving &, allegedly, Stradivarius violins & form. clogs (wood a poor conductor of heat) & gunpowder charcoal, for building scaffolding in Elizabethan times & (Virgil) for the first boats, the piles of Venice (durable under water), bark for tanning (reddens leather), medic. esp. gargle, many orn. cvs; *A. incana* (L.) Moench (white or grey a., N temp.); *A. japonica* (Thunb.) Steudel (Japanese a., NE As., Japan) – charcoal for gunpowder; *A. rubra* Bong. (*A. oregona*, Oregon or red a., W N Am.) – said to be best for fish-smoking, canoes; *A. viridis* (Chaix) DC (green a. (Eur.), Eur.)

Aloaceae Batsch ('Aloeaceae') = Asphodelaceae

Alocasia (Schott) G. Don f. Araceae (VII 4). 70 Indomal. to Aus. (13, R: Blumea 35(1991)499). Differs from *Colocasia* in having large sterile appendage & basal ovules. Some ed. rhiz., cult. orn. with many hybrids. *A. macrorrhizos* (L.) G. Don f. (giant taro, cunjevoi, Indomal.) – cult. for ed. rhiz. & orn., lvs to 3.02 x 1.92m, largest undivided lvs known, sap an antidote to *Dendrocnide gigas* stings; *A. veitchii* (Lindley) Schott (*A. picta*, Borneo) – housepl.

Alococarpum Riedl & Kuber. Umbelliferae (III 5). 1 Iran: *A. erianthum* (DC) Riedl & Kuber

aloe *Aloe* spp. (of Psalms = lign-aloes, q.v.); **Am. a.** *Agave americana*; **Bombay a.** *A. cantala*; **Cape a.** *Aloe ferox*; **Curaçao a.** *A. vera*; **partridge-breasted a.** *A. variegata*; **a. wood** *Aquilaria malaccensis* (= aloe (ahaloth) of Bible, 'aloexylum')

Aloe L. Asphodelaceae (Liliaceae s.l.). c. 365 trop. & esp. S Afr., Madag., Arabia, Canary Is. R: Excelsa 9(1980)57. Branched or unbranched trees to sessile pls with term. rosettes of succulent, non-fibrous (cf. *Agave*) lvs, usu. toothed or spiny at margin, & infls of yellow or red bird-poll. fls. Cult. orn. Hybrids with *Gasteria* spp. known. Yellow juice of lvs dried = drug (bitter) aloes, purgative (active principle an anthraquinone), esp. from *A. ferox* & *A. vera*. *A. bainesii* Dyer (S Afr.) – tree to 18m, though some with toxic hemlock alks, e.g. *A. ballyi* Reynolds (E Afr.); *A. ciliaris* Haw. (S Afr.) – runner-like stems, scrambling; *A. ferox* Miller (Cape aloe, S Afr.) – bird- & bee-poll., source of Cape aloes; *A. humilis* (L.) Miller (S Afr.) – sessile cult. orn., several cvs; *A. saponaria* (Aiton) Haw. (*A. latifolia*, S Afr.) – used in Afr. treatment of boils etc., form. for tanning; *A. variegata* L. (partridge-breasted aloe, S Afr.) – fav. housepl. with varieg. lvs; *A. vera* (L.) Burm.f. (*A. barbadensis*, Barbados or Curaçao a., SW Arabia, long natur. Medit., intr. warm Am.) – comm. source of aloes esp. in Carib., lvs used in shampooing, constituent of many cosmetics, mentioned in Babylonian texts, antibiotic barbaloin active against tubercle bacilli & long used in skin infections in Arabia, modern remedy for burns incl. sunburn

Aloeaceae Batsch = Asphodelaceae

aloes See under *Aloe*

Aloinopsis Schwantes (~ *Nananthus*). Aizoaceae (V). 12 W & C S Afr. Cult. orn.

Aloitis Raf. = *Gentiana*

Alomia Kunth. Compositae (Eup.-Alom.). 5 Mex. R: MSBMBG 22(1987)239

Alomiella R. King & H. Robinson. Compositae (Eup.-Ayap.). 2 Brazil. R: Phytol. 56(1984)256

Alona Lindley = *Nolana*

Alonsoa Ruíz & Pavón. Scrophulariaceae. 16 trop. Am., S Afr. (2). Some with elaiophores attractive to bees. Cult. orn. pot-pls, esp. *A. warscewiczii* Regel (Peru)

Alopecurus L. Gramineae (21d). 36 N temp. (Eur. 14), S Am. Foxtail, meadow grasses esp. *A. pratensis* L. (meadow f., blackgrass, Euras., natur. N Am.)

Alophia Herbert. Iridaceae (III 4). 5 warm Am. Cult. orn. as 'Eustylus'. 'Alophia' in cult. = *Herbertia*

Aloysia Palau. Verbenaceae. 37 Am. *A. citrodora* Palau (*A. triphylla*, *Lippia c.*, lemon verbena, Arg., Chile) – cult. orn., lemon flavouring in French liqueurs and used as a fragrant sedative tea in S Am., effective in pot-pourri

alpam root *Thottea* spp.

Alpaminia O. Schulz = *Weberbauera*

alpenrose *Rhododendron ferrugineum*

Alphandia Baillon. Euphorbiaceae. 3 New Guinea, Vanuatu, New Caled.

Alphitonia Reisseck ex Endl. Rhamnaceae. 6 Mal., Aus., W Pacific. *A. petriei* Braid & C. White (NE Aus.) – partly beetle-poll. Some timbers (red almond, red ash (Aus.)) inc. *A. ponderosa* Hillebrand (Hawaii) form. used for javelins etc., sinks in water; *A. excelsa* (Fenzl) Benth. (soap tree, Aus.) – lvs yield soap subs.

Alphonsea Hook.f. & Thomson. Annonaceae. 30 China, Indomal.

alpine campion *Silene suecica*; **a. rose** *Rhododendron ferrugineum*

Alpinia Roxb. Zingiberaceae. (Excl. *Pleuranthodium*) c. 200 warm As. to Pacific. R: EJB 49(1990)1. Cult. orn. & medic. antiseptic cineol). Rhiz. scented like ginger. Some monoec. *A. galanga* (L.) Sw. (galangal, Siamese ginger, trop. As.) – source of ess. oil, rhiz. a condiment, fls ed. Java; *A. officinarum* Hance similar (E & SE As., cult. China) – form. used in some Russian tea (Nastroika) & liqueurs; *A. zerumbet* (Pers.) B.L. Burtt & R.M. Sm. ('*A. speciosa*', shell ginger) with hexagonal rhiz. system uniformly exploiting environment, cult. Hawaii

Alposelinum Pim. = *Lomatocarpa*

Alrawia (Wendelbo) Persson & Wendelbo. Hyacinthaceae (Liliaceae s.l.). 2 Iran, Iraq

Alseis Schott. Rubiaceae (I 1). 20 Mex. to Brazil & Peru

Alseodaphne Nees (~ *Dehaasia*). Lauraceae (I). 50 Yunnan to W Mal. R: Candollea 28(1973)95. Light timbers

Alseuosmia Cunn. Alseuosmiaceae. 4 NZ, some hybridizing. R: NZJB 16(1978)271. Cult. orn.

Alseuosmiaceae Airy Shaw (~ Saxifragaceae). Dicots – Rosidae – Rosales. 3/8, W Pacif. R: Blumea 29(1984)387. Shrubs, *Pittosporum*-like in habit. Lvs spiral to subopp. or in 3, 4 (or 5)-verticillate pseudowhorls, simple with hairs in axils. Fls bisexual, often incl. cleistogamous ones heavily scented; K 5, rarely 4, 6 or 7; C(5) rarely (4) or (6–7), often crenate to erose; A 5, less often 4 or 6–7, attached to C or free, alt. with C-lobes, anthers introrse; $\overline{G}$ (2) sometimes only half inferior, with nectary-disk, ovules 2–∞ per locule. Fr. a 2-loc. berry with 1–∞ endospermous seeds

Genera: *Alseuosmia, Crispiloba, Wittsteinia*

Formerly considered allied to Caprifoliaceae, which have opp. lvs & valvate corolla-lobes with free stamens & different pollen, the constituent genera have been placed in Ericaceae & Gesneriaceae etc. Wood anatomy suggests affinity with Saxifragaceae

alsike *Trifolium hybridum*

Alsinaceae (DC) Bartling = Caryophyllaceae

Alsinidendron H. Mann (~ *Schiedea*). Caryophyllaceae (II 1). 4 (1 extinct) Hawaii. Shrubs & lianes

Alsinula Dostál = *Stellaria*

Alsmithia H. Moore. Palmae (V 4m). 1 Fiji (Tavenni). Close to *Clinostigma*

Alsobia Hanst. = *Episcia*

Alsodeiopsis Oliver. Icacinaceae. 11 trop. Afr. *A. zenkeri* Engl. a rheophyte. Roots of some spp. supposed aphrodisiac

Alsomitra (Blume) M. Roemer. Cucurbitaceae. 2 Indomal. *A. macrocarpa* (Blume) Cogn. seeds winged, to 120 mm across

Alsophila R. Br. = *Cyathea*

Alstonia R. Br. Apocynaceae. c. 43 Afr. (2), C Am. (3), China, Indomal., W Pacific (New Caled. 14 (R: Adansonia 16(1976)465)). Pagoda trees with whorled lvs & many alks, some ed. fr. R: Pac. Sci. 3(1949)133. Some useful timbers and latex, used to adulterate rubber or jelutong, some med. *A. congensis* Engl. (pattern wood, alstonia, ahun, awun, trop. Afr.) – soft white wood for boats, war-drums etc.; *A.constricta* F. Muell. (bitter bark, fever bark, E Aus.) – med. & tonic, bitters; *A. scholaris* (L.) R. Br. (dita bark, OW trop.) – long used med. as vermifuge, allegedly antimalarial, wood used for coffins & formerly slates (writing rubbed off with *Tetracera* lvs in Malaysia), hence specific name, now for gaudy ('dancer's') masks in Sri Lanka; *A. spatulata* Blume (W Mal.) – very light wood used for pith helmets, floats etc., med.

Alstroemeria L. Alstroemeriaceae (Liliaceae s.l.). 50 S Am. Lvs twisted at base during development so 'upper' surface facing downwards. Roots local starch sources. Cult. orn. esp. *A. aurea* Graham (*A. aurantiaca*, Peruvian lily, Chile (!), poss. not distinct from *A. versicolor* Ruíz & Pavón) but being replaced by 'Ligtu hybrids' (*A. ligtu* L. × *A. haemantha* Ruíz & Pavón, both Chile)

Alstroemeriaceae Dumort. (Liliaceae s.l.). Monocots – Liliidae – Liliales. 4/150 C & S Am.

esp. Andes. Erect or twining herbs with vessels & symp. rhiz., some roots tuberous. Lvs lanceolate to linear, usu. twisted at base during development so 'upper' surface faces downwards. Infls umbel-like helicoid cymes. Fls 3-merous, bisexual, ± reg.; P 3 + 3, similar or outer shorter, all often dotted, with nectaries at base of 2 or all inner; A 3 + 3 with longit. introrse dehiscence; $\overline{G}$ ((1)3) with ∞ anatropous ovules on axile (*Alstroemeria, Bomarea*) or parietal placentae. Fr. usu. loculicidal caps.; seeds ± globose. x = 8 or 9

Genera: *Alstroemeria, Bomarea, Leontochir, Schickedantzia*

Cult. orn. & some ed.

Altamiranoa Rose = *Villadia*

Altensteinia Kunth. Orchidaceae (III 3). 9 Andes

Alternanthera Forssk. Amaranthaceae (II 2). c.100 trop. & warm, esp. Am. Chaff-flower, joyweed, broad path (WI). Some ed. like spinach, others much used as trop. bedding plants for spectacular lvs. *A. ficoidea* (L.) Pal. (Mex. to Arg.) – widely cult. esp. 'Bettzickiana', with blotched yellow & red lvs. ('Jacob's coat'); *A. philoxeroides* (C. Martius) Griseb. (alligator weed, S Am.) – introd. N Am. as cult. orn. & as crayfish fodder now a weed rivalling water hyacinth, also in Aus.; *A. pungens* Kunth (S Am.) – natur. Aus., where lawn weed & harmful to stock; *A. sessilis* (L.) DC (mukunawanna, trop.) – ed. like spinach

Althaea L. Malvaceae. 12 Eur. (5) to NE Siberia, S Aus. (1! cf. *Lavatera*). *A. officinalis* L. (marsh mallow, Eur., natur. in US) – source of orig. marshmallow, roots medic. & used in salads, fibres form. used in paper-making

Althenia Petit. Zannichelliaceae. 2 Medit. to Turkey. R: Lagascalia 14(1986)102. *A. filiformis* Petit (W Med.) introd. S Afr.

Althoffia Schumann = *Trichospermum*

Altingia Noronha. Hamamelidaceae (IV; Altingiaceae). 8 Indomal. Male fls with naked stamens. Good timber, medic. *A. excelsa* Noronha (rasamala, SE As. to Mal.) – to 60 m, heavy wood for beams, yellow scented resin used in scents instead of styrax, lvs ed. in Java

Altingiaceae Horan. See Hamamelidaceae

alum root *Heuchera* spp. esp. *H. sanguinea*

aluminium plant *Pilea cadierei*

alunqua *Leichardtia australis*

alva marina *Zostera marina*

Alvaradoa Liebm. Picramniaceae. 5 trop. Am. R: Brittonia 5(1944)133

Alvesia Welw. Labiatae (VIII 4c). 3 trop. Afr.

Alvimia Calderón ex Söderstrom & Londoño. Gramineae. 3 Brazil. Climbers with fleshy fr. R: AJB 75(1988)819

Alvimiantha Grey-Wilson. Rhamnaceae. 1 Brazil: *A. tricamerata* Grey-Wilson

Alvordia Brandegee. Compositae (Helia.-Helia.). 4 Calif., Mex. R: PCAS 30(1964)157

alyce clover *Alysicarpus* spp. esp. *A. vaginalis*

Alyogyne Alef. (~ *Fugosia*). Malvaceae. 4 Aus. R: Austr. Pl. 4(1966)16. Cult. orn.

Alysicarpus Necker ex Desv. Leguminosae (III 9). 25 OW trop., 1 intr. NW. Alyce clover, good fodder esp. *A. vaginalis* (L.) DC (Indomal. natur. pantrop.) – as nutritious as alfalfa

Alyssoides Miller. Cruciferae. 3 S & SE Eur.(2), Turkey. Cult. orn.

Alyssopsis Boiss. Cruciferae. 1 Iran: *A. mollis* (Jacq.) O. Schulz

alyssum *Lobularia maritima*; **hoary a.** *Alyssum alyssoides*; **sweet a.** *L. maritima*

Alyssum L. Cruciferae. (Incl. *Ptilotrichum*) 190 Med. to Siberia (Eur. 70, Balkans 45 with 20 endemics; Turkey 90 with 50 endemics). Some cult. orn. Many accumulate nickel, 46 to a level of more than 1000 μg per g dry matter (all in sect. *Odontarrhena* & almost all E Med. & Turkey). *A. alyssoides* (L.) L. (hoary alyssum, Eur., natur. N Am.), *A. spinosum* L.(*Ptilotrichum s.*, S France), shrubby – cult. orn.

Alyxia Banks ex R. Br. Apocynaceae. 120 Indomal. to W Pac. (Madag. spp. = *Cabucala*). Some medic., also cult. for fls & bark to scent clothes. New Caled. spp. accumulate manganese, up to 1.15% dry weight in 1 sp.

Alzatea Ruíz & Pavón. Crypteroniaceae (Alzateaceae). 1 Costa Rica to Bolivia

Alzateaceae S. Graham = Crypteroniaceae

amacha drink prepared from *Hydrangea macrophylla* subsp. *serrata* in Japan

Amaioua Aublet. Rubiaceae (II 1). 25 trop. S Am.

Amalocalyx Pierre. Apocynaceae. 3 SE As.

Amalophyllon Brandegee. Scrophulariaceae. 1 Mex.

Amana Honda = *Tulipa*

Amanoa Aublet. Euphorbiaceae. 16 trop. Am. (13, R: Brittonia 42(1990)260), 3 trop. Afr. &

Madag. Some fish-disp. in Amazon

Amaraboya Linden ex Masters = *Blakea*

Amaracarpus Blume. Rubiaceae (IV 7). 60 Mal. to Micronesia

Amaracus Gled. = *Origanum*

amaranth (timber) *Peltogyne* spp.; **globe a.** *Gomphrena globosa*

Amaranthaceae Juss. Dicots – Caryophyllidae – Caryophyllales. 71/750 trop. & warm, few temp. Herbs (mostly) to climbers (e.g. *Stilbanthus*), shrubs or rarely trees (e.g. *Charpentiera*) with sec. thickening of concentric rings of vascular bundles, producing betalains but not anthocyanins nor tannins. Lvs spiral or opp., simple, usu. entire, often with Kranz anatomy. Fls bisexual or unisexual (pls dioec., polygamous or monoec.) small, solitary or (usu.) in cymose or thyrsoid infls often with bristly bracts and bracteoles, these sometimes conspicuously pigmented, wind- or insect-poll., rarely unisex. P 3–5 rarely 0–2, rarely connate at base, usu. falling with fr.; A usu. same no. as P & opp. them, usu. connate at base into a tube, often with teeth or lobes; nectary often a ring within tube; $\underline{G}$ 1-loc., with 1 style & 1 basal (rarely apical), campylotropous, bitegmic ovule, rarely several (e.g. *Celosia*) on a basal short placenta. Fr. an irreg. rupturing capsule, rarely a berry; seeds with abundant starchy perisperm, true endosperm ± absent. x = 8–13, 17+

Classification & principal genera:

I. **Amaranthoideae** (A 4-loc., 1–many ovules)
 1. **Celosieae** (ovules numerous to many): *Celosia*
 2. **Amarantheae** (ovule 1): *Amaranthus, Ptilotus*

II. **Gomphrenoideae** (A 2-loc., 1 ovule)
 1. **Pseudoplantageae** (fert. fls subtended by 1 or 2 spinous modified fls): *Pseudoplantago*
 2. **Gomphreneae** (fls not so): *Alternanthera, Gomphrena, Iresine, Pfaffia*

Probably should include Chenopodiaceae where only solitary ovules found

Some cult. orn. & grain crops (*Amaranthus*)

Amaranthus L. Amaranthaceae (I 2). 60 trop. & temp. (Eur: 2 native, many NW natur). Coarse ann. herbs – weeds, cult. orn., grain crops, some used like spinach (bhaji). *A. albus* L. (pigweed, US, natur. subcosmop.) – a tumbleweed; *A. caudatus* L. (love-lies-bleeding, Inca wheat, cat-tail, tumbleweed, trop., ? derived from *A. quitensis* Kunth, S Am.) – cult. orn., protein-rich seeds (high in lysine) & lvs ed. in Andes but displaced by colonists' cereals; *A. cruentus* L. (*A. paniculatus*, prob. a cultigen close to *A. hybridus*) – grain ed.; *A. hybridus* L. (inc. *A. hypochondriacus*, prince's feather) – ed. lvs & seeds (esp. form. Mex.), cult. orn. esp. var. *erythrostachys* Moq. with red spikes; *A. retroflexus* L. (N Am., natur. subcosmop. (pigweed)) – weed; *A. tricolor* L. (trop.) – cult. orn. (Joseph's coat, tampala), potherb

× **Amarcrinum** Coutts = *Amaryllis* × *Crinum*

× **Amaristetes** Hannibal = *Amaryllis* × *Cybistetes*. Poss. correct name for '× *Amarygia*' hybrids

Amaroria A. Gray = *Soulamea*

× **Amarygia** Cif. & Giac. = *Amaryllis* × *Brunsvigia*. Cult. orn. esp. × *A. parkeri* (W. Watson) H. Moore (× *A. bidwillii*) allegedly *Amaryllis belladonna* × *Brunsvigia* sp., though poss. truly × *Amaristetes*

Amaryllidaceae J. St-Hil. (Liliaceae s.l.). Monocots – Liliidae – Asparagales. 65/725 trop. & warm to Eur., esp. S Afr. & Andes. Perennial herbs with bulbs, rarely rhiz. (e.g. *Clivia, Scadoxus*); roots contractile, with vessels. Lvs usu. flat, sheathing at base, usu. glabrous. Infls term., umbel-like comprising condensed helicoid cymes. Fls 3-merous, bisexual, usu. reg. (irreg. in e.g. *Sprekelia*); P 3 + 3, sometimes atop a tube, corona present in *Narcissus*; A 3 + 3 (3 in *Zephyranthes*, up to 18 in *Gethyllis*) inserted at P bases, filament bases sometimes united by a staminal 'corona' (*Hymenocallis, Pancratium* etc.), anthers longit. dehiscent (apical pores in *Galanthus, Leucojum* etc.); $\overline{G}$ (3), 3-loc., each loc. with several to ∞ ± anatropous ovules with (0–)2 integuments. Fr. a capsule or berry

Principal genera: *Brunsvigia, Crinum, Cyrtanthus, Haemanthus, Hippeastrum, Hymenocallis, Stenomesson, Zephyranthes*

Form. placed with fams also with inferior G now believed to be parallelism

Many cult. orn. (all genera above, *Amaryllis, Ammocharis, Caliphruria, Cybistetes, Eucharis, Habranthus, Narcissus, Nerine, Proiphys, Sternbergia, Worsleya* etc.) incl. intergeneric hybrids

Amaryllis L. Amaryllidaceae (Liliaceae s.l.). 1. S Afr. (belladona lily: *A. belladonna* L.) – cult. orn. fls in autumn without lvs, signifying 'Pride' in Language of fls. '*Amaryllis*' of greenhouses etc. = hybrids of *Hippeastrum* (q.v.) and recent efforts to restore the generic name *Amaryllis* to them have failed; **blue a.** *Worsleya procera*

Amasonia L.f. Labiatae (II, Verbenaceae). 8 trop. Am. Lvs spirally arr.
Amatlania Lundell = *Ardisia*
amatungulu *Carissa bispinosa*
Amauria Benth. Compositae (Hele.-Per.). 3 SW N Am. R: Madrono 21(1972)516. Pappus 0
Amauriella Rendle = *Anubias*
Amauriopsis Rydb. = *Bahia*
Amauropelta Kunze = *Thelypteris*
amazakoué *Guibourtia ehie*
Amazon lily *Eucharis grandiflora*
ambal *Phyllanthus emblica*
ambarella *Spondias cytherea*
ambatch *Aeschynomene elaphroxylon*
Ambavia Le Thomas. Annonaceae. 2 Madagascar
Ambelania Aublet. Apocynaceae. 14 trop. S Am. R: AUWP 87-1(1987)23. Disk 0
amber fossil resins from largely extinct conifers esp. in Baltic region, occ. preserving insects caught in the fresh exudation; **a. bell** *Erythronium americanum*
Amberboa (Pers.) Less. (~ *Centaurea*). Compositae (Card.-Cent.). 6 Med. (Eur. 1) to C As. *A. moschata* (L.) DC (*C. moschata*, sweet sultan, E Med.) – cult. orn. annual with yellow, white, pink or purple scented fls
ambila *Pterocarpus angolensis*
Amblostoma Scheidw. = *Encyclia*
Amblyanthe Rauschert = *Dendrobium*
Amblyanthopsis Mez. Myrsinaceae. 2 Himal.
Amblyanthus A. DC. Myrsinaceae. 4 E Himal., New Guinea
Amblygonocarpus Harms. Leguminosae (II 3). 1 trop. Afr. savannas: *A. andongensis* (Oliver) Exell & Torre – timber (banga wanga), cult. orn.
Amblygonum (Meissner) Reichb. = *Polygonum*
Amblynotopsis J.F. Macbr. = *Antiphytum*
Amblynotus (A. DC) I.M. Johnston. Boraginaceae. 1 W Siberia to Mongolia
Amblyocalyx Benth. = *Alstonia*
Amblyocarpum Fischer & C. Meyer. Compositae (Inul.). 1 Caspian
Amblyolepis DC (~ *Hymenoxys*). Compositae (Hele.-Gai.). 1 Texas, Mex.: *A. setigera* DC
Amblyopappus Hook. & Arn. Compositae (Hele.-Bac.). 1 Calif., NW Mex., Peru, Chile
Amblyopetalum (Griseb.) Malme = *Oxypetalum*
Amblyopyrum Eig = *Aegilops*
Amblystigma Benth. Asclepiadaceae (III 1). 7 Bolivia, Arg. Corona 0
Ambongia Benoist. Acanthaceae. 1 Madagascar
Amborella Baillon. Amborellaceae. 1 New Caled.: *A. trichopoda* Baillon
Amborellaceae Pichon. Dicots – Magnoliidae – Magnoliales. 1/1 New Caled. Dioec. everg. shrub accumulating aluminium; wood without vessels. Lvs spiral, distichous later, simple, stipules 0. Fls in axillary cymose infls, unisexual, hypogynous to somewhat perigynous; P 5–8, spiral, weakly adnate at base; A $\pm\ \infty$ in several cycles, outer ones basally adnate to P, ± laminar, anthers dehiscing by slits, pollen without apertures; $\underline{G}$ 5 or 6, not closed at tips, ovule solit., marginal, anatropous. Fr. of separate drupe-like carpels, seeds with copious endosperm and minute basal embryo. 2n = 26
Only species: *Amborella trichopoda*
Orig. incl. in Monimiaceae. Virtual absence of ethereal oil cells confirms Magnoliales placement but cladistic study suggests move to Illiciales
Amboroa Cabrera. Compositae (Eup.-Crit.). 2 Bolivia, Peru. R: MSBMBG 22(1987)376
Amboyna wood *Pterocarpus* spp. esp. *P. indicus*
Ambrella Perrier. Orchidaceae (V 16). 1 Madagascar
ambrette seed *Abelmoschus moschatus*
ambrevade *Cajanus cajan*
Ambroma L.f. = *Abroma*
Ambrosia L. Compositae (Helia.-Amb.). (Incl. *Franseria*) 43 cosmop. (Eur. 1) esp. Am. R: JAA 45(1964)401. Ragweeds, imp. cause of hay-fever through copious pollen, molluscicides, fls used in liqueurs. Capitula 1-fld, unisexual; fr. encl. in involucre. *A. artemisiifolia* L. (hogbrake, N Am.) – poss. oil source; *A. maritima* L. (Medit.) – molluscicidal sesquiterpene lactones used on field scale, flavour in liqueurs; *A. peruviana* Willd. (warm Am.) – green dye in Peru; *A. psilostachya* DC (W N Am.) – medic. tea prepared by Indians, plant inhibits growth of nitrogen-fixing bacteria, algae & poss. some grasses through allelopathy; *A. tenuifolia* Sprengel – a fertility control in Uruguay

Ambrosina Bassi. Araceae (VIII 6). I Medit. (inc. Eur.): *A. bassii* L. – elaiosomes

Amburana Schwacke & Taubert. Leguminosae (III 2). 1–2 trop. S Am. Timber (amburana, umburana) & volatile oil from seeds, which contain coumarin

Ameghinoa Speg. Compositae (Mut.-Nass.). 1 arid Patagonia, Arg. Shrub

Amelanchier Medikus. Rosaceae (III). 33 N temp. (1 native in Eur.). R: BSBF 122(1975)243, CJB 68(1990)2231. Fr. ed. N Am., sarvis or servis berry eaten overripe (cf. *Mespilus*). Germ. of seeds of *A. arborea* (Michaux f.) Fern. (E N Am.) enhanced after passage through cedar waxwing. Cult. orn. (shad, service, juneberry) esp. *A. lamarckii* F. Schroeder (*A. laevis* of gardens, E N Am., natur. SE England, apomict poss. microsp. of hybrid origin) & *A. rotundifolia* (Lam.) Dum-Cours. (*A. ovalis*, *A. vulgaris*, snowy mespilus, C & S Eur. mts)

Amellus L. Compositae (Ast.-Ast.). 12 S Afr. R: MBSM 13(1977)579. *A. asteroides* (L.) Druce (*A. lychnitis*) – cult. orn.

Amentiferae old name for certain catkin-bearing families, now separated in modern systems, e.g. Salicaceae to Dilleniidae, Fagaceae to Hamamelidae

Amentotaxaceae Kudô & Yamam. = Cephalotaxaceae

Amentotaxus Pilger. Cephalotaxaceae. 4 China, SE As. R: Phytol.M 7(1984)13

American ash *Fraxinus americana*; **A. barberry** *Berberis canadensis*; **A. basswood** *Tilia americana*; **A. beech** *Fagus grandifolia*; **A. cherry** *Prunus serotina*; **A. chestnut** *Castanea dentata*; **A. cowslip** *Dodecatheon* spp.; **A. ebony** *Brya ebenus*; **A. elder** *Sambucus canadensis*; **A. elemi** *Bursera* spp.; **A. elm** *Ulmus americana*; **A. frogbit** *Limnobium spongia*; **A. grass** (WI) *Thysanolaena maxima*; **A. holly** *Ilex opaca*; **A. hornbeam** *Carpinus caroliniana*; **A. laurel** *Kalmia* spp.; **A. lime** *Tilia americana*; **A. lotus** *Nelumbo lutea*; **A. plane** *Platanus occidentalis*; **A. red gum** *Liquidambar styraciflua*; **A. senna** *Senna marilandica*; **A. spikenard** *Aralia racemosa, Maianthemum racemosum*; **A. walnut** *Juglans nigra*; **A. waterweed** *Elodea canadensis*; **A. whitewood** *Liriodendron tulipifera*; **A. wormseed** *Chenopodium ambrosioides*

Amerorchis Hultén (~ *Orchis*). Orchidaceae (IV 2). 1 N Am.: *A. rotundifolia* (Pursh) Hultén

Amerosedum Löve & D. Löve = *Sedum*

Amesiella Schltr. ex Garay. Orchidaceae (V 16). 1 Philipp.: *A. philippinensis* (Ames) Garay

Amesiodendron Hu. Sapindaceae. 1 China to W. Mal.: *A. chinense* (Merr.) Hu – furn. timber

Amethystanthus Nakai = *Isodon*

Amethystea L. Labiatae (II). 1 (*A. caerulea* L.) Turkey to Japan, cult. orn.

Amherstia Wallich. Leguminosae (I 4). 1 Burma, found only twice in wild. *A. nobilis* Wallich, cult. orn. with beautiful pink fls (orchid tree, pride of Burma). Lvs in limp flushes with brown spots, later becoming stiff and green (cf. *Brownea, Saraca*). Young lvs & fls ed.

Amianthemum A. Gray = *Zigadenus*

Amicia Kunth. Leguminosae (III 13). 7 trop. Am. *A. zygomeris* DC (Mex.) has lvs with sleep movements and protected by stipules in bud

amioki *Phyllanthus emblica*

Amischophacelus R. Rao & Kamm. = *Cyanotis*

Amischotolype Hassk. Commelinaceae (II 1d). 15 OW trop.

Amitostigma Schltr. Orchidaceae (IV 2). 10 India, E As.

Ammandra Cook (~ *Phytelephas*). Palmae (VI). 2 Colombia, Ecuador. R: OB 105(1991)41. Vegetable ivory (cf. *Phytelephas*)

Ammannia L. Lythraceae. 25 cosmop. (Eur. 2, Afr. 16, Am. 5 (R: JAA 66(1985)395), mostly wet places

Ammanthus Boiss. & Heldr. ex Boiss. = *Anthemis*

Ammi L. Umbelliferae (III 8). 3–4 Macaronesia, Med., W As. (Eur. 5). Medic. *A. majus* L. (Med.) – cult. for cut-flower trade; *A. visnaga* (L.) Lam. (S Eur., natur. N Am.) – cult. by Assyrians & ever since for med. esp. angina & asthma (visnaga), pedicels sold in Egypt for tooth-picks

Ammiopsis Boiss. = *Daucus*

Ammiaceae Small = Umbelliferae

Ammobium R. Br. Compositae (Gnap.-Cass.). 3 E Aus. R: OB 104(1991)83. *A. alatum* R. Br. cult. as 'everlasting' fl.

Ammobroma Torrey = *Pholisma*

× **Ammocalamagrostis** P. Fourn. Gramineae (*Ammophila* × *Calamagrostis*). × *A. baltica* (Schrader) P. Fourn. useful sand-binder

Ammocharis Herbert. Amaryllidaceae (Liliaceae s.l.). 4 trop. & S Afr. Alks. Cult. orn. bulbs

Ammochloa Boiss. Gramineae (17). 3 Med. (Eur. 1) to Mid. E. R: KBAS 13(1986)109

Ammocodon Standley = *Selinocarpus*

Ammodaucus Cosson & Durieu. Umbelliferae (III 3). 1 N to trop. Afr., Macaronesia: *A. leucotrichus* (Cosson & Durieu) Cosson & Durieu – often sold in markets, cult. as condiment and medic.

Ammodendron Fischer ex DC. Leguminosae (III 2). 6 C As.

Ammoides Adans. Umbelliferae (III 3). 2 Med. (Eur. 1)

ammoniacum, gum *Dorema ammoniacum*

Ammophila Host. Gramineae (21d). 3 E N Am., Eur. (1), N Afr. Hybrids with *Calamagrostis*. *A. arenaria* (L.) Link (marram, beach or mel grass – coastal Eur.) – sand-binder used for thatch, baskets, chair seats, brooms, mats etc., fibres found in 'seaballs' on Irish coast (cf. *Posidonia*)

Ammopiptanthus Cheng f. Leguminosae (III 30). 1–2 C As.

Ammopursus Small = *Liatris*

Ammoselinum Torrey & A. Gray. Umbelliferae (III 8). 3 N Am., temp. S Am.

Ammosperma Hook.f. Cruciferae. 1 N Afr.

Ammothamnus Bunge = *Sophora*

Amoebophyllum N.E. Br. = *Phyllobolus*

Amolinia R. King & H. Robinson (~ *Eupatorium*). Compositae (Eup.-Heb.). 1 Mex., Guatemala. R: MSBMBG 22(1987)401

Amomis O. Berg = *Pimenta*

Amomum Roxb. Zingiberaceae. 90 As. to Aus. (Borneo 25). Creeping aromatic rhiz., fls on separate scapes. *A. aculeatum* Roxb. (Indomal.) used to tranquillize wild bees (*Apis dorsata*) in Andamans before honey-collecting. Seeds of several spp. used like cardamom: *A. aromaticum* Roxb. (Bengal cardamom, India), *A. compactum* Sol. ex Maton (round c., Java), *A. maximum* Roxb. (Java c., Mal.) cult.

Amomyrtella Kausel. Myrtaceae (Myrt.). 1 N Arg.

Amomyrtus (Burret) Legrand & Kausel. Myrtaceae (Myrt.). 2 Chile. *A. luma* (Molina) Legrand & Kausel (cauchao, palo madrono) cult. orn.

Amoora Roxb. = *Aglaia*

Amoreuxia Moçiño & Sessé ex DC. Bixaceae (Cochlospermaceae). 3 SW US to C Am. R: BJ 101(1980)244. *A. palmatifida* Moçiño & Sessé ex DC ed. fr. & roots; *A. wrightii* A. Gray (SW US) – almost extinct

Amorpha L. Leguminosae (III 12). 15 N Am. R: Rhodora 77(1975)337. Wings & keel 0, standard folded round base of staminal tube. Cult. orn. *A. fruticosa* L. (bastard indigo) used as bedding by N Am. Indians

Amorphophallus Blume ex Decne. Araceae (VIII 5). 90 OW trop. Corms with solit. leaf produced after flowering. Cult. orn. curiosities for enormous lvs & infls or for ed. corms. Alks incl. coniine (!) must be removed by boiling. *A. bulbifer* (Roxb.) Blume (NE India) – bulbils on lvs; *A. konjac* K. Koch (*A. rivieri*, konjaku, Indonesia to Japan) – starch source as flour & comm. source of mannose for diabetic diets etc.; *A. paeoniifolius* (Dennst.) Nicolson (*A. campanulatus*, elephant yam, Indomal.) – ed. corms, 3rd after rice & maize as carbohydrate source in Indonesia; *A. titanum* Becc. (Sumatra) – carrion-beetle-poll. (eye-watering stench due to ? butyric aldehyde), leaf to 4.6 m tall & across, corm to 50 cm diam. & 50 kg and spathe to 2.4 m, growing 7.5 cm a day, often considered largest infl. of herbaceous (cf. *Corypha*) plants but *A. brooksii* Alderw. (Sumatra) alleged to have infl. incl. peduncle to 4.36 m (& corm to 70 kg)

Amorphospermum F. Muell. = *Niemeyera*

amourette *Brosimum guianense*

Ampalis Bojer ex Bureau = *Streblus*

Amparoa Schltr. Orchidaceae (V 10). 2 C Am.

Ampelamus Raf. (*Enslenia*; ~ *Cynanchum*). Asclepiadaceae (III 1). 1 N Am.: *A. albidus* (Nutt.) Britton

Ampelocalamus S.L. Chen, T.H. Wen & C.Y. Sheng (~ *Sinarundinaria*). Gramineae (1a). 2+ Himal. R: EJB 57 (1994)320

Ampelocera Klotzsch. Ulmaceae (II). 10 Mex. to Brazil. R: AMBG 76(1989)1087

Ampelocissus Planchon. Vitaceae. 100 trop. Some ed. fr. e.g. *A. abyssinica* (A. Rich.) Planchon (E Afr.) & local med.

Ampelodesma Pal. ex Benth. = *seq.*

Ampelodesmos Link. Gramineae (17). 1 Med. (inc. Eur.): *A. mauritanicus* (Poiret) T. Durand & Schinz (dis or diss grass) – grows with esparto & used for paper, fish-nets, ropes etc. locally, fodder when young

Ampelopsis Michaux. Vitaceae. 25 temp. & subtrop. Am. & As. Cult. orn. climbers with forked tendrils but no discoid tips (cf. *Parthenocissus* which is loosely referred to as '*Ampelopsis*' in hort.)

Ampelopteris Kunze = *Cyclosorus*

Ampelosycios Thouars. Cucurbitaceae. 3 Madagascar

Ampelosicyos = *praec.*

Ampelothamnus Small = *Pieris*

Ampelozizyphus Ducke. Rhamnaceae. 1 Brazil: *A. amazonicus* Ducke – bark a soap subs.

Amperea A. Juss. Euphorbiaceae. 8 Aus. esp. SW (6 endemic). R: AusJB 5(1992)1

Amphiachyris (DC) Nutt. Compositae (Ast.-Sol.). 2 C US, Texas. R: SB 4(1979)178

Amphianthus Torrey = *Bacopa*

Amphiasma Bremek. Rubiaceae (IV 1). 5–6 trop. & SW Afr.

Amphiblemma Naudin. Melastomataceae. 13 trop. W Afr. R: Adansonia 13(1973)429, 14(1974)469. Cult. orn.

Amphiblestra C. Presl = *Tectaria*

Amphibolia L. Bolus = *Eberlanzia*

Amphibolis Agardh. Cymodoceaceae. 2 off coasts of W & S Aus. Submerged marine aquatics forming 'meadows' imp. as fish-breeding grounds & origin of some 'seaballs' (cf. *Posidonia*); at least *A. antarctica* (Lab.) Asch. (to depths of 27 m) with submarine pollination (shorelines strewn with 1000s yellow-green empty anthers when male fls shed) & vivipary, fr. with 4-lobed 'comb' anchoring young seedling to substrate

Amphibologyne Brand. Boraginaceae. 1 Mex.

Amphibromus Nees (~ *Helictotrichon*). Gramineae (21b). 12 Aus. (10; R: Telopea 2(1986)715), NZ (1), S Am. (2)

Amphicarpa = *seq.*

Amphicarpaea Elliott ex Nutt. Leguminosae (III 10). 3 E As., N Am., Afr. R: SW Nat. 9(1964)207. Some have cleistogamous fls giving rise to subterr. fr. (cf. *Arachis*) esp. *A. bracteata* (L.) Fern. (*A. monoica*, hog peanut, N Am.) – seeds ed. Indians

Amphicarpon Raf. = *seq.*

Amphicarpum Kunth. Gramineae (34b). 2 SE US sandy pinewoods. *A. purshii* Kunth – geocarpic (fire-prone pine-barren habitat), self-fert. aerial (also cross-fert.) & subterr. fls

Amphicome (Royle) G. Don f. = *Incarvillea*

Amphidasya Standley. Rubiaceae (I 8). 7 trop. Am.

Amphidetes Fourn. = *Matelea*

Amphidoxa DC = *Gnaphalium*

Amphigena Rolfe = *Disa*

Amphiglossa DC. Compositae (Gnap.-Rel.). 5 S Afr. R: OB 104(1991)178

Amphilophium Kunth. Bignoniaceae. 5 C & trop. S Am.

Amphimas Pierre ex Harms. Leguminosae (III 2). 2 Afr. (Guineo-Congolian). Fleshy covering of seed prob. endocarp & not aril

Amphineurion (A. DC) Pichon = *Aganosma*

Amphineuron Holttum = *Cyclosorus*

Amphinomia DC = *Lotononis*

Amphiodon Huber = *Poecilanthe*

Amphiolanthus Griseb. (~ *Micranthemum*). Scrophulariaceae. 3 Cuba

Amphipappus Torrey & A. Gray. Compositae (Ast.-Sol.). 1 SW US. Spiny. R: AJB 30(1943)481

Amphipetalum Bacig. Portulacaceae. 1 Paraguay: *A. paraguayense* Bacig. R: Cand. 43(1988)409

Amphiphyllum Gleason. Rapateaceae. 2 Venez.

Amphipogon R. Br. Gramineae (25). 7 Aus. R: CNSWNH 1(1950)281

Amphipterum (Copel.) Copel. = *Hymenophyllum*

Amphipterygium Schiede ex Standley. Anacardiaceae (IV; Julianaceae). 4 Mex., Peru. *A. adstringens* (Schldl.) Standley (Mex.) – source of red dye

Amphirrhox Sprengel. Violaceae. 6 trop. Am.

Amphiscirpus Oteng-Yeboah (~ *Scirpus*). Cyperaceae. 1 W N Am., Arg.

Amphiscopia Nees = *Justicia*

Amphisiphon W. Barker. Hyacinthaceae (Liliaceae s.l.). 1 Cape: *A. stylosa* W. Barker.

Amphitecna Miers (~ *Dendrosicus*). Bignoniaceae. 18 trop. Am. R: FN 25(1980)50. *A. regalis* (Linden) A. Gentry (Mex.) – lvs to 1 m × 35 cm

Amphithalea Ecklon & Zeyher. Leguminosae (III 27). 20 Cape. R: OB 80(1985)

Amphitoma Gleason = *Miconia*

Amphoradenium Desv. = *Grammitis*

Amphoricarpos Vis. Compositae (Card.-Carl.). 3–4 SE Eur. (1) to Caucasus

Amphorocalyx Baker. Melastomataceae. 5 Madagascar

Amphorogyne Stauffer & Hürlimann. Santalaceae. 3 New Caled.

Amsinckia Lehm. Boraginaceae. 15 W US, W temp. S Am. R(N Am.): AJB 44(1957)529, JAA suppl. 1(1991)98. Distyly, alks; cult. orn.

Amsonia Walter. Apocynaceae. 20 N Am., Japan. R: AMBG 15(1928)379. Bluestar. Alks. Cult. orn. *A. hirtella* Standley (N Am.) – 2–5% rubber

Amydrium Schott. Araceae (III 2). 5 SE As. to Mal. R: Blumea 16(1968)123. Some cult. orn. ('*Epipremnopsis*')

Amyema Tieghem. Loranthaceae. 92 SE As. to Aus. (36, 32 endemic) & Samoa. R: Blumea 36(1992)293. *A. quandang* (Lindley) Tieghem (Aus.) only on *Acacia platycarpa* F. Muell. in Aus., reliant on just 2 bird spp., mistletoe bird & spiny-cheeked honeyeater, for disp., the latter getting most of energy requirement from the nectar & fr. (available all year) & pollinating

Amygdalus L. = *Prunus*

Amylotheca Tieghem. Loranthaceae. 4 SE As. to New Caled. (Aus. 2, 1 endemic). R: Blumea 38(1992)209

Amyrea Leandri. Euphorbiaceae. 2 Madagascar

Amyris P. Browne. Rutaceae. 40 trop. Am. (S 8; R: Cand. 46(1991)227). *A. balsamifera* L. (esp. Cuba) – timber (candlewood) also in incense & source of oil; *A. plumieri* DC (Mex.) – Yucatán elemi used in lacquers

Amyrsia Raf. = *Myrteola*

Amyxa Tieghem. Thymelaeaceae. 1 Borneo

Anabaena A. Juss. = *Romanoa*

Anabasis L. Chenopodiaceae (III 3). 42 Medit. (Eur. 4), C As. Alks (effective agent in *A. aphylla* L. (W As.) used as insecticide)

Anacampseros L. Portulacaceae. 22 S Afr. & Yemen, 1 S & E Aus. R: BJ 113(1992)471. Xerophytes with fleshy lvs; buds protected by bundles of hair (? stipules); *A. lanceolata* (Haw.) Sweet with 4 cotyledons. Cult. orn. esp. *A. telephiastrum* DC (S Afr.)

Anacampta Miers = *Tabernaemontana*

Anacamptis Rich. Orchidaceae (IV 2). 1 Eur., N Afr. to Iran: *A. pyramidalis* (L.) Rich. (pyramid orchid)

Anacantha (Iljin) Soják (*Modestia*). Compositae (Card.-Card.). 2 C As.

Anacaona Alain. Cucurbitaceae. 1 Hispaniola

Anacardiaceae Lindley. Dicots – Rosidae – Sapindales. 70/875 trop., subtrop., Med., temp. N Am. Trees, shrubs, lianes or rarely perennial herbs, with vertical resin-ducts (s.t. latex-channels) in bark, larger leaf-veins etc., the resin often allergenic. Lvs spiral, rarely opp. (e.g. *Bouea*) or whorled, pinnate or trifoliolate, less often simple, stipules 0 or vestigial (rare). Infls thyrsoid, terminal or axillary, rarely epiphyllous (females of *Campylopetalum, Dobinea*) often bisexual or usually unisexual fls (often with functionless parts of other sex), reg., usu. hypogynous. K & C usually 5-merous, s.t. more or fewer, valvate or imbricate, K usu. connate basally, P s.t. 0, K also rarely; A 10–5, rarely more or only 1 fertile, rarely connate basally, borne outside, on, or rarely in a nectary-disk, s.t. 5-lobed, or the disk a gynophore, $\underline{G}$, rarely (*Drimycarpus*) $\overline{G}$, (3–1) to (5) or rarely (12), pluriloc. or usually 1-loc., or rarely carpels discrete (then only one fertile), styles distinct or not; ovule 1 in each locule, anatropous, bitegmic or unitegmic. Fr. usually a drupe with ± resinous mesocarp; seeds with oily embryo, endosperm scanty or 0. x = 7–16

Classification & chief genera (after MNYBG 42(1987)29; *Pistacia* (dioecious, prob. naked fls, pollen features), *Blepharocarya* (opp. pinnate lvs, female fls with involucral cup like Fagaceae), *Campylopetalum* and *Dobinea* (toothed simple lvs, G 1, female fl. naked) are treated as distinct but related families by some authors: Pistaciaceae, Blepharocaryaceae and Podoaceae)

I. **Anacardieae** (usu. trees, lvs simple, A 5–10(–∞), G 1, 4–>6), trop.): 8 genera incl. *Buchanania, Gluta, Mangifera*

II. **Spondiadeae** (trees, shrubs (or lianes), lvs usu. pinnate, A = 2 x C, G ((1–)4 or 5(–12)), trop.): 17 genera incl. *Lannea, Spondias, Tapirira*

III. **Semecarpeae** (trees or shrubs, lvs simple, A = C, G (3), 1-loc. sunk in disk, trop. As. & Pacific): 5 genera incl. *Melanochyla, Semecarpus*

IV. **Rhoeae** (trees, shrubs (or lianes), lvs simple or pinnate, P sometimes 0 or 1 whorl, A variable, G (3)): 37 genera incl. *Ozoroa, Rhus, Schinus, Sorindeia, Trichoscypha*

V. **Dobineae** (shrubs or perennial herbs, lvs serrate, P 0 in female fls, A 8–10, G 1, trop. As.): *Campylopetalum, Dobinea*

The toxins from the resin seep out from the leaves during rain, though monkeys, at least, are immune to this. Alkali or antihistamine counteract the effect of the resin, but mango-eating & lacquered articles can cause a reaction in sensitive humans. Cult. orn. & fr. trees incl. sumachs (*Rhus*), cashew (*Anacardium*), pistachio (*Pistacia*), mango (*Mangifera*), *Bouea, Schinus, Spondias* etc., dyes (*Amphipterygium, Cotinus, Semecarpus*), timbers (*Antrocaryum, Astronium, Dracontomelum, Koordersiodendron, Lannea, Parishia, Swintonia* etc.), lacquer (*Gluta, Rhus*)

Anacardium L. Anacardiaceae (I). 11 trop. Am. R: MNYBG 42(1987)1, Brittonia 44(1992)331. Trees to geoxylic suffrutices with domatia. *A. occidentale* L. (cashew-nut, acajou (F.)) widely cult. In Mal. poll. by bee-flies (grasshopper parasites); G 1 giving kidney-shaped nut with hard acrid (oil irritant due to cardol & anacardic acid) pericarp around seed (promotion-nut, coffin-nail); pedicel swells into ed. pear-like body (cashew-apple, from which a drink, cajuado, is prepared in Brazil). Pericarp yields cashew-nutshell liquid (CNSL) used in brake-linings, clutches, plastic resins etc. Stem source of a gum like arabic (bookbinding – acajou gum) & bark an indelible ink from sap (held to be contraceptive); resin coll. euglossine bees for nest-building & by humans for tarring boats, fish-nets & for varnish. Seeds must be roasted & usu. sold salted. Molluscicidal alkylsalicylic acids used on field scale

Anacharis Rich. = *Elodea*

Anacheilum Reichb. ex Hoffmans. = *Epidendrum*

Anaclanthe N.E. Br. = *Babiana*

Anacolosa (Blume) Blume. Olacaceae. 16 OW trop. (As. 14, Madag. 1, Afr. 1). *A. frutescens* (Blume) Blume (*A. luzoniensis*, SE As. to C Mal.) – promising nut

Anacyclia Hoffsgg. = *Billbergia*

Anacyclus L. Compositae (Anth.-Anth.). 12 Med. (Eur. 4). R: BBMNHB 7(1979)83. Alks. Cult. orn. *A. pyrethrum* (L.) Lagasca (pellitory) medic. (*Radix Pyrethri*), mouthwashes, liqueur flavour etc.

Anadelphia Hackel. Gramineae (40g). 14 W to SE trop. Afr. R: KB 20(1966)275

Anadenanthera Speg. Leguminosae (II 3). 2 trop. Am. Cult. orn. *A. peregrina* (L.) Speg. seeds with alkaline ash give niopo or yopo, a potent hallucinogenic snuff (tryptamines & β-carbolines); *A. colubrina* (Vell. Conc.) Brenan used similarly esp. in Arg.

Anadendrum Schott. Araceae (III 1). 9 Indomal. Lvs used in curries

Anaectocalyx Triana ex Hook.f. Melastomataceae. 3 Venez. Fls 6-merous

Anagallidium Griseb. = *Swertia*

Anagallis L. Primulaceae. c. 28 Eur. (6), Afr. mts, S Am.(2), with 1 pantrop. Cult. orn.: *A. arvensis* L. (common or scarlet pimpernel, poor man's or shepherd's weather-glass, Eur. but widely natur.) – red fls close in dull or cold weather, form. medic., ssp. *caerulea* Hartman (*A. foemina*) has blue fls; *A. monelli* L. (blue p., Medit.); *A. tenella* (L.) L. (bog p., Eur.)

Anaglypha DC = *Gibbaria*

Anagyris L. Leguminosae (III 30). 1–2 Medit., Canaries. *A. foetida* L. form used as (dangerous) purgative, toxic

Anakasia Philipson. Araliaceae. 1 W New Guinea

Anamirta Colebr. Menispermaceae (IV). 1 Indomal.: *A. cocculus* (L.) Wight & Arn. (*A. paniculata*) – liane with furrowed bark and poisonous fr. (fish-berry, Indian berry, 'cocculus indicus') used as fish-poison, in beer to give bitter flavour & 'heady' character, & (dangerous) parasiticide (alks), also used (picrotoxin) in barbiturate poisoning & treating both head-lice & schizophrenia. Acarodomatia

Anamomis Griseb. = *Myrcianthes*

Ananas Miller. Bromeliaceae (3). 8 trop. Am. R: FN 14(1979)2051. *A. comosus* (L.) Merr. (*A. sativus*, pineapple, 'nana' of the Tupi Indians) a seedless cultigen, prob. derived from bird-poll. seeded spp. of Paraguay, but selected forms widely distrib. in pre-Columbian times. Terrestrial stem with terminal infl. which becomes a fleshy syncarp of 100–200 berry-like fr., bracts & axis, topped with a tuft of lvs, which can be used in propagation. First known to Europeans from Guadeloupe when Columbus landed in 1493, introduced to St Helena 1505, Philippines 1558, now a weed in some countries, & grown under glass in GB in early 1700s, when v. fashionable & the model for much garden statuary. Imp. crop in Hawaii (esp. spineless 'Smooth Cayenne'), Malaysia & Kenya for canning & juice & for fresh fr. export (esp. spiny 'Queen'). Fibre from lvs (crowa) strong & soft but diff. to extract, exported from Philippines & Taiwan to Spain for fine embroidery; *A. lucidus* Miller (curagua, NE S Am.) – fibre-pl.

Ananthocorus L. Underw. & Maxon = *Vittaria*

Anapalina N.E. Br. = *Tritoniopsis*

Anaphalioides auctt. = *Anaphalodes*

Anaphalis DC. Compositae (Gnap.-Cass.). 110 As., N Am. (natur. Eur.). Cult. orn. 'everlastings'. *A. margaritacea* (L.) Benth. ex C.B. Clarke (pearly e., N Am., natur. Eur. esp. on tips, where an early colonist) – form. medic., first Am. herb. pl. cult. in Eur.

Anaphaloides (Benth.) Kirpiczn. Compositae (Gnap.-Cass.). 4 NZ. Subdioec. subshrubs. R: OB 104(1991)99

Anaphyllopsis A. Hay (~ *Cyrtosperma*). Araceae (V 3). 3 trop. S Am. R: Aroid. 11(1988)25

Anaphyllum Schott. Araceae (V 3). 2 S India

Anarrhinum Desf. Scrophulariaceae. 8 Medit. (Eur. 5–6). R: D.A. Sutton, Rev. Antirrhineae (1988)249. Some cult. orn.

Anarthria R. Br. Restionaceae (Anarthriaceae). 6 SW Aus.

Anarthriaceae D. Cutler & Airy Shaw. See Restionaceae

Anarthrophyllum Benth. Leguminosae (III 28). 15 Andes. R: Darwiniana 18(1974)453

Anarthropteris Copel. Polypodiaceae (II 6). 1 NZ: *A. lanceolata* (Hook.f.) L. Moore. Cult. orn.

Anartia Miers = *Tabernaemontana*

Anaspis Rech.f. = *Scutellaria*

Anastatica L. Cruciferae. 1 Morocco to S Iran: *A. hierochuntica* L. (rose-of-Jericho, resurrection-pl.) – annual, lvs fall as seed matures, branches fold inwards & dead pl. is blown as a ball, dispersing seeds, branches opening out in moist conditions. Dead pls often sold as curiosities, opening & closing on wetting & drying (cf. *Selaginella lepidophylla*)

Anastrabe E. Meyer ex Benth. Scrophulariaceae. 1 S Afr. Elaiophores attractive to bees

Anastraphea D. Don = ? (Compositae)

Anastrophea Wedd. = *Sphaerothylax*

Anatropanthus Schltr. Asclepiadaceae (III 4). 1 Borneo: *A. borneensis* Schltr.

Anatropostylia (Plitm.) Kupicha = *Vicia*

anatto *Bixa orellana*

Anaueria Kosterm. (~ *Beilschmiedea*). Lauraceae (I). 1 Brazil: *A. brasiliensis* Kosterm.

Anax Ravenna = *Stenomesson*

Anaxagorea A. St-Hil. Annonaceae. 21 trop. Am., 3 Sri Lanka to W Mal. R: BJ 105(1984)73. Fr. explosive, thrown to 6 m. *A. rheophytica* Maas & Westra (Venez. Amazonia) – only rheophyte in Amazonia

Anaxeton Gaertner. Compositae (Gnap.-Cass.). 10 SW Cape. R: OB 104(1991)92

Ancana F. Muell. = *Meiogyne*

Ancathia DC. Compositae (Card.-Card.). 1 C As., China, Mongolia

Anchietea A. St-Hil. Violaceae. 8 trop. S Am. *A. salutaris* A. St-Hil. – liane with medic. root-bark

Anchistea C. Presl = *Woodwardia*

Anchomanes Schott. Araceae (V 6). 5 trop. Afr. Famine food

Anchonium DC. Cruciferae. 4 W & C As. R: Cand. 39(1984)715

anchovy pear *Grias cauliflora*

Anchusa L. Boraginaceae. c. 35 Eur. (24), N & S Afr., W As. Bugloss. Fls blue or yellow (e.g. *A. lutea* Moris, W Med.). Cult. orn. esp. *A. azurea* Miller (*A. italica*, Med., natur. N Am.), *A. capensis* Thunb. (Cape forget-me-not) & *A. officinalis* L. (Eur., W As.) – form. med. (melancholia etc.) & a vegetable, also rouge from roots (cf. *Alkanna*)

Ancistrachne S.T. Blake. Gramineae (34b). 3 Philippines, E Aus.

Ancistragrostis S.T. Blake = *Deyeuxia*

Ancistranthus Lindau. Acanthaceae. 1 Cuba

Ancistrocactus (Schumann) Britton & Rose = *Sclerocactus*

Ancistrocarpus Oliver. Tiliaceae. 5 trop. Afr.

Ancistrocarphus A. Gray (~ *Stylocline*). Compositae (Gnap.-Gnap.). 1 Calif.: *A. filagineus* A. Gray. R: OB 104(1991)175

Ancistrocarya Maxim. Boraginaceae. 1 Japan

Ancistrochilus Rolfe. Orchidaceae (V 11). 2 trop. Afr.

Ancistrochloa Honda = *Calamagrostis*

Ancistrocladaceae Planchon ex Walp. Dicots – Dilleniidae – Nepenthales. 1/12 OW trop. Lianes, branching sympodially, branch-tips hooked or twining, s.t. alks. Lvs spirally arranged, small waxy glands present; stipules minute, ephemeral. Fls small, in axillary or terminal spikes to panicles, often apparently dichot. cymes, regular exc. K, ± epigynous. K 5, adnate to ovary, imbricate, unequal, accrescent in fr.; C 5, s.t. connate at base, convolute, ± fleshy; A 10, 5 somewhat larger than others, rarely only 5, filaments ± connate at base & adnate to bases of C, anthers opening by longit. slits; $\overline{G}$ (3), semi-inf., apex free with 3 ± free styles, 1-loc. with 1 basal hemitropous bitegmic ovule. Fr. a nut crowned with sepals, floating in water; seed with hard, starchy ruminate endosperm

Only genus: *Ancistrocladus*

Form. associated with Dipterocarpaceae because of the winged fr. but now held to be allied to Nepenthaceae

Ancistrocladus Wallich. Ancistrocladaceae. 3 trop. Afr., 9 Indomal. Extracts active against HIV. *A. extensus* Wallich (SE As., W Mal.) – roots form. used in treatment of dysentery, young lvs as flavouring (Thailand)

Ancistrophora A. Gray (~ *Verbesina*). Compositae (Helia.-Verb.). 1 Cuba

Ancistrophyllum (G. Mann & H. Wendl.) H. Wendl. = *Laccosperma*

Ancistrorhynchus Finet. Orchidaceae (V 16). 14 trop. Afr.

Ancistrostylis Yamaz. Scrophulariaceae. 1 Laos

Ancistrothyrsus Harms. Passifloraceae. 1–2 W trop. S Am.

Ancrumia Harvey ex Baker. Alliaceae (Liliaceae s.l.). 1 Chile

Ancylacanthus Lindau = *Ptyssiglottis*

Ancylanthos Desf. Rubiaceae (III 2). 5 trop. Afr.

Ancylobothrys Pierre. Apocynaceae. 10 Afr., 1 extending to Madag. *A. petersiana* (Klotzsch) Pierre used against venereal disease on Mafia Is.

Ancylostemon Craib. Gesneriaceae. 11 China

anda-assy oil *Joannesia princeps*

Andersonia R. Br. Epacridaceae. 22 SW Aus. R: KB 16(1962)85

Anderssoniopiper Trel. = *Macropiper*

Andes rose *Befaria* spp. esp. *B. racemosa*

Andesia Hauman = *Oxychloe*

Andira Juss. Leguminosae (III 4). 30 trop. Am., W Afr.(1). Malodorous vertebrate (? bat) disp. drupes (andirá = bat in 'lingua geral' of Brazil). Timber, coffee-shade, bark medic. esp. vermifuge. *A. araroba* Aguiar (trop. Am.) – source of Goa powder, or araroba, derived from cavities in bark, medic.; *A. inermis* (Wright) DC (angelin, dog almond, Am.) – in dry forest fls biennially, in rain forest only 3 times in 12 yrs, timber (bastard mahogany, kuraru, partridge-wood) & planted for shelter belts in WI (cabbage-tree), bark medic., seeds emetic & anthelminthic

andiroba *Carapa guianensis*

andoung *Monopetalanthus heitzii*

Andrachne L. Euphorbiaceae. c. 25 trop. Am. (2), N Afr., Med. (Eur. 1), W Himal. Cult. orn.

Andradea Allemão. Nyctaginaceae (II). 1 SE Braz.: *A. floribunda* Allemão

Andrea Mez. Bromeliaceae (3). 1 C Braz.

Andreettaea Luer = *Pleurothallis*

Andresia Sleumer = *Cheilotheca*

Androcalymma Dwyer. Leguminosae (I 2). 1 Amaz. Brazil

Androcentrum Lemaire = *Bravaisia*

Androcera Nutt. = *Solanum*

Androchilus Liebm. ex Hartman. Orchidaceae. 1 Mexico

Androcorys Schltr. Orchidaceae (IV 2). 4 India, E As.

Androcymbium Willd. Colchicaceae (Liliaceae s.l.). 12 Med. (Eur. 2) to S Afr. Alks

Andrographis Wallich ex Nees. Acanthaceae. 20 trop. As. *A. paniculata* (Burm.f.) Nees (kariyat, India) – bitter stomachic & tonic, subs. for *Munronia pinnata*

Androlepis Brongn. ex Houllet. Bromeliaceae (3). 1 C Am.

Andromeda L. Ericaceae. 1–2 N temp. (Eur. 1). *A. polifolia* L. (bog rosemary, 'marsh andromeda') – cult. orn., lvs & twigs for tanning in Russia

Andromycia A. Rich. = *Asterostigma*

Andropogon L. Gramineae (40e). c. 100 trop. & warm (Eur. 1). R: HIP 37(1967)sub t. 3644. Dominant savanna genus. Thatching, erosion control & some fodder esp. *A. gayanus* Kunth (Gamba grass, trop. Afr.). Extrafl. nectaries

Andropterum Stapf. Gramineae (40d). 1 trop. Afr.

Andropus Brand = *Nama*

Androsace L. Primulaceae. c. 100 N temp. (Eur. 24). R: G.F. Smith & D.B. Lowe (1977) *Androsaces*. ± xerophytes, usu. heterostyled; differs from *Primula* in C-tube shorter than K & constricted at throat. Cult. orn. for rock gardens, 'rock-jasmine', esp. *A. carnea* L. (Eur.), *A. lanuginosa* Wallich & *A. sarmentosa* Wallich (Himal.)

Androsiphon Schltr. Hyacinthaceae (Liliaceae s.l.). 1 Karoo: *A. capense* Schltr. R: KM 7(1990)124

Androsiphonia Stapf (~ *Paropsia*). Passifloraceae. Spp? trop. Am.

Androstachyaceae Airy Shaw. See Euphorbiaceae

Androstachys Prain. Euphorbiaceae (Androstachyaceae). 1 SE trop. Afr.: *A. johnsonii* Prain (mecrusse, mzimbeet) – timber allegedly termite-proof. See also *Stachyandra*

Andostephanus Fern. Casas. Amaryllidaceae (Liliaceae s.l.). 1 Bolivia: *A. tarijensis* Fern. Casas.

Androstephium Torrey (~ *Bessera*). Alliaceae (Liliaceae s.l.). 2 Am. Cult. orn.

Androstylanthus Ducke = *Helianthostylis*

Androtium Stapf. Anacardiaceae (I). 1 Mal.

Androtrichum (Brongn.) Brongn. Cyperaceae. 3 E temp. S Am.

Androya Perrier. Buddlejaceae (Loganiaceae s.l.). 1 Madag.

Andruris Schltr. = *Sciaphila*

Andryala L. Compositae (Lact.-Hier.). c. 20 Med. (Eur. 5) to Canary Is.

Andrzeiowskia Reichb. Cruciferae. 1 Balkans to Caucasus: *A. cardamine* Reichb.

Anechites Griseb. Apocynaceae. 1 trop. Am. Alleged to improve memory

anegré *Pouteria* spp.

Aneilema R. Br. Commelinaceae (II). 62 warm (Afr. 60; Am. 1, also in Afr.). R: R. Faden (1975) *Biosyst. study of the genus A.*, SCB 76(1991). Some andromonoec.

Anelsonia J.F. Macbr. & Payson. Cruciferae. 1 W US

Anelytrum Hackel = *Avena*

Anemanthele Veldk. = *Stipa*

Anemarrhena Bunge. Anthericaceae (Liliaceae s.l.). 1 China: *A. asphodeloides* Bunge – med., molluscicidal steroid saponins, cult. orn.

Anemia Sw. Schizaeaceae. 100+ trop. & warm esp. Am. Basal pinnae fertile, erect, branched. R: ISCS 36(1962)349. Cult. orn. *A. tomentosa* (Savi) Sw. (Uruguay) – contraceptive

Anemocarpa P.G. Wilson. Compositae (Gnap.-Ang.). 3 Aus. R: Nuytsia 8(1992)452

Anemoclema (Franchet) W.T. Wang (~ *Anemone*). Ranunculaceae. 1 SW China

Anemone L. Ranunculaceae (II 2). 144 (exc. *Pulsatilla*) Euras. (Eur. 17), Sumatra, S & E Afr. (mts), N Am. to Chile. R: BJ 37(1906)172. Perianth petaloid, involucre in some resembling K. Tepals white, yellow, red or blue. Toxic: glycoside, ranunculin, hydrolyses to volatile toxic irritant oil, protoanemonin. Cult. orn., some med. ('Windflower' a confusion as generic name not from *anemos* (Gk. wind) but prob. Gk. from old Semitic Naaman (i.e. Adonis) whose blood held to have produced *Anemone coronaria* L.; in Mediaeval England '*anemone*' name for poppies). *A.* used as cut-fls orig. in Turkey but developed later in Italy usu. *A. coronaria*, prob. the Biblical 'lilies of the field', (domesticated c. 400 yrs; 'single' fls = 'de Caen') with its semidouble cvs (= 'St Brigid'), also *A. pavonina* Lam. (single fls = 'St Bavo') & *A. hortensis* L. (all Med.) & hybrid between last two: *A.* × *fulgens* Gay; spring spp. incl. *A. apennina* L. (S Eur.) – K 10–15 & *A. blanda* Schott & Kotschy (SE Eur. to Caucasus) – K <20, fr. head nodding; Japanese anemones of autumn are *A.* × *hybrida* Paxton (*A. vitifolia* Buch.-Ham. ex DC (Afghanistan to W China & Burma) × *A. hupehensis* Hort. Lemoine var. *japonica* (Thunb.) Bowles & Stearn (China, natur. Japan)) cvs esp. pure white 'Honorine Joubert'; *A. multifida* Poiret – NW bipolar distrib.; *A. nemorosa* L. (wood anemone, Euras.) – cult. orn., lvs & rhiz. imp. in bank-vole diet, nectar 0 but mainly bee-poll., ? ant-disp.

Anemonella Spach (~ *Anemone*). Ranunculaceae (II 2). 1 E N Am.: *A. thalictroides* (L.) Spach – ed. tubers; cult. orn.

Anemonidium (Spach) Löve & D. Löve = *Anemone*

Anemonopsis Siebold & Zucc. Ranunculaceae (I 2). 1 Japan (C Honshu): *A. macrophylla* Siebold & Zucc. – cult. orn.

Anemopaegma C. Martius ex Meissner. Bignoniaceae. 30 trop. Am. Lianes, some cult. orn. & aphrodisiac

Anemopsis Hook. & Arn. Saururaceae. 1 SW US, Mex.: *A. californica* (Nutt.) Hook. & Arn. – herb of wet places, infl. resembles single fl. Cult. orn., aromatic stock medic. & used as beads (Apache b.)

Anepsias Schott = *Rhodospatha*

Anerincleistus Korth. Melastomataceae. 30 S China & India to Philippines. R: PANSP 141(1989)29. Herbs & shrubs, some rheophytes incl. *A. rupicola* (Nayar) Maxw.

Anetanthus Hiern ex Benth. Gesneriaceae. 2 Brazil, Peru

Anethum L. Umbelliferae (III 10). 1 SW As.(?): *A. graveolens* L. (inc. *A. sowa*, dill, (E.) Indian d., natur. Euras., N Am.) – cult. since at least 400 BC, ann. herb with fennel-like flavour for fish & pickling gherkins, seeds & lvs in soup & salads, dill water used for infant flatulence

Anetium (Kunze) Splitg. Vittariaceae). 1 trop. Am.: *A. citrifolium* (L.) Splitg.

Aneulophus Benth. Erythroxylaceae. 2 W trop. Afr.

Aneurolepidium Nevski = *Leymus*

Angadenia Miers. Apocynaceae. 2 Florida, WI

Angasomyrtus Trudgen & Keigh. Myrtaceae. 1 SW Aus. Close to *Kunzea*

angel wings *Caladium* spp.

Angelica L. Umbelliferae (III 9). c. 110 N hemisph. (Eur. 8), NZ(?). Some orn., medic. & ed. *A. archangelica* L. (Euras., natur. GB – angelica) – lvs a vegetable & petioles, infl. axes etc. candied for cake-decoration etc., their form the inspiration for fluted (Doric) columns of Anc. Greece; tonic & flavouring in wines etc. esp. characteristic taste of Benedictine & Chartreuse, though root infusion taken 3 times a day alleged to create distaste for alcohol; *A. atropurpurea* L. (NE N Am.) – similar uses

angelin *Andira inermis*

angelique *Dicorynia guianensis*

Angelocarpa Rupr. = *Angelica*

Angelonia Bonpl. Scrophulariaceae. 25 trop. Am. Poll. by female oil-collecting bees (*Centris* spp.). Cult. orn. like *Alonsoa*

Angelphytum G. Barroso. Compositae (Helia.-Verb.). 14 Brazil, Arg., Paraguay. R: PBSW 97(1984)96

Angel's tears *Narcissus triandrus* 'Albus'; **a. trumpet** *Brugmansia* spp.

Angianthus Wendl. Compositae (Gnap.-Ang.). 17 S Aus. R: Muelleria 5(1983)153, OB 104(1991)131. Heads compound

angico gum *Parapiptadenia rigida*

Anginon Raf. (*Rhyticarpus*). Umbelliferae (III 8). 3–7 S Afr. R: NRBGE 45(1988)90

Angiopteridaceae Fée ex Bommer = Marattiaceae

Angiopteris Hoffm. Marratiaceae (Angiopteridaceae). (Incl. *Archangiopteris* & *Macroglossum*) 1 Madag., trop. As. to W Pacific: *A. evecta* (Forster f.) Hoffm., s.t. split into up to 200 microspp. Massive short stem with fronds to 3 m, sporangia distinct, annulus complicated. Oil used to perfume coconut-oil in Pacific, stem ed. India (starch) & basis of intoxicating drink

Angiospermae A. Braun & Doell (Magnoliophyta). 13 114/249 300 in 405 fams in 76 orders. Seed pls usu. with closed carpels (unclosed in certain Magnoliales, Resedaceae, Sterculiaceae etc.) & app. always with 'double fert.' where cells other than the egg unite during fertilization to give (s.t. short-lived) triploid endosperm, though diploid in at least some Onagraceae. The dominant group of pls on which civilization relies, trad. split into Dicotyledonae & Monocotyledonae (q.v.), & comprising the bulk of Spermatophyta (q.v.), themselves the bulk of vasc. pls (13 432/260 000 in 460 fams), the subject of this book. The origin of this group has vexed many but attempts to project the modern circumscription of the group into the past so as to judge fossils as 'true' angiosperms or not are muddleheaded: clearly the 'first' angiosperm would have been seen merely as an interesting 'gymnosperm' at the time. Similarly the ancestor of the group of plants to dominate the future world is presently seen as an 'angiosperm'. Pls resembling modern angiosperms at least in major part are known from the Triassic of Texas (*Sanmiguelia*), semi-aquatic pachycauls with plicate lvs & vessel-less xylem, app. monocot-like infls of fls with monosulcate pollen & closed carpels with app. bitegmic ovules

Angkalanthus Balf.f. Acanthaceae. 1 Socotra, 1 Transvaal

Angolaea Wedd. Podostemaceae. 1 Angola

Angolluma Munster = *Pachycymbium*

Angophora Cav. (~ *Eucalyptus*). Myrtaceae (Lept.). 13 E Aus. R: Telopea 2(1986)749. Differs from *Eucalyptus* in separate petals. Cult. orn. – gum myrtles, 'apples' (Aus.). *A. woodsiana* Bailey (NE Aus.) – poll. by 12 spp. of beetle

Angoseseli Chiov. Umbelliferae (III 3). 1 Angola

angostura *Galipea officinalis* bark used in bitters

Angostura Roemer & Schultes. Rutaceae. 38 trop. S Am. Fls sympetalous, zygomorphic; local med. in Venez. where few stomach disorders

Angostyles Benth. = *Angostylis*

Angostylidium (Muell. Arg.) Pax & K.Hoffm. = *Tetracarpidium*

Angostylis Benth. Euphorbiaceae. 2 trop. S Am.

Angraecopsis Kraenzlin. Orchidaceae (V 16). 15 trop. Afr., Masc. Cult. orn.

Angraecum Bory. Orchidaceae (V 16). 200 trop. Afr., Masc., Sri Lanka. R: KB 28(1973)496. Epiphytes, cult. orn. *A. fragrans* Thouars (Mauritius) – flavouring (faham) for a fine tea (thé de Bourbon), icecream etc. *A. arachnites* Schltr. (Madag.) – exclusively poll. by hawk-moth (*Panogena lingens* (Butler)) which also poll. other spp.; *A. sesquipedale* Thouars (Madag.) – spur to 45 cm, pollinating moth with corr. proboscis length predicted before discovery

angsana *Pterocarpus indicus*

Anguillaria R. Br. = *Wurmbea*

Anguloa Ruíz & Pavón. Orchidaceae (V 10). 10 trop. S Am. Cult. orn. ('baby in cradle' due to lip mobile when rocked)

Angylocalyx Taubert. Leguminosae (III 2). 7 trop. Afr.

Ania Lindley = *Tainia*

Aniba Aublet. Lauraceae (I). 41 Andes, Colombia to Bolivia R: FN 31(1982)18. Source of Brazilian sassafras & bois-de-rose oil. *A. coto* (Rusby) J.F. Macbr. (Bolivia) – medic. 'coto bark'; *A. perutilis* Hemsley (comino) – good timber; *A. rosaeodora* Duke (N S Am.) – 10–30 m trees extracted from Amazon alone, poll. stingless bees attracted by oil glands, disp. by toucans

aniègre *Pouteria* spp.

Anigozanthos Lab. Haemodoraceae. (Incl. *Macropidia*) 12 SW Aus. R: Aus. JB 25(1977)524, Aus. SB 4(1991)663. Cult. orn. with woolly fls (kangaroo-paw), inc. *A. humilis* Lindley (cat's-paw)

anime *Hymenaea verrucosa*

Aningeria Aubrév. & Pellegrin = *Pouteria*

Anisacanthus Nees. Acanthaceae. 8 SW US, Mex. Cult. orn. Oldest name is *Idanthisa*

Anisachne Keng = *Calamagrostis*

Anisadenia Wallich ex Meissner. Linaceae. 2 Himal. to C China. Approaches Plumbaginaceae

Anisantha K. Koch. See *Bromus*

Anisantherina Pennell = *Agalinis*

anise *Pimpinella anisum* (aniseed balls); **Chinese** or **star a.** *Illicium verum*; **Japanese star a.** *I. anisatum*; **purple** or **tree a.** *I. floridanum*

Aniseia Choisy. Convolvulaceae. 5 trop. Am. inc. *A. martinicensis* (Jacq.) Choisy pantrop.

Aniselytron Merr. = *Calamagrostis*

Aniserica N.E. Br. Ericaceae. 2 SW & S Cape

anisette *Foeniculum vulgare, Pimpinella anisum*

Anisocalyx L. Bolus = *Drosanthemopsis*

Anisocampium C. Presl = *Athyrium*

Anisochaeta DC. Compositae (Gnap.). 1 S Afr.: *A. mikanioides* DC. R: OB 104(1991)47

Anisochilus Wallich ex Benth. Labiatae (VIII 4c). 20 OW trop.

Anisocoma Torrey & A. Gray. Compositae (Lact.-Mal.). 1 SW US

Anisocycla Baillon. Menispermaceae (I). 3 trop. Afr., 3 Madag.

Anisodontea C. Presl. Malvaceae. 19 S Afr. R: GH 10(1969)215. Cult. orn.

Anisodus Link ex Sprengel (~ *Scopolia*). Solanaceae (1). 6 temp. E As. *A. stramoniifolius* (Wallich) G. Don f. – yak-fodder in Himal.

Anisomallon Baillon = *Apodytes*

Anisomeles R. Br. Labiatae (?VII). (Incl. *Epimeridi*) c. 12 OW trop. Some med. inc. *A. indica* (L.) Kuntze also eaten in sago cakes (Mal.), a good bee-pl.

Anisomeria D. Don. Phytolaccaceae (I). 2–3 Chile

Anisomeris C. Presl = *Chomelia*

Anisopappus Hook. & Arn. Compositae (Inul.). c. 30 trop. & S Afr., China (1). R: Kirkia 4(1964)45

Anisophyllea R. Br. ex Sabine. Anisophylleaceae. c. 30 trop. *A. laurina* R. Br. ex Sabine (W Afr.) – fr. ed. (monkey-apple)

Anisophylleaceae Ridley. Dicots – Rosidae – Rosales. 4/29 trop. Trees & shrubs, often accumulating aluminium. Lvs spirally arr., simple, usu. dimorphic; stipules 0. Fls in axillary racemes or in panicles on leafless shoots, usu. unisexual (monoecy), mostly 4-merous, epigynous. C entire to dissected or 0; A 8 in 2 whorls, opening by longitudinal slits; $\overline{G}$ (4), rarely (3), with separate styles; ovules 1 or 2 in each locule, pendulous. Fr. indehiscent, woody to drupaceous, sometimes winged; seeds 1–4 with no endosperm. n = 7, 8. R: AMBG 75(1988)1293

Genera: *Anisophyllea, Combretocarpus, Poga, Polygonanthus*

Affinities controversial and often placed near Rhizophoraceae, but differ in spirally arranged exstipulate lvs, distinct styles & pollen features (Cronquist); fl. morphology suggests interm. position between R. & Myrtales

Anisopoda Baker. Umbelliferae (III 8). 1 Madag.

Anisopogon R. Br. Gramineae (25). 1 SE Aus.: *A. avenaceus* R. Br.

Anisoptera Korth. Dipterocarpaceae. 11 Indomal. G semi-inf. Imp. timbers & veneers, though high silica levels blunt saws. *A. curtisii* Dyer ex King (W Mal.) – krabak; *A. laevis* Ridley (W Mal.) etc. – mersawa; *A. scaphulla* (Roxb.) Kurz (SE As. to W Mal.) – kaunghmu; *A. thurifera* (Blanco) Blume etc. – palosapis (Philippines), resin used to line tunnels in termite nests in N Moluccas by *Chalicodoma pluto* (world's biggest bee)

Anisopus N.E. Br. Asclepiadaceae (III 4). 1 trop. W Afr.: *A. mannii* N.E. Br.
Anisosciadium DC. Umbelliferae (III 1). 3 SW As.
Anisosepalum Hossain. Acanthaceae. 3 C Afr. R: BJBB 61(1991)127
Anisosperma A. Silva Manso = *Fevillea*
Anisostachya Nees = *Justicia*
Anisotes Nees. Acanthaceae. 19 trop. Afr., Madag. R: NJB 1(1981)623. Alks
Anisothrix O. Hoffm. Compositae (Gnap.). 2 SW Cape. R: OB 104(1991)46
Anisotoma Fenzl. Asclepiadaceae (III 5). 2 S Afr.
Anisotome Hook.f. Umbelliferae (III 8). 15 NZ, subAntarctic is. R: UCPB 33(1961)1. Dioec. Hybrids with *Aciphylla* spp.
Anisum Hill = *Pimpinella*
Ankyropetalum Fenzl. Caryophyllaceae (III 1). 4 E Med. to Iran. R: Wentia 9(1962)170
Anna Pellegrin. Gesneriaceae. 1 SE As.
Annamocarya A. Chev. = *Carya*
Anneliesia Brieger & Lueckel = *Miltonia*
Annesijoa Pax & K. Hoffm. Euphorbiaceae. 1 New Guinea
Anneslea Wallich. Theaceae. 4 Indomal., China. R: JAA 33(1952)79. G semi-inf. *A. fragrans* Wallich – source of beautifully marked timber
Annesorhiza Cham. & Schldl. Umbelliferae (III 8). 12–15 S Afr. *A. nuda* (Aiton). B.L. Burtt (*A. capensis*) – ed. roots
annatto *Bixa orellana*
Annickia Setten & Maas (*Enantia*). Annonaceae. 10 trop. Afr. (E Afr. 1). Some timbers esp. *A. chlorantha* (Oliver) Setten & Maas (*E. chlorantha*, African whitewood, W Afr.) & dyes. Alks
Annona L. Annonaceae. (Incl. *Raimondia*, excl. *Rollinia*, *Rolliniopsis*) 137 trop. Am. (129) & Afr. (incl. *Anonidium*, 8). R: AHB 10(1931)197. *A. coriacea* C. Martius (Brazil) fls heat up over 2 nights, *A. crassifolia* C. Martius (Brazil) fls 1 night then dropped, *A. cornifolia* A. St-Hil. (Brazil) fls over 1–2 nights, scarab-poll.; *A. sericea* Dunal (NE S Am.) – thick fleshy petals never fully opening, enclosing A & G, c.7 p.m. fl. temp. rises to 6°C above ambient with odour like chloroform & ether attractive to small chrysomelid beetles & flies which enter, stigmas then fall & anthers become erect & release pollen, stamens then fall off 1 by 1 & petals also to release insects. Fr. a large fleshy syncarp formed by amalgamation of pistils & receptacle, some fish-disp. (e.g. *A. hypoglauca* C. Martius – obligate) in Amazon, many ed. (custard, alligator or monkey apples). *A. cherimola* Miller (cherimoya, custard apple, Andes of Peru & Ecuador); *A. diversifolia* Saff. (ilama, C Am.); *A. glabra* L. (trop. Am. & W Afr.) – seeds with 6-10 layers of buoyancy cells in endotesta, sea-disp., roots used for corks; *A. muricata* L. (soursop. guanábana, graviola, trop. Am.) – domatia; *A. reticulata* L. (bullock's-heart, custard apple, sugar apple, trop. Am.) – inferior to *A. cherimola*; *A. scleroderma* Saff. (poshte, C Am.) – thick skin quality to be bred into other spp. to ease shipping problems; *A. squamosa* L. (sweetsop, custard apple, trop. Am.) – seeds with insecticides (acetogenins) inhibiting electron transport chain, local med. inc. in malaria, hybrids with *A. cherimola* grown
Annonaceae Juss. Dicots – Magnoliidae – Magnoliales. 112/2150 trop., *Asimina* extending N to Michigan. Trees (sometimes buttressed), shrubs or lianes, usu. with alks, resin canals & septate pith. Lvs simple, typically distichous and with glaucous or metallic sheen, stipules 0. Fls solitary or in various basically cymose infls, s.t. cauliflorous or even on underground suckers (some *Duguetia*), fragrant, often nodding, frequently opening before all parts fully developed, poll. by insects esp. beetles, which may be trapped inside, bisexual, rarely not so, hypogynous & commonly 3-merous, usu. brittle or fleshy. P usu. 3 whorls of 3, rarely more or fewer & rarely connate at base; A ∞, spirally arranged, rarely 3 or 6 cyclic or connate at base, anthers opening by longit. slits, pollen shed in monads, tetrads or even polyads, very varied. $\underline{G}$ ∞, rarely connate to form compound ovary (*Monodora*, *Isolona*); ovules 1–(usu.) ∞, anatr. or sometimes campylotr., bitegmic or rarely with a new integument intercalated between other 2. Fr. of distinct berries to dry indehiscent or explosive (*Anaxagorea*) or united into syncarp by development of receptacle; seeds bitegmic or tritegmic (e.g. *Cananga*, the third one lying between the others, a condition unique in angiosperms) often arillate, endosperm copious, ruminate, oily (& sometimes starchy). x = 7, 8, 9

Classification & large genera: although its limits are well-defined, the family is notoriously difficult to subdivide, though 2 subfams are commonly recognized, viz. Annonoideae – (carpels spirally arr. (rarely not), free (rarely not) incl. *Annona*, *Artabotrys*, *Goniothalamus*, *Guatteria*, *Polyalthia*, *Uvaria*, *Xylopia* & Monodoroideae (carpels cyclically arr., united into 1-loc. ovary with parietal placentation) – *Isolona*, *Monodora* (only genera), but this is now challeged & Kessler (in Kubitzki) proposes 14 'groups' (some with fuzzy margins), of which 1 corresponds to M.

Several *Annona* cult. for fruit (see also *Rollinia*) & others, e.g. *Cananga, Mkilua*, for scent (see also *Stelechocarpus*); timbers incl. spp. of *Annickia, Cleistopholis, Duguetia, Oxandra*; cult. orn. incl. *Polyalthia*

Anochilus (Schltr.) Rolfe = *Pterygodium*

Anoda Cav. Malvaceae. 23 S US & (esp.) Mex. to Bolivia, Arg. & Chile. R: Aliso 11(1987)485. Cult. orn.

Anodendron A. DC. Apocynaceae. 17 Indomal. to Japan. *A. candolleanum* Wight (Malay Pen.) – form. imp. fibre from this liane; *A. paniculatum* (Roxb.) A. DC (Sri Lanka) – fibre for fishing nets

Anodiscus Benth. Gesneriaceae. 1 Peru: *A. xanthophyllus* (Poepp.) Mansfeld (*A. peruvianus*), roadside weed. R: Selb. 6(1982)174

Anodopetalum A. Cunn. ex Endl. Cunoniaceae. 1 SW Tasmania: *A. biglandulosum* (Hook.) Hook.f. – trunk often bends to horizontal, branches forming thickets

Anoectocalyx Benth. = *Anaectocalyx*

Anoectochilus Blume. Orchidaceae (III 3). 35 trop. As. to W Pacif. Cult. orn. (jewel orchids), 'vogue' foliage stove-pls of mid-late C19

Anogeissus (DC) Wallich. Combretaceae. 8 OW trop. to Arabia. R: KB 33(1979)555. Timber, dyes, gums & medic. *A. acuminata* (DC) Wallich (India to SE As.) – comm. timber (yon); *A. latifolia* (DC) Wallich (India) – gatty gum, lvs produce a black dye & used in tanning (dhawa); *A. leiocarpa* (DC) Guillemin & Perrottet (trop. Afr.) – termite-proof wood, chewing-sticks in Nigeria, lvs source of yellow dye, plant a vermifuge for stock

Anogramma Link. Pteridaceae (III). 5 Azores & SW Eur.(1), OW trop. to NZ, trop. Am. Subterr. perennial prothallus in *A. leptophylla* (L.) Link, which has delicate annual sporophytes

Anoiganthus Baker = *Cyrtanthus*

anokye *Guibourtia ehie*

Anomacanthus R. Good (*Gilletiella*). Mendonciaceae (Acanthaceae). 1 Zaire, Angola

Anomalanthus Klotzsch (~ *Scyphogyne*). Ericaceae. 11 SW & S Cape

Anomalesia N.E. Br. = *Gladiolus*

Anomalluma Plowes = *Pseudolithos*

Anomalocalyx Ducke. Euphorbiaceae. 1 trop. S Am.: *A. uleana* (Pax) Ducke

Anomalostylus R. Foster = *Trimezia*

Anomanthodia Hook.f. (~ *Randia*). Rubiaceae (II 1). 6 Indomal. R: BMNHN 4,8(1986)275

Anomatheca Ker-Gawler (~ *Freesia*). Iridaceae (IV 3). 6 trop. & S Afr. Cult. orn. esp. *A. laxa* (Thunb.) Goldblatt (*A. cruenta, Lapeirousia laxa*). Differs from *L.* in round-based (not flat) corm with fibrous (not entire) tunics

Anomianthus Zoll. Annonaceae. 1 SE As., Java: *A. dulcis* (Dunn) J. Sinclair – liane with sweet fr.

Anomochloa Brongn. Gramineae (2). 1 Bahia, Brazil: *A. marantoidea* Brongn. R: SCB 68(1989)2. Forest grass of marantoid habit rediscovered 1976

Anomochloaceae Nakai. See Gramineae

Anomoctenium Pichon = *Pithecoctenium*

Anomopanax Harms = *Mackinlaya*

Anomospermum Miers. Menispermaceae (II). 6 trop. Am. R: MNYBG 22,2(1971)61

Anomostephium DC = *Wedelia*

Anomotassa Schumann. Asclepiadaceae (III 4). 1 Ecuador: *A. macranthus* Schumann

Anona Miller = *Annona*

Anonidium Engl. & Diels = *Annona*

Anoplocaryum Ledeb. Boraginaceae. 1 (?+) C As. (? – China)

Anopteris (Prantl) Diels. Pteridaceae (VI). 1 Bermuda, Greater Antilles: *A. hexagona* (L.) C. Chr. on limestone. Cult. orn.

Anopterus Lab. Grossulariaceae (Escalloniaceae). 2 SE Aus. Cult. orn.

Anopyxis (Pierre) Engl. Rhizophoraceae. 1–3 trop. Afr.

Anosporum Nees = *Cyperus*

Anotea (DC) Kunth. Malvaceae. 2 Mex.

Anotis DC = *Arcytophyllum*

Anotites E. Greene = *Silene*

Anplectrum A. Gray = *Diplectria*

Anredera Juss. Basellaceae. 10–15 warm Am. Lianes, cult. orn. esp. *A. cordifolia* (Ten.) Steenis (Madeira or mignonette vine, subtrop. S Am.) – natur. in S Eur. where cult. as vegetable

(*'Boussingaultia baselloides'*) & Indomal. to Aus. reproducing only by aerial tubers

Ansellia Lindley. Orchidaceae (V 9). 1–2 trop. & S Afr. Cult. orn.

Antarctic beech *Nothofagus antarctica*

Antegibbaeum Schwantes ex C. Weber (~ *Gibbaeum*). Aizoaceae (V). 1 C S Afr.

antelope bush *Purshia* spp.; **a. grass** *Echinochola pyramidalis*; **a. horn** *Asclepias viridis*; **a. orchid** *Dendrobium canaliculatum*

Antennaria Gaertner. Compositae (Gnap.-Cass.). 71 temp. (Eur. 6), warm (not Afr.). R: OB 104(1991)96. Small dioec. stoloniferous often apomictic (triploid to octoploid) herbs (pussy's toes) cult. rock gardens esp. *A. dioica* (L.) Gaertner (mountain everlasting, cat's-foot, cold N). *A. alpina* (L.) Gaertner (circumboreal) apomictic

Antenoron Raf. = *Persicaria*

Anteremanthus H. Robinson. Compositae (Vern.-Lych.). 1 Brazil

Anteriorchis E. Klein & Strack = *Orchis*

Anthacanthus Nees = *Oplonia*

Anthaenantia Pal. = *Anthenantia*

Anthaenantiopsis Mez ex Pilger. Gramineae (34b). 4 warm S Am. R: SB 18(1993)434

Antheliacanthus Ridley = *Pseuderanthemum*

Anthemis L. Compositae (Anth.-Anth.). c. 210 Eur. (62), Med. to Iran & E Afr. (1). Aromatic, medic., some cult. orn. *A cotula* L. (mayweed, Euras., cosmop. weed) – source of insecticide & alleged mouse-repellent, taints cows' milk; (*A. nobilis* L. = *Chamaemelum nobile*); *A. tinctoria* L. (dyer's or yellow chamomile, C & S Eur., natur. GB, N Am.) – source of yellow dye

Anthenantia Pal. (*Anthaenantia*). Gramineae (34d). 3 SE US pine barrens. R: SB 13(1988)589

Anthephora Schreber. Gramineae (34f). 12 trop. Afr., Arabia with *A. hermaphrodita* (L.) Kuntze ext. to Am. R: SB 13(1988)589

Anthericaceae J. Agardh (incl. Acanthochlamydaceae; Liliaceae s.l.). 29/500 subcosmop. esp. trop. & Aus. Subshrubs (*Alania, Borya*) to herbs with rhiz. & often vessels in scapes (anthroquinones 0, cf. Asphodelaceae). Lvs spirally arr. (distichous in *Caesia, Sowerbaea*), s.t. reduced & scapes effective photosynthetic organs (cf. Aphyllanthaceae), flat to terete. Infls simple to paniculate. Fls usu. reg.; P 3 + 3 sometimes with basal tube; A 3 + 3 (s.t. 3); G̲ (3), 3-loc., each loc. with 2–∞ usu. campylotropous ovules. Fr. a loculicidal caps.; seeds angled, often with elaiosomes (? arils)

Chief genera: *Anthericum* (shorn of *Echeandia* [& *Trachyandra* (Asphodelaceae!)]), *Chlorophytum, Thysanotus*

App. still heterogeneous with subshrubby genera poss. allied to Dasypogonaceae, while there are great app. similarities with certain Asphodelaceae. The subtle circumscription follows from the splitting of Liliaceae s.l.

Cult. orn. incl. spp. of *Anthericum, Arthropodium, Chlorophytum, Thysanotus*

Anthericopsis Engl. Commelinaceae (II 2). 1 trop. E Afr.

Anthericum L. Anthericaceae (Liliaceae s.l.). 50(?) S Afr. (3), Eur. (3). *A. liliago* L. (St Bernard's lily, S Eur.) – cult. orn. border pl. Afr. spp. with root-tubers = *Chlorophytum*; Am. spp. = *Echeandia*

Antherolophus Gagnepain = *Aspidistra*

Antheropeas Rydb. = *Eriophyllum*

Antheroporum Gagnepain. Leguminosae (III 6). 1 (or 4) China

Antherostele Bremek. Rubiaceae (I 9). 4 Philippines

Antherostylis C. Gardner = *Velleia*

Antherothamnus N.E. Br. Scrophulariaceae (Manuleae). 1 C & S Afr. *A. pearsonii* N.E. Br. R: O.M. Hilliard, *Manuleae* (1994)75

Antherotoma (Naudin) Hook.f. Melastomataceae. 2 trop. Afr., Madag.

Anthobembix Perkins = *Steganthera*

Anthobolus R. Br. Santalaceae. 3 Aus. R: MBMZ 213(1959)101. Ovary superior; fr. on coloured fleshy receptacle

Anthobryum Philippi = *Frankenia*

Anthocarapa Pierre. Meliaceae (I 6). 1 Aus., New Guinea to New Caled.: *A. nitidula* (Benth.) Mabb. R: Blumea 31(1985)132

anthocephalus *Neolamarckia cadamba*

Anthocephalus A. Rich. = *Breonia*. See also *Neolamarckia*

Anthocercis Lab. Solanaceae (2). 9 Aus. R: Telopea 2(1981)174. Some cult. orn.

Anthochlamys Fenzl. Chenopodiaceae (I 6). 2 SW & C As.

Anthochloa Nees & Meyen. Gramineae (19). 1 Chile & Peru

Anthochortus Nees. Restionaceae. 7 SW & S Cape. R: Bothalia 15(1985)484

Anthocleista Afzel. ex R. Br. Gentianaceae (Loganiaceae) s.l. 14 trop. Afr., Madag., Masc. Trees with big leaves (cabbage-trees) e.g. *A. vogelii* Planchon (W Afr.) sapling lvs to 2.5 m long

Anthoclitandra (Pierre) Pichon = *Landolphia*

Anthodiscus G. Meyer. Caryocaraceae. 10 trop Am. R: OB 92(1987)179. Fish-poisons

Anthodon Ruíz & Pavón. Celastraceae. 2 trop. Am.

Anthogonium Wallich ex Lindley. Orchidaceae (V). 1 E Himal. to SE As.: *A. gracile* Wallich ex Lindley. Cult. orn.

Antholyza L. = *Gladiolus*

Anthonotha Pal. (~ *Macrolobium*). Leguminosae (I 4). 28 trop. Afr. forests

Anthopteropsis A.C. Sm. Ericaceae. 1 C Am.

Anthopterus Hook. Ericaceae. 6 Andes. R: OB 92(1987)109

Anthorrhiza Huxley & Jebb. Rubiaceae (IV 7). 8 New Guinea. R: Blumea 36(1991)21. Ant-pls

Anthosiphon Schltr. Orchidaceae (V 10). 1 Colombia

Anthospermum L. Rubiaceae (IV 13). 40 Afr., Madag.

Anthostema A. Juss. Euphorbiaceae. 3 trop. Afr., Madag. Fls in cyathium like *Euphorbia* but have P

Anthotium R. Br. Goodeniaceae. 3 SW Aus. R: Nuytsia 7(1989)49

Anthotroche Endl. Solanaceae (2). 3 Aus.

Anthoxanthum L. Gramineae (21c). 18 Euras. (Eur. 6), Afr. mts., C Am. *A. odoratum* L. (sweet vernal grass, spring grass, Euras., widely natur.) – widespread tetraploid prob. hybrid of diploid N & S races, fodder (little food value) with strong scent (coumarin)

Anthriscus Pers. Umbelliferae (III 2). 10–12 Euras. (Eur. 7) to Afr. mts. *A. cereifolium* (L.) Hoffm. (chervil) – cult. since time of Pliny, lvs aniseed-flavoured for seasoning & salad, in *fines-herbes* (with parsley & chives) for omelettes etc.; *A. sylvestris* (L.) Hoffm. (cow-parsley, Queen Anne's lace, range of genus) – common roadside pl.

Anthurium Schott. Araceae (V 2). 700+ trop. Am. (e.g. Panamà 152). R(C Am.): AMBG 70(1984)211, SBM 14(1987). Terr. & epiphytic, in 19 sects (Aroid. 6(1983)85) incl. sect. *Pachyneurium* (114; R: AMBG 78(1991)539) usu. 'bird's nest' habit, sometimes ant-inhabited e.g. *A. gracile* (Rudge) Lindley. Fls bisexual with P, the spadix sometimes in brightly coloured spathe (usu. dull & small, here long-lived & poss. attractive to disp. birds), some with scent coll. male euglossine bees. Fr. a berry hanging from spadix by 2 threads from P when ripe, often brightly coloured. Some dried lvs used to perfume tobacco in S Am., many local med.; c. 30 spp. widely cult. orn. for foliage & spathes esp. *A. scherzerianum* Schott (C Am.) & *A. andraeanum* Linden ex André (Colombia) – flamingo fls. Numerous hybrids in cult., often involving several parental spp. inc. *A. amnicola* Dressler (Panamá) – scented mauve infls, form. (prob. now extinct) on boulders in streams. *A. armeniense* Croat (Guatemala) – lilac-scented white pollen; *A. harrisii* (R. Graham) G. Don f. (Brazil) – infls smell of red wine; *A. pendulifolium* N.E. Br. (trop. Am.) – lvs to 1.8 m long

Anthyllis L. Leguminosae (III 18). 20 Med., W Eur., Macaronesia, NE Afr. Some cult. orn. & fodder esp. *A. vulneraria* L. (kidney-vetch, lady's fingers, Eur. & Medit.) – complex sp. of some 35 sspp.

Antiaris Leschen. Moraceae (III). 1 OW trop.: *A. toxicaria* Leschen. (*A. africana*, upas-tree, ipoh) – timber in Afr. (ako) for canoes, plywood core, bark cloth for wrapping rubber, latex an adulterant of *Funtumia* rubber, arrow and ordeal poison (cardiac glycosides), fabled tree allegedly poisoning surroundings & fatal to approach

Antiaropsis Schumann. Moraceae (II). 1 New Guinea

Anticharis Endl. Scrophulariaceae. 14 Afr. to Mal.

Anticheirostylis Fitzg. = *Genoplesium*

Antidaphne Poeppig & Endl. Eremolepidaceae. 7 W trop. S Am. R: SMB 18(1988)17

Antidesma L. Euphorbiaceae (Stilaginaceae). c. 170 OW trop. & warm esp. As. (Afr. 10). Bird-disp. pink, red or black fr. Some rheophytes. High ploidy levels. Timber & fr. trees, cult. orn. *A. bunius* (L.) Sprengel (bignay, Chinese laurel, Indomal.) – fr., bitter to some but sweet to others, preserved, fls smell of powdered fish; *A. pulvinatum* Hillebrand (Hawaii) – domatia

Antigonon Endl. Polygonaceae (II). 2–3 C Am. Lianes with tendrils (infl. axes), cult. orn. esp. *A. leptopus* Hook. & Arn. (coral vine, corallita, Mex.) – bright pink fls, tubers ed. (nut-like flavour)

Antigramma C. Presl = *Asplenium*

Antillanorchis Garay. Orchidaceae (V 10). 1 Cuba

Antillia R. King & H. Robinson (~ *Eupatorium*). Compositae (Eup.-Crit.). 1 Cuba. R: MSBMBG 22(1987)304

Antimima N.E. Br. (~ *Ruschia*). Aizoaceae (V). 60 SW Namibia & S Afr.
Antinisa (Tul.) Hutch. = *Homalium*
Antinoria Parl. (~ *Aira*). Gramineae (21b). 2 Medit. R: KBAS 13(1986)132
Antiostelma (Tsiang & P.T. Li) P.T. Li (~ *Hoya*). Asclepiadaceae (III 4). 2 China, India. R: Novon 2(1992)218
Antiotrema Hand.-Mazz. Boraginaceae. 1 SW China
Antiphiona Merxm. Compositae (Inul.). 2 trop. & SW Afr.
Antiphytum DC ex Meissner. Boraginaceae. 10 Mex. & trop. Am. R: CGH 68(1923)48
Antirhea Comm. ex Juss. Rubiaceae (III 3). 36 Madag. & Mal. to Samoa (NW spp = *Pittoniotis* or *Stenostomum*). Dioec. Some timber trees
Antirrhinum L. Scrophulariaceae. 20 Med. (Eur. 19) esp. Iberia, natur. temp. R: D.A. Sutton, Rev. Antirrhineae (1988)67. Mouth of fl. closed and nectar accessible only to bees, which force an entry. *A. majus* L. (snapdragon, Medit.) – cult. orn., many cvs incl. tetraploids, peloric forms etc., ssp. *cirrhigerum* (Ficalho) Franco (SW Iberia) a scrambler
Antistrophe A. DC. Myrsinaceae. 5 Indomal.
Antithrixia DC. Compositae (Gnap.-Rel.). 1 Namaqualand: *A. flavicoma* DC. R: OB 104(1991)65. Shrub
Antizoma Miers. Menispermaceae (V). 2 arid S Afr. R: JSAB 46(1980)1
Antonella Caro = *Tridens*
Antongilia Jum. = *Dypsis*
Antonia Pohl. Strychnaceae (Loganiaceae s.l.). 1 S Am.
Antoniaceae Hutch. = Strychnaceae
Antonina Vved. = *Calamintha*
Antopetitia A. Rich. Leguminosae (III 19). 1 trop. Afr. mts. Close to *Ornithopus*
Antoschmidtia Boiss. = *Schmidtia*
Antrocaryon Pierre. Anacardiaceae (II). 8 trop. W Afr. Poss. oilseed. *A. micraster* A. Chev. & Guill. – timber for planks & furniture
Antrophora I.M. Johnston = *Lepidocordia*
Antrophyum Kauf. Vittariaceae. 50 trop. esp. Mal. Cult. orn.
Antunesia O. Hoffm. = *Vernonia*
añu *Tropaeolum tuberosum*
Anubias Schott. Araceae (VI 2). 7 C & W Afr. R: MLW 79-14(1979). Some rheophytes. Cult. orn. aquarium pls esp. *A. afzelii* Schott (Sierra Leone)
Anulocaulis Standley = *Boerhavia*
Anura (Juz.) Tscherneva = *Cousinia*
Anurosperma (Hook.f.) H. Hallier = *Nepenthes*
Anvillea DC. Compositae (Inul.). 2 N Afr., Middle East. R: NJB 2(1982)297
Anvilleina Maire = *praec.*
anyaran *Distemonanthus benthamianus*
Anychia Michaux = *Paronychia*
Aoranthe Somers (~ *Porterandia*). Rubiaceae (?I 8). 5 trop. Afr. R: BJBB 58(1988)47
Aorchis Vermeulen (~ *Habenaria*). Orchidaceae (IV 2). 1 Himal.
Aostea Buscal. & Muschler = *Vernonia*
Aotus Sm. Leguminosae (III 24). 15 Aus.
apa *Afzelia* spp.
Apache beads *Anemopsis californica*
Apacheria C. Mason. Crossosomataceae. 1 Arizona
Apalanthe Planchon = *Elodea*
Apalochlamys Cass. (~ *Cassinia*). Compositae (Gnap.-Cass.). 1 Aus.: *A. spectabilis* (Lab.) J.H. Willis. R: OB 104(1991)85
Apaloxylon Drake = *Neoapaloxylon*
Apama Lam. = *Thottea*
Apargidium Torrey & A. Gray = *Microseris*
Aparisthmium Endl. (*Conceveibum*). Euphorbiaceae. 1 trop. Am.
Apassalus Kobuski. Acanthaceae. 3 SE US, WI
Apatesia N.E. Br. Aizoaceae (V). 3 SW Cape. R: BJ 111(1990)479
Apatophyllum McGillivray. Celastraceae. 3 E Aus. R: Nuytsia 8(1992)193
Apatostelis Garay = *Stelis*
Apatzingania Dieterle. Cucurbitaceae. 1 Mex. Annual, fr. ripen in ground (cf. *Arachis*)
Apeiba Aublet. Tiliaceae. 10 trop. S Am. Timber trees. *A. tibourbou* Aublet (Brazil) – oil from seeds used in treatment of rheumatism.
Apera Adans. Gramineae (21d). 3 Eur. (3) to Afghanistan. Silky bents, cult. orn.

Apetahia Baillon. Campanulaceae. 3 Society Is., Marquesas, Rapa. Arborescent

Aphaenandra Miq. Rubiaceae (I 8). 2 SE As., W Mal.

Aphaerema Miers. Flacourtiaceae. 1 S Brazil

Aphanactis Wedd. Compositae (Helia.-Gal.). 8 trop. Am. R: BSAB 19(1980)35

Aphanamixis Blume. Meliaceae (I 5). 3 Indomal. to Solomons. R: Blumea 31(1985)136. *A. polystachya* (Wallich) R. Parker (throughout range) – timber (tasua), semi-drying oil from seeds, cult. orn., s.t. with hollow ant-infested shoots

Aphanandrium Lindau = *Neriacanthus*

Aphananthe Planchon. Ulmaceae (II). 5 Madag. (3), Indomal. to Japan & E Aus. (1), Mex. (1). *A. aspera* (Thunb.) Planchon (E As.) – fast-growing timber tree, lvs used like sandpaper

Aphandra Barfod. Palmae (VI). 1 Ecuador: *A. natalia* (Balslev & Henderson) Barford. R: OB 105(1991)44

Aphanelytrum Hackel. Gramineae (17). 1 W trop. S Am.: *A. procumbens* Hackel in montane forest

Aphanes L. Rosaceae. 20 Medit. (Eur. 5), Ethiopia, C As., Aus., Am. *A. arvensis* L. (parsley piert, Eur., W As., natur. N Am.) – thought to be useful in bladder probs as stimulates copious secretion of lithic acid

Aphania Blume = *Lepisanthes*

Aphanisma Nutt. ex Moq. Chenopodiaceae (I 1). 1 Calif.: *A. blitoides* Nutt. ex Moq.

Aphanocalyx Oliver. Leguminosae (I 4). 3 Afr. (Guineo-Congolian forest)

Aphanocarpus Steyerm. Rubiaceae (IV 7). 1 Venez.

Aphanococcus Radlk. = *Lepisanthes*

Aphanopetalum Endl. Cunoniaceae. 2 S Aus.

Aphanopleura Boiss. Umbelliferae (III 8). 6 C As. to Afghanistan

Aphanosperma Daniel. Acanthaceae. 1 NW Mex.: *A. sinaloense* (Leonard & Gentry) Daniel. R: AJB 75(1988)545

Aphanostelma Schltr. = *Metaplexis*

Aphanostemma St-Hil. (~ *Ranunculus*). Ranunculaceae (II 3). 1 S Brazil to Arg.

Aphanostephus DC. Compositae (Ast.-Ast.). 4 US, Mex. R: Phytol. 56(1984)81. *A. skirrhobasis* (DC) Trel. – cult. orn.

Aphanostylis Pierre = *Landolphia*

Aphelandra R. Br. Acanthaceae. 175 trop. Am. R: SCB 18(1975). Hummingbird-poll. Cult. housepls for showy bracts & foliage, esp. *A. squarrosa* Nees 'Louisae' with white veins, *A. scabra* (Vahl) Sm. (*A. deppeana*) – extrafl. nectaries in bracts attract ants protecting fr. (9 times unprotected ones matured) & *A. liboniana* Linden (Braz.) etc.

Aphelandrella Mildbr. Acanthaceae. 1 Peru

Aphelexis D. Don = *Edmondia*

Aphelia R. Br. Centrolepidaceae. 6 Aus.

Aphloia (DC) Bennett. Flacourtiaceae. 1 (polymorphic) trop. E Afr., Madag., Masc. Lvs used as tea in Masc.

Aphloiaceae Takht. = Flacourtiaceae

Aphoma Raf. = *Iphigenia*

Aphragmia Nees = *Ruellia*

Aphragmus Andrz. ex DC. Cruciferae. 6 Himal. to NE Siberia

Aphyllanthaceae Burnett (~ Anthericaceae; Liliaceae s.l.). Monocots – Liliidae – Asparagales. 1/1 W Medit. Herb with rhiz., vessels in roots, lvs reduced to sheaths, scapes photosynthetic. Fls 1 or 2(3) in clusters (? reduced panicles); P 3 + 3 with basal tube; A 3 + 3; $\underline{G}$ (3), 3-loc., each loc. with 1 anatropous ovule. Fr. a loculicidal capsule; seeds somewhat flattened

Only sp.: *Aphyllanthes monspeliensis*

Aphyllanthes L. Aphyllanthaceae (Liliaceae s.l.). 1 Portugal to Italy, N Afr.: *A. monspeliensis* L. Cult. orn. Rhizome with secondary thickening

Aphyllarum S. Moore = *Caladium*

Aphyllocladus Wedd. (~ *Hyalis*). Compositae (Mut.-Mut.). 5 Andes of S Bolivia, N Chile, NW Arg. Shrubs. R: Darw. 9(1957)367

Aphyllodium (DC) Gagnepain = *Hedysarum*

Aphyllon Michaux = *Orobanche*

Aphyllorchis Blume. Orchidaceae (V 1). 15 SE As., Indomal. Mycotrophs

Apiaceae Lindley. See Umbelliferae

Apiastrum Nutt. ex Torrey & A. Gray. Umbelliferae (III 5). 2 N Am.

Apinagia Tul. Podostemaceae. 50 trop. S Am. Incl. *Bladowia*, first described as a hepatic!

Apiopetalum Baillon. Araliaceae. 4 New Caled.

apio *Arracacia xanthorrhiza*

Apios Fabr. Leguminosae (III 10). 10 E As., N Am. *A. americana* Medikus (*A. tuberosa*, potato-bean, groundnut, N Am.) – ed. sweet tubers boiled or roasted imp. Indian food, occ. cult.

apitong *Dipterocarpus* spp. timber

Apium L. Umbelliferae (III 8). 25 temp. & warm (exc. Afr.; Eur. 5) esp. S Am. Some contraceptives (S Am.). *A. graveolens* L. (widespread) cult.: 2 forms – var. *dulce* (Miller) DC (celery) with blanched petioles eaten raw or cooked and fr. ('seed') used in flavouring & celery salt; var. *rapaceum* (Miller) DC (celeriac) with ed. turnip-like roots. Lvs in an Egyptian garland of 1200 BC; *A. leptophyllum* (Pers.) Benth. (? C Am.) – pantrop. annual weed leading to carrot-smelling milk from cows grazing on it

Aplanodes Marais. Cruciferae. 2 S Afr.

Aplectrum Torrey. Orchidaceae (V 8). 1 N Am.: *A. hyemale* Torrey – terr., mycotrophic when young, leaf 1 ephemeral, tuber med. & used in cementing earthenware ('puttyroot'), cult. orn.

Apleura Philippi = *Azorella*

Apluda L. Gramineae (40d). 1 Mauritius & Socotra to Taiwan & New Caled.: *A. mutica* L.

Apocaulon R. Cowan. Rutaceae. 1 Venezuela

Apochaete (C.E. Hubb.) J. Phipps = *Tristachya*

Apochiton C.E. Hubb. Gramineae (31d). 1 Tanzania

Apochoris Duby = *Lysimachia*

Apoclada McClure. Gramineae (1b). 4 Brazil. Bamboos. R: KBAS 13(1986)52

Apocopis Nees. Gramineae (40b). 15 Indomal., China. Racemes paired. R: KB 1952:101

Apocynaceae Juss. Dicots – Asteridae – Gentianales. Incl. Duckeodendraceae, 163/1850 mostly trop. & few temp. Usu. lianes, less often trees, shrubs or herbs or succulents with ubiquitous laticifer systems, glycosides & alks & sometimes unusual sec. thickening, internal phloem usu. present. Lvs simple, entire, opp. sometimes whorled or condensed (*Pachypodium*), rarely truly spiral, usu. with close parallel lateral veins. Fls in cymes or racemes or solit., bisexual, often showy, ± regular, usu. (4)5-merous exc. G: K (5) with imbricate lobes, C (5) usu. funnel- or salver-shaped, convolute or rarely valvate or imbricate, A 5 epipet., alt. with C (filaments without coronal appendages cf. Asclepiadaceae, q.v.) without pollinia, 5 nectary glands near ovary-base, s.t. confluent into disk, reduced or 0, $\underline{G}$ or semi-inf. (2(–8)) or 2 united by style, or more, 1–2-loc., with 2–∞ ovules in each locule, often pendulous, anatropous, unitegmic, style usu. simple with thickened head & ring of hairs below it. Fr. diverse (follicle, berry or drupe etc.), seeds flattened, often with crown of hairs, embryo straight, ± oily endosperm. x = 8–12+. JAA 70(1989)328

Classification & chief genera:

I. **Plumerioideae** (A free from style, anthers full of pollen, seeds usu. hairless) – *Alstonia, Alyxia, Aspidosperma, Rauvolfia, Tabernaemontana, Vinca*

II. **Apocynoideae** (A adnate to style-head, anthers empty at base, seeds hairy) – *Funtumia, Mandevilla, Nerium, Parsonsia, Strophanthus*

Asclepiadaceae, separated off by Robert Brown in 1810, should perhaps be returned to make a monophyletic fam.

Many sources of drugs & poisons, e.g. strophanthin (*Strophanthus*) in treatment of heart disease & as a cortisone precursor, see also *Acokanthera, Adenium, Catharanthus* (vincristine), *Cerbera, Picralema, Rauvolfia* (reserpine), *Voacanga*; many ornamentals incl. *Allamanda, Catharanthus, Mandevilla, Nerium, Pachypodium, Plumeria, Saba, Thevetia* & *Vinca*; timber from *Alstonia, Aspidosperma, Dyera, Gonioma, Ochrosia*, fibre from *Anodendron, Chonemorpha*, rubber from *Clitandra, Funtumia, Hancornia, Landolphia, Mascarenhasia, Urceola, Willughbeia*, dyes from *Wrightia*

Apocynum L. Apocynaceae. Incl. *Trachomitum* c. 12 S Russia (Eur. 3) to China, temp. Am. Seeds hairy. Cult. orn. (dogbane), form. medic. *A. cannabinum* L. (Ind. hemp, choctaw root) – bark a source of fibre for ropes, sails etc., root emetic & cardiac stimulant; *A. venetum* L. (*T.v.*, E Eur., W As.) – kendyr fibre locally used for sails and nets, seeds with useful floss

Apodandra Pax & K. Hoffm. Euphorbiaceae. 4 trop. S Am.

Apodanthaceae (R. Br.) Takht. = Rafflesiaceae

Apodanthera Arn. Cucurbitaceae. 15 warm Am. *A. undulata* A. Gray (melon-loco, SW N Am.) – minor oilseed

Apodanthes Poit. Rafflesiaceae (II 2). 1–7 trop. S Am.

Apodicarpum Makino. Umbelliferae (III 8). 1 E Japan

Apodiscus Hutch. Euphorbiaceae. 1 W Afr.

Apodocephala Baker. Compositae (Ast.-Ast.). 8 Madag. Woody

Apodolirion Baker. Amaryllidaceae (Liliaceae s.l.). 6 Cape to Transvaal. R: Willd. 15(1986)466. Cult. orn.

Apodostigma R. Wilczek. Celastraceae. 1 trop. Afr., Madag.: *A. pallens* (Oliver) R. Wilczek

Apodytes E. Meyer ex Arn. Icacinaceae. 1 OW trop. (*A. dimidiata* E. Meyer ex Arn. (white pear) – good timber), 1 Queensland

Apoia Merr. = *Sarcosperma*

Apollonias Nees (~ *Phoebe*). Lauraceae (I). 1 Macaronesia, 1 India

Apomuria Bremek. = *Psychotria*

Aponogeton L.f. Aponogetonaceae. 43 OW trop., S Afr. R: BB 33(137)(1985). Aquatics grown in aquaria, some ed. tubers. *A. distachyos* L.f. (Cape pondweed, water hawthorn, Cape asparagus, Cape) – cult. orn., fl. spikes ed. like spinach or pickled; *A. madagascariensis* (Mirbel) Bruggen (*A. fenestralis*, lace-leaf, Madag.) – (difficult) aquarium pl. with lace-like lvs due to patches of tissue between veins dying during leaf development

Aponogetonaceae J. Agardh. Monocots – Alismatidae – Najadales. 1/43 OW trop., S Afr. Perennial, glabrous hydrophytes with short rhiz. or corm & secretory canals, vessels in roots or 0. Lvs with parallel veins, floating or all submerged. Infl. emergent with spathe early lost, simple or 2–10-fid or contracted into a head. Fls small, bisex. usu. Tepals (1)2(–6) distinct, often persistent, when 1 with broad base & resembling a bract, or 0; A 3 + 3 (or ∞ in 3 or 4 whorls), pollen grains monosulcate; $\underline{G}$ 3–8, rarely more or fewer, with short styles, each loc. with (1)2–8(–14) basal, anatropous, bitegmic ovules. Fr. of distinct follicles, seeds without endosperm, embryo straight with single terminal cotyledon. x = 8
Only genus: *Aponogeton*
Distinct from Potamogetonaceae in coloured P and straight embryo

Apophyllum F. Muell. Capparidaceae. 1 NE Aus.: *A. anomalum* F. Muell.

Apoplanesia C. Presl. Leguminosae (III 12). 1 dry C Am.: *A. paniculata* C. Presl – cult. orn.

apopo *Lovoa trichilioides*

Aporocactus Lemaire = *Disocactus*

Aporosa Blume. Euphorbiaceae. c. 80 Indomal. (Borneo 30) to Solomon Is. Some timber & medic.

Aporosella Chodat & Hassler = *Phyllanthus*

Aporostylis Rupp & Hatch. Orchidaceae (V 1). 1 NZ

Aporrhiza Radlk. Sapindaceae. 6 trop. Afr.

Aporusa Blume = *Aporosa*

Aposeris Necker ex Cass. (~ *Hyoseris*). Compositae (Lact.-Hyp.). 1 C Eur.

Apostasia Blume. Orchidaceae (I). 8 trop. As. to Aus.

Apostasiaceae Lindley. See Orchidaceae

Apostates Lander. Compositae (Ast.-Ast.). 1 Rapa

Appendicula Blume. Orchidaceae (V 14). 50 trop. As. to Polynesia. *A. rupestris* Ridley (Malay Pen.) – rheophyte; *A. nuda* R. Br. (Malay Pen.) – treatment of diarrhoea

Appendicularia DC. Melastomataceae. 1 Guianas

Appertiella C. Cook & Triest. Hydrocharitaceae. 1 Madag.: *A. hexandra* C. Cook & Triest

apple usually *Malus* spp. esp. *M. pumila* long cult. in Eur. Many other fr. trees etc. also called apples (*Angophora* spp. (Aus.); that of the Garden of Eden perhaps *Strychnos*!): **akee a.** *Blighia sapida*; **alligator a.** *Annona* spp.; **Argyle a.** *Eucalyptus cinerea*; **balsam a.** *Momordica balsamina*; **a. banana** *Musa* × *paradisiaca*; **black** or **brush a.** *Planchonella australis*; **a. box** *E. bridgesiana*; **Chinese a.** *Malus prunifolia*; **cocky a.** *Planchonia careya*; **crab a.** *M. sylvestris*, but also applied to any naturalized seedling a. & some orn. flowering spp.; **custard a.** *Annona* spp.; **elephant a.** *Limonia acidissima*; **emu a.** *Owenia acidula*; **golden a.** *Aegle marmelos, Spondias cytherea*; **kangaroo a.** *Solanum aviculare, S. laciniatum*; **kei a.** *Dovyalis caffra*; **love a.** = tomato; **Malay a.** *Syzygium malaccense*; **mammee a.** *Mammea americana*; **a. mango** *Mangifera odorata*; **may a.** *Podophyllum peltatum*; **monkey a.** *Annona* spp., *Anisophyllea laurina, Strychnos* spp.; **oak a.** insect galls on *Quercus* spp.; **Otaheite a.** *Spondias cytherea*; **rose a.** *Syzygium jambos, S. malaccense*; **rose a., water** *S. aqueum*; **sand a.** *Parinari capensis*; **a. of Sodom** *Solanum linnaeanum* etc., *Calotropis procera*; **star a.** *Chrysophyllum cainito*; **sugar a.** *Annona* spp.; **thorn a.** *Datura stramonium*; **velvet a.** *Diospyros* spp.; **wood a.** *Limonia acidissima*

appleberry, purple a. *Billardiera longiflora*

Appunettia R. Good = *Morinda*

Appunia Hook.f. = *Morinda*

Aprevalia Baillon = *Delonix*

apricot *Prunus armeniaca*; **Briançon a.** *P. brigantina*; **Japanese a.** *P. mume*; **a. plum** *P. simonii*; **a.**

vine *Passiflora incarnata*

aprono *Mansonia altissima*

Aptandra Miers. Olacaceae (Aptandraceae). 5 trop. S Am. (4), Afr. (1). *A. spruceana* Miers (Brazil) – root-bark source of an oil

Aptandraceae Miers = Olacaceae

Aptandropsis Ducke = *Heisteria*

Aptenia N.E. Br. Aizoaceae (IV). 4 S Afr., natur. Aus. Cult. orn.

Apterantha C.H. Wright = *Lagrezia*

Apteria Nutt. Burmanniaceae. 3 trop. & warm Am.

Apterigia Galushko = *Thlaspi*

Apterokarpos Rizz. = *Loxopterygium*

Apteropteris (Copel.) Copel. = *Sphaerocionium*

Apterosperma H.T. Chang. Theaceae. 1 China: *A.oblata* H.T. Chang

Apterygia Baehni = *Sideroxylon*

Aptosimum Burchell. Scrophulariaceae. 20 Afr.

Apuleia C. Martius. Leguminosae (I 2). 1 NE Peru, SE Braz., Arg., variable: *A. leiocarpa* (J. Vogel) J.F. Macbr. (*A. praecox*) – tough timber; extrafl. nectaries

Apurimacia Harms. Leguminosae (III 6). 2–4 drier S Am.

Aquifoliaceae A. Rich. Dicots – Dilleniidae – Theales. 4/420 almost cosmop. (*Ilex*). Trees, usually small, or shrubs, evergreen usu. Lvs almost always spiral, rarely in pseudo-whorls (*Phelline*) or opp., often with resiniferous & laticiferous cells in mesophyll, stipules small or 0. Fls small, regular, hypogynous, usu. unisexual (pls dioecious), in axillary, supra-axillary or term. (*Sphenostemon*) fascicles, cymes or racemose, 4(–8)-merous: K imbricate, ± connate to 0; C ± connate at base, imbricate or valvate (*Phelline*), rarely 0; A usu. same no. as & alt. with C, often adnate to C (*Ilex*), s.t. more (12, *Sphenostemon*), ± laminar in *Sphenostemon*; $\underline{G}$ ((2–)4–6(–24)), style short or 0, ovule 1(2) per locule, pendulous, anatropous to s.t. hemitropous (*Phelline*) or ± campylotropous, unitegmic, funicle often with ventral protuberance poss. representing suppressed 2nd ovule. Fr. a drupe with as many pyrenes as carpels; seeds with small embryo near micropyle, endosperm copious, oily & proteinaceous, no starch. x = 9, 10

Genera: *Ilex*, *Nemopanthus*, *Phelline*, *Sphenostemon*

Phelline and *Sphenostemon* have been treated as separate families by some authors. *Ilex* incl. timber & orn. trees etc. (hollies). Prob. allied to Celastraceae but disk 0 & ovules usu. 1 per loc.

Aquilaria Lam. Thymelaeaceae (Aquilariaceae). 15 Indomal. *A. malaccensis* Lam. (*A. agallocha*, agarwood, aloewood, calambac, eaglewood, lign-aloes, Indomal.) – decaying heartwood saturated with a resin, the basis of incense, when distilled used in scent & medicine; fibre used for rope & textiles; the aloe (ahaloth) of Bible, 'aloexylum'

Aquilariaceae R. Br. = Thymelaeaceae

Aquilegia L. Ranunculaceae (III 1). 80 N temp. (Eur. 19). R: GH 7(1946)1. P with long spurs secreting nectar, coll. honeybees; A often >50 in whorls of 5. Alks. Cult. orn. (columbines, granny-bonnets) mostly hybrids, those with hooked spurs usu. derived from *A. vulgaris* L. (Eur., violet-blue fl. form native GB), those with long straight spurs from *A. coerulea* James & *A. chrysantha* A. Gray (N Am.); form. medic. incl. *A. formosa* Fischer (W N Am.) – seeds used to discourage headlice; *A. canadensis* L. (E N Am.) – scarlet fls visited by hummingbirds

Arabian coffee *Coffea arabica*; **A. violet** *Exacum affine*

Arabidella (F. Muell.) O. Schulz. Cruciferae. 6 Aus. R: TRSSA 89(1965)177

Arabidopsis Heynh. Cruciferae. (Incl. *Thellungiella*) 15–20 N temp. (Eur. 6) to trop. Afr. mts. *A. thaliana* (L.) Heynh. (thale cress, N temp.) – subject of genetic research (2n = 10, smallest known genome in vascular pls (haploid = 7×10^7 base-pairs, i.e. 50 times that in bacteria & 5 times that in yeast, life-cycle complete in 1 month)

Arabis L. Cruciferae. 180 N temp. (Eur. 35, N Am. 75 (60 endemic)), Med. to trop. Afr. mts. Cult. orn. herbs often mat-forming grown in rock-gardens etc. (rock or wall cresses) esp. *A. caucasica* Willd. ex Schldl. (*A. alpina* ssp. *caucasica*, *A. albida*, SE Eur. to Iran) – white fls, usu. wingless seeds (dry rocks), often confused with *A. alpina* L. (Eur. & Afr. mts, Himal., discovered on Skye 1887), rarely cult. (seeds winged, moist rocks), *A. glabra* (L.) Bernh. (tower-mustard, Euras., natur. N Am.); *A. pumila* Jacq. (Eur. mts) – prop. by leaves

Aracamunia Carnevali & I. Ramirez. Orchidaceae (III 3). 1 S Venez.: *A. liesneri* Carnevali & I. Ramirez

Araceae Juss. Monocots – Arecidae – Arales. 104 or 105/2550 mostly trop. & subtrop. with few temp. Scrambling shrubs or climbers with aerial roots, herbs (often enormous) with corms or tubers, rarely true epiphytes or free-floating aquatic (*Pistia*), usu. with bundles of raphides throughout shoot, often cyanogenic & s.t. with alks or other toxins; vessels mostly only in roots. Stems sympodial, rarely monopodial; roots mycorrhizal, without roothairs. Lvs spiral or distichous, variously parallel- or net-veined, usu. developing acropetally (like typical dicots). Infl. unbranched spadix (often smelly), usu. terminal & subtended by a ± prominent usu. coloured spathe. Fls many, small, without bracts, poll. by insects, esp. flies (rarely wind) when volatiles incl. ammonia, indole (both stimulating oviposition in carrion flies), skatole, trimethylamine etc., bisexual or unisexual (monoec. with males in upper spadix, rarely pseudo-dioec.). P of 4 or 6(8) distinct or connate tepals in 2 whorls, but reduced or 0 in unisexual fls or always 0; A (1–)4, 6 or 8(–12), sometimes ± connate, anthers opening by term. pores or slits or longit. slits; $\underline{G}$ ((2)3(–c. 50)) usu. with axile placentation, style short or 0, ovules 1–∞ per loc., often anatropous, bitegmic. Fr. usu. berry, rarely dry or leathery & splitting irreg. or whole spadix forming syncarp; seeds 1–∞, sometimes fleshy, embryo large sometimes with small 2nd cotyledon, linear, embedded in copious oily (& sometimes also starchy) endosperm or endosperm 0. x = 7–17 +. R: D. Bown (1988) *Aroids*; BJ 113(1991)396; Willd. 21(1991)35

Classification & chief genera N.B. *Acorus* now referred to own fam., to which subfam. I may belong:

I. **Gymnostachydoideae** (lvs grasslike with striate venation; spathe inconspic., fls bisexual: *Gymnostachys* (only)

II. **Pothoideae** (climbers; lvs entire with lateral veins all reticulate or at least those of 2nd & 3rd so, spathe persistent; 3 genera): *Pothos*

III. **Monsteroideae** (climbers or short-stemmed; lvs often perforate or pinnatifid with varied venation, spathe usu. deciduous; 12 genera in 4 tribes): *Epipremnum, Monstera, Rhaphidophora* (all with 'holes' in lvs, cf. *Monstera*), *Scindapsus, Spathiphyllum*

IV. **Calloideae** (land or marsh pl.; venation striate, fls usu. bisex.; N temp: *Calla* (only genus)

V. **Lasioideae** (land (often climbing) or marsh pls; lvs with reticulate venation; fls bisexual/unisexual, spathe rarely constricted, never deciduous; 23 genera in 8 tribes: *Cyrtosperma, Lysichiton, Symplocarpus, Zamioculcas*

VI. **Philodendroideae** (land or marsh pls; lvs with striate venation; fls unisex.; 18 genera (7 tribes): *Aglaonema, Dieffenbachia, Homalomena, Philodendron, Schismatoglottis, Typhonodorum, Zantedeschia*

VII. **Colocasioideae** (land or (rarely) marsh pls, tuberous or climbing.; lvs with sec. veins joining in a collective vein parallel to primaries, spathe usu. constricted, fls unisex., (A); 14 genera in 6 tribes): *Alocasia, Caladium, Colocasia, Syngonium, Xanthosoma*

VIII. **Aroideae** (land or marsh (some true aquatic) pls; tubers or rhizomes, lvs net-veined, spathe usu. constricted, fls usu. unisexual, A or (A); 32 genera in 6 tribes), several temp., some with epiphyll. infls: *Amorphophallus, Arisaema, Arisarum, Arum, Cryptocoryne, Dracunculus, Sauromatum, Typhonium*

IX. **Pistioideae** (free-living aquatic): *Pistia* (only)

Closely allied to Lemnaceae through Pistioideae and cladistic purists would include L. here. Only 10 genera in both NW & OW, those restricted to NW with some 1400 spp., those to As. 600 & to Afr. & Madag. only 100; many genera mono- or oligotypic with nearly half spp. in *Anthurium* & *Philodendron*. Relic *Protarum* in Seychelles but fam. absent from New Caled. & NZ (& Hawaii)

Of enormous economic importance as foodpls in tropics: *Alocasia, Amorphophallus, Colocasia, Cyrtosperma, Xanthosoma* – cocoyams etc., *Monstera* with ed. fr., *Calla & Arum* minor starch sources. Many familiar housepls incl. spp. of *Aglaonema, Anthurium, Caladium, Dieffenbachia* (poisonous), *Monstera, Philodendron* (poisonous) & *Spathiphyllum* & other cult. orn. esp. *Zantedschia* (journal: *Aroideana*)

Poll. often involves traps & lures for beetles & flies, the interior of the spathe of some trop. spp. concentrating light by reflection & refraction, besides 'window-panes' (cf. *Aristolochia*)

Arachis L. Leguminosae (III 13). 22 S Am. *A. hypogaea* L. (peanut, groundnut, monkeynut, earthnut, Braz.) – widely cult. for seeds crushed for oil for cooking, margarine, soap & cosmetics etc., the cake good for animal feed, raw or roasted for humans (though can cause urticaria), shells used for insulation (USDA Agric. Mon. 19(1954)1); prob. originated as allopolyploid hybrid between ann. & perenn. spp. of E Andes though an annual, fls bending down after fert., pedicel forcing young fr. underground to ripen; alleged remains in China 3300–2800 BC

Arachniodes Blume (~ *Dryopteris*). Dryopteridaceae (I 2). c. 60 trop. & warm esp. As. & Am. (Afr. 1, Madeira 1). Some cult. orn.

Arachnis Blume. Orchidaceae (V 16). 11 SE As., W Mal. Cult. orn. (scorpion orchids) esp. *A* × *maingayi* (Hook.f.) Schltr. 'Maggie Oei' (*A. flos-aeris* (L.) Reichb.f. × *A. hookeriana* Reichb.f. (Malay Pen.), natural & artificial hybrid) – most imp. cut-fl. export of Singapore

Arachnitis Philippi. Corsiaceae. 1 Chile. *A. uniflora* Philippi predicted to be poll. by fungus-gnats

Arachnocalyx Compton. Ericaceae. 2 S Afr.

Arachnothryx Planchon (~ *Rondeletia*). Rubiaceae (I 5). 80 trop. Am. R: ABAH 28(1982)68, 33(1987)301, 35(1989)309

Araeoandra Lefor = *Viviania*

Araeococcus Brongn. Bromeliaceae (3). 5 trop. Am. R: FN 14(1979)1505

Arafoe Pim. & Lavranova. Umbelliferae (III 8). I Caucasus: *A. aromatica* Pim. & Lavranova. R: BZ 74(1989)102

Aragoa Kunth. Scrophulariaceae. 8 Andes. R: Cand. 16(1991)301

Araiostegia Copel. = *Davallia*

Aralia L. Araliaceae (2). 36+ N Am., E As., Mal. Cult. orn., medic. & some ed. *A. chinensis* L. (E As.) – young lvs a vegetable; *A. cordata* Thunb. (Japan) – similar ('udo'); *A. hispida* Vent. (bristly sarsaparilla, N Am.); *A. nudicaulis* L. (wild s., N Am.); *A. racemosa* L. (American spikenard, N Am.) – rhiz. & roots medic.; *A. spinosa* L. (Hercules' club, E US) – bark medic. *A. merrillii* Shang (*A. scandens*, Mal.) – a prickly climber

Araliaceae Juss. Dicots – Rosidae – Apiales. 47/1325 trop. esp. Indomal., Am., few temp. Trees (usu. pachycaul, s.t. unbranched & hapaxanthic as *Harmsiopanax*), shrubs, lianes, woody epiphytes or rarely herbs, s.t. armed, usu. with secretory canals & multilacunar nodes. Lvs spiral, rarely opp. or whorls, ± stipules, usu. palmately or pinnately compound or lobed, s.t. to 2nd or 3rd degree, rarely simple (as *Meryta*). Infls term., rarely lateral, usu. umbels or heads arranged in panicles etc., rarely solit. Fls usu. bisexual, epigynous & 5-merous: K commonly small teeth to 0; C (3–)5(–12), rarely connate at base or forming calyptra, valvate or imbricate; A usu. 1 or 2 x C or ∞, anthers with longit. slits; $\overline{G}$ (2–5(–∞)), rarely $\underline{G}$, locules 1 (e.g. *Arthrophyllum*)–∞, with as many styles, sometimes ± connate, swollen basally & confluent with epigynous nectary-disk, ovules 1(2) per loc., pendulous, anatropous, unitegmic. Fr. a drupe with as many pyrenes as G or a berry, rarely a schizocarp with persistent carpophore like Umbelliferae (e.g. *Myodocarpus, Harmsiopanax*) or drupe with carpophore (*Stilbocarpus*); seeds with small embryo & oily endosperm. x = 11, 12(usu.), or more

Classification & chief genera (after Harms, no better alternative so far proposed):

1. **Schefflereae** (C valvate): *Fatsia, Hedera, Schefflera, Tetrapanax*
2. **Aralieae** (C ± imbricate, sessile with broad base): *Aralia, Panax*
3. **Mackinlayeae** (C valvate, shortly clawed): *Mackinlaya*

Fossils referred to Araliaceae known from Upper Cretaceous. Close to Umbelliferae, included by some authors, *Harmsiopanax* & *Astrotricha* having characters of both fams; also close to Cornaceae to which *Aralidium* and *Diplopanax* (form. here) have been moved Some medic. & drug pls (e.g. *Aralia, Eleutherococcus, Panax*), paper (*Tetrapanax*), timber (*Kalopanax*), but mostly orn. – *Aralia, Eleutherococcus, Fatsia, Hedera, Polyscias, Schefflera*

Aralidiaceae Philipson & Stone. See Cornaceae

Aralidium Miq. Cornaceae (Aralidiaceae). 1 W Mal. Largely intermediate between Cornaceae (esp. *Griselinia*) and Araliaceae, though app. not allied to C. on DNA data

Araliopsis Engl. Rutaceae. 3 trop. W Afr. Drupe with 4 pyrenes

Araliorhamnus Perrier = *Berchemia*

aramina fibre *Urena lobata*

Arapatiella Rizz. & A. Mattos. Leguminosae (I 1). 2 Brazil

arar *Tetraclinis articulata*

araroba *Andira araroba*

Aratitiyopea Steyerm. Xyridaceae. 1 Venez.

araucaria, oil of *Neocallitropsis pancheri*

Araucaria Juss. Araucariaceae. 18 SW Pacific (New Caled. 13), S Brazil to Chile (2). S Am. spp. with broad lvs – *A. araucana* (Molina) K. Koch (monkey-puzzle, Chile pine, Chile, where it is most imp. conifer) with lvs lasting 10–15 yrs (later marcescent & prickly) & ed. seeds (Chile nuts), a favourite architectural tree of the Victorians; *A. angustifolia* (Bertol.) Kuntze (Paraná pine, Brazilian p., S Brazil & nearby Arg.) – imp. timber, seeds ed. Broad lvs also in *A. bidwillii* Hook. (bunya-bunya pine, Queensland) – seeds ed. Narrow lvs in

A. hunsteinii Schumann (*A. klinkii*, klinki pine, New Guinea) – tallest trop. tree (88.9 m), wood imp. esp. for plywood when peeled 'green'. Needle lvs in *A. cunninghamii* D. Don (hoop pine, Moreton Bay pine, E Aus.) – timber & *A. heterophylla* (Salisb.) Franco (Norfolk Is. pine, Norfolk Is.) – planted on Ascension Is. for sailing-ship masts, popular conservatory plant when a seedling, lateral branches may be rooted but never produce a leader, continuing plagiotropic growth indefinitely. R: Phytol.M 7(1984)11

Araucariaceae Henkel & Hochst. Pinopsida. 3/39. S hemisph. (exc. Afr.) to SE As. Monoec. to dioec. evergreen trees with broad to needle lvs, sometimes pungent. Male cones cylindrical, the pollen without air-bladders, females usu. large, ± globose, taking 2–3 yrs to mature & disintegrating when seeds mature. Ligule in *Araucaria*, ± adnate to carpel, absent in others. Ovule 1, free (immersed in ligule in *Araucaria*). Cotyledons 2, occ. deeply 2-cleft. x = 13
Genera: *Agathis, Araucaria, Wollemia*
Differing from Pinaceae in lvs & ovule 1; fossils back to Triassic in both hemispheres
Important timber trees

Araujia Brot. Asclepiadaceae (III 1). 2–3 S Am. (1 natur. Aus.). *A. sericifera* Brot. (cruel plant, S Brazil) – cult. orn. climber with fls that hold proboscides of night-flying moths until daytime, fibre used in textiles

Arbelaezaster Cuatrec. (~ *Senecio*). Compositae (Sen.-Sen.) 1 Colombia: *A. ellsworthii* (Cuatrec.) Cuatrec. R: Cand. 15(1986)1

Arberella Söderstrom & Calderón (~ *Cryptochloa*). Gramineae (IV). 6 trop. Am.

arboloco *Montanoa quadrangularis*

arbor-vitae *Thuja* spp.; **American a.** *T. occidentalis*; **Chinese a.** *Platycladus orientalis*; **giant a.** *T. plicata*; **Japanese a.** *T. standishii*; **western a.** *T. plicata*

Arbulocarpus Tenn. = *Spermacoce*

Arbutus L. Ericaceae. c. 14 W N Am., W Eur. (2) to Med. Small trees & shrubs with red flaking bark & berries, cult. orn. *A. menziesii* Pursh (madroña, madrono, W N Am.) – tanbark & timber; *A. unedo* L. (strawberry tree, S Eur. & Ireland) – female gametophyte ready for fert. at pollination but first zygotic division 5 months, tanbark, fr. ed., preserved & flavour in liqueurs (esp. in Portugal), fls while previous yr's fr. ripening, part of Lusitanian element in British flora

Arcangelina Kuntze = *Tripogon*

Arcangelisia Becc. Menispermaceae (III). 2 SE As. to Mal. R: KB 32(1978)333. *A. flava* (L.) Merr. – germicidal dye from wood, root infusion medic. & abortifacient

Arceuthobium M. Bieb. Viscaceae. 31 N to C Am., WI, Medit. (Eur. 2, usu. on *Juniperus*), NE trop. Afr. (1), Sinohimal. to W Mal. R: USDA Agr. Handbk. 40(1972). Dwarf mistletoes (? always) on gymnosperms. Fr. explosive

Arceuthos Antoine & Kotschy = *Juniperus*

Archangel redwood or **A. yellow deal** *Pinus sylvestris*; **yellow a.** *Lamium galeobdolon*

Archangelica Wolf = *Angelica*

Archangiopteris Christ & Giesenh. = *Angiopteris*

Archboldia E. Beer & H.J. Lam = *Clerodendrum* (galled, *teste* De Kok)

Archboldiodendron Kobuski. Theaceae. 1 New Guinea

Archeria Hook.f. Epacridaceae. 4 Aus., NZ

Archiatriplex G.L. Chu. Chenopodiaceae (I 3). 1 Sichuan: *A. nanpinensis* G.L. Chu. R: JAA 68(1987)461

Archibaccharis Heering. Compositae (Ast.-Ast.). c. 30 Mex. to C Panamá. R: Phytologia 32(1975)81

Archiboehmeria C.J. Chen (~ *Boehmeria*). Urticaceae (III). 1 S China to SE As.: *A. atrata* (Gagnep.) C.J. Chen

Archiclematis (Tamura) Tamura (~ *Clematis*). Ranunculaceae (II 2). 1 Himal.

Archidendron F. Muell. Leguminosae (II 5). 94 Indomal. (esp. Borneo). R: OB 76(1984). G 1 – several, dark blue to black seeds hanging by funicles from red inner wall of carpel in many spp.

Archidendropsis I. Nielsen (~ *Albizia*). Leguminosae (II 5). 14 SW Pacific. R: BMNHN 4,5(1983)335

Archineottia S.C. Chen = *Neottia*

Archiphysalis Kuang (~ *Physalis*). Solanaceae. 3 E As.

Archirhodomyrtus (Niedenzu) Burret (~ *Rhodomyrtus*). Myrtaceae. 1 Aus., 5 New Caled.

Archontophoenix H.A. Wendl. & Drude. Palmae (V 4h). 3 E Aus. Cult. orn.: *A. alexandrae* (F. Muell.) H.A. Wendl. & Drude (Alexandra palm) & *A. cunninghamiana* (H.A. Wendl.) H.A. Wendl. & Drude (bangalow (palm), Illawara palm, piccabeen) – prot. in Aus.

Archytaea C. Martius. Guttiferae (II, Bonnetiaceae). 2 NE S Am.

Arcoa Urban. Leguminosae (I 1). 2 Hispaniola

Arctagrostis Griseb. Gramineae (17). 1 Arctic inc. Eur., marshy tundra

Arctanthemum (Tzvelev) Tzvelev. (~ *Chrysanthemum*). Compositae (Anth.-Art.). 3 Arctic & subarctic. Cult. orn.

Arcteranthis E. Greene (~ *Ranunculus*). Ranunculaceae (II 3). 1 NW N Am.

Arcterica Cov. = *Pieris*

Arctium L. Compositae (Card.-Card.). 11 OW temp. (Eur.4). R: Gorteria suppl. 3(1996). Burdocks, the burs being fr. heads covered with involucral bracts, which became hooked & woody after fert. & get attached to animal fur, clothing etc. & promote dispersal. *A. lappa* L. cult. for ed. roots (gobō) in Japan etc., young lvs ed. as salad in Scandinavia, Canada & Japan; dried root medic. esp. skin complaints (also *A. minus* (Hill) Bernh.), form. for gonorrhoea, now herbal treatment of arthritis, gout, eczema & anorexia nervosa

Arctogentia Löve = *Gentianella*

Arctogeron DC. Compositae (Ast.-Ast.). 1 C As. to China & Mongolia: *A. gramineus* (L.) DC

Arctomecon Torrey & Frémont. Papaveraceae (IV). 3 SW US. Persistent C. Cult. orn.

Arctophila (Rupr.) Anderss. (~ *Colpodium*). Gramineae (17). 1 Arctic

Arctopoa (Griseb.) Probat. = *Poa*

Arctopus L. Umbelliferae (I). 3 S Afr. *A. echinatus* L. roots medic.

Arctostaphylos Adans. Ericaceae. (Excl. *Comarostaphylis*) 50 W N Am. with 2 circumpolar (Eur. 2). R: JEMSS 56(1940)1. Mostly prostrate shrubs (palo blanco), cult. orn., fr. ed. – ground as meal or in drinks, seed germination enhanced by immersion in H_2SO_4. *A. alpina* (L.) Sprengel (black bearberry, circumpolar); *A. pungens* Kunth (manzanita, W N Am.) – to 3 m, orn. wood; *A. uva-ursi* (L.) Sprengel (bearberry, uva-ursi, circumpolar) – lvs used in tanning 'Russian leather', or N Am. smoking mixtures (kinnikinni(c)k) & in tea in Russia, as urinary antiseptic in UK since C13

Arctotheca Wendl. Compositae (Arct.-Arct.). 4 S Afr., 2 natur. Aus., 1 in Eur. (*A. calendula* (L.) Levyns – Capeweed)

Arctotis L. Compositae (Arct.-Arct.). 50 S Afr. to Angola. Cult. orn. (incl. *Venidium*)

Arctottonia Trel. Piperaceae. 3 C Am.

Arctous (A. Gray) Niedenzu = *Arctostaphylos*

Arcuatopterus Sheh & Shan (~ *Ferula*). Umbelliferae (III 10). 3 China. R: BBR 6,2(1986)11

Arcyosperma O. Schulz. Cruciferae. 1 Himalaya

Arcypteris Underw. = *Pleocnemia*

Arcytophyllum Willd. ex Schultes & Schultes f. Rubiaceae (IV 1). 15 trop. Am. mts. R: MNYBG 60(1990). Some weedy in OW ('*Anotis*')

Ardisia Sw. Myrsinaceae. 250 trop. & warm exc. Afr., rare Aus. R: PNASP 141(1989)268. Some med. and ed. fls or fr., cult orn. esp. *A. crenata* Sims (often confused with *A. crispa* (Thunb.) A. DC, which does not have crisped lvs) of NE India to Japan, undergrowth treelet with crimped lvs (poss. proteinaceous deposits or bacterial nodules responsible) & red fr. retained for several seasons

Ardisiandra Hook.f. Primulaceae. 3 trop. Afr. mts

Areca L. Palmae. 60 Indomal. Cult. orn. Alks. *A. catechu* L. (betelnut, pinang, areca nut), cultigen (? tetraploid orig. in Sulawesi) throughout trop. As. (prob. cult. by 'Hoabhinians' 8000–3000 BP): seed, cut into slices and usu. chewed in a wad of betel pepper (*Piper betle*) lvs with lime, causes saliva to turn red & promotes salivation & wellbeing (arecaine a mild narcotic), dulling appetite & medic. Prob. formerly used in some Eur. tooth-powders and poss. origin of myth of anthropophagous western invaders in New Guinea folklore. Bracts used as curd containers in Sri Lanka; brown dye from wood; closest ally: *A. concinna* Thwaites (Sri Lanka). *A. vestiaria* Giseke (C Mal.) – epidermis of young lvs for textiles

Arecaceae C.H. Schultz. See Palmae

Arecastrum (Drude) Becc. = *Syagrus*

Arechavaletaia Speg. = *Azara*

Aregelia Kuntze = *Nidularium*, Mez = *Neoregelia*

Aremonia Necker ex Nestler. Rosaceae. 1 SE Eur.: *A. agrimonioides* (L.) DC

Arenaria L. Caryophyllaceae (II 1). c. 150 N temp., consp. in mts & Arctic (Eur. 54). Sandworts. R: JLSB 33(1898)326. Cult. orn. & weeds. *A. bryophylla* Fern. (Himal.) – pl. at highest alt. (6180m) on Mt Everest

Arenga Lab. Palmae (V 1). c. 20 trop. As. (Mal. 15) to N Aus. Cult. orn., juicy fr. irritant. *A. microcarpa* Becc. (Mal.) – sago; *A. pinnata* (Wurmb) Merr. (*A. saccharifera*, sugar-palm, gomuti, ejow, Mal.) – widely cult., male spadices tapped for syrupy sap evaporated to produce palm sugar (jaggery), palm wine or toddy, (when distilled) arrack, sago from trunk, good fibre from leaf sheaths, hapaxanthic, infls appearing in descending order. *A. tremula* (Blanco) Becc. (Philipp.) – bud ed., narcotic

Arenifera Herre (~ *Psammophora*). Aizoaceae (V). 1 W S Afr.

arere *Triplochiton scleroxylon*

Arethusa L. Orchidaceae (V 11). 1 E N Am.: *A. bulbosa* L., cult. orn.

Aretiastrum (DC) Spach = *Valeriana*

Arfeuillea Pierre ex Radlk. Sapindaceae. 1 SE As.

argan = seq.

Argania Roemer & Schultes. Sapotaceae (III). 1 Morocco, Algeria, introd. Libya, natur. S Spain: *A. spinosa* (L.) Skeels (argan) – seed oil like olive oil for cooking, fr. ed. cattle, timber good, gum valuable

Argemone L. Papaveraceae (IV). c. 23 N & S Am., WI, Hawaii. R: MTBC 21(1958)1, Brittonia 13(1961)91. Alks. Cult. orn. annuals & 1 shrub, some medic. *A. mexicana* L. (Mexican poppy, prickly p., C Am., now pantrop. weed) – seeds a minor oil source though as contaminants in grain thought to cause glaucoma in India, latex a yellow dye in Peru

Argentipallium P.G. Wilson (~ *Ozothamnus*). Compositae (Gnap.-Ang.). 6 temp. Aus. R: Nuytsia 8(1992)455

Argeta N.E. Br. = *Gibbaeum*

Argillochloa W. Weber = *Festuca*

Argocoffeopsis Lebrun. Rubiaceae (II 1). 8 trop. Afr. R: BJBB 51(1981)361

Argomuellera Pax. Euphorbiaceae. 10 trop. Afr.

Argophyllaceae (Engl.) Takht. = Grossulariaceae

Argophyllum Forster & Forster f. Grossulariaceae (Escalloniaceae). 11 trop. Aus., New Caled.

Argopogon Mimeur = *Ischaemum*

Argostemma Wallich. Rubiaceae (IV 4). 100 OW trop.(Borneo 28, R: AMBG 76(1989)7). Some Mal. spp. with black lvs due to complete absorption of light by compact palisade cells full of chloroplasts

Argostemmella Ridley = *Argostemma*

Argusia Boehmer (~ *Tournefortia*). Boraginaceae. 4 Romania to Japan (1), trop.; halophytes. R: FNC 7(1976)108

Argyle apple *Eucalyptus cinerea*

Argylia D. Don. Bignoniaceae. 10 S Peru, Chile, Arg. Herbs

Argyll's tea-tree, Duke of *Lycium barbarum*

Argyranthemum Webb ex Schultz-Bip. Compositae (Anth.-Chrys.). 24 Macaronesia. R: BBMNH 5(1976)147. Cult. orn. (marguerites) incl. 'doubles' & hybrids, sometimes grown esp. as standards. *A. frutescens* (L.) Schultz-Bip. (Gran Canaria) – aggressive colonist in NZ

Argyreia Lour. Convolvulaceae. 90 Indomal. to Aus. (1). Few cult. orn. esp. *A. nervosa* (Burm.f.) Bojer (*A. speciosa*, elephant climber, India) – med. in India, dried fr. with accrescent K sold as 'wood rose' (smaller than *Merremia* spp.) in Hawaii

Argyrochosma (J. Sm.) Windham = *Cheilanthes*

Argyrocytisus (Maire) Raynaud (~ *Cytisus*). Leguminosae (III 31). 1 Morocco: *A. battandieri* (Maire) Raynaud, cult. orn.

Argyrodendron F. Muell. (~ *Heritiera*). Sterculiaceae. 7 Aus. Good timber (booyong, tulip oak)

Argyroderma N.E. Br. Aizoaceae (V). 10 S Afr. R: MSABH 15(1977)121. Cult. orn. succulents

Argyroglottis Turcz. (~ *Helichrysum*). Compositae (Gnap.-Gnap.). 1 N Aus.: *A. turbinata* Turcz. R: OB 104(1991)80

Argyrolobium Ecklon & Zeyher. Leguminosae (III 31). 70 S Afr. to trop. Afr. highlands, Madag., Med. (Eur. 2) to India. Alks. Mostly xerophytes, some with cleistogamous fls

Argyronerium Pitard = *Epigynum*

Argyrophanes Schldl. = *Chrysocephalum*

Argyrovernonia MacLeish (~ *Chresta*). Compositae (Vern.-Vern.). 2 Brazil

Argyroxiphium DC. Compositae (Hele.-Mad.). 5 Maui & Hawaii. Woody hapaxanthic monocaulous to polyxanthic branched pachycauls (silverswords), some high alt. R: Allertonia 4,1(1985)50. *A. sandwicense* DC fls after 15–50 yrs; *A. virescens* Hillebr. extinct (last seen 1945)

Argythamnia P. Browne. Euphorbiaceae. 17 trop. Am. R (subg.A.): GH 10(1966)1

arhar *Cajanus cajan*

Aria (Pers.) Host. Segregate of *Sorbus*, q.v.

Ariadne Urban. Rubiaceae (?). 2 E Cuba (serpentine)
Ariaria Cuervo = *Bauhinia*
aridan *Tetrapleura tetraptera*
Aridaria N.E. Br. = *Phyllobolus*
Aridarum Ridley. Araceae (VI 1). 7 Borneo, some rheophytes
Arikuryroba Barb. Rodr. = *Syagrus*
Arillastrum Pancher ex Baill. (~ *Stereocaryum*). Myrtaceae. 3 New Caled.
Ariocarpus Scheidw. Cactaceae (III 9). 6 S Texas, Mex. R: AJB 50(1963)724, 51(1964)144. Cult. orn. slow-growing
Ariopsis Nimmo. Araceae (VII 6). 1 Indomal.: *A. peltata* Nimmo
Ariosorbus Koidz. = *Sorbus*
Arisaema C. Martius. Araceae (VIII 6). 150 E Afr. & Arabia (10, R: KB 41(1986)261), trop. & E As. (Japan 42), W N Am. Some with long spadix appendages dispersing odours attractive to pollinators, some poss. poll. by snails, *A. utile* Hook.f. ex Schott (Himal.) by fungus-gnats. Cult. orn. (cobra-lily, dragon-arum, snail-fl.; R: Plantsman 3(1982)193), some medic. & with ed. corms, esp. *A. flavum* (Forssk.) Schott (Yemen to W China) in Arabia & *A. triphyllum* (L.) Torrey (Jack-in-the-pulpit, Indian turnip, E N Am.) – smallest infertile, larger reproduce as males, largest as females
Arisarum Miller. Araceae (VIII 6). 3 Med. (Eur. 2). R: Kew Mag. 7(1990)16. Cult. orn. *A. proboscideum* (L.) Savi (mouseplant, Italy, SW Spain) with long tail-like process on spathe, poll. by fungus-gnats (in spring – few fungi available!) like *Asarum* spp. (q.v.); *A. vulgare* Targ.-Tozz. (friar's cowl, Med.) – elaiosomes, corms a famine-food
Arischrada Pobed. = *Salvia*
Aristavena F. Albers & Butzin = *Deschampsia*
Aristea Aiton. Iridaceae (II). 50 trop. & S Afr., Madag. Cult. orn.
Aristeguietia R. King & H. Robinson. Compositae (Eup.-Crit.). 21 Andes. R: MSBMBG 22(1987)343
Aristeyera H. Moore = *Asterogyne*
Aristida L. Gramineae (28). c. 330 warm (Eur. 1). R: MRL 58, A & B (1929–33). *A. contorta* F. Muell. (Aus.) fr. round legs can immobilize sheep; *A. funiculata* Trin. & Rupr. (N Afr. to Pakistan) – florets in balls to 25 cm diam. disp. along ground by wind in Sudan; *A. pungens* Desf. (N Afr.) – famine grain
Aristocapsa Reveal & Hardman (~ *Centrostegia*). Polygonaceae (I 1). 1 Calif.: *A. insignis* (Curran) Reveal & Hardman. R: Phytol. 66(1989)84
Aristogeitonia Prain. Euphorbiaceae. 4 trop. Afr., Madag. R: KB 43(1988)627
Aristolochia L. Aristolochiaceae (II 2). 120 trop. & warm OW (Eur. 19 native; E & S As. 68, R: APS 27(1989)321). See also *Einomeia*, *Endotheca*, *Howardia*, *Isotrema* & *Pararistolochia*. Perianth (calyx) a flytrap attracting insects by smell to pollinate after sliding down a 'slip zone'. Their phototropism attracts them to windows at base of tube in some spp., so that they cannot escape until it expands later, by which time the stamens have matured & deposited pollen on them. In other spp. insects are detained until downward-pointing hairs wither & in yet others insects are attracted by production of heat (cf. *Arum*), though some can be regularly self-poll. All have poisonous alks or aristolochic acid. Many spp. with large fleshy funicle attractive to seed-dispersers. Many medic. (Gk. *aristos* = best, *lochia* = childbirth, from the curved form of the flower of *A. clematitis* recalling human foetus in right position prior to birth, the Doctrine of Signatures indicating that the plant would ease parturition (the pls are efficacious abortifacients), some cult. orn. *A. clematitis* L. (birthwort, Eur.) – abortifacient ± natur. GB; *A. petersiana* Klotzsch (E Afr.) toxic to stock, arrow poison
Aristolochiaceae Juss. Dicots – Magnoliidae – Aristolochiales. 12/475 trop. & warm esp. Am. Aromatic lianes, scramblers, shrubs or rhiz. herbs with alks or aristolochic acid. Lvs spiral, simple, s.t. lobed, true stipules 0. Fls solitary or in terminal or lateral racemes or cymes, bisexual, reg. or irreg., often smelling of rotting meat. P (prob. K) usu. (3) with S-shaped tube usu. & with 0–3 lobes or reg. & 3-lobed, often petaloid; C 0 or small, but 3 in *Saruma*; nectaries (patches of secretory hairs in C-tube) often present; A (5)6 – c. 40 (fewer in some *Aristolochia*) in 1 or 2 (to 4 in *Thottea*) whorls free or ± united with style to form gynostemium, anthers extrorse (rarely introrse); G (4–6), almost distinct in *Saruma* to inferior, 4–6-loc., s.t. with incomplete partitions; ovules ∞ per loc., bitegmic, usu. anatr. Fr. usu. septicidal capsule, sometimes with fleshy endocarp, many-seeded, rarely follicular (*Saruma*) or indehiscent & 1-seeded, dehiscing basipetally (*Thottea*) or acropetally (most *Aristolochia*); seeds with minute embryo & copious oily (sometimes also starchy) endosperm. x = 4–7, 12, 13, 16 poss. orig. 7

Classification & genera:

I. **Asaroideae** (herbs, not twining, fls not constricted between P & G): *Saruma, Asarum*

II. **Aristolochioideae** (woody or herbaceous, often climbing; fls constricted between P & G; 2 tribes): *Aristolochia , Asiphonia, Einomeia, Endotheca, Euglypha, Holostylis, Howardia, Isotrema, Parаristolochia, Thottea*

Very distinct family with *Saruma* showing app. primitive characters typical of magnoliid stock. Seed structure shows some affinity with Caricaceae

Cult. orn. (*Aristolochia, Asarum, Isotrema*), some medic. (*Aristolochia, Thottea*)

Aristopsis Guerra = *Aristida*

Aristotelia L'Hér. Elaeocarpaceae. 5 E Aus., NZ, Peru to Chile. R: KB 40(1985)491. Cult. orn. esp. *A. chilensis* (Molina) Stuntz (*A. macqui*, macqui, Chile – small ed. fr. preserved & used to colour wine) & *A. serrata* (Forster & Forster f.) W. Oliver (*A. racemosa*, NZ) – useful wood incl. inlay-work

Arizona poppy *Kallstroemia grandiflora*

Arjona Cav. Santalaceae. 10 temp. S Am. *A. tuberosa* Cav. with ed. tubers (macachi, Chile)

arjun *Terminalia arjuna*

Armatocereus Backeb. (~ *Lemaireocereus*). Cactaceae (III 7). c. 10 Colombia, Ecuador, Peru

Armeniaca Miller = *Prunus*

Armeria Willd. Plumbaginaceae (II). 100 N temp. (Eur. 43), Andes to Tierra del Fuego. Cult. orn. tufted perennials. After fert. K becomes a membranous organ aiding wind-disp. Common on coasts, mts etc. Most cvs derived from *A. maritima* (Miller) Willd. (range of genus (polymorphic), thrift, sea-pink) form. used in treatment of obesity

Armodorum Breda (~ *Arachnis*). Orchidaceae (V 16). 4 Indomal.

Armoracia P. Gaertner, Meyer & Scherb. Cruciferae. 4 E & SE Eur. (2) to Siberia, E N Am. (1). R: JAA 69(1988)160. *A. rusticana* P. Gaertner, Meyer & Scherb. (horse-radish, ? cultigen (or poss. native SW C As.) – clones with irreg. meiosis perhaps of hybrid origin, seeds not set) – root source (with salt, oil & vinegar) of relish for beef, oysters etc., prop. by root-cuttings for over 2000 yrs

Armouria Lewton = *Thespesia*

Arnaldoa Cabrera. Compositae (Barn.). 3 Peruvian Andes. Shrubs

Arnanthus Baehni = *Pichonia*

arnatto *Bixa orellana*

Arnebia Forssk. Boraginaceae. 25 Med. (Eur. 1), trop. Afr., Himal. Some, incl. *A. pulchra* (Roemer & Schultes) Edmondson (*A. echioides, Echioides longiflora*, prophet flower, Armenia, Cauc. to Iran), have black spots on C, fading as it matures. *A. benthamii* (G. Don f.) I.M. Johnston a pachycaul of the Himal.

Arnebiola Chiov. = *Arnebia*

Arnhemia Airy Shaw. Thymelaeaceae. 1 Aus.

Arnica L. Compositae (Hele.-Cha.). 32 N temp. & arctic (Eur. 2). R. Brittonia 4(1943)386. Cult. orn. incl. *A. montana* L. (arnica, mountain tobacco, C & N Eur.) – root & capitula medic. (tincture of a.), though *A. fulgens* Pursh (N Am.) more efficacious

Arnicastrum Greenman. Compositae (Hele.-Cha.). 2 Mex. R: SB 11(1986)277

Arnicratea Hallé. Celastraceae. 3 Indomal. R: BMNHN 4,6(1984)12

Arnocrinum Endl. & Lehm. Anthericaceae (Liliaceae s.l.). 3 SW Aus. R: FA 45(1987)246

Arnoglossum Raf. Compositae (Senec.-Tuss.). 7 E US. R: Phytologia 28(1974)294. New World '*Cacalia*' spp. *A. muhlenbergii* (Schultz-Bip.) H.Robinson (*C. m.*) – poss. oil source

Arnoseris Gaertner. Compositae (Lact.-Hier.). 1 Eur.: *A. pusilla* Gaertner (swine or lamb's succory)

Arnottia A. Rich. Orchidaceae (IV 2). 2 Mauritius

Aromadendron Blume = *Magnolia*

Aronia Medikus = *Photinia*

Arophyton Jum. Araceae (VIII 2). 7 Madag. R: BJ 92(1972)24

Aropsis Rojas = *Spathicarpa*

Arpitium Necker ex Sweet. Older name for *Endressia*

Arpophyllum Lex. Orchidaceae (V 13). 5 trop. Am. R: Orquidea 4(1974)16. Cult. orn.

Arrabidaea DC. Bignoniaceae. c. 50 trop. Am. 4-armed xylem in t.s. with 4 consp. ribs on bark of big lianes (phloem). Some cult. orn. esp. *A. rotundata* (DC) Schumann (Brazil); *A. chica* (Kunth) Verl. (chica, S Am.) – form. body cosmetic (red) prepared from lvs, dye-pl. (Peru) & to blacken teeth, local med.; many spp. toxic to stock

arracacha *Arracacia xanthorrhiza*

Arracacia Bancr. Umbelliferae (III 5). 55 trop. Am. *A. xanthorrhiza* Bancr. (arracacha, apio, Peruvian parsnip, N S Am.) – Andean root-crop with strong parsnip taste, locally replacing potatoes

arrack potable spirit distilled from sugary sap of palms, esp. spp. of *Arenga, Borassus, Caryota, Cocos, Corypha*

arrayán *Luma apiculata*

Arrhenatherum Pal. (~ *Helictotrichon*). Gramineae (21b). 6 Eur. (5), Med., N & W As. R: KBAS 13(1986)124. Like *H.* but florets dimorphic, though dimorphism varying within same panicle. *A. elatius* (L.) J. & C. Presl (false oat, French rye-grass, Eur.) – form. in seed mixtures for hay; sometimes with basal internodes swollen to 1.6 cm diam. – propagules (var. *bulbosum* (Willd.) St Amans, onion couch)

Arrhenechthites Mattf. Compositae (Senec.-Senec.). 6 mts of Celebes & New Guinea, Aus. alps. R: AMBG 43(1956)74

Arrhostoxylon Nees = *Ruellia*

Arrojadoa Britton & Rose. Cactaceae (III 3). 3 E Braz. R: Bradleya 6(1988)90, 7(1989)35. Cult. orn. Stem tubers in *A. dinae* Buining & Brederoo

Arrojadocharis Mattf. Compositae (Eup.-Gyp.). 2 E Brazil. R: MSBMBG 22(1987)118

arrow poisons Usu. mixtures of pl. toxins with adhesive gums & s.t. animal substances (often of magical significance only). In Afr. *Adenium* & *Acokanthera*, Am. *Strychnos* etc. (q.v.)

arrowgrass *Triglochin* spp.

arrowhead *Sagittaria sagittifolia* & other spp.

arrowroot *Maranta arundinacea* (**WI, St Vincent** or **Bermuda a.**). Other starches with small grains used as subs.: **Afr. a.** *Tacca leontopetaloides*, **Bombay a.** *Curcuma angustifolia;* **Braz. a.** *Manihot esculenta;* **Chinese a.** *Nelumbo nucifera;* **East Indian a.** *Curcuma angustifolia;* **Fiji a.** *Tacca leontopetaloides;* **Florida a.** *Zamia pumila;* **Guyana a.** *Dioscorea alata;* **Hawaiian a.** *Tacca leontopetaloides;* **Indian a.** *Curcuma angustifolia;* **Japanese a.** *Pueraria montana* var. *lobata;* **marble a.** *Myrosma cannifolia;* **Pará a.** *Manihot esculenta;* **Portland a.** *Arum maculatum;* **Queensland a.** *Canna indica;* **Rio a.** *Manihot esculenta;* **Tahiti a.** *Tacca leontopetaloides*

Arrowsmithia DC (~ *Macowania*). Compositae (Gnap.). 1 E Cape: *A. styphelioides* DC. R: OB 104(1991)51

arrow-weed *Tessaria* spp.

arrow-wood *Viburnum acerifolium, V. dentatum*

Artabotrys R. Br. Annonaceae. 100+ OW trop. (Afr. 31). Some climbing with recurved hooks (woody peduncles of term. infls, i.e. shoots sympodial). Alks. Some ed. fr. *A. hexapetalus* (L.f.) Bhand. (*A. odoratissimus, A. uncinatus*, India, Sri Lanka) – cult orn., a stimulant tea made from aromatic fls

Artanacetum (Rzazade) Rzazade = *Artemisia*

Artanema D. Don. Scrophulariaceae. 4 OW trop.

Artedia L. Umbelliferae (III 3). 1 E Med.

Artemisia L. Compositae (Anth.-Ast.). c. 350 N temp. (Eur. 55; China 170), W S Am., S Afr. (1); usu. dry areas. R: CN 25(1994)39. Aromatic shrubs & herbs (ragweeds): vermifuges, stimulants, culinary herbs, cult. orn. Charac. of Russian steppes. See also *Seriphidium*. Capitula small, wind-poll., causing hay-fever problems (in US 1 sq. mile emits 16 t in 2 wks (Aug.–Sept.). *A. abrotanum* L. (southernwood, lad's love, old man, origin unknown) – v. rarely flowering, medic., also tea; *A. absinthium* L. (absinthe, wormwood, Eur.) – a ketone (thujone) medic., pl.-pest control since time of Pliny (AD 77) & absinthe (harmful liqueur); *A. afra* Jacq. ex Willd. (trop. Afr. to E Cape) – local medic.; *A. annua* L. (Euras., natur. N Am.) – efficacious antimalarial (huanghuahaosu) in China, due to artemisin (sesquiterpene lactone) also phytotoxic, lvs used for burns; *A. arborescens* L. (Medit.) – hedgepl. in NZ; *A. cina* Berg ex Polj. (Levant wormseed, Turkestan) – medic., anthelmintic (santonin) known as santonica; *A. dracunculus* L. (tarragon, estragon, Euras.) – lvs for flavouring (terpineol), esp. t. vinegar used with fish; *A. glacialis* L. & *A. umbelliformis* Lam. (*A. laxa*, Alps) flavouring for genépi liqueur; *A. herba-alba* Asso (Med.) wormwood of the Bible; *A. lactiflora* Wallich (China) – cult. orn. herb. perenn.; *A. pontica* L. (Roman wormwood, SE Eur.) – flavour of vermouth; *A. tilesii* Ledeb. (Arctic) – medic. (Esquimaux), properties like codeine; *A. vulgaris* L. (mugwort, Euras.) – lvs a condiment, also used in magic & superstition in GB until C19

Artemisiastrum Rydb. = *Artemisia*

Artemisiella Ghafoor. Compositae (Anth.-Tan.). 1 Himal. to China

Artemisiopsis S. Moore. Compositae (Gnap.). 1 S trop. Afr.: *A. villosa* (O. Hoffm.) Schweick. R: OB 104(1991)55

Arthraerva (Kuntze) Schinz. Amaranthaceae (I 2). 1 SW Afr.

Arthragrostis Lazarides (~ *Panicum*). Gramineae (34b). 1 Queensland: *A. deschampsioides*

(Domin) Lazarides. R: Nuytsia 5(1984)285

Arthraxon Pal. Gramineae (40f). 20 warm As. to Aus. (2, 3 also pantrop.). R: Blumea 27(1981)255. *A. hispidum* (Thunb.) Makino (*A. ciliare*, trop. & warm OW, natur. Am.) – yellow dye, medic. in China

Arthrobotrya J. Sm. = *Teratophyllum*

Arthrocarpum Balf. f. Leguminosae (III 13). 2 C Somalia, 1 Socotra

Arthrocereus A. Berger. Cactaceae (III 4). 5 W & SE Brazil. Cult. orn.

Arthrochilus F. Muell. (~ *Spiculaea*). Orchidaceae (IV 1). 10 Aus. (10, 8 endemic), New Guinea. R: PRSQ 86(1975)155

Arthroclianthus Baillon. Leguminosae (III 9). 10 New Caled.

Arthrocnemum Moq. Chenopodiaceae (II 2). Excl. *Sarcocornia*, 2–3 coasts Medit. (Eur. 2), As. S Afr. Alks. Some ed.

Arthromeris (T. Moore) J. Sm. Polypodiaceae (II 2). 9 N India to S China & (?) Borneo

Arthrophyllum Blume. Araliaceae. 31 Indomal., W Pacific (New Caled. 10)

Arthrophytum Schrenk. Chenopodiaceae (III 3). 9 W & C As. *A. arborescens* Litv. (Turkestan) – valuable source of firewood, cult. in Sahara

Arthropodium R. Br. Anthericaceae (Liliaceae s.l.). 11 Madag. (1), Aus. (7), NZ (2), New Caled. (1). Cult. orn. (chocolate lilies, i.e. scented thus) esp. *A. cirrhatum* (Forst.f.) R. Br. (rock-lily, NZ)

Arthropogon Nees. Gramineae (34e). 7 trop. Am. R: Bradea 3(1982)303, Brittonia 38(1986)71

Arthropteris J. Sm. ex Hook.f. Oleandraceae). 12–15 OW trop. (esp. Madag. & New Guinea) to NZ & Juan Fernandez (1). Cult. orn.

Arthrosamanea Britton & Rose = *Albizia*

Arthrosolen C. Meyer = *Gnidia*

Arthrostemma Pavón ex D. Don. Melastomataceae. 4 trop. Am. Cult. orn.

Arthrostylidium Rupr. Gramineae (1b). 20 trop. Am. Climbing & scrambling bamboos, some prickly

Arthrostylis R. Br. Cyperaceae. 1 trop. Aus.

Artia Guillaumin. Apocynaceae. 7 China (1), Hainan (1), SE As. (1), New Caled. (4)

artichoke, Chinese or **Japanese** *Stachys affinis*; **common, globe** or **French a.** *Cynara scolymus*; **Jerusalem a.** *Helianthus tuberosus*

artillery plant *Pilea microphylla*

Artocarpus Forster & Forster f. Moraceae (II). 50 Indomal. R: JAA 40(1959)113, 298, 327, Blumea 22(1975)409. Timber & fr. trees. thick white latex used as a bird-lime; lvs simple, lobed or pinnate; stipules usu. large, conical, covering buds & leaving scars; monoec. with fls in heads, females swelling to become (often huge) fr. with large seeds embedded in head & covered with waxy or pulpy layer (enlarged K-tube), strap-shaped parts between seeds being undeveloped female fls. *A. altilis* (Z) Fosb. (*A. communis*, *A. incisus* breadfruit, C Mal. to Melanesia (seeded forms) – cult. esp. Pacific trop. fr. (starchy), cvs usu. seedless, orig. introd. from Bligh's 2nd voyage (1793); *A. chama* Buch.-Ham. (*A. chaplasha*, chaplash, India, Burma, Andamans, Nicobars) – timber; *A. elasticus* Reinw. ex Blume (SE As., W Mal.) – bark-cloth, lvs to 2 m in sapling stages; *A. heterophyllus* Lam. (jak, trop. As., ? native S India) – fast-growing (15 m in 3 yrs), huge fr. (barrel-shaped, to 90 cm long & 40 kg) on major branches or trunk, ed. raw or cooked, seeds ed. (jak-nuts), timber excellent for furniture & fuel (wood chips form. exclusive source of dye for Buddhist robes in SE As.), pollinating flies breed in male infls (cf. *Ficus*); *A. hirsutus* Lam. (S India) – boat-building ; *A. integer* (Thunb.) Merr. (chempedak) – fr. ed., seeds boiled; *A. lacucha* Roxb. ex Buch.-Ham. (*A. lakucha*, lacoocha, India to S China, SE As.) – ed. fr., timber good, seeds ed. fodder tree in Nepal; *A. nobilis* Thw. (Sri Lanka) – lvs crenate, seeds ed. boiled or fried; *A. odoratissimus* Blanco (marang, terap, Borneo, (? natur.) Philipp.) – ed., like jak; *A. tamaran* Becc. (Borneo) – loincloths

Artorima Dressler & G. Pollard. Orchidaceae (V 13). 1 Mex.

Artrolobium Desv. = *Coronilla*

arugula *Eruca vesicaria* ssp. *sativa*

Arum L. Araceae (VIII 6). 26 Eur. (8), Medit. R: P. Boyce (1993) *The genus A.* Alks. Cult. orn. incl. *A. maculatum* L. (cuckoo-pint, i.e. lively penis (Anglo-Saxon), jack-in-the-pulpit, lords-and-ladies, wake-robin, Eur.) – tubers leafless in first year, ed. cooked & source of Portland arrowroot (also *A. italicum* Miller (W Eur., natur. Arg.)), also form. used to starch linen (cypress powder); self-incompatible, spadix generates high temperature & stench attractive to flies, principally owl-midges, which are imprisoned by hairs (rudimentary male fls) at mouth of spathe, until male fls below them mature & hairs wither, allowing flies dusted with pollen to pass out to another infl. with mature female fls; *A. rupicola*

Boiss. (*A. conophalloides*, E Med., W As.) – poll. blood-sucking insects

arum-lily *Zantedeschia aethiopica*; **yellow a.-l.** *Z. elliottiana*

Aruncus L. Rosaceae. 1 N temp. & subarctic (incl. Eur.): *A. dioicus* (Walter) Fern. – cult. orn. (goat's-beard)

Arundina Blume. Orchidaceae (Arundineae). 1 Himal. to Pacific: *A. graminifolia* (D. Don) Hochr., to 3 m, cult. orn. R: OB 89(1986)16

Arundinaria Michaux. Gramineae (1a). (Incl. *Pleioblastus*) c. 50 As. esp. China & Japan, N Am. (1), S Am. (2). R: KBAS 13(1986)45. Frost-hardy bamboos flowering at infrequent intervals. Cult. orn. & for bamboo canes (many spp. trad. referred here now segregated in other genera, e.g. *Pseudosasa, Sasaella, Sinarundinaria, Thamnocalamus*). *A. amabilis* McClure (Tongking cane or bamboo, cultigen from China) – valued for split-cane fishing-rods; *A. gigantea* (Walter) Muehlenb. (giant or switch cane, SE US) – ed. young shoots, fr. eaten by Indians & early settlers; *A. viridistriata* (Regel) Nakai (Japan) – culms to 1.5 m

Arundinella Raddi. Gramineae (39). 47 warm. R: KB 10(1955)377, CJB 45(1967)1047

Arundo L. Gramineae (25). 3 Medit. (Eur. 1 + *A. donax* L. natur.), Taiwan. R: JAA 71(1990)156. *A. donax* L. (Med.), the 'reed shaken by the wind' of the Bible, used for 5000 yrs for pipe instruments (orig. Pan pipes), is the reed for clarinets & organ-pipes (Spanish cane); stems used for walking-sticks, fishing-rods, also source of cellulose in rayon-making & poss. useful for paper

Arundoclaytonia Davidse & R. Ellis. Gramineae. 1 Amazonas, Brazil: *A. dissimilis* Davidse & R. Ellis

arvi *Colocasia esculenta*

Arytera Blume. Sapindaceae. 25 Indomal. to E Aus. (10, 8 endemic). R: Blumea suppl. 9(1995)149

Asaemia (Harvey) Benth. = *Athanasia*

asafoetida *Ferula* spp. (q.v.)

asamela *Pericopsis elata*

Asanthus R. King & H. Robinson = *Steviopsis*

asarabacca *Asarum europaeum*

Asarca Lindley = *Gavilea*

Asarina Miller. Scrophulariaceae. (Excl. *Lophospermum* & *Maurandya*) 1 SW Eur.: *A. procumbens* Miller. Cult. orn.

Asarum L. Aristolochiaceae (I). (Incl. *Hexastylis*) c. 70 N temp. (Eur. 1; Japan 30). Cult. orn. & medic. Myrmechory in N Am. spp. Dark brown resiny-scented fls visited by flies, fungus-gnats etc. In latter, found in 8 sects. of genus, there are fungus-like appendages & gnats lay eggs but larvae cannot eat tissue. Cult. spp. incl. *A. canadense* L. (wild ginger, E N Am.) – ginger subs., rhiz. medic. & *A. europaeum* L. (asarabacca, Eur.) – form. medic. esp. after alcoholic excess (emetic), constituent of some snuffs

Ascarina Forster & Forster f. Chloranthaceae. 12 Madag. (1), Mal., Polynesia, NZ

Ascarinopsis Humbert & Capuron = *Ascarina*

Aschenbornia Schauer = *Calea*

Aschersoniodoxa Gilg & Muschler. Cruciferae. 3 Andes. R: SB 15(1990)

Aschistanthera C. Hansen. Melastomataceae. 1 Vietnam: *A. cristanthera* C. Hansen. R: NJB 7(1987)653

Asciadium Griseb. Umbelliferae (I). 1 Cuba

Ascidieria Seidenf. (~ *Eria*). Orchidaceae (V 14). 1 (?) Mal.: *A. longifolia* (Hook.f.) Seidenf. R: OB 89(1986)107

Ascidiogyne Cuatrec. Compositae (Eup.-Ager.). 2 Peru. R: MSBMBG 22(1987)152

Asclepiadaceae Medikus ex Borkh. (~ Apocynaceae). Dicots – Asteridae – Gentianales. 248/2800 trop. & warm (esp. Afr.), few temp. Lianes, scramblers, herbs, rarely shrubby or trees, s.t. succulent or with fleshy underground parts, with branched/unbranched laticifers throughout & various glycosides & alks; stem anatomy often unusual, internal phloem present. Lvs opp., whorled, rarely spiral, simple but often reduced; stipules usu. 0. Infls cymose, often umbelliform, rarely racemose. Fls usu. bisexual: K 5, imbricate or valvate, usu. ± connate at base; C (5), usu. with short tube & convolute; A epipet., 5 (Periplocoideae), or (5) in a short sheath round style, anthers in both coherent or connate around stylehead, in Asclepiadoideae & Secamonoideae A & style forming a columnar gynostegium. Fls usu. with a corona of appendages from base of filaments or also with a set from anthers, which open apically or longit., tetrasporangiate in Periplocoideae & Secamonoideae or bisporangiate, pollen-grains coherent into pollinia, which are extracted by translators (solidified secretions of anthers, stylehead or both, consisting of arms (retinacula) joined at middle by a 2-parted gland (corpusculum), one arm attached to pollinium(a) of 1 theca of 1 anther, the other to that (those) of adjoining theca of adjoining anther); G 2, united only by stylehead; ovules (1–)∞ on marginal placentas, pendulous, anatropous, unitegmic. Fr. of 2 distinct follicles, often only 1 developing; seeds usu. flattened & with term. head of long hairs, embryo straight, endosperm oily. x = 9–12

Classification & chief genera. 3 subfams (T 43(1994)201; I & III (incl. II) s.t. treated as separate fams):

I. **Periplocoideae** (pollen granular, in tetrads, transferred by spoon-shaped translator ending in a sticky disk; OW): 42 small genera incl. *Cryptostegia, Periploca*

II. **Secamonoideae** (pollen massed in pollinia; anthers 4-loc.): 5 genera form. incl. in III – *Secamone, Toxocarpus*

III. **Asclepiadoideae** (pollen massed in pollinia; anthers 2-loc.): often arranged in 4 tribes incl. II but whole fam. with its enormous number of monotypic/oligotypic genera in need of critical revision)

1. **Asclepiadeae** (88 genera) – *Asclepias, Astephanus, Cynanchum, Oxypetalum, Pachycarpus, Sarcostemma, Tylophora*
2. **Fockeeae** (2 genera) – *Fockea*
3. **Gonolobeae** (39 genera) – *Fischeria, Gonolobus, Matelea*
4. **Marsdenieae** (56 genera) – *Dischidia, Gymnema, Hoya, Marsdenia*
5. **Stapelieae** (42 genera) – *Brachystelma, Caralluma, Ceropegia, Heterostemma, Huernia, Orbea, Pachycymbium, Stapelia*

Closely related to Apocynaceae, in which it was at first included, but differing in specialized A, pollen-transfer system & gynostegium, though perhaps best returned to Apocynaceae

Elaborate insect-pollination, the corpuscula attaching to the legs (guided by grooves in the column) of nectar-seeking insects. Fls have disagreeable scent in many & are poll. by flesh-flies, some of which lay eggs in the fls of *Stapelieae* (up to 40 cm across). In *Ceropegia*, the C tube forms a trap in which flies are temporarily imprisoned, with hairs shrivelling to allow exit afterwards (cf. *Arum*) & often apparently very specific scents for particular pollinators. *Dischidia* spp. are ant-plants

Many cult. orn. esp. *Stapelieae* (R: T 40(1991)381) with 6 endemic genera & c. 160 endemic spp. in Karroo-Namib (succ.), *Marsdenia* (incl. stephanotis), *Ceropegia, Asclepias*. Fibres (*Asclepias*); seed-hairs used as floss (e.g. *Calotropis*); some poss. rubber (*Cryptostegia, Raphionacme*) and oil (*Asclepias*); some anti-sweeteners (*Gymnema*). Journal: *Asklepios*

Asclepias L. Asclepiadaceae (III 1). 100 N & C Am., esp. US, some natur. OW (Eur. 1: *A. syriaca* L., E N Am.!). R (N Am.): AMBG 41(1954)1, 261. Cult. orn. (milkweed, silkweed), shoots form. cooked N Am., rubber (milkweed r.), some weeds; cardiac glycosides but Monarch butterfly immune & becoming toxic to jays after feeding on A. *A. crispa* Bergius (S Afr.) – cardiac stimulant in Cape; *A. curassavica* L. (blood-flower, swallow-wort, matac, Indian root, bastard ipecacuanha, S Am., now pantrop. weed) – cult. orn., medic.; *A. eriocarpa* Benth. (Calif.) – fibre for cordage, latex for chewing-gum (Indians); *A. fruticosa* L. (*Gomphocarpus f.*, Med., Arabia, Afr., natur. Am. – latex used as hair-remover from hides in Yemen; *A. physocarpa* (E. Meyer) Schltr. (S Afr.) – introd. Hawaii as fibre-crop, now natur.; *A. speciosa* Torrey (Canada, W US) – roadside weed, 75% latex is a refinable oil, the rest suitable for food & cosmetics industries; *A. syriaca* form. cult. for fibre (stems), bee-forage & seed-floss (made up into vegetable silk), green fr. inverted & sold to tourists in Switzerland (budgerigar flowers); *A. tuberosa* L. (butterfly-weed, pleurisy-root, N Am.) – medic.

Asclepiodora A. Gray = *Asclepias*

Ascocentrum Schltr. ex J.J. Sm. Orchidaceae (V 16). 8 Himal. to Borneo. Cult. orn.

Ascochilopsis Carr. Orchidaceae (V 16). 1 Sumatra, Malay Pen.

Ascochilus Ridley. Orchidaceae (V 16). 5 W Mal.

Ascoglossum Schltr. (~ *Sarcochilus*). Orchidaceae (V 16). 2 New Guinea, Solomon Is.

Ascolabium Ying = *Ascocentrum*

Ascolepis Nees ex Steudel = *Lipocarpha*

Ascopholis C. Fischer = *Cyperus*

Ascotainia Ridley = *Tainia*

Ascotheca Heine. Acanthaceae. 1 trop. W Afr.

Ascyrum L. = *Hypericum*

Asemanthia (Stapf) Ridley = *Mussaenda*

Asemnantha Hook.f. Rubiaceae (III 4). 1 Mex.

Asepalum Marais. Cyclocheilaceae. 1 NE & E Afr.

ash *Fraxinus* spp., esp. *F. excelsior* (**common, Eur., French, Polish** or **Slavonian a.**); **alpine a.** *F. nigra, Eucalyptus delegatensis* (Aus.); **Am., Can.** or **white a.** *F. americana, F. nigra, F. pennsylvanica*; **black** or **brown a.** *F. nigra*; **blue a.** *F. quadrangularis*; **Blue Mt. a.** *Eucalyptus oreades*; **bumpy a.** *Flindersia schottiana*; **Canary a.** *Beilschmiedia bancroftii*; **Cape a.** *Ekebergia capensis*; **crow's a.** *Flindersia australis*; **flowering a.** *Fraxinus ornus*; **Japanese a.** *F. mandschurica*; **manna a.** *F. ornus*; **mountain a.** *Sorbus aucuparia, Eucalyptus regnans* (Aus.); **Oregon a.** *F. latifolia*; **prickly a.** *Orites excelsa*; **a. pumpkin** (or **gourd**) *Benincasa hispida*; **red a.** *F. pennsylvanica, Alphitonia* spp. (Aus.); **silver a.** *Flindersia schottiana*; **white a.** *Fraxinus americana, Eucalyptus fraxinoides* (Aus.); **yellow a.** *Emmenosperma alphitonioides, Eucalyptus luehmanniana*

Ashanti blood *Mussaenda erythrophylla*; **A. pepper** *Piper guineense*

ashplant *Leucophyllum frutescens*

Ashtonia Airy Shaw. Euphorbiaceae. 2 Malay Pen., Borneo

ashwagandha *Withania somnifera*

Asian mint *Persicaria odorata*

Asiasarum F. Maek. = *Asarum*

Asimina Adans. Annonaceae. 8 E N Am. (Eocene of London). R: Brittonia 12(1960)241. Alks. *A. reticulata* Chapman (dog apple, Seminole tea, Florida) – medic. tea; *A. triloba* (L.) Dunal (papaw, pawpaw, E US) – fr. (largest native US sp., to c. 400 g) ed. (can be baked in a 'bran bread'), buds naked but protected by rusty indumentum

Asiphonia Griff. (~ *Thottea*). Aristolochiaceae (II 1). 1 W Mal.: *A. piperiformis* Griff.

Askellia W. Weber = *Crepis*

Asketanthera Woodson. Apocynaceae. 4 trop. Am.

Askidiosperma Steudel. Restionaceae. 11 S & SW Cape. R: Bothalia 15(1985)431

asna *Terminalia alata*

asoka *Saraca asoca*

Aspalathus L. Leguminosae (III 28). 278 S Afr. esp. SW Cape. Often heath-like habit, cult. orn. *A. linearis* (Burm.f.) R. Dahlgren (*A. contaminatus*, Rooibos tea, W cape) – tea used by Hottentots: free of caffeine & low in tannin

Asparagaceae Juss. (incl. Herreriaceae & Ruscaceae; Liliaceae s.l.). Monocots – Liliidae – Asparagales. 6/150 OW excl. Aus., S Am. (*Herreria*) esp. arid. Shrubs or lianes with vessel elements, frequently with oxalate raphides & woody or annual aerial parts (sometimes prickly) arising from symp. rhiz. Leaf-like photosynthesizing organs intermediate between lvs & stems ('cladodes', 'phylloclades') s.t. bearing fls & often scale lvs. Fls app. in umbels or racemes, or solit., bisexual or unisexual; P 3 + 3 free or with basal tube, green to yellow or white, A 3 + 3, sometimes united into a column, G̲ (3), (1)3-loc. with axile placentation & (1)2–12 ovules per loc. Fr. a globose coloured berry, rarely a septicidal capsule; seeds often black, occ. winged

Genera: *Asparagus, Danae, Herreria, Herreriopsis, Ruscus, Semele*

Ruscaceae s.t. kept distinct (A united into column, G̲ (1)3 loc., each loc. with (1)2 ovules, seeds pale); Herreriaceae now considered a subfam. (fr. a capsule, seeds winged)

Cult. orn. & veg. (*Asparagus, Ruscus*)

Asparagopsis (Kunth) Kunth = *Asparagus*

Asparagus L. Asparagaceae (Liliaceae s.l.). (Incl. *Protasparagus*) 130–140 OW excl. Australasia. R (S Afr. incl. *Myrsiphyllum*): Bothalia 9(1966)31, 15(1984)77. Usu. rhizomes with photosynthetic infls, s.t. with longer-lived woody, thorny stems, some with molluscicidal steroid saponins. Cult. orn. & ed. (young shoots). *A. albus* L. (white a., W Med.) – ed., sold in markets in N Afr.; *A. asparagoides* (L.) W. Wight (*M. a.* S Afr.) – the smilax of florists; *A. cochinchinensis* (Lour.) Merr. (SE As.) – raw herb used in treatment of whooping cough in modern Chinese herbalism; *A. officinalis* L. (garden a., Eur. to N Afr.) – a. of commerce, prob. native in S Eur. but cult. since Ancient Greece, in GB native subsp. *prostratus* (Dumort.) Corb. (2n = 40, ?80) whereas cvs have 2n = 20, ground seeds said to be good subs. for coffee, & aerial parts suggested for high-quality paper-making; *A. setaceus* (Kunth) Jessop (*A. plumosus*, a. fern, S Afr.) – much used by florists in buttonholes etc., molluscicidal steroid saponins

asparagus, Bath *Ornithogalum pyrenaicum*; **a. bean** *Vigna unguiculata*; **Cape a.** *Aponogeton distachyos*; **a. fern** *Asparagus setaceus*; **a. lettuce** *Lactuca sativa*; **a. pea** *Psophocarpus tetragonolobus*; **white a.** *Asparagus albus*

Aspasia Lindley. Orchidaceae (V 10). 8 trop. Am. R: Brittonia 26(1974)333. Cult. orn.

Aspazoma N.E. Br. Aizoaceae (IV). 1 Cape

aspen *Populus tremula*, (N Am.) *P. tremuloides, P. grandidentata*

Asperella Humb. = *Hystrix*

Asperuginoides Rauschert (*Buchingera*). Cruciferae. 1 SW As.

Asperugo L. Boraginaceae. 1 Eur.: *A. procumbens* L. (madwort, so-called because its roots once a subs. for madder)

Asperula L. Rubiaceae (IV 16). c. 90 Euras. esp. Med. (Eur. 66, Turkey 41), Aus. (16, dioec. & poss. another genus). Cult. orn. incl. *A. cynanchica* L. (squinancywort, Euras. incl. GB), *A. tinctoria* L. (Medit.) – roots source of red dye

asphodel *Asphodelus aestivus*; **bog a.** *Narthecium ossifragum*; **giant a.** *Eremurus* spp.; **Scottish a.** *Tofieldia pusilla*; **yellow a.** *Asphodeline lutea*

Asphodelaceae Juss. (Incl. Aloaceae; Liliaceae s.l.). Monocots – Liliidae – Asparagales. 17/750 Eur. to C As., Afr. esp. S Afr. Pachycaul treelets to herbs when roots s.t. tuberous, occ. with vessels. Lvs often succ. Infls spikes to panicles with usu. leafless peduncles app. lateral as stem sympodial. Fls usu. bisexual, P 3 + 3 or (3 + 3), often somewhat zygomorphic, white to red or yellow, A 3 + 3, s.t. with hairs, G̲ (3), 3-loc. with axile placentation & 2–∞ anatropous to orthotropous ovules per loc. Fr. a loculicidal capsule; seeds usu. arillate, elongate & ovoid, winged in *Eremurus*. x = 7

Two subfams recog.: **Alooideae** (pachycaul trees to succ. rosette-pls, esp. S Afr.) – *Aloe, Gasteria, Haworthia,* & **Asphodeloideae** (herbaceous pls, though some *Bulbine, Kniphofia* & *Trachyandra* subshrubby, often temp.) – *Asphodeline, Asphodelus, Bulbine, Bulbinella, Eremurus, Kniphofia, Trachyandra*

Superficially resembling Hyacinthaceae (e.g. *Drimia*) & v. similar morphologically to Anthericaceae, such that field distinction often unclear, but lacking steroidal saponins though frequently with anthraquinones; even so, *Simethis* with the latter closely resembles certain Anthericaceae. The dismemberment of Liliaceae s.l. still causes practical difficulties

Many cult. orn.; *Aloe* medic. & in cosmetics etc.

Asphodeline Reichb. Asphodelaceae (Liliaceae s.l.). 14 Medit. (Eur. 3) to Caucasus. R: Candollea 42(1987)559. Cult. orn. incl. *A. lutea* (L.) Reichb. (yellow asphodel) – roots ed. like potatoes & *A. taurica* (Pallas) Endl. (E Med.)

Asphodelus L. Asphodelaceae (Liliaceae s.l.). 16 Medit. (Eur. 5) to Himal. Cult. orn. incl. *A. albus* Miller – source of alcohol, *A. fistulosus* L. (Canary Is., W Med.) – bulbs ed. Bedouin, noxious weed ('onion weed', but see also *Bulbine, Nothoscordum*) in Aus.; *A. aestivus* Brot. (*A. microcarpus*, asphodel, Medit.) – roots a source of a yellow dye for carpets (Egypt) and a gum used by bookbinders in Turkey (tchirish)

Aspicarpa Rich. Malpighiaceae. 12 warm Am.

Aspidiaceae Burnett = Dryopteridaceae

Aspidistra Ker-Gawler. Convallariaceae (Liliaceae s.l.). c. 10 E As. Mushroom-shaped pistil forms a lid over cavity formed by 6–8-lobed P; said to be snail-poll. *A. elatior* Blume (*'A.lurida'*, aspidistra, China) – v. tolerant housepl. incl. varieg. forms popular in C19 ('often regarded as a symbol of full middle class respectability' (Oxford English Dictionary))

Aspidocarya Hook.f. & Thomson. Menispermaceae (III). 1 NE India to SW China: *A. uvifera* Hook.f. & Thomson – fr. ed. R: KB 39(1984)101

Aspidogenia Burret = *Myrcianthes*

Aspidoglossum E. Meyer. Asclepiadaceae (III 1). 35 trop. & S Afr. R: KB 38(1984)633

Aspidogyne Garay. Orchidaceae (III 3). 30 trop. Am. R: Bradea 2(1977)200

Aspidonepsis Nicholas & Goyder (~ *Aspidoglossum*). Asclepiadaceae (III 1). 5 Drakensberg, S Afr. R: Bothalia 22(1992)24

Aspidophyllum Ulbr. = *Ranunculus*

Aspidopterys A. Juss. ex Endl. Malpighiaceae. 15–20 Indomal.

Aspidosperma C. Martius & Zucc. Apocynaceae. 80 trop. & S Am. Many alks, some antimicrobial. Good timber (peroba rosa), e.g. *A. excelsum* Benth. (paddlewood) for tool-handles etc., *A. polyneuron* Muell. Arg. (S & SE Brazil, overexploited) & *A. tomentosum* C. Martius & Zucc. (lemonwood, Brazil), bark (quebracha) used in tanning

Aspidostemon Rohwer & H. Richter (~ *Cryptocarya*). Lauraceae (I). 15+ Madag. R: BJ 109(1987)71

Aspidotis (Hook.) Copel. = *Cheilanthes*

Aspilia Thouars = *Wedelia*

Aspiliopsis Greenman = *Podachaenium*

Aspleniaceae Newman. Filicopsida. 1/c. 720 cosmop. Non-arborescent ferns incl. climbers & epiphytes. Fronds simple to multipinnate. Sori single, dorsal on the veins, indusium attached to side of vein, usu. long & narrow & with free tapering ends

Genus: *Asplenium*

See also Dryopteridaceae, Lomariopsidaceae form. incl. here

Aspleniopsis Mett. ex Kuhn = *Austrogramme*

Asplenium L. Aspleniaceae. Incl. *Camptosorus, Ceterach, Ceterachopsis, Diellia, Holodictyum, Loxoscaphe, Phyllitis, Pleurosorus, Schaffneria, Sinephropteris*, c. 720 subcosmop. (Eur. 31; 1 of few genera evenly trop. distrib.: 33% As., 22% Afr., 30% Am., 10% Aus. & Pacific; many is. endemics (Madag. 40, Hawaii, Tristan da Cunha, NZ, Réunion etc.); centres (sec.) of diversity in Appalachians, C Am. mts, Andes, Himal.). Hybrids & polyploids common, many cult. orn. inc. *A. adiantum-nigrum* L. (subcosmop.); *A. bulbiferum* G. Forster (SW Pacific) – viviparous; *A. ceterach* L. (rusty-back, Eur.); *A. marinum* L. (sea spleenwort, Eur.): *A. montanum* L. (mountain s., E US); *A. nidus* L. (bird's nest fern, OW trop.) – pioneering epiphyte forming a nest of simple fronds in which leaf-detritus etc. collects & roots grow, some forms with ed. young fronds for cooking; *A. rhizophyllum* L. (*Camptosorus r.*, N Am) – 'walking fern', cult. orn. with tips of fronds taking root & thus spreading pl., prothallus drought-resistant; *A. ruta-muraria* L. (wall-rue, w. spleenwort, Euras.); *A. scolopendrium* L. (Eur., hart's- tongue fern) – simple fronds, *A. trichomanes* L. (maidenhair spleenwort, subcosmop.). *A. acrobryum* Christ (New Guinea) – salt source (K^+, Na^+, Cl^-)

Asplundia Harling. Cyclanthaceae. 89 trop. Am. R: AHB 18(1958)139

Asplundianthus R. King & H. Robinson (~ *Eupatorium*). Compositae (Eup.-Crit.). 17 N Andes. R: MSBMBG 22(1987)346

Asraoa Joseph = *Wallichia*

assacu *Hura crepitans*

assagai wood *Curtisia dentata*

assai (palm) *Euterpe edulis*

Assam indigo *Strobilanthes cusia*

assié *Entandrophragma utile*

asta *Oxandra lanceolata*

Asta Klotzsch ex O. Schulz. Cruciferae. 2 Mex. R: CGH 214(1984)19

Astartea DC. Myrtaceae (Lept.). 8 Aus. Heterogeneous?

Astelia Banks & Sol. ex R. Br. Asteliaceae (Liliaceae s.l.). c. 25 Masc., New Guinea, Aus., NZ, Polynesia to Hawaii, Chile. Dioecious. R: KSVH III, 14(2)(1934)3. Cult. orn. inc. *A. nervosa* Hook.f. (bush flax, NZ) – berries ed. Maoris. Some locally useful fibres

Asteliaceae Dumort. (Liliaceae s.l.). Monocots – Liliidae – Asparagales. 4/35 Pacific, Masc. Herbs with tuberous rhiz., often dioec. Lvs spirally arr., linear to elliptic. Infl. of bracteate racemes or spikes; P 3 + 3 sometimes with basal tube, dull, A 3 + 3, $\underline{G}$ (3(4)), (1)3-loc., each loc. with 4–c. 15 anatropous ovules. Fr. a berry or a globose capsule (*Milligania*); seeds ovate, often angular

Genera: *Astelia, Collospermum, Milligania, Neoastelia*

The fam. which has some similarities with Asphodelaceae & also Phormiaceae may be better united with Xanthorrhoeaceae, while *Cordyline* recently put here from Agavaceae is in this work referred to Dracaenaceae, but it may be better with Anthericaceae, once again pointing up the difficulties raised by splitting up the Liliaceae too much & too soon

Astelma Schltr = *Papuastelma*

Astemma Less. = *Monactis*

Astenolobium Nevski = *Astragalus*

Astephania Oliver = *Anisopappus*

Astephanus R. Br. Asclepiadaceae (III 1). 30 Afr., Am. (Brazil 7, R: Bradea 4(1987)377)

aster (garden or **Chinese)** *Callistephus chinensis*; **Mexican a.** *Cosmos* spp.; **sea a.** *Tripolium vulgare*

Aster L. Compositae (Ast.-Ast.). Excl. *Tripolium, Galatella* c. 250 Am. (R: Brittonia 32(1980)230), Euras. (Eur. 15+ natur. N Am. spp.; Himal. 33 (R: NRBGE 26(1964)67)) & Afr. 17 (R: MBSM 11(1973)153). Form. incl. *T. vulgare* Besler ex Nees (*A. tripolium*), which prob. received name 'Michaelmas daisy' because of late flowering, the name transferred to the familiar garden pls derived from N Am. spp. after 1752, when the adoption of the Gregorian calendar caused Michaelmas Day to fall 11 days earlier to coincide with their flowering. Many cult. herb. perennials, the M. daisies (starworts or frost-flowers, US) in 5 major groups, 1 derived from *A. amellus* L. (Euras.), the rest from N Am. spp. esp. *A. novi-belgii* L. (E N Am. esp. coasts) – forms 20–150 cm, often natur. inc. complex hybrids. *A. oblongifolius* Nutt. (E US) – cult. orn., lvs scented; *A. spinosus* Benth. (C Am.) – rheophyte with spines

Asteraceae Martinov. See Compositae

× **Asterago** Everett = × *Solidaster*

Asteranthaceae Knuth. See Lecythidaceae

Asteranthe Engl. & Diels. Annonaceae. 2 trop. E Afr. R: BN 133(1980)53

Asteranthera Hanst. Gesneriaceae. 1 Chile: *A. ovata* (Cav.) Hanst. – woody climber, cult. orn.

Asteranthos Desf. Lecythidaceae (Asteranthaceae). 1 N Brazil: *A. brasiliensis* Desf. Bush; corona of staminal origin; bark purgative

Asteridea Lindley (~ *Athrixia*). Compositae (Gnap.-Ang.). 7 Aus. R: MBSM 16(1980)129

Asterigeron Rydb. = *Aster*

Asteriscium Cham. & Schldl. Umbelliferae (I 2c). 8 Chile, Arg. R: UCPB 33(1962)99

Asteriscus Miller. Compositae (Inul.). 3 Med. (Eur. 2.). R: NJB 5(1985)299. Cult. orn.

Asterochaete Nees = *Carpha*

Asterogyne H. Wendl. ex Hook.f. Palmae. 5 trop. Am. Cult. orn. *A. martiana* (H. Wendl.) Hemsley (C Am.) – crown acts as trap for nutrients which drip down trunk

Asterohyptis Epling. Labiatae (VIII 4a). 3 Mex. Gynodioec.

Asterolasia F. Muell. (~ *Urocarpus*). Rutaceae. 6 Aus.

Asterolinon Hoffsgg. & Link = *Lysimachia*

Asteromoea Blume. = *Kalimeris*

Asteromyrtus Schauer (~ *Melaleuca*). Myrtaceae. 7 S New Guinea, N Aus. R: AusSB 1(1988)373. Cult. orn.

Asteropeia Thouars. Theaceae (Asteropeiaceae). 5 Madag.

Asteropeiaceae (Syzyszyl.) Reveal & Hoogl. See Theaceae

Asterophorum Sprague. Tiliaceae. 1 Ecuador

Asteropsis Less. = *Podocoma*

Asteropterus Adans. = *Leysera*

Asteropyrum J.R. Drumm. & Hutch. Ranunculaceae (III 3). 2 China

Asterosedum Grulich = *Sedum*

Asterostemma Decne. Asclepiadaceae (III 4). 1 Java: *A. repandum* Decne

Asterostigma Fischer & C. Meyer. Araceae (VIII 6). 6 Brazil

Asterothamnus Novopokr. (~ *Aster*). Compositae (Ast.-Ast.). 7 C As. to China & Mongolia

Asterotricha V.V. Botsch. = *Pterygostemon*

Asterotrichion Klotzsch (~ *Plagianthus*). Malvaceae. 1 Tasmania

Asthenatherum Nevski = *Centropodia*

Asthenochloa Buese. Gramineae (40e). 1 Java, Philippines

Astianthus D. Don. Bignoniaceae. 1 C Am.

Astiella Jovet. Rubiaceae (IV 1). 1 Madag.

Astilbe Buch.-Ham. ex D. Don. Saxifragaceae. 12 E As., 2 Appalachians. Cult. orn. inc. hybrids. Often confused with *Aruncus* & *Filipendula* (Rosaceae) but those have many stamens & 3 to many separate pistils

Astilboides Engl. Saxifragaceae. 1 N China: *A. tabularis* (Hemsley) Engl.

Astiria Lindley. Sterculiaceae (? Malvaceae). 1 Masc.

Astoma DC = *seq.*

Astomaea Reichb. Umbelliferae (III 5). 2 E Med. to C As.

Astomatopsis Korovin = *Astomaea*

Astracantha Podlech (*Astragalus* subg. *Tragacantha*). Leguminosae (III 15). 250 Turkey to Iran. R: MBSM 19(1983)4. Spiny often cushion-forming shrubs, old petiole & rachis becoming woody & the leaflets sharp-tipped. Source of gum tragacanth (form. 'gum dragon') in W & C As. – tapped from stems or roots; hydrophilic & colloidal properties valuable in icecream, lotions, sizing etc. & in pharmaceuticals for suspending resinous tinctures & heavy insoluble powders, medicated creams, jellies etc.; several spp. involved but esp. *A. gummifera* (Lab.) Podlech (*Astragalus g.*), used since Anc. Greece, & in Iran as gum sarcocolla

Astragalus L. Leguminosae (III 15). (Excl. subg. *Tragacantha* Bunge = *Astracantha*) 1750 mainly N temp. (Eur. 133, China 287), esp. W N Am., C & W As. ext. to Chile, N India & trop. Afr. mts. Common in steppe, prairie etc., often ± spiny (goat's-thorn). Some ed. but some indicators of selenium associated with uranium ore e.g. in Utah, and some accumulators toxic to stock, 'locoweeds' (loco = Spanish for crazy) esp. in Midwest US and Canada. *A. boeticus* L. (Canary Is. to Iran) – seeds a coffee subs.; *A. canadensis* L. (N Am) – ed. roots; *A. crassicarpus* Nutt. (*A. caryocarpus*, W N Am., buffalo bean) – ed. pods; *A. fasciculifolius* Boiss. (Iran) – used as face-gloss by harem women; *A. glycyphyllos* L. (milk-vetch, fitsroot, Euras.) – fodder & herbal tea; *A. membranaceus* Bunge (temp. As.) – one of most commonly used herbs in modern Chinese herbalism; *A. sinicus* L. (E As.) – green manure for paddyland, medic. Some cult. orn.

Astranthium Nutt. Compositae (Ast.-Ast.). 11 S US & Mex. R: PMSUB 2(1965)429

Astrantia L. Umbelliferae (III I). 8–9 C & S Eur. (5), W As. Cult. orn. esp. *A. major* L. & *A. maxima* Pallas (masterworts). Whole infl. 'mimics' a single flower, the florets in *A. major* being bisexual but only functionally male at end of season.

Astrebla F. Muell. ex Benth. Gramineae (33a). 4 Aus. *A.* spp. esp. *A. pectinata* (Lindley) Benth. (Mitchell grass) – drought-resistant pasture grasses, form. Aboriginal grain, imp. food for wild budgerigars

Astrephia Dufr. = *Valeriana*

Astridia Dinter. Aizoaceae (V). 7 SW Afr. R: Bothalia 16(1986)205. Cult. orn.

Astripomoea Meeuse. Convolvulaceae. 3 S Afr. Cult. orn.

Astrocalyx Merr. Melastomataceae. 1 Philipp. R: Blumea 35(1990)75

Astrocaryum G. Meyer. Palmae (V 5e). 50 trop. Am. Fr. scatter-hoarded by agoutis, which s.t. peel off pericarp first (? thereby removing injurious invertebrate larvae) in Amaz. Fibre, oil, cult. orn. *A. jauary* C. Martius (Amaz.) – fr. used to catch fish though obligately disp. by characin fish; *A. murumuru* C. Martius (murumuru, Amazonia) – palm kernel oil (also other spp., guere); *A. tucuma* C. Martius (tucuma, Brazil etc.) – oil like *Cocos*; *A. tucumoides* Drude (awarra, NE S Am.) – lvs used in mats; *A. vulgare* C. Martius (trop. Am) – oil, fibre strongest in Amazon, poss. commercially viable, vitamin A in concs 3 times those of carrots

Astrocasia Robinson & Millsp. Euphorbiaceae. 5 trop. Am. R: SB 17(1992)311

Astrococcus Benth. Euphorbiaceae. 2 Brazil

Astrocodon Fed. = *Campanula*

Astrodaucus Drude (~ *Caucalis*). Umbelliferae (III 3). 2 Euras.

Astroloba Uitew. = *Haworthia*

Astroloma R. Br. (~ *Leucopogon*). Epacridaceae. 20+ Aus. Cult. orn. incl. *A. humifusum* (Cav.) R. Br. with sweet edible fr.

Astronia Blume. Melastomataceae. 59 Indomal. to Pacific (not Aus.). R: Blumea 35(1990)75

Astronidium A. Gray. Melastomataceae. 67 New Guinea, W Pacific to Society Is. R: Blumea 35(1990)115

Astronium Jacq. Anacardiaceae (IV). c. 15 trop. Am. Like *Rhus* but calyx winged, enveloping fr. Good timbers esp. *A. fraxinifolium* Schott (Brazil etc.) – Gonçalo alves, kingwood, locustwood, tigerwood, zebrawood & *A. urundeuva* (Allemão) Engl. (S Am.) – urunday, overexploited; *A. grande* Engl. (Brazil) – fls every 5 yrs

Astrophytum Lemaire. Cactaceae (III 9). 4+ Texas, Mex. Spineless cult. orn. esp. *A. asterias* (Zucc.) Lemaire, endangered in wild through overcollecting but easily raised from seed

Astrostemma Benth. = *Absolmsia*

Astrothalamus C. Robinson. Urticaceae (III). 1 W & C Mal.: *A. reticulatus* (Wedd.) C. Robinson

Astrotricha DC. Araliaceae. 16 Aus. (esp. SE, 1 in Hamersley Ranges, NW Aus.)

Astrotrichilia (Harms) J. Leroy. Meliaceae (I 4). c. 2 Madag.

Astydamia DC. Umbelliferae (III 10). 1 Canaries, Madeira

Astyposanthes Herter = *Stylosanthes*

Asyneuma Griseb. & Schenk. Campanulaceae. 50 Medit. (Eur. 4) to Cauc., E As. (1). Cult. orn.

Asystasia Blume. Acanthaceae. c. 70 OW trop. *A. gangetica* (L.) T. Anderson pantrop. weed, lvs ed. Mafia Is.

Asystasiella Lindau. Acanthaceae. 3 OW trop.

Ataenidia Gagnepain = *Phrynium*

Atalanthus D. Don = *Sonchus*

Atalantia Corr. Serr. Rutaceae. 11 trop. & E As. *A. monophylla* DC (India) – fr. an oil source, medic.

Atalaya Blume. Sapindaceae. 11 Afr. to Aus. *A. hemiglauca* (F. Muell.) F. Muell. (Aus.) – whitewood

Atalopteris Maxon & C. Chr. Older name for *Ctenitis*

Atamasco lily *Zephyranthes atamasco*

Atamisquea Miers ex Hook. & Arn. Capparidaceae. 1 Calif., temp. S Am.

Ataxipteris Holttum = *Ctenitis*

Ateixa Ravenna = *Sarcodraba*

Atelanthera Hook.f. & Thomson. Cruciferae. 1 C As. to Himal.: *A. perpusilla* Hook. f. & Thomson – some anthers monothecous

Ateleia (DC) Benth. Leguminosae (III 1). 16 trop. Am. R: Webbia 17(1962)153. C 1 (standard). *A. herbert-smithii* Pittier (C Am.) – dioec., wind-poll.

Atemnosiphon Leandri = *Gnidia*

Ateramnus P. Browne = *Sapium*

Athamanta L. Umbelliferae (III 8). 5–6 Medit. (Eur. 5–6). *A. cretensis* L. (Candy carrot) – flavouring for liqueurs

Athanasia L. Compositae (Anth.-Urs.). 40 S Afr. esp. Cape. R: OB 106(1991)5. Cult. orn.

Athenaea Sendtner. Solanaceae (1). 7 E Brazil. R: BSAB 26(1989)91

Atherandra Decne. Asclepiadaceae (I). 1 SE As., W Mal.

Atherolepis Hook.f. Asclepiadaceae (I). 3 SE As.

Atherosperma Lab. Monimiaceae (I 1; Atherospermataceae). 1 SE Aus.: *A. moschatum* Lab. V. pale wood for toys etc. Alks, strongly scented bark used as a tea (Tasmania)

Atherospermataceae R. Br. See Monimiaceae

Atherostemon Blume (~ *Atherandra*). Asclepiadaceae (I). 1 SE As.

Athertonia L. Johnson & B. Briggs. Proteaceae. 1 Queensland

Athrixia Ker-Gawler. Compositae (Inul.). 14 trop. & S Afr. R: MBSM 16(1980)46. *A. pinifolia* N.E. Br. (Natal) – rheophyte

Athroisma DC. Compositae (Hele.). 11 OW trop.

Athroostachys Benth. Gramineae (1b). 1 E Brazil: *A. capitata* (Hook.) Benth., clambering bamboo

Athrotaxis D. Don. Taxodiaceae. 3 Tasmania. R: Phytol.M 7(1984)15. Timber good (Tasmanian cedar) for general & cabinet work, esp. *A. selaginoides* D. Don (King William (Billy) pine). *A. cupressoides* D. Don – to 1000 yrs old, poss. useful in dendrochronology

Athyana (Griseb.) Radlk. Sapindaceae. 1 Paraguay, Arg.

Athyriaceae Alston = Dryopteridaceae

Athyrium Roth. Dryopteridaceae (II 1). 180 cosmop. (Eur. 2), esp. E & SE As. Close to *Diplazium*. Cult. orn. esp. *A. filix-femina* (L.) Roth (temp. N & Am., lady fern)

Athysanus E. Greene. Cruciferae. 1 W US

Atkinsia R. Howard = *Thespesia*

Atkinsonia F. Muell. Loranthaceae. 1 Blue Mts, SE Aus.: *A. ligustrina* (F. Muell.) F. Muell. – terr. shrub to 2 m parasitic on roots of diff. hosts

Atlanthemum Raynaud (~ *Helianthemum*). Cistaceae. I Medit. inc. Eur.: *A. sanguineum* (Lag.) Raynaud

Atlantic or Atlas cedar *Cedrus atlantica*; **A. ivy** *Hedera hibernica*

Atomostigma Kuntze. Rosaceae. 1 Brazil = ?

Atopostema Boutique = *Monanthotaxis*

Atractantha McClure. Gramineae (1b). 3–7 E Brazil. Climbing bamboos

Atractocarpa Franchet = *Puelia*

Atractocarpus Schltr. & K. Krause (~ *Randia*). Rubiaceae (II 1). c. 20 Aus., Lord Howe Is., New Caled.

Atractogyne Pierre. Rubiaceae (II 1). 3 W Afr.

Atractylis L. Compositae (Card.-Carl.). 14 Euras., Medit. (Eur. 4). R: BSBF 134 Lettres Bot. 1987(2)179. See also *Chamaeleon*

Atractylodes DC. Compositae (Card.-Carl.). 11 E As. R: BSBF 134 Lettres Bot. 1987(2)179. Imp. medic. pls in China inc. *A. lancea* (Thunb.) DC & *A. macrocephala* Koidz.

Atragene L. = *Clematis*

Atraphaxis L. Polygonaceae (II 4). 25 N Afr., SE Eur. (4) to Himal., E Siberia. Cult. orn., often spiny steppe-pls

Atrichantha Hilliard & B.L. Burtt. Compositae (Gnap.-Rel.). 1 SW Cape: *A. gemmifera* (L. Bolus) Hilliard & Burtt. R: OB 104(1991)74

Atrichodendron Gagnepain = ? *Lycium*

Atrichoseris A. Gray. Compositae (Lact.-Mal.). 1 SW US

Atriplex L. Chenopodiaceae (I 3). Incl. *Halimione* c. 300 temp. & warm (Eur. 24, Aus. 60). Salt-tolerant forage (saltbushes, e.g. *A. repanda* Philippi (Chile) – to 40 yrs old but over-exploited for sheep & goats; *A. vesicaria* Heward ex Benth., Aus.). Many ed. N Am. Indians, 1 widely cult.: *A. hortensis* L. (orache, As.) – lvs used like spinach, some cvs with coloured lvs (green after cooking)

Atropa L. Solanaceae (1). 4 W Eur. (2) to Himal. R: Flora 148(1960)84. *A. belladonna* L. (belladonna, deadly nightshade, Euras., Medit.) – all parts poisonous (atropine) to humans (not rabbits or pigs which can detoxify alks) though fr. said to taste nice (!), used medic. (in comm. cough mixtures etc.), dilates pupils (form. used by women to make eyes brighter & larger)

Atropanthe Pascher (~ *Scopolia*). Solanaceae (1). 1 China: *A. sinensis* (Hemsley) Pascher

Atroxima Stapf. Polygalaceae. 2 W & C Afr. R: MLW 77–18(1977)14

Attalea Kunth. Palmae (V 5c). 23 trop. Am. Allied to *Orbignya*. R: Fieldiana 38(1977)36. *A. colenda* (Cook) Balslev & A. Henderson (W Ecuador) – potential source of lauric acid; *A. funifera* C. Martius ex Sprengel (Bahia piassava, piassalba, coquilla, Brazil) – male when young & increasingly female as it reaches canopy, lvs (bases) sources of piassava fibre for brushes, brooms etc., nuts highly polished & used for carving etc.

attar of roses See *Rosa*

Atuna Raf. Chrysobalanaceae (4). 11 Indomal., W Pacific. R: PTRSB 320(1986)130. Distrib. ocean currents (? & squirrels, pigs). *A. excelsa* (Jack) Kosterm. (W Pacific) – hair dressing in Carolines, boat-caulking in Solomons, mashed seeds ed. with fish etc. in Moluccas

Atylosia Wight & Arn. = *Cajanus*

aubergine *Solanum melongena*

Aubregrinia Heine. Sapotaceae (IV). 1 trop. W Afr.: *A. taiensis* (Aubrév. & Pellegrin) Heine. R: T.D. Pennington, S. (1991)211

Aubrevillea Pellegrin. Leguminosae (II 3). 2 Guineo-Congolian forests

Aubrieta Adans. Cruciferae. c. 12 S Eur. (6) to Iran. R: [Q]BAGS 7(1939)157, 217. Cult. orn. esp. *A. deltoidea* (L.) DC (Sicily to Turkey) & hybrids of obscure parentage, the 'aubrietia' or 'aubretia' of gardens (*A.* x *cultorum* Bergmans)

Aucoumea Pierre. Burseraceae. 1 trop. W Afr.: *A. klaineana* Pierre (okoumé, Gaboon mahogany) – good timber & resin

Aucuba Thunb. Cornaceae (Aucubaceae). 3–4 Himal. to Japan. Prob. not allied to Cornaceae on DNA evidence. Cult. orn. evergreen dioec. shrubs esp. *A. japonica* Thunb. (E As., Japanese laurel) & its vulgar yellow-spotted cv., 'Variegata'

Aucubaceae J. Agardh. See Cornaceae

Audouinia Brongn. Bruniaceae. 1 S Afr.

Auerodendron Urban. Rhamnaceae. 8 WI. *A. pauciflorum* Alain (Puerto Rica) – reduced to 5 trees in wild

Augea Thunb. Zygophyllaceae. 1 S Afr. No endosperm

Augouardia Pellegrin. Leguminosae (I 4). 1 trop. W Afr.

Augusta Pohl. Rubiaceae (I 5). 1 E Brazil

Aulacocalyx Hook.f. Rubiaceae (II 4). 8 trop. Afr.

Aulacocarpus O. Berg = *Mouriri*

Aulacolepis Hackel = *Aniselytrum*

Aulacophyllum Regel = *Zamia*

Aulacospermum Ledeb. Umbelliferae (III 5). 12 Euras. (Eur. 1). R: BMOIPB 81,4 (1976)75, 5(1976)61, 88,6(1983)88

Aulandra H.J. Lam. Sapotaceae (II). 3 Borneo. Staminal tube. R: T.D. Pennington, S. (1991)152

Aulax Bergius. Proteaceae. 3 S Afr. Dioec. Cult. orn. R: SAJB 53(1987)464

Aulocaulis Standley = *Boerhavia*

Aulojusticia Lindau = *Justicia*

Aulonemia Goudot (~ *Arthrostylidium*). Gramineae (1a). 30 C & trop. S Am. R: SCB 9(1973)53, AMBG 77(1990)353. Montane forest bamboos, valuable forage for pack animals

Aulosepalum Garay. Orchidaceae (III 3). 4 Mex., Guatemala. R: HBML 28(1982)298

Aulosolena Kozo-Polj. = *Sanicula*

Aulospermum J. Coulter & Rose = *Cymopterus*

Aulostylis Schltr. (~ *Calanthe*). Orchidaceae (V 11). 1 New Guinea

Aulotandra Gagnepain. Zingiberaceae. 1 trop. W Afr., 5 Madag.

Aunt Lucy *Ellisia nyctelea*

Aureliana Sendt. (~ *Capsicum*). Solanaceae (1). 5 C S Am. R: Darw. 30(1990)95

Aureolaria Raf. Scrophulariaceae. 10 E US, 1 Mex. Hemiparasitic mostly on Fagaceae, but also Ericaceae. Sometimes cult. orn. incl. *A. flava* (L.) Farw. – iridoid (aucubin) sequestered by *Euphydryas phaeton* butterflies

auricula *Primula auricula*, *P.* x *pubescens* (usu.)

Auriculardisia Lundell = *Ardisia*

Aurinia Desv. (~ *Alyssum*). Cruciferae. c. 10 Eur. (9) & W As. Like *Alyssum* but infls usu. axillary & fl. buds spherical. Cult. orn. esp. *A. saxatilis* (L.) Desv. (gold-dust, C Eur. to Turkey) – rock-pl.

Australian blackwood *Acacia melanoxylon*; **A. bluebell** *Wahlenbergia gracilis*, **(honey)** *Echium plantagineum*; **A. cedar** *Toona ciliata*; **A. chestnut** *Castanospermum australe*; **A. currant** *Leucopogon* spp.; **A. daisy** *Vittadinia* spp.; **A. fuchsia** *Correa* & *Epacris* spp.; **A. grass-tree** *Xanthorrhoea* spp.; **A. heath** *Epacris* spp.; **A. honeysuckle** *Banksia* spp.; **A. kino** *Eucalyptus* *camaldulensis*; **A. maple** *Flindersia braylayana*; **A. mint** *Prostanthera* spp.; **A. pine** *Casuarina equisetifolia*; **A. red cedar** *Toona ciliata*; **A. rosewood** *Alectryon oleifolius*; **A. sandalwood** *Eucarya* spp.; **A. walnut** *Endiandra palmerstonii*

Australina Gaudich. Urticaceae (V). 1 SE Aus. & NZ, 1 Ethiopia & Kenya (cf. *Dietes, Pelargonium*). R: NJB 8(1988)53

Australopyrum (Tzvelev) Löve (~*Agropyron*). Gramineae (23). 3 E Aus., NZ

Australorchis Brieger = *Dendrobium*

Austrian pine *Pinus nigra*

Austroamericium Hendrych = *Thesium*

Austrobaileya C. White. Austrobaileyaceae. 1 Queensland: *A. scandens* C. White – fls smell of rotting fish, pollen like *Clavatipollenites* (Lower Cretaceous), young lvs on fast-growing shoots reflexed & acting as grapples, absence of phloem sieve-tubes now denied

Austrobaileyaceae (Croizat) Croizat. Dicots – Magnoliidae – Magnoliales. 1/1 NE Aus. Evergreen liane. Lvs opp. to subopp., simple; stipules 0. Fls large, solit., axillary or term. on leafy shoots, prob. fly-poll., hypogynous P 19–24, in compact spiral, sepaloid to petaloid; A 7–11, spirally arr., laminar, not differentiated into filament & anther, inner reduced to 9–16 staminodes, pollen monosulcate; G 10–13 free, ± spirally arr., style 2-lobed, ovules (4–)6–8(–13) in 2 series along ventral side of each carpel, anatropous, bitegmic. Fr. berry; seeds sarcotestal, embryo very small with abundant ruminate endosperm, germ. epigeal. 2n = 44

Only species: *Austrobaileya scandens* C. White

Pollen suggests affinity with Monimiaceae and laminar stamens, monosulcate pollen & several ovules per carpel suggest Magnoliales in the narrow sense of Cronquist, though unilacunar nodes, opp. lvs & climbing habit aberrant there, the first two features being compatible with his Laurales. An isolated plant poss. a relic of the group from which the numerically greater 2 orders s.t. recog. (here not) have arisen

Austrobassia Ulbr. = *Sclerolaena*

Austrobrickellia R. King & H. Robinson (~ *Eupatorium*). Compositae (Eup.-Alom.) 3 N S Am. R: MSBMBG 22(1987)253

Austrobuxus Miq. Euphorbiaceae. c. 10 Mal. to Fiji (Aus. 2)

Austrocactus Britton & Rose. Cactaceae (III 5). 5 Arg., Chile. Cult. orn.

Austrocedrus Florin & Boutelje (~ *Libocedrus*). Cupressaceae. 1 S Chile & Arg.: *A. chilensis* (D. Don) Florin & Boutelje (*L. chilensis*, Chilean cedar)

Austrocephalocereus Backeb. = *Micranthocereus*

Austrochloris Lazarides. Gramineae (33b). 1 Queensland

Austrocritonia R. King & H. Robinson (~ *Eupatorium*). Compositae (Eup.-Crit.). 4 Brazil. R: MSBMBG 22(1987)349

Austrocylindropuntia Backeb. = *Opuntia*

Austrocynoglossum M. Popov ex R. Mill (~ *Cynoglossum*). Boraginaceae. 1 E Aus.: *A. latifolium* (R. Br.) R. Mill. R: NRBGE 46(1989)43

Austrodolichos Verdc. Leguminosae (III 10). 1 Aus.

Austroeupatorium R. King & H. Robinson. Compositae (Eup.-Eup.). 13 trop. S Am. (esp. Brazil) to Uruguay. R: MSBMBG 22(1987)67

Austrofestuca (Tzvelev) Alexeev (~ *Festuca*). Gramineae (17). 4 Aus., NZ. R: Telopea 3(1990)601. Like a southern version of *Poa*

Austrogambeya Aubrév. & Pellegrin = *Chrysophyllum*

Austrogramme Fournier. Pteridaceae (III). 5 E Mal. to New Caled. R: FG 11(1975)61

Austroliabum H. Robinson & Brettell = *Microliabum*

Austromatthaea L.S. Sm. Monimiaceae (V 2). 1 NE Queensland

Austromuellera C. White. Proteaceae. 1 Queensland

Austromyrtus (Niedenzu) Burret. Myrtaceae (Myrt.). 37 E Aus. (19, endemic), Vanuatu, New Caled.

Austropeucedanum Mathias & Constance. Umbelliferae (III 10). 1 NW Arg.

Austrosteenisia Geesink (~ *Lonchocarpus*). Leguminosae (III 6). c. 3 N Aus., New Guinea

Austrosynotis C. Jeffrey (~ *Senecio*). Compositae (Sen.). 1 S trop. Afr.: *A. rectirama* (Baker) C. Jeffrey. R: KB 41(1986)878. Climbing by basally thickened petioles

Austrotaxus Compton. Taxaceae. 1 New Caled.: *A. spicata* Compton

Autranella A. Chev. (~ *Mimusops*). Sapotaceae (I). 1 W Afr.: *A. congolensis* (de Wild.) A. Chev.

Autrania Winkler & Barbey = *Jurinea*

autumn crocus *Colchicum autumnale*; **a. snowflake** *Leucojum autumnale*

Autumnalia Pim. Umbelliferae (III 10). 2 C As. R: BZ 74(1989)1485

Auxemma Miers. Boraginaceae (Ehretiaceae). 2 Brazil

Auxopus Schltr. Orchidaceae (V 5). 2 trop. Afr. Mycotrophic

avaram bark *Senna auriculata*

Avellanita Philippi. Euphorbiaceae. 1 Chile

Avellara Blanca & Díaz de la Guardia = *Scorzonera*

Avellinia Parl. = *Trisetaria*

Avena L. Gramineae (21b). c. 25 Eur., Medit. to Ethiopia. OW. R: Baum, B. (1977) *Oats*. Like *Helictotrichon* but annual & with smooth rounded glumes but *H. macrostachyum* intermediate; 2 weedy, 6–7 cult. spp. *A. sativa* L., cult. oats, derived from *A. sativa* ssp. *fatua* (L.) Thell. a weed of other cereals, domesticated c. 2000 BC & carried from Medit. to NW Eur. with weedy forms still evolving to become one of world's worst weeds. Basis of porridge, oat-cakes, oatmeal soap etc.; subsp. *sterilis* (L.) de Wet has awns which twist & untwist in moist surroundings & are s.t. used as 'flies' in fishing

Avenochloa Holub = *Helictotrichon*

avens, mountain *Dryas octopetala*; **water a.** *Geum rivale*; **wood** or **yellow a.** *G. urbanum*

Avenula (Dumort.) Dumort. = *Helictotrichon*

Averia Leonard = *Tetramerium*

Averrhoa L. Oxalidaceae. 2, (?) E Brazil or Mal., cult. pantrop. Lvs irritable. Fr. trees. *A. bilimbi* L. (bilimbing, bilimbi, camias, cucumber tree) – rather sour used for pickles, jams, jellies etc., borne on trunk & branches; *A. carambola* L. (carambola, caramba) – ribbed, axillary fr. pickled, or sweet forms eaten raw, transverse sections being called star-fruit, medic. Juice from both spp. will remove stains from hands, clothes & weapons

Averrhoaceae Hutch. = Oxalidaceae

Averrhoidium Baillon. Sapindaceae. 2 Brazil, Paraguay

Avetra Perrier. Dioscoreaceae. 1 E Madagascar

Avicennia L. Avicenniaceae (Verbenaceae s.l.). 4–7 warm – mangroves. Aerial roots projecting above mud at low tide, viviparous. Some good timber & tanbark. R: P.B. Tomlinson (1986) *Bot. Mangroves*: 186. *A. marina* Forssk. (OW) – timber, brown dye for batik

Avicenniaceae Endl. (~ Verbenaceae). Dicots – Asteridae – Lamiales. 1/4–7 warm. Mangrove trees & shrubs with erect pneumatophores. Lvs decussate, simple, exstipulate. Infls spicate to cymose, axillary or terminal. Fls bisexual, small; K 5 imbricate, ± connate at base, surounded by a 'pseudo-involucre' of bractlets, C 4 with basal tube, A 4 inserted in throat of tube, G̲ (2) with free central ± 4-winged placenta & 4 pendent orthotropous ovules. Fr. a capsule with somewhat fleshy exocarp & 2 valves, usu. 1-seeded; seeds without testa or endosperm, embryo ± viviparous

Genus: *Avicennia*

Form. incl. in Verbenaceae but s.t. considered allied to Dipterocarpaceae or Salvadoraceae (Moldenke)

Aviceps Lindley = *Satyrium*

Avignon berry *Rhamnus infectoria*

avocado (pear) *Persea americana*

avodiré *Turraeanthus africanus*

awari *Pterygota bequaertii*

awarra (palm) *Astrocaryum tucumoides*

awlwort *Subularia aquatica*

awun *Alstonia congensis*

awusa nut *Tetracarpidium conophorum*

Axinaea Ruíz & Pavón. Melastomataceae. 30 trop. Am. Close to *Meriana*. R: BTBC 63(1936)211

Axinandra Thwaites. Crypteroniaceae. 5 Sri Lanka (1), W Mal.

Axiniphyllum Benth. Compositae (Helia.-Mel). 5 Mex. R: Madroño 25(1978)46, 34(1987)164

Axonopus Pal. Gramineae (34b). 35 trop. & warm Am., Afr. (1). R: JAA suppl. 1(1991)291. Lawn-grasses (carpet grass). *A. fissifolius* (Raddi) Kuhlm. (*A. affinis*, C Am.) – sown in trop. pastures & *A. compressus* (Sw.) Pal. – lawn-grass widely natur.

Axyris L. Chenopodiaceae (I 3). 6 E Eur. (1) to Korea

ayahuasca *Banisteriopsis* spp. esp. *B. caapi*

ayan *Distemonanthus benthamianus*

Ayapana Spach (~ *Eupatorium*). Compositae (Eup.-Ayap.). 14 trop. Am. R: MSBMBG 22(1987)195. *A. triplinervis* (Vahl) R. King & H. Robinson a medicinal tea, cult. in Brazil (ayapana)

Ayapanopsis R. King & H. Robinson (~ *Eupatorium*). Compositae (Eup.-Ayap.). 15 Andes R: MSBMBG 22(1987)197

aye *Sterculia rhinopetala*
Ayenia L. Sterculiaceae. 68 S US to Arg. R: Op. Lill. 4(1960)1
Ayensua L.B. Sm. Bromeliaceae (1). 1 Venezuela
Aylacophora Cabrera. Compositae (Ast.-Ast.). 1 Arg.
Aylostera Speg. = *Rebutia*
Aylthonia N. Menezes = *Barbacenia*
Aynia H. Robinson. Compositae (Vern.-Vern.). 1 Peru
ayous *Triplochiton scleroxylon*
Aytonia L.f. = *Nymania*
Azadehdelia Braem = *Cribbia*
Azadirachta A. Juss. Meliaceae (I 2). 2 Indomal. R: FM I,12(1995)337. *A. excelsa* (Jack) Jacobs (Mal.) – timber, young shoots ed.; *A indica* A. Juss. (neem, nim, margosa, prob. native in Burma) – widely cult. & natur. OW trop., spread by bats, germ. enhanced by passage through baboon guts, one of world's most useful trees: fuelcrop easily grown on poor soils, shade-tree (50 000 in plains of Saudi Arabia for annual camp of 2 million Muslim Haj pilgrims), fodder, ed. fls & lvs, timber a mahogany subs., used in soaps, toothpaste, lotions, seeds an oil-source, lvs (incorporated in nests, ? reducing parasite loads, by house-sparrows in India) & seeds provide an insecticide (azadirachtin the most potent insect antifeedant known: 10 p.p.m. lethal to many Lepidoptera), medic. (skin diseases such as scabies; poss. antimalarial, postcoital contraceptive & for arthritis treatment) etc. etc. Nat. Research Council (1992) *Neem – a tree for solving global problems*
azalea *Rhododendron* spp. & hybrids (deciduous); **alpine** or **trailing a.** *Loiseleuria procumbens*; **spider a.** *Rhododendron stenopetalum*
Azalea L. = *Rhododendron*
Azanza Alef. = *Thespesia*
Azara Ruíz & Pavón. Flacourtiaceae (8). 10 S Am. R: BJ 98(1977)151. Lvs spirally arr. but 1 stipule often almost as large giving appearance of opp. lvs, *A. microphylla* Hook.f. (Chile, Arg.) with leaf, smaller 'stipule' & glandular smaller stipule-like structure interpreted as homoeotic replacement of 1 'stipule' by a 'leaf'. C 0, outer A without anthers. Cult. orn. Wood bitter
azarole *Crataegus azarolus*
Azilia Hedge & Lamond. Umbelliferae (III 10). 1 Iran.
Azima Lam. Salvadoraceae. 4 S Afr. to Hainan, Philippines, Lesser Sunda Is. Alks. Axillary thorns (prob. lvs of undeveloped shoot cf. Cactaceae). Crushed branches of *A. sarmentosa* (Blume) Benth. (Java) foetid
azobé *Lophira alata*
Azolla Lam. Azollaceae. 6 trop. & warm (2 natur. Eur., *A. filiculoides* Lam. in GB (native NW but in Eur. in earlier interglacials)). R: Webbia 31(1977)97. *A. pinnata* R. Br. (trop. As.) doubles weight in 7 days, used to control mosquitoes by blocking water-surface, as green manure & stock feed, in rice-fields may fix 50–150 kg ha^{-1} in 1–4 months
Azollaceae Wettst. (~ *Salviniaceae*). Filicopsida. 1/6 trop. & warm. Free-floating ferns with branched protostelic stems & simple deciduous chlorophyllous roots. Lvs sessile, alt. in 2 rows on dorsal side of stem, each divided into 2 lobes, the upper aerial & photosynthetic with large cavity containing mucilage & usu. colonies of *Anabaena azollae* Strasb., which is unique in using fructose to fix nitrogen to ammonia (passed to host, which returns amino-acids, proteins & ribonucleotides) even in dark, the lower with dorsal side immersed & usu. without chlorophyll. Growth of cotyledon leaf ruptures megasporangial wall allowing symbiont to reach developing lvs. Roots lose rootcap & resemble submerged lvs of *Salvinia*. Sporocarps 2(4) forming the lower lobe of 1st leaf of a lat. branch, the upper forming an involucre over them, each with micro- or megasporangia without a dehiscence mechanism; microspores held in massulae in each sporangium by hardened frothy mucilage, the massulae usu. covered with barbed hairs (glochidia); megasporangium with 1 spore, which becomes a floating female prothallus to be anchored to the massulae by the glochidia, the union sinking to the bottom for fert.
Only genus: *Azolla*
Azorella Lam. Umbelliferae (I 2b). Excl. *Bolax* 70 Andes to temp. S Am. (Arg. 30 – R: Darw. 29(1989)139), Falklands, Antarctic Is. Densely tufted cushion-pls, those in S Peru held to be up to 3000 yrs old. *A. monantha* Clos ex Gay ('*A. caespitosa*') forms the balsam bogs of the Falklands, an extract used medic.; *A. compacta* Philippi (*A. yareta*) & other spp. (yareta, Andes etc.) – imp. fuel-source burning with little smoke
Azorina Feer = *Campanula*
Aztec clover *Trifolium amabile*; **A. pine** *Pinus teocote*; **A. tobacco** *Nicotiana rustica*

Aztecaster Nesom. Compositae (Ast.). 2 Mex. Dioec. shrubs. R: Phytol. 75(1993)55

Aztekium Bödecker (~ *Strombocactus*). Cactaceae (III 9). 2 Mex. *A. ritteri* (Bödecker) Bödecker – slow-growing, threatened in wild, cult. orn., seeds arillate

Azukia Takah. ex Ohwi = *Dolichos*

Azureocereus Akers & J.H. Johnson = *Browningia*

B

babaco *Carica* × *pentagona*

babacú or **babassú palm** *Orbignya* spp. esp. *O. phalerata*

babai *Cyrtospermum merkusii*

Babbagia F. Muell. = *Osteocarpum*

Babcockia Boulos = *Sonchus*

Babiana Ker-Gawler. Iridaceae (IV 3). Incl. *Antholyza*, 63 S Afr., esp. Cape, 1 Socotra. R: JSAB suppl. 3(1959). Bird-poll. spp. with 'perches' in infls form. *Antholyza*; seeds black (usu. brown in Iridaceae). Cult. orn. *B. fragrans* (Jacq.) Ecklon (*B. plicata*, S Afr., baboon root) – corms ed. settlers

baboen *Virola surinamensis*

baboon root *Babiana fragrans*

babul bark *Acacia nilotica*

baby blue eyes *Nemophila menziesii*; **b. in cradle** *Anguloa* spp.; **b. corn** *Zea mays*; **b.'s breath** *Gypsophila paniculata*

bacaba palm *Oenocarpus distichus*

Baccaurea Lour. Euphorbiaceae. 80 Indomal., W Pacific. Several ed. fr. esp. *B. dulcis* (Jack) J. Voigt (Mal.), *B. motleyana* (Muell. Arg.) Muell. Arg. (rambai, Mal.) & B. *ramiflora* Lour. (*B. sapida*, rambai, lutqua, Indomal.) often cult.

Baccharidastrum Cabrera = *Baccharis*

Baccharidiopsis G. Barroso = *Baccharis*

Baccharis L. Compositae (Ast.-Ast.). c. 400 Am. (Chile). R: MBSM 21(1985)1. Some monoec., some dioec. shrubs, many leafless with winged or cylindrical photosynthetic stems. Some med., lvs ground for green dye & cult. orn. e.g. salt-tolerant *B. halimifolia* L. (consumption weed, groundsel bush, tree-groundsel, E N Am. saltmarshes) – noxious weed in NSW; *B. sarothroides* A. Gray (Calif.) – potential oilseed; *B. trimera* (Less.) DC (Uruguay) – fertility control; *B. viminea* DC (Calif.) – rheophyte, s.t. planted for erosion control

Baccharoides Moench (~ *Vernonia*). Compositae (Vern.-Vern.). 8 trop. Afr. *B. anthelmintica* (L.) Moench (*V. a.*) – used for skin problems ('bukchie').

bachelor's buttons usu. 'double-flowered' forms of e.g. *Achillea ptarmica*, *Bellis perennis*, *Centaurea cyanus*, *Ranunculus acris*

Bachmannia Pax. Capparidaceae. 2 S Afr.

Backhousia Hook. & Harvey. Myrtaceae (Lept.). 7 Aus., 1 New Guinea. *B. bancroftii* Bailey (Johnstone River hardwood, Queensland) – hard close-grained timber; *B. citriodora* F. Muell. (Queensland) – source of ess. oil

Baclea Fourn. Asclepiadaceae (III 3). 2 Brazil

bacon and eggs *Daviesia* spp.

Bacopa Aublet. Scrophulariaceae. 56 warm esp. Am. Mostly aquatic or paludal. R: PANSP 98(1946)83. Some cult orn. (see also *Sutera*) inc. *B. monnieri* (L.) Wettst. (pantrop., water hyssop) in aquaria, also medic. ('brahmi' in India)

Bactris Jacq. ex Scop. Palmae (V 5e). 239 trop. Am. Some fish-disp. in Amazon; ed. fr., oilseeds, wood (chonta) trad. source of bows, blowpipe darts, spears etc., cult. orn. *B. gasipaes* Kunth (*B. ciliata*, peach-palm, pejibay(e), pejivalle, prob. orig. Amazonian Peru) – insect-poll., also wind-, widely cult. S Am. for fleshy pulp & now comm. oilseed; *B. major* Jacq. (black roseau) – cult. orn.

bacu *Cariniana* spp., *Tieghemella heckelii*

Bacularia F. Muell. ex Hook.f. = *Linospadix*

bacury *Platonia esculenta*

badam *Terminalia procera*

badger's-bane *Aconitum lycoctonum*

badi *Nauclea diderrichii*

Badiera DC (~ *Polygala*). Polygalaceae. 15 trop. Am.

Badilloa R. King & H. Robinson (~ *Eupatorium*). Compositae (Eupat.-Crit.). 10 N Andes. R: MSBMBG 22(1987)350

badinjan (WI) *Solanum melongena*

Badula Juss. Myrsinaceae. 12 Masc. (3 extinct). *B. crassa* A. DC (Mauritius) a pachycaul treelet

Badusa A. Gray. Rubiaceae (I 7). 3 Palawan, New Guinea, W Pacific. R: Blumea 28(1982)145

Baeckea L. Myrtaceae (Lept.). 75 trop. As. (1) to Aus. (70, endemic), New Caled. *B. frutescens* L. (trop. As. to Aus.) – lvs in a tea, medic. & ess. oil for scents, soap etc.

bael fruit *Aegle marmelos*

Baeolepis Decne. ex Moq. = *Decalepis*

Baeometra Salisb. ex Endl. Colchicaceae (Liliaceae s.l.). 1 S Afr.: *B. uniflora* (Jacq.) Lewis – toxic to stock (alks)

Baeothryon Ehrh. ex A. Dietr. = *Eleocharis*

Baeria Fischer & C. Meyer = *Lasthenia*

Baeriopsis J. Howell. Compositae (Hele.-Bae). 1 Baja Calif.

Bafodeya Prance ex F. White (~ *Parinari*). Chrysobalanaceae (2). 1 W Afr. R: PTRSB 320(1988)98

Bafutia C. Adams (~ *Emilia*). Compositae (Sen.-Sen.). 1 Cameroun

bagac *Dipterocarpus grandiflorus*

Bagassa Aublet. Moraceae (II). 1 NE S Am.: *B. guianensis* Aublet – promising plantation tree

bagasse crushed *Saccharum* (q.v.)

bag-flower *Clerodendrum thomsoniae*

bag-pod *Glottidium vesicarium*

bagtikan *Parashorea malaanonan*

Bahama grass *Cynodon dactylon*; **B. hemp** *Agave sisalana*; **B. pitchpine** *Pinus caribaea*; **B. whitewood** *Canella winterana*

Bahia Lagasca. Compositae (Hele.-Cha.). 13 SW US, Mex., Chile. R: Rhodora 66(1964)67

Bahia fibre = piassava fibre; **B. grass** *Paspalum notatum*; **B. piassava** *Attalea funifera*; **B. rosewood** *Dalbergia nigra*; **B. wood** *Caesalpinia echinata*

Bahianthus R. King & H. Robinson (~ *Eupatorium*). Compositae (Eupat.-Gyp.). 1 NE Brazil. R: MSBMBG 22(1987)113

baib grass *Eulaliopsis binata*

Baikiaea Benth. Leguminosae (I 4). 4 trop. Afr. Most extensive decid. forests on Kalahari Sand in Zambesi basin. *B. plurijuga* Harms (C Afr.) – valuable timber (Zambesi redwood)

Baileya Harvey & A. Gray ex Torrey. Compositae (Hele.-Gai.). 3 SW US, Mex. R: Sida 15(1993)491. *B. multiradiata* Harvey & A. Gray ex Torrey – cult. orn.

Baileyoxylon C. White. Flacourtiaceae (4). 1 Queensland

Baillonella Pierre. Sapotaceae (I). 1 trop. W Afr.: *B. toxisperma* Pierre (djave) – ed. fat from seeds, good timber (moabi)

Baillonia Bocquillon. Verbenaceae. 1 S Am.

Baissea A. DC. Apocynaceae. c. 20 OW trop.

Baitaria Ruíz & Pavón (~ *Calandrinia*). Portulacaceae. Excl. *Montiastrum*, 22 S Am.

baitoa *Phyllostylon brasiliensis*

bajri *Pennisetum glaucum*

bakain *Melia azedarach*

Bakerantha L.B. Sm. = *Hechtia*

Bakerella Tieghem (~ *Taxillus*). Loranthaceae. 16 Madagascar

Bakeridesia Hochr. (~ *Abutilon*). Malvaceae. 13 C Am. R: GH 10(1973)446

Bakerolimon Lincz. (~ *Limonium*). Plumbaginaceae (III). 2 Peru, N Chile

Bakerophyton (Léonard) Hutch. = *Aeschynomene*

bakphul *Sesbania grandiflora*

baku *Tieghemella heckelii*

bakupari *Garcinia* sp. (*Rheedia brasiliensis*)

bakury *Platonia esculenta*

Balaka Becc. Palmae (V 4k). 7 Fiji (5), Samoa(2)

Balanitaceae Endl. See Zygophyllaceae

Balanites Del. Zygophyllaceae (Balanitaceae). 25 trop. Afr. to Burma. Oilseeds esp. drought-resistant *B. aegyptiaca* (L.) Del. (N trop. Afr. to E Med.) for soap, fr. ed., bark medic. (molluscicidal steroid saponins) & fibre, contraceptives from roots, wood useful (ship-building in Classical times); *B. maughamii* Sprague (menduro, E Afr.) – fr. ed., oilseed; *B. orbicularis* Sprague (kullam, Somalia) – gum resin & oilseed; *B. wilsoniana* Dawe & Sprague (trop. Afr.) – fr. eaten by elephant in Ghana

Balanopaceae Benth. Dicots – Dilleniidae – Buxales. 1/9 SW Pacific. Dioec. evergreen trees. Lvs dimorphic, spiral, the proximal on each branch scale-like, the distal normal & sometimes verticillate, all simple with vestigial stipules. Fls small, anemophilous, with 0 or vestigial perianth, males in axillary catkins – A (1–)3–6(–12), females solit., naked ovary subtended by many spiral deltoid bracts – G (2–3) with as many locules (sometimes not perfectly separated) & distinct bifid (sometimes forked again) styles & each with 2 nearly basal, anatropous, unitegmic ovules. Fr. a drupe in a persistent involucre, resembling an acorn, with 2 or 3 pyrenes; seeds with large green embryo in thin layer of endosperm. n = 21. R: Allertonia 2(1980)191

Only genus: *Balanops*

Apparently isolated family, considered by some to represent a 'prefloral' condition, by others to be extremely simplified. Form. placed near Fagaceae but now considered close to Buxaceae

Balanophora Forster & G. Forster. Balanophoraceae. 15 OW trop. R: DBA 28,1(1972)1. Acarpellate; seeds c. 7µg each. Known to parasitize at least 74 spp. in 35 families, *B. fungosa* Forster & G. Forster with at least 25 host spp. Wax from tubers used in torches in Java etc., as bird-lime in Thailand; some medic. (anti-asthmatic). Foxy smell associated with fly-poll.; *B. fungosa* monoec. with mousey nectar poll. beetles in N Queensland, infr. slowly wearing away & seeds disp. by rain (? insects, pigs)

Balanophoraceae Rich. Dicots – Rosidae – Santalales. 18/44 trop., subtrop. esp. upland forest, Med. Chlorophyll-less root-parasites with little host specificity, usu. no stomata or guard cells. Over-ground parts usu. fleshy club-shaped infls ('capitula'), fungus-like in appearance, pale yellow to brown, pink or purplish, bearing many fls, some of which are the smallest known; underground parts apparently modified root, tuber-like (vestigial roots in *Corynaea*), amorphous, up to the size of a baby's head, rarely with scale lvs (*Lophophytum*) entirely parasite tissue (e.g. *Dactylanthus*, NZ, *Helosis, Lophophytum, Scybalium*, Am.) or part host ('corpus intermedium'), a chimerical system unknown elsewhere in green pls, producing 'rhizomes' which grow through ground attacking new host-roots. Lvs spiral to whorled, scale-like, without stomata (exc. *Cynomorium*), or 0. Infls terminal, often developing inside 'tuber', rupturing its tissue which remains as a 'volva' at base (in *Chamydophytum* maturing completely before rupture), unbranched or with spiral branches (*Sarcophyte*). Fls unisexual (pls dioec. or monoec., when on sep. infls or together but males towards base, infl. in *Helosis* covered with startlingly geometric hexagonal scales, each of which is surrounded by 2 concentric rings of female fls, the males occupying the corners under the scales. Males diverse, P 0 or 3, 4(–8), discrete or connate at base, valvate, A 1 or 2 (where P 0) or same as P, anthers with 2, 4 or many locules, A in *Helosis* & *Scybalium* united at base with discrete anthers, in other genera A a tube tipped with pollen sacs; pollen not sculptured (exc. *Cynomorium*). Females v. reduced, P 0 or minute, epigynous in *Cynomorium*, cup-like in *Mystropetalon*; ovules, placentas & carpels often not easily recognizable but G ($\overline{G}$ in *Mystropetalon*) (2 or 3) or acarpellate (*Balanophora*), with 1 or 2 embryo-sacs without recognizable nucellus or integuments exc. *Cynomorium* where ovule unitegmic. Fr. indehiscent, s.t. surrounded by swollen 'pedicel' or perianth-tube (*Mystropetalon*) or aggregated to form fleshy multiple fr.; seed 1 with minute embryo embedded in endosperm. x = 8, 9, 12 etc.

Formerly split into several fams, now treated as tribes, 6 of them having a storage substance resembling starch, the *Balanophoroideae* having a waxy reserve (balanophorin). *Cynomorium* (Med. to Asia) often separated at family level (sculptured exine, ovule integument) but *Mystropetalon* very distinctive, notably in pollen which is unique in angiosperms in being triangular, square or pentagonal when viewed end on but almost always square when viewed from the side

Chief genera: most are monotypic, the largest is *Balanophora* (15 spp.)

Clearly simplified pls, it is difficult to decide whether the fam. represents a natural assemblage or comprises end-products of convergent evolution, though presently the first view is preferred, the fam. being derived from the less specialized parasitic Santalales, possibly near the ancestors of Olacaceae, though there are distant affinities with *Gunnera* (Haloragaceae)

App. fly-poll., though spp. of *Ombrophytum*, largely subterranean, prob. apomictic; juicy elaiosomes around fr. of *Mystropetalon* spp. attractive to ants which are disp. agents

Phallic infls. have suggested aphrodisiac qualities but *Balanophoroideae* provide waxes used in lighting

Balanops Baillon. Balanopaceae. 9 N Queensland, New Caled., Vanuatu, Fiji. R: Allertonia 2(1980)207

Balansaea Boiss. & Reuter = *Geocaryum*

balata *Manilkara bidentata, Ecclinusa balata*

balau *Shorea* spp. esp. *S. glauca, S. maxwelliana*; **red b.** *S. guiso, S. kunstleri*

Balaustion Hook. Myrtaceae (Lept.). 2 SW Aus.

Balbisia Cav. Geraniaceae (Ledocarpaceae). 8 S Am. Shrubby, ovary with many ovules per carpel

Balboa Planchon & Triana = *Chrysochlamys*

bald cypress *Taxodium distichum*

Baldellia Parl. (~ *Echinodorus*). Alismataceae. 2 W & S Eur. (2), N Afr.

baldmoney *Meum athamanticum*

Balduina Nutt. Compositae (Hele.-Gai). 3 SE US. R: Brittonia 27(1975)355

Balfourodendron Méllo ex Oliver. Rutaceae. 1 S Brazil: *B. riedelianum* (Engl.) Engl. – valuable timber (guatambu moroti), cult. orn.; alks

Baliospermum Blume. Euphorbiaceae. 6 India to Sumbawa. Drastic purgatives

balisier (WI) *Calathea lutea, Heliconia bihai*

Ballantinia Hook.f. ex E. Shaw. Cruciferae. 1 Victoria, Tasmania

ballart *Exocarpos* spp.

Ballochia Balf.f. Acanthaceae. 3 Socotra

balloon flower *Platycodon grandiflorus*; **b. pea** (S Afr.) *Sutherlandia frutescens*; **b. vine** *Cardiospermum halicacabum*

Ballota L. (~ *Marrubium, Stachys*). Labiatae (VI). 35 Eur. (7), Med., W As. *B. acetabulosa* (L.) Benth. (Med.) – fr. used as floating wick in olive oil lamps; *B. nigra* L. (black horehound, Euras., natur. US) – adulterant of *Marrubium vulgare*, ess. oil, form. medic.

Ballya Brenan = *Aneilema*

balm *Melissa officinalis*; **bastard b.** *M. melissophyllum*; **bee b.** *Monarda didyma*; **Canary b.** *Cedronella canariensis*; **b. of Gilead** *Abies balsamea, Commiphora gileadensis, Liquidambar orientalis, Populus balsamifera, P.* × *jackii* 'Gileadensis', *P. nigra*; **horse b.** *Collinsonia canadensis*; **lemon, sweet** or **tea b.** *Melissa officinalis*

Balmea Martinez. Rubiaceae (I 1). 1 Mex.

balmony *Chelone glabra*

Baloghia Endl. Euphorbiaceae. 12 E Aus. (3), New Caled., Norfolk Is. *B. lucida* Endl. (scrub bloodwood) – sap source of indelible paint

Balonga Le Thomas = *Uvaria*

Baloskion Raf. (~ *Restio*). Restionaceae. 12 E Aus.

balsa *Ochroma pyramidale*

balsam *Impatiens* spp.; **b. apple** *Momordica balsamina*; **b. bog** *Azorella caespitosa*; **Canada b.** *Abies balsamea*; **Copaiba b.** *Copaifera officinalis*; **b. fig** *Clusia rosea*; **b. fir** *A. balsamea*; **garden b.** *I. balsamina*; **gurjun b.** *Dipterocarpus* spp.; **Indian b.** *I. glandulifera*; **Mecca b.** *Commiphora opobalsamum*; **orange b.** *I. capensis*; **b. pear** *Momordica charantia*; **b. of Peru** *Myroxylon balsamum*; **b. poplar** *Populus balsamifera*; **rock b.** *Clusia, Peperomia*; **b. root** *Balsamorhiza sagittata*; **Tolu b.** *Myroxylon balsamum* var. *pereirae*; **b. tree** *Colophospermum mopane*; **umiry b.** *Humiria* spp.

Balsaminaceae Bercht. & J. Presl. Dicots – Rosidae – Geraniales. 2/850 trop. OW with few temp. Herbs (rarely subshrubby), subsucculent, nearly always glabrous, stems ± translucent. Lvs spiral, whorled or opp., simple, stipules 0 or a pair of petiolar glands. Fls bisexual, solit. or in cymes, zygomorphic, resupinate; K 3 (*Impatiens*) or 5 (*Hydrocera*), the app. lowermost petaloid with a slender spur-nectary; C 5 distinct in *Hydrocera*, ± connate in *Impatiens*; A 5, filaments connate at least above, anthers ± connate into calyptra over pistil, pollen sacs divided by trabeculae separating sporogenous tissue into islands; $\underline{G}$ (4 rare, 5) with axile placentation & 1 or 5 stigmas, ovules 1 per locule (*Hydrocera*) or ∞ (*Impatiens*), anatropous, bitegmic or rarely unitegmic. Fr. a berry-like drupe (*Hydrocera*) or explosively dehiscent loculicidal capsule, the valves twisting in dehiscence, the stone in the first eventually separating into 5 pyrenes; embryo straight, endosperm 0 (*Hydrocera*) or little. x = 6–11

Genera: *Hydrocera* (1 sp.), *Impatiens* (many cult. orn.)

Balsamita Miller = *Tanacetum*

Balsamocarpon Clos (~ *Caesalpinia*). Leguminosae (I 1). 1 Chile: *B. brevifolium* Clos (*Caesalpinia b.*, algarobilla)

Balsamocitrus Stapf. Rutaceae. 2 trop. E Afr.

Balsamodendron Kunth = *Commiphora*

Balsamorhiza Nutt. Compositae (Helia.-Verb.). 14 W N Am. R: AMBG 22(1935)119. Cult. orn. *B. sagittata* (Pursh) Nutt. (balsam root) – roots ed. N Am. Indians

Balthasaria Verdc. Theaceae. 2–3 trop. Afr.

Baltic redwood or **yellow deal** *Pinus sylvestris*; **B. whitewood** *Picea abies*

Baltimora L. Compositae (Helia.). 2 trop. Am. R: Fieldiana 36(1973)31

balucanat *Aleurites moluccana*

balustine flowers *Punica granatum*

Bambara groundnut *Vigna subterranea*

Bambekea Cogn. Cucurbitaceae. 2 trop. C & W Afr.

bamboo species of some 45 genera of grasses or woody culms of many used for building, pipes, walking-sticks, furniture etc., when split for mats, blinds, baskets, fans, hats, umbrellas, brushes etc. Also used for paper pulp. Culms to 37 m, ed. when young in many spp. (bamboo shoots, usu. *Phyllostachys* spp.), with deposits of silica in cell walls and fibrous when mature. Some flower annually, others hapaxanthic; Tai Khai-Chih (c. AD 460) *Chu Phu* [Treatise on bamboo] – first recorded botanical monograph of any plant group; **black b.** *Phyllostachys nigra*; **Calcutta b.** *Dendrocalamus strictus*; **common b.** *Bambusa vulgaris*; **fishpole b.** *P. aurea*; **heavenly b.** *Nandina domestica*; **madake b.** *Phyllostachys bambusoides*; **male b.** *B. bambos*; **b. palm** *Raphia* spp., *Chrysalidocarpus lutescens*; **spiny b.** *B. b.*; **Terai b.** *Melocanna baccifera*; **Tongking b.** *Arundinaria amabilis*; **umbrella b.** *Thamnocalamus spathaceus*; **b. vines** (N Am.) *Smilax* spp.

Bambusa Schreber. Gramineae (1b). 120 trop. & warm Am. (30) & As. Cult. orn., timber, pulp, bamboo shoots. *B. bambos* (L.) Voss (*B. arundinacea*, male or spiny bamboo, India) – to 37 m in Travancore (at Kew growing 91 cm in a day = 0.63 mm per minute!), hapaxanthic after 31–54 yrs with seeding over 5–6 yrs in India e.g. 1868–72, 1912–16, 1958–62, timber, ed. shoots, sacred in India, concretions of silica in stems (tabashir) medic.; *B. beecheyana* Munro (SE China) – cult. for shoots for canning; *B. multiplex* (Lour.) Räusch. (*B. glaucescens*, China) – widely grown hedging bamboo in SE As. etc.; *B. vulgaris* Schrader ex Wendl. (common bamboo, cultigen, widely grown) – pulp, construction, culms held during prayers in Sumatra to ward off spirits, ed. shoots, hapaxanthic after at least 150 yrs; *B. wrayi* Stapf (Malay Pen.) – clumps found in territories of orang asli as favoured for blowpipes

Bambusaceae Burnett = Gramineae

Bamiania Lincz. (~ *Cephalorhizum*). Plumbaginaceae (II). 1 Afghanistan: *B. pachycormum* (Rech.f.) Lincz.

Bamlera Schumann & Lauterb. = *Astronidium*

Bampsia Lisowski & Mielcarek. Scrophulariaceae. 2 Zaire

banak *Virola koschnyi*

banana *Musa* spp., cvs & hybrids; **apple b.** *M. × paradisiaca*

Bancroftia Porter ex Gaceta de Venez. = ? *Satureja*

Banara Aublet. Flacourtiaceae (7). 31 warm Am. R: FN 22(1980)83

bandakai (Ind.) = okra

Bandeiraea Welw. ex Benth. = *Griffonia*

baneberry *Actaea spicata*

banga wanga *Amblygonocarpus andogensis*

bangalay *Eucalyptus botryoides*

bangalow (palm) *Archontophoenix cunninghamiana*

banglang *Lagerstroemia speciosa*

Banisteria L. = *Heteropterys* + *Banisteriopsis*

Banisterioides Dubard & Dop = *Sphedamnocarpus*

Banisteriopsis C. Robinson. Malpighiaceae. 92 trop. Am. esp. Brazil. R: FN 30(1982). Mostly lianes. Some hallucinogens in S Am. esp. *B. caapi* (Griseb.) Morton (wild & cult. Amazonia) – bark infusion (ayahuasca (Peru), caapi (Brazil), yagé, yajé(Colombia)), lvs & bark sometimes smoked, psychotropic effects due to beta-carboline alks. *B. lutea* (Griseb.) Cuatrec. (Peru) – app. eglandular fls mimicking oil-producing M., visited by poll. bees

Banjolea S. Bowd. = *Nelsonia*

Banksia L.f. Proteaceae. 73 Aus. (73, 72 endemic, 59 endemic in SW Aus.), New Guinea (1). R: A.S. George (1986) *The B. Book*; Nuytsia 3(1981)239, 6(1988)309. Australian honeysuckles, nectar used as food by Aborigines. Evergreen with hard woody follicles encl. in woody infr. derived from bracts & bracteoles; seeds winged. Some poll. by honey possums; some, e.g. *B. ornata* F. Muell., have fire-dependent (the floral remnants acting as tinder) follicle dehiscence, stress between sclereids of different types in inner & outer layers allayed by a resin which is destroyed by fire – in other spp. the resin is chemically different & the fr. opens without fire. Some timbers, tanbarks & cult. orn. *B. aemula* R. Br. (wallum, E Aus.) – typical of sandy coastal heathlands; *B. grandis* Willd. (SW Aus.) – nectar eaten by aborigines

banyan *Ficus benghalensis*

baobab *Adansonia digitata*

Baolia H.W. Kung & G.L. Chu (~ *Chenopodium*). Chenopodiaceae (I 2). 1 China: *B. bracteata* H.W. Kung & G.L. Chu

Baphia Afzel. ex Lodd. Leguminosae (III 2). 45 trop. & S Afr., Madag. (1). R: KB 40(1985)291. *B. nitida* Afzel. ex Lodd. (cam-wood, trop. Afr.) – form. source of red dye, wood turning red from white on exposure to air, timber used for walking-sticks, violin bows etc., chewing-stick in Nigeria

Baphiastrum Harms. Leguminosae (III 2). 1–2 Guineo-Congolian forest. Lianes

Baphicacanthus Bremek. = *Strobilanthes*

Baphiopsis Benth. ex Baker. Leguminosae (III 1). 1 trop. Afr.

Baptisia Vent. Leguminosae (III 30). 17 E US. R: AMBG 27(1940)119. Alks. False indigo. *B. tinctoria* (L.) R. Br. etc. form. used as dye-plants. Cult. orn., some med. e.g. *B. lactea* (Raf.) Thieret (*B. leucantha*)

Baptistonia Barb. Rodr. = *Oncidium*

Baptorhachis W. Clayton & Renvoize. Gramineae (34d). 1 Mozambique

Baratranthus (Korth.) Miq. Loranthaceae. 4 Sri Lanka, W Mal.

barb grass *Hainardia cylindrica*

Barbacenia Vand. Velloziaceae. 104 S Am. R: SCB 30(1976)4

Barbaceniopsis L.B. Sm. Velloziaceae. 3 Andes. R: SCB 30(1976)37

barbadine *Passiflora quadrangularis*

Barbados almond *Terminalia catappa*; **B. cedar** *Juniperus bermudiana*; **B. cherry** *Malpighia glabra*; **B. gooseberry** *Pereskia aculeata*; **B. lily** *Hippeastrum puniceum*; **B. mastic** *Sideroxylon foetidissimum*; **B. pride** *Caesalpinia pulcherrima*; **B. snowdrop** *Habranthus tubispathus*

Barbara's buttons *Marshallia* spp.

Barbarea R. Br. Cruciferae. 20 N temp. (Eur. 10). Weeds (wintercress), esp. *B. vulgaris* R. Br. (yellow rocket) – noxious in US, & some ed. esp. *B. verna* (Miller) Asch. (*B. praecox*, land cress, American or Normandy c., W Med., Macaronesia, natur. elsewhere) – salad

barbary fig *Opuntia ficus-indica*; **b. nut** *Gynandriris sisyrinchium*

barbasco (S Am.) fish-poisoning plants esp. *Lonchocarpus* spp.

barbatimão *Stryphnodendron adstringens*

Barberetta Harvey. Haemodoraceae. 1 S Afr.

barberry *Berberis* spp.; **Alleghany** or **American b.** *B. canadensis*; **common** or **European b.** *B. vulgaris*

Barberton daisy *Gerbera jamesonii*

Barbeuia Thouars. Phytolaccaceae (V, Barbeuiaceae). 1 Madag.: *B. madagascariensis* Steudel – large liane, blackens on drying

Barbeuiaceae (H. Walter) Nakai = Phytolaccaceae

Barbeya Schweinf. Barbeyaceae. 1 NE Afr., Arabia

Barbeyaceae Rendle. Dicots – Hamamelidae – Urticales. 1/1 NE Afr., Arabia. Small dioec. tree. Lvs opp., simple, stipules 0. Fls small, wind-poll., reg. in short cymes without bracts or bracteoles: P 3 or 4, slightly connate at base; A 6–9(–12); $\underline{G}$ (1(–3)), ± connate, each carpel with 1 loc. & 1 pendulous, anatropous unitegmic ovule. Fr. a nut with accrescent P; embryo straight, endosperm 0

Only species: *Barbeya oleoides* Schweinf.

More-or-less distinct carpels & a primitive phloem type here suggest affinity of Urticales with Hamamelidae rather than Malvales if this genus is correctly placed though distinctive pollen & hairs suggest allocation to a monotypic Barbeyales

Barbieria DC = *Clitoria*

Barbosa Becc. = *Syragrus*

Barbosella Schltr. Orchidaceae (V 13). 15 trop. Am.

Barbrodria Luer (~ *Masdevallia*). Orchidaceae (V 13). 1 Brazil: *B. miersii* (Lindley) Luer, cult. orn.

Barcella (Trail) Drude (~ *Elaeis*). Palmae (V 5d). 1 Brazil

Barcelona nut *Corylus avellana*

Barclaya Wallich (*Hydrostemma*). Nymphaeaceae (Barclayaceae). 4 Indomal. Aquatics (3), *B. rotundifolia* Hotta (Borneo) terr.; ? allied to *Ondinea*. Some cult. aquaria. Seeds with hooked hairs adhering to coats of wild pigs in Mal.

Barclayaceae (Endl.) Li = Nymphaeaceae

Barcoo grass *Iseilema* spp.

bareet grass *Leersia hexandra*

barilla ash with high levels of sodium carbonate used in soap- & glass-making, derived from plants, esp. spp. of *Halogeton*, *Salsola*, *Suaeda*

Barjonia Decne. Asclepiadaceae (III 4). 6 Brazil. R: Rodriguesia 51(1979)7

Barkeria Knowles & Westc. (~ *Epidendrum*). Orchidaceae (V). 14 C Am. R: AOSB 42(1973)620. Cult. orn.

Barkerwebbia Becc. = *Heterospathe*

Barkleyanthus H. Robinson & Brettell (~ *Senecio*). Compositae (Sen.- Tuss.). 1 SW US

Barklya F. Muell. = *Bauhinia*

Barleria L. Acanthaceae. 250 trop. Many xerophytes with bracteolar thorns. Seeds with hairs which swell when wetted. Some med. & cult. orn. incl. *B. cristata* L. (? Mauritius, natur. trop.) – hedgepl. e.g. Christmas Is. (Ind. Ocean)

Barleriola Oersted. Acanthaceae. 6 WI

barley *Hordeum* spp. esp. *H. vulgare*; **meadow b.** *H. secalinum*

Barlia Parl. Orchidaceae (IV 2). 2 Med. (Eur. 1)

barna tree *Crateva religiosa*

Barnadesia Mutis ex L.f. Compositae (Barn.). 23 S Am. esp. trop. Andes. Trees & shrubs

Barnardiella Goldbl. Iridaceae (III 2). 1 W Cape

Barnebya Anderson & Gates. Malpighiaceae. 2 Brazil. R: Brittonia 33(1981)275

Barneoudia Gay. Ranunculaceae (II 2). 3 Chile, Arg.

Barnettia Santisuk = *Santisukia*

Barnhartia Gleason. Polygalaceae. 1 trop. S Am. Liane to tops of trees

barnyard grass = *Echinochloa crus-galli*

Barombia Schltr. = *Aerangis*

Barongia P.G. Wilson & Hyland. Myrtaceae. 1 N Queensland: *B. lophandra* P.G. Wilson & Hyland

Baronia Baker = *Rhus*

Baroniella Costantin & Gallaud. Asclepiadaceae (I). 4 Madag.

baros camphor *Dryobalanops aromatica*

Barosma Willd. = *Agathosma*

barrel, brown *Eucalyptus fastigiata*

barrenwort *Epimedium alpinum*

Barringtonia Forster & G. Forster. Lecythidaceae. 39 E Afr. (1), Madag. (2), trop. As. & Pacific. Seeds with saponins used as fish-poison like *Derris*; *B. novae-hiberniae* Lauterb. (W Pacific) – ed. seeds

Barringtoniaceae F. Rudolphi = *Lecythidaceae*

Barroetea A. Gray = *Brickellia*

Barrosoa R. King & H. Robinson (~ *Eupatorium*). Compositae (Eupat.-Gyp.). 10 trop. S Am. R: MSBMBG 22(1987)92

Barteria Hook.f. Passifloraceae. 1 trop. Afr.: *B. nigritana* Hook.f. Myrmecophilous, subsp. *nigritana* (Nigeria to Zaire) inhabited by small ants, subsp. *fistulosa* (Masters) Sleumer (Nigeria to Tanzania) by big ants. Plants without ants in Nigeria grow less well, ants deterring insect (& ? larger animal) grazing in myrmecophilous ones

Barthea Hook.f. Melastomataceae. 1 China, Taiwan

Bartholina R. Br. (~ *Holothrix*). Orchidaceae (IV 2). 3 Cape

Bartholomaea Standley & Steyerm. Flacourtiaceae (5). 2 C Am.

Bartlettia A. Gray. Compositae (Hele.-Cha). 1 SW N Am. R: SWN 8(1963)117

Bartlettina R. King & H. Robinson (~ *Eupatorium*). Compositae (Eupat.-Heb.). 37 trop. Am. R: MSBMBG 22(1987)403

Bartonia Muhlenb. ex Willd. Gentianaceae. 3–4 E N Am. R: Rhodora 61(1959)43. Mycotrophs with scale lvs & little chlorophyll

Bartonia Pursh ex Sims = *Mentzelia*

Bartschella Britton & Rose = *Mammilaria*

bartsia, alpine *Bartsia alpina* L.; **red b.** *Odontites vernus*; **yellow b.** *Parentucellia viscosa*

Bartsia L. Scrophulariaceae. 49 circumboreal (1), Eur. (2), Medit. (1), Afr. mts (2), Andes (45, endemic). R: OB 102(1990)5. Root-parasites, esp. on grasses; some Andean spp. hummingbird-poll.

barus camphor *Dryobalanops aromatica*

barwood *Pterocarpus erinaceus*, *P. soyauxii*

Barylucuma Ducke = *Pouteria*

Basanacantha Hook.f. = *Randia*

Basananthe Peyr. (*Tryphostemma*). Passifloraceae. c. 32 trop. & S Afr. R: Blumea 21(1974)327

Basedowia E. Pritzel. Compositae (Gnap.-Cass.). 1 C Aus.: *B. tenerrima* (F. Muell. & Tate) J. Black. R: OB 104(1991)85

Basella L. Basellaceae. 5: Madag. (3), E Afr. (1) & 1 pantrop. (*B. rubra* L. (*B. alba*), Malabar spinach or nightshade, Ceylon or Indian spinach) cult. potherb poss. not native in Am., climber with cleistogamous fls & fr. enclosed in P which becomes fleshy

Basellaceae Raf. Dicots – Caryophyllidae – Caryophyllales. 4/20 trop. & warm esp. Am. Perennial glabrous rhiz. herbs with fleshy, mucilaginous, annual climbing shoots & s.t. (*Ullucus*) tubers, without unusual sec. thickening or anthocyanins (betalains present). Lvs opp. to spiral, simple, entire, often succulent; stipules 0. Infls. term. or axillary spikes to panicles of small, regular, bisexual (s.t. functionally unisexual) fls; bracteoles 2, K (also considered bracteoles by some) 2, often coloured, s.t. adnate to base of petals; C 5 (s.t. considered K), imbricate with a basal tube or 0, persistent in fr.; A (4)5(–9) opp. C, adnate at base to C, anthers opening by term. slits or pores or longit. slits; annular nectary around base of A; $\underline{G}$ (3) becoming 1-locular with 1 basal, bitegmic ovule. Fr. a thin-walled nutlet enclosed in P; seeds with perisperm or 0, endosperm 0. x = 11, 12

Genera: *Anredera, Basella, Tournonia, Ullucus*

Related to Portulacaceae. *Basella* & *Ullucus* are foodpls, *Anredera* incl. some cult. orn.

Baseonema Schltr. & Rendle. Asclepiadaceae (I). 1 trop. E Afr.: *B. gregorii* Schltr. & Rendle

Bashania Keng f. & Yi = *Arundinaria*

Basigyne J.J. Sm. (~ *Dendrochilum*). Orchidaceae (V 12). 1 Sulawesi

basil *Ocimum basilicum* (**sweet b.**); **hoary b.** *O. canum*; **holy b.** *O. tenuiflorum*; **lemon b.** *O. basilicum*; **Thai b.** *O. tenuiflorum*; **b. thyme** *Clinopodium acinos*

Basilicum Moench. Labiatae (VIII 4b). 7 OW trop. to E Aus.

Basiphyllaea Schltr. Orchidaceae (V 13). 3 WI, Florida

Basisperma C. White. Myrtaceae (Lept.). 1 New Guinea

Basistelma Bartlett = *Cynanchum*

Basistemon Turcz. Scrophulariaceae. 8 trop. Am. R: SB 10(1985)125. Shrubs, elaiophores attractive to bees

Baskervilla Lindley. Orchidaceae (III 3). 7 Andes (6), Brazil (1)

basket flower *Hymenocallis narcissiflora*

basralocus *Dicorynia guianensis*

bass fibrous young bark esp. from *Tilia* spp.; **b. wood** *Tilia americana* etc.

bass-broom fibre = piassava

Basselinia Vieill. Palmae (V 4m). 11 New Caled. R: Allertonia 3(1984)355

Bassia All. Chenopodiaceae (I 4). Excl. *Chenolea*, 21 warm (Eur. 7). R: FR 89(1978)106 (incl. *Kochia*). *B. scoparia* (L.) Voss (*Kochia scoparia*, summer cypress, S Eur. to Japan) – cult. for broom-making (Japan), orn. foliage pl. for bedding with inconsp. (bee-poll.!) fls, widely natur.; '*K. s.* f. *trichophylla* (Schmeiss) Schinz & Thell.' (burning bush) – most commonly cult., lvs turning red in autumn

bassine fibre *Borassus flabellifer*

Bassovia Aublet = *Solanum*

basswood *Tilia americana*

bastard acacia *Robinia pseudoacacia*; **b. balm** *Melittis melissophyllum*; **b. box** *Polygala chamaebuxus*; **b. burr** *Xanthium spinosum*; **b. cedar** *Chukrasia tabularis, Soymida febrifuga*; **b. cinnamon** *Cinnamomum aromaticum*; **b. indigo** *Amorpha fruticosa*; **b. mahogany** *Eucalyptus botryoides*; **b. rosewood** *Synoum glandulosum*; **b. wild rubber** *Funtumia africana*

Bastardia Kunth. Malvaceae. 8 trop. Am.

Bastardiastrum (Rose) D. Bates. Malvaceae. 7 Mex. R: GH 11(1978)311, Aliso 11(1987)544

Bastardiopsis (Schumann) Hassler. Malvaceae. 1 S Am.

Basutica E. Phillips = *Gnidia*

bataan, b. mahogany *Shorea polysperma*

Bataceae C. Martius ex Perleb. Dicots – Dilleniidae – Capparidales. 1/2 Pacific & Am. coasts. Maritime shrubs. Lvs opp., simple, narrow, succulent; stipules minute. Fls unisexual (plants dioec. or monoec.), small, in strobiloid spikes, males initially enclosed in sac-like organ poss. representing K or pair of bracteoles, splitting down 1 side or into 4; P (? staminodes) 4, A 4 alt. with P, anthers opening by longitudinal slits. Female fls: P 0, $\underline{G}$ 2, 4-loc., with 1 anatropous bitegmic ovule in each and 2 stigmas. Fr. a sea-disp. drupe with 4 pyrenes; seeds without endosperm. 2n = 18

Only genus: *Batis*

Pollen referred here known from the Maestrichtian of Calif., making the fam. as old an 'identifiable' one as any in Dilleniidae or Caryophyllidae (Cronquist, who puts the fam. in an order of its own, Batales)

batai wood *Paraserianthes falcataria*

Batania Hatusima = *Pycnarrhena*

Bataprine Nieuw. = *Galium*

Batemannia Lindley. Orchidaceae (V 10). 45 trop. S Am. Cult. orn.
Batesanthus N.E. Br. Asclepiadaceae (I). 4 Afr.
Batesia Spruce ex Benth. Leguminosae (I 1). 1 Amazonia
Batesimalva Fryx. Malvaceae. 3 Mexico
Bathiaea Drake. Leguminosae (I 4). 1 Madagascar
Bathiorhamnus Capuron. Rhamnaceae. 2 Madagascar
Bathysa C. Presl. Rubiaceae (I 5). 10 Amazonia
Batidaceae auctt. = *Bataceae*
Batidaea (Dumort.) E. Greene = *Rubus*
Batiki blue grass *Ischaemum indicum*
batiputa *Ouratea parviflora*
Batis P. Browne. Bataceae. 2: *B. maritima* L. (saltwort, beachwort, Hawaii, SW US, WI, Atlantic S Am., dioec.), *B. argillicola* P. Royen (New Guinea, Queensland, monoec.) – occasionally used in salads, ash form. used in glass- & soap-making
Batocarpus Karsten. Moraceae (II). 4 trop. Am. R: BMNRB 37(1968)1
Batodendron Nutt. = *Vaccinium*
Batopedina Verdc. Rubiaceae (IV 1). 3 W & S trop. Afr.
Batrachium (DC) Gray = *Ranunculus*
Battandiera Maire = *Ornithogalum*
Baudouinia Baillon. Leguminosae (I 2). 4 Madagascar
Bauera Banks ex Andrews. Cunoniaceae. 4 temp. E Aus. Cult. orn.
Baueraceae Lindley = Cunoniaceae
Bauerella Borzi = *Acronychia*
Baueropsis Hutch. = *Cullen*
Bauhinia L. Leguminosae (I 3). 300 pantrop. (100 s.s.). Often lianes with flattened stems, e.g. *B. scandens* L. var. *anguina* (Roxb.) Ohashi (snake climber, India) curving in alt. directions at each node. Some dioec., also some heterostylous (otherwise unknown in fam.); some poll. by bats, others birds or insects; some with explosive fr. (seeds of *B. purpurea* L. (Indomal.) ejaculated to 15m), others, e.g. *B. binata* Blanco (SE As. to trop. Aus.) of coastal forest, with floating fr. Extrafl. nectaries. Many cult. orn. (butterfly tree, camel's foot): *B. variegata* L. (orchid tree, India & China) – lvs & pods ed., bark medic. & used in tanning, sacred to Buddhists; others used for cordage & food locally (e.g. *B. esculenta* Burchell (*Tylosema e.*, marama bean, morama b., S Afr.) – as much protein as soya & rich in oil, fl. buds of *B. purpurea* in Sikkim), also gum, dyes & timber
Baukea Vatke. Leguminosae (III 10). 1 Madagascar
Baumannia Schumann = *Knoxia*
Baumea Gaudich. (~ *Machaerina*). Cyperaceae. c. 30 Madag. to Pacific (Aus. 15)
Baumia Engl. & Gilg. Scrophulariaceae. 1 trop. Afr.
Baumiella H. Wolff = *Afrocarum*
bauno *Mangifera caesia*
bawchan seed *Cullen corylifolia*
Baxteria R. Br. ex Hook. Lomandraceae (Xanthorrhoeaceae s.l.). 1 SW Aus
bay berry *Myrica californica*; **b. laurel** *Laurus nobilis*; **b. rum** *Pimenta racemosa*; **b.wood** *Swietenia macrophylla*
bayonet grass *Scirpus paludosus*; **b. plant** *Aciphylla squarrosa*
Bdallophytum Eichler. Rafflesiaceae (III). 2 C Am.
bdellium resin from *Commiphora* spp.
beach heath *Hudsonia ericoides*
Beadlea Small = *Cyclopogon*
beadplant *Nertera granadensis*
beak rush *Rhynchospora* spp.
Bealea Scribner = *Muhlenbergia*
bean most widely cult. are spp. of *Phaseolus*; **adzuki b.** *Vigna angularis*; **asparagus b.** *V. unguiculata* subsp. *sesquipedalis*; **baked b.s** *P. vulgaris*; **Bengal b.** *Mucuna pruriens* var. *utilis*; **black b.** *Castanospermum australe*, *Kennedia nigricans*; **Boer b.** *Schotia* spp.; **bog b.** *Menyanthes trifoliata*; **borlotti b.** *P. vulgaris*; **broad b.** *Vicia faba*; **buffalo b.** *Astragalus crassicarpus*; **Burma b.** *P. lunatus*; **butter b.** *P. lunatus*; **Calabar b.** *Physostigma venenosum*; **canellini b.** *Phaseolus vulgaris*; **Cherokee b.** *Erythrina herbacea*; **cluster b.** *Cyamopsis tetragonolobus*; **b. curd** *Glycine max*, see also *Vigna*; **duffin b.** *P. lunatus*; **dwarf b.** *P. vulgaris*; **field b.** *Vicia faba*; **flageolet b.** *P. vulgaris*; **Florida velvet b.** *Mucuna pruriens* var. *utilis*; **French b.** *P. vulgaris*; **garbanzo b.** *Cicer arietinum*; **Goa b.** *Psophocarpus tetragonolobus*; **ground b.** *Macrotyloma geocarpum*; **haricot b.** *Phaseolus vulgaris*; **horse b.** *Vicia faba*;

hyacinth b. *Lablab purpureus;* **Indian b.** *Catalpa bignonioides;* **jack b.** *Canavalia ensiformis;* **jumbie b.** orn. seeds esp. of *Leucaena leucocephala* etc.; **kidney b.** *Phaseolus vulgaris;* **Lima b.** *P. lunatus;* **locust b.** *Ceratonia siliqua;* **lucky b.** spp. of *Abrus, Afzelia, Erythrina, Thevetia;* **maloga b.** *Vigna lanceolata;* **marama** or **morama b.** *Bauhinia esculenta;* **Mary's b.** *Merremia discoidesperma;* **mat** or **moth b.** *V. aconitifolia;* **mung b.** *V. radiata;* **navy b.** *Phaseolus vulgaris;* **nicker b.** *Caesalpinia bonduc;* **Nuñas b.** *P. vulgaris;* **ordeal b.** *Physostigma venenosum;* **pater-noster b.** *Abrus precatorius;* **pinto b.** *Phaseolus vulgaris,* **potato b.** *Apios americana;* **Queensland b.** *Entada phaseoloides;* **Rangoon b.** *P. lunatus;* **red b.** *Dysoxylum mollissimum;* **rice b.** *Vigna umbellata;* **sabre b.** *Canavalia ensiformis;* **scarlet runner b.** *P. coccineus;* **sea b.** *Entada gigas, Mucuna* spp.; **snake b.** *Vigna unguiculata* ssp. *sesquipedalis;* **snap b.** *P. vulgaris;* **soya b.** *Glycine max;* **b. sprouts** *Vigna radiata;* **string b.** *P. vulgaris;* **tepary b.** *P. acutifolius;* **Tonka b.** *Dipteryx odorata;* **velvet b.** *Mucuna pruriens* var. *utilis;* **walnut b.** *Endiandra palmerstonii;* **winged b.** *Psophocarpus tetragonolobus;* **yam b.** *Pachyrhizus erosus;* **wax b.** *Phaseolus vulgaris;* **yard-long b.** *Vigna unguiculata* subsp. *sesquipedalis*

bear *Hordeum vulgare*

bearberry *Arctostaphylos uva-ursi;* **black b.** *A. alpina*

beargrass *Yucca* spp., *Xerophyllum tenax*

bear's breech *Acanthus mollis;* **b.'s ear** *Primula auricula;* **b.'s foot** *Helleborus foetidus*

beard flower *Pogonia;* **b. grass** *Polypogon monspeliensis;* **b. tongue** *Penstemon* spp.

Beatsonia Roxb. = *Frankenia*

Beaucarnea Lemaire = *Nolina*

Beaufortia R. Br. Myrtaceae (Lept.). 17+ SW Aus.

Beaumontia Wallich. Apocynaceae. 9 India & China to Bali. R: AUWP 86-5(1986)3. Cult. orn. esp. *B. grandiflora* Wallich (Himal.)

Beauprea Brongn. & Gris. Proteaceae. 12 New Caled. R: FNC 2(1968)20

Beaupreopsis Virot. Proteaceae. 1 New Caledonia

Beautempsia Gaudich. = *Capparis*

beauty berry *Callicarpa americana;* **b. bush** *Kolkwitzia amabilis*

Beauverdia Herter = *Leucocoryne*

beaver poison *Cicuta maculata;* **b. wood** *Celtis occidentalis*

Bebbia E. Greene. Compositae (Helia.-Gal.). 2 SW N Am. R: Madrono 24(1977)112. Xerophytes.

Beccarianthus Cogn. (~ *Astronidium*). Melastomataceae. 22+ Borneo to New Guinea

Beccarinda Kuntze. Gesneriaceae. 8 Burma to Hainan

Beccariodendron Warb. = *Gonothalamus*

Beccariophoenix Jum. & Perrier. Palmae (V 5a). 1 E Madag., almost extinct. Tapped for sap

Becheria Ridley = mixture of *Ixora* & *Psychotria* spp.

Becium Lindley = *Ocimum*

Beckeropsis Figari & De Not. = *Pennisetum*

Beckmannia Host. Gramineae (21d). 2 N temp. (Eur. 2). Fodder; grains of *B. eruciformis* (L.) Host ed. Japan

Beckwithia Jepson = *Ranunculus*

Beclardia A. Rich. Orchidaceae (V 16). 2 Mascarenes

Becquerelia Brongn. Cyperaceae. 10 trop. S Am

Bedfordia DC (~ *Brachyglottis*). Compositae (Senec.-Tuss.). 3 SE Aus.

bedstraw *Galium* spp.; **hedge b.** *G. mollugo;* **lady's b.** *G. verum*

beech *Fagus* spp.; **African b.** *Faurea saligna;* **American b.** *Fagus grandifolia;* **Antarctic b.** *Nothofagus antarctica;* **black b.** *N. solanderi;* **blue b.** *Carpinus caroliniana;* **brown b.** *Cryptocarya glaucescens;* **Cape b.** *Myrsine melanophloeos;* **common** or **Eur. b.** *Fagus sylvatica;* **copper b.** *F. sylvatica* 'Atropunicea'; **Dawyck b.** *F. sylvatica* 'Dawyck'; **b. fern** *Phegopteris connectilis;* **Japanese b.** *Fagus crenata;* **myrtle b.** *N. cunninghamii;* **oriental b.** *F. sylvatica* ssp.*orientalis;* **red b.** *Nothofagus fusca, Flindersia* spp.; **roble b.** *N. obliqua;* **silver b.** *N. menziesii;* **southern b.** *N.* spp.; **Southland b.** *N. menziesii;* **Tasmanian b.** *N. cunninghamii;* **Turkish b.** *Fagus sylvatica* ssp. *orientalis*

beef apple *Manilkara zapota;* **b. plant** *Iresine herbstii;* **b. steak plant** *Acalypha wilkesiana;* **b. suet tree** *Shepherdia argentea;* **b.wood** *Casuarina equisetifolia, Grevillea striata, Stenocarpus salignus*

beer *Hordeum vulgare,* (**b., white**) *Triticum aestivum*

Beesia Balf.f. & W.W. Sm. Ranunculaceae (I 1). 1 N Burma, W & SW China

beet *Beta vulgaris*: cvs include beetroot, spinach beet & sugar beet

beetleweed *Galax urceolata*

beggar tick *Bidens* spp. esp. *B. pilosa;* **b. weed** *Desmodium tortuosum*

Begonia L. Begoniaceae. c. 900 trop. & warm esp. Am. R: SCB 60(1986)131. Monoec. herbs to pachycaul shrubs with fibrous or tuberous roots or rhiz. & upright, climbing or 0 stems, many epiphytic; lvs usu. consp. asymmetric (elephant ear) some with velvety surface appearing diff. colours at diff. light angles, e.g. *B. thaipingensis* King (Malay Pen.) blue-grey to gold; some cauliflory (*B. cauliflora* M. Sands (Borneo)), epiphyllous infls, branches & lvs in different spp., some propagated by leaf-cuttings; some with hooked or fleshy fr. or seeds (animal-disp.) or seeds of some with air-filled cells acting as balloons. Over 10 000 hybrids & cvs recorded, the genus divided hort. into fibrous-rooted, tuberous and rhizomatous though these categories prob. artificial, the genus being arranged botanically on ovary characters. Most cult. are fibrous-rooted *B.* Semperflorens-Cultorum Group (*B. cucullata* Willd. (*B. c.* var. *hookeri*, *B. semperflorens*, SE Brazil, NE Arg.) × *B. schmidtiana* Regel (Brazil)), tuberous *B.* × *tuberhybrida* Voss (cvs derived from crossing a number of Andean spp.) and rhizomatous *B. rex* Putzeys (Assam, though plants grown under this name are generally hybrids involving other spp. as well). Journal: *The Begonian*. Other cult. spp. incl. *B. grandis* Dryander (*B. evansiana*, E As.) – hardy in Britain, *B. masoniana* Irmscher (iron cross, New Guinea), *B. phyllomaniaca* C. Martius (? Brazil, ? *B. incarnata* Link & Otto (Mex.) × *B. manicata* Brongn. ex Cels (Mex.)) – lvs on stems & lvs; *B. socotrana* Hook.f. (Socotra) – bulbous, winter-flowering habit now bred into fibrous-rooted and tuberous hybrids (Rieger bs) grown comm. on huge scale as potpls. Some medic. & *B. muricata* Blume (*B. tuberosa*, Mal.) with ed. lvs

Begoniaceae Bercht. & J. Presl. Dicots – Dilleniidae – Violales. 2/c. 900 trop. & warm. R: SCB 60(1986). Succ. herbs or shrubs, s.t. climbing, usu. monoec. often with crassulacean acid metabolism, accumulating free organic acids (e.g. malic & oxalic) in cells. Lvs spiral, sometimes distichous, usu. asymmetric & simple, sometimes palmately lobed or compound; stipules often large, free. Infls usu. axillary, cymose; fls often irreg. P all petaloid, 2 sets of 5 (the outer larger) but more often fewer with 2 unlike valvate sets of 2 in males & a single imbricate set of 5 in females (10 in *Hillebrandia*), rarely connate at base; A 4 to ± ∞, originating centripetally, s.t. arranged on 1 side of fl., anthers opening by longit. slits or term. pores, the connective often elongated so that pollen-sacs well-separated; $\overline{G}$ (usu. 2 or 3, up to 6) with axile placentas or these not quite meeting, ovary v. often with (1–)3(–6) prominent wings, styles distinct or connate at base, bifid; ovules ∞, anatropous, bitegmic. Fr. usu. loculicidal capsule, rarely berry; seeds ∞, small with tiny straight embryo and almost 0 endosperm. x = 10–21+

Genera: *Begonia*, *Hillebrandia*

Usu. considered allied to Datiscaceae but pluriloc. ovary uncommon in Violales and centripetal stamens 'aberrant' in Dilleniidae

Cult. orn. (*Begonia*, q.v.)

Begoniella Oliver = *Begonia*

Beguea Capuron. Sapindaceae. 1 Madagascar

Behaimia Griseb. Leguminosae (III 6). 2 Cuba. *B. cubensis* Griseb. – good timber

Behnia Didr. Philesiaceae (Smilacaceae s.l.). 1 S Afr.

Behria E. Greene = *Bessera*

Behuria Cham. Melastomataceae. 3 S Brazil

Beilschmiedia Nees. Lauraceae (I). 250 trop. to Aus., NZ, C Chile. Some good timbers (bolly gums, Aus.) incl. *B. tarairi* (Cunn.) Kirk (taraire, NZ), *B. tawa* (Cunn.) Kirk (tawa, NZ) & *B. bancroftii* C. White (Canary ash, yellow walnut, Queensland); *B. kweo* (Mildbr.) Robyns (mkweo, Tanzania) – form. exported to Germany for luxury panelling etc. but supplies exhausted c. 1945; *B. roxburghiana* Nees (Himal.) – wood for house-building & tea-chests

Beirnaertia Louis ex Troupin. Menispermaceae. 1 trop. Afr.

Beiselia Forman. Burseraceae. 1 Mex.: *B. mexicana* Forman. G 10–12, each with 2 superposed ovules & fr. with 10–12-flanged columella

Bejaranoa R. King & H. Robinson (~ *Eupatorium*). Compositae (Eupat.-Gyp.). R: MSBMBG 22(1987)99. 2 trop. Am.

Bejaria Mutis ex L. Ericaceae. 15 trop. & warm Am. *B. racemosa* Vent. (SE US) & other spp. (Andes rose) like rhododendrons in vegetation

Bejaudia Gagnepain = *Myrialepis*

Belairia A. Rich. Leguminosae (III 13). 6 Cuba

Belamcanda Adans. Iridaceae (III 2). 1 E Russia, China, Japan, N India: *B. chinensis* (L.) DC (leopard lily) – spotted orange fls, seeds black (Irid. usu. brown), medic., cult. orn.

Belandra S.F. Blake = *Prestonia*

Belemia Pires. Nyctaginaceae (V?). 1 Brazil: *B. fucsioides* Pires

Belencita Karsten. Capparidaceae. 1 Colombia

beli *Aegle marmelos*
belian *Eusideroxylon zwageri*
Belicea Lundell = *Morinda*
Beliceodendron Lundell = *Lecointea*
bell flower *Campanula* spp.; **b. flower, Chilean** *Lapageria rosea;* **b. flower, climbing** *Littonia modesta;* **b. flower, ivy-leaved** *Wahlenbergia hederacea;* **b. pepper** *Capsicum* spp.; **Qualup b.** *Pimelea physodes;* **b. tree** *Halesia* spp.; **b. wort** *Uvularia* spp.
bella umbra *Phytolacca dioica*
belladonna *Atropa belladonna;* **b. lily** *Amaryllis belladonna*
Bellardia All. Scrophulariaceae. 1 Med. (incl. Eur.): *B. trixago* (L.) All. natur. Aus. & S Afr.
Bellardiochloa Chiov. = *Poa*
Bellendena R. Br. Proteaceae. 1 Tasmania
Bellevalia Lapeyr. (~ *Muscari*). Hyacinthaceae (Liliaceae s.l.). 45 Med. (Eur. 10) to Iran & N Afghanistan. Cult. orn. incl. *B. romana* (L.) Sweet (Roman hyacinth, Med.). R: PJBJS 1(1940)42, 131, 336
Bellida Ewart. Compositae (Gnap.-Ang.). 1 SW Aus.: *B. graminea* Ewart. R: Nuytsia 8(1992)367
Bellidastrum Scop. = *Aster*
Belliolum Tieghem = *Zygogynum*
Bellis L. Compositae (Ast.-Ast.). 8 Eur. (7), Medit. Daisies (daisy = day's eye because closing at night), some medic. Cult. orn. esp. forms of *B. perennis* L., a widespread lawn weed (= 'innocence' in 'Language of Fls'): 'Prolifera' with secondary heads in axils of involucral bracts (hen-and-chickens d.), 'double' forms with all ligulate florets (bachelor's buttons) etc.
Bellium L. Compositae (Ast.-Ast.). 4 Medit. Eur. (3). Some cult. orn.
Belloa Remy (~ *Lucilia*). Compositae (Gnap.-Gnap.). 9 Andes of Venez. to C Chile. R: BJLS 106(1991)189
Bellonia L. Gesneriaceae. 2 WI. Axillary thorns (? infls.) produced in dry localities, 0 in wet
bells of Ireland *Moluccella laevis*
Bellucia Necker ex Raf. (~ *Loreya*). Melastomataceae. 7 trop. Am. R: MNYBG 50(1989)5. Trees & shrubs; fls 5–8-merous, G 10–14(15)-loc. Fr. ed., *B. pentamera* Naudin (*B. axinanthera*) grown in Mal. as fr. & orn. tree
Bellynkxia Muell. Arg. = *Morinda*
Belmontia E. Meyer = *Sebaea*
Beloglottis Schltr. (~ *Spiranthes*). Orchidaceae (III 3). 8 warm Am. R: HBML 28(1982)302
Belonanthus Graebner = *Valeriana*
Belonophora Hook.f. Rubiaceae (? II 4). 6 trop. Afr. esp. W
Beloperone Nees = *Justicia*
Belostemma Wallich ex Wight (~ *Tylophora*). Asclepiadaceae (III 4). 2 India, China
Belosynapsis Hassk. (~ *Cyanotis*). Commelinaceae (II 1c). 4 As.
Belotia A. Rich = *Trichospermum*
Beltrania Miranda = *Enriquebeltrania*
bel-tree *Aegle marmelos*
Belvisia Mirbel. Polypodiaceae (II 3). 8 OW trop. R: Blumea 37(1993)511. Sporangia at leaf apex only. Cult. orn. epiphytes
Bemarivea Choux = *Tinopsis*
Bembicia Oliver. Flacourtiaceae (10). 1 Madag. $\overline{G}$; infl. strobiloid
Bembicidium Rydb. = *Poitea*
Ben, oil of see *Moringa oleifera*
Bencomia Webb & Berth. Rosaceae. (Incl. *Marcetella*) 7 Canary Is. R: BMac. 6(1980)71. Some cult. orn.
bendang *Agathis dammara*
Benedictella Maire = *Lotus*
Benedictine flavour due to *Angelica archangelica*
Beneditaea Tol. = *Ottelia*
Benevidesia Saldanha & Cogn. Melastomataceae. 1 SE Brazil
Bengal bean *Mucuna pruriens* var. *utilis;* **B. cardamom** *Amomum aromaticum;* **B. kino** *Butea monosperma;* **B. quince** *Aegle marmelos*
benge *Guibourtia arnoldiana*
Benguellia G. Taylor. Labiatae (VIII 4b). 1 Angola
beni seed *Polygala butyracea;* **black b. s.** *Hyptis spicigera*
Benin mahogany *Khaya* spp.; **B. walnut** *Lovoa trichilioides;* **B. wood** *K. grandifoliola*

Benincasa Savi. Cucurbitaceae. 1 trop. As., cultigen: *B. hispida* (Thunb.) Cogn. (wax or white gourd, ash pumpkin, petha), fr. coated with wax (vessel for scented coconut oil in Polynesia before Eur. contact), boiled as vegetable with curry, candied or pickled, local medic.; can grow at a rate of 2.3 cm in 3 hrs

Benitoa Keck = *Lessingia*

Benjamin, gum = benzoin; **stinking B.** *Trillium erectum*; **B. tree** *Ficus benjamina*

Benjaminia Mart. ex Benj. Scrophulariaceae. 1 trop. Am., aquatic

Benkara Adans. (~ *Randia*). Rubiaceae (II 1). 1 + ?, S India etc.

Bennettiodendron Merr. Flacourtiaceae (8). 3–4 Indomal., China

Benoicanthus Heine & A. Raynal. Acanthaceae. 2 Madagascar

Benoistia Perrier & Leandri. Euphorbiaceae. 2 Madagascar

Bensoniella Morton. Saxifragaceae. 1 NW Calif., SW Oregon: *B. oregona* (Abrams & Bacig.) Morton, cult. orn. like *Mitella*

bent grass *Agrostis* spp.; **black b.** *A. gigantea*; **brown b.** *A. canina*; **common** or **colonial b.** *A. capillaris*; **silky b.** *Cynosurus* spp.; **velvet b.** *A. canina*

Benthamantha Alef. = *Coursetia*

Benthamia A. Rich. Orchidaceae (IV 2). 26 Mascarenes

Benthamidia Spach = *Cornus*

Benthamiella Speg. Solanaceae (2). 12 S Patagonia. Cushion plants in open steppe. R: BN 133(1980)67

Benthamina Tieghem (~ *Amyema*). Loranthaceae. 1 E Aus.: *B. alyxifolia* (Benth.) Tieghem

Benthamistella Kuntze = *Buchnera*

Bentia Rolfe = *Justicia*

Bentinckia A. Berry ex Roxb. Palmae (V 4m). 2 India, Nicobar Is.

Bentinckiopsis Becc. = *Clinostigma*

Bentleya E. Bennett. Pittosporaceae. 2 SW Aus. R: BJLS 103(1990)309. Rhizomatous shrubs

benzoin (gum Benjamin) resinous balsams derived from *Styrax* spp.

Benzingia Dodson. Orchidaceae (V 10). 2 trop. S Am.

Benzonia Schum. Rubiaceae (?). 1 W Afr.

Bequaertia R. Wilczek (~ *Campylostemon*). Celastraceae. 1 trop. Afr.: *B. mucronata* (Exell) R. Wilczek

Bequaertiodendron De Wild. = *Englerophytum*

ber *Ziziphus mauritiana*

Berardia Villars. Compositae (Card.). 1 W Alps

Berberidaceae Juss. Dicots – Magnoliidae – Ranunculales. 15/680 N temp. to trop. mts. Trees, shrubs (some pachycaul &/or spiny at nodes) or perennial herbs, usu. glabrous, often with alks & tissues coloured yellow with berberine (an isoquinoline); vascular bundles often ± scattered, woody spp. often with broad medullary rays. Lvs spiral (opp. in *Podophyllum*), pinnate, ternate or simple (unifoliolate with articulation at base of leaflet in some *Berberis* spp.); stipules small or 0, though petiole often flared near base. Fls in racemes, cymes or solitary, bisexual, reg., (2, *Epimedium*)3(4)-merous. P (aborts in *Achlys*) usu. of 6 or 7(–9) series, typically outer 2 (? sepals but often considered bracts) sepaloid, often caducous, next 2 (? nectary-less C but often considered petaloid K) petaloid, inner 2 or 3 (? nectariferous C but often considered staminodes) usu. petaloid & nectariferous (not in *Diphylleia* nor *Podophyllum*); A (4–)6(–18) usu. same no. as nectariferous C, but s.t. 2 x & usu. opp. C, anthers usu. opening by 2 valves that lift up from base (longit. slits in *Nandina* and *Podophyllum*); G apparently 1 but often interpreted as derived from 3, ovules anatropous or hemitropous, bitegmic, commonly ∞ on a thickened marginal placenta, or 2 or (*Achlys*) 1, basal. Fr. usu. a berry, seldom dry, dehiscent or not; seeds often arillate, embryo small, endosperm abundant with oils, protein & s.t. hemicellulose. x = 6, 7, 8, 10

Classification & chief genera:

I. **Nandinoideae** (shrubs, A with longit. dehiscence): *Nandina*

II. **Berberidoideae** (shrubby or herbaceous, A with valvate (longit.) dehiscence):

1. **Leonticeae** (ovules 1–4; R: CJB 67(1989)2310): *Gymnospermium*
2. **Berberideae** (ovules several): *Berberis, Epimedium, Mahonia, Podophyllum*

Some genera v. close (e.g. *Berberis* & *Mahonia*, *Epimedium* & *Vancouveria*) but the fam. falls into distantly allied groups treated as separate fams by some authors, though the groups are in general more closely related to one another than they are to other fams. Poss. the assemblage is allied to Ranunculaceae, though *Nandina* with its 2–3-pinnate lvs & endotegmic seeds is somewhat isolated from the rest

Because of nuclear endosperm development fam. thought allied to Menispermaceae, Papaveraceae & Ranunculaceae but pollen does not resemble Ranunculaceae (exc.

Hydrastis) or Lardizabalaceae and that of *Mahonia* & *Berberis* is distinct from that of other genera here

Many cult. orn. esp. spp. of *Berberis, Epimedium* & *Mahonia*; *Podophyllum* is medic.

Berberidopsidaceae (Veldk.) Takht. = Flacourtiaceae

Berberidopsis Hook.f. Flacourtiaceae (1). 2 Chile (1), E Aus. (1). R: Blumea 30(1984)21. *B. corallina* Hook.f. (Chile, extinct in wild?) cult. orn.

Berberis L. Berberidaceae (II 2). 500+ Euras. (Eur. 2, China c. 200), N Afr., trop. Afr. mts, Am. R: JLSBot. 57(1961)1. Shrubs (barberries), usu. spiny, with alks & yellow wood (berberine) form. used in eye disease; fr. of many ed., wood a dyestuff; many cult. orn. (hybrids with *Mahonia* = × *Mahoberberis* C. Schneider) esp. as hedges. Simple lvs app. derived from leaflets of ancestors with pinnate lvs (cf. *Mahonia*); lvs of long shoots tripartite spines (transitions to 'true' lvs often present), short leafy & flowering shoots in their axils. Stamens sensitive, springing upwards on contact by insect & showering side of its head with pollen. Several spp. are alt. hosts of stem-rusts (first proposed by Banks (1805) but not accepted until de Bary (1865–6)) of wheat, oats, barley & rye & attempts at their eradication in US etc. have been made, though not a certain or indispensable phase in life-cycle of *Puccinia graminis* though the more dangerous *P. glumarum* (yellow rust) will not grow on *Berberis*, while in Russia removal not advocated as *B.* spp. imp. honey-pls. *B. aristata* DC (chitra, Nepal) – bitter tonic for fevers; *B. canadensis* Miller (American or Alleghany barberry, E N Am.) – ed. fr., alt. host for rusts; *B.* × *stenophylla* Lindley (*B. darwinii* Hook. (temp. S Am., introd. NZ where a pest) × *B. empetrifolia* Lam. (Chile, Arg.)) – common hedgepl.; *B. vulgaris* L. (common or Eur. barberry, Eur., natur. in Britain) – fine wood for turning, toothpicks, a dyestuff for silk, cotton, wool & leather & staining wood, form. a hair-dye, fr. preserved esp. a seedless form in a French jam, bark etc. form. medic.

Berchemia Necker ex DC. Rhamnaceae. c. 12 E Afr. to E As., W N Am. Cult. orn. twiners. *B. discolor* (Klotzsch) Hemsley (trop. & S Afr., Madag.) – fr. ed. Uganda.

Berchemiella Nakai. Rhamnaceae. 3 China, Japan. R: BBR 8,4(1988)119

bere (= **bear**) *Hordeum vulgare*

Berendtiella Wettst. & Harms. Scrophulariaceae. 4 C Am.

Berenice Tul. Campanulaceae. 1 Réunion

bergamot *Monarda* spp. esp. *M. didyma*; **b. orange** *Citrus* × *aurantium* (US), *C.* × *bergamia*

Bergenia Moench. Saxifragaceae. 6–8 temp. & subtrop. E As. R: APS 26(1988)20. Cult. orn. (elephant-ear, Siberian saxifrage) esp. as ground cover; many hybrids (high degree of self-incompatibility), most commonly *B. crassifolia* (L.) Fritsch (Siberia, Mongolia) – rarely sets seed. Some a source of tannin & medic., e.g. *B. ciliata* (Haw.) Sternb. (*B. ligulata*, W Pakistan to SW Nepal) – rhizomes a tea-leaf subs. in Kashmir

Bergeranthus Schwantes. Aizoaceae (V). 12 E Cape. Cult. orn.

Bergerocactus Britton & Rose. Cactaceae (III 8). 1 Calif. & Baja Calif. Fr. extrudes pulp & seeds; hybrids with *Pachycereus* & *Myrtillocactus* spp. in wild

Bergeronia M. Micheli. Leguminosae (III 6). 1 Paraguay, Arg.

Berghesia Nees. Rubiaceae (?). 1 Mex.

Bergia L. Elatinaceae. 24 warm (Aus. 10)

Berginia Harvey = *Holographis*

Berhautia Balle. Loranthaceae. 1 W Afr.

Berkheya Ehrh. Compositae (Arct.-Gort.). 75 trop. & S Afr. R: MBSM 3(1959)104

Berkheyopsis O. Hoffm. = *Hirpicium*

Berlandiera DC. Compositae (Helia.-Eng.). 4 S US, Mex. R: Brittonia 19(1967)285. Cult. orn. esp. *B. lyrata* Benth. – chocolate-scented

Berlinia Sol. ex Hook.f. & Benth. Leguminosae (I 4). 15 trop. Afr. Some timbers (abem, ekpogoi); an infusion of *B. globiflora* Hutch. & Burtt Davy used in ordeals in C Afr.

Berlinianche (Harms) Vatt. Rafflesiaceae (II 2). 2 trop. Afr. Parasitic on *Berlinia* & *Brachystegia* spp.

Bermuda arrowroot *Maranta arundinacea*; **B. buttercup** *Oxalis pes-caprae*; **B. cedar** *Juniperus bermudiana*; **B. grass** *Cynodon dactylon*

Bernardia Miller. Euphorbiaceae. 20 warm Am.

Bernardinia Planchon = *Rourea*

Berneuxia Decne. Diapensiaceae. 1 Himal.

Berniera Baillon = *Beilschmiedia*

Bernoullia Oliver. Bombacaceae. 2 trop. Am.

Berrisfordia L. Bolus (~ *Conophytum*). Aizoaceae (V). 1 W S Afr.

Berroa Beauverd. Compositae (Gnap.-Gnap.). 1 subtrop. S Am.: *B. gnaphalioides* (Less.) Beauverd. R: BJLS 106(1991)191

berry, Bay *Myrica californica*; **buffalo b.** *Shepherdia argentea*; **partridge b.** *Gaultheria procumbens*; **phenomenal b.** see *Rubus*; **purple apple b.** *Billardiera longiflora*; **yellow b.** *Rhamnus infectoria*

Berrya Roxb. Tiliaceae. 3–5 Indomal. *B. cordifolia* (Willd.) Burret (*B. ammonilla*, Trincomali wood, hamilla, S India, Sri Lanka) – valuable red timber e.g. for coconut arrack vats

Bersama Fres. Melianthaceae. 2 (1 v. polymorphic) trop. & S Afr.

bertam *Eugeissona tristis*

Berteroa DC. Cruciferae. 5 OW N temp. (Eur. 5). R: JAA 68 (1987)207

Berteroella O. Schulz. Cruciferae. 1 temp. E As.

Bertholletia Bonpl. Lecythidaceae. 1 trop. S Am.: *B. excelsa* Bonpl. (Brazil nut, Pará nut). Ligule (see fam.) pressed down in A & anthers available therefore only to big bees (*Xylocopa* spp. & female euglossines); self-sterile, fr. once trees c. 10 yrs old takes 14 months to mature, forming a large woody capsule; the operculum falls inwards & seeds with hard woody testa & oily endosperm (Brazil nuts of commerce) gnawed out by agoutis, which scatter-hoard them. Most in commerce (50 000 t per annum) collected from wild trees (logging now banned), the fr. being split with an axe; oil for foodstuffs & soap, hair-conditioner & cosmetics from seeds (60% oil, 17% protein) in commerce

Bertiera Aublet. Rubiaceae (inc. sed.). 55 trop. Am. & Afr. (41)

Bertolonia Raddi. Melastomataceae. 8 Brazil. Cult. orn. foliage pls

Bertya Planchon. Euphorbiaceae. c. 25 Aus. Resins

Berula Koch. Umbelliferae (III 8). 1 N temp. (inc. Eur.), E & S Afr.: *B. erecta* (Huds.) Cov. Aquatic, fatal to cattle in NSW

Beruniella Zak. & Nabiev = *Heliotropium*

Berzelia Brongn. Bruniaceae. 12 Cape. Cult. orn.

Beschorneria Kunth. Agavaceae. 7 C & S Mex. mts. R: Plantsman 10(1989)194. Lvs a soap subs., fls ed., cult. orn.

Besleria L. Gesneriaceae. 200 warm Am. Trees & shrubs

Bessera Schultes f. Alliaceae (Liliaceae s.l.). (Incl. *Behria*) 2–3. Calif., Mex. Cult. orn.

Besseya Rydb. Scrophulariaceae. 7 N Am. esp. Rockies. R: PPANS 85(1933)97. Catalpol (iridoid) sequestered by *Euphydryas anicia* butterflies. Cult. orn.

Beta L. Chenopodiaceae (I 1). 11–13 Eur. (6), Med. Selected forms of *B. vulgaris* L. subsp. *maritima* (L.) Arc. cult. since time of Assyrians, these derivatives of wild sea-beet being grouped as subsp. *vulgaris* (beetroot, sugarbeet, mangel-wurzel or mangold; inc. subsp. *cicla*, spinach beet or Swiss chards): all are biennial with sugar reserves in root, those of sugarbeet up to 20% of the weight (first factory in Silesia, 1801), those of beetroot with high concentrations of red betalains (used in herbal treatment of cancer, though 14% UK population cannot metabolize red betanin leading to beeturia (red urine); forms grown for their leaves & eaten as vegetables include those form. in subsp. *cicla*. The mangel-wurzel (UK record: 24.72 kg) is used as cattle-feed. The 'seeds' are actually coherent fr. enveloped in woody calyces

bété *Mansonia altissima*

betel (nut) *Areca catechu*

Betonica L. = *Stachys*

betony *Stachys officinalis*

betsa-betsa fermented cane-sugar, sometimes flavoured with *Nuxia congesta*, the national drink of Malagasy Republic

Betula L. Betulaceae (I). 35 N hemisphere (Eur. 4, birches). R: JAA 71(1990)32. Decid., monoec. trees & shrubs allied to *Alnus* but catkins shattering when ripe. Timber for furniture, plywood, skis etc., sap for sweetening, fermenting & shampoo, twigs for brooms & 'birching' schoolboys, birch bud oil form. giving characteristic smell to Russian leather. Betulinic acid in bark triggers cell death in melanomas but not other cell cultures. Birch bark 1800 yrs old (Afghanistan) bears oldest known Buddhist MSS. *B. alleghaniensis* Britton ('*B. lutea*', yellow b., E N Am.) – good timber; *B. lenta* L. (American black or cherry b., E N Am.) – timber, distilled bark gives oil high in methyl salicylate used medic. (wintergreen), 'birch beer' made from fermenting sap; *B. maximowicziana* Regel (Japanese b., Japan); *B. michauxii* Spach (NE Canada) – creeping shrub; *B. nana* L. (dwarf b., circumpolar); *B. nigra* L. (red b., E N Am.); *B. neoalaskana* Sarg. (*B. resinifera*, Alaska, Yukon) – triterpenoid papyriferic acid protects against depredation by snowshoe hares; *B. papyrifera* Marshall (paper b., N. Am.) – timber for turning, shoe-lasts, pegs & pulp, bark impervious to water & form. used for canoes, baskets, cups & wigwam covers by Indians; *B. pendula* Roth & *B. pubescens* Ehrh. (Eur. birches, poss. only subspecifically distinct, many hybrids) – reach northern limit of tree-growth, wood used for furniture esp. in

Scandinavia, forms with pretty markings (cause obscure) known as Karelian b. or masur, bark a famine food or eaten with sturgeon eggs in E As. as well as used for shoes, clothes etc., tar used for preserving leather & wood, lvs with alum give a green dye, with chalk a yellow one, oil medic., forms of *B. pendula* (silver b.) grown for ornament esp. weeping 'Youngii', shrubby variants in wild, e.g. *B. oycoviensis* Besser (SE Poland) perhaps of hybrid origin; *B. populifolia* Marshall (NE Am.) used as a lead indicator in Wisconsin; *B. schmidtii* Regel (E As.) – wood too dense to float in water; *B. utilis* D. Don (Himal.) – bark form. used as writing & packing material

Betulaceae Gray. Dicots – Hamamelidae – Fagales. (Incl. Corylaceae) 6/110 N temp., trop. mts. Decid. trees or shrubs, leptocaul, monoec., anemophilous, typically with ectotrophic mycorrhizae in roots. Lvs usu. spiral, simple, usu. toothed; stipules deciduous. Male fls in ± elongate pendulous catkins, females in pendulous or erect, short, often woody ones with dichasia subtended by bracts. K (0)1–6, scale-like; C 0; A same no. & opp. K or app. <18 (poss. a congested 3-flowered cymule), free or united at base, the pollen-sacs ± distinct (not in *Alnus*); $\overline{G}$ (2(3)) with ± distinct styles, 2(3)-loc. below, 1-loc. above, ovules axile, pendulous from near summit of partition, 1 or 2 each locule, anatropous, unitegmic or bitegmic (*Carpinus*); fert. chalazogamous after delayed growth of pollen-tube. Fr. a nut or 2-winged samara, in Coryloideae subtended or almost enclosed in 2 or 3 leafy bracts; seed usu. 1 with thin fleshy endosperm or 0, embryo with oily thickened cotyledons. x = 8, 14

Classification & genera:

I. **Betuloideae** (male fls. in groups of 3): *Alnus, Betula*

II. **Coryloideae** (males solit.): *Carpinus, Corylus, Ostrya, Ostryopsis*

These subfams have been treated as separate fams but recent serological study shows their unity (Cronquist). Pollen attrib. to Betulaceae known from Upper Cretaceous, wood attrib. to *Alnus* & *Carpinus* from Eocene

Timber, nuts & cult. orn.

Bewsia Goossens (~ *Leptochloa*). Gramineae (31d). 1 C & S Afr.

Beyeria Miq. Euphorbiaceae. 15 Aus. Turpentine bushes

Beyrichia Cham. & Schldl. = *Achetaria*

bezetta rubra dye from *Chrozophora tinctoria*

bhaji *Amaranthus* spp.

bhang *Cannabis sativa* subsp. *indica*

Bharbur grass *Eulaliopsis binata*

Bhesa Buch.-Ham. ex Arn. Celastraceae. 5 Indomal.

Bhidea Stapf ex Bor. Gramineae (40e). 2 India

Biarum Schott. Araceae (VIII 6). 15 Med. (Eur. 4). R: Aroid. 3(1980)24. *B. tenuifolium* (L.) Schott (S Eur.) – infls smellable at 20 m

bibolo *Lovoa trichilioides*

bicuhyba fat from seeds of *Bicuiba oleifera*

Bicuiba de Wilde (~ *Virola*). Myristicaceae. 1 Brazil: *B. oleifera* (Schott) de Wilde (*V. bicuhyba*, *V. oleifera*, bicuhyba) – seed oil for candles etc.

Bidaria (Endl.) Decne (~ *Gymnema*). Ascelpiadaceae (III 4). 10 Indomal.

Bidens L. Compositae (Helia.-Cor.). c. 240 cosmop., esp. Mex. (Eur. 3 + naturalized spp., Hawaii 19 all interfertile). Pappus of few mostly retrorsely barbed awns involved in fr. dispersal (even by migrating salamanders in Ontario). R: PFMB 16(1937)1, but boundary with *Coreopsis* unclear. Some cult. orn. e.g. *B. ferulifolia* (Jacq.) Sweet (S US to C Am.), many weeds (bur-marigold, cuckold, pitchfork, sticktight, tickseed, beggar tick (esp. *B. pilosa* L., trop., also known as blackjack, root exudates allelopathic to lettuce, beans, maize & sorghum) in US). *B. beckii* Torrey (N Am.) – heterophyllous aquatic; *B. cosmoides* (A. Gray) Sherff (Hawaii) – bird-poll. pendent capitula; others medic. locally e.g. *B. pilosa* against diarrhoea in Burundi

bidgee-widgee *Acaena* spp.

bidi (India) cheap cigarette rolled in a leaf, not paper

bidi-bidi *Acaena* spp.

Bidwillia Herb. = ? *Trachyandra*

Biebersteinia Stephan. Geraniaceae (Biebersteiniaceae). 5 Greece (? extinct) to C As. Seeds typically geraniaceous in anatomy

Biebersteiniaceae Endl. = Geraniaceae

Bienertia Bunge ex Boiss. Chenopodiaceae (III 2). 1 Eur. to W C As.: *B. cycloptera* Bunge ex Boiss.

Biermannia King & Pantl. Orchidaceae (V 16). 7 India, China

Bifora Hoffm. Umbelliferae (III 4). 3 Med. (Eur. 2) to C As.
Bifrenaria Lindley. Orchidaceae (V 10). 24 trop. Am. Cult. orn.
big *Hordeum vulgare*; **b. tree** *Sequoiadendron giganteum*
bigarade *Citrus* × *aurantium* Sour Orange Group
Bigelowia DC. Compositae (Ast.-Sol.). 2 SE US. R: Sida 3(1970) 451
Bignonia L. Bignoniaceae. 1 SE US: *B. capreolata* L. – cult. orn. liane with 4-armed 'anomalous' xylem & 4 conspicuous bark ribs (phloem)
Bignoniaceae Juss. Dicots – Asteridae – Scrophulariales. 109/750 mainly trop. esp. S Am. Trees, lianes, shrubs, rarely herbs, lianes usu. with unusual vascular structure. Lvs opp., sometimes whorled, rarely spiral, pinnate to 3-compound, less often simple or palmately compound, terminal leaflet s.t. a tendril; stipules 0. Fls bisexual, in cymes, racemes or solitary, usu. conspic. K (5), s.t. bilobed, or unlobed, rarely with a calyptra (*Lundia* spp.); C (5), often 2-lipped, rarely ± reg., imbricate or rarely valvate; A alt. with C but attached to tube, 4, in 2 pairs, the fifth (adaxial) staminodal or 0, rarely all 5 fertile (*Oroxylum*) or 2 fertile & 3 staminodal (*Catalpa*); pollen very diverse in structure; annular or cupular nectary-disk usu. round $\underline{G}$ (2) with 2-lobed stigma, 2-loc. with 2 axile placentas per loc., or 1-loc. with 2 or 4 ± intruded parietal placentas, or (*Tourrettia*) 4-locular with ovules uniseriate in each locule; ovules ∞ anatropous or hemitropous, unitegmic. Fr. a bivalved capsule, very often with a replum, rarely fleshy & indehiscent; seeds usu. flat, winged in capsules, endosperm 0. x = 20
Classification & chief genera (after Gentry):
Tecomeae (fr. dehiscent perpendicular to septum, placentation axile): *Jacaranda, Tabebuia*
Oroxyleae (fr. dehiscent parallel to septum, placentation axile, As.): *Oroxylum*
Bignonieae (fr. dehiscent, lianes, Am.): *Macfadyena*
Tourrettieae (fr. dehiscent, placentation axile, capsule spiny, Am.): *Tourrettia* (only)
Eccremocarpeae (fr. dehiscent, placentation parietal, wiry climbers, Andes): *Eccremocarpus* (only)
Coleeae (fr. indehiscent, lvs usu. pinnate, usu. bee-poll., Afr. & Madag.): *Colea, Phyllarthron*
Crescentieae (fr. indehiscent, bat-poll., lvs palmate &/or spiral): *Parmentiera*
[**Schlegelieae** (fr. indehiscent, bird- or insect-poll., lvs simple, opp.): here referred to Scrophulariaceae (Armstrong)]
Floral anatomy & seed morphology put Schlegelieae in Scrophulariaceae, where *Paulownia* is best placed (endosperm 0, floral anatomy, embryo & seed morphology)
Some valuable timbers (e.g. *Cybistax, Paratecoma, Tabebuia*) & many orn. street-trees (*Jacaranda, Spathodea*) & lianes (*Eccremocarpus, Macfadyena*)
Bijlia N.E. Br. Aizoaceae (V). 1 Cape: *B. cana* N.E. Br. – cult. orn.
Bikkia Reinw. Rubiaceae (I 7). 20 E Mal. to W Pacific
Bilacunaria Pim. & Tikhom. (~ *Hippomarathrum*). Umbelliferae (III 5). 4 SW As. R: FR 94(1983)151
bilberry *Vaccinium myrtillus*
Bilderdykia Dumort. = *Fallopia*
Bilegnum Brand = *Rindera*
bilimbi *Averrhoa bilimbi*
bilinga *Nauclea diderrichii*
bill, parrot's *Clianthus puniceus*
Billardiera Sm. Pittosporaceae. 30 Aus. Some with upright blue fls, others with pendent yellow-orange fls. Cult. orn. (apple berries) esp. *B. longiflora* Lab. (blueberry, purple apple berry)
Billbergia Thunb. Bromeliaceae (3). 56 trop. Am., esp. E Brazil. Many hybrids. Cult. orn. R: FN 14(1979)1975
Billia Peyr. Hippocastanaceae. 2 S Mex. to trop. S Am.
billian *Eusideroxylon zwageri*
Billieturnera Fryx. Malvaceae. 1 S Texas & NE Mex.
billion dollar grass *Echinochloa frumentacea*
Billy buttons *Craspedia uniflora*
Billya Cass. = *Petalacte*
billy-goat weed *Ageratum conyzoides*
bilsted *Liquidambar styraciflua*
bilum (New Guinea) bags made from *Gnetum* spp., *Pangium edule* etc.
bimble box *Eucalyptus populnea*

bindang *Agathis dammara*

bindweed *Calystegia, Convolvulus, Fallopia* spp.; **black b.** *Fallopia convolvulus;* **common** or **field b.** *Convolvulus arvensis;* **great** or **hedge b.** *Calystegia sepium;* **sea b.** *C. soldanella*

bine *Humulus lupulus*

Binotia Rolfe (~ *Gomesa*). Orchidaceae (V 10). 1 Brazil

bintangor *Calophyllum* spp.

binuang *Octomeles sumatrana*

Biolettia E. Greene = *Trichocoronis*

Biondia Schltr. Asclepiadaceae (III 1). 13 China

Biophytum DC. Oxalidaceae. 50 trop. Lvs pinnate, sometimes sensitive, e.g. *B. sensitivum* (L.) DC (OW trop.); seeds arillate. *B. petersianum* Klotzsch (OW trop.) – medic., e.g. for venereal disease in E Afr.

Bipinnula Comm. ex Juss. Orchidaceae (IV 1). 7 temp. S Am.

Bipontia S.F. Blake = *Soaresia*

birch *Betula* spp; **American black b.** *B. lenta;* **brown b.** *B. pubescens;* **canoe b.** *B. papyrifera;* **common** or **European b.** *B. pendula, B. pubescens;* **dwarf b.** *B. nana;* **Japanese b.** *B. maximowiczii;* **paper b.** *B. papyrifera;* **red b.** *B. nigra;* **silver b.** *B. pendula;* **WI b.** *Bursera simaruba;* **yellow b.** *Betula alleghaniensis*

bird cactus *Pedilanthus tithymaloides;* **b. cherry** *Prunus padus;* **b. pepper** *Capsicum annuum* var. *glabriusculum;* **b.'s eye** *Veronica* spp.; **b.'s eye maple** *Acer saccharum;* **b.'s eye primrose** *Primula farinosa;* **b.'s foot** *Ornithopus perpusillus:* **b.'s foot trefoil** *Lotus corniculatus;* **b.'s nest fern** *Asplenium nidus;* **b.'s nest orchid** *Neottia nidus-avis;* **yellow b.'s nest** *Monotropa hypopitys;* **b.'s tongue** *Ornithoglossum* spp.

bird-of-paradise *Strelitzia reginae*

biribá *Rollinia mucosa*

birthroot *Trillium erectum*

birthwort *Aristolochia* spp. esp. *A. clematitis*

Bisboeckelera Kuntze. Cyperaceae. 8 S Am.

Bischofia Blume. Euphorbiaceae (Bischofiaceae). 1 Indomal.: *B. javanica* Blume – timber tree, bark medic. & dyestuff for rattan baskets; 1 C & SE China: *B. polycarpon* (A. Léveillé) Airy Shaw

Bischofiaceae (Muell. Arg.) Airy Shaw. See Euphorbiaceae

Biscutella L. Cruciferae. 40 S & mid Eur. to Medit. R: Brittonia 38(1986)86. *B. laevigata* L. (SE Eur., widely natur., buckler mustard) – one of most polymorphic spp. in Eur. flora, cult. orn.

Biserrula L. (~ *Astragalus*). Leguminosae (III 15). 1 Med. to E Afr.: *B. pelecinus* L.

Bisglaziovia Cogn. Melastomataceae. 1 Brazil

Bisgoeppertia Kuntze. Gentianaceae. 2 WI

Bishopalea H. Robinson. Compositae (Vern.-Vern.). 1 Brazil

Bishopanthus H. Robinson. Compositae (Liab.). 1 Peru

Bishopiella R. King & H. Robinson. Compositae (Eupat.-Gyp.). 1 E Brazil. R: MSBMBG 22(1987)124

bishop's cap *Mitella* spp.; **b. weed** *Aegopodium podagraria*

Bishovia R. King & H. Robinson. Compositae (Eupat.-Crit.). 2 Arg., Bolivia. R: MSBMBG 22(1987)324

Bismarckia Hildebr. & H. Wendl. Palmae (I 3b). 1 W Madag. savannas: *B. nobilis* Hildebr. & H. Wendl. (*Medemia n.*) – lvs for thatch & basketry etc. & paper; trunk source of a type of sago

Bisquamaria Pichon = *Laxoplumeria*

bisselon *Khaya senegalensis*

Bistella Adans. = *Vahlia*

bistort *Persicaria bistorta*

Bistorta Adans. = *Persicaria*

Biswarea Cogn. Cucurbitaceae. 1 Himalaya

bitter almond *Prunus dulcis* var. *amara;* **b. aloes** *Aloe spp.;* **b. apple** *Citrullus colocynthis;* **b. bark** *Alstonia constricta;* **b.bush** *Adriana* spp.; **b.cress** *Cardamine* spp.; **b.cress, hairy** *C. hirsuta;* **b. nut** *Carya cordiformis;* **b. orange** *Citrus aurantium;* **b. root** *Lewisia rediviva;* **b. sweet** *Solanum dulcamara;* **b.weed** *Helenium amarum*

Bituminaria Heister ex Fabr. (~ *Psoralea*). Leguminosae (III 11). 2 Medit. *B. bituminosa* (L.) Fabr. – tar-smelling weed

Biventraria Small = *Asclepias*

Bivinia Jaub. ex Tul. (~ *Calantica*). Flacourtiaceae (9). E trop. Afr., Madag.: *B. jalbertii* Tul. –

borer-proof wood, protected Zimbabwe

Bivonaea DC. Cruciferae. 1 W Medit.: *B. lutea* (Biv.) DC

Bixa L. Bixaceae. 1 trop. Am.: *B. orellana* L. – cult. as living fence & tried for land rehabilitation for forestry in Amazonia, but esp. for orange colouring (bixin, a carotenoid) obtained from testa, the original Amerindian bodypaint, also effective insect-repellent (anatto, arnatto, roucou, urucu), form. imp. dyestuff now replaced by Congo red for fabric but still used in food esp. cheese, butter, margarine & chocolate (consumption doubled since 1950, synthetics banned) as it is almost tasteless & soaps & other skin prods. Extrafl. nectaries: seed production doubled in presence of attracted ants warding off predators

Bixaceae Kunth. Dicots – Dilleniidae – Malvales. 3/16 trop. Trees to rhiz. herbs with red or orange juice in secretory cells. Lvs palmate, palmately lobed to simple, spiral; petiole often with complex vascular anatomy; stipules present, in *Bixa* protecting term. buds. Fls bisexual, ± regular, in racemes to panicles; K 5 imbricate, deciduous, C5 imbricate or convolute, A ∞, centrifugal, often associated with 5(–10) trunk bundles, anthers with slits or pores; nectary-disk intrastaminal or A on it; $\underline{G}$ (2–5) with ± deeply intruded partitions meeting basally & apically, the placentation thus partly axile, partly parietal exc. *Bixa* where wholly parietal, style 1, ovules ∞, anatropous, bitegmic. Fr. a loculicidal capsule; seeds glabrous or woolly, embryo embedded in oily & proteinaceous or (*Bixa*) starchy endosperm. x = 6–8

Genera: *Amoreuxia, Bixa, Cochlospermum*

Cochlospermum & *Amoreuxia* are s.t. separated off as Cochlospermaceae, characterized by palmately lobed or palmate (not simple) lvs, fr. with 3–5 (not 2) valves and oily (not starchy) endosperm, but they are all more closely allied to one another than to other groups and are perhaps most closely allied to Flacourtiaceae, in which Bixaceae have been included by some workers, than to Malvales with which they have some resemblances (Cronquist)

Ornamentals, dyestuffs, kapok etc.

Bizonula Pellegrin. Sapindaceae. 1 trop. Afr.

Blabeia Baehni = *Pouteria*

Blaberopus A. DC = *Alstonia*

Blachia Baillon. Euphorbiaceae. 12 India, SE As., Philippines

black apple *Pouteria australis*; **b. bean** *Castanaspermum australe, Kennedia nigricans*; **b. beech** *Nothofagus solanderi*; **b.berry** *Rubus fruticosus* etc.; **b. bindweed** *Fallopia convolvulus*; **b. boy** *Xanthorrhoea* spp.; **b. brush** *Coleogyne ramosissima*; **b. bryony** *Tamus communis*; **b.bush** *Maireana pyramidata*; **b.butt** *Eucalyptus pilularis*; **b. cap** *Rubus occidentalis*; **b.currant** *Ribes nigrum*; **b. dammar** *Canarium* spp.; **b.-eyed Susan** *Thunbergia alata, Rudbeckia* spp.; **b. gin** *Kingia australis*; **b.grass** *Alopecurus pratensis*; **b. gum** *Nyssa sylvatica*; **b. haw** *Viburnum prunifolium*; **b. henna** *Indigofera tinctoria*; **b. jack** *Bidens pilosa*; **b. Jessie** *Pithecellobium unguis-cati*; **b. laurel** *Gordonia lasianthus*; **b. locust** *Robinia pseudoacacia*; **b. mulberry** *Morus nigra*; **b. mustard** *Brassica nigra*; **b. nightshade** *Solanum nigrum*; **b. oak** *Quercus velutina*; **b. pepper** *Piper nigrum*; **b. peppermint** *Eucalyptus amygdalina*; **b. raspberry** *Rubus occidentalis*; **b. roseau** *Bactris major*; **b. rosewood** *Dalbergia latifolia*; **B. Sea walnut** *Juglans regia*; **b. sloe** *Prunus umbellata*; **b. snakeroot** *Cimicifuga racemosa*; **b. stinkwood** *Ocotea bullata*; **b.thorn** *Prunus spinosa*; **b. walnut** *Juglans nigra*; **b. widow** *Geranium phaeum*; **b.wood, Afr.** *Dalbergia melanoxylon*; **b.wood, Aus.** *Acacia melanoxylon*; **b.wood, Bombay** or **Ind.** *D. latifolia*

Blackallia C. Gardner. Rhamnaceae. 2 W Aus.

Blackiella Aellen = *Atriplex*

Blackstonia Hudson. Gentianaceae. 5–6 Eur. (1), Medit. *B. perfoliata* (L.) Hudson (yellow-wort) in GB

bladder campion *Silene vulgaris*; **b. fern** *Cystopteris fragilis*; **b.nut** *Staphylea pinnata*; **b. pod** *Lesquerella* spp.; **b. seed** *Levisticum* spp., *Physospermum* spp.; **b. senna** *Colutea arborescens*; **b.wort** *Utricularia* spp.

blaeberry *Vaccinium myrtillus*

Blaeria L. = *Erica*

Blainvillea Cass. Compositae (Helia.-Verb.). 10 Am., some now pantrop. weeds

Blakea P. Browne. Melastomataceae. 100 trop. Am. Some hemiepiphytic spp. in cloud forest poll. by several rodent spp.; domatia. Fr. ed. Tanbark from *B. trinervis* Ruíz & Pavón (Guianas)

Blakeanthus R. King & H. Robinson (~ *Ageratum*). Compositae (Eup.-Ager.). 1 Guatemala, Honduras. R: MSBMBG 22(1987)149

Blakeochloa Veldk. = *Plinthanthesis*

Blakiella Cuatrec. Compositae (Ast.-Ast.). 1 Venez., Colombia

Blanchetia DC. Compositae (Vern.-Vern.). 1 NE Brazil: *B. heterotricha* DC, raises sweat in humans

Blanchetiastrum Hassler. Malvaceae. 1 Brazil

Blancoa Lindley (~ *Conostylis*). Haemodoraceae. 1 SW Aus.: *B. canescens* Lindley, bird-poll., self-incompatible. R: FA 45(1987)110

Blandfordia Sm. Blandfordiaceae (Liliaceae s.l.). 4 E Aus. R: FA 45((1987)175. Protected in wild; cult. orn. (Christmas bells); fls found in gut of first emu shot in Aus. (1788)

Blandfordiaceae Dahlgren & Clifford (~ Phormiaceae, Liliaceae s.l.). Monocots – Liliidae – Asparagales. 1/4 E Aus. Herb. perennials with distichous lvs & racemes of pendulous fls. P 3 + 3 with long tube; A 3 + 3 inserted within tube; G (3), stipitate, 3-loc., each loc. with 40–50 anatropous ovules. Fr. a septicidal caps.; seeds with short brown hairs

Only genus: *Blandfordia*

Although the stipitate ovary, septicidal caps. & hairy seeds are distinctive, there are similarities with *Xeronema* of W Pacific & the fam. may be better placed in Phormiaceae. The relationships of these genera, as so often in the 'old' Liliaceae are still imprecisely known

Blandibractea Wernham. Rubiaceae (I 5). 1 Brazil

Blandowia Willd. = *Apinagia*

blanket flower *Gaillardia pulchella* & other spp.; **b. leaf** *Verbascum thapsus*

Blastania Kotschy & Peyr. = *Ctenolepis*

Blastemanthus Planchon. Ochnaceae. 3 NE S Am. K 5 (+ 5 (+ 5)). R: BJ 113(1991)178

Blastocaulon Ruhl. Eriocaulaceae. 4 Brazil

Blastus Lour. Melastomataceae. 12 Assam to W Mal. R: BMNHN 4, 4 Adans. (1982)43

blazing star *Mentzelia laevicaulis, Chamaelirium luteum, Liatris* spp.

Bleasdalea F. Muell. ex Domin (*Turrillia*, ~ *Grevillea*). Proteaceae. 5 W Pacific, NE Aus. R: AJB 62(1975)138

Blechnaceae (C. Presl) Copel. Filicopsida. 9/200 cosmop. Usu. terr., s.t. tree-ferns, or climbing. Fronds usu. pinnate to pinnatifid. Sori continuous or not, on veins parallel to leaflet midrib, usu. with indusia opening towards midrib (acrostichoid in *Brainea*)

Classification & genera:

Blechnoideae (scales not peltate, petiole vasc. bundles forming U in cross-section): *Blechnum, Brainea, Doodia, Pteridoblechnum, Sadleria, Salpichlaena, Steenisioblechnum, Woodwardia*

Stenochlaenoideae (scales peltate, petiole vasc. bundles in 2 circles): *Stenochlaena*

Blechnum L. Blechnaceae (Blechnoideae). 150–200 subcosmop. esp. S hemisph. (Eur. 1: *B. spicant* (L.) Roth, hard fern). Fronds uniform or dimorphic, when the fertile ones v. reduced. Cult. orn. esp. *B. gibbum* (Lab.) Mett. (W Pacific), a tree-fern. *B. indicum* Burm.f. (As. to trop. Aus., Am.) – rhizome form. trad. foodstuff N & NE Aus. (bungwall); *B. minus* (R. Br.) Ettingh. (Aus., NZ) – ecdysteroids; *B. obtusatum* (Lab.) Mett. (New Caled., Vanuatu) – rheophyte

Blechum P. Browne. Acanthaceae. 6 trop. Am. Some diuretic

bleeding heart *Dicentra spectabilis, Clerodendrum thomsoniae*

bleedwood tree *Pterocarpus angolensis*

Bleekeria Hassk. = *Ochrosia*

Bleekrodia Blume (~ *Streblus*). Moraceae (I). 3 Madag., Malay Pen., Borneo. R: PKNAW C91(1988)359

Blennodia R. Br. Cruciferae. 2 Aus. R: TRSSA 89(1965)168

Blennosperma Less. Compositae (Sen.-Blenn.). 2 Calif., 1 Chile. R: Brittonia 16(1964)289

Blennospora A. Gray (~ *Calocephalus*). Compositae (Gnap.-Ang.). 2 S Aus. R: Muelleria 6(1987)354

Blepharandra Griseb. Malpighiaceae. 6 trop. Am.

Blepharidachne Hackel. Gramineae (31d). 4 W US & Arg. R: Brittonia 31(1979)446

Blepharidium Standley. Rubiaceae (I 1). 2 C Am.

Blephariglottis Raf. = *Platanthera*

Blepharipappus Hook. (~ *Layia*). Compositae (Hele.-Mad.). 1 W US

Blepharis Juss. Acanthaceae. 80 OW trop. to S Afr. & Med. Seed-hairs swell when wetted. Some seeds eaten in Afr. & some used in anthrax treatment

Blepharispermum Wight ex DC. Compositae (Hele.). 15 OW trop. Trees & shrubs, some timbers e.g. *B. hirtum* Oliver – planking in town-houses in Dhofar, Arabia, where endemic

Blepharistemma Wallich ex Benth. Rhizophoraceae. 1 SW India

Blepharitheca Pichon = *Cuspidaria*

Blepharizonia (A. Gray) E. Greene. Compositae (Hele.-Mad.). 1 Calif.

Blepharocalyx O. Berg. Myrtaceae (Myrt.). 25 warm S Am.

Blepharocarya F. Muell. Anacardiaceae (IV, Blepharocaryaceae). 2 NE Aus.

Blepharocaryaceae Airy Shaw = Anacardiaceae

Blepharodon Decne. Asclepiadaceae (III 1). c. 45 C & S Am., with 1 in N Am.

Blepharoneuron Nash. Gramineae (31d). 2 N Am. R: SB 15(1990)515

Blephilia Raf. Labiatae (VIII 2). 3 N Am. R: Rhodora 94(1992)1. Cult. orn.

blessed thistle *Centaurea benedicta*

Bletia Ruíz & Pavón. Orchidaceae (V 11). 30 trop. Am. Terr. cult. orn. Some med. esp. *B. purpurea* (Lam.) DC (widespread inc. in salt-spray zone of Costa Rica) in WI, where used against poisoning from fish. See also *Bletilla*

Bletilla Reichb.f. Orchidaceae (V 11). 9 temp. E As. Cult. orn. esp. *B. striata* (Thunb.) Reichb.f.

Blighia König. Sapindaceae. 4 trop. Afr. *B. sapida* König (W Afr.), the akee, grown for edible arils but the unripe fr., seeds & the raphe between aril & rest of seed poisonous with hypoglycin A, a non-protein amino-acid causing hypoglycaemia in man & other animals

Blighiopsis Veken. Sapindaceae. 1 trop. Afr.

blimbing = bilimbi

blind grass *Stypandra glauca.*; **b.-your-eyes** *Excoecaria agallocha*

blinks *Montia fontana*

Blinkworthia Choisy. Convolvulaceae. 3 Burma, trop. China

blite *Chenopodium* spp. esp. *C. bonus-henricus*

Blitum L. = *Chenopodium*

Blomia Miranda. Sapindaceae. 1 Mex.

blood berry *Rivina humilis*; **b. flower** *Scadoxus multiflorus, Asclepias curassavica, Calothamnus sanguineus*; **b. leaf** *Iresine* spp.; **b. orange** *Citrus sinensis*; **b. plum** *Haematostaphis barteri*; **b. root** *Sanguinaria canadensis, Haemodorum coccineum*; **b. wood** *Corymbia* spp. esp. *C. gummifera, Pterocarpus* spp.; **b. wood, scrub** *Baloghia lucida*; **b.wort** Haemodoraceae

Bloomeria Kellogg. Alliaceae (Liliaceae s.l.). 3 SW N Am. R: Madrono 12(1953)19. Cult. orn.

Blossfeldia Werderm. (~ *Parodia*). Cactaceae (III 5). 1 Andes of N Arg., Bolivia: *B. liliputana* Werderm. – a few mm diam.

Blotia Leandri. Euphorbiaceae. 5 Madagascar

Blotiella Tryon. Dennstaedtiaceae. 12 trop. Am. (1), Afr. to Masc. One of few fern genera strongly centred in Afr.

blue amaryllis *Worsleya procera*; **b. beard** *Salvia viridis*; **b.bell** (English) *Hyacinthoides non-scripta*, (Scottish) *Campanula rotundifolia*, (Am.) *Mertensia* spp., (Aus.) *Sollya* spp. (honey) *Echium plantagineum, Wahlenbergia* spp., (S Afr.) *Wahlenbergia* spp., *Gladiolus* spp., (NZ) *Wahlenbergia* spp.; **Spanish b.bell** *Hyacinthoides hispanica*; **b.bell creeper** *Sollya heterophylla*; **b.berry** *Vaccinum* spp., *Billardiera longiflora* (Aus.), *Dianella nigra* (NZ); **b.bottle** *Centaurea* spp.; **b. boys** *Pycnostachys urticifolia*; **b.bush** *Eucalyptus macrocarpa, Maireana sedifolia*; **b. buttons** *Succisa pratensis*; **b. cardinal flower** *Lobelia siphilitica*; **b. cohosh** *Caulophyllum thalictroides*; **b. corn** *Zea mays* cv.; **b. couch** *Digitaria didactyla*; **b. curls** *Trichostema* spp. esp. *T. lanatum*; **b. devil** *Eryngium ovinum*; **b.-eyed grass** *Sisyrinchium* spp.; **b.-eyed Mary** *Omphalodes verna*; **b. flag** *Iris versicolor*; **b. gem** *Craterostigma plantagineum*; **b. grama** *Chondrosum gracile*; **b.grass** *Poa* spp., *Festuca* spp.; **b. gum** *Eucalyptus globulus*; **b. gum, Sydney** *E. saligna*; **b. haze** *Selago spuria*; **b. jacket** *Tradescantia ohiensis*; **b. lace flower** *Trachymene coerulea*; **b. mahoe** *Hibiscus elatus*; **b. moor grass** *Sesleria caerulea*; **b. pine** *Pinus wallichiana*; **b. poppy** *Meconopsis* spp.; **b. sailor** *Cichorium intybus*; **b. star** *Amsonia* spp.; **b. stem** *Schizachyrium scoparium*; **b. toadflax** *Linaria canadensis*; **b. vine** *Clitoria ternatea*

bluet *Vaccinium* spp.; **mountain b.** *Centaurea montana*

bluets *Hedyotis caerulea*

Blumea DC. Compositae (Inul.). c.100 OW trop., S Afr., natur. Carib. (*B. viscosa* (Miller) Badillo (*B. aurita*)). R: Blumea 10(1960)176. ? Heterogeneous. *B. balsamifera* (L.) DC source of ai or ngai camphor (As.) used in food, medic. etc.; other spp. imp. medic. locally

Blumenbachia Schrader. Loasaceae. c. 6 S Am. Stinging hairs

Blumeodendron (Muell. Arg.) Kurz. Euphorbiaceae. 6 Andamans, Mal.

Blumeopsis Gagnepain. Compositae (Pluch.). 1 India to W Mal.

blush, maiden's *Sloanea australis*; **b.wood** *Hylandia dockrillii*

Blutaparon Raf. (~ *Philoxerus*). Amaranthaceae (II 2). 4 Ryukyu Is., Am., W Afr. R: T 31(1982)113

Blysmocarex Ivanova = *Kobresia*

Blysmopsis Oteng-Yeboah = *Blysmus*

Blysmus Panzer ex Schultes (~ *Scirpus*). Cyperaceae. 3 temp. Euras. (Eur. 2)

Blyttia Arn. (~ *Cynanchum*). Asclepiadaceae (III 1). 2 S Arabia to E Afr.

Blyxa Noronha ex Thouars. Hydrocharitaceae. 9 OW trop. R: AB 15(1983)1

boat lily *Tradescantia spathacea*

Bobartia L. Iridaceae (III 1). 12 Cape. R: OB 37(1974). Lvs sword-like to centric

Bobea Gaudich. (~ *Timonius*). Rubiaceae (III 3). 4 Hawaii. R: Man. Fl. Pl. Hawaii 2(1990)1114. Yellow wood form. for canoes, paddles etc. Domatia

Bocagea A. St-Hil. Annonaceae. 2 trop. Am. R: AHB 10(1931). Extinct?

Bocageopsis R. Fries. Annonaceae. 4 trop. Am.

Bocconia L. Papaveraceae (I). 9 warm Am. Apetalous, seeds arillate, alks. R: KB 1920: 275. Allied to *Macleaya*. *B. frutescens* L. (tree-celandine, trop. Am.), pachycaul, cult., natur. Java, latex used in treatment of warts (cf. *Chelidonium*) & to dye feathers

Bocoa Aublet (~ *Swartzia*). Leguminosae (III 1). 7 trop. S Am.

Bocquillonia Baillon. Euphorbiaceae. 14 New Caledonia

Boea Comm. ex Lam. Gesneriaceae. 17 China to trop. Aus. R: NRBGE 41(1984)413. Cult. orn. esp. *B. hygroscopica* F. Muell. (Queensland), a resurrection pl. retaining its chlorophyll when dry

Boeberastrum (A. Gray) Rydb. (~ *Dyssodia*). Compositae (Hele.-Pect.). 2 NW Mex.

Boeberoides (DC) Strother (~ *Dyssodia*). Compositae (Hele.-Pect). 1 Mex.

Boeckeleria T. Durand = *Tetraria*

Boehmeria Jacq. Urticaceae (III). c. 80 trop. & N subtrop. Monoec. or dioec. trees to herbs without stinging hairs (false nettles). *B. nivea* (L.) Gaudich. (ramie, China grass, trop. As.) cult. for longest, toughest & most silky of all known veg. fibres, used in rope etc., also Canton or Chinese linen & gas-mantles, but stems difficult to decorticate. Cult. in Med. & Calif. etc. (whitish lower leaf-surface); var. *tenacissima* (Gaudich.) Miq. (rhea, greenish) cult. in Mal. etc. Other spp. also tried for fibre & orn.

Boehmeriopsis Komarov = *Fatoua*

Boeica T. Anderson ex C.B. Clarke. Gesneriaceae. 9 China, SE As.

Boeicopsis Li = *Boeica*

Boenninghausenia Reichb. ex Meissner. Rutaceae. 1 Assam to C Japan. Cult. orn.

Boerhavia L. Nyctaginaceae (III). (Incl. *Aulocaulis*, *Commicarpus* & *Cyphomeris*) c. 50 warm (*B. chinensis* (L.) Rottb. (trop. OW) & allies with 10-ribbed (as opposed to 5-) fr. s.t. segregated as *C ommicarpus*). Weeds, local med. & veg. *B. diffusa* L. (OW) – antihelminthic in India, eaten by wild boar (*Sus scrofa*)

Boerlagea Cogn. Melastomataceae. 1 Borneo

Boerlagella Cogn. = ? *Pouteria*

Boerlagellaceae H.J. Lam. See Sapotaceae

Boerlagiodendron Harms = *Osmoxylon*

Boesenbergia Kuntze. Zingiberaceae. 30 Indomal.

bog asphodel *Narthecium ossifragum*; **b. bean** *Menyanthes trifoliata*; **b. myrtle** *Myrica gale*; **b. orchid** *Malaxis paludosa*; **b. pimpernel** *Anagallis tenella*; **b. rosemary** *Andromeda polifolia*; **b. rush, black** *Schoenus nigricans*; **b. spruce** *Picea mariana*; **b. violet** *Pinguicula vulgaris*

boga medaloa *Tephrosia candida*

Bognera Mayo & Nicolson. Araceae (VI 4). 1 Brazil: *B. recondita* (Madison) Mayo & Nicolson

Bogoria J.J. Sm. Orchidaceae (V 16). 4 Java, New Guinea

Bogotá tea *Symplocos theiformis*

Boholia Merr. Rubiaceae (? III 5). 1 C Mal.

bois de rose oil from *Aniba* spp.

Boisduvalia Spach (~ *Epilobium*). Onagraceae. 6 temp. WN & S Am. R: Brittonia 17(1965)238. Some cult. orn. incl. *B. densiflora* (Lindley) Bartl. (W N Am.)

Boissiera Hochst. ex Ledeb. = *Bromus*

Bojeria DC = *Inula*

bok choy *Brassica rapa* Chinensis Group

Bokhara clover *Melilotus albus*; **B. plum** *Prunus bokhariensis*

Bokkeveldia D. Mueller-Dombois & U. Mueller-Dombois (~ *Strumaria*). Amaryllidaceae (Liliaceae s.l.). 4 S Afr. R: BJ 107(1985)27

Bolandra A. Gray. Saxifragaceae. 2 NW US. R: BJLS 90(1985)57

Bolanosa A. Gray. Compositae (Vern.). 1 Mex.

Bolanthus (Ser.). Reichb. Caryophyllaceae (III 1). 8 Eur. (Eur. 6, 4 Greece eastwards) to Israel. R: Wentia 9(1962)157

Bolax Comm. ex Juss. (~ *Azorella*). Umbelliferae (I 2b). 4–5 temp. S Am.

Bolbidium Brieger = *Dendrobium*

Bolbitidaceae (Pichi-Serm.) Ching = Lomariopsidaceae

Bolbitis Schott. Lomariopsidaceae (Aspleniaceae s.l.). 44 trop. & warm esp. SE As. R: LBS 2(1977)123. *B. portoricensis* (Spreng.) Hennipman (Am.) & *B. quoyana* (Gaudich.) Ching (As.) – frond-tips rooting & forming new pls

Bolboschoenus (Asch.) Palla (~ *Scirpus*). Cyperaceae. 16 cosmop. (Aus. 4–5)

Bolbostemma Franquet. Cucurbitaceae. 2 China

boldo *Peumus boldus*

Boldoa Cav. ex Lag. (~ *Salvianthus*). Nyctaginaceae (I). 1 C Am.: *B. purpurascens* Cav. ex Lag.

Boleum Desv. Cruciferae. 1 E Spain: *B. asperum* (Pers.) Desv.

Bollea Reichb.f. Orchidaceae (V 10). 10 trop. S Am. Cult. orn.

Bolocephalus Hand.-Mazz. = *Dolomiaea*

Bolophyta Nutt. (~ *Parthenium*). Compositae (Helia.-Amb.). 3 W US. R: Phytol. 41 (1979) 486

Boltonia L'Hérit. Compositae (Ast.-Ast.). 5 C & E N Am. Cult. orn. esp. *B. asteroides* (L.) L'Hérit. (E US); some As. spp. used as leaf veg. in Japan

Bolusafra Kuntze. Leguminosae (III 10). 1 S Afr.

Bolusanthus Harms. Leguminosae (III 2). 1 S C Afr.: *B. speciosus* (Bolus) Harms, cult. orn. tree with hard ant- & borer-proof timber

Bolusia Benth. Leguminosae (III 28). 5 Afr. S of equator

Bolusiella Schltr. Orchidaceae (V 16). 6 trop. Afr.

bolwarra *Eupomatia* spp.

Bomarea Mirbel. Alstroemeriaceae (Liliaceae s.l.). c. 100 Mex. to trop. Am. Many climbers, cult. orn., some with ed. root tubers e.g. *B. edulis* (Tussac) Herbert (trop. Am.)

Bombacaceae Kunth. Dicots – Dilleniidae – Malvales. 26/250 trop. esp. Am. Trees, often very large but frequently with soft light wood and a thickened trunk with much parenchymatous tissue storing water. Lvs spiral, simple (venation palmate or pinnate) or palmate, deciduous, petiolar anatomy complex (often with included bundles), stipules deciduous. Fls usu. large, showy & often bat-poll., solitary or in short cymes, ± regular, often with epicalyx; K 5, s.t. basally connate, valvate with tufted glandular hairs (nectaries) at base; C 5, convolute, rarely 0; A 5–∞ (when centrifugal), adnate to base of C, generally connate into 5–15 bundles or with a tube with so many apically placed bundles, often some staminodal, anthers opening by longit. slits, pollen usu. smooth; $\underline{G}$ (2–5(–8)) with 1 simple or lobed style-head, ovary with 2–5(–8) locules, placentation axile, ovules 2+ per loc., anatropous, bitegmic. Fr. a loculicidal or septicidal capsule, rarely fleshy & indehiscent; seeds often arillate & often embedded in pithy or hairy ovary tissue, embryo often curved, endosperm 0 or scanty, oily. x = esp. 28, 36, 40

Principal genera: *Adansonia, Bombax, Ceiba, Durio, Matisia, Pachira, Quararibea*

The generic limits around *Bombax* are controversial. The fam. is v. closely allied to Malvaceae but its woodiness & (usu.) smooth pollen are used to keep them separate, though several genera are placed in one family by some authors & in the other or even Sterculiaceae by others, e.g. not incl. in B. here are *Chiranthodendron* (Sterc.), *Dicarpidium* (Sterc.), *Fremontodendron* (Sterc.), *Kydia* (Malv.), *Maxwellia* (Sterc.), while *Paradombeya* here in Sterc. & the tribe Durioneae similar to Sterc. in pinnate leaf venation, flat rather than folded cotyledons etc.

Baobabs (*Adansonia*), fibre-trees (*Ceiba*), fruit trees (*Durio*), balsa (*Ochroma*) & other timbers (esp. *Pachira*) & spectacular cult. orn. esp. *Ceiba*

Bombacopsis Pittier = *Pachira*

Bombax L. Bombacaceae. (Incl. *Rhodognaphalon*) 20 OW trop. R: BJBB 33(1963)84, 253. Sometimes spiny. Cult. orn., soft timber, ovary hairs a source of kapok. *B. brevicuspe* Sprague (W Afr.) – elephants strip off bark to 10 m in Ghana; *B. buonopozense* Pal. (W & C Afr.) – commercial timber; *B. ceiba* L. (*B. malabaricum*, simul, trop. As.) – silk cotton tree, inferior kapok, timber for matches, canoes etc., medic. (inc. constituent of supposed aphrodisiacs), red ed. fls

Bombay aloe *Agave cantala;* **B. ebony** *Diospyros montana;* **B. hemp** *Crotalaria juncea;* **B. mastic** *Pistacia atlantica;* **B. mix** see *Trachyspermum ammi*

bombway, white *Terminalia procera*

Bombycidendron Zoll. & Moritzi = *Hibiscus*

Bombycilaena (DC) Smoljan. Compositae (Gnap.-Gnap.). 3 Med. (Eur. 2) to Afghanistan. R: OB 104(1991)173

Bommeria Fourn. Pteridaceae (IV). 4 SW N & C Am. R: JAA 60(1979)445. Xeric & seasonal mts. Cult. orn.

bonace *Daphnopsis tinifolia*

Bonafousia A. DC = *Tabernaemontana*
Bonamia Thouars. Convolvulaceae. 45 trop. R: Phytol. 17(1968)121
Bonania A. Rich. Euphorbiaceae. 10 WI
Bonannia Guss. Umbelliferae (III 10). 1 SE Eur.
Bonatea Willd. (~ *Habenaria*). Orchidaceae (IV 2). 20 trop. & S Afr., Arabia
Bonatia Schltr. & K. Krause = *Tarenna*
bonavist *Lablab purpureus*
bonduc nut *Caesalpinia bonduc*
boneseed *Chrysanthemoides monolifera*
Bonetiella Rzed. Anacardiaceae (IV). 1 Mexico
Bongardia C. Meyer. Berberidaceae (II 4). 1 E Med. (Greece) to Afghanistan: *B. chrysogonum* (L.) Spach – lvs & rhiz. ed.
bongossi *Lophira alata*
Boninia Planchon. Rutaceae. 2 Bonin Is. 4-merous fls
Boninofatsia Nakai = *Fatsia*
Boniodendron Gagnepain. Sapindaceae. 1–2 China to SE As.
Bonnaya Link & Otto = *Lindernia*
Bonnayodes Blatter & Hallberg = *Limnophila*
Bonnetia C. Martius. Guttiferae (II, Bonnetiaceae). 29 trop. Am. Some pachycaul candelabriform shrubs
Bonnetiaceae (Bartling) Nakai. See Guttiferae
Bonniera Cordemoy. Orchidaceae (V 16). 2 Réunion
Bonnierella R. Viguier = *Polyscias*
Bonplandia Cav. Polemoniaceae. 2 Mex. Fls ± zygomorphic
bonsamdua *Distemonanthus benthamianus*
Bontia L. Myoporaceae. 1 WI, trop. S Am. Cult. orn.
Bonyunia Schomb. ex Progel. Strychnaceae (Loganiaceae s.l.). 4 S Am. R: ABN 18(1969)152
boobialla, boobyalla *Myoporum* spp.
boojum tree *Fouquieria columnaris*
boonaree, boonery *Alectryon oleifolius*
Boophane Herb. = seq.
Boophone Herb. Amaryllidaceae (Liliaceae). 6 S & E Afr. Alks, poisonous. *B. disticha* (L.f.) Herbert has haemanthine, medic. but also affects eyes
Boopis Juss. Calyceraceae. 13 Andes, Arg., S Brazil
booyong *Argyrodendron* spp.
boppel nut, red *Hicksbeachia pinnatifolia*
bopplefnut = macadamia nut
Boquila Decne. Lardizabalaceae. 1 Chile, Arg.: *B. trifoliolata* (DC) Decne. Dioec., occ. monoec.
borage *Borago officinalis*
Boraginaceae Juss. Dicots – Asteridae – Lamiales. 130/2300 Trees, shrubs, frequently herbs, rarely lianes, usu. with characteristic unicellular bristly hairs, often with alks & red alkannin (in roots). Lvs spiral (rarely with some opp.), simple, usu. entire, often with cystoliths &/or oxalate crystals, stipules 0. Fls usu. in cymes (scorpioid or helicoid), rarely solit. & axillary, usu. bisexual (gynodioecy in e.g. *Echium*, dioecy in *Cordia* etc.), ± regular, s.t. heterostylous; K (4)5(–8), free to connate, imbricate, rarely valvate; C ((4)5(6)), often salveriform, lobes imbricate or convolute, rarely valvate, the tube with hairy appendices between lobes in subfam. Boraginoideae; A (4)5(6) alt. with C, attached to tube, anthers opening by longit. slits; annular nectary-disk round G or 0; $\underline{G}$ (2) but (4, 5) in New Guinea spp. of *Trigonotis*, with twice as many compartments, each with 1 ovule, rarely 1 carpel suppressed; in subfam. Cordioideae G entire with terminal twice bifid style ripening to drupe usu. with 4-loc. stone; in subfam. Ehretioideae G entire to 4-lobed ripening to drupe with 2 2-seeded or 4 1-seeded stones or separating into 4 segments; in subfam. Heliotropioideae G entire or 4-lobed ripening into 2 (1)2-seeded or 4 1-seeded nutlets; in subfam. Boraginoideae G deeply 4-lobed with simple or bifid style ripening into (1–) 4 nutlets, in subfam. Wellstedioideae G 2-loc. ripening into 1- or 2-seeded capsule; ovules anatropous to hemitropous with single integument. Embryo with 2 cotyledons (deeply bifid in *Amsinckia*), straight; endosperm copious, oily to 0. x = 4–12. R: JAA suppl.1(1991)1
Classification & chief genera (for fr. differences see above):

Cordioideae (mainly trees, style term., 4-branched; mainly trop.): *Cordia*
Ehretioideae (mainly trees, style term., 2-branched): *Ehretia*
Heliotropioideae (commonly herbs, style term., undivided): *Heliotropium*,

Tournefortia

Boraginoideae (usu. herbs, style gynobasic):

Cynoglosseae (fls reg., style base ± conical, tips of nutlets not projecting above points of attachment): *Cynoglossum, Omphalodes*

Eritrichieae (fls reg., style base ± conical, tips of nutlets projecting above points of attachment): *Cryptantha, Eritrichium*; some genera with nutlets usu. tetrahedral s.t. excl. as Trigonotideae

Boragineae (fls reg., style base flat or slightly convex, nutlets with concave attachment surface): *Alkanna, Anchusa, Borago, Pulmonaria, Symphytum*

Lithospermeae (as above but attachment surface flat): *Arnebia, Cerinthe, Lithospermum, Myosotis* (contorted aestivation, sometimes excl. as monotypic tribe)

Echieae (fls zygomorphic; often incl. in praec.): *Echium*

Wellstedioideae (woody herbs, fls 4-merous, fr. a loculicidal capsule): *Wellstedia* (only)

Cordioideae with Ehretioideae and Wellstedioideae have been treated as separate fams, obscuring the close inter-relationship of this group, which is allied to Hydrophyllaceae. The woody tropical 'archaic' subfams are similar to woody Verbenaceae (i.e. Labiatae s.l.), which is why Cronquist places the families together in Lamiales, though similarities with Hydrophyllaceae suggest a position in Solanales. The parasitic Lennoaceae are perhaps best included here (cf. Orobanchaceae & Scrophulariaceae)

Timber and fr. trees (*Cordia*), dyeplants (*Alkanna* etc.), potherbs (*Symphytum* etc.) & cult. orn. (*Echium, Heliotropium, Myosotis* etc.)

Borago L. Boraginaceae. 3 Med., Eur. (2), As. Cult. orn. esp. *B. officinalis* L. (borage, Eur. & Med.) – bee-fodder, field-crop for oil (starflower o.) sold like evening primrose o. (seeds high in gamma-linoleic acid, unusual fatty-acid intermediate in biosynthesis of prostaglandins), flavouring for claret cup (replaced by cucumber), form. potherb

Borassodendron Becc. (~ *Borassus*). Palmae (I 3a). 2 Malay Pen., Borneo. R: Reinwardtia 8 (1972) 351

Borassus L. Palmae (I 3a). 11 OW trop. Dioec., lvs palmate. *B. aethiopum* C. Martius (E Afr.) – fls after 30–40 yrs (protected Zimbabwe), fr. ed. Uganda; *B. flabellifer* L. (Palmyra palm, India to Burma), cult., timber for rafters etc. (resists salt water), lvs for thatch and *olas* or writing-paper (perhaps the oldest form; a stylus used for writing), leaf-base fibre for brushes etc. (bassine or Palmyra fibre), split lvs used for weaving mats, baskets etc., fr. eaten roasted & infl. tapped for toddy from which sugar (jaggery), vinegar etc. prepared: 120 000 litres of toddy produced in lifetime of 1 palm; seeds surrounded by fibres worked as small orn. heads for sale in India; seedlings ed. & yield odiyal flour for breakfast food etc. when ground; according to an old Tamil song there are 801 uses in all; *B. sundaicus* Becc. a source of sugar in Java

Borderea Miégev. (~ *Dioscorea*). Dioscoreaceae. 2 Pyrenees. Tertiary relics; stems not twining, tuber napiform (cf. *Dioscorea*)

Boreava Jaub. & Spach. Cruciferae. 2 E Med.

borecole *Brassica oleracea* Acephala Group

Borinda Stapleton (~ *Fargesia*). Gramineae (1a). 8 Himal. R: EJB 51(1994)284

Borismene Barneby. Menispermaceae (III). 1 trop. S Am.: *B. japurensis* (C. Martius) Barneby

Borissa Raf. = *Lysimachia*

boriti poles *Rhizophora mucronata*

Borkonstia Ignatov = *Aster*

borlotti beans *Phaseolus vulgaris*

Borneacanthus Bremek. Acanthaceae. 6 Borneo

Borneo camphor *Dryobalanops aromatica*; **B. ironwood** *Eusideroxylon zwageri*; **B. mahogany** *Calophyllum inophyllum*; **B. rubber** *Willughbeia coriacea, W. firma*; **B. tallow** *Shorea palembanica*; **B. teak** *Intsia bijuga*

Borneodendron Airy Shaw. Euphorbiaceae (?). 1 N Borneo: *B. aenigmaticum* Airy Shaw – lvs & fls in whorls of 3, sap red (v. unusual in E.)

Bornmuellera Hausskn. Cruciferae. 6 Balkans (3), As. Minor. Nickel accumulators

Borodinia Busch. Cruciferae. 1 E Siberia

Borojoa Cuatrec. Rubiaceae (II 1). 8 trop. Am.

Boronella Baillon = *Boronia*

Boronia Sm. Rutaceae. 104 Aus. (c. 100 endemic, esp. SW), New Caled. Protected in Aus.; cult. orn. esp. *B. floribunda* Sieber ex Sprengel as cutfl. in Aus., *B. megastigma* Nees ex Bartling in Eur. for scent, this & other spp. grown comm. for that in Victoria

Borrachinea Lavy = *Borago*
Borreria G. Meyer = *Spermacoce*
Borrichia Adans. Compositae (Helia.-Verb.). 2 SE US, WI. R: AMBG 65(1978)681
Borsczowia Bunge. Chenopodiaceae (III 2). 1 C As.: *B. aralocaspica* Bunge
Borthwickia W.W. Sm. Capparidaceae. 1 Burma, Yunnan. Lvs opp.
Borya Labill. Anthericaceae (Liliaceae s.l.). 10 W Aus., Victoria, Queensland. Lvs of *B. nitida* Labill. revive after desiccation when orange
Borzicactella F. Ritter = *Cleistocactus*
Borzicactus Riccob. = *Cleistocactus* (some cult. spp. = *Oreocereus*)
Boschia Korth. = *Durio*
Boschniakia C. Meyer ex Bong. Orobanchaceae. 2 N & Arctic Russia, As. to Japan, NW N Am. R: ABY 9(1987)296. *B. rossica* (Cham. & Schldl.) B. Fedtsch. (N Euras., NW N Am.) – on *Alnus*, attractive to ants
Boscia Lam. Capparidaceae. 37 trop. & S Afr., Arabia (20). Some ed. fr., medic. & coffee subs. (seeds, roots)
Bosea L. Amaranthaceae (I 2). 3 Canary Is., Cyprus (not Turkey!), India
Bosistoa F. Muell. ex Benth. Rutaceae. 7 NE Aus. rain forests. R: JAA 58(1977)416
Bosleria Nelson = *Solanum*
Bosqueia Thouars ex Baillon = *Trilepisium*
Bosqueiopsis De Wild. & T. Durand. Moraceae (IV). 1 trop. Afr.
Bossera Leandri = *Alchornea*
Bossiaea Vent. Leguminosae (III 25). 50 temp. (esp. SW) Aus. Several spp. with flattened green stems & minute scale lvs, seedlings with more typical ones. Some cult. orn.
Boston fern *Nephrolepis exaltata* 'Bostoniensis'; **B. ivy** *Parthenocissus tricuspidata*
Bostrychanthera Benth. Labiatae (VI). 1 China
Boswellia Roxb. ex Colebr. Burseraceae. 19–20 dry trop. Afr. (esp. NE) & As. Fragrant resins used in incense & aromatherapy, esp. frankincense (gum olibanum) derived from *B. sacra* Flueckiger (Somalia, S Arabia) in classical times (in Roman times as valuable as gold), *B. papyrifera* (Del.) Hochst. (NE Nigeria to Ethiopia) in antiquity, largely *B. carteri* Birdw. & *B. frereana* Birdw. (African elemi, Somalia) today. *B. serrata* Roxb. ex Colebr. (consp. on Indian dry hills) – timber for charcoal, tea-chests etc.
Botelua Lagasca = *Bouteloua*
Bothriochilus Lemaire = *Coelia*
Bothriochloa Kuntze (~ *Dichanthium*). Gramineae (40c). c. 35 warm. Cult. for fodder
Bothriocline Oliver ex Benth. Compositae (Vern.-Vern.). c. 30 trop. Afr. (E Afr. 29; R: KB 43(1988)257), Madag.
Bothriospermum Bunge. Boraginaceae. 5 trop. & NE As.
Bothriospora Hook.f. Rubiaceae (inc. sed.). 1 trop. Am.: *B. corymbosa* (Benth.) Hook.f. – parasiticidal
Bothrocaryum (Koehne) Pojark. = *Cornus*
bo-tree *Ficus religiosa*
Botryarrhena Ducke. Rubiaceae (?III 1). 2 Brazil, Venezuela
Botrychiaceae Horan. = Ophioglossaceae
Botrychium Sw. Ophioglossaceae. c. 50 temp., polar, trop. mts. R: MTBC 19,2(1938)22. (Eur. 7 incl. *B. lunaria* (L.) Sw., moonwort). Habit of *Ophioglossum* but sterile lvs also lobed. *B. ternatum* (Thunb.) Sw. lvs said to be a veg. in Japan
Botryoloranthus (Engl. & K. Krause) Balle = *Oedina*
Botryomeryta R. Viguier = *Meryta*
Botryophora Hook.f. Euphorbiaceae. 1 SE As., W Mal.
Botrypus Michaux = *Botrychium*
Botryostege Stapf = *Cladothamnus*
Botschantzevia Nabiev. Cruciferae. 1 Kazakhstan
Bottegoa Chiov. Ptaeroxylaceae. 1 Somalia, SE Ethiopia, N Kenya: *B. insignis* Chiov.
Bottionea Colla = *Trichopetalum*
bottlebrush *Callistemon* spp.; **b. grass** *Hystrix* spp.
bottle-tree *Brachychiton* spp.
Boucerosia Wight & Arn. = *Caralluma*
Bouchardatia Baillon. Rutaceae. 3 New Guinea (2), E Aus. (1)
Bouchea Cham. (~ *Chascanum*). Verbenaceae. 9 trop. Am.
Bouchetia Dunal (~ *Salpiglossis*). Solanaceae. 3 S US to Brazil
Bouea Meissner. Anacardiaceae (I). 3 SE As., Mal. Lvs. opp. Fr. ed. esp. *B. macrophylla* Griffith (gandaria, plum mango, Mal.)

Bougainvillea Comm. ex Juss. Nyctaginaceae (IV). 18 C & trop. S Am. Fls arising from persistent coloured bracts in groups of 3 resembling a flower. Cult. orn. usu. spiny lianes to 25 m, esp. cvs of *B.* × *buttiana* Holttum & Standley (*B. glabra* Choisy (Brazil) × *B. peruviana* Bonpl. (Colombia to Peru)). Poll. by hummingbirds

Bougueria Decne. Plantaginaceae. 1 Andes

bouncing Bet *Saponaria officinalis*

bouquet, bridal *Poranopsis paniculata*

Bourasaha Thouars. Older name for *Burasaia*

bourbon *Zea mays*

Bourdaria A. Chev. = *Cincinnobotrys*

Bournea Oliver. Gesneriaceae. 2 SE China. R: EJB 49(1992)14

Bourreria P. Browne. Boraginaceae. c. 30 E Afr., Madag., Masc., warm Am. R: JAA suppl. 1(1991)59. Trees. *B. huanita* (La Llave & Lex.) Hemsley (huanita, Mex.) – scented fls used in tobacco, drinks etc.

Bousigonia Pierre. Apocynaceae. 2 SE As.

Boussingaultia Kunth = *Anredera*

Bouteloua Lagasca. Gramineae (33c). (Excl. *Chondrosum*) 24 Canada to Arg. esp. Mex. R: AMBG 66(1979)348. Pasture & cult. orn. grasses (C_4), esp. *B. curtipendula* (Michaux) Torrey (side-oats grama) – imp. forage & *B. gracilis* (Kunth) Griffiths (SW N Am., blue grama)

Boutiquea Le Thomas = *Neostenanthera*

Boutonia DC. Acanthaceae. 1 Madagascar

Bouvardia Salisb. Rubiaceae (I 1/IV 24). 20 trop. Am. R: AMBG 55(1968)1. Some heterostylous. Cult. orn., fls fragrant

Bouzetia Montr. Rutaceae. 1 New Caledonia

Bovonia Chiov. Labiatae (VIII 4). 1 Zaire

Bowdichia Kunth. Leguminosae (III 2). 4 trop. S Am. Strong timber (sucupira), e.g. *B. nitida* Spruce ex Benth. – wheel hubs & rims

Bowenia Hook. ex Hook.f. Boweniaceae (Zamiaceae s.l.). 2 NE Aus. R: MNYBG 57(1990)201. Lvs bipinnate, sold for decoration; underground stem with blue-green algal symbionts (*Anabaena*) ed. aborigines

Boweniaceae Stevenson (~ Stangeriaceae; Zamiaceae s.l.). Cycadopsida. 1/2 NE Aus. Subterranean stems with determinate ±bipinnate leaves with branched rachis (unique in cycads)

Only genus: *Bowenia*

Similarities to Zamiaceae s.s. considered parallelisms

bower plant *Pandorea jasminoides*

Bowiea Harvey ex Hook.f. Hyacinthaceae (Liliaceae s.l.). 1: S & E Afr.: *B. volubilis* Harvey ex Hook.f. – xerophyte with 1 or 2 evanescent lvs, bulb at ground-level & annual green, soon leafless, climbing 'stems' (infls) to 6 m & green fls. Cult. orn., bulb toxic & locally med.

Bowkeria Harvey. Scrophulariaceae. 5 S Afr. Cult. trees & shrubs with floral elaiophores attractive to bees

Bowlesia Ruíz & Pavón. Umbelliferae (I 2a). 15 S Am. R: UCPB 38(1965)1

Bowman's root *Gillenia trifoliata*

Bowringia Champ. ex Benth. Leguminosae (III 2). 2 W Afr., 1 E Madag., 1 S China

bowstring hemp *Sansevieria* spp.

bow-wood *Maclura pomifera*

box or **boxwood** *Buxus sempervirens*, (Aus.) *Eucalyptus* spp.; **Abassian b.** *B. sempervirens*; **apple b.** *E. bridgesiana*; **bimble b.** *E. populnea*; **Brisbane** or **brush b.** *Lophostemon confertus*; **Cape b.** *Gonioma kamassi*; **Colombian b.** *Casearia praecox*; **E. London b.** *B. macowanii*; **b. elder** *Acer negundo*; **Florida b.** *Schaefferia frutescens*; **Japanese b.** *Buxus microphylla*; **Maracaibo, Venezuelan** or **WI b.** *Casearia praecox*; **San Domingo b.** *Phyllostylon brasiliensis*; **boxthorn** *Lycium* spp.

Boyania Wurdack. Melastomataceae. 1 Guianas

Boykinia Nutt. Saxifragaceae. 9 E As., N Am., woodland. R: BJLS 90(1985)35. Cult. orn.

boysenberry *Rubus loganobaccus* cv.

Braasiella Braem, Lueckel & Russmann = *Oncidium*

Brabejum L. Proteaceae. 1 SW Cape: *B. stellatifolium* L. – seeds form. eaten roasted (wild almonds, chestnuts), coffee subs., supposed toxic

bracaatinga *Mimosa bracaatinga*

Brachanthemum DC (~ *Chrysanthemum*). Compositae (Anth.-Art.). 10 C As. to China

Brachiaria (Trin.) Griseb. Gramineae (34b). c. 100 warm, esp. Afr. (Eur. 1). Fodder, e.g. *B.*

arrecta (Durieu & Schinz) Stent (Tanner grass, Afr.), *B. brizantha* (A. Rich.) Stapf (palisade g., Afr.), *B. mutica* (Forssk.) Stapf (Pará or para g., trop.); *B. deflexa* (Schum.) Robyns (Afr. to India) – minor cereal in W Afr.

Brachionidium Lindley. Orchidaceae (V 13). 35 trop. Am. R: CJB 34(1956)160

Brachionostylum Mattf. (~ *Senecio*). Compositae (Sen.-Sen.). 1 New Guinea: *B. pullei* Mattf. – app. dioec., monopodial pachycaul

Brachistus Miers. Solanaceae (1). 3 C Am. R: AMBG 68(1981)226

Brachtia Reichb.f. Orchidaceae (V 10). 5 N S Am. R: Orquideologia 9(1974)5

Brachyachenium Baker = *Dicoma*

Brachyachne (Benth.) Stapf. Gramineae (33b). 9 trop. Afr., Java to Australasia. Rockpools & seasonally wet places; lvs of *B. patentiflora* (Stent & Rattray) C.E. Hubb. (Afr.) revive after reduction of water content to 5%

Brachyactis Ledeb. (~ *Aster*). Compositae (Ast.-Ast.). 5 N As., N Am.

Brachyandra Philippi = *Helogyne*

Brachyapium (Baillon) Maire = *Stoibrax*

Brachybotrys Maxim. ex Oliver. Boraginaceae. 1 Manchuria, E Siberia

Brachycarpaea DC. Cruciferae. 1 S Afr.

Brachycaulos Dixit & Panigr. Rosaceae. 1 Sikkim

Brachycereus Britton & Rose. Cactaceae (III 4). 1 Galápagos: *B. nesioticus* (Robinson) Backeb., on lava

Brachychaeta Torrey & Gray = *Solidago*

Brachychilum (Wallich) Petersen = *Hedychium*

Brachychiton Schott & Endl. Sterculiaceae. 31 Aus., New Guinea (2, 1 endemic). R: AusSB 1(1988)199. Bottle-trees (stems swollen); cult. orn. incl. hybrids: *B. acerifolius* (G. Don f.) Macarthur (flame-tree, NE Aus.); *B. populneus* (Schott & Endl.) R. Br. (NE Aus.) & *B. rupestris* (Lindley) Schumann (Queensland) cult. for fodder in E Aus.

Brachychloa S. Phillips. Gramineae (31d). 2 Mozambique, Natal. Sandy sites

Brachyclados Gillies ex D. Don. Compositae (Mut.-Mut.). 3 temp. S Am. Shrubs. Medic. (asthma)

Brachycodon Fed. = *Campanula*

Brachycome Cass. (*Brachyscome*). Compositae (Ast.-Ast.). 66 Aus. (62), NZ (3), New Guinea (1), New Caled. (2). R: PLSNSW 74(1949)97, 75(1950)122. Cult. orn. esp. *B. iberidifolia* Benth. (Swan River Daisy, W to S Aus.)

Brachycorythis Lindley. Orchidaceae (IV 2). 25 trop. & S Afr., trop. As. Incl. mycotrophs (*Schwartzkopffia*)

Brachycylix (Harms) R. Cowan. Leguminosae (I 4). 1 W Colombia

Brachycyrtis Koidz. = *Tricyrtis*

Brachyelytrum Pal. Gramineae (13). 3 N Am., E As. R: JAA 69(1988)253. Woodland grasses (? relicts of widespread Mid-Tertiary vegetation)

Brachyglottis Forster & Forster f. Compositae (Sen.-Tuss.). 29 NZ, Chatham Is., Tasmania. R: OB 44(1978)25. Prob. also in S Am. & Madag. Cult. orn. esp. shrubby Dunedin Group ('D. Hybrids'), usu. 'Sunshine' (*B. compacta* (Kirk) R. Nordenstam (NZ) × *B. laxifolia* (J. Buch.) R. Nordenstam (NZ), widely planted evergreen in towns ('*Senecio greyi*')

Brachyhelus (Benth.) Post & Kuntze = *Schwenckia*

Brachylaena R. Br. Compositae (Mut.-Mut.). c. 15 trop. & S Afr. to Madag. Trees & shrubs. *B. huillensis* O. Hoffm. (S and E Afr.) – endangered, strong termite-proof timber for floors, sleepers, figure-carving (muhugwe or muhuhu)

Brachylepis C. Meyer. Chenopodiaceae (II). 8 S Russia to C As.

Brachyloma Sonder. Epacridaceae. 7 Aus. Cult. orn.

Brachylophon Oliver. Malpighiaceae. 3 OW trop.

Brachymeris DC = *Phymaspermum*

Brachynema Benth. Olacaceae. 1 N Brazil

Brachyotum (DC) Hook.f. (*Alifana*). Melastomataceae. 50 S Am., paramo & subparamo. R: MNYBG 8(1953)343. Poll. by hummingbirds & *Diglossa* flower-piercers

Brachypeza Schltr. ex Garay. Orchidaceae (V 16). 7 Mal.

Brachypodium Pal. Gramineae (23). 17 temp. Euras. (Eur. 7; R: Boissiera 45(1991)), trop. Am. & mts. False bromes. Intergrades with *Elymus*; x = 5, 7, 9

Brachypremna Gleason = *Ernestia*

Brachypterum (Wight & Arn.) Benth. = *Derris*

Brachypterys A. Juss. Malpighiaceae. 3 trop. S Am., WI

Brachyscome Cass. Earlier name for *Brachycome*

Brachysema R. Br. Leguminosae (III 24). 8 SW Aus.

Brachysiphon A. Juss. Penaeaceae. 4 SW Cape

Brachystachyum Keng = *Semiarundinaria*

Brachystegia Benth. Leguminosae (I 5). 30 trop. Afr.: 8 in Guineo-Congolian forest, rest major components of decid. (miombo) woodlands of S C Afr. Some timbers (okwen) & barkcloth

Brachystele Schltr. Orchidaceae (III 3). 18 trop. Am. esp. Brazil

Brachystelma R. Br. Asclepiadaceae (III 5). (Incl. *Microstemma*) 100 Afr. with 1 India, New Guinea & Aus.: *B. glabriflorum* (F. Muell.) Schltr. (*Microstemma g.*) – ed. tubers sought by animals inc. man. R: R.A. Dyer, *Ceropegia, B. & Riocreuxia in S Afr.* (1983)12. Many cult. orn.

Brachystemma D. Don. Caryophyllaceae (II 1). 1 Himal.: *B. calycinum* D. Don

Brachystephanus Nees. Acanthaceae. 10 trop. Afr., Madagascar

Brachystigma Pennell = *Agalinis*

Brachythalamus Gilg = *Gyrinops*

Brachythrix Wild & Pope. Compositae (Vern.-Vern.). 5 C & E Afr. R: Kirkia 11(1978)25

Brachytome Hook.f. Rubiaceae (II 1). 4 S China, Indomal. R: Candollea 42(1987)351

Bracisepalum J.J. Sm. Orchidaceae (V 12). 2 Sulawesi. R: Blumea 28(1983)413, OM 1(1986)19

bracken *Pteridium aquilinum*

Brackenridgea A. Gray. Ochnaceae. 6–7 OW trop. Some yellow dyes

Bracteantha Anderb. & Haegi (~ *Helichrysum*). Compositae (Gnap.-Ang.). 5 Aus., natur. widely. R: OB 104(1991)102. Cult. orn. 'everlastings' esp. *B. bracteata* (Vent.) Anderb. & Haegi (*Helichrysum b.*, strawflower)

Bracteanthus Ducke = *Siparuna*

Bracteolanthus de Wit = *Bauhinia*

Bradburia Torrey & A. Gray = *Chrysopsis*

Bradea Standley ex Brade. Rubiaceae (IV 1). 5 Brazil

Braemea Jenny = *Houlletia*

Brahea C. Martius ex Endl. Palmae (I 1b). 12, limestones of C Am. R: GH 6(1943)177. Cult. orn., some ed. fr. & oil

brahmi *Bacopa monnieri*

Brainea J. Sm. Blechnaceae (Blechnoideae). 1 NE India to W Mal.: *B. insignis* (Hook.) J. Sm., dwarf fire-tolerant tree-fern with acrostichoid fronds

brake *Pteridium aquilinum*

bramble *Rubus* spp.

Brandegea Cogn. Cucurbitaceae. 1 SW N Am.

Brandella R. Mill (~ *Adelocaryum*). Boraginaceae. 1 NE Afr., Saudi Arabia

Brandisia Hook.f. & Thomson. Scrophulariaceae. 13 Burma, China

brandy See *Vitis vinifera*

brandy-bottle *Nuphar luteum*

Brandzeia Baillon = *Bathiaea* + *Albizia*

Brasenia Schreber. Cabombaceae. 1 trop. Am., Afr., India, temp. E As., Aus.: *B. schreberi* J. Gmelin. Submerged parts covered with mucilaginous jelly; A 12–18. R: Aqua-P. Sond. 3(1992)37. Cult. orn.; young lvs ed. Japan

brasiletto woods of spp. of *Caesalpinia* & *Peltophorum*

Brasilia G. Barroso (~ *Calea*). Compositae (Helia.-Mel.). 1 Brazil

Brasilicereus Backeb. (~ *Cereus*). Cactaceae (III 3). 2 E Brazil

Brasiliparodia = *Parodia*

Brasilocalamus Nakai = *Merostachys*

Brassaia Endl. = *Schefflera*

Brassaiopsis Decne & Planchon. Araliaceae. c. 20 Indomal.

Brassavola R. Br. Orchidaceae (V 13). 17 trop. Am. R: ANMW 79(1975)9. Cult. orn. incl. *B. nodosa* (L.) Lindley throughout genus range incl. salt-spray zone in C Am.

Brassia R. Br. Orchidaceae (V 10). 35 trop. Am. R (sect. *B.*): OD 43 (1979) 164. Cult. orn.

Brassiantha A.C. Sm. Celastraceae. 1 New Guinea

Brassica L. Cruciferae. 35 Euras. (Eur. 22), Medit. R: JAA 66(1985)288. Herbs, often pachycaul, s.t. woody, e.g. *B. somalensis* Hedge & Miller (Somalia) & Jersey longjacks cabbage – stems used as walking-sticks. Glucosinolates in crushed lvs broken down by myrosinase to bitter thiocyanates, nitriles etc.; crushed seeds mixed with 'must' of old wine = 'mustum ardens' i.e. mustard – flavour due to pungent allyl isothiocyanate released by enzyme within 10 mins of adding water to m. powder & 'hot' *p*-hydroxybenzyl isothiocyanate. Veg. ('greens' in GB), selected forms grown for seed, fl., stem or root. *B. carinata* A. Braun (Texsel greens, NE Afr.) – more protein-rich than spinach; *B. hirta* Moench (white mustard, Medit.) – mustard- & oil-producing seeds; *B. juncea* (L.) Czerniak. (Indian m., kai choy, rai, S & E As., natur. Eur.) – grown for spring greens, esp. 'Crispifolia' (var. *crispifolia*, *B. japonica*, Chinese m., Japanese greens) & seeds (Dijon mustard); *B. napus* L. (rape, colza, a cultigen) – Napobrassica Group (swede (introd. GB from Holland 1755), rutabaga) – ed. roots, Pabularia Group (Siberian kale) – curled bluish lvs eaten in winter, 'summer races' ('subsp. *oleifera*', oilseed rape) consp. yellow fields in GB, seed 40% oil – that (esp. 'Canola' (i.e. Can-adian oilseed, Canola oil)) low in erucic acid & glucosinolates, used in margarine, mayonnaise, salad & cooking oils & biodiesel (by esterification to RME (rape methyl ester) resembling fossil diesel with glycerine & animal feed as by-products) like soya, that with high levels used in lubricating jet-engines etc., source of colza oil (illuminant) & birdseed, hay-fever source (UK with 64 000 ha in 1978, 348 000 ha by 1988) diminishing racehorse performance, seedlings an ersatz subs. for slower-growing *Sinapis alba*; *B. nigra* (L.) Koch (black or brown m., Euras.) – the 'm. seed' of the Bible, source of a pungent m., seedlings = 'mustard' of 'm. & cress'; *B. oleracea* L. (coastal Eur.): – Acephala Group (kale, borecole, collards) – ed. lvs in cool season, some cvs cult. orn. with coloured lvs & others with tall woody stems used for walking-sticks & the lvs for cow fodder (e.g. Jersey longjacks in Channel Is.), Alboglabra Group (Chinese kale, kai lan) – oriental veg., Botrytis Group (broccoli, cauliflower) – ed. infls in compact heads, Capitata Group (cabbage) – large ed. term. bud, Savoy c. being a cv. with puckered lvs, Gemmifera Group ((Brussels) sprouts, selected c. 1750) – ed. compact lateral buds, Gongylodes Group (kohlrabi, knol-kohl) – ed. swollen stem, Italica Group (sprouting broccoli incl. calabrese) – like Botrytis but infl. not compacted into single head, Tronchuda Group – Portuguese cabbage; *B. rapa* L. (sarson, Eur.) – imp. oilseed in India, pest in NZ, Chinensis Group (*B. chinensis*, Chinese cabbage or mustard, pak-choi, bok (buk) choy, choi (choy) sum, Shantung cabbage) – ed. lvs, Pekinensis Group (*B. pekinensis*, pe-tsai, wong-bok, chihli, Chinese cabbage) – ed. lvs ± in head, Rapifera Group (turnip, neeps) – 1 of the oldest root-crops (UK record: 15.975 kg). Many long-cultivated, some known to Theophrastus (400 BC) & 12 distinct types listed in Pliny's time. Poss. ancestry of most from 3 diploid spp.: *B. nigra*, *B. oleracea* & *B. rapa*, *B. juncea* being *B. nigra* × *B. rapa*, *B. napus* being *B. oleracea* × *B. rapa* & *B. carinata* being *B. oleracea* × *B. nigra*, though kales may be derived from *B. cretica* Lam. & other spp. in E Medit. Sauerkraut is prepared from slightly fermented cabbage, coleslaw being cabbage dressed with mayonnaise or vinegar etc.; English mustard prepared from *B. hirta* with *B. nigra* (50:50) first ground to wettable dust 1720 and industrialized by Jeremiah Colman later

Brassicaceae Burnett. See Cruciferae

Brassiodendron C. Allen (~ *Endiandra*). Lauraceae (I). 1 E Mal., Aus.

Brassiophoenix Burret. Palmae (V 4k). 2 New Guinea. R: Principes 19 (1975) 100. Monoec.

brauna *Melanoxylum brauna*

Braunblanquetia Eskuche. Scrophulariaceae. 1 Argentina

Braunsia Schwantes. Aizoaceae (V). 4–6 Cape. Cult. orn.

Bravaisia DC. Acanthaceae. 3 trop. Am., 2 being mangrove trees. R: Proc. Calif. Acad. Sci. 45(1988)111

Bravoa Llave & Lex. = *Polianthes*

Braxireon Raf. = *Narcissus*

Braya Sternb. & Hoppe. Cruciferae. 20 N circumpolar, Alps (Eur. 3), C As., Himal. R: Pfl. 86 (IV 105) (1924)226

Brayopsis Gilg & Muschler (~ *Englerocharis*). Cruciferae. 3 Andes. R: NJB 8(1989)620. *B. gamosepala* Al-Shehbaz (Bolivia) – gamosepaly otherwise only known in Cruciferae in *Christolea*

Brayulinea Small = *Guilleminea*

Brazil or **Brazilian arrowroot** *Ipomoea batatas*; **B. cherry** *Eugenia* spp.; **B. copal** *Hymenaea courbaril*; **B. cress** *Acmella oleracea*; **B. nut** *Bertholletia excelsa*; **B. nutmeg** *Cryptocarya moschata*; **B. pepper** *Schinus* spp. esp. *S. terebinthifolius*; **B. pine** *Araucaria angustifolia*; **B. redwood** *Caesalpinia echinata*, *Brosimum rubescens*; **B. rosewood** *Dalbergia* spp.; **B. satin-wood** *Euxylophora paraensis*; **B. tea** *Ilex paraguariensis*; **B. tulipwood** *Dalbergia decipularis*; **b. wood** *Caesalpinia sappan*, *C. echinata* etc. etc.

Brazoria Engelm. ex A. Gray. Labiatae (VI). 3 S US

Brazzeia Baillon. Scytopetalaceae. 3 trop. Afr.

bread-and-cheese (GB) *Crataegus monogyna*

breadfruit *Artocarpus altilis*; **Nicobar b.** *Pandanus leram*

breadnut *Brosimum alicastrum*, *Pediomelum esculentum*

breadroot *Pediomelum esculentum*

Bredemeyera Willd. Polygalaceae. c. 20 trop. Am.
Bredia Blume. Melastomataceae. 30 E & SE As.
Breea Less. = *Cirsium*
Breitungia Löve & D. Löve = *Sedum*
Bremekampia Sreemad. = *Hapalanthodes*
Bremontiera DC (~ *Indigofera*). Leguminosae (III 8). 1 Réunion: *B. ammoxylon* DC. Almost extinct
Brenania Keay. Rubiaceae (II 1). 1 W Afr.
Brenesia Schltr. = *Pleurothallis*
Brenierea Humbert. Leguminosae (I 3). 1 Madag. Cladodes (unique in tribe), C 0, petaloid staminodes 5
Breonadia Ridsd. Rubiaceae (I 2). 1 trop. Afr. & Madag.: *B. salicina* (Vahl) Hepper & J.R. Wood (*B. microcephala*), rheophyte
Breonia A. Rich. ex DC. Rubiaceae (I 2). 5 Madag. R: Blumea 22(1975)544
Bretschneidera Hemsley. Bretschneideraceae. 1 China, Thailand, Vietnam: *B. sinensis* Hemsley
Bretschneideraceae Engl. & Gilg. Dicots – Dilleniidae – Capparidales. 1/1 SE As. Deciduous tree with mustard-oils & myrosin cells in bark & infl. Lvs spiral, pinnate; stipules 0. Fls bisexual, perigynous, slightly irreg., in terminal racemes; K (5), C 5, unequal, clawed at base, imbricate, A 8 attached to thin annular nectary-disk, anthers versatile, G (3–5), ovary pluriloc. with curved style, ovules 2 per locule, apical-axile & pendulous. Fr. a capsule; seeds red with large embryo, endosperm 0. 2n = 18
Only genus: *Bretschneidera*
Form. thought allied to Hippocastanaceae or Sapindaceae but differing in perigynous fls & myrosin cells etc.
Breviea Aubrév. & Pellegrin. Sapotaceae (IV). 1 W Afr.: *B. sericea* Aubrév. & Pellegrin. R: T.D. Pennington, Sapotaceae (1991)211
Brevoortia Alph. Wood = *Dichelostemma*
Brewcaria L.B. Sm., Steyerm. & H. Robinson. Bromeliaceae (1). 1 C Amazonas, Venez. R: AMBG 73(1986)714
Brewerina A. Gray = *Arenaria*
Brexia Noronha ex Thouars. Grossulariaceae (Brexiaceae). 1(?9) coastal lowlands E Afr., Madag., Seychelles. *B. madagascariensis* (Lam.) Ker with ed. fr. (E Afr.: mfukufuku)
Brexiaceae Loudon. See Grossulariaceae
Brexiella Perrier. Celastraceae. 2 Madagascar
Breynia Forster & Forster f. Euphorbiaceae. c. 25 China to New Caled. & Aus.(7). Some medic., cult. orn. esp. *B. nivosa* (W.G. Sm.) Small 'Roseapicta' (*B. disticha* f. *nivosa*, snow-bush, Pacific)
Brezia Moq. = *Suaeda*
briar root *Erica arborea*
Brickellia Elliot. Compositae (Eup.-Alom.). (Incl. *Barroetea*, *Phanerostylis*) 110 W US, Mex., C. Am. to Arg. R: MSBMBG 22(1987)221, 224. Some cult. orn.
Brickelliastrum R. King & H. Robinson. Compositae (Eup.-Alom.). 1 S US. R: MSBMBG 22(1987)228
bridal bouquet *Poranopsis paniculata*; **b. veil, Tahitian** *Gibasis pellucida*; **b. wreath** *Francoa sonchifolia*
Bridelia Willd. Euphorbiaceae. 60 OW trop. Some timbers, local medic., incl. for diabetes, tanbarks. *B.* spp. in Africa food of certain silkworms giving African wild or Anape silk
Bridgesia Bertero ex Cambess. Sapindaceae. 1 Chile
Briggsia Craib. Gesneriaceae. 29 E Himal., Burma, China. Some cult. orn. esp. *B. muscicola* (Diels) Craib (Yunnan & Bhutan to SE Tibet)
Briggsiopsis K.Y. Pan (~ *Briggsia*). Gesneriaceae. 1 China: *B. delavayi* (Franchet) K.Y. Pan
Brighamia A. Gray. Campanulaceae. 2 Hawaii. R: SB 14(1989)133. *B. insignis* A. Gray a pachycaul treelet
Brillantaisia Pal. Acanthaceae. c. 9 trop. Afr., Madag. 2 posterior A perfect (only genus in A. thus)
Brimeura Salisb. (~ *Hyacinthus*). Hyacinthaceae (Liliaceae s.l.). 2 S Eur. (both with disjunct distrib.). Cult. orn. esp. *B. amethystina* (L.) Chouard (Pyrenees & NW Yugoslavia)
brinjal *Solanum melanogena*
Briquetastrum Robyns & Lebrun = *Leocus*
Briquetia Hochr. Malvaceae. 5 warm Am.
Briquetina Macbr. = *Citronella*

Brisbane lily *Proiphys cunninghamii*
bristle-cone pine *Pinus longaeva*
Britoa O. Berg = *Campomanesia*
Brittenia Cogn. ex Boerl. Melastomataceae. 1 Sarawak: *B. subacaulis* Cogn. ex Boerl. Acaulescent herb; fls 5-merous, A all equal
brittlebush *Encelia farinosa*
Brittonella Rusby = *Mionandra*
Briza L. Gramineae (17). 20 temp. Euras. (Eur. 4), S Am. Quake or quaking grass, shaking grass. Cult. orn. esp. *B. maxima* L. (Med.), *B. media* L. (doddering-dillies, jiggle-joggles, Euras.)
Brizochloa V. Jir. & Chrtek = *Briza*
Brizula Hieron. = *Aphelia*
broad bean *Vicia faba*; **b. path** (WI) *Alternanthera* spp.
Brocchinia Schultes f. Bromeliaceae (1). 20 N S Am. R: FN 14(1974)437, AMBG 73(1986)702. Mostly terr. Some cult. orn. incl. *B. reducta* Bak. (Venez.) – resembles *Heliamphora* (q.v.) with bright green lvs forming cylinders, nectar-like odour unique in fam. but no nectar (unlike *Heliamphora*), poss. carnivore as trichomes on lvs can absorb leucine; some spp. have *Utricularia* pls in 'pitchers'
broccoli *Brassica oleracea* Botrytis Group; **sprouting b.** *B. oleracea* Italica Group
Brochoneura Warb. Myristicaceae. 3 E Madag. R: Adansonia 13(1973)205. Arils rudimentary
Brockmania W. Fitzg. = *Hibiscus*
Brodiaea Sm. Alliaceae (Liliaceae s.l.). 15 W N Am. R: UCPB 60(1971). Corms ed. raw or cooked by N Am. Indians. Cult. orn. but many plants sold as *B.* are spp. of *Tristagma*, *Triteleia* or *Dichelostemma*
Brodriguesia R. Cowan. Leguminosae (I 4). 1 E Brazil
Brombya F. Muell. (~ *Melicope*). Rutaceae. 2 Queensland. Ant-disp.
brome (grass) *Bromus* (*Anisantha, Bromopsis, Ceratochloa*) spp.; **false b.** *Brachypodium* spp.; **Hungarian b.** *Bromus inermis*; **soft b.** *B. hordaceus*
Bromelia L. Bromeliaceae (3). 48 trop. Am. R: FN 14(1979)1649. Terr. cult. orn. *B. pinguin* L. (WI, S Am.) – source of pinguin fibre, fr. ed.; *B. serra* Griseb. (Bolivia & Brazil to Arg.) – source of caraguata or chaguar fibre used for sacks, cordage & poss. useful for paper-making
Bromeliaceae Juss. Monocots – Zingiberidae – Bromeliales. 59/2400 trop. Am. (1 W trop. Afr.), few subtrop. Am. Terrestrial xeromorphic pachycauls to stemless epiphytic herbs, usu. with papain-like proteolytic enzymes. Lvs spiral, narrow, parallel-veined, entire or spinose-serrate, usu. with stalked peltate water-absorbing scales, usu. concave adaxially, channelling water to centre of rosette ('tank'). Fls bisexual or functionally unisexual, reg. or slightly not, 3-merous, hypogynous to epigynous, in spikes, racemes or heads with conspicuous coloured bracts, poll. by birds, insects, bats or rarely wind (*Navia*) or cleistogamous; K 3 green & herbaceous to ± petaloid, free or connate at base, C 3 free or connate at base, usu. brightly coloured, s.t. with paired basal nectary scales, A 3 + 3, often connate or adnate to P, anthers opening by longit. slits, $\underline{G}$ (3) to $\overline{G}$ (3) with term., often 3-fid style, ovules few to ± ∞ on axile placentas, anatropous or rarely campylotropous, bitegmic. Fr. a berry or less often a septicidal capsule, rarely multiple & fleshy (*Ananas*); seeds in capsules winged or plumose (due to splitting of testa starting at chalazal end but remaining attached at micropylar one), endosperm mealy, copious, the starch in compound grains. x = 8–28, often 25
Classification & chief genera:
1. **Pitcairnioideae** ($\underline{G}$, rarely half-inferior, fr. a capsule, seeds winged etc. but not with plumose crown, usu. terr.): *Dyckia, Hechtia, Navia, Puya*
2. **Tillandsioideae** (As above but seeds with plumose crown, lvs usu. entire, plants mostly epiphytic): *Guzmania, Tillandsia, Vriesea*
3. **Bromelioideae** ($\overline{G}$ (3), rarely only half-inferior; fr. a berry or rarely multiple fleshy, seeds without appendages, lvs usu. spiny-toothed, usu. epiphytes): *Aechmea, Ananas, Billbergia, Bromelia, Canistrum, Cryptanthus, Neoregelia, Nidularium*

Bromelioideae outnumber the other 2; Tillandsioideae difficult to assign to genera without flowers; Pitcairnioideae held to be most archaic, xeromorphy a 'pre-adaptation' to epiphytism, *Puya* being particularly primitive and *Navia* poss. a link to Rapateaceae which is represented in Africa also by 1 sp. (monotypic *Maschalocephalus*)
Epiphytes in rain forest and even deserts (on cacti), water economy marked by absorbing trichomes, succulence, dark CO_2 fixation, deciduous lvs etc. Some tank-forming spp. of *Brocchinia* in *Sphagnum* bogs resemble insectivorous plants & poss. tanks evolved for

nutritional rather than water stresses (*Catopsis* prob. also carnivorous). Tanks often habitats for nos. of insects & other animals and even spp. of *Utricularia* restricted to them; certain *Hyla* frogs block tanks with head thus reducing evaporation to advantage of both – 1 hibernates there in dry season

Fruit crops (*Ananas, Bromelia*), fibres (*Ananas, Bromelia, Chevaliera, Neoglaziovia, Puya, Tillandsia*) & very many cult. orn. housepls (*Flora neotropica* 14(1974–9), W. Rauh (1981), *Bromelien*; BBP 63(1988)403)

Bromelica (Thurb.) Farw. = *Melica*

Bromheadia Lindley. Orchidaceae (V 9). 12 Mal. to Aus. Fls drop in response to temp. change

Bromidium Nees & Meyen = *Agrostis*

Bromopsis (Dumort.) Fourr. = *Bromus*

Bromuniola Stapf & C. Hubb. Gramineae (24). 1 Tanzania to Angola

Bromus L. Gramineae (22). (Incl. *Anisantha, Bromopsis, Ceratochloa*) c. 100 temp. (Eur. 37); trop. mts. R: NRBGE 30 (1970)361. Some orn. & forage: bromes, easily confused with fescues but usu. with hairy tubular leafsheaths & subapical lemma awns. *B. catharticus* Vahl (*B. unioloides*, rescue grass, S Am.) – forage; *B. hordaceus* L. (*B. mollis*, soft b., Eur., natur. N Am.); *B. inermis* Leysser (Hungarian b., Euras., natur. N Am.) – hay; *B. interruptus* (Hackel) Druce (GB) – extinct in wild; *B. secalinus* L. (cheat or chess, Euras., Medit.); *B. mango* Desv. (Chile) – old cereal cult. Araucano Indians pre-Conquest; *B. rigidus* Roth. (ripgut, Euras.) – pungent callus & rough awn penetrate mouth, eyes & intestine; *B. tectorum* L. (*Anisantha t.*, downy chess, Medit.); *B. trinii* Desv. (Chile) – facultative cleistogamy in adverse conditions

Brongniartia Kunth. Leguminosae (III 23). 56 Am. *B. alamosana* Rydb. a poss. oilseed

Brongniartikentia Becc. Palmae (V 4m). 2 New Caled. R: Allertonia 3(1984)337

Brookea Benth. Scrophulariaceae. 3 Borneo

brooklime *Veronica beccabunga*

brookweed *Samolus valerandi*

broom *Cytisus, Genista, Spartium* etc. spp.; **common** or **Eur. b.** *Cytisus scoparius*; **coral b.** *Corallospartium crassicaule*; **Mt Etna b.** *Genista aetnensis*; **Spanish b.** *Spartium junceum*

broom-corn *Sorghum bicolor*

broomrape *Orobanche* spp.

Brosimopsis S. Moore = *Brosimum*

Brosimum Sw. Moraceae (IV). 13 trop. Am. R: FN 7(1972)161. Male fls (single stamens), female fls & fr. embedded in fleshy receptacle. *B. alicastrum* Sw. (breadnut, ramón) – lvs for fodder (drought-tolerant), seeds ed., latex potable, wood fine; *B. guianense* (Aublet) Huber (amourette, leopard wood, letterwood, Guianas) – for violin bows, turnery etc.; *B. rubescens* Taubert (*B. paraense*, cardinal wood, brazilwood, satiné, snakewood, Brazilian redwood) – furniture etc.; *B. utile* (Kunth) Oken (cow-tree) – latex potable & base for chewing-gum, barkcloth

Brossardia Boiss. Cruciferae. 1 Iran, Iraq

Broughtonia R. Br. Orchidaceae (V 13). 5 W I. R: Rhodora 86(1984)447. Cult. orn. like *Epidendrum*

Brousemichea Bal. = *Zoysia*

Broussa tea *Vaccinium arctostaphylos*

Broussaisia Gaudich. Hydrangeaceae. 1 Hawaii: *B. arguta* Gaudich. – fr. berry-like but inferior ovary

Broussonetia L'Hérit. ex Vent. Moraceae (I). 8 trop. & warm As.(7), Madag. (1). Dioec.: male fls in drooping catkins with explosive A (cf. *Urtica*, rare in Moraceae), females in globose heads, fr. a syncarp (cf. *Morus*) of orange-red drupelets. *B. kazinoki* Siebold (Korea) – paper in use in Japan by AD 610; *B. papyrifera* (L.) Vent. (paper mulberry, E As. early taken to Polynesia, natur. in N Am.) – tapa or kapa cloth & paper derived from inner bark of tree, often grown as coppice

Browallia L. Solanaceae. 2 trop. Am. A 4; fr. a capsule enclosed in K. Cult. orn., widely natur. esp. annual *B. americana* L. & perennial *B. speciosa* Hook., both with several cvs

brown barrel *Eucalyptus fastigiata*

Brownanthus Schwantes (~ *Psilocaulon*). Aizoaceae (IV). 8 S Afr. R: BJ 105(1985)316

Brownea Jacq. Leguminosae (I 4). 12 Costa Rica & WI to Peru. Fast-growing young shoots with pink or red lvs speckled white on flaccid stems later turn green & stiffen up (cf. *Saraca, Amherstia*). Several spp. with bright red bird-poll. heads of fls. with copious nectar; supposed contraceptives in NW Amazonia (? 'Doctrine of Signatures')

Browneopsis Huber. Leguminosae (I 4). 6 Panamá to Peru

Brownetera Rich. ex Tratt. = *Phyllocladus*

Browningia Britton & Rose (III 7). Cactaceae. 7 Andes of Bolivia, Peru, N Chile. Cult. orn. trees & shrubs incl. *B. viridis* (Rauh & Backeb.) F. Buxb. to 10 m

Brownleea Harvey ex Lindley. Orchidaceae (IV 3). 7 trop. & S Afr., Madag. R: JSAB 47(1981)13

Brownlowia Roxb. Tiliaceae. 25 SE As., wetter Mal. to Solomons

Brucea J. Miller. Simaroubaceae. 8 OW trop. Medic. esp. *B. amarissima* (Lour.) Gomes (*B. javanica*, Indomal.) for dysentery & worms, antimalarial quassinoids

Bruckenthalia Reichb. (~ *Erica*). Ericaceae. 1 SE Eur., As. Minor: *B. spiculifolia* (Salisb.) Reichb., cult. orn.

Brugmansia Pers. (~ *Datura*). Solanaceae (1). 14 S Am. esp. Andes. R: HBML 23 (1973)273. Perennial shrubs & trees with pendulous fls (*Datura* herbaceous with erect fls). Alks esp. scopolamine leading to hallucinations after violent intoxication. Cult. orn. (angel's trumpets) esp. *B.* × *candida* Pers. (*B. aurea* Lagerh. (Andes etc.) × *B. versicolor* Lagerh. (Ecuador), often labelled *B. arborea*); *B. arborea* (L.) Lagerh. (true, Ecuador & N Chile), *B. aurea* (many cvs, incl. *'Methysticodendron'*, veg. prop. by shamans) & *B. sanguinea* (Ruíz & Pavón) D. Don (Colombia to Chile) used as medic. & hallucinogens by Am. Indians

Bruguiera Sav. Rhizophoraceae. 6 E Afr. to Pacific (fossils similar in London Eocene, cf. *Nypa*). Mangroves. Conspicuous knee-roots but no aerial roots from branches (cf. *Rhizophora*). *B. gymnorhiza* (L.) Sav. – bird-poll. in Ryukyu Is. Tanbark, black dyes, timber & charcoal, chips for pulp & rayon; young seedlings of *B. sexangula* (Lour.) Poiret ed. Sulawesi

Bruinsmia Boerl. & Koord. Styracaceae. 1 Assam, Burma; 1 Mal. (not Malay Pen.)

Brunellia Ruíz & Pavón. Brunelliaceae. 52 trop. Am., esp. montane forests. R: FN 2(1970)

Brunelliaceae Engl. (~Cunoniaceae). Dicots – Rosidae – Rosales. 1/52 trop. Am. Evergreen trees, often densely tomentose, commonly dioec. or gynodioec. Lvs opp. or ternate, pinnate, trifoliolate to unifoliolate or simple with opp. leaflets entire to doubly-dentate; stipules small, deciduous, often more than 2. Fls small in cymes, ± hypogynous, reg.; K ((4)5(–8)), connate basally, valvate, persistent in fr., C 0; A in 2 whorls, each same number as K or inner with up to twice as many, inserted in notches of a nectary disk, anthers with longit. slits; $\underline{G}$ (2 or 3) up to same number as K, each with curved to almost circinate style with linear stigmatic surface, ovules 2 per carpel, collateral, pendulous, anatropous, bitegmic. Fr. a follicle, developing with stem pointing outwards or downwards, usu. densely tomentose & with long pointed trichomes; endocarp ± lignified; seeds 1 or 2 per carpel with shiny testa & corky 'aril', attached by funicle in dehisced fr., embryo large with flattened cotyledons & abundant mealy endosperm.

Only genus: *Brunellia*

Poss. best incl. in Cunoniaceae but differing in stigma, distinctive pubescence, fr. & also wood characters

Brunfelsia L. Solanaceae (1). c. 40 trop. Am. Alks. Cult. orn. & medic. Fls white fading to yellow or purple fading to white. *B. americana* L. (lady of the night, WI) – fragrant at night; *B. australis* Benth. (yesterday, today & tomorrow, trop. S Am.) – fls fading from purple to white; *B. grandiflora* D. Don (W S Am.) – lvs & bark hallucinogens in Amazon; *B. pauciflora* (Cham. & Schldl.) Benth. (yesterday, today & tomorrow, Brazil) – many cvs; *B. uniflora* (Pohl) D. Don (*B. hopeana*, Brazil & Venez.) – used against syphilis

Brunfelsiopsis (Urban) Kuntze = *Brunfelsia*

Brunia Lam. Bruniaceae. 7 S Afr. Cult. orn.

Bruniaceae Bercht. & J. Presl. Dicots – Rosidae – Rosales. 12/74 S Afr. esp. Cape. Shrubs to small trees, usu. ericoid & with long unicellular hairs. Lvs spiral, small, simple, often imbricate & with centric structure; stipules 0 or vestigial. Fls sessile in spikes or heads when s.t. involucrate & resembling Compositae, rarely solitary, usu. small, bisexual, reg. usu. epigynous; K (4)5, connate or not, imbricate; C (4)5, imbricate, rarely (*Lonchostoma*) connate at base with short tube; A same no. as C & alt. with C, anthers with longit. slits; intrastaminal disk s.t. present; G (2), rarely (3) (*Andouinia*) with as many locules but apparently 1 in *Berzelia* & *Mniothamnea*, ovules (1)2–4(–12) per locule, pendulous, anatropous, unitegmic. Fr. dry, often with persistent K, achene-like with 1 seed or 1- or 2-seeded carpels separating & opening along ventral suture; seeds small sometimes with aril, embryo small, straight, endosperm fleshy. 2n = 16

Chief genera: *Berzelia*, *Brunia*, *Raspalia*, *Staavia*, *Thamnea*

Prob. allied to woody relations of Saxifragaceae (Cronquist)

Some cut-fls

Brunnera Steven. Boraginaceae. 3 E Med. (Eur. 1) to W Siberia. *B. macrophylla* (Adams) I.M.

Johnston (Caucasus, W Siberia) – cult. orn.

Brunnichia Banks ex Gaertner. Polygonaceae (II 2). 3+ trop. Afr. Winged pedicel aids dispersal

Brunonia Sm. ex R.Br. Goodeniaceae (Brunoniaceae). 1 Aus.: *B. australis* Sm. ex R. Br. (pincushion) – cult. orn. 2n = 18, 36, 72. R: FA 35(1992)1

Brunoniaceae Dumort. See Goodeniaceae

Brunoniella Bremek. (~ *Ruellia*). Acanthaceae. 6 Aus. (5), New Guinea. R: JABG 9(1986)95

Brunsfelsia L. = *Brunfelsia*

Brunsvigia Heister. Amaryllidaceae (Liliaceae s.l.). c. 20 Afr. R: PL 6(1950)63. Much confused with *Amaryllis* etc. Alks

brush apple *Pouteria australis*; **b. box** *Lophostemon confertus*

Brussels sprouts *Brassica oleracea* Gemmifera Group

bruyère *Erica arborea*

Brya P. Browne. Leguminosae (III 9). 12 WI esp. Cuba. Spines in place of lvs on long shoots. *B. ebenus* (L.) DC a principal lumber tree of WI gives Jamaica or American ebony, cocus or cocos wood, the heartwood blackening with age, used for musical instruments & form. for door-handles etc.

Bryantea Raf. = *Neolitsea*

Bryanthus Gmelin. Ericaceae. 1 Kamchatka, Japan: *B. gmelinii* G. Don f.

Bryaspis Duvign. Leguminosae (III 13). 2 W trop. Afr.

Brylkinia F. Schmidt. Gramineae (20). 1 China, Japan

Bryobium Lindley = *Eria*

Bryocarpum Hook.f. & Thomson. Primulaceae. 1 E Himal.

Bryodes Benth. Scrophulariaceae. 3 Mascarenes

Bryomorphe Harvey. Compositae (Gnap.-Rel.). 1 S Afr.: *B. lycopodioides* (Walp.) Levyns. R: OB 104(1991)70

Bryonia L. Cucurbitaceae. c. 12 (many ill-defined) Euras., N Afr., Canary Is. *B. alba* L. marks northernmost limit of fam. (Scandinavia). R: KB 23(1969)441. Some medic. tubers & cult. orn. Some spp. dioec., some either dioec. or monoec. e.g. *B. alba* monoec. in N Eur., dioec. in *B. dioica* Jacq. (*B. cretica* subsp. *dioica*, white bryony, Euras., Medit.) – drastic purgative

bryony, black *Tamus communis*; **white b.** *Bryonia dioica*

Bryophyllum Salisb. = *Kalanchoe*

buaze *Securidaca longipedunculata*

Bubalina Raf. = *Burchellia*

Bubania Girard = *Limoniastrum*

Bubbia Tieghem = *Zygogynum*

bubinga *Copaifera salikounda, Guibourtia demeusei*

Bubon L. = *Athamanta*

Bubonium Hill = *Asteriscus*

Bucephalandra Schott. Araceae (VI 1). 3 Borneo. R: Aroid. 3(1980)134. *B. motleyana* Schott – rheophyte 1.5–8 cm tall

Bucephalophora Pau = *Rumex*

Buceragenia Greenman. Acanthaceae. 4 C Am.

Buchanania Sprengel. Anacardiaceae (I). 25 Indomal., W Pacific. G 4–6, 1 fert. *B. lanzan* Sprengel (Indomal.) – ed. oily seeds (cheronjee, Cuddapah almond, almondette) used like almonds

Buchenavia Eichler. Combretaceae. 20 trop. Am. *B. capitata* (Vahl) Eichler (WI) – hard timber for floors, boats, house-construction etc.

Buchenroedera Ecklon & Zeyher = *Lotononis*

Buchholzia Engl. Capparidaceae. 3 W Afr. *B. coriacea* Engl. – smelly elephant-disp. fr.

Buchingera Boiss. & Hohen. = *Asperuginoides*

Buchloe Engelm. Gramineae (33c). 1 N Am.: *B. dactyloides* (Nutt.) Engelm. (buffalo grass) monoec. or dioec. small creeping grass of western prairies, good fodder, sometimes a lawn-grass & for erosion control, trad. for Ind. grasshouses in Oklahoma

Buchlomimus Reeder, C. Reeder & Rzed. Gramineae (33c). 1 Mex.

Buchnera L. Scrophulariaceae. c. 100 warm (NW 16). Hemiparasites. *B. leptostachya* Benth. (trop. Afr.) – used for earache in Tanzania

Buchnerodendron Guerke. Flacourtiaceae (3). 2 trop. Afr. R: BJ 94(1974)289

Buchtienia Schltr. Orchidaceae (III 3). 3 trop. S Am. R: HBML 28(1982)304

buchu *Agathosma* spp.

Bucida L. Combretaceae. 3 S Florida, C Am., WI

Bucinellina Wiehler = *Columnea*

buck bean *Menyanthes trifoliata*; **b. berry** *Vaccinium* spp. (N Am.); **b.bush** *Salsola kali*; **b.eye** *Aesculus* spp.; **Mex. b.eye** *Ungnadia speciosa*; **yellow b. eye** *A. flava*; **b. spinifex** *Triodia longiceps*

Buckinghamia F. Muell. Proteaceae. 2 Queensland. R: Muelleria 6(1988)417

Bucklandia R. Br. ex Griffith = *Exbucklandia*

buckler fern *Dryopteris* spp.; **broad b.f.** *D. dilatata*; **narrow b. f.** *D. carthusiana*; **b. mustard** *Biscutella laevigata*

Buckleya Torrey. Santalaceae. 4 S US, Japan, C China. Deciduous, dioec. parasites. R: Castanea 47(1982)17. *B. distichophylla* (Nutt.) Torrey (S US) parasitic on *Tsuga* spp.

buckthorn *Rhamnus* spp.; **common** or **Eur. b.** *R. cathartica*; **alder b.** *R. frangula*; **Pallas's b.** *R. erythroxyloides*; **sea b.** *Hippophae rhamnoides*

buckwheat *Fagopyrum esculentum*; **Siberian, Tartary** or **Kangra b.** *F. tataricum*

Bucquetia DC. Melastomataceae. 3 trop. S Am.

Budawangia Telford (~ *Rupicola*). Epacridaceae. 1 NSW: *B. gnidioides* (Summerh.) Telford. R: Telopea 5(1992)231

Buddha's hand *Citrus medica* 'Fingered'

Buddleia = *Buddleja*

Buddleja L. Buddlejaceae (Loganiaceae s.l.). c. 100 warm esp. E As. Trees, shrubs, rarely herbs; wide range of pollen morphology, 2n = 38(?)–c. 456. Cult. orn. & some local med. *B. asiatica* Lour. (Indomal.) – strong freesia-like scent imperceptible to some (cf. *Freesia*), fish-poison; *B. davidii* Franchet (China) – many cvs, introd. GB 1890s & colonizing wasteground by 1930s being more attractive to butterflies etc. than any native pl., large associated fauna by 1980s; *B. salviifolia* (L.) Lam. (trop. & S Afr.) – useful timber

Buddlejaceae Wilhelm (Loganiaceae s.l.). Dicots – Asteridae – Scrophulariales. 7/120 trop. & warm. Trees or shrubs (rarely herbs: *Polypremum*), often with scales or branched hairs (never simple ones except s.t. within C). Lvs usu. opp. or whorled, simple. Fls bisexual, s.t. functionally dioec., 4(5)-merous; K with tube; C with tube & usu. imbricate lobes (valvate in *Peltanthera*); A attached to tube, alt. with C, anthers with longit. slits; G (2), 2-loc. (semi-inf. in *Polypremum*), style term. with 2-lobed stigma, ovules ∞ on thickened axile placentas, hemitropous or amphitropous, unitegmic. Fr. usu. septicidal capsule, rarely fleshy-indehiscent; seeds often winged, endosperm oily. x = 19 (11 in *Polypremum*)

Genera: Androya, Buddleja, Emorya, Gomphostigma, Nuxia, Peltanthera, Polypremum

The splitting up of Loganiaceae s.l. leads to tribe Buddlejieae, with pollen & other features (*Sanango* with somewhat irreg. C & only 4 functional A here referred to Gesneriaceae) recalling Scrophulariaceae, being recognized at fam. level. *Polypremum* clearly belongs in Loganiaceae s.l., but with the splitting here followed its position is unclear & would be best excluded from B.

Some cult. orn.

budgerigar flower *Asclepias syriaca*

Buergersiochloa Pilger. Gramineae (4). 1 New Guinea

Buesia (Morton) Copel. = *Hymenophyllum*

Buesiella C. Schweinf. (~ *Rusbyella*). Orchidaceae (V 10). 1

buffalo bean *Astragalus crassicarpus*; **b. berry** *Shepherdia argentea*; **b. bur** *Solanum rostratum*; **b. clover** *Trifolium reflexum*; **running b. c.** *Trifolium stoloniferum*; **b. currant** *Ribes odoratum*; **b. gourd** *Cucurbita foetidissima*; **b. grass** *Buchloe dactyloides, Panicum coloratum*, (Aus.) *Stenotaphrum secundatum*; **b. nut** *Pyrularia pubera*; **b. thorn** *Ziziphus mucronata*; **b. wood** *Burchellia bubalina*

buffel grass *Cenchrus ciliaris*

Bufonia L. Caryophyllaceae (II 1). 20 Medit. (Eur. 7), Canary Is. 4-merous. The allegation that Linnaeus deliberately altered the generic name from *Buffonia*, commemorating Buffon, as a malicious pun on *bufo* (= toad) is said to be unfounded (*J. Linn. Soc.* 2(1858)183)

Buforrestia C.B. Clarke. Commelinaceae (II 2). 3 Guianas, W Afr.

bugbane *Cimicifuga foetida*

bugle *Ajuga* spp.; **common** or **Eur. b.** *A. reptans*; **yellow b.** *A. chamaepitys*

bugles, red *Conostylis canescens*

bugloss *Echium plantagineum*; **viper's b.** *E. vulgare*

Buglossoides Moench = *Lithospermum*

bugseed *Corispermum* spp.

Buhsia Bunge. Capparidaceae. 1 Iran, Transcasp.

buk choy *Brassica rapa* Chinensis Group

bukchie *Baccharoides anthelmintica*

Bukiniczia Lincz. Plumbaginaceae (II). 1 Afghanistan, Pakistan: *B. cabulica* (Boiss.) Lincz.

Bulbine Wolf. Asphodelaceae (Liliaceae s.l.). 50 trop. & S Afr. (esp. SW Cape) diploids, Aus.(5) polyploids. Many medic. uses in S Afr.; cult. orn., weeds in Aus. ('onionweed', cf. *Asphodelus, Nothoscordum*). *B. frutescens* (L.) Willd. (E Cape) – shrubby cult orn.

Bulbinella Kunth. Asphodelaceae (Liliaceae s.l.). 21 S Afr. (16; R: SAJB 53(1987)431), NZ (6). Cult. orn. (cat-tail)

Bulbinopsis Borzi = *Bulbine*

Bulbocodium L. = *Colchicum*

Bulbophyllum Thouars. Orchidaceae (V 15). 1000 trop. (New Guinea 600). R: OM 2(1987)1 (Afr.72), 7(1993)1. Epiphytes with 1 or 2 lvs terminating each pseudobulb. Some sweetly scented & fly-poll. e.g. in clove-scented *B. macranthum* Lindley (Burma, Mal.) the 2 lateral sepals are directed upwards & meet near tips where flies land, holding on to outside of otherwise slippery sepals while licking surface. They work down sepals which are parted below such that flies can no longer straddle them: as they slip they clutch at the solid-looking tongue-like lip but as their weight is transferred to it, they are suddenly flung backwards & downwards, for the lip is pivoted. Two springy arms near tip of column embrace fly as lip returns to former position. The fly soon escapes but in so doing carries pollinia away on abdomen. In other spp. the flies are thrown head first against column & receive pollinia on head while others attract flesh-flies by bad smells etc. (after Proctor & Yeo). Several cult. orn. In *B. minutissimum* F. Muell. (Aus.), the pseudobulbs are hollow with stomata on inner surface

Bulbostylis Kunth (~ *Fimbristylis*). Cyperaceae. 100 trop. & warm (Aus. 7)

Bulbulus Swallen = *Rehia*

bull bay *Magnolia grandiflora*; **b. brier** *Smilax rotundifolia*; **b. horn thorn** *Acacia cornigera* etc.; **b. nettle** *Cnidoscolus stimulosus*; **b. oak** *Allocasuarina luehmannii*

bullace *Prunus × domestica* 'ssp. *institia*'

Bulleyia Schltr. Orchidaceae (V 12). 1 E Himal. to SW China

bullock's heart *Annona reticulata*

Bulnesia C. Gay. Zygophyllaceae. 9 S Am. R: Darw. 25(1984)299. Timber e.g. *B. arborea* (Jacq.) Engl. (Maracaibo lignum-vitae, verawood, Colombia, Venez.) – used like lignum-vitae & *B. sarmientoi* Lorentz ex Griseb. (Paraguay l.-v., palo santo, Paraguay, Brazil, SE Bolivia & Arg.) – wood distilled for fragrant oil of guaiac for soap as well as used like l.-v.; *B. retama* (Hook. & Arn.) Griseb. (Arg.) – common shrub, source of retamo wax used in shoe-polish etc.

bulrush prob. orig. (as in Bible) *Schoenoplectus lacustris* or *Vossia cuspidata* but name early transferred to *Typha* spp., Rubens's picture of Christ scourged shows this & all pictures of Moses in 'the bulrushes' also; **b. millet** *Pennisetum glaucum*

bulwaddy *Macropteranthes kekwickii*

bumble tree *Capparis mitchellii*

bumbo *Daniellia* spp.

Bumelia Sw. = *Sideroxylon*

bumpy ash *Flindersia schottiana*

buna *Fagus crenata*

bunch berry *Cornus canadensis*; **b. flower** *Melanthium virginicum*; **b. grass** *Schizachyrium scoparium*

Bunchosia Rich. ex Kunth. Malpighiaceae. 55 trop. Am. Fr. ed. in many spp. – 'marmelo' (Brazil)

Bungea C. Meyer. Scrophulariaceae. 2: SW As. (1), C As. & China (1)

bungwall *Blechnum indicum*

Bunias L. Cruciferae. 6 Med. (Eur. 3), As. Salad & fodder, e.g. *B. orientalis* L. (E Eur. & W As., adventive in C & W Eur.)

Buniella Schischkin = *Bunium*

Buniotrinia Stapf & Wettst. = *Ferula*

Bunium L. Umbelliferae (III 8). 48 Eur. (4) to N.Afr., (esp.) SW & C As. R: BMOIPB 93,1(1988)88. Seedlings with 1 cotyledon. *B. bulbocastanum* L. (Eur.) – tubers & lvs ed., form. much cult., medic.

bunya-bunya pine *Araucaria bidwillii*

Buphthalmum L. Compositae (Inul.). 2 Eur. (2), W As. Cult. orn. (ox-eye)

Bupleurum L. Umbelliferae (III 8). c. 180–190 Euras. (Eur. 39), N Afr., Canary Is., arctic N Am. (1), S Afr. (1). Shrubs & herbs with entire lvs., often parallel-veined (hare's ear). Some cult. orn. incl. *B. rotundifolium* L. (Eur., natur. N Am.) – thorow-wax, i.e. 'throw-wax' (through-grow) named because of perfoliate lvs

bur clover *Medicago* spp.; **b. cucumber** *Sicyos angulata*; **b. dock** *Arctium* spp.; **b. grass** *Cenchrus* spp.; **b. head** *Echinodorus* spp.; **b. marigold** *Bidens* spp.; **Noogoora b.** *Xanthium* spp.; **b. nut** *Tribulus* spp.; **b. oak** *Quercus macrocarpa*; **b. parsley** *Caucaulis platycarpos*; **b. reed** *Sparganium* spp.; **b. weed** *Medicago* spp., *Sparganium* spp., *Xanthium* spp. esp. *X. strumarium*

Burasaia Thouars (*Bourasaha*). Menispermaceae (III). 4 Madag. *B. madagascariensis* DC source of a bitter principle used in beer-making

Burbidgea Hook.f. Zingiberaceae. 5 Borneo. R: NRBGE 42(1985)262

Burchardia R. Br. Colchicaceae (Liliaceae s.l.). 5 SW, 1 ext. to E Aus. No colchicine!

Burchellia R. Br. Rubiaceae (II 1). 1 S Afr.: *B. bubalina* (L.f.) Sims, a shrub with v. hard wood (buffalo wood)

Burckella Pierre. Sapotaceae (II). 14 E Mal. to Tonga. *B. obovata* (Forst.f.) Pierre fr. ed. in Bismarcks

Burdachia C. Martius ex A. Juss. Malpighiaceae. 4 trop. S Am.

Burdekin plum *Pleiogynium timoriense*

burgan *Kunzea ericoides*

Burgesia F. Muell. (~ *Brachysema*). Leguminosae (III 24). 3 Aus.

Burgundy pitch from *Picea abies*

Burkartia Crisci. Compositae (Mut.-Nass.). 1 Patagonia: *B. lanigera* (Hook. & Arn.) Crisci

Burkea Hook. Leguminosae (I 1). 1 W & S Afr.: *B. africana* Hook. – source of a soluble gum & tanbark

Burkillanthus Swingle. Rutaceae. 1 W Mal.

Burkillia Ridley = *Burkilliodendron*

Burkilliodendron Sastry. Leguminosae (III 6). 1 Malay Pen.: *B. album* (Ridley) Sastry, known from 1 coll. from limestone in Perak, ? extinct

Burlemarxia Menezes & Semir (~ *Barbacenia*). Velloziaceae. 3 Brazil. R: T 40(1991)413

Burma padauk *Pterocarpus macrocarpus*

Burmabambus Keng f. = *Sinarundinaria*

Burmannia L. Burmanniaceae. 60 warm (Aus. 3)

Burmanniaceae Blume. Monocots – Liliidae – Liliales. 16/160 trop. N to Japan & E US, S to NZ. Small mycotrophs often with rhizomes or tubers & without chlorophyll (those with, sometimes with scalariform vessels). Lvs spiral, usu. scale-like & colourless to yellowish or reddish but sometimes green, simple. Fls terminal, 1 or in infls, bisexual, reg. to ± irreg., epigynous; P ((3 or)6), tubular or campanulate, outer valvate, inner smaller or 0, 3 (or 6) with elongate term. appendages, A 3(6) on tube, sessile or subsessile, s.t. connate by anthers forming tube round style & opening by lat. or longit. slits, connective often expanded, $\overline{G}$ (3), 1–3-loc., s.t. basally 3-, apically 1-loc., nectary glands on G or within, ovules ∞, anatropous, bitegmic, minute. Fr. a capsule, rarely fleshy, often winged, variously dehiscent; seeds minute with tiny embryo with as few as 4 cells when seed shed, endosperm ± 0. x = 6, 8

Principal genera: *Burmannia*, *Gymnosiphon*, *Thismia*

Burmeistera Karsten & Triana. Campanulaceae. c. 80 trop. S Am.

Burmese lacquer-tree *Gluta usitata*; **B. rosewood** *Pterocarpus indicus*

Burnatastrum Briq. = *Plectranthus*

Burnatia M. Micheli. Alismataceae. 1 trop. Afr.

burnet *Sanguisorba* spp. esp. *S. officinalis*; **b. rose** *Rosa spinosissima*; **b. saxifrage** *Pimpinella major*

Burnettia Lindley. Orchidaceae (IV 1). 1 SE Aus.: *B. cuneata* Lindley (lizard orchid)

burning bush *Bassia scoparia* 'f. *trichophylla*', *Dictamnus albus*

Burnsbaloghia Szlach. (~ *Schiedeella*). Orchidaceae. 1 Mex.: *B. diaphana* (Lindley) Szlach. R: FGP 26(1991)397

burr, bastard *Xanthium spinosum*; **Narrawa b.** *Solanum cinereum*; **Noogoora b.** *X. occidentale*

burrawang *Macrozamia spiralis*

Burretiodendron Rehder. Tiliaceae. 6 SW China, SE As. R: JAA 71(1990)375. *B. hsienmu* Chun & How (xianmu, SW China & N Vietnam) – valuable timber but vulnerable

Burretiokentia Pichi-Serm. Palmae (V 4m). 2 New Caled. R: Allertonia 3(1984)393

Burrielia DC = *Lasthenia*

burrograss *Scleropogon brevifolius*

Burroughsia Mold. = *Lippia*

burrowseed *Haplopappus tenuisectus*

Bursaria Cav. Pittosporaceae. 6 Aus. Usu. spiny, seeds 1 to several per fr. Cult. orn.

Bursera Jacq. ex L. Burseraceae. c. 50 trop. Am. Decid. aromatic polygamo-dioec. trees & shrubs, sources of resins, oils (linaloe, linaloa oil (Mex.)) & medic., esp. *B. simaruba* (L.) Sarg. (*B. gummifera*, WI birch, incense tree, S Florida, C Am., WI) – wood for fuel & matches, resin (American elemi, chibou) for varnish etc., used by Mayas for incense; *B. penicillata* (DC) Engl. (Indian lavender, Mex.) – much grown in India for ess. oil.

Burseraceae Kunth. Dicots – Rosidae – Sapindales. 17/540 trop. esp. Am. & NE Afr. Trees or shrubs with prominent resin ducts in bark. Lvs spiral, rarely opp., pinnate or trifoliolate, rarely unifoliolate; stipules rare. Fls in thyrses, rarely racemes or heads, bisexual or not, (trees often dioecious), reg., usu. hypogynous; K (3)4 or 5, usu. connate basally, C (3)4 or 5, rarely 0, both imbricate, rarely valvate; A 3–5 (+3–5), filaments rarely connate, without or within annular nectary-disk, anthers with longit. slits, staminodes often present in female fls. G ((2)3–5(–12 (*Beiselia*)) forming plurilocular ovary with terminal style, ovules pendulous on axile placenta, (1 or) 2 per locule, anatropous or hemitropous to campylotropous, bitegmic or rarely unitegmic. Fr. a drupe with 1–5 1-seeded stones or 1 stone with all seeds, rarely a 'pseudocapsule' opening to reveal 1-seeded nutlets; embryo oily with lobed or cleft cotyledons & almost 0 endosperm. x = 11, 13, 23

Classification & chief genera:

Protieae (drupe with 2–5 free or adhering but not fused parts): *Protium*, *Tetragastris*

Bursereae (drupe with endocarp completely fused, exocarp dehiscing by valves): *Boswellia*, *Bursera*, *Commiphora*

Canarieae (drupe with completely fused endocarp): *Canarium*, *Dacryodes*, *Haplolobus*, *Santiria*

Close to Anacardiaceae though the seeds show affinities with Laurales (Corner), i.e. Magnoliales; chemistry similar to Rutaceae & Meliaceae but limonoids prob. absent

Rain-forest as well as savanna trees: some timbers, ed. seeds & fr., incense & scents (incl. frankincense & myrrh), varnishes

Burseranthe Rizz. Euphorbiaceae. 1 Bahia

Burtonia R. Br. = *Gompholobium*

Burttdavya Hoyle. Rubiaceae (I 2). 1 trop. E Afr.

Burttia Baker f. & Exell. Connaraceae. 1 C Tanzania: *B. prunoides* Baker f. & Exell – seeds for poisoning animals. R: AUWP 89-6(1989)169

Buschia Ovcz. = *Ranunculus*

Buseria T. Durand = *Coffea*

bush flax *Astelia nervosa*, *Phormium tenax*; **b. lawyer** *Rubus australis*

Bussea Harms. Leguminosae (I 1). 6 trop. Afr., Madagascar

bussu palm *Manicaria saccifera*

Bustelma Fourn. = *Oxystelma*

Bustillosia Clos = *Asteriscium*

Busy Lizzie *Impatiens walleriana*

Butania Keng f. = *Yushania*

butcher's broom *Ruscus aculeatus*

Butea Roxb. ex Willd. Leguminosae (III 10). (Excl. *Meizotropis*, q.v.) 2 Indomal. *B. monosperma* (Lam.) Kuntze (*B. frondosa*, dhak, palas, flame of the forest, bastard teak, India to Burma) – gum (Bengal kino) astringent, seed-oil (muduga oil) vermifuge, fls give a red dye (tisso flowers), lvs stitched together as plates in restaurants in India, bark for cordage & sails, timber good under water also for charcoal; lac insects feed on it; it can be used to reclaim saline land & is one of the most beautiful of all flowering trees (fls bright orange-red), sacred to Brahmins in India

Buteraea Nees = *Strobilanthes*

Butia (Becc.) Becc. Palmae (V 5b). 8 S Am. R: Principes 23(1979)65. Cult. orn.

Butomaceae Mirb. Monocots – Alismatidae – Alismatales. 1/1 temp. Euras. Emergent aquatic perennial glab. herb with starchy rhizome. Lvs distichous, parallel-veined, linear, erect, ± triquetrous. Infl. axillary, cymose umbel with 3 bracts. Fls reg., bisexual, hypogynous; P 3 + 3, outer greenish, inner pink, A 3 prs + 3, anthers with longit. slits, pollen monosulcate, $\underline{G}$ 6 distally unclosed, connate at base in ring with short styles & nectaries basally, ovules ∞, over inner surface of carpels, anatropous, bitegmic. Fr. of separate follicles; embryo straight with single term. cotyledon & endosperm 0. x = 13 (small)

Only species: *Butomus umbellatus*, in many ways resembling Nymphaeaceae

Butomopsis Kunth (*Tenagocharis*). Alismataceae (Limnocharitaceae). 1 trop. OW: *B. latifolia* (D. Don) Kunth

Butomus L. Butomaceae. 1 temp. Euras. (natur. N Am.): *B. umbellatus* L. (flowering rush) –

cult. orn., rhiz. powdered for bread in N Euras.

butter and eggs *Linaria vulgaris*; **b.bur** *Petasites* spp.; **b.cup** *Ranunculus* spp.; **Bermuda b.cup** *Oxalis pes-caprae*; **dika b.** *Irvingia gabonensis*; **b.fruit** *Diospyros blancoi*; **illipe b.** *Shorea* spp.; **b.nut** *Juglans cinerea*; **shea b.** *Vitellaria paradoxa*; **b.wort** *Pinguicula* spp., esp. *P. vulgaris*

butterfly flower *Schizanthus* spp.; **b. lily** *Hedychium coronarium*; **b. orchid** *Platanthera* spp.; **b. pea** *Centrosema* spp., *Clitoria ternatea*; **b. tree** *Bauhinia* spp.; **b. weed** *Asclepias tuberosa*

button bush *Cephalanthus occidentalis*; **b. grass** *Gymnoschoenus sphaerocephalus*; **b. mangrove** *Conocarpus erectus*; **b.weed** *Spermacoce* spp.; **b.wood** *Platanus occidentalis*, *Conocarpus erectus*, *Laguncularia racemosa*

Buttonia McKen ex Benth. Scrophulariaceae. 3 trop. & S Afr.

Butumia G. Taylor. Podostemaceae. 1 Nigeria

Butyrospermum Kotschy = *Vitellaria*

Buxaceae Dumort. Dicots – Dilleniidae – Buxales. 4/70 nearly cosmop. Evergreen shrubs, trees or rarely herbs, often with steroid alks. Lvs opp., less often spiral, simple; stipules 0. Fls often in heads or spikes, small., reg., unisexual (plants monoec., rarely dioec.), P not petaloid, 2 + 2 (sometimes 5, or 3 + 3, 0 in pistillate fls of *Styloceras*), anthers with longit. slits, disk 0, $\underline{G}$ ((2)3(4)), primary locules divided into uniovulate cells in *Pachysandra* & *Styloceras*, ovules 2 per primary locule, pendulous, anatropous, bitegmic, often (at least) with an obdurator. Fr. a loculicidal dehiscent capsule, less often a drupe; seeds black, shiny, usually carunculate, embryo straight, endosperm firm, oily. x = 10, 14

Genera: *Buxus*, *Pachysandra*, *Sarcococca*, *Styloceras*

Here placed in Dilleniidae but seed anatomy suggests affinity with Hamamelidae, wood & stamens Pittosporaceae, & often put near Euphorbiaceae

Orn. & high quality timber

Buxiphyllum W.T. Wang & C.Z. Gao = *Paraboea*

Buxus L. Buxaceae. 50 W Eur. (2), Medit. to S Afr. (9 S of Sahara, R: KB 44(1989)296), temp. E As., WI (esp. Cuba 34), C Am. Monoec., terminal female surrounded by male fls. Fr. explosive, inner layer of pericarp separating from outer. Alks: lvs & seeds strongly purgative. Cult. orn. evergreen hedges withstanding pruning, topiary etc. esp. *B. sempervirens* L. (common or Eur. box, Abassian boxwood, Eur. & Medit.) – many cvs., wood hardest & heaviest of British trees, that used in rulers, musical instruments, croquet balls, inlay etc. since Anc. Egyptians (combs, flutes, furniture, lyres) & box balls still used for drawing lots for fixtures in F.A. Cup, now from Turkey, first used for wood engravings by Bewick c. 1800, now largely replaced by Venezuelan box, *Casearia praecox*; *B. macowanii* Oliv. (Cape b., E. London b.wood, E Cape, Natal); *B. microphylla* Siebold & Zucc. (Japanese b., ? E As.) – cult. orn. (many cvs) in Japan since 1450 but no wild pls found

Byblidaceae (Engl. & Gilg) Domin (Roridulaceae). Dicots – Dilleniidae – Ericales. 2/4 S Afr., Aus. Herbs & suffrutices with long-stalked oily or mucilaginous glands, also sessile ones in *Byblis*, the former trapping insects, the latter thought to digest them. Lvs spiral, sublinear; stipules 0. Fls solitary in axils (*Byblis*) or in term. racemes (*Roridula*), bisexual, hypogynous, reg.; K 5, imbricate, sometimes shortly connate basally, C 5, imbricate or convolute shortly connate basally, A 5 alt. with C, attached to C tube, anthers with term. pores or pore-like slits, $\underline{G}$ (2, *Byblis* or 3, *Roridula*) with simple style, ovules ± ∞ on axile placentas (*Byblis*) or 1–several per loc. & apical-axile (*Roridula*), anatropous, ? unitegmic. Fr. a loculicidal capsule; seeds with small straight embryo & copious endosperm

Genera: *Byblis*, *Roridula* – these differ so much as to be placed in separate families by some authors but they are closer to one another than to anything else (Cronquist) though together apparently closest to Pittosporaceae esp. *Cheiranthera*. Insectivory for *Roridula* is less firmly established than it is for *Byblis*. If the ovules are unitegmic an affinity with Nepenthales is a possibility (Cronquist)

Byblis Salisb. Byblidaceae. 2: N Aus. & New Guinea (1), SW Aus. (1). Fire-tolerant carnivores, *B. gigantea* Lindley (SW Aus.) dying back to tuber in summer (? fire necessary for germ.); capitate-glandular hairs do not move (cf. *Drosera*); ? buzz-poll., bees vibrating pollen out by flapping wings at right frequency

Byrsanthus Guillemin. Flacourtiaceae (5). 1 W Afr.

Byrsocarpus Schum. = *Rourea*

Byrsonima Rich. ex Kunth. Malpighiaceae. c. 130 trop. Am. Some fish-disp. in Amaz., some bird-disp. (unique in its subfam. but also in *Bunchosia* & *Malpighia*). Bark of some spp. for tanning, timber useful (surette), some ed. fr. (e.g. *B. crassifolia* (L.) Kunth (trop. Am. savannas) for icecream etc.) & cult. orn.

Byrsophyllum Hook.f. Rubiaceae (II 1). 2 India, Sri Lanka

Bystropogon L'Hérit. Labiatae (VIII 2). 10 Macaronesia

Bythophyton Hook.f. Scrophulariaceae. 1 Indomal., submerged aquatic

Byttneria Loefl. Sterculiaceae. 132 pantrop. R: Bonplandia 4(1976). Shrubs & lianes (*B. morii* L. Barnett & Dorr (Fr. Guiana) a tree), some myrmecophily

Byttneriaceae R. Br. = Sterculiaceae

by-yu *Macrozamia riedlei*

caa-ehe *Stevia rebaudiana*

caapi *Banisteriopsis caapi*

Caballeroa Font Quer = *Limoniastrum*

cabbage *Brassica oleracea* Capitata Group; **Chinese c.** *B. rapa* Chinensis Group & Pekinensis Group; **Kerguélen c.** *Pringlea antiscorbutica*; **Lundy c.** *Coincya wrightii*; **Moluccan c.** *Pisonia grandis* 'Alba'; **palm c.** bud of several palm spp., e.g. *Euterpe oleracea*, *Roystonea regia*, *Elaeis guineensis*, *Cocos nucifera*, eaten as salad fresh or tinned, its collection leading to death of tree; **Portuguese c.** *Brassica oleracea* Tronchuda Group; **c. rose** *Rosa × centifolia*; **Shantung c.** *Brassica rapa* Chinensis Group; **c. tree** name applied to many pachycaul trees with massive heads of lvs, e.g. spp. of *Anthocleista* & *Vernonia* (W Afr.), *Cussonia* (S Afr.), *Cordyline* (NZ), *Andira inermis* (WI)

cabelluda *Myrcia tomentosa*

Cabi Ducke = *Callaeum*

cabinet cherry *Prunus serotina*

Cabomba Aublet. Cabombaceae. 5 warm Am. Fanwort, fishgrass. R: NJB 11(1991)179. Aquarium pls popular as 'oxygenators'

Cabombaceae Rich. ex A. Rich. (~ Nymphaeaceae). Dicots – Magnoliidae – Nymphaeales. 2/6 trop. & warm temp. Aquatic herbs with rhiz. & elongate leafy stems; scattered vascular bundles & no cambium. Lvs dimorphic, in *Brasenia* spiral with floating elongate to peltate lamina, in *Cabomba* many or all opp. or whorled, submerged & deeply dissected; stipules 0. Fls solit., aerial & entomophilous, bisexual, hypogynous; K (2)3(4), C (2)3(4), A 3 or 6 (*Cabomba*), 18–36 (*Brasenia*) with slightly flattened filaments, staminodes 0, pollen usu. monosulcate, G (1)2–18 with terminal (*Cabomba*) or decurrent (*Brasenia*) styles, ovules (1)2 or 3, near the dorsal suture, anatropous, bitegmic (no aril). Fr. coriaceous, often dehiscent follicles; seeds small with little endosperm & copious perisperm, embryo small, dicotyledonous. 2n = 24, 80, 104

Genera: *Brasenia*, *Cabomba*

Differing from Nymphaeaceae in free carpels. S.t. referred to Ceratophyllaceae

Cabralea A. Juss. Meliaceae (I 6). 1 trop. Am. (variable): *C. canjerana* (Vell. Conc.) Mart. – imp. timber like *Cedrela*, sawdust a source of a red dye, bark used against fevers

Cabreriella Cuatrec. Compositae (Sen.-Tuss.). 2 Colombia. R: BSAB 19(1980)15

cabreuva oil *Myrocarpus frondosus*

Cabucala Pichon. Apocynaceae. 18 Madag. & Comoro Is. R: FMad. 169(1976)61. Some ed. fr. *C. striolata* Pichon alks have antiviral effect, esp. against influenza

cabuya fibre *Furcraea cabuya*

Cacabus Bernh. = *Exodeconus*

Cacalia L. OW spp. now referred to *Adenostyles* and *Parasenecio*, NW to *Arnoglossum*

Cacaliopsis A. Gray (~ *Senecio*). Compositae (Sen.-Tuss.). 1 W US

cacao = cocoa

Caccinia Savi. Boraginaceae. 6 W & C As.

cachana *Iostephane madrensis*

cachibou *Calathea lutea*

Cachrys L. Umbelliferae. (III 5). 3–4 Eur., N Afr.

Cacosmia Kunth. Compositae (Liab.). 3 Andes. R: BN 130(1977)279

Cactaceae Juss. Dicots – Caryophyllidae – Caryophyllales. 97/1400 NW, esp. hot & dry, OW trop. (*Rhipsalis* spp.). Xeromorphic trees or, most commonly, stem succulents, sometimes epiphytic, with crassulacean acid metabolism, accumulating organic acids & usually producing alks & always betalains; roots shallow, widespreading. Stem unbranched, columnar & sparsely branched or cushion-forming etc.; cuticle usually thick, shoots often photosynthetic for very many years, most often with spines, these in spiralled hairy areoles, those poss. representing axillary buds or short shoots with lvs or bud-scales replaced by spines often with tufts of short barbed irritant hairs (glochids); sieve-tubes with P-type plastids (globular protein crystalloid surrounded by ring of proteinaceous filaments). Lvs spirally arr., simple, succulent, or usu. very small & ephemeral or 0. Fls usu. solit. at areoles, rarely at branch-tips or in terminal cymes (*Pereskia*), often large & conspic., poll. by hummingbirds, bees, bats or hawkmoths, bisexual (rarely not), ± reg. P (∞), spirally arr. but not clearly divisible into K & C, united at base in an hypanthium, A ∞, centrifugally developed, spirally or in groups from hypanthium, anthers with longit. slits, nectary a ring within hypanthium, $\overline{G}$ (3–∞) but $\underline{G}$ weakly united in *Pereskia*, with single style & as many arms as G, ovules usu. ∞, basal in *Pereskia* (G partly partitioned) but usu. 3 or more parietal placentas in uniloc. G, campylotropous to rarely anatropous, bitegmic. Fr. usu. a berry, rarely dry-dehiscent; seeds (arillate in Opuntioideae) with usu. curved embryo & no true endosperm, perisperm present or 0, starchy. x = 11

Classification & principal genera:

I. **Pereskioideae** (lvs broad, glochids 0, fls term. or in term. cymes, seeds black, exarillate): *Maihuenia, Pereskia* (only)

II. **Opuntioideae** (lvs ± terete, deciduous, glochids present, seeds with pale bony aril or winged): *Opuntia, Pereskiopsis, Pterocactus, Quiabentia, Tacinga* (only)

III. **Cactoideae** (lvs 0 or v. small, glochids 0, seeds black or brown, exarillate): 9 tribes –

1. **Echinocereeae** (tree-like or shrubby, tube armed): *Echinocereus, Harrisia*
2. **Hylocereeae** (scandent or epiphytic, tube armed): *Disocactus, Epiphyllum, Hylocereus*
3. **Cereeae** (tree-like or shrubby, tube unarmed; R: Bradleya 7(1989)13): *Cereus, Melocactus, Pilosocereus*
4. **Trichocereeae** (tree-like, tube with numerous scales & hairy areoles): *Cleistocactus, Echinopsis, Gymnocalycium, Rebutia*
5. **Notocacteae** (like 4 but usu. small pls): *Copiapoa, Neoporteria, Parodia*
6. **Rhipsalideae** (epiphytic, stems usu. segmented): *Hatiora, Rhipsalis, Schlumbergera*
7. **Browningieae** (tree-like, v. spiny, fls lateral, usu. nocturnal): *Armatocereus*
8. **Pachycereeae** (poss. incl. 7): *Carnegiea, Cephalocereus*
9. **Cacteae** (pls usu. dwarf, usu. with many-ribbed, unjointed stems, fls diurnal): *Echinocactus, Ferocactus, Lophophora, Mammillaria*

The number of genera is perhaps still inflated by overfamiliarization in horticulture & overstressing of trivial features conspicuous therein

Xeromorphic features of thick cuticle & pachycaul structure of large volume to surface ratio, as well as weak secondary growth permitting extended retention of photosynthetic stem-surfaces, allow tolerance of extended periods of water-stress and diurnal temperature fluctuation. They are the NW stem-succulents, *Euphorbia* (unrelated) having similar role in Africa & Didiereaceae (same order) similar in Madag. The shallow root-system rapidly takes up whatever rainfall there is & CAM allows absorption of CO_2 at night for use in photosynthesis in day, both features 'pre-adapting' the group to epiphytism (cf. Bromeliaceae) while the spines retain air, condense dew & protect from grazing. Mainly animal-disp. esp. by birds, some by water or wind (*Pterocactus*). The Pereskioideae represent the archaic relics in the family, which is considered nearest to Phytolaccaceae (Cronquist)

Extremely commonly cult. as orn. potpls etc., others (esp. *Opuntia*) valued for fr. but others pestilential weeds in OW, where some used as hedges. The epiphytes incl. a group of cacti highly hybridized (cf. orchids) in cult. – 'epiphyllums', as well as *Rhipsalis*, whose OW distribution has been suspected as merely naturalized. Although typically of hot dry areas, the C. reach British Columbia & Patagonia & to 4000 m in the Andes

cactus Strictly a Cactacea but often ignorantly applied to succulent prickly plants in other families, e.g. *Agave*; **brittle c.** *Opuntia fragilis*; **Christmas c.** *Schlumbergera bridgesii*; **cholla c.** *Opuntia* spp.; **cochineal c.** *Opuntia* spp., *Nopalea cochenillifera*; **crab c.** *S. truncata*; **night-flowering c.** *Selenicereus grandiflorus*; **old man c.** *Cephalocereus senilis*

Cadaba Forssk. Capparidaceae. 30 OW trop. esp. Africa. Disk a tube; androphore & gynophore present. *C. farinosa* Forssk. (trop. Afr. to India) ed.

Cade, oil of *Juniperus oxycedrus*

Cadellia F. Muell. Surianaceae. 1 NE NSW: *C. pentastylis* F. Muell. Stipules

Cadetia Gaudich. Orchidaceae (V 15). 67 Papuasia. Cult. orn.

Cadia Forssk. Leguminosae (III 2). 6 Madag., 1 NE trop. Afr.: to S Arabia. R: ABN 19(1970)227. Adaxial C not always outermost

Cadiscus E. Meyer ex DC. Compositae (Sen.-Sen.). 1 S Afr.: *C. aquaticus* E. Meyer ex DC, aquatic

cadushi *Cereus repandus*

Caelospermum Blume (*Coelospermum*). Rubiaceae (IV 10). 7 SE As. to W Pacific. R: Blumea

33(1988)265

Caesalpinia L. Leguminosae (I 1). S.s. 15, s.l. (incl. *Mezoneuron*) c. 150 trop. (Mal. 21), warm Am., Namibia. Trees, shrubs or hook-climbers with extrafl. nectaries. Timbers (brasiletto, Brazil or Nicaragua wood, dyewoods) cult. orn. etc. *C. bonduc* (L.) Roxb. (*C. bonducella*, bonduc, nicker bean, trop.) – common drift seed on SW Br. coast etc. (seeds viable for at least 2½ years in seawater), used as beads; *C. coriaria* (Jacq.) Willd. (divi-divi, cascalote, trop. Am.) – pods with 40–45% tannin, giving light-coloured leather; *C. decapetala* (Roth) Alston (*C. sepiaria*, Mysore thorn, India) – spiny scrambler, used for hedging, noted as hallucinogenic in old Chinese herbals; *C. echinata* Lam. (Bahia wood, Braz. redwood, brazilwood, peach wood, Pernambuco wood, trop. Am.) – heartwood used for violin bows etc., also red dye source ('pau brasil', the origin of 'Brazil' because of amounts of dye exported thence to Portugal); *C. ferrea* C. Martius ex Tul. (leopard tree, E Brazil) – street-tree with spotty bark; *C. gilliesii* (Hook.) D. Dietr. (Arg., Uruguay) – cult. orn. with antitumour activity in seeds; *C. paraguariensis* (Parodi) Burkart (*C. melanocarpa*, S Am.) – lvs used for tanning (guayacán) – 21% tannin, forage with fr. available all year round; *C. pulcherrima* (L.) Sw. (Barbados pride, Paradise flower, peacock f., prob. native trop. As., now pantrop.) – cult. orn. medic. esp. laxative; *C. sappan* L. (sappanwood, Indomal.) – heartwood form. source red dye, cult. orn.; *C. spinosa* (Molina) Kuntze (deserts W S Am.) – pods an important tannin source in Peru; *C. violacea* (Miller) Standley (*C. brasiliensis*, C Am, WI) – cult. orn.

Caesalpiniaceae R. Br. = Leguminosae (I)

Caesaria Cambess. = *Viviania*

Caesia R. Br. Anthericaceae (Liliaceae s.l.). 11 S Afr., Aus. (8, 7 endemic) & New Guinea

Caesulia Roxb. Compositae (Asteroideae, tribe?). 1 NE India

cafta *Catha edulis*

caihuba *Virola surinamensis*

Cailliella Jacq.-Fél. Melastomataceae. 1 W Afr.

Caiophora C. Presl. Loasaceae. 65 S Am. Usu. with stinging hairs

Cairns satinwood *Dysoxylum pettigrewianum*

caja fruit *Spondias mombin*

Cajalbania Urban = *Poitea*

Cajanus DC. Leguminosae (III 10). (Incl. *Atylosia*) 37 OW trop. (Aus. 10) incl. 1 cultigen, *C. cajan* (L.) Millsp. (pigeon pea, Congo pea, arhar) apparently derived from arillate sp. in S As., where it may be crossed with local spp., though it has lost its aril, with large seeds & erect habit – widely cult for ed. seeds (dhal, red gram, catjang, ambrevade) & as cover crop, fuelwood, green manure, source of lac, silkworm foodpl. etc.

cajeput = cajuput

cajuado Brazilian drink made from cashew-apple

cajuput oil *Melaleuca cajuputi*

Cakile Miller. Cruciferae. 7 shores of Eur. (2), Medit., Arabia, Aus. (natur.), N Am. R: CGH 205(1974)1. Taproots form. powdered & mixed with other flour for bread by N Am. Indians – antiscorbutic famine-food; *C. maritima* Scop. – sea rocket

calaba *Calophyllum calaba*

Calabar bean *Physostigma venenosum*

calabash gourd *Lagenaria siceraria*; **c. nutmeg** *Monodora myristica*; **sweet c.** *Passiflora maliformis*; **tree c.** *Crescentia cujete*

calabrese *Brassica oleracea* Italica Group

calabura *Muntingia calabura*

Calacanthus T. Anderson ex Benth. Acanthaceae. 1 Indomal.

Caladenia R. Br. Orchidaceae (IV 1). 108 Aus. (104, 103 endemic), 1 ext. to Mal. & New Caled., NZ. Sexually attracted male thynnoid wasp-poll. in Aus. Nat. hybrids with *Glossodia* spp.; cult. orn.

Caladiopsis Engl. = *Chlorospatha*

Caladium Vent. Araceae (VII 1). 7 trop. S Am. R: Selb. 5(1981)367. Alks; some cyanogenic glycosides. Cult. orn. ('angel wings') for foliage, esp. cvs of *C. bicolor* (Aiton) Vent. (Brazil), sap irritant, cvs incl. '*C.* × *hortulanum* Birdsey', with almost all white lvs, 'vogue' foliage stove-pls of mid- to late C19; *C. lindenii* (André) Madison (*Xanthosoma l.*, Indian kale, C Panamá to Colombia) – varieg. potpl. pop. in US (var. *sylvestre* Grayum – wild form)

Calamagrostis Adans. Gramineae (21d). (Excl. *Deyeuxia*) c. 230 temp. (Eur. 14). R: FR 63(1960)229. Hybridity & apomixis common

calambac *Aquilaria malaccensis*

calamander wood *Diospyros quaesita*
calamint *Clinopodium* spp.; **common c.** *C. ascendens*; **wood c.** *C. menthifolia*
Calamintha Miller = *Clinopodium*
calamondin *Citrus × microcarpa*
Calamophyllum Schwantes = *Cylindrophyllum*
Calamovilfa (A. Gray) Hackel. Gramineae (31e). 4 N Am. R: Castanea 31(1966)145
Calamus L. Palmae (II 1d). c. 400 OW trop. (Afr. 1+) esp. Mal. (c. 280, Malaya 62). Usu. spiny climbers with long internodes, forming the canes of the rattan industry for furniture, basketry, mats, bridges etc., but some small, e.g. *C. minutus* Dransf. (Trengganu) only 50 cm tall. Often distal pinnae are backward-pointing spines, which act as grapples on surrounding vegetation with which the rattan grows up, the stem reaching 120 m (W Java) or may be up to 180 m, the rattans growing up again from the forest floor after the collapse of their supporting trees; rattan industry in Malaysia (worth $2 billion by 1977) based on *C. manan* Miq. (W Mal., endangered in wild) canes grow 3 m per season & *C. caesius* Blume (W Mal.) up to 4 m, 6 spp. used in India. *C. erinaceus* (Becc.) Dransf. (*C. aquatilis*, W Mal.) grows in mangrove or nearby; *C. muelleri* H.A. Wendl. & Drude (E Aus.) – chain-lengths used as standard measures by early Eur. surveyors; *C. pygmaeus* Becc. (Borneo) – infls root to form new pls; *C. scipionum* Lour. (W Mal.) – source of Malacca cane for walking-sticks, umbrella-handles etc. Some poss. fr. crops
calamus root *Acorus calamus*
Calanda Schumann. Rubiaceae (III 8). 1 trop. Afr.: *C. rubricaulis* Schumann. R: BJ 110(1989)546
Calandrinia Kunth. Portulacaceae. (Excl. *Baitaria*, *Cistanthe*, *Rumicastrum*). 60 Aus. (40), Am. Cult. orn. fleshy pls. Some ed.
Calandriniopsis Franz = *Baitaria*
Calanthe R. Br. Orchidaceae (V 11). c. 150 warm. Some with fls only openable by (large) euglossine bees. Cult. orn. All parts of pl. turn blue when damaged
Calanthea (DC) Miers (~ *Capparis*). Capparidaceae. 10 trop. Am.
Calantica Jaub. ex Tul. Flacourtiaceae (5). (Excl. *Bivinia*) 7 Madagascar
Calathea G. Meyer. Marantaceae. 300 trop. Am. Some ed. tubers, wax, cult. orn. foliage pls incl. *C. luciani* (Linden) N.E. Br. (trop. Am.) – lvs varieg., *C. splendida* (Lem.) N.E. Br. (Brazil) – lvs spotted yellow above, purple beneath, etc. Extrafl. nectaries; some e.g. *C. gymnocarpa* H. Kenn. (C Am.) with mass-flowerings; others e.g. *C. ovandensis* Matuda (Mex.) with extrafl. nectaries & with seeds dispersed by predatory *carnivorous* ants, which strip off arils – aril-less seeds grow best; in absence of ants, *Eurybia* sp. (Lepidopteran ant-tended herbivore) reduces seed prod. by 66% as opposed to 33%. *C. allouia* (Aublet) Lindley tubers (topee-tampo, topi-tamboo, topinambour) eaten like potatoes in WI; *C. lutea* (Aublet) G. Meyer (balasier, cachibou, WI) – lvs used for making & lining baskets, also promising source of wax (cauassú); *C. utilis* H. Kenn. (Ecuador) – thatch; *C. violacea* (Roscoe) Lindley – fl. buds & young shoots ed.
Calathiana Delarbre = *Gentiana*
Calathodes Hook.f. & Thomson. Ranunculaceae (I 1). 3 Himal. to Taiwan
Calathostelma Fourn. = *Ditassa*
Calatola Standley. Icacinaceae. 7 Mex. to Ecuador. *C. colombiana* Sleumer (Colombian Amaz.) – lvs chewed to blacken lips & teeth
Calcareoboea C.Y.Wu = *Platyadenia*
Calceolaria L. Scrophulariaceae. 388 trop. & S Am. Slipper flowers. Elaiophores attractive to bees. Some local med. but several cult. orn. esp. florists' potpl. (Herbeohybrida Group, *C. × herbeohybrida*, *C. × hybrida*), hybrids of *C. crenatiflora* Cav., *C. corymbosa* Ruíz & Pavón & *C. cana* Cav. (all Chile); some hardy. *C. andina* Benth. (Chile) – naphthoquinones effective against mites, aphids etc.
Calcitrapa Heister ex Fabr. = *Centaurea*
Calcitrapoides Fabr. = *Centaurea*
Caldcluvia D. Don. Cunoniaceae. 11 Chile (1), NZ (N Is., 1), trop. Aus. (2), Mal. (1), New Guinea (6), R: Blumea 25(1979)481. Aus. spp. good timber
Calderonella Söderstrom & H. Decker. Gramineae (24). 1 Panamá
Calderonia Standley = *Simira*
Caldesia Parl. (~ *Alisma*). Alismataceae. 4 OW trop. (1 ext. to Eur.)
Calea L. Compositae (Helia.-Mel.). c. 110 warm Am. *C. ternifolia* Kunth (*C. zacatechichi*) hallucinogen in Mex.; *C. urticifolia* (Miller) DC (C Am., weedy in OW) – medic. & intoxicant
Caleana R. Br. Orchidaceae (IV 1). (Incl. *Paracaleana*) 5 Aus. (5, 4 endemic), NZ. *C. major* R. Br. (duck orchid, E Aus.)

Calectasia R. Br. Xanthorrhoeaceae (Calectasiaceae). 2 S & W Aus.

Calectasiaceae Endl. = Xanthorrhoeaceae

Calendula L. Compositae (Cal.). c. 12 Medit. (Eur. 5) to Iran, Macaronesia. *C. officinalis* L. (common, pot or Scotch marigold, ruddles; origin unclear) – cult. orn. signifying 'grief' in Language of Fls, with many cvs, incl. 'Prolifera' with proliferated capitula (hen-and-chickens; also veg. shoots prod. from G bases of typical form in N India in spring), also medic. (fevers etc. in Portugal), effective against chilblains & warts, used in cosmetics & to colour butter & thicken soups, petals to garnish salads

Calepina Adans. Cruciferae. 1 Eur., Medit.: *C. irregularis* (Asso) Thell. R: JAA 66(1985)338

Calibanus Rose. Dracaenaceae (Agavaceae s.l.). 1 Mex.: *C. caespitosus* (Scheidw.) Rose with large tuber & few grass-like lvs

Calibrachoa Llave & Lex. = *Petunia*

× **Calicharis** Meerow. Hybrids between *Caliphruria* & *Eucharis* spp.

calico bush *Kalmia latifolia*

Calicorema Hook.f. Amaranthaceae (I 2). 2 trop. & S Afr.

Calicotome Link. Leguminosae (III 31). 2 Med. (inc. Eur.)

California allspice *Calycanthus occidentalis*; **C. buckeye** *Aesculus californica*; **C. laurel** *Umbellularia californica*; **C. lilac** *Ceanothus* spp.; **C. nutmeg** *Torreya californica*; **C. pepper** *Schinus molle*; **C. poppy** *Eschscholzia californica*; **C. redwood** *Sequoia sempervirens*

Caliphruria Herb. (~ *Eucharis*). Amaryllidaceae (Liliaceae s.l.). 4: W Colombia 3, Peru 1. R: AMBG 76(1989)212. Hybrids with *Eucharis* = × *Calicharis*

calisaya *Cinchona calisaya*

Calispepla Vved. Leguminosae (III). 1 C As.

Calla L. Araceae (IV). 1 N temp. bogs: *C. palustris* L., aquatic (water arum). P 0, fls biennially. Ed. starch from rhiz. (NB 'Calla' of florists is *Zantedeschia*)

Callaeum Small. Malpighiaceae. 10 trop. Am. R: SB 11(1986)335

Callerya Endl. (~ *Millettia*). Leguminosae (III 6). 19 SE & E As., Aus. R: Blumea 39(1994)1

Calliandra Benth. Leguminosae (II 5). (Excl. *Zapoteca*) 200 trop. Am., Madag., India. Pollen in polyads of 8 grains. Cult. orn. with heads of fls. esp. *C. tweedii* Benth. (*C.guildingii*, cunure, Brazil, natur. S US, WI) etc. *C. calothyrsus* Meissner (C Am.) – fuelcrop, coppices well in Indonesia, returning fertility to soil esp. in Java

Calliandropsis H. Hernández & Guinet (~ *Desmanthus*). Leguminosae (II 3). 1 Mex. (allied Madag. spp. = *Alantsilodendron*)

Callianthemoides Tamura. Ranunculaceae (II 3). 1 S Am.

Callianthemum C. Meyer. Ranunculaceae (II 1). 14 mts of Eur. (3) & C As. R: VKZGW 49(1899)316. Cult. orn. esp. *C. anemonoides* (Zahlbr.) Endl. (NE Alps)

Callicarpa L. Labiatae (I; Verbenaceae). c. 140 trop. & subtrop. Dioec. spp. on Bonin Is., females with non-germ. pollen perhaps poll. reward. Cult. orn. shrubs esp. *C. americana* L. (French mulberry, beautyberry, S N Am., WI), *C. bodinieri* A. Léveillé (China); others medic., fish-poisons etc.

Callicephalus C. Meyer (~ *Centaurea*). Compositae (Card.-Cent.). 1 SW to C As.

Callichilia Stapf. Apocynaceae. 7 trop. Afr. R: MLW 78–7(1978)1

Callichlamys Miq. Bignoniaceae. 1 trop. Am.

Callicoma Andrews. Cunoniaceae. 1 E Aus.: *C. serratifolia* Andr., – cult. orn., P 0.

Calligonum L. Polygonaceae (II 4). 80 Med. (Eur. 1). Shrubs used as sand-binders. *C. polygonoides* L. (W As. to India) – fls ed. on bread or cooked (India)

Callilepis DC. Compositae (Gnap.). 5 S Afr. R: OB 104(1991)46

Callipeltis Steven. Rubiaceae (IV 16). 3 Spain (1) & Egypt to Baluchistan

Callipteris Bory = *Diplazium*

Callirhoe Nutt. Malvaceae. 9 N Am. R: MNYBG 56(1990). Cult. orn., many with ed. roots, medic.

Callisia Loefl. Commelinaceae (II 1g). (Incl. *Cuthbertia*, *Hadrodemas*) 20 trop. Am. R: KB 41(1986)407, JAA 70(1989)117. Cult. orn. esp. *C. navicularis* (Ortgies) D. Hunt (*Tradescantia n.*, Mex.) with succ. lvs

Callistachys Vent. = *Oxylobium*

Callistemon R. Br. Myrtaceae (Lept.). 30 Aus. but app. merging with *Metrosideros* in New Caled. Bottlebrushes. Axis of infl. grows beyond head of fls to produce new lvs. Bird-poll. with conspic. stamens. Some rheophytes incl. *C. viminalis* (Gaertner) G. Don f. (NE Aus.) – oil anthelminthic *in vitro*, cult. orn. as are many spp. incl. *C. linearis* (Schrader & Wendl.) Sweet (E Aus.), *C. salignus* (Sm.) Sweet (S & E Aus.) etc.

Callistephus Cass. Compositae (Ast.-Ast.). 1 China: *C. chinensis* (L.) Nees, the aster or China a. of gardens (cult. China for 2000 yrs, introd. Eur. 1728) with fls from white & yellow to red, purple, & blue, the disk-florets often replaced by ray ones

Callisthene C. Martius. Vochysiaceae. 14 dry S & C Brazil, N Paraguay, E Bolivia. R: H.F. Martins, *O gênero C.* (1981)

Callistigma Dinter & Schwantes = *Mesembryanthemum*

Callistopteris Copel. = *Cephalomanes*

Callistylon Pittier = *Coursetia*

Callithauma Herb. = *Stenomesson*

Callitrichaceae Bercht. & J. Presl. Dicots – Asteridae – Callitrichales. 1/17 almost cosmop. Herbs, submerged aquatics with underwater poll. or amphibious or even terr. with aerial poll. Lvs decussate, to rosetted at branch-tips, usu. linear & bifid when submerged, linear to spatulate when floating or aerial; stipules 0. Fls unisexual, solit. in axils or rarely 1 male & 1 female per axil; P 0, A 1(2,3) in males, anthers with confluent longit. slits, G (2) in females with 2 elongate styles & each carpel divided into 2 locules, each with 1 pendulous, axile, anatropous, unitegmic ovule. Fr. dry, 4-lobed, each lobe winged or keeled, splitting into 4 1-seeded nutlets; seeds with oily endosperm. x = 3, 5+

Only genus: *Callitriche*

Callitriche L. Callitrichaceae. 17 almost cosmop. (Eur. 11 with 2 N Am. spp. natur.). Starworts. Some grown in aquaria incl. *C. stagnalis* Scop. (water s., Eur., Medit., Macaronesia), others v. sensitive to pollution & their performance can be used to predict presence of particular pollutants in S Germany

Callitris Vent. Cupressaceae. 14 Aus. R: Phytol. M 7(1984)16. Cypress pines. Lvs & cone-scales whorled; cones ripen in 1–2 yrs. Some good timbers & resins. *C. endlicheri* (Parl.) Bailey (*C. calcarata*, black cypress, SE Aus.) – source of Aus. sandarac used in varnishes; *C. rhomboidea* R. Br. ex A. Rich. & Rich. (Illawara pine, Oyster Bay p., Port Jackson p., SE Aus.), etc.

Callopsis Engl. Araceae (V 5). 1 Tanzania: *C. volkensii* Engl. – cult. orn.

Callostylis Blume = *Eria*

Calluna Salisb. Ericaceae. 1 Eur., As. Minor, natur. in N Am.: *C. vulgaris* (L.) Hull (common, white or Scottish heather, ling) – understorey shrub in woodland or more commonly maintained as major constituent of moors by burning & cutting in Scotland etc. for 'game' bird-shooting; poll. by bees (honey a major constituent of Drambuie) & other insects but in northerly sites wind-poll. or visited by thrips, *Ceratothrips ericae* (Haliday); baled shoots used in road-construction in New Forest etc., also for brooms; used like hops in Scotland in 'heather ale' for 4000 yrs (form. with *Filipendula ulmaria* & (to stop fermentation) *Osmunda regalis*), medic. & form. a yellow dye for wool; many cvs cult. orn.

Calocarpum Pierre = *Pouteria*

Calocedrus Kurz. Cupressaceae. 3 N Burma, SW China, Taiwan, Vietnam, Thailand, W N Am. *C. decurrens* (Torrey) Florin (incense cedar, W N Am.) – timber for shingles etc.; cult. orn. esp. fastigiate form in GB unknown in wild, though trees grown in Ireland & Italy broader (reason unknown)

Calocephalus R. Br. Compositae (Gnap.-Ang.). 11 temp. Aus. R: OB 104(1991)131. Cult. orn. See also *Leucophyta*

Calochilus R. Br. Orchidaceae (IV 1). 12 W Pacific (Aus. 11, 8 endemic). R: PLSNSW 71(1947)287. S Aus. spp. all self-poll.

Calochlaena (Maxon) R. White & M. Turner (~ *Culcita*). Dicksoniaceae (Thyrs.). 5 Philipp. & Java to Aus. R: AFJ 78(1988)86

Calochone Keay. Rubiaceae (II 1). 2 W Afr.

Calochortaceae Dumort.(~Liliaceae). Monocots – Liliidae – Liliales. 1/60 W N Am. Bulbous pls with few lvs & racemes of 3-merous bisexual fls. P 3 + 3, outer sepaloid, inner petaloid & clawed often fringed or ciliate on margins; A 3 + 3, anthers with longit. dehiscence; G3, 3-loc., style v. short, stigma 3-lobed. Capsules 3-angled or -winged, septicidal; seeds usu. flattened. x = 7-20

Only genus: *Calochortus*

Close to Liliaceae on DNA evidence but P distinctive. Perhaps linking with Colchicaceae

Calochortus Pursh. Calochortaceae (Liliaceae s.l.). c. 60 W N Am., British Columbia to Guatemala. Mariposa lilies. R: AMBG 27(1940)371. Bulbous pls, cult. orn., some ed. Fls white, yellow, red to purple, bluish or brownish

Calocrater Schumann. Apocynaceae. 1 Gabon

Calodecaryia J. Leroy. Meliaceae (I 1). 1–2 Madagascar

Calodendrum Thunb. Rutaceae. 2 E Afr., 1 ext. to Cape: *C. capense* (L.f.) Thunb. (Cape chestnut) – cult. orn. Seeds source of an oil for soap, timber useful

Calogyne R. Br. = *Goodenia*

Calolisianthus Gilg = *Irlbachia*

Calomeria Vent. *(Humea)*. Compositae (Gnap.-Ang.). 1 S Aus.: *C. amaranthoides* Vent. (*H. elegans)* – cult. orn. biennial (incense pl., Aus.), causing dermatitis in some. R: OB 104(1991)117

Caloncoba Gilg. Flacourtiaceae (3). 10 trop. Afr. R: BJ 94(1974)120. *C. echinata* (Oliver) Gilg source of gorli oil used like chaulmoogra oil in treatment of leprosy etc.

Calonyction Choisy = *Ipomoea*

Calopappus Meyen (~ *Nassauvia*). Compositae (Mut.). 1 Chilean Andes: *C. acerosus* Meyen. R: Cald. 15(1986)57. Shrub

Calophaca Fischer ex DC. Leguminosae (III 15). 5 Russia to C As. Cult. orn.

Calophanoides Ridley = *Justicia*

Calophyllum L. Guttiferae (III). 187 trop. (8 NW; most Indom.). R: JAA 61(1980)117. Timbers (bintangor) & oilseeds, cult. orn. *C. brasiliense* Cambess. (Santa Maria, galba, jacareuba, trop. Am.) – timber; *C. calaba* L. (calaba, Indomal.) – ed. fr., oilseed, timber; *C. inophyllum* L. (Alexandrian laurel, OW) – disp. by bats & sea, oilseed (domba oil, pinnay oil) med., illuminant, mixed with coconut oil to give Tongan Oil for massage, timber (Borneo mahogany) esp. for boats, bark medic., cult. orn.; *C. polyanthum* Wallich ex Choisy (India) – much exploited W Ghats for boat-masts & plywood; *C. tacamahaca* Willd. (Mauritius & Réunion) often confused with *C. inophyllum*

Calopogon R. Br. Orchidaceae (V 11). 4 N Am. Cult. orn.

Calopogonium Desv. Leguminosae (III 10). 6–8 NW trop., 1 (*C. mucunoides* Desv., widely grown ground-cover crop & green manure esp. under coconuts) extending to OW, isoflavonoid phytoalexins

Calopsis Pal. ex Desv. Restionaceae. 24 S Afr. R: Bothalia 15(1985)464

Calopteryx A.C. Sm. Ericaceae. 2 W trop. S Am.

Calopyxis Tul. Combretaceae. 23 Madagascar

Calorhabdos Benth. = *Veronicastrum*

Calorophus Lab. Restionaceae. 1 Tasmania

Caloscordum Herbert. (~ *Nothoscordum*). Alliaceae (Liliaceae s.l.). 1 C As. to China: *C. neriniflorum* Herbert. Infl. subtended by 1 (not 2 as in *N.*) bract

Calospatha Becc. Palmae (II 1d). 1 Malay Pen.: *C. scortechinii* Becc. Only palm genus restricted to Malay Pen.; thought extinct but rediscovered 1977, fr. ed.

Calostemma R. Br. Amaryllidaceae (Liliaceae s.l.). 1 E Aus.: *C. purpureum* R. Br. – P purple to yellow or white, alks, cult. orn.

Calostephane Benth. Compositae (Inul.). 6 trop. Afr., Madagascar

Calostigma Decne. = *Oxypetalum*

Calotesta Karis (~ *Metalasia*). Compositae (Gnap.-Rel.). 1 SW Cape: *C. alba* Karis. R: OB 104(1991)74

Calothamnus Lab. Myrtaceae (Lept.). 24 SW Aus. Like *Callistemon* with infl. axis continuing to produce lvs, but infls 1-sided. Cult. orn. esp. *C. quadrifidus* R. Br. – honeyeaters imp. pollinators & *C. sanguineus* Lab. (blood flower)

Calotis R. Br. Compositae (Ast.-Ast.). 26 SE As. (2), Aus. (24). R: PLSNSW 77(1952)146

Calotropis R. Br. Asclepiadaceae (III 1). 3 trop. & warm Afr. & India. R: NJB 11(1991)301. Bark fibre (madar, mudar), latex (alks) like gutta-percha, seed floss used like kapok. *C. gigantea* (L.) Aiton f. – multiple nectaries at petiole – lamina junction, lvs medic. & ed., fibre (wara, yercum), fls candied by Chinese in Java; *C. procera* (Aiton) Aiton f. now pantrop. (apple of Sodom) indicator of overgrazing in S Arabia, medic., poss. hydrocarbon source, roots used as chewsticks in Afr., strong fibre (akund fibre, French cotton), fine wood-ash form. used in gunpowder

Calpidochlamys Diels = *Trophis*

Calpocalyx Harms. Leguminosae (II 3). 11 W Afr. R: BMNHN 4,6(1974)297

Calpurnia E. Meyer. Leguminosae (III 26). 6 E Cape, Afromontane forests, S India. Pod winged. Cult. orn. esp. *C. aurea* (Aiton) Benth. (Afr.)

Caltha L. Ranunculaceae (I 1). 12 temp. (Eur. 1). R: Blumea 21(1973)119. Volatile protoanemonin irritant to skin & mucous membranes; alks. Cult. orn. esp. *C. palustris* L. (kingcups, marsh marigold, mayblob, N temp.) – molluscicidal saponins, fl. buds form. pickled like capers; *C. dionaeifolia* Hook.f. (S S Am.) & other S hemisph. spp. ±paired lobes to lvs, grows with *Drosera* & *Pinguicula* spp. but does not trap food; *C. introloba* F. Muell. (SE Aus.) – stabilizes soil in Snowy Mts, fls even under snow

caltrops *Tribulus terrestris*; **water c.** *Trapa natans*

calumba root *Jateorhiza palmata*; **false c. r.** *Coscinium fenestratum*

Caluera Dodson & Determan. Orchidaceae (V 10). 2 N S Am.

Calvaria Gaertner f. = *Sideroxylon*
Calvary clover *Medicago echinus*
Calvelia Moq. = *Suaeda*
Calvoa Hook.f. Melastomataceae. 18 trop. Afr. R: BMNHN Adans. 2(1981)123
Calycadenia DC (~ *Hemizonia*). Compositae (Hele.-Mad.). 11 W US
Calycacanthus Schumann. Acanthaceae. 1 New Guinea
Calycanthaceae Lindley. Dicots – Magnoliidae – Magnoliales. (Incl. Idiospermaceae) 4/10 trop. Aus., China, temp. N Am. Shrubs or small trees with aromatic bark; young stems with 4 inverted vasc. bundles in cortex. Lvs opp., simple, entire; stipules 0. Fls solit., term., bisexual, perigynous, beetle-poll. P 15–40, ± petaloid, spirally arr., A 5–30, spirally arr., ± ribbon-like with short or 0 filaments, connective extended beyond pollen-sacs with longit. slits, 10–15 staminodes, nectariferous, G 1–35 spirally arr. in hypanthial cup, each 1-ovulate, distal ovule abortive, proximal anatropous, bitegmic. Fr. of ∞ achenes in enlarged fleshy & oily, proteinaceous hypanthium; seeds endotestal, poisonous, cotyledons twisted, endosperm 0. x = 11, 12
Genera: *Calycanthus, Chimonanthus, Idiospermum, Sinocalycanthus*
Cult. orn., medic., spices & scent
Calycanthus L. Calycanthaceae. 2 SW & E US. R: Castanea 30(1965)63. Fls on leafy shoots (cf. *Chimonanthus*). Cult. orn.: *C. floridus* L. (Carolina allspice, E US) – bark form. medic., subs. for cinnamon; *C. occidentalis* Hook. & Arn. (Californian a., SW US)
Calycera Cav. Calyceraceae. c. 15 temp. S Am.
Calyceraceae R. Br. ex Rich. Dicots – Asteridae – Asterales. 6/55 C & S Am. Herbs with inulin. Lvs spirally arr., simple, entire to pinnately lobed; stipules 0. Fls bisexual, sometimes functionally unisexual, in centripetally developing involucrate heads, epigynous; K (4)5(6), C ((4)5(6)), reg. or not, lobes valvate, A same no. & alt. with lobes, attached near summit of tube, filaments ± connate, anthers opening by longit. slits & pollen released into anther-tube where pushed out by growth of style; $\overline{G}$ (2), uniloc., stigma capitate, ovule 1, pendulous, anatropous, unitegmic. Fr. achene-like with apical persistent K; seed with straight embryo & oily endosperm. x = 8, 15, 18, 21. R: NJB 12(1992)63
Genera: *Acarpha, Acicarpha, Boopis, Calycera, Gamocarpha, Moschopsis*
Lvs & inulin like Campanulaceae, filament-attachment, ovary & embryological features like Dipsacales, heads & pollen incl. presentation like Compositae though ovule basal there & chemistry different but cladistic analysis allies Calyceraceae with Compositae & Goodeniaceae. Cronquist suggests that Calyceraceae have an 'outmoded chemical arsenal' as defence & this might explain their lack of success compared with Compositae; most species occur in dry open vegetation
Calycobolus Willd. ex Schultes. Convolvulaceae. 30 trop. Afr. (25), Madag. (1), trop. Am. (4). R: BJBB 55(1985)29
Calycocarpum Nutt. ex Spach. Menispermaceae (III). R: JAA 45(1964)28. 1 E N Am.: *C. lyonii* (Pursh) A. Gray
Calycocorsus F.W. Schmidt = *Willemetia*
Calycogonium DC. Melastomataceae. 36 WI. Some timbers
Calycolpus Berg. Myrtaceae (Myrt.). 14 trop. Am.
Calycomis D. Don = *Acrophyllum*
Calycopeplus Planchon. Euphorbiaceae. 3 W Aus.
Calycophyllum DC. Rubiaceae (I 1). 6 trop. Am. 1 K-lobe often leaf-like. Some timbers esp. *C. candidissimum* (Vahl) DC (degami, dagame) – tool-handles & turnery, bows (lemonwood – US) & charcoal, *C. multiflorum* Griseb. (palo amarillo, Arg.)
Calycophysum Karsten & Triana. Cucurbitaceae. 5 NW trop. S Am.
Calycopteris Lam. Older name for *Getonia*
Calycorectes Berg. Myrtaceae (Myrt.). 17 trop. Am. (K)
Calycoseris A. Gray. Compositae (Lact.-Mal.). 2 SW N Am.
Calycosia A. Gray. Rubiaceae (IV 7). 5 Polynesia
Calycosiphonia Pierre ex Robbrecht. Rubiaceae (II 1). 2 trop. Afr.
Calycotropis Turcz. = ? *Polycarpaea*
Calyculogygas Krapov. Malvaceae. 1 Uruguay
Calydorea Herbert. Iridaceae (III 4). c. 10 temp. S Am.
Calylophus Spach. Onagraceae. 6 N Am. R: AMBG 64(1977)67. Close to *Oenothera*
Calymmanthera Schltr. Orchidaceae (V 16). 5 New Guinea
Calymmanthium F. Ritter. Cactaceae (III 7). 1 N Peru
Calymmatium O. Schulz. Cruciferae. 2 C As.
Calymmodon C. Presl. Grammitidaceae. 30 Sri Lanka to (esp. Borneo) Aus. & Polynesia. R:

PJS 34(1927)259

Calymmostachya Bremek. = *Justicia*

Calypso Salisb. Orchidaceae (V 8). 1 circumboreal, terrestrial: *C. bulbosa* (L.) Oakes – tubers form. ed. N Am., cult. orn.

Calyptocarpus Less. Compositae (Helia.-Verb.). 3 Texas to Guatemala

Calyptochloa C.E. Hubb. Gramineae (34b). 1 Queensland

Calyptostylis Arènes. Malpighiaceae. 1 Madagascar

Calyptraemalva Krapov. (*Calyptrimalva*). Malvaceae. 1 Brazil

Calyptranthes Sw. Myrtaceae (Myrt.). 130 trop. Am. Some fish-disp. in Amaz. Ed. fr.; lvs a spice. *C. luquillensis* Alain (Luquillo Mts, Puerto Rico) – 5 trees left

Calyptrella Naudin = *Graffenrieda*

Calyptridium Nutt. (~ *Cistanthe*). Portulacaceae. (Incl. *Spraguea*) 14 W N Am.

Calyptrimalva Krapov. = *Calyptraemalva*

Calyptrocalyx Blume (incl. *Paralinospadix*). Palmae (V 4j). 38 Moluccas (1), New Guinea

Calyptrocarya Nees. Cyperaceae. 8 trop. Am.

Calyptrochilum Kraenzlin. Orchidaceae (V 16). 10 trop. Afr.

Calyptrogenia Burret (*Neomitranthes*). Myrtaceae (Myrt.). 6 trop. Am.

Calyptrogyne H. Wendl. Palmae (V 6). 8 C Am. Some bat-poll. *C. ghiesbreghtiana* (Linden & H. Wendl.) H. Wendl. cult. orn. ± stemless

Calyptronoma Griseb. (~ *Calyptrogyne*). Palmae (V 6). 5 WI. Cult. orn.

Calyptrosciadium Rech.f. & Kuber. Umbelliferae (III 5). 1 Iran, Afghanistan

Calyptrotheca Gilg. Portulacaceae. 2 NE trop. Afr.

Calystegia R. Br. Convolvulaceae. c. 25 cosmop. (Eur. 3), bindweed. Some cult. orn., others pernicious weeds esp. *C. sepium* (L.) R. Br. – rhiz. ed. boiled in China, young lvs ed. India; *C. soldanella* (L.) Roemer & Schultes on seashores worldwide, seeds sea-disp. Differ from *Convolvulus* in solit. fls., usu. large bracts, fr. 1-loc. etc.

Calythropsis C. Gardner = *Calytrix*

Calytrix Lab. Myrtaceae (Lept.). 76 Aus. esp. SW. R: Brunonia 10(1987)1. Cult. orn. (fringe-myrtles) esp. *C. tetragona* Lab., heath-like

Camarea A. St-Hil. Malpighiaceae. 7 E S Am. R: Hoehnea 17,1(1990)1

Camarotea Scott-Elliot. Acanthaceae. 1 Madagascar

Camarotis Lindley = *Micropera*

camash *Camassia* spp., *Zigadenus* spp.; **common c.** *Camassia quamash*; **death c.** *Z. nuttallii*; **white c.** *Z. glaucus*

Camassia Lindley. Hyacinthaceae (Liliaceae s.l.). 5 N Am., 1 S Am. R: AMN 28(1942)712. Differs from *Scilla* in P-lobes 3-veined. Bulbs of N Am. spp. ed. esp. *C. quamash* (Pursh) E. Greene (*C. esculenta*), the camash or quamash of the Indians; others cult. orn. esp. *C. scilloides* (Raf.) Cory (E US)

Cambessedesia DC. Melastomataceae. 21 S Brazil

Camchaya Gagnepain. Compositae (Vern.-Vern.). 5 SE As. R: APG 23(1968)71

camel bush *Trichodesma zeylanicum*; **c.'s foot** *Bauhinia* spp.

Camelina Crantz. Cruciferae. 6–7 Eur. (4), Med. to C As. R: JAA 68(1987)234. *C. sativa* (L.) Crantz (gold-of-pleasure, false flax, Med. region) – domesticated in E or SE Eur., fibre, seeds for cagebirds, seed oil (cameline oil) like rape oil, imp. oil-crop in C & E Eur. until 1940s

Camelinopsis A. Miller. Cruciferae. 1 Iran, Iraq

Camellia L. Theaceae. c. 200 Indomal. (Mal. 1), E As. R: J.R. Sealy (1958) *A revision of the genus C.*, H.T. Chang & B. Bartholomew (1984) *Cs.* Tea, oilseeds, cult. orn. evergreens incl. hybrids, esp. *C. japonica* L. (Korea, Japan, Taiwan) – over 2000 named cvs with red, white or pink single or double (stamens petaloid) fls, seed oil (tsubaki oil) a hair-oil for Jap. women & cooking, lubricating & stamp-pad oil & soap in China; *C. oleifera* Abel (China) – source of comm. tea oil; *C. sasanqua* Thunb. (Japan) – oilseed lower quality than *C. japonica* & cult. orn.; *C. sinensis* (L.) Kuntze (tea, S & E As.) – long cult. by Chinese, now world's (& GB's: 180 000 t. per annum, i.e. more than N Am. & W Eur. together) most imp. caffeine drink (first in UK, in Holland 1610, public sale in London 1657 (Pepys's first cup 1660) but at £3.50 per lb. too dear for general drinking until C18 & in late C17 often adulterated with young lvs & shoots of *Uncaria gambier*), pls introd. to Java & India c. 1835 & later to Sri Lanka (thus responsible for transmigration of Indian Tamils thither); var. *assamica* (Masters) Kitam. with larger lvs native in wetter SE As., more suited to culture in Sri Lanka & Assam; trees pruned to table-top bushes & young shoots nipped off (finer qualities without expanded lvs), withered, rolled & fermented (except green tea, once thought to be a distinct sp.!), dried & sorted into grades, e.g. pekoe, souchong (black), gunpowder (green), s.t. compressed into bricks & form. transported by camel thus across C As. to Russia; stimulant due to alks (at least 6 incl. caffeine & theobromine), flavour s.t. added to with flower-petals (*Jasminum* etc.); seeds yield an oil & var. *sinensis* used as hedgepl. in US

Camelliaceae Dumort. = Theaceae

Camelostalix Pfitzer = *Pholidota*

Cameraria L. Apocynaceae. 6 trop. Am.

Camerunia (Pichon) Boit. = *Tabernaemontana*

camias *Averrhoa bilimbi*

Camissonia Link. Onagraceae. 62 Pacific Am. R: CUSNH 37(1969)161

Camoensia Welw. ex Benth. Leguminosae (III 2). 2 Gulf of Guinea. Lianes. *C. scandens* (Welw.) J.B. Gillett (*C. maxima*) cult. orn. with largest leguminous fl., to 20 cm across

camomile See chamomile

Campanea Decne = *Capanea*

Campanocalyx Valeton = *Keenania*

Campanula L. Campanulaceae. 300 N temp. esp. Medit. (Eur. 144; Turkey 95), trop. mts. Bellflowers: herbs, some hapaxanthic, rarely shrubby (e.g. *C. vidalii* H. Watson (*Azorina v.*, Azores)), many cult. orn. H.C. Crook (1951) *Campanulas*. Stamens dehisce in bud, depositing pollen on style-hairs; as fl. opens (but some cleistogamous e.g. *C. dimorphantha* Schweinf. (NE Afr., S & E As.) where seeds prod. more quickly in cleistogamous fls) A wither, except bases protecting nectar, style presenting pollen to insects; stigmas separate & receive external pollen, eventually curling right back & effecting self-poll. Fr. a capsule, if erect dehiscing distally, if pendent proximally, so that seeds escape only when pl. shaken, as by wind. *C. allionii* Villars (Alps) & *C. linifolia* Scop. (Alps) – constituents of Chartreuse; *C. rapunculus* L. (rampion, Euras.) – tap-root form. much eaten but old clones with parsnip-like roots app. lost by 1820s. Some almost ineradicable weeds, e.g. *C. rapunculoides* L. (Euras.) with brittle rhizomes; many beautiful garden plants incl.: *C. americana* L. (E N Am.) – poss. source of oil & rubber; *C. carpatica* Jacq. (Carpathians) – rock-pl.; *C. isophylla* Moretti (N Italy) – hanging baskets; *C. latifolia* L. (Euras.) – fls emetic; *C. medium* L. (Canterbury bell, S Eur.), 'Calycanthema' ('cup-and-saucer') with petaloid K forming saucer round C; *C. pyramidalis* L. (S C Eur.) – infls to 1.5 m; *C. robinsiae* Small (Florida) – aquatic annual thought extinct until 1982 when 2 populations found; *C. rotundifolia* L. (harebell, (Scottish) bluebell, N temp.) – polymorphic; *C. thyrsoides* L. (Alps) – 1 of only 2 spp. with yellow fls; *C. versicolor* Andrews (SE Eur.) – lvs ed. Greece

Campanulaceae Juss. Dicots – Asteridae – Asterales. 82/2000 cosmop. Mostly herbs but some shrubs & pachycaul trees with a network of laticifers in phloem & often medullary bundles or phloem, storing polysaccharides as inulin. Lvs simple, spiral, seldom opp. or whorled (*Ostrowskia*); stipules 0. Infls racemose to cymose (rarely epiphyllous – *Ruthiella*); fls bisexual (rarely not, e.g. dioec. spp. of *Lobelia*), epigynous to (rarely) perigynous (*Cyananthus*). K ((3–)5(–10)), imbricate or valvate with odd lobe posterior (appearing similar in Lobelioideae through resupination of fl. with anterior one), C ((3–)5(–10)), s.t. free, valvate ± reg. (Campanuloideae), irreg. in Lobelioideae with 3-lobed upper lip appearing as lower (resupination), other lip 2-lobed, A (3–)5(–10) connivent (Campanuloideae) or connate (Lobelioideae), forming anther-tube in which pollen shed & through which style with collecting hairs below initially adpressed stigmas grows, alt. with C, attached to annular epigynous nectary-disk or corolla-base, anthers separating after anthesis in Campanuloideae, pollen exine usually spinulose in Campanuloideae but reticulate in Lobelioideae, $\overline{G}$ (2–5) usu. (3) in Campanuloideae & (2) in Lobelioideae, rarely $\underline{G}$ 5 (*Cyananthus*), usu. with as many locules as G but in some Campanuloideae primary locules divided by partitions from carpellary midribs, in some Lobelioideae 1-loc. with 2 parietal placentas, ovules ∞ on axile, rarely parietal, placentas, anatropous, unitegmic. Fr. a berry or capsule, dehiscing variously; seeds small, s.t. winged, with straight embryo in oily endosperm (starchy in some *Wahlenbergia* (*Cephalostigma*)). x = 6–17

Classification & chief genera:

Campanuloideae (fls ± reg., anthers eventually usu. free): *Adenophora, Asyneuma, Campanula, Cyananthus, Phyteuma, Wahlenbergia*

Cyphioideae (fl. irreg., A sometimes united, anthers free): *Cyphia, Nemacladus* – all S Afr.

Lobelioideae (fl. irreg. usu.; anthers connate); *Burmeistera, Centropogon, Cyanea, Lobelia, Siphocampylus*

Some genera, e.g. *Lobelia*, show transition from fleshy to dry fr. & erect pachycaul treelets to creeping rhizomatous herbs with concomitant reduction in vessel-element length.

Woody spp. often considered 'anomalous' but occur widely scattered in the family (many endangered; 25% of Hawaiian endemics already extinct)
Many cult. orn., some ed. esp. *Adenophora, Phyteuma*

Campanulastrum Small = *Campanula*

Campanulorchis Brieger = *Eria*

Campanumoea Blume = *Codonopsis*

campeachy or **campeche wood** *Haematoxylum campechianum*

Campbellia Wight = *Christisonia*

Campecarpus H. Wendl. ex Becc. Palmae (V 4m). 1 New Caled.: *C. fulcitus* (Brongn.) Becc. R: Allertonia 3(1984)384

Campeiostachys Drobov = *Elymus*

Campelia Rich. = *Tradescantia*

Campestigma Pierre ex Costantin. Asclepiadaceae (III 4). 1 SE As.: *C. purpureum* Pierre ex Costantin

camphor *Cinnamomum camphora*; **baros, barus, Borneo** or **Sumatra c.** *Dryobalanops aromatica*; **E. Afr. c. wood** *Ocotea usambarensis*; **ngai c.** *Blumea balsamifera*; **c. plant** *Tanacetum balsamita*; **c. weed** *Pluchea camphorata*; **c. wood** *Tarchonanthus camphoratus*

Camphorosma L. Chenopodiaceae (I 4). 11 E Med. (Eur. 3), C As. *C. monspeliaca* L. shrubby, medic.

Campimia Ridley. Melastomataceae. 1 Malay Pen.: *C. wrayi* (King) Ridley. R: Willd. 17(1988)147

campion *Silene* spp.; **alpine c.** *S. suecica*; **bladder c.** *Silene vulgaris*; **moss c.** *S. acaulis*; **red c.** *S. dioica*; **rose c.** *S. coronaria*; **sea c.** *S. uniflora*; **white c.** *S. latifolia*

Campnosperma Thwaites. Anacardiaceae (IV). 10 trop. Am., Madag., Seychelles, Sri Lanka, SE As., Mal., W Pacific. Often forming monospecific stands in swamps. Timber for boxes, canoes etc. *C. coriaceum* (Jack) Steenis (Mal.) & *C. brevipetiolatum* Volkens (E Mal., Pacific) source of parasiticidal tigasco skin oil in Papua

Campomanesia Ruíz & Pavón. Myrtaceae (Myrt.). 80 S Am. Ed. fr. (Pará guava) esp. *C. guaviroba* (Berg) Kiaerskov (guabiroba, Brazil), some fish-disp. in Amaz.

Campovassouria R. King & H. Robinson (~ *Eupatorium*). Compositae (Eup.-Dis.). 1 Brazil: *C. bupleurifolia* (DC) R. King & H. Robinson. R: MSBMBG 22(1987)79

Campsiandra Benth. Leguminosae (I 1). 3 trop. Am. esp. Amazonia. Medic. *C. laurifolia* Benth. water-disp. (aerenchyma in seed-coat)

Campsidium Seemann. Bignoniaceae. 1 Chile, Arg.: *C. valdivianum* (Philippi) Bull – cult. orn.

Campsis Lour. Bignoniaceae. 2 E As. (*C. grandiflora* (Thunb.) Schumann, Chinese trumpet-flower medic. since ancient times; hexose-rich nectar for perching bird-poll.), E US (*C. radicans* (L.) Bureau, trumpet climber, creeper or vine; sucrose-rich nectar for humming-bird-poll.). Adventitious root-climbers like *Hedera*; extrafl. nectaries, attracted ants reducing herbivore attack. Cult. orn. incl. their hybrid *C.* × *tagliabuana* (Vis.) Rehder (nectar hexose-dominant!)

Camptacra N. Burb. (~ *Vittadinia*). Compositae (Ast.-Ast.). 2 New Guinea, trop. Aus. R: Brunonia 5(1982)11

Camptandra Ridley. Zingiberaceae. 3 W Mal.

Camptocarpus Decne. Asclepiadaceae (I). 11 Madag., Masc.

Camptodium Fée = *Tectaria*

Camptolepis Radlk. Sapindaceae. 1 E & NE Afr.

Camptoloma Benth. (~ *Sutera*). Scrophulariaceae (Manuleeae). 3 Canary Is., Somalia, S Yemen, Namibia. R: O.M. Hilliard, *The M.* (1994)80

Camptopus Hook.f. = *Psychotria*

Camptorrhiza Hutch. Colchicaceae (Liliaceae s.l.). 1 S Afr.: *C. strumosa* (Baker) Oberm. Alks

Camptosema Hook. & Arn. Leguminosae (III 10). 12 S Am.

Camptosorus Link = *Asplenium*

Camptostemon Masters. Bombacaceae. 2 C Mal., N Aus. A (∞)

Camptostylus Gilg. (*Cerolepis*). Flacourtiaceae (2). 3 trop. W & C Afr. R: BJ 94(1974)283

Camptotheca Decne. Cornaceae (Nyssaceae). 1 S & SE As.: *C. acuminata* Decne – cult. orn.

Campuloclinium DC (~ *Eupatorium*). Compositae (Eup.-Gyp.). 14 trop. Am. R: MSBMBG 22(1987)126

Campylandra Baker = *Tupistra*

Campylanthus Roth. Scrophulariaceae. 11 Macaronesia, NE Afr., Persian Gulf. R: NRBGE 38(1980)373, 40(1982)331, 45(1988)73. *C. salsoloides* (L.f.) Roth (Canary Is.) – cult. orn.

Campylocentrum Benth. Orchidaceae (V 16). 55 trop. Am. Some leafless

Campyloneurum C. Presl (~ *Polypodium*). Polypodiaceae (II 5). 20 trop. & warm Am. esp. Andes

Campylopetalum Forman. Anacardiaceae (V, Podoaceae). 1 Thailand: *C. siamense* Forman with epiphyllous female fls

Campylosiphon Benth. Burmanniaceae. 1 trop. S Am.

Campylospermum Tieghem = *Ouratea*

Campylostachys Kunth. Stilbaceae. 2 SW Cape

Campylostemon Welw. Celastraceae. 8+ trop. Afr. Links Celast. (s.s.) with form. segregated Hippocrateaceae

Campylotheca Cass. = *Bidens*

Campylotropis Bunge. Leguminosae (III 9). 65 As. Close to *Lespedeza*

Campynema Labill. Campynemataceae (Liliaceae s.l.). 1 Tasmania: *C. lineare* Lab. R: BMNHN4,8(1986)129. Leaf 1. Referred at diff. times to Amaryllidaceae, Colchicaceae, Hypoxidaceae, Iridaceae, Liliaceae or Melanthiaceae. See also *Campynemanthe*

Campynemataceae Dumort. (~ Melanthiaceae, Iridaceae). Monocots – Liliidae – Liliales. 2/4 SW Pacific. Rhiz. herbs with linear lvs and panicles of or solit. 6-merous fls. P 3 + 3; A 3 + 3 at base of P; $\underline{G}$ or half so, 1- or 3-loc. with parietal placentae & ∞ anatropous ovules. Fr. a 6-ribbed capsule with irreg. seeds.

Genera: *Campynema, Campynemanthe*

The fam. links Iridaceae & Melanthiaceae, showing how subtle the distinctions between such fams are

Campynemanthe Baillon. Campynemataceae (Liliaceae s.l.). 3 New Caled. R: BMNHN 4,8(1986)121

Camusiella Bosser = *Setaria*

camwood *Baphia nitida, Pterocarpus soyauxii*

Canaca Guillaumin = *Austrobuxus*

Canacomyrica Guillaumin. Myricaceae. 1 New Caled.: *C. monticola* Guillaumin, on ultramafics

Canacorchis Guillaumin = *Bulbophyllum*

Canada balsam *Abies balsamea*; **C. bluegrass** *Poa compressa*; **C. garlic** *Allium canadense*; **C. hemlock** *Tsuga canadensis*; **C. pitch** *Abies balsamea*

canaigre *Rumex hymenosepalus*

Cananga (DC) Hook.f. & Thomson. Annonaceae. 2 trop. As. to Aus. *C. odorata* (Lam.) Hook.f. & Thomson – cult. orn. & in Madag., Comoro Is. & Réunion, Philippines etc., fls the source of ylang-ylang or cananga oil (mature tree giving 9 kg fresh fls yielding 30 g oil per annum), a hair-dressing & constituent of Chanel 'No. 5', Revlon's 'Charlie' etc. (often mixed with pimento oil), P chewed with betel in Sri Lanka

Canarina L. Campanulaceae. 3 Canary Is. (1), trop. E Afr. (2). Cult. orn. with tubers; like *Campanula* but fls 6-merous & fr. a berry (ed.), lvs opp. or ternate & C yellow to red. *C. canariensis* (L.) Kuntze (Canary Is.) – bird-poll. by 'insectivorous' birds (no sunbirds in Canary Is.!)

Canarium L. Burseraceae. 77 trop. Afr. (2), Indomal. R: Blumea 9 (1959)275). Timbers (kedondong), resins & ed. seeds. (oily 'pili nuts') esp. SE As. e.g. *C. indicum* L. (ngali nut, Indomal.) – also source of Solomon nut oil used in tanning lotions & other cosmetics, *C. luzonicum* (Blume) A. Gray (Java almond, also source of Manila elemi for varnishes etc.) & fr. e.g. *C. album* (Lour.) Räusch. & *C. pimela* Leenh. (Chinese olives), those of *C. harveyi* Seemann widely sold in Polynesian markets. *C. euphyllum* Kurz (dhup, Indian white mahogany, Andamans); *C. schweinfurthii* Engl. (abé, abel, aielé, incense tree, trop. Afr.) – timber stained as mahogany subs., church incense in Uganda, fr. sold in markets so nat. distrib. unclear

Canary balm *Cedronella canariensis*; **c. creeper** *Tropaeolum peregrinum*; **c. grass** *Phalaris canariense*; **C. ivy** *Hedera algeriensis*; **C. pine** *Pinus canariensis*; **c. whitewood** *Liriodendron tulipifera*; **c. wood** *Morinda citrifolia*

Canavalia DC. Leguminosae (III 10). 51 trop. (Hawaii 6 endemic) esp. Am. R: Brittonia 16(1964)106. Alks. Green manure, stock feed & beans esp. *C. ensiformis* (L.) DC (jack bean, sword bean, sabre bean, Jamaican horse bean, trop. Am.) – young pods ed. though unripe seeds considered toxic; *C. gladiata* (Jacq.) DC (sword bean, poss. derived from *C. cathartica* Thouars, OW), similar; *C. rosea* (Sw.) DC (*C. maritima*) – common pantrop. beach pl.

Canbya Parry ex A. Gray. Papaveraceae (IV). 2 W N Am. Persistent C

Cancrinia Karelin & Kir. Compositae (Anth.-Can.). 30 C As. to Afghanistan

Cancriniella Tzvelev. Compositae (Anth.-Can.). 1 C As.

candelilla (wax) *Euphorbia antisyphilitica*

candied peel *Citrus medica* & other *C.* spp.

candle berry *Myrica cerifera, Aleurites moluccana;* **c. nut.** *A. moluccana;* **c. plant** *Senecio articulatus;* **c. tree** *Parmentiera cereifera;* **c. wood** *Amyris balsamifera*

Candollea Lab. = *Stylidium*

Candolleodendron R. Cowan. Leguminosae (III 1). 1 NE S Am.

Candy carrot *Athamanta cretensis;* **c. tuft** *Iberis* spp.

cane see rattan; **dumb c.** *Dieffenbachia* spp.; **Malacca c.** *Calamus scipionum* etc.; **rajah c.** *Eugeissona minor;* **Spanish c.** *Arundo donax;* **switch** or **giant c.** *Arundinaria gigantea;* **Tongking c.** *A. amabilis*

Canella P. Browne. Canellaceae. 1–2 S Florida, WI. *C. winterana* (L.) Gaertner – source of canella bark used as condiment & medic. (tonic & stimulant), a fish poison (Puerto Rico) & for flavouring tobacco; timber (Bahama whitewood)

Canellaceae Mart. Dicots – Magnoliidae – Magnoliales (? Illiciales). 5/13 trop. Afr. (E) & Am. Glabrous usu. aromatic trees (or shrubs), the oil-cells with terpenes in the parenchyma. Lvs spirally arr., simple, entire, often with pellucid dots; stipules 0. Fls bisexual, reg., solit. or in terminal or axillary racemes to panicles, or solit.; K (considered bracts by some) 3, leathery, imbricate, C (considered K by some) (4)5–12 in 1 or 2(–4) whorls and/or spirals (outer whorl sometimes considered K), imbricate, ± connate in *Canella* & *Cinnamosma*, A (6–12 (to 35 or 40 in *Cinnamodendron*)) forming a tube with extrorse anthers without, pollen monosulcate, $\underline{G}$ (2–6), 1-loc. with thick style & 2–6-lobed stigma, placentas parietal, 2–6 with 1 or 2 rows of 2–∞ ovules each, ovules hemitropous, bitegmic. Fr. a berry with 2 or more seeds with oily endosperm (ruminate in *Cinnamosma*). 2n = 22, 26, 28

Genera: *Canella, Cinnamodendron, Cinnamosma, Pleodendron, Warburgia*

Perhaps closest to Myristicaceae, the 2 from ancestors more like modern Magnoliaceae (Cronquist), though seeds like Winteraceae (Corner)

Some condiments & medic.

Canephora Juss. Rubiaceae (II 5). 5 Madagascar

canihua *Chenopodium pallidicaule*

canistel *Pouteria campechiana*

Canistrum C.J. Morren. Bromeliaceae (3). 10 E Brazil. R: FN 14(1979)1715. Some cult. orn. esp. *C. fragrans* (Linden) Mabb. (*C. lindenii*)

Canizaresia Britton = *Piscidia*

cankerberry *Coptis groenlandica*

Canna L. Cannaceae. 8–10 trop. Am. Cult. orn. esp. forms of *C. indica* L. (Indian shot, natur. throughout trop.) & commonly *C. × generalis* L. Bailey (fls yellow), '*C. × orchiodes* L. Bailey' (fls yellow & red) & other hybrids involving *C. flaccida* Salisb. (S US to Panamá); *C. edulis* Ker-Gawler (achira, Queensland arrowroot, tous-les-mois) prob. a form of *C. indica* selected for starchy ed. rhiz., the starch grains being suited to infants & invalids (cf. *Maranta*). Black seeds (some viable for c. 600 yrs) used as beads

Cannabaceae Martinov (~Urticaceae). Dicots – Dilleniidae – Urticales. 2/4 N temp. to SE As. Erect or twining (*Humulus*) dioec., rarely monoec., wind-poll. herbs with pyridine alks.; laticifers 0. Lvs opp., at least below, palmate(ly lobed, *Humulus*) with glandular hairs; stipules free, persistent. Infls basically cymose, females smaller, male fls – K 5, A 5 opp. K all in a spiral, female fls – K tubular (merely a ring in cultivars of *Cannabis*) round $\underline{G}$ (2), 1-loc. with 2 dry stigmas & 1 ovule, subapical, anatropous, bitegmic. Fr. an achene covered in K; seeds with curved or coiled (*Humulus*) embryo in oily endosperm. x = 8 (*Humulus*), 10 (*Cannabis*)

Genera: *Cannabis, Humulus*

Close to Moraceae

A fam. causing much human happiness (& misery)

Cannabidaceae See Cannabaceae

Cannabis L. Cannabaceae. 1 C As.: *C. sativa* L., annual to 8(–12) m, dioec. with sex chromosomes, though sex modifiable by environmental factors; v. variable & orig. wild material cannot be certainly identified but in cult. 2 races, subsp. *sativa*, more northerly cult. (in China for 4500 yrs, obligatory crop in Eliz. times in GB, where illegal since 1951) for fibre (hemp used for ropes, fibre-board, paper etc. since 4000 BC esp. in N & NE China where form. only fibre available, prob. used in first paper (AD 105) there) & subsp. *indica* (Lam.) E. Small & Cronq., more southerly cult. principally for psychotropic drugs (marijuana, marihuana (Mex.), pot (US, where allegedly the biggest cash crop worth $32 billion), dagga (S Afr.), kif (Morocco)), cannabis resin, which exudes from the glandular hairs & is used like opium (effects described 2736 BC by Chinese Emperor Shen Neng). In India, 3 common forms: ganja (dried unripe infrs), charas or churras (resin knocked off twigs, bark etc.) & bhang (largely mature lvs of wild pl.). Smoked ('weed') with or without tobacco ('skunk') in cigarettes ('joints') or taken as an intoxicating liquid formed from it (hashish; Arabic for 'hashishtaker' = root of word 'assassin'), in food or drink (e.g. in comm. beers in Netherlands) it has a stimulating & pleasantly exciting effect, relief from muscular sclerosis, cerebral palsy & glaucoma, though addictive & in excess can cause delirium & 'moral weakness and depravity' (Uphof). Seeds the source of hemp seed oil used in varnishes, food, soap, lip balm & fuel in Nazi tanks etc. & used as birdseed & to attract fishes

Cannaceae Juss. Monocots – Zingiberidae – Zingiberales. 1/8–10 trop. & warm Am. Glabrous herbs with starchy rhizomes. Lvs spirally arr. with a sheath passing into petiole, simple lamina with prominent midrib & numerous lateral veins but no ligule nor pulvinus. Infl. terminal often with short 2-flowered cymules axillary to principal bracts. Fls large, bisexual, obliquely orientated so that no organ is clearly median, K 3, spiral, persistent in fr., C 3 (1 smaller than others) joined in a basal tube with functional A 1 (middle of inner A cycle), petaloid with pollen-sac along more nearly median edge, & at least 1 staminode with (1)2(–4) additional ones, $\overline{G}$ (3), 3-loc. with axile placentation, style petaloid, ovules ± ∞ in each loc., anatropous, bitegmic. Fr. a caps. (? sometimes indehiscent); seed exarillate with straight embryo in thin starchy endosperm & copious hard, starchy perisperm. x = 9

Only genus: *Canna*

Petaloid staminode sometimes called labellum but not homologous with that in Zingiberaceae (Cronquist). Pollen shed on style in bud; insects land on staminode brushing stigma with foreign pollen & then the pollen already on style

Cannomois Pal. ex Desv. Restionaceae. 7 S & SW Cape. R: Bothalia 15(1985)480

cannonball tree *Couroupita guianensis*

Canola oil *Brassica napus* 'Canola'

Canotia Torrey. Celastraceae (Canotiaceae). 2 SW US. Leafless, poss. allied to *Acanthothamnus*. R: Brittonia 27(1975)119

Canotiaceae Airy Shaw = Celastraceae

Canscora Lam. Gentianaceae. 30 OW trop. Some medic.

Cansjera Juss. Opiliaceae. 3 Indomal. to Aus. R: Willdenowia 9(1979)43

cantala fibre *Agave cantala*

cantaloupe *Cucumis melo*

Canterbury bell *Campanula medium*

Canthium Lam. (*Plectronia*). Rubiaceae (III 2). 50 Afr., India (incl. Sri Lanka) 2. Some fine timbers & ed. fr.; some Afr. spp. with hollow ant-infested twigs. See also *Keetia*, *Multidentia*, *Psydrax*, *Pyrostria*

Cantleya Ridley. Icacinaceae. 1 Mal.: *C. corniculata* (Becc.) R. Howard – fragrant timber (sandalwood subs.), ed. fr.

Canton fibre *Musa* hybrid (*M. textilis* × ?); **C. linen** *Boehmeria nivea*

Cantua Lam. Polemoniaceae. 6 Andes. Shrubs & trees, some cult. orn.

Capanea Decne ex Planchon (*Campanea*). Gesneriaceae. 6 trop. Am. Herbs, shrubs or lianes, some cult. orn. & bat-poll. (fls 5 cm diam., green with purple spots)

Capanemia Barb. Rodr. Orchidaceae (V 10). 14 Brazil. R: Orquideologia 7(1972)215

caparrosa *Neea theifera*

Capassa Klotzsch = *Philenoptera*

Cape ash *Ekebergia capensis*; **C. asparagus** *Aponogeton distachyos*; **C. aster** *Felicia* spp.; **C. beech** *Myrsine melanophloeos*; **C. box** *Buxus macowanii*; **C. b.wood** *Gonioma kamassi*; **C. chestnut** *Calodendrum capense*; **C. cowslip** *Lachenalia aloides*; **C. ebony** *Euclea pseudebenus*, *Heywoodia lucens*; **C. figwort** *Phygelius capensis*; **C. gooseberry** *Physalis peruviana*; **C. honeysuckle** *Tecoma capensis*; **C. jasmine** *Gardenia augusta*; **C. lilac** (Aus.) *Melia azedarach*; **C. lily** *Crinum* spp.; **C. mahogany** *Trichilia emetica*; **C. pondweed** *Aponogeton distachyos*; **C. primrose** *Streptocarpus* spp.; **C. shamrock** *Oxalis* spp.; **C. spinach** *Emex australis*; **C. thorn** *Ziziphus mucronata*; **C.weed** *Arctotheca calendula*; **C. willow** *Salix mucronata*

caper *Capparis spinosa*; **c. spurge** *Euphorbia lathyris*

Caperonia A. St-Hil. Euphorbiaceae. 40 trop. Am., Afr. (6), Madagascar

Capillipedium Stapf (~ *Dichanthium*). Gramineae (40c). 14 E Afr., trop. As. to Aus. & New Caled. (1)

Capirona Spruce. Rubiaceae (I 1). 5 S Am. K like *Mussaenda*

Capitanopsis S. Moore. Labiatae (VIII 4C). 2 Madagascar

Capitanya Schweinf. ex Guerke. Labiatae (VIII 4c). 1 E Afr. K in fr. gives 'sail' for wind-disp.

Capitularia Valcken. = *Capitularina*

Capitularina Kern (~ *Chorizandra*). Cyperaceae. 2 Papuasia. 5-angled stems

Capnoides Miller (~ *Corydalis*). Fumariaceae (I 1). 1 N N Am.: *C. sempervirens* (L.) Borkh., natur. Norway, cult. orn. R: OB 88(1986)19

Capnophyllum Gaertner. Umbelliferae (III 10). 1 S Afr.: *C. africanum* (L.) Gaertner – natur. Aus. R: EJB 48(1991)189. See also *Krubera*

Capparaceae = Capparidaceae

Capparidaceae Juss. (Cruciferae s.l.; excl. Pentadiplandraceae & Physenaceae). Dicots – Dilleniidae – Capparidales. 39/650 warm, few temp. arid. Shrubs, herbs or rarely trees producing mustard-oil glucosides & (Capparidoideae) alks, s.t. with unusual sec. thickening of concentric type, roots s.t. with endotrophic mycorrhizae. Lvs spiral, rarely opp., simple, trifoliolate or usu. palmate (scale-like in *Koeberlinia*); stipules 0 or small, often glands or spiny. Fls usu. in racemes, rarely solit. & axillary, usu. bisexual, ± irreg., receptacle usu. prolonged into gynophore/androgynophore, K (2–)4(–6), often decussate, s.t. basally connate, C(2–)4(–6), rarely 0 or connate, alt. with K, often clawed, A 4 alt. with C but often 2 or all 4 of A primordia developing to give 6–∞ A, some staminodal, but A not tetradynamous like Cruciferae, anthers with longit. slits, nectary an extrastaminal ring or merely receptacular protrusion, G (2(–12)), 1-loc. with parietal placentas (s.t. ephemerally or never meeting, giving pluriloc. ovary), style 1, ovules (1–)∞ per placenta, campylotropous or rarely anatropous before fert., bitegmic. Fr. usu. stipitate, often a berry, s.t. segmented, or dry & siliquiform when indehiscent or valves falling to leave replum (v. rarely with partition like Cruciferae), rarely a nut or drupe; seeds often reniform with ± curved or folded oily embryo, s.t. with arils, endosperm little or 0 (rarely fleshy – *Oceanopapaver*), perisperm sometimes present. x = 8–17

Classification & chief genera:

Capparidoideae (usu. woody, fr. usu. a berry, at least without replum): *Boscia, Cadaba, Capparis, Maerua, Ritchiea*

Cleomoideae (more often herbaceous, fr. dehiscent, with replum): *Cleome, Cleomella*

These have been treated as different fams, the first 'reminiscent' of Flacourtiaceae, the second said to 'approach' Cruciferae (Cronquist). A number of genera do not fit either, e.g. *Pentadiplandra* & *Physena* which are here referred to separate fams

Some ed. fr. esp. *Capparis* (capers) & cult. orn. esp. *Cleome*

Capparidastrum Hutch. = *Capparis*

Capparis L. Capparidaceae. 250 trop. & warm (Eur. 1). Shrubs, scramblers or trees usu. with white or yellowish fls, some opening only at night (e.g. *C. lucida* (DC) Benth., Java to Aus.); some with stipular thorns; extrafl. nectaries. Some medic., others ed. fr. (high in vitamin C); *C. spinosa* L. (capers, Medit.) – cult. for fl. buds pickled as a relish, esp. in France, in sauce Tartare etc., ssp. *rupestris* (Sm.) Nyman (var. *inermis*) without stipular thorns; *C. mitchellii* (F. Muell.) Benth. (bumble tree, Aus.) – ed. fr.; *C. pittieri* Standley (Costa Rica) – poll. bats & sphingid moths

Capraria L. Scrophulariaceae. 4 warm Am. Lvs spiral; sometimes placed in Myoporaceae on pollen characters

Caprifoliaceae Juss. Dicots – Asteridae – Dipsacales. 15/420 Euras., Am., Aus., Medit., trop. Afr. mts. Shrubs or small trees, less often lianes or herbs. Lvs opp., simple (pinnate in *Sambucus*); stipules 0 or small (extrafl. nectaries in *Sambucus* & *Viburnum*). Fls bisexual, epigynous, constricted below calyx limb, usu. in cymose or mixed infls., K ((4)5), often small, lobes imbricate or open in bud, ± accrescent in fr., C ((4)5), usu. reg., lobes imbricate or valvate, tube often nectariferous, A (4)5 attached to tube alt. with lobes, or 4 (*Linnaea*) or 2 (*Carlemannia, Silvanthus*) even when P 5-merous, anthers with longit. slits, $\overline{G}$ (2–5(–8)), rarely semi-inf., with as many locules as G with axile placentation or partitions failing to meet in ovary apex or only 1 loc. fertile, style term. or stigma(s) subsessile, ovules 1–∞ per locule, pendulous, anatropous, unitegmic. Fr. a capsule, berry or drupe or dry & dehiscent; seeds often with small straight embryo & oily, fleshy endosperm. x = 8 or 9(–12)

Genera: *Abelia, Carlemannia, Diervilla, Dipelta, Heptacodium, Kolkwitzia, Leycesteria, Linnaea, Lonicera, Sambucus, Silvanthus, Symphoricarpos, Triosteum, Viburnum, Weigela*

Carlemannia & *Silvianthus* often segregated as Carlemanniaceae & *Sambucus* sometimes placed in monotypic family, though perhaps best united with *Viburnum* and Adoxaceae. Serological evidence suggests similarities with Cornales as well as more traditional view of affinity with Rubiaceae. A close relationship with Dipsacaceae & Valerianaceae suggests even uniting these fams

Horticulturally very imp. as most genera include fine hardy flowering shrubs or lianes

Capsella Medikus. Cruciferae. 5 temp. (Eur. 4), warm. *C. bursa-pastoris* Medikus (shepherd's purse) poss. allopolyploid (2n = 32) from *C. rubella* Reuter (Medit., 2n = 16) & *C. grandiflora* (Fauché & Chaub.) Boiss. (C Italy, W Greece, 2n = 16) now a cosmop. weed (cf. *Poa annua*), usu. self-poll., seeds attracting & trapping mosquito larvae; in Chinese medicine used in eye disease & dysentery, Herba Bursae Pastoris form. used as diuretic, febrifuge etc.; collected seeds found at Catal Huyuk (5950 BC) & in stomach of Tollund Man

Capsicodendron Hoehne = *Cinnamodendron*

Capsicum L. Solanaceae (1). 10 trop. Am. Fr. a many-seeded berry. 4 spp. widely cult. in trop. Am. (T 18(1969)277) & 2 elsewhere: *C. annuum* L. var. *annuum* (most of the cult. peppers) with 5 main groups of cvs, some of the first 3 also grown as orn.: 1. Cerasiforme Group (cherry p.) – fr. small, v. pungent, 2. Conoides Group (cone p.) – fr. usu. erect, ± conical, 3. Fasciculatum Group (red cone p.) – fr. erect, slender, red, clustered, v. pungent, 4. Grossum Group (bell p., green p., sweet p., pimento) – fr. large, thick-skinned, ± bell-shaped, with depression at base, scarcely pungent (the principal salad peppers, red, green or yellow), 5. Longum Group (Cayenne p., chilli p.) – fr. usu. drooping, to 30 cm long, v. pungent, the source of chilli powder, Cayenne pepper, paprika (esp. Medit. & in Hungarian goulash) also pickled e.g. fefferoni (Croatia); *C. annuum* var. *glabriusculum* (Dunal) Heiser & Pickersgill (var. *minimum*, bird pepper) incl. the wild or spontaneous forms in Am.; *C. frutescens* L. (a name much used for forms of *C. annuum*) is the source of Tabasco & other hot sauces. Capsicum used in treatment of rheumatism, neuralgia etc. (water with 1 part c. in 11 million distinctly pungent!), form. added to beer to give 'strength' or 'bite'

Captaincookia Hallé. Rubiaceae (II 2). 1 New Caled.: *C. margaretae* Hallé, a remarkable pachycaul of v. restricted distribution

capucin *Northea hornei*

capulin *Prunus serotina*

Capurodendron Aubrév. Sapotaceae (IV). 23 Madagascar

Capuronetta Markgraf = *Tabernaemontana*

Capuronia Lourteig. Lythraceae. 1 Madagascar

Capuronianthus J. Leroy. Meliaceae (III). 2 Madag. R: Adansonia 16(1976)174

Caracasia Szyszyl. Marcgraviaceae. 2 Venezuela. C free, A 3

Caragana Fabr. Leguminosae (III 15). 80 E Eur. (3), C As. to China. Cult. orn. & imp. fuel in treeless country. *C. arborescens* Lam. (Siberia, C As., Mongolia, Manchuria) – windbreak, bark used for ropes, young pods ed.; *C. chamlagu* Lam. (*C. sinica*, N China) – fls ed.; *C. spinosa* (L.) Hornem. (Mongolia, China) – spiny branches form. stuck on tops of walls (cf. broken glass) near Beijing

caraguata fibre *Bromelia serra*; *Eryngium pandanifolium*

Caraipa Aublet. Guttiferae (II). 21 trop. S Am. R: MNYBG 29(1978)97. Lvs spiral. Timber & medic. balsam, oils etc.

Carallia Roxb. Rhizophoraceae. 10 Madag., Indomal., trop. Aus. Some timber for furniture, flooring etc.

Caralluma R. Br. Asclepiadaceae (III 5). c. 56 Medit. (Eur. 1), Macaronesia to Somalia & NE Tanzania (Namibia 1?) to Burma. R: Bradleya 8(1990)10. Succulent cult. orn.; stems, fls & fr. ed. S Arabia

caramba or **carambola** *Averrhoa carambola*

Caramuri Aubrév. & Pellegrin = *Pouteria*

Carapa Aublet. Meliaceae (IV 3). 2–3 trop. Am. & Afr. *C. guianensis* Aublet (trop. Am.) – fls every 5 yrs, good timber (crabwood, bastard mahogany, andiroba, tallicona), seeds fish-bait & source of oil (andiroba) for lamps, soap- & candle-making, medic. (arthritis, throat infections in Braz.), insect-repellent & form. for shrinking heads

caraway *Carum carvi*

Cardamine L. Cruciferae. (Incl. *Cardaminopsis*, *Dentaria*) 200 temp. (Eur. 31) incl. trop. mts (Afr., New Guinea). R: JAA 69(1988)92. Some cult. orn. & weeds (bittercress) & some watercress subs. *C. bulbifera* (L.) Crantz (*Dentaria* b., Euras.) – cult. orn., axillary bulbils; *C. chenopodiifolia* Pers. (S Am.) – annual, amphicarpous, the geocarpic fr. with adv. roots on pedicel; *C. diphylla* (Michaux) Alph. Wood (E N Am.) – rhiz. ed.; *C. hirsuta* L. (hairy bittercress, N temp.) – noisome annual weed (usu. hairless!); *C. pratensis* L. (cuckoo flower, lady's smock, meadow cress, spinks, N temp.) – adv. buds on basal lvs, 2n = 16, 24 etc., 73–96

Cardaminopsis (C. Meyer) Hayek = *Cardamine*

cardamom *Elettaria cardamomum*; also loosely applied to *Aframomum* & *Amomum* spp. (q.v.)

Cardaria Desv. = *Lepidium*

Cardenanthus R. Foster. Iridaceae (III 4). c.8 S Am. incl. Andes
Cardenasiodendron F. Barkley. Anacardiaceae (IV). 1 Bolivia
Cardiacanthus Nees & Schauer = *Carlowrightia*
Cardiandra Siebold & Zucc. Hydrangeaceae. 2 E As. R: JJB 60(1985)139, 161. Rhiz. herbs; A ∞. Close to *Hydrangea*, but lvs spiral
cardinal climber *Ipomoea sloteri*; **c. flower** *Lobelia cardinalis*; **c. wood** *Brosimum rubescens*
Cardiochilos Cribb. Orchidaceae (V 16). 1 Malawi, S Tanzania
Cardiochlamys Oliver. Convolvulaceae. 2 Madagascar
Cardiocrinum (Endl.) Lindley. Liliaceae. 3 Himal., E As. Bulbs hapax. but reproducing by offsets, lvs cordate. Cult. orn. esp. *C. giganteum* (Wallich) Makino (Himal.) to 3.5 m tall
Cardiomanes C. Presl. Hymenophyllaceae (Card.). 1 NZ: *C. reniforme* (Forst.f.) C. Presl. Lamina some cells thick
Cardionema DC. Caryophyllaceae (I 2). 6 W N Am. to Chile
Cardiopetalum Schldl. Annonaceae. (Incl. *Froesiodendron*) 5 trop. S Am.
Cardiophyllarium Choux = *Doratoxylon*
Cardiopteridaceae Blume. Dicots – Rosidae – Celastrales. 1/2 SE As., Mal. R: FM 1,7(1972)93. Twining herbs with milky juice. Lvs spirally arr., cordate, lobed or not; stipules 0. Fls. bisexual or polygamous, small, in panicles grouped in forked cincinni, K ((4)5), lobes imbricate in bud, persistent, C ((4)5), lobes imbricate in bud, A (4)5 inserted in C tube alt. with lobes, anthers with longit. slits, disk 0, G̲ (2), 1-loc. with 2 styles (1 elongate persistent in fr., the other decid.) & 2 ovules, pendent. Fr. a 2-winged samara, ± stipitate; seed 1 with minute embryo & fleshy endosperm
Only genus: *Cardiopteris* (*Peripterygium*)
App. near Icacinaceae with v. similar pollen but considered like Convolvulaceae by Airy Shaw
Cardiopteris Wallich ex Royle. Cardiopteridaceae. 2 SE As., Mal. Lvs. used as vegetable
Cardiopterygaceae Tieghem = Cardiopteridaceae
Cardiospermum L. Sapindaceae. 14 trop. esp. Am. Climbers with inflated balloon-like frs, cult. orn. esp. *C. halicacabum* L. (balloon vine, trop. Am., widely natur.) – cult. in Amaz. for seeds worn in small bands by men to ward off snakebite, allegedly aphrodisiac in S As.; *C. grandiflorum* Sw. (heartseed, trop. Am. & Afr.) – black seeds used as beads (heart shape thereon), lvs a veg.
Cardioteucris C.Y. Wu = *Caryopteris*
cardol *Anacardium occidentale*
Cardonaea Aristeg., Maguire & Steyerm. = *Gongylolepis*
cardoon *Cynara cardunculus*
Cardopatium Juss. Compositae (Card.-Carl.). 2 Med. (Eur. 1)
Carduncellus Adans. Compositae (Card.-Cent.). 29 Med. (Eur. 4). R: AJBM 47(1990)29
Carduus L. Compositae (Card.-Card.). 91 Euras. (Eur. 48), Medit., E Afr. mts. R: MBSM 5(1963)139, (1964)279. Differs from *Cirsium* in minutely barbellate, non-plumose pappus. *C. nutans* L. (Euras., natur. N Am.), 'Scotch thistle' – ed. thick pith when boiled, dried fls form. used to curdle milk
Cardwellia F. Muell. Proteaceae. 1 Queensland: *C. sublimis* F. Muell. (silky oak) – fine cabinet wood
Carex L. Cyperaceae. c. 2000 cosmop. esp. temp. & cold (Eur. 180, China c. 500, Japan 202, NZ 73, N Am. 420) usu. wet places. R: CJB 68(1990)1405. Wind-poll. infls (s.t. unisexual) of female fl. reduced to naked G, enclosed in an utricle, &/or male of 3 stamens in axil of a glume, poss. representing a condensed spikelet of 3 1-staminate fls. Most primitive group is sect. *Vigneastra* with compound panicles of bisexual spikelets (trop. OW). *C. arenaria* L. (N temp.) on sand-dunes with habit of *Ammophila; C. capitata* L. & *C. macloviana* Urv. bipolar, *C. microglochin* Wahlenb. & *C. magellanica* Lam. poss. only subsp. distinct polar populations; some with elaiosomes & ant-disp. *C. brizoides* L. (Eur.) used as vegetable hair packing material, others somewhat medic. or used for hat-making etc.
Careya Roxb. Lecythidaceae (Barringtoniaceae). 4 trop. As. *C. arborea* Roxb. (patana oak) almost only tree sp. in grassy patanas of Sri Lanka, seeds ed., lvs used for silkworms
Carib grass *Eriochloa punctata*
Caribea Alain. Nyctaginaceae (III). 1 Cuba: *C. litoralis* Alain
Carica L. Caricaceae. 23 warm Am. R: V.M. Badillo, *Caricaceae* (1971)60. *C. papaya* L. (papaya, pawpaw, melon tree) – cult. throughout trop. for ed. fr. eaten fresh & tinned, ice-cream & chewing-gum flavour & for papain (an antibacterial protease, the only natur. pl. one in comm.) derived from scarifying unripe fr. & used as meat tenderizer (meat wrapped in lvs becomes tender too), to reduce cloudiness in beer, to shrinkproof wool & silk & in

control of termites & in upmarket toothpastes, lvs locally med.; fast-growing (everleafing (13–15 new lvs per month) & -flowering for 8–9 months) with pachycaul habit, palmate lvs & principal support in phloem fibres, usu. dioec. (male fls open for 1 day, females for 7) though sex changeable by damage, hormones etc., poll. by thrips in Afr., moths in Malaysia; *C. pubescens* Lenné & K. Koch (*C. candinamarcensis*, mountain pawpaw, Andes) – hardier & with smaller fr. usu. candied or preserved, hybrid with Ecuadorean *C. stipulata* Badillo (*C.* × *pentagona* Heilborn) the babaco of fr. stalls, cult. Aus. & NZ, all fls female so parthenocarpic & fr. seedless. *C. monoica* Desf. – fr. pulp dry but single-gene mutants with fleshy

Caricaceae Dumort. Dicots – Dilleniidae – Violales. 4/33 trop. & warm Am., trop. Afr. *(Cylicomorpha).* R: V.M. Badillo (1971) *Monografia de la familia C.* Trees, often sparsely branched & pachycaul, rarely prostrate herbs *(Jarilla)*, dioec., rarely monoec., with well-developed system of anastomosing, articulated laticifers. Lvs spirally arr., often large, palmately veined & lobed to palmate, rarely otherwise; stipules 0 or spine-like. Fls. axillary, solit. or in cymes, rarely bisexual, reg., K(5), P(5), the tube v. short in female fls, lobes convolute or valvate, A 5 or 5 + 5, attached to C-tube, distinct *(Carica)* or basally connate, anthers with longit. slits, $\underline{G}$ (5), 1-loc. with deeply intrusive parietal placentas or these meeting to give pluriloc. ovary with axile placentas, styles distinct, ovules ∞, anatropous, bitegmic with ± enlarged funicle. Fr. a large berry (pepo); seeds ∞ with gelatinous coat, straight embryo & oily, proteinaceous endosperm

Genera: *Carica, Cylicomorpha, Jacaratia, Jarilla*

Allied to Passifloraceae, particularly *Adenia*, some spp. of which are pachycaul. Presence of glucosinolates & cyanogenic glycosides suggests interm. position between Violales & Capparidales

Fr. trees, esp. *Carica papaya*

Carinavalva Ising. Cruciferae. 1 S Aus.: *C. glauca* Ising. R: Fl. S Aus. 1(1986)388

Cariniana Casar. Lecythidaceae. 15 trop. S Am. R: FN 21(1979)218. A somewhat 1-sided with small flies as poll. (see fam.). Fr. like *Lecythis* (monkey pots); valuable timber (abarco, albarco, bacu, jequitiba, jiquitiba) esp. *C. pyriformis* Miers (NE S Am.) – form. called Colombian mahogany & exported to Eur.

Carionia Naudin = *Medinilla*

Carissa L. Apocynaceae. 37 warm OW E to Aus. & New Caled. (1). Often with branch thorns: grown for hedging & tart fr. (conkerberry, congaberry in Aus.). *C. bispinosa* (L.) Brenan (amatungulu, S Afr.) – hedges, fr. ed.; *C. carandas* L. (karanda, Indomal.) – hedges, fr. pickled, sometimes called Christ's thorn; *C. grandiflora* (E. Meyer) A. DC (*C. macrocarpa*, Natal plum, S Afr.) – hedgepl., fr. large, sold in markets

Carissophyllum Pichon = *Tachiadenus*

Carlemannia Benth. Caprifoliaceae (Carlemanniaceae). 3 Indomal. mts

Carlemanniaceae Airy Shaw = Caprifoliaceae

Carlephyton Jum. Araceae (VIII 2). 3 Madag. R: BJ 92(1972)10

Carlesia Dunn. Umbelliferae (III 8). 1 E China: *C. sinensis* Dunn

Carlina L. Compositae (Card.-Carl.). 20 Eur. (13), Macaronesia, Medit., W As. R: FR 83(1972)213. *C. acaulis* L. (Eur., common in Alps) – typical form with sessile capitula, the bracts spreading out star-like in dry air to release cypselas, others with peduncles to 25 cm (subsp. *simplex* (Waldst. & Kit.) Nyman ('ssp. *caulescens*'))

Carlowrightia A. Gray. Acanthaceae. 23 SW US to Costa Rica, warm to arid. R: FN 34(1983), Brittonia 40(1988)245

Carludovica Ruíz & Pavón. Cyclanthaceae. 4 trop. Am. mainland. R: AHB 18(1958)127. Almost stemless, palm-like, source of fibres esp. *C. palmata* Ruíz & Pavón (C Am. to Bolivia) grown for hat manufacture (Panama hat palm, toquilla) – 6 young lvs per hat, of which 4 million exported (form. via Panamá) from Ecuador per annum); older lvs used for mats, baskets etc., also cult. orn. Other spp. for thatching & brooms

Carmenocania Wernham = *Pogonopus*

Carmichaelia R. Br. Leguminosae (III 16). 40 NZ, 1 Lord Howe Is. R: TRSNZ 75(1945)231. Photosynthetic flat branches without green lvs. Some cult. orn. esp. *C. enysii* Kirk & dwarfer *C. orbiculata* Colenso (NZ)

Carminatia Moçiño ex DC. Compositae (Eup.-Alom.). 3 SW US to El Salvador. R: PSE 110(1988)169

Carmona Cav. (~ *Ehretia*). Boraginaceae. 1 Indomal.: *C. retusa* (Vahl) Masam. (*E. buxifolia*, *E. microphylla*, Fukien tea) – cult. orn. esp. for hedging in Mal. & bonsai, natur. (bird-disp.) in Hawaii

Carnarvonia F. Muell. Proteaceae. 1 Queensland

carnation *Dianthus caryophyllus*

carnauba wax *Copernicia prunifera*

Carnegiea Britton & Rose. Cactaceae (III 8). 1 SW US, Mex.: *C. gigantea* (Engelm.) Britton & Rose (saguaro, giant cactus) – largest of all cacti, reaching 20 m tall, 60 cm thick, with candelabriform branching, 12 t in weight & alleged to live for 200 yrs, not thriving in cult.; poll. by birds & insects by day, bats by night; Indian house rafters in Arizona, fr. to 7.5 cm diam., ed., form. v. important as food & drink, still collected & used in ceremonies (see DP 2(1)(1980))

Carnegieodoxa Perkins = *Hedycarya*

caroá fibre *Neoglaziovia variegata*

carob *Ceratonia siliqua*

Carolina allspice *Calycanthus floridus*

Caropodium Stapf & Wettst. = *Grammosciadium*

Caropsis (Rouy & Camus) Rauschert (*Thorella* Briq.). Umbelliferae (III 8). 1 Eur.: *C. verticillato-inundata* (Thore) Rauschert

Carpacoce Sonder. Rubiaceae (IV 13). 7 S Afr.

Carpanthea N.E. Br. Aizoaceae (V). 1 SW Cape: *C. pomeridiana* (L.) N.E. Br. R: BJ 111(1990)478. Annual, cult. orn.

Carpentaria Becc. Palmae (V 4k). 1 N Aus.: *C. acuminata* (H.A. Wendl. & Drude) Becc.

Carpenteria Torrey. Hydrangeaceae (Philadelphaceae). 1 C Sierra Nevada, Calif: *C. californica* Torrey – cult. orn. evergreen

Carpesium L. Compositae (Inul.). 25 Euras. (Eur. 2), Indomal. to Aus. Pappus 0

carpet grass *Axonopus* spp.

Carpha Banks & Sol. ex R. Br. Cyperaceae. 4–5 Aus. (4), New Guinea, NZ, temp. S Am.

Carphalea Juss. Rubiaceae (IV 1). 10 trop. Afr. (3), Socotra (1), Madag. (6). R: BJBB 58(1988)271. K-lobes poll.-attractants, later for wind-disp.

Carphephorus Cass. Compositae (Eup.-Liat.). 4 SE US. R: MSBMBG 22(1987)277

Carphochaete A. Gray. Compositae (Eup.-Ager.). 7 SW N Am. R: Phytol. 64(1987)145

Carpinaceae Vest = Betulaceae

Carpinus L. Betulaceae (II, Carpinaceae). 26 N temp. (Eur. 2) to C Am. mts, esp. E As. R: JAA 71(1990)51. Decid. monoec. trees – good timber for turnery, tools etc. *C. betulus* L. (hornbeam, Eur. to Iran) – form. much used for mill cogwheels, ox-yokes etc., still imp. as mechanism between key & hammer in a piano, butchers' chopping-blocks, skittles, pulleys, riding-boot trees, draughtsmen, chessmen, dominoes, cult. orn. (several cvs) incl. as pleached hedging; *C. caroliniana* Walter (American h., blue beech, water b., E N Am.) – cult. orn.; *C. coreana* Nakai (Korea) – cult. orn., creeping

Carpobrotus N.E. Br. Aizoaceae (V). 25 SW Cape to Natal, Aus. (4), introd. Am. Cult. orn., some natur. esp. on cliffs in Eur. incl. *C. edulis* (L.) L. Bolus (pigface, Cape) – C yellow to purple, leaf-juice for dysentery, sore throats & smeared over newborn Hottentots, ed. fr. ('Hottentot fig') fresh, dried or in jam, though fr. of other Cape spp., e.g. *C. deliciosus* (L. Bolus) L. Bolus, *C. dulcis* L. Bolus (both ed. fresh) & *C. muirii* (L. Bolus) L. Bolus (ed. dried), tastier

Carpodetaceae Fenzl = Grossulariaceae

Carpodetus Forster & Forster f. Grossulariaceae (Escalloniaceae). 10 New Guinea, NZ (1)

Carpodinopsis Pichon = *Pleiocarpa*

Carpodinus R. Br. ex G. Don f. = *Landolphia*

Carpodiptera Griseb. (~ *Berrya*). Tiliaceae. 1 E Afr., 7 trop. Am.

Carpolepis (Dawson) Dawson (~ *Metrosideros*). Myrtaceae (Lept.). 3 New Caled. R: BMNHN 4,6 (1984)465

Carpolobia G. Don f. Polygalaceae. 4 trop. Afr. & Madag. R: MLW 77–18(1977)22. Some timber & ed. fr., *C. lutea* G. Don f. stem a chewing-stick in Nigeria

Carpolyza Salisb. Amaryllidaceae (Liliaceae s.l.). 1 S Afr.: *C. spiralis* (L'Hérit.) Salisb. R: BJ 107(1985)20 – cult. orn.

Carpotroche Endl. Flacourtiaceae (3). 11 trop. Am. R: FN 22(1980)31. Some parasiticidal oils used in skin disease

Carpoxylon H. Wendl. & Drude. Palmae (V 4). 1 Vanuatu. R: Principes 33(1989)68. Form. thought almost extinct

Carptotepala Mold. = *Syngonanthus*

Carramboa Cuatrec. See *Espeletia*

Carrichtera DC (~ *Vella*). Cruciferae. 1 Macaronesia to Iran, incl. Eur.: *C. annua* (L.) DC

Carrierea Franchet. Flacourtiaceae (8). 3 S & SW China, SE As.

Carrissoa Baker f. Leguminosae (III 10). 1 Angola (= ? *Eriosema*)

Carronia F. Muell. Menispermaceae (I). 3 New Guinea to NSW. R: KB 30(1975)94

carrot *Daucus carota*; **Candy c.** *Athamanta cretensis*

Carruanthus (Schwantes) Schwantes. Aizoaceae (V). 1–2 S Cape & Karoo. Cult. orn.

Carruthersia Seemann. Apocynaceae. 3 Philippines & Solomons (1), Fiji (2)

Carsonia E. Greene = *Cleome*

Carterella Terrell (~ *Bouvardia*). Rubiaceae (IV 1). 1 Baja Calif.: *C. alexanderae* (A. Carter) Terrell. R: Brittonia 39(1987)248

Carterothamnus R. King = *Hofmeisteria*

Carthamus L. Compositae (Card.-Cent.). 17 Medit. (Eur. 8) to C As. R. AJBM 47(1990)28. *C. lanatus* L. (saffron thistle, Medit. to C As.) – cosmop. weed; *C. tinctorius* L. (safflower, saffron thistle, cultigen poss. orig. Near E) – fls form. used in dyeing food & rouge (yellow & red) in decline after use of aniline dyes, potential oilseed crop (kurdee) with polyunsaturates for soft margarines & monounsaturate for frying – oil already used in easing Crohn's Disease, in cosmetics & aromatherapy, fr. used for poultry etc.

Cartiera E. Greene. Cruciferae. 6 W N Am.

Cartonema R. Br. Commelinaceae (I 1; Cartonemataceae). 6 Aru Is., trop. Aus.

Cartonemataceae Pichon = Commelinaceae

Carum L. Umbelliferae (III 8). c. 30 temp. (Eur. 5) & warm. *C. carvi* L. (caraway, Medit.) – cult. for fr. used as flavouring in bread, sauerkraut, cheese & 'seedcake' & liqueur (Kümmel), since time of Pliny

Carvalhoa Schumann. Apocynaceae. 1 E & SE Afr.: *C. campanulata* Schumann. R: AUWP 85-2(1985)47

Carvia Bremek. = *Strobilanthes*

Carya Nutt. Juglandaceae. 18 E N Am. to C Am., E As. (few). R: AMBG 65(1978)1080. Hickories – timber & nuts (pecans), strips of outer bark form. used in chair-seat-making & lacrosse sticks (now usu. plastic). *C. cordiformis* (Wangenh.) K. Koch (bitternut, E N Am.); *C. glabra* (Miller) Sweet (pignut, hognut, N Am.) – kernels usu. astringent; *C. illinoinensis* (Wangenh.) K. Koch (pecan, S US) – common dessert nut, first cv. selected 1846, now over 500 named, esp. thin-shelled ones, used like hazelnuts or walnuts in food, oil used in cosmetics etc., timber poor, Texas state tree (first to be designated – 1919); *C. ovata* (Miller) K. Koch, (N Am.), & *C. laciniosa* (Michaux f.) Loudon (E N Am.) & *C. tomentosa* (Poiret) Nutt. (mockernut, E N Am.) – nuts also used

Caryocar Allamand ex L. Caryocaraceae. 15 trop. Am. R: FN 12(1973), OB 92(1987)179. Bat-poll. Some timbers for ship-building. Drupes of some ed., others fish-poisons (saponins, prob. also effective against termites), seeds an oil source. *C. brasiliense* Cambess. (pequí, piquí, Brazil) – fr. & seeds local soap & cooking oil sources, comm. liquor from fr.; *C. amygdaliferum* Mutis (suarí, swarri nut) – pleasant-tasting fat; *C. glabrum* (Aublet) Pers. (soapwood, NE S Am.) – inner bark used for washing; *C. microcarpum* Ducke (Amaz.) – lvs repellent to leaf-cutting ants, fish-poison (saponins & tannins); *C. nuciferum* L. (Panamá to NE S Am.) – names & uses like *C. amygdaliferum*; *C. villosum* (Aublet) Pers. (NE S Am.) – names like *C. brasiliense*, also oilseed

Caryocaraceae Voigt. Dicots – Dilleniidae – Theales. 2/25 trop. Am. esp. Amazonia. R: FN 12(1973). Trees, rarely shrubs. Lvs opp. (*Caryocar*) or spirally arr., trifoliolate, ± dentate; stipules 0, 2 or 4, caducous. Fls in term. racemes, bisexual, reg., K 5(6), imbricate, basally connate in *Caryocar*, or reduced & lobed (*Anthodiscus*), C 5(6), imbricate, s.t. connate basally, connate above in *Anthodiscus*, forming calyptra, A ∞, shortly basally connate into ring or 5 bundles opp. C, inner ones s.t. without anthers, anthers with longit. slits, $\underline{G}$ (4–20) with as many locules & styles, each carpel with 1 axile, anatropous to orthotropous, bitegmic ovule. Fr. a drupe, the stone separating into 4 1-seeded pyrenes; seeds reniform with thin or 0 endosperm, embryo with large oily & proteinaceous spirally twisted hypocotyl & 2 small cotyledons; germination hypogeal

Genera: *Anthodiscus*, *Caryocar*

Close to Theaceae

All bat-poll. despite diff. habitats (savanna, rain forest etc.). Some ed. seeds (some toxic)

Caryodaphnopsis Airy Shaw (~ *Persea*). Lauraceae (I). 15 trop. As. (7) & Am. (8). R: ABY 13(1991)1. *C. theobromifolia* (A. Gentry) van der Werff & H. Richter (*P. t.*, Ecuador) – form. *the* 'mahogany' of Ecuador, now c. 12 trees left

Caryodendron Karsten. Euphorbiaceae. 3 trop. S Am. *C. orinocense* Karsten (tacay nut, Colombia) – seeds ed. roasted, poss. oil-crop

Caryolobis Gaertner = *Shorea*

Caryomene Barneby & Krukoff. Menispermaceae (II). 4 trop. Am. R: MNYBG 22, 2(1971)52

Caryophyllaceae Juss. Dicots – Caryophyllidae – Caryophyllales. 87/2300 cosmop. esp.

temp. & warm N hemisph. Herbs, rarely shrubs, lianes (e.g. *Alsinidendron*) or even small trees (*Sanctambrosia*), sometimes xeromorphic, usu. monoec., often with swollen nodes and sometimes with unusual sec. thickening with concentric rings of xylem & phloem; anthocyanins present, sieve-tube plastids typical of 'Centrospermae'. Lvs opp., rarely spirally arr., simple, entire; stipules 0 (exc. some Paronychioideae). Fls usu. reg., hypogynous (rarely not, as in *Scleranthus*, where perigynous), in dichasial cymes or solit. K ((4)5), completely free in some Alsinoideae, C 0 (many Paronychioideae), (4)5 (small & bifid in many Alsinoideae, clawed & usu. with large blade in Caryophylloideae), A (1–4) 5–10 in 1 or 2 whorls, sometimes basally adnate to C to form short tube that may be adnate to gynophore, or inserted at edge of nectary-disk around ovary, or adnate to K, anthers with longit. slits, G (2–5(+)) with ± united styles, often surmounting gynophore, 1-loc. distally but ± partitioned proximally, placental column reaching apex or not, ovules usu. ∞, sometimes >1, bitegmic, hemitropous to campylotropous (usu.). Fr. a capsule with as many or twice as many valves or apical teeth as styles, rarely indehiscent (berry, nutlet or achene); seeds small, usu. ornamented on testa, embryo usu. peripheral, curved round copious starchy perisperm, less often straight, endosperm little or 0. x = 5–19

Classification & chief genera:

I. **Paronychioideae** (stipules; C small or 0)
   1. **Polycarpeae** (lvs opp.; capsule): *Drymaria, Polycarpaea, Spergularia*
   2. **Paronychieae** (lvs opp.; nutlets; this & 3. sometimes separated as Illecebraceae): *Herniaria, Paronychia*
   3. **Corrigioleae** (lvs alt.): *Corrigiola*

II. **Alsinoideae** (stipules 0; K free)
   1. **Alsineae** (episepalous A usu. with nectariferous gland at base, styles free; capsule usu. with many seeds or nutlet): *Arenaria, Cerastium, Minuartia, Sagina, Stellaria*
   2. **Pycnophylleae** (styles at least basally connate, fr. indehiscent 1-seeded): *Pycnophyllum* (only)
   3. **Geocarpeae** (C 0, 5 vestigial episepalous 'staminodes'; capsule): *Geocarpon* (only)
   4. **Habrosieae** (C minute, styles free, G 2-ovulate; fr. indehiscent, 1-seeded): *Habrosia* (only)
   5. **Sclerantheae** (C 0, styles 2 free, G 1-ovulate, fr. indehiscent): *Scleranthus, Stellaria*

III. **Caryophylloideae** (stipules 0; (K))
   1. **Caryophylleae** (fls reg., styles 2(3); capsule with 4(6) teeth): *Acanthophyllum, Dianthus, Gypsophila, Saponaria*
   2. **Drypideae** (fls ± irreg., styles 3): *Drypis* (only)
   3. **Sileneae** (fls reg., styles 3–5, capsule with teeth = A or 2 x A): *Silene*

With Molluginaceae, differ from rest of order in anthocyanin prod. C supposed to be of staminodal origin (Cronquist)

Many cult. orn. herbs esp. *Dianthus*; many weeds esp. *Sagina, Stellaria*

Caryopteris Bunge. Labiatae (II, Verbenaceae). 6 E As. Cult. orn. esp. *C.* × *clandonensis* Rehder (*C. incana* (Thunb.) Miq. (China & Japan) × *C. mongholica* Bunge (N China)), a blue-flowered shrub often dying back in winter

Caryota L. Palmae (V 1). 12 Indomal. to trop. Aus. Only palms with bipinnate lvs, with fish-tail-like leaflets, flowering from top to bottom & dying, fr. filled with irritant crystals but disp. by animals, some cult. orn. e.g. *C. maxima* Blume (*C. aequatorialis*, Malay Pen.) – hapaxanthic to 35 m tall & fls with over A 100, *C. mitis* Lour. (Indomal.), 'fishtail palm', but prob. hybridize (taxonomy unclear). *C. urens* L. (kitul (kittool), toddy palm, Indomal.) – widely planted, somewhat 'weedy' in wild, a source of sago (famine food in e.g. Burma), palm-sugar, & toddy (7–14 l sap per day from 1 infl., up to 27 l per tree if more than 1 infl. tapped), timber & kitul (kittool) fibre (Ceylon piassava) derived from leaf-bases & used as a brush-fibre, young lvs ed., mature ones exported to Eur. for cut foliage

Caryotophora Leistner. Aizoaceae (V). 1 Cape

casabanana *Sicana odorifera*

Casabitoa Alain. Euphorbiaceae. 1 Hispaniola: *C. perfae* Alain

casana *Cyphomandra hartwegii*

Casasia A. Rich. Rubiaceae (II 1). 11 C Am., WI. G 1-loc. R: JAA 68(1987)176

Cascabela Raf. = *Thevetia*

Cascadia A.M. Johnson = *Saxifraga*

cascalote *Caesalpinia coriaria*

cascara (sagrada) *Rhamnus purshiana*
cascarilla bark *Croton eluteria*
Cascarilla (Endl.) Wedd. = *Ladenbergia*
Cascaronia Griseb. Leguminosae (III 4). 1 Bolivia, Argentina
Casearia Jacq. Flacourtiaceae (9). 160 trop. (75 Am. – R: FN 22(1980)280). Some timbers; main branches with scale lvs, short-lived twigs with foliage lvs; pellucid glands in lvs. *C. praecox* Griseb. (*Gossypiospermum praecox*, WI, Colombian or Maracaibo boxwood, zapatero, Venez., WI) – all trees in 1 population fl. at once & just for 1 day, principal boxwood largely replacing *Buxus sempervirens* in comm. for rules, veneers, carving, keyboards etc.; *C. sylvestris* Sw. (trop. Am.) – seeds a source of an oil like chaulmoogra oil; *C. tomentosa* Roxb. (*C. elliptica*, India) – fish poison, wood for carving
cashew nut *Anacardium occidentale*
Casimirella Hassler. Icacinaceae. 7 trop. Am. R: Brittonia 44(1992)166. Some with tubers ed. when cooked (*Humirianthera*)
Casimiroa Llave & Lex. Rutaceae. 5 C Am. highlands. Alks. *C. edulis* Llave & Lex. (white sapote, Mexican apple, Mex.) – cult. for ed. fr., bitter-sweet flavour, used in milk shakes, icecream etc., variable
cassabanana *Sicania odorifera*
Cassandra D. Don = *Chamaedaphne*
cassareep see *Manihot esculenta*
cassava *Manihot esculenta*
Cassebeera Kaulf. = *Doryopteris*
Casselia Nees & Mart. Verbenaceae. 12 trop. Am.
Cassia L. Leguminosae (I 2). Excl. *Chamaecrista* & *Senna* (segregation supported by floral ontogeny studies), q.v., c. 30 trop. (*Cassia* s.s. with 3 adaxial A sigmoidally curved; Afr. 10 (R: KB 43(1988)334)). Medic. & cult. orn. trees: *C. fistula* L. (pudding-pipe tree, purging cassia, Indian laburnum, golden shower, trop. As.) – widely cult., fr. to 60 cm with seeds embedded in laxative pulp, used against habitual constipation; *C. grandis* L.f. (horse c., trop. Am.) – cult. orn. with pink fls, medic.; *C. javanica* L. (Mal.) – cult. orn. with red fls, timber beautifully marked & used in house-building in Java, tanbark, fr. purgative, var. *indochinensis* Gagnepain (*C. nodosa*, Indomal.) with similar uses
cassia bark *Cinnamomum aromaticum*; **golden c.** *Chamaecrista fasciculata*; **horse c.** *Cassia grandis*; **purging c.** *C. fistula*; **ringworm c.** *Senna alata*; **tanner's c.** *S. auriculata*
Cassidispermum Hemsley = *Burckella*
cassie *Acacia farnesiana*
Cassine L. Celastraceae. c. 60 S Afr. Lvs opp. See also *Elaeodendron*
Cassinia R. Br. Compositae (Gnap.-Cass.). 21 Aus., NZ. R: OB 104(1991)90. Some cult. orn.
Cassinopsis Sonder. Icacinaceae. 4 Afr., Madagascar
Cassiope D. Don. Ericaceae. (Excl. *Harrimanella*) 11 Himal., circumboreal (Eur. 2). R: Plantsman 11(1989)106. Cult. orn. *C. tetragona* (L.) D. Don much used as fuel by Esquimaux
Cassipourea Aublet. Rhizophoraceae. (Excl. *Dactylopetalum*) c. 40 trop. Am., trop. & S Afr., Madag., Sri Lanka. R: KB 1925(1925)241. G. Alks. *C. elliottii* (Engl.) Alston (pillarwood, E Afr.) – commercial timber
cassumar (ginger) *Zingiber purpureum*
Cassupa Bonpl. = *Isertia*
Cassytha L. Lauraceae (I). 20 OW trop. esp. Aus. (15, 13 endemic). Alks. Parasites (dodder-laurels) with habit of *Cuscuta*, scale-like green lvs soon falling. *C. filiformis* L. a source of a brown dye in E Afr.
Cassythaceae Bartling ex Lindley = Lauraceae (I)
Castalis Cass. = *Dimorphotheca*
Castanadia R. King & H. Robinson = *Castenadia*
Castanea Miller. Fagaceae. 10 N temp. (Eur. 1 – *C. sativa* L. (common, sweet or Spanish chestnut, Medit. to Caucasus, introd. to GB by Romans – meal (pollenta) among relics of R. soldiers) cult. orn. (girth to 13.33 m in Dorset, greatest of any tree in GB but at 25 °C under long days merely a bush), for nuts (starchy, often sold roasted as in London's winter streets, the flour used in cooking esp. Italy, candied (marrons glacés)) & timber, esp. coppice for fencing & gates, also form. walking-sticks ('congo sticks'); bark used in tanning). R: A. Camus, *Les Châtaigniers* (1929)11. Entomophily; others planted incl.: *C. crenata* Siebold & Zucc. (Japanese chestnut) – timber & nuts; *C. dentata* (Marshall) Borkh. (American c., E N Am.) – form. important timber with best nuts (also used in a 'coffee' by Iroquois) & assoc. with Indian village-sites (planted), now almost extinct through chest-

nut blight introd. from As. in late 1800s, other (As.) spp. now tried in US to replace it in commerce; *C. pumila* (L.) Miller (chinquapin, E N Am.) – timber for railway sleepers, seeds ed.

Castanella Spruce ex Hook.f. = *Paullinia*

castanha de cutia *Couepia edulis*

Castanopsis (D. Don) Spach. Fagaceae. 110 trop. & warm As. For Am. spp. see *Chrysolepis*. No clear diffs from *Castanea*; seeds roasted like chestnuts; some cult. orn. *C. argentea* (Blume) A. DC (Indomal.) – timber, dyebark, ed. seeds; *C. cuspidata* (Thunb.) Schottky (Japan, Korea) – ed. seeds, much planted in Jap. parks & gardens, lvs form. used as rice 'bowls'

Castanospermum A. Cunn. ex Mudie. Leguminosae (III 2). (Incl. *Alexa*) 12 trop. Am., 1 NE Aus., W New Britain, New Caled., Vanuatu: *C. australe* A. Cunn. ex Mudie (black bean, Moreton Bay or Aus. chestnut) – coastal forest & beaches, seeds ed. (black beans roasted, poisonous if raw; screened as poss. AIDS vaccine as leucocyte movement to inflammation sites from blood vessels inhibited by castanospermine, poss. by interfering with the ability to bind with heparase, a key part of invasion), street tree, decorative timber

Castanospora F. Muell. Sapindaceae. 1 NE Aus.: *C. alphandii* (F. Muell.) F. Muell.

Castela Turpin. Simaroubaceae. 15 trop. & warm Am., Galápagos. Some med. R: TSDSNH 15,4(1968)31

Castelia Cav. (~ *Pitraea*). Verbenaceae. Spp.?

Castellanoa Traub (~ *Chlidanthus*). Amaryllidaceae (Liliaceae s.l.). 1 Argentina

Castellanosia Cárdenas = *Browningia*

Castellia Tineo. Gramineae (17). 1 Macaronesia, Medit. (inc. Eur.), N Afr. to Pakistan

Castelnavia Tul. & Wedd. Podostemaceae. 9 Brazil

Castenedia R. King & H. Robinson (*Castanedia*, ~ *Eupatorium*). Compositae (Eup.-Crit.). 1 Colombia. R: MSBMBG 22(1987)354

Castilla Sessé. Moraceae (III). 3 trop. Am. R: FN 7(1972)92. Form. imp. rubber sources now supplanted by *Hevea*. *C. elastica* Sessé (C Am. or Panama rubber, C Am.) – source of the rubber balls Columbus saw, *C. ulei* Warb. (uli, Amazon) etc.

Castilleja Mutis ex L.f. Scrophulariaceae. 200 (E N Am. 3, Euras. 5 (Eur. 3), C Am. 8, Andes 5, rest W N Am.) – Indian paintbrush. Hemiparasitic; upper lvs brightly coloured making infl. conspicuous – some cult. orn. *C. densiflora* (Benth.) Chuang & Heckard (*Orthocarpus d.*, W N Am.) – seeds disp. with fr. of introd. host-pl., *Hypochaeris glabra* L.; *C. integra* A. Gray (Mex.) – iridoids sequestered by *Euphydryas anicia* butterflies

castor oil *Ricinus communis*

Castratella Naudin. Melastomataceae. 1 NE S Am.

Casuarina L. Casuarinaceae. 17 SE As. to W Pacific (see also *Allocasuarina* & *Gymnostoma*). Timbers & fuelwoods – ironwood, she-oak, swamp oak, actinomycete root-symbionts (*Frankia*) fixing nitrogen; dioec. exc. *C. equisetifolia* L. (Australian pine, beefwood, jau, whistling pine, yar, Indomal. & widely planted e.g. long-established on E Afr. coast) – pioneer seashore tree good for hedging incl. topiary & windbreaks, timber used for shingles, fencing etc., burns with great heat ('best firewood in the world'), tanbark (Madag.). *C. cunninghamiana* Miq. (river oak, E Aus.) – protected sp.

Casuarinaceae R. Br. Dicots – Hamamelidae – Casuarinales. 4/95 Indomal., Aus. (esp.), W Pacific. Monoec. or dioec. evergreen trees & shrubs with drooping equisetoid twigs, tannins & roots often with nitrogen-fixing bacterial nodules. Lvs scale-like, up to 20 in sets of 4 in whorls, ± connate, forming toothed sheaths at each node; stipules 0. Fls anemophilous, P 1 or 2 in males (falling at anthesis) but 0 in females, so A 1 subtended by bract & 2 bracteoles, these males in whorled axillary spikes, G 2 subtended by an eventually woody bract & 2 bracteoles, 2-loc. but only anterior fertile, 2-branched style (winged in fr.) & 2(4) orthotropous, bitegmic ovules, these females in heads on short (usu.) lateral branches. Infr. a 'cone' of 1-seeded, winged nuts (samaras), initially enclosed in accrescent woody bracteoles, which separate at maturity appearing like a dehisced capsule; seed pendulous, endosperm 0, embryo (often more than 1) large, straight, oily. x = 8 –14 (? orig. 9). R: Telopea 3(1988)133

Genera: *Allocasuarina, Casuarina, Ceuthostoma, Gymnostoma*

Poss. allied to Betulaceae and Myricaceae, the simple fls representing reduction rather than primitive simplicity. The 'cones' of *Casuarina equisetifolia* are familiar on tropical beaches

Timbers esp. firewood

cat thyme *Teucrium marum*

Catabrosa Pal. Gramineae (17). 2 N temp. (Eur. 1), Chile. R: Darw. 23(1981)181. *C. aquatica*

(L.) Pal. (N temp.) – waterhair

Catabrosella (Tzvelev) Tzvelev = *Colpodium*

Catadysia O. Schulz. Cruciferae. 1 Peru

Catalepis Stapf & Stent. Gramineae (33d). 1 S Afr.

Catalonian jasmine *Jasminum officinale* f. *grandiflorum*

Catalpa Scop. Bignoniaceae. 11 E As., N Am. R: Candollea 13(1952)241. Cult. orn. & some timber. *C. bignonioides* Walter (Indian bean, SE US) – catalpol (iridoid) sequestered by sphingid *Ceratomia catalpae*, cult. orn. incl. golden 'Aurea', timber for railway sleepers etc.; *C. speciosa* (Warder ex Barney) Engelm. (catawba, cigar-tree, US) – similar uses, insect-damaged lvs produce more extrafl. nectar attracting insects which attack or remove eggs or larvae of the first herbivores; hybrids between As. & Am. spp. in cult.

Catamixis Thomson. Compositae (Mut.-Mut.). 1 NW Himal.

Catananche L. Compositae (Lact.-Cat.). 5 Medit. (Eur. 2). *C. caerulea* L. (cupid's dart, S Eur.) – cult. orn.

Catanthera F. Muell. Melastomataceae. 16 Sumatra, Borneo, New Guinea. R: Reinw. 10(1982)35. Ivy-like climbers

Catapodium Link (~ *Desmazeria*). Gramineae (17). 2 Eur. (2) & N Afr. to Iran. Like *D.* but lemmas glabrous

cataria *Nepeta cataria*

Catasetum Rich. ex Kunth. Orchidaceae (V 9). 100 trop. Am. Usu. epiphytes with 2 or 3 plicate lvs per pseudobulb and dimorphic male & female fls on separate infls (spp. with bisexual fls = *Clowesia*); pollinia violently ejected when column appendages touched. Female fls seldom produced in cult. & not always distinctive, though more produced in high light intensities. Closely related spp. with different fragrances attracting different pollinators. R: FR 30(1932)257, 31(1933)99. Many cult. orn.

Catatia Humbert. Compositae (Gnap.-Gnap.). 2 Madag. R: OB 104(1991)138

catawba *Catalpa speciosa*

catberry *Ribes grossularioides*

catbriar *Smilax* spp.

catchfly *Silene* spp.

catechu *Acacia catechu*

Catenularia Botsch. = *Catenulina*

Catenulina Soják. Cruciferae. 1 C As.

caterpillar plant *Spathicarpa sagittifolia*

Catesbaea L. Rubiaceae (in. sed.). 20 WI to Florida Keys. Spiny shrubs. *C. spinosa* L. (lily thorn, Spanish guava, Cuba) – cult., ed. fr.

catgut *Tephrosia virginiana*

Catha Forssk. ex Schreber. Celastraceae. 3 SW Arabia to S Afr. *C. edulis* (Vahl) Endl. (khat, miraa, qat, cafta) – cult. Ethiopia, Somalia, Yemen etc. for lvs chewed fresh by Moslems as daily stimulant, form. made into tea, fresh supplies airfreighted from E Afr. to Moslem countries

Catharanthus G. Don f. (~ *Vinca*). Apocynaceae. 6 Madag. (5), India & Sri Lanka (1). Differ from *Vinca* in 34 ways incl. sessile A without term. appendage, habit etc. *C. roseus* (L.) G. Don f. (Madag. periwinkle, old maid, Cayenne jasmine, pantrop. weed, orig. Madag.) – cult. orn. with at least 80 named alks notably vincristine & others which have retarding effect on progress of leukaemia discovered when tested for alleged effects in diabetes(!), also in Hodgkin's Disease, but 2 t of lvs needed for 1 g alks (= 6 wks' treatment for 1 child), see W.I. Taylor & N. Farnsworth (1975) *The C. alkaloids*

Cathaya Chun & Kuang (~ *Tsuga*). Pinaceae. 1 W China: *C. argyrophylla* Chun & Kuang – embryo & pollen like *Pinus*, wood anatomy & external morphology of female cones like *Picea*, wood like *Pseudotsuga*; discovered 1955 in Kwangsi & also known as fossil from Germany. R: NRBGE 45(1988)385

Cathayanthe Chun. Gesneriaceae. 1 Hainan: *C. biflora* Chun

Cathayeia Ohwi = *Idesia*

Cathcartia Hook.f. = *Meconopsis*

Cathedra Miers. Olacaceae. 11 trop. S Am.

Cathestecum J. Presl. Gramineae (33c). 5 S US to Guatemala. R: JWAS 27(1937)495

Cathormion (Benth.) Hassk. = *Albizia*

Catila Ravenna = *Calydorea*

cativo *Prioria copaifera*

catjang *Cajanus cajan*

catmint, catnep, catnip *Nepeta cataria* & other *N.* spp.

Catoblastus H. Wendl. Palmae (V 2b). 17 trop. S Am.

Catocoryne Hook.f. Melastomataceae. 1 Peru

Catoferia (Benth.) Benth. Labiatae (VIII 4). 4 C Am. R: KB 41(1986)299. A strongly exserted

Catopheria Benth. = *praec.*

Catophractes D. Don. Bignoniaceae. 1 trop. & S Afr.

Catopsis Griseb. Bromeliaceae (2). 19 trop. Am. R: FN 14(1977)1366. Cult. orn. incl. *C. berteroniana* (Schultes) Mez (S Florida to Braz.) – carnivore (absorbing organic matter through 'tank') at top of *Rhizophora* mangroves in Florida, infl. 90 cm above lvs so poll. insects clearly separated from prey (cf. *Cephalotus*)!

Catospermum Benth. = *Goodenia*

Catostemma Benth. Bombacaceae. 11 N S Am. R: AMBG 74(1987)636

Catostigma Cook & Doyle = *Catoblastus*

cat's claw *Acacia greggii*; **c. ear** *Hypochaeris* spp.; **c. eyes** *Dimocarpus longan* var. *malesianus*; **c. foot** *Antennaria dioica*; **c.paw** *Anigozanthos humilis*; **c. tail** *Amaranthus caudatus*, *Bulbinella* spp., *Typha* spp.; **c.t. grass** *Koeleria phleoides*, *Phleum* spp., *Rostraria cristata*; **c.t. millet** *Setaria pumila*; **c. whiskers** *Cleome gynandra*

Cattleya Lindley. Orchidaceae (V 13). 45 C & S Am. R: C.L. Withner (1988) *The Cs & their relatives* I. Epiphytes (mostly) with 1–3 thick lvs per pseudobulb & showy fls, much cult. & hybridized, the most familiar orchid of button-holes & bouquets being derived from large-flowered spp. with 1 leaf per pseudobulb. *C. bowringiana* O'Brien (C Am.), a 2-leaved sp. has up to 47 fls per spike (over 600 fls recorded in a season from 1 plant) its floriferousness leading to its being used in over 100 registered hybrids involving almost all the other spp. of *C.* as well as spp. in the closely allied *Epidendrum*, *Laelia* & *Sophronotis*; a terrestrial sp., it was overcollected in the wild but cultivation has saved it from extinction

Cattleyopsis Lemaire = *Broughtonia*

catuaba herbal med. based on diff. pls in diff. parts trop. Am., e.g. spp. of *Anemopaegma*, *Erythroxylum*, *Ilex*, *Micropholis*, *Secondatia*, *Tetragastris* (see KB 45(1990)186)

Catunaregam Wolf. Rubiaceae (II 1). 5–6 trop. Afr. to As. Incl. geoxylic suffrutex. *C. spinosa* (Thunb.) Tirv. (*Xeromphis s.*, C & S Afr., India, China) – molluscicidal saponins

cauassu *Calathea lutea*

Caucaea Schltr. Orchidaceae (V 10). 1 Colombia

Caucaliopsis H. Wolff = *Agrocharis*

Caucalis L. Umbelliferae. (III 3). 1 S Eur.: *C. platycarpos* L. (bur parsley) – casual in GB

Caucanthus Forssk. Malpighiaceae. 3 E & NE Afr. to Arabia

cauchao *Amomyrtus luma*

Caulanthus S. Watson. Cruciferae. 15 W US. *C. procerus* S. Watson ed. by Nevada Indians. R: AMBG 9(1922)283

Caularthron Raf. Orchidaceae (V 13). 3 trop. Am. Cult. orn. epiphytes incl. *C. bilamellatum* (Reichb.f.) R. Schultes (C Am. to Colombia) – facultative myrmecophyte with extrafl. nectar making up almost half diet of ants at some times of year

cauliflower *Brassica oleracea* Botrytis Group

Caulocarpus Baker f. = *Tephrosia*

Caulokaempferia Larsen. Zingiberaceae. 10 Himalaya to SE As.

Caulophyllum Michaux. Berberidaceae (II 1; Leonticaceae). 3 NE As., E N Am. (2). R: Rhodora 87(1985)463. *C. thalictroides* (L.) Michaux (blue cohosh, papoose root, squaw root, E N Am.) – dried rhiz. diuretic etc.; cult. orn. Ovary wall splits early to expose 2 large blue drupe-like seeds borne on sturdy funicles

Caulostramina Rollins. Cruciferae. 1 SW US

Caustis R. Br. Cyperaceae. 7 Aus. Protected spp. *C. blakei* Kükenthal ex S.T. Blake (NE Aus.) – dried & coloured to be sold in NSW as 'koala fern'

Cautleya Hook.f. Zingiberaceae. 5 Himal. Some cult. orn. like *Roscoea*

Cavacoa Léonard. Euphorbiaceae. 3 trop. Afr. R: BJBB 25 (1955)320

Cavalcantia R. King & H. Robinson (~ *Eupatorium*). Compositae (Eup.-Ager.). 2 Brazil. R: MSBMBG 22(1987)154

Cavanillesia Ruíz & Pavón. Bombacaceae. 3 trop. Am. *C. platanifolia* (Bonpl.) Kunth (cuipo, quipo, C Am.) – fr. mucilage improves water uptake & speeds germ., soft pith-like wood used for canoes & floating rafts of heavier timbers, washed up in Azores, poss. balsa subs.

Cavea W.W. Sm. & Small. Compositae (Asteroideae-tribe?). 1 E Himalaya

Cavendishia Lindley. Ericaceae. 100 trop. Am. R: FN 35(1983). Hummingbird-poll.; some cult. orn.

Cayaponia A. Silva Manso. Cucurbitaceae. 45 warm Am., 1 trop. W Afr. Bat-poll.? Some med. *C. glandulosa* (Poeppig & Endl.) Cogn. (Colombia) – crushed lvs & stems an insect repellent; *C. kathematophora* R. Schultes (? cultigen in Colombian Amaz.) – seeds for necklaces & anklets; *C. ophthalmica* R. Schultes (? cultigen in NW Amaz.) – treatment of conjunctivitis

cay-cay fat *Irvingia malayana*

Cayenne pepper *Capsicum annuum* Longum Group

Caylusea A. St-Hil. Resedaceae. 4 Cape Verde Is., N & E Afr. to India

Cayratia Juss. Vitaceae. 50 OW trop. Some ed. fr.

Ceanothus L. Rhamnaceae. 55 N Am. esp. W. R: M. Rensselaer & H.E. McMinn (1942) C. Calif. lilacs. Alks. Decid. & evergreen shrubs & small trees with nitrogen-fixing Actinomycetes (*Frankia*) in roots, sometimes spiny, the best blue-flowered shrubs for Br. gardens incl. hybrids esp. scented *C.* × *delilianus* Spach 'Gloire de Versailles' (*C. americanus* × *C. coeruleus* Lag., Mex. & Guatemala); *C. americanus* L. (E N Am.) – lvs fresh or dried used as a tea by N Am. Indians, also med. suggested remedy for lung-bleeding. In wild imp. constituents of chaparral

ceara rubber *Manihot glaziovii*

Ceballosia Kunkel = *Tournefortia*

Cecarria Barlow. Loranthaceae. 1 Philipp., Flores & Timor to N Queensland. Commemorates C.E. Carr, botanist

Cecropia Loefl. Cecropiaceae. c. 75 trop. Am. Dioec. fast-growing pioneer trees (some fish-disp. in Amaz.) with hollow septate twigs usu. inhabited by ants & with stilt-roots. The ants (*Azteca* spp. prob. with carnivorous ancestry) feed on food-bodies at petiole-bases (food is glycogen!) & attack other grazers esp. leaf-cutting ants. The hollow internodes are burrowed into by pregnant female, which raises brood therein. In spp. with waxy stems (difficult for leaf-cutters to climb) there are neither food-bodies nor a thin area at top of internode for *Azteca* penetration; moreover *C. peltata* L. (trumpet tree) in Puerto Rico has 98% of its trees without these symbiotic traits whereas in Trinidad & trop. Am. mainland all trees have them (cf. *Musanga*). *C. peltata* – pulp & cult. orn., replacing *Musanga cecropioides* in SW Cameroun (? relatively pest-free): *C. insignis* Liebm. (C Am.) – establishes only in gaps at least 215m^2 in forest; *C. sciadophylla* C. Martius (Brazil) – often left when forest felled for agric. as lvs a source of alkali for coca in Amaz. where limestone rare

Cecropiaceae C. Berg (~Urticaceae). Dicots – Hamamelidae – Urticales. 6/180 trop. Dioec. trees, shrubs or lianes, s.t. epiphytic with aerial &/or stilt roots & rather restricted latex system. Lvs spirally arr., simple & entire to deeply palmatifid, venation pinnate or palmate; stipules united. Infls cymose, the females sometimes ± capitate. K 2–4 or (2–4), A 1–4, filaments straight in bud, $\underline{G}$ 1 with 1 ± basal, bitegmic ± orthotropous ovule. Fr. a nutlet sometimes adnate to fleshy accrescent K, the seed with straight embryo ± endosperm. x = 7. R: Taxon 27(1978)39

Genera: *Cecropia, Coussapoa, Musanga, Myrianthus, Poikilospermum, Pourouma*

Form. treated as subfam. Conocephaloideae of Moraceae & later moved to Urticaceae, being intermediate between them

Pioneer trees with pith-like timbers

cedar *Cedrus* spp.; **c. acacia** *Paraserianthes toona*; **(E) Afr. c.** *Juniperus procera*; **Atlantic c.** *Cedrus atlantica*; **Aus. c.** *Toona ciliata*; **bastard c.** *Guazuma tomentosa, Soymida febrifuga*; **Bermuda c.** *Juniperus bermudiana*; **Burma** or **Moulmein c.** *T. ciliata*; **Chilean c.** *Austrocedrus chilensis*; **Clanwilliam c.** *Widdringtonia cedarbergensis*; **Cyprus c.** *C. brevifolia*; **c. elm** *Ulmus crassifolia*; **incense c.** *Calocedrus decurrens*; **Jap. c.** *Cryptomeria japonica*; **c. of Lebanon** *Cedrus libani*; **Mlanje c.** *Widdringtonia nodiflora*; **Port Orford** or **Oregon c.** *Chamaecyparis lawsoniana*; **red c.** *Juniperus virginiana* (US), *Toona ciliata* (Aus.); **Siberian c.** *Pinus cembra*; **Virginian pencil c.** or **red c.** *Juniperus virginiana*; **c. wattle** *Acacia terminalis*; **WI c.** *Cedrela odorata*; **western red c.** *Thuja plicata*; **white c.** *Melia azedarach, Calocedrus decurrens, Chamaecyparis thyoides, Chukrasia tabularis, Thuja occidentalis*, etc.; **yellow c.** *Chamaecyparis nootkatensis*

cedrat *Citrus medica*

Cedrela P. Browne. Meliaceae (IV 1). 8 trop. Am. R: FN 28(1981)360. Good timber esp. *C. odorata* L. (WI cedar) used in cigar-boxes, moth-proof chests etc.; domatia, moth-poll. in Costa Rica, natur. Medit., e.g. Stromboli. OW spp. now referred to *Toona*

Cedrelinga Ducke. Leguminosae (II 5). 1 Brazil

Cedrelopsis Baillon. Ptaeroxylaceae. 8 Madagascar. R: F Mad 107 bis (1991)97

cedron *Quassia cedron*

Cedronella Moench. Labiatae (VIII 2). 1 Macaronesia: *C. canariensis* (L.) Webb (*C. triphylla*, Canary balm) – shrubby, lvs 3-foliolate, form. used as a tea, cult. orn.

Cedronia Cuatrec. = *Picrolemma*

Cedrus Trew. Pinaceae. 4 (or 1–2) mts N Afr. to As. Cedars. Evergreen trees with long & short shoots of needle lvs, the short with capacity to develop into long. Infls solit. in position of short shoots; cone of closely enwrapped scales, each with 2 seeds, developing in 2–3 yrs. Cult. orn. esp. as specimen trees, when rapidly attaining an air of antiquity. *C. atlantica* (Endl.) Carr. (Atlantic or Atlas c., N Afr.) & 'Glauca', *C. brevifolia* (Hook.f.) Henry (Cyprus c., Cyprus), both sometimes considered as geographical sspp. (ssp. *atlantica* (Endl.) Battand. & Trabut & ssp. *brevifolia* (Hook.f.) Meikle, usu. differing in somewhat downy shoots & smaller lvs resp.) of *C. libani* A. Rich. (cedar of Lebanon, As. Minor, timber used in Solomon's temple but only 14 remnant groves from Tripolis to Sidon left) – all useful timber (allergenic to some) & some oils for scent etc. *C. deodara* (D. Don) G. Don f. (*C. libani* ssp. *deodara* (D. Don) Sell, deodar, Himal.) – reaching 12 m girth, an imp. timber tree in India. P. Maheshwari & C. Biswas (1970) *Cedrus*

Ceiba Miller. Bombacaceae. (Incl. *Chorisia*) 11+ trop. Am., 1 ext. to Afr. R: NRBGE 45(1988)134. Bat-poll. & *C. pentandra* (L.) Gaertner poll. by non-flying mammals in SE Peru rain forest. Seeds in hairs from carpel walls, the kapok of commerce esp. from *C. pentandra* (trop. Am. & Afr., where at 70 m the tallest tree in the Continent & revered as habitat of spirits) – the cult. plant app. derived from a savanna form crossed with a rain forest one in Africa (lemur-poll. in Madag.) & taken thence to Indonesia (prob. by AD 500) etc. (silk-cotton tree) – wind-disp., kapok used in stuffing mattresses, in insulation etc., seeds an oil-source, wood used for matches, canoes etc.; young fr. ed., shoots c. 25% protein dry weight & imp. food for red spider monkeys in C Am.; *C. insignis* (Kunth) Gibbs & Semir (*Chorisia i.*, *C. speciosa*, paina de seda, yachan, trop. Am.) – kapok used e.g. for arrow-proof jackets by Matico Indians, & silky cotton, cult. orn. tree with swollen often spiny trunk & white to yellow or red fls before lvs

Celaenodendron Standley. Euphorbiaceae. 1 Mexico

celandine *Ranunculus ficaria*; **greater c.** *Chelidonium majus*; **tree c.** *Bocconia frutescens*

Celastraceae R. Br. Dicots – Rosidae – Celastrales (incl. Hippocrateaceae). 88/1300 trop. & temp. (fewer). Trees, shrubs & lianes, usu. glabrous & with laticifers. Lvs spirally arr. or opp. (even in a single genus, e.g. *Cassine*), rarely much reduced (*Canotia*); stipules small or 0. Fls usu. small, in cymes, rarely racemes or infls coiled tendrils (*Lophopyxis*), rarely solit. & axillary, rarely unisexual, regular, hypogynous to occ. semi-epigynous, K (2–)5, imbricate (rarely valvate), s.t. with a basal tube, C (2–)5, usu. imbricate or valvate, rarely 0, A on, without, or within a nectary-disk, usu. alt. with C, seldom in 2 whorls or with set of staminodes opp. K, (2)3(4) or 5, when 3 aligned with sides of ovary, filaments s.t. ± connate at base, anthers extrorse or introrse, with longit. (transverse in *Hippocratea* & *Salacia*) slits, pollen often in tetrads or polyads, $\underline{G}$ (2–5), rarely half-inf. with as many locules as G (rarely all save 1 abortive) with axile placentation & as many stigmas from 1 style (free styles in *Goupia*), ovules (1)2–10(–∞, *Goupia*), anatropous, bitegmic (exotegmic). Fr. a berry, capsule, drupe or samara; seeds with arils, wings or angular or compressed, embryo with large cotyledons, endosperm ± oily or 0. x = 8, 12, 14

Chief genera: *Cassine*, *Celastrus*, *Crossopetalum*, *Euonymus*, *Hippocratea*, *Maytenus*, *Salacia*, *Tontelea*. Rather 'diversified and loosely knit' (Cronquist); various genera have been removed to separate fams at diff. times, esp. *Hippocratea* & *Salacia* with transverse anther dehiscence to Hippocrateaceae, though *Campylostemon* & *Sarawakodendron* intermediate in many characters; *Canotia* (minute lvs), *Goupia* (ovules ∞, styles free), *Lophopyxis* (infl. tendrils) & *Siphonodon* (locules divided horizontally into 2 superposed 1-ovulate locelli) have been put in monotypic fams

Cult. orn. esp. *Euonymus*, some medic. esp. *Maytenus*, some timbers & dyes, khat (*Catha*)

Celastrus L. Celastraceae. 32 trop. to warm temp. R: AMBG 42(1955)215. Cult. orn. lianes for pergolas etc. with fr. like *Euonymus*, some medic. *C. angulatus* Maxim. (NW & C China) & *C. glaucophyllus* Rehder & E. Wilson (W China) – insecticidal seed-oil & root-bark; *C. orbiculatus* Thunb. (*C. articulatus*, E As.) – natur. N Am.

celeriac *Apium graveolens* var. *rapaceum*

Celerina Benoist. Acanthaceae. 1 Madagascar

celery *Apium graveolens* var. *dulce*; **Chinese c.** *Oenanthe javanica*; **c.-top pine** *Phyllocladus aspleniifolius*

Celianella Jabl. Euphorbiaceae. 1 Venezuela

Celiantha Maguire. Gentianaceae. 3 Guayana Highland. R: MNYBG 32(1981)382

cellophane plant *Echinodorus berteroi*

Celmisia Cass. Compositae (Ast.-Ast.). 62 Aus. (3), NZ (59). R: NZJB 7(1969)400. Cult. orn., seeds not long-lived

Celome E. Greene = *Cleome*

Celosia L. Amaranthaceae (I 1). 45 warm Am. & Afr. Ovules 2–∞. *C. argentea* L. (trop.) – weedy polyploid complex largely octoploid (tetraploid in C India) – lvs used as spinach in Sri Lanka & E Afr. (mfungu); *C. cristata* L. (*C. argentea* var. *cristata*, cockscomb, a tetraploid cultigen of cvs app. derived from *C. argentea* with white to yellow, purple or red variously plumed (e.g. Plumosa Group) or fasciated (Cristata Group) etc. infls, much grown as potpls

Celsia L. = *Verbascum*

Celtis L. Ulmaceae (II). 100 trop. (most), temp. (Eur. 4). Hackberries, nettle-trees, sugarberries, often with ed. fr., a yellow dye from bark & timber good for charcoal. *C. australis* L. (lote-tree, Medit.) – widely planted street-tree in Medit. for shade, timber valuable, fr. the lotus of the Ancients (the lotophagi forgot their homeland); *C. occidentalis* L. (beaver wood, N Am.) – wood for fences & fuel; *C. zenkeri* Engl. (trop. Afr.) – powerful fetish tree in Ghana

celtuce *Lactuca sativa*

Cenarrhenes Labill. Proteaceae. 1 Tasmania

Cenchrus L. Gramineae (34f). (Incl. *Echinaria*) 30 warm & dry Am., Afr. & India (1 natur. Eur.). R: ISUJS 37(1963)259. Spikelet surrounded by involucre of sterile spikelets which are hardened & spiky in some spp. at maturity & act as burs in animal disp. (bur grasses, hedgehog grass), esp. troublesome in sheep-rearing areas of N Am. *C. ciliaris* L. (buffel grass, Afr. to India) – drought-tolerant forage (e.g. NSW) & soil stabilizer

Cenia Comm. ex Juss. = *Cotula*

Cenocentrum Gagnepain. Malvaceae. 1 SE As.

Cenolophium Koch. Umbelliferae (III 8). 1 Euras.: *C. denudatum* (Hornem.) Tutin

Cenolophon Blume = *Alpinia*

Cenostigma Tul. Leguminosae (I 1) 6. Brazil, Paraguay

Centaurea L. Compositae (Card.-Cent.). c. 500 Med. (Eur. (incl. *Cnicus*, excl. *Cheirolophus*) 218, Turkey 172) & Near E, N Euras. (few), trop. Afr., N Am. (2), Aus. (1). Herbs & some subshrubs; all fls tubular & bisexual or outer ones enlarged, ray-like & sterile. Many cult. orn. (bluebottle, knapweed; see also *Amberboa*). *C. calcitrapa* L. (star-thistle, Medit.) – long spiny involucral bracts, young stems ed. Egypt; *C. babylonica* (L.) L. (Turkey & Syria) – to 4 m, unbranched; *C. benedicta* (L.) L. (*Cnicus b.*, blessed thistle, Medit.) – form. imp. drug due to glucoside cnicin used in treatment of gout & as tonic, fr. an oil-source, with caruncle; *C. cyanus* L. (cornflower, bachelor's buttons, Euras.) – ann., form. common cornfield weed with many colour forms (blue, white, purple or pink) in cult.; *C. iberica* Trev. ex Sprengel (Medit.) – spiny, the 'thistle' of Genesis; *C. montana* L. (C Eur.) – common perennial cult.; *C. nigra* L. (hardheads, knapweed, Eur. natur. N Am.) – diuretic; *C. solstitialis* L. (St. Barnaby's thistle, S Eur.) – subcosmop. allelopathic weed noxious in Aus.

Centaurium Hill. Gentianaceae. 20 N hemisph., 1 ext. to Aus., another to Chile (Eur. 14). Some cult. orn. esp. *C. scilloides* (L.f.) Samp. (Azores, W Eur.) in rock gardens, others mostly ann. or bienn., some med.

Centaurodendron Johow. Compositae (Card.-Cent.). 3 (incl. *Yunquea*) Juan Fernandez. Pachycaul trees

Centauropsis Bojer ex DC. Compositae (Vern.-Vern.). 8 Madag. Shrubs

Centaurothamnus Wagenitz & Dittr. Compositae (Card.-Cent.). 1 Arabia

Centella L. Umbelliferae (I 1a). 40 warm, mostly S Afr., *C. asiatica* (L.) Urban (pantrop. to Chile, NZ etc.) grown as cover crop, lvs ed., extracts (largely from Madag.) used in western skin ointments to promote healing & in folk med. to treat leprosy, active agents being pentacyclic triterpenoid derivatives esp. asiaticoside, though depressant of central nervous system

Centema Hook.f. Amaranthaceae (I 2). 2 trop. Afr.

Centemopsis Schinz. Amaranthaceae (I 2). 3 trop. Afr. R: KB 36(1982)681

Centipeda Lour. Compositae (Asteroideae-tribe?). 6 Madag., Indomal., Pacific, Chile. Ground as snuff subs. (sneezeweed) in Aus. *C. minima* (L.) A. Br. & Asch. (*C. orbicularis*, trop. As. to Aus., widely introd.) – med. in China

centipede grass *Eremochloa ophiuroides*

Centosteca Desv. = *Centotheca*

Centotheca Desv. Gramineae (24). 1–4 OW trop. Lemma bristles disp. mechanism

Centradenia G. Don f. Melastomataceae. 6 Mex., C Am. R: JAA 58(1977)73. Opp. lvs of uneven sizes. Some cult. orn.

Centradeniastrum Cogn. Melastomataceae. 1 W trop. S Am.

Centranthera R. Br. Scrophulariaceae. 9 China to Aus. R: Ann. vol. ARBGC (1942)53

Centrantheropsis Bonati = *Phtheirospermum*

Centranthus DC. Valerianaceae. 9 Medit. & Eur. (8). R: BJLS 71(1976)211. C with spurred base with nectar, the C-tube with a partition separating the style from a passage of downward-pointing hairs leading to spur only penetrable by long-tongued insects. *C. ruber* (L.) DC (red valerian, Medit., Eur.) – cult. orn. natur. on walls in GB & Calif.

Centratherum Cass. Compositae (Vern.-Cent.) 2 trop. NW, Aus., Philipp. R: Rhodora 83(1981)14

Centrilla Lindau = *Justicia*

Centrochloa Swallen. Gramineae (34b). 1 Brazil

Centrogenium Schltr. = *Stenorrhynchos*

Centroglossa Barb. Rodr. Orchidaceae (V 10). 6 Brazil, Peru, Paraguay

Centrolepidaceae Endl. Monocots – Commelinidae – Restionales. 3/36 SE As., Mal., Pacific, temp. S Am. Tufted grass-or moss-like herbs with short erect rhiz. Lvs spirally arr. with ± well-defined open basal sheath, often with an adaxial ligule between it & lamina. Infl. a term. spike or head subtended by 1–several leafy bracts with vestigial or 2–many distichous bracts subtending individual fls or spikelets. Fls wind-poll. or selfed, P 0, A 1 or G 1 often in bisexual or female heads, ovule solitary, pendulous, orthotropous, bitegmic. Fr. small, dry, usu. dehiscing abaxially to release seed with copious starchy endosperm, 0 perisperm, embryo small, conical, incompletely developed with shield-like cotyledon capping or lying alongside endosperm. x = 10–13

Genera: *Aphelia*, *Centrolepis*, *Gaimardia* (*Hydatella* & *Trithuria* here treated as Hydatellaceae)

Mostly in nutrient-poor soils, on mts in trop.

Centrolepis Labill. Centrolepidaceae. 27 SE As., Mal. (mts), Aus. (19, endemic), NZ (3)

Centrolobium Mart. ex Benth. Leguminosae (III 4). 6 trop. Am. Legume spiny to 30 cm with large wing. *C. robustum* (Vell.) Mart. ex Benth. (zebrawood, Brazil) – timber for furniture, boat-building etc.

Centronia D. Don Melastomataceae. 15 C & W trop. Am., 1 Guianas

Centropetalum Lindley = *Fernandezia*

Centroplacus Pierre. Pandaceae. 1 W Afr.

Centropodia (R. Br.) Reichb. Gramineae (25). 4 Afr. to C As. R: KB 37(1982)658. Kranz anatomy

Centropogon C. Presl. Campanulaceae. 230 trop. Am. Herbs & shrubs, hummingbird-poll.; some cult. hanging baskets, some ed. fr.

Centrosema (DC) Benth. Leguminosae (III 10). 35 warm Am. Pedicel twists through 180° so that standard points downwards (cf. *Clitorea*). Some used as green manures (butterfly peas) esp. *C. plumieri* (Pers.) Benth. under rubber & coconuts. *C. carajasense* Cavalc. (Brazil) – fish-poison in Amazon

Centrosolenia Benth. = *Nautilocalyx*, see also *Episcia*

Centrospermae Eichler = Dicots – Caryophyllidae – Caryophyllales (see Cronquist's System, this volume p. **772**)

Centrostachys Wallich. Amaranthaceae (I 2). 1 N Afr., India, Java, Norfolk Is.

Centrostegia A. Gray ex Benth. (~ *Chorizanthe*). Polygonaceae (I 1). 1 SW N Am.: *C. thurberi* A. Gray ex Benth. R: Phytol. 66(1989)207

Centrostemma Decne = *Hoya*

Centrostigma Schltr. Orchidaceae (IV 2). 5 trop. Afr.

Centunculus L. = *Anagallis*

century plant *Agave americana*

Ceodes Forster & Forster f. = *Pisonia*

Cephaelis Sw. = *Psychotria*

Cephalacanthus Lindau. Acanthaceae. 1 Peru

Cephalanthera Rich. Orchidaceae (V 1). 14 temp. Euras. (Eur. 5). Helleborines. Rostellum 0; pollen germinates *in situ*. *C. rubra* (L.) Rich. (red helleborine) protected in GB; *C. austinae* (A. Gray) Heller (N Am.) & *C. calcarata* S.C. Chen & K.Y. Lang (Yunnan, coll. once) – mycotrophs

Cephalantheropsis Guillaumin. Orchidaceae (V 11). 6–8 SE As.

Cephalanthus L. Rubiaceae (III 6). 6 trop., N Am. R: Blumea 23(1976)179. Some rheophytes. *C. occidentalis* L. (button bush, N Am.) – cult. orn., medic. (laxative etc.)

Cephalaralia Harms. Araliaceae. 1 E Aus.: *C. cephalobotrys* (F. Muell.) Harms

Cephalaria Schrader ex Roemer & Schultes. Dipsacaceae. c. 65 Medit. (Eur. 14), to C As., Ethiopia, S Afr. Some cult. orn. *C. syriaca* (L.) Roemer & Schultes (Turkey to Iran & N Afr.) – imp. cornfield weed

Cephalipterum A. Gray (~ *Rhodanthe*). Compositae (Gnap.-Ang.). 1 W & S Aus.: *C. drummondii* A. Gray. R: Nuytsia 8(1992)417

Cephalobembix Rydb. = *Schkuhria*

Cephalocarpus Nees. Cyperaceae. 7 trop. S Am. Habit of *Dracaena*

Cephalocereus Pfeiffer. Cactaceae (III 8). 3 Mex. Cult. orn. esp. *C. senilis* (Haw.) Schumann (old man cactus, C Mex. where reaching 15 m). See also *Pilosocereus*

Cephalocleistocactus F. Ritter = *Cleistocactus*

Cephalocroton Hochst. Euphorbiaceae. 6 trop. Afr. to Sri Lanka. *C. cordofanus* Hochst. (trop. Afr.) – seeds ed., oil source (Sudan)

Cephalocrotonopsis Pax = *Cephalocroton*

Cephalodendron Steyerm. Rubiaceae (II 1). 2 Guayana Highland

Cephalohibiscus Ulbr. = *Thespesia*

Cephalomanes C. Presl. Hymenophyllaceae (Hym.). Incl. *Abrodictyum, Callistopteris, Davalliopsis, Nesopteris* & *Selenodesmium*, 60 trop. esp. OW. R(subg. C.): JJB 66 (1991)134

Cephalomappa Baillon. Euphorbiaceae. 5 S China, Mal.

Cephalopappus Nees & Mart. Compositae (Mut.-Nass.). 1 NE Brazil. Herb, pappus 0

Cephalopentandra Chiov. Cucurbitaceae. 1 NE trop. Afr.

Cephalophilon (Meissner) Spach = *Persicaria*

Cephalophyllum (Haw.) N.E. Br. Aizoaceae (V). 30 S Namibia & nearby. R: MIABH 22(1988)93. Cult. orn. leaf-succs. See also *Jordaaniella*

Cephalopodum Korovin. Umbelliferae (III 8). 2 C As.

Cephalorhizum Popov & Korovin. Plumbaginaceae (II). 2 C As.

Cephalorrhynchus Boiss. Compositae (Lact.-Lact.). c. 15 SE Eur. (1), SW As. to Himal.

Cephaloschefflera (Harms) Merr. = *Schefflera*

Cephalosorus A. Gray (~ *Angianthus*). Compositae (Gnap.-Ang.). 1 W Aus.: *C. carpesioides* (Turcz.) Short – annual with ectomycorrhizae. R: OB 104(1991)129

Cephalosphaera Warb. (~ *Brochoneura*). Myristicaceae. 1 trop. E Afr. mts: *C. usambarensis* (Warb.) Warb. (mtambara) – timber

Cephalostachyum Munro = *Schizostachyum*

Cephalostemon Schomb. Rapateaceae. (Incl. *Duckea*) 9 trop. S Am.

Cephalostigma A. DC = *Wahlenbergia*

Cephalotaceae Dumort. Dicots – Rosidae – Rosales. 1/1 SW Aus. Insectivorous herb with short rhiz. Lvs rosetted, the inner flat, simple, the outer being ground-level pitchers at first closed by lid, its lower surface & the distal inner surface of the pitcher slippery with overlapping, downwardly-directed projections from the epidermal cells. Multicellular glands on pitcher surfaces, petiole & lower surface of other lvs; flask-shaped ones within the pitcher esp. in bright-coloured cushion-like projections. Infl. arising from centre of rosette with apical racemes of dichasia of small reg. perigynous fls, the hypanthium appearing as a K-tube with K 6, valvate, as 'lobes', C 0, A 6 + 6, unequal, at summit of hypanthium above glandular setose disk, $\underline{G}$ 6 with circinnate styles, each G with 1(2) basal, erect, anatropous, bitegmic ovule. Fr. a hairy follicle, seeds with v. small straight embryo surrounded by copious fleshy endosperm; n = 10

Only sp.: *Cephalotus follicularis*; related to Crassulaceae & Saxifragaceae, though A structure diff.

Cephalotaxaceae Neger (~Taxaceae). Pinopsida. 2/10 E & SE As. Small resinous, dioec. evergreen trees or shrubs with opp. or whorled branches, dioec. (sometimes trees at first male, later female). Lvs linear, ± decussate. Female cones with a few decussate bracts, each subtending a pair of erect ovules on v. short secondary fertile short shoots, usu. only 1 ovule developing into a large olive-like seed, the outer fleshy layer around a stony one; male cones of 7–12 spirally arr. microsporophylls, each with 2 or 3 pollen-sacs, the pollen without air-bladders or any prothalial cells, the grains with only tube- & generative nuclei at pollination, which is followed by fertilization after the ensuing winter. Embryo large with 2 cotyledons; n = 12

Genera: *Amentotaxus, Cephalotaxus* (fossils also from Eur. & W N Am.)

Once thought related to *Torreya* (Taxaceae) & wood has some similarities, but the ovules are in cones & not borne terminally

Cephalotaxus Siebold & Zucc. ex Endl. Cephalotaxaceae. 6 E Himal. to Japan. R: Phytol. M 7(1984)19. Plum yews. Cult. orn. Lvs with 2 broad glaucous-green lines beneath. *C. harringtonia* (Forbes) K. Koch (var. *harringtonia* known only in cult.) & var. *drupacea* (Siebold & Zucc.) Koidz. (China, Japan) – most widely cult.

Cephalotomandra Karsten & Triana. Nyctaginaceae (VI). 1–3 C Am.

Cephalotus Labill. Cephalotaceae. 1 SW Aus.: *C. follicularis* Labill. Discovered by Robert

Brown in 1801; restricted to swampy coastal tracts between Donelly River & Cheyne Beach E of Albany. Pitchers develop from July to January; though plants can survive without insect food, the parallel with *Nepenthes* is remarkable even as far as to having a digestive-juice-resistant insect (gadfly, cf. mosquitoes in *N.*) and algae living in the pitchers; such pitchers arise by single gene mutants in e.g. *Codiaeum variegatum* (L.) Blume, but these do not have digestive glands; infl. to 60 cm so poll. insects separated from prey

Ceradenia L.E. Bishop. Grammitidaceae. 57 trop. Am. R: AFJ 78(1988)1

Ceranthera Elliott = *Dicerandra*

Ceraria Pearson & Stephens. Portulacaceae (I). 4 trop. & S Afr. Trees & shrubs

Cerastium L. Caryophyllaceae (III 1). c. 100 almost cosmop. (Eur. 58). Many weeds (chickweed, mouse-ear c. esp. *C. fontanum* Baumg. ssp. *vulgare* (Hartman) Greuter & Burdet (common m.e.)), some cult. orn. esp. *C. tomentosum* L. (snow-in-summer, Sicily, Italy) – rampant rock-pl.

Cerasus Miller = *Prunus*

Ceratandra Ecklon ex Bauer. Orchidaceae (IV 3). 2 SW to E Cape

Ceratandropsis Rolfe = *praec.*

Ceratanthus F. Muell. ex G. Taylor. Labiatae (VIII 4c). 10 SE As., Mal.

Ceratiola Michaux. Empetraceae. 1 SE US. Dioec., cult. orn.

Ceratiosicyos Nees. Achariaceae. 1 S Afr.

Ceratocapnos Durieu (~ *Corydalis*). Fumariaceae (I 2). 3 W Eur. (1) & Medit. R: OB 88(1986)38. Scrambling annuals with leaf tendrils

Ceratocarpus L. Chenopodiaceae (I 3). 2 E Eur. (1), temp. As.

Ceratocaryum Nees. Restionaceae. 5 SW & S Cape. R: Bothalia 15(1985)479

Ceratocentron Senghas. Orchidaceae (V 16). 1 Philippines

Ceratocephala Moench (~ *Ranunculus*). Ranunculaceae (II 3). 3 C Eur. (2), Medit. to NW China, NZ (1), natur. N Am.

Ceratocephalus Pers. = *praec.*

Ceratochilus Blume. Orchidaceae (V 16). 1 Mal.

Ceratochloa DC & Pal. = *Bromus*

Ceratocnemum Cosson & Bal. Cruciferae. 1 Morocco

Ceratogyne Turcz. Compositae (Ast.-Ast.). 1 W temp. Aus.: *C. obionoides* Turcz.

Ceratoides Gagnebin = *Ceratocarpus*

Ceratolacis (Tul.) Wedd. Podostemaceae. 1 Brazil

Ceratolobus Blume. Palmae (II 1d). 6 W Mal. R: KB 34(1979)1

Ceratominthe Briq. = *Xeropoma*

Ceratonia L. Leguminosae (I ?2). 2 Arabia & Somalia. Some mimosoid features. Relicts of Indomal. flora (flowers in autumn): *C. oreothauma* Hillc., J. Lewis & Verdc. (Oman & Somalia), notable for 3-colpate pollen & *C. siliqua* L. (carob, locust-bean (of the Bible, John the Baptist's 'husks that the swine did eat'), cultigen orig. from Oman?) – variable infl. type, ±bracteoles, variable nos of fl. parts & sex expression (1 sex usu. suppressed during development), variable pollen-grains, 'inverted' fr. etc.; fr. eaten & seeds disp. by bats (*Rousettus aegyptiacus*) pods (algaroba) full of juicy pulp containing sugar & gum (tragasol, a tragacanth subs.), used as fodder (e.g. for Wellington's cavalry in Peninsular War) & alcohol source, seeds a coffee subs., form sold as 'sweets' in GB & used as weights (? the original carats of jewellers), also yielding a diabetic flour also suitable for babyfood, timber for furniture etc., fls. semen-scented. See ITCJ 1(1980)15

Ceratopetalum Sm. Cunoniaceae. 5 New Guinea, E Aus. *C. apetalum* D. Don (coachwood) – light timber for furniture, veneers etc., characteristic caramel scent; *C. gummiferum* Sm. (Christmas bush, NSW) – cult. orn. (K red in fr.), protected sp.

Ceratophyllaceae Gray. Dicots – Magnoliidae – Ceratophyllales. 1/2–6 cosmop. Rootless floating submerged monoec. herbs, s.t. with root-like branches anchoring pl., vegetatively glabrous; stems usu. branched (only 1 per node). Lvs 1–4 times forked, rigid & often brittle in whorls of 3–10, with apical minute teeth & bristles; stomata & stipules 0. Fls unisexual, solit., extra-axillary, males & females usu. on alt. nodes; P 8–12, linear (s.t. interpreted as bracts), united at base, A (3–)10–20(–27), spirally arr. on flat receptacle, not clearly differentiated into filament & anther but connective extended into 2 points, extrorse, pollen smooth, inaperturate with no or reduced exine, $\underline{G}$ 1 with 1 orthotropous, unitegmic ovule. Fr. an achene with persistent spiny style & often other spines; seed with thin testa, 2 cotyledons from an annular common primordium, ± linear, radicle vestigial. x = 12

Only genus: *Ceratophyllum*

Form. considered reduced & specialized Nymphaeales, *Cabomba* providing 'a sort of link to the rest of the order' (Cronquist), but now thought a relictual group from pre-Dicot–Monocot split

Ceratophyllum L. Ceratophyllaceae. 2–6 variable (up to 30 recognized by some authors), cosmop. R: KB 40(1985)243. Hornworts, so-called from old translucent horny lvs. Plant decays as it grows apically, releasing laterals; winter buds not formed, the pl. sinking in autumn, rising in spring. Water-poll., the anthers breaking off & floating up through water, the pollen with same density as water. The floating veg. provides shelter for young fish but also for bilharzia-carrying snails & malarial mosquito larvae; it can rapidly choke waterways but, as in Arkansas, can be controlled using Chinese grass-carp. Both *C. demersum* L. (lvs 1- or 2-times forked, fr. usu. spiny) & *C. submersum* L. (lvs 3- or 4-times forked, fr. usu. smooth to warty) used as aquarium pls; often confused with *Myriophyllum* (Haloragidaceae), *Najas* (Najadaceae) or *Chara* (an alga), when sterile, but readily recognized by the forked lvs

Ceratophytum Pittier. Bignoniaceae. 1 Venezuela

Ceratopteris Brongn. Pteridaceae (II; Parkeriaceae). 4 trop. & warm. R: Brittonia 26(1974)139. Only homosporous aquatic (floating) ferns; fronds ed.: sterile ones simple to 3-pinnate, fertile ones larger & 4–5-pinnate; sporangia in 1–4 rows along veins. Some cult. aquarium pls. *C. cornuta* (Pal.) Lepr. (OW) – cult. & ed. Liberia; *C. pteridoides* (Hook.) Hieron. (trop. Am.) – covered 17 000 ha in a Surinam lake in 1966; *C. thalictroides* (L.) Brongn. (trop. As. & Pacific to S Japan) – much cult. in flooded rice-fields etc. as spring veg. (esp. Japan)

Ceratopyxis Hook.f. Rubiaceae (III 4). 1 W Cuba

Ceratosanthes Adans. Cucurbitaceae. 1 trop. Am.

Ceratosepalum Oliver = *Triumfetta*

Ceratostema Juss. Ericaceae. 23 S Am. mts, mostly E Ecuador. Some ed. fr.

Ceratostigma Bunge. Plumbaginaceae (I). 8 NE trop. Afr. (1), Tibet, China, SE As. Cult. orn. shrubby, often dying back in winter, esp. *C. plumbaginoides* Bunge (*Plumbago larpentiae*, W China, glabrous) & *C. willmottianum* Stapf (W China & Tibet, hairy lvs)

Ceratostylis Blume. Orchidaceae (V 14). c. 100 Indomal., W Pacific

Ceratotheca Endl. Pedaliaceae. 5 trop. & S Afr. R: MSB 25(1975)1. *C. sesamoides* Endl. (trop. Afr.) – cult. for seeds used like sesame; *C. triloba* (Bernh.) Hook.f. (S Afr.) – cult. orn., natur. N Am.

Ceratozamia Brongn. Zamiaceae (III). 10 Mex. to Belize. R: MNYBG 57(1990)201. *C. longifolia* Miq. (Mex.) male cone heats to 11.7 °C above ambient. Cult. orn. esp. *C. mexicana* Brongn. (Chiapas)

Cerbera L. Apocynaceae. 3–4 trop. coasts Indian & W Pacific Oceans. Poisonous trees & shrubs, frs common in drift, form. used as ordeal-poisons & for suicide. *C. manghas* L. (Seychelles to Pacific) – cult. orn. with fragrant white fls, wood used for vividly painted masks in S Sri Lanka (c.f. *Alstonia*)

Cerberiopsis Vieill. ex Pancher & Sébert. Apocynaceae. 3 New Caledonia. *C. candelabra* Vieill. ex Pancher & Sébert a tall branched tree of regular architecture but hapaxanthic; damaged younger pls can be found in flower

Cercestis Schott. Araceae (V 6). (Incl. *Rhektophyllum*) c. 10 W Afr. *C. afzelii* Schott – basket-making; cult. orn. esp. *C. mirabilis* (N.E. Br.) Bogner with irreg. perforated lvs

Cercidiphyllaceae Engler. Dicots – Hamamelidae – Trochodendrales. 1/2 E As. Dioec. tree with decid., simple lvs, palmately-veined & spiral on short shoots, pinnately-veined and opp. on long; stipules small, deciduous. Fls wind-poll. in term. infl. on short shoots maturing before or at same time as lvs; P 0, males in short raceme, 4 lower each subtended by 4-lobed bract, inner (upper) without, heads (individual fls diff. to recognize) with A 16–35 with latrorse anthers & longit. slits, females in pseudanthia with (2–)4(–8) K-like bracts, each subtending a naked carpel with decurrent 2-ridged stigma, ovules 15–30, in 2 rows, anatropous, bitegmic. Fr. of separate follicles; seeds flattened, winged, endosperm scanty, oily, embryo large, spatulate. 2n = 38

Only genus: *Cercidiphyllum*

Widely distrib. in N hemisph. in Tertiary. Seed anatomy & pollen etc. suggest placement in monotypic order

Cercidiphyllum Siebold & Zucc. Cercidiphyllaceae. 2 China, Japan. R: JAA 60(1975)367. *C. japonicum* Siebold & Zucc. – largest decid. tree in Japan (katsura), strong timber for house-interiors, furniture, etc., cult. orn. (though often inferior shrubby forms)

Cercidium Tul. = *Parkinsonia*

Cercis L. Leguminosae (I 3). 6 N temp. (Eur 1) to NE Mex. Orn. decid. trees with fls on branches & trunk (exc. *C. racemosa* Oliver, C China, with axillary racemes) before lvs expand; seeds with vestigial arils & lvs with pulvini at junction of lamina & petiole suggesting that leaf is terminal leaflet of an ancestral pinnate leaf; disjunct distr. (N Am., Medit. & E Asia) of v. homogeneous group of spp. not closely allied to any other genus, cauliflory (almost exclusively trop. trait) & arils etc. suggest C. a Tertiary relic cf. *Ceratonia siliqua*; fls superficially like Papilionoideae but standard ('back') C lies *inside* wings. *C. canadensis* L. (redbud, SE Canada to NE Mex.) – fls used in salads & pickles; *C. siliquastrum* L. (Judas tree, W Med. to E Bulgaria, Lebanon & Turkey) – cult. orn., legend that Iscariot hanged himself on it prob. a confusion between *Arbor Judae* & *Arbor Judaeae*, i.e. Judaea tree, as it was commonly cult. around Jerusalem), for in Medit. story associated with the fig (British legend has elder as the gibbet, 'Judas tree' in Kent, & edible fungus found on it called Jew's ear), serial buds in leaf axils

Cercocarpus Kunth. Rosaceae. 8 W & SW N Am. R: Brittonia 7(1950)91. Actinomycete root symbionts (*Frankia*) fix nitrogen. Wood for tool-handles etc., some cult. orn.

Cerdia Moçiño & Sessé ex DC. Caryophyllaceae (I 1). 4 Mexico

Cereus Miller. Cactaceae (III 3). 36 WI, E S Am. Formerly incl. many other ribbed columnar cacti esp. 'night-blooming cereus', now placed in segregate genera, e.g. *Hylocereus*, *Selenicereus* etc. Fls large, white; extrafl. nectaries; cult. orn., some ed. esp. *C. repandus* (L.) Miller (cadushi, WI) – despined young stems ed. Curaçao

ceriman *Monstera deliciosa*

Cerinthe L. Boraginaceae. 10 Eur. (4), Medit. *C. major* L. (honeywort, Medit.) – cult. orn.

Ceriops Arn. Rhizophoraceae. 3 trop. coasts of Indian & W Pacific Oceans (fossils like them in Eocene London Clay, cf. *Nypa*), inner mangrove; *C. tagal* (Perrottet) C. Robinson – timber the most durable of all mangroves, bark (45% tannin) used in tanning (e.g. *Crotalaria* nets in Sri Lanka) & a constituent of soga batik-dye

Ceriscoides (Hook.f.) Tirvengadum. Rubiaceae (II 1). 7 trop. As., Mal.

Ceriosperma (O. Schulz) Greuter & Burdet (~ *Nasturtium*). Cruciferae. 1 Syria

Cerochlamys N.E. Br. Aizoaceae (V). 1 Little Karoo: *C. pachyphylla* (L. Bolus) L. Bolus, cult. orn.

Cerolepis Pierre. Older name for *Camptostylus*

Ceropegia L. Asclepiadaceae (III 5). 160 Arabia (10, R: NRBGE 45(1988)287) warm Afr. incl. Canary Is., to Aus.(1). R: MSB 12(1957)5, R.A. Dyer, *C., Brachystelma & Riocreuxia in S Afr.* (1983)133. Usu. succulent twiners or subshrubs, some with tubers, others leafless & *Stapelia*-like. Fls held erect, the corolla often swollen at base, the whole acting as a poll. trap as in *Aristolochia* (q.v.): the C-tube is lined with downward-pointing hairs, in some heat is produced (cf. *Arum*) & flies are further attracted by smell, colour & sometimes long hairs at C-lobe tips flickering in breeze; once inside, they cannot escape until the tube hairs wither, when they leave with pollinia on their proboscides. Many cult. orn.; tubers of some eaten by humans (e.g. *C. bulbosa* Roxb. in Arabia & India) & other animals. *C. linearis* E. Meyer subsp. *woodii* (Schltr.) H. Huber (*C. woodii*, S Afr.) – much cult. in hanging baskets, lvs marbled, fls blackish, aerial tubers present

Cerosora (Baker) Domin. Pteridaceae (III). 3 Himal., Sumatra, Borneo. R: KB 13(1959)450

Ceroxylon Bonpl. Palmae (IV 2). c. 15 Andes. *C. alpinum* Bonpl. (*C. andicola*, wax palm, Colombia) – wax on trunk form. used for candles, wax matches, gramophone records; *C. quindiuense* (Karsten) H. Wendl. (Colombia) – at 60 m the tallest palm, national tree of Colombia, 10 kg wax per tree per annum; *C. utile* (Karsten) H. Wendl. (Colombia & Ecuador) found at 4000 m – highest record for a palm

Ceruana Forssk. Compositae (Ast.-Gran.). 1 Egypt, trop. Afr. Used in brooms; found in Egyptian tombs

Cervantesia Ruíz & Pavón. Santalaceae. 5 Andes

Cervaria Wolf (~ *Peucedanum*). Umbelliferae (III 10). 4 Euras.

Cervia Rodriguez ex Lagasca = *Rochelia*

Cespedesia Goudot. Ochnaceae. 1 trop. S Am.: *C. spathulata* (Ruíz & Pavón) Planchon

Cestrum L. Solanaceae. 175 trop. Am., (? Aus.). R: Candollea 6(1935)46, 7(1936)1. Alks. Berries. Many cult. orn. esp. for fragrant fls, e.g. *C. nocturnum* L. (lady of the night, WI); some med.; *C.* (?) *laevigatum* Schldl. a cannabis subs. in coastal Brazil; *C. parqui* L'Hérit. (S S Am.) – toxic to sheep & cattle

Ceterach Willd. = *Asplenium*

Ceterachopsis (J. Sm.) Ching = *Asplenium*

Ceuthocarpus Aiello. Rubiaceae (I 7). 1 E Cuba (serpentine)

Ceuthostoma L. Johnson (~ *Casuarina*). Casuarinaceae. 2 Palawan & Borneo to New Guinea. R: Telopea 3(1988)133

cevadilla *Schoenocaulon officinale*

Cevallia Lagasca. Loasaceae. 1 SW N Am. Connective with long process; G 1, ovule 1, pendulous

Ceylon cedar *Melia azedarach*; **C. ebony** *Diospyros ebenum*; **C. gooseberry** *Dovyalis hebecarpa*; **C. gurjun** *Dipterocarpus zeylanicus*; **C. mahogany** *M. azedarach*; **C. olive** *Elaeocarpus serratus*; **C. satinwood** *Chloroxylon swietenia*

Chaboissaea Fourn. (~ *Muehlenbergia*). Gramineae (31e). 4 Mex., Arg. (1). R: Madroño 39(1992)8

Chabrea Raf. (~ *Peucedanum*). Umbelliferae (III 10). 1 Eur.

Chacaya Escal. = *Discaria*

Chacoa R. King & H. Robinson (~ *Eupatorium*). Compositae (Eup.-Crit.). 1 Paraguay, Arg. R: MSBMBG 22(1987)332

Chadsia Bojer. Leguminosae (III 6). 17 Madag. Fls scarlet, some cauliflorous

Chaenactis DC. Compositae (Hele.-Cha.). 40 W N Am. R: CDH 3(1940)89. Some cult. orn. incl. *C. glabriuscula* DC (n = 6, Calif.) which has given rise by aneuploidy to *C. fremontii* A. Gray (n = 5) & *C. stevioides* Hook. & Arn. (n = 5)

Chaenanthe Lindley = *Diadenium*

Chaenomeles Lindley. Rosaceae (III). 4 E As. R: JAA 45(1964)302. CJB 68(1990)2232. 'Japonica' – 3 spp. have been hybridized & these & the parents but esp. *C. speciosa* (Sweet) Nakai (China) are grown as spring-flowering shrubs, often against walls, & are favourite bonsai subjects in Japan; the fr. can be made into preserves

Chaenorhinum (DC) Reichb. Scrophulariaceae. 21 Medit. (esp., Eur. 12), natur. temp. R: D.A. Sutton, Rev. Antirrhineae (1988)97; differs from *Antirrhinum* in spurred fls. Some cult. orn.

Chaerophyllopsis H. Boissieu. Umbelliferae (III 8). 1 W China: *C. huai* H. Boissieu

Chaerophyllum L. Umbelliferae (III 2). c. 35 N temp. (Eur. 12). *C. bulbosum* L. (turnip-rooted chervil, Eur., natur. US) – ed. carrot-like taproot, sometimes cult.

Chaetacanthus Nees. Acanthaceae. 4 S Afr.

Chaetachme Planchon. Ulmaceae (II). 1–4 trop. & S Afr., Madag.; wood sometimes used for musical instruments

Chaetacme Planchon = *Chaetachme*

Chaetadelpha A. Gray. Compositae (Lact.-Step.). 1 SW US. Zig-zag twigs

Chaetanthera Ruíz & Pavón. Compositae (Mut.-Mut.). 42 S Peru, Chile, Andes of Arg. R: RMLP (NS) 1(1937)87. Cushion-pls incl. *C. ramosissima* [D. Don ex] Taylor & Phillips (*C. tenella*, Chile)

Chaetanthus R. Br. (~ *Leptocarpus*). Restionaceae. 1–4 SW Aus.

Chaetium Nees. Gramineae (34b). 3 trop. Am.

Chaetobromus Nees. Gramineae (25). 3 SW Cape to Namaqualand

Chaetocalyx DC. Leguminosae (III 13). 12 trop. Am. Twiners

Chaetocarpus Thwaites. Euphorbiaceae. 19 trop. Am. (8 WI), Afr., As., W Mal. *C. castanicarpus* (Roxb.) Thwaites (trop. As., W Mal.) – good timber, young lvs ed.

Chaetocephala Barb. Rodr. = *Myoxanthus*

Chaetochlamys Lindau = *Justicia*

Chaetolepis (DC) Miq. Melastomataceae. 10 trop. Am. No appendages to connective

Chaetolimon (Bunge) Lincz. (~ *Acantholimon*). Plumbaginaceae (II). 3 C As.

Chaetonychia (DC) Sweet (~ *Paronychia*). Caryophyllaceae (I 2). 1 W Medit.: *C. cymosa* (L.) Sweet

Chaetopappa DC. Compositae (Ast.-Ast.). 10 (excl. *Pentachaeta*) SW N Am. R: Phytol. 64(1988)448

Chaetopoa C.E. Hubb. Gramineae (34f). 2 Tanzania

Chaetopogon Janchen (*Chaeturus*). Gramineae (21d). 1 Medit.

Chaetoptelea Liebm. (~ *Ulmus*). Ulmaceae (I). 1 C Am.: *C. mexicana* Liebm. – timber used for railway sleepers, cart-wheels etc.

Chaetosciadium Boiss. Umbelliferae (III 3). 1 E Med.: *C. trichospermum* (L.) Boiss.

Chaetoseris Shih (~ *Cicerbita*). Compositae (Lact.-Lact.). 18 Himal., China. R: APS 29(1991)398

Chaetospira S.F. Blake = *Pseudelephantopus*

Chaetostachydium Airy Shaw (~ *Psychotria*). Rubiaceae (IV 7). 1 New Guinea. Pachycaul

Chaetostichium C.E. Hubb. = *Oropetium*

Chaetostoma DC. Melastomataceae. 12 Brazil. G 3-loc.

Chaetosus Benth. = *Parsonsia*

Chaetothylax Nees = *Justicia*

Chaetotropis Kunth = *Polypogon*

Chaeturus Link = *Chaetopogon*
Chaetymenia Hook. & Arn. Compositae (Hele.). 1 Mex.
chaff-flower *Achyranthes* spp., *Alternanthera* spp.
chaguar fibre *Bromelia serra*
chahomilia *Salvia fruticosa*
Chailletia DC = *Dichapetalum*
chairmaker's rush *Scirpus americanus*
Chaiturus Willd. (~ *Leonurus*). Labiatae (VI). 1 W Eur. to C As.: *C. marrubiastrum* (L.) Reichb.
Chalarothyrsus Lindau. Acanthaceae. 1 Mexico
Chalcanthus Boiss. Cruciferae. 1 Iran mts, Afghanistan & C As.
Chalema Dieterle. Cucurbitaceae. 1 Mex.: *C. synanthera* Dieterle
Chalepophyllum Hook.f. Rubiaceae (I 5). 5 Venez., Guyana. 2 bracteoles; K unequal
chalice vine *Solandra maxima*
Chalybea Naudin (~ *Pachyanthus*). Melastomataceae. 1 trop. Am.
Chamabainia Wight. Urticaceae (III). 1–2 Indomal., Taiwan
Chamaeacanthus Chiov. = *Campylanthus*
Chamaealoe A. Berger = *Aloe*
Chamaeangis Schltr. Orchidaceae (V 16). 13 trop. Afr. to Masc.
Chamaeanthus Schltr. ex J.J. Sm. Orchidaceae (V 16). 2 Mal. & W Pacific
Chamaeanthus Ule = *Geogenanthus*
Chamaebatia Benth. Rosaceae. 2 Calif. & Baja Calif. Glandular, aromatic, evergreen shrubs. *C. foliolosa* Benth. (Sierra Nevada) – cult. orn. with nitrogen-fixing nodules
Chamaebatiaria (Brewer & Watson) Maxim. Rosaceae. 1 W N Am.: *C. millefolium* (Torrey) Maxim. – lvs 2-pinnate (cf. *Spiraea*), cult. orn.
Chamaecereus Britton & Rose = *Echinopsis*
Chamaechaenactis Rydb. Compositae (Hele.-Cha.). 1 SW US
Chamaeclitandra (Stapf) Pichon. Apocynaceae. 1 trop. Afr.: *C. henriquesiana* (Warb.) Pichon. R: BJBB 58(1988)165
Chamaecrista Moench (~ *Cassia*). Leguminosae (I 2). 265 trop. R: MNYBG 35(1982)636, (Afr. 36) KB 43(1988)335. Differs from *Cassia* s.s. in A all straight & *Senna* in 2 bracteoles.Trees, shrubs & herbs; some cult. orn. incl. *C. fasciculata* (Michaux) E. Greene (*Cassia* f., golden cassia, prairie senna, N Am.)
Chamaecrypta Schltr. & Diels = *Diascia*
Chamaecyparis Spach. Cupressaceae. 8 E As., N Am., rather local. R: Phytol. Monog. 7(1984)19. Monoec.; juvenile lvs sometimes needle-like whereas all adult lvs scale-like, adpressed; seeds winged, 2–5 per scale, maturing in 1 yr cf. *Cupressus* where many per scale taking 2 yrs. Commercial timbers & cult. orn. esp.: *C. lawsoniana* (A. Murray) Parl. (Lawson's cypress, Port Orford or Oregon cedar, NW N Am.) – one of most imp. timbers of Pacific NW & v. imp. cult. orn. (introd. GB 1854) with many cvs, incl. blue & golden forms, crushed foliage smells like parsley, lvs with white markings beneath; timber for house interiors incl. floors, boats, fences, sleepers, matches etc.; *C. nootkatensis* (D. Don) Sudw. (Nootka cypress, (Alaska) yellow cedar or cypress, Sitka cypress, W N Am.) – lvs without white markings, several cvs incl. drooping 'Pendula', a parent of × *Cupressocyparis leylandii*, timber similar to *C. lawsoniana*; *C. obtusa* (Siebold & Zucc.) Endl. (hinoki or Japanese cypress, Japan, Taiwan); *C. pisifera* (Siebold & Zucc.) Endl. (sawara cypress, Japan); *C. thyoides* (L.) Britton, Sterns & Pogg. (southern white cedar, E N Am.)
Chamaecytisus Link. Leguminosae (III 31). 30 Eur., Canary Is. Often incl. in *Cytisus*. *C. proliferus* (L.f.) Link ssp. *palmensis* (Christ) Kunkel (*C. palmensis*, tagasaste, Canary Is.) – fodder tree, introd. Pacific
Chamaedaphne Moench. Ericaceae. 1 N temp. (incl. Eur.): *C. calyculata* (L.) Moench – cult. orn., form. used as tea by N Am. Indians
Chamaedorea Willd. Palmae (IV 3). c. 100 trop. Am. R: D.R. Hodel (1992) *C. palms*. Dioec., unarmed, many tufted. Some ed. incl. fr., many cult. orn. with bamboo-like stems, surviving in dim light, *C. elegans* C. Martius (Mex., Guatemala) being the most common house-palm ('*Neanthe bella*') in SE US etc. (fr. ed. by Indians, the unopened spathes may be prepared like asparagus as may those of *C. graminifolia* H. Wendl. (? Guatemala), *C. sartorii* Liebm. ex C. Martius (Mex. to Honduras) & *C. tepejilote* Liebm. ex C. Martius (Mex. to Colombia), a sp. cult. for this delicacy esp. in Guatemala
Chamaegastrodia Makino & F. Maek. Orchidaceae (III). 1 Japan. Mycotrophic
Chamaegeron Schrenk. Compositae (Ast.-Ast.). 4 C As.
Chamaegigas Dinter (~ *Lindernia*). Scrophulariaceae. 1 Namibia : *C. intrepidus* Dinter. R: TSP 81(1992)328

Chamaegyne Süsseng. = *Eleocharis*

Chamaelaucium DC = *Chamelaucium*

Chamaele Miq. Umbelliferae (III 8). 1 Japan: *C. decumbens* (Thunb.) Makino

Chamaeleon Cass. (~ *Atractylis*). Compositae (Card.-Carl.). 2 Medit. R: BSBF 134 Lett. Bot. 1987,2:179. *C. gummifer* (L.) Cass. (*A. gummifera*) – adulterant of mastic, contains atractyloside, a toxic diterpene with effect like strychnine on mammals

Chamaelirium Willd. Melanthiaceae (Liliaceae s.l.). 1 E N Am.: *C. luteum* (L.) A. Gray (blazing star, fairy wand, unicorn root) – dried tubers medic., diuretic etc., cult. orn.

Chamaemeles Lindley. Rosaceae (III). 1 Madeira: *C. coriacea* Lindley. M's only endemic genus; G 1

Chamaemelum Miller (~ *Anthemis*). Compositae (Anth.-Art.). 6 Eur. (3), Medit., Canary Is. *C. nobile* (L.) All. (chamomile, S & W Eur., Medit.) – source of Oil of Roman Chamomile, a light blue (when fresh) oil distilled from heads particularly of a double-flowered form (single or 'German' c. considered inferior), used in flavouring liqueurs, as a tea, & esp. for hair shampoos (esp. blonde hair) & many other cosmetics; form. cult. as lawns with minimum maintenance & good drought tolerance (cf. grass) before mowing simplified through mechanization, still mixed with grass under heavy pressure (as at Buckingham Palace, London) esp. the sterile 'Treneague', the scented lvs giving off perfume when crushed, seats also made of this, though allegedly allergenic to some; fls used medic.

Chamaemespilus Medikus. Segregate of *Sorbus* (q.v.)

Chamaenerion Séguier = *Epilobium*

Chamaepentas Bremek. Rubiaceae (IV 1). 1 trop. E Afr.

Chamaepericlymenum Hill = *Cornus*

Chamaepus Wagenitz. Compositae (Gnap.-Gnap.). 1 Afghanistan: *C. afghanicus* Wagenitz. R: OB 104(1991)171

Chamaeranthemum Nees = *Chameranthemum*

Chamaeraphis R. Br. Gramineae (34f). 1 N Aus.

Chamaerhodiola Nakai = *Rhodiola*

Chamaerhodos Bunge. Rosaceae. 11 temp. C As. & N Am.

Chamaerops L. Palmae (I 1a). 1 W Med.: *C. humilis* L. Only Eur. mainland palm, clump-forming in maquis or ± arborescent in light woodland; cult. orn. (under which Goethe had inspiration) & lvs source of vegetable 'horse' hair (Algerian fibre, crin végétal) used in upholstery; young buds ed.

Chamaesaracha (A. Gray) Benth. Solanaceae (1). 7 N Am. R: Rhodora 75(1973)325. Prostrate herbs, some with ed. berries

Chamaesciadium C. Meyer (~ *Trachydium*). Umbelliferae (III 8). 1 W As.: *C. acaule* (M.Bieb.) Boiss.

Chamaescilla F. Muell. ex Benth. Anthericaceae (Liliaceae s.l.). 2 Aus. R: FA 45(1987)288

Chamaesium H. Wolff. Umbelliferae (III 8). 5 Himalaya, Tibet, W China

Chamaespartium Adans. = *Genista*

Chamaesphacos Schrenk ex Fischer & C. Meyer. Labiatae (VI). 1 C As., Iran, Afghanistan

Chamaesyce Gray = *Euphorbia*

Chamaexeros Benth. Xanthorrhoeaceae. 3 SW Aus. R: Nuytsia 2(1976)118

Chamaexiphion Hochst. ex Steudel = *Ficinia*

Chamarea Ecklon & Zeyher. Umbelliferae (III 8). 5 S Afr. R: EJB 48(1991)200, 261

chambala *Cyphomandra capanumensis*

Chambeyronia Vieill. Palmae (V 4h). 2 New Caled. R: Allertonia 3(1984)330

Chamelaucium Desf. Myrtaceae. 23 W Aus. Heath-like. 'Waxflowers'; *C. uncinatum* Schauer (Geraldton wax (flower)) – a cut-flower in Eur.

Chamelophyton Garay (V 13). Orchidaceae. 1 Venez., Guyana: *C. kegelii* (Reichb.f.) Garay

Chamelum Philippi = *Olsynium*

Chameranthemum Nees. Acanthaceae. 4 trop. Am. Cult. orn. with variegated lvs esp. *C. gaudichaudii* Nees (Brazil)

Chamerion (Raf.) Raf. = *Epilobium*

chamfuta *Afzelia quanzensis*

Chamguava Landrum (~ *Psidium*). Myrtaceae (Myrt.). 3 C Am. R: SB 16(1991)21

Chamira Thunb. Cruciferae. 1 S Afr. Basal lvs opp.

Chamissoa Kunth. Amaranthaceae (I 2). 2 warm Am. Aril

Chamissoniophila Brand. Boraginaceae. 2 S Brazil

chamomile (tea) *Chamaemelum nobile*; **corn c.** *Anthemis arvensis*; **scentless c.** *Tripleurospermum inodorum*; **stinking c.** *A. cotula*; **sweet c.** *C. nobile*; **wild c.** *Matricaria recutita*; **yellow** or **golden c.** *A. tinctoria*

Chamomilla Gray = *Matricaria*
Chamorchis Rich. Orchidaceae (IV 2). 1 Eur.
champak *Michelia champaca*
Champereia Griffith. Opiliaceae. 1 Indomal.: *C. manillana* (Blume) Merr. – veg., fr. ed., medic.
Championella Bremek. = *Strobilanthes*
Championia Gardner. Gesneriaceae. 1 Sri Lanka
chan *Shorea* spp.
chanal or **chañar** *Geoffroea decorticans*
Chandrasekharania Nair, Ramach. & Sreek. Gramineae (tribe ?). 1 Kerala: *C. keralensis* Nair, Ramach. & Sreek. R: Proc. Ind. Acad. Sci. (Pl. Sci.) 91(1982)79
Changium H. Wolff (~ *Conopodium*). Umbelliferae (III 5). 2, Tibet, E China
Changnienia Chien. Orchidaceae (V 8). 1 China = ?
Chapelieria A. Rich. ex DC. Rubiaceae (II 5). 2 Madagascar
chaplash *Artocarpus chama*
Chapmannia Torrey & A. Gray. Leguminosae (III 13). 1 Florida
Chapmanolirion Dinter = *Pancratium*
Chaptalia Vent. Compositae (Mut.-Mut.). c. 60 warm Am. R: Darw. 6(1944)505
Charadrophila Marloth. Scrophulariaceae. 1 S Afr.: *C. capensis* Marloth, long referred to Gesneriaceae
charas *Cannabis sativa* subsp. *indica*
Chardinia Desf. Compositae (Card.-Carl.). 1 W As.
chards *Cynara scolymus* (blanched summer shoots), *Tragopogon porrifolius* (young flowering shoots); **Swiss c.(s)** *Beta vulgaris* subsp. *vulgaris*
Chareis auctt. = *Felicia*
Charia C. DC = *Ekebergia*
Charianthus D. Don. Melastomataceae. 11 WI
Charidion Bong. = *Luxemburgia*
Charieis Cass. = *Felicia*
Chariessa Miq. = *Citronella*
charlock *Sinapis arvensis*
Charpentiera Gaudich. Amaranthaceae (I 2). 6 Hawaii, Austral Is., Cook Is. R: Brittonia 24(1972)283. Trees! Austral Is. c. 4500 km from Hawaii! Highly flammable dry timber lit & thrown over cliffs in pyrotechnical displays
Chartolepis Cass. = *Centaurea*
Chartoloma Bunge. Cruciferae. 1 C As.
Chartreuse Liqueur first made 1735, perfected to yellow form c. 1840, an infusion of over 100 herbs incl. *Angelica archangelica*, *Campanula allionii*, *C. linifolia*, *Gentiana bavarica*, *Saxifraga* sp. etc. manufactured by 3 monks & aged in 100 000 litre barrels
Chasallia Comm. ex Poiret = *Chassalia*
Chascanum E. Meyer (~ *Bouchea*). Verbenaceae. 30 Afr., Madag., Arabia to W India
Chascolytrum Desv. = *Briza*
Chascotheca Urban (~ *Securinega*). Euphorbiaceae. 2 WI
Chasechloa A. Camus = *Echinolaena*
Chaseella Summerh. Orchidaceae (V 15). 1 C & E Afr.: *C. pseudohydra* Summerh. R: OM 2 (1987)164
Chasmanthe N.E. Br. Iridaceae (IV 3). 3 SW Cape. R: SAJB 51(1985)253. Seeds bright orange (usu. brown in Irid.). Cult. orn. like *Gladiolus*
Chasmanthera Hochst. Menispermaceae (III). 2 trop. Afr. Some ed. roots
Chasmanthium Link. Gramineae (24). 6 E N Am. R: JAA 71(1990)176. Only temp. genus in tribe
Chasmatocallis R. Foster = *Lapeirousia*
Chasmatophyllum Dinter & Schwantes. Aizoaceae (V). 6 SW Afr., Cape. Cult. orn. shrubby succ.
Chasmopodium Stapf. Gramineae (40h). 2 W trop. Afr.
Chassalia Comm. ex Poiret. Rubiaceae (IV 7). 42 OW trop. esp. Madagascar
chaste tree *Vitex agnus-castus*
chats *Solanum tuberosum* (undersized tubers usu. fed to stock)
Chaubardia Reichb.f. Orchidaceae (V 10). 2 trop. S Am.
Chaubardiella Garay. Orchidaceae (V 10). 6 trop. Am.
Chauliodon Summerh. Orchidaceae (V 16). 1 trop. W Afr. Leafless
chaulmoogra oil *Hydnocarpus kurzii*

Chaunochiton Benth. Olacaceae. 5 trop. Am.
Chaunostoma J.D. Sm. Labiatae (VIII 2). 1 C Am.
Chavanessia A. DC = *Urceola*
chay root *Oldenlandia umbellata*
chaya *Cnidoscolus chayamansa*
Chaydaia Pitard = *Rhamnella*
chayote *Sechium edule*
chayotilla *Hanburia mexicana*
Chazaliella Petit & Verdc. Rubiaceae (IV 7). 25 trop. Afr. *C. letouzeyi* Robbrecht (Cameroun & Gabon) – usu. unbranched pachycaul
cheat or **chess** *Bromus secalinus*
checkerberry *Gaultheria procumbens*
Cheesemania O. Schulz. Cruciferae. 6–7 Tasmania (1), NZ (*C. wallii* (Carse) Allan prob. gynodioec.)
cheeseplant *Monstera deliciosa*
cheeses *Malva* & *Lavatera* spp. (frs resemble flat c.)
cheeseweed *Hymenoclea salsola*
cheesewood *Pittosporum undulatum*
Cheilanthes Sw. Pteridaceae (IV). 150 cosmop. (Eur. 7; Aus. 15 (R: Telopea 4(1991)509)), esp. Andes, Mex., S Afr. Usu. dry rocky sites – ferns with xeromorphic characters; some cult. orn.
Cheilanthopsis Hieron. = *Woodsia*
Cheiloclinium Miers. Celastraceae. 23 trop. Am.
Cheilophyllum Pennell. Scrophulariaceae. 8 WI
Cheiloplecton Fée = *Cheilanthes*
Cheilosa Blume. Euphorbiaceae. 1 W Mal.: *C. montana* Blume – fr. ed. & fermented. R: Blumea 38(1993)161
Cheilosoria Trevis. = *Pellaea*
Cheilotheca Hook.f. Ericaceae (Monotropaceae). 4 S & E As., Mal.
Cheiradenia Lindley. Orchidaceae (V 10). 1 NE S Am.
Cheiranthera A. Cunn. ex Brongn. Pittosporaceae. 5 SW & SE Aus.
Cheiranthus L. = *Erysimum*
Cheiridopsis N.E. Br. Aizoaceae (V). 23 S Afr. R: BJ 108(1987)567. Clump-forming succ., cult. orn.
Cheirodendron Nutt. ex Seemann. Araliaceae. 5 Hawaii, 1 Marquesas (*C. bastardianum* (Decne) Frodin)
Cheiroglossa C. Presl = *Ophioglossum*
Cheirolaena Benth. Sterculiaceae (? Malvaceae). 1 Mauritius
Cheirolophus Cass. (~ *Centaurea*). Compositae (Card.-Cent.). 20 SW Eur., N Afr., Canary Is.
Cheiropleuria C. Presl. Cheiropleuriaceae. 1 SE As. & Honshu to E Mal.: *C. bicuspis* (Blume) C. Presl
Cheiropleuriaceae Nakai. Filicopsida. 1/1 E As. Terrestrial; stem hairy with protostele, lvs dimorphous, sterile ones ± lobed, fertile ones acrostichoid with vein-branches serving sporangia & tetrahedral spores
Only species: *Cheiropleuria bicuspis* (Blume) C. Presl
Poss. allied to *Dipteris*, with primitive vasc. supply but advanced fertile fronds
Cheirorchis Carr = *Cordiglottis*
Cheirostemon Bonpl. = *Chiranthodendron*
Cheirostylis Blume. Orchidaceae (III 3). 15 OW trop. (Aus. 1)
Chelidonium L. Papaveraceae (I). 1 temp. & subarctic Euras. (natur. E US): *C. majus* L. (greater celandine, swallow-wort). R: FGP 17(1982)237. Differs from *Papaver* in stalked stigmas, arillate seeds etc. Alks. In GB in last interglacial (Ipswichian) but generally considered introduced by man in this one (Mediaeval) for medic. properties of orange latex long-used in eye disorders, classically mixed with fennel, wormwood, honey & a dash of human milk, certainly efficacious in treatment of warts & other skin disorders; mutant forms with epiphyllous infls or branches, double fls & the ragged 'var. *laciniatum* (Miller) Syme', a single-gene mutant with the effect of inhibiting intercalary growth of laminae & petals which arose at Heidelberg c. 1590, sometimes cult.
Chelonanthera Blume = *Pholidota*
Chelonanthus (Griseb.) Gilg = *Irlbachia*
Chelone L. Scrophulariaceae. 5–6 N Am. Cult. orn. (shellflowers, turtlehead) allied to *Penstemon*, esp. *C. glabra* L. (balmony, E N Am.) – medic., vermifuge, iridoids (catalpol

sequestered by *Euphydryas phaeton* butterflies, some Coleoptera & some Hymenoptera, which also sequester aucubin)

Chelonespermum Hemsley = *Burckella*

Chelonistele Pfitzer. Orchidaceae (V 12). 11 W Mal. (10 restricted to Borneo). R: OM 1(1986)23

Chelonopsis Miq. Labiatae (VI). 13 Kashmir, E Tibet, E As.

Chelyocarpus Dammer. Palmae (I 1a). 4 trop. S Am. R: Principes 16(1972)67

chempedak *Artocarpus integer*

chengal *Neobalanocarpus heimii*

Chennapyrum Löve = *Aegilops*

Chenolea Thunb. (~ *Bassia*). Chenopodiaceae (I 4). 4 Afr.

Chenopodiaceae Vent. (~Amaranthaceae). Dicots – Caryophyllidae – Caryophyllales. 103/1300 cosmop. esp. desert & semidesert areas. Shrubs, herbs, rarely small trees e.g. *Haloxylon* or climbers e.g. *Hablitzia*; stems often jointed, succulent or both, usu. with sec. growth of concentric rings of vascular bundles or alternating concentric rings of xylem & phloem. Betalains (not anthocyanins) present, plants often with C_4 photosynthesis, crassulacean acid metabolism & accumulating organic acids. Lvs spirally arr., rarely opp., simple, often succ. & with Kranz anatomy or ± reduced; stipules 0. Fls small, often green, 1–∞ in axils (these partial infls making up spikes, panicles or cymes), usu. reg., bisexual (plants rarely dioec. or monoec.); P (1–)5(6–8), sometimes basally connate or 0, A up to same no. as & opp. P, filaments s.t. connate at base, hypogynous, adnate to P or on annular disk, anthers with longit. slits, $\underline{G}$ (2,3(–5)), 1-loc. with ± connate styles (G half-inf. in *Beta*), ovule 1, basal, amphitropous to usu. campylotropous, bitegmic. Fr. usu. small nut, rarely with circumscissile dehiscence, often subtended by persistent P or bracteoles, s.t. (as in *Beta*) several maturing together with P to form multiple fr.; seed 1 with annular or spirally twisted or only slightly curved (*Dysphania*) embryo with 2(3) cotyledons, usu. ± surrounding starchy perisperm (sometimes 0), endosperm ± 0. x = (6–)9

Classification & chief genera:

I. **Chenopodioideae** (lvs well developed; infls usu. paniculate; embryo annular or only slightly curved)
   1. **Beteae** (fr. with persistent P & circumscissile lid): *Beta, Hablitzia*
   2. **Chenopodieae** (bladdery hairs often present; bracteoles 0): *Chenopodium, Dysphania* (form. put in own fam.)
   3. **Atripliceae** (fls usu. unisexual, males with K, females with 0, bracteoles accrescent in fr.): *Atriplex, Spinacia*
   4. **Camphorosmeae** (fls usu. ebracteolate in spikes or panicles; K persistent in fr.): *Bassia, Camphorosma*
   5. **Sclerolaeneae** (fls usu. solit. in leaf axils, P in fr. usu. hardened or succ.; Aus.): *Maireana, Sclerolaena*
   6. **Corispermeae** (infls spiciform; bracteoles 0, P often rudimentary): *Corispermum*

II. **Salicornioideae** (pls succ.; stems often articulate; fls usu. in 3s sunk in axes; embryo usu. annular or curved)
   1. **Halopeplideae** (lvs alt., amplexicaul): *Kalidium*
   2. **Salicornieae** (lvs usu. opp., connate): *Halosarcia, Salicornia*

III. **Salsoloideae** (pls leafy (linear) or succ.; fls usu. 1–3 in axils of bracts; embryo spiral)
   1. **Sarcobateae** (bracteoles 0, males with P 0): *Sarcobatus* (only)
   2. **Suaedeae** (bracteoles small): *Suaeda*
   3. **Salsoleae** (bracteoles surrounding fl. bud): *Anabasis, Halothamnus, Haloxylon, Nanophyton, Salsola*

IV. **Polycnemoideae** (shrubby with reg. sec. thickening to herbs; fls herm., solit., bracteolate
   1. **Polycnemeae**: *Nitrophila, Polycnemum*

Many halophytes (esp. *Halogeton, Salsola*) & also weeds of cultivation, others imp. constituents of desert floras as in W US & Aus.; imp. pl. products incl. spinach (*Spinacia*) & beet (*Beta*) besides *Agriophyllum, Cycloloma* & several *Chenopodium* spp. while *Arthrophytum* & *Haloxylon* are imp. firewood sources & *Nanophyton erinaceum* is efficacious in treatment of hypertension. Few cult. orn. e.g. *Bassia* (kochia)

Chenopodiopsis Hilliard. Scrophulariaceae. 3 S Afr. R: EJB 47(1990)339

Chenopodium L. Chenopodiaceae (I 2). c. 100 temp. (Eur. 25), *C. macrospermum* Hook.f. NW bipolar. Mostly weedy herbs, some grains (selection for increased fr. size, low dormancy & non-shattering infls (cf. cereals); saponins washed out before prep.), cult. orn. & medic. etc. *C. album* L. (goosefoot, fat-hen, N temp.) – form. a veg., ousted by spinach, fr. taken by poultry; *C. ambrosioides* L. (wormseed, American w., epazote, trop. Am., natur. elsewhere) – s.t. with allied spp. referred to *Teloxys*, cult. for medic. oil, a vermifuge, fr. also used to season rice, beans etc., fertility control in Uruguay; *C. bonus-henricus* L. (Good King Henry, all good, blite, Eur., natur. N Am.) – lvs like spinach, young shoots like asparagus; *C. graveolens* Willd. (yerba del zorillo, Mex.) – medic. & condiment; *C. oahuense* (Meyen) Aellen (Hawaiian goosefoot) – potherb; *C. pallidicaule* Aellen (canihua, Andes) – imp. grain; *C. quinoa* Willd. (quinoa, quinua, cultigen, Andes) – similar, both high in amino-acids but ousted by cereals introduced by colonists

chequers *Sorbus torminalis*

cherimoya *Annona cherimola*

Cherleria L. = *Minuartia*

cheronjee *Buchanania lanzan*

cherry *Prunus* spp., the ed. ones derived from *P. avium* & *P. cerasus*; **African c.** *Tieghemella heckelii*; **amarelle c.** *P. cerasus*; **American** or **black c.** *P. serotina*; **Barbados c.** *Malpighia glabra*; **bird c.** *P. padus*; **cabinet c.** *P. serotina*; **Cayenne c.** *Eugenia uniflora*; **choke c.** *P. virginiana*; **Cornelian c.** *Cornus mas*; **Duke c.** *P.* × *goudouinii*; **finger c.** *Rhodomyrtus macrocarpa*; **flowering c.** several *P.* spp. & hybrids but esp. *P. serrulata* cvs & *P.* × *yedoensis*; **Jamaican c.** *Muntingia calabura*; **c. laurel** *P. laurocerasus*; **Liberian c.** *Sacoglottis gabonensis*; **maraschino c.** *P. cerasus* 'Marasca'; **morello c.** *P. cerasus*; **c. pie** *Heliotropium arborescens*; **Pitanga c.** *Eugenia uniflora*; **c. plum** *P. cerasifera*; **prairie c.** *P. gracilis*; **St Lucie c.** *P. mahaleb*; **sand c.** *P. pumila*; **Surinam c.** *E. uniflora*; **sweet** or **wild c.** *P. avium*; **winter c.** *Physalis alkekengi*, *Solanum capicastrum*

Chersodoma Philippi. Compositae (Sen.-Sen.). 9 temp. S Am. R: RMLP 61(1946)343. Dioec.

chervil *Anthriscus cereifolium*; **turnip-rooted c.** *Chaerophyllum bulbosum*

Chesneya Lindley ex Endl. Leguminosae (III 15). 20 SW & C As. to Mongolia

Chesniella Boriss. = *Chesneya*

chess *Bromus secalinus*; **downy c.** *B. tectorum*

chestnut *Castanea sativa* (**common, Eur., Spanish** or **sweet c.**); **Amer. c.** *C. dentata*; **Aus. c.** *Castanospermum australe*; **Cape c.** *Calodendrum capense*; **China c.** *Sterculia monosperma*; **golden c.** *Chrysolepis chrysophylla*; **horse c.** *Aesculus hippocastanum*; **Jap. c.** *Castanea crenata*; **golden c.** *Castanopsis chrysophylla*; **Moreton Bay c.** *Castanospermum australe*; **c. oak** *Quercus prinus*; **Polynesian** or **Tahiti c.** *Inocarpus fagifer*; **c. vine** *Tetrastigma voinieranum*; **water c.** *Trapa natans*; **Chinese water c.** *Eleocharis dulcis*; **wild c.** (S Afr.) *Brabejum stellatifolium*

Chevaliera Gaudich. ex Beer (~ *Aechmea*). Bromeliaceae (3). 24 trop. Am. *C. magdalenae* (André) André (*A.m.*, pita, C & S Am.) – fibre from lvs for rope, twine & thread for sewing leather. R: Phytol. 66(1989)77

Chevalierella A. Camus. Gramineae (24). 1 Zaire. False petioles

Chevreulia Cass. Compositae (Gnap.-Gnap.). 5 S Am., Falkland Is., Tristan da Cunha. R: BJLS 106(1991)186

Chewing's fescue *Festuca rubra* ssp. *commutata*

chia seeds *Salvia* spp. esp. *S. columbariae* & *S. hispanica*

Chian turpentine *Pistacia terebinthus*

Chiangiodendron Wendt. Flacourtiaceae (4). 1 Mex.: *C. mexicanum* Wendt. R: SB 13(1988)435. Only non-OW Pangieae

Chiarinia Chiov. = *Lecaniodiscus*

Chiastophyllum (Ledeb.) A. Berger. Crassulaceae. 1 Caucasus: *C. oppositifolium* (Ledeb.) A. Berger (lamb's-tail), cult. orn. succ. with yellow fls in drooping infls

chibasa *Juncus* sp.

chibou *Bursera simaruba*

chica *Arrabidaea chica*

chick pea *Cicer arietinum*

chicken claws *Salicornia europaea*

chickrassy *Chukrasia tabularis*

chickweed *Stellaria* & *Cerastium* spp.; **common c.** *S. media*; **greater c.** *S. neglecta*; **mouse-ear c.** *Cerastium* spp.; **c. wintergreen** *Trientalis europaea*

chicle *Manilkara zapota*

Chiclea Lundell = *Manilkara*

chicory *Cichorium intybus*

Chidlowia Hoyle. Leguminosae (I 1). 1 trop. W Afr.

Chienia W.T. Wang = *Delphinium*

Chieniodendron Tsiang & P.T. Li = *Meiogyne*

Chigua Stevenson. Zamiaceae (III). 2 Colombia. R: MNYBG 57(1990)169
chiku *Manilkara zapota*
Chikusichloa Koidz. (~ *Leersia*). Gramineae (9). 3 China, Japan, Ryukyu Is., Sumatra
Chile, Chilean or **Chili cedar** *Austrocedrus chilensis*; **C. crocus** *Tecophilaea cyanocrocus*; **C. jasmine** *Mandevilla laxa*; **C. nut** *Gevuina avellana*; **C. pine** *Araucaria araucana*; **C. wine palm** *Jubaea chilensis*
Chileranthemum Oersted. Acanthaceae. 2 Mexico
Chiliadenus Cass. (~ *Jasonia*). Compositae (Inul.). 9 Medit. R: Webbia 94(1979)298
chilicote *Cucurbita foetidissima*
chil(l)ies *Capsicum annuum* Longum Group
Chiliocephalum Benth. (~ *Helichrysum*). Compositae (Gnap.-Gnap.). 2 Ethiopia. R: OB 104(1991)149
Chiliophyllum Philippi. Compositae (Ast.-Ast.). 3 temp. S Am.
Chiliotrichiopsis Cabrera. Compositae (Ast.-Ast.). 3 Argentina
Chiliotrichum Cass. Compositae (Ast.- Ast.). 7 temp. S Am. Evergreen shrubs; *C. diffusum* (Forst.f.) Kuntze commonly dominant in Fuegia, cult. orn.
chilito *Mammillaria* spp.
Chillania Roiv. = *Eleocharis*
Chilocardamum O. Schulz = *Sisymbrium*
Chilocarpus Blume. Apocynaceae. 15 Indomal.
Chiloglottis R. Br. Orchidaceae (IV 1). 18 Aus., 3 ext. to NZ. Sexually attractive to male thynnid wasp pollinators; some cult. orn.
Chilopogon Schltr. = *Appendicula*
Chilopsis D. Don. Bignoniaceae. 1 SW N Am.: *C. linearis* (Cav.) Sweet. Cult. orn. rheophyte, the branches used for baskets ('flowering willow')
Chiloschista Lindley. Orchidaceae (V 16). 18 Indomal. Many leafless
chilte *Cnidoscolus elasticus* & other *C.* spp.
Chimantaea Maguire, Steyerm. & Wurd. Compositae (Mut.-Mut.). 10 Venezuela, Guyana. Trees & shrubs, some unbranched pachcauls
Chimaphila Pursh. Ericaceae (Pyrolaceae). 4–5 Euras. (Eur. 1), N & trop. Am. Some cult. orn.; some med. esp. *C. umbellata* (L.) W. Barton (Euras.) for bladder problems
Chimarrhis Jacq. Rubiaceae (I 7). 14 trop. Am.
Chimborazoa H. Beck (~ *Paullinia*). 1 Ecuador: *C. lachnocarpa* (Radlk.) H. Beck. R: Brittonia 44(1992)306
Chimonanthus Lindley. Calycanthaceae. 6 China. R: JNTCFP 1984,2: 78. *C. praecox* (L.) Link (*C. fragrans*, wintersweet) – cult. orn. for fragrant winter fls produced before lvs (cf. *Calycanthus*) in China long used with linen, like lavender, though plants grown from seed may take 12–14 yrs to produce fls; beetle-poll. with A 5
Chimonobambusa Makino. Gramineae (1a). 10 Himal. (2), China, Japan. Bamboos, cult. orn., esp. *C. quadrangularis* (Fenzi) Makino (China) with culms square in TS & a flowering-cycle in excess of 100 yrs
Chimonocalamus Hsueh & Yi = *Sinarundinaria*
China or **Chinese anise** *Illicium verum*; **C. apple** *Malus prunifolia*; **C. artichoke** *Stachys affinis*; **C. aster** *Callistephus sinensis*; **C. banana** *Musa acuminata* 'Dwarf Cavendish'; **C. bellfl.** *Platycodon grandiflorus*; **C. berry** *Melia azedarach*; **C. box** *Murraya paniculata*; **C. briar** *Smilax bona-nox*; **C. cabbage** *Brassica rapa* Pekinensis Group & Chinensis Group; **C. chestnut** *Sterculia monosperma*; **C. coir** *Trachycarpus fortunei*; **C. crab apple** *Malus hupehensis*; **C. date** *Diospyros kaki*; **C. date-plum** *Ziziphus jujuba*; **C. foxglove** *Rehmannia elata*; **C. grass** *Boehmeria nivea*; **C. galls** insect galls on *Rhus javanica*; **C. gooseberry** *Actinidia chinensis*; **C. hat plant** *Holmskioldia sanguinea*; **C. houses** *Collinsia bicolor*; **C. jute** *Abutilon theophrasti*; **C. kale** *Brassica oleracea* Alboglabra Group; **C. lantern** *Physalis alkekengi*; **C. mustard** *B. juncea* 'Crispifolia'; **C. olive** *Canarium* spp.; **C. pink** *Dianthus chinensis*; **C. plum** *Prunus salicina*; **C. raisin** *Hovenia dulcis*; **C. root** *Smilax china*; **C. sumac** *Rhus javanica*; **C. tree** *Melia azedarach*; **C. white pine** *Pinus armandii*; **C. yew** *Taxus mairei*
chincherinchee *Ornithogalum thyrsoides*
Chingia Holttum = *Cyclosorus*
Chingiacanthus Hand.-Mazz. = *Isoglossa*
chingma *Abutilon theophrasti*
chinquapin *Castanea* spp. esp. *C. pumila*
Chiococca P. Browne. Rubiaceae (III 4). 6 trop. Am. incl. S Florida. Shrubs & lianes. *C. alba* (L.) A. Hitchc. (S Florida to Paraguay) – cult. orn. sometimes used for snakebite
Chiogenes Salisb. ex Torrey = *Gaultheria*

Chionachne R. Br. Gramineae (40j). 7 SE As., Indomal., E Aus., Polynesia. *C. cyathopoda* F. Muell. (Aus.) – good fodder

Chionanthus L. Oleaceae. (Incl. *Linociera*) 100 trop. (Afr. 9, Madag. 3) & subtrop., E As., E N Am. Some temp. spp. cult. orn. (fringe-flowers), esp. *C. virginicus* L. (old man's beard, E N Am.) – medic. bark

Chione DC. Rubiaceae (III 4). c. 15 C Am., WI

Chionocharis I.M. Johnston (~ *Eritrichium*). Boraginaceae. 1 Himalaya

Chionochloa Zotov. Gramineae (25). 21 SE Aus. (2 incl. Lord Howe Is.), 19 NZ. Tussock grasses intergrading with *Cortaderia* & *Danthonia*. *C. rigida* (Raoul) Zotov (NZ) – fire increases no. of tussocks in fl. & no. of infls per tussock

Chionodoxa Boiss. (~ *Scilla*). Hyacinthaceae (Liliaceae s.l.). 6 E Med. (Eur. 2). R: Naturk. Jahrb. Linz 21(1976)9. Cult. orn. esp. *C. siehei* Stapf (*C. luciliae* auctt. (true *C. luciliae* Boiss. (W Turkey) with only 1–3 fls per infl.), glory of the snow, W Turkey) – fls 4–12 blue with white centre (wild type), white or pink & *C. sardensis* Whittall ex Barr & Sugden (W Turkey) – fls all blue

Chionographis Maxim. Melanthiaceae (Liliaceae s.l.). 7 E As. Allied to *Chamaelirium*, P unequal

Chionohebe B. Briggs & Ehrend. (*Pygmaea*, ~ *Hebe*). Scrophulariaceae. 6 SE Aus., NZ

Chionolaena DC (~ *Gnaphaliothamnus*). Compositae (Gnap.-Gnap.). 17 Mex., S Am. R: OB 104(1991)91. Shrubs with revolute lvs

Chionopappus Benth. Compositae (Liab.). 1 Peru

Chionophila Benth. Scrophulariaceae. 2 Rocky Mts. *C. jamesii* Benth. cult. rock-pl.

× **Chionoscilla** J. Allen ex Nicholson. Hyacinthaceae (Liliaceae s.l.). *Chionodoxa* × *Scilla*

Chionothrix Hook.f. Amaranthaceae (I 2). 3 Somalia

chiquito *Combretum butyrosum*

chir pine *Pinus roxburghii*

Chiranthodendron Larréat. Sterculiaceae (? Bombacaceae). 1 Mex. & Guatemala: *C. pentadactylon* Larréat. K(5), C 0, A basally united into curved tube but apically separate as 5 exserted lobes, each with 2 linear, 1-celled anthers, the whole resembling a hand & thus a source of awe; poll. by perching birds & bats; fls used for eye disorders & piles

chiretta *Swertia* spp.

Chirita Buch.-Ham. ex D. Don. Gesneriaceae. 130 Indomal. (China 80). R: NRBGE 33(1974)123. Epiphyllous infls in sect. *Microchirita* C.B. Clarke; some cult. orn., e.g. *C. elphinstonia* Craib (Thailand) – annual with epiphyllous infls

Chiritopsis W.T. Wang (~ *Chirita*). Gesneriaceae. 7 China. R: EJB 49(1992)48

Chironia L. Gentianaceae. c. 15 subSaharan Afr., Madag. Some cult. orn. incl. *C. baccifera* L. (Christmas berry, S Afr.)

chironja citrus fr. believed to be a cross between grapefruit & orange

Chiropetalum A. Juss. = *Argythamnia*

Chirripoa Süsseng. = *Guzmania*

Chisocheton Blume. Meliaceae (1 6). 51 Indomal. to Vanuatu & trop. China. R: BBMNHB 6(1979)301. Most species with leaves (pinnate) with indeterminate growth; *C. pohlianus* Harms & *C. tenuis* P. Stevens (New Guinea) with infls borne on new apical growths of lvs; range of form from unbranched pachycauls to weeping leptocaul shrublets and timber trees, some myrmecophilous shoots; seeds arillate or sarcotestal, 1–2 whorls of C. *C. cumingianus* (C. DC) Harms (Indomal. to China) – fish-poison in New Guinea, seeds source of an oil for med. & lighting in Philipp.; similar oil from *C. macrophyllus* King (W Mal.) & that from *C. pentandrus* (Blanco) Merr. used as hair-oil in Philipp.

Chitonanthera Schltr. = *Octarrhena*

Chitonochilus Schltr. = *Agrostophyllum*

chitra *Berberis aristata*

Chittagong wood *Chukrasia tabularis*

chittam *Cotinus obovatus*

chives *Allium schoenoprasum*; **Chinese** or **garlic c.** *A. tuberosum*

Chlaenaceae Thouars = Sarcolaenaceae

Chlaenandra Miq. Menispermaceae (III). 1 New Guinea: *C. ovata* Miq.

Chlaenosciadium Norman. Umbelliferae (I 1a). 1 W Aus.: *C. gardneri* Norman

Chlamydacanthus Lindau (*Theileamea*). Acanthaceae. 4 trop. E Afr., Madagascar

Chlamydites J.R. Drumm. = *Aster*

Chlamydoboea Stapf = *Paraboea*

Chlamydocardia Lindau. Acanthaceae. 4 trop. W Afr.

Chlamydocarya Baillon. Icacinaceae. 6 trop. Afr.

Chlamydocola (Schumann) M. Bod. = *Cola*

Chlamydogramme Holttum (? = *Tectaria*). Dryopteridaceae (I 3). 2 New Guinea

Chlamydojatropha Pax & K. Hoffm. Euphorbiaceae. 1 trop. W Afr.

Chlamydophora Ehrenb. ex Less. Compositae (Anth.-Leuc.). 1 Eur., N Afr.

Chlamydophytum Mildbr. Balanophoraceae. 1 trop. W Afr.: *C. aphyllum* Mildbr. R: BJ 106(1986)367. Found on *Tessmannia* at 3 localities 1000 km apart

Chlamydostachya Mildbr. Acanthaceae. 1 trop. E Afr.

Chlamydostylus Baker = *Nemastylis*

Chlidanthus Herbert. Amaryllidaceae (Liliaceae s.l.). 1 Peruvian Andes: *C. fragrans* Herbert – cult. orn.

Chloachne Stapf = *Poecilostachys*

Chloanthaceae Hutch. (Dicrastylidaceae) = Labiatae (III)

Chloanthes R. Br. Labiatae (III; Chloanthaceae). 4 Aus.

Chloothamnus Buese = *Nastus*

Chlorantha Nesom et al. (~ *Boltonia*). Compositae (Ast.-Ast.). 1 S N Am. & C Am.: R: Phytol. 70(1991)371, 382

Chloraea Lindley. Orchidaceae (IV 1). 47 temp. S Am. R: Orquideologia 6(1971)231. Terrestrial

Chloranthaceae R. Br. ex Sims. Dicots – Magnoliidae – Magnoliales, 4/75 trop. & warm. Trees, shrubs or herbs (even annuals); vessels with scalariform end-plates (up to 200 crossbars!). Lvs opp., simple, petioles ± connate basally; stipules interpetiolar. Fls reduced, unisexual or not with 0–3 bracts, in crowded infls or (*Hedyosmum*) resembling a spike though perhaps solit. with a long spiral axis (Leroy), P 0 or weakly 3-fid K, A 1–5(–several 100 in *Hedyosmum* in Leroy's interpretation), usu. ± connate, lateral ones with only ½-anthers, in *Sarcandra* A 1 laminar with 2 separated pollen-sacs, anthers with longit. slits, pollen monosulcate (*Hedyosmum*) to multiaperturate, $\overline{G}$ 1 or half-inf., ovule 1, orthotropous, bitegmic. Fr. a berry or drupe, seeds with much oily, starchy endosperm, tiny embryo with 2 cotyledons. x = 8, 14, 15

Genera: *Ascarina, Chloranthus, Hedyosmum, Sarcandra*

Perhaps closest to Trimeniaceae, but Leroy (Taxon 32(1983)169) puts C. in a separate order, Chloranthales J. Leroy & argues that the app. strobiloid fl. of *Hedyosmum* is truly archaic, the fls of other genera being reduced from such a prototype, which he considers primitive in Angiosperms; others disagree (but the pigeonholing of the infl. type into pre-conceived categories (cf. Araceae, Pandanaceae) perhaps a sterile pursuit in any case)

Some teas (*Chloranthus*) & medic. locally imp.

Chloranthus Sw. Chloranthaceae. 18 Indomal., E As. *C. spicatus* (Thunb.) Makino (*C. inconspicuus*, E & SE As.) – fls to flavour tea in SE As., also medic.; *C. erectus* (Buch.-Ham.) Verdc. (*C. officinalis*, Mal.) – lvs used as tea in Java before *Camellia sinensis*, also febrifuge

Chloris Sw. Gramineae (33b). 40 trop. & warm. R: BYUSBB 1g, 2(1974)1. Some good pasture grasses esp. *C. gayana* Kunth (Rhodes grass, Afr., natur. Am.)

Chlorocalymma W. Clayton. Gramineae (34f). 1 Tanzania

Chlorocardium Rohwer, Richter & van der Werff (~ *Ocotea*). Lauraceae (inc. sed.). 2 trop. S Am. *C. rodiei* (Schomb.) Rohwer, Richter & van der Werff (*Nectandra r., Ocotea r., O. venenosa*, greenheart) – v. heavy timber, form. only one exploited in the forest, resistant to termites & borers so used for wharves, locks of Panama Canal etc., but v. difficult to work, arrow poison (alks close to D-tubocurarine; see *Chondodendron*)

Chlorocarpa Alston. Flacourtiaceae (4). 1 Sri Lanka

Chlorochorion Puff & Robbrecht (~ *Pentanisia*). Rubiaceae (III 8). 2 trop. Afr. R: BJ 110(1989)547. Name a translation of 'Verdcourt'

Chlorocrambe Rydb. Cruciferae. 1 W N Am.

Chlorocyathus Oliver = *Raphionacme*

Chlorogalum Kunth. Hyacinthaceae (Liliaceae s.l.). 5 W N Am. R: Madrono 5(1940)137. Cult. orn. bulbs incl. *C. pomeridianum* (DC) Kunth (N Calif.) – bulbs yield a lather usable as soap subs., outer scales v. fibrous

Chloroleucon (Benth.) Record = *Albizia*

Chloroluma Baillon = *Chrysophyllum*

Chloromyrtus Pierre = *Eugenia*

Chloropatane Engl. = *Erythrococca*

Chlorophora Gaudich. = *Maclura*. See also *Milicia*

Chlorophytum Ker-Gawler. Anthericaceae (Liliaceae s.l.). 215 OW trop. esp. Afr. & India (Madag. 1, Arabia 1, Socotra 1). Seeds thin, folded or flat, cf. angular small seeds of closely-allied *Anthericum*. Cult. orn. esp. *C. comosum* (Thunb.) Jacques (S Afr.) with white fls often replaced by young plants which weigh infl. axis down & take root; form usu. grown is 'Picturatum' (spider plant) with central yellow stripe on lvs to 30 cm, 'Variegatum' with creamy margins. *C. laxum* R. Br. (OW trop.) – tubers ed. S Arabia

Chlorosa Blume = *Cryptostylis*

Chlorospatha Engl. Araceae (VII 1). 16 trop. Am.

Chloroxylon DC. Rutaceae (Flindersiaceae). 1 S India, Sri Lanka: *C. swietenia* DC ((Ceylon) satinwood) – timber for furniture & veneers, though alks may irritate skin, gum useful

Choananthus Rendle = *Scadoxus*

cho-cho, choco *Sechium edule*

chocolate *Theobroma cacao*; **c. lily** *Arthropodium* spp.

Chodaphyton Minod = *Stemodia*

Chodsha-kasiana Rauschert = *Catenulina*

Choerospondias B.L. Burtt & A.W. Hill. Anacardiaceae (II). 1 NE India to N Thailand, SE China & Japan

choi sum *Brassica rapa* Chinensis Group

Choisya Kunth. Rutaceae. 7 SW N Am. *C. ternata* Kunth (SW Mex.) – hardy evergreen shrub with scented fls, poss. self-sterile & clonal in cult. (no fr. recorded, though *C.* 'Aztec Pearl' allegedly hybrid with *C. arizonica* Standley (Arizona))

chokeberry *Photinia* spp., *Prunus virginiana*; **red c.** *Photinia pyrifolia*

choko *Sechium edule*

cholla *Opuntia* spp. with cylindrical stems; **c. gum** from *O. fulgida* etc.

Chomelia Jacq. (~ *Tarenna*). Rubiaceae (III 3). Incl. *Anisomeris*, 20 trop. Am.

Chondradenia Maxim. ex F. Maek. = *Orchis*

Chondrilla L. Compositae (Lact.-Crep.). 25 temp. Euras. (Eur. 5 excl. *Calycocorsus*). *C. juncea* L. (skeleton weed) – lvs arranged in plane of the meridian, bad weed esp. in Aus., wiry stems damaging machinery

Chondrococcus Steyerm. = *Coccochondra*

Chondrodendron Ruíz & Pavón. Menispermaceae (I). 3 C & 7 trop. S Am. R: MNYBG 22,2(1971)5. Often large lianes. Root of *C. tomentosum* Ruíz & Pavón (Brazil & Peru) source of D-tubocurarine, a muscle-relaxant used in surgery & still (1982) not synthesized artificially, a constituent of curare arrow-poisons, root also medic. (pareira root, p. brava)

Chondropetalum Rottb. Restionaceae. 12 SW & S Cape. R: Bothalia 15(1985)427

Chondropyxis D. Cooke. Compositae (Gnap.-Ang.). 1 S Aus.: *C. halophila* D. Cooke. R: Fl. S Aus. 3(1986)1612

Chondrorhyncha Lindley. Orchidaceae (V 10). 24 trop. Am. Cult. orn.

Chondrostylis Boerl. Euphorbiaceae. 2 SE As., W Mal.

Chondrosum Desv. (~ *Bouteloua*). Gramineae (33c). 14 Canada to Arg. *C. gracile* Kunth (blue grama, N Am.) – valuable grazing

Chonemorpha G. Don f. Apocynaceae. 13 Indomal. R: KB 1947:47, 1948:68. *C. fragrans* (Moon) Alston (*C. macrophylla)* – cult. orn. liane with fragrant white fls to 8 cm diam.; bark source of water-resistant fibre used for fishing-nets

Chonocentrum Pierre ex Pax & K. Hoffm. Euphorbiaceae. 1 Amazonia

Chonopetalum Radlk. Sapindaceae. 1 trop. W Afr.

Chontalesia Lundell = *Ardisia*

Chordospartium Cheeseman. Leguminosae (III 16). 2 NE South Is., NZ. R: NZJB 23(1985)157. Leafless when mature; endangered: *C. stevensonii* Cheeseman not reproducing by seed in wild but seedlings look dead for 2 or more yrs

Choretrum R. Br. Santalaceae. 6 Aus. R: Fl. NSW 2(1991)150

Choriantha Riedl. Boraginaceae. 1 Iraq

Choricarpia Domin. Myrtaceae. 2 E Aus. *C. subargentea* (C. White) L. Johnson – ironwood

Choriceras Baillon (~ *Dissilaria* F. Muell.). Euphorbiaceae. 1 S New Guinea, NE Aus.

Chorigyne Eriksson. Cyclanthaceae. 7 Panamá & Costa Rica. R: NJB 9(1989)31

Chorilaena Endl. Rutaceae. 1 SW Aus.

Choriptera Botsch. (~ *Lagenantha*). Chenopodiaceae (III 3). 3 Somalia

Chorisandrachne Airy Shaw. Euphorbiaceae. 1 Peninsular Thailand

Chorisepalum Gleason & Wodehouse. Gentianaceae. 5 Guayana Highland. R: MNYBG 32(1981)342

Chorisia Kunth = *Ceiba*

Chorisis DC (~ *Lactuca*). Compositae (Lact.-Crep.). 1 coastal E As., Taiwan, Japan

Chorispora R. Br. ex DC. Cruciferae. 13 E Med. (Eur. 1), C As. *C. sabulosa* Cambess. (Himal.) – fr. ed.

Choristemon Williamson. Epacridaceae. 1 Victoria

Choristylis Harvey. Grossulariaceae (Iteaceae). 1 E trop. & S Afr.

Choritaenia Benth. Umbelliferae (I 2c). 1 S Afr.: *C. capensis* (Sonder) Burtt Davy

Chorizandra R. Br. Cyperaceae. 6–8 Aus.(5), New Caled. (1–3)

Chorizanthe R. Br. ex Benth. Polygonaceae (I 1). 50 dry W Am. (41 annuals – R: Phytol. 66(1989)100). Some with ochreae

Chorizema Labill. Leguminosae (III 24). 25 SW & (1) E Aus. R: AusSB 5(1992)249. Cult. orn. esp. *C. ilicifolium* Lab. (*C. cordatum*, flame pea, SW Aus.) – orange-red & purplish fls

Chortolirion A. Berger (~ *Haworthia*). Asphodelaceae (Liliaceae s.l.). 1 S Afr.: *C. angolense* (Baker) A. Berger

Chosenia Nakai = *Salix*

Choulettia Pomel = *Jaubertia*

Chouxia Capuron. Sapindaceae. 1 Madagascar

chow chow *Sechium edule*

chowlee *Vigna unguiculata*

choy sum *Brassica rapa* Chinensis Group

Chresta Vell. ex DC (~ *Eremanthus*). Compositae (Vern.-Vern.). 11 C Brazil. R: SB 10(1985)465

Christella A. Léveillé = *Cyclosorus*

Christensenia Maxon. Marattiaceae (Christenseniaceae). 2 Indomal. R: AFJ 83(1993)3. Lvs palmate; veins anastomosing; synangia circular. *C. aesculifolia* (Blume) Maxon (Assam, S China to Solomon Is. (not New Guinea)) – cult. orn.

Christenseniaceae Ching = Marattiaceae

Christia Moench. Leguminosae (III 9). 12 SE As., Indomal., Aus.

Christiana DC. Tiliaceae. 2 trop. S Am., trop. Afr., Madagascar

Christiopteris Copel. = *Christopteris*

Christisonia Gardner. Orobanchaceae. 17 SW China, SE As., Indomal. Roots parasitic on Acanthaceae & bamboos; infls emerge, mature & die in 2 weeks

Christmas bell *Blandfordia nobilis, B. punicea* (Aus.), *Sandersonia aurantiaca* (S Afr.); **C. berry** *Chironia baccifera* (S Afr.); **C. bush** *Ceratopetalum gummiferum, Prostanthera lasianthos* (Aus.), *Alchornea cordifolia* (W Afr.); **C. cactus** *Schlumbergera bridgesii;* **C. rose** *Helleborus niger;* **C. tree** *Picea abies* (GB but *Abies alba* used on Cont.), *Nuytsia floribunda* (Aus.), *Metrosideros excelsa* (NZ); **Nordmann** C. t. *Abies nordmanniana*

Christolea Cambess. Cruciferae. 13 C As., Himal., Kamchatka, Alaska. *C. himalayensis* (Camb.) Jafri (Himal.) – with *Ranunculus lobatus* the pl. at highest alt. (7756 m)

Christopher, Herb *Actaea spicata*

christophine *Sechium edule*

Christopteris Copel. Polypodiaceae (II 2/4). 2 Indomal. R: BJ 105(1984)1

Christ's thorn *Ziziphus spina-christi* or poss. *Paliurus spina-christi;* name also used for *Euphorbia milii, Carissa carandas* etc.

Chroesthes Benoist. Acanthaceae. 3–4 S China to SE As.

Chroilema Bernh. = *Haplopappus*

Chromolaena DC (~ *Eupatorium*). Compositae (Eup.-Prax.). 165 trop. & warm Am. R: MSBMBG 22(1987)383. x = 10. *C. odorata* (L.) R. King & H. Robinson (Siam weed, triffid weed) – invasive in OW, allelopathic effects on maize, toxic to stock, local med. (lvs anti-coagulant)

Chromolepis Benth. Compositae (Helia.-Verb.). 1 Mexico

Chromolucuma Ducke. Sapotaceae (IV). 2 NE S Am. R: T.D. Pennington, *Sapotaceae* (1991)216

Chronanthos (DC) K. Koch = *Cytisus*

Chroniochilus J.J. Sm. Orchidaceae (V 16). 6 Thailand, Malay Pen., Java, Fiji(!)

Chronopappus DC. Compositae (Vern.-Lych.). 1 Brazil

Chrozophora A. Juss. Euphorbiaceae. 6–12 Med. (Eur. 2), trop. Afr. to Thailand. *C. plicata* (Vahl) Sprengel (Afr., India) – fr. poss. oil-source for soap, also dye-source & purgative; *C. tinctoria* (L.) A. Juss. (Med.) – source of turn-sole dye (*bezetta rubra*, tournesol) used for colouring liqueurs, wine, pastries, linen & Dutch cheeses, properties known since antiquity

Chrysactinia A. Gray. Compositae (Helia.-Pect.). 6 SW N Am. R: Madrono 24(1977)129. Shrubs. *C. mexicana* A. Gray (Mex.) – medic.

Chrysactinium (Kunth) Wedd. (~ *Liabum*). Compositae (Liab.). 6 trop. S Am. R: SCB 54(1983)49

Chrysalidocarpus H. Wendl. Palmae (V 4e). c. 20 Madag., Comoro Is., Pemba. Monoec. with solit. or suckering trunks. Cult. orn. esp. *C. lutescens* H. Wendl. (*Areca madagascariensis*, bamboo or cane palm, Madag.) – slender clustered stems, anthelmintic in dogs, cosmetics exported; other spp. with ed. buds

Chrysallidosperma H. Moore = *Syagrus*

Chrysanthellum Rich. Compositae (Helia.-Car.). 13 S N & C Am. esp. Mex., Galápagos inc. 1 pantrop. weed, *C. americanum* (L.) Vatke

Chrysanthemoides Fabr. Compositae (Cal.). 2 E & S Afr. Fr. eaten & disp. by birds. *C. monolifera* (L.) Norl. (boneseed) – form. planting required in land-reclamation in Aus., now aggressive weed in NSW controlled biologically

Chrysanthemum L. Compositae (Anth.-Chrys.). s.s. 3 N Afr., Eur. (2). Annuals, cult.: *C. carinatum* Schousboe (? Morocco) – stems simple or forked; *C. coronarium* L. (Med.) – stems branched, lvs & fls ed. in China & Japan (chrysanthemum greens, shungiku); *C. segetum* L. (corn marigold, Eur., W As., natur. N Am.) – lvs ed. China. Plants form. included here are perennial & referred to the segregate genera *Dendranthema* ('chrysanthemums' of commerce, though it is to be hoped that proposals to refer them to this genus will prevail, in which case the above would become *Glebionis* spp.). See also *Argyranthemum*, *Leucanthemum*, *Tanacetum* etc.

chrysanthemum greens *Chrysanthemum coronarium*

Chrysanthoglossom Wilcox, Bremer & Humphries. Compositae (Anth.-Leuc.). 2 N Afr. R: BBMNHB 23(1993)143

Chrysitrix L. Cyperaceae. 2 Cape

Chrysobalanaceae R. Br. Dicots – Rosidae – Rosales. 17/460 trop. esp. Am. Trees or shrubs, sometimes geoxylic; alks & cyanogenic compounds unknown; seed fats unknown elsewhere in Rosales. Lvs spirally arr., simple, entire, with stipules. Fls in term. or axillary infls, rarely solit., ± strongly zygomorphic, small, bisexual or plants polygamous, perigynous with annular nectary in hypanthium below A; K (5), imbricate, C 5, imbricate, rarely 0, A(2–)8–20(–c. 300), s.t. some staminodal, filaments distinct, connate or connate in groups, in the more zygomorphic all on 1 side of hypanthium, anthers with longit. slits. $\underline{G}$ 3 but usu. 2 ± reduced, united by gynobasic style, often appearing $\underline{G}$ 1 eccentrically positioned in hypanthium, stigma simple or 3-lobed, ovules 2 per locule or single G subdivided to appear as 2 1-ovulate locules, erect (cf. Rosaceae), anatropous. Fr. a 1-seeded drupe; seeds with large embryo & 0 endosperm. n = 10, 11. R: PTRSB 320(1988)1

Classification & principal genera:

1. **Chrysobalaneae** (fls small, reg.): *Licania*
2. **Parinarieae** (fls irreg., A 7–17 posterior): *Parinari*
3. **Couepieae** (fls irreg., A (15)20 + usu. in ring): *Couepia*, *Maranthes*
4. **Hirtelleae** (fls strongly irreg., A 3–75 posterior): *Atuna*, *Hirtella*, *Magnistipula*

Stylobasium, once incl. with these genera, now referred to Surianaceae. Poss. referable to Dilleniidae – Theales. Some remarkable intercontinental distributions in *Hirtella* (all Am. save 1 in Afr. & Madag.), *Licania* (all Am. save 1 in Afr. & 2–3 Indomal.), *Maranthes* (all Afr. save 1 Am. closely allied to 1 As.) while *Chrysobalanus icaco* & (?) *Parinari excelsa* Sabine occur both in Am. & Afr.

Some oils, ed. seeds & fr. esp. *Chrysobalanus*, *Couepia* & *Licania*; timber

Chrysobalanus L. Chrysobalanaceae (1). 2 trop. Am., 1 (*C. icaco* L., cocoplum) extending to trop. Afr. & natur. Seychelles, Tanzania, Vietnam & Fiji. R: PTRSB 320(1988)80. *C. icaco* grown for ed. fr. usu. preserved (esp. NE SAM: 'icacos'); seed-oil used for candles etc. in W Afr.; in Benin inhabited by ants making nests from lvs & feeding on fr. & secretions, keeping off other intruders

Chrysobraya H. Hara. Cruciferae. 1 Himalaya

Chrysocephalum Walp. (~ *Helichrysum*). Compositae (Gnap.-Ang.). 8 Aus. R: OB 104(1991)119

Chrysochamela (Fenzl) Boiss. Cruciferae. 4 E Med. (Eur. 1) to Armenia & Iraq

Chrysochlamys Poeppig (~ *Tovomitopsis*). Guttiferae (III). 20 trop. Am.

Chrysochloa Swallen. Gramineae (33b). 4 trop. Afr.

Chrysochosma (J. Sm.) Kümmerle = *Cheilanthes*

Chrysocoma L. Compositae (Ast.-Ast.). 18 S Afr. esp. SW Cape. R: MBSM 17(1981)259. Some cult. orn.

Chrysocoryne Endl. = *Gnephosis*

Chrysocoryne Zoellner = *Leucocoryne*

Chrysocycnis Linden & Reichb.f. Orchidaceae (V 10). 3 Colombia & Ecuador

Chrysoglossum Blume. Orchidaceae (Collabinae). 6 Indomal., W Pacific

Chrysogonum L. Compositae (Helia.-Eng.). 1 SE US: *C. virginianum* L. – cult. orn., myrmechorous. R: Rhodora 79(1977)190

Chrysolaena H. Robinson (~ *Vernonia*). Compositae (Vern.-Vern.). 9 trop. S Am.

Chrysolepis Hjelmq. (~ *Castanopsis*). Fagaceae. 2 W N Am. *C. chrysophylla* (Hook.) Hjelmq. (golden chestnut) – timber, ed. fr.

Chrysoma Nutt. (~ *Solidago*). Compositae (Ast.-Sol.). 1 SE US

Chrysophae Kozo-Polj. = *Chaerophyllum*

Chrysophthalmum Schultz-Bip. Compositae (Inul.). 1 W As.

Chrysophyllum L. Sapotaceae (IV). 43 trop. Am., c. 15 Afr., c. 10 Madag., 2–3 Indomal. to Aus. R: T.D. Pennington, Sapotaceae (1991)216. Serial buds in leaf-axils of some spp. subseq. give rise to fls on old wood. Many ed. fr. esp. *C. cainito* L. (star-apple, fr. star-shaped in t.s., trop. Am.) & *C. oliviforme* L. (damson-plum, trop. Am.). Others with good timber, latex for bird-lime etc.

Chrysopogon Trin. Gramineae (40c). 26 warm esp. OW (1 Am., Eur. 1). *C. aciculatus* (Retz.) Trin. (As.) – intergrades with *Vetiveria*, a lawn-grass in trop.; *C. gryllus* (L.) Trin. (French whisk, Medit.) – used for brushes

Chrysopsis (Nutt.) Elliott (~ *Heterotheca*). Compositae (Ast.-Sol.). 10 SE US, esp. Florida, to Mex. & Bahamas. R: Rhodora 83(1981)323. Cult. orn. (golden aster)

Chrysoscias E. Meyer (~ *Rhynchosia*). Leguminosae (III 10). 6 S Afr.

Chrysosplenium L. Saxifragaceae. c. 60 Eur. (5), NE As., N Am., few in N Afr. & temp. S Am. R: JFSUTB 7(1957)1. K 4, C 0. Some cult. orn. incl. *C. oppositifolium* L. (golden saxifrage, Eur.) – emergency foodpl.

Chrysothamnus Nutt. Compositae (Ast.-Sol.). c. 15 SW N Am. R: AJB 53(1966)204. *C. nauseosus* (Pallas) Britton – gum form. chewed by Indians, poss. rubber source

Chrysothemis Decne. Gesneriaceae. 7 trop. Am. Cult. orn. tuberous herbs incl. *C. friedrichsthaliana* (Hanst.) H. Moore (C Am. & Andes) – ant-disp. (elaiosomes)

Chthonocephalus Steetz. Compositae (Gnap.-Ang.). 6 temp. Aus. R: OB 104(1991)128

Chuanminshen Sheh & Shan (~ *Peucedanum*). Umbelliferae (III 10). 1 China (cult.): *C. violaceus* Sheh & Shan

chuchupate *Ligusticum porteri*

Chucoa Cabrera. Compositae (Mut.-Mut.). 1 Peruvian Andes. Shrubs

chufa *Cyperus esculentus*

chuglam, white *Terminalia bialata*

Chukrasia A. Juss. Meliaceae (IV 2). 1 S China, Indomal.: *C. tabularis* A. Juss. Timber good (Chittagong wood, chickrassy, yinma, yonhin, Indian redwood, bastard cedar) & source of a gum. R: FM I,12(1995)354

chumprak *Heritiera cochinchinensis*

Chumsriella Bor = *Germainia*

Chunechites Tsiang. Apocynaceae. 1 SE China

Chunia H.T. Chang. Hamamelidaceae (II). 1 Hainan

Chuniophoenix Burret. Palmae (I 1c). 3 S China, Hainan, Vietnam

chupadilla *Cyrtocarpa procera*

Chuquiraga Juss. Compositae (Barn.). 20 Andes & Patagonia. R: Darw. 26(1985)219. Xeromorphic shrubs; thorns in axils

churnwood *Citronella moorei*

churras *Cannabis sativa* subsp. *indica*

Chusan palm *Trachycarpus fortunei*

Chusquea Kunth. Gramineae (1a). Incl. *Swallenochloa* c. 120 trop. Am., to snowline. R: SCB 9(1973)69, Brittonia 30(1978)303. Characteristic of cloud forest, often forming impenetrable thickets – tree-like, shrubby & climbing bamboos with monopodial, sympodial or mixed shoots, usu. solid pith. Cult. orn. esp. *C. culeou* Desv. (Chile) – to 6 m, hardy in GB

Chusua Nevski. Orchidaceae (IV 2). 17 N India to E As.

Chydenanthus Miers. Lecythidaceae. 1 Mal.: *C. excelsus* (Blume) Miers. R: Reinw. 10(1982)27

Chymsydia Albov (~ *Agasyllis*). Umbelliferae (III 9). 1–2 Transcaucasia

Chysis Lindley. Orchidaceae (V 11). 7 trop. Am. Cult. orn. epiphytes

Chytranthus Hook.f. Sapindaceae. 30 trop. Afr. Some ed. fr.

Chytroglossa Reichb.f. Orchidaceae (V 10). 3 Brazil

Chytroma Miers = *Lecythis* + *Eschweilera*

Chytropsia Bremek. = *Psychotria*

Cibirhiza Bruyns. Asclepiadaceae (III 2). 1 Oman: *C. dhofarensis* Bruyns – 1 of 4 Arabian endemic genera, local ed. esp. tubers. R: NRBGE 45(1988)51

Cibotarium O. Schulz = *Sphaerocardamum*

Cibotium Kaulf. Dicksoniaceae (Thyrs.). 11 trop. As. to Mal. (3), Am. (2), Hawaii (6). Cult. orn. tree-ferns & cut trunks used as flower-pots; stem-hairs used as styptic in As. C.

barometz (L.) J. Sm. (India & S China to W Mal.) – prostrate sp., the end of the trunk with bud covered in hairs was passed off as the 'Vegetable Lamb of Tartary' or Scythian lamb in C17; *C. glaucum* (Sm.) Hook. & Arn. (Hawaii) – pith eaten like breadfruit, the scales form. used & exported as 'pulu fibre' for packing & stuffing pillows etc.

Cicca L. = *Phyllanthus*

cicely, sweet *Myrrhis odorata*

Cicendia Adans. Gentianaceae. 1 Eur. & Med.; 1 Calif., W S Am.

Cicer L. Leguminosae (III 21). 40 C & W As. (36), Greece (1), Canary Is. & Morocco (1), Ethiopia (1) & *C. arietinum* L. (chick pea, garbanzo bean, poss. derived from *C. reticulatum* Ladiz. (sometimes treated as ssp.), SE Turkey c. 6500 BP). R: MLW 72–10(1972). Chick pea is the world's third pulse crop, after beans & peas, eaten fresh or dried, made into flour, coffee subs., fodder, the 'salted provender' of Isaiah

Cicerbita Wallroth (~ *Lactuca*). Compositae (Lact.-Lact.). 18 N temp. OW esp. mts (Eur. 4, damp shady places). *C. alpina* (L.) Wallroth (alpine sow-thistle, Eur.) – protected & cult. orn.

Ciceronia Urban. Compositae (Eup.-Crit.). 1 E Cuba (serpentine). R: MSBMBG 22(1987)1

Cichorium L. Compositae (Lact.). 7 Eur. (3), Med., Ethiopia. 2 salad pls: *C. endivia* L. (endive, poss. derived from subsp. *divaricatum* (Schousboe) Sell, coastal Medit.) – long cult. (Pliny mentions it; one of the 'bitter herbs' of the Passover) for lvs, usu. blanched; *C. intybus* L. (chicory, succory, witloof, blue sailor (Am.), Medit.) – fls open at 8 a.m., close at 4 p.m., lvs ed., usu. blanched, radicchio being a form with white-veined, red-purple lvs, also used for skin complaints & root as an adulterant or subs. of coffee, locust antifeedants

Ciclospermum Lagasca = *Cyclospermum*

Cicuta L. Umbelliferae (III 8). 8 N temp. (Eur. 1: *C. virosa* L., cowbane, water hemlock). All v. toxic, poss. the most violently poisonous (when eaten) of all N temp. plants, esp. in N Am. *C. maculata* L. (beaver poison)

cider see *Malus*; **c. tree** *Eucalyptus gunnii*

cidra *Cucurbita ficifolia*

Cienfuegosia Cav. Malvaceae. c. 26 trop. & warm Am. & Afr.

Cienkowskiella Kam = *Siphonochilus*

cigar flower *Cuphea ignea*; **c. tree** *Catalpa speciosa*

Cigarilla Aiello. Rubiaceae (I 1/IV 24). 1 Mexico

Cimicifuga Wernisch. (~ *Actaea*). Ranunculaceae (I 2). 18 N temp. (Eur. 1). Cult. orn. incl. *C. foetida* L. (bugbane, Siberia, E As.) – used to drive away bugs in Siberia, medic. in China, Eur. pls given this name usu. *C. europaea* Schipcz. (C Eur.); *C. racemosa* (L.) Nutt. (black cohosh or snakeroot, E N Am.) – dried rhiz. medic., cult pls with this name usu. *C. simplex* (DC) Turcz. (E As.)

Ciminalis Adans. = *Gentiana*

Cinchona L. Rubiaceae (I 1). 23 C Bolivia to N Colombia & Venez., 1 ext. to Costa Rica. R: MNYBG 80(1998). Trees & shrubs. Bark (Jesuits' bark, Peru b., druggists' b.) source of alks esp. antimalarial quinine (suppresses trophozoites, analgesic), still not completely superseded by synthetics & used in tonic water (bubbles speed entry of alcohol to blood so that gin & t. the most efficient means (short of injection) of so doing). Form. felled in forest before plantations established by colonial powers in As. esp. *C. calisaya* Wedd. ('*C. officinalis*', E Andes) – yellow bark, calisaya, Ledger bark, the main source of the high alk. yielding cvs of Indonesia, & *C. pubescens* Vahl (brown or red bark); true *C. officinalis* L. (S Ecuador) not important

Cincinnobotrys Gilg. Melastomataceae. 7 trop. Afr. R: Adansonia 16(1976)355

Cineraria L. (~ *Senecio*). Compositae (Sen.-Sen.). c. 30 trop. & (esp.) S Afr., Madag., SW Arabia. Garden 'cinerarias' now referred to *Pericallis* × *hybrida*

Cinna L. Gramineae (21d). 4 Am., temp. Euras. (Eur. 1). R: Sida 14(1991)581. A 1 or 2

Cinnabarinea F. Ritter = *Echinopsis*

Cinnadenia Kosterm. Lauraceae (inc. sed.). 2 Bhutan, Assam, Burma, Malay Pen.

Cinnagrostis Griseb. = *Calamagrostis*

Cinnamodendron Endl. Canellaceae. 5 trop. S Am., WI. *C. corticosum* Miers (WI) – bark a spice & a tonic

Cinnamomum Schaeffer. Lauraceae (I). (Incl. *Phoebe* p.p.) c. 350 E & SE As. to Aus., Fiji & Samoa, trop. Am. (60). R (p.p.): Ginkgoana 6(1986)1. Aromatic trees & shrubs, bark yielding products for flavouring, scent & medic. *C. aromaticum* Nees (*C. cassia*, cassia bark, cassia lignea, Chinese or bastard cinnamon, Burma) – long cult. China, bark used like cinnamon (one of the oldest spices), lvs distilled for oil used as flavouring; *C. burmanii* (Nees) Blume (Indomal.) – sold as spice 'Korintji cinnamon' in US; *C. camphora*

(L.) J. Presl (camphor, ho wood, China, Taiwan, Japan) – timber for cabinet work & distilled to give camphor (now synthesized artificially), used to keep off moths in wardrobes etc. & in liniments, aromatherapy oils etc.; *C. culitlawan* (L.) Kosterm. (*C. culilaban*, culilawan, China, Mal.) – bark gives spice, medic. etc.; *C. iners* Reinw. ex Blume (Indomal.) – bark a food-flavouring & tonic drink, in joss-sticks in Malay Pen.; *C. verum* J. Presl (*C. zeylanicum*, cinnamon, Sri Lanka & SW India) – cinnamon of commerce, though often adulterated with or replaced by *C. aromaticum* (coarser bark), cult. as low shrub & widely natur. (as in Seychelles), the bark sold as quills (quillings when broken) & used to flavour food (esp. stewed apples & pears in GB) & toothpaste, also as incense & medic.

cinnamon *Cinnamomum verum*; **Chinese c.** *C. aromaticum*; **c. fern** *Osmunda cinnamomea*; **Korintji c.** *C. burmanii*; **c. rose** *Rosa majalis*

Cinnamosma Baillon. Canellaceae. 3 Madag. *C. fragrans* Baillon (taggar) – scented wood exported via Zanzibar to Bombay for religious cerem.

cinquefoil *Potentilla* spp. esp. *P. reptans*

Cionidium T. Moore = *Tectaria*

Cionomene Krukoff = *Elephantomene*

Cionosicyos Griseb. Cucurbitaceae. 3 C Am., WI

Cionura Griseb. = *Marsdenia*

Cipadessa Blume. Meliaceae (I 4). 1 Indomal.: *C. baccifera* (Roth) Miq. R: FMI, 12(1995)57

Cipocereus F. Ritter. Cactaceae (III 3). 5 Brazil. R: Bradleya 9(1991)86

Cipum A. Rich. = ?

Cipura Aublet. Iridaceae (III 4). c. 6 trop. Am. Bulbs locally medic.

Cipuropsis Ule = *Vriesea*

Circaea L. Onagraceae. 7 N hemisph., woods (Eur. 2). Poll. syrphids & small bees, the first the more imp. the shadier the habitat; hybrids common but reprod. vegetatively & resemble good spp. R: AMBG 69(1982)804. Fl. 2-merous with 1 whorl of A; fr. with hooked bristles; *C. lutetiana* L. (enchanter's nightshade, E US, Euras.) – weed

Circaeaster Maxim. Circaeasteraceae. 1 NW Himal. to NW China: *C. agrestis* Maxim.

Circaeasteraceae Hutch. Dicots – Magnoliidae – Ranunculales. 1/1 NW Himal. to NW China. Annual herb with rosette of simple subdecussate lvs with open dichotomous venation at top of stem-like hypocotyl, stipules 0. Fls bisexual, reg., hypog., minute in term. fascicles; K 2 or 3, C 0, A (1)2(3), anthers with longit. slits bisporangiate, G̲ (1)2(3) each with 2 apical ovules, pendulous from ventral margin of G, orthotropous, unitegmic, only 1 maturing. Fr. an achene, seed with no testa but copious endosperm with suberized outer layer. 2n = 30

Only genus: *Circaeaster*

Kingdonia form. here has been put in a monotypic family or (as here) in Ranunculaceae. Pollen studies (AJB 69(1982)990) show similarities with herbaceous Berberidaceae & certain R. like *Trollius* and the affinities of this app. much reduced pl. still unclear. The open dichotomous venation has excited much morphological interest in the past even though such is found in petals of some *Ranunculus* spp.

Circaeocarpus C.Y. Wu = *Zippelia*

Circandra N. E. Br. (~ *Erepsia*). Aizoaceae (V). 1 SW S Afr. R: BBP 64(1989)473. Extinct?

Circassian seeds *Adenanthera pavonina*

cirio *Fouquieria columnaris*

Cirrhaea Lindley. Orchidaceae (V 10). 6 Brazil

Cirrhopetalum Lindley = *Bulbophyllum*

Cirsium Miller. Compositae (Card.-Card.). c. 250 N temp. (Eur. 60), c. 50 temp. Am. Few cult. orn.; some bad weeds, esp. *C. arvense* (L.) Scop. (swamp thistle, Eur.) – usu. dioec. & *C. vulgare* (Savi) Ten. (spear or bull t., Eur. & Med., natur. N Am.) – prob. 'true' t. of Scotland (though now applied to *Onopordum acanthium*); N Am. Indians eat inner stem of several spp. while the young shoots of *C. palustre* (L.) Scop. (Eur., Medit., W As.) eaten in salads & roots of *C. tuberosum* (L.) All. (Eur.) have been harvested & stored as food for winter

Cischweinfia Dressler & N. Williams. Orchidaceae (V 10). 7 trop. S Am.

Cissampelopsis (DC) Miq. Compositae (Sen.-Sen.). 10 trop. As.

Cissampelos L. Menispermaceae (V). 20 trop. R: Phytologia 30(1975)415. Alks. *C. pareira* L. (false pareira root, range of genus) – medic. esp. against snakebite & used as fertility control in Uruguay

Cissus L. Vitaceae. (Excl. *Cyphostemma*) c. 200 trop. & warm. Usu. lianes with tendrils; fls bisexual, C 4; berry usu. ined. Some cult. climbers & housepls (grape ivy), esp. *C. antarctica* Vent. (kangaroo vine, E Aus.); *C. quadrangularis* L. (OW trop., S Afr.) – succ.

liane with 4-winged stems, often ± leafless; lvs of many spp. ed. & medic. e.g. *C. gongylodes* (Baker) Planchon (Braz.) – cult. Kayapó (Amaz.) incl. cvs high in vitamins & nutrients, planted in 'abandoned' slash-and-burn, giving 40 yrs of ed. lvs & fr. before new clearance

Cistaceae Juss. Dicots – Dilleniidae – Violales (? Malvales). 8/175 temp. & warm esp. Medit. Shrubs or herbs; hairs often clustered as to seem stellate. Lvs opp. more rarely spiral or whorled, simple, s.t. with stipules. Fls bisexual, reg. (exc. K), hypog.; K 5 (2 outer often narrower than & s.t. adnate to inner 3) or 3, convolute, C 5 (3 in *Lechea*), convolute in opp. directions to K, seldom imbricate, often crumpled in bud & evanescent, A (3–) ± ∞ on or just outside annular nectary-disk, centrifugal, s.t. sensitive to touch, anthers with longit. slits, $\underline{G}$ ((3)5–10), 1-loc. with parietal placentation, the placentas ± deeply intruded & s.t. (as in *Cistus*) meeting to give discrete locules, style solit. to 0, stigma 1(3) s.t. lobed, ovules (1–)4–∞ on each placenta, orthotropous, rarely (as in *Fumana*) anatropous, bitegmic. Fr. a loculicidal capsule, seeds (1–)3–∞, usu. small with starchy endosperm & embryo usu. with flattened cotyledons & commonly curved into hook or ring, circinately coiled, or rarely straight. x = 5–11

Genera: *Atlanthemum, Cistus, Fumana, Halimium, Helianthemum, Hudsonia, Lechea, Tuberaria*

Related to Bixaceae & Flacourtiaceae, though chemistry does not support relationship with F. (Mears)

Many cult. orn. esp. spp. of *Cistus, Halimium* & *Helianthemum*; fragrant resin (*Cistus*)

Cistanche Hoffsgg. & Link. Orobanchaceae. 16 Med. (Eur. 2), Ethiopia to W India & NW China. *C. phelypaea* (L.) Cout. (SW Eur. & N Afr.) – eaten as asparagus by Tuareg

Cistanthe Spach (~ *Calandrinia*). Portulacaceae. (Excl. *Calyptridium* & *Spraguea*) 25 Am. (3 N). Some cult. orn. incl. *C. grandiflora* (Lindley) Schldl. (*Calandrinia d.*, Chile) – perennial grown as annual 1 m tall

Cistus L. Cistaceae. 18 Medit. (Eur. 16), Canary Is. R: Boissiera 4(1939)1. Although self-incompatible, pollen can germ. & obstruct stigma, but *C. ladanifer* & *C. salviifolius* with sensitive A moving from G to P reducing self-poll., *C. albidus* L. (alkaline soils) & *C. crispus* L. (acid) with style elongating above A by p.m. (fls last 1 day). Almost all cult. orn. shrubs incl. *C. ladanifer* L. (W Med.) – early colonist after fire (flammable as 0.1–0.2% leaf fresh wt. is alpha-pinene with flashpoint at 38 °C) & source of the resin ladanum coll. by dragging a kind of rake through shrubs (&, according to Pliny, goats' beards!) & used in scenting soap, deodorants, form. medic. & *C. creticus* L. (*C. incanus* subsp. *c.*, E Med.) – resin similar, prob. 'myrrh' of Genesis; resin from both obtained by boiling twigs when resin can be skimmed from water-surface; *C. salviifolius* L. (S Eur.) – lvs form. tea subs. in Greece

Cithareloma Bunge. Cruciferae. 3 Iran, C As.

Citharexylum L. Verbenaceae. c. 70 trop. Am. to Arg. Dioec. trees & shrubs superficially resembling cherries; fr. a berry-like drupe separating into 2 nutlets. Timber good (fiddlewood (meaning of Latin name), corruption of *bois-fidèle*) esp. *C. spinosum* L. (*C. fruticosum*, WI)

citrange, Troyer c. × *Citroncirus webberi*

Citriobatus A. Cunn. & Putterl. (~ *Pittosporum*). Pittosporaceae. 5 Aus. 1 ext. to Mal.

× **Citrofortunella** J. Ingram & H. Moore. *Citrus*

citron *Citrus medica*

× **Citroncirus** J. Ingram & H. Moore. Rutaceae. *Citrus* × *Poncirus*. Fr. trees esp. × *C. webberi* J. Ingram & H. Moore ((Troyer) citrange, *Citrus* × *aurantium* × *P. trifoliata*) – also a stock for other citrus fr. (resistant to tristeza virus)

Citronella D. Don. Icacinaceae. 21 Mal., Pacific, trop. Am. *C. gongonha* (Mart.) R. Howard (Brazil) used like maté; *C. moorei* (Benth.) R. Howard (churnwood, Aus.) – flanged trunks

citronella oil *Cymbopogon nardus*

Citropsis (Engl.) Swingle & Kellerman. Rutaceae. 8 trop. Afr. R: W.T. Swingle, *Botany of Citrus* (1943)302. Stock for citrus fr.

Citrullus Schrader. Cucurbitaceae. 4 trop. & S Afr., prob. As. Monoec. or dioec. lianes with branched tendrils. *C. colocynthis* (L.) Schrader (colocynth, bitter apple, 'vine of Sodom', 'gall'; cult. & natur. Medit. & India) – green throughout Iraq summer (deep root system), dried pulp purgative etc., cult. since time of Assyrians, rodent control since time of Columella; *C. lanatus* (Thunb.) Matsum. & Nakai (water melon, Afr., natur. Am.) – refreshing red-fleshed fr. sold in Med. etc., small white-fleshed cvs used for preserves, seeds used in soups etc. in China, excess fr. made into syrup in E Eur., ingredient of sun-lotions & other cosmetics

Citrus L. Rutaceae. (Incl. *Eremocitrus, Fortunella, Microcitrus*) c. 20 S & SE As, to E Aus. (6, R: Telopea 7(1998)333). R: H.J. Webber *et al.* (1943–8) *The C. Industry*. Evergreen trees & shrubs usu. spiny with simple lvs with a joint on the petiole showing derivation from trifoliolate (as *Poncirus*) and pinnate (as *Merrillia*) ones like those in many Rut.; subg. *Papeda* with winged 'petiole' as long as 'lamina', other spp. with smaller joints & some with 0. Fls usu. white, strongly fragrant, fr. a leathery-skinned berry (hesperidium), the rind insecticidal (source of bittering agents used in tonic water instead of quinine, esp. naringin half as bitter as q. but simply manipulated chemically to a dihydrochalcone 500 times as sweet as sucrose, related compounds now being developed as comm. sweeteners in US) with(3–)5–15(–18) locules (segments) filled with inflated hair-cells (fibre used in gut disease) full of juice, keeping seeds moist (& alleged to raise blood pressure, octopamine sometimes causing migraines), rich in vitamin C – 'limejuice' (actually usu. lemon, which is 3 times richer) form. used as antiscorbutic by British seamen (known as limers or 'limeys' in US). Acid oil droplets in juice vesicles of subg. *Papeda* make these ined. but subg. *Citrus* (R: Telopea 7(1997)167) gives most imp. fruit industry in warm countries esp. S US, Medit., Brazil, WI etc. based on *C.* spp. & hybrids. Long cult., imp. to Romans (fr. thought to come from tree with fragrant wood – 'citrus' (= *Tetraclinis*), the origins of some still obscure, but most in cult, are anc. apomictic hybrids (polyembryony first observed by Leeuwenhoek!) & selected cvs of these; hybrids still being synthesized incl with *Poncirus* spp. (= × *Citroncirus*) for new fr. crops & rootstocks (grafting in C. the earliest documented of all). *C.* × *aurantiifolia* (Christm.) Swingle (*C. maxima* × *C.* sp., lime) – for cooking & juice, oil from rind & from seeds used in soap etc., some cvs sweeter & much used in India; *C.* × *aurantium* L. (*C. maxima* × *C. reticulata*) Grapefruit Group (*C. paradisi*, a back-cross with female *C. maxima* made in C18 Barbados) – pop, breakfast fr. & source of juice, cvs incl. pink-fleshed mutants (usu.'Marsh Seedless' or 'Star Ruby') & NZ grapefruit with winged petioles & green cotyledons, also used in shampoos & other cosmetics, Sour Orange Group incl. 'Bouquet' ('bergamot' orange in US), bigarade & Seville oranges – upper surface of petals yield neroli oil for scent (chief ingredient of eau-de-Cologne, fr. used for marmalade, candied peel & in liqueurs (e.g. Curaçao, Cointreau), Sweet Orange Group (*C. sinensis*, raised in China) – most imp. citrus crop (66% of total), the pulp used fresh or for juice (diff. stereoisomers of limonene responsible for diffs between orange- & lemon-juice), rind for oil, seed-oil used in soap, extract from lvs, twigs & young fr. (petitgrain oil) used in aromatherapy, chewing sticks in Nigeria, many cvs incl. blood (red or red-streaked pulp) & navel (secondary fr. at stylar end, e.g. 'Baia' ('Washington Navel')) oranges, famous ones incl. 'Shamouti' ('Jaffa'), 'Valencia' with thin rind & c. 6 seeds, ortaniques (backcrosses with *C. reticulata*, Jamaica 1920s), Tangelo Group (*C.* × *tangelo*, grapefruit crossed with *C. reticulata*) – several cvs, in GB known incorrectly as uglis, such as 'Minneola', Tangor Group (*C. nobilis*, a backcross with *C. reticulata*) – easily removed rind; *C. australis* (Mudie) Planchon (*Microcitrus a.*, E Aus.) – fr. a passable 'lime'; *C.* × *bergamia* Risso (*C. aurantium* subsp. *b.*, *C.* × *a.* × *medica*, bergamot) – rind oil used in scent, hair-oil, tanning oil & Earl Grey tea (though allegedly carcinogenic); *C.* × *floridana* (J. Ingram & H. Moore) Mabb. (× *Citrofortunella floridana*, *Citrus* × *aurantiifolia* × *C. japonica*, limequat) – ed. fr., raised Florida 1909; *C. glauca* (Lindley) Burkill (*Eremocitrus g.*, desert kumquat, NE to S Aus.) – desert pl. resistant to cold, used as rootstock for citrus, fr. ed. used in drinks & preserves; *C. hystrix* DC (kaffir or makrut lime, SE As.) – 'lime leaves' of (esp. Thai) cuisine; *C. japonica* Thunb. (*Fortunella j.*, *C. margarita*, *F. m.*,, kumquat, ?S China) – fr. ed. complete with peel raw, preserved or candied; *C. limetta* Risso (limetta, sweetie, orig. unknown) – weak sweet or acid flavour, cult., Medit.; *C* × *limon* (L.) Osb. (*C medica* × *C.* sp., lemon) – rind yields lemon oil (flavour due to citral, less than 5% by weight), pulp gives lemon juice & citric acid used culinarily & med., many cvs; *C. maxima* (Burm.) Merr. (*C. decumana*, *C. grandis*, pomelo, pummelo, shaddock (after Capt. Chaddock, who introd. it to WI), pompelmous, pamplemousse, SE As.) – fr. to 8 kg with thick rind, source of bitter narinjin used in drinks & sweets, parent or many imp. frs; *C. medica* L. (citron, N India) – early cult. & spread to Medit. by Alexander but not in Bible (? seed in Cyprus 1200 BC), major source of candied peel, 'Etrog' (etrog, cedrat) used in Feast of Tabernacles 136 BC onwards, 'Fingered' (var. *sarcodactylis*, Buddha's hand) – fr. with finger-like processes; *C.* × *microcarpa* Bunge (*C* × *mitis*, × *Citrofortunellam.*, × *C. microcarpa*, *C. japonica* × *C. reticulata*, calamondin) – housepl. with fr. held all winter; *C. reticulata* Blanco (mandarin, tangerine, satsuma, subtrop. China) – rather hardy (satsumas the hardiest citrus) with small fr. introd. via Kew 1805, the segments separating from one another & peel readily, oil in shampoos & other cosmetics, ants used to capture insect pests in China (ant nests (*Oecophylla smaragdina*) sold for purpose) since AD 304 perhaps earliest use of biol. control (imported to US C20) with bamboo pole bridges between trees, cvs incl. 'Clementine' (clementine), Dancy' (common red tangerine of Florida), 'Owari' (seedless); *C. × taitensis* Risso (*C. × jambiri*, ?*C. medica* or *C. × limon × C. reticulata*, rough lemon) incl. 'Otaheite', a dwarf potted 'orange' esp. US Christmas flower trade. See also pomander

Cladanthus Cass. Compositae (Anth.-Art.). 1 S Spain, NW Afr.

Claderia Hook.f. Orchidaceae (? tribe). 1–2 Mal. R: OB 72(1983)17

Cladium P. Browne. Cyperaceae. 3: 2 N Am., 1 cosmop. exc. Am.: *C. mariscus* (L.) Pohl – paper-making in Danube, form. used for thatching in GB ('elk sedge' of Anglo-Saxons) but now rare there; *C. jamaicensis* Crantz (SE US to W India) – dominant in Everglades

Cladocarpa (H. St. John) H. St John = *Sicyos*

Cladoceras Bremek. Rubiaceae (II 2). 1 trop. E Afr.: *C. subcapitata* (Schumann & Krause) Bremek. R: BRSBB 117(1984)247

Cladochaeta DC (~ *Helichrysum*). Compositae (Gnap.-Gnap.). 2 Caucasus. R: OB 104(1991)150

Cladocolea Tieghem. Loranthaceae. 26 C Am. R: JAA 56(1975)272

Cladogelonium Leandri. Euphorbiaceae. 1 Madagascar

Cladogynos Zipp. ex Span. Euphorbiaceae. 1 SE As., Mal.

Cladomyza Danser. Santalaceae. 20 Borneo, Papuasia

Cladopus H. Möller. Podostemaceae. 5 E As., Mal.

Cladoraphis Franchet (~ *Eragrostis*). Gramineae (31d). 2 S Afr.

Cladostachys D. Don = *Deeringia*

Cladostemon A. Braun & Vatke. Capparidaceae. 1 S & SE Afr.

Cladostigma Radlk. Convolvulaceae. 2 NE trop. Afr. Ed. fr. R: NRBGE 8(1913)96

Cladothamnus Bong. = *Elliottia*

Cladrastis Raf. Leguminosae (III 2). 1 N Am., 5 E As. R: NRBGE 9 (1913)96. Yellow-wood esp. *C. lutea* (Michaux f.) K. Koch (Kentucky yellow-wood, yellow ash, SE US) – close-grained wood for gun-stocks, heartwood provides a yellow dye

Clandestinaria Spach = *Rorippa*

Clanwilliam cedar *Widdringtonia cedarbergensis*

Claoxylon A. Juss. Euphorbiaceae. 80 OW trop. A 10–200

Claoxylopsis Leandri. Euphorbiaceae. 3 Madag. R: KB 43(1988)642

Clappertonia Meissner. Tiliaceae. 2 trop. W Afr. R: FR 99(1988)267. Fibre plants esp. *C. ficifolia* (Willd.) Decne. – jute-like fibre

Clappia A. Gray. Compositae (Hele.-Flav.) 1 S US

Clara Kunth = *Herreria*

Clarisia Ruíz & Pavón. Moraceae (II). (Incl. *Sahagunia*) 3 trop. Am. Timber & ed. fr.

Clarkesia J.R.I. Wood. Acanthaceae. 1 Nepal to Thailand: *C. parviflora* (T. Anderson) J.R.I. Wood. R: EJB 51(1994)187

Clarkella Hook.f. Rubiaceae (inc. sed.). 2 Himalaya, Thailand

Clarkia Pursh. Onagraceae. 33 W N Am., S S Am. R: UCPB 20(1955)241. Cult. orn. ann. herbs esp. *C. amoena* (Lehm.) Nelson & Macbr. (coastal N Calif.) – the 'godetia' of gardens & *C. unguiculata* Lindley (Calif.) – the 'clarkia' of gardens with cvs of many colours incl. doubles

clary *Salvia sclarea*

Clastopus Bunge ex Boiss. Cruciferae. 1 Iran, Iraq

Clathrotropis (Benth.) Harms. Leguminosae (III 2). 6 trop. S Am.

Clausena Burm.f. Rutaceae. 23 OW trop., S Afr. R: W.T. Swingle, *The Botany of Citrus* (1943)158. Some ed. fr. esp. *C. dentata* (Willd.) M. Roemer (India) – taste like blackcurrants & *C. lansium* (Lour.) Skeels (wampi, S China) – cult. for lime-like fr., citrus rootstock. *C. heptaphylla* Wight & Arn. (SE As.) – carbazole alk. with antifungal activity

Clausenellia Löve & D. Löve = *Sedum*

Clausia Trotzky ex Hayek (~ *Hesperis*). Cruciferae. 5 E Eur. (1) to C As.

Clausospicula Lazarides. Gramineae (40C). 1 N Terr., Aus. R: AusSB 4(1991)391. Cleistogamous

Clavija Ruíz & Pavón. Theophrastaceae. 50 S Nicaragua to Braz., Hispaniola. R: OB 107(1991)1. Palm-like pachycauls often with cauliflory, most dioec., some gynodioec., androdioec. or polygamous

Clavinodum Wen = *Arundinaria*

claw, devil's *Harpagophytum procumbens*

Claytonia L. Portulacaceae. (Excl. *Montia*) 24 N Am., ext. to E As., 2 natur. Eur. Cotyledon 1; 'stemlvs' 1 pair (*Montia* several). *C. lanceolata* Pursh (E US) – corms ed. raw or cooked by

N Am. Indians; *C. perfoliata* Donn ex Willd. (*Montia p.*, winter purslane, N Am., natur. GB) – infls subtended by a perfoliate disc-like organ, ed. Some cult. orn. esp. *C. virginica* L. (spring beauty, E N Am.) – 2n = 12–almost 200 with diff. nos in same pl.; some with ant-disp. seeds

Claytoniella Yurtsev = *Montia*

Cleanthe Salisb. ex Benth. = *Aristea*

clearing nut *Strychnos potatorum*

clearweed *Pilea* spp.

cleavers *Galium aparine*

Cleghornia Wight. Apocynaceae. 4 Sri Lanka & SE As. to W. Mal. R: AUWP 88-6(1988)11

Cleidiocarpon Airy Shaw. Euphorbiaceae. 2 Burma, W China. *C. cavalieri* (Léveillé) Airy Shaw promoted in China as oil-crop

Cleidion Blume. Euphorbiaceae. 25 trop.

Cleisocentron Brühl. Orchidaceae (V 16). 1 Himalaya

Cleisomeria Lindley ex G. Don f. Orchidaceae (V 16). 2 SE As., Mal. R: OB 95(1988)131

Cleisostoma Blume (~ *Sarcochilus*). Orchidaceae (V 16). 80 Nepal to New Guinea & New Caled. (1). Cult. orn. incl. *C. striatum* (Reichb.f.) N.E. Br. (? incl. *C. javanicum*, Assam to W Mal.)

Cleistachne Benth. Gramineae (40c). 1 trop. Afr., India

Cleistanthus Hook. f. ex Planchon (~ *Bridelia*). Euphorbiaceae. c. 140 OW trop. Dried fr. of *C. collinus* (Roxb.) Benth. (India) used in criminal poisonings

Cleistes Rich. ex Lindley. Orchidaceae (Pogoniinae). 55 trop. Am., 1 NE US: *C. divaricata* (L.) Ames – hinged anther dispenses several (not just 1) pollen tetrad-masses to diff. insects

Cleistocactus Lemaire. Cactaceae (III 4). (Incl. *Borzicactus*) c. 45 C Peru to Bolivia & N Arg., Paraguay & Uruguay. Hummingbird-poll. Cult. orn. esp. *C. straussii* (Heese) Backeb. (S Bolivia) with 30–40 bristle-like spines to 2 cm long & 4 yellowish stouter spines per areole; *C. sepium* (Kunth) Roland-Gosselin (Peru) – ed. fr.

Cleistocalyx Blume (~ *Syzygium*). Myrtaceae (Myrt.). c. 21 Burma to Fiji. R: JAA 18(1937)322

Cleistochlamys Oliver. Annonaceae. 1 trop. E. Afr.: *C. kirkii* (Benth.) Oliver

Cleistochloa C.E. Hubb. Gramineae (34b). 3 NE Aus. Dry sandstone ridges. Cleistogamy

Cleistogenes Keng = *Kengia*

Cleistopholis Pierre ex Engl. Annonaceae. 3–4 trop. Afr. *C. patens* (Benth.) Engl. (otu) – comm. timber, bark for rope & mats, lvs & roots medic.

Clelandia J. Black = *Hybanthus*

Clematepistephium Hallé (~ *Epistephium*). Orchidaceae (V 4). 1 New Caled. Liane to 8 m long

Clematicissus Planchon. Vitaceae. 1 SW Aus. (Geraldton area)

Clematis L. Ranunculaceae (II 2). c. 295 N temp. (Eur. 10), S temp. (few), Oceania & trop. Afr. mts. R: APG 38(1987)36; C. Lloyd (1989) *C.*, ed. 3. Lianes, shrubs or herbs; lvs usu. opp. compound or simple, K petaloid, C 0, fr. an achene usu. with long feathery style. Imp. cult. orn. esp. lianes, the only woody Ranunculaceae of any size, some without interfascicular cambium (the sec. rays arise in the primary bundles), some with regular wood formation & some with extra thickenings in pith & cortex cells; petioles wrap round support & lignify. Linked to *Anemone* by *Clematopsis*. Several medic. (some spp. with acrid juice inflaming skin) & many hybrids in cult., the common garden plants being derived from *C. florida* Thunb. (E As.) – fls on old wood in summer, *C. patens* Morren & Decne. (China, Japan) – fls on old wood in spring, and esp. *C. × jackmanii* T. Moore (*C. lanuginosa* Lindley (? China, not known in wild) × *C. viticella* L. (Medit. & nearby As.), arose 1858 at Jackman's nurseries at Woking, England) – fls on new wood in summer & autumn. *C. afoliata* J. Buch. (NZ) – lvs reduced to petioles & petiolules, laminae developing only in young pl. or in shade; *C. armandii* Franchet (SW China) – evergreen cult. orn. liane; *C. flammula* L. (Euras.) – ed. shoots (cooked); *C. heracleifolia* DC (China) – commonly cult. herbaceous sp. with blue fls, hybrid with *C. vitalba* = *C. × jouiniana* Schneider; *C. montana* Buch.-Ham. ex DC (montana, Himal., W China) – rampant liane much planted for spring fls; *C. rehderiana* Craib (China) – cult. orn. liane with yellow, cowslip-scented, nodding fls; *C. vitalba* L. (old man's beard, 'traveller's joy', Euras., N Afr.) – may attain 30 m, lengths of stem a tobacco subs., young shoots ed. Eur., diuretic

Clematoclethra (Franchet) Maxim. Actinidiaceae. 1 W & C China: *C. scandens* (Franchet) Maxim. R: APS 27(1989)81

Clematopsis Bojer ex Hutch. = *Clematis*

Clemensiella Schltr. Asclepiadaceae (III 4). 1 Philipp.: *C. mariae* (Schltr.) Schltr.

clementine *Citrus reticulata* 'Clementine'

Cleobulia Mart. ex Benth. Leguminosae (III 10). 3 Brazil. R: Phytol. 38(1977)51

Cleomaceae Horan. = Capparidaceae

Cleome L. Capparidaceae. 150 trop. & warm (Eur. 3). A 6. Some medic. locally, others with ed. seeds, some cult. orn. *C. gynandra* L. (*Gynandropsis gynandra*, cat whiskers, OW trop. natur. Am.) – potherb; *C. sesquiorygalis* Naudin ex C. Huber (*C. hassleriana*), spider flower, SE Brazil to Arg.) – cult. orn. esp. cutfl., many cvs

Cleomella DC. Capparidaceae. 10 SW N Am. R: UWPSB 1(1992)29

Cleonia L. Labiatae (VIII 2). 2 W Med. (Eur. 1)

Cleophora Gaertner = *Latania*

Cleretum N.E. Br. Aizoaceae (V). 3 S Afr. R: BBP 61(1986)431

Clermontia Gaudich. Campanulaceae (Lob.). 22 Hawaii. Woody, some epiphytic with ed. fr.; latex used as birdlime. *C. arborescens* (H. Mann) Hillebrand visited by introd. white-eyes (*Zosterops japonica*, Japan), though orig. pollinators prob. extinct

Clerodendranthus Kudo = *Orthosiphon*

Clerodendrum L. Labiatae (II, Verbenaceae). Excl. *Kalaharia* & *Rotheca* (*Cyclonema*) c. 400 trop. & warm esp. E hemisph. Trees or shrubs, some scramblers; *C. laciniatum* Balf.f. (Rodrigues) – marked heterophylly; fls white, yellow, red, blue or violet; A project as a landing-stage for insects, later replaced by style rising from below as A wither & fall; K often brightly coloured & in some spp., e.g. *C. trichotomum* Thunb. (Japan), accrescent (in that sp. red & fleshy contrasting with blue fr. & evidently attractive to dispersing birds). Some medic. but name (Gk. for 'chance' & 'tree') refers to variable reports of efficacy. *C. fistulosum* Becc. (Borneo) has pithy stems inhabited by ants & the hollow ones of *C. capitatum* (Willd.) Schum. & Thonn. used as pipes & for tapping palm wine (W Afr.); those of *C. phyllomega* Steudel (W Mal.) with hollow internodes & thin-walled layers pierced & penetrated by ants. Many cult. orn. hardy & tender esp.: *C. bungei* Steudel (*C. foetidum*, China) – hardy shrub; *C. chinense* (Osb.) Mabb. (*C. fragrans*, *C. philippinum*, glory bower, Honolulu or Lady Nugent's rose, China to C Mal., natur. S US & pantrop.) – extrafl. nectaries with ants building caves over them, fls fragrant, pinkish, in cult. usu. double (wild pl. = var. *simplex* (Mold.) S.L. Chen), serious weed in WI, Samoa; *C. inerme* (L.) Gaertner (India to W Pacific, coastal) – widely introd. trop. as sand-binder, topiary in India, canework in Tonga; *C. paniculatum* L. (pagoda flower, SE As.) – tender with scarlet fls; *C. thomsoniae* Balf. (bleeding heart, bag flower, trop. W Afr.) – cult. greenhouses for white K & red C

Clethra L. Cyrillaceae (Clethraceae). 64 trop. Am., As.to Mal., N Am. (1), Madeira (1). R: BJ 87(1967)36. Some cult. orn. esp. *C. alnifolia* L. (bush pepper, E N Am.) & *C. arborea* Aiton (folhado, Madeira)

Clethraceae Klotzsch. = Cyrillaceae

Clevelandia Greene (~ *Orthocarpa*). Scrophulariaceae. 1 Baja Calif.: *C. beldingii* (Greene) Greene. R: SB 16(1991)663

Cleyera Thunb. Theaceae. 17 Himal. to Japan (1), Mex. to Panamá & WI (16). R: JAA 18 (1937)118, 22(1941)395. Some cult. orn. incl. *C. ochnacea* DC (Himal., Japan) – sacred tree of Shintoism (sakaki)

Clianthus Sol. ex Lindley. Leguminosae (III 15). 1 NE NZ: *C. puniceus* (G. Don f.) Lindley (parrot's bill), long cult. by Maoris & poss. native only in 2 areas in Urewawa Nat. Park. See also *Swainsona*

Clibadium L. Compositae (Helia.). 40 trop. Am. Some (e.g. *C. laxum* S.F. Blake) with fleshy cypselas. Fish-poisons (guaco) esp. *C. sylvestre* Baillon – extract stops the human heart reversibly

Clidemia D. Don. Melastomataceae. 117 trop. Am. Some ed. fr. (bush currants) & local med. Often in sec. veg. & weedy in Pacific (esp. *C. hirta* (L.) D. Don – 'Koster's curse' in Fiji). Differ from *Miconia* by usu. lateral infls. *C. hammelii* Almeida (C Am.) – *Ololaelaps* mites in domatia, otherwise unknown in fam.; *C. rubra* (Aublet) G. Don f. (Amaz.) & other spp. – cult. orn. shrubs

cliff brake *Pellaea* spp.

Cliffortia L. Rosaceae. 115 trop. & S Afr. esp. Cape mts. H. Weimarck (1934) *Monograph of the genus C.*

Cliftonia Banks ex Gaertner f. Cyrillaceae. 1 SE US: *C. monophylla* (Lam.) Britton & Sarg., a shrub of swamps of the coastal plain, its fls the forage for commerical honeybees

Climacoptera Botsch. = *Salsola*

climbing fern *Lygodium* spp.

Clinacanthus Nees. Acanthaceae. 2 S China to Mal. *C. nutans* (Burm. f.) Lindau (SE As.) – young lvs ed. Vietnam

Clinelymus (Griseb.) Nevski = *Elymus*

Clinopodium L. Labiatae. (Incl. *Acinos, Calamintha*) c. 20 temp. (Eur. 11). Cult. orn. incl. *C. acinos* (L.) Kuntze (*Acinos arvensis*, basil thyme, Eur., W As.), *C. menthifolium* (Host) Stace (*Calamintha sylvatica*, wood calamint, Eur.) & *C. ascendens* (Jordan) Samp. (common c., Eur.)

Clinosperma Becc. Palmae (V 4m). 1 New Caled.: *C. bracteata* (Brongn.) Becc.

Clinostemon Kuhlm. & Samp. = *Mezilaurus*

Clinostigma H. Wendl. Palmae (V 4m). 13 New Britain, Vanuatu, Fiji, Samoa, Micronesia to Bonin Is.

Clinostigmopsis Becc. = *Clinostigma*

Clintonia Raf. Convallariaceae (Liliaceae s.l.). 4 N Am., 1 E As. Wood lilies; cult. orn. esp. *C. andrewsiana* Torrey (C & N Calif.) & *C. borealis* (Aiton) Raf. (E N Am.) – young lvs a potherb & salad

Cliococca Bab. (~ *Linum*). Linaceae. 1 temp. S Am.

Clistax Mart. Acanthaceae. 2 Brazil

Clistoyucca (Engelm.) Trel. = *Yucca*

Clitandra Benth. Apocynaceae. 1 trop. Afr.: *C. cymulosa* Benth. – source of inferior (? medic.) rubber, fr. ed.

Clitandropsis S. Moore = *Melodinus*

Clitoria L. Leguminosae (III 10). 60 trop. esp. Am. (49). Fls inverted so that standard points downwards (cf. *Centrosema*) & A & pistil touch backs of visiting insects. Some cult. orn. climbers esp. blue-flowered *C. ternatea* L. (butterfly pea, blue p. or vine, prob. native trop. Am., now pantrop.) – fls for colouring rice, also like litmus paper, local medic.

Clitoriopsis R. Wilczek. Leguminosae (III 10). 1 Zaire, Sudan

Clivia Lindley. Amaryllidaceae (Liliaceae s.l.). 4 S Afr. Alks. Cult. orn. (kaffir lilies) esp. *C. miniata* (Lindley) Regel – stemless with fleshy roots & somewhat swollen leaf-bases like a bulb, many cvs incl. ones with coloured or varieg. lvs raised in Japan; *C. caulescens* R.A. Dyer has stem to 50cm

cloak fern *Notholaena* spp.

clock-vine *Thunbergia grandiflora*

Cloezia Brongn. & Gris. Myrtaceae (Lept.). 8 New Caledonia

Clonodia Griseb. Malpighiaceae. 2–3 trop. S Am. R: MNYBG 32(1981)203. Extrafl. nectaries

Clonostylis S. Moore = *Spathiostemon*

Closia Rémy = *Perityle*

clotbur *Xanthium spinosum*

cloudberry *Rubus chamaemorus*

clove gilliflower or **pink** *Dianthus caryophyllus, D. plumarius*

clover *Trifolium* spp.; **alsike c.** *T. hybridum*; **alyce c.** *Alysicarpus vaginalis*; **Aztec c.** *T. amabile*; **Bokhara c.** *Melilotus albus*; **Dutch c.** *T. repens*; **Hungarian c.** *T. pannonicum*; **Italian c.** *T. incarnatum*; **Japanese c.** *Kummerowia striata*; **Persian c.** *T. resupinatum*; **purple** or **red c.** *T. pratense*; **running buffalo c.** *Trifolium stoloniferum*; **strawberry c.** *T. fragiferum*; **subterranean c.** *T. subterraneum*; **Swedish c.** *T. hybridum*; **white c.** *T. repens*; **Uganda c.** *T. burchellianum* ssp. *johnstonii*; **zig-zag c.** *T. medium*

cloves *Syzygium aromaticum*; **Madagascar c.** *Cryptocarya* sp. (*Ravensara aromatica*)

Clowesia Lindley (~ *Catasetum*). Orchidaceae (V 9). 6 C Am. R: Selbyana 1(1975)134. Segregated from *Catasetum* on bisexual fls

clubmoss *Lycopodium* spp.

clubrush *Schoenoplectus lacustris*

Clusia L. Guttiferae (III). c. 145 trop. & warm Am. Dioec. trees (only genus of dicot. trees with crassulacean acid metabolism) & shrubs, c. 85 ± epiphytic, or stranglers with anastomosing aerial roots (cf. *Ficus*), e.g. *C. rosea* Jacq. (warm & trop. Am.) – to 20 m, at anthesis staminodes deliquesce into a mass attractive to bees, sticky resin coll. for their nests, that from seeds used as bird-lime & to caulk boats, used as iron indicator in Venez. & firebreak in Sri Lanka. Colour of sap useful in infrageneric classification; some medic. (rock balsam) & resin used for incense; in Mex. lvs of some spp. used as playing-cards; some apomictic; some cult. orn. incl. *C. major* (L.) Jacq. (copey, trop. Am.)

Clusiaceae Lindley = Guttiferae

Clusiella Planchon & Triana. Guttiferae (III). 1–4 Panamá & N S Am. Epiphytes

cluster bean *Cyamopsis tetragonolobus*

Clutia L. Euphorbiaceae. 70 Afr., Arabia (2). Some medic.

Clybatis Philippi = *Leucheria*

Clymenia Swingle. Rutaceae. 2 Bismarck Arch. R: PANSP 137(1985)223

Clypeola L. Cruciferae. 9 Medit. (Eur. 2). R: KB 1935:1

Clytostoma Miers ex Bur. Bignoniaceae. 9 trop. S Am. Cult. orn. lianes

Clytostomanthus Pichon (~ *Anemopaegma*). Bignoniaceae. 1 Ecuador, Brazil, Paraguay

Cnemidaria C. Presl = *Cyathea*

Cnemidiscus Pierre = *Glenniea*

Cneoraceae Vest. Dicots – Rosidae – Sapindales. 2/3 Cuba, Canary Is., Med. (1 each). Shrubs. Lvs spirally arr., simple, small; stipules 0. Fls axillary, 1-cymes, sometimes epiphyllous, small, bisexual, reg., hypog.; K 3 or 4, persistent, sometimes basally connate, C 3 or 4, imbricate, disk a columnar nectariferous androgynophore, anthers 3 or 4 with longit. slits, G (3 or 4) with axile placentas & term. style with lobed stigmas, ovules (1) 2 per locule, pendulous, amphitropous, bitegmic, when 2 ± separated by partition from carpellary midrib. Fr. a schizocarp, each segment with (1) 2 seeds with strongly curved embryo & copious fleshy, oily endosperm. x = 9

Genera: *Cneorum*, *Neochamaelea*

Apparently close to Zygophyllaceae & Rutaceae; limonoids present

Cneoridium Hook.f. Rutaceae. 1 S Calif.

Cneorum L. Cneoraceae. 1 Cuba, 1 W Med. inc. Eur.: *C. tricoccon* L. – violent purgative, cult. orn. Canary Is. sp. excluded (*Neochamaelea*) on pollen data etc.

Cnesmocarpon Adema (~ *Jagera*). Sapindaceae. 4 New Guinea, Aus. R: Blumea 38(1993)195. Fr. hairs irritant

Cnesmone Blume (*Cnesmosa*). Euphorbiaceae. 10 Assam to W Mal. Climbers

Cnestidium Planchon. Connaraceae. 3 trop. Am.

Cnestis Juss. Connaraceae. 13 trop. Afr., 1 Indomal. R: AUWP 89-6(1989)174. Lianes, rarely small trees, some poisonous

Cnicothamnus Griseb. Compositae (Mut.-Mut.). 2 Bolivia, Argentina

Cnicus L. = *Centaurea*

Cnidium Juss. = *Selinum*

Cnidocarpa Pim. Umbelliferae (III 8). 2 Caucasus, C As.

Cnidoscolus Pohl. Euphorbiaceae. c. 75 Am. Monoec. or rarely dioec. herbs, sometimes with stinging hairs. *C. chayamansa* McVaugh (trop. Am.) & other spp. (chaya) – lvs eaten like spinach; *C. elasticus* Lundell (chilte, Mex.) – source of rubber, latex 44–50% rubber; *C. marcgravii* Pohl (*Jatropha oligodon*, trop. Am.) – ed. fr. & oil for cooking etc.; *C. stimulosus* (Michaux) A. Gray (bull nettle, E N Am.) – stinging hairs; *C. urens* (L.) Arthur (trop. Am.) – urticating hairs & sticky latex deter grazers but in Costa Rica sphingid larvae graze hairs & constrict petiole preventing latex flow

coachwhip *Fouquieria splendens*

coachwood *Ceratopetalum apetalum*

Coaxana J. Coulter & Rose. Umbelliferae (III 8). 2 Mex. R: CUMH 11(1975)13

cob, Kentish *Corylus maxima*

Cobaea Cav. Polemoniaceae. c. 10 trop. Am. Lianes; cult. orn. as annuals in gardens esp. *C. scandens* Cav. (Mex.) growing to 8 m in 1 season. Lvs pinnate term. in branched tendril tipped with hooks, which prevent the spiral movements of the tendril from dragging a branch away before it has time to clasp its support. Fls protandrous, at first greenish cream with unpleasant smell (fly-poll.), later purplish with pleasant honey smell (bee-poll.) though prob. bat-poll. in wild; after anthesis, pedicel becomes twisted; seeds winged

Cobaeaceae D. Don = Polemoniaceae

Cobana Ravenna. Iridaceae (III 4). 2 Guatemala, Honduras

Cobananthus Wiehler (~ *Alloplectus*). Gesneriaceae. 1 Guatemala: *C. calochlamys* (J.D. Sm.) Wiehler – fls reg., dry capsule cf. fleshy fr. of *Alloplectus* & *Columnea*

cobnut *Corylus avellana*

coca *Erythroxylum coca*

cocaine *Erythroxylum coca*

Coccineorchis Schltr. = *Stenorrhynchos*

Coccinia Wight & Arn. Cucurbitaceae. 30 trop. & S Afr., 1 extending to Mal.: *C. grandis* (L.) J. Voigt – shoots & fr. eaten; *C. abyssinica* (Lam.) Cogn. cult. Ethiopia for ed. tubers

Coccochondra Rauschert (*Chondrococcus*). Rubiaceae (IV 7). 1 Guayana Highland

Coccocypselum P. Browne. Rubiaceae (IV 3). 20 trop. Am. Heterostyly

Coccoloba P. Browne. Polygonaceae (II 2). 120 trop. & warm Am. Trees, shrubs & lianes. Ed. fr., some cult. esp. *C. uvifera* (L.) L. (seaside grape, Jamaican kino, trop. Atlantic Am.) – pachycaul tree typical of littoral, fr. used for jelly. *C. caracasana* Meissner (trop. Am.) – assoc. with *Azteca* ants in Costa Rica

Cocconerion Baillon. Euphorbiaceae. 2 New Caledonia. Lvs whorled

Coccosperma Klotzsch (~ *Blaeria*). Ericaceae. 3 S & SW Cape

Coccothrinax Sarg. Palmae (I 1a). 47 WI, esp. Cuba. R: Selbyana 12(1991)91. Limestone & serpentine. Cult. orn.; lvs used for hats & basketry

Cocculus DC. Menispermaceae (V). 8 trop. & warm excl. S Am. & Aus. R (Mal.): KB 15 (1962)479. Alks. Dioec., usu. lianes, some medic., some ed. fr., some cult. orn. esp. *C. carolinus* (L.) DC (coral-beads, SE US)

cocculus indicus *Anamirta cocculus*

coccus wood *Brya ebenus*

Cochemiea (M. Brandegee) Walton = *Mammillaria*

cochineal red dye form. obtained from dried female c. insects grown on *Opuntia cochenillifera*, now prod. synthetically

Cochleanthes Raf. Orchidaceae (V 10). 15 trop. Am. Cult. orn. epiphytes

Cochlearia L. Cruciferae. c. 25 N temp. (Eur. 10). Two Turkish spp. accumulate nickel on serpentine. *C. officinalis* L. (scurvy grass, Eur.) – form. medic. (fashion for drinking extract in mornings in 1650s, cf. orange-juice today), salad (tarry flavour)

Cochleariella Zhang & Vogt (*Cochlearopsis*; ~ *Cochlearia*). Cruciferae. 1 China. R: ABY 7(1985)143

Cochleariopsis Löve & D. Löve = *Cochlearia*

Cochleariopsis Zhang = *Cochleariella*

Cochlianthus Benth. (~ *Apios*). Leguminosae (III 10). 2 Himalaya

Cochlidium Kaulf. = *Grammitis*

Cochlidosperma (Reichb.) Reichb. = *Veronica*

Cochlioda Lindley. Orchidaceae (V 10). 5 S Am. Cult. orn. epiphytes

Cochliostema Lemaire. Commelinaceae (II 1e). 2 Colombia to Bolivia. Epiphytic, bromeliad-like; anthers coiled, enclosed by filaments united in a stalked hood. Cult. orn.

Cochlospermaceae Planchon = Bixaceae

Cochlospermum Kunth. Bixaceae (Cochlospermaceae). 12 trop. R: BJ 101(1980)215. Mostly xeromorphic trees or shrubs, some with tuberous stems underground, often flowering when leafless in dry season. *C. religiosum* (L.) Alston (*C. gossypium*, silk-cotton tree, Burma, India) – flowers as an unbranched pole with term. infl., source of an insoluble gum (karaya, kutira), a subs. for tragacanth, fruit hairs stuffed in pillows said to induce sleep, cult. orn. for yellow fls esp. near temples in India

cockatoo bush *Myoporum insulare*

Cockaynea Zotov = *Hystrix*

Cockerellia Löve & D. Löve = *Sedum*

cockle, corn *Agrostemma githago*; **cow c.** *Vaccaria hispanica*; **white c.** *Silene latifolia*

cocklebur *Xanthium strumarium*

cockroach plant *Haplophyton crooksii*

cockscomb *Celosia cristata*

cocksfoot *Dactylis glomerata*

cockspur grass *Echinochloa crus-galli*; **c. thorn** *Acacia eburnea*

cocky apple *Planchonia careya*

coco or **cocoyam** *Colocasia esculenta*; **c. de macaco** *Orbignya* spp.; **c. de mer** *Lodoicea maldivica*; **c. grass** *Cyperus rotundus*; **c. plum** *Chrysobalanus icaco*

cocoa *Theobroma cacao*

cocobolo *Dalbergia* spp. esp. *D. retusa*

cocona *Solanum topiro*

coconut *Cocos nucifera*

Cocos L. Palmae (V 5b). 1 C Mal. (e.g. Samar, Philipp.) or Barrier Reef (?) now widely cult. & natur. pantrop.: *C. nucifera* L. (coconut) – v. imp. source of oil for margarine, soap, etc.; in Pacific, whole cultures hinge on it – lvs for shelter, weaving, 'grass skirts' of Hawaii (introd. from Gilbert Is.) etc., timber (porcupine wood) for building, fr. for food & drink. Monoec. insect-poll. (wind- supplementary) with (usu.) 1-seeded water-disp. drupe (inviable after 14 days in seawater, which retards germ.) with fibrous mesocarp (husk) which yields the fibre coir used for doormats, coconut matting, cord & rope (used by early Polynesians for flying kites), coir dust a by-product used for mulching & soil-less germination medium etc.; endocarp is hard with 3 pores, the seed adherent to it, endosperm hollow with spongy ed. embryo near base, when young with c. 500 ml of refreshing coconut milk used as a medium in plant physiology experiments; copra is the dried endosperm & oil (many cosmetics) is extracted by boiling or pressure, the refuse cake being used as stockfeed (70% world trade from Philipp.); desiccated c. is sliced & dried endosperm much used in confectionery (flavour due to alpha-nonalactone); glycerine a by-product so c. of strategic imp. until dynamite-based explosives superseded 1945. Apical bud of geriatric trees used for tinned palm-hearts (leads to tree death) & infl. axis is tapped for toddy, when evaporated producing jaggery (sugar), when fermented giving arrack & fermented further vinegar. Cult. in plantations (often imp. pollen source for honeybees (e.g. Sri Lanka) esp. near coasts, the polyphyletic dwarf cvs being highest yielding; origin obscure but forms in Atlantic basically uniform (& thus prone to catastrophic disease), the Am. ones prob. introd. from Cape Verde Is. 1499 & those poss. orig. via Mozambique. R: IBM 56 (1987)118. See also *Lytocaryum*

cocos or **cocus wood** *Brya ebenus*

cocozelle *Cucurbita pepo*

Codariocalyx Hassk. (~ *Desmodium*). Leguminosae (III 9). 2 SE As. to trop. Aus. *C. motorius* (Houtt.) Ohashi (*Desmodium gyrans*, telegraph plant, trop. As.) – long term. leaflet & 2 laterals moving by jerks in warmth, grown as curiosity under glass, cattle-fodder

Coddia Verdc. Rubiaceae (II 1). 1 S Afr.

Coddingtonia S. Bowd. = ? *Psychotria*

Codia Forster & Forster f. Cunoniaceae. (Inc. *Pullea*) 14 W Pacific

Codiaeum A. Juss. Euphorbiaceae. 15 Mal. to Pacific. *C. variegatum* (L.) Blume cult. for coloured lvs ('crotons'), long selected by people of Papuasia & W Pacific for ornament, now used for hedging also; many named cvs, some with twisted lvs, pitcher-lvs, or 2 blades separated by a length of midrib, or variously lobed, variegated or spotted, 'vogue' foliage pls of mid–late C19 in Eur.

Codiocarpus R. Howard. Icacinaceae. 2 Indomal.

codlin an early maturing, unstriped cooking apple; **c.s and cream** *Epilobium hirsutum*

Codon L. Hydrophyllaceae. 2 S Afr. 10–12-merous

Codonacanthus Nees. Acanthaceae. 2 NE India, S China & Japan

Codonanthe (Mart.) Hanst. Gesneriaceae. 20 trop. Am. Epiphytic subshrubs or lianes usually associated with aerial ant nests, often with extrafl. nectaries (red spots on lvs), seeds same size as ant eggs; fls small, poll. by hummingbirds (?) attracted by coloured lvs. Some cult. orn. incl. *C. crassifolia* (Focke) C. Morton (C Am. to Venez. & Peru) – liane, with ants, attracted by floral & extrafl. nectar, fr. pulp & arils, place seeds in walls of their nests: plant growth-rates slower away from nests. R: Baileya 19(1973)4

Codonanthopsis Mansf. (~ *Codonanthe*). Gesneriaceae. 6 Amaz.

Codonechites Markgraf = *Odontadenia*

Codonoboea Ridley = *Didymocarpus*

Codonocarpus Cunn. ex Endl. Gyrostemonaceae. 3 Aus.

Codonocephalum Fenzl = *Inula*

Codonochlamys Ulbr. Malvaceae. 2 Brazil

Codonopsis Wallich. Campanulaceae. c. 30 C & E As. to Mal. Usu. with tubers (some medic. e.g. *C. tangshen* Oliver (tang-shen, W China) – ginseng subs. or ed. e.g. *C. lanceolata* (Maxim.) Benth. (*C. ussuriensis*, Japan, Manchuria)); many cult. orn., esp. twining spp. with prettily marked C interior. R: QBAGS 48(1980)96

Codonorchis Lindley. Orchidaceae (IV 1). 3 S trop. & temp. S Am.

Codonosiphon Schltr. = *Bulbophyllum*

Codonostigma Klotzsch = *Scyphogyne*

Codonura Schumann = *Baissea*

Coelachne R. Br. Gramineae (35). 10 (closely allied) OW trop.

Coelachyropsis Bor = *Coelachyrum*

Coelachyrum Hochst. & Nees. Gramineae (31d). 8 trop. & S Afr. through Arabia to Pakistan

Coelanthum E. Meyer ex Fenzl. Molluginaceae. 3 SW Cape to Natal. R: JSAB 24(1958)48

Coelia Lindley. Orchidaceae (V 13). 5 C Am., WI. Cult. orn.

Coelidium J. Vogel ex Walp. (~ *Amphithalea*). Leguminosae (III 27). 20 SW Cape. R: BN 54(1980), NJB 7(1987)51

Coeliopsis Reichb.f. Orchidaceae (V 10). 1 C Am.: *C. hyacinthosma* Reichb.f. – cult. orn.

Coelocarpum Balf.f. Verbenaceae. 1 Socotra, 4 Madagascar

Coelocaryon Warb. Myristicaceae. 4 trop. Afr.

Coelococcus H. Wendl. = *Metroxylon*

Coeloglossum Hartman (*Satorkis*, ~ *Habenaria*). Orchidaceae (IV 2). 1 N temp.: *C. viride* (L.) Hartman (frog orchid)

Coelogyne Lindley. Orchidaceae (V 12). c. 100 Indomal., trop. China, W Pacific. Extrafl. nectaries; showy fls; allied to *Pleione* but epiphytic & without plicate lvs. Many cult. orn. esp. *C. cristata* Lindley (Himal.) form. much grown for winter fls for bouquets, plants to 2 m diam. recorded

Coelonema Maxim. Cruciferae. 1 SW China

Coeloneurum Radlk. Solanaceae (Goetzeaceae). 1 Hispaniola: *C. ferrugineum* (Sprengel) Urban

Coelophragmus O. Schulz (? = *Sisymbrium*). Cruciferae. 2 Mexico

Coelopleurum Ledeb. = *Angelica*

Coelopyrena Valeton. Rubiaceae (IV 7). 1 E Mal.

Coelorachis Brongn. (~ *Rottboellia*). Gramineae (40h). 21 trop. R: KB 24(1970)309

Coelospermum Blume = *Caelospermum*

Coelostegia Benth. Bombacaceae. 5 W Mal.

Coelostelma Fourn. = *Matelea*

Coespeletia Cuatrec. See *Espeletia*

Coffea L. Rubiaceae (II 3). 90 trop. Afr. to Masc. Poss. orig. in Kenya region before split-up of Gondwanaland (Leroy). Cult. for seeds (coffee beans), first coffee-house in GB 1650 ('Angel', High St., Oxford), esp. *C. arabica* L. (Arabian or arabica coffee, trop. Afr., allotetraploid (2n = 44)), the best, usu. grown at alt., timber used for furniture, *C. canephora* Pierre ex Fröhner (robusta or Congo c., trop. W Afr.) – often used in instant coffee, & *C. liberica* W. Bull ex Hiern (Liberian or Abeokuta c., trop. Afr.) – bitter flavour, often added to robusta in blends. Domatia; flowering in response to temp. drop in Mal.; fr. a drupe with 2 seeds; stimulatory effects (up to c. 3 espressos improving dexterity by c. 10%, calming & reducing depression) due to alks incl. caffeine (addictive; 150 mg per cup (tea 80), side effects after 200 mg, diuretic, excess leading to shaking, migraines etc.) & theobromine, over 700 compounds isolated but basis of flavour still unclear though alkylpyrazines, sulphur compounds & aliphatics imp.; powder in comm. hair-dyes; mostly produced in trop. Am. esp. Brazil, Colombia, WI etc. M.N. Clifford & K.C. Wilson (eds. 1985) *Coffee*, G. Wrigley (1988) *Coffee*

coffee *Coffea* spp.; **Abeokuta c.** *C. liberica*; **Arabian** or **arabica c.** *C. arabica*; **Congo c.** *C. canephora*; **Kentucky c.** *Gymnocladus dioica*; **Liberian c.** *C. liberica*; **robusta c.** *C. canephora*; **c. weed** *Senna occidentalis*

coffin nail *Anacardium occidentale*; **c. tree** or **wood** *Juniperus recurva, Persea nanmu*

cognac See *Vitis*

Cogniauxia Baillon. Cucurbitaceae. 1 trop. Afr.

Cogniauxiocharis (Schltr.) Hoehne = *Stenorrhynchos*

cogon grass *Imperata cylindrica*

cogwood *Ziziphus chloroxylon, Ceanothus* spp.

Cohnia Kunth = *Cordyline*

Cohniella Pfitzer. Orchidaceae. 1 C Am.

cohosh, black *Cimicifuga racemosa, Actaea spicata*; **blue c.** *Caulophyllum thalictroides*

cohune nut *Orbignya cohune*

coigue *Nothofagus dombeyi*

Coilocarpus F. Muell. ex Domin = *Sclerolaena*

Coilochilus Schltr. Orchidaceae (IV 1). 1 New Caledonia

Coilostigma Klotzsch (~ *Salaxis*). Ericaceae. 2 E Cape. R: Bothalia 17(1987)163

Coincya Rouy (*Hutera, Rhynchosinapis*). Cruciferae. 6 Eur (6), Medit. R: BJLS 102(1990)353. *C. wrightii* (O. Schulz) Stace (Lundy cabbage, Lundy Is., Bristol Channel) – 1 of v. few Br. endemics

Coinochlamys T. Anderson ex Benth. = *Mostuea*

Coix L. Gramineae (40k). 5 trop. As. *C. gigantea* Koenig ex Roxb. (Indomal.) – salt-source in New Guinea; *C. lacryma-jobi* L. (Job's tears, SE As.) – the sheath of the bract of the infl. a hollow pear-shaped organ containing the 1-flowered female spikelet, the 2 males projecting from it, this 'false fruit' with thin shell ed. (adlay) esp. India & Burma, or hard one used as beads (grey & shiny); aquatic forms grow up to 30 m long

Cojoba Britton & Rose (~ *Pithecellobium*). Leguminosae (II 5). 20 trop. Am.

coke *Erythroxylum coca*

Cola Schott & Endl. Sterculiaceae. c. 125 trop. Afr. Cola (kola) cult. trop. esp. OW for caffeine-containing seeds ('nuts'), which are chewed or used in cola drinks (largely supplanted by synthetics), notably *C. acuminata* (Pal.) Schott & Endl. (Abata cola) & to a lesser extent *C. anomala* Schumann (Bamenda c.), *C. nitida* (Vent.) Schott & Endl. & *C. verticillata* (Thonn.) A. Chév. (Owé c.)

Colania Gagnepain = *Aspidistra*

Colanthelia McClure & E.W. Sm. Gramineae (1a). 7 Brazil. R: SCB 9(1973)77. Bamboos

Colax Lindley = *Pabstia*

Colchicaceae DC (Liliaceae s.l.). Monocots – Liliidae – Liliales. 15/165 Aus. & S Afr. to W Eur. & W As. Geophytes without raphides but with starch-rich corms to tubers, s.t. with twining annual stems. Lvs dorsiventral, s.t. with tendrils. Fls reg., usu. bisexual; P 3 + 3, s.t. basally connate; A 3 + 3; G3, each loc. with several to ∞ anatropous ovules on axile placentas. Capsule loculicidal or septicidal with usu. globose seeds (ovoid in *Onixotis*)
Principal genera: *Androcymbium, Colchicum, Gloriosa, Wurmbea*
Kuntheria & *Schelhammera* here in Convallariaceae (q.v.) poss. referable to this fam. showing the somewhat provisional classification following from splitting Liliaceae s.l. *Iphigenia* & *Wurmbea* with disjunct distribs
Many cult. orn. in genera above, *Littonia* & *Sandersonia*

Colchicum L. Colchicaceae (Liliaceae s.l.). c. 65 Eur. (30 incl. *Bulbocodium* & *Merendera*), Medit. to Ethiopia, Somalia, C As. & N India. Many alks. Fls crocus-like but A 6 (3 in *Crocus*), many appearing in autumn after lvs wither; P-tube long (cf. *Crocus*) with ovary at ground level but in spring pedicel elongates & brings capsule above ground; many cult. orn. incl. hybrids & double-flowered forms. *C. autumnale* L. (autumn crocus, meadow saffron, naked boys, Eur. to N Afr.) – dried corms & seeds the source of medic. colchicum (*tinctura colchici* mentioned in Assyrian medic. texts & form. pain-killer esp. in gout) & alk. colchicine used in plant breeding as it causes a doubling of chromosomes by disorganizing the spindle-mechanism at mitosis when they usu. separate into 2 daughter nuclei

Coldenia L. Boraginaceae. 1 OW trop. & warm

Colea Bojer ex Meissner. Bignoniaceae. 20 Madag., Masc. Some pachycaul treelets with cauliflory

Coleactina Hallé. Rubiaceae (II 2). 1 trop. W Afr.

Coleanthera Stschegl. Epacridaceae. 3 W Aus.

Coleanthus Seidel. Gramineae (17). 1 temp. Euras., N Am.: *C. subtilis* (Tratt.) Seidel – umbel-like spikelet clusters

Colebrookea Sm. Labiatae (VII). 1 Pakistan to W China: *C. oppositifolia* Sm. – often gregarious, local medic., cult. orn.

Coleocarya S.T. Blake. Restionaceae. 1 NSW, Queensland: *C. gracilis* S.T. Blake

Coleocephalocereus Backeb. (~ *Cephalocereus*). Cactaceae (III 3). 6 E & SE Brazil. R: Bradleya 6(1988)91, 7(1989)35

Coleochloa Gilly. Cyperaceae. 7 trop. & warm Afr. (esp. E), Madag. Compressed stem, distich. deciduous lvs, 'ligule' of hairs & ventrally open leaf-sheaths superficially like Gramineae

Coleocoma F. Muell. Compositae (Pluch.). 1 NW Aus.

Coleogyne Torrey. Rosaceae. 1 SW US.: *C. ramosissimus* Torrey (black brush)

Coleonema Bartling & H.L. Wendl. Rutaceae. 8 SW to E Cape. R: JSAB 47(1981)401. Some cult. orn. like *Diosma*

Coleophora Miers = *Daphnopsis*

Coleostachys A. Juss. Malpighiaceae. 1 N trop. S Am.

Coleostephus Cass. Compositae (Anth.-Leuc.). 3 W Eur. (2), N Afr.

Coleotrype C.B. Clarke. Commelinaceae (II 1d). 9 SE Afr., Madagascar

coleslaw *Brassica oleracea*

Coleus Lour. = *Plectranthus*

colic root *Aletris farinosa*

Colignonia Endl. Nyctaginaceae (III). 6 Andes of Colombia to Arg. Shrubs, herbs & lianes to 15 m. R: NJB 8(1988)231

Collabium Blume. Orchidaceae (Collabiinae). 10 China to Mal. & Polynesia

Collaea DC (~ *Galactia*). Leguminosae (III 10). 4 S Am.

collards *Brassica oleracea* Acephala Group

Colletia Comm. ex Juss. Rhamnaceae. 5 S S Am. R: Parodiana 5(1989)279. In each axil 2 serial buds, upper develops into triangular thorn, lower into fls or a branch. Nitrogen-fixing actinomycete nodules on roots; some cult. orn. & timber. In *C. hystrix* Clos (*C. armata*, S Chile & Arg. to 2000 m) hawthorn-scented fls on spiny growths, lvs evanescent; *C. paradoxa* (Sprengel) Escal. (*C. cruciata*, S Brazil, Uruguay, Arg.) has flattened branches but spiny growths in juvenile & occ. adult phases; *C. spinosissima* J. Gmel. (Peru) – root saponins used as soap

Colletoecema E. Petit. Rubiaceae (IV 10). 1 trop. Afr.

Colletogyne Buchet. Araceae (VIII 2). 1 Madagascar

Colliguaja Molina. Euphorbiaceae. 3 temp. S Am. *C. odorifera* Molina (Chile) bark (colliguaji b.) – used as soap

colliguaji bark *Colliguaja odorifera*
collimamol *Luma apiculata*
Collinia (Liebm.) Oersted = *Chamaedorea*
Collinsia Nutt. Scrophulariaceae. c. 20 N Am., esp. W US. C 2-lipped functioning like a legume fl. Cult. orn. annuals esp. *C. bicolor* DC (*C. heterophylla* , Chinese houses, Calif.)
Collinsonia L. Labiatae (VIII 1). 5 E N Am. *C. canadensis* L. (horse balm, stone root) – form. medic.
Collomia Nutt. Polemoniaceae. 15 W N Am., Bolivia to Patagonia. R: AMN 31(1944)217. Differs from *Gilia* in that mature capsule does not rupture K; testa of some spp. mucilaginous when wet. Some cult. orn. esp. *C. grandiflora* Douglas ex Lindley (W N Am.), natur. in Eur., Aus.
Collospermum Skottsb. Asteliaceae (Liliaceae s.l.). 4 NZ (2), Fiji, Samoa. Dioec., epiphytic in NZ
Colobanthera Humbert. Compositae (Ast.-Gran.). 1 Madagascar
Colobanthium Reichb. = *Avellinia*
Colobanthus Bartling. Caryophyllaceae (II 1). 20 S Pacific (Aus., NZ, Kerguélen, New Amsterdam, temp. S Am. incl. Andes). C 0, A in 1 whorl. *C. quitensis* (Kunth) Bartling (*C. crassifolius*, trop. & S Am.) – with *Deschampsia antarctica* only angiosperms to penetrate Antarctic Circle
Colobogyne Gagnepain = *Acmella*
Colocasia Schott. Araceae (VII 4). 8 trop. As. Tuberous herbs with peltate lvs. Cult. ed. & orn. pls esp. *C. esculenta* (L.) Schott (cocoyam, taro (cognate with 'calo', 'cala' (e.g. *Caladium*) = dark or blue; -casia, kachu = tuber), dasheen, kalo (Pacific), keladi, talas (SE As.), aivi (India), imo (Japan)) – cult. for 10 000 ys As. (first in India?), first irrigation terraces in Asia with rice perhaps orig. a weed there, known in Medit. by classical times & prob. NW soon after 1492, young lvs (often blanched) & tubers ed. (boiled), starch (grains v. small) added to improve breakdown of 'biodegradable' plastics, many selected forms & cvs (84 in Hawaii alone) incl. 'var. *antiquorum*' (eddoes) – ed. small tubers, 'Fontanesii' (*C. violacea*) – triploid orn. with violet petioles & veins
colocynth *Citrullus colocynthis*
Cologania Kunth (~ *Amphicarpaea*). Leguminosae (III 10). 10 trop. Am. esp. Mex. R: Phytol. 73(1992)281
colomba root See calumba r.
Colombian boxwood *Casearia praecox*; **C. mahogany** *Cariniana pyriformis*
Colombiana Osp. = *Pleurothallis*
Colombobalanus Nixon & Crepet = *Trigonobalanus*
Colona Cav. Tiliaceae. 30 S China, SE As., Indomal.
colophony distillate from crude oleo-resin of *Pinus* spp.
Colophospermum Kirk ex Léonard. Leguminosae (I 4). 1 S trop. Afr.: *C. mopane* (Benth.) Léonard – dominates *mopane* woodlands of hot dry low-rainfall areas, wind-poll., lvs fold together in hottest time of day; resin makes it susceptible to fire; larvae of mopane worms (*Gonimbrasia* spp.) sought by Bushmen
Colorado grass *Panicum texanum*
Coloradoa Boissev. & C. Davidson = *Sclerocactus*
Colpias E. Meyer ex Benth. Scrophulariaceae. 1 S Afr. Elaiophores attractive to bees
Colpodium Trin. Gramineae (17). 3 (s.s.) or 19 (s.l.) Turkey & Caucasus to Nepal & E Siberia, Mts Kenya & Kilimanjaro, Lesotho mts. R: KBAS 13(1986)103. High alt. segregate of *Poa* with fewer florets etc.
Colpogyne B.L. Burtt. Gesneriaceae. 1 Madagascar
Colpoon P. Bergius = *Osyris*
Colpothrinax Griseb. & H. Wendl. Palmae (I 16). 2 C Am., Cuba. *C. wrightii* Griseb. & H. Wendl. ex Siebert & Voss (Cuba) – trunk greatly swollen in middle at maturity
Colquhounia Wallich. Labiatae (VI). c. 3 E Himal., SW China. Erect or twining herbs allied to *Stachys* but upper C lip entire or emarginate & shorter than lower. Some cult. orn. esp. *C. coccinea* Wallich, fls red, field-hedge in parts of Bhutan
coltsfoot *Tussilago farfara*
Colubrina Rich. ex Brongn. Rhamnaceae. 31 trop. & warm esp. Am. R: Brittonia 23(1971)2. Some cult. orn. incl. *C. arborescens* (Miller) Sarg. (C Am., WI) – source of snakebark, medic. & timber & *C. asiatica* (L.) Brongn. (trop. As.) – seeds long-viable in seawater, lvs ed., fr. a fish-poison & medic., bark & roots a soap subs.; *C. elliptica* (Sw.) Briz. & Stern (*C. reclinatus*, WI) – basis of a drink (mabee or mabi) in Puerto Rico & Haiti; *C. glandulosa* Perkins ('*C. rufa*', trop. Am.) – bark (saguaragy) used in treatment of fever in Brazil; *C. oppositifolia*

Brongn. ex H. Mann. (Hawaii) – hard wood used like metal by old Hawaiians

columba root See calumba r.

Columbiadoria Nesom (~ *Hesperodoria*). Compositae (Ast.-Sol.). 1 NW US: *C. hallii* (A. Gray) Nesom. R: Phytol. 71(1991)248

columbine *Aquilegia* spp., esp. *A. vulgaris*, & hybrids

Columellia Ruíz & Pavón. Columelliaceae. 2 N Andes

Columelliaceae D. Don. Dicots – Rosidae – Rosales. 1/2 Colombia to Bolivia. Trees or shrubs. Lvs opp., simple; stipules 0. Fls in few-flowered term. cymes or solit., bisexual, slightly irreg; K (4)5(8), valvate or weakly imbricate, C ((4)5(8)), tube v. short, lobes imbricate, A 2 attached to C near base, anthers with longit. slits, nectary-disk 0, $\overline{G}$ (2) with thick style & 2–4-lobed stigma & parietal placentae almost meeting to make 2-loc. ovary, ovules ∞, anatropous, unitegmic. Fr. a septicidal capsule with persistent K; seeds ∞ with small straight embryo & copious fleshy endosperm

Only genus: *Columellia*

The weakly sympetalous corolla has led some to ally C. with Asteridae but, despite the unitegmic ovules which also suggest such a disposition, the wood & other features suggest it belongs in the affinity of the woody relations of Saxifragaceae

Columnea L. Gesneriaceae. 75 trop. Am. Epiphytes & lianes, hummingbird-poll.; many spp. & hybrids cult. Lvs of a pair often unequal

Coluria R. Br. Rosaceae. 5 S Siberia, China. R: NRBGE 15(1925)48

Colutea L. Leguminosae (III 15). 28 Med. (Eur. 3) to China, Himal., E & NE Afr., mostly in dry mts. R: MB(P) 14(1963)3, AK 12(1967)33. Cult. orn. esp. *C. arborescens* L. (bladder senna, Med., natur. GB) – lvs with properties like *Senna italica* & used to adulterate it; fr. inflated

Coluteocarpus Boiss. Cruciferae. 1 SW As. mts: *C. vesicaria* (L.) Holmboe

Colvillea Bojer. Leguminosae (I 1). 1 Madag.: *C. racemosa* Bojer – A exserted ('shavingbrush syndrome'), C orange-red, much visited by parrots, cult. orn.

Colymbada Hill = *Centaurea*

Colysis C. Presl. Polypodiaceae (II 4). 30 Indomal. Sporangia in lines between main veins

colza *Brassica napus*

Comaclinium Scheidw. & Planchon (~ *Dyssodia*). Compositae (Hele.-Pect.). 1 C Am.

Comandra Nutt. Santalaceae. 1 N Am. (3 subspp.), Med. inc. Eur. (1 subsp.) – *C. umbellata* (L.) Nutt., parasitic on roots of other plants; fr. sweet, ed. R: MTBC 22(1965)1

Comanthera L.B. Sm. = *Syngonanthus*

Comanthosphace S. Moore. Labiatae (VII). 5 E As. R: BBMNHB 10(1982)68

Comarostaphylis Zucc. (~ *Arctostaphylos*). Ericaceae. 10 Mex. R: SB 12(1987)582

Comarum L. = *Potentilla*

Comastoma (Wettst.) Toyok. = *Gentiana*

Combera Sandw. Solanaceae (2). 2 temp. S Am. Allied to *Petunia*

Comborhiza Anderb. & Bremer. Compositae (Gnap.-Rel.). 2 S Afr. Geoxylic suffrutices

Combretaceae R. Br. Dicots – Rosidae – Myrtales. 20/500 trop. & warm esp. Afr. Trees or shrubs (sometimes geoxylic), often scandent. Internal phloem & intraxylary phloem often present. Lvs spirally aranged, opp. or whorled, simple, entire, leaf-base often with 2 gland-containing flask-shaped cavities at base; stipules minute or 0. Fls in racemes, spikes or heads, usu. small & bisexual & reg. & epigynous (half so in *Strephonema*), hypanthium often nectariferous within; K 4,5(–8) appearing as lobes of hypanthium, persistent, valvate or s.t. imbricate or v. small, C 4,5(–8) alt. with K, imbricate or valvate, often 0, A often twice K & bicyclic, outer s.t. reduced, 0, or rarely in pairs or triplets, rarely 3-cyclic, anthers usu. versatile, with longit. slits, epigynous disk often present, $\overline{G}$ (2–5), 1-loc. with term. style & 2(–6) ovules, pendulous, anatropous, bitegmic with zig-zag micropyle, an elaborate obturator often produced on funicle. Fr. 1-seeded, usu. indehiscent & water-disp. or drupaceous, generally ribbed, the ribs often wing-like, rarely dry & dehiscent; seeds without endosperm, embryo oily with 2 (or 3 in some *Terminalia* spp. of SE As.) folded or spirally twisted cotyledons (massive & hemispheric in *Strephonema*, united in some African spp. of *Combretum*). x = 7, 11–13

Chief genera: *Buchenavia, Calopyxis, Combretum, Quisqualis, Terminalia*

Many in savannas; *Conocarpus, Laguncularia* & *Lumnitzera* mangroves

Unicellular hairs of a type only otherwise found in some Myrtaceae

Timbers in above genera & *Anogeissus*, ed. seeds & fr. for tanning (*Terminalia*), several cult. orn. esp. spp. of *Combretum* & *Quisqualis*

Combretocarpus Hook.f. Anisophylleaceae. 1 wetter W Mal.: *C. rotundatus* (Miq.) Danser in peat-swamp forests (gregarious in Borneo), timber recommended for furniture

Combretodendron A. Chev. = *Petersianthus*

Combretum Loefl. Combretaceae. 250 trop. (excl. Aus.). Climbers or erect. Lvs spirally arr. or opp. (usu.). Cult. orn. lianes esp. *C. paniculatum* Vent. (trop. Afr.) – great length, often produces red fls when leafless; *C. butyrosum* (Bertol.f.) Tul. (trop. Afr.) – a butter-like substance (chiquito) from fr.; *C. imberbe* Wawra (trop. Afr.) – leadwood; others medic., dyes, poisons, scents & gums (Indian g.) imp. locally. *C. fruticosum* (Loefl.) Stuntz (trop. S Am.) – fls visited by non-flying mammals in SE Peru: 7 spp. primate by day & 3 at night; *C. molle* R. Br. ex G. Don f. (NE Afr.) – cytotoxic but eaten by healthy chimpanzees

Comesperma Labill. (~ *Bredemeyera*). Polygalaceae. 40 Aus. *C. volubile* Labill. – love(creeper)

Cometes L. Caryophyllaceae (I 2). 2 NE Afr. & Ethiopia to NW India, deserts

comfrey *Symphytum* spp.; **blue** or **Russian c.** *S.* × *uplandicum*

comino *Aniba perutilis*

Cominsia Hemsley. Marantaceae. 5 E Mal., Papuasia, Vanuatu

Comiphyton Floret. Rhizophoraceae. 1 Gabon to E Zaire, not N or S of 2°

Commelina L. Commelinaceae (II 2). c. 170 trop. & warm. Infls subtended by large leafy sheathing bract; fls short-lived, the upper 3 A sterile with cross-shaped anthers with juicy lobes pierced by bees for nectar; fr. dehiscent or not. Several cult. orn. as trop. ground cover, fls usu. blue; some with ed. tubers. 2n = 22, 24, 28, 30, 42, 44, 56, 60, 66, 90, 120, 150. *C. caroliniana* Walter (*C. hasskarlii*, India) – weedy pl. introd. SE US with rice seed; *C. communis* L. (China, Japan, natur. Eur., US) – 'Hortensis' with large fls source of blue dye (Japan); *C. erecta* L. (trop. & warm) – andromonoec.

Commelinaceae Mirbel. Monocots – Commelinidae – Commelinales. 39/640 trop. to subtemp. Herbs, usu. perennial, s.t. robust, or succ., or climbers (*Aetheolirion* etc.), almost always with 3-celled glandular microhairs, often with mucilage cells each containing bundle of calcium oxalate raphides (not *Cartonema*); stems swollen at nodes; vessel elements in all parts (only in roots in *Cartonema*); stems swollen at nodes; vessel elements in all parts (only in roots in *Cartonema*). Lvs spirally arr., simple with closed sheath with parallel-veined often ± succ. lamina s.t. separated from sheath by slender petiole, the blade halves rolled separately against midrib in bud, rarely plicate. Infls cymose, often breaking through subtending sheathing bract, rarely solit. or app. in racemes. Fls hypogynous, usu. bisexual, reg. to irreg.; K 3 usu. green (petaloid & coloured in e.g. *Dichorisandra*), rarely connate basally, C 3 ephemeral usu. blue or white, s.t. clawed or basally connate to form tube, alike or 1 diff. colour and/or ± reduced, A usu. 3 + 3 but s.t. 3 staminodal or not developed, rarely 1 (e.g. *Callisia*), filaments usu. slender often long-hairy, anthers basifixed or versatile often with expanded connective, with longit. slits (rarely apical (& basal) pores), G (3), 3-loc. or apically 1-loc., or 1 or 2 locs undeveloped or 0, style term., ovules 1–∞ per locule on axile placentas, orthotropous to anatropous, bitegmic. Fr. usu. loculicidal capsule, seldom indehiscent or fleshy; seeds with copious mealy endosperm & compound starch grains, usu. arillate, rarely winged (*Aetheolirion*), embryo s.t. with vestigial 2nd cotyledon opp. other. x = (4–)6–16(–29). R: T 40(1991)19

Classification & chief genera:

I. **Cartonematoideae** (shoots glandular-pubescent (glandular microhairs lacking), raphide-canals 0 or only next to veins; fls reg., yellow)
   1. **Cartonemateae** (perennials, raphide-canals 0): *Cartonema* (only; sometimes placed in own fam.
   2. **Triceratelleae** (annuals, raphide-canals next to veins): *Triceratella* (only)

II. **Commelinoideae** (shoots v. rarely glandular-pubescent (glandular microhairs almost always present, raphide canals present but never near veins; fls various but v. rarely both reg. & yellow. Tribes & subtribes separable on micro-characters in the main)
   1. **Tradescantieae** (7 subtribes): a (*Palisotinae*, Afr.) – *Palisota* (only); b (*Streptoliriinae*, usu. climbers, As.) – *Spatholirion*; c (*Cyanotinae*, OW trop.) – *Cyanotis*; d (*Coleotrypinae*, OW trop.) – *Amischotolype*; e (*Dichorisandrinae*, trop. Am.) – *Dichorisandra*; f (*Thyrsantheminae*, trop. & warm esp. C Am.) – *Tinantia*; g (*Tradescantiinae*, Am.) – *Callisia, Gibasis, Tradescantia, Tripogandra*
   2. **Commelineae**: *Aneilema, Commelina, Floscopa, Murdannia, Pollia*

Many cult. orn. esp. *Tradescantia* (incl. *Zebrina, Rhoeo*); some medic. & ed. (local spinach subs. in Pacific)

Commelinantia Tharp = *Tinantia*

Commelinidium Stapf = *Acroceras*

Commelinopsis Pichon = *Commelina*

Commersonia Forster & Forster f. Sterculiaceae. 14 SE As. to Aus. (14, 12 endemic), New

Caled. Fibre from bark. *C. bartramia* (L.) Merr. (Mal.) – termite-proof but soft building timber (Bismarcks)

Commicarpus Standley = *Boerhavia*

Commidendrum DC. Compositae (Ast.-Ast.). 4 (1 extinct) St Helena. Allied to pls in S Am. & Australasia. Form. common on island: pachycaul trees, extant spp. scarce: *C. spurium* (Forst.f.) DC with only 2 trees in wild (1993), *C. burchellii* Hemsley extinct but *C. rotundifolium* (Roxb.) DC form. reduced to 3 pls 1880s & extinct in wild 1986 but propagated by seed at Kew & 1000 trees re-established on is.

Commiphora Jacq. Burseraceae. 190 warm Afr. & Madag., Arabia to Sri Lanka, Mex. (2) & S Am. Mosquito-poll. at night? Fleshy outgrowths of stone not seed = pseudoarils. Resin exudes, the lumps used in medic., incense etc. (oleo-resins known as bdellium); live fences (long cuttings root); seedling roots ed., wood chewed as water source in Uganda. *C. myrrha* (Nees) Engl. (NE Afr. to Arabia) – gum = principal myrrh today imp. in Arab medic. but that of the Bible *C. guidotii*; *C. foliacea* Sprague (Arabia) – dye, charcoal for cleaning teeth, bark prep. in skin disease treatment; *C. gileadensis* (L.) C. Chr. (SW Arabia, Ethiopia & Sudan) – source of balm of Gilead (Mecca myrrh), form. medic., incense, scent; *C. guidotii* Chiov. (Somalia, Ethiopia) – scented myrrh; *C. kataf* (Forssk.) Engl. (N Kenya to S Arabia) – with allied spp. source of gum opopanax (300 t per annum), tick-repellent; *C. madagascarensis* Jacq. (Tanzania) – form. males cult. Mauritius & India for scent; *C. mukul* (Stocks) Engl. (India) – used to reduce blood cholesterol; *C. merkeri* Engl. (E Afr.) – oleo-resin poss. wound treatment; *C. wightii* (Arn.) Bhand. (Arabia to Ind. desert) – oleo-resin from trunk (guggul) imp. local medic. (arthritis etc.)

Commitheca Bremek. Rubiaceae (I 10). 1 trop. W Afr.

Comocladia P. Browne. Anacardiaceae (IV). 20 trop. Am.

Comolia DC. Melastomataceae. 22 trop. S Am.

Comoliopsis Wurdack. Melastomataceae. 1 Venez. R: ABV 14,3(1984)23

Comoranthus Knobloch. Oleaceae. 3 Madag., Comoro Is.

Comospermum Rauschert (*Alectorurus*; ~ *Anthericum*). Anthericaceae (Liliaceae s.l.). 1 Japan

Comparettia Poeppig & Endl. Orchidaceae (V 10). 10 trop. Am. Cult. orn. epiphytes

compass plant *Silphium laciniatum*; *Lactuca serriola*

Comperia K. Koch (~ *Orchis*). Orchidaceae (IV 2). 2 E Med. to Iran

Complaya Strother = *Thelechitonia*. But see SBM 33(1991)10

Compositae Giseke (Asteraceae). Dicots – Asteridae – Asterales. 1528/22 750 cosmop. (exc. Antarctica). Shrubs & herbs (esp. rhiz.), trees & climbers, often storing carbohydrate as polyfructosans (esp. inulin); articulated laticifers (mostly Lactuc.) or ± extensive resin-duct system present; sec. thickening well developed even in many herbaceous spp., s.t. in unusual configurations or with medullary &/or cortical bundles. Lvs spirally arr., less often opp., rarely whorled, simple, dissected or ± compound; stipules 0. Infls of 1–∞ dense heads (capitula) with 1–∞ sessile fls on common receptacle, nearly always subtended by an involucre of 1–several series of bracts (0 in *Psilocarphus*), opening in racemose sequence (mixed in *Espeletia*); capitula occ. aggregated into cymose secondary heads (over 40 genera in several tribes) s.t. with a sec. involucre; receptacle flat to conical or cylindrical, s.t. with a bract subtending each fl. (esp. Helia.) or bristly (esp. Card.). Fls epigynous, bisexual or some female, sterile or functionally male (radiate heads with marginal female or sterile ray-florets and central bisexual or functionally male disk-florets, ray-florets with tubular C prolonged into ± strap-shaped ligule often tipped with vestigial 3 C-lobes (other 2 ± absent), disk-florets with reg. tubular (4)5-lobed or toothed C; discoid heads with only disk-florets; disciform heads with central disk-florets & marginal female florets with eligulate C, or with only the latter; ligulate heads (mostly Lact.) of bisexual florets with C of 5 lobes; in Mutis. some or all florets with 2-lipped C, outer larger but marginal florets sometimes as in Lact. & some or all of central ones like disk-florets); K forming pappus on top of ovary or 0, of (1)2–∞ scales, awns, bristles or connate to form crown, A as many as C lobes, alt. with them & inserted in tube, usu. distinct, anthers usu. with short apical appendage & s.t. basal tails & pollen-sacs connate into tube, releasing pollen into tube through longit. slits to be pushed out by growth of style, $\overline{G}$ (2), 1-loc. with term. style, bifid, branches commonly separating after passage through anther-tube; nectary commonly a thickened scale or cup on top of G (nectar usu. amino-acids & hexoses), ovule 1, basal, erect, anatropous, unitegmic. Fr. a cypsela usu. with persistent pappus, rarely a drupe; embryo oily, straight with hypocotyl, endosperm 0 or v. thin peripheral. x = 2–19+ (? orig. 9). V.H. Heywood *et al.* (1977) *Biology & chemistry of the C.*; K. Bremer (1994) *Asteraceae*; Journal: Compositae Newsletter

Classification & chief genera (Bremer):

I. **Barnadesiodeae** (~Cich.-Mut.; capitula homogamous or heterogamous, discoid, disciform or radiate; florets reg., with char. long hairs, deeply 5-lobed or pseudobilabiate with a ± 4-lobed limb or rarely ligulate with a deeper split between 2 lobes, red to white or purple or yellow; anthers tailed; pollen variable; style ± shortly bilobed, never pilose; shrubs, small trees or herbs usu. with axillary spines – 9/92 S Am.): *Barnadesia, Dasyphyllum*

II. **Cichorioideae** (Lactucoideae; capitula usu. homogamous, ligulate, bilabiate or discoid; disk-florets usu. reg., deeply 5-lobed (if outer ones diff., then bilabiate or rarely true rays), purplish, pinkish or white, less often yellow; anthers usu. caudate; pollen variable; style-arms usu. with single stigmatic area on inner surface; shrubs, herbs & rarely trees; prob. heterogeneous – 6 tribes (some genera unassigned) – 392/6650 cosmop.):

1. **Mutisieae** (capitula bilabiate or ligulate, rarely not; latex-ducts 0, anthers usu. tailed, lvs spiral, rarely opp; trees (some unbranched pachycauls), lianes & herbs; heterogeneous 'grade' (R: OB 109(1991)5) – 77/950 trop. esp. S Am., 2 subtribes): *Achyrothalamus, Ainsliaea, Brachylaena, Chaetanthera, Chaptalia, Dicoma, Gerbera, Gochnatia, Mutisia, Onoseris, Tarchonanthus, Wunderlichia* (Mutisiinae); *Jungia, Leucheria, Nassauvia, Perezia, Trixis* (Nassauviinae)
   Cult. orn. – *Gerbera, Mutisia*; timber – *Brachylaena, Tarchonanthus*
2. **Cardueae** (capitula discoid, homogamous or with sterile outer florets; latex-ducts usu. 0; anthers acute, often tailed; pollen spiny not ridged, lvs spiral; usu. herbs, though *Centaurodendron* pachycaul trees – 82/2450 mostly Euras., 4 subtribes): *Echinops* (Echinopsidinae); *Atractylis, Carlina* (Carlininae); *Carduus, Cirsium, Cousinia, Cynara, Jurinea, Onopordum, Saussurea* (Carduinae); *Carduncellus, Carthamus, Centaurea, Serratula* (Centaureineae)
   Some ed. – *Carthamus* (oil, also dye), *Cynara* (cardoon, globe artichoke); cult. orn. – *Amberboa, Centaurea, Echinops, Onopordum, Silybum* (also med.); weeds – *Acroptilon, Carduus, Cirsium*
3. **Lactuceae** (capitula ligulate; latex ducts present; pollen usually ridged and spiny or spiny; lvs spiral; mostly herbs & some pachycaul trees e.g. *Dendroseris, Sonchus*) – 97/1530 cosmop. esp. N hemisp., 11 subtribes (*Cichorium* & *Scolymus* unassigned)): *Catananche* (Catananchinae); *Chondrilla, Crepis, Taraxacum, Youngia* (Crepidinae); *Cicerbita, Lactuca, Mycelis, Prenanthes* (Lactucinae); *Launaea, Sonchus* (Sonchinae); *Dendroseris* (Dendroseridinae); *Agoseris* (Microseridinae); *Stephanomeria* (Stephanomeriinae); *Malacothrix* (Malacothricinae); *Hieracium, Pilosella* (Hieraciinae); *Hypochaeris, Leontodon, Picris* (Hypochaeridinae); *Scorzonera, Tragopogon* (Scorzonerinae)
   Some vegetables – *Cichorium* (chicory, endive), *Lactuca* (lettuce), *Scorzonera, Taraxacum, Tragopogon* (salsify); cult. orn. – *Catananche*; rubber – *Taraxacum*; weeds – *Chondrilla, Crepis, Hypochaeris, Lactuca, Sonchus, Taraxacum*
4. **Vernonieae** (capitula discoid, rarely ligulate, homogamous, latex-ducts 0; anthers obtuse to acute, rarely tailed; pollen ridged &/or spiny; style-arms elongate; lvs spiral or opp.; shrubs, trees (sometimes large or pachycaul), lianes or herbs – 98/1530 mostly trop., 6 subtribes, prob. provisional): *Baccharoides, Bothriocline, Distephanus, Lepidaploa, Lessingianthus, Vernonanthera, Vernonia* (Vernoniinae); *Piptocarpha*, (Piptocarphinae); *Centratherum* (Centratherinae); *Eremanthus, Lychnophora* (Lychnophorinae); *Elephantopus* (Elephantopodinae); *Rolandra* (Rolandrinae)
   Cult. orn. – *Stokesia*; timber – *Eremanthus*; weeds – *Cyanthillium, Elephantopus, Ethulia*; *Pacourina* ed.
5. **Liabeae** (capitula radiate or discoid; latex-ducts 0; anthers acute or shortly tailed; pollen spiny; style-arms elongate; lvs opp. or whorled, trees & shrubs – 14/160 Am. esp. trop. (R: SCB 54(1983)1)): *Liabum, Munnozia, Sinclairia*
6. **Arctotideae** (Arctoteae, incl. Eremothamneae, prob. heterogeneous; capitula radiate, rarely discoid; latex-ducts usu. 0; anthers obtuse, acute, or shortly tailed; pollen spiny; lvs spiral; herbs or shrubs – 18/200 SW As., S Afr., 2 subtribes (some unassigned)): *Arctotis* (Arctotidinae); *Berkheya, Gazania* (Gorteriinae; R: MBSM 3(1959)71)
   Cult. orn. – *Arctotis, Gazania*

III. **Asteroideae** (capitula heterogamous, radiate or disciform, less often discoid; disk florets usu. with broad short lobes, usu. yellow; anthers basifixed; style-branches usu. with 2 distinct stigmatic areas; pollen spiny; latex ducts; 1119/16 000 cosmop. –

10 tribes (some 8 genera unassigned)):

1. **Inuleae** (excl. 2 & 3; anthers with tails often branched; style-branches with acute hairs; perennial herbs & shrubs (*Duhaldea* trees) – 38/480 OW esp. Euras.): *Anisopappus, Blumea, Buphthalmum, Inula, Pulicaria*
   Cult. orn. – *Inula, Telekia; Limbarda* ed.
2. **Plucheeae** (~Inuleae; anthers tailed or not; style & -branches with obtuse hairs; herbs, shrubs & trees – 28/180 trop. & warm incl. arid): *Epaltes, Laggera, Pluchea, Sphaeranthus*
3. **Gnaphalieae** (~Inuleae; involucral bracts papery, sometimes showy ('everlastings'); herbs (many rather dreary weeds), shrubs or shrublets (often ericoid) – 183/2050 cosmop. esp. S Afr. & Aus.; 5 subtribes (many genera unassigned); R: OB 104(1991)5): *Loricaria* (Loricariinae); *Metalasia, Oedera, Stoebe* (Relhaniinae); *Anaphalis, Antennaria, Ozothamnus* (Cassiniinae); *Angianthus, Bracteantha, Craspedia, Rhodanthe* (Angianthinae); *Filago, Gamochaeta, Gnapalium, Helichrysum, Leontopodium, Pseudognaphalium* (Gnaphaliinae)
   Cult. orn. – *Anaphalis, Antennaria, Bracteantha* ('helichrysum'), *Helichrysum, Leontopodium* (edelweiss), *Ozothamnus, Raoulia*
4. **Calenduleae** (involucral bracts in 1 or 2 series; receptacle naked; anthers acute, ± tailed; style-arms truncate with apical hairs; pappus 0; cypselas often curiously shaped; lvs spiral or opp.; herbs & shrubs – 8/85 Afr., SW As., Eur.): *Calendula, Osteospermum*, both cult. orn.; weeds – *Chrysanthemoides*
5. **Astereae** (involucral bracts in 1 or 2+ series; receptacle naked, rarely scaly; anthers obtuse, not tailed; style-arms with a shortly hairy triangular to lanceolate apical appendage; lvs usu. spiral; herbs, shrubs & some trees (e.g. *Apodocephala, Commidendrum, Melanodendron, Vernoniopsis*) – 178/2700 cosmop.; 3 subtribes): *Grangea* (Grangeinae); *Ericameria, Gutierrezia, Haplopappus, Machaeranthera, Pteronia, Solidago* (Solidagininae); *Archibaccharis, Aster, Baccharis, Brachycome, Celmisia, Conyza, Diplostephium, Erigeron, Felicia, Lagenophora, Microglossa, Olearia, Psiadia, Tetramolophium* (Asterinae)
   Many cult. orn. in above, *Bellis, Boltonia, Callistephus*; weeds – *Conyza*
6. **Anthemideae** (involucral bracts in 1–several series, usu. with thin dry transparent tips or margins; receptacle naked or scaly; anthers obtuse to acute, not tailed; style-arms truncate, fringed with short hairs; lvs spiral (rarely opp.), often much-divided, strongly scented; herbs & some shrubs (e.g. *Argyranthemum*) – 108/1725 cosmop. esp. N hemisph.; 12 subtribes): *Athanasia, Ursinia* (Ursiniinae); *Trichanthemis* (Cancriinae); *Tanacetopsis, Tanacetum* (Tanacetinae); *Inulanthera* (Gonosperminae); *Sclerorhachis* (Handeliinae); *Ajania, Artemisia, Dendranthema, Seriphidium* (Artemisiinae); *Achillea, Santolina* (Achilleinae); *Anthemis* (Anthemidinae); *Argyranthemum, Chrysanthemum* (Chrysantheminae); *Leucanthemum* (Leucantheminae); *Osmitopsis* (Thaminophyllinae); *Cotula, Leptinella, Pentzia* (Matricariinae)
   Cult. herbs – *Tanacetum* (costmary), *Artemisia* (absinthe & tarragon; also med.), *Chamaemelum* (chamomile), *Tanacetum* (incl. pyrethrum); cult. orn. – *Achillea, Anacyclus, Anthemis, Artemisia, Chrysanthemum, Dendranthema* (florist's chrysanthemum), *Leucanthemum, Santolina, Tanacetum, Ursinia*
7. **Senecioneae** (involucral bracts usu. in 1 row, often with outer series of reduced bracts; receptacle naked; style-branches usu. truncate, apically minutely hairy, less often variously appendaged; lvs spiral; shrubs, herbs (some succ.), trees & lianes) – 119/3225 cosmop., 3 subtribes: *Abrotanella* (Blennospermatinae, excl. by some workers); *Cineraria, Crassocephalum, Dendrorphorbium, Emilia, Euryops, Gynura, Kleinia, Monticalia, Othonna, Packera, Pentacalia, Senecio, Synotis* (Senecioninae); *Brachyglottis, Cremanthodium, Doronicum, Gynoxys, Ligularia, Parasenecio, Psacalium, Roldana, Sinosenecio, Tephroseris* (Tussilagininae)
   Cult. orn. – *Brachyglottis, Doronicum, Kleinia* (succ.), *Ligularia, Othonna* (succ.), *Pericallis* ('cineraria'), *Petasites, Senecio*; weeds – *Crassocephalum, Senecio* (some poisonous). Many pachycaul trees with similar candelabriform branching in diverse genera: *Dendrsenecio* (Afr. mts), *Brachyglottis* (Aus., NZ), *Kleinia* (dry Afr.), *Pittocaulon* (C Am.), *Pladaroxylon* (St Helena), *Robinsonia* (Juan Fernandez), *Telanthophora* (C Am.), etc., similar branching in Astereae (*Apodocephala, Psiadia, Vernoniopsis* in Madag.), in Cardueae (*Centaurodendron* on Juan Fernandez) & in Lactuceae (*Dendroseris* on Juan Fernandez) but also in small shrubby spp. in these tribes

8. **Helenieae** (~Heliantheae; receptacle without scales (cf. 9); involucral bracts in few rows; radiate fls usu. yellow; anthers pale or at least not black (cf. 9); usu. herbs (trees in *Dubautia*) with opp. lvs; 93/635 Am. esp. N, few OW warm, 8 subtribes (some genera unassigned) to be split further if 9 & 10 to be kept distinct): *Arnica* (Chaenactidinae); *Flaveria*, (Flaveriinae); *Pectis, Porophyllum, Tagetes, Thymophylla* (Pectidinae; Tageteae); *Hymenopappus* (Hymenopappinae); *Lasthenia* (Baeriinae); *Gaillardia, Helenium, Hymenoxys* (Gaillardiinae); *Perityle* (Peritylinae); *Dubautia, Madia* (Madiinae)
   Cult. orn. – *Helenium, Layia, Tagetes, Thymophylla*
9. **Heliantheae** (excl. 8; receptacle scaly; involucral bracts in few rows; anthers usu. blackened; shrubs, herbs (some aquatic), trees & lianes with strigose usu. opp. lvs; 196/2500 cosmop. esp. Am., 10 subtribes (many genera incl. *Montanoa* unassigned)): *Echinacea, Rudbeckia* (Rudbeckiinae); *Acmella, Zinnia* (Zinniinae); *Encelia, Espeletia, Flourensia, Perymenium, Verbesina, Wedelia* (Verbesininae); *Helianthus, Pappobolus, Scalesia, Simsia, Viguiera* (Helianthinae); *Galinsoga, Tridax* (Galinsoginae); *Calea, Melampodium* (Melampodiinae); *Bidens, Coreopsis, Cosmos, Dahlia* (Coreopsidinae); *Silphium* (Engelmanniinae); *Ambrosia, Parthenium, Xanthium* (Ambrosiinae); *Heptanthus* (Pinillosiinae)
   Cult. orn. – *Chrysogonum, Coreopsis, Cosmos, Dahlia, Echinacea, Encelia, Gaillardia, Helianthus, Ratibida, Rudbeckia, Sanvitalia, Tithonia, Wedelia, Zinnia*; some ed. – *Acmella, Galinsoga, Guizotia* & *Madia* (oil), *Helianthus* (oil & tubers) & *Smallanthus* (tubers); timber – *Montanoa*; rubber – *Parthenium*; weeds – *Bidens, Blainvillea, Chrysanthellum, Tridax, Xanthium*
10. **Eupatorieae** (form. in II; capitula discoid, homogamous; anthers obtuse to acute, not tailed; pollen spiny; style-arms elongate, club-shaped, papillose; lvs opp. or spiral; shrubs & herbs incl. aquatics) – 169/2400 mostly Am.; 16 subtribes (R: MSBMBG22(1987)1)): *Hofmeisteria* (Hofmeisteriinae); *Ageratina* (Oxylobinae); *Mikania* (Mikaniinae); *Trichocoronis* (Trichocoroninae); *Adenostemma* (Adenostemmatinae); *Ageratum, Stevia* (Ageratinae); *Fleischmannia* (Fleischmanniinae); *Chromolaena* (Praxelinae); *Campuloclinium, Trichogonia* (Gyptidinae); *Liatris* (Liatrinae); *Eupatorium* (much reduced but now being rebuilt; Eupatoriinae); *Symphyopappus* (Disynaphiinae); *Ayapana* (Ayapaninae); *Brickellia* (Alomiinae); *Critonia, Koanophyllum, Ophryosporus* (Critoniinae); *Bartlettina, Neomirandea* (Hebecliniinae)
   Cult. orn. – *Ageratum, Liatris*; sweetener – *Stevia*; medic. tea – *Ayapana*; weeds – *Adenostemma, Ageratina, Ageratum, Chromolaena, Gymnocoronis*

Most consp. in diversity in montane subtrop. & trop. areas with a tendency for tribal & subtribal specialization in ecology & habit, though within particular genera there may be great variation e.g. *Coreopsis* & *Erigeron* with aquatics to xerophytes; *Blepharispermum, Brachylaena, Eremanthus* & *Vernonia* include timber trees & *Piptocarpha* dominates the seasonally burnt landscapes of EC Brazil. The capitulum in animal-poll. spp. acts in many ways like a single flower (cf. Aizoaceae), the long-tubed fls of Cardueae being visited by bees & Lepidoptera, the yellow & white fls so common in the rest of the family being attractive to flies, beetles etc. At anthesis the pollen is forced out by the stigma, the floret being functionally male, later the stigmatic arms separate such that the floret is functionally female; often the stigmas then curl back to touch the pollen on the style thus effecting self-fert. A number of genera esp. *Hieracium* & *Taraxacum* have large numbers of apomictic lines ('microspecies'). Wind-poll. occurs in *Ambrosia, Artemisia* etc. It has been customary to argue that the ecological success in terms of spp. & individual nos. of the family is due to the capitulum system, the involucre acting like a calyx, the fr. wall like a testa, etc., but the pseudanthial head is 'duplicated in the small and unsuccessful family Calyceraceae' (Cronquist) while a similar pollen-presentation mechanism is found in *Brunonia* (Goodeniaceae s.l.; fls in cymose heads) and developed to varying degrees in Campanulaceae, other Goodeniaceae and part of Rubiaceae. Cronquist attributes the success of the C. to the defensive combination of polyacetylenes & the bitter sesquiterpene lactones followed by the development of other chemical repellents like the alks in Senecioneae, the latex system of Lactuceae, which do not have the polyacetylene-bearing resin system of the other tribes; the bad smell of the *Tagetes* group of Helenieae and the char. smells of Anthemideae are such repellents: many of these compounds make the pls imp. as sources of flavourings & insecticides

Recent cladistic & molecular work suggests the C. are closely allied to Calyceraceae, with which they share similar C venation, & Goodeniaceae all placed in an expanded Asterales with Campanulaceae & its allies (see AJB 82(1995)250). Contrary to received doctrine, the fam. is prob. primitively woody (cf. NP 73(1974)967) & perhaps of S Am. & Pacific origin (though paucity in New Caled. diff. to explain & fossils from mid-Jurassic China claimed), the early members being pachycaul with large discoid capitula of yellow fls & involucral bracts in several rows

Compared with other large fams, e.g. Gramineae, Leguminosae, the C. are of little value to Man except as ornamentals, the ed. ones having low levels of toxins or, as in lettuce, having had them selected out: some are insecticides & fish-poisons but many are noxious weeds, their fr. spread by wind (pappus) or animals (sticky pappus in *Adenostemma*, pappus of barbed bristles in *Bidens*, involucral bracts with hooked tips in *Arctium*, sticky bracts in *Sigesbeckia*, receptacle with hooks in *Xanthium*, etc.). With increasing clearance of native vegetation throughout the world, these aggressive toxic plants will inherit it

Compsoneura (A. DC) Warb. Myristicaceae. 14 trop. Am. *C. sprucei* (A. DC) Warb. – thrips-poll. in Costa Rica

Comptonanthus R. Nordenstam = *Ifloga*

Comptonella Baker f. Rutaceae. 8 New Caled. Usu. dioec.

Comptonia L'Hérit. Myricaceae. 1 E N Am.: *C. peregrina* (L.) J. Coulter sole survivor of (?) 12 Eocene spp. (some Eur.). Actinomycete root-symbionts fixing nitrogen. Differs from *Myrica* in pinnatifid lvs & usu. monoec. Lvs a tea for E Canadian Indians; local medic., cult. orn.

Comularia Pichon = *Hunteria*

Conamomum Ridley = *Amomum*

Conandrium (Schumann) Mez. Myrsinaceae. 2 E Mal. R: Blumea 33(1988)109

Conandron Siebold & Zucc. Gesneriaceae. 1 E China, Japan.: *C. ramondioides* Siebold & Zucc. – cult. orn.

Conanthera Ruíz & Pavón. Tecophilaeaceae (Liliaceae s.l.). 6 Chile. Some ed. bulbs

Conceveiba Aublet. Euphorbiaceae. 6 trop. Am. & Gabon (1)

Conceveibastrum Steyerm. = *praec.*

Conceveibum A. Rich. ex A. Juss. = *Aparisthmium*

Conchopetalum Radlk. Sapindaceae. 1 Madagascar

Conchophyllum Blume (~ *Dischidia*). Asclepiadaceae (III 4). 10 Mal.

Condalia Cav. Rhamnaceae. 18 warm Am. Some ed. fr., timber gives dye

Condaliopsis (Weberb.) Süsseng. = praec.

Condaminea DC. Rubiaceae (I 7). 3 Andes

condurango *Gonolobus cundurango*

Condylago Luer. Orchidaceae (V 13). 1 Colombia: *C. rodrigoi* Luer. R: MSB 24(1987)21

Condylidium R. King & H. Robinson (~ *Eupatorium*). Compositae (Eup.-Ayap.). 2 C. Am., WI Am. R: MSBMBG 22(1987)206

Condylocarpon Desf. Apocynaceae. 7 C Am. (1) to trop. S Am. R: AMBG 70(1983)149. Febrifuges

Condylopodium R. King & H. Robinson. Compositae (Eup.-Alom.). 4 Colombia. R: MSBMBG 22(1987)266

Condylostylis Piper = *Vigna*

cone flower *Echinacea* spp., *Ratibida* spp., *Rudbeckia* spp.; **c.sticks** *Petrophile* spp.

conessi *Holarrhena pubescens*

Confederate rose *Hibiscus mutabilis*

congaberry *Carissa* spp.

Congdonia Muell. Arg. = *Declieuxia*

Congea Roxb. Verbenaceae (Symphoremataceae). c. 7 SE As., W Mal. R: GBS 21(1966)259. Climbers, cult. orn. esp. *C. tomentosa* Roxb. (Burma, Thailand) with white to lilac tomentose leaf-like bracts subtending heads of white fls

Congo copal *Guibourtia demeusei*; **C. jute** *Urena lobata*; **C. pea** *Cajanus cajan*; **C. rubber** *Ficus lutea*; **C. stick** *Castanea sativa* coppice walking-stick; **C. wood** (Am.) *Lovoa trichilioides*

Congolanthus A. Raynal. Gentianaceae. 1 trop. Afr.

Conicosia N.E. Br. Aizoaceae (V). (Incl. *Herrea*) 2 Cape Prov. R: BJ 111(1990)482. Capsule opens hygroscopically & some seeds washed out by raindrops; it remains open when dry & many seeds shaken out & finally it breaks up into wind-disp. segments, each containing up to 2 seeds held in pocket-like folds (Rowley). Cult. orn.

Coniferae Juss., **Coniferales, Coniferopsida** See Pinopsida

Coniogramme Fée. Pteridaceae (IV). 20 OW trop. to Hawaii

Conioselinum Hoffm. Umbelliferae (III 8). 10 temp. Euras. (Eur. 1), E N Am.

Conium L. Umbelliferae (III 5). 6 temp. Euras. (Eur. 1), S Afr. (3). *C. maculatum* L. (hemlock,

Euras.), the only umbellifer with alks which may act as parts of coenzymes in oxid.–reduction processes, form. medic., v. poisonous (said to have killed Socrates, rodent control since C2) due to polyacetylenes esp. coniine paralysing respiratory system

conkerberry *Carissa* spp.

conkers *Aesculus hippocastanum*

Connaraceae R. Br. Dicots – Rosidae – Rosales. 12/180 trop. esp. OW. Trees (sometimes pachycaul & unbranched), shrubs or lianes commonly with solit. crystals of calcium oxalate in parenchyma cells & very often with mucilage-canals &/or tanniniferous secretory cavities; bark, fr. & seeds often very poisonous though agent unknown. Lvs spiral, pinnate to unifoliolate, s.t. reduced & hook-like, woody; stipules 0. Fls in racemes or panicles, small, usu. bisexual (pl. rarely dioec. as in *Ellipanthus*) & usu. heterostylous (some tristyly; *Connarus* in Afr. & As. with only medium- & long-styled fls), reg., ± hypogynous; K (4)5, s.t. basally connate, imbricate or valvate, v. often persistent around fr. base. C (4)5, s.t. basally connate, imbricate or rarely valvate, A 5 + 5, inner s.t. staminodes, anthers with longit. slits & 3-colporate or 3-colpate (4-colpate in *Jollydora*) pollen, disk 0 or small, usu. extrastaminal but receptacle s.t. nectariferous, G̲ 1 (3), 5 (7 or 8), often 5 with 4 abortive, s.t. ± connate basally, often not completely closed, each with term. style & capitate stigma & 2 marginal, collateral (cf. Leguminosae) ovules, ascending, anatropous to (often) hemitropous, bitegmic, usu. 1 abortive. Fr. usu. dry-dehiscent, opening along ventral or rarely both sutures or an indehiscent nut; seeds often arillate (cf. Sapindaceae) and/or sarcotestal with oily or 0 endosperm. x = 13, 14. R: AUWP 89-6(1989)131
Principal genera: *Agelaea, Cnestis, Connarus, Rourea*
Seeds like Sapindaceae but free carpels unlike Sapindales; exstipulate lvs inappropriate in Fabales so put in Rosales but isolated even there so s.t. put in own order Connarales (Goldberg)
Some timbers, local med. & poisons, tannins & fibres

Connarus L. Connaraceae. 77 trop. R: AUWP 89-6(1989)239. Some fish-poisons, anthelminthics, timber

Connellia N.E. Br. Bromeliaceae (1). 5 Venez. highlands. R: AMBG 73 (1986)690

Conobea Aublet. Scrophulariaceae. 7 Am.

Conocalyx Benoist. Acanthaceae. 1 Madagascar

Conocarpus L. Combretaceae. 2 trop. Am. & Afr. Mangroves. *C. erectus* L. (buttonwood, trop. Am., W Afr.) – tanbark & wood for charcoal

Conocephalus Blume = *Poikilospermum*

Conocliniopsis R. King & H. Robinson (~ *Eupatorium*). Compositae (Eup.-Gyp.). 1 Colombia, Venez., Brazil. R: MSBMBG 22(1987)97

Conoclinium DC (~ *Eupatorium*). Compositae (Eup.-Gyp.). 3 E US, Mex. R: MSBMBG 22(1987)130

Conomitra Fenzl (~ *Glossonema*). Asclepiadaceae (III 5). 1 Sudan: *C. linearis* Fenzl

Conomorpha DC = *Cybianthus*

Conopharyngia G. Don f. = *Tabernaemontana*

Conopholis Wallroth. Orobanchaceae. 2 SE US to Panamá. *C. americana* (L.) Wallroth (squawroot, US) – med.

conophor nut *Tetracarpidium conophorum*

Conophyllum Schwantes = *Mitrophyllum*

Conophytum N.E. Br. Aizoaceae (V). (Excl. *Ophthalmophyllum*) 85 S Afr. R: S.A. Hammer (1993) *The genus C.* Cult. orn. succ. with globose to oblong photosynthetic growths, new ones developing inside old, which dry to a thin shell protecting new in dry season

Conopodium Koch. Umbelliferae (III 8). 20 Euras., Medit. (Eur. 7). Cotyledon 1. Tuberous roots of *C. majus* (Gouan) Loret (earth-nut, pig-nut, Eur.) ed. roasted

Conospermum Sm. Proteaceae. 40 Aus. Smoke bushes; mooted for treatment of HIV, cult. orn.

Conostalix (Kraenzlin) Brieger = *Eria*

Conostegia D. Don. Melastomataceae. 43 trop. (esp. C) Am. Some ed. fr. esp. *C. xalapensis* (Bonpl.) D. Don (C Am.) – delicious flavour like blueberry, cult.

Conostephium Benth. Epacridaceae. 5 S Aus.

Conostomium (Stapf) Cuf. Rubiaceae (IV 1). 9 trop. E Afr.

Conostylidaceae (Pax) Takht. = Haemodoraceae

Conostylis R. Br. Haemodoraceae. (Excl. *Blancoa*) 45 SW Aus. R: FA 45(1987)57. Insect- and bird-poll.

Conothamnus Lindley. Myrtaceae (Lept.). 3 SW Aus.

Conradina A. Gray. Labiatae (VIII 2). 6 SE US. Some cult. orn.

Conringia Heister ex Fabr. Cruciferae. 6 Medit., Eur. (3) to C As. R: JAA 66(1985)348. *C. orientalis* (L.) Dumort. (Eur., Med.) – seeds source of cooking oil

Consolida Gray (~ *Delphinium*). Ranunculaceae (I 4). 43 Med. (Eur. 12) to C As. Annuals; differ from *Delphinium* in 2 upper C united and without 2 lower C. Cult. orn. (larkspur) esp. *C. ajacis* (L.) Schur ('*C. ambigua*', *Delphinium ajacis* auctt., Medit.) & *C. hispanica* (Costa) Greuter & Burdet (*C. orientalis*, *D. ajacis* auctt., Medit.) with abruptly beaked follicles; fls used to garland mummies in Egypt

Constantia Barb. Rodr. Orchidaceae (V 13). 4 Brazil

consumption weed *Baccharis halimifolia*

contrayerva *Aristolochia odoratissima*; **c. root** *Dorstenia contrajerva*

Convallaria L. Convallariaceae (Liliaceae s.l.). 3 N temp. (Eur. 1). R: FR 86(1975)543. Cult. orn. (lily-of-the-valley) esp. *C. majalis* L. (Eur.) – clones to 670+ yrs old, rhiz. medic., fls used in scent & in certain snuffs, poisonous (azetidine 2-carboxylic acid, non-protein amino-acid, interfering with proline chemistry in usu. being incorporated in proteins but proline t-RNA-synthetase discriminates against it) natur. N Am.

Convallariaceae Horan. (Liliaceae s.l.). Monocots – Liliidae – Asparagales. (Excl. Tricyrtidaceae, Uvulariaceae) 24/200 N temp., trop. As., Aus. Rhizomatous herbs with crystal raphides & basal lvs or aerial stems with spiral to verticillate lvs. Fls reg., ± 3-merous; P usu. 3, often connate at base; A (2)3(4) + (2)3(4) with anthers splitting introrsely longit.; $\underline{G}$ to $\overline{G}$, each loc. with 2–few anatropous to almost orthotropous ovules. Fr. usu. berry (capsules in *Liriope* & *Ophiopogon*), often brightly coloured & poisonous; seeds sometimes sarcotestal

Principal genera: *Aspidistra, Disporum, Maianthemum, Ophiopogon, Polygonatum, Streptopus, Tupistra*

Another volatile split from Liliaceae s.l. as s.t. taken to incl. Uvulariaceae (= Colchicaceae but *Disporum* & *Kuntheria* retained here) & *Tricyrtis* which links the latter to Liliaceae s.s., here including *Medeola*, s.t. referred to Trilliaceae, & *Streptopus*, s.t. put in Calochortaceae. Luzuriagaceae & Philesiaceae are dubiously distinct, though chromosome evidence supports separation; *Reineckia* forms link with Dracaenaceae. Clearly more needs to be done to find an acceptable working classification if Liliaceae s.l. are to be split up

Many cult. orn. in above genera, *Clintonia, Convallaria, Rohdea, Speirantha, Theropogon*

Convolvulaceae Juss. Dicots – Asteridae – Solanales. 56/1600 cosmop. esp. warm. Herbaceous climbers (always twining to right), sometimes parasitic, lianes to 30 m, herbs or shrubs or rarely trees (*Humbertia*); stems often with unusual sec. growth & usu. with articulated non-anastomosing latex-canals or cells & internal phloem, in *Cuscuta* without chlorophyll & attached to host by haustoria, the terr. root-system soon withering & internal phloem absent. Lvs spiral, simple, entire to lobed, scale-like in *Cuscuta*; stipules 0. Fls usu. in heads, dichasia or solit., often subtended by a pair of bracts, these s.t. enlarged & involucriform, bisexual (not in *Hildebrandtia*), usu. 5-merous ((3–)5 in *Cuscuta*, 4 in *Hildebrandtia*); K imbricate, ± connate, s.t. unequal, (C) commonly funnel-shaped, hardly lobed (obliquely irreg. in *Humbertia*), often induplicate-valvate & often convolute in bud, imbricate in *Cuscuta*, A attached at tube-base with filaments often unequal & anthers with longit. slits, usu. annular nectary-disk around ovary-base, $\underline{G}$ (2(3–5)) with as many locules (rarely 1) and free or united styles or 1, rarely united only by common style, ovules usu. 2 per carpel (∞ in *Humbertia*), basal, erect, anatropous, unitegmic. Fr. a loculicidal, circumscissile or irreg. dehiscing capsule, less often baccate, drupe or nut; embryo straight or curved with 2 plicate, often bifid cotyledons or these scarcely recognizable (*Cuscuta*), endosperm with oil, protein & carbohydrate. x = 7–15+

Classification & principal genera: excluding parasitic *Cuscuta*, Austin (AMBG 60(1973)349) recognizes 8 tribes & includes the arborescent *Humbertia* (s.t. excluded as Humbertiaceae) in Erycibeae with *Erycibe*, *Maripa* etc.; *Dichondra* & *Falkia* with deeply lobed ovaries are joined by *Nephrophyllum* (& ? *Wilsonia*) in Dichondreae, which has also been elevated to fam. level by some, while Cronquist keeps the allied *Cuscuta* in a separate fam. Other large genera are in Argyreieae (fr. indehiscent: *Argyreia*), Cresseae (fr. dehiscent, style ± bifid or 2: *Bonamia*), Convolvuleae (fr. dehiscent, style 1: *Calystegia, Convolvulus, Evolvulus, Jacquemontia*) & Ipomoeeae (fr. dehiscent or not, plicae or C sharply marked with 2 lateral veins: *Ipomoea*)

Most closely allied to Solanaceae, Boraginaceae & Polemoniaceae, though placed in Gentianales by Goldberg

Many cult. orn. esp. spp. of *Ipomoea, Calystegia, Convolvulus* & *Poranopsis*, dried fr. of *Argyreia* & *Merremia* sold as 'wood roses'; *Cladostigma* has ed. fr., the tubers of *Ipomoea batatas* are sweet potatoes, other *I.* spp. are leaf veg. & sources of hallucinogenic drugs &

purgatives

Convolvulus L. Convolvulaceae. c. 100 cosmop. esp. temp. (Eur. 23). Pollen smooth (that of *Ipomoea* spiny); see *Calystegia* for distinction. Bindweed; alks. Some cult. orn. esp. *C. tricolor* L. (S Eur.) & *C. gharbensis* Battand. & Pitard (Morocco) – annual herbs & *C. sabatius* Viv. (*C. mauritanicus*, Italy, N Afr.) – perennial. *C. arvensis* L. (common or field bindweed, cornbind, Euras., widely natur.) – pernicious weed with sweetly-scented fls visited by insects, var. *stonestreetii* Druce with 5-lobed C (cf. *Rhododendron stenopetalum*); *C. floridus* L.f. (Canary Is.) – woody, the roots a source of an essential oil; *C. scammonia* L. ((Levant) scammony, SW As.) – source of a drastic purgative

Conyza Less. Compositae (Ast.-Ast.). c. 60 temp. & warm (2 natur. Eur. incl. *C. canadensis* (L.) Cronq. (N Am.)). Some small trees, e.g. *C. vernonioides* (A. Rich.) Wild (trop. E Afr.)

Conyzanthus Tamamschjan = *Aster*

Conzattia Rose. Leguminosae (I 1). 2 Mexico

cooba *Acacia salicina*

coohoy nut *Floydia prealta*

Cooktown orchid *Dendrobium bigibbum*

coolabar, coolibah *Eucalyptus microtheca*; **c. grass** *Thellungia advena*

coolwort *Tiarella cordifolia*

Coombea P. Royen = *Medicosma*

coondi *Carapa guianensis*

Cooperia Herbert = *Zephyranthes*

Coopernookia Carolin. Goodeniaceae. 6 SW & SE Aus. R: FA 35(1992)80

copaiba balsam *Copaifera officinalis* etc.; **c. oil** *C. multijuga*

Copaifera L. Leguminosae (I 4). 30 trop. Am., Afr. (s.s. 4 Afr.), Mal. (1). Source of hard resins & oleo-resins (copals), used industrially, & timbers (ironwood) esp. *C. officinalis* (Jacq.) L. (trop. Am.) – copaiba balsam; *C. multijuga* Hayne (Amaz.) – oil (25 l over 6 months) from trunk can be used directly in diesel engines; *C. salikounda* Heckel (W Afr.) provides a veneer (bubinga)

copal hard resins form. much used in varnish & paint manufacture, esp. **Congo c.** dug up from ground (*Guibourtia demeusei*) & other semi-fossil c. from *Copaifera* spp. (W Afr.), *Hymenaea verrucosa* (**Zanzibar c.**, E Afr.); **Manila** or **East Ind. c.** is from *Agathis dammara*, **kauri c.** (NZ) from *A. australis*; **Madag. c.** *Hymenaea verrucosa*

copalchi bark *Croton niveus*

copalquía *Hintonia latiflora*, *Pachycormus discolor*

Copedesma Gleason = *Miconia*

Copelandiopteris Stone = *Pteris*

Copernicia Mart. ex Endl. Palmae (I 16). 29 WI (esp. Cuba) & S Am. savannas. R: GH 9(1961–3)1–232. Some cult. orn. incl. *C. prunifera* (Miller) H. Moore (*C. cerifera*, wax palm, carnauba w. p., NE Brazil where there are some 100 million trees) – source of carnauba wax used in shoe-polish, lipstick & carbon paper, form. for candles, gramophone records etc. (some other spp. less exploited), collected by beating off wax particles from young lvs; seeds ed.

copey *Clusia major*

Copiapoa Britton & Rose. Cactaceae (III 5). 20 Chile coastal deserts. R: CSJ 43(1981)49

copper-burr *Sclerolaena* spp.

copperweed *Iva acerosa*

coppice poles from stumps (stools) cut back on a regular basis of 5–15 yrs, long practised from Eur. to Himal., esp. with *Castanea sativa* & *Corylus avellana* in W Eur.; now *Tectona grandis* treated thus in trop. Poles used for stakes, fencing, hurdles etc. & form. in wattle-and-daub houses

copra *Cocos nucifera*

Coprosma Forster & Forster f. Rubiaceae (IV 13). 90 E Mal., Aus., Pacific. Dioec. shrubs & trees with extrafl. nectaries & domatia. R: BBPBM 132(1935)1. Some cult. orn. (v. variable), some dyes & med.; 15 in NZ are divaricate shrubs (51 spp. of div. shrubs in 23 families in NZ). *C. foetidissima* Forster & Forster f. a stinkwood (NZ)

Coptidipteris Nakai & Momose = *Dennstaedtia*

Coptis Salisb. Ranunculaceae (III 3). 15 N temp. R: JJB 24(1949)73. Alks. Cult. orn. herbs with rhiz. prod. yellow dye & some medic. esp. *C. trifolia* (L.) Salisb. (vegetable gold or gold thread, NE Am., Alaska) & *C. teeta* Wallich (Himal.), the latter over-exploited & only a few populations left

Coptocheile Hoffsgg. = ? Gesneriaceae

Coptophyllum Korth. Rubiaceae (IV 1). (Incl. *Jainia* & *Pomazota*) 14 India, W Mal.

Coptosapelta Korth. Rubiaceae (?I 1). 13 SE China to Mal. Some med.
Coptosperma Hook.f. = *Tarenna*
coquilla palm *Attalea funifera*
coquito palm *Jubaea chilensis*
coracan *Eleusine coracana*
coral beads *Cocculus carolinus*; **c. bean** *Erythrina* spp.; **c. bells** *Heuchera sanguinea*; **c. berry** *Symphoricarpos orbiculata*; **c. broom** *Corallospartium crassicaule*; **c. bush** or **plant** *Russelia equisetiformis*, *Templetonia retusa*; **c. creeper** *Antigonon leptopus*, *Kennedia* spp.; **c. fern** *Gleichenia* spp.; **c. flower** *Erythrina* spp.; **c. pea** *Kennedia* spp.; **c. root** *Corallorrhiza odontorhiza*, *Cardamine bulbifera*; **c. tree** *Erythrina* spp.; **c. vine** *Antigonon leptopus*
corallita *Antigonon leptopus*; **white c.** *Porana paniculata*
Corallocarpus Welw. ex Hook.f. Cucurbitaceae. 13 trop. Afr., Madag., India
Corallodiscus Batalin. Gesneriaceae. 18 Himal. to NW China & SE As.
Corallorhiza Châtel. = seq.
Corallorrhiza Rupp. ex Gagnebin. Orchidaceae (V 8). 15 N temp. (Eur. 1). Chlorophyll-less mycotrophs. Some cult. orn. esp. *C. odontorhiza* Nutt. (coral root, dragon claw, E N Am. to Guatemala) – fls commonly cleistogamous
Corallospartium Armstr. Leguminosae (III 16). 1 NZ (S Is.).: *C. crassicaule* (Hook.f.) Armstr. (coral broom)
Corbassona Aubrév. = *Niemeyera*
Corbichonia Scop. Molluginaceae. 2 SW Afr. (1), trop. Afr. to As. (1)
Corchoropsis Siebold & Zucc. Sterculiaceae. 3 E As. & Japan. Annuals
Corchorus L. Tiliaceae. 100 trop. Jute (gunny) obtained from phloem fibres; young shoots ed. Main source: *C. capsularis* L. (China, widely cult.) – can be grown inundated; *C. olitorius* L. (tossa jute, Jew's mallow, India) – grown more in uplands. Jute obtained by retting stalks, the fibre used in sacking, twine, carpeting & paper; young shoots ed. like spinach
cord grass *Spartina* spp. & hybrids
Cordanthera L.O. Williams. Orchidaceae. 1 Colombia
Cordeauxia Hemsley. Leguminosae (I 1). 1 NE Afr. *C. edulis* Hemsley (ye'eb or jeheb nut, Somalia, Ethiopia) – source of purplish dye, seeds ed. (taste somewhat like *Castanea sativa*)
Cordemoya Baillon. Euphorbiaceae. 1 Masc.
Cordia L. Boraginaceae. 320 trop. (WI c. 100 (80% endemic); perhaps splittable into *C.* s.s., *Gerascanthus* P. Browne & *Varronia* P. Browne on pollen & K characters etc.). R: JAA suppl. 1(1991)43. Homostyly, distyly, dioec.; hollow ant-infested domatia; trunk hissing when slashed; lvs used for sandpaper; branches used as self-tindering firesaws in Maikal Hills, India; cult. fast-growing timber in trop. & orn.; some ed. fr. *C. alliodora* (Ruíz & Pavón) Cham. (cyp, cypre, Ecuador laurel, salmwood, trop. Am.) – moth-poll., assoc. with *Azteca* ants in Costa Rica, plantation tree, fr. ed. Mex.; *C. africana* Lam. (*C. abyssinica*, mukumari, muringa, trop. Afr.) – timber; *C. collocca* L. (manjack, WI) – cherry-like fr., ed. & fed to fowls; *C. cylindristachya* Roemer & Schultes (trop. Am.) – hedgepl. in Mal. but weedy though rapidly controlled by beetle (*Schematiza cordiae*) feedling exclusively on lvs & wasp (*Eurytoma attiva*) laying eggs on young fr.; *C. dodecandra* DC (ziricote, C Am.) – fine dark timber, rough lvs used like sandpaper, cult. for ed. fr.; *C. gerascanthus* L. (prince wood, Spanish elm, WI) – timber; *C. goeldiana* Huber (freijo, Brazil) – comm. timber; *C. myxa* L. (Sudan teak, sebesten plum, India to Aus., natur. Afr.) – planted in Med., Calif. & trop. Am. for mucilaginous fr. used med. & as bird-lime, wood for cabinetwork; *C. sebestena* L. (geiger tree, trop. Am.) – cult. orn. with red fls., medic. fr.; *C. subcordata* Lam. (Indopacific strand) – buoyant fr. sea-disp.
Cordiaceae R. Br. ex Dumort. = Boraginaceae
Cordiglottis J.J. Sm. Orchidaceae (V 16). 7 W Mal.
Cordisepalum Verdc. Convolvulaceae. 1 SE As.
Cordobia Niedenzu. Malpighiaceae. 2 S Am.
Cordyla Lour. Leguminosae (III 1). 5 trop. Afr., Madag. C 0. *C. pinnata* (A. Rich.) Milne-Redh. (bush mango, W Afr.) – ed. fleshy pods
Cordylanthus Nutt. ex Benth. Scrophulariaceae. 18 W N Am. R: SBM 10(1988). Annuals. *C. wrightii* A. Gray lvs used for bleaching by Hopi Indians
Cordyline Comm. ex R.Br. Dracaenaceae (Agavaceae s.l.). 15 Australasia, Pacific, trop. Am. (1). Like *Dracaena* but 2 or more ovules per locule & no sec. tissue in roots. Cult. orn. 'cabbage-trees', fragments of trunk will regrow. *C. australis* (Forster f.) Endl. (palm lily, NZ) – form. eaten by moas, sudden decline in wild 1987 (? virus), lvs a source of fibre, cult. orn. esp. in tubs & at seaside, source of high fructose syrup comparable with that

used in food-processing; *C. fruticosa* (L.) Goeppert (*C. terminalis*, tanget (New Guinea), ti, trop. E As. to Polynesia) – many cvs with coloured lvs, often used for hedging, ritual, medic. & magic, trad. dress in New Guinea, wrapping food for cooking, crushed for soap for clothes in Amaz., roots ed. E Oceania

Cordyloblaste Henschel ex Moritzi = *Symplocos*

Cordylocarpus Desf. Cruciferae. 1 N Afr.

Cordylogyne E. Meyer. Asclepiadaceae (III 1). 1 S Afr.: *C. globosa* E. Meyer

Coreanaomecon Nakai = *Chelidonium*

Corema D. Don. Empetraceae. 2: *C. album* (L.) Steudel (Azores & Iberia), *C. conradii* (Torrey) Loudon (NE N Am.)

Coreocarpus Benth. Compositae (Helia.-Cor.). 9 SW N Am. esp. Mex. R: SB 14(1989)448

Coreopsis L. (~ *Bidens*). Compositae (Helia.-Cor.). 50 Am. R: FMNHB 11(1936)279; CN 27(1995)1. Cult. orn. esp. *C. basalis* (A. Dietr.) S.F. Blake (*C. drummondii*, Texas); *C. gigantea* (Kellogg) H.M. Hall (Calif.) – pachycaul with stems to 12.5 cm thick & *C. tinctoria* Nutt. (*Calliopsis t.*, E N Am.); some med. locally

Corethamnium R. King & H. Robinson (~ *Eupatorium*). Compositae (Eup.-Crit.). 1 Colombia. R: MSBMBG 22(1987)354

Corethrogyne DC. = *Lessingia*

coriander *Coriandrum sativum*

Coriandropsis H. Wolff = *Coriandrum*

Coriandrum L. Umbelliferae (III 4). 3 SW As. *C. sativum* L. (coriander) – long cult. for fr. ('seeds'; *coriandrum* one of the earliest words recognized in the deciphering of Linear B (Crete, C17–15 BC)), one of the 'bitter herbs' prescribed by Jews at Feast of the Passover, form. coated with sugar & sold as coriander comfits; form. added to weak beers, now most heavily used herb (India & Morocco biggest producers) – to flavour gin, confectionery, bread, curry powder & in some scents & soap, preserved in Tutankhamun's tomb 1325 BC

Coriaria L. Coriariaceae. 5 Mex. to Chile, W Med. (Eur. 1), Himal. to Japan & New Guinea, NZ & S Pacific. R: Rhodora 74(1972)242. Actinomycete root-symbionts. Some dyes, *C. myrtifolia* L. (W Med.) – lvs & bark for tanning, crushed fr. used as fly-poison; other cult. orn.; fr. poisonous, hallucinogenic

Coriariaceae DC. Dicots – Rosidae – Sapindales. 1/5 Euras., NZ, C & S Am. Trees to subshrubs, often with nitrogen-fixing root nodules. Branches & lvs frond-like. Lvs small, opp. or whorled, simple with palmate venation; stipules minute, caducous. Fls ± reg., in racemes, bisexual or plants polygamous; K 5, imbricate, C 5 becoming ± fleshy in fr., A 5 + 5, filaments of antepetalous A adnate to keel of C, anthers with longit. slits, $\underline{G}$ 5(10), s.t. basally united, each with long slender style & 1 pendulous, anatropous, bitegmic ovule. Fr. a head of achenes ± enclosed in fleshy C; seed rather compressed, embryo straight, oily; endosperm scanty or 0. x = 10

Only genus: *Coriaria*

On A structure & chromosome evidence recently returned to Sapindales though seed-coat anatomy (Corner) & other features suggest affinity with Ranunculaceae, but C. accumulate ellagic acid unlike rest of Magnoliidae. Wood structure suggests affinities with both groups & it may be that C. do indeed represent a transitional group pointing up the artificial nature of the 'higher' groupings of dicots, when such are 'nested' in the 'lower' ones

Coridaceae J. Agardh = Primulaceae

Coridothymus Reichb. (~ *Thymus*). Labiatae (VIII 2). Spp.?

Coris L. Primulaceae (Coridaceae). 1 (variable; recognized as 2 in Eur.) Med., Somalia

Corispermum L. Chenopodiaceae (I 6). 60 N temp. (Eur. 11, natur. Am.). Bugseed

Coristospermum Bertol. (~ *Ligusticum*). Umbelliferae (III 8). 3 Eur.

cork *Quercus suber*; **c. elm** *Ulmus thomasii*; **c. tree** *Ochroma pyramidale*, *Phellodendron* spp.; **c. wood** *Duboisia myoporoides* (Aus.), *Entelea arborescens* (NZ), *Hakea* spp. (Aus.), *Leitneria floridana* (N Am.), *Musanga cecropioides* (W Afr.), *Myrianthus arboreus* (Congo)

corkscrew grass *Stipa* spp.; **c. hazel** *Corylus avellana* 'Contorta'; **c. rush** *Juncus effusus* 'Spiralis'

Cormonema Reisseck ex Endl. = *Colubrina*

Cormus Spach Segregate of *Sorbus*, q.v.

corn grain commonly used in any particular territory, e.g. wheat in GB, maize in US; **baby c.** *Zea mays*; **c. bind** *Convolvulus arvensis*; **blue c.** *Z. mays*; **c. cockle** *Agrostemma githago*; **c. flakes, c. flour, Indian c., c. on the cob, pop c., sweet c.** *Zea mays*; **c. flower** *Centaurea cyanus*; **c. marigold** *Chrysanthemum segetum*; **c. poppy** *Papaver rhoeas*; **c.salad** *Valerianella* *locusta*; **c. spurrey** *Spergula arvensis*; **squirrel c.** *Dicentra canadensis*

Cornaceae (Dumort.) Dumort. Dicots – Rosidae – Cornales. (Incl. Nyssaceae) 14/120 N temp., rare in trop. & S temp. Trees, shrubs or rarely rhiz. herbs, s.t. accumulating cobalt or aluminium &/or inulin. Lvs opp., rarely spiral, simple, usu. entire (deeply pinnatifid in *Aralidium*); stipules 0 (exc. *Helwingia*). Fls usu. bisexual, reg., epigynous, small, in cymes or cymose heads (racemes in *Melanophylla*), these s.t. with consp. whorl of large petaloid bracts, that of *Helwingia* borne on leaf; K 4(5–7) small teeth or 0, s.t. forming a tube in male fls, C 4 or 5(–10) valvate (imbricate in *Griselinia* & *Melanophylla*), 0 in female fls, A as many as & alt. with C (rarely in 2 whorls), usu. attached to or around edge of epigynous disk, anthers with longit. slits, $\overline{G}$ (2–4(–9)) with as many locules (only 1 fert. in *Aralidium* & *Griselinia*) or 1-loc. & pseudomonomerous, style 1, term., or styles ± distinct, 1 ovule per locule, apical, pendulous, anatropous & bitegmic or rarely (*Curtisia*, *Mastixia*) epitropous, unitegmic. Fr. usu. a drupe with 1–5-loc. longit. grooved endocarp with 1 seed per locule (drupe with 2 pyrenes in *Kaliphora*), less often (*Aucuba*, *Griselinia*) a berry; seeds with small elongate embryo embedded in copious oily endosperm. x = 8–13, 19, 21, 22

Genera: *Aralidium*, *Aucuba*, *Camptotheca*, *Cornus*, *Curtisia*, *Davidia*, *Diplopanax*, *Griselinia*, *Helwingia*, *Kaliphora*, *Mastixia*, *Melanophylla*, *Nyssa*, *Torricellia*

Corokia (q.v.) is here excluded to Grossulariaceae (Escalloniaceae); *Diplopanax* has until lately been in Araliaceae, to which *Aralidium* has been referred, though this & all other genera exc. *Kaliphora*, whose place here has been questioned (? Melanophyllaceae), have been assigned to monotypic (*Camptotheca*, *Davidia* & *Nyssa* to Nyssaceae) families. Each of the segregates has some merit (esp. Aucubaceae, Griseliniaceae, Helwingiaceae & Melanophyllaceae) though on molecular data Nyssaceae are part of Cornaceae s.s. & the whole seems more closely interrelated than allied to any other family, so that Cronquist is followed here with the addition of Nyssaceae, though this arr. is clearly provisional. Moreover, Alangiaceae & certain Hydrangeaceae may also have to be included

Cult. orn. esp. *Cornus* & *Aucuba*, timber & ed. fr. (*Nyssa*)

cornel *Cornus* spp.

Cornelian cherry *Cornus mas*

Cornera Furt. = *Calamus*

Cornish moneywort *Sibthorpia europaea*

Cornopteris Nakai. Dryopteridaceae (II 1). 9 trop. As. R: APG 30(1979)101

Cornucopiae L. Gramineae (21d). 2 E Med. (Eur. 1) to Iraq. Fr. adhere to animals & alleged to burrow into soil (cf. *Heteropogon contortus*)

Cornuella Pierre = *Chrysophyllum*

Cornulaca Del. Chenopodiaceae (III 3). 6 Egypt to C As. *C. monacantha* Del. imp. camel fodder

Cornus L. Cornaceae. c. 65 N temp. (Eur. 4), rare S Am., Afr. (s.s. 4 Euras., Calif.; here incl. *Afrocrania*, *Benthamidia*, *Bothrocaryum*, *Chamaepericlymenum*, *Dendrobenthamia* (*Cynoxylon*), *Discocrania*, *Swida* (*Thelycrania*)), 15 red-fruited clearcut spp., c. 50 blue- (or white-)fruited less so: dogwoods (from 'dogs', i.e. skewers form. made from wood of C. spp.), cornels; trees, shrubs & rhiz. herbs (*Chamaepericlymenum*) with torn lvs hanging together with pulled-out xylem-thickenings ('magnetic lvs' cf. *Wimmeria*), some domatia, much cult. for bract-surrounded infls resembling simple fls, winter-bark colour etc. R: BR 54(1988)233. *C. alba* L. (C & E As. – cult. for red branches, esp. 'Sibirica' ('Westonbirt'); *C. canadensis* L. (*Chamaepericlymenum c.*, bunchberry, crackerberry, E As., N Am., Greenland) – rhiz., cult. orn. with conspic. bracts, pollen catapulted upwards in fl. by 'elbow springs' on filaments (not found in *Cornus* s.s.); *C. florida* L. (E N Am.) – cult. orn., molluscicidal steroid saponins; *C. kousa* (Miq.) Hance (E As.) – cult. orn. with consp. bracts & ed. infrs; *C. mas* L. (Cornelian cherry, C & S Eur., SW As.; '*C. femina*' of the herbals is *C. sanguinea*) – cult. orn. & for fr. used in jam or rob (syrup) & basis of alcoholic Vin de Cornoulle in France, fr. favoured by pigs (Ulysses & followers changed into ps by Circe after eating them); *C. nuttallii* Audubon (W N Am.) – timber for tools & cabinet-work, cult. orn. with large bracts; *C. obliqua* Raf. (E N Am.) – poss. polyphenol source; *C. sanguinea* L. (dogwood, swamp d., pegwood, Eur.) – wood for skewers, bobbins etc.

Cornutia L. Labiatae (I 1; Verbenaceae). c. 15 trop. Am. *C. pyramidata* L. (C Am., WI) – fr. used for blue ink (red with lime) & dyeing cloth

coro *Trichocline* spp.

Corokia Cunn. Grossulariaceae. 6 NZ, Rapa Is. Cult. evergreen shrubs with stomata on inner epidermis of inferior ovary esp. *C. cotoneaster* Raoul (NZ) – leaf colour in autumn (& size) like fr. so perhaps 'fr. flags' to attract dispersers. DNA data support removal from

Cornaceae

Corollonema Schltr. = ? *Mitostigma*

Coromandel wood *Diospyros quaesita*

Coronanthera Vieill. ex C.B. Clarke. Gesneriaceae. 10 New Caled., 1 Aus. Small trees & shrubs

Coronaria Guett. = *Silene*

Coronilla L. Leguminosae (III 19). 9 Atlantic is., Medit., Eur. (13). R: Willd. 19(1989)59. Cult. orn. with strongly-scented fls, esp. *C. valentina* L. (bastard senna, Medit.), also used for erosion control. See also *Hippocrepis*, *Securigera*

Coronopus Zinn. Cruciferae. 10 almost cosmop. (Eur. 3). Weeds (swine cress) esp. *C. squamatus* (Forssk.) Asch. (wart cress ? orig. S Am.) – ant-poll.

Corothamnus (Koch) C. Presl = *Genista*

Coroya Pierre = ? *Dalbergia*

Corozo Jacq. ex Giseke = *Elaeis*

Corpuscularia Schwantes (~ *Delosperma*). Aizoaceae (V). 3–5 SE Afr.

Correa Andrews. Rutaceae. 11 temp. Aus. R: TRSSA 85(1961)21. C(4). Cult. orn. (Australian fuchsia); all spp. hybridize. Lvs of *C. alba* Andrews have been used as tea

Correllia A.M. Powell. Compositae (Hele.-Per.). 1 S Mexico

Correlliana D'Arcy = *Cybianthus*

Corrigiola L. Caryophyllaceae (I 3). 11 cosmop. (Eur. 1–2). R: MBMHRU 285(1968)34. *C. litoralis* L. (strapwort, subcosmop.) – root used in scent & medic.

Corryocactus Britton & Rose. Cactaceae (III 5). 20 Bolivia, S Peru, N Chile. Columnar, cult. orn.

Corsia Becc. Corsiaceae. c. 25 Papuasia & Aus. (1)

Corsiaceae Becc. (~ Burmanniaceae). Monocots – Liliidae – Orchidales. 2/26 Chile & W Pacific. Chlorophyll-less mycotrophs with rhiz. or tubers. Lvs spiral, scale-like. Fls solit., term., irreg., bisexual or not, P petaloid, tubular basally, 3 + 3, posterior member of outer large & coloured enclosing other 5 linear-spathulate, A 6 on tube, filaments short, anthers with longit. slits, pollen monosulcate, $\overline{G}$ (3), 1-loc, with ± intruded 2-lobed placentas, style with 3 thick stigmas & ∞ tiny ovules. Fr. a capsule with 3 valves; seeds small, winged, with undifferentiated embryo & scanty endosperm. x = 9

Genera: *Arachnitis* (unisexual fls), *Corsia* (bisexual)

Corsican pine *Pinus nigra* subsp. *salzmannii*

Cortaderia Stapf. Gramineae (25). 24 S Am., NZ (4), New Guinea (1). Gynodioec. clump-forming coarse grasses, intergrading with *Chionochloa*; *C. selloana* is subdioec. & *C. bifida* Pilger apomictic. Cult. orn. (pampas grass) esp. *C. selloana* (Schultes & Schultes f.) Asch. & Graebner (*Gynerium argenteum*, Brazil, Arg., Chile) as specimen lawn pl., also comm. for dried infls (female larger) and in S Am. for paper, noxious weed in S Afr. (also *C. jubata* (Lemaire) Stapf (W trop. S Am.)), NSW & NZ

Cortesia Cav. Boraginaceae (Ehretiaceae). 2 temp. S Am.

Cortia DC. Umbelliferae (III 8). 8 C & S As.

Cortiella Norman. Umbelliferae (III 8). 3 C & S As.

Cortusa L. Primulaceae. c. 8 mts C Eur. (1) to N As. Cult. orn. esp. *C. matthioli* L. (Transylvania)

Corunastylis Fitzg. = *Genoplesium*

Coryanthes Hook. Orchidaceae (V 10). 33 trop. Am. R: TSP 83(1993). Epiphytes with extrafl. nectaries; massive pendent waxy fls with part of lip forming bucket into which drops of water are secreted from knobs on column; male bees of genus *Eulaema* attracted by strong scent (app. specific to each sp. preventing hybridization) scratch on area of tissue at the base of the lip to collect the liquid scent, which intoxicates their front tarsi such that the bees lose hold & fall into the bucket; they cannot scale the sides but must leave through a tunnel in which they deposit or collect the pollinia; a bee which took 45 mins to escape could not be induced to return as the scent was not replenished until next morning, suggesting that this would prevent self-fert. Some cult. orn.

Corybas Salisb. Orchidaceae (IV 1). c. 100 Indomal. to NZ (Aus. 20, 19 endemic), Polynesia & Japan with *C. macranthus* (Hook.f.) Reichb.f. (Macquarie Is., subAntarctic) most S distrib. of any orchid. R: PM 16 (1983) (74 E of Wallace's Line), KB 41(1986)575 (27 As. & W Mal.). Predicted to be poll. by fungus-gnats; some cult. orn. (helmet orchids)

Corycium Sw. Orchidaceae (IV 3). 14 S Afr. Elaiophores attractive to bees

Corydalis DC. Fumariaceae (I 1). (Excl. *Capnoides*, *Ceratocapnos*, *Pseudofumaria*, q.v.) c. 400 N temp. (Eur. 11; Sino-Himal. c. 280), trop. Afr. mts (1). R: AHB 17(1955)115, OB 88(1986)21. Herbs with tubers, 1 cotyledon & usu. pinnate (simple in *C. ludlowii* Stearn, Tibet) lvs;

tuber in *C. cava* (L.) Schweigger & Koerte (C Eur.) the main axis with depression on underside, each annual shoot arising from axil of scale leaf; that of *C. solida* (L.) Clairv. (*C. bulbosa*, Euras.) is a swollen current annual shoot. Fls transverse zygomorphic twisting through 90° to become vertical, 1 petal spurred & containing nectar from staminal nectary; inner petals united at tip enclose stigma & anthers, which are caused to emerge by bees alighting on & pushing down inner C (cf. Leguminosae). Many alks. Some ed. tubers in As. Many cult. orn. esp. the freq. confused *C. solida* & *C. cava* with pink to purple fls

Corylaceae Mirbel. See Betulaceae

Corylopsis Siebold & Zucc. Hamamelidaceae (I 2). 7 Bhutan to Japan. R: JAA 58(1977)382. Fls before lvs in spring, usu. in pendent spikes or racemes; cult. orn. esp. *C. pauciflora* Siebold & Zucc. & *C. spicata* Siebold & Zucc. (Japan)

Corylus L. Betulaceae (II; Corylaceae). 15 N temp. (Eur. 3). R: JAA 71(1990)61. Hazels or filberts; monoec. trees & shrubs with edible seeds in nuts; domatia. Anemophilous, the male fls in pendent catkins, the females in small red bud-like infls; poll. to fert. takes up to 4 months. Cult. ed. & orn. esp.: *C. americana* Walter (American hazel or filbert, E N Am.) – leafy involucre round nuts v. long; *C. avellana* L. (hazel, Euras.) – seriously threatened in GB by introduced (Am.) grey squirrels, form. much grown for coppice (q.v.), esp. for hurdles, firewood, legume-poles, wattle-and-daub, in GB 500 000 acres in 1905 reduced to c. 94 000 in 1965, diff. cvs of value for seeds esp. 'Barcelona' nuts selected from cobnuts (though 'Kentish cob' much grown actually a filbert (i.e. 'full beard' = long-husked nut), *C. maxima* Miller, SE Eur., poss. a cv. of *C. avellana*), the seeds used for oil for cooking, form. paint, soap etc., & for confectionery, esp. nut-chocolate, wood form. principal source of charcoal for gunpowder, can cause urticaria, 'Contorta' (corkscrew hazel, orig. plant (1863, ? somatic mutant) from Frocester, England) with twisted stems; hazel imp. in Eur. mythology, St Patrick supposed to have purged Ireland of poisonous snakes by brandishing a hazel wand & since time of Pliny favoured as a water-divining rod; *C. colurna* L. (Turkish hazel, SE Eur., SW As.) – cult. for nuts, wood form. used for spinning-wheels

Corymbia K. Hill & L. Johnson (~ *Eucalyptus*). Myrtaceae (Lept.). 113 Aus. (esp. N) & S New Guinea. R: Telopea 6(1995)185. Bloodwoods & ghost gums recog. by compound infls usu. ± urceolate fr. & usu. ± tesselate bark. Timbers & cult. orn. *C. calophylla* (Lindley) K. Hill & L. Johnson (*E. c.*, marri, SW Aus.) – v. large fr., ectomycorrhizae spread by red kangaroos (spores germ. after digestion); *C. citriodora* (Hook.) K. Hill & L. Johnson (*E. c.*, lemon-scented gum, Queensland) – source of a lemon-scented oil, structural timber, fuel crop; *C. ficifolia* (F. Muell.) K. Hill & L. Johnson (*E. f.*, SSW Aus) – widely cult. for showy panicles of red fls; *C. gummifera* (Gaertn.) K. Hill & L. Johnson (*E. g.*, bloodwood, E. Aus.) – sap red, one of first Aus. pls cult. GB (1771); *C. papuana* (F. Muell.) K. Hill & L. Johnson (*E. p.*, S New Guinea) & allied spp. in Aus.= ghost gum – bark white

Corymbium L. Compositae (Cichorioideae). 9 SW Cape. R: MIABH 23b(1990)631. Lvs narrow with parallel veins

Corymborkis Thouars. Orchidaceae (III 2). 8 Indomal. to Pacific. R: BT 71(1977)161

Corymbostachys Lindau = *Justicia*

Corynabutilon (Schumann) Kearney = *Abutilon*

Corynaea Hook.f. Balanophoraceae. 1 trop. Am. montane forests: *C. crassa* Hook.f. – vestigial roots on tuber

Corynanthe Welw. Rubiaceae (I 1). 8 trop. Afr. Alks.

Corynanthera J. Green. Myrtaceae. 1 W Aus.

Corynella DC. = *Poitea*

Corynemyrtus (Kiaerskov) Mattos = *Myrtus*

Corynephorus Pal. Gramineae (21b). 5 Eur. (4), Medit. to Iran. On dunes & other sands. Hair grass; basal awns with twisted column & clavate limb & ring of hairs at junction

Corynephyllum Rose = *Sedum*

Corynocarpaceae Engl. Dicots – Rosidae – Celastrales. 1/5 SW Pacific. Trees with bitter glucosides in bark & seeds. Lvs spiral, simple, entire, leathery; stipules intrapetiolar, decid. Fls in term. panicles, bisexual, reg.; K 5 quincuncial, C 5 imbricate, A 5 opp. & basally adnate to C all atop a short hypanthium, alt. with 5 nectaries (?staminodes) each bearing a petaloid scale (? C; ? staminode connective), anthers with longit. slits, 5 nectaries opp. A outside G, $\underline{G}$ 1 (rarely with second style) with 1 pendulous, anatropous, bitegmic ovule. Fr. a drupe; seeds with straight oily & starchy embryo, v. poisonous, endosperm 0. 2n = 44. R: BJLS 95(1987)9

Only genus: *Corynocarpus*

Affinities disputed (Goldberg suggests Rosales): wood anatomy suggests ally of Berberidaceae, bitter principle (karakin) said to be identical with hiptagin of *Hiptage* (Malpighiaceae), pollen like *Itea* (Saxifragaceae), so prob. best placed in its own order

Corynocarpus Forster & Forster f. Corynocarpaceae. 5 New Guinea, NE Aus. (2), New Caled., Vanuatu, NZ (poss. introd. by Maoris from Vanuatu & New Caled.: *C. laevigata* Forster & Forster f. (karaka nut, NZ etc.) – seeds ed. after roasting, a staple food of Maoris, cult. orn., fleshy part of fr. ed. raw, trunks used for canoes; natur. Hawaii after seeding Kauai by air in 1929)

Corynostylis Mart. Violaceae. 4 trop. Am. Some med. esp. *C. volubilis* L.B. Sm. & Fern. (Colombia) – powerful vermifuge

Corynotheca F. Muell. ex Benth. (~ *Caesia*). Anthericaceae (Liliaceae s.l.). 6 Aus. R: FA 45(1987)299

Corynula Hook. f. = *Leptostigma*

Corypha L. Palmae (I 1c). 6 trop. As. to Aus. Hapaxanthic; palmate lvs. Cult. esp. *C. utan* Lam. (*C. elata*, gebang) – infl. with 3–15 million functional fls & c. 250 000 fr. (biomass some 22% of a 44 yr-old tree recorded in cult.), source of toddy, sugar, vinegar, starch, seeds ed., petiole-fibre used in hats, ropes etc.; *C. umbraculifera* L. (talipot, cultigen (?) poss. derived from *C. utan*) – lvs (to 5.3 m across, petiole to 5 m long) used for thatching, umbrellas etc. & also for sacred Buddhist books (olas) written on strips (ready-ruled!) with metal stylus, infl. (after 20–30 yrs) to 8 m tall, the largest of any plant, followed 8 months later by fr. & further 4 months by death

Coryphantha (Engelm.) Lemaire. Cactaceae (III 9). 45 SW N Am. Cult. orn.

Coryphomia N. Rojas = *Copernicia*

Coryphopteris Holttum = *Thelypteris*

Coryphothamnus Steyerm. Rubiaceae (IV 7). 1 SE Venezuela

Corysanthes R. Br. = *Corybas*

Corythea S. Watson = *Acalypha*

Corytholoma (Benth.) Decne = *Sinningia*

Corythophora Knuth. Lecythidaceae. 4 Brazil

Corytoplectus Oersted (~ *Alloplectus*). Gesneriaceae. 10 trop Am.

Coscinium Colebr. Menispermaceae (IV). 2 Indomal., SE As. *C. fenestratum* (Gaertner) Colebr. (India, Sri Lanka) – seeds disp. bats & polecats; wood source of a turmeric-like dye, medic., rope for elephant-logging

Cosentinia Tod. = *Cheilanthes*

Cosmea Willd. = *Cosmos*

Cosmelia R. Br. Epacridaceae. 1 SW Aus.

Cosmianthemum Bremek. Acanthaceae. 8 W Borneo

Cosmibuena Ruíz & Pavón. Rubiaceae (I 1). c. 12 trop. Am. Usu. epiphytes

Cosmocalyx Standley. Rubiaceae (inc. sed.). 1 Mexico

Cosmos Cav. (~ *Bidens*). Compositae (Helia.-Cor.). 26 trop. & warm Am. esp. Mex. (*C. caudatus* Kunth commonly natur. OW). R: FMNHB 7(1932)401. Cult. orn. ('Mexican aster') esp. *C. bipinnatus* Cav. ('cosmea' of gardens, Mex.), natur. in Madag. & *C. atrosanguineus* (Hook.) Voss (Mex.) – fls colour & smell of chocolate

Cosmostigma Wight. Asclepiadaceae (III 4). 3 Hainan, Indomal.

Cossinia Comm. ex Lam. (*Cossignia*, *Cossignya*). Sapindaceae. 4 Masc., New Caled., E Aus., Fiji

Costaceae (Meissner) Nakai. See Zingiberaceae

Costaea A. Rich. = *Purdiaea*

Costantina Bullock. Asclepiadaceae (III 4). 1 SE As.: *C. inflexa* (Costantin) Bullock

Costarica L. Gómez = *Sicyos*

Costaricia Christ = *Dennstaedtia*

Costera J.J. Sm. Ericaceae. 9 Mal. Close to *Vaccinium*

costmary *Tanacetum balsamita*

Costularia C.B. Clarke. Cyperaceae. 20 S Afr., Indian Ocean, Mal. & New Caled. (where some woody)

Costus L. Zingiberaceae (Costaceae). 42 trop. to Aus. (c. 5). Staminodial labellum large, K & P rather small; floral mechanism like *Iris*; seeds arillate. Cult. orn., some local med.; *C. speciosus* (Koenig) Sm. (Indomal.) – diosgenin in rhiz. (commercial); *C. woodsonii* Maas (C Am.) – presence of ants attracted by extrafl. nectaries increases seed set 3-fold

costus root *Saussurea costus*

Cota Gay ex Guss. = *Anthemis*

Cotinus Miller (~ *Rhus*). Anacardiaceae (IV). 1 SE US, 1 S Eur. to China, 1 SW China. Cult.

orn. esp. *C. coggygria* Scop. (*Rhus cotinus*, smoke tree or bush, wig tree, Hungarian, Indian, Turkish, Tyrolean or Venetian sumac, S Eur. to China) – lvs used for tanning, sterile parts of infl. elongate & become hairy, the whole infl. falling & may be blown about, wood gives a dye ('young fustic', yellow), cvs incl. commonly cult. purple-lvd 'Purpureus'; *C. obovatus* Raf. (*C. americanus*, SE US) – chittamwood prod. orange dye

coto bark *Aniba coto*; **false c. b.** *Ocotea pseudocoto*

Cotoneaster Medikus. Rosaceae (III). 261 temp. OW (7 Eur.), many apomictic aggregates. R: CJB 68 (1990)2211. Close to *Crataegus*, on spp. of which many may be grafted, but without spines or lobed lvs. Many cult. orn. (but smelly due to trimethylamine (cf. *Ligustrum*)) evergreen & decid. with brightly coloured fr. (many retained through winter apparently unattractive to birds), esp. *C. horizontalis* Decne (W China) – apomictic with herring-bone sprays of foliage; most cult. spp. from China or Himal., e.g. *C. bacillaris* Wallich ex Lindley (Himal.) – walking-sticks, *C. conspicuus* Marquand (SE Tibet) – fr. untouched by birds in wild until April, *C. racemiflorus* (Desf.) Schldl. (Med. to China) – apomictic, source of a sweet manna-like subs. high in dextrose used in Iran & India; *C. cambricus* Fryer & Hylmö ('*C. integerrimus*') – apomict reduced to 4 pls at Llandudno, N Wales

Cotopaxia Mathias & Constance. Umbelliferae (III 8). 2 Colombia, Ecuador. R: Caldasia 14(1984)21

Cottea Kunth. Gramineae (29). 1 S US to C Mex., Ecuador to Arg.

Cottendorfia Schultes f. Bromeliaceae (1). 1 NE Brazil. R: FN 14(1974)212 (spp. 1–24 = *Lindmania*)

cotton *Gossypium* spp. & hybrids; **devil's c.** *Abroma augusta*; **c. grass** *Eriophorum* spp., *Imperata cylindrica*; **c. gum** *Nyssa aquatica*; **lavender c.** *Santolina chamaecyparissus*; **sea island c.** *Gossypium barbadense*; **c. sedge** *Eriophorum angustifolium*; **silk c.** *Ceiba pentandra*; **c. thistle** *Onopordum* spp.; **c.weed** *Otanthus maritimus*; **c. wood** *Populus* spp. esp. *P. deltoides*, *P. trichocarpa*

Cottonia Wight. Orchidaceae (V 16). 1 S India, Sri Lanka

Cotula L. Compositae (Anth.-Mat.). (Excl. *Leptinella*) c. 55 S hemisph. (esp. S Afr.) to N Afr. & Mex. some natur. N (e.g. Eur. 2 incl. *C. coronopifolia* L., S Afr.). *C. myriophylloides* Harvey (Cape) – obligate aquatic; some cult. orn. carpeting pls

Cotylanthera Blume. Gentianaceae. 4 Himal., SW China, C Mal., Polynesia. R: JJB 50(1975)321. Saprophytes

Cotyledon L. Crassulaceae. 9 S & E Afr. to Arabia (1). Succ. shrubs. Some cult. orn. (many pls described as *C.* belong in other genera like *Adromischus*, *Chiastophyllum*, *Dudleya*, *Echeveria*, *Tylecodon* & *Umbilicus*) esp. *C. orbiculata* L. (Cape & SW Afr.) – fresh leaf-juice allegedly beneficial in treatment of epilepsy but it & other spp. toxic

Cotylelobium Pierre. Dipterocarpaceae. 5 Sri Lanka, W Mal.

Cotylodiscus Radlk. = *Plagioscyphus*

Cotylolabium Garay = *Stenorrhynchos*

Cotylonychia Stapf. Sterculiaceae. 1 trop. Afr.

couch (grass) *Elytrigia repens*; (Aus.) *Cynodon dactylon*; **blue c.** *Digitaria didactyla*; **onion c.** *Arrhenatherum elatius* var. *bulbosum*

Couepia Aublet. Chrysobalanaceae (3). 67 trop. Am. exc. WI. R: FN 9(1972)202. Some rheophytes; hawkmoth-poll. (cf. allied *Hirtella*) with white fls open at night & copious nectar; some distrib. agoutis. *C. edulis* (Prance) Prance (castanha de cutia) an imp. local fruit tree, other spp. with ed. fr., *C. longipendula* Pilger (Brazil) – cv cult. nr. Manaus for ed. cotyledons & oil

cough root *Lomatium dissectum*

Coula Baillon. Olacaceae. 1 trop. W. Afr.: *C. edulis* Baillon (coula) – comm. mahogany subs., seeds (Gaboon nuts) ed. fresh, cooked or fermented

Coulterella Vasey & Rose. Compositae (Hele.). 1 Baja California

Coulterophytum Robinson. Umbelliferae (III 9). 5 Mexico

Couma Aublet. Apocynaceae. 15 NE S Am. Source of couma rubber (Brazil), that of *C. macrocarpa* Barb. Rodr. (sorva, lechi-caspi) a chewing-gum base; *C. guianensis* Aublet with ed. fr.

Coumarouna Aublet = *Dipteryx*

counter wood *Milicia excelsa*

Couratari Aublet. Lecythidaceae. 18 trop. S Am. Hood (ligule – see fam.) coiled under itself, concealing nectar which can be reached only by long-tongued female euglossine bees; wind-disp. winged seeds. Barkcloth

courbaril *Hymenaea courbaril*

courgette *Cucurbita pepo*

Couroupita Aublet. Lecythidaceae. 4 trop. Am. 1-sided ligule (see fam.) overlying stamens (with sterile pollen) & stigmas, *Xylocopa* bees entering for pollen & their backs rubbing stigmas & fertile stamens. *C. guianensis* Aublet (cannonball tree, Guianas) – infls on branches & trunk, fleshy, fruity, followed by woody caps. to 20 cm diam. with evil-smelling pulp, timber good

Coursetia DC. Leguminosae (III 7). 38 warm Am. R: SBM 21(1988)

Coursiana Homolle. Rubiaceae (I 1). 1–2 Madagascar

Courtoisia Nees = *Courtoisina*

Courtoisina Soják (~ *Cyperus*). Cyperaceae. 2 Afr., Madag., India

cous root *Lomatium ambiguum*

cousin mahoe *Urena lobata*

Cousinia Cass. Compositae (Card.-Card.). c. 500 E Medit. (Eur. 1) to C As. & W Himal.

Cousiniopsis Nevski. Compositae (Card.-Carl.). 1 C As.

Coussapoa Aublet. Cecropiaceae. 50 trop. S Am. R: FN 51(1990)16

Coussarea Aublet. Rubiaceae (IV 11). c. 100 trop. Am.

Coutaportla Urban (~ *Portlandia*). Rubiaceae (inc. sed.). 2 Mexico

Coutarea Aublet. Rubiaceae (I 1). 7 Mex. to Arg. Bark locally medic. esp. for malaria, notably *C. hexandra* (Jacq.) Schumann

Coutinia Vell. Conc. = *Aspidosperma*

Coutoubea Aublet. Gentianaceae. 5 trop S Am., WI. R: MNYBG 51(1989)19. Some medic.

Coveniella Tindale. Dryopteridaceae (I 3). 1 NE Queensland: *C. poecilophlebia* (Hook.) Tindale. R: GBS 39(1987)169

Covillea Vail = *Larrea*

cow bane *Cicuta virosa*; **c.berry** *Vaccinium vitis-idaea*; **c. cockle** *Vaccaria hispanica*; **c. itch** *Mucuna pruriens*; **c. parsley** *Anthriscus sylvestris*; **c. parsnip** *Heracleum sphondylium*; **c. pea** *Vigna unguiculata*; **c.slip** *Primula veris*; **c.slip, Amer.** *Dodecatheon* spp.; **c.slip, Cape** *Lachenalia* spp.; **c.slip, Virginian** *Mertensia virginica*; **c. tree** *Brosimum utile*; **c. wheat** *Melampyrum* spp.

Cowania D. Don ex Tilloch & Taylor (~ *Purshia*). Rosaceae. 5 SW N Am. *C. mexicana* D. Don – cucurbitacins poss. effective in tumour control

Cowiea Wernham (~ *Hypobathrum*). Rubiaceae (II 5). Spp.?

Coxella Cheeseman & Hemsley = *Aciphylla*

coyoli *Acrocomia mexicana*

crab apple *Malus sylvestris*, but also applied to any natur. seedling apples as well as some cult. for fls rather than fr.; **c.'s eyes** *Abrus precatorius*; **c. grass** *Digitaria, Eleusine* & *Panicum* spp.; **c. oil & wood** *Carapa guianensis*

Crabbea Harvey. Acanthaceae. 12 trop. & S Afr.

Cracca Benth. = *Coursetia*

crackerberry *Cornus canadensis*

Cracosna Gagnepain. Gentianaceae. 1 SE As.

Craibia Harms & Dunn. Leguminosae (III 6). 10 trop. Afr. Some good timber

Craibiodendron W.W. Sm. Ericaceae. 5 SE As. R: JAA 67(1986)441

Craigia W.W. Sm. & W.E. Evans. Tiliaceae. 1 SW China

Crambe L. Cruciferae. 20 Euras. (Eur. 8), Medit., Macaronesia, trop. Afr. mts. R: Pfl. 79(4, 105)(1919) 228. Cult.: *C. cordifolia* Steven (Caucasus) – lvs to 60 cm across, infl. to 2 m; *C. maritima* L. (sea kale, coasts W Eur. to SW As.) – grown for succ. blanched shoots, boiled & cooked; *C. hispanica* L. (Med.) – comm. oilseed

Crambella Maire. Cruciferae. 1 Morocco

cramp bark *Viburnum opulus*

cranberry *Vaccinium oxycoccos*, (Am.) *V. macrocarpon*

crane flower *Strelitzia reginae*

cranesbill *Geranium* spp.; **bloody c.** *G. sanguineum*; **meadow c.** *G. pratense*; **shining c.** *G. lucidum*

Cranichis Sw. Orchidaceae (III 3). 60 warm Am.

Craniolaria L. Pedaliaceae (Martyniaceae). 3 S Am. *C. annua* L. (N S Am.) – fleshy roots form. eaten, medic.

Craniospermum Lehm. Boraginaceae. 4 temp. As.

Craniotome Reichb. Labiatae (VI). 1 Himalaya

Cranocarpus Benth. Leguminosae (III 9). 3 Brazil

crape fern *Todea barbara*; **c. myrtle** *Lagerstroemia indica*

Craspedia Forster f. Compositae (Gnap.-Ang.). c. 11 temp. Aus., NZ. R: OB 104(1991)111. Cult. orn. esp. *C. uniflora* Forster f. (billy buttons)

Craspedolobium Harms. Leguminosae (III 6). 1 W China

Craspedophyllum (C. Presl) Copel. = *Hymenophyllum*
Craspedorhachis Benth. Gramineae (33b). 2 S trop. Afr.
Craspedosorus Ching & W.M. Chen. Thelypteridaceae. 1 Yunnan
Craspedospermum Airy Shaw = *Craspidospermum*
Craspedostoma Domke. Thymelaeaceae. 5 S Afr.
Craspidospermum Bojer ex A.DC. Apocynaceae. 1 Madagascar
Crassocephalum Moench. Compositae (Sen.). 24 warm Afr. to Yemen & Masc., *C. crepidioides* (Benth.) S. Moore (thickhead (Aus.)) an aggressive weed in Mal., now pantrop.
Crassula L. Crassulaceae. c. 200 almost cosmop. (Eur. 4, NW 13 (R: KB 39(1984)699)), esp. trop. & S (144, R: CBH 8(1977)) Afr. Succ. herbs, shrubs & treelets, divisible into 6 sections; many cult. orn. (incl. *Rochea*) esp. *C. arborescens* (Miller) Willd. (S Afr.) – familiar shrubby housepl., rarely flowering in cult.; *C. muscosa* L. (*C. lycopodioides*, S Afr.) – lvs arr. like clubmoss; *C.perfoliata* L. var. *falcata* (Wendl.) Toelken (Cape Prov., 'rochea') – lvs grey, sickle-shaped, 2-ranked, fls bright red; Many xeromorphic features, e.g. *C. perfoliata* with some epidermal cells swollen above rest into large bladders which meet over whole surface dead & air-filled when leaf mature, walls infiltrated with silica but *C. aquatica* (L.) Schönl. (N temp.) aquatic ann., while *C. pageae* Toelken (*Pagella archeri*, S Afr.) a liverwort-like ann. & *C. cloisiana* (Gay) Reiche (Arg.) also described in Podostemaceae; *C. helmsii* (Kirk) Cockayne (Aus., NZ) – outcompeting (introd.!) *Elodea* spp. in some UK sites & (1996) pestilential in 320 parks & nature reserves in England alone
Crassulaceae J. St-Hil. Dicots – Rosidae – Rosales. 33/1100 almost cosmop. esp. S Afr., exc. Aus. & W Pacific. Succ. shrubs & herbs, s.t. treelets, with crassulacean acid metabolism & often with red (anthocyanin) root-tips & crystals of calcium oxalate in parenchyma cells & unusual stem vasculature incl. cortical &/or medullary bundles. Lvs spiral to opp. or whorled, simple & usu. entire, often with hydathodes; stipules 0. Fls solit. or usu. in cymes, usu. bisexual, hypogynous to weakly perigynous usu. (3–)5(–6 or even ∞) -merous; K ± free, C usu. so, A usu. 2 × C in 2 whorls, less often same as & alt. with C, rarely basally connate, anthers with longit. slits, G as many as K or P, usu. free, with nectariferous appendage near base, ovules (1–)∞ per carpel, on submarginal placentas, anatropous, bitegmic. Fr. usu. head of follicles, capsule in *Diamorpha*; seeds small with ± copious oily proteinaceous endosperm. x = 4–22+ (2n to 540+)
Principal genera: *Crassula, Echeveria, Kalanchoe, Sedum, Sempervivum*
Often in arid habitats but also seen in rain forest & other moist sites (some *Crassula* spp. aquatic); closely allied to Saxifragaceae
Many cult. orn. esp. spp. of *Adromischus, Aeonium, Crassula, Echeveria, Kalanchoe* & *Villadia* under glass or as housepls, *Jovibarba, Sedum* & *Sempervivum* outside
+ **Crataegomespilus** Simon-Louis ex Bellair. Graft-hybrids between spp. of *Crataegus* & *Mespilus germanica*, known as Bronvaux medlars: layers of cells from 1 sp. overlying those of another but not as sharply distinguishable as formerly held
Crataegus L. Rosaceae (III). 186 (+ ?78) N temp. (Eur. 20) incl. many apomictic clones & hybrids form. considered spp. R: CJB 68(1990)2220. Usu. thorny decid. shrubs, hawthorns or 'thorns', the thorns in place of branches, the haws being pomes; some domatia. In GB *C. monogyna* Jacq. (open country, hedges) hybridizes with *C. laevigata* (Poiret) DC (woods), fls of both called may (smell of latter held to be like that of Great Plague, poss. origin in C19 of attribution of bad luck to bringing blossom indoors, though also associated with Virgin Mary), the first widely used for quickset or quickthorn hedging, young buds ed. ('bread and cheese' in country districts), dried fr. added to flour in Eur., wood a subs. for box; 'Biflora' ('Praecox', Glastonbury Thorn, poss. hybrid between them) flowering in winter as well as spring, legend of origin from staff of Joseph of Arimathea dating from 1714. Others cult. incl. *C. azarolus* L. (azarole, E Med.) – ed. apple-flavoured haws ('Medit. medlars'); *C. douglasii* Lindley (black haw, W N Am.) – fr. used in jellies; *C. pentagyna* Waldst. & Kit. (Chinese haw, E Eur. to Iran) – cult. fr. tree in China; *C. phaenopyrum* (L.f.) Medikus (Washington thorn, NE Am.); *C. pubescens* (Kunth) Steudel (Mexican hawthorn, Mex.) – fr. for stock; *C. stipulosa* (Kunth) Steudel (manzanilla, C Am.) – fr. ed.
× **Crataemespilus** Camus. Sexual hybrids between *Crataegus* & *Mespilus* spp.
Crataeva L. See *Crateva*
Crateranthus Baker f. Lecythidaceae (Napoleonaeaceae). 3 trop. W Afr.
Craterispermum Benth. Rubiaceae (III 7). 16 trop. Afr. & Madag. to Seychelles R: KB 28(1974)434
Craterocapsa Hilliard & B.L. Burtt. Campanulaceae. 4 C & S Afr.
Craterogyne Lanj. = *Dorstenia*

Craterosiphon Engl. & Gilg. Thymelaeaceae. 9 trop. Afr.
Craterostemma Schumann = *Brachystelma*
Craterostigma Hochst. Scrophulariaceae. 9 Yemen, trop. & S Afr., India. R: TSP 81(1992)85. Allied to *Lindernia*; some cult. orn. e.g. *C. plantagineum* Hochst. (E & S Afr., 'blue gem') – cold treatment in Uganda. See also *Crepidorhopalon*
Crateva L. Capparidaceae. 6 trop. Garlic pear. Some leaf veg. & cult. orn. esp *C. religiosa* Forster f. (barna, OW) in trop., medic. *C. benthamii* Eichler – obligate fish-disp. in Amaz.
Cratoxylum Blume. Guttiferae (I). 6 Indomal. R: Blumea 15(1967)453. Distyly in *C. formosum* (Jack) Dyer (medic.); *C. arborescens* (Vahl) Blume (gerongong) – soft wood for clogs, dayak drums, shingles
crattock *Ficus racemosa*
Cratylia Mart. ex Benth. Leguminosae (III 10). 5 S Am.
Cratystylis S. Moore. Compositae (Cichorioideae). 3 arid/semi-arid W & S Aus. Dioec. to subdioec. *C. conocephala* (F. Muell.) S. Moore imp. fodder pl.
craw-craw (plant) *Senna alata*
Crawfurdia Wallich = *Gentiana*
crazyweed *Oxytropis* spp.
Creaghiella Stapf = *Anerincleistus*
Creatantha Standley = *Isertia*
creeper, canary *Tropaeolum peregrinum*; **Virginia c.** *Parthenocissus quinquefolia*
creeping Jenny *Lysimachia nummularia*
Cremanthodium Benth. (~ *Ligularia*). Compositae (Sen.-Tuss.). 70 Himal., S China. R: JLSBot 48(1929)259. Cult. orn., fragrant fls
Cremaspora Benth. Rubiaceae (II 1). 3–4 trop. Afr., Comoro Is. *C. triflora* (Thonn.) Schumann (*C. africana*) – frs source of blue-black body-dye
Cremastogyne (H. Winkler) Czerep. = *Betula*
Cremastopus P. Wilson. Cucurbitaceae. 2 C Am.
Cremastosciadium Rech.f. = *Eriocycla*
Cremastosperma R.E. Fries. Annonaceae. 19 trop. S Am. R: AHB 10(1930)46
Cremastra Lindley. Orchidaceae (V 8). 2 E As. R: NJB 8(1987)197. *C. appendiculata* (D. Don) Mak. (*C. variabilis*, *C. wallichiana*) – roots used by Ainu for toothache
Cremastus Miers = *Arrabidaea*
Cremnophila Rose (~ *Sedum*). Crassulaceae. 2 Mex. R: CSJ 50(1978)139
Cremnophyton Brullo & Pavone. Chenopodiaceae (I 3). 1 Malta, Gozo, Sicily: *C. lanfrancoi* Brullo & Pavone. R: Cand. 42(1987)621
Cremnothamnus Puttock (~ *Ozothamnus*). Compositae (Gnap.). 1 NW Aus.: *C. thomsonii* (F. Muell.) Puttock. R: AusSB 7(1994)569
Cremocarpon Baillon. Rubiaceae (IV 7). 1 Comoro Is.
Cremolobus DC. Cruciferae. 7 Andes
Cremosperma Benth. Gesneriaceae. 23 N Andes
Crenea Aublet. Lythraceae. 2 trop. S Am., Trinidad. R: Caldasia 15(1986)121. Salt water
Crenias Sprengel (*Mniopsis*). Podostemaceae. 5 SE Brazil
Crenidium Haegi. Solanaceae (2). 1 W Aus.: *C. spinescens* Haegi
Crenosciadium Boiss. & Heldr. (~ *Opopanax*). Umbelliferae (III 10). 1 Turkey
Creochiton Blume. Melastomataceae. 6 C & E Mal.
creole tea *Sauvagesia erecta*
creosote bush *Larrea divaricata* subsp. *tridentata*
crepe flower *Lagerstroemia indica*; **c. jasmine** *Tabernaemontana divaricata*; **c. myrtle** *L. indica*
Crepidiastrum Nakai (~ *Ixeris*). Compositae (Lact.-Crep.). 15 E As.
Crepidomanes (C. Presl) C. Presl (~ *Trichomanes*). Hymenophyllaceae (Hym.). (Incl. *Gonocormus*, *Pleuromanes*, *Microtrichomanes*, *Reediella*, *Vandenboschia*) 120 Madag. to Japan & Tahiti, some Am. Buds on rachis of some spp.
Crepidopteris Copel. = praec.
Crepidorhopalon E. Fischer (~ *Craterostigma*). Scrophulariaceae. 24 trop. Afr. R: TSP 81(1992)126
Crepidospermum Hook.f. Burseraceae. 5 trop. Am. R: Brittonia 39(1987)51
Crepinella Marchal = *Schefflera*
Crepis L. Compositae (Lact.-Crep.). c. 200 N hemisph. (Eur. 70), S Afr., S Am. R: UCPB 21(1947)199. Mostly weedy with yellow fls (hawk's beard) incl. *C. capillaris* (L.) Wallr. (Eur.) & *C. vesicaria* L. (Eur.) – extracts inhibit *Staphylococcus aureus* growth; few cult. esp. *C. rubra* L. (E Eur.) – pink fls
Crescentia L. Bignoniaceae. 6 trop. Am. R: FN 25(1980)82. *C. cujete* L. (calabash tree) –

cauliflorous with gourd-like berries bearing nectaries thought to be ant-attractants, the ants warding off herbivores; woody pericarp used liked gourds as bowls, scoops etc. & form. with eye-holes used to camouflage swimming hunters who pulled down individual birds without disturbing the flock (Columbus); young fr. pickled (like walnuts), seeds ed. cooked & used to make a drink in Nicaragua; *C. amazonica* Ducke (Amaz.) – seeds rot in fr. unless disp. by characin fish; *C. portoricensis* Britton (W Puerto Rico) – known from 4 adults & some juveniles

cress *Lepidium sativum* (**common** or **garden c.**, c. of '**mustard & c.**', cvs incl. **Aus.** or **golden c.**), though often replaced by faster-growing *Brassica napus*; **American c.** *Barbarea verna*; **Brazilian c.** *Acmella oleracea*; **hoary c.** *Lepidium draba*; **Indian c.** *Tropaeolum majus*; **land c.** *B. verna*; **meadow c.** *Cardamine pratensis*; **Pará c.** *Acmella oleracea*; **penny c.** *Thlaspi arvense*; **rock c.** *Arabis* spp.; **shepherd's c.** *Teesdalia nudicaulis*; **violet c.** *Ionopsidium acaule*; **wall c.** *Arabis* spp.; **waterc.** *Rorippa nasturtium-aquaticum* & hybrids; **winter c.** *Barbarea verna*

Cressa L. Convolvulaceae. 1 trop. & warm inc. Eur.: *C. cretica* L. in arid saline sites, a tonic in Sudan

crested dog's-tail *Cynosurus cristatus*

Creusa P.V. Heath = *Crassula*

Cribbia Senghas (~ *Rangaeris*). Orchidaceae (V 16). 1–2 Afr.

Criciuma Söderstrom & Londoño (~ *Bambusa*). Gramineae (1b). 1 Brazil: *C. asymmetrica* Söderstrom & Londoño. Climber. R: AJB 74(1987)35

crin végétale *Chamaerops humilis*

Crinipes Hochst. Gramineae (25). 2 Sudan, Ethiopia, Uganda. R: KB 12(1957)54

Crinitaria Cass. = *Aster*

crinkle bush *Lomatia silaifolia*

Crinodendron Molina. Elaeocarpaceae. 4 temp. S Am. R: SB 16(1991)77. Cult. orn. esp. *C. hookerianum* C. Gay (Chile, 'Chile lantern-tree') – with red fls, the buds beginning to expand the season prior to anthesis

Crinonia Blume = *Pholidota*

Crinum L. Amaryllidaceae (Liliaceae s.l.). c. 120 trop. & warm. R: Herbertia 9 (1942)63. Bulbous pls with spirally arr. lvs & consp. fls maturing together. Alks. Many cult. orn. (Cape lilies), c. 10 spp. aquatic. *C. asiaticum* L. (trop. As.) – widely cult. trop., the seed with layer of cork over endosperm aiding water-disp. (fr. floating for 1–2 weeks, lvs a rheumatism compress in Sumatra; *C. flaccidum* Herbert (Darling lily, Aus.) & *C. kirkii* Baker (pyjama l., E Afr.) – cult. orn.; *C.* × *powellii* Baker (*C. bulbispermum* (Burm.f.) Milne-Redh. & Schweick. (S Afr.) × *C. moorei* Hook.f. (S Afr.)) – many cvs; some locally medic.

Crioceras Pierre. Apocynaceae. 1 Gabon to Angola

Criogenes Salisb. = *Cypripedium*

Criosanthes Raf. = *Cypripedium*

Crispiloba Steenis. Alseuosmiaceae. 1 Queensland

Cristaria Cav. Malvaceae. 75 temp. S Am.

Cristatella Nutt. (~ *Polanisia*). Capparidaceae. 2 C & S US

Critesion Raf. = *Hordeum*

Crithmum L. Umbelliferae (III 8). 1 Eur., maritime: *C. maritimum* L. (sea samphire) – lvs fleshy, form. much pickled (salty spicy taste)

Crithopsis Jaub. & Spach. Gramineae (23). 1 Libya & Crete to Iran

Critonia P. Browne (~ *Eupatorium*). Compositae (Eup.-Crit.). 43 trop. Am. R: MSBMBG 22(1987)295

Critoniadelphus R. King & H. Robinson (~ *Eupatorium*). Compositae (Eup.-Crit.). 2 C Am. R: MSBMBG 22(1987)299

Critoniella R. King & H. Robinson (~ *Eupatorium*). Compositae (Eup.-Crit.). 6 N Andes. R: MSBMBG 22(1987)341

Critoniopsis Schultz-Bip. Compositae (Vern.-Pip.). 28 S Am. esp. N Andes. R: Phytol. 46(1980)437

Crobylanthe Bremek. Rubiaceae (I 9). 1 Borneo

Crocanthemum Spach = *Halimium*

Crocidium Hook. Compositae (Sen.-Blen.). 1 NW N Am.

Crockeria E. Greene ex A. Gray = *Lasthenia*

Crocodeilanthe Reichb.f. & Warsc. = *Pleurothallis*

Crocopsis Pax = *Stenomesson*

Crocosmia Planchon (~ *Tritonia*). Iridaceae (IV 3). 9 trop. & S Afr. R: JSAB 50(1984)463. Like *Tritonia* but fr. globose, not oblong, each locule with 3 or more (not 1 or 2) seeds. Dried fls in water give strong smell of saffron, *C. aurea* (Hook.) Planchon source of yellow dye.

Cult. orn. (incl. *Curtonus*) esp. *C.* × *crocosmiiflora* (Burb. & Dean) N.E. Br. (montbretia, *C. aurea* × *C. pottsii* (Baker) N.E. Br., raised 1880 in France)

Crocoxylon Ecklon & Zeyher. Celastraceae. 2 S trop. & S Afr.

Crocus L. Iridaceae (IV 3). c. 80 Medit. (Eur. 43) to W China. R: B. Mathew (1982) *The C.* Corms with cormlets forming in axils of scales; peduncle & ovary subterr. (cf. *Colchicum*), A 3, style with 3 stigmatic branches, P closing at night & in dull weather, visited by bees & Lepidoptera for nectar produced by ovary; anthers dehisce outwards so that insects contact pollen while probing for nectar. 2 subgenera, subg. *Crocus* & subg. *Crociris* (Schur) B. Mathew (only *C. banaticus* Gay (*C. iridiflorus*)). Many cult. orn. esp. yellow-flowered *C. flavus* Weston (*C. aureus*, E Med., long but now rarely cult. & much in gardens sterile & big yellow-flowered 'Dutch Yellow' or 'Golden Yellow' hybrid with *C. angustifolius* Weston (*C. susianus*, SW Russia)) & white to lilac *C. vernus* (L.) Hill (Dutch c., C & S Eur.); *C. chrysanthus* (Herbert) Herbert (E Med.) is the common winter-flowering sp. with many cvs (some being hybrids with *C. biflorus* Miller (S Eur., W As.)); *C. nudiflorus* Sm. (Pyrenees, natur. in GB on former properties of Knights of St John of Jerusalem where introd. as saffron subs.) & *C. speciosus* M. Bieb. (E Med. to Iran) – commonly seen autumn-flowering spp.; *C. sativus* L. (saffron, a sterile autotriploid cultigen poss. selected from *C. cartwrightianus* Herbert (Greece)) – source of saffron (10^6 fls to give 10 kg dried spice) form. imp. dye (alpha crocin) in Ancient Greece derived from stigmas, now largely prepared in Spain (5000 ha yielding 4.7×10^9 stigmas, 47 t spice per annum) for cooking & colouring foods esp. Spanish rice, bouillabaisse, while saffron cakes & loaves trad. in Cornwall, 'karkom' (origin of word crocus) of Song of Solomon, prob. that featured on Minoan pottery & frescoes (1600 BC), app. reducing arteriosclerosis poss. explaining low levels of cardiovascular disease in Spain where consumption is high (richest known source of vitamin B_2)

crocus, autumn *Colchicum* spp. esp. *C. autumnale*; **Chilean c.** *Tecophilaea cyanocrocus*; **Dutch c.** *Crocus vernus*; **saffron c.** *C. sativus*

Crocyllis E. Meyer ex Hook.f. Rubiaceae (IV 12). 1 S Afr.

Croftia Small = *Carlowrightia*

Crofton weed *Ageratina adenophora*

Croizatia Steyerm. Euphorbiaceae. 3 Panamá, Venez. R: SB 12(1987)1

Cromapanax Grierson. Araliceae. 1 Bhutan: *C. lobatus* Grierson. R: EJB 48(1991)19

Cromidon Compton. Scrophulariaceae. 12 S Afr. R: EJB 47(1990)320

Croninia J. Powell (~ *Leucopogon*). 1 W Aus.: *C. kingiana* (F. Muell.) J. Powell. R: Nuytsia 9(1993)125

Cronquistia R. King = *Carphochaete*

Cronquistianthus R. King & H. Robinson (~ *Eupatorium*). Compositae (Eup.-Crit.). 29 N Andes. R: Phytol. 70(1991)158

crookneck *Cucurbita pepo* etc.

Croomia Torrey. Stemonaceae (Croomiaceae). 3 Japan, E US. Vascular system of discontinuous cylinders

Croomiaceae Nakai. See Stemonaceae

Croptilon Raf. (~ *Haplopappus*). Compositae (Ast.-Sol.). 3 SE US. R: Sida 9(1981)59

crosnes *Stachys affinis*

cross of Jerusalem or **Maltese c.** *Silene chalcedonica*; **c.wort** *Cruciata laevipes*

Crossandra Salisb. Acanthaceae. c. 50 trop. Afr. & As., Arabia, Madag. Seeds flat, covered with hairs or fringed scales making seed sticky when wet. Some cult. orn. greenhouse shrubs esp. *C. infundibuliformis* (L.) Nees (*C. undulifolia*, Afr., S India, Sri Lanka) – fls bright orange, allegedly aphrodisiac; *C. stenostachya* C.B. Clarke (trop. Afr.) – pollen grains 520 × 19 μm!

Crossandrella C.B. Clarke. Acanthaceae. 2 trop. Afr.

Crosslandia W. Fitzg. Cyperaceae. 1 W Aus.

Crossonephelis Baillon = *Glenniea*

Crossopetalum P. Browne. Celastraceae. Incl. *Myginda* c. 50 trop. Am.

Crossopteryx Fenzl. Rubiaceae (I 1). 1 trop. & S Afr. Timber & febrifuge

Crossosoma Nutt. Crossosomataceae. 2 SW N Am.

Crossosomataceae Engl. Dicots – Rosidae – Rosales. 3/11 SW N Am. Glabrous xeromorphic shrubs mostly growing on rhyolite. Lvs decid. or s.t. marcescent, simple, spirally arr. (opp. in *Apacheria*); stipules minute or 0. Fls solit., s.t. unisexual, shortly perigynous, the disk forming a nectary to which A attached or annular within A; K (3)4 or 5(6), C (3)4 or 5(6), white & imbricate, A to c. 50 in 3 or 4 whorls associated with c. 10 trunk bundles arising centrifugally or centripetally, in 2 whorls in *Apacheria* & some spp. of *Forsellesia* but ± reduced to 1 whorl (ante-C whorl lost) in rest of *F.*, anthers with longit. slits, G 1–5(–9) each with (1)2–∞ ovules on marginal placenta, amphitropous or campylotropous, bitegmic. Fr. follicular; seeds arillate with thin to copious oily endosperm. n = 12

Genera: *Apacheria, Crossosoma, Forsellesia*

Seed-coat anatomy like *Podophyllum*; aril & overall floral structure suggest Dilleniales, though Cronquist suggests the perigyny imp. in placement in Rosales: in both orders C. are isolated

Crossostemma Planchon ex Hook. Passifloraceae. 1 trop. Afr.

Crossostephium Less. (~ *Artemisia*). Compositae (Anth.-Art.). 1 E As.: *C. chinense* (L.) Makino (*C. artemisioides*) – hairs from young lvs coll. in China & used like cotton wool (moxa)

Crossostylis Forster & Forster f. Rhizophoraceae. 10 W Pacific

Crossothamnus R. King & H. Robinson (~ *Eupatorium*). Compositae (Eup.-Alom.). 2 Peru. R: MSBMBG 22(1987)262

Crotalaria L. Leguminosae (III 28). c. 600 trop. & subtrop. (Afr. & Madag. with 511, R: R.M. Polhill (1982) *C. in Afr. & Mad.*). Alks; fr. an inflated dehiscent legume (rattlepod (Aus.)). Fodders & fibres esp. *C. juncea* L. (Bombay, Madras, sann or sunn hemp, origin unknown) – fibre for cordage, canvas, fishing-nets (more durable than jute) & cigarette-papers; *C. burkeana* Benth. (S Afr.) – major cause of crotalism; *C. micans* Link (*C. anagyroides*, trop. Am.) – cult. OW as green manure; *C. spectabilis* Roth (Indomal.) – cult. for fodder in NW; *C. usaramoensis* Baker f. (S Am.) – green manure & cover crop in China

croton *Codiaeum* spp. esp. *C. variegatum*; **c. oil** *Croton tiglium*

Croton L. Euphorbiaceae. c. 750 trop. & warm (Afr. 50). R: Taxon 42(1993)793. Monoec. & dioec. herbs, shrubs, some rheophytes, & trees, usu. with stellate hairs; some may cause contact dermatitis while seeds of others suggested to promote tumours. Many alks. Some timbers, teas but esp. medic. (fever bark). *C. bonplandianus* Baillon (Brazil) – weedy in As. (introd. Chittagong 1897–8 & all over India in 90 yrs), leachate allelopathic to weedy associates; *C. eluteria* (L.) Wright (WI) – source of cascarilla bark used as tonic & bitters, flavouring for liqueurs & scenting tobacco (1–5% volatile oil incl. eugenol, vanillin etc.); *C. laccifer* L. (S & SE As.) – host-pl. for lac-prod. insects of imp. in varnish-making; *C. malambo* Karsten (malambo, Venez.) – bark med.; *C. megalocarpus* Hutch. (musine, E Afr.) – timber & shade-tree; *C. niveus* Jacq. (trop. Am.) – copalchi bark (Mex.) subs. for *C. eluteria*; *C. scouleri* Hook.f. (Galápagos) – capsules crushed by 2 spp. of finch & at least 1 seed always dropped, 2 other finch spp. forage but miss some, which can then germinate; *C. setigerus* Hook. (W N Am.) – natur. Aus. (dove-weed); *C. texensis* (Klotzsch) Muell. Arg. (skunkweed, N Am.) – tea & insecticide; *C. tiglium* L. (trop. As.) – source of croton oil, one of most purgative substances known, itself not tumour-inducing but when applied with a subeffective carcinogen it is & may account for high level of oesophageal cancer in China (*C. flavens* L. may do same in WI), phorbol myristate acetate the most active skin-irritant

Crotonogyne Muell. Arg. Euphorbiaceae. 1 trop. Afr.

Crotonogynopsis Pax. Euphorbiaceae. 1 trop. Afr.

Crotonopsis Michaux = *Croton*

crow('s) ash *Flindersia australis*; **c.berry** *Empetrum nigrum*; **c.foot** *Ranunculus* spp., *Dactyloctenium* spp. (N Am.); **c.f. grass** *Eleusine indica*; **c. poison** *Zigadenus densus*

crowa *Ananas comosus*

Crowea Sm. Rutaceae. 3 S Aus. R: Nuytsia 1(1970)15. Protected spp.

crown beard *Verbesina* spp.; **c. gum** *Manilkara chicle*; **c. imperial** *Fritillaria imperialis*; **c. of thorns** that of Christ prob. *Ziziphus spina-christi*, poss. *Sarcopoterium spinosum*, but now applied to *Euphorbia milii*

Crucianella L. Rubiaceae (IV 16). c. 30 Eur. (8), Medit. to Iran & C As.

Cruciata Miller. Rubiaceae (IV 16). 10 Eur. (5) to E Med. *C. laevipes* Opiz (crosswort) in GB

Crucicaryum Brand. Boraginaceae. 1 New Guinea

Cruciferae Juss. (Brassicaceae). Dicots – Dilleniidae – Capparales. 365/3250 cosmop. esp. temp. (particularly Med. to C As., W N Am.). Herbs, rarely shrubby when often pachycaul, with mustard-oil glucosides (glucosinolates effective in defence against bacteria, fungi, insects & mammals, sinigrin (glycoside) liberating allyl isothiocyanate, an injurious vesicant oil) & often cyanogenic, stem often with unusual vasculature; mycorrhizae rare. Lvs spiral, rarely opp., simple to pinnately dissected, rarely with articulated leaflets; stipules 0. Fls in usu. bractless racemes, rarely solit., bisexual (rarely dioec. or gynodioec.: see *Cheesemania, Hirschfieldia, Lepidium*) & usu. regular, hypogynous, receptacle rarely prolonged into gynophore (cf. Capparidaceae) but often with nectaries (s.t. forming ring around A or G); K 2 + 2 (v. rarely (K)), decussate, outer often gibbous basally, C 4 diagonal to K, imbricate or convolute, usu. with elongate claw, rarely 0 (e.g. *Pringlea*), A 2 short + 4 longer (tetradynamous), inner derived from only 2 primordia & s.t. basally connate in pairs (2–4 in *Lepidium*, up to 16 in *Megacarpaea*), anthers with longit. slits, G (2), ± style, ovary with 2 locules sep. by a thin replum connecting 2 parietal placentas, each with (1–) ∞ ovules in 2 rows sep. by replum, anatropous to campylotropous, bitegmic, endotestal (cf. exotegmic fibres of Capparidaceae); nectar hexose-rich. Fr. dry-dehiscent, usu. a siliqua (elongate) or silicula (short), valves falling to reveal replum; seeds with large oily embryo, folded, with the cotyledons often lying against radicle and 0 or little (often 1 layer of cells) endosperm. x = 5–12+

Principal genera: *Aethionema, Alyssum, Arabis, Biscutella, Cardamine, Descurainia, Draba, Erysimum, Heliophila, Lepidium, Lesquerella, Malcolmia, Matthiola, Rorippa, Thlaspi*

There have been several attempts at tribal arrangement of the genera but none is altogether satisfactory: features of indumentum, nectaries, fr. & embryo etc. have been tried, but the most often used system has many unnatural groupings & is not supported by seed structure studies. The family is relatively clear-cut but the limits of many genera are vague with hundreds of monotypic satellites: Krause suggested lumping the lot in a single genus *Crucifera*! Despite seed & other diffs, the C. are generally considered close to Capparidaceae in which they seem to be 'nested' such that fams should poss. be combined (cf. Umbelliferae & Araliaceae)

Many salads (cresses (*Barbarea, Lepidium, Sinapis*), rocket (*Eruca*)) & other vegetables incl. *Brassica, Raphanus* & of lesser imp. *Armoracia, Crambe, Rorippa* (watercress) & *Wasabia*, oilseeds (*Brassica, Descurainia*) & many cult. orn.: *Aethionema, Alyssum, Arabis, Aubrieta, Aurinia, Erysimum* (wallflowers), *Heliophila, Hesperis, Iberis, Ionopsidium, Lobularia, Lunaria, Malcolmia, Matthiola* (stock); *Isatis* gives woad & *Camelina* a fibre; weeds incl. spp. of *Arabidopsis, Capsella* & *Cardamine*

Although common in open dry habitats, C. include the aquatic *Subularia* & world alt. record (*Christolea*). R: J.G. Vaughan *et al.* (1976) *The biology & chemistry of the C.*

Cruckshanksia Hook. & Arn. Rubiaceae (IV 1). 7 Chile

Cruddasia Prain = *Ophrestia*

Crudia Schreber. Leguminosae (I 4). c. 55 trop., esp. riverine forests, some water-disp. in Amazon (air-filled cavity between cotyledons). R: BJBBuit. 18(1950)407

cruel plant *Araujia sericifera*

Crumenaria Mart. Rhamnaceae. 6 C Am. to Argentina

crummock *Sium sisarum*

Crunocallis Rydb. (~ *Montia*). Portulacaceae. 2 Aleutians to SW US & Minnesota

Crupina (Pers.) DC. Compositae (Card.-Cent.). 3 Med. (Eur. 2) & SW As. to China

Crusea Cham. & Schldl. Rubiaceae (IV 15). 13 SW N & C Am. R: MNYBG 22,4(1972)1

Cruzia Philippi = *Scutellaria*

cry-baby *Erythrina crista-galli*

Crybe Lindley. Orchidaceae. 1 C Am.

Cryophytum N.E. Br. = *Mesembryanthemum*

Cryosophila Blume. Palmae (I 1a). 8 W Mex. to N Colombia. Some cult. orn. often under name *Acanthorrhiza* – root-spines covering base of trunk. *C. argentea* Bartlett (escoba palm, C Am.) – often an indicator of *Swietenia macrophylla* presence

Cryphia R. Br. = *Prostanthera*

Cryphiacanthus Nees = *Ruellia*

Crypsinopsis Pichi-Serm. = *Selliguea*

Crypsinus C. Presl = *Selliguea*

Crypsis Aiton. Gramineae (31d). 8 Med. (Eur. 5) to N China, trop. Afr. R: BRCI 11D(1962)91. Seed extruded from fr.

Cryptadenia Meissner. Thymelaeaceae. 5 Cape

Cryptandra Sm. Rhamnaceae. 40 temp. Aus.

Cryptangium Schrader ex Nees = *Lagenocarpus*

Cryptantha Lehm. ex G. Don f. Boraginaceae. 100 W N Am. Cult. orn.

Cryptanthemis Rupp = *Rhizanthella*

Cryptanthopsis Ule = *Orthophytum*

Cryptanthus Osb. = *Clerodendrum*

Cryptanthus Otto & A. Dietr. Bromeliaceae. 21 E Brazil. R: FN 14(1979) 1586. Cult. orn. dwarf pls (earth-stars), esp. *C. zonatus* (Vis.) Beer & *C. bivittatus* (Hook.) Regel (only known in cult.), also hybrids incl. with spp. of *Billbergia*

Cryptarrhena R. Br. Orchidaceae (V 10). 4 trop. Am. R: Lindleyana 8(1993)163

Crypteronia Blume. Crypteroniaceae. 4 SE As., Mal. Young shoots of *C. paniculata* Blume (Mal.) eaten with rice, timber useful

Crypteroniaceae A. DC. Dicots – Rosidae – Myrtales. 3/10 trop. As., S Afr. & S Am. Trees often with quadrangular twigs & accumulating aluminium. Lvs opp., simple, entire with a continuous marginal vein; stipules minute. Fls in axillary racemes to panicles, v. small, bisexual (rarely trees dioec.), reg., often perigynous; K (4 or)5, valvate & often persistent, C 0 or (4 or)5 rudimentary, enveloping A (4 or 8)5 or 10, $\underline{G}$ or $\overline{G}$ (2–4(5)), 1–6-loc. with term. style, ovules 1, 2, 3 or ∞ per loc., anatropous on axile placentas. Fr. a loculicidal capsule, the valves often held together apically by persistent style; seeds usually small, flat with membranous wing & 0 endosperm

Possibly heterogeneous, genera: *Alzatea*, *Axinandra*, *Crypteronia*, *Dactylocladus* *Rhynchocalyx* excl. to Rhynchocalycaceae; *Alzatea* sometimes removed to own fam. See also under Melastomataceae

Cryptocapnos Rech.f. Fumariaceae (I 2). 1 Afghanistan: *C. chasmophytica* Rech.f. R: OB 88(1986)91

Cryptocarpus Kunth. Nyctaginaceae (I). 1 W S Am., Galápagos: *C. pyriformis* Kunth

Cryptocarya R. Br. Lauraceae (I). (Incl. *Ravensara*) c. 200 trop. & warm. Alks (touching bark of *C. glabella* Domin, poison walnut, Aus., may lead to hospitalization). *C. aromatica* (Becc.) Kosterm. (New Guinea) – essential oil from aromatic bark (massoy b.); *C. glaucescens* R. Br. (brown beech, jackwood, E Aus.) – good timber; *C. hornei* Gillespie (Polynesia) – timber for handicrafts in Tonga; *C. latifolia* Sonder (S Afr.) – source of ntonga nuts used locally for their oil; *C. moschata* Nees & Mart. (Brazil) – fr. (Brazilian nutmegs) used as spice; C. sp. (*Ravensara aromatica* Sonn., Madagascar cloves or nutmeg, Madag.) – seeds a spice, bark used in local rum-making

Cryptocentrum Benth. Orchidaceae (V 10). 19 trop. Am. R: BJ 97(1977)562

Cryptocereus Alexander = *Selenicereus*

Cryptochilus Wallich. Orchidaceae (V 14). 6 Himal. Bird-poll.?

Cryptochloa Swallen. Gramineae. 10 trop. Am. Bamboos, some ant-disp. (elaiosomes)

Cryptocodon Fed. (~ *Asyneuma*). Campanulaceae. 1 C As.

Cryptocoryne Fischer ex Wydler. Araceae (VIII 6). 50 Indomal. Marsh & water pl., often stoloniferous, with spathe-length governed by water-depth (open to air apically & entered by poll. beetles detained overnight), fr. underwater, seeds with cotyledons as floats dropped after 2 mins, some, e.g. *C. ciliata* Blume, viviparous; many cult. orn. aquarium pls esp. *C. spiralis* (Retz.) Wydler (Indian ipecacuanha, India) & *C. walkeri* Schott (Sri Lanka)

Cryptodiscus Schrenk ex Fischer & C. Meyer = *Prangos*

Cryptogramma R. Br. Pteridaceae (IV). 2 alpine & boreal (incl. Eur.). *C. crispa* (L.) R. Br. (parsley fern) – cult. orn.

Cryptogrammaceae Pichi-Serm. = Pteridaceae

Cryptogyne Hook.f. = *Sideroxylon*

Cryptolepis R. Br. Asclepiadaceae (I; Periplocaceae). c. 26 OW trop.

Cryptomeria D. Don. Taxodiaceae. 2+ China & Japan. *C. japonica* (L.f.) D. Don (Japanese cedar) – much grown for timber (sugi) in Japan, esp. used for boats, much cult. orn. with many cvs incl. dwarf forms derived from witches' brooms, the type natur. in Azores, where it (unlike trees in Japan) produces suckers; old wood buried in ground in Japan becomes dark green (jindai-sugi)

Cryptophoranthus Barb. Rodr. = *Pleurothallis*

Cryptophragmium Nees = *Gymnostachyum*

Cryptophysa Standley & J.F. Macbr. = *Conostegia*

Cryptopus Lindley. Orchidaceae (V 16). 4 Madagascar & Mascarenes

Cryptopylos Garay. Orchidaceae (V 16). 1 SE As., Sumatra: *C. clausus* (J.J. Sm.) Garay. R: OB 95(1988)266

Cryptorhiza Urban = *Pimenta*

Cryptosepalum Benth. Leguminosae (I 4). 11 trop. Afr. C 1, A 3

Cryptospora Karelin & Kir. Cruciferae. 3 C As.

Cryptostegia R. Br. Asclepiadaceae (I; Periplocaceae). 2 Madag. R: AusSB 4(1991)571. *C. grandiflora* R. Br. – cult. orn. liane with (Madag.) rubber comparable with that of *Hevea*, widely natur. trop., bad weed in Aus.

Cryptostemma R. Br. = *Arctotheca*

Cryptostephanus Welw. ex Baker. Amaryllidaceae (Liliaceae s.l.). 5 trop. Afr. Berry

Cryptostylis R. Br. Orchidaceae (IV 1). 20 Indomal. to W Pacific (Aus. 5) & Taiwan. Fls not resupinate. Aus. spp. with pseudocopulation (same ichneumon, *Lissopimpla semipunctata*). Cult. orn. incl. *C. subulata* (Lab.) Reichb.f. (Aus.) – recently self-introd. NZ

Cryptotaenia DC. Umbelliferae (III 8). 6 N temp. & trop. Afr. mts. *C. canadensis* (L.) DC (E As., N Am., natur. Austria) – cult. Japan for salad (Japanese parsley, mitsuba or mitzuba) & fried roots

Cryptotaeniopsis Dunn = *Pternopetalum*

Cryptothladia (Bunge) M. Cannon = *Morina*

Ctenanthe Eichler (~ *Myrosma*). Marantaceae. c. 20 Brazil, Costa Rica (1). Cult. orn. like *Calathea*

Ctenardisia Ducke. Myrsinaceae. 5 trop. Am. R: Wrightia 7(1982)42

Ctenitis (C. Chr.) C. Chr. (*Atalopteris*). Dryopteridaceae (I 3). c. 150 trop. & warm (not Aus.). R: Blumea 31(1985)1)

Ctenium Panzer. Gramineae (33b). 17 Am., Afr. (R: KB 16(1963) 471), Madag. Savannas. *C. aromaticum* (Walter) Wood (toothache grass, Virginia to Florida) – fls only after fire

Ctenocladium Airy Shaw = *Dorstenia*

Ctenolepis Hook.f. Cucurbitaceae. 2 trop. Afr. & India

Ctenolophon Oliver. Linaceae (Ctenolophonaceae). 1 trop. Afr., 1 Mal. Some timbers

Ctenolophonaceae (H. Winkler) Exell & Mendonça. See Linaceae

Ctenomeria Harvey (~ *Tragia*). Euphorbiaceae. 2 S Afr.

Ctenopaepale Bremek. = *Strobilanthes*

Ctenophrynium Schumann = *Saranthe*

Ctenopsis De Notaris = *Vulpia*

Ctenopteris Blume ex Kunze = *Grammitis*

Cuatrecasanthus H. Robinson (~ *Vernonia*). Compositae (Vern.-Pip.). 3 Ecuador, Peru

Cuatrecasasiella H. Robinson (~ *Luciliopsis*). Compositae (Gnap.-Gnap.). 2 Andes. R: BJLS 106(1991)185

Cuatrecasasiodendron Standley & Steyerm. Rubiaceae (I 5). 2 Colombia

Cuatrecasea Dugand = *Iriartella*

Cuatresia Hunz. Solanaceae (1). 9 trop. Am. R: OB 92(1987)73. Shrubs & trees

Cuba bast *Hibiscus elatus*; **C. hemp** *Furcraea hexapetala*; **C. lily** *Scilla peruviana*

Cubacroton Alain. Euphorbiaceae. 1 E Cuba

Cubanola Aiello. Rubiaceae (I 7). 2 Cuba, Hispaniola

Cubanthus (Boiss.) Millsp. (~ *Pedilanthus*). Euphorbiaceae. 3 Cuba, Hispaniola

cubebs *Piper cubeba*; **African c.** *P. clusii*; **Guinea c.** *P. guineense*

Cubilia Blume. Sapindaceae. 1 C Mal.: *C. cubili* (Blanco) Adelb. (*C. blancoi*) – lvs a veg. & seeds (kubili nuts) ed., cult. Java

Cubitanthus Barringer (~ *Anetanthus*). Gesneriaceae. 1 Braz.: *C. alatus* (Cham. & Schldl.) Barringer. R: JAA 65(1984)145

Cuchumatanea Seidenschnur & Beaman. Compositae (Helia.-Gal.). 1 Guatemala. Montane annual less than 1 cm tall

cuckold *Bidens* spp.

cuckoo flower *Cardamine pratensis*; **c. pint** *Arum maculatum*

Cucubalus L. = *Silene*

cucumber *Cucumis sativus*; **bitter c.** *Momordica charantia*; **bur c.** *Sicyos angulata*; **c. root** *Medeola virginiana*; **squirting c.** *Ecballium elaterium*; **c. tree** *Averrhoa bilimbi*, *Magnolia acuminata*

Cucumella Chiov. Cucurbitaceae. 6 trop. Afr., 1 India

Cucumeropsis Naudin. Cucurbitaceae. 1 trop. W Afr.: *C. mannii* Naudin – seeds (egusi) source of oil & protein

Cucumis L. Cucurbitaceae. 32 OW trop. R: J.H. Kirkbride (1993) *Biosystematic monograph of the genus C.* (5 cross-sterile spp. groups in 2 subg.) Cult. since earliest times. *C. anguria* L. (WI gherkin, cultigen poss. derived from *C. a.* subsp. *longipes* (Hook.f.) Greb. (Afr.)) – young fr. boiled or pickled; *C. melo* L. (melon, orig. (?) W Afr.) – 2 subspp.: subsp. *agrestis* (Naudin) Greb. (wild forms with inedible fr., incl. senat seed once exported from Sudan as oilseed) & subsp. *melo* with 7 major interfertile cv. groups incl. Cantalupensis Group (cantaloupe (true) with rough-warty but not netted skin, not in comm. prod. in Am.), Inodorus Group (honeydew melon with smooth rind, casaba m. with wrinkled, flesh crisp), Reticulatus Group (muskmelon, 'cantaloupe' (not true), rock m. (NSW), with ± strongly netted rind & musky orange flesh, the most imp. in commerce), & others largely grown as 'ornamental gourds' & cooking forms resembling small marrows; *C. humifructus* Stent (trop. & S Afr.) – fr. underground excavated & disp. by water-seeking aardvarks avoiding dangerous waterholes, also disp. by Bushmen; *C. metuliferus* Naudin (kiwano, horned or jelly melon, trop. & S Afr.) – fashionable dessert fr. grown in NZ (cf. *Actinidia deliciosa*); *C. sativus* L. (cucumber, Sino-Himal., v. rare in wild) – prob. cult. by 'Hoabhinian' culture 8000–3000 BP, now eaten raw or pickled, most 'gherkins' being immature small-fruited cvs, face-cleansing cosmetics, flavour due to an aldehyde (nona-2,6-dienal) with odour threshold of 0.0001 p.p.m.!

Cucurbita L. Cucurbitaceae. 13 spp. or groups trop. & warm Am. R: EB 44(3) suppl. (1990)58. Formed part of the squash/beans/maize culture of pre-Columbian Am. Monoec.; 2 genera of solit. bees (*Peponapis* & *Xenoglossa*) derive all food from nectar & pollen of C. spp. Genus prob. native in S as well as C Am., *C. moschata* most like original sp. & domesticated independently in C & S Am. while the rest as follows: *C. maxima* (S Am., from *C. andreana* Naudin, Arg. & Uruguay), *C. ficifolia* (C Am. S of Mex.), *C. argyrosperma* (Mex. S of M. City, perhaps derived from *C. sororia* L. Bailey, C Am.) & *C. pepo* (Mex. N of M. City, perhaps derived from var. *texana* (Scheele) Decker and/or *C. fraterna* L. Bailey perhaps domesticated independently in E US (subsp. *ovifera*) & Mex. (subsp. *pepo*)). The terms 'squash' & 'pumpkin' are used for more than 1 sp.: *C. argyrosperma* Hort. Huber (*C. mixta*, winter squash, pumpkin, cushaw, silverseed gourd); *C. ficifolia* Bouché (cidra, sidra (Spain), Malabar gourd) – made into kind of marmalade; *C. foetidissima* Kunth (buffalo gourd, chilicote) – drought-tolerant, seeds source of oil & protein, roots laxative; *C. maxima* Duchesne ex Lam. (autumn & winter squash & pumpkin) – fr. globose (to 2 m round & 90(–300) kg) to oblong, many cvs (incl. 'Turbaniformis', cult. orn. Turk's Cap gourd), anthelmintic in Malta; *C. moschata* (Duchesne) Poiret (pumpkin, winter squash) – a keeping-squash like *C. maxima*, anthelmintic (a carboxypyrrolidine called cucurbitine); *C. pepo* L. subsp. *pepo* (vegetable marrow, v. spaghetti, summer & autumn squash & pumpkin (non-keeping)) – one of oldest domesticated pls (9000 yrs BP Mex.), to 47.85 kg (UK record), seeds ('pumpkin nuts', pepitas) ed. & rich in zinc, many cvs (8 groups incl. 'Zucchini' (zucchini, courgettes) ed. immature, 'Summer Crookneck' – club-shaped with curved neck & subsp. *ovifera* (L.) Decker – the hard-shelled coloured orn. gourds), also vermifuge

Cucurbitaceae Juss. Dicots – Dilleniidae – Violales. 120/775 trop. & warm with few temp. Usu. juicy climbers or trailers, rarely somewhat woody or aborescent (*Dendrosicyos*), often with coiled tendrils, 1 at each node, these rarely spines or 0 (*Ecballium*), freq. with bitter purgative cucurbitacins; stem vascular bundles usu. in 2 cycles & nearly always bicollateral, sometimes otherwise unusual too. Lvs spirally arr., usu. palmately-veined or -lobed, often with extrafl. nectaries; stipules 0. Fls axillary, solit. or not, unisexual (plants monoec. or dioec.), v. rarely bisexual, usu. reg. & epigynous; K (3–)5(6), imbricate or open, C (3–)5(6) or ((3–)5(6)) usu. yellow or white ± valvate & often diff. in males & females, A essentially 5 attached to hypanthium (rarely around summit of ovary), alt. with C, but usu. reduced or displaced so that often apparently A 3 (2 with dithecal anthers, 1 with monothecal), free or connate, anthers with longit. slits, $\overline{G}$ ((2)3(–5)) with intruded parietal placentas, s.t. joined to form pluriloc. ovary, monomerous in Cyclanthereae, style solit. with 1–3(–5) usu. bilobed stigmas, or 2 or 3, each with bilobed stigma, ovules (1–)∞, anatropous, bitegmic. Fr. usu. a berry (when hard-walled termed a pepo), less often a capsule (s.t. fleshy & explosive), rarely samaroid; seeds (1–)∞, large, often flattened & even winged, with oily straight embryo & ± 0 endosperm. x = 7–14. (Jeffrey in BJLS 81(1980)233)

Classification & chief genera (Jeffrey):

I. **Zanonioideae** (free styles, seeds usu. winged, tendrils apically bifid, n = 8): *Fevillea* (& c. 16 others, mostly perennial & dioec.)

II. **Cucurbitoideae** (style 1, seeds unwinged, tendrils basally 2–7-fid or simple) – 8 tribes:

1. **Melothrieae** (receptable-tube long, similar in males & females, pollen usu. reticulate, fls small, A usu. free with simple thecae): *Cucumis, Dendrosicyos, Gurania, Melothria* (& c. 30 others)
2. **Schizopeponeae** (receptable-tube usu. short, if long in male then short in female, pollen reticuloid, petals unfringed, ovules 1 or 2): *Schizopepon* (only)
3. **Joliffieae** (receptable-tube as above, petals fringed): *Momordica, Telfairia, Thladiantha* (& 2 others)
4. **Trichosantheae** (receptable-tube long, pollen striate or verrucose, fls large, A united, with triplicate thecae) – *Trichosanthes* (& 9 others)
5. **Benincaseae** (receptable tube short, petals without fringes, ovules ∞, pollen reticulate): *Bryonia, Citrullus, Ecballium, Lagenaria, Luffa* (& c. 11 others)
6. **Cucurbiteae** (receptable-tube, petals & ovules as above (or 1–few), pollen

spinose): *Cucurbita* (& 11 others all NW exc. 1 sp. of *Cayaponia* (W Afr.))

7. **Cyclanthereae** (receptable-tube, petals & ovules as above, pollen smooth): *Cyclanthera, Marah* (& c. 10 others all with explosive fr.)
8. **Sicyoeae** (receptable-tube & petals as above, ovule 1, pendulous, pollen spinulose) – *Sechium, Sicyos* (& c. 4 others all NW to Hawaii)

All have frost-sensitive aerial parts & perennial spp. in temp. regions have tubers (e.g. *Bryonia*); 1 of 10 fams with elaiophores attractive to bees (*Momordica, Thladiantha*)
Presence of cucurbitacins supports affinity with Begoniaceae & Datiscaceae; seed-structure differs markedly from that in Passifloraceae or Caricaceae with which they have been allied: they have no members 'intermediate' between C. & another family
Of enormous economic importance as foodpls (gourds, melons, marrows, squashes, pumpkins etc.): *Benincasa, Citrullus, Cucumis, Cucurbita, Lagenaria, Momordica, Sechium, Trichosanthes* etc., also loofah (*Luffa*), oilseeds & medic. (*Citrullus, Cucumeropsis, Ecballium, Fevillea, Solena*); fls of many ed. too while others provide 'ornamental gourds' in cult.

Cucurbitella Walp. Cucurbitaceae. 1 S Am.

cudjoe wood *Jacquinia keyensis*

Cudrania Trécul = *Maclura*

cudweed *Filago* spp., *Gnaphalium* spp.

Cuenotia Rizz. Acanthaceae. 1 NE Brazil

Cuervea Triana ex Miers (~ *Hippocratea*). Celastraceae. 3 trop. Am., 1 W Afr.

Cufodontia Woodson = *Aspidosperma*

cuichunchulli *Hybanthus parviflorus*

cuipo *Cavanillesia platanifolia*

Cuitlauzenia Llave & Lex. (~ *Odontoglossum*). Orchidaceae (V 10). 1 Mex.: *C. pendula* Llave & Lex. – cult. orn.

Culcasia Pal. Araceae (V 7). 20 trop. Afr. esp. W & C

Culcita C. Presl. Dicksoniaceae (Thyrs.). (Excl. *Calochlaena*) 2 trop. Am., Azores to Iberia (1), Mal., Aus., New Caled.

Culcitaceae Pichi-Serm. See Dicksoniaceae

Culcitium Bonpl. (~ *Senecio*). Compositae (Sen.-Sen.). 15 Andes. Superficially resembling *Espeletia*

culilawan *Cinnamomum culitlawan*

Cullen Medikus (~ *Psoralea*). Leguminosae (III 11). 35 OW trop. *C. corylifolia* (L.) Medikus (*P. c.*, India & Sri Lanka) – principal cult. medic. pl. of S Arabia used in India against leprosy

Cullenia Wight = *Durio*

Cullumia R. Br. Compositae (Arct.-Gort.). 15 Cape to Karoo. R: MBSM 3(1959)271

Culver's root *Veronicastrum virginicum*

Cumarinia F. Buxb. = *Coryphantha*

cumbungi (reed) *Typha* spp.

cumin *Cuminum cyminum*; **black c.** *Nigella sativa*

Cumingia S. Vidal = *Camptostemon*

Cuminia Colla. Labiatae (VIII 2). 1 Juan Fernandez

Cuminum L. Umbelliferae (III 3). 4 Medit. to Sudan & C As. *C. cyminum* L. (cumin, Medit.) – cult. since time of Minoans (C13 BC), the frs ('seeds') used to flavour cheese, cakes, liqueurs & curry powder, largely replaced by caraway

Cumulopuntia F. Ritter = *Opuntia*

cundorangao *Gonolobus cundurango*

Cunila Royen ex L. (*Hedyosmos*). Labiatae (VIII 2). 15 E N Am. to Uruguay. *C. origanoides* (L.) Britton (*C. mariana*, American dittany, E N Am.) – culinary herb, cunila oil med.

cunjevoi *Alocasia macrorhizos*

Cunninghamia R. Br. ex Rich. Taxodiaceae. 2 E As. (Tertiary of C Eur.). R: Phytol. M 7(1984)21. Timber of *C. lanceolata* (Lambert) Hook. (*C. sinensis*, China) used for house-building, boats etc. & much used in reafforestation

Cunonia L. Cunoniaceae. 1 S Afr., 16 New Caled. (poss. best united with *Weinmannia*). Buds protected by stipules. *C. capensis* L. (S Afr.) – cult. orn., wood used for furniture etc.

Cunoniaceae R. Br. Dicots – Rosidae – Rosales. (Incl. Baueraceae) 19/340 mostly S hemisph. esp. Aus., New Guinea, New Caled. Trees, shrubs (sometimes unbranched) or climbers (e.g. *Aphanopetalum*), often accumulating aluminium. Lvs opp. or whorled, pinnate or trifoliolate, rarely simple; stipules often large & consp. (0 in *Bauera*), often interpetiolar & prs connate, commonly with small colleters. Fls small in heads, racemes to thyrses, rarely solit. in axils, reg., bisexual (rarely plants dioec. or polygamodioec. as in *Pancheria*), usu.

hypogynous; K (3)4–5(–10), imbricate or valvate, s.t. basally connate, C alt. with K, usu. smaller, sometimes 0, A usu. in 2 whorls, 8–10 or in 1 opp. K, rarely >20, with slender filaments longer than C & anthers s.t. versatile, with longit. slits & v. small pollen grains, shallow nectary-disk s.t. around G, G̲ (rarely half-inf.) (2(3–5)), with separate styles, rarely carpels ± distinct, ovules (1)2–∞ per locule on axile to apical-axile placentas, usu. anatropous, bitegmic with zig-zag micropyle. Fr. usu. a capsule, the carpels s.t. separating & opening ventrally at least apically, rarely drupe-like, nut-like or follicular; seeds small, winged or hairy with thin testa & usu. small straight embryo embedded in copious oily endosperm. n = 12, 15, 16. R: SB 17(1992)181

Principal genera: *Caldcluvia, Cunonia, Geissois, Pancheria, Schizomeria, Spiraeanthemum, Weinmannia*

Bauera (0 stipules, axillary fls) has been placed in its own fam. or assigned on fr. chars. to Saxifragaceae, Grossulariaceae, etc. but pollen & embryo characters suggest its being here. C. app. the woody equivalent of Saxifragaceae but also differ in opp. lvs. Brunelliaceae & Eucryphiaceae may also belong here (DNA data)

Few cult. orn.; some timber (*Ceratopetalum*)

cunure *Calliandra tweedii*

Cunuria Baillon = *Micrandra*

cup and saucer plant *Holmskioldia sanguinea*

Cupania L. Sapindaceae. c. 45 warm Am. Some timber (loblolly)

Cupaniopsis Radlk. Sapindaceae. 60 C Mal. to N & E Aus. (1, endemic: *C. anacardioides* (A. Rich.) Radlk., tuckeroo), New Caled., Fiji & Samoa. R: LBS 15 (1991)

Cuphea P. Browne. Lythraceae. 260 Am. Lvs opp. or whorled, often with 1 fl. at each node but origin is in axil of leaf below. Source of medium-chain triglycerides, subs. for palm kernel & coconut oil; several cult. orn. incl. hybrids, esp. *C. ignea* A. DC (cigar flower, Mex. & Jamaica)

Cupheanthus Seemann. Myrtaceae (Myrt.). 5 New Caled.

Cuphocarpus Decne & Planchon. Araliaceae. 1 Madag. Wood for musical instruments

Cuphonotus O. Schulz. Cruciferae. 2 Aus. R: CGH 205(1974)154

cupid's dart *Catananche caerulea*; **c. flower** *Ipomoea quamoclit*

cuprea bark *Remijia pedunculata, R. purdieana*

Cupressaceae Gray. Pinopsida. 20/125 subcosmop. R: Phytol. M. 7(1984)5. Monoec. or dioec. resinous trees & shrubs. Lvs decussate or in whorls of 3 or 4, in young pl. needle-like, usu. small & scale-like in mature pl. Fls small, solit., axillary or term. on short shoots (rarely males in axillary groups); cones term., woody, leathery or berry-like, cone-scales opp. or in whorls of 3, ovules usu. several per scale, erect; pollen without air-bladders, the nucleus of the grain acting as generative nucleus, there being no male prothallial cells, the first division delayed in many genera until after pollination. Seedlings usu. with 2(–6) cotyledons. n = 11

Genera: *Actinostrobus, Austrocedrus, Callitris, Calocedrus, Chamaeyparis, Cupressus, Diselma, Fitzroya, Fokienia, Juniperus, Libocedrus, Microbiota, Neocallitropsis, Papuacedrus, Pilgerodendron, Platycladus, Tetraclinis, Thuja, Thujopsis, Widdringtonia*. Half are basically N hemisph., half S, 8 are monotypic. Distinguished from Taxodiaceae & Pinaceae by opp. or whorled scale-like lvs (needle lvs in mature *Juniperus*)

Many imp. timbers esp. *Chamaecyparis, Cupressus, Juniperus, Thuja*; resins, flavourings & cult. orn.

× **Cupressocyparis** Dallimore. Cupressaceae (*Chamaecyparis* × *Cupressus*). Fast-growing conifers esp. × *C. leylandii* (Dallimore & Jackson) Dallimore ('leylandii', Leyland cypress grown from cones of *Cupressus macrocarpa* coll. 1888, 5 of the 6 resultant plants being named clones, pollen parent *Chamaecyparis nootkatensis*; cross now repeated) – prob. most widely planted shelter-belt tree in GB, reaching 20 m in 25 yrs, incl. colour-forms esp. golden 'Castlewellan'

Cupressus L. Cupressaceae. 13 N hemisph. (Eur. 2). Monoec., cone-scales woody, seeds maturing in 2 yrs. Cypresses. R: Phytol.M 7(1984)22. Many cult orn. & for timber (incl. shavings for hats in Japan): *C. arizonica* E. Greene (Arizona c., SW N Am.); *C. goveniana* Gordon (Californian c., Monterey County, Calif.); *C. lusitanica* Miller (Mexican c., C Am., long natur. in Eur.); *C. macrocarpa* Hartweg ex Gordon ('macrocarpa', Monterey c., M. County, Calif., now restricted to 2 groves in wild) – widely planted, a parent of × *Cupressocyparis leylandii*; *C. sempervirens* L. (Italian c., S Eur., SW As.) – cult. erect form (wild plant with spreading branches = f. *horizontalis* (Miller) Voss) familiar in Continental pictures, timber used for sarcophagi of Egyptians, statues of Greek gods & infusion as foot-bath long used to combat smelly feet and still used to scent soaps, coppices, young trees & new growth female but stressed trees more male; *C. torulosa* D. Don (Himalayan c., Himal.)

cupuaçu *Theobroma grandiflorum*

Cupulanthus Hutch. Leguminosae (III 24). 1 W Aus.

Cupularia Gordon & Gren. = *Dittrichia*

Cupuliferae A. Rich. = Betulaceae + Fagaceae, later restricted to F.

curagua *Ananas lucidus*

Curanga Juss. = *Picria*

curare form. arrow poisons esp. of *Strychnos toxifera* with *Chondrodendron tomentosum* (q.v.); see also *seq*.

Curarea Barneby & Krukoff. Menispermaceae (I). 4 trop. S Am. *C. toxicofera* (Wedd.) Barneby & Krukoff & *C. candicans* (Rich.) Barneby & Krukoff sources of curare; *C. tecunarum* Barneby & Krukoff an oral contraceptive in Brazil. R: MNYBG 22,2(1971)7

Curatella Loefl. Dilleniaceae. 2 trop. Am.

Curculigo Gaertner. Hypoxidaceae (Liliaceae s.l.). 10 trop. S hemisph. Sessile palm-like pls with plicate lvs; ovary loculi imperfect; some propagules epiphyllous; ? rat-disp. in Mal. Some cult. orn. & fibres e.g. *C. latifolia* Dryander (Indomal.) used for fishing-nets, fr. tasteless but sweetens taste of drinks for 1 hr after (unstable compound so comm. useless)

Curcuma L. Zingiberaceae. c. 40 trop. As. Spices & starch. *C. amada* Roxb. (mango ginger, India) – pickles; *C. angustifolia* Roxb. (India) – tubers yield Bombay or (East) Indian arrowroot; *C. aromatica* Salisb. (India) – dye; *C. caesia* Roxb. (India) – cosmetics & medic.; *C. longa* L. (*C. domestica*, turmeric, cultigen orig. India?) – dried & ground rhizome used in curry powder & orange or yellow dyes form. much used with silk & wool, incl. in carpets; *C. zedoaria* (Christm.) Roscoe (*C. zerumbet*, zedoary, NE India) – condiment or tonic, used in Ind. scents; others cult. orn.

Curcumorpha A. Rao & Verma. Zingiberaceae. 1 Himalaya

curlewberry *Empetrum nigrum*

currant *Vitis vinifera*; **blackc.** *Ribes nigrum*; **American blackc.** *R. americanum*; **buffalo c.** *R. aureum, R. odoratum*; **bush c.** *Clidemia* spp.; *Miconia* spp.; **c. bush** *Leptomeria* spp.; **flowering c.** *R. sanguineum*; **golden c.** *R. aureum*; **Missouri c.** *R. aureum*; **redc.** *R. rubrum*; **skunk c.** *R. glandulosum*; **whitec.** *R. rubrum* cv.

Curroria Planchon ex Benth. = Cryptolepsis

curry leaf *Murraya koenigii*

curse, Koster's *Clidemia hirta*; **Paterson's c.** *Echium plantagineum*

Curtia Cham. & Schldl. Gentianaceae. 10 Guianas to Uruguay

Curtisia Aiton. Cornaceae (Curtisiaceae). 1 S Afr.: *C. dentata* (Burm.f.) C.A. Sm. (*C. faginea*) – timber for spokes, furniture etc. (assagai wood). Prob. not referable to Cornaceae

Curtisiaceae (Harms) Takht. See Cornaceae

Curtonus N.E. Br. = *Crocosmia*

curua palm *Orbignya spectabilis*

Curupira G.A. Black. Olacaceae. 1 Brazil: *C. tefeensis* G.A. Black. Oilseed long-collected from Amaz. forests but not scientifically described until 1948

Cuscatlania Standley. Nyctaginaceae (III). 1 C Am.: *C. vulcanicola* Standley

cuscus *Vetiveria zizanioides*

Cuscuta L. Convolvulaceae (Cuscutaceae). c. 145 cosmop. (Eur. 17). R: MTBC 18(1932)113. Dodder, devil's guts, scald. Parasites with short-lived root-systems & often brightly-coloured chlorophyll-less thread-like twining stems (those of 1 pl. in Costa Rica totalling 500 m) linked by haustoria to host. Form. reduced yield of many crops, now largely controlled though some not restricted to 1 host sp.

Cuscutaceae Bercht. & J. Presl. See Convolvulaceae

cushaw *Cucurbita argyrosperma*

cush-cush *Dioscorea trifida*

cushion bush *Leucophyta brownii*

Cusickia M.E. Jones = *Lomatium*

Cusickiella Rollins (~ *Draba*). Cruciferae. 2 W US. R: JJB 63(1988)65

cusparia bark *Angostura* spp.

Cuspidaria DC. Bignoniaceae. 8 trop. Am.

Cuspidia Gaertner. Compositae (Arct.-Gort.). 1 S Afr. R: MBSM 3(1959)315

Cussonia Thunb. Araliaceae. c. 25 trop. & S Afr. to Masc. Pachycaul trees with soft wood, sometimes cult. ('cabbage-trees', 'umbrella-trees'), some molluscicidal saponins

custard apple *Annona reticulata, A. squamosa*

Cutandia Willk. Gramineae (17). 17 Medit. (Eur. 4) to Middle E. R: BJLS 76(1978)351

cutch *Acacia catechu*

Cuthbertia Small = *Callisia*

Cuttsia F. Muell. Grossulariaceae (Escalloniaceae). 1 E Aus.: *C. viburnea* F. Muell.

Cuviera DC. Rubiaceae (III 2). c. 20 trop. Afr. R: BSBF 106(1959)342. Some myrmecophilous with hollow swellings above nodes

Cwangayana Rauschert = *Aralia*

Cyamopsis DC. Leguminosae (III 8). 3 dry Afr. & Arabia. *C. tetragonolobus* (L.) Taubert (*C. psoraloides*, cluster bean, guar, guvar, gwar, cultigen poss. derived from *C. senegalensis* Guillemin & Perrottet, W Afr.) – much cult. esp. India for young pods as forage & seed-gum, in shampoos etc.

Cyanaeorchis Barb. Rodr. Orchidaceae (V 9). 2 Brazil

Cyanandrium Stapf. Melastomataceae. 4 Borneo

Cyananthus Wallich ex Benth. Campanulaceae. 19 Himal. R: APS 29(1991)25. Ovary superior! Some cult. rock-pl.

Cyanastraceae Engl. (~Tecophilaeaceae; Liliaceae s.l.). Monocots – Liliidae – Asparagales. 1/6 trop. Afr. Herbs with tuberous bases. Lvs basal with closed sheath & lanceolate to cordate blade, veins connected by consp. cross-veins, schizogenous oil channels throughout. Fls in raceme or panicle, regular, bisexual, P 3 + 3, petaloid, blue, shortly basally connate, A 3 + 3 adnate to P with anthers with term. pores or short slits, G (3), half-inf., 3-loc. with 2 anatropous, bitegmic ovules per loc. Fr. a capsule, often 1-seeded; seed with large term. cotyledon & large plumule, half-filled with large starchy & oily chalazosperm, endosperm 0. x = 11, 12

Only genus: *Cyanastrum* (perhaps only a forest-understorey offshoot of T.)

Cyanastrum Oliver. Cyanastraceae. 6 trop. Afr.

Cyanea Gaudich. (~ *Delissea*). Campanulaceae (Lobel.). 64 Hawaii. Many palm-like pachycauls, some with pinnatisect lvs; some with prickles poss. adapted to defence against flightless geese later exterminated by Polynesians

Cyanella L. Tecophilaeaceae (Liliaceae s.l.). 7 Afr. esp. SW Cape. R: SAJB 57(1991)34. Some cult. orn. incl. *C. hyacinthoides* L. – ed. corms poss. comm.

Cyanopsis Cass. = *Volutaria*

Cyanorchis Thouars = *Phaius*

Cyanoseris (Koch) Schur = *Lactuca*

Cyanostegia Turcz. Labiatae (III; Dicrastylidaceae). 5 C & W Aus. R: Brunonia 1(1978)45

Cyanothamnus Lindley = *Boronia*

Cyanothyrsus Harms = *Daniellia*

Cyanotis D. Don. Commelinaceae (II 1c). (Incl. *Amischophacelus*) 50 OW trop. Some cult. orn.

Cyanthillium Blume (~ *Vernonia*). Compositae (Vern.-Vern.). c. 25 OW trop., *C. cinerea* (L.) H. Robinson now pantrop. weed

Cyanus Miller = *Centaurea*

Cyathanthus Engl. = *Scyphosyce*

Cyathea Sm. Cyatheaceae. c. 620 trop & warm (Afr. 14), incl. *Alsophila* & *Sphaeropteris* (as subg., see KB 38(1983)167), *Cnemidaria* & *Schizocaena*. Tree-ferns esp. consp. in montane forest where often aggressive early colonists of sec. forest, marcescent fronds form skirts in some spp. poss. preventing climbers & epiphytes damaging apical buds; starchy pith of some used as food; all Aus. spp. protected; some cult. orn. esp. *C. arborea* (L.) Sm. (trop. Am.) – scales used in nest of Puerto Rico's endemic hummingbird, those of *C. mexicana* Schldl. & Cham. (Mex.) also used in nests; *C. macgregorii* F. Muell. & other spp. (New Guinea) withstand fires in otherwise treeless montane grasslands; *C. medullaris* (Forst.f.) Sw. (W Pacific) – 'pongoware' boxes etc. (NZ), mature pls each producing more than 15 billion spores (1.5 kg) per annum, cult. orn.; *C. tuyamae* H. Ohba (Volcano Is., Japan) – branches above ground level whereas other spp. have them underground; *C. usambarensis* Hieron. (E Afr.) – anthelmintic much used by German troops in World War I though excess leads to blindness

Cyatheaceae Kaulf. Filicopsida. 1/620 trop. & warm. Tree-ferns to 24 m tall with scaly young parts & lvs to 3-pinnate. Sori superficial variously indusiate or naked, differing from Polypodiaceae in having sporangia with complete, rather than incomplete, annulus & from Dicksoniaceae in having sori on lower surface of pinnules in forks of veins, rather than at margins on vein-tips

Only genus: *Cyathea*

Cyathobasis Aellen. Chenopodiaceae (III 3). 1 Turkey: *C. fruticulosa* (Bunge) Aellen

Cyathocalyx Champ. ex Hook.f. & Thomson. Annonaceae. 15 Indomal. Seeds tritegmic

Cyathochaeta Nees. Cyperaceae. 3 SW & E Aus.

Cyathocline Cass. Compositae (Ast.-Gran.). 3 trop. As. R: MBSM 15(1979)513

Cyathodes Labill. = *Styphelia*

Cyathogyne Muell.-Arg. = *Thecacoris*

Cyathomone S.F. Blake. Compositae (Helia.-Cor.). 1 Ecuador

Cyathophylla Bocquet & Strid. Caryophyllaceae (III 1). 1 Greece: *C. chlorifolia* (Poiret) Bocquet & Strid

Cyathopsis Brongn. & Gris (~ *Styphelia*). Epacridaceae. 1 New Caledonia

Cyathopus Stapf. Gramineae (21d). 1 Sikkim. Woods

Cyathoselinum Benth. (~ *Seseli*). Umbelliferae (III 8). 1 SE Eur.: *C. tomentosum* (Vis.) Benth.

Cyathostegia (Benth.) Schery. Leguminosae (III 2). 2 Peru, Ecuador

Cyathostelma Fourn. Asclepiadaceae (III 1). 2 Brazil

Cyathostemma Griff. Annonaceae. 8 Burma to Mal. Tonics, aphrodisiacs (alleged) etc.

Cyathula Blume. Amaranthaceae (I 2). 25 trop. Some drugs in China, incl. *C. prostrata* (L.) Blume (anthelmintic)

Cybebus Garay. Orchidaceae (III 3). 1 Colombia

Cybianthopsis (Mez) Lundell = *Cybianthus*

Cybianthus C. Martius. Myrsinaceae. 150 trop. Am. R: ABiV 10(1980)129

Cybistax C. Martius ex Meissner. Bignoniaceae. 3 trop. Am. Trees. *C. antisyphilitica* (C. Martius) C. Martius ex DC (Amazonian Brazil & Peru) – lvs used as blue dye for cloth; *C. donnell-smithii* (Rose) Seib. (primavera, C Am.) – imp. furniture timber; some cult. orn.

Cybistetes Milne-Redh. & Schweick. Amaryllidaceae (Liliaceae s.l.). 1 S Afr.: *C. longifolia* (L.) Milne-Redh. & Schweick., umbel falls off & is blown away scattering seeds, cult. orn., poss. parent of '× *Amarygia*' hybrids

cycad any member of Cycadopsida

Cycadaceae Pers. Cycadopsida. 1/17 E Afr. to Japan & Aus. Dioec. palm-like trees to 15 m with trunk clothed in leaf-bases. Pinnae with prominent mid-vein but no laterals, vernation circinately involute. Sporophylls spirally arr. in definite cones (male), the females leafy, toothed to deeply lobed, with large naked seeds terminally, the male cones are dropped & a lateral meristem takes over as leader while females are lateral

Only genus: *Cycas*

Cycadales Engler = the only order of Cycadopsida, q.v.

Cycadopsida (Cycadatae). 11/145. R: Brittonia 44(1992)220, Lyonia 2(4,5) (1986, 1989), MNYB 57(1990)8, 200. Gymnosperms, dioec., pachycaul, slow-growing with sec. thickening, tap-rooted & usu. unbranched. Lvs large, pinnate. Reproductive organs in cones (exc. female *Cycas*), term. or lateral; megasporophylls with sterile tips & 8–2 orthotropous ovules, seeds large; microsporophylls scale-like or peltate with pollen-sacs on abaxial side, sperm to 300 μm long (largest sperm known) with spiral band of flagella. Cotyledons 2. An archaic group recognized from the Triassic onwards & represented now by 4 fams: Boweniaceae (~ Stangeriaceae; fronds decompound), Cycadaceae (pinnae with midrib & no lateral veins), Stangeriaceae (midrib & lateral veins), Zamiaceae (numerous parallel or wavy, simple or forked veins running longitudinally)

The wood is centripetal, suggesting a relict condition from a protostelic ancestor. Most have a single persistent cambium but a succession of cambia give co-axial cylinders of secondary xylem & phloem in *Cycas* & some spp. of *Encephalartos* & *Macrozamia*; the wood of *Zamia* & *Stangeria* comprises scalariform tracheids. Apical meristems to 3000 μm across. Roots with cyanobacteria in special 'coralloid' much-branched lateral roots in at least 30 spp. In some at least sex is determined by X & Y chromosomes

Wind- and beetle-poll., weevils attracted by micropylar exudates (sugars & amino-acids), cf. nectar in angiosperms. Outer layer of seed fleshy. A group in decline since the Jurassic & now with relict distribution. All are cult. orn. & endangered through overcollecting; toxic due to carcinogens & neurotoxins but some yield sago (esp. *Cycas*)

Cycas L. Cycadaceae. 17 E Afr. to Japan & Aus. R: MNYBG 57(1990)201. Three-layered seed-coat but *C. circinalis* & allies have additional layer of spongy tissue allowing buoyancy. Cult. orn. & some ed. but destroy central nervous system leading to dementia, paralysis & death due to free amino-acid which scavenges Cu & Zn in CNS (Guam disease); cycasin (removed in cooking) converted to methylazomethanol, one of the most toxic carcinogens known. *C. circinalis* L. ('sago-palm', trop. As.) – lvs ed. Sri Lanka, sago from pith (also obtained from *C. beddomei* Dyer (? = *C. circinalis*) for flour & bread in India) & *C. revoluta* Thunb. (Japan) – dwarf, varieg. & cristate clones commonly cult. Japan); *C. media* R. Br. (NE Aus., New Guinea) – seeds ed. boiled, an aboriginal staple in Arnhemland

Cyclacanthus S. Moore. Acanthaceae. 2 SE As.

Cyclachaena Fres. ex Schldl. = *Iva*

Cycladenia Benth. Apocynaceae. 1 SW US

Cyclamen L. Primulaceae. 19 Eur. (8), Medit. to Iran & NE Somalia. R: Plantsman 13(1991)1. Swollen hypocotyls ('corms') to 100 yrs old; 1 cotyledon. Roots emerge from base of hypocotyl in e.g. *C. persicum*, from the top in e.g. *C. hederifolium* & all over in other spp. Reflexed C-lobes in all spp. (cf. *Dodecatheon*); fruiting scape of *C. persicum* extends & falls to ground, that of *C. hederifolium* & other spp. coils up spring-like before seed-release; seeds sticky, at least sometimes ant-disp. Cult. orn. esp. florist's cyclamen derived from *C. persicum* Miller (SE Eur., Tunisia & Aegean (not Iran!)), orig. with scented fls now with double, fringed or large fls, treated as annuals in the trade; other spp. grown outside esp. *C. hederifolium* Aiton (sowbread, S Eur. to SW As., natur. GB) – tuber to 20 cm diam., form. considered a cure for baldness when used as snuff

Cyclandrophora Hassk. = *Atuna*

Cyclanthaceae Poit. ex A. Rich. Monocots – Arecidae – Cyclanthales. 12/200 trop. Am. Perennial herbs to erect shrubs or lianes (± epiphytic). Lvs spiral or distichous with sheath, petiole & expanded blade, s.t. plicate, with parallel or parallel-pinnate venation & cross-veins, usu. bilobed or bifid, s.t. palmate or simple. Infls usu. axillary with 2–∞ large decid. spathes subtending spadix. Fls v. small, weevil-poll., much reduced in structure, unisexual, males & females (± embedded in axis) in same spadix, males with P 4–24, minute, s.t. connate in 1 or 2 series or merely an entire or toothed cup to 0 when fls ± confluent, A 6–∞ with filaments basally connate & anthers with longit. slits, females with P 4, small, s.t. connate & as many staminodes opp. & partly adnate to them, those often thread-like, G̲ (4), 1-loc. with 4 parietal or almost apical placentas & as many stigmas, rarely pseudomonomerous with 1 placenta & stigma (in *Cyclanthus* females confluent into rings round spadix, each ring with common ovular chamber & ∞ placentas & ovules), ovules ± ∞, anatropous, bitegmic. Fr. fleshy berry-like, often coalescent into multiple fr.; seeds up to ∞ with straight embryo and endosperm rich in oil, protein, often hemicellulose, rarely starch. x = 9–16. R: AHB 18(1958)1

Classification & genera:

I. **Carludovicoidae** (fls in spirally arr. groups; fr. spadix not screw-like; lvs bifid, palmate or simple): *Asplundia, Carludovica, Dianthoveus, Dicranopygium, Evodianthus, Ludovia, Pseudoludovia, Schultesiophytum, Sphaeradenia, Stelestylis, Thoracocarpus*

II. **Cyclanthoideae** (male & female fls in separate alt. whorls or part spirals; fr. spadix screw-like; lvs deeply bilobed): *Cyclanthus*

Apparently close to Palmae and Araceae but so distinct as to make up a separate order (Cronquist); insect-poll. recorded in *Asplundia, Carludovica, Cyclanthus* & *Evodianthus*; some ant-disp. fr.

Cult. orn. & fibre for hats (*Carludovica*)

Cyclanthera Schrader. Cucurbitaceae. 15 trop. Am. Anther-locules of A united into 2 ring-shaped locules around pistil; fr. usu. explosive but fleshy, the valves rolling back rapidly & expelling seeds in pulpy endocarp esp. in *C. brachystachya* (Ser.) Cogn. (*C. explodens*, N S Am.) – cult. orn.; *C. pedata* (L.) Schrader (trop. Am.) – fr. ed. Peru & Bolivia

Cyclantheropsis Harms. Cucurbitaceae. 2 trop. Afr.

Cyclanthus Poit. Cyclanthaceae. 1 trop. Am.: *C. bipartitus* Poit. – poll. by dynastine scarab beetles

Cyclea Arn. ex Wight. Menispermaceae (V). 29 China to Philipp. Alks. Some medic. locally esp. *C. peltata* (Lam.) Hook.f. & Thomson (Mal.)

Cyclobalanopsis Oersted = *Quercus*

Cyclocarpa Afzel. ex Baker. Leguminosae (III 13). 1 OW trop.

Cyclocarya Iljinsk. (~ *Pterocarya*). Juglandaceae. 1 E China

Cyclocheilaceae Marais (Verbenaceae s.l.). Dicots – Asteridae – Lamiales. 2/4 NE & E Afr. Shrubs. Lvs ± opp., simple; stipules 0. Fls solit., axillary, bracteate; bracteoles encl. bud, accrescent. K 0 or disc-like rim; C (5); A 2 + 2, filaments long-pilose, anthers often tailed. G̲, ± 2-loc. with 2–10 anatropous ovules on long funicles. Fr. a discoid capsule; seeds with 0 endosperm. R: KB 35(1981)805

Genera: *Asepalum, Cyclocheilon*

Cyclocheilon Oliver. Cyclocheilaceae (Verbenaceae s.l.). 3 NE Afr. R: KB 35(1981)806

Cyclocotyla Stapf. Apocynaceae 1 trop. W Afr. R: AUWP 85-2(1985)59

Cyclodium C. Presl. Dryopteridaceae (I 2). 10 trop. Am. R: AFJ 76(1986)56. 1 sp. a rheophyte

Cyclogramma Tag. = *Cyclosorus*

Cyclolepis Gillies ex D. Don. Compositae (Mut.-Mut.). 1 temp. S Am. Char. shrub of salty soils in N Patagonia

Cyclolobium Benth. Leguminosae (III 6). 5 trop. S Am.

Cycloloma Moq. Chenopodiaceae (I 4). 1 W & C N Am.: *C. atriplicifolium* (Roth) Coulter, seeds ed. by diff. Indian groups, natur. Eur., S Am.

Cyclonema Hochst. = *Rotheca*

Cyclopeltis J. Sm. Dryopteridaceae (I 3). c. 5 Indomal., 1 trop. Am.

Cyclophyllum Hook.f. (~ *Canthium*). Rubiaceae (III 2). 11 Vanuatu, New Caled.

Cyclopia Vent. Leguminosae (III 26). 20 SW & S Cape. R: Bothalia 6(1951)161. Stipules 0. Several spp. used for 'Cape' or bush teas, some comm.

Cyclopogon C. Presl. Orchidaceae (III 3). 55 Andes

Cycloptychis E. Meyer ex Sond. Cruciferae. 2 S Afr.

Cyclorhiza Shen & Shan (~ *Vicatia*). Umbelliferae (III 8). 1 SW China

Cyclosorus Link (*Meniscium*). Thelypteridaceae. (Incl. *Chingia, Christella, Goniopteris, Mesophlebeion, Pneumatopteris, Pronephrium, Pseudocyclosorus*) c. 600 trop. Some with trunks, some rheophytes

Cyclospermum Lag. (*Ciclospermum*). Umbelliferae (III 8). 3 C Am., WI. with *C. leptophyllum* (Pers.) Eichler natur. trop.

Cyclostachya Reeder & C. Reeder. Gramineae (33c). 1 Mexico

Cyclotrichium (Boiss.) Manden. & Scheng. Labiatae (VIII 2). 6 SW As. to Iran

Cycniopsis Engl. Scrophulariaceae. 3 trop. Afr.

Cycnium E. Meyer ex Benth. (~ *Escobedia*). Scrophulariaceae. 15 warm Afr. R: DBA 32,3(1978)1

Cycnoches Lindley. Orchidaceae (V 9). 23 trop. Am. Epiphytes with floral mechanism & dimorphism like *Catasetum*, producing more female infls in high light intensity. Some cult. orn.

Cydista Miers. Bignoniaceae. 4 trop. Am. Extrafl. nectaries. *C. diversifolia* Miers (Mex. to C Am.) – mass-flowering in Costa Rica

Cydonia Miller (~ *Pyrus*). Rosaceae (III). (Incl. *Pseudocydonia*) 2 W As., China. *C. oblonga* Miller (quince, long cult. poss. orig. wild in Turkey, now natur. in S Eur.) – fr. tart used as jelly or cooked with other fr. & often used as a stock for pears; in antiquity a love-token assoc. with Venus & poss. 'golden fruit' of the Hesperides; *marmelo* in Portuguese & Spanish & thus a contender for the 'original' marmalade. *C. sinensis* Thouin (*Pseudocydonia s.*, China) – cult. orn., large fr. used as air-freshener in China. C. differs from *Chaenomeles* in that latter has serrate lvs (not entire), decid. (not persistent) sepals & A 40–60 (not 15–25)

Cylicodiscus Harms. Leguminosae (II 3). 1 Guineo-Congolian forests: *C. gabunensis* (Taubert) Harms (denya, okan, Afr. greenheart) – comm. timber

Cylicomorpha Urban. Caricaceae. 2 trop. Afr. R: V.M. Badillo (1971) *Caric.*: 34. *C. parviflora* Urban (E Afr.) – hollow spiny trunk often tusked by elephant & with bees' nests within

Cylindrocarpa Regel. Campanulaceae. 1 C As.

Cylindrocline Cass. Compositae (Pluch.). 2 Mauritius

Cylindrokelupha Kosterm. = *Archidendron*

Cylindrolobus Blume = *Eria*

Cylindrophyllum Schwantes. Aizoaceae (V). 5 Cape. Cult. orn. with 4-ranked succ. lvs

Cylindropsis Pierre. Apocynaceae. 1 trop. W Afr.

Cylindropuntia (Engelm.) F. Knuth = *Opuntia*

Cylindropyrum (Jaub. & Spach) Löve = *Aegilops*

Cylindrosolenium Lindau. Acanthaceae. 1 Peru

Cylindrosperma Ducke = *Microplumeria*

Cymaria Benth. Labiatae (IV). 2 Indomal.

Cymatocarpus O. Schulz. Cruciferae. 3 Transcaucasia to C As.

Cymbalaria Hill (~ *Linaria*). Scrophulariaceae (Antirr.). 9 W Eur. (7), Medit. to Iran. R: D.A. Sutton, *Rev. Antirrhineae* (1988)158; differs from *Linaria* s.s. in axillary fls & palmately-veined lvs. Some cult. orn. creepers esp. *C. muralis* Gaertner f., Meyer & Scherb. (Kenilworth or Oxford ivy, ivy-leaved toadflax, mother-of-thousands, pennywort, wandering sailor, Eur. natur. GB (C17, said to have been introd. with sculpture from Italy & first known as Oxford weed) & N Am.) – fls negatively phototropic after pollination, developing fr. inserted in dark crannies; peloric fls produced in tissue-culture, not recorded in wild (cf. *Linaria*); f. *toutonii* (A. Chev.) Cuf. (France) with deeply lobed lvs (? virus)

Cymbaria L. Scrophulariaceae. 4 Eur. (1), C & E As.

Cymbidiella Rolfe. Orchidaceae (V 9). 3 Madag.: 2 epiphytic, 1 terrestrial (*C. flabellata* (Thouars) Rolfe, wasp-poll.)

Cymbidium Sw. Orchidaceae. 44 NW India to Japan & Aus. (3). R: D. du Puy & P. Cribb (1988) *The genus C.* Epiphytes (some mycotrophic) with extrafl. nectaries & large fls much cult. esp. for button-holes, *C. ensifolium* (L.) Sw. (trop. As. to Philipp.) is the Chien Lan or Fukien orchid, a pot-pl. in China for centuries, its fls infused as eye-treatment; *C. goeringii* (Reichb.f.) Reichb.f. (*C. virescens* (China) – fls salted & in hot water used as a drink & also preserved in plum vinegar. Some local medic. esp. for dysentery & diarrhoea

Cymbiglossum Halbinger = *Lemboglossum*

Cymbispatha Pichon = *Tradescantia*

Cymbocarpa Miers. Burmanniaceae. 2 trop. S Am., WI

Cymbocarpum DC ex C. Meyer. Umbelliferae (III 10). 3–4 SW As.

Cymbochasma (Endl.) Klok & Zoz = *Cymbaria*

Cymbolaena Smoljan. Compositae (Gnap.-Gnap.). 1 SW to C As.: *C. griffithii* (A. Gray) Wagenitz. R: OB 104(1991)173

Cymbonotus Cass. Compositae (Arct.-Arct.). 3 temp. Aus.

Cymbopappus R. Nordenstam. Compositae (Anth.-Mat.). 4 S Afr. R: BJLS 96(1988)308

Cymbopetalum Benth. Annonaceae. 27 Mex. to trop. S Am. R: SBM 40(1993)5. Poll. scarabs (*Cyclocephala* spp.); pollen of *C. odoratissimum* Barb.-Rodr. (Brazil) almost 350 μm diam. Petals of *C. penduliflorum* (Dunal) Baillon used with vanilla by Aztecs to flavour chocolate

Cymbopogon Sprengel. Gramineae (40f). 56 OW trop. & warm. R: Reinwardtia 9(1977)225, (1980)390. Many yield aromatic essential oils used in scent, medicine & flavouring. *C. citratus* (DC) Stapf (lemon grass, S India & Sri Lanka, cult. Florida) – serai of As. cooking, scent for soaps; *C. flexuosus* (Steudel) W. Watson (Malabar oil, Indomal.); *C. martinii* (Roxb.) W. Watson (ginger grass, palma-rosa, rosha, rusha, India, cult. Mal.) – 'geranium oil', used to flavour tobacco; *C. nardus* (L.) Rendle (citronella, mana grass, trop. As.) – used as tea, oil the source of insect-repellent; *C. schoenanthus* (L.) Sprengel (canal grass, N Afr. to N India) – oil medic. & adulterant of Otto of Roses in Middle E; *C. validus* (Stapf) Davy (S Afr.) – imp. thatching grass; *C. winterianus* Jowitt (cultigen) – cult. India for cheap scent

Cymbosema Benth. Leguminosae (III 10). 1 trop. Am.: *C. roseum* Benth. – fls a medic. tea for menstrual disorders in NW Amaz.

Cymbosetaria Schweick. = *Setaria*

Cymodocea König. Cymodoceaceae. 4 coasts of W Afr. & Canary Is. to Med. (Eur. 1), Indopacific. R: C. den Hartog (1970) *Sea grasses of the World* (1970)161

Cymodoceaceae N. Taylor. Monocots – Alismatidae – Najadales. 5/14 trop. & warm coastal shallows. R: C. den Hartog (1970) *Sea grasses of the World*. Glabrous, dioec. marine rhizomatous herbs without vessel-elements; stem sympodial, the app. axillary infls term. Lvs spiral, distichous or app. opp., linear, 3–∞-veined with open basal sheath & ligule at its junction with blade. Fls small, water-poll., solit., paired, in cymes or thyrses; P 0, anthers paired on common filament, with longit. slits & thread-like pollen grains to 1 mm long without exine, G 2, styles 2- or 3-fid, ovule 1, pendulous from locule apex, orthotropous, bitegmic. Fr. an achene, endosperm 0. x = 7 (*Cymodocea*)

Genera: *Amphibolis*, *Cymodocea*, *Halodule*, *Syringodium*, *Thalassodendron*

Prob. best included in Zannichelliaceae but differ in filamentous pollen & marine habitat

Cymophora Robinson = *Tridax*

Cymophyllus Mackenzie ex Britton & A. Brown. Cyperaceae. 1 S Appalachian Mts: *C. fraseri* (Andrews) Britton & A. Braun. R: JAA 68(1987)422. Differs from *Carex* in bractless culm each with just 1 leaf, without sheath, ligule or midrib, appearing after flowering

Cymopterus Raf. Umbelliferae (III 9). 32 W N Am. R: AMBG 17(1930)213. Several spp. (roots, lvs etc.) ed. (N Am. Indians)

Cynanchum L. Asclepiadaceae (III 1). 100 trop. & warm. Giant lianes, some cult. orn.

Cynapium Nutt. ex Torrey & A. Gray (~ *Ligusticum*). Umbelliferae (III 8). 1 N Am.: *C. apiifolium* Nutt. ex Torrey & A. Gray

Cynara L. Compositae (Card.-Card.). 8 Medit. (Eur. 7), Canary Is. Coarse thistle-like herbs cult. as veg. *C. cardunculus* L. (cardoon, S Eur.) – lvs blanched (chards) & ed., also root, now widespread natur. S Am. in pampas where fls used to curdle milk; *C. scolymus* L. (globe or French artichoke, cultigen prob. a form of *C. cardunculus*) – young fl. buds eaten (bases of involucral bracts & receptacle, florets discarded)

Cyne Danser. Loranthaceae. 6 Philipp. to New Guinea. R: Blumea 38(1993)101

Cynoctonum J. Gmelin = *Mitreola*

Cynodendron Baehni = *Chrysophyllum*

Cynodon Rich. Gramineae (33b). 8 trop. & warm (Eur. 1). R: T 19(1970)565. Hybrids with *Chloris*. Pasture & lawn-grasses (stargrasses) esp. *C. dactylon* (L.) Pers. (Bermuda grass,

Bahama or kweek grass, dhob, dhub or doob, warm & widely natur.) – lawn-grass in Aus. ('couch') sacred to Hindus (cattlefeed) & *C. transvaalensis* Davy (masindi, Uganda grass, S Afr., natur. US)

Cynoglossopsis Brand. Boraginaceae. 2 Somalia & Ethiopia. R: PSE 138(1981)283

Cynoglossum L. Boraginaceae. (Incl. *Paracynoglossum*) c. 75 temp. (Eur. 11) & warm esp. OW. Fr. of 4 nutlets covered with barbed prickles forming an animal-disp. bur. Alks. Few cult. orn. esp. *C. amabile* Stapf & J.R. Drumm. (Chinese forget-me-not, E As.) & *C. officinale* L. (hound's-tongue, Eur.) – form, medic. & young lvs used as salad

Cynoglottis (Guşuleac) Vural & Kit Tan (~ *Anchusa*). Boraginaceae. 2 SE Eur., SW As.

Cynometra L. Leguminosae (I 4). 70 trop. Some water-disp. in Amaz.; some timbers esp. *C. alexandri* C.H. Wright (ironwood, muhimbi, C Afr.); *C. cauliflora* L. (cultigen orig. E Mal.) – unripe pods ed. raw, cooked or pickled (nam-nam)

Cynomoriaceae (Agardh) Lindley. See Balanophoraceae

Cynomorium L. Balanophoraceae (Cynomoriaceae). 1 Med. incl. Eur. to S Arabia & Mongolia (also coll. NE Somalia): *C. coccineum* L. R: BJ 106(1986)374. Whole plant reddish brown to purplish black, parasitizing a range of salt-marsh pls; so unlike flowering pls as to be called *Fungus melitensis* (Maltese mushroom) in Middle Ages; ed. Bedouin, roots a condiment (Tuareg), valuable dysentery cure in Malta until 1860 with armed guards for pls, styptic & other local medic., ed. in Iraq

Cynorchis Thouars = *Cynorkis*

Cynorhiza Ecklon & Zeyher (~ *Peucedanum*). Umbelliferae (III 10). 4 S Afr.

Cynorkis Thouars. Orchidaceae (IV 2). 125 trop. & S Afr., Masc. Some apomicts, *C. uncata* Kraenzlin (E Afr.) with axillary bulbils; some cult. orn.

Cynosciadium DC. Umbelliferae (III 8). 2 N Am.

Cynosurus L. Gramineae (17). 5 Eur. (4), Medit. R: NBP 1964: 23. Pasture & hay grasses (silky bent) esp. *C. cristatus* L. (crested dog's-tail, Eur., natur. N Am.) – fodder, form. used for weaving mats & baskets

Cynoxylon (Raf.) Small = *Cornus*

cyp or **cypre** *Cordia alliodora*

Cypella Herbert. Iridaceae (III 4). c. 20 Mex. to Arg. Corms, lvs plicate; elaiophores attractive to bees. *C. aquatilis* Ravenna (Brazil) – submerged; other spp. cult. orn. esp. yellow-flowered *C. herbertii* (Lindley) Herbert (S Am.)

Cyperaceae Juss. Monocots – Commelinidae – Cyperales. 98/4350 cosmop. esp. temp. Perennial usu. rhiz. herbs (rarely annuals), s.t. lianoid (*Scleria*), shrubby & dracaenoid or vellozioid (*Afrotrilepis, Cephalocarpus*, some *Costularia, Gahnia, Microdracoides, Trilepis*), usu. with vessel-elements throughout vegetative parts; stems triangular (less often terete), usu. solid; roots with root-hairs (exc. *Eleocharis*), mycorrhiza rare. Lvs spiral, often 3-ranked with a (usu.) closed sheath & usu. long narrow blade (s.t. terete or even 0) with parallel veins; ligule at junction with sheath s.t. present. Fls small, usu. wind-poll. (some *Rhynchospora* (*Dichronema*) etc. insect-poll.), usu. unisexual (plants monoec., occ. dioec.), sessile in axils of spirally arr. or distich. bracts (scales), forming spikes or spikelets usu. in sec. infls, rarely solit. & term., very rarely a small bract between fl. & axis; P (1–)6(–∞) scales, bristles or 0, A (1–)3(6), anthers with longit. slits, pollen in pseudomonads (3 of the 4 nuclei formed at meiosis soon degenerating), $\underline{G}$ ((2)3(4)) with terminal style & stigma branches not always same no. as G, ovule 1, basal, anatropous, bitegmic. Fr. an achene; seed free from pericarp, with oily, ± starchy endosperm with outer proteinaceous layer. x = 5–60+. Journal: Cyperaceae Newsletter

Classification & chief genera (up to 5 subfams recog. in some treatments):

I. **Cyperoideae** (fls bisexual in many-flowered spikelets or solit. unisexual, ± P): *Cyperus, Eleocharis, Fimbristylis, Schoenoplectus, Scirpus*

II. **Rhynchosporoideae** (fls bisexual or unisexual ± P, in few-flowered spike-like racemes in spikes or heads): *Cladium, Mapania, Rhynchospora, Schoenus, Scleria*

III. **Caricoideae** (fls unisexual, naked, usu. in many-flowered spikes, females surrounded by a perigynium (utricle)): *Carex, Kobresia, Uncinia*

Controversy over generic limits & evolution of the infl., some contending that the bisexual fls in some genera are aggregates of unisexual. The ovules are unusual in being term. products of a meristem. C. are apparently related to Gramineae, but despite their frequency, particularly in cool wet habitats, are by no means so diversified: they are scarcely palatable to animals, incl. man & are of almost negligible economic importance when compared with G. Some are used for thatching, basketwork etc., *Cyperus* & *Lepidosperma* for paper-making, while spp. of *Eleocharis* & *Cyperus* have ed. tubers; few cult. orn.

Cyperochloa Lazarides & L. Watson. Gramineae (25). 1 SW Aus.: *C. hirsuta* Lazarides & L.

Watson. R: Brunonia 9(1987)215

Cyperus L. Cyperaceae. (Excl. *Kyllinga, Mariscus, Torulinium*) c. 300 trop. & warm (Eur. 27, Aus. 150 (50 endemic)). R: BSRBB 122(1989)103. Ann. or perenn. herbs with rhiz. *C. articulatus* L. (adrue, warm) – rhiz. medic.; *C. bulbosus* Vahl (OW trop. to Aus.) – ed. tubers; *C. corymbosus* Rottb.(OW trop.) – rhiz. a strong contraceptive in Brazil; *C. difformis* L. – trop. weed of rice e.g. in Aus. ('dirty Dora'); *C. esculentus* L. (W As. & Afr., now natur. NW) – weedy pest in US but var. *sativus* Boeck (tiger nut, earth almond, chufa, rush or Zulu nut) – rarely flowering, tubers ed. (rich in starch, sugar & fat) roasted, one of oldest crops in Egypt, made into flour or juice served as drink (Horchata de Chufa in Spain); *C. involucratus* Rottb. ('*C. alternifolius*', umbrella-plant, OW trop., widely natur.) – common housepl.; *C. iria* L. (warm OW) – grasshoppers fed it have underdeveloped ovaries (presence of juvenile hormone III); *C. laevigatus* L. (makaloa) form. much used in Hawaii for 'makaloa' mats for cloaks etc. esp. on Niihau with red sheaths of *Eleocharis calva*; *C. longus* L. ((sweet) galingale, Euras, incl. GB) – mooted for 'energy' production on poor soils, rhiz. (violet-scented) used in scent; *C. malaccensis* Lam. (Indomal.) – used for brushes & matting; *C. papyrus* L. (papyrus, C Afr. & Nile Valley (rare in upper parts), natur. in E Sicily) – dominates swamps of N Uganda forming species-poor trop. environment, 'sudd', poss. native in N Israel, form. leaf pith sliced into thin strips, laid side by side & another set over these at right angles then tapped so that a sheet (papyrus, same root as 'paper') was formed – used until c. C8 AD, examples 4000 years old extant, also used for sandals, ropes & boats (Moses in the bulrushes), recently the Ra series (transatlantic), tuberous rhiz. also ed.; *C. pulchellus* R. Br. (Aus.) – insect-poll.; *C. rotundus* L. (coco grass, nutgrass, trop., widely natur.) – 'the world's worst weed' through tubers, form. used in perfumery, local medic. in Germany & As. (antifebrile cyperene (also said to inhibit prostaglandin synthesis) & cyperinol in tubers), leaf extract allelopathic; *C. tegetiformis* Roxb. (As.) – cult. for fibre in China for mats etc.

Cyphacanthus Leonard. Acanthaceae. 1 Colombia

Cyphanthera Miers (~ *Anthocercis*). Solanaceae. 9 S Aus. (2 spp. may be hybrids between *C. albicans* (A. Cunn.) Miers & *Duboisia* spp.

cyphel *Minuartia sedoides*

Cyphia Bergius. Campanulaceae. 50 Afr. (esp. S), Cape Verde Is.

Cyphiaceae A. DC = *Campanulaceae*

Cyphisia Rizz. = *Justicia*

Cyphocalyx Gagnepain = *Trungboa*

Cyphocardamum Hedge. Cruciferae. 1 Afghanistan

Cyphocarpa (Fenzl) Lopr. = *Kyphocarpa*

Cyphocarpus Miers. Campanulaceae. 2 Chile

Cyphochilus Schltr. = *Appendicula*

Cyphochlaena Hackel. Gramineae (34b). 2 Madag. R: Adansonia 5(1965)411

Cyphokentia Brongn. Palmae (V 4m). 1 New Caled.: *C. macrostachya* Brongn. R: Allertonia 3(1984)352

Cypholepis Chiov. = *Coelachyrum*

Cypholophus Wedd. Urticaceae (III). 15 C Mal. to W Pacific. Some bark fibres for mats

Cypholoron Dodson & Dressler. Orchidaceae (V 10). 2 Ecuador. *C. frigidum* Dodson & Dressler, only 1.5 cm tall

Cyphomandra C. Martius ex Sendtner (~ *Solanum*). Solanaceae (1). 32 trop. Am. R: FN 63(1994)1. Distinction from *Solanum* not clear. Scent from anther connectives coll. male euglossine bees. Alks. Ed. fr. esp. *C. betacea* Cav. (*C. crassicaulis*, tree-tomato, tamarillo, S Am.) – long cult. Peru, & now widespread, for egg-shaped red fr. used for jelly etc., also ed. raw, *C. cajanumensis* (Kunth) Walp. ('*C. hartwegii*', casana, N Andes) – peach- & passion-fr.-flavoured & *C. capanumensis* Sendtner ex Walp. (chambala, Amaz.) – being tried in NZ; fr. juice of several spp. used as pottery paint

Cyphomeris Standley = *Boerhavia*

Cyphophoenix H. Wendl. ex Hook.f. Palmae (V 4m). 2 New Caled. (1), Loyalties (1). R: Allertonia 3(1984)380

Cyphosperma H. Wendl. ex Hook.f. Palmae (V 4m). 3 New Caled. (1), Fiji (2). R: Allertonia 3(1984)387

Cyphostemma (Planchon) Alston (~ *Cissus*). Vitaceae. c. 250 warm (Afr. 106). R: NS 16(1960)113. Some cult. orn. succ. esp. *C. juttae* (Dinter & Gilg) Descoings (elephant's foot, SW Afr.) – stem bottle-like to 1 m diam.

Cyphostigma Benth. Zingiberaceae. 1 Sri Lanka

Cyphostyla Gleason. Melastomataceae. 3 Colombia

Cyphotheca Diels. Melastomataceae. 1 Yunnan: *C. montana* Diels. R: NJB 10(1990)21

cypress *Cupressus* & *Chamaecyparis* spp.; **African c.** *Widdringtonia* spp.; **Arizona c.** *Cupressus arizonica*; **bald c.** *Taxodium distichum*; **black c.** *Callitris endlicheri*; **Calif. c.** *Cupressus goveniana*; **Himalaya c.** *C. torulosa*; **hinoki c.** *Chamaecyparis obtusa*; **Lawson's c.** *Chamaecyparis lawsoniana*; **Medit. c.** *Cupressus sempervirens*; **Mexican c.** *C. lusitanica*; **Monterey c.** *C. macrocarpa*; **Nootka c.** *Chamaecyparis nootkatensis*; **Patagonian c.** *Fitzroya cupressoides*; **c. pine** *Callitris* spp.; **Port Jackson c.** *Callitris rhomboidea*; **c. powder** French cosmetic with starch from *Arum maculatum*; **sawara c.** *Chamaecyparis pisifera*; **Sitka c.** *C. nootkatensis*; **southern c.** *Taxodium distichum*; **summer c.** *Bassia scoparia*; **swamp c.** *Taxodium distichum*; **yellow c.** *C. nootkatensis*

Cyprinia Browicz. Asclepiadaceae (I; Periplocaceae). 1 SE Turkey, Cyprus: *C. gracilis* (Boiss.) Browicz

Cypripediaceae Lindley = Orchidaceae (II)

Cypripedium L. Orchidaceae (II). 40 N temp. (Eur. 3). R: C. Cash *The slipper orchids* (1991)37. 'Lady's slipper' orchids, moccasin flowers (US), terr. with plicate lvs; contact irritant (cypripedin) from glands on shoots; labellum inflated, sac-like, column with 2 fert. anthers flanking gland-like staminode, pollen glutinous but not in pollinia; insects enter labellum but cannot return that way & are forced to pass out through opening at base, thereby brushing against stigma & then anthers. Some cult. orn., though most 'cypripediums', i.e. greenhouse epiphytes without plicate lvs, belong to *Paphiopedilum* & *Phragmipedium*. *C. calceolus* L. (Euras., varietally diff. races in N Am.) – grows c. 16 yrs before flowering, protected in GB though allegedly reduced to a single plant in Yorkshire through overcollecting for herbarium & garden, medic. E N Am.

Cypselea Turpin. Aizoaceae (II). 1 S Florida to Venez.: *C. humifusa* Turpin. R: JAA 51(1970)431

Cypselocarpus F. Muell. Gyrostemonaceae. 1 SW Aus.

Cypselodontia DC = *Dicoma*

Cyrilla Garden ex L. Cyrillaceae. 1 SE US to N S Am.: *C. racemiflora* L. – tree or shrub of wet ground with infls on last season's wood, imp. bee-tree for comm. honey

Cyrillaceae Endl. Dicots – Dilleniidae – Ericales. (Incl. Clethraceae) 4/78 warm As. & Am., Madeira (1). Glabrous to stellate-hairy trees or shrubs, sometimes accum. cobalt. Lvs spirally arr., simple, entire; stipules 0. Fls in racemes or panicles, each with a bract & often 2 bracteoles, bisexual, reg., hypogynous; K 5(6, 7), connate basally, imbricate, persistent & often accrescent in fr., C as many as & alt. with K, whitish, sometimes basally connate, imbricate or convolute, A twice C or same as & alt. with C (*Cyrilla*), anthers ± versatile, with longit. slits or apical pores & pollen-grains in monads, nectary-disk sometimes around G-base or that nectariferous, G (2–5), pluriloc. with axile placentas & ± style & 1–∞ ovules per locule, each pendulous from near tip, anatropous, unitegmic. Fr. indehiscent, drupaceous to capsule- or samara-like; seed often winged or with testa lost & straight embryo embedded in copious oily endosperm. x = 8, 10. R: CGH 186(1960)1

Genera: *Clethra*, *Cliftonia*, *Cyrilla*, *Purdiaea*

Clethraceae (∞ ovules) form. kept separate. Prob. closest to Ericaceae. Some cult. orn.

Cyrillopsis Kuhlm. Ixonanthaceae. 1 NE Brazil

Cyrilwhitea Ising = *Sclerolaena*

Cyrtandra Forster & Forster f. Gesneriaceae. c. 550 China (1), Mal. (Borneo 150, New Guinea 150) & Pacific. Many hybrids in Hawaii. Some tree-like, *C. dilatata* C.B. Clarke (Borneo) only rheophytic gesneriad

Cyrtandroidea F. Br. = *Cyrtandra*

Cyrtandromoea Zoll. Gesneriaceae. 11 SE As., Mal.

Cyrtandropsis Lauterb. = *Cyrtandra*

Cyrtanthera Nees = *Justicia*

Cyrtanthus Aiton. Amaryllidaceae (Liliaceae). 47 (incl. *Anoiganthus*, *Vallota*) trop. & S Afr. R: Herbertia 1939: 65. Cult. orn. (fire-lilies), like *Crinum* spp., esp. *C. elatus* (Jacq.) Traub (*V. speciosa*, Scarborough, George or Knysna lily)

Cyrtidiorchis Rauschert (*Cyrtidium*). Orchidaceae (V 10). 4 trop. Am. R: Taxon 31(1982)560

Cyrtidium Schltr. = *Cyrtidiorchis*

Cyrtocarpa Kunth. Anacardiaceae (II). 2 Mex. R: AMBG 78(1991)184. *C. procera* Kunth (chupadilla) – fr. ed., bark a soap subs.

Cyrtochilum Kunth = *Oncidium*

Cyrtococcum Stapf (~ *Panicum*). Gramineae (34b). 11 OW trop.

Cyrtocymura H. Robinson (~ *Vernonia*). Compositae (Vern.-Vern.). 8 trop. Am. esp. Brazil

Cyrtogonone Prain. Euphorbiaceae. 1 W Afr.
Cyrtomidictyum Ching = *Polystichum*
Cyrtomium C. Presl = *Polystichum*
Cyrtopodium R. Br. Orchidaceae (V 9). 30 trop. Am. Cult. orn. epiphytic or terr. *C. longibulbosum* Dodson & G. Romero (Ecuador) – pseudobulbs to 3.5 m tall!
Cyrtorchis Schltr. Orchidaceae (V 16). 16 trop. & S Afr. R: KB 14(1960)143. Hawkmoth-poll.
Cyrtorhyncha Nutt. (~ *Ranunculus*). Ranunculaceae (II 3). 1 W N Am.
Cyrtosia Blume (~ *Galeola*). Orchidaceae (V 4). 5 Indomal.
Cyrtosperma Griffith. Araceae (V 3). 11 Indomal., Oceania (similar seeds from Eocene of Br. Columbia). R: Blumea 33(1988)428. Some rheophytes, cult. orn. & ed. tubers esp. *C. merkusii* (Hassk.) Schott (*C. chamissonis*, *C. edule*, babai, swamp taro, W Pacific) – tubers to 60 kg (10 yrs old), slow-growing but good for stagnant & brackish swamps
Cyrtostachys Blume. Palmae (V 4i). 8 Mal. esp. New Guinea to Melanesia. Monoec. cult. orn. esp. *C. renda* Blume (*C. lakka*, sealing-wax palm, W Mal.) – widely planted in trop.
Cyrtostylis R. Br. (~ *Acianthus*). Orchidaceae (IV 1). 5 Aus. (4 endemic), NZ
Cyrtoxiphus Harms = *Cylicodiscus*
Cystacanthus T. Anderson = *Phlogacanthus*
Cysticapnos Miller. Fumariaceae (I). 5 S Afr. R: OB 88(1986)105
Cysticorydalis Fedde ex Ikonn. = *Corydalis*
Cystoathyrium Ching = *Athyrium*
Cystodiaceae Croft = Dicksoniaceae
Cystodiopteris Rauschert = *Cystodium*
Cystodium J. Sm. Dicksoniaceae. 1 Mal.: *C. sorbifolium* (Sm.) J. Sm. Stem prostrate
Cystopteris Bernh. Dryopteridaceae (II 1). c. 12 temp. & warm (esp. N (Eur. 6; GB 3) incl. trop. alpine. R: MTBC 21, 4(1963)1. Cult. orn. esp. *C. fragilis* (L.) Bernh. (bladder fern, N hemisph. to Chile)
Cystorchis Blume. Orchidaceae (III 3). 8 China, Mal. *C. aphylla* Ridley (Mal.) – mycotroph
Cystostemma Fourn. = *Sarcostemma*
Cystostemon Balf.f. Boraginaceae. 13 trop. Afr. to SW Arabia. R: NRBGE 40(1982)1
Cytinaceae (Brongn.) A. Rich. = Rafflesiaceae
Cytinus L. Rafflesiaceae (III). 8: 2 (subg. *Cytinus*) Canary Is. to Middle E (Eur. 2) on Cistaceae, monoec.; 6 (subg. *Hypolepis*) S Afr., Madag., dioec.
Cytisanthus Lang = *Genista*
Cytisophyllum Lang. Leguminosae (III 31). 1 W Med.: *C. sessilifolium* (L.) Lang
Cytisopsis Jaub. & Spach (= ? *Anthyllis*). Leguminosae (III 18). 2 E Med. & N Afr.
Cytisus Desf. Leguminosae (III 31). 33 Eur., N Afr., Canary Is., W As. Brooms. Explosive fls with more than 1 chance of cross-poll.: short A deposit pollen on undersurface of insect, long ones on back but preceded by style which may contact pollen deposited by another fl., then style grows round, such that stigma occupies position just above short stamens & is thus ready for pollen on undersurface of a later visitor. Lvs usu. reduced, stems photosynthetic; widely planted & natur. *C. scoparius* (L.) Link (*Sarothamnus s.*, common broom, Eur., natur. N Am.) – sand-binder, bee-forage, form. for fibre & dyes, prob. THE plantagenista (see *Genista*), alk. (sparteine); *C. striatus* (Hill) Rothm. (SW Eur.), widely planted & now natur. along motorways in GB. Some brooms forced for cut-flower trade under glass. See also *Argyrocytisus*
Cyttaranthus Léonard. Euphorbiaceae. 1 trop. Afr.
Czernaevia Turcz. = *Angelica*

D

dabéma, daboma *Piptadeniastrum africanum*
Daboecia D. Don. Ericaceae. 1 Ireland to Spain & Azores: *D. cantabrica* (Hudson) K. Koch (St Dabeoc's (*sic*) heath, fossils as far N as Shetland). G; bee-poll. cult. orn. evergreen
Dacrycarpus (Endl.) Laubenf. (~ *Podocarpus*). Podocarpaceae. 9 Burma to NZ. R: JAA 50(1969)315. Timber esp. *D. dacrydioides* (Rich.) Laubenf. (*Podocarpus d.*, kahikatea, NZ) – pulp for paper, fr. eaten by Maoris; *D. imbricatus* (Blume) Laubenf. (SE As. to W Pacific)
Dacrydium Sol. ex Forst. f. Podocarpaceae. 25 SE As. to NZ. Usu. dioec.; seeds arillate. R: Phytol.M 7(1984)25. Timber esp. *D. cupressinum* Sol. ex Forst. f. (rimu, red pine, NZ) – has mast years, *D. elatum* (Roxb.) Loudon (sempilor, Mal.). See also *Lagarostrobos*, *Lepidothamnus*
Dacryodes Vahl. Burseraceae. 40 trop. (c. 2 Am., 22 Afr., 15 As.). R: Blumea 7(1954)500. Some

timbers, ed. fr. pulp & resins. *D. belemensis* Cuatrec. (S Am.) – fr. used in drink by Colombian Indians; *D. edulis* (G. Don f.) H.J. Lam (trop. Afr.) – oily seeds ed.; *D. excelsa* Vahl (*D. hexandra*, WI) – source of WI elemi

Dacryotrichia Wild. Compositae (Ast.-Gran.). 1 Zambia: *D. robinsonii* Wild. R: GOB 1(1973)67

Dactyladenia Welw. (~ *Acioa*). Chrysobalanaceae (4). 27 trop. Afr. R: Brittonia 31(1979)483

Dactylaea Fedde ex H. Wolff. Umbelliferae (III 8). 2 C & E As.

Dactylaena Schrader ex Schultes & Schultes f. Capparidaceae. 6 WI, Brazil

Dactylanthaceae (Engl.) Takht. = Balanophoraceae

Dactylanthus Hook.f. Balanophoraceae. 1 NZ: *D. taylorii* Hook.f. R: APG 33(1982)96. Bat-poll.(?) smelly fls with copious nectar, attacked by introd. (Aus.) possums; seed fleshy (? aril). When boiled, parasite tissue is removed to expose host root – 'wood-rose'

Dactyliandra (Hook.f.) Hook.f. Cucurbitaceae. 1 Kenya, 1 disjunct distrib.: Namib (SW Afr.) & Rajasthan (India) deserts

Dactylicapnos Wallich (~ *Dicentra*). Fumariaceae (I 1). 10 Himal. to SE As. R: OB 88(1986)20

Dactyliophora Tieghem. Loranthaceae. 3 Papuasia, N Queensland. R: AusJB 22(1974)558

Dactylis L. Gramineae (17). 1–5 Euras. *D. glomerata* L. (cock's-foot) – cult. as meadow & pasture-grass, widely natur. N Am., S Afr., imp. cause of hay-fever; complex of tetraploid & diploid forms, latter incl. subsp. *aschersoniana* (Graebner) Thell. (*D. polygama* Horvat., C Eur.) prob. derived from subsp. *glomerata* by haplodiploidy

Dactylocardamum Al-Shehbaz. Cruciferae. 1 Peru: *D. imbricatifolium* Al-Shehbaz

Dactylocladus Oliver. Crypteroniaceae. 1 Borneo (+ ?1 New Guinea). *D. stenostachys* Oliver (jonkong) – imp. timber export of freshwater peatswamps used for boat-building & form. slats of Venetian blinds

Dactyloctenium Willd. Gramineae (31d). 10 warm. Crowfoot. *D. aegyptium* (L.) Willd. (Egyptian grass, Sahara & Sudan, widely natur. incl. N Am.) – sand-binder, fr. ed., locally medic. since time of Dioscorides; *D. australe* Steudel (Durban grass, S Afr.) – sand-binder, shade-tolerant lawn-grass

Dactylopetalum Benth. (~ *Cassipourea*). Rhizophoraceae. c. 15 trop. Afr., Madag.

Dactylopsis N.E. Br. Aizoaceae (IV). 1–2 Cape

Dactylorhiza Necker ex Nevski. Orchidaceae (IV 2). 30 temp. Euras. (Eur. 13), Alaska, Medit., Macaronesia. Spp. hybridize & reprod. isolation not clear-cut. Some tubers eaten Iran (Persian salep)

Dactylorhynchus Schltr. = *Bulbophyllum*

Dactylostalix Reichb.f. Orchidaceae (V 8). 1 Japan

Dactylostegium Nees = *Dicliptera*

Dactylostigma D. Austin = *Hildebrandtia*

Dactylostelma Schltr. = *Oxypetalum*

dadap *Erythrina subumbrans*

Dadjoua Parsa = ? Cruciferae

Daedalacanthus T. Anderson = *Eranthemum*

Daemia R. Br. = *Pergularia*

Daemonorops Blume. Palmae (II 1d). 114 Indomal. esp. W Mal. (New Guinea 1). Usu. spiny (alt. whorls of upward- & downward-pointing toothed flanges enclosing ant-inhabited passages – if disturbed ants beat on stem in unison providing an alarm alerting even elephants) dioec. rattans used for cane furniture, chimney-sweeps' brushes etc.; also resins ((Sumatran) dragon's blood) used in varnishes & form. medic. derived from fr. *D. verticillaris* (Griff.) Martius & *D. macrophylla* Becc. in Malay Penin. have ants' nests absorbing through-flow incl. that from débris accum. around apex so that nutrients passed to palms

Daenikera Hürl. & Stauffer. Santalaceae. 1 New Caledonia

daffodil *Narcissus pseudonarcissus* but also any large-flowered *N.* sp. or hybrid; **Peruvian d.** *Hymenocallis* spp.; **Tenby d.** *N. pseudonarcissus*

dagame *Calycophyllum candidissimum*

dagga *Cannabis sativa* subsp. *indica*

dahl *Cajanus cajan*

Dahlberg daisy *Thymophylla tenuiloba*

Dahlgrenia Steyerm. = *Dictyocarpum*

Dahlgrenodendron J. Merwe & Wyk (~ *Cryptocarya*). Lauraceae (I). 1 S Afr.: *D. natalense* (J. Ross) J. Merve & Wyk. R: SAJB 54(1988)80. Like *C.* but pollen unusual. Endangered

Dahlia Cav. Compositae (Helia.-Cor.). 29 mts of Mex. to Colombia. Tuberous roots; stems to 8 m, usu. unbranched, s.t. scrambling & epiphytic. R: Rhodora 71(1969)309, 367. Orig. grown for food (tubers, inulin) in C Am., introd. to Eur. where hybridization between *D.*

coccinea Cav. (Mex., fls yellow to red) & *D. pinnata* Cav. (Mex., fls purple) said to have given rise to garden dahlias (*D.* x *hortensis* Guillaumin), a mania in 1830s & 1840s; *D. excelsa* Benth. (incl. *D. imperialis* Roezl ex Ortgies, *D. arborea*, tree-dahlia, range of genus) – to 8 m, unbranched exc. infl., where up to 300 heads of fls, lvs 2–3-pinnate. Garden ds divided into 10 classes on capitulum form e.g. pompon d. with small heads of only ray-florets, single-flowered with disk-florets present (e.g. 'Coltness Gem' – much used dwarf bedder), cactus-flowered with only long ray-florets

Dahlstedtia Malme. Leguminosae (III 6). 1 S Brazil. Bird-poll.

dahoma *Piptadeniastrum africanum*

Dahomey rubber *Ficus lutea*

dahoon *Ilex cassine*

daikon *Raphanus sativus* 'Longipinnatus'

Dais L. Thymelaeaceae. 2 S Afr., Madag. Bark-fibres used for thread, string etc.

Daiswa Raf. = *Paris*

daisy *Bellis perennis* (but also used of many Compositae with radiate capitula); **Barberton d.** *Gerbera jamesonii*; **d.bush** *Olearia* spp.; **crown d.** *Chrysanthemum coronarium*; **Dahlberg d.** *Thymophylla tenuiloba*; **dog d.** *Leucanthemum vulgare*; **kingfisher d.** *Felicia bergeriana*; **Livingstone d.** *Dorotheanthus bellidiformis*; **Michaelmas d.** see *Aster*; **moon d.** *Leucanthemum vulgare*; **ox-eye d.** *L. vulgare*; **paper d.** (Aus.) *Bracteantha* spp. esp. *B. bracteata*, *Rhodanthe manglesii*; **Shasta d.** *L.* × *superbum*; **Singapore d.** *Wedelia trilobata*; **Swan River d.** *Brachycome iberidifolia*; **d. tree** *Olearia* spp. (Aus.), *Montanoa* spp. (Am.)

Daknopholis W. Clayton. Gramineae (33b). 1 E Afr., Madag., Aldabra

dakua *Agathis macrophylla*

Dalbergaria Tussac (~ *Columnea*). Gesneriaceae. 90 trop. Am.

Dalbergia L.f. Leguminosae (III 4). 100 trop. Trees, shrubs & lianes. Fr. flattened, indehiscent, some in Amaz. water-disp. Many timbers with dark colour & dense grain for furniture, musical instruments etc. (rosewood, Nicaragua wood, palisander (Brazil)). *D. cearensis* Ducke (kingwood, tulipwood, Brazil); *D. cochinchinensis* Pierre ex Laness. (trac, SE As.); *D. cultrata* R. Graham ex Ralph (Burma) – esp. for ploughs, resin red; *D. decipularis* Rizz. & Mattos (Brazilian tulipwood, Bahia) – valuable timber ('sebastião-de-arruda') known since C19 but tree not described until 1967; *D. granadillo* Pittier (granadillo, Mex.); *D. hainanensis* Merr. & Chun (huanghuali, Hainan) – wood for late C16/early C17 Chinese furniture, more recently combs etc.; *D. horrida* (Dennst.) Mabb. (India) – survives in only a few sacred groves in Kerala; *D. latifolia* Roxb. (Indian or Bombay blackwood or rosewood, Malabar r., black or E India r., S India); *D. melanoxylon* Guillemin & Perrottet (African blackwood, Mozambique ebony, trop. Afr.) – much used for carving e.g. Makonde; *D. nigra* (Vell. Conc.) Benth. (Brazilian rosewood, Bahia or Rio r., jacaranda, Brazil) – esp. for furniture, radio cabinets, pianos etc., now endangered sp.; *D. parviflora* Roxb. (Mal.) – liane with scented heartwood used in joss-sticks; *D. retusa* Hemsley (cocobolo, C Am.) – esp. for small work like knife-handles, chess-pieces, rosaries, buttons etc.; *D. sissoo* Roxb. ex DC (sheesham, sisso, shisham, India) – used for fancy trinket-boxes, fuelwood; *D. stevensonii* Standley (Honduras rosewood, Belize) – used for marimbas etc.

dalbergia, false *Pericopsis laxiflora*

Dalbergiella Baker f. Leguminosae (III 4). 3 trop. Afr.

Dalea L. Leguminosae (III 12). c. 160 Canada to Arg., esp. Mex. & Andes, in dry & desert areas. R: MNYBG 27(1977)135. Indigo bush. Some cult. orn.

Dalechampia L. Euphorbiaceae. 115 warm (OW 10, Afr. 6) esp. Am. (95). R: BJLS 105(1991)137. Mostly lianes; 9–10 male & 3 female fls + 2 bracts minic 1 fl, some with scent (trans-carvone oxide; also in *Catasetum* & other orchids) coll. male euglossines & fl. resins for bees' nests, visitors acting as pollinators; poll. by male euglossines evol. several times, that by female resin-collectors only once. *D. roezliana* Muell. Arg. (Mex.) – cult. orn. with 2 large pink or white outer bracts protecting fls, the females in a 3-flowered cyme protected by another bract lying below 4 more bracts with 9–14 male fls lying in front of a yellow cushion of rudimentary males sometimes secreting a resin; after flowering all that lying above the females is dropped. Some spp. with stinging glands

Dalembertia Baillon. Euphorbiaceae. 4 Mex. Male fl. of A 1 enclosed in K 1

Dalenia Korth. Melastomataceae. 3 Borneo. K-tube with calyptra

Dalhousiea Wallich ex Benth. Leguminosae (III 2). 1 W trop. Afr., 2 NE India, Bangladesh. R: BJB 4(1975)33

Dalibarda L. = *Rubus*

Dallachya F. Muell. = *Rhamnella*

dalli *Virola surinamensis*
Dallis grass *Paspalum dilatatum*
Dalzellia Wight. Podostemaceae. 4 trop. As. R: BMNHN 4,10(1988)71
Dalzielia Turrill. Asclepiadaceae (III 4). 1 W Afr.: *D. lanceolata* Turrill
damar minyak *Agathis dammara*
damask or **d. violet** *Hesperis matronalis*; **d. rose** *Rosa × damascena*
Damasonium Miller. Alismataceae. 5 Euras. (Eur. 1: *D. alisma* Miller, thrumwort), W N Am., S Aus.(1)
dame's rocket or **d.'s violet** *Hesperis matronalis*
damiana *Turnera diffusa*
dammar resins from trop. trees used in varnishes. See *Canarium, Hopea, Shorea, Vatica* & *Agathis*
Dammara Lam. = *Agathis*
Dammaropsis Warb. = *Ficus*
Dammera Lauterb. & Schumann = *Licuala*
Damnacanthus Gaertner f. Rubiaceae (IV 11). 6 E As. Usu. heterophyllous spiny shrubs, paired thorns = lateral shoots proximally or infl. shoots
Damnamenia Given (~ *Celmisia*). Compositae (Ast.-Ast.). 1 Auckland & Campbell Is.
Damnxanthodium Strother (~ *Perymenium*). Compositae (Helia.-Verb.). 1 Mex.: *D. calvum* (Greenman) Strother
Dampiera R. Br. Goodeniaceae. 66 Aus., esp. SW. R: Telopea 3(1988)183. Some cult. orn.
Damrongia Kerr ex Craib = *Chirita*
damson or **d. plum** *Prunus × domestica* 'ssp. *institia*'; **d. plum** (WI) *Chrysophyllum oliviforme*
Danae Medikus. Asparagaceae (Liliaceae s.l.; Ruscaceae). 1 SW As.: *D. racemosa* (L.) Moench (Alexandrian laurel) – phylloclades; cult. orn.
Danaea Sm. Marattiaceae. 30 poorly differentiated, trop. & warm Am.
Danaeaceae Agardh. See Marattiaceae
Danais Comm. ex Vent. Rubiaceae (I 1/IV 24). 40 Tanzania (1), Madag. & Masc. Some dyes & fibres
dandelion *Taraxacum officinale* etc.; **Russian d.** *T. bicorne*
Dandya H. Moore. Amaryllidaceae (Liliaceae s.l.). 4 Mex. R: ABM 18(1992)14
danewort *Sambucus ebulus*
Danguya Benoist. Acanthaceae. 1 Madagascar
Danguyodrypetes Leandri. Euphorbiaceae. 1 Madagascar
Daniellia Bennett. Leguminosae (I 4). 9 trop. Afr., forest & savannas. Bumbo; source of copals & timbers esp. *D. ogea* (Harms) Rolfe ex Holl. (faro, ogea) & *D. thurifera* Bennett (both W Afr.)
Dankia Gagnepain. Theaceae. 1 SE As.
Dansera Steenis = *Dialium*
Dansiea Byrnes. Combretaceae. 1 Queensland
danta *Nesogordonia papaverifera*
Danthonia DC. Gramineae (25). Incl. *Rytidosperma* 100 Eur. (2), Aus. (32) NZ (16), N & S Am. Wallaby grasses imp. for sheep in wool industry. *D. decumbens* (L.) DC (*Sieglingia d.*, heath grass, Eur., Medit.) – fls often cleistogamous
Danthoniastrum (Holub) Holub = *Metcalfia*
Danthonidium C. Hubb. Gramineae (25). 1 India
Danthoniopsis Stapf. Gramineae (39). 20 Afr. & Arabia to Pakistan. Fire-free sites
Danube grass *Phragmites australis*
dao *Dracontomelon dao*
Dapania Korth. Oxalidaceae. 1 Madag., 2 Mal. Lianes
Daphnandra Benth. Monimiaceae (I 2; Atherospermataceae). 6 New Guinea, E Aus. Alks
Daphne L. Thymelaeaceae. c. 50 Euras. (Eur. 17). C.D. Brickell & B. Mathew (1976) D. Shrubs often cult.; lvs usu. spiral (opp. in *D. genkwa* Siebold & Zucc., China, where used clinically as an effective & safe abortifacient); bark fibrous & used in Himal. for rope, paper & rayon (*D. odora* Thunb. (also cult. for local med., scent), *D. papyracea* Wallich ex Steudel (typewriter stencil paper), *D. genkwa* etc., Nepal paper from *D. bholua* Buch-Ham. ex D. Don); scent carnation-like in many spp. & attractive to Lepidoptera (also found in *Dianthus* (Caryophyllaceae), *Gymnadenia* (Orchidaceae), *Narcissus* (Amaryllidaceae) – all Eur., *Viburnum* (Caprifoliaceae) – E As., *Ribes* (Grossulariaceae) – N Am., *Petunia* (Solanaceae) – S Am.), the insects themselves producing similar scents: imp. constituents incl. indol & menthyl anthranilate; nectar produced at base of C-tube & accessible only to Lepidoptera & long-tongued bees; glycosides such as daphnin & an acrid resin (mezerein) give pls a

bitter burning taste when chewed – they are poisonous & some insect-repellent, form. medic. e.g. seed-oil of *D. gnidium* L. (Medit.) purgative; *D. laureola* L. (spurge laurel, Euras., natur. GB) – evergreen poll. principally by moths; *D. mezereum* L. (mezereon, Euras, protected in GB) – fls attractive to butterflies esp. Red Admiral & honeybees, fr. form. used as pepper subs. often with fatal consequences

Daphnimorpha Nakai. Thymelaeaceae. 2 Japan

Daphniphyllaceae Muell. Arg. Dicots – Dilleniidae – Buxales. 1/10 E As., Mal. Dioec. trees or shrubs with a unique type of alk. (daphniphylline group), often accumulating aluminium. Lvs spiral but s.t. so crowded at branch-tips as to appear whorled, simple, entire, pinnately-veined; stipules 0, Fls small, reg., hypogynous, in axillary racemes, each pedicel with a deciduous bract; K (0)2–6, ± imbricate, C 0, A 5–12 with anthers with longit. slits; female fls sometimes with staminodes, $\underline{G}$(2(–4)) with as many locules, styles united only basally, short, curved to circinate, ovules 2 per locule, apical-axile, pendulous, anatropous, bitegmic. Fr. a 1-seeded drupe; seed with v. small straight embryo in copious oily & proteinaceous endosperm, 2n = 32

Only genus: *Daphniphyllum*. Embryology confirms exclusion from Euphorbiaceae

Daphniphyllum Blume. Daphniphyllaceae. 10 (some polymorphic) China, through Indomal. to trop. Aus. R: Taiwania 11(1965)57, 12(1966)137. Some cult. orn. incl. *D. humile* Maxim. (*D. macropodum* var. *humile*, N Japan, Korea) – lvs used like tobacco by Ainu

Daphnopsis C. Martius (incl. *Coleophora*). Thymelaeaceae. 55 trop. Am. to E Arg. R: AMBG 46(1959)257, Phytol. 61(1986)361. *D. americana* (Miller) J.S. Johnston (WI) – inner bark used for cordage

Darbya A. Gray = *Nestronia*

dari *Sorghum bicolor*

Darling lily *Crinum flaccidum*; **D. pea** *Swainsona galegifolia*

Darlingia F. Muell. Proteaceae. 2 Queensland

Darlingtonia Torrey. Sarraceniaceae. 1 N Calif., SW Oregon: *D. californica* Torrey. Carnivorous herb differing from *Sarracenia* in having pitcher-hood with distinct flap ('fish-tail' covered in nectaries attracting insects which slide on footholdless detachable wax into pitchers), in bracts on scape & style apically 5-branched; first pair of pitchers usu. orientated N–S & taller than subsequent, all turned outwards; digestion by assoc. fauna as digestive glands 0

Darmera Voss ex Post & Kuntze (*Peltiphyllum*). Saxifragaceae. 1 N Calif., SW Oregon. *D. peltata* (Benth.) Post & Kuntze (*P. peltatum*, umbrella plant) – cult. orn. waterside pl. with long-petioled peltate lvs

darnel *Lolium temulentum*

Darniella Maire & Weiller = *Salsola*

Darwinia Rudge. Myrtaceae (Lept.). c. 45 Aus., esp. SW. Heath-like shrubs with oils used in perfumery; some cult. orn.

Darwiniera Braas & Lückel. Orchidaceae (V 10). 1 trop. Am.

Darwiniothamnus Harling. Compositae (Ast.-Ast.). 3 Galápagos. R: OB 91(1987)9

dasheen *Colocasia esculenta*

Dasiphora Raf. = *Potentilla*

Dasispermum Necker ex Raf. (~ *Heteroptilis*). Umbelliferae (III 8). 1 S Afr.: *D. suffruticosum* (Berg.) B.L. Burtt. R: NRBGE 45(1988)93

Dasistoma Raf. (~ *Seymeria*). Scrophulariaceae. 1 SE US

Dasoclema James Sincl. Annonaceae. 1 Thailand, W Mal.: *D. marginalis* (Scheffer) James Sincl.

Dasycephala (DC) Hook.f. = *Diodia*

Dasycondylus R. King & H. Robinson (~ *Eupatorium*). Compositae (Eup.-Gyp.). 8 Brazil. R: MSMMBG 22(1987)94

Dasydesmus Craib = *Oreocharis*

Dasylepis Oliver. Flacourtiaceae (2). 6 trop. Afr. R: BJ 92(1972)554

Dasylirion Zucc. Dracaenaceae (Agavaceae s.l.). 15 SW N Am. R: PAPS 50(1911)431. Stemless or tree-like dioec. pachycauls with linear, usu. spiny-edged lvs used for thatching & baskets, the polished leaf-bases sold as curios; sap used as a drink (sotol). Some cult. orn. esp. for 'sub-tropical' bedding; *D. wheeleri* S. Watson a poss. alcohol source

Dasymaschalon (Hook.f. & Thomson) Dalla Torre & Harms = *Desmos*

Dasynotus I.M. Johnson. Boraginaceae. 1 NW US

Dasyochloa Willd. ex Rydb. = *Erioneuron*

Dasyphyllum Kunth. Compositae (Barn.). 40 Chile. R: RMLP n.s. 9(1959)21. Shrubs & trees

Dasypoa Pilger = *Poa*

Dasypogon R. Br. Xanthorrhoeaceae (Dasypogonaceae). 3 SW Aus.

Dasypogonaceae Dumort. See Xanthorrhoeaceae

Dasypyrum (Cosson & Durieu) T. Durand. Gramineae (23). 2 Medit. incl. Eur. R: NJB 11(1991)135

Dasysphaera Volkens ex Gilg. Amaranthaceae (I 2). 4 E Afr.

Dasystachys Baker = *Chlorophytum*

Dasystachys Oersted = *Chamaedorea*

Dasystephana Adans. = *Gentiana*

Dasytropis Urban. Acanthaceae. 1 E Cuba (serpentine)

date *Phoenix dactylifera*; **Chinese d.** *Ziziphus jujuba*; **desert d.** *Balanites* spp,; **dwarf d.** *Phoenix reclinata*; **Indian d.** *Tamarindus indica*; **Trebizond d.** *Elaeagnus angustifolia*

date-plum *Diospyros* spp. esp. *D. kaki, D. lotus, D. virginiana*

Datisca L. Datiscaceae. 1 W N Am., 1 S As. Nitrogen-fixing actinomycete root-symbionts. *D. cannabina* L. (As. Minor to India) – dioec., cult. orn. foliage pl., source of a yellow dye form. much used for silk; *D. glomerata* (Presl) Baillon (W N Am.) – androdioec.

Datiscaceae Bercht. & J. Presl. Dicots – Dilleniidae – Violales. 3/4 As., Mal. to Aus., W N Am. Trees or perennial herbs. Lvs spiral, simple (trees) or pinnate; stipules 0. Fls usu. unisexual (plants dioec. but s.t. polygamous & androdioec. in *Datisca*), ± reg., in axillary spikes to panicles or slender term. leafy infls; male fls with K 4–8, s.t. connate basally, C 0 or 6–8 (*Octomeles*), A same no. as & opp. K or up to 25 with short (long in *Datisca*) anthers with longit. slits, ± pistillode; female (& bisexual) fls with K 3–8 on summit of G, C 0, ± functional A around G summit, G (3–8), 1-loc. with parietal placentas & distinct styles (bifid in *Datisca*), ovules anatropous, bitegmic. Fr. capsular, dehiscing apically between styles; seeds ∞ with v. small straight cylindrical oily embryo & very little or 0 endosperm. n = 11 (*Datisca*), 23 + (*Tetrameles*). Aliso 8(1973)49

Genera: *Datisca, Octomeles, Tetrameles*

Octomeles & *Tetrameles* are monotypic genera of soft-wooded trees & have been segregated from *Datisca* as a separate fam.; seed structure like Begoniaceae, though some allege affinity with Magnoliales

dattock *Detarium senegalense*

Datura L. (excl. *Brugmansia*). Solanaceae (1). 9 S N Am. but widely natur. Many alks. R: A. G. Avery *et al.* (1959) *The genus D.* Annuals with erect fls (cf. *Brugmansia*) & reg. candelabriform branching. *D. ceratocaula* Ortega (Mex.) – semi-aquatic, revered hallucinogen of Aztecs still in use (tropane alks as in all spp.); *D. innoxia* Miller (S N Am.) – sacred hallucinogen in SW N Am., cult. orn. (*D. meteloides*); *D. stramonium* L. (thorn-apple, Jimson or Jamestown weed, N Am. but now widely natur. incl. GB) – stramonium, a drug used in treatment of asthma, comprises dried lvs (alks – hyoscamine & hyoscine) gathered for 300 yrs in Eur., also intoxicant & hallucinogen used by Algonquin Indians (E US) as 'wysoccan' given to boys in adolescent rites over 18–20 days after which they are deemed adults; poll. by evening-flying moths (fls open c. 6 p.m., closing within 24 hrs); seeds remain dormant for many years; subject of intense genetic analysis, single gene mutants incl. pink-flowered forms (*D. tatula* – purple allele dominant over white) & ones with smooth capsules (*D. inermis* – smooth recessive to armed)

Daturicarpa Stapf = *Tabernanthe*

dau *Dipterocarpus costatus*

Daubentonia DC = *Sesbania*

Daubentoniopsis Rydb. = *Sesbania*

Daubenya Lindley. Hyacinthaceae (Liliaceae s.l.). 1 S Afr.: *D. aurea* Lindley

Daucosma Engelm. & A. Gray (~ *Discopleura*). Umbelliferae (III 8). 1 N Am.

Daucus L. Umbelliferae (III 3). c. 22 Eur. (10), Medit., SW & C As., trop. Afr., Aus., NZ & Am. *D. carota* L. complex (ed. carrots to 4.649 kg (GB record) & 2.8 m long), the eastern races with anthocyanin in root, the western with carotene, the eastern app. first domesticated in Afghanistan, the western ones being derived from yellow-rooted eastern ones (subsp. *sativus* (Hoffm.) Arc.), modern stocks deriving from a few C18 Dutch cvs; white-rooted & wild plants belong to subsp. *carota*. Biennial (wild pl. toxic) cult. for food & for animals, when roasted a coffee subs., fr. oil used in flavouring liqueurs, cosmetics etc. After poll. pedicels bend inwards but when frs ripen, spread out again, the burred mericarps adhering to animals; crossed with *D. capilliformis* to breed in resistance to carrot fly

Daumailia Arènes = *Urospermum*

Daumalia Airy Shaw = *Urospermum*

Dauphinea Hedge. Labiatae (VIII 4c). 1 Madagascar

Davallia Sm. Davalliaceae. (Incl. *Araiostegia, Humata, Parasorus* & *Scyphularia*) 34 W Med. (1, incl. Eur.), Himal. & N Japan to Aus. & Tahiti, Afr. & Madag. (2). R: Blumea 39(1994)151. Epiphytes often cult. in hanging baskets or on 'fern-balls' esp. *D. canariensis* (L.) Sm. (hare's-foot fern, SW Med. & Canary Is.), *D. mariesii* Veitch (E As.), etc.

Davalliaceae Mett. ex Frank. Filicopsida. 4/44 trop. & warm OW. Mostly epiphytes with dorsiventral scaly rhiz. & simple to articulate lvs; sori at ends of veins, or dorsal, usu. with pouch-shaped indusium (excl. Oleandraceae). R: Blumea 39(1994)154
Genera: *Davallia, Davallodes, Gymnogrammitis, Leucostegia*

Davalliopsis Bosch = *Cephalomanes*

Davallodes (Copel.) Copel. Davalliaceae. 7 Mal. R: Blumea 37(1992)176

Daveaua Willk. ex Mariz. Compositae (Anth.-Mat.). 1 Portugal, Morocco.

Davidia Baillon. Cornaceae (Davidiaceae). 1 SW China: *D. involucrata* Baillon (dove-tree, ghost-tree). Rare native of *Quercus-Prunus-Corylus* forests, with scented young lvs & large flakes in bark; the 'flower' comprises a condensed infl. of C-less male fls surrounding 1 female reduced to an ovary with a ring of aborted stamens (cf. *Euphorbia*), the whole being subtended by 2(3) large bracts to 30 cm long; the solit. fr. has 6–10 seeds & after drop, the outer layers or pericarp decay & inner ones dehisce to reveal seeds which all germinate at once, 1 eventually crowding out the others, possibly a mechanism to pay off hungry browsers for even the cotyledons have axillary buds which can develop if the apex is grazed off; cult. orn. hardy tree

Davidiaceae (Harms) Li. See Cornaceae

Davidsea Söderstrom & R. Ellis = *Schizostachyum*

Davidson's plum *Davidsonia pruriens*

Davidsonia F. Muell. Davidsoniaceae. 2–3 NE Aus. *D. pruriens* F. Muell. – fr. ed. (preserves)

Davidsoniaceae Bange. Dicots – Rosidae – Rosales, 1/2–3 NE Aus. Small trees with spiral lvs to 1m long, pinnate with toothed leaflets & distally tooth-winged rachis, when young covered with bright red somewhat irritant hairs; stipules large & consp., palmately-veined. Fls bisexual, reg., hypogynous, in ± axillary panicles; K 4 or 5, connate in a tube as long as lobes, valvate, P 0, A 8 or 10 alt. with as many nectary-scales, anthers versatile with apical pores becoming longit. slits, G (2), 2-loc, with distinct styles, ovules apical-axile, c. 5–7 per loc., pendulous, anatropous, bitegmic. Fr. red-velvety becoming plum-like, indehiscent, (1)2-seeded with fleshy mesocarp, the leathery endocarp forming (1)2 flattened pyrenes; endosperm 0
Only genus: *Davidsonia*, poss. allied to Cunoniaceae but with spiral lvs & 0 endosperm

Daviesia Sm. Leguminosae (III 24). 120 Aus., esp. SW. Bitter peas or 'bacon-and-eggs'; *D. arborea* W. Hill (queenwood, NE Aus.) – tree to 14 m

Davilia Mutis = *Llagunea*

Davilla Vand. Dilleniaceae. 20 trop. Am. 2 inner K accrescent in fr., forming leathery casing for it. Most primitive spp. the most restricted

dawn redwood *Metasequoia glyptostroboides*

day lily *Hemerocallis* spp.

Dayaoshania W.T. Wang. Gesneriaceae. 1 E Guangxi, China: *D. cotinifolia* W.T. Wang

dead man's finger *Orchis mascula*; **d.nettle** *Lamium* spp.; **red d.n.** *L. purpureum*; **white d.n.** *L. album*

deadly nightshade *Atropa belladonna*

deal, red or **(Baltic) yellow** *Pinus sylvestris*; **white d.** *Picea abies*

Deamia Britton & Rose = *Selenicereus*

death-camus *Zigadenus* spp.

Debesia Kuntze = *Chlorophyllum*

Debregeasia Gaudich. Urticaceae (III). 4 NE Afr., trop. & warm As. R: KB 43(1988)673, 44(1989)702. Monoec. & dioec. trees & shrubs with achenes enclosed in fleshy K & aggregated into spherical syncarps. Useful fibres & ed. fr. e.g. *D. longifolia* (Burm.f.) Wedd. ('*D. edulis*', janatsi, yanagi, India to Japan & Philipp.) – cult. orn. & *D. saeneb* (Forssk.) Hepper & Wood (*D. salicifolia, D. hypoleuca*, NE Afr. to Tibet & Bhutan)

Decabelone Decne = *Tavaresia*

Decachaena (Hook.) Lindley = *Gaylussacia*

Decachaeta DC. Compositae (Eup.). 7 Mex. & Guatemala. R: MSBMBG 22(1987)406

Decagonocarpus Engl. Rutaceae. 1 Amazonia

Decaisnea Hook.f. & Thomson. Lardizabalaceae. 1 E Himal. to C China: *D. insignis* (Griffith) Hook.f. & Thomson (*D. fargesii*) – pachycaul shrub, not climber like rest of family; primitive wood with scalariform perforation plates & pitting; fr. ed., cult. orn.

Decaisnina Tieghem. Loranthaceae. 25 Java & Philipp. to trop. Aus., Tahiti & Marquesas. R: Blumea 38(1993)70

Decalepidanthus Riedl. Boraginaceae. 1 Pakistan: *D. sericophyllus* Riedl – for 70 yrs known only from type (rediscovered 1971)
Decalepis Wight & Arn. Asclepiadaceae (I; Periplocaceae). 4 India
Decalobanthus Oostr. Convolvulaceae. 1 Sumatra
Decamerium Nutt. = *Gaylussacia*
Decanema Decne. = *Cynanchum*
Decanemopsis Costantin & Gallaud. = ? *Sarcostemma*
Decaphalangium Melchior = *Clusia*
Decaptera Turcz. Cruciferae. 1 Chile
Decarya Choux. Didiereaceae. 1 SW Madag.: *D. madagascariensis* Choux
Decarydendron Danguy. Monimiaceae (V 1). 3 Madag. R: AMBG 72(1985)77
Decaryella A. Camus. Gramineae (33d). 1 Madag.
Decaryia Choux = *Decarya*
Decaryochloa A. Camus. Gramineae (1b). 1 Madag.: *D. diadelpha* A. Camus
Decaschistia Wight & Arn. Malvaceae. 18 India, SE As. to Aus. Some promising fibres
Decaspermum Forster & Forster f. Myrtaceae (Myrt.). 30 Indomal. to W Pacific (Aus. 1). *D. parviflorum* (Lam.) A.J. Scott (Mal.) – bee-poll., cryptically dioec. with male anthesis 20 mins before female
Decastelma Schltr. = *Oxypetalum*
Decastylocarpus Humbert. Compositae (Vern.-Vern.). 1 Madag.: *D. perrieri* Humbert
Decatoca F. Muell. Epacridaceae. 1 New Guinea
Decatropis Hook.f. Rutaceae. 2–3 Mex. & Guatemala
Decazesia F. Muell. Compositae (Gnap.-Ang.). 1 W Aus.: *D. hecatocephala* F. Muell. R: OB 104(1991)127
Decazyx Pittier & S.F. Blake. Rutaceae. 1 Honduras, 1 Mex. R: BSBM 43(1982)1
Deccania Tirvengadum (~ *Randia*). Rubiaceae (II 1). 1 India
Decemium Raf. = *Hydrophyllum*
Deckenia H. Wendl. ex Seemann. Palmae (V 4n). 1 Seychelles. Palm cabbage prized; massive leaf-litter preventing regeneration
Declieuxia Kunth. Rubiaceae (IV 7). 27 trop. Am. savannas. R: MNYBG 28, 4(1976)1
Decodon Gmelin. Lythraceae. 1 E N Am.: *D. verticillatus* (L.) Elliott (fossils in Euras.) – aquatic with *Hippuris* habit; alks; cult. orn.
Decorsea R. Viguier. Leguminosae (III 10). 4 Afr., Madagascar
Decorsella A. Chev. Violaceae. 1 trop. W Afr. Seeds still developing after capsule dehiscence
Decumaria L. Hydrangeaceae. 2–3 E As., E N Am. Lianes with aerial rootlets, sometimes cult. orn.
Decussocarpus Laubenf. = *Retrophyllum*
Dedeckera Rev. & J. Howell. Polygonaceae (I 1). 1 Calif.: *D. eurekensis* Rev. & J. Howell. R: Phytol. 66(1989)238. Fls in summer so monopolizing poll. insects as rest of desert pls 'dormant'
Deeringia R. Br. Amaranthaceae (I 1). 11 OW trop. *D. amaranthoides* (Lam.) Merr. (trop. As.) – young lvs ed., juice with vinegar & onion inhaled to clear nose
Deeringothamnus Small. Annonaceae. 2 Florida. R: Britt. 12 (1960)273
degami *Calycophyllum candidissimum*
Degeneria I. Bailey & A. C. Sm. Degeneriaceae. 2 Fiji. R: JAA 69(1988)277. *D. vitiensis* I. Bailey & A.C. Sm. – poll. by Coleoptera
Degeneriaceae I. Bailey & A.C. Sm. Dicots – Magnoliidae – Magnoliales. 1/2 Fiji. Large glabrous tree with vessel-elements. Lvs spirally arr., simple, entire; stipules 0. Fls solit. on long supra-axillary pedicels, bisexual, reg., hypogynous; K 3, C 12–25 in 3–5 whorls (? or spiral), larger, A ∞ often 20–30 in spiral, laminar, 3-nerved, the 4 microsporangia paired & embedded in abaxial surface between veins with pollen monosulcate, staminodes s.t. present, G 1, largely open at anthesis with stigmatic surface along margins, ovules c. 20–32, laminar, in 1 row near margin of G, anatropous, bitegmic with consp. funicular obturator. Fr. thick ± fleshy but with hard exocarp, poss. dehiscent; seeds ∞ with orange-red sarcotesta, s.t. on dangling funicles (cf. Magnoliaceae), embryo v. small, with 3(4) cotyledons in copious oily ruminate endosperm. 2n = 24

Only genus: *Degeneria*

Primitive features incl. laminar stamens, unsealed G & 3(4) cotyledons but G 1 & vessels more advanced than Magnoliaceae & Winteraceae resp.
Degenia Hayek. Cruciferae. 1 NW Yugoslavia: *D. velebitica* (Degen) Hayek
Degranvillea Determann. Orchidaceae (III 1). 1 Fr. Guiana
Deguelia Aublet = *Derris*

Dehaasia Blume. Lauraceae (I). 35 SE As. to New Guinea. R: BJ 93(1973)427

Deherainia Decne. Theophrastaceae. 2 C Am. R: NJB 9(1989)20. *D. smaragdina* Decne with large green fls, foetid; at anthesis A lie close to stigma but later spring away, the anthers tipped with a fibrous deposit & between lobes are crystal deposits in connective, such also found in pollen: function of movements & crystals not understood but crystal-form a useful character in separating several genera in family

Deianira Cham. & Schldl. Gentianaceae. 7 trop. Am. R: AJBRJ 21(1977)45

Deidamia Thouars. Passifloraceae. 5 Madagascar

Deilanthe N.E. Br. = *Nananthus*

Deinacanthon Mez = *Bromelia*

Deinanthe Maxim. Hydrangeaceae. 2 C China & Japan. Coarse cult. orn. herbs

Deinbollia Schum. & Thonn. Sapindaceae. 40 warm Afr. & Madag.

Deinocheilos W.T. Wang. Gesneriaceae. 2 China

Deinostema Yamaz. (~ *Gratiola*). Scrophulariaceae. 2 E As.

Deinostigma W.T. Wang & Z.Y. Li. Gentianaceae. 1 Vietnam: *D. poilanei* (Pellegrin) W.T. Wang & Z.Y. Li. R: APS 30(1992)356

Deiregyne Schltr. Orchidaceae (III 1). 7 Mex., Guatemala. R: HBML 28(1982)311

Deiregynopsis Rauschert = *Aulosepalum*

Dekinia M. Martens & Galeotti = *Agastache*

Delaetia Backeb. = *Neoporteria*

Delairea Lemaire (~ *Senecio*). Compositae (Sen.-Sen.). 1 S Afr.: *D. odorata* Lemaire (*S. mikanioides*), climber natur. in England & Calif., serious weed in Hawaii, fls scented

Delamerea S. Moore. Compositae (Pluch.). 1 N Kenya: *D. procumbens* S. Moore

Delaportea Thorel ex Gagnepain = *Acacia*

Delarbrea Vieill. Araliaceae. 6 Mal. (1), Queensland (1), New Caled. R: Allertonia 4,3(1986)1. Cult. orn. pachycauls

Delavaya Franchet. Sapindaceae. 1 SW China

Delia Dumort. = *Spergularia*

Delilia Sprengel. Compositae (Helia.). 3 trop. Am. (1: *D. biflora* (L.) Kuntze, natur. W Afr., Cape Verde Is.), Galápagos (2)

Delissea Gaudich. Campanulaceae. 9 Hawaii. R: W.L. Wagner *et al.* (1990) Man. Fl. Pl. Hawaii: 467. Woody

Delonix Raf. Leguminosae (I 1). (Incl. *Aprevalia*) 12 trop. Afr., Madag. (most), India. *D. regia* (Hook.) Raf. (flamboyant, peacock flower, flame-tree, (gul) mohur, Madag., where v. rare & rediscovered there only in 1932) – widely planted street-tree with scarlet fls & lvs with up to 1000 leaflets

Delopyrum Small = *Polygonella*

Delosperma N.E. Br. Aizoaceae (V). 155 S Afr. to Arabia. Succ. shrubs or herbs; some cult. orn.

Delostoma D. Don. Bignoniaceae. 4 Andes

Delphinacanthus Benoist = *Pseudodicliptera*

Delphinium L. Ranunculaceae (I 4). c. 320 N temp. (Eur. 25; R (As.): JAA 48(1967)249, 49(1968)73, 233) to trop. Afr. mts (3–4, R: JAA 48(1967)31, 476). K 5, 1 spurred, C 2 or 4 smaller, upper pair with spurs entering K spur where there is nectar; bee-poll. Many alks & v. toxic; many cult. orn. (R: C. Edwards (1990) *Ds.*). herbaceous perennials (cf. *Consolida*) with complex ancestry involving *D. elatum* L. (Eur.), *D. grandiflorum* L. (As.), etc. (= *D.* × *cultorum* Voss), some 'double', those with red or pink fls involving red-flowered Am. spp. *D. nuttallianum* Pritzel ex Walp. (W N Am.) – toxic to cattle (delphinine) but sheep immune & used to clear pasture; *D. semibarbatum* Bien. ex Boiss. (*D. zalil*, *zalil*, Iran) – yellow fls source of a dye for silk & cotton; *D. staphisagria* L. (stavesacre, S Eur., SW As.) – seeds form. used to control ectoparasites, rats & ants in W Eur. since 1788

Delphyodon Schumann = *Parsonia*

Delpinophytum Speg. Cruciferae. 1 Patagonia

Delpya Pierre ex Radlk. = *Sisyrolepis*

Delpydora Pierre. Sapotaceae (IV). 2 trop. W Afr. R: T.D. Pennington, *S.* (1991)227

Deltaria Steenis. Thymelaeaceae. 1 New Caledonia

Deltocheilos W.T. Wang = *Chirita*

Demavendia Pim. Umbelliferae (III 10). 1 SW & C As.

Demidium DC = *Gnaphalium*

Demosthenesia A.C. Sm. Ericaceae. 12 Andes

Dendranthema (DC) Des Moul. (~ *Chrysanthemum*). Compositae (Anth.-Art.). 37 Eur. (2), C & E As. R: APG 29(1978)168. Garden 'chrysanthemums' (vulgarly 'mums') derived from

D. × *grandiflora* (Ramat.) Kitam. (*C. morifolium*), prob. a complex hybrid group raised in China from *D. indica* (L.) Des Moul. (E As., fls yellow, form. used as insect-fumigant in China & still for sore eyes in herbalism) & other spp.; in China first yellow, then white then purple cvs – 500 by 1630 (monograph written), introd. Eur. 1688 but not established until 1789 (France), s.t. trained as standards on 'artemisia' stocks (Fortune). Korean chrysanthemums are hybrids between this & *D. zawadskii* (Herbich) Tzvelev (*C. erubescens*, Euras.), originally with 'single' heads with 1 row of ligulate florets. All are imp. cut-fls & form. assoc. with nobility in China (Mah Jong tiles; the sign of the Japanese Mikado – (sun symbol, royal stock descended from Sun Goddess) with 16-floreted 'double' (16 outer, 16 inner) exclusive to royal household, 14-floreted 'single' used by royal princes. It is to be hoped that the Latin name, *Chrysanthemum* (q.v.) can be restored for these

Dendriopoterium Svent. = *Sanguisorba*

Dendrobangia Rusby. Icacinaceae. 3 trop. S Am.

Dendrobenthamia Hutch. = *Cornus*

Dendrobium Sw. Orchidaceae (V 15). c. 900 trop. & warm As. to Aus. (71) & Pacific. R (sect. *Oxyglossum*): NRBGE 46(1989)161. Epiphytes with extrafl. nectaries; alks. Many cult. orn. (over 100 in Eur. incl. *D. bigibbum* Lindley (Cooktown orchid, N Aus.), *D. canaliculatum* R. Br. (antelope orchid, N Aus.), *D. crumenatum* Sw. (pigeon orchid, Indomal.) – fls 9 days after cold snap in wild, *D. nobile* Lindley (Himal. to China)) – many cvs, *D. speciosum* Sm. (rock lily, r. orchid, E Aus.) & *D. taurinum* Lindley (Philipp.) – in wild always on *Pterocarpus* spp.; some locally medic., others used for basket-work & bangles etc. (stems) & gum of some Aus. spp. used to fix body colour of Aborigines in N Aus.

Dendrocacalia (Nakai) Tuyama. Compositae (Sen.-Tuss.). 1 Bonin Is.: *D. crepidifolia* (Nakai) Tuyama – shrub

Dendrocalamopsis (Chia & Fung) Keng f. = *Bambusa*

Dendrocalamus Nees. Gramineae (1b). c. 35 India & Sri Lanka to China & Philipp. R: KBAS 13(1986)54. Huge clump-forming woody bamboos, *D. giganteus* Munro (giant bamboo, Burma) to 35 m the largest grass known, growing at c. 46 cm a day at best, stems used for buckets & rafts, when split for chopsticks; *D. asper* (Schultes f.) Heyne (SE As., Malaya) – widely cult., as are other spp., for ed. shoots, most imp. thus in Thailand (phai-tong); *D. strictus* (Roxb.) Nees (male bamboo, Calcutta b., India) – solid stems used for construction, paper pulp, charcoal etc., cult. Hawaii, S N Am., WI etc.

Dendrocereus Britton & Rose = *Acanthocereus*

Dendrochilum Blume. Orchidaceae (IV 12). 100 SE As. & Mal. Cult. orn. epiphytes

Dendrochloa C.E. Parkinson = *Schizostachyum*

Dendrocnide Miq. Urticaceae (I). 37 Indomal., Pacific. R: GBS 25(1969)1. Stinging trees (gympie, 'stingers', bad weeds of regrowth in N Aus.) with painful effects (pollen also irritant; *Alocasia macrorrhizos* alleged antidote), *D. excelsa* (Wedd.) Chew (trop. Aus.) the tallest to 40 m with 2 m buttresses (cf. *Obetia*). Some fibres incl. *D. sinuata* (Blume) Chew (Himal.) for ropes

Dendroconche Copel. = *Microsorum*

Dendrocousinsia Millsp. = *Sebastiania*

Dendroglossa C. Presl = *Leptochilus*

Dendrokingstonia Rauschert (*Kingstonia*). Annonaceae. 1 W Mal.: *D. nervosa* (Hook.f. & Thomson) Rauschert

Dendroleandria Arènes = *Helmiopsiella*

Dendrolobium (Wight & Arn.) Benth. Leguminosae (III 9). 12 trop. As., Indian Ocean, Aus.

Dendromecon Benth. Papaveraceae (II). 1–2 SW N Am. Evergreen shrubs with arils, cult. orn.

Dendromyza Danser. Santalaceae. 7 SE As., Indomal.

Dendropanax Decne & Planchon. Araliaceae. 60 trop. Am. (50), E As., Mal. Some cult. orn. & timber esp. *D. arboreus* (L.) Decne & Planchon (trop. Am.)

Dendropemon (Blume) Reichb. (~ *Phthirusa*). Loranthaceae. 20 WI

Dendrophorbium (Cuatrec.) C. Jeffrey. Compositae (Sen.-Sen.). c. 50 WI, trop. S Am. Woody

Dendrophthoe C. Martius. Loranthaceae. 30 OW trop. (Aus. 6)

Dendrophthora Eichler. Viscaceae. 54 trop. Am. R: Wentia 6(1961)1

Dendrophylax Reichb.f. Orchidaceae (V 16). 6 WI. Leafless epiphytes

Dendrosenecio (Hedb.) R. Nordenstam (~ *Senecio*). Compositae (Sen.-Sen.). 4 C & E Afr. R: KB 34(1973)53. Pachycaul 'giant groundsels' in montane forest & forming groves above treeline (some to c. 250 yrs old), dwarf & creeping spp. in swamps (cf. *Espeletia*, *Lobelia*)

Dendroseris D. Don. Compositae (Lact.-Dend.). 11 Juan Fernandez. Palm-like pachycauls allied to S Am. *Hieracium* spp.

Dendrosicus Raf. = *Amphitecna*

Dendrosicyos Balf.f. Cucurbitaceae. 1 S Arabia, Socotra: *D. socotranus* Balf.f. – only aborescent cucurbit, now threatened by goats though protected by the ± goat-proof *Cissus subaphylla* (Balf.f.) Planchon

Dendrosida Fryxell. Malvaceae. 2 S Mex. R: Brittonia 23(1971)237

Dendrosipanea Ducke. Rubiaceae (I 5). 3 N S Am.

Dendrostellera (C. Meyer) Tieghem = *Diarthron*

Dendrostigma Gleason = *Mayna*

Dendrothrix Esser (~ *Sapium*). Euphorbiaceae. 3 trop. S Am. R: Novon 3(1993)245

Dendrotrophe Miq. Santalaceae. 5 Indomal., S China, SE As., trop. Aus.

Denea Cook = *Howea*

Denekia Thunb. (~ *Amphidoxa*). Compositae (Gnap.). 1 S Afr.: *D. capensis* Thunb. R: OB 104(1991)55

Denhamia Meissner. Celastraceae. 9–10 trop. Aus.

Denisonia F. Muell. = *Pityrodia*

Denisophytum R. Viguier = *Caesalpinia*

Denmoza Britton & Rose. Cactaceae (III 4). 1 W & NW Argentina

Dennettia Baker f. Annonaceae. 1 Nigeria: *D. tripetala* Baker f.

Dennstaedtia Bernh. Dennstaedtiaceae. 45 trop. to warm temp. Some forming thickets on poor soils, *D. glauca* (Cav.) Looser green manure in Peru; some cult. orn.

Dennstaedtiaceae Lotsy. Filicopsida. 16/370 cosmop. Usu. creeping rhizomes usu. with protostele or solenostele usu. articulate scales; scaleless lvs small to large & much-divided. Sori term. on a vein but often on a commissure joining 2 to many vein ends ± indusium. (incl. Lindsaeaceae, excl. Monachosoraceae)

Classification & principal genera:

Saccolomatoideae (dictyostele; sori term. single veins): *Saccoloma* (only)

Dennstaedtioideae (usu. solenostele; sori on 1 to many vein ends): *Dennstaedtia, Hypolepis, Pteridium*

Lindsaeoideae (usu. protostele; sori usu. on commissures): *Lindsaea, Odontosoria*

Some cult. orn. – *Blotiella, Dennstaedtia, Hypolepis, Microlepia*; some weedy (*Hypolepis, Paesia* but esp. *Pteridium* (bracken))

Dentaria L. = *Cardamine*

Dentella Forster & Forster f. Rubiaceae (IV 1). 10 Indomal., Aus., New Caled. (1)

Dentoceras Small = *Polygonella*

denya *Cylicodiscus gabunensis*

deodar *Cedrus deodara*

Depacarpus N.E. Br. = *Meyerophytum*

Depanthus S. Moore. Gesneriaceae. 1 New Caledonia

Deparia Hook. & Grev. (~ *Athyrium*). Dryopteridaceae (II 1). 34 trop. OW (Afr. 1) to Hawaii, N Am. (1). R: JFSUTB III, 13(1984)375

depgul *Lancea tibetica*

Deplanchea Vieill. Bignoniaceae. 5 Mal., Aus., New Caled. Trees. *D. tetraphylla* (R. Br.) F. Muell. (New Guinea, NE Aus.) – bird-poll.

Deppea Cham. & Schldl. Rubiaceae (IV 5). 25 C & S Mex. (most) to C Am., SE Brazil (1). R: Allertonia 4(1988)389

Deprea Raf. (~ *Physalis*). Solanaceae (1). 8 trop. Am.

Derenbergia Schwantes = *Conophytum*

Derenbergiella Schwantes = *Mesembryanthemum*

Dermatobotrys Bolus. Scrophulariaceae. 1 S Afr.: *D. saundersii* Bolus – epiphytic shrub on palms & *Calodendrum capense*

Dermatocalyx Oersted = *Schlegelia*

Derosiphia Raf. Older name for *Podocaelia*

Derris Lour. Leguminosae (III 6). 40 SE As. to N Aus., 1 extending to E Afr. & W Pacific in mangrove (*D. trifoliata* Lour.). Sources of rotenones, efficacious as insecticides & fish-poisons, esp. *D. elliptica* (Roxb.) Benth. (tuba-root, derris root) – chewed in suicide in New Ireland, *D. malaccensis* (Benth.) Prain (trop. As.) & *D. trifoliolata* Lour. (*D. uliginosa*, OW trop.), cult. lianes in trop. for derris powder; some cult. orn.

Derwentia Raf. (~ *Parahebe*). Scrophulariaceae. 8 SE Aus. R: Telopea 5(1992)258

Desbordesia Pierre ex Tieghem. Irvingiaceae (Simaroubaceae s.l.). 1 trop. W Afr. Fr. a samara

Deschampsia Pal. Gramineae (21b). (Incl. *Vahlodea*) c. 40 temp. & cool, esp. N (Eur. 6). Hairgrasses. Tufted growth with rough lvs esp. *D. cespitosa* (L.) Pal. (tussock grass, N hemisph.) – coarse fodder; *D. antarctica* Desv. (S S Am.) – most southerly fl. pl. known (68° 21′ S, Refuge Is.); *D. flexuosa* (L.) Trin. (N temp, S S Am.) – bipolar distrib.

Descurainia Webb & Berth. Cruciferae. 40 temp. & cool N (Eur. 1), S Afr. R (N Am.): AMN 22(1939)481. Minor oilseeds esp. *D. pinnata* (Walter) Britton (N Am.); *D. sophia* (L.) Webb ex Prantl (flixweed, Euras., natur. N Am.) – seeds used like mustard, attract & trap mosquito larvae

Desdemona S. Moore = *Basistemon*

desert pea (Sturt('s)) *Swainsona formosa*; **d. rose** *Adenium obesum*; **Sturt's d. r.** *Gossypium sturtianum*

Desfontainia Ruíz & Pavón. Desfontainiaceae (Loganiaceae s.l.). 1 Andes (Costa Rica to Cape Horn): *D. spinosa* Ruíz & Pavón – lvs holly-like but opp., fls scarlet & yellow poll. green-backed firecrown hummingbird, cult. orn.; source of yellow dye for cloth; tea from lvs medic. & allegedly hallucinogenic

Desfontainiaceae Endl. (Loganiaceae s.l.). Dicots – Asteridae – (?) Dipsacales. 1/1 Andes. Excluded from Loganiaceae because of fr. (berry) etc.
Genus: *Desfontainia* (q.v.)

Desideria Pampan. = *Christolea*

Desmanthodium Benth. Compositae (Helia.). 8 Mex. & C Am. mts

Desmanthus Willd. Leguminosae (II 3). 24 warm Am. R: SBM 38(1993). Mostly herbs. *D. pernambucanus* (L.) Thell. – pantrop. weed

Desmaria Tieghem = *Loranthus*

Desmazeria Dumort. Gramineae (17). (Excl. *Catapodium*) 1 Med. (incl. Eur.). R: BJ 94(1974)556. Fern grass

Desmidorchis Ehrenb. = *Caralluma* (but see Adansonia 19(1980)322)

Desmodiastrum (Prain) Praminik & Thoth. = *Alysicarpus*

Desmodium Desv. Leguminosae (III 9). 450 warm esp. E As., Brazil, Mex. (excl. *D. gyrans* = *Codariocalyx*). R: Ginkgoana 1(1973)1 (As. spp.). Legumes fall into 1-seeded units spread on animals, trousers etc. (beggarweed). Alks; some medic. locally; some fodders & green manure; some light timbers (*Ougeinia*) e.g. *D. oojeinense* (Roxb.) Ohashi (Himal.). *D. gangeticum* (L.) DC (OW trop.) – fibres allegedly valuable for paper-making; *D. triflorum* (L.) DC (trop.) – form. used for lawns by whites in Bismarcks

Desmogymnosiphon Guinea. Burmanniaceae. 1 trop. W Afr.

Desmoncus C. Martius. Palmae (V 5e). c. 65 trop. Am. Lianes superficially like *Calamus*. Some ed. fr. & cult. orn. incl. *D. orthacanthos* C. Martius (*D. major*, picmoc, Trinidad to E S Am.) – stems used for basketry

Desmopsis Safford. Annonaceae. 17 Mexico to Cuba

Desmos Lour. Annonaceae. 30 Indomal. to W Pacific. Fr. a head of berries, some ed.; infl.-axes act as grapples in climbing. *D. chinensis* Lour. (Assam & S China to Philipp.) grown as living fence in Java

Desmoscelis Naudin. Melastomataceae. 1 trop. S Am.

Desmoschoenus Hook.f. = *Scirpus*

Desmostachya (Stapf) Stapf (*Stapfiola*). Gramineae (31d). 1 N Afr. & Middle E to India & SE As.: *D. bipinnata* (L.) Stapf – used for cheap ropes, baskets & whips used as bird-scarers (hieroglyphic sign of Ancient Egyptians), ceremonial armbands in Hindu funerals

Desmostachys Miers. Icacinaceae. 7 trop. Afr. & Madagascar

Desmotes Kallunki (~ *Erythrochiton*). Rutaceae. 1 Oiba Is., Panamá: *D. incomparabilis* (Riley) Kallunki

Desmothamnus Small = *Lyonia*

Desplatsia Bocquillon. Tiliaceae. 7 trop. W Afr.

Detarium Juss. Leguminosae (I 4). 3 W Afr. forests & Sudanian savanna. *D. senegalense* J. Gmelin (dattock, tallow-tree, W Afr.) – grey timber mahogany subs.; fr. (round indehiscent & drupe-like) ed.

determa *Ocotea rubra*

Dethawia Endl. Umbelliferae (III 8). 1 Pyrenees: *D. tenuifolia* (DC) Godron

Detzneria Schltr. ex Diels. Scrophulariaceae. 1 New Guinea: *D. tubata* Diels (= ? *Chionohebe*)

Deuterocohnia Mez. Bromeliaceae (1). 7 S Am. R: FN 14(1974)231. Terrestrial shrubs with extrafl. nectaries

Deuteromallotus Pax & K. Hoffm. Euphorbiaceae. 1 Madagascar

Deutzia Thunb. Hydrangeaceae. c. 60 temp. As. to Philipp., C Am. mts. Seeds small, winged. R: T.I. Zaikonnikova (1966) *Deitsii-Dekorativnye Kustarniki* (see Baileya 19 (1975)133). Many (50 in US) cult. orn. shrubs esp. hybrids between v. similar spp. & cvs

Deutzianthus Gagnepain. Euphorbiaceae. 1 SE As.

Deverra DC (*Pituranthos*). Umbelliferae (III 8). 7 OW desert & semi-desert. *D. triradiata* Hochst. ex Boiss. (Arabia) – lvs with 0.6–1.7% dry wt. furanocoumarins leading to photosensitization in grazers

Devia Goldbl. & Manning. Iridaceae (IV 3). 1 W Karoo

devil, blue *Eryngium ovinum*; **Mexican d.** *Ageratina adenophora*

Devillea Tul. & Wedd. Podostemaceae. 1 Brazil

devil's bit *Succisa pratensis*; **d.'s cotton** *Abroma augusta*; **d.'s guts** *Cuscuta* spp., *Equisetum* spp.; **d.'s paintbrush** *Pilosella aurantiaca*

dewberry *Rubus* spp. esp. (GB) *R. caesius*, but applied in US to trailing blackberries & *R. vitifolius*

dewdrop, golden *Duranta erecta*

Dewevrea M. Micheli. Leguminosae (III 6). 1–2 W Afr.

Dewevrella De Wild. Apocynaceae. 1 trop. Afr.: *D. cochliostema* De Wild. R: AUWP 85-2(1985)67

Dewildemania O. Hoffm. Compositae (Vern.-Vern.). 3 trop. Afr.

Dewindtia De Wild. = *Cryptosepalum*

Deyeuxia Clarion ex Pal. (~ *Calamagrostis*). Gramineae (20c). c. 40 temp. Australasia

dhaincha *Sesbania bispinosa*

dhak *Butea monosperma*

dhal *Cajanus cajan*

dhani *Coriandrum sativum*

dhawa *Anogeissus latifolia*

Dhofaria A. Miller. Capparidaceae. 1 Oman: *D. macleishii* A. Miller. R: NRBGE 45(1988)55. Allied to Aus. *Apophyllum*!

dhop or **dhub** *Cynodon dactylon*

dhup *Canarium euphyllum*; **red d.** *Parishia insignis*

dhupa fat *Vateria indica*

Diacalpe Blume = *Peranema*

Diacarpa Sim = *Atalaya*

Diachyrium Griseb. = *Sporobolus*

Diacidia Griseb. Malpighiaceae. 12 trop. Am. R: MNYBG 32(1981)61

Diacranthera R. King & H. Robinson. Compositae (Eup.-Gyp.). 2 Brazil. R: MSBMBG 22(1987)94

Diacrodon Sprague. Rubiaceae (IV 15). 1 Brazil

Diadenium Poeppig & Endl. Orchidaceae (V 10). 2 W trop. S Am.

Dialium L. Leguminosae (I ?2). c. 40 trop. (Am. 1). C 0–2, A 2(3), fr. ± globose indehiscent 1–2-seeded; extrafl. nectaries. *D. guineense* Willd. (trop. W Afr.) & *D. ovoideum* Thw. (Sri Lanka) – velvet tamarinds, orange-red ed. fr. pulp; other spp. with ed. fr. incl. *D. indum* L. (Mal.) – good timber

Dialyanthera Warb. = *Otoba*

Dialyceras Capuron. Ochnaceae (Diegodendraceae). 3 Madagascar

Dialypetalanthaceae Rizz. & Occh. Dicots – Rosidae – Rosales. 1/1 Brazil. Tree with tomentose twigs. Lvs opp., simple, entire; stipules intrapetiolar, large, pairs laterally connate basally. Infl. a term. thyrse of bisexual, reg. fls each with 2 bracteoles. K 2 + 2, C 2 + 2, A (16–)18(–25) in 2 whorls, filaments basally connate, anthers with term. pores, nectary-disk a ring at apex of $\overline{G}$ (2), 2-loc. with 1 style, ovules ∞, anatropous, bitegmic borne on axile placentas. Fr. a septifragal capsule with persistent K; seeds ∞, elongate, fusiform with straight embryo in thin oily endosperm

Only species: *Dialypetalanthus fuscescens* Kuhlm.

Similar to Rubiaceae in lvs, stipules etc. but free C & A ∞ better suggest Myrtales, though Cronquist places it in 'the rather amorphous' Rosales

Dialypetalanthus Kuhlm. Dialypetalanthaceae. 1 E Brazil

Dialypetalum Benth. Campanulaceae. 5 Madag. Pithy shrubs

Dialytheca Exell & Mendonça. Menispermaceae (III). 1 Angola: *D. gossweileri* Exell & Mendonça

Diamena Ravenna = *Anthericum*

Diamorpha Nutt. Crassulaceae. 1 SE US: *D. smallii* Britton, on granite outcrops, easily manipulated experimental pl., ? ant-poll.

Diandriella Engl. = *Homalomena*

Diandrochloa De Winter = *Eragrostis*

Diandrolyra Stapf. Gramineae (4). 7 SE Brazil. Bamboos

Diandrostachya (C. Hubb.) Jacq.-Fél. = *Loudetiopsis*

Dianella Lam. ex Juss. Phormiaceae (Liliaceae s.l.). 20 trop. E Afr., Indomal., Madag., W Pacific. R: MBMUZ 163(1940)5. Some cult. orn. incl. epiphytes; berries blue, a dye-source in Hawaii, *D. nigra* Colenso the blueberry of NZ; *D. caerulea* Sims (NSW) – 1 of first 6 Aus. pls introd. UK (1771); *D. ensifolia* (L.) DC (range of genus) – germ. after 3 months (embryo immature until then)

Dianellaceae Salisb. = Phormiaceae

Dianthoseris Schultz-Bip. ex A. Rich. Compositae (Lact.-Crep.). 1 E & NE Afr. highlands: *D. schimperi* Schultz-Bip. ex A. Rich.

Dianthoveus Hammel & Wilder. Cyclanthaceae. 1 SW Colombia to N Ecuador: *D. cremnophilus* Hammel & Wilder

Dianthus L. Caryophyllaceae (III 1). c. 300 Euras. (Eur. 115) to Afr. mts (few), N Am. (1). Pinks. Strongly-scented fls visited by butterflies (cf. *Daphne*). Many cult. orn. esp. rock-pls & many hybrids (over 30 000 cvs recorded), some form. used in mulling ale. *D. arboreus* L. (SE Eur.) – shrub to 1 m, poss. the pink in the murals at Knossos; *D. armeria* L. (Deptford pink, Eur. to Iran); *D. barbatus* L. (Sweet William, Eur. mts, natur. China & N Am.); *D. caryophyllus* L. (carnation, clove pink, ? Medit.) – source of oil for soap & scent, cult. by Moors in Valencia 1460, now comm. cut-fls (trad. in Liguria), though some carnations are hybrids (= *D.* × *allwoodii* Hort. Allwood) with *D. plumarius*, with variously coloured fls (e.g. picotee, orig. speckled, now edged. with diff. colour), subject of florists' societies C17 & 18 (pinks in C18); *D. chinensis* L. (Chinese p., C & E China) – esp. 'Heddewigii' (Japanese p.) flowering in first year from seed; *D. deltoides* L. (maiden p., meadow p., Eur.); *D. gratianopolitanus* Villars (*D. caesius*, Cheddar p., Eur. incl. GB where protected) – poll. butterflies, diurnal hawkmoths, diurnal & nocturnal noctuids (nectar mostly sucrose but high in amino-acids); *D. plumarius* L. (clove p., C Eur.) – origin of most 'pinks', involved in carnation breeding

Diapensia L. Diapensiaceae. 4 Himal. & W China, circumboreal (1: *D. lapponica* L., arctic-alpine not discovered in Scotland before 1951, now protected). R: Taxon 32(1983)419

Diapensiaceae Lindley. Dicots – Dilleniidae – Theales. 6/15 arctic & N temp. to Himal. R: Taxon 32(1983)417. Subshrubs or perennial herbs, mycorrhizal. Lvs spirally arr., stipules 0. Fls solit. or in racemes, reg., bisexual, K 5 or (5), imbricate, C (5), almost distinct in *Galax*, imbricate or convolute, A usu. 5 alt. with C & attached to tube with 5 staminodes (0 in *Diapensia* & *Pyxidanthera*) opp. C, all sometimes connate, this tube adnate to C-tube & falling with it in *Galax*, anthers with longit. (transverse in *Pyxidanthera*) slits, nectary a weakly developed ring at base of ovary or 0, $\underline{G}$(3), 3-loc., with 3-lobed stigma, ovules several–∞, anatropous to hemitropous or campylotropous, unitegmic. Fr. a 3-valved loculicidal or denticidal capsule; seeds small with copious fleshy endosperm around ± straight embryo. x = 6

Genera: *Berneuxia, Diapensia, Diplarche, Galax, Pyxidanthera, Shortia*

Often included in Ericales but differ in A, pollen not in tetrads, shed from slits & not term. pores. An app. relict group (sometimes placed in own order, Diapensiales), with low chromosome numbers & a marked inability for seedling regeneration

Some cult. orn. esp. ground cover (*Galax*)

Diaperia Nutt. = *Filago*

Diaphananthe Schltr. Orchidaceae (V 16). 20 trop. Afr. R: KB 14(1960)140

Diaphanoptera Rech.f. (~ *Acanthophyllum*). Caryophyllaceae (III 1). 6 NE Iran to Afghanistan

Diaphractanthus Humbert. Compositae (Vern.-Vern.). 1 Madag.: *D. homolepis* Humbert

Diarrhena Pal. Gramineae (12). 4 E As., N Am. (2, R: BTBC 118(1991)128). ? Relics of wide-spread Tertiary woodland

Diarthron Turcz. Thymelaeaceae. 19 Eur. (1) to W As. R: NRBGE 40(1982)216

Diascia Link & Otto. Scrophulariaceae. 38 S Afr. *D. barberae* Hook.f. cult. ann. with pink fls. Melittid bees (*Rediviva* spp.) with long legs orientated by translucent 'windows' at base of upper C lip (ultraviolet-absorbing cf. surroundings) harvest oils from trichomal elaio-phores at apices of long spurs of fls by rubbing forelegs against trichomes & poll. (co-evolution)

Diaspananthus Miq. = *Ainsliaea*

Diaspasis R. Br. Goodeniaceae. 1 SW Aus.: *D. filifolia* R. Br. R: FA 35(1992)146

Diastatea Scheidw. Campanulaceae. 7 Mex. (most) to Peru

Diastella Salisb. Proteaceae. 7 S Afr. R: JSAB 42(1976)185

Diastema Benth. Gesneriaceae. 20 trop. Am. Herbs with scaly underground rhizomes (like *Achimenes*); some cult. orn.

Diateinacanthus Lindau = *Odontonema*

Diatenopteryx Radlk. Sapindaceae. 2 S Am.

Diblemma J. Sm. = *Microsorum*

Dibrachionostylus Bremek. Rubiaceae (IV 1). 1 trop. E Afr.

Dicarpellum (Loesener) A.C. Sm. = *Salacia*

Dicarpidium F. Muell. Sterculiaceae. 1 Aus.

Dicarpophora Speg. Asclepiadaceae (III 1). 1 Bolivia: *D. mazzuchii* Speg.

Dicella Griseb. Malpighiaceae. 6 trop. S Am.

Dicellandra Hook.f. Melastomataceae. 3 trop. W Afr. R: Adansonia 14(1974)77

Dicellostyles Benth. Malvaceae. 1 Sri Lanka

Dicentra Benth. Fumariaceae (I 1). 12 As. & N Am. R: Brittonia 13(1961)1, OB 88(1986)8. Alks. Fls pendulous, each outer C with large basal pouch, inner C spoon-shaped cohering at tips so as to form a hood covering A & G; bees hang on pend. fl. & probe for nectar in the pouches & in so doing push aside the hood & touch stigma on which there is usu. pollen from same fl. Cult. orn. & local medic. *D. canadensis* (Goldie) Walp. (squirrel-corn, E N Am.) – many small yellow tubers, fls whitish; *D. formosa* (Haw.) Walp. – common in gdns, natur. GB; *D. scandens* (D. Don) Walp. (Sino-Himal.) – climber to 4.5 m with leaflet-tendrils, fls yellowish; *D. spectabilis* (L.) Lemaire (bleeding hearts, Dutchman's breeches, Japan) – most commonly cult. 'old-fashioned' (not introd. until 1846) – red fls

Dicerandra Benth. Labiatae (VIII 2). 8 SE US. R: SB 14(1989)197

Diceratella Boiss. Cruciferae. 7 NE trop. Afr. (5), Iran (2)

Diceratostele Summerh. Orchidaceae (III 1). 1 trop. Afr.

Dicerma DC. Leguminosae (III 9). 1 trop. As. to Aus.: *D. biarticulatum* (L.) DC

Dicerocaryum Bojer. Pedaliaceae. 3 S trop. Afr. R: MSB 25(1975)1

Diceroclados C. Jeffrey & Y.L. Chen. Compositae (Sen.-Tuss.). 1 China

Dicerospermum Bakh.f. = *Poikilogyne*

Dicerostylis Blume. Orchidaceae (III 3). 3 Taiwan to Mal.

Dichaea Lindley. Orchidaceae (V 10). 55 trop. Am. Cult. orn.

Dichaelia Harvey = *Brachystelma*

Dichaetanthera Endl. Melastomataceae. 34 trop. Afr. (7), Madagascar (27)

Dichaetaria Nees ex Steudel. Gramineae (25). 1 India, Sri Lanka

Dichaetophora A. Gray. Compositae (Ast.-Ast.). 1 S US, N Mex.: *D. campestris* A. Gray. R: Wrightia 1(1946)90.

Dichanthelium (Hitchc. & Chase) Gould = *Panicum*

Dichanthium Willemet. Gramineae (40c). 20 OW trop. (Eur. 2 natur.). R (sect. D.): BSAB 12(1968)206

Dichapetalaceae Baillon. Dicots – Rosidae – Celastrales. 3/160 trop. (& S Afr. 1). Trees, shrubs or lianes, usu. v. poisonous (fluoroacetic acid & pyridine alks) & with char. unicellular hairs with warty papillae. Lvs spirally arr., simple, entire; stipules usu. caducous. Fls in axillary to petiolar or epiphyllous cymes, usu. bisexual & reg. (± irreg. in *Tapura*) with articulated pedicels; K (4)5, imbricate, s.t. basally connate, C (4)5, usu. 2-lobed or bifid, imbricate, rarely with basal tube, A(4)5, alt. with C, rarely 3 with 2 staminodes, anthers with longit. slits, basal nectary gland opp. each C (confluent when tube present), $\underline{G}$ to $\overline{G}$ (2, 3(4)), pluriloc. with term. style, lobed or rarely distinct styles, ovules 2 per loc., apical-axile, pendulous, anatropous. Fr. a drupe with 1(–3)-loc. stone usu. with 1 seed per loc., exocarp s.t. splitting; seed with large straight oily embryo & no endosperm, often with caruncle. 2n = 20, 24

Genera: *Dichapetalum* with satellites, *Stephanopodium* & *Tapura*

Prob. best moved to Euphorbiales or Linales

Some cult. orn. & sources of poisons

Dichapetalum Thouars. Dichapetalaceae. 124 trop. (Indomal. 19, trop. Afr. & Madag. 86, Am. 19). R: AUWP 86-3(1986). Many extremely poisonous & used to kill wild pigs, monkeys & rats in Afr. (fluoroacetic acid disrupts tricarboxylic acid cycle of respiration); *D. cymosum* (Hook.) Engl. of high veldt of S Afr. begins growth before veldt grasses & is therefore eaten by cattle leading to 'gifblaar' poisoning & death. Many spp. with epiphyllous infls

Dichasianthus Ovcz. & Yanusov = *Torularia*

Dichazothece Lindau. Acanthaceae. 1 E Brazil

Dichelachne Endl. Gramineae (21d). 9 E Mal., Aus., NZ, introd. Hawaii. R: Blumea 22(1974)5, NZJB 20(1982)303. Some pasture grasses; *D. crinita* (L.) Hook.f. (Aus. to Pacific) – stems for paper-making

Dichelostemma Kunth (~ *Brodiaea*). Alliaceae (Liliaceae s.l.). 5 N Am. R: Four Seasons

9,1(1991)24. Cult. orn. esp. *D. idamaia* (Alph. Wood) E. Greene (*Brevoortia i., Brodiaea i.*, Calif., Oregon) – scarlet fls; *D. volubile* (Kellogg) A. Heller (California) – fls pink, stem flexuous & twining to 1.5 m

Dicheranthus Webb. Caryophyllaceae (I 2). 1 Canary Is.: *D. plocamoides* Webb

Dichilanthe Thwaites. Rubiaceae (III 3). 2 Sri Lanka, Borneo

Dichiloboea Stapf = *Trisepalum*

Dichilus DC. Leguminosae (III 28). 5 S Afr. R: SAJB 54(1988)182. Alks

Dichocarpum W.T. Wang & Hsiao (~ *Isopyrum*). Ranunculaceae (III 2). 20 Himal., E As. R: APS 26(1988)249

Dichodon (Reichb.) Reichb. = *Cerastium*

Dichoglottis Fischer & C. Meyer = *Gypsophila*

Dichondra Forster & Forster f. Convolvulaceae. 9 trop. & warm (1 natur. Eur. incl. Cornwall). R (partial): Brittonia 13(1961)346. *D. micrantha* Urban (E As., etc.) – grown as lawn-grass subs.

Dichondraceae Dumort. = Convolvulaceae

Dichondropsis Brandegee = *Dichondra*

Dichopogon Kunth (~ *Arthropodium*). Anthericaceae (Liliaceae s.l.). 6 Aus., 1 ext. to New Guinea. R: FA 45(1987)345

Dichorisandra Mikan. Commelinaceae (II 1e). 25 trop. S Am., 1 extending to C Am. & WI. Infls racemes or thyrses; seeds arillate. Some cult. orn. housepls esp. *D. reginae* (L. Linden & Rodigas) W. Ludw. (*Tradescantia r.*, Peru)

Dichosciadium Domin. Umbelliferae (I 2b). 1 Aus.: *D. ranunculaceum* (Hook.f.) Domin

Dichostemma Pierre. Euphorbiaceae. 3 trop. Afr.

Dichotomanthes Kurz. Rosaceae (III). 1 SW China: *D. tristaniicarpa* Kurz. Differs from *Cotoneaster* in fleshy K around dry fr.

Dichroa Lour. Hydrangeaceae. 12 China to Mal. R: APS 25(1987)388. Alks. *D. febrifuga* Lour. used against fever

Dichrocephala L'Hérit. ex DC. Compositae (Ast.-Gran.). 4 OW trop. R: MBSM 15(1979)491

Dichromanthus Garay. Orchidaceae (III 1). 1 Mex., Guatemala

Dichromena Michaux = *Rhynchospora*

Dichromochlamys Dunlop. Compositae (Ast.-Ast.). 1 Aus.: *D. dentatifolia* (F. Muell.) Dunlop. R: JABG 2(1980)235

Dichrospermum Bremek. = *Spermacoce*

Dichrostachys (DC) Wight & Arn. Leguminosae (II 3). 12 trop. OW (1 pantrop., Aus. 1) esp. Madag. C 0, basal fls in spike sterile with 10 staminodes, term. ones bisexual with 10 short A; some with stipular thorns e.g. *D. cinerea* (L.) Wight & Arn. (Afr., India) – used as a thorny hedge but now a pest in e.g. Cuba

Dichrotrichum Reinw. ex Vriese = *Agalmyla*

Dickasonia L.O. Williams. Orchidaceae (V 12). 1 Burma: *D. vernicosa* L.O. Williams

Dickinsia Franchet. Umbelliferae (I 1a). 1 SW China: *D. hydrocotyloides* Franchet

Dicksonia L'Hérit. Dicksoniaceae. 20–25 trop. Am., St Helena, Mal. (mts, esp. New Guinea), Aus., New Caled., NZ. Differs from *Cyathea* in indumentum having only hairs & no scales on young parts. All Aus. spp. protected. Some cult. orn. greenhouse tree-ferns esp. *D. antarctica* Labill. (Aus.) to 15 m, pith a starch source for aborigines

Dicksoniaceae (C. Presl) Bower (~ Cyatheaceae; incl. Culcitaceae, Thyrsopteridaceae, excl. Lophosoriaceae). Filicopsida. 6/20 trop. Am., Mal., SW Pacific, St Helena. Tree-ferns, trunk prostrate in *Cystodium*, with young parts without scales; fronds 2- or more-pinnate; sori term. at ends of veins, indusium with modified lobe of frond segment giving a cup- or box-shaped structure

Genera: *Calochlaena, Cibotium, Culcita, Cystodium, Dicksonia, Thyrsopteris*

App. relict group (many fossils referred here) diff. from Cyatheaceae in marginal sori

Dicladanthera F. Muell. Acanthaceae. 2 W Aus. R: JABG 9(1986)171

Diclidanthera C. Martius. Polygalaceae. 8 trop. S Am.

Diclidantheraceae J. Agardh = Polygalaceae

Diclinanona Diels. Annonaceae. 3 E Peru, W Brazil

Dicliptera Juss. Acanthaceae. c. 150 trop. & warm. Some cult. orn. e.g. *D. dodsonii* Wassh. (W Ecuador, only 1 wild pl. known) – lianoid; *D. javanica* Nees (Mal.) – normal & cleistogamous fls

Diclis Benth. Scrophulariaceae. 10 trop. & S Afr., Madagascar

Dicoelia Benth. Euphorbiaceae. 2–3 W Mal.

Dicoelospermum C.B. Clarke. Cucurbitaceae. 1 S India

Dicoma Cass. Compositae (Mut.-Mut.). 65 trop. & S Afr., Madag., Socotra, India (1). Trees to herbs

Dicoria Torrey & A. Gray. Compositae (Helia.-Amb.). 5 SW N Am. Fr. winged, pappus present

Dicorynia Benth. Leguminosae (I 2). 1–2 trop. S Am. A 2. *D. guianensis* Amshoff (angelique, basralocus) – timber

Dicoryphe Thouars. Hamamelidaceae (I 1). 15 Madag., Comoro Is. Some rheophytes

Dicotyledonae DC (Magnoliopsida). Angiospermae. 10 463/193 500 here arranged in 321 fams. Seedlings usu. with 2 cotyledons. Lvs usu. net-veined. Fls often 5-merous. Monocotyledonae much less diversified. See the modified Cronquist System – (this volume, p. **772**)

Dicraeanthus Engl. Podostemaceae. 4 W Afr. *D. africanus* Engl. eaten as salad

Dicraeia Thouars = *Podostemum*

Dicraeopetalum Harms. Leguminosae (III 2). 1 SE Ethiopia, S Somalia, NE Kenya

Dicranocarpus A. Gray. Compositae (Helia.-Cor.). 1 SW N Am.: *D. parviflorus* A. Gray – some fr. without pappus

Dicranoglossum J. Sm. Polypodiaceae (II 5). 6 trop. Am.

Dicranolepis Planchon. Thymelaeaceae. 15 trop. Afr.

Dicranopteris Bernh. Gleicheniaceae (Gleich.). 12 trop. esp. Mal. & S temp. Forming dense tangles particularly on poor soils & after fires. *D. pectinata* (Willd.) Underw. (trop. Am.) – only solenostele in G. *D. linearis* (Burm.f.) Underw. (trop. As.) – stems used for fish-traps (last 2 yrs in saltwater), plaiting, chair-seats, hut-walls, mats (e.g. for tea-shade) etc., & split as pens for writing Arabic, veg. in Sumatra

Dicranopygium Harling. Cyclanthaceae. 49 trop. Am. R: AHB 18(1958)274

Dicranostegia (A. Gray) Pennell = *Cordylanthus*

Dicranostigma Hook.f. & Thomson. Papaveraceae (I). 3–5 Himal. W China. Alks. Cult. orn. herbs

Dicranostyles Benth. Convolvulaceae. 15 trop. S Am. R: AMBG 60(1975)385. Lianes to 30 m

Dicraspidia Standley. Tiliaceae. 1 C Am.

Dicrastylidaceae J.L. Drumm. ex Harvey (= Chloanthaceae). See Labiatae (III)

Dicrastylis J.L. Drumm. ex Harvey. Labiatae (III; Chloanthaceae). 28 Aus. R: Brunonia 1(1978)437, JABG 14(1991)85

Dicraurus Hook.f. = *Iresine*

Dicrocaulon N.E. Br. Aizoaceae (V). 12 Cape

Dictamnus L. Rutaceae. 1 C & S Eur. to N China: *D. albus* L. (*D. fraxinella*, burning bush, dittany) – herb. perennial with aromatic lvs & irreg. fls with 10 upward-pointing A in term. racemes, the stems covered with glands which release a volatile inflammable oil which may be lit without harming the plant; causes allergy in some

Dictyandra Welw. ex Hook.f. Rubiaceae (II 2). 2 trop. W Afr. R: PSE 145(1984)105. K large

Dictyanthus Decne = *Matelea*

Dictymia J. Sm. Polypodiaceae (II 4). 2–3 Aus. & New Guinea to Fiji. *D. brownii* (Wikström) Copel. (W Pacific) – cult. orn. with crassulacean acid metabolism

Dictyocaryum H. Wendl. Palmae (V 2a). 1–2 trop. S Am.

Dictyochloa (Murb.) Camus = *Ammochloa*

Dictyocline T. Moore = *Cyclosorus*

Dictyodroma Ching (~ *Diplazium*). Dryopteridaceae (II 1). 2–3 E & SE As.

Dictyolimon Rech.f. (~ *Limonium*). Plumbaginaceae (II). 4 Afghanistan to India

Dictyoloma A. Juss. Rutaceae. 2 Peru, Brazil

Dictyoneura Blume. Sapindaceae. 2 Mal.

Dictyophleba Pierre. Apocynaceae. 5 trop. Afr. R: BJBB 59(1989)207

Dictyophragmus O. Schulz. Cruciferae. 2 Peru, Arg. R: Novon 1(1991)71

Dictyophyllaria Garay. Orchidaceae (V 4). 1 trop. Am.

Dictyosperma H. Wendl. & Drude. Palmae (V 4m). 1 Masc.: *D. album* (Bory) R. Scheffer, now rare in wild through over-exploitation of the cabbage, but cult. orn. in tropics

Dictyospermum Wight (~ *Aneilema*). Commelinaceae (II 2). 5 As.

Dictyostega Miers. Burmanniaceae. 2 trop. Am.

Dictyoxiphium Hook. = *Tectaria*

Dicyclophora Boiss. Umbelliferae (III 1). 1 Iran: *D. persica* Boiss.

Dicymanthes Danser = *Amyema*

Dicymbe Spruce ex Benth. Leguminosae (I 4). 16 trop. S Am., esp. Amazonia

Dicymbopsis Ducke = *Dicymbe*

Dicypellium Nees & C. Martius. Lauraceae (I). 2 E Amaz. R: BJ 110(1988)168. *D. caryophyllaceum* (C. Martius) Nees & C. Martius – bark sold in quills (*Cassia caryophyllata*) smells like cloves (oil has 95% eugenol), used as flavouring & with lvs in a stimulant tea, timber good

Dicyrta Regel = *Achimenes*

Didelotia Baillon. Leguminosae (I 5). 9 Guineo-Congolian forests, Afr. R: Blumea 12(1964)209

Didelta L'Hérit. Compositae (Arct.). 2 SW Afr. R: MBSM 3(1959)304

Didesmandra Stapf. Dilleniaceae. 1 Borneo

Didesmus Desv. Cruciferae. 2 E Med. (Eur. 1)

Didiciea King & Prain. Orchidaceae (V 8). 2 E Himal., Japan

Didierea Baillon. Didiereaceae. 2 S & SW Madagascar

Didiereaceae Radlk. Dicots – Caryophyllidae – Caryophyllales. 4/11 Madagascar. Spiny pachycaul xerophytes, usu. dioec. (rarely gynodioec.), with betalains (not anthocyanins); wood-formation reg. Spiny short axes in axils of lvs of long axes. Phloem sieve-tube plastids of P-type with central globular protein crystalloid & subperipheral ring of proteinaceous filaments. Lvs spirally arr., entire, small; stipules 0. Fls in term. ± dichasial cymes with 2 involucral bracts; K 2, petaloid (but derived from bracts?), C 2 + 2 (derived from K?), A (6)8(10), the filaments adnate to outside of annular nectary, staminodes often in female fls, G ((2)3(4)) with single style & (2)3(4)-lobed stigma & as many loc., but only 1 fert. with 1 basal, erect, campylotropous, bitegmic ovule. Fr. dry, indehiscent, usu. 3-angled, encl. in involucral bracts; seeds with small funicular aril & large curved or folded embryo with almost 0 endosperm or perisperm. 2n = c. 150, 190–200. R: FMad. 121(1963)

Genera: *Alluaudia, Alluaudiopsis, Decaryia, Didierea*

Serology, sieve-tube plastid features, betalains & pollen ornamentation place D. in the 'Centrospermae', which is supported by some being successfully grafted on Cactaceae (*Pereskia*)

Didiplis Raf. = *Lythrum*

Didissandra C.B. Clarke. Gesneriaceae. 30 India, China

Didonica Luteyn & Wilbur. Ericaceae. 4 Panamá & Costa Rica. R: SB 16(1991)587

Didymaea Hook.f. Rubiaceae (IV 16). 5 C Am. ? Apogamous

Didymanthus Endl. Chenopodiaceae (I 5). 1 W Aus.: *D. roei* Endl. R: FA 4(1984)218

Didymaotus N.E. Br. Aizoaceae (V). 1 Karoo: *D. lapidiformis* (Marloth) N.E. Br. – cult. orn.

Didymelaceae Leandri. Dicots – Dilleniidae – Buxales. 1/2 Madag. Dioec. evergreen, glabrous trees with spirally arr. simple lvs; stipules 0. Fls small, C 0, males in panicles, P 0, A 2, the filaments connate & anthers with longit. slits; females often in racemes (rarely solit. or in 3s), s.t. with K 1–4, scale-like, G 1 with 1 hemitropous, bitegmic ovule. Fr. a drupe; seed without endosperm

Genus: *Didymeles*

Affinities debatable (see Cronquist) but steroid alks like Buxaceae

Didymeles Thouars. Didymelaceae. 2 Madagascar

Didymia Philippi = *Mariscus*

Didymiandrum Gilly. Cyperaceae. 3 trop. S Am.

Didymocarpus Wallich ex Buch.-Ham. Gesneriaceae. c. 180 Madag. (2), China (31), Indomal., SE As. to trop. Aus. Long narrow capsules split to form gutters down which raindrops carry away tiny seeds. Some cult. orn. Sect. *Codonoboea* (Ridley) Kiew (4 Malay Pen.; R: Blumea 35(1990)71) – fls epiphyllous (cf. *Chirita*). *D. geitleri* A. Weber (Pahang) – bright yellow style an anther dummy; *D. platypus* C.B. Clarke (Malay Pen.) – produces 4 lvs a year, each lasting 22 months

Didymocheton Blume = *Dysoxylum*

Didymochlaena Desv. Dryopteridaceae (I 2). 1–2 trop. *D. truncatula* (Sw.) J. Sm. – rhizome erect, fronds to 2 m, cult. orn.

Didymochlamys Hook.f. Rubiaceae (inc. sed.). 2 Panamá to Venez. Epiphytes

Didymocistus Kuhlm. Euphorbiaceae. 1 trop. S Am.

Didymodoxa Wedd. (~ *Australina*). Urticaceae (V). 2 S & E trop. Afr. to N Ethiopia. R: NJB 8(1988)45

Didymoecium Bremek. Rubiaceae (IV 10). 1 Sumatra

Didymoglossum Desv. = *Trichomanes*

Didymopanax Decne & Planchon = *Schefflera*

Didymophysa Boiss. Cruciferae. 2 Iran to C As. & Himal.

Didymoplexiella Garay. Orchidaceae (V 5). 1 Borneo

Didymoplexis Griffith. Orchidaceae (V 5). 10 OW trop. exc. W Afr., India. Mycotrophic

Didymopogon Bremek. Rubiaceae (I 9). 1 Sumatra

Didymosalpinx Keay. Rubiaceae (II 1). 3 trop. Afr.

Didymosperma H. Wendl. & Drude ex Hook.f. = *Arenga*

Didymostigma W.T. Wang (~ *Chirita*). Gesneriaceae. 1 S China: *D. obtusum* (C.B. Clarke) W.T. Wang

Didymotheca Hook.f. = *Gyrostemon*

Didyplosandra Wight ex Bremek. = *Strobilanthes*

Diectomis Kunth = *Anadelphia*

Dieffenbachia Schott. Araceae (VI 4). c. 20 trop. Am. Dumb canes. Stout herbs with female fls with consp. staminodes. Many cult. orn. housepls (esp. C19 vogue) incl. *D. seguine* (Jacq.) Schott, the orig. 'dumb cane' used to torture slaves – chewing a portion of stem leads to speechlessness in adults & death in children or pets due to an obscure proteinaceous poison & irritant raphides (believed to cause temp. sterility & use in Russian concentration camps proposed by Nazis); many cvs derived from *D. seguine* & *D. maculata* (Lodd.) G. Don f. (? = *D. seguine*; narrower lvs, often irreg. heavily white-spotted); *D. longispatha* Engl. & Krause (C Am.) – poll. by 9 scarab spp.

Diegodendraceae Capuron. See Ochnaceae

Diegodendron Capuron. Ochnaceae (Diegodendraceae). 1 Madag.

Dielitzia Short. Compositae (Gnap.-Ang.). 1 Aus.: *D. tysonii* Short. R: Muelleria 7(1989)103

Diellia Brackenr. = *Asplenium*

Dielsantha F. Wimmer. Campanulaceae. 1 trop. Afr.

Dielsia Gilg. Restionaceae. 1 Aus.

Dielsiocharis O. Schulz. Cruciferae. 1 Iran

Dielsiochloa Pilger. Gramineae (21b). 1 Peru, Bolivia, Arg.: *D. floribunda* (Pilger) Pilger – high-alt. grassland

Dielsiothamnus R.E. Fries. Annonaceae. 1 trop. E Afr.: *D. divaricatus* (Diels) R.E. Fries

Dielytra Cham. & Schldl. = *Dicentra*

Dienia Lindley = *Malaxis*

Dierama K. Koch. Iridaceae (IV 3). 44 E Afr. mts to Cape. R: O.M. Hilliard & B.L. Burtt (1991) *Dierama*. Specially prot. in Zimbabwe; some cult. orn. with pendent fls (wandflowers) esp. *D. pendulum* (L.f.) Baker

Diervilla Miller. Caprifoliaceae. 2–3 E N Am. Cult. orn. decid. shrubs

Dieterlea Lott. Cucurbitaceae. 1 Mex. R: Britt. 38(1986)407

Dietes Salisb. ex Klatt. Iridaceae (III 2). 6 E & S Afr. (5). Lord Howe Is. (1!). R: AMBG 68(1981)132. Some cult. orn.

Dieudonnaea Cogn. = *Gurania*

Diflugossa Bremek. = *Strobilanthes*

Digastrium (Hackel) A. Camus = *Ischaemum*

Digera Forssk. Amaranthaceae (I 2). 1 OW trop.: *D. muricata* (L.) C. Martius (*D. arvensis*) – potherb in Sudan, fodder for stock

Digitacalia Pippen. Compositae (Sen.-Tuss.). 5 Mex. R: Phytol. 69(1990)150

Digitalis L. Scrophulariaceae. c. 19 Medit., Eur. (12) to C As. R: BJ 79(1960)222. Foxgloves. Cult. orn. & medic. esp. as source of digitalis (cardiac glycoside) derived from *D. purpurea* & *D. lanata* & used as cardiac stimulant since 1785 (Withering) but not analysed until 1933: *D. grandiflora* Miller (*D. ambigua*, Euras.) – yellow fls; *D. lanata* Ehrh. (Austrian d., C & SE Eur., natur. N Am.) – source of digitalis; *D. mertonensis* Buxton & C. Darl. (*D. grandiflora* × *D. purpurea* L., a true-breeding tetraploid); *D. purpurea* (common f., fairy fingers, polymorphic centred on Medit., homogeneous N Eur. populations expanded from Iberia post-glacial) – principal source of digitalis (copious draughts form. taken to induce intoxication; the minute seeds coll. by children S England as part of 'War Effort' in World War II), contains loliolide, a potent ant-repellent, sign of insincerity in 'Language of Fls', many cvs incl. 'Monstrosa' (? orig. Leiden Botanic Garden before 1842) with reg. term. fl.

Digitaria Haller. Gramineae (34d). c. 220 trop. & warm (Eur. 3, Aus. 44, N Am. 22). R: J.T. Henrard (1950) *Monogr. D.*; Blumea 21(1973)1. Crab or finger grasses, many weedy esp. *D. abyssinica* (A. Rich.) Stapf (*D. scalarum*) in E Afr. *D. decumbens* Stent (pangola grass, S Afr.) – pasture-grass in S US; *D. didactyla* Willd. (blue couch, trop. Afr.) – lawn-grass; *D. exilis* Stapf (W Afr., ? cultigen) – a staple crop (acha, fonio, fundi, hungry rice) tolerant of poor soils; *D. iburua* Stapf (iburu, W Afr.) – eaten like millet

Digitariella de Winter = *Digitaria*

Digitariopsis C. Hubb. = *Digitaria*

Diglyphosa Blume. Orchidaceae (Collabiinae). 12 Indomal.

Dignathe Lindley = *Leochilus*

Dignathia Stapf. Gramineae (33d). 5 trop. E Afr., W India

Digomphia Benth. Bignoniaceae. 3 Guayana Highland & nearby

Digoniopterys Arènes. Malpighiaceae. 1 Madagascar

Diheteropogon (Hackel) Stapf (~ *Andropogon*). Gramineae (40f). 5 trop. & S Afr. R: KB 20(1966)73

Dijon mustard *Brassica juncea*

dika (fat) *Irvingia gabonensis*

Dilatris Bergius. Haemodoraceae. 4 SW Cape

Dilepis Süsseng. & Merxm. = *Flaveria*

Dilkea Masters. Passifloraceae. 2–5 trop. S Am. A 6 basally connate

dill or **(E) Indian d.** *Anethum graveolens*

Dillenia L. Dilleniaceae. 60 Indian Ocean, Indomal., Aus. (1). R: Blumea 7(1952)1. Cult. orn. esp. *D. indica* L. (India to C Mal., water-disp.) – fr. ed. (curry, jellies), also shampoo, & *D. suffruticosa* (Griffith) Martelli (Mal.) – evergrowing shoots, seldom out of flower; *D. pentagyna* Roxb. (China, Indomal.) – decid., charcoal source; other spp. minor timbers, ed. fr. locally medic. etc. Most primitive spp. the most restricted

Dilleniaceae Salisb. Dicots – Dilleniidae – Dilleniales. 12/300 trop. & warm esp. Australasia. Trees to subshrubs or even herbs (*Acrotrema*) or lianes, with usu. spirally arr. simple lvs, rarely lobed or scale-like; stipules 0 or wing-like & adnate to petiole. Fls solit. or in racemose or cymose infls, yellow or white, usu. bisexual, reg., usu. without nectar; K (3–)5(–20), spirally imbricate, persistent, C (2–)5, imbricate, often crumpled in bud, A ± ∞, but s.t. few or 1, often asymmetrically placed but always assoc. with 5–15 stamen-trunks, the members of each originating centrifugally or all A so, anthers with longit. slits or apical pores, G̲ (1–) several (–20), rarely in 2 whorls, ± conduplicate & s.t. not fully sealed, rarely connate forming pluriloc. ovary with distinct styles, 1–∞ anatropous to amphitropous or campylotropous bitegmic ovules with zig-zag micropyle per loc. Fr. dry-dehiscent or indehiscent & then incl. in ± fleshy calyx (*Dillenia*); seeds endotestal with funicular aril (sometimes laciniate or even ± 0) & very small straight embryo in copious oily proteinaceous endosperm. x = 4, 5, 8–10, 12, 13

Chief genera: *Dillenia, Doliocarpus, Hibbertia, Tetracera*

Many primitive features incl. ± distinct conduplicate carpels, some not completely closed, ± ∞ A & wood with scalariform vessel-elements but differ markedly from Magnoliidae in chemistry (almost no alks, no ethereal oils etc.)

Some timbers, ed. fr. & cult. orn. (esp. *Dillenia*)

dillon bush *Nitraria billardierei*

Dillwynia Sm. Leguminosae (III 24). 24 Aus. Some cult. orn. (parrot pea)

Dilobeia Thouars. Proteaceae. 1 Madagascar

Dilochia Lindley. Orchidaceae (Arundinae). 6 Mal.

Dilodendron Radlk. Sapindaceae. 3 trop. Am. R: AMBG 74(1987)533. *D. costaricense* (Radlk.) A. Gentry & Steyerm. – ed. seeds

Dilomilis Raf. Orchidaceae (V 13). 4 WI (Hispaniola 3 (R: Moscosoa 5(1989)231)) to Brazil

Dilophia Thomson. Cruciferae. 5 C As.

Dilophotriche (C. Hubb.) Jacq.-Fél. Gramineae (39). 3 Senegal to Ivory Coast

Dimerandra Schltr. (~ *Epidendrum*). Orchidaceae (V 13). 8 trop. Am.

Dimeresia A. Gray. Compositae (Hele.). 1 W US: *D. howellii* A. Gray – heads 2-flowered, each fl. with bract

Dimeria R. Br. Gramineae (40e). 40 Madag. (3), India to Aus. R: KB 7(1953)553. Spikelets 1-flowered

Dimerocostus Kuntze. Zingiberaceae (Costaceae). 2 trop. Am.

Dimerostemma Cass. Compositae (Helia.-Verb.). 12 trop. S Am. R: PBSW 97 (1984)618

Dimetra Kerr. Verbenaceae. 1 Thailand. Geoxylic suffrutex, ? extinct

Dimitria Ravenna = *Sisymbrium*

Dimocarpus Lour. (~ *Litchi*). Sapindaceae. 5 SE As. to Aus. *D. longan* Lour. (*Euphoria l.*, *Nephelium l.*, longan or longyen, dragon's eyes, India, Sri Lanka) – cult. China & Mal. for juicy fr., exported dried from China, var. *malesianus* Leenh. being mata kuching or cat's eyes

Dimorphandra Schott. Leguminosae (I 1). 25 trop. Am. Intermediate between Leg. I & II

Dimorphanthera (Drude) F. Muell. ex J.J. Sm. Ericaceae. 68 Mal. esp. New Guinea (cf. *Satyria* in trop. Am.). R: FM I 6(1967)885, 9(1982)563

Dimorphocalyx Thwaites. Euphorbiaceae. 12 Indomal. to Aus.

Dimorphocarpa Rollins (~ *Dithyrea*). Cruciferae. 4 N Am.

Dimorphochloa S.T. Blake = *Cleistochloa*

Dimorphocoma F. Muell. & Tate. Compositae (Ast.-Ast.). 1 C Aus.: *D. minutula* F. Muell. & Tate

Dimorpholepis (G. Barroso) R. King & H. Robinson = *Grazielia*

Dimorphorchis Rolfe (~ *Arachnis*). Orchidaceae (V 16). 2 Borneo. Cult. orn.

Dimorphosciadium Pim. (~ *Pachypleurum*). Umbelliferae (III 8). 1 C As.: *D. gayoides* (Regel & Schmalh.) Pim.

Dimorphostemon Kitag. = *Sisymbrium*

Dimorphotheca Moench. Compositae (Calend.). (Incl. *Castalis*) 19 S & trop. Afr. R: M. Norlindh, *Stud. Calend.* 1(1943)38. Disk cypselas straight, compressed, with 2 thick wings, ray cypselas incurved, 3-angled to nearly cylindrical, usu. sharply tubercled or wrinkled, pappus 0. Cult. orn. annuals to shrubby perennial esp. *D. sinuata* DC ('*D. aurantiaca*', *D. calendulacea*, 'sun marigolds') but many plants cult. as *D.* belong to *Osteospermum*; *D. cuneata* (Thunb.) Less. a poss. oilseed for dry country

Dinacria Harvey ex Sonder = *Crassula*

Dinebra Jacq. (~ *Leptochloa*). Gramineae (31d). 3 trop. E Afr. & Madag. to W India

Dinema Lindley = *Encyclia*

Dinemagonum A. Juss. Malpighiaceae. 1 Chile: *D. gayanum* A. Juss. R: SB 14(1989)419

Dinemandra A. Juss. ex Endl. Malpighiaceae. 1 Peru, Chile: *D. ericoides* A. Juss. ex Endl. R: SB 14(1989)422

Dinetopsis Roberty = *Porana*

Dinetus Buch.-Ham. ex Sweet (~ *Porana*). Convolvulaceae. 6 trop. As. R: Novon 3(1993)198

Dinizia Ducke. Leguminosae (II 3). 1 Brazil, Guyana

Dinklageanthus Melchior ex Mildbr. = *Dinklageodoxa*

Dinklagiella Mansf. Orchidaceae (V 16). 2 trop. W Afr.

Dinklageodoxa Heine & Sandw. Bignoniaceae. 1 Liberia

Dinocanthium Bremek. = *Pyrostria*

Dinochloa Buese. Gramineae (1b). c. 25 Burma to Philipp. R: KB 36(1981)613. Scrambling bamboos with zig-zag culms & fleshy fr.

Dinophora Benth. Melastomataceae. 2 trop. W Afr.

Dinoseris Griseb. = *Hyaloseris*

Dintera Stapf. Scrophulariaceae. 1 trop. Afr.

Dinteracanthus C.B. Clarke ex Schinz = *Ruellia*

Dinteranthus Schwantes. Aizoaceae (V). 56 S Afr. Stemless mat-forming succ., cult. orn. with fls opening p.m.

Dioclea Kunth. Leguminosae (III 10). c. 30 trop. Am., a few OW incl. *D. hexandra* (Ralph) Mabb. (*D. reflexa*, pantrop.), a drift seed from Carib. in Carolinas, successfully sea-disp. to W Afr.

Diodeilis Raf. = *Clinopodium*

Diodella (Torrey & A. Gray) Small = *Diodia*

Diodia L. Rubiaceae (IV 15). c. 30 trop. & warm Am. & Afr., *D. serrulata* (Pal.) G. Taylor (*D. maritima*) common to both

Diodontium F. Muell. (~ *Glossogyne*). Compositae (Helia.-Cor.). 1 N Aus.

Dioecrescis Tirvengadum (~ *Gardenia*). Rubiaceae (II 1). 1 India

Diogenesia Sleumer. Ericaceae. 13 Andes. R: NRBGE 36(1978)251

Diogoa Exell & Mendonça. Olacaceae. 1 trop. Afr.

Dioicodendron Steyerm. Rubiaceae (I 7). 2 NW trop. S Am.

Dion Lindley = *Dioon*

Dionaea Sol. ex Ellis. Droseraceae. 1 SE US (pine barrens): *D. muscipula* Ellis (Venus' fly-trap), carnivore attracting by ultraviolet light absorbing leaf-margins secreting carbohydrates. Lvs in basal rosette, spathulate, the petioles winged & lamina of 2 hinged lobes fringed with 14–20 teeth, interlocking when trap closed, & with 3–7(+) sensitive hairs on upper surface: the lobes fold together on stimulation of 2 of the triggers (or 1 twice) within 20–40 secs, the mechanism said to be similar to that in *Mimosa*, i.e. hydraulic, the loss of turgor due to an increase in cell-wall plasticity with trap cells losing in 1–3 secs 30% of their ATP used for rapid transport of protons from motor cells, action potential predominantly dependent on extracellular concentration of calcium. Once closed, it seems that the trap actually squashes the engulfed insect prey before digestive enzymes from reddish surface glands (visible to naked eye) are secreted (prey excretions & movement (struggling against triggers) promoting this), breakdown products stimulating trap to narrow further, mucilage from peripheral glands sealing it like a gasket: opening is by real growth & several cycles possible. Much coll. & exported ('protected') for the insectivorous plant-trade, though readily propagated as almost every part of the plant is capable of producing adventitious buds seen even in infl. Fls' initiation apparently stimulated by trap-activity. Intolerant of thalloid liverwort competition

Dionaeaceae Raf. = Droseraceae

Dioncophyllaceae (Gilg) Airy Shaw. Dicots – Dilleniidae – Nepenthales. 3/3 trop. Afr. Lianes or shrubs with hooked or cirrhose leaf-tips & peltate scales & char. multicellular glands secreting acid insect-trapping mucilage; stems with successively produced irreg. arranged vascular bundles. Lvs spirally arr., simple, midrib prolonged & forked into hooks or tendrils; stipules 0. Fls bisexual, reg., in cymes; K 5 small, s.t. basally connate, valvate or open, peristent, C 5, convolute, A 10–30, anthers with longit. slits, $\underline{G}$ (2)or(5), 1-loc. with parietal placentas, styles 2 or 5 distinct to united, stigmas capitate or feathery, ovules ∞, anatropous, bitegmic. Fr. a loculicidal capsule, opening before ripe, ovules on thickened funicles; seeds large, winged all round, with ± discoid embryo & copious starchy endosperm. 2n = 36 (*Triphyophyllum*)
Genera: *Dioncophyllum*, *Habropetalum*, *Triphyophyllum*
Form. incl. in Flacourtiaceae; associated with Ancistrocladaceae by Cronquist
Triphyophyllum is certainly carnivorous

Dioncophyllum Baillon. Dioncophyllaceae. 1 trop. W Afr.

Dionycha Naudin. Melastomataceae. 2 Madag. *D. bojeri* Naudin a source of a black dye for silk

Dionychastrum Fernandes & R. Fernandes. Melastomataceae. 1 trop. E Afr.

Dionysia Fenzl. Primulaceae. 44 mts C As., N Iraq, Iran (25 of which 22 endemic), Afghanistan. R: C. Grey-Wilson (1989) *The genus D.* Tufted alpine garden pls allied to *Primula*

Dioon Lindley. Zamiaceae (II). 10 Mex. & C Am. R: MNYBG 57(1990)203. *D. edule* Lindley (Mex.) – average annual growth 0.76 mm, male cone 10 °C above ambient when ripe, seeds ed. if cooked

Diora Ravenna. Anthericaceae (Liliaceae s.l.). 1 Peruvian Andes: *D. cajamarcaensis* (Poelln.) Ravenna. R: OB 92(1987)189

Dioscorea L. Dioscoreaceae, c. 850 trop. & warm (Eur. 1, for others see *Borderea*). Yams (see D.G. Coursey (1967) *Y.*). Annual twining stems from tubers; alks (OW where x = 10 (NW, x = 9)); extrafl. nectaries. Tubers arise variously: in *D. batatas* by lateral hypertrophy of hypocotyl, in others from hypertrophy of internodes above cotyledon, in *D. pentaphylla* L. etc. from internode above cotyledon as well as hypocotyl, in *D. villosa* L. etc. as fleshy rhizomes. Many cult. trop. esp. W Afr. for starch (fufu), their culture intimately related to life & ceremonial of many peoples; others are collected or cult. for manufacture of steroidal hormones of use as oral contraceptives. *D. alata* L. (white yam, water y., Guyana arrowroot, trop. As.) – most widely grown with many cvs with 3x to 8x (all ploidy levels), tubers to 50 kg & 2 m long; *D. batatas* Decne (Chinese y. or potato, temp. E As.); *D. bulbifera* L. (air potato, acom (y.), Otaheite y. or potato, aerial y., As.) – axillary bulbils, some cvs ed.; *D.* × *cayenensis* Lam. (yellow y., Guinea y., *D. abyssinica* Hochst. ex Kunth (Afr.) × ?); *D. elephantipes* (L'Hérit.) Engl. (elephant's-foot, hottentot-bread, S Afr.) – tuber derived from 1st internode, projecting out of soil with thick covering of cork & annual shoots in wet season, cult. orn., & emergency food; *D. esculenta* (Lour.) Burkill (potato-yam, E Asia) – clones with 4x, 6x, 9x, 10x; *D. trifida* L.f. (cush-cush, yampee, yampi yam, trop. Am.) – each plant produces several tubers. Yams have been domesticated by different cultures independently in diff. parts of the world. Diosgenin, a precursor of progesterone, cortisone etc., is yielded in commercial quantities by a number of spp., coll. of which in Africa is licensed: *D. floribunda* M. Martens & Galeotti (Mex.) produces 10% by weight. *D. daemona* Roxb. (Indomal.) – tubers used to kill fish & pigs in Sumatra

Dioscoreaceae R. Br. Monocots – Liliidae – Dioscoreales. (Incl. Trichopodaceae) 8/880 trop. & warm with few N temp. Herbs usu. dioec. with twining, or rarely erect or spiny, shoots arising from a fleshy starchy rhizome or a tuber derived from lower internodes &/or hypocotyl; raphides usu. present, also steroidal saponins & often lactone alks; vessel-elements usu. present throughout, the vascular bundles in 2 dissimilar whorls or 1 whorl of alternating types & no sec. thickening exc. in tuber. Lvs spirally arr., rarely opp., usu. with distinct petiole & lamina, entire to less often lobed or compound, usu. with 3–13 curved-convergent main veins & often with nectaries or mucilaginous pits s.t. with nitrogen-fixing bacteria. Fls in spikes to panicles, rarely bisexual, reg., epigynous; P 6, usu. basally connate with short tube, nectaries commonly present, A 3 + 3, but inner s.t. staminodal or obs., filaments s.t. shortly connate, attached to P-tube base, anthers with longit. slits, $\overline{G}$ (3), 3-loc. with axile placentation & ± distinct styles & 2–∞ anatropous, bitegmic ovules per loc. Fr. a denticidal capsule, often 3-angled or -winged, rarely (*Tamus*) a berry or (*Trichopus* & *Rajania*) indehiscent & samaroid; seeds mostly winged with small embryo, subterm. plumule & a broad lateral cotyledon s.t. with a rudimentary 2nd one, embedded in copious endosperm with oil, protein & hemicellulose. x = 9(?), 10, 12, 14 +

Genera: *Avetra, Borderea, Dioscorea* (95% of the spp.), *Nanarepenta, Rajania, Stenomeris, Tamus, Trichopus*

Avetra, Stenomeris & *Trichopus* have bisexual fls; *Trichopus* has a short stem with only 1 leaf & it, like *Tamus* (berries), has been segregated into separate fam.

Dioscorea is an important carbohydrate & steroid source

Dioscoreophyllum Engl. Menispermaceae (III). 3 trop. Afr. *D. cumminsii* (Stapf) Diels (W Afr.) – fr. source of a sweetener, monellin (protein) 9000 times as sweet as sucrose tried for low-calorie food & drinks

Diosma L. Rutaceae. 28 SW Cape. R: JSAB 48(1982)329. Heath-like, some cult. orn.

Diosphaera Buser = *Trachelium*

Diospyros L. Ebenaceae. c. 475 trop. (Am. 80; Afr. 94, Madag. c. 100, As. 200). Ebonies. Timbers (favoured furn. wood in Ancient Egypt) & fr. trees etc., *D. ebenum* Koenig (Ceylon ebony, India, Sri Lanka) being the ebony of commerce, though other spp. also used, e.g. *D. abyssinica* (Hiern) F. White (trop. Afr.) – for shuttles for weaving sisal, *D. celebica* Bakh. Macassar ebony, Sulawesi), *D. haplostylis* Boivin (Madagascar e., Madag.); *D. montana* Roxb. (Bombay ebony, Indomal.); other timbers incl. marblewood or zebra-wood with streaked or marbled figure esp. from *D. marmorata* R. Parker, *D. oocarpa* Thwaites (Sri Lanka, Andamans) & calamander or Coromandel wood (*D. quaesita* Thwaites, Sri Lanka) greyish-brown with black bands much used in Sheraton furniture. Those cult. for fr. (date-plums, velvet apples) incl. *D. blancoi* A. DC (*D. discolor*, mabola, butterfruit, C Mal.), *D. digyna* Jacq. ('*D. ebenaster*', black sapote, C Am., natur. As.), *D. kaki* L.f. (Japanese persimmon, kaki, E As.) – much grown in Japan & China (where dried & also a sugar-source), derived by polyploidy from *D. roxburghii* Carr. & cvs selected for smooth-skinned fr. & frost-hardiness, *D. lotus* L. (As.) – fr. eaten fresh, dried or bletted, with wild form ('var. *brideliifolia*') in Philipp., *D. virginiana* L. (persimmon, SE US) – wood also useful (golf-club heads in US). *D. affinis* Thw. (India, Sri Lanka) – wood for carving in Sri Lanka; *D. decandra* Lour. (SE As.) – topiary & bonsai in Buddhist temples (Thailand); *D. major* (Forst.f.) Bakh. (Pacific) – ed. fr. in Tonga; *D. melanoxylon* Roxb. (India, Sri Lanka) – lvs used as cigarette-papers; *D. malabarica* (Desr.) Kostel. (*D. embryopteris*, gaub, trop. As.) – fr. sticky & used for caulking boats; *D. mespiliformis* Hochst. ex A. DC (trop. Afr.) – medic., ed. fr., timber; *D. mollis* Griffith (makua or ma-plua, Thailand) – fr. a source of a black dye for silk; *D. mweroensis* F. White (C & SE Afr.) – fish-poison and anti-bilharzia agent; *D. natalensis* (Harvey) Brenan (S Afr.) – some rheophytic forms

Diostea Miers. Verbenaceae. 3 S Am.

Diotacanthus Benth. = *Phlogacanthus*

Diothonea Lindley. = *Epidendrum*

Diotocranus Bremek. = *Mitrasacmopsis*

Dipanax Seemann = *Tetraplasandra*

Dipcadi Medikus. Hyacinthaceae (Liliaceae s.l.). c. 30 Medit. (Eur. 1), Afr., SW As. (India 9; R: BNHS 75(1978)50). Dreary cult. orn.

Dipelta Maxim. Caprifoliaceae. 4 China. Cult. orn. like *Weigela* but A 4 & C 2-lipped, esp. *D. floribunda* Maxim. (W & C China)

Dipentodon Dunn. Dipentodontaceae. 1 S China & adjacent Burma

Dipentodontaceae Merr. Dicots – Dilleniidae – Violales. 1/1 S China & Burma. Small tree with long vessel-elements with scalariform endplates. Lvs spirally arr., simple, toothed; stipules small, deciduous. Fls in globose axillary, pedunculate umbels, at first subtended by 4 or 5 involucriform bracts, regular, hypogynous; pedicels jointed in middle, K 5–7, connate at base, valvate, C similar & alt. with K, valvate, A as many as & alt. with C, anthers with longit. slits, nectaries (? staminodes) opp. C, $\underline{G}$ (3) with simple style & term. stigma, 1-loc., but with partial partitions basally, ovules 6 on top of free-central placenta. Fr. a tardily dehiscent capsule with 1 seed

Only species: *Dipentodon sinicus* Dunn

Form. incl. in Rosidae (Santalales)

Diphalangium Schauer. Alliaceae (Liliaceae s.l.). 1 Mexico

Diphasia Pierre. Rutaceae. 6 trop. Afr., Madagascar

Diphasiastrum Holub. See *Lycopodium*

Diphasiopsis Mendonça. Rutaceae. 2 trop. E Afr.

Diphasium C. Presl ex Rothm. = *Lycopodium*

Diphelypaea Nicolson = *Phelypaea*

Dipholis A. DC = *Sideroxylon*

Diphylax Hook.f. Orchidaceae (IV 2). 1 NE India, China

Diphyllarium Gagnepain. Leguminosae (III 10). 1 SE As.

Diphylleia Michaux. Berberidaceae (II 2). 1 E N Am. (*D. cymosa* Michaux), 1 C & N Japan (*D. grayi* F. Schmidt), 1 W China (*D. sinensis* H.L. Li – rhizomes locally medic.)

Diphysa Jacq. Leguminosae (III 7). 15 trop. Am. *D. robinioides* Benth. – fuelwood

Dipidax Salisb. = *Onixotis*

Diplachne Pal. = *Leptochloa*

Diplacrum R. Br. (~ *Scleria*). Cyperaceae. 6 trop.

Dipladenia A. DC = *Mandevilla*

Diplandra Hook. & Arn. = *Lopezia*

Diplandrorchis S.C. Chen = *Neottia*

Diplarche Hook.f. & Thomson. Diapensiaceae. 2 E Himal., SW China

Diplarpea Triana. Melastomataceae. 1 Colombia

Diplarrhena Labill. Iridaceae (III 1). 2 SE Aus. *D. moraea* Labill., cult. orn.

Diplasia Rich. Cyperaceae. 2 trop. Am., 1 SE As.

Diplaspis Hook.f. Umbelliferae (I 2b). 2 Tasmania

Diplatia Tieghem. Loranthaceae. 3 trop. Aus.

Diplaziopsis C. Chr. (~ *Diplazium*). Dryopteridaceae (II 1). 3–4 E As., Polynesia

Diplazium Sw. (~ *Athyrium*). Dryopteridaceae (II 1). c. 400 trop. (Afr. few) & N temp. (Eur. 2); incl. *Allantodia*). Cult. orn. incl. *D. esculentum* (Retz.) Sw. (trop. & E As. to Polynesia, natur. Florida) – young shoots a vegetable in Sikkim, in curry in Sri Lanka etc.

Diplazoptilon Ling (~ *Dolomiaea*). Compositae (Card.-Card.). 2 SW China

Diplectria (Blume) Reichb. Melastomataceae. 11 SE As., Mal. R: Blumea 24(1978)405

Diplobryum C. Cusset. Podostemaceae. 1 Vietnam

Diplocarex Hayata = *Carex*

Diplocaulobium (Reichb.f.) Kraenzlin. Orchidaceae (V 15). 94 Mal. (esp. New Guinea) to W Pacific. *D. pentanema* (Schltr.) Kraenzlin (Papuasia) – yellow fibres from stems used in otherwise black armbands in New Ireland

Diplocentrum Lindley. Orchidaceae (V 16). 2 India

Diploclisia Miers. Menispermaceae (V). 2 Indomal., China, SE As.

Diplocyatha N.E. Br. = *Orbea*

Diplocyclos (Endl.) Post & Kuntze. Cucurbitaceae. 4 trop. Afr., 1 extending to trop. As. (*D. palmatus* (L.) C. Jeffrey, cult. orn.)

Diplodiscus Turcz. Tiliaceae. 7 Sri Lanka (1), W Mal. *D. paniculatus* Turcz. (Philipp.) – Baroba nut, seeds ed.

Diplodium Sw. = *Pterostylis*

Diplofatsia Nakai = *Fatsia*

Diploglottis Hook.f. Sapindaceae. 10 E Mal., NE Aus. (8 endemic), New Caled. *D. australis* (G. Don f.) Radlk. (*D. cunninghamii*, Aus.) – ed. arils used in jam

Diplokeleba N.E. Br. Sapindaceae. 2 trop. S Am.

Diploknema Pierre. Sapotaceae (II). 10 Indomal. (Yunnan 1). Seeds of *D. butyracea* (Roxb.) H.J. Lam (India–SW China) yield a fat used in soap etc.

Diplolabellum F. Maek. Orchidaceae (V 8). 1 Korea. Mycotrophic, coll. once (= ?)

Diplolaena R. Br. Rutaceae. 6–8 W Aus.

Diplolegnon Rusby = *Corytoplectus*

Diplolepis R. Br. Asclepiadaceae (III 1). 2 China

Diplolophium Turcz. Umbelliferae (III 8). 5–7 trop. Afr.

Diplomeris D. Don. Orchidaceae (IV 2). 2 Himalaya

Diplomorpha Meissner = *Wikstromia*

Diploon Cronq. Sapotaceae (III). 1 trop. S Am.: *D. cuspidatum* (Hoehne) Cronq. R: T.D. Pennington, *S.* (1991)180

Diplopanax Hand.-Mazz. Cornaceae. 1 China: *D. stachyanthus* Hand.-Mazz. congeneric with Tertiary fossils referred to *Mastixia* (cf. *Metasequoia*)

Diplopeltis Endl. Sapindaceae. 5 NW Aus.

Diplopilosa Dvořák = *Hesperis*

Diplopogon R. Br. Gramineae (25). 1 SW Aus., wet

Diploprora Hook.f. Orchidaceae (V 16). 2 trop. As. R: OB 95(1988)39

Diplopterygium (Diels) Nakai. Gleicheniaceae (Gleich.). 25 trop. & warm As. to Hawaii, 1 trop. Am. (*D. bancroftii* (Hook.) A.R. Sm.). Thicket-formers but lvs bipinnatifid, not pseudodichotomous

Diplopterys A. Juss. Malpighiaceae. 4 trop. Am. R: FN 30(1982)208. Close to *Banisteriopsis*. *D. cabrerana* (Cuatrec.) Gates yields an hallucinogenic drug (tryptamines) s.t. mixed with *Banisteriopsis caapi*

Diplora Baker = *Asplenium*

Diplorhynchus Welw. ex Ficalho & Hiern. Apocynaceae. 1 trop. & S Afr.

Diplosoma Schwantes. Aizoaceae (V). 2 Cape, limestone. R: BBP 63(1980)375

Diplospora DC (~ *Tricalysia*). Rubiaceae (II 1). 10+ Indomal. R: Blumea 35(1990)297 (Aus. spp. = *Tarenna*)

Diplostephium Kunth. Compositae (Ast.-Ast.). 90 trop. Andes, 1 Costa Rica

Diplostigma Schumann. Asclepiadaceae (III 1). 1 E Afr.: *D. canescens* Schumann

Diplotaenia Boiss. (~ *Peucedanum*). Umbelliferae (III 10). 2 Turkey, Iran. *D. cachyridifolia* Boiss. – a source of jatamansi (cf. *Nardostachys*)

Diplotaxis DC (*Dyplotaxis*). Cruciferae. 27 Eur. (11), Medit. to NW India. R: Pfl. 70(4,105)(1919)149. *D. erucoides* (L.)DC (SW Eur.) – 'white rocket', *D. muralis* (L.) DC (S & C Eur.) – 'wall rocket'

Diplotropis Benth. Leguminosae (III 2). 7 Amazonia. Commercial timbers ('sucupira')

Diplusodon Pohl. Lythraceae. 74 Brazil. R: Bradea 5(1989)205

Diplycosia Blume. Ericaceae. 99 Mal. (many restricted to Mt Kinabalu). R: Reinwardtia 4(1957)119

Dipodium R. Br. Orchidaceae (V 9). 20 Mal. to New Caled. Some cult. orn. incl. *D. squamatum* (Forster f.) Sm. (*D. punctatum*, SW Pacific)

Dipogon Liebm. Leguminosae (III 10). 1 S trop. Afr., cult. elsewhere: *D. lignosus* (L.) Verdc.

Dipoma Franchet. Cruciferae. 2 SE Tibet, SW China

Diposis DC. Umbelliferae (I 2c). 3 temp. S Am.

Dipsacaceae Juss. Dicots – Asteridae – Dipsacales (excl. Morinaceae). 11/290 Euras. & Afr. (esp. Medit.). Herbs or shrubs. Lvs opp. or whorled, simple, toothed, or lobed; stipules 0. Fls usu. in dense cymose (mixed or even racemose in *Cephalaria*), involucrate heads with naked to bracteate receptacle, bisexual, epigynous, usu. with epicalyx (s.t. adnate to G) but 0 in some *Cephalaria* spp. & *Succisa* spp.; K ((0)4 or 5) or with 10 teeth or bristles, C (4 or 5), lobes imbricate, A 4 attached to C-tube, alt. with lobes, anthers with longit. slits, $\overline{G}$ (2), 1 obs. so that G 1-loc. with 1 style, a nectary around its base, & 1 pendulous, anatropous, bitegmic ovule. Fr. a cypsela enclosed in epicalyx (except apically) & usu. topped with persistent K; seeds with large straight embryo & rather scanty, fleshy, oily endosperm. x = 5–10

Principal genera: *Cephalaria*, *Knautia*, *Pterocephalus*, *Scabiosa*

Morinaceae are excluded (fls in axillary verticillasters, functional A 2, alks, 0, X = 17 & pollen & anther differences) but are closely allied. D. are close to Valerianaceae & perhaps best incl. in Caprifoliaceae s.s. Generic limits around *Scabiosa* debatable

Some cult. orn. esp. *Knautia*, *Lomelosia* & *Scabiosa*; teasels from *Dipsacus*

Dipsacus L. Dipsacaceae. 15 Euras. (Eur. 8), Medit., trop. Afr. & Sri Lanka (1) mts. Lvs basally connate, forming rainwater-collecting troughs, which may prevent insects climbing up to fls. Rigid bracts act as fr. catapults when animals pass by. In *D. sativus* (L.) Honck. (cultigen ? derived from *D. ferox* Lois., Medit.), these are the effective part of the head (teasel) form. used in raising nap on cloth (bracts downward-pointing cf. wild pls); *D. fullonum* L. (*D. sylvestris*) much used in dried fl. arrangements like other spp. cult. orn. e.g. *D. pilosus* (shepherd's rod, Euras.); other spp. medic. in As.

Dipteracanthus Nees = *Ruellia*

Dipteranthemum F. Muell. = *Ptilotus*

Dipteranthus Barb. Rodr. Orchidaceae (V 10). 2 trop. S Am.

Dipteridaceae (Diels) Seward & Dale. Filicopsida. 1/8 Indomal. to Pacific. Terr. ferns with fronds with 2 flabellate halves.

Only genus: *Dipteris*.

Many similar Jurassic fossils (*Dictyophyllum*, *Hausmannia*) of cosmop. distrib.

Dipteris Reinw. Dipteridaceae. 8 As. to Polynesia. *D. lobbianum* (Hook.) Moore (Mal.) – facultative rheophyte

Dipterocarpaceae Blume. Dicots – Dilleniidae – Malvales. 16/680 trop. esp. Mal. Trees, usu. resinous (char. branching r. canals in Dipterocarpoideae) & often with ectotrophic mycorrhizae. Lvs spirally arr., simple, leathery; stipules sometimes protecting bud, persistent or deciduous. Fls usu. bisexual, reg., nodding, scented, in usu. axillary panicles, racemes or rarely cymes; K 5, sometimes valvate in fr. but usu. imbricate, often with basal tube, persistent (when 2–5 usu. greatly enlarging into wing-like lobes in fr.), C 5, contorted, sometimes basally connate, A 5–110 in 1–3 whorls or ± irreg., initiated centrifugally, typically served by 10 trunk-bundles, ± hypogynous (on an androgynophore in Monotoideae), filaments distinct or ± basally connate, anthers basifixed (appearing versatile in Monotoideae and Pakaraimaeoideae), with longit. slits & connective usu. consp.

prolonged, G, rarely semi-inf. ((2 or)3(–5)), pluriloc. with axile placentation & term. style & 2(–4) anatropous, bitegmic ovules per locule. Fr. dry indehiscent, 1-seeded with woody pericarp; seeds without endosperm or dormancy, cotyledons often folded & enclosing radicle. x = 6, 7, 10, 11 (FM I, 9(1982)237)

Classification & genera:

Dipterocarpoideae (wood, lvs & G with resin-ducts, anthers basifixed, 2 or 3 K becoming accrescent in fr., G 2–3-loc., each with 2 ovules, trop. As., Mal.; Tertiary of Mt Elgon, E Afr.): *Anisoptera, Cotylelobium, Dipterocarpus, Dryobalanops, Hopea, Neobalanocarpus, Parashorea, Shorea, Stemonoporus, Upuna, Vateria. Vateriopsis, Vatica*

Monotoideae (wood, G & often lvs without resin-ducts, anthers basiversatile, K equally accrescent in fr., papery, G 3(–5)-loc., each with 2 ovules, C longer than K, trop. Afr. & Madag. (1 sp.): *Marquesia, Monotes*

Pakaraimaeoideae (as Monotoideae but P shorter than K, G 4(5)-loc., each with 4 ovules, Guianas): *Pakaraimaea*

The last 2 have been referred to Tiliaceae & the 2nd to a separate fam. by some authors. Seed & floral vasculature studies suggest affinity with Malvales & esp. Sarcolaenaceae. They are richest in Mal. (10/386 with 267 spp. in Borneo alone) & char. show gregarious flowering poss. initiated by periods of high irradiation though reserves must also be built up beforehand. Fruiting assoc. with pig (*Sus barbatus*) migrations; pigs mate when trees fl. & gestation = time to fr. ripening). They are the char. trees of Indomal., being the principal emergents & the basis of the export timber-trade, *Shorea* being the most imp. genus of all in Mal. Much is used as plywood. Resins are also used in varnishes & *Shorea* frs yield a comm. ed. fat

Dipterocarpus Gaertner f. Dipterocarpaceae. 69 Indomal. (Tertiary fossils in Afr.). Large stipules protect buds. Light timber, which absorbs preservative, much used for sleepers & heavy construction (keruing (Mal.) or gurjun (Ind.), apitong or 'bagac' (Philipp.) or yang (Burma)) e.g. *D. costatus* Gaertner f. (dau, Andamans to Negri Sembilan), *D. grandiflorus* (Blanco) Blanco (bagac, Andamans to Philipp.), *D. tuberculatus* Roxb. (eng, SE As.), *D. zeylanicus* Thwaites (hora, Ceylon gurjun, Sri Lanka). Oleoresins (gurjun oil or balsam) form. much exported from India for slow-drying varnishes. *D. oblongifolius* Blume (neram, S Thailand, Malay Pen., Borneo) – char. rheophyte of fast-flowing rivers ('neram' rivers)

Dipterocome Fischer & C. Meyer. Compositae (? Cal.). 1 E Med. to Afghanistan: *D. pusilla* Fischer & C. Meyer

Dipterocypsela S.F. Blake. Compositae (Vern.-Vern.). 1 Colombia: *D. succulenta* S.F. Blake

Dipterodendron Radlk. = *Dilodendron*

Dipteronia Oliver. Aceraceae. 2 C & S China. Fr. winged all round

Dipteropeltis Hallier f. Convolvulaceae. 2 trop. W Afr.

Dipterostele Schltr. = *Stellilabium*

Dipterygium Decne. Capparidaceae. 1 Egypt to Pakistan. Methylglucosinolates present like Cruciferae; fr. a samara

Dipteryx Schreber. Leguminosae (III 3). 10 trop. Am. Bat-disp. *D. odorata* (Aublet) Willd. (Tonka bean) – fls every 5 yrs, fragrant seeds cured in rum used for scenting tobacco & snuff; *D. oleifera* Benth. (ebor or eboe) – similar seeds

Diptychandra Tul. Leguminosae (I 1). 3 Brazil, Bolivia, Paraguay

Diptychocarpus Trautv. Cruciferae. 1 Eur. to C As.: *D. strictus* (Fischer) Trautv.

Dipyrena Hook. Verbenaceae. 1 temp. S Am. Pyrenes 2, 2-loc.

Dirachma Schweinf. ex Balf.f. Geraniaceae (?; Dirachmaceae). 1 Socotra: *D. socotrana* Schweinf. ex Balf.f., a tree reduced to 30 individuals by 1967; 1 C Somalia

Dirachmaceae Hutch. See Geraniaceae

Dirca L. Thymelaeaceae. 2 N Am. Leatherwood. K almost 0, C 0, A 8, G 1-loc., flexible shoots used for baskets, the bark for rope, cult. orn. esp. *D. palustris* L. (moosewood)

Dirhamphis Krapov. Malvaceae. 1 Bolivia, Paraguay

Dirichletia Klotzsch = *Carphalea*

dirty Dora *Cyperus difformis*

Disa P. Bergius. Orchidaceae (IV 3). 99 trop. & S Afr., Arabia, Madag., Masc. R(p.p.): CBH 9(1981)1. Terrestrial, often with conspic. fls. esp. *D. uniflora* P. Bergius (butterfly-poll. (like *D. ferruginea* (Thunb.) Sw., though *D. filicornis* (L.f.) Thunb. nectarless & poll. mason-bees), 10 cm across) from S Afr. & cult. orn. incl. hybrids

Disaccanthus E. Greene. Cruciferae. 6 W N Am.

Disakisperma Steudel = *Leptochloa*

Disanthus Maxim. Hamamelidaceae (II). 1 E China, S Japan: *D. cercidifolius* Maxim. – cult. orn. (autumn colour) with small paired fls, C with 2 basal nectaries & 5 small staminodes. R: Cathaya 3(1991)1

Discalyxia Markgraf = *Alyxia*

Discaria Hook. Rhamnaceae. 8 S Am. (5), Aus. (2), NZ (1). R: BSAB 22(1983)301. Spiny shrubs allied to *Colletia*, those with explosive fr. expelling seeds to 2.4 m, those without have floating fr. Some cult. orn.

Dischidanthus Tsiang. Asclepiadaceae (III 4). 1 S China, SE As.: *D. urceolatus* (Blume) Tsiang

Dischidia R. Br. Asclepiadaceae (III 4). 80 Indomal. to Aus. & W Pacific (New Caled. 1). Epiphytes with adventitious roots & with fleshy wax-covered lvs.; some local medic. *D. major* (Vahl) Merr. (*D. rafflesiana* Wallich, India to Aus.) also has pitcher-shaped lvs (stomata inside, those outside prob. closed). c. 10 cm deep, into each of which an adventitious root grows from stem or petiole near it; detritus, largely coll. by ants, accumulates in pitchers, as does rainwater thus made available to roots

Dischidiopsis Schltr. Asclepiadaceae (III 4). 9 E Mal.

Dischisma Choisy. Globulariaceae. 11 SW Cape to Namibia. R: MBSM 15(1979)94

Dischistocalyx T. Anderson ex Benth. ('*Distichocalyx*'). Acanthaceae. 20 trop. Afr. Terrestrial, later becoming climbers & epiphytes

Disciphania Eichler. Menispermaceae (III). c. 25 trop. Am. R: MNYBG 20(1970)124

Discipiper Trel. & Stehlé = *Piper*

Discocactus Pfeiffer. Cactaceae (III 4). 9 E S Am. R: CSJ 43(1981)37 (up to 36 spp. recognized by some growers)

Discocalyx (A.DC) Mez (~ *Tapeinosperma*). Myrsinaceae. 50 C Mal. (Philipp. 31) to Polynesia. *D. dissectus* Kaneh. & Hatusima (New Guinea) subherbaceous with much divided chamomile-like lvs

Discocapnos Cham. & Schldl. (~ *Fumaria*). Fumariaceae (I 2). 1 SW & S Cape: *D. mundtii* Cham. & Schldl.

Discocarpus Klotzsch. Euphorbiaceae. 5 Brazil, Guyana

Discoclaoxylon (Muell. Arg.) Pax & K. Hoffm. (~ *Claoxylon*). Euphorbiaceae. 4 trop. Afr. (3 restricted to Gulf of Guinea is.)

Discocleidion (Muell. Arg.) Pax & K. Hoffm. Euphorbiaceae. 3 China, Ryukyus

Discocnide Chew. Urticaceae (I). 1 Mex., Guatemala: *D. mexicanus* (Liebm.) Chew

Discocrania (Harms) Král = *Cornus*

Discoglypremna Prain. Euphorbiaceae. 1 trop. Afr.

Discolobium Benth. Leguminosae (III 13). 8 S Am.

Discophora Miers. Icacinaceae. 2 Panamá to trop. S Am.

Discopleura DC = *Ptilimnium*

Discopodium Hochst. Solanaceae. 1 trop. Afr. mts: *D. penninervium* Hochst.

Discospermum Dalzell (~ *Diplospora*). Rubiaceae (V 16). 6 Indomal. R: Blumea 35(1990)300

Discovium Raf. Cruciferae. Oldest name for *Lesquerella*

Discyphus Schltr. Orchidaceae (III 1). 1 trop. Am.

Diselma Hook.f. Cupressaceae. 1 Tasmania: *D. archeri* Hook.f. – foliage resembling *Microcachrys tetragona*

Disepalum Hook.f. Annonaceae. (Incl. *Enicosanthellum*) 9 SE As. to W Mal. R: Brit. 41(1989)356. 2-merous

Disocactus Lindley. Cactaceae (III 2). (Incl. *Aporocactus*, *Heliocereus*, × *Heliochia* & *Nopalxochia*) 16 C Am., WI (2), 1 ext. to N S Am. R: Bradleya 9(1991)86. Epiphytes, cult. orn. allied to *Epiphyllum*, esp. hybrid *D.* × *hybridus* (Geel) Barthlott (× *Heliochia* 'Ackermannii', '*N. ackermannii*'), *D. speciosus* (Cav.) Barthlott (*Heliocereus s.*, Mex.) × *D. phyllanthoides* (Haw.) Barthlott (*Nopalxochia p.*, S Mex.) – large red to pink fls (most of the 'epiphyllums' of comm. (R: S.E. Haselton (1946) *E. handbook*)) & *D. flagelliformis* (L.) Barthlott (*Aporocactus f.*, rat's-tail, Mex.) – crimson fls

Disparago Gaertner. Compositae (Gnap.-Rel.). 9 S Afr. R: JSAB 2(1936)95, Bothalia 21(1991)158

Disperis Sw. Orchidaceae (IV 3). 75 trop. & S Afr., Masc., Indomal. Only OW orchids with oil-secreting fls (at least 17 spp. in S Afr.) visited by poll. *Rediviva* bees (cf. *Diascia*). Some cult. orn. epiphytes & terr. spp.

Disphyma N.E. Br. Aizoaceae (V). 4 Cape, Aus. (1). Cult. orn. incl. *D. crassifolium* (L.) L. Bolus (pigface, SW Cape)

Disporopsis Hance. Convallariaceae (Liliaceae s.l.). 4–5 trop. E As., Philipp.

Disporum Salisb. ex D. Don. Convallariaceae (Liliaceae s.l.). 10 E As., Indomal., W N Am. Some cult. orn. (fairy-bells) incl. *D. lanuginosum* (Michaux) Nicolson (EC US) – myrmechory

diss or **dis grass** *Ampelodesmos mauretanicus*

Dissanthelium Trin. Gramineae (21b). 16 Peru & Bolivia (esp. high Andes) ext. to Calif. R: Phytol. 11(1965)361

Dissilaria F. Muell. ex Baillon. Euphorbiaceae. 1–2 NE Aus.

Dissocarpus F. Muell. Chenopodiaceae (I 5). 4 Aus. R: FA 4(1984)226

Dissochaeta Blume. Melastomataceae. 20 Indomal.

Dissochondrus (Hillebrand) Kuntze. Gramineae (34b). 1 Hawaii

Dissomeria Hook.f. ex Benth. Flacourtiaceae (5). 1 trop. W Afr.

Dissothrix A. Gray. Compositae (Eup.-Alom.). 1 Brazil. R: MSBMBG 22(1987)251

Dissotis Benth. Melastomataceae. c. 100 trop. & S Afr. Some cult. orn. shrubs & herbs

Disteganthus Lemaire. Bromeliaceae (3). 2 NE S Am. R: FN 14(1979)1764

Distemonanthus Benth. Leguminosae (I 2). 1 trop. W Afr.: *D. benthamianus* Baillon (anyaran, ayan, bongassi, Nigerian satinwood) – comm. timber

Distephanus Cass. (~ *Vernonia*). Compositae (Vern.-Vern.). c. 40 OW trop. esp. Madag.

Disterigma (Klotzsch) Niedenzu. Ericaceae. c. 25 trop. Andes. Some ed. fr. reaching local markets

Distichella Tieghem = *Dendrophthora*

Distichia Nees & Meyen. Juncaceae. 3 Andes. R: NJB 6(1986)151

Distichlis Raf. Gramineae (31c). 5 Am., Aus. (1). R: RAA 22(1955)86. Alkali grass. *D. spicata* (L.) E. Greene (N Am.) used for binding sandy soil

Distichocalyx Benth. = *Dischistocalyx*

Distichochlamys M. Newman. Zingiberaceae. 1 Vietnam. R: EJB 52(1995)65

Distichoselinum García Martín & Silvestre (~ *Elaeoselinum*). Umbelliferae (III 12). 1 Iberia: *D. tenuifolium* (Lag.) García Martín & Silvestre. R: Lagascalia 13(1985)232

Distichostemon F. Muell. Sapindaceae. 6 Aus. R: Austrobaileya 2(1984)57

Distictella Kuntze. Bignoniaceae. 10 trop. S Am. & Tobago. Domatia

Distictis C. Martius ex Meissner. Bignoniaceae. 9 Mex., WI

Distoecha Philippi = *Hypochaeris*

Distomocarpus O. Schulz = *Rytidocarpus*

Distrianthes Danser. Loranthaceae. 3 New Guinea

Distyliopsis Endress. Hamamelidaceae (I 4). 6 Burma to Mal., SE As. & Taiwan

Distylium Siebold & Zucc. Hamamelidaceae (I 4). 12 Indomal. (10), C Am. (2). Domatia; some rheophytes & cult. orn. esp. *D. racemosum* Siebold & Zucc. (Japan) – fine-grained wood used for furniture & art

Distylodon Summerh. Orchidaceae (V 16). 1 Uganda: *D. comptus* Summerh.

Disynaphia Hook. & Arn. ex DC (~ *Eupatorium*). Compositae (Eup.-Dis.). 16 S Am. R: MSBMBG 22(1987)77

Disynstemon R. Viguier. Leguminosae (III 6). 1 Madagascar

dita bark *Alstonia scholaris*

Ditassa R. Br. Asclepiadaceae (III 1). 75 S Am.

Ditaxis Vahl ex A. Juss. = *Argythamnia*

Ditepalanthus Fagerl. Balanophoraceae. 1 Madagascar

Dithrix (Hook.f.) Schlechter = *Habenaria*

Dithrydanthus Garay. Orchidaceae (III 1). 1 Mex.

Dithyrea Harvey. Cruciferae. 3 SW N Am.

Dithyridanthus Garay = *Schiedeella*

Dithyrostegia A. Gray (~ *Angianthus*). Compositae (Gnap.-Ang.). 2 SW Aus. R: Muelleria 7(1989)106, OB 104(1991)114

Ditrichospermum Bremek. = *Strobilanthes*

Ditta Griseb. Euphorbiaceae. 2 WI

dittander *Lepidium latifolium*; (in old Herbals = dittany)

dittany *Origanum dictamnus*; *Dictamnus albus*; **American d.** *Cunila origanoides*; **false d.** *Ballota acetabulosa*

Dittoceras Hook.f. Asclepiadaceae (III 4). 3 E Himal., Thailand

Dittostigma Philippi = *Nicotiana*

Dittrichia Greuter (~ *Inula*). Compositae (Inul.). 2 Med. incl. Eur. introd. S Am.

Diuranthera Hemsley = *Chlorophytum*

Diuris Sm. Orchidaceae (IV 1). 1 Timor, 39 Aus. Bee-poll., many wild hybrids; *D. maculata* Sm. (SE Aus.) fls mimic those of *Daviesia* spp. & *Pultenaea scabra* R. Br. (Leguminosae). Some cult. orn. (donkey orchids)

divi-divi *Caesalpinia coriaria*

Diyaminauclea Ridsd. Rubiaceae (I 2). 1 Sri Lanka

Dizygostemon (Benth.) Radlk. ex Wettst. Scrophulariaceae. 1 Brazil

Dizygotheca N. E. Br. = *Schefflera*

Djaloniella P. Taylor. Gentianaceae. 1 trop. W Afr.

djave *Baillonella toxisperma*

Djinga Cusset. Podostemaceae. 1 Cameroun

Dobera Juss. Salvadoraceae. 2 trop. E Afr., S Arabia to NW India

Dobinea Buch.-Ham. ex D. Don. Anacardiaceae (V; Podoaceae). 2 E Himal., S China. Female fls epiphyllous

dock *Rumex* spp.; **patience d.** or **spinach d.** *R. patientia*; **tanner's d.** *R. hymenosepalus*; **water d.** *R. hydrolapathum*; **yellow d.** *R. crispus*

Dockrillia Brieger = *Dendrobium*

Docynia Decne. Rosaceae (III). 2 Himal. to SE As. R: CJB 68(1990)2233. Trees allied to *Cydonia*, some cult. orn. & ed. fr. sometimes used in China to speed bletting of persimmons

Docyniopsis (C.K. Schneider) Koidz. (~ *Malus*). Rosaceae (III). 4 E As.

Dodartia L. Scrophulariaceae. 1 S Russia, W As.

dodder *Cuscuta* spp.; **d.-laurel** *Cassytha* spp.

doddering-dillies *Briza media*

Dodecadenia Nees (~ *Litsea*). Lauraceae (II). 1 S. Himal.: *D. grandiflora* Nees

Dodecahema Rev. & Hardham (~ *Centrostegia*). Polygonaceae (I 1). 1 Calif.: *D. leptoceras* (A. Gray) Rev. & Hardham. R: Phytol. 66(1989)86

Dodecastigma Ducke. Euphorbiaceae. 3 trop. S Am.

Dodecatheon L. Primulaceae. 13 N Am. (all but 1 in W), E Siberia (1). American cowslip, shooting-star. R: CDH 4(1953)73, Plantsman 13(1991)157. Allied to *Primula* but C reflexed & self-poll. poss. Many cult. orn. esp. *D. meadia* L. (E US) & *D. pulchellum* (Raf.) Merr. ('*D. pauciflorum*', N Am.)

dodo cloth *Tabernaemontana pachysiphon*

Dodonaea Miller. Sapindaceae. 68 trop. & warm esp. Aus. (61, 59 endemic). Wind-poll. usu. viscid shrubs & trees, some cult. orn. & medic. or fodders. *D. viscosa* Jacq. (*D. angustifolia*, akeake (NZ), hop-bush, trop. & warm) – cult. orn. & winged fr. allegedly used as hops by early settlers in Aus., molluscicidal saponins

Dodonaeaceae Link = Sapindaceae

Dodsonia Ackerman. Orchidaceae (V 10). 2 Ecuador

Doellia Schultz-Bip. ex Walp. (~ *Blumea*). Compositae (Pluch.). 2 Arabia, Afr.

Doellingeria Nees (~ *Aster*). Compositae (Ast.-Ast.). c. 7 E As., N Am.

Doerpfeldia Urban. Rhamnaceae. 1 Cuba

dog apple *Asimina reticulata*; **d.bane** *Apocynum* spp.; **d. daisy** *Leucanthemum vulgare*; **d.'s mercury** *Mercurialis perennis*; **d. nettle** *Urtica urens*; **d. plum** *Ekebergia capensis*; **d. rose** *Rosa canina*; **d. senna** *Senna italica*; **d.'s tail grass, crested** *Cynosurus cristatus*; **d.'s tooth violet** *Erythronium dens-canis*; **d. violet** *Viola canina*; **d.wood** *Cornus* spp.; **d.wood, swamp** *C. sanguinea*

Dolichandra Cham. Bignoniaceae. 1 S Brazil, Paraguay, Arg.: *D. cynanchoides* Cham. – cult. orn. liane with red fls

Dolichandrone (Fenzl) Seemann. Bignoniaceae. 10 E Afr. (1), Indomal. (5), trop. Aus. (3), Indopacific (1). Trees with white fls opening at night, very fragrant, bat-poll. (cf. allied bird-poll. *Markhamia*); cult. orn. incl. *D. spathacea* (L.f.) Schumann (As. to New Caled.) – mangrove with sea-disp. seeds with corky wings

Dolichlasium Lagasca (~ *Trixis*). Compositae (Mut.-Nass.). 1 Arg. Andes: *D. lagascae* Gillies ex D. Don

Dolichocentrum (Schltr.) Brieger = *Dendrobium*

Dolichochaete (C. Hubb.) J. Phipps = *Tristachya*

Dolichodelphys Schumann & A. Krause. Rubiaceae (II 1). 1 Peru

Dolichoglottis R. Nordenstam (~ *Senecio*). Compositae (Sen.-Tuss.). 2 NZ

Dolichokentia Becc. = *Cyphokentia*

Dolicholobium A. Gray. Rubiaceae (I 1). 28 Philipp. to Fiji. R: Blumea 29(1983)251. Some rheophytes incl. *D. rheophilum* M. Jansen (New Guinea)

Dolicholoma D. Fang & W.T. Wang. Gesneriaceae. 1 SW Guangxi, China: *D. jasminiflorum* D. Fang & W.T. Wang

Dolichometra Schumann. Rubiaceae (IV 1). 1 E Afr. (Usambaras)

Dolichopetalum Tsiang. Asclepiadaceae (III 4). 1 China: *D. kwangsiense* Tsiang

Dolichopsis Hassler. Leguminosae (III 10). 2–3 Paraguay & Arg.

Dolichopterys Kosterm. = *Lophopterys*

Dolichorhynchus Hedge & Kit Tan. Cruciferae. 1 N Saudi Arabia: *D. arabicus* Hedge & Kit Tan

Dolichorrhiza (Pojark.) Galushko (~ *Senecio*). Compositae (Sen.-Sen.). 4 Caucasus, Iran

Dolichos L. Leguminosae (III 10). 60 OW trop. Usu. climbers. Some cult. orn. & fodder; others with ed. pods & seeds. *D. kilimandscharicus* Taub. (SC & E Afr.) – molluscicidal saponins. See also *Lablab*

Dolichostachys Benoist. Acanthaceae. 1 Madagascar

Dolichostegia Schltr. Asclepiadaceae (III 4). 1 Philipp.: *D. boholensis* Schltr.

Dolichostemon Bonati. Scrophulariaceae. 1 SE As.

Dolichothrix Hillard & B.L. Burtt. Compositae (Gnap.-Rel.). 1 Cape: *D. ericoides* Hilliard & B.L. Burtt. R: OB 104 (1991)71

Dolichoura Brade. Melastomataceae. 1 Brazil

Dolichovigna Hayata = *Vigna*

Doliocarpus Rolander. Dilleniaceae. 40 trop. Am. Some lianes with potable water if cut

Dollinera Endl. = *Desmodium*

Dolomiaea DC. Compositae (Card.-Card.). (Incl. *Vladimiria*) 14 Tibet, Himal.

domba oil *Calophyllum inophyllum*

Dombeya Cav. Sterculiaceae. 225 Afr. to Mascarenes (Madag. 187). Small trees & shrubs, some cult. orn. incl. the hybrid *D.* × *cayeuxii* André (*D. burgessiae* Gerrard ex Harvey (S & E Afr.) × *D. wallichii* (Lindley) B.D. Jackson (Madag.)); other spp. locally important as fibre-sources

Domeykoa Philippi. Umbelliferae (I 2c). 4 Peru, Chile. R: UCPB 33(1962)173

Dominella F. Wimmer. Campanulaceae. 1 trop. S Am.

Domingoa Schltr. Orchidaceae (V 13). 3 WI

Dominia Fedde = *Uldinia*

Domkeocarpa Markgraf = *Tabernaemontana*

Domohinea Leandri. Euphorbiaceae. 1 Madagascar

Donatia Forster & Forster f. Donatiaceae. 2 Tasmania, NZ, subAntarctic S Am. Dominate Fuegian bogs

Donatiaceae Chandler. Dicots – Asteridae – Asterales. 1/2 S S Am. (1), NZ & Tasmania (1). Dwarf cushion-pls storing inulin. Lvs spirally arr., linear, hairy in axils; stipules 0. Fls solit., term., bisexual, reg., epigynous; K (5–7), C 5–10, imbricate, A 2 or 3 within epigynous nectary-disk at summit of G, filaments distinct, anthers with longit. slits, $\overline{G}$ (2 or 3) with as many loc. & free styles, ovules ∞ on axile placentas. Fr. turbinate, dry, indehiscent; seeds few with minute embryo & copious oily endosperm

Genus: *Donatia*

Often associated with Stylidiaceae but distinct C has suggested that it has affinities with Saxifragaceae though pollen features typical of the Campanulales

Donax Lour. Marantaceae. 6 Indomal. to Vanuatu. Split stems used for basketwork & fish-traps etc. locally

Donella Pierre ex Baillon = *Chrysophyllum*

Donepea Airy Shaw = *Douepea*

dong (nut) *Santalum acuminatum*

Doniophyton Wedd. Compositae (Barn.). 2 Andes & Patagonia. Annuals

donkey orchid *Diuris* spp.

Donnellsmithia J. Coulter & Rose. Umbelliferae (III 5). 15–20 Mex., C Am.

Dontostemon Andrz. ex C. Meyer. Cruciferae. 8 C to E As.

doob grass *Cynodon dactylon*

Doodia R. Br. Blechnaceae (Blechn.). c. 12 Sri Lanka to Hawaii & Easter Is. Some cult. orn.

doon *Shorea* spp.

Doona Thwaites = *Shorea*

Dopatrium Buch.-Ham. ex Benth. Scrophulariaceae. 7 trop. As. to Aus., *D. junceum* (Roxb.) Buch.-Ham. ex Benth. a ricefield weed natur. in N Am.

Dora, dirty *Cyperus diffusus*

Doratoxylon Thouars ex Hook.f. Sapindaceae. 1 Mascarenes

Dorema D. Don. Umbelliferae (III 10). 12 C & SW As. R: BMOIPB 93,2(1988)89. *D. ammoniacum* D. Don (Iran to India) – principal source of gum ammoniacum used in scent, incense & medic., obtained from insect punctures etc.

Doricera Verdc. (~ *Pyrostria*). Rubiaceae (II 2). 1 Rodrigues: *D. trilocularis* (Balf.f.) Verdc. – consp. foliar dimorphism

Doritis Lindley (~ *Phalaenopsis*). Orchidaceae (V 16). 2 Indomal. R: OB 95(1988)31. Cult. orn.

Dorobaea Cass. (~ *Senecio*). Compositae (Sen.-Sen.). 3 Andes

Doronicum L. Compositae (Sen.-Tuss.). c. 40 temp. Euras. (Eur. 12), Medit. Some medic. & cult. orn. (leopard's bane), the common early-flowering pls being hybrids: *D.* × *willdenowii* Rouy (? *D. pardalianches* L. (incl. *D. cordatum*, W Eur.) × *D. plantagineum* L. (W Eur.)) & *D.* × *excelsa* (N.E. Br.) Stace (? *D. pardalianches* × *D. plantagineum* × *D. columnae* Ten. ('*D. cordatum*', SE Eur.)) with more cordate basal lvs

doronoki *Populus maximowiczii*

Dorothea Wernham = *Aulacocalyx*

Dorotheanthus Schwantes. Aizoaceae (V). 5 S Afr. R: BBP 61(1986)441. Cult. orn. esp. *D. bellidiformis* (Burm.f.) N.E. Br. (*Mesembryantheum b., Cleretum b., M. criniflorum*, Livingstone daisy, Cape) – fr. opens on wetting

Dorstenia L. Moraceae (IV). 105 trop. (As. 1, Afr. c. 60, Am. c. 45). Shrubs & herbs, monoec., fls on a flat or hollowed receptacle, fr. ejected 'as one might flip away a bit of soap between finger & thumb' (Willis). Medic. esp. *D. brasiliensis* Lam. (S Am.) – fertility control in Uruguay, *D. contrajerva* L. (trop. Am.) – febrifuge in Costa Rica, rhiz. used to flavour cigarettes

Dorvalia Hoffsgg. = *Fuchsia*

Doryalis Warb. = *Dovyalis*

Doryanthaceae Dahlgren & Clifford (Liliaceae s.l.). Monocots – Liliidae – Asparagales. 1/2 E Aus. Giant rosette pls with spiral lvs. Infl. a term. thyrse. Fls ± reg.; P 6; A 3 + 3 with elongate anthers dehiscing longit.; $\overline{G}$ 3 with simple style, septal nectaries & several–∞ anatropous ovules per loc. Fr. a loculicidal capsule

Only genus: *Doryanthes*

Many similarities with Hemerocallidaceae

Doryanthes Correa. Doryanthaceae (Liliaceae s.l.). 2 E Aus. Pachycaul pls with strap-shaped lvs, protected, sometimes cult. esp. *D. excelsa* Correa (gymea lily)

Dorycnium Miller = *Lotus*. But see BJ 31(1901)314

Dorycnopsis Boiss. = *Anthyllis*

Doryopteris J. Sm. Pteridaceae (IV). 25 trop. & warm esp. SE Brazil. R: CGH 143(1942)1. Some cult. orn.

Doryphora Endl. Monimiaceae (I 2; Atherospermataceae). 2 NE Aus. Alks. *D. sassafras* Endl. has fragrant wood used for furniture, insect-proof boxes etc.

Dorystaechas Boiss. & Heldr. ex Benth. Labiatae (VIII 2). 1 W As.

Dorystephania Warb. = *Sarcolobus*

Doryxylon Zoll. Euphorbiaceae. 1 C Mal.

Dossifluga Bremek. = *Strobilanthes*

Dossinia Morren. Orchidaceae (III 3). 1 Borneo

double coconut *Lodoicea maldivica*

Douepea Cambess. Cruciferae. 1 NW India

Dougal grass *Isolepis cernua*

Douglas fir *Pseudotsuga menziesii*

Douglasia Lindley (~ *Androsace*). Primulaceae. 8 N Am. R: CJB 70(1992)594. Cult. orn. rock-pls

doum (dom) palm *Hyphaene thebaica*

Douradoa Sleumer. Olacaceae. 1 Brazil

dove flower or **orchid** *Peristeria elata*; **d. tree** *Davidia involucrata*; **d.-weed** *Croton setigerus*

Dovea Kunth (~ *Restio*). Restionaceae. 1 SW Cape: *D. macrocarpa* Kunth. R: Bothalia 15(1985)435

Dovyalis E. Meyer ex Arn. Flacourtiaceae (8). 15 warm Afr., Sri Lanka (1). Dioec. shrubs & trees, some armed with axillary spines. R: BJ 92(1972)64. *D. caffra* (Hook.f. & Harvey) Hook.f. (kei apple, S & E Afr.) – hedgepl. with ed. fr. used in jelly, marmalade etc.; *D. hebecarpa* (Gardner) Warb. (Ceylon gooseberry, kitembilla, Sri Lanka) – ed. fr., added to arrack with sugar to give a sherry subs.

down tree *Ochroma pyramidale*

Downingia Torrey. Campanulaceae. 11 W N Am., 1 extending to Chile. R: MTBC 19,4(1941)1. Elongated pedicel-like ovary. Some cult. orn.

downy chess *Bromus tectorum*

Doxantha Miers = *Macfadyena*

Doxanthemum D. Hunt. Bignoniaceae. 1 N Am. = ?

Doyerea Grosourdy. Cucurbitaceae. 2 trop. Am.

Draba L. Cruciferae. c. 300 N temp. (Eur. 44) & boreal, S Am. mts. Largest genus in fam.

Some cult. orn. rock-garden pls incl. *D. aizoides* L. (C & SE Eur., SW Wales (? natur.)). *D. cacuminum* Ekman – one of few Scandinavian endemics, an octoploid that has arisen at least 3 times from *D. norvegica* Gunnerus (hexaploid) & a diploid sp. See also *Erophila*

Drabastrum (F. Muell.) O. Schulz. Cruciferae. 1 SE Aus.: *D. alpestre* (F. Muell.) O. Schulz

Drabopsis K. Koch. Cruciferae. 1 SW & C As.: *D. nuda* (Bél.) Stapf. R: PCUH 7-8(1977)295

Dracaena Vand. ex L. Dracaenaceae. (Agavaceae s.l.). 60 trop. OW to Canary Is., 1 C Am., 1 Cuba. Shrubs & trees with extrafascicular cambium like *Cordyline* but ovule 1 per loc. & sec. tissues in roots. Resins from stems of some spp. a source of dragon's blood used in varnishes & photo-engraving, that of *D. cinnabari* Balf.f. (Socotra) prob. that known to the Ancients & used to stain horn to resemble tortoiseshell. Extrafl. nectaries. Several cult. orn. incl. *D. draco* (L.) L. (dragon-tree, Canary Is.) – source of dragon's-blood, slow-growing, branching every 10 yrs, prob. after flowering, one 20 m tall & c. 12 m girth in 1868 alleged to have been 6000 yrs old but in 1971 none more than 365 years old alive; *D. fragrans* (L.) Ker-Gawler (trop. Afr.) – commonly seen, particularly varieg. cvs. Some used as live fences in trop.

Dracaenaceae Salisb. (~Agavaceae). Monocots – Liliidae – Asparagales. 6/220 SW N Am. & trop. esp. Afr. Pachycaul trees to rhizomatous succ. (*Sansevieria*) or tuberous (*Calibanus*) pls with sec. thickening & with vessels in lvs & roots; steroidal saponins (e.g. diosgenin) & s.t. resins. Lvs linear to ovate or terete. Infls axillary racemes or panicles. Fls bisexual or not (pls s.t. dioec.), articulated; P 6, A 3 + 3 inserted at base of lobes with introrse longit. dehiscent anthers; $\underline{G}$ (3) (app. 1-loc. in *Dasylirion*) with septal nectaries, simple style & 3-lobed or capitate stigma, each loc. with 1 (*D.* s.s.) or 2 anatropous ovules. Fr. a fleshy, woody or dry 'berry'

Genera: *Calibanus, Cordyline, Dasylirion, Dracaena, Nolina, Sansevieria*

Many cult. orn.; some fibres & resins

Dracocephalum L. Labiatae (VIII 2). c. 45 Euras. (Eur. 3), Medit., N Am. (1). Cult. orn. (dragonhead) allied to *Nepeta* but K 2-lipped & 15-veined, upper A longer than lower

Dracontioides Engl. Araceae (V 3). 1 SE & E Brazil

Dracontium L. Araceae (V 3). 23 trop. Am. Sympodial rhizome gives 1 leaf & infl. (in *D. gigas* (Seemann) Engl. 3 m & 1.5 m tall resp.); leaf with 3 major lobes, the laterals developing pseudo-dichotomously initially; fls bisexual with P. *D. polyphyllum* L. tubers ed.; some medic. esp. for snakebite cure (Amaz.) incl. *D. trianae* Engl. cult. NW Amaz.

Dracontomelon Blume. Anacardiaceae (II). 8 SE As., Indomal. to Fiji. Some timbers (New Guinea, Pacific or Papuan walnut) esp. *D. dao* (Blanco) Merr. & Rolfe (*D. mangiferum*, dao, Indomal.) – 'paldao' used for veneers & also matches, fr. ed., fls used as flavouring

Dracophilus (Schwantes) Dinter & Schwantes (~ *Juttadinteria*). Aizoaceae (V). 3–4 S Afr.

Dracophyllum Labill. Epacridaceae. c. 48 Aus., New Caled. & esp. NZ. Trees & shrubs with lvs like monocots. Some cult. orn.

Dracopsis Cass. (~ *Rudbeckia*). Compositae (Helia.-Rudb.). 1 N Am: *D. amplexicaulis* (L.) Less.

Dracosciadium Hilliard & B.L. Burtt. Umbelliferae (III 8). 2 Natal. R: NRBGE 43(1986)220

Dracula Luer (~ *Masdevallia*). Orchidaceae (V 13). 100+ trop. Am. R: MSB 46(1993)1. Cult. orn.

Dracunculus Miller. Araceae (VIII 6). 2 (excl. *Helicodiceros*) Medit. incl. Eur. Cult. orn. tuberous herbs with malodorous appendage & poll. mechanism like *Arum*. *D. vulgaris* Schott (Medit.) poisonous & avoided by grazing animals (dragon arum), depicted in Minoan paintings

dragon arum *Dracunculus vulgaris*, (N Am.) *Arisaema* spp.; **d. claw** *Corallorrhiza odontorhiza*; **d. gum** *Astracantha* spp. esp. *A. gummifera*; **d. head** *Dracocephalum* spp.; **d. tree** *Dracaena draco*

dragon's blood Several reddish resins used in varnishes etc., orig. from *Dracaena cinnabari* &, later, *D. draco*, more recently from *Daemonorops* spp.; **d. eyes** *Dimocarpus longan*

Drakaea Lindley. Orchidaceae (IV 1). 4 SW Aus. *Ophrys*-like insect mimics

Drake-Brockmania Stapf. Gramineae (31d). 2 E & NE Afr.

Drambuie Liqueur made from whisky & honey derived from bees visiting *Calluna vulgaris*

Draperia Torrey. Hydrophyllaceae. 1 California

Drapetes Lam. Thymelaeaceae. (Excl. *Kelleria*) 1 Fuegia & Falkland Is.: *D. muscosa* Lam.

Dregea E. Meyer. Asclepiadaceae (III 4). 3 warm OW. *D. abyssinica* (Hochst.) Schumann (trop. Afr.) – lvs cooked Uganda; *D. volubilis* (L.f.) Hook.f. (*Wattakaka v.*, India to W Mal.) – eaten with curry

Dregeochloa Conert (~ *Danthonia*). Gramineae (25). 2 S Afr. R: SBi 47(1966)335

Drejera Nees. Acanthaceae. 4 trop. Am.

Drejerella Lindau = *Justicia*

Drepananthus Maingay ex Hook.f. (~ *Cyathocalyx*). Annonaceae. 10 Mal.

Drepanocarpus G. Meyer = *Machaerium*

Drepanocaryum Pojark. Labiatae (VIII 2). 1 C As.

Drepanostachyum Keng f. = *Sinarundinaria*

Drepanostemma Jum. & Perrier = *Sarcostemma*

Dresslerella Luer. Orchidaceae (V 13). 8 C & NW S Am. R: MSB 26(1988)1. Allied to *Pleurothallis*

Dressleria Dodson. Orchidaceae (V 9). 5 C Am. R: Selbyana 1(1975)130. Fls bisexual (cf. *Catasetum*)

Dressleriella Brieger = *Jacquiniella*

Dresslerioopsis Dwyer = *Lasianthus*

Dresslerothamnus H. Robinson. Compositae (Sen.-Sen.). 4 C Am. to Colombia. R: SB 14(1989)382. Lianes

Driessenia Korth. Melastomataceae. 14 W (11 Borneo endemics) & C Mal. R: NJB 5(1985)335

Drimia Jacq. ex Willd. (incl. *Urginea*). Hyacinthaceae (Liliaceae s.l.). 120 S Eur. (3), Afr., As. Many medic. & poisonous (cardiac glycosides) esp. *D. maritima* (L.) Stearn (*U. maritima*, squill, sea onion, Medit.) – insect- & wind-poll., diff. forms used as cardiac stimulant etc. & as rat-poison (specific to rodents, used since time of Theophrastus; other animals vomit), grown comm. in US since 1946, hung as amulet outside houses in Greece even today, bulbs resistant to maquis fires; *D. indica* (Roxb.) Jessop (trop. As. & Afr.) used as subs. *D. noctiflora* (Batt. & Trabut) Stearn (N W Afr.) – night-flowering with tepals reflexed like *Cyclamen*

Drimiopsis Lindley & Paxton. Hyacinthaceae (Liliaceae s.l.). c. 15 Afr. S of Sahara. R: JSAB 38(1972)157

Drimycarpus Hook.f. Anacardiaceae (III). 2+ Indomal. $\overline{G}$. *D. luridus* (Hook.f.) Ding Hou (W Mal.) – v. allergenic fr.

Drimys Forster & Forster f. Winteraceae. 6 trop. Am. (bisexual fls), 5 C Mal. to Tahiti (usu. dioec.; *D. piperita* Hook.f. v. variable with 39 local 'entities' recognized). Poll. by wide range of Diptera; P with distinct K & C. *D. lanceolata* (Poiret) Baillon (*D. aromatica*, Tasmania) – dried fr. a pepper subs., planted as hedges in Ireland; *D. winteri* Forster & Forster f. (Winter's bark, Mex. to Tierra del Fuego) – v. variable, first known as a 'medicine very powerful against the scurvy' introd. by Capt. John Winter in 1578, also stomachic

Droceloncia Léonard. Euphorbiaceae. 1 trop. Afr.

Droguetia Gaudich. Urticaceae (V). 7 trop. & warm. Afr. to Java. R: NJB 8(1988)36, 10(1990)431

Droogmansia De Wild. Leguminosae (III 9). 5 SC Afr. (1 variable), W Afr. (4), merging with *Tadehagi*

dropwort *Filipendula vulgaris*; **water d.** *Oenanthe crocata*

Drosanthemopsis Rauschert (~ *Drosanthemum*). Aizoaceae (V). 2 W S Afr.

Drosanthemum Schwantes. Aizoaceae (V). 100 SW Afr., Cape & Namaqualand (most). Succ. shrubs; cult. orn.

Drosera L. Droseraceae. c. 110 cosmop. (Eur. 3) esp. S hemisph. (SW Aus. 68) in wet places. R: T 43(1994)583. Insectivorous herbs (sundews, e.g. less than 1 ha of *D. anglica* Hudson (N temp.) trapped c. 6 m *Pieris rapae* in one morning!) with rhiz. or s.t. scrambling stems (*D. gigantea* Lindley (SW Aus.) to 1 m tall, *D. erythrorrhiza* Lindley (SW Aus.) to over 50 yrs old) & round to linear palisade-less leaf-blades with gland-tipped red or greenish hairs capable of movement when irritated (bending to centre of leaf in 3–20 mins, the surface voltage dropping when touched & the receptor potential correlated with intensity of stimulus, above a certain threshold a series of short electrical pulses towards the base of the tentacle giving the action potential leading to the bending) & of holding (trapping mucilage c. 4% solution of acidic polysaccharide often suffocating insects through spiracles) & digesting insects, which in some spp. at least promotes flowering. Some cult. as curiosities (over 100 grown); some with cleistogamous fls, others with epiphyllous propagules. Some locally medic. (bronchial complaints) due to quinones, rootstocks ed. S W Aus. aborigines

Droseraceae Salisb. Dicots – Dilleniidae – Nepenthales. 4/110 cosmop. Insectivorous herbs (subshrubby in *Drosophyllum*), often with basal rosette of lvs & cyanogenic. Lvs spirally arr. to whorled, often circinate in bud, simple, either with irritable gland-tipped hairs (*Drosera, Drosophyllum*) or an active trap (*Aldrovanda, Dionaea*); stipules often present. Fls bisexual, reg., hypogynous with marcescent K, C & A, solit. (*Aldrovanda*) or in cymes; K

(4)(5–8), ± basally connate, imbricate, C same no., convolute, A (4)5(10–20), connate basally in *Dionaea* with v. variable pollen (tetrads in all but *Drosophyllum* where monads), G (3(5)), 1-loc., with distinct often bifid styles (united in *Dionaea*) & (3–)∞ ovules (anatropous, bitegmic) on parietal placentas or a basal one. Fr. a loculicidal capsule (rarely indehiscent); seeds (3–)∞, fusiform with short straight embryo embedded in copious endosperm rich in starch, oil & protein. x = 6–17+

Genera: *Aldrovanda, Dionaea, Drosera, Drosophyllum*

Non-moving mucilage traps (*Drosophyllum*, which is remarkably similar to *Byblis*) to slow but perceptible movements (*Drosera*), from dry habitats (*Drosophyllum*) to a wide range incl. wet (*Drosera*) to total submergence (*Aldrovanda*)

Drosophyllaceae Chrtek et al. = Droseraceae

Drosophyllum Link. Droseraceae. 1 Portugal, S Spain, Morocco: *D. lusitanicum* (L.) Link – somewhat woody calcifuge colonist with marcescent lvs. Insect-trapping like *Drosera* (mucilage) but the hairs are not irritable & leaf-tips face out (reverse circinnation); used to treat conjunctivitis

Drudeophytum J. Coulter & Rose = *Tauschia*

druggists' bark *Cinchona* spp.

Drummondita Harvey (~ *Philotheca*). Rutaceae. 5 Aus. R: Nuytsia 1(1971)206, 9(1993)96

drumsticks *Isopogon* spp., *Moringa oleifera*

Drusa DC. Umbelliferae (I 2a). 1 Canary Is. & Somalia: *D. glandulosa* (Poiret) Boran.

Dryadanthe Endl. = *Sibbaldia*

Dryadella Luer (~ *Masdevallia*). Orchidaceae (V 13). c. 25 trop. Am.

Dryadodaphne S. Moore. Monimiaceae (I 2). 3 New Guinea, Queensland

Dryadorchis Schltr. Orchidaceae (V 16). 3 New Guinea

Dryandra R. Br. Proteaceae. 66 SW Aus.: R: R.M. Sainsbury(1985) *A field guide to D.*

Dryas L. Rosaceae. 2 Arctic-alpine. R: SBT 53(1959)507. Mountain avens. Nitrogen-fixing nodules (actinomycetes); lvs char. as fossil in cool periods. Cult. orn. rock-pls esp. *D. octopetala* L.

Drymaria Willd. ex Schultes. Caryophyllaceae (I 1). 48 W US to Patagonia (46), Galápagos (1) with *D. cordata* (L.) Roemer & Schultes pantrop. R: AMBG 48(1961)173. Some with elaiophores attractive to poll. male euglossine bees

Drymoanthus Nicholls. Orchidaceae (V 16). 3 NE Aus. , NZ (2)

Drymocallis Fourr. ex Rydb. = *Potentilla*

Drymoda Lindley. Orchidaceae (V 15). 2 Burma, Thailand. Laos. R: OB 89(1986)167

Drymoglossum C. Presl = *Pyrrosia*

Drymonia C. Martius. Gesneriaceae. c. 110 trop. Am. Mostly lianes with salt-shaker anthers. Some cult. orn. incl. *D. serrulata* (Jacq.) C. Martius – decid. liane with dichogamous fls (male 1st day, female next) with more nectar than any other bee-poll. pl., *Epicharis* bees being poll. but oil sticking pollen grains together also coll. non-poll. *Trigona bees*

Drymophila R. Br. (~ *Luzuriaga*). Convallariaceae (Liliaceae s.l.). 2 E & SE Aus. R: FA 45(1987)156

Drymophloeus Zipp. Palmae (V 4h). c. 15 Moluccas to Solomon Is. Monoec. graceful cult. orn.

Drymotaenium Makino. Polypodiaceae (II 3). 1 India, Japan, Taiwan: *D. miyoshianum* (Makino) Makino

Drynaria (Bory) J. Sm. Polypodiaceae (II 1). 15 OW trop. esp. China. Epiphytes. Lvs usu. dimorphic, the sterile erect, short & broad, soon becoming dry & collecting humus, the fertile deeply lobed or pinnate, petiolate; (? ant-attracting) nectaries reported

Drynariopsis (Copel.) Ching = *Aglaomorpha*

Dryoathyrium Ching = *Lunathyrium*

Dryobalanops Gaertner f. Dipterocarpaceae. 7 Mal. R: FM I,9(1982)371. Timber (Brunei teak, kapur) a popular light brown hardwood; *D. aromatica* Gaertner f. (*D. sumatrensis*; & other spp.) source of camphor (baros, barus, Borneo or Sumatra c.) form. much exported from N Sumatra & Johore since C6 to Arabs (mentioned by Marco Polo), the crystals collected from splits in the bole; fr. boiled as vegetable

Dryopetalon A. Gray. Cruciferae. 5 SW N Am. R: CDH 3(1941)199, CGH 214(1984)20. Petals 5–7-lobed

Dryopoa Vickery (~ *Festuca*). Gramineae (17). 1 SE Aus.: *D. dives* (F. Muell.) Vickery

Dryopolystichum Copel. Dryopteridaceae (I 3). 1 Papuasia: *D. phaeostigma* (Cesati) Copel.

Dryopsis Holttum & Edwards (~ *Ctenitis*). Dryopteridaceae (I 3). 26 Indomal. mts

Dryopteridaceae Herter (~ Aspleniaceae). Filicopsida. (Incl. Athyriaceae, Onocleaceae,

Woodsiaceae – oldest name). 47/1700 cosmop. esp. temp. & montane. Terr. (rarely epiphytic) pls with erect or creeping stems, dictyostelic, scaly. Fronds simply pinnate (simple) to decompound. Sporangia usu. in ± orbicular sori

Classification & principal genera (after Holttum):

I Dryopteridoideae (petiole with at least 3 vasc. bundles; spores achlorophyllous; 3 tribes): *Rumohra* (1. Rumohreae); *Arachniodes, Dryopteris, Polybotrya, Polystichum, Stigmatopteris* (2. Dryopterideae); *Ctenitis, Dryopsis, Tectaria* (3. Tectarieae)

II Athyrioideae (petiole with 2 vasc. bundles; spores sometimes chlorophyllous; 2 tribes): *Athyrium, Deparia, Diplazium, Woodsia* (1. Physematieae); *Matteuccia, Onoclea* (2. Onocleeae)

Some cult. orn.; *Diplazium* ed.

Dryopteris Adans. Dryopteridaceae (I 2). c. 225 subcosmop. (Eur. 19) but rare in lowland trop. (Aus. 1, NZ 0). R: BBMNHB 14(1986). Shield ferns, buckler f. Much allopolyploidy. Some cult. orn. esp. *D. dilatata* (*D. austriaca*, N temp.) incl. 'Florists' fern', 'fern' used in soap, oil in spa massage & *D. filix-mas* (L.) Schott (male fern, N temp.) – one of oldest known vermifuges which paralyses tapeworms, which may then be removed by purgatives but dangerous as it paralyses voluntary muscles of patients & is now replaced by quinacrine, app. used in silk reeling in Ancient China; buckler ferns incl. *D. carthusiana* (Villars) H.P. Fuchs (narrow b.f.) & *D. dilatata* (Hoffm.) A. Gray (broad b.f.)

Drypetes Vahl. Euphorbiaceae. c. 200 trop., E As., S Afr. Some timbers. *D. caustica* (Cordemoy) Airy Shaw now reduced to 2 trees on Mauritius & 12 on Réunion; *D. gossweileri* S. Moore (W & C Afr.) – elephant-disp.; *D. pellegrinii* Leandri (W Afr.) – bark local medic., roots for chewing-sticks (Ghana); *D. pendula* Ridley (Malay Pen.) – ant-pl. with hollow twigs

Drypis L. Caryophyllaceae (III 2). 1 S Eur. & Lebanon: *D. spinosa* L., cult. orn. herb

Duabanga Buch.-Ham. Sonneratiaceae. 2 Indomal. in rain forest (cf. *Sonneratia* in mangrove); timber of *D. grandiflora* (DC) Walp. used for tea-chests

Duabangaceae Takht. = Sonneratiaceae

Dubardella H.J. Lam = *Pyrenaria*

Dubautia Gaudich. Compositae (Hele.-Mad.). (Incl. *Raillardia*) 21 Hawaii. R: Allertonia 4,1(1985)62. Trees, shrubs & lianes; lvs with parallel veins

Duboisia R. Br. Solanaceae (2). 3 Aus., New Caled. Alks incl. atropine, a commercial source & form. used as emu poison. *D. hopwoodii* (F. Muell.) F. Muell. (Aus.) – masticatory (pituri) of C Aus. Aborigines; *D. myoporoides* R. Br. (Aus., New Caled.) – alk. comm. source for insecticide & medic. drugs (hyoscine) & timber (corkwood) for carving

Duboscia Bocquillon. Tiliaceae. 3 trop. W Afr.

Dubouzetia Pancher ex Brongn. & Gris (~ *Crinodendron*). Elaeocarpaceae. 10 Moluccas to New Caled. R: KB 42(1987)796

Dubyaea DC. Compositae (Lact.-Crep.). 10 Himal., W China. R: MTBC 19,3(1940)8

Ducampopinus A. Chev. = *Pinus*

Duchesnea Sm. = *Potentilla*

duck meat *Lemna* spp.; **d. orchid** *Caleana major*; **d. plant** *Sutherlandia frutescens*; **d.weed** *Lemna* spp. esp. *L. minor*

Duckea Maguire = *Cephalostemon*

Duckeanthus R.E. Fries. Annonaceae. 1 trop. S Am.: *D. grandiflorus* R.E. Fries

Duckeella Porto & Brade. Orchidaceae (Pogoniinae). 3 trop. S Am.

Duckeodendraceae Kuhlm. = Apocynaceae

Duckeodendron Kuhlm. Apocynaceae (Duckeodendraceae). 1 Amazonian Brazil: *D. cestroides* Kuhlm. Wood anatomy suggests affinity with Solanaceae

Duckera F. Barkley = *Rhus*

Duckesia Cuatrec. Humiriaceae. 1 Amazonian Brazil

Ducrosia Boiss. Umbelliferae (III 11). 3 Egypt to NW India. R: NRBGE 34(1975)190

Dudleya Britton & Rose. Crassulaceae. c. 40 SW N Am. Glabrous succ.; some cult. orn.

duffin bean *Phaseolus lunatus*

Dufrenoya Chatin. Santalaceae. 14 Indomal.

Dugaldia Cass. (~ *Helenium*). Compositae (Hele.-Gai.). 3 US & C Am. R: Brittonia 26(1974)385

Dugandia Britton & Killip = *Acacia*

Dugesia A. Gray. Compositae (Helia.-Eng.). 1 Mex.: *D. mexicana* (A. Gray) A. Gray. R: Brittonia 26(1974)385

Duggena Vahl = *Gonzalagunia*

Duguetia A. St-Hil. Annonaceae. 82 trop. Am. R: AHB 12(1934)28. Fr. united with fleshy receptacle to give false fr. Some timbers esp. *D. quitarensis* Benth. (Jamaica & Cuba lancewood). *D. rhizantha* (Eichler) Huber has subterr. rhizome with scale lvs & aerial flowering shoots

Duhaldea DC (~ *Inula*). Compositae (Inul.) 13 Iran to Himal. & E Afr. Trees to herbs

Duidaea S.F. Blake. Compositae (Mut.-Mut.). 4 Venez. & Guyana

Duidania Standley. Rubiaceae (I 1). 1 Venezuela

Dukea Dwyer = *Raritebe*

Dulacia Vell. (~ *Liriosma*). Olacaceae. 14 trop. Am.

Dulichium Pers. Cyperaceae. 1 N Am.: *D. arundinaceum* (L.) Britton

Dulongiaceae J. Agardh = Grossulariaceae

Dumasia DC. Leguminosae (III 10). 9 OW trop.

dumb cane *Dieffenbachia* spp. esp. *D. seguine*

Dumori butter *Tieghemella heckelii*

Dunalia Kunth (~ *Acnistus*). Solanaceae (1). 7 Andes

Dunbaria Wight & Arn. Leguminosae (III 10). 16 trop. As. to Aus.

Dunkeld larch *Larix × marschlinsii*

Dunnia Tutcher. Rubiaceae (I 1/IV 24). 2 India, China. R: Blumea 24(1978)367

Dunniella Rauschert = *Pilea*

Dunstervillea Garay. Orchidaceae (V 10). 1 Venezuela

Duosperma Dayton. Acanthaceae. 12 trop. & S Afr.

Duparquetia Baillon. Leguminosae (I 2). 1 trop. W Afr. Fls seem to mimic those of orchids but poll. mechanism unknown

Duperrea Pierre ex Pitard. Rubiaceae (II 2). 2 India, China, SE As.

Dupontia R. Br. (~ *Colpodium*). Gramineae (17). 1 Arctic incl. Eur.

Durandea Planchon. Linaceae. 15 C Mal. to Fiji

Duranta L. Verbenaceae. 17 Carib. to S Am., natur. S & E As. R: Sida 10(1984)308. Shrubs & trees, sometimes spiny. Some cult. orn. incl. *D. erecta* L. (*D. repens*, pigeon-berry, golden dewdrop, Florida to Brazil, natur. S US)

Durban grass *Dactyloctenium australe*

Duriala (R. Anderson) Ulbr. = *Maireana*

durian *Durio zibethinus*

Durio Adans. Bombacaceae. 28 Burma to W Mal. Bat-poll. trees with lvs in 2 ranks; some cauliflorous. *D. zibethinus* Murray (durian, W Mal.) – each fl. produces 0.63 ml watery nectar a night, highly esteemed fr., usu. malodorous but with arils tasting of caramel, banana, vanilla etc. with slight onion tang, much sought by animals incl. humans early in morning after fall of the spiny capsule up to some kg in weight; seeds ed. roasted

durma mats *Phragmites australis*

durmast oak *Quercus petraea*

Duroia L.f. Rubiaceae (II 1). 20 trop. Am. Myrmecophilous: in *D. petiolaris* Hook.f. & *D. hirsuta* (Poeppig & Endl.) Schumann (app. allelopathic growing in monospecific stands in rain forest; caustic bark used to make short-lived tattoos), stem below infl. hollow & with 2 longit. slits as ant-doors; *D. saccifera* Hook.f. has ant-houses on lvs

durra *Sorghum bicolor*

Durringtonia R. Henderson & Guymer. Rubiaceae (IV 13). 1 trop. E Aus.: *D. paludosa* R. Henderson & Guymer – dioec. herb. pl. of swamps. R: KB 40(1985)97

durum (wheat) *Triticum turgidum* Durum group

Duschekia Opiz = *Alnus*

Duseniella Schumann. Compositae (Barn.). 1 Arg., arid Patagonia: *D. patagonica* (Dusén) Schumann – annual

Dussia Krug & Urban ex Taubert. Leguminosae (III 2). 2 trop. Am.

dusty miller *Senecio cineraria*

Dutaillyea Baillon. Rutaceae. 2 New Caled. R: BMNHN 4,6(1984)29

Dutch clover *Trifolium repens*; **D. elm** *Ulmus × hollandica*; **D. iris** *Iris × hollandica*; **D. man's breeches** *Dicentra spectabilis*; **D. man's pipe** *Isotrema macrophyllum*

Duthiastrum De Vos. Iridaceae (IV 3). 1 S Afr.

Duthiea Hackel. Gramineae (21a). 3 Afghanistan to China. R: KB 8(1954)547

Duthiella De Vos = *Duthiastrum*

Duvalia Haw. Asclepiadaceae (III 5). 12 S Afr., 5 Somalia to Arabia. R: MIABH 23b(1990)95. Cult. orn. leafless succ.

Duvaliandra M. Gilbert (~ *Caralluma*). Asclepiadaceae (III 5). 1 Socotra: *D. dioscoridis* (Lavranos) M. Gilbert. R: Bradleya 8(1990)29

Duval-Jouvea Palla = *Cyperus*

Duvaucellia S. Bowdich = *Kohautia*
Duvernoia E. Meyer ex Nees (*Duvernoya*) = *Justicia*
Duvigneaudia Léonard. Euphorbiaceae. 1 trop. Afr.
duzhong *Eucommia ulmoides*
Dyakia Christenson. Orchidaceae (V 16). 1 Borneo: *D. hendersoniana* (Reichb.f.) Christenson. R: OD 50(1986)63.
Dybowskia Stapf = *Hyparrhenia*
Dyckia Schultes f. Bromeliaceae (1). 107 trop. S Am. esp. S. R: FN 14(1974)500. Terr. & saxicolous cult. orn. with extrafl. nectaries
Dyera Hook.f. Apocynaceae. 2–3 W Mal. *D. costulata* (Miq.) Hook.f. (jelutong) – comm. lightweight hardwood, chicle obtained by tapping used as chewing-gum (now superseded by synthetics)
Dyerophytum Kuntze (~ *Plumbago*). Plumbaginaceae (I). 3 S Afr., Socotra, Arabia, India
dyer's chamomile *Anthemis tinctoria*; **d. greenweed** *Genista tinctoria*; **d. rocket** or **weld** *Reseda luteola*
Dymondia Compton (~ *Arctotis*). Compositae (Arct.-Arct.). 1 S Afr.: *D. margaretae* Compton
Dypsidium Baillon = *Neophloga*
Dypsis Noronha ex C. Martius. Palmae (V 4e). c. 20 Madag. Incl. some of the tiniest palms like *D. hildebrandtii* (Baillon) Becc. only 30 cm tall
Dyschoriste Nees. Acanthaceae. c. 65 trop. & warm. R(Am.): AMBG 15(1928)9
Dyscritogyne R. King & H. Robinson = *Steviopsis*
Dyscritothamnus Robinson. Compositae (Helia.-Gal.). 2 Mexico
Dysodiopsis (A. Gray) Rydb. (~ *Hymenatherum*). Compositae (Hele.-Pect.). 1 S US
Dysolobium (Benth.) Prain (~ *Vigna*). Leguminosae (III 10). 4 SE As. R: Blumea 30(1985)363
Dysophylla Blume = *Pogostemon*
Dysopsis Baillon. Euphorbiaceae. 1 Andes, Juan Fernandez
Dysosma Woodson (~ *Podophyllum*). Berberidaceae (II 2). 7 E As.
Dysoxylum Blume. Meliaceae (I 6). 80 Indomal. (Mal. 50, R: FM I, 12(1995)61 to NZ (1) & Tonga. Timber: *D. acutangulum* Miq. (Mal.) – coffins; *D. fraserianum* (A. Juss.) Benth. (Aus. mahogany or rosewood, Aus.) – fragrant wood for turnery etc.; *D. loureiroi* Pierre (SE As.) – sandalwood-like scented timber used for coffins & joss-sticks; *D. mollissimum* Bl. (*D. forsteri*, *D. muelleri*, China to Aus.) – red timber for cabinet-work etc. esp. in Tonga (red bean, Aus.), form. largest trees in Java; *D. pettigrewianum* Bailey (Cairns satinwood, E Mal. to trop. Aus. & Solomon Is.); *D. spectabile* (A. Juss.) Hook.f. (kohekohe, NZ) – similar, winter-flowering, poll. tuis (birds); *D. angustifolium* King (Malay Pen.) – rheophyte, ? fish-disp., though fr. allegedly toxic to mammals
Dyspemptemorion Bremek. = *Justicia*
Dysphania R. Br. Chenopodiaceae (I 2, Dysphaniaceae). 17 Aus., NZ. R: Nuytsia 4(1983)180
Dysphaniaceae Pax = Chenopodiaceae
Dyssochroma Miers = *Markea*
Dyssodia Cav. Compositae (Hele.-Pect.). Excl. *Thymophylla* (q.v.) etc. 4 US to Guatemala. R: UCPB 48(1969)1. Some cult. orn.
Dystaenia Kitag. (~ *Ligusticum*). Umbelliferae (III 8). 2 Korea, Japan.
Dystovomita (Engl.) D'Arcy. Guttiferae (III). 3 trop. Am.

E

eagle fern *Pteridium aquilinum*; **e. wood** *Aquilaria malaccensis*
Earina Lindley. Orchidaceae (V 13). 7 W Pacific
early Nancy *Wurmbea dioica*
earth almond *Cyperus esculentus*; **e. chestnut** *Lathyrus tuberosus*; **e. nut** *Arachis hypogaea*, *Conopodium majus*; **e. stars** *Cryptanthus* spp.
East African cedar *Juniperus procera*; **E. A. sandalwood** *Osyris tenuifolia*; **E. Indian arrowroot** *Curcuma angustifolia*, *Tacca leontopetaloides*; **E. I. dill** *Anethum graveolens*; **E. I. rosewood** *Dalbergia latifolia*; **E. London boxwood** *Buxus macowanii*
Easter cactus *Hatiora gaertneri*; **E. lily** *Lilium longiflorum* var. *eximium*
Eastwoodia Brandegee. Compositae (Ast.-Sol.). 1 SW N Am.
Eatonella A. Gray. Compositae (Helia.). 1 SW N Am.
eau de Cologne principal ingredient *Citrus* × *aurantium*; **e. de Créole** *Mammea americana*
eba *Lophira alata*
Ebandoua Pellegrin = *Jollydora*
Ebenaceae Guerke. Dicots – Dilleniidae – Theales. 2/485 trop. & warm, few temp. Trees,

shrubs or rarely geoxylic suffrutices, often dioec., s.t. with extrafl. nectaries, s.t. with black heartwood. Lvs usu. spirally arr., simple, entire; stipules 0. Fls small, reg., solit. or in cymose, rarely thyrsoid clusters, often with well-developed staminodes or pistillodes; K (3–7), persistent, often accrescent in fr., C (3–7), lobes contorted, A 2–100+, epipetalous or borne on receptacle, often with 2 anthers per filament, usu. with longit. slits, rarely apical pores, connective often larger than anthers, $\underline{G}$ (2–5(+)), rarely $\overline{G}$, pluriloc., each loc. with 2 pendulous, anatropous, bitegmic ovules & ± divided by a false septum, styles ± distinct. Fr. a berry, rarely tardily dehiscent; seeds large, with thin testa, hard, sometimes ruminate endosperm with oil & hemicellulose, & straight or slightly curved embryo with flat, leafy, usu. emergent cotyledons, photosynthetic or not. x = 15

Genera: *Diospyros, Euclea*

Timbers & fr.-trees: ebonies etc., persimmons & date-plums

Ebenus L. Leguminosae (III 17). 18 Medit. (Eur. 2) to Baluchistan

Eberhardtia Lecomte. Sapotaceae (I). 3 S China, SE As., Sabah. R: T.D. Pennington, S. (1991)145

Eberlanzia Schwantes (V). Aizoaceae (V). 12 SW Afr. Spiny (fr. pedicels)

ebony *Diospyros* spp. esp. *D. ebenum*; **American** or **Jamaican e.** *Brya ebenus*; **black e.** *Euclea pseudebenus*; **Bombay e.** *D. montana* etc.; **Cape e.** *Heywoodia lucens*; **e. heart** *Elaeocarpus bancroftii*; **Macassar e.** *D. celebica*; **Madagascar e.** *D. haplostylis*

ebor or **eboe** *Dipteryx oleifera*

Ebracteola Dinter & Schwantes. Aizoaceae (V). 5 SW Afr. R: Bothalia 16(1986)218

Ecastaphyllum P. Browne = *Dalbergia*

Ecballium A. Rich. Cucurbitaceae. 1 Medit.: *E. elaterium* (L.) A. Rich. (squirting cucumber) – monoec. ± trailing herb with fr. which falls when ripe & pericarp contracts, reducing turgidity, such that seeds in watery fluid are ejected explosively through basal hole; fr. used as purgative (elaterium) & anti-inflammatory (cucurbitacin B)

Ecbolium Kurz = ? *Justicia*

Ecclinusa C. Martius. Sapotaceae (IV). 11 trop. S Am. to Trinidad. Some, esp. *E. balata* Ducke (Brazil), sources of inferior balata

Eccoilopus Steudel = *Spodiopogon*

Eccoptocarpha Launert. Gramineae (34b). 1 S trop. Afr.

Eccremis Baker = *Excremis* Willd.

Eccremocactus Britton & Rose = *Weberocactus*

Eccremocarpus Ruíz & Pavón. Bignoniaceae. 5 Peru & Chile. Fr. a capsule with valves remaining apically united after dehiscence. Climbers with tendrils & sensitive petioles. *E. scaber* Ruíz & Pavón (Chile) – cult. orn. with orange (cvs with yellow or red) fls

Ecdeiocolea F. Muell. Restionaceae (Ecdeiocoleaceae). 1 SW Aus.

Ecdeiocoleaceae Cutler & Airy Shaw. See Restionaceae

Ecdysanthera Hook. & Arn. = *Urceola*

Echeandia Ortega. Anthericaceae (Liliaceae s.l.). 80 SW US to Peru (S Am. spp. poss. generically distinct)

Echetrosis Philippi = *Parthenium*

Echeveria DC. Crassulaceae. c. 150 warm Am. esp. Mex. R: E. Walther (1972) *E.* Succ. herbs & shrubs with term. leaf-rosettes & axillary (cf. *Cotyledon*) infls. Many cult. orn. spp. & hybrids, often used for formal bedding-displays (floral clocks etc.) when infls removed; propagated from stem or leaf-cuttings or, in some spp., fragments of infl. scape

Echidnium Schott = *Dracontium*

Echidnopsis Hook.f. Asclepiadaceae (III 5). 19 trop. NE Afr. to Arabia. R: Bradleya 6(1988)1. Some cult. orn. leafless succ.

Echinacanthus Nees. Acanthaceae. 4 Himal. & China. R: EJB 51(1994)186

Echinacea Moench. Compositae (Helia.-Rudb.). 9 E US. R: UKSB 48(1968)113. Cult. orn. (cone flowers) esp. *E. angustifolia* DC – root the most widely used Indian med. of Plains, patent med. 1870 now shown to have antiviral effects & stimulate prod. of white blood-cells & *E. purpurea* (L.) Moench – rhizomes used in Indian medicine, both with capitula to 15 cm diam.

Echinaria Desf. = *Cenchrus*

Echinocactus Link & Otto. Cactaceae (III 9). 6 SW N Am. Ribbed, often large cult. orn. often with long spines (to 8 cm in *E. polycephalus* Engelm. & Bigelow), incl. *E. horizonthalonius* Lemaire – pulp eaten & *E. platyacanthus* Link & Otto (C Mex.) – sacred in pre-Columbian culture, now used in confectionery, medic. etc.

Echinocaulon (Meissner) Spach = *Persicaria*

Echinocephalum Gardner = *Melanthera*

Echinocereus Engelm. Cactaceae (III 1). 47 SW N Am. R: KMMS 1(1985). All in cult. (hedgehog cacti). *E. enneacanthus* Engelm. & *E. pectinatus* (Scheidw.) Engelm, with ed. (strawberry-flavoured) fr., the spine-clusters easily removed

Echinochloa Pal. Gramineae (34b). c. 35 warm (Eur. 1). R: JJB 19(1966)277. *E. colona* (L.) Link (jungle rice, Shama millet, trop. & warm) – weed, esp. in rice; *E. crus-galli* (L.) Pal. (cockspur) – bad weed in Am.; *E. frumentacea* Link (barnyard grass or millet, jungle rice, OW, natur. US, where grown for fodder – billion dollar grass) – cereal cropping 6 weeks after sowing, prob. derived from *E. colona*; *E. oryzoides* (Ard.) Fritsch (origin unknown) – paddy-rice mimic weed; *E. pyramidalis* (Lam.) A. Hitchc. & Chase (antelope grass, trop. & S Afr., Madag.) – fodder, locally used flour; *E. turneriana* (Domin) J. Black (channel millet, Aus.) – promising forage & grain crop; *E. utilis* Ohwi & Yab. (Jap. or Sanwa millet) – cereal in China & Japan, poss. derived from *E. crus-galli*

Echinocitrus Tanaka = *Triphasia*

Echinocodon Hong. Campanulaceae. 1 China

Echinocoryne H. Robinson (~ *Vernonia*). Compositae (Vern.-Vern.) 6 Brazil

Echinocystis Torrey & A. Gray. Cucurbitaceae. 1 N Am.: *E. lobata* (Michaux) Torrey & A. Gray – climbing herb with tubers, sometimes used medic. by Am. Indians, cult. orn.

Echinodorus Rich. ex Engelm. Alismataceae. 48 Am., Afr. R: K. Rataj (1975) *Studie ČSAV* 2. Cult. orn. aquarium pls (burhead, sword plant) esp. *E. berteroi* (Sprengel) Fassett (cellophane pl., N Am. to WI) – submerged lvs membranous, ribbon-like

Echinofossulocactus Lawrence = *Echinocactus*

Echinoglochin (A. Gray) Brand = *Plagiobothrys*

Echinolaena Desv. (= ?*Panicum*). Gramineae (34b). 8 trop. Am., Madag. ? Heterogeneous

Echinomastus Britton & Rose = *Sclerocactus*

Echinopaepale Bremek. = *Strobilanthes*

Echinopanax Decne & Planchon ex Harms = *Oplopanax*

Echinopepon Naudin (~ *Echinocystis*). Cucurbitaceae. 12 Am.

Echinophora L. Umbelliferae (III 1). 9 Medit. (Eur. 2) to Iran, 1 carpel aborts; umbel with 1 bisexual fl. surrounded by males, the spiny pedicels to the latter enclosing fr.

Echinopogon Pal. Gramineae (21d). 7 Aus., NZ, New Guinea. R: HIP 33(1935)t.3261

Echinops L. Compositae (Card.-Ech.). c. 120 Eur. (12), Medit. to C As. & trop. Afr. mts. Alks. Spherical heads comprising ∞ 1-flowered capitula, each with an involucre, bee-poll. Cult. orn. robust herb. pls, esp. *E. bannaticus* Rochel ex Schrader ('*E. ritro*', SE Eur.) – blue fls, (globe thistles)

Echinopsis Zucc. Cactaceae (III 4). (Incl. *Lobivia* & *Trichocereus*) c. 50–100 S Am. Cult. orn. ribbed cacti with diurnal or nocturnal fls incl. *E. chiloensis* (Colla) Friedrich & Rowley (*E. chilensis*, *T. chiloensis*, Chile) – fr. ed., made into drinks & *E. pachanoi* (Britton & Rose) Friedrich & Rowley (Ecuador, Peru) – hallucinogenic (mescaline); *E. pasacana* (Ruempler) Friedrich & Rowley (*T. pasacana*, W Arg., S Bolivia) – vasc. system (closed) used for boxes, church beams, doors & fences

Echinopterys A. Juss. Malpighiaceae. 3 Mex. Mericarps spiny

Echinosophora Nakai = *Sophora*

Echinospartum (Spach) Fourr. (~ *Genista*). Leguminosae (III 31). 3 SW Eur.

Echinostephia (Diels) Domin. Menispermaceae (V). 1 SE Queensland

Echiochilon Desf. Boraginaceae. (Incl. *Leurocline*, *Tetraedrocarpus*) 17 arid NE Afr. & Arabia to Iran & Baluchistan. R (s.s.): JAA 38(1957)261. Some firewood sources

Echiochilopsis Caball. = *Echiochilon*

Echioides Ortega = *Arnebia*

Echiostachys Levyns = *Lobostemon*

Echitella Pichon = *Mascarenhasia*

Echites P. Browne. Apocynaceae. 6 Carib.

Echium L. Boraginaceae. 60 Macaronesia (28, 27 endemic), Eur. (18), W As., N & S Afr. R: JAA suppl. 1(1991)137. Alks; some form. medic. incl. *E. vulgare* L. (viper's bugloss, Euras., pernicious weed in N Am., NZ where *E. candicans* L. (Madeira) – shrub to 2.5 m also aggressive colonist). Cult. orn., esp. shrubby & pachycaul spp. of Macaronesia, some unbranched & with infls to 4 m tall, also *E. plantagineum* L. (*E.lycopsis*, bugloss, Euras.) – biennial, pernicious weed (Paterson's curse) but emergency fodder (Salvation Jane) & imp. honeybee-pl. (for 'Aus. bluebell honey') in E Aus. (introd. 1843, weedy by c. 1890); *E. nervosum* Dryander (Madeira) – visited by lizards, most other spp. bee-poll.; *E. wildpretii* Pearson ex Hook.f. (Canary Is.) – red fls & copious nectar attractive to birds though colour in ultaviolet & shape attractive to bees

Eclipta L. Compositae (Helia.). 4 warm. Pappus 0. *E. alba* (L.) Hassk. (*E. prostrata*, Am., introd. OW) – blackish dye for hair, tattooing etc. in India, source of thiophene derivatives active against nematodes

Ecliptostelma Brandegee. Asclepiadaceae (III 4). 1 Mex.: *E. molle* Brandegee

Ecpoma Schumann. Rubiaceae (I 8). 1 trop. Afr.

Ectadiopsis Benth. = *Cryptolepis*

Ectadium E. Meyer. Asclepiadaceae (I; Periplocaceae). 3 S Afr. R: SAJB 56(1990)113

Ectinocladus Benth. = *Alafia*

Ectopopterys W.R. Anderson. Malpighiaceae. 1 Colombia, Peru: *E. soejartoi* W.R. Anderson. R: CUMH 14(1980)11

Ectotropis N.E. Br. = *Delosperma*

Ectozoma Miers = *Markea*

Ectrosia R. Br. Gramineae (31d). 11 trop. Aus. to New Guinea (1). R: HIP (1936) t. 3312

Ectrosiopsis (Ohwi) Jansen. Gramineae (31d). 1 Aus., New Guinea, Carolines

Ecuador laurel *Cordia alliodora*

Edanyoa Copel. = *Bolbitis*

Edbakeria R. Viguier = *Pearsonia*

eddoes *Colocasia esculenta* cvs

edelweiss *Leontopodium alpinum*; **NZ e.** *Leucogenes* spp.

Edgaria C.B. Clarke. Cucurbitaceae. 1 E Himal.

Edgeworthia Meissner. Thymelaeaceae. 3 China, Japan & (? introd.) SE US (Georgia). Shrubs with bark used for high-class paper (for currency etc.) esp. *E. papyrifera* Siebold & Zucc. (paperbush, mitsumata, China; long cult. Japan) & *E. gardneri* (Wallich) Meissner (Nepal, Sikkim); cult. orn.

edinam *Entandrophragma angolense*

Edisonia Small = *Matelea*

Edithcolea N.E. Br. Asclepiadaceae (III 5). 1 E & NE Afr. to Socotra & Arabia: *E grandis* N.E. Br., cult. orn.

Edithea Standley = *Omiltemia*

Edmondia Cass. (~ *Helichrysum*). Compositae (Gnap.-Gnap.). 3 Cape. R: OB 104(1991)153

Edraianthus (A. DC) DC (~ *Wahlenbergia*). Campanulaceae. 24 SE Eur. (9) to Caucasus. R: GBIUS 26(1973)1. Like *Wahlenbergia* but capsule with irreg. dehiscence & lvs linear, elongate. Cult. orn. rock-pls esp. *E. graminifolius* (L.) DC (Eur.)

Eduardoregelia Popov = *Tulipa*

Edwardsia Salisb. = *Sophora*

eelgrass *Zostera marina*

Eenia Hiern & S. Moore = *Anisopappus*

Efulensia C.H. Wright (~ *Deidamia*). Passifloraceae. 2 trop. Afr.

efwatakala grass *Melinis minutiflora*

Eganthus Tieghem = *Minquartia*

Egenolfia Schott = *Bolbitis*

Egeria Planchon (~ *Elodea*). Hydrocharitaceae. 2 subtrop. S Am. R: Darwiniana 12(1961)293

egg fruit or **e. plant** (incl. **Thai e. p.**) *Solanum melongena*

Eggelingia Summerh. Orchidaceae (V 16). 3 trop. Afr.

eggs and bacon *Lotus corniculatus*

eglantine *Rosa rubiginosa*

Egleria Eiten. Cyperaceae. 1 Amazonian Brazil

Eglerodendron Aubrév. & Pellegrin = *Pouteria*

Egletes Cass. Compositae (Ast.-Ast.). 10 trop. Am. R: Lloydia 12(1949)239, 248

egusi *Cucumeropsis mannii*

Egyptian bean *Nelumbo nucifera*; **E. grass** *Dactyloctenium aegyptium*; **E. lotus** *Nymphaea lotus*; **E. onion** *Allium cepa* Proliferum Group

Ehretia P. Browne. Boraginaceae. 75 trop. & warm (NW 3, R: AMBG 76(1989)1059). Some timbers incl. *E. acuminata* R. Br. (Aus.) – koda wood. Some medic. incl. *E. philippensis* A. DC (Philipp.) – effective in treatment of diarrhoea

Ehretiaceae C. Martius = Boraginaceae

Ehrharta Thunb. Gramineae (11). 35 S Afr. (25, 1 ext. to Ethiopia), Masc., Indonesia to NZ. Some pasture grasses; *E. villosa* (L.f.) Schultes f. (S Afr.) – sand-binder in NSW

Eichhornia Kunth. Pontederiaceae. 7 trop. Am. Rhizomatous aquatics with floating or submerged lvs. *E. azurea* (Sw.) Kunth (Brazil) – fls dimorphic; *E. crassipes* (C. Martius) Solms-Laub. (water-hyacinth, trop. Am.) – trimorphic, heterostylous, though broken down in some populations in Costa Rica. *E. crassipes* is a free-floater with swollen petioles containing aerenchyma, the laminas raised above water-level & acting as sails; 2 parents can give

30 offspring by vegetative budding in 23 days & 1200 in 4 months, a yield of 470 t/ha. The plant, orig. introd. as an ornamental, has spread to choke many trop. waterways & is natur. in Portugal; when grown on sewage it can yield 800 kg dry matter/ha/day, which in Indonesia is harvested for pigs & has been suggested as a methane source; *E. paniculata* (Sprengel) Solms-Laub. (trop. Am.) – tristylous though monomorphic in Jamaica (morphs self-compatible?)

Eichlerago Carrick = *Prostanthera*

Eichleria Progel = *Rourea*

Eichlerodendron Briq. = *Xylosma*

Eigia Soják (*Stigmatella*). Cruciferae. 1 E Med.

Einadia Raf. Chenopodiaceae (I 2). 4 Aus., 2 NZ. R: Nuytsia 4(1983)199

einkorn *Triticum monococcum*

Einomeia Raf. (~ *Aristolochia*). Aristolochiaceae (II 2). 36 Mex. to Carib. *E. bracteolata* (Retz.) Klotzsch (*A.b.*) – in Sri Lanka visited by swallowtail butterflies able to cope with toxins like monarchs in S Am.

Eionitis Bremek. = *Oldenlandia*

Eirmocephala H. Robinson. Compositae (Vern.-Vern.). 3 C Am., Andes to Bolivia

Eisocreochiton Quis. & Merr. = *Creochiton*

Eitenia R. King & H. Robinson (~ *Eupatorium*). Compositae (Eup.-Prax.). 2 Brazil. R: MSBMBG 22(1987)395

Eizia Standley. Rubiaceae (I 5). 1 Mexico

ejow *Arenga pinnata*

Ekebergia Sparrman. Meliaceae (I 4). 4 trop. & S Afr. Some timbers esp. *E. capensis* Sparrman (Cape ash, dog plum, trop. & S Afr.)

ekhimi *Piptadeniastrum africanum*

ekki *Lophira* spp.

Ekmania Gleason. Compositae (Vern.-Pip.). 1 Cuba

Ekmanianthe Urban. Bignoniaceae. 2 Cuba, Hispaniola

Ekmaniocharis Urban = *Mecranium*

Ekmanochloa A. Hitchc. Gramineae (4). 2 E Cuba (serpentine). R: KBAS 13(1986)65. Caespitose; blades of *E. subaphylla* A. Hitchc. ± suppressed, the culms photosynthetic

ekpogoi *Berlinia* spp.

ekra, ekar *Saccharum bengalense*

Elachanthemum Ling & Y.R. Ling (~ *Artemisia*). Compositae (Anth.-Art.). 1 China

Elachanthera F. Muell. = *Asparagus*

Elachanthus F. Muell. Compositae (Ast.-Ast.). 2 temp. Aus.

Elacholoma F. Muell. & Tate. Scrophulariaceae. 1 C Aus.: *E. hornii* F. Muell. & Tate

Elachyptera A.C. Sm. Celastraceae. 7 trop. Afr., Madag., Am.

Elaeagia Wedd. Rubiaceae (I 5). 10 trop. Am. *E. utilis* Wedd. (Colombia) – pasto lacquer

Elaeagnaceae Juss. Dicots – Rosidae – Proteales. 3/45 temp. & warm N hemisph. to trop. As. & Aus. Shrubs to small trees, often armed, usu. with nitrogen-fixing root nodules (Actinomycetes); indumentum of scales or stellate hairs. Lvs simple, entire, spirally arr. (opp. in *Shepherdia*); stipules 0. Fls bisexual (s.t. pls dioec. or polygamo-dioec.), reg., perigynous with hypanthium usu. constricted above G (cupulate to ± flat in male fls), solit. or in small umbels; K (2)4(6) valvate lobes on hypanthium, often petaloid, C 0, A in throat alt. with & same no. as K (*Elaeagnus*) or 2 × K alt. & opp. (*Hippophae* & *Shepherdia*), anthers with longit. slits, nectary-disk (lobed) also in throat, $\overline{G}$ 1 with 1 style & 1 basal, anatropous, bitegmic ovule with funicular obturator. Fr. drupe- or berry-like, the achene enveloped by, but free from, persistent hypanthium which becomes fleshy often with a bony inner layer; seed with straight embryo & 2 fleshy oily & proteinaceous cotyledons & ± 0 endosperm. x = 6, 10, 11, 13

Genera: *Elaeagnus*, *Hippophae*, *Shepherdia*

Pls mostly of steppe & coasts; some cult. orn. & ed. fr.

Elaeagnus L. Elaeagnaceae. c. 40 Eur. (1 natur.), As., N Am. (1). Nitrogen-fixing Actinomycete root-symbionts. Alks. Some cult. orn. shrubs (oleaster) esp. *E. angustifolia* L. (Russian olive, Trebizond date, SE Eur., W As.), *E.* × *ebbingei* Boom (*E.* × *submacrophylla* Serv., *E. macrophylla* Thunb. (Korea, Japan) × *E. pungens*), *E. multiflora* Thunb. (Japan, China) – fr. ed., *E. pungens* Thunb. (China & Japan) – common variegated forms & *E.* × *reflexa* Morren & Decne (*E. pungens* × *E. glabra* Thunb. (China, Japan) – scrambling hedgepl. widely used in NZ, now escaped; many spp. locally imp. as fr., e.g. *E. latifolia* L. (*E. conferta*, S & E As.) in Sikkim, swamping creeper in forest canopy of Krakatoa & *E. multiflora* Thunb. (Japan) – goumi

Elaeis Jacq. Palmae (V 5d). 2 trop. Am. (1) & Afr. (1). Monoec. palms which hybridize in cult. *E. guineensis* Jacq. (oil palm, W Afr., now widely planted in trop., natur. Malay Pen.) – poll. weevils (now introd. to As.) suppl. by wind, most imp. source of oil for margarine; oil from pericarp (palm oil) used for soap, candles etc., oil from rest is palm kernel oil used in many cosmetics etc.; palm-oil 'diesel' taxis in some trop. cities; oil yield c. 3475 kg/ha/yr. *E. oleifera* (Kunth) Cortés (Am. oil palm, trop. Am.) – lesser imp.

Elaeocarpaceae Juss. ex DC. Dicots – Dilleniidae – Violales. 9/540 trop. & warm exc. Cont. Afr. Trees or shrubs, often with alks. Lvs usu. spirally arr., simple; stipules persistent or not. Fls usu. bisexual, reg., without epicalyx, in racemose or cymose infls; K (3)4, 5(–11), usu. valvate, s.t. basally connate, C (3)4 or 5, usu. free, often apically fringed, usu. valvate, rarely 0, A ∞, originating centrifugally & often ± in 5 antesepalous groups on a ± definite disk or enlarged receptacle forming an androgynophore, anthers with apical slits or pores or short longit. lateral slits, connective often ± conspic. prolonged, $\underline{G}$ ((1–)2–∞) with as many locs & 1 style, each loc. with 2–∞ anatropous, bitegmic ovules with zig-zag micropyle. Fr. a capsule (often armed) or a drupe; seeds with much oily, proteinaceous endosperm, often arillate, & with straight or J–U-shaped embryo. x = 12, 14, 15

Genera: *Aceratium, Aristotelia, Crinodendron, Dubouzetia, Elaeocarpus, Peripentadenia, Sericolea, Sloanea, Vallea*

Usu. considered allied to Tiliaceae but seed-structure militates against this; *Muntingia*, form. included here, is referred to Flacourtiaceae 'highly diversified' cf. 'the more narrowly limited family' E. (Cronquist) while *Petenaea* has been removed to Tiliaceae

Some cult. orn. (*Crinodendron* etc.) & some locally imp. timbers & fr.

Elaeocarpus L. Elaeocarpaceae. c. 360 trop. & warm OW exc. Afr. Some rheophytes, heteroblasty in NZ; domatia; cult. orn. & ed. fr. *E. bancroftii* F. Muell. (ebony heart, Aus.) – Karanda nuts (Queensland); *E. bifidus* Hook. & Arn. (Hawaii) – form. bark for cordage, wood for frames of grass huts; *E. dentatus* (Forster & Forster f.) Vahl (NZ) – bark source of a blue-black dye; *E. hookerianus* Raoul (pohaka, NZ) – juvenile divaricate; *E. reticulatus* Sm. (E Aus.) – lignotubers; *E. serratus* L. (Indomal.) – fr. (Ceylon olive) used in curry; *E. sphaericus* (Gaertner) Schumann (*E. ganitrus*, olive nut, India to W Mal.) – seeds used as beads; *E. tectorius* (Lour.) Poiret (*E. robustus* Roxb., Indomal.) – a village fr. tree

Elaeodendron Jacq.f. (~ *Cassine*). Celastraceae. 15 OW trop., S Afr., C Am., WI. Seeds exarillate (cf. *Euonymus*)

Elaeoluma Baillon. Sapotaceae (IV). 4 trop. Am.

Elaeophora Ducke = *Plukenetia*

Elaeophorbia Stapf (~ *Euphorbia*). Euphorbiaceae. 4 trop. & S Afr. Trees like *Euphorbia* but with drupes. *E. drupifera* (Thonn.) Stapf (W Afr.) – ordeal-poison in Ivory Coast, rubbed into eyes (danger to cornea)

Elaeopleurum Korovin = *Seseli*

Elaeoselinum Koch ex DC. Umbelliferae (III 12). (Excl. *Distichoselinum, Margotia*) 4 Medit. (Eur. 3). R: Lagascalia 13(1985)213

Elaeosticta Fenzl (~ *Scaligeria*). Umbelliferae (III 8). (Incl. *Muretia*) 24 Eur. (2) to C As. R: BMOIPB 81,6(1976)92

Elaphandra Strother. Compositae (Helia.-Verb.) 10 trop. Am. R: SBM (1991)17

Elaphanthera Hallé (~ *Exocarpos*). Santalaceae. 1 New Caledonia

Elaphoglossum Schott ex J. Sm. Lomariopsidaceae. c. 400 trop. & warm (Eur. 1: *E. semicylindricum* (Bowdich) Benl (Azores, Madeira), Aus. 1, NZ 0) to Japan esp. Am. (e.g. Venez. 98). R: AFJ 70(1980)47. Epiphytes with simple, tongue-like fronds without indusia, some with epiphyllous bud propagules. Some cult. orn.

Elasis D. Hunt. Commelinaceae (II 1f). 1 Ecuador

Elateriopsis Ernst. Cucurbitaceae. 5 S Am.

Elateriospermum Blume. Euphorbiaceae. 1 S Thailand, Malay Pen.: *E. tapos* Blume seeds ed. when prussic acid removed by boiling, also used in a game like conkers

elaterium *Ecballium elaterium*

Elatinaceae Cambess. Dicots – Dilleniidae – Guttiferales. 2/34 temp. & (esp.) trop. Herbs to suffrutices of wet places, often creeping & rooting, resinous. Lvs opp. or whorled, simple, entire to toothed; stipules small. Fls small, reg., bisexual, axillary, solit. or in cymes; K 2–5(6), s.t. connate, C 2–5(6), imbricate, persistent, A same as or 2 × C, anthers with longit. slits, $\underline{G}$ (2–5) with distinct styles & as many loc. but partitions not reaching apex in some *Bergia* spp. Ovules ∞, on axile placentas, anatropous, bitegmic, sometimes with zig-zag micropyle. Fr. a septicidal capsule; seeds with straight or curved embryo & 0 endosperm. x = 6, 9

Genera: *Bergia, Elatine*

Prob. allied to Guttiferae but exceptional in alliance in being largely aquatic & embryology recalls Myrtales

Elatine L. Elatinaceae. 10 trop. & temp. (Eur. 8). Waterwort. Some grown in aquaria, incl. *E. alsinastrum* L. (Medit., Euras.) with unbranched stems & whorls (unique in fam.) of lvs ('*Hippuris* syndrome')

Elatinoides (Chav.) Wettst. = *Kickxia*

Elatostema Forster & Forster f. Urticaceae (II). (Incl. *Pellionia*) c. 300 trop. OW to NZ (1). R: FRBeih. 83, 1-2(1935–6). Fr. ejected by staminodes; some apogamous. Some cult. orn. for coloured lvs & stems. *E. elegans* Winkler (New Guinea) – rheophytic

Elatostematoides Robinson = *Elatostema*

Elattospermum Soler. = *Breonia*

Elattostachys (Blume) Radlk. Sapindaceae. c.20 Mal. to Aus. (4) & W Pacific

Elburzia Hedge. Cruciferae. 1 NW Iran

Elcomarhiza Barb. Rodr. = *Marsdenia*

elder *Sambucus nigra*; **American e.** *S. canadensis*; **blue e.** *S. caerulea*; **box e.** *Acer negundo*; **dwarf e.** *S. ebulus*; **poison e.** *Rhus vernix*; **stinking e.** *S. pubens*

elecampane *Inula helenium*

Elegia L. Restionaceae. 35 S Afr. R: Bothalia 15(1985)418

Eleiodoxa (Becc.) Burret. Palmae (II 1d). 1 W Mal.: *E. conferta* (Griff.) Burret – sarcotesta a tamarind subs. (with raw prawns = 'umei' (Sarawak))

Eleiosina Raf. Older name for *Sibiraea*

Eleiotis DC. Leguminosae (III 9). 2 India

elemi oleo-resins from various pls form. much used in varnishes, printing inks & ointments. **African e.** *Canarium schweinfurthii*; **American e.** *Bursera simarouba*; **Brazilian e.** *Protium heptaphyllum*; **Carana e.** *P. carana*; **E Afr. e.** *Boswellia frereana*; **Manila e.** *C. luzonicum*; **WI e.** *Dacryodes excelsa*; **Yucatan e.** *Amyris plumieri*

Eleocharis R. Br. Cyperaceae. c. 120 cosmop. (Eur. 15). Spike rushes. *E. calva* Torrey (kohekohe, N Am., Hawaii) – red basal sheath used in Niihau mats (see *Cyperus laevigatus*); *E. dulcis* (Burm.f.) Henschel ('Tuberosa', *E. tuberosa*, (Chinese) water chestnut, OW trop.) – cult. China etc. in flooded fields drained before harvest of tubers or corms which comprise the principal white crunchy vegetable in Chinese food like chop suey; lvs of other spp. used for matting & for women's skirts in New Guinea

Eleogiton Link = *Isolepis*

Eleorchis F. Maek. Orchidaceae (V 11). 1 Japan

elephant apple *Limonia acidissima*; **e. climber** *Argyreia nervosa*; **e. ear** *Begonia* spp.; **e. foot** *Dioscorea elephantipes*; **e. grass** *Pennisetum purpureum*; **e. tree** *Pachycormus discolor*

Elephantomene Barneby & Krukoff. Menispermaceae (II). 1 NE S Am.: *E. eburnea* Barneby & Krukoff – disjunct distrib., Fr. Guiana & Peru/Brazil border

Elephantopus L. Compositae (Vern.-Ele.). c. 30 trop. & warm. *E. scaber* L. a bad weed in warm regions

Elephantorrhiza Benth. Leguminosae (II 3). 9 trop. & S Afr. R: Bothalia 11(1974)247

Elettaria Maton. Zingiberaceae. 7 India to W Mal. Infls on prostrate shoots with scale lvs arising from rhizomes. *E. cardamomum* (L.) Maton (cardamom, India) – widely cult. in As. for spicy seeds used in medic. & cooking, imported to Eur. since Roman period: still used in meat dishes such as hamburgers, also a masticatory in As.

Elettariopsis Baker. Zingiberaceae. 8 SE As., W Mal.

Eleusine Gaertner. Gramineae (31d). c. 9 Afr. (8), S Am. (1: *E. tristachya* (Lam.) Kunth). R (Afr.): KB 27(1972)251. *E. coracana* (L.) Gaertner (finger millet, coracan, kurakkan, ragi, tetraploid ? orig. E Afr. from *E. indica* (L.) Gaertner subsp. *africana* (Kenn.-O'Byrne) S. Phillips (tetraploid race of pantrop. weed toxic to stock (crowsfoot grass)) – imp. grain crop in Afr. & India, cult. in Ethiopia (earliest Afr. agriculture) 3rd Millenium BC, fermented for alcoholic drinks, seeds viable for more than 10 yrs without weevil damage; others weeds (crabgrass) incl. *E. indica* (L.) Gaertner (yardgrass, trop. OW, now pantrop.)

Eleutharrhena Forman. Menispermaceae (I). 1 China, Assam: *E. macrocarpa* (Diels) Forman. R: KB 30(1975)98

Eleutherandra Slooten. Flacourtiaceae (2). 1 Mal.

Eleutheranthera Poit. Compositae (Helia.-Verb.). 1 trop. Am., natur. OW

Eleutheranthus Schumann = *Eleuthranthes*

Eleutherine Herbert. Iridaceae (III 4). 2 trop. Am.

Eleutherococcus Maxim. (*Acanthopanax*). Araliaceae. 30 E As., Himal., Mal. Some cult. orn., usu. prickly trees & shrubs, few timbers. *E. senticosus* (Rupr. & Maxim.) Maxim. (Siberian ginseng, NE As.) – source of a tonic (used by Russian athletes at the Moscow Olympics,

1984)

Eleutheropetalum (H. Wendl.) H. Wendl. ex Oersted = *Chamaedorea*

Eleutherospermum K. Koch. Umbelliferae (III 5). 1 Caucasus, S W As.

Eleutherostigma Pax & K. Hoffm. = *Plukenetia*

Eleutherostylis Burret. Tiliaceae. 1 New Guinea

Eleuthranthes F. Muell. = *Opercularia*

Elgon olive *Olea capensis* ssp. *welwitschii*

Elide Medikus. Older name for *Myrsiphyllum* = *Asparagus*

Eliea Cambess. Guttiferae. 1 Madagascar

Eligmocarpus Capuron. Leguminosae (I 2). 1 SE Madag. (v. local)

Elingamita Baylis. Myrsinaceae. 1 NZ (Three Kings Group): *E. johnsonii* Baylis – c. 12 left in 1970

Elionurus Kunth ex Willd. (*Elyonurus*). Gramineae (40h). 15 trop. Afr. to Sind, Am., Aus. (1). R: KB 32(1978)665. Some essential oils (unexploited)

Elisena Herbert = *Urceolina*

Elissarrhena Miers = *Anomospermum*

Elizabetha Schomb. ex Benth. Leguminosae (I 4). 11 trop. S Am. R: PKNAW C79(1976)323

Elizaldia Willk. Boraginaceae. 5 W Medit. (Eur. 1–? extinct)

elk nut *Pyrularia pubera*; **e. sedge** *Cladium mariscus*

elkshorn fern *Platycerium* spp.

Ellangowan poison bush *Eremophila deserti*

Elleanthus C. Presl. Orchidaceae (V 13). 115 trop. Am. Some cult. orn.

Elleimataenia Kozo-Polj. = *Osmorhiza*

Ellenbergia Cuatrec. Compositae (Eup.-Ager.). 1 Peru. R: MSBMBG 22(1987)159

Ellertonia Wight = *Kamettia*

Elliottia Muhlenb. ex Elliott. Ericaceae. (Incl. *Cladothamnus*, excl. *Tripetaleia*) 2 Alaska, NW & SE N Am. R: JJB 63(1988)163. Some cult. orn.

Ellipanthus Hook.f. Connaraceae. 7 E Afr. coast & Madag. (2, R: AUWP 89-6(1989)268) to Mal.

Ellipeia Hook.f. & Thomson. Annonaceae. 5 W Mal.

Ellipeiopsis R.E. Fries. Annonaceae. 2 India to SE As.

Ellisia L. Hydrophyllaceae. 1 N Am.: *E. nyctelea* (L.) L. ('Aunt Lucy')

Ellisiophyllaceae Honda = Scrophulariaceae

Ellisiophyllum Maxim. Scrophulariaceae. 1 India to Japan, Taiwan to E New Guinea: *E. pinnatum* (Benth.) Makino

elm *Ulmus* spp.; **American e.** *U. americana*; **Camperdown e.** *U. glabra* 'Camperdown'; **cedar e.** *U. crassifolia*; **cork e.** *U. thomasii*; **Dutch e.** *U.* × *hollandica*; **English e.** *U. procera*; **hickory e.** *U. thomasii*; **Huntingdon e.** *U.* × *vegeta*; **Japanese e.** *U. davidiana, Zelkova serrata*; **rock e.** *U. thomasii, Milicia excelsa*; **Scotch e.** *U. glabra*; **water e.** *Planera aquatica*; **white e.** *U. americana*; **WI e.** *Guazuma ulmifolia*; **wych e.** *U. glabra*

Elmera Rydb. (~ *Heuchera*). Saxifragaceae. 1 Mts of Washington: *E. racemosa* (S. Watson) Rydb. – cult. orn. like *Heuchera* but petals cleft & scapes leafy

Elmerrillia Dandy. Magnoliaceae (I 2). 4 Mal. Some timbers. R: Blumea 31(1985)100

Elodea Michaux. Hydrocharitaceae. 12 Am. Submerged aquatics. Grown in ponds etc. but often becoming weedy, e.g. *E. canadensis* Michaux (Canadian pondweed or waterweed, American w., N Am.) – at depths to 14 m in Canada, dioec., introd. Br. Is. 1834 & rapidly spread by vegetative reproduction (only females) until 1880s when in decline & now scarce, introd. NZ 1870s & still a pest with stems to 6 m long & dry matter per sq.m greater than any other aquatic macrophyte, male fls (P 6, A 9) break off as buds & float to surface where they open, females (P 6, 3 staminodes, $\overline{G}$ (3)) with ovaries extending to surface where pollinated; *E. nuttallii* (Planchon) St John (N Am.) – introd. Br. Is. 1966 & still spreading & displacing *E. canadensis* as stronger-growing, though not pestilential; *E. callitrichoides* (Rich.) Caspary (*E. ernstiae*, S Uruguay, NE Arg.) – natur. Eur. incl. GB (1948). See also *Crassula* & *Lagarosiphon*

Eloyella Ortíz = *Phymatidium*

Elsholtzia Willd. Labiatae (VIII 1). c. 35 temp. OW incl. trop. mts. *E. ciliata* (Thunb.) Hylander (*E. cristata*, C & E As., natur. Eur., N Am.) – cult. orn. & used for hangovers in Japan; *E. haichowensis* Y.Z. Sun (China) used as copper indicator. R: BBMNHB 10(1982)69

Elsiea F.M. Leighton = *Ornithogalum*

Eltroplectris Raf. = *Stenorrhynchos*

Elvasia DC. Ochnaceae. 10 trop. S Am. R: BJ 113(1991)171

Elvira Cass. = *Delilia*

Elymandra Stapf. Gramineae (40g). 6 trop. Afr., 1 also in Brazil. R: KB 20(1966)287

Elymus L. Gramineae (23). (Incl. *Festucopsis*) c. 150 N temp. (Eur. 24) esp. As. *E. farctus* (Viv.) Meld. (*E. junceus*, Russian wild rye, Euras.) – sand-binder esp. in saline areas. See also *Elytrigia*, *Leymus*

Elyonurus Willd. = *Elionurus*

Elythranthera (Endl.) A.S. George. Orchidaceae (IV 1). 2 W Aus.

Elytranthe (Blume) Blume. Loranthaceae. 10 SE As., Mal.

Elytraria Michaux. Acanthaceae. 17 trop. & warm

Elytrigia Desv. (~ *Elymus*). Gramineae (23). 5 Euras. *E. repens* (L.) Nevski (*Elymus r.*, *Agropyron r.*, couch, twitch or witch grass, Euras.) – bad weed of cult. with tough rhiz.

Elytropappus Cass. Compositae (Gnap.-Rel.). 8 S Afr. R: JSAB 1(1935)89. Poss. heterogeneous. *E. rhinocerotis* (L.f.) Less. (rhinoceros bush) – char. pl. of Karoo, where it ousts more desirable fodder pls

Elytrophorus Pal. Gramineae (25). 4 OW trop., wet places

Elytropus Muell. Arg. Apocynaceae. 1 Chile

Elytrostachys McClure. Gramineae (1b). 2 C & N S Am. R: SCB 9(1973)79. Bamboos

Embadium J. Black. Boraginaceae. 3 S Aus. R: TRSAS 89(1965)285

Embelia Burm.f. Myrsinaceae. c. 100 trop. & warm OW. Some ed. & medic. esp. *E. ribes* Burm.f. (Indomal.) – liane with fr. used to adulterate pepper, as tapeworm treatment etc., fed to rats induce infertility; *E. micrantha* A. DC (Mauritius) – favoured kidney-stone remedy & now v. rare; *E. philippensis* A. DC (Philipp.) – ed. fr. & lvs, cordage

Embergeria Boulos (~ *Sonchus*). Compositae (Lac.-Son.). 1 NZ

embero *Lovoa trichilioides*

Emblemantha Stone. Myrsinaceae. 1 Sumatra: *E. urnulata* Stone. R: PANSP 140(1988)275

emblic *Phyllanthus emblica*

Emblica Gaertner = *Phyllanthus*

Emblingia F. Muell. Polygalaceae (Emblingiaceae). 1 W Aus. Autogamous

Emblingiaceae (Pax) Airy Shaw. See Polygalaceae

Embolanthera Merr. Hamamelidaceae (I 1). 2 SE As., Philippines

Embothrium Forster & Forster f. Proteaceae. 8 C & S Andes. *E. coccineum* Forster & Forster f. (Chilean firebush, Arg. & Chile) – variable shrub of open places from coast to treeline, some cvs hardy & cult. esp. 'Norquinco Valley' (Arg. Andes)

Embreea Dodson (~ *Stanhopea*). Orchidaceae (V 10). 1 Colombia, Ecuador

embul *Musa × paradisiaca*

Emelianthe Danser. Loranthaceae. 2 trop. E Afr.

emeri *Terminalia ivorensis*

Emex Campderá. Polygonaceae. 2 Medit. (Eur. 1), S Afr., natur. Aus. Fr. surrounded by persistent P (3 spiny). *E. spinosa* (L.) Campderá (*E. australis*) medic. in S Afr., weedy elsewhere ('Cape spinach', Aus.)

Emicocarpus Schumann & Schltr. Asclepiadaceae (III 1). 1 SE Afr.: *E. fissifolius* Schumann & Schltr.

Emilia Cass. Compositae (Sen.-Sen.). 100 OW trop. (3 now pantrop. weeds). Some cult. orn. ('Flora's paintbrush')

Emiliella S. Moore (~ *Emilia*). Compositae (Sen.-Sen.). 5 Angola, Zambia. R: GOB 2(1975)85

Eminia Taubert. Leguminosae (III 10). 4 trop. Afr. R: BJBB 53(1983)153. Roots used in brewing (amylase)

Eminium (Blume) Schott. Araceae (VIII 6). 8 E Med. to C As. R: Willdenowia 20(1991)49. Cult. orn. with infrs often developing below ground & growing to surface before seed-disp.

Emmenanthe Benth. Hydrophyllaceae. 1 SW N Am.: *E. penduliflora* Benth. – seeds germ. only in presence of charred wood, cult. orn. annual with pale yellow or pink fls

Emmenopterys Oliver. Rubiaceae (I 1). 2 Burma, China, Thailand

Emmenosperma F. Muell. Rhamnaceae. 3 Aus., New Caled. *E. alphitonioides* F. Muell. (Aus.) – 'yellow ash'

Emmeorhiza Pohl ex Endl. Rubiaceae (IV 15). 1 trop. S Am.

emmer *Triticum turgidum* Dicoccon Group

Emmotum Desv. Icacinaceae. 12 trop. S Am.

Emodiopteris Ching & S.K. Wu = *Dennstaedtia*

Emorya Torrey. Buddlejaceae (Loganiaceae s.l.). 1 Texas & N Mexico

Empedoclesia Sleumer = *Orthaea*

Empetraceae Bercht. & J. Presl (~ Ericaceae). Dicots – Dilleniidae – Ericales. 3/5 N temp., temp. S Am. Evergreen usu. dioec. shrubs with mycorrhizae. Lvs spirally arr., crowded (sometimes in pseudo-whorls), ericoid, apparently rolled with an abaxial groove & basal pulvinus; stipules 0. Fls 1–3, axillary or in small term. heads, small, reg. usu. unisexual; P 3–6, ± recog. as 2 or 3 imbricate K & C, or 3 or 4 all alike (*Corema*), A 2 (*Ceratiola*) or 3(4), alt. with C when C distinct from K, anthers inverted in ontogeny with longit. slits & no appendages, pollen in tetrads, nectary-disk 0, G (2–9), pluriloc. with axile-basal placentas & ± deeply cleft style, 1 erect, anatropous to ± campylotropous, unitegmic ovule per loc. Fr. a drupe with 2–9 pyrenes, sometimes juicy; seeds with thin testa & slender elongate embryo in copious fleshy oily endosperm. X = 13

Genera: *Ceratiola*, *Corema*, *Empetrum*

Disjunct distributions in *Corema & Empetrum*. Some cult. orn. & ed. fr.

Empetrum L. Empetraceae. 2 N temp. (Eur. 1) & Arctic, S Andes, Falkland Is., Tristan da Cunha. Diploids usu. dioec.; monoec. & bisexual pls app. derived. *E. nigrum* L. (crowberry, curlew berry, N temp.) – ed. black fr.

emping *Gnetum gnemon*

Emplectanthus N.E. Br. Asclepiadaceae (III 4). 2 S Afr.

Emplectocladus Torrey = *Prunus*

Empleuridium Sonder & Harvey. Rutaceae. 1 S Afr.

Empleurum Sol. ex Aiton. Rutaceae. 2 S Afr. R: JSAB 50(1984)427. Lvs used like buchu (*Agathosma* spp.)

Empodisma L. Johnson & Cutler. Restionaceae. 2 Aus., NZ

Empodium Salisb. Hypoxidaceae (Liliaceae s.l.). 10 S Afr.

Empogona Hook.f. = *Tricalysia*

emu apple *Owenia acidula*; **e. bush** *Eremophila* spp.

Enallagma (Miers) Baillon = *Amphitecna*

Enantia Oliver = *Annickia*

Enantiophylla J. Coulter & Rose. Umbelliferae (III 9). 1 C Am.

Enarganthe N.E. Br. (~ *Ruschia*). Aizoaceae (V). 1 W S Afr.

Enarthrocarpus Labill. Cruciferae. 5 E Med. (Eur. 2), N Afr.

Enaulophyton Steenis. Melastomataceae. 2 Borneo incl. *E. lanceolatum* Steenis, a rheophyte

Encelia Adans. Compositae (Helia.-Verb.). 15 SW N Am., Chile, Peru, Galápagos. R: PAAAS 49(1913)358. Shrubs or herbs, some cult. orn. incl. *E. farinosa* A. Gray (brittle-bush, SW N Am.) – stem resin form. used as incense by Spanish

Enceliopsis (A. Gray) Nelson. Compositae (Helia.-Verb.). 4 SW US

Encephalartos Lehm. Zamiaceae (I). 46 trop. & S Afr. R: MNYBG 57(1990)203. Poll. curculionids. Stems a source of sago (cf. *Cycas*); cult. orn. incl. *E. woodii* Sander (Natal) – extinct in wild, propagated from single male known

Encephalosphaera Lindau. Acanthaceae. 2 trop. S Am.

enchanter's nightshade Orig. applied to *Mandragora* spp. but transferred to *Circaea* spp.

Encheiridion Summerh. = *Microcoelia*

Encholirium C. Martius ex Schultes & Schultes f. Bromeliaceae (1). 14 E S Am. R: FN 14(1974)191, 604

Enchosanthera King & Stapf ex Guillaumin = *Creochiton*

Enchylaena R. Br. Chenopodiaceae (I 5). 2 Aus. R: FA 4(1984)213

Encopella Pennell. Scrophulariaceae. 1 Cuba

Encyclia Hook. (~ *Epidendrum*). Orchidaceae (V 13). 235 trop. Am. Some cult. orn. epiphytes

Endadenium Leach (~ *Monadenium*). Euphorbiaceae. 1 Angola

Endertia Steenis & De Wit. Leguminosae (I 4). 1 Borneo

Endiandra R. Br. Lauraceae (I). (? 1 Assam,) c. 100 Mal. to Aus. (38, endemic) & New Caled. (6). Some timbers esp. *E. palmerstonii* C. White (Queensland or Aus. walnut, walnut bean, NE Aus.) – wood varieg. black, green, brown, pink etc., valuable & form. much used for panelling, seeds viable for 1.5 yrs

endive *Cichorium endivia*

Endlicheria Nees. Lauraceae (I). 40 trop. Am. R: RTBN 34(1937)500. Some fish-disp. in Amazon

Endocaulos C. Cusset. Podostemaceae. 1 Madagascar

Endocellion Turcz. ex Herder (~ *Petasites*). Compositae (Sen.-Tuss.). 2 Siberia. R: FGP 7(1972)393

Endocomia Wilde (~ *Horsfieldia*). Myristicaceae. 4 Indomal. R: Blumea 30(1984)179

endod *Phytolacca dodecandra*

Endodesmia Benth. Guttiferae (III). 1 trop. W Afr.

Endolepis Torrey (~ *Atriplex*). Chenopodiaceae (I 3). 3 US to N Mex.

Endomallus Gagnepain = *Cajanus*

Endonema A. Juss. Penaeaceae. 2 S Afr.
Endopappus Schultz-Bip. (~ *Chrysanthemum*). Compositae (Anth.-Mat.). 1 N Afr.
Endopleura Cuatrec. Humiriaceae. 1 Amazonian Brazil
Endosamara Geesink (~ *Millettia*). Leguminosae (III 6). 1–2 Indomal. Lianes
Endosiphon T. Anderson ex Benth. = *Ruellia*
Endospermum Benth. Euphorbiaceae. 12–13 SE As. to Fiji. Myrmecophily with stem-domatia, ants keeping off other plants, esp. *E. moluccanum* (Teijsm. & Binn.) Becc. (Moluccas) therefore called 'Arbor Regis' by Rumphius. Some timbers
Endosteira Turcz. = *Cassipourea*
Endostemon N.E. Br. Labiatae (VIII 4b). c. 17 trop. & S Afr. to India
Endotheca Raf. (~ *Aristolochia*). Aristolochiaceae (II 2). 2 C & E N Am. *E. serpentaria* (L.) Raf. (*A. serpentaria*, Virginian snakeroot) – rhiz. shape suggestive of snakes (Doctrine of Signatures) & actually efficacious in treatment of snakebite!
Endresiella Schltr. = *Trevoria*
Endressia Gay. Umbelliferae (III 8). 2 Pyrenees, N Spain. (*Arpitium* Necker ex Sweet app. correct name)
Endusa Miers ex Benth. = *Minquartia*
Endymion Dumort. = *Hyacinthoides*
Enemion Raf. (~ *Isopyrum*). Ranunculaceae (III 1). 6 NE As., N Am.
eng *Dipterocarpus tuberculatus*
Engelhardia Leschen. ex Blume = seq.
Engelhardtia Leschen. ex Blume. Juglandaceae. 5 Himal. to Mal. R: AMBG 65(1978)1076. Some timber & tanbarks. *E. roxburghiana* Wallich (*E. chrysolepis*, Indomal.) – dihydro-flavonoid sweeteners suggested in tumour treatment
Engelmannia Torrey & A. Gray ex Nutt. Compositae (Helia.-Eng.). 1 N Am.: *E. peristenia* (Raf.) Goodm. & C. Lawson (*E. pinnatifida*), cult. orn.
Englerastrum Briq. (~ *Rabdosia*). Labiatae (VIII 4c). 20 trop. Afr.
Englerella Pierre = *Pouteria*
Engleria O. Hoffm. Compositae (Ast.-Sol.). 2 Angola, Namibia
Englerina Tieghem (~ *Tapinanthus*). Loranthaceae. Spp.?
Englerocharis Muschler. Cruciferae. 2 Andes. R: NJB 8(1989)623
Englerodaphne Gilg = *Gnidia*
Englerodendron Harms. Leguminosae (I 5). 1 E Afr. (Usambaras)
Englerophytum K. Krause (*Bequaertiodendron*). Sapotaceae (IV). 5–10 trop. Afr. R: T.D. Pennington, *S.* (1991)47
Engysiphon G. Lewis = *Geissorhiza*
Enhalus Rich. Hydrocharitaceae. 1 Indomal. to W Pacific: *E. acoroides* (L.f.) Royle. Marine. The female fls float horiz. at low tide (associated with spring tides) & catch the males which break off & float (cf. *Elodea*); as tide rises, fls stand vertically & pollen sinks (heavier than water) on to stigmas; testa bursts when seed ripens & frees embryo. Fibre for fishing nets in E Mal.; seeds ed.; chief food of dugong
Enhydra DC = *Enydra*
Enhydrias Ridley = *Blyxa*
Enicosanthellum Tien Ban = *Disepalum*
Enicosanthum Becc. Annonaceae. 16 Sri Lanka to W Mal.
Enicostema Blume. Gentianaceae. 3 C Am. & WI (1), Afr. to Lesser Sunda Is. (1), Madag. (1). R: Adansonia 9(1969)57. Alks. Medic. locally
Enkianthus Lour. Ericaceae. 13 Himal. to Japan. R: ABY 4(1982)355. Cult. orn. decid. shrubs with pagoda form
Enkleia Griffith. Thymelaeaceae. 3 Andamans, SE As., Mal.
Ennealophus N.E. Br. (~ *Trimezia*). Iridaceae (III 4). 5 S Am. Elaiophores attractive to bees
Enneapogon Desv. ex Pal. Gramineae (29). 28 warm (Eur. 1). R (p.p.): KB 22(1968)393
Enneastemon Exell = *Monanthotaxis*
Enneatypus Herzog. Polygonaceae. 1 Bolivia = ?
Enochoria Baker f. = *Schefflera* + *Meryta*
Enriquebeltrania Rzed. (*Beltrania*). Euphorbiaceae. 1 Mexico
Ensete Horan. Musaceae. 7 OW trop. R: KB 1947(1947)97. Hapaxanthic banana-like herbs. *E. ventricosum* (Welw.) Cheesman (Ethiopia) – fl. heads & seeds eaten cooked, seeds for beads in N Uganda
Enslenia Nutt. = *Ampelamus*
Entada Adans. Leguminosae (II 3). c. 30 trop. Am & Afr. to Aus. Trees & lianes, some sea-disp., seeds germ. in N Is., NZ after 21 months in sea, carved as snuff-boxes. *E. africana*

Guillemin & Perr. (W Afr.) – fls to 2 cm but fr. to 80+ cm; *E. gigas* (L.) Fawcett & Rendle (*E. scandens*, sea bean, nicker or Mackay b., trop Am. & Afr.) – liane with legumes to 1.8m long, seeds flat & brown to 5cm across (common drift seeds in W Br.), lvs a vegetable & seeds ed. roasted, fibre for nets, sails etc., saponin source; *E. phaseoloides* (L.) Merr. (*E. pursaetha*, Queensland bean, trop. Afr. to Aus.) similar liane to 75m long but legumes (to 200 × 13 cm) not spirally twisted, seeds used for hair-washing in As.

Entandrophragma C. DC. Meliaceae (IV 2). 11 trop. Afr. Timber (sapele, subs. for true Am. mahoganies incl. *E. candollei* Harms (omu), *E. angolense* (Welw.) C. DC (edinam, gedu nohor), *E. cylindricum* (Sprague) Sprague (aboudikro), *E. utile* (Dawe & Sprague) Sprague (utile, assié))

Entelea R. Br. Tiliaceae. 1 NZ.: *E. arborescens* R. Br. – timber half the weight of cork (corkwood) used for fishing-net floats etc. by Maoris

Enterolobium C. Martius. Leguminosae (II 5). 5 trop. Am. Close to *Albizia*. Cult. orn., timber & legumes for livestock (though some in Brazil cause lesions (photosensitive)) esp. *E. cyclocarpum* (Jacq.) Griseb., dispersed by modern (introd.) horses suggesting extinct horse-like herbivores were orig. dispersal agents, but also by peccaries which eat seeds, crushing some & spitting out rest there being a variation in seed weight (the heavier the harder) correlated with crushability which ensues that peccaries both disperse seeds yet eat some as a 'reward', bark & fr. used as soap subs.

Enteropogon Nees. Gramineae (33b). 11 trop. savanna

Enterosora Baker (~ *Grammitis*). Grammitidaceae. 9 trop. Am., Afr. (1). R: SB 17(1992)345

Enterospermum Hiern = *Tarenna*

Entolasia Stapf. Gramineae (34b). 5 trop. Afr. to E Aus. & New Caled. (1)

Entomophobia de Vogel (~ *Pholidota*). Orchidaceae (V 12). 1 Borneo: *E. kinabaluensis* (Ames) de Vogel. R: OM 1(1986)41. Fls almost completely closed

Entoplocamia Stapf. Gramineae (31a). 1 SW Afr.

Enydra Lour. Compositae (Helia.-Mel.). 10 warm. Wet places

Eomatucana F. Ritter = *Matucana*

Eomecon Hance. Papaveraceae (I). 1 E China: *E. chionantha* Hance – arils, cult. orn.

Eosanthe Urban. Rubiaceae (inc. sed.). 1 E Cuba (serpentine)

Epacridaceae R. Br. (~ Ericaceae). Dicots – Dilleniidae – Ericales. 31/375 Australasia with few Indomal. & S Am. Small trees (often pachycaul) or shrubs (often ericoid), mycorrhizal. Lvs xeromorphic, usu. spirally arr., often crowded, simple & usu. sessile, often small & ericoid, sometimes with sheathing base; venation palmate or subparallel (parallel & monocot-like in *Richea* & *Dracophyllum*, pinnate in *Prionotes*); stipules 0. Fls usu. bisexual, reg., solit., axillary or in bracteate racemes, bibracteolate & s.t. with sepaloid bracteoles, usu. 5-merous; K (4)5, persistent, C ((4)5) with valvate to imbricate lobes, A (2, 4)5 usu. attached to tube & alt. with lobes, anthers usu. bisporangiate & monothecal (tetrasporangiate & dithecal in *Prionotes*), becoming inverted during development, with longit. slits & no appendages, pollen in tetrads or pseudomonads, (staminodal) nectary scales alt. with A or connate (0 in *Sprengelia*), $\underline{G}$ (4 or 5 (10)), pluriloc. with axile placentas or 1-loc. with deeply intruded ones, s.t. some locs empty & only 1–4 carpels fertile, style usu. hollow, ovules (1) several–∞ on each placenta, anatropous, unitegmic. Fr. a loculicidal capsule with 1–∞ seeds or a drupe (usu. fleshy) with 1–5 pyrenes; seeds with thin testa (1 layer of cells) & straight cylindrical embryo embedded in copious oily endosperm. x = 4–14

Classification & principal genera:

Richeoideae (stems with annular leaf-scars): 3 genera incl. *Dracophyllum* & *Richea* – mostly small trees

Epacridoideae (stems without annular leaf-scars) – 5 tribes of mostly ericoid shrubs:

Cosmelieae (lvs sheathing): 3 genera (all Aus.) incl. *Andersonia* & *Sprengelia*

Oligarrheneae (lvs not sheathing; A 2): *Oligarrhena* (only)

Needhamielleae (as above, A 5, C lobes rolled): *Needhamiella* only

Styphelieae (as above, C lobes valvate, style attenuate from G, fr. usu. drupe): 16 genera incl. *Leucopogon*, *Styphelia* (generic limits debatable)

Epacrideae (as above, style in deep depression at G apex, fr. a loculicidal capsule): 7 genera incl. *Epacris*, *Lebetanthus*, *Prionotes*; (*Wittsteinia* is referred to Alseuosmiaceae)

The pinnate venation of *Prionotes* resembles Ericaceae, a family of which this fam. is the austral counterpart, its species forming heath-like moorland: in Mal. E. are found growing with *Erica* spp. Most are insect-poll., though some are visited by honey-eaters &

parakeets. They are separable from Ericaceae in appendage-less anthers with slits, rather than pores, as well as the open leaf venation

Some cult. orn.

Epacris Cav. Epacridaceae. 40 SE Aus. (38), NZ (2). Australian heath. Sweet- (dianthus-) scented

Epallage DC = *Anisopappus*

Epaltes Cass. Compositae (Pluch.). 14 trop. (not Aus.). Lvs usu. decurrent; pappus 0

Eparmatostigma Garay (~ *Saccolabium*). Orchidaceae (V 16). 1 India

epazote *Chenopodium ambrosioides*

Eperua Aublet. Leguminosae (I 4). 15 NE S Am. R: SCB 28(1975). Timber (wallaba) ± resistant to decay in water, that of *E. falcata* Aublet (flagelliflory, bat-poll.) much used for roof shingles. *E. purpurea* Benth. – ashes added to clay for pots; this & other spp. held to encourage growth of thick hair

Ephedra L. Ephedraceae. c. 40 Medit. (Eur. 4) to China, 14 W US & Mex., 13 Andes. Mostly shrubby switch-pls, 1 ± arborescent, some lianoid, many with rhiz. & underground buds. Wind-poll. though some visited by insects seeking nectar on outside of 'P'; double fert. (without triploid endosperm) in *E. nevadensis* S. Watson (W N Am.). Alks incl. ephedrine (not all spp.) medic. in China for 5000 yrs & still used in treatment of asthma, sinusitis etc. & stimulant (abused by athletes); *E. trifurca* Torrey ex S. Watson (Mormon tea, SW N Am.) – form. imp. tea; some cult. orn. e.g. *E. gerardiana* Wallich (Himal.) – fleshy berry-like 'P' ed. but acid

Ephedraceae Dumort. Gnetopsida. 1/65 N hemisph. & S Am. Xeromorphic equisetoid shrubs, climbers or small trees, usu. dioec.; sec. xylem ring porous. Lvs opp. or in whorls of 3 (4), scale-like, evanescent (stems photosynthetic). Female infl. a short shoot with 2–8 prs (or whorls) of bracts, the lowermost sterile, sometimes becoming swollen & juicy, & 1–3 fls, each a nucellus encl. by 2 layers (integuments & bract envelope); male fls subtended by a bract, 'P' 2-lipped, microsporangiophore ('stamen') 1 central, forked or even 3. Female gametophyte develops from lowermost of a linear tetrad of spores produced by megaspore mother-cell & is cellular prothallus with (1)2 or 3 archegonia with necks to 40 cells long; male gametophyte a prothallus with a tube nucleus, sperm cell, 1 sterile cell & 2 prothallial cells (cf. Pinopsida); pollen tube releases 2 nuclei, 1 uniting with egg nucleus, the 2nd with another but no embryo or nutritive tissue (cf. Angiosperms) results from latter. Zygote divides 3 times & all (but usu. only 3–5) can become embryos, each with 2 cotyledons. Seeds solit. or paired forming syncarp with 2 prs of bracts, membranous & winged or fleshy & coloured. 2n = 14, 28

Only genus: *Ephedra*

V. diff. from Gnetaceae with which it is generally associated

Ephedranthus S. Moore. Annonaceae. 5 trop. S Am. R: AHB 10(1931)175

Ephemerantha P. Hunt & Summerh. = *Flickingeria*

Ephippiandra Decne. Monimiaceae (V 1). 6 Madag. R: AMBG 72(1985)81

Ephippianthus Reichb.f. Orchidaceae (V 8). 1 Korea, Japan, Sakhalin

Ephippiocarpa Markgraf = *Callichilia*

Epiblastus Schltr. Orchidaceae (V 14). 20 E Mal., Polynesia

Epiblema R. Br. Orchidaceae (IV 1). 1 SW Aus.: *E. grandiflorum* R. Br.

Epicampes J. Presl = *Muhlenbergia*

Epicharis Blume = *Dysoxylum*

Epicion Small = *Metastelma*

Epiclastipelma auctt. = seq.

Epiclastopelma Lindau. Acanthaceae. 2 trop. E Afr.

Epicranthes Blume = *Bulbophyllum*

Epidanthus L.O. Williams = *Epidendrum*

Epidendropsis Garay & Dunsterv. = *Epidendrum*

Epidendrum L. Orchidaceae (V 13). c. 800 trop. Am. Epiphytes much cult. (see also *Encyclia*, *Barkeria* sometimes incl. here). Extrafl. nectaries. *E. ibaguense* Kunth flowers all year round, the fls with 0 nectar & app. mimicking *Lantana camara* L. & *Asclepias curassavica* L. in Panamá, attracting monarch butterflies

Epidryos Maguire. Rapateaceae. 3 Colombia, Panamá

Epifagus Nutt. Orobanchaceae. 1 N Am.: *E. virginiana* (L.) Barton (beech drops) on *Fagus grandifolia*

Epigaea L. Ericaceae. 3: *E. asiatica* Maxim. (Japan), *E. repens* L. (E US) – cryptic dioecy (app. gynodioec.), mymechory, *E. gaultherioides* (Boiss.) Takht. (Caucasus & E As. Minor), all cult. orn. creeping evergreen shrubs

Epigeneium Gagnepain (~ *Dendrobium*). Orchidaceae (V 15). 12 As. & Mal. Cult. orn. epiphytes

Epigynum Wight. Apocynaceae. 14 SE As., Indomal.

Epilasia (Bunge) Benth. Compositae (Lact.-Scor.). 3 W & C As. to China

Epilobium L. Onagraceae. 165 temp. (Eur. 24 native & natur.) esp. W N Am. (incl. *Zauschneria*), arctic & trop. mts. Willow-herbs. Herbs or subshrubs with spirally arr., opp. or whorled lvs, some weedy & hybridizing e.g. *E. parviflorum* Schreber (Euras., Med.) with A 4 shorter than style (pollen for cross-poll.), 4 longer, which curl back giving autogamy; seeds with chalazal tuft of hairs making seeds buoyant in air; some bird-poll. (*Zauschneria* – Calif. fuchsia). Cult. orn. incl. *E. angustifolium* L. (*Chamaenerion a.*, *Chamerion a.*, rose-bay willow-herb, fireweed, wickup, N temp.) – distr. greatly increased in GB (highland populations poss indig.; lowland from Am. or Eur. in 20C, poss. through increase in habitats (bomb-sites etc.), autogamy almost impossible (highly protandrous: dichogamy first described by Sprengel from this pl.), roots live for 20 yrs, lvs ed. as greens by N Am. Indians & used by them & in Russia for tea, pollen concentrates gold, excellent honey; *E. brunnescens* (Cockayne) Raven & Engelhorn (*E. nerteroides*, NZ) – natur. NW England; *E. hirsutum* L. (codlins and cream, Euras., N Afr.) – natur. N Am.

Epilyna Schltr. = *Elleanthus*

Epimedium L. Berberidaceae (II 2). 44 N Afr., N Italy to Caspian (2), W Himal., NE As., Japan. R: JLSBot. 51(1938)409. K 4(unequal & in 2 prs) + 4 petaloid, C 4 flat or extended into pouches, A 4 (but see fam. for various interpretations of structure). Fls pendulous, protogynous, anthers later bending up over stigma & dehiscing, followed by style elongation carrying stigma among anthers & poss. of self-poll.; seeds with membranous arils, ant-disp. In modern Chinese herbalism used for impotence, paralysis of legs & high blood pressure in elderly women, *E. sagittatum* (Sieb. & Zucc.) Maxim. (C China, natur. Japan) for hypertension. Many cult. orn. spp. & hybrids much confused in cult. incl. *E. alpinum* L. (barrenwort, S Eur., natur. in N) & *E.* × *youngianum* Fischer & C. Meyer (*E. diphyllum* Lodd. (S Japan) × *E. grandiflorum* Morren (E As.)) orig. Japan, other spontaneous hybrids occurring in gdns. *E. elatum* Morren & Dcne (W Himal.) – effective mosquito-repellent

Epimeredi Adans. = *Anisomeles*

Epinetrum Hiern = *Albertisia*

Epipactis Zinn. Orchidaceae (V 1). 22 N temp. (Eur. 9) to trop. Afr., Thailand & Mex. Helleborines

Epiphyllanthus A. Berger = *Schlumbergera*

Epiphyllum Haw. Cactaceae (III 2). 15 trop. Am. Epiphytes with large fragrant fls & extrafl. nectaries; cult. orn. incl. *E. crenatum* (Lindley) G. Don f. (Mex. to Honduras) – diurnal fls, parent of many hybrids incl. intergeneric ones, though 'epiphyllums' of hort. usu. = *Disocactus* × *hybridus*

Epipogium S. Gmelin ex Borkh. Orchidaceae (V 5). 3 temp. Euras. (*E. aphyllum* Sw., ghost orchid), OW trop.to Aus. Leafless mycotrophs with branched rhiz. & 0 roots; endotrophic mycorrhiza. *E. roseum* (D. Don) Lindley (OW trop. to Aus.) – seeds disp. a few days after fls emerge

Epipremnopsis Engl. = *Amydrium*

Epipremnum Schott. Araceae (III 2). c. 15 SE As. to W Pacific (alleged fossils in Oligocene of N Egypt). Lianes, some medic. & cult. orn. esp. *E. pinnatum* (L.) Engl. (Indomal. to W Pacific) – like *Monstera deliciosa* with perforated lvs, many cvs esp. 'Aureum' (? orig. Solomon Is., 'money plant' – rarely flowering so owners of fl. pls considered 'in the money'), irreg. variegated, widely planted in trop.

Epiprinus Griffith. Euphorbiaceae. 5–6 Assam to W Mal.

Epirixanthes Blume = *Salomonia*

Epischoenus C.B. Clarke. Cyperaceae. 8 S Afr.

Episcia C. Martius. Gesneriaceae. 9 trop. Am. R: Selbyana 5(1978)25, 6(1982)183. Stolons (unique in fam.). Cult. orn. spp. & hybrids

Episcothamnus H. Robinson = *Lychnophoriopsis*

Epistemma D.V Field & J. Hall. Asclepiadaceae (I; Periplocaceae). 3 W Afr.

Epistephium Kunth. Orchidaceae (V 4). 14 trop. Am. Some to 180 cm tall; lvs net-veined like *Smilax*; toothed cup-shaped organ at summit of G, persistent in fr.

Epitaberna Schumann = *Heinsia*

Epithelantha A. Weber ex Britton & Rose. Cactaceae (III 9). 1 (variable) SW N Am.: *E. micromeris* (Engelm.) Britton & Rose

Epithema Blume. Gesneriaceae. 22 OW trop. (Afr. 1: *E. tenue* C.B. Clarke, W Afr.) – epiphyllous fls on leaf 3 (cf. *Streptocarpus*)

Epitriche Turcz. (~ *Angianthus*). Compositae (Gnap.-Ang.). 1 SW Aus.: *E. demissus* (A. Gray) Short. R: Muelleria 5(1983)181

Epixiphium (Engelm. ex A. Gray) Munz. Scrophulariaceae (Antirr.). 1 SW N Am.

Eplingia L.O. Williams = *Trichostema*

Equisetaceae Michaux ex DC. Equisetopsida. 1/15 almost cosmop. Rhiz. herbs with aerial hollow, jointed stems impregnated with silica, s.t. unbranched. Lvs a series of teeth united by a sheath, usu. without chlorophyll (stems photosynthetic). Sporangia in term. cones, sometimes on special unbranched stems, each with whorls of sporangiophores without bracts, each peltate & bearing 5–10 sporangia which originate on outer surface but during ontogeny are carried underneath peltate head. Spores with 4 spathulate bands which are hygroscopic ('haptera') coiling & uncoiling when humidity changes. Prothalli male or hermaphrodite (when female first), so that selfing is possible, though proportions of male to hermaphrodite can be manipulated by differing light intensities or crowding; antherozoids spirally coiled & multiflagellate. n = 108

Only genus: *Equisetum*

Poss. related to Calamitaceae but ancestors ('*Equisetites*') were herbaceous as well as arborescent, the 2 growing together in the Carboniferous, but the reduced lvs app. derived from larger dichotomizing structures (Sporne)

Usu. in marshes, wet woods, on riverbanks etc.

Equisetopsida (Equisetatae, Sphenopsida). 1/15 in 1 fam. Vascular spore-pls with roots, stems & whorled lvs. Some with sec. thickening. Sporangia thick-walled, homosporous (or heterosporous) usu. borne in a reflexed position on sporangiophores in whorls; antherozoids multiflagellate (Sporne). Three extinct orders incl. Calamitales which were a dominant group of the swamp-forests of the Carboniferous, and the extant Equisetales comprising the family Equisetaceae

Equisetum L. Equisetaceae. 15 almost cosmop. (exc. Australasia; Eur. 10). Horsetails. Alks incl. nicotine but toxicity due to thiaminase breaking down the vitamin thiamine (like silica poss. antiherbivores (? dinosaurs)); poss. treatment of Alzheimer's Disease; extracts in cosmetics; affinity for gold in solution & concentrate it more than any other pl. but indicators because only 0.25 g of gold per kg of stems or rhizomes. Up to 13 m long (*E. giganteum* L., but stems only 2 cm thick & no sec. thickening so a sprawler). Several bad weeds (devil guts) & stems s.t. used for scouring (s. rush) & polishing (form. for pewter, & lime wood-carvings by Grinling Gibbons), others medic. (form. used to remove white spots in fingernails) or cones ed. (e.g. *E. arvense* L. (temp., natur. Aus.) in Japan) after boiling, *E. arvense* shoots eaten like asparagus by Romans who also dried them as a tea & a thickening powder, *E. telmateia* Ehrh. (N temp.) cones juicy & sweet; *E. telmateia* provides a black fibre for Vancouver Indians' baskets; *E. sylvaticum* L. (N temp.) – a yellow dye extracted from dried stems in Norway

Eragrostiella Bor. Gramineae (31d). 5 E Afr., Sri Lanka to N Aus. R: CHA 22(1976)1

Eragrostis Wolf. Gramineae (31d). c. 300 temp. (Eur.9) & trop. R: Preslia 24(1952)281. *E. reptans* Nees (Brazil) – dioec.; pericarp of many spp. peeled away readily when moist. Some cult. orn. & fodder incl. *E. curvula* (Schrader) Nees (African or weeping lovegrass, S Afr.) – planted to stabilize soil in tea estates; *E. tef* (Zucc.) Trotter (*E. abyssinica*, t'ef or teff, NE Afr., poss. derived from *E. pilosa* (L.) Palib.) – also cult. for ed. seeds, straw used in brick manufacture (e.g. this or other *E.* spp. in pyramid bricks (3359 BC)). The Afr. spp. *E. hispida* Schumann, *E. nindensis* Ficalho & Hiern & *E. paradoxa* Launert have lvs which revive after desiccation to 6–12% water-content

Eranthemum L. Acanthaceae. 30 trop. As. Cult. orn. foliage pls closely allied to *Pseuderanthermum*

Eranthis Salisb. Ranunculaceae (I 1). 8 Euras. R: APG 38(1987)96. Tuberous toxic herbs much cult. for early spring fls (winter aconites) esp. *E. hyemalis* (L.) Salisb. (S Eur., natur. N Eur. & Am.); *E. cilicica* Schott & Kotschy (E Medit.) often considered conspecific but alleged hybrid, *E. × tubergenii* Bowles is sterile with aborted anthers & has more carpels than either. Consp. P app. K, C being nectaries, though organs intermediate between these latter & A found in *E. × tubergenii* (cf. *Nymphaea*)

Erato DC (~ *Liabum*). Compositae (Liab.). 4 trop. Am. R: SCB 54(1983)52

Erblichia Seemann (~ *Piriqueta*). Turneraceae. 1 C Am., 4 Madag. R: Adansonia 19(1979)459

Ercilla A. Juss. Phytolaccaceae (I). 1–2 Chile. Temp. rain forest; adventitious clinging roots like *Hedera*; coloured P prob. K. Cult. orn. esp. *E. volubilis* A. Juss.

Erdisia Britton & Rose = *Corryocactus*

Erechtites Raf. Compositae (Sen.-Sen.). 15 Am. introd. OW (Eur. 1). R: AMBG 43(1956)1. Alks

Eremaea Lindley. Myrtaceae (Lept.). 16 SW Aus. R: Nuytsia 9(1993)137

Eremaeopsis Kuntze = *Eremaea*

Eremalche E. Greene (~ *Malvastrum*). Malvaceae. 4 W US

Eremanthus Less. Compositae (Vern.-Lych.). 18 Brazil, esp. arid cerrado. R: AMBG 74(1987)265. Trees & shrubs; some timber (*Vanillosmopsis*)

Eremia D. Don. Ericaceae. 7 S & SW Cape. R: Bothalia 12(1976)29

Eremiastrum A. Gray = *Monoptilon*

Eremiella Compton. Ericaceae. 1 S Afr.

Eremiopsis N.E. Br. = *Eremia*

Eremitis Doell. Gramineae (5). 7 E Brazil. 3 types of infl., 1 burying itself like a peanut

Eremobium Boiss. Cruciferae (I). 3 N Afr. to Arabia

Eremoblastus Botsch. Cruciferae. 1 W Kazakhstan to C As.: *E. caspicus* Botsch.

Eremocarpus Benth. = *Croton*

Eremocarya E. Greene = *Cryptantha*

Eremocaulon Söderstrom & Londoño (~ *Bambusa*). Gramineae (1b). 1 Brazil: *E. aureofimbriatum* Söderstrom & Londoño. R: AJB 74(1987)37

Eremocharis Philippi. Umbelliferae (I 2 c). 9 Chile, Peru. R: UCPB 33(1962)153

Eremochion Gilli = *Horaninovia*

Eremochloa Buese. Gramineae (40h). 9 Indomal. to Aus. R: KB 7(1952)309. *E. ophiuroides* (Munro) Hackel (centipede grass, SE As.) – used for erosion control & (SE US) as lawn-grass

Eremocitrus Swingle = *Citrus*

Eremocrinum M.E. Jones. Anthericaceae (Liliaceae s.l.). 1 Utah

Eremodaucus Bunge. Umbelliferae (III 5). 1 Caucasus to C As. & Afghanistan

Eremodraba O. Schulz. Cruciferae. 2 Peru, Chile. R: AMBG 77(1990)602

Eremogeton Standley & L.O. Williams. Scrophulariaceae. 1 Mex., Guatemala: *E. grandiflorus* (A. Gray) Standley & L.O. Williams. R: Sida 11(1985)167

Eremohylema Nelson = *Pluchea*

Eremolaena Baillon. Sarcolaenaceae. 2 E Madagascar

Eremolepidaceae Tieghem ex Nakai. Dicots – Rosidae – Santalales. 3/11 trop. Am. Hemiparasitic, epiphytic, dioec. to monoec. shrubs on trees, ± green & with thickened haustoria promoting host to engage in growing a dual organ at contact point, often with epicortical roots. Lvs usu. spirally arr., simple, entire; stipules 0. Fls small, in bracteate spikes or catkins (females sometimes solit.), 2–4-merous, males with P (0) 2–4, females 2–4, A opp. P with anthers with longit. slits, $\overline{G}$ (3–5), 1-loc. (half-inf. in *Antidaphne*), with 2 ovules embedded in base, each comprising embryo-sac without clearly differentiated nucellus or integument. Fr. a berry; seed 1 without a testa, surrounded or tipped with viscid tissue, the embryo embedded in chlorophyllous endosperm. n = 10, 13. R: SBM 18(1988)1

Genera: *Antidaphne, Eubrachion, Lepidoceras*

Form. considered the primitive group of Viscaceae but poss. closer to Opiliaceae (Kuijt) or rep. 'diverse aerial-parasitic members of the Santalaceae' (Cronquist)

Eremolepis Griseb. = *Antidaphne*

Eremolimon Lincz. (~ *Limonium*). Plumbaginaceae (II). 7 C As.

Eremomastax Lindau. Acanthaceae. 1 trop. Afr. (? + Madag.), variable. Seeds often with toothed scales, spreading when wetted

Eremonanus I.M. Johnson = *Eriophyllum*

Eremopanax Baillon = *Arthrophyllum*

Eremopappus Takht. = *Centaurea*

Eremophea P.G. Wilson. Chenopodiaceae (I 5). 2 Aus. R: FA 4(1984)224

Eremophila R. Br. Myoporaceae. 206 Aus. esp. W, 1 ext. to NZ. Some timber & tanbark (e.g. *E. oppositifolia* R. Br., emu bush), trad. medic. & ceremonial for Aboriginals but many toxic to stock (e.g. *E. deserti* (Benth.) Chinnock (Ellangowan poison bush, Aus.)), some weedy

Eremophyton Bég. Cruciferae. 1 Algeria, Sahara

Eremopoa Rosch. Gramineae (17). 4 E Med. (Eur. 1) to W China

Eremopogon Stapf = *Dichanthium*

Eremopyrum (Ledeb.) Jaub. & Spach. Gramineae (23). 4 Morocco & Eur. (2) to W China. R: NJB 11(1991)271

Eremosemium E. Greene = *Grayia*

Eremosis (DC) Gleason (~ *Vernonia*). Compositae (Vern.-Pip.). 20 Mex. & C Am.

Eremosparton Fischer & C. Meyer. Leguminosae (III 15). 3 SE Russia (1) to C As., sandy deserts

Eremospatha (G. Mann & H. Wendl.) H. Wendl. Palmae (II 1a). 12 trop. Afr. Herm. rattans

Eremostachys Bunge. Labiatae (VI). c. 5 v. variable (some recog. up to 60) Eur. (1) to C As. (where approach *Phlomis*)

Eremosynaceae Dandy = Saxifragaceae

Eremosyne Endl. Saxifragaceae (Eremosynaceae). 1 SW Aus.

Eremothamnus O. Hoffm. Compositae (Arct.). 1 coastal S Namibia: *E. marlothianus* O. Hoffm. R: T 43(1994)36

Eremotropa Andres = *Monotropastrum*

Eremurus M. Bieb. Asphodelaceae (Liliaceae s.l.). c. 45 Eur. (2), alpine W & C As. R: MAISSP 8,23(1909)1. Foxtail lilies, giant asphodels. Columnar infls to 3.5 m, of white, pink or yellow fls with tepals withering before A & G mature, followed by capsules of winged seeds (wind-disp.). Typical of dry open, almost barren habitats of W As. to subalpine grasslands of Afghanistan & W China. Lvs of *E. spectabilis* M. Bieb. (Crimea to Pakistan) ed. Several cult. orn. esp. hybrids & forms of *E. stenophyllus* (Boiss. & Buhse) Baker (SW As.) with yellow fls & *E. olgae* Regel (Turkestan) with white (the hybrid between them = *E.* × *isabellinus* P.L. Vilm. (Shelford hybrids))

Erepsia N.E. Br. Aizoaceae (V). 27 SW Cape. R: BBP 64(1989)391. Shrubs, some cult. orn. (incl. *Kensitia*, *Semnanthe*)

Ergocarpon C. Towns. Umbelliferae (III 1). 1 Iraq/Iran border: *E. cryptanthum* (Rech.f.) C. Towns.

Eria Lindley. Orchidaceae (V 14). c. 500 Indopacific. Epiphytes, some cult. orn.

Eriachaenium Schultz-Bip. Compositae (Mut.-Nass.). 1 S S Am.

Eriachne R. Br. Gramineae (37). 40 Sri Lanka to (mostly) Aus. *E. mucronata* R. Br. used as lead indicator in Queensland

Eriadenia Miers = *Mandevilla*

Eriandra P. Royen & Steenis. Polygalaceae. 1 Papuasia. Tree

Eriandrostachys Baillon = *Macphersonia*

Erianthecium L. Parodi. Gramineae (17). 1 Uruguay & S Brazil

Erianthemum Tieghem. Loranthaceae. 12 trop. & S Afr.

Erianthus Michaux = *Saccharum*

Eriastrum Wooton & Standley. Polemoniaceae. 14 SW US. R: Madrono 8(1945)65. Some cult. orn. like *Gilia* but K-lobes unequal

Eriaxis Reichb. f. Orchidaceae (V 4). 3 New Caledonia

Erica L. Ericaceae. (Incl. *Blaeria*, *Ericinella*, *Philippia*) 735 S Afr. (most, esp. SW Cape), trop. Afr. mts, Madag., Medit., Macaronesia, Eur. (16). Heaths. Shrubs & small trees with endotrophic mycorrhiza; some cyanogenic. Fls bell-shaped, pendulous & visited esp. by bees searching for nectar secreted by disk, bird-poll. spp. in Cape with significantly thicker stems. Many fynbos spp. germ. enhanced by smoke. *E. cinerea* L. & *E. tetralix* (crossed-leaved h.) cover great areas of drier & wetter moorland resp. in Eur. Many cult. orn. esp. form. in greenhouses incl. *E. arborea* L. (tree h., Medit. to trop. Afr. mts) – woody nodules at ground level trad. used to make briar ('bruyère') pipes, *E. ciliaris* L. (Dorset h., W Eur.), *E. erigena* R. Ross (Irish h., W Eur.), *E. carnea* L. (*E. herbacea*, Eur.) – lime-tolerant winter-flowering, *E. lusitanica* Rudolphi (SW Eur., natur. SW England) – major weed in many parts of NZ, *E. tetralix* L. (Eur.) – source of yellow dye, *E. vagans* L. (Cornish h., W Eur.)

Ericaceae Juss. (incl. Monotropaceae, Pyrolaceae). Dicots – Dilleniidae – Ericales. 107/3400 cosmop. exc. deserts & usu. montane in trop., scarce in Australasia (cf. Epacridaceae). Shrubs & trees (s.t. pachycaul), more rarely lianes or (sub)herbaceous, s.t. epiphytic, mycorrhizal & often on acid soils or even achlorophyllous (Monotropoideae). Lvs simple, often ericoid, spirally arr., opp. or whorled; stipules 0. Fls usu. bisexual, ± reg., usu. in bracteate racemes (s.t. solit.) with 2 bracteoles; K (3–)5(–7), valvate or imbricate, persistent, C ((3–)5(–7)) usu. a tube with convolute or imbricate lobes, s.t. free, A in 2 whorls, usu. 2 × C, less often up to 20 or (some of) whorl opp. C absent, s.t. attached at tube-base & rarely united, anthers becoming inverted during ontogeny with term. pores, less often slits, appendages often present, pollen grains in tetrads (monads in e.g. *Enkianthus*), nectary-disk intrastaminal, surrounding & often attached to G, $\underline{G}$ to $\overline{G}$ ((2–)5(–10)), pluriloc. (1-loc. in *Scyphogyne* etc. with 1 pendulous ovule), style hollow, placentation axile basally, parietal apically or all axile, ovules (1–)∞ on each placenta, anatropous to ±

campylotropous, unitegmic. Fr. a capsule (loculicidal or septicidal), berry or drupe (esp. Vaccinioideae), rarely a nut; seeds ± ∞, usu. small, sometimes winged, with straight or short cylindric or spathulate embryo embedded in copious oily & proteinaceous endosperm. x = (8–)12 or 13(–23)

Classification & chief genera (after Stevens, BJLS 64(1971)1, though this does not satisfy cladistic criteria (Brittonia 45(1993)99)):

Rhododendroideae (Lvs revolute, infls usu. term., C usu. other than urceolate, decid., anther & filament appendages 0, usu. viscin threads in pollen tetrads, G̲, fr. usu. septifragal or septicidal capsule): 7 tribes incl. Rhodoreae (*Rhododendron*), Bejarieae (*Bejaria*), Epigaeae (*Epigaea*), Phyllodoceae (*Kalmia, Loiseleuria, Phyllodoce*), Daboecieae (*Daboecia*)

Ericoideae (Lvs v. revolute with abaxial channel, infls variable in position, C gamopetalous, persistent, appendages if present flattened spurs, viscin threads absent, G̲, fr. usu. loculicidal capsule): 2 or 3 tribes incl. Ericeae (*Erica*) & Calluneae (*Calluna*)

Vaccinioideae (characters not in same combinations as above but a v. variable group without viscin threads & with superior to inf. ovaries & fr. a berry, drupe or loculicidal capsule): 5 tribes incl. Arbuteae (*Arbutus, Arctostaphylos*), Enkiantheae (*Enkianthus*), Andromedeae (*Gaultheria, Leucothoe, Lyonia, Pieris*) & Vaccinieae (*Agapetes, Cavendishia, Gaylussacia, Vaccinium*)

Pyroloideae (herbaceous, G imperf. 5-loc.; sometimes treated as separate fam., Pyrolaceae): *Pyrola*

Monotropoideae (chlorophyll-less; sometimes treated as separate fam., Monotropaceae; R: WJB 33(1975)1): *Cheilotheca, Monotropa*

Wittsteinioideae referred to Alseuosmiaceae. The Empetraceae & Epacridaceae may eventually be incl. here on cladistic grounds

Imp. cult. orn. esp. *Rhododendron* (incl. azaleas) & *Erica*, also *Arbutus, Arctostaphylos, Cassiope, Elliottia, Gaultheria, Pieris* etc. Almost all require acidic conditions (all Br. spp. exc. *Arbutus unedo* for example); some fr. trees or bushes esp. *Vaccinium*, through others e.g. *Kalmia* & others in its subfam. poisonous. Although char. of moorlands of N temp. they are most diversified in Cape (18 endemic genera & 650 endemic spp.), the Andes & (esp. *Rhododendron*) Sino-Himal. & New Guinea

Ericameria Nutt. (~ *Haplopappus*). Compositae (Ast.-Sol.). 27 SW & W N Am. R: Phytol. 68(1990)144

Ericentrodea S.F. Blake & Sherff. Compositae (Helia.-Cor.). 6 Andes. R: Novon 3(1993)77

Erichsenia Hemsley. Leguminosae (III 24). 1 W Aus.

Ericinella Klotzsch = *Erica*

Erigenia Nutt. Umbelliferae (III 5). 1 E N Am.: *E. bulbosa* (Michaux) Nutt. – tuberous, one of the first spring-flowers

Erigeron L. Compositae (Ast.-Ast.). (Excl. *Trimorpha*) c. 150 cosmop. esp. N Am. (Eur. 15 incl. introd. *E. karvinskianus* DC (*E. mucronatus*, C Am.), apomict natur. on walls trop. & temp.). Many cult. orn. (fleabanes) incl. *E. annuus* (L.) Pers. (daisy f., N Am., natur. Eur.) & a number of complex hybrids used as herb. perennials, many derived from *E. speciosus* (Lindley) DC (NW US mts); *E. heteromorphus* Robinson (Mex.) – aquatic

erima (New Guinea) *Octomeles sumatrana*

erimado *Ricinodendron heudelotii*

Erinacea Adans. Leguminosae (III 31). 1 W Medit. (incl. Eur.): *E. anthyllis* Link (*E. pungens*, hedgehog broom) – cult. orn. shrublet with branch-thorns

Erinna Philippi. Alliaceae (Liliaceae s.l.). 1 Chile. A 3 with 3 staminodes

Erinocarpus Nimmo ex J. Graham. Tiliaceae. 1 SW India: *E. nimmonii* J. Graham. Androphore; fr. spiny; bark-fibre for rope

Erinus L. Scrophulariaceae. 2 N Afr., Pyrenees & Alps (*E. alpinus* L.) – cult. orn. rock-pls

Erioblastus Nakai ex Honda = *Deschampsia*

Eriobotrya Lindley. Rosaceae (III). 26 Himal. to E As. & W Mal. R: CJB 68(1990)2233. Cult. orn. evergreen trees & shrubs esp. *E. japonica* (Thunb.) Lindley (loquat, Japanese medlar, nispero, China & Japan) – widely cult. in subtrop. for yellow acid fr. eaten raw or in jams etc.

Eriocaulaceae Pal. ex Desv. Monocots – Commelinidae – Eriocaulales. 9/1100 trop. & warm esp. Am., few temp. Perennial (usu.) small herbs without sec. thickening, growing in wet places; vessel-elements in all veg. organs. Lvs in a dense basal spiral, parallel-veined, grass-like but without a well differentiated basal sheath. Infl. a dense centripetally flowering, usu. grey or white, head on a scape, the receptacle naked, hairy or with chaffy bracts subtending fls. Fls small wind- or insect-poll., without nectaries exc. for glands in petal-tips of *Eriocaulon*, (2- or) 3-merous, often reg., unisexual, the sexes mixed or females marginal, rarely plants dioec.; K 2 or 3, sometimes with basal tube or forming spathe, C (0) 2 or 3, sometimes with a tube, males with filaments adnate to tube or often with stipe-like androphore at top of which C & filaments diverge, A 2 or 4 (2-merous fls), 3 or 6((or 1)3 + -merous fls) opp. C when 2 or 3, anthers with longit. slits, G (2 or 3), often stipitate, with as many loc. & term. style with as many branches, each loc. with 1 ventral–apical, pendulous, orthotropous, bitegmic ovule. Fr. a loculicidal capsule; seeds with small lenticular embryo forming a cap over the copious starchy endosperm at micropylar end

Principal genera: *Eriocaulon, Leiothrix, Paepalanthus* & *Syngonanthus* (see Flora 177(1985)335 for poss. arrangement)

Apparently closely allied to Xyridaceae

Infls of *Syngonanthus* spp. used as 'everlastings'

Eriocaulon L. Eriocaulaceae. c. 400 trop. & warm incl. c. 30 in Japan & c. 8 in N Am. incl. *E. aquaticum* (Hill) Druce (*E. septangulare*, pipewort) also in Ireland & Hebrides. Some spp. with trunks to 80 cm. *E. australe* R. Br. (E Aus., New Guinea) – salt-source in New Guinea

Eriocephalus L. Compositae (Anth.-Mat.). c. 26 S Afr. Some cult. orn. & fodder

Eriocereus (A. Berger) Riccob. = *Harrisia*

Eriochilus R. Br. Orchidaceae (III 1). 6 Aus.

Eriochiton (R. Anderson) A.J. Scott = *Maireana*

Eriochlamys Sonder & F. Muell. Compositae (Gnap.-Ang.). 1 Aus.: *E. behrii* Sonder & F. Muell. R: OB 104(1991)129

Eriochloa Kunth. Gramineae (34b). c. 30 trop. & warm (Eur. 1). Fodders incl. *E. punctata* (L.) Desv. (*E. polystachya*, Carib grass, WI) in S US

Eriochrysis Pal. Gramineae (40a). 7 trop. Am. & Afr., India (1)

Eriocnema Naudin. Melastomataceae. 1 Brazil

Eriocoelum Hook.f. Sapindaceae. 10 trop. Afr.

Eriocycla Lindley. Umbelliferae (III 8). 7 N Iran to W China, alpine

Eriodes Rolfe (*Tainiopsis*). Orchidaceae (V 11). 1 NE India to Vietnam: *E. barbata* (Lindley) Rolfe. R: OB 89(1986)65

Eriodictyon Benth. Hydrophyllaceae. 8 SW N Am. Some cult. orn. shrubs incl. *E. californicum* (Hook. & Arn.) Torrey (Calif., Oregon) – lvs used medic. (yerba santa) & as a tea

Erioglossum Blume = *Lepisanthes*

Eriogonella Goodman = *Chorizanthe*

Eriogonum Michaux. Polygonaceae (I 1). 240 W & S N Am. R: Phytol. 66(1989)267. Shrubs, herbs & cushion-pls differing from most P. in having no ocreae but cymose umbels or heads of fls (often compound heads). Several cult. orn. & ed. lvs & roots; *E. ovalifolium* Nutt. used as silver indicator in Montana

Eriolaena DC. Sterculiaceae. 17 Indomal., China, SE As.

Eriolobus (DC) M. Roemer (~ *Malus*). Rosaceae (III). 1 E Med.: *E. trilobatus* (Poiret) M. Roemer

Eriolopha Ridley = *Alpinia*

Erioneuron Nash (~ *Tridens*). Gramineae (31d). 5 S US, Mex., Arg. R: AJB 48(1961)565

Eriope Humb. & Bonpl. ex Benth. Labiatae (VIII 4a). 30 trop. & warm S Am. R: HIP 28,3(1978)1. *E. crassipes* Benth. (Braz. savannas) – explosive poll. mechanism (like *Ulex*) associated with bee-visits

Eriophoropsis Palla = seq.

Eriophorum L. Cyperaceae. 12 N temp. (Eur. 7) & Arctic, S Afr. (1). Typically on wet moorland. Female fls each with P of bristles which grow into long hairs after fert. & act to disperse fr.; sometimes used to stuff pillows (cotton grass)

Eriophyllum Lagasca. Compositae (Helia.-Bae.). 11 W N Am. R: UCPB 18(1937)69. Some cult. orn.

Eriophyton Benth. Labiatae (VI). 1 Himalaya

Eriopidion Harley (~ *Eriope*). Labiatae (VIII 4a). 1 Braz. & Venez.: *E. strictum* (Benth.) Harley – disjunct distrib. R: HIP 28,3(1978)1

Eriopsis Lindley. Orchidaceae (inc. sed.). 4 trop. Am. Epiphytes, some cult. orn.

Erioscirpus Palla = *Eriophorum*

Eriosema (DC) Reichb. Leguminosae (III 10). 130 trop. & warm. Some ed. & med.

Eriosemopsis F. Robyns. Rubiaceae (III 2). 1 S Afr.

Eriosolena Blume. Thymelaeaceae. 5 E Himal. & SW China to Java

Eriosorus Fée. Pteridaceae (III). 25 trop. Am. (Andes). R: CGH 200(1970)54

Eriospermaceae Endl. (Liliaceae s.l.). Monocots – Liliidae – Asparagales. 1/c. 100 subSaharan Afr. Tuberous herbs with 1–5 ovate to linear lvs. Infls racemes; fls reg., small; P 3 + 3, connate at base, sometimes dimorphic; A 6 with versatile anthers; G̲ (3), each loc. with 2–6 axile ovules, style with small stigma. Fr. a loculicidal capsule; seeds oval to comma-shaped, densely hairy

Only genus: *Eriospermum*

In many details approaching Tecophilaeaceae to which Cyanastraceae undoubtedly allied. Uniting all these would, however, lead to the collapsing of more of the present classification of Liliaceae s.l.

Eriospermum Jacq. ex Willd. Eriospermaceae (Liliaceae s.l.). c. 100 subSaharan Afr. Some with epiphyllous lvs

Eriosphaera Less. = *Galeomma*

Eriostemon Sm. Rutaceae. 35 Aus., New Caled. (1). R: Nuytsia 1(1970)19, 4(1982)47. Cult. orn. (waxflowers, wax plants)

Eriostrobilus Bremek. = *Strobilanthes*

Eriostylos C. Towns. Amaranthaceae (I 2). 1 Somalia

Eriosyce Philippi. Cactaceae (III 5). 2 Chile. Cult. orn. allied to *Echinocactus* but with axillary spines at top of fr.

Eriosynaphe DC. Umbelliferae (III 10). 1 SE Russia to C As.: *E. longipes* (Fischer & Sprengel) DC

Eriotheca Schott & Endl. Bombacaceae. 19 trop. Am. R: BJBB 33(1963)124

Eriothrix Cass. = *Eriotrix*

Eriothymus (Benth.) Reichb. (~ *Hedeoma*). Labiatae (VIII 2). 1 Brazil

Erioxylum Rose & Standley. Malvaceae. 2 W Mexico

Eriotrix Cass. Compositae (Sen.-Sen.). 2 Réunion. Shrubs

Erisma Rudge. Vochysiaceae. 16 trop. S Am. R: ABN 3(1954)462. Seeds a source of tallow for candles, soap etc., those of *E. japura* Spruce ex Warm. collected & stored for lean times in Amaz.

Erismadelphus Mildbr. Vochysiaceae. 2 trop. W Afr. R: ABN 1(1953)594

Erismanthus Wallich ex Muell. Arg. Euphorbiaceae. 2 SE As. to Hainan & W Mal.

Erithalis P. Browne. Rubiaceae (III 4). 10 Florida, WI

Eritrichium Schrader ex Gaudin. Boraginaceae. c. 30 N temp. (Eur. 4). Some cult. orn. (American forget-me-nots) esp. *E. nanum* (L.) Gaudin (Alps)

Erlangea Schultz-Bip. Compositae (Vern.). 5 trop. Afr.

Ermaniopsis Hara. Cruciferae. 1 Nepal

Ernestia DC. Melastomataceae. 16 trop. S Am.

Ernestimeyera Kuntze = *Alberta*

Ernodea Sw. Rubiaceae (IV 15). 9 SE US, WI

Erocallis Rydb. (~ *Lewisia*). Portulacaceae. 1 W N Am.

Erodiophyllum F. Muell. Compositae (Ast.-Ast.). 2 W & S Aus.

Erodium L'Hérit. Geraniaceae. c. 60 Eur. (34), Medit. to C As., temp. Aus. (3) & S trop. S Am. Awn twists into corkscrew & is hygroscopic; mericarp with sharp point with backward-pointing hairs. Fr. falls & in damp awn untwists, lengthens & drives fr. into soil (cf. *Stipa*). Cult. orn. (storksbills) with cincinnal umbel (cf. *Geranium*) incl. *E. cicutarium* (L.) L'Hérit. (Eur., Medit., natur. N & S Am. where used as forage) – form. root a source of a dye in Hebrides; others with ed. roots or lvs, some weedy

Erophila DC (~ *Draba*). Cruciferae. 10 Eur. (2), Medit. Whitlow-grass (alleged to cure whitlow = inflammation around the nails). *E. verna* (L.) Chevall. comprises a number of selfing pure lines (jordanons) form. considered spp. (Jordan)

Erpetion Sweet = *Viola*

Errazurizia Philippi. Leguminosae (III 12). 1 coastal Chile, 3 SW US deserts. R: MNYBG 27(1977)13

Ertela Adans. = *Monniera*

eru *Gnetum africanum*

Eruca Miller. Cruciferae. 3 Medit. (Eur. 1), NE Afr. R: JAA 66(1985)324. *E. vesicaria* (L.) Cav. subsp. *sativa* (Miller) Thell. (garden or salad rocket, jamba, Medit.) – salad greens, oilseed subs. for rape in India

Erucaria Gaertner. Cruciferae. 6 E Med. (Eur. 1), Arabia, Iran

Erucastrum C. Presl. Cruciferae. 20 Macaronesia, Medit., C & S Eur. (4). R: JAA 66(1985)305

Ervatamia (DC) Stapf = *Tabernaemontana*

Erxlebenia Opiz = *Pyrola*

Erycibe Roxb. Convolvulaceae. 67 Indomal. to trop. Aus. & Japan. C-lobes bifid & sessile, conical or subglob. 5–10-rayed stigma unique in fam. *E. ramiflora* Hallier f. (Sumatra) – cauliflorous; *E. stenophylla* Hoogl. (Borneo) – rheophyte

Erycina Lindley. Orchidaceae (V 10). 2 Mex. R: Schlechteriana 2(1991)115. Cult. orn.

Erymophyllum P.G. Wilson (~ *Rhodanthe*). Compositae (Gnap.-Ang.). 5 W Aus. R: OB 104(1991)105

Eryngiophyllum Greenman = *Chrysanthellum*

Eryngium L. Umbelliferae (II 1). 230-250 trop. & temp. (exc. trop. & S Afr.; Eur. 25). Lvs spiny-toothed, simple & monocot-like or lobed; fls in bracteate heads with K often longer than C, bee-poll.; many ed. or cult. orn. *E. agavifolium* Griseb. ('*E. bromeliifolium*', Arg.) & *E. pandanifolium* Cham. & Schldl. (*E. lassauxii*, S Am.) – cult. orn. with pandan-like lvs from Pampas, *E. pandanifolium* lvs source of caraguata fibre; *E. campestre* L. (eryngo, Eur., (?) natur. sandy fields in GB) – candied roots form. used as 'kissing comfits', also med.; *E. foetidum* L. (trop. Am.) – lvs pickled in Sikkim; *E. giganteum* M. Bieb. (Caucasus) – cult. orn. ('Miss Willmott's ghost') visited by wasps; *E. maritimum* L. (sea holly, Eur., natur. N Am.) – roots candied (? eringoes of Falstaff), form. valued as tonic; *E. ovinum* A. Cunn. ('*E. rostratum*', Chile) – weedy in Aus. ('blue devil')

Erysimum L. Cruciferae. (Incl. *Cheiranthus, Syrenia*) c. 200 Medit., Eur. (58), As. Some weeds, e.g. *E. cheiranthoides* L. (treacle mustard, Eur., natur. N Am.), & cult. orn. esp. *E. cheiri* (L.) Crantz (*Cheiranthus c.*, wallflower, prob. hybrid origin of 2 or more Aegean spp., cultigen natur. GB) – spring-bedding, scented perennial with yellow, brown, red or pink fls form. medic. & used in scent; also alpine or fairy wallflowers esp. 'Siberian wallflower', *E.* × *marshallii* (Henfrey) Bois ('*E.* × *allionii*', prob. *E. humile* Pers. (*E. decumbens*, SW Alps) × *E. perofskianum* Fischer & C. Meyer (Caucasus to Afghanistan)) differing in its bilobed stigma; *E. asperum* (Nutt.) DC (western wallflower, W & C N Am.)

Erythea S. Watson = *Brahea*

Erythradenia (Robinson) R. King & H. Robinson. Compositae (Eup.-Heb.). 1 Mex. R: MSBMBG 22(1987)408

Erythraea Borkh. = *Centaurium*

Erythranthera Zotov (~ *Rytidosperma*). Gramineae (25). 3 Aus. (2), NZ

Erythrina L. Leguminosae (III 10?). 112 warm (As. 12, Afr. 31, Am. c. 70). R: Lloydia 37(1974)332. Affinities obscure. Prob. all bird-poll. red or orange fls, some Mal. spp. with long peduncles acting as perches for non-hovering sunbirds, many Am. spp. without & with smaller tubular C-tubes sticking out assoc. with long-billed hummingbird poll. (short-billed ones merely thieves) & with high sucrose & low amino-acid content in nectar, whereas spp. visited by passerines have fls twisted back towards peduncle & low sucrose & high amino-acid contents; insufficient calorific value in nectar in fls open on any tree in 1 day to sustain a single hummingbird's energy requirements so outcrossing inevitable; nectar of Mal. spp. 'sweet, if somewhat bitter, watery' (Corner) & as fls open in dry season in monsoon climates an imp. water-source for birds (& squirrels); some with ant-infested twigs. Cult. as shade trees & orn. (coral trees); fls cooked & eaten; seeds used as beads. Extrafl. nectaries attractive to ants, which act as guards; many alks (some hypnotic incl. those of *E. americana* & *E. variegata*). *E. americana* Miller (C Mex.) – fls ed.; *E. caffra* Thunb. (kaffir boom, SE Afr.) – red seeds with black hilum used as beads, wood v. light; *E. crista-galli* L. (warm S Am.) – fls inverted with v. small wings, keel forming nectar sac at base, bird-poll. often so nectariferous as to be called 'cry-baby' in Louisiana, grown under glass; *E. herbacea* L. (Cherokee bean, S N Am.) – seeds red with black line from hilum used as beads; *E. mildbraedii* Harms (trop. Afr.) – thorns used for making 'rubber stamps'; *E. mitis* Jacq. (*E. umbrosa*, mortel, Venez.) – shade tree for woody crops; *E. subumbrans* (Hassk.) Merr. (*E. lithosperma*, dadap, Indomal.) – coffee-shade; *E. sandwicensis* Degener (~ *E. tahitensis* Nad., *E. monosperma*, Hawaii) – fl. colour polymorphic in populations (orange, yellow, white or pale green), toxic to 8 introd. spp. of bruchid beetles, timber for outriggers & form. surfboards & fish-floats; *E.* × *sykesii* Barneby & Krukoff – cult. orn. sterile (? orig. Aus.) with nectar sugar suggesting hybrid between OW & NW bird-poll. spp. (? *E. coralloides* A. DC (S US, Mex.) × *E. lysistemon* Hutch. (S Afr.)); *E. variegata* L. (*E. indica*, E Afr. to Pacific) – poplar-like (esp. fastigiate 'Tropical Coral') coffee-shade, wind-break & crop-support, face-powder prepared from wood in Thailand

Erythrocephalum Benth. Compositae (Mut.-Mut.). 12 trop. E Afr. Herbs with scaly pappus

Erythrochiton Nees & C. Martius. Rutaceae. 7 trop. Am. R: Brittonia 44(1992)123. K coloured, (C). *E. hypophyllanthus* Planchon & Linden (Colombia & Venez.) – known only from type coll., unbranched treelet with fls borne on abaxial surface of lvs

Erythrochlamys Guerke. Labiatae (VIII 4b). 5 trop. E Afr.
Erythrococca Benth. Euphorbiaceae. 50 trop. & S Afr., S Arabia
Erythrocoma E. Greene = *Geum*
Erythrodes Blume. Orchidaceae (III 3). 60 trop. As., Mal., New Caled. (1), Am. Some cult. orn. terrestrial
Erythronium L. Liliaceae. c. 20 temp. N Am., Euras. (1). R (p.p.): Madrono 3(1935)58. P reflexed. Many cult. orn. with membranous-coated corms, some hybrids, incl. *E. americanum* Ker-Gawler (amberbell, trout lily, E N Am.) – medic., fls yellow; *E. californicum* Purdy (fawn lily, N California) – fls whitish; *E. dens-canis* L. (dog's-tooth violet, Euras.) – corms source of a starch used for vermicelli & cakes in Japan, also eaten with reindeer or cow milk in Mongolia & Siberia, lvs ed. boiled, cult. orn. with pink fls; *E. grandiflorum* Pursh (W US) – corms dug up & eaten by bears in Canada
Erythropalaceae Planchon ex Miq. = Olacaceae
Erythropalum Blume. Olacaceae (Erythropalaceae). 1 Indomal.: *E. scandens* Blume
Erythrophleum Afzel. ex G. Don f. Leguminosae (I 1). 9 OW trop. (Afr. 4, Madag. 1, E As. & Mal. 3, Aus. 1). Alks. *E. suaveolens* (Guillemin & Perrottet) Brenan (*E. guineense*, trop. Afr.) – 3–5 axillary buds, timber (missanda), bark (sassy) source of poison for fishes & arrows & form. much used ordeal poison for 'trying' criminals; other spp. used similarly
Erythrophysa E. Meyer ex Arn. Sapindaceae. 7–9 Ethiopia (1), NW Cape (1), Transvaal (1), W Madag. (4–6). R: MMNHP 19(1969)17. Petiole often winged
Erythrophysopsis Verdc. = praec.
Erythropsis Lindley ex Schott & Endl. = *Firmiana*
Erythrorchis Blume (~ *Galeola*). Orchidaceae (V 4). 3 Indomal. to Aus. (1)
Erythrorhipsalis A. Berger = *Rhipsalis*
Erythroselinum Chiov. Umbelliferae (III 10). 1 NE trop. Afr.
Erythrospermum Lam. Flacourtiaceae (2). 4 Mauritius to Fiji
Erythroxylaceae Kunth. Dicots – Rosidae – Linales. 4/240 trop. esp. Am. Glabrous trees & shrubs often with alks incl. cocaine. Lvs spirally arr. (opp. in *Aneulophus*), simple, entire; stipules intrapetiolar. Fls small, reg., usu. bisexual, 5-merous, often heterostylous, solit. or in axillary fascicles; K a tube with imbricate or valvate lobes, C imbricate & usu. with adaxial ± basal appendages, disk 0, A 10 usu. forming a tube, anthers with longit. slits, $\underline{G}$ ((2 or) 3) with as many locs & styles (± connate), ovule 1(2) in 1 fert. loc., axile, pendulous, anatroprous to hemitropous, bitegmic. Fr. a 1-seeded drupe; seed with straight embryo in copious (rarely 0) starchy endosperm. x = 12
Genera: *Aneulophus, Erythroxylum, Nectaropetalum, Pinacopodium*
Close to Linaceae
Erythroxylum a source of narcotics etc.
Erythroxylum P. Browne. Erythroxylaceae. 230 trop. esp. Am. (180) & Madag. (Aus. 3). Branches often covered with distichous scales (vestigial lvs); lvs often with longit. folds & pale broad band along centre; alks with medic. properties. *E. coca* Lam. (coca, E Andes) – cv. ('var. *ipadu*') perhaps brought to Amaz. from Andean highlands propagated vegetatively by men (cassava by women) as hedgepl. & for lvs, dried & powdered & mixed with lime the daily masticatory of W S Am. Indians maintaining blood glucose levels despite poor diets; source of cocaine ('coke'), the oldest anaesthetic, a debilitating addictive narcotic causing euphoria, indifference to pain & tiredness, increased alertness & enhancing sexual desire ('girl'), sniffed in lines from flat surface through straw, also dissolved & injected, effect lasting c. 15–40 mins; crack = freebase cocaine in raisin-sized bits ('rocks', freebasing = cocaine hydrochloride dissolved in water & heated with another agent to 'free' cocaine) with instant euphoria wearing off after c. 15 mins, form. with cola in non-alcoholic drinks (1885, no longer used) & 'Coca des Incas' tonic 'wine' and 'Vin Mariani' sanctioned by the Pope; *E. novogranatense* (Morris) Hieron. (Amaz.) also cult. for alks. Other spp. with useful timber
Escallonia Mutis ex L.f. Grossulariaceae (Escalloniaceae). 39 S Am. esp. Andes. Evergreen shrubs & trees widely cult. & often hybridized. R: KNAWC 58,2(1968). *E. rubra* (Ruíz & Pavón) Pers. (incl. *E. macrantha*, Chile) – variable sp. used as hedgepl. in SW England & Ireland
Escalloniaceae R. Br. ex Dumort. See Grossulariaceae
Eschscholzia Cham. Papaveraceae (II). c. 12 W N Am. Concave receptacle; (K) falling as a cap; in dull weather each petal rolls longit. encl. some stamens; valves of fr. curl spirally & fr. explodes; alks. Cult. as orn. ann. esp. *E. californica* Cham. (California poppy, Calif. (state flower)) – chlorophyllous seed-coat with stomata, cvs with white or pink fls as well as yellow wild form. colourless latex mildly narcotic & used by Indians against tooth-

ache, forming ± pure stands in NZ; *E. mexicana* E. Greene used as copper indicator in Arizona

Eschweilera C. Martius ex DC. Lecythidaceae. 90 trop. Am. incl. Mex. (1). Ligule (see fam.) pressed down on A & bearing staminodes & nectar available only to big bees which can force way in, i.e. *Xylocopa* & female euglossines; seed sessile. *E. itayensis* Knuth – bark ash 'salt' used to coat pellets of hallucinogenic *Virola* paste in NW Amaz.; *E. odorata* Miers – seeds coll., ed.; *E. ovalifolia* (DC) Niedenzu – obligate fish-disp. in Amaz.

Esclerona Raf. = *Xylia*

Escobaria Britton & Rose (~ *Coryphantha*). Cactaceae (III 9). 15 SW N Am. to S Canada & Cuba. R: BCSJ 4(1986)36

Escobedia Ruíz & Pavón. Scrophulariaceae. c. 6 trop. Am. Roots used for dyeing

Escontria Rose. Cactaceae (III 8). 1 S Mex.: *E. chiotilla* (Schumann) Rose – fr. ed., sold in markets

Esenbeckia Kunth. Rutaceae. c. 20 trop. Am. Bark of some spp. used like angostura, others medic. (gasparillo, WI)

Esfandiaria Charif & Aellen = *Anabasis*

Eskemukerjea Malick & Sengupta = *Fagopyrum*

Esmeralda Reichb.f. (~ *Arachnis*). Orchidaceae (V 16). 2 Himal. to SE As. Cult. orn.

Espadaea A. Rich. Solanaceae (Goetzeaceae). 1 Cuba: *E. amoena* A. Rich.

esparto grass *Stipa tenacissima*

Espejoa DC (~ *Jaumea*). Compositae (Hele.-Cha.). 1 Mex., C Am.

Espeletia Mutis ex Bonpl. Compositae (Helia.-Verb.). 88 (recently split into 7 genera largely on habit etc.: *Carramboa*, *Coespeletia*, *Espeletia* s.s., *Espeletiopsis*, *Libanothamnus*, *Ruilopezia*, *Tamania*). Char. pachycaul pls of paramo (cf. *Dendrosenecio*) with dense rosettes of spirally arr. lvs (cf. tribe!), often densely pubescent; some wind-poll., *E. schultzii* Wedd. (Venez.) visited by hummingbirds

Espeletiopsis Cuatrec. See *Espeletia*

espercet *Hedysarum coronarium*

Espostoa Britton & Rose. Cactaceae (III 4). c. 10 Peru & Brazil. Cult. orn. like *Cephalocereus* but fls scaly & hairy

Espostoopsis Buxb. Cactaceae (III 4). 1 N Bahia

essia *Petersianthus macrocarpus*

Esterhazya Mikan. Scrophulariaceae. 5 Bolivia, Brazil. R: Brittonia 37(1985)195

Esterhuysenia L. Bolus (~ *Lampranthus*). Aizoaceae (V). 1 W S Afr.

estragon *Artemisia dracunculus*

Esula (Pers.) Haw. = *Euphorbia*

Etaballia Benth. Leguminosae (III 4). 1 trop. S Am.

eteng *Pycnanthus angolensis*

Etericius Desv. Rubiaceae. 1 Guyana = ?

Ethulia L.f. Compositae (Vern.-Vern.). 19 trop. Afr. (15 endemics), Indomal. R: KB 43(1988)165. *E. conyzoides* L.f., medic. in Afr., a ricefield weed

Etlingera Giseke (~ *Amomum*). Zingiberaceae. c. 60 Indomal. *E. elatior* (Jack) R.M. Sm. (torch ginger, Mal.) – fls used in curries, cult. orn.

etrog *Citrus medica* 'Etrog'

Euadenia Oliver. Capparidaceae. 3 trop. Afr.

Euanthe Schltr. = *Vanda*

Euaraliopsis Hutch. = *Brassaiopsis*

Eubotryoides (Nakai) H. Hara = *Leucothoe*

Eubotrys Nutt. = *Leucothoe*

Eubrachion Hook.f. Eremolepidaceae. 2 WI, S Am. R: SBM 18(1988)43

eucalypt *Eucalyptus* spp.; also *Angophora* & *Corymbia* ssp.

Eucalyptopsis C. White. Myrtaceae (Lept.). 3 E Mal.

Eucalyptus L'Hérit. Myrtaceae (Lept.). c. 600+ E Mal. (few), Aus. See also *Corymbia* (bloodwoods). Evergreen trees most with distinct juvenile phase with diffs in shape, position & colour of lvs, the waxy surface of juvenile lvs preventing adult tortoise beetles getting a foothold & grazing; fls (some beetle-poll.) with an operculum thought to be derived from K & C through intercalary growth, leaving vestigial lobes around rim, dropped at anthesis, A ∞, often brightly coloured; seeds usu. ∞, small. Most char. genus of Aus. landscape, varying from dwarf shrubs with lignotubers to 250 yrs old & coppice shoots (mallees) to some of the tallest trees known (*E. regnans* F. Muell. at least 97 m nr. Melbourne & one felled Gippsland, Victoria (c. 1872) held to be 132.5 m tall & the biggest tree of all time, though even 152.4 m claimed). Over 200 spp. have been introduced elsewhere & dominate the scenery of parts of California, E Afr., Sri Lanka, Portugal etc.: they are fast-growing (*E. gunnii* Hook.f. is the fastest tree hardy in GB & other spp. in plantation in Kenya are more productive than even estuarine grasses) & the most imp. dicot. plantation trees worldwide, though aggressive & invasive in e.g. Cape Province, S Afr. Besides timber, they are sources of oils (medicinal) & tannins & are grown as orn. & for cut foliage. There are 5 main groups recognizable by their bark: gums (smooth & decid.), boxes (rough but fibrous), peppermints (finely fibrous), stringybarks (long fibrous) & ironbarks (hard, rough-fissured & dark). L.D. Pryor (1976) *Biology of eucalypts*; S. Kelly (1983) *Eucalypts*, ed. 2. *E. acmenoides* Schauer (white mahogany, yellow stringybark, E Aus.) – timber; *E. amygdalina* Labill. (black peppermint, Tasmania) – a source of Aus. kino; *E. astringens* Maiden (W Aus.) – bark (mallet b., 40–50% tannin) form. much used for tanning shoe-leather; *E. botryoides* Sm. (bangalay, bastard mahogany, swamp m., SE Aus.) – timber imp. in ship-building; *E. brassiana* S.T. Blake (Cape York red gum, NE Aus.) – fuel crop; *E. bridgesiana* R. Baker (apple box, E Aus.); *E. camaldulensis* Dehnh. (*E. rostrata*, (river or Murray) red gum, Aus.) – most widespread sp., now more planted than *E. globulus* elsewhere, prob. from Eur. introd. from SE Aus. (1803), to at least 950 yrs old, principal source of Aus. kino; *E. capitellata* Sm. (brown stringybark, NSW) – general construction; *E. cinerea* F. Muell. ex Benth. (Argyle apple, SE Aus.); *E. cornuta* Labill. (yate, Aus.) – tough timber; *E. crebra* F. Muell. (red ironbark, NE Aus.); *E. deglupta* Blume (kamarere, E Mal.) – general construction, much planted in Mal., fuel crop; *E. delegatensis* R. Baker (alpine ash, Tasmanian oak, SE Aus.) – imp. timber tree; *E. diversicolor* F. Muell. (karri, W Aus.) – v. hard timber ± resistant to termites & teredo; *E. dives* Schauer (SE Aus.) – imp. source of E. oil; *E. dumosa* A. Cunn. ex Oxley (S & SE Aus.) – source of E. oil, also a manna called lerp or larap; *E. fastigiata* Deane & Maiden (brown barrel, SE Aus.) – timber; *E. forrestiana* Diels (W Aus.) – 'fuchsia mallee'; *E. fraxinoides* Deane & Maiden ((Aus.) white ash, SE Aus.); *E. globoidea* Blakely (white stringybark, NSW) – timber for heavy construction; *E. globulus* Labill. (blue gum, fever tree, Victoria & Tasmania) – form. most widely cult. sp. throughout world, natur. in Calif. etc., major source of E. oil in Spain, imp. firewood crop, good bee-forage, pulp etc. cut juvenile twigs in floristry, buds & fr. solit. (rare in *E.*); *E. gomphocephala* DC (tuart, SW Aus.) – one of Aus. heaviest & strongest timbers form. used for railway sleepers etc.; *E. grandis* W. Hill ex Maiden (rose gum, NE Aus.) – construction timber; *E. gunnii* Hook.f. (cider gum, c. tree, Tasmania) – potable sweet sap, hardiest in GB, growing up to 1.5 m per annum; *E. luehmanniana* F. Muell. (yellow ash, NSW); *E. macrorhyncha* F. Muell., (red stringybark, S & E Aus.) – lvs a source of rutin; *E. marginata* Sm. (jarrah, W Aus.) – principal timber of W Aus. much exported; *E. microcorys* F. Muell. (tallow wood, E Aus.) – valuable timber for telegraph poles etc.; *E. microtheca* F. Muell. (coolibah (as in 'Waltzing Matilda'), coolibar, desert box, N Aus.) – one of strongest & hardest of all timbers, good fuel; *E. moluccana* Roxb. (*E. hemiphloia*, yellow box, E Aus.) – timber for railway sleepers etc.; *E. muelleriana* Howitt (yellow stringybark, S & E Aus.) – timber for general outdoor use; *E. oreades* R. Baker (Blue Mt. ash, E Aus.) – timber for barrels, cabinet work etc.; *E. paniculata* Sm. (grey ironbark, E Aus.) – timber v. hard; *E. pauciflora* Sieber ex A. Sprengel (white Sally, E & SE Aus.), subsp. *niphophila* (Maiden & Blakely) L. Johnson & Blaxell (snow gum); *E. pilularis* Sm. (blackbutt, E Aus.) – timber for poles, ship-decking etc.; *E. populnea* F. Muell. (bimble box, NE Aus.); *E. redunca* Schauer (wandoo, SW Aus.) – tough timber, tanbark; *E. regnans* F. Muell. (Aus. mountain ash or oak, SE Aus.) – a major v. tall timber-tree in even-aged stands, when fruiting producing c. 12×10^5 seeds/ha annually, seeds germinate after fire when predators (esp. ants) killed & mother trees incinerated giving even-aged stand; *E. resinifera* Sm. (red mahogany, E Aus.) – timber & kino source; *E. robusta* Sm. (swamp or white mahogany, E Aus.) – timber for shingles, ship-building etc.; *E. saligna* Sm. (Sydney blue gum, saligna g., NSW) – general construction; *E. salmonophloia* F. Muell. (salmon gum, SW Aus.) – durable timber; *E. sideroxylon* Cunn. ex Woolls (mugga, SE Aus.) – form. exploited for tannin; *E. tereticormis* Sm. (forest river gum, E Aus.) – fuel crop; *E. viminalis* Labill. (white gum, S & SE Aus.) – timber for shingles etc., manna, widely cult.

Eucarpha (R. Br.) Spach (~ *Knightia*). Proteaceae. 2 New Caled.

Eucarya T. Mitchell ex Sprague & Summerh. = *Santalum*

Eucephalus Nutt. = *Aster*

Euceraea C. Martius. Flacourtiaceae (9). 1 N Amaz.: *E. nitida* C. Martius

Euchaetis Bartling & Wendl. Rutaceae. 23 S & SW Cape. R: JSAB 47(1981)157

Eucharis Planchon & Linden (~ *Urceolina*). Amaryllidaceae (Liliaceae s.l.). (Incl. *Caliphruria*) 17 with 2 nat. hybrids trop. Am. R: AMBG 76(1989)170. A on margin of corona; alks. Cult.

orn. esp. *E.* × *grandiflora* Planchon & Linden (often sold as *E. amazonica*, sterile hybrid, poss. involving *E. sanderi* Baker (W Colombia) & *E. moorei* (Baker) Meerow (Andes of Ecuador & Peru), Amazon or star lily, Andes of Colombia & Peru) – large scented white fls; *E. amazonica* Linden ex Planchon (NE Peru) – emetic tea prepared from pl. incl. bulbs

Euchilopsis F. Muell. Leguminosae (III 24). 1 SW Aus.: *E. linearis* (Benth.) F. Muell.

Euchiton Cass. (~ *Gnaphalium*). Compositae (Gnap.-Gnap.). c. 20 Aus., NZ, New Guinea, E As. R: OB 104(1991)166

Euchlaena Schrader = *Zea*

Euchlora Ecklon & Zeyher = *Lotononis*

Euchorium Ekman & Radlk. Sapindaceae. 1 W Cuba

Euchresta Bennett. Leguminosae (III 29). 4 E As., Java (1). R: APS 30(1992)43

Euclasta Franchet. Gramineae (40c). 2 Afr., India, trop. Am.

Euclea L. Ebenaceae. 12 trop. Afr. to Arabia & Comoro Is. Lvs spirally arr., opp. or whorled; fr. ed. *E. latidens* Stapf (trop. Afr.) – seeds for beads in Karamoja (Uganda); *E. pseudebenus* E. Meyer (black ebony, S Afr.) – timber for furniture etc.; *E. schimperi* (A. DC) Dandy (Arabia) – timber, ed. fr., insecticidal smoke & dye from bark

Euclidium R. Br. Cruciferae. 2 E Eur. (1: *E. syriacum* (L.) R. Br.) to India

Euclinia Salisb. (~ *Randia*). Rubiaceae (II 1). 3 trop. Afr., Madag.

Eucnide Zucc. Loasaceae. 14 SW N Am. R: JAA 48(1967)56. Like *Mentzelia* but C with basal tube & G with 4 or 5 placentas, *E. bartonioides* Zucc. (W Texas to Mex.) – cult. orn. biennial with yellow fls

Eucodonia Hanst. (~ *Achimenes*). Gesneriaceae. 2 C Am. R: Selbyana 1(1976)389

Eucomis L'Hérit. Hyacinthaceae (Liliaceae s.l.). 10 trop. (1) & S Afr. R: Plantsman 12(1990)129. Fls in term. bracteate raceme topped with cluster of sterile leafy bracts – 'pineapple lilies' incl. some cult. orn. esp. *E. autumnalis* (Miller) Chitt. (*E. undulata*) – lvs undulate & *E. comosa* (Houtt.) Wehrh. (*E. punctata*) – lvs with purple spots (both S Afr.)

Eucommia Oliver. Eucommiaceae. 1 (?) China: *E. ulmoides* Oliver poss. no longer known in wild state, much cult. in China; timber for furniture & fuel; aucubin-containing bark (duzhong) a Chinese tonic & for arthritis, solid latex yields a gutta percha which can be seen as strands if leaf broken across, used for lining oil pipelines & insulating electric cables, also tooth-fillings

Eucommiaceae Engl. Dicots – Hamamelidae – Eucommiales. 1/1 China. Dioec. wind-poll. tree with articulated laticifers in phloem & cortex & scattered latex-cells elsewhere. Lvs spirally air., simple, toothed, decid.; stipules 0. Fls solit. & shortly pedicellate in axils of bracts racemosely arr. on proximal section of a distally leafy shoot, reg.; P 0, males with A 5–12 with short filaments & apically prolonged connective, anthers with longit. slits, females with G (2), 1-loc., flattened with short style & 2 unequal stigmas & 2 collateral, pendulous anatropous unitegmic ovules, 1 aborting. Fr. a samara; seed with large embryo embedded in copious endosperm. 2n = 34

Only species: *Eucommia ulmoides*

Variously associated with Urticales, Hamamelidales (pollen similar) or even Magnoliales, morphologically nearest U. & supposed to be allied to Ulmaceae but embryological features differ as do ovules 2 (? primitive), stipules 0 & unitegmic ovules (? advanced). Cronquist considers an origin from Hamamelidales independent of that of Urticales

Eucorymbia Stapf. Apocynaceae. 1 Borneo

Eucosia Blume. Orchidaceae (III 3). 2 Java, New Guinea

Eucrosia Ker-Gawler. Amaryllidaceae (Liliaceae s.l.). 7 Andes. R: SB 12(1987)460. Lvs stalked; ? butterfly-poll.; some cult. orn.

Eucryphia Cav. Eucryphiaceae. 6 SE Aus. (4, endemic), Chile. R: Plantsman 5(1983)169. Evergreen trees or shrubs cult. for white fls esp. (allegedly lime-tolerant) *E.* × *nymansensis* Bausch (*E. cordifolia* Cav. (ulmo, Chile: commercial timber & tannin source, allegedly lime-tolerant) × *E. glutinosa* (Poeppig & Endl.) Baillon (Chile)); *E. lucida* (Labill.) Baillon (leatherwood, Tasmania) – good timber, cult. orn., honey sold UK; *E. moorei* F. Muell. (plumwood, SE Aus.) – useful timber

Eucryphiaceae Endl. Dicots – Rosidae – Rosales. 1/5 SE Aus., Chile. Evergreen trees & shrubs; sieve-tubes with P-type plastids with single polygonal protein-crystalloid. Lvs pinnate or simple, opp.; stipules interpetiolar, caducous with large colleters; term. buds sticky. Fls solit., axillary, reg., bisexual; K 4(5), imbricate, leathery, apically coherent & falling as a cap, C 4(5), imbricate, white, A ∞ multiseriate, originating centripetally from a limited number of trunk-bundles on an enlarged often domed receptacle, anthers versatile with longit. slits, G (4–14(–18)), pluriloc. with ± basally united persistent styles & ovules on axile placentas, ∞ anatropous, bitegmic & biseriate in each loc. Fr. a leathery or woody septicidal capsule, carpels separating from axis in fr. & dehiscing ventrally; seeds few, pendulous, flattened & winged with ± copious endosperm & embryo with large foliaceous cotyledons. n = 15, 16

Only genus: *Eucryphia*

App. closest to (& best included in) Cunoniaceae, another austral group

Eucrypta Nutt. Hydrophyllaceae. 2 SW N Am.

Eudema Humb. & Bonpl. Cruciferae. 6 Andes. R: JAA 71(1990)100

Eudianthe (Reichb.) Reichb. = *Silene*

Eugeissona Griffith. Palmae (II 1b). 6 Malay Pen. (2), Borneo (4). Branching due to unequal development of products of dichotomizing apex; dioec.; C woody; sago, pollen ed. *E. minor* Becc. (Borneo) – stilt-roots form. favoured for umbrella-handles & walking-sticks ('rajah canes'); *E. tristis* Griffith (bertam, Malay Pen.) – lvs used for thatch etc., the plant forming dense groves app. antagonistic to regeneration of dipterocarps; *E. utilis* Becc. (Borneo) – starch from trunk the basic sago source for Penans of Sarawak

Eugenia L. Myrtaceae (Myrt.). c. 550+ trop. esp. Am. (Aus. 1, E New Guinea 1 with woody caps.). Differs from *Syzygium* (to which many spp. form. here now referred) in a number of characters, *E.* usu. having cotyledons united, seed-coat smooth & free of pericarp & infls being racemes of pedicellate fls. Some fish-disp. in Amazon. Many with ed. fr. (Brazilian cherries) & cult. *E. dombeyi* Skeels (*E. brasiliensis* Lam. non Aublet, grumichama, S Brazil) – fr. eaten fresh, in pies, candied etc.; *E. luschnathiana* (O. Berg) B.D. Jackson (pitomba, Brazil) – aromatic fr. for jelly; *E. uniflora* L. (Surinam cherry, pitanga (name also applied to other spp.); trop. Am.) – widely cult. trop. as hedgepl., for ed. fr. used in jellies etc., crushed lvs insect-repellent; *E. uvalha* Cambess. (uvalha, Brazil) – aromatic fr. ed. used in drinks

Euglypha Chodat & Hassler. Aristolochiaceae (II 2). 1 N Arg., Paraguay: *E. rojasiana* Chodat & Hassler

Euhesperida Brullo & Furnari. Labiatae (VIII 2). 1 Libya & Cyrenaica: *E. linearifolia* Brullo & Furnari. R: Webbia 34(1979)433

Euklisia Rydb. ex Small (~ *Streptanthus*.) Cruciferae. 10 W US

eulalia *Miscanthus sinensis*

Eulalia Kunth. Gramineae (40a). 30 OW trop. (Aus. 4). *E. aurea* (Bory) Kunth (*E. fulva*, sugar grass, SE As. to Aus.) – palatable to stock

Eulaliopsis Honda. Gramineae (40a). 2 Afghanistan, India, China, Taiwan, Philipp. *E. binata* (Retz.) C. Hubb. (baib, Bharbur or sawai grass, India) – imp. in paper-making & also locally for cordage

Euleria Urban. Anacardiaceae (II). 1 Cuba

Eulophia R. Br. Orchidaceae (V 9). c. 200 trop. Usu. terr., some mycotrophic. Tubers of some spp. provide Indian salep

Eulophidium Pfitzer = *Oeceoclades*

Eulophiella Rolfe. Orchidaceae (V 9). 2 Madag. R:OD36(1972)120. Cult. orn.

Eulychnia Philippi. Cactaceae (III 5). 6 Chile, Peru. Cult. orn.

Eumorphia DC. Compositae (Anth.-Urs.). 6 S Afr. R: NJB 5(1986)538

Eunomia DC = *Aethionema*

Euodia Forster & Forster f. (*Evodia*). Rutaceae. (Incl. *Zieridium*) 6 New Guinea & NE Aus. to Pacific. *E. hortensis* Forster & Forster f. (W Pacific) – abortifacient. See also *Melicope* & *Tetradium*

Euonymopsis Perrier (*Evonymopsis*). Celastraceae. 4 Madagascar

Euonymus L. (*Evonymus*). Celastraceae. 177 N temp. (Eur. 4) esp. As., Aus. R: KB 1951(1951)210. Decid. or evergreen trees & shrubs, rarely creeping & rooting, fr. a 3–5-valved brightly coloured capsule dehiscing to expose bird-disp. scarlet to orange arillate seeds; seeds of many spp. with cardiotoxic glycosides. Many cult. orn. esp. ones with corky outgrowths on stems & bright autumn foliage or evergreen lvs. *E. alatus* (Thunb.) Siebold (E As.) – branches with corky wings; *E. atropurpureus* Jacq. (Indian arrow wood, C & E N Am.); *E. europaeus* L. (spindle-tree, Eur., W As.) – indicator of hedges 100+ yrs old, host for aphid *Myzus persicae*, charcoal form. used in gunpowder & yellow dye from seeds form. used to colour butter, timber form. used for skewers, spindles, toothpicks & violin bows (pegwood), poss. use as insecticide; *E. fortunei* (Turcz.) Hand.-Mazz. (C & W China) – evergreen with many cvs incl. varieg. ones; *E. hamiltonianus* Wallich (Himal. to Japan) – cult. orn. incl. var. *maackii* (Rupr.) Komarov (N China to Korea) used as box subs. ('paich-ha'); *E. japonicus* Thunb. (Japan) – evergreen widely planted, many cvs incl.

varieg. ones offered as housepls. *E. globularis* Ding Hou (Aus.) links with *Brassiantha* & *Hedraianthera*

Eupatoriadelphus R. King & H. Robinson = *Eupatorium*

Eupatoriastrum Greenman. Compositae (Eup.-Crit.). 4 Mex., C Am.

Eupatorina R. King & H. Robinson (~ *Eupatorium*). Compositae (Eup.-Crit.). 1 Hispaniola. R: MSBMBG 22(1988)308

Eupatoriopsis Hieron. Compositae (Eup.-Prax.). 1 Brazil. R: MSBMBG 22(1987)388

Eupatorium L. Compositae (Eup.). 45 E US, Euras. (Eur. 1). R: MSBMBG 22(1987)64. Form. incl. some 1200 spp. also from trop., now segregated into many genera e.g. *Ayapana*. Some cult. orn. incl. *E. cannabinum* L. (hemp agrimony, Eur. & Med. to C As.) – visited by Lepidoptera, form. medic. ('ague weed') as was *E. purpureum* L. (Joe-pye weed, E N Am.) – 'gravel root' for urinary problems

Euphlebium Brieger = *Dendrobium*

Euphorbia L. Euphorbiaceae. c. 2000 cosmop. (Eur. 105; Turkey 91), esp. warm. Poss. largest angiosperm genus. Monoec. or dioec. herbs (c. 1300), some geophytic, succ. (esp. Afr., Madag.), shrubs & trees with milky latex, sometimes carcinogenic; stems often spiny, the plants cactoid (but latex & paired spines distinguish them at once), lvs spiral, opp. or whorled; fls in cyathia without P, the males reduced s.t. to A 1 (jointed, rep. pedicel, below), surrounding 1 female comprising G (3) mature before males, the whole surrounded by a 'whorl' of green P-like bracts & between them 4 horn-shaped glands prob. rep. stipules of bracts (in allied *Anthostema* both male & females have P); explosive fr., seeds disp. ants, wind, birds etc. The succ. spp. occupy the same position in Afr. as the Cactaceae in Am. & a series can be drawn up showing increasing xeromorphy assoc. with aridity: lvs normal & photosynthesizing; shoot not water-storing stems which do not photosynthesize (e.g. *E. bupleurifolia* Jacq., S Afr.) but lvs falling in dry season or stem fleshy & green, the lvs present only in wet season (e.g. *E. neriifolia* L., India); where the lvs are abortive & soon fall, the principal photosynthesis being due to the stems, these may be thin & cylindrical but branched (*E. tirucalli*) or flattened (*E. xylophylloides* Brongn. ex Lemaire, Madag., cf. *Schlumbergera*) or a stout central stem with a number of thinner apical branches (e.g. *E. caput-medusae* L., S Afr.), the branches covered with cushion-like papillae; in other spp. these scarcely separate in development giving a ridged stem resembling the Cereeae of Cactaceae (e.g. *E. polygona* Haw., S Afr.) & the stem may be almost spherical like *Echinocactus* (e.g. *E. globosa* (Haw.) Sims, S Afr.). The infls in some leafy spp. are surrounded by brightly coloured bracts, app. attractive to pollinating birds as in *E. pulcherrima* ('poinsettia'). The latex of all spp. is toxic & can bring out skin allergies; it has been used to stupefy fish. Many cult. as succ. curiosities or as hardy border pls etc., while *E. tirucalli* L. (trop. & S Afr.) is a promising source of hydrocarbons for fuel (in latex, almeidena, which may be tapped) & is used for charcoal (incl. in fireworks in India) & much planted around burial-grounds (tirucalli). *E. amygdaloides* L. (wood spurge, Eur., SW As.) – seeds can wait more than 125 yrs for forest clearing before germ. in England, ssp. *robbiae* (Turrill) Stace (Mrs Robb's bonnet) – cult. ground-cover introd. from E Med. by Mrs Robb in her hatbox; *E. antisyphilitica* Zucc. (candelilla, SW N Am.) – waxy exudate refined for use in polishes & creams for leather, furniture, babies etc. (form. for gramophone records), mixed with rubber for electric insulation materials, also in waterproofing fabrics, paper etc., when mixed with paraffin used as candles; *E. balsamifera* Aiton (NW Afr., Arabia, Canary Is.) – young shoots ed. boiled, latex a bird-lime, that of ssp. *adoensis* (Deflers) Bally a valuable glue in Arabia; *E. characias* L. (S Eur.) – commonly cult. hardy shrub, ssp.*veneta* (Willd.) Litard (*E. wulfenii*) larger eastern subsp.; *E. cyparissias* L. (cypress spurge, Eur., natur. N Am.) – ground-cover becoming weedy, form. used as cosmetic in Ukraine; *E. esula* L. (Eur.) – 'hay' can yield 4 times more energy per annum than wheat straw; *E. fischeriana* Steudel (*E. pallasii*) – antitumour drug in Chinese med. for over 2000 yrs; *E. hirta* L. (C Am., now pantrop. weed) – latex effective bee & scorpion sting treatment (Uganda); *E. intisy* Drake (Madag.) – form. imp. source of (intisy) rubber now almost exterminated; *E. ipecacuanhae* L. (ipecacuanha spurge, Carolina i., E US) – form. used as emetic; *E. lancifolia* Schldl. ('ixbut', C Am.) – natural increaser of milk-flow in women; *E. lathyris* L. (caper spurge, mole plant, Eur., natur. N Am.) – 8% dry wt. terpenes can be 'cracked' to give fuel oil, fls visited by wasps in UK, carunculate seeds toxic but fr. used as caper subs. (poor), plant incorrectly believed to be mole-repellent; *E. marginata* Pursh (snow-on-the-mountain, N Am.) – glands with white petal-like appendages, garden-pl. but latex corrosive, seed oil (also that of *E. heterophylla* L. (warm Am.)) app. superior to linseed; *E. milii* Des Moul. (*E. bojeri*, crown of thorns, Madag.) – spiny shrub with bright red or yellow bracts often grown as housepl. esp. var. *splendens* (Bojer ex

Hook.) Ursch & Leandri with brilliant red bracts; *E. peplus* L. (petty spurge, Eur.) – bad weed of cult.; *E. pulcherrima* Willd. ex Klotzsch (poinsettia, Mex.) – basis of multi-million pound Christmas industry, rooted cuttings being flowered under short days & growth retardants, e.g. B9, to produce the bright red bracts on small pl. for housepl. sale, some cvs with white bracts, used for hedging & vegetable in trop., latex used as depilatory in Mex., red dye from bracts, extrafl. nectaries. Euphorbia J.

Euphorbiaceae Juss. Dicots – Dilleniidae – Euphorbiales. 313/8100 cosmop. (few in Amazon basin) exc. Arctic. Trees, shrubs, lianes or herbs (even free-floating as in *Phyllanthus fluitans*), s.t. succ., monoec. or dioec. with stems & lvs often with specialized cells or tubes of milky or coloured latex; internal phloem s.t. present. Lvs usu. simple (3-foliolate in e.g. *Bischofia*) & spirally arr., s.t. opp. or whorled, s.t. compound, with pinnate or palmate venation, s.t. v. reduced; stipules large & protecting term. but s.t. reduced to merely glands or 0. Fls usu. reg. in basically cymose infls, s.t. v. reduced & in bisexual pseudanthia; P usu. reg. inconspicuous, sometimes basally connate or 0, A 5– ∞, s.t. fewer or 1, s.t. basally connate, anthers with longit. slits or rarely apical pores, nectary-disk of discrete or united segments s.t. present without or within A, $\underline{G}$ ((2)3(4–∞)), pluriloc. with distinct styles or style with bifid or more-branched branches, rarely G pseudomonomerous, each loc. with 1 or 2 pendulous, apical-axile, anatropous or hemi-tropous bitegmic ovules, the nucellus often protruding through micropyle & contact with obturator (placental) roofing the micropyle & forming a passageway for pollen-tubes. Fr. often a capsular schizocarp, the mericarps separating from persistent columella & opening adaxially to release seeds, or a drupe, samara or berry; seeds often with a caruncle around micropyle & with straight or curved embryo embedded in copious oily, rarely 0, endosperm, often with poisonous proteins. x = 6–14+

Classification & principal genera: 'The classification of the family needs drastic overhauling' (Airy Shaw) & the limits are unclear, a number of peripheral genera having been segregated as distinct fams by some authors (e.g. *Androstachys*, *Antidesma* (Stilaginaceae), *Bischofia*) not followed here, though the allied Buxaceae & Pandaceae are kept up here. A recent arrangement within the fam. (Webster) is:

I. **Phyllanthoideae** (2 ovules per loc., no milky latex; lvs spirally arr., simple, stipulate, fls ± C, seeds without caruncles, endosperm copious to 0, x usu. = 12 or 13): 13 tribes incl. Amanoeae (*Actephila*), Antidesmeae (*Antidesma*), Drypeteae (*Drypetes*), Phyllantheae (*Phyllanthus*, *Securinega*), Bischofieae (*Bischofia*)

II. **Oldfieldioideae** (as above, but lvs sometimes opp. or whorled, simple or palmate, stipules small or 0, C 0, seeds often with caruncles, endosperm usu. copious): 4 tribes incl. Petalostigmateae (*Androstachys*)

III. **Acalyphoideae** (1 ovule per loc., latex absent; lvs often with petiolar or laminar glands, seeds ± caruncle, endosperm usu. copious, x usu. = 8–11): 19 tribes incl. Clutieae (*Clutia*), Chrozophoreae (*Chrozophora*), Alchorneeae (*Alchornea*), Acalypheae (*Acalypha, Macaranga, Mallotus, Mercurialis, Ricinus, Trewia*), Dalechampieae (*Dalechampia*), Pereae (*Pera*)

IV. **Crotonoideae** (as above but latex reddish or yellowish to milky, rarely 0, innocuous, laticifers articulate or not, lvs often palmately veined, lobed or compound, indumentum simple or often stellate, usu. C present): 11 tribes incl. Micrandreae (*Hevea, Micrandra*), Manihoteae (*Manihot, Cnidoscolus*), Joannesieae (*Jatropha, Ricinodendron*), Codiaeae (*Codiaeum*), Aleuritideae (*Aleurites, Neoboutonia*), Crotoneae (*Croton*)

V. **Euphorbioideae** (as above but latex whitish, often caustic or poisonous, laticifers inarticulate (rarely 0), lvs simple (rarely lobed), pinnately veined or 3-veined, indumentum of simple hairs or 0, C 0): 5 tribes incl. Hippomaneae (*Excoecaria, Hippomane, Sapium*), Hureae (*Hura*), Euphorbieae (*Anthostema, Euphorbia, Monadenium, Pedilanthus*)

The different basic chromosome nos of the 1st 2 tribes have suggested that they represent a separate line independent of the last 3 and that the family is diphyletic. The seed structure of the first suggests affinities with Flacourtiaceae, Celastraceae & Violaceae, whereas the others show more affinity with Bombacaceae, Malvaceae, Sterculiaceae & Tiliaceae. *Euphorbia* (q.v.) represents the most florally advanced group in the family

Imp. crops incl. rubber (*Hevea, Manihot, Micrandra, Sapium*), cassava (*Manihot*) & other vegetables (*Cnidoscolus*), oils both industrial & medic. (*Aleurites, Baliospermum, Cleidocarpon, Croton, Joannesia, Plukenetia, Ricinus, Sapium, Schinzophyton, Tetracarpidium, Trewia*), fruit-trees (*Antidesma, Baccaurea, Phyllanthus*), timber (*Antidesma, Chaetocarpus, Hevea, Heywoodia, Ricinodendron, Uapaca*), dye-plants (*Chrozophora, Mallotus*), hydrocarbon sources (*Euphorbia*) & many cult. orn. (*Acalypha, Breynea, Codiaeum, Euphorbia, Monadenium, Pedilanthus, Ricinus*)

Euphoria Comm. ex Juss. = *Litchi*

Euphorianthus Radlk. (~ *Diploglottis*). Sapindaceae. 1 Indomal.: *E. euneurus* (Miq.) Leenh.

Euphrasia L. Scrophulariaceae. c. 170 with many microspp. (Eur. 46). Hemiparasites, some reg. autogamous. *E. officinalis* L. (s.s. = *E. rostkoviana* Hayne, eyebright, Eur.) – form. used in eye disease

Euphronia C. Martius & Zucc. Vochysiaceae. 3 N trop. S Am. R: AMBG 74(1987)89

Euphroniaceae Marc.-Bert. = Vochysiaceae

Euphrosyne DC. Compositae (Helia.-Amb.). 1 Mexico. Aquatic

Euplassa Salisb. Proteaceae. 25 trop. Am.

Euploca Nutt. = *Heliotropium*

Eupomatia R. Br. Eupomatiaceae. 2 E New Guinea, E Aus. R: Fl. NSW 1 (1990)124. Staminodes with food-bodies, musky fragrance & sticky exudate attractive to poll. weevils (*Elleschodes*), overarching & protecting them from predators. *E. laurina* R. Br. (Aus.) – fls last less than 1 day, poll. *Elleschodes hamiltonii* T. Blackburn only, fr. ed.

Eupomatiaceae Endl. Dicots – Magnoliidae – Magnoliales. 1/2 New Guinea, E Aus. Trees & rhiz. wiry shrubs; sieve-tube plastids of P type. Lvs spirally arr. (distichous), simple, entire; stipules 0. Fls 1–3 in axils or term. on longer shoots, reg., bisexual, receptacle urceolate; P 0, calyptra (? bract) attached to rim of receptacle falling to expose A 20–100 near rim of receptacle, outer ones with short, broad laminar base & well-defined anther & prolonged thickened connective, staminodes c. 40–80, fleshy & ± petaloid, all forming continous spiral & basal united to form synandrium, $\underline{G}$ 13–70 spirally arr. on receptacle, ± connate at margins, but unclosed, style 0, stigma feathery, each carpel with 2–11 anatropous bitegmic ovules. Fr. subglobose berry-like aggregate with carpels laterally coalescent so as to appear sunk in fleshy receptacle; seeds 1 or 2 per carpel, with v. small embryo embedded in oily ruminate endosperm. 2n = 20

Only genus: *Eupomatia*

Euptelea Siebold & Zucc. Eupteleaceae. 2 Assam to SW & C China (1), Japan (1). R: JAA 27(1946)175. Cult. orn.

Eupteleaceae Willhelm. Dicots – Hamamelidae – Trochodendrales. 1/2 E As. Small subglabrous, decid. trees. Lvs spirally arr., simple, toothed; stipules 0. Fls wind- to insect-poll., long-pedicellate, solit. in axils of 6–12 closely crowded bracts of vegetative shoots, bisexual or sometimes some male; P 0, A 6–19 in 1 whorl with short filaments & long red anthers & prolonged connectives, $\underline{G}$ 8–31 incompletely closed with decurrent stigma, style 0, each carpel with 1–3(4) ± marginal, anatropous, bitegmic ovules. Fr. a head of small samaras with papery pericarp; seeds with tiny embryo & poorly differentiated cotyledons embedded in oily & proteinaceous endosperm. 2n = 28

Only genus: *Euptelea* (? fossils in Oregon)

Carpel structure similar to Winteraceae (Magnoliales)

Eureiandra Hook.f. Cucurbitaceae. 10 trop. Afr. to Socotra

Euroschinus Hook.f. Anacardiaceae (IV). 6 Mal. (1), Aus. (1). New Caled. (4). *E. falcatus* Hook.f. (Aus.) – second-rate timber for cheap furniture etc.

Eurotia Adans. = *Axyris*

Eurya Thunb. Theaceae. c. 70 trop. & warm As. & W Pacific. R: BSBF 42(1895)151. Dioec. trees & shrubs, some cult. orn. *E. acuminata* DC (Indomal.) is an aluminium-accumulator; some locally used timbers

Euryalaceae J. Agardh = Nymphaeaceae

Euryale Salisb. Nymphaeaceae (Euryalaceae). 1 China to N India & Japan: *E. ferox* Salisb. – like *Victoria* but lvs with flat margins, smaller fls & all A fertile; cleistogamous fls over longer period & setting more seeds than chasmogenous ones. Fossils known from Eur. Cult. by Chinese for ed. seeds & rhizomes for 3000 yrs (for nut), now 'puffed' by roasting & sold in markets in India

Eurybiopsis DC (~ *Minuria*). Compositae (Ast.-Ast.). 1 trop. Aus.: *E. macrorhiza* DC. R: Brunonia 5(1982)10

Eurycarpus Botsch. Cruciferae. 1 W Tibet

Eurycentrum Schltr. Orchidaceae (III 3). 7 Papuasia

Eurychone Schltr. Orchidaceae (V 16). 2 trop. Afr. Cult. orn.

Eurycles Salisb. ex Schultes & Schultes f. = *Proiphys*

Eurycoma Jack. Simaroubaceae. 3 Indomal. A5, staminodes 5. *E. longifolia* Jack (Indomal.) – locally medic.

Eurycorymbus Hand.-Mazz. Sapindaceae. 1 S China, Taiwan

Eurydochus Maguire & Wurd. = *Gongylolepis*
Eurylobium Hochst. Stilbaceae (Verbenaceae s.l.). 2 SW Cape
Eurynotia R. Foster = *Ennealophus*
Euryops (Cass.) Cass. Compositae (Sen.-Sen.). 97 S Afr. to Arabia & Socotra. R: OB 20(1968)1. Some cult. orn. esp. *E. acraeus* M.D. Henderson (S Afr., often confused with *E. evansii* Schltr. not in cult.)
Eurypetalum Harms. Leguminosae (I 4). 3 Cameroun to Gabon. Some cabinet timbers
Eurysolen Prain. Labiatae (VII/VI). 1 SE As., Mal.
Eurystemon Alexander = *Heteranthera*
Eurystyles Wawra. Orchidaceae (III 1). 10 trop. Am. R: Brittonia 37(1985)160
Eurytaenia Torrey & A. Gray. Umbelliferae (III 8). 2 N Am.
Euscaphis Siebold & Zucc. Staphyleaceae. 1 temp. E As.: *E. japonica* (Thunb.) Kanitz – cult. orn. decid. tree
Eusideroxylon Teijsm. & Binnend. Lauraceae (I). 1 Sumatra to Borneo: *E. zwageri* Teijsm. & Binnend. (belian, billian, Borneo ironwood) – ± pure stands in Sumatra, heavy construction timber sinking in water, much exported. See also *Potoxylon*
Eusiphon Benoist. Acanthaceae. 3 Madagascar
Eustachys Desv. (~ *Chloris*). Gramineae (33b). 10 trop. Am. & Afr., S Afr., savannas
Eustegia R. Br. Asclepiadaceae (III 1). 5 Cape
Eustephia Cav. Amaryllidaceae (Liliaceae s.l.). 6 Peru, Arg. Alks; cult. orn.
Eusteralis Raf. = *Pogostemon*
Eustigma Gardner & Champ. Hamamelidaceae (I 3). 2 S China, SE As. C minute, fleshy with 2 basal nectaries abaxially, A subsessile, $\overline{G}$ (2) with large stigmas
Eustoma Salisb. Gentianaceae. 3 S N Am. to N S Am. R: SWNat. 2(1957)38. *E. grandiflorum* (Raf.) Shinn. (N Am). – showy annual or biennial with purple fls widely cult. potpl. & cut-fl. ('prairie gentian', 'lisianthus') since early 1980s, many cvs
Eustrephus R. Br. Philesiaceae (Smilacaceae s.l.). 1 polymorphic S & E New Guinea, E Aus., New Caled.: *E. latifolius* R. Br. (wombat berry) – cult. orn. R: Muelleria 6(1987)367
Eustylis Engelm. & A. Gray = *Alophia*
Eutaxia R. Br. Leguminosae (III 24). 8–9 Aus. (esp. W). Cult. orn. esp. *E. microphylla* (R. Br.) C.H. Wright & Dewar
Euterpe C. Martius. Palmae (V 4f.). 30 trop. Am. Monoec. palms, a principal source of palm-hearts esp. *E. edulis* C. Martius (assai, açai, Brazil) & *E. oleracea* C. Martius (Brazil to Venez. & Guyana); some fish-disp. Amazon, cult. orn.
Eutetras A. Gray. Compositae (Hele.-Per.). 2 Mex. R: SW Nat 11(1966)118
Euthamia (Nutt.) Cass. Compositae (Ast.-Sol.). 8 N Am. R: Rhodora 83(1981)551. *E. graminifolia* (L.) Nutt. (*Solidago g.*) – poss. rubber source
Euthemis Jack. Ochnaceae. 2 SE As. to Borneo. R: BJ 113(1991)180
Eutheta Standley = *Melasma*
Euthryptochloa T. Cope. Gramineae (21d). 1 SW China
Euthystachys A. DC. Stilbaceae. 1 S Afr.
Eutrema R. Br. Cruciferae. 15 Arctic (Eur. 1), C & E As., SW US (1)
Euxylophora Huber. Rutaceae. 1 Amaz. Brazil: *E. paraensis* Huber (Brazilian satinwood) – clear yellow hard timber for flooring, furniture etc.
Euzomodendron Cosson. Cruciferae. 1 S Spain: *E. bourgaeanum* Cosson
Evacidium Pomel. Compositae (Gnap.-Gnap.). 1 Sicily & N Afr.: *E. discolor* (DC) Maire. R: OB 104(1991)173
Evandra R. Br. Cyperaceae. 2 SW Aus.
Evax Gaertner = *Filago*
Evea Aublet = *Psychotria*
Everardia Ridley. Cyperaceae. 12 Venezuela, Guyana
everlastings usu. coloured heads of *Bracteantha bracteata* but many other materials also used esp. spp. of *Limonium, Rhodanthe, Syngonanthus* etc.; **mountain e.** *Antennaria dioica*; **pearly e.** *Anaphalis margaritacea*
Eversmannia Bunge. Leguminosae (III 7). 1 SE Russia, N Iran & C As.: *E. subspinosa* (DC) B. Fedtsch.
Evodia Lam. = *Euodia*
Evodianthus Oersted. Cyclanthaceae. 1 trop. Am.: *E. funifer* (Poit.) Lindm. – pendulous aerial roots used for basket-weaving in Ecuador
Evodiella van der Linden. Rutaceae. 2 New Guinea, Queensland. R: Blumea 32(1987)195
Evodiopanax (Harms) Nakai = *Gamblea*
Evolvulus L. Convolvulaceae. 98 warm & trop. Am., 2 extending to OW incl. *O. alsinoides*

(L.) L. – only pl. on some Sri Lanka serpentines

Evonymopsis Perrier = *Euonymopsis*

Evonymus L. = *Euonymus*

Evota (Lindley) Rolfe = *Ceratandra*

Evotella Kurzweil & Linder. Orchidaceae (IV 3). 1 SW & S Cape. R: PSE 175(1991)215

Evrardia Gagnepain = *Evrardianthe*

Evrardianthe Rauschert (*Evrardia*; ~ *Hetaeria*). Orchidaceae (III 1). 1 SE As. Mycotrophic

Evrardiella Gagnepain (~ *Aspidistra*). Convallariaceae (Liliaceae s.l.). 1 SE As.

Ewartia Beauverd. Compositae (Gnap.-Cass.). 4 SE Aus. R: OB 104(1991)95

Ewartiothamnus Anderb. (~ *Ewartia*). Compositae (Gnap.-Gnap.). 1 NZ: *E. sinclairii* (Hook.f.) Anderb. R: OB 104 (1991)94

Exaculum Caruel = *Cicendia*

Exacum L. Gentianaceae. 65 OW trop. (As. 21, Socotra & Oman 4, Afr. 2, Madag. 38). R: OB 84(1985). Style turned to 1 side: diff. fls with it turned both ways found on same plant (enantiostyly). Some cult. orn. esp. *E. affine* Balf. f. ex Regel (Socotra, Dhofar) – often seen as flowering pot-plants ('Arabian' or 'Persian violets'), sometimes with 'double' fls

Exallage Bremek. = *Hedyotis*

Exandra Standley = *Simira*

Exarata A. Gentry. Bignoniaceae. 1 Ecuador & Colombia: *E. chocoensis* A. Gentry. R: SB 17(1992)503

Exarrhena R. Br. = *Myosotis*

Exbucklandia R.W. Br. Hamamelidaceae (II). 2 E Himalaya & S China, Malay Pen. & Sumatra. Large stipules folded against one another & protecting young axillary but or infl.; fls in heads in groups of 4 sunk in axis. *E. populnea* (R. Br. ex Griffith) R.W. Br. (E Himal. to Malay Hills) – multiple branches at nodes arising from branching of trace to primary axillary bud; timber useful, lvs quiver like aspen

Excavatia Markgraf = *Ochrosia*

excelsior short thin curled wood shavings, esp. from *Populus* spp., used for stuffing mattresses etc.

Excentrodendron H.T. Chang & Miau = *Burretiodendron*

Excoecaria L. Euphorbiaceae. 40 OW trop. to Pacific. *E. africana* Muell. Arg. (African sandalwood, S & SE Afr.) – good timber; *E. agallocha* L. (blind-your-eye(s), Indomal.) – latex blinding, lvs turn bright red before falling; *E. cochinchinensis* Lour. (SE As.) – cult. orn., medic.

Excremis Willd. Phormiaceae (Liliaceae s.l.). 1 Peru, Colombia. (A) thick

Exechostylus Schumann = *Pavetta*

Exellia Boutique (~ *Monanthotaxis*). Annonaceae. 1 trop. Afr.: *E. scammopetala* (Exell) Boutique

Exellodendron Prance. Chrysobalanaceae (2). 5 trop. S Am. R: FN 9(1972)195

Exoacantha Labill. Umbelliferae (III 3). 1 SW As.

Exocarpos Labill. Santalaceae. 26 SE As., Mal. to Hawaii (Aus. 10, 9 endemic; New Caledonia 5). R: MBMUZ 213(1959)117. Some Aus. timbers (ballart) for cabinet-making

Exocarya Benth. Cyperaceae. 1 E New Guinea, NE Aus.: *E. sclerioides* (F. Muell.) Benth.

Exochogyne C.B. Clarke. Cyperaceae. 4 trop. S Am.

Exochorda Lindley. Rosaceae. 4 C As., China. Cult. orn. *Spiraea*-like shrubs; fr. of 5 carpels flattened & arranged starwise

Exodeconus Raf. (~ *Physalis*). Solanaceae (1). 6 S Am., Galápagos

Exogonium Choisy = *Ipomoea*

Exohebea R. Foster = *Tritoniopsis*

Exolobus Fourn. = *Gonolobus*

Exomiocarpon Lawalrée. Compositae (Helia.-Verb.). 1 Madagascar

Exomis Fenzl ex Moq. Chenopodiaceae (I 3). 2 SW Cape & dry S Afr.

Exorhopala Steenis. Balanophoraceae. 1 Malay Pen.

Exorrhiza Becc. = *Clinostigma*

Exospermum Tieghem = *Zygogynum*

Exostema (Pers.) Bonpl. Rubiaceae (I 1). c. 45 trop. Am. esp. WI. Some rheophytes; febrifugal alks in bark (Jamaica b.). *E. caribaeum* (Jacq.) Schultes (princewood, C Am., WI) – wood for turnery etc.

Exostyles Schott. Leguminosae (III 1). 2 SE Brazil

Exothea Macfad. Sapindaceae. 3 Florida, WI, C Am.

Exotheca Andersson. Gramineae (40g). 1 trop. Afr. & Vietnam. Odd distrib. poss. reflexion of coastal trading going back to antiquity

eyan *Lovoa trichilioides*

eyebright *Euphrasia* spp. esp. *E. officinalis*

Eylesia S. Moore = *Buchnera*

eyong *Sterculia oblonga*

Eysenhardtia Kunth. Leguminosae (III 12). 10 C Am. *E. polystachya* (Ortega) Sarg. (*E. amorphoides*, kidneywood, lignum-nephriticum, SW N Am.) – wood chips placed in water against a black background produce peacock blue phosphorescence, form. imported for medic.

Ezosciadium B.L.Burtt (*Trachysciadium* auctt.) Umbelliferae (III 8). 1 S Afr.: *E. capense* (Ecklon & Zeyher) B.L. Burtt. R: EJB 48(1991)207

F

Faba Miller = *Vicia*

Fabaceae Lindley = Leguminosae

Faberia Hemsley = *Prenanthes*

Fabiana Ruíz & Pavón. Solanaceae (2). 25 warm temp. S Am. *F. imbricata* Ruíz & Pavón (pichi, Chile) – cult. orn. shrub, medic.

Fabrisinapis C. Towns. = *Hemicrambe*

Facelis Cass. Compositae (Gnap.-Gnap.). 3 S Am., *F. retusa* (Lam.) Schulz-Bip. introd. OW. R: BJLS 106(1991)191

Facheiroa Britton & Rose (~ *Espostoa*). Cactaceae (III 4). 3 NE Brazil

Factorovskya Eig = *Medicago*

Fadenia Aellen & C. Towns. Chenopodiaceae (III 3). 1 Ethiopia, Kenya: *F. zygophylloides* Aellen & C. Towns. R: NJB 11(1991)315

Fadogia Schweinf. Rubiaceae (III 2). c. 45 trop. Afr. Some with large red, ? bird-poll. fls; some ed. fr.

Fadogiella F. Robyns. Rubiaceae (III 2). 2 (1 known only from (destroyed!) type specimen) trop. Afr.

Fadyenia Hook. = *Tectaria*

Fagaceae Dumort. Dicots – Hamamelidae – Fagales. 8/700 cosmop. excl. trop. & S Afr. Strongly tanniferous monoec. (rarely dioec.) trees & shrubs; roots often with ectotrophic mycorrhizae. Lvs simple to deeply lobed, spirally arr., rarely opp. or whorled; stipules decid. Infls pendulous, rarely erect or variously reduced, even to solit fls. Fls usu. wind-poll. (insects in *Castanea*), inconspicuous, males in ± reduced dichasia in catkins or heads (1–3-flowered dichasia in *Nothofagus*), P (4–)6(7–9), small, scale-like, sometimes basally connate or almost 0, A (4–)6–12(–90), anthers with longit. slits; females 1–7(–15) at base of male infls or in diff. axils, individually or collectively involucrate, ± 6–12 staminodes, $\overline{G}$ ((2)3) or (6(7–12)) with styles & locules as many as carpels but septa not reaching to apex, each loc. with 2 axile, anatropous, bitegmic (*Nothofagus* unitegmic) ovules. Fr. usu. a nut with stony or leathery pericarp (rarely ± a samara), subtended by accrescent involucre (cupule); seed 1 with large straight starchy or oily embryo & 0 endosperm. x = (11)12(13)
Genera: *Castanea, Castanopsis, Chrysolepis, Fagus, Lithocarpus, Nothofagus, Quercus, Trigonobalanus*
Prominent & often dominant in angiosperm forests of N & to a lesser extent S hemispheres (oaks, beeches) & montane forests of Mal. Delayed fert. (cf. Pinopsida, Hamamelidae). The char. cupules have been thought derived from the coalescence of 3-lobed extensions of the pedicel beneath each fl., though it has been more convincingly argued that it is derived from P; within the fam., increased enclosure of fr. by cupule has led to decrease in ecol. imp. of former; cupule reduction elsewhere has increased its imp. (BR 59(1993)81). Source of some of the most imp. timbers – oak, beech, chestnut, also cork (*Quercus suber*), seeds (up to 46% oil, by wt.) for human (*Castanea*) & stock (*Fagus, Quercus*) & wild pig (*Castanopsis, Lithocarpus* – leading to spectacular pig migrations in 'masting' years) consumption; many cult. orn.

Fagara L. = *Zanthoxylum*

Fagaropsis Mildbr. ex Siebenl. Rutaceae. 1 trop. & NE Afr.: *F. angolensis* (Engler) Dale (mafu)

Fagerlindia Tirvengadum (~ *Randia*). Rubiaceae (II 1). 6 India to W Mal.

Fagonia L. Zygophyllaceae. 30 Med. (Eur. 1), SW As. & NE trop. Afr. to NW India, SW Afr., SW N Am. (6)

Fagopyrum Miller (~ *Polygonum*). Polygonaceae (II 5). 8 As. & E Afr. Fls like *Polygonum* but heterostyled. Grown for fr. ('seed'), the seeds with floury endosperm (buckwheat) esp. *F. esculentum* Moench (C or N As.) – flour much used in US for pancakes, imp. bee-pl. for comm. honey, source of rutin, seeds of other spp. inhibited from germinating for period after cropping b.; *F. tataricum* (L.) Gaertner (Siberian, Tartary or Kangra b., India) – flour used locally. Both spp. perhaps derived from *F. dibotrys* (D. Don) Hara (*F. cymosum*, N India to China)

Fagraea Thunb. Gentianaceae. 35 Sri Lanka & S India to China, Mal. & W Pacific. Form. assigned to Loganiaceae s.l. Often epiphytic. Cult. orn. with scented fls, some medic. & locally used timber. *F. auriculata* Jack (Burma to Philipp.) – tree to epiphyte, fls to 30 cm across

Faguetia Marchand. Anacardiaceae (IV). 1 Madagascar

Fagus L. Fagaceae. 10 N temp. (Eur. 1). Beeches. Decid. monoec. trees with imp. timber, nuts & cult. orn. e.g. *F. crenata* Blume (Japanese b., buna, Japan) – imp. forest tree, etc. but esp. *F. grandifolia* Ehrh. (American b., E N Am.) – differing from Eur. b. in longer, coarsely serrate lvs with more veins, uses similar, distilled wood for creosote & *F. sylvatica* L. ((Eur.) b., Eur. incl. ssp. *orientalis* (Lipsky) Greuter & Burdet (oriental or Turkish b., C Greece to SE Russia)) – to 300 yrs old in Carpathians, in pure stands in mts of Eur., also (artificially) in areas where oak removed & b. encouraged for timber for chairs etc. (e.g. 'Windsor' chairs in Chilterns, England), casting dense shade with little undergrowth beyond mycotrophs like *Lathraea* spp. etc., masts at irreg. intervals with males in drooping heads & fr. of 1 or 2 brown 3-angled 'nuts' in a prickly involucre, toxic saponins, straight-grained timber used for kitchen utensils, turnery etc. as well as furniture, seeds produce an oil, timber form. source of creosote, cult. orn. esp. as hedges when lvs retained (marcescent) as in young trees, many cvs incl. 'Atropunicea' (copper b.) – first coll. in German forests. 'Dawyck' (Fastigiata, Dawyck b.) – fastigiate, 'Laciniata' – deeply cut lvs etc., 'Pendula' (weeping b.), a specimen of the wild type grown at 3000 m in Java for 60 yrs was a densely branched evergreen shrub with annual rings but no fls, while the allied *Castanea sativa* produced fls regularly

faham tea *Angraecum fragrans*

Fahrenheitia Reichb.f. & Zoll. ex Muell. Arg. (~ *Ostodes*). Euphorbiaceae. 4 S India, Sri Lanka, W Mal. to Bali

Faidherbia A. Chev. (~ *Acacia*). Leguminosae (II 5). 1 warm Afr. to E Med.: *F. albida* (Del.) A. Chev. – drops lvs in wet season & gets new ones in dry, valuable fodder tree, source of gum, tannin etc.

Faika Philipson. Monimiaceae (V 2). 1 W New Guinea

fair maids of France (or **Kent**) *Ranunculus aconitifolius*

fairy bells *Disporum* spp.; **f. fingers** *Digitalis purpurea*; **f. flax** *Linum catharticum*; **f. wand** *Chamaelirium luteum*

Falcaria Fabr. Umbelliferae (III 8). 1 C Eur., Medit., W & C As.: *F. vulgaris* Bernh.

Falcata J. Gmelin = *Amphicarpaea*

Falcataria (I. Nielsen) Barneby & Grimes. See *Paraserianthes*

Falcatifolium Laubenf. (~ *Podocarpus*). Podocarpaceae. 5 Mal. to New Caled. R: Phytol.M 7(1984)29

Falckia Thunb. = *Falkia*

Falconeria Hook.f. = *Kashmiria*

Falkia L.f. (*Falckia*). Convolvulaceae. 3 Afr.

Fallopia Adans. (~ *Polygonum*). Polygonaceae (II 4). Excl. *Reynoutria* 9 N temp. (Eur. 2). *F. baldschuanica* (Regel) Holub (*F. aubertii*, Russian vine, W China & Tibet) – rampant liane planted as a screen

Fallugia Endl. Rosaceae. 1 SW N Am.: *F. paradoxa* (Tilloch & Taylor) Endl. – cult. orn. shrub, poss. oilseed

Falona Adans. = *Cynosurus*

false acacia *Robinia pseudoacacia*; **f. oat** *Arrhenatherum elatius*; **f. sarsaparilla** *Hardenbergia violacea*; **f. spikenard** *Maianthemum racemosum*

Famatina Ravenna. Amaryllidaceae (Liliaceae s.l.). 3 Andes of Chile & Arg.

Fanninia Harvey. Asclepiadaceae (III 1). 1 S Afr.: *F. caloglossa* Harvey

fanwort *Cabomba* spp.

Faradaya F. Muell. Labiatae (II; Verbenaceae). 3 N Borneo, New Guinea, Aus., Polynesia. *F. splendida* F. Muell. (N Borneo to N Aus.) – fish-poison in NE Queensland (koie-yan)

Faramea Aublet. Rubiaceae (IV 11). 125 trop. Am. Pollen dimorphic

Farfugium Lindley. Compositae (Sen.-Tuss.). c. 3 E As. Cult. orn. esp. *F. tussilagineum* (Burm.f.) Kitam. (*Ligularia tussilaginea*)

Fargesia Franchet = *Thamnocalamus*

farkleberry *Vaccinium arboreum*

Farmeria Willis ex Trimen. Podostemaceae. 2 SW India, Sri Lanka

faro *Daniellia ogea*

Faroa Welw. Gentianaceae. 17 trop. Afr. R: GOB 1(1973)69

Farquharia Stapf. Apocynaceae. 1 S Nigeria

Farrago W. Clayton. Gramineae (33d). 1 Tanzania

Farringtonia Gleason = *Siphanthera*

Farsetia Turra. Cruciferae. 25 Morocco to NW India, Tanz. mts. R: SBU 25,3(1986)

Fascicularia Mez. Bromeliaceae (3). 5 Chile. Usu. terr. incl. *F. pitcairniifolia* (Verlot) Mez – cult. orn. natur. Eur. is.

fat hen *Chenopodium album*

Fatoua Gaudich. Moraceae (I). 2 Madag. to E As., N Aus. & New Caled.

× **Fatshedera** Guillaumin. Araliaceae. *Fatsia* × *Hedera*. × *F. lizei* (C.-Cochet) Guillaumin (*Fatsia japonica* 'Moseri' × *Hedera helix* cv. (allegedly ssp. *hibernica* though n = 44–49 & that of *F. japonica* 12 or 24)), an hybrid arising in Lizé Frères nursery, Nantes, 1910, but not repeated, with floral structure intermediate between parents & sterile but with large stipular outgrowths found in neither though typical of other genera in A., cult. orn.

Fatsia Decne & Planchon. Araliaceae (1). Incl. *Boninofatsia*, 3 E As. *F. japonica* (Thunb.) Decne & Planchon – cult. orn. evergreen for bold effect (name misrendering of Jap. yatsude as '*fatsi*')

Faucaria Schwantes. Aizoaceae (V). 33 E Cape & Karoo. Cult. orn. succ. with 4-ranked lvs, usu. long-toothed esp. *F. tigrina* (Haw.) Schwantes (tiger's-jaws, Cape)

Faucherea Lecomte (~ *Manilkara*). Sapotaceae (I). 11 Madagascar

Faujasia Cass. Compositae (Sen.-Sen.). 4 Réunion

Faujasiopsis C. Jeffrey. Compositae (Sen.-Sen.). 3 Mascarenes

Faurea Harvey. Proteaceae. 15 trop. & S Afr., Madag. Some timber esp. *F. saligna* Harvey (African beech, trop. & S Afr.), also yields tannin & medic. locally

Fauria Franchet = *Nephrophyllidium*

Favargera Löve & D. Löve = *Gentiana*

Fawcettia F. Muell. = *Tinospora*

Faxonanthus Greenman. Scrophulariaceae. 1 Mexico

Faxonia Brandegee. Compositae (Helia.-Gal.) 1 Baja California

feaberry *Ribes uva-crispa*

feather flower *Verticordia* spp.; **f. grass** *Stipa* spp., *Leptochloa* spp.; **f. hyacinth** *Muscari comosum* 'Plumosum'; **parrot's f.** *Myriophyllum aquaticum*

Feddea Urban. Compositae (Ast. tribe?). 1 E Cuba (serpentine)

Fedia Gaertner. Valerianaceae. 3 Medit. (Eur. 2). R: NM 54(1990)3, AJBM 48(1991)157. *F. cornucopiae* (L.) Gaertner (horn of plenty) – cult. orn. & as salad pl.

Fedorovia Yakovlev. See *Ormosia*

Fedtschenkiella Kudr. = *Dracocephalum*

Feea Bory = *Trichomanes*

Feeria Buser (~ *Trachelium*). Campanulaceae. 1 Morocco

Fegimanra Pierre. Anacardiaceae (I). 2 trop. Afr.

Feijoa O. Berg = *Acca*

Feldstonia Short. Compositae (Gnap.-Ang.). 1 W Aus.: *F. nitens* Short. R: OB 104(1991)132

Felicia Cass. Compositae (Ast.-Ast.). 83 S Afr. with few extending to trop. Afr. & Arabia. R: MBSM 9(1973)195. Some cult. orn. esp. *F. amelloides* (L.) Voss & *F. bergeriana* (Sprengel) O. Hoffm. (kingfisher daisy) – blue fls, S Afr.

Feliciadamia Bullock (~ *Miconia*). Melastomataceae. 1 W Afr.

Felipponia Hicken = *Mangonia*

Felipponiella Hicken = *Mangonia*

Femeniasia Susanna. Compositae (Card.-Cent.). 1 Balearics. Spiny shrub

Fendlera Engelm. & A. Gray. Hydrangeaceae (Philadelphaceae). 2–3 SW N Am.

Fendlerella A.A. Heller. Hydrangeaceae (Philadelphaceae). 3 SW US

Feneriva Airy Shaw = *Polyalthia*

Fenerivia Diels = *Polyalthia*

Fenestraria N.E. Br. Aizoaceae (V). 1 SW Afr.: *F. rhopalophylla* (Schltr & Diels) N.E. Br. – stemless clump-forming succ. with erect club-shaped lvs, the truncate tips with transparent windows, which, in the wild, are all of the pl. above ground, light being focused by them to internal photosynthetic tissues; cult. orn.

Fenixia Merr. Compositae (Helia.-Verb.). 1 Philippines

fennel *Foeniculum vulgare*; **Florence f.** *F. vulgare* var. *azoricum*; **f. flower** *Nigella sativa*; **giant f.** *Ferula communis*; **hog's f.** *Peucedanum officinale*; **water f.** *Oenanthe aquatica*

fenugreek *Trigonella foenum-graecum*

Fenzlia Endl. = *Myrtella*

Ferdinandea Pohl = *Ferdinandusa*

Ferdinandusa Pohl. Rubiaceae (I 1). 20 trop. Am. Some local febrifuges

Feretia Del. Rubiaceae (II 5). 2 trop. Afr. R: KB 34(1979)368. Seeds a coffee subs.

Fergania Pim. (~ *Ferula*). Umbelliferae (III 10). 1 C As.: *F. polyantha* (Korovin) Pim.

Fergusonia Hook.f. Rubiaceae (?IV 7). 1 S India, Sri Lanka

fern, air (US) – *Sertularia argentea* L., a zoophyte dyed green & resembling plastic *Ceratophyllum*; **beech f.** *Phegopteris connectilis*; **bird's nest f.** *Asplenium nidus*; **bladder f.** *Cystopteris fragilis*; **bristle f.** *Trichomanes* spp.; **cinnamon f.** *Osmunda cinnamomea*; **eagle f.** *Pteridium aquilinum*; **elk's horn f.** *Platycerium* spp.; **filmy f.** Hymenophyllaceae; **f. grass** *Desmazeria* spp.; **hart's tongue f.** *Asplenium scolopendrium*; **holly f.** *Polystichum* spp.; **kangaroo f.** *Phymatosorus diversifolius*; **koala f.** *Caustis blakei*; **lady f.** *Athyrium* spp.; **maidenhair f.** *Adiantum* spp.; **male f.** *Dryopteris filix-mas*; **marsh f.** *Acrostichum* spp.; **ostrich f.** *Matteuccia* spp.; **parsley f.** *Cryptogramma crispa*; **royal f.** *Osmunda* spp.; **shield f.** *Dryopteris* spp.; **stag's horn f.** *Platycerium* spp.; **f. tree** *Filicium decipiens*, *Jacaranda* spp.; **walking f.** *Asplenium rhizophyllum*

Fernaldia Woodson. Apocynaceae. 4 Mex., C Am. *F. pandurata* (A. DC) Woodson (C Am.) – ed. fls exported to US

Fernandezia Ruíz & Pavón. Orchidaceae (V 10). 9 trop. Am.

Fernandoa Welw. ex Seemann. Bignoniaceae. 14 OW trop. (Afr. 4, Madag. 3, S China & SE As. to Sumatra 7)

Fernelia Comm. ex Lam. Rubiaceae (II 5). 4 Masc. R: KB 37(1983)551

Fernseea Baker. Bromeliaceae (3). 2 Brazil. R: Bradea 3(1983)343

Ferocactus Britton & Rose. Cactaceae (III 9). 23 SW N Am. R: Bradleya 2(1984)19. Stems candied; cult. orn. large ovoid ribbed cacti esp. *F. wislizeni* (Engelm.) Britton & Rose with largest spines to 10cm, hooked; *F. histrix* (DC) G. Lindsay (N C Mex.) – fls (sometimes pickled) & fr. ed., sacred in pre-Columbian times, local medic.

Feronia Corr. Serr. = *Limonia*

Feroniella Swingle. Rutaceae. 3 SE As., Java. R: W. Swingle, *Bot. Citrus* (1943)468. *F. lucida* (R. Scheffer) Swingle (Java) – locally cult. for fr.

Ferraria Burm. ex Miller. Iridaceae (III 2). 10 trop. (1) & S Afr. R: JSAB 45(1979)295. Cult. orn. though fls 'malodorous & fugacious'

Ferreirea Allemão = *Sweetia*

Ferreyanthus H. Robinson & Brettell = *Ferreyranthus*

Ferreyranthus H. Robinson & Brettell. Compositae (Liab.). 7 Peru & S Ecuador (1). R: SCB 54(1983)27. Trees & shrubs form. common on slopes along roads

Ferreyrella S.F. Blake. Compositae (Eup.-Ager.). 1 Peru. R: MSBMBG 22(1987)163

Ferrocalamus Xue & Keng f. = *Indocalamus*

Ferula L. Umbelliferae (III 10). 172 Medit. (Eur. 8) to C As. Drugs & gums & some statuesque garden pls with large lvs. Asafoetida, a gum-resin sometimes used in veterinary med., derived from *F. assa-foetida* L. (W Iran), *F. foetida* (Bunge) Regel (E Iran etc.) & *F. narthex* Boiss. (Afghanistan), form. used as condiment ('food of the gods') in Iran & more widely as medic.; gum galbanum also medic. derived from *F. gummosa* Boiss. (*F. galbaniflua*, Iran) & *F. rubricaulis* Boiss.; sagapenum another resin medic. from diff. spp., also 'silphion' prob. now extinct but form. of great imp. in Cyrenaica as a source of an aromatic gum, & sumbul from *F. sumbul* Hook.f. (C As.) – scent & incense; cult. orn. incl. *F. communis* L. (giant fennel, S Eur. to Syria) – scape tipped with a pine cone = thyrse of 'bacchantes', pith burns slowly leaving vasc. tissue, Prometheus bringing fire to Earth in it

Ferulago Koch. Umbelliferae (III 10). 43 Medit. (Eur. 9) to C As. R: FR 100(1989)119

Ferulopsis Kitagawa. Umbelliferae (III 10). 2 N & C As.

fescue *Festuca* spp.; **bearded f.** *Vulpia ambigua*; **Chewing's f.** *F. rubra* ssp. *commutata*; **meadow f.** *F. pratensis*; **red f.** *F. rubra*; **sheeps's f.** *F. ovina*

Festuca L. Gramineae (17). 450 temp., trop. mts. R: KBAS 13(1986)94. Fescues, bluegrass. Lvs roll inwards when dry. Pasture- & lawn-grasses, some dioec., alpine ones often viviparous. *F. brevipila* Tracey ('*F. longifolia*', W Eur.), *F. rubra* L. (red. f., N temp.) esp. ssp. *commutata* Gaudin (*F. nigrescens*, Chewing's f.) & *F. filiformis* Pourret (*F. tenuifolia*, Eur.) – imp. lawn-grasses; *F. ovina* L. (sheep's f., temp. N OW) & *F. pratensis* Hudson (meadow f., Euras.) – pasture-grasses

Festucella Alexeev = *Austrochloa*

Festucopsis (C. Hubb.) Meld. = *Elymus*

×**Festulolium** Asch. & Graebner. Gramineae. *Festuca* × *Lolium*

feterita *Sorghum bicolor* Caudatum Group

fetterbush *Pieris floribunda*

fever bark *Alstonia constricta, Croton* spp.; **f. bush** *Garrya elliptica, Ilex verticillata; Lindera* spp.; **f.few** *Tanacetum parthenium;* **f. tree** *Acacia xanthophloea, Eucalyptus globulus, Pinckneya bracteata, Zanthoxylum capense;* **f. wort** *Triosteum perfoliatum*

Fevillea L. Cucurbitaceae. 7 trop. Am. A 5 all alike. Water-disp. purgative & emetic seeds with oil content greater than any other dicot (55%), those of *F. pedatifolia* (Cogn.) Jeffrey used as candles in Peru

Fezia Pitard. Cruciferae. 1 Morocco

Fibigia Medikus. Cruciferae. 10 E Medit. (Eur. 3) to Afghanistan. Some cult. orn. herbs esp. *F. lunarioides* Sweet (Aegean)

Fibraurea Lour. Menispermaceae (IV). 2 Assam. SE As., Philipp., Borneo, Sulawesi. R: KB 40(1985)546. Alks. *F. recisa* Pierre yields an antibacterial used in Chinese medicine ('huangteng'), also a yellow dye

Ficalhoa Hiern. Theaceae. 1 trop. Afr.

Ficinia Schrader. Cyperaceae. 60 trop. & S Afr. (50 restricted to Cape)

Ficoidaceae Juss. = Aizoaceae

Ficus L. Moraceae (V). c. 750 trop. & warm esp. Indomal. to Aus. (c. 500, R: GBS 21(1965)1; Afr. 105 (R: Kirkia 13(1990)253), Am. c. 150). Figs. Dioec. or monoec. trees, shrubs & root-clinging lianes, epiphytes & stranglers (with coalescing roots, some 'individuals' comprising more than one genome!), occ. podagric semi-succ. (e.g. *F. palmeri* S. Watson, Baja Calif.), unbranched pachycaul treelets to tall trees with slender twigs & small lvs or rheophytes, with milky latex; roots to 120 m deep recorded, clogging drains & engulfing buildings; lvs simple (*F. otophora* Corner & Guillaumin (New Caled.) with separate pinnae, *F. hirta* Vahl (NE India to W Mal.) palmate), spirally arr. to distichous or opp. with stipules enveloping bud but soon lost. Fls borne in a globose, oblong or pyriform receptacle (fig, syconium) with a small apical opening (ostiole) closed by overlapping bracts, 2 or 3 fls to several thousand in largest figs, male, female & gall fls (short-styled sterile females): all 3 in monoec. figs (subg. *Pharmacosycea, Sycomorus, Urostigma*) or males & galls in gall pl., the females in seed pl. in dioec. figs (subg. *Ficus*, mainly As. & Pacific, the males usually around ostiole); males with P (0)1–7, A 1–7; infls protogynous & attractive (? pheromones) to species-specific gall-wasps (Hymenoptera – Agaonidae), though *F. ottoniifolia* (Miq.) Miq. (S trop. Afr.) occupied by diff. spp. in forest & open country, the pregnant females (usu. more than 1 per syconium) forcing a passage through the ostiole, often losing limbs (& prob. any attached fungal spores too) in the process, removing pollen from 'pollen pockets' & poll. (? 1 grain per stigma), laying eggs in short-styled fl. ovaries in which the ovules are stimulated only to produce endosperm on which the grubs feed, then dying; the CO_2 content is higher within the fig than without & is believed in some spp. to control the hatching of the grubs, it in turn being controlled by the activity of bacteria, in turn controlled by populations of nematodes e.g. in *F. religiosa* early male phase has 10% CO_2, 10% 0_2 & some ethylene so females inactive; as seeds mature, the emerging male wasps (wingless with reduced legs & eyes) fertilize & then help out of their fls the females which leave the fig through holes cut by the males (which then die, in *F. religiosa* the change in atmosphere rousing the females), past the male fls which dust them with pollen put into the pockets & taken to the next fig. There are also a number of parasitic Hymenoptera which do not pollinate, ovipositing through syconium wall, some also taking pollen. 30–80% ovules are killed. Syconia (aborted if not poll.) borne in axils, on branches or trunks even at ground level in diff. spp., while others are geocarpic, the figs borne on underground stolons to 10 m long & buried up to 10 cm deep. The frs are drupes borne in the fleshy receptacle, which may be brightly coloured & are the principal food of many birds & mammals esp. bats & primates (incl. human) which are the disp. agents, though some fish-disp. in Amaz., protected by latex with ficin (a protease) until ripe

The latex of some spp. is used as birdlime & vermifuge, meat-tenderizer & chill-proofing agent in beer, while others provide bark cloth (e.g. *F. thonningii* Blume (trop. Afr., where often planted as shade), *F. lutea* Vahl (*F. nekbudu*, trop. & S Afr. to Seychelles)), though in Mex. largely replaced by *Trema micrantha* Blume. Many others cult. orn. (few artificial hybrids made), sources of rubber, fibres, paper, timber, medic. & are revered in religion (considered haunts of spirits etc. in diff. parts of world) & figuring consp. in mythology (e.g. Homer). *F. benghalensis* L. (*F. indica*, banyan (from 'banians' the traders seen resting below such trees), India, Pakistan) – 2 sorts of gall fls in July as opposed to Nov. flowering, small fls with *Blastophaga* wasps, large with *Apocrypta*, initially epiphytic, crown spreading by aerial roots (root-hairs 0, cortex & pericycle thick, periderm with chloroplasts & lenticels present) dropping down to become accessory trunks, such that a plant may occupy some hectares e.g. 200 yr-old over 1.6 ha at Calcutta, 412 m circumference with 100 subsid. trunks & 1775 proproots (Alexander's army said to have sheltered beneath one; at least their views of 'roots' & 'stems' were shaken), timber & fibre, a sacred tree of the Hindus, 'Krishnae' with cup-shaped lvs (10% of seedlings said to come true); *F. benjamina* L. (Benjamin tree, weeping or Java fig, Indomal.) – often initially epiphytic & sometimes a strangler, dropping roots all round 'host' tree & shading out its crown to smother it completely & become a living shell round the dead tree, cult. orn. with drooping branches; *F. carica* L. (common fig) – decid. tree (allied to *F. palmata* Forssk. of NE Afr. to India but with diff. wasp poll.) prob. native in SW As. but early spread to Medit. where Egyptians were cultivating it 4000 BC, poss. present in pre-pottery A Neolithic level of Jericho: much grown are Adriatic fs with no male fls & parthenocarpic while the Smyrna figs require poll. achieved by suspending infls of 'wild' caprifigs from which the fig-wasps, *Blastophaga psenes* L., are about to emerge on the branches, now, with almonds, olives & carobs typical of the anthropogenic landscape of the W Medit., where many grown for eating fresh or drying, extract a laxative ('syrup of figs'), latex used against warts; *F. elastica* Roxb. ex Hornem. (rubber plant, India rubber tree, Indomal., ? extinct in wild) – large tree with buttresses usu. grown as juvenile in houses, form. imp. source of rubber (India, Indian or Assam r.); *F. exasperata* Vahl (trop. Afr., Yemen to India) – cytotoxic but eaten by healthy chimpanzees; *F. lutea* (*F. vogelii*, trop. & S Afr. to Seychelles) – source of Abbo, Congo or Dahomey rubber; *F. lyrata* Warb. ex De Wild. & Durand (fiddle-leaf f., trop. Afr.) – juveniles cult. as housepls; *F. macrophylla* Desf. ex Pers. (Moreton Bay f., E Aus.) – to 60 m, but usu. smaller in cult. where used as street & shade tree (as at Sydney); *F. microcarpa* L.f. (Indomal.) – introd. C Am. & now potential pest as wasp has caught up; *F. pseudopalma* Blanco (Philipp.) – unbranched pachycaul cult. orn.; *F. pumila* L.('*F. repens*', climbing f., Vietnam to Japan) – liane with aerial roots which secrete a gummy exudate & absorb fluid constituent to leave root cemented to support, at top of which it becomes tree-like, fr. used for jelly in China (okgue), though some cvs e.g. 'Minima' are juvenile forms much cult. in conservatories etc.; *F. racemosa* L. (crattock, Indomal.) – timber, shade, ed. fr.; *F. religiosa* L. (peepul, pipal, pipul, bo-tree, India to SE As.) – decid. fast-growing tree, usu. an epiphyte when young & splits host (not a strangler) with lvs with v.long drip-tips containing vestigial veins representing an ancestral large lamina here unexpanded, sacred to Hindus & Buddhists (Buddha had the true insight beneath one; the Anuradhapura, Sri Lanka tree grown from a sprig of that tree brought from India 288 BC, the gall-wasps soon introd.), lvs used for miniature paintings, fibre form. used for paper in Burma, host for lac insects & some silkworms; *F. rubiginosa* Desf. ex Vent. (Port Jackson fig, NSW) – cult. street tree, natur. NZ as wasp has now been blown from Aus. as has that of *F. macrophylla* (1993); *F. semicordata* Buch.-Ham. ex Sm. (Nepal to SE As.) – imp. fodder crop; *F. sycomorus* L. (sycamore (of the Bible), mulberry fig, trop. & S Afr.), brought to Medit. in cult. for ed. figs (parthenocarpic clones, maturing when syconia slashed) & timber used for Pharaohs' sarcophagi (now extinct in wild in Egypt where first domesticated); *F. tinctoria* Forster f. (*F. globosa*, Indomal. to Pacific) – cult. orn. strangler with coalescing roots, first through union of epidermal hairs, then compression of adjacent cortex, at periphery rays prod. parenchyma eventually uniting & some becoming cambial cells linking existing cambia so that continuous ring of vasc. cambium reorganized & giving rise to more secondary tissues; *F. virens* Aiton (Indomal.) – cult. for shade, young shoots pickled in Sikkim

fiddle dock *Rumex pulcher*; **f. greens** or **f. heads** *Matteuccia struthiopteris*; **f.-leaf** *Ficus lyrata*; **f.wood** *Citharexylum* spp., *Petitia domingensis*, *Vitex* spp.

Fiebrigia Fritsch = *Gloxinia*

Fiebrigiella Harms. Leguminosae (III 13). 1 Bolivia

Fiedleria Reichb. = *Petrorhagia*

Fieldia Cunn. Gesneriaceae. (Incl. *Lenbrassia*) 1 Queensland, 1 SE Aus. *F. australis* Cunn. (SE Aus.) – anisophyllous climber allied to *Mitraria* (Chile), usu. on tree-fern trunks

fig *Ficus carica*; **climbing f.** *F. pumila*; **fiddle-leaf f.** *F. lyrata*; **Hottentot f.** *Carpobrotus edulis*; **Indian f.** *Opuntia* spp.; **Java f.** *F. benjamina*; **mulberry** or **sycamore f.** *F. sycomorus*; **Port Jackson f.** *F. rubiginosa*; **weeping f.** *F. benjamina*

figwort *Scrophularia* spp.; **Cape f.** *Phygelius capensis*; **water f.** *S. auriculata*

Fijiian kauri *Agathis macrophylla*

Filaginella Opiz = *Gnaphalium*

Filago L. Compositae (Gnap.-Gnap.). (Incl. *Evax*) 41 Euras. (Eur. 24), N Am., N Afr. R: OB

104(1991)170

Filarum Nicolson. Araceae (VIII 4). 1 Peru

filbert *Corylus* spp.

Filetia Miq. Acanthaceae. 8 Sumatra, Malay Peninsula

Filicium Thwaites ex Hook.f. Sapindaceae. 3 OW trop. *F. decipiens* (Wight & Arn.) Thw. – fern tree (E Afr., India) with fern-like lvs

Filicopsida Ferns. 223/8550 in 33 fams. Sporophytes with roots, stems & spirally arr. lvs (megaphylls) often markedly compound, protostelic, solenostelic or dictyostelic, sometimes polycyclic; some with restricted sec. thickening. Sporangia homosporous or heterosporous borne term. on axis or on fronds (marginal or superficial abaxially); antherozoids multiflagellate. From tree ferns (Cyatheaceae) to delicate filmy ferns (Hymenophyllaceae) & free-floating aquatics (Azollaceae, Salviniaceae), mostly rhiz. perennials, e.g. Aspleniaceae, Dryopteridaceae, Polypodiaceae, Pteridaceae. Not as biochemically variable as angiosperms but with a range of toxic sec. compounds: condensed tannins & other phenolics, thiaminase, phytoecdysones, cyanogenic glycosides & sesquiterpenes

Filifolium Kitam. (~ *Artemisia*). Compositae (Anth.-Art.). 1 NE As.

Filipedium Raiz. & Jain = *Capillipedium*

Filipendula Miller. Rosaceae. 10 N temp. (Eur. 2). *F. ulmaria* (L.) Maxim. (meadowsweet, Euras., natur. N Am.) – form. medic. (salicylic acid compounds basis of efficacy in arthritis treatment etc.), fragrant oil, cult. orn.; *F. vulgaris* Moench (dropwort, Euras.) – form. medic. esp. renal diseases, cult. orn.

Fillaeopsis Harms. Leguminosae (II 3). 1 W & SW Afr., forest margins. *F. discophora* Harms – easily polished timber used for wood-block floors etc.

filmy ferns Hymenophyllaceae

Fimbriella Farw. ex Butzin = *Platanthera*

Fimbristemma Turcz. = *Matelea*

Fimbristylis Vahl (*Abildgaardia*, excl. *Bulbostylis*). Cyperaceae. c. 250 warm (Eur. 3; Aus. 85). Some copper indicators (Queensland) & fibres esp. *F. umbellaris* (Lam.) Vahl (*F. globulosa*, trop. As.) – cult. for weaving

Findlaya Hook.f. = *Orthaea*

Finetia Gagnepain = *Anogeissus*

finger cherry *Rhodomyrtus macrocarpa*; **dead man's f.** *Orchis mascula*; **fairy f.s** *Digitalis purpurea*; **f. grasses** *Digitaria* spp.

Fingerhuthia Nees. Gramineae (31b). 1–2 Arabia, Afghanistan, S Afr. Disjunct distrib. like allied *Tetrachne*

Finlaysonia Wallich. Asclepiadaceae (I; Periplocaceae). c. 3 Indomal. to Aus.

finocchio *Foeniculum vulgare* var. *azoricum*

Finschia Warb. Proteaceae. 4 Papuasia & W Pacific. *F. chloroxantha* Diels (Papuasia) – ed. seeds, planted Vanuatu

Fintelmannia Kunth = *Trilepis*

Fioria Mattei = *Hibiscus*

fiorin *Agrostis stolonifera*

fique *Furcraea* spp.

fir *Abies* spp.; **alpine f.** *A. lasiocarpa*; **balsam f.** *A. balsamea*; **Caucasian f.** *A. nordmanniana*; **Douglas f.** *Pseudotsuga menziesii*; **Fraser f.** *A. fraseri*; **giant f.** *A. grandis*; **Greek f.** *A. cephalonica*; **Himalayan f.** *A. pindrow*; **Japanese f.** *A. firma*; **noble f.** *A. procera*; **Norway f.** *Pinus sylvestris*; **red f.** *A. magnifica*; **silver f.** *A. alba*; **white f.** *A. grandis*

fire bush *Pyracantha coccinea*, *Hamelia patens*; **Chilean f. bush** *Embothrium coccineum*; **f. cracker plant** *Russelia equisetiformis*; **f. pink** *Silene virginica*; **f. thorn** *Pyracantha coccinea*; **f. tree** *Nuytsia floribunda*; **f. weed** *Epilobium angustifolium*, *Senecio madagascariensis* (Aus.); **f.wheel tree** *Stenocarpus sinuatus*

Firmiana Marsili. Sterculiaceae. 12 OW trop., E Afr. eastwards. R: Reinwardtia 4(1957)281, 5(1960)384, Blumea 34(1989)117. Ovary wings in wind-disp. Some light timbers & cult. orn. esp. *F. simplex* (L.) W. Wight (E As.) – foliage pl. with lvs to 30 cm across, v. common street-tree in Japan

Fischeria DC. Asclepiadaceae (III 3). 16 trop. Am. R: SB 11(1986)229

fish berries *Anamirta cocculus*; **f. grass** *Cabomba* spp.

Fissendocarpa (Haines) Bennet = *Ludwigia*

Fissenia Endl. = *Kissenia*

Fissicalyx Benth. Leguminosae (III 4). 1 Venez., Guyana

Fissistigma Griffith. Annonaceae. 60 OW trop.

fitches *Nigella sativa* seeds

Fitchia Hook.f. Compositae (Helia.-Cor.). 7 Polynesia. R: UCPB 29, 1(1957). Trees

fitsroot *Astragalus glycyphyllos*

Fittingia Mez. Myrsinaceae. 5 New Guinea (? & New Britain). R: Blumea 33(1988)94. Pachycaul treelets

Fittonia Coëm. Acanthaceae. 2 Peru. Leaf infusions used for toothache in NW Amaz.; cult. orn. housepls with white or coloured leaf-veins, esp. *F. verschaffeltii* (Lemaire) van Houtte

Fitzalania F. Muell. Annonaceae. 1 trop. E Aus.: *F. heteropetala* (F. Muell.) F. Muell.

Fitzroya Hook.f. ex Lindley. Cupressaceae. 1 S Chile & S Arg.: *F. cupressoides* (Molina) I.M. Johnston (*F. patagonica*, alerce, Patagonian cypress) – timber esp. for shingles

Fitzwillia Short. Compositae (Gnap.-Ang.). 1 W Aus.: *F. axilliflora* (Ewart & J.W. White) Short. R: OB 104(1991)132

Flabellaria Cav. Malpighiaceae. 1 trop. Afr.

Flabellariopsis R. Wilczek. Malpighiaceae. 1 trop. Afr.

Flacourtia Comm. ex L'Hérit. Flacourtiaceae (8). c. 15 trop. & S Afr. to Fiji. Fr. ± ed., some medic. esp. *F. indica* (Burm. f.) Merr. (*F. ramontchi*, Madagascar or Governor's plum, ramontchi, OW trop.) – fr. for jelly, medic.; *F. inermis* Roxb. (lovi-lovi, origin uncertain) – fr. variable, for jelly; *F. jangomas* (Lour.) Räuschel (another cultigen); *F. rukam* Zoll. & Moritzi (rukam, Mal.)

Flacourtiaceae Rich. Dicots – Dilleniidae – Violales. (Incl. Aphloiaceae, Berberidopsidaceae, Kiggelariaceae & Lacistemataceae; excl. Plagiopteraceae) 86/875 trop., few subtemp. Shrubs, less often trees, often cyanogenic, s.t. (as in New Caled.) accumulating nickel. Lvs spirally arr., rarely opp. or whorled, simple, sometimes gland-dotted; stipules often caducous; ± accessory buds. Fls solit., axillary or epiphyllous (e.g. *Mocquerysia*, *Phyllobotryon*, *Phylloclinium*), or in usu. cymose infls, or (*Bembicia*) in cone-like pseudanthia, usu. bisexual, ± hypogynous ($\overline{G}$ in *Bembicia*); K 3–8(–15), usu. imbricate, sometimes basally connate or accrescent in fr., C (0)3–8(–15), imbricate, alt. with K or spirally arr. & not distinct from K, nectaries various, s.t. a disk within or without A or discrete glands there, s.t. a basal scale on each petal, A (4–)∞, centrifugal, s.t. in groups opp. petals, s.t. some staminodal, anthers with longit. slits (term. pores in *Kiggelaria*), s.t. with prolonged connective, corona rare (cf. Passifloraceae) & intrastaminal (e.g. *Trichostephanus*), $\underline{G}$ (2–10), app. 1 in *Aphloia*, $\overline{G}$ in *Bembicia*, 1-loc. with parietal or almost basal placentas or placentas ± deeply intruded so that ovary effectively pluriloc., distinctly so & with axile placentas rarely (e.g. *Prockia*), styles ± united, stigmas distinct, 2–∞ anatropous to orthotropous, bitegmic ovules per placenta. Fr. a berry, less often a loculicidal capsule or drupe; seeds often arillate, embryo straight (horseshoe-shaped in *Aphloia*) in often copious oily proteinaceaous endosperm, often with fatty acids of the cyclopentenoid (chaulmoogric) series. x = 10–12. R: Aliso 12(1988)29

Classification & principal genera (after Lemke):

10 tribes but some genera of uncertain disposition & many mono- or oligo-typic: (1) Berberidopsideae (*Berberidopsis*); (2) Erythrospermeae (*Dasylepis*); (3) Oncobeae (*Caloncoba, Lindackeria*); (4) Pangieae (*Hydnocarpus, Pangium, Ryparosa*); (5) Homalieae (*Calantica, Homalium*); (6) Scolopieae (*Scolopia*); (7) Prockieae (*Banara*); (8) Flacourtieae *Azara, Dovyalis, Flacourtia, Ludia, Tisonia, Xylosma*; (9) Casearieae (*Casearia, Lunania, Samyda*); (10) Bembiceae (*Bembicia*)

Least advanced group of Violales with 2 distinctive seed-types & affinities with several other fams in the order, most clearly with Passifloraceae to which *Paropsia* & its allies are referred here. *Hasseltia, Hasseltiopsis, Macrohasseltia, Pleuranthodendron* & *Prockia* (all 7) s.t. referred to Tiliaceae to which *Muntingia* is here returned; *Dankia* is moved to Theaceae while *Lacistema* & *Lozania* (= Lacistemataceae incl. here) poss. not even in Dilleniidae

Some timbers (*Casearia, Kiggelaria, Scolopia*), fr. trees (*Dovyalis, Flacourtia, Pangium*), oils used in skin disease like leprosy (*Hydnocarpus* but also *Caloncoba, Carpotroche, Gynocardia*) & cult. orn. esp. *Azara* & *Berberidopsis*

flag *Iris* spp.; **blue** or **poison f.** *I. versicolor*; **f. root** or **sweet f.** *Acorus calamus*; **water** or **yellow f.** *I. pseudacorus*

Flagellaria L. Flagellariaceae. 4 OW trop. Stems used for basketry, fish-traps etc. Young lvs of *F. indica* L. (As.) used as shampoo

Flagellariaceae Dumort. Monocots – Commelinidae – Restionales. (Incl. Hanguanaceae) 2/5–6 OW trop. Glabrous, cyanogenic herbs or lianes without sec. growth, stems arising from sympodial rhizomes accumulating sucrose & not starch, without axillary buds but with apically-dichotomizing meristems; cell-walls silicified; vessel-elements throughout vegetative body. Lvs spirally arr. with closed sheath, short petiole & 0 ligule, lamina parallel-veined, circinate in bud, apically cirrhose at maturity acting as a tendril. Fls small, bisexual, or pls dioec., reg., 3-merous, (?) wind-poll., in term. bracteate panicles; P 3 + 3, white or greenish, A 6 with longit. slits, G (3), 3-loc. with 3 styles, sometimes weakly connate basally & stigmatic ± throughout length, 1 orthotropous (? or anatropous) ovule per loc. on axile placenta. Fr. a drupe with 1 pyrene & usu. 1 or 2 seeds with copious endosperm with starch in simple grains; embryo small, undifferentiated, capping endosperm. 2n = 38

Genera: *Flagellaria, Hanguana*

Flagenium Baillon. Rubiaceae (II 5). 2 trop. W Afr., 4 Madagascar

flamboyant *Delonix regia*; **yellow f.** *Peltophorum pterocarpum*

flame flower *Ixora coccinea*; **f. of the forest** (Afr.) *Mussaenda erythrophylla*, (India) *Butea monosperma*; **Nandi f.** *Spathodea campanulata*; **f. pea** *Chorizema ilicifolium*; **f. tree** (Aus.) *Brachychiton acerifolius, Nuytsia floribunda*, (Afr.) *S. campanulata*, (WI) *Delonix regia*; **f. vine** *Pyrostegia venusta*

flamingo flower *Anthurium andraeanum* & *A. scherzerianum*

Flanagania Schltr. = *Cynanchum*

flannel flower *Actinotus helianthi*; **f. plant** *Verbascum thapsus*

Flaveria Juss. Compositae (Hele.-Flav.). 21 Am. (esp. SW N Am.), Aus. 1 incl. 2 pantrop. weeds. R: AMBG 65(1978)590. Genus with both Kranz & non K. leaf-anatomy syndromes

flax *Linum usitatissimum*; **blue f.** *L. narbonense*; **bush f.** *Astelia nervosa*; **fairy f.** *L. catharticum*; **false f.** *Camelina sativa*; **flowering f.** *L. grandiflorum*; **NZ f.** *Phormium tenax*

fleabane *Pulicaria* spp., *Erigeron* spp., *Pluchea* spp.; **common f.** *Pulicaria dysenterica*; **daisy f.** *E. annuus*

fleawort *Tephroseris integrifolia*

Fleischmannia Schultz-Bip. Compositae (Eup.-Flei.). 80 N Am., W S Am. R: MSBMBG 22(1987)285

Fleischmanniopsis R. King & H. Robinson. Compositae (Eup.-Crit.). 5 Mex., C Am. R: MSBMBG 22(1987)310

Flemingia Roxb. ex Aiton f. Leguminosae (III 10). 30 OW trop. Some dyes. *F. vestita* Benth. ex Baker a root-crop (souphlong) in N India

fleur-de-lis or fleur-de-lys *Iris* × *germanica* 'Florentina' or *I. pseudacorus* (cf. ASNB 13,10(1989)1)

Fleurya Gaudich. = *Laportea*

Fleurydora A. Chev. Ochnaceae. 1 Guinea: *F. felicis* A. Chev. R: BJ 113(1991)172

Flexanthera Rusby. Rubiaceae (I 7). 2 Colombia, Bolivia

Flickingeria A. Hawkes. Orchidaceae (V 15). 70 trop. As. to Oceania

Flinders grass *Iseilema* spp.

Flindersia R. Br. Rutaceae (Flindersiaceae). 16 E Mal., E Aus. (14, 11 endemic), New Caled. Timber-trees (yellow-wood, red beech etc.) with tubercled capsules of winged seeds. R: JAA 50(1969)481, 56(1975)243. Alks. Timbers: *F. australis* R. Br. (crow('s) ash, Aus. teak, E Aus.); *F. brayleyana* F. Muell. (Queensland or Aus. maple) – beetle-poll. in N Queensland, high quality timber; *F. collina* Bailey & *F. maculosa* (Lindley) Benth. (leopard tree, l.-wood, E Aus.); *F. pimentaliana* F. Muell. (silkwood, New Guinea, NE Queensland); *F. schottiana* F. Muell. (*F. pubescens*, silver or bumoy ash, NE & E Aus.); *F. xanthoxyla* (Hook.) Domin (*F. oxleyana*, Long Jack, E Aus.) – timber & yellow dye

Flindersiaceae (Engl.) C. White ex Airy Shaw = Rutaceae

flixweed *Descurainia sophia*

Floerkea Willd. Limnanthaceae. 1 N Am.

flopper *Kalanchoe daigremontiana*

floradora *Marsdenia floribunda*

Flora's paintbrush *Emilia* spp.

Florence whisk *Sorghum bicolor*

flores de palo Loranthaceae esp. *Phoradendron* spp.

Florestina Cass. Compositae (Hele.-Cha.). 8 S US, Mex., Guatemala. R: Brittonia 15(1963)27

Florida arrowroot *Zamia pumila*; **F. boxwood** *Schaefferia frutescens*; **F. moss** *Tillandsia usneoides*; **F. trema** *Trema micrantha*; **F. velvet bean** *Mucuna pruriens* var. *utilis*

florist's fern *Dryopteris dilatata*

Floscaldasia Cuatrec. Compositae (Ast.-Ast.). 1 Colombia

Floscopa Lour. Commelinaceae (II 2). 20 trop. & warm

Flosmutisia Cuatrec. Compositae (Ast.-Ast.). 1 Colombia

Flourensia DC. Compositae (Hel.-Verb.). 13 S N Am., 18 S Am. (disjunct). R: Fieldiana Bot. n.s. 16(1984)

flowering ash *Fraxinus ornus*; **f. cherry** *Prunus serrulata* & hybrids; **f. currant** *Ribes sanguineum*; **f. rush** *Butomus umbellatus*

Floydia L. Johnson & B. Briggs. Proteaceae. 1 NE Aus.: *F. prealta* (F. Muell.) L. Johnson & B. Briggs – nut (coohoy n.) ed. roasted. R: FlNSW 2(1991)65

Flueckigera Kuntze = *Ledenbergia*

Flueggea Willd. (~ *Securinega*). Euphorbiaceae. 14 trop. R: Allertonia 3(1984)259. *F. neowawraea* W. Hayden (Hawaii) – reduced to 41 trees by 1982; *F. virosa* (Roxb.) J. Voigt (*S. v.*, trop. OW) – fr. eaten on Mafia Is., E Afr., seeds germ. best after passage through baboon (? & human) gut

fly honeysuckle *Lonicera xylosteum*; **f. orchid** *Ophrys insectifera*; **shoo f.** *Nicandra physalodes*; **Venus's f. trap** *Dionaea muscipula*

Flyriella R. King & H. Robinson. Compositae (Eup.-Alom.). 6 SW N Am. R: Sida 11(1986)300

foam flower *Tiarella cordifolia*

Fockea Endl. Asclepiadaceae (III 2). 6 trop. & S Afr. R: Asclepios 40(1987)69. Water-storing tubers with aerial twining stems; cult. orn. esp. *F. crispa* (Jacq.) Schumann (Karoo) – 'oldest living potpl.' (1989, potted Vienna c. 1801)

Foeniculum Miller. Umbelliferae (III 8). 4–5 As., natur. widely. *F. vulgare* Miller (*F. officinale*, fennel), rep. by 2 vars in cult.: var. *azoricum* (Miller) Thell. (Florence f., finocchio) – swollen leaf-bases ed. when blanched, & var. *dulce* Battand. & Trabut – fr. larger, cult. for ess. oils (esp. *trans*-anethole) for flavouring (e.g. anisette, ouzo, pastis raki) & medic. since Mycenaean times (C13 BC)

Foetidia Comm. ex Lam. Lecythidaceae (Foetidiaceae). 1 E Afr., 14 Madag., 2 Masc.

Foetidiaceae (Niedenzu) Airy Shaw = Lecythidaceae

fog, Yorkshire *Holcus lanatus*

Fokienia A. Henry & H. Thomas. Cupressaceae. 1(2) SE China, SE As.: *F. hodginsii* (Dunn) A. Henry & H. Thomas – pe-mou oil used in scent-making extracted by distillation of roots

Foleyola Maire. Cruciferae. 1 W Sahara

folhado *Clethra arborea*

Folotsia Costantin & Bois (~ *Cynanchum*). Asclepiadaceae (III 1). 5 Madag.

fonio *Digitaria exilis*

Fonkia Philippi = *Gratiola*

Fontainea Heckel. Euphorbiaceae. 6 New Guinea, NE Aus., Vanuatu, New Caled. R: Austrobaileya 2(1985)112

Fontanesia Labill. Oleaceae. 2 Sicily, W As., China. Cult. orn. shrubs

Fontquera Maire = *Perralderia*

fool's parsley *Aethusa cynapium*

Forbesina Ridley = *Eria*

Forchhammeria Liebm. Capparidaceae. 10 Calif. to C Am. & WI

Forcipella Baillon. Acanthaceae. 5 Madagascar

Fordia Hemsley. Leguminosae (III 6). 18 SE As., W Mal. R: Blumea 36(1991)191. Ramiflorous & cauliflorous trees to 30 m. *F. rheophytica* (Buijsen) Dasuki & Schot ('*F. angustifoliola*') a rheophyte of rapids in Sarawak

Fordiophyton Stapf. Melastomataceae. 10 S China, SE As.

Forestiera Poiret. Oleaceae. 15 Am. (esp. SW N Am.). Some cult. orn. incl. *F. acuminata* (Michaux) Poiret (N Am.) – timber good for turnery

Forficaria Lindley (*Herschelia*, ~ *Disa*). Orchidaceae (IV 3). 16 trop. & S Afr. R: Bothalia 13(1981)365 ('*H*'.)

Forgesia Comm. ex Juss. Grossulariaceae (Escalloniaceae). 1 Réunion

forget-me-not *Myosotis* spp.; **American f.** *Eritrichium* spp.; **Cape f.** *Anchusa capensis*; **Chatham Is. f.** *Myosotidium hortensia*; **Chinese f.** *Cynoglossum amabile*; **garden f.** *Myosotis sylvatica*; **water f.** *M. scorpioides*

Formania W. Sm. & Small. Compositae (Anth.-Art.). 1 SW China

Formanodendron Nixon & Crepet = *Triganobalanus*

Formosia Pichon = *Anodendron*

Forrestia A. Rich. = *Amischotolype*

Forsellesia E. Greene = *Glossopetalon*

Forsskaolea L. Urticaceae (V). 6 Canary Is., SE Spain (1), Afr., Arabia, India. R: NJB 8(1988)34. Bark of *F. tenacissima* L. (S Spain to Sahara & India) used for rope

Forstera L.f. Stylidiaceae. 5 NZ, Tasmania (1)

Forsteronia G. Meyer. Apocynaceae. c. 50 trop. Am. Rubber of little value

Forsythia Vahl. Oleaceae. 7 SE Eur. (1), E As. (6). Decid. shrubs with winged seeds (& heterostylous fls before lvs emerge), much cult. esp. *F.* × *intermedia* Zabel (*F. suspensa* (Thunb.) Vahl (China, long cult. Japan; pith solid at nodes) × *F. viridissima* Lindley (China, pith hollow at nodes)) esp. 'Spectabilis', one of the most widely planted of all shrubs, from which colchicine-induced tetraploids & back-crossed triploids have been raised; fr. wall of *F. suspensa* medic. in China

Forsythiopsis Baker = *Oplonia*

Fortunatia J.F. Macbr. = *Camassia*

Fortunearia Rehder & E. Wilson. Hamamelidaceae (I 3). 1 C & E China: *E. sinensis* Rehder & E. Wilson – cult. orn. decid. shrub with male fls & bisexual fls in separate racemes

Fortunella Swingle = *Citrus*

Fortuynia Shuttlew. ex Boiss. Cruciferae. 2 Iran, Afghanistan, Baluchistan

Fosterelia Airy Shaw = *Fosterella*

Fosterella L. B. Sm. Bromeliaceae (1). 13 C Am. & W S Am. R: FN 14(1974)199

Fosteria Molseed. Iridaceae (III 4). 1 Mexico

Fothergilla L. Hamamelidaceae (I 4). 2 E N Am. R: Arnoldia 31(1971)89. Cult. orn. decid. shrubs with consp. white filaments, C 0

fountain plant *Russelia equisetiformis*; **f. tree** *Spathodea campanulata*

Fouquieria Kunth. Fouquieriaceae. 11 arid SW N Am. R: Aliso 7(1972)439. *F. columnaris* (Kellogg) Kellogg (*Idria c.*, boojum tree, cirio, Baja Calif.) – succ. columnar trunk to 20 m & 360 yrs old (R.R. Humphrey (1974) *The boojum & its home*); *F. splendens* Engelm. (coachwhip, ocotillo) – planted as spiny hedge, latex poss. rubber source

Fouquieriaceae DC. Dicots – Dilleniidae – Violales. 1/11 SW N Am. Woody or succ. xeromorphic spiny shrubs & small trees. Lvs spirally arr., small, simple, those of long shoots each arising from a decurrent ridge, basal part of petiole marcescent as a spine, in axils of which are short shoots with clustered spineless lvs. Fls hypogynous, ± reg.; K 5 imbricate, persistent, 2 outer often larger, C (5) forming a tube or salveriform corolla with imbricate lobes, A 10–18(–23), in single whorl on receptacle, exserted but those opp. K sometimes larger, anthers with longit. slits, $\underline{G}$ (3) with 1 branched style, placentation basally axile but otherwise parietal with deeply intruded placentas, which meet as fr. matures forming a central column & may even appear axile or free-central, ovules (6)14–18(–20), anatropous, bitegmic. Fr. a loculicidal capsule; seeds winged with straight embryo in thin (or 0) oily proteinaceous endosperm. x = 12. R: Aliso 7(1972)439

Only genus: *Fouquieria* (incl. *Idria*)

Prob. to be excluded from Dilleniidae

four-o'clock plant *Mirabilis jalapa*

Fourraea Greuter & Burdet = *Arabis*

Foveolaria Ruíz & Pavón = *Styrax*

Foveolina Källersjö (~ *Pentzia*). Compositae (Anth.-Mat.). 5 S Afr. R: BJLS 96(1988)316. Annuals

fox and cubs *Pilosella aurantiaca*; **f.bane** *Aconitum vulparia*; **f. berry** *Vaccinium vitis-idaea*; **f.glove** *Digitalis* spp. esp. *D. purpurea*; **f.g., Chinese** *Rehmannia elata*; **f. grape** *Vitis labrusca*; **f. nut** *Euryale ferox*; **f.tail grass** *Alopecurus pratensis*; **f.t.g., giant** *Setaria magna*; **f.t.g., green** *S. viridis*; **f.t.g., yellow** *S. pumila*; **f.t. lily** *Eremurus* spp.; **f.t., marsh** *Alopecurus geniculatus*; **f.t. millet** *S. italica*

Fragaria L. Rosaceae. c. 12 N temp. (Eur. 2). Chile. R: CJB 40(1962)869. Strawberries. Herbs with rooting runners & fleshy usu. red receptacles (active flavour principles unknown: synthetics = ethyl 1-methyl 2-phenylglycate) bearing achenes over surface, causing urticaria in some (fr. & lvs). Commercial strawberries (UK record: 231g a fruit) usu. *F. ananassa* (Weston) Lois et al. (octoploid (2n = 56), *F.* × *magna*, *F. chiloensis* Duchesne (coastal Alaska to Calif., S Am.) × *F. virginiana* Duchesne (E N Am.), raised early 1800s ('Keen's Seedling') or earlier in France), esp. 'Royal Sovereign' (1892), some cvs e.g. 'Little Scarlet' much used for jam), in C19 picked early morning in Kent & turned into jam for mid-day sales in London; *F. moschata* (Duchesne) Duchesne (*F. muricata*, hautbois, Eur.) – 'fr.' small & largely replaced by *F. ananassa*; *F. vesca* L. (wild s., N temp.) – small 'fr.' used in fancy pastries etc. esp. in France & C Eur., app. incl. continuously flowering 'alpine' s. (*F. semperflorens* (Duchesne) Staudt) & var. *monophylla* Duchesne with simple lvs & transitions to 3-foliolate

Fragariopsis A. St-Hil.= *Plukenetia*

Fragosa Ruíz & Pavón = *Azorella*

Frailea Britton & Rose. Cactaceae (III 5). 15 S Am. Cult. orn.

Franciscodendron Hyland & Steenis (~ *Hildegardia*). Sterculiaceae. 1 Queensland: *F. laurifolia* (F. Mueller) Hyland & Steenis. R: Brunonia 10(1987)211

Francoa Cav. Saxifragaceae (Francoaceae). 1 Chile (polymorphic): *F. sonchifolia* Cav. ('*F. ramosa*', bridal wreath, wedding fl.), cult. orn. herb

Francoaceae A. Juss. = Saxifragaceae

Francoeuria Cass. = *Pulicaria*

frangipani *Plumeria rubra*; **native f.** *Hymenosporum flavum*

Frangula Miller = *Rhamnus*

Frankenia L. Frankeniaceae. 80 temp. (Eur. 6, Aus. 50, Am. 14 – R: SBM 17(1987)1) & subtrop. salty habitats. Herbs or subshrubs with wiry branches & usu. inrolled hairy lvs. Some medic., poisons or ashes a salt-source; some cult. rock-pls (sea heath esp. *F. laevis* L., Eur.). *F. salina* (Molina) I.M. Johnston (*F. grandifolia*, yerba reuma, SW N Am.) – shrub used in treatment of rheumatism

Frankeniaceae Desv. Dicots – Dilleniidae – Violales. 2/81 mostly temp. & subtrop. saline habitats. Herbs or shrubs. Lvs opp., simple & often ericoid with inrolled margins & often with salt-excreting glands; stipules 0. Fls axillary, solit. or cymose, bibracteolate, usu. bisexual, reg.; K (4–7), basally connate with short induplicate-valvate lobes, C 4–7, imbricate with long claws, A 4–7(–24) usu. 3 + 3, ± basally connate, anthers with longit. slits, $\underline{G}$ ((2)3(4)), 1-loc. with as many parietal placentas & 1 style with distinct stigmas, each placenta with (1)2–6(–∞) anatropous, bitegmic ovules. Fr. a loculicidal capsule, enclosed in persistent K; seeds with central straight embryo in abundant starchy endosperm. n = 10, 15

Genera: *Frankenia*, *Hypericopsis*

Allied to Tamaricaceae but lvs opp.

frankincense *Boswellia* spp. esp. *B. sacra*; **f. fern** *Mohria caffrorum*; **f. pine** *Pinus taeda*

Franklandia R. Br. Proteaceae. 2 SW Aus.

Franklinia Bartram ex Marshall. Theaceae. 1 SE Georgia, USA (now extinct in wild): *F. alatamaha* Marshall last seen in wild in 1803, cult. orn. decid. tree, can hybridize with *Gordonia lasianthus*

Franseria Cav. = *Ambrosia*

Frantzia Pittier = *Sechium*

Frasera Walter. Gentianaceae. 15 N Am. Some cult. orn. incl. *F. speciosa* Douglas ex Griseb. (Rocky Mts) – sporadic synchronous flowering every 2–4 yrs (fls preformed at least 3 yrs before anthesis) suggesting that massive seed-crops necessary to overcome pressure of seed-eaters; some locally medic. as tonics

Fraser's fir *Abies fraseri*

Fraunhofera C. Martius. Celastraceae. 1 Brazil

Fraxinus L. Oleaceae. 65 N temp. (Eur. 5) a few extending to trop. R: Pfl. 4,243(1920)9. Ash. Timber trees usu. with pinnate lvs (some domatia) & 1-seeded winged samaras esp. *F. excelsior* L. (common or Eur. a., Eur., SW As.) – serial axillary buds, fls in short racemes before lvs with P 0, A 2, G 2, anemophilous, with male, female & bisexual fls in diff. combinations on diff. trees, elastic timber (French, Polish or Slavonian a.) form. for wheels & still for shooting-brake timbers (e.g. Morris Minor, also carbody frames of Morgan cars), tool-handles & sports goods e.g. tennis racquets, polo mallets, billiard cues, hockey-sticks (handle = cane), cricket stumps & croquet-mallet handles, the toughest of Br. timbers, bark form. used for fevers, lvs a cattle-fodder in Scandinavia. Other spp. with similar uses & cult. orn. incl. *F. americana* L. (American, Canadian or white a., N Am.) incl. 'Ascidia' with lvs pitcher-shaped at base; *F. caroliniana* Miller (water a., E N Am.); *F. chinensis* Roxb. (China) – source of 'Chinese insect white wax' extruded by feeding insects on pollarded trees (insects form. coll. from *Ligustrum lucidum* & transported by humans in 25 kg loads for over 300 km to trees, where each kg of insects gave up to 5 kg of wax in 6 mm layers on lvs after 100 days), used in Chinese candles, for coating pills & high quality paper & for polishing jade, soapstone etc.; *F. latifolia* Benth. (Oregon a., W N Am.); *F. mandschurica* Rupr. (Japanese a., NE As.); *F. nigra* Marshall (Amer., black, brown, Can. or white a., E N Am.); *F. ornus* L. (flowering or manna a., S Eur., W As.) – white corolla in fragrant insect-poll. fls produced at same time as lvs, cult. in Sicily & Calabria for manna sugar or syrup which exudes from branches when damaged by insects & is a mild laxative, mite-galls with organs intermediate between stems & lvs; *F. pennsylvanica* Marshall (*F. pubescens*, Amer., Can., green, red or white a., E N Am.) – imp. urban forestry in Canadian prairies esp. to replace elms

Fredolia (Bunge) Ulbr. = *Anabasis*

Freesia Ecklon ex Klatt. Iridaceae (IV 9). 11 S Afr. esp. Cape. R: JSAB 48(1982)39. Tunicate corms & 2-ranked lvs with fragrant fls, the basis of imp. cut-flower industry, hybridization in earnest since 1898, florists' freesias being referred to *F.* × *hybrida* L. Bailey with blood of *F. refracta* (Jacq.) Klatt (fls greenish-yellow, v. fragrant), *F. alba* (G.L. Meyer) Gumbleton (fls white, v. strong scent), *F. corymbosa* (Burm.f.) N.E. Br. (esp. forms with small pink fls, '*F. armstrongii*') & now with many named cvs incl. double ones & tetraploids. The ability to smell f. varies & some find it difficult to detect any scent at all. *Anomatheca* prob. referable here

Fregea Reichb.f. = *Sobralia*

freijo *Cordia goeldiana*

Fremontia Torrey = *Fremontodendron*

Fremontodendron Cov. Sterculiaceae. 3 SW N Am. R: SB 16(1991)3. Cult. orn. ± evergreen shrubs & trees with irritant brown hairs & petaloid K, C 0 & A united basally in a tube, esp. 'California Beauty', the hybrid between *F. californicum* (Torrey) Cov. (1 sold in 1850s for £37.80!) & *F. mexicanum* Davidson, & allegedly hardier than either

French bean *Phaseolous vulgaris*; **F. berries** *Rhamnus infectoria*; **F. cotton** *Calotropis procera*; **F. honeysuckle** *Hedysarum coronarium*; **F. jujube** *Ziziphus jujuba*; **F. lavender** *Lavandula stoechas*; **F. marigold** *Tagetes patula*; **F. millet** *Panicum miliaceum*; **F. mulberry** *Callicarpa americana*; **F. ryegrass** *Arrhenatherum elatius*; **F. turpentine** *Pinus pinaster*; **F. whisk** *Chrysopogon gryllus*

Frerea Dalzell. Asclepiadaceae (III 5). 1 NW India: *F. indica* Dalzell. R: Bradleya 8(1990)9

Fresenia DC = *Felicia*

Freya Badillo. Compositae (Helia.). 1 Venezuela

Freycinetia Gaudich. Pandanaceae. 175 Sri Lanka to NZ (1) & Polynesia. R: Blumea 16(1968)361. Usu. lianes, dioec. (some monoec. in Aus.) with brightly coloured fleshy bracts but fr. a syncarp of berries (drupes in *Pandanus*), some poll. bats attracted by fleshy bracts, e.g. *F. baueriana* Endl. (Norfolk Is., NZ) – bats extinct or rare in NZ & prob. poll. & disp. by introd. Aus. possum, *F. arborea* Gaudich. (Hawaii) by *Zosterops japonica* (white-eye, introd. 1929), the orig. pollinators (drepanids) now extinct; *F. reineckei* Warb. (Samoa) produces occ. bisexual spikes & fls. Leaf-fibre used for weaving, skirts etc.

Freyera Reichb. = *Geocaryum*

Freylinia Colla. Scrophulariaceae. 4 trop. & S Afr. Cult. orn. fragrant shrubs

Freyliniopsis Engl. = *Manuleopsis*

Freziera Willd. Theaceae. (Incl. *Patascoya*) c. 42 trop. Am. Dioec. with more females than males (unlike most trop. trees). *F. forerorum* A. Gentry (E Panamá) with v. asymmetric lvs

Friar's balsam tincture largely derived from *Styrax* spp.; **f.'s cowl** *Arisarum vulgare*

Fridericia C. Martius. Bignoniaceae. 1 S Brazil

Friesodielsia Steenis. Annonaceae. 55 trop. W Afr. (7), Indomal. ? Heterogeneous. Some lianes

frijoles *Phaseolus vulgaris*

fringe flower or **tree** *Chionanthus* spp. esp. *C. virginicus*; **f.-myrtle** *Calytrix* spp. esp. *C. tetragona*; **f.d lily** *Thysanotus* spp.

Frithia N.E. Br. Aizoaceae (V). 1 Pretoria area, S Afr.: *F. pulchra* N.E. Br. – stemless succ. herb. with club-shaped lvs with apical windows through which light passes to photosynthetic tissues within, cult. orn.

Fritillaria L. Liliaceae. 100 W Eur. & Medit. (Eur. 24) to E As. R: HIP 39(1980)265. Bulbous with many alks & 1 or more fleshy scales & sometimes rice-grain-like bulblets; seeds many, flat. Many cult. orn. & few ed. etc. *F. camschatcensis* (L.) Ker ('black sarana', E As., coastal NW Am.) – bulbs ed.?, fls purple-black; *F. cirrhosa* D. Don (Himal.) – lvs with tendril-like tips (cf. *Gloriosa, Littonia*), bulb pounded with orange juice & sugar for chest ailments; *F. imperialis* L. (crown imperial, Iran to N India) – fls in whorl below term. tuft of bracts, whole pl. with fox or skunk smell, a favourite of the Dutch Masters, once medic. & starch source, ed. cooked (imperialine a heart poison); *F. meleagris* L. (snake's head, guinea fl., Eur., W As., natur. GB where protected though not recorded before 1732) – P chequered shades of red & purple (Latin: *fritillus* = dicebox, which were form. chequered thus), form. called leper-lily as fl. resembles bells lepers were obliged to carry, cult. orn. since Tudor times, seeds water-disp.; *F. persica* L. (Cyprus to Iran) – bulb with single massive scale invested with remnants of older ones; *F. roylei* Hook. (Himal.) – bulb medic. in China (pei-mu) used like *F. cirrhosa*

fritillary *Fritillaria* spp.

Fritzschia Cham. Melastomataceae. 1 Brazil

Froelichia Moench. Amaranthaceae (II 2). 18 warm Am., Galápagos. Lateral fls sterile & developing wings serving to aid dispersal

Froelichiella R.E. Fries. Amaranthaceae (II 2). 1 Brazil

Froesia Pires. Quiinaceae. 4 trop. S Am. R: Novon 4(1994)246

Froesiochloa G.A. Black. Gramineae (4). 2 trop. S Am.

Froesiodendron R.E. Fries = *Cardiopetalum*

frog-bit *Hydrocharis morsus-ranae;* **American f.-b.** *Limnobium spongia;* **f. orchid** *Coeloglossum viride*

Frolovia (DC) Lipsch. = *Saussurea*

Frommia H. Wolff. Umbelliferae (III 8). 1 S C Afr.

Frondaria Luer (~ *Pleurothallis*). Orchidaceae (V 13). 1 C Colombia to C Bolivia: *F. caulescens* (Lindley) Luer

Froriepia K. Koch. Umbelliferae (III 8). 2 Turkey to Iran

frost flowers *Aster* spp.; **f. grape** *Vitis vulpina;* **f.weed** *Helianthemum canadense*

Fryxellia D. Bates (~ *Anoda*). Malvaceae. 1 SW N Am.

Fuchsia L. Onagraceae. 105 C & S Am., Tahiti (1), NZ (4). R: PCAS IV, 25(1943)1, AMBG 69(1982)1. Shrubs (some with tubers) & trees with opp. or whorled lvs & usu. pendulous bird-poll. fleshy fls & ed. berries. A few locally medic. but many cult. orn. esp. hybrids grown in hanging baskets & as standards e.g. *F.* × *hybrida* Hort. ex Siebold & Voss (prob. involving *F. magellanica* & *F. fulgens* DC (Mex.) but *F. coccinea* Dryander (S Brazil), *F. microphylla* Kunth & *F. thymifolia* Kunth (both Mex.) prob. involved in ancestry of modern hybrids). *F. excorticata* (Forster & Forster f.) L.f. (NZ) – decid. gynodioec. tree a target for introd. possums; *F. magellanica* Lam. (S Chile, Arg.) – forms (e.g. 'Riccartonii') grown for hedging in Azores, Isle of Man & Ireland, where ± natur., weedy in Hawaii; *F. procumbens* R. Cunn. (NZ) – prostrate, fls erect

fuchsia, Californian *Epilobium* spp.

Fuernrohria K. Koch. Umbelliferae (III 4). 1 Caucasus, Armenia

Fuerstia T.C.E. Fries. Labiatae (VIII 4b). 6 trop. Afr.

Fuertesia Urban. Loasaceae. 1 Hispaniola

Fuertesiella Schltr. Orchidaceae (III 3). 1 WI

fufu (W Afr.) mashed or pounded starchy food, orig. *Dioscorea* spp.

Fuirena Rottb. Cyperaceae. 30 warm (Eur. 1)

Fukien tea *Carmona retusa*

Fulcaldea Poiret. Compositae (Barn.). 1 W trop. S Am. Shrub

fuller's herb *Saponaria officinalis;* **f. teasel** *Dipsacus sativus*

Fumana (Dunal) Spach. Cistaceae. c. 9 Eur., N Afr. Cult. orn. like *Helianthemum* but outer A sterile

Fumaria L. Fumariaceae (I 2). 50 Eur. (39), Medit. to C As. & Himal. (1), trop. E Afr. highlands (1). R: OB 88(1986)42. Fumitory. Usu. autogamous annuals, mostly polyploid, with fls like *Corydalis;* many scramble with sensitive petiolules; caruncles; alks. *F. officinalis* L. (Eur., Medit. to Iran) – form. yellow dye source

Fumariaceae Bercht. & Presl (~ Papaveraceae). Dicots – Magnoliidae – Papaverales. (Excl. Pteridophyllaceae) 17/530 N temp. with few trop. Afr. mts & S Afr. Herbs often with swollen underground parts, sometimes scramblers. Lvs spirally arr. to subopp. usu. ± dissected; stipules 0. Fls (1) few to several, bisexual, irreg. (± reg. in *Hypecoum*); K 2, bract-like, often ± peltate, not encl. bud, caducous, C 2 + 2 in all but *H.*, 1 or both outer with basal spur or pouch & inner ± connate over stigmas at tip, A 6 in 2 bundles opp. outer C (4 free in *H.*), dimorphic with middle ones with 4 microsporangia & 2 pollen-sacs, others with 2 & 1 resp., nectaries often at base of A, $\underline{G}$ (2), 1-loc. with 1–∞ anatropous to campylotropous bitegmic ovules on parietal placentas. Fr. usu. a capsule, often with replum, longit. dehiscent, rarely breaking into l-seeded segments or a nut; seeds usu. arillate with small embryo & abundant oily endosperm. Seedlings s.t. with only 1 cotyledon (some *Corydalis* & *Dicentra* spp.). x = (6–)8(+)

Classification & principal genera:

I. **Fumarioideae** (fls irreg., A 6, 2 tribes):
  1. **Corydaleae** (fr. usu. many-seeded capsule): *Corydalis, Dicentra*
  2. **Fumarieae** (fr. a nut or few-seeded capsule): *Fumaria, Rupicapnos*

II. **Hypecoideae** (fls ± reg., A 4): *Hypecoum*

Cult. orn. esp. *Corydalis* & *Dicentra;* some weeds esp. *Fumaria*

Fumariola Korsh. Fumariaceae (I 2). 1 C As.: *F. turkestanica* Korsh. R: OB 88(1986)42

fumitory *Fumaria* spp.

Funastrum Fourn. = *Sarcostemma*

fundi *Digitaria exilis*

Funifera Leandro ex C. Meyer. Thymelaeaceae. 3–4 Brazil

Funkia Sprengel = *Hosta*

Funkiaceae Horan. = Agavaceae

Funkiella Schltr. Orchidaceae (III 1). 4 Mex., Guatemala. R: FFG 35(1991)19

Funtumia Stapf. Apocynaceae. 2 trop. Afr. R: MLW 81–16(1981). Alks. Rubber sources esp. *F. elastica* (Preuss) Stapf (Lagos or Iré r.), comm. valuable tree in regrowth of forest after felling timber trees

Furarium Rizz. = *Oryctanthus*

Furcaria (DC) Kostel. = *Hibiscus*

Furcraea Vent. Agavaceae. c. 20 trop. Am. R: RMBG 18(1907)25. Pachycaul succ. ± stem with large lvs in rosettes & term. infls even larger than in *Agave*; usu. hapaxanthic with bulblets in infls; some imp. fibres (fique) & cult. orn. esp.: *F. foetida* (L.) Haw. (*F. gigantea*, Mauritius hemp, N S Am.) – comm. cult. in Mauritius, St Helena etc. (firebreak in Sri Lanka) & *F. hexapetala* (Jacq.) Urban (*F. cubensis*, ? poss. not distinct from *F. foetida*, Cuba hemp, Cuba, Haiti) – both for twine, cordage, sacking etc. Others less used incl. *F. cabuya* Trel. (cabuya, C Am.) – ropes, hammocks etc.

Furtadoa Hotta (~ *Homalomena*). Araceae (VI 1). 2 Sumatra

furze *Ulex* spp.

Fusaea (Baillon) Saff. Annonaceae. 2 trop. S Am. R: AHB 12(1934)207, (1937)279

Fusispermum Cuatrec. Violaceae. 2 Colombia

fustic *Maclura tinctoria*; **young f.** *Cotinus coggygria*

futi, futui *Jacaranda copaia*

G

Gaboon chocolate *Irvingia gabonensis*; **G. mahogany** *Aucoumea klaineana*

Gabunia Schumann ex Stapf = *Tabernaemontana*

Gaertnera Lam. Rubiaceae (IV 7). 30 OW trop. G

Gaertnera Schreber = *Hiptage*

gage See *Prunus*

Gagea Salisb. Liliaceae. 70 temp. Euras. (Eur. 23). Axillary bulbils develop if poll. fails. R: PL 14(1958)124, 15(1959)151, 16(1960)163. 2 in Br. Is. (*G. lutea* (L.) Ker-Gawler (yellow star-of-Bethlehem, Eur. to Himal., where major food of Himal. snowcock) & *G. bohemica* (Zauschner) Schultes & Schultes f. (C & S Eur.), a relict pop. recently discovered in Welsh mts)

Gagnebina Necker ex DC. Leguminosae (II 3). 4–5 Madag. & W Indian Ocean. R: KB 41(1986)463

Gagnepainia Schumann. Zingiberaceae. 3 SE As.

Gagria Král = *Pachyphragma*

Gahnia Forster & Forster f. Cyperaceae. 40 E As., Mal., Aus. (22, 20 endemic), Pacific. R: Bot.Arch. 40(1940)151. Some woody; fr. exposed by attachment to marcescent A bases thrust outwards, their tips being retained by surrounding bracts

Gaiadendron G. Don f. Loranthaceae. 1 trop. Am mts: *G. punctatum* (Ruíz & Pavón) G. Don f. – terr., treelike in S Am., epiphytic in C Am.

Gaillardia Foug. Compositae (Helia.). 28 N & temp. S (2) Am. Cult. orn. (blanket flowers, Indian b.) esp. perennial *G.* × *grandiflora* Van Houtte (*G. aristata* Pursh (N Am.) × *G. pulchella* Foug. (N Am.), natur. W N Am.) & annual *G. pulchella*

Gaillonia A. Rich. ex DC = *Neogaillonia*

Gaimardia Gaudich. Centrolepidaceae. 3 New Guinea, Tasmania, NZ, Antarctic S Am.

Galactia P. Browne. Leguminosae (III 10). 140 warm esp. Am. (Aus. 3). Latex (rare in L.)

Galactites Moench. Compositae (Card.-Card.). 3 Canary Is. & Medit. (Eur. 2)

Galactophora Woodson. Apocynaceae. 7 trop. S Am.

Galagania Lipsky = *Muretia*

galanga *Kaempferia* spp., *Alpinia* spp.

galangal *Alpinia galanga*, *A. officinarum*

Galanthus L. Amaryllidaceae (Liliaceae s.l.). c. 12 Eur. (4) to Iran. R: DTYB 32(1966)62. Snowdrops (an allusion to Mediaeval earring fashions, not weather). Alks. P 3 + 3 shorter (cf. *Leucojum* where all same length), inner with nectaries, visited by bees which touch stigma when grasping pendulous fl. & are showered with pollen when probing for nectar. Many cult. orn. (30 million exported annually from Turkey, *G. elwesii* Hook. f. (Balkans, W Turkey) much depleted along Med. coast, yet can be readily prop. by 'chipping' from bulb fragments) esp. *G. nivalis* L. (Eur., natur. GB), 'Scharlokii' with 2 fls per peduncle, others with 'double' fls; *G. reginae-olgae* Orph. (*G. corcyrensis*, *G. nivalis* ssp. *r.-o.*, SE Eur.) – flowering autumn, lvs revolute

Galatella (Cass.) Cass. (~ *Aster*). Compositae (Ast.-Ast.). c. 25 Euras.

Galax Sims. Diapensiaceae. 1 SE US: *G. urceolata* (Poiret) Brummitt ('*G. aphylla*', beetleweed) – cult. ground cover

Galaxia Thunb. Iridaceae (III 2). 15 Cape (winter rainfall region). R: JSAB 45(1979)385

galba *Calophyllum brasiliense*

galbanum *Ferula galbaniflua, F. gummosa, F. rubricaulis*

Galbulimima Bailey. Himantandraceae. 2 E Mal., NE Aus. Alks; fr. taken by fruit-pigeons. *G. belgraveana* (F. Muell.) Sprague (New Guinea) when taken with lvs of *Homalomena* spp. leads to violent intoxication & deep sleep with hallucinations

Gale Duhamel = *Myrica*

gale, sweet *Myrica gale*

Galeana Llave & Lex. Compositae (Hele.-Hym.). 3 C Am.

Galeandra Lindley & Bauer. Orchidaceae (V 9). 25 C & S Am. Cult. orn. terr. & epiphytic incl. *G. devoniana* Lindley – flood-tolerant epiphyte in igapó of Amaz.

Galearia Zoll. & Moritzi. Pandaceae. 6 SE As. to Solomon Is.

Galearis Raf. Orchidaceae (IV 2). 12 India, Tibet, temp. N Am., Greenland

Galeatella (F. Wimmer) Degener & I. Degener = *Lobelia*

Galega L. Leguminosae (III 15). 6 Euras. (Eur. 1), E Afr. mts. Alks. *G. officinalis* L. (goat's rue, Eur. to Iran) – fodder (also *G. orientalis* Lam., Cauc.) & cult. orn., form. medic., stem-fibre used for paper

Galenia L. Aizoaceae (I). 15 Namibia & W S Afr. R: JSAB 22(1956)88. Heterogeneous?

Galeobdolon Adans. = *Lamium*

Galeola Lour. Orchidaceae (V 4). 10 Madag., Indomal. to Aus. Lianoid mycotrophs to several m. *G. septentrionalis* Reichb.f. (Japan) – endozoochory

Galeomma Rauschert (*Eriosphaera*). Compositae (Gnap.-Gnap.). 2 Cape. R: OB 104(1991)135

Galeopsis L. Labiatae (VI). 10 temp. Euras. (Eur. 9). Hemp nettles. *G. tetrahit* L. (2n = 32, Eur.) = *G. speciosa* Miller (2n = 16, Eur.) × *G. pubescens* Besser (2n = 16, Eur.), first allotetraploid to be resynthesized (the triploid hybrid back-crossed with *G. p.*)

Galeottia A. Rich. (*Mendoncella*). Orchidaceae (V 10). 11 trop. Am. R: Lindleyana 3(1988)221. Cult. orn.

Galeottiella Schltr. (~ *Brachystele*). Orchidaceae (III 1). 5 Mex., Guatemala

Galianthe Griseb. = *Spermacoce*

galingale *Cyperus longus*

Galiniera Del. Rubiaceae (II 5). 2 trop. Afr., Madagascar

Galinsoga Ruíz & Pavón. Compositae (Helia.-Gal.). 13 temp. & subtrop. C & S Am. R: Rhodora 79(1977)319. *G. parviflora* Cav. & *G. quadriradiata* Ruíz & Pavón (*G. ciliata*), diploid & polyploid resp., weeds now almost cosmop., cooked as veg. in SE As.

Galipea Aublet. Rutaceae. 14 trop. Am. Alks. Bark of *G. officinalis* Hancock produces angostura used in pink gins etc.; others medic.

Galitzkya Botsch. = *Alyssum*

Galium L. Rubiaceae (IV 16). c. 300 cosmop. (Eur. 145, Turkey 101). Slender herbs with usu. square stems with whorled lvs & leaflike 'stipules' (bedstraws), some with coumarin when dried (allegedly not in living pls) & used to keep with linen or as mattress-stuffing, esp. *G. odoratum* (L.) Scop. (*Asperula o.*, woodruff, Euras., Medit.) – flavouring for drinks, snuff etc., locally medic., cult. orn.; *G. aparine* L. (goose-grass, cleavers, sticky bobs, sticky willy, sweethearts, Euras., S S Am.) – scrambler with reflexed hooks on stems & schizocarps which are animal-disp. & form. used as coffee subs. in Ireland; *G. circaezans* Michaux (N Am.) – myrmechory; *G. mollugo* L. (hedge b., Eur., natur. N Am.) – occ. cult.; *G. verum* L. (lady's b., Eur. to Iran, natur. N Am.) – yellow fls smelling of urine, attractive to flies etc., form. used as styptic & to curdle milk in cheese-making in Eng., medic., roots yield a red dye

gall *Quercus pubescens*

Gallardoa Hicken. Malpighiaceae. 1 Argentina

gallberry *Ilex glabra*

Gallesia Casar. Phytolaccaceae (II). 1 Peru, Brazil: *G. integrifolia* (Speg.) Harms. Pl. smells of garlic; fr. a samara. Locally medic. for worms etc.

Gallienia Dubard & Dop. Rubiaceae (II 5). 1 Madagascar

Galopina Thunb. Rubiaceae (IV 13). 4 SW Cape to S Malawi

Galphimia Cav. Malpighiaceae. 10 Texas to Arg. Cult. orn. shrubs

Galpinia N.E. Br. Lythraceae. 3 S Afr.

Galtonia Decne. Hyacinthaceae (Liliaceae s.l.). 4 Drakensberg, S Afr. R: NRBGE 45(1988)95. Cult. orn. bulbous pls esp. *G. candicans* (Baker) Decne (summer hyacinth, Berg lily)

Galvezia Dombey ex Juss. Scrophulariaceae (Antirr.). 4 coastal Peru, Ecuador, Galápagos. R:

D.A. Sutton, *Rev. Antirr.* (1988)514

gama grass *Tripsacum dactyloides*

gamalu *Pterocarpus marsupium*

Gamanthera van der Werff & Endress. Lauraceae (I). 1 Costa Rica

Gamanthus Bunge (~ *Halanthium*). Chenopodiaceae (III 3). 5 E Med. to C As. & Afghanistan

gamar *Gmelina arborea*

gamba grass *Andropogon gayanus*

Gambelia Nutt. (~ *Antirrhinum*). Scrophulariaceae (Antirr.). 4 SW N Am. R: D.A. Sutton, Rev. Antirr. (1988)510. Shrubs, some juncoid; *G. speciosa* Nutt. poll. hummingbirds

Gambeya Pierre = *Chrysophyllum*

Gambeyobotrys Aubrév. = *Chrysophyllum*

Gambia pods *Acacia nilotica*

gambier or **gambir** *Uncaria gambier*

Gamblea C.B. Clarke. Araliaceae. (Incl. *Evodiopanax*) 4 E Himal. to Japan & W Mal.

gamboge *Garcinia* spp. esp. *G. xanthochymus*

Gamocarpha DC. Calyceraceae. 6 temp. S Am.

Gamochaeta Wedd. (~ *Gnaphalium*). Compositae (Gnap.-Gnap.). 52, some weedy cosmop. esp. S Am. R: OB 104(1991)155

Gamochaetopsis Anderb. & Freire (~ *Lucilia*). Compositae (Gnap.-Gnap.). 2 subAntarctic S Andes. R: BJLS 106(1992)186

Gamolepis Less. = *Steirodiscus*

Gamopoda Baker = *Rhaptonema*

Gamosepalum Hausskn. = *Alyssum*

Gamosepalum Schltr. = *Aulosepalum*

Gamotopea Bremek. Rubiaceae (IV 7). 5 trop. S Am.

gampi *Wikstroemia sikokiana*

Gamwellia Baker f. = *Gleditsia*

gandaria *Bouea macrophylla*

gang-flower *Polygala vulgaris*

ganja *Cannabis sativa* subsp. *indica*

Ganophyllum Blume. Sapindaceae. 1 trop. W Afr. (*G. giganteum* (A. Chev.) Hauman – timber & ed. fr.), 1 Andamans, Vietnam, Malay Pen., Sumatra, Java, Philipp., New Guinea (*G. falcatum* Blume – timber, seeds yield an oil)

Gantelbua Bremek. = *Hemigraphis*

Ganua Pierre ex Dubard = *Madhuca*

Garaventia Looser. Alliaceae (Liliaceae s.l.). 1 Chile

Garaya Szlach. (~ *Cyclopogon*). Orchidaceae (III 3). 1 Brazil: *G. atroviridis* (Barb. Rodr.) Szlach.

Garayella Brieger = *Chamelophyton*

garbanzo beans *Cicer arietinum*

Garberia A. Gray. Compositae (Eup.-Liat.). 1 SE US. R: MSBMBG 22(1987)280

Garcia Rohr. Euphorbiaceae. 2 Mex. (cult. trop. Am.). Yield an oil like tung oil; fls consp. (K (2 or 3), C 8–12 pinkish, A 60+), cult. orn.

Garcibarrigoa Cuatrec. (~ *Senecio*). Compositae (Sen.-Sen.). 1 Colombia, Ecuador: *G. telembina* (Cuatrec.) Cuatrec. R: Candollea 15(1986)7

Garcilassa Poeppig. Compositae (Helia.-Helia.). 1 Costa Rica to Peru & Bolivia

Garcinia L. Guttiferae (III). (Incl. *Rheedia*, *Tripetalum*) 200 trop. OW esp. As., S Afr. Polygamous usu. slow-growing trees & shrubs (some rheophytic) with usu. nocturnal highly scented fls & berries with (often ed.) fleshy endocarp around seeds, some parthenocarpic; resins give pigments incl. gamboge; some waxes & timbers. Some wild spp. & also *G. mangostana* in cult. apomictic (adventive embryony). *G. cambogia* Desr. (India) – pink deeply lobed endocarp, pericarp dried for fish curry; *G. cowa* Roxb. ex DC (Indomal.) – endocarp & pericarp good flavour; *G. indica* (Thouars) Choisy (India) – ed. fat (kokam, kokum or Goa butter) from seeds, pericarp used in flavouring curries; *G. kola* Heckel (W & C Afr.) – chewing-sticks; *G. mangostana* L. (allopolyploid female (2n = 88–90) = *G. hombroniana* Pierre (2n = 48) × *G. malaccensis* Hook.f. (2n = ?42), mangosteen, Mal.) – delicious endocarp, one of the best trop. frs only productive there; *G. mannii* Oliver (W Afr.) – antibacterial chewing-stick in Cameroun; *G. pedunculata* Roxb. ex Buch.-Ham. (Bengal) – large ed. fr.; *G. sopsopia* (Buch.-Ham.) Mabb. (*G. paniculata*, E Himal.) – fr. cherry-sized, mangosteen-flavoured; *G. xanthochymus* Hook.f. ex T. Anderson (N India) & other spp. when tapped yield gamboge used in watercolours & dyeing (e.g. Buddhist priests' robes); *G.* sp. (*Rheedia brasiliensis*, bakupari, Brazil) – fr. sold in markets

gardener's garters *Phalaris arundinacea* var. *picta*

Gardenia Ellis. Rubiaceae (II 1). c. 60 trop. & warm OW. Shrubs & trees with opp. or whorled lvs (3s, app. 2 prs with 1 reduced to a scale) & usu. large scented (like *Polianthes*) white or yellow fls. Form. much cult. under glass for button-holes, esp. *G. augusta* (L.) Merr. (*G. florida*, Cape jasmine, China) esp. double-flowered forms, used to scent tea etc., fr. source of a yellow dye, used in treatment of influenza & colds in modern Chinese herbalism. Other spp. medic. & app. insecticidal, minor sources of timber & dyes

Gardeniopsis Miq. Rubiaceae (?II 4). 1 W Mal.

Gardneria Wallich. Strychnaceae (Loganiaceae s.l.). 5 India & C Japan to Java. R: BJBB 32(1962)431. Alks

Gardnerina R. King & H. Robinson. Compositae (Eup.-Ager.). 1 Brazil (known only from 1 specimen). R: MSBMBG 22(1987)157

Gardnerodoxa Sandw. Bignoniaceae. 1 Brazil

Gardoquia Ruíz & Pavón = *Satureja*

Garhadiolus Jaub. & Spach (~ *Rhagadiolus*). Compositae (Lac.-Hyp.). 4 S W As. to China

gari *Manihot esculenta*

Garidella L. (~ *Nigella*). Ranunculaceae (I 3). 2 S Eur. (2) to C As. C longer than K unlike in *Nigella*

garlic *Allium sativum*; **Canadian g.** *A. canadense*; **g. chives** *A. tuberosum*; **crow** or **false g.** *A. vineale*; **field g.** *A. oleraceum*; **hedge g.** *Alliaria petiolata*; **g. pear** *Crateva* spp.; **wild** or **wood g.** *A. ursinum*

garnet berry *Ribes rubrum*

Garnieria Brongn. & Gris. Proteaceae. 1 New Caledonia

Garnotia Brongn. Gramineae (39). 29 S & E As. to Pacific. R: KB 27(1972)515

Garnotiella Stapf = *Asthenochloa*

Garrettia Fletcher. Labiatae (IV; Verbenaceae). 1 Thailand, Java

Garrya Douglas ex Lindley. Garryaceae. 13 Washington to Panamá, WI (1). R: CGH 209(1978). Cult. orn. shrubs esp. *G. elliptica* Douglas ex Lindley (Calif. to Oregon), form. medic. (feverbush), bark with at least 5 alks incl. delphinine otherwise known only from *Aconitum* & *Delphinium*

Garryaceae Lindley. Dicots – Rosidae – Cornales. 1/13 W N & C Am. Dioec. evergreen trees & shrubs with highly toxic alks. Lvs decussate though traces arising at diff. levels, simple; stipules 0. Fls small, wind-poll., 1–3 in axils of decussate bracts in usu. pendent catkins; males with P 4, apically connate, bract-like, A 4 alt. with P, anthers with longit. slits, females with P 0 or 2 appendages near styles, $\overline{G}$ (2(3)), 1-loc. with distinct styles & 2(3) pendulous, anatropous, unitegmic ovules. Fr. a 2-seeded berry dry & tardily dehiscent at maturity; seed 1(2) with small linear embryo in copious oily endosperm with reserves of hemicellulose. x = 11

Only genus: *Garrya*

Poss. allied to Cornaceae, particularly *Aucuba* (pollen similar & both have petroselinic acid as major fatty acid in seeds – otherwise unknown in Cornales) but chloroplast DNA evidence against this

Cult. orn. & medic. locally

Garuga Roxb. Burseraceae. 4 Himal., Indomal. to W Pacific. R: Blumea 7(1953)459, 498. Some locally used timber

Garuleum Cass. Compositae (Cal.). 8 S Afr. Pappus 0

Gasoul Adans. = *Mesembryanthemum*

gasparillo *Esenbeckia* spp.

Gasteranthus Benth. (~ *Besleria*). Gesneriaceae. 40 trop. Am. R: Selbyana 1(1975)150. One group poll. by hummingbirds, others by bees

Gasteria Duval. Asphodelaceae (Aloaceae). 14 S Afr. R: Excelsa 9(1979)29. Mostly stemless succ. with rosettes of lvs, much cult. & many variants form. regarded as spp.

Gastonia Comm. ex Lam. Araliaceae. 10 E Afr., Madag., Masc. (3), Mal., Papuasia. Trees allied to *Polyscias*; *G. spectabilis* (Harms) Philipson (New Guinea) at 40 m the tallest araliad, timber for light carpentry

Gastranthus Moritz ex Benth. = *Stenostephanus*

Gastridium Pal. Gramineae (21d). 3 Canaries, W Eur. (1), Medit. to Iran. Glumes persistent on axis

Gastrochilus D. Don (~ *Saccolabium*). Orchidaceae (V 16). 50 Himal. & E As. to Japan. R: AOSB 54 (1985)1111. Cult. orn.

Gastrococos Morales (~ *Acrocomia*). Palmae (V 5e). 1 Cuba (limestones): *G. crispa* (Kunth) H. Moore

Gastrocotyle Bunge. Boraginaceae. 2 E Med. to C As. & NW India

Gastrodia R. Br. Orchidaceae (V 5). 35 E As., Indomal. to NZ (Aus. 7, 6 endemic). Some mycotrophs, with tubers roasted & eaten by aborigines; *G. sesamoides* R. Br. (potato orchid, S & E Aus., NZ); *G. elata* Blume (SE As.) – dried & powdered a common cure for headaches in China

Gastrolepis Tieghem. Icacinaceae. 1 New Caledonia

Gastrolobium R.Br. Leguminosae (III 24). 40 W Aus. Produce fluoroacetate (in *G. bilobum* up to 2.65 g per kg) highly toxic to stock ('heart-leaf') as blocks Krebs Cycle at citrate stage, though some marsupials ± resistant. *G. bilobum* R. Br. – ectomycorrhizae spread by red kangaroo, spores germ. after digestion; *G. calycinum* Benth. (York road poison) & *G. spinosum* Benth. (prickly poison) – v. toxic to stock

Gastrolychnis (Fenzl) Reichb. = *Silene*

Gastronychia Small = *Paronychia*

Gastropyrum (Jaub. & Spach) Löve = *Aegilops*

Gastrorchis Schltr. = *Phaius*

gaub tree *Diospyros malabarica*

Gaudichaudia Kunth. Malpighiaceae. c. 10 Mex. to Bolivia

Gaudinia Pal. Gramineae (21b). 4 Azores & Medit. (Eur. 3). Spikelet many-flowered

Gaudiniopsis (Boiss.) Eig = *Ventenata*

× **Gaulnettya** Marchant = *Gaultheria*

Gaultheria Kalm ex L. Ericaceae. (Incl. *Pernettya*) 134 Mal. (24), E As. (c. 33), Aus. & NZ (14), N Am. (5), C & S Am. (c. 58). R: BJLS 106(1991)229. Many with methyl salicylate, the basis of medic. wintergreen. Some cult. orn. shrubs incl. *G. mucronata* (L.f.) Hook. & Arn. (*P. mucronata*, S S Am.) – grown for game cover, potentially wind-poll., some forms bisexual (most wild ones functionally but not structurally dioec.); *G. procumbens* L. (checkerberry, partridge berry, E N Am.) – medic. & refreshing tea for E Canad. Indians, source of orig. wintergreen (now extracted from *Betula lenta*), & *G. shallon* Pursh (salal, shallon, W N Am.) – natur. in GB; *G.* × *wisleyensis* Marchant ex Middleton ('× *Gaulnettya wisleyensis*', the distinction from *Pernettya* (fr. of *G.* a capsule, that of *P.* a berry) not tenable) – hybrids between *G. mucronata* & *G. shallon*, spontaneous in S GB where parents natur., incl. 'Wisley Pearl'

× **Gaulthettya** Camp = *Gaultheria*

Gaura L. Onagraceae. 21 N Am. R: MTBC 23(1972)1. Fr. nut-like. Some cult. cut-fls

Gaurella Small = *Oenothera*

Gauropsis C. Presl = *Clarkia*

Gaussia H. Wendl. Palmae (IV 3). 4 C Am., WI, on limestone. R: SB 11(1986)145. Stems with swollen 'bellies'

gauze tree *Lagetta lagetto*

Gavarretia Baillon. Euphorbiaceae. 1 Amaz. Brazil. G 2-loc.

Gavilea Poeppig. Orchidaceae (IV 1). 11 temp. S Am. R: BSAB 6(1956)73

gay feather *Liatris* spp.

Gaya Kunth. Malvaceae. 20 trop. Am., 3 NZ. Epicalyx 0

Gayella Pierre = *Pouteria*

Gaylussacia Kunth. Ericaceae. 48 N & S (most) Am. R: BJ 86(1967)309. Huckleberries. Fr. ed. esp. in pies, notably *G. baccata* (Wangenh.) K. Koch (N Am.). Cult. orn. incl. *G. brachycera* (Michaux) Torrey & Gray (E US) – 13 000 yr-old clone of 40 ha in Pennsylvania (biggest clone of any sp. known). Distinguished from *Vaccinium* in ovary with 10 divisions because of outgrowths in the 5 carpels

Gayophytum A. Juss. Onagraceae. 9 temp. W N Am. & W S Am. R: Brittonia 16(1964)343

Gazachloa Phipps = *Danthoniopsis*

Gazania Gaertner. Compositae (Arct.-Gort.). 17 trop. (1) & S Afr. R: MSBM 3(1959)364. Milky latex. Cult. orn. esp. *G. rigens* (L.) Gaertner & *G. pectinata* (Thunb.) Hartweg (*G. pinnata*) – both S Afr. & many hybrids

gean *Prunus avium*

Geanthus Reinw. = *Etlingera*

gear *Papaver somniferum*

Gearum N.E. Br. Araceae (VIII 3). 1 Brazil

gebang (palm) *Corypha utan*

gedu nohor *Entandrophragma angolense*

geebung *Persoonia* spp.

Geesinkorchis de Vogel. Orchidaceae (IV 12). 2 Borneo. R: OM 1(1986)43

geiger tree *Cordia sebestena*

Geigeria Greiss. Compositae (Inul.). 28 trop. & S Afr. R: MBSM 1(1953)251

Geijera Schott. Rutaceae. 8 N New Guinea, E Aus. (5, 4 endemic), New Caled. Alks. *G. parviflora* Lindley (E Aus.) – a fodder tree (sheep bush, wilga)

Geissanthus Hook.f. Myrsinaceae. 35 W trop. S Am.

Geissaspis Wight & Arn. Leguminosae (III 13). 3 warm As.

Geissois Labill. Cunoniaceae. (Incl. *Lamanonia*) c. 25 Aus., New Caled., Vanuatu, Fiji, S Am.

Geissolepis Robinson. Compositae (Ast.-Ast.). 1 Mex.

Geissoloma Lindley ex Kunth. Geissolomataceae. 1 S Afr.

Geissolomataceae Endl. Dicots – Rosidae – Celastrales. 1/1 S Afr. Xeromorphic evergreen shrub accumulating aluminium. Lvs opp., simple; stipules petiolar, minute. Fls term. with 3 prs of bracts on axillary shoots, bisexual, reg., 4-merous; P 4 petaloid, A 4 + 4 adnate to ovary-base, anthers with longit. slits, disk 0, G (4), 4-loc. each with slender style, the stigmas distally connivent, & 2 pendulous, anatropous, bitegmic ovules. Fr. a loculicidal capsule enveloped by persistent P; seeds 1 per loc. with elongate central embryo & scant endosperm

Only species: *Geissoloma marginatum* (L.) A. Juss. of the Lageberg Mts, Cape Province

Form. associated with Myrtales but absence of internal phloem etc. makes this unlikely; wood studies support alliance with Bruniaceae & Grubbiaceae

Geissomeria Lindley. Acanthaceae. 15 Mex. to trop. S Am.

Geissopappus Benth. = *Calea*

Geissorhiza Ker-Gawler. Iridaceae (IV 3). 84 S Afr. esp. Cape (winter rainfall). R: AMBG 72(1985)277, Novon 5(1995)156. Some cult. orn. like *Ixia* but related to *Hesperantha*

Geissospermum Allemão. Apocynaceae. 5 Brazil. Some medic. bark used as febrifuge: alks. R: AMBG 71(1984)1077

Geitonoplesium Cunn. ex R. Br. Philesiaceae (Smilacaceae s.l.). 1 C & E Mal. to Fiji (polymorphic): *G. cymosum* (R. Br.) R. Br. R: Muelleria 6(1987)368

Gelasia Cass. = *Scorzonera*

Gelasine Herbert. Iridaceae (III 4). 4 subtrop. S Am. Elaiophores attractive to bees

Geleznowia Turcz. Rutaceae. 1 SW Aus.

Gelibia Hutch. = *Polyscias*

Gelidocalamus Wen = *Indocalamus*

Gelsemiaceae (G. Don f.) Struwe & Albert (~ Loganiaceae). Dicots – Asteridae – Rubiales. 2/10 N to trop. Am., Afr., E As. Shrubs & lianes with heterostylous fls recently split off from Loganiaceae because of imbricate C, twice-dichotomously divided stigmas, latrorse anthers & flattened seeds

Genera: *Gelsemium*, *Mostuea*

Heterostyly & complex indole alks like Rubiaceae

Local stimulants etc.

Gelsemium Juss. Gelsemiaceae (Loganiaceae s.l.). 2 S US to Guatemala, 1 SE As. to W Mal. Allspice jasmine. Alks. *G. elegans* (Gardner & Champ.) Benth. (Indomal.) – alk. used in murder & suicide; *G. sempervirens* (L.) J. St-Hil. (Am.) – cult. orn. liane, medic. (neuralgia, migraine etc.)

Gemmaria Salisb. (~ *Hessea*). Amaryllidaceae (Liliaceae s.l.). 10 S Afr. R: BJ 107(1985)23

Gendarussa Nees = *Justicia*

genépi absinthe-type liqueur flavoured with *Artemisia glacialis* & *A. umbelliformis*

Genianthus Hook.f. Asclepiadaceae (II). 10 Indomal.

Geniosporum Wallich ex Benth. Labiatae (VIII 4b). 25 trop. Afr., Madag., S China, SE As.

Geniostemon Engelm. & A. Gray. Gentianaceae. 2 Mexico

Geniostoma Forster & Forster f. Geniostomaceae (Loganiaceae). 52 Mal. to NZ, Tahiti & Japan. R: Blumea 26(1980)245. Herm. but self-incompatible

Geniostomaceae Struwe & Albert (~Loganiaceae). Dicots – Asteridae – Gentianales. 2/67 Mal. to Japan, NZ & Hawaii. Recently split from Loganiaceae s.s. (similarities seen as parallelisms) because of anthers with apical appendages, porate pollen & capsules with persistent fleshy placentas & decid. valves like some Apocynaceae

Genera: *Geniostoma*, *Labordia*

genip, genipa, genipapo *Genipa americana*

Genipa L. Rubiaceae (II 1). 7 trop. Am. Some fish-disp. in Amaz. *G. americana* L. (genip, Mex. to Brazil) – tree used as live-fences, timber useful, fr. rather unpalatable but source of a body dye (iridoid, genipin, turning black with protein, e.g. skin) for Am. Indians & used in drinks (genipapo)

Genista L. Leguminosae (III 31). 87 Eur. (58), Canary Is., Medit. to W As. R: NRBGE 27(1966)11 & BSB 2,45(1972)269 (*Teline*) – brooms, 'planta-genista' origin of 'Plantagenets' name prob. *Cytisus scoparius*; alks; explosive fls showering poll. insects with pollen. Shrubs, sometimes spiny, differing from *Cytisus* etc. in style curved only near apex. Cult. orn. & dyepls. *G. aetnensis* (Biv.) DC (*Cytisanthus a.*, Mt Etna broom, Sicily, Sardinia) – to 7 m, fls scented; *G. anglica* L. (petty whin, needle furze, W Eur.) – spiny cult. orn.; *G. tinctoria* L. (dyer's greenweed, Eur., W As., natur. N Am.) – fls yield a yellow dye &, when mixed with woad, Kendal green; *G. tridentata* L. (*Chamaespartium* t., W Iberia) – medic. tea in S Portugal

Genistidium I.M. Johnston. Leguminosae (III 7). 1 Mexico

Genlisea A. St-Hil. Lentibulariaceae. 19 trop. Am., trop. & S. Afr., Madag. Rosette-pls with 2 kinds of leaf, thread-like forked & chlorophyll-less ones = traps (soil particles within suggest prey are sucked in); roots 0; fr. like a globe dehiscing along equator & to certain extent tropics

Gennaria Parl. Orchidaceae (IV 2). 1 Macaronesia, W Med. incl. Eur.: *G. diphylla* (Link) Parl.

Genoplesium R. Br. (~ *Prasophyllum*). Orchidaceae (IV 1). 40 Aus. (39, 38 endemic), New Caled., NZ. Midge orchids

gentian *Gentiana* spp.; **field g.** *Gentianella campestris*; **marsh g.** *Gentiana pneumonanthe*; **prairie g.** *Eustoma grandiflorum*; **yellow g.** *G. lutea*

Gentiana L. Gentianaceae. 361 (15 sects) temp. (Eur. 27, As. 312 – China 247 (SW mts 190), Aus. 1) & Arctic, usu. montane elsewhere but absent from Afr. exc. Morocco. R: BBMNHB 20(1990)169. Usu. perennial (annuals incl. *G. lilliputiana* C.J. Webb (NZ) – to 20 mm with 1 4-merous fl. & *G. prostrata* Haenke (Alps, N Am., Andes, S S Am.) – bipolar distrib.) herbs, often tufted, with blue to purple or red, yellow or white fls; roots with bitter glucosides yield medic.; cult. orn. esp. *G. acaulis* L. (*G. kochiana*, Alps & Pyrenees) – fls dark blue to 5 cm tall, poll. by bees, other spp. by butterflies, while yellow-flowered *G. lutea* L. (yellow gentian, Eur., W As.) by short-tongued insects (nectar accessible), lvs resembling *Veratrum* spp., comm. source of g. root used as tonic & in flavouring liqueurs & Suze; *G. pneumonanthe* L. (marsh g., Eur., N As.) – fls a source of a blue dye; *G. prostrata* Haenke – bipolar; *G. verna* L. (Ireland, GB (Teesdale), C Eur. to As. mts) – poss. survived the glaciation in GB (pollen evidence)

Gentianaceae Juss. Dicots – Asteridae – Gentianales. (Incl. Saccifoliaceae & Loganiaceae p.p.) 78/1225 cosmop. but esp. temp. & subtrop. & in trop. mts. Small trees, some pachycaul, shrubs & (usu.) herbs, often with mycorrhiza (s.t. chlorophyll-less & mycotrophic), usu. accumulating bitter iridoid substances & often with internal phloem. Lvs opp., seldom whorled or even spirally arr. (some *Swertia* spp.), simple (scale-like in mycotrophic *Voyria, Voyriella* etc.); stipules 0. Fls solit. or in cymose (rarely racemose) infls, usu. bisexual & reg.; K (4 or 5(–12)) with lobes imbricate (sometimes valvate or open), rarely reduced or 0 when tube 2-cleft, rarely K free, C (4 or 5(–12)) with ± elongate tube & usu. convolute lobes (with long thread-like projection in *Urogentias*) & often with scales or nectary-pits within, A as many as & alt. with lobes, attached to tube, rarely some staminodal or wanting, anthers with longit. slits (rarely term. pores as in *Exacum*), nectary-disk or glands usu. around $\underline{G}$ (2), 1-loc. with parietal placentas s.t. ± deeply intruded & bifid, seldom 2-loc. with axile placentas or 1-loc. with free-central placenta, style term. with entire or 2-lobed stigma (0 in *Lomatogonium*, where stigmas decurrent along ovary-sides), ovules ± ∞, anatropous, unitegmic. Fr. a septicidal capsule, rarely a berry; seeds usu. with small, straight embryo in copious oily endosperm but in chlorophyll-less genera seeds tiny with undifferentiated embryo & scanty endosperm. x = 5–13+

Principal genera (generic limits controversial): *Canscora, Exacum, Fagraea, Gentiana, Gentianella, Halenia, Lisianthus, Macrocarpaea, Sebaea, Swertia, Voyria*

Incl. Loganiacaeae tribe Potalieae, first moved here in 1856, with bilobed placentas, xanthones & unique iridoids

Cult. orn. esp. *Eustoma* ('lisianthus'), *Exacum, Gentiana, Sabatia*, & medic. esp. *Canscora, Gentiana*. Some timber (*Fagraea*)

Gentianella Moench (~ *Gentiana*). Gentianaceae. 125 temp. (exc. Afr.; Eur. 22). Cult. orn. & some med., differing from *Gentiana* in lacking small lobes or pleats between C-lobes, in sessile G & rounded wingless seeds. *G. campestris* (L.) Boerner (felwort, field gentian, Eur.); *G. quinquefolia* (L.) Small (ague weed, E N Am.)

Gentianodes Löve & D. Löve = *Gentiana*

Gentianopsis Ma. Gentianaceae. 16–25 N temp. As. & Am. Cult. orn.

Gentianothamnus Humbert. Gentianaceae. 1 Madagascar

Gentingia Johansson & Wong (~ *Prismatomeris*). Rubiaceae (IV 10). 1 NW Malay Pen. R: Blumea 33(1988)351

Gentlea Lundell = *Ardisia*

Gentrya Breedlove & Heckard = *Castilleja*

Genyorchis Schltr. Orchidaceae (V 15). 6 trop. Afr.

Geoblasta Barb. Rodr. (~ *Chloraea*). Orchidaceae (IV 1). 1 S Am.

Geocarpon Mackenzie. Caryophyllaceae (II 3). 1 SW Missouri, SE Arkansas

Geocaryum Cosson (incl. *Huertia*). Umbelliferae (III 8). 13–15 E Medit. R: L. Engstrand (1977) *Biosystematics & taxonomy of G.* 2 Greek spp. already extinct

Geocaulon Fern. Santalaceae. 1 Alaska, Canada, NE US

Geocharis (Schumann) Ridley. Zingiberaceae. 4 W Mal.

Geochorda Cham. & Schldl. = *Bacopa*

Geococcus J.L. Drumm. ex Harvey. Cruciferae. 1 semi-arid Aus.: *G. pusillus* J.L. Drumm. ex Harvey. R: TRSSA 89(1965)231

Geodorum Jackson. Orchidaceae (V 9). 10 Indomal., W Pacific (Aus. 1). *G. nutans* (C. Presl) Ames (Taiwan to Philipp.) – source of a strong adhesive used in musical instruments; *G. densiflorum* (Lam.) Schltr. (*G. pictum*, Aus.) – eaten by aborigines (yeenga)

Geoffraya Bonati = *Lindernia*

Geoffroea Jacq. Leguminosae (III 4). 3 S Am. *G. decorticans* (Hook. & Arn.) Burkart (*Gourliea d.*, chanal, chañar, Arg., Bolivia, Chile, Peru, Uruguay) imp. fodder, pods ed., medic., dyepl., timber

Geogenanthus Ule. Commelinaceae (II 1e). 4 trop. S Am. Cult. orn. esp. *G. poeppigii* (Miq.) Faden (*G. undatus*, seersucker pl., Brazil, Peru) – lvs with undulating surface

Geohintonia Glass & FitzMaurice. Cactaceae (III 9). 1 NE Mex. on gypsum outcrops

Geomitra Becc. = *Thismia*

Geonoma Willd. Palmae (V 6). c. 75 trop. Am. R: VKNAW II,18(1968)82. Monoec., unarmed forest palms incl. some rheophytes incl. *G. linearis* Burret (Colombia, Ecuador)

Geopanax Hemsley = *Schefflera*

Geophila D. Don. Rubiaceae (IV 7). c. 20 trop.

Georgia bark tree *Pinckneya bracteata*

Geosiridaceae Jonker = Iridaceae

Geosiris Baillon. Iridaceae (II; Geosiridaceae). 1 Madag.: *G. aphylla* Baillon

Geostachys (Baker) Ridley. Zingiberaceae. 19 SE As., W Mal.

Geraea Torrey & A. Gray. Compositae (Helia.-Verb.). 2 SW N Am. Cult. orn.

Geraldton wax (flower) *Chamelaucium uncinatum*

Geraniaceae Juss. Dicots – Dilleniidae – Malvales. 11/700 temp., few trop. Herbs or shrubs (s.t. pachycaul), usu. with aromatic oils in multicellular capitate-glandular hairs. Lvs spirally arr., less often opp., usu. lobed, compound or dissected pinnately or palmately, rarely (e.g. *Dirachma, Viviania*) simple & ± entire; stipules usu. present. Fls in cymes, less often solit. & axillary, usu. bisexual, reg. (irreg. in *Pelargonium*); K (4)5, imbricate or less often valvate, s.t. basally connate or forming a lobed tube (in *Pelargonium* adaxial sepal with spurred nectary; *Balbisia, Dirachma* & *Wendtia* with epicalyx), C (0, 4)5(8), imbricate, rarely (e.g. *Balbisia, Dirachma*) convolute, nectary-glands alt. with P, around A (0 in *Pelargonium*), A 5 + 5 but some or all of outer staminodal, rarely (*Monsonia, Sarcocaulon*) 5 + 5 + 5 or 8 (*Dirachma*), filaments ± basally connate, anthers with longit. slits, G̲ ((2, 3)5 (8 in *Dirachma*), pluriloc. with axile placentas & 1 style with distinct stigmas (styles distinct in *Biebersteinia*) & (Geranieae) elongating persistent column (beak), ovules anatropous to campylotropous, bitegmic, usu. 2 per loc., superposed & at least upper pendulous (1 per loc. in *Biebersteinia* & s.t. *Wendtia* & *Rhynchotheca*; 1 erect one per loc. in *Dirachma*; 2 rows of ovules per loc. in *Balbisia*). Fr. in Geranieae usu. 5 1-seeded mericarps separating acropetally from beak & often opening to release seed, in other genera a loculicidal capsule (*Balbisia, Viviania, Wendtia*) or mericarps without a beak (*Biebersteinia, Dirachma, Rhynchotheca*); seeds (exotegmic like Guttiferae) with straight or usu. curved embryo & little or 0 endosperm (copious & oily in *Viviania*). x = 7–14

The Geranieae (*Erodium, Geranium, Monsonia, Pelargonium*, etc. all with some spp. with spiny petioles) are closely related, the other genera less closely allied & have been segregated as discrete fams on the characteristics outlined above: poss. Vivianiaceae & Dirachmaceae worthy of segregation but Cronquist keeps them here pending evidence of a better position

Cult. orn. (*Geranium, Pelargonium, Sarcocaulon* etc.), forage (*Erodium*), essential oils (*Pelargonium*) & locally used med., dyes etc.

Geraniopsis Chrtek = *Geranium*

geranium (of house-pls) *Pelargonium* spp.; **g. oil** ('East India' or 'Turkish') *Cymbopogon martinii, Geranium* or *Pelargonium* spp.

Geranium L. Geraniaceae. 300 temp. (Eur. 38), montane tropics (Mal. 15, 13 endemic;

Hawaii 6 endemic incl. *G. arboreum* A. Gray to 3950 m – red-flowered, bird-poll. (unique in fam.)). Ann. or perennial rabbit-proof herbs & sometimes (e.g. Macaronesia) pachycaul shrublets; infls of 1 fl. or a pair as units in dichasial cymes or cincinni; fr. explosive often all mericarps opening at once & ejecting seeds. Many cult. orn. (cranesbills) esp. for ground-cover; R: P. Yeo (1985) *Hardy Gs. G. lucidum* L. (shining c., Eur. & Medit. to Himal.) – weed; *G. macrorrhizum* L. (Eur.) – Zdravetz oil from hairs; *G. nepalense* Sweet (As. mts) – dye a subs. for *Rubia cordifolia* L. from roots also used in tanning; *G. phaeum* L. (black or mourning widow, W & C Eur.) – blackish fls; *G. pratense* L. (meadow c., Euras.); *G. robertianum* L. (Herb Robert, N temp.) – form. medic.; *G. wallichianum* D. Don (Himal.) – root with 25–32% tannin used in tanning & dyeing

Gerardia Benth. = *Agalinis*

Gerardia L. = *Stenandrium*

Gerardiina Engl. Scrophulariaceae. 2 trop. & S Afr. R: FFG 31/2(1987)3

Gerardiopsis Engl. = *Anticharis*

Gerascanthus P. Browne. See *Cordia*

Gerbera L. Compositae (Mut.-Mut.). 30 NE to S Afr. with *G. piloselloides* Cass. W Afr. to China. R: OB 78(1985). Herbs. Cult. orn. greenhouse rosette-pls esp. *G. jamesonii* Bolus ex Adlam (Barberton or Transvaal daisy, Transvaal)

Germainia Bal. & Poitr. Gramineae (40b). 9 Assam & SE As. to NE Aus. R: TFB 6(1972)29

germander *Teucrium* spp. esp. *T. chamaedrys*; **g. speedwell** *Veronica chamaedrys*; **wood g.** *T. scorodonia*

gero *Pennisetum glaucum*

geronggong *Cratoxylum arborescens*

Geropogon L. Compositae (Lact.-Scor.). 1 Eur., Medit., Turkey: *G. hybridus* (L.) Schultz-Bip. R: Lazaroa 9(1986)31

Gerrardanthus Harvey ex Hook.f. Cucurbitaceae. 5 trop. & S Afr. *G. macrorhiza* Harvey ex Hook.f. (Natal) – spherical succ. stem to 50 cm diam.

Gerrardina Oliver. Flacourtiaceae (5). 2 E trop. & S Afr.

Gerritea Zuloaga, Morrone & Killeen. Gramineae (34b). 1 Bolivia: *G. pseudopetiolata* Zuloaga, Morrone & Killeen. R: Novon 3(1993)213

Gesneria L. Gesneriaceae. (Incl. *Rhytidophyllum*) 70 trop. Am. R: SCB 29(1976)43, Selbyana 6(1982)200. Some with pseudostipules; mucilage from lvs scented & (?) attractive to bat pollinators. Some cult. orn.

Gesneriaceae Rich. & Juss. ex DC. Dicots – Asteridae – Scrophulariales. 141/2900 trop. (esp. OW – over 60% genera & spp., China (R: EJB 49(1992)5) with 56 genera (28 endemic); R(NW): Selbyana 6(1983)1), few Tertiary relics in temp. Euras. Usu. herbs, s.t. epiphytic, or shrubs, s.t. lianes, rarely trees. Lvs opp., rarely whorled or spirally arr., s.t. members of a pair unequal, usu. simple (rarely pinnatifid); stipules 0. Fls solit., axillary or in usu. cymose infls, sometimes epiphyllous, bisexual; K 5 or usu. (5), lobes usu. valvate, C (5), irreg., & usu. 2-lipped & often spurred, or almost reg., lobes imbricate, adaxial ones usu. enclosed, A 4 (posterior 1 absent) alt. with C & attached to tube, anthers all connivent or in pairs, less often A 5 (*Ramonda*, some *Sinningia* spp.) or 2, staminodes 1–3 in place of missing A, anthers with longit. slits, nectary-disk often at base of $\overline{\text{G}}$ (2) or $\underline{\text{G}}$ (2) with term. style, 1-loc. & with 2 parietal placentas ± intruded & bifurcate, s.t. meeting (e.g. *Monophyllaea*) so G 2-loc., ovules ∞, anatropous, unitegmic. Fr. a usu. loculicidal (or septicidal) capsule, less often a berry; seeds ∞, small with straight embryo in oily endosperm (Gesnerioideae; 0 in Cyrtandroideae) & 1 (esp. C.) or 2 cotyledons. x = 4–17+

Classification & principal genera:

I. **Cyrtandroideae** (Cotyledons becoming unequal after germ. & 1 s.t. becoming the only photosynthetic organ, forming a lamina by intercalary growth, $\underline{\text{G}}$; OW exc. for 1 *Rhynchoglossum* sp.): 5 tribes incl. Trichosporeae (*Aeschynanthus*), Didymocarpeae (*Boea, Briggsia, Didissandra, Chirita, Haberlea, Ramonda, Saintpaulia, Streptocarpus*), Cyrtandreae (*Cyrtandra*), Klugieae (*Monophyllaea, Rhynchoglossum*)

II. **Gesnerioideae** (Cotyledons equal after germ.; $\underline{\text{G}}$ or ± $\overline{\text{G}}$; NW + SE Australasia): 11 tribes, 5 with $\underline{\text{G}}$ incl. Beslerieae (*Besleria*), Columneeae (*Alloplectus, Codonanthe, Columnea, Drymonia* (lianes), *Nautilocalyx*), Coronanthereae (*Coronanthera* (trees)), Mitrarieae (*Fieldia* (Aus.), *Mitraria*) & 6 with ± $\overline{\text{G}}$ incl. Gloxinieae (*Achimenes, Gloxinia, Smithiantha*), Gesnerieae (*Gesneria, Rhytidophyllum, Sanango*), Kohlerieae (*Kohleria*), Sinningieae (*Sinningia*)

Close to Scrophulariaceae, of which G. are in some ways the trop. counterparts but usu. have 1-loc. ovary. The related Andean & Aus. genera of Mitrarieae fit rather unhappily in either subfam. & have been raised to subfam. status by some authors. Many genera esp. in NW show transitions from insect- to bird-poll. & *Columnea* (NW) is paralleled by *Aeschynanthus* in being bird-poll. epiphytes; some have extrafl. nectaries inc. *Codonanthe* epiphytic on ants' nests; 1 of the 10 fams with elaiophores attractive to bees (*Drymonia*) Some locally imp. med. but internationally imp. as cult. orn. esp. *Saintpaulia, Sinningia* (gloxinias), *Streptocarpus* but also *Achimenes, Columnea, Haberlea, Nematanthus, Ramonda* etc. (H.E. Moore (1957) *African violets, gloxinias & their relatives*)

Gesnouinia Gaudich. Urticaceae (IV). 2 Canary Is. Monoec. shrubs & trees. *G. arborea* (L.f) Gaudich – cult. orn. tree to 7m

Gethyllis L. Amaryllidaceae (Liliaceae). 32 S Afr. R: Willdenowia 15(1986). Some ed. fr.

Gethyum Philippi. Alliaceae (Liliaceae s.l.). 1 Chile

Getonia Roxb. (*Calycopteris*). Combretaceae. 1 Indomal. Climber, like *Bucida* but lvs opp.

Geum L. Rosaceae. c. 40 temp. (Eur. 12) & cold. R: MB 4(1957)1. Avens. Cult. orn. esp. forms of *G. chiloense* Balbis ex Ser. (Chile) – fls red, e.g. 'Mrs Bradshaw'; *G. rivale* L. (water a., N temp.) – glycosides medic., nodding bee-poll. fls found in less disturbed habitats than *G. urbanum* L. (wood a., Herb Bennet (*herba benedicta*), Eur., W As., Medit.) with erect yellow fls, roots clove-scented used for flavouring beer etc., locally medic., but where they meet hybrid swarms form; both have hooked achenes promoting animal-disp., the hook an accrescent Z-shaped style, of which the upper piece lost after fert.

Geunsia Blume = *Callicarpa*

Gevuina Molina. Proteaceae. 1 Chile, Arg. (OW spp. = *Bleasdalea*): *G. avellana* Molina (Chile nut) – cult. orn., useful timber, seeds ed., flavour reminiscent of hazel-nuts

Ghaznianthus Lincz. (~ *Acantholimon*). Plumbaginaceae (II). 1 Afghanistan: *G. rechingeri* (Freitag) Lincz.

gherkin *Cucumis sativus*; **WI g.** *C. anguria*

Ghikaea Volkens & Schweinf. Scrophulariaceae. 1 trop. Afr.

Ghinia Schreber = *Tamonea*

ghost gum *Corymbia* spp.; **Miss Willmott's g.** *Eryngium giganteum*; **g. orchid** *Epipogium aphyllum*; **g. tree** *Davidia involucrata*

Giadotrum Pichon = *Cleghornia*

giam *Hopea* spp.

giant bamboo *Dendrocalamus giganteus*; **g. cactus** *Carnegiea gigantea*; **g. fennel** *Ferula communis*; **g. hogweed** *Heracleum mantegazzianum*; **g. taro** *Alocasia macrorrhizos*

Gibasis Raf. Commelinaceae (II 1g). 11 trop. Am. esp. Mex. R: KB 41(1986)107. Cult. orn. esp. *G. pellucida* (Martens & Galeotti) D. Hunt (Tahitian bridal veil, Mex.)

Gibasoides D. Hunt. Commelinaceae (II 1f). 1 Mexico

Gibbaeum Haw. ex N.E. Br. Aizoaceae (V). 15 Cape. G.C. Nel (1953) *The G. handbook*. Cult. orn. succ. with shoots like *Conophytum*

Gibbaria Cass. Compositae (Cal.). 2 Cape

Gibbesia Small = *Paronychia*

Gibbsia Rendle. Urticaceae (III). 2 W New Guinea

Gibraltar mint *Mentha pulegium*

Gibsoniothamnus L. O. Williams. Scrophulariaceae. 14 Mex. to C Am. Epiphytic shrubs

Gifola Cass. = *Filago*

Gigantochloa Kurz ex Munro. Gramineae (1b). 18 Indomal. R: Reinwardtia 10(1987)291. Giant bamboos; fls of *G. albociliata* (Munro) Munro (Burma) fragrant & visited by meliponid bees. *G. apus* (Schultes f.) Munro (Burma) & other spp. with ed. young shoots (esp. in Java, where long cult.)

Gigasiphon Drake = *Bauhinia*

Gigliolia Becc. = *Areca*

Gilberta Turcz. (~ *Myriocephalus*). Compositae (Gnap.-Ang.). 1 Aus.: *G. tenuifolia* Turcz. R: Nuytsia 8(1992)419

Gilbertiella Boutique = *Monanthotaxis*

Gilbertiodendron Léonard. Leguminosae (I 4). 26 trop. W Afr. Some good timber. *G. dewevrei* (De Wild.) Léonard can form single-dominant forests in C Afr.

Gilead, balm of *Abies balsamea, Commiphora gileadensis, Liquidambar orientalis, Populus balsamifera, P.* × *jackii* 'Gileadensis', *P. nigra*

Gilesia F. Muell. (~ *Hermannia*). Sterculiaceae. 1 C Aus.: *G. biniflora* F. Muell.

Gilgiochloa Pilger. Gramineae (39). 1 trop. Afr.

Gilia Ruíz & Pavón. Polemoniaceae. c. 25 NW esp. W N Am. Some cult. orn. esp. annuals but most spp. removed to other genera e.g. *Ipomopsis, Linanthus*

Giliastrum (Brand) Rydb. = *Gilia*

Gilibertia Ruíz & Pavón = *Dendropanax*
Gilipus Raf. = ? *Myrica*
Gillbeea F. Muell. Cunoniaceae. 2 Queensland, New Guinea
Gillenia Moench (*Porteranthus*). Rosaceae. 2 E N Am.: *G. stipulata* (Willd.) Baillon (American ipecacuanha, SE US) – glycosides in root & bark form. medic.; *G. trifoliata* (L.) Moench (Bowman's root, E N Am.)
Gillespiea A.C. Sm. Rubiaceae (IV 7). 1 Fiji
Gilletiella De Wild. & T. Durand = *Anomacanthus*
Gilletiodendron Vermoesen. Leguminosae (I 4). 5 trop. Afr.
Gilliesia Lindley. Alliaceae (Liliaceae). 3 Chile
gilliflower (gyllofer, gilofre, corruptions of caryophyllum, i.e. clove) *Dianthus* spp.; also *Erysimum* & *Matthiola* spp.
Gilmania Cov. Polygonaceae (I 1). 1 Death Valley area, Calif.: *G. luteola* (Cov.) Cov. R: Phytol. 66(1989)243
Gilruthia Ewart. Compositae (Gnap.-Ang.). 1 W Aus.: *G. osbornii* Ewart & White. R: OB 104(1991)125
gin flavour due to *Juniperus communis*
Ginalloa Korth. Viscaceae. 8 Indomal.
gingelly *Sesamum orientale*
ginger *Zingiber officinale*; **g. bread palm**. *Hyphaene thebaica*; **g.b. plum** *Neocarya macrophylla*; **cassumar g.** *Z. purpureum*; **g. grass** *Cymbopogon martinii*; **Japanese g.** *Z. mioga*; **g. lily** *Hedychium* spp.; **mango g.** *Curcuma amada*; **g. mint** *Mentha* × *gracilis*; **Mioga g.** *Z. mioga*; **shell g.** *Alpinia zerumbet*; **Siamese g.** *Alpinia galanga*; **torch g.** *Etlingera elatior*; **wild g.** *Asarum canadense*
Gingidia Dawson (*Gingidium*). Umbelliferae (III 8). 10 Aus. (3), NZ (6). R: NZJB 4(1966)84. Gynodioec. Hybrids, also with *Aciphylla* spp.
Gingidium Forster & Forster f. = *Gingidia*
Ginkgo L. Ginkgoaceae. 1: *G. biloba* L. maidenhair tree – dioec. (males with earlier leaf-fall & usu. less spreading form), flowering when 20 yrs old, mycorrhizal, a relic in E China much cult., to 1000 yrs old & bole to 9 m circumference in temples, few pests, tolerates pollution & salt; wild hill pops with basal (following damage) & aerial (in old age) outgrowths used for bonsai. Timber & street-tree though females are objectionable as fallen seeds stink of rancid butter & make a mess, these (ginkgo nuts, when nauseous layer removed leaving ed. female gametophyte) roasted or consumed with bird's-nest soup, also an oil-source though causing dermatitis in sensitive people; some insecticidal compounds, v. imp. in preventing blood-flow while ginkgoilides used in Parkinson's Disease. App. identical fossils 200 million yrs old, ± continuous with fossil *G. adiantoides* (Unger) Heer, though 6 spp. in Cretaceous, treated as living pteridosperm by some theorists. 'Ohazuki' (Epiphylla) with peduncles adnate to petioles
Ginkgoaceae Engl. Ginkgoopsida. 1/1 extant, China (at least 6 genera extinct). Decid. dioec. trees with resinous, slightly mucilaginous wood, strap-shaped (extinct) or fan-shaped lvs all with open dichotomous venation on long & short shoots. Ovules on peduncles in axils of lvs or scale-lvs of short shoots, 2(–4+) per peduncle with fleshy collars around bases; meiosis leads to linear tetrad of spores, lowermost giving gametophyte with chlorophyll & 2 or 3 archegonia. Microsporangiophores borne on catkin-like axis, the gametophytes with 2 prothallial cells, a tube nucleus, a sterile cell & a sperm cell, released as pollen which has 2 large motile sperms with spiral bands of flagella. Fert. may be up to 4 months after poll. & after the ovule has fallen; embryo with 2 cotyledons. n = 12 (incl. XY sex system, XX being female, XY male)
Extant genus: *Ginkgo*
Ginkgoopsida 1/1 extant. Gymnosperms, trees with lvs with dichot. venation. Ovules 2–10, term. on axillary branching axes. Seed large with outer fleshy & inner stony layers. Microsporangiophores in catkins with 2–12 pendulous microsporangia; sperms with spiral bands of flagella
Families: Ginkgoaceae, Trichopityaceae (extinct, ? lvs circular in cross-section)
Ginoria Jacq. Lythraceae. 14 Mexico & WI
ginseng *Panax pseudoginseng*; **American g.** *P. quinquefolius*; **Siberian g.** *Eleutherococcus senticosus*
Giraldiella Dammer = *Lloydia*
Girardinia Gaudich. Urticaceae (I). 2 warm & trop. OW. R: KB 36(1981)143. 1 E & NE Afr.: *G. bullosa* (Steudel) Wedd. – bark fibre for sewing. 1 warm & trop. OW: *G. diversifolia* (Link) I. Friis (*G. palmata*, Nilgiri nettle) – bark-fibre made into cloth, ropes, bowstrings, in dry

season in Java deer eat the stinging lvs just before fighting for mates

Girgensohnia Bunge ex Fenzl. Chenopodiaceae (III 3). 3 Eur. (1) to C As. Saline semi-deserts

girigiri *Sphenostylis stenocarpa*

girl *Erythroxylum coca*

Gironniera Gaudich. Ulmaceae (II). 6 Indomal. to Pacific. Timber used locally for tea-chests & matches

Gisekia L. Phytolaccaceae (?; Gisekiaceae). 1 Afr., 1 OW trop. (*G. pharnacioides* L., ed.)

Gisekiaceae (Endl.) Nakai = ? Phytolaccaceae

Gitara Pax & K. Hoffm. Euphorbiaceae. 2 trop. Am.

Githopsis Nutt. Campanulaceae. 4 W N Am. R: SB 8(1983)436

Giulianettia Rolfe = *Glossorhyncha*

Givotia Griffith. Euphorbiaceae. 4 E Afr. (1), Madag. (2), India (1). Dioec. *G. rottleriformis* Griffith (India) – timber soft & light for figures & toys etc., cult. orn.

Gjellerupia Lauterb. Opiliaceae. 1 New Guinea

Glabraria L. = *Brownlowia*

gladdon *Iris foetidissima*

Gladiolimon Mobayen. Plumbaginaceae (II). 1 Afghanistan: *G. speciosissimum* (Aitch. & Hemsley) Mobayen

Gladiolus L. Iridaceae (IV 3). (Incl. *Acidanthera* & sunbird-poll. spp. form. called *Anomalesia*, *Petamenes* (*Homoglossum*) & *Oenostachys*) c. 195 Eur. (6), Medit., E Med., trop. Afr. mts but esp. S Afr. (R: JSAB suppl. 10(1972), Madag. (8; R: BMNHN 4,11(1990)235); (afrikanders, bluebell (S Afr.)), gladiolus being applied to *Acorus* or *Iris pseudacorus* in mediaeval GB. Florists' gladiolus (*G.* × *hortulanus* L. Bailey) with complex ancestry involving *G. cardinalis* Curtis, *G. carneus* Delaroche, *G. dalenii* Geel (*G. natalensis*, Afr., Madag.), *G. oppositiflorus* Herbert, *G. papilio* Hook.f. (*G. purpureoauratus*), *G. saundersii* Hook.f. & *G. tristis* L. (all S Afr.), early hybrids known as *G.* × *colvillei* Sweet (1823, *G. cardinalis* × *G. tristis*, leading to 'Nanus Hybrids', *G.* × *insignis* Paxton, 1834) & *G.* × *gandavensis* Van Houtte (1837, *G. dalenii* × *G. oppositiflorus*), over 30 000 cvs derived from 8 spp., record pl. 2.55 m tall; other spp. cult. incl. *G. communis* L. (S Eur., N Afr., widely natur. e.g. Scilly Is. – 'whistling jacks'); *G. callianthus* Marais (*Acidanthera bicolor*, Ethiopia to E Afr.) – scented fls, esp. 'Murieliae' (*A. murieliae*); *G. illyricus* Koch (Eur. to Turkey, widely natur. incl. GB, where protected); *G. italicus* Miller (S Eur. to C As., Macaronesia & Afghanistan) – perhaps the 'hyacinth' of antiquity; *G. ukambanensis* (Baker) Marais (Tanzania to Arabia) – corm v. imp. wild food in Dhofar, sought also by red-legged partridges; *G. watsonioides* Baker (trop. Afr. mts) – up to snow-line

Gladiopappus Humbert. Compositae (Mut.-Mut.) 1 Madag. Subshrub

gladwin, gladwyn (, stinking) *Iris foetidissima*

Glandonia Griseb. Malpighiaceae. 3 trop. S Am. R: MNYBG 32(1981)135

Glandularia J. Gmelin (~ *Verbena*). Verbenaceae. Spp.? S Am. Cult. orn. herbs esp. *G.* × *hybrida* (Groenl. & Ruempler) Nesom & Pruski (*V.* × *hybrida*, *V.* × *hortensis*, complex hybrid group involving S Am. spp. incl. *G. incisa* (Hook.) Tronc., *G. peruviana* (L.) Sm., *G. platensis* (Sprengel) Schnack & Covas) – garden verbena

Glaphyropteridopsis Ching = *Cyclosorus*

Glaphyropteris (Fée) Fée = *Cyclosorus*

Glaribraya H. Hara. Cruciferae. 1 Nepal

glasswort *Salicornia* spp., *Salsola* spp.

Glastaria Boiss. Cruciferae. 1 SW Turkey, Syria, Iraq

Glastonbury thorn *Crataegus monogyna* 'Biflora'

Glaucidiaceae Tamura = Paeoniaceae

Glaucidium Siebold & Zucc. Paeoniaceae (Glaucidiaceae). 1 Japan: *G. palmatum* Siebold & Zucc. – cult. orn. with embryology & cytology like *Paeonia*

Glaucium Miller. Papaveraceae (I). c. 23 Eur. (2), SW & C As. R: FR 89(1979)499. Many alks. *G. flavum* Crantz (sea poppy, horned p., Eur., Medit., natur. N Am.) – seed-oil for illumination & soap, form. medic., cult. orn.

Glaucocarpum Rollins. Cruciferae. 1 SW US

Glaucocochlearia (O. Schulz) Pobed. = *Cochlearia*

Glaucosciadium B.L. Burtt & P. Davis. Umbelliferae (III 10). 1 S Turkey, Cyprus

Glaux L. Primulaceae. 1 N temp. coasts: *G. maritima* L. (sea milkwort) – hibernating shoot with root produced in axil of seedling, which dies, gives rise vegetatively to new pl. for some yrs before flowering; shoots fleshy; C 0, K petaloid

Glaziocharis Taubert = *Thismia*

Glaziophyton Franchet. Gramineae (1a). 1 E Brazil mt. tops: *G. mirabile* Franchet. Reed-like sprouting typical bamboo twigs after burning

Glaziostelma Fourn. = *Tassadia*

Glaziova Bureau. Bignoniaceae. 1 Brazil. Adhesive disks at tendril tips (cf. *Parthenocissus*)

Glazioviantlius G. Barroso (~ *Chresta*). Compositae (Vern.-Vern.). 2 Brazil

Gleadovia Gamble & Prain. Orobanchaceae. 6 W Himal., W China

Gleasonia Standley (~ *Henriquesia*). Rubiaceae (I 4). 5 trop. S Am.

Glebionis Cass. Corr. name for cult. annual spp. of *Chrysanthemum* if that name applied to garden chrysanthemums

Glechoma L. Labiatae (VIII 2). c. 10 temp. Euras. (Eur. 2). *G. hederacea* L. (ground ivy, alehoof, Eur., natur. N Am.) – form. medic. tea & added to beer on long voyages

Glechon Sprengel. Labiatae (VIII 2). 6 Brazil, Paraguay

Gleditsia L. Leguminosae (I 1). 14 (2–3 E N Am., 1 S Am., 1 Caspian, rest India & Japan to New Guinea). Trees (honey-locusts) usu. with stout branched thorns in axils, arising from uppermost of serial buds; some used as hedges, others timber, shade & cult. orn. Spp. from NE Arg. & China v. similar though separated for 60 million yrs. Tepals scarcely different from one another, spirally arr., floral parts inconstant in number; lvs pinnate or bipinnate often on same tree (cf. constancy of *Astragalus*). Cult. orn. esp. *G. triacanthos* L. (OW temp.) – many cvs, some thornless trees become thorny after 40–55 yrs

Glehnia Schmidt ex Miq. Umbelliferae (III 9). 2 NE As., W N Am. *G. littoralis* Schmidt ex Miq. – cough treatment in modern Chinese herbalism

Gleichenia Sm. Gleicheniaceae (Gleich.). 10 S Afr., Masc., Mal. to NZ. Leaf-segments v. small. See also *Dicranopteris*

Gleicheniaceae (R. Br.) C. Presl. Filicopsida. (Incl. Stromatopteridaceae) 5/125 trop. & warm, S temp. Terr. often thicket-forming ferns of open ground; rhizomes usu. protostelic, creeping, dichotomizing. Rachis (pinnate to) bipinnate or pseudodichotomously branching to 7 m. Sori of 2–15 pear-shaped sporangia borne abaxially without indusia. Gametophytes massive, slow-growing, with endotrophic mycorrhiza when old, & v. large antheridia (to 100 μm diam.) with several hundred antherozoids

Classification & genera:

Gleichenioideae (frond at least 1-forked): *Dicranopteris, Diplopterygium, Gleichenia, Sticherus*

Stromatopteridoideae (frond 1-pinnate): *Stromatopteris*

Fossils back to Mesozoic & poss. Carboniferous

Some used for weaving

Glekia Hilliard (~ *Phyllopodium*). Scrophulariaceae. 1 Lesotho, E Cape: *G. krebsiana* (Benth.) Hilliard. R: NRBGE 45(1988)482. Commem. G.L.E. Krebs (cf. *Afgekia*)

Glenniea Hook.f. Sapindaceae. 8 trop. Afr., Madag., Indomal. R: Blumea 22(1977)411

Glia Sonder (~ *Annesorhiza*). Umbelliferae (III 8). 2 Macaronesia, St Helena, 1 SW Cape: *G. prolifera* (Burm.f.) B.L. Burtt (*G. capensis*). R: NRBGE 45(1988)198

Glinus L. Molluginaceae. c. 12 warm (Eur. 1, Aus. 3) & trop. Some potherbs

Glionettia Tirv. (~ *Randia*). Rubiaceae (I 5). 1 Seychelles: *G. sericea* (Baker) Tirv. R: BMNHN 4,6(1984)197

Gliopsis Rauschert = *Rutheopsis*

Gliricidia Kunth. Leguminosae (III 7). 4 trop. Am. *G. sepium* (Jacq.) Walp. – explosive fr. dehiscence with seeds ejected to 40 m; grown widely as cocoa shade incl. in OW, also as living fence, termite-proof (? coumarin) building timber in C Am., firewood crop (10–15 t/ha/yr dry matter in W Afr.), green manure in Sri Lanka; seeds or powdered bark a poison for rats & mice; fls fried & eaten in C Am.

Glischrocaryon Endl. (~ *Loudonia*). Haloragidaceae. 4 S & SW Aus. Looks like *Haloragis* but pollen like *Gunnera*. R: BAIM 10(1975)150

Glischrocolla (Endl.) A. DC. Penaeaceae. 1 S Afr.

Glischrothamnus Pilger. Molluginaceae. 1 NE Brazil

Globba L. Zingiberaceae. 35 E As., Indomal. Lower cymes usu. replaced by bulbils

globe amaranth *Gomphrena globosa*; **g. artichoke** *Cynara scolymus*; **g. flower** *Trollius* spp.; **g. mallow**, scarlet *Sphaeralcea coccinea, S. augustifolia* spp. *cuspidata*; **g. thistle** *Echinops* spp.

Globimetula Tieghem. Loranthaceae. 14 trop. Afr. Sunbirds probe edges of specialized C segments causing reflex, further probes splitting C tube & A coil inwards explosively

Globularia L. Globulariaceae. 22 Cape Verde Is., Canary Is., Eur. (15), As. Minor. R: BJ 69(1938)318, BAGS 35(1967)305. Herbs & subshrubs, cult. orn. esp. rock garden pls

Globulariaceae DC. Dicots – Asteridae – Scrophulariales. 9/230 Eur., W As., Afr. Shrubs & herbs with distinctive glandular hairs (single stalk-cell & head of 2(4) cells, cf. Scrophulariaceae where they are different). Lvs spirally arr. (opp. in *Globulariopsis*),

simple, small, entire; stipules 0. Fls bisexual, ± strongly irreg., in spikes, involucrate heads or compound corymbs; K ((2)3–5), lobed or adnate to bract, rarely 2-lipped, C (4 or 5), ± irreg., often 2-lipped with imbricate lobes, A alt. with C & attached to tube, usu. 4, the 5th (posterior) 0 or rarely a staminode or perfect, rarely A 2, anther thecae apically confluent at maturity opening by 1 distal slit, nectary-disk often at base of G (2) with term. style, 2-loc., or 1 scarcely developing or 1-loc. or pseudomonomerous, ovules anatropous, unitegmic, 1(2) per loc., pendulous or, where 2, upper erect. Fr. of 2 unequal nutlets or 1 or achene in persistent K; seeds with straight embryo in oily endosperm. x = 8
Principal genera: *Globularia, Selago, Walafrida*
Form. considered to comprise only *Globularia* & *Poskea*, these have been joined by most of the Selagineae of Scrophulariaceae (Selaginaceae; *Hebenstretia* returned), to which they have generally been held to be related: this move makes G. & Scrophulariaceae more clear-cut fams (Cronquist) but see *Glumicalyx*
Some cult. orn.

Globulariopsis Compton. Globulariaceae. 1 S Afr. (only 2 localities): *G. wittebergensis* Compton. R: SAJB 56(1990)477

Globulostylis Wernham = *Cuviera*

Glochidion Forster & Forster f. Euphorbiaceae. 300 Madag. (few), to W Pacific (many), trop. Am. (few). Some local timbers & medic. (alks)

Glochidocaryum W.T. Wang. Boraginaceae. 1 NW China

Glochidotheca Fenzl (*Turgeniopsis*). Umbelliferae (III 3). 1 Eur., W As.

Glockeria Nees = *Habracanthus*

Gloeocarpus Radlk. Sapindaceae. 1 Philipp. R: Blumea 35(1991)389

Gloeospermum Triana & Planchon. Violaceae. 12 trop. Am. Endosperm 0

Glomera Blume. Orchidaceae (V 13). 50 Mal. to W Pacific. R: FGP 10(1974)81

Glomeropitcairnia Mez. Bromeliaceae (2). 2 Carib. R: FN 14(1977)1388. Up to 20 litres water recorded in one 'tank'

Gloneria André = *Psychotria*

Gloriosa L. Colchicaceae (Liliaceae s.l.). 1–9 OW trop. *G. superba* L. (v. variable esp. Somalia) – flame lily. Scrambler of forest edge & scrub esp. in abandoned cult. with tendrils at leaf-tips (cf. *Littonia*) & pedicels adnate for 2 nodes above the point of emergence of their vasc. bundles from the stem; the fls have reflexed P, are nodding & with spreading A & style projecting from fl. horizontally yet close to one another so that seed-set by selfing possible in absence of pollinators; alks incl. colchicine, brittle underground tubers poisonous (used for suicide in India); several selected forms with specific names (& diff. ploidies) in cult., the best poss. *'G. rothschildiana'* (octoploid)

glory bower *Clerodendrum chinense*; **g. bush** *Tibouchina urvilleana*; **g. of the snow** *Chionodoxa forbesii*; **g. of the sun** *Leucocoryne ixioides*; **g. pea** *Clianthus* & *Swainsona* spp.

Glossanthis Polj. = *Trichanthemis*

Glossarion Maguire & Wurd. Compositae (Mut.-Mut.). 2 Venez., Guyana. R: Brittonia 41(1989)39. Shrubs

Glossocalyx Benth. Monimiaceae (IV; Siparunaceae). 4 trop. W Afr. 1 leaf of each pair rep. by midrib only

Glossocardia Cass. Compositae (Helia.-Cor.). 12 SE As. to Pacific, introd. Afr. R: Blumea 35(1991)466

Glossocarya Wallich ex Griffith. Labiatae (II; Verbenaceae). 9 Indomal., Aus.(R: JABG 13(1990)17)

Glossochilus Nees. Acanthaceae. 2 S Afr.

Glossodia R. Br. Orchidaceae (IV 1). 2 Aus. (wax-lip (orchid))

Glossogyne Cass. = *Glossocardia*

Glossolepis Gilg = *Chytranthus*

Glossonema Decne. Asclepiadaceae (III 1). 4 trop. Afr. to As. R: KB 37(1982)344

Glossopappus Kunze. Compositae (Anth.-Leuc.). 1 SW Eur., N Afr.

Glossopetalon A. Gray (*Forsellesia*). Crossosomataceae. 8 W US

Glossopholis Pierre = *Tiliacora*

Glossorhyncha Ridley (~ *Glomera*). Orchidaceae (V 13). 80 Mal. & W Pacific. R: FGP 9(1974)82. Some cult. orn.

Glossostelma Schltr. Asclepiadaceae (III 1). 2–3 trop. & S Afr.

Glossostemon Desf. Sterculiaceae. 1 Iran, Iraq, Arabia: *G. bruguieri* Desf. – moghat root, sold in bazaars, medic.

Glossostigma Wight & Arn. Scrophulariaceae. 6 Aus., 3 ext. to NZ & 1 of them to Afr. & India. Minute short-lived herbs ('mud-mats')

Glossostipula Lorence (~ *Randia*). Rubiaceae (II 1). 2 Mex., Guatemala. R: Candollea 41(1986)453

Glottidium Desv. (~ *Sesbania*). Leguminosae (III 7). 1 SE US: *G. vesicarium* (Jacq.) Harper (bag-pod) – poisonous to stock

Glottiphyllum Haw. ex N.E. Br. Aizoaceae (V). 16 Karoo. Cult. orn. dwarf succ. with tongue-shaped or ± cylindrical succ. lvs in 2 or 4 ranks

gloxinia *Sinningia speciosa*

Gloxinia L'Hérit. Gesneriaceae. 15 trop. Am. R: Selbyana 1(1976)385. Bird- & bee-poll. groups of spp. (cf. *Gasteranthus* & *Sinningia*), male euglossines coll. scents

Gluema Aubrév. & Pellegrin. Sapotaceae (I). 1 trop. W Afr.: *G. ivorensis* Aubrév. & Pellegrin. R: T.D. Pennington, S. (1991)144

Glumicalyx Hiern. Scrophulariaceae. 6 S & SE Afr. R: NRBGE 35(1977)155. Links Manuleeae with Selagineae (here partly in Globulariaceae)

Gluta L. Anacardiaceae (I). 30 Madag. (1), Indomal. (incl. *Melanorrhoea*). *G. renghas* L. (rengas, Mal.) – timber useful though can cause dermatitis, seeds ed. roasted; *G. laccifera* (Pierre) Ding Hou (*M. l.*, SE As.) – lacquer and *G. usitata* (Wallich) Ding Hou (*M. u.*, theet-see, thitsi, trop. As.) – source of Burmese lacquer

Glutago Comm. ex Raf. = *Oryctanthus*

Glyceria R. Br. Gramineae (18). 40 temp. (Eur. 10) esp. N Am. R: AJB 36(1949)155. Luscious pasture-grasses for cows; waterfowl favour fr. form. used by N Am. Indians; those of *G. fluitans* (L.) R. Br. (sweet or manna (because fls shower down when shaken) grass, N temp.) the basis of 'manna croup'

Glycine Willd. Leguminosae (III 10). (Incl. *Neonotonia*) 18 As. to Aus. (15). R: EB 35(1981)275. Cleistogamous infls in axils of lower lvs of most spp., some on rhiz. *G. max* (L.) Merr. (soy or soya bean, prob. selected in NE China c. C11 BC (known to Europeans since C16) from *G. soja* Siebold & Zucc. (C & E As., Taiwan), its weedy form being *G. gracilis* Skvortzov) with seeds with 35–40% animal-like protein, richest of all pl. foods, first cult. in US 1924, now imp. oilseed (35% of US oil & fats), fodder & cosmetics; flour used in biscuits & confectionery, icecream etc., a milk made from seeds much promoted; seeds cooked with roasted wheat & fermented by *Aspergillus oryzae* give soy(a) sauce used in Worcestershire sauce; bean curd (tofu, tou-fou, tau-foo) cooked, mashed, strained & precipitated with gypsum, much used in China & SE As.; alleged to reduce incidence of breast cancer; *G. wightii* (Arn.) Verdc. – at germ. releasing canavanine inhibiting growth of e.g. lettuce

Glycosmis Corr. Serr. Rutaceae. 43 Indomal. (1 natur. Afr. & Am.). R: PANSP 137(1985)1. Alks. *G. parviflora* (Sims) Little (*G. citrifolia*, SE As.) introd. 1788 via England to Jamaica & now spread throughout S Am.; *G. pentaphylla* (Retz.) Corr. Serr. (Indomal.) – ed. fr., twigs used as toothpicks in India; *G. perakensis* Naray. (Malay Pen.) – rheophyte

Glycoxylon Ducke = *Pradosia*

Glycydendron Ducke. Euphorbiaceae. 1 Amazonia

Glycyrrhiza L. Leguminosae (III 15). 18 Euras. (Eur. 5) with few in Aus., N Am. & temp. S Am. Saponins; oligoglycosides incl. glycyrrhizin 50 times sweeter than sugar (but liquorice aftertaste), slows tooth decay. Many spp. local sources of liquorice incl. *G. glabra* L. (Medit. to C As.) – cult. (esp. Russia, Spain, Middle E) for rhizomes, a source of liquorice used in confectionery, cough mixtures, lozenges & other medic., in plug tobacco & in brewing stout, shoe-polish. Fibre for plastics & fibreboard in US; medic. for sore throats & food poisoning in modern Chinese herbalism

Glycyrrhizopsis Boiss. & Bal. = *Glycyrrhiza*

Glyphaea Hook.f. Tiliaceae. 2 trop. Afr. *G. brevis* (Sprengel) Monachino – chewing-stick in Nigeria

Glyphochloa W. Clayton. Gramineae (40h). 8 C & S India. R: KB 35(1981)814

Glyphosperma S. Watson = *Asphodelus*

Glyphostylus Gagnepain. Euphorbiaceae. 1 SE As.

Glyphotaenium (J. Sm.) J. Sm. = ? *Enterosora*

Glyptocarpa Hu = *Pyrenaria*

Glyptocaryopsis Brand = *Plagiobothrys*

Glyptopetalum Thwaites. Celastraceae. c. 20 Indomal.

Glyptopleura Eaton. Compositae (Lact.-Mal.). 2 W US

Glyptostrobus Endl. Taxodiaceae. 1 SE China: *G. pensilis* (Staunton) K. Koch (*G. lineatus*) – cult. orn., poss. no longer wild. Like *Taxodium* but lvs trimorphic

Gmelina L. Labiatae (I; Verbenaceae). 2 trop. Afr., Mascarenes, 39 E As., Indomal., Aus. (New Caled. 2). R: Phytol. 55(1984)308, 424, 460 etc. Some light & medium timbers (grey teak) esp. *G. arborea* Roxb. (yamar, yemani, India) – utility timber & firewood crop (up to

30m^3/ha/annum)

Gnaphaliothamnus Kirpiczn. (~ *Gnaphalium*). Compositae (Gnap.-Cass.). 1 Mexico: *G. salicifolium* (Bertol.) Anderb. R: OB 104(1991)92

Gnaphalium L. Compositae (Gnap.-Gnap.). c. 50 cosmop. (Eur. 1+ *G. undulatum* L. (S Afr.) natur.; 19 Afr. & Madag.). R: OB 104(1991)167. Cudweeds

Gnaphalodes A. Gray = *Actinobole*

Gnephosis Cass. Compositae (Gnap.-Ang.). 8 temp. Aus. R: OB 104(1991)130

Gnetaceae Blume. Gnetopsida. 2/29 trop. Usu. lianes, less often trees or shrubs with sec. thickening (single cambium in trees, successive ones in lianes, like angiosperms), vessels in xylem, & sec. phloem with companion cells as well as sieve-tubes (unlike most gymnosperms, but derived from diff. initials unlike angiosperms). Lvs decussate, evergreen, with broad lamina & reticulate venation. Fls in whorls in spike-like infls, the whorls subtended by fleshy collars (pl. dioec. to bisexual s.t. with males & females in same infl.); females with nucellus with 3 coats (? 3 integuments or 2 + P), the innermost extending beyond others as micropylar tube & at maturity middle one stony, outer one fleshy; males with 1 microsporangiophore with 1 or 2 sporangia surrounded by a tubular P, the pollen grains with 3 nuclei (? tube nucleus, sterile cell & sperm cell). Pollen drawn down micropyle by droplet mechanism; no archegonia are formed & poss. any prothallial nucleus can act as egg. Several pollen-tubes may penetrate prothallus giving many zygotes & as there are many prothalli per seed & suspensors may branch giving multiple embryos, polyembryony is of a high order but only 1 usu. reaches maturity in seed
Genera: *Gnetum*, *Vinkiella*

Gnetopsida (Gnetatae). Gymnosperms. 4/95 in 3 fams. Trees, shrubs, lianes or stumpy turnip-like pls. Lvs opp. or whorled, simple, strap-like (Welwitschiaceae), angiosperm-like (Gnetaceae) or scales (Ephedraceae). Sec. xylem with vessels. Pl. usu. dioec. (except some Gnetaceae) with fls unisexual in compound strobili ('inflorescences'); females with 1 erect ovule, nucellus enclosed in 2 or 3 coats, micropyle projecting as a long tube; males with 'P'; fert. through pollen-tubes with 2 male nuclei. Embryo with 2 cotyledons
Families: Ephedraceae, Gnetaceae, Welwitschiaceae
Poss. this group represents a grade of sophistication rather than a phyletic unit, each fam. attaining features somewhat like those characteristic of angiosperms; fls in compound strobili with 'P' in males & several coats (cf. integuments) around nucellus; vessels long denied as truly similar to those in angiosperms ('remarkable example of convergent evolution') but indistinguishable microscopically; angiosperm-like lvs in Gnetaceae. Beyond *Drewria* (Cretaceous) & some pollen records there is no fossil known of this group but as so many 'angiosperm' features are here, other fossils may have been pigeonholed as angiosperms

Gnetum L. Gnetaceae. 28 Indomal., Amazonia (7), trop. W Afr. (2). R: P. Maheshwari & R. Vasil (1961) *G.* Some water-disp., fish-disp. etc. Sources of fibre & ed. seeds esp. *G. gnemon* L. (melindjo, Indomal.) – tree cult. for ed. young lvs & seeds ('Fructus Beretinus' (Beretina, Philippines) brought to Eur. from Drake's voyage, 1580) cooked & roasted, excellent crackers (emping) made from flour, fibre from bark (inner bark from several spp. New Guinea ('tulip') for bilum bags); *G. africanum* Welw. (eru, W. & C Afr.) – liane with ed. lvs & tuber; *G. costatum* Schumann (Papuasia) – seeds, lvs & fls for potherb

Gnidia L. (*Lasiosiphon*). Thymelaeaceae. 140 trop. & S Afr. to Arabia (1), Madag. to W India & Sri Lanka. *G. glauca* Steudel (E Afr.) – source of high-quality paper

Goa bean *Psophocarpus tetragonolobus*; **G. butter** *Garcinia indica*; **G. ipecacuanha** *Naregamia alata*; **G. powder** *Andira araroba*

Goadbyella R. Rogers = *Microtis*

goat's beard *Tragopogon pratensis*, *Aruncus dioicus*; **g. nut** *Simmondsia chinensis*; **g. rue** *Galega officinalis*, *Tephrosia virginiana*; **g. thorn** *Astragalus* spp.; **g. weed** *Ageratum conyzoides*; **g. willow** *Salix caprea*

gobbo *Abelmoschus esculentus*

gobō *Arctium lappa*

Gochnatia Kunth. Compositae (Mut.-Mut.). 68 trop. & warm Am., SE As. mts (2). R: RMP 12(1971)1

Godetia Spach = *Clarkia*

Godmania Hemsley. Bignoniaceae. 2 trop. Am. Trees

Godoya Ruíz & Pavón. Ochnaceae. 2 Colombia, Peru & Bolivia, 1–2000 m. R: BJ 113(1991)173. Lvs simple

Godwinia Seemann = *Dracontium*

Goebelia Bunge ex Boiss. = *Sophora*

Goeldinia Huber = *Allantoma*

Goerziella Urban = *Amaranthus*

Goethalsia Pittier. Tiliaceae. 1 C Am. to Colombia

Goethartia Herzog = *Pouzolzia*

Goethea Nees (? = *Pavonia*). Malvaceae. 2 Brazil. Serial axillary buds, some delayed development so infls on branches; epicalyx red, longer than K & C, which is white & tube-like & through which styles & later stamens emerge (reverse of most M.). Cult. orn. shrubs

Goetzea Wydler. Solanaceae (Goetzeaceae). 2 Puerto Rico, Hispaniola. *G. elegans* Wydler reduced to fewer than 50 pls (Puerto Rico)

Goetzeaceae Miers ex Airy Shaw. See Solanaceae

goflo (Canary Is.) *Pteridium aquilinum*

Golaea Chiov. Acanthaceae. 1 Somalia

gold fruit *Citrus* spp.; **g. of pleasure** *Camelina sativa*; **g. thread** or **vegetable g.** *Coptis trifolia*

Goldbachia DC. Cruciferae. 8 Eur. (1) to temp. As.

gold-dust *Aurinia saxatilis*

golden apple *Spondias cytherea*; **g. aster** *Chrysopsis* spp.; **g. cassia** *Chamaecrista fasciculata*; **g. chain** *Laburnum anagyroides*; **g. chestnut** *Chrysopsis chrysophylla*; **g. club** *Orontium aquaticum*; **g. currant** *Ribes aureum*; **g. dewdrop** *Duranta erecta*; **g. drop** *Onosma frutescens*; **g. feather** *Tanacetum parthenium* 'Aureum'; **g. gram** *Vigna radiata*; **g. larch** *Pseudolarix kaempferi*; **g. rain** *Laburnum anagyroides*, *Koelreuteria paniculata*; **g. rod** *Solidago* spp.; **g. samphire** *Limbarda crithmoides*; **g. seal** *Hydrastis canadensis*; **g. shower** *Pyrostegia venusta*, *Cassia fistula*; **g. thread** *Coptis trifolia*; **g. walnut, Nigerian** *Lovoa trichilioides*; **g. willow** *Salix alba* 'Vitellina'; **g. yew** *Taxus baccata* 'Aurea'

goldenaster *Ionactis* spp.

goldfussia *Strobilanthes anisophylla*

Goldfussia Nees = *Strobilanthes*

goldilocks *Ranunculus auricomus*

Goldmanella Greenman. Compositae (Helia.-Cor.). 1 C Am.

Goldmania M. Rose ex M. Micheli. Leguminosae (II 3). 1 Mex. & C Am., 1 Paraguay & Arg.

Golionema S. Watson = *Olivaea*

Gomara Ruíz & Pavón = *Sanango*

Gomaranthus Rauschert = *Sanango*

gombo *Abelmoschus esculentus*

Gomesa R. Br. Orchidaceae (V 10). 13 S Am. esp. Brazil. Cult. orn. epiphytes

Gomidesia O. Berg = *Myrcia*

Gomortega Ruíz & Pavón. Gomortegaceae. 1 S Chile: *G. keule* (Molina) I.M. Johnson – intoxicating fr., form. imp. narcotic in Chile

Gomortegaceae Reiche. Dicots – Magnoliidae – Magnoliales. 1/1 S Chile. Tree, evergreen, with ethereal oil-cells in lvs & young stems. Lvs opp., simple, entire; stipules 0. Fls in racemes, bisexual, P (5–)7(–10), spirally arr. to 3-merous whorls, inner ones smaller; A 7–13 spirally arr., outer 1–3(4) tepaloid with imperf. anthers, others with filaments & anthers dehiscing from base upwards by 2 valves, (1–)3(4) staminodes between A & $\overline{G}$ ((2)3(–5)), 2- or 3-loc., style with 2(3) branches, each loc. with 1 pendulous, anatropous ovule. Fr. a drupe, yellow, ed., usu. 1-loc. & 1-seeded; seed with large embryo in oily endosperm. 2n = 42

Only species: *Gomortega keule*. Allied to Monimiaceae

Gomphandra Wallich ex Lindley. Icacinaceae. 33 SE As. to Solomon Is.

Gomphia Schreber = *Ouratea*

Gomphichis Lindley. Orchidaceae (III 3). 23 Andes, Brazil, Costa Rica

Gomphocalyx Baker. Rubiaceae (IV 15). 1 Madagascar

Gomphocarpus R. Br. = *Asclepias*

Gomphogyne Griffith. Cucurbitaceae. 2 E Himal. to C China & SE As.

Gompholobium Sm. Leguminosae (III 24). (Incl. *Burtonia*) c. 30 Aus. Wedge peas

Gomphostemma Wallich ex Benth. Labiatae (VI). 30 SE As., Mal. Rain forest. Some medic. esp. *G. javanicum* (Blume) Benth. (SE As., W & C Mal.) said to have antitumour activity

Gomphostigma Turcz. Buddlejaceae (Loganiaceae s.l.). 2 S Afr. R: MLW 77-8(1977)15. 1 a rheophyte, other in Karoo

Gomphotis Raf. = *Thryptomene*

Gomphrena L. Amaranthaceae (II 2). (Incl. *Philoxerus*) c. 120 trop. & warm Am., Aus., natur. OW trop. Cult. orn. for bedding & everlastings esp. *G. globosa* L. (globe amaranth, OW trop.) – fl. heads subtended by 2 or 3 purple, orange, rose, white or varieg. leafy bracts

gomuti palm *Arenga pinnata*
gonagra *Rumex hymenosepalus*
Gonatanthus Klotzsch = *Remusatia*
Gonatopus Hook.f. ex Engl. Araceae (V 4). 5 trop. & S Afr. *G. boivinii* (Decne) Engler – tubers & fr. toxic to dogs, man & birds
Gonatostylis Schltr. Orchidaceae (III 3). 1 New Caledonia
Gonçalo alves *Astronium fraxinifolium*
Gongora Ruíz & Pavón. Orchidaceae (V 10). 50 trop. Am. Cult. orn. epiphytes with extrafl. nectaries
Gongrodiscus Radlk. Sapindaceae. 2 New Caledonia
Gongronema (Endl.) Decne. Asclepiadaceae (III 4). 15 OW trop.
Gongrospermum Radlk. Sapindaceae. 1 Philippines
Gongrostylus R. King & H. Robinson (~ *Eupatorium*). Compositae (Eup.-Ayap.). 1 C Am. R: MSBMBG 22(1987)201. Epiphytic liane
Gongrothamnus Steetz = *Distephanus*
Gongylocarpus Schldl. & Cham. Onagraceae. 2 Mex., C Am. G eventually sunk in receptacle
Gongylolepis Schomb. Compositae (Mut.-Mut.). 15 trop. S Am. (14 in Guayana Highland, 11 endemics of Venez. G.). R: AMBG 76(1989)997. Trees & shrubs
Gongylosciadium Rech.f. (~ *Pimpinella*). Umbelliferae (III 8). 1 Turkey, Caucasus, Iran
Gongylosperma King & Gamble. Asclepiadaceae (I; Periplocaceae). 2 Malay Pen.
Gonioanthela Malme. Asclepiadaceae (III 1). 6 S Brazil
Goniocaulon Cass. Compositae (Card.-Cent.). 1 trop. Afr., India
Goniochilus M. Chase (~ *Hybochilus*). Orchidaceae (V 10). 1 Costa Rica: *G. leochilinus* (Reichb.f.) M. Chase. R: CUMH 16(1987)124
Goniocladus Burret. Palmae (V 4m). 1 Fiji: *G. petiolatus* Burret – seen only once, in 1937 (= ? *Physokentia*)
Goniodiscus Kuhlm. Celastraceae. 1 Brazil
Goniolimon Boiss. Plumbaginaceae (II). 20 Russia (Eur. 11) to Mongolia, NW Afr. Like *Limonium* but styles hairy & stigmas capitate; some cult. orn.
Gonioma E. Meyer. Apocynaceae. 1 S Afr., 1 Madag. Alks. *G. kamassi* E. Meyer (kamassi, S Afr.) – dense wood (Cape boxwood) used for engraving etc., exported as 'boxwood'
Goniophlebium C. Presl (~ *Polypodium*). Polypodiaceae (II 5). 5–10 trop. As. to Fiji (? & Am.). Some cult. orn.
Goniopteris C. Presl = *Cyclosorus*
Goniorrhachis Taubert. Leguminosae (I 4). 1 SE Brazil
Gonioscypha Baker. Convallariaceae (Liliaceae s.l.). 2 Himal., SE As.
Goniostemma Wight = *Toxocarpus*
Goniothalamus (Blume) Hook.f. & Thomson. Annonaceae. 50–100 Indomal. Some ed. fr. (grape-flavoured); *G. wightii* Hook.f. & Thomson bark yields a strong fibre
Gonocalyx Planchon & Linden ex Lindley. Ericaceae. 9+ C Am. R: SB 15(1990)747
Gonocarpus Thunb. (~ *Haloragis*). Haloragidaceae. 41 Aus. (36) & NZ extending to SE As. & Japan. R: BAIM 10(1975)164
Gonocaryum Miq. Icacinaceae. 9 or 10 Indomal. to Taiwan
Gonocormus Bosch = *Crepidomanes*
Gonocrypta Baillon = *Pentopetia*
Gonocytisus Spach. Leguminosae (III 31). 3 E Med. (Eur. 1). Like *Spartium* but lvs 3-foliolate
Gonolobus Michaux. Asclepiadaceae (III 3). 100 Am. *G. cundurango* Triana (condurango, cundurango, trop. Am.) – bitters, form. medic.; *G. edulis* Hemsley (guayato, Costa Rica) – fr. ed.
Gonopyrum Fischer & C. Meyer = *Polygonella*
Gonospermum Less. Compositae (Anth.-Gon.). 4 Canary Is.
Gonostegia Turcz. (~ *Pouzolzia*). Urticaceae (III). 5 SE As. to N Aus.
Gontscharovia Boriss. (~ *Satureja*). Labiatae (VIII 2). 1 C As.
Gonyanera Korth. = *Acranthera*
Gonypetalum Ule = *Tapura*
Gonystylaceae Gilg = Thymelaeaceae
Gonystylus Teijsm. & Binnend. Thymelaeaceae. 20 Indomal., Pacific. Seeds large, arillate, *G. bancanus* (Miq.) Kurz (ramin, W Mal.) – peat swamp-forest tree with knee-roots & inner bark with irritant fibres, lightweight comm. timber used for dowelling, walking canes, etc., much exported
Gonzalagunia Ruíz & Pavón. Rubiaceae (I 8). 15 trop. Am.
Good King Henry *Chenopodium bonus-henricus*

Goodallia Benth. Thymelaeaceae. 1 Guyana. (NB *Goodallia* Bowdich is an older name for another pl.)

Goodenia Sm. Goodeniaceae. (Incl. *Calogyne, Catospermum, Neogoodenia, Symphyobasis*) 179 Australasia (Aus. 178 mainly endemic, some ext. to SE As. with 1 endemic Java). G 1-loc. apically, ± 2-loc. basally

Goodeniaceae R. Br. Dicots – Asteridae – Campanulales. 12/400 trop. & warm, mostly Aus. Sappy shrubs, herbs or even trees with simple to stellate hairs (multicellular with term. cell papillate in *Brunonia*), often storing inulin & poisonous; laticifers 0. Lvs spirally arr. (rarely opp. or whorled); stipules 0. Fls bisexual, solit. or in heads, racemes or cymes; K ((3–)5), lobed, s.t. v. reduced, C (5), 2-labiate or 1-labiate (i.e. adaxial bifid to base), lobes valvate, A 5 alt. with & attached to tube of C or free, anthers with longit. slits, connivent or connate, style growing up through & presenting pollen to insects, intrastaminal nectaries s.t. present, $\overline{G}$ (2), s.t. half-inf. or (*Velleia*) $\underline{G}$, (1)2-loc. (4-loc. in *Scaevola porocarya* F. Muell., 2 locs & ovules reduced almost to common condition in other *S.* spp.), ovules 1–∞ per loc., mostly erect or ascending on axile placentas, anatropous & unitegmic. Fr. usu. a capsule, s.t. a drupe or nut; seeds usu. flat, s.t. winged, with straight embryo embedded in copious oily endosperm. x = 7–9

Principal genera: *Dampiera, Goodenia, Leschenaultia, Scaevola, Velleia*

Brunonia allied to *Leschenaultia-Anthotium-Dampiera* group. Fam. allied to Calyceraceae & Compositae

Some cult. orn.

Goodia Salisb. Leguminosae (III 25). 1–3 SW & SE Aus.

Goodmania Rev. & Ertter (*Gymnogonum*). Polygonaceae (I 1). 1 California & Nevada: *G. luteola* (C. Parry) Rev. & Ertter. Allied to *Eriogonum*

Goodyera R. Br. Orchidaceae (III 3). 55 cosmop. (Eur. 1: *G. repens* (L.) R. Br.)

googa *Owenia acidula*

gooseberry *Ribes uva-crispa*; **American g.** *R. cynosbati*; **Barbados g.** *Pereskia aculeata*; **Cape g.** *Physalis peruviana*, **Ceylon g.** *Dovyalis hebecarpa*; **Chinese g.** *Actinidia deliciosa*; **Otaheite g.** *Phyllanthus acidus*; **prickly g.** *Ribes cynosbati*

goosefoot *Chenopodium album*

goosegrass *Galium aparine*

Gooringia F. Williams = *Arenaria*

gopher wood see *Pinus*

Gorceixia Baker. Compositae (Vern.-Pip.). 1 SE Brazil

Gordonia Ellis. Theaceae. 70 SE As., warm N. Am. Cult. orn. evergreen trees & shrubs esp. *G. axillaris* (Ker-Gawler) Endl. (Taiwan to Vietnam (greenhouse)) & *G. lasianthus* (L.) Ellis (loblolly bay, black laurel, SE US) – cabinet wood

Gorgonidium Schott. Araceae (VIII 3). 3 Peru, Bolivia, N Arg. R: BJ 109(1988)529

gorli oil *Caloncoba echinata*

Gormania Britton = *Sedum*

Gorodkovia Botsch. & Karav. Cruciferae. 1 NE Siberia

gorse *Ulex europaeus*; **dwarf g.** *U. minor, U. gallii*

Gorteria L. Compositae (Arct.-Gort.). 3 SW Cape to Namibia. R: MBSM 3(1959)319, 11(1973)91. *G. diffusa* Thunb. – dark marks on ray-florets poss. mimicking beetles

Gosela Choisy. Globulariaceae. 1 S Afr.: *G. eckloniana* Choisy. R: SAJB 56(1990)477

Gossweilera S. Moore. Compositae (Vern.-Vern.). 2 Angola

Gossweilerochloa Renvoize = *Tridens*

Gossweilerodendron Harms. Leguminosae (I 4). 2 Gulf of Guinea forests. *G. balsamiferum* (Vermoesen) Harms (agba, tola wood) – timber for furniture etc., copals

Gossypianthus Hook. = *Guilleminea*

Gossypioides Skovsted ex J.B. Hutch. Malvaceae. 2 trop. Afr., Madag.

Gossypiospermum (Griseb.) Urban = *Casearia*

Gossypium L. Malvaceae. 39 warm temp. to trop. R: P.A. Fryxell (n.d.) *Nat. Hist. Cotton tribe*: 37 (OW 12, R: KB 42(1987)337). Cotton: seeds covered with long hairs (single-celled, 3000 times longer than wide), which when dried are flat & can be spun (200 million per kg), & short hairs (fuzz) used for felt; oil extracted & (C Am.) flour (high protein), incaparina; seed-cake valuable cattle-feed. Cotton used for cotton wool, thread, carpets etc., oil in cooking & soap powders; possible male contraceptive (China) as gossypol in small quantities inhibits sperm formation; petals source of yellow dyes (India); over 1000 diff. uses claimed for waste in former USSR, incl. growth promoters, concrete plasticizers, food preservatives, etc. Fls visited by bees & (Am.) hummingbirds; extrafl. nectaries. Complex & controversial history; diploid *G. arboreum* L. (tree cotton cult. Pakistan 1800

BC & taken to Middle E in 1st millenium BC) more variable than diploid *G. herbaceum* L. (domesticated E Afr.) in Afr., independently selected from diff. ancestors, though their genome claimed as also appearing in diff. NW cottons which (comm. ones) are tetraploid esp. *G. barbadense* L. (*G. peruvianum*, Sea Island c.), the one widely planted in S US. Some cult. orn. incl. *G. sturtianum* J.H. Willis (Sturt('s) desert rose, Aus.) – emblem of N Terr., Aus.

Gouania Jacq. Rhamnaceae. 50–70 trop. & warm (Am. 15, Afr. 2, Madag. & Ind. Ocean 5, As. 4, New Caled. 1). Saponins lead to use as shampoo etc. – *G. lupuloides* (L.) Urban (trop. Am.) used as chewing-stick gives soapy mouthwash when chewed; some with watch-spring tendrils

Gouinia Fourn. ex Benth. Gramineae (31d). 12 trop. Am. R: BTBC 88(1961)143. Allied to *Silentvalleya* (India), *Lophacme* (Afr.)

Gouldia A. Gray = *Hedyotis*

Gouldochloa Valdés R., Morden & Hatch = *Chasmanthium*

goumi *Elaeagnus multiflora*

Goupia Aublet. Celastraceae (Goupiaceae). 3 Guyana, N Brazil. *G. glabra* Aublet (Guyana) – fine timber for outdoor use

Goupiaceae Miers. See Celastraceae

gourd See Cucurbitaceae, *Benincasa, Cucumis, Cucurbita, Lagenaria, Trichosanthes*, also *Crescentia*; **silverseed g.** *Cucurbita argyrosperma*

gourd-tree *Adansonia gregorii*

Gourliea Gillies ex Hook. & Arn. = *Geoffroea*

goutweed *Aegopodium podagraria*

Govenia Lindley. Orchidaceae (V 9). 20 trop. Am. Some cult. orn. terr.

governor's plum *Flacourtia indica*

gowan Scottish word for various spp. with yellow fls, e.g. *Ranunculus* & *Taraxacum* spp.

Goyazia Taubert. Gesneriaceae. 2 Brazil. R: Selbyana 1(1976)392

Goyazianthus R. King & H. Robinson. Compositae (Eup.-Alom.). 1 Brazil. R: MSBMBG 22(1987)258

Goydera Liede. Asclepiadaceae (III 1). 1 Somalia: *G. somaliensis* Liede. R: Novon 3(1993)265

Grabowskia Schldl. Solanaceae (1). 6 S Am. Cult. orn. spiny divaricately branched shrubs

Graciela Rzed. = *Strotheria*

Graderia Benth. Scrophulariaceae. 5 Afr. & Socotra

Graellsia Boiss. Cruciferae. 7 Morocco (1), Turkey to Pakistan. R: BJLS 102(1990)17

Graffenrieda DC. Melastomataceae. 44 trop. Am.

Grafia A. Hawkes = *Phalaenopsis*

Grafia Reichb. Umbelliferae (III 5). 1 Eur.

Grahamia Gillies ex Hook. & Arn. Portulacaceae. 1 Arg. R: FGP 23(1988)217

Grajalesia Miranda. Nyctaginaceae (VI). 1 Mex.: *G. fasciculata* (Standley) Miranda

gram, black *Vigna mungo*; **golden** or **green g.** *V. radiata*; **horse g.** *Macrotyloma uniflorum*; **red g.** *Cajanus cajan*

grama or **gramma** *Bouteloua* spp.; **blue g.** *Chondrosum gracile*; **side-oats g.** *B. curtipendula*

Gramineae Juss. (Poaceae). Monocots – Commelinidae – Cyperales. 668/9500 cosmop. but esp. trop. & N temp. sub-arid. Usu. perennial & often rhizomatous herbs or (bamboos) ± woody & tree-like but without sec. thickening; cell-walls, esp. epidermis ± strongly silicified, vessel-elements usu. in all vegetative organs; stems usu. terete & with hollow internodes; roots often with root-hairs but often with endomycorrhizae also. Lvs distichous (spirally arr. in e.g. *Micraira*), never 3-ranked, with usu. open sheath & elongate lamina usu. with basal meristem & pair of basal auricles (narrowed to a petiolar base above sheath in many bamboos; ligule usu. adaxial at junction of lamina & sheath, rarely 0. Fls usu. wind-poll., usu. bisexual, in 1–∞-flowered spikelets in spike-like to panicle-like sec. infls; spikelets usu. with pair of subopp. bracts (glumes) & 1–several distichous florets often on zig-zag rhachilla, the florets usu. comprising a pair of subopp. subtending scale-like bracts (lemma & palea), 2 or 3 small lodicules (? P, the upper bract (palea) sometimes interpreted as derived from P, up to 6 or more lodicules in Bambusoideae), A (1–)3 or 6 (esp. Bambusoideae, where up to >100 in *Ochlandra*), anthers elongate, basifixed but deeply sagittate so as to appear versatile, with longit. slits & nearly smooth pollen grains, G̲ (2 (3 in Bambusoideae)), 1-loc. with 2(3) stigmas, often large & feathery, ovule 1, orthotropous to almost anatropous etc., (1)2-tegmic. Fr. (caryopsis) usu. enclosed in persistent lemma & palea, usu. dry-indehiscent, integuments adnate to pericarp, the seed rarely falling free of these accessory structures such as when pericarp becomes mucilaginous when wet & expelling the seed on drying out, fleshy in some bamboos (e.g. *Alvimia*

& *Olmeca* (NW), *Dinochloa, Melocalamus, Melocanna, Ochlandra* (OW)); embryo straight with well-developed plumule covered by a closed cylindrical coleoptile, radicle with a similar coleorhiza, & enlarged lateral cotyledon (scutellum), all peripheral to copious starchy endosperm usu. with proteinaceous tissue & s.t. also oily, rarely (*Melocanna*) 0. x = 2–23+. R: KBAS 13(1986)

Classification & principal genera:

Although still in a state of flux (see e.g. BR 55(1989)141), G. can be subdivided into 6 subfams & 40 tribes on features of spikelets, anatomy, physiology & caryology (Clayton & Renvoize):

I. **Bambusoideae** (Biotropica 11(1979)161; tribes with similar leaf anatomy & clear links with other fams, suggesting this group to be the most archaic); trop. esp. forest & aquatic habitats, 13 tribes:

1. **Bambuseae** (mostly tree-like & shrubby bamboos to 40 m; some with 3 lodicules, A 6 & 3 stigmas; some with periodic flowering as rarely as every 120 yrs; mainly plants of trop. forests & mostly tetraploids (*Dendrocalamus* hexaploids). Many hundreds of uses in paper-making, construction, engineering & handicrafts, ed. young shoots, cult. orn.): some 50 genera in 3 subtribes (a Arundinariinae (*Arundinaria, Chusquea, Sasa, Sinarundinaria*), b Bambusinae (*Bambusa, Dendrocalamus, Phyllostachys*), c Melocaninae (*Schizostachyum*)
2. **Anomochloeae** (herbaceous, forest pl. resembling Marantaceae but with bambusoid leaf anatomy & ligular hairs): *Anomochloa* (only)
3. **Streptochaeteae** (herbaceous; 1-flowered spikelets. 3 lodicules to 2 cm long): *Streptochaeta* (only)
4. **Olyreae** (creeping herbs or 'rambling canes' with broad laminas; spikelets 1-flowered, unisexual; largely S Am.): *Olyra*
5. **Parianeae** (herbs with A 2–30, insect-poll.): *Pariana*
6. **Phareae** (like Olyreae but veins not running parallel to leaf midrib; rain forest): *Leptaspis, Pharus*
7. **Phaenospermateae** (herbaceous; 1-flowered spikelets; lodicules 3, A 3): *Phaenosperma* (only)
8. **Streptogyneae** (herbaceous, several-flowered spikelets): *Streptogyna* (only)
9. **Oryzeae** (herbs; spikelets 1-flowered with glumes scarcely developed, in *Oryza* 2 sterile lemmas simulating them; trop. & warm temp. esp. in swamps): *Leersia, Oryza* (rice), *Zizania*
10. **Phyllorachideae** (suffrutescent; spikelets 2-flowered; lodicules 2, A 3 or 6): *Humbertochloa*
11. **Ehrharteae** (rice-like spikelets but on open hillsides): *Ehrharta* (only)
12. **Diarrheneae** (herbaceous with 2–5-flowered spikelets & ellipsoid knobbed fr. with 2 term. stigmas): *Diarrhena* (only)
13. **Brachyelytreae** (like 13. but 1-flowered spikelets): *Brachyelytrum* (only)

II. **Pooideae** (incl. Stipoideae; Taxon 29(1980)664; differs from other subfams in cryptic chars. (incl. no epidermal microhairs (exc. 14 & 15 poss. misplaced here), caryology etc.); lodicules 2 or 3; largely N temp. but also on trop. mts): 10 tribes:

14. **Nardeae** (microhairs; infl. a unilateral raceme): *Nardus* (only)
15. **Lygeae** (microhairs; infl. a single spikelet): *Lygeum* (only)
16. **Stipeae** (tussocky or reed-like, spikelets 1-flowered, lemmas usu. with term. hygroscopic awn): *Piptochaetium, Stipa*
17. **Poeae** (incl. Seslerieae; infls spikes, racemes or non-capitate panicles, usu. 1–many fert. fls per spikelet (variable), lemmas usu. herbaceous, awns usu. term. & straight, rarely 0, lodicules usu. free; caryopses sometimes adherent to lemma &/or palea): *Briza, Cynosurus, Dactylis, Festuca, Lolium, Poa, Puccinellia, Sesleria, Vulpia*
18. **Hainardieae** (infl. a single bilateral raceme, 1- or 2–6)-flowered): *Pholiurus*
19. **Meliceae** (sheath margins connate; infls racemes or panicles, 1 or 1-several fert. florets per spikelet (variable), lemmas herbaceous, awns dorsal or 0, lodicules often connate, glabrous): *Glyceria, Melica*
20. **Brylkinieae** (sheath margins connate; infl. a raceme, lodicules free): *Brylkinia* (only)
21. **Aveneae** (incl. Agrostideae; infls racemes or non-capitate panicles; usu. 1–several fert. florets per spikelet, glumes usu. 2, the lower up to 11-veined, lemmas hard or herbaceous, awns dorsal or from term. notch, usu. twisted or angled; 4 subtribes): a Duthieinae (*Duthiea*), b Aveninae (*Arrhenatherum, Avena*

(oats), *Deschampsia, Helictotrichon, Koeleria, Trisetum*, c Phalaridinae (*Hierochloe, Phalaris*), d Alopecuridinae (*Agrostis, Alopecurus, Calamagrostis, Deyeuxia, Phleum*

22. **Bromeae** (auricles usu. 0; sheath margins connate; infl. a panicle, 1–30 florets per spikelet, lemmas herbaceous, awns 1 or more per lemma, dorsal, lodicules glabrous): *Bromus*
23. **Triticeae** (incl. Brachypodieae; auricles usu. present; sheath margins usu. free; infl. usu. a spike, 1-several (variable) fls per spikelet, lemmas hard or herbaceous, awns term. (or 0), lodicules usu. hairy): *Aegilops, Brachypodium, Elymus, Elytrigia, Hordeum* (barley), *Leymus, Secale* (rye), *Triticum* (wheat)

III. **Centothecoideae** (1 tribe similar in some respects to both Bambusoideae & Panicoideae but distinctive embryo):

24. **Centotheceae** (broad-leaved herbs with panicles of 1–several-flowered spikelets, mainly rain forest): *Centotheca, Chasmanthium*

IV. **Arundinoideae** (unspecialized group largely defined by absence of specialized characters & considered relics of ancient line from which non-bambusoid grasses derived; mainly trop. esp S), 4 tribes:

25. **Arundineae** (incl. Danthonieae; perhaps not homogeneous; robust grasses with panicles of plumose spikelets): *Arundo, Cortaderia* (much cult. orn.), *Danthonia, Phragmites* (pulp), *Rytidosperma*
26. **Thysanolaeneae** (spikelets with dimorphic florets disarticulating below pedicel; poss. best inc. in 25): *Thysanolaena* (only)
27. **Micraireae** (moss-like with spirally arr. lvs (unique), Aus.): *Micraira* (only)
28. **Aristideae** (panicles of 1-flowered needle-like spikelets; lemma with 3-branched awn; trop.): *Aristida, Stipagrostis*

V. **Chloridoideae** (leaf anatomy with 'Kranz syndrome' associated with C_4 photosynthesis app. adapted to high light intesities; spikelets shattering at maturity; mainly trop., 5 tribes):

29. **Pappophoreae** (panicles with several-flowered uniform spikelets; lemma 9–11-nerved): *Enneapogon*
30. **Orcuttieae** (panicles or racemes of uniform several-flowered spikelets; lemma 13–15-nerved): *Orcuttia*
31. **Eragrostideae** (incl. Sporoboleae; panicles or racemes with several-flowered spikelets & 1–3-nerved lemmas; trop. with many colonist spp.): 5 subtribes: a Triodiinae (Aus.; *Triodia*), b Uniolinae (*Uniola*), c Monanthochloinae (halophytes; *Distichlis*), d Eleusininae (*Eleusine, Eragrostis, Leptochloa, Tridens, Tripogon*), e Sporoboloinae (*Muhlenbergia, Sporobolus*)
32. **Leptureae** (spikelets embedded in rachis; maritime): *Lepturus* (only)
33. **Cynodonteae** (incl. Chlorideae, Zoysieae; racemes of spikelets with 1 fert. floret ± sterile ones; trop. & subtrop. savannas, 4 subtribes): a Pommereullinae (*Astrebla*), b Chloridinae (*Chloris, Cynodon, Spartina*), c Boutelouinae (*Bouteloua*), d Zoysiinae (R: KB 28(1973)37; *Zoysia*)

VI. **Panicoideae** (leaf anatomy usu. with Kranz syndrome; spikelets 2-flowered, the lower male or sterile; mainly trop., 7 tribes):

34. **Paniceae** (spikelets falling entire, lemma & palea of upper floret enclosing fr.; pantrop., 7 subtribes): a Neurachininae (Aus.; *Neurachne*), b Setariinae (*Axonopus, Brachiaria, Echinochloa, Eriochloa, Panicum, Paspalidium, Paspalum, Sacciolepis, Setaria*), c Melinidinae (R: BB 138(1988); *Melinis*), d Digitariinae (*Digitaria*), e Arthropogoninae (*Arthropogon*), f Cenchrinae (*Cenchrus, Pennisetum*), g Spinificinae (*Spinifex*)
35. **Isachinae** (like 34. but 2 fertile disarticulating florets): *Isachne*
36. **Hubbardieae** (like 35. but paleas 0): *Hubbardia* (only, now extinct)
37. **Eriachneae** (like 34. & 35. but awned lemmas in bisexual florets with Kranz anatomy): *Eriachne*
38. **Steyermarkochloeae** (aerenchymatous, of obscure relationship): *Steyermarkochloa* (only)
39. **Arundinelleae** (spikelets with typical disarticulation; emergent panicles often with juvenile spikelets; trop. savannas esp. OW): *Arundinella, Loudetia*
40. **Andropogoneae** (infls of fragile racemes (sometimes paniculate) with usu. paired spikelets; trop. savannas, 11 subtribes): a Saccharinae (*Eulalia, Imperata, Miscanthus, Saccharum* (sugar-cane)), b Germainiinae (*Apocopsis*), c Sorghinae (*Bothriochloa, Sorghastrum, Sorghum*), d Ischaeminae (*Ischaemum*), e Dimeriinae (*Dimeria*), f Andropogoninae (*Andropogon, Cymbopogon, Schizachyrium*), g

Anthistiriinae (*Hyparrhenia, Themeda*), h Rottboelliinae (*Coelorachis*), i Tripsacinae (*Tripsacum, Zea* (maize)), j Chionachininae (*Chionachne*), k Coicinae (*Coix*)

A very natural family, with fossils back to Palaeocene/Eocene boundary (Tennessee), but with considerable variation from the lamina-less sheaths of *Spartochloa* to laminas 5 m long in *Neurolepis*, from the solit. 1-flowered spikelet of *Aciachne* to the 2 m plumes of *Gynerium*, from the familiar dry 'seeds' to the fleshy berries several cm long of *Melocanna*, from annuals a few mm tall to 40 m bamboos; the stereotyped pooids typical of the N temp. are misleadingly uniform. It is argued that from bambusoid ancestors arose Arundineae & Stipeae, giving rise to savanna regimes of burning & grazing, expanding grasslands as major ecosystems, these then ousted by emergence of C_4 tribes. Panicoideae occupied mesic & 'climax' environments, Chloridoideae the pioneering & stressful ones, with Arundineae surviving esp. in S, the Stipeae giving rise to Pooideae (Clayton)

The combination of basal shoots (tillers) & intercalary meristems allows grasses to tolerate burning & grazing which eliminate their competitors & hold up successional sequences to forest; in turn, saliva from herbivorous mammals appears to contain growth-factors stimulatory to the grass. Silica app. grazing tolerant adaptation (cf. *Equisetum*) but, after fires in mills etc., lumps of 'glass' formed form. thought to be 'thunderbolts' & the cause rather than result of fire! Most spp. are wind-poll., the pollen viable for less than a day though of immense effect on hay-fever sufferers. Some are apomictic or cleistogamous (e.g. wheat), while others have bulbils in infls (esp. Arctic spp.); vivipary reported in *Melocanna*. At least 80% have a polyploid ancestry & these have been partially analysed in wheat, *Spartina anglica* etc., though reduction in chromosome number via 'polyhaploidy' recorded. Vegetative spread of clones estimated to be 1000 yrs old known in *Festuca*. Many are wind-disp. or transported ectozoically (awns, hooks etc.), the awns s.t. hygroscopic & screwing the fr. into the ground, some fleshy & endozoic; some whole infls acting as 'tumbleweeds'

Often confused with sedges & rushes by laymen, G. differ from Juncaceae (P 6, fr. a 1–3-loc. capsule with 3 to ∞ seeds) & Cyperaceae (1 scale beneath each floret, stems usu. 3-angled & solid) in a number of clear-cut characters (A. Arber (1934) *The Gramineae*; C.E. Hubbard (1984), *Grasses*, ed. 3)

Most major civilizations are based on the triploid endosperm of G. (AMBG 68(1981)87: wheat, barley, oats, rye etc. in Euras., millets & tef in parts of Afr., rice in E As., maize in C Am. (wheat, rice & maize provide more than 50% of calories consumed by humans); only the Maoris have become (form.) international power without, being based on sweet potato) & on animals raised on G. as forage as well as bamboos as building materials in many trop. societies. Other imp. products include sugar-cane & aromatic oils (esp. *Cymbopogon* spp.) for soap, cooking etc., thatching & weaving materials, sand-binders (*Ammophila* etc.) & toxic metal-tolerant colonist grasses (*Agrostis* spp. etc.) for reclamation of derelict land; others are sources of paper (*Arundo, Eulaliopsis, Leymus, Stipa* etc. spp.), ed. bamboo shoots, beads (*Coix*), reeds for wind instruments (*Arundo donax* & other spp.), fishing-rods, lawn-grasses (esp. *Festuca* spp. in temp., *Cynodon dactylon* etc. in trop.) & cult. orn. e.g. *Cortaderia, Hakonechloa, Melinis, Miscanthus, Phalaris, Zea* & many bamboos. Many are bad weeds (*Elytrigia repens, Imperata cylindrica, Nassella trichotoma, Avena sativa* ssp. *fatua, Poa annua* etc. & others rep. fire-climax vegetation of a noxious type e.g. *Heteropogon contortus, Imperata cylindrica*

Grammadenia Benth. = *Cybianthus*

Grammangis Reichb.f. Orchidaceae (V 9). 2 Madag. Cult. orn. epiphytes

Grammatophyllum Blume. Orchidaceae (V 9). 12 Mal. to Polynesia. Some locally medic. & supposed aphrodisiac; *G. speciosum* Blume with infls to 3 m of up to 100 fls c. 15 cm diam., spectacular cult. orn. epiphyte

Grammatopteridium Alderw. = *Selliguea*

Grammatotheca C. Presl (~ *Lobelia*). Campanulaceae. 1 S Afr., 1 (? native) Aus.

Grammitidaceae Newman. Filicopsida. 4/450 trop. & Aus., esp. cloud forest on trop. mts. Epiphytes with solenosteles & ± hairy fronds with free or reticulate venation; indusia 0; spores green, usu. trilete & s.t. multicellular before release

Genera: *Acrosorus, Calymmodon, Grammitis, Scleroglossum*

Anarthropteris & *Loxogramme* form. incl. here now referred to Polypodiaceae in which whole group form. incl.

Grammitis Sw. Grammitidaceae. (incl. *Adenophorus, Ctenopteris, Prosaptia, Xiphopteris*) c. 400 trop. (New Guinea 64, R: Blumea 29(1983)13) esp. As. & Am., warm & S temp. (s.s. 150

trop. & Aus. (NZ 9, R: NZJB 14(1976)85)). *G. succinea* L.D. Gomez (Oligocene amber of Dominican Republic) – only grammitid fossil known

Grammosciadium DC. Umbelliferae (III 2). 7 E Med.

Grammosolen Haegi. Solanaceae (2). 2 S Aus. R: Telopea 2(1981)178

Grammosperma O. Schulz. Cruciferae. 1 Patagonia

granadilla *Passiflora quadrangularis*; **purple g.** *P. edulis*; **sweet g.** *P. ligularis*; **yellow g.** *P. laurifolia*

granadillo *Dalbergia granadillo*

Grandidiera Jaub. Flacourtiaceae (3). 1 trop. E Afr.

Grangea Adans. Compositae (Ast.-Gran.). 10 trop. & warm Afr., Madag., 1 (*G. maderaspatana* (L.) Poiret) warm Asia. R: MBSM 15(1979)450

Grangeopsis Humbert. Compositae (Ast.-Gran.). 1 Madag. R: MBSM 15(1979)524

Grangeria Comm. ex Juss. Chrysobalanaceae (I). 2 Madag., Mauritius (1). R: PTRSB 320(1988)83

granite gooseberry *Ribes curvatum*

granny-bonnets *Aquilegia* spp.

Grantia Boiss. = *Iphiona*

grape *Vitis vinifera*; **bush g.** *V. acerifolia*; **bullace g.** *V. rotundifolia*; **canyon g.** *V. arizonica*; **cat g.** *V. palmata*; **chicken g.** *V. vulpina*; **fox g.** *V. labrusca*, *V. rotundifolia*; **frost g.** *V. vulpina*; **g. hyacinth** *Muscari* spp.; **g. ivy** *Cissus* spp.; **mountain g.** *V. monticola*; **Oregon** or **Rocky Mt. g.** *Mahonia aquifolium*; **sand g.** *V. rupestris*; **seaside g.** *Cocoloba uvifera*; **skunk g.** *V. labrusca*; **tree g.** *C. uvifera*; **g.-vine** (largely US) *Vitis vinifera* (N.B. 'on the g.-v.' = 'on the g.-v. [cf. bush] telegraph' (US Civil War))

grapefruit *Citrus* × *aurantium* Grapefruit Group

Graphandra Imlay. Acanthaceae. 1 Thailand

Graphardisia (Mez) Lundell = *Ardisia*

Graphephorum Desv. Gramineae (21b). 3 N & C Am.

Graphistemma (Benth.) Benth. Asclepiadaceae (III 1). 1 Hong Kong: *G. pictum* (Benth.) B.D. Jackson

Graphistylis R. Nordenstam (~ *Senecio*). Compositae (Sen.-Sen.). 8 S Brazil

Graphorkis Thouars. Orchidaceae (V 9). 2–3 trop. Afr. to Masc. Cult. orn. incl. *G. lurida* (Sw.) Kuntze (Afr.) – flowering throughout year, often epiphytic on palms esp. *Hyphaene*

grapple plant *Harpagophytum procumbens*

Graptopetalum Rose. Crassulaceae. 12 SW N Am. Cult. orn. (inc. hybrids with *Echeveria* spp.) succ. with rosettes of succ. lvs esp. *G. paraguayense* (N.E. Br.) Walther (W Mex.!)

Graptophyllum Nees. Acanthaceae. 10 Aus., SW Pacific. Shrubs often with coloured or spotted lvs; *G. pictum* (L.) Griffith (? New Guinea) much cult. in trop. as foliage pl. with purplish or green lvs marked with yellow, local medic.

grass See Gramineae; (slang) *Cannabis sativa*; (skirts) *Cocos nucifera*; **g. of Parnassus** *Parnassia palustris* **g. tree** *Xanthorrhoea* spp.; **g. wrack** *Zostera marina*

Grastidium Blume = *Dendrobium*

Gratiola L. Scrophulariaceae. c. 20 temp., trop. mts (Eur. 2). Dried pl. of *G. officinalis* L. (N temp.) form. medic. ('hedge hyssop')

Gratwickia F. Muell. Compositae (Gnap.-Ang.). 1 Aus.: *G. monochaeta* F. Muell. R: OB 104(1991)117

Grauanthus Fayed. Compositae (Ast.-Gran.). 2 trop. Afr.

gravel root *Eupatorium purpureum*

Gravesia Naudin. Melastomataceae. 110 Afr. (5), Madagascar

Gravesiella Fernandes & R. Fernandes = *Cincinnobotrys*

Gravisia Mez = *Aechmea*

Grayia Hook. & Arn. (*Eremosemium*). Chenopodiaceae (I 3). 2 W US

Grazielia R. King & H. Robinson (~ *Eupatorium*). Compositae (Eup.-Dis.). 10 S Am. R: MSBMBG 22(1987)79

Grazielodendron Lima (~ *Pterocarpus*). Leguminosae (III 4). 1 Brazil: *G. rio-docensis* Lima. R: Bradea 3(1983)401

greasewood *Sarcobatus vermiculatus*

green briar *Smilax* spp.; **g. gram** *Vigna radiata*; **Kendal g., dyer's g. weed** *Genista tinctoria*; **g. rose** *Rosa chinensis* 'Viridiflora'

Greenea Wight & Arn. Rubiaceae (I 5). 7–8 SE As. to W Mal.

Greenella A. Gray = *Gutierrezia*

greengage *Prunus* × *domestica* 'ssp. *italica*'

greenheart *Chlorocardium rodiei*; **Afr. g.** *Cylicodiscus gabunensis*

greenhood *Pterostylis* spp.
Greeneocharis Guerke & Harms = *Cryptantha*
Greeniopsis Merr. Rubiaceae (I 1). 6 Philippines
Greenmania Hieron. = *Unxia*
Greenmaniella W. Sharp. Compositae (Helia.-Mel.). 1 Mexico
Greenovia Webb & Berth. Crassulaceae. 4 Canary Is. Cult. orn. succ.
greens (or g.tuff, GB) *Brassica* spp. – lvs used as vegetables; **chrysanthemum g.** *Chrysanthemum coronarium*; **Japanese g.** *B. juncea* 'Crispifolia'; **sour g.** *Rumex venosus*; **Texsel g.** *B. carinata*
Greenwayodendron Verdc. Annonaceae. 2 trop. Afr.
Greenwoodia Burns-Balogh. Orchidaceae (III 1). 1 Mexico
Greigia Regel. Bromeliaceae. 26 C & NW S Am. R: FN 14(1979)1629. *G. sphacellata* (Ruíz & Pavón) Regel (Chile) – fr. ed.
Grenacheria Mez. Myrsinaceae. 10 Mal.
grenadine *Punica granatum*
Greslania Balansa. Gramineae (1b). 4 New Caledonia
Grevea Baillon. Grossulariaceae (Montiniaceae). 2 trop. E Afr., Madag.
Grevillea R. Br. ex J. Knight. Proteaceae. 261 Sulawesi (1), New Guinea (3), New Caled. (3, endemic), Vanuatu, Aus. (256, 254 endemic). R: D.J. McGillivray (1993) *G.*, Nuytsia 9(1993)237. Fls protandrous, stigma carrying mature pollen in position to be collected by pollinators (sometimes birds; *G. leucopteris* Meissner (W Aus.) nocturnal scarabs; *G. myosodes* McGillivray (NW Aus.) poss. mammal-poll., smells mousy; also effectively poll. by humans sucking nectar!) before stigma ripe; ant-disp. in *G. pteridifolia* J. Knight (N Aus.) & some other spp., seed-wing high in lipid & protein, rest of seed with cyanide. Some timbers & cult. orn. esp. *G. robusta* Cunn. ex R. Br. (silky oak, E Aus.) – timber for cabinet-work etc., cult. as coffee-shade, a tree to 40m much used when a seedling as a housepl., street-tree as at Kathmandu; *G. striata* R. Br. (beefwood, Aus.) – cabinet-work (Aus.)
Grewia L. Tiliaceae. 150 warm OW. Some Afr. spp. with seeds disp. endozoically via elephants; some domatia. (Aus.) twigs with chewed ends used as paint-brushes; others yield locally ed. fr. (c. 50% diet of some 'hunter-gatherers' in C Tanzania at some times of year), timbers, fibres & medic. e.g. *G. crenata* (Forster f.) Schinz & Guillaumin (Tonga) – fibre form. for textiles & cordage, fr. ed., stems for friction-lighting fires, spears etc.; *G. asiatica* L. (phassa, Himal.) – ed. fr., juice for drinks in India; *G. paniculata* Roxb. ex DC (Indomal.) – local medic., wood distillate high in acetone
Greyia Hook. & Harvey. Greyiaceae. 3 S Afr. Some cult. orn.
Greyiaceae Hutch. Dicots – Rosidae – Rosales. 1/3 S Afr. Shrubs & small trees. Lvs spirally arr., decid., simple, crenately-lobed & dentate (*Pelargonium*-like), petiole with sheathing decurrent base; stipules 0. Fls bisexual, reg., 5-merous in racemes; K 5, imbricate, persistent, C 5, imbricate, red, nectary-disk extrastaminal, well-developed with 10 small staminodes alt. with A 5 + 5 appearing as 1 series, anthers with longit. slits, $\underline{G}$ ((4)5(6)), 5-lobed with axile placenta & term. style & ∞ ovules, anatropous, bitegmic in 2 rows in each loc. In fr. carpels separate from central axis & dehisce ventrally; seeds ∞, small with copious endosperm & small straight embryo. x = 16, 17
Only genus: *Greyia*
Form. incl. in Melianthaceae (A structure similar) but simple lvs & ∞ ovules clearly against this
Grias L. Lecythidaceae. 6 Panamá to Peru. Locally medic. (NW Amaz.); ed. fr. high in vitamins esp. *G. cauliflora* L. (anchovy pear, C Am., Jamaica) – lvs to 110 × 28 cm
Griegia auct. = *Greigia*
Grielum L. Neuradaceae. 5 S Afr.
Griffinia Ker-Gawler. Amaryllidaceae (Liliaceae s.l.). 7 Brazil. Cult. orn. esp. *G. liboniana* Morren (C Brazil)
Griffithella (Tul.) Warm. Podostemaceae. 1 India (W Ghats). Shoots polymorphic – cup- or wine-glass-shaped, creeping or erect, diverse in size & shape
Griffithia Wight & Arn. = *Benkara*
Griffithsochloa Pierce. Gramineae (33c). 1 Mexico
Griffonia Baillon. Leguminosae (I 3). 4 trop. W Afr.
Grimaldia Schrank = *Chamaecrista*
Grimmeodendron Urban. Euphorbiaceae. 2 WI
Grindelia Willd. Compositae (Ast.-Sol.). c. 55 WN & S Am. Some cult. orn. coarse herbs & evergreen shrubs e.g. *G. glutinosa* (Cav.) C. Martius (S Am.) & local medic. esp. *G.*

camporum E. Greene (Calif.) – grown for resin prod. like rosin & *G. squarrosa* (Pursh) Dunal (NW Am., natur. GB & Aus.) – used for burns, poison ivy dermatitis etc.

Grisebachia Klotzsch. Ericaceae. 8 SW Cape. R: Bothalia 13(1980)65

Grisebachianthus R. King & H. Robinson (~ *Eupatorium*). Compositae (Eup.-Crit.). 8 E Cuba. R: MSBMBG 22(1987)328

Grisebachiella Lorentz = *Astephanus*

Griselinia Forster & Forster f. Cornaceae (Griseliniaceae). 7 NZ & Chile. R: Brittonia 45(1993)261. Dioec. shrubs & trees, often littoral. Some cult. orn. & timber for boats, railway-sleepers (esp. *G. littoralis* (Raoul) Raoul, NZ). Not allied to Cornaceae on DNA data

Griseliniaceae (Wangenh.) Takht. See Cornaceae

Grisollea Baillon. Icacinaceae. 2 Madag., Seychelles

Grisseea Bakh.f. = *Parsonsia*

groats usu. oats, wheat or buckwheat

Grobya Lindley. Orchidaceae (V 9). 3 Brazil. Cult. orn. epiphytes

Groenlandia Gay (~ *Potamogeton*). Potamogetonaceae. 1 W Eur. & N Afr. to SW As.: *G. densa* (L.) Fourr.

gromwell *Lithospermum officinale, Mertensia maritima*

Gronophyllum R. Scheffer. Palmae (V 41). 33 E Mal., Aus. (1). R: Principes 29(1985)129

Gronovia L. Loasaceae. 2 trop. Am. A 5, staminodes 0, G 1

Grosourdya Reichb.f. Orchidaceae (V 16). 10 SE As., W Mal.

Grossera Pax. Euphorbiaceae. 11 trop. Afr., Madagascar

Grossheimia Sosn. & Takht. = *Centaurea*

Grossularia Miller = *Ribes*

Grossulariaceae DC. Dicots – Rosidae – Rosales. 24/330 cosmop. Trees & shrubs, s.t. armed, s.t. accumulating aluminium, s.t. cyanogenic. Lvs usu. spirally arr. (opp. in e.g. *Polyosma*), simple, though s.t. deeply lobed, pinnately- or palmately-veined & v. often with hydathodes; stipules usu. 0 (present in *Brexia, Itea, Phyllonoma, Pterostemon* & then usu. small, rarely consp. & basally adnate to petiole in some *Ribes* spp.). Fls ± reg., usu. bisexual (or pl. dioec.), often with saucer-shaped to tubular hypanthium, in term. or axillary racemes (less often panicles or small umbels or solit. in upper axils or (*Phyllonoma*) borne on lvs); K ((3–)5(–9)) persistent, imbricate or valvate as lobes on hypanthium or with tube, s.t. ± petaloid, C as many & alt. with K (or 0), imbricate, valvate or convolute, A usu. opp. K, s.t. with staminodes alt. with A or 2nd series of A, anthers with longit. slits, disk (intrastaminal) often present, $\underline{G}$ (2,3(–7)) or $\overline{G}$, pluriloc., with axile placentas or 1-loc. with ± intruded parietal placentas, rarely (*Tetracarpaea*) carpels free, & distinct or basally connate styles & ∞ anatropous unitegmic (bitegmic in *Ribes*) ovules. Fr. a capsule or berry (follicles in *Tetracarpaea*); seeds ∞, often arillate, endosperm with oil & protein & s.t. hemicellulose. x = 8, 9, 11, 12, 17, 30

Principal genera: *Argophyllum, Escallonia, Itea, Polyosma, Ribes*

Form. incl. in a variable Saxifragaceae, G. have been recognized as rep. up to 8 families (incl. Argophyllaceae, Escalloniaceae (most genera), Iteaceae, Montiniaceae & Pterostemonaceae) so that this arr. is a compromise, though the genera are app. closely interrelated (Cronquist) – cf. Cornaceae, though increasing DNA evidence suggests Argophyllaceae (*Argophyllum, Corokia*) should be moved to Asteridae – Asterales while *Ribes* is rather isolated from rest of fam.

Ribes spp. yield fr. (gooseberries, currants) & cult. orn. Spp. of *Corokia, Escallonia* & *Itea* also cult. orn.

Grosvenoria R. King & H. Robinson (~ *Eupatorium*). Compositae (Eup.-Crit.). 4 Andes. R: MSBMBG 22(1987)352

ground elder *Aegopodium podagraria*; **g. ivy** *Glechoma hederacea*; **g. pine** *Ajuga chamaepitys*

groundnut *Arachis hypogaea*, (Am.) *Apios americana*; **Bambara g.** *Vigna subterranea*

groundsel *Senecio vulgaris*; **g. bush** *Baccharis halimifolia*; **tree** or **giant g.** *Dendrosenecio* spp., *B. halimifolia*

Grubbia P. Bergius. Grubbiaceae. 3 SW & S Cape. R: JSAB 43(1977)115

Grubbiaceae Endl. Dicots – Dilleniidae – Ericales. 1/3 Cape. Ericoid shrubs. Lvs opp., simple, revolute; stipules 0 but leaf-bases joined by transverse ridge. Fls bisexual, reg., small in compact axillary cymes; K 4, small, valvate, C 0, A 8, 4 weakly adnate to K & longer than rest, anthers inverted & adnate to filament & with longit. slits, pollen in monads, $\overline{G}$ (2) with apical disk & style with simple or bilobed stigma, at first ± 2-loc., later 1-loc. with central placenta with 2 apical, pendulous, unitegmic, anatropous ovules. Fr. a 1-seeded drupe, those of an infl. forming a compact cluster looking like a cypress cone;

seed with thin testa & long straight embryo in oily proteinaceous endosperm
Only genus: *Grubbia*
Affinities obscure (Cronquist) & prob. to be excluded from Dilleniidae

gru-gru *Acrocomia totai*
gruie *Owenia acidula*
grumichama *Eugenia dombeyi*
Grumilea Gaertner = *Psychotria*
Grypocarpha Greenman = *Philactis*
guaba *Inga vera*
guabiroba *Campomanesia guaviroba*
Guacamaya Maguire. Rapateaceae. 1 Colombia, Venezuela
guacimilla *Trema micrantha*
guaco *Clibadium* spp.
Guadua Kunth = *Bambusa*
Guaduella Franchet. Gramineae (1a). 6 trop. Afr. R: KB 16(1962)247, 37(1983)660. Resemble *Aframomum* & easily mistaken
guaiac *Bulnesia sarmientoi*
Guaiacum L. Zygophyllaceae. 6 warm Am. Evergreen trees & shrubs of dry areas, sources of lignum-vitae much used in seawater etc. (hardest of comm. timbers, S.G. 1.333, with fibre-layers diagonally opposed & lots of resin so self-lubricating & therefore good for pestles etc.), from which medic. resin guaiacum is obtained by heating; timber used for pulleys, bowls for bowling, etc. esp. from *G. officinale* L. & *G. sanctum* L., the former almost exterminated for medic. use in venereal disease by natives from 1450, introd. Spain 1501 & in Eur. used with mercury in treatment of syphilis; g. lozenges ('Plummer's pills') for sore throats
Guaicaia Maguire = *Glossarion*
guaje *Leucaena esculenta*
Guamatela J.D. Sm. Rosaceae. 1 C Am.
Guamia Merr. = *Meiogyne*
guanacaste *Enterolobium cyclocarpum*
Guapeba B.A. Gomes = *Pouteria*
Guapira Aublet (~ *Pisonia*). Nyctaginaceae (VI). c. 70 trop. Am.
guar *Cyamopsis tetragonolobus*
guaraná *Paullinia cupana*
Guardiola Cerv. ex Bonpl. Compositae (Helia.). 10 SW N Am.
Guarea Allam. ex L. Meliaceae (I 6). 40 trop. Am. (36; R: FN 28(1981)255), Afr. Some moth-poll. incl. *G. rhopalocarpa* Radlk. (C Am.) – 5 flowering periods (4–20 wks) per yr. Some timbers esp. *G. cedrata* (A. Chev.) Pellegrin & *G. thompsonii* Sprague & Hutch. (pale mahogany subs., trop. Afr.) & *G. guidonia* (L.) Sleumer (WI redwood, alligator w., 'acajou', trop. Am.), also medic. locally. Most spp. with lvs with indefinite growth (cf. *Chisocheton*)
guatambu moroti *Balfourodendron riedelianum*
Guatemala grass *Tripsacum fasciculatum*
Guatteria Ruíz & Pavón. Annonaceae. (Incl. *Guatteriella*, *Guatteriopsis*) 279 trop. Am. R: AHB 12(1939)108, 291, PSE 148(1985)20. Alks; *G. scandens* Ducke one of few climbing A. in Am. Some local timbers & fibres, *G. modesta* Diels oral contraceptive in Peru
Guatteriella R.E. Fries = *Guatteria*
Guatteriopsis R.E. Fries = *Guatteria*
guava *Psidium* spp. esp. *P. guajava*; **black g.** *Guettarda argentea*; **Chilean g.** *Ugni molinae*; **Costa Rican g.** *P. friedrichsthalianum*; **Pará g.** *Campomanesia* spp.; **pineapple g.** *Acca sellowiana*; **purple g.** *P. cattleianum*; **Spanish g.** *Catesbaea spinosa*; **strawberry g.** *P. cattleianum*
guaxima *Urena lobata*
guayacán *Caesalpinia paraguariensis*
Guayania R. King & H. Robinson (~ *Eupatorium*). Compositae (Eup.-Heb.). 6 Guayana Highland. R: MSBMBG 22(1987)410
Guaymasia Britton & Rose = *Caesalpinia*
guayato *Gonolobus edulis*
guayule *Parthenium argentatum*
guayusa *Ilex guayusa*
Guazuma Miller. Sterculiaceae. 4 trop. Am. R: Ceiba 1(1951)193. *G. ulmifolia* Lam. (bastard cedar, WI elm) – light timber for boats, barrels, fuelwood etc.; fr. passes through cattle which act as dispersal agents suggesting that prob. orig. disp. by extinct megafauna of Pleistocene

gubgub *Vigna unguiculata*

Gueldenstaedtia Fischer. Leguminosae (III 15). (Incl. *Tibetia*) 14 Sino-Himal. to Siberia. R: Candollea 18(1962)137, BBLNEFI 5(1979)48

guelder rose *Viburnum opulus*

guere palm *Astrocaryum* spp.

Guerkea Schumann = *Baissea*

Guernsey lily *Nerine sarniensis*

Guerreroia Merr. = *Glossocardia*

Guettarda L. Rubiaceae (III 3). c. 80 Vanuatu, New Caled., trop. Am. with 1, *G. speciosa* L., common on trop. coasts. Some ed. fr. esp. *G. argentea* Lam. (black guava, Am.) & locally used timbers

Guettardella Champ. ex Benth. = *Antirhea*

Guevaria R. King & H. Robinson (~ *Eupatorium*). Compositae (Eup.-Ager.). 4 Ecuador & Peru. R: MSBMBG 22(1987)161

guggul *Commiphora wightii*

Guibourtia Bennett. Leguminosae (I 4). 16–17 trop. Afr. (14), Am. Timbers & copal esp. from *G. arnoldiana* (De Wild. & T. Durand) Léonard (benge, C Afr.); *G. demeusei* (Harms) Léonard (African rosewood, bubinga, Congo copal, trop. Afr.) & *G. copallifera* Bennett (Sierra Leone c., trop. Afr.); *G. ehie* (A. Chev.) Léonard (amazakoué, anokye, trop. Afr.)

Guichenotia Gay. Sterculiaceae. 7 SW Aus. R: Nuytsia 8(1992)319

Guiera Adans. ex Juss. Combretaceae. 1 N trop. Afr.

Guihaia Dransf., S.K. Lee & F.N. Wei. Palmae (I 1a). 2 S China & N Vietnam. R: Principes 29(1985)9

Guildford grass *Romulea rosea*

Guilelma auctt. = *Bactris*

Guilfoylia F. Muell. (~ *Cadellia*). Surianaceae. 1 NE Aus.: *G. monostylis* (Benth.) F. Muell. Stipules

Guilielma C. Martius = *Bactris*

Guillainia Vieill. = *Alpinia*

Guillauminia Bertrand = *Aloe*

Guilleminea Kunth. Amaranthaceae (II 2). 2 C Am. (*G. densa* (Willd.) Moq. weedy in OW)

Guillenia E. Greene. Cruciferae. 4 SW US

Guillonea Cosson (~ *Laserpitium*). Umbelliferae (III 12). 1 S Spain

Guinea corn *Sorghum* spp.; **g.(-)flower** *Fritillaria meleagris*, (Aus.) *Hibbertia* spp.; **G. grains** *Aframomum melegueta*; **G. grass** *Panicum maximum*; **G. peach** *Sarcocephalus latifolius*; **G. pepper** *Xylopia aethiopica*; **G. yam** *Dioscorea* × *cayenensis*

Guioa Cav. Sapindaceae. 64 Thailand to Samoa. R: P.L. van Welzen (1990) *Guioa* Cav. (Sapindaceae). Some ed. arils

Guiraoa Cosson. Cruciferae. 1 Spain: *G. arvensis* Cosson

guisaro *Psidium guineense*

Guizotia Cass. Compositae (Helia.-Mel.). 6 warm Afr. R: BT 69(1974)1. *G. abyssinica* (L.f.) Cass. (ramtil, Niger seed, Ethiopia) – cult. (esp. India) for oilseed fed to cagebirds & oil used for cooking & paint

gul mohur *Delonix regia*

Gularia Garay = *Schiedeella*

Gulubia Becc. Palmae (V 4l). 9 E Mal. (5) to Fiji. R: Principes 26(1982)159

Gulubiopsis Becc. = *Gulubia*

gum *Eucalyptus* spp., also *Angophora* & *Corymbia* spp.; **Aden g.** *Acacia* spp.; **almond g.** *Prunus dulcis*; **American red g.** *Liquidambar styraciflua*; **Amrad g.** *A. nilotica*; **Amritsar g.** *A. modesta*; **Amanis g.** *Hymenaea courbaril*; **angico g.** *Parapiptadenia rigida*; **g. arabic** *A. senegal* etc.; **Ashanti g.** *Terminalia* spp.; **Barbary g.** *A.* spp.; **black g.** *Nyssa sylvatica*; **blue g.** *Eucalyptus globulus*; **Cape g.** *A. karroo*; **Cape York red g.** *E. brassiana*; **carob seed g.** *Ceratonia siliqua*; **cashew g.** *Anacardium occidentale*; **cherry g.** *Prunus cerasus*; **g. dragon** *Astracantha* spp.; **E Afr. g.** *Acacia drepanolobium*; **E Indian g.** *A.* spp.; **forest red g.** *E. tereticornis*; **gatty g.** *Anogeissus latifolia*; **ghost g.** *Corymbia* spp.; **karaya** or **kutira g.** *Cochlospermum religiosum, Sterculia urens*; **khair g.** *A. catechu*; **kolhol g.** *A. senegal*; **lemon-scented g.** *Corymbia citriodora*; **locust g.** *Ceratonia siliqua*; **mesquite g.** *Prosopis glandulosa*; **Morocco g.** *A.* spp.; **Murray red** or **river red g.** *E. camaldulensis*; **rose g.** *E. grandis*; **saligna g.** *E. saligna*; **salmon g.** *E. salmonophloia*; **snow g.** *E. pauciflora* ssp. *niphophila*; **sweet g.** *Liquidambar styraciflua*; **Sydney blue g.** *E. saligna*; **talh g.** *A. seyal*; **Tartar g.** *Sterculia cinerea*; **g. tragacanth** *Astracantha* spp.; **g. tree** *Eucalyptus* spp.; **water g.** *Tristania neriifolia*; **yellow g.** *Xanthorrhoea* spp. esp. *X. resinosa*

gumbo *Abelmoschus esculentus*
Gumillea Ruíz & Pavón = *Picramnia*
gundabluey *Acacia victoriae*
Gundelia L. Compositae (Arct.-Gort.). 1 Cyprus & Turkey to Iran: *G. tournefortii* L. – a steppe-pl. tumbleweed with ed. fr. & young infls, molluscicidal saponins. R: T 43(1994)37
Gundlachia A. Gray. Compositae (Ast.-Sol.). 10 Bahamas to Curaçao & Haiti
Gunillaea Thulin. Campanulaceae. 2 trop. Afr. & Madagascar
Gunnarella Senghas. Orchidaceae (V 16). 20 New Caledonia
Gunnera L. Gunneraceae. c. 40 trop. & S Afr., Mal., Tasmania, NZ, Antarctic is., Hawaii, S Am. Creeping herbs to pachycaul ± stemless herbs with some of the largest lvs known; cult. orn. esp. *G. tinctoria* (Molina) Mirbel (*G. chilensis*, Chile) – lvs to 1.5 m across, peeled young petioles ed. Chile, & *G. manicata* Linden (*G. brasiliensis*, S Brazil) – lvs to 3 m across by waterside; other spp. creeping & grown as rock-pls; *G. petaloidea* Gaudichaud (Hawaii) – mycorrhizal as well as bacterial symbionts
Gunneraceae Meissner. Dicots – Rosidae – Haloragidales. 1/40 S trop. & S hemisph. Terr., often pachycaul herbs with stems & adventitious roots harbouring symbiotic *Nostoc punctiforme* (intracellular infection starting in 2 glands below cotyledons in seedlings) or *Chlorococcus* colonies in warts entered through hydathodes or mucilage glands; stem polystelic. Lvs spirally arr., orbicular or ovate (–peltate), with large median axillary scale (? stipules). Infl. a term. or upper axillary panicle of small epigynous fls, the basal ones often female, upper ones male & middle ones bisexual or pl. dioec.; K 2(3) or almost 0, small, valvate, C 0(2), A 1 or 2 with short filaments & anthers with longit. slits, $\overline{G}$ (2), 1-loc. with 2 term. styles & 1 pendulous, anatropous, bitegmic ovule. Fr. a drupe; seed with v. small embryo embedded in copious oily endosperm. x = (?11) 12, 17, 18
Only genus: *Gunnera*
Form. incl. in Haloragidaceae but differing in habit, A, G, embryology & chemistry etc. (Cronquist)
Gunnessia P. Forster. Asclepiadaceae (III 4). 1 Queensland: *G. pepo* P. Forster
Gunniopsis Pax (~ *Aizoon*). Aizoaceae (I). 14 W, S & C Aus. R: JABG 6(1983)133
gunny *Corchorus* spp.
gunpowder plant *Pilea microphylla*
Gurania (Schldl.) Cogn. (~ *Psiguria*). Cucurbitaceae. c. 75 trop. Am. Monoec. but changing sex so as to appear dioec.; some with tough bird-poll. fls like *P.*
Guraniopsis Cogn. Cucurbitaceae. 1 Peru
gurjun or **g. oil** *Dipterocarpus* spp.
Gustavia L. Lecythidaceae. 41 trop. Am. R: FN 21(1979)128. Reg. fls (some self-sterile) poll. trigonid bees. Some with ed. fr. sold in Colombian markets; some timbers esp. *G. augusta* L. (stinkwood), some medic.; bark ash 'salt' used to coat pellets of hallucinogenic (*Virola*) paste in NW Amaz.
Gutenbergia Schultz-Bip. Compositae (Vern.-Vern.). 25 trop. Afr.
Guthriea Bolus. Achariaceae. 1 S Afr.
Gutierrezia Lagasca. Compositae (Ast.-Sol.). (Incl. *Xanthocephalum*) 27 W N (16, R: SB 10(1985)7) & warm S Am. *G. sarothrae* (Pursh) Britton & Rusby (*X. sarothrae*, S N Am.) – protein with antitumour activity
gutta percha *Palaquium* spp. esp. *P. gutta*
Guttiferae Juss. (Clusiaceae; incl. Bonnetiaceae). Dicots – Dilleniidae – Guttiferales. 45/1370 trop. with Hypericoideae also N temp. Trees (sometimes pachycaul), shrubs, lianes & herbs with yellow or otherwise brightly coloured resinous juice in schizogenous secretory canals & cavities. Lvs spiral, opp. or whorled, simple & usu. entire & often with many slender lateral veins & resin-cavities appearing as pellucid dots (e.g. Hypericoideae); stipules 0. Fls in term. cymes, rarely solit., bisexual or not, reg., hypogynous, often with bracteoles (s.t. passing into K), K imbricate, C (2)3–6(–14), imbricate or convolute, s.t. basally connate, A ∞ centrifugally dev. from few trunk-bundles, or grouped in 2–5 centrifugal bundles opp. & often adnate to C when many staminodal, or fewer (as few as 3), anthers with longit. slits, $\underline{G}$ ((1–)3–5(–20+)), with as many locules or 1-loc. (intruded placentas not reaching centre) & as many styles, ± basally connate or 1 style with lobed or peltate stigma, placentation axile (rarely parietal on intruded placentas with (1)2–∞ anatropous to hemitropous bitegmic ovules per carpel. Fr. a drupe or berry or septicidal capsule; seeds exotegmic, often arillate, with straight or curved embryo & 0 endosperm. x = 7–10
Classification & principal genera (TSSTIF 21(1985)61):
I. **Hypericoideae** (lvs opp., styles 3 or 5; 3 tribes):

1. **Cratoxyleae** (C never yellow; capsule or berry (then interstaminal glands absent)): *Cratoxylum*
2. **Hypericeae** (C yellow or orange): *Hypericum*
3. **Vismieae** (drupe or berry (then interstaminal glands present): *Psorospermum, Vismia*

II. **Bonnetioideae** (lvs spiral with styles 1, 3 or 5, or lvs opp. with 1, 2, 4 or 6 styles; fls bisexual; capsule; 3 tribes):
1. **Bonnetieae** (pollen with circular endoaperture): *Bonnetia*
2. **Kielmeyereae** (pollen with transversely elongated aperture; placentas paired): *Kielmeyera*
3. **Caraipeae** (pollen as in 2.; placenta solit.): *Caraipa*

III. **Calophylloideae** (Clusioideae; as II. but fls unisexual, bisexual (then fr. not a capsule) or polygamous; 7 tribes):
1. **Calophylleae** (placenta basal): *Calophyllum, Mammea, Mesua*
2. **Endodesmieae** (placenta apical): *Endodesmia*
3. **Allanblackieae** (placenta parietal): *Allanblackia*
4. **Moronbeae** (placenta axile, fls bisexual): *Montrouziera, Pentadesma, Symphonia*
5. **Garcinieae** (placenta axile, fls usu. unisexual, fr. a berry): *Garcinia*
6. **Clusieae** (as 5. but fr. a capsule): *Clusia*
7. **Tovomiteae** (as 5. but fr. a 'capsular drupe or capsular berry'): *Tovomita, Tovomitopsis*

Classification still somewhat fluid

Many timbers (*Calophyllum, Caraipa, Cratoxylum, Mesua, Montrouziera, Platonia*), drugs & dyes from bark (*Calophyllum, Harungana, Vismia*), gums, pigments & resins from stems (esp. *Garcinia*, gamboge), drugs from lvs (*Harungana, Hypericum*), ed. fr. (*Garcinia* incl. mangosteen, *Mammea, Platonia*), oilseeds (*Allanblackia, Calophyllum, Garcinia, Mammea, Pentadesma*) & cult. orn. (esp. *Hypericum*)

Gutzlaffia Hance = *Strobilanthes*

guvar *Cyamopsis tetragonolobus*

Guya Frapp. ex Cordem. = *Drypetes*

Guyania Airy Shaw = *Guayania*

Guyonia Naudin. Melastomataceae. 2 trop. W Afr.

Guzmania Ruíz & Pavón. Bromeliaceae (2). 131 trop. Am. R: FN 14(1977)1275, 1401. Terr. or epiphytic; some cult. orn. incl. hybrids with *Vriesea* spp.

gwar *Cyamopsis tetragonolobus*

Gymapsis Bremek. = *Strobilanthes*

gymea lily *Doryanthes excelsa*

Gyminda (Griseb.) Sarg. Celastraceae. 4 SN & C Am., WI

Gymnacanthus Nees. Acanthaceae. 1 Mexico

Gymnachne L. Parodi = *Rhombolytrum*

Gymnacranthera Warb. Myristicaceae. 7 Indomal. R: Blumea 31 (1986)451. Some oilseeds used in candle-manufacture

Gymnadenia R. Br. (~ *Habenaria*). Orchidaceae (IV 2). 10 NE Am. & temp. Euras. (Eur. 2, incl. *G. conopsea* (Willd.) R. Br. (scented orchid, Euras. to Japan))

Gymnadeniopsis Rydb. = *Platanthera*

Gymnanthemum Cass. = *Vernonia*

Gymnanthera R. Br. Asclepiadaceae (I; Periplocaceae). 4 Mal. to Aus. 2, 1 endemic). Alks

Gymnanthes Sw. (*Ateramnus* auctt.). Euphorbiaceae. 15 SN & C Am., WI

Gymnarrhena Desf. Compositae (? Anthem.). 1 Medit., 5 W As.

Gymnartocarpus Boerl. = *Parartocarpus*

Gymnaster Kitam. = *Miyamayomena*

Gymnema R. Br. Asclepiadaceae (III 4). 25 OW trop. to Aus. & S Afr. Some emetics; sweet taste sensation blocked by chewing *G. sylvestre* (Retz.) Sm. (OW), gymnemic acid even overcoming miraculin in *Synsepalum* spp.

Gymnenopsis Costantin. Asclepiadaceae (III 4). 2 SE As.

Gymnocactus Backeb. = *Turbinicarpus*

Gymnocalycium Pfeiffer ex Mittler. Cactaceae (III 4). c. 50 S Am. R: J. Pilbeam (1994) *G. A collector's guide*. Alks incl. mescaline. Many cult. orn. globose cacti

Gymnocarpium Newman. Dryopteridaceae (II 1). 6 N temp. (Eur. 3), Indomal. (1), Taiwan (1). R: ABF 15(1971)101. Some cult. orn. incl. *G. dryopteris* (L.) Newman (oak fern, N temp.)

Gymnocarpos Forssk. = *Paronychia*

Gymnochilus Blume. Orchidaceae (III 3). 3 Mascarenes

Gymnocladus Lam. Leguminosae (I 1). 5 E N Am. (1), E & SE As. (4, 2 v. rarely coll.). R: JAA 57(1976)91. Dioec. or polygamous pachycaul trees with bipinnate lvs & fr. opening along parietal suture like a follicle (primitive). *G. dioica* (L.) Koch (*G. canadensis*, chicot, Kentucky coffee tree, E US) – cult. orn. with durable timber & with seeds roasted as coffee-subs., form. planted by Indians in New York State

Gymnocondylus R. King & H. Robinson (~ *Eupatorium*). Compositae (Eup.-Ayap.). 1 Brazil. R: MSBMBG 22(1987)206

Gymnocoronis DC. Compositae (Eup.-Aden.). 5 C & S Am. R: MSBMBG 22(1987)62. All aquatic, *G. spilanthoides* (D. Don) DC (S Am.) – noxious weed in NSW ('Senegal tea')

Gymnodiscus Less. Compositae (Sen.-Sen.). 2 SW Cape to Namaqualand

Gymnogonum Parry = *Goodmania*

Gymnogrammitidaceae Ching = Davalliaceae

Gymnogrammitis Griffith (~ *Davallia*). Davalliaceae. 1 E Himal. & S China to SE As.: *G. dareiformis* (Hook.) Ching – cult. orn. epiphyte. R: Blumea 37 (1992)186

Gymnolaena (DC) Rydb. Compositae (Hele.-Pect.). 3 Mex. R: Sida 3(1967)110

Gymnolomia Kunth = *Wedelia*

Gymnopentzia Benth. Compositae (Anth.-Urs.). 1 S Afr.: *G. bifurcata* Benth. R: NJB 5(1986)538

Gymnopetalum Arn. Cucurbitaceae. 3 Indomal., S China. Some locally ed. & med.

Gymnophragma Lindau. Acanthaceae. 1 NE New Guinea

Gymnophyton Clos. Umbelliferae (I 2c). 6 Andes of Chile & Arg. R: UCPB 33(1962)137

Gymnopodium Rolfe. Polygonaceae (II 1). 3 C Am. Shrubs yielding good charcoal

Gymnopogon Pal. Gramineae (33b). c. 15 warm Am., India to Thailand (1). R: ISUJS 45(1971)319

Gymnopoma N.E. Br. = *Skiatophytum*

Gymnopteris Bernh. = *Hemionotis*

Gymnoschoenus Nees. Cyperaceae. 2 Aus. *G. sphaerocephalus* (R. Br.) Hook.f. (SE Aus.) – distinctive ('button grass') plains esp. W Tasmania

Gymnosciadium Hochst. = *Pimpinella*

Gymnosiphon Blume. Burmanniaceae. (Incl. *Ptychomeria*) c. 50 trop.

Gymnosperma Less. Compositae (Ast.-Sol.). 1 S US to C Am.

Gymnospermae Lindley, **gymnosperms** 86/840 in 17 fams. A grouping used to designate those seed plants not considered to be angiosperms, differing from those in not having the seeds enclosed in carpels & not having a double fert. char. of a. Of gymnosperms there are 4 extant groups: Cycadopsida (cycads), Ginkgoopsida, Gnetopsida, Pinopsida (conifers, incl. Taxopsida (yews)) – possibly not closely interrelated. The extinct seed-ferns (pteridosperms) prob. gave rise to most of these & to the angiosperms independently

Gymnospermium Spach (~ *Leontice*). Berberidaceae (II 1; Leonticaceae). 8 Eur. (1) to E As. R: BZ 55(1970)191

Gymnosphaera Blume = *Cyathea*

Gymnosporia (Wight & Arn.) Hook.f. = *Maytenus*

Gymnostachys R. Br. Acoraceae (?). 1 E Aus.: *G. anceps* R. Br. – lvs form. for cordage

Gymnostachyum Nees. Acanthaceae. 30 India to C Mal.

Gymnostemon Aubrév. & Pellegrin. Simaroubaceae. 1 trop. W Afr.

Gymnostephium Less. Compositae (Ast.-Ast.). 7 S Afr.

Gymnosteris E. Greene. Polemoniaceae. 2 W US. R: AMN 31(1944)230

Gymnostoma L. Johnson (~ *Casuarina*). Casuarinaceae. 18 Mal. to W Pacific

Gymnostyles Juss. = *Soliva*

Gymnotheca Decne. Saururaceae. 2 SW China

gympie *Dendrocnide* spp.

Gynandriris Parl. (~ *Iris*). Iridaceae (III 2). 9 Medit. (2, incl. Eur.), S Afr. R: BN 133(1980)239. Cult. orn. with fugacious fls incl. *G. sisyrinchium* (L.) Parl. (Barbary nut, Medit. to Pakistan, poss. spread by agric.) – fls open p.m.

Gynandropsis DC = *Cleome*

Gynatrix Alef. (~ *Plagianthus*). Malvaceae. 1 E Aus.: *G. pulchella* (Willd.) Alef.

Gynerium Willd. ex Pal. Gramineae (25). 1 trop. Am.: *G. sagittatum* (Aublet) Pal. (uva grass) – giant reed to 10 m, stems used for arrow-shafts, laths etc., lvs for weaving hats, mats etc., young shoots used for shampoo

Gynocardia R. Br. Flacourtiaceae (4). 1 Assam & Burma: *G. odorata* R. Br. – seeds yield an oil similar to, but less efficacious than, chaulmoogra oil (cf. *Hydnocarpus*), fr. used as fish-poison

Gynochthodes Blume. Rubiaceae (IV 10). (Incl. *Tetralopha*) c. 20 SE As. to Pacific

Gynocraterium Bremek. Scrophulariaceae. 1 trop. S Am.

Gynoglottis J.J. Sm. Orchidaceae (V 12). 1 Sumatra

Gynopachis Blume (~ *Randia*). Rubiaceae (II 1). 9 Mal. R: BMNHN 4,8(1986)280

Gynophorea Gilli. Cruciferae. 1 Afghanistan

Gynophyge Gilli = *Agrocharis*

Gynostemma Blume. Cucurbitaceae. 2 Indomal., E As.

Gynotroches Blume. Rhizophoraceae. 1 Burma to Mal., Caroline & Solomon Is.

Gynoxys Cass. Compositae (Sen.-Tuss.). 60 C Am. to Peru. Shrubs & trees

Gynura Cass. Compositae (Sen.-Sen.). c. 40 OW trop. Cult. orn., medic. & some ed. *G. aurantiaca* (Blume) DC (Java, Sulawesi, widely natur.) – foliage pl. with velvety-purple hairy lvs; *G. bicolor* (Willd.) DC (Himal.) – cult. orn. 'Okinawa spinach'; *G. japonica* (Thunb.) Juel (E As.) – cult. for foliage & tubers; *G. pinnatifida* DC (China) – medic. in modern Chinese herbalism; *G. pseudochina* (L.) DC (OW trop.) – source of medic. 'China root'

Gypothamnium Philippi. Compositae (Mut.-Mut.). 1 N Chile. Shrub

Gypsacanthus Lott, Jaramillo & Rzed. Acanthaceae. 1 Mex.: *G. nelsonii* Lott, Jaramillo & Rzed. R: BSBM 46(1984)47

Gypsophila L. Caryophyllaceae (III 1). 150 temp. Euras. (Eur. 27), Egypt, Aus. & NZ (1, ? introd.). R: Wentia 9(1962)4. Allied to *Saponaria* but with many small fls with short tubes. Some medic. & cult. orn. esp. *G. elegans* M. Bieb. (Ukraine–Iran) & *G. paniculata* L. (baby's breath, C Eur. to C As.) – much planted, the latter esp. for wedding bouquets etc. with cvs incl. 'doubles' esp. 'Bristol Fairy'; *G. patrinii* Ser.(E Russia to temp. As.) – used as copper indicator in Russia; *G. struthium* Loefl. (Spain) – used form. as soap in Medit.; *G. rokejeka* Del. (Egypt, E Med.) used in halva with sesame seeds & honey

gypsywort *Lycopus europaeus*

Gyptidium R. King & H. Robinson (~ *Eupatorium*). Compositae (Eup.-Gyp.). 2 Brazil, Arg. R: MSBMBG 22(1987)89

Gyptis (Cass.) Cass. (~ *Eupatorium*). Compositae (Eup.-Gyp.). 7 trop. S Am. R: MSBMBG 22(1987)87

Gyranthera Pittier. Bombacaceae. 2 Panamá, Venez. *G. caribensis* Pittier has extreme anisocotyly

Gyrinops Gaertner. Thymelaeaceae. 9 Sri Lanka (1), Laos (1), E Mal. *G. walla* Gaertner (wallapatta, Sri Lanka) – light timber & bark fibre used for ropes

Gyrinopsis Decne = *Aquilaria*

Gyrocarpaceae Dumort. = Hernandiaceae

Gyrocarpus Jacq. Hernandiaceae (Gyrocarpaceae). 3 trop. & warm. R: BJ 89(1969)181. Alks. *G. americanus* Jacq. (trop.) – light timber for toys, rafts etc.

Gyrocaryum Valdés. Boraginaceae. 1 Spain

Gyrocheilos W.T. Wang (~ *Didymocarpus*). Gesneriaceae. 4 S China. R: EJB 49(1992)54

Gyrodoma Wild. Compositae (Ast.-Gran.). 1 Mozambique: *G. hispida* (Vatke) Wild. R: Kirkia 9(1974)294

Gyrogyne W.T. Wang (~ *Stauranthera*). Gesneriaceae. 1 S China: *G. subaequifolia* W.T. Wang

Gyroptera Botsch. = *Choriptera*

Gyrostelma Fourn. = *Matelea*

Gyrostemon Desf. Gyrostemonaceae. 12 Aus. (11 W Aus.). R: BJ 106(1985)108

Gyrostemonaceae Endl. Dicots – Dilleniidae – Capparidales. 5/18 Aus. Trees or shrubs with mustard-oil glucosides but without myrosin-cells. Lvs spirally arr., simple, entire, often ± succ.; stipules minute or 0. Fls in spikes, racemes or solit. axillary, small, unisexual (pls usu. dioec.) with enlarged receptacle; P usu. discoid or cupular, ± lobed, persistent in fr., A 6–∞, in 1 or more whorls developing centripetally, filaments v. short or 0, anthers with longit. slits, $\underline{G}$ (2–)±∞ adnate to central column, forming compound ovary with as many locs as carpels (1 in *Cypselocarpus*), the column often apically expanded with short distinct styles, placentation axile with 1 campylotropous ovule per loc. Fr. dry, each carpel opening dorsally &/or ventrally & separating from column, or indehiscent (*Cypselocarpus*, *Tersonia*); seeds basally arillate with peripheral embryo curved around copious oily endosperm. x = 14, 15. R: BJ 106(1985)112

Genera: *Codonocarpus*, *Cypselocarpus*, *Gyrostemon*, *Tersonia*, *Walteranthus*

All genera with indehiscent fr. restricted to temp. W Aus. Embryology suggests alliance with Resedaceae

Gyrostipula J. Leroy. Rubiaceae (I 2). 2 Madag. & Comoro Is.

Gyrotaenia Griseb. Urticaceae (I). 6 WI

H

H *Papaver somniferum* (heroin)

Haageocactus Backeb. = *seq.*

Haageocereus Backeb. Cactaceae (III 4). 5–10 deserts of Peru & N Chile to Bolivia. Cult. orn. ribbed cylindrical cacti

Haarera Hutch. & E. A. Bruce = *Erlangea*

Haastia Hook.f. Compositae (Asteroideae–tribe?). 3 NZ incl. *H. pulvinaris* Hook.f. forming large cushions in subalpine & alpine veg. (cf. *Raoulia*, vegetable sheep)

Habenaria Willd. Orchidaceae (IV 2). c. 600 pantrop. & subtrop. Terr., some cult. orn., some in Ayurvedic medic. in India. See also *Coeloglossum, Gymnadenia, Platanthera*

Haberlea Friv. Gesneriaceae. 1 Balkans: *H. rhodopensis* Friv. (incl. *H. ferdinandi-coburgii*, a fine cult. form known from 1 locality in Bulgaria), a Tertiary relic like *Jancaea* & *Ramonda*

Hablitzia M. Bieb. Chenopodiaceae (I 1). 1 Cauc.: *H. tamnoides* M. Bieb. – perennial subterr. stem with annual climbing shoots with sensitive petioles

Habracanthus Nees. Acanthaceae. 60 trop. Am. (Colombia 34, R: KB 43(1988)24)

Habranthus Herbert. Amaryllidaceae (Liliaceae s.l.). 10 temp. S Am. Like *Zephyranthes* but P zygomorphic. Some cult. orn. esp. *H. robustus* Herbert ex Sweet (Arg., S Brazil) & *H. tubispathus* (L'Hérit.) Traub (Barbados snowdrop, warm S Am.)

Habrochloa C. Hubb. Gramineae (31d). 1 C Afr.

Habroneuron Standley. Rubiaceae (? I 5). 1 Mex.: *H. radicans* (Wernham) S. Darwin

Habropetalum Airy Shaw. Dioncophyllaceae. 1 Sierra Leone

Habrosia Fenzl. Caryophyllaceae (II 4). 1 W As.: *H. spinulifera* (Ser.) Fenzl

Hachettea Baillon. Balanophoraceae. 1 New Caled.: *H. austrocaledonica* Baillon. R: APG 33(1982)95

hackberry *Celtis* spp.

Hackelia Opiz. Boraginaceae. 45 N temp., C & S Am. R: MNYBG 26, 1(1976)121

Hackelochloa Kuntze. Gramineae (40h). 2 trop.

hackmatack *Larix laricina, Populus balsamifera*

Hacquetia Necker ex DC. Umbelliferae (II 1). 1 C Eur.: *H. epipactis* (Scop.) DC – cult. orn. rock-pl. with short-stalked umbels surrounded by petal-like green bracts

Hadrodemas H. Moore = *Callisia*

Haeckeria F. Muell. (~ *Calomeria*). Compositae (Gnap.-Cass.). 4 SE Aus. R: OB 104(1991)186

Haegiela Short & P. Wilson (~ *Epaltes*). Compositae (Gnap-Ang.). 1 Aus.: *H. tatei* Short & P. Wilson. R: Muelleria 7(1990)259

Haemacanthus S. Moore = *Satanocrater*

Haemanthus L. Amaryllidaceae (Liliaceae). 21 Cape. R: JSAB supp. 12(1984). Cult. orn. (blood lilies (incl. *H. albiflos* Jacq.!); trop. spp. referred to *Scadoxus*). Alks; 2n = 16 (reduced from *Scadoxus* condition where 18)

Haematocarpus Miers (*Baterium*). Menispermaceae. 2 E Himal. to Sulawesi. R: KB 26(1972)419, 30(1975)81

Haematodendron Capuron. Myristicaceae. 1 Madagascar

Haematostaphis Hook.f. Anacardiaceae (II). 2 trop. W Afr. *H. barteri* Hook.f. (blood plum) – fr. ed.

Haematostemon (Muell. Arg.) Pax & K. Hoffm. Euphorbiaceae. 2 trop. S Am.

Haematoxylum L. Leguminosae (I 1). 3 trop. Am. (2), Namibia (1). Timber (Nicaragua wood) & dyes esp. *H. brasiletto* Karsten (peachwood, Am.) & *H. campechianum* L. (logwood, campeachy wood or campeche, Mex. & WI) – dark heartwood source of haematoxylin used in microscopical preparations & as dye ('logwood chips', rasped by prisoners in C17 Holland) & in ink, timber for furniture, cult. orn. & used as spiny hedge, fls good bee-forage

Haemodoraceae R. Br. Monocots – Liliidae – Liliales. (Incl. Lanariaceae) 14/100 trop. & N Am., S Afr., Aus. (7/84, most in SW), New Guinea. Perennial herbs with tubers or short rhiz. & usu. raphides & typical red pigment (phenalenones) in roots & rhiz. (not in *Lanaria*); vessels usu. restricted to roots, s.t. also stem, rarely 0; vasc. bundles in whorls in *Lophiola* (2 in stem, 1 in rhiz.) or scattered. Lvs all basal with sheathing base & linear parallel-veined lamina. Fls bisexual, reg. to irreg., 3-merous, hypogynous to epigynous, in racemes, panicles, cymes or cymose umbels, which are usu. long-hairy; persistent P 6 to (6) in 1 (Conostyloideae) or 2 whorls, tube straight or curved, A 3 or 6, free or adnate to tube, anthers basifixed or versatile, with longit. slits, $\underline{G}$ to $\overline{G}$ (3), 3-loc. with axile placentation & septal nectaries, 1–∞ anatropous to orthotropous, bitegmic ovules per loc. Fr. a loculicidal capsule with (1–)several–∞ seeds or nut-like (*Phlebocarya*); embryo small with term. cotyledon & lateral plumule embedded in copious starchy endosperm with oil, protein & hemicellulose. x = 4–8, 15+. R: AMBG 77(1990)722

Classification & principal genera:

Haemodoroideae (P in 2 whorls, tube ± 0, A 3 or 6): *Haemodorum*, *Wachendorfia*

Conostyloideae (P in 1 whorl, tube often long & curved, A 6, fls always long-hairy): *Anigozanthos*, *Conostylis*

Allied to Pontederiaceae but variously placed near Iridaceae or Liliaceae. The red alpha-phenylphenalenones are not found in any other organisms

Some cult. orn. incl. cut-fls (*Anigozanthos* (kangaroo-paws)), dyes (*Lachnanthes*). Natural hybrids frequent

Haemodorum Sm. Haemodoraceae. 20 Aus., 1 ext. to New Guinea. R: FA 45(1987)136. *H. corymbosum* Vahl (E Aus.) & other spp. – roasted rhiz. ed. (Aborigines)

Haenianthus Griseb. Oleaceae. 2 WI. R: CJB 69(1991)489

hagberry or **hegberry** *Prunus padus*, *P. avium*

Hagenbachia Nees & C. Martius. Anthericaceae (Liliaceae s.l.). 6 trop. Am. R: NJB 7(1987)255. Form. referred to Haemodoraceae

Hagenia J. Gmelin. Rosaceae. 1 E Afr. highlands: *H. abyssinica* (Bruce) J. Gmelin – female fls (koso or kousso) used as taenicide

hag-taper (i.e. hedge-taper, high-taper) *Verbascum thapsus*

Hagsatera González (~ *Epidendrum*). Orchidaceae (V 13). 2 Mexico

haiari *Lonchocarpus* spp.

Hainania Merr. Tiliaceae. 1 Hainan

Hainardia Greuter. Gramineae (18). 1 Medit. incl. Eur.: *H. cylindrica* (Willd.) Greuter

hair grass spp. of *Aira*, *Cornynephorus*, *Deschampsia*, *Koeleria*, *Muhlenbergia* & *Trisetum* etc.

Haitia Urban. Lythraceae. 1 Hispaniola

Haitiella L. Bailey = *Coccothrinax*

Hakea Schrader. Proteaceae. 140 Aus. esp. SW. Evergreen xerophytes (needlebush, needlewood) with divided lvs only when young in most xeromorphic spp. (cf. *Acacia*), others resembling plastic foliage; like *Grevillea* but seeds with long term. wings. Some timbers (corkwood) & cult. orn., others used in reclamation of arid areas, e.g. *H. salicifolia* (Vent.) B.L. Burtt in Spain, though some invasive weeds e.g. *H. suaveolens* R. Br. (W Aus.) & *H. sericea* Schrader & Wendl. (*H. acicularis*, SE Aus.) in Cape fynbos, also natur. NZ, Iberia

Hakoneaste F. Maek. = *Ephippianthus*

Hakonechloa Makino ex Honda (~ *Phragmites*). Gramineae (25). 1 Japan: *H. macra* (Munro) Honda – varieg. cvs cult. as pot-pls in Japan

Halacsya Doerfler. Boraginaceae. 1 W Balkans (serpentine)

Halacsyella Janchen = *Edraianthus*

Halanthium K. Koch. Chenopodiaceae (III 3). 3 W & C As.

Halarchon Bunge. Chenopodiaceae (III 3). 1 Afghanistan: *H. vesiculosus* (Moq.) Bunge

Haldina Ridsd. Rubiaceae (I 2). 1 S & SE As.: *H. cordifolia* (Roxb.) Ridsd. (haldu)

haldu *Haldina cordifolia*

Halenbergia Dinter = *Mesembryanthemum*

Halenia Borkh. Gentianaceae. c. 70 Euras. mts (3; Eur. 1), Am. esp. Andes (C Am. 6, R: BTBC 111(1984)366). Many with cleistogamous fls, few cult. orn.

Halerpestes E. Greene (~ *Ranunculus*). Ranunculaceae (II 3). c. 10 As., Am.

Halesia Ellis ex L. Styracaceae. 5 E China & E N Am., $\overline{G}$; fr. with 2 or 4 longit. wings. Cult. orn. decid. shrubs (bell trees) esp. *H. carolina* L. (*H. tetraptera* , E US) – snowdrop tree, silver bell, opossumwood, with white drooping fls

halfa *Stipa tenacissima*

Halfordia F. Muell. Rutaceae. 4 New Guinea, E Aus., New Caled. Some timber

Halgania Gaudich. Boraginaceae. 18 Aus.

× **Halimiocistus** Janchen. Cistaceae. Hybrids between *Cistus* & *Halimium* spp. Cult. orn.

Halimione Aellen = *Atriplex*

Halimiphyllum (Engl.) Boriss. Zygophyllaceae. 5 C As.

Halimium (Dunal) Spach. Cistaceae. 9 Medit. (Eur. 9). Cult. orn. shrubs like *Helianthemum*, hybridizing with *Cistus* spp. (= × *Halimiocistus*)

Halimocnemis C. Meyer. Chenopodiaceae (III 3). 19 Eur. (1) to C As.

Halimodendron Fischer ex DC. Leguminosae (III 15). 1 Eur. & Turkey to C As. *H. halodendron* (Pallas) Voss, salt-steppe shrub with persistent spine-tipped leaf-rachides, cult. orn.

Halimolobos Tausch. Cruciferae. 19 W Am. (Rockies to Andes). R: CDH 3(1943)241

Hallea J. Leroy (~ *Mitragyna*). Rubiaceae (I 1). 3 trop. Afr. R: Adansonia 15(1975)66

hallelujah *Oxalis acetosella*

Halleria L. Scrophulariaceae. 4 trop. & S Afr., Madag. *H. lucida* L. (African honeysuckle, Afr.) – cult. orn. shrub with scarlet drooping fls & allegedly ed. fr.

Hallia Thunb. Leguminosae (III 11). 9 S Afr.

Hallieracantha Stapf = *Ptyssiglottis*

Hallianthus H. Hartman (~ *Leipoldtia*). Aizoaceae (V). 1 SW Cape: *H. planus* (L. Bolus) H. Hartman. R: BJ 104(1983)143

Halmoorea Dransf. & Uhl = *Orania*

Halocarpus Quinn (~ *Dacrydium*). Podocarpaceae. 3 NZ. R: Phytol.M 7(1984)31. *H. biformis* (Hook.) Quinn (manoao) – timber used locally for building, railway-sleepers etc.

Halocharis Moq. Chenopodiaceae (III 3). 13 SW & C As.

Halocnemum M. Bieb. Chenopodiaceae (II 2). 1 C Med. incl. Eur. to C As.: *H. strobilaceum* (Pallas) M. Bieb. Alks

Halodule Endl. Cymodoceaceae. 6 shallow trop. seas. R: VKNAW II, 59(1970)1. Pollen in rafts reaching stigmas (cf. *Halophila*, *Lepilaena*, *Ruppia*)

Halogeton C. Meyer. Chenopodiaceae (III 3). 9 Med. (Eur. 1) to C As. Toxic weeds in W US (sodium oxalate). *H. glomeratus* (M. Bieb.) C. Meyer (SE Russia) – ant-poll.; *H. sativus* (L.) Moq. (W Med.) – form. cult. & burnt for base-rich ash (barilla)

Halopegia Schumann. Marantaceae. 6 trop. OW E to Java

Halopeplis Bunge ex Ung.-Sternb. Chenopodiaceae (II 1). 3 S W Eur. (1), E Medit., Cauc.

Halophila Thouars. Hydrocharitaceae. 8 trop. coasts WI, Indian & Pacific Oceans. R: C. den Hartog (1970) *Sea grasses of the world*: 238. Pollen in rafts reaching stigmas (cf. *Halodule*, *Lepilaena*, *Ruppia*). *H. stipulacea* (Forssk.) Asch. spread from Indian Ocean & Red Sea through Suez Canal (opened 1869) to Malta etc. & still spreading

Halophytaceae Soriano (~ Chenopodiaceae). Dicots – Caryophyllidae – Caryophyllales. 1/1 temp. Arg. Succ. monoec. herb. Lvs spirally arr.; stipules 0. Fls in racemes; P 4 membranous, 0 in females; A 4; $\underline{G}$, 1-loc., 1-ovulate; style with 3 stigmas. Fr. a thin-walled nutlet encl. in axial tissue which hardens to form a syncarp encl. a few nutlets; seed with annular embryo

Only genus: *Halophytum*

Excl. from Chenopodiaceae: sieve-tube plastids diff. while cuboid pollen resembles that of some *Basella* spp.

Halophytum Speg. Halophytaceae (Chenopodiaceae s.l.). 1 Patagonia: *H. ameghinoi* (Speg.) Speg.

Halopyrum Stapf. Gramineae (31d). 1 Indian Ocean coasts

Haloragaceae = seq.

Haloragidaceae R. Br. Dicots – Rosidae – Haloragidales. 9/145 cosmop. but esp. S hemisph. Shrubs or small trees (*Haloragodendron*) but usu. aquatic or amphibious herbs; cortex commonly with air-cavities; in aquatics, vasc. system reduced (s.t. to single central fibrovascular strand). Lvs spirally arr., opp. or whorled, v. varied in form; stipules 0. Fls solit. & axillary or in term. spikes to panicles, usu. small & unisexual, epigynous, reg., (3)4-merous, wind-poll., 2-bracteolate; K valvate, persistent in fr., C often larger or 0, A 4 + 4 or 4(3) with usu. short filaments & rather large anthers with longit. slits, $\overline{G}$ ((2)3 or 4) with as many locules (sometimes partitions ± absent) & distinct feathery styles, 1 anatropous, bitegmic pendulous ovule per loc. Fr. a nut or drupe (a schizocarp in *Myriophyllum* & *Vinkia*); seeds with straight cylindrical embryo in ± copious oily endosperm. x = 7 (often). R: BAIM 10(1975)

Principal genera: *Gonocarpus*, *Haloragis*, *Myriophyllum*

Poss. close to Saxifragaceae. *Gunnera* is excluded as Gunneraceae

Some cult. aquatics in aquaria esp. *Myriophyllum* spp.

Haloragis Forster & Forster f. Haloragidaceae. 27 Aus. (22), New Caled (1), NZ (1), Rapa (1), Juan Fernandez (2). R: BAIM 10(1975)64. Desert ephemerals to obligate aquatics. *H. micrantha* (Thunb.) Siebold & Zucc. (SE As. to Aus.) – aquatic natur. W Galway (Ireland)

Haloragodendron Orch. Haloragidaceae. 5 S Aus. (1 almost extinct). R: BAIM 10(1975)140. Trees & shrubs to 3 m

Halosarcia P.G. Wilson. Chenopodiaceae (II 2). 23 Aus., 1 extending to Mal. R: Nuytsia 3(1980)25

Halosciastrum Koidz. (~ *Cymopterus*). Umbelliferae (III 8). 1 E As.: *H. melanotilingia* (Boissieu) Pim. & Tichomirov

Halosicyos Mart. Crov. Cucurbitaceae. 1 N Argentina

Halostachys C. Meyer ex Schrenk. Chenopodiaceae (I). 1 SE Russia to C As.: *H. belangeriana* (Moq.) Botsch.

Halothamnus Jaub. & Spach (~ *Salsola*). Chenopodiaceae (III 3). 21 Middle E to Afghanistan

& Somalia. R: OB 143(1993)

Halotis Bunge (~ *Halimocnemis*). Chenopodiaceae (III 3). 2 C As. to Iran & Afghanistan

Haloxanthium Ulbr. = *Atriplex*

Haloxylon Bunge. Chenopodiaceae (III 3). (Incl. *Hammada*) c. 25 W Med. (Eur. 1) to Iran, Mongolia, Burma & SW China. Steppe-pls with alks & jointed twigs (saxaul) in deserts E of Caspian & Aral Seas to China; imp. fuel & for stock-pens, charcoal & fodder for camels, sheep etc., dune-stabilizers esp. *H. persicum* Bunge ex Boiss. & Buhse (C As.) to 6 m tall & bole to 20 cm diam. with timber used for general carpentry & source of charcoal, introd. Mongolia for dune-stabilizing; *H. aphyllum* (Minkw.) Iljin (widespread) – similar & a promising firewood crop

Halphophyllum Mansf. = *Gasteranthus*

Hamadryas Comm. ex Juss. Ranunculaceae (II 3). 6 Antarctic S Am. Dioec.

Hamamelidaceae R. Br. Dicots – Hamamelidae – Hamamelidales. 30/95 widespread but chiefly subtrop. esp. E As. Shrubs or trees, often with stellate hairs. Lvs usu. spirally arr. (distichous), simple, often palmately-lobed; stipules usu. present. Fls bisexual or unisexual, usu. reg., ± epigynous, s.t. crowded (& considered by some to be s.t. pseudanthial) but usu. in spikes or heads, wind- or insect-poll.; K (0)4 or 5(–7), small, distinct to connate, C usu. (0)4 or 5 small & narrow, A (1–)4 or 5(–24), alt. with C developing centripetally (*Matudaea*) or centrifugally (*Fothergilla*), anthers with valves or longit. slits, connective usu. extended, G (2 (or 3)) united at least basally, placentation usu. axile, styles distinct, each loc. with 1(2–∞) ± pendulous, anatropous (orthotropous in Altingioideae), bitegmic ovules, often all but 1 aborting. Fert. often delayed until long after poll. (cf. Pinopsida). Fr. a woody capsule, septicidal (& loculicidal); seeds with thick hard testa & large straight embryo in oily & proteinaceous endosperm. x = 8, 12, 15, 16

Classification & principal genera:

I. **Hamamelidoideae** (male fls in male infls s.t. not discrete, others clearly separate; locules with 1 or 2 ovules; 4 tribes):
   1. **Hamamelideae**: *Dicoryphe, Hamamelis, Trichocladus*
   2. **Corylopsideae**: *Corylopsis* (only)
   3. **Eustigmateae**: *Eustigma*
   4. **Fothergilleae** (incl. Disanthoideae): *Distylium, Sycopsis*

II. **Exbucklandioideae** (pls polygamo-monoec. with fls in capitula; A 10–14; stipules broad, enclosing young shoot): *Exbucklandia*

III. **Rhodoleioideae** (fls bisexual in 5–10-flowered capitulum subtended by bracts so as to resemble single fl.): *Rhodoleia*

IV. **Altingioideae** (Liquidambaroideae; pls dioec., male infls with P 0, females globose heads with P of numerous scales): *Altingia, Liquidambar*

12–14 of the genera are monotypic; fam. loosely knit & Altingioideae often recognized as discrete fam., though clearly allied. App. relics of an ancient group, *Disanthus* having most archaic features

Timbers (*Altingia, Exbucklandia, Liquidambar*), extracts medic. & used in scent (*Hamamelis, Liquidambar*), cult. orn. (*Corylopsis, Hamamelis, Liquidambar, Parrotia, Sycopsis* etc.)

Hamamelis L. Hamamelidaceae (I 1). 5–6 E N Am., E As. Witch hazels. Cult. orn. shrubs (As. spp. fl. in winter, Am. spp. in autumn when leafy) esp. *H.* × *intermedia* Rehder (*H. japonica* Siebold & Zucc. (Japan) × *H. mollis* Oliver (China)) & *H. virginiana* L. (E N Am.) – bark & lvs source of medic. witch hazel for bruises, haemorrhoids, varicose veins etc. & grown comm. in England for eye-lotions, form. used as divining-rod (witch/wych an Old Engl. term for pliant branches), used as grafting stock for other spp. though seeds take up to 2 years to germinate; all spp. have strongly-scented fls

Hamatocactus Britton & Rose = *Thelocactus*

Hamelia Jacq. Rubiaceae (IV 5). 16 trop. Am. Interpetiolar stipules small, decid.; extrafl. nectaries continue to secrete for up to 21 days after fall of C. Cult. orn. shrubs with red or yellow fls, esp. hummingbird-poll. *H. patens* Jacq. (fire bush) often confused with *Ixora* spp. – ed. fr. used in a fermented drink, tanbark

Hamilcoa Prain. Euphorbiaceae. 1 trop. W Afr.

hamilla *Berrya cordifolia*

Hamiltonia Roxb. = *Spermadictyon*

Hammada Iljin = *Haloxylon*

Hammarbya Kuntze = *Malaxis*

Hammatolobium Fenzl. Leguminosae (III 19). 2 Medit. (Eur. 1)

Hampea Schldl. Malvaceae. 21 trop. Am. R: P. Fryxell (n.d.) *Nat. Hist. Cotton tribe*: 72

Hanabusaya Nakai. Campanulaceae. 2 Korea

Hanburia Seemann. Cucurbitaceae. 2 Mex., Guatemala. Fr. explosive incl. in *H. mexicana* Seemann (chayotilla)
Hanceola Kudô. Labiatae (VIII 4). 3 China
Hancockia Rolfe. Orchidaceae (V 11). 1 E & SE As.: *H. uniflora* Rolfe. R: OM 6(1992)63
Hancornia B.A. Gomes. Apocynaceae. 1 Brazil: *H. speciosa* B.A. Gomes – source of mangabeira rubber & fr. used in marmalade & to flavour icecream, sherbet etc.
hand, Buddha's *Citrus medica* 'Fingered'
Handelia Heimerl. Compositae (Anth.-Han.). 1 C As. to China
Handeliodendron Rehder = *Sideroxylon*
Handroanthus Mattos = *Tabebuia*
Hanghomia Gagnepain & Thénint. Apocynaceae. 1 SE As.: *H. marseillei* Gagnepain & Thénint – liane with roots of ceremonial significance, burnt as incense etc. in Laos
Hanguana Blume. Flagellariaceae (Hanguanaceae). 1 or 2 Sri Lanka, SE As. & Mal. The alleged 2 spp. have identical male pl., but females have diff. sized fr.
Hanguanaceae Airy Shaw. = Flagellariaceae
Haniffia Holttum. Zingiberaceae. 2 Peninsular Thailand & Malaya
Hannafordia F. Muell. Sterculiaceae. 4 C Aus.
Hannoa Planchon = *Quassia*
Hannonia Braun-Blanquet & Maire. Amaryllidaceae (Liliaceae s.l.). 1 NW Afr.
Hansenia Turcz. (~ *Ligusticum*). Umbelliferae (III 8). 1 N & C As.
Hanslia Schindler = *Desmodium*
Hansteinia Oersted = *Habracanthus*
Hapaline Schott. Araceae (VII 2). 5 Indomal., SE As.
Hapalochilus (Schltr.) Senghas = *Bulbophyllum*
Hapalorchis Schltr. Orchidaceae (III 1). 9 warm Am. R: HBML 29(1982)326
Haplanthodes Kuntze (~ *Andrographis*). Acanthaceae. 4 India. R: BBSI 23(1983)198
Haplanthoides Li = *Andrographis*
Haplanthus Nees = *Andrographis*
Haplocalymma S.F. Blake = *Viguiera*
Haplocarpha Less. Compositae (Arct.-Arct.). 8 E Afr. to Ethiopia
Haplochilus Endl. = *Bulbophyllum*
Haplochorema Schumann (~ *Boesenbergia*). Zingiberaceae. 1 Sumatra, 3 Borneo. R: NRBGE 44(1987)211
Haploclathra Benth. Guttiferae (II). 3 Amazonia. R: MNYBG 22,4(1972)129. Some locally used timbers (red)
Haplocoelum Radlk. Sapindaceae. 7 trop. Afr.
Haplodictyum C. Presl = *Cyclosorus*
Haplodypsis Baillon = *Dypsis*
Haploesthes A. Gray. Compositae (Helia.-Flav.). 3 SW N Am. R: Wrightia 5(1975)108
Haplolobus H.J. Lam. Burseraceae. 22 Mal. & Pacific
Haplolophium Cham. Bignoniaceae. 4 Brazil, Peru. R: Novon 2(1992)163
Haplopappus Cass. Compositae (Ast.-Sol.). (Excl. N Am spp. = *Ericameria, Tonestus*) c. 70 S Am. esp. Chile. R: CIWP 389(1928)1. Some cult. orn.
Haplopetalon A. Gray = *Crossostylis*
Haplophandra Pichon = *Odontadenia*
Haplophloga Baillon = *Dypsis*
Haplophragma Dop = *Fernandoa*
Haplophyllophorus (Brenan) Fernandes & R. Fernandes = *Cincinnobotrys*
Haplophyllum A. Juss. (~ *Ruta*). Rutaceae. 66 Medit. (Eur. 8) to NW Afr., Arabia & E Siberia. R: HIP 40(1-3)(1986). Alks. *H. tuberculatum* (Forssk.) A. Juss. (N Afr. & Middle E) – imp. local medic.
Haplophyton A. DC. Apocynaceae. 3 SW N Am. Alks (insecticidal) esp. *H. cimicidium* A. DC & *H. crooksii* (L. Benson) L. Benson (cockroach plant)
Haplorhus Engl. Anacardiaceae (IV). 1 Peru
Haplormosia Harms. Leguminosae (III 2). 1 W Afr., swamp forest: *H. monophylla* (Harms) Harms – timber (haplormosia, idewa)
Haplosciadium Hochst. Umbelliferae (III 8). 1 E & NE Afr.: *H. abyssinicum* Hochst. – geocarpic rosette-pl.
Haploseseli H. Wolff & Hand.-Mazz. = *Physospermopsis*
Haplosphaera Hand.-Mazz. Umbelliferae (III 8). 2 S & E As.
Haplostachys Hillebrand. Labiatae (VI). 5 (4 extinct) Hawaii. *H. haplostachya* (A. Gray) St John reduced to 1 pop.

Haplostephium C. Martius ex DC. = *Lychnophora*
Haplostichanthus F. Muell. Annonaceae. 6 C Mal. to Queensland: R: Blumea 39(1994)215
Haplothismia Airy Shaw. Burmanniaceae. 1 S India
Haplotrichion P.G. Wilson (~ *Waitzia*). Compositae (Gnap.-Ang.). 2 Carnarvon Dist., W Aus. R: Nuytsia 8(1992)422
Haptanthus Goldberg & C. Nelson. Family? 1 Honduras: *H. hazlettii* Goldberg & C. Nelson. R: SB 14(1989)16. Fr. unknown
Haptocarpum Ule. Capparidaceae. 1 E Brazil
Haradjania Rech.f. = *Myopordon*
Haraella Kudô. Orchidaceae (V 16). 1 Taiwan
Harbouria J. Coulter & Rose. Umbelliferae (III 8). 1 SW US: *H. trachypleura* (A. Gray) J. Coulter & Rose
hard fern *Blechnum spicant*; **h. heads** *Centaurea nigra*
Hardenbergia Benth. Leguminosae (III 10). 3 Aus. Allied to *Kennedia*. Cult. orn. esp. *H. violacea* (Schneev.) Stearn (*H. monophylla*, false sarsparilla, E Aus.) with 1-foliolate lvs
Harding grass *Phalaris aquatica*
Hardwickia Roxb. Leguminosae (I 4). 1 semi-arid India: *H. binata* Roxb. – heaviest of Indian timbers used for construction & orn. work, bark used for tanning & as fibre & for sails & paper, resin a wood preservative, lvs used as fodder
harebell *Campanula rotundifolia*, (Aus. & NZ) *Wahlenbergia* spp.
hare's ear *Bupleurum* spp.
hare's-foot fern *Davallia canariensis*
harewood *Acer pseudoplatanus*
Harfordia E. Greene & C. Parry. Polygonaceae (I 2). 1 Baja Calif., Mex.: *H. macroptera* (Benth.) E. Greene & C. Parry
haricot bean *Phaseolus vulgaris*
Harlanlewisia Epling = *Scutellaria*
harlequin flower *Sparaxis* spp.
Harleya S.F. Blake. Compositae (Vern.-Vern.). 1 Mex., C Am.
Harleyodendron R. Cowan. Leguminosae (III 1). 1 E Brazil
harmal *Peganum harmala*
Harmandia Pierre ex Baillon. Olacaceae. 1 SE As., Mal.
Harmandiella Costantin. Asclepiadaceae (III 4). 1 SE As.: *H. cordifolia* Costantin
Harmsia Schumann. Sterculiaceae. 3 trop. Afr.
Harmsiodoxa O. Schulz. Cruciferae. 3 Aus. R: TRSSA 89(1965)204
Harmsiopanax Warb. Araliaceae. 3 Mal. Some hapaxanthic (unique in fam.). Fr. with 2 mericarps like Umbelliferae but leaf-base, habit & petal-shape typical of Araliaceae. R: FM I, 9(1979)9. *H. ingens* Philipson (New Guinea) to 18 m, spiny, hapaxanthic
Harnackia Urban. Compositae (Hele-Pect.). 1 E Cuba: *H. bisecta* Urban – liane on serpentine. R: Madrono 24(1977)137
Haronga Thouars = *Harungana*
Harpachne Hochst. ex A. Rich. Gramineae (31d). 2 N & NE Afr.
Harpagocarpus Hutch. & Dandy = *Fagopyrum*
Harpagonella A. Gray. Boraginaceae. 1 SW N Am.
Harpagophytum DC ex Meissner. Pedaliaceae. 2 S Afr., Madag. R: MSABH 13(1970)15. *H. procumbens* DC ex Meissner (devil's claw, grapple plant) has fr. with large woody grapples c. 2.5 cm long, pointed & barbed, so that it is animal-disp. (&, like *Xanthium*, a nuisance to shepherds) but it can cause grazing animals to become lame or even starve as their jaws may become locked by them; used as mouse-traps in Madag., also medic., for rheumatism in W Eur.
Harpalyce Sessé & Moçiño ex DC. Leguminosae (III 23). 24 Mex., Cuba & Brazil
Harpanema Decne. Asclepiadaceae (I; Periplocaceae). 1 Madag.: *H. acuminatum* Decne
Harpephyllum Bernh. ex K. Krause. Anacardiaceae (II). 1 S Afr.: *H. caffrum* Bernh. ex K. Krause (kaffir plum) – fr. used in jelly & local medic. Dioec.(?) cult. orn. flowering when unbranched
Harperia W. Fitzg. Restionaceae. 1+ SW Aus.
Harperocallis McDaniel. Melanthiaceae (Liliaceae s.l.). 1 Florida
Harpochilus Nees (~ *Justicia*). Acanthaceae. 3 Brazil
Harpochloa Kunth. Gramineae (33b). 2 C & S Afr.
Harpullia Roxb. Sapindaceae. 37 Indomal., trop. Aus. (8) & New Caled. (1). R: Blumea 28(1982)1. Some cult. orn. trees incl. *H. pendula* Planchon ex F. Muell. (Aus. tulipwood, NE Aus.) – timber beautifully marked black to yellow used in cabinet-making

Harrimanella Cov. (~ *Cassiope*). Ericaceae. 2 Arctic & subarctic. Cult. orn. shrublets differing from C. in spiral lvs & term. infls

Harrisella Fawcett & Rendle (~ *Campylocentrum*). Orchidaceae (V 16). 3 trop. Am.

Harrisia Britton. Cactaceae (III 1). c. 20 Florida, WI, S Am. Cult. orn. slender ribbed cacti with caffeine. *H. martinii* (Labouret) Britton & Rose (Arg.) declared a noxious weed in C Afr.

Harrisonia R. Br. ex A. Juss. Rutaceae. 3–4 OW trop. *H. abyssinica* Oliver (trop. Afr.) – root used for swollen testicles on Mafia Is.

Harrysmithia H. Wolff. Umbelliferae (III 8). 2 China

Harthamnus H. Robinson = *Plazia*

Hartia Dunn = *Stewartia*

Hartleya Sleumer. Icacinaceae. 1 New Guinea. Allied to *Gastrolepis*

Hartliella E. Fischer (~ *Lindernia*). Scrophulariaceae. 4 Zaire (Shaba). R: TSP 81(1992)204

Hartmannia Spach = *Oenothera*

Hartogia Thunb. ex L.f. = *Hartogiella*

Hartogiella Codd. Celastraceae. 1 Cape

Hartogiopsis H. Perrier. Celastraceae. 1 Madagascar: *H. tribulocarpa* (Baker) H. Perrier

hart's tongue (fern) *Asplenium scolopendrium*

Hartwegiella O. Schulz = *Mancoa*

Hartwrightia A. Gray ex S. Watson. Compositae (Eup.-Gyp.). 1 Georgia, Florida. R: MSBMBG 22(1987)282

Harungana Lam. Guttiferae (I). 1 trop. Afr. to Mauritius: *H. madagascariensis* Lam. ex Poiret – resin yellow turning red on exposure; wood easily worked; lvs medic.

Harveya Hook. Scrophulariaceae. 40 trop. & S Afr., Masc. Some root-parasites

Haselhoffia Lindau = *Physacanthus*

hashish *Cannabis sativa* subsp. *indica*

Hasseanthus Rose = *Dudleya*

Hasselquistia L. = *Tordylium*

Hasseltia Kunth. Flacourtiaceae (7; ? Tiliaceae). 3 trop. Am. R: FN 22(1980)73

Hasseltiopsis Sleumer. Flacourtiaceae (7; ? Tiliaceae). 1 C Am.

Hasskarlia Baillon = *Tetrorchidium*

Hasslerella Chodat = *Polypremnum*

Hasteola Raf. (~ *Synosma*). Compositae (Sen.-Sen.). 2 N Am. R: SB 19(1994)211

Hastingsia (Durand) S. Watson (~ *Schoenolirion*). Hyacinthaceae (Liliaceae s.l.). 4 W N Am. R: Madrono 38(1991)135

Hatiora Britton & Rose. Cactaceae (III 5). (Incl. *Rhipsalidopsis*) 4 E & SE Brazil. Cult. orn. epiphytes like *Rhipsalis* but fls apical, esp. *H. gaertneri* (Regel) Barthlott (*R. gaertneri*, Easter cactus, E Brazil) – fls bright red, *H. rosea* (Lagerh.) Barthlott (SE Brazil) – fls pink, & their hybrid with range of fl. colours

Hatschbachiella R. King & H. Robinson (~ *Eupatorium*). Compositae (Eup.-Eup.). 2 S Brazil. R: MSBMBG 22(1987)71

haulm aerial stems of potatoes, beans, peas etc.

Haumania Léonard. Marantaceae. 3 trop. Afr.

Haumaniastrum Duvign. & Plancke. Labiatae (VIII 4b). 23 trop. Afr.

Hausa potato *Plectranthus rotundifolius*

Haussknechtia Boiss. Umbelliferae (inc.sed.). 1 SW Iran: *H. elymaitica* Boiss. – umbels globose, not seen since 1860s

hautbois or **hautboy** *Fragaria moschata*

Hauya Moçiño & Sessé ex DC. Onagraceae. 2 C Am. Woody

Havardia Small. Leguminosae (II 5). 15 warm Am., 3 As.

Havetia Kunth. Guttiferae (III). 1 Colombia

Havetiopsis Planchon & Triana. Guttiferae (III). 5 trop. Am.

haw fr. of *Crataegus* spp. esp. *C. monogyna*; **black h.** *C. douglasii*, *Viburnum prunifolium*

Hawaiian goosefoot *Chenopodium oahuense*

Hawkesiophyton Hunz. Solanaceae (1). 3 trop. S Am.

hawk's beard *Crepis* spp.; **h.bit** *Leontodon* spp.; **h.weed** *Hieracium* spp.

Haworthia Duval. Asphodelaceae (Aloaceae). 87 dry S Afr. R: M.B. Bayer (1982) *The new H. handbook*; C.L. Scott (1985) *The genus H.* Stemless or short-stemmed rosette-pls with succ. warty lvs, much cult. as orn.

hawthorn *Crataegus* spp. esp. *C. monogyna*; **Chinese h.** *C. pentagyna*; **Mexican h.** *C. pubescens*; **water h.** *Aponogeton distachyos*

Haya Balf.f. Caryophyllaceae (I 1). 1 Socotra: *H. obovata* Balf.f.

Hayataella Masam. Rubiaceae (?IV 1). 1 Taiwan

Haydonia R. Wilczek = *Vigna*

Haylockia Herbert. Amaryllidaceae (Liliaceae s.l.). 6 Andes. Fls projecting from soil like *Colchicum*

Haynaldia Schur = *Dasypyrum*

hayrattle *Rhinanthus minor*

Hazardia E. Greene (~ *Haplopappus*). Compositae (Ast.-Sol.). 13 SW N Am. R: Madrono 26(1979)105

hazel *Corylus* spp. esp. *C. avellana*; **American h.** *C. americana*; **corkscrew h.** *C. avellana* 'Contorta'

Hazomalania Capuron (~ *Hernandia*). Hernandiaceae. 1 W Madag.: *H. voyroni* Capuron

Hazunta Pichon = *Tabernaemontana*

headache tree *Premna serratifolia*

heart-leaf *Gastrolobium* spp.

heart's ease *Viola tricolor*; **h.pea** or **seed** *Cardiospermum grandiflorum*

heath *Erica* spp.; **Australian h.** *Epacris* spp.; **Cornish h.** *Erica vagans*; **cross-leaved h.** *E. tetralix*; **Dorest h.** *E. ciliaris*; **h. grass** *Danthonia decumbens*; **Irish h.** *E. erigena*; **St Dabeoc's h.** *Daboecia cantabrica*; **sea h.** *Frankenia* spp.; **tree h.** *E. arborea*

heather (**common** or **Scottish** or **white**) *Calluna vulgaris*

heaven, tree of *Ailanthus altissima*

Hebanthe C. Martius = *Pfaffia*

Hebe Comm. ex Juss. (~ *Veronica*). Scrophulariaceae. 90 Australasia (esp. NZ), Papuasia, temp S Am. R: D. Chalk (1988) *Hs & Parahebes*. Differs from *Veronica* in being mostly woody, in opp. lvs & dehiscent capsule. Trees & shrubs sometimes cult. as hedges & many cult. orn. incl. hybrids, swarms forming readily in wild & cult., several cvs derived from crosses with *H. speciosa* (Cunn.) Cockayne (NZ) prob. incl. *H.* × *franciscana* (Eastw.) Souster (*H. elliptica* (Forster f.) Pennell (NZ to S Am.) × *H. speciosa*), the shrub used as hedging in SW GB, *H. salicifolia* (Forster f.) Pennell (S NZ to S Am.), the hardiest sp.; some with small adpressed lvs resembling conifers etc. e.g. *H. cupressoides* (Hook.f.) Cockayne & Allan & *H. lycopodioides* (Hook.f.) Cockayne & Allan (both NZ). Some with young lvs used as cure for diarrhoea & dysentery

Hebea L. Bolus = *Tritoniopsis*

Hebecladus Miers = *Jaltomata*

Hebeclinium DC (~ *Eupatorium*). Compositae (Eup.-Heb.). 20 trop. Am. R: MSBMBG 22(1987)399

Hebecoccus Radlk. = *Lepisanthes*

Hebenstretia L. Scrophulariaceae. 25 trop. & S Afr. R: MBSM 15(1979)1, 18(1982)183. Some cult. orn.

Hebepetalum Benth. = *Roucheria*

Heberdenia Banks ex A. DC. Myrsinaceae. 1 Mex., 1 Macaronesia

Hebestigma Urban. Leguminosae (III 7). 1 Cuba: *H. cubense* (Kunth) Urban

Hecastocleis A. Gray. Compositae (Mut.-Mut.). 1 SW US. Shrub

Hecatactis F. Muell. ex Mattf. = *Keysseria*

Hecatostemon S.F. Blake. Flacourtiaceae (9). 1 trop. S Am.

Hechtia Klotzsch. Bromeliaceae (1). 51 C Am. R: FN 14(1974)577. Dioec., habit of *Agave* or *Yucca*; some cult. orn.

Hecistopteris J. Sm. Vittariaceae. 1 trop. Am.: *H. pumila* (Sprengel) J. Sm.

Heckeldora Pierre. Meliaceae (I 6). 1–4 trop. W Afr.

Hectorella Hook.f. = *Lyallia*

Hectorellaceae Philipson & Skip. (~ Portulacaceae). Dicots – Caryophyllidae – Caryophyllales. 1/2 S Is., NZ & Kerguélen Is. Cushion-forming shrubs with taproots. Lvs small, spirally arr.; stipules 0. Fls solit., axillary, insect- or self-poll.; bracteoles 0–3; K 2 opp.; C 4 or 5(6), free or basally connate; A 3–5(6) alt. with C, anthers with longit. slits; $\underline{G}$ 1-loc. with few ovules. Capsule; seed kidney-shaped with smooth hard testa & embryo curved round perisperm. 2n = 96

Genus: *Lyallia*

Hecubaea DC = *Helenium*

Hedbergia Molau (~ *Bartsia*). Scrophulariaceae. 1 trop. Afr mts: *H. abyssinica* (Benth.) Molau. R: NJB 8(1988)193

Hedeoma Pers. Labiatae (VIII 2). 38 SW N Am., S Am. R: Sida 8(1980)218. Some cult. incl. *H. pulegioides* (L.) Pers. (American pennyroyal, Am.) – dried lvs home medic.

Hedera L. Araliaceae (1). 4–11 Eur. (1–3), Medit. to E As. Ivies. R: P.Q. Rose (1980) *Ivies*.

Woody lianes with distinct juvenile & mature stages, discussed by Theophrastus, the first with usu. lobed lvs & rooting stems, the mature with rootless flowering shoots with elliptic lvs, this 'phase-change' assoc. with increase in nuclear size & DNA content. Perhaps not generically distinct from *Schefflera* (Frodin). *H. helix* L. (Eur., Medit., W As.) – fls v. late, poll. by wasps & moths etc., v. nutritious fr. ripens over winter & imp. food-source for nestling birds in spring (tropical 'behaviour'), molluscicidal saponins (toxic); wood used as boxwood subs., young twigs form. source of dyes, form. used in treatment of corns & considered to counteract effects of alcohol so figuring in Bacchus's chaplet & as a sign for a tavern; signifying 'fidelity' in 'Language of Fls'. Species boundaries not clear-cut, many cvs selected for ground-cover, climbers & pot-pls (sometimes grafted on × *Fatshedera* to give 'standards', incl. many varieg. forms. *H. algeriensis* Hibb. ('*H. canariensis*', Canary ivy, N Afr.) esp. 'Gloire de Marengo' (Variegata) – housepl., *H. colchica* (K. Koch) Hibb. (Persian i., Cauc. to N Iran) – lvs v. large, *H. helix* ssp. *hibernica* (Kirchner) D. McClint. ('Hibernica', *H. hibernica*, Irish i., Iberia, W France, W Br., Ireland) – tetraploid (ssp. *helix* diploid; poss. allotetraploid involving another sp. from Morocco), lvs larger than type, 'Conglomerata' etc. – adult forms prop. vegetatively, 'Congesta' – stems erect with 2-ranked juvenile lvs, & many varieg. cvs, *H. rhombea* (Miq.) Bean (Japanese i., Korea, Japan) – esp. 'Variegata'

Hederopsis C.B. Clarke = *Macropanax*

Hederorkis Thouars (~ *Bulbophyllum*). Orchidaceae (V 13). 2 Masc. R: Adansonia 16(1976)225

hedge bedstraw *Galium mollugo*; **h. mustard** *Sisymbrium officinale*; **h. parsley** *Torilis japonica*; **h. woundwort** *Stachys sylvatica*

hedgehog broom *Erinacea anthyllis*; **h. cactus** *Echinocereus* spp.; **h. grass** *Cenchrus* spp.; **h. holly** *Ilex aquifolium* 'Ferox'

Hedinia Ostenf. Cruciferae. 1 C As. to NW Himal.: *H. tibetica* (Thomson) Ostenf.

Hediniopsis Botsch. & Petrovsky. Cruciferae. 1 E As.

Hedraianthera F. Muell. Celastraceae. 1 E Aus.: *H. porphyropetala* F. Muell.

Hedranthera (Stapf) Pichon = *Callichilia*

Hedstromia A.C. Sm. Rubiaceae (IV 7). 1 Fiji

Hedyachras Radlk. = *Glenniea*

Hedycarya Forster & Forster f. Monimiaceae (V 1). 11 W Pacific (New Caled. 9), Aus., NZ. R: Adansonia 18(1978)25, BMNHN 4,5(1983)247. Dioec. trees & shrubs incl. rheophytes (e.g. *H. rivularis* Guillaumin, New Caled.); *H. angustifolia* Cunn. (Aus.) – timber for cabinet-work

Hedycaryopsis Danguy = *Ephippiandra*

Hedychium J. Koenig. Zingiberaceae. 50 Madag. (?), Indomal., Himal. Robust herbs with rhiz. & showy fls with long tubes & narrow free P-lobes, larger staminodes (lip 2-lobed) & with stigmas projecting just beyond anther. Many cult. orn. (ginger lilies). *H. coronarium* J. Koenig (butterfly lily, trop. As., widely natur. trop. Am.) – fls v. fragrant, stems poss. paper-source; *H. gardnerianum* Ker (Himal.) – pest in Hawaii etc.; *H. spicatum* Buch.-Ham. ex Sm. (India) – rhiz. much used in perfumery e.g. abir (q.v.)

Hedyosmos Mitch. = *Cunila*

Hedyosmum Sw. Chloranthaceae. 45 trop. Am. (R: FN 40(1988)1; WI 5, R: JAA 69(1988)51), SE As. (1). Infl. resembling a spike or poss. a solit. strobiloid fl. with A several × 100 on spiral axis; pollen monosulcate. Some medic.

Hedyotis L. Rubiaceae (IV 1). c. 250 trop. & warm esp. OW. Trees, lianes, shrubs (some rheophytes) & herbs; some cult. orn. esp. *H. caerulea* (L.) Hook. (bluets, innocence, *Houstonia c.*, N Am.); *H. terminalis* (Hook. & Arn.) W.C. Wagner & Herbst (Hawaii) – prob. most polymorphic pl. sp. in Hawaii. See also *Oldenlandia*

Hedypnois Miller. Compositae (Lact.-Hyp.). 2 Macaronesia to Iran. (Eur. 2), *H. rhagadioloides* (L.) F.W. Schmidt widely introd.

Hedysarum L. Leguminosae (III 17). 100 N temp. (Eur. 18), Medit. R: AHP 19(1902)183. Some cult. orn. incl. *H. alpinum* L. (circumboreal) – roots eaten raw, boiled or roasted by Esquimaux & N Am. Indians; *H. coronarium* L. (French honeysuckle, W Med., natur. rest of S Eur.) – sometimes cult. as fodder (espercet, sulla)

Hedyscepe H. Wendl. & Drude. Palmae (V 4h). 1 Lord Howe Is: *H. canterburyana* (C. Moore & F. Muell.) H. Wendl. & Drude (*Kentia c.*, umbrella palm) – cult. orn.

Hedythyrsus Bremek. Rubiaceae (IV 1). 2 trop. Afr.

Heeria Meissner. Anacardiaceae (IV). (Excl. *Ozoroa*) 1 SW Cape

Hegemone Bunge ex Ledeb. = *Trollius*

Hegnera Schindler = *Desmodium*

Heimerliodendron Skottsb. = *Pisonia*

Heimia Link. Lythraceae. 3 (closely allied) S US to Arg. Alks. Cult. orn. shrubs used in fertility control & crushed lvs, fermented & drunk, as mildly intoxicating hallucinogen (everything seems to be yellow)

Heinsenia Schumann (~ *Aulacocalyx*). Rubiaceae (II 4). 1 trop. Afr.

Heinsia DC. Rubiaceae (I 8). 4–5 trop. Afr. Some myrmecophily; some ed. fr., timber

Heisteria Jacq. Olacaceae. 35 trop. Am., Afr. (3). R: FN 38(1984)42. Some timber. *H. latifolia* Standley (*H. olivae*, trop. Am.) – psychoactive (scopolamine); *H. spruceana* Engl. (trop. Am.) – leaf infusions used to relieve swollen limbs

Hekistocarpa Hook.f. Rubiaceae (IV 1). 1 trop. W Afr.

Heladena A. Juss. Malpighiaceae. 6 trop. & warm Am.

Helcia Lindley. Orchidaceae (V 10). 4 Colombia

Heldreichia Boiss. Cruciferae. 4 As. Minor to Afghanistan

Helenium L. Compositae (Hele.-Gai.). 40 Am. Cult. orn. esp. cvs of *H. autumnale* L. (N Am.). *H. amarum* (Raf.) Rock (bitterweed, N Am.) – sesquiterpene lactones toxic to herbivores & used as antifeedant against Colorado beetle; *H. hoopesii* A. Gray (sneezeweed, W US, like *H. amarum* in spreading as weed) – causes 'spewing sickness' in stock

Heleochloa Host ex Roemer = *Crypsis*

Heliabravoa Backeb. = *Polaskia*

Heliamphora Benth. Sarraceniaceae. 5 (closely allied) Guayana Highland. R: Preslia 64(1992)219. Subshrubby to herbaceous pitcher pls like *Sarracenia* but G 3-loc., C 0. Ants & other arthropods main food, attracted by sarracenin secreted by special glands, though spiders s.t. take them before they fall in; in low light, only weakly pouched laminas produced

Heliamphoraceae Chrtek et al. = Sarraceniaceae

Helianthella Torrey & A. Gray. Compositae (Helia.-Verb.). 8 W N Am. Extrafl. nectaries. Some cult. orn.

Helianthemum Miller. Cistaceae. c. 110 Eur. (30) to Sahara, NE Afr. to C As., N & S Am. Rock-roses. Cult. orn. shrublets esp. forms & hybrids of *H. nummularium* (L.) Miller (*H. chamaecistus*, Eur.), also *H. canadense* (L.) Michaux (frostweed, E US), *H. canum* (L.) Hornem. (Eur., Medit.) etc.

Helianthopsis H. Robinson = *Pappobolus*

Helianthostylis Baillon. Moraceae (IV). 2 Amazonia

Helianthus L. Compositae (Helia.-Helia). 50 N Am. R: MTBC 22, 3(1969)1. Sunflowers. *H. annuus* L. (common s., US) – introd. Eur. 1510, second only to soyabean as oil crop in 1970s with 2/3 prod. in former USSR, oil (seeds 25–32% oil: high in polyunsaturates esp. linoleic acid) used in cooking oil, margarine, paints etc., cult. orn. annual to 3 m (UK record = 7.17 m; 'bonsai' specimens flowering at 5.6 cm!) with capitulum to 30 cm diam. or more (ray-florets sterile; signifying 'haughtiness' because of size in 'Language of Fls'), seeds ed. usu. salted, oil-cake used as fodder, pericarp made into fuel-logs in Canada; *H. tuberosus* L. (Jerusalem (prob. corruption of girasole, i.e. sunflower in Italian) artichoke, N Am.) – ed. tubers with sweet taste due partly to fructose from inulin, good carbohydrate food for diabetics (sweeter on molar basis than glucose from starch), eaten by N Am. Indians before Eur. contact, introd. OW 1616, boiled, also livestock feed, bred with *H. annuus* increasing latter's disease resistance; other (rabbit-proof) cult. orn. incl. *H. microcephalus* Torrey & A. Gray (? = *H. parviflorus* Hornem., E US) & *H.* × *multiflorus* L. (2n = 51, allotriploid between *H. annuus* (2n = 34) & *H. decapetalus* L. (2n = 68, C & SE US) arising in Eur. cult.)

Helicanthes Danser. Loranthaceae. 1 India

Helichrysopsis Kirpiczn. (~ *Gnaphalium*). Compositae (Gnap.-Gnap.). 1 trop. E Afr.: *H. septentrionale* (Vatke) Hilliard. R: OB 104(1991)153

Helichrysum Miller. Compositae (Gnap.-Gnap.). c. 600 warm OW (Eur. 14) esp. S Afr., not Aus. (for 'everlastings' see *Bracteantha*). Polyphyletic? Many cult. orn. herbs & shrubs incl. *H. forskahlii* (J. Gmelin) Hilliard & B.L. Burtt (*H. fruticosum*, E & S Afr.), *H. petiolare* Hilliard & B.L. Burtt ('*Gnaphalium lanatum*', S Afr., natur. in hedges in Portugal) – orn. fol. esp. in hanging baskets. *H. italicum* (Roth) G. Don.f.(Medit.) – oil said to have antiviral activity; *H. serpyllifolium* (Berg.) Pers. (S Afr.) – lvs used as tea (Hottentot t.)

Helicia Lour. Proteaceae. c. 100 Indomal., Pacific. *H. diversifolia* C. White (Queensland) – source of helicia nuts

Helicilla Moq. = ? *Suaeda*

Heliciopsis Sleumer. Proteaceae. 7 Indomal.

Helicodiceros Schott (~ *Dracunculus*). Araceae (VIII 6). 1 W Med. is.: *H. muscivorus* (L.f.)

Engl. – cult. orn. with stinking hairy infl. likened to dead animal's anus (Bown), app. attractive to poll.

Heliconia L. Musaceae (Heliconiaceae). c. 100–200 trop. Am. (R (subg. *Stenochlamys* – 42): OB 82(1985))& Moluccas to Fiji & Samoa (6, R: Allertonia 6,1(1990)). R: F. Berry & W.J. Kress (1991) *H.: an identification guide*. Hummingbird-poll. in Am., some poll. bats in OW. Cult. orn. banana-like pls (lobster-claw). *H. bihai* (L.) L. (balisier, trop. Am.) – wild pl. prob. not cult., though many cvs referred to it, poss. paper source

Heliconiaceae Nakai = Musaceae subfam. Heliconioideae

Helicostylis Trécul. Moraceae (III). 7 trop. Am. R: FN 7(1972)75. Hallucinogenic bark used in ceremonial witchcraft in Guianas

Helicteres L. Sterculiaceae. 40 trop. As. & Am. Shrubs & trees (some bat-poll.) with useful bark fibre e.g. kaivum (*H. isora* L., As. – fr. medic.)

Helicteropsis Hochr. Malvaceae. 2 Madagascar

Helictonema Pierre (~ *Hippocratea*). Celastraceae. 1 W Afr.: *H. velutinum* (Afzel.) Hallé. R: BMNHN 4,5(1983)20

Helictotrichon Besser. Gramineae (21b). (Incl. *Amphibromus* (glabrous G), *Avenula*) 100 temp. Euras. (Eur. 39), trop. mts to S temp. Oatgrass

Helietta Tul. Rutaceae. 7 trop. Am. to Calif. (N Am. 3 – R: Brittonia 36(1984)455). *H. parvifolia* A. Gray (Texas, Mex.) – effective insecticide against Mexican flies

Helinus E. Meyer ex Endl. Rhamnaceae. 5 trop. & S Afr. (3), Madag. (1), NW India (1)

Heliocarpus L. Tiliaceae. 10 trop. Am. *H. donnell-smithii* Rose (Mex.) – source of fibre for hammocks etc., bark beaten to give paper, soft wood used for floats, bottle-stoppers etc.

Heliocarya Bunge = *Caccinia*

Heliocauta Humphries. Compositae (Anth.-Tan.). 1 Morocco

Heliocereus (A. Berger) Britton & Rose = *Disocactus*

× **Heliochia** G. Rowley = *Disocactus*

Heliomeris Nutt. Compositae (Helia.-Helia). 6 N Am. R: PIAS 88(1979)364

Heliophila Burm. f. ex L. Cruciferae. 72 S Afr. Some cult. orn. herbs & shrubs esp. *H. longifolia* DC with blue fls reminiscent of flax. *H. glauca* DC is a shrub to 2 m, *H. scandens* Harvey a woody climber to 3 m

Heliopsis Pers. Compositae (Helia.-Zinn.). 15 upland trop. Am. R: OJS 57(1957)171. *H. helianthoides* (L.) Sweet (N Am.) cult. orn., like *H. longipes* (A. Gray) S.F. Blake source of a promising insecticide (scabrin)

Heliosperma (Reichb.) Reichb. = *Silene*

Heliostemma Woodson. Asclepiadaceae. 1 Mexico

heliotrope *Heliotropium* spp. esp. *H. arborescens*; **winter h.** *Petasites fragrans*

Heliotropium L. Boraginaceae. c. 250 trop. & temp. (Eur. 10, W As. c. 80, S Am. c. 80). R: JAA suppl. 1(1991)74. Alks; some not eaten by locusts in E Afr. even though common in swarming places of young stages. Cult. orn. & locally imp. medic. *H. amplexicaule* Vahl (S Am.) – cult. orn. & fert. control; *H. arborescens* L. (*H. peruvianum*, heliotrope, cherry pie, Peru) – cult. orn., in S Eur. used in scent

Helipterum DC = *Helichrysum*; for Aus. spp. see *Rhodanthe*, *Syncarpha*

Helixanthera Lour. Loranthaceae. 25 trop. Afr. to Sulawesi. *H. parasitica* Lour. (Indomal.) – fr. ed. Philippines

Helixyra Salisb. ex N.E. Br. = *Gynandriris*

Hellalia Král = *Sedum*

hellebore *Helleborus* spp.; **black h.** *H. niger*; **false h.** *Veratrum* spp.; **green h.** *H. viridis*; **stinking h.** *H. foetidum*; **white h.** *V. album*

helleborine *Epipactis* spp.; *Cephalanthera* spp.

Helleborus L. Ranunculaceae (I 1). 21 Eur. (11), Medit., As., limestones. R: B. Mathew (1989) *Hellebores*. Cardiac glycosides v. poisonous with burning taste. Rhiz. with aerial shoots taking several yrs to flower; carpels slightly connate basally, C prob. represented by nectaries, K coloured; seeds with elaiosome along raphe, attractive to ants (? & snails) which act as disp. agents. Many cult. orn. for winter & spring fls esp.: *H. foetidus* L. (setterwort, stinkwort, stinking h., bear's-foot, W & S Eur.) – green fls seen as yellow by bees, alleged to find smelly fls (cf. *Crocus & Cytisus*) attractive, form. used as dangerous cathartic & veterinary medic.; *H. lividus* Aiton ssp. *corsicus* (Briq.) Fourn. (*H. argutifolius*, Corsica, Sardinia); *H. niger* L. (Christmas rose, black (referring to cut surface of rhiz.) h., Alps & Apennines) – fls white to pink becoming green after fert., rodent & bird pest control since C1 in W Eur.; *H. orientalis* Lam. (Lenten rose, Greece & Turkey) – fls cream, some hybrids with *H. niger*; *H. thibetanus* Franchet (Tibet) – germination hypogeal (rest of genus epigeal); *H. vesicarius* Aucher (S Turkey) – mature follicle 3-winged inflated to 8 cm diam.,

? wind-disp.; *H. viridis* L. (green h., W & C Eur., natur. N Am.)

Hellenocarum H. Wolff (~ *Carum*). Umbelliferae (III 8). 3 Eur., SW As.

Helleria Fourn. = *Festuca*

Helleriella A. Hawkes (~ *Platyglottis*). Orchidaceae (V 13). 2 C Am.

Hellerochloa Rauschert = *Festuca*

Hellmuthia Steudel (~ *Scirpus*). Cyperaceae. 1 Cape

helmet flower *Scutellaria* spp.; **h. orchid** *Corybas* spp.; **policeman's h.** *Impatiens glandulifera*

Helmholtzia F. Muell. Philydraceae. 1 New Guinea, 2 E Aus.

Helminthocarpon A. Rich. = *Lotus*

Helminthostachys Kaulf. Ophioglossaceae. 1 Sri Lanka, Himal. to Queensland & New Caled.: *H. zeylanica* (L.) Hook. – dorsiventral rhiz. & 2-ranked lvs on upper surface, fert. spikes with lateral sporangiophores of globose sporangia, ed. in salads (Philipp.), alleged antimalarial

Helminthotheca Zinn (~ *Picris*). Compositae (Lac.-Hyp.). 4 SE Eur. to Iran, *H. echioides* (L.) Holub widely introd. R: T 24(1975)111

Helmiopsiella Arènes. Sterculiaceae. 4 Madag. R: BMNHN 4,10(1988)69

Helmiopsis Perrier. Sterculiaceae. 9 Madagascar

Helmontia Cogn. Cucurbitaceae. 1 Brazil, Guyana

Helogyne Nutt. Compositae (Eup.-Alom.). 8 Andes. R: MSBMBG 22(1987)264

Helonema Süsseng. = *Eleocharis*

Helonias L. Melanthiaceae (Liliaceae s.l.). (Incl. *Heloniopsis*, *Ypsilandra*) 10 Tibet to E As., E US (*Helonias* s.s.). Cult. orn. with fragrant pink or purplish fls with blue anthers incl. *H. bullata* L. (E US) & *H.* sp. (*Heloniopsis o.*, *H. japonica*, Japan, Korea) – epiphyllous propagules

Heloniopsis A. Gray = praec.

Helonoma Garay. Orchidaceae (III 3). 2 Venezuela, Guyana

Helosaceae (Schott & Endl.) Rev. & Hoogl. = Balanophoraceae

Helosciadium Koch = *Apium*

Helosis Rich. Balanophoraceae (Helosaceae). 3 trop. Am. *H. cayanensis* (Sw.) Sprengel (*H. guyanensis*) – used NW Amaz. as styptic (blood-red! Cf. 'Doctrine of Signatures')

Helwingia Willd. Cornaceae (Helwingiaceae). 3 E As. Dioec. shrubs with epiphyllous infls initiated adjacent to leaf-axil on base of leaf primordium; DNA data suggest not related to C. *H. japonica* (Thunb.) Dietr. (China & Japan) – cult. orn., lvs ed.

Helwingiaceae Decne. See Cornaceae

Helxine Req. = *Soleirolia*

Hemandradenia Stapf. Connaraceae. 2 W & C Afr. R: AUWP 89-6(1989)275

Hemarthria R. Br. Gramineae (40h). 12 warm OW (Eur. 1, Aus. 1)

Hemerocallidaceae R. Br. (Liliaceae s.l.). Monocots – Liliidae – Asparagales. 1/15 C Eur. to E As. Glabrous rhizomatous herbs with swollen roots with vessels. Lvs linear. Fls in helicoid cymes on bracteate scapes; P 3 + 3 yellow to red atop a tube; A 3 + 3, the anthers with longit. dehiscence; (G 3), 3-loc., each locule with many anatropous ovules. Fr. a loculicidal capsule. Seeds subglobose to angled

Only genus: *Hemerocallis*

Hemerocallis L. Hemerocallidaceae (Liliaceae s.l.). c. 15 C Eur. (1) to China & (esp.) Japan. R: A.B. Stout (1934) *Daylilies*. Cult. orn. with many hybrids (daylilies, spiderlilies) esp. derived from *H. fulva* (L.) L. (As., early natur. in Eur., later US) – fls orange, scentless, commonly a self-sterile triploid ('Europa') & *H. lilioasphodelus* L. (*H. flava*, E Siberia to Japan) – fls yellow, fragrant; dried fls of *H. fulva* used as food-flavouring in China & Japan, form. medic.

Hemiadelphis Nees = *Hygrophila*

Hemiandra R. Br. Labiatae (III). 8 SW Aus.

Hemiangium A.C. Sm. = *Semialarium*

Hemianthus Nutt. = *Micranthemum*

Hemiarrhena Benth. = *Lindernia*

Hemiarthron (Eichler) Tieghem = *Psittacanthus*

Hemibaccharis S.F. Blake = *Archibaccharis*

Hemiboea C.B. Clarke. Gesneriaceae. 21 S China & S Japan to SE As. R: APS 25(1987)81, 220

Hemiboeopsis W.T. Wang (~ *Lysionotus*). Gesneriaceae. 1 SE Yunnan, Laos: *H. longisepala* (Li) W.T. Wang. R: ABY 6(1984)397

Hemicarpha Nees = *Lipocarpha*

Hemichaena Benth. Scrophulariaceae. 1 C Am.

Hemichlaena Schrader = *Ficinia*

Hemichroa R. Br. Chenopodiaceae (IV 1). 3 Aus.
Hemicicca Baillon = *Phyllanthus*
Hemicrambe Webb. Cruciferae. 2 Morocco, Socotra. *H. fruticosa* (C. Towns.) Gómez-Campo (*Fabrisinapis f.*, Socotra) – known only from 2 pls
Hemicrepidospermum Swart = *Crepidospermum*
Hemicyatheon (Domin) Copel. = *Hymenophyllum*
Hemidesmus R. Br. Asclepiadaceae (I; Periplocaceae). 1 S India, SE As., Mal. *H. indicus* (L.) Sm.– used in local medic.
Hemidictyum C. Presl. Dryopteridaceae (II 1). 1 trop. Am.: *H. marginatum* (L.) C. Presl (*Diplazium limbatum*)
Hemidiodia Schumann = *Diodia*
Hemieva Raf. = *Suksdorfia*
Hemifuchsia Herrera. Onagraceae. 1 Peru = ?
Hemigenia R. Br. Labiatae (III). 37 Aus. Lvs in whorls of 3
Hemigramma Christ = *Tectaria*
Hemigraphis Nees. Acanthaceae. 90 trop. As. to New Caled. (2). Some cult. orn. groundcover
Hemilophia Franchet. Cruciferae. 2 SW China
Hemimeris L.f. Scrophulariaceae. 4 S Afr. Elaiophores attractive to bees
Hemimunroa L. Parodi = *Munroa*
Hemionitidaceae Pichi-Serm. = Pteridaceae
Hemionitis L. Pteridaceae (IV). (As. sp. excl.) 7 trop. Am. Lvs dimorphic. Cult. orn.
Hemiorchis Kurz. Zingiberaceae. 3 Indomal. Orchid-like
Hemipappus K. Koch = *Tanacetum*
Hemiphora (F. Muell.) F. Muell. Labiatae (III; Dicrastylidaceae). 1 W Aus.
Hemiphragma Wallich. Scrophulariaceae. 1 W Himal. to Assam: *H. heterophyllum* Wallich – cult. orn. tufted herb
Hemiphylacus S. Watson. Asphodelaceae (Liliaceae s.l.). 1 N Mexico
Hemipilia Lindley. Orchidaceae (IV 2). 16 Himal., E As., Thailand
Hemipogon Decne. Asclepiadaceae (III 1). 10 S Am.
Hemiptelea Planchon (~ *Zelkova*). Ulmaceae (I). 1 N China & Korea: *H. davidii* (Hance) Planchon – cult. orn. spiny decid. tree or shrub with fr. only half-encircled by wing
Hemiscleria Lindley. Orchidaceae. 1 Ecuador, Peru
Hemiscolopia Slooten. Flacourtiaceae (6). 1 Indomal.
Hemisiphonia Urban = *Micranthemum*
Hemisorghum C. Hubb. ex Bor. Gramineae (40c). 2 trop. As.
Hemisphaerocarya Brand = *Oreocarya*
Hemisteptia Bunge (~ *Saussurea*). Compositae (Card.-Card.). 1 India, E As., Aus.: *H. lyrata* (Bunge) Fischer & C. Meyer
Hemistylus Benth. Urticaceae (IV). 4 trop. S Am.
Hemitelia R. Br. = *Cyathea*
Hemithrinax Hook.f. = *Thrinax*
Hemitomes A. Gray. Ericaceae (Monotropaceae). 1 W US
Hemitria Raf. Older name for *Phthirusa*
Hemizonella (A. Gray) A. Gray = *Madia*
Hemizonia DC. Compositae (Hele.-Mad.). 30 Calif. & Baja Calif. Extrafl. nectaries
Hemizygia (Benth.) Briq. Labiatae (VIII 4b). 28 trop. & S Afr.
hemlock *Conium maculatum*; **water h.** *Cicuta virosa*
hemlock (spruce) *Tsuga* spp.; **Canada h.** *T. canadensis*; **Carolina h.** *T. caroliniana*; **eastern h.** *T. canadensis*; **Japanese h.** *T. sieboldii*; **western h.** *T. heterophylla*; **white h.** *T. canadensis*
hemp *Cannabis sativa*; **African h.** *Sparrmannia africanum*; **Ambari h.** *Hibiscus cannabinus*; **Bahama h.** *Agave sisalana*; **Bombay h.** *Crotalaria juncea*; **bowstring h.** *Sansevieria* spp.; **Chinese h.** *Abutilon theophrasti*; **Cuba h.** *Furcraea hexapetala*; **Deccan h.** *Hibiscus cannabinus*; **Indian h.** *Cannabis sativa*; **Madras h.** *Crotalaria juncea*; **Manila h.** *Musa textilis*; **Mauritius h.** *Furcraea foetida*; **h. nettle** *Galeopsis* spp.; **NZ h.** *Phormium tenax*; **Queensland h.** *Sida rhombifolia*; **Russian h.** *Cannabis sativa*; **sisal h.** *Agave sisalana*; **sunn** or **sann h.** *Crotalaria juncea*
Hemsleya Cogn. ex F.B. Forbes & Hemsley. Cucurbitaceae. 30 E As. Allied to *Gomphogyne*. Cucurbitacins poss. effective in tumour control
hen-and-chickens *Jovibarba sobolifera*, *Bellis perennis* 'Prolifera', proliferating captula forms of *Calendula officinalis*
henbane *Hyoscyamus niger*

henbit *Lamium amplexicaule*
Henckelia Sprengel = *Didymocarpus*
henequen *Agave fourcroydes*
Henleophytum Karsten. Malpighiaceae. 1 Cuba
henna *Lawsonia inermis*; **black h.** *Indigofera tinctoria*; **neutral h.** *Ziziphus jujuba*
Hennecartia Poisson. Monimiaceae (V 3). 3 S Brazil, Paraguay, NE Arg. Pollen received on a 'hyperstigma'
Henonia Moq. Amaranthaceae (I 1). 1 Madag.: *H. scoparia* Moq.
Henoonia Griseb. (*Bissea*). Solanaceae (Goetzeaceae). 1 Cuba
Henophyton Cosson & Durieu = *Oudneya*
Henrardia C. Hubb. Gramineae (23). 2 Turkey & Iran to C As.
Henricia Cass. = *Psiadia*
Henricksonia B. Turner. Compositae (Helia.-Cor.). 1 Mexico
Henriettea DC. Melastomataceae. (Incl. *Henriettella, Llewelynia*) 67 trop. S Am.
Henriettella Naudin = praec.
Henriquezia Spruce ex Benth. Rubiaceae (I 4; Henriqueziaceae). 7 Amaz. Brazil
Henriqueziaceae Bremek. = Rubiaceae
Henrya Nees ex Benth. (~ *Tetramerium*). Acanthaceae. 2 C Am. R: CUMH 17(1990)99
Henryettana Brand = *Antiotrema*
Hensmania W. Fitzg. Anthericaceae (Liliaceae s.l.). 3 SW Aus. R: FA 45(1987)249
Hepatica Miller (~ *Anemone*). Ranunculaceae (II 2). 7 N temp. (Eur. 2). R: BAGS 58(1990)144. Cult. orn. esp. *H. acutiloba* DC (N Am.) – seeds dispersed by ants, & *H. nobilis* Schreber (*H. triloba, Anemone h.*, Euras.) – supposed medic.
Heppiella Regel. Gesneriaceae. 4 Andes. R: SB 15(1990)720
Heptacodium Rehder. Caprifoliaceae. 1 C & E China: *H. miconioides* Rehder. R: NSPV 1965(1965)230
Heptanthus Griseb. Compositae (Helia.-Pin.). 7 Cuba
Heptaptera Margot & Reuter. Umbelliferae (III 5). 6 E Medit. (Eur. 4), SW As. R: NRBGE 31(1971)91
Heracleum L. Umbelliferae (III 11). 65 N temp. (Eur. 8), trop. mts. Coarse biennials & perennials. *H. mantegazzianum* Sommier & Levier (giant hogweed, Cauc., natur. Eur. (GB 1893), US) – to 3 m with umbels to 1 m across, causing phytodermatitis, sensitizing skin to ultraviolet radiation, in bright sunlight (same but lesser effect in carrots & parsnips etc.), cult. orn.; *H. sphondylium* L. (hogweed, N temp.) – form. pigfood, medic. & used in liqueurs & basis of alcoholic drink in E Eur., subsp. *montanum* (Gaudin) Briq. (*H. lanatum*, cow parsnip) – roots, fls & stems ed. N Am., fr. used as spice in Sikkim
Herb Bennet *Geum urbanum*; **H. Christopher** *Actaea spicata*; **H.-of-Grace** *Ruta graveolens*; **H. Paris** *Paris quadrifolia*; **H. Patience** *Rumex patientia*; **H. Robert** *Geranium robertianum*; **willowh.** *Epilobium* spp.
Herbertia Sweet. Iridaceae (III 4). 4 temp. S Am., S US. R: NJB 9(1989)55. Cult. orn.
Herbstia Sohmer. Amaranthaceae (I 2). 1 Brazil
Hercules' club *Aralia spinosa, Zanthoxylum clava-herculis*
Herderia Cass. Compositae (Vern.-Vern.). 1 trop. W Afr.
Hereroa (Schwantes) Dinter & Schwantes. Aizoaceae (V). 30 S Afr. Cult. orn. succ. hummock-formers or shrubs
Hericinia Fourr. = *Ranunculus*
Herissantia Medikus = *Abutilon*
Heritiera Dryander. Sterculiaceae. 30 trop. Afr., Indomal. to Aus. & New Caled. (1). R: Reinw. 4(1959)465. Monoec. trees with large buttresses. Some comm. timbers incl. menkulang (Mal.). *H. littoralis* Dryander (OW) – timber for ship-building e.g. masts for dhows in E Afr. & *H. utilis* (Sprague) Sprague (niangon, W Afr.)
Hermannia L. Sterculiaceae. 100+ trop. & warm esp. S Afr. Some cult. orn. with honey-scented fls
Hermas L. Umbelliferae (I 2c). 7 S Afr. R: EJB 48(1991)211
Hermbstaedtia Reichb. Amaranthaceae (I 1). 14 trop. & S Afr. R: KB 37(1982)83
Hermidium S. Watson = *Mirabilis*
Herminiera Guillemin & Perrottet = *Aeschynomene*
Herminium L. Orchidaceae (IV 2). 30 temp. Euras. (Eur. 1: *H. monorchis* (L.) R. Br., musk orchid), Thailand, C Mal.
Hermodactylus Miller (~ *Iris*). Iridaceae (III 2). 1 S France to Middle E: *H. tuberosa* (L.) Miller (snake's head iris) – cult. orn. with G 1-loc. & outer P plum-purple, velvety, tubers form. medic. in Eur.

Hernandia L. Hernandiaceae. 22 trop. esp. Indopacific. R: BJ 89(1969)122. Many alks. Primitive spp. the most restricted in distr. Monoec. trees; some timbers for canoes etc. esp. *H. nymphaeifolia* (C. Presl) Kubitzki (*H. peltata*, *H. ovigera* auctt., jack-in-the-box, OW trop. coasts) – leaf-extract a painless depilatory, street-tree

Hernandiaceae Bercht. & J. Presl. Dicots – Magnoliidae – Magnoliales. 5/57 trop. Trees, shrubs or lianes with alks & scattered spherical ethereal oil-cells that may unite as mucilage cavities. Lvs spirally arr., simple (s.t. 3-lobed) or palmate; stipules 0. Fls small, reg., epigynous, bisexual or not (plants then polygamous, monoec., rarely dioec.) in cymes; P 4–8 or 3 or 4(–6) + 3 or 4(–6), usu. imbricate, A 3–5(–7), filaments often with nectary appendages (cf. Lauraceae), anthers with longit. valves, $\overline{G}$ 1 with elongate style & 1 pendulous, anatropous, bitegmic ovule. Fr. dry-indehiscent, often winged or in accrescent involucre derived from 2 or 3 connate bracteoles; seeds without endosperm, embryo with large, folded, wrinkled or lobed oily cotyledons. x = 15, 20. R: BJ 89(1969)78
Genera: *Gyrocarpus, Hazomalania, Hernandia, Illigera, Sparattanthelium*
The 1st & last often treated as separate subfam. or fam.
Some timbers

Herniaria L. Caryophyllaceae (I 2). 48 Eur. (17) & Afr. to India, 1 N Arg. & Bolivia. C 0 or rudimentary. *H. glabra* L. (rupturewort, herniary, Eur. to C As.) – medic. (diuretic etc.)

herniary *Herniaria glabra*

Herodotia Urban & E. Ekman. Compositae (Sen.-Tuss.). 3 Hispaniola

Herpestis Gaertner f. = *Bacopa*

Herpetacanthus Nees. Acanthaceae. 10 Panamá to Brazil

Herpetophytum (Schltr.) Brieger = *Dendrobium*

Herpetospermum Wallich ex Hook.f. Cucurbitaceae. 1 Himal., China

Herpolirion Hook.f. Anthericaceae (Liliaceae s.l.). 1 SE Aus., NZ: *H. novae-zelandiae* Hook.f. R: FA 45(1987)242

Herpysma Lindley. Orchidaceae (III 3). 1 India to Philippines

Herpyza Sauvalle. Leguminosae (III 10). 1 W Cuba

Herrania Goudot. Sterculiaceae. 20 trop. S Am. Monkey-disp. in Amaz. *H. camargoana* R. Schultes (Amazonia) – ground seeds used as a condiment on meat

Herrea Schwantes = *Conicosia*

Herreanthus Schwantes. Aizoaceae (V). 1 Namaqualand: *H. meyeri* Schwantes

Herreria Ruíz & Pavón. Asparagaceae (Herreriaceae; Liliaceae s.l.). 7 S Am.

Herreriaceae Endl. = Asparagaceae

Herreriopsis Perrier. Asparagaceae (Herreriaceae; Liliaceae s.l.). 2 Madag.

Herrickia Wooton & Standley = *Aster*

Herschelia Lindley = *Forficaria*

Herschelianthe Rauschert = *Forficaria*

Hertia Less. (~ *Othonna*). Compositae (Sen.-Sen.). 10 SW As., N & S Afr.

Herya Cordemoy = *Pleurostylia*

Hesiodia Moench = *Sideritis*

Hesperaloe Engelm. Agavaceae. 3 SW N Am. R: CSM 23(1978)56. Cult. orn. stemless herbs forming grassy clumps

Hesperantha Ker-Gawler. Iridaceae (IV 3). 65–70 subSaharan Afr. R: SAJB 50(1984)15. Cormous herbs like *Ixia*; some cult. orn.

Hesperelaea A. Gray. Oleaceae. 1 NW Mexico

Hesperethusa M. Roemer = *Naringi*

Hesperevax (A. Gray) A. Gray (~ *Filago*). Compositae (Gnap.-Gnap.). 3 W US. R: SB 17(1992)293

Hesperhodos Cockerell = *Rosa*

Hesperidanthus (Robinson) Rydb. = *Schoenocrambe*

Hesperis L. Cruciferae. 25 Eur. (13), Medit. to Iran, C As., W China. R: FR 84(1973)259. Biennial or short-lived perennial herbs; *H. matronalis* L. (dame's violet or d.'s rocket, (sweet) r., damask (v.), C & S Eur., natur. N Eur. & Am.) – old-fashioned garden pl. with fragrant fls, seeds crushed to give an oil

Hesperocallidaceae Traub = Agavaceae

Hesperocallis A. Gray. Agavaceae (Funkiaceae, Liliaceae s.l.). 1 SW US deserts: *H. undulata* A. Gray – bulbous herb with fragrant fls, cult. orn.

Hesperochiron S. Watson. Hydrophyllaceae. 2 SW N Am. Cult. orn. stemless herbs

Hesperochloa (Piper) Rydb. = *Festuca*

Hesperocnide Torrey. Urticaceae (I). 2 Calif., Hawaii

Hesperodoria E. Greene (~ *Haplopappus*). Compositae (Ast.-Sol.). 2 SW US. R: Phytol.

71(1991)245

Hesperogreigia Skottsb. = *Greigia*

Hesperolaburnum Maire. Leguminosae (III 31). 1 Morocco

Hesperolinon (A. Gray) Small. Linaceae. 12 Calif., 1 extending to Oregon & Mex. R: UCPB 32(1961)235

Hesperomannia A. Gray. Compositae (Mut.-Mut.). 4 Hawaii. Trees; only Mut. in Hawaii

Hesperomecon E. Greene (~ *Platystigma*). Papaveraceae (III). 1 W N Am.: *H. linearis* (Benth.) E. Greene. R: UKSB 47(1987)25

Hesperomeles Lindley. Rosaceae (III). 11 C Am. to Peru. R: CJB 68(1990)2230

Hesperonia Standley = *Mirabilis*

Hesperopeuce (Engelm.) Lemmon (~ *Tsuga*). Pinaceae. 1 W N Am.: *H. mertensiana* (Bong.) Rydb. R: NRBGE 45(1988)387

Hesperoscordum Lindley = *Muilla*

Hesperoseris Skottsb. = *Dendroseris*

Hesperothamnus Brandegee = *Millettia*

Hesperoxiphion Baker. Iridaceae. 5 Andes, Colombia. R: BN 132(1979)466

Hesperoyucca (Engelm.) Baker = *Yucca*

Hesperozygis Epling. Labiatae (VIII 2). 8 Mex. to Brazil

Hessea Herbert. Amaryllidaceae (Liliaceae s.l.). 14 S Afr. R: BJ 107(1985)37

Hetaeria Blume. Orchidaceae (III 3). 27 OW trop.

Heterachne Benth. Gramineae (31d). 3 N Aus. R: HIP (1935) t.3283

Heteracia Fischer & C. Meyer. Compositae (Lact.-Crep.). 2 SW As. to China

Heteradelphia Lindau. Acanthaceae. 1 São Tomé (W Afr.)

Heteranthelium Hochst. ex Jaub. & Spach. Gramineae (23). 1 Turkey to Pakistan

Heteranthemis Schott (~ *Chrysanthemum*). Compositae (Anth.-Anth.). 1 SW Eur., N Afr.: *H. viscidehirta* Schott – cult. orn. annual

Heteranthera Ruíz & Pavón. Pontederiaceae. 12 trop. & warm Afr., Am. extending to N Am. Some with merely submerged linear lvs, others with orbicular floating ones, some with both. Cult. orn. in aquaria

Heteranthia Nees & C. Martius. Solanaceae (?, 2). 1 Brazil

Heteranthoecia Stapf. Gramineae (35). 1 trop. Afr.

Heteraspidia Rizz. = *Justicia*

Heteroaridarum Hotta. Araceae (VI 1). 1 Sarawak: *H. borneense* Hotta – rheophyte

Heteroarisaema Nakai = *Arisaema*

Heterocalycium Rauschert = *Cuspidaria*

Heterocarpha Stapf & C. Hubb. = *Drake-Brockmania*

Heterocaryum A. DC = *Lappula*

Heterocentron Hook. & Arn. Melastomataceae. 6 Mexico & C Am. A dimorphic, some merely attractants for insects. Some cult. orn. incl. some aggressive spp. e.g. *H. subtriplinervium* (Link & Otto) A. Braun (*H. macrostachyum*, Mex.), widely natur. in trop.

Heterochaenia A. DC. Campanulaceae. 3 Masc. Pachycaul treelets

Heterochiton Graebner & Mattf. = *Herniaria*

Heterocodon Nutt. Campanulaceae. 2 China (*Homocodon*). 1 W N Am.

Heterocoma DC. Compositae (Vern.-Vern.). 1 Brazil

Heterocondylus R. King & H. Robinson (~ *Eupatorium*). Compositae (Eup.-Ayap.). 13 C & S Am., esp. Brazil. R: MSBMBG 22(1987)204

Heterocypsela H. Robinson. Compositae (Vern.-Vern.). 1 Brazil

Heterodendrum Desf. = *Alectryon*

Heteroderis (Bunge) Boiss. Compositae (Lact.-Crep.). 1 SW & C As. to Pakistan

Heterodraba E. Greene. Cruciferae. 2 W N Am.

Heterogaura Rothr. Onagraceae. 1 W US

Heterogonium C. Presl (~ *Tectaria*). Dryopteridaceae (I 3). 20 Mauritius, SE As., Mal. R: Kalikasan 4(1975)205. Form bybrids with *T.* spp.

Heterolamium C.Y. Wu. Labiatae (VIII). 1 China

Heterolepis Cass. Compositae (Arct.). 3 Cape to Karoo

Heterolobium Peter = *Gonatopus*

Heteromeles M. Roemer (~ *Photinia*). Rosaceae (III). 1 Calif.: *H. arbutifolia* (Lindley) M. Roemer (toyon, tollon) – char. of chaparral, used like holly in decorations & origin of name Hollywood, Calif.

Heteromera Pomel (~ *Chrysanthemum*). Compositae (Anth.-Mat.). 1 N Afr.

Heteromma Benth. Compositae (Ast.-Ast.). 3 S Afr. mts

Heteromorpha Cham. & Schldl. Umbelliferae (III 8). 8 trop. & S Afr. to Yemen. Trees. *H.*

arborescens (Sprengel) Cham. & Schldl. (*H. trifoliata*) – tree, shrubby or subherbaceous forms, locally medic.

Heteropanax Seemann. Araliaceae. 2 India, S China

Heteropappus Less. Compositae (Ast.-Ast.). c.20 C & E As. Some cult. orn. bienn. herbs

Heteropetalum Benth. Annonaceae. 1 Venez., Brazil: *H. brasiliense* Benth. R: AHB 10(1930)73

Heteropholis C. Hubb. Gramineae (40h). 5 C Afr. to Aus. R: GBS 36(1983)137

Heterophragma DC. Bignoniaceae. 2 India, SE As.

Heterophyllaea Hook.f. Rubiaceae (I 1/IV 24). (Incl. *Teinosolen*) 8 Bolivia, Arg.

Heteroplexis C.C. Chang. Compositae (Ast.-Ast.). 2 China

Heteropogon Pers. Gramineae (40g). 6 trop. & warm Afr., S Eur. (1) etc. *H. contortus* (L.) Roemer & Schultes (trop.& warm) – good grazing when young but awns with hygroscopic action painful to stock & man, fire-climax veg. of much of Madag.

Heteropsis Kunth. Araceae (III 3). 13 trop. S Am. *H. spruceanum* Schott (Brazil) – aerial roots for lashing poles in house construction

Heteropteris Fée = *Neurodium*

Heteropterys Kunth (*Heteropteris* Kunth). Malpighiaceae. 120 Mex. to Arg., W Afr. (1). Fr. a samara. Some locally used fibres

Heteroptilis E. Meyer ex Meissner = *Dasispermum*

Heteropyxidaceae Engl. & Gilg = Myrtaceae

Heteropyxis Harvey. Myrtaceae (Lept.; Heteropyxidaceae). 3 C & S Afr. R: MBSM 10(1971)222

Heterorhachis Schultz-Bip. ex Walp. Compositae (Arct.-Gort.). 1 SW Cape: *H. aculeata* (Burm.f.) Roessler – cult. orn. shrub

Heterosciadium Lange ex Willk. = *Daucus*

Heterosmilax Kunth (~ *Smilax*). Smilacaceae. 11 E As., SE As., W Mal. R: Brittonia 36(1984)184

Heterospathe R. Scheffer. Palmae (V 4m). 32 C & E Mal., Papuasia (New Guinea 16), Micronesia. Monoec., unarmed feather-palms. Some cult. orn. incl. *H. elata* R. Scheffer (Philippines, Ambon) – fr. chewed like betel, petioles & lvs used for basketry & hat-making etc.

Heterosperma Cav. Compositae (Helia.-Cor.). 5 SW US to C Am.

Heterostachys Ung.-Sternb. Chenopodiaceae (II 2). 2 C & S Am.

Heterostemma Wight & Arn. Asclepiadaceae (III 5). c. 40 Indomal., W Pacific

Heterostemon Desf. Leguminosae (I 4). 7 trop. Am. esp. Upper Amazon. R: PKNAW, C 79(1976)42. Extrafl. nectaries

Heterostipa (Elias) Barkworth (~ *Stipa*). Gramineae (16). 4 N Am. R: Phytol. 74(1993)15

Heterothalamus Less. Compositae (Ast.-Ast.). 8 S Am.

Heterotheca Cass. Compositae (Ast.-Sol.). c.25 S N Am. Cult. orn. but some referred to *Chrysopsis* & *Pityopsis*

Heterotis Benth. = *Dissotis*

Heterotoma Zucc. Campanulaceae. 1 Mex. & C Am.: *H. lobelioides* Zucc. (Mexico) – cult. orn. R: SB 15(1990)296

Heterotrichum DC. Melastomataceae. 10 trop. Am. Cult. orn. shrubs, some with ed. fr.

Heterotristicha Tobler = *Tristicha*

Heterotropa Morren & Dcne = *Asarum*

Heterozostera (Setch.) Hartog (~ *Zostera*). Zosteraceae. 1 coasts of temp. Aus. & Chile to 31 m deep: *H. tasmanica* (Asch.) Hartog

Heuchera L. Saxifragaceae. 55 N Am. (E, 7). R: MSPS 2(1936)1. Cult. orn. tufted pls (alum-root) for ground-cover, esp. *H. sanguinea* Engelm. (coralbells, SW N Am.) & its hybrids

× **Heucherella** Wehrh. Saxifragaceae. *Heuchera* × *Tiarella*. Hybrids known only in cult. esp. × *H. tiarelloides* (Lemoine) Wehrh. ex Stearn (*Heuchera* × *brizoides* Hort. ex Lemoine (garden hybrids involving *H. sanguinea* Engelm. & other spp.) × *Tiarella cordifolia* L.) – sterile

Hevea Aublet. Euphorbiaceae. 12 Amazon basin. Fr. explodes disp. seeds to 20+ m, seeds water-disp. but mainly destroyed by fish, e.g. piranhas at floodtime. *H. brasiliensis* (A. Juss.) Muell. Arg. (Pará rubber) – domatia; seeds stay afloat up to 2 months; trunk the source of best natural rubber (so-named because it would rub out pencil-marks) & most of that planted in OW esp. Mal. Latex tapped by making sloping incisions in bark & exudate collected in suspended cups, coagulated with acid & pressed into sheets etc. – literally thousands of uses; the last 'tappings' or waste called almeidina in Angola; old logs used for chipboard, plywood & furniture in Thailand & Sri Lanka, now exported to Eur. as chopping-boards etc. The OW provenances selected from Amazonia some of the best known but reduced to a small number of successful trees when introduced via Kew

& Sri Lanka to Mal., where Tamils imported from India as tappers
Hewardia J. Sm. = *Adiantum*
Hewittia Wight & Arn. Convolvulaceae. 1 OW trop.: *H. malabarica* (L.) Suresh (*H. sublobata*)
Hexachlamys O. Berg (~ *Eugenia*). Myrtaceae (Myrt.). 4 temp. S Am.
Hexacyrtis Dinter. Colchicaceae (Liliaceae s.l.). 1 SW Afr.
Hexadesmia Brongn. = *Scaphyglottis*
Hexaglottis Vent. Iridaceae (III 2). 6 W Cape. R: AMBG 74(1987)542
Hexalectris Raf. Orchidaceae (V 11). 7 S US, Mex. Mycotrophs
Hexalobus A. DC. Annonaceae. 4 trop. & S Afr., Madagascar
Hexaneurocarpon Dop = *Fernandoa*
Hexapora Hook.f. = *Micropora*
Hexaptera Hook. = *Menonvillea*
Hexapterella Urban. Burmanniaceae. 2 trop. S Am. R: AMBG 76(1989)956
Hexaspermum Domin = ? *Phyllanthus*
Hexaspora C. White. Celastraceae. 1 N Queensland
Hexastemon Klotzsch = *Eremia*
Hexastylis Raf. = *Asarum*
Hexatheca C.B. Clarke. Gesneriaceae. 3 Borneo. R: NRBGE 46(1989)54
Hexisea Lindley = *Scaphyglottis*
Hexopetion Burret = *Astrocaryum*
Hexuris Miers = *Peltophyllum*
Heynea Roxb. ex Sims (~ *Trichilia*). Meliaceae (I 4). 2 India & S China to W Mal. R: FM 1,12(1995)41
Heynella Backer. Asclepiadaceae (III 4). 1 Java: *H. lactea* Backer – rare
Heywoodia Sim. Euphorbiaceae. 1 E Afr., S Afr. (disjunct): *H. lucens* Sim (Cape ebony)
Heywoodiella Svent. & Bramw. = *Hypochaeris*
hiba *Thujopsis dolabrata*
Hibbertia Andrews. Dilleniaceae. c. 115 Madag. (1), Mal. (2), Aus. (c. 110), New Caled., Fiji (1). Mostly ericoid or climbing shrubs, some with phylloclades. A varied, from 200 not in obvious groups to ∞ in 15 bundles to A 1. Some cult. orn. with yellow fls (guinea-fl.). *H. conspicua* (Harvey) Gilg (W Aus.) – buzz-poll. by bees; *H. scandens* (Willd.) Hoogl. (snake climber, Aus.) – cult.
Hibiscadelphus Rock (~ *Hibiscus*). Malvaceae. 7 Hawaii, extinct (4) or endangered. R: Novon 5(1995)183. Trees & shrubs. *H. wilderianus* Rock extinct & only 1 tree of it ever known; *H. giffordianus* Rock reduced to 1 tree by 1930 & known now only in cult.
Hibiscus L. Malvaceae. c. 300 warm temp. (Eur. 2) to trop. (excl. *Ablemoschus*). Fls white to red, yellow or even bluish, usu. with basal maroon spots on petals & extrafl. nectaries (even in originally ant-less Hawaii), differing from *Abelmoschus* (okra etc.) in thick K not longit. split nor connate with bases of C & A to fall as a unit; alks. Fibres, medic. & many cult. orn., (rose) mallows – *H. cannabinus* L. (kenaf, Ambari or Deccan hemp, Bimlipatum jute, trop. Afr., long cult. India & SE Eur.) – fibre like jute, seed-oil for illumination in Afr.; *H. elatus* Sw. (blue mahoe, Cuban bast, Jamaica & Cuba) – bark-fibre for ropes & hat-making, timber for gunstocks, cabinet-making etc.; *H. macrophyllus* Roxb. ex Hornem. (India to Java) – fibre-pl. in SE As., timber for building; *H. mutabilis* L. (Confederate rose, China); *H. rosa-sinensis* L. (China rose, shoe-flower, unknown in wild & poss. anc. hybrid involving several spp., but widely cult. trop. & warm & under glass in temp.) – many cvs, fls used for shining shoes in India; *H. sabdariffa* L. (roselle, Jamaica sorrel, ? trop. Afr., but now natur. pantrop.) – fibre used for rope, fleshy red calyx used in drinks, jellies, lvs used like spinach; *H. schizopetalus* (Dyer) Hook.f. (trop. E Afr.) – petals deeply laciniate, poss. an old cv. or parent of *H. rosa-sinensis*; *H. syriacus* L. (rose-of-Sharon (Am.), E As.) – many cvs, the common hibiscus of gardens (temp.); *H. tiliaceus* L. (trop.) – leaf starch like sago, imp. fibre for cordage, mats, sails, nets, exported to Eur., firewood & erosion control
Hickelia A. Camus. Gramineae (1b). 1 Madag.: *H. madagascariensis* A. Camus – climber
Hickenia Lillo = *Oxypetalum*
hickory *Carya* spp.; **Australian h.** *Acacia implexa*; **h. elm** *Ulmus thomasii*
Hicksbeachia F. Muell. Proteaceae. 2 NE Aus. R: Telopea 3(1988)231. *H. pinnatifolia* F. Muell. (red boppel nut, rose nut) – seed ed.
Hicoria Raf. = *Carya*
Hicriopteris C. Presl = *Dicranopteris*
Hidalgoa Llave & Lex. Compositae (Helia.-Cor.). 6 Mex., C Am.
Hieracium L. Compositae (Lact.-Hier.). c. 90 (c. 1000 microspp., excl. *Pilosella*); *H. umbellatum* L. (N temp.) diploid sexual, rest being triploid or tetraploid apomicts, temp. (excl.

Aus.), trop. mts. Hawkweeds. Allied to *Crepis* & linked to it via S Am. spp.

Hieris Steenis. Bignoniaceae. 1 Penang (Malay Pen.)

Hiernia S. Moore. Scrophulariaceae. 1 Angola, Namibia

Hierobotana Briq. Verbenaceae. 1 Ecuador, Peru

Hierochloe R. Br. (~ *Anthoxanthum*). Gramineae (21c). 20 temp. (Eur. 7) exc. Afr., Arctic, trop. mts. Close to *Anthoxanthum, H. odorata* (L.) Wahlenb. (holy grass, N temp.) with similar strong scent of coumarin, form. strewn on church floors (introd. to Scotland by Prussian monks for this; cf. *Myrrhis*), burned as incense in New Mexico & used to scent clothes; *H. horsfieldii* (Kunth) Maxim. (Java) – good fodder, form. for stock of princes in Java

Hieronima Allemão = *Hyeronima*

Hieronymiella Pax. Amaryllidaceae (Liliaceae s.l.). 1 Arg.

Hieronymusia Engl. = *Suksdorfia*

Hilaria Kunth. Gramineae (33c). 10 S US to Guatemala. R: JWAS 46(1956)311. Some cult. incl. *H. mutica* (Buckley) Benth. (tobosa grass, SW US)

hildaberry See *Rubus*

Hildebrandtia Vatke ex A. Braun. Convolvulaceae. 9 Afr., Madag. 2 sepals accrescent in fr.

Hildegardia Schott & Endl. Sterculiaceae. 8 Cuba (1), OW trop.

Hildewintera Ritter = *Cleistocactus*

Hillebrandia Oliver. Begoniaceae. 1 Hawaii. P 10 in female fls

Hilleria Vell. Conc. Phytolaccaceae (II). 3 S Am., 1 ext. to Afr., Madag. & Masc.

Hillia Jacq. Rubiaceae (I 3). 24 trop. Am. Epiphytic shrubs with solit. term. fls & alks

Hilliardia R. Nordenstam (~ *Matricaria*). Compositae (Anth.-Mat.). 1 Natal: *H. zuurbergensis* (Oliver) R. Nordenstam. R: OB 93(1987)147

Hilliella (O. Schulz) Zhang & Li = *Cochlearia*

Himalayacalamus Keng f. = *Thamnocalamus*

Himalrandia Yamaz. Rubiaceae (II 4). 3 Himalaya

Himantandraceae Diels. Dicots – Magnoliidae – Magnoliales. 1/2 or 3 E Mal. to N Aus. Large aromatic trees with alks, young parts densely covered with fimbriate peltate scales. Lvs spirally arr. (distichous), simple, entire, gland-dotted; stipules 0. Fls large, bisexual, solit. (2 or 3) in axils; K (2), calyptrate, 1 enclosing other (prob. bracts), C 3–23 (prob. staminodes) spirally arr., linear, A 13–130, spirally arr., not differentiated into filament & anther, 1 pair of sporangia on each side, each with single longit. slit, c. 13–22 spirally arr. staminodes between A & G, pollen-grains monosulcate, G (6)7–10(–28) spirally arr. closed, each with ovary & style, weakly connate (more fully in fr.), with 1(2) pendulous, anatropous ovules. Fr. gall-like, a syncarp of coalesced carpels; seeds small with copious, oily non-ruminate endosperm. 2n = 24

Only genus: *Galbulimima*

Himantochilus Anderson ex Benth. = *Anisotes*

Himantoglossum Koch. Orchidaceae (IV 2). 2 Eur. (1: *H. hircinum* (L.) Sprengel, lizard orchid), N Afr., E Med. *H. hircinum* visited by solit. bees but otherwise unattractive to insects, distr. slowly spreading northwards in GB (? warming)

Himantostemma A. Gray = *Matelea*

Himatanthus Willd. ex Schultes (~ *Plumeria*). Apocynaceae. 13 S Am. R: Bradea 5, suppl. (1991)

Hindsia Benth. ex Lindley. Rubiaceae (I 1/IV 24). 8 trop. S Am.

hinoki *Chamaecyparis obtusa*

Hinterhubera Schultz-Bip. ex Wedd. Compositae (Ast.-Ast.). 9 Andes

Hintonella Ames. Orchidaceae (V 10). 1 Mexico

Hintonia Bullock. Rubiaceae (inc. sed.). 4 C Am. *H. latifolia* (DC) Bullock – used for malaria & other fevers (quinine)

Hionanthera Fernandes & A. Diniz. Lythraceae. 1–2 S Tanzania, Mozambique, Zimbabwe

Hippeastrum Herbert. Amaryllidaceae (Liliaceae s.l.). 76 trop. Am., W Afr. (1). Cult. orn. bulbous pls ('amaryllis') esp. hyrbids of complicated parentage involving Am. spp. esp. *H. aulicum* (Ker-Gawler) Herbert, *H. elegans* (Sprengel) H. Moore, *H. puniceum* (Lam.) Urban (*H. equestre*, Barbados or fire lily, perhaps the true *Amaryllis belladonna*), *H. reginae* (L.) Herbert (also in W Afr.), *H. reticulatum* (L'Hérit.) Herbert, *H. striatum* (Lam.) H. Moore (*H. rutilum*), much cult. for selling at Christmas as dormant bulbs to grow

Hippeophyllum Schltr. Orchidaceae (V 7). 6 Mal.

Hippia L. Compositae (Anth.-Mat.). 8 SW & S Cape. R: KB 1918:176

Hippobroma G. Don f. (~ *Laurentia*). Campanulaceae. 1 WI, natur. pantrop.: *H. longiflora* (L.) G. Don f. – toxic latex

Hippobromus Ecklon & Zeyher. Sapindaceae. 1 S Afr.

Hippocastanaceae A. Rich. (~ Sapindaceae). Dicots – Rosidae – Sapindales. 2/15 N temp. to N S Am. & SE As. Trees & shrubs. Lvs opp., 3–11-palmate; stipules 0. Infl. term. thyrses or racemes. Fls large, irreg., bisexual (exc. esp. some apical ones male); K (5), almost distinct in *Billia*, C 4 or 5, unequal, clawed, imbricate, disk small, extrastaminal & often 1-sided, A 5 + (0) 1–3, anthers dehiscing longit. G̲ ((2)3(4)), pluriloc. with term. style & simple stigma, 2 superposed, anatropous to orthotropous, bitegmic ovules per loc. Fr. a loculicidal capsule, usu. 1-loc. & 1-seeded by abortion; seed large with hard testa & v. large hilum (funicle held in placenta; obturator adnate to ovule) & large, curved often starchy ovule with 1 cotyledon larger than other, endosperm 0. x = 20. R (Am.): Brittonia 9(1957)145

Genera: *Aesculus* (decid.), *Billia* (evergreen)

V. close to Sapindaceae but palmate lvs (though some *Aesculus* mutants have pinnate lvs) & large seeds char. *Aesculus* (horse-chestnuts, buck-eyes) cult. orn., some timber etc.

Hippocratea L. Celastraceae. 100 trop. Twining shrubs s.t. used as 'rope' bridges in Afr.

Hippocrateaceae Juss. = Celastraceae

Hippocrepis L. Leguminosae (III 19). 30 Eur. (10), W As., Medit. R: Willdenowia 19(1989)59. Some fodders incl. *H. comosa* L. (horseshoe vetch, Eur.) – cult. orn. with scented fls esp. *H. emerus* (L.) Lassen (scorpion senna [jointed legumes suggesting scorpion tail & therefore sting cure – 'Doctrine of Sigantures'], Eur., Turkey)

Hippodamia Decne = *Solenophora*

Hippolytia Polj. (~ *Tanacetum*). Compositae (Anth.-Tan.). 19 C As. to N China. R: APS 17, 4(1978)70

Hippomane L. Euphorbiaceae. 5 Mex., WI. G 6–9-loc.; fr. a drupe; alks. *H. mancinella* L. (manchineel) – coastal tree cult. as windbreak, latex in eyes can blind (form. thought of like upas – death to sleepers below)

Hippomarathrum Link = *Cachrys*

Hippophae L. Elaeagnaceae. 3 temp. Euras. (Eur. 1). R: ABF 8(1971)177. Alks; actinomycete root-symbionts fixing nitrogen. *H. rhamnoides* L. (sea buckthorn, sallow thorn, Eur. to N China) – small dioec. tree or shrub of sandy coasts & shingle-banks etc. in mts, some rheophytic forms; bracteoles form hood over A in wet weather & separate on drying out so that pollen may be blown away; fr. ed. – sauce for meat or fish in Eur., with milk or cheese in C As.; wood suitable for turning; yellow dye; oil in cosmetics

Hippotis Ruíz & Pavón. Rubiaceae (inc. sed.). 12 trop. Am.

Hippuridaceae Vest. Dicots – Asteridae – Callitrichales. 1/1 cosmop. Aquatic perennial herb with symp. rhiz. & erect emergent stems, storing carbohydrate as stachyose; vasc. system of stems an axile strand of xylem surrounded by phloem, cortex with a system of intercellular spaces. Lvs small, in whorls of (4–)6–12(–16), linear, entire; stipules 0. Fls wind-poll., small, solit., axillary, bisexual or s.t. at least some unisexual, when females above males; K (2–4), a lobed rim atop G, C 0, A 1 atop G, style lying between pollen-sacs, $\overline{G}$ pseudomonomerous with term. style stigmatic throughout length, 1-loc. with 1 apical, pendulous, anatropous, unitegmic ovule, the embryo with large haustorial suspensor. Fr. an achene or drupe with thin fleshy exocarp; embryo elongate, straight, endosperm thin, proteinaceous. X = (? 8 or) 16

Only species: *Hippuris vulgaris*

Form. allied with Haloragidaceae (Rosidae) but ovule studies etc. support placement in Asteridae (Cronquist), though some suggest Cornaceae as allied

Hippuris L. Hippuridaceae. 1 cosmop., with ecological races in Arctic & Baltic: *H. vulgaris* L. (mare's tail) – young lvs eaten by Esquimaux. The char. habit (*Hippuris* syndrome) seen in unrelated genera: *Decodon* & *Rotala* (Lythraceae), *Elatine* (Elatinaceae), *Pogogyne* (Labiatae) etc. etc.

Hiptage Gaertner. Malpighiaceae. 20–30 trop. As. to Fiji. Fr. a 3-winged samara. Some cult. orn. incl. *H. benghalensis* (L.) Kurz (Indomal.) – cult. in trop. for fragrant fls, bonsai in Thailand, locally medic. (insecticidal)

Hiraea Jacq. Malpighiaceae. 40 trop. Am.

Hirpicium Cass. Compositae (Arct.-Gort.). 12 trop. & S Afr. R: MBSM 3(1959)333

Hirschfeldia Moench. Cruciferae. 2 Medit. (Eur. 1: *H. incana* (L.) Lagr.-Fossat (hoary mustard) – gynodioec.), Socotra

Hirschia Baker = *Iphiona*

Hirtella L. Chrysobalanaceae (4). 103 trop. Am., E Afr. & Madag. (1). Pink or purple butterfly-poll. dayfls (cf. *Couepia*). Bark of *H. americana* L. (cf. *Licania*) baked with clay gives heat-resistance to pottery cooking-pots; *H. carbonaria* Little (N S Am.) – charcoal; fresh fr. of *H. pendula* Lam. used as ear pendants on St Lucia

Hispaniella Braem = *Oncidium*

Hispidella Barnadez ex Lam. Compositae (Lact.-Hier.). 1 Iberian Pen.
Histiopteris (J. Agardh) J. Sm. Dennstaedtiaceae. 8 trop. Terr. *H. incisa* (Thunb.) J. Sm. – cult. orn.; in wild, lvs scramble or climb, the lamina to 3 m long on petiole to 2 m
Hitchcockella A. Camus. Gramineae (1a). 1 Madagascar
Hitchenia Wallich. Zingiberaceae. 3 Indomal. Some used like arrowroot & in glue
Hitcheniopsis (Baker) Ridley = *Scaphochlamys*
Hitoa Nad. = *Ixora*
Hladnikia Reichb. (~ *Grafia*). Umbelliferae (III 5). 1 Slovenia
ho wood *Cinnamomum camphora*
hoary cress *Lepidium draba*
hobblebush *Viburnum lantanoides*
Hochreutinera Krapov. Malvaceae. 2 Mexico, temp. S Am.
Hochstetteria DC = *Dicoma*
Hockinia Gardner. Gentianaceae. 1 E Brazil
Hodgkinsonia F. Muell. Rubiaceae (III 4). 1 E Aus.
Hodgsonia Hook.f. & Thomson. Cucurbitaceae. 1 Indomal.: *H. macrocarpa* (Blume) Cogn. (*H. heteroclita*) – inside of seeds eaten by Sikkimese; opium roasted in seed-oil
Hodgsoniola F. Muell. Anthericaceae (Liliaceae s.l.). 1 SW Aus.: *H. junciformis* (F. Muell.) F. Muell. R: FA 45(1987)306
Hoehnea Epling. Labiatae (VIII 2). 4 S Brazil to N Argentina
Hoehneella Ruschi. Orchidaceae (V 10). 2 Brazil
Hoehnelia Schweinf. = *Ethulia*
Hoehnephytum Cabera. Compositae (Sen.-Sen.). 3 Brazil
Hoffmannia Sw. Rubiaceae (IV 5). c. 45 Mex. to Arg. Some cult. orn. foliage plants under glass
Hoffmanniella Schltr. ex Lawalrée. Compositae (Helia.-Verb.). 1 C Afr., Cameroun
Hoffmannseggella H. Jones = *Laelia*
Hoffmannseggia Cav. Leguminosae (I 1). 28 SW US to Chile (25), S Afr. (3). Tubers of some SW US spp. ed. roasted
Hofmeisterella Reichb.f. Orchidaceae (V 10). 1 Ecuador
Hofmeisteria Walp. Compositae (Eup.-Hofm.). (Incl. *Carterothamnus*) 12 Mex. R: MSBMBG 22(1987)453
hog brake *Pteridium aquilinum, Ambrosia artemisiifolia*; **h. fennel** *Peucedanum officinale*; **h. gum** *Metopium toxiferum, Moronobea* spp.; **h. nut** *Carya glabra*; **h. plum** *Spondias* spp. esp. *S. mombin, Symphonia globulifera, Ximenia americana*; **h.weed** *Heracleum sphondylium*; **h.w., giant** *H. mantegazzianum*
Hohenackeria Fischer & C. Meyer. Umbelliferae (III 6). 2 E Eur. (2), Cauc., N Afr.
Hohenbergia Schultes f. Bromeliaceae (3). 41 trop. Am. R: FN 14(1979)1731. Some cult. orn. terrestrial or epiphytic
Hohenbergiopsis L.B. Sm. & Read. Bromeliaceae (3). 1 Guatemala, ? Mex.
Hoheria Cunn. Malvaceae. 5 NZ. R: Plantsman 5(1983)178. Cult. orn. small trees with heteroblasty & white fls esp. *H. lyallii* Hook.f. & *H. populnea* Cunn. (bark for cordage, wood for cabinet-making, locally medic.)
Hoita Rydb. = *Orbexilum*
Holacantha A. Gray = *Castela*
Holalafia Stapf = *Alafia*
Holarrhena R. Br. Apocynaceae. 4 trop. Afr., Indomal. R: MLW 81-2(1981). Many alks. Decid. trees & shrubs, imp. locally medic., cult. orn. esp. *H. pubescens* (Buch.-Ham.) G. Don.f. (*H. antidysenterica*, conessi, kurchi, Indomal.) – bark (Tellichery b.) effective against dysentery, (*H. febrifuga*, kumbanzo, trop. Afr.) – febrifuge, timber for spoons etc.
Holboellia Wallich (~ *Stauntonia*). Lardizabalaceae. 10 Himal., China. K petaloid, C nectaries. Cult. orn. monoec. lianes with fleshy indehiscent fr. (some ed.) & black seeds
Holcoglossum Schltr. Orchidaceae (V 16). 9 Taiwan, S Japan. R: NRBGE 44(1987)251
Holcolemma Stapf & C. Hubb. (~ *Setaria*). Gramineae (34b). 4 E Afr. & India, Aus. R: KB 32(1978)773
Holcosorus T. Moore = *Selliguea*
Holcus L. Gramineae (21b). 8 Eur. (8), Medit. to Middle E. Intergrading with *Deschampsia*. *H. lanatus* L. (Yorkshire fog, creeping soft grass, Euras., natur. N Am. etc.) – pasture-grass
hold-me-tight *Achyranthes indica*
Holigarna Buch.-Ham. ex Roxb. Anacardiaceae (III). 8 Indomal. $\overline{G}$. Some timber
Hollandaea F. Muell. Proteaceae. 2 E Aus.
Hollermayera O. Schulz. Cruciferae. 1 Chile

Hollisteria S. Watson. Polygonaceae (I 1). 1 C Calif.: *H. lanata* S. Watson. R: Phytol. 66(1989)210

Hollrungia Schumann. Passifloraceae. 1 E Mal., Papuasia

holly *Ilex* spp. esp. *I. aquifolium* & *I.* × *altaclerensis*; **American h.** *I. opaca*; **h. fern** *Polystichum* spp.; **h. grape** *Mahonia repens*; **Japanese h.** *I. crenata*; **h. oak** *Quercus ilex*; **sea h.** *Eryngium maritimum*. See also *Heteromeles*

hollyhock *Alcea rosea*

holm (e) = holly

Holmbergia Hicken. Chenopodiaceae (I 2). 1 Uruguay, Paraguay, Arg.: *H. tweedii* (Moq.) Speg.

Holmesia Cribb = *Angraecopsis*

Holmgrenanthe Elisens. Scrophulariaceae (Antirr.). 1 SW N Am.

Holmskioldia Retz. Labiatae (V; Verbenaceae). 1 Himal.: *H. sanguinea* Retz. (Chinese hat plant, cup-and-saucer plant) – orn. with red fls. now pantrop. cult. See also *Karomia*

Holocalyx M. Micheli. Leguminosae (III 2). 1 trop. S Am.: *H. balansae* M. Micheli. Timber

Holocarpa Baker = *Pentanisia*

Holocarpha E. Greene. Compositae (Hele.-Mad.). 4 California

Holocheila (Kudô) S. Chow (~ *Teucrium*). Labiatae (IV). 1 SW China

Holocheilus Cass. Compositae (Mut.-Nass.). 6 S Brazil, Paraguay, Uruguay, N & C Arg. R: RMLP n.s. 11(1968)1. Perennial herbs

Holochlamys Engl. Araceae (III 4). 1 New Guinea

Holodictyum Maxon = *Asplenium*

Holodiscus (K. Koch) Maxim. Rosaceae. 8 W N Am. to Colombia. R: BTBC 70(1943)275. Cult. orn. shrubs

Holographis Nees. Acanthaceae. 10 Mex., arid & semi-a. R: JAA 64(1983)129

Hologyne Pfitzer = *Coelogyne*

Hololachna Ehrenb. (~ *Reaumuria*). Tamaricaceae. 2 C As.

Hololeion Kitam. (~ *Hieracium*). Compositae (Lact.-Hier.). 3 E As.

Hololepis DC = *Vernonia*

Holopogon Komarov & Nevski = *Neottia*

Holoptelea Planchon. Ulmaceae (I). 2 trop. Afr. (1), India (1). Timber

Holopyxidium Ducke = *Lecythis*

Holostachyum (Copel.) Ching = *Aglaomorpha*

Holostemma R. Br. Asclepiadaceae (III 1). 1–2 Indomal., China

Holosteum L. Caryophyllaceae (II 1). 3–4 temp. Euras. (Eur. 1). Mouse-ear

Holostyla DC = *Caelospermum*

Holostylis Duchartre. Aristolochiaceae (II 2). 1 C & SE Brazil: *H. reniformis* Duchartre

Holostylon F. Robyns & Lebrun. Labiatae (VIII 4c). 3–4 trop. Afr.

Holothrix Rich. ex Lindley. Orchidaceae (IV 2). 55 trop. & S Afr., Arabia

Holozonia E. Greene (~ *Lagophylla*). Compositae (Hele.-Mad.). 1 W US

Holstia Pax = *Neoholstia*

Holstianthus Steyerm. Rubiaceae (I 5). 1 Guayana Highland: *H. barbigularis* Steyerm. 11 other R. genera restricted to G. H.

Holtonia Standley = *Elaeagia*

Holttumiella Copel. = *Taenitis*

Holtzea Schindler = *Desmodium*

Holubia Oliver. Pedaliaceae. 1 S Afr.

Holubia Löve & D. Löve = *Gentiana*

Holubogentia Löve & D. Löve = *Gentiana*

holy basil *Ocimum tenuiflorum*; **h. clover** *Onobrychis viciifolia*; **h. flax** *Santolina* spp.; **h. grass** *Hierochloe odorata*; **h. thistle** *Silybum marianum*

Holzneria Speta (~ *Chaenorhinum*). Scrophulariaceae (Antirr.). 2 SW As. R: BJ 103(1982)16

Homalachne Kuntze = *Holcus*

Homalanthus A. Juss. (*Omalanthus*). Euphorbiaceae. 35 trop. As. to Aus. *H. populifolius* Graham (New Guinea, trop. Aus.) – widely cult. & natur., source of black dye for staining rattan goods etc.

Homalium Jacq. Flacourtiaceae (5). c. 180 trop. & warm (Afr. (R: BJBB 43(1973)239) & Madag. 59, Mal. 23, Am. 3). K &/or C accrescent & forming wings in fr. *H. guillainii* (Vieill.) Briq. (New Caled.) – accumulates nickel to 14% leaf ash dry weight & carries a moss which accumulates chromium to 5000 µg/g (20 times that in itself). Some timbers for building, boats etc., *H. tomentosum* (Vent.) Benth. (SE As., Mal.) – source of Moulmein lancewood

Homalocalyx F. Muell. Myrtaceae (Lept.). (Incl. *Wehlia*) 11 Aus. R: Brunonia 10(1987)139

Homalocarpus Hook. & Arn. (~ *Bowlesia*). Umbelliferae (I 2a). 6 Chile. R: UCPB 38(1965)58

Homalocheilos J.K. Morton = *Isodon*

Homalocladium (F. Muell.) L. Bailey = *Muehlenbeckia*

Homalodiscus Bunge ex Boiss. = *Ochradenus*

Homalomena Schott. Araceae (VI 1). 140 trop. As. & Am. Some cult. orn. & locally used poisons & medic. etc. *H. peekelii* Engl. (New Guinea) – scented pl. worn round neck in New Ireland

Homalopetalum Rolfe. Orchidaceae (V 13). 4 Jamaica, Cuba

Homalosciadium Domin. Umbelliferae (I 1a). 1 SW Aus.: *H. homalocarpum* (F. Muell.) H. Eichler

Homalosorus Small ex Pichi-Serm. = *Diplaziopsis*

Homalospermum Schauer (~ *Leptospermum*). Myrtaceae (Lept.). Myrtaceae. 1 SW Aus. coastal swamplands: *H. firmum* Schauer. R: Telopea 2(1983)381

Homeria Vent. Iridaceae (III 2). (Incl. *Sessilistigma*) 32 S Afr. R: AMBG 68(1981)413. Fly- & bee-poll. cult. orn. pls like *Ixia* with tunicated corms esp. *H. breyniana* (L.) G. Lewis (*H. collina*); some with bulbils in axils of lower lvs. Many noxious weeds in Aus.; homeridin with digitalis-like effects on heart

hominy *Zea mays*

Homochroma DC = *Mairia*

Homocodon Hong = *Heterocodon*

Homoglossum Salisb. = *Gladiolus*

Homognaphalium Kirpiczn. (~ *Gnaphalium*). Compositae (Gnap.-Gnap.). 1 N Afr.: *H. pulvinatum* (Del.) Fayed & Zareh. R: OB 104(1991)149

Homogyne Cass. (~ *Petasites*). Compositae (Sen.-Tuss.). 3 Eur. mts. *H. alpina* (L.) Cass. (C Eur.) – cult. orn., natur. in GB, lactones insect-antifeedant

Homolepis Chase. Gramineae (34b). 3 trop. Am.

Homollea Arènes. Rubiaceae (? II 2). 3 Madagascar

Homolliella Arènes. Rubiaceae (? II 2). 1 Madagascar

Homonoia Lour. Euphorbiaceae. 2 SE As., Mal. Rheophytes. *H. riparia* Lour. planted as erosion retardant in Sumatra, rope made from bark in S China, skin-medic. in Mal.

Homopholis C. Hubb. Gramineae (34d). 2 E Aus.

Homopogon Stapf = *Trachypogon*

Homoranthus Cunn. ex Schauer. Myrtaceae (Lept.). (Incl. *Rylstonea*) 7 E Aus. R: AusSB 4(1991)19. Essential oil suggested for scent-making; some cult. orn. shrubs

Homozeugos Stapf. Gramineae (40b). 5 trop. Afr. R: GOB 1(1973)11

Honckenya Ehrh. (*Honkenya*). Caryophyllaceae (II 1). 1–2 N temp. (Eur. 1), circumpolar, S Patagonia. R: NSL 20(1960)142. *H. peploides* (L.) Ehrh. (sea purslane) – sandy coasts

honesty *Lunaria annua*

honewort *Trinia glauca*

honey flower *Melianthus major, Lambertia formosa*; **h. locust** *Gleditsia* spp.; **h. palm** *Jubaea chilensis*; **h.suckle** *Lonicera* spp.; **African h.s.** *Halleria lucida*; **Cape h.s.** *Tecoma capensis*; **coral h.s.** *L. sempervirens*; **fly h.s.** *L. xylosteum*; **French h.s.** *Hedysarum coronarium*; **Jamaica h.s.** *Passiflora laurifolia*; **swamp h.s.** *Rhododendron viscosum*; **trumpet h.s.** *L. sempervirens*; **h. wort** *Cerinthe major*

Honolulu rose *Clerodendrum chinense*

Hoodia Sweet ex Decne. Asclepiadaceae (III 5). 13 SW trop. & S Afr. R: Excelsa 7(1977)75. Cult. orn. succ. with flat or cup-shaped, almost lobeless C

× **Hoodiopsis** Lückh. Asclepiadaceae. Natural hybrid between *Hoodia* & *Stapelia* spp.

Hookerochloa Alexeev = *Austrofestuca*

hoop pine *Araucaria cunninghamii*

hop *Humulus lupulus*; **h.-bush** *Dodonaea* spp.; **h. clover** *Medicago lupulina*; **h. hornbeam** *Ostrya* spp.; **h. tree** *Ptelea trifoliata*; **h. trefoil** *Trifolium campestre*

Hopea Roxb. (~ *Shorea*). Dipterocarpaceae. 102 Indomal. Resin of *H. papuana* Diels (New Guinea) coll. megachilid bees & inhibits growth of pollen-associated fungi. Some form. imp. as sources of resin (damar mata kuching) used in linoleum & paints; some timbers (giam) esp. *H. odorata* Roxb. (thingwa, SE As.) etc., this sp. also a street-tree, & *H. mengerawan* Miq. (merawan, Sumatra). Poll. by thrips; some spp. triploid & apomictic (emergent & understorey spp.). *H. centipeda* Ashton (Borneo) – rheophyte; *H. ponga* (Dennst.) Mabb. (*H. wightiana*, ilapongu, India) – shoots often with echinate galls initiated by *Mangalorea hopeae* Takagi so that larvae are protected by a spiny 'fruit' reminiscent of those in other fams (no spines in Dipterocarpaceae otherwise!), similar forming in other insect-attacked genera in this & allied fams

Hopkinsia W. Fitzg. Restionaceae. 2 SW Aus.

Hoplestigma Pierre. Hoplestigmataceae. 2 W C Afr.

Hoplestigmataceae Gilg. Dicots – Dilleniidae – Violales. 1/2 W C Afr. Trees with large spirally arr. simple lvs; stipules 0. Fls bisexual, reg. in term. bractless cymes; K globose irreg. splitting into 2–4 lobes, C (11–14) with short tube, lobes imbricate in 2–4 irreg. series, A ± ∞ (c. 20–35) in c. 3 irreg. series attached to base of C, anthers with longit. slits, G (2), 1-loc. with 2 intruded forked parietal placentas & deeply bifid style & 2 pendulous, anatropous, unitegmic ovules per placenta. Fr. a drupe with leathery pericarp & hard endocarp; seeds with nearly straight embryo with elongate hypocotyl & expanded cotyledons but scanty endosperm

Only genus: *Hoplestigma*

Infl., C, G & ovules suggest affinity with Boraginaceae but A ∞ aberrant in Asteridae; C, A & no. of ovules suggest Ebenaceae but parietal placentation unknown there. Placed in Violales though unitegmic ovules aberrant (Cronquist) & prob. to be excluded from Dilleniidae

Hoplophyllum DC. Compositae (Arct.). 2 NW Cape, S Afr. R: Taxon 43(1994)36

Hoppea Willd. Gentianaceae. 2 India

hora *Dipterocarpus zeylanicus*

Horaninovia Fischer & C. Meyer. Chenopodiaceae (III 3). 7 SW & C As.

Hordelymus (Jessen) Harz. Gramineae (23). 1 Eur., N Afr. to Caucasus

Hordeum L. Gramineae (23). c. 20 N temp. (Eur. 8 + cultigens). Barley. Spikelets in 3s on axis forming dense spike, fls of central or lateral spikelets often aborted. *H. vulgare* L. subsp. *spontaneum* (K. Koch) Körn. (*H. spontaneum*, W As.) with brittle rachis & husked grains first harvested c. 9000 BC, giving rise to 2-rowed subsp. *distichon* (L.) Koern. (*H. distichon* L.) with non-brittle rachis, which mutated to subsp. *vulgare* (6-rowed or '4-rowed' b. with central spikelet sterile – 'bere' (= bear), 'big' (Scotland)) all spikelets of each triad fertile, a single recessive gene giving 6-rowed ears from 2-rowed; predominantly selfed (2n = 14), hardiest forms cult. to 70° N (Norway), somewhat tolerant of saline soils; cereal grown by Anc. Egyptians & found in Swiss lake-dwellings, now mostly used in malting when germinated in water, kiln-dried & used as substrate for yeasts in beer- (first successfully bottled c. 1736, £8347m spent on it (cf. £4051m on bread) in GB in 1985, 29m pints per day drunk in GB 1988, i.e. 108 l per head per annum (10 times wine consumption)) & whisky-making (first recorded as distilled 1494 (by a friar), now 4m bottles a day made in Scotland) with diff. cvs (form. 'Maris Otter' in old Brit. ales now replaced by bigger-yielding but lesser-tasting ones; 'Golden Promise' in whisky), malt-extract used in proprietary spreads, med. etc.; pearl b. used in soups & stews is grain worn to ± spherical shape; b. water is a watery solution used as soft drink & medic. Some cult. orn. – *H. jubatum* L. (N temp.), some weeds esp. *H. secalinum* Schreber (meadow b., Eur., N Afr.); *H. murinum* L. subsp. *leporinum* (Link) Arc. (*H. leporinum*, Medit.) – seeds 200 yrs old from adobe in SW N Am. found to be viable

horehound *Marrubium vulgare*; **black h.** *Ballota nigra*; **water h.** *Lycopus europaeus*

Horichia Jenny. Orchidaceae (V 10). 1 Panamá: *H. dressleri* Jenny

Horkelia Cham. & Schldl. Rosaceae. 17 W N Am. R: Lloydia 1(1938)79. Some cult. orn. allied to *Potentilla*

Horkeliella (Rydb.) Rydb. (~ *Horkelia*). Rosaceae. 3 N Am.

Hormathophylla Cullen & T. Dudley = *Alyssum* s.l.

Hormidium (Lindley) Heynh. = *Encyclia*

Horminum L. Labiatae (VIII 2). 1 S Eur. mts. *H. pyrenaicum* L. – cult. orn. with showy purplish-blue fls

Hormocalyx Gleason = *Myrmidone*

Hormuzakia Guşul. = *Anchusa*

hornbeam *Carpinus betulus*; **American h.** *C. caroliniana*; **hop h.** *Ostrya* spp.

horn-nut *Trapa natans*; **h.-of-plenty** *Fedia cornucopiae*; **h.wort** *Ceratophyllum* spp.

Hornea Baker. Sapindaceae. 1 Mauritius

horned melon *Cucumis metalliferus*; **h. poppy** *Glaucium flavum*; **h. pondweed** *Zannichellia palustris*; **h. rampion** *Phyteuma* spp.

Hornschuchia Nees. Annonaceae. 6 E Brazil

Hornstedtia Retz. Zingiberaceae. 24 Indomal. (Borneo 8) to Aus. *H. rumphii* (Sm.) Valeton (Mal.) – fr. ed.; *H. scyphifera* (König) Steudel (Mal.) – runners 'walk' on 1 m high adventitious roots in Malay Pen.

Hornungia Reichb. Cruciferae. 2 W Eur. (2) to Med.

Horridocactus Backeb. = *Neoporteria*

horse brush *Tetradymia* spp.; **h. chestnut** *Aesculus* spp. esp. *A. hippocastanum*; **h. gram** *Macrotyloma uniflorum*; **h. hair, vegetable** *Chamaerops humilis*; **h. mint** *Mentha longifolia*; **h.radish** *Armoracia rusticana*; **h.r. tree** *Moringa oleifera*; **h.shoe vetch** *Hippocrepis comosa*; **h.tail** *Equisetum* spp.

Horsfieldia Willd. Myristicaceae. 100 Indomal. to Aus. R: GBS 37(1985)115, 38(1985)55,185, 39(1986)1. *H. iryaghedhi* (Gaertner) Warb. (Indomal.) – fine timber, fls v. fragrant & suggested for scent-making, oilseed used for candle-making

Horsfordia A. Gray. Malvaceae. 4 SW N Am.

Horstrissea Greuter, Gerstb. & Egli. Umbelliferae (III 8). 1 Crete: *H. dolinicola* Greuter, Gerstb. & Egli. R: Willdenowia 19(1990)389

Horta Vell. Conc. = *Clavija*

hortensia *Hydrangea macrophylla*

Hortia Vand. Rutaceae. 9 trop. S Am. Alks

Hortonia Wight Monimiaceae (I). 3 Sri Lanka

Horvatia Garay. Orchidaceae (V 10). 1 Peru

Horwoodia Turrill. Cruciferae. 1 Arabia, Iraq

Hosackia Benth. ex Lindley = *Lotus*

Hosea Ridley. Labiatae (II; Verbenaceae). 1 Sarawak

Hoseanthus Merr. = *Hosea*

Hoshiapuria Hajra, Daniel & Philcox = *Rotala*

Hosiea Hemsley & E. Wilson. Icacinaceae. 2 W & C China, Japan

Hoslundia Vahl. Labiatae (VIII 4c). 2–3 trop. Afr. *H. opposita* Vahl (trop. Afr.) – source of vanilla-scented oil

Hosta Tratt. Agavaceae (Hostaceae; Liliaceae s.l.). c. 25 China, Korea & Japan (15). R: Plantsman 3(1981)20; W.G. Schmid (1991) *The Genus H.* Cult. orn. perennial herbs incl. hybrids & varieg. pls with short rhizomes (plantain lilies), long cult. Japan & many cvs not assignable to wild spp., many not producing viable seeds, *H. ventricosa* Stearn (China) apomictic

Hostaceae Mathew = Agavaceae

hot water plant *Achimenes grandiflora*

Hottarum Bogner & Nicolson. Araceae (VI). 5 Borneo. Some rheophytes

Hottea Urban. Myrtaceae (Myrt.). 5 Hispaniola

Hottentot bread *Dioscorea elephantipes*; **H. fig** *Carpobrotus edulis*; **H. tea** *Helichrysum serpyllifolium*

Hottonia L. Primulaceae. 2 W N Am. (1), Eur. & W As. (1). Floating aquatics with finely dissected submerged lvs; heterostylous aerial fls. *H. palustris* L. (water violet, Euras.) – cult. in aquaria

Houlletia Brongn. Orchidaceae (V 10). 10 trop. Am. Cult. orn. epiphytes

Houmiri, Houmiriaceae See *Humiria*, Humiriaceae

hound's-tongue *Cynoglossum officinale*

house leek *Sempervivum* spp. esp. *S. tectorum*; **h. lime** *Sparrmannia africana*

houses, Chinese *Collinsia bicolor*

Houssayanthus Hunz. Sapindaceae. 3 Venez. to Arg. R: Candollea 42(1987)805

Houstonia L. = *Hedyotis*

Houttuynia Thunb. Saururaceae. 1 Japan S to mts of Nepal & Java: *H. cordata* Thunb. – parthenogenetic & almost completely male-sterile as microspores degenerate, cult. orn., shoots eaten as veg. in China

Hovea R. Br. Leguminosae (III 25). 12 SW & E Aus. Alks. Some cult. orn. (purple pea)

Hovenia Thunb. Rhamnaceae. 2 E As. Close to *Colubrina*, decid. trees incl. *H. dulcis* Thunb. (Chinese or Japanese raisin (tree), China, Korea, Japan) – pedicels fleshy & sweet, turning red after frost: coll. from wild & also used in med. esp. for hangovers, timber valuable

Hoverdenia Nees. Acanthaceae. 1 Mexico

Howardia Klotzsch (~ *Aristolochia*). Aristolochiaceae (II 2). 150 trop. & warm Am. *H. fimbriata* (Cham.) Klotzsch (*A.f.*, Uruguay) – fertility control; *H. glaucescens* (Kunth) Klotzsch (*A.g.*, S Am.) – source of yellow pareira used as diuretic etc.; *H. grandiflora* (Sw.) Klotzsch (*A. g.*, WI) – perianth to 20 cm across with 'tail' to 60 cm

Howea Becc. (*Howeia*). Palmae (V 4j). 2 Lord Howe Is. (Aus.). Cult. orn. monoec. palms esp. *H. forsteriana* (C. Moore & F. Muell.) Becc. (kentia) – to 18 m but commonly grown as a sessile housepl.

Howeia Becc. = *praec.*

Howellia A. Gray. Campanulaceae. 1 W N Am.: *H. aquatilis* A. Gray – aquatic s.t. with

Hippuris 'syndrome'

Howelliella Rothm. Scrophulariaceae (Antirr.). 1 E Calif.: *H. ovata* (Eastw.) Rothm.

Howethoa Rauschert = *Lepisanthes*

Howittia F. Muell. Malvaceae. 1 SE Aus.: *H. trilocularis* F. Muell.

Hoya R. Br. Asclepiadaceae (III 4). c. 70 Indomal. to Pacific. Cult. orn. root-climbers. twiners or sprawling shrubs (wax flowers, w. plant) esp. *H. carnosa* (L.f.) R. Br. (S China to Aus.) & *H. lanceolata* Wallich ex D. Don subsp. *bella* (Hook.) Kent (*H. bella*, Burma); some spp. with accrescent infl. axes; some local medic.

Hoyella Ridley. Asclepiadaceae (III 4). 1 Sumatra: *H. rosea* Ridley

Hua Pierre ex De Wild. Huaceae. 1 trop. Afr.

Hua Wang *Paeonia suffruticosa*

Huaceae A. Chev. Dicots – Dilleniidae – Malvales. 2/3 trop. Afr. Shrubs & trees with garlic smell. Lvs spirally arr., simple, entire; stipules present, caducous in *Afrostyrax*. Fls solit. or in clusters, axillary, small, bisexual, reg.; K 5, valvate (*Hua*), or (3–5) with irreg. lobes (*Afrostyrax*), C (4)5, induplicate-valvate, A (8)10, G̲ (5), 1-loc. with term. style & 1 (*Hua*) or (4–)6 (*Afrostyrax*) basal, erect, anatropous, bitegmic ovules. Fr. indehiscent (*Afrostyrax*) or dehiscent, 5-valved (*Hua*); seed 1(2) with straight embryo & garlic-scented copious endosperm

Genera: *Afrostyrax, Hua*

huanghuahaosu *Artemisia annua*

huanghuali *Dalbergia hainanensis*

Huanaca Cav. Umbelliferae (I 2b). 4 Patagonia & Chile. R: Kurtziana 6(1971)7

huangteng *Fibraurea recisa*

huanita *Bourreria huanita*

Huarpea Cabrera. Compositae (Barn.). 1 S Andes (Arg.). Perenn. herb

Hubbardia Bor. Gramineae (36). 1 W India: *H. heptaneuron* Bor of waterfalls – extinct because of damming rivers

Hubbardochloa Auq. (~ *Muehlenbergia*). Gramineae (31e). 1 C Afr. mts

Huberia DC. Melastomataceae. 6 Brazil, 1 ext. to Ecuador

Huberodendron Ducke. Bombacaceae. 5 trop. Am.

Huberopappus Pruski. Compositae (Vern.-Pip.). 1 Venez.: *H. maigualidae* Pruski. R: Novon 2(1992)19

Hubertia Bory (~ *Senecio*). Compositae (Sen.-Sen.). 25 Madag., Masc.

huckleberry *Gaylussacia* spp., *Vaccinium* spp., *Solanum melanocerasum*

Hudsonia L. Cistaceae. 1 (variable) NW Am.: *H. ericoides* L. (beach heath) – cult. orn.

Hueblia Speta = *Chaenorhinum*

Huernia R. Br. Asclepiadaceae (III 5). 64 trop. & S Afr., S Arabia. R: Excelsa Tax. Ser. 4(1988). Cult. orn. dwarf succ.

Huerniopsis N.E. Br. Asclepiadaceae (III 5). 2 S Afr. Cult. orn. succ.

Huertea Ruíz & Pavón. Staphyleaceae. 4 WI to Peru

Huetia Boiss. = *Geocaryum*

Hughesia R. King & H. Robinson (~ *Eupatorium*). Compositae (Eup.-Crit.). 1 Peru. R: MSBMBG 22(1987)361

Hugonia L. Linaceae. 32 OW trop. Lower branches of infl. rep. by conspic. climbing-hooks

Hugoniaceae Arn. = Linaceae

Hugueninia Reichb. (~ *Descurainia*). Cruciferae. 1 S Eur. mts: *H. tanacetifolia* (L.) Reichb.

Huilaea Wurd. Melastomataceae. 4 Colombia

Hulemacanthus S. Moore. Acanthaceae. 1 New Guinea

Hullettia King ex Hook.f. Moraceae (II). 2 S Burma & Thailand, Malay Pen., Sumatra. R: JAA 41(1960)334

Hulsea Torrey & A. Gray. Compositae (Hele.-Cha.). 7 W N Am. R: Brittonia 27(1975)228. Cult. orn. aromatic lvs

Hulteniella Tzvelev. Compositae (Anth.-Art.). 1 Arctic Euras., N Am.

Hulthemia Dumort. = *Rosa*

hulver (archaic) *Ilex aquifolium*

Humata Cav. = *Davallia*

Humbertacalia C. Jeffrey. Compositae (Sen.-Sen.). 8 Madag., Masc.

Humbertia Comm. ex Lam. Convolvulaceae (Humbertiaceae). 1 Madag.

Humbertiaceae Pichon = Convolvulaceae

Humbertianthus Hochr. Malvaceae. 1 Madagascar

Humbertiella Hochr. Malvaceae. 6 Madag. R: BMNHN 4,12(1990)7

Humbertina Buchet = *Arophyton*

Humbertiodendron Leandri. Trigoniaceae. 1 Madagascar

Humbertioturraea J. Leroy. Meliaceae (I 1). 7 Madag. R: KB 44(1989)369

Humbertochloa A. Camus & Stapf. Gramineae (10). 2 trop. E Afr., Madag.

Humblotiella Tard. = *Lindsaea*

Humblotiodendron Engl. = *Vepris*

Humboldt willow *Salix chilensis*

Humboldtia Vahl. Leguminosae (I 4). 6 S India, Sri Lanka. Flowering shoots with extrafl. nectaries & hollow obconical internodes, in the top of each of which, opp. leaf, is a slit to an ant-inhabited cavity

Humboldtiella Harms = *Coursetia*

Humea Sm. = *Calomeria*

Humeocline A. Anderb. (~ *Calomeria*). Compositae (Gnap.-Gnap.). 1 Madag.: *H. madagascariensis* (Humbert) A. Anderb. R: OB 104(1991)139

Humiria Aublet. Humiriaceae. 4 trop. S Am. R: CUSNH 35(1961)87. Bat-disp. Some locally used med. & timbers, some of which (cf. *Aquilaria*) beautifully scented & used as incense once attacked by fungi; bark a source of 'umiry-balsam'; exocarp of some spp. e.g. *H. balsamifera* Aublet, ed.

Humiriaceae A. Juss. (Houmiriaceae). Dicots – Rosidae – Linales. 8/50 trop. S Am. to Costa Rica & W Afr. (1). Trees & shrubs. Lvs evergreen, spirally arr. to distichous, simple, ± entire; stipules caducous, tiny or 0. Fls bisexual, ± reg., in cymes; K (5), lobes imbricate, 2 outer often smaller than others, rarely 0, C 5, thick, convolute or imbricate, usu. 3–5-veined, whitish (rarely red), A 10–30 (s.t. 5 groups of 3 opp. K & 5 opp. C; ∞ in bundles in *Vantanea*) with filaments forming a tube, s.t. some without anthers (staminodes), anthers with expanded prolonged connective, nectary-disk intrastaminal, free or adnate to G, usu. cupulate to tubular, with lobes, or s.t. of 10–20 distinct scales, G ((4)5(–7)), pluriloc. with axile placentas & 1 style, s.t. 1-loc. apically (partitions not reaching summit), 1 or 2 (superposed) pendulous, anatropous, bitegmic ovules per loc. Fr. a drupe with usu. pluriloc. stone, s.t. with resinous secretory cavities & adapted to water-disp., the stone with as many valves as G, 1 or more being pushed off at germination; seeds 1 or 2 with slightly curved embryo & copious oily endosperm. x = 12. R: CUSNH 35(1961)25

Genera: *Duckesia, Endopleura, Humiria, Humiriastrum, Hylocarpa, Sacoglottis, Schistostemon, Vantanea*

Some timber etc. (*Humiria*)

Humirianthera Huber = *Casimirella*

Humiriastrum (Urban) Cuatrec. Humiriaceae. 12 C Am. to SE Brazil. R: CUSNH 35(1961)122. Local medic.

Humularia Duvign. Leguminosae (III 13). 40 trop. Afr. Some roots, smoked like tobacco, stimulatory

Humulus L. Cannabaceae. 3 N temp. Twining dioec. climbers, male fls in loose axillary panicles, females in short bracteate spikes, cone-like at maturity, each bract with 2 fls. *H. japonicus* Siebold & Zucc. (Japanese hop, temp. E As.) – cult. orn.; *H. lupulus* L. (hop, N temp.) – used in brewing beer thanks to (bacteriostatic) resinous & bitter substances (incl. alks like codeine & morphine) in female infls, when grown trained up wires as in Kent, where the stems (bines) are cut down when the hops coll., propagated by cuttings of rhiz., introd. from Flanders early C15 as beer keeps better than ale (though strongly opposed at first), lager (orig. Bavaria) brewed at cool temps (lager yeasts active then), fibre for coarse cloth in Sweden & paper, Romans ate shoots like asparagus

Hunaniopanax C.J. Qi & T.R. Cao. Araliaceae. 1 Hunan: *H. hypoglaucus* C.J. Qi & T.R. Cao. R: APS 26(1988)47

Hunga Pancher ex Prance. Chrysobalanceae (2). 11 New Guinea (3), New Caled. & Loyalty Is. (8, 3 on serpentine). R: PTRSB 320(1988)102

hungry rice *Digitaria exilis*

Hunnemannia Sweet. Papaveraceae (II). 2 E Mex. R: Phytol. 73(1992)330. *H. fumariifolia* Sweet – alks, cult. orn. like *Eschscholzia* but sepals separate

Hunteria Roxb. Apocynaceae. 4 trop. Afr. (3), Indomal. (1). Alks

Huntleya Bateman ex Lindley. Orchidaceae (V 10). 10 trop. Am. Cult. orn. epiphytes

huntsman's cup or **horn** *Sarracenia purpurea*

Hunzikeria D'Arcy. Solanaceae (2). 3 warm Am.

Huodendron Rehder. Styracaceae. 6 S China, SE As.

Huon pine *Lagarostrobos franklinii*

Huperzia Bernh. (~ *Lycopodium*). Lycopodiaceae. c. 300 subcosmop. (Eur. 1: *H. selago* (L.) Schrank & C. Martius (*L. selago*). R: OB 92(1987)163. Like *Lycopodium* but sporophylls not apical; many hybrids in N Am., reprod. by gemmae (? wind-disp.)

Hura L. Euphorbiaceae. 2 trop. Am. Monoec. trees with irritant latex. *H. crepitans* L. (huru) – widely cult. trop., fr. 5–20-loc. with explosive dehiscence, expelling seeds to 14 m, form. used wired together as sand-boxes before blotting-paper used, latex a fish-poison alleged to kill even anacondas, timber (assacu)

huru *Hura crepitans*

Husnotia Fourn. = *Ditassa*

Hutchinsia R. Br. = *Pritzelago*

Hutchinsiella O. Schulz. Cruciferae. 1 W Tibet

Hutchinsonia Robyns (~ *Rytigynia*). Rubiaceae (III 2). 2 trop. W Afr.

Hutera Porta = *Coincya*

Huthamnus Tsiang. Asclepiadaceae (III 4). 1 SW China: *H. sinicus* Tsiang

Huthia Brand. Polemoniaceae. 2 Peru

Huttonaea Harvey. Orchidaceae (IV 3). 5 S Afr. Elaiophores attractive to bees

Huttonella Kirk = *Carmichaelia*

Huxleya Ewart. Labiatae (II; Verbenaceae). 1 N Aus.: *H. linifolia* Ewart & B. Rees. R: JABG 13(1990)35

Huynhia Greuter = *Arnebia*

hyacinth *Hyacinthus orientalis*; see also *Gladiolus italicus*; **h. bean** *Lablab purpureus*; **feather h.** *Muscari comosum* 'Plumosum'; **grape h.** *M.* spp.; **Roman h.** *Bellevalia romana*; **summer h.** *Galtonia candicans*; **tassel h.** *M. comosum*; **water h.** *Eichhornia crassipes*

Hyacinthaceae Batsch ex Borckh. (~ Liliaceae). Monocots – Liliidae – Asparagales. 41/770 subcosmop. esp. Medit. & S Afr. Bulbous (rhizomatous) herbs with rosette of lvs or 0. Infl. usu. a spike or raceme, or infl. photosynthetic, branched & ±leafless. P 6, sometimes with basal tube; A 6; (G3), each loc. with 1–many anatropous ovules. Fr. a loculicidal capsule; seeds black, subglobose to flattened & even winged

Principal genera: *Albuca, Bellevalia, Bowiea, Camassia, Chionodoxa, Dipcadi, Drimia, Eucomis, Galtonia, Hyacinthoides, Hyacinthus, Lachenalia, Ledebouria, Muscari, Ornithogalum, Puschkinia, Scilla, Veltheimia*

Facies v. like Asphodelaceae; the position of the rhiz. *Schoenolirion* as well as *Chlorogalum* debated

Many cult. orn.; some *Camassia* & *Ornithogalum* spp. ed.; scent from *Hyacinthus*

Hyacinthella Schur (~ *Hyacinthus*). Hyacinthaceae (Liliaceae s.l.). 16 SE Eur. (3), SW As. R: Candollea 36(1981)513, 37(1982)157. Some cult. orn.

Hyacinthoides Heister ex Fabr. (*Endymion*). Hyacinthaceae (Liliaceae s.l.). 3–4 W Eur. (3), N Afr. Bluebells. Splash-cup disp. mechanism. Cult. orn. ('Constancy' in 'Language of Fls') esp. *H. hispanica* (Miller) Rothm. (Spanish bluebell, SW Eur.) & *H. non-scripta* (L.) Rothm. (*Scilla nutans*, (English) bluebell, W Eur.) – bulbs form. source of glue for book-binding & fixing arrow-flights & of starch for linen etc.; these form hybrid swarms in cult.

Hyacinthus L. Hyacinthaceae (Liliaceae s.l.). 3 W & C As. R: BN 127(1974)297. Hyacinth (cf. *Gladiolus italicus*) v. ancient name, pre-Greek (cf. *Crocus*) & non-Indoeuropean, orig. a name of a god assoc. with spring. *H. orientalis* L. (hyacinth, NE Medit., natur. Eur.) – introd. to W Eur. via Turkey & Venice to Padua flowering 1562, 351 cvs offered by one nurseryman in 1753, 2000 known in C19 but by 1996 only 220 (60 in comm.); cvs of different colours & fasciated rachis, diploids 1560–1850, triploids 1700–1900, higher ploidies 1900–1950, 'double' fls selected end C17 Haarlem with some cvs realizing £200 a bulb (cf. *Tulipa*); subject of C18 Florists' societies ('sorrow' in 'Language of Fls'); may be propagated by gouging out base of bulbs which stimulates growth of new bulblets therein; fls used in scent-making in S France (competing with synthetic phenyl acetaldehyde)

Hyaenanche Lambert. Euphorbiaceae. 1 S Afr. Lvs verticillate

Hyalea (DC) Jaub. & Spach = *Centaurea*

Hyalis D. Don ex Hook. & Arn. (~ *Plazia*). Compositae (Mut.-Mut.). 2 S Bolivia, Paraguay & Arg. Shrubs

Hyalisma Champ. = *Sciaphila*

Hyalocalyx Rolfe. Turneraceae. 1 trop. E Afr., Madagascar

Hyalochaete Dittrich & Rech. f. (~ *Jurinea*). Compositae (Card.-Card.). 1 Afghanistan, Pakistan

Hyalochlamys A. Gray (~ *Angianthus*). Compositae (Gnap.-Ang.). 1 SW Aus.: *H. globifera* A. Gray. R: OB 104(1991)129

Hyalocystis Hallier f. Convolvulaceae. 1 trop. Afr.

Hyalolaena Bunge. Umbelliferae (III 8). 10 SW & C As.

Hyalopoa (Tzvelev) Tzvelev = *Colpodium*

Hyalosepalum Troupin = *Tinospora*
Hyalosema (Schltr.) Rolfe = *Bulbophyllum*
Hyaloseris Griseb. Compositae (Mut.-Mut.). 7 Bolivia, Arg. R: Kurtziana 7(1973)195. Shrubs
Hyalosperma Steetz. Compositae (Gnap.-Ang.). 9 Aus. R: OB 104(1991)124
Hyalotricha Copel. = *Campyloneurum*
Hyalotrichopteris Wagner = *Campyloneurum*
Hybanthus Jacq. Violaceae. 150 trop. & warm. R: BJ 67(1936)437. Alks. *H. floribundus* (Lindley) F. Muell. (Aus.) – nickel hyperaccumulator. Some medic. locally esp. *H. calceolaria* (L.) Oken (white ipecacuanha, trop. Am.) used like ipecacuanha & *H. parviflorus* (L.f.) Baill. (cuichunchulli, warm Am.); *H. prunifolius* (Roemer & Schultes) G. Schulze (trop. Am.) – understorey shrub induced to flower by 12 mm rain after dry spell, leading to mass-flowering
Hybochilus Schltr. Orchidaceae (V 10). 1 Costa Rica: *H. inconspicuus* (Kraenzlin) Schltr. R: CUMH 16(1987)120
Hybosema Harms = *Gliricidia*
Hybosperma Urban = *Colubrina*
Hybridella Cass. Compositae (Helia.). 2 Mexico
Hydatella Diels. Hydatellaceae. 5 Aus. (4), NZ (1)
Hydatellaceae Hamann (~Centrolepidaceae). Monocots – Commelinidae – Hydatellales. 2/8–9 Aus., NZ. Small tufted annual aquatics with vessels only in roots. Lvs spirally arr., slender, subterete to flattened but centric internally. Infl. a term. head (unisexual in *Hydatella*, bisexual in *Trithuria*) with 2–∞ bracts each with 1 or several fls in axis; fls water-poll., naked, unisexual of A 1 or pseudomonomerous G with 2 or 3 (*Trithuria*) or 5–10 (*Hydatella*) filamentous structures (? styles) & 1 pendulous, anatropous, bitegmic ovule. Fr. small dry, indehiscent (*Hydatella*) or opening by 3 valves (*Trithuria*); seed 1 with starchy perisperm but almost 0 endosperm & minute scarcely differentiated embryo
Genera: *Hydatella, Trithuria*
Form. placed in Centrolepidaceae but Cronquist suggests affinity with Commelinales
Hydnocarpus Gaertner. Flacourtiaceae (4). 40 Indomal. (Borneo 13). R: BJ 69(1936)1. Main branches with only scale lvs; short-lived twigs with foliage lvs. Several spp. sources of oil (chaulmoogra o.) form. much used in treatment of eczema, leprosy & other skin conditions esp. *H. pentandra* (Buch.-Ham.) Oken (*H. laurifolia, H. wightiana*, India), *H. kurzii* (King) Warb. (SE As.), *H. castanea* Hook.f. & Thomson (kalaw, Burma), *H. venenata* Gaertner (lucraban, India, Sri Lanka)
Hydnophytum Jack. Rubiaceae (IV 7). 52 Indomal. (esp. New Guinea). Epiphytes, some with ant-inhabited 'tubers', like *Myrmecodia*, but c. 40% not inhabited by ants though other animals bring in nutrients (Jebb), havens for small frogs; fr. spread by birds on to bark
Hydnora Thunb. Hydnoraceae. c. 5 Arabia, trop. & S Afr., Madag., Masc., arid & semi-a. *H. africana* Thunb. & *H. triceps* Drège ex Meyer restricted to *Euphorbia* spp., others on *Acacia* & other legume spp.; beetle-poll. Ed. fr. esp. *H. africana* (Arabia, Afr.) – sought by foxes & humans, *H. esculenta* Jum. & Perrier (Madag.), *H. johannis* Becc. (Afr.) – sought by monkeys, infls break through asphalt to flower, imp. fuel in Sudan, medic.
Hydnoraceae Agardh. Dicots – Magnoliidae – Aristolochiales. 2/7 drier trop. & S Afr., drier S Am. Leafless chlorophyll-less terr. parasites with rhizome-like root-body from which arise haustorial unbranched roots parasitizing host-roots. Fls arising singly & endogenously from root-body, often partially buried, large, fleshy, smelly, beetle-poll., bisexual, reg., epigynous; P ((2)3 or 4(5)), basally connate, fleshy, valvate, abaxial surfaces cracked & brown, adaxial white to red, (A) with filaments ± 0, same no. as & opp. P but with many elongate bisporangiate pollen-sacs with longit. slits, forming a lobed ring on hypanthium (*Hydnora*) or connate anthers forming a dome (*Prosopanche*, where fleshy staminodes below alt. with them), $\overline{G}$ (3 or 4(5)), 1-loc. but becoming filled with accrescent placentas (laminar & covered with ∞ ovules, in 3 groups in *Prosopanche*, parietal but not meeting at centre), ovules orthotropous, unitegmic, the integument scarcely discernible in *Prosopanche*. Fr. with woody pericarp & ed. flesh; seeds ∞, minute, with thin layer of perisperm & well-developed endosperm (arabinose) enclosing minute undifferentiated embryo
Genera: *Hydnora, Prosopanche*. Near Rafflesiaceae but fls bisexual
Hydrangea L. Hydrangeaceae. 23 Himal. to Japan & Philipp., Am. R: PCAS 20(1957)147; M. Haworth-Booth (1984) *The Hs*, ed. 5. Erect or climbing, evergreen or decid. shrubs with small fertile bisexual fls., often with showy sterile fls with C-like K surrounding them (in cult. forms, these s.t. comprise all the infl.). Colour of fls dependent on capacity to absorb

Al^{3+} which cause pigments to go from red to blue; this is made difficult on limy soils where 'blue' forms go pink though the addition of iron reverses this. During development, the stomata of K disintegrate as lobes green & lose char. colours. Dried roots source of hydrangin, a diaphoretic & diuretic alk. – some spp. medic. & steamed lvs of *H. macrophylla* (Thunb.) Ser. subsp. *serrata* (Thunb.) Makino (Japan, Korea) used in a drink, amacha, in Japan, though *H.* spp. can poison man & stock. Many cult. orn. forms esp. derived from *H. macrophylla* (hortensia) – typical var. with all sterile fls but wild var. *normalis* E. Wilson (Japan) with both fertile & sterile; *H. anomala* D. Don is a climbing (aerial rootlets) sp. of which subsp. *petiolaris* (Siebold & Zucc.) McClint. (Japan, Taiwan) is often cult. as wall pl., sterile fls inverting as inner fls mature into fr. Other cult. incl. *H. paniculata* Siebold (E Aus.) – wood for umbrella-handles

Hydrangeaceae Dumort. Dicots – Rosidae – Rosales. 17/190 N temp. to Mal. Trees, shrubs, lianes or rhiz. herbs (*Cardiandra, Deinanthe, Kirengeshoma*), often accum. aluminium. Lvs opp. (rarely whorled or spiral (*Cardiandra*)), simple; stipules 0. Fls bisexual, reg. or the marginal ones sterile, irreg. with enlarged K (polygamo-dioec. in *Broussaisia*), in cymes (often corymbiform to paniculiform); K (4 or 5(–12)), lobes valvate or imbricate, C 4 or 5(–12), valvate, imbricate to convolute, A (1)2–several × C (to 200 in *Carpenteria*), when ∞ arising centripetally, filaments sometimes basally connate, G ((2)3–5(–12)), half to fully inf. (superior in *Carpenteria, Fendlera, Fendlerella, Jamesia, Whipplea*), pluriloc. or 1-loc. with intruded parietal placentas & distinct (sometimes basally connate) styles (rarely 1) & intrastaminal nectary-disk at apex of G, (1–)∞ anatropous, unitegmic ovules on each placenta. Fr. a capsule or berry; seeds with straight embryo in fleshy endosperm. x = 13–18+

Principal genera: *Deutzia, Hydrangea, Philadelphus* (with allies sometimes segregated as distinct fam.). On DNA data, perhaps best included in Cornaceae s.s.

Many cult. orn. shrubs

Hydranthelium Kunth = *Bacopa*

Hydrastidaceae Augier ex Martinov. See Ranunculaceae

Hydrastis L. Ranunculaceae (V; Hydrastidaceae). 1 C & E N Am.: *H. canadensis* L. (golden seal, yellow or turmeric root) – rhiz. used in prep. a tonic & form. a yellow dye, the sp. much reduced through over-exploitation

Hydriastele H. Wendl. & Drude. Palmae (V 4l). 8 New Guinea, NE Aus. Slender monoec. palms incl. rheophytes; weevil-poll.

Hydrilla Rich. Hydrocharitaceae. 1 OW, introd. to SE US & C Am.: *H. verticillata* (L.f.) C. Presl, allied to *Elodea*. Submerged aquatic, male fls floating to surface, opening & floating to females; sometimes cult. in aquaria, pops in US monoec. & female

Hydrobryopsis Engl. = *Hydrobryum*

Hydrobryum Endl. Podostemaceae. 10 S India, E Nepal, Assam, China & S Japan

Hydrocera Blume ex Wight & Arn. (*Tytonia*). Balsaminaceae. 1 Indomal.: *H. triflora* (L.) Wight & Arn. K 5, C 5, fr. a 5-seeded berry-like drupe (cf. *Impatiens*). Semi-aquatic, poss. with water-disp. seeds; fls used to dye fingernails in India

Hydrocharis L. Hydrocharitaceae. 3–6 OW (Eur. 1). Floating aquatics often (erroneously) reported as rootless. *H. morsus-ranae* L. (frog-bit, Eur., E As., natur. N Am.) – cult. orn. dioec. pl. with horizontal runners which form new pl. at tips, overwintering as buds from stolons

Hydrocharitaceae Juss. Monocots – Alismatidae – Hydrocharitales. 15/80 cosmop. Aquatics (marine or freshwater, usu. perenn.), submerged or partly emergent, s.t. free-floating; vessels only in roots. Lvs spirally arr., opp. or whorled, often ± sheathing, subtending axillary scales & s.t. with distinct lamina, the margins often with thick-walled prickle-hairs (pl. usu. otherwise glabrous). Fls reg. (slightly irreg. in *Vallisneria*), usu. unisexual, solit. or in few-flowered cymes, infls subtended by (1)2 distinct ± connate bracts forming a spathe; K 3, C 3 attached to G or hypanthium, or C (P) 0, A 2 or 3 to ∞ in 1–∞ 3-merous whorls, s.t. paired opp. K, when ∞ developing centripetally, some s.t. nectariferous staminodes, anthers with longit. slits, pollen grains globose, united in thread-like chains in marine genera (*Halophila, Thalassia*), male fls often released from submerged infls & floating on surface (*Elodea, Enhalus, Hydrilla, Lagarosiphon, Vallisneria*), though pollen alone does this in some *Elodea* spp., $\overline{G}$ ((2)3–6(–20)), 1-loc. often with ± deeply intruded partial partitions, styles shortly basally connate, ovules anatropous (orthotropous), bitegmic. Fr. a septicidal capsule, usu. opening irreg., submerged; seeds several–∞, endosperm 0 (scanty in *Ottelia*), embryo straight. x = 7–12

Classification & principal genera (Dandy):

I. **Hydrocharitoideae** (sometimes Vallisnerioideae & Hydrilloideae separated; usu. freshwater, poll. at or above surface, K & C usu. present): *Blyxa, Elodea,*

Lagarosiphon, Ottelia, Enhalus (marine). Some insect-poll.

II. **Thalassioideae** (marine, poll. below surface, P of 1 whorl, styles 6–12, bifid): *Thalassia*

III. **Halophiloideae** (marine, poll. below surface, P of 1 whorl, styles 2–5 entire): *Halophila*

Many monotypic genera. Of freshwater genera, *Ottelia* & *Vallisneria* pantrop., most of others OW. Najadaceae may belong here

Some cult. orn. esp. *Vallisneria*; *Ottelia* ed.; some introd. have become weeds (see *Elodea, Lagarosiphon*)

Hydrochloa Pal. = *Luziola*

Hydrocleys Rich. Alismataceae (Limnocharitaceae). 5 S Am. R: Phytol. 57(1985)421. Aquatics with milky latex, *H. nymphoides* (Willd.) Buchenau (*Limnocharis humboldtii*, water poppy) – cult. orn. with large yellow fls

Hydrocotylaceae (Drude) Hylander = Umbelliferae

Hydrocotyle L. Umbelliferae (Ia; Hydrocotylaceae). c. 130 cosmop. (Eur. 1, Aus. 55 mostly endemic). Creeping perennial herbs often with peltate lvs; some cult. orn. ground-cover (navelwort, pennywort) esp. *H. sibthorpioides* Lam. (As., but widely natur.), *H. bonariensis* Lam. (S Am.) – lvs to 120 mm across, aggressive lawn-weed in Aus.; lvs of *H. javanica* Thunb. a fish-poison in Indonesia

Hydrodea N.E. Br. = *Mesembryanthemum*

Hydrodyssodia B.L. Turner (~ *Hydropectis*). Compositae (Helia.). 1 Mex.: *H. stevensii* (McVaugh) B.L. Turner. R: Phytol. 65(1988)132. Aquatic annual

Hydrogaster Kuhlm. Tiliaceae. 1 Brazil

Hydroidea Karis (~ *Atrichanthum*). Compositae (Gnap.). 1 SW Cape: *H. elsiae* (Hilliard) Karis. R: OB 104(1991)74

Hydrolea L. Hydrophyllaceae. 11 trop. R: Rhodora 90(1988)169. Semiaquatics; lvs spirally arr., some with axillary thorns; autogamous

Hydrolythrum Hook.f. (~ *Rotala*). Lythraceae. 1 Indomal.

Hydromystria G. Meyer = *Limnobium*

Hydropectis Rydb. Compositae (Helia.-Pect.). 1 Mex.: *H. aquatica* (S. Watson) Rydb. – aquatic

Hydrophilus Linder (~ *Leptocarpus*). Restionaceae. 1 SW Cape: *H. rattrayi* (Pillans) Linder. R: Bothalia 15(1985)484

Hydrophylax L.f. Rubiaceae (IV 15). 1 coasts of India, Sri Lanka, Thailand: *H. maritima* L.f. Dune-colonist; fr. corky, indehiscent

Hydrophyllaceae R. Br. Dicots – Asteridae – Solanales. 18/270 subcosmop. (exc. Aus.) esp. dry W N Am. Herbs or shrubs, s.t. pachycaul (e.g. *Wigandia*) often with glandular hairs & odorous, or rough-hairy. Lvs spirally arr. or at least some opp., simple to pinnatisect or pinnate (rarely palmate); stipules 0. Fls bisexual, (4)5(10–12 in *Codon*)-merous, in (often helicoid) cymes, rarely solit.; (K) divided s.t. to base, lobes imbricate, (C) ± reg. with imbricate (convolute) lobes, A as many as & alt. with C, attached to C-tube, usu. with small scale on either side, nectary-disk usu. 0, $\underline{G}$ (2), half-inf. in some *Nama* & *Wigandia* spp. etc., usu. 1-loc. with term. style ± bifid (to base in *Nama* spp., undivided in *Romanzoffia*) & 2 parietal ± intruded, s.t. meeting placentas (G then 2-loc. & placentation axile) each with 2–∞ anatropous or amphitropous, unitegmic ovules. Fr. a capsule, sometimes irreg. dehiscent or indehiscent; seeds with straight spatulate or sometimes linear embryo in ± fleshy oily endosperm. x = 5–13+

Principal genera: *Hydrolea, Nama, Nemophila, Phacelia* (many genera mono- or oligo-typic)
Cult. orn. esp. spp. of *Hesperochiron, Nemophila, Phacelia, Romanzoffia, Wigandia*; some ed. (*Hydrophyllum*)

Hydrophyllum L. Hydrophyllaceae. 8 N Am. (W & E spp. separated). R: AMN 27(1942)710. Fls protandrous with the A scales united to C-tube forming nectar-tubes which bees probe. Some form. eaten by Am. Indians as greens, esp. *H. virginianum* L. (E N Am.) – cult. orn.

Hydrostachyaceae Engl. Dicots – Asteridae – Callitrichales. 1/22 S trop. & S Afr., (esp.) Madag. Submerged perennial dioec. (monoec.) aquatics with tubers & no vessels. Lvs basal, entire to 3-pinnatifid, ligulate basally & often covered with small scaly or fringed appendages, a small membranous intrapetiolar stipule present. Fls sessile in bract axils of dense spike term. a central scape, small, unisexual; P 0, males with A 1 with short filament & extrorse anther, females with $\underline{G}$ (2), 1-loc. with 2 parietal placentas & 2 elongate persistent styles, s.t. basally connate, with ∞ anatropous, unitegmic ovules. Fr. a fissuricidal capsule with ∞ tiny seeds with 0 endosperm

Only genus: *Hydrostachys*

Poss. allied to Plantaginaceae & Scrophulariaceae

Hydrostachys Thouars. Hydrostachyaceae. 22 S trop. & S Afr., (esp.) Madag. R: Adansonia 13(1973)76. Flower when water level drops; bases attached to rocks; leaf morphology v. variable within spp.

Hydrostemma Wallich = *Barclaya*

Hydrothauma C. Hubb. Gramineae (34b). 1 Zambia: *H. manicatum* C. Hubb. – aquatic annual with air-canals

Hydrothrix Hook.f. Pontederiaceae. 1 E Brazil. Submerged annuals with cleistogamous fls (outer stamen fertile, inner 1 or 2 staminodal)

Hydrotriche Zucc. Scrophulariaceae. 4 Madag. R: Adansonia 19(1979)145. *H. hottoniiflora* Zucc. – 'Hippuris' syndrome

Hyeronima Allemão (*Hieronima*). Euphorbiaceae. 36 trop. Am. (S Am. 10, R: BJ 111(1990)297). Some timbers for cabinet-work

Hygea Hanst. Gesneriaceae. 1 Chile

Hygrochilus Pftitzer (~ *Arachnis*). Orchidaceae (V 16). 1 SE As.: *H. parishii* (Veitch & Reichb. f.) Pftitzer. R: OB 95(1988)138

Hygrochloa Lazarides. Gramineae (34b). 2 N Aus. Aquatics (? allied to *Paspalidium*)

Hygrophila R. Br. Acanthaceae. c. 25 trop., wet places. Some with sticky seeds disp. on birds' feet; some cult. orn. as submerged aquatics in aquaria, esp. *H. difformis* (L.f.) Blume (*Synnema triflorum*, water wisteria, India to Thailand) – often a weed of ricefields; *H. schulli* (Buch.-Ham.) M.R. & S.M. Almeida (India, Sri Lanka) – used like spinach

Hygroryza Nees. Gramineae (9). 1 India, Sri Lanka. Floating

Hylaeanthe A. Jonker & Jonker. Marantaceae. 4 trop. S Am.

Hylandia Airy Shaw. Euphorbiaceae. 1 Queensland: *H. dockrillii* Airy Shaw (blushwood)

Hylandra Löve = *Arabidopsis*

Hylebates Chippindale. Gramineae (34d). 2 trop. Afr.

Hylenaea Miers. Celastraceae. 3 trop. Am.

Hyline Herbert. Amaryllidaceae (Liliaceae). 2 Brazil

Hylocarpa Cuatrec. Humiriaceae. 1 Amazonian Brazil

Hylocereus (A. Berger) Britton & Rose. Cactaceae (III 2). c. 16 trop. Am. Usu. climbing with aerial roots, stems usu. 3-angled, fls nocturnal & white, rarely red. Fr. of some ed. (pitaya); some widely cult. esp. *H. undatus* (Haw.) Britton & Rose (origin unknown) – one of the 'night-blooming cereus' with fragrant white fls & ed. fr. (sold in markets)

Hylocharis Miq. = *Oxyspora*

Hylodendron Taubert. Leguminosae (I 4). 1 Gulf of Guinea

Hylomecon Maxim. (~ *Chelidonium*). Papaveraceae (I). 3 temp. E As. Arils; alks. Cult. orn. esp. *H. japonica* (Thunb.) Prantl & Kündig, incl. 'f. *dissectum*' (cf. *Chelidonium*)

Hylomyza Danser = *Dufrenoya*

Hylophila Lindley. Orchidaceae (III 3). 6 Malay Pen., New Guinea

Hylotelephium H. Ohba. See *Sedum*

Hymenachne Pal. (~ *Panicum*). Gramineae (34b). 5 trop. As. & Am., swamps. R: SB 13(1988)595. Some fodders. *H. amplexicaulis* (Rudge) Nees forms part of floating islands in Amazon

Hymenaea L. Leguminosae (I 4). 16 trop. Am. & Afr. (incl. *Trachylobium*). Fr. ed. & sold in markets; copals form. much used for varnish etc. up to 3 kg, the best quality subfossil, esp. from *H. courbaril* L. (WI locust, Brazilian copal, anami gum, trop. Am.) – heavy timber like mahogany (courbaril); *H. verrucosa* Gaertner (*Trachylobium* v., E Afr. to Seychelles) – source of Madag. copal

Hymenandra (A. DC) Spach. Myrsinaceae. 8 Indomal. R: GBS 43(1991)1

Hymenanthera R. Br. = *Melicytus*

Hymenatherum Cass. = *Thymophylla*

Hymenidium Lindley (~ *Pleurospermum*). Umbelliferae (III 5). 1 Himal.: *H. densiflorum* Lindley

Hymenocallis Salisb. Amaryllidaceae (Liliaceae s.l.). 30–40 warm Am. Filaments basally united by a consp. tube (corona) larger than P. R: KB 1954: 201. Some with chlorophyllous seed integuments; alks – some used locally medic. Some cult. orn. (Peruvian daffodil, ismene) esp. *H.* × *festalis* (Worsley) Schmarse (*H. longipetala* (Lindley) Macbr. (Peru) × *H. narcissiflora*) & *H. narcissiflora* (Jacq.) Macbr. (basket flower, Andes of Peru & Bolivia)

Hymenocardia Wallich ex Lindley. Euphorbiaceae (Hymenocardiaceae). 6–7 trop. & S Afr., 1 SE As. to Sumatra. Some dyes. Close to Ulmaceae

Hymenocardiaceae Airy Shaw. See Euphorbiaceae

Hymenocarpos Savi. Leguminosae (III 18). 1 Medit. (incl. Eur.), W As.: *H. circinnatus* (L.)

Savi

Hymenocephalus Jaub. & Spach (~ *Centaurea*). Compositae (Card.-Cent.). 1 Iran

Hymenochlaena Bremek. = *Strobilanthes*

Hymenoclea Torrey & A. Gray. Compositae (Helia.-Amb.). 3 SW N Am. R: Brittonia 25(1973)243. *H. salsola* Torrey & A. Gray ex A. Gray (cheeseweed) – crushed lvs cheese-smelling

Hymenocnemis Hook.f. Rubiaceae (IV 7). 1 Madagascar

Hymenocoleus Robbrecht. Rubiaceae (IV 7). 12 trop. Afr. R: BJBB 45(1975)273, 47(1977)8

Hymenocrater Fischer & C. Meyer. Labiatae (VIII 2). 12 SW As.

Hymenodictyon Wallich. Rubiaceae (I 1). 20 OW E to Sulawesi. Alks. *H. orixense* (Roxb.) Mabb. (*H. excelsum*, trop. As.) – soft timber for tea-boxes, school slates etc., bark a febrifuge

Hymenoglossum C. Presl. Hymenophyllaceae (Hym.). 1 Chile & Juan Fernandez: *H. cruentum* (Cav.) C. Presl

Hymenogyne Haw. Aizoaceae (V). 2 Cape. R: BJ 111(1990)481

Hymenolaena DC. (~ *Pleurospermum*). Umbelliferae (III 5). 6 Himal. to C As.

Hymenolepis Cass. (~ *Athanasia*). Compositae (Anth.-Urs.). 7 S Afr. R: NJB 5(1986)517

Hymenolobium Benth. Leguminosae (III 4). c. 12 trop. S Am.

Hymenolobus Nutt. Cruciferae. 5 Eur. (2), Medit., C As., Aus., N Am., Chile

Hymenolophus Boerl. Apocynaceae. 1 Sumatra: *H. romburghii* Boerl. – rubber source

Hymenolyma Korovin = *Hyalolaena*

Hymenonema Cass. Compositae (Lact.-Cat.). 2 Greece

Hymenopappus L'Hérit. Compositae (Hele.-Hym.). 14 S N Am. R: Rhodora 58(1956)163. Roots of some spp. form. chewed as medic. by Indians

Hymenophyllaceae Link. Filicopsida. 10/600 trop. & temp. Filmy ferns. Usu. epiphytic with scale-less rhiz., vasc. tissue usu. in protostele, only 1 tracheid (in section) or 0 in some spp., some rootless. Lvs 5 mm–60 cm long but lamina only 1 cell thick (exc. *Cardiomanes*) with open venation & 0 stomata. Sporangia on receptacles continuous with vein-tips; indusia tubular or 2-lobed. Pls of ever-humid habitats esp. cloud forests of trop. mts, streamsides etc.

Classification & principal genera:

Cardiomanoideae (petioles >7 cm; lamina usu. of 4 cell-layers): *Cardiomanes* (only)

Hymenophylloideae (petioles smaller; lamina usu. 1 cell thick (exc. veins)): form. 2 polymorphous genera recognized – *Hymenophyllum* & *Trichomanes* but their limits still controversial, principal segregates presently recognized incl. *Cephalomanes, Crepidomanes, Sphaerocionium*

Hymenophyllopsidaceae Pichi-Serm. Filicopsida. 1/8 NE S Am. Small delicate usu. terr. ferns with solenostelic scaly rhiz. Fronds like *Hymenophyllum*, stomata 0; sori term. on vein with pouch-like indusium

Genus: *Hymenophyllopsis*

Affinities obscure

Hymenophyllopsis Goebel. Hymenophyllopsidaceae. 8 Roraima to W Venezuela

Hymenophyllum Sm. Hymenophyllaceae (Hym.). (Incl. *Amphipterum, Craspedophyllum, Hemicyatheon, Meringium, Myriodon*) c. 250 trop. & S temp., N temp. only in Eur. (2) & Japan. Differs from *Trichomanes* (s.l.) in thread-like (not 2–4 mm across), glabrous (not hairy) rhizome & 2-lipped (not tubular) indusium

Hymenophysa C. Meyer = *Lepidium*

Hymenopogon Wallich = *Neohymenopogon*

Hymenopyramis Wallich ex Griffith. Labiatae (I; Verbenaceae). 6 India to China & SE As.

Hymenorchis Schltr. Orchidaceae (V 16). 10 Philipp. to New Caled. (1) esp. New Guinea (6)

Hymenosporum R. Br. ex F. Muell. Pittosporaceae. 1 E Aus., New Guinea: *H. flavum* (Hook.) F. Muell. Cult. orn. tree with fragrant fls ('native honeysuckle') & winged seeds.

Hymenostachys Bory = *Trichomanes*

Hymenostegia (Benth.) Harms. Leguminosae (I 4). 16 Gulf of Guinea

Hymenostemma Kunze ex Willk. Compositae (Anth.-Leuc.). 1 Spain, Morocco

Hymenostephium Benth. = *Viguiera*

Hymenothrix A. Gray. Compositae (Hele.-Cha.). 5 SW N Am. R: Brittonia 14(1962)101

Hymenoxys Cass. Compositae (Hele.-Gai.). 28 W N Am. to Arg. Some cult. orn. but others toxic weeds of rangeland (sesquiterpene lactones killing many millions of sheep & goats per annum) esp. *H. odorata* DC & *H. richardsonii* (Hook.) Cockerell (pingue) in Texas. The latter & other spp. form. used as chewing-gum by Indians

Hyobanche L. Scrophulariaceae. 7 S Afr.

Hyophorbe Gaertner. Palmae (IV 3). 5 Masc. Unarmed, monoec. R: GH 11(1978)212. Some cult. orn. incl. *H. lagenicaulis* (L. Bailey) H. Moore, reduced (1984) to 9 trees in wild (Round Is., Mauritius); *H. amaricaulis* C. Martius reduced (1988) to 1 tree in a botanic garden

Hyoscyamus L. Solanaceae (1). 15 W Eur. (5) & N Afr. to Somalia, SW & C As. Highly toxic alks. *H. niger* L. (henbane, temp. Euras., natur. N Am.) – seed dormancy up to at least 100 yrs, form. cult. as source of alkaloidal drugs used as hypnotic & narcotic from dried lvs, one of poisons used by Dr Crippen, used to control pl.-pests in Eur. since C2; other spp. used locally

Hyoseris L. Compositae (Lact.-Hyp.). 5 Medit. incl. Eur. (3)

Hyospathe C. Martius. Palmae (V 4f). 2 trop. Am. R: NJB 9(1989)189

Hypacanthium Juz. Compositae (Card.-Card.). 3 C As.

Hypagophytum A. Berger. Crassulaceae. 1 Ethiopia

Hyparrhenia Andersson ex Fourn. Gramineae (40g). 55 Afr., Madag., (with few in) trop. Am. & As., 1 ext. to Medit. R: KB Add. ser. 2(1969)1. *H. filipendula* (Hochst.) Stapf – source of paper pulp of moderate quality; *H. hirta* (L.) Stapf (OW) – grown in SW N Am. for erosion control; *H. rufa* (Nees) Stapf (trop. Afr.) – good fodder (jaragua grass)

Hypecoaceae (Dumort.) Willk. & Lange = Fumariaceae

Hypecoum L. Fumariaceae (II; Hypecoaceae). 20 Med. (Eur. 6) to C As. & N China. R: NJB 10(1990)129. Alks. Fls 2-merous; inner C 3-sect, middle lobe encl. A. In *H. procumbens* L. (Medit.) pollen shed into these lobes which close up before stigma is receptive; when pressed by insects the lobes open & dust the visitors with pollen

Hypelate P. Browne. Sapindaceae. 1 Florida, WI. White timber

Hypelichrysum Kirpiczn. = *Pseudognaphalium*

Hypenanthe (Blume) Blume = *Medinilla*

Hypenia (Benth.) Harley (~ *Hyptis*). Labiatae (VIII 4a). 24 trop. & warm Am. R: BJLS 98(1988)91. Internodes with waxy scales so ants cannot climb to rob nectar ('greasy pole syndrome')

Hyperacanthus E. Meyer ex Bridson (~ *Gardenia*). Rubiaceae (II 1). 2 SE Afr. R: KB 40(1985)275

Hyperaspis Briq. = *Ocimum*

Hyperbaena Miers ex Benth. Menispermaceae (I). 19 warm Am. R: Brittonia 33(1981)81

Hypericaceae Juss. = Guttiferae

Hypericophyllum Steetz. Compositae (Hele.-Cha.). 7 trop. Afr.

Hypericopsis Boiss. Frankeniaceae. 1 S Iran

Hypericum L. Guttiferae (I). c. 370 temp., trop. mts (Eur. 56 with 5 from N Am. natur.). R: BBMNHB 5(1977)293, 8(1981)55, 12(1985)163, 16(1987)1, 20(1990)1. Trees, shrubs or herbs with opp. gland-dotted lvs; some apomictic; extrafl. nectaries; fr. usu. a capsule but berries in 4 distinct parts of genus. Some locally medic. (diuretic etc.) in OW (comm. as 'Hypercal Ointment' (with *Calendula officinalis*) & some cosmetics) & Am., temp. & trop., but toxic (quinones) esp. to sheep; many cult. orn. (St John's wort) esp. *H. androsaemum* L. (tutsan, Eur., W As., N Afr.) – fr. a berry, diuretic shrubs; *H. calycinum* L. (rose of Sharon, SE Bulgaria, N Turkey) – cult. for ground-cover in shade; *H.* 'Hidcote' (prob. *H. calycinum* × *H.* × *cyathiflorum* N. Robson 'Gold Cup' (*H. addingtonii* N. Robson (SW China) × *H. hookerianum* Wight & Arn. (India to Nepal & Thailand) prob. arising Hidcote, Glos., UK)) – sterile familiar garden shrub hypericum now virus-infected; *H.* × *inodorum* Miller (*H. elatum, H. hircinum* L. (Medit. to SW Saudi Arabia) × *H. androsaemum* but without goat odour of *H. hircinum*); *H. perforatum* L. (Eur. & Medit. to C China, natur. W N Am. (Klamath weed), where controlled biologically) – poisonous to stock through photosensitization esp. in sunny countries e.g. Aus., SW US, Iraq, first noted to affect white but not black sheep by Cirillo (1787)

Hypertelis E. Meyer ex Fenzl. Molluginaceae. 9 S Afr. (8, 1 ext. to trop. Afr., Madag.), St Helena (1)

Hyperthelia W. Clayton. Gramineae (40g). 7 trop. Afr., H. dissoluta (Steudel) W. Clayton introd. S Am. R: KB 20(1967)438. Grains of *H. edulis* (C. Hubb.) W. Clayton coll. for food

Hyphaene Gaertner. Palmae (I 3b). c. 10 Afr., Madag., Arabia & India. Dioec., often forked by true dichotomy; bee-poll.; seeds disp. by elephants & baboons. All used by humans, *H. petersiana* Klotzsch basis of imp. basket-weaving industry in Botswana; some cult. orn. incl. *H. thebaica* (L.) C. Martius (doum or dom palm, gingerbread palm, Nile region) – fibrous part of fr. considered to taste like gingerbread, endocarp used for buttons (veg. ivory subs.), lvs used for mats, ropes, paper & fuel

Hypobathrum Blume. Rubiaceae (II 5). 20 Indomal. Some rheophytes

Hypocalymma (Endl.) Endl. Myrtaceae (Lept.) 14 SW Aus. Some cult. orn. shrubs
Hypocalyptus Thunb. Leguminosae (III 27). 3 S & SW Cape
Hypochaeris L. Compositae (Lact.-Hyp.). c.60 Eur. (9), As., N Afr., (esp.) S Am. Some weeds (cat's-ear) & cult. orn. *H. oligocephala* (Svent. & Bramwell) Lack (*Heywoodiella o.*, NW Tenerife) – suffrutex
Hypochoeris L. = praec.
Hypocylix Woloszczak = *Salsola*
Hypocyrta C. Martius = *Nematanthus*
Hypodaphnis Stapf. Lauraceae (I). 1 trop. W Afr.: *H. zenkeri* (Engl.) Stapf
Hypodematiaceae Ching = Dryopteridaceae
Hypodematium Kunze. Dryopteridaceae (II 1). 4 OW warm esp. limestone
Hypoderris R. Br. ex Hook. Dryopteridaceae (I 3). 1 trop. Am.: *H. brownii* J. Sm. ex Hook. – cult. orn., s.t. with epiphyllous plantlets
Hypodiscus Nees. Restionaceae. 15 S & SW Cape, Namaqualand. R: Bothalia 15(1985)488
Hypoestes Sol. ex R. Br. Acanthaceae. c. 40 OW trop. Some cult. orn. foliage pot-pls esp. *H. phyllostachya* Baker ('*H. sanguinolenta*', polka-dot plant, Madag.). *H. verticillaris* (L.f.) Roemer & Schultes (trop. & warm Afr., Arabia) – source of 2 antineoplastic alks (hypoestestafins)
Hypogomphia Bunge. Labiatae (VI). 4 C As., Afghanistan
Hypogon Raf. = *Collinsonia*
Hypogynium Nees = *Andropogon*
Hypolaena R. Br. Restionaceae. 8 Aus.
Hypolepis Bernh. Dennstaedtiaceae. c. 40 trop. & warm. Sori protected by small reflexed margins of frond (no indusium). Some cult. orn., *H. sparsiora* (Schrader) Kuhn (Afr., Madag.) – weedy
Hypolobus Fourn. = ? *Matelea*
Hypolytrum Rich. ex Pers. Cyperaceae. 40 trop. & warm
Hypophyllanthus Regel = *Helicteres*
Hypopitys Hill = *Monotropa*
Hypoxidaceae R. Br. (Liliaceae s.l.). Monocots – Liliidae – Asparagales. 8/220 S hemisph. Herbs (some pachycaul) with corms or rhiz. Lvs usu. tristichous, linear to lanceolate, often plicate, hairy; roots with vessels. Infls spikes to umbel-like clusters or solit. fl. Fls reg.; P 3 + 3 s.t. with short tube, usu. yellow or white (red in *Rhodohypoxis*); nectaries 0; A 3 + 3 (A 3 in *Pauridia*), anthers with longit. dehiscence; $\overline{G}$ (3), 3-loc. (1-loc. in *Empodium*), each loc. with several anatropous to hemianatropous ovules. Fr. a capsule with remains of P or fleshy; seeds small, globose

Genera: *Curculigo, Empodium, Hypoxidia, Hypoxis, Molineria, Pauridia, Rhodohypoxis, Spiloxene*

Some cult. orn.; some fibres (*Curculigo*)

Hypoxidia Friedmann (~ *Hypoxis*). Hypoxidaceae. 2 Seychelles. R: BMNHN 4,6(1984)453. Fls to 12 cm diam.
Hypoxis L. Hypoxidaceae. c. 150 warm & trop., esp. S hemisph. Some cult. orn. with white or yellow fls.
Hypsela C. Presl. Campanulaceae. 4 Aus., NZ, S Am. Cult. orn. creeping pls esp. *H. reniformis* (Kunth) C. Presl (S Am.)
Hypselandra Pax & K. Hoffm. = *Boscia*
Hypselodelphys (Schumann) Milne-Redh. Marantaceae. 4 trop. Afr.
Hypseloderma Radlk. = *Camptolepis*
Hypseocharis Remy. Oxalidaceae (Hypseocharitaceae). 8 Andes. Loculicidal capsule
Hypseocharitaceae Wedd. = Oxalidaceae
Hypseochloa C. Hubb. Gramineae (21d). 2 Mt Cameroun & Tanzania
Hypserpa Miers. Menispermaceae (V). 9 Indomal. to Polynesia. *H. nitida* Miers (Indomal.) – local medic., mountaineers' ropes in SE As.
Hypsophila F. Muell. Celastraceae. 3 NE Aus.
Hyptianthera Wight & Arn. Rubiaceae. 2 N India, Thailand
Hyptidendron Harley (~ *Hyptis*). Labiatae (VIII 4a). 16 trop. & warm Am. R: BJLS 98(1988)93. *H. arboreum* (Benth.) Harley (NE S Am.) – to 18 m tall
Hyptis Jacq. Labiatae (VIII 4e). Excl. *Hypenia* & *Hyptodendron*, c. 300 warm & trop. Am. with few weedy spp. OW. R: RMLP 7(1949)153. Explosive poll. mech. with middle of C lobes holding A & style under tension until set off by insects (leading to cross-poll.) or wind (to self-poll.). Some locally used medic. & seeds for food (e.g. *H. emoryi* Torrey, SW N Am.) or oil (e.g. *H. spicigera* Lam. (black sesame or beni seed, Am. & Afr.), stems also provide a

fibre): *H. suaveolens* (L.) Poit. (S Am.) locally medic. & poss. anticancer agent

Hyrtanandra Miq. = *Gonostegia*

hyssop *Hyssopus* spp. esp. *H. officinalis*; (of the Bible) *Origanum syriacum*; *Agastache* spp.; **anise h.** *A. foeniculum*; **hedge h.** *Gratiola officinalis*; **water h.** *Bacopa monnieri*

Hyssopus L. Labiatae (VIII 2). 5 S Eur. (1), Medit. to C As. *H. officinalis* L. (hyssop, S Eur.) – medic., oil for liqueurs, antibacterial

Hysterionica Willd. Compositae (Ast.-Ast.). 10 S Brazil, Uruguay, Arg.

Hystrichophora Mattf. Compositae (Vern.-Vern.). 3 Tanzania. *H. macrophylla* Mattf., known only from 1 specimen. R: KB 43(1988)249

Hystrix Moench. Gramineae (23). 9 N Am., temp. As. Some cult. orn. N Am. for bouquets (bottle-brush grass)

I

i *Inocarpus fagifer*

Ianthe Salisb. = *Spiloxene*

Ibarrea Lundell = *Ardisia*

Ibatia Decne. = *Matelea*

Iberidella Boiss. = *Aethionema*

Iberis L. Cruciferae. 40 Eur. (19), Medit. Candytuft. Fls in umbelliform clusters which elongate with age, outer 2 C longer than inner. Usu calcicoles. Some locally medic. but imp. cult. orn. esp. *I. amara* L. (W Eur.) & *I. umbellata* L. (florist's c., S Eur.) – annuals & *I. sempervirens* L. (Medit.) – evergreen perennial rock-pl.

Ibervillea E. Greene. Cucurbitaceae. 4 SW N Am. *I. sonorae* (S. Watson) E. Greene with swollen caudex to several kg

Ibetralia Bremek. = *Alibertia*

Ibicella Eselt. Pedaliaceae (Martyniaceae). 3 warm & trop. S Am. *I. lutea* (Lindley) Eselt. (natur. Calif., Aus., S Afr.) – carnivorous with viscid lvs entangling gnats & flies

ibogo *Tabernanthe iboga*

Iboza N.E. Br. = *Tetradenia*

iburu *Digitaria iburua*

Icacina A. Juss. Icacinaceae. 6 trop. Afr. *I. oliviformis* (Poiret) Raynal (*I. senegalensis*, W Afr.) – pyrophyte, tubers (to 50 kg) & seeds can provide a flour

Icacinaceae (Benth.) Miers. Dicots – Rosidae – Celastrales. 52/300 trop., few temp. Trees & shrubs, s.t. scrambling, or lianes, s.t. accum. aluminium & producing alks; stems often with unusual anatomy, s.t. with interxylary phloem. Lvs spirally arr. (opp. in *Iodes*), simple, entire to toothed; stipules 0. Fls reg., bisexual (rarely pl. polygamous to dioec.), (3)4 or 5(6)-merous in usu. axillary infls; pedicel articulated with fl., K a tube with imbricate (rarely valvate) lobes, C usu. valvate, free or ± connate, rarely 0, A same no. as C or K with filaments free or on C-tube alt. with lobes, anthers with longit. slits, disk s.t. present, $\underline{G}$ ((2)3(–5)) with term. style, pseudomonomerous in *Gomphandra*, *Gonocaryum* & *Phytocrene*, usu. only 1 loc. fertile (3 ovuliferous in *Emmotum*) with (1)2 anatropous, unitegmic ovules pendulous, back-to-back, from top of G, with funicular thickening near micropyle. Fr. often a 1-seeded drupe, rarely samaroid; seeds with straight or curved embryo & well-developed oily endosperm or 0. x = 10, 11

Principal genera: *Alsodeiopsis*, *Citronella*, *Gomphandra*, *Iodes*, *Phytocrene*, *Pyrenacantha*, *Stemonurus*, *Urandra*. Many monotypic

Timber (*Apodytes*, *Cantleya*, *Pennantia*), some local foods & medic.

Icacinopsis Roberty = *Dichapetalum*

icaco, icacos *Chrysobalanus icaco*

Icacorea Aublet = *Ardisia*

ice-plant *Mesembryanthemum crystallinum*

Iceland poppy *Papaver nudicaule*

Ichnanthus Pal. Gramineae (34b). 39 trop. Am. with 1 pantrop. R: SB 7(1982)85

Ichnocarpus R. Br. Apocynaceae. Incl. *Micrechites*, 12 Indomal. to S China & trop. Aus. R: Blumea 39(1994)73. Some rubber

Ichthyothere C. Martius. Compositae (Helia.). 18 trop. S Am. Some fish-poisons (active principle a polyacetylene – ichthyothereol)

ichu *Stipa ichu*

Icianthus E. Greene = *Euklisia*

Icma Philippi = ? (Compositae)

Icomum Hua = *Aeollanthus*

Idahoa Nelson & Macbr. Cruciferae. 1 W US
Idanthisa Raf. = *Anisacanthus*
Idenburgia Gibbs = *Sphenostemon*
Idertia Farron = *Ouratea*
Idesia Maxim. (*Cathayeia*). Flacourtiaceae (8). 1 China, Japan: *I. polycarpa* Maxim. – cult. orn. dioec. or polygamous tree
idewa *Haplormosia monophylla*
idigbo *Terminalia ivorensis*
Idiopteris T. Walker = *Pteris*
Idiospermaceae S.T. Blake = Calycanthaceae
Idiospermum S.T. Blake. Calycanthaceae (Idiospermaceae). 1 Queensland: *I. australiense* (Diels) S.T. Blake – mature twigs with vessel elements with simple perforation plates, those of young twigs with scalariform; embryos to 6.5 cm across & 100 g (largest known?) found in stomachs of poisoned cattle in Queensland & thus rediscovered (1971) but still a threatened rain-forest tree
Idiothamnus R. King & H. Robinson (~ *Eupatorium*). Compositae (Eup.-Crit.). 4 S Am. R: MSBMBG 22(1987)335
Idria Kellogg = *Fouquieria*
Ifloga Cass. Compositae (Gnap.-Gnap.). 6 Medit. (Eur. 1), Macaronesia, S Afr. R: OB 104(1991)154
Ighermia Wiklund (~ *Asteriscus*). Compositae (Inul.). 1 SW Morocco
Ignatius bean *Strychnos ignatii*
Iguanura Blume. Palmae (V 4m). 18 W Mal. R: GBS 28(1976)191, KB 34(1979)143
igusa *Juncus effusus*
Ihlenfeldtia Hartmann (~ *Cheiridopsis*). Aizoaceae (V). 2 Cape. R: BJ 114(1992)29
Ikonnikovia Lincz. Plumbaginaceae (II). 1 C As., NW China: *I. kaufmanniana* (Regel) Lincz.
ilama *Annona diversifolia*
ilang-ilang = ylang-ylang
ilapongu *Hopea ponga*
ilb *Ziziphus spina-christi*
Ildefonsia Gardner = *Bacopa*
Ileostylus Tieghem. Loranthaceae. 1 NZ
ilex *Quercus ilex*
Ilex L. Aquifoliaceae. c. 400 cosmop. (Eur. 3) esp. trop. & temp. As. & Am. Dioec. or polygamo-dioec., often evergreen trees & shrubs (some epiphytes, e.g. *I. baasii* Stone & Kiew (Borneo) – tubers) with spirally arr. (opp. in some Bornean spp.) lvs & drupes; some domatia. Hollies, with alks incl. caffeine & theobromine: cult. orn. (many hybrids & other cvs), timber & stimulants; R: Holly Soc. Am., *The Holly handbook*. Most commonly cult. are *I. aquifolium* & *I.* × *altaclerensis*, which differs in spines pointing towards leaf-apex. *I.* × *altaclerensis* (Loudon) Dallimore (*I.* × *altaclarensis*, hybrids ('altacs') between *I. aquifolium* & *I. perado* Aiton (Tenerife, Gomera)) incl. varieg. forms like 'Golden King' (female) & 'Camelliifolia' with usu. entire spineless lvs; *I. aquifolium* L. ((Eur.) holly, Eur. & Medit.) – vern. name (arch. *hulver, holm(e)* from Anglo-Saxon *holegn* (i.e. holy tree when tradition of decorating houses with it in winter, prob. assoc. with Romans' Saturnalia, was taken over by Christians for Christmas) – thorns & red drupes added to attraction in Christian eyes, much cult., with many cvs incl. 'Ferox' (hedgehog h., male) with rows of spines on adaxial surfaces of lvs, & form. the most used hedgepl. for formal gardening since Tudor times, some cvs with entire thornless lvs like adult forms of the type – app. this is usu. only manifest in maturity (cf. *Hedera*), timber white (colour preserved if timber cut in winter & dried before hot weather) used for veneers, inlay, musical instruments, etc. or stained black as ebony subs., burns when felled (little water), foliage also flammable, form. pollarded for fodder in GB; *I. anomala* Hook. & Arn. (Hawaii) – wood form. for saddle-trees, canoe decoration & anvils for kapa beating; *I. cassine* L. (dahoon, SE US, Cuba) – lvs used as a tea, cult. orn.; *I. crenata* Thunb. (Japanese h., Japan, Korea) – much cult. in Japan incl. as bonsai; *I. glabra* (L.) A. Gray (inkberry, E N Am.) – imp. bee-plant; *I. guayusa* Loes. (guayusa, E Colombia, Ecuador, W Peru) – lvs a stimulant tea (caffeine); *I. opaca* Aiton (American h., C & E US) – similar uses to *I. aquifolium*, many cvs; *I. paraguariensis* A. St-Hil. ((yerba) maté, Brazilian or Paraguay tea, Paraguay & adjacent parts of Argentina & Brazil, Uruguay) – locally imp. tea made from lvs of cult. & wild pls; *I. verticillata* (L.) A. Gray (feverbark, E N Am.) – lvs a tea subs., cult. orn. with many cvs; *I. vomitoria* Aiton (yaupon, SE US. Mex., natur. Bermuda) – dried lvs a tea for N Am. Indians
Iliamna E. Greene = *Sphaeralcea*

iliau *Wilkesia gymnoxiphium*

Iljinia Korovin ex Komarov (~ *Haloxylon*). Chenopodiaceae (III 3). 1 C As.: *I. regelii* (Bunge) Komarov

Illawara palm *Archontophoenix cunninghamiana*; **I. pine** *Callitris rhomboidea*

Illecebraceae R. Br. = Caryophyllaceae

Illecebrum L. Caryophyllaceae (I 2). 1 Canary Is., W Eur., Medit.: *I. verticillatum* L.

Illiciaceae (DC) A.C. Sm. Dicots – Magnoliidae – Illiciales. 1/42 SE As., SE US to Hispaniola. Glabrous aromatic evergreen trees & shrubs with scattered ethereal oil-cells. Lvs spirally arr. (often crowded as to appear whorled), simple, entire; stipules 0. Fls small, solit. (or 2 or 3 together), axillary or supra-a., bisexual, reg.; P (7–)12–30(–33) often not clearly differentiated into C & K, often in spiral series, the outer bract- or sepal-like, others more petaloid or reduced & transitional to A, A (4–)c. 50, in (1–)several spiral series with short thick filaments & anthers with longit. slits & prolonged connectives, $\underline{G}$ (5–)7–15(–21) in a tight spiral, often compressed & narrowed to unsealed style, each carpel with 1 ventrally near-basal anatropous, bitegmic ovule. Fr. a head of often radiating 1-seeded follicles; embryo v. small, endosperm copious, oily. x = 13, 14

Only genus: *Illicium* – some comm. oils

Illicium L. Illiciaceae. 42 India to Korea & W Mal., SE N Am. to Hispaniola. R: Sargentia 7(1947)10. First insecticidal fumigants known (China, C2 BC); some cult. orn. (beetle-poll.) & comm. oils etc. *I. anisatum* L. (Japanese anise, Korea & Japan) – toxic seeds used to kill fish, locally medic., branches used to decorate Buddhist graves; *I. floridanum* Ellis (purple or tree a., SE US) – cult. orn.; *I. verum* Hook.f. (star a., Chinese a., SE China, NE Vietnam) – unripe fr. a culinary spice (main sweet constituent = *trans*-anethol) esp. in liqueurs when distilled, also medic.

Illigera Blume. Hernandiaceae. 18 OW trop. R: BJ 89(1969)157

illipe *Shorea macrophylla* & other spp.; *Madhuca* spp.

ilomba *Pycnanthus angolensis*

Iltisia S.F. Blake = *Microspermum*

Ilysanthes Raf. = *Lindernia*

Imantina Hook.f. = *Morinda*

Imbralyx Geesink = *Fordia*

Imbricaria Comm. ex Juss. = *Mimusops*

imbu *Spondias* spp.

imbuya *Phoebe porosa*

Imeria R. King & H. Robinson (~ *Eupatorium*). Compositae (Eup.-Crit.). 2 Venez. R: MSBMBG 22(1987)357

Imerinaea Schltr. Orchidaceae (V 13). 1 Madagascar

Imitaria N.E. Br. (~ *Gibbaeum*). Aizoaceae (V). 1 S Afr.: *I. muirii* N.E. Br. – cult. orn.

immortelles = everlastings

imo *Colocasia esculenta*

Impatiens L. Balsaminaceae. 850 trop. & N temp. esp. India (N & C Am. 6, Eur. 1). R (Afr. 109: C. Grey-Wilson (1980) *I. of Afr.*). Balsams, jewel-weed, touch-me-not (allusion to explosive 5-valved capsule with fleshy pericarp, outer layers of which are v. turgid &, when ripe, a touch can set off the valves, rolling up inwards, starting at base, & scattering seeds) – tender succ. herbs much grown as bedding & housepls with resupinate fls, insect or bird-poll., & extrafl. nectaries, some Indian spp. with 3-sporangiate anthers; chromosomes can be counted in pollen-grains on herbarium sheets app. at a resting stage at first mitotic interphase until released from pl.; some epiphytes. *I. balsamina* L. (garden balsam, C & S India, – cult., fls used to dye fingernails; *I. capensis* Meerb. (orange b., N Am.(!)) – cult., natur. Eur.; *I. glandulifera* Royle (policeman's helmet, Himal.) – natur. in N Am. & Eur. incl. GB where v. aggressive streamside herb growing to 3 m in 1 season & ousting native vegetation; *I. hawkeri* W. Bull (Papuasia) – v. variable & now fashionable as pot-pl. ('New Guinea Hybrids'), salt-source in New Guinea; *I. parviflora* DC (Siberia) – scruffy annual ± natur. GB where a favourite subject for physiologists; *I. walleriana* Hook.f. (busy lizzie, sultan's fl., Tanzania to Mozambique) – ubiquitous pot-pl. but not natur. in GB as no seeds set, cf. bee-poll. *I. glandulifera*, many cvs

Impatientella Perrier = *Impatiens*

Imperata Cirillo. Gramineae (40a). 8 trop. & warm. *I. cylindrica* (L.) Pal. (lalang, alang-alang, cogon or cotton grass, kunai (New Guinea), OW trop. to Medit.) – v. bad weed of burnt-over forest, abandoned pasture etc., suggested for reclamation & for thatching, hat-making etc., palatable to few animals incl. water buffalo

Imperatoria L. (~ *Peucedanum*). Umbelliferae (III 10). 3 Eur., N Afr.

imphee *Sorghum bicolor*
Inca wheat *Amaranthus caudatus*
incaparina *Gossypium* spp.
Incarvillea Juss. Bignoniaceae. 11–14 Himal., C & E As. (Eur. in Oligocene). R: NRBGE 23(1961)303. Herbs with tuberous or woody roots & alks. Cult. orn. hardy pls esp. *I. delavayi* Bureau & Franchet (purple fls) & *I. mairei* (A. Léveillé) Grierson (*I. grandiflora*, crimson fls) from China
incense *Boswellia* spp.; **Cayenne i.** *Protium guianense*; **i. cedar** *Calocedrus decurrens*; **i. plant** *Calomeria amaranthoides*; **i. tree** *P. heptaphyllum*
India rubber tree *Ficus elastica*
Indian almond *Terminalia catappa*; **I. arrow wood** *Euonymus atropurpureus*; **I. bean-tree** *Catalpa bignonioides*; **I. beech** *Millettia* sp. (*Pongamia pinnata*); **I. berry** *Anamirta paniculata*; **I. blanket** *Gaillardia* spp.; **I. corn** *Zea mays*; **I. cress** *Tropaeolum majus*; **I. dill** *Anethum graveolens*; **I. fig** *Opuntia ficus-indica*; **I. gum** spp. of *Acacia*, *Anogeissus*, *Combretum*, *Limonia*; **I. hemp** *Apocynum cannabinum*; **I. horse-chestnut** *Aesculus indica*; **I. kale** *Caladium lindenii*; **I. laburnum** *Cassia fistula*; **I. lavender** *Bursera penicillata*; **I. liquorice** *Abrus precatorius*; **I. madder** *Oldenlandia umbellata*, *Rubia cordifolia*; **I. mallow** *Abutilon theophrasti*; **I. millet** *Pennisetum glaucum*; **I. mulberry** *Morinda citrifolia*; **I. olive** *Olea europaea* ssp. *cuspidata*; **I. paint-brush** *Castilleja* spp.; **I. pipe** *Monotropa uniflora*; **I. potato** *Ipomoea pandurata*; **I. root** *Asclepias curassavica*; **I. shot** *Canna indica*; **I. strawberry** *Potentilla indica*; **I. tobacco** *Lobelia inflata*; **I. willow** *Polyalthia longifolia* (fastigiate form)
Indigastrum Jaub. & Spach = *Indigofera*
indigo *Indigofera* spp.; **Assam i.** *Strobilanthes cusia*; **bastard i.** *Amorpha fruticosa*; **Chinese green i.** *Rhamnus* spp.; **false i.** *Amorpha* & *Baptisia* spp.; **Java i.** *I. arrecta*, *I. tinctoria*
Indigofera L. Leguminosae (III 8). c. 700 trop. & warm. Indigo. Form. imp. dye-pls cult. in India & Sumatra esp. *I. arrecta* Hochst. ex A. Rich. (trop. Afr.) & *I. tinctoria* L. (SE As.), the natural product almost replaced by aniline dyes (still in hair dyes – black henna), though some spp. still grown as green manure & fodders, esp. *I. spicata* Forssk. (trop. OW) but dogmeat made from horses fed on it hepatotoxic (indospicine); *I. pentaphylla* L. (India) – lvs sold in market as sour component for curry; some cult. orn. incl. *I. heterantha* Wallich ex Brandis (*I. gerardiana*, NW Himal.) – decid. shrub, fls boiled in milk as a tonic in Kashmir
Indobanalia A.N. Henry & Roy. Amaranthaceae (I 2). 1 SW India
Indocalamus Nakai. Gramineae (1a). c. 10 S China. *I. tesselatus* (Munro) Keng with flowering cycle of at least 115 yrs
Indochloa Bor = *Euclasta*
Indocourtoisia Bennet & Raiz. = *Courtoisina*
Indofevillea Chatterjee. Cucurbitaceae. 1 Assam: *I. khasiana* Chatterjee
Indokingia Hemsley = *Gastonia*
Indoneesiella Sreemad. Acanthaceae. 2 India
Indopiptadenia Brenan. Leguminosae (II 3). 1 India, Nepal: *I. oudhensis* (Brandis) Brenan
Indopoa Bor. Gramineae (31d). 1 India. Caryopsis needle-like in narrow pocket along keel of lemma
Indopolysolenia Bennet (*Polysolenia*). Rubiaceae (I 8). 2 E Himal., Burma
Indorouchera Hallier f. Linaceae. 3 Nicobar & Andaman Is., SE As. to W Mal.
Indosasa McClure. Gramineae (1a). 12 S China & Vietnam. R: APS 21(1983)60
Indosinia J.E. Vidal. Ochnaceae. 1 S Vietnam. R: BJ 113(1991)183
Indotristicha P. Royen (~ *Tristicha*). Podostemaceae. 2 India. R: BMNHN 4,10(1988)173
Indovethia Boerl. = *Sauvagesia*
Inezia E. Phillips. Compositae (Anth.-Tham.). 2 S Afr.
Inga Miller. Leguminosae (II 5). c. 350 trop. & warm Am. Only pinnate-leaved II. Some fish-, some monkey-disp. Amazon; some with foliar nectaries visited by ants which remove other insects. Some planted as crop-shade or orn., others with ed. sweet pulp in fr., those of *I. feuillei* DC offered by Atahualpa (last Inca Emperor) to Pizarro, who killed him anyway. Some timbers, fuelwood etc. e.g. *I. laurina* (Sw.) Willd. (Spanish oak, WI), *I. vera* Willd. (guaba) – fls open afternoon & poll. lepidoptera & birds (sucrose-rich nectar), at night nectar hydrolysed (? micro-organisms) to hexoses & with sour smell attractive to bats; *I. acrocephala* Steudel (Amaz.) – red latex used to fix dyes for tourist souvenirs
Inhambanella (Engl.) Dubard. Sapotaceae (I). 1 W & 1 SE trop. Afr. R: T.D. Pennington, *S.* (1991)140
inkberry *Phytolacca americana*, *Ilex glabra*

inkweed *Phytolacca octandra*
innocence *Hedyotis caerulea*
Inobulbon (Schltr.) Schltr. & Kränzlin = *Dendrobium*
Inocarpus Forster & Forster f. Leguminosae (III 4). 3 Mal., Pacific. Leaf, fl. & fr. v. 'unleguminous' in aspect. *I. fagifer* (Parkinson) Fosb. (*I. edulis*, i, Polynesian, Tahiti or O'taheite chestnut, Pacific) – seeds ed. raw or roasted
inoy *Poga oleosa*
inside-out-flower *Vancouveria hexandra*
Intsia Thouars. Leguminosae (I 4). 3 trop. As., coasts of Indian & Pacific Oceans. Timber esp. *I. bijuga* (Colebr.) Kuntze (kwila, esp. abundant on Kabara so centre of canoe-building in Fiji & much of Tonga) & *I. palembanica* Miq. (*Afzelia p.*, merbau, Borneo or Malacca teak, Mal.)
Intybellia Cass. = *Crepis*
Inula L. Compositae (Inul.). c. 90 temp. & warm OW (Eur. 18 (excl. *Limbarda*)). Herbs (often statuesque) with alks. Some medic., cult. orn., dyes etc.: *I. britannica* L. (Euras.) – cult. orn. with lactone antifeedant effective on flour-beetles; *I. conyzae* (Griess.) Meikle (*I. conyza*, ploughman's spikenard, Eur., Med.): *I. helenium* L. (elecampane (enula campana), C As., natur. Eur., N Am., Japan etc.) – root (*Radix helenii*, Enulae) form. medic. in skin & chest disease, also candied & used to flavour absinthe, cult. as root vegetable at time of Pliny; *I. magnifica* Lipsky (E Cauc.) & *I. racemosa* Hook.f. (W Himal., cult. as medic. pl. in India; a lactone (isoalantodiene) potent plant-growth regulator) – robust herb. perennials to 2–3 m
Inulanthera Källersjö (~ *Athanasia*). Compositae (Anth.-Gon.). 10 S Afr. R: NJB 5(1986)539
Inulopsis (DC) O. Hoffm. = *Podocoma*
Inversodicraea Engl. = *Ledermanniella*
Iocenes R. Nordenstam (~ *Senecio*). Compositae (Sen.-Sen.). 1 Chile, Argentina
Iochroma Benth. Solanaceae (1). 15 trop. S Am. Shrubs & trees, some local medic. in Amaz. & cult. orn. esp. *I. cyaneum* (Lindley) Green (*I. tubulosum*), NW S Am.) with deep blue tubular fls (other spp. with yellow, scarlet or white fls)
Iodanthus (Torrey & A. Gray) Steudel (*Oclorosis*). Cruciferae. 1 E N Am., 3 Mex. R: JAA 69(1988)116
Iodes Blume (*Ioedes*). Icacinaceae, 28 OW trop.
Iodina Miers (*Jodina*). Hook. & Arn. ex Meissner. Santalaceae. 1 S Brazil, Uruguay, Arg.: *I. rhombifolia* (Hook. & Arn.) Reisseck – tanning material
iodine bush *Mallotonia gnaphalodes*
Iodocephalus Thorel ex Gagnepain. Compositae (Vern.-Vern.). 1 Laos
Ioedes Blume = *Iodes*
Iogeton Strother. Compositae (Helia.-Verb.). 1 Panamá
Ionacanthus Benoist. Acanthaceae. 1 Madagascar
Ionactis E. Greene (~ *Aster*). Compositae (Ast.-Sol.). 5 N Am. R: Brittonia 44(1992)247. Goldenasters
Ionidium Vent. = *Hybanthus*
Ionopsidium Reichb. (*Jonopsidium*). Cruciferae. 5 Medit. incl. Eur. *I. acaule* (Desf.) Reichb. (violet cress, Portugal) – minute cult. orn. ann. for paving
Ionopsis Kunth. Orchidaceae (V 10). 3 trop. & warm Am. Cult. orn. epiphytes
Iostephane Benth. Compositae (Helia.-Helia.). 4 Mex. R: Madrono 30(1983)34. *I. madrensis* (S. Watson) Strother (cachana) – roots medic.
ipecacuanha *Psychotria ipecacuanha*; **American i.** *Gillenia stipulata*; **bastard i.** *Asclepias curassavica*; **Carolina i.** *Euphorbia ipecacuanhae*; **false i.** *Richardia scabra*; **Goa i.** *Naregamia alata*; **Indian i.** *Cryptocoryne spiralis*; **i. spurge** *E. ipecacuanhae*; **white i.** *Hybanthus calceolaria*
Ipheion Raf. = *Tristagma*
Iphigenia Kunth. Colchicaceae (Liliaceae s.l.). 5 Afr., 2 Madag., 6 India with 1 ext. to Aus. (? NZ 1, perhaps referable to *Wurmbea*). Alks
Iphigeniopsis F. Buxb. = *Iphigenia*
Iphiona Cass. Compositae (Inul.). 11 Middle E & NE Afr. to C As. R: NJB 5(1985)169
Iphionopsis A. Anderb. (~ *Iphiona*). Compositae (Inul.). 2 NE & E Afr., Madag. R: NJB 5(1985)51
ipoh *Antiaris toxicaria*
Ipomoea L. Convolvulaceae. (Incl. *Mina*, *Pharbitis*) c. 650 trop. & warm temp. (Am. 327 (R: T 45(1996)3); Eur. 2 incl. *I. sagittata* Poiret, trop. Am. & Eur.). Climbing herbs or shrubs (even trees incl. some pioneers in Mex. arid areas), some with tubers, others caudiciform

succ. (e.g. *I. bolusiana* Schinz, S & trop. Afr., Madag.), differing from *Convolvulus* in stigma not linearly divided; extrafl. nectaries. Alks. Some ed. tubers, fls, shoots etc., drugs & cult. orn. (morning glory). *I. alba* L. (*I. bona-nox*, moonflower, trop. Am., widely natur.) – cult. for large white scented night fls, calyces a curry veg. in Sri Lanka; *I. aquatica* Forssk. (kangkong, water spinach, OW) – cult. for ed. shoots esp. white-stemmed form in e.g. Hong Kong; *I. batatas* (L.) Lam. (sweet potato, 'yam' (US), Brazilian arrowroot, cultigen spread from C Am. ('kumar' in Peru) to Polynesia ('kumara') by ? Spanish & thence to NZ by Maoris (staple crop, no grains), now widely cult. in trop. & warm incl. Japan where v. imp.) – hexaploid (*I. trifida* (Kunth) G. Don f. = feral forms) poss. derived from *I. leucantha* Jacq. (diploid, ? *I. trichocarpa* (Kunth) G. Don f. × *I. lacunosa* L., trop. Am.), swollen tubers ed. starch & alcohol source, 110m t per annum in 1980s (more than fifth world root-crop, many cvs, early Hawaiians having 230 but all exc. 24 now lost, genes from '*I. trifida*' being introduced to stocks to improve nematode resistance; *I. coccinea* L. (*Quamoclit c.*, US) – fls scarlet; *I. imperati* (Vahl) Griseb. (*I. stolonifera*, trop. & warm) – sand-binder; *I. lobata* (Llave & Lex.) Thell. (trop. Am.) – lvs palmately lobed, cult. orn. with red bird-poll. fls fading to yellow; *I. leptophylla* Torrey (W N Am.) – ants attracted to extrafl. nectaries on lvs & K permit seed production increase × 100; *I. nil* (L.) Roth (Jap. morning glory, trop., introd. As. by Portuguese early 1500s) – much used in plant physiology (*Pharbitis n.*), cult. orn., esp. fashionable in Jap. Edo period being much depicted on screens, seeds allegedly purgative; *I. obscura* (L). Ker Gawler (OW trop.) – potherb in Sri Lanka; *I. orizabensis* (J. Pellett.) Steudel (Mex.) – scammony root, the source of a resin, ipomoea, a drastic purgative; *I. pandurata* (L.) G. Meyer (Indian potato, p. vine, N Am.) – large tuberous root, pedicel nectaries attractive to *Crematogaster* & other ants so fr. survival inceased 3–4 times, cult. orn. with white fls; *I. pes-caprae* (L.) R. Br. (trop. beaches) – sand-binder; *I. purga* (Wender.) Hayne (jalap, Mex.) – dried tubers with high levels of purging resin, ipomoea; *I. purpurea* (L.) Roth (morning glory, trop. Am. (? orig. Mex.), natur. N Am. etc.) – cult. orn., *I. quamoclit* L. (cupid flower, trop. Am.) – fls scarlet, lvs deeply dissected pinnately, cult. orn., weed of Aus. sugar-cane; *I. sloteri* (House) Oostr. (cardinal climber) – tetraploid derived from *I. coccinea* × *I. quamoclit* (*I.* × *multifida* (Raf.) Shinners), cult. orn.; *I. spathulata* Hall. f. (E Afr.) – root infusion used against eye disease; *I. tricolor* Cav. (*I. rubrocaerulea*, Mex. & C Am.) – ergoline alks the effective hallucinogenic agent in seeds, cult. orn. esp. 'Heavenly Blue' (reddish buds but increasing pH turns C blue); *I. tuboides* Degener & Oostr. (Hawaii) – extrafl. nectaries but no native ants in Hawaii (phylogenetic inertia?); *I. violacea* L. (*I. macrantha*, *I. tuba*, coastal & pantrop.) – floating seeds with air cavity, cult. orn.

Ipomopsis Michaux. Polemoniaceae. 26 W N Am. & Florida, S Am. (1: Arg. & Chile). R: Aliso 3(1956)351. Like *Gilia* but stems with well-developed lvs. Some cult. orn. herbs

Ipsea Lindley. Orchidaceae (V 11). 2 India, Sri Lanka. R: KB 42(1987)937

Iranecio R. Nordenstam (~ *Senecio*). Compositae (Sen.-Sen.). 16 SE Eur. (1) to Iran. R: Fl. Iranica 164(1989)53

Irania Hadač & Chrtek = *Fibigia*

Iré rubber *Funtumia elastica*

Irenella Süsseng. Amaranthaceae (II 2). 1 Ecuador

Irenepharsus Hewson. Cruciferae. 3 SE Aus. R: JABG 6(1982)1

Iresine P. Browne. Amaranthaceae (II 2). c. 80 trop. & temp. esp. Am. & Aus. (? orig. S Am.). Some locally medic., others cult. for orn. foliage (blood-leaf) esp. *I. herbstii* Hook. (S Am.) – lvs purplish-red or green, with yellow veins

Iriartea Ruíz & Pavón. Palmae (V 2a). 1–2 trop. Am. R: GH 9(1963)275. Stem withers basally to be supported by aerial roots, allowing tree to move away from obstacles (cf. *Socratea*); some roots spiny; bee-poll. *I. ventricosa* C. Martius (paxiuba) has thickened 'belly' halfway up trunk

Iriartella H. Wendl. Palmae (V 2a). 2 trop. S Am.

Iridaceae Juss. Monocots – Liliidae – Liliales. 82/1700 cosmop. but esp. S Afr., E Medit., C & S Am. Usu. geophytic herbs (symp. rhiz., corms or bulbs), less often (S Afr.) evergreen (*Aristea*, *Bobartia*, *Dietes*, *Dierama*, *Pillansia*) or even shrubby (*Klattia*, *Nivenia*, *Witsenia*), rarely annual (3 spp. of *Sisyrinchium* – Cronquist); usu. with crystals of calcium oxalate in some cells, alks 0 but s.t. pl. poisonous, storing starch or fructosans; vessels in roots (in aerial organs also, in *Sisyrinchium*), commonly mycorrhizal & without root-hairs. Lvs usu. distichous, parallel-veined, with sheathing base & narrow blade (rarely with petiole & expanded blade). Fls bisexual, reg. or irreg. in term. infls or solit., subtended by a bract, the infl. by 1 or 2 expanded bladeless sheaths forming a spathe; P 3 + 3 alike or not, often united by basal tube, nectaries septal or at base of P or on inner P (e.g. *Tigridia*) or base of

A (*Iris*) or 0 (*Sisyrinchium*), A 3 (2 in *Diplarrhena*) opp. outer P, filaments often united by basal tube, anthers extrorse with longit. slits (apical pores in *Cobana*), $\overline{G}$ (3), 3-loc. with axile placentation ($\underline{G}$ in *Isophysis*, 1-loc. with parietal placentation in *Hermodactylus*) & term. 3-lobed style, the branches s.t. subdivided or often expanded & petaloid with stigma on outer side of branch, (1–) ±∞ anatropous, bitegmic ovules per loc. Fr. a loculicidal capsule; seeds sometimes arillate or sarcotestal with rather small linear embryo and fleshy endosperm with reserves of hemicellulose, protein & oil (usu. 0 starch). x = 3–19+. C. Innes (1985) *The World of I.*

Classification & principal genera (AMBG 77(1990)620):

I. **Isophysidoideae** ($\underline{G}$, Tasmania): *Isophysis* (only)

II. **Nivenioideae** (incl. Geosiridaceae; some woody or mycotrophic, fugacious blue fls, Afr., Madag., New Guinea, Aus.): *Aristea, Nivenia, Patersonia*

III. **Iridoideae** (fugacious fls, perigonal nectaries, style long-branched below A level; 4 tribes):

1. **Sisyrinchieae** (style-branches extending between rather than opp. A; esp. S hemisph.): *Bobartia, Sisyrinchium*
2. **Irideae** (P divided into limb & claw, flattened style-branches & crests tepaloid; esp. OW): *Galaxia, Gyandriris, Homeria, Iris, Moraea*
3. **Mariceae** (like 2. but style-branches merely thickened; C & S Am.): *Neomarica, Trimezia*
4. **Tigridieae** (like 3. but lvs plicate; C & S Am.): *Cypella, Tigridia*

IV. **Ixioideae** (basal-rooting corms, infls of sessile fls in spikes or panicles, P-tubes, OW esp. S Afr.; 3 tribes):

1. **Pillansieae** (panicle, Cape): *Pillansia* (only)
2. **Watsonieae** (spike, deeply-divided style-branches, trop. & S Afr.: *Lapeirousia, Watsonia*
3. **Ixieae** (style diff., OW esp. S Afr.): *Babiana, Crocus, Dierama, Freesia, Geissorhiza, Gladiolus, Hesperantha, Ixia, Romulea, Tritonia, Tritoniopsis*

Allied to Liliaceae which usu. have A 6

Many v. imp. garden pls esp. in commerce as cut-fls etc.: *Crocosmia* (montbretia), *Crocus, Freesia, Gladiolus, Iris, Ixia, Sparaxis, Tigridia* & limited uses as spices (*Crocus*), dyes (*Crocosmia*) & scents (*Iris*); some S Afr. spp. (esp. *Romulea, Sparaxis, Watsonia*) now serious weeds in Aus.

Iridodictyum Rodionenko = *Iris*

Iridosma Aubrév. & Pellegrin. Simaroubaceae. 1 W trop. Afr.

Iris L. Iridaceae. (III 2). c. 210 Euras. (Eur. 30), N Afr., N Am. Outer P 3 the falls narrowed basally, sometimes bearded, inner 3 the standards usu. erect, A 3 free at base of falls, style-branches 3, petal-like covering A; just above the anther on the outer side of the style is a little flap whose upper surface is the stigma. Bees, attracted by the flag-like falls & s.t. scent, enter the fl. for nectar when pollen from other fls is brushed against the stigma; when they retreat, carrying newly deposited pollen, they close the flap. thereby separating the anther from the stigma & preventing self-poll. (similar is found in *Viola* but here the fl. comprises 3 functional units). Seeds often arillate or flattened (wind-disp.). Fls of many colours & much cult. (flags), some v. anc. hybrids; 6 subgg. recog.: *Iris* (I.) with rhiz. & distinctly bearded falls (incl. bearded (pogon i., many cvs), oncocyclus & regelia i. & their hybrids (Regeliocyclus i.), *Limniris* (Li.) with rhiz. & beardless falls (incl. 'Pacific Coast Irises' (PCIs) or Calif. i., many hybrids) & *Hexagonae* (Louisiana i.), waterside spp. & hybrids of S US for warm countries), *Nepalensis* (e.g. *I. decora* Wallich (*I. nepalensis*, Himal.) with dahlia-like tuberous roots, *Xiphium* (X.) with bulbs & fibrous roots (incl. xiphion (English, Dutch & Spanish i.) usu. on alk. soils), *Scorpiris* with bulbs & usu. thick fleshy roots when dormant (comprises juno i.), *Hermodactyloides* (H.) (R: Davis & Hedge Festschrift (1989)83) with bulbs with fibrous netted tunics (reticulata i.). B. Mathew (1981) *The Iris*; F. Köhlein (1987) *The I.*, ed. 2. The common flag iris (I.) of gardens is of complex history involving the Eur. spp. *I. pumila* L. (E Eur. to Urals), *I. lutescens* Lam. (S Eur.), *I. variegata* L. (C & SE Eur.) etc. & referred to as *I. × germanica* L., of which group 'Florentina' (*I. florentina*) poss. used to decorate Sphinx & known to Thutmose III (1501–1447 BC), rhiz. ground & used as scent of violets or medic. (orris, i.e. iris, root), as fixative in pot-pourri & for powdering wigs & hair in C18, the fl. poss. the orig. fleur-de-lis (i.e. -Louis); the whole group has 2n = 24–c. 64 with increasing pl. size. Other cult. spp. incl.: *I. × albicans* Lange (I., Yemen, natural sterile hybrid of *I. × germanica* group) – cult. in Mohammedan graveyards, fls white; *I. atrofusca* Baker (I., Israel) – ordinary sword-lvs on deep soils but in heavily grazed areas linear lvs like the toxic *Asphodelus aestivus* in the area (? mimicry); *I.*

clarkei Hook.f. (Li., Himal.) – dried lvs fodder for yaks & horses; *I. confusa* Sealy (Li., W China) & *I. wattii* Baker ex Hook.f. (Li., Assam & China) – stems elongate, carrying up lvs before flowering: *I. danfordiae* (Baker) Boiss. (H., Turkey) – yellow-flowered reticulata of early spring; *I. ensata* Thunb. (Li., *I. kaempferi*, Japanese i., E As.) – red-purple fls, long cult. Japan, many cvs incl. tetraploids & hybrids with *I. pseudacorus*; *I. foetidissima* L. (Li., gladdon, (stinking) gladwin, gladwyn or iris, S & W Eur. to N Afr.) – bruised lvs smell of meat ('roast beef plant'), rhiz. steeped in ale form. an efficient purge; *I. graminea* L. (Li., NE Spain to Cauc.) – fls smell of ripe plums; *I. latifolia* (Miller) Voss (X., *I. xiphioides*, English i., Spain & Pyrenees) – falls deep blue-purple; *I. pseudacorus* L. (Li., yellow flag, water f. (orig. 'gladiolus' in Mediaeval GB), W Eur., Medit. to Iran, natur. N Am.) – rhiz. & seeds form. medic., rhiz. form. a source of black dye & ink, seeds a coffee subs., fl. poss. orig. fleur-de-lis; *I. reticulata* M. Bieb. (H., Caucasus) – violet-scented fls in early spring, many hybrids with *I. histrioides* (G. Wilson) S. Arn. (C N Turkey); *I. sibirica* L. (Li., C Eur. to Siberia) – cult. ('Siberian') hybrids with *I. sanguinea* Hornem. ex Donn (C As. to Japan); *I. spura* L. (Li., Eur., Medit.) – v. variable in wild, over 70 cvs; *I. unguicularis* Poiret (Li., *I. stylosa*, Algerian i., Algeria to E Medit.) – winter-flowering, seeds covered with glistening sessile glands attractive to disp. agents (ants); *I. versicolor* L. (Li., blue f., E US) – dried rhizomes medic.; *I. xiphium* L. (X., Spanish i., W Medit.) – a parent of the bulbous Dutch i. (*I.* × *hollandica* Hort.) grown as cut-fls, derived also from *I. filifolia* Boiss. & *I. tingitana* Boiss. & Reuter (both W Medit.) etc. (poss. *I. latifolia*)

See also *Gynandriris, Hermodactylus*

iris, Algerian *Iris unguicularis*; **Cape i.** *Moraea* spp.; **Dutch i.** *I.* × *hollandica*; **English i.** *I. latifolia*: **Japanese i.** *I. ensata*; **snake's-head i.** *Hermodactylus tuberosa*; **Spanish i.** *I. xiphium*; **stinking i.** *I. foetidissima*

Irish ivy *Hedera helix* subsp. *hibernica*; **I. yew** *Taxus baccata* 'Stricta'

Irlbachia C. Martius. Gentianaceae. 17 trop. S Am. R: PKNAW C88(1985)406

Irmischia Schldl. = *Metastelma*

iroko *Milicia excelsa*

iron grass *Lomandra* spp.

ironbark *Eucalyptus* spp.; **grey i.** *E. paniculata*; **red i.** *E. crebra*; **scrub i.** *Dysoxylum pettigrewianum*

ironweed (US) *Vernonia* spp.

ironwood *Choricarpia subargentea, Cynometra alexandri; Eusideroxylon zwageri, Mesua ferrea, Ostrya virginiana, Parrotia persica, Xylia xylocarpa* & also spp. of *Casuarina, Copaifera, Olea*; **desert i.** *Olneya tesota*

Irvingbaileya R. Howard. Icacinaceae. 1 Queensland

Irvingia Hook.f. Irvingaceae (Ixonanthaceae s.l.; Simaroubaceae s.l). 3 trop. Afr., 1 Mal. Fr. ed., seeds used for soap & as wax source, those of *I. gabonensis* (O'Rorke) Baillon (dika nuts, W trop. Afr.) giving d. butter, the paste from the mashed kernels being d. bread or Gaboon chocolate, fr. smelly & disp. elephants; *I. malayana* Oliver ex Bennett (SE As., W Mal.) – hard-wooded relic in cleared forest, seeds yielding cay-cay fat

Irvingiaceae (Engl.) Exell and Mendonça (Ixonanthaceae s.l.; Simaroubaceae s.l.). Dicots – Rosidae – Sapindales. 3/8 trop. OW. Excl. from Simaroubaceae (q.v.) because of mucilage cavities in stem, stipules, leaf & fr. anatomy, phytochemistry

Genera: *Desbordesia, Irvingia, Klainedoxa*

Affinities with Erythroxylaceae, Linaceae, Malpighiaceae ... suggested

Irwinia G. Barroso. Compositae (Vern-Vern.). 1 Brazil

Iryanthera Warb. Myristicaceae. 24 trop. S Am. to Panamá. Bark of some sources of hallucinogenic pastes & medic.

Isabelia Barb. Rodr. Orchidaceae (V 13). 2 Brazil

Isachne R. Br. Gramineae (35). 50 trop. & warm esp. trop. As. *I. globosa* (Thunb.) Kuntze (*I. australis*) (As. to Aus.) – fodder

Isaloa Humbert = *Barleria*

Isalus J. Phipps = *Tristachya*

Isandra F. Muell. = *Symonanthus*

Isandraea Rauschert = *Symonanthus*

isano oil *Ongokea gore*

Isanthus Michaux = *Trichostema*

isaño *Tropaeolum tuberosum*

Isatis L. Cruciferae. c. 30 Eur. (5–9), Medit. to C As. *I. tinctoria* L. (woad, SW As., ? SE Eur.) – form. cult. for lvs to be dried, powdered & fermented (smelly) to give v. fast blue dye (indigotin) from lvs form. used for sailors' clothes, policemen's uniforms & students' gowns at Christ's Hospital (a London school, hence 'Blue-coat Boys'), displaced by indigo (1631) & synthetics (1890) but factory in England until 1930s

Ischaemum L. Gramineae (40d). c. 65 warm & trop. esp. As. *I. rugosum* Salisb. (trop.) – bad weed of ricefields

Ischnea F. Muell. Compositae (Sen.-Blenn.). 5 New Guinea. Pappus 0

Ischnocarpus O. Schulz. Cruciferae. 1 NZ

Ischnocentrum Schltr. = *Glossorhyncha*

Ischnochloa Hook.f. = *Microstegium*

Ischnogyne Schltr. Orchidaceae (V 12). 1 China

Ischnolepis Jum. & Perrier. Asclepiadaceae (I, Periplocaceae). 1 Madag.: *I. tuberosa* Jum. & Perrier

Ischnosiphon Koern. Marantaceae. 35 trop. Am. R: OB 43(1977). Stalks used as pipes for blowing snuff up noses in Amaz. *I. arouma* (Aublet) Koern. (tirite) – lvs used for basketry

Ischnostemma King & Gamble. Asclepiadaceae (III 1). 1 Mal. to trop. Aus. coasts: *I. selangoricum* King & Gamble

Ischnurus Balf.f. = *Lepturus*

Ischyrolepis Steudel (~ *Restio*). Restionaceae. 49 S Afr. R: Bothalia 15(1985)397

Iseia O'Don. (~ *Ipomoea*). Convolvulaceae. 1 trop. S Am.

Iseilema Andersson. Gramineae (40g). 20 Indomal. to Aus. R: HIP 33(1935)t.3286. Barcoo or Flinders grass

Isertia Schreber. Rubiaceae (I 8). 13 trop. Am. Some cult. orn. shrubs; soap

Isidorea A. Rich. ex DC. Rubiaceae (I 7). 20 WI

Isidrogalvia Ruíz & Pavón (~ *Tofieldia*). Melanthiaceae (Liliaceae s.l.). 6 S Am. R: SB 16(1991)270, Brittonia 44(1992)368

Isinia Rech.f. = *Lavandula*

Iskandera N. Busch. Cruciferae. 1 C As.

Islaya Backeb. = *Neoporteria*

Ismelia Cass. (~ *Chrysanthemum*). Compositae (Anth.-Anth.). 1 Morocco: *I. carinata* (Schousboe) Sch.-Bip. (*C. carinatum*) – cult. orn.

Ismene Salisb. ex Herbert = *Hymenocallis*

Isoberlinia Craib & Stapf ex Holland. Leguminosae (I 4). 5 trop. Afr. Imp. components of miombo woodlands in S trop. Afr.

Isocarpha R. Br. Compositae (Eup.-Ayap.). 5 trop. & warm Am. R: MSBMBG 22(1987)215. *I. oppositifolia* (L.) Cass. used as a tonic in Cuba

Isochilus R. Br. Orchidaceae (V 13). 10 trop. Am. Cult. orn. epiphytes

Isochoriste Miq. = *Asystasia*

Isocoma Nutt. (~ *Haplopappus*). Compositae (Ast.-Sol.). 16 SW N Am. R: Phytol. 70(1991)70

Isodendrion A. Gray. Violaceae. 4 Hawaii (1 extinct: *I. pyrifolium* A. Gray (form. on all is.), 3 rare). Shrubs

Isodesmia Gardner = *Chaetocalyx*

Isodictyophorus Briq. (~ *Plectranthus*). Labiatae (VIII 4c). 1 W Afr.

Isodon (Benth.) Spach (*Rabdosia*, ~ *Plectranthus*). Labiatae (VIII 4c). 96 trop. & warm As., Afr. (1). R: JAA 69(1988)289

Isoetaceae Reichb. Lycopsida. 1/150 cosmop. exc. Pacific is. Terr. or aquatic pls with lvs ± linear, flat to terete, arising from corm-like rhiz., apically tapered, basally spathulate, in bases of outermost of which few large megaspores with the many microspores at bases of inner ones

Genus: *Isoetes*

Isoetes L. Isoetaceae. (Incl. *Stylites*) c. 150 cosmop. (Eur. 11) exc. Pacific is. Quillworts. Rosette-pls with roots dichot. branched or tussock-forming with elongate dichotomously branching stems & unbranched roots ('*Stylites*'). Crassulacean acid metabolism like xerophytes as daytime carbon limitation in oligotrophic aquatic habitats, but terr. temp. spp. without even when submerged, though trop. alpine ones e.g. *I. andicola* (Amstutz) Gomez (*Stylites a.*, Andean Peru to 4750 m, discovered 1940) have variable levels of CAM & are unique in terr. pls as stomata 0 & overall structure comparable with Cretaceous *Nathorstia* Richter & ultimately Lepidodendrales (Sporne). Spores said to be dispersed in excreta of earthworms. Some cult. orn. Spp. difficult to identify, usu. requiring microscopic examination of spores

Isoetopsis Turcz. Compositae (Gnap.-Ang.). 1 temp. Aus.: *I. graminifolia* Turcz. – female florets 3-lobed with zygomorphic throat. R: OB 104(1991)123

Isoglossa Oersted. Acanthaceae. 50 OW trop. to Arabia (As. 8, R: NJB 5(1985)1)). Some W Afr. spp. on 9-yr flowering cycles

Isolepis R. Br. (~ *Scirpus*). Cyperaceae. 70 temp. esp Aus. (29, 15 endemic) & Afr., trop. (esp. mts). *I. cernua* (Vahl) Roemer & Schultes (*Scirpus c.*, Eur., Medit.) – grown in tube & sold as 'optical fibre pl.' or 'Dougal grass'

Isoleucas Schwartz. Labiatae (VI). 1 Arabia

Isoloma J. Sm. = *Lindsaea*

Isoloma Benth. ex Decne. = *Kohleria*

Isolona Engl. Annonaceae. 19 trop. Afr., Madagascar

Isomacrolobium Aubrév. & Pellegrin = *Anthonotha*

Isomeris Nutt. = *Cleome*

Isometrum Craib. Gesneriaceae. 13 China. R: ABY 8(1986)36

Isonandra Wight (~ *Palaquium*). Sapotaceae (II). 10 S India, Sri Lanka

Isonema R. Br. Apocynaceae. 3 W Afr.

Isopappus Torrey & A. Gray = *Haplopappus*

Isophysis T. Moore ex Seemann. Iridaceae (I). 1 Tasmania. G!

Isoplexis (Lindley) Loudon (~ *Digitalis*). Scrophulariaceae. 3 Macaronesia. R: BJ 79(1980)218. Shrubby. Cult. orn. incl. *I. canariensis* (L.) G. Don f. (Canary Is.) – bird-poll. *I. isabellina* (Webb & Berth.) Masf. – pollen found on heads of insectivorous birds (no sunbirds there)

Isopogon R. Br. ex J. Knight. Proteaceae. 30+ Aus. esp. SW (25). Like *Petrophile* but bracts fall before fr. Some cult. orn. (drumsticks)

Isopyrum L. Ranunculaceae (III 1). 4 Euras. (Eur. 1). R: NRBGE 28(1968)272. Alks. Some cult. orn. delicate herbs

Isostigma Less. Compositae (Helia.-Cor.). 11 subtrop. S Am., on campos. R: BG 81(1926)241

Isotheca Turrill. Acanthaceae. 1 Trinidad

Isotoma (R.Br.) Lindley = *Laurentia*

Isotrema Raf. (~ *Aristolochia*). Aristolochiaceae (II 2). 50 temp. & trop. As. to Sumatra, N & C Am. Cult. orn. incl. *I. macrophyllum* (Lam.) C. Reed (*A. macrophylla*, '*A. durior*', *A. sipho*, Dutchman's pipe, E N Am.) – rapid climber

Isotria Raf. Orchidaceae (Pogoniinae). 2 E US

Isotropis Benth. Leguminosae (III 24). 14 Aus.

isphaghul seeds *Plantago ovata*

istle = ixtle

ita palm *Mauritia flexuosa*

Italian alder *Alnus cordata*; **I. cypress** *Cupressus sempervirens*; **I. millet** *Setaria italica*; **I. poplar** *Populus nigra*; **I. rye** *Lolium multiflorum*; **I. senna** *Senna italica*; **I. whisk** *Sorghum bicolor* Technicum Group

Itaobimia Rizz. = *Riedeliella*

Itasina Raf. (*Thunbergiella*). Umbelliferae (III 8). 1 S Afr.: *I. filifolia* (Thunb.) Raf. R: NRBGE 45(1988)93

Itatiaia Ule = *Tibouchina*

Itaya H. Moore. Palmae (I 1a). 1 Peru, Brazil

Itea L. Grossulariaceae (Iteaceae). 15 Himal. to Japan & W Mal., E N Am. (1). Cult. orn. evergreen trees & shrubs esp. *I. ilicifolia* Oliver (China) & *I. virginica* L. (E US); *I. riparia* Collett & Hemsley (Burma) – rheophyte

Iteaceae J. Agardh = Grossulariaceae

Iteadaphne Blume (~ *Lindera*). Lauraceae (II). 2 SE As. to W Mal.

Iti Garnock-Jones & P. Johnson. Cruciferae. 1 SW NZ: *I. lacustris* Garnock-Jones & P. Johnson. R: NZJB 25(1987)603

Itoa Hemsley. Flacourtiaceae (8). 2 S China, trop. As.

Ituridendron De Willd. = *Omphalocarpum*

Itysa Ravenna = *Calydorea*

Itzaea Standley & Steyerm. Convolvulaceae. 1 C Am.

Iva L. Compositae (Helia.-Amb.). 15 N Am. to WI. R: KUSB 41(1960)783. *I. acetosa* (Nutt.) R. Jackson (*Oxytenia a.*, copperweed) – toxic to stock in SW US; *I. annua* L. – cypsela kernels eaten by N Am. Indians since prehistoric times, poss. oil source; *I. xanthifolia* Nutt. (N Am.) – introd. Eur. where imp. hay-fever pl.

Iva wine or **liqueur** flavoured with *Achillea erba-rota* subsp. *moschata*

Ivania O. Schulz. Cruciferae. 1 N Chile

Ivanjohnstonia Kazmi (~ *Cynoglossum*). Boraginaceae. 1 NW Himalaya

Ivesia Torrey & A. Gray (~ *Potentilla*). Rosaceae. 30 W N Am. R: Lloydia 1(1938)111

Ivodea Capuron. Rutaceae. 6 Madagascar

ivory nut *Phytelephas macrocarpa*; **vegetable i.** *Phytelephas* spp., *Palandra aequatorialis* (best quality); **i.wood** *Siphonodon australis*

ivy *Hedera* spp. esp. *H. helix*; **Boston i.** *Parthenocissus tricuspidata*; **Canary i.** *H. canariensis*; **grape i.** *Cissus* spp.; **Irish i.** *H. helix* subsp. *hibernica*; **Japanese i.** *H. rhombea*; **Kenilworth or Oxford i.** *Cymbalaria muralis*; **Persian i.** *H. colchica*; **poison i.** *Rhus radicans*

Ixanthus Griseb. Gentianaceae. 1 Canary Is.

ixbut *Euphorbia lancifolia*

Ixerba Cunn. Grossulariaceae (Brexiaceae). 1 N NZ

Ixeridium (A. Gray) Tzvelev (~ *Ixeris*). Compositae (Lact.-Crep.). c. 15 E & SE As. to New Guinea

Ixeris Cass. Compositae (Lact.-Crep.). c. 20 E & SE As.

Ixia L. Iridaceae (IV 3). c. 50 trop. (1) & S Afr., natur. W Aus. R: JSAB 28(1962)45, SAJB 51(1985)66. Cult. orn. corms ((African) corn lilies) esp. hybrids involving *I. maculata* L. (Cape)

Ixianthes Benth. Scrophulariaceae. 1 S Afr.: *I. retzioides* Benth. – rheophyte

Ixiochlamys F. Muell. & Sonder (~ *Podocoma*). Compositae (Ast.-Ast.). 4 Aus. R: JABG 2(1980)241

Ixiolaena Benth. Compositae (Gnap.). 8 Aus. R: OB 104 (1991)58. Heterogeneous?

Ixiolirion Herbert. Tecophilaeaceae (Ixioliriaceae, Amaryllidaceae; Liliaceae s.l.). 4 W & C As. Cult. orn. bulbs esp. *I. tataricum* (Pallas) Herbert (*I. ledebourii*, *I. montanum*, SW & C As., Kashmir) – fls blue

Ixioliriaceae (Pax) Nakai = Tecophilaeaceae

Ixocactus Rizz. (~ *Cladocolea*). Loranthaceae. 6 Colombia & Venez. R: SB 16(1991)

Ixodia R. Br. Compositae (Gnap.-Cass.). 2 Aus. R: OB 104(1991)87

Ixodonerium Pitard. Apocynaceae. 1 SE As.

Ixonanthaceae (Benth.) Exell & Mendonça. Dicots – Rosidae – Linales. 4/21 trop. Trees & shrubs with spirally arr. simple entire to toothed lvs; stipules small or 0. Fls bisexual, ± reg., ± hypogynous, in racemose to cymose infls, often 5-merous; K imbricate, s.t. with basal union, C imbricate or convolute, A 5–20 with widened filaments, free or adnate to annular to cupular intrastaminal nectary-disk, anthers with longit. slits, $\underline{G}$ ((2–)5), pluriloc. with axile-apical placentas & term. style, sometimes divided into locelli (cf. Linaceae) & apically 1-loc.; style & filaments folded in bud; 1 or 2 pendulous, anatropous bitegmic ovules per loc. Fr. a septicidal (sometimes also loculicidal) capsule with or without persistent central column; seeds arillate or winged, endosperm little or 0

Genera: *Allantospermum*, *Cyrillopsis*, *Ixonanthes*, *Ochthocosmus*

Some genera now excl. as Irvingiaceae (via Simaroubaceae)

Ixonanthes Jack. Ixonanthaceae. 3 SE As., Mal. R: Blumea 26(1980)191. Bark of *I. icosandra* Jack used for tanning fishing-nets in Malay Pen.

Ixophorus Schldl. Gramineae (34b). 1 Mexico

Ixora L. Rubiaceae (II 2). c. 300 trop. (Afr. 30). Fls long-tubular, white, yellow, orange, pink or red, poll. by (?) Lepidoptera. Shrubs or trees, some rheophytic, some cult. orn. greenhouse shrubs esp. *I. coccinea* L. (flame flower, S India) – fls red

Ixorhea Fenzl. Boraginaceae. 1 Arg. Andes

Ixtlania M.E. Jones = *Justicia*

ixtle or **ixtli fibre** *Agave* spp.; **Jaumave i.** *A. funkiana*; **tula i.** *A. lecheguilla*. See also *Yucca*

Izabalaea Lundell = *Agonandra*

J

Jablonskia Webster (~ *Securinega*). Euphorbiaceae. 1 N trop. S Am.: *J. congesta* (Muell. Arg.) Webster. R: SB 9(1984)232

jaborandi Name for several pungent aromatic S Am. pl. which promote saliva when chewed, esp. Piperaceae & Rutaceae (notably *Pilocarpus* spp. which provide the drug jaborandi (dried lvs))

Jaborosa Juss. Solanaceae (1). 23 temp. S Am. (Chile 21, 11 endemics). R: Kurtziana 19(1987)77

jaboticaba *Myrciaria* spp. esp. *M. cauliflora*

jaca = jak

Jacaima Rendle = *Matelea*

Jacaranda Juss. Bignoniaceae. 34 trop. Am. Shrubs & trees usu. with bipinnate lvs (fern-trees); 2 or 3 buds per axil. Some timber (palisander), e.g. *J. copaia* (Aublet) D. Don (futi, futui, NE S Am.) & pulp; some medic. esp. bark; cult. orn. esp. as street-trees, esp. *J. mimosifolia* D. Don (*J. ovalifolia*, NW Arg.). 2 or 3 buds per axil

Jacaratia A. DC. Caricaceae. 7 trop. Am. R: V.M. Badillo (1971) *Caric.*: 38. Locally eaten fr. *J. dolichaula* (Donn.Sm.) Woodson (S Mex. to Panamá) – sphingid-poll., dioec. with females resembling male infl. as petals (5) resemble male fl. buds & lobed stigma like a female fl., but no nectar reward!

jacareuba *Calophyllum brasileense*

Jacea Miller = *Centaurea*

jack = jak; **j.ass clover** *Wislizenia refracta*; **j.**(i.e. jakes = latrine)**-by-the-hedge** *Alliaria petiolata*; **j.-in-the-box** *Hernandia nymphaeifolia*; **j.-in-the-pulpit** *Arum maculatum, Arisaema* spp.; **j. fruit** = jak; **long j.** *Triplaris* spp.; **j. pine** *Pinus banksiana*; **j.wood** *Cryptocarya glaucescens*

Jackia Wallich = *Jackiopsis*

Jackiopsis Ridsd. Rubiaceae (inc. sed.). 1 W Mal.

jacks, whistling *Gladiolus communis*

Jacksonia R. Br. ex Sm. Leguminosae (III 24). 40 Aus. Cult. orn. incl. *J. scoparia* R. Br. (stinkwood, E Aus.) – burning wood foetid

Jacmaia R. Nordenstam (~ *Senecio*). Compositae (Sen.-Sen.). 5 Jamaica, Costa Rica

jacobaea *Senecio elegans*

Jacobean lily *Sprekelia formosissima*

Jacobinia Nees ex Moricand = *Justicia*

Jacobsenia L. Bolus & Schwantes (~ *Drosanthemum*). Aizoaceae (V). 2 W S Afr. Cult. orn. shrublets

Jacob's coat *Alternanthera ficoidea* 'Bettzickiana', *Acalypha wilkesiana*; **J.'s ladder** *Polemonium* spp. esp. *P. caeruleum*

Jacquemontia Choisy. Convolvulaceae. 80–100 trop. & warm, esp. Am. (Aus. 1: *J. pannosa* (R. Br.) Mabb.). Some cult. orn. lianes

Jacquesfelixia J. Phipps = *Danthoniopsis*

Jacqueshuberia Ducke. Leguminosae (I 1). 6 Brazil, Colombia. Fls expose partly joined filaments & pollen sticky with viscin threads at night – bat-poll.

Jacquinia L. Theophrastaceae. 35 C Am. & Carib. Some fr. used to poison fish; some cult. orn. shrubs incl. *J. arborea* Vahl (*J. barbasco*, WI) – fish-poison, yellow & brown seeds used as beads; *J. keyensis* Mez (Florida, WI) – source of cudjoe wood; *J. pungens* A. Gray (Mex.) – leafing in dry season, lvs spine-tipped

Jacquiniella Schltr. Orchidaceae (V 13). 12 trop. Am. Cult. orn. epiphytes

jade plant *Crassula ovata*

Jadunia Lindau. Acanthaceae. 1 New Guinea

Jaegeria Kunth. Compositae (Helia.-Gal.). 6 Mexico to Uruguay, Galápagos. R: Phytol. 55(1984)243

Jaeschkea Kurz. Gentianaceae. 3 Himalaya

Jagera Blume. Sapindaceae. 4 E Mal. to Aus. Some unbranched pachycauls; fr. hairs irritant. Some fish-poisons, e.g. *J. protorhus* (A. Rich.) Radlk. (NE Aus., New Caled.) – saponins (also foaming agents)

jaggery crude sugar from palms or sugar-cane

Jahnia Pittier & S.F. Blake = *Turpinia*

jak or **jack (fruit)** *Artocarpus heterophyllus*

jalap *Ipomoea purga*; **Brazilian j.** *Merremia tuberosa*; **false j.** *Mirabilis jalapa*; **Indian j.** *I. turpethum*

Jaimehintonia B. Turner. Alliaceae. 1 Mex.: *J. gypsophila* B. Turner. R: Novon 3(1993)86

Jainia Balakr. = *Coptophyllum*

Jalcophila Dillon & Sagást. Compositae (Gnap.-Gnap.). 3 Andes of Bolivia, Peru, Ecuador. R: BJLS 106(1991)185

Jaliscoa S. Watson. Compositae (Eup.-Oxy.). 3 Mex. R: MSBMBG 22(1987)440

Jaltomata Schldl. (~ *Saracha*). Solanaceae (1). (Incl. *Hebecladus*) 30 trop. & warm Am.

jam tree *Muntingia calabura*

Jamaica bark *Exostema* spp.; **J. cherry** *Muntingia calabura*; **J. dogwood** *Piscidia piscipula*; **J. ebony** *Brya ebenus*; **J. honeysuckle** *Passiflora laurina*; **J. horse bean** *Canavalia ensiformis*; **J. pepper** *Pimenta dioica*; **J. plum** *Spondias mombin*; **J. sarsaparilla** *Smilax regelii*; **J. sorrel** *Hibiscus sabdariffa*

Jamaicella Braem = *Oncidium*

jamba *Eruca vesicaria* subsp. *sativa*

jamberberry *Physalis philadelphica*

jambolan *Syzygium cumini*

Jambosa Adans. = *Syzygium*

jambu *Syzygium* spp. esp. *S. jambos*; **wax j.** *S. samarangense*

Jamesbrittenia Kuntze (~ *Sutera*). Scrophulariaceae (Manuleeae). 83 trop. & S Afr., 1 N Afr. to India. R: O. Hilliard (1994) *The M. a tribe of S. J. atropurpurea* (Benth.) Hilliard (*S. a.*., S Afr.) – fls a saffron subs.; *J. fodina* (Wild) Hilliard (*S. f.*, Zimbabwe) – ash has 15.3% nickel & 4.8% chromium by weight

Jamesia Torrey & A. Gray. Hydrangeaceae. 2 W N Am. R: Brittonia 41(1989)335. *J. americana* Torrey & A. Gray – cult. orn. shrub with pink fls

Jamesianthus S.F. Blake & Sherff. Compositae (Hele.-Cha.). 1 S US

Jamesonia Hook. & Grev. Pteridaceae (III). 19 C Am., N Andes, SE Braz. Xerophytes with fronds of indefinite apical growth. R: CGH 191(1962)109

Jamestown weed *Datura stramonium*

Janakia Joseph & Chandra = *Decalepis*

janatsi *Debregeasia longifolia*

Jancaea Boiss. (*Jankaea*). Gesneriaceae. 1 Mt Olympus: *J. heldreichii* (Boiss.) Boiss. – Tertiary relic (cf. *Haberlea, Ramonda*), cult. orn. Hybrids synth. with *Ramonda myconi* (Pyrenees)

Jankaea Boiss. = praec.

Janotia J. Leroy. Rubiaceae (I 2). 1 Madagascar

Jansenella Bor. Gramineae (39). 1 India & Sri Lanka

Jansonia Kippist. Leguminosae (III 25). 1 SW Aus. Lvs simple, opp.

Janusia A. Juss. ex Endl. Malpighiaceae. (Incl. *Schwannia*) 18 Calif. to Arg.

Japan or **Japanese alder** *Alnus japonica*; **J. anemone** *Anemone* × *hybrida*; **J. apricot** *Prunus mume*; **J. arrowroot** *Pueraria montana* var. *lobata*; **J. artichoke** *Stachys affinis*; **J. birch** *Betula maximowicziana*; **J. cedar** *Cryptomeria japonica*; **J. chestnut** *Castanea crenata*; **J. clover** *Kummerowia striata*; **J. galls** *Rhus javanica*; **J. ginger** *Zingiber mioga*; **J. greens** *Brassica juncea* 'Crispifolia'; **J. honeysuckle** *Lonicera japonica*; **J. hop** *Humulus japonica*; **J. hyacinth** *Ophiopogon* spp.; **J. iris** *Iris ensata*; **J. knotweed** *Fallopia japonica*; **J. lacquer** *Rhus verniciflua*; **J. larch** *Larix kaempferi*; **J. laurel** *Aucuba japonica*; **J. lilac** *Syringa reticulata*; **J. lime** *Tilia japonica*; **J. maple** *Acer palmatum*; **J. medlar** *Eriobotrya japonica*; **J. millet** *Echinochloa utilis*; **J. morning glory** *Ipomoea nil*; **J. parsley** *Cryptotaenia canadensis*; **J. pear** *Pyrus pyrifolia*; **J. pepper** *Zanthoxylum piperitum*; **J. persimmon** *Diospyros kaki*; **J. privet** *Ligustrum japonicum*; **J. raisin-tree** *Hovenia dulcis*; **J. toad-lily** *Tricyrtis hirta*; **J. wax tree** *Rhus succedanea*; **J. yew** *Taxus cuspidata*

Japonasarum Nakai = *Asarum*

Japonolirion Nakai (~ *Tofieldia*). Melanthiaceae. 1 Japan: *J. osense* Nakai

japonica *Chaenomeles speciosa*

jaragua grass *Hyparrhenia rufa*

Jaramilloa R. King & H. Robinson (~ *Eupatorium*). Compositae (Eup.-Oxy.). 2 N Colombia. R: MSBMBG 22(1987)450

Jarandersonia Kosterm. Tiliaceae. 3 Borneo

Jardinea Steudel = *Phacelurus*

Jarilla Rusby. Caricaceae. 1 Mexico. Herb

jarrah *Eucalyptus marginata*

Jasarum Bunting. Araceae (VII 1). 1 Guyana, Venez.: *J. steyermarkii* Bunting – only submerged aroid in S Am.

Jasione L. Campanulaceae. c. 20 Eur. (9), Medit., SW As. Some cult. orn. incl. *J. montana* L. (sheep's-bit, Eur.) – fls in dense heads, blue

Jasionella Stoy. & Stefanoff = *Jasione*

jasmine *Jasminum* spp.; **allspice j.** *Gelsemium* spp.; **Arabian j.** *Jasminum sambac*; **Cape j.** *Gardenia augusta*; **Catalonian j.** *J. officinale* f. *grandiflorum*; **Cayenne j.** *Catharanthus roseus*; **common j.** *J. officinale*; **crepe j.** *Tabernaemontana divaricata*; **Italian j.** *J. humile* 'Revolutum'; **Madag. j.** *Marsdenia floribunda*; **rock j.** *Androsace* spp.; **Spanish j.** *J. officinale* f. *grandiflorum*; **white j.** *J. officinale*; **winter j.** *J. nudiflorum*; **yellow j.** *J. humile*, *J. mesnyi*

Jasminocereus Britton & Rose. Cactaceae (III 7). 1 Galápagos: *J. thouarsii* (A. Weber) Backeb. – cult. orn. tree to 8 m with bole to 30 cm diam.

Jasminochyla (Stapf) Pichon = *Landolphia*

Jasminum L. Oleaceae. c. 200 trop. to (few) temp. OW (China 46 – R: BBR 4,1(1984)88), Am. (1). Jasmine, jessamine (arch.). Decid. & evergreen shrubs & lianes with scented white, yellow or pink fls used in scent-making & perfuming tea, the oil (c. £5000 per kg in 1993) called malatti (India); many cult. orn. esp. *J. beesianum* Forrest & Diels (W China) – fls pale red; *J. humile* L. (yellow j., Himal.) esp. 'Revolutum' (Italian j.) – fragrant yellow fls; *J. mesnyi* Hance (*J. primulinum*, yellow j., W China) – fls yellow; *J. nudiflorum* Lindley (winter j., China) – yellow fls in winter; *J. odoratissimum* L. (Macaronesia) – yellow fls used in scent-making; *J. officinale* L. (common j., Cauc. to SW China, long cult. China) – white fls used in scent-making esp. f. *grandiflorum* (L.) Kobuski (*J. grandiflorum*, Catalonian or Spanish j.) – more robust but less hardy form taken by Moors to Spain; *J. paniculatum* Roxb. (China) – fls used to scent tea; *J. parkeri* Dunn (NW India) – cult. orn. dwarf shrub endangered in wild; *J. polyanthum* Franchet (Yunnan) – much cult. pot-pl. with v. fragrant fls pinkish without; *J. sambac* (L.) Aiton (Arabian j., zambac, ? India) – fls (mohle flowers) used to scent tea, incl. double-flowered 'Grand Duke of Tuscany'

Jasonia Cass. Compositae (Inul.). 1 Spain. R: Webbia 34(1979)296

Jateorhiza Miers. Menispermaceae (III). 2 trop. Afr. *J. palmata* (Lam.) Miers (*J. columba*, calumba, colomba, columba) – liane the source of *Radix Columba*, a tonic

Jatropha L. Euphorbiaceae. 175 trop. & warm (Afr. 70), N Am. R: UCPB 74(1979). Monoec. or dioec. trees, shrubs & herbs incl. annuals; alks. *J. aconitifolia* Miller (C Am.) – lvs boiled & eaten by Mayans; *J. cuneata* Wiggins & Rollins (Mex.) – stems for basket-making; *J. curcas* L. (physic nut, purging nut, pulza, trop. Am.) – cult. as boundary hedge, (foaming) chewing-stick in Nigeria, oil for candle- & soap-making, high-octane fuel trials in Sri Lanka, purgative, seed provides short-lived taper when lodged in cleft stick, shown to have antitumour activity; other spp. cult. orn. esp. *J. podagrica* Hook. (C Am.) – stem grotesquely swollen, fls red, or locally medic. & tanbarks (e.g. *J. spathulata* (Ortega) Muell. Arg. for tanning leather & red dye in Mex., SW US)

jau *Casuarina equisetifolia*

Jaubertia Guillemin. Rubiaceae (IV 12). (Incl. *Choulettia*) 16 N Afr. to C As., Socotra

jaumave fibre *Agave funkiana*

Jaumea Pers. Compositae (Hele.-Flav.). 2 W N Am., S S Am.

Jaundea Gilg = *Rourea*

Java almond *Canarium luzonicum*; **J. cardamom** *Amomum maximum*; **J. fig** *Ficus benjamina*; **J. grass** *Polytrias indica*; **J. indigo** *Marsdenia tinctoria*; **J. plum** *Syzygium cumini*

Javorkaea Borh. & J.-Komlódi (~ *Rondeletia*). Rubiaceae (I 5). 1 Honduras: *J. hondurensis* (Donn. Sm.) Borh. & J.-Komlódi. R: ABH 29(1983)16

Jedda Clarkson. Thymelaeaceae. 1 Aus.: *J. multicaulis* Clarkson – cryptogeal germ. (only T. & only Aus. dicot thus), resistant to fire & grazing

Jefea Strother. Compositae (Helia.-Verb.). 5 S US to Guatemala. R: SBM 33(1991)22

Jeffersonia Barton. Berberidaceae (II 2; Podophyllaceae). 2 NE As., E N Am. R: KB 1920:242. Cult. orn. herbs incl. *J. diphylla* (L.) Pers. (E N Am.) – local med.

Jeffreya Wild. Compositae (Ast.-Ast.). 1 Zambia, Tanzania: *J. palustris* (O. Hoffm.) Wild. R: Kirkia 9(1974)295

jeheb nut *Cordeauxia edulis*

Jehlia Rose = *Lopezia*

Jejosephia A.N. Rao & Mani (~ *Trias*). Orchidaceae (V 15). 1 NE India. R: JETB 7(1985)216

jelutong *Dyera costulata*

Jenkinsia Griffith = *Miquelia*

Jenmaniella Engl. Podostemaceae. 7 NE S Am.

Jensenobotrya Herre. Aizoaceae (V). 1 SW Afr.: *J. lossowiana* Herre – cult. orn. dwarf shrub

Jepsonia Small. Saxifragaceae. 1–2 S Calif. R: Brittonia 21(1969)286. *J. parryi* (Torrey) Small – cult. orn. herb with lvs in spring, withering before fls appear in autumn; corm-like rhiz.

jequerity seeds *Abrus precatorius*

jequitiba *Cariniana* spp.

Jerdonia Wight. Scrophulariaceae. 1 S India

jereton *Schefflera morototoni*

Jericho, rose of *Anastatica hierochuntica*, *Selaginella lepidophylla*

jerry-jerry *Ammannia multiflora*

Jersey long jacks *Brassica oleracea* Acephala Group

Jerusalem artichoke *Helianthus tuberosus*; **J. cherry** *Solanum pseudocapsicum*; **J. cross** *Silene chalcedonica*; **J. rye** *Triticum turgidum* Polonicum Group; **J. sage** *Phlomis fruticosa*; **J. thorn** *Parkinsonia aculeata*

jessamine *Jasminum* spp.

Jessa H. Robinson & Cuatrec. Compositae (Sen.-Sen.). 3 C Am. R: Novon 4 (1994)49

Jessenia Karsten = *Oenocarpus*

Jesuits' bark *Cinchona officinalis*; **J. nut** *Trapa natans*; **J. tea** *Otholobium glandulosum*

jewel orchid *Anoectochilus* spp; **j.weed** *Impatiens* spp.

Jew's apple *Solanum melongena*; **J. mallow** *Corchorus olitorius*; **J. plum** *Spondias cytherea*

jicama *Pachyrrhizus erosus*

jiggle-joggles *Briza media*

Jimson weed *Datura stramonium*

jiquitiba *Cariniana* spp.
Joannesia Vell. Euphorbiaceae. 2 Brazil. Monoec. trees. *J. princeps* Vell. (araranut tree, S Braz. coast) – source of anda-assy oil, purgative & for skin disease, timber good
Jobinia Fourn. Asclepiadaceae (III 4). 3–4 trop. S Am.
jobo *Spondias mombin*
Job's tears *Coix lacryma-jobi*
Jodina Hook. & Arn. ex Meissner = *Iodina*
Jodrellia Baijnath. Asphodelaceae (Liliaceae s.l.). 3 C to NE Afr.
Joe-pye weed *Eupatorium purpureum*
Johannesteijsmannia H. Moore. Palmae (I 1b). 4 W Mal. R: GBS 26(1972)63. Some cult. orn. with large scarcely-divided lvs used as thatch
Johnny-go-to-bed [at noon] *Tragopogon pratensis*; **J.-jump-up** *Viola tricolor*
Johnson grass *Sorghum halepense*
Johnsonia R. Br. Anthericaceae (Liliaceae s.l.). 5 SW Aus. R: FA 45(1987)242
Johnstone River hardwood *Backhousea bancroftii*
Johnstonella Brand = *Cryptantha*
Johrenia DC. Umbelliferae (III 10). 15 Eur. (1), SW & C As.
Johreniopsis Pim. Umbelliferae (III 10). 4 Caucasus, SW As.
joint *Cannabis sativa* ssp. *indica*
Joinvillea Gaudich. ex Brongn. & Gris. Joinvilleaceae. 2 W Mal., Solomon & Caroline Is. to Hawaii
Joinvilleaceae A.C. Sm. & Toml. (~Flagellariaceae). Monocots – Commelinidae – Restionales. 1/2 W Mal. & Pacific is. Coarse erect herbs, with unbranched hollow (except nodes) stems & sympodial rhiz. without sec. growth; cell-walls ± strongly silicified like Gramineae. Lvs spirally arr., with open sheath, grassy lamina, ligule & pair of auricles at base of lamina, which is plicate in bud & parallel-veined. Fls bisexual, 3-merous, reg., in term. much-branched bracteate panicles with caducous bracteoles; P 3 + 3, imbricate, s.t. basally connate, A 6 with basifixed anthers with longit. slits, $\underline{G}$ (3), 3-loc. with term. styles sometimes basally connate & 1 orthotropous pendulous ovule per loc. on axile placenta. Fr. ± 3-quetrous drupe with 1 endocarp & 1–3 seeds with copious starchy endosperm capped by small undifferentiated embryo. R: T 19(1970)887
Only genus: *Joinvillea*, form. incl. in Flagellariaceae (q.v.)
jo-jo weed *Soliva sessilis*
jojoba *Simmondsia chinensis*
Jollydora Pierre ex Gilg. Connaraceae. 3 E Nigeria to Angola. R: AUWP 89-6(1989)284. Usu. unbranched treelets (Corner's Model)
jonkong *Dactylocladus stenostachys*
Jonopsidium Reichb. = *Ionopsidium*
jonquil *Narcissus jonquilla*
Joosia Karsten. Rubiaceae (I 1). 7 W S Am. R: Brittonia 27(1975)251
Jordaaniella H. Hartman (~ *Cephalophyllum*). Aizoaceae (V). 4 SW Cape. R: BJ 104(1983)321. Cult. orn. mat-formers
Joseanthus H. Robinson (~ *Vernonia*). Compositae (Vern.-Pip.). 5 Ecuador, Colombia
Josephinia Vent. Pedaliaceae. 3–4 Kenya & Somalia (1), trop. & arid Aus. (2–3), 1 ext. to Mal.
Joseph's coat *Amaranthus tricolor*
Joshua tree *Yucca* spp. esp. *Y. brevifolia*
Jossinia Comm. ex DC = *Eugenia*
Jostaberry See *Ribes* hybrids
Jouvea Fourn. Gramineae (31c). 2 Baja Calif. to Panamá. R: BTBC 66(1939)315
Jovellana Ruíz & Pavón. Scrophulariaceae. 6 NZ, Chile. Some cult. orn. like *Calceolaria*
Jovetia Guédès. Rubiaceae (II 5). 2 Madagascar
Jovibarba Opiz (~ *Sempervivum*). Crassulaceae. 6 Eur. R: ABH 34(1988)221. C 6 or 7, fringed (C 9–20 not fringed in *S.*). No natural hybrids (cf. *S.*). Cult. orn. esp. *J. heuffelii* (Schott) Löve with new rosettes arising by division of old ones, & not as offsets, & *J. globifera* (L.) J. Parnell (*J. sobolifera*) with offsets ('hen and chickens')
jowar *Sorghum bicolor*
joyweed *Alternanthera* spp.
Juania Drude. Palmae (IV 2). 1 Juan Fernandez: *J. australis* (C. Martius) Hook.f. – cult. orn. dioec. palm
Juanulloa Ruíz & Pavón. Solanaceae (1). 8 trop. Am. R: Kurtziana 21(1991)209. Epiphytes, ? bird-poll.
Jubaea Kunth. Palmae (V 5b). 1 coastal C Chile (excl. *Paschalococos* q.v.): *J. chilensis* (Molina)

Baillon (*J. spectabilis*, Chilean wine-palm, coquito, honey-palm) – massive bole tapped or felled for sap (up to 300 litres per tree) for treacle (palm honey) so that tree now rare in wild, but widely cult.; seeds ed. R: IBM 56(1987)120

Jubaeopsis Becc. Palmae (V 5b). 1 Transkei: *J. caffra* Becc. R: IBM 56(1987)122

Jubelina A. Juss. (~ *Diplopterys*). Malpighiaceae. 5 trop. Am. R: CUMH 17(1990)21

Judas tree *Cercis siliquastrum*

Judas's bag *Adansonia digitata*

Juelia Aspl. = *Ombrophytum*

Jugastrum Miers = *Eschweilera*

Juglandaceae DC ex Perleb. Dicots – Hamamelidae – Juglandales. 8/63 temp. & warm N hemisph. to S Am. & Mal. R: AMBG 65(1978)1058. Monoec. or dioec. wind-poll. trees & shrubs, aromatic. Lvs spirally arr. (opp. in *Alfaroa* & *Oreomunnea*), (trifoliolate or) pinnate with aromatic peltate glands; stipules 0. Fls small, unisexual, in axils of bracts of catkins, s.t. in term. panicles, these unisexual or not; C 0, males with K (1–)4(5) ± adnate to bractlets, or K (& s.t. bractlets) 0, A (3–)5–40(–100+) in 1 or 2 or more whorls, with short filaments & anthers with longit. slits, females with bract & bracteoles often united & forming an involucre becoming a husk in fr., K 4 teeth or 0 (e.g. *Carya*), $\overline{G}$ (2(3)) with free styles s.t. basally united, rarely stigmas sessile, 2(3)-loc. below (4–8 with extra partitions in some), 1-loc. apically, with 1 orthotropous (morphologically truly apical) unitegmic ovule. Fr. a nut or samara, or drupe-like ('tryma'), the soft husk (involucre) splitting to release bony pericarp; seed 1, ± 0 endosperm & embryo with 4-lobed massive cotyledons. x = 16

Genera: *Alfaroa, Carya, Cyclocarya, Engelhardtia, Juglans, Oreomunnea, Platycarya, Pterocarya*

Samaroid fr. with epigeal germ. seeds, drupe-like with hypogeal

Timber: *Carya* (hickories), *Juglans* (walnut); ed. nuts: *Carya* (pecan) & *Juglans*; cult. orn.

Juglans L. Juglandaceae. 21 Medit. (Eur. 1) to E As., N Am. to Andes. R: AMBG 65(1978)1071. Walnuts, differing from *Carya* in chambered pith in twigs. Ed. seeds, timber & cult. orn. Fr. drupe-like with hard pericarp splitting into 2 'boats' along midribs of carpels. Produce juglone, an allelopathic red crystalline compound, in lvs, bark & fr.; some domatia. *J. ailantifolia* Carrière (*J. sieboldiana*, Japanese w., Japan) – seeds ed.; *J. cinerea* L. (butternut, NE US) – ed. seeds, timber for furniture; *J. mandschurica* Maxim. (Manchurian w., NE China, Korea) – timber; *J. neotropica* Diels (*J. honorei*, S Am.) – ed. seeds, timber for cabinet-work; *J. nigra* L. (black or Am. w., E US, natur. C Eur.) – seeds ed. (used in confectionery in US), timber for furniture, gunstocks etc.; *J. regia* L. ((English or Black Sea) w., SE Eur., W–C As. & W China (? origin), natur. US) – walnuts of commerce, timber 'the foremost cabinet wood of N Am.' much used in C18 for furniture (ousted by mahogany less vulnerable to woodworm), gunstocks & as veneer, seed-oil used in salad oil (usu. from France), paint & soap, dye from husk used to stain floors, fertility symbol in Pliny's time, plant-pest control since time of Evelyn (1664), some hardier forms known as Carpathian w., hybrid with *J. nigra* (*J.* × *intermedia* Carr.) breeds true

jujube *Ziziphus jujuba, Z. mauritiana* (**Indian j.**)

Julbernardia Pellegrin. Leguminosae (I 4). 10 trop. Afr. (4 Guineo-Congolian forest, 6 major constituents of Sudano-Zambesian woodlands – miombo)

Julianiaceae Hemsley = Anacardiaceae (IV)

Julocroton C. Martius = *Croton*

Julostylis Thwaites. Malvaceae. 2 Sri Lanka, S India. A 10–20, G (2), seeds oblong to reniform

jumbie beans Various leguminous seeds esp. *Abrus precatorius, Adenanthera pavonina, Erythrina* spp., *Leucaena* spp., *Ormosia* spp.

Jumellea Schltr. Orchidaceae (V 16). 60 Afr. (2), Madag., Masc.

Jumelleanthus Hochr. Malvaceae. 1 Madagascar

jumping beans See *Sapium* & *Sebastiania* spp.

Juncaceae Juss. Monocots – Commelinidae – Juncales. 7/430 temp. & cold + trop. mts. Herbs (pachycaul shrubs in *Prionium*) often with starchy symp. rhiz. (rarely annual) rarely with mycorrhiza; vessels in all veg. organs. Stems often not extending above ground except in fl., photosynthetic. Lvs spirally arr., usu. basal, simple, parallel-veined with sheath & usu. flat to channelled terete, folded, or centric lamina (rarely 0), sheath often with apical auricles often ± confluent to form adaxial ligule. Fls usu. bisexual (rarely pl. dioec.) & wind-poll. (no nectaries when insect-poll.), small, (solit.) in heads or cymes; P (2)3 + (2)3 (or P 3) usu. greenish to blackish (rarely white or yellow), A 3 + 3 (or 3 + 0 or 2) with basifixed anthers with longit. slits & pollen in tetrads, $\underline{G}$ (3), 1- or 3-loc. (parietal or axile placentas (3 basal ovules in *Luzula*)), ovules 3–∞, anatropous, bitegmic, morphologically term. Fr. a loculicidal capsule (rarely indehisc.); embryo small, straight, embedded in starchy endopserm. x = 3–36

Principal genera: *Juncus, Luzula, Oxychloe, Prionium,* with greatest generic diversity in S hemisph.

Apparently closer to Gramineae & Cyperaceae than form. held

Some matting & chair-seats (*Juncus*); cult. orn.

Juncaginaceae Rich. Monocots – Alismatidae – Alismatales. 4/20 temp. & cold. Perennial (usu.), often rhizomatous herbs of wet places with elongate secretory canals, often cyanogenic; vessels confined to roots. Lvs spirally arr., mostly basal with ligule at junction of open sheath & slender (0 in *Maundia*) lamina, s.t. not distinct. Fls small, wind-poll., bisexual or not, in term. bractless spikes or racemes (in *Lilaea*, each leaf-axil with 1 or 2 female fls & spike with male & bisexual fls); P 3 + 3 (2 + 2; 3; *Lilaea* with P 1 in males & bisexual, 0 in females), A (3, 4)6(8; 1 in *Lilaea*), anthers with longit. slits G̲ 6 (alt. ones sterile or 0; 4 in *Maundia* & *Tetroncium*), usu. adnate to central axis but distinct at maturity (free in some *Triglochin* ('*Cycnogeton*'), 1-loc. (but prob. G 3 in *Lilaea* where style filiform to 30 cm), 1 erect, bitegmic, anatropous (pendulous, apical, orthotropous in *Maundia*) ovule per loc. Fr. a follicle (dry-indehiscent in *Lilaea*); seed 1 without endosperm, embryo straight. x = 6, 8, 9

Genera: *Lilaea* (often placed in its own fam.), *Maundia, Tetroncium, Triglochin*

Some *Triglochin* spp. ed., some poisonous (hydrogen cyanide)

Juncellus C.B. Clarke = *Scirpus*

Juncus L. Juncaceae. c. 300 cosmop. (Eur. 53 (*J. planifolius* R. Br. (S Pacific) natur. Ireland & Hawaii), Aus. 47 (31 endemic) with 21 natur.) but rare in trop. Rushes. R: BJ 12(1890)167. Usu. symp. rhiz. producing 1 leafy shoot a year; lvs grassy, needle-like or centric, infls often appearing lateral on a leaf-like cylindrical stem, but actually bract takes over term. position. Some medic. e.g. chibasa (Colombia) – roots used in skin infections. Some small e.g. *J. bufonius* L. (toad rush, cosmop.) others robust & used for matting (in Egypt since Neolithic) & chair-seats esp. *J. effusus* L. (Euras., N Am., Aus. & NZ) in Japan where cult. for weaving into standard floor-matting of Japanese house (igusa), some variants cult. orn. incl. 'Spiralis' with spiral stems found in Ireland, W Scotland & Surrey; pith of several spp. used as rush-lights before candles common; *J. planifolius* R. Br. (Aus., NZ, S Am.) – natur. Hawaii, W Ireland; *J. rigidus* Desf. (pantrop.) – used for making writing instruments in Egypt to 300 BC

Juneberry *Amelanchier* spp.

Junellia Moldenke (*Thryothamnus* Philippi). Verbenaceae. 47 S Am. R: Darw. 29(1989)382

Jungia L.f. (*Trinacte*). Compositae (Mut.-Nass.). 26 Mex. to Andes

jungle rice *Echinochloa colona, E. frumentacea*

juniper *Juniperus* spp. esp. *J. communis*, (in Bible) *Retama raetam*; **African j.** *J. procera*; **plum j.** *J. drupacea*; **Rocky Mt. j.** *J. scopulorum*; **Sierra j.** *J. occidentalis*

Juniperus L. Cupressaceae. 50 N hemisph. (Eur. 9) to trop. Afr. mts & WI. Junipers. R: Phytol.M 7(1984)31. Timbers, cult. orn. & flavourings: monoec. or dioec. trees & shrubs, some with needle-like lvs, some with scale-like ones at maturity; females cones with 3–8 fleshy coalescing scales becoming berry-like & bird-disp., the seeds hard & unwinged. *J. bermudiana* L. (Bermuda or Barbados cedar, Bermuda) – timber form. for pencils; *J. chinensis* L. (temp. E As.) – v. many orn. cvs; *J. communis* L. ((common) juniper, N temp.) – shrub with sweet aromatic fr. (200 t coll. wild pls in C Eur. imported to GB annually) used to flavour gin (= genever, Dutch for juniper), liqueurs & eaten with meat, many orn. cvs incl. creeping forms & slow-growing ones esp. 'Compressa'; *J. drupacea* Labill. (plum j., E Med.) – cone ed.; *J. excelsa* M. Bieb. (E Medit. to C As.) – fr. used medic.; *J. horizontalis* Moench (N Am.) – many creeping cvs; *J. occidentalis* Hook. (Sierra j., W US) – wood for fencing, fr. ed.; *J. oxycedrus* L. (Medit.) – heartwood distilled to give parasiticidal oil of Cade; *J. phoenicea* L. (Medit., Canary Is.) – said to have been used to build Mogador in Morocco; *J. procera* Hochst. ex Endl. (~ *J. excelsa*, E Afr. cedar, E Afr. mts, SE Arabia) – timber for pencils etc.; *J. sabina* L. (savin, C & S Eur. to W As.) – young twigs medic. & insecticidal; *J. squamata* D. Don (India to Taiwan) – lvs & twigs burnt in Sikkim temples; *J. virginiana* L. (red cedar, NE N Am.) – wood for insect-proof chests, pencils, oil for scenting soap, cult. orn. (many cvs) but alt. host for apple rust

Juno Tratt. = *Iris*

Junopsis W. Schulze = *Iris* (subg. *Nepalenses*)

Juno's tears *Verbena officinalis*

Jura turpentine *Picea abies*

Jurinea Cass. Compositae (Card.-Card.). c. 200 C & S Eur. (17), NW Afr.(1), SW & esp. C As. Some cult. orn. incl. *J. mollis* (L.) Reichb. – when lepidopterans attack rosette, multiple rosettes & therefore more fls & seeds produced but mammalian attack leads to only lateral heads which are less successful at producing seeds

Jurinella Jaub. & Spach (~ *Jurinea*). Compositae (Card.-Card.). 4 SW As.

Juruasia Lindau. Acanthaceae. 2 Brazil

Jussiaea L. = *Ludwigia*

Justago Kuntze = *Cleome*

Justenia Hiern = *Bertiera*

Justicia L. Acanthaceae. (Incl. *Adhatoda*, *Monechma*) c. 600 trop. & warm, temp. N Am. R: KB 43(1988)551, 592. Herbs & shrubs. *J. heterocarpa* T. Anders. (trop. Afr.) – heterocarpy otherwise unknown in fam. Cult. orn. esp. *J. adhatoda* L. (*A. vasica*, adhatoda, India, Sri Lanka) – lvs medic., wood used for beads, good fuel; *J. brandegeeana* Wassh. & L.B. Sm. (*Beloperone guttata*, *Drejerella guttata*, shrimp-plant, lobster-plant, Mex., natur. Florida) – v. popular housepl.; *J. pectoralis* Jacq. (Brazil) – hallucinogenic snuff

jute *Corchorus* spp.; **American**, **Chinese** or **Manchurian j.** *Abutilon theophrasti*; **Bimlipatum j.** *Hibiscus cannabinus*; **Congo j.** *Urena lobata*

Juttadinteria Schwantes. Aizoaceae (V). 5 S Afr. Cult. orn. succ.

Juzepczukia Chrshan. = *Rosa*

K

Kabulia Bor & C. Fischer. Caryophyllaceae (I). 1 Afghanistan: *K. akhtarii* Bor & C. Fischer

kadam *Neolamarckia cadamba*

Kadenia Lavrova & Tikh. (~ *Selinum*). Umbelliferae (III 8). 2 Euras.

Kadsura Juss. Schisandraceae. 22 India to Japan, W Mal., Moluccas. R: Sargentia 7(1947)156. Twining shrubs with unisexual fls, parts spirally arr. *K. japonica* (L.) Dunal (E As.) – cult. dioec. with red fr.; *K. scandens* (Blume) Blume (Mal.) – fr. ed. (sour)

Kadua Cham. & Schldl. = *Hedyotis*

Kaempferia L. Zingiberaceae. c. 50 India to S China & Mal. R: NRBGE 38(1980)1. Cult. orn. pot-pls incl. *K. galanga* L. (SE As) – rhiz. a spice & scent (galanga, name referring to other pls too), & *K. rotunda* L. (SE As.) – flavouring & medic.

Kafirnigania Kamelin & Kinzik. Umbelliferae (III 10). 1 C As.

kaffir boom *Erythrina caffra*; **k. lily** *Clivia* spp., *Schizostylis coccinea*; **k. plum** *Harpephyllum caffrum*

Kageneckia Ruíz & Pavón. Rosaceae. 3 Peru, Chile. Dioec. trees. *K. lanceolata* Ruíz & Pavón (Peru) – black dye from lvs

kagné butter *Allanblackia oleifera*

kahikatea *Dacrycarpus dacrydiodes*

kahua bark *Terminalia arjuna*

kai choy *Brassica rapa*; **k. lan** *Brassica oleracea* Alboglabra Group

Kaieteuria Dwyer = *Ouratea*

Kailarsenia Tirvengadum (~ *Randia*). Rubiaceae (II 1). 6 SE As.

Kairoa Philipson. Monimiaceae (V 2). 1 E New Guinea

Kairothamnus Airy Shaw. Euphorbiaceae. 1 New Guinea

kaivum fibre *Helicteres isora*

Kajewskia Guillaumin = *Veitchia*

Kajewskiella Merr. & Perry. Rubiaceae (? I 7). 2 Solomon Is. R: Blumea 25(1979)283

kaki *Diospyros kaki*

kaku oil *Lophira* spp.

Kalaharia Baillon (~ *Clerodendrum*). Labiatae (II; Verbenaceae). 1 trop. & S Afr.: *K. uncinata* (Schinz) Moldenke

Kalakia Alava. Umbelliferae (III 11). 1 Iran: *K. marginata* (Boiss.) Alava

kalamet *Mansonia gagei*

Kalanchoe Adans. (incl. *Bryophyllum*). Crassulaceae. 125 OW trop. esp. Afr. & Madag., S Am. (1). R: BHB II 7(1907)869, 8(1908)17. Succ. shrubs & herbs s.t. hapaxanthic; 3 sections incl. sect. *Bryophyllum* with plantlets borne in leaf-margins & infls. Favoured for pl. physiology experiments; *K. petitiana* A. Rich. (Ethiopia) – facultative crassulacean acid metabolism, shifting from C_3 as leaves age or there is water stress. Cult. orn. *K. blossfeldiana* Poelln. (Madag.) occasionally, but commonly grown pot-pls for red fls in winter are hybrids (poss. with *K. flammea* Stapf (Somalia), *K. pumila* Baker (Madag.) etc.); *K. beharensis* Drake (Madag.) – tree-like to 6 m; *K. daigremontiana* Hamet & Perrier (flopper, Madag.) – commonly cult. sp. with plantlets at margin of blotched lanceolate lvs; *K. delagoensis* Ecklon & Zeyher (*K. tubiflora*, Madag., S Afr.) – lvs linear with apical plantlets, widely cult. & weedy; *K. pinnata* (Lam.) Pers. (? origin Madag., but widely natur. in trop.) – commonly cult. sp. with pinnate lvs & plantlets ('air plant')

Kalappia Kosterm. Leguminosae (I 2). 1 Sulawesi

kalaw *Hydnocarpus castanea*

Kalbfussia Schultz-Bip. = *Leontodon*

Kalbreyera Burret = *Geonoma*

Kalbreyeracanthus Wassh. (*Syringidium*; ~ *Habracanthus*). Acanthaceae. 2 Colombia. R: Brittonia 37(1985)199

Kalbreyeriella Lindau. Acanthaceae. 3 Panamá, Colombia

kale *Brassica oleracea* Acephala Group; **Chinese k.** *B. oleracea* Alboglabra Group; **Indian k.** *Caladium lindenii*; **sea k.** *Crambe maritima*; **Siberian k.** *B. napus* Pabularia Group

Kalidiopsis Aellen = *Kalidium*

Kalidium Moq. Chenopodiaceae (II 1). 5 Medit. (Eur. 2) to C As.

Kalimeris Cass. (~ *Aster*). Compositae (Ast.-Ast.). 10 C & E to SE As. Some cult. orn.

Kalimpongia Pradhan = *Dickasonia*

Kaliphora Hook.f. Cornaceae (Melanophyllaceae). 1 Madag.

Kallstroemia Scop. Zygophyllaceae. 17 trop. & warm Am. R: CGH 198(1969)41. *K. grandiflora* Torrey & A. Gray (Arizona poppy, SW N Am.) – conspic. desert annual

Kalmia L. Ericaceae. 7 N Am., Cuba (2 natur. Eur.). R: Rhodora 76(1974)315; R.A. Jaynes (1988) *K.* C with 10 pouches in which anthers held under tension like bows until insect probing for nectar releases them & is showered with pollen. American laurel; lvs poisonous. Cult. orn. usu. evergreen shrubs esp. *K. angustifolia* L. (sheep laurel, pig l., lambkill), *K. latifolia* L. (mountain l., calico bush) & *K. polifolia* Wangenh. (bog l.) – all E N Am.

Kalmiella Small = *Kalmia*

Kalmiopsis Rehder. Ericaceae. 1 Oregon: *K. leachiana* (L. Henderson) Rehder – cult. orn. rock pl., protected in wild

kalo *Colocasia esculenta*

kalonji *Nigella sativa*

Kalopanax Miq. (~ *Eleutherococcus*). Araliaceae. 1 temp. E As.: *K. septemlobus* (Thunb.) Koidz. (*Acanthopanax ricinifolius*, *K . pictus*) – cult. orn. decid. tree, timber (sen)

Kalopternix Garay & Dunsterv. = *Epidendrum*

kamala *Mallotus philippensis*

kamarere *Eucalyptus deglupta*

kamassi *Gonioma kamassi*

Kamettia Kostel. (*Ellertonia*). Apocynaceae. 1 S India: *K. caryophyllata* (Roxb.) Nicolson & Suresh

Kamiesbergia Snijman = *Hessea*

Kampmannia Steudel = *Cortaderia*

Kampochloa W. Clayton. Gramineae (33b). 1 S trop. Afr.

Kanahia R. Br. Asclepiadaceae (III 1). 2 trop. E Afr., Arabia. R: NJB 6(1986)787

Kanaloa Lorence & K. Wood. Leguminosae (II 3). 1 Hawaii: *K. kahoolawensis* Lorence & K. Wood. R: Novon 4(1994)137

Kandaharia Alava. Umbelliferae (III 11). 1 Afghanistan: *K. rechingerorum* Alava

Kandelia (DC) Wight & Arn. Rhizophoraceae. 1 E As., W Mal: *K. candel* (L.) Druce – tanbark

kanga butter *Pentadesma butyracea*

kangaroo apple *Solanum aviculare, S. laciniatum*; **k. fern** *Phymatosorus pustulatus*; **k. grass** *Themeda* spp. esp. *T. australis*; **k. paw** *Anigozanthos* spp.; **k. thorn** *Acacia paradoxa*; **k. vine** *Cissus antarctica*

kangkong *Ipomoea aquatica*

Kania Schltr. (~ *Metrosideros*). Myrtaceae. 6 Philippines (2), New Guinea (4). R: Blumea 28(1982)177, KB 45(1990)205

Kanimia Gardner = *Mikania*

Kanjarum Ramam. = *Strobilanthes*

kanluang *Nauclea orientalis*

kan-non-chiku *Rhapis excelsa*

Kansas gay-feather *Liatris spicata*; **K. thistle** *Solanum rostratum*

Kantou Aubrév. & Pellegrin = *Inhambanella*

Kaokochloa De Winter. Gramineae (29). 1 SW Afr.

kaolang *Sorghum* spp. (China)

Kaoue Pellegrin = *Stachythyrsus*

kapa *Broussonetia papyrifera*

kapok *Ceiba pentandra*; see also *Bombax* spp.

kapong *Tetrameles nudiflora*
kapur *Dryobalanops* spp. esp. *D. aromatica*
karaka *Corynocarpus laevigata*
Karamyschewia Fischer & C. Meyer = *Oldenlandia*
karanda *Carissa carandas*; **k. nuts** *Elaeocarpus bancroftii*
karanja *Millettia* sp. (*Pongamia pinnata*)
Karatas Miller = *Bromelia*
Karatavia Pim. & Lavrova. Umbelliferae (III 8). 1 C As.
karaya gum *Cochlospermum religiosum*
Kardanoglyphos Schldl. Cruciferae. 1 Andes
karela, karella *Momordica charantia*
Karelinia Less. Compositae (Pluch.). 1 Russia & NE Iran to Mongolia
Kariba weed *Salvinia molesta*
Karimbolea Descoings. Asclepiadaceae (III 1). 1 Madag.: *K. verrucosa* Descoings
Karina Boutique. Gentianaceae. 1 Zaire
kariyat *Andrographis paniculata*
Karnataka P.K. Mukherjee & Constance. Umbelliferae (III 8). 1 S India: *K. benthamii* (C.B. Clarke) P.K. Mukherjee & Constance. R: Brittonia 38(1986)145
karo *Pittosporum crassifolium*
Karomia Dop. Labiatae (II; Verbenaceae). 1 SE As., 8 Afr. & Madag. R: GOB 7(1985)36
karri *Eucalyptus diversifolia*
Karroochloa Conert & Türpe = *Rytidosperma*
Karvandarina Rech.f. Compositae (Card.-Cent.). 1 Iran, Pakistan
Karwinskia Zucc. Rhamnaceae. 16 SW US to Bolivia & WI. *K. calderonii* Standley (Salvador) – fine timber; *K. humboldtiana* (Roemer & Schultes) Zucc. (SW N Am.) – fr. ed. but seeds toxic, paralyzing motor nerves, used in treatment of tetanus in Mex.
Kaschgaria Polj. Compositae (Anth.-Art.). 2 C As. to W China
Kashmiria Hong. Scrophulariaceae. 1 Himal.: *K. himalaica* (Hook.f.) Hong
ķataka *Strychnos potatorum*
Katherinea A.D. Hawkes = *Epigeneium*
katon *Sandoricum koetjape*
katsura *Cercidiphyllum japonicum*
Kaufmannia Regel. Primulaceae. 1–2 C As.
Kaulfussia Nees = *Felicia*
kaunghmu *Anisoptera scaphula*
Kaunia R. King & H. Robinson (~ *Eupatorium*). Compositae (Eup.-Oxy.). 14 S Am. R: MSBMBG 22(1987)448
kauri *Agathis* spp.; **Fijian k.** *A. macrophylla*
kava *Piper methysticum*, (NZ) *Macropiper excelsum*
kawaka *Libocedrus plumosa*
kawa-kawa *Macropiper excelsum*
kawal *Senna obtusifolia*
Kayea Wallich = *Mesua*
keaki *Zelkova serrata*
Kearnemalvastrum D. Bates. Malvaceae. 2 Mexico to Colombia
Keayodendron Leandri. Euphorbiaceae. 1 trop. Afr.
Keckiella Straw (~ *Penstemon*). Scrophulariaceae. 7 N Am.
kedam *Neolamarckia cadamba*
Kedarnatha P.K. Mukherjee & Constance. Umbelliferae (III 8). 1 Himal. R: Brittonia 38(1986)147
kedondong *Canarium* spp.
Kedrostis Medikus. Cucurbitaceae. 23 OW trop. Often with swollen caudiciform bases
Keenania Hook.f. Rubiaceae (I 8). 5 Assam, SE As. Poss. incl. *Myrioneuron*
Keerlia DC = *Chaetopappa*
Keetia E. Phillips (~ *Canthium*). Rubiaceae (III 2). 40 trop. & S Afr. R: KB 41(1986)965
Kefersteinia Reichb.f. Orchidaceae (V 10). 36 trop. Am.
Kegeliella Mansf. Orchidaceae (V 10). 3 trop. Am.
kei apple *Dovyalis caffra*
Keiria Bowdich = ? *Cissus*
Keiskea Miq. Labiatae (VIII 1). 5 E As. R: BBMNHB 10(1982)70
Keithia Benth. = *Rhabdocaulon*
Keithia Sprengel = ? Capparidaceae

keladi *Colocasia esculenta*

Kelissa Ravenna. Iridaceae (III 4). 1 temp. S Am.

Kelleria Endl. (~ *Drapetes*). Thymelaeaceae. 5 Mt Kinabalu (1), New Guinea, Aus. (2), NZ. R: AusSB 3(1990)609. ? Beetle-poll.

Kelleronia Schinz (~ *Tribulus*). Zygophyllaceae. 10 Somalia

Kelloggia Torrey ex Benth. Rubiaceae (IV 12). 2 China (1), SW US (1)

Kelly grass *Rottboellia cochinchinensis*

Kelseya S. Watson ex Rydb. (~ *Luetkea*). Rosaceae. 1 W US: *K. uniflora* (S. Watson) Rydb. – cult. rock-pl.

kempas *Koompassia malaccensis*

Kemulariella Tamaschjan = *Aster*

kenaf *Hibiscus cannabinus*

kenda *Macaranga peltata*

Kendal green dye made from fls of *Genista tinctoria* with woad

Kendrickia Hook.f. Melastomataceae. 1 S India, Sri Lanka. Root-climber

kendyr fibre *Apocynum venetum*

Kengia Packer (*Cleistogenes*). Gramineae (31d). c. 10 S Eur. (2), Turkey to temp. E As.

kenguel seed *Silybum marianum*

Kengyilia C. Yen & J.-L. Yang = *Elymus*

Kennedia Vent. Leguminosae (III 10). 16 Aus., New Guinea. Showy lianes with red to almost black fls, some cult. orn. (coral peas) incl. *K. nigricans* Lindley (black bean, W Aus.) – fls almost black

Kenopleurum Candargy = *Thapsia*

Kensitia Fedde = *Erepsia*

kentia *Howea forsteriana*

Kentiopsis Brongn. Palmae (V 4h). 1 New Caled.: *K. oliviformis* (Brongn. & Gris) Brongn. R: Allertonia 3(1984)324

Kentish cob *Corylus maxima*

Kentranthus Raf. = *Centranthus*

Kentrochrosia Schumann & Lauterb. = *Kopsia*

Kentrosiphon N.E. Br. = *Gladiolus*

Kentrothamnus Süsseng. & Overk. Rhamnaceae. 1 Bolivia, Arg.

Kentucky blue grass *Poa pratensis*; **K. coffee tree** *Gymnocladus dioica*; **K. yellowwood** *Cladrastis lutea*

keppel *Stelechocarpus burahol*

Keracia (Cosson) Calest. = *Hohenackeria*

keratto *Agave* spp.

Keraudrenia Gay. Sterculiaceae. 7 Madag. (1), New Guinea, Aus.

Keraymonia Farille. Umbelliferae (III 5). 3 Nepal

Kerbera Fourn. = *Melinia*

Kerguélen cabbage *Pringlea antiscorbutica*

Kerianthera Kirkbride. Rubiaceae (I 7). 1 Amaz. Braz.: *K. preclara* Kirkbride. R: Brittonia 37(1985)109

Kerigomnia P. Royen = *Octarrhena*

Kermadecia Brongn. & Gris. Proteaceae. 4 New Caled. R: AJB 62(1975)135. *K. rotundifolia* Brongn. & Gris – timber for general use

kermek *Limonium gmelinii* & *L. latifolium*

Kernera Medikus (~ *Cochlearia*). Cruciferae. 1 C & S Eur. mts. *K. saxatilis* (L.) Sweet – cult. orn. rock-pl.

kerong *Pongamia pinnata*

kerosene weed *Ozothamnus ledifolius*

Kerria DC. Rosaceae. 1 temp. E As: *K. japonica* (L.) DC – cult. orn. suckering shrub, usu. double-flowered form

Kerriochloa C. Hubb. Gramineae (40d). 1 SE As.: *K. siamensis* C. Hubb.

Kerriodoxa Dransf. Palmae (I 1c). 1 peninsular Thailand: *K. elegans* Dransf. R: Principes 27(1983)3. One of 12 most endangered pl. spp. on Phuket

Kerriothyrsus C. Hansen. Melastomataceae. 1 Laos: *K. tetrandrus* (Nayar) C. Hansen. R: Willdenowia 17(1988)153

Kerstingiella Harms = *Macrotyloma*

keruing *Dipterocarpus* spp.

ketaki *Pandanus tectorius*

Keteleeria Carrière. Pinaceae. 3–7 S China, Taiwan, SE As. (Eur. in Tertiary). R: NRBGE

46(1989)81. Some timber & cult. orn.

kewda *Pandanus fascicularis*

keyaki *Zelkova serrata*

Keyserlingia Bunge ex Boiss. = *Sophora*

Keysseria Lauterb. = *Lagenophora*

Khadia N.E. Br. Aizoaceae (V). 5 Transvaal. Drink made from roots (khadi)

Khasiaclunea Ridsd. Rubiaceae (I 2). 1 E As.

khat *Catha edulis*

Khaya A. Juss. Meliaceae (IV 2). c. 7 trop. Afr. & Madag. African mahogany & used as subs. for *Swietenia*. Bark locally medic. Esp. imp. in commerce are *K. grandifoliola* C. DC (Benin mahogany, B.wood, W & C Afr.), *K. ivorensis* A. Chev. & *K. senegalensis* (Desr.) A. Juss. (bisselon, W Afr.) & *K. madagascariensis* Jum. & Perrier (Madagascar m., E Madag.)

khayer *Acacia catechu*

khesara *Lathyrus sativus*

khus-khus *Vetiveria zizanioides*

kiaat *Pterocarpus angolensis*

Kibara Endl. Monimiaceae (V 2). 43 Nicobar Is., Mal. to trop. Aus. R: Blumea 30(1985)389

Kibaropsis Vieill. ex Jérémie (~ *Hedycarya*). Monimiaceae (V 1). 1 New Caledonia

Kibatalia G. Don f. Apocynaceae. 15 SE As. to Philipp. R: AUWP 86-5(1986)36. Wood for shoes etc.

Kibessia DC = *Pternandra*

Kickxia Dumort. (~ *Linaria*). Scrophulariaceae (Antirr.). 47 Medit. (Eur. 5) to S C & SW As., W Afr. is. R: D.A. Sutton *Rev. Antirrhineae* (1988)169. Fluellen. Weedy: *K. elatine* (L.) Dumort. (Eur., Medit.) caused abandonment of c. 1500 acres barley SW of Dayton, Calif.

kidney bean *Phaseolus vulgaris*; **k. fern** *Trichomanes* spp.; **k. vetch** *Anthyllis vulneraria*; **k. wood** *Eysenhardtia polystachya*

Kielmeyera C. Martius. Guttiferae (II). c. 30 S Brazil. Characteristic of campos. *K. coriacea* C. Martius (pau santo) – ground bark a subs. for ground cork

kif *Cannabis sativa* subsp. *indica*

Kigelia DC. Bignoniaceae. 1 (variable) trop. Afr. (fossil wood in Libya): *K. africana* (Lam.) Benth. (*K. pinnata*, sausage-tree) – infls of claret, bat-poll. fls. hanging on long peduncles, fr. gourd-like, locally medic. (purgative)

Kigelianthe Baillon = *Fernandoa*

Kiggelaria L. Flacourtiaceae (4). 1 trop. & S Afr.: *K. africana* L. (*K. dregeana*, wild peach) – good timber (Natal mahogany)

Kiggelariaceae Link = Flacourtiaceae

Kiharopyrum Löve = *Aegilops*

Kikuyu grass *Pennisetum clandestinum*

Killipia Gleason. Melastomataceae. 4 Colombia

Killipiella A.C. Sm. = *Disterigma*

Killipiodendron Kobuski. Theaceae. 1 Colombia

Kilmarnock willow *Salix caprea* 'Pendula'

Kinepetalum Schltr. = *Tenaris*

king cup *Caltha palustris*; **k.fisher daisy** *Felicia bergeriana*; **K. William (Billy) pine** *Athrotaxis selaginoides*; **k. wood** *Astronium fraxinifolium*, *Dalbergia cearensis*

Kingdonia Balf.f. & W. Sm. Ranunculaceae (II 2; Circaeasteraceae; Kingdoniaceae). 1 W & N China

Kingdoniaceae A.S. Foster ex Airy Shaw. See Ranunculaceae (II)

Kingella Tieghem = *Trithecanthera*

Kinghamia C. Jeffrey (~ *Gutenbergia*). Compositae (Vern.-Vern.). 5 trop. W Afr. R: KB 43(1980)274

Kingia R. Br. Xanthorrhoeaceae. 1 SW Aus.: *K. australis* R. Br. (black gin) – pachycaul tree with mantle of concealed aerial roots under marcescent leaf-bases; as the stem dies from the base after 300+yrs, roots keep contact with ground (cf. *Iriartea*); oldest specimen known c. 650 yrs old; flowering stimulated by fire

Kingianthus H. Robinson. Compositae (Helia.- Verb.). 2 Ecuador. R: Phytol. 58(1978)415

Kingidium P. Hunt = *Phalaenopsis*

Kingiodendron Harms (~ *Oxystigma*). Leguminosae (I 4). 6 Indomal. R: Blumea 18(1970)46, KB 32(1977)244

Kingstonia Hook.f. & Thomson = *Dendrokingstonia*

Kinia Raf. = ? *Ottelia*

kinnikinni(c)k *Arctostaphylos uva-ursi*

kino astringent resin-like substance from tapped trees; **African k.** *Pterocarpus erinaceus;* **Australian k.** *Eucalyptus* spp. esp. *E. camaldulensis* & *E. resinifera;* **Bengal k.** *Butea monosperma;* **E. Indian** or **Malabar k.** *P. marsupium;* **Jamaican k.** *Coccoloba uvifera*. Used medic. & for tanning locally

Kinostemon Kudô = *Teucrium*

Kinugasa Tatew. & Sûto = *Paris*

Kionophyton Garay = *Stenorrhynchos*

Kippistia F. Muell. (~ *Minuria*). Compositae (Ast.-Ast.). 1 S Aus.: *K. suaedifolia* F. Muell. R: Nuytsia 3(1980)215

Kirengeshoma Yatabe. Hydrangeaceae. 1 E China, Korea & Japan: *K. palmata* Yatabe – cult. orn. herb

kiri wood *Paulownia tomentosa*

Kirilowia Bunge. Chenopodiaceae (I 4). 2 C & SW As.

Kirkbridea Wurd. Melastomataceae. 1 Colombia

Kirkia Oliver. Kirkiaceae. (Simaroubaceae s.l.). 5 trop. & S Afr. R: KB 35(1981)829. Some cult. orn. incl. *K. acuminata* Oliver (trop. Afr.) – timber & *K. wilmsii* Engl. (pepper-tree, S Afr.)

Kirkiaceae (Engl.) Takht. (Simaroubaceae s.l.). Dicots – Rosidae – Sapindales. 2/6 trop. & S Afr., Madag. Form. subfam. of Simaroubaceae excl. now on absence of quassinoids & DNA work

Genera: *Kirkia, Pleiokirkia*

Kirkianella Allan (~ *Launaea*). Compositae (Lact.-Son.). 1 NZ

kisidwe *Allanblackia floribunda*

Kissenia R. Br. ex Endl. Loasaceae. 2: 1 S Arabia, Somalia, Ethiopia, 1 SW Afr.

Kissodendron Seemann = *Polyscias*

Kita A. Chev. = *Hygrophila*

Kitagawia Pim. (~ *Peucedanum*). Umbelliferae (III 10). 5 C As.

Kitaibela Willd. Malvaceae. 1 Lower Danube: *K. vitifolia* Willd. – cult. orn.; 1 E Med. R: BJ 109(1987)59

Kitamuraea Rauschert = *Aster*

Kitchingia Baker = *Kalanchoe*

kite tree *Nuxia floribunda*

kitembilla *Dovyalis hebecarpa*

Kitigorchis Maek. = *Oreorchis*

kittool or **kitul** *Caryota urens*

kiwano *Cucumis metuliferus*

kiwi (fruit) *Actinidia deliciosa*

Kjellbergia Bremek. = *Strobilanthes*

Kjellbergiodendron Burret. Myrtaceae (Lept.). 1+ Sulawesi, New Guinea

Klaineanthus Pierre ex Prain. Euphorbiaceae. 1 trop. W Afr.

Klaineastrum Pierre ex A. Chev. = *Memecylon*

Klainedoxa Pierre ex Engl. Irvingiaceae (Simaroubaceae s.l.). 3 trop. Afr. R: AMBG 71(1984)166. *K. gabonensis* Pierre ex Engl. (trop. W Afr.) – seeds ed., timber valuable

Klamath weed *Hypericum perforatum*

Klaprothia Kunth. Loasaceae. 2 trop. S Am. R: SB 15(1990)671

Klasea Cass. = *Serratula*

Klattia Baker. Iridaceae (II). 3 Cape. Shrubby

Kleinhovia L. Sterculiaceae. 1 trop. As. to Aus.: *K. hospita* L. – commonly cult. (in trop.) orn. tree with red fls

Kleinia Miller (~ *Senecio*). Compositae (Sen.-Sen.). c. 40 Canary Is., trop. & S Afr., Madag., Arabia (incl. *Notonia*), S India, Sri Lanka. R (partial): HIP 39,4(1988)1. Cult. orn. succ. with red, yellow or white fls esp. *K. neriifolia* Haw. (*S. kleinia*, Canary Is.) – pachycaul treelet with pseudo-dichotomous branching

Kleinodendron L.B. Sm. & Downs = *Savia*

Klemachloa R. Parker = *Dendrocalamus*

Klingia Schönl. = *Gethyllis*

klinki pine *Araucaria hunsteinii*

Klossia Ridley. Rubiaceae (? IV 2). 1 Malay Peninsula

Klotzschia Cham. Umbelliferae (I2b). 3 Brazil

Klugia Schldl. = *Rhynchoglossum*

Klugiodendron Britton & Killip = *Abarema*

Kmeria (Pierre) Dandy. Magnoliaceae. 2 S China, SE As.

knapweed *Centaurea nigra*

Knautia L. Dipsacaceae. 60 Eur. (48), Medit. Some cult. orn. like *Scabiosa*, incl. *K. arvensis* (L.) Coulter (field scabious, Eur. to N Afr.) with heads of fls, the more towards the periphery the more the C is extended outwards; stigmas of a head ripen together; fatty outgrowth at fruit-base attractive to animal dispersers

knawel *Scleranthus annuus*

Knema Lour. Myristicaceae. 90 Indomal. R: Blumea 25(1979)321. 27(1981)223, 32(1987)115. Some medic. seed-oils (some e.g. *K. glauca* (Blume) Petermann (W Mal.) form. for lighting) & locally used timbers

Knightia Sol. ex R. Br. Proteaceae. (Excl. *Eucarpha*) 1 NZ: *K. excelsa* Sol. ex R. Br. (rewa-rewa) – beautifully figured timber for cabinet-making

Kniphofia Moench. Asphodelaceae (Aloaceae). 65 Arabia (1), trop.(22) & S (46) Afr. mts, Madag. (1). Red-hot pokers, torch lilies. R: (S Afr.) Bothalia 9(1968)363, (trop. Afr.) KB 28(1973)465. Cult. orn. spp. & hybrids (fls in winter in NZ) with heads of pendent red or yellow fls visited by sunbirds; bees entering some fls in Brit. gardens may be trapped & unable to leave cylindrical C. *K. caulescens* Baker ex Hook.f. (E S Afr.) – rosettes of lvs term. stout stems to 30 cm; *K. uvaria* (L.) Oken (S Afr.) – 1 of first Cape pls introd. Eur. though the commonly cult. pls today are complex hybrids & cvs (*K.* × *praecox* Baker etc.), natur. Victoria

knobthorn or **knobwood** *Zanthoxylum* spp. esp. *Z. capense*

knol-kohl = kohlrabi

Knorringia (Czuk.) Tsvelev = *Persicaria*

knotweed or **knotgrass** *Fallopia*, *Persicaria* & *Polygonum* spp. esp. *P. aviculare*; **Japanese k.** *F. japonica*

Knowltonia Salisb. Ranunculaceae (II 2). 8 E (1) & S Afr. R: BN 53(1979). Allied to *Anemone*, but fleshy drupelets

Knoxia L. Rubiaceae (III 8). 7 Indomal., 2 Afr.

Koanophyllon Arruda (~ *Eupatorium*). Compositae (Eup.-Crit.). c. 120 trop. Am. R: MSBMBG 22(1987)314

Kobresia Willd. Cyperaceae. 35 N temp. (Eur. 2). R: Pfl. IV,20(1909)33. Close to *Carex*. Imp. pasture pls in alpine Russia

Kochia Roth = *Bassia*

Kochummenia Wong (~ *Randia*). Rubiaceae (II 1). 2 Malay Pen. R: MNJ 38(1984)31

Kodalyodendron Borh. & Acuña. Rutaceae. 1 E Cuba (serpentine). Allied to *Amyris*

kodo millet *Paspalum scrobiculatum*; **k. wood** *Ehretia acuminata*

Koeberlinia Zucc. Capparidaceae. 1 SW N Am.: *K. spinosa* Zucc. (allthorn) – leafless xerophyte

Koeberliniaceae Engl. = Capparidaceae

Koechlea Endl. = *Ptilostemon*

Koehneola Urban. Compositae (Helia.-Pin.). 1 E Cuba (serpentine)

Koehneria S.A. Graham, Tobe & Baas (~ *Pemphis*). Lythraceae. 1 S Madag.: *K. madagascariensis* (Baker) S.A. Graham, Tobe & Baas. Pollen diff. from *Pemphis*

Koeiea Rech.f. = *Prionotrichon*

Koeleria Pers. Gramineae (21b). c. 35 temp. (Eur. 13). Hair grass. Spp. intergrading

Koellensteinia Reichb.f. Orchidaceae (V 10). 17 trop. Am.

Koellikeria Regel. Gesneriaceae. 1 trop. Am.: *K. erinoides* (DC) Mansf. R: Selbyana 6(1982)174

Koelpinia Pallas. Compositae (Lact.-Scor.). 5 N Afr. to E As. (Eur. 1)

Koelreuteria Laxm. Sapindaceae. 3 China (2), Taiwan (1). R: JAA 57(1976)129 Capsule large & bladdery & may be blown as a seed-disp. mechanism. *K. paniculata* Laxm. (golden rain, pride of India, China & Korea, natur. Japan) – cult. orn. tree esp. as street-tree in C Eur., etc., fls medic. & source of yellow dye in China, seeds used as beads, poss. oil source

Koelzella Hiroe = *Prangos*

Koenigia L. (~ *Persicaria*). Polygonaceae (II 5). 6 Arctic & N Eur. mts, temp. E As., Himalayas, *K. islandica* L. circumpolar to Tierra del Fuego

Koernickanthe Andersson. Marantaceae. 1 trop. Am. R: NJB 1(1981)240

Kohautia Cham. & Schldl. (~ *Oldenlandia*). Rubiaceae (IV 1). 60 OW trop. Some weedy like *K. tenuis* (S. Bowdich) Mabb. (*K. senegalensis*, W Afr.)

kohekohe *Dysoxylum spectabile*, (Hawaii) *Eleocharis calva*

Kohleria Regel. Gesneriaceae. 17 trop. Am. (Colombia 14, 9 endemic). R: SCB 79(1992). Shrubs & herbs, some with scaly runners, or aerial rhiz. in place of infls. Many cult. orn. incl. numerous hybrids

Kohlerianthus Fritsch = *Columnea*

kohlrabi *Brassica oleracea* Gongylodes Group
Kohlrauschia Kunth (~ *Petrorhagia*). Caryophyllaceae (III 1). 5 Medit. & W As.
koie-yan *Faradaya splendida*
Koilodepas Hassk. Euphorbiaceae. 10 S India to Hainan & New Guinea
kokam or **kokum butter** *Garcinia indica*
kokerite *Maximiliana maripa*
Kokia Lewton. Malvaceae. 4 Hawaii. R: P.A. Fryxell, *Nat. Hist. Cotton Tribe* (n.d.) 79. Some extinct (1) or only surviving in botanic gardens, e.g. *K. cookei* Degener reduced to 1 pl. in Eur., now being propagated by leaf-fragments & callus. *K. drynarioides* (Seemann) Lewton with extrafl. nectaries, though no native ants in Hawaii (? 'phylogenetic inertia')
kokko *Albizia lebbeck*
kokoon *Kokoona zeylanica*
Kokoona Thwaites. Celastraceae. 8 Indomal. *K. zeylanica* Thwaites (kokoon, Sri Lanka) – seed-oil effective as leech-deterrent
kokrodua *Pericopsis alata*
koksaghyz *Taraxacum bicorne*
kokum *Garcinia indica*
kola *Cola* spp.
Kolkwitzia Graebner. Caprifoliaceae. 1 C China: *K. amabilis* Graebner (beautybush) – floriferous shrub like *Abelia* cult. esp. in US
Kolobochilus Lindau. Acanthaceae. 2 Costa Rica
Kolobopetalum Engl. Menispermaceae (III). 4 trop. Afr.
Kolowratia C. Presl = *Alpinia*
Kolpakowskia Regel = *Ixiolirion*
Komaroffia Kuntze (~ *Nigella*). Ranunculaceae (I 3). 2 Iran & C As. Differs from *N.* in entire or palmate lvs
Komarovia Korovin. Umbelliferae (? tribe). 1 C As.
kombo *Pycnanthus angolensis*
Kompitsia Costantin & Gallaud = *Pentopetia*
Konantzia Dodson & N. Williams. Orchidaceae (V 10). 1 W Ecuador
konjaku *Amorphophallus konjac*
Koompassia Maingay. Leguminosae (I 2). 3 Mal. *K. excelsa* (Becc.) Taubert (tualang, W Mal.) – at 84 m a tree in Sarawak the tallest trop. angiosperm – timber splits too easily to be valuable; *K. malaccensis* Maingay (kempas, W Mal.) – timber good
Koordersiodendron Engl. Anacardiaceae (II). 1 Philipp. to New Guinea: *K. pinnatum* (Blanco) Merr. – timber good
Kopsia Blume. Apocynaceae. 25 SE As. to W Mal., Carolines. Alks. Some cult. orn. trees & shrubs
Kopsiopsis (G. Beck) G. Beck. Orobanchaceae. 2 W N Am.
korakaha *Memecylon umbellatum*
Korolkowia Regel = *Fritillaria*
Korovinia Nevski & Vved. = *Galagania*
Korshinskia Lipsky. Umbelliferae (III 5). 5 SW & C As.
Korthalsella Tieghem. Viscaceae. 30 N trop. Afr., Madag., Masc., Himal. to Japan & NZ. Explosive fr. (not bird-disp.)
Korthalsia Blume. Palmae (II 1c). 27 Indomal. (25 Mal.). R: KB 36(1981)163. Rattans. Herm.; hapaxanthic shoots branching high in canopy; suckers
Kosmosiphon Lindau. Acanthaceae. 1 trop. W Afr.
koso or **kousso** *Hagenia abyssinica*
Kosopoljanskia Korovin. Umbelliferae (III 4). 2 C As.
Kosteletzkya C. Presl (*Thorncroftia*). Malvaceae. 30 N Am., trop. & S Afr., Madagascar
Kostermansia Soeg.-Reks. Bombacaceae. 1 Malay Peninsula
Kostermanthus Prance. Chrysobalanaceae (4). 2 W Mal. R: PTRSB 320(1988)149
Koster's curse *Clidemia hirta*
kosumba *Schleichera oleosa*
Kotchubaea Regel ex Hook.f. = *Kutchubaea*
kotibé *Nesogordonia papaverifera*
Kotschya Endl. Leguminosae (III 13). 30 trop. Afr.
kowhai *Sophora tetraptera*
Koyamacalia H. Robinson & Brettell = *Parasenecio*
Koyamaea W. Thomas & Davidse. Cyperaceae. 1 Venez. & Brazil: *K. neblinensis* W. Thomas & Davidse. R: SB 14(1989)189. Own tribe

Kozlovia Lipsky. Umbelliferae (III 2). 1 Afghanistan
kozo *Broussonetia kazinoki*
krabak *Anisoptera spp. esp. A. curtisii*
Kraenzlinella Kuntze = *Pleurothallis*
Krameria L. Krameriaceae. 15 SW US to Arg. & Chile, esp. dry. Local medic. (rhatany root) & dyes. *K. lanceolata* Torrey (SW US) – hemiparasite with haustoria
Krameriaceae Dumort. Dicots – Rosidae – Polygalales. 1/15 warm Am. Hemiparasitic shrubs or trees or herbs. Lvs spirally arr., simple (–trifoliolate), entire; stipules 0. Fls bisexual, not resupinate, solit. & axillary or in term. racemes; K (4)5, imbricate, the 3 outer often larger than 2 inner & ± enclosing flower, C (4)5, the 3 adaxial long-clawed, 2 lower smaller, broad, thick, sessile often lipid-secreting glands (nectary-disk 0), A (3)4 alt. with upper C (5th rarely below & sterile) with thick filaments sometimes basally connate or adnate to C & anthers with 1 or 2 term. pores or short slits, G pseudomonomerous (2), 1 carpel reduced & empty, the other with 2 collateral pendulous anatropous bitegmic ovules. Fr. dry-dehiscent, 1-seeded, usu. armed with barbed bristles or spines; seeds with straight embryo & 0 endosperm. x = 6
Only genus: *Krameria*
App. closest to Polygalaceae though trifoliolate *K. cytisoides* Cav. (Mex.) reminiscent of Leguminosae (I), another assumed ally, but wood anatomy, serology & G (2) support affinity with P.
Female *Centris* bees coll. saturated fatty-acids from the lipid-gland C for larvae; they also visit Malpighiaceae for the same
Krapfia DC (~ *Ranunculus*). Ranunculaceae (II 3). 8 Andes
Krapovickasia Fryx. Malvaceae. 4 C & S Am. R: Brittonia 30(1978)454
Krascheninnikovia Gueldenst. (*'Eurotia'*). Chenopodiaceae (I 3). 8 Med. (Eur. 1: *K. ceratoides* (L.) Gueldenst. – boron indicator in Russia), temp. As., W N Am. Some used as lice-killing shampoo in N Am.
Krasnovia Popov. Umbelliferae (III 2). 1 C As.
Krassera Schwartz = *Anerincleistus*
Krausella H.J. Lam = *Pouteria*
Krauseola Pax & K. Hoffm. Caryophyllaceae (I 1). 1 N Kenya & S Ethiopia, 1 Mozambique
Kraussia Harvey. Rubiaceae (II 5). 3 trop. & S Afr.
Krebsia Harvey = *Stenostelma*
Kremeria Durieu = *Leucanthemum*
Kremeriella Maire. Cruciferae. 1 NW Afr.
Kreodanthus Garay. Orchidaceae (III 3). 6 trop. Am.
kretek clove-flavoured cigarette
Kreysigia Reichb. = *Schelhammera*
Krigia Schreber. Compositae (Lact.-Mic.). 7 N Am. R: Brittonia 44(1992)172. Some cult. orn. with yellow or orange fls
krobonko *Telfairia occidentalis*
Krokia Urban = *Pimenta*
Krubera Hoffm. (~ *Capnophyllum*). Umbelliferae (III 10). 1 Medit. incl. Eur.: *K. peregrina* (L.) Hoffm.
Krugia Urban = *Marlierea*
Krugiodendron Urban. Rhamnaceae. 1 WI: *K. ferreum* (Vahl) Urban – bark & roots medic.
Krukoviella A.C. Sm. Ochnaceae. 1 Peru, Brazil: *K. scandens* A.C. Sm. R: BJ 113(1991)174
Krylovia Schischkin (~ *Aster*). Compositae (Ast.-Ast.). 4 C As., China
Kryptostoma (Summerh.) Geer. = *Habenaria*
Kubitzkia van der Werff. Lauraceae (I). 1 NE S Am.: *K. macrantha* (Kosterm.) van der Werff (*'Systemonodaphne geminiflora'*). R: BJ 110(1988)161
Kudoacanthus Hosok. Acanthaceae. 1 Taiwan
Kudrjaschevia Pojark. Labiatae (VIII 2). 4 C As.
kudzu vine *Pueraria montana* var. *lobata*
kümmel *Carum carvi*
Kuhitangia Ovcz. = *Acanthophyllum*
Kuhlhasseltia J.J. Sm. Orchidaceae (III 3). 6 Mal.
Kuhlmannia J. Gómes = *Pleonotoma*
Kuhlmanniella Barroso = *Dicranostyles*
Kuhnia L. = *Brickellia*
kukui nut *Aleurites moluccana*
kullam nut *Balanites orbicularis*

kumara *Ipomoea batatas*

kumbanzo *Holarrhena pubescens*

kumbuk *Terminalia arjuna*

Kumlienia E. Greene (~ *Ranunculus*). Ranunculaceae (II 3). 1 W N Am.

Kummerowia Schindler. Leguminosae (III 9). 2 As., N Am. *K. striata* (Thunb.) Schindler (*Lespedeza* s., Japanese clover, E As., widely natur. SE US) – fodder

kumquat *Citrus japonica*; **desert k.** *C. glauca*

kunai *Imperata* spp.

Kundmannia Scop. Umbelliferae (III 8). 2 S Eur., Medit.

Kunhardtia Maguire. Rapateaceae. 1 Venezuela

Kuniwatsukia Pichi-Serm. = *Athyrium*

Kunkeliella Stearn. Santalaceae. 4 Canary Is.

Kunstleria Prain. Leguminosae (III 6). 8 Kerala (1), W & C Mal. (7). R: Blumea 38(1994)465. Lianes, rarely coll.

Kuntheria Conran & Clifford (~ *Schelhammera*). Convallariaceae. 1 N Queensland: *K. pedunculata* (F. Muell.) Conran & Clifford. R: FA 45(1987)417

Kunzea Reichb. Myrtaceae (Lept.). 36 Aus. (36, 35 endemic), NZ (1: *K. ericoides* (A. Rich.) J. Thompson (burgan) – timber & medic.). Some cult. orn. heath-like shrubs like *Leptospermum* but A longer than C

kurakkan *Eleusine coracana*

kurara *Andira inermis*

kurchi bark *Holarrhena pubescens*

kurdee *Carthamus tinctorius*

kurrat *Allium ampeloprasum*

Kurzamra Kuntze. Labiatae (VIII 2). 1 temp. S Am.

kussum oil *Schleichera oleosa*

Kutchubaea Fischer ex DC (*Kotchubaea*). Rubiaceae (II 1). 11 trop. S Am. R: MNYBG 10,1(1963)212

kuth *Saussurea costus*

kutira gum *Cochlospermum religiosum*

kutki *Picrorhiza kurrooa*

kweek grass *Cynodon dactylon*

kwei *Osmanthus fragrans*

kweme nut *Telfairia pedata*

kwila *Intsia bijuga*

Kydia Roxb. Malvaceae. 1 (variable) Sikkim to SE As.: *K. calycina* Roxb. – monoec. tree with bark used for coarse ropes

kyetpaung *Urceola esculenta*

Kyllinga Rottb. (~ *Cyperus*). Cyperaceae. 40 warm esp. Afr., Madag.

Kyllingiella R. Haines & Lye. Cyperaceae. 3–4 trop. E Afr.

kyor *Sagittaria sagittifolia*

Kyphocarpa (Fenzl) Lopr. (*Cyphocarpa*). Amaranthaceae (I 2). 3–4 S trop. & S Afr.

Kyrsteniopsis R. King & H. Robinson (~ *Eupatorium*). Compositae (Eup.-Alom.). 4 Mex. R: MSBMBG 22(1987)243

L

Labatia Sw. = *Pouteria*

labdanum = ladanum

Labiatae Juss. (Lamiaceae). Dicots – Asteridae – Lamiales. S.l. (incl. Chloanthaceae, Verbenaceae p.p.) 251/6700 cosmop.; s.s. cosmop. but esp. Medit. to C As., v. rare in NZ. S.s.: shrubs (rarely trees to 18 m – *Hyptis*) or herbs, often with short-stalked epidermal glands containing ethereal oils & commonly storing carbohydrate as stachyose; young stems often 4-angled. Lvs opp. (s.t. whorled or even spirally arr. as in *Icomum*), simple (rarely pinnate); stipules 0. Fls bisexual (pl. s.t. gynodioec.), usu. bracteolate & in verticillasters (compact axillary cymes) or single fls in axils; K usu. persistent & with 5 teeth or lobes or bilabiate, fleshy & berry-like in fr. in *Hoslundia*, C (5), (1) 2-labiate or less often ± reg. with 4 lobes (e.g. *Mentha*) where 1 app. rep. 2, A 4 or 2 (+ 2 staminodes), attached to tube, anthers with longit. slits, connective s.t. such that pollen-sacs are sep. (e.g. *Salvia*), usu. nectary-disk (nectar sucrose-rich) annular to anterior at base of $\underline{G}$ (2) s.t. on gynophore, each carpel longit. divided in 2, style often bifid apically, 1 erect, anatropous to hemitropous unitegmic ovule with funicular obturator in each of 4 lobes. Fr. of

(1–)4 1-seeded nutlets with tough pericarp or a drupe (e.g. *Prasium*); seed with straight embryo & little or 0 oily endosperm. x = 5–11+

S.l. incl. many trees & lianes (see R.M. Harley et al. (eds) *Advances in Labiate Science* (1992))

Classification & chief genera:

I. **Viticoideae** (Verbenaceae p.p.): *Callicarpa, Gmelina, Premna, Vitex*

II. **Teucrioideae** (Verbenaceae p.p.): *Aegiphila, Caryopteris, Clerodendrum, Oxera, Rotheca, Teucrium*

III. **Chloanthoideae** (Chloanthaceae): *Dicrastylis, Hemigenia, Tectona* (teak)

IV. **Ajugoideae**: *Ajuga*

V. **Scutellarioideae**: *Holmskioldia, Scutellaria, Tinnea*

VI. **Lamioideae**: *Ballota, Lamium, Leonotis, Leucas, Melittis, Molucella, Phlomis, Physostegia, Sideritis, Stachys*

VII. **Pogostemonoideae**: *Pogostemon*

VIII. **Nepetoideae** (4 tribes):

1. **Elsholtzieae**: *Elsholtzia, Perilla*
2. **Mentheae**: *Agastache, Clinopodium, Dracocephalum, Hyssopus, Lycopus, Melissa, Mentha, Micromeria, Monarda, Nepeta, Origanum, Perovskia, Prunella, Rosmarinus, Salvia, Satureja, Thymus*
3. **Lavanduleae**: *Lavandula*
4. **Ocimeae** (3 subtribes): *Eriope, Hypenia, Hyptis* (a, Hyptidinae); *Acrocephalus, Basilicum, Ocimum, Orthosiphon* (b, Ociminae); *Aeollanthus, Hoslundia, Isodon, Plectranthus* (c, Plectranthinae)

Some genera are not readily placed here & some are v. isolated in fam., *Tetrachondra* (ditypic) sometimes being placed in its own fam. near Callitrichaceae. Trad. distinction from Verbenaceae (usu. without ess. oils & 4-lobed G) untenable

Labiatae s.s. usu. insect-poll. pls of open country (*Gomphostemma* in rainforest), lipped fls acting as landing-stages for pollinators: in most temp. spp. lower lip (C 3) acting as such, the upper usu. hooded & protecting A, while in trop. spp. upper lip of C 4 with A lying along lower (C 1) lip or ascendant from it; in *Teucrium* lower lip is 5-lobed & upper 0, A often completely exposed. Pollinators incl. birds as well as insects (incl. Lepidoptera) – see *Salvia* for most intricate mech.; explosive systems in *Aeollanthus* & *Hyptis*, where A held under tension by enfolding lobes of C such that arrival of insect on lower lip releases A & pollinator is dusted with grains. Some disp. by modifications of K – persistent & bladdery or hooks formed by teeth (wind- & animal-disp. respectively). Volatile oils incl. terpenes which are allelopathic in some chaparral spp. but also the basis of many of their uses to man

A fam. long-recognized because of its medic. & culinary value, many still of great importance as flavourings & scents: *Lavandula* (lavender), *Mentha* (mint), *Ocimum* (basil), *Origanum* (oregano, marjoram), *Rosmarinus* (rosemary), *Salvia* (sage), *Satureja* (savory), *Thymus* (thyme), also *Pogostemon* (patchouli) & *Perilla* (used in printing-inks & paints) etc.; tubers of *Stachys* (Chinese artichoke) & *Plectranthus* (Hausa potato) ed.; many cult. orn. in above genera & *Ajuga, Clinopodium, Horminum, Lamium, Moluccella, Monarda, Nepeta, Perovskia, Physostegia, Plectranthus, Prostanthera, Scutellaria, Westringia* etc. etc. & 'coleus' now referable to *Plectranthus*

The tribes form. in Verbenaceae incl. imp. timbers: *Gmelina, Petitia, Premna, Tectona* (teak), *Vitex*; cult. orn.: *Callicarpa, Caryopteris, Clerodendrum, Holmskioldia, Rotheca*

Labichea Gaudich. ex DC. Leguminosae (I 2). 14 Aus. R: Muelleria 6(1985)22

Labidostelma Schltr. = *Matelea*

Labisia Lindley. Myrsinaceae. 6 W Mal., New Guinea (1). Subherbaceous with creeping stems; blue fr. poss. attractive to pheasants. Some medic. locally (childbirth, gonorrhoea)

lablab *Lablab purpureus*; also used for 'greens' eaten with rice in trop. As.

Lablab Adans. (~ *Dolichos*). Leguminosae (III 10). 1 (prob.) trop. Afr., widely cult.: *L. purpureus* (L.) Sweet (*L. niger, Dolichos l.*, lablab, hyacinth bean, bonavist, prob. derived from wild subsp. *uncinatus* Verdc.) – pods & seeds ed., green manure & fodder

Labordia Gaudich. (~ *Geniostoma*). Geniostomaceae (Loganiaceae s.l.). 15 Hawaii. R: Man. Fl. Pl. Hawaii (1990)852

Labourdonnaisia Bojer. Sapotaceae (I). 3 Masc., ? S Madag. Timber hard

Labrador tea *Rhododendron tomentosum*

Labramia A. DC (~ *Manilkara*). Sapotaceae (I). 8 Madag.

+ **Laburnocytisus** C. Schneider. Leguminosae. Graft hybrid between *Laburnum* & *Cytisus*. + *L. adamii* (Poit.) C. Schneider derived (once, 1826) accidentally from graft on *Laburnum* *anagyroides* of *C. purpureus* Scop. (Austria), the chimaera (epidermis *C.*, rest *L.*) often breaking down so that there are branches of 'pure' *L.* or (rarely) *C.* Cult. curiosity

Laburnum Fabr. Leguminosae (III 31). 2 SC & SE Eur. Cult. orn. trees & shrubs esp. *L. anagyroides* Medikus (golden chain or rain, C & S Eur.) – long-lived (tree planted at Leiden in 1601, 17 m tall in 1937), all parts poisonous (cytisine (alk.), seeds can be fatal to children), timber hard, used as ebony subs. esp. in inlays, musical instruments etc., also *L. alpinum* (Miller) Bercht. & J. Presl (Scotch l., S Eur.) & their hybrid *L.* × *watereri* (Wettstein) Dippel ('*L.* × *vossii*') with longer racemes

laburnum, Indian *Cassia fistula*; **Scotch l.** *Laburnum alpinum*

lac insect resin, secreted by *Laccifer lacca* esp. (in India) on *Butea monosperma*, *Cajanus cajan*, *Ficus religiosa*, *Schleichera oleosa* & *Ziziphus mauritiana*, the original shellac

Lacaena Lindley. Orchidaceae (V 10). 2 C Am. R: Die Orchidee 30(1979)55. Cult. orn.

Lacaitaea Brand. Boraginaceae. 1 E Himalaya

Lacandonia E. Martínez & C.H. Ramos = *Triuris*

Lacandoniaceae E. Martínez & C.H. Ramos = Triuridaceae

Laccodiscus Radlk. Sapindaceae. 4 W Afr.

Laccopetalum Ulbr. (~ *Ranunculus*). Ranunculaceae (II 3). 1 Peru. *L. giganteum* (Wedd.) Ulbr. – coarse herb with pachycaul stems & large yellowish green fls

Laccospadix Drude & H. Wendl. (~ *Calyptrocalyx*). Palmae (V 4j). 1 NE Queensland: *L. australasica* Drude & H. Wendl.

Laccosperma (G. Mann & H. Wendl.) Drude (*Ancistrophyllum*). Palmae (II 1a). 7 trop. Afr. R: KB 37(1982)455. Rattans – hapaxanthic, herm.

lace, Queen Anne's *Anthriscus sylvestris*

lace-bark *Lagetta lagetto*; **l.-b. pine** *Pinus bungeana*

lace-leaf *Aponogeton madagascariensis*

lacewood *Platanus* × *hispanica* veneers

Lachanodes DC (~ *Senecio*). Compositae (Sen.-Sen.). 1 St Helena: *L. arborea* (Roxb.) R. Nordenstam (*S. redivivus*) – tree, reduced to 10 pls in wild but now in cult., allies in S Am. & Australasia

Lachemilla (Focke) Rydb. = *Alchemilla*

Lachenalia Jacq.f. ex Murray. Hyacinthaceae (Liliaceae s.l.). 101 S Afr. R: Plantsman 8(1986)129. Cape cowslip esp. *L. aloides* (L.f.) Asch. & Graebner – cult. orn., many cvs, & hybrids with other spp. – pendent orange or yellow fls bird.-poll. while *L. unifolia* Jacq. has pink or purplish fls visited by bees

Lachnaea L. Thymelaeaceae. 29 SW Cape

Lachnagrostis Trin. = *Agrostis*

Lachnanthes Elliott. Haemodoraceae. 1 Massachusetts to Florida, Cuba: *L. caroliniana* (Lam.) Dandy (*L. tinctoria*) – root form. medic. (chest disease) & source of red dye

Lachnocapsa Balf.f. Cruciferae. 1 Socotra

Lachnocaulon Kunth. Eriocaulaceae. 7 SE US & Cuba (2)

Lachnoloma Bunge. Cruciferae. 1 C As., Iran

Lachnophyllum Bunge. Compositae (Ast.-Ast.). 2 W to C As.

Lachnopylis Hochst. = *Nuxia*

Lachnorhiza A. Rich. (~ *Vernonia*). Compositae (Vern.-Vern.). 1 W Cuba

Lachnosiphonium Hochst. = *Catunaregam*

Lachnospermum Willd. Compositae (Gnap.-Rel.). 3 SW Cape to Namaqualand. R: OB 104(1991)74

Lachnostachys Hook. Labiatae (III; Verbenaceae, Dicrastylidaceae). 6 S Aus. R: Brunonia 1(1978)643

Lachnostoma Kunth. = *Matelea*

Lachnostylis Turcz. Euphorbiaceae. 2 S Cape

Lacistema Sw. Flacourtiaceae (Lacistemataceae). 11 trop. Am. Species delimitation difficult, fls v. small. R: FN 22(1980)183

Lacistemataceae C. Martius = Flacourtiaceae

Lacmellea Karsten. Apocynaceae. c. 35 trop. S Am. Some rubber & ed. fr. *L. lactescens* (Kuhlm.) Markgraf – powdered lvs chewed as coca subs. in NW Amaz.

lacoocha *Artocarpus lacucha*

Lacostea Bosch = *Trichomanes*

Lacosteopsis (Prantl) Nakaike = *Crepidomanes*

lacquer (SE As.) *Gluta laccifera*; **Burmese l.** *G. usitata*; **Chinese** or **Japanese l.** *Rhus verniciflua*; **pasto l.** *Elaeagia utilis*

Lactoridaceae Engl. Dicots – Magnoliidae – Magnoliales. 1/1 Juan Fernandez (Cretaceous pollen from S Afr. referred here too). Shrub, polygamo-dioec., with ethereal oil-cells in parenchyma; nodes swollen. Lvs distichous, gland-dotted, simple, small; stipules large, adnate to petiole, forming sheath around branch. Fls small, axillary, 1–4-flowered cymes; K 3, C 0, A 3 + 3, narrowly laminar with well-separated pollen-sacs & v. short connective (inner ones s.t. staminodes), pollen grains in tetrads, monosulcate, G̲ 3, ± connate basally, with short style & decurrent stigma, each with 4–8 anatropous bitegmic ovules on intruded marginal placenta. Fr. follicular, beaked; seeds with small embryo in copious oily endosperm. 2n = 40(?)

Only species: *Lactoris fernandeziana*

Isolated sp. placed in 'loosely defined' Magnoliales (Cronquist) here incl. Laurales, poss. near Annonaceae, but sometimes put in own order, its wood anatomy suggesting Piperales

Lactoris Philippi. Lactoridaceae. 1 Masatierra, Juan Fernandez: *L. fernandeziana* Philippi. 12 individuals left in 1962 due to depredations by goats & more recently rampant *Rubus* spp.

Lactuca L. Compositae (Lact.-Lact.). c. 75 cosmop. esp. N temp. (Eur. 17). *L. sativa* L. (lettuce, cultigen ? orig. E Medit. from *L. serriola*) – cos (upright heads of lvs with broad midribs) & cabbage (globose heads of lvs) types known to Ancient Persians & Pliny mentions 11 distinct sorts; unlike wild spp. the bitter sesquiterpene lactones lactucin & lactupicrin absent: these have sedative properties & *L. virosa* L. (opium lettuce, C & S Eur.) form. cult. for this (lactucarium), used as home remedy for coughs etc.; *L. sativa* – the salad green of temp. regions with many cvs incl. forms with thick main stem ed. (asparagus lettuce, celtuce (stem l.)), 'Grand Rapids' a form in which the red/far-red reactions of phytochrome orig. investigated, aphrodisiac to Anc. Egyptians but opp. for Anc. Greeks. Other spp. ed. locally incl. *L. serriola* L. (*L. scariola*, prickly l., Eur. now subcosmop. weed) – seeds the source of Egyptian lettuce-seed oil used in foods, the lvs tending to be held edgewise upwards (compass plant)

lactucarium *Lactuca virosa*

Lactucella Naz. (~ *Lactuca*). Compositae (Lact.-Lact.). 1 SW & C As. to W China

Lactucosonchus (Schultz-Bip.) Svent. (~ *Sonchus*). Compositae (Lact.-Son.). 1 Canary Is.

Lacunaria Ducke. Quiinaceae. 12 trop. Am.

ladanum *Cistus ladanifer* & *C. incanus*

Ladenbergia Klotzsch. Rubiaceae (I 1). (Incl. *Cascarilla*) c. 55 trop. Am. Alks in bark used like quinine but less efficacious

Ladino clover *Trifolium repens* f. *lodigense*

lad's love *Artemisia abrotanum*

lady fern *Athyrium filix-femina*; **L. Nugent's rose** *Clerodendrum chinense*; **l. of the night** *Brunfelsia americana*, *Cestrum nocturnum*

Ladyginia Lipsky. Umbelliferae (III 10). 3 SW & C As. R: EJB 49(1992)215

lady's bedstraw *Galium verum*; **l. comb** *Scandix pecten-veneris*; **l. fingers** *Anthyllis vulneraria*, *Abelmoschus esculentus*; **l. mantle** *Alchemilla vulgaris*; **l. slipper** *Cypripedium* spp.; **l. smock** *Cardamine pratensis*; **l. tresses** *Spiranthes spiralis*

Laelia Lindley. Orchidaceae (V 13). 69 trop. Am. R: C.L. Withner (1990) *The Cattleyas & their relatives* II. Cult. orn. epiphytes with extrafl. nectaries, esp. *L. anceps* Lindley (Mex., Honduras) – many cvs & hybrids (incl. some with C. spp. etc.); mucilaginous paste from pseudobulbs of *L. speciosa* (Kunth) Schltr. (Mex.) used to make images of animals, skulls etc. on All Saints Day & 'Day of the Dead' (Uphof)

Laeliopsis Lindley & Paxton = *Broughtonia*

Laennecia Cass. (~ *Conyza*). Compositae (Ast.-Ast.). 16 SW N Am. to Bolivia. R: Phytol. 68(1990)205

Laestadia Kunth ex Less. Compositae (Ast.-Ast.). 6 trop. Andes, WI. Pappus 0

Laetia Loefl. ex L. Flacourtiaceae (9). 10 trop. Am. R: FN 22(1980)237. Local medic.

Lafoensia Vand. Lythraceae. 10 trop. Am. Trees & shrubs with 8–16-merous fls. *L. punicifolia* DC (Mex. to N S Am.) – tree with yellow fls turning to red, cult. orn., source of yellow dye

Lafuentea Lagasca. Scrophulariaceae. 2 S Spain (1), Morocco. Shrubby

Lagarinthus E. Meyer = *Schizoglossum*

Lagarosiphon Harvey. Hydrocharitaceae. 9 trop. Afr. & Madag. R: BJBB 53(1983)441. Male fls released; A 3, staminodes 2 or 3. *L. major* (Ridley) Wager (S Afr.) – grown in aquaria ('*Elodea crispa*'), pest in NSW, NZ

Lagarosolen W.T. Wang (~ *Chirita*). 1 SE Yunnan: *L. hispidus* W.T. Wang. R: ABY 6(1984)11

Lagarostrobos Quinn (~ *Dacrydium*). Podocarpaceae. 1 Tasmania, 1 NZ. R: AusJB 30(1982)316. Cult. orn. with lax seed-cones (cf. *Dacrydium*) incl. *L. franklinii* (Hook.f.) Quinn (*D. franklinii*, Huon pine, Tasmania) – timber, trees to over 2200 yrs old so poss. use in dendrochronology (used for endgrain blocks in printmaking by Margaret Preston), H.p. oil in medic. soaps

Lagascea Cav. Compositae (Helia.-Helia.). 9 Mex., C Am., R: Fieldiana 38(1978)75. Synflorescence of 1-flowered capitula, unique in Heliantheae

Lagedium Soják (~ *Mulgedium*). Compositae (Lact.-Lact.). 1 temp. Euras.

Lagenandra Dalz. Araceae (VIII 6). 12 S India, Sri Lanka. R: MLW 78–13(1978). Some aquarium pls (allied to *Cryptocoryne*)

Lagenantha Chiov. Chenopodiaceae (III 3). 1 Somalia: *L. nogalensis* Chiov.

Lagenanthus Gilg = *Lehmanniella*

Lagenaria Ser. Cucurbitaceae. 6 trop. Afr., Madag. (1) with 1 extending to rest of trop. Night-flowering. *L. siceraria* (Molina) Standley (calabash or bottle gourd) – 1 subsp. in Afr. & Am., 1 in As. & Pacific perhaps domesticated independently though that in E Polynesia poss. disp. from Am., that in OW known to Pliny & prob. cult. 'Hoabhinian' culture 8000–3000 yrs ago; young fr. ed., locally medic. (purgative), mature fr. with tough pericarp used for flasks, cups, dippers, penis-sheaths (New Guinea) etc., many diff. shaped forms cult.

Lagenifera Cass. = *Lagenophora*

Lagenocarpus Nees. Cyperaceae. 34 trop. Am.

Lagenocypsela Svenson & Bremer. Compositae (Ast.-Ast.). 2 New Guinea. R: Aus. SB 7 (1994)265

Lagenophora Cass. (*Lagenifera*). Compositae (Ast.-Ast.). (Incl. *Keysseria*) 30 Indomal., Pacific, trop. Am. (6). R: Blumea 14(1966)285

Lagerstroemia L. Lythraceae. 53 trop. As. to Aus. R: GBS 24(1969)185. Trees with A 15–200. Cult. orn. & some good timber; alks. *L. hypoleuca* Kurz (Andaman pyinama, Andamans) – timber; *L. indica* L. (crape or crepe myrtle, crepe flower, China) – widely cult. trop. & warm, many cvs with white, pink or purple fls with A 36–42 & potent psychoactive properties; *L. microcarpa* Wight (*L. lanceolata*, India) – reddish timber for many uses incl. bridges, coffee-boxes, furniture, tanbark good; *L. speciosa* (L.) Pers. (*L. flos-reginae*, pride-of-India, queen-flower, pyinma, India & China to Aus.) – cult. orn. with purple or white fls with A 130–200, medic., timber (banglang) for railway sleepers

Lagetta Juss. Thymelaeaceae. 5 WI. *L. lagetto* (Sw.) Nash (*L. lintearia*, lace-bark, gauze tree) – stretched inner bark a reticulated lace-like material used ornamentally & as textile in Jamaica

lagetto *Lagetta lagetto*

Laggera Schultz-Bip. ex Benth. (~ *Blumea*). Compositae (Pluch.) 17 OW trop.

Lagoa T. Durand. Asclepiadaceae (III 1). 1 Brazil: *L. calcarata* (Dcne) Baillon

Lagochilopsis Knorr. = *seq.*

Lagochilus Bunge ex Benth. Labiatae (VI). 35 C As. to Iran & Afghanistan. Alks

Lagoecia L. Umbelliferae (II 2). 1 Medit. (incl. Eur.): *L. cuminoides* L. – 1 of the 2 loc. aborts, fr. a cumin subs.

Lagonychium M. Bieb. = *Prosopis*

Lagophylla Nutt. Compositae (Hele.-Mad.). 5 W N Am.

Lagopsis (Benth.) Bunge = *Marrubium*

Lagos rubber *Funtumia elastica*

Lagoseriopsis Kirpiczn. = ? *Launaea*

Lagoseris M. Bieb. = *Crepis*

Lagotis Gaertner. Scrophulariaceae. 20 E Eur (2), N & C As. to Cauc., Himal. & W China

Lagrezia Moq. (~ *Celosia*). Amaranthaceae (I 2). 12 Madag., Indian Ocean

Laguna Cav. Older name for *Abelmoschus*

Lagunaria (DC) Reichb. Malvaceae. 1 E Aus. (? natur.), Norfolk Is., Lord Howe Is.: *L. patersonia* (Andrews) G. Don f. – cult. orn. evergreen tree, fr. irritant

Laguncularia Gaertner f. Combretaceae. 2 trop. Am. & W Afr. Mangroves. *L. racemosa* (L.) Gaertner f. (button wood, trop. Am.) – tanbark, timber for external use

Lagurus L. Gramineae (21d). 1 Medit. (incl. Eur.): *L. ovatus* L. (natur. Channel Is.) – cult. orn. for dried displays

Lagynias E. Meyer ex F. Robyns. Rubiaceae (III 2). 4 trop. E & S Afr. R: BJBB 11(1928)312

Lahia Hassk. = *Durio*

lalang *Imperata cylindrica*

Lalldhwojia Farille. Umbelliferae (III 10). 2 Himalaya

Lallemantia Fischer & C. Meyer. Labiatae (VIII 2). 5 Turkey to C As. & Himal. Some oils, potherbs & local medic.

Lamanonia Vell. Conc. = *Geissois*

Lamarchea Gaudich. Myrtaceae (Lept.). 2 W Aus.

Lamarckia Moench. Gramineae (17). 1 Medit. incl. Eur.: *L. aurea* (L.) Moench – cult. orn., weedy in Aus.

Lambertia Sm. Proteaceae. 9 SW & 1 E Aus. Some cult. orn. incl. *L. formosa* Sm. (honey flower, E Aus.)

lamb-kill *Kalmia angustifolia*

lamb's ears *Stachys byzantina*; **l. lettuce** *Valerianella locusta*; **l.-tail** *Chiastophyllum oppositifolium*

lambswool (i.e. lamasool) Ale with roasted apples, sugar & spice drunk on 31 October (also in wassail in e.g. Sussex)

Lamechites Markgraf = *Ichnocarpus*

Lamellisepalum Engl. = *Sageretia*

Lamiacanthus Kuntze = *Strobilanthes*

Lamiaceae Martinov. See Labiatae

Lamiastrum Heister ex Fabr. = *Lamium*

Lamiodendron Steenis (~ *Fernandoa*). Bignoniaceae. 1 New Guinea

Lamiophlomis Kudô = *Phlomis*

Lamium L. Labiatae (VI). c. 40 N Afr., Euras. (Eur. 13). Dead nettles, non-flowering stems superficially resembling *Urtica* spp. & referred to as 'urtica' in med. *L. album* L. (white dead nettle, Euras., natur. E N Am.) – lvs cooked like spinach, fls large, white, bee-poll.; *L. amplexicaule* L. (henbit, Eur., Med. to Iran, natur. N Am.) – produces cleistogamous fls in spring & autumn, can cause 'staggers' in stock; *L. galeobdolon* (L.) L. (*Lamiastrum g., G. luteum*, yellow archangel, W Eur. to Iran) – cult. orn. incl. varieg. cvs & *L. maculatum* L. (Eur., Medit.) – over 20 cvs esp. varieg. for ground-cover; *L. purpureum* L. (red. d. n., Eur., Med.) – locally medic.

Lamottea Pomel (~ *Carduncellus*). Compositae (Card.-Cent.). 8 Med. esp. W (Eur. 2). R: AJBM 47(1990)27

Lamourouxia Kunth. Scrophulariaceae. 28 Mexico to Peru. R: SCB 6(1971)

Lampas Danser. Loranthaceae. 1 N Borneo: *L. elmeri* Danser. R: Blumea 38(1993)108

Lampaya Philippi = *Lampayo*

Lampayo Philippi ex Murillo. Verbenaceae. 3 Bolivia, Chile, Argentina

Lamprachaenium Benth. Compositae (Vern.-Vern.). 1 India

Lampranthus N.E. Br. Aizoaceae. (Incl. *Oscularia*) c. 180 S Afr., Aus. (1). Prob. heterogeneous. Cult. orn. subshrubs, esp. hybrids (many complex)

Lamprocaulos Masters = *Elegia*

Lamprocephalus R. Nordenstam (~ *Senecio*). Compositae (Sen.-Sen.). 1 SW Cape

Lamprococcus Beer (~ *Aechmea*). Bromeliaceae (3). 13 trop. Am. R: Phytol. 66(1989)70

Lamproconus Lemaire = *Pitcairnia*

Lamprolobium Benth. Leguminosae (III 25). 1 N Queensland. Alks

Lamprophragma O. Schulz = *Pennellia*

Lamprothamnus Hiern. Rubiaceae (II 5). 1 trop. E Afr.

Lamprothyrsus Pilger. Gramineae (25). 3 Bolivia to Arg.

Lamy butter *Pentadesma butyracea*

Lamyra (Cass.) Cass. = *Ptilostemon*

Lamyropappus Knorr. & Tamamschjan. Compositae (Card.-Card.). 1 C As.

Lamyropsis (Charadze) Dittr. Compositae (Card.-Card.). 8 Medit. (Eur. 2), SW As.

Lanaria Aiton. Haemodoraceae. 1 S Afr.: *L. lanata* (L.) Durand & Schinz. Phenalenone typical of H. absent

Lanariaceae H. Huber ex Dahlgren = Haemodoraceae

Lancea Hook.f. & Thomson. Scrophulariaceae. 2 Tibet, China. *L. tibetica* Hook.f. & Thomson (depgul, Tibet) – roots roasted & smoked with tobacco as narcotic in Ladakh

lancewood Elastic woods used for fishing-rods, bows etc. e.g. *Duguetia quitarensis, Oxandra lanceolata*; **degame l.** *Calycophyllum candidissimum*; **Moulmein l.** *Homalium tomentosum*; **NZ l.** *Pseudopanax crassifolius*

Lancisia Fabr. = *Cotula*

Landolphia Pal. Apocynaceae. 60 trop. Am., Afr., Madag. Many lianes with hook-tendrils like *Strychnos* spp.; fr. (some ed.) a berry of acid pulp (seed-hairs), many yield a rubber (Madagascar or African r.) e.g. *L. gummifera* (Lam.) Schumann (vahy, Madag.)

land cress *Barbarea verna*

Landtia Less. = *Haplocarpha*

Lanessania Baillon = *Trymatococcus*

Langebergia A. Anderb. (~ *Petalacte*). Compositae (Gnap.-Cass.). 1 S Afr.: *L. canescens* (DC)

A. Anderb. R: OB 104(1991)93. Shrub

Langlassea H. Wolff = *Prionosciadium*

Langloisia E. Greene (~ *Gilia*). Polemoniaceae. 1 W N Am. deserts: *L. setosissima* (Torrey & A. Gray) E. Greene. R: Madrono 33(1986)167

langsat *Lansium domesticum*

Langsdorffia C. Martius. Balanophoraceae. 3 Madag. (1), New Guinea (1), trop. Am. (1): *L. hypogaea* C. Martius – wax made into candles

Lanium (Lindley) Benth. = *Epidendrum*

Lankesterella Ames = *Stenorrhynchos*

Lankesteria Lindley. Acanthaceae. 7 trop. Afr., Madagascar

Lannea A. Rich. Anacardiaceae (II). 40 trop. Afr., Indomal. (1: *L. coromandelica* (Houtt.) Merr. (*L. grandis*, *Odina wodier*) – timber for external use, furniture etc., gum used for sizing paper & cloth, confectionery, tanbark also powdered as tooth-powder, roadside tree often pollarded). Roots roasted or boiled (v. starchy) in Uganda; *L. schweinfurthii* (Engl.) Engl. (E & S Afr.) – ed. fr.; *L. stuhlmannii* (Engl.) Engl. (trop. Afr.) – roots a 'wool' used for life-belts etc. ('flotite' in NZ); *L. welwitschii* (Hiern) Engl. (trop. Afr.) – elephants strip off bark in pieces to 10 m long (Ghana)

Lansium Corr. Serr. Meliaceae (I 5). 3 Mal. R: Blumea 31(1985)140. *L. domesticum* Corr. Serr. (langsat) – ed. arils, a market fr. with a no. of distinctive cvs, some, at least, apomictic; pericarp used for incense in Java

Lantana L. Verbenaceae. 150 trop. Am., trop. & S Afr. (few). Shrubs & herbs, often armed, some with ed. fr. Alks. Some hedgepls, others aggressive weeds esp. *L. camara* L. (trop. Am.) – serious weed (toxic triterpenoids (lantadenes), verbascoside, an inhibitor of protein kinase C) in parts of US, Hawaii (where biologically controlled), New Caled. etc., obligate outbreeder, fls open yellow when visited by thrips (in India), changing to orange once poll., diploid to hexaploid

Lantanopsis C. Wright ex Griseb. Compositae (Helia.). 3 Cuba, Hispaniola

lantern (s), Chinese *Physalis alkekengi*; **l. tree, Chile** *Crinodendron hookeranum*

Lanugia N.E. Br. = *Mascarenhasia*

lapacho(l) *Tabebuia* spp. esp. *T. impetiginosa*

Lapageria Ruíz & Pavón. Philesiaceae (Smilacaceae). 1 S Chile & nearby Arg.: *L. rosea* Ruíz & Pavón (Chilean bellflower) – national fl. of Chile ('copihue'), cult. orn. liane with pendulous red (or white) fls, berry allegedly ed. Has been crossed with *Philesia magellanica* J. Gmelin

Lapeirousia Pourret. Iridaceae (IV 2). c. 36 subSaharan Afr. esp. SW Coast. R (trop. Afr. – 16): AMBG 77(1990)430. Some cult. orn. See also *Anomatheca*

Laphamia A. Gray = *Perityle*

Lapidaria (Dinter & Schwantes) N.E. Br. Aizoaceae (V). 1 SW Afr.: *L. margaretae* (Schwantes) N.E. Br. – cult. orn. succ. herb

Lapiedra Lagasca. Amaryllidaceae (Liliaceae s.l.). 2 W Med. (Eur. 1)

Lapithea Griseb. = *Sabatia*

Laplacea Kunth = *Gordonia*

Laportea Gaudich. (*Fleurya*). Urticaceae (I). 21 trop. & warm, temp. E As., E N Am. R: GBS 25(1969)111. See also *Dendrocnide*. Some with stings e.g. *L. mooreana* (Hiern) Chew (trop. Afr.), though those of *L. alatipes* Hook.f. (trop. & S Afr.) not a deterrent to gorillas which eat shoots. Some lvs used like spinach; some with strong fibres esp. *L. canadensis* (L.) Wedd. (wood nettle, E N Am.) – fibre comparable with ramie, v. strong

Lappula Gilib. Boraginaceae. 40 temp. Euras. (Eur. 6 incl. *L. squarrosa* (Retz.) Dumort. (*L. myosotis*)), 5 N Am. R: JAA suppl. 1(1991)108. Fr. hooked; in *L. squarrosa* fls change from white to red & blue

Lapsana L. Compositae (Lact.-Crep.). 9 temp. OW (Eur. 1: *L. communis* L., nipplewort)

laran *Neolamarckia cadamba*

larch *Larix* spp., see also *Abies procera*; **common** or **European l.** *L. decidua*; **Dunkeld l.** *L.* × *marschlinsii*; **golden l.** *Pseudolarix kaempferi*; **hybrid l.** *L.* × *marschlinsii*; **Japanese l.** *L. kaempferi*; **tamarack** or **American l.** *L. laricina*

Lardizabala Ruíz & Pavón. Lardizabalaceae. 1–2 Chile. Monoec. to dioec. evergreen lianes. *L. biternata* Ruíz & Pavón – cult. orn., fr. ed., good bark fibre

Lardizabalaceae R. Br. Dicots – Magnoliidae – Ranunculales. (Incl. Sargentodoxaceae) 8/45 Himal. to SE As. & Taiwan, C Chile. Pachycaul shrubs (*Decaisnea*) & lianes (monoec. or dioec.; *D.* polygamous). Lvs spirally arr., palmate (pinnate in *Decaisnea*), rarely simple (*Sargentodoxa*); stipules usu. 0. Fls small, 3-merous, reg., in usu. drooping racemes from scaly axillary buds; K 3 + 3(–8) (3 in *Akebia*), usu. petaloid & imbricate or outer valvate, C

3 + 3, smaller, nectariferous or 0, A (3)6(–8) opp. C, filaments ± basally connate, anthers with longit. slits, the pollen-sacs ± embedded in thickened connective with short term. appendage, G 3 (or 6–12) in 1(–5) whorls of 3 ± 6 staminodes (pistillodes s.t. in male fls), carpels with term. ± sessile stigma, sometimes (at least *Akebia*) conduplicate & unclosed, with (few–)∞ anatropous to campylotropous or orthotropous bitegmic ovules. Fr. a head of berries or fleshy follicles, often ed., the pericarp (at least in *Decaisnea*) with latex-system; seeds with short straight embryo in copious oily endosperm s.t. also with carbohydrate. 2n = 28, 30, 32

Genera: *Akebia, Boquila, Decaisnea, Holboellia, Lardizabala, Sargentodoxa, Sinofranchetia, Stauntonia*

Major disjunction in distr. Most archaic genus the pachycaul *Decaisnea* (Cronquist), the most advanced the dioec. Chilean genera

Cult. orn. & some locally ed. fr.

Larentia Klatt = *Alophia*

Laretia Gillies & Hook. Umbelliferae (I 2c). 2 Chilean Andes

Larix Miller. Pinaceae. 9 cool N hemisphere (Eur. 2). R: NSPV 9(1972)4; Phytologia M 7(1984)38. Larches. Much like *Cedrus* but lvs decid. & cones mature in 1 yr, protogynous. Timber & turpentine. *L. decidua* Miller (Common or European l., Eur.) – timber for telegraph poles, pit-props, boat-building, shingles, fencing & gates, also nurse for broadleaf trees, distilled resin the source of Venice turpentine, sugar-like manna form. medic. in bronchial conditions, bark has 8–9% tannin & used for tanning leather esp. in former USSR: *L. kaempferi* (Lamb.) Carrière (*L. leptolepis*, Japanese l., Japan); *L. laricina* (Duroi) K. Koch (tamarack, hackmatack, Am. l., N Am.) – uses similar to those of *L. decidua*; *L.* × *marschlinsii* Coaz (*L.* × *eurolepis*, *L. decidua* × *L. kaempferi*, Dunkeld or hybrid l.) – much planted hybrid larch resistant to l. canker, seed from seed orchards of parental spp. best coll. from *L. decidua* as *L. kaempferi* fls first & with more pollen so has higher percentage of pure *L. kaempferi* offspring

larkspur *Consolida* spp.

Larrea Cav. (*Covillea*). Zygophyllaceae. 5 SW N Am., S Am., desert disjuncts esp. *L. divaricata* Cav. with subsp. *divaricata* in S Am. & subsp. *tridentata* (DC) Felger & Lowe (*L. tridentata*, creosote bush) in SW N Am., where suckering clones to 7.8m radius estimated at 11 700 yrs old (Mojave desert); twigs steeped in boiling water yield antiseptic lotion, fl. buds pickled & eaten like capers

Larsenia Bremek. = *Strobilanthes*

Lasallea E. Greene = *Aster*

Laseguea A. DC = *Mandevilla*

Laser Borkh. ex P. Gaertner, Meyer & Scherb. Umbelliferae. 1 C & S Eur., W As.

Laserpitium L. Umbelliferae (III 12). 35 Canary Is., Medit. (Eur. 13) to SW As. Fr. & roots locally medic., liqueur flavourings etc.

Lasersia Liben = *Synsepalum*

Lasia Lour. Araceae (V 3). 2 Indomal. (1 only known from Bogor botanic garden, Java). R: Blumea 33(1988)459. *L. spinosa* (L.) Thwaites (*L. aculeata*) – a potherb like spinach, rhizome cooked, curried in Sri Lanka

Lasiacis (Griseb.) A. Hitchc. Gramineae (34b). 20 warm Am., Madag. R: AMBG 65(1978)1133. At maturity glumes & lower lemma turn black & epidermis fills with oil globules attractive to frugivorous birds, the tough upper floret not being digested by disp. agents

Lasiadenia Benth. Thymelaeaceae. 2 trop. S Am. R: Brittonia 38(1986)114. *L. rupestris* Benth. used to remove warts in Amaz.

Lasiagrostis Link = *Stipa*

Lasianthaea DC (~ *Zexmenia*). Compositae (Helia.-Verb.). c.12 Arizona to Venez. (esp. Mex.). R: MNYBG 31, 2(1979)1, Phytol. 65(1988)359, 66(1989)496. Cult. orn.

Lasianthera Pal. Icacinaceae. 1 trop. W Afr.

Lasianthus Jack. Rubiaceae (IV 10). c. 170 trop. Afr. (c. 15), Indomal. (Malay Pen. 54) to Aus., WI (1)

Lasiarrhenum I.M. Johnston. Boraginaceae. 2 Mexico

Lasimorpha Schott = *Lasiomorpha*

Lasiobema (Korth.) Miq. = *Bauhinia*

Lasiocarpus Liebm. Malpighiaceae. 4 Mexico

Lasiocaryum I.M. Johnston. Boraginaceae. 7 C As., Himalaya

Lasiocephalus Willd. ex Schldl. (*Aethiolaena*; ~ *Culcitium*). Compositae (Sen.-Sen.). 10 Colombia, Ecuador, Peru

Lasiochlamys Pax & K. Hoffm. Flacourtiaceae. 13 New Caled. R: Blumea 22(1974)124

Lasiochloa Kunth = *Tribolium*

Lasiocladus Bojer ex Nees. Acanthaceae. 5 Madagascar

Lasiococca Hook.f. Euphorbiaceae. 3 E Himalaya, Hainan, Malay Peninsula

Lasiocoma Bolus = *Euryops*

Lasiocorys Benth. = *Leucas*

Lasiocroton Griseb. Euphorbiaceae. 4 WI

Lasiodiscus Hook.f. (~ *Colubrina*). Rhamnaceae. 9 trop. Afr., Madagascar

Lasiolaena R. King & H. Robinson (~ *Eupatorium*). Compositae (Eup.-Gyp.). 5 E Brazil. R: MSBMBG 22(1987)120

Lasiomorpha Schott. Araceae (V 3). 1 W & C trop. Afr. R: Blumea 33(1988)465

Lasiopetalum Sm. Sterculiaceae. 30–35 Aus.

Lasiopogon Cass. Compositae (Gnap.-Gnap.). 8 S Afr., 1 ext. to N Afr. & Middle E. R: OB 104(1991)155

Lasiorrhachis (Hackel) Stapf = *Saccharum*

Lasiosiphon Fres. = *Gnidia*

Lasiospermum Lagasca. Compositae (Anth.-Urs.). 4 S Afr., Egypt

Lasiostelma Benth. = *Brachystelma*

Lasiurus Boiss. Gramineae (40h). 1 Mali to NE India: *L. scindicus* Henrard – valuable fodder in subdesert

Lastarriaea Rémy (~ *Chorizanthe*). Polygonaceae (I 1). 3 Calif., N & C Chile. R: Phytol. 66(1989)213

Lasthenia Cass. Compositae (Helia.-Bae.). 17 Pacific N Am., Chile. R: UCPB 40(1966)1. Some cult. orn. (*Baeria*)

Lastrea Bory = *Thelypteris*

Lastreopsis Ching. Dryopteridaceae (I 3). c. 40 trop. & S temp. (Am. 6) esp. Aus.

Latace Philippi = *Leucocoryne*

Latania Comm. ex Juss. Palmae (I 3a). 3 Masc. R: Fl. Masc. 189(1984)6. Dioec. fan-palms. Cult. orn. in trop. esp. *L. lontaroides* (Gaertner) H. Moore (*L. borbonica*, Réunion)

Lateristachys Holub = *Lycopodiella*

Lateropora A.C. Sm. Ericaceae. 2 Panamá

Lathraea L. Scrophulariaceae. 7 temp. Euras. (Eur. 3). R: E. Heinricher (1931) *Monogr. der Gattung L. L. squamaria* L. (toothwort, Eur. to Himal.) parasitic on roots of *Fagus, Corylus* etc. with thick rhiz. bearing 4 rows of tooth-like scale lvs, each hollow with glands in side-chambers in which dead insects sometimes found, in S France apparently sometimes mycotrophic; *L. clandestina* L. (S & W Eur.) – cult. parasite with brilliant fls growing on *Populus, Salix* etc. & natur. in GB (esp. in C Cambridge)

Lathraeocarpa Bremek. Rubiaceae (IV 9). 2 Madagascar

Lathriogyna Ecklon & Zeyher = *Amphithalea*

Lathrophytum Eichler. Balanophoraceae. 1 Brazil (nr. Rio)

Lathyrus L. Leguminosae (III 20). 160 N temp. (most; Eur. 54), trop. E Afr. mts, temp. S Am. R; NRBGE 41(1983)209. Vetchlings; fls yellow, blue, red etc.. Close to *Vicia* but most distinguishable by their winged stems & parallel veins in lvs but one section only separable on characters of style-pubescence: a tuft of these hairs brushes the pollen out of the apex of the keel where the anthers shed it. Climbing spp. usu. with branched tendrils; in *L. aphaca* L. (yellow v., Eur. & Med. to Afghanistan) leaf rep. by tendril, photosynthesis carried out by v. large stipules; in *L. nissolia* L. (Eur. & Med., natur. N Am.) there are phyllodes with parallel veins (cf. *Acacia*). Many cult. orn. (incl. *Orobus*) & some fodders etc. *L. cicera* L. (Medit.) & *L. clymenum* L. (Medit.) – fodders, but eating seeds can lead to lathyrism (paralysis of legs); *L. grandiflorus* Sibth. & Sm. (S Eur.) – perennial 'everlasting' pea with 2 or 3 rose-purple fls per peduncle; *L. japonicus* Willd. (*L. maritimus*, beach pea, sea p., circumpolar) – seeds viable for 4–5 yrs in seawater, a coffee subs. for Esquimaux; *L. latifolius* L. (*L. sylvestris* subsp. *l.* C & S Eur., natur. N Am.) – perennial 'everlasting' pea with several to many white to purple fls per peduncle; *L. linifolius* (Reichard) Baessler (*L. macrorrhizus, L. montanus*, W & C Eur.) – tubers ed. like potatoes, form. used to flavour some whiskies; *L. odoratus* L. (sweet pea, Crete, Sicily & S Italy) – cult. since anc. times for large scented fls, *the* cut flower of the Edwardians, v. many cvs (autogamous); most 'scent' for soap etc. now synthetic; *L. pratensis* L. (common or meadow v., Eur. & Medit. to Afghanistan); *L. sativus* L. (Indian, Riga or dogtooth pea, khesari) – fodder & used like *Cajanus cajan*, cultigen, cult. early Neolithic & perhaps first crop domesticated in Eur. but beta-N-oxalyl-amino-alanine causing motorneurone disease in C India; *L. tingitanus* L. (Tangier pea, W Medit.) – green manure; *L. tuberosus* L. (earth chestnut, Euras., Medit.) –

tubers ed., fls form. distilled for scent

Latipes Kunth = *Leptothrium*

Latouchea Franchet. Gentianaceae. 1 E China

Latrobea Meissner. Leguminosae (III 24). 6 SW Aus.

Latua Philippi. Solanaceae. (1). 1 S Chile: *L. pubiflora* (Griseb.) Baillon – heteroblastic, spiny, woody pl. to 10 m, with alks incl. hyoscyamine & scopolamine, form. used as malevolent hallucinogen, leading to permanent madness; fish-poison

lauan lightweight timbers of *Parashorea* & *Shorea* spp.; **red l.** *Shorea negrosensis* etc.; **white l.** *S. contorta, Parashorea malaanonan*

Laubertia A. DC. Apocynaceae. 6 trop. Am.

Laumoniera Nooteboom = *Brucea*

Launaea Cass. Compositae (Lact.-Son.). 30 Canaries to S Afr. & E As. (Eur. 6), WI. *L. sarmentosa* (Willd.) Kuntze a pl. char. of trop. sandy beaches

Lauraceae Juss. Dicots – Magnoliidae – Magnoliales. 52/2850 trop. & warm esp. SE As. & Brazil. Aromatic usu. evergreen trees & shrubs (*Cassytha* a parasitic twiner), usu. with spherical ethereal oil-cells in parenchymatous tissues. Lvs spirally arr. (rarely whorled or opp.), simple (lobed in e.g. *Sassafras*, scale-like in *Cassytha*), often coriaceous, venation usu. pinnate; stipules 0. Fls bisexual (or pl. polygamous to dioec.) with well-developed hypanthium like a K-tube (epigyny in *Hypodaphnis*), reg., usu. 3-merous & small, in axillary usu. thyrsoid infls (rarely solit.); P usu. 3 + 3, sepaloid, A in 4 whorls of 3, the innermost often sterile or 0 & s.t. 1 or 2 of the other 3 also, whorl 3 usu. with pair of glands at base, anthers usu. opening by (1)2 or 4 valves from base upwards (small pores in *Micropora*), $\underline{G}$ ($\overline{G}$ in *Hypodaphnis*) 1 (?(3)), 1-loc. with 1 large ± apical, anatropous, bitegmic ovule. Fr. a 1-seeded berry (rarely dry-indehiscent), often enclosed in persistent accrescent fleshy to woody hypanthium; seeds endotestal with large straight oily embryo (s.t. also starchy), endosperm 0. x = 12

Classification & chief genera (after Rohwer):

1. **Perseeae** (infl. thyrsoid, umbels without involucres): *Alseodaphne, Aniba, Beilschmiedia, Cassytha, Cinnamomum, Cryptocarya, Dehaasia, Endiandra, Endlicheria, Eusideroxylon, Licaria, Ocotea, Persea, Phoebe*
2. **Laureae** (umbels surrounded by involucre of bracts): *Actinodaphne, Laurus, Lindera, Litsea, Neolitsea*

Uncertain position: *Cinnadenia, Chlorocardium*

The dodder-like *Cassytha*, despite its macro-morphological features was correctly placed in L. by Robert Brown in 1810; unlike other parasitic groups, it has a normal well-developed embryo. The fam. has imp. relationships with birds (food/disp.)

Several genera in Macaronesia represent relict laurel-forest trees: *Apollonias, Laurus, Ocotea, Persea*

Many spices & flavourings (*Aniba, Cinnamomum* (camphor, cinnamon), *Cryptocarya, Laurus, Licaria, Lindera, Litsea, Sassafras*), some medic.; timbers (*Aniba, Beilschmiedia, Caryodaphnopsis, Chlorocardium, Dicypellium, Endiandra, Eusideroxylon, Nectandra, Ocotea, Persea, Umbellularia*); fr. of *Persea americana* the avocado pear

Lauradia Vand. = *Sauvagesia*

laurel Orig. *Laurus nobilis* but the l. crown of Caesar *Ruscus hypoglossum* & the l. of shrubberies is usu. *Prunus laurocerasus*; **Alexandrian l.** *Calophyllum inophyllum*; **bay l.** *L. nobilis*; **bog l.** *Kalmia polifolia*; **Californian l.** *Umbellularia californica*; **cherry l.** *Prunus laurocerasus*; **Chilean l.** *Laurelia aromatica*; **Chinese l.** *Antidesma bunius*; **Ecuador l.** *Cordia alliodora*; **Indian l.** *Terminalia alata*; **Japanese l.** *Aucuba japonica*; **mountain l.** *K. latifolia*; **pig** or **sheep l.** *K. angustifolia*; **Portugal l.** *P. lusitanica*; **spurge l.** *Daphne laureola*

Laurelia Juss. Monimiaceae (I 2; Atherospermataceae). 1 NZ, 1 Chile & Peru. Polygamous to dioec. trees with alks. *L. novae-zelandiae* Cunn. (pukatea, NZ) – timber & medic.; *L. sempervirens* (Ruíz & Pavón) Tul. (*L. aromatica, L. serrata*, Chile & Peru) – fr. a spice (Peruvian nutmegs, tepa)

Laureliopsis Schodde. Monimiaceae (I 1). 1 S Arg., S Chile

Laurembergia P. Bergius. Haloragidaceae. 4 trop. & warm

Laurentia Adans. (*Solenopsis*). Campanulaceae. 25 W N Am., Medit., S Afr. See also *Hippobroma*

Laurocerasus Duhamel = *Prunus*

Laurophyllus Thunb. Anacardiaceae (IV). 1 S Afr.

Laurus L. Lauraceae (II). 1–2 Medit. (1), Macaronesia (1: *L. azorica* (Seub.) Franco (*L. canariensis*)). R: Bot. JLS 68(1974)51. Alks. *L. nobilis* L. ((true or bay) laurel, sweet bay or b. tree) – domatia, lvs used to flavour food, fr. in veterinary medic., aromatherapy, lvs the original crown of l. (resting on your ls.; baccalaureat (Bachelor of Arts etc.), poet laureate) of Greece

laurustinus *Viburnum tinus*

Lautembergia Baillon. Euphorbiaceae. 3 Madagascar, Mauritius

Lauterbachia Perkins. Monimiaceae (V ?2). 1 New Guinea (known only from one gathering, now destroyed)

lavandin *Lavandula* × *intermedia*

Lavandula L. Labiatae (VIII 3). 30 Atlantic Is., Medit. (Eur. 7) to Somalia & India. Lavenders. R: JLSBot. 51(1937)153. Protandrous bee-poll. fls sources of nectar for good honey; aromatic oil extracted today in GB & S France largely from *L.* × *intermedia* Emeric ex Lois. (spontaneous hybrid (lavandin, 'English lavender') SW Eur.: *L. angustifolia* Miller (*L. officinalis*, *L. vera*, (orig.) English l., Medit.) × *L. latifolia* Medik. (Medit.), much cult., many orn. cvs incl. dwarfs) & *L. latifolia*, also *L. stoechas* L. (French l., Spanish l., Medit., Portugal) for scent, lavender water & medic., in porcelain-painting; fls used in sachets (l. bags) to scent linen, though lavender = 'distrust' in 'Language of Fls'; oil (1000 t per annum (1993) in commerce) antiseptic & has been used empirically since anc. times, in cosmetics today

Lavatera L. Malvaceae. c. 25 Macaronesia, Medit. (Eur. 11) to NW Himal., C As., Aus. (1), Calif. Herbs & shrubby trees with soft wood, differing from *Malva* in that epicalyx segments basally united though not all spp. show this! (Curious disjunct distribution may reflect weak taxonomy). Some medic., others fast-growing cult. orn. with flattened fr. ('cheeses') like *Malva*. *L. arborea* L. (tree mallow, Eur., natur. SW US) – maritime & *L. thuringiaca* L. ('*L. olbia*', C & SE Eur.) – perennial, many cvs; *L. trimestris* L. (Medit.) – annual

Lavauxia Spach = *Oenothera*

lavender *Lavandula* spp. esp. *L. angustifolia*, *L.* × *intermedia* (**English l.**); **French** or **Spanish l.** *L. stoechas*; **Indian l.** *Bursera penicillata*; **sea l.** *Limonium* spp. esp. *L. vulgare*

Lavigeria Pierre. Icacinaceae. 1 trop. W Afr.

Lavoisiera DC. Melastomataceae. 46 Brazil

Lavoixia H. Moore (~ *Clinostigma*). Palmae (V 4m). 1 New Caled.: *L. macrocarpa* H. Moore. R: Allertonia 3(1984)334. Only 4 left (Mt Panié) 1980

Lavradia Roemer = *Sauvagesia*

Lavrania Plowes. Asclepiadaceae (III 5). 5 S Afr.

Lawia Griffith ex Tul. = *Dalzellia*

Lawrencella Lindley (~ *Helichrysum*). Compositae (Gnap.-Ang.). 2 Aus. R: Nuytsia 8(1992)369

Lawrencia Hook. (~ *Plagianthus*). Malvaceae. 12 Aus. (W Aus. 11). R: Nuytsia 5(1984)201. Some dioecy

Lawsonia L. Lythraceae. 1 N Afr., trop. OW: *L. inermis* L. Powdered lvs (henna) a green powder staining fingernails (hands & feet of brides in Sumatra – lasts up to 6 months) etc. red, with indigo hair glossy blue-black, much used in E; Egyptian mummies swathed in cloth dyed with it; fls used to scent oil etc. & distilled for similar; bark medic.; cult. orn.

Lawson's cypress *Chamaecyparis lawsoniana*

Laxmannia R. Br. Anthericaceae (Liliaceae s.l.). 13 Aus. R: FA 45(1987)254

Laxoplumeria Markgraf. Apocynaceae. 3 E Peru, Brazil

Layia Hook. & Arn. ex DC. Compositae (Hele.-Mad.). c. 15 Calif. *L. platyglossa* (Fischer & C. Meyer) A. Gray (*L. elegans*, tidy-tips) – consp. early summer field flower with yellow rayflorets tipped with white, much-cult. annual

Lazarum A. Hay. Araceae (VIII 6). 1 N Aus. R: BJLS 109 (1992)427

Leachia Plowes = *Trichocaulon*

Leachiella Plowes = praec.

lead tree *Leucaena leucocephala*; **l.wort** *Plumbago* spp.

Leandra Raddi. Melastomataceae. 175 trop. Am. Trees, shrubs & herbs; *L. salicina* (DC) Cogn. (Brazil) – rheophyte

Leandriella Benoist. Acanthaceae. 1 Madagascar

Leaoa Schltr. & Porto = *Scaphyglottis*

leatherwood *Eucryphia lucida*

Leavenworthia Torrey. Cruciferae. 8 SE US. R: JAA 69(1988)118. Limestone glades

Lebeckia Thunb. Leguminosae (III 28). c. 35 S Afr.

Lebetanthus Endl. (~ *Prionotes*). Epacridaceae. 1 S S Am.

Lebetina Cass. = *Adenophyllum*

Lebronnecia Fosb. Malvaceae. 1 Marquesas: *L. kokioides* Fosb. – known from only 1 tree & seedlings when described

Lebrunia Staner. Guttiferae (III). 1 trop. Afr.
Lebruniodendron Léonard. Leguminosae (I 4). 1 Gulf of Guinea
Lecananthus Jack. Rubiaceae (I 8). 2 W Mal.
Lecaniodiscus Planchon ex Benth. Sapindaceae. 3 trop. Afr. *L. cupanioides* Planchon ex Benth. – chewing-stick in Nigeria
Lecanium C. Presl = *Trichomanes*
Lecanolepis Pichi-Serm. = *Trichomanes*
Lecanopteris Reinw. Polypodiaceae (II 4). 13 Mal. R: GBS 45(1994)293. Epiphytes with fleshy ant-inhabited rhiz. *L. spinosa* Jermy & Walker (Sulawesi) – spines used by ants in runways
Lecanorchis Blume. Orchidaceae (V 4). 20 Indomal., Japan
Lecanosperma Rusby = *Heterophyllaea*
Lecanthus Wedd. Urticaceae (II). 1 OW trop.: *L. peduncularis* (Royle) Wedd.
Lecardia Poisson ex Guillaumin = *Salaciopsis*
Lecariocalyx Bremek. Rubiaceae (IV 7). 1 Borneo
Lechea L. Cistaceae. 17 Am: R: CGH 121(1938)
Lechenaultia R. Br. = *Leschenaultia*
Lecocarpus Decne. Compositae (Helia.-Mel.). 3 Galápagos. R: Madrono 20(1969)255
lecheguilla *Agave lecheguilla*
Lecointea Ducke. Leguminosae (III 1). 5 trop. Am. R: BZ 61(1976)1307
Lecokia DC. Umbelliferae (III 5). 1 Crete to Iran
Lecomtea Koidz. = *Cladopus*
Lecomtedoxa (Engl.) Dubard. Sapotaceae (I). 5 Gabon. R: T.D. Pennington, *S.* (1991)143. Seeds a source of cooking oil
Lecomtella A. Camus. Gramineae (34b). 1 Madagascar
Lecontea A. Rich. ex DC = *Paederia*
Lecythidaceae Poit. Dicots – Dilleniidae – Lecythidales (?Theales). 20/285 trop. esp. S Am. rain forests. Trees or shrubs, often pachycaul. Lvs spirally arr. commonly at branch-tips, simple, entire or toothed; stipules small & caducous or 0. Fls bisexual, reg. or not, epigynous (or half so) with hypanthium s.t. extended beyond G, often large, showy & ephemeral, insect- or bat-poll. with much nectar, axillary, solit. or in term. or axillary panicles, fascicles or racemes sometimes from old wood; K (2–)4–6(–12), imbricate (valvate in some OW genera, rarely connate & forming calyptra), C 4–6, imbricate or 0, A (10–)∞(to c. 1200) symmetrical in several centrifugal series basally connate on a basal ring, or this ring asymmetrical & extended on 1 side into a flat ligule s.t. curved over G as a hood (the stamens thereon staminodal), in fls with C 0 outer A sterile & connate forming a corona, intrastaminal disk scarcely developed in NW genera, enlarged & ± covering G in OW, anthers opening by longit. slits (or apical pores), $\overline{G}$ (2–6) with a term. style & as many loc. as G with axile placentation (basal in *Eschweilera*) & 1–∞ anatropous, bitegmic ovules per loc. Fr. a capsule with distal operculum, often v. large (monkey-pots) or a drupe or berry; seeds often nut-like, winged or often with funicular aril, endosperm usu. 0 (well developed in *Asteranthos*), embryo large oily & proteinaceous, the cotyledons s.t. rudimentary & hypocotyl much-thickened. x = 13 (Planchonioideae), 16 (Napoleonaeoideae), 17 (Lecythidoideae)

Classification & chief genera:

Four subfams sometimes treated as separate but closely related fams:

I. **Lecythidoideae** (C present, A in several whorls, ± basally united, fr. a berry or capsule, trop. Am.): 9 genera incl. *Cariniana, Couroupita, Eschweilera, Gustavia, Lecythis*
II. **Napoleonaeoideae** (Napoleonaeaceae, Asteranthaceae; C 0, outer A sterile forming corona or disk, fr. a caps. with persistent K or large berry): *Asteranthos, Crateranthus, Napoleonaea*
III. **Foetidioideae** (Foetidiaceae; C 0, staminal disk large, fr. a drupe): *Foetidia* (OW)
IV. **Planchonioideae** (Barringtoniaceae; C present, fr. a usu. 1-seeded berry or dry 4-winged indehiscent capsule, OW): 6 genera incl. *Barringtonia, Petersianthus, Planchonia*

Tradi. placed in Myrtales, L. differ in embryology, bitegmic ovules, spirally arr. lvs etc.; ovule characters suggest Dilleniidae esp. near Theales (Cronquist). The pachycaul genus *Gustavia* is considered the most archaic (Cronquist)

The range of flower & fruit form associated with diff. poll. & disp. mechanisms, the last including water, air & a range of animals. In Amaz., poll. from open-flowered *Gustavia* (small & middle-sized bees) to those with increasing ligule (*Couroupita* (q.v.) etc.) to *Couratari* with hidden nectar available only to female euglossines

Ed. fr. or seeds (*Bertholletia* – Brazil nuts, *Grias, Gustavia, Lecythis*), timber (*Bertholletia, Careya, Cariniana, Couratari, Couroupita, Lecythis*), cult. orn.

Lecythis Loefl. Lecythidaceae. 25 trop. Am. R: FN 21(1979)197. Like *Couroupita* but A of ligule (see fam.) sterile; fr. a 'monkey-pot' form. used to trap monkeys, which grab sugar inside cannon-ball-like caps. (secured) & cannot withdraw extended fist. Some timbers & ed. oily seeds (paradise nuts) esp. *L. zabucajo* Aublet (Brazil, Guianas) – sapucaia nuts, with delicate flavour suitable for chocolates, & other spp.; *L. minor* Jacq. & *L. ollaria* L. have toxic seeds which if eaten lead to temporary loss of hair & nails & nausea caused by the selenium analogue of the amino-acid cystathionine

Leda C.B. Clarke = *Isoglossa*

Ledebouria Roth (~ *Scilla*). Hyacinthaceae (Liliaceae s.l.). c. 30 trop. & S Afr., India. R: JSAB 36(1970)233. Cult. orn. with blotched or striped lvs, some with toxic bulbs

Ledebouriella H. Wolff. Umbelliferae (III 8). 2 C As.

Ledenbergia Klotzsch ex Moq. (*Flueckigera*). Phytolaccaceae (III). 2 Mex. to Venez.

Ledermanniella Engl. Podostemaceae. 43 trop. Afr. R: BMNHN 4,5(1983)361, 6(1984)249

Ledger bark *Cinchona calisaya*

Ledocarpaceae Meyen. See Geraniaceae

Ledothamnus Meissner. Ericaceae. 9 Guayana Highland. R: MNYBG 29(1978)141, ABV 14,1(1983)170

Ledum L. = *Rhododendron*

Leea Royen ex L. Leeaceae. 34 OW trop. (Afr. & Madag. 2, Indomal. 32). Some locally medic. & cult. orn. foliage pls

Leeaceae (DC) Dumort. Dicots – Rosidae – Rhamnales 1/34 OW trop. R: Blumea 22(1974)57. Erect shrubs, some pachycaul & unbranched, & herbs. Lvs spirally arr. (or opp.), pinnate to 4-pinnate or simple; stipules with petiole s.t. basally auricled or flanged. Fls bisexual, reg., (4)5-merous, in large cymose often umbelliform infls opp. lvs; K a short tube with valvate teeth; C & A basally adnate forming a tube, C-lobes valvate, A opp. C, filaments connate distal to C – A tube, the A tube often with lobes alt. with A & s.t. extended near middle to form pendulous membrane, anthers introrse, G̲ s.t. sunk in disk or disk 0, incompletely divided into 4 or 6(8) app. locules (? 2 or 3(4) locules with 'false' partitions) with term. style & 1 basal, erect, anatropous, bitegmic ovule per ovary chamber. Fr. a pluriloc. berry; embryo linear, small, endosperm ruminate, oily & proteinaceous

Only genus: *Leea*

Differs from Vitaceae in erect habit & 0 tendrils & in G characters

leek *Allium ampeloprasum* Porrum Group; **lady's l.** *A. cernuum*; **l. orchid** *Prasophyllum* spp.; **round-headed l.** *A. sphaerocephalon*; **sand l.** *A. scorodoprasum*

Leersia Sw. Gramineae (9). 17 trop. & warm temp. (Eur. 1). R: SBi 46(1965)129, ISJS 44(1969)215. Marsh grasses allied to *Oryza* but sterile lemmas 0, used as fodder in As. esp. *L. hexandra* Sw. (bareet grass, trop.). *L. oryzoides* (L.) Sw. (N temp.) has cleistogamous fls

Leeuwenbergia Letouzey & Hallé. Euphorbiaceae. 2 trop. Afr.

Lefebvrea A. Rich. Umbelliferae (III 10). 6 trop. & SW Afr.

Legazpia Blanco = *Torenia*

Legenere McVaugh. Campanulaceae. 1 Calif., 1 Chile. Name anagram of E.L. Greene

Legnephora Miers. Menispermaceae (V). 5 New Guinea, NE Aus. Alks

Legousia Durande (*Specularia*). Campanulaceae. 15 Eur. (5), Medit. Some cult. orn. annuals incl. *L. hybrida* (L.) Delarbre (Venus's looking-glass, Eur. to N Afr.)

Legrandia Kausel. Myrtaceae (Myrt.). 1 Chile

Leguminosae Juss. (Fabaceae). Dicots – Rosidae – Fabales. 643/18 000 cosmop. (I & II mostly trop.). Trees, shrubs incl. lianes, & herbs, sometimes spiny, often with root nodules containing nitrogen-fixing bacteria (25–30% of I, 60–70% of II, c. 95% of III) & frequently with non-protein amino-acids in seeds &/or vegetative parts as well as alks. Lvs pinnate (rarely palmate), bipinnate, unifoliolate, trifoliolate or simple, s.t. phyllodic or reduced to a tendril (some *Lathyrus*), usu. spirally arr., the petiole & leaflets with basal pulvini often controlling orientation & 'sleep movements'; stipules present, s.t. large or represented by spines (s.t. ant-inhabited in *Acacia*) or prickles, leaflets s.t. with stipellules. Fls usu. bisexual, reg. or not, hypogynous to perigynous, usu. in racemes, spikes or heads, K 5 or ((3–)5(6)), when a tube with valvate (rarely imbricate) lobes (II), or often ± bilabiate (III), C (0–)5 irreg. (adaxial small & lying within its laterals in I or large (standard) & outside them (resupinate occasionally) in III) or (3–)5(6) s.t. basally connate, reg. & usu. valvate (II), A usu. twice P (to ∞ in I & II), distinct or ± connate, alike or not, s.t. some staminodal, forming a sheath around G in III, often coloured & long-exserted in II, anthers usu. with longit. slits (pores in some I), nectary often a ring on receptacle around G, G̲ 1 (2–16 in some II), each with (1)2–∞ anatropous to campylotropous, bitegmic ovules often with zig-zag micropyles. Fr. usu. dry & dehiscent down both sutures (a legume), occ. breaking up into 1-seeded sections (lomentum) or indehiscent & samariod or a drupe; seeds hard, exotestal, s.t. with an elongate funicle and an aril (often merely a vestige), rarely sarcotestal or winged, embryo straight to curved with 0 or little endosperm (much in *Prosopis*). x = 5–14. R.M. Polhill & P.H. Raven (eds, 1981) *Advances in legume systematics* 1; M. Crisp & J. Doyle (eds, 1995) *idem* 7

Classification & chief genera:

Sometimes treated as 3 fams, the legumes are most often arranged as below but the wood-anatomy would support amalgamation of I (which is paraphyletic & basal to II) & III:

I. **Caesalpinioideae** (L. Watson & M.J. Dalwitz (1983) *The genera of L. C.*; fls usu. irreg., C imbricate in bud, free or some united; adaxial C overlapped by laterals, K free (exc. in 3), radicle usu. straight; lvs bipinnate or pinnate, rarely unifoliolate or simple, usu. trees (s.t. forming pure stands in Afr. forests) & shrubs of trop.): 153/2175. 4 tribes (diff. to recognize):
   1. **Caesalpinieae** (stipules interpetiolar or 0, rarely herbs; some genera transitional to other subfams, the tribe defined by negative characters & the base group for the whole family, comprising relics & complexes undergoing rapid speciation): widely distrib. esp. trop., incl. *Acrocarpus* (plantation), *Caesalpinia* (timber, cult. orn.), *Covillea* (cult. orn.), *Cordeauxia* (ed. seeds), *Delonix* (cult. orn.), *Gleditsia* (timber), *Gymnocladus* (timber, cult. orn.), *Haematoxylum* (dyes), *Melanoxylon*, *Pterogyne* & *Recordoxylon* (timber), *Tachigali*
   2. **Cassieae** (Sclerolabieae; stipules lateral or 0, anthers with slits or pores): trop. & warm, divisible into 5 subtribes, incl. *Cassia*, *Ceratonia* (carobs), *Chamaecrista*, *Dialium*, *Koompassia* (timber; tallest dicot. in trop.), *Senna* (medic.)
   3. **Cercideae** (lvs with palmate venation & 2 halves with independent nyctitropic movement, whether free or not, on a single pulvinus unique in fam.; trees, shrubs or lianes; 2 subtribes): *Bauhinia*, *Cercis*
   4. **Detarieae** (incl. Amherstieae; stipules intrapetiolar, ectotrophic mycorrhiza typical but rare elsewhere in L.; trees to suffrutices esp. consp. in African woodlands (miombo) e.g. *Brachystegia*, *Isoberlinia*, *Julbernardia*): *Afzelia* (timber), *Amherstia* (cult. orn.), *Baikiaea* (timber), *Brownea*, *Copaifera* (copals), *Cynometra* (ed.), *Daniellia* (copals), *Detarium*, *Guibourtia* (copals), *Intsia* & *Peltogyne* (timber), *Saraca*, *Schotia* & *Tamarindus* (ed. fr.). *Macrolobium* + 22 genera incl. *Brachystegia* & allies of trop. Afr. & Am. excluded lately as Macrolobieae (bracteoles protecting fl. buds, K well-developed)

II. **Mimosoideae** (fls reg., C valvate in bud (excl. *Dinizia*), often with basal tube; lvs bipinnate or, less often, pinnate or phyllodic; trees, shrubs & some herbs incl. aquatics (*Neptunia*) esp. trop. & warm): 64/2950. 5 tribes often with uncommon amino-acids in seeds (but not canavanine); Journal: Bull. Int. Group Study M.:
   1. **Parkieae** (trees; K-lobes with valvate aestivation): *Parkia* (ed.) & *Pentaclethra* (only), both with disjunct distribs.
   2. **Mimozygantheae** – 1 sp. (of *Mimozyganthus*) of unclear relationships
   3. **Mimoseae** (poss. the basal group, more than half spp. in *Mimosa*, some 15 genera mono- or di-typic; trop. & warm esp. S Am. & trop. Afr.): *Adenanthera* (beads), *Anadenanthera* (hallucinogens), *Cylicodiscus* (timber), *Desmanthus*, *Dinizia* (v. tall), *Entada* (v. long), *Leucaena* (ed. fr., pulp), *Mimosa*, *Neptunia*, *Piptadenia*, *Prosopis* (fodder etc.), *Xylia* (timber)
   4. **Acacieae** (gum exudations common, filaments usu. free (cf. 5); trees or shrubs incl. climbers, rarely herbs): *Acacia* & *Faidherbia* (only; timber, gums, tanbarks, fodder)
   5. **Ingeae** (like 4 (with which it should poss. be united) but filaments united; trop. esp. Am. & As. to Aus.): *Albizia* (timber), *Archidendron*, *Calliandra*, *Enterolobium*, *Inga*, *Paraserianthes* (timber), *Pithecellobium*

III. **Papilionoideae** (Faboideae; fls like I but adaxial petal (standard) outside laterals (wings; exc. Swartzieae & Sophoreae p.p.), K basally united; radicle usu. curved; no bipinnate lvs or compound pollen grains as found in other subfams; the most familiar legumes with the standarized papilionoid flower assoc. with insect- (& bird-) poll., able to synthesize quinolizidine alks & isoflavones as well as non-protein amino-acids incl. canavanine, found from rain forests to edges of deserts with greatest diversity of form in Brazilian planalto, Mex., E Afr., Madag. & Sino-

Himal.; spectacular radiations of a few basic stocks in Mediterranean, Cape & Aus.): 426/12 150. 31 tribes:

1. **Swartzieae** (polyphyletic; often sep. as a subfam. or incl. in I, but close to 2; (usu. trees; adaxial C not outside wings, A usu. free; fr. various, usu. trop.): *Cordyla, Swartzia*
2. **Sophoreae** (paraphyletic, not sharply divided from I or rest of III: trees, shrubs or rarely herbs; fls reg. to papilionoid, A free or not; fr. various; canavanine 0; mostly trop., ext. to higher latitudes in *Sophora*): *Cadia, Camoensia* (largest L. fl.), *Castanospermum* (medic., ed. seeds), *Cladrastis, Myroxylon* (balsam), *Ormosia, Pericopsis* (afrormosia), *Sophora*
3. **Dipteryxeae** (trees; lvs paripinnate; K with 2 enlarged upper lobes; balsams & gums, trop. S Am.): *Dipteryx*
4. **Dalbergieae** (trees, lianes & shrubs with hard timber, rarely suffrutices; gum & phenolics abundant; trop. (all rep. Am. exc. *Dalbergiella* & *Inocarpus*): *Andira, Dalbergia* (rosewood), *Machaerium* & *Pterocarpus* (all timber), *Inocarpus* (ed. seeds)
5. **Abreae** (woody lianes or shrubs, lvs paripinnate, A 9, pantrop.): *Abrus* (only)
6. **Millettieae** (Tephrosieae; R. Geesink LBS 8(1984); paraphyletic & overlapping in some characters with other tribes esp. Galegeae where gradual change in vasculature associated with herbaceous habit therein; trees, lianes or shrubs, mainly trop.): *Derris* & *Lonchocarpus* (fish poisons & insecticides – rotenones), *Millettia, Tephrosia, Wisteria* (cult. orn.; shows similarities (loss of a chloroplast-DNA inverted repeat) with 6 temp. tribes: 15, 16, 17, 20, 21, 22, but 19 & 20 do not have it & are similar to other trop. M. & 10)
7. **Robinieae** (incl. Sesbanieae; trees, shrubs & herbs with canavanine in seeds, centred in trop. Am. but extending to rest of Am. with *Sesbania* pantrop.; R: Crisp & Doyle, *op. cit.*: 158): 12 genera incl. *Gliricidia, Robinia* (cult. orn.)
8. **Indigofereae** (herbs, less often shrubs or small trees, s.t. with flattened or even leafless branches; mainly trop. Afr. but *Indigofera* widespread warm; R: Crisp & Doyle, *op. cit.*: 217): 7 genera incl. *Cyamopsis* (ed.), *Indigofera, Phylloxylon* (timber)
9. **Desmodieae** (shrubs or often herbs, s.t. trees; canavanine frequent; mainly trop. esp. India to China): *Alysicarpus, Brya* (timber), *Codariocalyx, Desmodium, Uraria*
10. **Phaseoleae** (polyphyletic; dextrarotatory twining, prostrate or erect herbs, occ. shrublets, rarely trees; lvs usu. 3-foliolate; cosmop. esp. warm, the most imp. economically): *Cajanus* (pigeon pea), *Canavalia* (sword bean), *Dioclea, Erythrina, Flemingia* (root-crop), *Glycine* (soya), *Lablab, Mucuna, Phaseolus* (beans), *Rhynchosia, Vigna*
11. **Psoraleeae** (small trees or shrubs, rarely herbs; rarely trop.): *Cullen, Psoralea*
12. **Amorpheae** (small trees or shrubs or herbs; canavanine 0, NW): *Amorpha, Dalea*
13. **Aeschynomeneae** (mostly shrubs & herbs; warm): *Aeschynomene, Arachis* (ground nut), *Kotschya, Ormocarpum, Smithia, Stylosanthes* (fodder), *Zornia*
14. **Adesmieae** (herbs or shrubs of montane & temp. S Am.): *Adesmia* (only)
15. **Galegeae** (herbs or shrubs usu. with canavanine, principally N temp.): *Astracantha, Astragalus, Caragana, Clianthus, Colutea, Glycyrrhiza* (liquorice), *Swainsona*
16. **Carmichaelieae** (small trees, shrubs or subshrubs mostly with ribbed or flattened branches & canavanine; NZ region): *Carmichaelia*
17. **Hedysareae** (herbs or subshrubs with 1-seeded fr. or lomenta, sometimes with canavanine; N temp. to Horn of Afr.): *Ebenus, Hedysarum* (fodder), *Onobrychis* (sainfoin)
18. **Loteae** (herbs or small shrubs with canavanine; mainly N temp.): *Anthyllis, Lotus*
19. **Coronilleae** (like 18 but lomenta; OW esp. Medit. & Afr.): *Coronilla* & *Hippocrepis*
20. **Vicieae** (annual or perennial herbs with canavanine, largely temp., imp. economically): *Lathyrus* (sweet pea), *Lens* (lentil), *Pisum* (pea), *Vicia* (vetch)
21. **Cicereae** (perennial & annual herbs, often spiny; E Medit. to W As.): *Cicer* (chick pea) only (poss. best in 20.)
22. **Trifolieae** (annual or perennial herbs usu. with canavanine; largely Euras.): *Medicago, Ononis, Trifolium* (clover)
23. **Brongniartieae** (trees, shrubs or subshrubs; Am.): *Brongniartia, Harpalyce, Poecilanthe* (only)
24. **Mirbelieae** (usu. shrubs sometimes with canavanine; Aus. (esp. SW) with *Gompholobium* ext. to New Guinea): *Daviesia, Jacksonia, Pultenaea, Oxylobium*

25. **Bossiaeeae** (shrubs; quinolizidine alks or canavanine; Aus.): *Bossiaea* & *Hovea*
26. **Podalyrieae** (shrubs or small trees; quinolizidine alks in *Podalyria*; Cape; R: Crisp & Doyle, *op. cit.*: 304): *Cyclopia*, *Podalyria*
27. **Liparieae** (shrubs with canavanine or quinolizidine alks; Cape; R: Crisp & Doyle, *op. cit.*: 304): 5 genera incl. *Amphithalea*, *Liparia*
28. **Crotalarieae** (shrubs or herbs sometimes with quinolizidine & pyrrolizidine alks; mostly Afr.; R: Crisp & Doyle, *op. cit.*: 305): 11 genera incl. *Aspalathus*, *Crotalaria*, *Lebeckia*, *Lotononis*, *Rafnia*
29. **Euchresteae** (shrubs with quinolizidine alks; Indomal.): *Euchresta* (only)
30. **Thermopsideae** (shrubs or perennial herbs usu. with 3-foliolate lvs & often with lupine alks; N temp.): *Baptisia*, *Thermopsis*
31. **Genisteae** (shrubs, herbs or small trees, some armed or switch pls; mostly Eur. & Afr. + *Lupinus*): all but *Lupinus* making up the perplexing *Cytisus–Genista* complex, incl. *Laburnum* & *Ulex*

Almost a third of all spp. in 6 genera: *Acacia*, *Astragalus*, *Cassia*, *Crotalaria*, *Indigofera* & *Mimosa*, all characteristic of open or disturbed vegetation. Apparently most closely allied to Sapindales, some genera, esp. *Ceratonia*, *Dialium* & *Gymnocladus*, with features reminiscent of S. The defence mechanisms of the L. varied & complex, showing an increasing sophistication in the more advanced tribes: I & II have ants, tannins & terpenoids while III have many chemicals incl. diverse alks, non-protein amino-acids & isoflavonoids, the most advanced tribes with phytoalexins rather than stored compounds. Some of these compounds are important in tanning, insecticides etc. (*Acacia, Derris* etc.). The presence of *Rhizobium* bacterial nodules makes many imp. in land improvement & as green manures (*Leucaena*, *Medicago*, *Onobrychis* etc.) & valuable fodder (*Trifolium*)

Some aquatics (*Neptunia*, *Aeschynomene*) with aerenchyma useful in making floats etc., some xerophytes & many climbers: leaf-tendrils in *Lathyrus*, *Vicia* etc., stem-tendrils in *Bauhinia*, hooks in *Acacia*, others with twining stems (*Phaseolus*); lianes often with unusual anatomy. Thorns may be branches as in *Gleditsia* or stipules as in *Acacia*, where s.t. ant-inhabited. Lvs v. small in e.g. *Ulex*, *Stauracanthus*, though seedlings with typical 3-foliolate or 3-lobed lvs, sometimes phyllodic (*Acacia*, *Lathyrus*, *Phylloxylon*), when small the stems photosynthetic (*Cytisus*, *Carmichaelia* – where flattened). Reg. lvs usu. with 'sleep' movements, the leaflets folding together upwards, downwards etc.; in *Mimosa*, they are sensitive to touch & assume sleep position rapidly, while in *Codariocalyx* the lateral leaflets move continuously at high temperatures

Poll. in III principally by insects, the adaxial standard being the 'flag', the lateral wings encl. the other 2 petals forming a keel in which A & G lie. Nectar accumulates around base of G & on either side of the base of the free 10th stamen is an opening leading to it: moderately long-tongued insects like bees can reach it. Bees land on the wings depressing them with their weight while probing for nectar under the standard: the keel is depressed & the stigmas emerge, collecting any pollen already on the insect, rapidly followed by the stamens. The stigmas & stamens may return to the keel & the whole process can be repeated as in *Trifolium*; or, they are under tension such that a visit leads to explosive poll. which cannot be repeated as in *Medicago* & *Ulex*; or there is a piston mechanism squeezing pollen in small quantities out of the keel-tip, requiring several insect visits as in *Lotus*, *Lupinus*, *Ononis*; or this latter may be achieved by a brush of hairs as in *Lathyrus* & *Vicia*. Cleistogamy quite frequent. Some fls also brought to ground level or below where fr. ripens (geocarpy) as in *Arachis*, *Macrotyloma*, *Medicago*, *Tephrosia*, *Vigna*. Legumes often opening explosively, s.t. inflated as in *Colutea*; seed-coats usu. hard, making good beads & weights (*Abrus*, *Adenanthera*, *Afzelia*, *Caesalpinia*, *Delonix*, *Erythrina*)

Most imp. fam. for food-pls esp. pulses (beans, peas, gram) & oil (soya, groundnut) but also tanbarks, timber, copals, gums, insecticides & cult. orn. (see above), form. dyes & kino & still medic. (*Cassia*, *Senna*), rivalling Gramineae in the world economy

Lehmanniella Gilg. Gentianaceae. 4 Panamá, Colombia, Peru. R: PKNAW, C 88(1985)411

Leibergia J. Coulter & Rose = *Lomatium*

Leibnitzia Cass. Compositae (Mut.). 4 S & E As. R: NJB 8(1988)67. Herbs. *L. anandria* (L.) Turcz. (E As.) – cleistogamous in autumn

Leiboldia Schldl ex Gleason (~ *Vernonia*). Compositae (Vern.-Vern.). 1 Mex.: *L. serrata* (D. Don) Gleason. R: BJ 108(1987)225

Leichardtia R. Br. (~ *Marsdenia*). Asclepiadaceae (III 4). 6 Aus. (5), Fiji (1). Fr. ed., *L. australis* R. Br. (alunqua) an aboriginal food-source

Leichhardt (pine) *Nauclea orientalis*

Leichhardtia F. Muell. = *Phyllanthus*

Leidesia Muell. Arg. (~ *Seidelia*). Euphorbiaceae. 3 S Afr.

Leiophaca Lindau = *Whitfieldia*

Leiophyllum Hedwig f. Ericaceae. 1 E N Am.: *L. buxifolium* (Bergius) Elliott (sand myrtle) – cult. orn. shrub. R: SB 16(1991)529

Leiopoa Ohwi = *Festuca*

Leiospora (C. Meyer) A. Vassiljeva. Cruciferae. 2 C As.

Leiothrix Ruhl. Eriocaulaceae. 37 S Am. (Brazil 35 endemic)

Leiothylax Warm. Podostemaceae. 3 trop. Afr. R: Adansonia 20(1980)202

Leiphaimos Schldl. & Cham. = *Voyria*

Leipoldtia L. Bolus. Aizoaceae (V). 10 SW Cape, Karoo to Namibia. Cult. orn. succ. shrublets

Leisostemon Raf. Scrophulariaceae. 1 N Japan, Kamchatka

Leitgebia Eichler = *Sauvagesia*

Leitneria Chapman. Simaroubaceae (Leitneriaceae). 1 SE US (also Eocene of London): *L. floridana* Chapman (corkwood) – light timber (lighter than cork) used for floats for fishing-nets

Leitneriaceae Benth. = Simaroubaceae

Leleba Nakai = *Bambusa*

Lellingeria A.R. Sm. & R.C. Moran. Grammitidaceae. 60 trop. Afr., Madag., trop. Am., Hawaii. R: AFJ 81(1991)72

Lelya Bremek. Rubiaceae (IV 1). 1 trop. Afr.

Lemaireocereus Britton & Rose = *Pachycereus*. See also *Stenocereus*

Lembertia E. Greene (~ *Eatonella*). Compositae (Hele.-Bae.). 1 Calif.: *L. congdonii* (A. Gray) E. Greene. R: Novon 1(1991)122

Lembocarpus Leeuwenb. Gesneriaceae. 1 Guiana: *L. amoenus* Leeuwenb. Tuber producing 1 leaf & 1 infl. a season

Lemboglossum Halb. (*Cymbiglossum*). Orchidaceae (V 10). 14 C Am.

Lembotropis Griseb. = *Cytisus*

Lemeea P.V. Heath = *Aloe*

Lemmaphyllum C. Presl. Polypodiaceae (II 3). 6 Mal. Cult. orn. epiphytes

Lemmonia A. Gray = *Nama*

Lemna L. Lemnaceae. 7 cosmop. esp. Arctic regions. R: IBM 34(1965)16. Floating (*L. trisulca* L. (temp.) at depths to 14 m in Canada) plant-body, photosynthetic with single root & conspic. rootcap. Duckweed or duckmeat esp. *L. minor* L., eaten by wildfowl & tried as fodder for farm animals

Lemnaceae Martinov. Monocots – Arecidae – Arales. 4/25 cosmop. R: IBM 34(1965)1, Blumea 18(1970)355. Small to minute free-floating thalloid plants, glabrous, with 0–several unbranched roots & 0 xylem (tracheids in *Spirodela* roots), s.t. accumulating manganese. Thallus ± fleshy globular to membranous flat & linear, s.t. with visible xylem-less 'veins'; reproductive pouches 2, 1 marginal, the other marginal or on upper surface, with vegetative budding of new pl. within. Infls (rare) in 1 marginal pouch, when 2(3) male & 1 female fls with small membranous spathe, or in upper pouch, when 1 male & 1 female with no spathe; males with P 0, A 1, anthers with longit. or transverse slits; females with G pseudomonomerous, 1-loc. with short term. style & 1–7 bitegmic ovules at first orthotropous but becoming hemitropous, campylotropous or anatropous after fert. Fr. a utricle with 1–4 seeds; embryo large, straight, sometimes radicle 0, endosperm 0 or starchy (etc.) & sheathing embryo. x = 5, 8, 10, 11, 21

Genera: *Lemna, Spirodela, Wolffia, Wolffiella, Wolffia* incl. the smallest angiosperms. L. derived from Araceae (neotenous), the free-floating *Pistia* with reduced spathe showing intermediate structure

Food for wildfowl & fish, some grown on dairy waste-water & fed to cattle (*Spirodela*), *Wolffia* used to test levels of herbicide in water

lemon *Citrus × limon*; **l. balm** *Melissa officinalis*; **l. basil** *Ocimum basilicum*; **l. grass** *Cymbopogon citratus*; **l. mint** *Mentha × piperita* 'Citrata'; **l.-scented gum** *Corymbia citriodora*; **l. verbena** *Aloysia citrodora*; **l. wood** *Aspidosperma tomentosum, Calycophyllum candidissimum, Pittosporum eugenioides, Xymalos monospora*

Lemoorea Short. Compositae (Gnap.-Ang.). 1 Aus.: *L. burkittii* (Benth.) Short. R: Muelleria 7(1989)112

Lemphoria O. Schulz = *Arabidella*

Lemurella Schltr. Orchidaceae (V 16). 4 Madagascar

Lemurodendron Villiers & Guinet. Leguminosae (II 3). 1 Madag.: *L. capuronii* Villiers & Guinet. R: BMNHN 4,11(1989)3

Lemuropisum Perrier (~ *Caesalpinia*). Leguminosae (I 1). 1 SW Madagascar

Lemurorchis Kraenzlin. Orchidaceae (V 16). 1 Madagascar

Lemurosicyos Keraudren. Cucurbitaceae. 1 Madagascar

Lemyrea (A. Chev.) A. Chev. & Beille (~ *Coffea*). Rubiaceae (? II 5). 3 Madagascar

Lenbrassia G. Gillett = *Fieldia*

Lencymmoea C. Presl. Myrtaceae (?). 1 Burma

Lennea Klotzsch. Leguminosae (III 7). 5 C Am.

Lennoa Llave & Lex. Lennoaceae. 1 C Mex., Colombia & Venez.: *L. madreporoides* Llave & Lex. R: SB 11(1986)539. Stamens of 2 lengths

Lennoaceae Solms-Laub. (~Boraginaceae). Dicots – Asteridae – Lamiales. 2/4 SW US to Venez. R: SB 11(1986)531. Chlorophyll-less, fleshy root-parasites with spirally arr. scale lvs. Fls bisexual, ± reg., (4)5–9(10)-merous, in pyramidal to discoid heads; K narrow, ± distinct, (C) with imbricate or induplicate-valvate lobes, A as many as & alt. with C-lobes, filaments attached above middle of C-tube, anthers with longit. slits, nectary-disk 0, G (5–16) with term. style & ± lobed stigma, each locule divided by a partition into 2 locelli, each with 1 anatropous, unitegmic ovule. Fr. a fleshy capsule, tardily irregularly circumscissile; seeds small with undifferentiated globose embryo embedded in copious starchy endosperm. x = 9

Genera: *Lennoa, Pholisma*

Floral, pollen & embryological characters indicate relationship with Hydrophyllaceae & Boraginaceae, the features of the gynoecium leading Cronquist to ally L. with the latter. Plants of deserts (mainly), the succ. underground stems of *Pholisma* form. much prized as food

Lenophyllum Rose. Crassulaceae. 5–6 SW N Am. Cult. orn. succ.

Lens Miller. Leguminosae (III 20). 4 Medit., W As., Afr. (1: *L. ervoides* (Brign.) Grande) incl. 2: cultigens: *L. nigricans* (M. Bieb.) Godron & *L. culinaris* Medikus (*L. esculenta, Ervum l.*, lentil, masur, cultigen with cleistogamous fls app. derived from *L. orientalis* (Boiss.) Schmalh. (*L. culinaris* ssp. *o.*, E Medit. to Iraq) – one of earliest crops cult. (form. with oats & barley, threshed together & separated by throwing – cereals go further) for seeds, split into cotyledons for sale, & used in soups (the 'mess of pottage' for which Esau traded his birthright) or as flour, foliage a valuable forage for animals. R: NRBGE 43(1986)491

Lent lily *Narcissus pseudonarcissus*

Lenten rose *Helleborus orientalis*

Lentibulariaceae Rich. Dicots – Asteridae – Scrophulariales. 3/245 cosmop. Insectivorous plants of wet places, sometimes rootless & free-floating, with stalked &/or sessile glands. Lvs simple, spirally arr. in rosettes in *Pinguicula* & *Genlisea*, which has tubular trap-lvs also; stipules 0; stem in *Utricularia* often with spirally arr. or whorled simple or dissected photosynthetic appendages bearing bladders with a trap-mechanism capturing small animals (see *Utricularia*). Fls bisexual, in a bracteate raceme (rarely 1-flowered) or solit. & term. without bracts (*Pinguicula*); K (4 or 5), lobed or ± 2-cleft, C (5), 2-labiate & ± 5-lobed, lobes imbricate, the lower lip basally spurred, A 2 (anterior) borne on C-tube, nectary-disk 0, G (2), 1-loc. with free-central placentation & ± sessile unequally 2-lobed stigma & 2–∞ anatropous, unitegmic ovules. Fr. usu. a capsule opening by 2–4 valves or irreg., or circumscissile (indehiscent & 1-seeded in some *Utricularia* (*Biovularia*) spp.); embryo scarcely differentiated, endosperm 0. x = 6, 8, 9, 11, 21

Genera: *Genlisea, Pinguicula, Utricularia*

App. derived from Scrophulariaceae but differing in placentation & carnivory, which is achieved variously: *Pinguicula* with glandular hairs & inrolling lvs, *Genlisea* with bottle-like pitchers with bands of hairs & digestive glands, *Utricularia* with bladder-traps

Some *Utricularia* spp. are weedy in ricefields; some cult. orn.

lentil *Lens culinaris*

lentisco, lentisk *Pistacia lentiscus*

Lenzia Philippi. Portulacaceae. 1 Chile

Leocereus Britton & Rose. Cactaceae (III 4). 1 E Brazil: *L. bahiensis* Britton & Rose. R: Bradleya 8(1990)107. Caffeine synthesized, cult. orn.

Leochilus Knowles & Westc. Orchidaceae (V 10). 10 trop. Am. R: SBM 14(1986). Cult. orn. epiphytes

Leocus A. Chev. Labiatae (VIII 4c). 1 trop. W Afr.

Leonardendron Aubrév. = *Anthonotha*

Leonardoxa Aubrév. Leguminosae (I 4). 3 Guineo-Congolian forests. R: Adansonia 8(1968)177. *L. africana* (Baillon) Aubrév. – foliar nectaries, swollen petioles excavated & occupied by ants (*Petalomyrex phylax*) which are app. restricted to tree & patrol young lvs

& protect from other herbivores necessary for survival of shoot to maturity, but *Catalaucus mckeyi* ants exlude them & do not patrol so parasitic on system

Leonia Ruíz & Pavón. Violaceae. 6 trop. S Am. *L. glycycarpa* Ruíz & Pavón used for bird-lime in Brazil

Leonohebe Heads = *Hebe*

Leonotis (Pers.) R. Br. Labiatae (VI). 15 trop. Afr., 1 extending to Am. & As. (*L. nepetifolia* (L.) R. Br., natur. in US – young infls cooked Uganda, infusion a lice-remover). *L. ocymifolia* (Burm.f.) Iwarsson (*L. leonurus*, S Afr.) – locally medic. & cult. orn.

Leonticaceae Bercht. & J. Presl. See Berberidaceae

Leontice L. Berberidaceae (II 1; Leonticaceae). 3 SE Eur. (1) to N Afr. & C As. Alks. Fr. a papery wind-dispersed bladder. Locally medic. & countering opium, *L. leontopetalum* L. an agric. weed

Leontochir Philippi. Alstroemeriaceae (Liliaceae s.l.). 1 Chile. Placentation parietal

Leontodon L. Compositae (Lact.-Hypo.). c. 50 temp. Euras. (Eur. 27) to Medit. & Iran. Often weedy in lawns (hawkbit) like *Taraxacum* spp.

Leontonyx Cass. = *Helichrysum*

Leontopodium (Pers.) R. Br. Compositae (Gnap.-Gnap.). 58 Eur. (1) to Burma & China, esp. mts. R: OB 104(1991)134. *L. alpinum* Cass. (edelweiss, Eur. mts), the familiar tufted woolly herb with small fl. heads crowded into dense cymes surrounded by consp. bract-like lvs, beloved of alpinists & cult. orn. as are a no. of other spp.; some dioec.

Leonuroides Rauschert = *Panzerina*

Leonurus L. Labiatae (VI). (Excl. *Chaiturus*) 3 temp. Euras. (Eur. 1). Alks. *L. cardiaca* L. (motherwort, Eur., Medit.) – form. locally imp. medic. & source of green dye, cult. orn.

leopard bane *Doronicum pardalianches*; **l. flower** *Belamcanda chinensis*; **l. plant** *Ligularia dentata*; **l. tree** *Caesalpinia ferrea, Flindersia collina*; **l. wood** *Brosimum aubletii, F. collina, F. maculosa*

Leopoldia Parl. = *Muscari*

Leopoldinia C. Martius. Palmae (V 4c). 4 Colombia, Venez., Brazil. Some fish-disp. in Amaz. *L. piassaba* Wallace (piassaba palm) – source of Pará piassava, a valuable fibre collected in the wild, used like *Musa textilis* in ropes, also for brushes & brooms

Lepanthes Sw. Orchidaceae (V 13). 460 trop. Am. Few cult. orn. usu. dwarf epiphytes

Lepanthopsis Ames. Orchidaceae (V 13). 39 trop. Am. R: MSB 39(1991)1

Lepargochloa Launert = *Loxodera*

Lepechinella Airy Shaw = *Lepechiniella*

Lepechinia Willd. Labiatae (VIII). (Incl. *Sphacele*) 55 warm Am., Hawaii (1)

Lepechiniella Popov. Boraginaceae. 9 C As.

Lepeostegeres Blume. Loranthaceae. 13 W Mal. to New Guinea. R: Blumea 38 (1993)175

Lepianthes Raf. = *Piper*. See also *Pothomorphe*

Lepidacanthus C. Presl = *Aphelandra*

Lepidagathis Willd. Acanthaceae. 100 trop. & warm. *L. longiflora* Wight (W Mal.) – common forest pl. never found in fr., lvs fall if stem shaken

Lepidaploa (Cass.) Cass. (~ *Vernonia*). Compositae (Vern.-Vern.). 116 trop. Am.

Lepidaria Tieghem. Loranthaceae. 12 W Mal.

Lepiderema Radlk. Sapindaceae. 8 N Aus. (6, R: Austrobaileya 1(1982)488), New Guinea (2, R: Blumea 36(1991)235)

Lepidesmia Klatt (~ *Ayapana*). Compositae (Eup.-Ayap.). 1 Cuba, Colombia, Venez. coastal: *L. squarrosa* Klatt. R: MSBM-BG 22(1987)213

Lepidium L. Cruciferae. (Incl. *Cardaria*) c. 140 cosmop. esp. temp. (Eur. 20, Aus. 35 endemic + 8 natur.). Pepperwort, peppercress. R: NDSNG 41(1906)1. Seeds slimy; alks. Some cult. for salads etc. esp. *L. sativum* L. (Egypt & W As., cress) – usu. eaten at cotyledon (characteristic 3-lobed) or seedling stage with mustard sown 4 days later; *L. draba* L. (*Cardaria d.*, hoary cress, Medit., Euras.) – pungent seeds form. a pepper subs., natur. GB & US spreading by root suckers (cf. *Armoracia*); *L. latifolium* L. (dittander, Eur., Medit., N Afr.) – cult. salad pl. of Ancient Greeks, form. medic. & veterinary (camels); *L. meyenii* Walp. (maca, Andes) – form. cult. salad veg.; *L. sisymbrioides* Hook.f. (S NZ) – dioec.

Lepidobolus Nees. Restionaceae. 3 S Aus.

Lepidobotryaceae Léonard (Oxalidaceae s.l.). Dicots – Rosidae – ? Sapindales. 2/2 trop. Afr. & Am. R: Novon 3(1993)408. Trees with unifoliolate, articulate lvs, lately removed from Oxalidaceae because of bitter (not sour or acidic) substances in lvs & bark, unclawed petals, disk, G 2 or 3 (rather than 5) septicidal or irreg. (not loculicidal) dehiscent capsules, collateral rather than superposed ovules, obturators and 0 endosperm

Genera: *Lepidobotrys, Ruptiliocarpon*

Similarities to Sapindaceae but also Euphorbiaceae

Lepidobotrys Engl. Lepidobotryaceae (Oxalidaceae s.l.). 1 trop. Afr.

Lepidocaryum C. Martius. Palmae (II 2). 9 trop. S Am. Dioec. *L. tessmannii* Burret – thatch in Amaz.

Lepidoceras Hook.f. Eremolepidaceae (Viscaceae s.l.). 2 Peru to Chile. R: SBM 18(1988)48

Lepidocordia Ducke. Boraginaceae (Ehretiaceae). 2 N trop. S Am. R: AJB 77(1990)548

Lepidogrammitis Ching = *Lemmaphyllum*

Lepidogyne Blume. Orchidaceae (III 3). 3 Mal.

Lepidolopha Winkler. Compositae (Anth.-Tan.). 6 C As.

Lepidolopsis Polj. Compositae (Anth.-Han.). 1 C As., Iran, Afghanistan

Lepidonia S.F. Blake (~ *Vernonia*). Compositae (Vern.-Vern.). 7 Mex. & C Am. R: BJ 108(1987)225

Lepidopetalum Blume. Sapindaceae. 6 Andamans & Nicobars, Sumatra, Philipp. to Bismarck Arch. R: Blumea 36(1992)439

Lepidopharynx Rusby = *Hippeastrum*

Lepidophorum Necker ex DC. Compositae (Anth.-Leuc.). 1 Portugal, Spain

Lepidophyllum Cass. Compositae (Ast.-Ast.). 1 Patagonia

Lepidorrhachis (H. Wendl. & Drude) Cook. Palmae (V 4m). 1 Lord Howe Is.

Lepidospartum (A. Gray) A. Gray. Compositae (Sen.-Tuss.). 3 SW US

Lepidosperma Labill. Cyperaceae. 60 Mal., Aus. (55, 53 endemic), New Caledonia, NZ. *L. gladiatum* Labill. (sword sedge, Aus.) – sand-binder & material for paper-making

Lepidostemon Hook.f. & Thomson. Cruciferae. 1 E Himalaya

Lepidostephium Oliver (~ *Athrixia*). Compositae (Gnap.). 2 S Afr. R: OB 104(1991)49

Lepidostoma Bremek. Rubiaceae (I 9). 1 Sumatra

Lepidothamnus Philippi (~ *Dacrydium*). Podocarpaceae. 2 NZ, 1 S Chile. R: AusJB 30(1982)316. *L. laxifolius* (Hook.f.) Quinn (mountain rimu, NZ mts) – planted to check erosion

Lepidotis Pal. ex Mirbel = *Lycopodium*

Lepidotrichilia (Harms) J. Leroy. Meliaceae (I 4). 4 Madag. (3), E Afr. (1)

Lepidotrichum Velen. & Bornm. = *Aurinia*

Lepidozamia Regel. Zamiaceae (I). 2 NE Aus. R: MNYBG 57(1990)204. *L. hopei* Regel (NE Queensland) – at 20 m the tallest cycad

Lepilaena J. L. Drumm. ex Harvey (~ *Althenia*). Zannichelliaceae. 6 Aus. (4) to S Pacific is. Pollen in rafts reaching stigma, cf. *Halodule*, *Halophila*, *Ruppia*

Lepinia Decne. Apocynaceae. 3 Trobriand, Caroline & Solomon Is., Tahiti. R: BM(T) 48(1934)528

Lepiniopsis Valeton. Apocynaceae. 2 C & E Mal. (1), Micronesia (1)

Lepionurus Blume. Opiliaceae. 1 Indomal.

Lepironia Rich. Cyperaceae. 5 Madag. to Polynesia. *L. articulata* (Retz.) Domin (*L. mucronata*, Madag. to Polynesia) – cult. esp. China for fibre used in junk sails & mats for packing tobacco, rubber, kapok etc.

Lepisanthes Blume. Sapindaceae. 24 OW trop. R: Blumea 17(1969)33. Some cult. orn. trees, some with ed. fr. grown as *Aphania* or *Erioglossum* spp. e.g. *L. fruticosa* (Roxb.) Leenh. (luna nut, Mal.)

Lepismium Pfeiffer (~ *Rhipsalis*). Cactaceae (III 6). 16 E Bolivia & Arg., a few ext. to Brazil. R: Bradleya 5(1987)99. Cult. orn. like *Rhipsalis* but without char. branching pattern & usu. spiny

Lepisorus (J. Sm.) Ching (~ *Pleopeltis*). Polypodiaceae (II 3). (Incl. *Paragramma*) c. 40 trop. & warm OW. Some cult. orn.

Lepistemon Blume. Convolvulaceae. 10 OW trop.

Lepistemonopsis Dammer. Convolvulaceae. 1 trop. E Afr.

Leporella A.S. George. Orchidaceae (IV 1). 1 SW & S Aus.: *L. fimbriata* (Lindley) A.S. George – pheromones attract male flying ants (*Myrmecia urens*) for pseudocopulation poll.

leprosy gourd *Momordica charantia*

Leptacanthus Nees = *Strobilanthes*

Leptactina Hook.f. Rubiaceae (II 2). 25 trop. & S Afr. R: PSE 145(1984)105

Leptadenia R. Br. Asclepiadaceae (III 5). 4 OW trop. *L. pyrotechnica* (Forssk.) Decne (Afr. to India) – fr. & twigs eaten by Bedouin

Leptagrostis C. Hubb. Gramineae (25). 1 Ethiopia

Leptaleum DC. Cruciferae. 1 E Med. incl. Eur. to C As. & Baluchistan

Leptaloe Stapf = *Aloe*

Leptarrhena R. Br. Saxifragaceae. 1 W N Am.: *L. amplexifolia* (Sternb.) Ser. (*L. pyrifolia*) – cult.

orn. rhiz. herb

Leptaspis R. Br. Gramineae (6). 5 OW trop. Panicle tough or shed intact

Leptaulus Benth. Icacinaceae. 6 trop. Afr., Madagascar

Lepterica N.E. Br. = *Scyphogyne*

Leptinella Cass. (~ *Cotula*). Compositae (Anth.-Mat.). 33 New Guinea, Aus., NZ, subAntarctic is., S Am. R: NZJB 25(1987)99. Cult. orn. carpeting pls esp. *L. squalida* Hook.f. (*C. squalida*, NZ)

Leptobaea Benth. Gesneriaceae. 2 E Himalaya, 1 Borneo

Leptocallisia (Benth.) Pichon = *Callisia*

Leptocanna Chia & Fung = *Schizostachyum*

Leptocarpha DC. Compositae (Helia.-Verb.). 1 Chile

Leptocarpus R. Br. (excl. *Calopsis*). Restionaceae. 16 Indomal. (3), Aus.–NZ (12), Chile (1). *L. chilensis* (Gay) Masters (Chile) – mat-manufacturing

Leptocarydion Hochst. ex Stapf. Gramineae (31d). 1 E & S Afr. Links *Leptochloa* & *Trichoneura*

Leptoceras Fitzg. = *Leporella*

Leptocereus (A. Berger) Britton & Rose. Cactaceae (III 1). 13 Cuba, Hispaniola, Puerto Rico

Leptochilus Kaulf. Polypodiaceae (II 4). 3 Indomal.

Leptochiton Sealy. Amaryllidaceae (Liliaceae s.l.). 2–3 Andes

Leptochloa Pal. Gramineae (31d). (Incl. *Diplachne*) 27 trop. & warm Am. & Aus. Some fodders. *L. scabra* Nees (Brazil) – bad weed of rice

Leptochloopsis Yates = *Uniola*

Leptocionium C. Presl = *Hymenophyllum*

Leptoclinium (Nutt.) Benth. Compositae (Eup.-Alom.). 1 Brazil. R: MSBMBG 22(1987)258

Leptocodon (Hook.f.) Lemaire (~ *Codonopsis*). Campanulaceae. 1 Himal. Pedicels concrescent with infl. axis

Leptocoryphium Nees = *Anthaenantia*

Leptodactylon Hook. & Arn. Polemoniaceae. 12 W N Am. Some cult. orn. shrubs

Leptodermis Wallich. Rubiaceae (IV 12). 30 Himalaya to Japan

Leptoderris Dunn. Leguminosae (III 6). 20 trop. Afr.

Leptodesmia (Benth.) Benth. Leguminosae (III 9). 5 Madagascar, 1 India

Leptofeddea Diels = *Leptoglossis*

Leptoglossis Benth. Solanaceae (2). 7 Peru, Argentina. Xerophytes

Leptoglottis DC = *Schrankia*

Leptogonum Benth. Polygonaceae (II 1). 1 Hispaniola: *L. domingense* Benth. R: NJB 10(1990)487. Small tree

Leptogramma J. Sm. = *Stegnogramma*

Leptolaena Thouars. Sarcolaenaceae. (Incl. *Xerochlamys*) 16 Madagascar

Leptolepia Mett. ex Diels (~ *Microlepis*). Dennstaedtiaceae. 1 NZ (New Guinea, Aus.)

Leptolepidium Hsing & S.K. Wu = *Cheilanthes*

Leptoloma Chase = *Digitaria*

Leptomeria R. Br. Santalaceae. 17 Aus. esp. SW. *L. acida* R. Br. (26 mg vitamin C per 100 g fr.) & other spp. (currant bushes) with fr. used in preserves

Leptomischus Drake. Rubiaceae (IV 1). 1 SE As.

Leptonema A. Juss. Euphorbiaceae. 2 Madagascar

Leptonychia Turcz. Sterculiaceae (?). 45 OW trop. (SE As. 3, R: Blumea 32(1987)443). Seeds very different from rest of S.

Leptonychiopsis Ridley = *Leptonychia*

Leptopharyngia (Stapf) Boit. = *Tabernaemontana*

Leptopharynx Rydb. = *Perityle*

Leptophoenix Becc. = *Gronophyllum*

Leptophyllochloa Caldéron = *Koeleria*

Leptoplax O. Schulz = *Peltaria*

Leptopteris C. Presl. Osmundaceae. 6 New Guinea, Aus., New Caled., Polynesia, NZ. Cult. orn. esp. *L. superba* (Colenso) C. Presl (NZ) with plume-like fronds to 1.2 m. *L. wilkesiana* (Brack.) Christ (W Pacific) – tree fern to 133 yrs old on Fiji

Leptopus Decne. Euphorbiaceae. 20 W Himalaya to Aus.

Leptopyrum Reichb. Ranunculaceae (III 1). 1 W Siberia to E As.: *L. fumarioides* (L.) Reichb. – cult. orn. annual, natur. Eur.

Leptorhabdos Schrenk. Scrophulariaceae. 1 Cauc. & Iran to C As. & Himalaya

Leptorhynchos Less. Compositae (Gnap.-Ang.). 10 temp. Aus. R: OB 104(1991)120

Leptorumohra (H. Itô) H. Itô = ? *Dryopteris*

Leptosaccharum (Hackel) A. Camus = *Eriochrysis*

Leptoscela Hook.f. Rubiaceae (IV 1). 1 E Brazil

Leptosema Benth. (~ *Brachysema*). Leguminosae (III 24). 8 Aus.

Leptosiphon Benth. = *Linanthus*

Leptosiphonium F. Muell. (~ *Ruellia*). Acanthaceae. 10 Papuasia

Leptosolena C. Presl (~ *Alpinia*). Zingiberaceae. 1 Philippines

Leptospermopsis S. Moore = *Leptospermum*

Leptospermum Forster & Forster f. Myrtaceae (Lept.). 79 SE As. to NZ (1, Aus. 77 (75 endemic)). R: Telopea 3(1989)301. Evergreen shrubs (some rheophytes) & trees, some beetle-poll.; cult. orn. & lvs used for tea (tea trees, Aus.). *L. petersonii* Bailey (*L. citratum*, E Aus.) – source of lemon-scented oil & comm. grown in Kenya & Guatemala; *L. scoparium* Forster & Forster f. (manuka, E Aus., NZ) – most commonly cult. sp. (natur. in Scilly Is.), timber & tea

Leptostachya Nees (~ *Justicia*). Acanthaceae. 1 Indomal.: *L. wallichii* Nees. R: NJB 5(1986)469

Leptostigma Arn. (~ *Nertera*). Rubiaceae (IV 13). (Incl. *Corynula*) 6: SE As. 1, NZ 1, W S Am. 4. R: APG 33(1982)73

Leptostylis Benth. Sapotaceae (IV). 8 New Caledonia. R: FNC 1(1967)20

Leptotaenia Nutt. ex Torrey & A. Gray = *Lomatium*

Leptoterantha Louis ex Troupin. Menispermaceae (III). 1 trop. Afr.: *L. mayumbense* (Exell) Troupin

Leptotes Lindley. Orchidaceae (V 13). 3 trop. S Am. Cult. orn. like *Cattleya* with lvs terete or subterete; fr. of *L. bicolor* Lindley (Brazil) locally used to flavour icecream

Leptothrium Kunth. Gramineae (33d). 2 Caribbean & Senegal to Pakistan

Leptothyrsa Hook.f. Rutaceae. 1 Amazon: *L. sprucei* Hook.f. R: Candollea 45(1990)376

Leptotriche Turcz. (*Gnaphosis* auctt.). Compositae (Gnap.). 12 Aus. R: OB 104(1991)128

Leptunis Steven = *Asperula*

Lepturella Stapf = *Oropetium*

Lepturidium A. Hitchc. & Ekman (~ *Brachyachne*). Gramineae (33b). 1 Cuba: *L. insulare* A. Hitchc. & Ekman – salt flats

Lepturopetium Morat. Gramineae (33b). 1 New Caled., Marshall Is., Cocos Is. ? *Chloris* × *Lepturus* spp.

Lepturus R. Br. Gramineae (32). 8 coasts of E Afr. to Polynesia & Aus.

Lepuropetalaceae (Engl.) Nakai. See Saxifragaceae

Lepuropetalon Elliott. Saxifragaceae (Lepuropetalaceae). 1 SE US, Mex., C Chile, Uruguay: *L. spathulatum* (Muhlenb.) Elliott. Winter annual c. 2 cm tall, one of the smallest terr. herbs, growing on powerlines & in cemeteries etc.

Lepyrodia R. Br. Restionaceae. 19 Aus. For NZ sp. see *Sporadanthus*

Lepyrodiclis Fenzl. Caryophyllaceae (II 1). 3 W As. R: ANMW 61(1937)74

Lerchea L. Rubiaceae (IV 1). 8 Sumatra, Java. R: Blumea 32(1987)91

Lereschia Boiss. (~ *Cryptotaenia*). Umbelliferae (III 8). 1 S Eur.

Leretia Vell. Conc. = *Mappia*

Leroya Cavaco = *Pyrostria*

Leroyia Cavaco = *Pyrostria*

lerp manna from *Eucalyptus dumosa*

Lescaillea Griseb. Compositae (Hele.-Pect.). 1 W Cuba (serpentine). R: Madrono 24(1977)138. Liane

Leschenaultia R. Br. (*Lechenaultia*). Goodeniaceae. 26 Aus. (SW 20 endemics), 1 ext. to New Guinea. R: FA 35(1992)17. Some cult. orn.

Lesliea Seidenf. Orchidaceae (V 16). 1 Thailand: *L. mirabilis* Seidenf. R: OB 95(1988)190

Lespedeza Michaux. Leguminosae (III 9). 40 temp. N & S Am., trop. & E As., Aus. R (sect. *Macrolespedeza*): S. Akiyama (1988) *Rev. of the genus L. sect. M.* C sometimes 0 & fls cleistogamous; herbs & shrublets with v. small 1-seeded indehiscent legumes cult. for forage & as green manures; alks. *L. bicolor* Turcz. (E As.) & *L. thunbergii* (DC) Nakai (? hybrid cultigen, E As.) – cover crops & fuel

Lesquerella S. Watson (*Discovium*, ~ *Alyssum*). Cruciferae. 40 N Am. Bladder pod. R (partial): R.C. Rollins & E.A. Shaw (1973) *The genus L. in N. America*. *L. fendleri* (A. Gray) S. Watson (S W N Am.) – cult. instead of castor oil for hydroxy fatty-acids

Lesquereuxia Boiss. & Reuter. Scrophulariaceae. 1 E Med.: *L. syriaca* Boiss. & Reuter – root-parasite on *Hedera helix*, *Castanea sativa* etc.

Lessertia DC. Leguminosae (III 15). 50 S (few trop.) Afr.

Lessingia Cham. Compositae (Ast.-Sol.). (Incl. *Corethrogyne*) 14 SW N Am. R: UCPB 16 (1929) 1. Some cult. orn.

Lessingianthus H. Robinson (~ *Vernonia*). Compositae (Vern.-Vern.). c. 100 S Am. esp. Brazil

Letestua Lecomte. Sapotaceae (I). 1 W trop. Afr.: *L. durissima* (A. Chev.) Lecomte

Letestudoxa Pellegrin. Annonaceae. 2 W trop. Afr. Lianes

Letestuella G. Taylor. Podostemaceae. 1 W trop. Afr.

Lethedon Sprengel. Thymelaeaceae. 10 New Caled., (?1) Vanuatu, 1 NE Aus.

Lethia Ravenna. Iridaceae. 1 (?) Brazil: *L. umbellata* (Klatt) Ravenna. R: NJB 6(1986)585

letterwood *Brosimum guianense*

Lettowianthus Diels. Annonaceae. 1 trop. E Afr.: *L. stellatus* Diels

lettuce, asparagus l. *Lactuca sativa*; **lamb's l.** *Valerianella locusta*; **opium l.** *L. virosa*; **prickly l.** *L. serriola*; **stem lettuce** *L. sativa*; **l. tree** *Pisonia grandis*; **wall l.** *Mycelis muralis*; **water l.** *Pistia stratiotes*

Leucactinia Rydb. Compositae (Hele.-Pect.). 1 Mexico

Leucadendron R. Br. Proteaceae. 79 S Afr. R: CBH 3(1972). When fr. matures, P splits into 4 parts united round stigma & acting as wing in dispersal. *L. argenteum* (L.) R. Br. has lvs covered with soft silky hairs – much exploited for bookmarks, mats, sometimes painted

Leucaena Benth. Leguminosae (II 3). 22 Texas to Peru. Some polyploids; at least 13 spp. & hybrids with ed. fr. (unripe), seeds (also fl. buds & galls, esp. Mex.) in local markets. Livestock fodder, green manure, firewood crops, soil conservation; spp. grown beyond natural ranges & hybridizing with indigenous ones (Hughes). *L. esculenta* (Moçiño & Sessé) Benth. (guaje, Mex.) – seeds eaten with salt but in others toxic mimosine leads to loss of hair, infertility etc.; *L. leucocephala* (Lam.) De Wit (*L. glauca*, *L. latisiliqua*, lead tree, allopolyploid (?) arising in Mex., natur. in OW) – firewood crop, green manure, pulp & forage (toxic to non-ruminants causing hair loss, e.g. horses' manes & tails) in trop., seeds (flat, brown) used as beads (jumbie beans), now mooted as charcoal source & for energy (1 million barrels of oil per annum from 12 000 ha), anthelminthic in Sumatra

Leucampyx A. Gray = *Hymenopappus*

Leucanthemella Tzvelev (~ *Tanacetum*). Compositae (Anth.-Leuc.). 2 SE Eur. (1: *L. serotina* (L.) Tzvelev – cult. orn. with white or red ray-florets), E As.

Leucanthemopsis (Giroux) Heyw. (~ *Tanacetum*). Compositae (Anth.-Leuc.). 9 S Eur., Morocco (1). R: AIBC 32,2(1975)175

Leucanthemum Miller (~ *Chrysanthemum*). Compositae (Anth.-Leuc.). 33 Eur. (12) esp. mts, N As. Cult. orn. esp. *L.* × *superbum* (J. Ingram) Kent (Shasta daisy, *L. maximum* (Ramond) DC (Pyrenees) × *L. lacustre* (Brot.) Samp. (SW Eur.) though poss. involving *Nipponanthemum nipponicum* (Maxim.) Kitam.) esp. 'double-flowered' 'Esther Read' & *L. vulgare* Lam. (*C. l.*, moon, dog or ox-eye daisy, marguerite, Euras., natur. N Am.) – home remedy for catarrh, ed. (just about)

Leucas R. Br. Labiatae (VI). 150 Afr. & Arabia to Indomal. R (Afr. Arabia): SBNA 341(1980)1. Some locally medic.

Leucaster Choisy. Nyctaginaceae (II). 1 SE Brazil: *L. caniflorus* (C. Martius) Choisy

Leucelene E. Greene = *Chaetopappa*

Leucheria Lagasca. Compositae (Mut.-Nass.). 46 Andes, Chile, Patagonia, Falkland Is. R: Darwiniana 20(1976)9. Herbs

Leuchtenbergia Hook. (~ *Ferocactus*). Cactaceae (III 9). 1 N Mex.: *L. principis* Hook. – cult. orn. with v. long tubercles

Leucobarleria Lindau = *Neuracanthus*

Leucoblepharis Arn. (~ *Blepharispermum*). Compositae (Hele.). 1 India: *L. subsessilis* (DC) Arn. R: BJ 112(1990)183

Leucocalantha Barb. Rodr. Bignoniaceae. 1 Amazonian Brazil

Leucocarpus D. Don. Scrophulariaceae. 1 trop. Am.: *L. perfoliatus* (Kunth) Benth. – cult. orn. like *Mimulus*

Leucochrysum (DC) P.G. Wilson. Compositae (Gnap.-Ang.). 5 S Aus. R: Nuytsia 8(1992)439

Leucocodon Gardner. Rubiaceae (I 8). 1 Sri Lanka

Leucocorema Ridley = *Trichadenia*

Leucocoryne Lindley. Alliaceae (Liliaceae s.l.). 12 Chile. R: AMHNV 5(1972)9. Bulbous pls cult. orn. like *Ixia* esp. *L. ixioides* (Hook.) Lindley (glory of the sun)

Leucocrinum Nutt. ex A. Gray. Agavaceae (Hostaceae). 1 SW US: *L. montanum* Nutt. ex A. Gray – cult. orn. with fragrant white fls

Leucocroton Griseb. Euphorbiaceae. 27 Cuba, Hispaniola

Leucocyclus Boiss. (~ *Anacyclus*). Compositae (Anth.-Art.). 1 Turkey

Leucogenes Beauverd (~ *Leontopodium*). Compositae (Gnap.-Gnap.). 2 NZ. R: OB 104(1991)134. Hybrids with *Raoulia* spp. in wild. Cult. orn. silvery tomentose herbs (NZ edelweiss) resembling *Leonotopodium*

Leucoglossum Wilcox et al. = *Mauranthemum*

Leucohyle Klotzsch. Orchidaceae (V 10). 3 trop. Am.

Leucojum L. Amaryllidaceae (Liliaceae s.l.). 10 Eur. (8) to Morocco & Iran. Snowflakes. R: F.C. Stern (1956) *Snowdrops & snowflakes*, Plantsman 14(1992)72. Alks. x = 7–9, 11. App. derived from west of a complex forced south in Pleistocene, the eastern part becoming *Galanthus* (x = 12), differing in having inner P segments shorter than outer. Cult. orn. esp. *L. aestivum* L. (summer s., Eur. to Iran, natur. in valleys of Shannon & Thames) – seeds with outer layer (? aril) of air-filled tissue allowing water-disp.; *L. autumnale* L. (autumn s., Medit.); *L. vernum* L. (spring s., C Eur., natur. GB) – seeds with caruncle (? aril) attractive to ants, allegedly disp. agents

Leucolophus Bremek. = *Urophyllum*

Leucomphalos Benth. ex Planchon. Leguminosae (III 2). 1 W Afr.

Leuconotis Jack. Apocynaceae. 10 Mal. Trees, poss. rubber sources

Leucopholis Gardner = *Chionolaena*

Leucophrys Rendle = *Urochloa*

Leucophyllum Bonpl. Scrophulariaceae. 12 SW N Am. R: Sida 11(1985)107. *L. frutescens* (Berland.) I.M. Johnson (ashplant, Texas & Mex.) – grown as low hedgepl. or as a lawn locally

Leucophysalis Rydb. Solanaceae (1). 9 N Am. (2), Bhutan to E As. (9)

Leucophyta R. Br. (~ *Calocephalus*). Compositae (Gnap.-Ang.). 1 Aus.: *L. brownii* Cass. (*Calocephalus b.*, cushion bush) – cult. orn. R: OB 104(1991)125

Leucopoa Griseb. = *Festuca*

Leucopogon R. Br. (~ *Styphelia*). Epacridaceae. 120 Mal., Aus. esp. SW. Fls term. or in upper axils; other (c. 70) spp. referred to diff. genera. Some with ed. fr. ('Aus. currant') disp. by birds, *L. malayanus* Jack (Mal.) – inner bark waterproofs canoes, lvs & roots medic.

Leucopsis (DC) Baker = *Noticastrum*

Leucoptera R. Nordenstam. Compositae (Anth.-Mat.). 3 SW Cape to Namaqualand

Leucorchis E. Meyer = *Pseudorchis*

Leucosalpa Scott-Elliott. Scrophulariaceae. 3 Madagascar

Leucosceptrum Sm. Labiatae (VII). 3 Himalaya to China

Leucosidea Ecklon & Zeyher. Rosaceae. 1 S Afr.

Leucospermum R. Br. Proteaceae. 46 SW Cape to Zimbabwe (excl. *Vexatorella*). R: JSAB supp. 8(1972). *L. conocarpodendron* (L.) Buek (Cape) – ant-disp. (elaiosomes). Some cult. orn. shrubs

Leucosphaera Gilg. Amaranthaceae (I 2). 1 S trop. Afr.

Leucospora Nutt. = *Stemodia*

Leucostegane Prain. Leguminosae (I 4). 2 Malay Peninsula, Borneo

Leucostegia C. Presl. Davalliaceae. 2 C Himal. & S China, S India to E Polynesia. R: Blumea 37(1992)184. Terr. & epiphytic cult. orn.

Leucosyke Zoll. & Moritzi. Urticaceae (III). 35 Mal. to Polynesia. R: BJ 73(1943)191. *L. capitellata* (Poiret) Wedd. (Taiwan to Java & New Guinea) – fibre for ropes, medic. locally

Leucothoe D. Don. Ericaceae. (Excl. *Agarista*) 8 Himal., Japan, N Am. Cult. orn. shrubs esp. *L. fontanesiana* (Steudel) Sleumer ('*L. catesbaei*', SE US) – locally med.

Leunisia Philippi. Compositae (Mut.-Nass.). 1 C Chile. Shrub

Leurocline S. Moore = *Echiochilon*

Leutea Pim. Umbelliferae (III 10). 6 SW & C As.

Leuzea DC (~ *Centaurea*). Compositae (Card.-Cent.). 3 Medit. (excl. *Stemmacantha*)

Levant galls *Quercus pubescens*; **L. garlic** *Allium ampeloprasum*; **L. madder** *Rubia peregrina*; **L. scammony** *Convolvulus scammonia*; **L. storax** *Liquidambar orientalis*

Levenhookia R. Br. Stylidiaceae. 8 S Aus. Shoe-shaped labellum embracing column springs downwards if touched

lever wood *Ostrya virginiana*

Levieria Becc. Monimiaceae (V 1). 7 E Mal., N Aus. R: Blumea 26(1980)373

Levisticum Hill. Umbelliferae (III 8). 1 E Medit.: *L. officinale* Koch (lovage, bladderseed, natur. Eur. & N Am.) – cult. since time of Pliny, lvs may be blanched like celery, fr. & root used as flavouring in liqueurs etc.

Levya Bureau ex Baillon = *Cydista*

Lewisia Pursh. Portulacaceae. 22 W N Am. Thick starchy roots of some spp. ed. esp. *L. rediviva* Pursh (bitter root) – cult. orn. rock-pl. which can be grown after 2 yrs drought or even boiling. R: BAGS 34(1966)

Leycephyllum Piper = *Rhynchosia*

Leycesteria Wallich. Caprifoliaceae. 6 W Himal. to SW China. R: KB 1932:161. *L. formosa*

Wallich (W China, Himalaya & E Tibet) – cult. orn. thicket-forming shrub with verticillate simple lvs & consp. bracts, natur. Eur. incl. GB, fr. attractive to gamebirds

leylandii × *Cupressocyparis leylandii*

Leymus Hochst. (~ *Elymus*). Gramineae (23). c. 40 N temp. (Eur. 7), 1 Arg. R (N Am.): AJB 71(1984)609. Steppes, often saline, alkaline or dunes; some useful sand-binders esp. *L. arenarius* (L.) Hochst. (lyme grass, Euras.) – cult. in Japan for ropes, mats & paper

Leysera L. (~ *Astropteris*). Compositae (Gnap.-Rel.). 3 S Afr. with 1 Medit. to SW As. R: BN 131(1978)369

Lhotskya Schauer = *Calytrix*

Lhotzkyella Rauschert = *Matelea*

Liabellum Rydb. = *Sinclairea*

Liabum Adans. Compositae (Liab.). 38 C Am., WI, Andes. R: SCB 54(1983)31

Liatris Gaertner ex Schreber. Compositae (Eup.-Gyp.). 43 E N Am. R: Rhodora 48(1946)165, 216, 273, 331, 393. Cult. orn. perennial herbs (gayfeather) esp. *L. spicata* (L.) Willd. (Kansas gayfeather, E N Am.), locally medic.

Libanothamnus Ernst. See *Espeletia*

Libanotis Haller ex Zinn = *Seseli*

Liberatia Rizz. = *Lophostachys*

Liberbaileya Furt. = *Maxburretia*

Libertia Sprengel. Iridaceae (III 1). 9 Aus. (2), New Guinea, NZ, Andes. Cult. orn. herbs with fibrous roots & white or blue fls

Libocedrus Endl. Cupressaceae. (Excl. *Austrocedrus, Papuacedrus*) 5 New Caled. (3), NZ (2). R: Phytol.M 7(1984)40. Timber esp. *L. bidwillii* Hook.f. (pahautea, NZ) & *L. plumosa* (D. Don) Sarg. (*L. doniana*, kawaka, NZ)

Librevillea Hoyle. Leguminosae (I 4). 1 Gabon

Libyella Pampan. Gramineae (17). 1 N Afr.: *L. cyrenaica* (Durand & Baratte) Pampan. – annual 2–5 cm tall

Licania Aublet. Chrysobalanaceae (1). 193 trop. Am., Afr. (1), As. (2–3). R: FN 9(1972)21. PTRSB 320(1988)86. Some fish-disp. Amaz.; *L. michauxii* Prance (SE US) – geoxylic suffrutex with spreading underground 'trunk' to 30 m & aboveground shoots to 30 cm. Timbers, oil (oiticica o.), esp. for illumination, & ed. fr., used for pots (silica grains). *L. arborea* Seemann (Am.) – oil; *L. cecidophora* Prance (NC Peru) – leaf galls used in necklaces & capes (only insect galls used for ornament!); *L. elaeosperma* (Mildbr.) Prance & F. White (*Afrolicania e.*, poyok, W Afr.) – seeds yield a drying oil form. subs for tung oil (paints etc.); *L. octandra* (Roemer & Schultes) Kuntze (*L. utilis*, pottery tree of Pará, NE S Am.) – powdered bark mixed with clay makes heat-resistant pots; *L. tomentosa* (Benth.) Fritsch (Am.) – cult. for fr.

Licaria Aublet. Lauraceae (I). 40 trop. Am. Some oilseeds & local medic. esp. *L. pucheri* (Ruíz & Pavón) Kosterm. (puchurin nut)

Lichtensteinia Cham. & Schldl. Umbelliferae (III 8). 7 S Afr., 1 St Helena. R: EJB 48(1991)221

licorice See liquorice

Licuala Thunb. Palmae (I 1b). 108 SE As. to Vanuatu & Aus. Palms with bisexual fls & palmate lvs much grown in trop. as cult. orn.; lvs used to improve opium burning; stems for walking sticks ('Penang lawyers', i.e. loyak or loyar (Malay names for the palms))

Lidbeckia Berg. Compositae (Anth.-Tham.). 2 S Afr.

Liebigia Endl. = *Chirita*

Lietzia Regel. Gesneriaceae. 1 Brazil: *L. brasiliensis* Regel & Schmidt – bat-poll.

Lifago Schweinf. & Muschler. Compositae (Gnap.). 1 Morocco, Algeria

Ligaria Tieghem (~ *Loranthus*). Loranthaceae. 2+ C Brazil. R: Brittonia 42(1990)66

Ligeophila Garay. Orchidaceae (III 3). 8 trop. Am.

Lightfootia L'Hérit. = *Wahlenbergia*

Lightia Schomb. = *Euphronia*

Lightiodendron Rauschert = *Euphronia*

lightwood *Acacia implexa, Ceratopetalum apetalum*

Ligia Fasano = *Thymelaea*

lign-aloes *Aquilaria malaccensis*

Lignariella Baehni. Cruciferae. 2 Tibet, Himalaya. R: JETB 3(1982)974

Lignocarpa Dawson. Umbelliferae (III 8). 2 NZ. Gynodioec.

lignum *Muehlenbeckia* spp. esp. *M. florulenta*

lignum-nephriticum *Eysenhardtia polystachya*

lignum-vitae *Guaiacum* spp.; **Maracaibo l.** *Bulnesia arborea*; **Paraguay l.** *B. sarmentoi*; **Queensland l.** *Premna lignum-vitae*

Ligularia Cass. Compositae (Sen.-Tuss.). (Excl. *Cremanthodium*) 125 temp. Euras. (China most). Herb. perennial cult. orn. esp. *L. dentata* (A. Gray) H. Hara (*L. clivorum*, leopard pl., China & Japan), its cvs & hybrids. See also *Farfugium*

Ligusticella J. Coulter & Rose. = *Podistera*

Ligusticopsis Leute. Umbelliferae (III 8). 14 temp. E As.

Ligusticum L. Umbelliferae (III 8). 40-50 circumboreal (Eur. 7). ? Heterogeneous. Some ed. & medic. *L. canbyi* J. Coulter & Rose (N Am.) – still used by Flathead Indians for colds & sore throats; *L. porteri* J. Coulter & Rose (chuchupate, Mex.) – root tea medic.; *L. scoticum* L. (E N Am., Eur.) – potherb like celery

Ligustrina Rupr. = *Syringa*

Ligustrum L. Oleaceae. c. 40 Eur. & N Afr. (1), E & SE As. to Aus. Privets. R: BJ Beib. 132 (1924)19. Decid. & evergreen shrubs & small trees with a heavy scent, offensive to many due to undertone of ammonia because of presence of trimethylamine (esp. after rain as in *Cotoneaster* & *Sorbus*, Rosaceae) giving overall fishy smell not found in related *Syringa* & *Jasminum* spp. but tainting honey of bees feeding on *L.* nectar; extrafl. nectaries; weedy when introd. to NZ. Commonly planted as hedges, form. *L. vulgare* L. (common or European p., Eur., Medit., natur. N Am.) since at least 1548, but chiefly now *L. ovalifolium* Hassk. (Japan), though the 'golden' p. is app. cv. Aureum of *L. vulgare*, all tolerant of city pollution; others incl. *L. japonicum* Thunb. (Japanese p., Korea, Japan) – evergreen & *L. lucidum* Aiton f. (glossy p., China, Korea) – a common street tree in S Eur., comm. insect wax coll. in China; *L. sinense* Lour. (Chinese p., China); *L. vulgare* fr. toxic (some child fatalities recorded), form. source of dyes. *L. indicum* (Lour.) Merr. (*L. wallichii*, Himal. to SE As.) – with honey, extract a popular cold cure in modern Chinese herbalism

Lijndenia Zoll. & Moritzi (~ *Memecylon*). Melastomataceae. 12 OW trop. (Madag. 8, Indomal. 3, Afr. 2. R: NJB 2(1982)121

lilac *Syringa* spp. esp. *S. vulgaris*; **Calif. l.** *Ceanothus* spp.; **Cape l.** (Aus.) *Melia azedarach*; **Himalaya l.** *S. emodi*; **Indian** or **Persian l.** *M. azedarach*; **Rouen l.** *S.* × *persica*

Lilaea Bonpl. Juncaginaceae (Lilaeaceae). 1 Rocky Mts, Mex., Andes: *L. scilloides* (Poiret) Hauman (*L. subulata*) – lvs for thatch & brooms in S Am. (natur. Portugal & Aus.)

Lilaeaceae Dumort. See Juncaginaceae

Lilaeopsis E. Greene. Umbelliferae (III 8). 13 Am., Aus., NZ, 1 natur. Portugal. R: SBM 6(1985)

Liliaceae Juss. Monocots – Liliidae – Liliales. S.l.: 288/4950 cosmop. esp. dryish temp. & warm. Perennial often geophytic herbs, usu. with alks, & often with starchy rhizome, bulb or corm, vessels usu. only in roots, which usu. have no root-hairs. Lvs simple, usu. annual, spirally arr., rarely opp. or whorled, usu. narrow & parallel-veined. Fls usu. bisexual & reg., insect-poll. & (2)3(4)-merous, solit. or in spikes, racemes, panicles or umbels; P usu. in 2 petaloid whorls (outer narrower in e.g. *Calochortus* & *Trillium*), s.t. with basal tube, s.t. a corona (app. s.t. derived from P, s.t. A), nectaries septal or at base of P or A, or 0, A as many as P (rarely 3 or up to 12), s.t. connate by filaments or on P-tube, anthers with longit. slits (rarely term. pores), $\underline{G}$ or $\overline{G}$ ((2)3(4)) with axile or deeply intruded parietal placentation, with as many stigmas as carpels (or 1) on 1(3) styles & (1–)∞ anatropous to campylotropous or rarely orthotropous (uni-)bi-tegmic ovules. Fr. usu. a capsule, rarely a berry or like a nut; seeds with arils, sarcotestas, wings etc., embryo ± linear embedded in copious endosperm with reserves of protein, oil & hemicellulose. x = 3–27+

Classification: 'The relationship between subdivisions (subfamilies or tribes) and between them and other families are not always clear and there is a wide range of opinion on how to classify the genera of this and related families' (Jessop); nonetheless, the botanical & now hort. (e.g. Plantsman 11(1989)89) literature has taken to the splitting. In this book, readers can use either system; for those wishing to keep a narrowly circumscribed Liliaceae, see e.g. Alliaceae ($\underline{G}$, umbels), Alstroemeriaceae ($\overline{G}$, racemes or heads), Amaryllidaceae ($\overline{G}$, cymose esp. umbels), Aphyllanthaceae ($\underline{G}$, lvs rep. by sheaths), Hypoxidaceae ($\overline{G}$, racemes), Liliaceae s.s. ($\underline{G}$, racemes), Tecophilaeaceae (G semi-inf., anthers usu. with term. pores), Trilliaceae ($\underline{G}$, perianth ± differentiated into C & K) etc. & System (p. 780)

Principal genera: *Albuca, Allium, Alstroemeria, Androcymbium, Anthericum, Asparagus, Bellevalia, Bomarea, Bulbine, Calochortus, Chlorophytum, Colchicum, Crinum, Cyrtanthus, Dipcadi, Drimia, Eremurus, Eriospermum, Fritillaria, Gagea, Hippeastrum, Hymenocallis, Hypoxis, Lachenalia, Ledebouria, Lilium, Maianthemum, Muscari, Nerine, Nothoscordum, Ophiopogon, Ornithogalum, Polygonatum, Scilla, Spiloxene, Thysanotus, Trachyandra, Trillium, Tulipa, Wurmbea, Zephyranthes*

Many of these are familiar cult. orn. as are spp. of *Agapanthus, Amaryllis, Asphodelus, Aspidistra, Brodiaea, Brunsvigia, Camassia, Chionodoxa, Clivia, Convallaria, Erythronium, Eucharis, Eucomis, Galanthus, Gloriosa, Haemanthus, Hemerocallis, Hosta, Hyacinthoides, Hyacinthus, Leucocoryne, Leucojum, Liriope, Narcissus, Pancratium, Proiphys, Rhodohypoxis, Ruscus, Scadoxus, Semele, Sprekelia, Sternbergia, Tricyrtis, Tristagma, Triteleia, Uvularia, Veratrum*, so familiar that small differences in ovary position due to intercalary meristems (e.g. $\overline{G}$ & P-tube in Amaryllidaceae) early led to a separation of related plants into different alliances. An arrangement on cladistic principles, R.M.T. Dahlgren *et al.* (1985) *The families of monocotyledons*, puts many scattered groups back together but, at the same time, splinters the whole group & its allies into a large number of (often small) fams and revolutionizes the classification of the group with respect to fams trad. kept far apart, 'It is certainly of no advantage to unite Liliaceae [as circumscribed narrowly by them] with Asphodelaceae, Hypoxidaceae, Tecophilaeaceae, and Trilliaceae, if at the same time the Liliaceae are kept distinct from Orchidaceae, Alstroemeriaceae or Iridaceae'. Their arrangement has many problem genera & fams which may be placed in one order or another and there is no similar system for Dicotyledonae, yet, not even those allegedly allied to Liliaceae. s.l. Nevertheless, Liliaceae s.l. as circumscribed above, are accommodated in 'Superorder Liliiflorae', comprising 5 orders (Dioscoreales, Asparagales, Melanthiales, Burmanniales & Liliales) – the bulk of them in Asparagales (30 fams) & Liliales (7 fams) differing in a suite of characters of which 'taken singly none... is sufficient for distinguishing the families into different orders, but in conjunction they seem to be of great significance'. Of the alliances included in Asparagales from Liliaceae s.l., are Alliaceae (incl. Agapanthaceae), Amaryllidaceae, Anthericaceae, Aphyllanthaceae, Asparagaceae, Asteliaceae, Blandfordiaceae, Convallariaceae, Dianellaceae (incl. *Phormium* from Agavaceae), Doryanthaceae, Eriospermaceae, Funkiaceae, Hemerocallidaceae, Herreriaceae, Hyacinthaceae, Hypoxidaceae, Ixioliriaceae, Ruscaceae & Tecophilaeaceae, with Luzuriagaceae & Philesiaceae (split from Smilacaceae), Agavaceae (as A. s.s. with Dracaenaceae & Nolinaceae), Hanguanaceae, Xanthorrhoeaceae (as X. s.s. with Calectasiaceae & Dasypogonaceae) & Cyanastraceae; those incl. in Liliales are Alstroemeriaceae, Calochortaceae, Colchicaceae, Tricyrtidaceae (= Uvulariaceae) & Liliaceae s.s., with Geosiridaceae, Iridaceae & the orchids; Melanthiaceae are taken to include Petrosaviaceae &, with Campynemataceae, make up the order Melanthiales, while Trilliaceae are included in Dioscoreales. In the forthcoming volume for Kubitzki's *The fams & genera of vasc. pls*, there are considerable modifications of this (see Appendix) so that those following a narrow view of Liliaceae will now find Aloeaceae in Asphodelaceae, Calectasiaceae in Dasypogonaceae, Dianellaceae in Phormiaceae, Funkiaceae (Hostaceae) incl. in Agavaceae, Geosiridaceae in Iridaceae, Hanguanaceae in Flagellariaceae, Ixioliriaceae in Tecophilaeaceae, Nolinaceae in Dracaenaceae, Ruscaceae in Asparagaceae etc. etc.; Burmanniales & Melianthales are merged with Liliales. For further concerns here, see comments at end of each segregate fam.

Liliaceae s.s. (10/350 N hemisph. esp. SW As. to China; bulbous pls, $\underline{G}$) comprise merely *Cardiocrinum, Erythronium, Fritillaria, Gagea, Lilium, Lloydia, Nomocharis, Notholirion, Tricyrtis* & *Tulipa*, though Calochortaceae are incl. by some, *Tricyrtis* excl. by others, underlining even with this rump the problems associated with the splitting approach. Clearly the last word has not been said

Besides imp. cult. orn. (see above), L. s.l. provide edible products (*Allium* – onions, garlic, leeks etc., *Asparagus, Brodiaea, Camassia, Fritillaria, Ornithogalum, Ruscus, Tupistra*), though many are v. toxic (*Baeometra, Convallaria, Gloriosa, Ornithogalum, Veratrum*); medic. incl. *Allium, Colchicum, Drimia* & minor products incl. dyes (*Asphodelus*), glue (*Hyacinthoides*), scent (*Muscari* etc.)

Liliopsida. See Monocotyledonae

Lilium L. Liliaceae. 100 N temp. (Eur. 10) to Philippines. Lilies. Bulbs scaly, s.t. stoloniferous or rhizomatous, many ed.; aerial stems unbranched, s.t. with roots above the bulbs ('stem-rooting l.'). Allied to *Fritillaria* & linked by *Notholirion* & *Nomocharis*, much cult. with many hybrids, *L. candidum* in GB in C15 but only sp. known to Shakespeare. H.B.D. Woodcock & W.T. Stearn (1950) *Lilies of the world*. *L. auratum* Lindley (golden-rayed l. of Japan, Honshu) – strongly scented fls to 30 cm diam., bulbs eaten in Japan; *L. brownii* Miellez. (Burma, China) – cough treatment in modern Chinese herbalism; *L. bulbiferum* L. (orange l., C & E Eur.) – long cult., an emblem of the Orangemen in N Ireland; *L. canadense* L. (Canada l., E N Am.); *L. candidum* L. (Madonna l., Bourbon l., Balkans (? natur.), Israel & Lebanon) – long cult. for white fls used in scent-making (500 kg yield

300 g pure essence), figured in Cretan frescos 5000 yrs old, poss. the Rose of Sharon of the Bible & cult. since at least 1500 BC & more recently associated with the Virgin Mary, figuring in many religious pictures, overwintering basal lvs prod. autumn, no fr. in cult.; *L. catesbaei* Walter (leopard l., SE US); *L. formosanum* A. Wallace (Taiwan) – flowering within a few months from seed; *L. lancifolium* Thunb. (*L. tigrinum*, tiger lily, E China, Korea, Japan, natur. N Am.) – bulbs eaten in Japan; *L. longiflorum* Thunb. var. *eximium* (Courtois) Baker (Easter l., Bermuda l., November l. (Aus.), Japan) – fragrant white fls; *L. martagon* L. (Turk's-cap l., Eur. to Mongolia, natur. GB) – most widespread sp., bulbs ed.; *L. michauxii* Poiret (Carolina l., SE US); *L. regale* E. Wilson (royal l., W China); *L. superbum* L. (Turk's-cap l., E US)

lilli-pilli, lilly(-)pilly *Acmena smithii*

lily *Lilium* spp. but **l. of the field** is *Anemone coronaria* or poss. *Sternbergia lutea*; **African l.** *Agapanthus* spp.; **arum l.** *Zantedeschia aethiopica*; **atamasco l.** *Zephyranthes atamasco*; **Amazon l.** *Eucharis × grandiflora*; **belladonna l.** *Amaryllis belladonna*; **Berg l.** *Galtonia candicans*; **Bermuda l.** *Lilium longiflorum* var. *eximium*; **blood l.** *Haemanthus* spp.; **boat l.** *Tradescantia spathacea*; **Bourbon l.** *L. candidum*; **Brisbane l.** *Proiphys cunninghamii*; **Canadian l.** *L. canadense*; **Carolina l.** *L. michauxii*; **Chinese sacred l.** *Narcissus tazetta*; **chocolate l.** *Arthropodium* spp.; **climbing l.** *Gloriosa superba*; **cobra l.** *Arisaema* spp.; **corn l.** *Ixia* spp., *Clintonia borealis*; **Darling l.** *Crinum flaccidum*; **day l.** *Hemerocallis* spp.; **Easter l.** *L. longiflorum* var. *eximium*; **Eucharist l.** *Eucharis grandiflora*; **fawn l.** *Erythronium californicum*; **fire l.** *Cyrtanthus* spp.; *Hippeastrum puniceum*; **flame l.** *Gloriosa superba*; **fringed l.** *Thysanotus* spp.; **George l.** *C. elatus*; **glory l.** *G. superba*; **golden(-rayed) l. of Japan** *L. auratum*; **Guernsey l.** *Nerine sarniensis*; **gymea l.** *Doryanthes excelsa*; **impala l.** *Adenium multiflorum*; **Jacobean l.** *Sprekelia formosissima*; **Kaffir l.** *Schizostylis coccinea*; **Knysna l.** *Cyrtanthus elatus*; **Lent l.** *Narcissus pseudonarcissus*; **leopard l.** *Lilium catesbaei, Belamcanda chinensis*; **Madonna l.** *L. candidum*; **Malta l.** *Sprekelia formosissima*; **mariposa l.** *Calochortus* spp.; **May l.** *Maianthemum bifolium*; **November l.** *L. longiflorum*; **orange l.** *L. bulbiferum*; **palm l.** *Cordyline australis*; **peace l.** *Spathiphyllum* spp.; **Peruvian l.** *Alstroemeria aurea*; **plantain l.** *Hosta* spp.; **Poor Knight's l.** *Xeronema callistemon*; **pyjama l.** *Crinum kirkii*; **queen l.** *Phaedranassa* spp.; **rain l.** *Zephyranthes carinata*; **rock l.** *Arthropodium cirrhatum, Dendrobium speciosum*; **royal l.** *Lilium regale*; **spider l.** *Hymenocallis* spp.; **St Bernard's l.** *Anthericum liliago*; **St Bruno's l.** *Paradisea liliastrum*; **St John's l.** *Clivia miniata*; **Scarborough l.** *Cyrtanthus elatus*; **Snowdon l.** *Lloydia serotina*; **star l.** *Eucharis × grandiflora*; **swamp l.** *Zephyranthes* spp.; **l. thorn** *Catesbaea spinosa*; **tiger l.** *Lilium lancifolium*; **torch l.** *Kniphofia* spp.; **trout l.** *Erythronium americanum*; **Turk's-cap l.** *L. martagon, L. superbum*; **veld l.** *Crinum* spp.; **voodoo l.** *Sauromatum venosum*; **wood l.** *Clintonia* spp., *Trillium grandiflorum*

lily of the valley *Convallaria majalis*; **l. o. t. v. tree** *Clethra arborea*

Lima bean *Phaseolus lunatus*

Limacia Lour. Menispermaceae (V). 3 Burma to W Mal. R: KB 1957:447

Limaciopsis Engl. Menispermaceae (V). 1 trop. Afr.: *L. loangensis* Engl.

Limatodis Blume = *Calanthe*

limay Several dipterocarp woods (trade name in Philippines)

limba *Terminalia superba*

Limbarda Adans. (~ *Inula*). Compositae (Inul.). 1 Eur. & Medit.: *L. crithmoides* (L.) Dumort. (*Inula c.*, golden samphire) – lvs sometimes used as greens

lime (tree) *Tilia* spp.; **American l.** or basswood *T. americana*; **European** or **common l.** *T. × europaea*; **house l.** *Sparrmannia africana*; **Japanese l.** *T. japonica*; **silver** or **weeping l.** *T. tomentosa* 'Petiolaris'; **small-leaved l.** *T. cordata*

lime (fr.) *Citrus × aurantiifolia*; **l. berry** *Triphasia trifolia*; **Chinese l.** *T. trifolia*; **kaffir** or **makrut l.** *C. hystrix*; **myrtle l.** *T.* spp.

limequat *Citrus × floridana*

limetta *Citrus limetta*

Limeum L. Molluginaceae. 20 S Afr. (most) to trop. Afr., Arabia & Pakistan. R: MBSM 2(1956)133

Limnalsine Rydb. (~ *Montia*). Portulacaceae. 1 W N Am.

Limnanthaceae R. Br. Dicots – Dilleniidae – Capparidales. 2/8 temp. N Am. Small delicate sub-succ. annuals of moist places, often with myrosin cells & mustard oils. Lvs spirally arr., pinnatisect to pinnate; stipules 0. Fls reg., bisexual, solit, on long axillary pedicels, (4)5-merous (*Limnanthes*) or 3-merous (*Floerkea*); K ± distinct, valvate, C distinct, convolute, A twice C in 2 whorls or as many as & alt. with C (*Floerkea* s.t.), filaments opp. K with basal nectary gland, $\underline{G}$ 2 or 3 (*Floerkea*), (4)5 (*Limnanthes*), united by gyno-basic style, deeply lobed into globular segments, style cleft or with lobed stigma, each locule with 1 basal, anatropous, unitegmic ovule. Fr. separating into indehiscent 1-seeded mericarps; embryo straight with unusual fats, endosperm 0. X = 5

Genera: *Floerkea, Limnanthes*

Form. allied with Geraniaceae. Embryology like Boraginaceae

Limnanthes is cult. orn. & poss. oilseed

Limnanthemum S. Gmelin = *Nymphoides*

Limnanthes R. Br. Limnanthaceae. 7 W N Am. esp. Calif. R: UCPB 25(1952)455. Potential oilseeds with long-chain fatty-acids esp. *L. alba* Hartweg ex Benth. with wax like jojoba, but also *L. douglasii* R. Br. – conspic. spring fl. in the wild, cult. orn. (poached egg fl.) with white-tipped yellow C

Limnas Trin. (~ *Alopecurus*). Gramineae (21d). 2 C As. to NE Siberia

Limniboza R. Fries = *Geniosporum*

Limnobium Rich. Hydrocharitaceae. (Incl. *Hydromystria*) 1 SE US: *L. spongia* (Bosc) Steudel (*H. laevigatum*, American frogbit) – floating herb with spongy layer in abaxial leaf-surface & stolons, cult. orn. (only female pl. in Eur.), root-hairs used to demonstrate protoplasmic streaming. R: Rhodora 94(1992)124

Limnocharis Bonpl. Alismataceae (Limnocharitaceae). 1 trop. Am.: *L. flava* (L.) Buchenau grown & natur. in SE As. & Mal. – eaten like spinach, also pig-fodder

Limnocharitaceae Takht. ex Cronq. = Alismataceae

Limnocitrus Swingle = *Pleiospermium*

Limnodea L. Dewey. Gramineae (21d). 1 S US. Prairie

Limnophila R. Br. Scrophulariaceae. 37 OW trop. R: KB 24(1970)101, (Aus. 5) AusJB 33(1985)367. Aromatic marsh herbs or aquatics sometimes grown in aquaria, some locally medic. & for flavouring food

Limnophyton Miq. Alismataceae. 4 OW trop.

Limnopoa C. Hubb. Gramineae (35). 1 S India: *L. meeboldii* (Fischer) C. Hubb. – forming mats on water-surface

Limnosciadium Mathias & Constance. Umbelliferae (III 8). 2 S C US

Limnosipanea Hook.f. Rubiaceae (I 6). 7 Panamá to trop. S Am. *L. spruceana* Hook.f. (Amazon) only truly aquatic Rubiacea, with habit of *Hippuris*

Limodorum Boehmer. Orchidaceae (V 1). 1 Medit. incl. Eur. to Iran: *L. abortivum* (L.) Sw. – leafless mycotroph; 4 lateral stamens s.t. fertile

Limonia L. (*Feronia*). Rutaceae. 1 India to Java: *L. acidissima* L. (*Feronia elephantum*, elephant apple, wood a.) – elephants eat lvs, useful timber, gum (Indian g.) used in glue, water-colours etc., fr. ed. esp. when strained to remove seeds (also in jam) & used as soap subs. in Japan, *Citrus* may be grafted on to it

Limoniaceae Ser. = Plumbaginaceae

Limoniastrum Fabr. (~ *Limonium*). Plumbaginaceae. (II, Limoniaceae). 9 Medit. (Eur. 1) to N Somalia

Limoniopsis Lincz. Plumbaginaceae (II). 2 Turkey to Caucasus

Limonium Miller (*Statice*). Plumbaginaceae (II, Limoniaceae). c. 350 cosmop. esp. maritime & arid N hemisph. (Eur. 87). Herbs or subshrubs with persistent K around capsule; pollen finely or coarsely reticulate, stigmas with rounded or prominent papillae (those with coarse p. & rounded s. or with fine p. & prominent s. self-incompatible); monomorphic (coarse p. & prominent s.) self-compatible or apomicts e.g. *L. binervosum* (G.E. Sm.) Salmon agg. (Atlantic Eur.). *L. carolinianum* (Walter) Britton (E US) – 9–44% dry wt. soluble phenolics deterrent to grazing Canada geese. Cult. orn. esp. for cut-fls & everlastings (statice, sea lavender, marsh rosemary; see also *Psylliostachys*), some esp. *L. gmelinii* (Willd.) Kuntze (E Eur. to Siberia) & *L. latifolium* (Sm.) Kuntze (E Eur.; stellate indumentum) with roots (kermek) used for tanning & others, in Uruguay, used for fertility control. *L. arborescens* (Brouss.) Kuntze (Canary Is.) – shrubby, rare in wild but commonly cult.; *L. sinuatum* (L.) Miller (Medit.) – K blue-violet, C white & *L. vulgare* Miller (W Eur., N Afr.) – K pale purple, C blue-purple – commonly seen 'everlastings'; *L. suffruticosum* (L.) Kuntze – used as a boron indicator in former USSR; *L. thouinii* (Viv.) Kuntze (Medit.) – cult. orn.

Limosella L. Scrophulariaceae. 11 cosmop. (Eur. 3). R: BJ 66(1934)488. *L. aquatica* L. (mudwort, N temp.) – multiplies by runners

Linaceae DC ex Perleb. Dicots – Rosidae – Linales. 14/250 cosmop. Trees, lianes, shrubs & herbs, sometimes cyanogenic. Lvs spirally arr. to opp., simple, entire; stipules decid., often small, sometimes glands or 0. Fls bisexual, usu. reg. & 5-merous in cymose or racemose infls; K imbricate, distinct to basally connate, sometimes ± dissimilar, C distinct, convolute, often clawed, 2–5 extrastaminal nectary glands or a disk (*Ctenolophon*), A alt. with C or opp. (*Anisadenia*) or 10 (15), when sometimes unequal, or alt. with staminodes when 5, filament bases connate into a tube, anthers with longit. slits, G (2–5), pluriloc. with axile or apical-axile placentas & sometimes 1-loc. apically (partitions not reaching apex), styles distinct or 1 deeply cleft, each locule with 2 pendulous, anatropous, bitegmic ovules, sometimes separated by incomplete septa from ovary wall & often with placental obturators. Fr. a septicidal capsule, drupe, nut (*Ctenolophon*) or pair of 1-seeded mericarps (*Anisadenia*); seeds with straight oily embryo & little or 0 endosperm (thick in *Indorouchera*). x = 6–11+

Principal genera: *Durandea, Hesperolinon, Hugonia, Linum, Roucheria*

A trad. view of the L. is taken, some authors (incl. Cronquist) segregating *Hugonia, Ctenolophon* & allies as a separate fam., s.t. allied with Celastraceae; *Anisadenia* supposed to show features char. of Malpighiaceae or to be transitional to Plumbaginaceae; *Roucheria* etc. not clearly distinct from Erythroxylaceae

Some timbers & locally eaten fr., cult. orn. esp. *Linum*, which provides linseed oil & flax

Linanthastrum Ewan = *Linanthus*

Linanthus Benth. Polemoniaceae. 35 W N Am., Chile. Usu. annuals, some cult. orn.

linaloe or **linaloa oil** *Bursera* spp.

Linaria Miller. Scrophulariaceae (Antirr.). 150 Euras. (Eur. 77), Medit., widely natur. temp. R: D.A. Sutton, Rev. Antirrh. (1988)260. Toadflax; some locally medic. esp. in treatment of haemorrhoids; cult. orn. (fls blue, purple or yellow) incl. *L. vulgaris* Miller (common t., butter-and-eggs, Euras., natur. N Am.) – fls yellow & orange, 'Peloria' being a form with reg. fls & 5 spurs due to linked multiple alleles (cf. *Cymbalaria*)

Linariantha B.L. Burtt & R.M. Sm. Acanthaceae. 1 Borneo

Linariopsis Welw. Pedaliaceae. 2 W & SW trop. Afr.

Linconia L. Bruniaceae. 2 Cape

Lindackeria C. Presl. Flacourtiaceae (3). 14 trop. Afr. (8, R: BJ 94(1974)311), trop. Am. (6, R: FN 22(1980)11)

Lindauea Rendle. Acanthaceae. 1 Somalia

Lindbergella Bor. Gramineae (17). 1 Cyprus

Lindelofia Lehm. (~ *Cynoglossum*). Boraginaceae. 11 C As. to Himal. Alks; some cult. orn.

linden *Tilia* spp.

Lindenbergia Lehm. Scrophulariaceae. 15 OW trop.

Lindenia Benth. Rubiaceae (I 5). 3 C Am. (1), Fiji (1), New Caled. (1). R: JAA 57(1976)426. Rheophytes in middle of shallow watercourses; transpacific distrib.! Cult. orn. with long white fls

Lindeniopiper Trel. = *Piper*

Lindera Thunb. Lauraceae (II). 100 trop. & temp. As., Aus. (1), E N Am. (3, fever bush, wild allspice). Aromatic trees. *L. benzoin* (L.) Blume (spice bush, E N Am.) – bush with lvs form. used as a tea & fr. as an allspice subs., bark a febrifuge, poss. oil source; *L. praecox* (Siebold & Zucc.) Blume (*Parabenzoin p.*, Japan) – seed-oil used as illuminant

Lindernia All. Scrophulariaceae. (Incl. *Ilysanthes*) 80 warm. esp. OW (Eur. 1, Afr. 38 – R: TSP 81(1992)214)

Lindheimera A. Gray & Engelm. Compositae (Helia.-Eng.). 1 SW N Am.: *L. texana* A Gray & Engelm. (Texas star) – cult. orn. annual. R: Sida 15(1993)533

Lindleya Kunth. Rosaceae. 2 Mexico. Capsule

Lindmania Mez (~ *Cottendorfia*). Bromeliaceae (1). 37 Guayana Highland. R: AMBG 73(1986)690

Lindneria T. Durand & Lubbers (*Pseudogaltonia*). Hyacinthaceae (Liliaceae s.l.). 1 S Afr.: *L. clavata* (Masters) Speta. R: BJ 106(1985)125

Lindsaea Dryander ex Sm. Dennstaedtiaceae (Lindsaeoideae). 150 trop. to Aus., New Caledonia (10) & Japan. *L. integra* Holttum (N Mal.) – rheophyte

Lindsaeaceae Pichi-Serm. See Dennstaedtiaceae

Lindsayella Ames & C. Schweinf. Orchidaceae. 1 Panamá

Lindsayomyrtus B. Hyland & Steenis. Myrtaceae (Lept.). 1 E Mal. & Queensland: *L. racemoides* (Greves) Craven

linen cloth made from *Linum usitatissimum*; **China** or **Canton l.** that from *Boehmeria nivea*

ling *Calluna vulgaris*

Lingelsheimia Pax = *Drypetes*

Lingnania McClure = *Bambusa*

lingue *Persea lingue*

Linkagrostis Romero García, Blanca & Morales Torres = *Agrostis*

Linnaea L. Caprifoliaceae. 1 circumpolar (incl. Eur.): *L. borealis* L. (twinflower) – trailing evergreen cult. orn.

Linnaeobreynia Hutch. = *Capparis*

Linnaeopsis Engl. Gesneriaceae. 3 trop. E Afr.

Linocalix Lindau = *Justicia*

Linociera Sw. ex Schreber = *Chionanthus*

Linodendron Griseb. Thymelaeaceae. 3 Cuba

Linospadix H. Wendl. Palmae (V 4j). 11 NE Aus., New Guinea. Small cult. orn. palms with pinnate or bifid lvs incl. *L. monostachya* (C. Martius) H. Wendl. (NE Aus.) – stems used for walking-sticks

Linostoma Wallich ex Endl. Thymelaeaceae. 6 Indomal. to trop. Aus.

Linosyris Cass. (~ *Aster*). Compositae (Ast.-Ast.). 10 Euras.

linseed oil *Linum usitatissimum*

Lintonia Stapf. Gramineae (33a). 2 trop. E Afr.

Linum L. Linaceae. c. 180 temp. & subtrop. esp. Medit. (Eur. 36). Flax. Fls red, yellow, blue or white, some heterostyled; seeds with mucilaginous testa, swelling on wetting. *L. pratense* (Norton) Small (S US) selfed – after C falls, K moves to force A on to stigma. Comm. flax derived from retting stems of *L. usitatissimum* L. (cultigen with non-dehiscent capsules & big seeds derived from subsp. *bienne* (Miller) Stankevich (*L. angustifolium*, *L. bienne*), W Eur., Medit.) & used for textiles (linen – decline due to cotton imports), thread, carpets, twine, canvas, paper etc., the fibres having great tensile strength, that left after removal = tow form. used for ropes & sacks; seeds of other cvs yield linseed oil (known in archaeology from 8000 BC), a drying oil used in food-processing, paints, varnish, printing inks, water-proofing, soap etc., also medic., seed-cake good stock feed; increasing fertilizer can alter permanently amounts of DNA in cells (Lamarckian evolution!). Others locally medic. e.g. *L. catharticum* L. (purging or fairy f., Eur. & Medit.) – form. purgative & diuretic, & fibres, though *L. marginale* A. Cunn. (Aus.) suspected of cyanide poisoning of stock; many cult. orn. esp. *L. flavum* L. (C & S Eur.) – fls yellow, *L. grandiflorum* Desf. (flowering f., N Afr.) – fls red, *L. perenne* L. (Eur. & N Am.) & allied *L. narbonense* L. (blue f., Medit.) – fls blue

Liparia L. Leguminosae (III 27). (Incl. *Priestleya*) 14 SW Cape fynbos. R: T 43(1994)577. Bird-, bee- & ? mammal-poll.

Liparis Rich. (*Stichorkis*) Orchidaceae (V 7). 350 cosmop. (not NZ; Eur. 1: *L. loeselii* (L.) Rich. (fen orchid, N temp.) – regularly autogamous, the success of seed-set increased 4 times by rain through pressure on anther cap pushing pollinia on to stigma or by cohesion of water droplet). Terr. orchids, a few (dingy) cult. orn.

Liparophyllum Hook.f. Menyanthaceae. 1 Tasmania & NZ. Homostylous

Lipocarpha R. Br. Cyperaceae. 8 trop. & warm (N Am. 5, R: JAA 68(1987)409) except Pacific

Lipochaeta DC (~ *Wollastonia*). Compositae (Helia.-Verb.). 20 Hawaii. R: Rhodora 81(1979)291. Chromosome evidence suggests origin from *Wedelia*

Lipostoma D. Don = *Coccocypselum*

Lippia L. Verbenaceae. c. 200 trop. Afr. & Am., widely natur. *L. carviodora* Meikle (NE Afr.) – dioec. (rare in fam.). *L. plicata* Bak. (E Afr.) – insecticidal, eaten by healthy chimps. Some with lvs used as tea, others as 'oregano' in C & S Am. (esp. *L. graveolens* Kunth (S N to C Am.) & *L. micromera* Schauer (WI to N S Am.)). Cult. orn. incl. *L. dulcis* Trev. (C Am.) – hernandulcin 800 times as sweet as sucrose, known to Aztecs

Lipskya (Koso-Polj.) Nevski. Umbelliferae (III 4). 1 C As.

Lipskyella Juz. Compositae (Card.-Card.). 1 C As.

Liquidambar L. Hamamelidaceae (IV; Altingiaceae). 5 E Med., E As., SE N & C Am. Monoec. decid. trees; C 0. Valuable timber & aromatic balsam (storax) used medic. & in scent; cult. orn. with spectacular autumn colours. *L. formosana* Hance (S China, Taiwan) – silkworms fed on it yield 'Marvello hair'; *L. orientalis* Miller (As. Minor) – source of Levant storax, the balm (of Gilead) of the Bible; *L. styraciflua* L. (sweet gum, American red g., Connecticut to C Am.) – street-tree as in NSW, timber (satin walnut, bilsted) for cabinet-work & veneers etc., storax medic., antiseptic, used in skin disease

liquorice *Glycyrrhiza glabra*

Liriodendron L. Magnoliaceae (II 1). 2 E N Am. (1), China (1). Alks; lactone antifeedant affects gypsy moth. *L. tulipifera* L. (tulip tree or t. poplar, E N Am.) – food-bodies at bases of some tepals, cult. orn. (a tree at Wheathampstead Herts, GB planted 1581 still alive 1987), timber (yellow poplar, (American or canary) whitewood) used for cabinet-work, shingles, clapboards, riding-boot lasts etc. form. for Indian canoes & for cigar- & bible-boxes for white settlers; *L. chinense* (Hemsley) Sarg. (C China, N Indochina) – cult. orn.

differing in smaller yellow-green (rather than orangeish) fls, separated for 10–16m yrs but still interfertile (hybrids (1970) = '*L.* × *chinamerica*')

Liriope Lour. Convallariaceae (Liliaceae s.l.). 5 Japan, China & Vietnam. Cult. orn. ground-cover esp. *L. muscari* (Decne) L. Bailey (Japan, China) incl. fasciated forms; some locally medic. & supposed aphrodisiac

Liriosma Poeppig = *Dulacia*

Liriothamnus Schltr. = *Trachyandra*

Lisaea Boiss. (~ *Turgenia*). Umbelliferae (III 3). 3 E Medit., SW As.

Lisianthius P. Browne = *seq*.

Lisianthus P. Browne. Gentianaceae. 30 trop. Am. Woody & semi-woody, some cult. orn. R: JAA 53(1972)76. See also *Eustoma*

Lissanthe R. Br. (~ *Styphelia*). Epacridaceae. 6 Aus.

Lissocarpa Benth. Lissocarpaceae. 3–4 trop. S Am.

Lissocarpaceae Gilg. Dicots – Dilleniidae – Theales. 1/3–4 trop. S Am. Trees without stellate or peltate indumentum. Lvs spirally arr., simple, entire; stipules 0. Fls bisexual, reg., bibracteolate, in small cymes; K (4), lobes imbricate, C (4), lobes convolute in bud, corona 8-toothed, A 8, weakly connate, borne on C tube below middle, anthers with longit. slits, $\overline{G}$ (4), pluriloc. with axile placentas & term. style sometimes with 4-lobed stigma, each loc. with 2 pendulous ovules. Fr. indehiscent, rather fleshy, with 1 or 2 seeds; embryo straight, endosperm copious

Genus: *Lissocarpa*. Form. incl. in Ebenaceae but differing in presence of corona, inf. ovary etc.

Lissochilus R. Br. = *Eulophia*

Lissospermum Bremek. = *Strobilanthes*

Listera R. Br. Orchidaceae (V 1). 20 N temp. & boreal (Eur. 2). Twayblade (orchids). *L. ovata* (L.) R. Br. (twayblade, Eur. to Siberia) grows 13–15 yrs before flowering; the rostellum when touched splits explosively ejecting a viscous fluid which glues the pollinia to the visiting insect; higher percentages of B chromosomes at higher latitudes & in wetter habitats

Listia E. Meyer = *Lotononis*

Listrobanthes Bremek. = *Strobilanthes*

Listrostachys Reichb. f. Orchidaceae (V 16). 2 trop. Afr., (?) Réunion

Litanthus Harvey. Hyacinthaceae (Liliaceae s.l.). 1 S Afr.

Litchi Sonn. Sapindaceae. 1 trop. China to W Mal.: *L. chinensis* Sonn. with (Blumea 24(1978)398) 3 recog. subspp.: *philippensis* (Radlk.) Leenh. (Philippines, wild) poss. ancestor of the cult. *javensis* Leenh. (Java) & *chinensis* (litchi, lychee, *Nephelium litchi*) of which 2 forms grown – 'mountain' used as a stock or grown in mts (small prickly fr. more like wild pl.) & 'water' grown best in monsoon climates, the litchi of commerce eaten fresh, tinned or dried, the edible flesh being an aril

Lithachne Pal. Gramineae (4). 4 trop. Am. Bamboos. *L. humilis* Söderstrom has 'sleep' movements with the lvs hanging vertically & not held upwards as in most bamboos

Lithobium Bong. Melastomataceae. 1 Brazil

Lithocarpus Blume. Fagaceae. 100+ Indomal., fossils in W N Am. R: A. Camus (1952–4) *Les chênes* 3: 511

Lithocaulon Bally = *Pseudolithos*

Lithococca Small ex Rydb. = *Heliotropium*

Lithodora Griseb. Boraginaceae. 7 NW France & W Med. (Eur. 7) to As. Minor. Cult. orn. shrublets esp. *L. prostrata* (Loisel.) Griseb. (*Lithospermum prostratum*, W Medit.)

Lithodraba Boelcke = *Xerodraba*

Lithophila Sw. Amaranthaceae (II 2). 1 WI, 2 Galápagos

Lithophragma (Nutt.) Torrey & A. Gray. Saxifragaceae. 9 W N Am. R: UCPB 37(1965). Some cult. orn. like *Tellima*

Lithophytum Brandegee = *Plocosperma*

Lithops N.E. Br. Aizoaceae (V). 37 S Afr. R: Excelsa 3(1973)41.; D.T. Cole (1988) *L.* Living stones. Glabrous succ. with annual growths of 1 or more obconical bodies composed of a pair of fleshy lvs on top of a dual column; leaf apices flattened & often with transparent 'windows' where aqueous tissue of the body of the leaf meets the surface; the solit. fl. produced between the lvs terminates season's growth after which a new shoot develops in the axil of 1 or both lvs; new shoots withdraw moisture from old lvs which remain as withered sheaths. Plants well camouflaged in their stony surroundings, the colour variations being due to orange-red chromoplasts in the epidermal & subepidermal cells as well as betalain pigments in the latter; in those with 'windows', light penetrates to chloroplasts inside lvs; in areas of v. high light intensity, pl. has app. protective layer of calcium oxalate. Many cult. orn. incl. hybrids

Lithospermum L. Boraginaceae. (Incl. *Buglossoides*) c. 45 temp. exc. Aus. (Eur. 8). R: JAA suppl. 1(1991)119. Some dyes (*L. arvense* L., Euras., bastard alkanet, etc.) & cult. orn. incl. *L. canescens* (Michaux) Lehm. (yellow puccoon, N Am.) – red dye used as face-paint by Indians; *L. officinale* L. (gromwell, Euras., natur. N Am.) – lvs used as a tea; *L. ruderale* Douglas ex Lehm. (W N Am.) – medic. & contraceptive (inspiration for perfecting oral contraceptives). See also *Lithodora*

Lithostegia Ching. Dryopteridaceae (I 2). 1 E Himal. to SW China: *L. foeniculacea* (Hook.) Ching

Lithraea Miers ex Hook. & Arn. Anacardiaceae (IV). 3 S Am. *L. caustica* (Molina) Hook. & Arn. (Chile) – severe dermatitis with painful swellings lasting many days

Litogyne Harvey (~ *Epaltes*). Compositae (Pluch.). 1 trop. & S Afr.

Litosanthes Blume. Rubiaceae (IV 7). 20 Indomal. R: Candollea 44(1989)209

Litothamnus R. King & H. Robinson (~ *Eupatorium*). Compositae (Eup.-Gyp.). 2 Brazil. R: MSBMBG 22(1987)110

Litrisa Small (~ *Carpephorus*). Compositae (Eup.-Gyp.). 1 SE US. R: MSBMBG 22(1987)273

Litsea Lam. Lauraceae (II). c. 400 warm & trop. esp. As. & Aus. Dioec. trees & shrubs with alks; *L. rheophytica* Kosterm. (Sarawak) – rheophyte. Some timbers. *L. cubeba* (Lour.) Pers. (Himal. to SE As.) – fr. ed. & med., silkworms reared on lvs, effective insecticide in seed-oil; *L. glutinosa* (Lour.) Robinson (Himal., SE As.) – bark & fr. medic.; *L. monopetala* (Roxb.) Pers. (Himal.) – Imp. fodder tree

Littaea Tagl. = *Agave*

Littledalea Hemsley. Gramineae (22). 3 C As. to W China. Close to *Bromus* but leaf-sheaths with free margins

Littonia Hook. Colchicaceae (Liliaceae s.l.). 5 trop. Arabia & NE trop. Afr., 2 S Tanzania (1) to S Afr. Climbing like *Gloriosa*; alks. *L. modesta* Hook. (climbing bellflower, S Afr.) – cult. orn. with orange fls

Littorella P. Bergius. Plantaginaceae. 3: 1 N Am., 1 S Am., 1 Eur. & Azores: *L. uniflora* (L.) Asch. (*L. lacustris*, shoreweed) – land form with rosettes of ± flattened lvs with protogyny, 2 sessile female fls flanking 1 male (wind-poll.); submerged form with larger centric lvs (often confused with *Isoetes*) & runners (fls 0), taking up carbon through lvs & roots (cf. *Isoetes*)

Litwinowia Woronow = *Euclidium*

live forever *Sedum telephium*; **l. long** *S.* spp.

living stones *Lithops* spp.

Livingstone daisy *Dorotheanthus bellidiformis*

Livistona R. Br. Palmae (I 1b). 28 NE Afr., Arabia, Ryukyus, Indomal. to Aus. Fanpalms with bisexual fls & v. varied ecol. Cult. orn. & locally eaten bud (cabbage palm) esp. *L. australis* (R. Br.) C. Martius (E Aus.) – protected in wild

lizard orchid *Himantoglossum hircinum*, (Aus.) *Burnettia cuneata*; **l. plant** *Kalanchoe delagoensis*

Llagunoa Ruíz & Pavón. Sapindaceae. 3 W trop. S Am.

Llavea Lagasca. Pteridaceae (IV). 1 trop. Am.: *L. cordifolia* Lagasca – cult. orn.

Llerasia Triana. Compositae (Ast.-Sol.). 14 Colombia, Bolivia. R: Biotropica 2(1970)39. Trees, lianes, shrubs

Llewelynia Pittier = *Henriettea*

Lloydia Salisb. ex Reichb. Liliaceae. 12 temp. Euras. (Eur. 1), W N Am. *L. serotina* (L.) Reichb. (Snowdon lily, range of genus) – cult. orn., protected pl. in GB

Loasa Adans. Loasaceae. 105 Mex. to S Am. esp. mts. Herbs & subshrubs usu. with stinging hairs & yellow (usu.), white or red fls facing downwards; petals boat-shaped, protecting A, nectaries large & conspic. Some cult. orn.

Loasaceae Juss. ex DC. Dicots – Dilleniidae – Violales. 14/260 Am. & (*Kissenia*) Afr., Arabia. Herbs, s.t. climbing, shrubs or even small trees with coarse silicified (& often calcified) hairs, s.t. stinging, often gland-tipped. Lvs simple & often lobed, spirally arr. or opp.; stipules 0. Fls bisexual, reg., solit. or in cymes; K (4)5(–7), convolute or imbricate, persistent, C (4)5(–7) or 10 (when incl. 5 petaloid staminodes), distinct or lobes of a tube, induplicate-valvate in bud, A (10–)∞ centripetal (Mentzelioideae), centrifugal (Loasoideae) or 5 (Gronovioideae), distinct or with basal tube or in antepetalous bundles, s.t. the anthers almost sessile on C-tube, some often petaloid or nectariferous staminodes, anthers with longit. slits, $\overline{G}$ (3–5(–7)), 1-loc. with parietal placentas, often ± deeply intruded (rarely pluriloc. with axile placentas or (Gronovioideae) pseudomonomerous

with 1 pendulous, apical ovule), each placenta with 1–∞ anatropous to hemitropous, unitegmic ovules. Fr. a caspule, rarely dry-indehiscent, seeds with straight or curved embryo in copious oily endosperm or endosperm ± 0 (Gronovioideae). X = 7–15 +
Principal genera: *Blumenbachia, Caiophora, Eucnide, Loasa, Mentzelia*
Prob. not Dilleniidae; the unitegmic seeds suggest Asteridae but A ∞ & free petals, etc. led Cronquist to keep L. out. Possibly heterogenous (see A) & affinities unclear
Some cult. orn. herbs

Lobanilia Radcl-Sm. (~ *Claoxylon*). Euphorbiaceae. 7 Madag. R: KB 44(1989)334

Lobelia L. Campanulaceae (Lobelioideae). c. 300 trop. & warm esp. Am., few temp. (Eur. 2). R: Pfl. IV 267b(1943)408, (1953)775. Incl. *Pratia* (fleshy berry) as there are intermediates e.g. *L. angulata* Forster f. (Am., As.). in Sulawesi showing transition to typical capsule, natur. Scotland. Herbs, shrubs & pachycaul trees (Giant L.), some aquatics (*L. dortmanna* L., water l., W Eur., N N Am. – roots take up free CO_2 (cf. *Isoetes*)) & succ., some dioec. (e.g. *L. irrigua* R. Br. – islands nr. Tasmania, almost extinct but saved by cult.). Fls twisted through 180°; style pushes through the anther-tube forcing out the pollen, after which stigmas separate, exposing receptive surface free of 'home' pollen (cf. Compositae). Many alks. The pachycaul spp. of the Afr. mts, like *Dendrosenecio* spp. there, are adapted to the diurnal fluctuation in temp. above the tree-line: their massiveness acts as a thermal buffer & the hollow infl. axes have chemicals causing ice crystals to form at highish temperatures thus protecting the rest of the pl.; the lvs close over bud at night & provide a haven for many invertebrates; pith eaten by gorillas which make nests in the pl.; fls bird-poll. Many cult. orn. incl. *L. cardinalis* L. (cardinal fl., N Am.) – fls scarlet, more alks than *L. inflata*; *L. erinus* L. (S Afr.) – the common bedding sp. with many cvs esp. 'Pendula' used in hanging baskets; *L. inflata* L. (Indian tobacco, N Am.) – cult. for lvs used medic. esp. in chest conditions; *L. siphilitica* L. (blue cardinal fl., E US); *L. tupa* L. (Chile) – psychoactive & poss. hallucinogenic, certainly narcotic & medic. esp. in treatment of toothache but overpowering smell can cause sickness

Lobeliaceae Bonpl. = Campanulaceae (Lobelioideae)

Lobivia Britton & Rose = *Echinopsis*

loblolly bay *Gordonia lasianthus*; **l. magnolia** *Magnolia grandiflora*; **l. pine** *Pinus taeda*; **l. tree** *Cupania* spp.

Lobostemon Lehm. Boraginaceae. 28 S Afr. *L. fruticosus* (L.) Buek (Cape) – dressings for wounds & sores

Lobostephanus N.E. Br. = *Emicocarpus*

lobster claw *Heliconia* spp.; **l. plant** *Justicia brandegeeana*

Lobularia Desv. Cruciferae. 4 Cape Verde & Canary Is., Medit. (Eur. 2) to Arabia. R: OB 91(1987)5. *L. maritima* (L.) Desv. ((sea or sweet) alyssum, S Eur.) – commonly grown white-flowered edging pl.

Lochia Balf.f. Caryophyllaceae (I 2). 2 Socotra, Abd al Kuri, Oman (1), Somalia (1)

Lockhartia Hook. Orchidaceae (V 10). 24 trop. Am. Cult. orn. (R: AOSB 43(1974)399) epiphytes with small fls

locoweed *Astragalus* spp. (also *Oxytropis* spp.)

locust (bean) Several legumes etc. with large mealy pods esp. spp. of *Astronium, Byrsonima, Ceratonia, Gleditsia, Hymenaea, Parkia, Robinia* etc. (l. of Bible = *Ceratonia siliqua*); **African l.** *Parkia filicoidea* etc.; **black l.** *R. pseudoacacia*; **l. bean gum** *Ceratonia siliqua*; **black l.** *Robinia pseudoacacia*; **honey l.** *Gleditsia* spp.; **WI l.** *Hymenaea courbaril*; **l. wood** *Astronium fraxinifolium*

lodgepole pine *Pinus contorta* var. *latifolia*

lodh bark *Symplocos racemosa*

Lodoicea Comm. ex Labill. Palmae (I 3a). 1 Seychelles: *L. maldivica* (J. Gmelin) Pers. (coco-de-mer, double coconut) – dioec. palm to 30m (male) & 200–350 yrs old, with v. large lvs (petiole to 8m long, lamina c. 4 × 2 m) & male infl. to 2 m long, fls to 10 cm diam., slow-growing & now restricted to reserves, its conservation first mooted 1864; fr. takes 6 yrs to develop & contains 1(–3) 2-lobed seeds in a bony endocarp, the largest seeds (50 cm long) of any pl., killed by seawater & first known washed up on other islands & believed to be from the sea (Seychelles then unknown), their suggestive shape (like female hindquarters) indicating the devil's work or, at least, aphrodisiacal qualities. An evolutionary dead-end on a small granite fragment of Gondwanaland, rarely cult. but endocarp used as bowls etc.

Loefgrenianthus Hoehne. Orchidaceae (V 13). 1 Brazil

Loeflingia L. Caryophyllaceae (I 1). 7 N Am., Medit. (Eur. 3)

Loerzingia Airy Shaw. Euphorbiaceae. 1 Sumatra

Loeselia L. Polemoniaceae. 9 California to Venezuela. Fl. ± irreg.

Loeseliastrum (Brand) Timbrook (~ *Langloisia*). Polemoniaceae. 2 W N Am. deserts. R: Madrono 33(1986)170

Loesenera Harms. Leguminosae (I 4). 4 trop. W Afr.

Loeseneriella A.C. Sm. (~ *Hippocratea*). Celastraceae. 16 OW trop.

Loewia Urban. Turneraceae. 3 NE trop. Afr.

loganberry *Rubus loganobaccus*; **S African l.** = 'Youngberry'

Logania R. Br. Loganiaceae. 25 Aus. (22), New Caledonia (1), NZ (1)

Loganiaceae Martinov s.l. (see below). Dicots – Asteridae – Gentianales. 29/570 trop. to temp. (few). R: Pfl. ed. 2, 28b1(1980), incl. Buddlejaceae, Retziaceae). Trees, shrubs, lianes & herbs, s.t. accum. aluminium & often with alks & internal phloem. Lvs opp. (rarely spiral or ternate etc.), simple, entire to lobed; stipules present, or lines linking petiole-bases or 0, petioles s.t. with an ocrea. Fls usu, bisexual, sometimes heterostylous (V), reg. or not, solit. or usu. in cymes; K 4 or 5 free or not, valvate or imbricate, C (4 or 5(–16 (*Anthocleista, Potalia*)), variously shaped, the lobes valvate, imbricate or contorted in bud, A inserted on C-tube & usu. alt. with lobes, sometimes fewer (*Sanango, Usteria*), anthers basifixed, versatile with longit. slits, disk only in *Sanango*, G ((1)2(–4)) s.t. half-inf. with as many loc., the partitions s.t. incomplete apically, with term. style s.t. apically lobed & seldom branched & 2–∞ anatropous to hemitropous or amphitropous unitegmic ovules on axile placentas. Fr. a capsule, berry or drupe (*Neubergia*); seeds 1–∞ sometimes winged, with fleshy, starchy or horny endosperm surrounding straight (slightly curved in *Gardneria*) embryo. x = 6–12, 19

Classification and genera: 10 tribes –

I. **Spigelieae** (herbaceous, connate leaf-sheaths & stipules, usu. half-inf. ovary): *Mitrasacme, Mitreola, Polypremum, Spigelia*

II. **Loganieae** (ochreae frequent, fls s.t. unisexual, C imbricate or contorted): *Geniostoma, Labordia, Logania* (wood anatomy like Oleaceae)

III. **Strychneae** (C fleshy, valvate): *Gardneria, Neubergia, Strychnos* – some similarities with Apocynaceae & Rubiaceae

IV. **Plocospermeae** (capsule with few large narrow hair-tufted seeds): *Plocospermum* – much like Apocynaceae

V. **Gelsemieae** (heterostyly): *Gelsemium, Mostuea*

VI. **Antonieae** (C valvate, seeds winged): *Antonia, Bonyunia, Norrisia, Usteria*

VII. **Buddlejeae** (indumentum s.t. stellate, lvs s.t. not opp., pollen like Scrophulariaceae, X = 7, 19): *Androya, Buddleja, Emorya, Gomphostigma, Nuxia, Peltanthera, Sanango*

VIII. **Retzieae** (lvs acicular, usu. opp.): *Retzia*

IX. **Potalieae** (glabrous, fleshy often large fls, berries): *Anthocleista, Fagraea, Potalia*

X. **Desfontainieae** (lvs holly-like, berry): *Desfontainia*

Although fam. 'may represent a relict from common ancestors' (Leeuwenberg *et al.*), the fam. has lately been split up (Cladistics 10(1994)206): *Spigelia*, III & VI = Strychnaceae, *Geniostoma* & *Labordia* = Geniostemaceae, IV = Plocospermataceae, V = Gelsemiaceae, VIII has been assigned to Stilbaceae, IX to Gentianaceae, X = Desfontainiaceae (see those fams), but the position of *Polypremum* is still uncertain. Loganiaceae s.s. (3/70 trop. & warm, subshrubs & herbs) comprises only *Logania, Mitrasacme* & *Mitreola*

Logfia Cass. (~ *Filago*). Compositae (Gnap.-Gnap.). 9 Eur. (6), Medit. to Afghanistan. R: OB 104(1991)171

logwood *Haematoxylum campechianum*

Loheria Merr. Myrsinaceae. 6 E Mal. Pachycaul treelets

Loiseleuria Desv. Ericaceae. 1 N circumpolar: *L. procumbens* (L.) Desv. (alpine azalea) – cult. orn. shrublet with revolute lvs & reg. protogynous fls with A 5

Lojaconoa Bobrov = *Trifolium*

lolagbola *Oxystigma oxyphyllum*

Loliolum Krecz. & Bobrov (~ *Vulpia*). Gramineae (17). 1 E Medit. to C As.

loliondo *Olea capensis* subsp. *welwitschii*

Lolium L. Gramineae (17). 8 Euras. (Eur. 5). R: USDATB 1392(1968). Ryegrass; spikelets in 2-ranked heads. Valuable fodder & lawn-grasses; spp. interfertile & intergrade, will hybridize with *Festuca arundinacea* Schreber & allies. *L. multiflorum* Lam. (*L. italicum, L. perenne* subsp. *multiflorum*, Italian r., introd. to GB c. 1830 from C & S Eur., crossed with *L. perenne*) – fodder; *L. perenne* L. (perennial r., Eur., natur. N Am.) – contains loliolide a potent ant-repellent, much used for playing-fields & lawns, prob. the first grass deliberately sown for pastures, artificial autotetraploids more competitive than diploids; *L.*

temulentum L. (darnel, the 'tares' of the Bible, Euras.) – fr. infected with fungal mycelia form. mixed with barley to make an intoxicating beer

Lomagramma J. Sm. Lomariopsidaceae. c. 20 Indomal. to Pacific (not Aus.); *L. guianensis* (Aublet) Ching perhaps separable

Lomandra Labill. Xanthorrhoeaceae (Lomandraceae). c. 50 New Guinea (2), Aus., New Caled. (1). Dioec. 'iron grasses'. P sepaloid or inner petaloid

Lomandraceae Lotsy = Xanthorrhoeaceae

Lomanodia Raf. = *Astronidium*

Lomariopsidaceae Alston (~ Aspleniaceae). Filicopsida. 6/525 trop. & warm (few). Scaly rhizomes/stems with dorsiventral dictyosteles. Petiole with U-shaped arr. of vasc. bundles. Lamina divided to simple; fert. fronds usu. contracted with sporangia usu. covering fertile parts abaxially

Genera: *Bolbitis, Elaphoglossum, Lomagramma, Lomariopsis, Teratophyllum, Thysanosoria*

Allied to Dryopteridaceae

Lomariopsis Fée. Lomariopsidaceae. 45 trop. (Aus. 1). Thick rhizomes climbing trees to 15 m

Lomatia R. Br. Proteaceae. 12 Aus. (8), S Am. Like *Grevillea* but fr. with several seeds. *L. myricoides* (Gaertner f.) Domin (SE Aus.) – rheophyte. Some cult. orn. incl. *L. hirsuta* (Lam.) Diels (Peru, Chile, Ecuador, Arg.) – timber for furniture; *L. silaifolia* (Sm.) R. Br. (E Aus.) – protected in wild

Lomatium Raf. Umbelliferae (III 10). 74 W N Am. R: AMBG 25(1938)225. Many with ed. roots esp. *L. ambiguum* (Nutt.) J. Coulter & Rose (cous root) – eaten raw or pounded & made into cakes or biscuits; *L. dissectum* (Nutt.) Mathias & Constance (cough root) – boiled root medic.

Lomatocarpa Pim. Umbelliferae (III 8). 3 SW & C As., Afghanistan

Lomatogoniopsis T.N. Ho & S.W. Liu (~ *Lomatogonium*). Gentianaceae. 1 China

Lomatogonium A. Braun. Gentianaceae. 18 temp. Euras. (Eur. 2). R: APS 30(1992)289

Lomatophyllum Willd. (~ *Aloe*). Asphodelaceae (Aloaceae). 12 Madagascar, Mascarenes

Lomatozona Baker. Compositae (Eup.-Prax.). 4 Brazil. R: MSBMBG 22(1987)390. Lvs opp., pappus basally connate

Lombardochloa Roseng. & Arrill. = *Briza*

Lombardy poplar *Populus nigra* 'Italica'

Lomelosia Raf. (*Trematostelma*; ~ *Scabiosa*). Dipsacaceae. 40 Medit. to Cauc. R: Willdenowia 15(1985)72. Cult. orn. incl. *L. caucasica* (M. Bieb.) Greuter & Burdet (*Scabiosa c.*, Cauc.) – cult. orn. herbaceous perennial scabious with pale blue fls

Lonas Adans. Compositae (Anth.-Mat.). 1 SW Medit. incl. Eur.: *L. annua* (L.) Vines & Druce – cult. orn. ann. with yellow fls

Lonchitis L. Dennstaedtiaceae. 2 trop. Am., Afr., Madagascar

Lonchocarpus Kunth. Leguminosae (III 6). 130 trop. Am., *L. sericeus* (Poiret) DC reaching W Afr. Many used as fish-poisons (barbasco, haiari) in S Am. (3 p.p.m. eliminates piranha but not other spp.) & sources of rotenone (2–4% by wt.) for insecticides esp. *L. nicou* (Aublet) DC (timbo, Guyana); some timbers esp. *L. sericeus* (savonette)

Lonchophora Durieu = *Matthiola*

Lonchostephus Tul. Podostemaceae. 1 Brazilian Amazon

Lonchostoma Wikström. Bruniaceae. 5 SW Cape

Londesia Fischer & C. Meyer = *Kirilowia*

London plane *Platanus × hispanica*; **L. pride** *Saxifraga × urbium*

long jack *Flindersia xanthostyla, Triplaris* spp.; **l. purples** *Orchis mascula*

longan or **longyen** *Dimocarpus longan*

Longetia Baillon = *Austrobuxus*

Lonicera L. Caprifoliaceae. 180 N hemisphere (Eur. 16) to Mex. & Philipp. R: RMBG 14(1903)27. Honeysuckle. Usu. decid. shrubs & lianes, opp. lvs sometimes connate. Paired fls s.t. discrete with separate fr., s.t. at summit of a common ovary or even more united. Spp. with short-tubed fls bee-poll., long-tubed (e.g. *L. hildebrandiana* Collett & Hemsley (Burma) with fls to 15(–18) cm long) v. fragrant poll. by hawkmoths: fls open in evening just after anthers have dehisced & moved into horizontal position, acting as landing-stage for moths with the style projecting beyond it; later the stamens wither & droop, the style taking up the horizontal position when the fl. has turned from white to yellow. Many cult. orn. spp. & hybrids incl. *L. caprifolium* L. (Euras.) – fls white or purplish, fragrant, liane locally medic., though increasingly replaced in cult. by hybrid with *L. etrusca* Santi (Medit.) = *L. × italica* Schmidt ex Tausch ('*L. × americana*') – decid., true *L. × americana* (Miller) K. Koch (*L. etrusca × L. implexa* Sol.) being evergreen; *L. ciliosa* (Pursh)

Poiret (N Am.) – plant boiled to make shampoo (Flathead Indians); *L. fragrantissima* Lindley & Paxton (China, not found wild) – ± decid. shrub with white fragrant fls in winter; *L. japonica* Thunb. (Japanese h., E As., natur. & serious weed in US) – ± evergreen liane, many cvs incl. varieg.; *L. nigra* L. (C & S Eur. mts) – shrub with molluscicidal saponins; *L. nitida* E. Wilson (China) – evergreen shrub with small lvs, much used as hedging ('Ernest Wilson') & game-cover, rarely fertile exc. 'Fertilis' etc.; *L. periclymenum* L. (woodbine, Eur. & Medit.); *L. xylosteum* L. (fly h., Euras., locally natur. in GB)

Loniceroides Bullock. Asclepiadaceae (III 4). 1 Brazil: *L. harrisonae* Bullock

loofah *Luffa aegyptiaca*

loosestrife, purple *Lythrum salicaria*; **yellow l.** *Lysimachia* spp. esp. *L. vulgaris*

Lopez root *Toddalia asiatica*

Lopezia Cav. Onagraceae. 21 Mex. & C Am. R: AMBG 60(1973)478. C 4, upper 2 ascending with 1 or 2 nectar-drop-like tubercles at base, A 2 with the upper fertile & enfolded by expanded petaloid one; at first insects alight on A, which later grow up out of the way leaving style in their position; in *L. coronata* Andrews (cult. orn.) there is explosive pollen-shedding in that the 2 stamens are under tension & the arrival of insect releases them. Interstaminal nectaries in bird-poll. spp.; chemical stimulants & the pseudonectaries above in fly-poll.

Lophacme Stapf. Gramineae (31d). 2 S C Afr.

Lophanthera A. Juss. Malpighiaceae. 4 Brazil. R: MNYBG 32(1981)36. *L. lactescens* Ducke – cult. orn. tough street-tree with panicles of 300–500 fls

Lophanthus Adans. (~ *Nepeta*). Labiatae (VIII 2). Spp.?

Lophatherum Brongn. Gramineae (24). 2 E As., Indomal. to trop. Aus. Root tubers; false petioles; awns of sterile lemmas involved in disp.

Lophiocarpus Turcz. Phytolaccaceae (III). 4 S Afr. Sieve-tube plastid type confirms placing in P. & not Chenopodiaceae

Lophiola Ker-Gawler. Melanthiaceae (Lophiolaceae; Liliaceae s.l.). 1 E N Am.: *L. aurea* Ker-Gawler – form. placed in Haemodoraceae

Lophiolaceae Nakai = Melanthiaceae

Lophira Banks ex Gaertner f. Ochnaceae. 2 trop. W Afr. R: BJ 113(1991)170. Antibacterial flavonoids (lophirones) & inhibitors of tumour promotors. Seeds yield an oil (meni oil, kaku, zawa) used for cooking, on hair & as soap, chewing-sticks; timber (ekki, ironwood, Afr. oak) esp. (rain forest) *L. alata* Banks ex Gaertner f. (azobé, bongossi, eba) superior to reinforced concrete for wharves, dams & locks & used (untreated) as railway sleepers (*L. lanceolata* Tieghem ex Keay in savanna)

Lophocarpinia Burkart. Leguminosae (I 1). 1 Paraguay, Argentina

Lophocereus (A. Berger) Britton & Rose = *Pachycereus*

Lophochlaena Nees = *Pleuropogon*

Lophochloa Reichb. = *Rostraria*

Lophogyne Tul. Podostemaceae. 2 E C Brazil

Lopholaena DC. Compositae (Sen.-Sen.). 19 trop. & S Afr. R: TRSSA 21(1934)221, NJB 11(1991)79

Lopholepis Decne. Gramineae (33d). 1 S India, Sri Lanka

Lophomyrtus Burret. Myrtaceae (Myrt.). 2 NZ

Lophopappus Rusby. Compositae (Mut.-Nass.). 6 Andes. Shrubs

Lophopetalum Wight ex Arn. Celastraceae (Lophopetalaceae). 18 Indomal. to trop. Aus. *L. multinervium* Ridley (Mal.) – tree with consp. kneeroots in swamp forests; *L. toxicum* Loher (Philippines) – arrow poisons (digitaloid gylcosides) from bark; other spp. with bark (oily) making good firelighters

Lophophora J. Coulter. Cactaceae (III 9). 1–2 S Texas, N & E Mex. R: Brittonia 21 (1969)299. Spineless cacti with turnip-like roots & alks, the button-like crowns mescal buttons containing the hallucinogenic peyote. *L. williamsii* (Salm-Dyck) J. Coulter, used for over 7000 yrs, contains 30 alks, mescaline the active one, leading to brilliantly coloured hallucinations & a feeling of weightlessness

Lophophytaceae Horan. = Balanophoraceae

Lophophytum Schott & Endl. Balanophoraceae. 3 warm S Am. R: FN 23(1980)45. *L. mirabile* Schott & Endl. – pollen tricolporate & hexacolporate (tetrahedral)

Lophopogon Hackel. Gramineae (40a). 2 India. R: HIP 37(1967)t.3648. Glumes 3-toothed

Lophopterys A. Juss. Malpighiaceae. 3 trop. S Am.

Lophopyxidaceae (Engl.). H. Pfeiffer = Celastraceae

Lophopyxis Hook.f. Celastraceae (Lophopyxidaceae). 1 Mal., W Pacific: *L. maingayi* Hook.f. – split stems used for tying thatch

Lophoschoenus Stapf = *Costularia*
Lophosciadium DC = *Ferulago*
Lophosoria C. Presl. Lophosoriaceae. 1 trop. Am.: *L. quadripinnata* (Gmelin) C. Chr. – small tree-fern, the stipe-base hairs & rhiz. hairs found as linings in nests made of *Cyathea* in Mex.
Lophosoriaceae Pichi-Serm. Filicopsida. 1/1 trop. Am. Terr. fern with stem to a few m. tall, solenostelic, apex hairy; lamina bi-(–tri)pinnate with sori superficial without indusia. n = 65
Genus: *Lophosoria*
Prob. allied to Cyatheaceae & Dicksoniaceae
Lophospatha Burret = *Salacca*
Lophospermum D. Don ex R.Taylor (~ *Asarina*). Scrophulariaceae (Antirr.). 8 N & C Am. R: D.A. Sutton, Rev. Antirr. (1988)497. Some cult. orn. scramblers
Lophostachys Pohl. Acanthaceae. 15 trop. Am.
Lophostemon Schott (~ *Tristania*). Myrtaceae (Lept.). 4 N & E Aus., S New Guinea. Cult. orn. street-trees in trop. esp. *L. confertus* (R. Br.) P.G. Wilson & Waterhouse (red, brush or Brisbane box, E Aus.) – timber, varieg. form in cult.
Lophostigma Radlk. Sapindaceae. 2 Bolivia, Peru, Ecuador. R: SB 18(1993)379
Lophostoma (Meissner) Meissner. Thymelaeaceae 4 N trop. S Am.
Lophothecium Rizz. = *Justicia*
Lophotocarpus T. Durand = *Sagittaria*
Lopimia C. Martius (~ *Pavonia*). Malvaceae. 2 trop. Am.
Lopriorea Schinz. Amaranthaceae (I 2). 1 E Afr.
loquat *Eriobotrya japonica*; **wild l.** (W Afr.) *Uapaca kirkiana*
Loranthaceae Juss. Dicots – Rosidae – Santalales. 68 (1 extinct)/900 trop. & temp. (esp. S). Photosynthetic hemiparasites, typically brittle shrublets on tree-branches, less often terrestrial shrubs, lianes or even (*Atkinsonia, Macrosolen*) shrubs or (*Nuytsia*) trees on host roots. Haustorium 1, or several at ends of epicortical roots, plant rarely *Cuscuta*-like, the haustoria usu. promoting gall-like host growth. Stem often dichasial but nodal constrictions 0. Lvs opp. or ternate, simple, entire, rarely scale-like; stipules 0. Fls usu. bisexual, usu. reg., consp. & frequently red or yellow, insect- or bird-poll., in dichasia s.t. resembling heads, racemes, umbels etc.; K a toothed or lobed rim or cup at summit of G, C (3–)5, 6(–9) often with a basal tube equally or unequally cleft, valvate & s.t. nectariferous basally, A as many as & opp. & adnate to C, anthers with longit. slits, disk s.t. present, $\overline{G}$ (0 or) (3 or 4), usu. 1-loc. with c. 4–12 ovules comprising 8-nucleate egg-sac without obvious nucellus or integument. Fr. usu. a berry or drupe with latex & 1(–3) seeds, rarely dry-indehiscent; seed without testa but ± covered with viscous material & often with more than 1 embryo, at least s.t. without obvious radicle but with 2 cotyledons eventually becoming united, endosperm copious, starchy (derived from primary endosperm nuclei). x = 8–12 (reducing series from 12, polyploidy v. rare)
Principal genera: *Amyema, Cladocolea, Decaisnina, Dendrophthoe, Helixanthera, Macrosolen, Phthirusa, Psittacanthus, Scurrula, Struthanthus, Tapinanthus, Taxillus* (many form. referred to *Loranthus*)
Viscaceae (q.v.; poss. close to Santalaceae) form. included, but prob. L. closer to Olacaceae. An old southern fam. (like Proteaceae & Restionaceae). Some may be troublesome in plantation crops in trop., esp. spp. of *Phthirusa*; some used as bird-lime, some dried for orn.
Loranthus Jacq. Loranthaceae. 1 Eur.: *L. europaeus* Jacq. – parasitic on Fagaceae. Trop. spp. now assigned to other genera but some Mal. spp. prob. referable here
Lordhowea R. Nordenstam (~ *Senecio*). Compositae (Sen.-Sen.). 1 Lord Howe Is. Treelet
lords and ladies *Arum maculatum*
Lorentzia Griseb. = *Pascalia*
Lorentzianthus R. King & H. Robinson (~ *Eupatorium*). Compositae (Eup.-Crit.). 1 Bolivia, Arg. R: MSBMBG 22(1987)330
Lorenzochloa Reeder & C. Reeder = *Ortachne*
Loretoa Standley = *Capirona*
Loreya DC. Melastomataceae. (Incl. *Myriaspora*) 13 trop. S Am. R: MNYBG 50(1989)30. *L. collata* Wurdack – fr. ed., imp. food for tapirs & wild pigs
Loricalepis Brade. Melastomataceae. 1 Brazil
Loricaria Wedd. Compositae (Gnap.-Lor.). 19 Andes. R: OB 104(1991)62
Lorinseria C. Presl = *Woodwardia*
Loropetalum R. Br. ex Reichb. Hamamelidaceae (I 1). 2–3 Himal., China, Japan. *L. chinense*

(R. Br.) Oliver – cult. orn. evergreen

Lorostelma Fourn. = *Stenomeria*

Lorostemon Ducke. Guttiferae (III). 3 Brazil

lote-tree *Celtis australis*

Lotononis (DC) Ecklon & Zeyher. Leguminosae (III 28). (Incl. *Buchenroedera*) 120 S Afr. (most), ext. to Medit. (Eur. 2) & India

lotus *Ziziphus lotus*; **sacred l.** *Nelumbo nucifera*

Lotus L. Leguminosae (III 18). 100 N temp. (Eur. 30, Macaronesia 27). Incl. *Dorycnium* & *Tetragonolobus*; R: BJ 25(1898)166, 31(1901)314. Piston poll. mechanism (see fam.) with style receptive only after abrasion, promoting cross-poll. Some cult. orn. & for forage. *L. berthelotii* Lowe ex Masf. (Canary Is.) – fls scarlet; *L. corniculatus* L. (bird's foot trefoil, eggs & bacon, Euras., Medit.) – fodder but southern populations often highly cyanogenic (pack-animals of the Sudan Campaign (1896) died from eating it); *L. jolyi* Batt. (Sahara) – imp. toxic pl.

Loudetia Hochst. ex Steudel. Gramineae (39). 26 trop. & S Afr., Madag., S. Am. (1). *L. esculenta* C. Hubb. (Sudan) – grain ed.

Loudetiopsis Conert (~ *Tristachya*). Gramineae (39). 11 trop. W Afr., S Am.

Loudonia Lindley = *Glischrocaryon*

Louiseania Carrière = *Prunus*

Louisiella C. Hubb. & Léonard. Gramineae (34b). 1 Sudan, Zaire

louro *Ocotea* spp.; **l. inamui** *O. cymbarum*; **l. preto** *O.* spp.; **red l.** *O. rubra*

Lourteigia R. King & H. Robinson (~ *Eupatorium*). Compositae (Eup.-Gyp.). 9 Colombia & Venez. R: MSBMBG 22(1987)132

Lourtella S. Graham, Baas & Tobe. Lythraceae. 1 Peru: *L. resinosa* S. Graham, Baas & Tobe. R: SB 12(1987)519

lousewort *Pedicularis* spp. esp. *P. sylvatica*

Louteridium S. Watson. Acanthaceae. 6 Mexico to C Am.

Louvelia Jum. & Perrier = *Ravenea*

lovage *Levisticum officinale*; **Scotch l.** *Ligusticum scoticum*

Lovanafia M. Pelt. = *Dicraeopetalum*

love apple *Lycopersicon esculentum*; **l. (creeper)** *Comesperma volubile*; **African** or **weeping l.grass** *Eragrostis curvula*; **l.-in-idleness** *Viola tricolor*; **l.-in-a-mist** *Nigella damascena*; **l.-lies-bleeding** *Amaranthus caudatus*

lovi-lovi *Flacourtia inermis*

Lovoa Harms. Meliaceae (IV 2). 2 trop. Afr. *L. trichilioides* Harms (African or Nigerian golden walnut, Benin w., apopo, bibolo, emero, eyan, tigerwood, Congo wood) – much used for furniture as subs. for mahogany

Lowiaceae Ridley. Monocots – Zingiberidae – Zingiberales. 1/7 Indomal. Glabrous perennial herbs with sympodial rhizomes & vessels only in roots. Lvs distichous, with sheathing base, petiole & expanded simple lamina, rolled in bud, with parallel-pinnate venation though some veins not reaching tip. Fls bisexual, orchid-like, bracteate, in axillary cymes, foul-smelling; P 6 petaloid, the lateral 2 inner small, anterior 1 an elliptic or spatulate labellum, the 3 outer narrow & free, A 5 (0 opp. labellum), anthers with longit. slits, $\overline{G}$ (3), extending into slender hypanthium-like neck, 3-loc. with term. style & 3 stigmas & ∞ anatropous (?) bitegmic ovules on axile placentas. Fr. a capsule with ∞ seeds with 3-lobed arils & starchy reserves. x = 9

Genus: *Orchidantha* (cult. orn.). Close to Musaceae

Loxanthera (Blume) Blume. Loranthaceae. 1 W Mal.: *L. speciosa* Blume. R: Blumea 38(1993)114

Loxanthocereus Backeb. = *Cleistocactus*

Loxocalyx Hemsley. Labiatae (VI). 2 China

Loxocarpus R. Br. Gesneriaceae. 15 Malay Peninsula, Java

Loxocarya R. Br. Restionaceae. 7 SW Aus.

Loxococcus H. Wendl. & Drude. Palmae (V 41). 1 Sri Lanka

Loxodera Launert. Gramineae (40h). 5 trop. Afr.

Loxodiscus Hook.f. Sapindaceae. 1 New Caledonia

Loxogrammaceae Ching ex Pichi-Serm. = Polypodiaceae

Loxogramme (Blume) C. Presl. Polypodiaceae (II 6; Loxogrammaceae). 33 Indomal., Pacific (1), Afr. (4), C Am. (1)

Loxoma Garay = *Smithsonia*

Loxoma auctt. See *Loxsoma*

Loxomorchis Rauschert (*Loxoma*) = *Smithsonia*
Loxonia Jack. Gesneriaceae. 3 W Mal. R: PSE 127(1977)201
Loxoptera O. Schulz. Cruciferae. 1 Peru
Loxopterygium Hook.f. Anacardiaceae (IV). 5 trop. S Am. *L. sagotii* Hook.f. (Guyana) – general purpose timber
Loxoscaphe T. Moore = *Asplenium*
Loxostemon Hook.f. & Thomson. Cruciferae. 1 E Himal. to SW China: *L. pulchellus* Hook.f. & Thomson
Loxostigma C.B. Clarke. Gesneriaceae. 4 E Himal. R: NRBGE 34(1975)103
Loxostylis A. Sprengel ex Reichb. Anacardiaceae (IV). 1 S Afr.
Loxothysanus Robinson. Compositae (Helia.-Hym.). 3 E Mex. R: Wrightia 5(1974)45
Loxsoma R. Br. ex A. Cunn. Loxsomataceae. 1 N NZ: *L. cunninghamii* R. Br. ex A. Cunn.
Loxsomataceae C. Presl. Filicopsida. 2/5 NZ, trop. Am. Creeping hairy rhizomes with solenosteles; fronds like *Davallia*; sori like *Trichomanes*
Genera: *Loxsoma, Loxsomopsis*
Affinities obscure; Jurassic fossils (*Stachypteris*) referred here
Loxsomopsis Christ. Loxsomataceae. 1–4 trop. Am. Isolated populations in Costa Rica & Colombia to Bolivia
loyak, loyar *Licuala* spp.
lozane *Tephrosia macropoda*
Lozanella Greenman. Ulmaceae (II). 2 trop. Am.
Lozania S. Mutis. Flacourtiaceae (Lacistemataceae). 4 trop. Am. R: FN 22(1980)201
luan See lauan
Lubaria Pittier. Rutaceae. 1 Venezuela
lucerne *Medicago sativa*; **Paddy's l.** *Sida rhombifolia*
Lucilia Cass. Compositae (Gnap.-Gnap.). (Excl. *Belloa*) 8 S Am. R: OB 104(1991)158
Luciliocline A. Anderb. & Freire (~ *Lucilia*). Compositae (Gnap.-Gnap.). 5 Andes of Peru, Bolivia, Arg. R: BJLS 106(1991)187
Luciliopsis Wedd. = *Chaetanthera*. See also *Cuatrecasasiella*
Lucinaea DC. Rubiaceae (IV 6). 25 Mal., New Caledonia
Luckhoffia A. White & B. Sloane = *Hoodia* × *Stapelia* natural hybrid
lucky bean *Abrus precatorius, Afzelia quanzensis, Erythrina* spp.; **l. nut** *Thevetia peruviana*
lucraban *Hydnocarpus venenata*
Luculia Sweet. Rubiaceae (I 1). 5 Himal. & Yunnan. Cult. orn. evergreen shrubs
Lucuma Molina = *Pouteria*
Lucya DC. Rubiaceae (IV 1). 1 WI
Ludekia Ridsd. Rubiaceae (I 2). 2 Borneo, Philippines. R: Blumea 24(1978)335
Ludia Comm. ex Juss. Flacourtiaceae (8). 23 E Afr., Madag., Masc. R: Adansonia 12(1972)79
Ludisia A. Rich. Orchidaceae (III 3). 1 SE As., Malay Peninsula
Ludovia Brongn. Cyclanthaceae. 3 trop. Am. Climbers with female fls (P 0) sunken
Ludwigia L. (*Ludvigia*). Onagraceae. 82 cosmop. (Eur. 1) esp. Am. R: Reinwardtia 6(1963)327. Herbaceous & floating to woody & erect pls (*L. anastomosans* (DC) Hara (Brazil) – tree) of wet places, many producing aerenchyma facultatively, *L. repens* (L.) Sw. (trop.) in water producing erect spongy roots reaching to water surface as well as ordinary roots. Some cult. orn. esp. in aquaria; *L. hyssopifolia* (G. Don f.) Exell (OW trop. weed) – seeds buoyant but also adhere to birds etc., those near base of capsule shed with corky endosperm but those near apex free, pl. stocked by Chinese herbalists, source of black dye
Lueddemannia Reichb.f. Orchidaceae (V 10). 1 W trop. S Am.
Luehea Willd. Tiliaceae. 15 trop. Am. *L. divaricata* C. Martius (Brazil) – timber for general construction; *L. seemannii* Planchon & Triana (C & N S Am.) – moth-poll.
Lueheopsis Burret. Tiliaceae. 7–9 trop. S Am.
Luerella Brass = *Masdevallia*
Luerssenia Kuhn ex Luerssen = *Tectaria*
Luerssenidendron Domin = *Acradenia*
Luetkea Bong. Rosaceae. 1 W N Am.: *L. pectinata* (Pursh) Kuntze – high mt. cult. orn. rock-pl.
Luetzelburgia Harms. Leguminosae (III 2). 6 Brazil. R: Rodriguesia 42(1977)7
Luffa Miller. Cucurbitaceae. 7 trop. (4 OW). Fr. & seeds medic. Am. *L. acutangula* (L.) Roxb. (As.) – young fr. (sing-kwa, sinqua (melon)) ed.; *L. aegyptiaca* Miller (*'L. cylindrica'*, loofah, vegetable sponge, OW) – bleached vasc. system of mature fr. the loofah of bathrooms
Lugoa DC (~ *Gonospermum*). Compositae (Anth.-Gon.). 1 Canary Is.
Lugonia Wedd. = *Philibertia*

Luina Benth. (? = *Tetradymia*). Compositae (Sen.-Tuss.). 2 NW N Am.

Luisia Gaudich. Orchidaceae (V 16). 40 trop. As. to Japan & Polynesia (Mal. 9, R: Reinw. 10(1988)383)

Lulia Zardini. Compositae (Mut.-Mut.). 1 Brazil: *L. nervosa* (Less.) Zardini. R: BSAB 19(1982)254

lulo *Solanum quitoense*

Luma A. Gray. Myrtaceae (Myrt.). 4 Chile, Arg. Cult. orn. shrubs incl. *L. apiculata* (DC) Burret (arrayán, collimamol, palo colorado)

lumbang oil *Aleurites moluccana*

Lumnitzera Willd. Combretaceae. 2 E Afr. to Pacific: *L. racemosa* Willd. with white fls borne throughout tree ('Modèle d'Attims'), butterfly-poll.; *L. littorea* (Jack) J. Voigt with red somewhat zygomorphic fls borne on outside of canopy ('Modèle de Scarronne'), bird-poll. Mangroves with sea-disp. fr. Timber

Lunania Hook. Flacourtiaceae (9). 14 trop. Am. R: FN 22(1980)207

Lunaria L. Cruciferae. 3 C & SE Eur. R: JAA 68(1987)190. Silicula with satiny paper-white septum. Cult. orn. esp. *L. annua* L. (honesty, penny flower, S Eur., natur. N Eur., N Am.) – dried infrs for winter decoration, young roots allegedly ed.

Lunasia Blanco. Rutaceae. 10 Borneo & E Mal. Many alks

Lunathyrium Koidz. = *Deparia*

Lundellia Leonard = *Holographis*

Lundellianthus H. Robinson (~ *Lasianthaea*). Compositae (Helia.-Verb.). 8 C Am. R: SB 14(1989)544

Lundia DC. Bignoniaceae. 12 trop. Am. Some with calyptrate K

lungwort *Pulmonaria officinalis*; **sea l.** *Mertensia maritima*

lunumidella *Melia azedarach*

lupin(e) *Lupinus* spp.; **Egyptian l.** *L. albus*; **pearl l.** *L. mutabilis*; **Russell l.** *L. × regalis*; **tree l.** *L. arboreus*; **white l.** *L. albus*; **yellow l.** *L. luteus*

Lupinophyllum Hutch. = *Tephrosia*

Lupinus L. Leguminosae (III 31). 200 Andes, Rocky Mts (Calif. 82), Medit. (Eur. 6), trop. Afr. highlands, E S Am. Lupins. Poll. mechanism with piston action (see fam.); many alks. Cult. orn., green manure & fodder with palmate lvs, most commonly seen 'Russell lupins' (*L. × regalis* Bergmans, introd. 1937) prob. hybrids between *L. arboreus* & *L. polyphyllus* (? & other spp.), with 1- or 2-coloured fls. *L. albus* L. (*L. termis*, Egyptian or white l., derived from subsp. *graecus* (Boiss. & Spruner) Franco & Sylva, Aegean) – cattle food, form. used as flour by Egyptians, Greeks & Romans, seeds ed. boiled to remove alks or roasted & s.t. ground as coffee subs.; *L. angustifolius* L. (Medit.) – some strains with alks to 2.5% dry wt., some with 0 (single gene diff.), distinguishable by sheep; *L. arboreus* Sims (tree l., Calif.) – shrubby pl. used in land restoration, 180 kg N /ha/yr fixed on China clay tips in Cornwall; *L. arcticus* S. Watson (N Am.) – seeds from 8000 (? 13 000) BC germ. 1966); *L. latifolius* Lindley ex J. Agardh (W US) – carpeted devastated forest-floors after Mt St Helens eruption (1980); *L. luteus* L. (yellow l., Medit.) – green manure, seeds a coffee subs.; *L. mutabilis* Sweet (tarwi, tarhui, pearl lupin, Andes) – high alt. pulse crop (cvs without toxic quinolizidine alks); *L. perennis* L. (E N Am.) – fodder pl.; *L. polyphyllus* Lindley (W N Am.) – form. cult. orn. cvs

Luronium Raf. Alismataceae. 1 Eur.: *L. natans* (L.) Raf. – cult. orn. rootless floating herb

Luteidiscus H. St. John = *Tetramolopium*

lutqua *Baccaurea ramiflora*

Lutzia Gand. = *Alyssoides*

Luvunga Wight & Arn. Rutaceae. 12 Indomal. R: PANSP 137(1985)221. Locally med.

Luxemburgia A. St-Hil. Ochnaceae. 17 Brazil. R: BJ 113(1991)176

Luziola Juss. Gramineae (9). (Incl. *Hydrochloa*) 12 Am. R: AMBG 52(1965)472. Wet places

Luzonia Elmer. Leguminosae (III 10). 1 Philippines

Luzula DC. Juncaceae. 115 cosmop. (Aus. 15, 12 endemic) esp. temp. Euras. (Eur. 31). R: BMOIPB 95,6(1990)70. Usu. hairy with flat lvs (wood rushes). Some incl. *L. campestris* (L.) DC (Euras., natur. N Am.) with ant-disp. seeds with juicy outgrowths

Luzuriaga Ruíz & Pavón. Luzuriagaceae (Philesiaceae). 4 Peru to Tierra del Fuego, NZ. R: Willdenowia 17(1988)170. Some cult. orn.

Luzuriagaceae Lotsy (~Philesiaceae, Smilacaceae). Monocots – Liliidae – Asparagales. 1/4 S Am., NZ. Separated from Philesiaceae as twiners with poricidal anthers & black seeds but limits controversial & here most genera form. believed allied referred to Convallariaceae or retained in Smilacaceae etc.
Genus: *Luzuriaga*

Another mess in the Liliidae complex (see Liliaceae)

Lyallia Hook.f. Hectorellaceae. (Incl. *Hectorella*) 2 S Is. NZ & Kerguélen Is. Tertiary relicts

Lyauteya Maire = *Cytisopsis*

Lycapsus Philippi. Compositae (Helia.-Per.). 1 Desventuradas Is. (Chile)

Lycaste Lindley. Orchidaceae (V 10). 49 trop. Am. R: J.A. Fowlie (1970) *The genus L.* Cult. orn. epiphytes esp. *L. skinneri* (Lindley) Lindley (*L. virginalis*, Mex. to Belize) – many cvs

lychee *Litchi sinensis*

Lychniothyrsus Lindau. Acanthaceae. 5 Brazil

Lychnis L.=*Silene*

Lychnodiscus Radlk. Sapindaceae. 8 trop. Afr.

Lychnophora C. Martius. Compositae (Vern.-Lych.). 26 Brazil. Some coniferoid xerophytes

Lychnophoriopsis Schultz-Bip. (~ *Lychnophora*). Compositae (Vern.-Lych.). 4 Brazil

Lycianthes (Dunal) Hassler (~ *Solanum*). Solanaceae (1). 200 trop. Am., E As. *L. moziniana* (Poiret) Bitter (Mex. mts) – weedy (?) cultigen of Aztecs ('tlanochtle') with fr. rich in vitamin C

Lycium L. Solanaceae. 100 warm temp. (Eur. 3, Aus. 1) esp. Am. (45, R: AMBG 19(1932)179). Shrubs, often thorny (boxthorn) s.t. used as hedging, some becoming noxious weeds e.g. *L. ferocissimum* Miers (S Afr.) in Aus. *L. barbarum* L. (Duke of Argyll's tea-tree, SE Eur. to China) – medic. in modern Chinese med., natur. GB esp. on walls; *L. pallidum* Miers (W N Am.) – fr. a delicacy in SW US

Lycocarpus O. Schulz. Cruciferae. 1 S Spain: *L. fugax* (Lag.) O. Schulz

Lycochloa Samuelsson. Gramineae (19). 1 Syria. ? Derived from *Schizachne*

Lycomormium Reichb.f. Orchidaceae (V 10). 5 trop. S Am.

Lycopersicon Miller (~ *Solanum*). Solanaceae. 7 W S Am., Galápagos. Tomatoes (love-apples). *L. esculentum* Miller (*L. lycopersicum*, Andes) – comm. tomato eaten as salad veg. (UK record – 2.537 kg), cooked & tinned (esp. irreg. 'Italian' t.), puree or ketchup, soup or juice, or green in chutney & pickle, the seeds yielding a cooking oil also used in soap with the resultant cake as cattle food, many cvs incl. small-fruited var. *cerasiforme* (Dunal) A. Gray (var. *leptophyllum*, cherry t.); seeds pass through herbivores incl. man & germinate in dung, sewage farms etc. Other spp. with tiny fr., e.g. *L. pimpinellifolium* (Jusl.) Miller (currant t., Andes) – fr. c. 1 cm diam.

Lycopodiaceae Pal. ex Mirbel. Lycopsida. 4/380 cosmop. R: OB 92(1987)153. Evergreen herbs, sometimes epiphytic. Stems dichotomously branched, protostelic. Lvs scale- or needle-like with 1 vasc. strand. Sporangia in term. strobili or in axils of lvs; spores all alike. Prothalli monoec. with mycorrhiza
Genera: *Huperzia* , *Lycopodiella*, *Lycopodium* (clubmosses) & *Phylloglossum*

Lycopodiastrum Holub = *Lycopodium*

Lycopodiella Holub (~ *Lycopodium*). Lycopodiaceae. 40 moist temp. (Eur. 2) & trop. R: OB 92(1987)174. Fertile interspecific hybrids in wild

Lycopodium L. Lycopodiaceae. (Incl. *Diphasiastrum*) 40 trop. & temp. (Eur. 5; see fam. for segregate genera). R: OB 92(1987)170. Clubmosses. Some with alks (only pteridophytes thus). Some cult. orn., others used for stuffing upholstery, basket-, bag- & fishing-net-making; spores (l. powder) form. used in sound experiments in physics, those of *L.* (*Diphasiastrum*) *alpinum* L. (N temp. & arctic) dye wool yellow & those of *L. clavatum* L. (N temp.) v. flammable & form. used in fireworks & stage-lighting, also medic., hair powder & dusting agent on some condoms (vegetable sulphur) though some humans are allergic & develop granulosis resembling syphilis!

Lycopsida (Lycopodiatae). Clubmosses. 6/1230 in 3 fams (usu. put in monotypic orders). Clubmosses & allies. Sporophytes with roots & usu. spirally arr. microphylls. Sporangia thick-walled, homo- or heterosporous, borne on a sporophyll or associated with one; antherozoids bi- or multiflagellate. Five orders (2 extinct – Protolepidodendrales appearing in Devonian & Lepidodendrales typically arborescent plants of the coal-forests of the Upper Carboniferous) with the extant Lycopodiales (Lycopodiaceae), Isoetales (Isoetaceae) & Selaginellales (Selaginellaceae)

Lycopsis L. = *Anchusa*

Lycopus L. Labiatae (VIII 2). 4 N temp. (Eur. 2), Aus. R: AMN 68(1962)95. *L. europaeus* L. (gipsywort, water horehound, Euras., natur. N Am.) – roots used to stain face brown

Lycoris Herbert. Amaryllidaceae (Liliaceae s.l.). 11 China & Japan to Burma. Cult. orn. bulbous herbs

Lycoseris Cass. Compositae (Mut.-Mut.). 11 trop. Am. Dioec. shrubs

Lycurus Kunth. Gramineae (31e). 3 warm & trop. Am. R: Parodiana 4(1986)267. *L. setosus* (Nutt.) C. Reeder in SW N Am. & S Bolivia to NW Arg.

Lydenburgia N. Robson = *Catha*

Lygeum Loefl. ex L. Gramineae (15). 1 Medit. incl. Eur. *L. spartum* L. (albardine) – fibre used for mats, sails & ropes etc., one of the 'esparto' grasses (cf. *Ampelodesmos, Stipa*)

Lyginia R. Br. Restionaceae. 1 SW Aus.: *L. barbata* R. Br. Filaments ± connate

Lygisma Hook.f. Asclepiadaceae (III 4). 3 SE As.

Lygodesmia D. Don. Compositae (Lact.-Step.). 7 Am. esp. W N Am. R: SBM 1(1980)

Lygodisodea Ruíz & Pavón = *Paederia*

Lygodium Sw. Schizaeaceae. 40 trop. & warm (E US 1). Climbing ferns with fronds of determinate growth in young stages & homologous ones with dichotomizing apices with indeterminate, giving rise to twining axes used for basketry, fish-traps, mats & yarn

Lygos Adans. = *Retama*

Lymanbensonia Kimnach = *Lepismium*

Lymania Read (~ *Araeococcus*). Bromeliaceae (3). 6 trop. Am. R: Bradea 4(1987)396

lyme grass *Leymus arenarius*

Lyonia Nutt. Ericaceae. 35 E & SE As., E N Am., Mex., WI. R: JAA 62(1981)129,315. Some cult. orn. shrubs

Lyonothamnus A. Gray. Rosaceae. 1 is. off S Calif.

Lyonsia R. Br. = *Parsonsia*

Lyperanthus R. Br. Orchidaceae (IV 1). 5 Aus. (4 endemic), NZ. New Caledonian spp. = *Megastylis*

Lyperia Benth. (~ *Sutera*). Scrophulariaceae. 6 S Afr. R: O.M. Hilliard (1994) *The Manuleeae*: 212

Lyrocarpa Hook. & Harvey. Cruciferae. 2 California. R: CDH 3(1941)169

Lyroglossa Schltr. = *Stenorrhynchos*

Lyrolepis Rech.f. = *Carlina*

Lysiana Tieghem. Loranthaceae. 8 Aus. G 4-loc. with axile placentas

Lysicarpus F. Muell. Myrtaceae (Lept.). 1 Queensland

Lysichiton Schott. Araceae. (V 1). 1 Kamchatka, Sakhalin & Kuriles: *L. camtschatcensis* (L.) Schott; 1 W N Am.: *L. americanus* Hultén & H. St John (skunk cabbage) – poll. *Pelecomalius testaceum* (staphylinid beetle) seeking food & mating site, locally medic. & eaten; both cult. orn. waterside pls, the first with white, the second yellow spathes but hybridize (poss. only subspp.)

Lysichlamys Compton = *Euryops*

Lysiclesia A.C. Sm. = *Orthaea*

Lysidice Hance. Leguminosae (I 4). 1 S China, Vietnam: *L. rhodostegia* Hance – cult. orn. shrub

Lysiloma Benth. Leguminosae (II 5). 9 trop. Am. Some mahogany-like timbers esp. sabicu (*L. sabicu* Benth., WI) used in boat-building

Lysimachia L. Primulaceae. c. 150 temp. (esp. Himal.; China 134, Eur. 14, incl. *Asterolinon*) & warm. Herbs, rarely shrubs, elaiophores attractive to bees, some locally medic., some cult. orn. *L. clethroides* Duby (NE to SE As.) – lvs a local condiment; *L. minorcensis* Rodr. (Minorca) – extinct in wild; *L. nemorum* (yellow pimpernel, W & C Eur.) – form. medic.; *L. nummularia* L. (creeping Jenny, moneywort, Eur., natur. E N Am.) – form. medic. & tea; *L. vulgaris* L. (yellow loosestrife, Euras.) – garden pl.

Lysinema R. Br. Epacridaceae. 5 (excl. *Woollsia*) SW Aus.

Lysionotus D. Don. Gesneriaceae. 30 Himal. to Japan (China 28). Epiphytic shrubs, some cult. orn.

Lysiosepalum F. Muell. Sterculiaceae. 2 SW Aus.

Lysiostyles Benth. Convolvulaceae. 1 trop. S Am.

Lysiphyllum (Benth.) de Wit = *Bauhinia*

Lysipomia Kunth. Campanulaceae. 27 high Andes

Lythraceae J. St Hil. Dicots – Rosidae – Myrtales. (Excl. Punicaceae & Sonneratiaceae) 27/600 trop. with few temp. Herbs, less often shrubs or trees, often with alks & usu. with internal phloem. Lvs opp. (rarely whorled or spirally arr.), simple; stipules vestigial or 0. Fls bisexual, often heterostylous, solit., fascicled in axils or term. racemes, reg. or not, with consp. hypanthium s.t. spurred or with epicalyx, (3)4–,6–,8(–16)-merous; K valvate lobes of hypanthium, C free attached at summit of or within hypanthium, crumpled in bud, A usu. twice K or C & in 2 whorls inserted in hypanthium (rarely fewer (1 in *Rotala*) or numerous & centrifugal as in *Lagerstroemia*), anthers with longit. slits, G̲ (2–4(–6)), pluriloc., but partitions s.t. not reaching apex, rarely pseudomonomerous with 1 ovule, surrounded by annular nectary disk & term. by style, with (2–)±∞ anatropous bitegmic ovules with zig-zag micropyles per loc. Fr. usu. a capsule, dehiscing variously; seeds (1–)±∞ s.t. winged, ± 0 endosperm, embryo straight, oily. x = 5–11 (? orig. 8)

Principal genera: *Ammania, Cuphea, Diplusodon, Lagerstroemia, Lythrum, Nesaea, Rotala*

Distyly (e.g. *Lythrum, Nesaea, Pemphis*) app. derived from tristylous condition; some aquatics (e.g. *Decodon, Rotala*) with '*Hippuris* syndrome'

Many dye-pls esp. henna (*Lawsonia*) & *Lafoensia*, timber from *Lagerstroemia* & *Physocalymma* spp.; many cult. orn. esp. *Cuphea, Lagerstroemia* & *Lythrum*; some hallucinogens (*Heimia, Lagerstroemia*)

Lythrum L. Lythraceae. 36 cosmop. (Eur. 13, incl. *Peplis*). Herbs with C 6, A 12, in tristylous fls – styles of 3 lengths on diff. pls promoting outcrossing, Darwin having demonstrated that fewest viable seeds formed by selfing; the longest stamens have largest pollen-grains which are deposited by insects on longest styles with largest stigmatic papillae. This is condition in *L. salicaria* L. (purple loosestrife, OW (diploid Aus., tetraploid Eur.), natur. N Am.) – cult. orn. In some, tristyly has broken down, distyly believed to have evolved polyphyletically for some have populations with fls equivalent to the long-& short-styled forms, others to the long- & mid-styled. *L. portula* (L.) D. Webb (*Peplis p.*, Eur. to Cauc.) – water purslane

Lytocaryum Tol. (~ *Syragrus*). Palmae (V 5b). 3 SE Brazil. R: IBM 56(1987)109. *L. weddellianum* (H. Wendl.) Tol. (*Cocos w., Microcoelum w.*) – much cult. housepl.

M

Maackia Rupr. Leguminosae (III 2). 8 E As. Some cult. orn. decid. trees. R: NRBGE 8(1913)99

Maba Forster & Forster f. = *Diospyros*

Mabea Aublet. Euphorbiaceae. 50 trop. Am. Water-disp. seeds nearly destroyed by fish. *M. occidentalis* Benth. app. poll. by red woolly opossum & bats (only E. thus)

mabee or **mabi bark** *Colubrina elliptica*

mabola *Diospyros blancoi*

Mabrya Elisens. Scrophulariaceae (Antirr.). 5 SW N Am. R: D.A. Sutton, *Rev. Antirrhineae* (1988)490

maca *Lepidium meyenii*

macachi *Arjona tuberosa*

Macadamia F. Muell. Proteaceae. 12 E Mal., Aus. (10, endemic), New Caled. (pachycaul). Seeds ed. (macadamia nuts, Queensland nuts, bopplefnuts (Aus.)) esp. *M. integrifolia* Maiden & Betche (Queensland) – principal sp. grown, & *M. tetraphylla* L. Johnson (Queensland, NSW) – fr. dehiscing on tree, both confused with *M. ternifolia* F. Muell. (rarely cult.) & grown with one another, also their hybrid. Sold either in endocarp & then cracked like almonds, or shelled, roasted & salted, poss. the most delicious of all nuts but expensive; oil used in cosmetics. Much cult. in Calif., Hawaii & Malawi; *M. hildebrandtii* Steenis (Sulawesi) – planted in Sumatra as fire-lane tree in *Pinus* plantations

Macairea DC. Melastomataceae. 22 trop. S Am. R: MNYBG 50(1989)54

Macaranga Thouars. Euphorbiaceae. 280 OW trop. Dioec. or rarely monoec. trees & shrubs, many typical of sec. forest & some with extrafl. nectaries & ant-inhabited twigs: seedlings of *M. triloba* Muell. Arg. (W Mal.) 7–10 cm tall already with hollows in stipules & stems; some with lvs to 60 cm diam. e.g. *M. mappa* (L.) Muell. Arg. (*M. grandifolia*, Philippines, cult. orn.); some locally medic. *M. peltata* (Roxb.) Muell. Arg. (Kenda, S India, Sri Lanka) – lvs used for steaming jaggery etc.

Macarenia P. Royen. Podostemaceae. 1 Colombia

Macarisia Thouars. Rhizophoraceae. 7 Madagascar

macaroni *Triticum turgidum* Durum Group

Macarthuria Huegel ex Endl. Molluginaceae. 6–7 SW & SE Aus.

Macartney rose *Rosa bracteata*

macary bitter *Picramnia antidesma*

macassà *Aeollanthus suaveolens*

Macassar ebony *Diospyros celebica*; **M. oil** *Schleichera oleosa*

macaw bush *Solanum mammosum*; **m. palm** *Aiphanes erosa*

Macbridea Elliott ex Nutt. Labiatae (VI). 2 SE US. Antedated by *M.* Raf.

Macbrideina Standley. Rubiaceae (I 5). 1 Peru

Macdougalia Heller = *Hymenoxys*

mace *Myristica fragrans*

Macfadyena A. DC. Bignoniaceae. 3–4 trop. Am. Lianes, cult. orn. esp. *M. unguis-cati* (L.) A. Gentry (*Bignonia u., Doxantha u.*, Mex. to Arg.) – yellow fls, fr. to 30 cm long, mass-flowering in dry but not wet regions, 2n = 40, 80 (only example of polyploidy in *M.*)

Macgregoria F. Muell. Stackhousiaceae. 1 E Aus.: *M. racemigera* F. Muell.

Machaeranthera Nees (excl. *Xylorhiza*; ~ *Aster*). Compositae (Ast.-Sol.). 36 W N Am. R: Phytol. 68(1990)439. Some cult. orn. herbs; *M. venusta* (M.E. Jones) Cronq. & Keck used as selenium & uranium indicator

Machaerina Vahl. Cyperaceae. 45 trop. & warm esp. Aus.

Machaerium Pers. Leguminosae (III 4). 120 trop. Am., 1 extending to W Afr. coast. Many lianes with recurved stipular thorns; some timbers (palisander) & locally medic.

Machaerocarpus Small = *Damasonium*

Machaerocereus Britton & Rose = *Stenocereus*

Machairophyllum Schwantes. Aizoceae (V). 10 Cape. Cult. orn. cushion-formers to 120 cm diam.

Machaonia Bonpl. Rubiaceae (III 3). 30 trop. Am.

Machilus Nees = *Persea*

Mackay bean *Entada gigas*

Mackaya Harvey. Acanthaceae. 1 S Afr.: *M. bella* Harvey – cult. orn. shrub

Mackeea H. Moore. Palmae (V 4h). 1 New Caled.: *M. magnifica* H. Moore. R: Allertonia 3(1984)327. Tallest palm in New Caled., allied to *Actinokentia*

Mackenziea Nees = *Strobilanthes*

Mackinlaya F. Muell. Araliaceae (3). 5 E Mal., W Pacific. R: BBMNHB 1(1951)3. 2-celled ovary & other features like Umbelliferae. Alks

Macleania Hook. Ericaceae. 45 trop. Am. Some cult. orn. & ed. fr. sold in local markets

Macleaya R. Br. (~ *Bocconia*). Papaveraceae (I). 2 temp. E As. Alks. Cult. orn. herbs esp. *M. cordata* (Willd.) R. Br. (plume poppy, China & Japan) to 2.5 m with small fls (C 0) & *M.* × *kewensis* Turrill, its hybrid with *M. microcarpa* (Maxim.) Fedde (C China)

Maclura Nutt. Moraceae (I). (Incl. *Cudrania, Plecospermum*, excl. *Chlorophora* p.p.) 12 Indomal. to Aus., Afr. Am. R: PKNAWC 89(1986)243. *M. pomifera* (Raf.) C. Schneider (Osage orange, bow wood, Arkansas & Texas) – dioec. spiny tree with male fls in racemes & females in heads, fr. a yellow syncarp of achenes enclosed in fleshy P on a fleshy receptacle, timber used for bows, war-clubs, railway sleepers, fence-posts, widely planted in US as living fence in C19 & now natur. with roots source of yellow dye & lvs used to feed silkworms; *M. tinctoria* (L.) Steudel (*Chlorophora t.*, fustic, trop. Am.) – heartwood provides the yellow, brown or green dye, fustic, bark medic.; *M. tricuspidata* Carrière (*Cudrania t.*, China, Korea) – fr. ed., cult. as hedgepl. in US

Maclurodendron Hartley. Rutaceae. 6 SE As. to Mal. R: GBS 35(1982)1

Maclurolyra C. Calderón & Söderstrom (~ *Olyra*). Gramineae (4). 1 Panamá: *M. tecta* C. Calderón & Söderstrom

Macnabia Benth. ex Endl. = *Nabea*

Macodes (Blume) Lindley. Orchidaceae (III 3). 14 Mal. to Papuasia

Macoubea Aublet. Apocynaceae. 2 trop. Am. R: MMNHN 30(1985)170. Some ed. fr.

Macowania Oliver. Compositae (Gnap.). 12 S Afr. (10), Ethiopia & Yemen (2). R: NRBGE 34(1975)260

Macphersonia Blume. Sapindaceae. 8 trop. E Afr., Madag.

macqui *Aristotelia chilensis*

Macrachaenium Hook.f. Compositae (Mut.-Nass.). 1 S Am. Andes. Herb in *Nothofagus* forest

Macradenia R. Br. Orchidaceae (V 10). 12 Florida & C Am. to trop. S Am. Cult. orn. epiphytes

Macraea Hook.f. (~ *Lipochaeta*). Compositae (Helia.-Verb.). 1 Galápagos

Macranthera Nutt. ex. Benth. Scrophulariaceae. 1 SE US

Macranthisiphon Bureau ex Schumann. Bignoniaceae. 1 coastal Ecuador & Peru

Macroberlinia (Harms) Hauman = *Berlinia*

Macrobia (Webb & Berth.) Kunkel = *Aichryson*

Macrobriza (Tzvelev) Tzvelev = *Briza*

macrocarpa *Cupressus macrocarpa*

Macrocarpaea (Griseb.) Gilg. Gentianaceae. 30 trop. Am. Shrubs & trees, s.t. pachycaul, to 5 m

Macrocaulon N.E. Br. = *Carpanthea*

Macrocentrum Hook.f. Melastomataceae. 15 N trop. S Am.

Macrochaetium Steudel = *Tetraria*

Macrochlaena Hand-Mazz. = *Nothosmyrnium*

Macrochloa Kunth = *Stipa*

Macrochordion de Vriese (~ *Aechmea*). Bromeliaceae (3). 10 trop. Am. R: Phytol. 66(1989)77

Macroclinidium Maxim. (~ *Pertya*). Compositae (Mut.-Mut.). 3 Japan. Herbs
Macroclinium Barb. Rodr. = *Notylia*
Macrocnemum P. Browne. Rubiaceae (I 1). 20 C Am. to Colombia
Macrococculus Becc. Menispermaceae (I). 1 New Guinea: *M. pomiferus* Becc. R: KB 26(1972)418
Macrodiervilla Nakai = *Weigela*
Macroditassa Malme = *Ditassa*
Macroglena (C. Presl) Copel. = *Cephalomanes*
Macroglossum Copel. = *Angiopteris*
Macrohasseltia L.O. Williams. Flacourtiaceae (7; ? Tiliaceae). 1 C Am.
Macrolenes Naudin ex Miq. Melastomataceae. 15 Thailand, Mal. *M. muscosa* (Blume) Bakh.f. (Mal.) – liane with ed. fr. & shoots, local medic.
Macrolepis A. Rich. = *Bulbophyllum*
Macrolobium Schreber. Leguminosae (I 4). c. 70 trop. Am. Extrafl. nectaries
Macromeria D. Don. Boraginaceae. 10 Mexico to S Am.
Macropanax Miq. Araliaceae. 14 E Himal. to W Mal. R: JNTCFP 1985, 1:29
Macropelma Schumann = *Sacleuxia*
Macropeplus Perkins. Monimiaceae (V 2). 1 E Brazil
Macropetalum Burchell ex Decne. Asclepiadaceae (III 5). 1 S Afr.: *M. burchellii* Decne.
Macropharynx Rusby. Apocynaceae. 5 trop. S Am.
Macropidia J.L. Drumm. ex Harvey = *Anigozanthos*
Macropiper Miq. (~ *Piper*). Piperaceae. 9 Pacific. R: BJLS 71(1975)1. Shrubs & small trees. *M. excelsum* (Forster f.) Miq. (kawa-kawa, pepper-tree, NZ) – locally imp. medic. & supposed aphrodisiac (juvenile hormone analogue in roots!), cult. orn. to 6 m
Macroplectrum Pfitzer = *Angraecum*
Macropodanthus L.O. Williams. Orchidaceae (V 16). 6+ Mal.
Macropodiella Engl. Podostemaceae. 5 W trop. Afr. R: Adansonia 17(1977)293
Macropodina R. King & H. Robinson (~ *Eupatorium*). Compositae (Eup.-Gyp.). 3 S Brazil, Arg. R: MSBMBG 22(1987)28
Macropodium R. Br. Cruciferae. 2 C As. & Sakhalin, Japan
Macropsychanthus Harms ex Schumann & Lauterb. (~ *Dioclea*). Leguminosae (III 10). 4 Philippines, New Guinea, Micronesia
Macropteranthes F. Muell. Combretaceae. 3 N Aus. *M. kekwickii* F. Muell. (bulwaddy) forms dense thickets in N Aus.
Macroptilium (Benth.) Urban. Leguminosae (III 10). 8 trop. Am. R: Boissiera 28(1978)151
Macrorhamnus Baillon = *Colubrina*
Macrorungia C.B. Clarke = *Anisotes* + *Metarungia*
Macrosamanea Britton & Rose (~ *Albizia*). Leguminosae (II 5). 8 trop. Am.
Macroscepis Kunth. Asclepiadaceae (III 3). 8 trop. Am.
Macrosciadium Tikh. & Lavrova. Umbelliferae (III 8). 2 SW As., Cauc.
Macroselinum Schur (~ *Peucedanum*). Umbelliferae (III 10). 1 S Eur., Cauc.
Macrosiphonia Muell. Arg. Apocynaceae. 10 trop. Am. Cult. orn. like *Mandevilla* but fls opening in evening
Macrosolen (Blume) Reichb. Loranthaceae. 25 S & SE As. to New Guinea. *M. parasiticus* (L.) Danser (S India, Sri Lanka) – bushes to several m tall, trunks to 15 cm diam.
Macrosphyra Hook.f. Rubiaceae (II 1). 3 trop. Afr. R: BJBB 28(1958)27
Macrostegia Nees = *Vitex*
Macrostelia Hochr. Malvaceae. 3 Madagascar, 1 Queensland
Macrostigmatella Rauschert = *Eigia*
Macrostylis Bartling & Wendl. (~ *Euchaetis*). Rutaceae. 10 S & SW Cape. R: JSAB 47(1981)373
Macrosyringion Rothm. = *Odontites*
Macrothelypteris (H. Itô) Ching. Thelypteridaceae. 10 Natal & Masc. to Hawaii. R: Blumea 17(1969)25. Mostly with 3-pinnatifid fronds & small indusia. Cult. orn. esp. *M. torresiana* (Gaudich.) Ching (Masc. to Japan & Polynesia) – natur. Am.
Macrotomia DC ex Meissner = *Arnebia*
Macrotorus Perkins. Monimiaceae (V 2?). 1 SE Brazil
Macrotyloma (Wight & Arn.) Verdc. Leguminosae (III 10). 24 OW trop. R: HIP 38,4(1982). Pulses esp. *M. uniflorum* (Lam.) Verdc. (horse gram, OW trop.) – fodder etc. *M. geocarpum* (Harms) Maréchal & Baudet (*Kerstingiella g.*, ground bean, W Afr.) – ed. seeds, fr. buried like *Arachis*
Macroule Pierce. See *Ormosia*
Macrozamia Miq. Zamiaceae (I). 15 Aus. (13 E, 1 C, 1 SW: *M. riedlei* (Gaudich.) Gardner –

possums effect disp., seed form. trad. food 'by-yu'). R: PLSNSW 84(1959)87, MNYBG 57(1990)204. Dioec. cult. orn. with alks; seeds ed. locally (Queensland nut) if soaked & pounded or baked. *M. spiralis* (Salisb.) Miq. (NSW) – source of good quality arrowroot & ed. seeds (burrawang)

Macvaughiella R. King & H. Robinson (~ *Eupatorium*). Compositae (Eup.-Ager.). 4 C Am. R: SB 16(1991)639

Madagascar cardamom *Aframomum angustifolium*; **M. clove** *Cryptocarya* sp. (*Ravensara aromatica*); **M. copal** *Hymenaea verrucosa*; **M. ebony** *Diospyros haplostylis*; **M. jasmine** *Marsdenia floribunda*; **M. mahogany** *Khaya madagascariensis*; **M. nutmeg** *C.* sp. (*R. aromatica*); **M. periwinkle** *Catharanthus roseus*; **M. plum** *Flacourtia indica*; **M. rubber** *Cryptostegia*, *Landolphia*, *Marsdenia* & *Mascarenhasia* spp.

Madagaster Nesom (~ *Aster*). Compositae (Ast.-Ast.). 5 Madag. R: Phytol. 75(1993)94

madar *Calotropis* spp.

Madarosperma Benth. = *Tassadia*

Maddenia Hook.f. & Thomson. Rosaceae. 4 Himal., China. Dioec. trees & shrubs

madder *Rubia tinctorum*; **field m.** *Sherardia arvensis*; **Indian m.** *R. cordifolia*; **Levant** or **wild m.** *R. peregrina*

Madeira marrow *Sechium edule*; **M. vine** *Anredera cordifolia*

Madhuca Buch.-Ham. ex J. Gmelin. Sapotaceae (II). (Incl. *Ganua*) 100 Indomal., esp. W Mal., to Aus. R: Blumea 7(1953)364, 10(1966)1. Timbers & oilseeds (the original illipe nuts used for margarine etc.) esp. *M. longifolia* (Koenig) Macbr. (mahwa, mahua, mowa, moa, mi, India) – fls also ed. (rich in nectar), seed-cake ('mahwa meal') used as worm-killer in lawns; *M. pallida* (Burck) Baehni (*Ganua p.*, Mal.) – yields a gutta-purcha

Madia Molina. Compositae (Helia.-Mad.). 18 W N Am., Chile. Tarweeds. Oilseeds esp. *M. sativa* Molina (Chile) used as olive oil subs., cult. in Eur.

Madras hemp *Crotalaria juncea*; **M. thorn** *Pithecellobium dulce*

madroña, madrono (laurel) *Arbutus menziesii*

madwort *Asperugo procumbens*

Maerua Forssk. Capparidaceae. 50 trop. & S Afr. to India. Some locally used timbers (incl. toothpicks) & ed. lvs. Fr. a lomentum-like berry, that of *M. crassifolia* Forssk. ed. Sahara

Maesa Forssk. Myrsinaceae. 100 OW trop. to Japan & Aus. Some locally medic. Some cult. orn. e.g. *M. indica* (Roxb.) Sweet (India). *M. lanceolata* Forssk. (trop. & S Afr., Madag.) – toxic, the fr. an effective bactericide

Maesobotrya Benth. Euphorbiaceae. 20 trop. Afr.

Maesopsis Engl. Rhamnaceae. 1 trop. Afr.: *M. eminii* Engl. (musizi) – comm. timber

mafoureira or **mafurra** *Trichilia emetica*

mafu *Fagaropsis angolensis*

Maga Urban = *Montezuma*

Magadania Pim. & Lavrova. Umbelliferae (III 8). 2 NE As.

Magallana Cav. Tropaeolaceae. 1 temp. S Am. R: OB 108(1991)125

Magdalenaea Brade. Scrophulariaceae. 1 SE Brazil

Magnistipula Engl. Chrysobalanaceae (4). 11 trop. Afr. & Madagascar (2). R: BJBB 46(1976)281. Some with ant-infested stem domatia. *M. butayei* de Wild. (trop. Afr.) – imp. in rain-making ceremonies

Magnolia L. Magnoliaceae (I 1). (Incl. *Talauma*) c. 100 Himal. to Japan & W Mal., E N Am. to trop. Am. Trees & shrubs with alks much cult. (incl. hybrids) for spectacular white, pink, purple or yellow fruity-scented, beetle-poll. fls with congested fr. & often sarcotestal seeds (some spp. form. referred to *Talauma* with ± indehiscent fr.). R (temp. spp.): N. Treseder (1978) *Magnolias*. Some timbers (e.g. for trad. Jap. shoes) & medic. barks. *M. acuminata* (L.) L. (cucumber tree, E N Am.) – timber for flooring etc., fr. purplish red to 10 cm long; *M. grandiflora* L. (bull bay, loblolly m., SE US) – evergreen usu. grown against a wall in GB, flowering when 20 yrs old from seed (faster from cuttings), several cvs incl. 'Goliath' with fls to 30 cm diam. & hybrids with *M. virginiana* (Freeman hybrids incl. 'Exmouth'); *M. heptapeta* (Buc'hoz) Dandy (*M. conspicua*, *M. denudata*, yulan, E & S China) – cult. orn. decid. tree, fls ed.; *M. kobus* DC (*M. praecocissima*, Japan) – cult. orn. decid. tree, timber used for engraving, matches etc. in Japan; *M. officinalis* Rehder & E. Wilson (cultigen, ? orig. C China) – bark used as tonic in Chinese medic.; *M. × soulangiana* Soul.-Bod. (*M. heptapeta* × *M. quinquepeta* (Buc'hoz) Dandy (*M. liliiflora*, pollen parent, temp. China), a deliberate cross c. 1820) – most commonly planted m. with many cvs; *M. stellata* (Sieb. & Zucc.) Maxim. ('*M. tomentosa*', Japan) – free-flowering small sp.; *M. virginiana* L. (*M. glauca*, E N Am.) – decid. or evergreen cult. orn., timber used for broom-handles etc.

Magnoliaceae Juss. Dicots – Magnoliidae – Magnoliales. 7/165 trop. to warm temp. esp. N.

Trees & shrubs usu. with alks & always with oil-cells in parenchyma. Lvs spirally arr., simple, entire (lobed in *Liriodendron*); stipules large, enclosing term. bud, decid. Fls large, usu. term. & solit., bisexual (unisexual in *Kmeria*), reg., usu. with long receptacle; P spiral or in 3 or more whorls, 6–18, often ± all petaloid, A ∞ spirally arr., originating centripetally, often ± strap-shaped with 4 paired microsporangia embedded in surface (adaxial; abaxial in *Liriodendron*) & ± elongate connective, pollen monosulcate, G (2–)∞, conduplicate & s.t. not completely closed but with ± distinct style & term. stigma & 2(–±∞) anatropous, bitegmic ovules on marginal placenta. Fr. a follicle or indehiscent & berry-like or samaroid (*Liriodendron*), with carpels growing together s.t. forming fleshy syncarp; seeds usu. large, endotestal with sarcotesta, suspended by fibrils from vasc. bundle of raphe in spp. with dehiscent fr., embryo v. small with suspensor in copious oily, proteinaceous endosperm. x = 19

Classification & genera (after Nooteboom):

I. **Magnolioideae** (lvs entire, sarcotesta, fr. not a samara; 2 tribes):
   1. **Magnolieae** (sympodial, fls term. on twigs): *Kmeria, Magnolia, Manglietia, Pachylarnax*
   2. **Michelieae** (monopodial, infls on axillary short shoots): *Elmerrillia, Michelia*

II. **Liriodendroideae** (lvs 2–10-lobed, testa adhering to endocarp, fr. a samara): *Liriodendron*

Both subfams show an E As./E N Am. (–trop. Am.) distribution. The family has many primitive features but has no staminodes intermediate between P & A & the sarcotestal seed achieves the same ecological effect as a primitive arillate seed dangling from the funicle rather than the vasc. bundle of the raphe

Many magnificent cult. orn. & some timbers (*Liriodendron, Magnolia, Michelia*)

Magnoliophyta. See Angiospermae

Magnoliopsida. See Dicotyledonae

Magodendron Vink. Sapotaceae (V). 2 E New Guinea: R: Blumea 40(1995)91

Magonia A. St-Hil. Sapindaceae. 1 Paraguay, Brazil: *M. pubescens* A. St-Hil.

maguey *Agave cantala*

Maguireanthus Wurd. Melastomataceae. 1 Guyana

Maguireocharis Steyerm. Rubiaceae (I 1). 1 Guayana Highland

Maguireothamnus Steyerm. Rubiaceae (I 5). 2 Venezuela

Magydaris Koch ex DC. Umbelliferae (III 5). 2 Medit. (inc. Eur.)

Mahafalia Jum. & Perrier = *Cynanchum*

mahaleb (cherry) *Prunus mahaleb*

Maharanga A. DC (~ *Onosma*). Boraginaceae. 10 E Himal. to SW China, Thailand (1)

Mahawoa Schltr. Asclepiadaceae (III 1). 1 Sulawesi: *M. montana* Schltr

× **Mahoberberis** C. Schneider. Hybrids between *Berberis* & *Mahonia* spp.

mahoe (WI) *Thespesia populnea*; (NZ) *Melicytus ramiflorus*

mahogany *Swietenia* spp. esp. *S. mahagoni* (**Cuban m.**), (Afr.) *Khaya* spp. Many medium coloured medium-weight timbers have been called m. (see KB 1936:193) incl.: **Australian m.** *Dysoxylum fraserianum*; **bastard m.** *Eucalyptus botryoides*; **Bataan** or **Philippine m.** *Shorea polysperma*; **Benin m.** *Khaya* spp.; **Borneo m.** *Calophyllum inophyllum*; **Burma m.** *Pentace burmanica*; **Cape m.** *Trichilia emetica*; **Ceylon m.** *Melia azedarach*; **cherry m.** *Tieghemella heckelii*; **Colombian m.** *Cariniana pyriformis*; **East India m.** *Pterocarpus dalbergioides*; **Gaboon m.** *Aucoumea klaineana*; **Indian white m.** *Canarium euphyllum*; **Natal m.** *Kiggelaria africana*; **red m.** *Afzelia quanzensis, Eucalyptus resinifera*; **white m.** *E. acmenoides, E. robusta*

maholtine *Wissadula amplissima*

Mahonia Nutt. Berberidaceae (II 2). c. 100 Himal. to Japan & Sumatra, N & C Am. R: JLSBot. 57(1961)296. Like *Berberis* but thornless shrubs & pachycaul treelets with pinnate thorny lvs & fascicled racemes or panicles of yellow scented fls; alt. hosts for cereal rusts, cult. orn. incl. ground-cover, some ed. fr.; alks. *M. aquifolium* (Pursh) Nutt. (NW N Am., natur. Eur., 'Oregon grape') – berries cooked by N Am. Indians, planted as game cover in Eur., locally medic.; *M.* × *media* C. Brickell (*M. japonica* (Thunb.) DC (China, long cult. Japan) × *M. lomariifolia* Takeda (Burma, W China), accidental cross in Ireland in 1950s) – commonly planted winter-flowering pachycauls, several cvs esp. 'Charity'; *M. repens* (Lindley) G. Don f. (holly grape, W N Am.) – cult. orn., locally medic., fr. used in jellies, drinks etc.; *M. swaseyi* (Buckley) Fedde (Texas) – promising fruit-bush

Mahurea Aublet. Guttiferae (II). 2 trop. S Am. R: MNYBG 29(1978)134

mahwa or **mahua** *Madhuca longifolia*

Mahya Cordemoy = *Lepechinia*

Maianthemum G. Weber ex Wigg. Convallariaceae (Liliaceae). (Incl. *Smilacina*) 27 N temp., Himal., C Am. R (Am.): JAA 67(1986)371. Fls 2- or 3-merous; *M. paludicola* LaFrankie (Costa Rica) – rhiz. upright. Cult. orn. esp. *M. bifolium* (L.) F.W. Schmidt (May lily, Euras.) & *M. racemosum* (L.) Link (*Smilacina r.*, false, American or wild spikenard, N Am.)

maiden pink *Dianthus deltoides*; **m's blush** *Sloanea australis*

maidenhair fern *Adiantum* spp. esp. *A. capillus-veneris*; **m. tree** *Ginkgo biloba*

Maidenia Rendle. Hydrocharitaceae. 1 NW Aus. Male fls float to surface & float towards females

maidu *Pterocarpus marsupium*

Maieta Aublet. Melastomataceae. 2 trop. Am. Heterophyllous; some have bladdery outgrowths of lvs inhabited by ants

maigyee *Strobilanthes cusia*

Maihuenia (F. Weber) Schumann. Cactaceae (I). 2 S Chile, S Arg. Cult. orn. winter-hardy in S Eur.

Maihueniopsis Speg. = *Opuntia*

Maillardia Frappier & Duchartre = *Trophis*

Maillea Parl. = *Phleum*

Maingaya Oliver. Hamamelidaceae (I 1). 1 Penang, Perak (Malay Pen.)

maire *Nestegis* sp.

Maireana Moq. (*'Kochia'*). Chenopodiaceae (I 5). 57 Aus. R: FA 4(1984)179. Char. of dry plains of C & S Aus. esp. *M. sedifolia* (F. Muell.) P.G. Wilson (bluebush, common on Nullabor) & *M. pyramidata* (Benth.) P.G. Wilson (blackbush, S Aus.)

Mairetis I.M. Johnston. Boraginaceae. 1 Canary Is., Morocco

Mairia Nees. Compositae (Ast.-Ast.). 14 S & SW Cape

maize *Zea mays*

Majidea J. Kirk ex Oliv. Sapindaceae. 5 trop. Afr., Madagascar

Majorana Miller = *Origanum*

makalao *Cyperus laevigatus*

makoré *Tieghemella heckelii*

makrut lime *Citrus hystrix*

makua *Diospyros mollis*

Malabaila Hoffm. Umbelliferae (III 11). 8–10 E Med. (Eur. 3) to C As. & Iran. Some locally ed.

Malabar gourd *Cucurbita ficifolia*; **M. kino** *Pterocarpus marsupium*; **M. nightshade** *Basella alba*; **M. oil** *Cymbopogon citratus*; **M. rosewood** *Dalbergia latifolia*; **M. tallow** *Vateria indica*

Malacantha Pierre = *Pouteria*

Malacca cane *Calamus* spp. esp. *C. scipionum*; **M. teak** *Afzelia rhomboidea*, *Intsia palembanica*

Malaccotristicha Cusset & G. Cusset. Podostemaceae. 1 Malay Pen.: *M. malayana* (Dransf. & Whitmore) Cusset & G. Cusset. R: BMNHN 4,10(1988)174

Malachra L. Malvaceae. 8 warm & trop. Am., some natur. OW. Epicalyx s.t. 0. *M. capitata* (L.) L. – source of excellent jute-like fibre

Malacocarpus Fischer & C. Meyer (~ *Peganum*). Zygophyllaceae. 1 C As.

Malacocera R. Anderson. Chenopodiaceae (I 5). 4 Aus. R: JABG 2(1980)139

Malacomeles (Decne) Engl. Rosaceae (III). 3 Texas to Guatemala. R: CJB 68(1990)2234

Malacothamnus E. Greene. Malvaceae. 20 SW US, Mex., Chile (1). R: LWB 6(1951)113. Cult. orn. tree-like shrubs

Malacothrix DC. Compositae (Lact.-Mal.). 16 W N Am. R: AMN 58(1957)494, Madrono 16(1962)258

Malacurus Nevski = *Elymus*

Malagasia L. Johnson & B. Briggs. Proteaceae. 1 Madagascar

malagueta pepper *Aframomum melegueta*

Malaisia Blanco = *Trophis*

malambo *Croton malambo*

Malanea Aublet. Rubiaceae (III 3). 21 trop. Am. Lianes

Malania Chun & S.K. Lee. Olacaceae. 1 S China. Oil-pl.

malatti *Jasminum* spp.

Malaxis Sol. ex Sw. Orchidaceae. c. 300 subcosmop. (Eur. 2 (*Hammarbya*, *Microstylis*), Afr. 12). Terr. herbs (adder's mouth, US). *M. paludosa* (L.) Sw. (*Hammarbya p.*, bog orchid, N temp.) – foliar embryos at leaf-tips effect veg. reproduction

Malay apple *Syzygium malaccense*

Malcolmia R. Br. Cruciferae. c. 30 Medit. (Eur. 11) to Afghanistan. R: OB 33(1972)1, SBT 66(1972)239. *M. maritima* (L.) R. Br. (Virginia stock, Albania & Greece) – cult. orn. annual with reddish to white fls

male bamboo *Dendrocalamus strictus*; **m. fern** *Dryopteris filix-mas*

Malea Lundell. Ericaceae. 1 Mexico

Malephora N.E. Br. Aizoaceae (V). 8–9 S Afr. Cult. orn. succ. shrubs used for erosion control in Calif.

Malesherbia Ruíz & Pavón. Malesherbiaceae. 27 S Peru, N Chile, W Arg. R: Gayana 16(1967)1

Malesherbiaceae D. Don. Dicots – Dilleniidae – Violales. 1/27 S Am. Herbs, s.t. woody, often with foetid glandular hairs. Lvs simple (s.t. pinnatifid), spirally arr.; stipules 0. Fls bisexual, reg., with long hypanthium (s.t. curved), solit. or in racemose infls; K 5, valvate, C 5 valvate, toothed corona, A 5 with longit. slits & $\underline{G}$ (3 or 4), 1-loc. with parietal placentas, both borne on central androgynophore, styles free, ovules ∞, anatropous. Fr. a capsule within persistent hypanthium; seeds with oily endosperm & no aril
Genus: *Malesherbia*
Close to Passifloraceae & Turneraceae (arillate)

Malinvaudia Fourn. = *Matelea*

Malleastrum (Baillon) J. Leroy. Meliaceae (I 4). 16 Madag., Comoro Is., Aldabra

mallee *Eucalyptus* spp. with pl. forming a thicket of stems; **fuchsia m.** *E. forrestiana*

Malleola J.J. Sm. & Schltr. Orchidaceae (V 16). c. 30 SE As., Mal.

Malleostemon J. Green. Myrtaceae (Lept.). 6 SW Aus. R: Nuytsia 4(1983)295

mallet bark *Eucalyptus astringens*

Mallinoa J. Coulter = *Ageratina*

Mallophora Endl. Labiatae (III; Dicrastylidiaceae). 2 W Aus.

Mallophyton Wurd. Melastomataceae. 1 Venezuela

Mallotonia (Griseb.) Britton (~ *Tournefortia*). Boraginaceae. 1 Florida, Mex., WI: *M. gnaphalodes* (L.) Britton (iodine bush)

Mallotopus Franchet & Savat. (~ *Arnica*). Compositae (Hele.-Cha.). 1 NE As.

Mallotus Lour. Euphorbiaceae. c. 140 Indomal. to E As., Fiji & E Aus., trop. Afr. with Madag. (2). Some light timbers, medic. & dyes of local imp. esp. *M. philippensis* (Lam.) Muell. Arg. (kamala, NW Himal. to Aus.) – red dye from fr.

mallow spp. of *Abutilon, Hibiscus, Lavatera, Malva* & other Malvaceae; **common m.** *M. sylvestris*; **curled m.** *M. verticillata* var. *crispa*; **Egyptian m.** *M. parviflora*; **Indian m.** *A. theophrasti*; **Jew's m.** *Corchorus olitorius*; **marsh m.** *Althaea officinalis*; **musk m.** *M. moschata*; **rose m.** *Hibiscus* spp.; **scarlet globe m.** *Sphaeralcea coccinea, S. angustifolia* ssp. *cuspidata*; **tree m.** *L. arborea*

Malmea R. Fries. Annonaceae. 18 trop. Am. R: AHB 10(1930)37, (1931)318

Malmeanthus R. King & H. Robinson (~ *Eupatorium*). Compositae (Eup.-Crit.). 3 Brazil, Uruguay, Arg.. R: MSBMBG 22(1987)359

maloga bean *Vigna lanceolata*

Malope L. Malvaceae. 3 Medit. (Eur. 2). Annuals with 3 leafy involucral bracts & many 1-seeded mericarps app. arranged in superposed whorls. *M. trifida* Cav. (Spain, N Afr.) – cult. orn. with white to purple fls

Malortiea H. Wendl. = *Reinhardtia*

Malosma (Nutt.) Raf. = *Rhus*

Malouetia A. DC. Apocynaceae. 25 trop. Am. & Afr. (4, R: AUWP 85-2(1985)70). Alks. Some fish-poisons & allegedly hallucinogenic narcotics sometimes fatal. *M. tamaquarina* A. DC common in Amaz., toxic but eaten by the pajuíl bird (*Nothocrax urumutum* (Spix)), which is often domesticated & flesh eaten all year, though bones in fruiting season (Mar–June) poison dogs

Malouetiella Pichon = *Malouetia*

Malperia S. Watson. Compositae (Eup.-Alom.). 1 Calif. & Baja Calif. R: MSBMBG 22(1987)235

Malpighia L. Malpighiaceae. c. 40 trop. Am. esp. Carib. Trees & shrubs, s.t. with stinging hairs; some with cleistogamous fls. Some cult. orn. incl. *M. glabra* L. (*M. punicifolia*, Barbados cherry, acerola, trop. Am.) – fr. ed., high in vitamin C & used in syrups, jam etc.

Malpighiaceae Juss. Dicots – Rosidae – Polygalales. 67/1100 trop. & warm esp. S Am. Small trees, shrubs or lianes often with unusual sec. growth & s.t. with alks. Lvs usu. opp., simple & entire, very often with 2 large fleshy glands on petiole or abaxial surface; stipules s.t. large & united. Fls usu. bisexual & bilaterally symmetrical to ± reg., 5-merous, borne on jointed 2-bracteolate pedicels in racemes of cymes; receptacle ± convex, K imbricate s.t. basally connate, very often with pair of conspic. abaxial glands at base, C imbricate, often clawed, with ciliate to fringed margins, A (1)2(3) whorls (some often sterile), filaments ± basally connate, anthers with longit. slits (rarely term. pores), disk 0,

G ((2)3(–5)), pluriloc. with axile placentas & ± distinct styles & 1 anatropous to hemitropous bitegmic ovule per loc. Fr. often a schizocarp of winged to nut-like mericarps, s.t. a nut or drupe; seeds with large straight to circinate oily embryo with ± 0 endosperm. x = 6, 9–12+

Principal genera: *Acridocarpus, Banisteriopsis, Byrsonima, Heteropterys, Hiptage, Hiraea, Malpighia, Mascagnia, Stigmaphyllon*

Intermediate between Polygalales & Linales (Cronquist) but app. closest to Vochysiaceae & Trigoniaceae in P. App. with most archaic genera in Guayana Highland, the NW genera visited by oil-gathering bees such as *Centris* spp. absent from OW, where char. oil-glands of K are vestigial or 0 (Anderson in Cronquist) but in 41 genera (only 10 fams with them) & known from mid-Eocene fossils. Bird-disp. seems to have arisen independently in unrelated genera (*Bunchosia, Byrsonima, Malpighia*)

Ed. fr. (*Bunchosia, Malpighia*), hallucinogens (*Banisteriopsis, Diplopterys*), timber (*Byrsonima*) & some cult. orn.

Malpighiantha Rojas. Malpighiaceae. 2 Argentina (= ?)

malt (extract) *Hordeum vulgare*

Maltebrunia Kunth. Gramineae (9). 5 trop. & S Afr., Madag. R: HIP 36(1962)t.3595

Maltese cross *Silene chalcedonica*

malukang fat *Polygala butyracea*

Malus Miller. Rosaceae (III). 55 N temp. (Eur. 5). R: CJB 68(1990)2234. Apples. Sometimes incl. in *Pyrus* but differing in ± pubescent lvs (for distinctly 3-lobed *M. trilobata* (Labill.) Schneider see *Eriolobus t.*), acute leaf-margin teeth, yellow pollen (purple in *Pyrus*), gritty sclereid tissue absent from fr., etc. Fr. & orn. trees, *M. pumila* Miller (*M. communis, M. domestica*, '*M. sylvestris* (L.) Miller ssp. *mitis* (Wallroth) Mansfield' C As.) the most imp. temp. fr. tree of over 2000 cvs (those from pips = 'pippins', orig. small green-fruited cvs = 'codlins', a pop. C13 (& long after) apple being 'Costard', hence term 'costermonger'), derived from variable wild populations form. called *M. sieversii* (Lindley) M. Roemer, millions of 'wild' apples known from Neolithic & Bronze Age Switzerland, the Lake-dwellers storing them dried; fr. (flavour due to a single compound) used as dessert (most widely grown is 'Red Delicious') or cooked or for juice, s.t. fermented to cider (hard c. in US), some 70% in GB, or distilled to apple brandy (calvados) esp. in Normandy whence cider-making introd. GB c. 1630 ('Redstreak' in Herefordshire, which county was producing 1.5 m gallons per annum in 1890s), the vinegar culinary & a cure-all; good eating cvs incl. 'Cox's Orange Pippin' (chance seedling (like 'Bramley's Seedling'), 'Golden Delicious' (orig. 'Mullins' Yellow Seedling') & 'Granny Smith' (orig. c. 1830). The fr. (pome) an enlarged receptacle, containing the inf. ovary (core) with seeds (pips), to 1.357 kg (UK record); fls protogynous, visited by bees; extrafl. nectaries present. Escaped forms ('*M. pumila* Miller', '*M. sylvestris* (L.) Miller') often known as crab apples, a name also used for cult. orn. spp. & hybrids. Cult. spp. incl.: *M. baccata* (L.) Borkh. (E As.) – fr. ed. fresh, dried or preserved; *M. floribunda* van Houtte – pop. early-flowering tree prob. of hybrid origin; *M. hupehensis* (Pampan.) Rehder (*M. theifera*, China, Assam) – cult. orn. with early fls & small orn. fr., lvs used as tea; *M. prunifolia* (Willd.) Borkh. (Chinese a., NE As.) – poss. ancestor of some orchard apples

Malva L. Malvaceae. c. 40 Eur. (12), Medit., temp. As., trop. Afr. mts. Mallows, the fr. known as cheeses (shape reminiscent of some types of c.). Nectar secreted in receptacle covered with hairs preventing rain-wash & access by short-tongued insects, the fl. v. protandrous with A in centre at first, the styles later lengthening to occupy this position as A curve back & down; in small-flowered spp. the styles later curve down & twist amongst A effecting self-poll. Cult. orn. & some leaf vegetables etc. *M. moschata* L. (musk mallow, Eur., N Afr., natur. N Am.) cult. orn.; *M. nicaeensis* All. (Euras., Medit., natur. N Am.) & *M. parviflora* L. (Egyptian m., Medit. to Afghanistan) imp. ed. mallows used since 6000 BC, viable seeds of latter recovered from 200-yr.-old adobe in Calif. & Mex.; *M. pusilla* Sm. (*M. rotundifolia*, p.p., Euras.) – locally medic., fibre-crop for small farms on Amaz. varzéa, seeds germ. after sealing in cannister for 100 yrs (Dr Beal's experiment); *M. sylvestris* L. (common m., Eur., Medit., natur. N Am.) – fr. ed.; *M. verticillata* L. 'Crispa' (curled m., Euras.) – cult. salad pl.

Malvaceae Juss. Dicots – Dilleniidae – Malvales. 111/1800 cosmop. esp. trop. Herbs, shrubs & some trees usu. with stellate hairs & parenchyma typically with scattered mucilage-cells & -cavities or -canals. Lvs spirally arranged, simple to ± dissected & usu. palmately-veined; stipules usu. present. Fls usu. bisexual, often with epicalyx, solit. & axillary or in cymes; K 5 valvate, s.t. basally united, C 5 often adnate to base of A-tube, convolute (or imbricate), A (5–)∞ initiated centrifugally & assoc. with limited no. of trunk-bundles, connate in a tube for most of length, anthers with longit. slits & usu. spinulose pollen, G (1–)5(–∞) with as many (twice in Ureneae) styles as carpels, ± basally united, & as many locules with axile placentation & 1–∞ anatropous to campylotropous bitegmic ovules with zig-zag micropyles. Fr. a loculicidal capsule, schizocarp, rarely berry or samara; seeds exotegmic, cotyledons folded, endosperm oily & proteinaceous, copious to 0. x = 6–17, 20+

Classification & principal genera:

Gossypieae (G(3–5) with as many connate styles, 5-dentate staminal column, capsule & gossypol glands, R: BG 129(1968)303): *Gossypium*

Hibisceae (G(5) with 5 free styles, otherwise like G. but gossypol glands 0): *Hibiscus, Kosteletzkya*

Decaschistieae (G(6–)10, otherwise like H. exc. fr. a capsule eventually breaking up into segments at maturity): *Decaschistia* (only)

Ureneae (Like H. but fr. a schizocarp & twice as many styles as carpels): *Pavonia, Sida*

Malveae (Like H. but fr. a schizocarp, G 3–∞, apex of staminal column antheriferous): *Abutilon, Lavatera, Malva, Sphaeralcea, Wissadula*

V. close to Bombacaceae & some genera assigned differently by authors, *Montezuma* here in B., *Cenocentrum* here in M. etc., but pollen-grain sculpturing (spiny in M., smooth in B.) supports usu. separation of fams, in which most woody genera (but cf. *Plagianthus*) are in B. *Astiria* & *Cheirolaena* (Sterculiaceae) poss. referrable here. Carpels in vertical rows in *Malope* & allies reduced to 1 whorl in other tribes though Ureneae have 10 carpels initiated but the upper 5 abort except for styles

Many cult. orn. & fibre-pls esp. cotton (*Gossypium*), but also *Abutilon, Decaschistia, Hibiscus, Kydia, Malachra, Plagianthus, Sida, Urena* & *Wissadula*, the most familiar cult. orn. being mallows (*Malva, Lavatera* etc.), hollyhocks (*Alcea*), *Abutilon, Hibiscus, Malope, Sidalcea* etc. Some food-pls typically v. mucilaginous like okra (*Abelmoschus*) & *Malva* while *Hoheria* & *Thespesia* yield timber & *Sphaeralcea* used in hair conditioners. Spp. of *Hibiscadelphus, Kokia* & *Lebronnecia*, woody island pls of Pacific, extinct or almost so

Malvastrum A. Gray. Malvaceae. 14 trop. & warm. R: Rhodora 84(1982)1

Malvaviscus Fabr. Malvaceae. 3 trop. Am. R: AMBG 29(1942)183. Fr. a berry. Cult. orn. in trop. where natur. esp. *M. arboreus* Cav. with bright red C, hummingbird-poll.

Malvella Jaub. & Spach (~ *Sida*). Malvaceae. 4 Am., Medit. (Eur. 1). R: SWNat. 19(1974)97

Mamillopsis C.J. Morren ex Britton & Rose = *Mammillaria*

Mammea L. Guttiferae (III). c. 50 trop. (Am. 1, Afr. 1, Madag. 20, 27 Indomal. to Pacific). Trees; some monkey-disp. in Amaz. *M. americana* L. (mammee apple, WI) – cult. for ed. fr. & fls used in liqueur-making (eau de Créole), seeds toxic to fish, chicks & some insects (insecticidal substituted coumarins, also in *M. africana* Sabine, trop. Afr.); *M. suriga* (Roxb.) Kosterm. (India) – cult. orn., timber, fls to dye silk, fr. ed.

mammee (apple) *Mammea americana*; **m. zapote** *Pouteria sapota*

Mammillaria Haw. Cactaceae (III 9). c. 150 SW US to Colombia & Venezuela esp. Mex. R: CSJ 43(1981)41 (incl. *Dolicothele*); J. M. Soc. Low, often tuft-forming cacti v. commonly cult. orn., some with hooked spines (e.g. *M. bocasana* Poselger, C Mex.) or sometimes unisexual fls (e.g. *M. dioica* M. Brandegee, SW Calif. & Baja Calif.) or with long soft hair-like spines (e.g. *M. hahniana* Werderm., C Mex.); some with apices with true dichotomies (e.g. *M. parkinsonii* Ehrenb. & *M. perbella* Hildm. ex Schumann (C Mex.). Fr. ed. (chilitos)

Mammilloydia F. Buxb. (~ *Mammillaria*). Cactaceae (III 9). 1 NE Mex.

mammoth tree *Sequoiadendron giganteum*

mamoncillo *Melicoccus bijugatus*

Mamorea Sota = *Thismia*

man orchid *Aceras anthropophorum*

mana grass *Cymbopogon nardus*

Mananthes Bremek. = *Justicia*

Manaosella J. Gómes. Bignoniaceae. 1 Brazil

manchineel *Hippomane mancinella*

Manchurian jute *Abutilon theophrasti*; **M. walnut** *Juglans mandschurica*

Mancoa Wedd. Cruciferae. 10 Mexico, Andes

mandarin (orange) *Citrus reticulata*

Mandenovia Alava. Umbelliferae (III 11). 1 Caucasus

Mandevilla Lindley. Apocynaceae. 114 trop. Am. Usu. lianes, many form. cult. orn. greenhouse pls (*Dipladenia*) with some hybrids, incl. Chilean jasmine (*M. laxa* (Ruíz & Pavón)

Woodson, Bolivia & N Arg.). Local medic., latex against warts in NW Amazonia, that of *M. vanheurckii* (Muell. Arg.) Markgraf antifungal (Colombian Amaz.)

mandioca *Manihot esculenta*

Mandragora L. Solanaceae (1). 6 Medit. (Eur. 2) to Himal. Mandrake. Alks; thick tuberous roots with fancied resemblance to human form, esp. *M. officinarum* L. (the mandrake of western literature assoc. with many myths (see ASNB 13,11(1991)49), allegedly screaming when pulled from ground & deafening human gatherer so that dogs were supposed to extract them, S Eur.) – contains hyoscyamine, form. medic., e.g. as anaesthetic up to 1846 (introd. of ether); *M. turcomanica* Mizg. – now restricted to SW Turkmenistan ('white mandragora') but poss. the 'soma' (khaoma) of ancient India & Iran

mandrake *Mandragora* spp. esp. *M. officinarum*

Manekia Trel. Piperaceae (Peperomiaceae). 1 Hispaniola (WI)

Manettia Mutis ex L. Rubiaceae (I 1/IV 24). c. 80 trop. Am. R: JB 57 suppl. (1918)1. Lianes & twining herbs, some cult. orn. esp. *M. cordifolia* C. Martius (Bolivia to Arg. & Peru) – ipecacuanha adulterant, fls bright red, & *M. luteorubra* Benth. (*M. inflata*, Paraguay & Uruguay) – fls yellow-tipped. In Peru lvs of some spp. chewed, blackening (? & preserving) teeth, also febrifugal

Manfreda Salisb. (~ *Agave*). Agavaceae. 22 SE US to Mex. R: CSM 30(1985)56. Rhizomes produce suds; some cult. orn.

mangel-wurzel *Beta vulgaris*

Mangenotia Pichon. Asclepiadaceae (I). 1 trop. W Afr.: *M. eburnea* Pichon

mange-tout *Pisum sativum* var. *macrocarpon*

Mangifera L. Anacardiaceae (I). c. 40–60 Indomal. Mango. R: Lloydia 12(1949)73; A.J.G.H. Kostermans & J.-M. Bompard (1993) *The mangoes*. *M. indica* L. (mango, cultigen (introd. Afr. by AD 1000) with v. many cvs) – cross-poll. (between diff. cvs) required, the drupe an excellent fr., also used in chutney, pickles, squashes, the ground seed a source of a flour, the timber for floor-boards, tea-chests etc. (L.B. Singh (1960) *The Mango*); 25 other spp. with ed. fr. e.g. *M. pajang* Kosterm. (bambangan, Borneo) with skin peeled like a banana & *M. caesia* Jack (bauno, binjai, W Mal.); *M. odorata* Griff. (apple m., cultigen) – grows better than *M. indica* in everwet sites, flour used in Javanese delicacies (do dol); some grown as shade-trees etc.

mangium *Acacia mangium*

Manglietia Blume (~ *Magnolia*). Magnoliaceae (I 1). 25 E Himal. & S China to Mal.

Manglietastrum Law = *Magnolia*

mango *Mangifera* spp. esp. *M. indica*; **apple m.** *M. odorata*; **bush m.** *Cordyla pinnata*; **plum m.** *Bouea macrophylla*

mangold *Beta vulgaris*

Mangonia Schott. Araceae (VIII 3). 1 Brazil

mangosteen *Garcinia mangostana*

mangrove Woody pls growing in muddy swamps inundated by tides, e.g. spp. of *Aegiceras, Avicennia, Bruguiera, Ceriops, Conocarpus, Kandelia, Laguncularia, Lumnitzera, Pelliciera, Rhizophora, Sonneratia, Xylocarpus*; **m. bark** bark of these pls used in comm. leather-tanning, though leather turned intense red by it

Manicaria Gaertner. Palmae (V 4b). 4 trop. Am. *M. saccifera* Gaertner (bussu palm) – source of sago in Venez., lvs used as sails, the spathe used as a hat, the temiche cap of the Lower Amazon, seeds an oil source

Manihot Miller. Euphorbiaceae. 98 trop. & warm Am. R: FN 13(1973)19. Monoec. trees, shrubs & herbs. *M. esculenta* Crantz (*M. utilissima*, cassava, manioc, mandioca, tapioca, gari (W Afr.), cultigen) – shrubby tree with large tuberous roots rather immune to insect attack because of high levels of cyanide, a reliable crop (cult. by women in Amaz., cf. coca) on somewhat impoverished soils with over 100 cvs with differing amounts of cyanide ('sweet' ones with glycosides only in bark of tubers), which is removed by squeezing the ground tuber in water & by evaporation during drying; cassava meal (Brazilian arrowroot) & tapioca used in soups, puddings etc., a glue form. used on postage stamps, sugar, alcoholic drinks & acetone all derived from it, the toxic juice evaporated being cassareep used for preserving meat & in certain table sauces, wood suitable for chip- & particle-board, extrafl. nectaries attract red ants which prevent pl. being climbed by beans which only then climb tougher maize in the maize–beans–cassava system of the Kayapó people of S Am.; *M. glaziovii* Muell. Arg. (Brazil) – source of Ceara or Manicoba rubber & oilseeds

Manihotoides D. Rogers & Appan. Euphorbiaceae. 1 Mexico

Manila copal *Agathis dammara*; **M. elemi** *Canarium luzonicum*; **M. hemp** *Musa textilis*; **M.**

maguey *Agave cantala*

Manilkara Adans. Sapotaceae (I). 65 trop. (As. to Pacific 15, Afr. & Madag. 20, Am. 30). Milky latex & fr. of commercial imp. esp. *M. bidentata* (A. DC) A. Chev. (balata, trop. Am.) – source of non-elastic rubber used in machine-belts, boot-soles, largely from wild trees; *M. chicle* (Pittier) Gilly (crown gum, trop. Am.) – subs. for *M. zapota*; *M. hexandra* (Roxb.) Dubard (palu, India) – hard timber; *M. obovata* (Sabine & G. Don f.) J.H. Hemsley (African pear, W Afr.); *M. zapota* (L.) P. Royen (sapodilla (plum), chiku, chicle, naseberry, beef apple, Mexico & C Am.) – tapped like rubber but only once every 2–3 yrs, the chewing-gum of the Aztecs, the tree being encouraged by the Mayas such that its presence in many sites a reflection of the practices of that lost culture, fr. ed.

Manilkariopsis (Gilly) Lundell = *Manilkara*

Maniltoa R. Scheffer. Leguminosae (I 4). 20–25 Indomal. to Aus. esp. New Guinea

manio *Podocarpus nubigenus, P. salignus*

manioc *Manihot esculenta*

Manisuris L. = *Rottboellia*

manjack *Cordia collocca*

manna Various ed. materials, some of pl. origin & usu. sweet, often exudations following insect attack, like honeydew but set hard; that of Bible could have been lichen; modern comm. sources incl. *Fraxinus ornus* (**m. ash**) & it is coll. from *Tamarix* spp., *Hammada salicornica* (poss. Biblical m.), *Larix decidua* etc.; **m. croup** *Glyceria fluitans*

Mannagettaea H. Sm. Orobanchaceae. 3 E Siberia to W China

Manniella Reichb.f. Orchidaceae (III 3). 1 trop. W Afr.

Manniophyton Muell. Arg. Euphorbiaceae. 1 trop. Afr. (C): *M. africanum* Muell. Arg. (gasso nut) – liane with fibres used for ropes & nets

manoao *Halocarpus biformis*

Manochlamys Aellen = *Exomis*

Manoelia Bowdich = *Withania*

Manongarivea Choux = *Lepisanthes*

Manostachya Bremek. Rubiaceae (IV 1). 3 trop. Afr.

Manotes Sol. ex Planchon. Connaraceae. 4 humid trop. Afr. R: AUWP 89-6(1989)294

Manothrix Miers. Apocynaceae. 2 Brazil

Mansoa DC. Bignoniaceae. 15 trop. S Am. *M. alliacea* (Lam.) A. Gentry etc. widely used indig. medic., cult. orn.; *M. standleyi* (Steyerm.) A. Gentry (Ecuador) – garlicky leaf decoction used as febrifuge

Mansonia J.R. Drumm. ex Prain. Sterculiaceae. 5 W, C, E Afr., Assam, Burma – v. disjunct, *M. gagei* J.R. Drumm. ex Prain (kalamet, Burma) – fragrant wood used as cosmetic; *M. altissima* (A. Chev.) A. Chev. (W Afr.) – timber (aprono, bété) used for furniture

Mantalania Capuron ex J. Leroy. Rubiaceae (II 1). 2–3 Madag.

Mantisalca Cass. (~ *Centaurea*). Compositae (Card.-Cent.). 4 Medit. (Eur. 1)

Mantisia Sims. Zingiberaceae. 3 Indomal.

manuka *Leptospermum scoparium*

Manulea L. Scrophulariaceae. 74 S Afr. R: O.M. Hilliard (1994) *The Manuleeae*: 291. Some cult. orn. annuals & shrublets

Manuleopsis Thell. Scrophulariaceae. 1 SW Afr.: *M. dinteri* Thell. R: O. Hilliard (1994) *The Manuleeae*: 78

manzanilla *Crataegus stipulosa*

manzanita *Arctostaphylos pungens*

manzanote *Olmediella betschleriana*

Maoutia Wedd. Urticaceae (III). 15 Indomal., Polynesia. Female fls with P 0. *M. puya* (Hook.) Wedd. (Himal. to Burma) – fibre for cloth & sails

Mapania Aublet. Cyperaceae. 73 trop. (exc. Madag.). D.A. Simpson (1992) *Revision of the genus M.* Some resemble pandans; some in Malay Pen. allegedly rat-disp. *M. palustris* (Steudel) Fernandez-Villar (Indomal.) – used for weaving mats & baskets

Mapaniopsis C.B. Clarke. Cyperaceae. 1 N Brazil

maple *Acer* spp.; **Australian m.** *Flindersia brayleyana*; **field** or **hedge m.** *A. campestre*; **great** or **Scottish m.** *A. pseudoplatanus*; **Japanese m.** *A. palmatum*; **Norway m.** *A. platanoides*; **Queensland m.** *F. brayleyana*; **red m.** *A. rubrum*; **rock, sugar** or **striped m.** *A. saccharum* (**fiddle-back** or **bird's-eye m**. from its burrs)

Mapouria Aublet = *Psychotria*

Mappia Jacq. Icacinaceae. 5 trop. Am.

Mappianthus Hand.-Mazz. Icacinaceae. 2 S China, Borneo

Maprounea Aublet. Euphorbiaceae. 4 trop. Am. (2), Afr. (2)

maqui *Aristotelia chilensis*

Maquira Aublet. Moraceae (III). 5 trop. Am. R: FN 7(1972)64. Arrow poisons (cardenolides). *M. sclerophylla* (Ducke) C. Berg (Amaz. Brazil) – hallucinogenic snuff from fr.

Maracanthus Kuijt = *Oryctina*

maracuja, maracuya *Passiflora quadrangularis*

Marah Kellogg. Cucurbitaceae. 7 W N Am. R: Madrono 13(1955)113. Tendrilled monoec. climbers with large tubers. Some cult. orn. incl. *M. macrocarpus* (E. Greene) E. Greene (S Calif. & Baja Calif.) – seeds source of a red dye

Marahuacaea Maguire (~ *Amphiphyllum*). Rapateaceae. 1 Venez.: *M. schomburgkii* (Maguire) Maguire. R: ABV 14,3(1984)16

marama bean *Bauhinia esculenta*

marang *Artocarpus odoratissimus*

Maranta L. Marantaceae. c. 20 trop. Am. R(subg. M.: 16): NJB 6(1986)729). *M. arundinacea* L. (arrowroot, Bermuda or St Vincent a., C Am.) – rhizome the source of arrowroot, a readily digestible (small-grained) starch used for infants & invalids esp. in treatment of diarrhoea, also used as face-powder & coating on carbon-less paper used for computer print-out, poss. fuel-alcohol source & fibre for tear-resistant paper bags, not cult. pre-Columbian Am., mostly grown at St Vincent, WI; other spp. cult. orn. housepls esp. *M. leuconeura* C. J. Morren (prayer plant, Brazil) notably var. *kerchoviana* C. J. Morren with light green lvs marked with a row of dark brown blotches on either side of midrib

Marantaceae R. Br. Monocots – Zingiberidae – Zingiberales. 29/535 trop. esp. Am. Rhizomatous perennial herbs, some with sublianoid stems, the rhizomes sympodial usu. starchy, vessels usu. only in roots. Lvs ± distichous with open sheath, distinct & sometimes winged petiole & simple lamina rolled from 1 side in bud & often patterned, with distinct pulvinus at lamina base, venation pinnate-parallel with prominent midrib. Fls bisexual, basically 3-merous in thyrses, s.t. on a separate shoot from rhizome; K 3, not petaloid, C 3 with a basal tube, usu. white, with 1 petal often hood-like & larger, 1 functional stamen (posterior member of inner whorl), petaloid bearing 1 pollen-sac on 1 edge, staminodes (2)3 or 4, petaloid but small (2 from inner A whorl, other (1)2 the laterals of outer), 1 of the inner forming a labellum over pistil before anthesis, other often a landing-stage for insects, $\overline{G}$ (3), 3-loc. (2 often empty or obsolete) with term. style & septal nectaries at ovary summit, each loc. with 1 almost basal, anatropous to campylotropous bitegmic ovule. Fr. a capsule or berry; seeds with micropylar operculum & basal aril (in capsules), embryo linear & usu. curved to plicate in starchy perisperm. x = 4–14+

Principal genera: *Calathea, Ctenanthe, Ischnosiphon, Maranta, Monotagma, Phrynium*

Pollen released from the anther is deposited in a subapical cavity on the outside of the style; when an insect forces its head between the 2 inner staminodes searching for nectar, it pushes the labellum & releases the style until then under tension, receiving the pollen on its back (cf. Leguminosae); on euglossine bees deposited where they cannot groom & remove it

Arrowroot (*Maranta, Myrosma*), wax (*Calathea*), basket-weaving (*Donax, Ischnosiphon, Phrynium*), v. sweet polypeptides (*Thaumatococcus*) & some cult. orn. esp. *Calathea* (also ed. tubers & fls), *Ctenanthe* & *Maranta*

Maranthes Blume. Chrysobalanaceae (3). 12 trop. (Afr. 10, trop. As. & Pacific 1, Panamá 1 (closely allied to As. sp.)). R: PTRSB 320(1988)121. *M. aubrevillei* (Pellegrin) Prance (W Afr.) – toothed lvs (v. rare in C.), bubbling & hissing slash; *M. corymbosa* Blume (As., Pacific) – imp. timber in Solomon Is.; *M. polyandra* (Benth.) Prance (W Afr.) – bat-poll., form. favoured for charcoal by Nigerian blacksmiths

Marantochloa Brongn. ex Gris. Marantaceae. c. 3 trop. Afr.

marara *Pseudoweinmannia lachnocarpa*

maraschino cherry or **marasco** *Prunus cerasus* 'Marasca'

Marasmodes DC. Compositae (Anth.-Mat.). 4 SW Cape. R: BJLS 96(1988)306

Marathrum Bonpl. Podostemaceae. 25 trop. Am.

Marattia Sw. Marattiaceae. 60 trop., temp. S hemisph. Fronds to 5 m long. Synangia paired along each side of a vein, dehiscing inwards. *M. fraxinea* Sm. (*M. salicina*, trop. Afr., As. to NZ) – cult. orn., the thick stem form. much eaten by Maoris & Aborigines

Marattiaceae Bercht. & J. Presl. Filicopsida. (Incl. Angiopteridaceae, Christenseniaceae & Danaeaceae) 4/100 trop. & warm. Large terr. ferns with erect but short stout stems, s.t. markedly dorsiventral. Fronds often v. large, simple to 1-or more-pinnate. Sori intramarginal on lower surface with free sporangia or these in synangia; indusium 0. Spores all of 1 type giving rise to monoec. prothalli, potentially rather long-lived resembling anthoceroid liverwort

Genera: *Angiopteris* (= Angiopteridaceae – fronds pinnate, veins free, sporangia separate), *Marattia* (= Marattiaceae s.s. – fronds pinnate, veins free, synangia paired, stem erect, radially symmetrical), *Danaea*, (= Danaeaceae – as M. but synangia single, stem dorsiventral), *Christensenia* (Christenseniaceae (Kaulfussiaceae) – fronds palmate, veins reticulate), the fam. sometimes treated as an order comprising the above segregate fams. Generally considered the most primitive surviving group of ferns with similar fossils from Upper Carboniferous

Some cult. orn. & ed. stems etc. (*Angiopteris, Marattia*)

marble wood *Diospyros* spp. esp. *D. marmorata* & *D. oocarpa*

Marcania Imlay. Acanthaceae. 1 Thailand

Marcelliopsis Schinz. Amaranthaceae (I 2). 3 S Afr.

Marcetella Svent. = *Bencomia*

Marcetia DC. Melastomataceae. 23 trop. S Am.

Marcgravia L. Marcgraviaceae. 45 trop. Am. Climbing epiphytes, the climbing shoots with small 2-ranked, round lvs adpressed to surface climbed & covering adventitious roots, the pl. later producing (cf. *Hedera*) pendulous shoots with leathery lanceolate lvs & the capacity to revert to the climbing condition; the pendulous shoots are tipped with dense racemes of green fls with stalked pitcher-like nectaries (bracts). Infls of *M. rectiflora* Triana & Planchon (WI, cult. orn. liane to 10 m) erect with pendulous nectaries, while in other spp. infls pendulous; some fls cleistogamous, others allegedly poll. by birds, bees or lizards, while *M. myriostigma* Triana & Planchon (Brazil) app. bat-poll.

Marcgraviaceae Juss. ex DC. Dicots – Dilleniidae – Theales. 5/108 trop. Am. Glabrous lianes or epiphytes with clinging roots, rarely erect shrubs or trees. Lvs simple, entire, often dimorphic, those on juvenile rooting shoots sessile & oriented distichously, those on flowering ones petiolate & spirally arr.; stipules 0. Fls bisexual, reg. in term., often pendulous, racemes, spikes or umbels, often bird-poll. though some cleistogamous, bracts pitcher-like, saccate or spurred nectaries; K 4 or 5, s.t. basally connate, C 4 or 5 ± basally connate & s.t. (*Marcgravia*) distally so, forming a calyptra (decid.), A 3–40, the filaments ± basally connate & sometimes adnate to C, anthers with longit. slits, $\underline{G}$ (2–8) with simple or lobed stigma, 1-loc. becoming pluriloc. by intrusion of placental partitions with numerous anatropous, bitegmic ovules on resultant axile placentas. Fr. tardily dehiscent with ∞ small seeds with straight or weakly curved embryo in little or 0 endosperm. Starch app. 0

Genera: *Caracasia, Marcgravia, Norantea, Ruyschia, Souroubea*

Allied to Theaceae

mare's tail *Hippuris vulgaris*

Maresia Pomel. Cruciferae. 6 Medit. (Eur. 1) to Caspian & S Iran

Mareya Baillon. Euphorbiaceae. 3 trop. Afr.

Mareyopsis Pax & K. Hoffm. Euphorbiaceae. 1 W trop. Afr.

Margaranthus Schldl. Solanaceae (1). 1 SW US to C Am.: *M. solanaceus* Schldl.

Margaretta Oliver. Asclepiadaceae (III 1). 1 trop. Afr.: *M. rosea* Oliver

Margaritaria L.f. Euphorbiaceae. 13 trop. R: JAA 60(1979)407. *M. discoidea* (Baillon) Webster (trop. Afr.) – fr. ed. Mafia Is.

Margaritolobium Harms. Leguminosae (III 6). 1 Venezuela

Margaritopsis Sauvalle. Rubiaceae (IV 7). 3 Cuba, Hispaniola

Margelliantha Cribb. Orchidaceae (V 16). 4 NE Zaire & Tanz. R: KB 34(1979)329

Marginaria Bory = *Polypodium*

Marginariopsis C. Chr. = *Pleopleltis*

margosa *Azadirachta indica*

Margotia Boiss. (~ *Elaeoselinum*). Umbelliferae (III 12). 1 W Medit.: *M. gummifera* (Desf.) Lange. R: Lagascalia 13(1985)228

marguerite *Leucanthemum* spp. esp. *L. vulgare, Argyranthemum* spp.

Margyricarpus Ruíz & Pavón (excl. *Tetraglochin*). Rosaceae. 1 Andes: *M. pinnatus* (Lam.) Kuntze (*M. setosus*, pearl fruit) – cult. orn. rock-pl. with white berries, used in fertility control in Uruguay

Marianthus Huegel ex Endl. = *Billardiera*

marigold, common or **pot** *Calendula officinalis*; **African m.** *Tagetes erecta*; **bur m.** *Bidens tripartita*; **corn m.** *Chrysanthemum segetum*; **French m.** *T. patula*; **marsh m.** *Caltha palustris*; **Scotch m.** *Calendula officinalis*

marihuana, marijuana *Cannabis sativa* subsp. *indica*

Marila Sw. Guttiferae (II). 15 trop. Am.

Marina Liebm. Leguminosae (III 12). 38 Mex., ext. to SW US & C Am. R: MNYBG 27(1977)55

marionberry See *Rubus*

Maripa Aublet. Convolvulaceae. 19 warm & trop. Am. esp. N S Am. Lianes to 30 m. R: AMBG 60(1973)357. Some ed. fr.

Mariposa (Alph. Wood) Hoover = *Calochortus*

mariposa lily *Calochortus* spp.

Mariscopsis Chermezon = *Queenslandiella*

Marisculus Goetgb. = *Alinula*

Mariscus Vahl = *Cyperus*

marjoram, sweet m. *Origanum majorana*; **pot m.** *O. onites*

Markea Rich. Solanaceae (1). 18 trop. Am. Epiphytic or scandent. *M. neurantha* Hemsley – bat-poll. in Costa Rica

Markhamia Seemann ex Baillon. Bignoniaceae. 10 trop. Afr. (most) & As. Some cult. orn. & timber

marking nut *Semecarpus anacardium*

× **Markleya** Bondar = *Maximiliana* × *Orbignya*

Marlierea Cambess. Myrtaceae (Myrt.). 50 trop. S Am. Fr. ed.

Marlieriopsis Kiaerskov = *Blepharocalyx*

marlock *Eucalyptus* spp.

Marlothia Engl. = *Helinus*

Marlothiella H. Wolff. Umbelliferae (III 8). 1 S Afr.: *M. gummifera* H. Wolff. R: EJB 48(1991)225

Marlothistella Schwantes = *Ruschia*

marmalade plum *Pouteria sapota*

Marmaroxylon Killip = *Zygia*

Marmoritis Benth. (*Phyllophyton*, *Pseudolophanthus*; ~ *Nepeta*). Labiatae (VIII 2). 5 Pakistan to China

Marojejya Humbert. Palmae (V 4o). 2 NE Madag.

maroola plum *Sclerocarya birrea* subsp. *caffra*

Marquesia Gilg. Dipterocarpaceae. 3–4 trop. Afr.

marram grass *Ammophila arenaria*

marri *Corymbia calophylla*

marrow *Cucurbita pepo*

Marrubium L. Labiatae (VI). 30 Eur. (12). Medit., As. Some cult. orn. incl. *M. vulgare* L. ((white) horehound, Euras., Medit., Macaronesia) – form. much-used medic. herb as tea, in sweets & in liqueurs

Marsdenia R. Br. Asclepiadaceae (III 4). (Incl. *Stephanotis*) c. 100 trop. & warm. Usu. twining shrubs. Cult. orn. esp. *M. floribunda* (Brongn.) Schltr (*Stephanotis f.*, stephanotis, Madagascar jasmine, floradora, Madag.) – liane grown for large white waxy fls used in wedding corsages etc. *M. castillonii* Lillo ex Meyer (Arg., Paraguay) – roots ed.; *M. erecta* R. Br. (SE Eur. to Iran) – latex blisters skin; *M. hamiltonii* Wight (India) – ed. fr.; *M. tenacissima* (Roxb.) Moon (Himal., NE India) – source of fibre for rope etc.; *M. tinctoria* R. Br. (Java indigo, Mal.) – stimulates appetite, form. cult. for indigo-like dye; others medic., poisonous, some Madag. spp. a source of Madagascar rubber

marsh betony *Stachys palustris*; **m. fern** *Acrostichum* spp.; **m. grass** *Spartina* spp.; **m. mallow** *Althaea officinalis*; **m. marigold** *Caltha palustris*; **m. rose** *Orothamnus zeyheri*; **m. rosemary** *Ledum* spp., *Limonium* spp.

Marshallfieldia J.F. Macbr. = *Adelobotrys*

Marshallia Schreber. Compositae (Helia.-Gai.). 7 US. R: CGH 181(1957)41. Cult. orn. (Barbara's buttons) esp. *M. grandiflora* Beadle & Boynton (E N Am.)

Marshalljohnstonia Henricksen. Compositae (Lact.). 1 Mexico

Marsilea L. Marsileaceae. 50–70 trop. (esp. Afr.) & temp. (Eur. 4). R: SBi 49(1968)273, (Am.) SBM 11 (1986)1. Lvs petiolate with 4 clover-like lobes floating in deep water, held above water in shallow. Some Aus. spp. with v. drought-resistant sporocarps living up to 100 yrs; on water uptake, gelatinous interior swells, splitting sporocarp & worm-like mass exudes carrying sori with it leading to germ. of spores & fertilization. Sporocarps (nardoo) of Aus. spp. esp. *M. drummondii* A. Braun ground & eaten by Aborigines & early white settlers

Marsileaceae Mirbel. Filicopsida. (Incl. Pilulariaceae) 3/55–75 trop. & temp. Small amphibious or aquatic heterosporous pls with creeping rhizomes with solenosteles. Fronds circinate, borne in 2 dorsal rows. Micro- & mega-sporangia together in hard sporocarps (not produced when pls submerged) attached to frond-base or in axil, the sporocarps perhaps tightly folded pinnae with a number of elongate sori, each covered with a membranous indusium. Spores can be dormant for up to 100 yrs. Male gametophyte of only 9 cells incl. 1 prothallial cell

Genera: *Marsilea* (fronds 4-lobed), *Pilularia* (frond without lobes), *Regnellidium* (frond with 2 lobes). *Pilularia* is s.t. placed in a fam. of its own, the 2 fams comprising Marsileales. Young *Marsilea* pls have unlobed fronds like *Pilularia*

Marsippospermum Desv. Juncaceae. 3 temp. S Am., Falkland Is., NZ & is.

Marssonia Karsten = *Napeanthus*

Marsypianthes C. Martius ex Benth. Labiatae (VIII 4a). 6 trop. Am.

Marsypopetalum R. Scheffer. Annonaceae. 1 W Mal.: *M. pallidum* (Blume) Kurz

Marthella Urban. Burmanniaceae. 1 Trinidad

Marticorenia Crisci. Compositae (Mut.-Nass.). 1 Chile. Shrub

Martinella Baillon. Bignoniaceae. 2 trop. Am. *M. obovata* (Kunth) Bureau & Schumann used & cult. by Am. Indians for eye troubles

Martinezia Ruíz & Pavón = *Prestoea* + *Aiphanes*

Martiodendron Gleason. Leguminosae (I 2). 4 trop. S Am.

Martretia Beille. Euphorbiaceae. 1 W & C Afr.: *M. quadricornis* Beille. R: BJBB 59(1989)319

Martynia L. Pedaliaceae (Martyniaceae). 1 Mex.: *M. annua* L. Fr. s.t. pickled; sticky lvs used to remove lice from fowls

Martyniaceae Horan. See Pedaliaceae

marupa *Quassia amara*

Maruta (Vass.) Gray = *Anthemis*

marvel of Peru *Mirabilis jalapa*

Marvello hair Gut from Chinese silkworms fed on *Liquidambar formosana*

Maryland pinkroot *Spigelia marilandica*

Mary's bean *Merremia discoidesperma*

Mascagnia (DC) Colla. Malpighiaceae. 50 Mexico to Argentina

Mascarena L. Bailey = *Hyophorbe*

Mascarenhasia A. DC. Apocynaceae. 12 Madag. with 1 ext. to Afr. R: FMadag. 169(1976)248. Rubber sources (Madagascar r.)

Maschalocephalus Gilg & Schumann. Rapateaceae. 1 trop. W Afr.

Maschalocorymbus Bremek. = *Urophyllum*

Maschalodesme Schumann & Lauterb. Rubiaceae (II 5). 2 New Guinea

Masdevallia Ruíz & Pavón. Orchidaceae (V 13). c. 380 trop. Am. highlands. R: MSB 16(1986)1. Cult. orn. epiphytes with K with elongated tails. Some attract poll. flesh-flies through colour & smell of bad meat but there is no 'reward'. Sect. Chimaeroideae predicted to be poll. by fungus-gnats

masindi *Cynodon transvaalensis*

Masoala Jum. Palmae (V 4). 1 NE Madag.: *M. madagascariensis* Jum. – now reduced to 3 trees in wild

Massangea C.J. Morren = *Guzmania*

Massartina Maire = *Elizaldia*

Massia Bal. = *Eriachne*

Massonia Thunb. ex L.f. Hyacinthaceae (Liliaceae s.l.). 8 dry S Afr. R: JSAB 42(1976)406

massoy bark *Cryptocarya aromatica*

Massularia (Schumann) Hoyle. Rubiaceae (II 1). 1 trop. Afr.: *M. acuminata* (G. Don f.) Hoyle – chewing-stick in Nigeria

mast fallen fr. of beech, oak etc. form. much used as food for pigs etc. Right to be permitted to allow animals into woodland to eat it known as pannage; some woodlands (e.g. in Domesday Book) measured by amount of mast rather than area

Mastersia Benth. Leguminosae (III 10). 2 Assam (1), C Mal. (1). R: Blumea 30(1984)77. *M. bakeri* (Koord.) Backer (Mal.) – a cover-pl. in plantations

Mastersiella C. Benedict. Restionaceae. 3 SW & S Cape. R: Bothalia 15(1985)481

masterwort *Astrantia major* & *A. maxima*

mastic *Pistacia lentiscus*; **American m.** *Schinus molle*; **Barbados m.** *Sideroxylon foetidissimum*; **Bombay m.** *P. atlantica*

Mastichodendron (Engl.) H.J. Lam = *Sideroxylon*

Mastigosciadium Rech.f. & Kuber. Umbelliferae (III 5). 1 Afghanistan: *M. hysteranthum* Rech.f. & Kuber

Mastigostyla I.M. Johnston. Iridaceae (III 4). c. 16 Peru, Arg.

Mastixia Blume. Cornaceae (Mastixiaceae). 13 Indomal. R: Blumea 23(1976)51. Some used as plywood in India

Mastixiaceae Calestani See Cornaceae

Mastixiodendron Melchior. Rubiaceae (? III 4). 7 E Mal. to Fiji. R: JAA 58(1977)349. Polypetalous corollas; some spp. with semi-sup. ovaries

masur *Lens culinaris;* **m. wood** *Betula pendula* & *B. pubescens*

mat grass *Nardus stricta*

mata kucing, m. kuching *Dimocarpus longan* var. *malesianus*

matac *Asclepias curassavica*

matai *Prumnopitys taxifolia*

Matara tea *Senna auriculata*

matarique *Psacalium decompositum*

Matayba Aublet. Sapindaceae. 45 trop. Am. Some timbers & medic. oilseeds

maté *Ilex paraguariensis*

Matelea Aublet. Asclepiadaceae (III 3). c. 200 trop. & warm Am.

mathers *Vaccinium ovalifolium*

Mathewsia Hook. & Arn. Cruciferae. 6 Peru, Chile

Mathiasella Constance & C. Hitchc. Umbelliferae (III 10). 1 Mex.: *M. bupleuroides* Constance & C. Hitchc.

Mathieua Klotzsch = *Eucharis*

Mathurina Balf.f. Turneraceae. 1 Rodrigues: *M. penduliflora* Balf.f. – heterophylly like many Rodrigues endemics

Matisia Bonpl. Bombacaceae. 25+ trop. S Am. *M. cordata* Bonpl. (trop. S Am.) – fr. ed., pulp rich in vitamins A & C

Matonia R. Br. Matoniaceae. 2 Mal., exposed ridges & high mts. R: Blumea 38(1993)167. Frond with initial dichotomy, followed by each half undergoing reg. series of unequal dichotomies such that growing point curved back parallel to petiole; the pinnae forming a fan-like frond on surface of rachis away from petiole; pinnae pinnatifid. Fossil *Matonidium* with similar char. frond v. widespread in Triassic

Matoniaceae C. Presl. Filicopsida. 2/4 Mal. Terr. ferns with creeping rhizome with 2 co-axial cylinders of vasc. tissue surrounding central solid stele. Fronds 3-chotomous-pedate or pseudodichotomous; veins free or ± anastomosing. Sori superficial, consisting of a small number of sporangia in a ring around a receptacle which is continuous with the stalk of an umbrella-shaped indusium

Genera: *Matonia, Phanerosorus.* Relict group, fossils back to Lower Mesozoic referred here

Matricaria L. (*Chamomilla*). Compositae (Anth.-Mat.). (Excl. *Tripleurospermum*) 7 Euras. (Eur. 3 incl. *M. recutita* (L.) Rauschert, wild chamomile, & *M. discoidea* DC (*M. matricarioides*, pineapple weed, As., (?) natur. N Am., whence introd. GB, fr. spread on vehicle tyres, in Am. medic. tea & insecticide), N Afr.

Matsumurella Mak. = *Lamium*

Matteuccia Tod. Dryopteridaceae (II 2). 2 N temp. (Eur. 1). Fronds for food; dimorphic – sterile ones forming leafy cup round deeply bipinnatifid fertile with pinnule edges contracted round sori. *M. struthiopteris* (L.) Tod. (ostrich fern, cool Euras.) – cult. orn., the young fronds ed. if steamed or tinned. *M. intermedia* C. Chr. = *M. struthiopteris* × *Onoclea orientalis* (Hook.) Hook.

Mattfeldanthus H. Robinson & R. King. Compositae (Vern.-Vern.). 3 Brazil

Mattfeldia Urban. Compositae (Sen.-Sen.). 1 Hispaniola

Matthaea Blume. Monimiaceae (V 2). 6 Mal. R: Blumea 28(1972)77

Matthiola R. Br. Cruciferae. 55 Macaronesia, W Eur. (10), Medit. Stock. R: MHB 1, 18(1900)9. Cult. orn. herbs (sometimes subshrubby), esp. *M. incana* (L.) R. Br. (Brompton s., S Eur., natur. Calif.) – highly-scented fls grown for cutting, 'Annua' (ten week stock), a fast-maturing form used for bedding; *M. longipetala* (Vent.) DC (S Eur.) esp. subsp. *bicornis* (Sm.) P. Ball (*M. bicornis*, night-scented s., E Medit., natur. SW N Am.) – heavy scent in evening

Mattiastrum (Boiss.) Brand = *Paracaryum*

mattipaul *Ailanthus malabarica*

Matuccana Britton & Rose (~ *Oreocereus*). Cactaceae (III 4). 6–7 Peru

Matudacalamus Maekawa = *Aulonemia*

Matudaea Lundell. Hamamelidaceae (I 4). 2 C Mexico to Honduras

Matudanthus D. Hunt. Commelinaceae (II 1f). 1 Mexico

Matudina R. King & H. Robinson (~ *Eupatorium*). Compositae (Eup.-Heb.). 1 Mex. R: MSBMBG 22(1987)412

Maughaniella L. Bolus = *Diplosoma*

Mauloutchia Warb. (~ *Brochoneura*). Myristicaceae. 6 E Madagascar. R: Adansonia 13(1973)209. Aril rudimentary

Maundia F. Muell. Juncaginaceae. 1 E Aus.: *M. triglochinoides* F. Muell. Lamina 0; ovule orthotropous, apical, pendulous

Maurandella (A. Gray) Rothm. (~ *Asarina*). Scrophulariaceae (Antirrh.). 1 SW N Am.

Maurandya Ortega (~ *Asarina*). Scrophulariaceae (Antirrh.). 2 W & S Am. R: D.A. Sutton, Rev. Antirrhineae (1988)484

Mauranthemum Vogt & Oberprieler (*Leucoglossum*). Compositae (Anth.-Leuc.). 4 W Med. R: Taxon 44 (1995)377. Cult. orn. esp. *M. paludosum* (Poiret) Vogt & Oberprieler

Mauria Kunth. Anacardiaceae (IV). 10 Andes

Mauritia L.f. Palmae (II 2). 3 trop. Am. Dioec.; often in vast stands. *M. flexuosa* L.f. (ita palm, N S Am., Trinidad) – water-disp. in Amaz., sources of food (sago, oil, wine), fibre, thatch, etc.

Mauritiella Burret (~ *Mauritia*). Palmae (II 2). 14 N S Am.

Mauritius hemp *Furcraea foetida*

Maurocenia Miller. Celastraceae. 1 S Afr.: *M. frangularia* (L.) Miller (*M. capensis*)

Mausolea Polj. (~ *Artemisia*). Compositae (Anth.-Art.). 1 C As., Iran

maw seed *Papaver somniferum*

mawah *Pelargonium* spp. & hybrids

Maxburretia Furt. Palmae (I 1a). 3 Thailand & Malay Pen., limestone hills. R: GH 11(1978)187

Maxia O. Nilsson (~ *Montia*). Portulacaceae. 1 W N Am.

Maxillaria Ruíz & Pavón. Orchidaceae (V 2). 420 trop. (e.g. Costa Rica 82) & warm Am. Some with 0 pseudobulbs e.g. *M. valenzuelana* (A. Rich.) Nash – iris-like lvs in fans. Cult. orn. epiphytes (incl. *Ornithidium*) with false pollen on lip – protein &/or starch – in *M. rufescens* Lindley & other spp. Several spp. with elaiophores attractive to bees

Maximiliana C. Martius. Palmae (V 5c). 1 NE S Am. & Trinidad: *M. maripa* (Correa) Drude (*M. regia*, kokerite, NE S Am., Trinidad) – monoec., cult. orn., oilseeds & thatch

Maximowiczia Khokhr. = *Scirpus*

Maximowicziella Khokr. = *Scirpus*

Maxonia C. Chr. Dryopteridaceae (I 2). 1 WI, S Am.: *M. apiifolia* (Sw.) C. Chr.

Maxwellia Baillon. Sterculiaceae. 1 New Caledonia. ? Tiliaceae

may *Crataegus* spp.; **m. apple** *Podophyllum peltatum*; **m. blob** *Caltha palustris*; **m. weed** *Anthemis cotula*

Mayaca Aublet. Mayacaceae. 4 trop. Am., Angola. R: NS Paris 14(1952)234

Mayacaceae Kunth. Monocots – Commelinidae – Commelinales. 1/4 trop. Am., Angola. *Lycopodium*-like freshwater aquatics with vessels in roots & stems; longit. air-channels throughout vegetative body, hairy only in leaf-axils. Lvs sessile, without sheaths, spirally arr., narrow & often apically bifid. Fls bisexual, reg., term. but app. axillary due to sympodial stem-growth, aerial but without nectaries; K 3 green, valvate, C 3 whitish, imbricate, A 3 alt. with C, anthers with apical pores or short slits & monosulcate pollen, $\underline{G}$ (3), 1-loc. with term. style & 3 parietal placentas with several–∞ bitegmic, orthotropous ovules. Fr. a loculicidal capsule (often immersed on recurved pedicel at maturity); embryo small, forming a cap on starchy proteinaceous endosperm beneath an operculum at micropylar end

Genus: *Mayaca*

Mayanaea Lundell = *Orthion*

Mayna Aublet. Flacourtiaceae (3). 6 trop. Am. R: FN 22(1980)22. Local med.

Mayodendron Kurz = *Radermachera*

mayten *Maytenus boaria*

Maytenus Molina. Celastraceae. (Incl. *Gymnosporia*) c. 200 trop. to warm (Madeira 1: *M. umbellata* (R. Br.) Mabb. (*M. dryandri*), S Spain 1: *M. senegalensis* subsp. *europaeus* (Boiss.) Rivas Martínez). *M. acuminata* (L.f.) Loes. (trop. & S Afr.) – fine silky threads from broken leaf edges due to polyisoprenes from laticifer-like cells (not xylem thickenings as in *Cornus*). Alks incl. caffeine. *M. boaria* Molina (mayten, Chile) – lvs a febrifuge; *M. officinalis* Mabb. (*M. ilicifolia*, S Am.) – medic. (? antitumoural pristimerin) & fert. control; *M. senegalensis* (Lam.) Exell (Spain & N Afr. to Bangladesh) – locally medic. in Afr., extracts have cytotoxic effect on some cancers

Mazaea Krug & Urban (*Neomazaea*). Rubiaceae. 2 Cuba (serpentine), Haiti

Mazus Lour. Scrophulariaceae. 10–15 As. to Aus., NZ. R: Brittonia 8(1954)29. Mat-forming ground-cover pls esp. *M. reptans* N.E. Br. ('*M. japonicus*', '*M. rugosus*', Himal.) – natur. in lawns in Scotland, US, Jamaica

mazzard *Prunus avium*

Mazzettia Iljin = *Vladimiria*

mbocaya *Acrocomia totai*

mboga *Sesuvium portulacastrum*

mbura *Parinari curatellifolia*

Mcvaughia W.R. Anderson. Malpighiaceae. 1 Bahia. Prob. correctly *Macvaughia*

meadow crane(')s-bill *Geranium pratense*; **m. fescue** *Festuca pratensis*; **m. foxtail** *Alopecurus pratensis*; **m.grass** *Poa* spp.; **annual m.g.** *P. annua*; **common m.g.** *P. pratensis*; **rough m.g.** *P. trivialis*; **wood. m.g.** *P. nemoralis*; **m. pink** *Dianthus deltoides*; **m. rue** *Thalictrum* spp.; **m. saffron** *Colchicum autumnale*; **m. saxifrage** *Saxifraga granulata*; **m.sweet** *Filipendula ulmaria*

mealberry *Arctostaphylos uva-ursi*

mealies *Zea mays*

Mearnsia Merr. = *Metrosideros*

Mecardonia Ruíz & Pavón (~ *Bacopa*). Scrophulariaceae. 10 warm Am. R: Candollea 42(1987)431

Mecca galls *Quercus pubescens*

Mechowia Schinz. Amaranthaceae (I 2). 2 S trop. Afr.

Mecodium (Copel.) Copel. = *Hymenophyllum*

Mecomischus Cosson ex Benth. Compositae (Anth.-Art.). 2 Morocco, Algeria

Meconella Nutt. ex Torrey & A. Gray. Papaveraceae (III). 3 W N Am.

Meconopsis Viguier. Papaveraceae (IV). c. 50 Himal. to W China, 1 W Eur. (*M. cambrica* (L.) Viguier (Welsh poppy) prob. not congeneric)). R: G. Taylor (1934) *An account of the genus M*, J.L.S. Cobb (1989) *M*. Alks. Cult. orn. with blue, reddish or yellow fls, esp. 'blue poppies' incl. *M. betonicifolia* Franchet (*M. baileyi*, China) & *M.* × *sheldonii* G. Taylor (*M. betonicifolia* × *M. grandis* Prain (Himal.)), some hapaxanthic incl. *M. horridula* Hook.f. & Thomson (to 6000 m in Himal. & China)

Mecopus Bennett. Leguminosae (III 9). 1 Indomal.

Mecranium Hook.f. Melastomataceae. 23 WI. R: SBM 39(1993). Trees & shrubs

mecrusse *Androstachys johnsonii*

Medemia Wuerttemb. Palmae (I 3b). 1 N trop. Afr.: *M. argun* (C. Martius) Wuerttemb. (Egypt, Sudan) – endangered through overuse as matting (not seen living since 1964); fr. used as offerings in Ancient Egyptian tombs, ed., esp. toothsome after burying

Medeola L. Convallariaceae (Liliaceae s.l.; Trilliaceae). 1 E N Am.: *M. virginiana* L. (cucumber root) – rhizomes crisp & ed. like cucumber

Medeolaceae (S. Watson) Takht. = Convallariaceae

Mediasia Pim. (~ *Seseli*). Umbelliferae (III 8). 1 SW & C As.

Medicago L. Leguminosae (III 22). (Incl. *Trigonella* p.p.) 85 Eur. (43), Medit., Ethiopia, S Afr. R: CJB 67(1989)3260. Medick, burweed. Fodder & green manures, some cult. orn. *M. arborea* L. (tree m., moon trefoil, S Eur.) – shrubby to 3.5 m, cult. orn.; *M. hypogaea* E. Small (*Factorovskya aschersoniana*, SE Med. to Iraq) – geocarpic annual, the 1 mm gynophore extending to 10 cm underground; *M. intertexta* (L.) Miller (*M. echinus*, calvary clover, Medit.) – fr. twisted & spiny like crown of thorns; *M. lupulina* L. (black medick, hop clover, nonsuch, yellow trefoil, Euras., natur. E Afr., N Am.) – fodder pl., the (?) original shamrock; *M. polymorpha* L. (*M. nigra*, OW) – viable seeds 200 yrs old from adobe in Calif. & Mex.; *M. sativa* L. (lucerne, alfalfa, SW As., natur. Eur., N Am.) – successfully domesticated autotetraploid (2n = 4x = 32, few diploids known) selected by horse-raising culture ('alfalfa' Arabic from old Irani 'an aspo-asti' (= horsefodder) introd. Eur. 2400 BC (Persian-Greek Wars)), long-cult. for fodder & silage, poss. oilseed, lvs a comm. source of chlorophyll, seeds grown like beansprouts

medick *Medicago* spp.; **black m.** *M. lupulina*

Medicosma Hook.f. Rutaceae. 22 New Guinea (1), E Aus. (6, endemic), New Caled. 15. R: AusJB 33(1985)27. Ant-disp. through persistent placental endocarp

Medinilla Gaudich. Melastomataceae. c. 400 trop. Afr. to Pacific (Borneo 48, R: Blumea 35(1990)5; Philippines 80, R: Blumea 40(1995)113). Some cult. orn. pot-pls with large lvs esp. *M. magnifica* Lindley (Philippines) – 4-winged stems

Mediocalcar J.J. Sm. Orchidaceae (V 14). 20 New Guinea. Polynesia

Mediusella (Cavaco) Dorr (~ *Leptolaena*). Sarcolaenaceae. 1 Madag.

medlar *Mespilus germanica*; **Bronvaux m.** + *Crataegomespilus* cvs; **Japanese m.** *Eriobotrya japonica*; **Medit. m.** *Crataegus azarolus*

Medusagynaceae Engl. & Gilg. Dicots – Dilleniidae – Theales. 1/1 Seychelles. R: PSE 171(1990)27. Glabrous small tree with cortical vasc. bundles. Lvs simple & opp.; stipules 0. Fls bisexual or male (andromonoecy), reg., in small term. panicles, foetid; K 5, basally connate, imbricate, decid., C 5 imbricate, pinkish-white, A ∞, anthers with longit. slits, $\underline{G}$ (17–25), with as many locs & styles forming a 'crown', each loc. with 1 ascending & 1 descending bitegmic, anatropous ovules. Fr. a capsule opening from the base septicidally, the carpels diverging from the central column like the spokes of an opening umbrella & opening ventrally; seeds exotegmic, winged

Only species: *Medusagyne oppositifolia* Baker

Affinities unclear

Medusagyne Baker. Medusagynaceae. 1 Mahé (Seychelles): *M. oppositifolia* Baker, thought extinct (not seen between 1908 & 1970) but c. 12 pls (2 populations) found at c. 250 m, the sp. now spread in cult. by seed. R: Kew Mag. 6(1989)166

Medusandra Brenan. Medusandraceae. 2 trop. W Afr. Fr. relished by baboons & parrots

Medusandraceae Brenan. Dicots – Rosidae – Santalales. 2/9 trop. W Afr. Trees with secretory canals throughout. Lvs simple, spirally arr., abaxially with hairs of a unique (thick-based) type, the petiole with apical pulvinus (? lvs unifoliolate); stipules small, decid. Fls bisexual, reg., in axillary catkin-like racemes with minute caducous bracts; K 5, ± basally connate, open in bud & ± accrescent in fr., C 5 small, imbricate, A 5 opp. C, anthers opening by longit. recurving valves, staminodes 5 opp. K, disk 0, $\underline{G}$ (3(4)) with distinct short styles & 1 loc. with 6(–8) anatropous ovules pendulous from top of placental column. Fr. a capsule with 1 large pendulous seed with straight embryo embedded in copious somewhat ruminate endosperm

Genera: *Medusandra, Soyauxia*

Variously placed near Olacaceae, Flacourtiaceae or Euphorbiaceae

Medusanthera Seemann. Icacinaceae. 4–5 Mal. to W Pacific

Meeboldia H. Wolff. Umbelliferae (inc. sed.). 1 NW Himalaya

Meeboldina Süsseng. (~ *Leptocarpus*). Restionaceae. 1 SW Aus.: *M. denmarkica* Süsseng.

Meehania Britton. Labiatae (VIII 2). 1 E As., 1 E US: *M. cordata* (Nutt.) Britton. – cult. ground cover

Meehaniopsis Kudô = *Glechoma*

Megacarpaea DC. Cruciferae. 7 Eur. (1) to C As., Himal. & China. Some statuesque hapaxanthic pachycauls with A 16 incl. *M. polyandra* Benth. (Himal.) – young lvs ed.

Megacaryon Boiss. = *Echium*

Megacodon (Hemsley) H. Sm. Gentianaceae. 1 Himal.: *M. stylophorus* (C.B. Clarke) H. Sm. – cult. orn. perennial to 2 m

Megadenia Maxim. Cruciferae. 2 C As., China

Megalachne Steudel. Gramineae (17). 2 Juan Fernandez

Megalastrum Holttum (~ *Ctenitis*). Dryopteridaceae (I 3). 30 trop. Am., Afr. (1) to Masc. R: GBS 39(1987)161

Megaleranthis Ohwi. Ranunculaceae (I 1). 1 S Korea

Megalochlamys Lindau. Acanthaceae. 3 SW Afr.

Megalodonta E. Greene. Compositae (Helia.-Cor.). 1 N Am.: *M. beckii* (Torrey) E. Greene, an aquatic with entire aerial lvs & dissected submerged ones, v. sensitive to pollution. R: AB21(1985)99

Megalopanax Ekman (~ *Aralia*). Araliaceae. 1 Cuba

Megaloprotachne C.E. Hubb. Gramineae (34d). 1 S & S trop. Afr.

Megalorchis Perrier. Orchidaceae (IV 2). 1 Madagascar

Megalostoma Leonard. Acanthaceae. 1 C Am.

Megalostylis S. Moore. Euphorbiaceae. 1 Amazonian Brazil

Megalotheca F. Muell. = Aus. spp. form. in *Restio*

Megalotus Garay. Orchidaceae (V 16). 1 Philippines

Megaphrynium Milne-Redh. Marantaceae. 5 trop. Afr.

Megaphyllaea Hemsley = *Chisocheton*

Megasea Haw. = *Bergenia*

Megaskepasma Lindau. Acanthaceae. 1 Venez.: *M. erythrochlamys* Lindau – cult. for brilliantly coloured bracts

Megastachya Pal. (~ *Eragrostis*). Gramineae (24). 1 Afr. & Madag.

Megastigma Hook.f. Rutaceae. 2 Mexico, C Am.

Megastoma Cosson & Durieu = *Ogastemma*

Megastylis (Schltr.) Schltr. Orchidaceae (IV 1). 8 New Caled., 1 E Aus.

Megatritheca Cristóbal. Sterculiaceae. 2 trop. Afr.

Megistostegium Hochr. Malvaceae. 3 Madagascar

Megistostigma Hook.f. Euphorbiaceae. 3–4 SE As. to W Mal.

Meiandra Markgraf = *Alloneuron*

Meineckia Baillon. Euphorbiaceae. 19 trop. Am., SW & NE trop. Afr., Socotra, S Arabia, Madag., S India & Sri Lanka, Assam. R: ABN 14(1965)323

Meiocarpidium Engl. & Diels. Annonaceae. 1 W trop. Afr.: *M. lepidotum* Engl. & Diels
Meiogyne Miq. Annonaceae. 9 Indomal. to Pacific. R: Blumea 38(1994)487
Meiomeria Standley = *Chenopodium*
Meionandra Gauba = *Valantia*
Meiostemon Exell & Stace. Combretaceae. 2 SE trop. Afr. (1), Madag. (1)
Meiracyllium Reichb.f. Orchidaceae (V 13). 2 C Am. Cult. orn. epiphytes
Meizotropis J. Voigt (~ *Butea*). Leguminosae (III 10). 2 Himal.
mel grass *Ammophila arenaria*
Melachone Gilli = *Amaracarpus*
Meladenia Turcz. = *Cullen*
Meladerma Kerr. Asclepiadaceae (I; Periplocaceae). 3 Thailand
Melaleuca L. Myrtaceae (Lept.). 220 Indomal. (few), Aus. (215, 210 endemic; 120 W Aus.), Pacific. Fls in heads or spikes, the infl. axis often extending as a leafy shoot (cf. *Callistemon*), the fls with consp. long stamens in 5 bundles opp. petals. Cult. orn. timber & medic. oil (paperbarks – Aus.). *M. cajuputi* Powell (Burma to Aus.) – distilled for cajuput oil used like eucalyptus oil in cough sweets etc., timber good; *M. quinquenervia* (Cav.) S.T. Blake (*'M. leucadendra'*, E Aus., SE New Guinea, New Caled.) – source of niaouli oil like cajuput, the trees forming almost pure stands after the native vegetation of New Caled. is destroyed by fire, weedy pest in S Florida
Melampodium L. Compositae (Helia.-Mel.). 37 trop. & warm Am. esp. Mex. R: Rhodora 74(1972)1. *M. americanum* L. (Mex. to Guatemala) – insecticidal lactone effective against army worm
Melampyrum L. Scrophulariaceae. 35 N temp. (Eur. 24). R: FR 23(1926)159, 385, 24(1927)127. Cow-wheats. Hemiparasites. Seeds of *M. arvense* L. (Euras.) taint wheat so form. pls & even seeds were removed from fields in S England
Melananthus Walp. Solanaceae (2). 5 trop. Am. G with 1 ovule
Melancium Naudin. Cucurbitaceae. 1 E & S Brazil
Melandrium Roehl. = *Silene*
Melanocenchris Nees. Gramineae (33c). 3 Chad & NE trop. Afr. to India & Sri Lanka. R: BBSI 16(1974)141. No OW relations
Melanochyla Hook.f. Anacardiaceae (III). 25 Mal. R: FM I, 8(1978)490
Melanocommia Ridley = *Semecarpus*
Melanodendron DC. Compositae (Ast.-Ast.). 1 St Helena: *M. integrifolium* DC, a tree; allies in Australasia, S Am.
Melanodiscus Radlk. = *Glenniea*
Melanolepis Reichb.f. ex Zoll. Euphorbiaceae, 1–2 Taiwan & SE As. to Pacific
Melanoloma Cass. = *Centaurea*
Melanophylla Baker. Cornaceae (Melanophyllaceae). 8 Madag. Not allied to Cornaceae according to DNA evidence
Melanophyllaceae Takht. ex Airy Shaw. See Cornaceae
Melanopsidium Cels ex Colla (~ *Billotia*). Rubiaceae (II 1). Spp?
Melanorrhoea Wallich = *Gluta*
Melanosciadium Boissieu. Umbelliferae (III 8). 1 W China: *M. pimpinelloides* Boissieu
Melanoselinum Hoffm. (~ *Thapsia*). Umbelliferae (III 12). 4 Macaronesia
Melanospermum Hilliard (~ *Phyllopodium*). Scrophulariaceae. 6 C & S Afr. R: NRBGE 45(1988)482
Melanoxylon Schott = *Melanoxylum*
Melanoxylum Schott. Leguminosae (I 1). 1 Amazonia: *M. brauna* Schott – timber used for bridges etc. (brauna)
Melanthera J.P. Rohr. Compositae (Helia.-Verb.). 20 trop. (exc. Far East)
Melanthiaceae Batsch. ex Borkh. (incl. Petrosaviaceae, excl. Campynemataceae; ~ Liliaceae). Monocots – Liliidae – Liliales. 21/c. 120 N hemisph. to SE As. & S Am. Perennial herbs, s.t. pachycaul (*Veratrum*), with rhizomes & spiral to distichous lvs. Infls spikes or racemes (to panicles) of usu. bisexual ± hypogynous (cf. C.) fls. P 3 + 3, s.t. basally connate; A 3 + 3 (–12 in *Pleea*); G (3), 3-loc., s.t. apically free, each loc. with 2–∞ ovules. Fr. loculicidal or septicidal capsule; seeds usu. rounded & winged or with term. appendages
Principal genera: *Helonias, Narthecium, Schoenocaulon, Tofieldia, Veratrum, Zigadenus*
Some cult. orn. & local medic.: *Aletris, Chamaelirion, Veratrum*
Melanthium L. (~ *Zigadenus*). Melanthiaceae (Liliaceae s.l.). 5 N Am. Some cult. orn. rhizomatous herbs incl. *M. virginicum* L. (bunchflower, E US)
Melasma P. Bergius. Scrophulariaceae. 5 S Afr. (3), Am. (2)

Melasphaerula Ker-Gawler. Iridaceae (IV 3). 1 S Afr.: *M. ramosa* (L.) N.E. Br. (*M. graminea*) – cult. orn. like *Sparaxis* but G acutely 3-angled

Melastoma L. Melastomataceae. c. 70 Indomal. to Pacific (Aus. 1). Some cult. orn. esp. *M. malabathricum* L. (Indomal.) – locally medic., fr. sweet, staining mouth black like *Vaccinium*, a common plant of regrowth app. everflowering, one form rheophytic

Melastomastrum Naudin. Melastomataceae. 6 trop. Afr. R: BMNHN 3°, 270 Bot. 14(1974)49

Melastomataceae Juss. Dicots – Rosidae – Myrtales. 188/4950 trop. & warm esp. S Am. Shrubs & herbs, less often trees or lianes, often with 4-angled stems, s.t. myrmecophilous, often accumulating aluminium & with internal phloem as well as cortical and/ or pith vascular bundles. Lvs opp. (s.t. anisophyllous or 1 suppressed), rarely whorled, simple, often with 3–9 prominent sub-parallel veins; stipules rare (e.g. *Astronidium*). Fls usu. bisexual & without nectar (present in *Blakea*), insect-poll., in cymes, usu. ± perigynous, reg. exc. A, (3)4 or 5(–10)-merous; K lobes valvate or a rim on hypanthium, s.t. a calyptra, C usu. free, convolute in bud, A usu. 2 whorls, often dimorphic, filaments often twisted at anthesis bringing anthers to 1 side of fl., these with single term. pore or less often 2 pores or longit. slits, connective often with appendages, G ((2)3–5(–15)) with as many loc. (s.t. 1-loc. through partitions not developing), with term. style & (1–)∞ anatropous (or campylotropous) bitegmic ovules with zig-zag micropyle on usu. axile placentas per loc. Fr. a loculicidal or septifragal capsule or berry; seed usu. small, endosperm 0, the cotyledons often unequal. x = 7–18+

Classification & principal genera:

According to Blumea 27(1981)463, Crypteroniaceae exc. *Rhynchocalyx* (Rhynchocalycaceae) form a subfam. here, the others being:

Memecyloideae (inc. Astronioideae in part, s.t. treated as sep. fam.; some with buzz-poll.; fr. a 1–5-seeded berry, embryo large): *Memecylon, Mouriri, Pternandra*

Melastomatoideae (inc. Astronioideae in part; fr. many-seeded, embryo small): *Astronia, Blakea, Brachyotum, Clidemia, Conostegia, Dissotis, Gravesia, Leandra, Medinilla, Melastoma, Miconia, Microlicia, Monochaetum, Osbeckia, Ossaea, Sonerila, Tibouchina, Tococa, Topobea*

Memecyloideae often considered to approach Myrtaceae; wood anatomy (fibres with bordered (not simple) pits; included phloem etc.) like Myrt., also Lythraceae & Onagraceae. Details of A app. associated with poll. mechanisms imp. in generic & specific distinction. 15 genera with elaiophores attractive to bees (only 10 fams thus). Fleshy fr. distrib. mainly by birds but also marsupials, monkeys, bats & other mammals, turtles & other reptiles, & fish

Some timber (*Astronia*), ed. fr. (*Bellucia* & *Conostegia* cult., *Heterotrichum*), dyes (*Dionycha*, *Memecylon* etc.) & cult. orn. esp. *Bertolonia, Dissotis, Medinilla, Melastoma, Memecylon, Rhexia, Sonerila, Tibouchina*, some becoming weedy, e.g. *Clidemia, Heterocentron, Melastoma*

Melchiora Kobuski = *Balthasaria*

Melhania Forssk. Sterculiaceae. 60 trop. OW. Epicalyx with 3 large segments

Melia L. Meliaceae. (I 2). 3 OW trop. R: GBS 37(1984)49. *M. azedarach* L. (As. to Aus.) – timber tree (*M. dubia, M. composita*, white cedar, bakain, lunumidella, Ceylon cedar or mahogany) used for construction, bark & lvs medic. & insecticidal (antifeedant triterpenoids), fr. used as beads, early-flowering forms domesticated in China & India (Persian or Indian or Cape (Aus.) lilac, chinaberry, 'syringa', Pride of India) & widely natur. in NW, Afr. & Medit. where cult. orn.

Meliaceae Juss. Dicots – Rosidae – Sapindales. 51/565 trop. & subtrop. (few). Trees, often pachycaul, rarely shrubs or suckering shrublets, dioec., polygamous, monoec. or with only bisexual fls; bark bitter & astringent. Lvs pinnate to bipinnate, unifoliolate or simple, spirally arr. (rarely decussate) with usu. entire leaflets, s.t. spiny; stipules 0. Fls if unisexual often with rudiments of opp. sex, in spikes to thyrses, axillary to supra-axillary, cauliflorous or even epiphyllous, reg.; K (2)3–5(–7), s.t. transitional to bracteoles, usu. atop a tube, imbricate or s.t. almost closed when basally circumscissile, C 3–7(–14) in 1(–2) whorls, s.t. basally connate, imbricate or convolute or valvate, A usu. atop a tube with 3–19(–30) anthers in 1(2) whorls, nectary usu. a disk usu. around ovary, $\underline{G}$ ((1)2–6(–20)) with as many locules & usu. axile placentation (1–loc. with intruded parietal placentas in *Heckeldora* etc.) with 1–∞ bitegmic anatropous, campylotropous or orthotropous ovules per loc. Fr. a capsule, berry or drupe; seeds winged & then attached to woody columella, or with corky outer layers, or with fleshy sarcotesta or aril or a combination of both, or none of these, usu. without endosperm. x = 10–14+. R: Blumea 22(1975)419

Classification & principal genera:

I. **Melioideae** (buds naked; fr. a capsule, berry or drupe with unwinged seeds): 7 tribes:
   1. **Turraeeae** (lvs often simple, fr. usu. a capsule): 6 genera incl. *Munronia, Turraea* & *Nymania* (sometimes treated in a separate tribe)
   2. **Melieae** (lvs pinnate or bipinnate, fr. a drupe): *Azadirachta, Melia*
   3. **Vavaeeae** (trees with *Terminalia*-branching, lvs simple, fr. a berry): *Vavaea*
   4. **Trichilieae** (lvs 1-foliolate to pinnate, fr. a capsule, berry or drupe): 10 genera incl. *Owenia, Trichilia*
   5. **Aglaieae** (lvs 1-foliolate to pinnate, fr. a berry or capsule): 5 genera incl. *Aglaia* & *Lansium*
   6. **Guareeae** (lvs pinnate, sometimes with indefinite growth from apical 'pseudogemmula', fr. usu. a capsule): 10 genera incl. *Chisocheton, Dysoxylum, Guarea*
   7. **Sandoriceae** (lvs 3-foliolate, fr. a drupe): *Sandoricum*

II. **Quivisianthoideae** (buds naked, fr. a dry capsule with winged seeds): *Quivisianthe*

III. **Capuronianthoideae** (buds naked, lvs decussate; fr. dry with seeds with corky sarcotesta): *Capuronianthus*

IV. **Swietenioideae** (buds usu. with scale lvs, fr. a woody capsule with central columella & winged seeds or columella rudimentary & seeds with woody or corky sarcotesta): 3 tribes:
   1. **Cedreleae** (A 5 free, seeds winged at 1 or both ends): *Cedrela, Toona*
   2. **Swietenieae** (A 8–10 atop a tube, seeds winged): 9 genera incl. *Entandrophragma, Khaya, Lovoa, Swietenia*
   3. **Xylocarpeae** (A 8–10 atop a tube, seeds unwinged with corky or woody outer layer): *Carapa, Xylocarpus*

Subfams II & III monotypic & restricted to Madag.; the recently described *Neomangenotia* assigned to a 5th subfam. & also from Madagascar referable to *Commiphora*. Char. limonoids allied to those in Rutaceae

Imp. high-quality timbers esp. *Cedrela, Dysoxylum, Entandrophragma, Khaya, Lovoa, Melia, Swietenia* (mahogany), *Toona*; fr. trees (*Lansium, Sandoricum*), insecticides (*Azadirachta, Melia*), tanbark (*Xylocarpus*), medic. (*Munronia*), oilseeds etc.

Melianthaceae Bercht. & Presl. Dicots – Rosidae – Sapindales. 2/8 trop. & S Afr. Trees & shrubs. Lvs pinnate, spirally arr.; stipules connate, intrapetiolar. Fls usu. bisexual, resupinate through twisting of pedicel, in racemes; K 5 (2 s.t. united), imbricate, sometimes basally connate, C 5 imbricate & unequal or 1 abortive, nectary-disk extrastaminal, A 4 or 5, s.t. basally connate, alt., anthers with longit. slits, $\underline{G}$ (4(5)), pluriloc., each loc. with 1 (*Bersama*) or 2–5 (*Melianthus*) erect to pendulous, anatropous, bitegmic ovules. Fr. a loculicidal capsule; seeds usu. 1 or 2 per loc., with small straight embryo & copious oily or sometimes starchy endosperm, arillate in *Bersama*. n = 18, 19

Genera: *Bersama, Melianthus*. Greyiaceae poss. referable here (again) on A structure. App. close to Sapindaceae but seed structure reminiscent of Lardizabalaceae

Melianthus L. Melianthaceae. 6 S Afr. Shrubs with rather pachycaul stems, foetid when bruised, posterior sepal saccate or spurred. *M. major* L. (honeyflower) – cult. orn. for 'subtrop.' bedding, fls rich in nectar (black, taken by sunbirds), locally medic.

Melica L. Gramineae (19). 80 temp. (Eur. 10) exc. Aus. R: KBAS 13(1986)113. Melick. some cult. orn.; some forest grasses incl. *M. uniflora* Retz. (wood melick, Eur., Medit.)

melick *Melica* spp.; **wood m.** *M. uniflora*

Melichrus R. Br. (~ *Styphelia*). Epacridaceae. 4 Aus.

Melicocca L. = *Melicoccus*

Melicoccus P. Browne. Sapindaceae. 2 trop. Am. *M. bijugatus* Jacq. (mamoncillo) – fr. tree, seeds ed. roasted, bark decoction used against dysentery

Melicope Forster & Forster f. Rutaceae. 150 Indomal. to Aus. (14–16), NZ & Hawaii (47). Alks

Melicytus Forster & Forster f. Violaceae. (Incl. *Hymenanthera*) 8–9 Solomon Is., E Aus., NZ, Norfolk Is., Fiji. Fls almost reg.; ovules ∞ per loc.; berry. Dioec. trees & shrubs; *M. ramiflorus* Forster & Forster f. (NZ etc.) – cult. orn., the timber form. used for charcoal in gunpowder

Melientha Pierre. Opiliaceae. 2 Yunnan & SE As. to Philipp. *M. suavis* Pierre (SE As. to Philipp.) – wood for charcoal, fr. ed.

melilot *Melilotus* spp.; **white m.** *M. albus*; **yellow m.** *M. officinalis*

Melilotus Miller. Leguminosae (III 22). 20 temp. & subtrop. Euras. (Eur. 16), N Afr., Ethiopia. R: BJ 29(1901)660, CJPS 49(1969)1. Melilot. Fragrant herbs grown as green manure, forage crops (though a glucoside of o-hydroxycinnamic acid gets converted to dicoumarol, an anticlotting factor if it reaches blood of sheep or cattle) or bee-pls esp. *M. albus* Medikus (white m., Bokhara clover, Euras., natur. N Am., ? white-flowered form of *M. officinalis*) & *M. officinalis* (L.) Medikus (yellow m., Euras., natur. N Am.)

melindjo *Gnetum gnemon*

Melinia Decne. Asclepiadaceae (III 1). 8 S Am. R: Darw. 30(1990)279

Melinis Pal. Gramineae (34c). (Incl. *Rhynchelytrum*) 22 trop. & S Afr. R: BB 138(1988)50. *M. minutiflora* Pal. (molasses grass, efwatakala g., trop. Afr.) – widely introd. for forage, whole pl. sweet-smelling, said to be insect-repellent; *M. repens* (Willd.) Zizka (*R. r.*, *Tricholaena rosea*, Natal grass, trop. & S Afr.) – cult. pasture grass natur. US, cult. orn.

Meliosma Blume. Sabiaceae (Meliosmaceae). 55 trop. As. (15) & Am. (40; Tertiary fossils in Euras. & N Am.). Some timbers & cult. orn. with fragrant fls

Meliosmaceae Endl. See Sabiaceae

Melissa L. Labiatae (VIII 2). 3 Eur. (1) to Iran & C As. *M. officinalis* L. (balm, lemon or sweet or tea b., S Eur., natur. N Am. etc.) – cult. for lemon-scented lvs used as seasoning & medic. (mild sedative & antidepressant, terpenoids acting on involuntary nervous system & relieving spasm in smooth muscle), in scents & liqueurs, lvs in empty skeps attract bee-swarms as same terpenoids as in Nasonov glands of honeybees

Melissitus Medikus = *Trigonella*

Melittacanthus S. Moore. Acanthaceae. 1 Madagascar

Melittis L. Labiatae (VI). 1 Eur.: *M. melissophyllum* L. (bastard balm)

Mellera S. Moore. Acanthaceae. 4–5 warm Afr.

Mellichampia A. Gray ex S. Watson = *Cynanchum*

Melliniella Harms. Leguminosae (III 9). 1 trop. W Afr. Close to *Alysicarpus*

Melliodendron Hand.-Mazz. Styracaceae. 3 S & SW China

Mellissia Hook.f. Solanaceae (2). 1 St Helena: *M. begoniifolia* (Roxb.) Hook.f. – extinct since 1875, allies in S Am.

Melloa Bureau. Bignoniaceae. 1 trop. Am.

Melocactus Link & Otto. Cactaceae (III 3). 31 W Mex. to S Peru (E Brazil 14, 11 endemic). R: Bradleya 9(1991)1. Term. fl.-bearing structure permanently distinct from rest. Cult. orn. (melon cactus) esp. Braz. spp. & *M. intortus* (Miller) Urban (*M. communis*, Turk's-cap c., WI) to 1 m tall, seeds removed from fr. & disp. by fireants

Melocalamus Benth. (~ *Dinochloa*). Gramineae (1b). 2 India & S China to SE As. Climbers with fleshy fr.

Melocanna Trin. (~ *Schizostachyum*). Gramineae (1c). 2 Indomal. Some vivipary. *M. baccifera* (Roxb.) Kurz (*M. bambusoides*, Terai bamboo, NE India, Burma) – stems used for allegedly white-ant resistant building material, fr. a berry size of an avocado, ed. baked; flowering on a 7–51 yr cycle

Melochia L. Sterculiaceae. 54 trop. esp. Am. R: CUSNH 34(1967)191. Some locally medic. & veg.; *M. pyramidata* L. (trop. & warm Am.) – locally used fibres

Melodinus Forster & Forster f. Apocynaceae. 75 Indomal. & Pacific. Alks.

Melodorum Lour. Annonaceae. 4–5 trop. Afr. (1), SE As. to W Mal. Some ed. fr.

Melolobium Ecklon & Zeyher. Leguminosae (III 28). 20 S Afr.

melon *Cucumis melo*; **m. cactus** *Melocactus* spp.; **horned m.** *Cucumis metalliferus*; **m.-loco** *Apodanthera undulata*; **rock m.** *C. melo* Reticulata Group; **m. tree** *Carica papaya*; **water m.** *Citrullus lanatus*

Melosperma Benth. Scrophulariaceae. 1 Chile. Large seeds

Melothria L. Cucurbitaceae. 10 NW. Some ed. fr. & local medic.

Melothrianthus Mart. Crov. Cucurbitaceae. 1 Brazil

Melpomene A.R. Sm. & R.C. Moran (~ *Grammitis*). Grammitidaceae. 20 trop. Afr. & Am. R: Novon 2(1992)426

Memecylaceae DC = Melastomataceae

Memecylanthus Gilg & Schltr. = *Wittsteinia*

Memecylon L. Melastomataceae (Memecylaceae). c. 250 OW trop. (Borneo 27, R: OB 69(1983)5). Some rheophytes; elaiophores attractive to bees. Some timbers & cult. orn., some ed. fr. *M. umbellatum* Burm.f. (Indomal.) – fr. ed., lvs yield yellow dye

Memora Miers. Bignoniaceae. 27 trop. Am. Four-armed 'anomalous' sec. xylem with 4 complementary phloem ribs in bark of these lianes

Menabea Baillon. Asclepiadaceae (II; Periplocaceae). 1 Madag.: *M. venenata* Baillon – ordeal-poison

Menais Loefl. Boraginaceae. 1 S Am.

Mendoncella A. Hawkes = *Galeottia*

Mendoncia Vell. ex Vand. Mendonciaceae (Acanthaceae). 60 trop. Am. & Afr., Madag.

Mendonciaceae Bremek. (~ Acanthaceae). Dicots – Asteridae – Scrophulariales. 2/60 trop. Am. & Afr. Twining shrubs with jointed stems & unusual sec. growth; cystoliths 0. Lvs opp., simple, entire; stipules 0. Fls bisexual, solit. & axillary or in term. racemes, with 2 large bracteoles; K minute or shortly lobed, C (5) ± irreg., A 4 attached to C-tube & alt. with lobes, paired upper 1 a staminode or 0, cupular nectary-disk around $\underline{G}$ (2), 1–2-loc. with 2 collateral (?) unitegmic ovules per loc., funicles not developed. Fr. a drupe with 1(2)-loc. endocarp & 1 or 2 seeds; endosperm 0, cotyledons twice folded

Genera: *Anomacanthus*, *Mendoncia*. Considered a subfam. of Acanthaceae by some (not really separable on wood anatomy; *Pseudocalyx* like M. but fr. dry) but cystoliths & jaculators 0, poss. close to *Thunbergia*

Mendoravia Capuron. Leguminosae (I 2). 1 SE Madag., local

menduro *Balanites maughamii*

Menendezia Britton = *Tetrazygia*

Menepetalum Loes. Celastraceae. 6 New Caledonia

meni oil *Lophira* spp.

Meniscium Schreber. Older name for *Cyclosorus*

Meniscogyne Gagnepain (~ *Elatostema*). Urticaceae (II). 2 SE As.

Menisorus Alston = *Cyclosorus*

Menispermaceae Juss. Dicots – Magnoliidae – Ranunculales. 72/450 trop. & warm, few temp. Usu. dioec. lianes & scandent shrubs, rarely trees or herbs, usu. with bitter sesquiterpenoids (poisonous) & alks, the stems often with unusual sec. thickening & becoming flattened. Lvs simple (3-foliolate in *Syntriandrum* etc.), spirally arr., rarely lobed but with palmate venation; stipules usu. 0. Infls racemes, panicles or cymose heads, fls rarely solit., rarely on old wood. Fls ± reg., usu. dull, 3-merous often with 2 whorls of K, C & A; K usu. free, imbricate or valvate (1–)6(–12+), C often 6 but s.t. more or fewer or even 0, A (2–3, 6(–40) often opp. C, filaments s.t. ± connate, anthers with longit. (rarely transverse) slits & usu. introrse, females with $\underline{G}$ (1)3(6–30) in 1 or more whorls, often held on a gynophore, each with 2 (1 soon aborting) submarginal, pendulous, hemitropous to amphitropous bi- or unitegmic ovules. Fr. a head of drupes or nuts, usu. ± curved, the term. style often appearing ± basal; seed often horseshoe-shaped with straight to curved embryo & oily proteinaceous endosperm, s.t. ruminate, or scanty or 0. x = 11–13, 19, 25

Classification & principal genera:

I. **Pachygoneae** (Tiliacoreae, Triclisieae incl. Hyperbaeneae & Peninantheae); endosperm 0): *Chondrodendron*, *Hyperbaena*, *Tiliacora*, *Triclisia*

II. **Anomospermeae** (endosperm strongly ruminate; cotyledons not foliaceous, appressed): *Abuta*

III. **Tinosporeae** (endosperm weakly ruminate; cotyledons foliaceous, divaricate): *Disciphania*, *Tinospora*

IV. **Fibraureae** (incl. Coscinieae; endosperm not ruminate; cotyledons thin, foliaceous, divaricate): *Coscinium*, *Fibraurea*

V. **Menispermeae** (endosperm not ruminate; cotyledons subcarnose, appressed): *Cissampelos*, *Cocculus*, *Stephania*

App. close to Lardizabalaceae but L. have compound lvs & copious endosperm

Many medic., esp. in preparation of curare or fish-poisons, sweeteners or contraceptives: *Abuta*, *Anamirta*, *Chondrodendron*, *Cissampelos*, *Curarea*, *Dioscoreophyllum*, *Fibraurea*, *Jateorhiza*, *Pachygone*, *Pycnarrhena*, *Stephania*, *Tinospora*; dyes: *Coscinium*; few cult. orn. lianes: *Cocculus*, *Menispermum*, while some rain-forest spp. gigantic

Menispermum L. Menispermaceae (V). 2–4 E N Am. & E As. *M. canadense* L. (moonseed, yellow parilla, E N Am.) – cult. orn. liane with medic. rhizome

Menkea Lehm. Cruciferae. 6 Aus. R: CGH 200(1970)175

mengkulang *Heritiera* spp.

Menodora Bonpl. Oleaceae. c. 25 warm Am., S Afr. (3). R: AMBG 19(1932)87

Menonvillea R. Br. ex DC. Cruciferae. 30 Peru, Chile, Arg. Shrublets to annuals; gynophore present

Mentha L. Labiatae (VIII 2). 25 temp. OW (Eur. 10 with 4 common hybrids). Mints. C subreg., 4-merous. V. aromatic herbs long-cult. as flavourings, the Romans using *M. aquatica* L. (waterm., Euras., Medit.), most common now being *M. spicata* L. (spearm., *M. longifolia* × *M. suaveolens*) & *M. suaveolens* Ehrh. (apple or woolly m., S & W Eur., incl. 'Variegata', pineapple m.), a hairier pl., used in mint sauce & jelly, *M.* × *piperita* L. (*M. spicata* × *M. aquatica*, peppermint), the flavouring for chocolate, crème-de-menthe, tea, icecream, toothpaste etc., incl. 'Citrata' (nm. *citrata*, eau-de-cologne m., bergamot, lemon or orange m.), 'Officinalis' (nm. *officinalis*, white p.) & 'Piperita' (nm. *piperita*, black p. used in tea), while an imp. source of menthol for cigarettes etc. is *M. arvensis* L. (field m., Euras.) var. *piperascens* Malinv. (Japanese m.). Others cult. incl. *M.* × *gracilis* Sole ('*M.* × *gentilis*', *M. spicata* × *M. arvensis*, ginger or Scotch m.) – cult. for spearmint oil; *M. longifolia* (L.) Hudson (horse m., Eur., Medit.) – often confused with *M. spicata*; *M. pulegium* L. (pennyroyal, Eur., W As.) – used in soap etc., a form being Gibraltar m., medic.; *M.* × *villosa* Hudson (*M. spicata* × *M. suaveolens*) esp. 'Alopecuroides' (nm. *alopecuroides*, Bowles's m.) – rumoured to be the best

Mentocalyx N.E. Br. = *Gibbaeum*

Mentodendron Lundell = *Pimenta*

Mentzelia L. Loasaceae. 60 warm Am. R: AMBG 21(1934)103. No stinging hairs; outer A s.t. staminodes; shrubs & herbs, some cult. orn. esp. *M. laevicaulis* (Douglas) Torrey (blazing star, W N Am.) & *M. lindleyi* Torrey & A. Gray (bartonia, C Calif.) – scented yellow fls opening in evening

Menyanthaceae Bercht. & Presl. (~ Gentianaceae). Dicots – Asteridae – Asterales. 5/40 cosmop. Aquatics or helophytes with intercellular canals & spaces in stems, where vasc. bundles often scattered. Lvs simple (3-foliolate in *Menyanthes*), cordate or reniform to linear, spirally arr.; petiole sheathing at base, stipules 0 or wing-margins of petiole. Fls bisexual, reg., usu. 5-merous, solit. or in various infl. types, K sometimes connate, C with tube & valvate to imbricate lobes (margins fringed or crested), filaments attached to tube alt. with lobes, anthers with longit. slits, s.t. scales (? staminodes) alt. with filaments, nectary-disk often around G (2), superior to half-inf., 1-loc. with term. style & ∞ anatropous, unitegmic ovules on the 2 parietal placentas. Fr. a capsule, variously dehiscent or a berry; seeds with linear embryo in copious firm oily endosperm. x = 9, 17

Genera: *Liparophyllum*, *Menyanthes*, *Nephrophyllidium*, *Nymphoides*, *Villarsia*

Form. assoc. with Gentianaceae but G. have opp. lvs, internal phloem etc. & Cronquist places M. in Solanales but cladistic work places M. in an enlarged Asterales

Some locally medic. or ed., some cult. orn. & bad weeds

Menyanthes L. Menyanthaceae. 1 circumboreal: *M. trifoliata* L. (bogbean, or buckbean) – self-incompatible heterostylous fls, rhizome medic. (trad. cure for arthritis in Germany), lvs used as hop subs. in beer, rhizome powdered for a bread by Esquimaux & in N Euras., mitsugashiwalactone attractive to cats

Menziesia Sm. Ericaceae. 7 N Am. (2), E As. (7). Some cult. orn. decid. shrubs

Meopsis (Calest.) Kozo-Polj. = *Daucus*

meranti *Shorea* spp.

merawan *Hopea mengerawan*

merbau *Intsia bijuga*, *I. palembanica*

Merciera A. DC. Campanulaceae. 4 SW Cape. G 1-loc. with basal ovule

Mercurialis L. Euphorbiaceae. 8 Medit. & temp. Euras. (Eur. 7) to N Thailand. Mercury. Dioec. with extrafl. nectaries; lvs opp.; G (2). Some dye-sources incl. *M. perennis* L. (dog's m., Euras., Medit.) – carpeting rhiz. herb often char. of disturbed woodland; *M. annua* L. (annual m., Eur., Medit.) – seeds with caruncles, weed of gardens

mercury *Mercurialis* spp.; **annual m.** *M. annua*; **dog's m.** *M. perennis*

Merendera Ramond = *Colchicum*

Meresaldia Bullock = *Metastelma*

Meriandra Benth. Labiatae (VIII 2). 2 Ethiopia, Himalaya

Meriania Sw. Melastomataceae. 74 trop. Am. Connective spurred

Merianthera Kuhlm. Melastomataceae. 3 Brazil

Mericarpaea Boiss. Rubiaceae (IV 16). 1 W As.

Mericocalyx Bamps = *Otiophora*

Meringium C. Presl = *Hymenophyllum*

Meringogyne H. Wolff = *Angoseseli*

Meringurus Murb. = *Gaudinia*

Merinthopodium J.D. Sm. = *Markea*

Merinthosorus Copel. = *Aglaomorpha*

Merismostigma S. Moore = *Caelospermum*

Meristotropis Fischer & C. Meyer = *Glycyrrhiza*

Merope M. Roemer (~ *Atalantia*). Rutaceae. 1 trop. As.

Merostachys Sprengel. Gramineae (1b). 40 trop. Am. Bamboos flowering in 11–34 yr cycles; some climbers

Merremia Dennst. ex Endl. Convolvulaceae. c. 70 trop. Some serious weeds in plantations, some cult. orn. & locally medic. & ed. *M. discoidesperma* (J.D. Sm.) O' Don. (Mary's bean, C

Am.) – most widespread of all drift-seeds; *M. tuberosa* (L.) Rendle (Brazilian jalap, trop. Am. but now natur. pantrop.) – cult. orn. with yellow fls, dried fr. with accrescent K = 'wood rose' of dried floral decorations

Merrillanthus Chun & Tsiang. Asclepiadaceae (III 1). 1 Hainan: *M. hainanensis* Chun & Tsiang

Merrillia Swingle. Rutaceae. 1 Burma to W Mal.: *M. caloxylon* (Ridley) Swingle – pinnate lvs with rachis joints like petiole of *Citrus* spp., fr. with large amounts of eupatorin (also in *Eupatorium* spp.) with high antitumour activity

Merrilliodendron H.L. Kaneh. Icacinaceae. 1 Philippines & W Pacific

Merrilliopanax H.L. Li. Araliaceae. 3 E Himal. to SW China. R: BMNHN 4,5(1983)289

Merrittia Merr. (~ *Blumea*). Compositae (Pluch.). 1 Philippines

mersawa *Anisoptera* spp. esp. *A. laevis*

Mertensia Roth. Boraginaceae. 45 N temp. (Eur. 1, N Am. 24 – R: JAA suppl. 1(1991)88) to Mex. & Afghanistan. R: AMBG 24(1937)17. Bluebells (N Am.). Cult. orn. rock-pls incl. *M. maritima* (L.) Gray (gromwell, sea lungwort, oysterleaf, N temp. coasts) – rhizome eaten by Esquimaux, & *M. virginica* (L.) Link (Virginian b. or cowslip, E N Am.)

Merumea Steyerm. Rubiaceae (I 1/IV 24). 2 Guayana Highland

Merwia B. Fedtsch. = *Ferula*

Merwiopsis Sophina = *Pilopleura*

Merxmuellera Conert = *Rytidosperma*

Meryta Forster & Forster f. Araliaceae. 30 NZ, Aus., New Caled. (11), Pacific is. Dioec. pachycauls, lvs simple; some cult. orn.

Mesadenella Pabst & Garay = *Stenorrhynchos*

Mesadenus Schltr. = *Brachystele*

Mesanthemum Koern. Eriocaulaceae. 12 trop. Afr., Madagascar

Mesanthophora H. Robinson. Compositae (Vern.-Vern.). 1 C Paraguay: *M. brunneri* H. Robinson – on limestone. R: Novon 2(1992)169

mescal *Lophophora williamsii*; see also *Agave*

Mesechites Muell. Arg. Apocynaceae. 10 trop. Am.

Mesembryanthemum L. Aizoaceae (IV). 25 Medit. (Eur. 2), Arabia, drier S Afr., S Aus., Calif., W S Am., Atlantic Is. (incl. *Cryophytum*). *M. crystallinum* L. (*Cryophytum c.*, ice-pl., Cape, natur. in Calif., Medit. etc.) – lvs covered with glistening papillae, cult. bedding pl. with lvs eaten like spinach. Many succ. & cult. orn. form. referred here now in *Carpobrotus, Conophytum, Dorotheanthus* etc.

Mesoglossum Halb. Orchidaceae (V 10). 1 Mexico

Mesogyne Engl. Moraceae (III). 1 trop. Afr.

Mesomelaena Nees (~ *Gymnoschoenus*). Cyperaceae. 5 SW Aus. R: Telopea 2(1981)181

Mesona Blume. Labiatae (VIII 4b). 2–3 SE As., Mal. *M. palustris* Blume (Java) – cooling drink & flavouring

Mesophlebion Holttum = *Cyclosorus*

Mesoptera Hook.f. = *Psydrax*

Mesopteris Ching = *Amphineuron*

Mesosetum Steudel. Gramineae (34b). 32 trop. Am. R: Brittonia 2(1937)363

Mesospinidium Reichb.f. Orchidaceae (V 10). 7 trop. Am. R: Orquideologia 8(1973)165

Mesostemma Vved. = *Stellaria*

Mespilus L. Rosaceae. 1 Arkansas (*M. canescens* Phipps) & 1 SE Eur. to C As.: *M. germanica* L. (medlar) – fr. ed. esp. after frost & slightly rotted (bletted) when malic acid reduced & sugar increased, poss. cult. Assyrians & Babylonians, poss. a Roman introd. to GB (seed found at Silchester); can be grafted on hawthorn & can form graft- as well as sexual hybrids

mespilus, snowy *Amelanchier rotundifolia*

mesquite *Prosopis glandulosa*

Messerschmidia Hebenstr. = *Argusia*

messmate *Eucalyptus* spp. with stringy bark

Mestoklema N.E. Br. ex Glen. Aizoaceae (V). 6 Cape. R: Bothalia 13(1981)454

Mesua L. Guttiferae (III). 40 Indomal. *M. ferrea* L. (ironwood, na) – timber v. hard used for railway sleepers in S India, form. for lances, the tree sacred in India, fls used in medic. & cosmetics & to scent the stuffing of pillows

Mesyniopsis W. Weber = *Linum*

Mesynium Raf. = *Linum*

Metabolos Blume = *Hedyotis*

Metabriggsia W.T. Wang. Gesneriaceae. 2 W Guangxi, China. R: EJB 49(1992)30

Metadina Bakh.f. Rubiaceae (I 2). 1 Indomal.
Metaeritrichium W.T. Wang = *Eritrichium*
Metalasia R. Br. Compositae (Gnap.-Rel.). 52 S Afr. esp. SW Cape. R: OB 99(1989)1. Lvs s.t. twisted spirally
Metalepis Griseb. (~ *Cynanchum*). Asclepiadaceae (III 3). 2 WI
Metanarthecium Maxim. = *Aletris*
Metanemone W.T. Wang. Ranunculaceae (II 2). 1 Yunnan
Metapetrocosmea W.T. Wang (~ *Petrocosmea*). Gesneriaceae. 1 Hainan: *M. peltata* (Merr. & Chun) W.T. Wang
Metaplexis R. Br. Asclepiadaceae (III 1). 6 E As. *M. japonica* (Thunb.) Makino – stems for rope, seeds & lvs medic., floss a cotton-floss subs. (China)
Metapolypodium Ching = *Goniophlebium*
Metaporana N.E. Br. (~ *Bonamia*). Convolvulaceae. 5 Socotra (1), E Afr. (1), Madag. (3). R: JAA 71(1990)252
Metarungia Baden (*Macrorungia* auctt.). Acanthaceae. 3 trop. & S Afr. R: NJB 1(1981)143, KB 34(1980)638
Metasequoia Miki ex Hu & W.C. Cheng. Taxodiaceae, 1 C China: *M. glyptostroboides* Hu & W.C. Cheng (dawn redwood) – cult. orn. decid. (branchlets) tree to 45m or more, discovered in 1945 & seed coll. 1947 introd. to Western gardens in 1948. Genus first described from fossils as *Taxites* Brongn. (1828) & as *Metasequoia* (1941), from modern pls in 1948. R: BR 42(1976)215
Metasocratea Dugand = *Socratea*
Metastachydium Airy Shaw ex C.Y. Wu & Li. Labiatae (VI). 1 C As.
Metastelma R. Br. (~ *Cynanchum*). Asclepiadaceae (III 1). 100 trop. & warm Am.
Metastevia Grashoff. = *Stevia*
Metathelypteris (H. Itô) Ching = *Thelypteris*
Metatrophis F. Br. Moraceae. 1 Polynesia = ?
Metaxya C. Presl. Metaxyaceae. 1 trop. S Am.: *M. rostrata* (Kunth) C. Presl
Metaxyaceae Pichi-Serm. Filicopsida. 1/1 trop. S Am. Rhizome creeping, solenostelic & fronds 1-pinnate with sori dorsal on the veins (sometimes 2 or 3 on same vein – unique in modern ferns)
Genus: *Metaxya*
App. allied to *Lophosoria*
Metcalfia Conert. Gramineae (21b). 2 Mex., Balkans & Cauc. R: KBAS 13(1986)122
Meteoromyrtus Gamble. Myrtaceae. (Myrt.). 1 S India
Meterostachys Nakai (~ *Sedum*). Crassulaceae. 1 S Japan & S Korea
Metharme Philippi ex Engl. Zygophyllaceae. 1 N Chile
Methysticodendron R. Schultes = *Brugmansia*
Metopium P. Browne. Anacardiaceae (IV). 3 Florida, Mex., WI. Purging resins esp. hog gum from *M. toxiferum* (L.) Krug & Urban (Florida)
Metrodorea A. St-Hil. Rutaceae. 5–6 Brazil
Metrosideros Banks ex Gaertner. Myrtaceae (Lept.). (Excl. *Carpolepis*) 50 E Mal., Pacific (NZ 10), S Afr. (1: *M. angustifolia* Sm., only capsular Myrtacea in Afr.). Trees, shrubs (incl. rheophytes) & woody climbers app. merging with *Callistemon* in New Caled. Some cult. orn. & timbers esp.: *M. polymorpha* Gaudich. (*M. collina*, Hawaii) – wood for construction & form. for idols, (exc. for young lvs) resistant to volcanic fumes (SO_2 to 100 p.p.m.) as stomata close; *M. excelsa* Sol. ex Gaertner (Christmas tree (NZ), NZ) – locally medic.; *M. robusta* Cunn. (rata, NZ) – sometimes epiphytic, timber for telegraph-poles etc.
Metroxylon Rottb. Palmae (II 1c). 5 E Mal. Sago palms. R: Principes 30(1986)170. Trunks hapaxanthic but, in most, suckers develop from base; fr. takes 3 yrs to ripen. Trunks felled when infl. appears & sago removed from stem by crushing & washing, esp. *M. sagu* Rottb. (natur. W Pacific to SE As.) spineless or ('*M. rumphii* C. Martius') prickly – petioles used for walls in Indonesia. Some 'seeds' used as buttons
Mettenia Griseb. = *Chaetocarpus*
Metteniusa Karsten. Icacinaceae. 3 NW trop. S Am.
Metternichia Mikan. Solanaceae (1). 1 E Brazil: *M. principis* Mikan
meu *Meum athamanticum*
Meum Miller. Umbelliferae (III 8). 3 N Afr. & Eur. incl.: *M. athamanticum* Jacq. (baldmoney, meu, spignel) – cult. orn., form. cult. for ed. roots & medic.
Mexacanthus T. Daniel. Acanthaceae. 1 W Mexico
Mexerion Nesom (~ *Gnaphalium*). Compositae (Gnap.-Gnap.). 2 Mex. R: Phytol. 68(1990)247
Mexianthus Robinson. Compositae (Eup.-Crit.). 1 Mex.: *M. mexicanus* Robinson. R: Sida

13(1989)340

Mexican apple *Casimiroa edulis*; **M. devil** *Ageratina adenophora*; **M. hat** *Rudbeckia* spp.; **M. mahogany** *Swietenia humilis*; **M. poppy** *Argemone mexicana*

Mexicoa Garay (~ *Oncidium*). Orchidaceae (V 10). 1 Mex.: *M. ghiesbreghtiana* (A. Rich. & Galeotti) Garay – cult. orn. diminutive

Meximalva Fryx. Malvaceae. 2 Mexico

Mexipedium V. Albert & M. Chase (~ *Phragmipedium*). Orchidaceae (II). 1 Mex.: *M. xerophyticum* (Soto, Salazar & Hagsater) V. Albert & M. Chase. R: Lindleyana 7(1992)172

Meyenia Nees (~ *Thunbergia*). Acanthaceae. 1 India, Sri Lanka

Meyerophytum Schwantes. Aizoaceae (V). 1 Cape: *M. meyeri* (Schwantes) Schwantes

Meyna Roxb. ex Link. Rubiaceae (III 2). 11 trop. Afr. & Comoro Is. to SE As.

mezcal = mescal

mezereon *Daphne mezereum*

Mezia Schwacke ex Niedenzu. Malpighiaceae. 1 trop. Am.: *M. includens* (Benth.) Cuatrec. – imp. local medic. vine

Meziella Schindler (~ *Haloragis*). Haloragidaceae. 1 SW Aus.: *M. trifida* (Nees) Schindler, known only from type-specimen until 1990 when a pop. discovered

Mezilaurus Kuntze ex Taubert. Lauraceae (I). 20 trop. S Am. to Costa Rica (1). R: AMBG 74(1987)158

Mezleria C. Presl = *Lobelia*

Mezobromelia L.B. Sm. Bromeliaceae (2). 5 Colombia, Ecuador. R: FN 14(1977)1364

Mezochloa Butzin = *Alloteropsis*

Mezonevron Desf. = *Caesalpinia*

Mezzettia Becc. Annonaceae. 4 W Mal. to Moluccas. R: Blumea 35(1990)217

Mezzettiopsis Ridley (~ *Orophea*). Annonaceae. 1 W Mal.: *M. creaghii* Ridley

mfukufuku *Brexia madagascariensis*

mfungu *Celosia argentea*

mi *Madhuca longifolia*

Mibora Adans. Gramineae (21d). 2 W Eur., NW Afr. *M. minima* (L.) Desf. – cult. orn. annual, sometimes flowering when less than 1 cm tall

Michaelmas daisy *Aster* spp. & hybrids

Michauxia L'Hérit. Campanulaceae. 7 SW As. Fls 6–10-merous. *M. campanuloides* L'Hérit. (E Medit.) – cult. orn. biennial with white fls

Michelia L. Magnoliaceae (I 2). 30 China, trop. As. to S Japan & W Mal. Like *Magnolia* but fls axillary & gynophore between A & G. Alks. Some timbers & cult. orn. esp. *M. champaca* L. (champak, sapu, Himal.) – cult. around Hindu & Jain temples, ess. oil from fls used in scent-making, timber for tea-boxes, furniture, carving & fuel, bark a febrifuge, lvs used to feed silkworms; *M. doltsopa* Buch.-Ham. ex DC (E Himal. to W China) – timber good

Michelsonia Hauman (? = *Tetraberlinia*). Leguminosae (I 4). 1 E Zaire: *M. microphylla* (Troupin) Hauman – forming almost pure stands

Micholitzia N.E. Br. = *Dischidia*

Miconia Ruíz & Pavón. Melastomataceae. c. 1000 trop. Am., 1 W Afr. Some with hollow ant-infested stems. Some ed. fr. (bush currants); some yellow dyes. Many related spp. in same habitat, 20 in Trinidad with complementary fruiting times throughout yr & same dispersing birds. *M. albicans* (Sw.) Steudel (trop. Am.) – disp. (Brazil savanna) by vole-like rat, *Bolomys lasiurus*, only small rodent-disp. known; *M. argentea* (Sw.) DC (trop. Am.) – pioneer establishing in gaps greater than 102 m^2 & becoming canopy emergent; *M. calvescens* Blume ('*M. magnifica*', trop. Am., natur. Tahiti (penetrating native forest) & Sri Lanka) – lvs to 70 cm long

Micractis DC (~ *Sigesbeckia*). Compositae (Helia.-Mel.). 3 trop. Afr., Madag.

Micraeschynanthus Ridley. Gesneriaceae. 1 Malay Peninsula

Micraira F. Muell. Gramineae (27). 13 trop. Aus. R: Brunonia 2(1979)76, Nuytsia 5(1984)290. Moss-like pls with spiral phyllotaxis, lvs reviving after desiccation

Micrandra Benth. Euphorbiaceae. 14 trop. S Am. Some rubber ('M.' r.), local medic.

Micrandropsis Rodr. Euphorbiaceae. 1 Amazon

Micrantha Dvořák = *Hesperis*

Micranthemum Michaux. Scrophulariaceae. (Incl. *Hemianthus*, *Hemisiphonia*) 14 Am., WI. Aquatics s.t. cult. in aquaria

Micrantheum Desf. Euphorbiaceae. 3 Aus.

Micranthocereus Backeb. Cactaceae (III 3). 9 C & E Brazil

Micranthus (Pers.) Ecklon. Iridaceae (IV 2). 3 SW Cape

Micrargeria Benth. Scrophulariaceae. 4–5 trop. Afr., India

Micrargeriella R. Fries. Scrophulariaceae. 1 trop. E Afr.
Micrasepalum Urban. Rubiaceae (IV 15). 2 Cuba, Hispaniola
Micrechites Miq. Apocynaceae. 20 Indomal., S China. Some rubber
Microberlinia A. Chev. Leguminosae (I 4 or III 1). 2 Gulf of Guinea. Timber (zebrawood, zebrano, zingana) esp. *M. brazzavillensis* A. Chev.
Microbiota Komarov (~ *Thuja*). Cupressaceae. 1 E Siberia: *M. decussata* Komarov
Microbriza Parodi ex Nicora & Rugolo (~ *Briza*). Gramineae (17). 2 Brazil to Arg.
Microcachrys Hook.f. Podocarpaceae. 1 Tasmania: *M. tetragona* (Hook.) Hook.f. – seeds encl. in scarlet aril
Microcala Hoffsgg. & Link = *Cicendia*
Microcalamus Franchet. Gramineae (34b). 1 trop. W Afr.: *M. barbinodis* Franchet – superficially like a bamboo
Microcalia A. Rich. = *Lagenophora*
Microcardamum O. Schulz. Cruciferae. 1 temp. S Am.
Microcarpaea R. Br. = *Peplidium*
Microcaryum I.M. Johnston. Boraginaceae. 4 C As. to W China
Microcasia Becc. = *Bucephalandra*
Microcephala Pobed. Compositae (Anth.-Mat.). 4 Iran, C As. R: MBSM 12(1976)655
Microchaete Benth. = *Senecio*
Microcharis Benth. = *Indigofera*
Microchlaena Ching = *Athyrium*
Microchloa R. Br. Gramineae (33b). 4 Afr., with 1 pantrop. Lvs of *M. caffra* Nees & *M. kunthii* Desv. (Afr.) revive after desiccation to 3–9% of water content. R: SBi 47(1966)291
Microcitrus Swingle = *Citrus*
Microcnemum Ung.-Sternb. Chenopodiaceae (II 2). 1 Medit. (incl. Eur.) to Cauc.: *M. coralloides* (Loscos & Pardo) Buen
Micrococca Benth. Euphorbiaceae. 12 OW trop.
Microcodon A. DC. Campanulaceae. 4 S Afr.
Microcoelia Lindley. Orchidaceae (V 16). 23 trop. & S Afr., Madag. R: SBU 23,4(1981)1. Leafless epiphytes with photosynthetic roots
Microcoelum Burret & Potztal = *Lytocaryum*
Microconomorpha (Mez) Lundell = *Cybianthus*
Microcorys R. Br. Labiatae (III). 17 SW Aus.
Microcos L. Tiliaceae. 53 Indomal., Fiji. *M. nervosa* (Lour.) S.Y. Hu (S China) – imp. local medic., exported to US
Microculcas Peter = *Gonatopus*
Microcybe Turcz. Rutaceae. 3 Aus.
Microcycas A.DC. Zamiaceae (III). 1 W Cuba: *M. calocoma* (Miq.) A. DC – almost extinct
Microdactylon Brandegee = *Matelea*
Microderis DC = *Leontodon*
Microdesmis Hook.f. ex Hook. Pandaceae. 10 trop. Afr., SE As., W Mal. *M. puberula* Hook.f. ex Hook. (trop. Afr.) – flexible hard timber for tool-handles, combs etc.
Microdon Choisy. Globulariaceae. 5 SW Cape
Microdracoides Hua. Cyperaceae. 1 trop. Afr.: *M. squamosa* Hua – pachycaul with woody trunk & term. panicle of fls
Microglossa DC. Compositae (Ast.-Ast.). 10 Afr., Madag., Masc., trop. As. Usu. scandent or twining shrubs. *M. afzelii* O. Hoffm. (trop. Afr.) – local medic. esp. for chests
Microgonium C. Presl = *Trichomanes*
Microgramma C. Presl (~ *Polypodium*). Polypodiaceae (II 5). 15 trop. Am., 1 Afr., Madag. & Masc. Cult. orn. epiphytes
Microgyne Less. = *Microgynella*
Microgynella Grau (*Microgyne*). Compositae (Ast.-Ast.). 1 Arg., Brazil, Uruguay
Microgynoecium Hook.f. Chenopodiaceae (I 3). 1 Tibet: *M. tibeticum* Hook.f.
Microholmesia Cribb = *Angraecopsis*
Microlaena R. Br. (~ *Ehrharta*). Gramineae (11). 10 Aus., NZ, Oceania
Microlagenaria (C. Jeffrey) A. M. Lu & J. Q. Li (~ *Thladiantha*). Cucurbitaceae. 1 trop. Afr.
Microlecane Schultz-Bip. ex Benth. = *Bidens*
Microlepia C. Presl. Dennstaedtiaceae. 45 OW trop. (Afr. 1) to Japan & NZ. Close to *Dennstaedtia*; some with fronds several m long, supported by other pls, some rheophytes. Some cult. orn. incl. *M. speluncae* (L.) Moore (OW trop., natur. Am.)
Microlepidium F. Muell. (~ *Capsella*). Cruciferae. 2 Aus. R: CGH 205(1974)158
Microlepis (DC) Miq. Melastomataceae. 4 S Brazil

Microliabum Cabrera. Compositae (Liab.). 6 C Bolivia to Arg. R: SB 15(1990)738
Microlicia D. Don. Melastomataceae. 100 trop. S Am.
Microlobius C. Presl = ? (Leguminosae)
Microloma R. Br. Asclepiadaceae (III 1). 10 S Afr. R: BJ 112(1991)453
Microlonchoides Candargy = *Jurinea*
Microlonchus Cass. = *Centaurea*
Microlophopsis Czerep. = ? *Serratula*
Micromeles Decne = *Sorbus* (*Aria*)
Micromelum Blume. Rutaceae. 9 Indomal., Pacific. R: W.T. Swingle, *Bot. Citrus* (1943)139
Micromeria Benth. (~ *Satureja*). Labiatae (VIII 2). 90 temp. (Canary Is. 15 endemic, Eur. 21) to WI. Some cult. orn. rock-pls. *M. chamissonis* (Benth.) E. Greene (*Satureja douglasii*, yerba buena, W US) – fragrant lvs a local tea
Micromonolepis Ulbr. = *Monolepis*
Micromyrtus Benth. Myrtaceae (Lept.). 22 Aus.
Micromystria O. Schulz = *Arabidella*
Micronoma H. Wendl. Palmae. 1 E Peru = ?
Micronychia Oliver. Anacardiaceae (IV). 5 Madagascar
Micropapyrus Süsseng. = *Rhynchospora*
Microparacaryum (Riedl) Hilger & Podlech = *Paracaryum*
Micropeplis Bunge = *Halogeton*
Micropera Lindley (*Camarotis*). Orchidaceae (V 16). 17 Indomal. to Aus. (New Caled. 1)
Micropholis (Griseb.) Pierre. Sapotaceae (IV). 38 trop. Am. R: FN 52(1990)172. Some timbers
Microphyes Philippi. Caryophyllaceae (I 1). 3 Chile
Microphysa Schrenk. Rubiaceae (IV 16). 1 C As.
Microphysca Naudin = *Tococa*
Microphytanthe (Schltr.) Brieger = *Dendrobium*
Micropleura Lagasca. Umbelliferae (I 1a). 2 Colombia, Chile
Microplumeria Baillon. Apocynaceae. 1 Amazonian Brazil
Micropolypodium Hayata (~ *Xiphopteris*). Grammitidaceae. 30 trop. Am. & Mal. R: Novon 2(1992)419
Micropora Hook.f. (*Hexapora*). Lauraceae (I). 1 Malay Pen.: *M. curtisii* (Hook.f.) Hook.f. – coll. once on Penang Hill
Micropsis DC. Compositae (Gnap.-Gnap.). 8 Brazil, Arg., Uruguay. R: OB 104(1991)158
Micropterum Schwantes = *Cleretum*
Micropus L. Compositae (Gnap.-Gnap.). 1 W Medit. inc. Eur. to Iran: *M. supinus* L. R: OB 104(1991)174
Micropyropsis Romero-Zarco & Cabezudo. Gramineae (17). 1 S Spain
Micropyrum (Gaudin) Link. Gramineae (17). 3 C Eur. (2), Medit.
Microrphium C.B. Clarke. Gentianaceae. 1 W Mal.
Microsaccus Blume. Orchidaceae (V 16). 11 Indomal.
Microschoenus C.B. Clarke = *Juncus*
Microsciadium Boiss. Umbelliferae (III 8). 1 As. Minor
Microsechium Naudin. Cucurbitaceae. 2 Mexico
Microsemia E. Greene = *Streptanthus*
Microseris D. Don. Compositae (Lact.) 14 W N Am., Chile (1), Aus. & NZ (1). R (partial): CDH 4(1955)207. *M. scapigera* Schultz-Bip. (murnong, Aus.) – tuber ed. roasted
Microsisymbrium O. Schultz. Cruciferae. 6 C As. to Afghanistan & W Himal., NW Am. (1)
Microsorum Link (*Microsorium*, ~ *Polypodium*). Polypodiaceae (II 4). 60 OW trop. to NZ. Some cult. orn. epiphytes. *M. scandens* (Forst.f.) Tindale (E Aus., NZ) – climbs ivy-like
Microspermum Lagasca. Compositae (Eup.-Ager.). (Incl. *Iltisia*) 18 C Am. R: MSBMBG 22(1987)184
Microstaphyla C. Presl = *Elaphoglossum*
Microstegium Nees. Gramineae (40a). 15 OW trop. & warm esp. Afr. R: KB 7(1962)209
Microsteira Baker. Malpighiaceae. 25 Madag. Polygamo-dioec.
Microstelma Baillon = *Gonolobus*
Microstemma R. Br. = *Brachystelma*
Microstephanus N.E. Br. = *Pleurostelma*
Microsteris E. Greene (~ *Phlox*). Polemoniaceae. 1 (variable) W N & S Am.
Microstigma Trautv. = *Matthiola*
Microstrobilus Bremek. = *Strobilanthes*
Microstrobos J. Garden & L. Johnson. Podocarpaceae. 2 E & SE Aus. R: CNSWH 1(1959)316.

M. fitzgeraldii (F. Muell.) J. Garden & L. Johnson (Blue Mts) – dioec. (app. monoec. in cult.), ? water-poll., only 440 pls left in 7 waterfalls, protected

Microstylis (Nutt.) Eaton = *Malaxis*

Microtatorchis Schltr. Orchidaceae (V 16). 45 New Guinea, W Pacific. Some leafless

Microtea Sw. Phytolaccaceae (III). 9 trop. Am. Often placed in Chenopodiaceae

Microterangis (Schltr.) Senghas. Orchidaceae (V 16). 6 Madag., Masc.

Microthelys Garay = *Brachystele*

Microtis R. Br. Orchidaceae (IV 1). 11 E As. & Mal. to NZ & W Pacific (Aus. 10 (all in W Aus., R: JABG 13(1990)49), 8 endemic). Onion orchids (single onion-like leaf). Fls green, some fragrant, stems & tubers emitting strong scent too; if not visited then self-poll., many apomicts. *M. parviflora* R. Br. (Aus., NZ) & other spp. ant-poll.

Microtoena Prain. Labiatae (VI). 25 Himal. to W China, natur. in Mal. *M. patchouli* (C.B. Clarke) C.Y. Wu & Hsuan (*M. cymosa*, Chinese patchouly, trop. As.) – oil used like patchouly

Microtrichia DC = *Grangea*

Microtrichomanes (Mett.) Copel. = *Crepidomanes*

Microtropis Wallich ex Meissner. Celastraceae. 70 trop. amphipacific

Microula Benth. Boraginaceae. 30 Himal. to W China (most)

Mida Cunn. ex Endl. Santalaceae. 1 NZ, 1 Juan Fernandez

midge orchid *Genoplesium* spp.

Miersia Lindley. Alliaceae (Liliaceae s.l.). 5 Bolivia, Chile

Miersiella Urban. Burmanniaceae. 4 trop. S Am.

Miersiophyton Engl. = *Rhigiocarya*

Migandra Cook = *Chamaedorea*

mignonette *Reseda odorata*; **m. vine** *Anredera cordifolia*; **wild m.** *R. lutea*

Mikania Willd. Compositae (Eup.-Mik.). c. 430 trop. (OW 9, R: BJ 103(1982)211). R: MSBMBG 22(1987)419. Lianes, *M. dentata* Sprengel (*M. ternata*, S Brazil) grown as hanging-basket pl.; *M. cordata* (Burm.f.) Robinson used as antimalarial in Afr., headache cure in Sumatra

Mikaniopsis Milne-Redh. (~ *Cissampelopsis*). Compositae (Sen.-Sen.). c. 15 trop. Afr.

Mila Britton & Rose (~ *Echinopsis*). Cactaceae (III 4). 1 C Peru: *M. caespitosa* Britton & Rose – cult. orn. variable dwarf sp.

Mildbraedia Pax. Euphorbiaceae. 3–4 trop. Afr.

Mildbraediochloa Butzin = *Melinis*

Mildbraediodendron Harms. Leguminosae (III 1). 1 trop. Afr.

Mildella Trevis. = *Cheilanthes*

milfoil *Achillea millefolium*; **water m.** *Myriophyllum* spp.

Milicia Sim (~ *Chlorophora*). Moraceae (I). 2 trop. Afr. *M. excelsa* (Welw.) C. Berg (*Chlorophora e.*, iroko, odum, mvule, African teak, rock elm, counter wood) – timber good for furniture etc. (but allergenic to some), termite-proof, used as oak & teak subs., now endangered through logging, plantations with bananas in Uganda, best not grown in pure stands (galls)

Milium L. Gramineae (16). 4 N temp. OW (Eur.2), E N Am. Spikelets 1-flowered; x = 4, 5, 7, 9. *M. effusum* L. – cult orn. perenn. (millet grass)

Miliusa Leschen. ex A. DC. Annonaceae. 40 Indomal. to Aus.

milk thistle *Silybum marianum*, *Sonchus* spp.; **m. tree** *Brosimum* & *Manilkara* spp. etc.; **m. vetch** *Astragalus* spp.; **m.weed** *Asclepias* spp.; **m.wort** *Polygala* spp.; **sea m.wort** *Glaux maritima*

Milla Cav. Alliaceae (Liliaceae s.l.). 6 SW N Am. to Guatemala. R: GH 8(1953)278. Some cult. orn.

Milleria L. Compositae (Helia.-Mel.). 2 Mex. to C Am. & Peru

millet, African *Pennisetum glaucum*; **barnyard m.** *Echinochloa frumentacea*; **bulrush m.** *P. glaucum*; **cat-tail m.** *Setaria italica*; **channel m.** *E. turnerana*; **common or French m.** *Panicum miliaceum*; **foxtail** or **German m.** *S. italica*; **m. grass** *Milium effusum*; **great m.** *Sorghum bicolor*; **Hungarian m.** *Setaria italica*; **Indian m.** *Pennisetum americanum*; **Italian m.** *S. italica*; **Japanese m.** *E. utilis*; **Kodo m.** *Paspalum scrobiculatum*; **little m.** *Panicum sumatrense*; **milo m.** *Sorghum bicolor*; **pearl m.** *Pennisetum glaucum*; **proso** or **Russian m.** *Panicum miliaceum*; **Sanwa m.** *E. utilis*; **Shama m.** *E. colona*; **spiked m.** *Pennisetum glaucum*; **m. spray** (birdseed) *Setaria italica*

Millettia Wight & Arn. Leguminosae (III 6). c. 90 OW trop. (incl. *Pongamia*). Some fish & bilharzia-snail poisons, cult. orn. & timbers. *M.* sp. (*P. pinnata*, Indian beech, karanja, kerong, saw, thinwin, Indomal.) – oil in skin treatment etc., fuelwood; *M. grandis* (E.

Meyer) Skeels (*M. caffra*, S Afr.) – timber for walking-sticks (umzimbeet); *M. laurentii* De Wild. (wenge, trop. Afr.) & *M. stuhlmannii* Taubert (panga-panga, SE trop. Afr.) – comm. timbers; *M. nieuwenhuisii* J.J. Sm. (Borneo) – climbing ant-plant; *M. thonningii* (Schum. & Thonn.) Baker (W Afr.) – chewing-stick in Nigeria

Milligania Hook.f. Asteliaceae (Liliaceae s.l.). 5 Tasmania. R: FA 45(1987)169

Millingtonia L.f. Bignoniaceae. 1 SE As. & Mal.: *M. hortensis* L.f. – cult. orn. with fragrant fls & fr. to 30 cm long; lvs a poor subs. for opium in cigarettes; timber suggested for tea-chests

Millotia Cass. Compositae (Gnap.). (Incl. *Toxanthes*) 16 temp. Aus. R: AusJB 8 (1995)1

Miltianthus Bunge. Zygophyllaceae. 1 C As.

Miltitzia A. DC. Hydrophyllaceae. 8 W N Am.

Miltonia Lindley. Orchidaceae (V 10). 9 trop. Am. Cult. orn. epiphytes

Miltoniodes Brieger & Lückel = *Oncidium*

Miltoniopsis God.-Leb. (~ *Miltonia*). Orchidaceae (V 10). 5 trop. Am.

Milula Prain. Alliaceae (Liliaceae s.l.). 1 E Himal. Infl. a raceme

Mimetanthe E. Greene = *Mimulus*

Mimetes Salisb. Proteaceae. 13 SW & S Cape. R: JSAB 50(1984)171. Complex compound infls of bird-poll. fls

Mimetophytum L. Bolus = *Mitrophyllum*

Mimophytum Greenman. Boraginaceae. 1 Mexico

Mimosa L. Leguminosae (II 3). 480 trop. & warm esp. Am. (461). R: MNYBG 65(1991). Herbs, shrubs, lianes & trees often with stipular thorns. Some bad weeds of sugar-cane etc. esp. *M. diplotricha* C. Wright (*'M. invisa'*, trop. Am.), *M. pigra* Humb. & Bonpl. ex Willd. (trop.) & *M. pudica* L. (sensitive plant, trop. Am. but widely natur.) – lvs bipinnate with 4 sec. petioles & marked 'sleep movements' at night where leaflets fold together & petioles droop, a condition reached v. rapidly if pl. touched or shaken, subjected to sudden temp. change, high hydrostatic pressure, chemical agents etc., poss. a mechanism reducing transpiration or deterring grazers etc. the stimulus app. passed through phloem of pulvinus, which has living wood-fibres, & poss. implemented by contractile proteins or migration of ions (action potential poss. a chloride spike) through membranes leading to drop in turgor of pulvinus cells; cult. orn. curiosity & sand-binder. *M. scabrella* Benth. (*M. bracaatinga*, Brazil) – locally imp. fuelwood crop; *M. tenuiflora* (Willd.) Poiret (*M. hostilis*, trop. & warm Am.) – potent drink rich in tryptamines form. prepared from roots & taken before battles, source of tepescohuite (Mex.) used in treatment of burns. Florist's mimosa & 'mimosa bark' of tanners derived from *Acacia* spp.

Mimosaceae R. Br. = Leguminosae (II)

Mimosopsis Britton & Rose = *Mimosa*

Mimozyganthus Burkhart. Leguminosae (II 2). 1 Arg., W Paraguay: *M. carinatus* (Griseb.) Burkart

Mimulicalyx Tsoong. Scrophulariaceae. 2 China

Mimulopsis Schweinf. Acanthaceae. 30 trop. Afr., Madag. Some with gregarious flowering, *M. solmsii* Schweinf. (W Afr.) – hapaxanthic after 9 yrs

Mimulus L. Scrophulariaceae. 150 S Afr., As., (esp.) Am. Monkey flowers. R: AMBG 11(1924)99. Shrubs (e.g. *M. aurantiacus* Curtis (Oregon, Calif.) to 1.2 m) & herbs; stigma sensitive to contact such that it closes up after being touched by visiting insect. Many cult. orn. & some natur. esp. yellow-flowered *M. guttatus* Fischer ex DC (Alaska to Mex.) – densely glandular pubescent lvs to 15 cm, fls to 6 cm, *M. luteus* L. (yellow m.f., Chile) – usu. glabrous lvs to 2.5 cm, fls to 2.5 cm, & their hybrid *M. × hybridus* Hort. ex Siebert & Voss (*M. × robertsii*), the common upland pl. in GB with many orn. cvs; *M. cupriphilus* Macnair (Calif.) – close to *M. guttatus*, restricted to 2 copper mines; *M. moschatus* Douglas ex Lindley (musk plant, N Am.) – much cult. cottage-pl. in C19 for fragrance but since c. 1914 surviving clones in cult. scentless

Mimusops L. Sapotaceae (I). 41 trop. Afr. (20), Madag. (15), Masc. (4), Seychelles (1), Indomal. (1: *M. elengi* L. – cult. for fragrant fls). See also *Baillonella*, *Manilkara*, *Tieghemella*

Mina Llave & Lex. = *Ipomoea*

Minasia H. Robinson. Compositae (Vern.-Lych.). 3 Brazil

mind-your-own-business *Soleirolia soleirolii*

Mindium Adans. = *Michauxia*

minjiri *Senna siamea*

Minkelersia M. Martens & Galeotti = *Phaseolus*

minneola *Citrus × aurantium* Tangelo Group

Minquartia Aublet. Olacaceae. 3 N trop. S Am.

mint *Mentha* spp. & hybrids esp. *M. spicata*; **apple m.** *M. suaveolens*; **bergamot m.** *M.* × *piperita* 'Citrata'; **Bowles's m.** *M.* × *villosa* 'Alopecuroides'; **cat m.** *Nepeta* spp.; **eau-de-cologne m.** *M.* × *piperita* 'Citrata'; **field m.** *M. arvensis*; **Gibraltar m.** *M. pulegium*; **ginger m.** *M.* × *gracilis*; **horse m.** *M. longifolia*; **Japanese m.** *M. arvensis* var. *piperascens*; **lemon** or **orange m.** *M.* × *piperita* 'Citrata'; **pepperm.** *M.* × *piperita*; **black p.m.** *M.* × *p.* 'Piperita'; **white p.m.** *M.* × *p.* 'Officinalis'; **pineapple m.** *M. suaveolens* 'Variegata'; **Scotch m.** *M.* × *gracilis*; **spearm.** *M. spicata*; **water m.** *M. aquatica*; **woolly m.** *M. suaveolens*

mintbush *Prostanthera* spp.

Minthostachys (Benth.) Spach. Labiatae (VIII 2). 12 Andes

Minuartia L. Caryophyllaceae (II 1). c. 100 Arctic to Mex., Ethiopia & Himal. (Eur. 53), 1 Chile. Some cult. orn. like *Arenaria*, (incl. *Cherleria* – *M. sedoides* (L.) Hiern (*C. s.*, mossy cyphel, Eur. mts incl. Scotland – often dioec., female usu. C 0) & *M. stricta* (Sw.) Hiern (Teesdale sandwort, Arctic & N Eur. mts) – protected in GB)

Minuria DC. Compositae (Ast.-Ast.) 11 C & SE Aus. R: Nuytsia 3(1980)221, 6(1987)63

Minuriella Tate = *Minuria*

Minurothamnus DC = *Heterolepis*

Mionandra Griseb. Malpighiaceae. 1 Bolivia, Paraguay, Arg.

Miquelia Meissner. Icacinaceae. 8 Indomal. *M. caudata* King (Malay Pen., Borneo) – climber with potable water in stem

miraa *Catha edulis*

mirabelle *Prunus* × *domestica* 'ssp. *institia*'

Mirabilis L. Nyctaginaceae (III). 54 warm Am. (esp. SW N), Himal. (1!). Fls 1–several in K-like involucre of 5 bracts, K C-like, persistent in fr., some spp. (*Oxybaphus*) with enlarged papery involucre in fr. Some cult. orn. incl. night-flowering *M. jalapa* L. (four-o'clock, marvel of Peru, false jalap, Mex.) – cult. as annual but tuberous root to c. 20 kg containing purgative trigenollin when perennial, fls in water give a crimson dye used for tinting seaweed cakes & jellies in China, cosmetic powder from ground seeds used in Japan; *M. multiflora* (Torrey) A. Gray (S US) – root coll. for food since Prehistoric times, antitumour activity in extract

miraculous berry *Synsepalum dulciferum*; **m. fruit** *Thaumatococcus daniellii*

Miraglossum Kupicha (~ *Schizoglossum*). Asclepiadaceae (III 1). 7 S Afr. R: KB 38(1984)625

Mirandaceltis Sharp = *Gironniera*

Mirandea Rzed. Acanthaceae. 2 Mexico. R: CUMH 15(1982)171

Mirbelia Sm. Leguminosae (III 24). 20 Aus.

Miricacalia Kitam. (~ *Ligularia*). Compositae (Sen.-Tuss.). 1 Japan

miro *Prumnopitys ferruginea*

Miscanthidium Stapf = *Miscanthus*

Miscanthus Andersson. Gramineae (40a). c. 20 OW trop. S Afr., E As. Hybridize with *Saccharum* spp.; mooted as biofuel. Some local uses for thatch or brooms esp. *M. ecklonii* (Nees) Mabb. (*Miscanthidium capense*, S Afr.); some cult. orn. esp. *M. sinensis* Andersson (eulalia, E As.) – commonly grown varieg. cvs

Mischarytera (Radlk.) H. Turner (~ *Arytera*). Sapindaceae. 3 New Guinea, N Aus. R: Blumea suppl. 9(1995)

Mischobulbum Schltr. (~ *Tainia*). Orchidaceae (V 11). 8 Indomal. R: OM 6(1992)64

Mischocarpus Blume. Sapindaceae. 15 SE As. to Aus. (9, 7 endemic). R: Blumea 23(1977)251. Some ant-infested domatia

Mischocodon Radlk. = *Mischocarpus*

Mischodon Thwaites. Euphorbiaceae. 1 S India, Sri Lanka. Lvs verticillate

Mischogyne Exell. Annonaceae. 1 trop. Afr.

Mischophloeus R. Scheffer = *Areca*

Mischopleura Wernham ex Ridley = *Sericolea*

Misodendraceae J. Agardh (Myzodendraceae). Dicots – Rosidae – Santalales. 1/8 temp. S Am. Dioec. hemiparasitic shrublets mostly on *Nothofagus* spp., with thickened haustorial region & stout twigs, the apex aborting annually & laterals growing out to give pseudo-2 or 3-chotomous shoots. Lvs small, s.t. scale-like, spirally arr.; stipules 0. Fls small in racemes or spikes, males with P 0, A 2 or 3 around small nectary-disk, anthers with term. slits, females with P 3, basally connate & adnate, as are alt. accrescent staminodes, to $\underline{G}$ (3), 1-loc. with 3 stigmas & 3 pendulous ovules, not differentiated into nucellus & integuments, hanging from free-central placenta. Fr. an achene or nut with accrescent feathery staminodes (wind-disp.); seed without testa, embryo straight in oily green endosperm

Genus: *Misodendrum*

Misodendron G. Don f. = *Misodendrum*

Misodendrum Banks ex DC. Misodendraceae. 8 S Am. S of 33°, mainly on Andean *Nothofagus* spp. R: Parodiana 1(1982)245

Misopates Raf. (~ *Antirrhinum*). Scrophulariaceae (Antirrh.). 8 Medit. (Eur. 3) to Cape Verde Is., Ethiopia & NW India. R: D.A. Sutton, Rev. Antirrh. (1988)145. *M. orontium* (L.) Raf. (weasel's snout, Euras.) – natur. N Am., GB

Miss Wilmott's ghost *Eryngium giganteum*

missanda *Erythrophleum suaveolens*

mistletoe *Viscum album*, name applied to any Loranthaceae or Viscaceae; **American m.** *Phoradendron* spp. esp. *P. leucarpon*; **m. cactus** *Rhipsalis* spp. esp. *R. baccifera*; **dwarf m.** *Arceuthobium* spp.; **WI m.** *Phthirusa caribaea*

mistol *Ziziphus mistol*

Mitchell grass *Astrebla* spp. esp. *A. pectinata*

Mitchella L. Rubiaceae (inc. sed.). 1 N temp., 1 N Am., 1 Japan. Heterostyly. *M. repens* L. (partridge berry, N Am.) – lvs medic.

Mitella L. Saxifragaceae. 20 N Am., E As. Mitrewort, bishop's cap. Some cult. orn. herbs

mitnan *Thymelaea hirsuta*

Mitolepis Balf.f. = *Cryptolepis*

Mitophyllum E. Greene. Cruciferae. 1 California

Mitostemma Masters. Passifloraceae. 3 trop. S Am.

Mitostigma Decne. Asclepiadaceae (III 1). 20 S Am.

Mitracarpum auctt. = *Mitracarpus*

Mitracarpus Zucc. ex Schultes & Schultes f. Rubiaceae (IV 15). 30 trop. Am., 1 introd. OW: *M. hirtus* (L.) DC

Mitragyna Korth. Rubiaceae (I 1). 10 OW trop. R: Blumea 24(1978)56. Many alks. Some timbers (nazingu, Uganda) esp. *M. parvifolia* (Roxb.) Korth. (S As.) – subs. for *Chloroxylon swietenia* & *M. stipulosa* (DC) Kuntze (abura, W Afr.) – furniture, barrels etc., lvs medic. Some spp., e.g. *M. speciosa* (Korth.) Haviland (Mal.) with lvs smoked like, & alleged to be more dangerous than, opium

Mitranthes O. Berg. Myrtaceae (Myrt.). 11 trop. Am

Mitrantia P.G. Wilson & Hyland. Myrtaceae. 1 N Queensland: *M. bilocularis* P.G. Wilson & Hyland. R: Telopea 3(1988)264

Mitraria Cav. Gesneriaceae. 1 Chile: *M. coccinea* Cav. – cult. orn. liane allied to *Fieldia* (Aus.)

Mitrasacme Labill. Loganiaceae. 40 Indomal., E As., Aus. (most), New Caled., NZ. Reminiscent of *Hedyotis* (Rubiaceae)

Mitrasacmopsis Jovet. Rubiaceae (IV 1). 1 E Afr. & Madagascar

Mitrastemma Makino. Rafflesiaceae (I; Mitrastemmataceae). 1 SE As. to Mal. & Japan; 1 Mex. & C Am. R: ('*Mitrastema*'): Blumea 38 (1993)221. On Fagaceae

Mitrastemmataceae Makino = Rafflesiaceae (I)

Mitrastemon Makino = *Mitrastemma*

Mitrastylus Alm & T.C.E. Fries = *Erica*

Mitratheca Schumann = *Oldenlandia*

Mitrella Miq. (~ *Fissistigma*). Annonaceae. 5 Mal.

Mitreola L. Loganiaceae. 6 trop. R: MLW 74–23(1974)

Mitrephora Hook.f. & Thomson. Annonaceae. 40 SE As., W Mal.

mitrewort *Mitella* spp.; **false m.** *Tiarella* spp.

Mitriostigma Hochst. Rubiaceae (II 1). 5 trop. & S Afr.

Mitrocereus (Backeb.) Backeb. = *Pachycereus*

Mitrophyllum Schwantes. Aizoaceae (V). 6 Cape. R: BJ 97(1976)339. Cult. orn. succ. pls

Mitropsidium Burret = *Psidium*

mitsuba or **mitzuba** *Cryptotaenia canadensis*

mitsumata *Edgeworthia papyrifera*

Mitwabachloa Phipps = *Zonotriche*

Miyakea Miyabe & Tatew. Ranunculaceae. 1 Sakhalin

Miyamayomena Kitam. (~ *Aster*). Compositae (Ast.-Ast.). 6 E As.

Mizonia A. Chev. = *Pancratium*

mkani fat *Allanblackia stuhlmannii*

Mkilua Verdc. Annonaceae. 1 trop. E Afr.: *M. fragrans* Verdc. – fls used in making scent for Swahili women & Arabs

mkweo *Beilschmiedia kweo*

Mnesithea Kunth. Gramineae (40h). 1 Indomal.

Mniochloa Chase. Gramineae (4). 2 Cuba

Mniodes (A. Gray) Benth. Compositae (Gnap.-Lor.). 5 Peru. R: FBA 1(1954)1. Dioec. cushion-formers

Mniopsis C. Martius = *Crenias*

Mniothamnea (Oliver) Niedenzu. Bruniaceae. 2 S Afr.

moa *Madhuca longifolia*

moabi *Baillonella toxisperma*

Moacroton Croizat. Euphorbiaceae. 7 Cuba (serpentine)

Mobilabium Rupp. Orchidaceae (V 16). 1 Queensland

mobola (plum) *Parinari curatellifolia*

moccasin flower *Cypripedium* spp.

mock orange *Philadelphus* spp. esp. *P. coronarius*

mocker nut *Carya tomentosa*

Mocquerysia Hua. Flacourtiaceae (6). 1 trop. Afr. Epiphyllous infls

Modestia Charadze & Tamamschjan = *Anacantha*

Modiola Moench. Malvaceae. 1 Am.: *M. caroliniana* (L.) G. Don f.

Modiolastrum Schumann. Malvaceae. 7 S Am. R: Brittonia 32(1980)484

Moehringia L. (~ *Arenaria*). Caryophyllaceae (II 1). c. 25 N temp. (Eur. 25)

Moenchia Ehrh. Caryophyllaceae (II 1). 3 W & C Eur. (3), Medit.

Moerenhoutia Blume (~ *Malaxis*). Orchidaceae (III 3). 10 Pacific

Moghania J. St-Hil. = *Flemingia*

moghat root *Glossostemon bruguieri*

Mogoltavia Korovin. Umbelliferae (III 8). 2 C As.

Mohavea A. Gray. Scrophulariaceae (Antirrh.). 2 SW US. R: D.A. Sutton, Rev. Antirrh. (1988)482. Cult. orn. annuals

mohle flowers *Jasminum sambac*

Mohria Sw. Schizaeaceae. 7 trop. & S Afr. Margins of fronds rolled back over solit. sporangia. *M. caffrorum* (L.) Desv. (frankincense fern, trop. & S Afr., Madag., Masc.) – cult. orn. with scented fronds

mohur *Delonix regia*

molasses uncrystallizable sugars (syrup) from raw sugar; **m. grass** *Melinis minutiflora*

Moldenhawera Schrader. Leguminosae (I 1). 6 Brazil, Venezuela

Moldenkea Traub = *Hippeastrum*

Moldenkeanthus Morat = *Paepalanthus*

mole plant *Euphorbia lathyrus*

Molinadendron Endress. Hamamelidaceae (I 4). 3 Mex., C Am.

Molinaea Comm. ex Juss. Sapindaceae. 10 Madag., Masc.

Molineria Colla (~ *Curculigo*). Hypoxidaceae. 7 Indomal. Cult. orn. esp. *M. capitulata* (Lour.) Herbert (Indomal.) – housepl. in Aus.

Molineriella Rouy = *Periballia*

Molinia Schrank. Gramineae (25). 2–4 temp. Euras. (Eur. 1: *M. caerulea* (L.) Moench (purple moor grass) – cult. orn. esp. 'Variegata', catches insects between paleae in a trap like *Dionaea* (lodicules shrink slowly but if punctured by insect snap shut, though app. not absorbing nutrients). Often diff. to distinguish from allopatric *Eragrostis*

Moliniopsis Hayata = *Molinia*

Mollera O. Hoffm. = *Calostephane*

Mollia C. Martius. Tiliaceae. 18 trop. S Am.

Mollinedia Ruíz & Pavón. Monimiaceae (V 2). 90 trop. Am.

Molluginaceae Raf. (~ Caryophyllaceae). Dicots – Caryophyllidae – Caryophyllales. 13/130 trop. & warm esp. S Afr. Herbs (often weedy; rarely shrubs), often with unusual sec. thickening giving concentric rings of vasc. bundles or alt. rings of phloem & xylem & app. always with anthocyanins & not betalains. Lvs simple, scarcely succ., at least s.t. with Kranz syndrome, opp. to spirally arr. or whorled; stipules small & decid. or 0 (larger in *Pharnaceum*). Fls usu. bisexual, reg., often small, solit. or in cymes; K (4 – *Polpoda*) 5, rarely basally connate (*Coelanthum*), C (? staminodes) 5 small or 0, rarely a tube, A (2–)4 or 5(–10 or more), usu. basally connate, anthers with longit. slits, nectary a ring around, or on lower part of, $\underline{G}$ (2–5(–more)), usu. with distinct styles & at least basally pluriloc., placentation usu. axile with (1–)∞ campylotropous (almost anatropous) bitegmic ovules. Fr. usu. septifragal capsule, often with persistent K; seeds usu. reniform, with embryo curved around perisperm (endosperm 0), sometimes with arils. x = 9

Genera: *Adenogramma, Coelanthum, Corbichonia, Glinus, Glischrothamnus, Hypertelis, Limeum, Macarthuria, Mollugo, Pharnaceum, Polpoda, Psammotricha, Suessenguthiella*

Form. incl. in Aizoaceae but scarcely succ.; $\underline{G}$, & free sepals & anthocyanins present – a classic example of chemotaxonomic studies leading to recognition of family status

Mollugo L. Molluginaceae. c. 35 trop. & warm (Eur. 2). Some potherbs esp. *M. pentaphylla* L. (As.) – medic.

Molongum Pichon. Apocynaceae. 3 trop. S Am. R: AUWP 87-1(1987)

Molopanthera Turcz. Rubiaceae (I 1). 1 E Brazil

Molopospermum Koch. Umbelliferae (III 5). 1 W Medit.: *M. peloponnesiacum* (L.) Koch. R: Bauhinia 10(1992)75

Moltkia Lehm. Boraginaceae. 3 N Italy to N Greece, 3 SW As. Cult. orn. rock-pls esp. *M.* × *intermedia* (Froebel) J. Ingram (*M. petraea* (Tratt.) Griseb. (C. Greece to Dalmatia) × *M. suffruticosa* (L.) Brand (N Italy)) like *Lithospermum* but C-throat not crested & nutlets bent

Moltkiopsis I.M. Johnston. Boraginaceae. 1 N Afr. to Iran

moluccana *Paraserianthes falcataria*

Moluccella L. Labiatae (6). 4 Medit. (Eur. 1) to NW India. *M. laevis* L. (bells of Ireland, shell flower, E Medit.) – cult. orn. with large green bell-like K, grown as an everlasting

moly *Allium moly*

mombin, red *Spondias purpurea*; **yellow m.** *S. mombin*

Mommsenia Urban & Ekman. Melastomataceae. 1 Hispaniola

Momordica L. Cucurbitaceae. 45 OW trop. Usu. monoec. scramblers with elaiophores attractive to bees & bitter fr., ed. when cooked esp. *M. balsamina* L. (balsam apple, Mal., natur. trop.) – fr. foul-smelling but 'red jelly' & seeds ed., lvs & stems camel fodder; *M. charantia* L. (balsam pear, bitter cucumber, leprosy gourd, trop., natur. SE US) – immature fr. (karel(l)a) a favourite of Pakistanis in UK, extract has effect like insulin; *M. cochinchinensis* (Lour.) Sprengel (India & Japan to New Guinea) – roots produce a lather for clothes-washing

Mona O. Nilsson (~ *Montia*). Portulacaceae. 1 high mts Venez., Colombia

Monachather Steudel (~ *Danthonia*). Gramineae (25). 1 Aus.: *M. paradoxa* Steudel

Monachosoraceae Ching (~ Dennstaedtiaceae). Filicopsida. 1/3 Himal. to Mal. Terr. ferns with glandular-hairy dictyostelic stem. Frond 1-pinnate to 4-pinnate; rachis gemmiferous; sori term., indusium 0. n = 56

Only genus: *Monachosorum*

Monachosorum Kunze. Monachosoraceae (Adiantaceae s.l.). c. 3 Himal. to S China & Mal.

Monachyron Parl. = *Melinis*

Monactis Kunth. Compositae (Helia.-Verb.). c. 9 trop. S Am. R: Phytologia 34(1976)34

Monadenia Lindley. Orchidaceae (IV 3). 16 trop. & S Afr., Madag.

Monadenium Pax. Euphorbiaceae. 50+ trop. E, SW & C (few) Afr. R: P.R.O. Bally (1961) *The genus M.* Succ. monoec. pls, cult. orn.

Monandriella Engl. = *Ledermanniella*

Monanthes Haw. Crassulaceae. 13 Canary & Salvage Is., Morocco. Cult. orn. dwarf succ.

Monanthochloe Engelm. Gramineae (31c). 2 S US, Mex., WI, Arg. R: Kurtziana 5(1969)369. Seashores in N, saltpans in S Am.

Monanthocitrus Tanaka. Rutaceae. 4 Borneo, New Guinea. R: PANSP 140(1988)272

Monanthotaxis Baillon. Annonaceae. 56 trop. Afr. & Madag. Some lianes

monarch of the East *Sauromatum venosum*

Monarda L. Labiatae (VIII 2). 12 N Am. A 2. R: UCPB 20 (1942)147. Cult. orn., teas & herbs etc. esp. *M. didyma* L. (Oswego tea, bee balm, bergamot, E N Am.) – eaten with meat, oil used as pomade (*M. citriodora* Cerv. ex Lagasca (C US to N Mex.) with greatest oil yield by wt.), *M. fistulosa* L. (N Am.) & their hybrids; *M. menthifolia* Graham (*M. fisulosa* var. *menthifolia*, N Am.) – poss. comm. source of geraniol; others medic.

Monardella Benth. Labiatae (VIII 2). 19 W N Am. R: AMBG 12(1925)1. Some cult. orn. & medic. teas esp. *M. odoratissima* Benth.

Monarrhenus Cass. Compositae (Pluch.). 2 Mauritius

Monarthrocarpus Merr. = *Desmodium*

Mondia Skeels. Asclepiadaceae (I; Periplocaceae). 2 trop. Afr.

Monechma Hochst. = *Justicia*

Monelytrum Hackel ex Schinz. Gramineae (33d). 1 SW Afr.

Monenteles Labill. = *Pterocaulon*

Monerma Pal. = *Lepturus*, see also *Hainardia*

Moneses Salisb. ex Gray. Ericaceae (Pyrolaceae). 1 cool N temp.: *M. uniflora* (L.) A. Gray

money plant *Epipremnum pinnatum* 'Aureum'; **m.wort** *Lysimachia nummularia*; **Cornish m.w.** *Sibthorpia europaea*

mongongo *Schinziophyton rautanenii*

Moniera Loefl. = *Monniera*

Monilaria (Schwantes) Schwantes. Aizoaceae (V). 6 Cape. R: MSABH 14(1973)49. Some cult. orn. dwarf succ.

Monimia Thouars. Monimiaceae (VI 2). 3 Mauritius & Réunion. R: AMBG 72(1985)68

Monimiaceae Juss. Dicots – Magnoliidae – Magnoliales. 34/440 trop. & warm, esp. S. R: NJB 7(1987)25, 8(1988)25. Evergreen trees, shrubs or lianes often accumulating aluminium & with alks.; twigs often flattened at nodes. Lvs usu. opp., simple, often with gland-dots; stipules 0. Fls small, usu. unisexual, reg. to oblique, usu. perigynous with concave hypanthium, (solit. or) in cymose infls., s.t. cauliflorous; K 2 + 2, fleshy, decussate, C 7–20 (–more) or P not diff. into K & C, reduced or 0, A ∞ in 1 or 2 series with short filaments & basal (? staminodes) nectaries, anthers with longit. slits or valves, staminodes around G (1–)few–∞ with short styles & term. stigmas, carpels s.t. sunk in receptacle, with 1 anatropous bitegmic or (Siparunoideae) unitegmic apical & pendulous (Hortonioideae, Monimioideae) or basal & erect (Atherospermatoideae, Siparunoideae) ovule. Fr. a head of drupes or nuts, often enclosed in hypanthium; seeds endotestal with copious oily endosperm, cotyledons 2(4). x = 18–22, 39, 43

Classification & principal genera:

I. **Hortonioideae** (fls bisexual, inner P petaloid; fr. a drupe): *Hortonia* (only)

II. **Atherospermatoideae** (fls bisexual or unisexual, P sepaloid or petaloid; fr. a plumose nutlet; 2 tribes):

1. **Atherospermateae** (leaf-hairs centrifixed, inner staminodes elongating in fr.): *Atherosperma, Laureliopsis* (only)
2. **Laurelieae** (leaf-hairs basifixed, inner staminodes not enlarging): *Daphnandra, Laurelia*

III. **Siparunoideae** (fls unisexual, P sepaloid; anthers opening by valves): *Siparuna* (only)

IV. **Glossocalycoideae** (fls unisexual, P obscure, fr. a drupe): *Glossocalyx* (only)

V. **Mollinedioideae** (fls unisexual, P sepaloid, filaments with appendages, anthers opening by slits, fr. a drupe; 3 tribes):

1. **Hedycaryeae** (anthers with longit. slits; receptacle splitting): *Hedycarya, Tambourissa*
2. **Mollinedieae** (anthers with longit. slits; receptacle opening by abscission of upper part): *Kibara, Mollinedia, Steganthera*
3. **Hennecartieae** (anthers with equatorial slit): *Hennecartia* (only)

VI. **Monimioideae** (pls dioec.; P sepaloid (inner petaloid in *Peumus*); filaments without appendages (except *Palmeria*), anthers opening by slits, fr. a drupe; 3 tribes):

1. **Palmerieae** (lianes): *Palmeria* (only)
2. **Monimieae** (trees & shrubs, P sepaloid): *Monimia* (only)
3. **Peumeae** (tree, P petaloid): *Peumus* (only)

16 genera monotypic.'Somewhat heterogeneous' (Cronquist). Some with almost closed hypanthium (cf. *Ficus*), pollen germinating on a mucilaginous plug closing the opening Some timbers (*Doryphora, Hedycarya, Peumus, Xymalos*), scents & medic., *Laurelia* the source of Peruvian nutmegs

Monimiastrum Guého & A.J. Scott. Myrtaceae. 5 Mauritius

Monimopetalum Rehder. Celastraceae. 1 China

Monium Stapf = *Anadelphia*

Monizia Lowe (~ *Melanoselinum*). Umbelliferae (III 12). 1 Madeira: *M. edulis* Lowe

monkey apple *Annona* spp., *Anisophyllea laurina, Strychnos* spp.; **m. bread** *Adansonia digitata*; **m. cocoa** *Theobroma angustifolium*; **m. comb** *Pithecoctenium* spp.; **m. flower** *Mimulus* spp.; **m. nut** *Arachis hypogaea*; **m. pot** *Lecythis* spp., *Cariniana* spp.; **m. puzzle (tree)** *Araucaria araucana*

monk's rhubarb *Rumex pseudoalpinus*

monkshood *Aconitum* spp.

Monniera Loefl. (*Ertela*). Rutaceae. 2 trop. S Am. R: Candollea 45 (1990)369

Monnina Ruíz & Pavón. Polygalaceae. 150 New Mexico to Chile, esp. Andes. One of G 2 usu. rudimentary; fr. a drupe, rarely samaroid. *M. salicifolia* Ruíz & Pavón (Peru) – fr. source of a blue dye

Monocardia Pennell = *Bacopa*

Monocarpia Miq. Annonaceae. 1 Thailand, W Mal.: *M. euneura* Miq.

Monocelastrus Wang & Tang = *Celastrus*

Monochaetum (DC) Naudin. Melastomataceae. 45 upland trop. Am. A dimorphic

Monochasma Maxim. ex Franchet & Savat. Scrophulariaceae. 4 E As.

Monochilus Fischer & C. Meyer. Labiatae (II; Verbenaceae). 1 Brazil

Monochoria C. Presl. Pontederiaceae. 6 NE Afr. to Japan & Aus. (4). *M. hastata* (L.) Solms-Laub. (Indomal.) & *M. vaginalis* (Burm.f.) Kunth (Mal.) – all save roots relished as veg., locally medic. *M. korsakowii* Regel & Maack (C & E As.) – natur. N Black Sea by 1932, N Italian ricefields by 1985

Monocladus Chia, Fung & Y.L. Yang = *Bambusa*

Monococcus F. Muell. Phytolaccaceae (II). 1 E Aus., New Caled., Vanuatu: *M. echinophorus* F. Muell.

Monocosmia Fenzl (~ *Calandrinia*). Portulacaceae. 1 Chile, Arg.

Monocostus Schumann. Zingiberaceae (Costaceae). 1 E Peru

Monocotyledonae DC. (Liliopsida). Angiospermae. 2651/55 800 here arr. in 84 fams. Usu. pls without reg. sec. thickening, with 1 'cotyledon' (better a prophyll as no trace of 'second' one), parallel-veined lvs & 3-merous fls. No parasites. Range of seed-size from dust-like in orchids to the biggest seeds known, in *Lodoicea* (Palmae). See modified Cronquist System (this volume p. 779) & R. Dahlgren *et al.* (1985) *The families of the Monocotyledons*

Monocyclanthus Keay. Annonaceae. 1 trop. W Afr.: *M. vignei* Keay

Monocymbium Stapf. Gramineae (40g). 3 trop. & S Afr.

Monodia S. Jacobs. Gramineae (31a). 1 W Aus.: *M. stipoides* Jacobs. R: KB 40(1985)659

Monodiella Maire. Gentianaceae. 1 Sahara = ?

Monodora Dunal. Annonaceae. 15 trop. Afr. Fleshy syncarps, though in *M. crispata* Engl. & Diels, at least, a single term. carpel & not the syncarp it resembles. *M. myristica* (Gaertner) Dunal (calabash nutmeg) – seeds used like nutmeg, as rosary beads & locally medic.

Monogereion G. Barroso & R. King. Compositae (Eup.-Ayap.). 1 NE Brazil. R: MSBMBG 22(1987)210

Monogramma Comm. ex Schkuhr. Vittariaceae. 7 Madag. to Mal. Epiphytes, minutest of all ferns, mature fronds 3 mm–2 cm long

Monogramme *auctt.* = praec.

Monolena Triana. Melastomataceae. 15 trop. S Am. Forest-floor herbs, seeds distrib. heavy rain

Monolepis Schrader. Chenopodiaceae (I 2). 6 N & E As., N Am., temp. S Am., introd. Aus. Some form. eaten

Monolopia DC. Compositae (Hele.-Bae.). 4 Calif. R: Novon 1(1991)122. *M. major* DC – cult. orn. annual

Monomeria Lindley. Orchidaceae (V 15). 2 E Himalaya, SE As.

Monopera Barringer (~ *Angelonia*). Scrophulariaceae. 2 S Am. R: Brittonia 35(1983)111

Monopetalanthus Harms. Leguminosae (I 4). 15 Guineo-Congolian forest. *M. heitzii* Pellegrin (andouong, Gabon) – timber

Monopholis S.F. Blake = *Monactis*

Monophrynium Schumann. Marantaceae. 3 C Mal.

Monophyllaea R. Br. Gesneriaceae. 20 S Thailand & Mal. R: NRBGE 37(1978)1. Pls resembling single lvs rooted at foot of petioles & bearing infls at junction of 'lamina' & petiole, the 'lamina' to 60 cm long; poss. derived from a system with caulescent growth & anisophylly as in some Acanthaceae

Monophyllanthe Schumann. Marantaceae. 1 Guyana

Monophyllorchis Schltr. Orchidaceae (V 3). 2 Colombia & Ecuador

Monoplegma Piper = *Oxyrhynchus*

Monoporus A. DC. Myrsinaceae. 8 Madagascar

Monopsis Salisb. Campanulaceae. 18 trop. & S Afr. Some cult. orn., like *Lobelia* but stigma-lobes filiform & recurved, esp. *M. debilis* (L.f.) C. Presl (*'M. simplex'*, S Afr.) natur. W Aus.

Monopteryx Spruce ex Benth. Leguminosae (III 2). 2–3 Brazil & Venez. R: Brittonia 36(1984)48. Large trees with big buttresses of mythological imp. in NW Amaz. *M. angustifolia* Spruce ex Benth. (Amazonia) – seeds coll. & stored for lean times

Monoptilon Torrey & A. Gray. Compositae (Ast.-Ast.). 2 SW N Am. deserts

Monopyle Moritz ex Benth. Gesneriaceae. 23 trop. Am.

Monopyrena Speg. Older name for *Junellia*

Monorchis Séguier. Older name for *Herminium*

monos plum *Myrcianthes umbellulifera*

Monosalpinx Hallé. Rubiaceae (II 1). 1 trop. W Afr.

Monoschisma Brenan = *Pseudopiptadenia*

Monosepalum Schltr. (~ *Bulbophyllum*). Orchidaceae (V 15). 3 E Mal.

Monostachya Merr. = *Rytidosperma*

Monostylis Tul. = *Apinagia*

Monotaceae (Gilg) Takht. = Dipterocarpaceae
Monotagma Schumann. Marantaceae. 20 trop. S Am.
Monotaxis Brongn. Euphorbiaceae. 10 Aus.
Monotes A. DC. Dipterocarpaceae. c. 26 trop. Afr., Madag. (1) & Tertiary Eur. *M. elegans* Gilg (C S Afr.) – timber
Monotheca A. DC. = *Sideroxylon*
Monothecium Hochst. Acanthaceae. 3 trop. Afr. to S India
Monothrix Torrey = *Perityle*
Monotoca R. Br. Epacridaceae. 11 Aus.
Monotrema Koern. Rapateaceae. 3–5 Colombia, Venez., Brazil
Monotropa L. Ericaceae (Monotropaceae). 2 N temp. (Eur. 1). *M. hypopitys* L. (yellow bird's-nest, N temp.) – yellowish stems, *M. uniflora* L. (Indian pipe, Himal., Japan, N & C Am.) – locally medic.
Monotropaceae Nutt. See Ericaceae
Monotropanthum Andres = *seq.*
Monotropastrum Andres (~ *Cheilotheca*). Ericaceae (Monotropaceae). 2 E As. R: WJB 33(1975)1
Monotropsis Schwein. ex Elliott. Ericaceae (Monotropaceae). 1 N Am.
Monrosia Grondona = *Polygala*
Monsonia L. Geraniaceae. 25 Afr., Madag., SW As. R: MLW 79–9(1979)
Monstera Adans. Araceae (III 2). 25 trop. Am. R: CGH 207(1977)3. Large epiphytic lianes with entire, perforated or pinnatifid lvs & often long corky aerial roots; holes in lvs caused by local slowing of growth of lamina, drying & splitting as rest expands. Several cult. as housepl. when young esp. *M. deliciosa* Liebm. (ceriman, (Swiss) cheeseplant, Mex. to Panamá) – introd. 1840 but rare in wild, with ed. spadix (to 15 °C above ambient when receptive) with banana & pineapple flavour though eaten unripe disagreeable (calcium oxalate crystals). Some with dopamine
Montagueia Baker f. = *Polyscias*
Montamans Dwyer. Rubiaceae (IV 7). 1 Panamá
montana *Clematis montana*
Montanoa Cerv. Compositae (Helia.). 25 trop. Am. R: MNYBG 36(1982). Tree daisy. Pachycaul treelets & trees, some cult. orn., *M. quadrangularis* Schultz-Bip. (*M. moritziana*, arboloco, C Am.) – timber used locally (telegraph-poles), lvs to 90 cm; others medic.
montbretia *Crocosmia* × *crocosmiiflora*
Montbretiopsis L. Bolus = *Tritonia*
Monteiroa Krapov. Malvaceae. 5 trop. S Am.
Monterey pine *Pinus radiata*
Montezuma Moçiño & Sessé ex DC. Bombaceae. 1 Puerto Rico: *M. speciosissima* Moçiño & Sessé ex DC – cult. orn., street-tree in S Florida, timber for furniture
Montia L. Portulacaceae. 10 temp. (Eur. 1, Aus. 2), wet places. *M. fontana* L. ((water) blinks, N temp., trop. mts, SE Aus.) – annual of wet places, fls cleistogamous in dull weather or when inundated, lvs eaten as salad esp. in France
Montiastrum (A. Gray) Rydb. (~ *Montia*). Portulacaceae. 4 W N Am., E Siberia
Monticalia C. Jeffrey. Compositae (Sen.-Sen.). c. 60 C & S Am. (Andes)
Montinia Thunb. Grossulariaceae (Montiniaceae). 1 S Afr.
Montiniaceae Nakai. See Grossulariaceae
Montiopsis Kuntze (~ *Baitaria*). Portulacaceae. 18 temp. S Am. esp. Chile. R: Phytol. 74(1993)273
Montrichardia Crüger. Araceae (V 8). 1 trop. S Am.: *M. arborescens* (L.) Schott – pachycaul to 7 m, v. fragrant bee-poll. infls with large starchless pollen, water-disp. in Amaz., locally medic., spadix ed.
Montrouziera Pancher ex Planchon & Triana. Guttiferae (III). 5 New Caled. Principal exploited timber of New Caled.
Monttea C. Gay. Scrophulariaceae. 3 Chile. Inside anterior lip oils providing poll. reward for centridine bees (cf. oil-collecting bees in *Diascia*)
Monvillea Britton & Rose = *Acanthocereus*; see also *Cereus*
mooley plum *Owenia acidula*
mooli *Raphanus sativus* 'Longipinnatus'
moon beam *Tabernaemontana divaricata*; **m. daisy** *Leucanthemum vulgare*; **m. flower** *Ipomoea alba*; **m. seed** *Menispermum canadense*; **m. trefoil** *Medicago arborea*; **m.wort** *Botrychium lunaria*
Moonia Arn. Compositae (Helia.-Cor.). 1 Sri Lanka, S India. Allied to *Dahlia*. R: Brittonia

27(1975)97

moorgrass, blue *Sesleria caerulea*; **purple m.** *Molinia caerulea*

Mooria Montr. = *Cloezia*

mooseberry *Viburnum lantanoides*; **moosewood** *Acer pensylvanicum, Dirca palustris, V. lantanoides*

mopane *Colophospermum mopane*

Mopania Lundell = *Manilkara*

Moquinia DC. Compositae (Vern.-Vern.). 2 E Brazil. R: T 43(1994)41. Dioec. shrubs & trees

Moquiniella Balle. Loranthaceae. 1 S Afr.

Mora Benth. Leguminosae (I 1). 6 trop. Am. V. tall emergent trees with large seeds, 1 per fr., those of *M. oleifera* (Triana) Ducke the largest dicot. seed (to 1 kg, cf. *Lodoicea*) with the largest embryo of any pl., drifting in currents; *M. paraensis* Ducke (Brazil) – air-filled cavity between cotyledons promoting water-disp. in Amaz. Some timbers for ship-building etc.

Moraceae Link. Dicots – Hamamelidae – Urticales. 38/1100 trop. & warm, few temp. Monoec. or dioec. trees, shrubs, lianes (incl. stranglers) & herbs, usu. with laticifers with milky latex (0 in *Fatoua*) & s.t. alks. Lvs spirally arr. or opp., usu. simple, often with cystoliths & cell walls with silica or calcium carbonate; stipules present, s.t. minute. Fls small, wind-poll. (insect-poll. in *Ficus* etc.), in axillary infls with the axis often thickening to form a head or invaginated receptacle (almost closed in syconia of *Ficus*); P (0–4) or 5(–10) ± basally connate, s.t. in 2 whorls, A (1–4(–6) as many as & opp. P, $\underline{G}$ or $\overline{G}$ (2(3)) with 1 often rudimentary so 1(2)-loc. with usu. 2 style(arm)s & 1 ± apical, anatropous to hemitropous or campylotropous bitegmic ovule. Fr. a drupe (s.t. with dehiscent exocarp), P &/or receptacle often becoming fleshy; seeds with straight or curved embryo with often unequal cotyledons in fleshy & oily or 0 endosperm. x = 7+. R: GBS 19(1962)187

Classification & principal genera:

I. **Moreae** (infl. unisexual or, at least, not discoid, female 1-flowered or racemose, male fls with P 3–5, A 3–5 inflexed in bud): *Broussonetia, Fatoua, Maclura, Streblus, Trophis*

II. **Artocarpeae** (like I., female infls with several to many fls, A straight in bud); *Artocarpus, Sorocea*

III. **Castilleae** (like I., but female infl. a stout spike or head, male similar with involucre, male fls usu. without pistillodes): *Antiaris, Castilla, Perebea, Naucleopsis, Pseudolmedia*

IV. **Dorstenieae** (incl. Brosimeae; infls mostly bisexual, not enclosed): *Brosimum, Dorstenia*

V. **Ficeae** (fls in invaginated receptacles, sterile female fls with gall-wasps): *Ficus*

Cannabaceae & Cecropiaceae kept distinct

Imp. fr. trees (*Artocarpus, Brosimum, Ficus, Morus*), rubber (*Castilla, Ficus*), timber (*Brosimum, Ficus, Maclura, Milicia*), dyes (*Maclura*), paper (*Broussonetia, Ficus*), arrow poisons – cardenolides (*Antiaris, Maquira, Naucleopsis*); cult. orn. (*Broussonetia, Ficus, Maclura*)

Moraea Miller. Iridaceae (III 2). 120 trop. (c. 24) & S Afr. (over half in, & most of those restricted to, winter rainfall region). R: AKBG 14(1986). Outer integument becomes fleshy as seed ripens; many toxic to stock, some cult. orn. (peacock lilies) incl. *M. ramossisima* (L.f.) Druce (*M. ramosa*, S & W Cape) – roots metamorphosed into thorns

morama bean *Bauhinia esculenta*

Morangaya G. Rowley = *Echinocereus*

Moratia H. Moore (~ *Cyphokentia*). Palmae (V 4m). 1 New Caled.: *M. cerifera* H. Moore – adaxial cuticle 4 times as thick as epidermal cells, seeds take 1½ yrs to germinate. R: Allertonia 3(1984)349

Morelia A. Rich. ex DC. Rubiaceae (II 1). 1 trop. Afr.

Morelotia Gaudich. Cyperaceae. 2 Hawaii, NZ

Morenia Ruíz & Pavón = *Chamaedorea*

Moreton Bay chestnut *Castanospermum australe*; **M. B. fig** *Ficus macrophylla*; **M. B. pine** *Araucaria cunninghamii*

Morettia DC. Cruciferae. 4 N Afr. to Arabia

Morgania R. Br. = *Stemodia*

Moricandia DC. Cruciferae. 7 Medit. (Eur. 3) to Pakistan

Moriera Boiss. Cruciferae. 1 C As. & Iran to Afghanistan

Morierina Vieill. Rubiaceae (I 7). 2 New Caledonia

Morina L. Morinaceae. (Incl. *Cryptothladia*) 10 Balkans to Sino-Himal. R: BBMNHB

12(1984)12, 15. *M. persica* L. (Balkans) – seeds eaten like rice

Morinaceae Raf. (~ Valerianaceae). Dicots – Asteridae – Dipsacales. 2/13 Balkans to Himal. & China. R: BBMNHB 12(1984)1. Perennial herbs with persistent leaf-bases. Lvs opp. or whorled, often spiny; stipules 0. Fls bisexual, zygomorphic, in verticillasters (sometimes subcapitate); epicalyx of 4 united bracteoles, calyx cupular, 2-lobed or with oblique mouth, C ± 2-lipped, the 2 lateral lobes of anterior lip overlapping median & overlapped by 2 posterior in bud, A 4, 1 pair above the other near C mouth, or 2 fertile + 2 staminodes in tube, anther-lobes ± unequal with longit. slits, $\overline{G}(3)$, 1-loc. with 1 pendulous ovule & slender style. Fr. dry & indehiscent encl. in epicalyx & topped by persistent K. n = 17

Genera: *Acanthocalyx*, *Morina*

Considered close to Dipsacaceae but differing in verticillasters, epicalyx, aestivation, 6-veined ovary, palynology & embryology

Some cult. orn.

Morinda L. Rubiaceae (IV 10). (Excl. *Rennellia*) 80 OW trop., few Am. Fr. a head of ± coherent drupes in succ. enlarged calyces; some dye-sources esp. *M. citrifolia* L. (Indian mulberry, Kenya, Indomal.) – domatia present, yields a permanent red dye (Turkey red; *Symplocos* sp. used as mordant) for batik, source of canarywood, fr. allegedly toxic, source of foetid insecticidal hair-oil; *M. lucida* Benth. (trop. Afr.) – chewing-stick in Nigeria, diabetes treatment in Zaire; *M. titanophylla* Petit (Zaire, Uganda) – pachycaul

Morindopsis Hook.f. Rubiaceae (II 5). 2 SE As.

Moringa Rheede ex Adans. Moringaceae. 12 semi-arid Afr. to As. R: KB 40(1985)1. Seeds imp. water purifiers. *M. oleifera* Lam. (horse-radish tree, NW India) – alks, fls bird-poll., spinning seeds with 3 equidistant wings, oilseed (Ben nut, oil of Ben used in artist's paints, salad oil & soap), pods ed. (drumsticks), roots smell of horse-radish, timber suited for pulp for cellophane & rayon, fuelwood crop; *M. ovalifolia* Dinter & A. Berger (S Afr.) – trunk water-storing

Moringaceae Martinov. Dicots – Dilleniidae – Capparidales. 1/12 semi-arid Afr. to As. Decid. trees with mustard-oil glucosides. Lvs spirally arr., 2–3-pinnate with opp. leaflets; stipules 0, glands at base of petioles & pinnae. Fls bisexual, ± irreg., in axillary panicles or thyrses, with cupular hypanthium lined with a nectary-disk apically free; K 5, spreading or reflexed, unequal, imbricate, C 5 imbricate with the outermost usu. largest & inner 2 smallest, A 5 opp. C & alt. with staminodes, filaments around disk margin, anthers with longit, slits, $\underline{G}$ ((2)3(4)), 1-loc. on short gynophore with term. hollow style, parietal placentas, each with 2 rows of anatropous, bitegmic ovules with zig-zag micropyles. Fr. woody loculicidal capsule, elongate pod without replum, explosively dehiscent; seeds 3-winged (less often wingless) with straight oily embryo ± without endosperm. x = 14

Genus: *Moringa*

The seed-structure suggests affinities with Theales (Corner)

Morisia Gay. Cruciferae. 1 Corsica & Sardinia: *M. monanthos* (Viv.) Asch. (*M. hypogaea*) – cult. orn. sessile herb which buries fr.

Morisonia L. Capparidaceae. 4 Carib. & trop. S Am.

Morithamnus R. King, H. Robinson & G. Barroso (~ *Eupatorium*). Compositae (Eup.-Gyp.). 2 E Brazil. R: MSBMBG 22(1987)115

Moritizia DC ex Meissner. Boraginaceae. 5 trop. Am.

Morkillia Rose & Painter. Zygophyllaceae. 2 Mexico

Mormodes Lindley. Orchidaceae (V 9). 60 trop. Am. R: Selbyana 2(1978)149. Column twisted to 1 side of claw of lip; pollinia violently ejected if insect touches joint of anther to column. Cult. orn. epiphytes

Mormolyca Fenzl. Orchidaceae (V 10). 7 trop. Am. Cult. orn. epiphytes

Mormon tea *Ephedra trifurca*

morning glory *Ipomoea purpurea*

Morolobium Kosterm. = *Archidendron*

Morongia Britton = *Mimosa*

Moronobea Aublet. Guttiferae (III). 8–10 trop. S Am. Latex a source of adhesive & flammable resins (hog gum)

Morrenia Lindley. Asclepiadaceae (III 1). 2 S Am. R: JAA 70(1989)477. *M. odorata* (Hook. & Arn.) Lindley – pestliential climber of citrus in Florida

Morrisiella Aellen = *Atriplex*

morrison *Verticordia* spp.

Morsacanthus Rizz. Acanthaceae. 1 Brazil

mortel *Erythrina mitis*

mortiña *Vaccinium mortinia*

Mortonia A. Gray. Celastraceae. 8 S N Am.

Mortoniella Woodson. Apocynaceae. 1 C Am.

Mortoniodendron Standley & Steyerm. Tiliaceae. 5 C Am. Arils

Mortoniopteris Pichi-Serm. = *Crepidomanes*

Morus L. Moraceae (I). c. 12 temp. & warm, trop. Afr. Decid. monoec. to dioec. trees – mulberries. Male fls in catkins, females in pseudo-spikes, wind-poll.; fr. a bird-disp. juicy syncarp of fleshy P with the same *ecological* effect as *Rubus* (q.v.). Form. used for barkcloth in Mex. (superseded by *Trema micrantha*); cult. orn. & fr. trees esp. *M. alba* L. (white m., C & E China, natur. Eur. & N Am.) – used for coughs, colds & sore eyes in modern Chinese herbalism, the food-pl. of silkworm *Bombyx mori* in China, coppiced to produce readily harvestable lvs, James I's 1610 plan for silk sufficiency in GB mistaken in that *M. nigra* planted (incl. 4 acres at what is now Buckingham Palace, London) instead of *M. alba*! 'Macrophylla' (*M. bombycis, M. kagayamae*) street tree in Medit.; *M. australis* Poiret (temp. E As.) – fr. eaten in Sikkim; *M. nigra* L. (black or common m., ? derived from *M. alba*) – cult. for millennia, introd. GB C16 for juicy fr. (causing illness if eaten unripe, also source of rouge) & timber used for furniture, snuff-boxes, inlay etc., can be propagated from cuttings 2.5 m long; *M. rubra* L. (red m., E US) – cult. orn., timber

Moscharia Ruíz & Pavón. Compositae (Mut.-Nass.). 2 Chile. R: CGH 205(1974)163. Annuals, cult. orn.

moschatel *Adoxa moschatellina*

Moschopsis Philippi. Calyceraceae. 8 Chile, Patagonia

Mosdenia Stent. Gramineae (33d). 1 Transvaal

Mosheovia Eig = *Scrophularia*

Mosiera Small (~ *Myrtus*). Myrtaceae. 3+ C Am. 4-merous fls

Mosla (Benth.) Buch.-Ham. ex Maxim. Labiatae (VIII 1). 10 E As., Mal.

mosquito bush *Ocimum* spp.; **m. wood** *Mosquitoxylum jamaicense*

Mosquitoxylum Krug & Urban. Anacardiaceae (IV). 1 Jamaica: *M. jamaicense* Krug & Urban (mosquito wood) – timber for building

moss campion *Silene acaulis*; **clubm.** *Lycopodium* spp.; **m. pink** *Polemonium* spp.

Mossia N.E. Br. Aizoaceae (V). 1 Transvaal, Lesotho

mossy cyphel *Minuartia sedoides*

Mostacillastrum O. Schulz = *Sisymbrium*

Mostuea Didr. Gelsemiaceae (Loganiaceae s.l.). 8 trop. Afr. (7), S Am. (1). R: MLW 61-4(1961). *M. batesii* Baker (*M. stimulans*, W Afr.) – stimulant keeping revellers awake through all-night dances

Motandra A. DC. Apocynaceae. 10 trop. W Afr.

moth mullein *Verbascum blattaria*

mother-in-law's tongue *Sansevieria trifasciata*

mother-of-thousands *Saxifraga stolonifera, Cymbalaria muralis, Soleirolia soleirolii*

Motherwellia F. Muell. Araliaceae. 1 NE Aus.

motherwort *Leonurus cardiaca*

Motleyia J.T. Johansson. Rubiaceae (IV 10). 1 NW Borneo. R: Blumea 32(1987)149

Moullava Adans. Leguminosae (I 1). 1 S India

Moultonia Balf.f. & W.W. Sm. = *Monophyllaea*

Moultonianthus Merr. Euphorbiaceae. 1 Sumatra, Borneo

Mount Etna broom *Genista aetnensis*

mountain ash *Sorbus aucuparia*; **Australian m. a.** *Eucalyptus regnans*; **m. avens** *Dryas octopetala*; **m. pawpaw** *Carica pubescens*; **m. pride** *Spathelia sorbifolia*; **m. rimu** *Lepidothamnus laxifolius*

Mourera Aublet. Podostemaceae. 6 N S Am.

Mouretia Pitard. Rubiaceae (IV 1). 1 SE As.

Mouriri Aublet. Melastomataceae. (Memecylaceae). 81 trop. Am. R: FN 15(1976)33. Elaiophores attractive to *Centris* bees with 'buzz'-poll.; some fish-disp. in Amaz. Some ed. fr. sold in markets

mourning widow *Geranium phaeum*

mouse ear *Myosotis* spp. (Am.), (**chickweed**) *Cerastium* spp., *Holosteum* spp.; **common m. e.** *C. fontanum* ssp. *vulgare*; **m. plant** *Arisarum proboscideum*; **m. tail** *Myosurus minimus*

Moussonia Regel. Gesneriaceae. 11 C Am. R: Selbyana 1(1975)22

Moutabea Aublet. Polygalaceae. 10 trop. Am.

moutan *Paeonia suffruticosa*

moxa *Crossostephium chinense*

Moya Griseb. = *Maytenus*

moya grass *Pennisetum hohenackeri*
Mozartia Urban = *Myrcia*
mpesi *Trema orientalis*
Mrs Robb's bonnet *Euphorbia amygdaloides* ssp. *robbiae*
Msuata O. Hoffm. Compositae (Vern.-Vern.). 1 trop. Afr.
mtambara *Cephalosphaera usambarensis*
Muantijamvella J. Phipps = *Tristachya*
Muantum Pichon = *Beaumontia*
Mucizonia (DC) A. Berger. Crassulaceae. 2 W Medit. (Eur. 2)
Mucoa Zarucchi (~ *Ambelania*). Apocynaceae. 2 trop. S Am. R: AUWP 87-1(1987)40
Mucronea Benth. (~ *Chorizanthe*). Polygonaceae (I 1). 2 Calif. R: Phytol. 66(1989)202
Mucuna Adans. Leguminosae (III 10). 100 trop. (Pacific is. 11, R: KB 45(1990)1). Lianes often with legumes covered in irritant hairs; some fodders (buffalo bean) & cult. orn. Alks; L-dopa in seeds used in treatment of Parkinson's Disease. Bat-poll. spp. in both O & NW. *M. bennettii* F. Muell. (New Guinea) – fls scarlet, pergola pl.; *M. pruriens* (L.) DC (cow itch, trop. As., widely natur.) – irritant hairs form. used as vermifuge, var. *utilis* (Wight) Burck (*M. deeringiana*, Bengal bean, (Florida) velvet b.) with glabrous pods & used as fodder & cover-crop
mudar fibre *Calotropis* spp.
Mudgee wattle *Acacia spectabilis*
mud-nut *Glossostigma* spp.
muduga oil *Butea monosperma*
mudwort *Limosella aquatica*
Muehlenbeckia Meissner. Polygonaceae (II 2). (Incl. *Homalocladium*) 23 New Guinea, Aus. (14), NZ, S Am. Dioec. climbers or creeping pls with wiry stems, *M. florulenta* Meissner (*M. cunninghamii*, lignum) forming dense thickets in Aus. Some cult. orn. esp. *M. axillaris* (Hook.f.) Walp. (Aus., NZ) – creeping shrublet; *M. adpressa* (Labill.) Meissner (Aus.) – climber with ed. 'berries' (fleshy calyx) used in pies etc.; *M. australis* (Forster f.) Meissner & *M. complexa* (A. Cunn.) Meissner (both NZ) – form. eaten by moas; *M. platyclada* (F. Muell.) Meissner (*Homalocladium p.*, Solomon Is.) – myrmecophilous, cult. orn. shrub with jointed stems usu. leafless in fl.
Muellera L.f. = *Lonchocarpus*
Muelleranthus Hutch. Leguminosae (III 25). 3 Aus.
Muellerargia Cogn. Cucurbitaceae. 2 Madag. (1), E Mal. to Torres Strait (1)
Muellerina Tieghem (~ *Phrygillanthus*). Loranthaceae. 4 E Aus.
Muellerolimon Lincz. (~ *Limonium*). Plumbaginaceae (II). 1 W Aus.: *M. salicorniaceum* (F. Muell.) Lincz
mugga *Eucalyptus sideroxylon*
mugongo *Schinziophyton rautanenii*
mugwort *Artemisia* spp.
muhimbi *Cynometra alexandri*
Muhlenbergia Schreber. Gramineae (31e). (Incl. *Epicampes*) c. 160 trop. & warm Am. (esp S N), S As. (8). Some fodders ('hair grass'); some with roots used for exported brooms, esp. *M. macroura* (Kunth) A. Hitchc. (*E. m.*, zakaton, Mex.)
muhugwe or **muhuhu** *Brachylaena huillensis*
Muilla S. Watson ex Benth. Alliaceae (Liliaceae s.l.). 5 SW N Am. Cult. orn. with fibre-covered corms. R: Aliso 10(1984)623
Muiria N.E. Br. Aizoaceae (V). 1 Little Karoo: *M. hortenseae* N.E. Br. Cult. orn. leaf-succ.
Muiriantha C. Gardner. Rutaceae. 1 SW Aus.
Mukdenia Koidz. Saxifragaceae. 2 N China, Manchuria, Korea
Mukia Arn. Cucurbitaceae. 4 OW trop.
mukumari *Cordia africana*
mukunawanna *Alternanthera sessilis*
mula or **muli** *Raphanus sativus* 'Longipinnatus'
mulberry *Morus* spp. esp. *M. nigra* (**common** or **black m.**); **m. fig** *Ficus sycomorus*; **Indian m.** *Morinda citrifolia*; **paper m.** *Broussonetia papyrifera*; **red m.** *Morus rubra*; **white m.** *M. alba*
mulga *Acacia aneura*
Mulgedium Cass. (~ *Lactuca*). Compositae (Lact.-Lact.). 15 Euras.
Mulinum Pers. Umbelliferae (I 2c). 20 S Andes
mullein *Verbascum* spp.; **common m.** *V. thapsus*; **dark** or **black m.** *V. nigrum*; **moth m.** *V. blattaria*; **m. pink** *Silene coronaria*; **white m.** *V. lychnitis*
Multidentia Gilli. Rubiaceae (III 2). 11 trop. (E 8) Afr. R: KB 42(1987)645

Muluorchis J.J. Wood = *Tropidia*
mume *Prunus mume*
mums *Dendranthema ×grandiflorum*
Munbya Pomel = *Psoralea*
Mundulea (DC) Benth. Leguminosae (III 6). 15 Madag. (1 ext. to OW trop.), S Afr. (1). *M. sericea* (Willd.) A. Chev. (OW trop.) – fish-poison (rotenone)
mung (beans) *Vigna radiata*
muninga *Pterocarpus angolensis*
munjeet *Rubia cordifolia*
Munnozia Ruíz & Pavón. Compositae (Liab.). 43 Andes. R: SCB 54(1983)54
Munroa Torrey. Gramineae (31d). 5 Am. Disjunct – (1 W US, 4 SW S Am.). R: BANCC 52(1978)229
Munroidendron Sherff. Araliaceae. 1 Hawaii
Munronia Wight. Meliaceae (I 1). 10 Indomal. Shrublets, some medic. esp. *M. pinnata* (Wall.) Theob.
Muntafara Pichon = *Tabernaemontana*
Muntingia L. Tiliaceae. 1 trop. Am.: *M. calabura* L. (calabura, jam tree (Sri Lanka), Jamaican cherry) – flowers continuously with each fl. lasting 1 day (C falling p.m.), ovules do not develop until poll., pedicels elongating to present fls for bee-poll., later lowered below lvs, the ed. fr. disp. by birds & bats, so natur. in trop. As.; A 10–100, the pistil reduced at high nos & possibility of fr. formation proportionately reduced; firewood crop
Munzothamnus Raven (~ *Stephanomeria*). Compositae (Lact.-Mal.). 1 San Clemente (Channel Is. off Calif.)
Muraltia DC. Polygalaceae. 115 trop. & S Afr. (100 endemic in Cape). R: JSAB supp. 2(1954)
Murbeckiella Rothm. Cruciferae. 5 SW Eur. (4) & Algeria, 1 Caucasus
Murchisonia Brittan (~ *Thysanotus*). Anthericaceae (Liliaceae s.l.). 2 S & W Aus. R: FA 45(1987)340
Murdannia Royle. Commelinaceae (II 2). c. 50 trop. & warm
Muretia Boiss. = *Elaeosticta*
Murianthe (Baillon) Aubrév. = *Manilkara*
Muricaria Desv. Cruciferae. 1 N Afr.
Muriea Hartog = *Manilkara*
muringa *Cordia africana*
murnong *Microseris scapigera*
murraim berries *Tamus communis*
Murray red gum *Eucalyptus camaldulensis*
Murraya Koenig ex L. Rutaceae. 4 Indomal. to Pacific. R: W.T. Swingle, *Bot. Citrus* (1943)192. Spineless trees & shrubs. Alks. *M. koenigii* (L.) Sprengel (curry leaf, India, Sri Lanka) – lvs *always* used in Indian curries, carabazole alks with antifungal activity, oil used in soap industry; *M. paniculata* (L.) Jack (*M. exotica*, Chinese box, SE As. to Aus.) – flowers several times a year, wood for cutlery handles & walking-sticks, wood & roots powdered to give sweet-scented thanaka powder used by Thai & Burmese women, cult. orn. (incl. bonsai), leaf coumarin inhibits rat thyroid activity, trad. Chinese med. anti-implantation agent (yuehchukene) an indole derivative
Murtonia Craib = *Desmodium*
murumuru *Astrocaryum murumuru*
Musa L. Musaceae. 35 trop. As. Bananas. Like *Ensete* but with distinct petiole between blade & sheath & not hapaxanthic while A usu. 5 (not 6). Those spp. with pendent infls bat-poll., the fls functional for only 1 night, those with erect infls self-poll. or poll. by sunbirds. The wild forms have arillate seeds surrounded by simple trichomes in mucilage, though the fr. of *M. lasiocarpa* Franchet (Yunnan) a pl. c. 60 cm tall, rather dry; ed. bananas have degenerate ovules in a yellow mass derived from pericarp. Some with seeds germ. only after 6 months (cf. *Ravenala*), unusual in rain-forest pls. Perhaps first grown in China as fibre-pls for cloth (still in S China for gunny sacks); seen by Alexander the Great (327 BC), cult. Medit. c. AD 650, W Afr. C15 (before Europeans), Carib. 1516 (introd. by Portuguese), widely exported since use of refrigerated ships (1901). Fr. a berry, the clonal parthenocarpic forms ed. & a major carbohydrate source, the dessert bananas derived from triploid forms of *M. acuminata* Colla (Indomal.) esp. 'Dwarf Cavendish' – widely cult. the bat-poll. (0.63 ml watery nectar per fl. per night) wild pl. (seeding form known also from Pemba, E Afr.) from wetter areas than *M. balbisiana* Colla (Indomal.), a more disease-resistant sp. with which hybrids have been formed – *M. ×paradisiaca* L. (plantain) – the major cooking bananas used in trop. incl. the sour embul (apple banana) exported

by air from Sri Lanka to UK. R: N.W. Simmonds (1966) *The bananas*, ed. 2. Prod. 1 leaf c. every 10 days for 7–9 months, then flowers. Flavour due to mixture of 2 aliphatic esters with eugenol (aromatic phenol), the immature fr. c. 7% tannin, though, like cheese, can promote migraines in some (tyramine affecting blood vessels to brain), pulp used in shampoos etc. in Eur. Other spp. locally eaten, their lvs also wax & fibre-sources, esp. *M. textilis* Née (? cultigen, Philipp. – abacá, Manila hemp) – cables, twine, tea-bags, 'paper' walls of Japanese houses, & 'Canton fibre', less good & allegedly from a hybrid between *M. textilis* & an unknown sp. Many cult. orn. incl. *M. basjoo* Siebold & Zucc. ex Iinuma (Ryukyus) – Japanese favourite, fibre for textiles etc. *M. ingens* Simmonds (New Guinea) to 15 m tall with 'trunk' to 89 cm diam.

Musaceae Juss. Monocots – Zingiberidae – Zingiberales. 6/200 trop. Glabrous woody pls with erect stems, s.t. compressed &/or truly dichotomous (Strelitzioideae), or large pachycaul herbs with massive subterranean corm & hapaxanthic shoots (Musoideae), or perennial herbs with sympodial rhiz. (Heliconioideae) without sec. growth; roots with scattered vessel-elements & phloem strands. Aerial shoots (pseudostems) in Musoideae largely the rolled sheathing leaf-bases, the lvs simple & spirally arr. or distichous (Strelitzioideae, Heliconioideae) with long petiole & expanded lamina rolled from 1 side to other in bud, & with prominent midrib & pinnate-parallel venation, the laterals incurved only at margin & between which tears may form under windy conditions. Infl. bracteate, term. or lateral, with spirally arr. or distichous (Strelitzioideae, Heliconoideae) leathery keeled bracts, each subtending few-flowered cymes of bisexual (unisexual in Musoideae, the apical males) flowers. Fls irreg. with much nectar from septal nectaries (visited by birds, bats & insects); P 6, petaloid in 2 distinct whorls but outer & 2 inner merely teeth or lobes of a P-tube split along 1 side, the 3rd (adaxial) inner free (Musoideae) or lateral inner larger than median (connivent in *Strelitzia*, q.v.) or the adaxial outer distinct & the rest ± connate as 5-toothed boat-shaped vessel (Heliconioideae), A 5 or 6, the 6th opp. free P but often a staminode or 0, anthers with longit. slits, $\overline{G}(3)$, 3-loc., with term. style & ∞ (1 in Heliconioideae) anatropous, bitegmic ovules per loc. Fr. a loculicidal capsule with arillate seeds (Strelitzioideae) or fleshy berry with few–∞ seeds with rudimentary arils (Musoideae) or schizocarp splitting into (2)3 1-seeded fleshy mericarps with exarillate seeds (Heliconioideae) & straight or (*Ensete*) curved embryo in copious starchy & mealy endosperm & perisperm. x = 7–11, 16, 17

Classification & genera:

Strelitzioideae (Strelitziaceae; lvs distichous, fls bisexual; capsule with arillate seeds): *Strelitzia, Phenakospermum, Ravenala*

Musoideae (lvs spirally arr.; fls unisexual in infl. arising from underground corm and growing up through pseudostem; berry with seeds with rudimentary arils): *Ensete, Musa*

Heliconioideae (Heliconiaceae; lvs distichous; fls bisexual; schizocarp with exarillate seeds): *Heliconia*

Ensete, Musa (bananas) – imp. fruit & fibre pls; cult. orn.

Musanga R. Br. Cecropiaceae. 2 trop. Afr. R: BJBB 46(1976)496. V. like *Cecropia* but without ants & poss. derived thus. *M. cecropioides* R. Br. (umbrella tree) – cult. orn. fast-growing (stilt-roots allow germ. on temporary humps or tree-trunks rich in nutrients in clearings, disappearing to leave bole on stilts, reaching 24 m in 15–20 yrs & soon dying) but being ousted by introd. allied *Cecropia peltata* (? fewer pests) with light timber (corkwood) used for floats & rafts, isothermic ceilings, long thin fibres suitable for paper, charcoal for floor-polish

Muscadinia (Planchon) Small = *Vitis*

Muscari Miller. Hyacinthaceae (Liliaceae s.l.). 30 Eur. (13), Medit., W As. R: Lily Yearbook 29(1965)126. Grape hyacinths. Alks incl. colchicine. Cult. orn. & fls used in scent-making incl. *M. armenaicum* Leichtlin ex Baker (SE Eur. to Cauc.) – lvs developing in autumn, but very commonly *M. neglectum* Guss. ex Ten. (*M. atlanticum*, '*M. racemosum*', Eur. Medit.) & *M. botryoides* (L.) Miller (C & SE Eur.); others incl. *M. comosum* (L.) Miller (tassel hyacinth, Medit.) – bulbs ed., 'Plumosum' ('Monstrosum', feather h.) – branched infls of sterile fls, P thread-like, *M. muscarimi* Medikus (*M. moschatum*, SW Turkey) – long-cult. for scent-making

Muscarimia Kostel. ex Losinsk. = *Muscari*

muscatel *Vitis vinifera*

Muschleria S. Moore. Compositae (Vern.-Vern.). 1 Angola

muscovado raw or unrefined cane-sugar

Musella (Franchet) Li = *Musa*

Museniopsis (A. Gray) J. Coulter & Rose = *Tauschia*

Musgravea F. Muell. Proteaceae. 2 Queensland

musine *Croton megalocarpus*

Musineon Raf. Umbelliferae (III 5). 3 W N Am.

musizi *Maesopsis eminii*

musk (plant) *Mimulus moschatus*; **m. mallow** *Malva moschata*; **m. weed** *Myagrum perfoliatum*; **m.wood** *Alangium villosum*

muskit = mesquite

Mussaenda L. Rubiaceae (I 8). 100 OW trop. (not Aus.). Erect or scrambling shrubs, some locally medic. & several spectacular cult. orn. esp. *M. erythrophylla* Schum. & Thonn. (Ashanti blood, flame of the forest, trop. W Afr.) – the enlarged sepal bright red, such sepals making the infls of many *M.* spp. conspicuous (cf. *Euphorbia*); *M. kingdon-wardii* Jayaw. (N Burma) – rheophyte

Mussaendopsis Baillon (I 1). Rubiaceae. 2 W Mal. Durable timber

Mussatia Bureau ex Baillon. Bignoniaceae. 2 trop. Am.

Musschia Dumort. Campanulaceae. 2 Madeira. Pachycaul treelets with dull fls, *M. aurea* (L.f.) Dumort. visited by nectar-seeking lizards; fr. a capsule with tranverse slits between vasc. ribs

mustard *Brassica nigra* (**m. seed** of Bible, though poss. *Salvadora persica*, **black m.**), *Sinapis alba*, *B. napus* (m. of '**m. & cress**'); **ball m.** *Neslia paniculata*; **brown m.** *B. nigra*; **buckler m.** *Biscutella laevigata*; **Chinese m.** *B. juncea* 'Crispifolia', *B. oleracea* Chinensis Group; **Dijon m.** *B. juncea*; **garlic m.** *Alliaria petiolata*; **hedge m.** *Sisymbrium officinale*; **hoary m.** *Hirschfeldia incana*; **Indian m.** *B. juncea*; **Mithridate m.** *Thlaspi arvense*; **tower m.** *Arabis glabra*; **treacle m.** *Erysimum cheiranthoides*; **tumble m.** *S. altissimum*; **white m.** *B. hirta*

Mutellina Wolf (~ *Ligusticum*). Umbelliferae (III 8). 3 Euras.

Mutisia L.f. Compositae (Mut.-Mut.). 59 Andes from Colombia to S Arg. & Chile, SE Brazil, Paraguay, Uruguay & NE Arg. R: OL 13(1965)1. Shrubs & lianes, some cult. orn. Lvs often term. by a tendril. *M. acuminata* Ruíz & Pavón (Peru to Bolivia) – hummingbird-poll. in altiplano of Peru though nectar hexose-rich (typical of Compositae)

mvule *Milicia excelsa*

Myagrum L. Cruciferae. 1 Medit. & C Eur. to India: *M. perfoliatum* L. – introd. Aus., where troublesome (musk weed)

myall *Acacia pendula*

Mycelis Cass. Compositae (Lact.-Lact.). 1 temp. Euras., N W Afr. (rare): *M. muralis* (L.) Dumort. (wall lettuce))

Mycerinus A.C. Sm. Ericaceae. 5 Guayana Highland. R: MNYBG 29(1978)175

Mycetia Reinw. Rubiaceae (I 8). 25 Indomal.

Myginda Jacq. = *Crossopetalum*

Myladenia Airy Shaw. Euphorbiaceae. 1 Thailand

Myllanthus R. Cowan = *Raputia*

Myodocarpus Brongn. & Gris. Araliaceae. 12 New Caledonia

Myonima Comm. ex Juss. (~ *Ixora*). Rubiaceae (II 2). 4 Mauritius, Réunion. R: KB 37(1983)555

Myoporaceae R. Br. Dicots – Asteridae – Scrophulariales. 3/235 Aus. (most), Pacific; Indian Ocean, trop. Am. (few). Small trees & shrubs with scattered secretory cavities (not in excl. *Oftia*, which is odd in having internal phloem). Lvs simple, usu. gland-dotted, spirally arr. (rarely opp.). Fls bisexual, reg. or not, solit. or in axillary cymes; K (5), lobes imbricate or open, C (5) often 2-labiate, lobes imbricate, A 4 with upper posterior 0 or staminode, rarely 5, alt. with C-lobes on C-tube, anthers with longit. slits, $\underline{G}$ (2) with term. style, 2-loc., each loc. with (1)2 pendulous, anatropous, unitegmic ovules from near summit, or 4–8 superposed in pairs, or the loc. subdivided into 4–10 uniovulate compartments. Fr. a drupe or separating into 1-seeded drupe-like segments; seeds with ± straight embryo in scanty or 0 endosperm. x = 27

Genera: *Bontia*, *Eremophila*, *Myoporum*

(*Oftia* & *Ranopisoa* referred to Scrophulariaceae)

Some timber (*Eremophila*, *Myoporum*) & locally eaten fr.

Myopordon Boiss. Compositae (Card.-Card.). 6 SW & C As.

Myoporum Sol. ex Forster f. Myoporaceae. c. 28 Aus. (most), Mauritius, E As., E Mal., NZ, Hawaii. Some drought-tolerant pls (boobialla, boobyalla) used as shelter-belts in S Eur. e.g. *M. tenuifolium* Forster f. (*M. acuminatum*, Aus.), some toxic to stock; some locally coll. ed. fr. & manna. Cult. orn. incl. *M. insulare* R. Br. (cockatoo bush, Aus.), *M. laetum* Forster f. (ngaio, NZ) & *M. sandwicense* A. Gray (Hawaii) – timber

Myoschilos Ruíz & Pavón. Santalaceae. 1 Chile

Myosotidium Hook. Boraginaceae. 1 Chatham Is. (NZ): *M. hortensia* (Decne.) Baillon (*M. nobile*, Chatham Is. forget-me-not) – cult. orn.

Myosotis L. Boraginaceae. c. 100 temp. (Eur. 41) incl. trop mts. R: JAA suppl. 1(1991)159. Forget-me-nots (scorpion grass). C with scales almost closing throat, where s.t. a coloured ring honey-guide; fl. often changing from pink to blue at anthesis. Cult. orn. small herbs esp. *M. scorpioides* L. (water f., Euras., natur. N Am.) – K without hooked hairs, but usu. *M. sylvatica* Ehrenb. ex Hoffm. (garden f., Euras.) – K with hooked hairs, many cvs incl. pink ones

Myosoton Moench. Caryophyllaceae (II 1). 1 temp. Euras.: *M. aquaticum* (L.) Moench

Myosurus L. Ranunculaceae (II 3). 6 temp. esp. Am. (Eur. 2 incl. *M. minimus* L. (mouse tail, Euras., Afr. Aus.)) – receptacle elongate

Myoxanthus Poepp. & Endl. (~ *Pleurothallis*). Orchidaceae (V 13). 47 trop. Am. R: MSB 44(1992)1

myrabolans = myrobalans

Myracrodruon Allemão & M. Allemão = *Astronium*

Myrceugenella Kausel = *Luma*

Myrceugenia O. Berg. Myrtaceae. 40 Juan Fernandez, Chile & Arg. & (1000 km away) SE Brazil. R: FN 29(1981), Brittonia 36(1984)161

Myrcia DC ex Guillemin. Myrtaceae (Myrt.). c. 250 trop. Am. K ± free; 2 ovules per loc. Some fish-disp. in Amaz. *M. splendens* (Sw.) DC (Amaz.) – bark extract used to paint insides of gourds black; *M. tomentosa* (Aublet) DC – ed. fr. (cabelluda)

Myrcialeucus N. Rojas = *Eugenia*

Myrcianthes O. Berg. Myrtaceae. (Myrt). 50 trop. Am. *M. umbellulifera* (Kunth) Alain (monos plum, WI) – ed. fr.

Myrciaria O. Berg. Myrtaceae (Myrt.). 40 trop. Am. Like *Eugenia* but K-tube extended above ovary. Many ed. fr. esp. *M. cauliflora* (C. Martius) O. Berg (jaboticaba, S Brazil, where much cult. for fr.)

Myrciariopsis Kausel = *Myrciaria*

myrhh See myrrh

Myriactis Less. Compositae (Ast.-Ast.). 12 Cauc. to Japan & New Guinea. Pappus 0

Myrialepis Becc. Palmae (II 1e). 1 Sumatra, Malay Pen.: *M. paradoxa* (Kurz) Dransf. – hapaxanthic rattan

Myrianthemum Gilg = *Medinilla*

Myrianthus Pal. Cecropiaceae. 7 trop. Afr. R: BJBB 46(1976)472. *M. arboreus* Pal. with ed. fr. & timber (corkwood) lighter than cork

Myriaspora DC. (~ *Loreya*). Melastomataceae. 1 trop. S Am. Cauliflory; fls with calyptra

Myrica L. Myricaceae. c. 55 subcosmop. (not Medit., Aus.; Eur. 2). Shrubs usu. with nitrogen-fixing actinomycetes in roots; some wax sources, ed. fr. & cult. orn. incl. *M. californica* Cham. (bay berry, coastal W US) & *M. cerifera* L. (wax myrtle, candleberry, tallow shrub, E US) – wax on fr. surface rich in palmitic acid removed in boiling water made into candles by early settlers & used in the scented bayberry soap; *M. esculenta* Buch.-Ham. ex D. Don (Indomal.) – fr. ed., bark for fish-poison & yellow dye; *M. faya* Aiton (Macaronesia, natur. S Portugal) – introd. to Hawaii for afforestation or for wine-making by Portugese labourers in 1880s, an invasive pest since 1944 though fixing 10–20 kg nitrogen/ha/yr (no N-fixing native pls in Hawaii); *M. gale* L. (*Gale belgica*, (sweet) gale, bog myrtle, N Am., NW Eur., NE Siberia) – form. medic. & source of yellow dye, insecticidal (increases no. of oil-glands when attacked), lvs used to flavour & improve foaming of beer, a common pl. of wet heaths with capacity to change sex from yr to yr & with morphologically term. ovules; *M. javanica* Blume (Java) – fr. ed., wood excellent as fuel & charcoal, planted for land reclamation; *M. rubra* Siebold & Zucc. (E As.) – cult. China for ed. fr.

Myricaceae Martinov. Dicots – Hamamelidae – Juglandales. 3/55 subcosmop. Aromatic trees & shrubs with indumentum of long colourless unicellular hairs & peltate usu. yellow multicellular glands; roots usu. with nitrogen-fixing bacteria (? not *Canacomyrica*). Lvs simple, spirally arr.; stipules in *Comptonia*. Fls small, wind-poll., usu. unisexual in spikes (simple or compound); P 0 (present in *Canacomyrica*), males often with 2 bracteoles as well as bract, A (2–)4(5, 6 in *Canacomyrica* where at summit of G in bisexual fls, –20), fewest in most acropetal fls, anthers with longit. slits, females with 2 bracteoles (sometimes K-like) & G (2), 1-loc., with ± distinct styles & 1 basal erect orthotropous, unitegmic ovule. Fr. a drupe or nutlet, sometimes with accrescent bracteoles; seed with straight embryo & ± 0 endosperm. x = 8

Genera: *Canacomyrica*, *Comptonia*, *Myrica*

Some waxes, timber & ed. fr.

Myricanthe Airy Shaw. Euphorbiaceae. 1 New Caledonia

Myricaria Desv. Tamaricaceae. Incl. *Tamaricaria* 10 temp. Euras. (Eur. 1). *M. elegans* Royle (*T.e.*, C & S As.) – rheophyte

Myriocarpa Benth. Urticaceae (III). 18 trop. Am.

Myriocephalus Benth. Compositae (Gnap.-Ang.). 8 temp. Aus. R: OB 104(1991)115. Prob. heterogeneous. Compound heads of 1–9-flowered capitula

Myriocladus Swallen. Gramineae (1a). 13 sandstone tablelands of Venez. R: SCB 9(1973)94. Bamboos resembling bromeliads on poles

Myriodon (Copel.) Copel. = *Hymenophyllum*

Myrioneuron R. Br. ex Hook.f. = *Keenania*

Myriophyllum L. Haloragidaceae. 60 cosmop. (Eur. 3) esp. Aus. (36, 31 endemic). R: Brunonia 8(1985)173. Water milfoil. Mostly submerged aquatics like *Ceratophyllum* but lvs pinnate & not dichot.; with aerial wind-poll. fls & overwintering buds. Some cult. in aquaria esp. *M. aquaticum* (Vell. Conc.) Verdc. (*M. brasiliense*, parrot's feather, trop. Am., natur. Afr. & US)

Myriopteron Griffith. Asclepiadaceae (I; Periplocaceae). 1 NE India to W Mal.: *M. paniculatum* Griffith

Myriostachya (Benth.) Hook.f. Gramineae (31d). 1 S India,, Sri Lanka to SE As.

Myripnois Bunge. Compositae (Mut.-Mut.). 1 N China. Gynodioec. shrubs

Myristica Gronov. Myristicaceae. 72 trop. As. to Aus. (3, R: Blumea 36(1991)183). Dioec.; some domatia. *M. fragrans* Houtt. (nutmeg, Moluccas) – comm. nutmeg cult. in Mal. & Grenada (WI), seedlings sexed by colour reaction with ammonium molybdate, when ground the seed used as flavouring for milk puddings, biscuits etc. though excess is toxic (myristicin), hallucinatory & addictive with reputation as aphrodisiac; the fimbriate aril is mace used in flavouring for fish, doughnuts etc.; oil medic. also used to flavour toothpaste, cigarettes etc.; *M. inspida* R. Br. (trop. Aus.) – beetle-poll.

Myristicaceae R. Br. Dicots – Magnoliidae – Magnoliales. 19/400 trop. – usu. lowland rain forest. Dioec. or monoec. evergreen trees usu. with aromatic tissues containing spherical ethereal oil-cells often with the phenolic, myristicin; bark typically with coloured sap when slashed. Lvs simple, entire, spirally arr. & often distichous, often gland-dotted; stipules 0. Fls in usu. axillary cymes or racemes; P ((2)3(–5)) with valvate lobes, A 2–∞ with ± united filaments & often laterally connate anthers with longit. slits & monosulcate or inaperturate pollen grains, $\underline{G}$ 1 with unclosed carpel, stigma rarely with style, & 1 nearly basal, ± anatropous, bitegmic ovule. Fr. fleshy to leathery, usu. dehiscent along 2 sutures to reveal pendulous endotestal seed with conspicuous aril, copious oily & usu. ruminate endosperm & small embryo, the cotyledons sometimes basally connate. x = 9, 21, 25

Principal genera: *Gymnacranthera, Horsfieldia, Iryanthera, Knema, Myristica, Virola*; a very uniform archaic group

Male fls have no pistil, females no staminodes. According to Corner the seeds are the most primitive surviving; they are very large & the attractive arils are taken by birds, the toxic seed dropped

Some oils & spices esp. nutmeg & mace (*Myristica*), timber (esp. *Horsfieldia, Virola*) & hallucinogens (esp. *Virola*)

Myrmechis (Lindley) Blume. Orchidaceae (III 3). 6 E As., W Mal.

Myrmecodia Jack. Rubiaceae (IV 7). 26 Mal. to Fiji. R: Blumea 37(1993)271. Epiphytes with swollen stems ('tubers') penetrated by numerous interconnecting galleries & chambers inhabited by ants, the system forming independent of them in the hypocotyl & deriving from the activity of phellogens producing cork layers which split. Ants, which patrol the stems collecting nectar, bring nutrient-rich material (decay facilitated by fungi) into the tuber enhancing the mineral regime for the plant in a nutrient-poor habitat but deter pollinators (so pls largely selfed) & disp. agents. Seeds spread on to bark by birds, also removed by ants & planted in their runways

Myrmeconauclea Merr. Rubiaceae. 3 Mal. R: Blumea 24(1978)342. Some rheophytes incl. *M. strigosa* (Korth.) Merr., unique in being myrmecophilous too

Myrmecophila (Christ) Nakai = *Lecanopteris*

Myrmecophila Rolfe (~ *Schomburgkia*). Orchidaceae (V 13). 6 trop. Am.

Myrmecopteris Pichi-Serm. = *Lecanopteris*

Myrmecosicyos C. Jeffrey. Cucurbitaceae. 1 Kenya: *M. messorius* C. Jeffrey – found around holes of harvester ants in Rift Valley

Myrmedoma Becc. = *seq.*

Myrmephytum Becc. Rubiaceae. 8 Philippines, Sulawesi, W New Guinea. Ant-pls. R: Blumea 36(1991)43. Epiphytes

Myrmidone C. Martius. Melastomataceae. 2 trop. S Am. Heterophylly

myrobalans *Terminalia* spp. esp. *T. chebula, Prunus cerasifera*

Myrocarpus Allemão. Leguminosae (III 2). 4 trop. S Am. R: Phytologia 23(1972)401. Some timber & oil used in scent-making

Myrosma L.f. Marantaceae. 15 trop. Am. *M. cannifolia* L.f. is marble arrowroot

Myrosmodes Reichb.f. (~ *Aa*). Orchidaceae (III 3). 9 trop. Am.

Myrospermum Jacq. Leguminosae (III 2). 2–3 trop. Am. Lvs dotted; fr. winged, indehiscent

Myrothamnaceae Niedenzu. Dicots – Hamamelidae – Hamamelidales. 1/2 S Afr. & Madag. Small dioec. xeromorphic aromatic glabrous shrubs. Lvs opp., flabellate, plicate, venation not reticulate; resin- or oil-ducts & stipules present. Fls reg., wind-poll. in erect spikes; P 0–4 esp. in term. fls, A 3–8 alt. with P when present, anthers with prolonged connective & longit. slit & pollen shed in tetrads, G (3 or 4), 3- or 4-loc. with distinct recurved styles & axile placentation & rather numerous anatropous bitegmic ovules per loc. Fr. a capsule, the carpels separating apically & opening ventrally; seeds ∞ with thin testa & copious oily endosperm. 2n = 20

Genus: *Myrothamnus*

Myrothamnus Welw. Myrothamnaceae. 2 Kenya, S trop. Afr., Madag. *M. flabellifolia* Welw. (S Afr.) a resurrection pl. rapidly reviving after rain when the folded blackish fragile lvs green up, for the mitochondria (? & plastids) separated from rest of cell contents during desiccation (up to 2 yrs) by barriers perforated on rehydration

Myroxylon L.f. Leguminosae (III 2). 2–3 trop. Am., natur. OW. *M. balsamum* (L.) Harms (Venez. to Peru) – source of balsam of Tolu used in ointments & for flavouring cough syrups etc., var. *pereirae* (Royle) Harms – source of balsam of Peru, form. much valued medic. now with similar uses; both with good timber for furniture

myrrh *Commiphora* spp. esp. *C. myrrha*, (Genesis) *Cistus creticus*; **Abyssinian m.** *C. madagascariensis*; **garden m.** *Myrrhis odorata*; **Mecca m.** *C. opobalsamum*; **scented m.** *C. guidotii*

Myrrhidendron J. Coulter & Rose. Umbelliferae (III 10). 5 C Am. to Colombia

Myrrhinium Schott. Myrtaceae (Myrt.). 3 trop. S Am. R: Loefrgrenia 80(1983)1. *M. atropurpureum* Schott with A 4–8, fr. ed.

Myrrhis Miller. Umbelliferae (III 2). 1 Eur.: *M. odorata* (L.) Scop. (sweet cicely, garden myrrh) – form. potherb, prob. widely introd. for strewing on church floors in mediaeval GB, home remedy & flavouring for brandy (*trans*-anethole the main sweet constituent); only male fls at end of flowering so that last (protandrous) bisexual fls pollinated

Myrrhoides Heister ex Fabr. = *Physocaulis*

Myrsinaceae. R. Br. Dicots – Dilleniidae – Primulales. 33/1225 trop. & warm with few temp. OW. Trees, shrubs, lianes & few subherbaceous (esp. *Discocalyx* spp.), usu. with secretory ducts or cavities with resins. Lvs spirally arr., simple & usu. entire with gland dots & dashes or glandular-hairy; stipules 0. Fls usu. reg. & bisexual, small in usu. ebracteolate infls. K (3)4 or 5(6), often basally connate, lobes imbricate, convolute or valvate, C same but rarely free, A opp. lobes & usu. adnate to tube with filaments s.t. basally connate & anthers with longit. slits or apical pores, G (half-inf. in *Maesa*) (3–5(6)), 1-loc. with term. style & few to many anatropous to almost campylotropous bitegmic (unitegmic in *Aegiceras*) ovules. Fr. a berry or drupe (1-seeded capsule in *A.*), 1-seeded (several in *Maesa*); seeds with ± straight embryo in oily endosperm (0 in *Aegiceras*). x = 10–13, 23

Classification & principal genera:

1. **Maesoideae** (G half-inf.; fr. many-seeded): *Maesa* (only)
2. **Myrsinoideae** (G; fr. 1-seeded): *Ardisia, Cybianthus, Discocalyx, Embelia, Geissanthus, Myrsine, Oncostemum, Parathesis, Stylogyne, Tapeinosperma*

Some cult. orn. (*Ardisia, Myrsine*) & local medic.; *Aegiceras* in mangrove

Myrsine L. Myrsinaceae. s.s. 5 Azores, Afr. & As. (s.l. incl. *Rapanea, Suttonia* 150 trop. & warm). Some timbers incl. *M. melanophloeos* (L.) Sweet (*Rapanea m.*, Cape beech, trop. & S Afr.); *M. africana* L. (Azores to China) – cult. orn.

Myrsiphyllum Willd. = *Asparagus*

Myrtaceae Juss. Dicots – Rosidae – Myrtales. 129/4620 trop. & warm + temp. Aus. Trees & shrubs with abundant, scattered secretory cavities &, characteristically, internal phloem in pith; usu. with ectotrophic mycorrhizae. Lvs usu. opp., simple & often leathery; stipules rudimentary or 0. Fls usu. bisexual, reg. with hypanthium extended beyond ovary, rarely perigynous, with conspic. bracts, K, C or A attractive to animals esp. birds, in complex infls (solit. in *Myrtus communis*); K (3)4 or 5(6), often imbricate & free or not, s.t. v.

reduced or splitting at anthesis or forming calyptra, C similar (a calyptra in *Eucalyptus*), A usu. ∞ developing centripetally on rim of hypanthium, free or basally united into 4 or 5 groups each supplied by 1 vasc. trunk-bundle, less often 2 × K or C in 2 whorls or single whorl opp. K (*Heteropyxis*), anthers with longit. slits or term. pores, nectary-disk lining prolonged hypanthium or at apex of G (2–5(–16)) with as many loc. or rarely pseudo-monomerous & term. style (stigma sessile in *Psiloxylon*), placentation axile, each loc. with 2–∞ anatropous to campylotropous, bitegmic ovules per loc. though embryo often apomictic (developing from nucellus after degeneration of zygote). Fr. a (1–)few(–∞)-seeded berry, capsule, or drupe or nut; seeds mesotestal, often polyembryonous initially, with ± 0 endosperm, the embryos various with cotyledons s.t. curved or spiralled. x = (6–9–)11(12)

Classification & chief genera:

(See PLSNSW 102(1979)157, Taxon 29(1980)587)

The arrangement is as yet unfixed, though recent work recognizes a number of generic groups in 13 alliances (tribes) in the 2 trad. recog. subfams, Myrtoideae (6) & Leptospermoideae (7, incl. Heteropyxidoideae & Chamelaucioideae recog. as separate subfams by some workers); Psiloxyloideae (s.t. incl. in H., s.t. kept as separate fam. (unisexual fls, A 10(–12), G̲ & large lobed sessile stigma – only *Psiloxylon*, Masc.); the other subfams:

I. **Leptospermoideae** (fr. usu. dehiscent, pericarp usu. dry, lvs spirally arr. or opp., SE As. to Pacific with (v. few) outliers in S Afr. & Chile e.g. *Tepualia*): *Baeckea, Callistemon, Calothamnus, Calytrix, Corymbia, Darwinia, Eucalyptus, Kunzea, Leptospermum, Melaleuca, Metrosideros, Tristaniopsis, Verticordia, Xanthostemon*

II. **Myrtoideae** (fr. indehiscent, pericarp usu. succ., lvs opp.; major concentrations in trop. Am., SE As., E Aus., Pacific): *Austromyrtus, Blepharocalyx, Calyptranthes, Campomanesia, Decaspermum, Eugenia, Marlierea, Myrceugenia, Myrcia, Myrcianthes, Myrciaria, Plinia, Psidium, Rhodamnia, Syzygium, Xanthomyrtus*

Timber (*Corymbia, Eucalyptus*, which characterize much of Aus. veg., *Syncarpia*), ed. fr. (*Acca, Eugenia, Psidium* (guava), *Myrciaria, Myrteola, Syzygium, Ugni* etc.), spices & medic. oils (*Eucalyptus, Pimenta* (allspice, bay rum), *Leptospermum, Melaleuca, Myrtus, Syzygium* (cloves)) & cult. orn. esp. *Acmena, Callistemon, Calothamnus, Chamelaucium* (cut-fl.), *Eucalyptus, Leptospermum, Myrtus, Verticordia*

Myrtama Ovcz. & Kinsik. = *Myricaria*

Myrtastrum Burret. Myrtaceae. (Myrt.). 1 New Caledonia

Myrtekmania Urban = *Pimenta*

Myrtella F. Muell. Myrtaceae (Myrt.). 9 New Guinea, Aus.

Myrteola O. Berg. Myrtaceae (Myrt.). 3 S S Am. & trop. Am. mts. R: SB 13(1988)120. Ed. fr.

Myrtillocactus Console. Cactaceae (III 8). 4 Mex., Guatemala. Tree-like or shrubby. Fr. ed. like *Vaccinium myrtillus*

myrtle *Myrtus communis*; **m. beech** *Nothofagus cunninghamii*; **bog m.** *Myrica gale*; **crape m.** *Lagerstroemia indica*; **m. lime** *Triphasia trifolia*; **sand m.** *Leiophyllum buxifolium*; **strawberry m.** *Ugni molinae*; **Tasmanian m.** *Nothofagus cunninghamii*; **wax m.** *Myrica cerifera*; **willow m.** *Agonis* spp.

Myrtopsis Engl. Rutaceae. 8 New Caledonia

Myrtus L. Myrtaceae. (s.s.) 2 Medit. & N Afr. *M. communis* L. (myrtle, Medit., origin unclear) – lvs, fragrant fls & fr. rich in oil, the 'Eau d'Ange' of scent-making, also medic., wood for walking-sticks, furniture etc., bark & roots yield tannin used on finest leathers of Russia & Turkey, cult. orn. with several cvs, long associated with ritual & ceremony

Mysore thorn *Caesalpinia decapetala*

Mystacidium Lindley. Orchidaceae (V 16). 9 E & S Afr. Cult. orn.

Mystropetalon Harvey. Balanophoraceae. 1 SW Cape: *M. thomii* Harvey – on Proteaceae, pollen unique in being triangular, square or pentagonal end-on, but square from the side; elaiosomes around fr. attractive to ants which disperse seeds

Mystroxylon Ecklon & Zeyher (~ *Cassine*). Celastraceae. 3 trop. & S Afr.

Mytilaria Lecomte. Hamamelidaceae (II). 1 Kwangsi & Laos. G (2) semi-inf., partly sunk in fleshy axis

Myuropteris C. Chr. = *Colysis*

Myxopappus Källersjö (~ *Pentzia*). Compositae (Anth.-Mat.). 2 S Afr. R: BJLS 96(1988)314. Annuals

Myxopyrum Blume. Oleaceae. 4 Indomal. R: Blumea 29(1984)499

Myzodendraceae J. Agardh = Misodendraceae

Myzodendron Banks & Sol. ex R. Br. = *Misodendrum*

Myzorrhiza Philippi = *Orobanche*
mzimbeet *Androstachys johnsonii*

N

na *Mesua ferrea*
Nabaluia Ames. Orchidaceae (V 12). 3 N Borneo. R: OM 1(1986)47
Nabalus Cass. (~ *Prenanthes*). Compositae (Lact.-Lact.). 15 N Am., E As. Some cult. orn.
Nabea Lehm. ex Klotzsch (*Macnabia*). Ericaceae. 1 S Afr.
Nablonium Cass. = *Ammobium*
Nacrea Nelson = *Anaphalis*
Nageia Gaertner (*Decussocarpus*, ~ *Podocarpus*). Podocarpaceae. 5 Indomal. R: NRBGE 45(1988)381. Some cult. orn. incl. *N. wallichiana* (C. Presl) Kuntze, only conifer native in S India
Nageliella L.O. Williams. Orchidaceae (V 13). 2 C Am. & Venezuela. Cult. orn. epiphytes esp. *N. purpurea* (Lindley) L.O. Williams
Nagelocarpus Bullock. Ericaceae. 1 SW Cape
Naias Adans. = *Najas*
Naiocrene (Torrey & A. Gray) Rydb. = *Montia*
Najadaceae Juss. (~ Hydrocharitaceae). Monocots – Alismatidae – Najadales. 1/c. 32 cosmop. Monoec. or dioec. freshwater aquatics rooted but with reduced vessel-less xylem in branched stems. Lvs ± opp. to app. whorled, narrow, 1-veined, somewhat sheathing at base & often subtending 2 small axillary scales; stomata 0. Fls water-poll., small, usu. 1 per axil, males almost always with spathe-like set of scales & flask-shaped inner involucre (?P), A 1, basal, with sessile irreg. dehiscing anther, females with 0 or inconsp. involucre adnate to $\underline{G}$ 1, 1-loc. with 2–4 long stigmas & 1 basal, erect, anatropous, bitegmic ovule. Fr. indehiscent with thin pericarp; seed with straight embryo & 0 endosperm. x = 6, 7
Genus: *Najas*. Poll. like *Zostera* but pollen spherical
Najas L. Najadaceae. c. 32 cosmop. (Eur. 4 + 2 natur., Aus. 8). *N. tenuifolia* R. Br. found in mudpools at 60 °C in Java; some bad weeds in ricefields but good green fertilizer & valuable fish-food (e.g. in water 5 m deep for tilapia, to 14 m in Canada), also as packing material & local food (Hawaii)
naked boys or **ladies** *Colchicum autumnale*
nal *Phragmites australis*
Naletonia Bremek. = *Psychotria*
Nama L. Hydrophyllaceae. 45 SW US & trop. Am., Hawaii (1)
Namacodon Thulin. Campanulaceae. 1 SW Afr. Unique in fam. as capsule has septicidal dehiscence
Namaquanthus L. Bolus. Aizoaceae (V). 1 Cape
Namaquanula D. Mueller-Doblies & U. Mueller-Doblies. Amaryllidaceae (Liliaceae s.l.). 1 Namaqualand. R: BJ 107(1985)20
Namation Brand. Scrophulariaceae. 1 Mexico
Namibia (Schwantes) Schwantes (~ *Juttadinteria*). Aizoaceae (V). 1 SW Namibia
nam-nam *Cynometra cauliflora*
Nananthea DC. Compositae (Anth.-Anth.). 1 Corsica, Sardinia
Nananthus N.E. Br. Aizoaceae (V). 7 S Afr. Cult. orn. dwarf succ.
Nanarepenta Matuda. Dioscoreaceae. 1 Mexico
Nancy, early *Wurmbea dioica*
Nandi flame *Spathodea campanulata*
Nandina Thunb. Berberidaceae (I; Nandinaceae). 1 India to Japan: *N. domestica* Thunb. (heavenly bamboo) – much cult. orn. evergreen shrub in Japan, lvs 2- or 3-pinnate
Nandinaceae Horan. See Berberidaceae
Nani Adans. = *Xanthostemon*
Nanking cherry *Prunus tomentosa*
nanmu wood *Persea nanmu*
Nannoglottis Maxim. Compositae (Ast.-Ast.). 9 W China
Nannorrhops H. Wendl. Palmae (I 1c). 1 Arabia to Pakistan: *N. ritchieana* (Griffith) Aitch. – bushy hapaxanthic unarmed, ed. fr. & bud, cult. orn. with fibre used for rope, lvs for basketry etc.
Nannoseris Hedb. = *Dianthoseris*
Nannothelypteris Holttum = *Cyclosorus*

Nanochilus Schumann. Zingiberaceae. 1 Sumatra

Nanocnide Blume. Urticaceae (I). 2 E As.

Nanodea Banks ex Gaertner f. Santalaceae. 1 temp. S Am.

Nanodes Lindley = *Epidendrum*

Nanolirion Benth. = *Caesia*

Nanophyton Less. Chenopodiaceae (III 3). 3 Eur. to SW & C As. *N. erinaceum* (Pallas) Bunge – piperidine derivatives used in treatment of hypertension

Nanostelma Baillon = *Tylophora*

Nanothamnus Thomson. Compositae (Inul.). 1 Bombay

Nanuza L.B. Sm. & Ayensu. Velloziaceae. 1 Brazil

Napaea L. Malvaceae. 1 E to C US: *N. dioica* L. – cult. orn. with useful fibre from bark

Napeanthus Gardner. Gesneriaceae. 16 trop. Am. R: ABN 7(1958)340

Napier grass *Pennisetum purpureum*

Napoleonaea Pal. Lecythidaceae (Napoleonaeaceae). 8 trop. W Afr. R: BJBB 41(1971)363. Some locally medic. e.g. *N. vogelii* Hook. & Planchon – chewing-stick in Nigeria; *N. imperialis* Pal. – cult. orn.

Napoleonaeaceae A. Rich. = Lecythidaceae

nara or **narras** *Acanthosicyos horridus*

naranjilla *Solanum quitoense*

Naravelia Adans. (~ *Clematis*). Ranunculaceae (II 2). 7 Indomal. R: APG 37(1986)106. Leaflet tendrils

Narcissus L. Amaryllidaceae (Liliaceae s.l.). 27 Eur. (26), Medit. R: J.W. Blanchard (1990) *N.: a guide to wild daffodils*. Corona well developed and free of A. Daffodils. Some bee-poll. e.g. *N. pseudonarcissus*, some butterfly-poll. e.g. *N. jonquilla*, some both e.g. *N. triandrus*. Cult. orn. spp. & hybrids with many alks; imp. as cut-fls (5300 ha in UK, more than anywhere else) & in scent-making; many garden pls (10 000 cvs) of complex origin & frequently ± natur., there being a botanical classification for wild forms & a horticultural one based on fl. proportions e.g. Trumpet narcissi (Division 1, corona as long as or longer than P etc.), Large- (Division 2, corona ⅓ to as long as P) e.g. 'Carlton' the most commonly grown cv. (registered 1927 & now some 9450 million bulbs (c. 350 000 tons) & perhaps the biggest genetic 'individual' in the world, but cf. *Saccharum*) & Small-cupped (Division 3, less than ⅓ as long) narcissi derived from *N.* × *incomparabilis*, etc. Cult. spp. (signifying egotism in 'Language of Fls') incl. *N. asturiensis* (Jordan) Pugsley ('*N. minimus*', C & N Spain, N Portugal) – only 6–12 cm tall; *N. bulbocodium* L. (W Medit.) – v. variable even within wild populations, the corona the conspicuous part of the fl., P v. reduced, some tall swamp populations in S Portugal; *N. cyclamineus* DC (NW Spain & Portugal) – P strongly reflexed; *N.* × *incomparabilis* Miller (*N. pseudonarcissus* × *N. poeticus*) – many cvs (Hort. divisions 2 & 3); *N. jonquilla* L. (jonquil, SW Eur. natur. elsewhere) – lvs rush-like, fls v. fragrant used in scent-making; *N.* × *medioluteus* Miller (*N.* × *biflorus*, *N. poeticus* × *N. tazetta*) – natural hybrid (S France) usu. with 2 fls per infl., P white; *N. papyraceus* Ker-Gawler (paper white, W Medit.) – grown forced for winter fls, poss. a form of *N. tazetta*; *N. poeticus* L. (pheasant's-eye, France to Greece) – fragrant white fl. with v. short red-tipped corona, used in scent-making; *N. pseudonarcissus* L. (*N. obvallaris*, wild daffodil, Lent lily, Tenby d., Eur.) – 'wild' in parts of GB, parent of many hybrids & commonly natur.; *N. serotinus* L. (Medit.) – fls white & *N. viridiflorus* Schousboe (SW Spain, Morocco) – fls green, night-flowering, (?) moth-poll.; both autumn-flowering without lvs, the scapes photosynthetic & doubling in length after flowering; *N. tazetta* L. (Medit.) – fls 4–8, white with pale yellow corona, long cult., one sterile form early taken to China (? c. AD 700 via Iran), the 'Chinese sacred lily' (poss. back-cross with *N.* × *incomparabilis*) v. popular at New Year, 'Grand Soleil d'Or' yellow, scented cut-fl. in UK; *N. triandrus* L. (Angel's [after a tired collector, A. Gancedo] tears, Spain, Portugal) – fls nodding, white or corona yellowish, some pops tristylous

nard *Nardostachys grandiflora*; **n. grass** *Nardus stricta*

nardoo *Marsilea drummondii*

Nardophyllum (Hook. & Arn.) Hook. & Arn. Compositae (Ast.-Ast.). 7 S Andes. R: NMBA 17(1954)55

Nardosmia Cass. = *Petasites*

Nardostachys DC. Valerianaceae. 1 Himal.: *N. grandiflora* DC ('*N. jatamansi*', nard, spikenard) – rhizomes with oil form. prized in salves in Roman society, medic. & in scent-making etc. (cf. *Diplotaenia*). Close to *Patrinia*

Narduretia Villar = *Vulpia*

Narduroides Rouy. Gramineae (18). 1 Medit. incl. Eur.

Nardurus (Bluff, Nees & Schauer) Reichb. = *Vulpia*

Nardus L. Gramineae (14). 1 Eur. & W As.: *N. stricta* L. (mat grass, nard g.) on drier moors, poor grazing

Naregamia Wight & Arn. (~ *Turraea*). Meliaceae (I 1). 1 SW trop. Afr., 1 (closely allied) India: *N. alata* Wight & Arn. (Goa ipecacuanha) – locally medic. (alks)

Narenga Bor = *Saccharum*

Nargedia Beddome. Rubiaceae (II 5). 1 Sri Lanka: *N. macrocarpa* (Thw.) Bedd.

Naringi Adans. Rutaceae. 1 Indomal.: *N. crenulata* (Roxb.) Nicolson

narinjin *Citrus maxima*

Narrawa burr *Solanum cinereum*

Narthecium Hudson. Melanthiaceae (Liliaceae s.l.). 7 N temp. (Eur. 3). *N. ossifragum* (L.) Hudson (bog asphodel, Eur.) – iris-like pl. with yellow nectarless fls, a subs. for saffron in Shetland & C17 hair-dye in Lancashire

Narvalina Cass. Compositae (Helia.-Cor.). 4 trop. Am.

naseberry *Manilkara zapota*

nashi (pear) *Pyrus pyrifolia*

Nashia Millsp. Verbenaceae. 6 WI. Lvs used as a tea

Nassauvia Comm. ex Juss. Compositae (Mut.-Nass.). 39 S Andes. R: Darw. 24(1982)283. Shrubs

Nassella (Trin.) E. Desv. (~ *Stipa*). Gramineae (16). 79 warm & trop. Am. esp. Andes. R: T 39(1990)597. *N. trichotoma* (Nees) Arechav. (S Am.) – natur. trop. As. to Aus., noxious weed in NSW ('serrated tussock')

Nastanthus Miers = *Acarpha*

Nasturtiicarpa Gilli = *Calymmatium*

Nasturtiopsis Boiss. Cruciferae. 2 N Afr. to Arabia

nasturtium *Tropaeolum majus*

Nasturtium R. Br. = *Rorippa*

Nastus Juss. Gramineae (1b). 18 Madag., Réunion, Sumatra to Solomon Is. Tree-like & scramblers in montane forest

Natal grass *Melinis repens*; **N. mahogany** *Kiggelaria africana*; **N. plum** *Carissa grandiflora*

Nathaliella B. Fedtsch. Scrophulariaceae. 1 C As.

Natocrene (Torrey & A. Gray) Rydb. Portulacaceae. 2 W N Am.

Natsiatopsis Kurz. Icacinaceae. 1 Burma

Natsiatum Buch.-Ham. ex Arn. Icacinaceae. 1 E Himal. to SE As.

Nauclea L. Rubiaceae (I 2). 10 OW trop. R: Blumea 24(1978)325. *N. diderrichii* (de Wild.) Merr. (badi, bilinga, opepe, W Afr.) – timber resistant to borers & used for harbour work & yam mortars, medic. but the alk. a cumulative heart-poison; *N. orientalis* (L.) L. (*N. cordata*, kanluang, Leichhardt (pine), Indomal.), etc. – trade timber, fr. ed.

Naucleaceae (DC) Wernham = Rubiaceae

Naucleopsis Miq. Moraceae (III). 20–25 trop. Am. R: FN 7(1972)104. Arrow-poisons (cardenolides)

Naudinia Planchon & Linden. Rutaceae. 1 Colombia. (C)

Naufraga Constance & Cannon. Umbelliferae (I). 1 Majorca: *N. balearica* Constance & Cannon. Closest relations in NZ & Chile

Nauplius (Cass.) Cass. (~ *Asteriscus*). Compositae (Inul.). 8 Macaronesia, N Afr. R: NJB 7(1987)1. Cult. orn. esp. shrubby *N. sericeus* (L.f.) Cass. (*A. s.*, Canary Is.) to 1 m

Nautilocalyx Linden. Gesneriaceae. 50+ trop. Am. R: Selbyana 5(1978)29. Cult. orn.

Nautochilus Bremek. = *Orthosiphon*

Nautonia Decne. Asclepiadaceae (III 1). 1 S Brazil: *N. nummularia* Decne

Nautophylla Guillaumin = *Logania*

naval stores *Pinus* spp. (rosin & turpentine)

Navarretia Ruíz & Pavón. Polemoniaceae. 30 W N Am., Chile & Arg. (1). Annuals with various capsule dehiscence, some irreg., *N. filicaulis* (A. Gray) E. Greene (W N Am.) with both loculicidal & septicidal dehiscence. Cult. orn. incl. *N. squarrosa* (Eschsch.) Hook. & Arn. (skunkweed, W N Am., natur. GB) – foetid

navelwort *Umbilicus rupestris*; *Hydrocotyle* spp.

Navia Schultes f. Bromeliaceae (1). 106 N S Am. R: FN 14(1974)451, AMBG 73(1986)703. Wind-poll. Poss. link to Rapateaceae

Nayariophyton Paul (~ *Kydia*). Malvaceae. 1 E Himal. to Yunnan: *N. zizyphifolium* (Griffith) Long & A.G. Miller (*N. jujubifolium*). R: BJ 110(1988)43

nazingu *Mitragyna* spp.

Nealchornea Huber. Euphorbiaceae. 1 Upper Amazon: *N. yapurensis* Huber – fish-poison

Neamyza Tieghem = *Peraxilla*
Neanotis W. Lewis. Rubiaceae (IV 1). 28 trop. As. to Aus. R: AMBG 53(1966)32. Some smelling of faeces but lvs used as vegetable
Neanthe P. Browne = ? Leguminosae
Neanthe Cook = *Chamaedorea*
Neatostema I.M. Johnston. Boraginaceae. 1 Macaronesia to Medit. (inc. Eur.) & Iraq: *N. apulum* (L.) I.M. Johnston
Nebelia Necker ex Sweet (~ *Brunia*). Bruniaceae. 6 S Afr.
Neblinaea Maguire & Wurd. Compositae (Mut.-Mut.). 1 Venez., Guyana. Shrub
Neblinantha Maguire. Gentianaceae. 2 Guayana Highland. R: MNYBG 51(1989)29
Neblinanthera Wurd. Melastomataceae. 1 Venezuela
Neblinaria Maguire = *Bonnetia*
Neblinathamnus Steyerm. Rubiaceae (I 5). 3 Venezuela
Necepsia Prain. Euphorbiaceae. 3 trop. Afr., Madag. R: BJBB 56(1986)179
Nechamandra Planchon = *Lagarosiphon*
Neckia Korth. = *Sauvagesia*
Necramium Britton = *Sagraea*
Necranthus Gilli. Orobanchaceae (Scrophulariaceae s.l.). 1 Turkey
Nectandra Rolander ex Rottb. Lauraceae (I). 120 trop. Am. R: FN 60(1993). Some hard, heavy timbers (silverballi in Guyana) like *Ocotea; N. elaiophora* Barb.-Rodr. (Amaz.) – volatile oil a kerosene subs., also used in skin disease; *N. pichurim* (Kunth) Mez (pichurim or purchury bean, Brazil) – seeds medic. See also *Aniba*
nectarberry See *Rubus*
nectarine *Prunus persica* var. *nucipersica*
Nectaropetalum Engl. Erythroxylaceae. 6 trop. & S Afr.
Nectaroscordum Lindley (~ *Allium*). Alliaceae (Liliaceae s.l.). 2 S E Eur., W As. Differs from *A.* in repugnant smell & sharply keeled leaf-sheaths
Nectouxia Kunth. Solanaceae. (1). 1 Mex.: *N. formosa* Kunth. R: Solanaceae Newsl. 2,3(1984)15
Neea Ruíz & Pavón. Nyctaginaceae (VI). 83 S Florida to Bolivia. Alks incl. caffeine: *N. parviflora* Poeppig & Endl. (Peru) – dental preservative (blackening teeth), *N. theifera* Oersted (Brazil) – lvs used as tea (caparrosa), source of comm. black dye
Needhamiella L. Watson. Epacridaceae. 1 SW Aus.: *N. pumilio* (R. Br.) L. Watson
needle, Adam's *Yucca* spp. esp. *Y. filamentosa*; **n.bush** or **n. wood** *Hakea* spp.; **n. furze** or **whin** *Genista anglica*
neem *Azadirachta indica*
Neeopsis Lundell. Nyctaginaceae (VI). 1 Guatemala: *N. flavifolia* (Lundell) Lundell
neeps *Brassica rapa* Rapifera Group
Neeragrostis Bush = *Eragrostis*
Neesenbeckia Levyns. Cyperaceae. 1 S Afr.
Neesia Blume. Bombacaceae. 8 W Mal. Fr. with irritant hairs lost only when the skin is sloughed off. Light timbers. *N. altissima* (Blume) Blume – dried fr. hung above doors in Sumatra to ward off spirits
Neesiochloa Pilger. Gramineae (31d). 1 NE Brazil
Negria F. Muell. Gesneriaceae. 1 Lord Howe Is.: *N. rhabdothamnoides* F. Muell. – tree
Negripteridaceae Pichi-Serm. = Pteridaceae
Negripteris Pichi-Serm. = *Cheilanthes*
negrita *Sphaeralcea* spp.
negro pepper *Xylopia aethiopica*
Negundo Boehmer ex Ludwig = *Acer*
Neillia D. Don. Rosaceae. 11 E Himal. to China & W Mal. R: Adansonia 3(1963)141. Some cult. orn. shrubs like *Spiraea*
Neisosperma Raf. = *Ochrosia*
Nelia Schwantes. Aizoaceae (V). 1 Namaqualand
Nelmesia Veken. Cyperaceae. 1 trop. Afr.
Nelsia Schinz. Amaranthaceae (I 2). 2 S trop. & S Afr.
Nelsonia R. Br. Acanthaceae. 1 OW trop.: *N. canescens* (Lam.) Sprengel. R: Willdenowia 14(1984)397
Nelsonianthus H. Robinson & Brettell (~ *Senecio*). Compositae (Sen.-Tuss.). 1 Mex., 1 Guatemala – epiphytic shrub
Nelumbium Juss. = *Nelumbo*
Nelumbo Adans. Nelumbonaceae. 1 S & E N Am., WI to Colombia (*N. lutea*), 1 Lower Volga,

S & SE As. to trop. Aus. (*N. nucifera*), sometimes treated as subspp. R: BBP 68(1994)421. *N. lutea* (Willd.) Pers. (*N. pentapetala*, water chinquapin, American lotus) – fls yellow, seeds & rhizome ed.: *N. nucifera* Gaertner (*N. speciosa*, sacred lotus, Egyptian bean) – fls red, sacred in India, Tibet & China, being the 'padma' from which lotus motif of Asia derived, in Hindu religion considered to have sprung from navel of the god Vishnu & to have given birth to Brahma (sacred colour red) creator of the world, introd. to Egypt c. 500 BC but no longer in Nile, receptacle (often seen in Chinese food) heating & volatilizing odour attractive to pollinators (cf. Araceae), when dried a source (in China) of antihaemorrhagic quercetin; 'seeds' (fr.) viable for several hundred years in river-muds; much grown in E for rhizomes (source of Chinese arrowroot) which can grow to 20 m in 1 season, also cult. orn. with red- or white-flowered (incl. 'doubles' much used in temples) scented cvs. R: P. Swindells, *Waterlilies* (1983)153

Nelumbonaceae A. Rich. Dicots – Magnoliidae – Nelumbonales. 1/2 E As. & E N Am. Aquatic rhizomatous herbs with alks; articulated laticifers present, vascular bundles 'scattered', vessels only in roots. Lvs peltate, concave, usu. held above water on long petioles; phyllotaxy unique, leaf-primordia in 3s with 1 scale-leaf on underside of rhizome 1 on top next to 1 foliage-leaf, the lower orig. wrapped around term. bud but split on growth, the upper scale-leaf wrapped around base of petiole & a basal stipule; branches arise from axils of foliage lvs. Fl. solit., ebracteolate, from axil of upper scale-leaf, held above water, beetle-poll., bisexual; P c. 22–30, spirally arr., outer 2 K-like, rest in ± 2 series, the outer less consp. than inner, A c. 200–300, spirally arr. with slender filaments, 4 introrse-latrorse pollen-sacs on narrow laminar connective & term. thermogenic appendage, staminodes 0, pollen-grains 3-colpate, $\underline{G}$ 12–40 in 2–4 ± distinct whorls, individually sunken in large spongy receptacle, each with 1(2) ventral-apical, anatropous, bitegmic ovule(s). Fr. comprising separate hard-walled nuts loose in accrescent receptacle; seed 1 without perisperm & ± 0 endosperm, cotyledons 2 (arising as separate lobes of annular primordium) basally connate, forming a sheath around green plumule; radicle non-functional. 2n = 16

Genus: *Nelumbo*

Form. incl. in Nymphaeales but app. isolated, electron-microscopic work suggesting no affinity with N. or Magnoliales, but aporphin alks in cuticular waxes suggest alliance with Ranunculales

Seeds ed.

Nemacaulis Nutt. Polygonaceae (I 1). 1 SW N Am.: *N. denudata* Nutt.

Nemacladus Nutt. Campanulaceae. 10 SW N Am.

Nemaluma Baillon = *Pouteria*

Nemastylis Nutt. Iridaceae (III 4). 5 S US to C Am. Some cult. orn.

Nematanthus Schrader. Gesneriaceae. 26 S Am. R: Candollea 39(1984)297. Epiphytic shrubs, cult. orn. housepls (*Hypocyrta* spp.) with many cvs

Nematolepis Turcz. Rutaceae. 1 SW Aus.

Nematopoa C. Hubb. Gramineae (25). 1 Zimbabwe

Nematopteris Alderw. = *Scleroglossum*

Nematosciadium H. Wolff (~ *Arracacia*). Umbelliferae (III 5). 1 Mexico

Nematostemma Choux = *Cynanchum*

Nematostylis Hook.f. (~ *Alberta*). Rubiaceae (III 5). 1 Madag.: *N. anthophylla* (A. Rich.) Baillon. R: BJBB 54(1984)348

Nematuris Turcz. = *Ditassa*

Nemcia Domin (~ *Oxylobium*). Leguminosae (III 24). c. 40 Aus. esp. SW. R: Nuytsia 9(1993)223

Nemesia Vent. Scrophulariaceae. 65 trop. (few) & S Afr. Cult. orn. annuals derived from *N. strumosa* Benth. & *N. versicolor* E. Meyer ex Benth. (S Afr.), many cvs

Nemopanthus Raf. Aquifoliaceae. 2 E N Am. R: JAA 55(1974)436. Like *Ilex* but C free & K 0 or early lost. *N. mucronatus* (L.) Trel. cult. for orn. fr.

Nemophila Nutt. Hydrophyllaceae. 11 W & SE (2) N Am. R: UCPB 19(1941)341. Cult. orn. annuals esp. *N. menziesii* Hook. & Arn. (baby-blue eyes, Calif.) – fls blue with white centres

Nemosenecio (Kitam.) R. Nordenstam (~ *Senecio*). Compositae (Sen.- Tuss.). 6 China, Japan. R: KB 39(1984)262

Nemuaron Baillon. Monimiaceae (I 2; Atherospermataceae). 1 New Caledonia

Nemum Desv. (~ *Scirpus*). Cyperaceae. 5 trop. Afr.

Nenax Gaertner. Rubiaceae (IV 13). 9 S Afr.

Nenga H. Wendl. & Drude. Palmae (V 4l). 5 SE As. to Mal. R: Principes 27(1983)55

Nengella Becc. = *Gronophyllum*
Neoabbottia Britton & Rose = *Leptocereus*
Neoalsomitra Hutch. Cucurbitaceae. 12 Indomal. to Aus. & W Pacific
Neoancistrophyllum Rauschert = *Laccosperma*
Neoapaloxylon Rauschert (*Apaloxylon*). Leguminosae (I 4). 2 Madagascar
Neoastelia J. Williams (~ *Astelia*). Asteliaceae (Liliaceae s.l.). 1 NE NSW: *N. spectabilis* J. Williams. R: FA 45(1987)173
Neoathyrium Ching & Z.R. Wang = *Cornopteris*
Neoaulacolepis Rauschert = *Calamagrostis*
Neobaclea Hochr. Malvaceae. 2 temp. S Am.
Neobakeria Schltr. = *Massonia*
Neobalanocarpus Ashton. Dipterocarpaceae. 1 Peninsular Thailand, Malay Pen.: *N. heimii* (King) Ashton (chengal) – durable timber for boats & houses, hardest & finest in Malaysia
Neobartlettia R. King & H. Robinson = *Bartlettina*
Neobartlettia Schltr. Orchidaceae. 6 trop. S Am.
Neobassia A.J. Scott. Chenopodiaceae (I 5). 3 Aus. R: FA 4(1984)221
Neobathiea Schltr. Orchidaceae (V 16). 5 Madagascar. Cult. orn.
Neobaumannia Hutch. & Dalz. = *Knoxia*
Neobeguea J. Leroy. Meliaceae (IV 2). 3 Madag. R: Adansonia 16(1976)170
Neobenthamia Rolfe. Orchidaceae (V 13). 1 E Afr.: *N. gracilis* Rolfe – cult. orn. straggly
Neobertiera Wernham. Rubiaceae. 1 Guyana = ?
Neobesseya Britton & Rose = *Escobaria*
Neoblakea Standley. Rubiaceae (?III 3). 1 Venezuela
Neobolusia Schltr. (~ *Brachycorythis*). Orchidaceae (IV 2). 4 trop. E & S Afr.
Neobouteloua Gould (~ *Chondrosum*). Gramineae (33c). 1 Argentina & Chile
Neoboutonia Muell.-Arg. Euphorbiaceae. 3 trop. Afr. Gap colonists
Neobracea Britton. Apocynaceae. 8 Cuba, Bahamas
Neobreonia Ridsd. Rubiaceae (I 2). 1 Madagascar
Neobrittonia Hochr. Malvaceae. 1 Mexico
Neobuchia Urban. Bombacaceae. 1 Hispaniola
Neobuxbaumia Backeb. (~ *Carnegiea*). Cactaceae (III 8). 7 Mexico
Neobyrnesia J.A. Armstr. Rutaceae. 1 N Aus.
Neocabreria R. King & H. Robinson (~ *Eupatorium*). Compositae (Eup.-Crit.). 5 trop. S Am. R: MSMMBG 22(1987)371
Neocaldasia Cuatrec. = ? (Compositae)
Neocallitropsis Florin. Cupressaceae. 1 E New Caled.: *N. pancheri* (Carrière) Laubenf. – oil for scenting soaps etc. marketed as 'oil of araucaria'
Neocalyptrocalyx Hutch. = *Capparis*
Neocarya (DC) Prance ex F. White (~ *Parinari*). Chrysobalanaceae (2). 1 W Afr.: *N. macrophylla* (Sabine) Prance ex F. White (gingerbread plum) – fr. ed. (potential fr. tree in N Aus.), timber good. R: PTRSB 320(1988)105
Neocentema Schinz. Amaranthaceae (I 2). 1 Somalia, 1 Tanzania
Neochamaelea (Engl.) Erdtman (~ *Cneorum*). Cneoraceae. 1 Canary Is. Fls 4-merous & pollen diff. from *C.*
Neocheiropteris Christ. Polypodiaceae (II 4). 5–10 Indomal. to Japan
Neochevalierodendron Léonard. Leguminosae (I 4). 1 Gabon
Neocinnamomum H. Liu (~ *Cinnamomum*). Lauraceae (I). 6 S China to SE As.
Neoclemensia Carr. Orchidaceae (V 5). 1 Borneo: *N. spathulata* Carr – mycotroph coll. only once
Neocogniauxia Schltr. Orchidaceae (V 13). 2 WI
Neocollettia Hemsley. Leguminosae (III 9). 1 Burma, C Java. Geocarpic with strong roots on gynophore poss. pulling fr. down into earth
Neoconopodium (Kozo-Polj.) Pim. & Kljuykov. Umbelliferae (III 2). 1 India. R: FR 90(1981)373
Neocouma Pierre. Apocynaceae. 2 N S Am. R: AUWP 87-1(1987)
Neocracca Kuntze = *Coursetia*
Neocryptodiscus Hedge & Lamond = *Prangos*
Neocuatrecasia R. King & H. Robinson (~ *Eupatorium*). Compositae (Eup.-Gyp.). 9 C Andes. R: MSBMBG 22(1987)108
Neocussonia Hutch. = *Schefflera*
Neodeutzia Small = *Deutzia*
Neodielsia Harms = ? *Astragalus*

Neodissochaeta Bakh.f. = *Dissochaeta*

Neodistemon Babu & A.N. Henry. Urticaceae (III). 1 Indomal.

Neodonnellia Rose = *Tripogandra*

Neodregea C. Wright. Colchicaceae (Liliaceae). 1 S Afr.

Neodriessenia Nayar. Melastomataceae. 6 Borneo. R: BJ 106(1985)1

Neodryas Reichb.f. Orchidaceae (V 10). 6 trop. S Am.

Neodunnia R. Viguier = *Millettia*

Neodypsis Baillon. Palmae (V 4e). 14 Madag. Like *Chrysalidocarpus* but seeds with ruminate endosperm; lvs twisted so as to appear in 3 ranks

Neoeplingia Ramam., Hiriart & Medrano. Labiatae (VIII 2). 1 Mex.: *N. leucophylloides* Ramam., Hiriart & Medrano. R: BSBM 43(1982)61

Neoescobaria Garay = *Helcia*

Neofabricia J. Thompson (~ *Leptospermum*). Myrtaceae. 3 N Queensland. R: Telopea 3(1989)291

Neofinetia Hu (~ *Holcoglossum*). Orchidaceae (V 16). 1 E As.: *N. falcata* (Thunb.) Hu – cult. orn.

Neofranciella Guillaumin. Rubiaceae (II 1). 1 New Caledonia

Neogaerrhinum Rothm. Scrophulariaceae. 2 SW N Am. R: D.A. Sutton, Rev. Antirrh. (1988) 479

Neogaillonia Lincz. (*Gaillonia*). Rubiaceae (IV 12). 17 NW Afr. to C As. R: NS 10(1973)226

Neogardneria Schltr. ex Garay. Orchidaceae (V 10). 1 NE trop. Am.

Neoglaziovia Mez. Bromeliaceae (3). 2 E Brazil. R: FN 14(1979)2036. *N. variegata* (Arruda) Mez – source of caroa fibre used for nets & suggested for paper & artificial silk

Neogleasonia Maguire = *Bonnetia*

Neogoezia Hemsley. Umbelliferae (III 5). 5 Mex. R: OB 92(1987)59

Neogontscharovia Lincz. (~ *Acantholimon*). Plumbaginaceae (II). 3 C As., Afghanistan

Neogoodenia C. Gardner & A.S. George = *Goodenia*

Neoguillauminia Croizat. Euphorbiaceae. 1 New Caledonia

Neogunnia Pax & K. Hoffm. = *Gunniopsis*

Neogyna Reichb.f. (~ *Coelogyne*). Orchidaceae (V 12). 1 N India & SW China to SE As.: *N. gardneriana* (Lindley) Reichb.f. R: OB 89(1980)75

Neohallia Hemsley (~ *Justicia*). Acanthaceae. 1 S Mexico

Neoharmsia R. Viguier. Leguminosae (III 2). 2 W & NW Madagascar

Neohemsleya Penn. Sapotaceae (III). 1 Tanz.: *N. usambarensis* Penn. R: T.D. Pennington, *S.* (1991) 175

Neohenricia L. Bolus. Aizoaceae (V). 1 S Afr.

Neohenrya Hemsley = *Tylophora*

Neohintonia R. King & H. Robinson = *Koanophyllum*

Neoholstia Rauschert (*Holstia*). Euphorbiaceae. 1 trop. Afr.

Neohouzeaua A. Camus = *Schizostachyum*

Neohuberia Ledoux = *Eschweilera*

Neohumbertiella Hochr. = *Humbertiella*

Neohusnotia A. Camus = *Acroceras*

Neohymenopogon Bennet (*Hymenopogon*). Rubiaceae (I 1/IV 24). 3 Himal. R: IF 107(1981)436

Neohyptis J.K. Morton. Labiatae (VIII 4c). 1 Angola

Neojatropha Pax = *Mildbraedia*

Neojeffreya Cabrera (~ *Conyza*). Compositae (Pluch.). 1 Afr., Madagascar

Neojobertia Baillon. Bignoniaceae. 1 NE Brazil

Neokoehleria Schltr. Orchidaceae (V 10). 7 Peru

Neolabatia Aubrév. = *Pouteria*

Neolamarckia Bosser ('*Anthocephalus*'). Rubiaceae (I 2). 2 Indomal. to trop. Aus. R: BMNHN 4,6(1984)247. Fast-growing colonist trees. *N. cadamba* (Roxb.) Bosser ('*Anthocephalus chinensis*', kadam, kedam, laran) – first planted Indonesia 1933 but growth sensitive to soil moisture, timber used for disposable chopsticks in Malaysia, matchboxes, tea-chests & considered for pulp

Neolauchea Kraenzlin = *Isabelia*

Neolaugeria Nicolson. Rubiaceae (III 3). 5 WI. R: Brittonia 31(1979)119

Neolehmannia Kraenzlin = *Epidendrum*

Neolemonniera Heine. Sapotaceae (I). 5 W trop. Afr.

Neolepisorus Ching = *Neocheiropteris*

Neoleptopyrum Hutch. = *Leptopyrum*

Neoleroya Cavaco = ? *Pyrostria*
Neolindenia Baillon = *Louteridium*
Neolitsea Merr. Lauraceae (II). 100 Indomal. to E As. & Aus. Alks. Some medic. & oil-sources esp. *N. sericea* (Blume) Koidz. (*Litsea glauca*, China & Japan) – oil for burning & soap; *N. zeylanica* (Nees) Merr. (Indomal.) – fr. used as 'peas' in peashooters by Malay boys
Neolloydia Britton & Rose. Cactaceae (III 9). 14 Texas & Mex. R: Bradleya 4(1986)1. Cult. orn.
Neolophocarpus Camus = *Schoenus*
Neolourya L. Rodrigues = *Peliosanthes*
Neoluederitzia Schinz. Zygophyllaceae. 1 SW Afr.
Neoluffa Chakrav. = *Siraitia*
Neomacfadya Baillon = *Arrabidaea*
Neomandonia Hutch. = *Tradescantia*
Neomangenotia J. Leroy = *Commiphora*
Neomarica Sprague (~ *Trimezia*). Iridaceae (III 3). 12 trop. Am. Cult. orn. herbs with fugacious fls & orange seeds (usu. brown in I.), sometimes viviparous with plantlets in old infls
Neomartinella Pilger. Cruciferae. 1 China
Neomazaea Krug & Urban = *Mazaea*
Neomezia Votsch = *Deherainia*
Neomicrocalamus Keng f. (~ *Racemobambos*). Gramineae (1b). 4 Mal. R: EJB 51(1994)324
Neomillspaughia S.F. Blake (~ *Podopterus*). Polygonaceae (II 2). 2 C Am.
Neomirandea R. King & H. Robinson (~ *Eupatorium*). Compositae (Eup.-Heb.). 27 Mex. to Ecuador. R: MSBMBG 22(1987)415. Small trees, shrubs & perennial herbs, often epiphytes
Neomitranthes Legrand = *Calyptrogenia*
Neomolinia Honda & Sakis. = *Diarrhena*
Neomoorea Rolfe. Orchidaceae (V 10). 1 Panamá to Ecuador: *N. wallisii* (Reichb.f.) Schltr. – cult. orn. epiphyte
Neomortonia Wiehler. Gesneriaceae. 3 C Am. R: Selbyana 6(1982)194
Neomuellera Briq. = *Plectranthus*
Neomyrtus Burret. Myrtaceae (Myrt.). 1 NZ
Neonauclea Merr. Rubiaceae (I 2). 65 Indomal. to China (1: *N. tsaiana* S.Q. Zow – timber). R: Blumea 34(1989)177. Some rheophytes, some myrmecophytes (W of New Guinea)
Neonelsonia J. Coulter & Rose. Umbelliferae (III 5). 2 Mex. to N S Am.
Neonicholsonia Dammer (~ *Prestoea*). Palmae (V 4f). 1 C Am.
Neonotonia Lackey = *Glycine*
Neopalissya Pax = *Necepsia*
Neopallasia Polj. (~ *Artemisia*). Compositae (Anth.-Art.). 3 C As. to China. R: APS 18(1980)86
Neopanax Allan = *Pseudopanax*
Neoparrya Mathias. Umbelliferae (III 8). 2 SW US
Neopatersonia Schönl. Hyacinthaceae (Liliaceae s.l.). 3 S & SW Afr.
Neopaulia Pim. & Kljuykov = *Paulita*
Neopaxia O. Nilsson (~ *Montia*). Portulacaceae. 1 mts SE As. to NZ
Neopentanisia Verdc. Rubiaceae (III 8). 2 S trop. Afr. R: BJ 110(1989)548
Neopetalonema Brenan = *Gravesia*
Neophloga Baillon. Palmae (V 4e). 30 Madagascar
Neopicrorhiza Hong (~ *Picrorhiza*). Scrophulariaceae. 1 E Himal.: *N. scrophulariifolia* (Pennell) Hong – locally medic. esp. febrifuge
Neopilea Leandri = *Pilea*
Neoplatytaenia Geld. = *Semenovia*
Neopometia Aubrév. = *Pradosia*
Neoporteria Britton & Rose. Cactaceae (III 5). 25 S Peru, Chile, W Arg. Cult. orn. globose ribbed cacti
Neopreissia Ulbr. = *Atriplex*
Neopringlea S. Watson. Flacourtiaceae (5). 3 Mex., Guatemala. R: SB 8(1983)430
Neoptychocarpus Buchheim. Flacourtiaceae (9). 3 trop. S Am.
Neoraimondia Britton & Rose. Cactaceae (III 7). 2 Peru, N Chile, Bolivia. Cult. orn. columnar cacti
Neorapinia Mold. = *Vitex*
Neoraputia Emmerich = *Raputia*
Neorautanenia Schinz. Leguminosae (III 10). 3 S trop. Afr. *N. brachypus* (Harms) C.A. Sm.

(*Dolichos seineri*) – giant swollen caudex to 150 kg; *N. mitis* (A. Rich.) Verdc. alleged to kill bilharzia-carrying snails

Neoregelia L.B. Sm. Bromeliaceae (3). 78 trop. & warm S Am. R: FN 14(1979)1533. Infls sunk in centre of rosette with brightly coloured inner lvs. Many cult. orn. spp. & hybrids esp. *N. farinosa* (Ule) L.B. Sm. (Brazil) with crimson inner lvs

Neoregnellia Urban. Sterculiaceae. 1 Cuba, Hispaniola

Neorhine Schwantes = *Rhinephyllum*

Neorites L.S. Sm. Proteaceae. 1 NE Aus.

Neoroepera Muell. Arg. & F. Muell. Euphorbiaceae. 2 NE Aus.

Neorosea Hallé = *Tricalysia*

Neorudolphia Britton. Leguminosae (III 10). 1 WI. Close to *Rhodopis*

Neosabicea Wernham = *Manettia*

Neoschimpera Hemsley = *Amaracarpus*

Neoschischkinia Tzvelev = *Agrostis*

Neoschroetera Briq. = *Larrea*

Neoschumannia Schltr. Asclepiadaceae (III 5). 1 trop. W Afr.: *N. kamarunensis* Schltr.

Neosciadium Domin. Umbelliferae (I 1a). 1 SW Aus.: *N. glochidiatum* (Benth.) Domin. R: OBZ 115(1968)28

Neoscortechinia Pax. Euphorbiaceae. 6 Burma, Nicobar Is., Mal., Solomon Is. R: Blumea 39(1994)301

Neosepicaea Diels. Bignoniaceae. 4 Moluccas, New Guinea, Queensland

Neosinacalamus Keng f. = *Dendrocalamus*

Neosloetiopsis Engl. = *Streblus*

Neosparton Griseb. Verbenaceae. 4 temp. S Am.

Neosprucea Sleumer = *Hasseltia*

Neostapfia Davy (~ *Anthocloa*). Gramineae (30). 1 Calif. Spring pools

Neostapfiella A. Camus. Gramineae (33b). 3 Madag. R: NS 11(1943)189

Neostenanthera Exell. Annonaceae. 5 trop. Afr., S Am.

Neostrearia L.S. Sm. Hamamelidaceae (I 1). 1 NE Aus.: *N. fleckeri* L.S. Sm.

Neostricklandia Rauschert = *Phaedranassa*

Neotainiopsis Bennet & Raiz. = *Eriodes*

Neotatea Maguire = *Bonnetia*

Neotchihatchewia Rauschert (*Tchihatchewia*). Cruciferae. 1 Armenia: *N. isatidea* (Boissier) Rauschert – cult. orn.

Neotessmannia Burret. Tiliaceae. 1 Peru

Neothorelia Gagnepain. Capparidaceae. 1 SE As.

Neothymopsis Britton & Millsp. = *Thymopsis*

Neotina Capuron. Sapindaceae. 2 Madagascar

Neotinea Reichb.f. Orchidaceae ((IV 2). 2 Macaronesia to W Eur. & Medit. incl. GB (1: *N. maculata* (Desf.) Stearn (*N. intacta*))

Neotorularia Hedge & Léonard (*Torularia*). Cruciferae. 12 Medit. (Eur. 2) to C As. & Afghanistan

Neotreleasea Rose = *Tradescantia*

Neotrewia Pax & K. Hoffm. Euphorbiaceae. 1 C Mal.

Neottia Guett. Orchidaceae (V 1). 9 temp. Euras. (Eur. 1: *N. nidus-avis* (L.) Rich., bird's-nest orchid) – mycotrophs

Neottianthe (Reichb.) Schltr. Orchidaceae (IV 2). 7 temp. Euras. (Eur. 1), Himalaya

Neotuerckheimia J.D. Sm. = *Amphitecna*

Neoturczaninowia Polj. Umbelliferae. Spp.? S Am. (= ?)

Neotysonia Dalla Torre & Harms. Compositae (Gnap.-Ang.). 1 SW Aus. (Mt Narryer): *N. phyllostegia* (F. Muell.) P.G. Wilson. R: OB 104(1991)114

Neo-urbania Fawcett & Rendle = *Maxillaria*

Neo-uvaria Airy Shaw. Annonaceae. 2 W Mal.

Neoveitchia Becc. Palmae (V 4m). 1 Fiji: *N. storckii* (H.A. Wendl.) Becc. – almost extinct

Neowawraea Rock = *Flueggea*

Neowerdermannia Frič (~ *Gymnocalycium*). Cactaceae (III 5). 2 S Am. Cult. orn.

Neowilliamsia Garay = *Epidendrum*

Neowimmeria Degener & I. Degener = *Lobelia*

Neowollastonia Wernham = *Melodinus*

Neowormia Hutch. & Summerh. = *Dillenia*

Neoxythece Aubrév. & Pellegrin = *Pouteria*

Neozenkerina Mildbr. = *Staurogyne*

Nepa Webb = *Stauracanthus*

Nepal paper *Daphne bholua*

Nepenthaceae Bercht. & J. Presl. Dicots – Dilleniidae – Nepenthales. 1/82 Madag. & Seychelles to Aus. & New Caled. Shrubby dioec. insectivorous pls, often climbing &/or epiphytic; cortical &/or medullary vasc. bundles often present. Lvs spirally arr., comprising ± distinct winged petiole, strap-shaped lamina & term. tendril by which pl. climbs, its apex usu. in form of a pitcher with a lid projecting over mouth; multicellular nectar-glands & peltate hydathodes on stem & lvs with digestive glands in pitchers partly filled with digestive fluid; stipules 0. Fls small, reg., in racemes or thyrses; K (3)4, imbricate & usu. free, with nectar glands within, C 0, A (4–)8–25, the filaments united into central column, pollen grains in tetrads, G ((3)4) with as many loc. & ± sessile stigma & ∞ anatropous, bitegmic ovules in many rows on axile placentas. Fr. a loculicidal capsule; seeds ∞ filiform with straight embryo in starchy as well as oily & proteinaceous endosperm

Genus: *Nepenthes*

Nepenthes L. Nepenthaceae. 82 Madag. (2), Seychelles (1), Sri Lanka (1), Assam (1) to N Queensland & New Caled. (Borneo c. 28; pollen from Eur. Tertiary). Pitcher plants; sea level (some ±halophytic to 3500 m (Mt Kinabalu, Borneo), up to 25 spp. at 1 locality in Mal., 1 Mal. sp. myrmecophytic. Pitchers develop as intercalary invaginations below the tendril tips, the lips growing out below them, the basal pitchers capturing creeping arthropods, the later-formed upper ones (s.t. diff.-shaped) flying insects. Arthropods attracted by the nectar, amines & bright pitcher colour, work down pitcher on to slippery cuticles (several thousand wax scales per cell like tiles snapping off stalks when touched) & slide into the base of pitcher where they decompose & their nutrients absorbed, 10 pitchers found to contain remains of 1994 arthropods (150 spp.). Some pitchers to 35 cm long & 18 cm across (*N. rajah* Hook.f., Mt Kinabalu) large enough to hold 2 litres of fluid (potable & ? medic. before pitcher opens), as well as fungi, slime-moulds, protozoans, desmids, diatoms, rotifers, oligochaetes, crustaceans, larvae of mosquitoes & flies, even tadpoles; above fluid level spiders catching falling prey. Some stems locally used for twine etc.; many cult. orn. (incl. complex hybrids raised in C19)

Nepeta L. Labiatae (VIII 2). c. 250 temp. Euras. (Eur. 24), N Afr., trop. Afr. mts. Usu. dry habitats; gynodioecy. Catmint, some spp. irresistible to cats. Rabbit-proof cult. orn. esp. *N. cataria* L. (catnip, catnep, SW & C As.) – lvs medic. & used as a tea, also psychedelic effects like cannabis ('cataria'), nepetalactone (an iridoid sequestered by *Romalea guttata* grasshoppers) sensitizing genetically inherited response in Felidae so that cats behave as with cannabis (cf. *Actinidia*, *Boschniakia*, *Menyanthes*, *Teucrium*, *Valeriana*) but similar activity found in urine of tomcats so plant prod. poss. mimicking a pheromone assoc. with courtship behaviour; *N.* × *faassenii* Bergmans ex Stearn (*N. racemosa* Lam. (*N. mussinii*, Caucasus, Iran) × *N. nepetella* L. (W Med.)) – common catmint of gardens, sterile pl. often confused with *N. racemosa*

Nephelaphyllum Blume. Orchidaceae (V 11). 17 Indomal.

Nephelea Tryon = *Cyathea*

Nephelium L. Sapindaceae. 22 Indomal. Some fruit trees esp. *N. lappaceum* L. (rambutan, Mal.) – apomictic, wild seeds only germ. after passage through monkey, pericarp form. source of black dye & *N. ramboutan-ake* (Labill.) Leenh. (*N. mutabile*, pulasan, W Mal.) – many cvs, pulp (sarcotesta) ed.

Nephelochloa Boiss. (~ *Eremopoa*). Gramineae (17). 1 Turkey

Nephopteris Lellinger. Pteridaceae (III; Adiantaceae s.l.). 1 C Colombia: *N. maxonii* Lellinger – at 3000 m. R: AFJ 56(1966)180

Nephradenia Decne. Asclepiadaceae (III 4). 10 trop. Am.

Nephrangis (Schltr.) Summerh. Orchidaceae (V 16). 1 trop. Afr.

Nephrocarpus Dammer = *Basselinia*

Nephrocarya Candargy = *Nonea*

Nephrodesmus Schindler (~ *Arthroclianthus*). Leguminosae (III 9). 6 New Caledonia

Nephrodium Michaux = *Dryopteris*

Nephrolepidaceae Pichi-Serm. (~Davalliaceae). Filicopsida. 1/30 trop. & warm. Terr. to epiphytic ferns (some annuals) with ± erect dictyostelic rhiz. & runners at first protostelic, s.t. with tubers. Frond pinnate; vein ends with hydathodes. Sori term. on vein-branches, usu. ± round; indusium usu. reniform to suborbicular-subpeltate. n = 41

Genus: *Nephrolepis*

Nephrolepis Schott. Nephrolepidaceae (Davalliaceae s.l.). 30 trop. & warm. Cult. orn. (sword ferns) esp. *N. cordifolia* (L.) C. Presl (trop.) – widely natur., *N. exaltata* (L.) Schott (trop.) – many cvs incl. 'Bostoniensis' (Boston fern), *N. falcata* (Cav.) C. Chr. (Indomal.) esp. 'Furcans' – pinnae bifurcate & sometimes even subdivided, cult. Pacific for leis as in Hawaii, *N. hirsutula* (Forst.f.) C. Presl (Indomal., Pacific) – dominant in sec. grasslands in e.g. Niue

Nephromeria (Benth.) Schindler = *Desmodium*

Nephropetalum Robinson & Greenman = *Ayenia*

Nephrophyllidium Gilg (*Fauria*). Menyanthaceae. 1 N Japan, NW N Am.: *N. crista-galli* (Hook.) Gilg – cult. orn. bog pl.

Nephrophyllum A. Rich. Convolvulaceae. 1 Ethiopia

Nephrosperma Balf.f. Palmae (V 4n). 1 Seychelles: *N. vanhoutteanum* (Van Houtte) Balf.f. – cult. orn.

Nephthytis Schott. Araceae (V 6). 1 Borneo, 7 trop. Afr.(!)

Nepsera Naudin. Melastomataceae. 1 trop. Am.

Neptunia Lour. Leguminosae (II 3). 11 trop. & warm esp. Aus. & Am. R: AusJB 14(1966)379. Sensitive lvs like *Mimosa*, those of *N. plena* (L.) Benth. (trop. & warm Am.) with living wood-fibres in pulvinus; *N. prostrata* (Lam.) Baillon (*N. aquatica*, *N. oleracea*, trop.) – floating stems with aerenchyma, locally eaten as potherb

neram *Dipterocarpus oblongifolius*

Neraudia Gaudich. Urticaceae (III). 5 Hawaii

Neriacanthus Benth. Acanthaceae. 4 trop. Am.

Nerine Herbert. Amaryllidaceae (Liliaceae s.l.). 22 S Afr. R: Plant Life 23(1967)suppl. Cult. orn. bulbs with many alks, one group flowering whilst lvs photosynthesizing, other when they have withered. *N. bowdenii* Will. Watson – fls after lvs die, hardiest sp.; *N. sarniensis* (L.) Herbert (Guernsey lily) – described from Guernsey & alleged to have arrived in ballast of ship from Japan but prob. presented to islanders by shipwrecked sailors from Cape in 1650s; many hybrids & a N. Society

Nerisyrenia E. Greene. Cruciferae. 11 S N Am. R: Rhodora 80(1978)159, Phytol. 75(1993)231

Nerium L. Apocynaceae. 1 Medit. (incl. Eur.) to Cape Verde Is. & Japan: *N. oleander* L. (oleander, rose-bay) – rheophyte with bright but nectarless fls poll. by deceit, seeds water-disp., cult. orn., v. toxic due to cardiac glycosides incl. oleandrin (rat poison but gazelles & hyrax immune), in humans 1 leaf a potentially lethal dose, death within 24 hrs

Nernstia Urban. Rubiaceae (I 7). 1 Mexico

neroli *Citrus × aurantium* Sour Orange Group

Nerophila Naudin. Melastomataceae. 1 trop. W Afr.

Nertera Banks & Sol. ex Gaertner. Rubiaceae (IV 13). 15 S China to Java, Aus., NZ, Society Is., Hawaii, temp. S Am., Tristan da Cunha. *N. granadensis* (L.f.) Druce (*N. depressa*, bead-plant, Aus., NZ, S Am.) – cult. orn. ground cover or pot-pl. with orange bead-like fr.

Nervilia Comm. ex Gaudich. Orchidaceae (V 6). 65 trop. & warm OW (Arabia, Afr., Madag. 16). Some cult. orn. terr.

Nesaea Comm. ex Kunth. Lythraceae. 56 trop. & S. Afr. (27, R: Bothalia 21(1991)35), S India, Sri Lanka, Aus., S Am.

Nesiota Hook.f. Rhamnaceae. 1 St Helena: *N. elliptica* (Roxb.) Hook.f. – 1 tree found alive in 1978 (12 in 1875), allies in Afr., Masc.

Neslia Desv. Cruciferae. 2 SE Eur. (1: *N. paniculata* (L.) Desv. (ball mustard)), Medit., SW As. R: FR 64(1961)11

Nesocaryum I.M. Johnston. Boraginaceae. 1 Desventuradas Is. (Chile)

Nesocodon Thulin. Campanulaceae. 1 Mauritius

Nesogenaceae Marais (~ Verbenaceae). Dicots – Asteridae – Lamiales. 1/8 E Afr. to Pacific. Herbs with simple & multicelluar unbranched hairs and glandular hairs. Lvs opp., simple; stipules 0. Infl. axillary, of 1 or more superposed cymes, each 1–several-flowered or in bracteate spikes. K (4)5-lobed, accrescent; C ((4)5), lobes imbricate; A 2 long, 2 short; G, 2-loc., each loc. with 1 axile erect ovule. Fr. indehiscent with crustaceous pericarp; seeds with endosperm. R: KB 35(1981)799

Genus: *Nesogenes*

Nesogenes A. DC. Nesogenaceae (Verbenaceae s.l.). 8 Tanz., Madag., Indian & Pacific Oceans. R: KB 35(1981)799, 38(1983)37

Nesogordonia Baillon. Sterculiaceae. 18 trop. Afr., Madag. *N. papaverifera* (A. Chév.) Capuron (*Cistanthera p.*, danta, kotibé, W Afr.) – hard reddish timber used for veneers, plywood & flooring

Nesohedyotis (Hook.f.) Bremek. Rubiaceae (IV 1). 1 St Helena. Allies in Indomal.

Nesoluma Baillon. Sapotaceae (III). 3 Polynesia

Nesomia B. Turner (~ *Ferreyrella*). Compositae (Eup.-Ayap.). 1 Mex.: *N. chiapensis* B. Turner.

R: Phytol. 7(1991)208

Nesopteris Copel. = *Cephalomanes*

Nesothamnus Rydb. = *Perityle*

Nesphostylis Verdc. Leguminosae (III 10). 1 trop. Afr., 1 Burma

Nestegis Raf. Oleaceae. 5 Norfolk Is. & NZ (1), NZ (3), Hawaii (1). Some timber (maire); *N. sandwicensis* (A. Gray) Degener, I. Degener & L. Johnson (Hawaii) – form. imp. for tool-handles & fuel

Nestlera Sprengel (~ *Relhania*). Compositae (Gnap.-Rel.). 1 S Afr.

Nestoria Urban = *Pleonotoma*

Nestronia Raf. = *Buckleya*

nettle *Urtica* spp. & other pls with stinging hairs; **bull n.** *Cnidoscolus stimulosus;* (**common**) **stinging n.** *U. dioica;* **dead n.** *Lamium* spp.; **dog n.** *U. urens;* **false n.** *Boehmeria* spp.; **giant n.** *Dendrocnide* spp.; **hedge n.** *Stachys* spp.; **hemp n.** *Galeopsis* spp.; **Nilgiri n.** *Girardinia diversifolia;* **Roman n.** *U. pilulifera;* **small n.** *U. urens;* **n. tree** *Celtis* spp. esp. *C. australis;* **wood n.** *Laportea canadensis*

Nettoa Baillon = *Grewia*

Neuburgia Blume. Strychnaceae (Loganiaceae s.l.). 10–12 E Mal. to Pacific. Fr. a drupe

Neumanniaceae 'Tieghem' = Flacourtiaceae

Neuontobotrys O. Schulz. Cruciferae. 2 Chile, Argentina

Neuracanthus Nees. Acanthaceae. 20 trop. Afr. & Arabia to India

Neurachne R. Br. Gramineae (34a). 6 Aus. R: CQH 13(1972)

Neuractis Cass. = *Glossocardia*

Neurada L. Neuradaceae. 1 E Medit. to Indian desert: *N. procumbens* L.

Neuradaceae Link (Grielaceae; ~ Rosaceae). Dicots – Dilleniidae – ? Malvales. 3/9 Afr. deserts & Medit. to Indian desert. Annual prostrate tomentose herbs with mucilage-ducts in pith. Lvs toothed to pinnatifid, spirally arr., some with stipules. Fls bisexual, reg., solit. on axillary pedicels, ± epigynous; K 5, valvate, C 5 imbricate or convolute, A 10 on hypanthium with unique bipolar pollen with 3(4) pores at each end, G (10, ? derived from 5), ± inferior, pluriloc. but 2–4 ± reduced or ovules not maturing, with distinct styles & 1 apical-axile, pendulous, anatropous, bitegmic ovule. Fr. indehiscent, dry; seeds without endosperm. n = 6

Genera: *Grielum, Neurada, Neuradopsis*

Close to Rosaceae but distinct by pollen & G

Neuradopsis Bremek. & Oberm. Neuradaceae. 3 SW Afr.

Neurocallis auctt. = *Nevrocallis*

Neurocalyx Hook. Rubiaceae (IV 4). 5 Sri Lanka, 1 ext. to S India

Neurodium Fée (*Paltonium*). Polypodiaceae (II 5). 1 C Am.: *N. lanceolatum* (L.) Fée

Neurolaena R. Br. Compositae (Helia.-Mel.). 13 trop. Am. R: PSE 140(1982)119, Phytologia 58(1985)497

Neurolakis Mattf. Compositae (Vern.-Vern.). 1 Cameroun

Neurolepis Meissner. Gramineae (1a). 9 trop. Am., high alts. R: SCB 9(1973)97. Lvs to 5 m long, 30 cm wide. *N. aristata* (Munro) A. Hitchc., bamboo, to snowline in Andes, at 4500 m forming impenetrable 'carizal'; *N. pittieri* McClure – mass-flowering on 5 yr cycle

Neurolobium Baillon = *Diplorhynchus*

Neuromanes Trevis. = *Trichomanes*

Neuropeltis Wallich. Convolvulaceae. 4 trop. As., 9 trop. Afr. Epiphyllous infls

Neuropeltopsis Oostr. Convolvulaceae. 1 Borneo

Neurophyllodes (A. Gray) Degener = *Geranium*

Neuropoa Clayton (~ *Poa*). Gramineae (17). 1 Aus.: *N. fax* (Willis & Court) Clayton – annual

Neurosoria Mett. ex Kuhn = *Cheilanthes*

Neurotheca Salisb. ex Benth. Gentianaceae. 1 S Am., trop. Afr.: *N. loeselioides* Oliver

Neustruevia Juzep. = *Pseudomarrubium*

neutral henna *Ziziphus jujuba*

Neuwiedia Blume. Orchidaceae (I; Apostasiaceae). 8 Mal.

Neves-armondia Schumann = *Pithecoctenium*

Nevillea Esterh. & Linder (~ *Restio*). Restionaceae. 2 SW Cape. R: Bothalia 15(1985)482

Neviusia A. Gray. Rosaceae. 1 Alabama, 1 N Calif. (fossils in Canadian Eocene). R: Novon 2(1992)285. *N. alabamensis* A. Gray (cliff-faces above Black Warrior River, Alabama) – cult. orn. decid. shrub

Nevrocallis Fée (*Neurocallis*). Pteridaceae (VI). 1 trop. Am.: *N. praestantissima* Fée

Nevrodium Fée = *Neurodium*

Nevskiella Krecz. & Vved. = *Bromus*

New Guinea teak *Vitex cofassus*
New Zealand flax *Phormium tenax*
Newbouldia Seemann ex Bureau. Bignoniaceae. 1 trop. W Afr.: *N. laevis* (Pal.) Bureau – cult. orn. used as living fence, medic.
Newcastelia F. Muell. Labiatae (III; Chloanthaceae). 12 trop. Aus. R: Brunonia 1(1978)589
Newtonia Baillon. Leguminosae (II 3). 14 trop. Afr. (Am. spp. referable to new genus). R: BJBB 60(1990)119
Neyraudia Hook.f. Gramineae (31d). 2 OW trop. R: JAA 71(1990)164
ngai (camphor) *Blumea balsamifera*
ngaio *Myoporum laetum*
ngali nut *Canarium indicum*
Nialel Adans. = *Aglaia*
niangon *Heritiera utilis*
niaouli oil *Melaleuca quinquenervia*
Nicandra Adans. Solanceae. 1 Peru: *N. physalodes* (L.) Scop. – cult. orn. annual alleged to keep off flies (shoo-fly), with alks, weedy & natur. in Euras., trop. Am. & US; some seeds lack 1 chromosome & these can remain dormant for 30 yrs
Nicaragua wood *Caesalpinia, Dalbergia* & *Haematoxylum* spp.
Nichallea Bridson. Rubiaceae (II 2). 1 trop. Afr.
nicker bean *Caesalpinia bonduc, Entada gigas*
Nicobariodendron Vasudeva & Chakrab. Celastraceae. 1 Nicobar Is. R: JETB 7(1985)513
Nicodemia Ten. = *Buddleja*
Nicolaia Horan. = *Etlingera*
Nicolasia S. Moore. Compositae (Pluch.). 7 SW trop. Afr. R: MBSM 2 (1954)1
Nicolletia A. Gray. Compositae (Helia.-Pect.). 3 S N Am. R: Sida 7(1978)369
Nicolsonia DC = *Desmodium*
Nicotiana L. Solanaceae (1). 67 Am., S Pacific, Aus. (17, R: JABG 3(1981)1), SW Afr. (*N. africana* Merxm.). R: T.H. Goodspeed (1954) *The genus N.* (CB 16); herbs & a few shrubs; alks (10% of carbon metabolism directed to their prod.). Tobacco. Smoking tobacco from *N. tabacum* L. (cultigen orig. in trop. Am. in pre-Columbian period, prob. *N. sylvestris* Speg. (Arg.), *N. tomentosiformis* Goodspeed (Bolivia) & *N. otophora* Griseb. (Arg.) in ancestry) – in Amaz. form. only medic. & rarely smoked though 'cigars' there now 90 cm long, grown as annual esp. in US, trop. Am., Medit., S trop. Afr. & China (largest producer & consumer in 1980s), lvs coll. when yellowing, dried & fermented ('cured') in diff. ways to give tobacco for cigars, cigarettes (trade worth £9916m in GB alone in mid-1980s) etc. & variously flavoured; rolled lvs for outer surface of cigars & bidi cigarettes; ground & flavoured for snuff; mixed with molasses etc. for chewing tobacco (Cavendish plug or cake tobacco). Different cvs for diff. purposes e.g. Virginia t. for cigarettes, Havana t. – a large-lved form for cigars; also source of comm. nicotine, the stimulatory alk., used for insecticides etc. since 1664; a principal source of taxation since time of James I & cause of lung cancer but increasing efficiency & reducing stress; glands often trapping insects. *N. alata* Link & Otto (temp. S Am.) but usu. *N.* × *sanderae* Hort. Sander ex Will. Watson (*N. alata* × *N. forgetiana* Hort. Sander ex Hemsley (Brazil)) – cult. orn. flowering t. or t. plant, usu. with fragrant fls in evening; *N. glauca* Graham (S Bolivia to N Arg., natur. US & S Eur.) – cult. orn. shrubby tree, introd. Aus. where hybridizing with native spp.; *N. rustica* L. (wild or Aztec t., cult. in pre-Columbian times in Mexico & E N Am., *N. paniculata* L. (S Am.) × *N. undulata* Ruíz & Pavón) – orig. tobacco introd. to Eur. & that smoked by Raleigh etc., first cult. Virginia 1612 (John Rolfe), pl. smaller & hardier, now grown for insecticides in Euras., source of citric acid, etc.
nicuri *Syagrus coronata*
Nidema Britton & Millsp. (~ *Epidendrum*). Orchidaceae (V 13). 2 trop. Am. Cult. orn.
Nidorella Cass. Compositae (Ast.-Ast.). 13 trop. & S Afr. R: BSB 43(1969)209
Nidularium Lemaire. Bromeliaceae (3). 35 E S Am. R: FN 14(1979)1604. Cult. orn. epiphytes with prickly lvs & sessile infls
Niebuhria DC = *Maerua*
Niederleinia Hieron. = *Frankenia*
Niedzwedzkia B. Fedtsch. = *Incarvillea*
Niemeyera F. Muell. (~ *Chrysophyllum*). Sapotaceae (IV). (Incl. *Sebertia*) 20 Queensland, New Caledonia
Nierembergia Ruíz & Pavón. Solanaceae (2). 23 Mexico to Chile, esp. Arg. Some with elaiophores attractive to bees. Cult. orn. esp. *N. hippomanica* Miers var. *violacea* Millán ('*N. caerulea*', Arg.) – C violet-blue; *N. repens* Ruíz & Pavón (*N. rivularis*, Andes & warm temp.

S Am.) – fls white

Nietneria Klotzsch ex Benth. Melanthiaceae (Liliaceae s.l.). 1 Venezuela, Guyana

Nigella L. Ranunculaceae (I 3). (Excl. *Garidella*) 20 Euras. (Eur. 14), Medit. R: PSE 142(1983)71. Annuals with usu. finely pinnate lvs & fls with coloured K & 2-lipped petals with hollow nectariferous claws; fr. of ± united follicles making a capsule. Cult. orn. esp. *N. damascena* L. (love-in-a-mist, Medit.); *N. sativa* L. (fennel flower, Medit.) – cult. since Assyrian times for seeds (fitches, kalanji, black cumin), preserved in Tutankhamun's tomb, used to flavour bread (e.g. 'nan' bread, N India) & cakes, medic. (oil antibacterial)

Niger seed *Guizotia abyssinica*

Nigerian golden walnut *Lovoa trichilioides*

night-flowering cactus *Selenicereus grandiflorus, Epiphyllum hookeri*

nightshade, black *Solanum nigrum*; **deadly n.** *Atropa belladonna*; **enchanter's n.** *Circaea lutetiana*

Nigritella Rich. Orchidaceae (IV 2). 2 Eur.

Nigromnia Carolin = *Scaevola*

nikau palm *Rhopalostylis sapida*

Nikitinia Iljin. Compositae (Card.-Card.). 1 Turkmenistan to Iran

Nile grass *Acroceras macrum*

Nilgirianthus Bremek. = *Strobilanthes*

nim *Azadirachta indica*

Nimiria Prain ex Craib = *Acacia*

ninde *Aeollanthus myrianthus*

niopo snuff *Anadenanthera peregrina*

niové *Staudtia stipitata*

Nipa Thunb. = *Nypa*

Niphaea Lindley. Gesneriaceae. 5 trop. Am.

Niphidium J. Sm. Polypodiaceae (II 5). 10 trop. Am. R: AFJ 62(1972)101

Niphogeton Schldl. Umbelliferae (III 8). 18 N Andes

nipplewort *Lapsana communis*

Nipponanthemum Kitam. Compositae (Anth.-Leuc.). 1 Japan: *N. nipponicum* (Maxim.) Kitam. – cult. orn.

Nipponobambusa Muroi = *Sasa*

Nipponocalamus Nakai = *Arundinaria*

Nirarathamnos Balf.f. Umbelliferae (III 8). 1 Socotra: *N. asarifolius* Balf.f.

nirre *Nothofagus antarctica*

nispero *Eriobotrya japonica*

Nispero Aubrév. = *Manilkara*

Nissolia Jacq. Leguminosae (III 13). 13 trop. Am. esp. Mex. R: CUSNH 32(1956)173

Nitraria L. Nitrariaceae (Zygophyllaceae s.l.). 6 salt deserts of Sahara & S Russia (Eur. 1) to Afghanistan & E Siberia, SE Aus. (1: *N. billardierei* DC (dillon bush) – salt-rich pericarp eaten by emus, germ. improved (mammals less effective)). R: BZ 50(1965)1053. Some locally eaten fr. & soda sources

Nitrariaceae Bercht. & J. Presl (Zygophyllaceae s.l.). Dicots – Rosidae – Sapindales. 1/6 Sahara, Euras., SE Aus. Shrubs often pubescent & spiny. Lvs simple, semi-succ., spirally arr.; stipules v. small. Fls reg., app. bisexual but often functionally unisexual, in dichasial cymes; K (5), persistent; C 5(6), hooded, induplicate-valvate; A 10–15 without appendages, anthers ± basifixed; $\underline{G}$(3), 3-loc. with 1 pendulous ovule per loc., style short, stigmas 3. Fr. a drupe with fleshy exocarp & bony endocarp, 1-seeded; endosperm 0
Genus: *Nitraria*

Nitrophila S. Watson. Chenopodiaceae (IV 1). 6–7 SW N & temp. S Am.

Nivellea Wilcox, Bremer & Humphries. Compositae (Anth.-Leuc.). 1 Morocco

Nivenia Vent. Iridaceae (II). 5 Cape mts. Shrubby; heterostyly (? or androdioecy). R: SBT 34(1940)355. Some cult. orn.

Niveophyllum Matuda = *Hechtia*

Noaea Moq. Chenopodiaceae (III 3). 3 Eur. (1) & W As.

Noahdendron Endress, Hyland & Tracey. Hamamelidaceae (I 1). 1 N Queensland: *N. nicholasii* Endress, Hyland & Tracey. R: BJ 107(1985)369

noble fir *Abies procera*

Nocca Cav. = *Lagascea*

Noccaea Moench = *Thlaspi*

Nodocarpaea A. Gray. Rubiaceae (IV 15). 1 Cuba

Nodonema B.L. Burtt. Gesneriaceae. 1 W Afr.

Nogalia Verdc. (~ *Heliotropium*). Boraginaceae. 1 Somalia & S Arabia: *N. drepanophylla* (Baker) Verdc. R: KB 43(1988)431

Nogo Baehni = *Lecomtedoxa*

Nogra Merr. Leguminosae (III 10). 3 As. (prob. heterogeneous)

Noisettia Kunth. Violaceae. 1 Peru, Brazil, Guyana

Nolana L. ex L.f. Nolanaceae. 18 W S Am. from N Chile to S Peru, Galápagos (1). R: FN 26(1981). Often fleshy-leaved shore-pls, some cult. orn. grown as annuals

Nolanaceae Dumort. (~ Solanaceae (3)). Dicots – Asteridae – Solanales. 1/18 W S Am. R: FN 26(1981). Shrubs & herbs with glandular hairs. Lvs simple, ± succ. or ericoid, spirally arr. but in unequal pairs on same side in infls (as in some Solanaceae); stipules 0. Fls bisexual, 5-merous, axillary, solit.; K tubular-campanulate, lobes usu. imbricate, s.t. unequal, C campanulate, plicate between lobes, sometimes ± 2-lipped, A unequal with 3 longer, filaments attached to C tube alt. with lobes, anthers with longit. slits, nectary-disk often around $\underline{G}$ ± united, s.t. in multiple ranks of 5, each carpel with 1–several anatropous to hemitropous, unitegmic ovules. Fr. a head of ± distinct nutlets with 1–several seeds; embryo curved or spiral, in copious, oily endosperm. x = 12

Genus: *Nolana* (incl. *Alona*)

Nolina Michaux. Dracaenaceae (Agavaceae s.l.). 30 SW N Am. R: PAPS 50(1911)412. Dioec. pachycaul xerophytes with linear lvs. Some cult. orn. incl. *N. longifolia* (Schultes & Schultes f.) Hemsley (zacate, Mex.) – lvs used for brooms, basketry & thatching, & *N. microcarpa* S. Watson (SW N Am.) – used similarly

Nolinaceae Nakai = Dracaenaceae

Nolletia Cass. Compositae (Ast.-Ast.). 10 S & N Afr. to Eur. (1)

Noltea Reichb. Rhamnaceae. 1 S Afr:. *N. africana* (L.) Endl. – cult. orn.

Nomaphila Blume = *Hygrophila*

Nomismia Wight & Arn. = *Rhynchosia*

Nomocharis Franchet. Liliaceae. 7 W China, SE Tibet & NE Burma (with 1 extending to N Assam), 3770–4300 m. R: BJLS 87(1983)285. Cult. orn. intermediate between *Lilium* & *Fritillaria*

Nomosa I.M. Johnson. Boraginaceae. 1 Mexico

Nonatelia Aublet = *Palicourea*

nonda *Parinari nonda*

Nonea Medikus. Boraginaceae. 35 Medit. (Eur. 9, dry places)

none-so-pretty *Saxifraga* × *urbium*

nongo *Albizia grandibracteata*

Noogoora burr *Xanthium occidentale* (= *X. strumarium* s.l.)

Nopalea Salm-Dyck = *Opuntia*

Nopalxochia Britton & Rose = *Disocactus*

Norantea Aublet. Marcgraviaceae. 35 trop. Am. *Philodendron*-like pls with fls all fertile. *N. guianensis* Aubl. has red nectar-spurs looking like fr. visited by birds which poll., the rest of the fl. dull so potential beak damage reduced

Nordmann (Christmas tree) *Abies nordmanniana*

Norfolk Island pine *Araucaria heterophylla*

Normanbokea Klad. & F. Buxb. = *Neolloydia*

Normanboria Butzin = *Acrachne*

Normanbya F. Muell. ex Becc. (~ *Ptychosperma*). Palmae (V 4k). 1 Queensland: *N. normanbyi* (W. Hill) L. Bailey

Normandia Hook.f. Rubiaceae (IV 13). 1 New Caledonia

Normandy cress *Barbarea verna*

Noronhia Stadman ex Thouars. Oleaceae. 40 Madag. & Comore Is.

Norrisia Gardner. Strychnaceae (Loganiaceae s.l.). 2 W Mal.

Northea Hook.f. = seq.

Northia Hook.f. Sapotaceae (I). 1 Seychelles: *N. hornei* (Hartog) Pierre (capucin) – endosperm 0

Northiopsis Kaneh. = *Manilkara*

Norway maple *Acer platanoides*; **N. spruce** *Picea abies*

Nosema Prain. Labiatae (VIII 4b). 6 SE As., Mal.

Nostolachma T. Durand. Rubiaceae (II 1). 10 Indomal.

Notanthera (DC) G. Don f. (~ *Loranthus*). Loranthaceae. 1 temp. S Am.

Notaphoebe Griseb. = *Nothaphoebe*

Notechidnopsis Lavranos & Bleck (~ *Echidnopsis*). Asclepiadaceae (III 5). 2 S Afr. R: CSJAm. 57(1985)255

Notelaea Vent. Oleaceae. 9 E Aus. R: JAA 49(1968)333. Hard timbers. *N. longifolia* Vent. has lignotubers

Nothaphoebe Blume (~ *Persea*). Lauraceae (I). 40 Indomal. Alks

Nothapodytes Blume. Icacinaceae. 5 E As., Mal. R: BMNHN 4,7(1985)82

Nothoalsomitra Telford (~ *Alsomitra*). Cucurbitaceae. 1 SE Queensland

Nothobaccharis R. King. & H. Robinson. Compositae (Eup.-Crit.). 1 Peru. R: MSBMBG 22(1987)324

Nothocalais (A. Gray) E. Greene (~ *Microseris*). Compositae (Lact.-Mic.). 4 C & W N Am.

Nothocestrum A. Gray. Solanaceae (1). 4 Hawaii, two endangered spp. R: Man. Fl. Pl. Hawaii 2(1990)1262

Nothochelone (A. Gray) Straw. Scrophulariaceae. 1 W N Am.

Nothochilus Radlk. Scrophulariaceae. 1 Brazil

Nothocissus (Miq.) Latiff. Vitaceae. 1 W Mal.

Nothocnide Blume ex Chew (*Pseudopipturus*). Urticaceae. 5 Mal. to Solomon Is. R: GBS 24(1969)361

Nothodoritis Tsi. Orchidaceae (V 16). 1 China. R: APS 27(1989)58

Nothofagaceae Kuprian. = Fagaceae

Nothofagus Blume. Fagaceae. 35 New Guinea, New Caled., temp. Aus., NZ, temp. S Am. R: BJLS 105(1991)68. Southern beeches. At anthesis ovules merely bulges in ovary & do not develop until fert. 9–10 weeks later. Imp. timber trees, in S hemisph. second only to eucalypts; some fast-growing spp. promoted in Eur. for timber or to 'replace elm in the landscape', after the depredations of Dutch Elm disease, or as cult. orn. incl. *N. antarctica* (Forster f.) Oersted (nirre, Antarctic beech, Chile, Arg.) – decid.; *N. cunninghamii* (Hook.) Oersted (myrtle beech, Tasmanian b., T. m., SE Aus.) – evergreen, timber for furniture, flooring etc.; *N. dombeyi* (Mirbel) Oersted (coigue, Chile) – decid., timber imp.; *N. fusca* (Hook.f.) Oersted (red b., NZ) – evergreen, timber for railway sleepers; *N. menziesii* (Hook.f.) Oersted (silver b., Southland b., NZ) – evergreen; *N. obliqua* (Mirbel) Blume (roble b.) & *N. nervosa* (Philippi) Dmitri & Milano (*N. procera*, ?*N. alpina*, rauli) – both Chile & Arg., decid., with their hybrid the most promising in GB; *N. solanderi* (Hook.f.) Oersted (black b., NZ) – evergreen, timber for general construction

Notholaena R. Br. (~ *Cheilanthes*). Pteridaceae (IV). 25 warm (esp. SW US to Mex.) to trop. Am. Cloak ferns. Usu. xerophytes. Some cult. orn.

Notholirion Wallich ex Boiss. Liliaceae. 4 Afghanistan to W China. Cult. orn. bulbous pls

Nothomyrcia Kausel = *Myrceugenia*

Nothopanax Miq. = *Polyscias*

Nothopegia Blume. Anacardiaceae (III). 7 Sri Lanka, India

Nothopegiopsis Lauterb. = *Semecarpus*

Nothoperanema (Tag.) Ching. Dryopteridaceae (I 2). c. 2–3 OW trop.

Nothophlebia Standley = *Pentagonia*

Nothoruellia Bremek. & Nannenga-Bremek. = *Ruellia*

Nothosaerva Wight. Amaranthaceae (I 2). 1 trop. Afr. & Mauritius to trop. As.

Nothoscordum Kunth. Alliaceae (Liliaceae s.l.). 20 Am. (*N. borbonicum* Kunth, i.e. '*N. inodorum*', poss. *N. gracile* (Aiton) Stearn × *N. entrerianum* Ravenna arising nr. Buenos Aires, natur. Eur., Aus. ('onion weed') etc.). *Allium*-like but odourless & style term.; adventitious embryony from nucellar tissue; few cult. orn.

Nothosmyrnium Miq. Umbelliferae (III 8). 2 E As.

Nothospondias Engl. Simaroubaceae. 1 W trop. Afr.

Nothostele Garay. Orchidaceae (III 3). 1 Brazil

Nothotsuga Hu ex Page (~ *Tsuga*). Pinaceae. 1 China: *N. longibracteata* (W.C. Cheng) Page. R: NRBGE 45(1988)390

Noticastrum DC. Compositae (Ast.-Sol.). 20 trop. S Am. R: RMP 13(1985)313

Notiosciadium Speg. Umbelliferae (I). 1 Argentina

Notobasis (Cass.) Cass. (~ *Cirsium*). Compositae (Card.-Card.). 1 Medit. (incl. Eur.) to C As.

Notobuxus Oliver = *Buxus*

Notocactus (Schumann) Frič = *Parodia*

Notoceras R. Br. Cruciferae. 1 Canary Is., Medit., incl. Eur., to NW India

Notochaete Benth. Labiatae (VI). 1 Himalaya

Notochloe Domin. Gramineae (25). 1 Blue Mts, NSW: *N. microdon* (Benth.) Domin

Notodanthonia Zotov = *Danthonia*

Notodon Urban = *Poitea*

Notodontia Pierre ex Pitard = *Spiradiclis*

Notonerium Benth. = *Heliotropium*

Notonia DC = *Kleinia*
Notoniopsis R. Nordenstam = *Kleinia*
Notopleura (Hook.f.) Bremek. = *Psychotria*
Notopora Hook.f. Ericaceae. 5 E Venezuela. R: MNYBG 29(1978)158
Notoptera Urban = *Otopappus*
Notopterygium Boissieu. Umbelliferae (III 8). 2 China. R: APS 13,3(1975)83
Notoseris Shih (~ *Prenanthes*). Compositae (Lact.-Lact.). 12 China S of Yangtze. R: APS 25(1987)196
Notospartium Hook.f. Leguminosae (III 16). 3 S Is., NZ
Notothixos Oliver. Viscaceae. 8 Sri Lanka to SE Aus. R: Brunonia 6(1983)25
Notothlaspi Hook.f. Cruciferae. 2 NZ. *N. rosulatum* Hook.f. (penwiper pl.) – lvs at first covered with hairs, the flannel-textured rosette of them supposedly much like Victorian penwiper
Nototriche Turcz. Malvaceae. 100 S Am. Infls epiphyllous
Nototrichium (A. Gray) Hillebrand (~ *Achyranthes*). Amaranthaceae (I 2). 2 Hawaii. R: Man. Fl. Pl. Hawaii 1(1990)193. Small trees & shrubs
Notoxylinon Lewton (~ *Cienfugosia*). Malvaceae. 8 Aus.
Notylia Lindley. Orchidaceae (V 10). (Incl. *Macroclinium*) c. 75 trop. Am. Small. cult. orn. epiphytes with extrafl. nectaries & fragrant fls
Nouelia Franchet. Compositae (Mut.-Mut.). 1 SW China. Shrub
Nouettea Pierre. Apocynaceae. 1 SE As.
November lily *Lilium longiflorum*
Novenia Freire (~ *Gnaphalium*). Compositae (tribe?). 1 Andes: *N. acaulis* (Benth.) Freire & Hellwig (*N. tunariense*). R: BSAB 24(1986)295
Novosieversia F. Bolle. Rosaceae. 1 arctic NW Am., W & arctic Siberia to Kamchatka
Nowickea Martínez & McDonald. Phytolaccaceae (I). 2 Mex. R: Brittonia 41(1989)399. Gynophores. (Monstrous forms of *Phytolacca* spp.?)
Noyera Trécul = *Perebea*
ntonga nut *Cryptocarya latifolia*
Nucularia Battand. Chenopodiaceae (III 3). 1 Algeria, Sahara: *N. perrinii* Battand.
Nudilus Raf. = *Forestiera*
Nuihonia Dop = *Craibiodendron*
Numaeacampa Gagnepain = *Codonopsis*
Nuñas bean *Phaseolus vulgaris*
Nuphar Sm. Nymphaeaceae. 16 N temp. & cold (Eur. 2). R: P. Swindells, *Waterlilies* (1983)152. Waterlilies with alks, yellow or purplish hypogynous fls usu. held above water & large berries maturing above water, breaking off & splitting into carpels, the seeds without arils (cf. *Nymphaea*) but slimy pericarp with air-bubbles aiding dispersal. Some cult. orn. esp. *N. luteum* (L.) Sm. (brandy-bottle, yellow waterlily, E N Am., WI, Euras., Medit.) – fls with alcoholic smell attractive to beetles; *N. polysepalum* Engelm. (N N Am.) – seeds used for a flour by Klamath Indians
Nurmonia Harms = *Turraea*
nut technically a hard 1-seeded brittle fr., in common parlance any edible 'kernel' or even a pod with edible seeds (e.g. monkey n.) – E.A. Menninger (1977) *Edible nuts of the world*; **arara n.** *Joannesia princeps*; **Areca n.** *Areca catechu*; **Australian chestn.** *Castanospermum australe*; **Barbary n.** *Gynandriris sisyrinchium*; **Barcelona n.** *Corylus avellana*; **Baroba n.** *Diplodiscus paniculatus*; **Ben n.** *Moringa oleifera*; **betel n.** *Areca catechu*; **bladder n.** *Staphylea* spp.; **bread n.** *Brosimum alicastrum*; **Brazil n.** *Bertholletia excelsa*; **butter n.** *Juglans cinerea*; **candle n.** *Aleurites moluccana*; **cashew n.** *Anacardium occidentale*; **Chile n.** *Araucaria araucana*, *Gevuina avellana*; **chufa n.** *Cyperus esculentus*; **clearing n.** *Strychnos potatorum*; **cocon.** *Cocos nucifera*; **cohune n.** *Orbignya cohune*; **coohoy n.** *Floydia prealta*; **coquilla n.** *Attalea funifera*; **dika n.** *Irvingia gabonensis*; **double cocon.** *Lodoicea maldivica*; **earthn.** *Conopodium majus*; **fox n.** *Euryale ferox*; **Gaboon n.** *Coula edulis*; **gasso n.** *Manniophyton africanum*; **n. grass** *Cyperus rotundus*; **ground n.** *Arachis hypogaea*; **gru-gru n.** *Acrocomia totai*; **helicia n.** *Helicia diversifolia*; **illipe n.** *Shorea*, *Madhuca* spp.; **inoi n.** *Poga oleosa*; **ivory n.** *Phytelephas macrocarpa*; **jack n.** *Artocarpus heterophyllus*; **jojoba n.** *Simmondsia chinensis*; **Jamaica cobn.** *Omphalea triandra*; **Jesuit's n.** *Trapa natans*; **karaka n.** *Corynocarpus laevigata*; **kaya n.** *Torreya nucifera*; **keranda n.** *Elaeocarpus bancroftii*; **kola n.** *Cola* spp.; **kubili n.** *Cubilia cubili*; **ling n.** *Trapa bicornis*; **Lunan n.** *Lepisanthes fruticosa*; **macadamia n.** *Macadamia* spp.; **manketti n.** *Ricinodendron rautanenii*; **marking n.** *Semecarpus anacardium*; **monkey n.** *Arachis hypogaea*; **ngali n.** *Canarium indicum*; **nicuri palm n.** *Syagrus coronata*; **olive n.** *Elaeocarpus sphaericus*; **owusa n.** *Plukenetia conophora*; **oyster n.** *Telfairia pedata*; **paradise n.** *Lecythis*

spp.; **pean.** *Arachis hypogaea*; **physic n.** *Jatropha curcas*; **pign.** *Conopodium majus*; **pili n.** *Canarium* spp.; **pistachio n.** *Pistacia vera*; **quandong n.** *Eucarya acuminata*; **Queensland n.** *Macadamia* spp., *Macrozamia* spp.; **red boppel n.** or **rose n.** *Hicksbeachia pinnatifolia*; **rush n.** *Cyperus esculentus*; **sapucaia n.** *Lecythis* spp.; **shea n.** *Vitellaria paradoxa*; **snake n.** *Ophiocaryon paradoxum*; **soap n.** *Sapindus* spp.; **Solomon n. oil** *Canarium indicum*; **swarri n.** *Caryocar nuciferum*; **tacy n.** *Caryodendron orinocense*; **tagua n.** *Phytelephas macrocarpa*; **tallow n.**; *Ximenia americana*; **tiger n.** *Cyperus esculentus*; **n.wood** *Terminalia arostrata*; **yeheb n.** *Cordeauxia edulis*; **Zulu n.** *Cyperus esculentus*

nutmeg *Myristica fragrans*; **Brazilian n.** *Cryptocarya moschata*; **calabash n.** *Monodora myristica*; **false n.** *Pycnanthus angolensis*; **Madagascar n.** *Cryptocarya* sp. (*Ravensara aromatica*); **Peruvian n.** *Laurelia sempervirens*; **W African n.** *Monodora myristica*

Nuttallanthus D. Sutton (~ *Linaria*). Scrophulariaceae (Antirrh.). 4 N Am., W S Am. R: D.A. Sutton, Rev. Antirrh. (1988)455

nux vomica *Strychnos nux-vomica*

Nuxia Comm. ex Lam. Buddlejaceae (Loganiaceae s.l.). 15 S Arabia to trop. Afr., Masc. & S Afr. R: MLW 75–8(1975). *N. floribunda* Benth. (kite tree, trop. & S Afr.) – cult. orn. tree; *N. verticillata* Comm. ex Lam. (Madag.) – used as flavouring for betsa-betsa

Nuytsia R. Br. ex G. Don f. Loranthaceae. 1 W Aus.: *N. floribunda* (Labill.) R. Br. ex G. Don f. (fire-tree, flame-tree, Christmas tree) – tree to 12 m, parasitic on roots of grass, with brilliant orange bird-poll. fls, dry wind-disp. fr. with broad wings; cotyledons 2–4, unequal

nyala tree *Xanthocercis zambesiaca*

nyatoh *Palaquium* spp.

Nyctaginaceae Juss. Dicots – Caryophyllidae – Caryophyllales. 30/390 trop. & warm esp. Am., few temp. Trees, shrubs & herbs producing betalains but not anthocyanins, often with unusual sec. growth with concentric rings of vasc. bundles or alt. rings of xylem & phloem. Lvs usu. opp., simple, often unequal; stipules 0. Fls usu. bisexual, often in cymes, s.t. subtended by large & s.t. coloured involucre & when reduced giving a 1-flowered pseudanthium with K-like involucre & C-like K; P tubular with (3–)5(–8) valvate to plicate lobes, often C-like, A (1–) as many as K (–40) with filaments of unequal length, s.t. basally connate, anthers with longit. slits, annular nectary-disk often around $\underline{G}$ 1 with long slender style & 1 basal campylotropous (-hemitropous) (uni- or) bitegmic ovule. Fr. an achene or nut, often encl. in persistent base of P-tube (anthocarp, often glandular & animal-disp.); seed with large usu. curved embryo with ± copious starchy perisperm, endosperm 0 or a cap over radicle. x = 10, 13, 17, 29, 33+

Classification & principal genera:

I. **Boldoeae** (lvs spirally arr.; involucre 0, bracteoles 0, style filiform): *Boldoa*
II. **Leucastereae** (lvs spirally arr.; involucre 0, ±bracteoles; style thick or 0): *Reichenbachia*
III. **Nyctagineae** (lvs opp. to whorled; cotyledons 2): *Boerhavia, Mirabilis, Selinocarpus*
IV. **Abronieae** (lvs opp.; cotyledon 1)
V. **Bougainvilleeae** (lvs spirally arr.; involucre of coloured bracts): *Bougainvillea*
VI. **Pisonieae** (lvs usu. opp.; P inconspic.): *Guapira, Neea, Pisonia*

Some ed. & medic. (*Mirabilis, Neea, Pisonia*) but imp. cult. orn. esp. *Bougainvillea* & *Mirabilis*

Nyctaginia Choisy. Nyctaginaceae (III). 1 SW N Am.: *N. capitata* Choisy – cult. orn. herb with deep red fls

Nyctanthes L. Oleaceae. 2 India to Java. *N. arbor-tristis* L. a holy tree of India where the fallen corollas are swept up each morning; the fragrant fls opening at night are a source of scent & a fugitive orange dye (cheap saffron subs.)

Nycticalanthus Ducke. Rutaceae. 1 Amazonia

Nyctocalos Teijsm. & Binnend. Bignoniaceae. 3 Indomal.

Nyctocereus (A. Berger) Britton & Rose = *Peniocereus*

Nylandtia Dumort. Polygalaceae. 1 S Afr.: *N. spinosa* (L.) Dumort. R: SAJB 57(1991)229

Nymania Lindb. (*Aitonia*). Meliaceae (I 1). 1 S Afr.: *N. capensis* (Thunb.) Lindb. – shrub with inflated pink to purple capsules, which are blown about (cf. *Physalis*), break up & scatter seeds

Nymphaea L. Nymphaeaceae. c. 50 cosmop. (Eur. 4). R: PCIW 4(1905), P. Swindells, *Waterlilies* (1983)152, (subg. *Hydrocallis* (nocturnal fls, G united) 14) SMB 16(1987). Waterlilies. Herbs with rhizomes or tubers, the fls solit. on pedicels with phyllotactic spirals independent of those of lvs, which usu. float on water-surface & have stomata, cuticle & palisade on adaxial surface. Fls often floating, some night-flowering, poll. by beetles, Hymenoptera & syrphid flies; K 4, C ∞ – outer 4 alt. with K & next 4 with those, the 8 being basal members of spirally arr. C eventually passing into A ∞ continuing spirals, C & A inserted on 10–20-loc. G sunk in receptacle with radiating apical stigmas; fr. a large berry with ∞ seeds with spongy aril entrapping air-bubbles, the fr. maturing under water, dehiscing to release seeds which float until aril decays, when they sink & germinate (cf. *Nuphar*). Rhizomes & seeds of some spp. ed., many cult. orn. & (many of those) hybrids of obscure parentage. *N. alba* L. (Euras., Medit.) – common white-flowered (open most of day), in shallow water in Ukraine with smaller lvs & fls ('*N. minoriflora*', app. environmentally induced); *N. ampla* (Salisb.) DC (warm & trop. Am.) – narcotic of Mayans; *N. nouchali* Burm.f. var. *caerulea* (Savigny) Verdc. (*N. caerulea*, N & C Afr.) & *N. lotus* L. (Egypt, trop. & S Afr., Madag., hotsprings of NW Romania) – found as wreaths on mummies of Rameses II etc. back to 2000 BC, the original Egyptian lotus, *Nelumbo nucifera* Gaertner not introd. before 500 BC; *N. nouchali* var. *caerulea*, a narcotic in Ancient Egypt, & other spp. have coleoptiles & mesocotyls; *N. micrantha* Guillemin & Perrottet (W Afr.) – plantlets form in umbo of decaying lvs, prob. a parent of *N.* × *daubenyana* Baxter ex Daubeny, which arose in the Oxford Botanic Garden, *N. nouchali* Burm.f. (*N. stellata*, OW trop.) prob. the other; *N. pubescens* Willd. (Indomal. to trop. Aus.) – night lotus with white fls, devoted to Ciwa (sacred colour white, cf. *Nelumbo*); *N. tetragona* Georgi (*N. pygmaea*, Finland & N Russia to Japan, N Am.) – smallest sp. in cult. much used in hybridization

Nymphaeaceae Salisb. Dicots – Magnoliidae – Nymphaeales. (Incl. Barclayaceae) 6/75 cosmop. Aquatic herbs with rhizomes or tubers, often with alks; vessels 0; root-hairs from specialized cells, like many monocots, & scattered vascular bundles like them or a single ring. Lvs spirally arr. on rhizome, usu. rounded cordate & floating; stipules median-axillary or 0. Fls solit. in axils or extra-axillary, bisexual, reg., aerial, insect-poll., hypo- to epigynous; K 4–6 (–12), s.t. petaloid (*Nuphar*), C (0–)8–∞ inserted around top of ovary, usu. free, often passing into A 14–∞, spirally arr. & usu. free & laminar with elongate microsporangia or transitional to forms with differentiated filaments & anthers with longit. slits, pollen grains usu. monosulcate, G (3–) 5–35, ± united into compound superior to inferior ovary with ∞ scattered anatropous or rarely orthotropous bitegmic ovules. Fr. berry-like ± (irreg.) dehiscent through swelling of mucilage within; seeds arillate, with hooked hairs or nude with 2 cotyledons, wholly distinct or arising as lobes from common primordium (that appearing to be 1 is a true leaf); endosperm scanty, perisperm copious. X = 12–29+

Genera: *Barclaya* (*Hydrostemma*), *Euryale*, *Nuphar*, *Nymphaea*, *Ondinea*, *Victoria*

The fam. has many characteristics of monocots; were N. the first angiosperms studied in detail, it is likely that our overall classification would be different – as it is, criteria used to keep other homogeneous fams separate are inappropriate in that they would shatter this group, much in the same way that the characters which separate fams in modern classifications prob. represent differences at the specific or even population level in the ancestral groups of angiosperms (Taxon 33(1984)77). So inferior or superior ovaries, C present or 0, laminar or 'orthodox' stamens, orthotropous or anatropous ovules, scattered or ringed vasc. bundles & varied chemical features are all found here. Moreover, spiral floral parts typical of early 'angiosperms' (how an early angiosperm is recognizable from its 'gynmosperm' ancestor is a moot point) fossils (Nature 319(1986)723) & long assumed to represent a primitive condition, as well as arillate seeds, also held to be primitive (Corner), are here. The sometimes C-less *Ondinea* approaches *Barclaya* in a number of features, so that the retention of Barclayaceae in Nymphaeaceae is supported. The group is a specialized one ecologically & has penetrated the temp. zones through a 'loophole' in that the rhizomes of perennial spp. persist in the buffered habitat of pond- & river-bottoms; they have irregular morphology, not 'conforming' to morphological 'rules'. N. demonstrate a range of interesting disp. features incl. air-filled arils (*Nymphaea*), pericarp with air-bubbles (*Nuphar*), minute hooked hairs (*Barclaya*) – associated with water or pig-dispersal, etc. Despite their advanced aquatic habit, the plants have retained beetle- & other insect-poll., another allegedly primitive feature, while their pollen is also of a primitive type. Overall, there is compelling evidence that this group is a successful, specialized relic of the stock which existed before the monocotyledons were recognizably distinct from dicotyledons – it is tempting, then, to imagine that that stock, presumably a terr. one, had large insect-poll. flowers with spirally arr. parts & monosulcate pollen, animal-disp. arillate seeds & 'irregular' morphology & anatomy

Euryale, *Nuphar*, *Nymphaea* & *Victoria* yield ed. rhizomes or seeds & are cult. orn.

Nymphoides Séguier (*Limnanthemum*). Menyanthaceae. 20 cosmop. (Eur. 1: *N. peltata* Kuntze (water-fringe, Euras.)). The floating lvs are part of infl. axis (cf. *Nymphaea*); some with air-filled seed-hairs for disp., others (Aus.) disp. by ants. Some ed. tubers & medic.

seeds; cult. orn.

Nypa Steck. Palmae (III). 1 India to Ryukyu & Solomon Is. (natur. W Afr., Panamá): *N. fruticans* Wurmb (nipa palm) – a char. pl. of mangrove swamps with creeping rhizomes with truly dichotomizing apices & fronds to 10 m tall, infls with 'irreg.' positions, infr. heads breaking up into char. obovoid angular fibrous sea-disp. fr., the seeds app. germinating before release, the plumule assisting in detachment. Fossils known from Cretaceous & Eocene of much of OW incl. England & poss. showing the distrib. of pl. around Tethys Sea, though poss. a drift-seed from further south. Lvs used for thatch, cigarette papers, basketry etc., the drosophilid-poll. infls tapped for sugar etc. (gula malacca, an ingredient of 'three palms pudding' with coconut milk & sago), immature seeds ed.

Nypaceae Brongn. ex Le Maout & Decne = Palmae

Nyssa L. Cornaceae (Nyssaceae). 8 S N Am. (5, R: Sida 15(1992)323), C Am. (1), China (1), Indomal. (1). Some timbers & cult. orn. decid. trees (tupelo) with bright autumn colour, esp. *N. aquatica* L. (cotton gum, SE US) – good bee-pl., swamp tree standing floods of 2 m or more, water-disp.; *N. ogeche* Bartram ex Marshall (ogeechee lime, SE US) – fr. ed. when preserved; *N. sinensis* Oliver (Chinese tupelo, C China); *N. sylvatica* Marshall (black gum, cotton g., pepperidge, SE N Am.) – cobalt indicator, the principal tupelo timber used for wharves, veneers, pulp etc.

Nyssaceae Juss. ex Dumort. = Cornaceae

Nyssanthes R. Br. Amaranthaceae (I 2). 2 E Aus. P 4, 2 inner smaller

O

oak *Quercus* spp., in GB usu. *Q. robur* or *Q. petraea* (**English**, **French**, **Polish** or **Slavonian o.**); **African o.** *Oldfieldia africana*, *Lophira* spp.; **(American) red o.** *Q. rubra*, *Q. falcata* etc.; **(A.) white o.** *Q. alba*, *Q. prinus*; **o. apple** gall on *Q.* spp.; **Australian o.** *Eucalyptus regnans*; **bear o.** *Q. ilicifolia*; **Brazilian o.** *Posoqueria latifolia*; **bull o.** *Allocasuarina luehmannii*; **Californian o.** *Q. lobata*; **Ceylon o.** *Careya arborea*, *Schleichera oleosa*; **chestnut o.** *Q. prinus*; **cork o.** *Q. suber*; **evergreen** or **holly o.** *Q. ilex*; **o. fern** *Gymnocarpium dryopteris*; **holm o.** *Q. ilex*; **iron o.** *Q. obtusifolia*; **jack o.** *Q. marilandica*; **Japanese o.** *Q. mongolica* etc.; **Kermes o.** *Q. coccifera*; **live o.** *Q. virginiana*; **Lucombe o.** *Q.* × *pseudosuber*; **manna o.** *Q. cerris*; **Oregon white o.** *Q. garryana*; **patana o.** *Careya arborea*; **poison o.** *Rhus radicans*; **possum o.** *Q. nigra*; **post o.** *Q. obtusiloba*; **Quebec o.** *Q. alba*; **river o.** *Casuarina cunninghamiana*; **satin o.** *Alloxylon* spp.; **scarlet o.** *Q. coccinea*; **she-o.** *Casuarina* spp.; **shin o.** *Q. gambelii*; **shingle o.** *Q. imbricaria*; **silky o.** *Grevillea robusta*, *Cardwellia sublimis*; **Spanish o.** *Inga laurina*; **swamp o.** *Casuarina* spp.; **Tasmanian o.** *Eucalyptus delegatensis*; **tulip o.** *Argyrodendron* spp.; **Turkey o.** *Q. cerris*

Oakes-amesia C. Schweinf. & P. Allen = *Sphyrastylis*

Oakesiella Small = *Uvularia*

oat *Avena sativa*; **black o. grass** *Stipa avenacea*; **false o.** *Arrhenatherum elatius*; **o. grass** *Avena*, *Arrhenatherum* & *Helictotrichon* spp.; **swamp o.** *Trisetum pennsylvanicum*; **water o.** *Zizania* spp.; **wild o.** *Avena sativa* subsp. *fatua*; **yellow o.** *T. flavescens*

Oaxacania Robinson & Greenman = *Hofmeisteria*

obada *Ficus vogelii*

obeche *Triplochiton scleroxylon*

obedient plant *Physostegia virginiana*

Oberonia Lindley. Orchidaceae. 300 OW trop. (India 41, R: OM 4(1990)). Some locally medic., few cult. orn.

Obetia Gaudich. Urticaceae (I). 8 trop. & S Afr., Madag., Masc. R: KB 38(1983)221. Trees with stinging hairs

Obione Gaertner = *Atriplex*

Oblivia Strother (~ *Otopappus*). Compositae (Helia.-Verb.). 2 trop. Am. R: SB 14(1989)541

Obolaria L. Gentianaceae. 1 E N Am. Mycotroph (cf. *Voyria*), purplish green with scale lvs

Obolinga Barneby. Leguminosae (II 5). 1 Hispaniola: *O. zanonii* Barneby. R: Brittonia 41(1989)167

Obregonia Frič. Cactaceae (III 9). 1 NE Mex.: *O. denegrii* Frič – cult. orn. like *Leuchtenbergia* but tubercles short

oca *Oxalis tuberosa*; **o. quina** *Ullucus tuberosus*

Oceanopapaver Guillaumin = *Corchorus*

Ochagavia Philippi. Bromeliaceae (3). 3 Chile incl. Juan Fernandez. R: FN 14(1979)1527. *O.*

carnea (Beer) L.B. Sm. & Looser (*O. lindleyana*, coastal C Chile) – cult. orn., natur. Scilly Is.

Ochanostachys Masters. Olacaceae. 1 W Mal.: *O. amentacea* Masters – fr. ed., timber for furniture etc.

Ochlandra Thwaites. Gramineae (1c). 7 Sri Lanka, India. R: KBAS 13(1986)57. Bamboos; lodicules 1–15 & to 1.5 cm long, A 15–120, fr. fleshy; some used for pulp in India. *O. scriptoria* (Dennst.) C. Fischer (flowering annually), *O. travancorica* (Bedd.) Gamble take over from cardamom under trees, preventing forest regeneration leading to 'brakes' in S India, *O. t.* favoured by elephants

Ochna L. Ochnaceae. 86 OW trop. K coloured, persistent; G 3–12 with common style, which falls after fert., when receptacle becomes fleshy & each carpel becomes an oil-rich drupelet attractive to birds. Cult. orn. trees & shrubs, some with ed. fr.

Ochnaceae DC. Dicots – Dilleniidae – Theales. 28/370 trop. esp. Brazil. R: BJ 113(1991)105. Evergreen trees & shrubs, few herbs, usu. glabrous, young stems usu. with cortical & s.t. medullary vasc. bundles. Lvs usu. simple & with v. many parallel lateral veins; petioles with siphonosteles; stipules present. Fls bisexual, ± reg., in term. panicles etc.; K (3–)5(–12), imbricate, often persistent, some s.t. larger than rest, C (4)5(–10) convolute (rarely imbricate), A (5,10–)∞ (s.t. in 3–5 whorls), associated with 5 trunk-bundles with members arising centripetally, often eccentrically positioned & s.t. on long androgynophore, anthers usu. with term. pores, staminodes s.t. internal to A or a tube or lobed disk around $\underline{G}$ (1)2–15, united at least by a common style, the ovary ± partitioned with ± distinctly axile placentation or so deeply lobed that carpels appear distinct, when receptacle often enlarging in fr., each loc. with 1–∞ anatropous to almost campylotropous bitegmic (unitegmic in *Lophira*) ovules. Fr. various, often of distinct 1-seeded drupelets, s.t. a capsule, nut or berry; seeds often winged, the endosperm oily & proteinaceous or 0. x = 12, 14 (?7)

Principal genera: *Luxemburgia*, *Ochna*, *Ouratea*, *Sauvagesia*

Fam. rather 'loosely knit' (Cronquist) & diff. authors split off subgroupings as separate fams, esp. Sauvagesiaceae, which is v. distinct in its seed anatomy, suggesting a position near Flacourtiaceae & Violaceae; the monotypic *Diegodendron* (endosperm 0, Madag., now joined with Madagascan Sphaerosepalaceae prob. best excl. to Malvales) & *Strasburgeria* (K 8–10, New Caled.; removed here) are also somewhat distinctive, but even their removal leaves a rather diverse group

Some timber (*Lophira*) & cult. orn. (*Ochna*, *Ouratea*)

Ochoterenaea F. Barkley. Anacardiaceae (IV). 1 Colombia

Ochotonophila Gilli. Caryophyllaceae (III 1). 1 Afghanistan: *O. allochrusoides* Gilli

Ochradenus Del. Resedaceae. 6 Middle E to Socotra, Libya, Pakistan. R: NRBGE 41(1984)491

Ochreata (Lojac.) Bobrov = *Trifolium*

Ochreinauclea Ridsd. & Bakh.f. Rubiaceae (I 2). 2 Indomal. R: Blumea 24(1978)331

Ochrocarpos Thouars = *Garcinia*

Ochrocephala Dittr. Compositae (Card.-Cent.). 1 N trop. Afr. savanna

Ochroma Sw. Bombacaceae. 1 trop. Am.: *O. pyramidale* (Lam.) Urban (*O. lagopus*, balsa, down tree, cork t.) – fast-growing tree of clearings, seeds requiring v. high temps for germ., world's lightest comm. timber used for insulation, model aeroplanes, architect's models, form. for life-belts etc.; fls bat-poll., to 12 × 8 cm, seed embedded in hairs (cf. *Ceiba*)

Ochropteris J. Sm. Pteridaceae (VI). 1 Madag., Masc.: *O. pallens* (Sw.) J. Sm.

Ochrosia Juss. Apocynaceae. (Incl. *Neisosperma*) 30 Masc. & Seychelles to Aus. & W Pacific. R: AUWP 87-5(1987)47. Some timber (yellow wood), medic. & dyes, fr. of *O. elliptica* Labill. (pokosola, trop. Aus. to New Caled.) with ed. seeds

Ochrosperma Trudgen (~ *Baeckea*). Myrtaceae. 3 E Aus. R: Nuytsia 6(1987)9.

Ochrothallus Pierre ex Baillon = *Niemeyera*

Ochthephilus Wurd. Melastomataceae. 1 Guyana

Ochthocharis Blume. Melastomataceae. 2 trop. Afr., 5 trop. As. R: KB 36(1981)13

Ochthochloa Edgew. (~ *Eleusine*). Gramineae (31d). 1 NE Afr., Arabia, Sind

Ochthocosmus Benth. Ixonanthaceae. 15 trop. S Am. (7, esp. Guayana Highland, R: Brittonia 32(1980)128), trop. Afr. (8)

Ochthodium DC. Cruciferae. 1 E Medit.

Ocimum L. Labiatae (VIII 4b,). 150 warm temp. & trop. esp. Afr. (incl. *Becium*). Many aromatic herbs & shrubs with thymol-containing ess. oils used in flavouring etc., esp. *O. basilicum* L. (*O. citriodorum*, (sweet or lemon) basil, OW trop.) – c. 100 t oil prod. per annum, chopped lvs used in casseroles, sauces, pizza & liqueurs esp. in Medit., local tonic, flavour is linalool, main sweet constituent is estragol (phenylpropanoid), juvenile hormone analogue in lvs; *O. canum* Sims (hoary b., OW trop. & warm) – potherb in India, medic. in Sudan; *O. gratissimum* L. (*O. suave*, *O. urticifolium*, S As., trop. Afr.) – oil contains eugenol, trad. medic. in Afr., infusion a louse-remover in N Uganda; *O. selloi* Benth. used for fertility control in Uruguay & *O.* spp. lvs burnt as mosquito- repellents; *O. tenuiflorum* L. (*O. sanctum*, holy or Thai b., OW trop.) – herb sacred to Hindus & found nr. every Hindu house in India

Oclorosis Raf. Older name for *Iodanthus*

Ocotea Aublet. Lauraceae (1). 350 trop. & warm Am. with few Macaronesia (1), trop. & S Afr. (7) to Madag. (50) & Masc. R: MIABH 20(1986)87. Alks. *O. insularis* (Meissner) Mez (*O. pedalifolia*, trop. Am.) – hollow ant-infested stems. Timbers (louro in S Am.) esp. *O. cymbarum* Kunth (*O. barcellensis*, l. inamui, Brazil) – fragrant bark (Orinoco sassafras) form. comm., *O. bullata* (Burchell) Baillon (black stinkwood, S Afr.), *O. rubra* Mez (determa, red l., NE S Am.); *O. usambarensis* Engl. (E Afr. camphorwood, trop. E Afr.). *O. quixos* (Lam.) Schmidt (Ecuadorian Andes) – cinnamaldehyde presence inspired members of 1514 voyage of Orellana so that they thought they saw female warriors whom they called Amazonians, hence the name of the river. See also *Chlorocardium*

ocotillo *Fouquieria splendens*

Octamyrtus Diels. Myrtaceae (Myrt.). 6 Moluccas, New Guinea

Octarrhena Thwaites. Orchidaceae (V 14). (Incl. *Chitonanthera*) c. 45 Sri Lanka, Mal., New Caled. (2), Polynesia

Octerium Salisb. = *Deidamia*

Octoceras Bunge. Cruciferae. 1 C As. to Iran & Afghanistan

Octoclinis F. Muell. = *Callitris*

Octoknema Pierre. Olacaceae. 5 trop. Afr.

Octoknemaceae Soler. = Olacaceae

Octolepis Oliver. Thymelaeaceae. 6 trop. Afr.

Octolobus Welw. Sterculiaceae. 5 trop. Afr.

Octomeles Miq. Datiscaceae (Tetramelaceae). 1 Mal. (not Malay Pen., Java): *O. sumatrana* Miq. (binuang, erima (New Guinea)) – soft pale timber good for canoes, weather-boarding, plywood, veneers etc. but large crown diff. silviculturally but in plantation in New Guinea

Octomeria R. Br. Orchidaceae (V 13). 135 trop. Am. esp. Brazil. Some cult. orn. epiphytes

Octomeron F. Robyns. Labiatae (VIII 4b). 1 trop. Afr.

Octopoma N.E. Br. (~ *Ruschia*). Aizoaceae (V). 8 S Afr.

Octospermum Airy Shaw. Euphorbiaceae. 1 New Guinea

Octotheca R. Viguier = *Schefflera*

Octotropis Beddome. Rubiaceae (? III 5). 2 S India, Burma

Ocyroe Philippi = *Nardophyllum*

Oddoniodendron De Wild. Leguminosae (I 4). 2 Gulf of Guinea

Odicardis Raf. (*Oligospermum* Hong; ~ *Veronica*). Scrophulariaceae. 2 Cauc., E Afr. R: OB 75(1984)56, T 38(1989)139

Odixia Orch. (~ *Ixodia*). Compositae (Gnap-Cass.). 2 Tasmania. R: OB 104(1991)87

odiyal flour *Borassus flabellifer*

odoko *Scottellia coriacea*

Odoniella K. Robertson (~ *Jacquemontia*). Convolvulaceae. 2 trop. Am. R: Brittonia 34(1982)417

Odontadenia Benth. Apocynaceae. 30 trop. Am. Some cult. lianes; flea-, louse- & mosquito-repellents in Colombia

Odontanthera Wight. Asclepiadaceae (III 1). 1 Middle E to Somalia

Odontelytrum Hackel. Gramineae (34f). 1 Yemen & E Afr.: *O. abyssinicum* Hackel – in running water

Odontitella Rothm. = *Odontites*

Odontites Ludwig. Scrophulariaceae. 30 W & S Eur. (15), Medit. to W As. Hemiparasites esp. *O. vernus* (Bellardi) Dumort. (red bartsia, Euras.)

Odontocarya Miers. Menispermaceae (III). 30 trop. Am. R: MNYBG 202(1970)82. Some locally med. NW Amazonia, *O. asarifolia* Barneby ed. in Chaco

Odontochilus Blume = *Anoectochilus*

Odontocline R. Nordenstam (~ *Senecio*). Compositae (Sen.-Sen.). 6 Jamaica. Trees & climbers

Odontoglossum Kunth (~ *Oncidium*). Orchidaceae (V 10). c. 140 trop. Am. R: L. Bockemühl (1989) *O.* Epiphytes & saxicolous pls of uplands, c. 20 spp. in cult. in cool greenhouses

(1150 guineas paid for 1 form of *O. crispum* Lindley (Colombia) in 1906), many cvs & parents of intergeneric hybrids with *Brassia* & *Oncidium* spp. etc.

Odontonema Nees = *Justicia*

Odontonemella Lindau = *Mackaya*

Odontophorus N.E. Br. Aizoaceae (V). 3 S Afr. R: BJ 97(1976)161

Odontophyllum Sreemad. = *Aphelandra*

Odontorrhynchus Correa. Orchidaceae (III 3). 5 S Am. R: HBML 28(1982)340

Odontosoria Fée. Dennstaedtiaceae (Lindsaeoideae). (Incl. *Sphenomeria*) 22 trop. extending to Korea & Florida. Lvs of indefinite growth forming spiny thickets (cf. *Dicranopteris*). Cult. orn. incl. *O. chinensis* (L.) J. Sm. (*S. chinensis*, E As. to Polynesia) – can be woody, red dye obtained from fronds

Odontospermum Necker ex Schultz-Bip. = *Asteriscus*

Odontostelma Rendle. Asclepiadaceae (III 1). 1 S trop. Afr.: *O. welwitschii* Rendle

Odontostomum Torrey. Tecophilaeaceae (Liliaceae s.l.). 1 California

Odontotrichum Zucc. = *Psacalium*

Odosicyos Keraudren (~ *Ampelosicyos*). Cucurbitaceae. 1 Madagascar

odum *Milicia excelsa*

Odyendea Pierre ex Engl. = *Quassia*

Odyssea Stapf. Gramineae (13d). 2 trop. & SW Afr. & Red Sea coasts. Xerophytes

Oeceoclades Lindley (~ *Angraecum*). Orchidaceae (V 9). 31 trop. Am. (1), Afr. & W Indian Ocean. R: HBML 24(1976)249. Some cult. orn. incl. *O. maculata* (Lindley) Lindley (trop. Afr., Am.) – spreading weed, autogamous with the movement of pollinia to stigma effected by rain

Oecopetalum Greenman & C. Thompson. Icacinaceae. 3 Mexico & C Am.

Oedematopus Planchon & Triana. Guttiferae (III 6). 20 trop. Am.

Oedera L. Compositae (Gnap.-Rel.). 18 S Afr. Shrubs

Oedibasis Kozo-Polj. Umbelliferae (III 8). 4 C As.

Oedina Tieghem = *Dendrophthoe*

Oemleria Reichb. (*Osmaronia*). Rosaceae. 1 W N Am.: *O. cerasiformis* (Hook. & Arn.) Landon (oso-berry) – cult. orn. early-flowering shrub, dioec., with fragrant fls & ed. fr. Like *Prunus* but G 5 free

Oenanthe L. Umbelliferae (III 8). 40 N temp. (Eur. 13), Indomal. to Aus. & trop. Afr. mts. Toxic polyacetylene hydrocarbons; some with tubers e.g. *O. divaricata* (R. Br.) Mabb. (Madeira). Water dropwort esp. *O. crocata* L. (Eur.) – poisonous (most toxic UK pl.) & used to stupify fish & *O. aquatica* (L.) Poiret (water fennel) – poisonous. *O. javanica* (Blume) DC (Indomal.) – imp. vegetable in Taiwan & Mal., eaten with rice. Cult. orn. incl. *O. pimpinelloides* L. (Eur. to SW As.) – natur., forming almost pure stands in NZ.

Oenocarpus C. Martius. Palmae (V 4f). (Incl. *Jessenia*) 9 trop. S Am. *O. bataua* C. Martius (*J. bataua*, *J. polycarpa*, pataua, seje) – form. imp. oil like olive o., 18–24% of pericarp (only 1% of kernel) but 40% more protein than soya, spines used as darts in blowpipes; *O. distichus* C. Martius (bacaba palm) – fr. an oil-source

Oenosciadium Pomel = *Oenanthe*

Oenostachys Bullock = *Gladiolus*

Oenothera L. Onagraceae. 124 Am. (esp. temp.), many natur. elsewhere (e.g. GB where hybridizing). Evening primroses. Endosperm diploid; chromosomes of some species in rings, with the rings passed through to next generation without any recombination, combinations of rings giving new genotypes sometimes with startling morphological features: an ignorance of the mechanism led de Vries to argue that these were mutations (he also had polyploids as well) & that mutations of great magnitude were the stuff of evolution, where Darwin had argued for the accumulation of small ones. Cult. orn. with fls often opening & scented in evening, pollinated by moths. Seeds a source of gamma-linolenic acid imp. in production of fatty acids & prostaglandins & used (evening primrose oil; £36m- worth sold in 1993) in treatment of premenstrual tension etc. though efficacy questioned as placebos found to be as good (more gammalinolenic acid in borage anyway), so some grown as crops incl. *O. glazioviana* Micheli ex C. Martius (*O. erythrosepala*, W N Am.), *O. biennis* L. (originated in Eur.) – seeds germinated after 80 but not 90 years of Dr Beal's experiment set up in 1879, roots & lvs ed.

Oenotheridium Reiche = *Clarkia*

Oenotrichia Copel. (~ *Microlepia*). Dennstaedtiaceae. 2 New Caledonia

Oeonia Lindley. Orchidaceae (V 16). 5 Madag., 1 ext. to Masc. R: BMNHN 4,11(1989)157

Oeoniella Schlechter. Orchidaceae (V 16). 5 Mascarenes

Oerstedella Reichb.f. (~ *Epidendrum*). Orchidaceae (V 13). 32 trop. Am. Cult. orn.

Oerstedianthus Lundell = *Ardisia*

Oerstedina Wiehler. Gesneriaceae. 3 C Am.

Oestlundorchis Szlach. Orchidaceae (III 3). 11 Mex., Guatemala. R: FFG 35(1992)23

Ofaiston Raf. Chenopodiaceae (III 3). 1 S Russia to C As.: *O. monandrum* (Pallas) Moq.

ofram *Terminalia superba*

Oftia Adans. Scrophulariaceae. 3 S Afr. R: BN 124(1971)451; see *Ranopisoa* for Madag. sp. Excl. from Myoporaceae (wood features etc.)

Ogastemma Brummitt (*Megastoma*). Boraginaceae. 1 Canary Is. & N Afr.: *O. pusillum* (Bonnet & Barratte) Brummitt – red dye used as cosmetic in N Afr.

Ogcodeia Bur. = *Naucleopsis*

ogea *Daniellia ogea*

ogeechee lime *Nyssa ogeche*

'ohelo berry *Vaccinium reticulatum*

ohia *Syzygium malaccense*

Oianthus Benth. = *Heterostemma*

oil grass *Cymbopogon* & *Vetiveria* spp., q.v.; **o. palm** *Elaeis guineensis*

Oiospermum Less. Compositae (Vern.-Cent.). 1 NE Brazil

Oistanthera Markgraf = *Tabernaemontana*

Oistonema Schltr. Asclepiadaceae (III 4). 1 Borneo: *O. dischidioides* Schltr.

oiticica *Licania* spp.

okan *Cylicodiscus gabunensis*

okari *Terminalia kaernbachii*

Okenia Schldl. & Cham. Nyctaginaceae (III). 1–2 Florida to Mexico & Nicaragua

ok-gue *Ficus pumila*

Okinawa spinach *Gynura bicolor*

Okoubaka Pellegrin & Normand. Santalaceae. 2 trop. Afr.

okoumé *Aucoumea klaineana*

okra *Abelmoschus esculentus*

Olacaceae Martinov (incl. Erythropalaceae; ~ Santalaceae). Dicots – Rosidae – Santalales. 27/180 trop., S Afr. Usu. evergreen root-parasites, trees, lianes & shrubs. Lvs simple, entire, alt.; stipules 0. Fls small, bisexual (or pl. dioec.), reg., hypogynous (–epi- or perigynous), in axillary panicles, racemes or heads; K ± cupular & ± 3–6-toothed, often accrescent in fr., C 3–6, usu. valvate & s.t. basally united (a long tube in *Schoepfia* etc.), disk annular or glands alt. with C outside A or around G, A same no. & opp. C or 2–5 times as many in 1 whorl with some staminodes, filaments sometimes basally united or adnate to C, G ((2)3(–5)) with term. style & usu. basally 2–5-loc. & apically 1-loc., each loc. with 1 pendulous, usu. anatropous, bitegmic or unitegmic ovule from top of free-central (or axile) placenta. Fr. a usu. 1-seeded drupe or nut often encl. by accrescent K; seeds with small embryo near tip of copious oily (& s.t. starchy) endosperm, thin testa, cotyledons 2–6. x = 19, 20

Principal genera: *Anacolosa, Heisteria, Olax, Schoepfia* (12 genera monotypic)

The considerable variation has led to recognition of a number of segregate families, all kept close together & here re-included

Some timbers (*Coula, Heisteria, Scorodocarpus, Ximenia*), ed. fr. (*Coula, Ochanostachys, Ximenia*) & lvs (*Olax, Strombosia*), oils (*Aptandra, Ongokea, Ximenia*), local spices & medic.

Olax L. Olacaceae. 25 OW trop., S Afr. Some lvs & fr. smell of garlic (cf. *Scorodocarpus*) – local condiments, lvs of *O. zeylanica* L. (Sri Lanka) ed.; *O. nana* Wallich ex Benth. (W Himal.) – annual shoots from woody stock; *O. subscorpioidea* Oliver (W Afr.) – chewing-stick in Nigeria

old maid *Catharanthus roseus*; **o. man** *Artemisia abrotanum*; **o. m.'s beard** *Clematis* spp. esp. *C. vitalba, Chionanthus virginicus, Tillandsia usneoides*; **o. m. cactus** *Cephalocereus senilis*

Oldenburgia Less. Compositae (Mut.-Mut.). 4 Cape. R: SAJB 53(1987)493. Subshrubs (2), cushion-pls (2)

Oldenlandia L. (~ *Hedyotis*). Rubiaceae (IV 1). 300 trop. esp. Afr. Shrubs & herbs. Dyes esp. from *O. umbellata* L. (*Hedyotis puberula* (G. Don f.) Arn., chay, Indian madder, Indomal.) – roots boiled yield yellow colour turning red in alkali & used for cloth esp. turbans

Oldenlandiopsis Terrell & W. Lewis (~ *Oldenlandia*). Rubiaceae (IV 1). 1 Carib.: *O. callitrichoides* (Griseb.) Terrell & W. Lewis. R: Brittonia 42(1990)185

Oldfieldia Benth. & Hook. Euphorbiaceae. 4 trop. Afr. R: BJBB 26(1956)338. Timber from *O. africana* Benth. & Hook. (African oak or teak)

Olea L. Oleaceae. 20 OW trop. & warm temp. Olives. Evergreen trees & shrubs with extrafl. nectaries. *O. europaea* L. (olive, complex from Macaronesia (subsp. *cerasiformis* (Batt. &

Trabaut) Ciferri) to Sind, Himal. & S Afr.) – subsp. *europaea* (incl. var. *sylvestris*, oleaster, spontaneous from seeds of cvs in Med.) cultigen of Medit. with large drupe with oily mesocarp, prob. derived from subsp. *cuspidata* (DC) Ciferri (subsp. *africana*, *O. cuspidata*, Indian o., Arabia, As.) with small drupes with thin mesocarp, certainly in cult. N of Dead Sea since 3700–3600 BC for oil & fresh or preserved drupes; expressed oil used as salad oil (best = virgin oil) & for preserving tinned sardines, the fr. used in cooking or preserved in brine & eaten as appetiser sometimes stuffed with red pepper etc., black or green, many cvs (large or small etc.); 1m t oil per annum used in cooking, lubrication, lighting & soap, a green form in Greece being made of pyrene oil from the 'stones', also medic. (bowel regulation & poss. insulin subs., a spoonful before an evening's drinking helping to prevent drunkenness), twigs used as toothbrushes to promote healthy gums in Saudi Arabia, source of manna in Dhofar; foliage an ancient sign of good will – 'olive branch'. Good timber esp. *O. europaea* subsp. *cuspidata* & *O. capensis* L. (*O. laurifolia*, trop. & S Afr.) incl. subsp. *welwitschii* (Knobloch) Friis & P.S. Green (Elgon olive, loliondo, S & E trop. Afr.) & other spp. (ironwood) – heaviest wood known with specific gravity 1.49

Oleaceae Hoffsgg. & Link. Dicots – Asteridae – Scrophulariales. 24/600 subcosmop. esp. As. Trees & shrubs, sometimes lianoid, usu. with peltate secretory hairs. Lvs opp. (spiral in *Jasminum*), simple, pinnate, 1- or 3-foliolate; stipules 0. Fls reg., usu. bisexual, in basically cymose infls or solit.; K 4–15-lobed, valvate (0 in *Fraxinus*), C (4(–12)), lobes imbricate, valvate or convolute, s.t. ± distinct or 0 (*Fraxinus*), A 2(4) attached to C-tube, anthers with longit. slits, s.t. a disk around $\underline{G}$ (2), 2-loc. with term. style, each loc. with (1)2(–4, ∞) anatropous or amphitropous unitegmic ovules in axile placentas. Fr. a capsule, berry, drupe or samara; seed with straight embryo in oily or 0 endosperm. x = 10, 11, 13–4, 23–4

Classification & chief genera (CNSWNH 2(1957)395):

Oleoideae (C in 4s (5s or 6s) or 0; ovules 2 per loc., pendulous) – 2 tribes: **Fraxineae** (fr. a samara) – *Fraxinus* (only), **Oleeae** (fr. a drupe or berry or 2-loc. capsule) – *Chionanthus, Ligustrum, Noronhia, Olea, Osmanthus, Phillyrea, Syringa*

Jasminoideae (C 4–12; ovules 1, 4 or ∞ per loc.) – 4 tribes:

Jasmineae (fr. a capsule or berry): *Jasminum, Menodora*

Fontanesieae (fr. indehiscent, compressed with surrounding wing): *Fontanesia* (only)

Forsythieae (fr. a tough capsule or indehiscent compressed with surrounding wing): *Abeliophyllum, Forsythia*

Myxopyreae (fr. fleshy): *Myxopyrum* (only)

Trad. associated with Gentianales, but A 2 & absence of internal phloem anomalous there, so Cronquist allies O. with Scrophulariales where he places Buddlejaceae (form. incl. in Loganiaceae, Gentianales), both fams with C 4, reg. otherwise rare in the S. Serologically O. form a link between the orders

Many timbers (*Olea, Fraxinus, Nestegis, Notelaea*), olives (*Olea*), scents (*Jasminum, Osmanthus, Syringa*) & many cult. orn. hardy shrubs & hedging pls (*Abeliophyllum, Chionanthus, Fontanesia, Forsythia, Fraxinus, Jasminum, Ligustrum* (privet), *Osmanthus, Phillyrea, Syringa* (lilac)): several of these hardy genera show disjunct Asiatic– European distributions – *Forsythia, Ligustrum, Osmanthus, Syringa* (KB 26(1972)487)

oleander *Nerium oleander*; **yellow o.** *Thevetia peruviana*

Oleandra Cav. Oleandraceae. 40 trop. Fronds simple with sori near midrib; some with erect stem-branches, forming thickets (almost unique in ferns)

Oleandraceae (J. Sm.) Pichi-Serm. (~ Davalliaceae). Filicopsida. 3/55 trop. & warm. Terr. to epiphytic, sometimes scandent, ferns. Rhiz. dictyostelic, with peltate scales. Frond simple or pinnate. Sori usu. term. on veins & round with reniform to suborbicular or 0 indusium. n = 40, 41

Genera: *Arthropteris, Oleandra, Psammiosorus*

Nephrolepis excl. on anatomical grounds

Oleandropsis Copel. = *Selliguea*

Olearia Moench. Compositae (Ast.- Ast.). 75 Aus., 25 New Guinea & NZ. Tree daisy, daisy bushes. Trees, shrubs & herbs, many cult. orn. (R: JRHS 90(1965)207,245) incl. *O. albida* Hook.f. (NZ) – 2n = 400+! & *O. avicenniifolia* (Raoul) Hook.f. (akeake, S Is., NZ) & a few timber trees

oleaster *Elaeagnus* spp., *Olea europaea*

Oleicarpon Airy Shaw = *Dipteryx*

Oleiocarpon Dwyer = *Dipteryx*

Olfersia Raddi (~ *Polybotrya*). Dryopteridaceae (I 2). 1 trop. Am.: *O. cervina* (L.) Kunze – cult. orn. terr. R: AFJ 76(1987) 161

Olgaea Iljin. Compositae (Card.-Card). 16 C As. to N China
olibanum *Boswellia* spp.
Oligactis (Kunth) Cass. (~ *Liabum*). Compositae (Liab.). 12 trop. Am. R: SCB 54(1983)37
Oligandra Less. = *Lucilia*
Oliganthemum F. Muell. = *Allopterigeron*
Oliganthes Cass. Compositae (Vern.-Vern.). 9 Madag. R: FMad 189(1960)182. Treelets & shrubs
Oligarrhena R. Br. Epacridaceae. 1 SW Aus.
Oligobotrya Baker = *Maianthemum*
Oligocarpus Less. (~ *Osteospermum*). Compositae (Cal.). 1 S Afr.
Oligoceras Gagnepain. Euphorbiaceae. 1 SE As. Fr. ed.
Oligochaeta (DC) K. Koch (~ *Centaurea*). Compositae (Card.-Cent.). 4 SW & C As., India. R: VGIETHZ 37(1962)315
Oligocladus Chodat & Wilczek. Umbelliferae (II). 2 Argentina
Oligocodon Keay. Rubiaceae (II 1). 1 trop. W Afr.
Oligolobos Gagnepain = *Ottelia*
Oligomeris Cambess. Resedaceae. 2 dry N temp., 1 S Afr.
Oligoneuron Small (~ *Solidago*). Compositae (Ast.). 6 N Am. R; Phytol. 75(1993)26. *O. rigida* (L.) Small – poss. rubber source
Oligophyton Linder. Orchidaceae (IV 2). 1 Zimbabwe
Oligospermum Hong = *Odicardis*
Oligostachyum Wang & Ye = *Arundinaria*
Oligothrix DC. Compositae (Sen.-Sen.). 1 Cedarberg Mts, S Afr.
Olinia Thunb. Oliniaceae. 8 E & S Afr., St Helena
Oliniaceae Harvey & Sonder. Dicots – Rosidae – Myrtales. 1/8 E & S Afr. Trees & shrubs with simple hairs & internal phloem next to pith. Lvs simple, entire, opp.; stipules rudimentary. Fls small, bisexual, reg. with hypanthium extended beyond ovary, in cymes; K a 4- or 5-lobed or -toothed rim on hypanthium, C 4 or 5, imbricate alt. with K, scale-like staminodes alt. with C, s.t. small staminodes alt. with those & A with last, anthers with longit. slits $\overline{G}$ ((3) 4 or 5), with axile placentation, term. style & as many locs as G, each with 2(3) superposed hemitropous bitegmic ovules. Fr. a drupe with 1 pluriloc. endocarp, each loc. with 1 seed; cotyledons spirally twisted or irreg. folded, endosperm 0. x = ?10
Genus: *Olinia*
Olivaea Schultz-Bip. ex Benth. Compositae (Ast.-Sol.). 2 Mex. R: Brittonia 15(1963)86
olive *Olea europaea*; **Californian o.** *Umbellularia californica*; **Ceylon o.** *Elaeocarpus serratus*; **Chinese o.** *Canarium* spp.; **Elgon o.** *O. capensis* subsp. *welwitschii*; **Indian o.** *O. europaea* subsp. *cuspidata*; **Java o.** *Sterculia foetida*; **Russian o.** *Elaeagnus* spp.
Oliveranthus Rose = *Echeveria*
Oliverella Tieghem = *Tapinanthus*
Oliveria Vent. Umbelliferae (III 8). 1 Syria to Iran
Oliveriana Reichb.f. Orchidaceae (V 10). 4 Colombia
olivillo *Aextoxicon punctatum*
Olmeca Söderstrom. Gramineae (1a). 2 Mex., 1 on acid & 1 on alkaline soil. Bamboos with fleshy fr.
Olmedia Ruíz & Pavón = *Trophis*
Olmediella Baillon. Flacourtiaceae (8). 1 C Am.: *O. betschleriana* (Goeppert) Loes. – grown as a park tree in C Am. & in Eur. botanic gardens, (?) extinct in wild
Olmedioperebea Ducke = *Maquira*
Olmediophaena Karsten = *Maquira*
Olneya A. Gray (~ *Coursetia*). Leguminosae (III 7). 1 SW N Am.: *O. tesota* A. Gray (desert ironwood, tesota) – seeds form. eaten by Am. Indians
olona *Touchardia latifolia*
Olsynium Raf. (~ *Sisyrinchium*). Iridaceae (III 1). 12 Am. R: SB 15(1990)507. Some cult. orn.
Olymposciadium H. Wolff = *Aegokeras*
Olyra L. Gramineae (4). 23 trop. Am. R: SCB 69(1989)2. Forest grasses. *O. latifolia* L. a weed spread to Afr. & Madag.
Omalanthus A. Juss. = *Homalanthus*
Omalocarpus Choux = *Deinbollia*
Omalotes DC= *Tanacetum*
Omalotheca Cass. (~ *Gnaphalium*). Compositae (Gnap.). 8 Euras. R: Phytol. 68(1990)241
Omania S. Moore. Scrophulariaceae. 1 Arabia
Ombrocharis Hand.-Mazz. Labiatae (VIII). 1 China

Ombrophytum Poeppig ex Endl. Balanophoraceae. 4 warm S Am., Galápagos. R: FN 23(1980)55. Largely subterr. & ? apomictic

Omeiocalamus Keng f. = *Arundinaria*

Omiltemia Standley. Rubiaceae (IV 5). 4 Mexico. R: SB 9(1984)410

omixochitl *Polianthes tuberosa*

Omoea Blume. Orchidaceae (V 16). 2 Java, Philippines

omoto *Rohdea japonica*

Omphacomeria (Endl.) A. DC. Santalaceae. 1 SE Aus.: *O. acerba* (R. Br.) A. DC

Omphalea L. Euphorbiaceae. 20 trop. (Afr. 1). *O. triandra* L. (Jamaican cobnut, trop. Am.) & other spp. with seeds ed. after cooking

Omphalocarpum Pal. Sapotaceae (V). 6 W & C Afr. R: T.D. Pennington, *S.* (1991)260. Local oilseeds

Omphalodes Miller. Boraginaceae. 30 temp. Euras. (Eur. 8), Mex. (6, R: Sida 13(1988)25). Cult. orn. herbs esp. *O. verna* Moench (blue-eyed Mary, Eur.)

Omphalogramma (Franchet) Franchet. Primulaceae. 15 Himal., W China. R: NRBGE 20(1949)125. Cult. orn. like *Primula* but fls bractless & seeds winged

Omphalolappula Brand. Boraginaceae. 1 temp. Aus.: *O. concava* (F. Muell.) Brand

Omphalopappus O. Hoffm. Compositae (Vern.-Vern.). 1 Angola

Omphalophthalma Karsten = *Matelea*

Omphalopus Naudin. Melastomataceae. 1 Sumatra, Java, New Guinea

Omphalotrix Maxim. Scrophulariaceae. 1 NE As.

Omphalotrigonotis W.T. Wang = *Trigonotis*

omu *Entandrophragma candollei*

Ona Ravenna = *Olsynium*

Onagraceae Juss. Dicots – Rosidae – Myrtales. 18/650 cosmop. esp. temp & warm Am. Herbs & shrubs, rarely trees to 30 m often with epidermal oil-cells, usu. with internal phloem. Lvs whorled, opp. or spirally arr., simple, entire to pinnatifid; stipules s.t. present. Fls usu. bisexual & reg. & often 4-merous, solit. or in spikes to panicles, usu. with long hypanthium nectariferous within (bird or insect (not beetle)-poll. or reg. selfed (apomixis 0)); K often valvate lobes on hypanthium, C usu. same as K, valvate, imbricate or convolute & often clawed (rarely 0), A within hypanthium or on disk, often in 2 whorls, s.t. reduced to A 2, anthers with longit. slits, pollen in monads or tetrads with viscin threads in groups, G a compound ovary with as many locs as K or partitions imperfect so placentation axile or parietal, each loc. with (1–)several–∞ anatropous, bitegmic ovules. Fr. a loculicidal capsule, berry or nut; seeds usu. ∞ with straight oily embryo & endosperm 0 (diploid initially). x = (6) 7 (–18, orig. ?11)

Principal genera: *Camissonia, Clarkia, Epilobium, Fuchsia, Gaura, Lopezia, Ludwigia, Oenothera*

Fuchsia with woody habit, fleshy fr. & unspecialized placentation held to most resemble ancestral 0.

Many cult. orn. esp. *Clarkia, Epilobium, Fuchsia, Ludwigia, Oenothera*, though species of *E., O.* & *Circaea* can be weedy

Oncella Tieghem. Loranthaceae. 4 trop. Afr.

Oncidium Sw. Orchidaceae (V 10). 680 trop. Am. to temp. S Am. R: Bradleya 1(1974)398. Cult. orn. epiphytes with extrafl. nectaries, some parents of intergeneric hybrids with spp. of *Macradenia* etc.

Oncinema Arn. Asclepiadaceae. 1 Cape: *O. roxburghii* Arn.

Oncinocalyx F. Muell. Labiatae (II; Verbenaceae). 1 E Aus.: *O. betchei* F. Muell. R: JABG 14(1991)77

Oncinotis Benth. Apocynaceae. 7 Afr. (6), Madag. (1). R: AUWP 85–2(1985)5

Oncoba Forssk. Flacourtiaceae (3). 4 trop. Afr. Some cult. orn. incl. *O. spinosa* Forssk. (trop. Afr.) – ed. pulp

Oncocalamus (G. Mann & H. Wendl.) Hook.f. Palmae (II 1h). 5 trop. Afr. Hapaxanthic rattans

Oncocalyx Tieghem (~ *Loranthus*). Loranthaceae. 7 trop. Afr., Arabia. R: NJB 5(1985)222

Oncocarpus A. Gray = *Semecarpus*

Oncodostigma Diels = *Meiogyne*

Oncosiphon Källersjö (~ *Pentzia*). Compositae (Anth.-Mat.). 7 S Afr. esp. Atlantic coast. R: BJLS 96(1988)310. Annuals. Some cult. orn.

Oncosperma Blume. Palmae (V 4n). 5 Sri Lanka to Moluccas. V. spiny monoec. palms with true dichotomous branching, s.t. cult. orn. incl. *O. tigillarium* (Jack) Ridley (*O. filamentosum*, SE As., Mal.) – young lvs a veg., timber for flooring, underwater piles etc.

Oncostemma Schumann. Asclepiadaceae. 1 São Tomé (W Afr.): *O. cuspidatum* Schumann

Oncostemum A. Juss. Myrsinaceae. 100 Madagascar, Mascarenes

Oncostylus (Schldl.) F. Bolle = *Geum*

Oncotheca Baillon. Oncothecaceae. 2 New Caledonia

Oncothecaceae Kobuski ex Airy Shaw. Dicots – Dilleniidae – Theales. 1/2 New Caled. R: BMNHN Adansonia 3(1981)305. Glabrous trees & shrubs. Lvs spirally arr. at branch-tips, simple, leathery, pinnate venation obscure, sometimes with minute glandular teeth distally; stipules 0. Fls small, regular, greenish, shortly pedicellate in axillary thyrses with angular rachides, pedicels with 1 bract & 2 sepaloid bracteoles; K 5, free, ± quincuncial before anthesis, persistent in fr., C (5) quincuncial in bud, A 5 attached to C, alt. with lobes, filaments v. short, anthers basifixed, extrorse with prolonged connective & longit. slits, G̲ (5) 5-loc. & -grooved with 5 short free styles & 1 or 2 collateral, pendulous, anatropous, unitegmic ovules per loc. Fr. a compressed drupe, ± woody with 5-loc. stone, each loc. with 1 or 0 seed with copious endosperm. n = 25
Genus: *Oncotheca*. Rather isolated taxonomically

Ondetia Benth. Compositae (Inul.). 1 SW Afr.

Ondinea Hartog. Nymphaeaceae. 1 W Aus.: *O. purpurea* Hartog. Herb in temp. pools; usu. without petals

Ongokea Pierre. Olacaceae. 1 W trop. Afr.: *O. gore* (Hua) Pierre (*O. klaineana*) – seeds yield a drying oil (isano o.)

onion *Allium cepa*; **o. couch** *Arrhenatherum elatius* var. *bulbosum*; **Egyptian o.** *A. cepa* Proliferum Group; **o. grass** *Romulea* spp.; **Japanese bunching o.** *A. fistulosum*; **multiplier** or **potato o.** *A. cepa* Aggregatum Group; **o. orchid** *Microtis* spp.; **sea o.** *Drimia maritima*; **tree o.** *A. cepa* Proliferum Group; **o. weed** (Aus.) *Asphodelus fistulosus*, *Nothoscordum borbonicum*; **Welsh o.** *A. fistulosum*

Onira Ravenna. Iridaceae (III 4). 1 temp. S Am.

Onixotis Raf. (*Dipidax*). Colchicaceae (Liliaceae s.l). 2 SW Cape & Namaqualand. R: Taxon 29(1980)600: *O. punctata* (L.) Mabb. & *O. stricta* (Burm. f.) Wijnands (*O. triquetra*) s.t. cult. orn.

Onobrychis Miller. Leguminosae (III 17). 130 Euras. (Eur. 23), Ethiopia. R: PFSUM 56 (1925) 1, 57 (1926)1. *O. viciifolia* Scop. (sainfoin, holy clover, As., natur. Eur.) – cult. for fodder, good bee-pl.

Onoclea L. Dryopteridaceae (II 2; Woodsiaceae). 2 N temp.: *O. orientalis* (Hook.) Hook. – hybridizes with *Matteuccia struthiopteris* & *O. sensibilis* L. – cult. orn.

Onocleaceae Pichi-Serm. = Dryopteridaceae

Onocleopsis Ballard. Dryopteridaceae (II 2: Woodsiaceae). 1 Mexico, C Am.

Onohualcoa Lundell = *Mansoa*

Ononis L. Leguminosae (III 22). 75 Eur. (49), Medit., Canary Is., Ethiopia & Iran. Restharrow. R: BBC 49, 2(1932)381. Shrubs & herbs, often with thorny lateral branches; fls with piston mechanism at first, later like *Trifolium* (see fam.). Some cult. orn. incl. *O. spinosa* L. (*O. campestris*, Eur.) – locally medic.

Onopordum L. Compositae (Card.-Card.). c. 60 Eur. (13). Medit. & W As. Coarse prickly biennial (to triennial) herbs with spiny decurrent lvs & usu. spiny involucral bracts. Some cult. orn. incl. *O. acanthium* L. (cotton thistle, the modern-day 'Scotch t.', Eur. to C As.) & *O. nervosum* Boiss. ('*O. arabicum*', Spain, Portugal) to 3 m tall, fls of the first form. used to adulterate saffron

Onoseris Willd. Compositae (Mut.-Mut.). 32 Mex. to Andes. R: JAA 25(1944)349. Herbs & shrubs

Onosma L. Boraginaceae. 150 Medit. (Eur. 33, Turkey 88) to Himal. & China. Some dyes esp. red dye used like alkanet from *O. echioides* L. (Medit. to W Himal.), some cult. orn. incl. *O. frutescens* Lam. (golden drop, E Medit.) etc.

Onosmodium Michaux. Boraginaceae. 5 N Am. R: JAA Suppl. 1(1991)133

Onuris Philippi. Cruciferae. 5 Chile, Patagonia. R: Parodiana 3(1984)53

Onus Gilli = *Mellera*

Onychium Kaulf. Pteridiaceae (III). 8 NE Afr. & Iran to New Guinea (esp. China). Some cult. orn. incl. *O. siliculosum* (Desv.) C. Chr. (Indomal.) – juice from crushed lvs alleged to prevent baldness

Onychopetalum R. Fries. Annonaceae. 4 Brazil

Onychosepalum Steudel. Restionaceae. 2 SW Aus.

Oonopsis (Nutt.) E. Greene (~ *Haplopappus*). Compositae (Ast.- Sol.). 3 C US

Oophytum N.E. Br. (~ *Conophytum*). Aizoaceae (V). 2 W S Afr.

Oosterdyckia Boehmer = *Cunonia*

Oparanthus Sherff. Compositae (Helia.-Cor.). 3 Rapa, Polynesia. R: FB 38 (1977)63. Trees allied to *Petrobium* (St Helena!)

opepe *Nauclea diderrichii*

Opercularia Gaertner. Rubiaceae (IV 13). 18 Aus. (11 in W Aus.). Bonsai subjects

Operculicarya Perrier (~ *Lannea*). Anacardiaceae (II). 4 Madagascar

Operculina Silva Manso. Convolvulaceae. 15 trop. *O. turpethum* (L.) Silva Manso (turpeth root, turpethum, E Afr. to Pacific) – roots drastic purgative

Ophellantha Standley. Euphorbiaceae. 2 C Am.

Ophidion Luer. Orchidaceae (V 13). 4 C Am. to Andes. R: Selbyana 7(1982)79

Ophiobotrys Gilg (~ *Osmelia*). Flacourtiaceae (9). 1 W trop. Afr.

Ophiocaryon Endl. Sabiaceae (Meliosmaceae). 7 trop. S Am. incl. *O. paradoxum* Schomb. ex Hook. (snakenut) – imported to Eur. as curiosity, embryo coiled & visible like snake

Ophiocephalus Wiggins. Scrophulariaceae. 1 Baja Calif.: *O. angustifolius* Wiggins

Ophiochloa Filgueiras, Davidse & Zulonga. Gramineae. 1 Brazil: *O. hydrolithica* Filgueiras, Davidse & Zulonga – on serpentine. R: Novon 3(1993)360

Ophiocolea Perrier. Bignoniaceae. 5 Madagascar, Comoro Is.

Ophioderma (Blume) Endl. = *Ophioglossum*

Ophioglossaceae (R. Br.) C. Agardh. Filicopsida. 3/75 cosmop. Small homosporous herbs, some trop. spp. epiphytic; roots mycorrhizal, without root-hairs. Lvs solit. or few, without circinate vernation, comprising succ. sterile often entire blade & a long-stalked spike or panicle with sorus-less sporangia sunk in it. Prothalli subterr., colourless, mycotrophic. Polyploidy to v. high levels

Genera: *Botrychium* (but see GBS 40(1987)1 for cladistic splitting), *Helminthostachys*, *Ophioglossum*

Held to be allied to the progymnosperm line (periderm, circular bordered-pitted tracheids & non-circinate vernation etc. suggest alliance with cycads!). Some locally eaten & medic.

Ophioglossum L. Ophioglossaceae. 25–30 subcosmop. (Eur. 3). R: MTBC 19, 2(1938)111. Adder's-tongue, snake-tongue. Fertile blade usu. a spike with 2 rows of sporangia. 2n = 140–1440 (*O. reticulatum* L. (trop.) – highest chromosome no. known). Some ed. & locally medic., few cult. orn. long-lived & slow-growing incl. *O. palmatum* L. (SE As., Madag., Réunion, trop. & warm Am.) – bizarre epiphyte with coarse palmatifid frond bearing pendent spikes of sporangia

Ophiomeris Miers = *Thismia*

Ophionella Bruyns = *Pectinaria*

Ophiopogon Ker-Gawler. Convallariaceae (Liliaceae s.l.). c. 20 Indomal. to Himal. & Japan. Japanese hyacinth. Cult. orn. evergreen turf-forming pls esp. *O. japonicus* (L.f.) Ker-Gawler (Japan, Korea) – fls white to lilac not prod. in trop. lowlands where often grown; fr. blue, pea-sized; tuberous roots ed.

Ophiorrhiza L. Rubiaceae (IV 2). 150 Indomal. *O. tomentosa* Jack (Mal.) s.t. viviparous

Ophiorrhiziphyllon Kurz. Acanthaceae. 5 SE As.

Ophiuros Gaertner f. Gramineae (40h). 4 NE trop. Afr. to S China & Aus.

Ophrestia H.M. Forbes. Leguminosae (III 10). 13 trop. OW

Ophryococcus Oersted. Rubiaceae (I 8). 1 C Am.

Ophryosporus Meyen. Compositae (Eup.-Crit.). 37 S Am. R: MSBMBG 22(1987)363

Ophrypetalum Diels. Annonaceae. 1 trop. E Afr.: *O. odoratum* Diels

Ophrys L. Orchidaceae (IV 2). 25 Eur. (20), W As., N Afr. R: E. Nelson (1962) *Gestaltwandel ... Monographie ... Gattung O.* Terrestrial herbs with swollen tubers. Lip often resembling large insect entering fl. Insects attracted by cyclic sesquiterpene alcohols & hydrocarbons (aliphatic acetates esp. octyl a. in *O. speculum* Link (*O. vernixia*, Medit.), prob. chemically mimetic of female bee sex hormones in *O. lutea* Cav. (Medit.) at least) are males, which emerge before females & attempt to copulate with the fl., leaving sperm & taking pollinia for a second fl. (cf. *Caladenia*, *Chiloglottis*, *Cryptochilus*, *Leporella*). Some cult. orn. & named (e.g. spider orchids) after fancied resemblance to insects or spiders, not their pollinators. *O. apifera* Hudson (bee orchid, W & C Eur.) – usu. hapaxanthic after 5–8 yrs veg. growth, accounting for great fluctuations in annual nos. of (usu. self-poll. at least in N) flowering pls, mutant form with deformed lip known as wasp o.; *O. insectifera* L. (fly o., Eur.) etc. Some tubers form. used as salep in Turkish delight

Ophthalmoblapton Allemão. Euphorbiaceae. 3 Brazil

Ophthalmophyllum Dinter & Schwantes (~ *Conophytum*). Aizoaceae (V). 15 S W Afr. Photosynthetic organs with translucent window-spots

Opilia Roxb. Opiliaceae. 2 OW trop. R: Willdenowia 12(1982)161

Opiliaceae (Benth.) Valeton (~ Olacaceae). Dicots – Rosidae – Santalales. 10/32 trop. Usu. evergreen root-parasitic trees & shrubs, sometimes lianoid. Lvs simple, alternate; stipules 0. Fls small, usu. bisexual (dioec. in *Agonandra* & *Gjellerupia*), in axillary or cauliflorous spikes to panicles or umbels; K small, cupular ± 4 or 5 small lobes or teeth (not accrescent), C 4 or 5, sometimes basally connate, valvate (usu. 0 in female fls), A opp. C, s.t. on C or C-tube, anthers with longit. slits, disk of free or ± connate nectaries around G (2–5) with simple or 0 style, sunk in disk, 1-loc. with 1 pendulous (erect in *Agonandra*), anatropous, unitegmic ovule (integument s.t. not recognizable). Fr. a drupe; seeds with small embryo with (2) 3 or 4 cotyledons in copious oily starchy endosperm. n = 10

Genera: *Agonandra* (Am., basal ovule, dioec.), *Cansjera*, *Champereia*, *Gjellerupia*, *Lepionurus*, *Melientha*, *Opilia*, *Pentarhopalopilia*, *Rhopalopilia*, *Urobotrya*

Some ed. fr. (*Champereia*, *Melientha*)

Opisthiolepis L.S. Sm. Proteaceae. 1 Queensland

Opisthocentra Hook.f. Melastomataceae. 1 N Brazil

Opisthopappus Shih (~ *Chrysanthemum*). Compositae (Anth.-Tan.). 2 China. R: APS 17, 3(1979)110

Opithandra B.L. Burtt. Gesneriaceae. 8+ E As.

opium *Papaver somniferum*

Opizia J. Presl. Gramineae (33c). 1 S Mexico, Cuba: *O. stolonifera* J. Presl

Oplismenopsis L. Parodi. Gramineae (34b). 1 Uruguay, Arg.: *O. najada* (Hackel & Arech.) L. Parodi – floating

Oplismenus Pal. Gramineae (34b). 7 trop. & warm. R: PM 13(1981). Forest shade, some with sticky awns for disp. *O. hirtellus* (L.) Pal. (trop. Am.) – fr. in comm. bird food in US, 'Variegatus' ('*Panicum variegatum*') – cult. greenhouse hanging-basket pl.

Oplonia Raf. Acanthaceae. 19 trop. Am. (Peru 1, WI 12) & Madag. (5), presumed extinct in Afr. R: BBMNHB 4(1971)259, ABASH 23(1977)303. Heterostyly

Oplopanax (Torrey & A. Gray) Miq. Araliaceae. 3 NW N Am., Japan. Prickly decid. treelets, cult. orn. esp. *O. horridus* (Sm.) Miq. (N Am.) – 'devil's club'

Opoidea Lindley (~ *Peucedanum*). Umbelliferae (inc. sed.). 1 Iran

opopanax *Opopanax chironium*, *Commiphora* spp., esp. *C. kataf*, *Acacia farnesiana*

Opopanax Koch. Umbelliferae (III 10). 3 Balkans (Eur. 2) to Iran. *O. chironium* Koch (Medit.) – source of gum opopanax used in scent-making & form. medic.

Opophytum N.E. Br. = *Mesembryanthemum*

opossum wood *Halesia carolina*

Opsiandra Cook = *Gaussia*

optical fibre plant *Isolepis cernua*

Opuntia Miller. Cactaceae (II). (Incl. *Nopalea*) c. 200 Massachusetts & Br. Columbia to Galápagos & Straits of Magellan (range of fam. in Am.). Chollas (cylindrical-stemmed spp.) & spp. with flattened joints, with usu. early decid. lvs & many minute irritant glochids often with larger spines; alks incl. mescaline. Some ed. fr. (prickly pears) & sources of cochineal (e.g. *O. cochenillifera* (L.) Miller (*Nopalea c.*, cochineal pl., cultigen orig. Mex.) – long cult. trop. Am. but rarely fls in cult.), i.e. dried bodies of mealy bugs feeding on pls form. coll. for preparation of carmine (scarlet dye now synthesized artificially), others for living fences, spineless forms for forage in bad times; some introd. to OW & natur., esp. *O. aurantiaca* Lindley (tiger pear, Uruguay & adjacent Arg.) to S Afr. & E Aus. (& S Am.), occupying range-land to detriment of grazing (c. 25m ha in E Aus. by 1925, advancing 100 ha per hr), now largely controlled by larvae of moths introduced from native habitats where usu. controlled thus – an early success of 'biological control'. Many cult. orn.; 4 subg. (*Brasiliopuntia* Schumann – tree-like, main stem unjointed, S Brazil to N Arg.; *Consolea* (Lemaire) A. Berger – tree-like with stems often bearing flattened joints at first, WI & Florida; *Cylindropuntia* Engelm. – tree-like or prostrate with stems globose to cylindrical, N & S Am.; *Opuntia* (prickly pear, tuna) – stems of flattened joints, prostrate to tree-like, N & S Am.); commonly seen spp. incl. *O. articulata* (Pfeiffer) D. Hunt (W Arg.) – some cvs with long papery spines (*O. diademata*, *O. papyracantha*); *O. compressa* (Salisb.) Macbride (C & E US) – hardy England, natur. Switzerland; *O. ficus-indica* (L.) Miller (Indian or Barbary fig, origin Mex.) – widely cult. for fr. exported to Eur. etc., natur. in Medit.; *O. floccosa* Salm Dyck (mts of Peru, Bolivia) – cushion-forming at 4000 m; *O. fulgida* Engelm. (SW N Am.) – source of cholla gum; *O. microdasys* (Lehm.) Pfeiffer (C & N Mex.) – commonly cult. sp. with flattened joints with many closely set yellow glochids

orache *Atriplex hortensis*

orange *Citrus × aurantium*; **bergamot o.** *C. × aurantium* (US), *C. × bergamia*; **mandarin o.** *C. reticulata*; **mock o.** *Philadelphus* spp.; **Osage o.** *Maclura pomifera*; **satsuma o.** *C. reticulata*; **Seville** or **sour o.** *C. × aurantium* Sour Orange Group

Orania Zipp. Palmae (V 4a). 19 Madag. (3), Mal. to trop. Aus. R: Lyonia 1(1980)211. Poisonous. *O. disticha* Burret (New Guinea) – lvs distichous

Oraniopsis (Becc.) Dransf., A. Irvine & Uhl. Palmae (IV 2). 1 N Queensland: *O. appendiculata* (Bailey) Dransf., A. Irvine and Uhl. R: Principes 29(1985)57

Orbea Haw. (~ *Stapelia*). Asclepiadaceae (III 5). 20 trop. & S Afr. R: Excelsa Tax. Ser. 1(1978)4. Cult. orn. succ. with stinking mottled fls attractive to flies & succ. fanciers esp. *O. variegata* (L.) Haw. (*Stapelia v.*, Cape, natur. S Aus.) – 1 of first Cape pls known in Eur. (drawing sent back in 1624)

Orbeanthus Leach. Asclepiadaceae (III 5). 2 S Afr. Cult. orn.

Orbeopsis Leach (~ *Caralluma*). Asclepiadaceae (III 5). 10 trop. & S Afr. R: Excelsa 1(1972)61

Orbexilum Raf. Leguminosae (III 11).(Excl. *Pediomelum*) 25 Am., esp. N

Orbignya C. Martius ex Endl. Palmae (V 5c). 20 trop. Am. R: Phytologia 36(1977)89. Slow-growing, often large, palms with valuable palm kernel oil, the most imp. source for indigenous peoples & used in Western aromatherapy & cosmetics (aguassú, babassu, babacu, esp. *O. phalerata* C. Martius (*O. barbosiana*, Brazil; coco de macaco) – poll. beetles & s.t. wind, source of coal-like fuel, methanol from endocarp & mesocarp, plastics, animal feed etc.; *O. cohune* (C. Martius) Standley (*Attalea c.*, cohune nut, C Am.) – source of oil, lvs also ed.; *O. cuatrecasana* Dugand (táparos, Colombia) – seeds ed.; *O. spectabilis* (C. Martius) Burret (curua, Brazil) – oil & thatch

Orchadocarpa Ridley. Gesneriaceae. 1 Malay Peninsula

orchid, bee *Ophrys apifera*; **bird's-nest o.** *Neottia nidus-avis*; **bog o.** *Malaxis paludosa*; **butterfly o.** *Platanthera* spp.; **donkey o.** *Diurus* spp.; **duck o.** *Caleana major*; **early purple o.** *Orchis mascula*; **fen o.** *Liparis loeselii*; **fly o.** *Ophrys insectifera*; **frog o.** *Coeloglossum viride*; **green-winged o.** *Orchis morio*; **Jersey o.** *Orchis laxiflora*; **lady's slipper o.** *Cypripedium* spp. esp. *C. calceolus*; **l.'s tresses o.** *Spiranthes* spp.; **leek o.** *Prasophyllum* spp.; **lizard o.** *Himantoglossum hircinum*, (Aus.) *Burnettia cuneata*; **man o.** *Aceras anthropophorum*; **midge o.** *Genoplesium* spp.; **military o.** *O. militaris*; **monkey o.** *O. simia*; **musk o.** *Herminium monorchis*; **onion o.** *Microtis* spp.; **poor man's o.** *Schizanthus* spp.; **potato o.** *Gastrodium sesamoides*; **pyramid o.** *Anacamptis pyramidalis*; **rock o.** *Dendrobium speciosum*; **soldier o.** *O. militaris*; **spider o.** *Ophrys* spp.; **sweet-scented o.** *Gymnadenia conopsea*; **o. tree** *Amherstia nobilis*, *Bauhinia* spp.; **twayblade o.** *Listera* spp.; **wax-lip o.** *Glossodia* spp.

Orchidaceae Juss. (incl. Apostasiaceae & Cypripediaceae). Monocots – Liliidae – Orchidales. 788/18 500 cosmop. (e.g. New Guinea 133/2300). Perennial mycotrophic epiphytic (the great majority) or terr. herbs, rarely lianes (e.g. *Clematepistephium*, *Vanilla*) or annuals (see *Zeuxine*), s.t. without chlorophyll, v. rarely (Aus.) completely subterr. or rheophytic (e.g. *Appendicula*), always with raphides in some cells & often with mucilage-cells & alks, frequently with crassulacean acid metabolism & generally with roots with multi-layered velamen, in terr. spp. often swollen into tubers, or stems forming corms or rhizomes; stems of epiphytic spp. often thickened to form a pseudobulb with adventitious roots. Lvs entire, spirally arr., distichous, rarely opp. or whorled, s.t. scale-like, often ± fleshy & basally sheathing. Fls usu. bisexual, 3-merous, epigynous, irreg., in racemes or panicles to solit., often resupinate; P usu. petaloid though outer 3 s.t. greenish, the median app. adaxial (truly abaxial) s.t. diff. from others or 2 or 3 basally connate, inner 3 with app. abaxial (truly adaxial) one (exc. Apostasioideae) usu. larger & diff. colour from laterals forming labellum (lip), the laterals often like outer P, nectaries various (s.t. hollow spur from base of labellum, or a cup on or embedded in G, or extrafl. etc.), A 1(–3) all truly abaxial opp. labellum, when 1 united with style forming gynostemium (column) & truly median stamen of outer whorl, other 2 laterals of inner s.t. staminodal with vasc. strands of adaxial stamens in gynostemium, anthers with longit. slits, pollen grains solit. in Apostasioideae & Cypripedioideae, in tetrads & pollinia in rest with 1–6 pollinia per pollen-sac, each often with slender tip or caudicle, $\overline{G}$ (3), 1-loc. (3-in Apostasioideae) with marginal placentas & usu. gynostemium subtended by enlarged stigma-lobe (rostellum) to which caudicles often attached & from which a sticky viscidium is removed when pollinia taken by pollinators, ovules anatropous, (uni-) bitegmic, minute, ∞, development triggered by poll. & fertilization often delayed (up to 6 months). Fr. usu. a capsule with 3(6) longit. slits but apically & basally closed (fissuricidal); seeds minute & ∞ (to several million) with minute undeveloped embryo, endosperm formation arrested at 2–4(–16)-nucleate stage, only testa usu. persisting. Seeds usu. germinating only in presence of appropriate fungus, when forming a protocorm with basal rhizoids, no radicle &

usu. no cotyledon, the protocorm eventually giving rise to apical lvs. x = 6–29+. R: R.L. Dressler (1993) *Phylogeny & classification of the orchid fam.*

Classification & chief genera (Dressler; though a small no. of genera & tribes unplaced in this system):

I. **Apostasioideae** (lvs spirally arr., plicate; fls weakly irreg., s.t. resupinate (*Neuwiedia*), A 2 or 3, pollen in monads & pollinia 0, some selfing; Indomal.): *Apostasia* & *Neuwiedia* (only)

II. **Cypripedioideae** (usu. terr.; lvs spirally or distichously arr., s.t. plicate; fls resupinate, labellum slipper-shaped, A 2 (inner whorl) present with median outer A a staminode, true pollinia rare; N temp. to trop. exc. Afr.): *Cypripedium, Paphiopedilum, Phragmipedium, Selenipedium*

III. **Spiranthoideae** (3 tribes):
   1. **Diceratostleae** (terr. with reed-like stems, pollinia 4): *Diceratostele* (only)
   2. **Tropidieae** (as 1 but pollinia 2): *Tropidia*
   3. **Cranichideae** (as 2 but soft herbaceous lvs; 4 subtribes): *Anoectochilus, Cranichis, Cyclopogon, Erythrodes, Goodyera, Spiranthes, Zeuxine*

IV. **Orchidoideae** (survive adverse periods as dormant tuberoids, lvs soft herbaceous, pseudocopulation frequent, (?) orig. southern but now the common O. of N; 3 tribes rather subtly distinguished):
   1. **Diurideae** (trop. As. to (esp. Aus.), S Am.; 10 subtribes): *Chloraea, Caladenia, Corybas, Prasophyllum, Pterostylis, Rhizanthella*
   2. **Orchideae** (Afr., N hemisph.; 2 subtribes): *Anacamptis, Cynorkis, Dactylorhiza, Habenaria, Holothrix, Ophrys, Orchis, Peristylus, Platanthera*
   3. **Diseae** (Afr. to As.; 4 subtribes): *Disa, Disperis, Satyrium*

V. **Epidendroideae** (usu. epiphytes with pseudobulbs, often with distichous fleshy lvs & lateral infls but exceptions to all these; the bulk of the fam. arr. in 16 subtly distinguished tribes, most of these arr. into 'phylads' by Dressler, l.c.):
   1. **Neottieae** (usu. terr. without pseudobulbs; 2 subtribes): *Cephalanthera, Epipactis, Listera, Neottia*
   2. **Palmorchideae** (terr. with reed-like stem): *Palmorchis* (only)
   3. **Triphoreae** (terr. or mycotrophic): *Triphora*
   4. **Vanilleae** (mycotrophic &/or lianoid, 3 subtribes): *Epistephium, Galeola, Vanilla*
   5. **Gastrodieae** (leafless mycotrophs, poss. distinct subfam., 3 subtribes): *Didymoplexis, Epipogium, Gastrodia*
   6. **Nervilieae** (terr. with globose corm): *Nervilia* (only)
   7. **Malaxideae**: *Liparis, Malaxia, Oberonia*
   8. **Calypsoeae**: *Corallorrhiza*
   9. **Cymbidieae**: *Catasetum, Cymbidium, Cyrtopodium, Eulophia, Grammatophyllum, Mormodes*
   10. **Maxillarieae**: *Batemania, Dichaea, Gongora, Lycaste, Maxillaria, Miltonia, Odontoglossum, Oncidium, Rodriguezia, Stanhopea, Telipogon, Zygopetalum*
   11. **Arethuseae**: *Bletia, Calanthe, Phaius, Phocoglottis, Spathoglottis*
   12. **Coelogyneae**: *Coelogyne, Dendrochilum, Pholidota*
   13. **Epidendreae**: *Cattleya, Elleanthus, Encyclia, Epidendrum, Laelia, Lepanthes, Pleurothallis, Polystachya, Scaphyglottis, Sobralia, Stelis, Trichosalpinx*
   14. **Podochileae**: *Appendicula, Ceratostylis, Eria, Phreatia*
   15. **Dendrobieae**: *Bulbophyllum, Cadetia, Dendrobium, Diplocaulobium*
   16. **Vandeae**: *Aerangis, Aerides, Angraecum, Cleisostoma, Gastrochilus, Jumellea, Luisia, Phalaenopsis, Taeniophyllum, Thrixspermum, Trichoglottis, Tridactyle, Vanda*

Distinct from all other monocots in reduction of adaxial A & presence of ∞ endospermless seeds & gynostemium (cf. Asclepiadaceae & Stylidiaceae) but prob. derived from terr. Lilialean type with A 6. Mycorrhizas (non-specific) incl. strains of *Armillaria* & *Rhizoctonia* pathogenic in other pls, suggesting they were originally thus in orchids too (Smith & Douglas). Found in driest deserts to highest mts, from few mm to several m tall, but no aquatics save rheophytes, nor halophytes (but see *Bletia, Brassavola*), poll. by bees, wasps, flies (e.g. *Acianthus*), ants (e.g. *Microtis*), beetles, birds (e.g. *Cryptochilus?, Sacoila*), bats & poss. frogs, which transfer pollinia, often with quick-setting 'glue' (? polyisoprene) or explosive systems projecting them up to 60 cm from fl.; the pollinators often attracted by scent (from elaiophores as found in only 9 other fams) & males even carrying out pseudocopulation (*Ophrys* (q.v.), *Caladenia, Chiloglottis, Cryptostylis, Leporella, Trichoceros*), though it is unlikely that any sp. is poll. by just 1 animal sp., while the fam. is notorious for occurrence of natural hybrids & their synthesis in horticulture (up to

20 spp. in 5 genera combined in 1 pl.), comm. hybrids entering trade at rate of c. 150 a month (Hunt). Sometimes insects imprisoned (see *Coryanthes*, *Porroglossum*) & s.t. intoxicated, but pollen never offered (a third of orchids offer no reward at all, merely deception), though pollen-like 'pseudopollen' (some *Polystachya* spp.) or nutritive oils offered though some spp. mimic other fls & there is no 'reward', while others attract egg-laying Diptera by imitating carrion (e.g. *Bulbophyllum* spp.), fungal fr. bodies (*Corybas*) or even aphids; some insects coll. pheromones, hold territory or roost in fls & incidentally poll. them

Fr. usu. dry (indehiscent with seeds escaping as it rots in *Galeola* etc.), with hygroscopic hairs often between seeds, which are expelled by irreg. movements of hairs when wetted. The small seeds have allowed O. to compete with spore-pls (esp. ferns) in the rain forest canopy

Despite the huge size of this fam., it is of almost no significance to the Common Man in most of the world, though millions of pounds are expended in cultivating the more showy tropical spp., hybridizing them & selling them as cut-fls, while they have an aura of the exotic & even temp. spp. are sought out & dug up, such that many are endangered (all epiphytic & lithophytic spp. in Aus. now protected): fortunately many are now propagated by micro-techniques from meristems. Most often seen in florists' are spp. & hybrids of *Arachnis*, *Cattleya*, *Cymbidium*, *Odontoglossum*, *Paphiopedilum*, *Papilionanthe*, *Phalaenopsis* & *Vanda*. Salep used in Turkish delight form. gathered from *Dactylorhiza*, *Eulophia*, *Ophrys* & *Orchis* spp., flavourings from *Angraecum*, *Leptotes* & *Vanilla*, while *Dendrobium* spp. used in Aus. aboriginal body-paint & *Geodorum* spp. provide a strong gum for musical instruments

Orchidantha N.E. Br. Lowiaceae. 7 S China to W Mal.

Orchipedum Breda. Orchidaceae (III 3). 1 Malay Peninsula, Java, Philippines

Orchis L. Orchidaceae (IV 2). 33 N temp. (Eur. 23) to SW China & India. Terr. orchids sometimes cult., dried starchy tubers form. used medic. & culinarily (salep) as in Turkish delight, their shape reminiscent of testicles (Gk. *orchis*) so assoc. with potency, *O. mascula* (L.) L. (early purple orchid, dead man's finger, Eur.) being Shakespeare's 'long purples'. Many Br. spp. rare & endangered, *O. militaris* L. (military or soldier orchid, Eur., Medit.) – orchinol (a phytoalexin) in tubers) & *O. simia* L. (monkey o., Eur. Medit.) being protected spp.; other Br. spp. incl. *O. laxiflora* Lam. (Jersey o., Channel Is. & Belgium to Medit.) & *O. morio* L. (green-winged o., Euras.). *O. caspia* Trautv. (E Medit.) – nectarless sp. attracting insects by deceit in mimicking nectariferous fls of *Asphodelus* & *Bellevalia* spp. etc. in same habitat in Israel

Orcuttia Vasey. Gramineae (30). 5 Calif. R: BTBC 68(1941)149. Dormant in drought yrs but flowering in spring pools when flooding adequate

Oreacanthus Benth. Acanthaceae. 4 C Afr. R: KB 37(1982)467. *O. mannii* Benth. – hapaxanthic after 9 yrs growth

Oreanthes Benth. Ericaceae. 4 Andes of Ecuador

Orectanthe Maguire. Xyridaceae (Abolbodaceae). 1 Venezuela

Oregandra Standley. Rubiaceae (*inc. sed.*). 1 Panamá

oregano *Origanum* spp. esp. *O. vulgare*, (Am.) *Lippia* spp. esp. *L. graveolens*

Oreiostachys Gamble = *Nastus*

Oreithales Schldl. Ranunculaceae (II 2). 1 Andes from Ecuador to Bolivia

Oreobambos Schumann. Gramineae (1b). 1 trop. E Afr.: *O. buchwaldii* Schumann – bamboo

Oreoblastus Susl. = *Christolea*

Oreobliton Durieu. Chenopodiaceae (I 1). 1 Algeria, Tunisia: *O. thesioides* Durieu

Oreobolopsis Koyama & Guag. Cyperaceae. 1 Bolivia: *O. tepalifera* Koyama & Guag. R: Darw. 28(1988)79

Oreobolus R. Br. Cyperaceae. 8 Pacific (Aus. 5, 4 endemic). R: BJLS 96(1988)119

Oreocalamus Keng = *Chimonobambusa*

Oreocallis R. Br. Proteaceae. 2 Peru, Ecuador (OW spp. = *Alloxylon*)

Oreocarya E. Greene = *Cryptantha*

Oreocereus (A. Berger) Riccob. (~ *Borzicactus*). Cactaceae (III 4). 6–7 W S Am. Cult. orn. esp. *O. celsianus* (Salm Dyck) Riccob. (*B. celsianus*, NW Arg., Bolivia) & *O. trollii* (Kupper) Backeb. (*B. trollii*, S Bolivia, N Arg.) with long matted hairs

Oreocharis Benth. Gesneriacae. 27 China to SE As. R: APS 25(1987)264

Oreochloa Link. Gramineae (17). 4 S Eur. R: OBC 3(1946)239

Oreochorte Kozo-Polj. = *Anthriscus*

Oreochrysum Rydb. (~ *Solidago*). Compositae (Ast.-Sol.). 1 WN Am.: *O. parryi* (A. Gray) Rydb. R: Phytol. 75(1993)334

Oreocnide Miq. Urticaceae (III). (Incl. *Villebrunea*) 15 China & Japan, Indomal. Shrubs, s.t. scandent, & trees. *O. integrifolia* (Gaud.) Miq. (trop. As.) – useful fibre

Oreocome Edgew. (~ *Selinum*) Umbelliferae (III 8). 1 Himalaya

Oreodendron C. White. Thymelaeaceae. 1 Queensland

Oreodoxa auctt. = *Roystonea*

Oreogrammitis Copel. = *Grammitis*

Oreograstis Schumann = *Carpha*

Oreoherzogia W. Vent = *Rhamnus*

Oreoleysera Bremer (~ *Leysera*). Compositae (Gnap.-Rel.). 1 SW Cape: *O. montana* (Bolus) Bremer. R: OB 194(1991)67

Oreoloma Botsch. Cruciferae. 4 China. 4: BZ 65(1980)425

Oreomitra Diels. Annonaceae. 1 New Guinea: *O. bullata* Diels

Oreomunnea Oersted. Juglandaceae. 2 C Am. R: AMBG 59(1972)298

Oreomyrrhis Endl. Umbelliferae (III 5) . 23 Taiwan, N Borneo, W Pacific (Aus. 7), C & S Am. R: UCPB 27(1955)347

Oreonana Jepson. Umbelliferae (III 5). 3 California mts. R: Madrono 26(1979)133

Oreonesion A. Raynal. Gentianaceae. 1 trop. W Afr.

Oreopanax Decne. & Planchon. Araliaceae. c. 80 trop. Am. Some cult. orn. hairy trees & shrubs esp. *O. capitatus* (Jacq.) Decne. & Planchon grown as juvenile epiphyte

Oreophysa (Boiss.) Bornm. Leguminosae (III 15). 1 Iran mts: *O. microphylla* (Jaub. & Spach) Browicz (*O. triphylla*)

Oreophyton O. Schulz. Cruciferae. 1 E & NE Afr. mts

Oreopolus Schldl. = *Cruckshanksia*

Oreoporanthera Hutch. Euphorbiaceae. 1 NZ

Oreopteris Holub = *Thelypteris*

Oreorchis Lindley. Orchidaceae (V 8). 9 Himalaya to Japan

Oreoschimperella Rauschert (*Schimperella*). Umbelliferae (III 8). 3 Yemen, Ethiopia, Kenya mts

Oreoselinum Miller (~*Peucedanum*). Umbelliferae (III 10). 1 Eur.

Oreosolen Hook.f. Scrophulariaceae. 3 Himalaya

Oreosparte Schltr. Asclepiadaceae (III 4). 1 Sulawesi: *O. celebica* Schltr.

Oreosphacus Leyb. = *Satureja*

Oreostemma E. Greene (~ *Aster*). Compositae (Ast.-Ast.). 3 W US

Oreostylidium S. Berggren. Stylidiaceae. 1 NZ

Oreosyce Hook.f. Cucurbitaceae. 2 trop. Afr., Madagascar

Oreothyrsus Lindau = *Ptyssiglottis*

Oreoxis Raf. Umbelliferae (III 8). 4 W N Am. *O. alpina* (A. Gray) J. Coulter & Rose – cult. orn. alpine

Oresitrophe Bunge. Saxifragaceae. 1 NE China

Orestias Ridley. Orchidaceae (V 7). 3 trop. Afr.

Orias Dode = *Lagerstroemia*

Oricia Pierre. Rutaceae. 8 trop. & S Afr.

Oriciopsis Engl. Rutaceae. 1 trop. W Afr.

oriental beech *Fagus sylvatica* subsp. *orientalis*; **o. spruce** *Picea orientalis*

Origanum L. Labiatae (VIII 2). 36 Euras. (Eur. 13), Medit. R: LBS 4(1980)1. Dwarf shrubs or herbs with ess. oils, some cult. as potherbs esp. *O. dictamnus* L. (dittany, Greece & Crete) – cult. orn., form. med.; *O. majorana* L. (*Majorana hortensis*, (sweet) marjoram, Medit., Turkey but now widely natur.) – cult. herb for flavouring (methyl chavicol) meat & sausages, ess. oil used in tinned meat, form. medic.; *O. onites* L. (pot m., SE Eur., E Medit.) – inferior; *O. vulgare* L. (oregano, Eur. to C As., natur. E US) – dried lvs much used in pizza etc., source of red dye. *O. syriacum* L. (Near E) – hyssop of the Bible (e.g. at Crucifixion). A Turkish sp. used against rheumatism as containing corvacrone, a prostaglandin synthetase inhibitor & this a painkiller

Orinoco sassafras *Ocotea cymbarum*

Orinus A. Hitchc. Gramineae (31d). 2 + China. Desert dunes at high alts. R:KB 6(1951)453

Orites R. Br. Proteaceae. 8 temp. E Aus., 3 S Am. Some timbers & cult. orn. e.g. *O. excelsa* R. Br. (prickly ash, E Aus.)

Oritrephes Ridley = *Anerincleistus*

Oritrophium (Kunth) Cuatrec. Compositae (Ast.-Ast.). 15 Andes, Mex. (1)

Orixa Thunb. Rutaceae. 1 China, Korea, Japan: *O. japonica* Thunb. – cult. orn. dioec. shrub with alks & scented lvs used for hedging in Japan

Orlaya Hoffm. Umbelliferae (III 3). 3 SE Eur. (3) to C As.

Orleanesia Barb. Rodr. Orchidaceae (V 13). 8 trop. S Am.
Ormenis (Cass.) Cass. = *Chamaemelum*
Ormocarpopsis R. Viguier. Leguminosae (III 13). 5 Madagascar
Ormocarpum Pal. Leguminosae (III 13). 20 trop. & warm OW
Ormoloma Maxon. Dennstaedtiaceae (Lindsaeoideae). 1–2 trop. Am.
Ormopteris J. Sm. = *Pellaea*
Ormopterum Schischkin. Umbelliferae (III 8). 1 C As.
Ormosciadium Boiss. Umbelliferae (III 11). 1 E As. Minor
Ormosia Jackson. Leguminosae (III 2). 100 E S Am., E As. to NE Aus. Alks; seeds orn. & used as beads, some resembling *Abrus precatorius*, & others cult. orn. or useful timber esp. *O. krugii* Urban (WI) & *O. monosperma* (Sw.) Urban (WI to NE Venez.). Five segregates based on diff. dispersal mechanisms: *Fedorovia, Macroule, Placolobium, Ruddia, Trichocyamos*; R (China – 35 spp.): APS 22(1984)117. *O. nobilis* Tul. (Amaz.) – fine roots grow under damaged bark & coalesce, repairing it
Ormosiopsis Ducke = *Ormosia*
Ormosolenia Tausch (~ *Peucedanum*). Umbelliferae (III 10). 1 E Medit.: *O. alpina* (Schultes) Pim. R: EJB 49(1992)214
Ornichia Klack. (~ *Chironia*). Gentianaceae. 3 R: BMNHN4, 8 (1986) 195
Ornithidium R. Br. = *Maxillaria*
Ornithoboea Parish ex C.B. Clarke. Gesneriaceae. 8 SE As. R: NRBGE 22 (1958)287
Ornithocarpa Rose. Cruciferae. 2 Mexico
Ornithocephalus Hook. Orchidaceae (V 10). 28 trop. Am. Some cult. orn.
Ornithochilus (Lindley) Wallich ex Benth. Orchidaceae (V 16). 3 India to China & Thailand. R: OB 95(1988)42
Ornithogalum L. Hyacinthaceae (Liliaceae s.l.). c. 200 Euras. (Eur. 34), Afr. (S Afr. 54 – R: Bothalia 12(1978)330). Alks incl. colchicine, cardiac glycosides. Cult. orn. bulbous pls, some toxic, others ed., incl. *O. angustifolium* Bor ('*O. umbellatum*', star of Bethlehem, GB etc. with 2n = 18, 27, 36, the true *O. umbellatum* L. (*O. divergens*, lowland Eur., Middle E.) with 2n = 36, 45, 54) – bulb s.t. eaten; *O. pyrenaicum* L. (Bath asparagus, Medit., natur. N Eur.) – young infls eaten like asparagus; *O. thyrsoides* Jacq. (chincherinchee, S Afr.) – populations of diploids, triploids, hexaploids & aneuploids, winter cut-fl. exported to Eur. (name said to sound like the scapes knocking against one another in breeze), toxic (even seeds) to stock, noxious weed in S Aus.
Ornithoglossum Salisb. Colchicaceae (Liliaceae s.l.). 8 trop. & S Afr. R: BN 64(1982) – 'bird's-tongue', with alks toxic to stock
Ornithophora Barb. Rodr. = *Sigmatostalix*
Ornithopus L. Leguminosae (III 19). 6 temp. S Am. (1), Atlantic is., Eur. (4), Medit. to W As. Bird's foot (esp. *O. perpusillus* L., Eur.); *O. sativus* Brot. (serradella, Medit.) – good fodder
Ornithostaphylos Small = *Arctostaphylos*
Orobanchaceae Vent. (~ Scrophulariaceae). Dicots – Asteridae – Scrophulariales. 15/210 N hemisph. esp. temp. & subtrop. OW. Herbaceous root-parasites without chlorophyll, the stems usu. fleshy, the radicle becoming a haustorium penetrating host-root. Lvs usu. spirally arr. scales with often disorganized stomata; stipules 0. Fls bisexual, (solit. as in *Phelipaea* or in term. bracteate racemes or spikes; K ((1–)4 or 5), segments open or valvate, C (5), ± 2-labiate, often curved, lobes imbricate (adaxial ones internal), A 4 with adaxial 1 staminodal or 0, attached to C & alt. with lobes, anthers with longit. slits, G (2(3)) with thin style & 2–4-lobed stigma, 1-loc. with (2)4(6) intruded parietal placentas, often branched, with ∞ anatropous unitegmic ovules. Fr. a loculicidal capsule each of the 2(3) valves typically with 2 placentas; seeds minute (c. 1 m/g), with undifferentiated embryo in oily endosperm. x = 12, 18–21

Principal genera: *Aeginetia, Christisonia, Cistanche, Orobanche*

V. close to hemiparasitic Scrophulariaceae but trad. segregated, *Lathraea* being placed in one or other

Some locally eaten or pests; parasitism rarely restricted to 1 host sp.; seeds wind-disp.

Orobanche L. Orobanchaceae. 150 temp. (Eur. 45) & warm esp. N. Broomrapes. Some restricted to particular fams as hosts, rarely to 1 sp. as in *O. hederae* Duby on Irish ivy (W Eur. & Medit., but also on *Fatsia japonica* in cult.), known as broomrapes because form. thought to be outgrowths of brooms etc., 'rapum' being a knob or tuber. Seeds (0.0001 mg in *O. arenaria* Borkh. (*O. ionantha*, Euras.)) germ. giving spiral filaments carrying testa apically, 'searching' habitat in a spiral movement for hosts; pl. often annual with tuber developing as swelling behind tip lying in contact with host-root, the tuber becoming nodulated & covered with papillae, one of which penetrates host as far as xylem; thereafter host & parasite tissue difficult to discern but parasite certainly has own vessel-elements; from near junction, infl. arises as bud & pushes above ground (even through tarmac). Some infls eaten by Amerindians (cf. *Pholisma*)

Orobus L. = *Lathyrus*

Orochaenactis Cov. Compositae (Hele.-Cha.). 1 California

Orogenia S. Watson. Umbelliferae (III 5). 2 W N Am.

Orontium L. Araceae (V 1). 1 E N Am.: *O. aquaticum* L. (golden club) – (?) bee-poll. cult. orn. aquatic, starch in rhizome & seeds ed. once boiled

Oropetium Trin. Gramineae (31d). 6 OW trop. *O. capense* Stapf (Afr.) lvs revive after reduction to 8% water content

Orophea Blume. Annonaceae. 41 India & S China to Moluccas. R: Blumea 33(1988)1. *O. katschallica* Kurz (Andamans) – sap used to repel *Apis dorsata* bees during honey-collecting

Orophochilus Lindau. Acanthaceae. 1 Peru

Orostachys (DC) Sweet (~ *Sedum*). Crassulaceae. 10 Eur. (2) & temp. As. Some cult. orn. with hapaxanthic leaf-rosettes often with offsets withering to compact winter-buds of callose lvs which grow basally in spring to give callus-tipped lvs (s.t. spiny)

Orothamnus Pappe ex Hook. Proteaceae. 1 S Afr.: *O. zeyheri* Pappe ex Hook. (marsh rose)

Oroxylum Vent. Bignoniaceae. 1 Indomal.: *O. indicum* (L.) Kurz – sparsely branched pachycaul tree of clearings & roadsides, with 2–3 pinnate lvs, stinking, bat-poll. fls with copious nectar opening at night & fr. to 1 m long; young lvs cooked as vegetable, bitter bark medic., dye used in rattan basketry in Sarawak

Oroya Britton & Rose (~ *Oreocerus*). Cactaceae (III 4). 1–2 Peru. R: Ashingtonia 1(1975)136

Orphanidesia Boiss. & Bal. = *Epigaea*

Orphandodendron Barneby & Grimes. Leguminosae (I 1). 1 NW Colombia: *O. bernalii* Barneby & Grimes. R: Brittonia 42(1990)249

Orphium E. Meyer. Gentianaceae. 1 S Afr.: *O. frutescens* (L.) E. Meyer – cult. orn. shrub with pink fls

orpine *Sedum telephium*

Orrhopgium Löve = *Aegilops*

orris root *Iris × germanica* 'Florentina'

Ortachne Nees ex Steudel. Gramineae (16). 3 Costa Rica to Peru, Patagonia. Dwarf grasses of montane woodland glades. R: KBAS 13(1986)86

ortanique *Citrus × aurantium* Sweet Orange Group

Ortegia L. Caryophyllaceae (I 1). 1 Spain & Portugal: *O. hispanica* L.

Ortegocactus Alexander. Cactaceae (III 9). 1 SE Mex.: *O. macdougallii* Alexander. Intermediate between *Mammillaria* & *Escobaria*

Ortgiesia Regel (~ *Aechmea*). Bromeliaceae (3). 22 trop. Am. R: Phytol. 66(1989)72

Orthaea Klotzsch. Ericaceae. 31 trop. Am. R: NJB 7(1987)31

Orthandra Burret = *Mortoniodendron*

Orthantha (Benth.) Wettst. = *Odontites*

Orthanthera Wight. Asclepiadaceae (III 5). 4 India & Nepal (1), S trop. Afr. (3)

Orthechites Urban = *Secondatia*

Orthilia Raf. Ericaceae (Pyrolaceae). 1 circumboreal: *O. secunda* (L.) House

Orthion Standley & Steyerm. Violaceae. 3 C Am.

Orthiopteris Copel. = *Saccoloma*

Orthocarpus Nutt. Scrophulariaceae. 9 W Am. R: SB 17(1992)560

Orthoceras R. Br. Orchidaceae (IV 1). 2 E Aus., New Caled., NZ

Orthoclada Pal. Gramineae (24). 1 trop. Am., 1 SE trop. Afr. Lvs with 'false petioles'

Orthodon Benth. = *Mosla*

Orthogoneuron Gilg. Melastomataceae. 1 trop. Afr.

Orthogynium Baillon. Menispermaceae (tribe?). 1 Madagascar (?). Female unknown

Orthomene Barneby & Krukoff. Menispermaceae (II). 4 trop. Am. R: MNYBG 22,2(1971)79

Orthopappus Gleason = *Elephantopus*

Orthophytum Beer. Bromeliaceae (3). 22 E Brazil. R: Plantsman 13(1991)180. Some cult. orn.

Orthopichonia H. Huber. Apocynaceae. 6 trop. W Afr. R: AUWP 89-4(1989)29

Orthopogon R. Br. = *Oplismenus*

Orthopterum L. Bolus (~ *Faucaria*). Aizoaceae (V). 2 E Cape

Orthopterygium Hemsley. Anacardiaceae (IV; Julianaceae). 1 Peru. Winged pedicel aids disp.

Orthoraphium Nees = *Stipa*

Orthosia Decne. Asclepiadaceae (III 1). 20 trop. Am.

Orthosiphon Benth. Labiatae (VIII 4b). 40 OW trop. *O. aristatus* (Blume) Miq. (Mal.) – cult. diuretic with high levels of potassium salts & a glycoside

Orthosphenia Standley. Celastraceae. 1 Mexico

Orthotactus Nees = *Justicia*

Orthotheca Pichon = *Xylophragma*

Orthothylax (Hook.f.) Skottsb. = *Helmholtzia*

Orthrosanthus Sweet. Iridaceae (III 1). 9 SW Aus., S & C Am.

Orthurus Juz. = *Geum*

Orumbella J. Coulter & Rose = *Podistera*

Orvala L. = *Lamium*

Orychophragmus Bunge. Cruciferae. 2 C As., China

Oryctanthus (Griseb.) Eichler (*Glutago*). Loranthaceae. 10 trop. Am. R: BJ 95(1976)478

Oryctes S. Watson. Solanaceae (1). 1 SW US

Oryctina Tieghem = *Oryctanthus*

Orygia Forssk. = *Corbichonia*

Oryza L. Gramineae (9). c. 18 trop. (Mal. 7, R: Blumea 32(1987)169). Rice, esp. *O. sativa* L., poss. derived from *O. rufipogon* Griffth in As., though prob. several centres of domestication & arising as a selected weed (c. 5000 BC Lower Yangtze, Thailand, N India) in flooded *Colocasia* fields; now principal carbohydrate source of As., the unhusked grain known as paddy; many cvs grouped as lowland (grown under inundation) & upland (dry) r., some glutinous & used for puddings, others not so & poss. the world's most imp. food-pl. eaten with e.g. curries, & in Eur. as paella & risotto, but often without the husks (pearl r.) when much protein & vitamin removed, & often 'polished'; the husks & polishings (shude(s)) an imp. cattle food; flour used in cooking & breakfast foods etc. or as comm. starch; fermented to give beer & eventually sake; straw imp. fodder & used for 'sea-grass matting' (Aus.). *O. glaberrima* Steudel ('red rice', W Afr.) – locally cult. (in Niger Delta for 3500 yrs), weedy in rice but reduced by early flooding after seeding

Oryzetes Salisb. = *Hygrophila*

Oryzidium C. Hubb. & Schweick. Gramineae (34b). 1 Zambia to SW Afr.: *O. barnardii* C. Hubb. & Schweick. – floating

Oryzopsis Michaux (~ *Piptatherum*). Gramineae (16). 4–35 N temp. & subtrop. (esp. Middle E). Some grains (mountain rice) for N Am. Indians. *O. miliacea* (L.) Asch. & Schweinf. (Medit.) – germ. much improved after heating seed to 90 °C (fire-adapted)

Osa Aiello (~ *Hintonia*). Rubiaceae (I 7). 1 Costa Rica

Osage orange *Maclura pomifera*

Osbeckia L. Melastomataceae. 50 OW trop. (As. 31, R: Ginkgoana 4(1977)1)

Osbertia E. Greene (~ *Haplopappus*). Compositae (Ast.-Sol.). 2 WN & C Am. R: Phytol. 71(1991)132

Osbornia F. Muell. Myrtaceae (Myrt.). 1 C Mal. to NE Aus.: *O. octodonta* F. Muell. in mangrove

Oschatzia Walp. (~ *Azorella*). Umbelliferae (I 2b). 2 Aus.

Oscularia Schwantes (~ *Lampranthus*). Aizoaceae (V). 1–3 SW S Afr.

Oserya Tul. & Wedd. Podostemaceae. 7 trop. Am. R: ABN 3(1954)216

osier *Salix* spp.

Osmadenia Nutt. = *Hemizonia*

Osmanthus Lour. Oleaceae. 15 As., S US (1–2). R: NRBGE 22(1958)439. Cult. orn. evergreen shrubs & trees with extrafl. nectaries, esp. *O.* × *burkwoodii* (Hort. Burkw. & Skipw.) P. Green (× *Osmarea burkwoodii*, *Osmanthus decorus* (Boiss. & Bal.). Kasapl. (*Phillyrea d.*, Caucasus, Lazistan) × *O. delavayi* Franchet (W China) – often considered an 'intergeneric hybrid' (faulty taxonomy!)); *O. fragrans* Lour. (kwei, prob. E As., but long cult. in As.) – male infls used to flavour tea & confectionery; *O. heterophyllus* (G. Don f.) P. Green (Taiwan, Japan) – cult. orn. with holly-like foliage. See also *Nestegis*

× **Osmarea** C. Curtis. See *Osmanthus*

Osmaronia E. Greene = *Oemleria*

Osmelia Thwaites. Flacourtiaceae (9). 3 Mal., 1 Sri Lanka

Osmiopsis R. King & H. Robinson (~ *Eupatorium*). Compositae (Eup. -Prax.). 1 Haiti. R: MSBMBG 22(1987)397

Osmites L. = *Relhania*

Osmitopsis Cass. Compositae (Anth.-Tham.). 9 SW Cape. R: BN 125(1972)9, 129(1976)21

Osmoglossum (Schltr.) Schltr. (~ *Palumbina*). Orchidaceae (V 10). 7 trop. Am.

Osmorhiza Raf. Umbelliferae (III 2). 10 Am., E As. Some ed. roots & medic. R: AMBG 71(1984)1128. Some cult. orn. incl. *O. longistylis* (Torrey) DC (W N Am.) – main sweet constituent is *trans*-anethol

Osmoxylon Miq. Araliaceae. 50 C & E Mal. (R: Blumea 23(1976)99), Taiwan & W Pacific. *O. borneense* Seemann (Borneo) – rheophyte

Osmunda L. Osmundaceae. c. 10 temp. (Eur. 1), trop. E As. Cult. orn. (royal ferns) & locally eaten, esp. *O. regalis* L. (N temp. to Afr. & S Am.) with stock like small tree-fern, & *O. cinnamomea* L. (cinnamon fern, fiddleheads, Am. & E As.); fibre used for orchid-growing, hairs around young fronds used with wool in Japan to make a textile for raincoats. *O. regalis* used in brewing Celtic heather ale (spores contain a thiaminase destroying vitamin B, & therefore yeast so stopping fermentation, a prehistoric 'Camden tablet'!)

Osmundaceae Bercht. & J. Presl. Filicopsida. 3/22 trop. & temp. Stems erect & enveloped in persistent frond-bases & coarse roots; vasc. system of separate xylem strands. Fronds 1–2-pinnate or -pinnatifid. Sporangia not arr. in sori, on underside of fronds or on specialized fronds or pinnae (*Osmunda*) s.t. forming a kind of panicle at frond-tip; sporangia short-stalked with poorly-developed lateral annulus & many spores

Genera: *Leptopteris, Osmunda, Todea*

Fossils back to Permian referred here. Cult. orn.

Osmundastrum C. Presl = *Osmunda*

oso-berry *Oemleria cerasiformis*

Ossaea DC. Melastomataceae. 91 trop. Am.

Ossiculum Cribb & van der Laan. Orchidaceae (V 16). 1 Cameroun

Ostenia Buchenau = *Hydrocleys*

Osteocarpum F. Muell. (*Babbagia*; ~ *Threlkeldia*). Chenopodiaceae (I 5). 5 temp. Aus. R: FA 4(1984)231

Osteomeles Lindley. Rosaceae (III). 3 China to Hawaii. R: CJB 68(1990)2230. Some cult. orn. shrubs with pinnate lvs & pyrenes

Osteophloeum Warb. Myristicaceae. 1 Amazonia: *O. platyspermum* (A. DC) Warb. – bark a source of hallucinogenic paste, also medic.

Osteospermum L. Compositae (Cal.). (Excl. *Tripteris*) c. 45 S Afr. to Arabia, St Helena (1). Shrubs & herbs; some cult. orn. as '*Dimorphotheca*' esp. *O. barberae* (Harvey) Norlindh & *O. ecklonis* (DC) Norlindh (S Afr.)

Ostericum Hoffm. = *Angelica*

Ostodes Blume. Euphorbiaceae. 4 E Himal. to Borneo. *O. paniculata* Blume (W Indomal.) – lvs used for wrapping, latex for gum in Sikkim

Ostrearia Baillon. Hamamelidaceae (I 1). 1 Queensland: *O. australiana* Baillon. R: FA 3(1989)1

ostrich fern *Matteuccia struthiopteris*

Ostrowskia Regel. Campanulaceae. 1 Turkestan: *O. magnifica* Regel – thicket-forming to 2.5 m with tubers & whorled lvs, C to 15 cm diam., difficult cult. orn.

Ostrya Scop. Betulaceae (II; Carpinaceae). 5 N temp. to C Am. R: JAA 71(1990)57. Hop-hornbeam (nutlets encl. in bladder-like green involucres). Timbers esp. *O. carpinifolia* Scop. (S Eur., As. Minor) & *O. virginiana* (Miller) K. Koch (ironwood, leverwood, E N Am.), irritant hairs on fr. scales of former irritant

Ostryocarpus Hook.f. Leguminosae (III 6). 6 trop. OW

Ostryopsis Decne. Betulaceae (II; Carpinaceae). 2 E Mongolia & SW China. Shrubs; A 4–6

Oswego tea *Monarda didyma*

Osyridicarpos A. DC. Santalaceae. 6 trop. & S Afr.

Osyris L. Santalaceae. 6–7 Medit. (Eur. 2) & Afr. (incl. *Colpoon*) to India. *O. tenuifolia* Engl. (E Afr. sandalwood, trop. E Afr.) – timber like true sandalwood, oil used in scent-making. Some tanbark

Otacanthus Lindley. Scrophulariaceae. 4 Brazil

Otachyrium Nees. Gramineae (34b). 7 S Am. R: SCB 57(1984)1

Otaheite apple *Syzygium malaccense, Spondias* spp.; **O. chestnut** *Inocarpus fagifer*; **O. gooseberry** *Phyllanthus acidus*; **O. myrtle** *Securinega durissima*; **O. potato** *Dioscorea bulbifera*

Otanthera Blume. Melastomataceae. 15 Nicobars, Mal. to N Aus.

Otanthus Hoffsgg. & Link. Compositae (Anth.-Art.). 1 coasts W Eur. to Near East: *O. maritimus* (L.) Hoffsgg. & Link (cottonweed) – cult. orn. herb with felted lvs & rayless yellow capitula

Otatea (McClure & E.W. Sm.) Calderón & Söderstrom = *Sinarundinaria*

Oteiza Llave (~ *Calea*). Compositae (Helia.-Helia.). 3 Mexico, Guatemala

Othake Raf. = *Palafoxia*

Otherodendron Makino = *Microtropis*

Otholobium Stirton (~ *Psoralea*). Leguminosae (III 11). 36 E & S Afr., S Am. *O. glandulosum*

(L.) Grimes (*P. g.*, Jesuit's tea, temp. S Am.)

Othonna L. Compositae (Sen.-Sen.). c. 120 trop. & S Afr., Aus. Xeromorphic shrubs & herbs with tubers; ray-florets fertile, disk-florets sterile. Cult. orn. succ. pls esp. *O. capensis* L. Bailey ('*O. crassifolia*', S Afr.) – shrubby with long succ. lvs & yellow fls, much grown in hanging-baskets & (in Calif.) as ground-cover, ± everblooming

Othonnopsis Jaub. & Spach = *Othonna*

Otilix Raf. = *Lycianthes*

Otiophora Zucc. Rubiaceae (IV 1). 20 trop. Afr. & Madag. R: GOB 1(1973)25

Otoba (A. DC) Karsten. Myristicaceae. 6 trop. Am. Source of otoba fat used in soap-making

Otocalyx Brandegee. Rubiaceae. 1 Mexico = ?

Otocarpus Durieu. Cruciferae. 1 Algeria

Otocephalus Chiov. = *Calanda*

Otochilus Lindley. Orchidaceae (V 12). 4 E Himal. to SE As. R: BT 71(1976)8

Otoglossum (Schltr.) Garay & Dunsterv. (~ *Odontoglossum*). Orchidaceae (V 10). 7 N S Am.

Otomeria Benth. Rubiaceae (IV 1). 8 trop. Afr., Madagascar

Otonephelium Radlk. Sapindaceae. 1 India

Otopappus Benth. Compositae (Helia.-Verb.). 14 Mex. & C Am., WI. R: SB 8(1983)185

Otophora Blume = *Lepisanthes*

Otoptera DC (~ *Vigna*). Leguminosae (III 10). 2 Afr.

Otospermum Willk. Compositae (Anth.-Mat.). 1 SW Eur., NW Afr.

Otostegia Benth. Labiatae (VI). 20 NE trop. Afr. to C As.

Otostylis Schltr. Orchidaceae (V 10). 3 trop. S Am., Trinidad

Ottelia Pers. Hydrocharitaceae. 21 trop. & warm, esp. OW (NW 1). R: AB 18(1984)263. Submerged aquatics incl. *O. alismoides* (L.) Pers. (China & Japan to Aus. & NE Afr.) – cult. orn., green veg. in Mal., natur. Italy

otto of roses See *Rosa*

Ottoa Kunth. Umbelliferae (III 5). 1 Mexico

Ottochloa Dandy. Gramineae (34b). 6 Indomal. to N Aus. R: Blumea 4(1941)530

Ottonia Sprengel = *Piper*

Ottoschmidtia Urban. Rubiaceae (III 3). 2 Cuba, Hispaniola

Ottoschulzia Urban. Icacinaceae. c. 3 WI, Guatemala

Ottosonderia L. Bolus (~ *Ruschia*). Aizoaceae (V). 1 W S Afr.

otu *Cleistopholis patens*

Oubanguia Baillon. Scytopetalaceae. 3 trop. Afr.

Oudneya R. Br. Cruciferae. 1 Algeria

Ougeinia Benth. = *Desmodium*

Ouratea Aublet. Ochnaceae. 150 trop. (incl. *Gomphia*). Some timbers, oils & local med. esp. *O. parviflora* (A. St-Hil.) Engl. (batiputa, Brazil) – seed-oil medic. & *O. serrata* (Gaertner) Robson (*G. serrata*, Indomal.) – lvs, roots & twigs medic. *O. amplectens* (Stapf) Engl. (W Afr.) – litter-coll. rosette of lvs

ouricuri (palm) *Syagrus coronata*

Ourisia Comm. ex Juss. Scrophulariaceae. 27 Andes (12, R: Parodiana 4(1986)239), NZ (14), Tasmania (1). Some cult. orn. rock-pls

Ourisianthus Bonati = *Artanema*

Outreya Jaub. & Spach. Compositae (Card.-Card.). 1 SW As.

ouzo Flavour due to some or all of *Foeniculum vulgare*, *Pimpinella anisum*, *Pistacia lentiscus*

Ovidia Meissner. Thymelaeaceae. 4 temp. S Am. Cult. orn. shrubs like *Daphne*

Owenia F. Muell. Meliaceae (I 4). 5 E Aus. Disp. emus. *O. acidula* F. Muell. (gooya, gruie, emu apple, mooley or sour plum) – fr. with refreshing pulp; *O. cepiodora* F. Muell. – timber tree reduced to 31 trees by 1984

owala *Pentaclethra macrophylla*

Oxalidaceae R. Br. Dicots – Rosidae – Geraniales. (Excl. Lepidobotryaceae) 6/775 trop. to temp. (few). Small trees, shrubs & esp. herbs with tubers or bulbs, usu. accum. oxalates. Lvs spirally arr., pinnate, palmate but often 3-foliolate, rarely 1-foliolate or phyllodic, often with leaflets folded together at night ('sleep movements'); stipules usu. 0. Fls bisexual, reg., 5-merous, usu. with tristyly & s.t. cleistogamous when C 0, (solit. or) in axillary cymes on peduncles; K 5, imbricate, C 5 s.t. basally weakly connate, convolute (rarely imbricate), A 10 in ± 2 whorls with outer usu. with shorter filaments, all of which basally connate & s.t. 5 without anthers (A 5 + 5 + 5 in *Hypseocharis*), outer often with basal nectaries, anthers with longit. slits, $\underline{G}$ ((3–)5) with axile placentas & discrete styles (1 in *Hypseocharis*), each loc. with (1) 2–several ± pendulous, anatropous or hemitropous, bitegmic ovules. Fr. a loculicidal capsule or berry (*Averrhoa*, *Sarcotheca*); seeds often with basal aril involved in expulsion from capsule, embryo embedded in usu. copious oily endosperm. x = (5–)7(–12)

Genera: *Averrhoa, Biophytum, Dapania, Hypseocharis, Oxalis, Sarcotheca*

Lepidobotrys excl., *Hypseocharis* retained though s.t. treated as monotypic fam.; the woody *Averrhoa* & *Sarcotheca* with the lianoid *Dapania* form. made another fam., reflecting an obsession going back to the Anc. Greeks of the essential nature of the diffs between woody & herbaceous pls: this has done much to retard the progress of botany & was the crux of Hutchinson's system

Some fr. trees (*Averrhoa*), ed. tubers (*Oxalis*), cult. orn. & weeds (*Oxalis, Biophytum*)

Oxalis L. Oxalidaceae. 700 cosmop. esp. S Am. & Cape (Cape shamrock (s.t. distrib. as true s. in US; S Afr. c. 200 – R: JSAB suppl. 1(1944)). Herbs & subshrubs, s.t. succ. or even aquatic (e.g. *O. natans* L.f. (S Afr.) – endangered sp.). Some with distyly derived from tristyly; many with tubers making them weedy pests esp. *O. pes-caprae* L. (*O. cernua*, Bermuda buttercup, soursob, S Afr.) – tristylous with diploids & tetraploids in S Afr., short-styled pentaploids common in Medit. (introd. Sicily 1796, Iran by 1928, Pakistan 1940s) & Aus. but rare in S Afr., bulbs ± ed. (distrib. by pole cats etc.) though toxic to stock in Aus. & NZ (continued grazing by sheep leading to kidney damage), widely natur. in trop. & warm with spontaneous double fls incl. Eur. where only 2 native spp. but 10 S Afr. & S Am. spp. widely natur. Some cult. orn. & some ed. but rapidly spreading vegetatively or by seed (arils turn inside-out expelling seeds from capsule explosively); nectar of some spp. with pH 1.6 (most usu. pH 5.6–5.9), some fly-poll., e.g. *O. acetosella* L. (wood sorrel, Euras.) – form. used instead of *Rumex* spp. as 'sorrel'; *O. corniculata* L. (origin unknown) – cosmop. weed with yellow fls often seen on greenhouse floors, lvs ed. India as scurvy cure; *O. debilis* Kunth (*O. corymbosa*, S Am.) – sparsely hairy lvs with translucent dots a tamarind subs. & *O. latifolia* Kunth (Mex. to Peru) – hairless lvs without dots, both cosmop. weeds with pinkish fls; *O. magellanica* Forster f. (New Guinea, Aus., NZ, Chile) – cult. orn. (s.t. pest), fls white; *O. tetraphylla* Cav. (*O. deppei*, Mexico) – cult. orn. with 4-foliolate lvs & red fls, tubers ed. (form. cult. in Eur. for them); *O. tuberosa* Molina (*O. crenata*, oca, Andes) – long cult. as root veg. in Peru with tubers white, yellow or red (the latter 2 not flowering); other spp. locally medic. or lvs eaten

Oxandra A. Rich. Annonaceae. 30 trop. Am. R: AHB 10(1931)153. *O. lanceolata* (Sw.) Baillon (WI) – timber (asta, lancewood).

Oxanthera Montr. Rutaceae. 4 New Caled. R: W.T. Swingle, Bot. Citrus (1943)229

Oxera Labill. Labiatae (II; Verbenaceae). 20 New Caled., Vanuatu (1). Lianes & pachycaul treelets allied to *Faradaya*

ox-eye *Buphthalmum* spp.; **o. daisy** *Leucanthemum vulgare*

Oxford & Cambridge bush *Rotheca myricoides* 'Ugandensis'; **O. ivy** or **weed** *Cymbalaria muralis*; **O. ragwort** *Senecio squalidus*

oxlip *Primula elatior*

oxtongue *Picris* spp. esp. *P. echioides*

Oxyanthera Brongn. = *Thelasis*

Oxyanthus DC. Rubiaceae (II 1). 40 Afr. Some timber

Oxybaphus L'Hérit. ex Willd. = *Mirabilis*

Oxycarpha S.F. Blake. Compositae (Helia.-Verb.). 1 Venezuela

Oxycaryum Nees (~ *Scirpus*). Cyperaceae. Spp.?

Oxyceros Lour. (~ *Randia*). Rubiaceae (II 1). Spp.?

Oxychlamys Schltr. = *Aeschynanthus*

Oxychloe Philippi. Juncaceae. 7 C & S Andes

Oxychloris Lazarides (~ *Chloris*). Gramineae (33b). 1 Aus.: *O. scariosa* (F. Muell.) Lazarides – dry savanna, adventive in Switzerland. R: Nuytsia 5(1984)283

Oxycoccus Hill = *Vaccinium*

Oxydendrum DC. Ericaceae. 1 E US: *O. arboreum* (L.) DC (sourwood, tree sorrel) – decid. tree to 25 m with acid-tasting lvs alleged to slake thirst

Oxygonum Burchell ex Campderá. Polygonaceae (II 4). 30 trop. & S Afr., Madag. (1). R: KB 1(1957)145

Oxygraphis Bunge. Ranunculaceae (II 3). 5 temp. As.

Oxygyne Schltr. Burmanniaceae. 2 W trop. Afr., 1 Japan (!)

Oxylaena Benth. ex Anderb. (~ *Macowania*). Compositae (Gnap.). 1 S Afr.: *O. acicularis* (Benth.) Anderb. R: OB 104(1991)53

Oxylobium Andrews. Leguminosae (III 24). (Excl. *Podolobium*) 15 Aus. esp. SW. Shaggy peas; 5 produce fluoroacetate (highly toxic except to some resistant marsupials) blocking Krebs Cycle at citrate stage, *O. parviflorum* Benth. (SW Aus.) with 2.5 g/kg

Oxylobus (DC) A. Gray. Compositae (Eup.-Oxy.). 6 Mex. to Venez. R: MSBMBG 22(1987)436
Oxyosmyles Speg. Boraginaceae. 1 Argentina
Oxypappus Benth. (*Stelmanis*). Compositae (Helia.- Bae.). 2 Mexico
Oxypetalum R. Br. Asclepiadaceae (III 1). 80–100 trop. Am.
Oxyphyllum Philippi. Compositae (Mut.-Nass.). 1 N Chile. Shrub
Oxypolis Raf. Umbelliferae (III 10). 7 N Am.
Oxyrhachis Pilger. Gramineae (40h). 1 trop. E Afr., Madag.: *O. gracillima* (Baker) C.E. Hubb. – upland bogs
Oxyrhynchus Brandegee. Leguminosae (III 10). 1 New Guinea, 3 C Am.
Oxyria Hill. Polygonaceae (II 3). 1 arctic to mts of Euras. & California: *O. digyna* (L.) Hill – like *Rumex* but K 4, cult. orn. rock-pl. with ed. lvs
Oxysepala Wight = *Bulbophyllum*
Oxyspora DC. Melastomataceae. 24 Indomal. to S China. R: GBS 35(1982)216. Some cult. orn.
Oxystelma R. Br. (~ *Sarcostemma*). Asclepiadacene. 4 trop. OW. *O. esculentum* (L.f.) Sm. (*S. secamone*, As.) – local medic., fr. eaten in times of scarcity
Oxystigma Harms. Leguminosae (I 4). 5 trop. Afr. *O. oxyphyllum* (Harms) Léonard (*Pterygopodium o.*, lolagbola, tchitola, W Afr.) – comm. timber
Oxystophyllum Blume = *Dendrobium*
Oxystylis Torrey & Frémont. Capparidaceae. 1 SW US
Oxytenanthera Munro. Gramineae (2c). 1 trop. Afr.: *O. abyssinica* (A. Rich.) Munro – bamboo with 7–21-yr flowering cycle, culms used in boat-making, tapped for wine in Tanzania
Oxytenia Nutt. = *Iva*
Oxytheca Nutt. Polygonaceae (I 1). 7 W N Am., temp. S Am., deserts. R: Brittonia 32(1980)70
Oxytropis DC. Leguminosae. 300 N temp. (Eur. 24) esp. C As. R: MAISP VII, 22-1(1874)1, (N Am.) PCAS IV, 27 (1952)177. Crazyweed or locoweed (cf. *Astragalus*), harmful to stock; some cult. orn. *O. deflexa* (Pallas) DC (N temp.) – pops (Finmark, Norway (500 pls), As. (5000 km away) & N Am.) separated by glaciations
Oyedaea DC. Compositae (Helia.-Verb.). 14 trop. Am.
Oyster Bay pine *Callitris rhomboidea*; **o. leaf** *Mertensia maritima*; **mock o.** *Tragopogon porrifolius*; **o. nut** *Telfairia pedata*; **o. plant** *T. porrifolius*, (Aus.) *Acanthus spinosus*; **Spanish o. p.** *Scolymus hispanicus*; **vegetable o.** *T. porrifolius*
Ozoroa Del. (~ *Heeria*). Anacardiaceae (IV). 40 trop. Afr. Some locally ed. fr. & medic.
Ozothamnus R. Br. (~ *Helichrysum*). Compositae (Gnap.- Cass). 50 Aus., New Caled., NZ. R: AusJB 6(1958)229, OB 104 (1991)87. *O. ferrugineus* (Labill.) Sweet (SE Aus.) – tree to 5 m. Cult. orn. evergreen shrubs & perennial herbs e.g. *O. ledifolius* (DC) Hook.f. (kerosene weed, Tasmania) – scented exudate highly flammable & *O. rosmarinifolius* (Labill.) Sweet (SE Aus.)

P

Pabellonia Quezada & Martic. = *Leucocoryne*
Pabstia Garay (*Colax*). Orchidaceae (V 10). 5 Brazil. Cult. orn.
Pabstiella Brieger & Senghas = *Pleurothallis*
Pachecoa Standley & Steyerm. Leguminosae (III 13). 1 C Am. (poss. introd. (or relict) in Venez.): *P. prismatica* (Sessé & Moçiño) Schubert. R: Brittonia 37(1985)78
Pachira Aublet. Bombacaceae. (Incl. *Bombacopsis*) c. 20 trop. Am. Some light timbers & cult. orn. incl. *P. aquatica* Aublet (trop. Am. estuaries) – seeds ed.
Pachites Lindley. Orchidaceae (IV 3). 2 SW Cape
Pachyacris Schltr. ex Bullock = *Xysmalobium*
Pachyanthus A. Rich. Melastomataceae. 16 Cuba, Hispaniola & Colombia (1)
Pachycarpus E. Meyer. Asclepiadaceae (III 1). 40 trop. & S Afr.
Pachycentria Blume. Melastomataceae. 8 Burma, Mal.
Pachycereus (A. Berger) Britton & Rose. Cactaceae (III 8). 12 SW US & Mex. Cult. orn. tree-like cacti incl. *P. pecten-aboriginum* (Engelm.) Britton & Rose, *P. pringlei* (S. Watson) Britton & Rose – fr. ed., seeds ground into flour & *P. schottii* (Engelm.) D. Hunt (*Lophocereus s.*) – toxic to most *Drosophila* spp. living in rotten trunks but not *D. pachea* resistant to alks & reliant on a steroid precursor of moulting hormone
Pachycladon Hook.f. Cruciferae. 2 NZ mts
Pachycormus Cov. ex Standley. Anacardiaceae (IV). 1 Baja Calif.: *P. discolor* (Benth.) Cov. ex Standley (elephant tree, copalquin) – stem swollen. R: CJSAm 63(1991)35
Pachycornia Hook.f. Chenopodiaceae (II 2). 1 Aus.: *P. triandra* (F. Muell.) J. Black. R: FA 4(1984)308. Like *Salicornia*

Pachyctenium Maire & Pampan. Umbelliferae (inc. sed.). 1 N Ethiopia

Pachycymbium Leach (~ *Caralluma*). Asclepiadaceae (III 5). 32 Arabia, trop. & S Afr. R: Bradleya 8(1990)20. Cult. orn. succ.

Pachydesmia Gleason = *Miconia*

Pachyelasma Harms. Leguminosae (I 1). 1 W Afr. rain forest

Pachygone Miers. Menispermaceae. 10 China & Indomal. to Pacific. Some used to stupefy fish & against vermin in Aus.

Pachylaena D. Don ex Hook. & Arn. Compositae (Mut.-Mut.). 2 Andes of Chile & Arg. Herbs

Pachylarnax Dandy. Magnoliaceae (I 1). 2 Assam to W Mal. Fr. a woody loculicidal capsule. R: Blumea 31(1985)97. *P. pleiocarpa* Dandy (Assam) – timber for cabinet-making

Pachylecythis Ledoux = *Lecythis*

Pachyloma DC. Melastomataceae. 6 Brazil. Appendages behind connective

Pachymitus O. Schulz. Cruciferae. 1 Aus.: *P. cardaminoides* (F. Muell.) O. Schulz. R: TRSSA 89(1965)226

Pachynema R. Br. ex DC. Dilleniaceae. 7 NW Aus. R: AusSB 5(1992)477. Phyllodes; stamens thickened basally

Pachyneurum Bunge. Cruciferae. 1 C As.

Pachypharynx Aellen = *Atriplex* (galled)

Pachyphragma (DC) Reichb. = *Thlaspi*

Pachyphyllum Kunth. Orchidaceae (V 10). 35 W trop. S Am.

Pachyphytum Link, Klotzsch & Otto. Crassulaceae. 12 Mex. Cult. orn. succ. like *Echeveria* (intergeneric hybrids = × *Pachyveria*) but each petal with 2 scale-like appendages within

Pachyplectron Schltr. Orchidaceae (III 3). 2 New Caledonia

Pachypleuria (C. Presl) C. Presl = *Humata*

Pachypleurum Ledeb. (~ *Ligusticum*). Umbelliferae (III 8). 2 Euras.

Pachypodanthium Engl. & Diels. Annonaceae. 3 trop. W Afr. Some timber

Pachypodium Lindley. Apocynaceae. 13 Madag. (most), S & SW Afr. R: CSJAm 44(1972)7. Some cult. orn. succ. incl. *P. succulentum* (L.f.) Sweet (S Afr.) ranging in size from pachycaul treelets of dry forest to mound-forming spp. of mts

Pachyptera DC ex Meissner = *Mansoa*

Pachypterygium Bunge. Cruciferae. 4 C As. to Afghanistan & Iran

Pachyrhizus Rich. ex DC. Leguminosae (III 10). 4–5 NW trop. R: NJB 8(1988)167. Cult. for ed. tubers & starch esp. *P. erosus* (L.) Urban (jicama, yam bean, C Am.) cult. in pre-Columbian times for tubers eaten cooked or raw (as in Mal. where eaten with fermented prawn sauce), seeds toxic

Pachyrhynchus DC = *Lucilia*

Pachysandra Michaux. Buxaceae. 3 E As. (2), E US (1). R: BBAS 33(1992)201. Alks. Monoec. shrubby herbs cult. as ground-cover esp. *P. terminalis* Siebold & Zucc. (Japan) in shade

Pachystachys Nees. Acanthaceae. 12 trop. Am. Cult. orn. greenhouse shrubs esp. *P. lutea* Nees (Peru) with consp. yellow bracts

Pachystegia Cheeseman (~ *Olearia*). Compositae (Ast.-Ast.). 3 NZ. R: NZJB 25(1987)144. *P. insignis* (Hook.f.) Cheeseman – cult. orn. shrub

Pachystela Pierre ex Radlk. = *Synsepalum*

Pachystele Schltr. = *Scaphyglottis*

Pachystelma Brandegee = *Matelea*

Pachystigma Hochst. Rubiaceae (III 2). 10 warm Afr.

Pachystoma Blume. Orchidaceae (V 11). 6 China & Indomal. to W Pacific

Pachystrobilus Bremek. = *Strobilanthes*

Pachystroma Muell. Arg. Euphorbiaceae. 1 S Brazil

Pachystylidium Pax & K. Hoffm. Euphorbiaceae. 1 India to C Mal.

Pachystylus Schumann. Rubiaceae (II 2). 2 New Guinea

Pachythamnus (R. King & H. Robinson) R. King & H. Robinson (~ *Eupatorium*). Compositae (Eup.-Oxy.). 1 C Am.: *P. crassirameus* (Robinson) R. King & H. Robinson. R: MSBMBG 22(1987)442. Fat stems leafless in fl.

Pachytrophe Bureau = *Streblus*

× **Pachyveria** Hort. Haage & Schmidt. Crassulaceae. Hybrids between spp. of *Echeveria* & *Pachyphytum*. Cult. orn. succ.

Packera Löve & D. Löve (~ *Senecio*). Compositae (Sen.- Sen.). c. 65 N Am., Siberia. Some cult. orn. esp. alpines

paco-paco *Wissadula spicata*

Pacouria Aublet = *Landolphia*

Pacourina Aublet. Compositae (Vern.-Vern.). 1 trop. Am.: *P. edulis* Aublet – aquatic with sessile heads & ed. lvs

padauk *Pterocarpus* spp.; **Andaman p.** *P. dalbergioides*; **Burma p.** *P. macrocarpus*; **W African p.** *P. soyauxii*

Padbruggea Miq. = *Callerya*

paddlewood *Aspidosperma excelsum*

paddy *Oryza sativa*; **P.'s lucerne** *Sida rhombifolia*

padma *Nelumbo nucifera*

padri *Stereospermum colais*

Padus Miller = *Prunus*

Paederia L. Rubiaceae (IV 12). 30 trop. Lianes with faecal smell (sulphur group released on damage) but lvs of *P. scandens* (Lour.) Merr. (SE As., Mal.) used as a vegetable

Paederota L. (~ *Veronica*). Scrophulariaceae. 2 S Eur. incl. *P. bonarota* (L.) L. (E Alps) – cult. orn. herb

Paederotella (Wulff) Kem.-Nat. (~ *Veronica*). Scrophulariaceae. 1–3 As. Minor, Caucasus

Paedicalyx Pierre ex Pitard = *Xanthophytum*

Paeonia L. Paeoniaceae. 33 temp. Euras. (Eur. 10), W N Am. (2). Peonies. R: F.C. Stern (1946) *A study of the genus P.* Perennial herbs with rhizomes & thickened tuberous roots (sects *Paeonia* (Spain to Japan) & *Onaepia* (2 W N Am.)) or subpachycaul shrubs (sect. *Moutan*, tree ps, 4 W China & Tibet) mainly in calcareous woodland; many cult. orn. (signifying shame & bashfulness in 'Language of Fls'), herb. forms largely derived from *P. lactiflora* Pallas (Tibet to China & Siberia) though some cvs involving other parents, cult. since 900 BC, tree ps from *P. suffruticosa* Andrews (moutan, Bhutan to Tibet & China) but many other spp. now involved (publ. of American P. Society); some form. medic. *P. lutea* Delavay ex Franchet (*P. delavayi* var. *lutea*, W China) & esp. var. *ludlowii* F. Stern & G. Taylor (Tibet) – common yellow-flowered tree p.; *P. suffruticosa* in cult. since C7 becoming a 'rage' a century later when grafted on to herbaceous stocks, its red roots much prized but flowering pl. discarded once petals fell (Hua Wang, the King of Flowers), medic. esp. modern Chinese herbalism (period pains, high blood pressure etc.); *P. mascula* (L.) Miller (S & E Eur. to As. Minor) – natur. Steep Holm in Bristol Channel since at least 1803, capsules with red (viable) & black (inviable) seeds

Paeoniaceae Raf. Dicots – Dilleniidae – Dilleniales. 2/34 temp. Euras., W N Am. Herbs & subpachycaul shrubs with vessels with scalariform endplates (cf. Ranunculaceae). Lvs spirally arr. & binately lobed or dissected; stipules 0. Fls solit., term., large, at least sometimes beetle-poll., bisexual, ± reg. with ± concave receptacle & usu. continuous phyllotactic spiral from lvs & bracts to K, C, A-trunks & G; K (3)4, 5(–7), leathery, persistent, or C-like (*Glaucidium*), C 0 (*Glaucidium*), 5–8(–13) with 3 or more vasc. strands like K, A ∞ with dichotomizing vascular strands derived from 5 basal trunks, members of each group usu. maturing centrifugally, anthers with longit. slits, G often surrounded by nectary-disk, (2)3–5(–15) with expanded ± subsessile stigma & several – ∞ marginal, anatropous, bitegmic ovules with massive outer integument. Fr. a head of follicles; seeds mesotestal, large with funicular aril (*Paeonia*) or compressed & winged (*Glaucidium*), with copious oily endosperm. X = 5

Genera: *Glaucidium*, *Paeonia*

Position of fam. controversial (Rosidae lately suggested), some characters (centrifugal stamens, pollengrain sculpturing, arils etc.) like Dilleniaceae but other features like Ranunculaceae: the forcing of P. into either order demonstrates the artificiality of the system, the fam. clearly retaining features in concert reminiscent of the angiosperms when not clearly divisible into the groups now recognized. *Glaucidium* has embryological & cytological features like *Paeonia* but has generally been treated as a fam. of its own – the fr. & seed differences characterize diff. dispersal syndromes – the arillate for birds, the winged for wind, such diversity being common in many fams. The group is prob. best treated separately from other orders in the present system as the seeds show affinities with Cucurbitaceae & Convolvulaceae (Corner), not Dilleniaceae, but as this is controversial (see KB 38(1983)87), it is maintained in its place in the Cronquist system for convenience. Cult. orn.

paeony *Paeonia* spp.

Paepalanthus Kunth. Eriocaulaceae. 485 trop. Am. (? + 1 Afr.). Some to 180 cm in Amaz. savannas

Paesia J. St-Hil. Dennstaedtiaceae. 12 Mal. to Polynesia, NZ & trop. Am. Like *Hypolepis* but with true indusium. *P. scaberula* (A. Rich.) Kuhn (NZ) – troublesome weed in disturbed habitats (cf. *Pteridium*)

Pagaea Griseb. = *Irlbachia*

Pagamea Aublet. Rubiaceae (IV 7). 24 trop. S Am. R: MNYBG 12,3(1965)270, Brittonia 41(1989)129. Local medic. esp. *P. coriacea* Spruce ex Benth. – bark used to restore mobility resulting from attacks of unknown cause

Pagameopsis Steyerm. Rubiaceae (IV 7). 2 Venez. R: AMBG 74(1987)106

Pagella Schönl. = *Crassula*

Pagesia Raf. = *Mecardonia*

Pagetia F. Muell. = *Bosistoa*

Pagiantha Markgraf = *Tabernaemontana*

pagoda flower *Clerodendrum paniculatum*; **p. tree** *Plumeria rubra, Sophora japonica*

pahautea *Libocedrus bidwillii*

paich-ha *Euonymus hamiltonianus* var. *maackii*

paigle *Primula veris, Ranunculus* spp.

paina de seda *Ceiba insignis*

paintbrush, devil's *Pilosella aurantiaca*; **Flora's p.** *Emilia* spp.; **Indian p.** *Castilleja* spp.

Painteria Britton & Rose = *Havardia*

Paivaea O. Berg = *Campomanesia*

Pajanelia DC. Bignoniaceae. 1 Indomal.: *P. longifolia* (Willd.) Schumann – good timber

Pakaraimaea Maguire & Ashton. Dipterocarpaceae. 1 Guayana Highland: *P. dipterocarpacea* Maguire & Ashton. Some features of Tiliaceae

pak-choi *Brassica rapa* Chinensis Group

Paladelpha Pichon = *Alstonia*

Palaeocyanus Dostál = *Centaurea*

Palafoxia Lagasca. Compositae (Hele.-Cha.). 12 S US, Mex. R: Rhodora 78(1976)567. Some cult. orn. ann. herbs

palamut *Quercus macrolepis* & other spp.

Palandra Cook = *Phytelephas*

Palaquium Blanco. Sapotaceae (II). 110 Taiwan & Indomal. to Samoa. Timber (nyatoh) & gutta-percha – rubbery substance which softens on heating, used in dentistry & obtained by ringing or felling trees, form. *P. gutta* (Hook.) Baillon usu. used, but other spp. exploited (see also *Payena*) but now ± superseded by synthetics. Some oilseeds

palas *Butea monosperma*

Palaua Cav. Malvaceae. 15 Andes

paldao *Dracontomelon dao*

Paleaepappus Cabrera. Compositae (Ast.-Ast.). 1 Patagonia

Palenia Philippi = *Heterothalamus*

Paleodicraeia C. Cusset. Podostemaceae. 1 Madagascar

Palhinhaea Franco & Vasc. = *Lycopodiella*

Paliavana Vell. ex Vand. Gesneriaceae. 2 Brazil. Cult. orn., bat- & bee-poll. (cf. bird- in *Vanhouttea*, though *Sinningia* similar but including bird-poll. *Rechsteineria*!)

Palicourea Aublet. Rubiaceae (IV 7). 200+ trop. Am. Local medic. esp. emetics in Amaz., some cult. orn.

Palimbia Besser ex DC. Umbelliferae (III 10). 3 S & E Russia to C As.

palisade grass *Brachiaria brizantha*

palisander timbers of Braz. spp. of *Dalbergia, Jacaranda* & *Machaerium*

Palisota Reichb. Commelinaceae (II 1a). 18 trop. Afr. Stout subpachycaul herbs with brightly coloured berries; some cult. orn. esp. *P. barteri* Hook. (W & C Afr.) – fls open 4–6 a.m. in wild & *P. hirsuta* (Thunb.) K. Schum. (W & C Afr.) – andromonoec.

Paliurus Miller. Rhamnaceae. 8 S Eur. (1) to Japan. *P. spina-christi* Miller (Christ's thorn, S Eur. to N China) – tree to 7 m with stip. thorns, 1 straight, 1 curved (other spp. with 2 straight) & fr. with horizontal wing, hedgepl.

Pallasia Klotzsch = *Wittmackanthus*

Pallenis (Cass.) Cass. = *Asteriscus*

palm Any sp. of Palmae, but in rural dists of GB = *Salix* spp.; **American oil p.** *Elaeis oleifera*; **Alexandra p.** *Archontophoenix alexandrae*; **bacaba p.** *Oenocarpus distichus*; **bamboo p.** *Chrysalidocarpus lutescens*; **bangalow p.** *A. cunninghamiana*; **betel p.** *Areca catechu*; **Black roseau p.** *Bactris major*; **p. cabbage** = **p. hearts**; **cabbage p.** *Livistona australis*; *Roystonea oleracea*; **c. palmetto palm** *Sabal palmetto*; **cane p.** *Chrysalidocarpus lutescens*; **carnauba wax p.** *Copernicia prunifera*; **coco-de-mer p.** *Lodoicea maldivica*; **coconut p.** *Cocos nucifera*; **cohune p.** *Orbignya cohune*; **coquito p.** *Jubaea chilensis*; **corozo p.** *Elaeis oleifera*; **date p.** *Phoenix dactylifera*; **double coconut p.** *Lodoicea maldivica*; **doum p.** *Hyphaene thebaica*; **ejow p.** *Arenga pinnata*; **fishtail p.** *Caryota mitis*; **gingerbread p.** *H. thebaica*; **gomuti p.** *A.*

pinnata; **gru-gru p.** *Acrocomia totai*; **p. hearts** – buds of spp. of *Acrocomia, Bactris, Euterpe, Roystonea* & *Sabal* spp. etc. but esp. geriatric *Cocos nucifera*; **honey p.** *Jubaea chilensis*; **ivory p.** *Phytelephas macrocarpa*; **Japanese peace p.** *Rhapis excelsa*; **kentia p.** *Howea forsteriana*; **kitul (kittool) p.** *Caryota urens*; **nikau p.** *Rhopalostylis sapida*; **nipa p.** *Nypa fruticans*; **oil p.** *Elaeis guineensis*; **palmetto p.** *Sabal* spp.; **palmyra p.** *Borassus flabellifer*; **Panama-hat p.** *Carludovica insignis*; **paxiuba p.** *Socratea exorrhiza*; **peach p.** *Bactris gasipaes*; **piassaba p.** *Attalea funifera, Leopoldinia piassaba*; **raffia p.** *Raphia farinifera*; **rattan p.** *Calamus* spp. etc.; **royal p.** *Roystonea regia*; **sago p.** *Metroxylon sagu*; **saw palmetto p.** *Serenoa repens*; **sealing-wax p.** *Cyrtostachys renda*; **sugar p.** *Arenga pinnata*; **tagua p.** *Phytelephas macrocarpa*; **talipot p.** *Corypha umbraculifera*; **thatch p.** *Thrinax parviflora*; **toddy p.** *Caryota urens*; **traveller's p.** *Ravenala madagascariensis*; **vegetable ivory p.** *Phytelephas macrocarpa*; **wax p.**, **(Colombian)** *Ceroxylon alpinum*, **(carnauba)** *Copernicia prunifera*; **windmill p.** *Trachycarpus fortunei*; **wine p.** *Jubaea chilensis*

palma-Christi *Ricinus communis*; **p. fibre** *Samuela carnerosana*; **p. rosa** *Cymbopogon martinii*

Palmae Juss. (Arecaceae). Monocots – Arecidae – Arecales. 201/2650 trop. (Afr. few) & warm. Evergreen trees, unbranched (or dichotomously branched), usu. erect, sometimes miniature &/or suckering or slender & lianoid (rattans) to 150 m with clear internodes; sec. thickening diffuse without new vasc. tissue, vasc. bundles closed, numerous, usu. with silicified fibres; raphides, polyphenols, s.t. alks present; roots with mycorrhizae & 0 root-hairs; vessels throughout. Lvs spirally arr. (rarely distichous or tristichous), usu. in term. rosettes & often v. large, with basal sheath, tubular but often splitting at maturity, petiole, lamina pinnate (feather palms) or palmate (fan p.), less often entire or 2-pinnate (*Caryota*), simple initially & often splitting during development into V-shaped (induplicate) or Λ-shaped (reduplicate) leaflets from plicate condition like Cyclanthaceae; ligule a prolongation of sheath, s.t. surrounding stem, or 0. Infls usu. axillary (simple to) paniculate, the branchlets often thick & spadix-like, the peduncle with prophyll & 1–several spathes. Fls usu. small & ± sessile, usu. insect-poll. & unisexual (monoec. & dioec. common) or bisexual, 3-merous, ± reg.; P often 3 + 3, leathery or fleshy, green to yellow, red or white, the outer often smaller usu. imbricate, inner usu. valvate in males, imbricate in females, rarely 2 + 2 or spirally arr. (10) or 0, A (3)6–950+, usu. 3 + 3, filaments s.t. connate &/or adnate to P, anthers latrorse with longit. slits & often monosulcate pollen grains; staminodes free, forming a cup or 0, $\underline{G}$ or 0 ($\underline{G}$(1–)3(–10), pluriloc., or pseudomonomerous with only 1 fertile loc., s.t. with septal nectaries, stigmas sessile or atop free or ± united styles, each loc. with 1 anatropous to hemitropous, campylotropous or orthotropous, bitegmic ovule. Fr. usu. a fleshy or fibrous drupe, rarely ± dehiscent (*Astrocaryum, Socratea* spp.); seed 1(–10), endosperm usu. v. oily, with protein & hemicellulose, s.t. ruminate. x = 13–18 (polyploidy in 1 *Areca* sp., also *Voaniola* (2n = 596+)). R: N.W. Uhl & J. Dransfield (1987) *Genera palmarum*

Classification & chief genera:

Six subfams:

I. **Coryphoideae** (lvs usu. induplicately palmate (pinnate in 2.); fls pedicellate to sessile, borne singly or in clusters but not in triads of 2 male & 1 female, often apocarpous, A 6–24; some in dry country or temp., 3 tribes):
   1. **Corypheae** (lvs palmate, bisexual or fls not dimorphic; 4 subtribes): *Chamaerops, Coccothrinax, Rhapis, Thrinax, Trachycarpus* (a, Thrinacinae); *Copernicia, Licuala, Livistona, Pritchardia, Serenoa, Washingtonia* (b, Livistoninae); *Corypha* (c, Coryphinae); *Sabal* (d, Sabalinae)
   2. **Phoeniceae** (lvs pinnate): *Phoenix* (only)
   3. **Borasseae** (lvs palmate; dioec.; 2 subtribes): *Borassus, Lodoicea* (a, Lataniinae); *Hyphaene* (b, Hyphaeninae)

II. **Calamoideae** (Lepidocaryoideae; erect or scandent, s.t. suckering; lvs reduplicate, (entire-) pinnate or (rarely) palmate; fls singly or in prs, pls with bisexual fls, polygamous, monoec. or dioec., A 6(–70), G (3) covered with reflexed imbricate scales; fr. usu. with 1 sarcotestal seed; mostly As., 2 tribes):
   1. **Calameae** (lvs pinnate; OW (exc. 1 *Raphia* sp.), 8 subtribes): *Eremospatha* (a, Ancistrophyllinae); *Eugeissona* (only; b, Eugeissoninae); *Korthalsia, Metroxylon* (only; c, Metroxylinae); *Calamus, Daemonorops, Salacca* (d, Calaminae); *Plectocomia* (e, Plectocomiinae); *Pigafetta* (only; f, Pigafettinae); *Raphia* (only; g, Raphiinae); *Oncocalamus* (only; h, Oncocalaminae)
   2. **Lepidocaryeae** (lvs palmate, NW): *Lepidocaryum, Mauritiella*

III. **Nypoideae** (dichot. branched creeping stem in mangrove; lvs pinnate with reduplicate leaflets; infl. with term. head of female fls & lateral branches of males, A 3 with filaments united in a column, G 3(4), 1 or more becoming water-disp. fr. with fibrous mesocarp, As.): *Nypa fruticans* (only) – s.t. referred to separate fam.

IV. **Ceroxyloideae** (lvs pinnate, reduplicate, infls with several peduncular bracts, fls solit. or in rows, G 3-ovulate; 3 tribes):
  1. **Cyclospatheae** (proximal fls bisexual): *Pseudophoenix* (only)
  2. **Ceroxyleae** (dioec., fls stalked): *Ceroxylon, Ravenea*
  3. **Hyophorbeae** (monoec. or dioec., fls sessile): *Chamaedorea, Hyophorbe*

V. **Arecoideae** (monoec. or dioec., usu. with reduplicately paripinnate (or bipinnate (1.)), fls in triads of central female flanked by males; 6 tribes):
  1. **Caryoteae** (lvs bipinnate or pinnate, leaflets praemorse, OW): *Arenga, Caryota*
  2. **Iriarteeae** (infls with more than 2 peduncular bracts, fls not sunken in pits; 2 subtribes): *Iriartea, Socratea* (a, Iriarteinae); *Catoblastus* (b, Wettiniinae)
  3. **Podococceae** (as 2. but fls sunk in pits): *Podococcus* (only)
  4. **Areceae** (usu. 1 peduncular bract, G usu. pseudomonomerous; the bulk of palms arr. in 15 subtribes): *Orania* (a, Oraniinae); *Manicaria* (only; b, Manicariinae); *Leopoldinia* (only; c, Leopoldiniinae); *Reinhardtia* (only; d, Malortieinae); *Chysalidocarpus, Dypsis, Neodypsis, Neophloga* (e, Dypsidinae); *Euterpe, Oenocarpus, Prestoea* (f, Euterpeinae); *Roystonea* (only; g, Roystoneinae); *Rhopalostylis* (h, Archontophoenicinae); *Cyrtostachys* (only; i, Cyrtostachydinae); *Calyptrocalyx, Howea* (j, Linospadicinae); *Ptychosperma, Veitchia* (k, Ptychospermatinae); *Areca, Gronophyllum, Nenga, Pinanga* (l, Arecinae); *Heterospathe, Iguanura* (m, Iguanurinae); *Oncosperma* (n, Oncospermatinae); *Sclerosperma* (o, Sclerospermatinae)
  5. **Cocoeae** (like 4. but G 3-loc.; 5 subtribes): *Beccariophoenix* (only; a, Beccariophoenicinae); *Butia, Cocos, Jubaea, Lytocaryum, Syagrus* (b, Butiinae); *Attalea, Orbignya, Scheelea* (c, Attaleinae); *Elaeis* (d, Elaeidinae); *Acrocomia, Aiphanes, Astrocaryum, Bactris, Desmoncus* (e, Bactridinae)
  6. **Geonomeae** (like 4. but fls sunk in pits): *Geonoma*

VI. **Phytelephantoideae** (dioec. with reduplicately paripinnate lvs with persistent petioles, male infls usu. dense spikes, females usu. globose heads, A ∞, female fls with spirally arr. P 7 + 7 & ∞ staminodes, G (5–10); Am.; R: OB 105(1991)5): *Phytelephas* (vegetable ivory)

Often char. of habitats being typical understorey pls of rain forest in As. & Am. (few in Afr.) & in canopy as lianes (rattans) but also high in Andes (*Ceroxylon*), in dry country & OW mangrove (*Nypa*), to 44° N (*Chamaerops*). The apical bud is usu. well-protected from predators by leaf-bases, spines or poss. toxic tissues: this the palm-heart of commerce & its removal leads to death of the trunk

App. an archaic group with the Afr. & Am. genera retaining the most primitive characters (Whitmore), palms are intimately associated with human societies in trop. & warm countries esp. *Cocos* in Pacific, *Phoenix* in N Afr., *Metroxylon, Arenga* & *Caryota* in As. etc. A no. of spp. so long cult. as to be unknown wild: *Cocos nucifera* (coconut), *Corypha umbraculifera* (talipot), *Phoenix dactylifera* (date). Many imp. international commodities esp. oils (*Acrocomia, Bactris, Cocos, Elaeis, Orbignya*), sago (*Metroxylon*), fibres (*Astrocaryum, Attalea, Caryota, Chamaerops, Cocos, Leopoldinia, Raphia, Sabal, Serenoa*), waxes (*Ceroxylon, Copernicia*), fr. (*Cocos, Phoenix*) & palm-hearts, drugs (*Areca*), rattan furniture, some timber & many cult. orn. street-trees & house-pls. E.J.H. Corner (1966) *Natural history of Palms*; *Principes* (journal)

Palmerella A. Gray = *Lobelia*

Palmeria F. Muell. Monimiaceae (VI 1). 15 New Guinea (12), Aus. (3, end.). R: Blumea 28(1982)85

Palmervandenbroekia Gibbs = *Polyscias*

palmetto *Sabal palmetto*; **saw p.** *Serenoa repens*

palmiet *Prionium serratum*

palmilla *Yucca schidigera*

palmiste *Roystonea oleracea*

palmito palm-heart

Palmolmedia Ducke = *Naucleopsis*

Palmorchis Barb.-Rodr. Orchidaceae (V 2). 12 trop. Am.

palmyra *Borassus flabellifer*

palo amarillo *Calycophyllum multiflorum*; **p. colorado** *Luma apiculata*; **p. madrono** *Amomyrtus luma*; **p. santo** *Bulnesia sarmientoi*; **p. verde** *Parkinsonia* spp.

palosapis *Anisoptera thurifera* & other spp.

Paloue Aublet. Leguminosae (I 4). 4 trop. Am.
Paloveopsis Cowan. Leguminosae (I 4). 1 NE S Am.
palsywort *Primula veris*
palta *Persea americana*
Paltonium C. Presl = *Nevrodium*
palu *Manilkara hexandra*
Palumbina Reichb.f. Orchidaceae (V 10). 1 Guatemala: *P. candida* (Lindley) Reichb.f. – cult. orn.
Pamburus Swingle (~ *Atalantia*). Rutaceae. 1 S India, Sri Lanka
pameroon bark *Trichilia moschata*
Pamianthe Stapf. Amaryllidaceae (Liliaceae s.l.). 3 N Andes. R: Brittonia 36(1984)22. Bulb with long false neck. *P. peruviana* Stapf (Peru) – cult. orn. with scented fls
pampas grass *Cortaderia selloana*
pampelmousse *Citrus grandis*
Pamphalea Lagasca. Compositae (Mut.-Nass.). 6 subtrop. & temp. S Am. Herbs; pappus 0
Pamphilia C. Martius ex A. DC. Styracaceae. 3 Brazil
Pampletantha Bremek. (~ *Pauridiantha*). Rubiaceae (I 10). 5 trop. Afr. R: BJ 71(1941)217
Panama-hat plant *Carludovica insignis*
Panamanthus Kuijt (~ *Struthanthus*). Loranthaceae. 1 Panamá. Allied to *Gaiadendron*
Panax L. Araliaceae (2). 6 N Am., E As. Glabrous herbs with rhizomes & thick roots used medic. (ginseng; active principles glycosides) in E & thought to affect membrane transport of steroids, esp. *P. pseudoginseng* Wallich (sanchi, Korea, Manchuria), used in cancer treatment in China, but also *P. ginseng* C. Meyer (sang, Korea, NE China) now v. rare in wild, roots fetching up to $10 000, & (American g.) *P. quinquefolius* L. (E N Am.)
Pancheria Brongn. & Gris. Cunoniaceae. 26 New Caled. Accumulate nickel
Panchezia Montr. = *Pancheria*
Pancicia Vis. (~ *Pimpinella*). Umbelliferae (III 8). 1 SE Eur.
Pancovia Willd. Sapindaceae. 12 trop. Afr. Some timber & ed. fr.
Pancratium L. Amaryllidaceae (Liliaceae s.l.). 16 Canary Is., W Afr., Medit. (Eur. 2), Namibia. *P. maritimum* L. (sea daffodil, Medit.) – figured in Minoan bronze (1560 BC), with other spp. cult. orn. like *Hymenocallis* but seeds angular (not globose or oblong)
Panda Pierre. Pandaceae. 1 trop. W Afr.: *P. oleosa* Pierre – seed-oil used locally for cooking; alks; large pyrene distrib. only by elephants
Pandaca Noronha ex Thouars = *Tabernaemontana*
Pandacastrum Pichon = *Tabernaemontana*
Pandaceae Engl. & Gilg (~ Euphorbiaceae). Dicots – Rosidae – Euphorbiales. 4/18 OW trop. Dioec. trees. Lvs simple, distichous on branches of determinate growth, often without axillary buds though such a bud between branch & axis; stipules small. Fls small, reg., usu. 5-merous, in axillary fascicles (*Microdesmis*) or cymes (*Centroplacus*) or in term. (*Galearia*) or cauliflorous (*Panda*, *Galearia*) thyrses; K free or connate, C valvate or imbricate, A 5, 5 + 5, 10 or 15, anthers in longit. slits, disk 0 (present in *Centroplacus* female), $\underline{G}$ (2–5) with 1 (2 in *Centroplacus*) anatropous (orthotropous in *Panda*) bitegmic ovule per loc. Fr. a drupe (capsule in *Centroplacus*) with bony endocarp with as many loc. & seeds as G; seeds without caruncles, oily endosperm copious. 2n = 30

Genera: *Centroplacus*, *Galearia*, *Microdesmis*, *Panda*

Stony fr. not found in Euphorbiaceae, though 'throw-away' branches found in e.g. *Phyllanthus*

Oilseeds (*Panda*)

pandan wangi *Pandanus amaryllifolius*
Pandanaceae R. Br. Monocots – Arecidae – Pandanales. 3/875 OW trop. to NZ. Dioec. (bisexual infls or fls in *Freycinetia*) usu. pachycaul trees & shrubs or (*Freycinetia*) lianes with clasping aerial roots, s.t. epiphytic; prop-roots usu. at base of stem; primary thickening, often with compound vasc. bundles; vessels throughout vegetative body; branching sympodial, apical meristems forming infls (s.t. infls only on lateral branches in *Pandanus*). Lvs simple, glabrous, in 3 (4 in *Sararanga*) ranks appearing as spirals ('screw-pines') because of spiral growth of stem, bases sheathing, blades usu. elongate (to 5 m), usu. xeromorphic with parallel veins & marginal spines (also on midrib). Fls v. small, ∞, in term. panicles (*Sararanga*) or in 1–several racemosely arranged spadices subtended by coloured spathes, homologies obscure – pedicellate & bracteate with irreg. 3–4-lobed cup (?P), A ∞ with fleshy filaments, G 10–80, 1-seeded with sessile stigmas, in forked double row, in *Sararanga*; in *Pandanus* & *Freycinetia* P 0 & distinction between A & G of individual fls obscured in development from app. reg. initials (in *F.*, A grouped around vestigial

pistillode, G (1–12), 1-loc. with several ovules & often with basal staminodes; in *P.*, A arranged in phalanges, often with branched filaments & elongated connectives, G 2–30 in phalanges, often incompletely closed, each with 1(– few) ovules); ovules anatropous, bitegmic. Fr. – berries (*Freycinetia*) or drupes, in heads, drupes with 12–80 pyrenes in *Sararanga*, monocarpellary with 1 seed or 'polydrupes' (connate carpels of phalanges) with united or separate endocarps; seeds small with copious oily (starchy in *Freycinetia*) endosperm, strophiole (from raphe) s.t. present. x = 30 (some aneuploidy)

Classification & genera:

1. **Pandanoideae** (arborescent, G 1-ovulate): *Pandanus, Sararanga*
2. **Freycinetioideae** (lianoid, G multi-ovulate): *Freycinetia*

Trad. associated with Palmae but strikingly diff.; fam. in all continents (exc. Aus.) in Upper Cretaceous, now OW only

Some ed. fr., scent-making, flavouring, thatch & cult. orn. (*Pandanus*)

pandani, pandanni, pandanny *Richea pandanifolia*

Pandanus Parkinson. Pandanaceae. 700 OW trop. (Mal. c. 450, Madag. c. 100, Afr. only 26). Screw-pines (all Aus. spp. protected); fr. fibrous, cooked & eaten, disp. by sea, freshwater, turtles, fish, birds & bats; R: BJ 94(1974)466. Usu. erect with stilt-roots, some maritime, others rheophytes, in mountain forests or epiphytes; in subg. *Vinsonia* sect. *Acanthostyla* (13 spp. Madag. swamps) monopodial trunk to 15 m with huge term. rosette of lvs & lateral horizontal forked branches with infls & smaller lvs, a growth-form unique in angiosperms, the apical lvs of saplings to 10 m long & 36 cm wide; wind- or insect-poll. Lvs used for basketry & hat-making, mats (Tonga), thatch, umbrellas (Solomon Is.) etc., fibres from stilt-roots used for chair-seats, cordage etc.; some flavourings for food esp. *P. amaryllifolius* Roxb. (pandan wangi) – usu. sterile (? orig. Moluccas), scented lvs used to flavour rice, jellies etc. in Mal.; *P. fascicularis* Lam. (? = *P. tectorius*) – male spadices distilled for kewda (perfume) esp. in Orissa (30 million infls annually) to flavour food, tobacco, soap, hair-oil etc.; *P. julianettii* Martelli (New Guinea) – imp. local fr. tree; *P. kaida* Kurz (? cultigen, SE As.) – planted on paddy-field 'bunds' in Sri Lanka, pollen insect-repellent; *P. leram* Jones ex Fontana (Nicobar breadfruit, E Indian Ocean) – fr. ed.; *P. tectorius* Sol. ex Parkinson (*P. odoratissimus*, ketaki, OW maritime) – males form 30 branches & fl. annually, females c. 16 & biennially (Fiji), on atolls disp. by landcrabs eating fr. & discarding seed, often cult. for ed. fr., scent from male fls, lvs for matting & basketry, as living hedge & cult. orn., a staple on some Pacific atolls, where black dye prepared from roots; *P. utilis* Bory (Madag.) – lvs much used for mats etc.; *P. veitchii* Hort. Veitch ex Masters & T. Moore (Polynesia) – sp. most commonly cult. as housepl., lvs varieg., fls & fr. unknown (? *P. tectorius* cv.)

Panderia Fischer & C. Meyer. Chenopodiaceae (I 4). 1 SW & C As.

Pandiaka (Moq.) Hook.f. Amaranthaceae (I 2). 12 trop. & S Afr. R: KB 34(1980)425

Pandorea (Endl.) Spach. Bignoniaceae. 6 E Mal., Aus., New Caled. Evergreen lianes, cult. orn. esp. *P. jasminoides* (Lindley) Schumann (bower plant, Aus.) & *P. pandorana* (Andrews) Steenis (wonga-wonga vine, Mal. & W Pacific) – wiry branches straightened over fire by C Aus. aborigines for spear-shafts

panga-panga *Millettia stuhlmannii*

Pangium Reinw. Flacourtiaceae (4). 1 Mal.: *P. edule* Reinw. – fr. to 30cm, seeds with hydrogen cyanide removed by boiling, ed. (selected cvs low in cyanide), oil for illumination & soap, bony seed-coats used for rattles, door-'curtains' etc., bark used for string bags ('bilum') in New Guinea & an antimalarial paste in Sumatra; seeds washed up on shores of Netherlands (poss. aided by humans!)

pangola grass *Digitaria decumbens*

panic grass *Panicum* spp.

Panicum L. Gramineae (34b). (Incl. *Dichanthium*) 500+ trop. to warm temp. (Eur. 2 + 2 natur.). Panic or crab grass & millets; some fodders, grains & cult. orn. esp. OW. R: JAA suppl. 1(1991)224. Some myrmechory (N Am.). *P. gilvum* Launert (Afr.) – weed in NSW suspected of photosensitizing sheep; *P. hemitomon* Schultes (pifine grass, N Am.) – prairie fodder; *P. maximum* Jacq. (guinea grass, Afr., natur. Am.) – most imp. lowland trop. Am. cult. forage, facultative apomict, tetraploid races commonest; *P. miliaceum* L. (proso millet, common, French or Russian m., cultigen (?) domesticated C As.) – flour, alcoholic drinks, pig-food, cult. E & C Eur. 5th millenium BC; *P. molle* Sw. (water grass, trop. Am.) – bad weed; *P. sonorum* L. (sauwi, Afr.) – domesticated; *P. sumatrense* Roth ex Roemer & Schultes (little millet, sama, Mal.) – minor grain imp. in S India, subsp. *psilopodum* (Trin.) de Wet (*P. psilopodum*) alleged progenitor; *P. texanum* Buckley (Colorado grass, Texas millet, S N Am.) – fodder

panirband *Withania coagulans*

Panisea (Lindley) Steudel. Orchidaceae (V 12). 7 NE India to SE As. R: NJB 7(1987)511

Panopsis Salisb. Proteaceae. 20 trop. Am. Some timbers

pansy *Viola* spp. esp. *V.* × *wittrockiana*; **field p.** *V. arvensis*

Pantacantha Speg. Solanaceae (1). 1 Patagonia: *P. ameghinoi* Speg. Allied to *Lycium*

Pantadenia Gagnepain. Euphorbiaceae. 1 SE As.

Pantathera Philippi = *Megalachne*

Pantlingia Prain = *Stigmatodactylus*

Panulia (Baillon) Kozo-Polj. = *Apium*

Panurea Spruce ex Benth. Leguminosae (III 2). 1 Colombia, Brazil

Panzeria Moench = *Panzerina*

Panzerina Soják (*Leonuroides*). Labiatae (VI). 6 W Siberia to Mongolia. R: Taxon 31(1982)559

Paolia Chiov. = *Coffea*

papain *Carica papaya*

papapsco *Acer* spp.

Papaver L. Papaveraceae (IV). 80 Eur. (26), As., Cape Verde Is. (1), S Afr. (1: *P. aculeatum* Thunb., allied to E Med. spp., (?) introd. Aus.), W N Am. Poppies. V. many alks. Fls in bud nodding through asymmetric growth of pedicel, later rectified at anthesis; fr. a capsule roofed by persistent stigmas, seeds protected thus & shaken out through pores beneath them. Many cult. orn. & for alks esp. *P. somniferum* L. (opium p., = ssp. *somniferum* (2n = 22) cultigen poss. derived from ssp. *setigerum* (DC) Arc. (*P. setigerum*, 2n = 44, SW As., but these two effectively 'triploid' & hexaploid of pl. similar to *P. glaucum* Boiss. & Hausskn. (2n = 14) & *P. gracile* Boiss. (2n = 28)), E Med.) – ssp. *somniferum* (C 20–80 mm long!) early cult. W Med. but now esp. Iran to China, most imp. drug pl. unmatched by synthetics, opium being dried latex obtained from immature capsules by lancing them, & containing c. 25 diff. alks esp. morphine (9–17%, available in pure form by 1806) – powerful analgesic & potentially addictive narcotic (suppressing emotional activity & making pain, hunger, discomfort, fear & anxiety tolerable; blood vessel dilation giving pleasant warm feeling), thebaine, codeine, narceine, narcotine & papaverine (alleged to promote prolonged penile erections (when injected!)), source of heroin ('gear', 'smack', 'scag', 'H'; 'chasing the Dragon' = inhaling smoke from heated powder through tube) & form. used in many patent medicines etc., production in 1971 some 1450 t (legal) & c. 1200 t (illicit), c. 700 t from the Golden Triangle; seeds (maw s.) without opium, used on bread etc. in baking (high incidence of glaucoma in p. seed eaters) & as birdseed; ssp. *hortensis* (Hussenot) Corb. – source of poppy oil used in artists' paints & salad oil, soap etc.; *P. bracteatum* Lindley (W As.) – poss. comm. source of codeine (prepared from thebaine in this sp. allied to *P. orientale* which has no morphine); *P. nudicaule* L. (Iceland p., Arctic & N temp. As.) – some opium, many cvs with diff. fl. colours; *P. pseudo-orientale* (Fedde) Medw. ('*P. orientale*', oriental p., SW As.) – the coarse perennial p. of herbaceous borders; *P. radicatum* Rottb. (N Eur., W As.) – with *Salix arctica* the most northerly plant (83° N); *P. rhoeas* L. (red, common, corn or field p., Euras., N Am.) – orig. E Med. & poss. selected by anthropogenic habitats up to NW Eur., prob. 'flowers of the field' of Isaiah, the p. of Flanders in World War I & Poppy Day in GB, petals a source of red dye used in some wines & medicines, also medic., selected cult. orn. cvs being 'Shirley ps' (Rev. W. Wilks C19)

Papaveraceae Juss. Dicots – Magnoliidae – Ranunculales. 23/230 mostly N temp. Herbs, subshrubs or pachycaul treelets with many isoquinoline alks in articulated laticifers (or latex-cells); vasc. bundles in 1, 2 or more rings. Lvs usu. spirally arr., (entire–) lobed or dissected; stipules 0. Fls usu. large, reg., bisexual, hypogynous (perigynous in *Eschscholzia*), solit. (less often in cymes etc.); K 2(3), sometimes basally connate, rather asymmetrical, encl. bud & often caducous, C usu. 2 × K in 2 whorls (0 in *Macleaya*), often 2 + 2, 3 + 3(–16), imbricate & crumpled in bud, A (4–6 in *Meconella*) ± ∞, developing centripetally in multiples of K, nectaries 0, $\underline{G}$ (2+), 1-loc. ± style, the stigmas often connate to form ± discoid roof but separate atop ± ∞ carpels in 1 whorl in *Platystemon*, ovules (1–)∞, anatropous to amphitropous or ± campylotropous, bitegmic, often on parietal placentas s.t. meeting to form pluriloc. ovary with axile placentation. Fr. a capsule with longit. valves or pores (follicle with lomentum-like units in *Platystemon*); seeds s.t. arillate, with elongate embryo in copious oily endosperm. x = (5)6 or 7(8–11, 19)

Classification & chief genera:

I. **Chelidonioideae** (hairs multicellular; pollen grains tricolpate or polyporate; G 2(+), seeds usu. arillate): *Bocconia, Glaucium*

II. **Eschscholzioideae** (hairs unicellular; pollen grains polycolpate; G 2): *Eschscholzia*

III. **Platystemonoideae** (hairs multicelluar-multiseriate; pollen grains tricolpate; G many): *Meconella*

IV. **Papaveroideae** (hairs multicelluar-multiseriate; pollen grains usu. tricolpate; G (many): *Argemone, Meconopsis, Papaver*

Most of above & *Chelidonium, Eomecon, Macleaya, Roemeria, Romneya* & *Sanguinaria* cult. orn. & often medic. esp. *Papaver somniferum* (opium); some oilseeds

Fumariaceae prob. best here as subfam. Fumarioideae (R: OB 88(1986)5)

papaw *Carica papaya, Asimina triloba;* **mountain p.** *C. pubescens*

papaya *Carica papaya*

paper birch *Betula papyrifera;* **p. bush** *Edgeworthia papyrifera;* **p. daisy** *Bracteantha* spp. esp. *B. bracteata, Rhodanthe manglesii;* **p. mulberry** *Broussonetia papyrifera*

paperbark *Melaleuca* spp., *Streblus asper*

Paphia Seemann = *Agapetes*

Paphinia Lindley. Orchidaceae (V 10). 12 trop. Am. R: Die Orchidee 29(1978)207. Cult. orn. epiphytes

Paphiopedilum Pfitzer. Orchidaceae (II). 64 trop. As. to Papuasia. R: C. Cash, *The slipper orchids* (1991)73. Cult. orn. epiphytes or terr. herbs without pseudobulbs often grown as 'Cypripedium' but lvs folded & not plicate as in C. Many hybrids in cult.

Papilionaceae Giseke = Leguminosae (III)

Papilionanthe Schltr. (~ *Vanda*). Orchidaceae (V 16). 10 Himal. to Mal. Cult. orn. scramblers esp. 'Miss Joaquim' (*P. hookeriana* (Reichb.f.) Schltr. (SE As. to W Mal.) × *P. teres* (Roxb.) Schltr. (Himal. to SE As.)

Papilionopsis Steenis = *Desmodium* (trick-pl. of *D.* infl. inserted in a Burmanniacea (cf. *Actinotinus, Stalagmitis*)

Papillilabium Dockr. Orchidaceae (V 16). 1 E Aus.: *P. beckleri* (Benth.) Dockrill

papoose root *Caulophyllum thalictroides*

Pappagrostis Rosch. = *Stephanachne*

Pappea Ecklon & Zeyher. Sapindaceae. 1 Dhofar, trop. E to S Afr.: *P. capensis* Ecklon & Zeyher (wild plum) – ed. fr., seed-oil & timber

Papperitzia Reichb.f. Orchidaceae (V 10). 1 Mex.: *P. leiboldii* (Reichb.f.) Reichb.f. – cult. orn.

Pappobolus S.F. Blake Compositae (Helia.-Helia.). (Incl. *Helianthopsis*) 38 Colombia, Ecuador, Peru, Bolivia. R: SBM 36(1992)

Pappophorum Schreber. Gramineae (29). 8 S US to Arg. R: SB 14(1989)356

Pappothrix (A. Gray) Rydb. = *Perityle*

paprika *Capsicum annuum* Longum Group

Papuacalia Veldk. Compositae (Sen.-Tuss.). 13 New Guinea mts. R: Blumea 36(1991)168. Trees & shrubs

Papuacedrus Li (~ *Libocedrus*). Cupressaceae. 1 Moluccas & New Guinea: *P. papuana* (F. Muell.) Li – timber for housing. R: BM 12(1995)66

Papuaea Schltr. Orchidaceae (III 3). 1 New Guinea

Papualthia Diels = *Haplostichanthus*

Papuanthes Danser. Loranthaceae. 1 New Guinea

Papuapteris C. Chr. = *Polystichum*

Papuastelma Bullock. Asclepiadaceae (III 4). 1 New Guinea: *P. secamonoides* (Schltr.) Bullock

Papuechites Markgraf. Apocynaceae. 1 New Guinea

Papuodendron C. White = *Hibiscus*

Papuzilla Ridley = *Lepidium*

papyrus *Cyperus papyrus*

Pará cress *Acmella oleracea;* **P. grass** *Brachiaria mutica;* **P. nut** *Bertholletia excelsa;* **P. rubber** *Hevea brasiliensis*

Parabaena Miers. Menispermaceae (III). 6 Indomal. R: KB 39(1984)103

Parabarium Pierre ex C. Spire = *Urceola*

Parabeaumontia Pichon = *Vallaris*

Parabenzoin Nakai = *Lindera*

Paraberlinia Pellegrin = *Julbernardia*

Parabignonia Bur. ex Schumann. Bignoniaceae. 1 NE Brazil

Paraboea (C.B. Clarke) Ridley. Gesneriaceae. c. 80 SE As. to W Mal. R: NRBGE 41(1984)422. Calcium salts accumulate in lvs leading to necrotic blisters

Parabouchetia Baillon. Solanaceae (2). 1 Brazil: *P. brasilensis* Baillon ex Wettst. – known only from orig. coll. & poss. not in right fam.

Paracaleana Blaxell = *Caleana*

Paracalia Cuatrec. Compositae (Sen.-Tuss.). 2 Peru, Bolivia. Scandent shrubs

Paracalyx Ali. Leguminosae (III 10). 6 OW trop. R: USK 5(1968)93
Paracarpaea (Schumann) Pichon = *Arrabidaea*
Paracaryopsis (H. Riedl) R. Mill (~ *Cynoglossum*). Boraginaceae. 3 India. R: EJB 48(1991)56
Paracaryum (A. DC) Boiss. Boraginaceae. 9 (?70) Medit. to C As. R: PSE 148(1985)296. Some cult. orn.
Paracautleya R.M. Sm. Zingiberaceae. 1 S India
Paracephaelis Baillon. Rubiaceae (?II 2). 1 Madagascar
Paraceterach (F. Muell.) Copel. (~ *Gymnopteris*). Pteridaceae (IV). 7 S Eur., Macaronesia to Queensland. R: AFJ 76(1987)186
Parachampionella Bremek. = *Strobilanthes*
Parachimarrhis Ducke. Rubiaceae (I 7). 1 Amazonia
Paracoffea (Miq.) J. Leroy = *Psilanthus*
Paracolpodium (Tzvelev) Tzvelev = *Colpodium*
Paracorynanthe Capuron. Rubiaceae (I 1). 2 Madag. R: Adansonia 18(1978)159
Paracroton Miq. = *Fahrenheitia*
Paracryphia Baker f. Paracryphiaceae. 1 New Caledonia
Paracryphiaceae Airy Shaw. Dicots – Dilleniidae – Theales. 1/1 New Caled. Rather small tree with unicellular hairs; vessel-elements with up to 200 cross-bars in perforation plates. Lvs simple, toothed, subverticillate; stipules 0. Fls bisexual or male, ∞ in compound spikes; P 4 decussate resembling larger outer bract encl. inner 3 bracteoles, caducous. A usu. 8, filaments in male fls somewhat flattened, anthers basifixed with longit. slits, G (8–15) with as many distinct stigmas & 4 anatropous, unitegmic ovules in single row on axile placenta of each loc. Fr. a capsule with carpels separating from a central column except at its apex & opening ventrally (cf. Medusagynaceae); seeds small, winged, the straight embryo in copious endosperm
Only species: *Paracryphia alticola* (Schltr.) Steenis
Position controversial. S-type plastids like those in Dilleniaceae, Theaceae & Actinidiaceae
Paractaenum Pal. (*Parectenium*). Gramineae (34b). 1 Aus.: *P. novae-hollandiae* Pal.
Paracynoglossum Popov = *Cynoglossum*
Paradavallodes Ching = *Davallia*
Paraderris (Miq.) Geesink = *Derris*
Paradina Pierre ex Pitard = *Mitragyna*
Paradisanthus Reichb.f. Orchidaceae (V 10). 4 trop. S Am.
paradise flower *Caesalpinia pulcherrima*; **grains of p.** *Aframomum melegueta*; **p. nut** *Lecythis* spp.; **p. tree** *Quassia glauca*
Paradisea Mazzucc. Asphodelaceae (Liliaceae s.l.). 2 S Eur. Cult. orn. esp. *P. liliastrum* (L.) Bertol. (St Bruno's lily, mts of S Eur.)
Paradolichandra Hassler = *Parabignonia*
Paradombeya Stapf. Bombacaceae. 5 Burma, SW China
Paradrymonia Hanst. Gesneriaceae. 40+ trop. Am.
Paradrypetes Kuhlm. Euphorbiaceae. 2 W Amaz. & E Brazil. R: SB 17(1992)74
Parafaujasia C. Jeffrey. Compositae (Sen.-Sen.). 2 Mascarenes
Parafestuca Alexeev (? ~ *Festuca*). Gramineae (17). 1 Madeira
Paragelonium Leandri = *Aristogeitonia*
Paragenipa Baillon. Rubiaceae (II 5). 1 Madagascar: *P. lancifolia* (Baker) Tirv. & Robbrecht. R: NJB 5(1985)458
Paraglycine F.J. Herm. = *Ophrestia*
Paragoldfussia Bremek. = *Strobilanthes*
Paragonia Bur. Bignoniaceae. 1 trop. Am.: *P. pyramidata* (Rich.) Bur. – local med. in NW Amaz.
Paragophyton Schumann = *Spermacoce*
Paragramma (Blume) T. Moore = *Lepisorus*
Paragrewia Gagnepain ex R. Rao = *Leptonychia*
Paraguay tea *Ilex paraguariensis*; **P. palm** *Acrocomia totai*
Paragulubia Burret = *Gulubia*
Paragynoxys (Cuatrec.) Cuatrec. Compositae (Sen.-Tuss.). 15 NW trop. S Am. Pachycaul trees
Parahancornia Ducke. Apocynaceae. 7 trop. S Am. R: Novon 1(1991)42
Parahebe W. Oliver. Scrophulariaceae. 30 New Guinea, Aus. (1), NZ. Like *Hebe* but with compressed capsule etc. Some cult. orn. rock-pls esp. *P. catarractae* (Forster f.) W. Oliver (NZ)

Parahemionitis Panigr. (~ *Hemionitis*). Pteridaceae (IV). 1 India to C Mal.: *P. arifolia* (Burm.f.) Panigr.

Parahyparrhenia A. Camus. Gramineae (40g). 5 trop. W Afr., India, Thailand. R: KB 20(1967)434

Paraia Rohwer, Richter & van der Werff. Lauraceae (I). 1 Amaz. R: AMBG 78(1991)392

Paraixeris Nakai = *Ixeris*

Parajaeschkea Burkill = *Gentianella*

Parajubaea Burret. Palmae (V 5b). 2 Ecuador, Bolivia & Colombia, high alts (some 10 months dry). R: Brittonia 42(1990)92. Cult. orn. unarmed monoec.

Parajusticia Benoist = *Justicia*

Parakaempferia A. Rao & Verma. Zingiberaceae. 1 Assam: *P. synantha* A. Rao & Verma. R: BBSI 11(1971)206

parakeelya *Calandrinia* spp.

Parakibara Philipson. Monimiaceae (V 2). 1 Moluccas

Paraknoxia Bremek. Rubiaceae (III 8). 1 trop. Afr.: *P. parviflora* (Stapf & Verdc.) Bremek. R: BJ 110(1989)550

Parakohleria Wiehler. Gesneriaceae. 20 Andes, Colombia

Paralabatia Pierre = *Pouteria*

Paralamium Dunn. Labiatae (VI). 1 SW China

Paralbizzia Kosterm. = *Archidendron*

Paralepistemon Lejoly & Lisowski. Convolvulaceae. 2 S trop. Afr. R: BTBC 118(1991)267

Paraleptochilus Copel. = *Colysis*

Paraligusticum Tichom. (~ *Ligusticum*). Umbelliferae (III 8). 1 Altai Mts

Paralinospadix Burret = *Calyptrocalyx*

Paralstonia Baillon = *Alyxia*

Paralychnophora MacLeish (*Sphaerophora*). Compositae (Vern.- Lych.). 2 Brazil

Paralyxia Baillon = *Aspidosperma*

Paramachaerium Ducke. Leguminosae (III 4). 5 trop. Am. R: Brittonia 33(1981)435

Paramacrolobium Léonard. Leguminosae (I 4). 1 trop. Afr.: *P. coeruleum* (Taub.) Léonard

Paramammea Leroy = *Mammea*

Paramansoa Baillon = *Arrabidaea*

Paramapania Uittien (~ *Mapania*). Cyperaceae. Spp.?

Paramelhania Arènes. Sterculiaceae. 1 Madagascar

Parameria Benth. Apocynaceae. 3 SE As., Mal. Some medic., fibres & rubber

Parameriopsis Pichon = praec.

Paramichelia Hu = *Michelia*

Paramicropholis Aubrév. & Pellegrin = *Micropholis*

Paramicrorhynchus Kirpiczn. = *Launaea*

Paramignya Wight (~ *Atalantia*). Rutaceae. 12 Indomal. R: Swingle, *Bot. Citrus* (1943)253. *P. monophylla* Wight (India) – citrus stock

Paramitranthes Burret = *Siphoneugena*

Paramoltkia Greuter (~ *Moltkia*). Boraginaceae. 1 Balkans: *P. doerfleri* (Wettst.) Greuter & Burdet. R: Willd. 11(1981)38

Paramomum Tong. Zingiberaceae. 1 Yunnan. R: ABY 7(1985)309

Paramongaia Velarde. Amaryllidaceae (Liliaceae s.l.). 1 Peru: *P. weberbaueri* Velarde (Peru) – cult. orn. like *Pamianthe* but without false neck on bulb

Paramyrciaria Kausel. Myrtaceae (Myrt.). 6 S Am. R: Cand. 46(1991)512

Paramyristica de Wilde = *Myristica*

Paraná pine *Araucaria angustifolia*

Paranecepsia R.-Sm. Euphorbiaceae. 1 SE Afr.

Paranephelium Miq. Sapindaceae. 4 SE As., W Mal. R: Blumea 29(1984)425. *P. macrophyllum* King (SE As. to Malay Pen.) – form. cult. for seed-oil used in lamps & skin problems

Paranephelius Poeppig. Compositae (Liab.). 7 Peru, Bolivia, Arg. R: SCB 54(1983)45

Paraneurachne S.T. Blake. Gramineae (34a). 1 Aus.

Paranneslea Gagnepain = *Anneslea*

Paranomus Salisb. Proteaceae. 18 S Afr. R: CBH 2(1970)

Parantennaria Beauverd. Compositae (Gnap.-Cass.). 1 E Aus. mts: *P. uniceps* (F. Muell.) Beauverd. R: OB 104(1991)95

Parapantadenia Capuron = *Pantadenia*

Parapentapanax Hutch. = *Pentapanax*

Parapentas Bremek. Rubiaceae (IV 1). 3–4 trop. Afr., (?) 1 Madag.

Paraphalaenopsis A. Hawkes (~ *Phalaenopsis*). Orchidaceae (V 16). 4 W Borneo. R: H.R.

Sweet, *Genus Phalaenopsis* (1986)118. Cult. orn.

Paraphlomis Prain. Labiatae (VI). 8 E As., Mal.

Parapholis C. Hubb. (~ *Pholiurus*). Gramineae (18). 6 Eur. to S As.

Paraphyadanthe Mildbr. = *Caloncoba*

Parapiptadenia Brenan. Leguminosae (II 3). 3 trop. S Am. *P. rigida* (Benth.) Brenan source of angico gum used like g. arabic

Parapiqueria R. King & H. Robinson (~ *Eupatorium*). Compositae (Eup.- Ayap.). 1 Brazil. R: MSBMBG 22(1987)210

Parapodium E. Meyer. Asclepiadaceae (III 1). 3 S Afr.

Paraprenanthes Chang ex Shih (~ *Lactuca*). Compositae (Lact.- Lact.). 11 E (esp. China) & SE As. R: APS 26(1988)418

Paraprotium Cuatrec. = *Protium*

Parapteroceras Averyanov (~ *Saccolabium*). Orchidaceae (V 16). 3 Indomal. R: NJB 10(1990)485

Parapteropyrum A.J. Li (~ *Pteropyrum*). Polygonaceae (II 4). 1 Tibet

Parapyrenaria H.T. Chang = *Pyrenaria*

Paraquilegia J.R. Drumm. & Hutch. Ranunculaceae (III 1). 5 W Iran to Himal. & W China. Like *Isopyrum* but follicles several, *Semiaquilegia* but staminodes 0. *P. anemonoides* (Willd.) Ulbr. (*P. grandiflora*) – cult. orn.

Pararchidendron I. Nielsen (~ *Pithecellobium*). Leguminosae (II 5). 1 Java & Aus.: *P. pruinosum* (Benth.) I. Nielsen. R: BMNHN 4, 6(1984)379

Parardisia Nayar & Giri = *Ardisia*

Pararistolochia Hutch. & Dalziel (~ *Aristolochia*). Aristolochiaceae (II 2). 18 OW trop. (Afr. 9). R: Adansonia 17(1977)465

Parartocarpus Baillon. Moraceae (II). 4 Mal. Some ed. seeds (those of *P. venenosus* (Zoll. & Mor.) Becc. ed. after soaking in seawater for some days)

Pararuellia Bremek. Acanthaceae. 5 Mal.

Parasamanea Kosterm. = *Albizia*

Parasarcochilus Dockr. = *Pteroceras*

Parasassafras Long (~ *Litsea*). Lauraceae (II). 2 Himal. to W China & Burma

Parascheelea Dugand = *Orbignya*

Parascopolia Baillon = *Lycianthes*

Paraselinum H. Wolff. Umbelliferae (III 8). 1 Peru

Parasenecio W.W. Sm. & Small. Compositae (Sen.-Tuss.). 60+ Russia, E As. ('*Cacalia*' of As.), Alaska. R (as *Koyamacalia*): Phytologia 27(1973)270

Paraserianthes I. Nielsen (~ *Albizia*). Leguminosae (II 5). 4 Mal. to trop. Aus. & Solomon Is. Timbers esp. *P. falcatoria* (L.) I. Nielsen (*A. falcatoria*, *A. moluccana* (with most spp. lately distinguished as *Falcataria* – *F. moluccana* (Miq.) Barneby & Grimes), batai wood, moluccana, sau, Mal.) – fast-growing (to 5 m height increase (record = 10.74 m in 13 months in Sabah (1975), the world's fastest growing tree) & 15 cm diam. increase per annum) for tea-chests & fuel & *P. toona* (Bailey) I. Nielsen (red siris, cedar acacia, Aus.) – resistant to termites, used for carving in Sri Lanka

Parashorea Kurz. Dipterocarpaceae. 14 S China, SE As., Mal. Timbers ((white) lauan or seraya) esp. *P. malaanonan* (Blanco) Merr. (bagtikan, Borneo, Philippines) – major veneer; *P. stellata* Kurz (thingadu, Burma) – imp. timber

Parasicyos Dieterle. Cucurbitaceae. 1 Guatemala

Parasilaus Leute. Umbelliferae (III 5). 1 SW & C As.

Parasitaxus Laubenf. (~ *Podocarpus*). Podocarpaceae. 1 New Caled.: *P. ustus* (Vieill.) Laubenf. – parasitic (root-graft, no haustoria but vasc. bundles penetrating to host cambium) on *Falcatifolium taxoides* (Brongn. & Gris) Laubenf.; only parasitic gymnosperm known

Paraskevia W. Sauer & G. Sauer. Boraginaceae. 1 Greece: *P. cesalina* (Fenzl & Friedrichsthal) W. Sauer & G. Sauer. R: Phyton 20(1980)285

Parasorus Alderw. = *Davallia*

Parasponia Miq. Ulmaceae (II). 5 Mal., W Pacific. *Rhizobium* root-nodules

Parastemon A. DC. Chrysobalanaceae (1). 2 Nicobar Is. to New Guinea. R: PTRSB 320(1988)95. *P. urophyllus* (A. DC) A. DC – hard timber for construction, boats etc.

Parastrephia Nutt. Compositae (Ast.-Ast.). 5 Andes

Parastriga Mildbr. Scrophulariaceae. 1 trop. Afr.

Parastrobilanthes Bremek. = *Strobilanthes*

Parastyrax W.W. Sm. Styracaceae. 1 Burma

Parasympagis Bremek. = *Strobilanthes*

Parasyringa W.W. Sm. = *Ligustrum*

Paratecoma Kuhlm. Bignoniaceae. 1 Brazil: *P. peroba* (Rec.) Kuhlm. (peroba) – imp. building & furniture timber, now threatened with extinction

Paratephrosia Domin = *Tephrosia*

Parathelypteris (H. Itô) Ching = *Thelypteris*

Paratheria Griseb. Gramineae (34f). 2 W Afr., Madag., trop. Am.

Parathesis (A. DC) Hook.f. Myrsinaceae. 84 trop. Am. R: CTRFBS 5 (1966), Phytol. 55(1984)237

Parathyrium Holttum = *Deparia*

Paratriaina Bremek. Rubiaceae (IV 8). 1 Madagascar

Paravallaris Pierre = *Kibatalia*

Paravitex Fletcher. Labiatae (I, Verbenaceae). 1 Thailand

Pardanthopsis (Hance) Lenz (~ *Iris*). Iridaceae (III 2). 1–2 Mongolia, N China. *P. dichotoma* (Pallas) Lenz can be crossed with *Belamcanda chinensis*

Pardoglossum Barbier & Mathez. Boraginaceae. 6 W Med. (Eur. 1). R: Candollea 28(1973)281

Parduyna Salisb. = *Schelhammera*

Parectenium auctt. = *Paractaenum*

pareira root or **p. brava** *Chondrodendron tomentosum*; **false p. r.** *Cissampelos pareira*; **white p. r.** *Abuta rufescens*; **yellow p. r.** *Aristolochia glaucescens*

Parenterolobium Kosterm. = *Albizia*

Parentucellia Viv. Scrophulariaceae. 2 Med. to SW As. incl. Eur. e.g. *P. viscosa* (L.) Caruel (yellow bartsia)

Parepigynum Tsiang & P.T. Li. Apocynaceae. 1 Yunnan

Parhabenaria Gagnepain. Orchidaceae (IV 2). 1 SE As. = ?

Pariana Aublet. Gramineae (5). 30 Amazon, Costa Rica. R: JLSBot. 50(1936)337. Some spp. transitional to *Eremitis*. Lvs of *P. lunata* Nees used to pack gold & platinum dust

Parietaria L. Urticaceae (IV). c. 10 (diff. to identify) subcosmop. (Eur. 6). Fls bisexual in axillary cymes though first fl. female & last are males, the bisexuals v. protogynous with style emerging from bud, stamens developing explosively later by which time style dropped so fls appear male. Achenes of some spp. attractive to ants. *P. judaica* L. (*P. diffusa*, pellitory of the wall, W & S Eur.) – wall weed in GB, serious in Sydney, NSW; *P. officinalis* L. (C & S Eur.) – form. used as laxative & for urinary problems

parilla, yellow *Menispermum canadense*

Parinari Aublet (*Parinarium*). Chrysobalanaceae (2). 44 trop. (Am. 18, Afr. 6, As. & Pacific 21). R: PTRSB 320(1988)110. Frs disp. by bats, elephants (e.g. *P. excelsa* Sabine, trop. Afr.), baboons, pigeons, rheas, emus, agoutis & fish. Some timber & ed. fr. *P. capensis* Harvey (sand apple, trop. & S Afr.) – geoxylic suffrutex in poor grasslands of C Afr.; *P. curatellifolia* Planchon ex Benth. (trop. Afr.) – ed. fr. (mbura, mobola) & timber for poles etc. (Livingstone buried beneath one); *P. nonda* F. Muell. (nonda, N Aus.) – fr. ed. See also *Neocarya*

Paripon Voigt. Palmae. 1 = ?

Paris L. Trilliaceae (Liliaceae s.l.). c. 20 OW temp. (incl. *Daiswa*. R: Brittonia 35(1983)255); Eur. 1: *P. quadrifolia* L. (herb Paris), form. medic. with whorl of 4 (or more) net-veined lvs on aerial shoots produced irreg. (not annually) from monopodial rhizomes & 4-merous fls attractive to pollinating carrion-flies, though stamens grow on to stigmas if crossing fails; fr. a poisonous berry with seeds without sarcotesta. R: Plantsman 9(1987)81, 10(1988)167. Some cult. orn. esp. *P. polyphylla* Sm. (*Daiswa* p., Himal.) – largest pl. chromosomes known

Parishella A. Gray. Campanulaceae. 1 California

Parishia Hook.f. Anacardiaceae (IV). 5 Indomal. *P. insignis* Hook.f. (dhup) – trade timber in Andamans

Pariti Adans. = *Hibiscus*

Parkeriaceae Hook. Older name for Pteridaceae

Parkia R. Br. Leguminosae (II 1). 30 trop. (As. 10, Afr. 3, Madag. 1, Am. 16). Fls in dense heads lasting 1 night; entomophilous spp. (Am.) with all fls fertile (no nectar), others with term. nectariferous ones or with fertile ones separated from basal staminodal ones by zone of nectariferous; heads pendulous or erect, staminodal fls sometimes forming a fringe (? landing-stage for bats, OW spp. of which landing head upwards, Am. ones head downwards); red heads seen as black against sky in most spp., yellow ones as white against foliage (bats colour-blind); seeds disp. by mammals & birds; mealy pods ed. (locusts) esp. *P. biglobosa* (Jacq.) G. Don f. (Afr., introd. WI) & *P. filicoidea* Welw. ex Oliver (African locust,

Afr.), seeds of first boiled & fermented in W Afr., tannin from pods used on floor-surfaces; *P. speciosa* Hassk. (petai, W Mal.) – imp. 'jungle fr.' in Mal., ed. pods imparting garlic scent to eaters

Parkinsonia L. Leguminosae (I 1). (Incl. *Cercidium*) 29 drier Am. (25), S Afr. (1), NE Afr. (3). Consp. along water-courses in desert SW US (palo verde). *P. aculeata* L. (Jerusalem thorn, trop. & warm Am.) – cult. orn. spiny tree, phloem fibre suitable for mixed paper-pulp; *P. microphylla* Torrey (SW N Am.) – some cells survived after 250 yrs as herbarium material!

Parlatoria Boiss. Cruciferae. 2 SW As.

Parma violets *Viola odorata* cv.

Parmentiera DC. Bignoniaceae. 9 C Am. to NW Colombia. R: FN 25(1980)97. Cult. orn. trees incl. *P. alata* (Kunth) Miers (C Am.) – bat-poll. though pollen & nectar stolen by social bees, horse- (& poss. extinct megafauna-) dispersed, timber, seeds prepared as cooling drink, pericarp as cups; *P. cerifera* Seemann (candle-tree, Panamá) – fodder tree with fr. to 1.3 m long, borne on trunk; *P. aculeata* (Kunth) L.O. Williams (*P. edulis*, C Am.) – fr. ed. (poor)

Parnassia L. Parnassiaceae (Saxifragaceae s.l.). 15 N temp. & Arctic (Eur. 1: *P. palustris* L. (grass of Parnassus, N temp.)). Some cult. orn. herbs with solit. term. white fls, nectarless staminodes forming glistening knobs (false nectaries) attractive to poll. insects

Parnassiaceae Martinov (~ Saxifragaceae). Dicots – Dilleniidae – Theales. 1/15 N hemisphere. Prob. best excl. from Saxifragaceae, though that fam. & allies need careful study Genus: *Parnassia*; *Lepuropetalon* s.t. incl. here

Parochetus Buch.-Ham. ex D. Don. Leguminosae (III 22). 1 mts of trop. Afr. & As. to Java: *P. communis* Buch.-Ham. ex D. Don – some cleistogamous fls; cult. orn. rock-pl. & in hanging baskets

Parodia Speg. Cactaceae (III 5). (Excl. *Blossfeldia*) c. 50 S Am. Cult. orn. small globose cacti

Parodianthus Tronc. Verbenaceae. 1 Argentina

Parodiochloa C. Hubb. = *Poa*

Parodiodendron Hunz. Euphorbiaceae. 1 Argentina

Parodiodoxa O. Schulz. Cruciferae. 1 N Argentina mts

Parodiolyra Söderstrom & Zuloaga (~ *Olyra*). Gramineae (4). 3 trop. Am. R: SCB 69(1989)64

Parolinia Webb. Cruciferae. 5 Canary Is.

Paronychia Miller. Caryophyllaceae (I 2). (Excl. *Chaetonychia*, incl. *Gymnocarpos*) 110 cosmop. esp. Medit. (Eur. 17), Turkey, SE US & Peru to Bolivia (not native Aus.). R: MBMHRH 285(1968)64. Tufted herbs with small axillary fls concealed by stipules; some cult. orn. rock pls incl. *P. argentea* Lam. (Medit.) – form. medic.; *P. decandra* (Forssk.) Rohw. & Urmi-König (*G. d.*, Macaronesia to Pakistan) – much used as camel fodder

Paropsia Noronha ex Thouars. Passifloraceae. 4 trop. Afr., 6 Madag., 1 E Mal. Transitional to Flacourtiaceae

Paropsiopsis Engl. Passifloraceae. 7 trop. W Afr. Allied to *Paropsia*

Paropyrum Ulbr. = *Isopyrum*

Parosela Cav. = *Dalea*

Paroxygraphis W.W. Sm. Ranunculaceae (II 3). 1 E Himalaya

Parquetina Baillon. Asclepiadaceae (I, Periplocaceae). 1 W trop. Afr.: *P. gabonica* Baillon – local medic., stuns fish

Parramatta grass *Sporobolus africanus*

Parrotia C. Meyer. Hamamelidaceae (I 4). 1 SW Caspian: *P. persica* (DC) C. Meyer (ironwood) – C 0, A 5–7, fls bisexual; cult. orn. with flaking bark & bright autumn colour, hybrid with *Sycopsis sinensis* formed in Switzerland c. 1950

Parrotiopsis (Niedenzu) C. Schneider. Hamamelidaceae (I 4). 1 Himalaya: *P. jacquemontiana* (Decne.) Rehder – cult. orn. like *Parrotia* but A c. 15–24 & fl. heads surrounded by large white bracts, alleged to prevent regeneration of coniferous forest; twigs used in basket-making

parrot's bill *Clianthus speciosus*; **p.'s feather** *Myriophyllum aquaticum*

Parrya R. Br. Cruciferae. 25 N temp. (Eur. 1). Cult. orn. rock- pls esp. *P. nudicaulis* (L.) Regel with scented violet fls

Parryella Torrey & A. Gray. Leguminosae (III 12). 1 Mexico

Parryodes Jafri. Cruciferae. 1 S Tibet, E Himalaya

Parryopsis Botsch. Cruciferae. 1 Tibet

Parsana Parsa & Maleki = *Laportea*

parsley *Petroselinum crispum*; **bur p.** *Caucalis platycarpos*; **cow p.** *Anthriscus sylvestris*; **p. fern** *Cryptogramma crispa*; **fool's p.** *Aethusa cynapium*; **Hamburg p.** *P. crispum* 'Tuberosum'; **hedge p.** *Torilis japonica*; **Italian p.** *P. crispum* var. *neapolitanum*; **p. piert** *Aphanes arvensis*;

turnip-rooted p. *P. crispum* 'Tuberosum'; **water p.** *Oenanthe sarmentosa*

parsnip *Pastinaca sativa*; **cow p.** *Heracleum sphondylium*; **Peruvian p.** *Arracacia xanthorhiza*; **water p.** *Sium* spp.

Parsonsia R. Br. Apocynaceae. (Incl. *Lyonsia*) 82 E As., Indomal. to W Pacific (Aus. 22). Some locally medic.; fr. of *P. edulis* (G. Benn.) Guillaumin (New Caled.) ed.

Parthenice A. Gray. Compositae (Helia.-Amb.). 1 SW N Am.

Parthenium L. Compositae (Helia.-Amb.). 16 N Am., WI. R: CGH 172(1950)1. Cypselas released with an involucral bract & 2 adjacent disk florets (sterile). *P. argentatum* A. Gray (guayule, Texas & N Mex.) – 20% by weight rubber, cult. emergency rubber pl. as in World War II in US; *P. fruticosum* Less. (N Am.) – lactones inhibiting growth of insect larvae; *P. hysterophorus* L. (N Am.) – pollen inhibits fr. set in other spp. through allelopathy on stigmas, serious dermatitis through sesquiterpene lactone (parthenin) can be fatal

Parthenocissus Planchon. Vitaceae. 10 temp. As., N Am. Decid. lianes with tendrils often tipped with adhesive disks, much cult. ('Virginia creeper', 'ampelopsis') esp. *P. quinquefolia* (L.) Planchon (true V. c., NE US to Mex.) – leaflets 5, bark medic., phloem eaten, boiled stalk water used as cooking syrup by E Canadian Indians, less often seen in Eur. than *P. tricuspidata* (Siebold & Zucc.) Planchon (Boston ivy, C China to Japan) – lvs simple to 3-lobed, many cvs

partridge berry *Gaultheria procumbens*; **p.-breasted aloe** *Aloe variegata*; **p. wood** *Andira inermis*

Parvatia Decne. = *Stauntonia*

Parvisedum R.T. Clausen = *Sedella*

Parvotrisetum Chrtek = *Trisetaria*

Pasaccardoa Kuntze. Compositae (Mut.-Mut.). 4 trop. & S Afr. Annuals

Pascalia Ortega (~ *Wedelia*). Compositae (Helia.-Verb.). 2 S Am.

Paschalococos Dransf. (~ *Jubaea*). Palmae (V 5b). 1 Easter Is.: *P. disperta* Dransf. – extinct when described from subfossil endocarps

Pascopyrum A. Löve = *Elymus*

Pasithea D. Don. Anthericaceae (Liliaceae s.l.). 1 Chile

Paspalidium Stapf = *Setaria*. But see JAA suppl. 1(1991)305

Paspalum L. Gramineae (34b). c. 330 trop. & warm esp. Am. R: KBAS 13(1986)287, JAA suppl. 1(1991)277. Char. of pampas, campos etc., *P. pyramidale* Nees growing to 15 m in Amazonia. Many imp. fodders & some grains: *P. dilatatum* Poiret (Dallis grass, S Am., natur. US & S Afr.) – fodder; *P. distichum* L. & *P. vaginatum* Sw. (warm coasts) – stabilizing saltmarsh; *P. notatum* Fluegge (Bahia grass, trop. Am., introd. US) – fodder, some forms as lawn grass, erosion control in Afr.; *P. scrobiculatum* L. (Kodo millet, rice grass, India) – livestock feed

pasque-flower *Pulsatilla vulgaris*

Passacardoa Wild = *Pasaccardoa*

Passaea Adans. = *Ononis*

Passerina L. Thymelaeaceae. 18 S Afr. C 0, G 1-loc. Cult. orn. heath-like shrubs

Passiflora L. Passifloraceae. 430 trop. & warm Am., 20 Indomal. & Pacific (Afr. 0). R: FMBot. 19(1938)1. Passion flowers: lianes (*P. arborea* Sprengel (Colombia) a shrub or tree to 10 m) with axillary tendrils & entire or lobed lvs (sometimes central lobe not developing so lvs crescentic or bilobed), usu. extrafl. nectaries on petioles; alks (fertility control in Uruguay). Co-evolved (in S Am.) with heliconiid butterflies, species of which have 'partitioned' host resources allowing co-existence of several spp. per plant. Vernacular name bestowed by Catholic missionaries in S Am. as 'Calvary Lesson' – 3 styles = nails of the Crucifixion, A 5 = 5 wounds, corona = crown of thorns, K 5 + C 5 = apostles (less Peter & Judas), lobed lvs & tendrils = hands & scourges of Christ's persecutors. Many cult. orn. incl. hybrids & some imp. trop. fr. (passion fruit) – ed. part being arillate pulp used in drinks & ices; susceptible to nematodes. *P. adenopoda* DC (trop. Am.) – lvs with sharp recurved trichomes puncturing cuticle of heliconiid caterpillars & killing them through haemorrhage; *P.* 'Banana Poka' (? *P. mollissima* × *P. cumbalensis* (Karsten) Harms (Ecuador) – serious weed in Hawaii (almost 4000 ha); *P.* × *belotii* Pépin (*P.* × *alatocaerulea*, *P. alata* Dryander (E Brazil, NE Peru) × *P. caerulea*) – fls used in scent-making; *P. caerulea* L. (blue p.f., Brazil to Arg.) – ± hardy in GB; *P. edulis* Sims (purple granadilla, Brazil) – cult. for fr. in Aus., Mex., Hawaii (f. *edulis* with fls open a.m., fr. purple & f. *flavicarpa* Degener with fls open p.m., fr. yellow), US for juice etc.; *P. foetida* L. (trop. Am., widely natur. OW trop.) – ed. fr., but lvs (cyanogenic) refused even by starving horses; *P. incarnata* L. (apricot vine, E N Am.) – ed. fr. long cult. by Indians; *P. laurifolia* L. (yellow granadilla, Jamaica honeysuckle, trop. Am.) – widely cult. for ed. fr.; *P. ligularis* Juss. (sweet granadilla, trop. Am.) –

fr. superior to that of *P. laurifolia*; *P. maliformis* L. (trop. Am.) – cult. for grape-flavoured juice; *P. mollissima* (Kunth) L. Bailey (banana passion fruit, trop. Am.) – fr. ed.; *P. quadrangularis* L. (granadilla, maracuja, ? orig.) – large yellow fr. to 30 cm, eaten as veg. when immature, flavour for yoghurt etc.

Passifloraceae Juss. ex DC. Dicots – Dilleniidae – Violales. 17/575 trop. & warm temp. esp. Am. Lianes with axillary tendrils (? derived from infls), or (*Paropsia* & allies) shrubs & trees, often with unusual secondary growth & alks. Lvs spirally arr., entire or lobed palmately (compound in *Deidamia*), usu. with nectaries on petiole; stipules usu. small decid., or 0; axillary bud often aborting or growing into infl. or tendril, vegetative branches forming from accessory bud. Fls reg., usu. bisexual (solit. or) in cymes, with flat to tubular hypanthium (fls hypogynous in *Paropsia* etc.), often with elongate androgynophore; K (3–)5(–8) sometimes basally connate, imbricate, persistent, C as many as & alt. with K (or 0), imbricate, with corona of 1 or more rows of filaments or scales around A (4)5(–∞), usu. alt. with C, free or on gynophore (connate in tube around G in *Androsiphonia*), anthers with longit. slits, often (staminodal) disk around G, $\underline{G}$ (2)3(–5), 1-loc. with parietal placentas but styles usu. only basally connate (1 in *Barteria* & *Crossostemma*), ± ∞ anatropous (or orthotropous) bitegmic ovules usu. with long funicles. Fr. a berry or capsule; seeds with fleshy apical aril & bony testa, straight embryo embedded in copious oily endosperm. x = 6, 9–11

Principal genera: *Adenia, Basananthe, Passiflora*

Close to Flacourtiaceae & linked to them by *Paropsia* & allies, trees with corona but with pollen & fibre-tracheids of Passifloraceae, though chemistry much like *Pangium* of Flacourtiaceae; 'it seems useful to draw the arbitrary distinction between the P. & F. on the basis of the presence of the corona, rather than on growth-habit' (Cronquist)

Some fr. (*Passiflora*) & cult. orn.

passion flower *Passiflora* spp.; **blue p.f.** *P. caerulea*

passion fruit *Passiflora* spp. esp. *P. edulis*; **banana p. f.** *P. mollissima*

Pastinaca L. Umbelliferae (III 11). 14 temp. Euras. (Eur. 4). Alks; furanocoumarin toxic to insects. *P. sativa* L. (parsnip) – biennial with ed. taproot (UK record: 3.62 m long) with best flavour when frosted, allegedly useful in arthritis treatment

Pastinacopsis Golosk. Umbelliferae (III 11). 1 C As.

pastis *Pimpinella anisum* (also *Foeniculum vulgare*)

pasto lacquer *Elaeagia utilis*

Pastorea Tod. ex Bertol. = *Ionopsidium*

Patagonula L. Boraginaceae (Ehretiaceae). 2 Brazil, Arg. Timber for furniture, also source of dyes to colour *Cedrela* wood to appear like *Swietenia*

patana oak *Careya arborea*

Patascoya Urban = *Freziera*

pataua *Oenocarpus bataua*

patchouli or **patchouly** *Pogostemon cablin*; **Chinese p.** *Microtoena patchouli*; **Indian p.** *P. heyneanus*

Patellaria J. Williams & Ford-Lloyd = *Beta*

Patellifolia A.J. Scott, Ford-Lloyd & J. Williams = *Beta*

Patellocalamus W.T. Lin = *Ampelocalamus*

Patersonia R. Br. Iridaceae (II). 19 Sumatra, Borneo & New Guinea, Aus. (17 endemic). R: BJBB 44(1974)41. Some cult. orn.

Paterson's curse *Echium plantagineum*

patience *Rumex patientia*

Patima Aublet = *Sabicea*

Patinoa Cuatrec. Bombacaceae. 4 trop. S Am. Fish-poison

Patis Ohwi = *Stipa*

Patosia Buchenau = *Oxychloe*

Patrinia Juss. Valerianaceae. 15 Eur. (1) to Himal. & E As. Some cult. orn. rock-pls esp. *P. triloba* Miq. (*P. palmata*, Japan)

Pattalias S. Watson = *Cynanchum*

pattern wood *Alstonia congensis*

pau santo *Kielmeyera coriacea*

Paua Caball. = *Andryala*

Pauia Deb & Dutta. Solanaceae (1). 1 Assam: *P. belladonna* Deb & Dutta

Pauldopia Steenis. Bignoniaceae. 1 NE India to SE As.

Paulia Korovin = *Paulita*

Paulita Soják. Umbelliferae (III 8). 3 C As.

Paullinia L. Sapindaceae. 194 trop. Am., *P. pinnata* also in Afr. R: Pfl. IV 165,1(1931)219. Lianes with watch-spring tendrils & alks incl. caffeine & theobromine, some fish-disp. in Amazon. *P cupana* Kunth (guaraná, S Am.) – comm. fizzy soft drink prepared with seeds (c. 5m bottles a day in Brazil by 1980, 'Tai' in US), also dried as a tea (4.3% caffeine, i.e 3–5 times that in coffee) &, with cassava, principal alcohol of Mato Grosso, locally medic., tonic (allegedly aphrodisiac) imported to UK & used in chewing-gum there; *P. pinnata* L. (Am. & Afr.) – arrow & fish-poison, stems for cordage & roots for chewing-sticks (Nigeria); *P. yoco* R. Schultes & Killip (S Am. esp. Colombia) – caffeine-high drink (yoco)

Paulownia Siebold & Zucc. Scrophulariaceae. 6 E As. R: QJTM 12(1959)1. Decid. trees with winged seeds suggesting Bignoniaceae but placed in S. because of floral anatomy & seed-structure esp. endosperm. V. similar in aspect to *Catalpa* (Bignoniaceae). Cult. orn. esp. *P. tomentosa* (Thunb.) Steudel (*P. imperialis*, China) – fls strongly scented to some but faint to others (cf. *Freesia*), cabinet wood much used in Japan (kiri) for musical instruments; *P.* × *taiwaniana* T. Hu & H.T. Chang (*P. kawakamii* Itô (S China, Taiwan) × *P. fortunei* (Seem.) Hemsley (China, Japan) – fast-growing timber tree in S Am.

Paulseniella Briq. = *Elsholtzia*

Pauridia Harvey. Hypoxidaceae (Liliaceae s.l.). 2 SW Cape

Pauridiantha Hook.f. Rubiaceae (I 10). 25 trop. Afr.

Paurolepis S. Moore = *Gutenbergia*

Pausandra Radlk. Euphorbiaceae. 12 trop. Am.

Pausinystalia Pierre ex Beille. Rubiaceae (I 1). 13 trop. W Afr. *P. johimbe* (Schumann) Beille – aphrodisiac, etc. (alks), yohimbine shown to increase sexual motivation in rats

Pavetta L. Rubiaceae (II 2). 400 OW trop. R: FR 37(1934)1. Lvs with nitrogen-fixing bacterial nodules; some cult. orn. shrubs & small trees

Pavieasia Pierre. Sapindaceae. 1 SE As.

Pavonia Cav. Malvaceae. 150 trop. & warm. Some cult. orn. esp. *P.* × *gledhillii* Cheek (*P. makoyana* C.J. Morren × *P. multiflora* Juss. (Brazil)); some fibres used locally

pawpaw = papaw

Paxia Gilg = *Rourea*

Paxistima Raf. Celastraceae. 2 N Am. R: Sida 14(1990)231. Cult. orn. evergreen shrubs

paxiuba palm *Iriartea ventricosa*

Paxiuscula Herter = *Argythamnia*

Payena A. DC. Sapotaceae (II). 15 W Mal., Mindanao. *P. leerii* (Teijsm. & Binnend.) Kurz (W Mal.) – source of a gutta-percha (g. sundek)

Payera Baillon. Rubiaceae (IV 1). 1 Madagascar

Paypayrola Aublet. Violaceae. 7 trop. S Am. Trees with A-tube

pea *Pisum sativum*; **asparagus p.** *Psophocarpus tetragonolobus*; **p.-bean** *Phaseolus vulgaris*; **bitter p.** *Daviesia* spp.; **blue p.** *Hovea* spp., *Clitoria ternatea*, *Psoralea pinnata*; **butterfly p.** *C. ternatea*; **chick p.** *Cicer arietinum*; **Congo p.** *Cajanus cajan*; **coral p.** *Abrus precatorius*, *Kennedia* spp.; **cow p.** *Vigna unguiculata*; **Darling p.** *Swainsona galegifolia*; **(Sturt's) desert p.** *S. formosa*; **dogtooth p.** *Lathyrus sativus*; **dun p.** *Pisium sativum* var. *arvense*; **everlasting p.** *L. latifolius*, *L. grandiflorus*; **field p.** *P. sativum* var. *arvense*; **flame p.** *Chorizema ilicifolium*; **glory p.** *Clianthus puniceus*, *Swainsona formosa*; **grey p.** *P. sativum* var. *arvense*; **Indian p.** *Lathyrus sativus*; **maple, mutter** or **partridge p.** *P. sativum* var. *arvense*; **parrot p.** *Dillwynia* spp.; **pigeon p.** *Cajanus cajan*; **prince's p.** *Psophocarpus tetragonolobus*; **purple p.** *Hovea* spp.; **Riga p.** *L. sativus*; **rosary p.** *Abrus precatorius*; **sea p.** *L. japonicus*; **shaggy p.** *Oxylobium* spp.; **snow p.** = mange-tout; **Spanish p.** *Cicer arietinum*; **sugar p.** *Pisum sativum* var. *macrocarpum*; **sweet p.** *L. odoratus*; **Tangier p.** *L. tingitanus*; **winged p.** *Psophocarpus tetragonolobus*

peace-lily *Spathiphyllum* spp.

peach *Prunus persica*; **African p.** *Sarcocephalus latifolius*; **p. palm** or **nut** *Bactris gasipaes*; **wild p.** *Kiggelaria africana*; **p. wood** *Haematoxylum brasiletto*, *Caesalpinia echinata*

peacock flower *Caesalpinia pulcherrima*, *Delonix regia*, *Moraea* spp.; **p. tiger flower** *Tigridia pavonia*

peanut *Arachis hypogaea*; **hog p.** *Amphicarpaea bracteata*

pear *Pyrus communis* etc.; **African p.** *Manilkara obovata*; **avocado, alligator** or **aguacate p.** *Persea americana*; **balsam p.** *Momordica charantia*; **Chinese p.** *Pyrus pyrifolia*; **garlic p.** *Crateva* spp.; **Japanese** or **nashi p.** *P. pyrifolia*; **prickly p.** *Opuntia* spp.; **thorn p.** *Scolopia ecklonii*; **tiger p.** *Opuntia aurantiaca*; **white p.** *Apodytes dimidiata*; **Whitty p.** *Sorbus domestica*; **woody p.** *Xylomelum* spp.

Pearcea Regel. Gesneriaceae. 1 Ecuador: *P. hypocyrtiflora* (Hook.) Regel. R: Selbyana 6(1982)175

pearl barley *Hordeum vulgare*; **p. fruit** *Margyricarpus pinnatus*; **p. lupin** *Lupinus mutabilis*; **p. millet** *Pennisetum glaucum*; **p.wort** *Sagina procumbens* & other spp.

Pearsonia Dümmer. Leguminosae (III 28). 12 trop. & S Afr., Madag. (1) R: KB 29(1974)383. *P. metallifera* Wild accumulates up to 15.3% nickel & 4.8% chromium (in ash)

pecan nut *Carya illinoinensis*

Pechuel-Loeschea O. Hoffm. (~ *Pluchea*). Compositae (Inul.). 1 SW Afr. to Zimbabwe

Peckoltia Fourn. = *Matelea*

Pecluma Price. Polypodiaceae (II 5). 25 warm Am. R: AFJ 73(1983)109

Pecteilis Raf. Orchidaceae (IV 2). 4 trop. & E As. Cult. orn. terr.

Pectinaria Haw. Asclepiadaceae (III 5). 3 S Afr. Cult. orn. dwarf succ. herbs. R: CSJ 43(1981)62

Pectis L. Compositae (Hele.-Pect.). c. 100 warm & trop. Am., Galápagos. *P. papposa* Harvey & A. Gray (SW US) – fls used to flavour meat etc.

Pectocarya DC ex Meissner. Boraginaceae. 10 W Am.

Pedaliaceae R. Br. Dicots – Asteridae – Scrophulariales. 17/85 trop. & warm esp. coasts & arid. Herbs (aquatic in *Trapella*) or shrubs with indumentum of short-stalked hairs with head of 4 or more mucilage-filled cells & usu. storing stachyose rather than starch. Lvs opp. (upper s.t. spirally arr.), simple; stipules 0. Fls bisexual, solit. or in dichasia with 1 or 2 extrafloral nectaries (abortive fls) in axils of bracts at pedicel base; K (4)5 usu. forming lobed tube, C (5), irreg. s.t. with basal spur, the limb often oblique or ± 2-labiate, A alt. with lobes & attached to tube, posterior 1 & s.t. anterior 2 staminodal or latter 0, anthers with longit. slits, nectary-disk often around $\underline{G}$ (2) ($\overline{G}$ in *Trapella*) with term. style, 2-loc. with axile placentas, s.t. subdivided by partitions from carpellary midribs, or 1-loc. with intruded parietal placentas ± reaching centre or 8 1-ovulate locelli (*Josephinia*), with 1–∞ anatropous, unitegmic ovules. Fr. a loculicidal capsule or drupe or nut, often armed with horns, hooks or prickles, or winged; seeds with straight embryo in 0 or little oily endosperm. x = 8, 13–15, 18

Classification & chief genera (after Cronquist):

- **Pedalioideae** (fls axillary; $\underline{G}$, placentation axile; OW land-pls): *Harpagophytum, Pterodiscus, Rogeria, Sesamothamnus, Sesamum, Uncarina*
- **Martynioideae** (Martyniaceae (distinct on pollen evidence); infls term.; $\underline{G}$, placentation parietal; NW land-pls): *Craniolaria, Martynia, Proboscoidea*
- **Trapelloideae** ($\overline{G}$, As. aquatics): *Trapella* (only)

Ibicella carnivorous

Some cult. orn., oilseeds (*Ceratotheca, Proboscoidea, Sesamum* (sesame)), ed. fr. (*Proboscoidea*), local medic., while animal-disp. armed fr. of *Harpagophytum* & *Uncaria* used as mouse-traps

Pedaliodiscus Ihlenf. = *Pedalium*

Pedalium Royen ex L. Pedaliaceae. 1 OW trop: *P. murex* L. – lvs a veg., seeds locally medic.

Peddiea Harvey. Thymelaeaceae. 12 trop. (Tanz. 6, in E arc mts) & S Afr.

Pedicellarum Hotta (~ *Pothos*). Araceae (II). 2 Borneo incl. *P. paiei* Hotta coll. only twice. R: APG 27(1976)61. Fls pedicellate

Pedicularis L. Scrophulariaceae. 350+ N hemisp. (esp. mts of C & E As. (China allegedly 350), Eur. 54), 1 in Andes. Louseworts esp. *P. sylvatica* L. (Eur.) – presumed to become lice on sheep contacting them. Hemiparasites with alks. Roots of *P. lanata* Cham. & Schldl. eaten by N Am. Indians

Pedilanthus Necker ex Poit. (~ *Euphorbia*). Euphorbiaceae. 14 S N Am. to trop. Am. R: CGH 182(1957)1. Monoec. succ. shrubs with cyathia. Some cult. orn. esp. *P. tithymaloides* (L.) Poit. (bird cactus, Florida to S Am.) – used as hedgepl. & in cemeteries

Pedilochilus Schltr. Orchidaceae (V 15). 15 Papuasia

Pedinogyne Brand = *Trigonotis*

Pedinopetalum Urban & H. Wolff. Umbelliferae (III 8). 1 Hispaniola

Pediocactus Britton & Rose. Cactaceae (III 9). 6 SW US. R: CSJ 53(1981)17. Cult. orn. small cacti

Pediomelum Rydb. (~ *Orbexilum*). Leguminosae (III 11). 25 S N Am. incl. *P. esculentum* (Pursh) Rydb. (breadnut)

Pedistylis Wiens. Loranthaceae. 1 S Afr.

Peekelia Harms = *Cajanus*

Peekeliopanax Harms = *Gastonia*

peel, candied *Citrus medica* & other spp.

peelu extract *Salvadora persica*

peepul *Ficus religiosa*

Peersia L. Bolus = *Rhinephyllum*

Pegaeophyton Hayek & Hand.-Mazz. Cruciferae. 3 C As. & Himal. to W China

Peganaceae (Engl.) Takht. = Zygophyllaceae

Peganum L. Zygophyllaceae. 5–6 Medit. (Eur. 1) to Mongolia, S N Am. Alks. *P. harmala* L. (harmal, Medit. to As.) – alks incl. harmine (used as 'truth drug' by Nazis; also found in *Banisteriopsis*) & harmalol so avoided by all stock save camels, locally medic. esp. in eye-disease, rheumatism, Parkinson's Disease etc., burnt as intoxicant in C As. since (?) 5th millenium BC, seeds hallucinogenic (images reflected in local art-styles (?) incl. Persian carpets & concept of 'flying carpets' cf. 'flying broomsticks' in Eur. perhaps due to henbane) & sexually stimulatory (carbolines), source of an oil & of a dye (Turkey red) used for dyeing carpets & the hats known as tarbooshes

Pegia Colebr. Anacardiaceae (II). 3 E As., Mal.

Pegolettia Cass. Compositae (Inul.). 9 Afr., Arabia & Middle E with *P. senegalensis* Cass. to India (? Java). R: Cladistics 2(1986)158

pegwood *Cornus sanguinea, Euonymus europaeus*

Pehria Sprague. Lythraceae. 1 Colombia, Venezuela

pei-mu *Fritillaria roylei*

Peixotoa A. Juss. Malpighiaceae. 28 Brazil. R: CUMH 15(1982)1. Some apomixis

pejibaye, pejivalle *Bactris gasipaes*

Pelagatia O. Schulz = *Weberbauera*

Pelagodendron Seemann (~ *Randia*). Rubiaceae (II 1). 1 New Guinea, 2 New Caled., 1 Fiji. R: Candollea 46(1991)251

Pelagodoxa Becc. Palmae (V 4m). 1 Marquesas: *P. henryana* Becc. – large bifid lvs, now v. rare

Pelargonium L'Hérit. Geraniaceae. 280 trop. (few) & S (most) Afr. with 2 in E Medit. to Iraq (allied to S Afr. spp.), & 1 in each of S Arabia, St Helena, Tristan da Cunha & S India, Aus. 6, NZ 1 (*P. inodorum* Willd.). 'Geraniums' of greenhouses but differing from *Geranium* in irreg. fls & 5–7 of A 10 fertile, the rest just filaments; shrubs & herbs (some annual e.g. *P. apetalum* P. Taylor, C Afr.), s.t. with tuberous roots, or succ. when often with stipular spines & reduced lvs. R: D. Clifford (1970) *Ps*; J.J. van der Walt et al. (1977–88) *Ps of southern Africa* 1–3; (sect. P.) Bothalia 15(1985)345. V. imp. house-pls & bedding-pls esp. *P.* × *hortorum, P.* × *domesticum, P. peltatum* & spp. with scented lvs due to essential oils thought to be deterrent to animal-grazers though some used as basis for scent & soaps (geranium oil, see Arnoldia 34(1974)104 for spp. & scents), the oils reminiscent of lemon, peppermint, pennyroyal, nutmeg, strawberry, mint, camphor, apple, ginger, rue etc., those of *P. graveolens* L'Hérit. (*P. asperum*), *P. odoratissimum* (L.) L'Hérit. & *P. quercifolium* (L.) L'Hérit. (all S Afr.) being coll. comm. (mawah oil) in France, Algeria & Réunion (where 'Rosé' (*P. capitatum* (L.) L'Hérit. × *P. radens* H. Moore (*P. radula*), both S Afr.) much grown). *P. crispum* (Bergius) L'Hérit. (SW Cape) – lemon-scented; *P.* × *domesticum* L. Bailey (regal p. or geranium, group of cvs derived from *P. cucullatum* (L.) L'Herit. (*P. angulosum*), *P. grandiflorum* (Andrews) Willd. (S Afr. shrubs) & other spp.); *P. endlicherianum* Fenzl (Turkey) – hardy in GB; *P.* × *hortorum* L. Bailey (zonal p. or g., complex hybrid cultigen involving S Afr. *P. inquinans* (L.) L'Hérit. & *P. zonale* (L.) L'Hérit., their hybrid, *P.* × *hybridum* (L.) L'Hérit. (poss. correct name for *P.* × *hortorum*) back-crossed with parents) – familiar 'geranium' with many ('zonal') cvs incl. varieg. & even haploid ('Kleine Liebling' with n = 9, roots diploid) ones, much used for bedding e.g. scarlet-flowered 'Paul Crampel' (raised by Lemoine); *P. monoliforme* E. Meyer ex Harvey (S Afr.) – succ. still sprouting 7 months after being pressed; *P. peltatum* (L.) L'Hérit. (ivy-leaved p. or g., S Afr.) – familiar hanging basket geranium; *P. pinnatum* (L.) L'Hérit. (S Afr.) – taproot-like tuber to 25 cm, lvs pinnate

Pelatantheria Ridley. Orchidaceae (V 16). 5 India to Taiwan & W Mal. R: OB 95(1988)115

Pelea A. Gray = *Melicope*

Pelecostemon Leonard. Acanthaceae. 1 Colombia

Pelecyphora Ehrenb. Cactaceae (III 9). 2 NE Mex. R: EGF 3(1989)279

Pelexia Poit. ex Lindley. Orchidaceae (III 3). 73 warm & trop. Am. R: HBML 28(1982)342

Peliosanthes Andrews. Convallariaceae (Liliaceae s.l.). 1 Indomal.: *P. teta* Andrews – blue fr. attractive to (?)pheasant dispersers

Peliostomum E. Meyer ex Benth. Scrophulariaceae. 7 trop. & S Afr.

Pellacalyx Korth. Rhizophoraceae. 7–8 Indomal.

Pellaea Link. Pteridaceae (IV). 35 trop. & warm (Eur. 1) to Canada, esp. SW N Am. R (sect. P.): AMBG 44(1957)125. Cult. orn. ferns of rocks (cliff brake, US)

Pellegrinia Sleumer. Ericaceae. 5 Andes

Pellegriniodendron Léonard. Leguminosae (I 4). 1 Guineo-Congolian forest

Pelletiera A. St-Hil. Primulaceae. 2: 1 S Am., 1 Macaronesia

Pelliceria Planchon & Triana = *Pelliciera*

Pelliciera Planchon & Triana ex Benth. Pellicieraceae. 1 Costa Rica to Colombia (Pacific shores) & Carib. coast (rare), form. circum-Carib.: *P. rhizophorae* Planchon & Triana – contraction of range poss. due to salinity changes (intolerant of more than 3.7%)

Pellicieraceae (Triana & Planchon) Bullock. (~ Theaceae). Dicots – Dilleniidae – Theales. 1/1 C Am. Mangrove-tree. Lvs spirally arr., simple, leathery, sessile & decurrent; stipules 0. Fls bisexual, reg., axillary, solit., each with 2 red decid. petaloid bracts larger than C; K 5, petaloid, red, unequal, imbricate, C 5 elongate, pink to white, caducous, A 5 alt. with C & adpressed to alt. grooves of 10-grooved G with anthers adpressed to style with longit. slits & long connective, G (2) with tapering style, 2-loc. (partition sometimes only partial), each loc. with 1 campylotropous ovule (or 2nd loc. empty or 0). Fr. dry, leathery, onion-shaped with beak, ridged & grooved, with 1 1-seeded loc.; embryo with broad fleshy cotyledons & 0 endosperm

Only species: *Pelliciera rhizophorae* Triana & Planchon

Prob. to be excl. from Dilleniidae

Pellionia Gaudich. = *Elatostema*

pellitory *Anacyclus pyrethrum*; **p. of the wall** *Parietaria judaica*

Pelma Finet = *Bulbophyllum*

Pelozia Rose = *Lopezia*

Peltaea (C. Presl) Standley (~ *Pavonia*). Malvaceae. 4 trop. Am.

Peltandra Raf. Araceae (VI 7). 3 N Am. Allied to *Typhonodorum*. R: JEMSS 90(1975)137. *P. virginica* (L.) Kunth – starchy rhizomes form. imp. Indian food (roasted)

Peltanthera Benth. Buddlejaceae (Loganiaceae s.l.). 1 trop. Am. mts

Peltapteris Link = *Elaphoglossum*

Peltaria Jacq. Cruciferae. 6 E Medit. (Eur. 2) to Iran & C As. Some cult. orn. *P. emarginata* (Boiss.) Hausskn. (Greece, serpentines) has nickel up to 1% dry weight of lvs

Peltariopsis (Boiss.) N. Busch. Cruciferae. 2 Caucasus, N Iran

Peltastes Woodson. Apocynaceae. 7 trop. Am.

Pelticalyx Griff. Annonaceae (?). 1 India: *P. argentea* Griffith

Peltiphyllum (Engl.) Engl. = *Darmera*

Peltoboykinia (Engl.) H. Hara (~ *Boykinia*). Saxifragaceae. 1–2 Japan. R: BJLS 90(1985)31. *P. tellimoides* (Maxim.) H. Hara – cult. orn. coarse herb with peltate lvs

Peltobractea Rusby = *Peltaea*

Peltocalathos Tamura. Ranunculaceae (II 3). 1 S Afr.

Peltodon Pohl. Labiatae (VIII 4a). 5 Paraguay, Brazil

Peltogyne Vogel. Leguminosae (I 4). 23 trop. Am. esp. Amazonia. R: AA 6(1) suppl. (1976)1. Some water-disp. fr. in Amaz. Timber (amaranth) for panelling going black on contact with water (purple heart) esp. *P. paniculata* Benth.

Peltophoropsis Chiov. = *Parkinsonia*

Peltophorum (Vogel) Benth. Leguminosae (I 1). 8 trop. Timber (brasiletto) & shade trees; some cult. orn. esp. *P. pterocarpum* (DC) K. Heyne (*P. ferrugineum*, yellow flamboyant, Indomal. seashores) – bark medic. & a constituent of soga, a yellow-brown dye used in batik

Peltophyllum Gardner (*Hexuris*; ~ *Triuris*). Triuridaceae. 2 trop. S Am.

Peltostigma Walp. Rutaceae. 4 C Am., WI

Pelucha S. Watson. Compositae (Pluch.). 1 Mexico

peluskins *Pisum sativum* var. *arvense*

pe-mou oil *Fokienia hodginsii*

Pemphis Forster & Forster f. Lythraceae. 1 OW trop.: *P. acidula* Forster & Forster f. on coasts esp. eroding beaches, distylous having lost 'mid-type'; wood for knife-handles (Cocos Is.). See also *Koehneria*

Penaea L. Penaeaceae. 4 SW Cape. R: OB 29(1971). Endosperm tetraploid. Sources of gums tasting like liquorice, locally medic.

Penaeaceae Sweet ex Guillemin. Dicots – Rosidae – Myrtales. 7/21 Cape. Usu. small shrubs, often ericoid, ± glabrous, usu. with internal phloem. Lvs simple, entire, decussate; stipules minute to 0. Fls bisexual, reg. perigynous, solit. in axils of upper lvs with hypanthium & K often coloured like a corolla; K 4 valvate lobes of hypanthium, C 0, A 4 attached to hypanthium & alt. with K, anthers with longit. slits & much-expanded laminar connective, disk 0, G (4) with term. style & axile-basal &/or -apical or median-axile placentation, each loc. with 2–4 pendulous or erect anatropous, bitegmic ovules. Fr. a loculicidal capsule in persistent hypanthium; seeds usu. 1 per loc., with ± 0 endosperm,

embryo with large hypocotyl & v. small cotyledons. n = 11 or 12
Genera: *Brachysiphon, Endonema, Glischrocolla, Penaea, Saltera, Sonderothamnus, Stylapterus*
Locally medic.

Penang lawyer *Licuala* spp.

pencil cedar *Juniperus virginiana*

penda *Xanthostemon oppositifolius*

Penelopeia Urban. Cucurbitaceae. 1 Hispaniola

Penianthus Miers. Menispermaceae (I). 4 W & C Afr. R: BJBB 53(1983)41

Peniocereus (A. Berger) Britton & Rose. Cactaceae (III 1). (Incl. *Nyctocereus*) 20 SW N Am. to C Am. Cult. orn. day- or night-flowering cacti; fr. ed. esp. *P. serpentinus* (Lagasca & Rodriguez) N. Taylor (*N. serpentinus*, Mex.)

Peniophyllum Pennell = *Oenothera*

Pennantia Forster & Forster f. Icacinaceae. 3 Aus., NZ, Norfolk Is. *P. baylisiana* (W. Oliver) Baylis (Three Kings Is., NZ) – found 1945, only females; *P. corymbosa* Forster & Forster f. (NZ) – hard timber used for tool-handles, cabinet-work & form. (Maoris) for fire-making by friction

Pennellia Nieuw. Cruciferae. 9 Mex. (7), SW US, Guatemala, Bolivia, N Arg.

Pennellianthus Crosswh. = *Penstemon*

Pennilabium J.J. Sm. Orchidaceae (V 16). 11 Thailand to Java

Pennisetum Rich. Gramineae (34f). 130 trop. & warm (Eur. 1). R: JAA suppl. 1(1991)205. Fodders, lawn-grasses & some grains: *P. clandestinum* Hochst. ex Chiov. (Kikuyu grass, trop. E & NE Afr.) – apomictic tetraploid, pasture grass used for erosion control & as lawns; *P. glaucum* (L.) R. Br. (*P. americanum, P. typhoides*, bulrush or African or Indian or pearl or spiked millet, bajri, gero, OW cultigen) – food-crop also used in beer-making; *P. hohenackeri* Hochst. ex Steudel (moya grass, E Afr. to India) – suggested for paper-making; *P. polystachyon* (L.) Schultes (OW trop.) – fodder & hay but bad weed in Aus.; *P. purpureum* Schum. (elephant or Napier grass, Afr.) – fodder & paper

penny cress *Thlaspi arvense*; **p. flower** *Lunaria annua*; **p. pies** *Umbilicus rupestris*

pennyroyal *Mentha pulegium*; **American p.** *Hedeoma pulegioides*

pennywort *Hydrocotyle* spp., *Cymbalaria muralis, Sibthorpia europaea, Umbilicus rupestris*

Penstemon Schmidel. Scrophulariaceae. 250 N Am. (esp. W), NE As. (1). Shrubs & perennial herbs with posterior staminode bent down to lower side of C, naked or bearded. Sucrose-rich or -dominant nectars in primarily hummingbird-poll. spp., hexose-rich or -dominant ones in insect-poll. spp. even in closely related spp. pairs. Many cult. orn. esp. in US with over 130 named cvs, some treated as annuals incl. *P. barbatus* (Cav.) Roth (SW N Am.) – fls pink, catalpol (an iridoid) sequestered by *Neoterpes graefiaria* moths, *P. spectabilis* Thurber ex A. Gray (SW N Am.) – fls blue & purple, poll. pseudomasarid wasps & *P. virgatus* A. Gray (SW US) – fls violet, catalpol sequestered by *Poladryas arachne* butterflies & *Meris alticola* moths

Pentabothra Hook.f. = *Vincetoxicum*

Pentabrachion Muell.-Arg. = *Microdesmis*

Pentacalia Cass. (~ *Senecio*). Compositae (Sen.-Sen.). c. 200 C & S Am. R: Phytol. 40(1978)37, 49(1981)241

Pentacarpaea Hiern = *Pentanisia*

Pentace Hassk. Tiliaceae. 25 Burma to W Mal. R: PLPH 87(1964). Some timbers esp. *P. burmannica* Kurz (thitka, Burmese mahogany) – light timber

Pentaceras Hook.f. Rutaceae. 1 E Aus.: *P. australe* (F. Muell.) Benth. – some alks with anti-tumour activity

Pentachaeta Nutt. (~ *Chaetopappa*). Compositae (Ast.-Ast.). 6 SW N Am. R: UCPB 65(1973)32

Pentachlaena Perrier. Sarcolaenaceae. 2 Madagascar

Pentachondra R. Br. Epacridaceae. 4 S Aus., 1 ext. to NZ (*P. pumila* (Forster & Forster f.) R. Br.) – cult. orn. dwarf shrub

Pentaclethra Benth. Leguminosae (II 1). 2 trop. Am. & Afr. Oilseeds: *P. macroloba* (Willd.) Kuntze (*P. filamentosa*, pracaxi fat, S Am.) – explosive fr. ejecting toxic (alks & free amino-acids) seeds to 10 m, wood rotting after only 10–20 yrs, & v. similar *P. macrophylla* Benth. (owala oil, Afr.); useful timber & locally medic.

Pentacme A. DC = *Shorea*

Pentacrostigma K. Afzel. Convolvulaceae. 1 Madagascar

Pentactina Nakai (~ *Spiraea*). Rosaceae. 1 N Korea: *P. rupicola* Nakai – cult. orn. decid. shrub

Pentacyphus Schltr. Asclepiadaceae (III 1). 1 Peru

Pentadenia (Planchon) Hanst. = *Columnea*

Pentadesma Sabine. Guttiferae (III). 5 trop. Afr. R: BJBB 41(1971)430. *P. butyracea* Sabine (tallow tree, W Afr.) – seeds source of Sierra Leone, Kanga or lamy butter, used for cooking, soap, margarine & candles

Pentadiplandra Baillon. Pentadiplandraceae (Capparidaceae s.l.). 1 trop. W Afr.

Pentadiplandraceae Hutch. & Dalz. (~ Capparidaceae). Dicots – Dilleniidae – Capparidales. 1/1 W Afr. Shrubby climber with short axillary racemes lately re-excluded from C.
Only genus: *Pentadiplandra*

Pentadynamis R. Br. = *Crotalaria*

Pentaglottis Tausch. Boraginaceae. 1 SW Eur.: *P. sempervirens* (L.) L. Bailey (alkanet) – cult. orn. herb with blue fls

Pentagonanthus Bullock = *Raphionacme*

Pentagonia Benth. Rubiaceae (inc. sed.). 20 trop. Am. Lvs v. large, often bright red; local medic. *P. gigantifolia* Ducke (Amaz. Peru) – unbranched pachycaul

Pentagramma Yatsk., Windham & Wollenw. (~*Pityrogramma*). Pteridaceae (III). 2 N Am.

Pentalepis F. Muell. (~ *Chrysogonum*). Compositae (Helia.- Eng.). 3 W Aus. R: AusSB 6(1993)149

Pentalinon Voigt (*Urechites*). Apocynaceae. 2 Florida & C Am. to WI. R: JAA 70(1989)383

Pentaloncha Hook.f. Rubiaceae (inc. sed.). 3 trop. W Afr.

Pentameris Pal. Gramineae (25). 5 S Afr. mt. grasslands

Pentamerista Maguire. Tetrameristaceae. 1 N S Am.

Pentanema Cass. (~ *Inula*). Compositae (Inul.). 18 C As. to Sri Lanka, natur. elsewhere

Pentanisia Harvey. Rubiaceae (III 8). 15 trop. Afr. *P. ouranogyne* S. Moore (NE & E Afr.) – arrow-poison in N Uganda

Pentanopsis Rendle. Rubiaceae (IV 1). 1 Somalia

Pentanura Blume. Asclepiadaceae (I, Periplocaceae). 2 Burma, Sumatra

Pentapanax Seemann. Araliaceae (2). 14 S China, Indomal.

Pentapeltis (Endl.) Bunge (~ *Xanthosia*). Umbelliferae (I 16). 1 W Aus.

Pentapera Klotzsch = *Erica*

Pentapetes L. Sterculiaceae. 1 Indomal.

Pentaphalangium Warb. = *Garcinia*

Pentaphragma Wallich ex G. Don f. Pentaphragmataceae. 25 Mal. *P. horsfieldii* (Miq.) Airy Shaw – produces 2 lvs a year, each lasting 33 months; lvs used like spinach

Pentaphragmataceae J. Agardh. Dicots – Asteridae – Asterales. 1/25 Mal. Perennial ± succ. herbs often with branched hairs; alks & latex-system 0. Lvs simple, spirally arr., usu. asymmetric at base; stipules 0. Fls usu. bisexual, in dense sympodial ± axillary helicoid cymes with usu. conspic. bracts; K 5 unequal, imbricate, persistent, C ((4)5 or (4)5, valvate, usu. fleshy, A alt. with C attached to tube or when C free to top of G, anthers with longit. slits, $\overline{G}$ (2 or 3), 2- or 3-loc. separated from hypanthium by nectariferous pits, style without collecting-hairs, axile placentas with ∞ anatropous, unitegmic ovules. Fr. a berry with apical persistent P; seeds minute with copious starchy endosperm
Genus: *Pentaphragma*
Differing from Campanulaceae in absence of pollen-presenting mechanism and poss. diff. pollen-types

Pentaphylacaceae Engl. (~Theaceae). Dicots – Dilleniidae – Theales. 1/1 SE As. & Mal. Small tree accum. aluminium. Lvs simple, entire, spirally arr.: stipules 0. Fls bisexual, small, 5-merous with 2 persistent bracteoles adpressed to K in racemes often with sterile leafy tips; K free, imbricate, persistent, C free, imbricate, A alt. with C, the filaments expanded esp. at middle, anthers with term. pores with valves, $\underline{(G)}$, 5-loc. with stout persistent style & 2 pendulous, anatropous to ± campylotropous bitegmic ovules on axile placentas. Fr. a ± dehiscent woody capsule with persistent central axis (like Theaceae); seeds ± winged with horseshoe-shaped embryo in scanty endosperm
Only species: *Pentaphylax euryoides* Gardner & Champ.

Pentaphylax Gardner & Champ. Pentaphylacaceae. 1 S China to Sumatra

Pentaplaris L.O. Williams & Standley. Tiliaceae. 1 C Am.

Pentapleura Hand-Mazz. Labiatae (VIII 2). 1 Kurdistan

Pentapogon R. Br. Gramineae (21d). 1 SE Aus.: *P. quadrifidus* (Labill.) Baillon

Pentaptilon E. Pritzel. Goodeniaceae. 1 SW Aus.: *P. careyi* (F. Muell.) E. Pritzel. R: FA 35(1992)297

Pentarhaphia Lindley = *Gesneria*

Pentarhopalopilia (Engl.) Hiepko (~ *Rhopalopilia*). Opiliaceae. 2 W & S C Afr. & E Afr. coast, 2 Madag. R: BJ 108(1987)280

Pentarrhaphis Kunth (~ *Bouteloua*). Gramineae (33c). 3 C Am.

Pentarrhinum E. Meyer. Asclepiadaceae (III 1). 3 trop. & S Afr.

Pentas Benth. Rubiaceae (IV 1). 34 Afr. to Arabia & Madag. R: BJBB 23(1953)237. Shrubs & herbs, some cult. orn. esp. *P. lanceolata* (Forssk.) Deflers (E Afr. to S Arabia)

Pentasacme Wallich ex Wight & Arn. Asclepiadaceae (III 5). 4 Indomal. R: Blumea 36(1991)109. Rheophytes. Lvs smoked like marijuana in Mal.

Pentaschistis (Nees) Spach. Gramineae (25). c. 65 Afr. (esp. S; trop. Afr. to Yemen 6), Madag. 3

Pentascyphus Radlk. Sapindaceae. 1 Guyana

Pentaspadon Hook.f. Anacardiaceae (IV). 6 SE As. to Papuasia. Utility timbers. *P. motleyi* Hook.f. – stem a source of oil used in treating skin-disease, esp. ringworm

Pentaspatella Gleason = *Sauvagesia*

Pentastelma Tsiang & Li. Asclepiadaceae (III 1). 1 Hainan

Pentastemona Steenis. Pentastemonaceae. 2 Sumatra. R: Blumea 36(1991)245. Only 5-mery in monocots. *P. sumatrana* Steenis – vivipary in apical part of infl.

Pentastemonaceae Duyfjes (~ Stemonaceae). Monocots – Liliidae – Haemodorales. 1/2 Sumatra. Like Stemonaceae but monopodial, lvs without pulvinus, infls racemose, fls usu. unisexual, P 5, A 5, $\overline{G}$ etc.

Only genus: *Pentastemona* of everwet forests, cf. Stemonaceae in seasonal habitats

Pentastemonodiscus Rech.f. Caryophyllaceae (II 5). 1 Afghanistan: *P. monochlamydeus* Rech.f.

Pentasticha Turcz. = *Fuirena*

Pentatherum Náb. = *Agrostis*

Pentathymelaea Lecomte. Thymelaeaceae. 1 E Tibet

Pentatrichia Klatt. Compositae (Gnap.). 4 S Afr. R: OB 104(1991)43

Pentatropis R. Br. ex Wight & Arn. Asclepiadaceae (III 1). 2 warm OW, 3 Aus.

Penthoraceae Rydb. ex Britton = Saxifragaceae

Penthorum L. Saxifragaceae (Penthoraceae). 1–3 E & SE As., E N Am. *P. sedoides* L. (N Am.) – medic. Perhaps better in its own fam.

Pentodon Hochst. Rubiaceae (IV). 2 warm & trop. Am., Afr., Arabia, Seychelles

Pentopetia Decne. Asclepiadaceae (I, Periplocaceae). 10 Madag. Some rubber sources

Pentopetiopsis Costantin & Gallaud = praec.

Pentossaea W. Judd (~ *Sagraea*). Melastomataceae. 7 trop. Am.

Pentstemon auctt. = *Penstemon*

Pentstemonacanthus Nees. Acanthaceae. 1 Brazil

Pentzia Thunb. Compositae (Anth.-Mat.). 23 Morocco & Algeria (2), S Afr. See also *Oncosiphon*

penwiper plant *Notothlaspi rosulatum*

peony *Paeonia* spp.; **tree p.** *P.* sect. *Moutan* esp. *P. suffruticosa*

Peperomia Ruíz & Pavón. Piperaceae (Peperomiaceae). 1000 trop. & warm esp. Am. (Afr. 17). R: Loefgrenia 82(1983)1, JAA suppl. 1(1991)357. Mostly small ± succ. herbs, many epiphytic with water-storing tissue in upper leaf-epidermis (facultative crassulacean acid metabolism dependent on developmental & environmental conditions); lvs spiral to decussate or verticillate; endosperm octoploid. Many cult. orn. housepls tolerant of little light (rock balsam), esp. *P. argyreia* C.J. Morren (trop. Am.) – lvs silver-striped, *P. caperata* Yuncker (Brazil) – lvs strongly plicate-bullate (several cvs) & *P. velutina* Linden (Ecuador) – lvs silver & red. *P. berteroana* Miq. ssp. *berteroana* on Juan Fernandez, ssp. *tristanensis* (Christoph.) Valdebenito on Tristan da Cunha 5000km away; *P. foliiflora* Ruíz & Pavón & *P. haenkeana* Opiz (trop.Am.) – epiphyllous infls; *P. leptostachya* Hook. & Arn. (W Pacific) – locally medic., cult. orn.; *P. pellucida* (L.) Kunth (trop., orig. Am.) – oil antagonistic to growth of *Helminthosporium oryzae* (brown spot) in rice

Peperomiaceae A.C. Sm. = Piperaceae

Pepinia Brongn. ex André (~ *Pitcairnia*). Bromeliaceae (1). 44 trop. Am. R: SB 13(1988)297

pepino *Solanum muricatum*

pepita *Cucurbita pepo*

Peplidium Del. Scrophulariaceae. 7 Aus. (6 endemic), 1 ext. to India & N Afr.

Peplis L. = *Lythrum*

Peplonia Decne. Asclepiadaceae (III 1). 3 Brazil

Peponidium Baillon ex Arènes. Rubiaceae (III 2). 20 Madag., Comoro Is.

Peponium Engl. Cucurbitaceae. 20 Afr., Madagascar

Peponopsis Naudin. Cucurbitaceae. 1 Mexico

pepper *Piper nigrum*; **African p.** *Xylopia aethiopica*; **Ashanti** or **Benin p.** *P. guineense*; **bell p.** *Capsicum annuum* Grossum Group; **betel p.** *P. betle*; **bird p.** *C. a.* var. *glabriusculum*; **black**

p. *P. nigrum*; **bush p.** *Clethra alnifolia*; **cayenne p.** *C. a.* Longum Group; **cherry p.** *C. a.* Cerasiforme Group; **chilli p.** *C. a.* Longum Group; **cone p.** *C. a.* Conoides Group; **Ethiopian p.** *Xylopia aethiopica*; **green p.** *C. a.* Grossum Group; **Guinea p.** *Aframomum melegueta*; **Indian long p.** *P. longum*; **Jamaican p.** *Pimenta dioica*; **Japanese p.** *Zanthoxylum piperitum*; **Java p.** *Piper cubeba*; **Kawa p.** *P. methysticum*; **long p.** *P. longum*; **Madagascar p.** *P. nigrum*; **malagueta** or **melegueta p.** *Aframomum melegueta*; **negro p.** *Xylopia aethiopica*; **red** or **sweet p.** *C. a.* Grossum Group; **Sichuan p.** *Zanthoxylum piperitum*; **p. tree** *Schinus molle, Kirkia wilmsii, Macropiper excelsum*; **wall p.** *Sedum acre*; **water p.** *Persicaria hydropiper*; **W African black p.** *Piper clusii*; **white p.** *P. nigrum*

peppercress *Lepidium* spp.

pepperidge *Nyssa sylvatica*

peppermint *Mentha × piperita*; name used for some *Eucalyptus* spp. e.g. *E. amygdalifolia* (**black p.**)

pepperwood *Umbellularia californica*

pepperwort *Lepidium* spp.

pequí *Caryocar brasiliense*; **p.-á** *C. villosum*

Pera Mutis. Euphorbiaceae (Peraceae). 40 trop. Am.

Peracarpa Hook.f. & Thomson. Campanulaceae. 1 Indomal.

Peraceae Klotzsch = Euphorbiaceae

Perakanthus F. Robyns. Rubiaceae (III 2). 1 Malay Peninsula

Perama Aublet. Rubiaceae (inc. sed.). 9 trop. Am. R: Brittonia 29(1977)191

Peranema D. Don. Dryopteridaceae (I 2). 3–4 Indomal.

Peranemataceae (C. Presl) Ching = Dryopteridaceae

Peraphyllum Nutt. Rosaceae (III). 1 W US: *P. ramosissimum* Nutt. – cult. orn. shrub

Peratanthe Urban. Rubiaceae (IV 13). 2 Cuba, Hispaniola

Peraxilla Tieghem. Loranthaceae. 2 NZ. R: AusJB 14(1966)429

Perdicium L. Compositae (Mut.-Mut.). 2 SW Cape. R: NJB 5(1986)543. Herbs

Perebea Aublet. Moraceae (III). 9–10 trop. Am. R: FN 7(1972)60. Copious latex brilliant yellow

Peregrina W.R. Anderson (~ *Janusia*). Malpighiaceae. 1 Brazil & Paraguay: *P. linearifolia* (A. St.-Hil.) W.R. Anderson. R: SB 10(1985)303. Geoxylic suffrutex?

Pereilema J. Presl. Gramineae (31e). 3 trop. Am.

pereira, yellow *Howardia glaucescens*

Perella (Tieghem) Tieghem = *Peraxilla*

Perenideboles Goyena. Acanthaceae. 1 Nicaragua

Pereskia Miller. Cactaceae (I). 16 trop. Am. (disjunct: Carib. & SE trop. S Am.). R: MNYBG 41(1986)1. Leafy trees, shrubs or lianes often used as stocks for cacti-grafting; some cult. orn. esp. *P. aculeata* Miller (trop. Am.) – liane to 10 m with recurved spines, cult. for ed. fr., lvs ed. as are those of *P. grandifolia* Haw. ('*P. bleo*', Brazil)

Pereskiopsis Britton & Rose. Cactaceae (II). 9 Mex. & C Am. Allied to *Opuntia* but habit & lvs like *Pereskia*

Perezia Lagasca. Compositae (Mut.-Nass.). 32 Andes from Colombia to S Patagonia, S Brazil, Paraguay, Uruguay & NE Arg. R: CGH 199(1970)1. Herbs. N Am. spp. referred to *Acourtia*

Pereziopsis J. Coulter = *Onoseris*

Pergularia L. Asclepiadaceae (III 1). 3–5 Afr. to India. *P. daemia* (Forssk.) Chiov. (trop. & S Afr. to Arabia, India, Sri Lanka) & *P. tomentosa* L. (N & trop. Afr. to India) – latex smeared on hides which are buried so removing hair in Arabia, locally medic.

Periandra C. Martius ex Benth. Leguminosae (III 10). 6 Brazil & Hispaniola. R: Loefgrenia 59(1973). Some with roots used as liquorice subs.

Perianthomega Bur. ex Baillon. Bignoniaceae. 1 C Brazil. Petiole of compound leaf twining

Periarrabidaea Samp. Bignoniaceae. 2 Amazonia

Periballia Trin. (~ *Deschampsia*). Gramineae (21b). 2–4 Medit. (Eur. 1–3). Each sp. diff. chromosome no.

Periboea Kunth. Hyacinthaceae (Liliaceae s.l.). 2 SW Cape

Pericalia Cass. = *Roldana*

Pericallis D. Don (~ *Senecio*). Compositae (Sen.-Sen.). 15 Macaronesia. R: OB 44(1978)15. Florist's cinerarias (*P. × hybrida* R. Nordenstam) derived in GB from *P. cruenta* (L'Hérit.) Bolle, *P. lanata* (L'Hérit.) R. Nordenstam (*S. heritieri*) & poss. other spp., now naturalizing in Calif., many cvs

Pericalymma (Endl.) Endl. (~ *Leptospermum*). Myrtaceae (Lept.). 1 coastal SW Aus.: *P.*

ellipticum (Endl.) Schauer. R: Telopea 2(1983) 381

Pericalypta Benoist. Acanthaceae. 1 Madagascar

Pericampylus Miers. Menispermaceae (V). 4–6 Indomal. *P. glaucus* (Lam.) Merr. – basketry, eye medic.

Perichasma Miers = *Stephania*

Perichlaena Baillon (~ *Fernandoa*). Bignoniaceae. 1 Madagascar

Pericome A. Gray. Compositae (Hele.-Per.). 4 SW N Am. R: SWNat 18(1973)335

Pericopsis Thwaites. Leguminosae (III 2). 3 trop. Afr. (*Afrormosia*), 1 Sri Lanka to Micronesia. R: BJBB 32(1962)213. Valuable timbers esp. *P. elata* (Harms) van Meeuwen (afrormosia, asamela, kokrodua) & *P. laxiflora* (Baker) van Meeuwen (false dalbergia) in W Afr.

Perictenia Miers = *Odontadenia*

Perideridia Reichb. Umbelliferae (III 8). 13 N Am. R: UCPB 55(1969)1. Herbs of wet places with ed. tubers esp. *P. gairdneri* (Hook. & Arn.) Mathias form. much eaten by Indians & early settlers

Peridictyon Seberg, Frederiksen & Baden (~ *Festucopsis*). Gramineae (23). 1 Balkans: *P. sanctum* (Janka) Seberg, Frederiksen & Baden. R: Willd. 21(1991)96

Peridiscaceae Kuhlm. (~ Bixaceae). Dicots – Dilleniidae – Malvales. 2/2 trop. Am. Trees. Lvs simple, large, spirally arr., leathery & basally 3-veined; stipules intrapetiolar, decid. Fls bisexual, small, reg., with large persistent bracteoles, in axillary fascicles or groups of racemes; K 4–7 imbricate, C 0, A ∞ on or around outside of large fleshy lobed annular to cupular disk with ± connate filaments & small anthers with longit. slits, $\underline{G}$ (3 or 4), 1-loc. sunk in disk (*Peridiscus*) or not, with distinct styles & 6–8 pendulous ovules. Fr. a 1-seeded drupe; small embryo lying alongside copious horny endosperm

Genera: *Peridiscus*, *Whittonia*

Differing from Flacourtiaceae in apical placentation etc.

Peridiscus Benth. Peridiscaceae. 1 Amazonian Brazil, Venezuela

Periestes Baillon = *Hypoestes*

Periglossum Decne. = *Cordylogyne*

Perilepta Bremek. = *Strobilanthes*

Perilimnastes Ridley = *Anerincleistus*

Perilla L. Labiatae (VIII 1). c. 6 India to Japan. *P. frutescens* (L.) Britton (Himal. to E As.) – much cult. in E As. & SE Eur. for orn. (incl. garnish in Japan) & oil (yegoma) like linseed o. used to waterproof paper, as in umbrellas, in paints & printing inks etc., medic. in modern Chinese herbalism; natur. in Ukraine

Perillula Maxim. Labiatae (VIII 1). 1 Japan

Periomphale Baillon = *Wittsteinia*

Peripentadenia L.S. Sm. Elaeocarpaceae. 2 N Aus. R: KB 42(1987)809

Peripeplus Pierre. Rubiaceae (IV 7). 1 W trop. Afr.

Periphanes Salisb. = *Hessea*

Periploca L. Asclepiadaceae (I, Periplocaceae). 11 Medit. (Eur. 2), E As., trop. Afr. R: AK 11(1966)5. Some locally medic. incl. *P. aphylla* Decne. (Egypt & Iran to India) – fls ed., cordage, fodder & firewood; *P. graeca* L. (silk vine, SE Eur., W As.) – cult. orn. decid. twiner to 10 m; *P. visciformis* (Vatke) Schumann (NE Afr., Arabia) – fodder & dye-pl.

Periplocaceae Schltr. See Asclepiadaceae

Periptera DC. Malvaceae. 5 Mexico. R: BSBM 33(1974)39

Peripterygia (Baillon) Loes. Celastraceae. 1 New Caledonia

Peripterygium Hassk. = *Cardiopteris*

Perispermum Degener = *Bonamia*

Perissocarpa Steyerm. & Maguire. Ochnaceae. 2 Venez. R: BJ 113(1991)172

Perissocoeleum Mathias & Constance. Umbelliferae (III 10). 4 Colombia

Perissolobus N.E. Br. = *Machairophyllum*

Peristeranthus Hunt. Orchidaceae (V 16). 1 NE Aus.: *P. hillii* (F. Muell.) Hunt

Peristeria Hook. Orchidaceae (V 10). 15 trop. Am. Cult. orn. epiphytes esp. *P. elata* Hook. (Costa Rica to Venez.) with white waxy fls, national fl. of Panamá

Peristrophe Nees. Acanthaceae. 15 trop. OW. Some cult. orn. greenhouse pls

Peristylus Blume (~ *Herminium*). Orchidaceae (IV 2). 70 China to Pacific

Peritassa Miers. Celastraceae. 14 trop. S Am., Tobago

Peritoma DC = *Cleome*

Perittostema I.M. Johnston. Boraginaceae. 1 Mexico

Perityle Benth. Compositae (Helia.). c. 63 (incl. *Laphamia*) SW N Am. with 1 in Peru & Chile

periwinkle, greater *Vinca major*; **lesser p.** *V. minor*; **Madagascar p.** *Catharanthus roseus*

Pernambuco wood *Caesalpinia echinata*

Pernettya Gaudich. = *Gaultheria*

Pernettyopsis King & Gamble (~ *Diplycosia*). Ericaceae. 2 Malay Pen., Borneo. R: BJLS 85(1982)2

peroba *Paratecoma peroba*; **p. rosa** *Aspidosperma* spp.

Peronema Jack. Labiatae (II; Verbenaceae). 1 S Burma to Sumatra & Borneo: *P. canescens* Jack – timber for house-building

Perotis Aiton. Gramineae (33d). 10 OW trop. Glumes awned

Perotriche Cass. = *Stoebe*

Perovskia Karelin. Labiatae (VIII 2). 7 NE Iran to NW India. Cult. orn. shrubby herbs esp. *P. atriplicifolia* Benth. (Afghanistan, Pakistan) & 'Blue Spire' (*P. a.* × *P. abrotanoides* Karelin (Afghanistan, W Himal.), 'P. Hybrida')

Perplexia Iljin = *Jurinea*

Perralderia Cosson. Compositae (Inul.). 3 NW Afr. R: BJLS 102(1990)166

Perralderiopsis Rauschert = *Iphiona*

Perriera Courchet. Simaroubaceae. 1 Madagascar

Perrieranthus Hochr. = *Perrierophytum*

Perrierastrum Guillaumin. Labiatae (VIII 4c). 1 Madagascar

Perrierbambus A. Camus. Gramineae (1a). 2 Madagascar

Perrieriella Schltr. (~ *Oeonia*). Orchidaceae (V 16). 1 Madagascar

Perrierodendron Cavaco. Sarcolaenaceae. 1 Madagascar

Perrierophytum Hochr. Malvaceae. 10 Madagascar

Perrierosedum (A. Berger) H. Ohba (~ *Sedum*). Crassulaceae. 1 Madag.

Perrottetia Kunth. Celastraceae. 15 China, Mal., Aus. & C Am. Anomalous in C. in exotegmic fibres in seeds (cf. Turneraceae)

Persea Miller. Lauraceae (I). 200 trop. As. & Am. R (NW): MNYBG 14(1966)1). Alks. Cult. orn., some ed. fr. & timber. *P. americana* Miller (avocado (pear), aguacate, alligator pear, palta,? C Am.) – cult. since 8000 BC for highly nutritious fr., many cvs (Guatemalan, Mexican & W Indian), rich in oils (some comm. aromatherapy oils) & vitamins A, B & E & potassium, the most delicious of all salad veg., also eaten as dessert fr., much cult. S Africa, Israel, Calif. (75% there = 'Fuerte', a cross between Guatemalan & Mexican cvs); *P. borbonia* (L.) Sprengel (SE US swamps) – timber; *P. indica* (L.) Sprengel (Macaronesia!) – distinct genus?, cult. orn. S US; *P. lingue* Nees (Chile) – tanbark & comm. timber; *P. nanmu* Oliver (*Machilus n.*, *Phoebe n.*, coffin wood, nanmu, China) – timber esp. for coffins

Persian berries *Rhamnus infectoria*; **P. clover** *Trifolium resupinatum*; **P. insect powder** *Tanacetum coccineum*; **P. lilac** *Melia azedarach*

Persicaria Miller (~ *Polygonum*). Polygonaceae (II 5). Incl. *Aconogonon* & *Bistorta* c. 150 subcosmop. Some cult. orn. but some cause dermatitis in stock in Aus. Usu. herbs, s.t. aquatic (*P. amphibia* (L.) Gray & *P. hydropiper* (L.) Opiz (water pepper)). *P. bistorta* (L.) Samp. (*Polygonum b.*, bistort, Euras.) – dried rhizome medic., an ingredient of 'Easter Ledger Pudding' in Lake Dist. of GB at Easter poss. through legend of its aiding conception & retention of child; *P. maculosa* Gray (*Polygonum persicaria*, redleg, redshank, N temp.) – bad agric. weed; *P. mollis* (D. Don) Gross (Himal.) – young shoots a veg. in Sikkim; *P. odorata* (Lour.) Soják (*Polygonum o.*, rau ram, Asian or Vietnamese mint, SE As.) – potherb, other spp. also ed. & medic.; *P. vacciniifolia* (Wallich) Ronse Decraene & *P. affinis* (D. Don) Ronse Decraene (Himal.) – cult. orn. ground-cover; *P. vivipara* (L.) Ronse Decraene (*Polygonum v.*, N temp.) – bulbils in place of fls at base of infl.

persimmon (Am.) *Diospyros virginiana*; **Japanese p.** *D. kaki*

Persoonia Sm. Proteaceae. 90 Aus. Some cult. orn. shrubs & trees (geebung, snodgollion, snot-goblin, snottygobble)

Pertusadina Ridsd. (~ *Adina*). Rubiaceae (I 2). 4 E As., Mal. R: Blumea 24(1978)353. *P. eurhyncha* (Miq.) Ridsd. (*Adina rubescens*, W Mal.) with fenestrated trunk

Pertya Schultz-Bip. Compositae (Mut.-Mut.). 15 Afghanistan to Japan. Shrubs & herbs

Peru or **Peruvian balsam** *Myroxylon balsamum* var. *pereirae*; **P. bark** *Cinchona* spp.; **P. cotton** *Gossypium barbadense*; **P. daffodil** *Hymenocallis* spp.; **marvel of P.** *Mirabilis jalapa*; **P. nutmeg** *Laurelia sempervirens*

Perulifera A. Camus = *Pseudechinolaena*

Perymeniopsis H. Robinson = seq.

Perymenium Schrader. Compositae (Helia.-Verb.). 41 Mex. to Peruvian Andes. R: (Mex. & C Am., 33): Allertonia 1,4(1978)1

Pescatoria Reichb.f. Orchidaceae (V 10). 16 trop. Am. R: OD 32(1968)86. Cult. orn. epiphytes without pseudobulbs; lvs plicate

Peschiera A. DC = *Tabernaemontana*

Pessopteris Underw. & Maxon = *Niphidium*
Petagnaea Caruel (*Petagnia*). Umbelliferae (II 2). 1 Sicily
Petagnia Guss. = *Petagnaea*
Petagomoa Bremek. = *Psychotria*
petai *Parkia speciosa*
Petalacte D. Don (*Billya*). Compositae (Gnap.-Cass.). 3 Cape. R: OB 104(1991)93
Petaladenium Ducke. Leguminosae (III 2). 1 Rio Negro (Brazil)
Petalidium Nees. Acanthaceae. 35 trop. & S Afr., W Himal. & W India (1)
Petalocentrum Schltr. = *Sigmatostalix*
Petalochilus R. Rogers = *Caladenia*
Petalodiscus Baillon = *Savia*
Petalolophus Schumann. Annonaceae. 1 NE New Guinea: *P. megalopus* Schumann
Petalonyx A. Gray. Loasaceae. 5 SW N Am. Hairs barbed. Allied to Cornaceae s.s. on DNA data
Petalostelma Fourn. = *Cynanchum*
Petalostemon Michaux = *Dalea*
Petalostigma F. Muell. Euphorbiaceae. 7 Aus. (6), New Guinea. *P. pubescens* Domin (quinine bush, q. tree, Queensland) – disp. by emus, voided & endocarp explodes in sun, the seeds taken by ants attracted by elaiosomes
Petalostylis R. Br. Leguminosae (I 2). 2–3 arid Aus. R: Muelleria 6(1986)212. Alks
Petamenes Salisb. ex J.W. Loudon = *Gladiolus*
Petasites Miller. Compositae (Sen.-Tuss.). 19 N temp. (Eur. 9). Butterbur. R: FGP 7(1972)382. Usu. ± dioec. stoloniferous perennial herbs with large lvs. Some cult. orn. esp. *P. fragrans* (Villars) C. Presl (? correctly *P. pyrenaicus* (Loefl.) G. López, winter heliotrope, C Medit., natur. GB) – lvs last until second season, vanilla-scented fls in spring, only male clones known; *P. hybridus* (L.) Gaertner, Meyer & Scherb. (Euras.) – females (heads of 150 female fls with 1–3 males which produce nectar) not uncommon in N & N Midlands of GB, rare elsewhere in GB (elsewhere in Eur. male exists alone only where sp. introduced) used since Middle Ages as anticonvulsive, the active principle (petasin, a terpene) 14 times as effective as papaverine, discovered in 1950s, also febrifuge etc.; *P. japonicus* (Siebold & Zucc.) Maxim. (China, Korea, Japan, females rarely natur. GB) – petioles to 2 m tall with laminas to 1.5 m across on Sakhalin Is., but clones in cult. smaller, petioles used like rhubarb or as vegetable by Japanese & fl.-buds used as condiment etc.
Petastoma Miers = *Arrabidaea*
Petchia Liv. Apocynaceae. 1 Sri Lanka: *P. ceylanica* (Wight) Livera
Petelotiella Gagnepain. Urticaceae (II). 1 SE As.: *P. tonkinensis* Gagnepain – coll. only once
Petenaea Lundell. Tiliaceae. 1 Guatemala
Peteniodendron Lundell = *Pouteria*
Peteravenia R. King & H. Robinson (~ *Eupatorium*). Compositae (Eup.-Crit.). 5 Mex., C Am. R: Phytol. 21(1971)394, MSBMBG 22(1987)339
Peteria A. Gray. Leguminosae (III 7). 4 SW N Am.
Petermannia F. Muell. Philesiaceae (Smilacaceae s.l.). 1 NSW: *P. cirrosa* F. Muell.
Petermanniaceae Hutch. See Philesiaceae
Peterodendron Sleumer (~ *Poggea*). Flacourtiaceae (3). 1 trop. E Afr.: *P. ovatum* (Sleumer) Sleumer
Petersianthus Merr. (*Combretodendron*). Lecythidaceae. 1 trop. W Afr., 1 Philippines. *P. macrocarpus* (Pal.) Liben (essia, W Afr.) – imp. comm. timber
petha *Benincasa hispida*
Petimenginia auctt. = *Petitmenginia*
Petinotia Léonard. Cruciferae. 1 Iran, Pakistan: *P. purpurascens* (Boiss.) Léonard. R: BJBB 50(1980)228
petitgrain oil *Citrus × aurantium*
Petitia Jacq. Labiatae (I, Verbenaceae). 2 WI. *P. domingensis* Jacq. (fiddlewood (corruption of *bois fidèle*)) – good timber for poles, furniture etc.
Petitiocodon Robbrecht (~ *Didymosalpinx*). Rubiaceae (II 1). 1 Nigeria & Cameroun: *P. parviflorus* (Keay) Robbrecht. R: BJBB 58(1988)116
Petitmenginia Bonati. Scrophulariaceae. 2 S China, SE As.
Petiveria L. Phytolaccaceae (II). 1 trop. & warm Am.: *P. alliacea* L. – locally medic. R: JAA 66(1985)30
Petiveriaceae C. Agardh = Phytolaccaceae (II)
Petkovia Stefanoff = *Campanula*
Petopentia Bullock (~ *Tacazzea*). Asclepiadaceae (I, Periplocaceae). 1 Natal & Transkei: *P.*

natalensis (Schltr) Bullock. R: SAJB 56(1990)393

Petradoria E. Greene. Compositae (Ast.-Sol.). 3 SW US

Petraeovitex Oliver. Labiatae (II, Verbenaceae). 7 wetter Mal., NZ, Pacific. R: GBS 21(1965)215. *P. multiflora* (Sm.) Merr. (Mal.) – liane to 30 m with infls to 70 cm

Petrea L. Verbenaceae. 30 trop. Am. R: FR 48(1938)161. Cult. orn. decid. trees, shrubs & lianes esp. *P. volubilis* L. – liane to 10 m with pale lilac to purple fls

Petrina J. Phipps = *Danthoniopsis*

Petrobium R. Br. Compositae (Helia.-Cor.) 1 St Helena: *P. arboreum* (Forster f.) Sprengel – pachycaul tree, dioec. or (?) gynodioec., allies in Polynesia

Petrocallis R. Br. Cruciferae. 1 mts S Eur.: *P. pyrenaica* (L.) R. Br. – cult. orn. rock-pl. See also *Elburzia*

Petrocodon Hance. Gesneriaceae. 1 C & S China: *P. dealbatus* Hance

Petrocoma Rupr. = *Silene*

Petrocoptis A. Braun. Caryophyllaceae (III 3). 5 Pyrenees. R: FE ed. 2, 1(1993)190. Some cult. orn. rock-pls

Petrocosmea Oliver. Gesneriaceae. 27 S China (24, NE India, Burma, Thailand, Vietnam, mts. Some cult. orn. R: ABY 7(1985)49

Petroedmondia Tamamsch. Umbelliferae (III 5). 1 SW As.

Petrogenia I.M. Johnston = *Bonamia*

Petrollinia Chiov. = *Inula*

Petromarula Vent. ex Hedwig f. Campanulaceae. 1 Crete: *P. pinnata* (L.) A. DC

Petronymphe H. Moore. Alliaceae (Amaryllidaceae, Liliaceae s.l.). 1 Mex.: *P. decora* H. Moore – cult. orn.

Petrophila R. Br. = seq.

Petrophile R. Br. ex Knight (~ *Isopogon*). Proteaceae. 42 Aus. Some cult. orn. shrubs (cone-sticks)

Petrophyton Rydb. = seq.

Petrophytum (Nutt.) Rydb. Rosaceae. 3 W N Am. Cult. orn. rock-pls

Petroravenia Al-Shehbaz. Cruciferae. 1 Arg.: *P. eseptata* Al-Shehbaz. R: Novon 4(1994)191

Petrorhagia (Ser.) Link (~ *Velezia*). Caryophyllaceae (III 1). 28 Canary Is. & Medit. (Eur. 18) to Kashmir. R: BBMNHB 3(1964)119. Some cult. orn.

Petrosavia Becc. Melanthiaceae (Petrosaviaceae). 2 E As., W Mal. R: JJB 59(1984)106

Petrosaviaceae Hutch. = Melanthiaceae

Petroselinum Hill. Umbelliferae (III 8). 2 Eur. (2), Medit. Parsley. *P. crispum* (Miller) A.W. Hill (garden p., Eur., W As.) – one of richest sources of vitamin C, cult. for lvs used as garnish, salads & in meat-dishes, sausages etc. (flavour due to terpineol; dried fr. medic.; cult. pl. is a form with curled lvs compared with wild one; 'Tuberosum' (Hamburg or turnip-rooted p.) – ed. root, lvs hardier than type; var. *neapolitanum* Danert (Italian p.) – lvs not curled

Petrosimonia Bunge. Chenopodiaceae (III 3). 11 SE Eur. (7) to C As.

pe-tsai *Brassica rapa* Chinensis Group

Petteria C. Presl. Leguminosae (III) 31. 1 Balkans: *P. ramentacea* (Sieber) C. Presl – seeds toxic

pettigree or **pettigrue** *Ruscus aculeatus*

petty spurge *Euphorbia peplus*; **p. whin** *Genista anglica*

Petunga DC = *Hypobathrum*

Petunia Juss. Solanaceae (2). (Excl. *Calibrachoa*, 7–15) 35 trop. (esp. Brazil) & warm S Am. R: KSVH 46,5(1911)1. Cult. orn. bedding-pls esp. many (over 200 incl. striped, picotee & double) cvs of *P.* × *hybrida* (Hook.) Vilm. (cultigen app. derived from *P. axillaris* (Lam.) Britton, Sterns & Pogg. × *P. integrifolia* (Hook.) Schinz & Thell. ('*P. violacea*')); at least 1 sp. in Ecuador source of hallucinogen inducing sense of flying or levitation

Peucedanum L. Umbelliferae (III 10). 100–120 temp. Euras. (Eur. 29), trop. & S Afr. mts. Prob. heterogeneous. Many diff. spp. used medic., in Eur. *P. officinale* L. (hog fennel) used veterinarily. *P. kerstenii* Engl. (trop. E Afr. mts) on Ruwenzori with woody erect stem to 2 m topped with rosette of lvs said to be gorilla food on Virunga Volcanoes; *P. palustre* (L.) Moench (Eur. to C As.) – food of swallowtail butterfly

Peucephyllum A. Gray. Compositae (Hele.-Cha.). 1 SW US, N Mex.

Peumus Molina. Monimiaceae (VI 3). 1 Chile: *P. boldus* Molina (boldo) – wood hard, bark a dye-source, fr. ed., lvs used as a digestive tea after meals like coffee (alks incl. boldine)

peyote or **peyotl** *Lophophora williamsii*

Peyritschia Fourn. = *Trisetum*

Peyrousea DC = *Schistostephium*

Pezisicarpus Vernet. Apocynaceae. 1 SE As.

Pfaffia C. Martius. Amaranthaceae (II 2). 33 trop. S Am. Mostly shrubby some with basis for patented antitumour compounds. *P. iresinoides* Spreng. (Brazil) – ecdysteroids

Pfeiffera Salm-Dyck = *Lepismium*

Phacelia Juss. Hydrophyllaceae. 150 W N Am. (most), E US, S Am. Bee-poll. with anthers turning inside-out at maturity. Many cult. orn. esp. in US, *P. tanacetifolia* Benth. (W N Am.) cult. as bee-fodder in Eur.

Phacellanthus Siebold & Zucc. Orobanchaceae. 2 E As.

Phacellaria Benth. Santalaceae. 7 SE As.

Phacellothrix F. Muell. Compositae (Ast.-Ast.). 1 E Mal. & E trop. Aus.

Phacelophrynium Schumann. Marantaceae. 9 Nicobar Is., Mal.

Phacelurus Griseb. Gramineae (40h). 9 warm OW (Eur. 1). R: KB 33(1978)175

Phacocapnos Bernh. = *Cysticapnos*

Phaeanthus Hook.f. & Thomson. Annonaceae. 12 Indomal. Alks

Phaedranassa Herbert. Amaryllidaceae (Liliaceae s.l.). 9 Andes. R: PL 25(1969)55. Cult. orn. (queen lilies)

Phaedranthus Miers = *Distictis*

Phaenanthoecium C. Hubb. Gramineae (25). 1 NE trop. Afr. mts

Phaenocoma D. Don. Compositae (Gnap.-Rel.). 1 SW Cape: *P. prolifera* (L.) D. Don – cult. orn. R: OB 104(1991)70

Phaenohoffmannia Kuntze = *Pearsonia*

Phaenosperma Munro ex Benth. Gramineae (7). 1 Assam to Japan: *P. globosum* Munro

Phaeocephalus S. Moore = *Hymenolepis*

Phaeoneuron Gilg = *Ochthocharis*

Phaeonychium O. Schulz. Cruciferae. 3 C As. to Himalaya

Phaeopappus (DC) Boiss. (*Tomanthea*) = *Centaurea*

Phaeoptilum Radlk. Nyctaginaceae (III). 1 SW Afr.: *P. spinosum* Radlk. – *Lycium* habit

Phaeosphaerion Hassk. = *Commelina*

Phaeostemma Fourn. = *Matelea*

Phaeostigma Muldashev. Compositae (Anth.-Art.). 3 China

Phagnalon Cass. Compositae (Gnap.). 43 Canary Is. & Medit. (Eur. 6) to Arabia & Middle E. R: OB 104(1991)53

Phainantha Gleason. Melastomataceae. 4 trop. S Am.

Phaiophleps Raf. = *Olsynium*

phai-tong *Dendrocalamus asper*

Phaius Lour. Orchidaceae (V 11). 45 Indomal. to S China, trop. Afr. & Aus. (3), New Caledonia (4). Cult. orn. terr. orchids esp. *P. tankervilliae* (L'Hérit.) Blume (Himal. to Aus.) with infl. of brown & yellow fls to 1 m

Phalacrachena Iljin (~ *Centaurea*). Compositae (Card.-Cent.). 2 SE Russia (1) to Caucasus & C As.

Phalacraea DC (~ *Piqueria*). Compositae (Eup.-Ager.). 4 Colombia, Ecuador & Peru. R: MSBMBG 22(1987)147

Phalacrocarpum (DC) Willk. (~ *Chrysanthemum*). Compositae (Anth.- Leuc.). 2 Spain & Portugal

Phalacroloma Cass. = *Erigeron*

Phalacroseris A. Gray. Compositae (Lact.-Mic.). 1 California

Phalaenopsis Blume. Orchidaceae (V 16). (Incl. *Kingidium*; R: OB 95(1988)182) 44 Himal. to Aus. R: H.R. Sweet (1980) *The genus P.*, MNJ 36(1982)17. Alks. Cult. orn. epiphytes or chasmophytes incl. hybrids. *P. taenialis* (Lindley) Christenson & U.C. Pradhan (*Doritis t.*, *K. t.*, NW Himal.) – leafless

Phalaris L. Gramineae (21c). c. 20 Eur. (7, 1 natur.), Medit., N As., N & S Am. Alks. R: ISJS 36(1961)1. Fodders etc., *P. aquatica* & *P. arundinacea* with tryptamine-based alks gramine & hordemine, gramine at 0.01% a feeding stimulant but at higher conc. repellent: *P. arundinacea* L. (N temp.) – valuable hay & fodder when young, erosion-control, 'Picta' (var. *picta* L.) a varieg. form cult. orn. (ribbon grass); *P. canariensis* L. (canary grass, W Medit., widely natur.) – canary seed of commerce; *P. aquatica* L. (*P. tuberosa*, Toowoomba canary grass, Medit.) – fodder esp. in Aus., incl. 'var. *stenoptera*' (Harding grass, origin obscure) – forage

Phaleria Jack. Thymelaeaceae. 20 Indomal., W Pacific

Phalocallis Herbert = *Cypella*

phalsa *Grewia asiatica*

Phanera Lour. = *Bauhinia*

Phanerodiscus Cavaco. Olacaceae. 1 Madagascar

Phaneroglossa R. Nordenstam (~ *Senecio*). Compositae (Sen.-Sen.). 1 SW Cape mts

Phanerogonocarpus Cavaco = *Tambourissa*

Phanerophlebia C. Presl = *Polystichum*

Phanerophlebiopsis Ching = *Arachniodes*

Phanerosorus Copel. Matoniaceae. 1 Sarawak, 1 is. S of Moluccas & off W New Guinea, on limestone. Fronds pendent, lateral branches often tipped with dormant apex

Phanerostylis (A. Gray) R. King & H. Robinson = *Brickellia*

Phania DC. Compositae (Eup.-Ager.). 2 Cuba. R: MSBMBG 22(1987)145

Phanopyrum (Raf.) Nash = *Panicum*

Pharbitis Choisy = *Ipomoea*

Pharnaceum L. Molluginaceae. 20 S Afr. R: JSAB 24(1959)18

Pharus P. Browne. Gramineae (6). 8 warm Am. R: KBAS 13(1986)67

Phaseolus L. Leguminosae (III 10). 36 trop. & warm Am. Beans (see also *Vigna* – stipules with basal appendages, etc.). Many cult. edible seeds &/or pods esp.: *P. acutifolius* A. Gray (S N Am.) esp. 'var. *latifolius* G. Freeman' (tepary bean) – drought- resistant; *P. coccineus* L. (scarlet runner b., C Am.) – runner beans much cult. GB; *P. flavescens* Piper (Andes) – promising but perennial; *P. lunatus* L. (incl. *P. limensis*, butter or Lima or Burma or duffin or Rangoon b., trop. S Am.) – imp. ed. beans; *P. vulgaris* L. (kidney or French or flageolet or dwarf or string or snap or wax or haricot b., frijoles, pea-b., trop. Am. poss. derived from *P. aborigineus* Burkart, Andes) – predominantly self-poll., one of most imp. legumes giving ed. young pods (e.g. cannellini beans) as well as dried seeds (e.g. borlotti & pinto beans, Nuñas beans being cooked in hot oil like popcorn) allegedly promoting hypoglycaemia in humans, with tomato sauce the ubiquitous baked beans (navy beans – mainly US navy) of commerce (first tinned 1911 when a luxury but now 1m tins consumed a day (5 kg per person per annum) in UK) allegedly effective in lowering cholesterol levels, & forage, resistant to cowpea weevils & adzuki bean weevils as 1% tissue by weight is alpha-amylase inhibitor inhibiting insect-gut digestion leading to insect starvation (gene introduced thence to peas via *Agrobacterium* genetic engineering)

Phaulopsis Willd. Acanthaceae. 20 trop. Afr., Masc., India

Phaulothamnus A. Gray. Achatocarpaceae. 1 Calif. & Texas to N Mex.: *P. spinescens* A. Gray

pheasant's eye *Adonis annua*; **p. e. narcissus** *Narcissus poeticus*

Phebalium Vent. Rutaceae. 49 Aus. (48 endemic), NZ (1). R: Nuytsia 1(1970)60. Alks

Phegopteris (C. Presl) Fée. Thelypteridaceae. 3: 1 N temp., 1 E N Am., 1 Mal. R: Blumea 17(1969)9. *P. connectilis* (Michaux) Watt (*Thelypteris p.*, beech fern, N temp.) – cult. orn.

Pheidochloa S.T. Blake. Gramineae (37). 1 New Guinea, 1 trop. Aus. R: Blumea 19(1971)61

Pheidonocarpa Skog = *Gesneria*

Phelipaea Desf. = *Phelypaea*

Phellinaceae (Loes.) Takht. See Aquifoliaceae

Phelline Labill. Aquifoliaceae (Phellinaceae). 12 New Caled.

Phellocalyx Bridson. Rubiaceae (II 1). 1 Malawi, Mozambique & Tanz.

Phellodendron Rupr. Rutaceae. 10 E As. Alks. Dioec. decid. trees – some cult. orn. (cork trees), locally medic. & useful timber

Phellolophium Baker. Umbelliferae (III 8). 1 Madagascar

Phellopterus Benth. = *Glehnia*

Phelpsiella Maguire. Rapateaceae. 1 Venezuela

Phelypaea L. (*Diphelypaea, Phelipaea*). Orobanchaceae. 4 Medit. (Eur. 2). Some emergency foods

Phenakospermum Endl. Strelitziaceae. 1 E trop. S Am.: *P. guyanense* (Rich.) Miq. – term. infl. (hapaxanthic), arils bright red & hairy

Phenax Wedd. Urticaceae (III). 12 trop. Am., some natur. trop. As.

phenomenal berry see *Rubus*

Pherolobus N.E. Br. = *Dorotheanthus*

Pherotrichis Decne. Asclepiadaceae (III 3). 2 Mexico

Phialacanthus Benth. Acanthaceae. 5 Himal. to Malay Pen.

Phialanthus Griseb. Rubiaceae (III 4). 18 WI

Phialodiscus Radlk. = *Blighia*

Phidiasia Urban = *Odontonema*

Philacra Dwyer. Ochnaceae. 4 Venez. & N Brazil. R: BJ 113(1991)177

Philactis Schrader. Compositae (Helia.-Zinn.). 4 C Am. R: Brittonia 21(1969)322

Philadelphaceae Martinov. See Hydrangeaceae

Philadelphus L. Hydrangeaceae (Philadelphaceae). 65 N temp. (Eur. 1). R: JAA 35(1954)275, 36(1955)52, 325, 37(1956)15. Mock orange; shrubs with cucumber-scented lvs & strongly protogynous white, usu. strongly-scented, fls, regrettably known as syringa (*Syringa* = lilac). Water scented with the fls independently used by ancient Parthians (OW) & in Mex. (*P. mexicanus* Schldl.) in pre-Columbian times. Many cult. orn. incl. hybrids, esp. *P. coronarius* L. (Eur., SW As.) – poss. oil-source, v. fragrant, many cvs & *P. microphyllus* A. Gray (SE US to C Mexico) & their hybrid *P.* × *lemoinei* Lemoine as well as its hybrid with *P. coulteri* S. Watson (Mex.) = *P.* × *purpureomaculatus* Lemoine with many cvs such as 'Sybille'

Philbornea H. Hallier. Linaceae. 2 Philippines, Borneo

Philenoptera Hochst. ex A. Rich. (~ *Lonchocarpus*). Leguminosae (III 6). 15 trop. Afr., 4 S Am.

Philesia Comm. ex Juss. Philesiaceae (Smilacaceae s.l.). 1 S Chile: *P. magellanica* J. Gmelin – evergreen shrub with petioled 1-veined revolute lvs & red fls, cult. orn.

Philesiaceae Dumort. (incl. Petermanniaceae, excl. Luzuriagaceae; ~ Smilacaceae). Monocots – Liliidae – Asparagales. 6/6 C Mal. to Pacific, S S Am. Shrubs & lianes. Lvs spirally arr., entire with pinnate or parallel veins & net-like transverse venation. Fls bisexual, reg., solit. or in panicles, axillary. P 3 + 3, the outer sometimes sepaloid; A 3 + 3, s.t. basally united; $\underline{G}$ to $\overline{G}$, 1-loc.; ovules anatropous. Fr. a berry with few to ∞ seeds with oily endosperm. n = 15, 19

Genera: *Behnia, Eustrephus, Geitonoplesium, Lapageria, Petermannia, Philesia*

Some cult. orn.

Philgamia Baillon. Malpighiaceae. 4 Madagascar

Philibertia Kunth (~ *Sarcostemma*). Asclepiadaceae (III 1). 40 trop. Am.

Philippia Klotzsch = *Erica*

Philippiamra Kuntze = *Cistanche*

Philippiella Speg. Caryophyllaceae (I 2). 1 Patagonia: *P. patagonica* Speg. R: MBMHRU 285(1968)398

Philippinaea Schltr. & Ames = *Orchipedum*

Phillyrea L. Oleaceae. 4 Madeira, Medit. (Eur. 2) to N Iran. Some locally used timber; some cult. orn. evergreen trees & shrubs esp. *P. latifolia* L. (S Eur., SW As.) – often mistaken for evergreen oak

Philodendron Schott. Araceae (VI 1). 350–400 trop. Am. R: KB 45(1990)37. Usu. epiphytic lianes with entire to deeply lobed lvs, some scarab-poll with resin gluing pollen to beetles. Some locally ed. fr. & medic. (berberine-type alks reputedly bactericidal & fungicidal); many cult. orn. incl. hybrids but many v. poisonous to children & pets fermented lvs of *P. craspedodendrum* R.E. Schultes used as fish-poison in NW Amazon), the most familiar being *P. bipinnatifidum* Endl. (SE Brazil) – stems to 10 cm diam., spadix to 17 °C above ambient (lipid metabolism) & *P. scandens* K. Koch & Sello (trop. Am.)

Philodice C. Martius. Eriocaulaceae. 2 Brazil

Philoglossa DC. Compositae (Liab.). 5 S Am. R: Phytologia 26(1973)381, SCB 54(1983)59

Philotheca Rudge. Rutaceae. 6 Aus.

Philoxerus R. Br. = *Gomphrena*

Philydraceae Link. Monocots – Liliidae – Haemodorales. 3/6 Aus. to SE As. & S Japan. Perennial herbs with rhizome, tuber or thickened stem-base; vessels (with scalariform end-plates) only in roots, no sec. growth. Lvs parallel-veined, basal ones distichous, rest spirally arr., usu. with basal sheath. Fls bisexual, zygomorphic, in axils of bracts in simple or branched spike, P outer 2 (lateral), small, other (upper) united with inner upper 2 to give upper lip with 3 prominent veins (& sometimes teeth), lower inner also forming a lip, A 1 s.t. with filament united with inner P, anther introrse to extrorse with longit. slits, pollen s.t. in tetrads, $\underline{G}$ (3), 3-loc. with axile placentas or 1-loc. with intruded parietal placentas & term. style, ovules ∞ anatropous, bitegmic. Fr. usu. loculicidal capsule; seeds usu. ± ∞, embryo with term. cotyledon & lateral plumule embedded in starchy, oily, proteinaceous endosperm. x = 8, 17

Genera: *Helmholtzia, Philydrella, Philydrum*

App. close to Pontederiaceae, though some similarities with Burmanniaceae

Philydrella Caruel. Philydraceae. 2 SW Aus. R: FA 45(1987)42

Philydrum Banks ex Sol. Philydraceae. 1 E & SE As., Mal to Aus.: *P. lanuginosum* Banks ex Sol. R: FA 45(1987)41

Philyra Klotzsch. Euphorbiaceae. 1 S Brazil, Paraguay

Philyrophyllum O. Hoffm. Compositae (Gnap.). 1 S Afr.: *P. schinzii* O. Hoffm. R: OB 104(1991)43

Phinaea Benth. Gesneriaceae. 9 Mex. to Colombia & WI. G semi-inf.; hairy herbs with scaly rhizomes. *P. multiflora* C. Morton (Mex.) – cult. orn. with white fls

Phippsia (Trin.) R. Br. Gramineae (17). 3 circumpolar (Eur. 2). Hybrids with *Puccinellia* spp.

Phitopis Hook.f. Rubiaceae (inc. sed.). 2 Peru. Densely hairy trees

Phlebiophragmus O. Schulz. Cruciferae. 1 Peru

Phlebocarya R. Br. Haemodoraceae. 3 SW Aus. G 1-loc. in fr. R: FA 45(1987)128

Phlebodium (R. Br.) J. Sm. = *Polypodium*

Phlebolobium O. Schulz. Cruciferae. 1 Falkland Is.

Phlebophyllum Nees = *Strobilanthes*

Phlebotaenia Griseb. (~ *Polygala*). Polygalaceae. 3 Cuba, Puerto Rico

Phlegmatospermum O. Schulz. Cruciferae. 4 Aus. R: CGH 205(1974)150

Phleum L. Gramineae (21d). 15 N temp. (Eur. 11), temp. S Am. R: KBAS 13(1986)142. Cat-tail grass. Cult. forage grasses esp. *P. pratense* L. (timothy, Euras., natur. US) – grazing & hay, many cvs, first widely grown in US (introd. *Timothy* Hansen (1720) & later reintrod. UK), common cause of hay-fever

Phloeophila Hoehne & Schltr. = *Pleurothallis*

Phloga Noronha ex Hook.f. Palmae (V 4e). 2 Madag. Leaflets whorled

Phlogacanthus Nees. Acanthaceae. 15 Indomal. A 2. Some cult. greenhouse shrubs esp. *P. thyrsiformis* (Hardw.) Mabb. (*P. thyrsiflorus*, N India) – C orange

Phlogella Baillon = *Dypsis*

Phlojodicarpus Turcz. ex Ledeb. Umbelliferae (III 8). 4 Siberia

Phlomidoschema (Benth.) Vved. (~ *Stachys*). Labiatae (VI). 1 Iran to W Himal.

Phlomis L. Labiatae (VI). c. 100 Medit. (Eur. 12) to C As. & China, usu. dry stony habitats. Some cult. orn. coarse herbs esp. *P. fruticosa* L. (Jerusalem sage, Medit. W to Sardinia)

Phlomoides Moench = praec.

Phlox L. Polemoniaceae. 67 N Am., NE As. (1). R: MAM 3(1955). Annual or perennial herbs or shrubs with A of unequal length or unequally inserted in C tube. Many handsome cult. orn. (signifying 'agreement' in 'Language of Fls') esp. *P. drummondii* Hook. (E Texas) – annual with many cvs of diff. fl. colour, *P. paniculata* L. (E N Am.) – common perennial garden phlox with scented fls, *P. subulata* L. (NE US) – mat-forming rock-pl.

Phlyctidocarpa Cannon & W.L. Theob. Umbelliferae (III 8). 1 SW Afr.

Phoebanthus S.F. Blake. Compositae (Helia.-Verb.). 2 N Am.

Phoebe Nees (~ *Ocotea*). Lauraceae (I). 100 Indomal. Alks. Am. spp. referred to *Cinnamomum*

Phoenicanthus Alston. Annonaceae. 2 Sri Lanka. R: Rev. Handbk Fl. Ceylon 5(1985)23

Phoenicaulis Nutt. Cruciferae. 1 W N Am.: *P. cheiranthoides* Nutt. – cult. orn. rock-pl.

Phoenicophorium H. Wendl. Palmae (V 4n). 1 Seychelles: *P. borsigianum* (K. Koch) Stuntz

Phoenicoseris (Skottsb.) Skottsb. = *Dendroseris*

Phoenicosperma Miq. = *Sloanea*

Phoenix L. Palmae (I 2). 17 trop. & warm Afr. & As.; 1 Eur.: *P. theophrasti* Greuter (~*P. dactylifera*) – Crete (& SW Turkey), figured in Minoan frescoes, endangered sp. allied to *P. dactylifera* L. (date palm, cultigen, though poss. wild in NE Sahara & Arabia where trees with small ined. fr., form spontaneous hybrids with *P. reclinata* & *P. sylvestris*). R: O. Beccari, Malesia 3(1890)345. Dioec. (males heterozygous, females homozygous recessive) trees (only palms with induplicate pinnate lvs, lower leaflets spines), imp. sources of fr., sugar & cult. orn. (many hybridize in cult.), most imp. *P. dactylifera* cult. since 4000 BC, ? insect- or wind-poll., each tree producing up to 700 kg fr. annually (60–70% sugar), staple of nomadic peoples in Arabia & N Afr. (soft drupes eaten locally, semi-dry exported, e.g. Deglet Noor, dry used locally & stored), comm. crop in Iraq, N Afr. & Calif., dried or preserved with sugar & used as dessert fr., in puddings etc., lvs for matting, thatch etc.; in Israel old leaf-bases harbour an epiphytic angiosperm flora of 32 spp., ants long used to control pests in Arabia (noted by Forssk. (1755) as first 'biological control' in Eur. literature), seeds & pollen contain oestrone. *P. acaulis* Roxb. (Assam to Burma) – fr. chewed like betel in Sikkim; *P. canariensis* Hort. ex Chabaud (Canary Is.) – widely cult. in S Eur. towns, the trunk broader than date-palm; *P. paludosa* Roxb. (Bengal to Malay Pen.) – mangrove; *P. reclinata* Jacq. (dwarf date palm, trop. Afr.) – large hedgepl. because of vigorous suckering habit, overexploited in NE Kenya thus endangering Tana River crested mangabey which is heavily reliant on it; *P. roebelenii* O'Brien ex C. Roebelen (Laos) – elegant pot-pl.; *P. sylvestris* (L.) Roxb. (India) – imp. source of palm sugar & toddy

Pholidia R. Br. = *Eremophila*

Pholidocarpus Blume. Palmae (I 1b). 6 Mal. *P. macrocarpus* Becc. to 40 m tall in Johore, corky fr. prob. elephant-disp.

Pholidostachys H. Wendl. ex Hook.f. (~ *Calyptrogyne*). Palmae (V 6). 3+ trop. Am.

Pholidota Lindley ex Hook. Orchidaceae (V 12). 29 Indomal. R: OM 3(1988)1. Some cult. orn. epiphytes with wand-like spiralling racemes clothed in bracts

Pholisma Nutt. ex Hook. Lennoaceae. (Inc. *Ammobroma*) 3 SW N Am. R: SB 11(1986)543.

Perennials with 4–10-merous fls, in sand on roots esp. of *Croton* or *Hymenoclea* spp. *P. sonorae* (A. Gray) Yatskievych (SW US) – succ. underground stems used as food by Arizona Indians

Pholistoma Lilja (~ *Nemophila*). Hydrophyllaceae. 3 SW N Am. *P. auritum* (Lindley) Lilja (Calif.) – cult. orn. annual with lavender to blue or violet fls

Pholiurus Trin. Gramineae (18). 1 SE Eur. & C As. (saline soils): *P. pannonicus* (Host) Trin.

Phonus Hill = *Carthamus*. But see AJBM 47(1990)26

Phoradendron Nutt. Viscaceae. 190 Am. esp. trop. Alks. *P. leucarpon* (Raf.) Reveal & M. Johnston (*P. flavescens*, *P. serotinum*, N Am.) – the principal mistletoe of commerce in US, toxic amines. Dried growths of *P.* spp. (& other Loranthaceae) sold as 'flores de palo' (wooden fls) in C Am.

Phormiaceae J. Agardh (~ Agavaceae). Monocots – Liliidae – Asparagales. 7/30 warm S, esp. Aus. Subshrubs & rhiz. herbs without sec. thickening. Lvs distichous, linear, often tough; vessels in roots. Infl. term. panicle; fls bisexual, ± reg. P 3 + 3, almost free, outer usu. slightly smaller; A 3 + 3 rarely basally united, anthers with apical pores; G (3), 3-loc., each loc. with 4–∞ anatropous (epitropous) ovules. Fr. a capsule or berry, seeds black (phytomelan). x = 8

Genera: *Agrostocrinum, Dianella, Excremis, Phormium, Stypandra, Thelionema, Xeronema*

Close to Doryanthaceae? Poss. Hemerocallidaceae referable here? Poss. Dianellaceae distinct as app. close to Anthericaceae & Asphodelaceae: further problems from the dismemberment of Liliaceae s.l.

Phormium a fibre source; some cult. orn.

Phormium Forster & Forster f. Phormiaceae (Agavaceae s.l.). 2 NZ, Norfolk Is. Lvs of *P. tenax* Forster & Forster f. (NZ) source of NZ flax or hemp (bush flax), long-used by Maoris for textiles, cordage & nets, cult. orn. incl. cvs with red lvs, poll. honey-eaters; *P. cookianum* Le Jolis (*P. colensoi*, NZ) – similiar cult. orn. but smaller

Phornothamnus Baker = *Gravesia*

Photinia Lindley. Rosaceae (III). 65 Himal. to Japan & Sumatra, E N & C Am. R: CJB 68(1990)2236. Some rheophytes; others (incl. *Aronia*, *Stranvaesia*) cult. orn. evergreen & decid. shrubs esp. *P. pyrifolia* (Lam.) K. Robertson & J. Phipps (*A. arbutifolia*, red chokeberry, NE Am.) & *P. glabra* (Thunb.) Maxim. (Japan) – imp. hedgepl.

Photinopteris J. Sm. = *Aglaomorpha*

Phragmanthera Tieghem (~ *Tapinanthus*). Loranthaceae. 6+ trop. E Afr.

Phragmipedium Rolfe. Orchidaceae (II). c. 15 trop. Am. R: C. Cash, Slipper orchids (1991)137. Cult. orn. terr. orchids incl. *P. caudatum* (Lindley) Rolfe with petals to 60 cm long!

Phragmites Adans. Gramineae (25). 3 cosmop. (Eur. 1). R: JAA 71(1990)156. *P. australis* (Cav.) Steudel (*P. communis*, reed, nal (India), Danube grass, cosmop.) – forms floating fens as at mouth of Danube, where harvested, the pulp used for paper, cellophane, cardboard & synthetic textiles, fibre-board, fuel, alcohol, insulation & fertilizer, its growth some of the most rapid of any pl.; grains consumed by N Am. Indians, young shoots by Japanese, lvs used for mats (durma m.) in India. Culms used like quill pens by Anc. Egyptians

Phragmocarpidium Krapov. Malvaceae. 1 Brazil

Phragmorchis L.O. Williams. Orchidaceae (V 16). 1 Philippines

Phragmotheca Cuatrec. Bombacaceae. 5 trop. S Am. R: Brittonia 43(1991)73

Phreatia Lindley. Orchidaceae (V 14). 160 NE India to Taiwan & Aus. (esp. New Guinea)

Phrissocarpus Miers = *Tabernaemontana*

Phrodus Miers. Solanaceae (2). 1 Chile (Atacama & Coquimbo): *P. microphyllus* (Miers) Miers. R: Kurtziana 19(1987)69

Phryganocydia C. Martius ex Bur. Bignoniaceae. 3 trop. Am. Cult. orn. rampant lianes esp. *P. corymbosa* (Vent.) Schumann with fragrant pink or purple fls

Phrygilanthus Eichler = *Notanthera*

Phryma L. Phrymaceae (Verbenaceae). 1–2 India to Japan, E N Am.

Phrymaceae Schauer (~ Verbenaceae). Dicots – Asteridae – Lamiales. 1/1–2 India to Japan, E N Am. Recently resplintered from the Labiatae–Verbenaceae complex

Genus: *Phryma*

Phryna (Boiss.) Pax & K. Hoffm. = *Phrynella*

Phryne Bubani = *Lycocarpus* + *Murbeckiella*

Phrynella Pax & K. Hoffm. Caryophyllaceae (III 1). 1 Turkey: *P. ortegioides* (Fischer & C. Meyer) Pax & K. Hoffm.

Phrynium Willd. Marantaceae. 20 OW trop. Local basketry

Phtheirospermum Bunge ex Fischer & C. Meyer. Scrophulariaceae. 7 E As.

Phthirusa C. Martius (= *Hemitria*). Loranthaceae. (Excl. *Dendropemon*) 40 trop. Am. *P. caribaea* (Krug & Urban) Britton & P. Wilson (WI mistletoe, Carib.) – pest on woody crops. *P. pyrifolia* (Kunth) Eichler (S Am.) – acarpellate

Phuodendron (Graebner) Dalla Torre & Harms = *Valeriana*

Phuopsis (Griseb.) Hook.f. Rubiaceae (IV 16). 1 Caucasus, E Turkey, NW Iran: *P. stylosa* (Trin.) B.D. Jackson (*Crucianella s.*) – cult. orn. annual

Phycella Lindley = *Hippeastrum*

Phygelius E. Meyer ex Benth. Scrophulariaceae. 2 S Afr. R: Plantsman 9(1988)233. Cult. orn. subshrubs esp.*P. capensis* E. Meyer ex Benth. (Cape figwort) – pendent scarlet fls & *P.* × *rectus* Coombes (*P. capensis* × *P. aequalis* Harvey ex Hiern (fls pink)) – many cvs

Phyla Lour. (~ *Lippia*). Verbenaceae. c. 11 trop. & warm. Some cult. orn. ground-cover esp. *P. nodiflora* (L.) E. Greene

Phylacium Bennett. Leguminosae (III 9). 2 SE As. to NE Aus. R: Blumea 24(1978)485

Phylactis Schrader = *Philactis*

Phylica L. Rhamnaceae. 150 S Afr. (140 endemics in Cape), Madag. & Tristan da Cunha (incl. Gough Is. where ant-disp. *P. arborea* Thouars not regenerating as ants gone & introd. mice eat seeds). Mostly heath-like shrubs with revolute lvs

Phyllacantha Hook.f. = *Phyllacanthus*

Phyllacanthus Hook.f. Rubiaceae (inc. sed.). 1 W Cuba (prob. extinct)

Phyllachne Forster & Forster f. Stylidiaceae. 4 Tasmania, NZ, temp. S Am.

Phyllactis Pers. = *Valeriana*

Phyllagathis Blume. Melastomataceae. 47 S China to W Mal. Apical part of 'leaf' of *P. scortechinii* King & other spp. (Malay Pen.) falls, the basal part bears infls & buds

Phyllanoa Croizat. Euphorbiaceae. 1 Colombia

Phyllanthera Blume. Asclepiadaceae (I, Periplocaceae). 2 W Mal.

Phyllanthodendron Hemsley = *Phyllanthus*

Phyllanthus L. Euphorbiaceae. c. 600 trop. & warm (Afr. c. 100). Monoec. or dioec. trees, shrubs & herbs often with flattened leaf-like organs bearing fls on margins or branches with lvs arranged so that branches appear like pinnate lvs; alks; some nickel accumulators (New Caled.) & rheophytes. Some ed. fr., cult. orn. etc. (e.g. for fish-poisons & insecticides in Amazon). R (WI): JAA 37–39(1956–8) var. pp. *P. acidus* (L.) Skeels (*P. distichus*, Otaheite gooseberry, orig. S Am.?, natur. widely) – fr. used in pickles & preserves; *P. amarus* Schum. & Thonn. (? orig. trop. Am., now pantrop. weed) – extracts reduce or eliminate detectable hepatitis B virus surface antigen in humans; *P. buxifolius* (Blume) Muell. Arg. (orig.?) – trop. cult. orn.; *P. emblica* L. (emblic, ambal, amioki, trop. As.) – animal-disp. the drying endocarp splitting explosively to release seeds, ed. fr. (acid at first then sweet, saliva breaking down glycoside but rich in vitamin C & minerals), medic., tanbark, fuelwood & dyes; *P. fluitans* Muell. Arg. (S Am.) – floating aquatic like *Salvinia*, unique in fam.; *P. muellerianus* (Kuntze) Exell (trop. Afr.) – chewing-stick in W Afr.; *P. nummulariifolius* Poiret (trop. & S Afr. to Seychelles) – stems for basket-making in Uganda; *P. piscatorum* Kunth (NE S Am.) – fish-poison in Colombia; *P. niruri* L. (WI) & other spp. locally medic.

Phyllapophysis Mansf. = *Catanthera*

Phyllarthron DC. Bignoniaceae. 13 Madag., Comoro Is. Lvs resembling linear series of articulated segments, app. rachis & petiole wings of ancestral compound leaf (cf. *Citrus*)

Phyllis L. Rubiaceae (IV 13). 2 Macaronesia

Phyllitis Hill = *Asplenium*

Phyllitopsis Reichst. = *Asplenium*

Phyllobaea Benth. = seq.

Phylloboea Benth. Gesneriaceae. 3 China, Burma, Malay Pen.

Phyllobolus N.E. Br. (*Sphalmanthus*). Aizoaceae (IV). (Incl. *Sceletium*) 25 S Namibia, W & C S Afr. Dried fermented lvs of some spp. chewed leading to intoxication due to mesembrine, like cocaine; local soporific

Phyllobotryon Muell. Arg. Flacourtiaceae (6). 3 trop. Afr. R: BJBB 51(1981)426. Infls epiphyllous

Phyllobotryum Muell. Arg. = praec.

Phyllocactus Link = *Epiphyllum*

Phyllocara Guşul. = *Anchusa*

Phyllocarpus Riedel ex Tul. Leguminosae (I 4). 2 trop. Am.

Phyllocephalum Blume. Compositae (Vern. – Vern.). 3 India, Java. R: Rhodora 83(1981)9

Phyllocharis Diels = *Ruthiella*

Phyllocladaceae Bessey. Pinopsida. 1/5 Philippines & Borneo to New Guinea, Tasmania,

NZ. Allied to Podocarpaceae but with complex lateral branch-systems poss. similar to those of the earliest gymnosperms arising in axils of scale-lvs, often lobed or dentate ('phylloclades') & bearing scales on edges; monoec. or dioec., the fls in similar positions to those of phylloclades, each scale of female cones bearing 1 ovule; seeds with basal cup-shaped arils (unique)

Genus: *Phyllocladus*

Phyllocladus Mirbel. Phyllocladaceae. 5 Borneo & N Luzon to New Guinea, Tasmania, NZ. Fossils in S S Am., W Antarctica, NZ, S Aus.; migration to Mal. prob. recent (Keng). Cult. orn. v. slow-growing evergreens & timber (celery pines). R: JAA 59(1978)249. Sylleptic growth giving photosynthetic units with 3 orders of branching, lvs needlelike in seedlings but ephemeral & non-photosynthetic in adults. Bark of NZ spp. esp. *P. trichomanoides* D. Don (tanekaha) form. used for red dye by Maoris, timber valuable. *P. aspleniifolius* (Lab.) Hook.f. (celery-top pine, Tasmania) – imp. timber, trees up to 780 yrs old so poss. use in dendrochronology

Phylloclinium Baillon. Flacourtiaceae (6). 2 trop. Afr. Epiphyllous infls

Phyllocomos Masters = *Anthochortus*

Phyllocosmus Klotzsch = *Ochthocosmus*

Phyllocrater Wernham (~ *Oldenlandia*). Rubiaceae (IV 1). 1 Borneo

Phylloctenium Baillon. Bignoniaceae. 2 Madagascar

Phyllodium Desv. (~ *Desmodium*). Leguminosae (III 9). 6 S As., N Aus.

Phyllodoce Salisb. Ericaceae. 5 circumpolar, N temp. (Eur. 1). R: Plantsman 10(1988)88. Some cult. orn. evergreen heathers esp. *P. caerulea* (L.) Bab. (circumpolar, N temp.) – a protected sp. in GB, rest in mts W N Am. W to Japan. Hybrids with spp. of *Kalmiopsis* & *Rhodothamnus*

Phyllogeiton (Weberb.) Herzog = *Berchemia*

Phylloglossum Kunze. Lycopodiaceae. 2 Aus., NZ. R: OB 92(1987)170. *P. drummondii* Kunze – dies back to a tuber in hot summer months

Phyllomelia Griseb. Rubiaceae (*inc. sed.*). 1 W Cuba (serpentine)

Phyllonoma Willd. ex Schultes. Grossulariaceae (Phyllonomaceae). 4 Mex. to NW Bolivia. R: Brittonia 29(1977)69. Infls epiphyllous, those of *P. ruscifolia* Willd. ex Schultes (*P. integerrima*), at least, initiated on leaf primordia & not 'adnate' etc.; *P. laticuspis* (Turcz.) Engl. has reputation in Mex. as remedy for smallpox

Phyllonomaceae Small = Grossulariaceae

Phyllophyton Kudô = *Marmoritis*

Phyllopodium Benth. (~ *Polycarena*). Scrophulariaceae (Man.). 115 S Afr. esp. W & SW Cape. R: O. Hilliard, *Manuleeae* (1994)429

Phyllorachis Trimen. Gramineae (10). 1 S trop. Afr.

Phylloscirpus C.B. Clarke. Cyperaceae. 1 Argentina

Phyllosma L. Bolus ex Schltr. Rutaceae. 2 SW Cape mts

Phyllospadix Hook. Zosteraceae. 5 Japan, W N Am. R: C. den Hartog (1970) *Seagrasses of the world*, 5. Some locally eaten roots

Phyllostachys Siebold & Zucc. Gramineae (1b). 55 Himal. to Japan. Largest & most frequently cult. hardy bamboos forming thickets by rhizomatous spread, some used for paper, timber, fishing-rods & ed. shoots; the stripped stems of *P. nigra* (Lindley) Munro (black bamboo, China, long cult. Japan) & other spp. the w(h)angee canes used as walking-sticks, umbrella-handles, musical instruments, furniture etc. *P. aurea* (Carr.) M. Riv. & Riv. (fish-pole bamboo, SE China, natur. Japan) – stems for fishing-rods, plant-stakes etc., young shoots ed.; *P. bambusoides* Siebold & Zucc. (madake b., China, long cult. Japan) – most imp. timber b. in E, ed. shoots frequently seen in Chinese food, in plantation pulp prod. twice that of pine, cult. orn. to 25 m tall (fl. cycle 120 yrs); *P. dulcis* McClure (C China) – most imp. ed. b. of C China; *P. edulis* (Carr.) Houz. (*P. pubescens*, Japan introd. China 1746 & now 50% of Chinese bamboo forests) – major source of ed. bamboo shoots esp. those tinned for export from China & Japan

Phyllostegia Benth. Labiatae (VI). 1 Tahiti, 27 Hawaii. R: W.L. Wagner, Man. Fl. Pl. Hawaii (1990)810

Phyllostelidium Beauverd = *Baccharis*

Phyllostemonodaphne Kosterm. Lauraceae (I). 1 SE Brazil: *P. geminiflora* (Mez) Kosterm. R: BJ 110(1988)167

Phyllostylon Capanema ex Benth. Ulmaceae (I). 2 trop. Am. R: Sida 15(1992)263. Fine-grained yellow wood often ebonized esp. *P. brasiliensis* Capanema ex Benth. (baitoa, San Domingo boxwood)

Phyllota (DC) Benth. Leguminosae (III 24). 10 SW & E Aus.

Phyllotrichum Thorel ex Lecomte. Sapindaceae. 1 SE As.

Phylloxylon Baillon. Leguminosae (III 8). 5 Madag. Hard timber

Phylogyne Salisb. ex Haw. = *Narcissus*

Phylohydrax Puff (~ *Hydrophylax*). Rubiaceae (IV 15). 2 E Afr., Madag. R: PSE 154(1986)343

Phymaspermum Less. Compositae (Anth.-Urs.). 18 S Afr. R: NJB 5(1986)535

Phymatarum Hotta. Araceae (VI 1). 2 Borneo. R: MCSUK B32(1965)29

Phymatidium Lindley. Orchidaceae (V 10). 10 Brazil

Phymatocarpus F. Muell. Myrtaceae (Lept.). 2 W Aus.

Phymatosorus Pichi-Serm. (*'Phymatodes'*). Polypodiaceae (II 4). 12 trop. & warm OW. R: Webbia 28(1978)457. Cult. orn. incl. *P. pustulatus* (Forster f.) Large, Braggins & P. Green (*Microsorum diversifolium, Phymatodes d.*, kangaroo fern, E Aus., NZ), *P. scolopendria* (Burm.f.) Pichi-Serm. (OW) – oil used to flavour coconut oil

Phymosia Desv. Malvaceae. 8 C Am. R: Madrono 21(1971)153. Some cult. orn. trees & shrubs

Phyodina Raf. = *Callisia*

Physacanthus Benth. Acanthaceae. 5 trop. Afr.

Physaliastrum Makino = *Leucophysalis*

Physalidium Fenzl = *Graellsia*

Physalis L. Solanaceae (1). 80 cosmop. (Eur. 1, Aus. 1) esp. Am. Cult. orn. & some ed. fr. R (partial): Rhodora 69(1967)82, 203, 319. Alks. *P. alkekengi* L. (Chinese lantern, winter cherry, C & S Eur., W As. to Japan, incl. *P. franchetii*) – cult. orn. 'everlasting' with vermilion inflated calyces (toxic) around ± ed. berry (calyces skeletonizing as frs redden), form. medic.; *P. peruviana* L. (Cape gooseberry, trop. S Am.) – fr. small, ed.; *P. philadelphica* Lam. (*P. ixocarpa*, tomatillo, jamberry, Mex., natur. E N Am.) – fr. yellow to purple, ed.; *P. pruinosa* L. (strawberry tomato, E N Am.) – common cult. 'husk tomato', fr. yellow, ed.; *P. viscosa* L. (Am.) – natur. Aus., noxious weed in Victoria

Physandra Botsch. = *Salsola*

Physaria (Nutt.) A. Gray. Cruciferae. 14 W N Am. R: Rhodora 41(1939)392. Some cult. orn. rock-pls

Physena Noronha ex Thouars. Physenaceae (Capparidaceae s.l.). 2 Madagascar

Physenaceae Takht. (~ Capparidaceae). Dicots – 1/2 Madag. Lately split off from Capparidaceae of which limits still controversial
Genus: *Physena*

Physetobasis Hassk. = *Holarrhena*

physic nut *Jatropha curcas*

Physinga Lindley = *Epidendrum*

Physocalymma Pohl. Lythraceae. 1 trop. S Am.: *P. scaberrima* Pohl – timber red with darker markings (tulipwood), used for furniture, oil used in scent-making

Physocalyx Pohl. Scrophulariaceae. 2 Brazil

Physocardamum Hedge. Cruciferae. 1 E Turkey

Physocarpus (Camb.) Maxim. Rosaceae. 10 N Am., NE As. Cult. orn. shrubs with inflated follicular fr. esp. *P. opulifolius* (L.) Maxim. (C & E N Am.)

Physocaulis (DC) Tausch (*Myrrhoides*). Umbelliferae (III 2). 1 Medit. incl. Eur.

Physoceras Schltr. Orchidaceae (IV 2). 7 Madagascar

Physochlaina G. Don f. Solanaceae (1). 6 China. Alks. Some cult. orn. perennial herbs

Physodium C. Presl = *Melochia*

Physogyne Garay = *Schiedeella*

Physokentia Becc. Palmae (4m). 7 New Britain, Solomon Is., Vanuatu, Fiji. R: Principes 13(1969)120

Physoleucas Jaub. & Spach = *Leucas*

Physoplexis (Endl.) Schur (~ *Phyteuma*). Campanulaceae. 1 Alps: *P. comosa* (L.) Schur (*Phyteuma c.*) – cult. orn. rock-pl.

Physopsis Turcz. Labiatae (III, Verbenaceae). 2 W Aus.

Physoptychis Boiss. Cruciferae. 2 E Turkey, NW Iran

Physopyrum Popov. Polygonaceae. 1 C As.

Physorhynchus Hook. Cruciferae. 2 S Iran to NW India

Physosiphon Lindley = *Pleurothallis*

Physospermopsis H. Wolff. Umbelliferae (III 5). 17–18 C to E As.

Physospermum Cusson. Umbelliferae (III 5). 2 temp. Euras. (Eur. 2). Bladderseed

Physostegia Benth. Labiatae (VI). 12 N Am. R: CGH 211(1982)1. Cult. orn. herb. pls esp. *P. virginiana* (L.) Benth. (obedient plant) – if fls moved sideways, they stay put – catalepsy, the fls not springing back because of rigidity of bracts & friction between trichomes of bract & those of calyx & pedicel

Physostelma Wight = *Hoya*

Physostemon C. Martius = *Cleome*

Physostigma Balf. Leguminosae (III 10). 4 trop. Afr. Alks. *P. venenosum* Balf. (Calabar or ordeal bean, trop. W Afr., introd. Am. & As.). Ordeal poison (physostigmine) an alk. used in ophthalmic med., inhibiting acetylcholine esterase interfering with nerve impulse transmission

Physothallis Garay = *Pleurothallis*

Physotrichia Hiern. Umbelliferae (III 8). 10 trop. Afr.

Phytelephas Ruíz & Pavón. Palmae (VI). (Incl. *Palandra*) 4 trop. Am. R: OB 105(1991)48. Dioec. unarmed short-stemmed palms with A 900+ & aggregated fr., the 'cellulose' (2 mannans) endosperm v. hard (vegetable ivory) used for billiard-balls, buttons, chessmen, dice etc., esp. that of *P. macrocarpa* Ruíz & Pavón (ivory nuts, tagua), though that of *P. aequatorialis* Spruce (*Palandra a.*) the best quality

Phyteuma L. Campanulaceae. 40 Medit., Eur. (24 excl. *Physoplexis*), As. Horned rampion. Fls in heads, tips of C forming tube with anthers inside, the style pushing pollen out where insects take it; later style emerges & stigmas open while C separate & recurve (cf. Compositae). Some cult. orn. rock- & border-pls incl. *P. orbiculare* L. (Alps) – root s.t. ed. in salad etc. (rampion)

Phytocrene Wallich. Icacinaceae. 11 SE As., Mal. Lianes with large vessel-elements – sources of potable water when cut

Phytolacca L. Phytolaccaceae (I). 25 trop. & warm. Trees, shrubs & herbs with alks., some poisonous, some used as veg., cult. orn. etc., medic. (antiviral proteins used against leukaemia): *P. acinosa* Roxb. (incl. *P. esculenta*, E As., natur. trop. As.) – young lvs & shoots a veg.; *P. americana* L. (*P. decandra*, pokeweed, pigeonberry, inkberry, N Am.) – cult. orn., dye from berries used to colour ink, wine, sweets, potherb (boiled water to be discarded – toxic), kills snails which carry bilharzia; *P. dioica* L. (trop. S Am.) – dioec. evergreen tree with thick trunk & wide-spreading surface roots, cambia successive formed in layers external to earlier ones, planted in S Eur. etc. as shade-tree (bella umbra); *P. dodecandra* L'Hérit. (endod, trop. & S Afr., Madag.) – v. poisonous, molluscicidal saponins (triterpenes) used in field trials; '*P. electrica*' Lévy (Nicaragua) – alleged to have given shock to its collector (1876); *P. octandra* L. (inkweed, NZ) – introd. Aus.

Phytolaccaceae R. Br. Dicots – Caryophyllidae – Caryophyllales. 18/65 trop. & warm esp. Am. R: Pfl. 4(1960)83. Trees, shrubs, lianes or usu. herbs, often glabrous & somewhat succ., producing betalains & not anthocyanins; sec. growth with concentric rings of vasc. bundles; sieve-tubes with P-type plastids incl. central globular (polyhedral in *Stegnosperma*) protein crystalloid & a sub-peripheral ring of proteinaceous filaments. Lvs simple, spirally arr. (opp. in *Gisekia*); stipules minute or 0 (spines in *Seguieria*). Fls usu. reg., usu. bisexual, small in spikes to panicles, rarely in cymes or solit.; P 4 or 5(–10), s.t. basally connate, A (2–)4(–∞), when ∞ developing centrifugally, anthers with longit. slits, G̲ (1)2–17, ± connate but with distinct styles & as many loc. as G, each loc. with 1 usu. basal, campylotropous bitegmic ovule. Fr. a drupe or nut (or capsule in *Barbeuia*), carpels often separating; seeds with embryo curved around ± copious hard or starchy perisperm (true endosperm 0) & sometimes an aril (*Barbeuia*)

Classification & chief genera:

I. **Phytolaccoideae** (P 5, G 3–17): *Ercilla, Phytolacca*
II. **Rivinoideae** (P 4(5 in *Seguieria*)), G 1: *Hilleria, Rivina*
III. **Microteoideae** (P usu. 5, G (2)): *Microtea*
IV. **Agdestidioideae** (Agdestidaceae; P 4, G (3)4): *Agdestis* (only)
V. **Barbeuioideae** (Barbeuiaceae; P 5, G 2, fr. a capsule): *Barbeuia* (only)

'Rather loosely knit' (Cronquist), *Agdestis* & *Barbeuia* (s.t.) treated as segregate monotypic fams, while Achatocarpaceae, Gyrostemonaceae & Stegnospermataceae here treated separately incl. in P. by some workers

Some cult. orn., locally medic., dye-pls, potherbs etc. (esp. *Phytolacca*)

Phytomorula Kofoid = *Acacia*. Described as an alga, *P. regularis* Kofoid is known to be pollen-grains!

Piaggiaea Chiov. = *Wrightia*

Piaranthus R. Br. Asclepiadaceae (III 5). 5 Namibia, S Afr. Cult. orn. succ.

piassava or **piassaba** Fibre from various palms used as bristles of scrubbing-brushes, brooms, mechanical road-sweepers (bass broom fibre) etc. **African p.** *Raphia hookeri, R. palma-pinus*; **Bahia p.** *Attalea funifera*; **Ceylon p.** *Caryota urens*; **Pará p.** *Leopoldinia piassaba*

Picardaea Urban. Rubiaceae (I 7). 2 Cuba, Hispaniola

piccabeen *Archontophoenix cunninghamiana*

Picconia A. DC. Oleaceae. 2 Macaronesia

Picea A. Dietr. Pinaceae. c. 40 cool N hemisph. (Eur. 2). R: Phytol. M 7(1984)43. Spruce. Monoec. evergreens differing from *Abies* in persistent raised leaf-bases & pendulous woody female cones with persistent scales; cones mature in 1 yr. Imp. softwoods used in chipboard, hardboard, as pulp, cellophane & rayon. *P. abies* (L.) Karsten (*P. excelsa*, Norway spruce, N & C Eur., native in GB in last interglacial) – common Christmas tree in GB, timber ('white deal', Baltic whitewood) & pulp, bark used in tanning (Germany), source of Jura turpentine & Burgundy pitch used in wound-dressings, form. much used for telegraph-poles & (because of resonance) violins & cellos, young shoots & lvs basis of spruce beer, many cult. orn. cvs incl. dwarfs esp. 'Clanbrasiliana' c. 2 m tall; *P. breweriana* S. Watson (S Oregon & N Calif.) – striking drooping branches; *P. glauca* (Moench) Voss (*P. alba*, white s., N N Am.) – timber & pulp; *P. jezoensis* (Sieb. & Zucc.) Carr. (NE As. to Japan) – needles mixed with those of *Abies sachalinensis* to yield Japanese pine needle oil; *P. mariana* (Miller) Britton, Sterns & Pogg. (bog or (Canadian) black s., N N Am.) – timber & pulp, resin used as chewing-gum (spruce g.), boiled branches giving s. beer, trunk & twigs accum. gold & platinum, the bark arsenic; *P. orientalis* (L.) Petermann (oriental s., Caucasus, NE Turkey) – many cvs incl. dwarfs; *P. rubens* Sarg. (red s., N N Am.) – timber for pulp & planks (form. ships), major source of spruce gum; *P. sitchensis* (Bong.) Carr. (Sitka s., W N Am.) – introd. GB 1834, much planted in Scottish highlands, timber & pulp

pichi *Fabiana imbricata*

Pichi-Sermollia Neto = *Areca*

Pichleria Stapf & Wettst. = *Zosima*

Pichonia Pierre (~ *Lucuma*). Sapotaceae (IV). 6 Papua New Guinea, Solomon Is., New Caled. R: T.D. Pennington, *S.* (1991)228

pichurim *Nectandra pichurim*

pick-a-back plant *Tolmiea menziesii*

pickerel weed *Pontederia cordata*, (E Anglia) *Potamogeton natans*

Pickeringia Nutt. ex Torrey & A. Gray. Leguminosae (III 30). 1 SW N Am.: *P. montana* Nutt. ex Torrey & A. Gray – xerophytic shrub of chapparal spreading by underground stems & rapidly re-establishing after fire

picmoc *Desmoncus orthacanthus*

Picnomon Adans. (~ *Cirsium*). Compositae (Card.-Card.). 1 Medit. (incl. Eur.), SW As.

picotee *Dianthus caryophyllus*

Picradeniopsis Rydb. ex Britton = *Bahia*

Picraena Lindley = *Picrasma*

Picralima Pierre. Apocynaceae. 1 W trop. Afr.: *P. klaineana* Pierre – bark a febrifuge, seeds used as quinine subs. (alks)

Picramnia Sw. Picramniaceae (Simaroubaceae s.l.). 45 trop. Am. *P. antidesma* Sw. (macary bitter, WI); others locally medic., dyes etc.

Picramniaceae (Engl.) Fernando & Quinn (~ Simaroubaceae). Dicots – Rosidae – ?Sapindales. 2/50 trop. Am. R: T 44(1990)177. Dioec. trees & shrubs. Lvs pinnate, spirally arr.; stipules 0. Infls axillary (term. or cauliflorous) racemes or panicles. Fls 3–5(6)-merous; K connate basally, lobes imbricate or valvate; C s.t. 0 in males, small & imbricate in females; A alt. with K; G (2 or 3), 1–3-loc., each loc. with 2 ovules, styles short. Fr. a berry or samaroid capsule; seeds planoconvex to narrowly ellipsoid with 0 endosperm

Genera: *Alvaradoa, Picramnia* (placed in separate subfams)

Simaroubaceae s.s. have uniovulate locules

Picrasma Blume. Simaroubaceae. 6 trop. Am., 2 Indomal. Stipules (rare in fam.), alks. *P. excelsa* (Sw.) Planchon (*Aeschrion e.*, WI) – source of quassia chips, form. used in bitters & as insecticide

Picrella Baillon = ? *Helietta*

Picria Lour. (*Curanga*). Scrophulariaceae. 1 Indomal.

Picridium Desf. = *Reichardia*

Picris L. Compositae (Lact.-Hypo.). c. 40 Eur. (12), Medit. (most), As., N Aus., Afr. mts. Ox-tongue (esp. *P. echioides* L. (Medit., natur. elsewhere) – wide range of insects trapped by grapple-hook hairs & *P. hieracioides* L. (Euras., widely natur.))

Picrodendraceae Small = Euphorbiaceae

Picrodendron Griseb. Euphorbiaceae (Picrodendraceae). 1 WI: *P. baccatum* (L.) Krug & Urban. R: JAA 65(1984)105. Allied to Oldfieldioideae

Picrolemma Hook.f. Simaroubaceae. 3 E Peru, Amazonian Brazil. Some with hollow ant-infested stem domatia; some fish-poisons

Picrorhiza Royle ex Benth. Scrophulariaceae. 1 W Himal.: *P. kurrooa* Royle ex Benth. (*P.*

scrophulariiflora, kutki) – locally medic. esp. febrifuge. R: OB 75(1984)56

Picrosia D. Don. Compositae (Lact.-Mic.). 2 warm S Am.

Picrothamnus Nutt. (~ *Artemisia*). Compositae (Anth.-Art.). 1 W US

picrotoxin *Anamirta cocculus*

Pictetia DC. Leguminosae (III 13). 6 WI

Pieris D. Don. Ericaceae. 7 E As., E N Am. R: JAA 63(1982)103. Cult. orn. evergreen shrubs & trees incl. *P. floribunda* (Pursh) Benth. (fetterbush, SE US)

Pierranthus Bonati. Scrophulariaceae. 1 SE As.

Pierrodendron Engl. = *Quassia*

Pierrina Engl. Scytopetalaceae. 1 W trop. Afr.

Pietrosia Nyár. = *Andryala*

pifine grass *Panicum hemitomon*

pig balsam *Tetragastris balsamifera*; **p. laurel** *Kalmia angustifolia*; **p. nut** *Conopodium majus, Carya glabra, Simmondsia chinensis*; **p.weed** *Amaranthus albus, A. retroflexus* etc.

Pigafetta (Blume) Becc. Palmae (II 1f). 1 (?+) E Mal.: *P. filaris* (Giseke) Becc. – fast-growing colonist to 50 m with small seeds

pigeon bean *Vicia faba* var. *equina*; **p.berry** *Phytolacca americana, Duranta erecta*; **p. grass** *Setaria pumila* & other spp.; **p. orchid** *Dendrobium crumenatum*; **p. pea** *Cajanus cajan*

pigface *Carpobrotus edulis, Disphyma* spp. esp. *D. crassifolium*

piggyback plant *Tolmiea menziesii*

pignoli, pignon *Pinus pinea*

Pilea Lindley. Urticaceae (II). 200+ trop. & warm excl. Australasia. Monoec. & dioec. herbs without stinging hairs; some greenhouse pls (clearweed) esp.: *P. cadierei* Gagnepain & Guillaumin (aluminium plant, Vietnam) – lvs with silvery stripes, *P. involucrata* (Sims) C.H. Wright & Dewar (trop. Am.) – many cvs & *P. microphylla* (L.) Liebm. (*P. muscosa*, artillery pl., gunpowder pl., trop. Am., widely natur. Afr. & As. & Balkans) – anthers eject pollen explosively, mordant in New Britain; *P. peperomioides* Diels (Yunnan) – elongated stems with succ. lvs like *Peperomia*, widely cult. pot-pl. introd. to W via Norway

Pileanthus Labill. Myrtaceae (Lept.). 3 SW Aus.

Pileostegia Hook.f. & Thomson (~ *Schizophragma*). Hydrangeaceae. 4 E As. Evergreen lianes. *P. viburnoides* Hook.f. & Thomson (Himal.) – cult. orn.

Pileus Ramírez = *Jacaratia*

pilewort *Ranunculus ficaria*

Pilgerochloa Eig = *Ventenata*

Pilgerodendron Florin (~ *Libocedrus*). Cupressaceae. 1 S S Am. N to 40° S: *P. uviferum* (D. Don) Florin – timber useful

pili nut *Canarium luzonicum, C. ovatum*

Pilidiostigma Burret. Myrtaceae. 5 NE Aus.

Piliocalyx Brongn. & Gris. Myrtaceae (Myrt.) 8 New Caled., Fiji

Piliostigma Hochst. = *Bauhinia*

Pillansia L. Bolus. Iridaceae (IV 1). 1 SW Cape

pillarwood *Cassipourea elliottii*

pillwort *Pilularia globulifera* & other spp.

Piloblephis Raf. (~ *Satureja*). Labiatae (VIII 2). Spp.?

Pilocarpus Vahl. Rutaceae. 22 trop. Am. Alks. Source of medicinal jaborandi, active alk. being pilocarpine

Pilocereus Lemaire = *Cephalocereus*

Pilocopiapoa F. Ritter = *Copiapoa*

Pilocosta Almeda & Whiffin. Melastomataceae. 5 trop. Am. R: Novon 3(1993)311

Pilophyllum Schltr. = *Chrysoglossum*

Pilopleura Schischkin. Umbelliferae (III 8). 2 C As.

Pilosella Hill (~ *Hieracium*). Compositae (Lact.-Hier.). c. 20–80 (dependent on sp. concept) Euras., N Afr. Sexual spp. (cf. *Hieracium*) or partially apomictic, most hybridizing freely. Some cult. orn. but invasive (stolons) esp. *P. aurantiaca* (L.) F.W. Schultz & Schultz-Bip. (devil's paintbrush, fox and cubs, Eur., natur. N Am.) – florets orange-red

Piloselloides (Less.) C. Jeffrey ex Cufod. = *Gerbera*

Pilosocereus Byles & G. Rowley (~ *Cephalocereus*). Cactaceae (III 3). 35 Mex. to trop. S Am. esp. Brazil. Cult. orn. shrubs & trees to 10 m

Pilosperma Planchon & Triana. Guttiferae (III). 2 Colombia

Pilostemon Iljin. Compositae (Card.-Card.). 2 C As. to China

Pilostigma Tieghem = *Amyema*

Pilostyles Guillemin. Rafflesiaceae (II 2). 1 Iran, 2 SW Aus., 1 trop. Afr., c. 7 S Calif. to trop. Am. The pls parasitize all 3 subfams of Leguminosae & are internal save fls; *P. haussknechtii* Boiss. (SW As.) on *Astragalus* produces paired fls at base of host lvs

Pilothecium (Kiaerskov) Kausel = *Myrtus*

Pilularia L. Marsileaceae. 3–6 N temp. (Eur. 2), Aus., NZ, Ethiopian mts, W S Am.. Pillwort esp. *P. globulifera* L. (Eur.) – grows at lake-edges, lvs filiform, entire, with pea-shaped 4-chambered sporocarp, each chamber with a sorus bearing both macro- & micro-sporangia, rhizome & roots with 'vessels'. *P. minuta* Dur. ex A. Br. (SW Eur.) – one of smallest of all ferns

Pimelea Banks & Sol. Thymelaeaceae. 108 NZ (17) & Aus. (90 endemic; W Aus. 45, R: Nuytsia 6(1988)129) with few ext. to Timor & Lord Howe Is. Fls in term. heads often surrounded by involucre of leafy & often coloured bracts (*P. physodes* Hook. (W Aus.) = Qualup bell). Some cult. orn. evergreen shrubs (riceflowers); some with bark used as sources of twine by early settlers; some toxic to stock in Aus.

Pimelodendron Hassk. Euphorbiaceae. 6–8 Mal. *P. amboinicum* Hassk. (Moluccas) – ed. seeds, bark a purgative, milky latex used as varnish

Pimenta Lindley. Myrtaceae (Myrt.). 2–5 trop. Am. Aromatic trees. *P. dioica* (L.) Merr. (*P. officinalis*, allspice, pimento, Jamaica pepper, C Am., WI) – cryptically dioec., spice reminiscent of cinnamon, nutmeg & cloves etc. derived from unripe fr. used as flavouring (as in Benedictine & Chartreuse) & medic. etc.; *P. racemosa* (Miller) J. Moore (*P. acris*, bay rum tree, trop. Am.) – introd. Pacific, oil form. distilled with rum, used in scent- & soap-making (flavour due to eugenol)

Pimentelia Wedd. Rubiaceae (I 1). 1 Peru

pimento *Pimenta dioica*

Pimia Seemann. Sterculiaceae. 1 Fiji

pimpernel (**common** or **scarlet**) *Anagallis arvensis*; **blue p.** *A. monelli*; **bog p.** *A. tenella*; **yellow p.** *Lysimachia nemorum*

Pimpinella L. Umbelliferae (III 8). 150 Euras. (Eur. 16), Afr. Some spices & cult. orn. esp. *P. anisum* L. (anise, Greece to Egypt) – cult. since 2000 BC, food- & drink-flavouring (absinthe, anis, anisette, ouzo, pastis, raki), distilled oil medic., familiar as aniseed balls; *P. major* (L.) Hudson & *P. saxifraga* L. – burnet saxifrage, Eur.

pin cushion *Brunonia australis*

Pinacantha Gilli. Umbelliferae (inc. sed.). 1 Afghanistan

Pinaceae Sprengel ex Rudolphi. Pinopsida. 12/220 N temp. S to C Am., WI, Sumatra & Java. Usu. evergreen, resinous monoec. trees (few prostrate shrubs) with usu. opp. or whorled branches & scaly, often resinous, term. buds. Lvs linear, spirally arr. Male cones small, herbaceous, with spirally arr. microsporophylls each with 2 pollen-sacs abaxially; female cones usu. woody, with spirally arr. scales, each usu. with 2 ovules adaxially & subtended by a ± united bract; seeds usu. 2 per scale & usu. winged. Embryo with several cotyledons

Genera: *Abies, Cathaya, Cedrus, Hesperopeuce, Keteleeria, Larix, Nothotsuga, Picea, Pinus, Pseudolarix, Pseudotsuga, Tsuga*

Most imp. sources of timber & pulp, turpentine, resins, cult. orn., ed. seeds etc.

Pinacopodium Exell & Mendonça. Erythroxylaceae. 2 trop. Afr.

pinang *Areca catechu*

Pinanga Blume. Palmae (V 4l). 120 Indomal., SE As. Usu. undergrowth palms with up to 9 spp. growing sympatrically on limestone in Borneo; beetle-poll. (pollen v. variable) though in *P. simplicifrons* (Miq.) Becc. (Sumatra) & *P. cleistantha* Dransf. (Malay Pen.) infls enclosed in prophyll so no obvious pollinator access. Some rheophytes in Borneo; some masticatories like betel, a few cult. orn. small pls

Pinaropappus Less. Compositae (Lact.-Mal.). 6 S N Am. R: CUMH 9(1972)371

Pinarophyllon Brandegee. Rubiaceae (IV 5). 2 C Am.

pinaster *Pinus pinaster*

Pinckneya Michaux. Rubiaceae (I 7). 1 SE US: *P. bracteata* (Bartram) Raf. (*P. pubens*, fever tree, Georgia bark tree) – 1 or more K lobes rose-coloured 'flags' to 7 cm long, bitter bark medic., used against malaria in Am. Civil War (cinchonin). R: JAA 68(1987)142

Pinda P.J. Mukherjee & Constance (~ *Heracleum*). Umbelliferae (III 11). 1 W Ghats, S India: *P. concanensis* (Dalzell) P.K. Mukherjee & Constance. R: KB 41(1986)224

pine *Pinus* spp.; **Aleppo p.** *P. halepensis*; **American pitch p.** *P. palustris*; **Australian p.** *Casuarina equisetifolia*; **Austrian p.** *P. nigra* subsp. *nigra*; **Bhutan** or **blue p.** *P. wallichiana*; **Brazilian p.** *Araucaria angustifolia*; **bristle-cone p.** *P. longaeva*; **bunya-bunya p.** *Araucaria bidwillii*; **Canadian red p.** *P. resinosa*; **Caribbean pitch p.** *P. caribaea*; **celery p.** *Phyllocladus* spp.; **celery-top p.** *P. aspleniifolius*; **Chile p.** *A. araucana*; **chir p.** *Pinus roxburghii*; **Corsican**

p. *P. nigra* subsp. *salzmannii*; **cypress p.** *Callitris* spp.; **digger p.** *P. sabiniana*; **golden p.** *Pseudolarix kaempferi*; **hoop p.** *Araucaria cunninghamii*; **Huon p.** *Lagarostrobos franklinii*; **Illawara p.** *Callitris rhomboidea*; **jack p.** *Pinus banksiana*; **King Billy** or **William p.** *Athrotaxis selaginoides*; **klinki p.** *Araucaria hunsteinii*; **Leichhardt p.** *Nauclea orientalis*; **loblolly p.** *P. taeda*; **lodgepole p.** *P. contorta* ssp. *latifolia*; **long-leaf p.** *P. palustris*; **maritime p.** *P. pinaster*; **Monterey p.** *P. radiata*; **Norfolk Is. p.** *A. heterophylla*; **Oregon p.** *Pseudotsuga menziesii*; **Oyster Bay p.** *Callitris rhomboidea*; **Paraná p.** *A. angustifolia*; **parasol p.** *Sciadopitys verticillata*; **pitch p.** *Pinus palustris*; **ponderosa p.** *P. ponderosa*; **Port Jackson p.** *Callitris rhomboidea*; **radiata p.** *P. radiata*; **Scots p.** *P. sylvestris*; **screwp.** *Pandanus* spp.; **slash p.** *Pinus elliotii*; **southern p.** *P. echinata, P. elliotii, P. palustris, P. taeda*; **stone p.** *P. pinea*; **sugar p.** *P. lambertiana*; **umbrella p.** *P. pinea*; **western white p.** *P. monticola*; **Weymouth** or **white p.** *P. strobus*; **Wollemi p.** *Wollemia nobilis*; **yellow p.** *P. echinata*

pineapple *Ananas comosus*; **p. guava** *Acca sellowiana*; **p. lily** *Eucomis* spp.; **p.(-scented) sage** *Salvia elegans*; **p. weed** *Matricaria discoidea*

Pineda Ruíz & Pavón. Flacourtiaceae (7). 1 Andes of Ecuador & Peru, above tree-line: *P. incana* Ruíz & Pavón – lvs a source of black dye

Pinelia Lindley (*Pinelianthe*, ~ *Homalopetalum*). Orchidaceae (V 13). 3 Brazil

Pinelianthe Rauschert = *Pinelia*

Pinellia Ten. Araceae (VIII 6). 6 China, Japan. Some cult. orn. incl. *P. tripartita* (Blume) Schott (S Japan) – poss. self-poll.

piney varnish *Vateria indica*

pingue *Hymenoxys richardsonii*

Pinguicula L. Lentibulariaceae. 46 Am., Medit. & few circumboreal (Eur. 12). R: BB 127/8(1966)1. Small carnivorous herbs with lvs in rosettes & usu. viscid above (secretory glands), trapping small insects as in *P. vulgaris* L. (butterwort – alleged to coagulate milk (poss. by micro-organisms associated with mucilage rather than digestive enzymes of pl.), bog violet, N N Am., Euras. exc. NE) where leaf-margins curl over insects & sessile glands secrete enzymes which digest prey after mucilage glands collapse & epidermal cells below them lose turgor bringing insects in touch with sessile glands

pinguin fibre *Bromelia pinguin*

Pinillosia Ossa. Compositae (Helia.-Pin.). 1 Cuba, Hispaniola

pink *Dianthus* spp. esp. *D. plumarius* (app. from **pinkster**, i.e. Whitsun (Dutch), now used for the colour); **Cheddar p.** *D. gratianopolitanus*; **Chinese p.** *D. chinensis*; **clove p.** *D. caryophyllus*; **Deptford p.** *D. armeria*; **p.-eye** *Tetratheca* spp.; **Japanese p.** *D. chinensis* 'Heddewigii'; **maiden p.** *D. deltoides*; **mullein p.** *Silene coronaria*; **rose p.** *Sabatia* spp.; **sea p.** *Armeria maritima*

pinnay oil *Calophyllum inophyllum*

Pinophyta. See Gymnospermae

Pinopsida (Coniferae, Pinatae). 70/598 in 9 fams, trop. to cold. Branching woody pls often with long & short shoots; sec. wood of tracheids. Resin-canals in lvs, cortex & s.t. wood. Lvs spirally arr. or opp. (whorls in *Sciadopitys*), usu. needle-like or scale-like. Female cones with bract-scales united with ovuliferous scales each bearing usu. several to 2 ovules. Males with many scale-like microsporophylls with 2–∞ pollen-sacs. Embryo with 2–∞ cotyledons. Fams: Araucariaceae, Cephalotaxaceae, Cupressaceae, Phyllocladaceae, Pinaceae, Podocarpaceae, Sciadopityaceae, Taxaceae & Taxodiaceae Imp. timber trees & cult. orn. incl. 'dwarf conifers' (mutants or diseased clones)

Pinosia Urban. Caryophyllaceae. 1 Cuba

pinto beans *Phaseolus vulgaris*

Pintoa C. Gay. Zygophyllaceae. 1 Chile

Pinus L. Pinaceae. 93 N temp. (Eur. 11) & S to C Am., Sumatra & Java. Pines. R: N.T. Mirov (1967) *The genus P.*, Phytol.M 7(1984)47. Evergreen monoec. trees with scale lvs soon lost & long needle lvs in clusters of (1)2–5, each year's growth rep. by unbranched long shoot term. in a series of buds, 1 carrying on the axial growth, the rest forming a whorl of branches in the next season. Male cones with many fertile scales, each with 2 pollen-sacs abaxially, each grain with 2 air-bladders; female cones take 2–3 yrs to mature following arrival of pollen grain which is brought into contact with nucellus by drying up of mucilage & then does not develop complete pollen tubes for another yr. Pines amongst the most imp. timbers in temp. (incl. merchant ships of classical Medit. & perhaps the 'gopher wood' of the Ark ('kapher' = resinous wood)) & trop. regions – in the latter, *P. patula* & *P. caribaea* the most planted of all trees; cult. for timber, pulp & resinous products (incl. constituents of comm. cough mixtures in Eur.) like turpentine & rosin (together = naval stores, when crude = colophony, used in cigarette flavourings & as soldering flux but allergenic fumes cause asthma in some 20% of operatives) esp. from *P. palustris* & *P. pinaster*, rosin being used to size paper etc.; wood tar & pitch obtained by destructive distillation, esp. from *P. sylvestris* ('Stockholm tar'); some have ed. seeds (pignons) & evolved with bird-dispersers, or manna & many cult. orn. Although helpful in identification, the no. of needles in a cluster is not a good phylogenetic marker & the genus is divided into 2 groups which are marked by the no. of vasc. bundles in the needles, viz. sect. *Strobus*, soft pines (timber soft, resin little) having 1 & sect. *Pinus* (relatively hard wood & frequently much resin) having 2. *P. armandii* Franchet (Chinese white p., temp. C & W China) – cheap furniture; *P. banksiana* Lamb. (*P. divaricata*, jack p., E N Am.) – pulp; *P. bungeana* Zucc. ex Endl. (lace-bark p., C & N China) – cult. orn. with flaky bark; *P. canariensis* C. Sm. (Canary p., Canary Is.) – app. closely allied to *P. roxburghii* 8000 km away in Himal.; *P. caribaea* Morelet (Caribbean or Bahamas pitch p., C Am. & WI) – hard timber much planted in lowland trop., high incidence of pls ± unbranched ('foxtails') in plantation esp. in aseasonal climates; *P. cembra* L. (Siberian cedar, Euras.) – lvs for turpentine, timber for building, seeds ed. (mast-fruiting over 1000s of km^2; *P. cembroides* Zucc. (SW N Am.) – seeds locally imp. as food; *P. contorta* Douglas ex Loudon ssp. *latifolia* (Engelm.) Critchf. (lodge-pole p., W N Am.) – imp. plantation tree introd. GB 1853/4; *P. densiflora* Siebold & Zucc. (Korea, Japan) – typical Japanese sp. with many cvs; *P. echinata* Miller (yellow or southern p., SE US) – hard wood; *P. edulis* Engelm. (SW US) – distrib. piñon jays, prod. irregularly abundant nutritious food in which seasons bird testes grow & remain big throughout season so breeding in late winter & late summer; *P. elliotii* Engelm. (slash or southern p., SE US); *P. halepensis* Miller (Aleppo p., Medit.) – flavours retsina, turpentine; *P. kesiya* Royle ex Gordon (*P. khasya*, mts of SE As.) – turpentine; *P. koraiensis* Sieb. & Zucc. (NE As. to Japan) – to 350 yrs old; *P. lambertiana* Douglas ex Taylor & Phillips (sugar p., W N Am.) – timber, damaged heartwood yields sweet laxative material, seeds ed.; *P. longaeva* D. Bailey (Calif., Nevada & Utah) – ancient 'bristle-cone pines' confused with shorter-lived *P. aristata* Engelm. (Rocky Mts) to 2000 yrs old, *P. longaeva* to 5000 yrs old known (one 4900 yrs old felled in 1964) & by extrapolation on dead specimens (4000 yrs to decay), material known to be 8200 yrs old now available for correcting carbon-dating; *P. monophylla* Torrey & Frémont (*P. cembroides* var. *m.*, SW US) – 1-needled; *P. monticola* Douglas ex D. Don (western white p., NW Am.) – valuable timber; *P. mugo* Turra (C Eur., Balkans) – source of pumilio pine oil used in inhalants, soaps etc.; *P. nigra* J.F. Arnold (Austrian p., Eur. & SW As.), subsp. *nigra* (E Eur.) & esp. subsp. *salzmannii* (Dunal) Franco (ssp. *laricio*, incl. var. *corsicana* (Loudon) Hyl. (var. *maritima*), Corsican p., W Medit.) – widely planted; *P. palustris* Miller (longleaf or pitch or southern p., SE US) – remains in dwarf 'grass stage' until exposed to heat of a forest fire, source of timber, pulp, & turpentine, imp. in preparation of dressings in medic., varnishes, printing inks, soap, polish, sealing-wax, fireworks, plastics etc. etc.; *P. parviflora* Siebold & Zucc. (S & C Japan) – typical sp. in Jap. landscape, timber for building, shingles, sculptures; *P. patula* Schldl. & Cham. (Mex. to Nicaragua) – widely planted in trop.; *P. pinaster* Aiton (maritime p., pinaster, Medit.) – much grown in Les Landes, imp. source of turpentine ('French t.'), natur. in S GB; *P. pinea* L. (stone or umbrella p., N Medit.) – cult. for ed. seeds (pine nuts, pignons, pignoli) much used in salads, found in Roman middens in GB where it cannot be grown, sugar-coated like almonds at Easter in Portugal, cones used as altar fuel by R.; *P. ponderosa* Douglas ex Lawson & P. Lawson (ponderosa p., W & S N Am.) – valuable timber; *P. pungens* Lamb. (E N Am.) – cones retained 15–20 yrs; *P. radiata* D. Don (Monterey or radiata p., SW N Am.) – almost extinguished by disease in Calif. but forming biggest cult. forests in world (NZ) & planted in Medit. climates for timber, 1% of all seedlings with truly dichotomizing apices; *P. resinosa* Aiton (red p., NE N Am.) – timber & tanbark; *P. roxburghii* Sarg. (*P. longifolia*, chir, Himalaya) – turpentine, charcoal for Chinese fireworks; *P. sabiniana* Douglas (digger p., California) – seeds ed.; *P. strobus* L. (white or Weymouth p., E N Am.) – imp. timber tree, form. for ship-masts supplied from New England, so lack of supply in American War of Independence a factor in conflict as GB ship-masts weaker; *P. sylvestris* L. (Scots pine, S. or Norway fir, Euras.) – relict short-leaved trees in old Caledonian forests of Scotland (ssp. *scotica* (P. Schott) E. Warb. (var. *s.*)), trees of lowlands being long-leaved introductions from Eur., escaped from plantations, a tree producing 1000m pollen grains a yr, one of hardest timbers in pines ((Archangel or Baltic) yellow or red deal, Archangel or B. redwood) used for sleepers, telegraph poles & other outside work besides joists & pitprops, for hardboard, pulp, cellophane & plastics, the pine tar antiseptic, the oil from lvs used in men's colognes etc. (pollen contains traces of human sex hormones!); *P. taeda* L. (loblolly p., frankincense or southern p., SE US) – timber & pulp; *P. teocote* Schiede ex Schldl. & Cham. (Aztec p., Mex.) – turpentine etc.; *P.*

torreyana C. Parry ex Carr. (S Calif.) – seeds ed.; *P. wallichiana* A.B. Jackson (? = *P. dicksoniana* Forbes, Bhutan or blue p., Himal.) – 5-leaved

Pinzona C. Martius & Zucc. Dilleniaceae. 1–2 trop. Am.

Piofontia Cuatrec. = *Diplostephium*

Pionocarpus S.F. Blake = *Iostephane*

Piora J. Koster. Compositae (Ast.-Ast.). 1 New Guinea

pipal *Ficus religiosa*

pipe, Dutchman's *Isotrema macrophyllum*; **p. tree** *Syringa vulgaris*; **p.wort** *Eriocaulon aquaticum*

Piper L. Piperaceae. c. 2000 trop. R (NW): BBMNHB 19(1989)117, 20(1990)193. Dioec. or monoec. (NW hermaphr.) shrubs incl. rheophytes, lianes & small trees often with swollen nodes; odour pungent; fr. a 1-seeded drupe with thin mesocarp. Fls of some Costa Rican spp. visited by insects; others there produce food-bodies in ant-hollowed petioles – ants keep off other spp. of liane & bring nutrients absorbed by pl., in *P. cenocladium* C. DC the food bodies produced in response to ant presence (unique) but larvae of *Phyllobaenus* beetles eat ants & can occupy stems when food bodies still produced; many bat-disp. NW. Extracts shown to have antifertility & insecticidal effects. Cult. orn. & many spices esp. *P. nigrum* L. (pepper, Madagascar p., S India, Sri Lanka) – more consumed worldwide than any other spice though having hypoglycaemic effect in humans (piperine an alk. but pungency from piperidine, chavicine & piperittine (amides) confined to & with limited distrib. in genus), ground unripe fr., 'peppercorns', giving black pepper, with pericarp removed giving white, a liane (monoec. or dioec., hermaphr. forms selected) much cult. Mal. up poles or wires, mentioned by Theophrastus (C4 BC) & much of former wealth of Venice & Genoa due to its trade. Other imp. spp. at least locally incl. *P. aduncum* L. (S Am.) – fr. ed. in Puerto Rico, introd. C19 to W Mal. where a consp. roadside weedy shrub; *P. angustifolium* Lam. (matico, trop. S Am.) – treatment of skin complaints in Ecuador; *P. betle* L. (betel p., cultigen, orig. ? Indomal.) – dioec., widely cult. (since at least 'Hoabinhian' culture 8000–3000 BP) for fresh lvs made up into a packet or plug containing betel nut & slaked lime as masticatory, roots etc. locally medic. (antiseptic essential oils); *P. clusii* C. DC (W Afr. black p., Afr. cubebs) – used like *P. cubeba* L.f. (cubebs, Java p., Indonesia) – cult. in E & WI for dried unripe fr. used medic. & to flavour cigarettes; *P. guineense* Schum. & Thonn. (Guinea c., Ashanti or Benin p., trop. Afr.) – used similarly; *P. longum* L. ((Indian) long p., trop. E Himal.) – seasoning & medic., seeds fed to rats induce infertility; *P. methysticum* Forster f. (kava, kawa p., yang(g)ona, ? cultigen derived from *P. wichmannii* C. DC, New Guinea to Vanuatu) – dioec. shrub with roots a source of narcotic sedative drink (effect due to a lactone) form. prepared by children & women chewing them & spitting out pulp to be diluted & fermented, cult. esp. Fiji & W Pacific

Piperaceae Giseke. Dicots – Magnoliidae – Piperales. 8/3000 trop. Shrubs, s.t. lianes or epiphytes, herbs or small trees; aromatic, usu. with spherical ethereal oil cells, often with alks; vasc. bundles usu. scattered like those of monocots but with intrafascicular cambium, the outermost often becoming continuous by cambial growth. Lvs simple, usu. spirally arr., petiolate. Fls v. small, unisexual or not, in axils of small peltate bracts on dense fleshy spikes; P 0, A 2–6 but often 3 + 3, the filaments often free, the anthers with longit. slits (1 in *Peperomia*, 2 in *Piper*), G̲ 1 (*Peperomia*), (3) or (4) (*Piper*), 1-loc. with 1 orthotropous, ± basal, erect bitegmic (unitegmic in *Peperomia*) ovule. Fr. a berry or drupe; seed 1, endotegmic, with scanty endosperm & copious starchy perisperm, the embryo minute & hardly differentiated when seed ripe. 2n = 8–c. 128 (x = ?11 (*Peperomia*) & ?12 (*Piper*))

Genera: *Arctottonia, Macropiper, Manekia, Peperomia, Piper, Sarcorhachis, Trianaeopiper, Zippelia*

Piper & *Peperomia* are huge genera, though the no. of spp. is prob. far fewer than the no. of binomials in use. The second differs in A 2, G 1 & is s.t. put with its allies in a separate fam. Spices & stimulants (*Macropiper, Piper*) incl. some jaborandi, house-pls (*Peperomia*)

Piperanthera C. DC = *Peperomia*

Piperia Rydb. (~ *Habenaria*). Orchidaceae (IV 2). 4 N Am. R: BJLS 75(1978)245

Pippenalia McVaugh. Compositae (Sen.-Tuss.). 1 Mexico

Piptadenia Benth. Leguminosae (II 3). 15 trop. Am. Some timber etc., see also *Anadenanthera, Parapiptadenia* & *Piptadeniastrum*

Piptadeniastrum Brenan. Leguminosae (II 3). 1 C Afr. forests: *P. africanum* (Hook.f.) Brenan (agboin, dabéma, daboma, dahoma, ekhimi) – trade timber used for planking

Piptadeniopsis Burkart. Leguminosae (II 3). 1 Paraguay

Piptanthus Sweet. Leguminosae (III 30). 2 Himal. R: Brittonia 32(1980)281. Alks. *P. nepalensis* (Hook.) Sweet (*P. laburnifolius*) – cult. orn. sappy shrub with yellow fls

Piptatherum Pal. = *Oryzopsis*

Piptocalyx Oliver ex Benth. = *Trimenia*

Piptocarpha R. Br. Compositae (Vern.-Pip.). 45 trop. Am.

Piptochaetium J. Presl. Gramineae (16). 30 US to Arg. Steppes

Piptocoma Cass. Compositae (Vern.-Pip.). 3 WI. R: Rhodora 83(1981)77. Shrubs

Piptolepis Schultz-Bip. Compositae (Vern.-Lych.). 9 Brazil

Piptophyllum C. Hubb. Gramineae (25). 1 Angola

Piptoptera Bunge. Chenopodiaceae (III 3). 1 C As.: *P. turkestanica* Bunge

Piptospatha N.E. Br. Araceae (VI 1). 10 W Mal.

Piptostachya (C. Hubb.) J. Phipps = *Zonotriche*

Piptostigma Oliver. Annonaceae. 14 trop. Afr.

Piptothrix A. Gray. Compositae (Eup.-Oxy.). 5 Mex., Guatemala. R: MSBMBG 22(1987)438

Pipturus Wedd. Urticaceae (III). 30 Masc. to Polynesia. Some bark used as fibre or bark-cloth in Pacific

pipul *Ficus religiosa*

Piqueria Cav. Compositae (Eup.-Agar.). 8 trop. R: MSBMBG 22(1987)168. *P. trinervia* Cav. (tabardillo, Mex., C Am., Haiti) – pioneering allelopathic (monoterpenoids) weed, cult. orn. for winter fls ('stevia')

Piqueriella R. King & H. Robinson (~ *Eupatorium*). Compositae (Eup.-Ager.). 1 Brazil. R: MSBMBG 22(1987)166

Piqueriopsis R. King (~ *Microspermum*). Compositae (Eup.-Agar.). 1 Mex. R: MSBMBG 22(1987)166

piquí *Caryocar brasiliense, C. villosum*

Piranhea Baillon. Euphorbiaceae. 1–2 NE trop. S Am.

Piratinera Aublet = *Brosimum*

Piresia Swallen. Gramineae (4). c. 7 Venez. & Trinidad to Brazil. R: Brittonia 34(1982)203. Fern-like; infls borne in leaf-litter layer

Piresodendron Aubrév. ex Le Thomas & Leroy = *Pouteria*

Piriadacus Pichon = *Cuspidaria*

Pirinia Král. Caryophyllaceae (I 1). 1 SW Bulgaria: *P. koenigii* Král

Piriqueta Aublet. Turneraceae. 21 trop. Am. & Afr.

Pironneava Gaudich. ex Regel = *Hohenbergia*

Piscaria Piper = *Croton*

Piscidia L. Leguminosae (III 6). 7 C Am. to WI & Florida. *P. piscipula* (L.) Sarg. (*P. erythrina*, Jamaica dogwood, Mex., Florida, C Am., WI) – hard timber for boat-building etc., charcoal, roots yield a fish poison

Pisonia L. Nyctaginaceae (VI). 40 trop. & warm esp. Am. Trees & shrubs with uni-or bisexual fls. Glandular anthocarps ensnare or disable birds which may die as a result. *P. grandis* R. Br. (Madag.–Polynesia, esp. atolls & islets with guano) figures in wayang (shadow-puppet) shows in Java (where v. rare) & form. coll. only for coronation of Sultan of Solo, some cvs cult. ed. lvs (lettuce tree) incl. some albino ones – Moluccan cabbage, 'Alba'

Pisoniella (Heimerl) Standley. Nyctaginaceae (III). 1 warm Am.: *P. arborescens* (Lag. & Rodrigues) Standley

pistachio *Pistacia vera*; **p. galls** *P. terebinthus* etc.

Pistacia L. Anacardiaceae (IV). 9 Medit. (Eur. 3), As. & Mal., S US & C Am. R: PJBJ 5(1952)187. Resins, oils & ed. seeds: *P. atlantica* Desf. (*P. mutica*, Macaronesia to Pakistan) – tan-galls, gum (Bombay mastic) used in varnishes, ripe fr. coll. Near E in Neolithic & Bronze Age & still in some markets; *P. chinensis* Bunge (Afghanistan to China & Philippines) – young shoots a Chinese veg.; *P. lentiscus* L. (lentisco, lentisk, mastic, Medit.) – Sardinian warbler only found near fruiting shrubs of this sp., resin used for chewing since time of Theophrastus, also used for varnishes esp. for oil-pictures, in quelling halitosis & as filler for caries, ingredient in ouzo, 160–170 t per annum from male pls on Chios; *P. terebinthus* L. (terebinth, Medit.) – source of tan-galls & form. (Chian) turpentine; *P. vera* L. (pistachio, Iran to C As., much cult. in Med. & US) – domesticated C As., ed. seeds used as dessert, in icecream & confectionery

Pistaciaceae C. Martius ex Perleb. See Anacardiaceae (IV)

Pistaciovitex Kuntze = *Vitex*

Pistia L. Araceae (IX). 1 trop. (? orig. Lake Victoria): *P. stratiotes* L. (water lettuce) – free-floating stoloniferous rosette-pl. with feathery roots & lvs with 'sleep-movements' (moving together at night) & covered with short depressed hairs giving a water-repellent surface; larva of a mosquito sp. lives anchored to root & obtains air by piercing roots with tail; known since time of Pliny (AD 77) & often a serious pest in reservoirs etc., the fragmenting colonies soon occupying open water, used locally for pig-fodder, against warts in Amazonia, etc. Genus linking the neotenous Lemnaceae to Araceae

Pistorinia DC (~ *Cotyledon*). Crassulaceae. 2 W Medit. (2 Eur.)

Pisum L. Leguminosae (III 20). 2 (excl. *Vavilovia*) Medit. (Eur. 1), W As. *P. sativum* L. (garden pea, selected from 'steppe' type (*P. humile*), Medit. to Iran) – self-poll. in bud, cult. since c. 7000 BC (as long as wheat & barley, var. *pumilio* Meikle one of extant crops of Neolithic agriculture), seeds used as veg., fresh, tinned, frozen or dried or as flour; var. *arvense* (L.) Poiret (field, dun, grey, mutter or partridge p., peluskins) – seeds usu. for animal feed, pls as fodder & green manure; var. *macrocarpon* Ser. (sugar p., mange-tout, snow p.) – pods without fibrous inner layer, the whole unripe fr. ed., shoots ed. Aus. Peas used by Mendel to demonstrate the laws of inheritance, his 'wrinkled' peas being due to a lack of one form of 'starch-branching' enzyme

pita fibre *Chevaliera magdalenae*

pitanga *Eugenia uniflora*

Pitardia Battand. ex Pitard. Labiatae (VIII 2). 2 NW Afr.

Pitavia Molina. Rutaceae. 1 Chile

pitaya ed. cactus fr. esp. of spp. of *Hylocereus*, *Stenocereus* & *Opuntia*

Pitcairnia L'Hérit. Bromeliaceae (1). (Excl. *Pepinia*) c. 320 trop. Am., W Afr. (1: *P. feliciana* A. Chev., only non-Am. bromeliad). R: FN 14(1974)244, 605, AMBG 73(1986)700. Mostly terr., some epiphytic or saxicolous. Some cult. orn. usu. with spiny lvs

pitch, Burgundy *Picea abies*; **p. pine** *Pinus palustris* & other spp.

pitcher plants See *Cephalotus*, *Darlingtonia*, *Heliamphora*, *Nepenthes*, *Sarracenia*

pitchfork *Bidens* spp.

pith plant *Aeschynomene* spp.

Pithecellobium C. Martius. Leguminosae (II 5). 37 trop. & warm Am. Some water-disp. in Amazon esp. *P. inaequale* (Kunth) Benth. with air-filled cavity between cotyledons. *P. dulce* (Roxb.) Benth. (Madras thorn, Manila tamarind, C Am.) – introd. As. as shade-tree & thorny hedge, fr. ed. (pulp made into a 'lemonade'), seed-oil for soap etc., firewood crop, tanbark; *P. unguis-cati* (L.) Benth. (black Jessie, S Florida to N S Am.) – spiny shrub or tree with ed. fr.

Pithecoctenium C. Martius ex Meissner. Bignoniaceae. 3 trop. Am. Local medic. & cult. orn. lianes (monkey comb) with white fls esp. *P. crucigerum* (L.) A. Gentry (*P. echinatum*)

Pithecoseris C. Martius ex DC. Compositae (Vern. - Lych.). 1 N Brazil. *Echinops*-like

Pithocarpa Lindley. Compositae (Gnap.-Ang.). 4 SW Aus. R: OB 104(1991)107

pitomba *Eugenia luschnathiana*

pitpit *Saccharum edule* (lowland New Guinea), *Setaria palmifolia* (upland)

Pitraea Turcz. Verbenaceae. 1 S Am.

Pittierothamnus Steyerm. = *Amphidasya*

Pittocaulon H. Robinson & Brettell (~ *Senecio*). Compositae (Sen. -Tuss.). 5 Mex. & C Am. R: Phytol. 26(1973)451. Trees & shrubs incl. *P. praecox* (Cav.) H. Robinson & Brettell (*S. p.*, Mex.) – cult. orn.

Pittoniotis Griseb. (~ *Antirhea*). Rubiaceae (III 3). Spp.? (NW *Antirhea* spp.)

Pittosporaceae R. Br. Dicots – Rosidae – Rosales. 9/200 trop. & warm OW, esp. Aus. (6 endemic genera, 2 others ext. to Mal., 1 OW trop. – *Pittosporum*). Trees, shrubs, lianes, s.t. spiny; schizogenous secretory canals present. Lvs simple, subentire, leathery, spirally arr.; stipules 0. Fls usu. bisexual & reg., solit. or in corymbs or thyrses, each with 2 bracteoles; K 5 sometimes basally connate, imbricate, C 5 usu. basally connate forming ± distinct tube with imbricate lobes, A 5 alt. with C, s.t. weakly connate basally, anthers with longit. slits or term. pores, $\underline{G}$ (2(3–5)), usu. 1-loc. with simple style & several to usu. ± ∞ anatropous to ± campylotropous, unitegmic ovules on each usu. parietal placenta. Fr. a loculicidal capsule or berry; seeds often in viscid pulp (aril; wing in *Hymenosporum*) & with 2–5 cotyledons at base of copious oily proteinaceous endosperm. x = 12. R: Nuytsia 5(1984)405

Genera: *Bentleya*, *Billardiera*, *Bursaria*, *Cheiranthera*, *Citriobatus*, *Hymenosporum*, *Pittosporum*, *Pronaya*, *Sollya*

Cult. orn. esp. *Pittosporum*

Pittosporopsis Craib. Icacinaceae. 1 SE As.

Pittosporum Banks ex Sol. Pittosporaceae. c. 150 trop. & S Afr. (Afr. & Arabia 5, R: KB 42(1987)319) to NZ (R (S & E As.): JAA 32(1951)263, 303; (Aus. & NZ): AMBG 43(1956)87)

& Pacific (R: Allertonia 1(1977)73). Evergreen trees & shrubs much cult. in trop. & Medit. climates, esp. as street-trees; some local timbers etc.: *P. crassifolium* Banks & Sol. ex Cunn. (karo, NZ natur. Scilly Is.), *P. eugenioides* Cunn. (lemonwood, NZ), *P. napaulense* (DC) Rehder & Wilson (*P. verticillatum* Wallich, India) – local med., *P. resiniferum* Hemsley (Borneo & Philippines) – seed-oil (c. 30% terpenes) for illumination, fuel-oil & medic.; *P. stenopetalum* Baker (*P. verticillatum* Bojer, Madag.) – evil-smelling sap said to ward off spirits; *P. tenuifolium* Gaertner (NZ) – cult. orn. with many cvs, foliage pl. for fl. arranging; *P. tobira* (Thunb.) Aiton f. (tobira, China, Japan) – cult. esp. near sea; *P. undulatum* Vent. (E Aus.) – oil extracted from fragrant fls, wood used in golf-clubs (cheesewood), aggressive weed-tree in disturbed forests of Jamaica & Aus.

Pituranthos = *Deverra*

pituri *Duboisia hopwoodii*

Pitygentias Gilg = *Gentianella*

Pityopsis Nutt. (~ *Heterotheca*). Compositae (Ast.-Sol.). 8 E US to C Am. Some cult. orn. rock-pls

Pityopus Small. Ericaceae (Monotropaceae). 1 W US

Pityphyllum Schltr. Orchidaceae (V 10). 4 Colombia

Pityranthe Thwaites = *Diplodiscus*

Pityrodia R. Br. Labiatae (III, Dicrastylidaceae). 41 Aus. R: JABG 2(1979)1

Pityrogramma Link. Pteridaceae (III). 16 trop. Am. (Andes 10), SW & W US (1), Afr. & Madag. (4–5). R: CGH 189(1962)52. Terr. ferns with white or yellow waxy powder on abaxial surface of fronds, some cult. orn. esp.: *P. calomelanos* (L.) Link (trop. Am., natur. pantrop.) – first colonizer of erupted volcanoes in Mex., spore-prints used as face-paint in New Guinea & *P. triangularis* (Kaulf.) Maxon (W N Am.) – black petioles used in Indians' basketware

Placea Miers. Amaryllidaceae (Liliaceae s.l.). 5 Chile

Placocarpa Hook.f. Rubiaceae (III 4). 1 Mexico

Placodiscus Radlk. Sapindaceae. 15 trop. Afr. (esp. W)

Placolobium Miq. See *Ormosia*

Placopoda Balf.f. Rubiaceae (IV 1). 1 Socotra

Placospermum C. White & Francis. Proteaceae. 1 Queensland: *P. coriaceum* C. White & Francis

Pladaroxylon (Endl.) Hook.f. (~ *Senecio*). Compositae (Sen.-Sen.). 1 St Helena: *P. leucadendron* (DC) Hook.f. – pachycaul tree with reg. pseudo-dichot. candelabriform branching, allies in S Am. & Australasia

Plaesianthera (C.B. Clarke) Liv. = *Hygrophila*

Plagiantha Renvoize (~ *Panicum*). Gramineae (34b). 1 Brazil (Bahia)

Plagianthus Forster & Forster f. Malvaceae. 2–3 NZ. Cult. orn. trees & shrubs incl. *P. regius* (Poit.) Hochr. (*P. betulinus*) – bark a fibre-source & raffia subs.

Plagiobasis Schrenk. Compositae (Card.-Cent.). 1 C As.

Plagiobothrys Fischer & C. Meyer. Boraginaceae. c. 70 W Am. (Calif. 39, 18 endemic), E As. (1), Aus. (4). R: JAA suppl. 1(1991)95

Plagiocarpus Benth. Leguminosae (III 25). 1 N Aus.: *P. axillaris* Benth. R: Muelleria 7(1992)421

Plagioceltis Mildbr. ex Baehni = *Ampelocera*

Plagiocheilus Arn. ex DC. Compositae (Ast.-Ast.). 7 S Am. R: Phytol. 26(1973)159

Plagiochloa Adamson & Sprague = *Tribolium*

Plagiogyria (Kunze) Mett. Plagiogyriaceae. c. 30 E As. to New Guinea, Am. (6). Forest ferns on mountain ridges

Plagiogyriaceae Bower. Filicopsida. 1/c. 30 E As., Am. Terr. ferns with dictyostelic rhiz. without indumentum (v. rare in ferns). Lvs pinnate, when young bathed in mucilage through which pneumatathodes protrude. Sporangia on enlarged distal part of veins of very narrow pinnae so fertile fronds appearing acrostichoid, annulus complete, oblique
Genus: *Plagiogyria*

Plagiolirion Bak. = *Urceolina*

Plagiolophus Greenman. Compositae (Helia.-Verb.). 1 Yucatán (Mex.)

Plagiopetalum Rehder = *Anerinocleistus*

Plagiopteraceae Airy Shaw (~ Flacourtiaceae). Dicots – Dilleniidae – Violales. 1/1 E & SE As. Liane recently separated from Flacourtiaceae; pollen reminiscent of Elaeocarpaceae
Genus: *Plagiopteron*

Plagiopteron Griffith. Plagiopteraceae (Flacourtiaceae s.l.). 1 China, S Burma & Thailand. R: ABY 12(1990)126

Plagiorhegma Maxim. = *Jeffersonia*
Plagioscyphus Radlk. Sapindaceae. 2 Madagascar
Plagiosetum Benth. Gramineae (34b). 1 Aus.: *P. refractum* (F. Muell.) Benth.
Plagiosiphon Harms. Leguminosae (I 4). 5 Guineo-Congolian rain forest
Plagiospermum Oliver = *Prinsepia*
Plagiostachys Ridley. Zingiberaceae. 10 W Mal. (Borneo 7)
Plagiostyles Pierre. Euphorbiaceae. 1 trop. Afr.
Plagiotheca Chiov. = *Isoglossa*
Plagius L'Hérit. ex DC (~ *Chrysanthemum*). Compositae (Anth.- Leuc.). 1 Balearics, Tunisia, Morocco: *P. flosculosus* (L.) Alava & Heyw. – cult. orn.
Plakothira Florence. Loasaceae. 1 Marquesas: *P. frutescens* Florence – woody. R: BMNHN 4e,7(1985)239
Planaltoa Taubert. Compositae (Eup.-Alom.). 2 Brazil. R: MSBMBG 22(1987)260
Planchonella Pierre = *Pouteria*
Planchonia Blume. Lecythidaceae (Barringtoniaceae). 5 Andamans to N Aus. *P. careya* (F. Muell.) Knuth (trop. Aus., New Guinea, cockey apple) – ed.; lvs of *P. valida* (Blume) Blume (W Mal.) turn red before falling
plane *Platanus* spp., (of Isaiah) *Viburnum tinus*; **American p.** *P. occidentalis*; **common** or **London p.** *P.* × *hispanica* 'Acerifolia'; **oriental p.** *P. orientalis*
planer tree *Planera aquatica*
Planea Karis (~ *Metalasia*). Compositae (Gnap.-Rel.). 1 Cape: *P. schlechteri* (L. Bolus) Karis. R: OB 104(1991)73
Planera J. Gmelin. Ulmaceae (I). 1 SE US: *P. aquatica* (Walt.) J. Gmelin (water elm, planer tree)
Planichloa Simon = *Ectrosia*
Planodes E. Greene = *Sibara*
Planotia Munro = *Neurolepis*
Plantaginaceae Juss. Dicots – Asteridae – Plantaginales. 3/275 cosmop. Herbs or less often shrubs, s.t. pachycaul, s.t. with alks &/or medullary vasc. bundles. Lvs simple, usu. spirally arr., venation parallel; stipules 0. Fls usu. small & reg. & bisexual (*Littorella* (poss. not distinct) monoec.; *Bougueria* gynomonoec.), usu. in pedunculate bracteate heads or spikes without bracteoles & wind-poll.; K ((3 or) 4), C ((3 or) 4), membranous, imbricate, A (1 or 2 – *Bougueria*, 3) 4 alt. with C & attached to it, anthers versatile with longit. slits, <u>G</u> (2) usu. with 2-lobed stigma, 2-loc. with 1–40 anatropous to hemitropous unitegmic ovules on axile placenta in each loc. (*Plantago*; 1-loc. with 1 basal ovule in others). Fr. a circumscissile capsule or (*Bougueria* & *Littorella*) achene in persistent K; seeds with usu. straight embryo in copious translucent endosperm. x = 4–12+
Genera: *Bougueria, Littorella, Plantago*
App. quite close to Scrophulariaceae
Laxatives & weeds (*Plantago*)
Plantago L. Plantaginaceae. c. 270 cosmop. (Eur. 35, Aus. 24 native). Plantains, ribwort. R: Pfl. IV, 269(1937)39, 440. Herbs & few shrubs e.g. *P. sempervirens* Crantz (*P. cynops*, C & S Eur.) & pachycaul shrublets in e.g. Hawaii but many weeds esp. *P. lanceolata* L. (ribwort p., Euras., natur. temp.) – iridoids sequestered by some butterflies, *P. major* L. (common p. Euras., widely natur.) – fr. & lvs used for influenza & colds in modern Chinese herbalism, 'Rosularis' a cult. orn. teratologous mutant with head of lvs instead of fls (rose plantain), & *P. media* L. (hoary p., Euras.) – ± insect-poll., as well as herbs of salt-marsh (*P. maritima* L. – sea p.) & sea-cliffs (*P. coronopus* L. – buck's-horn p.) – form. cult. as salad veg., in Eur.; alks. Seeds of some spp. mucilaginous when wetted – efficient laxatives esp. *P. afra* L. ('*P. psyllium*', psyllium, Medit. to India); *P. ovata* Forssk. (isphagul, Medit. to India) – used in treatment of dysentery, bowel regulation etc. (isphagul seeds)
plantain *Plantago* spp., (trop.) *Musa* × *paradisiaca* etc.; **buck's-horn p.** *P. coronopus*; **common p.** *P. major*; **hoary p.** *P. media*; **p. lily** *Hosta* spp.; **ribwort p.** *P. lanceolata*; **rose p.** *P. major* 'Rosularis'; **sea p.** *P. maritima*; **water p.** *Alisma plantago-aquatica*
Plarodrigoa Looser = *Cristaria*
Platanaceae Lestib. Dicots – Hamamelidae – Hamamelidales. 1/c. 8 N hemisph. Decid., monoec. trees with branched hairs & bark consp. scaling irreg. Lvs simple, palmately-lobed & -veined (toothed & pinnately veined in 1 sp.); stipules usu. conspic. & united around twig; petiole base encircling axillary bud. Fls small, reg. in dense unisexual wind-poll. heads in pendulous strings; K 3 or 4(–7) without vasc. tissue, s.t. basally united, C 3 or 4(–7) minute, alt. with K in males, 0 in females, A as many as & opp. K, anthers ± sessile with longit. slits & connective with prolonged apical appendage (3 or 4 staminodes often in females), <u>G</u> (3–)5–8(9) in 2 or 3 whorls, distally unclosed (s.t. vestigial G in males) with

1(2) pendulous, orthotropous or slightly hemitropous, bitegmic ovule. Frs hairy (accrescent P) achenes or nutlets in globose heads; seeds wind-disp., with slender straight embryo in scanty oily, proteinaceous endosperm. x = 7

Genus: *Platanus* (planes) – timber

Heads app. derived from main branches of a panicle

Platanocephalus Crantz = *Nauclea*

Platanthera Rich. (~ *Habenaria*). Orchidaceae (IV 2). 40 N hemisph. (Eur. 5, E As. 50). Butterfly orchids. *P. obtusata* (Pursh) Lindley (N Am.) – poll. by mosquitoes in Alaska

Platanus L. Platanaceae. c. 8 N hemisph. (planes): 1 SE As. – *P. kerrii* Gagnepain, lvs unlobed, venation pinnate (also seen in first lvs of spring shoots in other spp.), female fls in heads of 10–12 (primitive wood features); 1 SE Eur. to N Iran – *P. orientalis* L. (oriental p.); 1 NE N Am. – *P. occidentalis*. L. (Am. p., buttonwood); SW N Am. R: AJB 60(1973)678. Some timbers for furniture & pulp but esp. cult. orn. esp. London p. (*P.* × *hispanica* Miller ex Münchh. 'Acerifolia' (*P.* × *acerifolia*, *P. hybrida*, allegedly an hybrid of *P. orientalis* & *P. occidentalis* raised at Oxford Botanic Garden from seed received from Montpellier, but poss. only a cv. of *P. orientalis* from Turkey) – much planted in towns (most common tree in City of Westminster & many London squares e.g. Berkeley Square (1789), Cavendish Square (1810)) as withstands pollution though hairs from lvs & fr. can cause bronchial problems, veneer sold as 'lacewood', seeds fertile

Platea Blume. Icacinaceae. 5 Mal.

Plateilema (A. Gray) Cockerell. Compositae (Hele.-Gai.). 1 Mexico

Plathymenia Benth. Leguminosae (II 3). 4 Brazil, Arg. Yellowish timber prized for cabinet-work & parquet flooring esp. *P. reticulata* Benth. (vinhatico, Brazil)

Platonia C. Martius. Guttiferae (III). 1–2 Guyana, Brazil. *P. esculenta* (Arruda) Rickett & Stafleu (*P. insignis*, bacury, bakury) – parrot-poll.?, large yellow fr. ed., seeds source of oil for candles & soap, timber good

Platostoma Pal. Labiatae (VIII 4b). 5 OW trop.

Platyadenia B.L. Burtt. Gesneriaceae. 1 Sarawak

Platyaechmea (Bak.) L.B. Sm. & Kress (~ *Aechmea*). Bromeliaceae (3). 18 trop. Am. R: Phytol. 66(1989)73. Cult. orn. esp. room-pls e.g. *P. fasciata* (Lindley) L.B. Sm. & Kress (*A. f.*, urn pl., Brazil)

Platycalyx N.E. Br. (~ *Eremia*). Ericaceae. 1 S Cape

Platycapnos (DC) Bernh. Fumariaceae (I 2). 3 Macaronesia, W Medit. esp. Spain (Eur. 3). R: OB 88(1986)39

Platycarpha Less. Compositae (Arct.). 3 S Afr.

Platycarpum Bonpl. Rubiaceae (I 4; Henriqueziaceae). 10 N trop. S Am.

Platycarya Siebold & Zucc. Juglandaceae. 3 C & E China, Korea, Japan, Vietnam. R: AMBG 65(1978)1069. *P. strobilacea* Siebold & Zucc. (C & S China) – cone-like infrs of winged nutlets in axils of stiff lanceolate bracts; cult. orn., bark & fr. sources of black dye used in E for textiles, nets, etc.; *P. simplicifolia* G.-R. Long (China) – lvs simple!

Platycaulos Linder. Restionaceae. 8 S & SW Cape. R: Bothalia 15(1985)435

Platycelyphium Harms. Leguminosae (III 2). 1 drier E & NE Afr.

Platycentrum Naudin = *Leandra*

Platycerium Desv. Polypodiaceae (I). 17 trop. esp. OW (Afr. & Madag. 6, As. 9, Andes of Peru & Bolivia 1). Elk's horn or stag's horn fern. Ferns of rocks, or epiphytes with short rhizome & dimorphic fronds: sterile ± entire, shield-like, marcescent fronds clasping support & in some spp. e.g. *P. grande* (Fée) C. Presl (Mal. to Aus.) & *P. superbum* De Joncheere & Hennipman (E Aus.) accumulating humus within where roots ramify, while in *P. bifurcatum* (Cav.) C. Chr. (SE As. to Aus. & Pacific), commonly cult. sp., no such accumulation, the only humus being supplied by decaying host bark & old sterile fronds; fertile fronds pendulous & much branched, covered with stellate hairs & bearing sporangia in clusters but not sori abaxially. All Aus. spp. protected; some cult. orn. housepls incl. *P. angolense* Welw. ex Baker (trop. Afr.) with hygroscopic movements of marcescent lvs protecting growing-points during drought

Platychaete Boiss. = *Pulicaria*

Platycladus Spach (~ *Thuja*). Cupressaceae. 1 China & Korea: *P. orientalis* (L.) Franco (*Biota o.*, *T. o.*, Chinese arbor-vitae) – distinct from *T.* in female cone-scales fleshy when young & in wingless seeds, cult. orn. with many cvs incl. dwarf ones

Platycodon A. DC. Campanulaceae. 1 NE As.: *P. grandiflorus* (Jacq.) A. DC (balloon flower, Chinese bellflower) like *Campanula* but with apically dehiscent 5-valved capsule, cult. orn. esp. dwarf 'Apoyama' (Japan), used for influenza & colds in modern Chinese herbalism

Platycoryne Reichb.f. (~ *Habenaria*). Orchidaceae (IV 2). 17 trop. Afr., Madag.
Platycraspedum O. Schulz. Cruciferae. 1 E Tibet
Platycrater Siebold & Zucc. Hydrangeaceae. 1 Japan: *P. arguta* Sieb. & Zucc. – differing from *Hydrangea* only in greater no. of stamens
Platycyamus Benth. Leguminosae (III 6). 2 Brazil & Peru
Platydesma H. Mann. Rutaceae. 4 Hawaii. Alks
Platyglottis L.O. Williams. Orchidaceae (V 13). 1 Panamá
Platygyna Merc. (~ *Tragia*). Euphorbiaceae. 7 Cuba
Platygyria Ching & S.K. Wu = *Neocheiropteris*
Platykeleba N.E. Br. Asclepiadaceae (III 1). 1 Madag.: *P. insignis* N.E. Br.
Platylepis A. Rich. Orchidaceae (III 3). 10 trop. Afr. to Seychelles
Platylobium Sm. Leguminosae (III 25). 4 E Aus. R: Muelleria 5(1983)127
Platylophus D. Don. Cunoniaceae. 1 Cape: *P. trifoliatus* (L.f.) D. Don – white timber for furniture, picture-framing etc.
Platymiscium Vogel. Leguminosae (III 4). 20 trop. Am. Some with ant-infested stem domatia. Timber for kitchen-knife handles, brush-backs, turnery etc. esp. *P. pinnatum* (Jacq.) Dugand (roble, Colombia)
Platymitra Boerl. Annonaceae. 2 W Mal. R: Blumea 33(1988)471
Platypholis Maxim. Orobanchaceae. 1 Bonin Is. (Japan)
Platypodanthera R. King & H. Robinson (~ *Eupatorium*). Compositae (Eup.-Gyp.). 1 E Brazil. R: MSBMBG 22(1987)106
Platypodium Vogel. Leguminosae (III 4). 1–2 trop. Am.
Platypterocarpus Dunkley & Brenan. Celastraceae. 1 trop. E Afr. (W Usambaras): *P. tanganyikensis* Dunkley & Brenan – winged dehiscent fr. (only Afr. C. thus), ? extinct
Platyrhiza Barb. Rodr. Orchidaceae (V 10). 1 Brazil
Platyrhodon Hurst = *Rosa*
Platysace Bunge. Umbelliferae (I 1a). 25 Aus.
Platyschkuhria (A. Gray) Rydb. Compositae (Hele.-Cha.). 1 SW US. R: Brittonia 23(1971)269
Platysepalum Welw. ex Baker. Leguminosae (III 6). 12 trop. Afr.
Platyspermation Guillaumin. Grossulariaceae (Escalloniaceae). 1 New Caled. $\overline{G}$ or semi-inferior (form. in Rutaceae)
Platystele Schltr. Orchidaceae (V 13). 80 trop. Am. R: MSB 38(1990)1, 39(1991)147, 44(1992)112. *P. jungermannioides* (Schltr.) Garay (C Am.) – fls when less than 1 cm tall. Some cult. orn.
Platystemma Wallich. Gesneriaceae. 1 Himal.: *P. violoides* Wallich
Platystemon Benth. Papaveraceae (III). 1 W N Am.: *P. californicus* Benth. – cult. orn. with mainly opp. lvs & alks. R: UKSB 47(1967)25
Platystigma Benth. = *Meconella*
Platytaenia Kuhn = *Taenitis*
Platytaenia Nevski & Vved. = *Semenovia*
Platytheca Steetz. Tremandraceae. 2 SW Aus.
Platythelys Garay. Orchidaceae (III 3). 8 warm Am. R: Bradea 2(1977)196
Platythyra N.E. Br. = *Aptenia*
Platytinospora (Engl.) Diels. Menispermaceae (III). 1 W trop. Afr.: *P. buchholzii* (Engl.) Diels
Platyzoma R. Br. Pteridaceae (I, Platyzomataceae). 1 N & NE Aus.: *P. microphyllum* R. Br. – terr. xerophyte,juveniles with filiform fronds also in zones on adults; heterospory unique in Filicopsida with megaspores twice size of microspores, the latter germ. as filamentous prothalli bearing antheridia, the former giving spathulate prothalli with archegonia when young & antheridia later if fert. has not occurred (cf. *Equisetum*)
Platyzomataceae Nakai = Pteridaceae (I)
Plazia Ruíz & Pavón. Compositae (Mut.-Mut.). Incl. *Harthamus*, 3 S Andes, Arg. Shrubs
Plecosorus Fée = *Polystichum*
Plecospermum Trécul = *Maclura*
Plecostachys Hillard & B.L. Burtt (~ *Helichrysum*). Compositae (Gnap.-Gnap.). 2 SW Cape to Natal. R: OB 104(1991)140
Plectaneia Thouars. Apocynaceae. 13 Madag. R: FMad. 169(1976)114. Some rubber
Plectis Cook = *Euterpe*
Plectocephalus D. Don = *Centaurea*
Plectocomia C. Martius ex Blume. Palmae (II 1e). 16 trop. As. to W Mal. R: Kalikasan 10(1981)1. Hapaxanthic rattans of little economic imp.
Plectocomiopsis Becc. Palmae (II 1e). 5 S Burma & Thailand, W Mal. R: KB 37(1982)244. Hapaxanthic dioec. rattans

Plectorrhiza Dockr. Orchidaceae (V 16). 3 E Aus., 1 Lord Howe Is.

Plectrachne Henrard. Gramineae (31a). 16 Aus. Pungent lvs dominate much of the veg. of Aus. ('spinifex')

Plectranthus L'Hérit. Labiatae (VIII 4c). (Incl. *Coleus* & *Solenostemon*) c. 200 trop. & warm OW. Herbs & shrubs, s.t. succ.; ed. tubers & cult. orn. esp.: *P. amboinicus* (Lour.) Sprengel (*Coleus a.*, ? Afr., spread to As.) – flavouring, medic. & shampoo; *P. barbatus* Andr. (*C. b.*, OW trop.) – hedgepl. in Kenya, infusion a lice-remover in N Uganda, diterpene (forskolin) potential drug for hypertension, glaucoma, asthma etc.; *P. esculentus* N.E. Br. (trop. & S Afr.) – ed. tubers; *P. oertendahlii* T.C.E. Fries (Natal) – cult. in hanging baskets; *P. rotundifolius* (Poiret) Sprengel (*C. r.*, *Solenostemon r.*, Hausa potato, OW) – tubers eaten like potatoes; *P. scutellarioides* (L.) R. Br. (*C. blumei*, *S. s.*, coleus, ? New Guinea) – cult. orn. foliage pot-pl., the variegated lvs with purple dominant over green & both over patterned which is expressed only in homozygous condition – some are true chimaeras with green–yellow–green etc. while others with dark-edged yellowish lvs are apparently due to reversible hormonal bleaching initiated from the mesophyll, another dominant char.

Plectrelminthus Raf. Orchidaceae (V 16). 12 trop. W Afr.: *P. caudatus* (Lindley) Summerh. – cult. orn.

Plectritis DC. Valerianaceae. 4 W N Am., 1 Chile. R: CDH 5(1959)119

Plectrocarpa Gillies ex Hook. & Arn. Zygophyllaceae. 3 temp. S Am.

Plectroniella F. Robyns. Rubiaceae (III 2). 2 trop. Afr.

Plectrophora H. Focke. Orchidaceae (V 10). 8 NE S Am., Trinidad

Pleea Michaux. Melanthiaceae (Liliaceae s.l.). 1 SE US

Plegmatolemma Bremek. = *Justicia*

Pleiacanthus (Nutt.) Rydb. = *Lygodesmia*

Pleiadelphia Stapf = *Elymandra*

Pleioblastus Nakai = *Arundinaria*

Pleiocardia E. Greene. Cruciferae. 6 California

Pleiocarpa Benth. Apocynaceae. 3 trop. Afr. Alks

Pleiocarpidia Schumann. Rubiaceae (I 9). 27 W Mal. R: RTBN 37(1940)198

Pleioceras Baillon. Apocynaceae. 3 trop. W Afr.

Pleiochiton Naudin ex A. Gray. Melastomataceae. 7 S Brazil

Pleiococca F. Muell. = *Acronychia*

Pleiocoryne Rauschert (*Polycoryne*). Rubiaceae (II 1). 1 trop. W Afr.

Pleiocraterium Bremek. Rubiaceae (IV 1). 4 S India, Sri Lanka, Sumatra

Pleiogynium Engl. Anacardiaceae (II). 2–3 C Mal. to Pacific. *P. timoriense* (DC) Leenh. (*P. cerasiferum*, *Owenia c.*, Burdekin plum, C Mal. to Pacific) – imp. timber in Tonga where seeds strung into skirt-like garments, cult. orn. street-tree in Afr. etc., fr. used in jam & jellies

Pleiokirkia Capuron. Kirkiaceae (Simaroubaceae s.l.). 1 Madagascar

Pleiomeris A. DC. Myrsinaceae. 1 Macaronesia

Pleione D. Don. Orchidaceae (V 12). 16 Nepal to Taiwan & Thailand. R: P. Cribb & I. Butterfield (1988) *The genus P.* Terr. orchids with corm-like pseudobulbs; cult. orn. almost hardy incl. hybrids esp. *P. bulbocodioides* (Franchet) Rolfe (China) & *P. hookeriana* (Lindley) Rollison (Himal.)

Pleioneura (C. Hubb.) J. Phipps = *Danthoniopsis*

Pleioneura Rech.f. Caryophyllaceae (III 1). 1 C As. to W Himal.

Pleiosepalum Hand.-Mazz. Rosaceae. 1 SW China

Pleiospermium (Engl.) Swingle. Rutaceae. 5 S India & Sri Lanka (1), Sumatra (1), Java (1), Borneo (2). R: W.T. Swingle, *Bot. Citrus* (1943)274

Pleiospilos N.E. Br. Aizoaceae (V). 4 Cape. R: BJ 106(1986)472. Cult. orn. stemless succ. resembling stones & with fls often coconut-scented

Pleiostachya Schumann. Marantaceae. 3 C Am., Ecuador

Pleiostachyopiper Trel. = *Piper*

Pleiostemon Sonder = *Flueggea*

Pleiotaenia J. Coulter & Rose = *Polytaenia*

Pleiotaxis Steetz. Compositae (Mut.-Mut.). 26 trop. Afr. Shrubs & herbs. R: KB 21(1967)180

Plenasium C. Presl = *Osmunda*

Plenckia Reisseck. Celastraceae. 4 S Am.

Pleocarphus D. Don (~ *Jungia*). Compositae (Mut.-Nass.). 1 N Chile. Shrub

Pleocaulus Bremek. = *Strobilanthes*

Pleocnemia C. Presl. Dryopteridaceae (I 3). 19 Burma & Mal. to Fiji. R: KB 29(1974)341

Pleodendron Tieghem. Canellaceae. 1 Hispaniola, Puerto Rico

Pleogyne Miers. Menispermaceae (I). 1 trop. E Aus.: *P. australis* Benth. – alks. R: KB 30(1975)88

Pleomele Salisb. = *Dracaena*

Pleonotoma Miers. Bignoniaceae. 14 trop. Am.

Pleopeltis Humb. & Bonpl. ex Willd. (~ *Polypodium*). Polypodiaceae (II 5). 10 trop. Am., 1 ext. to India (*P. macrocarpa* (Willd.) Kaulf.). Fronds often with peltate scales

Plerandra A. Gray = *Schefflera*

Plesioneuron (Holttum) Holttum = *Cyclosorus*

Plesmonium Schott = *Amorphophallus*

Plethadenia Urban. Rutaceae. 2 Cuba, Hispaniola

Plethiandra Hook.f. Melastomataceae. 7 W Mal.

Plettkea Mattf. Caryophyllaceae (II 1). 4 Andes (Peru at 4000–5000 m!)

Pleurandropsis Baillon = *Asterolasia*

Pleuranthemum (Pichon) Pichon = *Hunteria*

Pleuranthodendron L.O. Williams. Flacourtiaceae (7). 1 trop. Am.

Pleuranthodes Weberb. Rhamnaceae. 2 Hawaii

Pleuranthodium (Schumann) Ridley (~ *Alpinia*). Zingiberaceae. 23 New Guinea, trop. Aus. R: EJB 48(1991)63

Pleuraphis Torrey = *Hilaria*

Pleuricospora A. Gray. Ericaceae (Monotropaceae). 2 W N Am.

Pleurisanthes Baillon. Icacinaceae. 5 trop. S Am.

pleurisy-root *Asclepias tuberosa*

Pleuroblepharis Baillon = *Crossandra*

Pleurobotryum Barb. Rodr. = *Pleurothallis*

Pleurocalyptus Brongn. & Gris. Myrtaceae (Lept.). 1 New Caledonia

Pleurocarpaea Benth. Compositae (Vern.-Vern.). 2 trop. Aus. R: JABG 14(1991)93

Pleurocitrus Tanaka = *Citrus*

Pleurocoffea Baillon = *Coffea*

Pleurocoronis R. King & H. Robinson (~ *Eupatorium*). Compositae (Eup.-Alom.). 3 SW N Am. R: MSBMBG 22(1987)237

Pleuroderris Maxon = *Tectaria*

Pleurogyna Eschsch. ex Cham. & Schldl. = *Lomatogonium*

Pleurogynella Ikonn. = *Swertia*

Pleuromanes C. Presl = *Crepidomanes*

Pleuropappus F. Muell. (~ *Angianthus*). Compositae (Gnap.-Ang.). 1 S Aus.: *P. phyllocalymnus* F. Muell. R: OB 104(1991)130

Pleuropetalum Hook.f. Amaranthaceae (I 1). 4 trop. Am., Galápagos

Pleurophora D. Don. Lythraceae. 7 S Am.

Pleurophragma Rydb. Cruciferae. 4 W US

Pleurophyllum Hook.f. Compositae (Ast.-Ast.). 3, islands S of NZ

Pleuropogon R. Br. (~ *Glyceria*). Gramineae (19). 5 W US with 1 circumpolar. R: AJB 28(1941)358

Pleuropterantha Franchet. Amaranthaceae (I 2). 3 NE trop. Afr.

Pleuropteropyrum Gross = *Persicaria*

Pleurosoriopsis Fomin. Pteridaceae/Grammitidaceae. 1 China, Japan: *P. makinoi* (Maxim.) Fomin

Pleurosorus Fée = *Asplenium*

Pleurospa Raf. = *Montrichardia*

Pleurospermopsis Norman. Umbelliferae (III 5). 1 Sikkim

Pleurospermum Hoffm. Umbelliferae (III 5). S.s. 3 temp. Euras. (Eur. 2), s.l. 45 (? heterogeneous)

Pleurostachys Brongn. Cyperaceae. 50 S Am.

Pleurostelma Baillon. Asclepiadaceae (III 1). 2 E Afr. to Madag. & Aldabra

Pleurostima Raf. (~ *Barbacenia*). Velloziaceae. 25 trop. S Am. R: RBB 3(1980)37, B de B 8(1980)65

Pleurostylia Wight & Arn. Celastraceae. 3–4 trop. & S Afr., S As., N Aus., New Caled. Timber good for joists

Pleurothallis R. Br. Orchidaceae (V 13). 1120 trop. Am. R: MSB 20(1986)1. Some cult. orn. epiphytes

Pleurothallopsis Porto & Brade = *Octomeria*

Pleurothyrium Nees (~ *Ocotea*). Lauraceae (I). 45 trop. Am. Some with ant-infested stem domatia

Plexipus Raf. = *Chascanum*

Plicosepalus Tieghem. Loranthaceae. (Incl. *Tapinostemma*) 5 trop. Afr.

Plinia L. Myrtaceae (Myrt.). 30 trop. Am.

Plinthanthesis Steudel (*Blakeochloa*, ~ *Danthonia*). Gramineae (25). 3 SE Aus. R: T 30(1978)478

Plinthus Fenzl. Aizoaceae (I). c. 4 S Afr. R: JSAB 27(1961)147

Plocama Aiton. Rubiaceae (IV 2). 1 Canary Is.: *P. pendula* Aiton – subdioec., saurochory

Plocaniophyllon Brandegee = *Deppea*

Plocoglottis Blume. Orchidaceae (V 11). 40–45 Mal. to Pacific. Terr. Lip connected to lower part of column by 2 elastic flanges. On flower opening lip is forced down by lateral K stretching elastic 'bands'; light touch of lip by insect releases trigger & lip springs up to column

Plocosperma Benth. Plocospermataceae (Loganiaceae s.l.). 1 Mex., Guatemala

Plocospermataceae Hutch. (~ Loganiaceae). Dicots – Asteridae – Scrophulariales. 1/1 C Am. Excl. from Loganiaceae s.s.; capsule with few narrow hair-tufted seeds
Genus: *Plocosperma*

Ploiarium Korth. Guttiferae (II; Bonnetiaceae). 3 SE As. to New Guinea. R: JAA 31(1950)201. Some timber.

ploughman's spikenard *Inula conyzae*

Plowmania Hunz. & Subils (~ *Brunfelsia*). Solanaceae. 1 Mex.: *P. nyctaginoides* (Standley) Hunz. & Subils

Pluchea Cass. Compositae (Pluch.). c. 40 warm. Fleabane. Herbs & shrubs, some weedy, e.g. *P. camphorata* (L.) DC (camphor weed, N Am.); *P. indica* (L.) Less. (Indomal.) – lvs ed., locally medic., hedge-pl.

Plukenetia L. Euphorbiaceae. 16 trop. (Am. 11, R: SB 18(1993)575). *P. conophora* Muell. Arg. (trop. Afr.) – liane cult. for oilseeds used in cooking etc. (owusa nut)

plum *Prunus × domestica*; **Alleghany p.** *P. alleghaniensis*; **apricot p.** *P. simonii*; **beach p.** *P. maritima*; **Bokhara p.** *P. bokhariensis*; **Burdekin p.** *Pleiogynium timoriense*; **Canada p.** *Prunus nigra*; **chickasaw p.** *P. angustifolia*; **coco p.** *Chrysobalanus icaco*; **date p.** *Diospyros kaki*; **Davidson's p.** *Davidsonia pruriens*; **gingerbread p.** *Neocarya macrophylla*; **governor p.** *Flacourtia indica*; **hog p.** *Spondias* spp.; **Japanese p.** *P. salicina*; **Java p.** *Syzygium cumini*; **kaffir p.** *Harpephyllum caffrum*; **Madagascar p.** *F. indica*; **p. mango** *Bouea macrophylla*; **marmalade p.** *Pouteria sapota*; **mobola p.** *Parinari curatellifolia*; **monkey p.** *Ximenia caffra*; **mooley p.** *Owenia acidula*; **Natal p.** *Carissa grandiflora*; **Oklahoma p.** *Prunus gracilis*; **sapodilla p.** *Manilkara zapota*; **Sebesten p.** *Cordia myxa*; **sour p.** *Owenia acidula*; **Spanish p.** *Spondias purpurea*; **spiny p.** *Ximenia americana*; **wild p.** *Pappea capensis*; **w. goose p.** *Prunus hortulana*, *P. munsoniana*; **p. wood** *Santalum lanceolatum*; **p. yew** *Cephalotaxus* spp.

Plumbagella Spach (~ *Plumbago*). Plumbaginaceae (I). 1 C As.: *P. micrantha* (Ledeb.) Spach – cult. orn. annual

Plumbaginaceae Juss. Dicots – Caryophyllidae – Plumbaginales. 27/730 cosmop. esp. maritime. Shrubs, lianes or, usu., (perennial) herbs; stems often with cortical &/or medullary bundles or alt. rings of concentric xylem & phloem; anthocyanins not betalains present. Lvs simple, entire to lobed, spirally arr.; stipules usu. 0; foliage usu. with scattered chalk-glands exuding water & calcium salts. Fls reg., bisexual, 5-merous, often with heterostyly & in panicles or cymose heads (Staticoideae) or racemes (Plumbaginoideae); K (5) forming 5- or 10-ribbed tube, often petaloid & membranous, C usu. (5), often persistent, lobes convolute, A 5 opp. C, adherent to C in Armerioideae, anthers with longit. slits, pollen often dimorphic in Armerioideae, G̲ (5), 1-loc. with distinct styles (Armerioideae) or 1 apically lobed, & 1 basal, anatropous, bitegmic ovule (no synergids in *Plumbago* & 3 other genera) on slender funicle. Fr. (usu.) an achene, s.t. a circumscissile capsule or one with apical valves, encl. in persistent K; seed with straight embryo in copious or 0 starchy endosperm. x = usu. 6–9

Classification & chief genera:

I. **Plumbaginoideae** (infl. a spike, raceme or head): *Ceratostigma*, *Plumbago*

II. **Staticoideae** (Armerioideae; thyrses with circinnate partial infls): *Acantholimon*, *Armeria*, *Limonium* (from which many small genera have lately been split)

True affinities unclear. Conspic. saltmarsh or maritime genera incl. *Limonium & Armeria*; *Aegialitis* grows in mangrove. Heteromorphic incompatibility systems broken down in some monomorphic *Limonium* spp.; some apomicts

Cult. orn. incl. spp. of *Acantholimon*, *Armeria*, *Ceratostigma*, *Limonium* ('statice' – 'everlasting'), *Plumbago*

Plumbago L. Plumbaginaceae (I). 24 trop. & warm (Eur. 1). Leadwort (*P. europaea* L. (W

Medit. to C As.) form. used to treat eye disease, a side-effect being the skin going lead-coloured). Endosperm tetraploid. Some locally medic., several cult. orn. esp. *P. auriculata* Lam. (*P. capensis*, S Afr.) – evergreen greenhouse scandent shrub with blue fls; *P. indica* L. (*P. rosea*, SE As.) – fruit unknown, pl. always in man-made habitats (? mutant of an As. sp.)

plumcot plum × apricot cross

Plumeria L. Apocynaceae. (Excl. *Himatanthus*) c. 17 trop. Am. Pachycaul trees & shrubs with regularly pseudo-dichotomizing candelabriform branching & fragrant fls. Cult. orn. esp. *P. rubra* L. (frangipani, pagoda tree, temple t. or flower, Mex. to Panamá) – mass-flowering & poll. by hawkmoths fooled by strong scent & colour (no nectar!), fls offered in Buddhist temples, bark purgative & *P. obtusa* L. (WI) – differing in lvs with rounded tips etc.; propagated by cuttings when the large apex & lvs typical of young seedlings are regained – in the mature tree, these get smaller with each branching of the axis (apoxogenesis)

Plummera A. Gray (~ *Hymenoxys*). Compositae (Hele.-Gai.). 2 Arizona

Plumosipappus Czerep. = *Centaurea*

plumwood *Eucryphia moorei*

Plutarchia A.C. Sm. Ericaceae. 10 Ecuador & Colombia

Pneumatopteris Nakai = *Cyclosorus*

Pneumonanthe Gled. = *Gentiana*

Poa L. Gramineae (17). 200+ temp. & cold (Eur. 45 + 4 stabilized hybrids, Aus. c. 40), trop. mts. Perennial (c. 15 annual), s.t. dioec., some apomixis; v. uniform usu. distinguishable from *Festuca* in having keeled lemma & round hilum. Meadow grass, (US) blue g. – imp. forage grasses in pasture & some lawn gs. *P. annua* L. (annual m. g., Eur.) – weed poss. a tetraploid hybrid between *P. infirma* Kunth (Eur., Medit.) & *P. supina* Schrader (Euras.) or either & another sp., now cosmop. cool climates; *P. arachnifera* Torrey (Texas b.g., W US) – cult. winter fodder & lawns; *P. compressa* L. (Canada b.g., N Am.) – fodder in N Am.; *P. nemoralis* L. (wood m.g., Euras., Medit., introd. N Am.) – pasture & hay; *P. pratensis* L. (common m.g., Euras., Medit., introd. N Am. (Kentucky blue grass) late C17 or early C18) – lawns & pasture; *P. trivialis* L. (rough m.g., Euras., Medit., introd. N Am.) – meadows & pastures

Poaceae Caruel = Gramineae

poached egg flower *Limnanthes douglasii*

Poacynum Baillon = *Apocynum*

Poaephyllum Ridley. Orchidaceae (V 14). 9 Mal.

Poagrostis Stapf. Gramineae (25). 1 S Afr.

Poarium Desv. Scrophulariaceae. 20 trop. Am.

Pobeguinea Jacq.-Fél. = *Anadelphia*

Pochota Goyena = *Pachira*

Poculodiscus Danguy & Choux = *Plagioscyphus*

pod mahogany *Afzelia quanzensis*

Podachaenium Benth. ex Oersted. Compositae (Helia.-Zinn.). 2 Mex., Costa Rica. R: SB 7(1982)481. Trees & shrubs. *P. eminens* (Lag.) Schultz-Bip. – cult. orn. with candelabriform branching

Podadenia Thwaites (~ *Ptychopyxis*). Euphorbiaceae. 1 Sri Lanka

Podaechmea (Mez) L.B. Sm. & Kress (~ *Aechmea*). Bromeliaceae (3). 5 trop. Am. R: Phytol. 66(1989)70

Podagrostis (Griseb.) Scribner & Merr. = *Agrostis*

Podalyria Willd. Leguminosae (III 26). 22 Cape to Natal. Alks; some cult. orn. shrubs

Podandra Baillon = *Amblystigma*

Podandrogyne Ducke. Capparidaceae. 10 C Am. to Andes

Podangis Schltr. Orchidaceae (V 16). 1 trop. Afr.: *P. dactyloceras* (Reichb.f.) Schltr – cult. orn.

Podanthus Lagasca. Compositae. (Helia.-Verb.). 2 Chile, Argentina

Podistera S. Watson. Umbelliferae (III 8). 4 W N Am.

podo timber of E Afr. spp. of Podocarpaceae

Podoaceae Baillon ex Franchet. See Anacardiaceae (V)

Podocaelia (Benth.) Fernandes & R. Fernandes (= *Derosiphia*). Melastomataceae. 1 trop. W Afr.

Podocalyx Klotzsch. Euphorbiaceae. 1 Amazonia

Podocarpaceae Endl. Pinopsida. 17/168 S hemisph. to Japan & C Am., trop. Afr. mts. R (excl. *Phyllocladus*): Phytol.M 7(1984)4; AusJB 30(1982)319. Evergreen, usu. dioec. resinous trees (root-parasite in *Parasitaxus*). Lvs spirally arr. (decussate in *Microcachrys*), linear to scale-like. Male cones catkin-like with many bracts, each bearing 2 pollen sacs, pollen grains with (0)2(3) air-bladders; female cones with 1–many bracts, each or only 1 with 1 ovule, & s.t. a peduncle (receptacle) which may become dry or fleshy; mature cones drupe-like, rarely cone-like or with usu. 1 protruding seed s.t. seated on an aril-like fleshy growth attractive to bird-dispersers (0 in *Microstrobos*); cotyledons 2. x = 9–19

Genera: *Acmopyle, Afrocarpus, Dacrycarpus, Dacrydium, Falcatifolium, Halocarpus, Lagarostrobos, Lepidothamnus, Microcachrys, Microstrobos, Nageia, Parasitaxus, Podocarpus, Prumnopitys, Retrophyllum, Saxegothaea, Sundacarpus*

Cult. orn. & imp. timbers (*Dacrycarpus, Dacrydium, Halocarpus, Podocarpus, Prumnopitys*)

Podocarpium (Benth.) Y.C. Yang & S.H. Huang (non Unger) = *Desmodium*

Podocarpus L'Hérit. ex Pers. Podocarpaceae. 94 S temp. through trop. highlands to WI & Japan. R: Blumea 30(1985)251. Lvs usu. narrow & flat; female cones with only 2–4 scales, of which only 1 or 2 bearing ovules; seeds drupe-like on a fleshy red or purple 'receptacle'; some S Am. spp. with ant-infested stem domatia. Cult. orn. & many imp. timbers (white pine, podo, (African) yellow wood) esp.: *P. macrophyllus* (Thunb.) D. Don (S China, Japan) – some forms much used for hedges in Japan; *P. nivalis* Hook. (NZ) – hardy in GB as rock-pl.; *P. nubigenus* Lindley (S Chile, SW Arg.) & *P. salignus* D. Don (S Chile) – manio; *P. totara* G. Benn. ex D. Don (totara, NZ) – war canoes & general construction; *P. urbanii* Pilger (yacca, Jamaica). See also *Afrocarpus, Dacrycarpus, Nageia, Prumnopitys, Retrophyllum*

Podochilopsis Guillaumin = *Adenoncos*

Podochilus Blume. Orchidaceae (V 14). 60 Indomal., SW China, W Pacific

Podochrosia Baillon = *Rauvolfia*

Podococcus G. Mann & H. Wendl. Palmae (V 3). 1 trop. Afr.

Podocoma Cass. Compositae (Ast.-Ast.). 10 Brazil, Argentina

Podocytisus Boiss. & Heldr. Leguminosae (III 31). 1 Balkans, Turkey: *P. caramanicus* Boiss. & Heldr. – cult. orn., unarmed

Podogynium Taubert = *Zenkerella*

Podolasia N.E. Br. Araceae (V 3). 1 W Mal. R: Blumea 33(1988)463

Podolepis Labill. Compositae (Gnap.-Ang.). 20 Aus. R: PLSNSW 81(1957)245, OB 104(1991)120. Cult. orn. 'everlastings'

Podolobium R. Br. (~ *Oxylobium*). Leguminosae (III 24). 6 SW Aus. R: M.D. Crisp & J.J. Doyle, Adv. Legume Syst. 7(1995)279

Podonephelium Baillon. Sapindaceae. 4 New Caledonia

Podopetalum F. Muell. = *Ormosia*

Podophania Baillon = *Hofmeisteria*

Podophorus Philippi. Gramineae (17). 1 Juan Fernandez

Podophyllaceae DC = Berberidaceae

Podophyllum L. Berberidaceae (II 2; Podophyllaceae). c. 5 E N Am. (1) & Himal. to E As. Rhizomes long-known as medic.; flesh of 'berry' mainly derived from placenta: *P. hexandrum* Royle (*P. emodi*, Himal.) & *P. peltatum* L. (may apple, American mandrake, E N Am.) – major disp. in Delaware forests is eastern box turtle, ripe fr. ed. (toxic unripe), long used medic. by N Am. Indians for warts & now against testicular cancer

Podopterus Bonpl. Polygonaceae (II 2). 3 Mexico & Guatemala

Podorungia Baillon. Acanthaceae. 1 Madagascar

Podosorus Holttum = *Microsorum*

Podosperma Labill. = *Podotheca*

Podospermum DC = *Scorzonera*

Podostelma Schumann. = *Pleurostelma*

Podostemaceae Rich. ex Kunth. Dicots – Rosidae – Podostemales. (incl. Tristichaceae) 47/280 trop. (esp. Am. & As.), few temp. Moss-like herbs of stony rivers, usu. submerged & often annual, producing aerial fls & fr. at low water, thallus often lichenoid, with 0 vessels or even 0 xylem; sieve-plates recorded only in *Tristicha* & *Marathrum*; primary root 0; lvs (when discernible) entire to dissected with 0 axillary buds, the thallus held by root-like haptera. Fls bisexual, reg. or not, small, solit. or in (often spiciform) cymes, wind- or insect-poll, or cleistogamous, subtended or encl. by 2 bracteoles (spathe-like & encl. up to 20 fls in Podostemoideae); P 2 or 3(–5) ± connate (? K, Tristichoideae) or 2–∞ free or a small annular scale or 0 (Podostemoideae), A – rarely 1, in up to several whorls with filaments usu. basally connate, pollen grains in diads or monads, G ((1)2(3)) with as many loc. & ± basally connate styles & (2–)±∞ anatropous, bitegmic ovules on thickened axile placentas. Fr. a septicidal capsule with usu. ∞ v. small seeds, often with mucilaginous testa, straight embryo & 0 endosperm (nor double fert.). x = 10

Principal genera: *Apinagia, Hydrobryum, Ledermanniella, Marathrum, Podostemum, Rhyncholacis, Sphaerothylax*

Some developmental processes in the thallus lead to photosynthetic organs which are leaf-stem intermediates ('fuzzy morphology'). App. nearest to Crassulaceae (cf. aquatic tendency in *Crassula*) but raised to Podostemopsida by some

Some locally eaten as salad

Podostemopsida Cusset & G. Cusset. See Podostemaceae

Podostemum Michaux. Podostemaceae. c. 18 trop. to N Am. *P. ceratophyllum* Michaux (N Am.) – functional pl. from meristem at base of hypocotyl reaching flowering size in 2 months, some cleistogamous populations; poss. indicator of clean streams in US

Podostigma Elliott = *Asclepias*

Podotheca Cass. Compositae (Gnap.). 6 temp. Aus. R: Muelleria 7(1989)39

Podranea Sprague. Bignoniaceae. 2 trop. & S Afr. Cult. orn. lianes like *Pandorea* esp. *P. ricasoliana* (Tanf.) Sprague (S Afr.)

Poecilandra Tul. Ochnaceae. 3 N trop. S Am. R: BJ 113(1991)178

Poecilanthe Benth. Leguminosae (III 23). 8 trop. S Am.

Poecilocalyx Bremek. Rubiaceae (I 10). 2 trop. Afr.

Poecilochroma Miers = *Saracha*

Poecilolepis Grau (~ *Aster*). Compositae (Ast.-Ast.). 2 Cape

Poeciloneuron Beddome. Guttiferae (III). 2 S India. Locellate anthers. *P. indicum* Beddome (W Ghats, S India) – locally monospecific forests, timber for electricity poles

Poecilostachys Hackel. Gramineae (34b). 20 Madag., trop. Afr. Forest understorey

Poellnitzia Uitew. (~ *Haworthia*). Asphodelaceae (Aloaceae). 1 SW Cape: *P. rubriflora* (L. Bolus) Uitew. – poll. by mites, cult. orn. succ.

Poeppigia C. Presl. Leguminosae (I 1?). 1 trop. Am.: *P. procera* (Sprengel) C. Presl

Poga Pierre. Anisophylleaceae. 1 W trop. Afr.: *P. oleosa* Pierre (afo, inoy or inoi nut, poga) – timber & oilseed

Poggea Guerke. Flacourtiaceae (3). 4 trop. W & C Afr. R: BJ 94(1974)296

Pogogyne Benth. Labiatae (VIII 2). 6 W N Am. R: PCAS IV, 20(1931)105. Aquatics with 'Hippuris syndrome'

Pogonachne Bor. Gramineae (40d). 1 Bombay: *P. racemosa* Bor

Pogonanthera Blume. Melastomataceae. 1–4 Mal.

Pogonarthria Stapf. Gramineae (31d). 4 trop. & S Afr. R: SBi 47(1966)303

Pogonatherum Pal. Gramineae (40a). 3 trop. As. A 1 or 2. *P. paniceum* (Lam.) Hackel – cult. orn.

Pogonia Juss. Orchidaceae (Pogoniinae). 2 temp. As., N Am. Cult. orn. terr. orchids (beard flower) with bearded lip, esp. *P. ophioglossoides* (L.) Ker-Gawler (N Am.)

Pogoniopsis Reichb.f. Orchidaceae (Pogoniinae). 2 Brazil. Mycotrophic

Pogonochloa C. Hubb. Gramineae (33b). 1 S trop. Afr.

Pogonolepis Steetz (~ *Angianthus*). Compositae (Gnap.-Ang.). 2 Aus. R: Muelleria 6(1986)237

Pogonolobus F. Muell. = *Coelospermum*

Pogononeura Napper. Gramineae (31d). 1 trop. E Afr.: *P. biflora* Napper

Pogonophora Miers ex Benth. Euphorbiaceae. 1–2 trop. S Am., 1 trop. Afr.

Pogonopus Klotzsch. Rubiaceae (I 7). 2–3 trop. Am. Alks. *P. speciosus* (Jacq.) Schumann (C Am.) – cult. orn. with brightly-coloured enlarged K-lobe

Pogonorrhinum Betsche = *Linaria*

Pogonotium Dransf. Palmae (II 1d). 3 W Mal.

Pogostemon Desf. Labiatae (VII). (Incl. *Eusteralis*) 96 temp. As., Indomal. to Aus. R: BBMNHB 10(1982)71. Alks. *P. cablin* (Blanco) Benth. (patchouli, patchouly, Indomal.) – oil used in scenting cosmetics, shampoos, in insecticides & leech-repellent & to soothe menstrual pains but in commerce often replaced by *P. heyneanus* Benth. (Indian p., Indomal.); *P. mutamba* (Hiern) G. Taylor (S trop. Afr.) – starchy ed. tubers; *P. parviflorus* (Benth.) Benth. (E As.) – imp. honey pl.

Pohlidium Davidse, Söderstrom & R. Ellis. Gramineae (24). 1 Panamá: *P. petiolatum* Davidse, Söderstrom & R. Ellis

Pohliella Engl. Podostemaceae. 2 W Afr.

Poicilla Griseb. = *Matelea*

Poicillopsis Schltr. ex Rendle = *Matelea*

Poidium Nees = *Poa*

Poikilacanthus Lindau. Acanthaceae. 6 trop. Am.

Poikilogyne Baker f. Melastomataceae. 20 Borneo, New Guinea. R: GBS 35(1982)223

Poikilospermum Zipp. ex Miq. Cecropiaceae. 20 E Himal. to Mal. R: GBS 20(1963)1. Myrmecophily

Poilanedora Gagnepain. Capparidaceae. 1 SE As.

Poilaniella Gagnepain. Euphorbiaceae. 1 SE As.

Poilannammia C. Hansen. Melastomataceae. 4 Vietnam. R: BMNHN 4, 9(1987)263

Poinciana L. = *Caesalpinia*

poinsettia *Euphorbia pulcherrima*

Poinsettia J. Graham = *Euphorbia*

pointvetch *Oxytropis* spp.

Poiretia Vent. Leguminosae (III 13). 11 trop. Am.

poison ivy or **oak** *Rhus radicans.* Names used loosely for other *R.* spp. in N Am.; **prickly p.** *Gastrolobium spinosum*; **p. walnut** *Cryptocarya glabella*

Poissonia Baillon = *Coursetia*

Poitea Vent. Leguminosae (III 7). (Incl. *Sabinea*) 12 WI. R: SBM 37(1993). *P. carinalis* (Griseb.) Lavin (Dominica) – cult. orn. tree with bright red fls

Poivrea Comm. ex DC = *Combretum*

Pojarkovia Askerova. Compositae (Sen.-Sen.). 1 Cauc.: *P. stenocephala* (Boiss.) Askerova

pokaka *Elaeocarpus hookerianus*

poke or **p.weed** *Phytolacca americana*

pokosola *Ochrosia elliptica*

Polakia Stapf = *Salvia*

Polakowskia Pittier = *Sechium*

Polanisia Raf. (~ *Cleome*). Capparidaceae. 6 N Am. Like *Cleome* but A (8–)12–16(–27), some cult. orn.

Polaskia Backeb. (~ *Lemaireocereus*). Cactaceae (III 8). 2 S Mex. Links *Stenocereus* & *Myrtillocactus*. Fr. ed.

Polemannia Ecklon & Zeyher. Umbelliferae (III 8). 3 SE Afr. mts. R: EJB 48(1991)242

Polemanniopsis B.L. Burtt (~ *Polemannia*). Umbelliferae (*inc. sed.*). 1 W Cape: *P. marlothii* (H. Wolff) B.L. Burtt. R: NRBGE 45(1988)498. Woody; mericarps heteromorphic

Polemoniaceae Juss. Dicots – Asteridae – Solanales. 20/290 Am. (esp. W N), Euras. Shrubs, lianes (*Cobaea*), small trees (*Cantua*) or, usu., herbs, usu. storing carbohydrate as inulin & often stinking, xylem usu. forming continuous ring. Lvs simple to pinnatisect, palmatisect or pinnate, spirally arr. (opp. in *Linanthus* & *Phlox*, whorled in *Gymnosteris*); stipules 0. Fls bisexual, (4)5(6)-merous, (solit. or, usu.,) in head-like cymes; (K) (K in *Cobaea*), lobes s.t. unequal, (C) reg. or ± 2-labiate, the lobes convolute in bud, A attached to C-tube alt. with lobes, s.t. at diff. levels, anthers with longit. slits, annular nectary-disk usu. around G ((2)3(4)) with as many loc. & axile placentas & term. style with as many stigma-lobes, each loc. with 1–∞ anatropous to hemitropous unitegmic ovules. Fr. a capsule usu. loculicidally dehiscent (septicidally in *Cobaea*), sometimes ± indehiscent; seeds 1–∞, often mucilaginous when wetted, with ± straight embryo in (scanty–) copious oily endosperm. x = 9 (sometimes reduced to 6)

Principal genera: *Cobaea* (sometimes segregated as monotypic fam.), *Collomia*, *Eriastrum*, *Gilia*, *Ipomopsis*, *Leptodactylon*, *Linanthus*, *Navarretia*, *Phlox*, *Polemonium*

App. close to Hydrophyllaceae, with v. diverse pollen structure & sculpturing app. not correlated with diverse poll. systems (bees from which poll. by hummingbirds, flies & beetles seems to have arisen independently in several genera) incl. bats (*Cobaea*) & Lepidoptera. The trop. genera usu. woody pls with large to medium-sized C & winged seeds with little or 0 endosperm whereas temp. ones mostly herbaceous with medium to small C & wingless seeds with endosperm

Many cult. orn.

Polemonium L. Polemoniaceae. 25 N temp. (Eur. 3) to Mex., Chile (2). R: UCPB 23(1950)209. Cult. orn. herbs (moss pink, N Am.) esp. *P. caeruleum* L. (Jacob's ladder, Greek valerian, Euras.)

polenta *Zea mays*

Polevansia de Winter. Gramineae (33b). 1 SW Afr.

Polhillia Stirton. Leguminosae (III 28). 6 SW Cape. R: SAJB 52(1986)167

Polianthes L. Agavaceae. 2 Mex. R: CUSNH 8(1903)8. Herbs with thick bulb-like bases. *P. tuberosa* L. (tuberose, unknown in wild) – fls gardenia-scented waxy-white, cult. in preColumbian Mexico ('omixochitl') & added to chocolate as flavouring, now grown for v. fragrant fls (often double form), the oil used in scent-making

Poliomintha A. Gray. Labiatae (VIII 2). 7 SW N Am. R: Sida 5(1972)8, Phytol. 74(1993)164. *P. incana* (Torrey) A. Gray – locally used potherb & fls used for flavouring

Poliophyton O. Schulz. Cruciferae. 1 SW N Am.

Poliothyrsis Oliver. Flacourtiaceae (8). 1 China: *P. sinensis* Oliver – cult. orn. decid. tree with winged seeds

polka-dot plant *Hypoestes phyllostachya*

Pollalesta Kunth. Compositae (Vern.-Pip.). 16 trop. & warm Am., Philipp. (introd.?). R: Rhodora 83(1981)398. Trees & shrubs to 30 m

Pollia Thunb. Commelinaceae (II 2). 26 OW trop. (Aus. 2), Panamá (1). Blue fr. attractive to pheasants in Mal. forest undergrowth

Pollichia Sol. Caryophyllaceae (I 2). 1 trop. & S Afr., Arabia: *P. campestris* Sol.

Polliniopsis Hayata = *Microstegium*

Polpoda C. Presl. Molluginaceae (?). 2 SW & W Cape. R: JSAB 21(1955)93

Polyachyrus Lagasca. Compositae (Mut.-Nass.). 7 Peru, Chile (W Andean slope). R: Gayana 26(1974)3. Herbs

Polyadoa Stapf = *Hunteria*

Polyalthia Blume. Annonaceae. 100 OW trop. (Madag. 18 (R: BMNHN 4,12(1990)113; Afr. spp. = *Greenwayodendron*), Aus. 5 (4 endemic)). *P. cerasoides* (Roxb.) Bedd. (India, SE As.) – timber for boats & carpentry; *P. hypogaea* King (Malay Pen.) with fls on runners at ground-level; *P. longifolia* (Sonn.) Thw. (S India, Sri Lanka) – riparian, fastigiate form a common garden & roadside tree in India & Sri Lanka ('Indian willow'); *P. micrantha* (Hassk.) Boerl. – living fence in Java

Polyandra Leal. Euphorbiaceae. 1 Brazil

Polyandrococos Barb. Rodr. Palmae (V 5b). 2 Brazil. *P. caudescens* (C. Martius) Barb. Rodr. – cult. orn. unarmed monoec. palm with ed. fr.

Polyanthina R. King & H. Robinson (~ *Eupatorium*). Compositae (Eup.-Ayap.). 1 trop. Am. along Andes to Bolivia: *P. nemorosa* (Klatt) R. King & H. Robinson. R: MSBMBG 22(1987)199

polyanthus *Primula × polyantha*

Polyarrhena Cass. (~ *Felicia*). Compositae (Ast.-Ast.). 4 SW Cape. R: MBSM 7(1970)347

Polyaster Hook.f. Rutaceae. 1–2 Mexico

Polyaulax Backer = *Meiogyne*

Polybactrum Salisb. = *Pseudorchis*

Polybotrya Humb. & Bonpl. ex Willd. Dryopteridaceae (I 2). 35 trop. Am. esp. Andes. R: INHSB 34(1987)1. Large climbing ferns with dimorphic fronds

Polycalymma F. Muell. & Sonder (~ *Myriocephalus*). Compositae (Gnap.-Ang.). 1 Aus.: *P. stuartii* F. Muell. & Sonder. R: OB 104(1991)115

Polycardia Juss. Celastraceae. 9 Madag. Epiphyllous infls in *P. aquifolium* Tul., *P. phyllanthoides* (Lam.) DC etc.

Polycarena Benth. Scrophulariaceae (Man.). 17 SW Cape. R: O. Hilliard, *Manuleeae* (1994)395. See also *Phyllopodium*

Polycarpaea Lam. Caryophyllaceae (I 1). 50 trop. & warm esp. OW. Some locally medic. in China, *P. spirostylis* F. Muell. (Aus.) a copper indicator

Polycarpon L. Caryophyllaceae (I 1). 16 cosmop. (Eur. 2). *P. tetraphyllum* (L.) L. (Eur. & Medit., widely natur.) – four-leaved allseed

Polycephalium Engl. Icacinaceae. 2 trop. Afr.

Polyceratocarpus Engl. & Diels. Annonaceae. 7 trop. Afr., esp. W

Polychilos Breda = *Phalaenopsis*

Polychrysum (Tzvelev) Kovalevsk. Compositae (Anth.-Han.). 1 C As., Afghanistan

Polyclathra Bertol. Cucurbitaceae. 1 (variable) C Am.

Polyclita A.C. Sm. Ericaceae. 1 Bolivia

Polycnemum L. Chenopodiaceae (IV 1). 7–8 C & S Eur. (4), Medit. to C As.

Polycodium Raf. ex E. Greene = *Vaccinium*

Polycoryne Keay = *Pleiocoryne*

Polyctenium E. Greene. Cruciferae. 1 NW Am.

Polycycliska Ridley = *Lerchea*

Polycycnis Reichb. f. Orchidaceae (V 10). 15 trop. Am. Cult. orn.

Polygala L. Polygalaceae. c. 500 subcosmop. (not NZ; Eur. 33). Milkwort (once thought to increase cow milk yields). Trees, shrubs & herbs with irreg. fls (inner 2 K petaloid (wings); C 3–5, often united, the lowermost (keel) often crested), the stamens & pistil emerging when insect alights (cf. Leguminosae), & arillate (s.t. reduced to hairs etc.) seeds. Some cult. orn. & oilseeds etc.: *P. arillata* D. Don (Himal.) – roots fermented for alcoholic drinks in Nepal; *P. butyracea* Heckel (Afr., not known wild) – source of fibre used in sacking & of Beni-seed from which malukang butter made; *P. chamaebuxus* L. (C Eur. to Italy) – cult.

orn. evergreen shrubby rock-pl. esp. var. *grandiflora* Neilr. ('Atropurpurea') with purple wings & C yellow; *P.* × *dalmaisiana* Hort. (*P. oppositifolia* L. (S Afr.) × *P. myrtifolia* L. (S Afr.)) – shrub with almost continuously prod. purplish or rose fls, cult. Medit. climates; *P. senega* L. (senega root, N Am.) – dried root medic. esp. for snakebite; *P. sibirica* L. (E Eur. to As.) – 'root' with liquorice used to treat depression, irritability & insomnia in modern Chinese herbalism; *P. vulgaris* L. (gang flower, Eur. & Medit.) – wings red, white or blue (red crossed with white gives blue), pollinator unknown, *Lasius niger* take seeds (with elaiosomes) to 2 m; others locally medic. & in comm. cough mixtures etc.

Polygalaceae Hoffmsgg. & Link. Dicots – Rosidae – Polygalales. (Incl. Emblingiaceae, excl. Xanthophyllaceae) 17/950 subcosmop. but not W Pacific. Trees, shrubs, lianes (e.g. *Barnhardtia*) & herbs, s.t. parasitic (*Salomonia*) & often with extrafl. nectaries. Xylem usu. in closed ring even in herbs. Lvs usu. spirally arr., simple & entire; stipules usu. 0, s.t. a pair of glands or spines. Fls bisexual, 2-bracteolate, usu. strongly irreg. & hypogynous, in spikes, racemes or panicles: K 5, rarely basally connate or 2 lower ones so, the 2 inner often petaloid, C 3 or 5 often adnate to A to form tube, when 3 the 2 upper ones + lower median boat-shaped & often apically fringed, A 4 + 4 or 10 or 3–7 usu. basally connate with anthers with apical pores to longit. slits, annular disk s.t. around $\underline{G}$ (2–5(–8)) with axile placentas (s.t. pseudomonomerous & 1-loc.) with term. style, often curved & 2-lobed (1 lobe stigmatic, other ending in tuft of hairs), each loc. with 1 pendulous, epitropous, anatropous to hemitropous bitegmic ovule. Fr. a loculicidal capsule, nut, samara or drupe; seeds arillate (caruncle) or at least hairy with 0–copious oily, proteinaceous endosperm. x = 5–11+

Principal genera: *Bredemeyera, Comesperma, Monnina, Moutabea, Polygala, Securidaca* (*Xanthophyllum* here kept in own fam.)

Fls superficially like Leguminosae (III) but prob. derived from Malpighiaceae–Vochysiaceae complex; Emblingiaceae (K (5), C 2) incl. here

Some cult. orn. & medic. etc. (*Polygala*)

Polygaloides Haller = *Polygala*

Polygonaceae Juss. Dicots – Caryophyllidae – Polygonales. 46/1100 ± cosmop., esp. (N) temp. Trees, shrubs, lianes or often herbs often with unusual vascular structure & app. swollen nodes, producing anthocyanin & not betalain. Lvs simple, usu. entire, usu. spirally arr.; stipules usu. conspic. & often united as a scarious sheath (ocrea) around stem, or ± 0 (e.g. *Eriogonum*). Fls usu. bisexual (or pl. dioec.), reg., 2-, 3- (orig.?) or 5-merous, small, often with pseudopedicel above articulation with pedicel often in involucrate fascicles subtended by persistent ocreola in simple or branched infls; P (2–6) with minute tube, green to ± petaloid, often 3 + 3 but not recognizably K & C, or 5 but always arising spirally, persistent & often accrescent in fr., A (2,)3 + 3, 8(9–), arising in front of P (1 or 2 usu. per P), filaments s.t. basally connate, often of 2 lengths, anthers usu. versatile, introrse with longit. slits, pollen grains v. variable, annular nectary-disk around G or nectaries between A, $\underline{G}$ ((2)3(4)), 1-loc. with ± united styles & 1 basal, orthotropous (–anatropous) bi-(or ± uni-)tegmic ovule. Fr. often 3-gonous achene or nut, sometimes encl. in persistent P often forming a wing (wind-disp.) or fleshy hypanthium; seed with straight to curved usu. eccentric embryo in starchy & oily (s.t. ruminate) endosperm. x = 7–13

Classification & principal genera:

I. **Eriogonoideae** (ocrea 0; 2 tribes; R: Phytol. 66(1989)266):
   1. **Eriogoneae** (infl. involucre tubular or series of 3 to many bracts): *Chorizanthe, Eriogonum*
   2. **Pterostegieae** (involucre a single bract encl. mature achene): *Harfordia*

II. **Polygonoideae** (ocrea present; 5 tribes):
   1. **Triplareae** (trees & shrubs, often dioec., P 3 + 3): *Ruprechtia, Triplaris*
   2. **Coccolobeae** (trees, shrubs & lianes, P 5): *Antigonon, Coccoloba, Muehlenbeckia*
   3. **Rumiceae** (herbs, P (2)3 + (2)3): *Rheum, Rumex*
   4. **Polygoneae** (shrubs & herbs, P 5, outer often winged): *Atraphaxis, Calligonum, Fallopia, Oxygonum, Polygonum*
   5. **Persicarieae** (herbs, P 5, outer rarely winged; R: BJLS 98(1988)321): *Fagopyrum, Persicaria*

Absence of betalains, ± 0 perisperm, S-type sieve-tube plastids & 3-mery set P. off from Caryophyllales to which P. are clearly allied. Pentamery is derived from 3-mery here as there is a 'compound tepal' (Cronquist) with (frequently) 2 midveins rather than 1 in the 5-merous, suggesting amalgamation of 3 + 3

Some ed. products (*Coccoloba, Fagopyrum* (buckwheat), *Persicaria, Rheum, Rumex*), timber (*Triplaris*), charcoal (*Gymnopodium*), tanning material (*Rumex*), cult. orn. (*Antigonon*,

Eriogonum, Fallopia, Muehlenbeckia, Persicaria, Polygonum, Rheum) & some bad weeds (*Emex, Persicaria, Polygonum* – knotweeds, *Reynoutria, Rumex* – docks)

Polygonanthus Ducke. Anisophylleaceae. 2 Amazonia

Polygonatum Miller. Convallariaceae (Liliaceae s.l.). 55 N temp. (Eur. 5) esp. SW China. Herbs with robust horiz. rhizomes, locally used as food & medic. but widely cult. orn. (Solomon's seal i.e. mender (use in treatment of bruises)); usu. woodland pl. (*P. hookeri* Baker also in high alt. Himal. grasslands, cult. orn. dwarf sp.) visited by bees esp. solit. *Anthophora* spp. Most commonly cult. is *P.* × *hybridum* Bruegger (*P. odoratum* (Miller) Druce (2n = 20, Euras.) × *P. multiflorum* (L.) All. (2n = 18, Euras.)) – usu. sterile (2n = 19 with usu. 8 bivalents & 1 trivalent at meiosis, though it, like its parents, may have up to 2n = 30)

Polygonella Michaux (~ *Polygonum*). Polygonaceae (II 4). 9 E N Am. R: Brittonia 15(1963)177. Sandy & scrub sites

Polygonum L. Polygonaceae (II 4). 20 N temp. (*P. maritimum* L. also in S S Am.). Annual herbs (knotweeds) usu. insect-poll. Some bad weeds; for cult. orn. see *Fallopia, Persicaria* & *Reynoutria*. *P. aviculare* L. (knotweed, knotgrass, Euras., widely natur.) – form. widely used medic. esp. as tea in treatment of asthma etc., now bad weed resistant to herbicide in many places

Polylepis Ruíz & Pavón. Rosaceae. 15 Andes. R: SCB+ 43(1979)1. Shrubs & trees to 5000 m in mts, the highest alt. reached by an arborescent angiosperm genus

Polylophium Boiss. Umbelliferae (III 12). 2 W As.

Polylychnis Bremek. Acanthaceae. 2 NE S Am.

Polymeria R. Br. Convolvulaceae. 7 Aus. (6 endemic) to E Mal. & New Caledonia

Polymita N.E. Br. (~ *Ruschia*). Aizoaceae (V). 2 NW S Afr.

Polymnia L. Compositae (Helia.-Mel.). 2 E N Am. R: Phytol. 39(1978)49

Polyosma Blume. Grossulariaceae (Escalloniaceae). 60 E Himalaya to trop. Aus. & New Caledonia (5)

Polyosmaceae Blume = Grossulariaceae

Polyotidium Garay. Orchidaceae (V 10). 1 Colombia

Polyphlebium Copel. (~ *Trichomanes*). Hymenophyllaceae (Hym.). 1 Aus., NZ: *P. venosum* (R. Br.) Copel.

Polypleurella Engl. Podostemaceae. 1 Thailand

Polypleurum (Tul.) Warm. Podostemaceae. 5 trop. W Afr. (1), India, Sri Lanka

Polypodiaceae Bercht. & J. Presl. Filicopsida. (Excl. Grammitidaceae) 33/700 cosmop. esp. trop. Usu. terr. or epiphytic with dorsi-ventral dictyostelic rhizome covered with peltate scales. Fronds sometimes dimorphic, simple, lobed to pinnate (bipinnate, pedate etc.) with sori usu. round, spreading along veins or on v. diff. fronds (acrostichoid); indusium 0

Classification & principal genera:

I. **Platycerioideae** (fronds with stellate hairs): *Platycerium, Pyrrosia*

II. **Polypodioideae** (scales &/or hairs but never stellate; 6 tribes):

1. **Drynarieae** (fronds dimorphic): *Aglaomorpha, Drynaria*
2. **Selligueeae** (fronds monomorphic, stem scales opaque): *Selliguea*
3. **Lepisoreae** (as 2. but scales clathrate, exospore thick): *Belvisia, Lepisorus*
4. **Microsoreae** (as 3. but exospore thin): *Colysis, Lecanopteris, Microsorum, Phymatosorus*
5. **Polypodieae** (as 3. but scales sometimes opaque, exospore sometimes thin): *Pecluma, Polypodium*
6. **Loxogrammeae** (all but roots lacking internal sclerenchyma): *Loxogramme*

Some ant-pls (*Lecanopteris, Solanopteris*); some ed. & medic. esp. *Polypodium*, & some cult. orn. esp. *Aglaomorpha, Drynaria, Microsorum, Phymatosorus, Platycerium, Polypodium*

Polypodiastrum Ching = *Goniophlebium*

Polypodiodes Ching = *Gonophlebium*

Polypodiopteris C. Reed. Polypodiaceae (II 5). 3 Borneo. R: Blumea 39(1994)365. Montane forests

Polypodium L. Polypodiaceae (II 5). Incl. *Phlebodium*, c. 150 cosmop. (Eur. 1–4) esp. trop. Am. Polypody. Some cult. orn. (though many now referred to other genera e.g. *Goniophlebium, Microgramma* & *Pleopeltis*) incl. *P. aureum* L. (*Phlebodium a.*, trop. Am.) – many cvs & *P. vulgare* L. (Euras.) – hardy epiphyte form. used to flavour tobacco (liquorice taste) & containing ostadin, a steroid saponin, 3000 times as sweet as sucrose, in small amounts – one of sweetest of all known compounds; closely allied *P. glycyrrhiza* D. Eaton (N Am.) with liquorice-flavoured rhizomes eaten by N Am. Indians, while similar *P. virginianum* L. (E As., E N Am.) does not have sweet taste, but it & *P. vulgare* etc. have phytoecdysones (25 mg per 2.5 g of rhiz. (same as 0.5 t silkworms!))

polypody *Polypodium* spp. esp. *P. vulgare*

Polypogon Desf. Gramineae (21d). 10 warm temp. (Eur. 3), trop. mts. Hybrids with *Agrostis* spp. *P. monspeliensis* (L.) Desf. (beard-grass, Eur.) – cult. orn., natur. Am., Aus.

Polypompholyx Lehm. = *Utricularia*

Polyporandra Becc. Icacinaceae. 1 E Mal., Melanesia: *P. scandens* Becc. – a liane with lvs cooked with taro

Polypremum L. Fam.? (Scrophulariales; Loganiaceae s.l.). 1 warm Am.

Polypsecadium O. Schulz. Cruciferae. 1 C Andes

Polypteris Nutt. = *Palafoxia*

Polyradicion Garay (~ *Dendrophylax*). Orchidaceae (V 16). 4 Florida & WI. Leafless epiphytes with chlorophyllous roots & fls to 12 cm diam. *P. lindenii* (Lindley) Garay (S Florida, WI) – cult. orn. with fragrant white fls

Polyrhabda C. Towns. Amaranthaceae (I 2). 1 Somalia

Polyschemone Schott, Nyman & Kotschy = *Silene*

Polyscias Forster & Forster f. Araliaceae. 150 OW trop. esp. New Caled. Unarmed shrubs & trees (all A. with pinnate lvs & articulated pedicels), often with reg. pseudo-di(–5)-chotomous candelabriform branching (umbrella trees), some with cryptic dioec.; incl. *Nothopanax*. Many used in W Pacific as cult. orn. esp. varieg. forms as hedges & locally as veg., esp. cvs of *P. fruticosa* (L.) Harms (? W Pacific) & *P. guilfoylei* (W. Bull) L. Bailey (? E Mal.) – rarely flowering in cult. *P. scutellaria* (Burm.f.) Fosb. (? E Mal.) – source of scent & medic.; other spp. used to stupefy fish

Polysolen Rauschert = *Indopolysolenia*

Polysolenia Hook.f. = *Indopolysolenia*

Polyspatha Benth. Commelinaceae (II 2). 3 trop. W Afr.

Polysphaeria Hook.f. Rubiaceae (II 5). 20 trop. Afr., Madag., Comoro Is. R: KB 35(1980)97

Polystachya Hook. Orchidaceae (V 13). 150 trop. & warm. Epiphytes with swollen stems or pseudobulbs; some with false pollen rich in protein & or starch on lip as pollinator attractant. Some cult. orn.

Polystemma Decne. = *Matelea*

Polystemonanthus Harms. Leguminosae (I 4). 1 W Afr.

Polystichopsis (J. Sm.) Holttum = *Arachniodes*

Polystichum Roth. Dryopteridaceae (I 2). Incl. *Cyrtomium* & *Phanerophlebia*, 200 cosmop. (Eur. 4). Terr. woodland ferns, some cult. orn. (holly f.) incl. *P. falcata* (L.f.) Diels (S Afr. to Hawaii, rare in trop. lowlands but natur. GB, Azores, Netherlands, Sydney Harbour, US etc.) – tough greenhouse fern & *P. munitum* (Kaulf.) C. Presl (W N Am.) – peeled rhizome with banana taste

Polytaenia DC. Umbelliferae (III 10). 2 N Am.

Polytaenium Desv. = *Antrophyum*

Polytaxis Bunge (~ *Jurinea*). Compositae (Card.-Card.). 2 C As.

Polytepalum Süsseng. & Beyerle. Caryophyllaceae (I 1). 1 Angola: *P. angolense* Süsseng. & Beyerle

Polytoca R. Br. Gramineae (40j). 2–6 Indomal. Ant-disp. (elaiosome). *P. macrophylla* Benth. (Mal.) – cattle fodder

Polytrema C.B. Clarke = *Ptyssiglottis*

Polytrias Hackel (~ *Eulalia*). Gramineae (40a). 1 SE As.: *P. indica* (Houtt.) Veldk. (*P. amaura*, Java grass, Mal.) – lawn-grass escaping pantrop.

Polyura Hook.f. Rubiaceae (IV 1). 1 Assam

Polyxena Kunth. Hyacinthaceae (Liliaceae s.l.). 5–6 Cape. Cult. orn. dwarf bulbous pls

Polyzygus Dalz. Umbelliferae (III 11). 1 S India

pomade (pomatum) scented ointment esp. for hair, orig. alleged to contain apple pulp

Pomaderris Labill. Rhamnaceae. 55 SE As. (1), Aus. (53, 50 endemic), NZ. Some NZ spp. triploid. Cult. orn. shrubs & trees incl. *P. apetala* Labill. (Aus.) – ? introd. NZ as used for canoe skids & *P. elliptica* Labill. (Tasmania) – locally medic.

pomander orig. a mixture of aromatic subs. moulded into balls & suspended as preservatives against infection, latterly an orange stuck with cloves

Pomatocalpa Breda. Orchidaceae (V 16). 35 China to Mal., Aus. & Polynesia. Cult. orn.

Pomatosace Maxim. Primulaceae. 1 NW China

Pomatostoma Stapf = *Anerincleistus*

pomatum = pomade

Pomax Sol. ex DC. Rubiaceae (IV 13). 1 Aus.: *P. umbellata* (Gaertner) A. Rich.

Pomazota Ridley = *Coptophyllum*
pomegranate *Punica granatum*
pomelo *Citrus maxima*
pomerac *Syzygium malaccense*
Pometia Forster & Forster f. Sapindaceae. 2 Indomal. R: Reinwardtia 6(1962)109. *P. pinnata* Forster & Forster f. (Mal.) – v. variable, fr. ed. (much valued in Bismarcks & Tonga), timber locally used, local medic.
Pommereschea Wittm. Zingiberaceae. 2 Burma
Pommereulla L.f. Gramineae (33a). 1 S India, Sri Lanka
pompelmous *Citrus maxima*
pomion *Cucurbita pepo*
pompon See *Dahlia*
Ponapea Becc. = *Ptychosperma*
Poncirus Raf. Rutaceae. 1 C & N China: *P. trifoliata* (L.) Raf. – cult. orn. spiny decid. tree with 3-foliolate lvs showing that the 'joint' of *Citrus* lvs rep. vestiges of pinnate construction, fl. buds overwintering on leafless shoots (cf. *Citrus*); used as hedges & stock for *Citrus* with which it hybridizes (see × *Citroncirus*), fr. fragrant but acid & with little flesh (suitable for marmalade). A second sp. (evergreen) from Yunnan poss. a citrange
pondweed *Potamogeton* spp.; **Canadian p.** *Elodea canadensis*; **Cape p.** *Aponogeton distachyos*
Ponera Lindley. Orchidaceae (V 13). 11 trop. Am.
Ponerorchis Reichb.f. = *Gymnadenia*
Pongamia Vent. = *Millettia*
Pongamiopsis R. Viguier. Leguminosae (III 6). 3 Madagascar
pongoware *Cyathea medullaris*
Pontederia L. Pontederiaceae. 5 NW. R: Rhodora 75(1973)426. Tristyly. *P. cordata* L. (pickerel weed, wampee, E N Am. to Carib.) – pollen trimorphism but short-style morph rarely legit. poll. N Am., among pollinators a bee (*Dufourea novae-angliae* Robertson) which visits no other pl., its emergence coinciding with anthesis of *P. c.*; cult. orn. with ed. 1-seeded fr.
Pontederiaceae Kunth. Monocots – Liliidae – Haemodorales. 6/32 trop. & warm (esp. Am.) to N temp. (few). Hydrophytes, s.t. free-floating, occ. annual, with sympodial branching, vegetatively glabrous. Lvs usu. with sheath, distinct (s.t. inflated with aerenchyma) petiole & expanded lamina with parallel curved-convergent veins (filiform in *Hydrothrix*), in basal rosettes, distichous or spirally arr. along stem. Roots with vessel-elements (often in stem too). Fls bisexual, reg. or not, often tristylous, solit. or in term. racemes, spikes or panicles subtended by a sheath, usu. insect-poll.; P 3 + 3 (4 in *Scholleropsis*), petaloid, ± basally connate, A (1 with 2 staminodes – *Hydrothrix*) 3 + 3 or 3 ± staminodes, adnate to P-tube, anthers with longit. slits or term. pores (*Monochoria*), $\underline{G}$ (3), 3-loc. with axile placentas or 1-loc. with intruded parietal placentas (pseudomonomerous in *Pontederia*), usu. with septal nectaries & 1 style & (1 pendulous (*Pontederia*)–)∞ anatropous, bitegmic ovules. Fr. a loculicidal capsule (nut in *Pontederia*), maturing underwater; seeds longit. ribbed with red-coloured tegmen, axile cylindrical embryo in copious starchy endosperm with outer aleurone layer. x = 8, 14, 15
Genera: *Eichhornia, Heteranthera, Hydrothrix, Monochoria, Pontederia, Scholleropsis*
Some bad weeds esp. *Eichhornia*, ed. parts (*Monochoria, Pontederia*) & cult. orn. esp. *Eichhornia, Heteranthera* & *Pontederia*
Ponthieva R. Br. Orchidaceae (III 3). 53 trop. & warm Am.
pontianak, black *Shorea* spp.
poor man's orchid *Schizanthus* spp.; **p. m. weather glass** *Anagallis arvensis*
popinac *Acacia farnesiana*
poplar *Populus* spp.; (US), p. (yellow) used for *Liriodendron tulipifera*; **balsam p.** *P. balsamifera*; **black p.** *P. nigra*: **Canada p.** *P. balsamifera*; **grey p.** *P.* × *canescens*; **Italian** or **Lombardy p.** *P. nigra* 'Italica'; **Manchester p.** *P. nigra* ssp. *betulifolia*; **white p.** *P. alba*
Popoviocodonia Fed. (~ *Adenophora*). Campanulaceae. 2 E Siberia, Sakhalin
Popoviolimon Lincz. Plumbaginaceae (II). 1 C As.: *P. turcomanicum* (Lincz.) Lincz.
Popowia Endl. Annonaceae. 30 trop. As. to Aus. Some locally used for scent & ed. fr.
poppy *Papaver* spp.; **blue p.** *Meconopsis* spp.; **Californian p.** *Eschscholzia californica*; **common, corn** or **Flanders p.** *P. rhoeas*; **horned p.** *Glaucium flavum*; **Iceland p.** *P. nudicaule*; **p. mallow** *Callirhoe papaver*; **Matilija p.** *Romneya coulteri*; **Mexican p.** *Argemone mexicana*; **opium p.** *P. somniferum*; **oriental p.** *P. pseudoorientale*; **plume p.** *Macleaya cordata*; **prickly p.** *Argemone mexicana*; **Shirley p.** *P. rhoeas* cvs; **water p.** *Hydrocleys nymphoides*; **Welsh p.** *Meconopsis cambrica*

Populina Baillon. Acanthaceae. 2 Madagascar

Populus L. Salicaceae. 35 N temp. (Eur. 3) with *P. ilicifolia* (Engl.) Rouleau (*Tsavo i.*, allied to *P. euphratica*) endangered sp. at low alts in E Afr. Dioec. trees – poplars (name from their growing around public squares & meeting-places in Horace), cottonwoods. Fls in catkins before lvs expand, wind-poll. Fast-growing trees for shelter-belts, pulp, comm. coppice-fuel etc., esp. hybrids, being fastest-growing trees in N temp., though suckers & roots often troublesome in drains, paving or house-walls (recommended to be planted at least 40m away); timber odourless & of low flammability, used for brake-blocks, oast-house floors, matches, veneers, chips etc. esp. *P. alba* L. (abele, white p., Euras., N Afr.) – its resistance to splitting making it suitable for matchboxes, punnets & trugs, *P. balsamifera* L. (balsam p., tacamahac, hackmatack, balm of Gilead, N N Am., temp. As.) – plywood, boxes, pulp, excelsior, *P.* × *canadensis* Moench (*P.* × *euroamericana*, *P. deltoides* (introd. France 1870s) × *P. nigra*) – fast-growing cvs imp. in match industry like 'Serotina' widely planted, 'Regenerata' tolerant of urban pollution & often pollarded; *P.* × *canescens* (Aiton) Sm. (grey p., *P. alba* × *P. tremula*), *P. deltoides* Bartram ex Marshall (cottonwood, SE US), *P. euphratica* Olivier (Arabia to Himal.) – orig. 'weeping willow', *P. grandidentata* Michaux (Canadian aspen, E N Am.), *P. heterophylla* L. (swamp cottonwood, E US) – timber for interior joinery, *P.* × *jackii* Sarg. (*P.* × *gileadensis*, poss. *P. candicans* Aiton, *P. balsamifera* × *P. deltoides*) esp. 'Gileadensis' (balm of Gilead) – suckering & known only as female; *P. maximowiczii* A. Henry (doronoki, NE As.) – timber for matches & pulp, *P. nigra* L. (black p., Euras., rare in GB where 1000 left though in C19 imp. for making wagons, clothes-pegs etc.) – timber & pulp, apical buds a source of resin used medic. (some 'balm of Gilead' coll. Eur.) & sought by bees for propolis, 'Italica' (var. *italica*, Lombardy or Italian p.) – fastigiate male mutant of As. form, planted for centuries (1758 introd. GB) but prone to collapse ('Plantierensis' better & now preferred), ssp. *betulifolia* (Pursh) Wettst. (Manchester p.) – much planted in Am., *P. tremula* L. (aspen, Euras., N Afr.) – trembling lvs due to versatile petiolar anatomy, seeds viable a few days, germ. in 6, timber imp. in match industry & gunpowder charcoal, *P. tremuloides* Michaux (Canadian a., N Am.) – clones to 10 000 yrs old, root-suckers only mode of reprod. in northern prairies, timber & medic., *P. trichocarpa* Torrey & A. Gray ex Hook. (cottonwood, W N Am.) – most commonly planted 'balsam poplar'

Porana Burm.f. Convolvulaceae. 3 trop. As. See also *Poranopsis*

Porandra Hong = *Amischotolype*

Poranthera Rudge. Euphorbiaceae. 10 Aus. (8–9 endemic), NZ

Poranopsis Roberty (~ *Porana*). Convolvulaceae. 3 trop. As. R: Novon 3(1993)200. *P. paniculata* (Roxb.) Roberty (*Porana p.*, bridal bouquet, N India & Burma) – cult. liane with white fls

Poraqueiba Aublet. Icacinaceae. 3 trop. S Am. Bat-disp. in Amazon. Locally eaten fr. *P. sericea* Tul. – lvs for treating dysentery, seed-oil for frying fish, a cultigen poss. derived from *P. guianensis* Aublet

Porcelia Ruíz & Pavón. Annonaceae. 7 trop. Am. R: SBM 40(1993)89. Scarab-poll. *P. saffordiana* Rusby (N Bolivia) – fr. to 29 kg, ed.

porcupine wood *Cocos nucifera*

Porlieria Ruíz & Pavón. Zygophyllaceae. 6 Mex., Andes. Some locally medic. & cult. orn. shrubs incl. *P. hygrometra* Ruíz & Pavón (Andes) – leaflets fold together at night

Porocystis Radlk. Sapindaceae. 2 trop. S Am.

Porodittia G. Don f. ex Kraenzlin = *Stemotria*

Porolabium Tang & F.T. Wang. Orchidaceae (IV 2). 1 Mongolia = ?

Porophyllum Adans. Compositae (Hele.-Pect.). 28 warm Am. R: UKSB 48(1969)225. Some v. smelly

Porospermum F. Muell. = *Delarbrea*

Porpax Lindley. Orchidaceae (V 14). 11 trop. As. R: BT 72(1977)1

Porphyrocoma Scheidw. ex Hook. = *Justicia*

Porphyrodesme Schltr. Orchidaceae (V 16). 3 New Guinea

Porphyroglottis Ridley. Orchidaceae (V 9). 1 W Mal.

Porphyroscias Miq. = *Angelica*

Porphyrospatha Engl. = *Syngonium*

Porphyrostachys Reichb.f. Orchidaceae (III 3). 2 Andes

Porphyrostemma Benth. ex Oliver. Compositae (Pluch.). 4 trop. Afr.

Porroglossum Schltr. Orchidaceae. 30 Andes. R: MSB 24(1987)25, 26(1988)108, 31(1989)124, 39(1991)147. Cult. orn. incl. *P. echidnum* (Reichb.f.) Garay (*Masdevallia muscosa*, Colombia) – prob. poll. by Diptera, which settle on a ridge on distal triangular part of lip; the ridge is sensitive & lifts, carrying the insect up into funnel formed by united sepal-bases so that it can only escape through tube between lip & column where pollinia & stigma are, the 'trap' remaining closed for c. 30 mins

Porrorhachis Garay. Orchidaceae (V 16). 2 Mal.

Port Jackson *Acacia saligna*; **P. J. fig** *Ficus rubiginosa*; **P. J. willow** *A. saligna*

Portea Brongn. ex K. Koch. Bromeliaceae (3). 7 E Brazil. R: FN 14(1979)2038

Portenschlagiella Tutin. Umbelliferae (III 8). 1 Adriatic

Porterandia Ridley. Rubiaceae (II 1). 9–10 W Mal. See also *Aoranthe* for Afr. spp.

Porteranthus Britton = *Gillenia*

Porterella Torrey (~ *Solenopsis*). Campanulaceae. 1 W N Am.

Porteresia Tateoka (~ *Oryza*). Gramineae (9). 1 India, Burma

Portlandia P. Browne. Rubiaceae (I 7). 6 Jamaica. R: JAA 60(1979)99

Portuguese cabbage *Brassica oleracea* Tronchuda Group

Portulaca L. Portulacaceae. 40 trop. & warm (Eur. 1). R: FR 37(1934)240. Succ. trailing usu. annual herbs with semi-inf. G & A 8–∞, s.t. sensitive to contact. Many selfing lines (jordanons) form. described as spp. now referred to *P. oleracea* L. (polyploid cosmop. weed); *P. oleracea* subsp. *sativa* (Haw.) Čelak (purslane) – potherb & locally medic. Some cult. orn. esp. *P. grandiflora* Hook. (N S Am.) – tender annual with short-lived pink, red, yellow or white fls incl. doubles (Lemoine at Nancy 1852)

Portulacaceae Juss. Dicots – Caryophyllidae – Caryophyllales. 32/380 cosmop. esp. W Am. Small trees, shrubs to herbs, often succ. with betalains & no anthocyanins; sieve-tubes with P-type plastids with central globular crystalloid & ring of proteinaceous filaments. Lvs spirally arr. or opp., simple & s.t. (e.g. *Portulaca*) with Kranz anatomy & C_4 photosynthesis; stipules(?) scarious, tufts of hairs or 0. Fls usu. reg. & bisexual, solit. or in heads or other infls; K 2(3) or up to 9 in *Lewisia*, usu. persistent, C (2–)5(–18, *Lewisia*) often ephemeral, s.t. basally connate, imbricate, A as many as & opp. C, or 1 or ∞ & in bundles, s.t. basally adnate to C, anthers with longit. slits, nectary ring or nectaries around A, $\underline{G}$ to $\overline{G}$ (2,3(–9)) with distinct styles or lobed style, eventually 1-loc. with free-central placenta with 1 or 2–∞ campylotropous or amphitropous or s.t. anatropous bitegmic ovules. Fr. a capsule, rarely indehiscent; seeds often shiny & strophiolate with peripheral ± curved embryo around abundant starchy perisperm (endosperm 0). x = 4–42+. R: AusJB 35(1987)406, FR 101(1990)237

Principal genera: *Anacampseros, Calandrinia, Cistanthe, Claytonia, Lewisia, Montia, Portulaca, Rumicastrum, Talinum* (*Hectorella* segregated as sep. fam.)

Unlike most Caryophyllales, P. have no unusual sec. thickening

Few ed. (*Portulaca*) & some cult. orn. in genera above & *Portulacaria*

Portulacaria Jacq. Portulacaceae. 2 S Afr. *P. afra* (L.) Jacq. – cult. orn. succ. shrub useful as fodder in S Afr.

Posadaea Cogn. (~ *Cucumeropsis*). Cucurbitaceae. 1 trop. Am.

posh-te *Annona scleroderma*

Posidonia C. Koenig. Posidoniaceae. 1 cosmop., 8 Aus. Tapeweeds. R: AB 20(1984)267. Fr. a drupe with spongy pericarp, free-floating but eventually dehiscent, seed with ventral wing remaining attached to young pl. for 1–2 yrs after germ. Stems used in glass-packing & fibre in coarse sacking etc., dead pl. fibres s.t. seen on seashores round & worn – posidonia balls of which tons bulldozed from Medit. beaches to 'clean' them (used to fertilize fields)

Posidoniaceae Hutch. (~ Potamogetonaceae). Monocots – Alismatidae – Najadales. 1/9 Medit. & Aus. coasts. Marine perennial glabrous herbs with flattened rhizomes & erect stems term. in lvs; vessels & stomata 0. Lvs distichous with open persistent sheath, ligule & linear blade; intravaginal scales at nodes. Fls small, bisexual, protandrous, water-poll., in (?) cymes like term. spikes with 2–4 bracts; P 3 scales or 0, A 3 (4) with sessile anthers with ± laminar connectives & longit. slits, pollen grains filamentous (pollen tubes) with 0 exine, $\underline{G}$ 1 with irreg. sessile stigma & 1 orthotropous pendulous ovule. Fr. floating & detached, eventually dehiscent; seed with straight embryo & 0 endosperm

Genus: *Posidonia*. *P. oceanica* (L.) Del. (Medit.) – to depths of 40 m

Marine forms of Potamogetonaceae to which fam. they should be moved

Poskea Vatke. Globulariaceae. 2 Somalia, Socotra

Posoqueria Aublet. Rubiaceae (II 1). 12 trop. Am. Cult. orn. shrubs & trees with fragrant fls incl. *P. latifolia* (Rudge) Roemer & Schultes (Brazilian oak, trop. Am.) – when anthers touched, pollen expelled several cm towards pollinator moth & C-tube closes; in fls with pollen removed tube reopens allowing cross-poll.; fr. ± ed., timber for walking-sticks, powdered fls used to repel fleas, bark yields blood-clotting agent used after wound from poison arrows in Amazon & recently tried as AIDS vaccine. *P. grandiflora* Standley (C Am.) – fls every 3 yrs, fr. taking 32 months to mature

possumwood *Quintinia sieberi*

Postia Boiss. & Blanche = *Rhanteriopsis*

Postiella Kljuykov. Umbelliferae (III 5). 1 Turkey

pot *Cannabis sativa* subsp. *indica*

Potalia Aublet. Gentianaceae. 1 trop. S Am.: *P. amara* Aublet – used against snakebite in Amazonia

Potaliaceae C. Martius = Gentianaceae

Potameia Thouars. Lauraceae (I). 30 Madag. (20), As. (10)

Potamoganos Sandw. Bignoniaceae. 1 NE S Am.

Potamogeton L. Potamogetonaceae. c. 90 cosmop. (Eur. 22). Pondweeds; often hybridizing (as in GB); R (N Am.): Rhodora 45(1943)57, 119, 171. Some with floating as well as submerged lvs; some with overwintering rhizomes, or tubers borne on branches or winter buds. Nitrogen-fixing bacteria in rhizosphere; useful manure (lime in lvs & animal protein also attached). Some cult. orn. in aquaria incl. *P. natans* L. (temp. & warm) – floating lvs in E Anglia form. believed to give rise to young pikes, hence local name pickerel weed, rhiz. possible starch-source

Potamogetonaceae Reichb. Monocots – Alismatidae – Najadales. 2/90 cosmop. Freshwater perennial glabrous herbs rooted in substrate (roots with vessel-elements) with creeping sympodial rhizomes & erect leafy shoots. Lvs spirally arr., with basal open sheath, ligule & parallel-veined lamina, linear or expanded atop a petiole in floating ones, s.t. attached some way down sheath which appears as a stipule. Fls bisexual, small, usu. wind-poll., reg., 4-merous in rather fleshy bractless (bracts s.t. initiated but not developed) spikes held a little above water; P 4 valvate, fleshy, with short claws, A 4 opp. & adnate to claws, anthers sessile with longit. slits, $\underline{G}$ 4 alt. with A, each with term. style or sessile stigma & 1 orthotropous bitegmic ovule often becoming campylotropous or anatropous at maturity. Fr. a head of drupelets, usu. floating (pericarp aerenchymatous); endosperm 0. x = (?)7, 13–15

Genera: *Groenlandia, Potamogeton* (*Ruppia* s.t. incl.)

Posidoniaceae & Zosteraceae, marine derivatives of this fam., should be incl.

Potamophila R. Br. Gramineae (9). 1 N NSW: *P. parviflora* R. Br. on gravel banks in streams

Potaninia Maxim. Rosaceae. 1 Mongolia. Epicalyx; K, C & A 3; G 1

Potarophytum Sandw. Rapateaceae. 1 NE S Am.

potato *Solanum tuberosum*; **air p.** *Dioscorea bulbifera*; **p. bean** *Apios americana*; **Chinese p.** *D. batatas*; **p. crisps** *S. tuberosum* esp. 'Record'; **p. fern** *Solanopteris bifrons*; **Hausa p.** *Plectranthus rotundifolius*; **p. onion** *Allium cepa* Aggregatum Group; **p. orchid** *Gastrodia sesamoides*; **sweet p.** *Ipomoea batatas*; **p. tree** *Solanum macranthum*; **p. vine** *I. pandurata*, *S. jasminoides*; **p. yam** *Dioscorea esculenta*

Potentilla L. Rosaceae. c. 500 N temp. (Eur. 75) & boreal, few S temp. Herbs & shrubs with bractlets (epicalyx) alt. with K; some cult. orn. esp. herb. perennials *P. atrosanguinea* D. Don (W & C Himal.) cvs, *P. nepalensis* Hook.f. (Pakistan to C Nepal) & their hybrids, locally medic. & ed. R: BB 71(1908). *P. anserina* L. (silverweed, N temp.) – roots ed.; *P. erecta* (L.) Räusch. (tormentil, Euras.) – dried rhiz. medic.; *P. fruticosa* L. (widdy, circumpolar, relict in GB) – common shrubby sp. in gardens, parent of many cvs; *P. lineata* Trev. (*P. fulgens*, Himal.) – cult. orn.; *P. palustris* (L.) Scop. (Eur., N Am.) – crossed with a strawberry to give 'Serenata' with large ed. fr. (should be in a separate hybrid genus); *P. reptans* L. (cinquefoil, Euras., natur. N Am.) – locally medic; *P. rivalis* Nutt. ex Torrey & A. Gray (N Am.) – cross-poll. by bees or self-poll. by thrips; *P. sterilis* (L.) Garcke (barren strawberry, Eur., Medit.)

Poteranthera Bong. Melastomataceae. 2 trop. S Am. Fls 5-merous

Poterium L. = *Sanguisorba*

Pothoidium Schott. Araceae (II). 1 C Mal.: *P. lobbianum* Schott – no spathe; fibre for fish-traps in Philippines

Pothomorphe Miq. = *Piper*

Pothos L. Araceae (II). 50 trop. OW. Usu. lianes with lvs in 2 rows, P 4–6, in bisexual fls. Some cult. orn., local medic. *P. rumphii* Schott (E Mal., Pacific) – lvs form. shredded to make 'pubic aprons' in New Ireland

Pothuava Gaudich. (~ *Aechmea*). Bromeliaceae (3). 21 trop. Am. R: Phytol. 66(1989)75

Potoxylon Kosterm. (~ *Eusideroxylon*). Lauraceae (I). 1 Borneo: *P. melagangai* (Sym.) Kosterm.

pottery tree of Pará *Licania octandra*

Pottingeria Prain. Celastraceae (Pottingeriaceae). 1 Assam to NW Thailand. Venation like *Cinnamomum*

Pottingeriaceae (Engl.) Takht. = Celastraceae

Pottsia Hook. & Arn. Apocynaceae. 4 India to Java

Pouchetia A. Rich. ex DC. Rubiaceae (II 5). 6 trop. Afr.

Poulsenia Eggers. Moraceae (II). 1–2 trop. Am. Young lvs imp. protein source for red spider-monkeys. Bark fibre used for cloth, mats etc.

pounce *Tetraclinis articulata*

Pounguia Benoist = *Whitfieldia*

Poupartia Comm. ex Juss. Anacardiaceae (II). 12 trop. Some ed. fr.

Pourouma Aublet. Cecropiaceae. 25+ trop. Am. R: FN 50(1990)116. Some ed. fr. esp. *P. cecropiifolia* C. Martius (uvilla, Brazil) – excellent grape flavour

Pourthiaea Decne = *Photinia*

Pouteria Aublet. Sapotaceae. (Incl. *Planchonella*) c. 200 trop. Am., 120 As. to Pacific, 5 Afr. R: T.D. Pennington, S. (1991)184. Trees & shrubs incl. some rheophytes, some fish-disp. Amazon. Alks. Some timber (anegré, aniègre) & ed. fr. (abiu) esp. *P. cainito* (Roemer & Schultes) Radlk. (trop. Am.); *P. campechiana* (Kunth) Baehni (canistel, C Am., WI) – ed. pulp & *P. sapota* (Jacq.) H. Moore & Stearn (sapote, marmalade plum, mammee zapote, C Am.) – imp. Carib. fr., fresh or preserved. *P. australis* (R. Br.) Baehni (*Planchonella* a., black or brush apple, trop. Aus.) – timber

Pouzolzia Gaudich. Urticaceae (III). 70 trop. Fibres esp. for fishing-nets – *P. mixta* Solms (*P. hypoleuca*, tingo, trop. Afr.) & *P. sanguinea* (Blume) Merr. (*P. viminea*, Mal.). *P. tuberosa* Wight (Indomal.) – ed. tubers

Poveodaphne Burger (~ *Ocotea*). Lauraceae (I). 1 Costa Rica: *P. quadriporata* Burger. R: Brittonia 40(1988)276

poyok *Licania elaeosperma*

Pozoa Lagasca. Umbelliferae (I 2b). 2 Andes of Chile & Arg. R: UCPB 33(1962)131

pracaxi fat *Pentaclethra macroloba*

Pradosia Liais. Sapotaceae (IV). 23 trop. S Am., 1 ext. to C Am.

Praecereus F. Buxb. = *Cereus*

Praecitrullus Pang. Cucurbitaceae. 1 India: *P. fistulosus* (Stocks) Pang. – cultigen

Prainea King ex Hook.f. Moraceae (II). 4 Mal. R: JAA 40(1959)30

prairie gentian *Eustoma grandiflorum*; **p. senna** *Chamaecrista fasciculata*

Pranceacanthus Wasshausen (~ *Juruasia*). Acanthaceae. 1 Amaz. Brazil: *P. coccineus* Wasshausen. R: Brittonia 36(1984)1

Prangos Lindley (~ *Cachrys*). Umbelliferae (III 5). 38 Euras., Medit. R: Boissiera 26(1977)

Praravinia Korth. Rubiaceae (I 9). 50 C Mal.

Prasium L. Labiatae (VI). 1 Medit. incl. Eur.: *P. majus* L. Drupes

Prasophyllum R. Br. Orchidaceae (IV 1). 60 Aus. (58, 57 endemic), NZ. Leek orchids; fls not resupinate, mainly wasp-poll. Some cult. orn.

Pratia Gaudich. = *Lobelia*

Pravinaria Bremek. Rubiaceae (I 9). 2 Borneo

Praxeliopsis G. Barroso. Compositae (Eup.-Prax.). 1 Brazil. R: MSBMBG 22(1987)392

Praxelis Cass. (~ *Eupatorium*). Compositae (Eup.-Prax.). 15 trop. Am. R: MSBMBG 22(1987)380

prayer plant *Maranta leuconeura*

Premna L. Labiatae (I; Verbenaceae). c. 50 trop. & warm OW. Alks. Pyrogenous forms form. distinguished as *Pygmaeopremna*. Some locally medic. & some timbers esp. *P. serratifolia* L. (*P. integrifolia*, headache tree, Indomal. to Pacific) – timber beautifully marked & used for cutlery handles, *P. lignumvitae* (Schauer) Pieper (Queensland lignum-vitae, NE Aus.)

Prenanthella Rydb. Compositae (Lact.-Step.). 1 SW N Am. deserts: *P. exigua* (A. Gray) Rydb. R: Brittonia 24(1972)223

× **Prenanthenia** Svent. = *Prenanthes* × *Sventenia*

Prenanthes L. Compositae (Lact.-Lact.). 30 N temp. (Eur. 1) to Afr. mts. *P. subpeltatum* Stebb. (Afr. mts) – climber; other spp. locally medic.

Prenia N.E. Br. = *Phyllobolus*

Prepodesma N.E. Br. = *Aloinopsis*

Prepusa C. Martius. Gentianaceae. 5 Brazil

Prescotia Lindley = seq.

Prescottia Lindley (*Prescotia*). Orchidaceae (III 3). 21 trop. Am.

Preslia Opiz = *Mentha*

Prestelia Schultz-Bip. ex Benth. & Hook.f. (~ *Eremanthus*). Compositae (Vern.-Vern.). 1 Brazil

Prestoea Hook.f. Palmae (V 4f). 11 C Am. R: AMBG 75(1988)203. Palm-hearts

Prestonia R. Br. Apocynaceae. 65 trop. Am. Lianes with alks, some hallucinogens
Preussiella Gilg. Melastomataceae. 2 trop. W Afr. R: Adansonia 16(1976)405
Preussiodora Keay. Rubiaceae (II 1). 1 trop. W Afr.
Priamosia Urban. Flacourtiaceae (8). 1 Hispaniola. Allied to *Xylosma*
prickly ash *Orites excelsa*; **p. pear** *Opuntia* spp.; **p. poison** *Gastrolobium spinosum*; **p. poppy** *Argemone mexicana*
Pridania Gagnepain = *Pycnarrhena*
pride of Burma *Amherstia nobilis*; **p. of India** *Lagerstroemia speciosa, Koelreuteria paniculata*
Priestleya DC = *Liparia*
Prieurella Pierre = *Chrysophyllum*
primavera *Cybistax donnell-smithii*
primrose *Primula vulgaris*; **bird's-eye p.** *P. farinosa*; **Cape p.** *Streptocarpus* spp.; **Chinese p.** *P. sinensis*; **evening p.** *Oenothera* spp.
Primula L. Primulaceae. 425 N hemisph. (Eur. 33, China 294), incl. Ethiopia & trop. As. mts to Java & New Guinea, S S Am. (*P. magellanica* Lehm. (*P. farinosa* var. *magellanica*)). R: W.W. Smith, H.R. Fletcher & G. Forrest (reprinted 1977) *The genus P.*, G.K. Fenderson, *Synoptic guide ... P.* (1986); arranged in 37 sects. Herbs with rhizomes & term. infls, s.t. a series of whorls as in *P. verticillata*, an umbel in *P. veris* where 'stalk' is peduncle & fls erect until anthesis when pendent but becoming erect again in fr., or reduced peduncle hidden in basal rosette of lvs as in *P. vulgaris* where 'stalk' is an elongated pedicel. Heterostyly, with pl. either bearing fls with short styles & anthers at mouth of fl. ('thrum') or long styles with anthers down inside fl. ('pin'), the diff. morphs with diff. pollen grains & stigmatic surfaces, the whole system promoting outcrossing when pollinators (bees & Lepidoptera) receive pollen in a position appropriate to depositing it on stigmas in the fls of second morph; homostyles, where system has broken down, incl. *P. scotica* Hook. (Scotland, one of v. few British endemic pls, now in decline; 2n = 54) a derivative of *P. farinosa* L. (Euras., 2n = 18, 36), also populations of *P. prolifera* Wallich in Java, the pl. in Himal. being heterostyled & *P. eximea* Greene a widespread colonist in Alaska & nearby Russia derived from heterostylous *P. tschuktschorum* Kjellm. (Bering Strait). Many spp. with 'farina', a wax from glands on lvs, abaxial surfaces of which frequently bearing glands with primin, a flavone allergenic to c. 6% humans (0.001 mg may suffice), leading to frequently delayed dermatitis & further complications. Many cult. orn. hardy & greenhouse pls, the latter esp. *P. obconica* Hance (China) with umbels of fls with notched petals, *P. sinensis* Sabine ex Lindley (*P. praenitens*, Chinese primrose, China) with whorls of large fls & *P. malacoides* Franchet (China) with whorls of small fls & less allergenic than *P. obconica*, all treated as annuals for winter flowering; most familiar hardy pls incl. *P. vulgaris* Hudson (primrose, Eur.) – many cvs incl. doubles & 'Hose in Hose' with one corolla inside another & 'Jack in the Green' with leafy K & *P.* × *polyantha* Miller (polyanthus, known since 1660, prob. complex hybrid group involving *P. vulgaris*, *P. elatior* (L.) Hill (oxlip, Eur. to Iran) – indicator of woods >100 yrs old in E England & *P. veris*) – many cvs. Other cult. orn. incl. *P. allionii* Lois. (Alps) – many cvs & hybrids; *P. auricula* L. (auricula, bear's ears, Alps) – largely calcicole, fls yellow, form. much cult. as 'florist's flower' esp. in mining communities in GB, esp. forms with farina & prob. some of hybrid origin (*P.* × *pubescens* Jacq., other parent = *P. hirsuta* All. (*P. rubra*, C Alps, Pyrenees, calcifuge, fls pink with white eye), form. medic., first cult. Vienna c. mid C16, like polyanthus the subjects of florists' societies esp. in C17 & C18 with cvs costing up to £20 by 1682 (cf. *Tulipa*) & those with green-edged C selected by mid C18, *P. sieboldii* C.J. Morren (NE As., Japan) a similar cult in C18 Japan; *P. denticulata* Sm. (Himal.) – 'drumsticks' of pink or purple yellow-eyed fls; *P. farinosa* L. (bird's-eye primrose, sub-boreal & boreal OW with closely allied pl. in N Am.); *P. florindae* Kingdon-Ward (SE Tibet) – 'cowslip' over 1m tall; '*P. kewensis*' (enigmatic greenhouse pl. alleged to be allopolyploid (2n = 36) derived from sterile hybrid *P.* × *kewensis* Will. Watson (2n = 18), a cross between *P. verticillata* Forssk. (Ethiopia, S Arabia, 2n = 18) & *P. floribunda* Wallich (Himal., 2n = 18) but sterile & with 'thrum' fls) – the pl. now grown as '*P. kewensis*' allegedly arising as a cross between a shoot bearing a 'pin' fl. & the original (the chance of the double recessive 'pin' arising as a somatic mutation on double dominant 'thrum' plants is so slim that the story is prob. more complex); *P. marginata* Curtis (Alps) – many cvs & hybrids with 'auriculas'; *P. soldanelloides* Watt (E Himal.) – minute with nodding fls; *P. veris* L. (cowslip (*primula veris* = Lat. firstling of spring), paigle or palsywort (arch.), Eur.) – form. medic. incl. tea from lvs, fls (fragrant) form. used in cowslip wine; *P. vialii* Delavay ex Franchet (China) – dense spikes of small pinkish fls resembling an *Orchis* sp.; also many hybrids of complex ancestry incl. Candelabra Group ('C. Hybrids', sect. *Proliferae*, E As.), Petiolares Group ('P. Hybrids',

Himalaya), Pruhonicensis Group ('P. Hybrids', *P. juliae* Kuzn. (Cauc.) × spp. in sect. 'Vernales' (*Primula*)) esp. 'Wanda'

Primulaceae Batsch ex Borkh. Dicots – Dilleniidae – Primulales. 22/825 subcosmop. but esp. N hemisph. Herbs, rarely subshrubby. Lvs spirally arr., opp. or whorled, usu. simple & ± toothed (pinnatisect in *Hottonia*), often basal; stipules 0. Fls bisexual, reg. (irreg. in *Coris*), (3–)5(–9)-merous, often heterostylous, in umbels, panicles, heads or solit.; (K) often persistent, (C) with imbricate or convolute lobes (C 3 in *Pelletiera*; C 0 in *Glaux*), A opp. C & attached in tube, anthers introrse, with longit. slits or term. pores, staminodes alt. with A s.t. present, G̲ (5), half-inf. in *Samolus*, 1-loc. with rudimentary partitions at base, 1 style & (5–)∞ hemitropous or anatropous, bitegmic ovules on free-central placenta (campylotropous & unitegmic in *Cyclamen*). Fr. a capsule (s.t. circumscissile), rarely indehiscent, with (1–) ± ∞ seeds; embryo linear or short, straight (1 cotyledon in *Cyclamen*), embedded in copious starchless endosperm with reserves of oil, protein & amylose. x = 5, 8–15, 17, 19, 22

Classification & principal genera:

Primuleae (G̲, C imbricate, capsule with valves): *Androsace, Dionysia, Dodecatheon, Hottonia, Primula, Soldanella*

Cyclamineae (G̲, C convolute, capsule with valves, seeds unitegmic & monocot., tubers): *Cyclamen* (only)

Lysimachieae (G̲, C convolute, capsule with valves or circumscissile & bitegmic seeds): *Anagallis, Lysimachia*

Samoleae (G half-inf.): *Samolus* (only)

Corideae (fls irreg., subshrubby): *Coris* (only; s.t. segregated as sep. fam.)

One of 10 fams with elaiophores attractive to bees (*Lysimachia*)

Many cult. orn. in above-mentioned genera, *Cortusa* etc.

Primularia Brenan = *Cincinnobotrys*

Primulina Hance. Gesneriaceae. 1 China (N Guangdong): *P. tabacum* Hance – leaf glands said to smell of tobacco

Prince Albert's yew *Saxegothaea conspicua*; **p.'s feather** *Amaranthus hybridus* var. *erythrostachys*

Princea Dubard & Dop = *Triainolepis*

princewood *Cordia gerascanthus, Exostema caribaeum*

Principina Uittien. Cyperaceae. 1 Principé (W Afr.)

Pringlea T. Anderson ex Hook.f. Cruciferae. 1 Kerguélen & Crozet Is. (S Indian Ocean): *P. antiscorbutica* R. Br. ex Hook.f. (Kerguélen cabbage) – pachycaul giant cabbage with lat. infls of usu. wind-poll. C-less fls, though s.t. (allegedly in shaded sites) with C, app. s.t. also visited by thrips; no septum in G. Pollen evidence supports theory of its being a relic of Tertiary flora; antiscorbutic

Pringleochloa Scribner. Gramineae (33c). 1 Mexico

Prinsepia Royle. Rosaceae. 4 Himal. to N China & Taiwan. Thorny decid. shrubs with lat. styles & differing from *Prunus* in having chambered pith in stems. Some oilseeds & cult. orn. esp. *P. sinensis* (Oliver) Oliver (E As.) with yellow fls

Printzia Cass. Compositae (Gnap.). 6 Afr. R: MBSM 16(1980)108. Heterogeneous?

Prionanthium Desv. Gramineae (25). 3 Cape. R: Bothalia 18(1988)143. Annuals on seasonally wet flats

Prionium E. Meyer. Juncaceae. 1 S Afr. mt. streams: *P. serratum* (L.f.) E. Meyer (palmiet) – pachycaul aloe-like pl. to 2 m with stems clothed in marcescent leaf-bases & runners

Prionophyllum K. Koch = *Dyckia*

Prionopsis Nutt. (~ *Haplopappus*). Compositae (Ast.-Sol.). 1 C & S US

Prionosciadium S. Watson. Umbelliferae (III 9). 8 Mexico

Prionostemma Miers. Celastraceae. 5 trop. Am., Afr., India

Prionotaceae Hutch. = Epacridaceae

Prionotes R. Br. Epacridaceae. 1 Tasmania: *P. cerinthoides* R. Br. – scrambling epiphyte, cult. orn.

Prionotrichon Botsch. & Vved. Cruciferae. 4 C As. to Afghanistan

Prioria Griseb. Leguminosae (I 4). 1 trop. Am.: *P. copaifera* Griseb. (cativo) – resins used in euglossine bees' nests, timber for coarse construction, veneer etc.

Prismatocarpus L'Hérit. Campanulaceae. 30 trop. (1) & S esp. SW Afr.

Prismatomeris Thwaites. Rubiaceae (IV 10). 15 Sri Lanka to S China & W Mal. R: OB 94(1987)5

Pristiglottis Cretz. & J.J. Sm. Orchidaceae (III 3). 13 Sumatra to Samoa

Pristimera Miers. Celastraceae. 26 trop. R: BMNHNAdans. 3, 1(1981)8

Pritchardia Seemann & H. Wendl. Palmae (I 1b). 25 Fiji, Samoa, Tonga, Tuamotus, Hawaii (19). Fan palms with bisexual fls; some cult. orn. & many endangered spp. known only from 1 tree in cult.

Pritchardiopsis Becc. (~ *Licuala*). Palmae (I 1b). 1 New Caled.: *P. jeanneneyi* Becc. – almost extinct save 3 small pops on serpentine through overexploitation of palm cabbage (apical bud). R: Allertonia 3(1984)317

Pritzelago Kuntze (*Hutchinsia*). Cruciferae. 1 C & S Eur. mts: *P. alpina* (L.) Kuntze (*H. a.*) – cult. orn. rock-pl. R: Willdenowia 15(1985)68

Priva Adans. Verbenaceae. 20 trop. & warm. R: FR 41(1936)1

privet *Ligustrum* spp., esp. *L. vulgare* (**common** or **European p.**) & *L. ovalifolium*; **Chinese p.** *L. sinense*; **glossy p.** *L. lucidum*; **golden p.** *L. vulgare* 'Aureum'; **Japanese p.** *L. japonicum*

Proboscidea Schmidel. Pedaliaceae (Martyniaceae). 9 warm Am. Cult. orn. & ed. fr. esp. *P. louisianica* (Miller) Thell. (warm US, introd. Aus.) – fr. interferes with shearing of sheep & can get in eyes or even clamp mouth of sheep, but young fr. pickled & pl. grown comm. for this; strips from endocarp used by N Am. Indians in black or dark designs in basketry; some oilseeds esp. *P. parviflora* (Wooton) Wooton & Standley (S N Am.) & *P. triloba* (Schldl. & Cham.) Decne (Mex.) – poss. crops

Prochnyanthes S. Watson. Agavaceae. 1 Mex.: *P. mexicana* (Zucc.) Rose – used medic. & in washing clothes etc.

Prockia P. Browne ex L. Flacourtiaceae (7). 3 trop. Am. (1 restricted to Venez.)

Prockiopsis Baillon. Flacourtiaceae (3). 1 Madagascar

Procopiania Guşul. = *Symphytum*

Procris Comm. ex Juss. Urticaceae (II). 16 OW trop. R: FR 45(1938)179, 257

Proferea C. Presl = *Sphaerostephanos*

Proiphys Herbert (*Eurycles*). Amaryllidaceae (Liliaceae s.l.). 3 Mal. to trop. & E Aus. R: FA 45(1987)376. 'Seeds' actually bulbils. Some cult. orn. incl. *P. amboinensis* (L.) Herbert (Mal. to Aus.) & *P. cunninghamii* (Lindley) Mabb. (Brisbane lily, SE Queensland, NSW)

Prolobus R. King & H. Robinson. Compositae (Eup.-Gyp.). 1 Bahia (Brazil). R: MSBMBG 22(1987)100

Prolongoa Boiss. (~ *Chrysanthemum*). Compositae (Anth.-Leuc.). 1 Spain: *P. hispanica* G. López & Jarvis

Promenaea Lindley. Orchidaceae (V 10). 14 Brazil. Cult. orn. epiphytes

Prometheum (A. Berger) H. Ohba (~ *Sedum*). Crassulaceae. 2 Euras. R: JFSUTB III, 12(1978)168

Pronaya Huegel. Pittosporaceae. 1 SW Aus.: *P. fraseri* (Hook.) E. Bennett

Pronephrium C. Presl = *Cyclosorus*

prophet flower *Arnebia pulchra*

propolis resin coll. from buds of poplars, horse-chestnut etc. by bees used to fix combs in hives (bee-glue)

Prosanerpis S.F. Blake = *Clidemia*

Prosaptia C. Presl = *Grammitis*

Prosartes D. Don = *Disporum*

Proscephaleium Korth. (~ *Psychotria*). Rubiaceae (IV 7). 1 Java

Proserpinaca L. Haloragidaceae. 2–3 E N Am. to WI. Fls 3-merous. *P. palustris* L. (N Am. to WI) – sometimes cult. in aquaria

proso millet *Panicum miliaceum*

Prosopanche Bary. Hydnoraceae. 2 Paraguay, Argentina

Prosopidastrum Burkart. Leguminosae (II 3). 1 Mex., 1 Patagonia

Prosopis L. Leguminosae (II 3). 44 warm Am. (most), SW As., Afr. Usu. spiny trees or shrubs of value in dry country for browse, agro-forestry, ed. fr. etc. R: JAA 57(1976)219, 450. Alks. *P. chilensis* (Molina) Stuntz (algaroba, Arg. & Chile) – fast-growing in dry conditions, locally cult. for cattle fodder & sweet-tasting legumes; *P. cineraria* (L.) Druce (*P. spicigera*, Iran to India) – ed. fr., bark fibre & local timber; *P. glandulosa* Torrey (algaroba, mesquite, SW N Am.) – timber, fuel, gum; *P. pallida* (Willd.) Kunth (S Am., natur. Hawaii (derived from 1 pl. 1828) & Aus.) – salt-tolerant, seeds disp. animals (endozoic), legume ed. & source of sweet syrup used in drinks; *P. tamarugo* F. Philippi (N Chile) – only tree to survive in arid salt flats of N C., where ann. rainfall 70 mm, water-table at 20 m & salt 0.5 m thick; *P. velutina* Wooton (SW N Am.) – a million fls per tree, attracting hundreds of insect spp. incl. 60 spp. of solitary bee

Prosopostelma Baillon. Asclepiadaceae (III 1). 3 Madagascar

Prosphytochloa Schweick. Gramineae (9). 1 S Afr.: *P. prehensilis* (Nees) Schweick. – climbing to 10 m by wind bringing leaf-tips with retrorse scabrid surface in contact with supports

Prostanthera Labill. Labiatae (III). c. 100 Aus. (R (sect. *Klanderia*, 15): JABG 6(1984)207). Many natural hybrids. Some cult. orn. (mintbush)

Prosthecidiscus F.D. Sm. Asclepiadaceae (III 3). 1 C Am.: *P. guatemalensis* F.D. Sm.

Protarum Engl. Araceae (VII 3). 1 Seychelles: *P. sechellarum* Engl.

Protasparagus Oberm. = *Asparagus*

Protea L. Proteaceae. c. 100 trop. (high alts, 21) & S Afr. (82 (R: J.P. Rourke (1980) *The Ps. of S Afr.*), 69 endemic in SW Cape, *P. gaguedi* J.F. Gmel. from Ethiopia to Namibia). Fls in large heads with much nectar (sugar bushes esp. *P. repens* (L.) L. (*P. mellifera*, Cape – visited by Cape Sugar Bird, good bee-pl.); some cult. orn. & used as long-lasting cut-fls esp. *P. cynaroides* (L.) L. (Cape) with heads to 20 cm across

Proteaceae Juss. Dicots – Rosidae – Proteales. 77/1600 trop. & subtrop. esp. S hemisph. (most in Aus. & S Afr.). R: BJLS 70(1975)83. Evergreen trees & shrubs usu. with 3-celled hairs, often accum. aluminium; roots without mycorrhizae but often with short lateral ('proteoid') roots releasing organic acids making otherwise inaccessible phosphorus available for absorption. Lvs usu. spirally arr., simple (usu.) to pinnate or bipinnate, often xeromorphic; stipules 0. Fls usu. bisexual (or pl. monoec. or dioec.), reg. or not, protandrous, poll. by insects, birds or marsupials, 4-merous, solit. or paired in axil of bract, in racemes, umbels or involucrate heads or primary 2-flowered infls in sec. racemes; K valvate, often petaloid, usu. with basal tube s.t. cleft most on 1 side or 3 connate & 1 free, C app. an annular ± 4-lobed nectary disk or (2–)4 scales or glands alt. with K, or 0, A opp. K with broad filaments usu. adnate to K & anthers with longit. slits & 1 theca often sterile & often elongated connective, $\underline{G}$ 1, s.t. not closed, with elongate style s.t. a pollen-presenter & 1 or 2 (– ± ∞) marginal (anatropous to) hemitropous or amphitropous (or orthotropous) bitegmic ovules. Fr. a follicle, nut, achene or drupe, often 1-seeded when indehiscent; seeds often winged, embryo straight & oily with 2(–8 in *Persoonia*) cotyledons & usu. 0 endosperm. x = 5, 7, 10–13 (? orig. 7), the chromosomes s.t. v. large

Classification & principal genera:

I. **Persoonioideae** (*sensu* Johnson & Briggs; fls solit. in axils, style not a pollen-presenter, ovules orthotropous, chromosomes large; 2 tribes, As. to Aus.): *Bellendena, Persoonia*

II. **Proteoideae** (fls solit. in axils, chromosomes small at 4x & higher levels, fr. indehiscent, 1(2)-seeded; 3 tribes, OW): *Adenanthos, Conospermum, Leucadendron, Leucospermum, Protea*

III. **Sphalmioideae** (fls solit. in axils of bracts of raceme, andromonoec., reg., style not a pollen-presenter, ovules 2 anatropous, follicles with winged seeds, chromosomes small, NE Aus.): *Sphalmium* (only)

IV. **Carnarvonioideae** (partly digitate, partly pinnately divided lvs, fls loosely grouped but not paired, reg., bisexual, style not a pollen-presenter, ovules 2 hemitropous, follicles with winged seeds, chromosomes small, NE Aus.): *Carnarvonia* (only)

V **Grevilleoideae** (fls paired; 7 tribes, mostly Aus., some Am., few Afr.): *Banksia, Dryandra, Euplassa, Grevillea, Hakea, Helicia, Macadamia, Roupala*

App. primitively rain-forest trees, the xeromorphic features sec. & of parallel origin in Aus., S Afr., New Caled. & to lesser extent S Am. The geoflorous habit assoc. with mammalian poll. & burning also in parallel in *Dryandra* (SW Aus.) & *Protea* (S Afr.). Some cult. orn. esp. *Banksia, Embothrium, Gevuina, Grevillea, Persoonia* etc., some timber (esp. *Cardwellia, Grevillea, Knightia, Orites, Panopsis* & *Stenocarpus*), medic. (*Conospermum*) & ed. seeds, esp. *Macadamia* but also *Brabejum, Finschia, Floydia, Gevuina* & *Hicksbeachia*; some weedy e.g. *Hakea* in S Afr.

Proteopsis C. Martius & Zucc. ex Schultz-Bip. Compositae (Vern.- Lych.). 2 S Brazil campos

Protium Burm.f. Burseraceae. 85 trop. Am., Madag., Mal. Source of resins (used by euglossine bees in their nest-making) esp. in trop. Am. esp. *P. carana* (Humb.) Marchand (Carana elemi) & *P. heptaphyllum* (Aublet) Marchand (Brazilian e., incense tree); *P. javanicum* Burm.f. (Mal.) – durable timber

Protoceras Joseph & Vajravelu. Orchidaceae (V 16). 1 S India

Protocyrtandra Hosok. = *Cyrtandra*

Protogabunia Boit. = *Tabernaemontana*

Protolirion Ridley. (~ *Petrosavia*). Melanthiaceae (Liliaceae s.l.). Spp.?

Protomarattia Hayata = *Angiopteris*

Protomegabaria Hutch. Euphorbiaceae. 2 trop. W Afr. Noisy slash

Protorhus Engl. Anacardiaceae (1V). 2 S Afr., 20 Madag. (seed-structure suggests perhaps a separate genus)

Protoschwenkia Soler. Solanaceae (2). 1 Bolivia: *P. mandonii* Soler.

Protowoodsia Ching = *Woodsia*

Proustia Lagasca. Compositae (Mut.-Nass.). 3 Bolivia, Chile & Arg. R: RMLP 11(1968)23. Shrubs

Provancheria B. Boivin (*Provencheria*) = *Cerastium*

Prumnopitys Philippi (~ *Podocarpus*). Podocarpaceae. 8 Costa Rica to Venez., S Chile, NZ, New Caled. R: Blumea 24(1978)189, Phytol.M 7(1984)68. Timber. *P. andina* (Endl.) Laubenf. (Chile) – seed ed. (no resins); *P. ferruginea* (D. Don) Laubenf. (miro, NZ) – timber for cabinet-work, gum medic.; *P. taxifolia* (D. Don) Laubenf. (*Podocarpus spicatus*, matai, NZ) – timber for flooring, railway sleepers etc.

prune *Prunus domestica*

Prunella L. Labiatae (VIII 2). 4 N temp. (Eur. 4), NW Afr. *P. vulgaris* L. (selfheal, Euras.) – form. medic. esp. for sore throat; some ground-cover etc.

Prunus L. Rosaceae. 200+ temp. (esp. N; Eur. 17), trop. mts. Decid. & evergreen trees & shrubs, s.t. spiny, sometimes with domatia or extrafl. nectaries (e.g. on leaves of *P. serotina* which attract *Formica obscuripes* ants repelling caterpillar attack, the nectar flow ceasing once caterpillars too big to be killed by ants); imp. fr. trees – plums, cherries, almonds, peaches, apricots etc., cult. orn. & gums. Many hybrids incl. most 'flowering cherries' most common cvs (e.g. the ubiquitous fastigiate 'Amanogawa' of suburbia) grouped as *P.* Sato-zakura Group. Subg. s.t. recog. as separate genera: *Prunus* s.l. (*P. spinosa* & allies), *Cerasus* Miller (*P. avium* & allies), *Armenaica* Scop. (*P. armenaica*), *Persica* Miller (*P. persica* & allies), *Amygdalus* L. (*P. dulcis* & allies), *Padus* Miller (*P. padus* & allies) & *Laurocerasus* Duham. (*P. laurocerasus* & allies). *P. africana* (Hook.f.) Kalkman (*Pygeum a.*, red stinkwood, trop. & S Afr.) – comm. treatment for prostitis, timber form. used for wagons etc.; *P. alleghaniensis* Porter (Alleghany plum, E N Am.); *P. angustifolia* Marshall (Chickasaw plum, SE US) – cult. ed. fr.; *P. armeniaca* L. (apricot, N China where cult. since 2000 BC) – fr. (rich in vitamin A) eaten fresh (10 monoterpenes active in flavour compared with peach) or tinned, juice, brandy & liqueurs, seed-oil used in cosmetics etc., old cvs incl. 'Moor Park' mentioned by Jane Austen (*Mansfield Park*); *P. avium* (L.) L. (gean, mazzard, wild cherry, hagberry, Euras.) – self-compatible, 2n = 16 (cf. *P. cerasus*), source of sweet cherries, timber for veneers, walking-sticks etc.; *P. brigantina* Villars (Briançon apricot, SE France) – seed-oil scented; *P. cerasifera* Ehrh. (myrobalan, mirabelle, cherry plum, C As. to Balkans (or cultigen?)) – small ed. fr., often candied, stock for plums; *P. cerasus* L. (morello or amarelle cherry, origin unclear, widely natur.) – self-incompatible, 2n = 32, source of sour cooking cherries, 'Marasca' (var. *marisco*, maraschino cherry or marasco) used in liqueurs, & crossed with *P. avium* to give Duke cherries (*P.* × *gondouinii* (Poit. & Turpin) Rehder), lvs a tea-source, gum form. used in cotton printing (cherry gum); *P.* × *domestica* L. (plum, gage, ancient cultigen, poss. orig. SW As. but hexaploid & of complex background poss. involving *P. cerasifera* & *P. spinosa*) – the familiar plum with many cvs used fresh (form. much dried), tinned, or as juice, in liqueurs etc., timber used for turnery, gum to flavour cider, but plums in US may be native Am. spp. or (or hybrids with) *P. salicina* Lindley (China, Korea) – all diploids, dried fr. esp. of *P.* × *domestica* being prunes, an inf. type being the quetsche(n) plum, while the Victoria plum is a self-fertile form found in a garden in Sussex in 1840, 'ssp. *institia*' (*P. institia*) – sweet forms = 'mirabelle', others being damson (bullace) – usu. preserved, 'ssp. *italica*' = greengage, hybrids with apricots = 'plumcots'; *P. dulcis* (Miller) D. Webb (*Amygdalus communis*, almond, W As.) – ed. seed for dessert & confectionery, var. *dulcis* sweet (sweetness a result of a dominant gene) the most grown of all 'nuts', ground for macaroons & cakes or coated in sugar ('sugared a.s'), var. *amara* (DC) H. Moore – bitterer, source of almond oil used medic. (e.g. earache, allegedly keeping cholesterol levels low) & as hair-conditioner, both much cult. in Medit., Calif., Aus. etc., the 2nd s.t. as stock for first, endocarps ground for fibre used in clay-tile-making in SW Eur.; *P. gracilis* Engelm. & A. Gray (Oklahoma plum, Arkansas to Texas) – fr. ed. & form. much dried by Indians for winter use; *P. hortulana* L. Bailey (wild goose plum, C & SE US) – several cvs with ed. fr.; *P. laurocerasus* L. (cherry laurel, SE Eur., SW As.) – the laurel of suburbia, usu. clipped as hedge but a tree to 14 m with many cvs, lvs with alks & the glycoside prulaurasine which is broken down to release HCN by an enzyme when lvs damaged, so used in 'killing' bottles by entomologists, glands at base of lamina extrafl. nectaries visited by ants poss. homologous with lateral leaflets of pinnate-leaved Rosaceae; *P. lusitanica* L. (Portugal laurel, SW Eur., Macaronesia) – evergreen resembling bay; *P. mahaleb* L. (mahaleb (cherry), St Lucie c., Euras., introd. N Am.) – timber for turnery & tobacco pipes, lvs a flavouring incl. in tobacco; *P. maritima* Marshall (beach plum, E N Am.) – ed. fr. & sand-binder; *P. mume* Siebold & Zucc. (mume, Japanese apricot, SW Japan) – winter-flowering 'plum blossom' of Japan much grown & illus. there, used as bonsai & ed. fr.; *P. munsoniana* W. Wight & Hedr. (wild goose plum, N Am.) – ed. fr., several cvs; *P. nigra* Aiton (Canada p., E N Am.) – ed. fr., some cvs, planted by Iroquois in New York State; *P. padus* L. (bird cherry, hagberry, Eur. to Japan) – timber for interior work, boats etc.; *P. persica* (L.) Batsch (peach, China, poss. cultigen derived from *P. davidiana* (Carr.) N.E. Br. (China), early natur. N Am. prob. via Florida in C16 but in Carolinas & Georgia before English arrived) – flavour due to undecalactone alone (cf. apricot), next to apple world's most widely grown tree-fr., dwarf forms produced by budding on *P. tomentosa* etc., many cvs for fr. eaten fresh ('free-stoned'), dried, tinned ('cling-stoned'), source of juice, brandy, ground seeds in certain Italian sweetmeats (e.g. 'Amaretto'), var. *nucipersica* C. Schneider (var. *nectarina*, nectarine) with smooth fr. skin; *P. pumila* L. (sand cherry, NE US) – shrub sometimes prostrate (*P. depressa*) forming mats to 2 m across; *P. salicina* Lindley (Japanese plum, China, Korea) – ed. fr., see *P.* × *domestica*; *P. serotina* Ehrh. (American or black or cabinet or rum cherry, capulin, E N Am.) – timber for cabinet-making, bark form. medic., flavouring for rum & brandy; *P. serrulata* Lindley (NE As.) – many cvs, prob. contributing towards the bulk of the Jap. flowering cherries ('*P. lannesiana*'), the wild form (var. *spontanea* (Maxim.) E. Wilson) being the Jap. cherry of poetry & painting; *P. simonii* Carr. (apricot plum, N China, not known wild) – bitter but ed. fr.; *P. spinosa* L. (sloe, blackthorn, Eur., W As.) – spiny shrub or small tree with fr. used in liqueurs incl. sloe gin, etc., wood for turnery, hay-rake teeth, shillelaghs etc., lvs form. tea subs., fr. eaten by birds, badgers & foxes in GB – disp. agents; *P. subhirtella* Miq. (Japan) – winter-flowering esp. 'Autumnalis' & 'A. Rosea' & hybrids with other spp.; *P. tenella* Batsch (dwarf Russian almond, C Eur. to Siberia) – dwarf suckering cult. orn.; *P. tomentosa* Thunb. (Nanking cherry, temp. E As.) – sweet ed. fr.; *P. virginiana* L. (chokecherry, E N Am.) – fr. ed. cooked, bark medic.; *P.* × *yedoensis* Matsum. (most frequently cult. flowering cherry in Japan) – poss. hybrid between *P. speciosa* (Koidz.) Ingram (? = *P. serrulata*, Japan) & *P. subhirtella*

pry *Tilia cordata*

Przewalskia Maxim. Solanaceae (1). 2 C As.

Psacadopaepale Bremek. = *Strobilanthes*

Psacaliopsis H. Robinson & Brettell (~ *Senecio*). Compositae (Sen.-Tuss.). 5 Mex., Guatemala

Psacalium Cass. (~ *Senecio*). Compositae (Sen.-Tuss.). 41 C Am. R: Phytol. 27(1973)254. *P. decompositum* (A. Gray) H. Robinson & Brettell (matarique, Mexico) – roots medic.

Psammagrostis C. Gardner & C. Hubb. Gramineae (31d). 1 W Aus. Pericarp free

Psammetes Hepper. Scrophulariaceae. 1 trop. W Afr.

Psammiosorus C. Chr. Oleandraceae. 1 NW Madag.: *P. paucivenius* (C. Chr.) C. Chr.

Psammisia Klotzsch. Ericaceae. 55 trop. Am.

Psammochloa A. Hitchc. Gramineae (16). 1 Gobi Desert: *P. villosa* (Trin.) Bor. R: KB 6(1951)196. Allied to *Stipa*

Psammogeton Edgew. Umbelliferae (III 8). 7 SW As. R: BDBG 69(1956)227

Psammomoya Diels & Loes. Celastraceae. 2 SW Aus.

Psammophora Dinter & Schwantes. Aizoaceae (V). 2–3 S Afr. Cult. orn. mat-forming succ. pls, the stems s.t. underground

Psammotropha Ecklon & Zeyher. Molluginaceae. 11 SE Afr. with 2 ext. to trop. Afr. R: JSAB 25(1959)51

Psathura Comm. ex Juss. Rubiaceae (IV 7). 8 Madag. & Masc.

Psathyranthus Ule = *Psittacanthus*

Psathyrostachys (Boiss.) Nevski (~ *Hordeum*). Gramineae (23). 8 E Med. (Eur. 1) to C As. R: NJB 11(1991)3. *P. fragilis* (Boiss.) Nevski crossed with *Hordeum vulgare* L. leads to chromosome elimination resulting in haploid (*Hordeum*) tillers; *P. juncea* (Fischer) Nevski (Russian wild rye)

Psathyrotes A. Gray. Compositae (Hele.-Gai.). 3 SW N Am. R: Madrono 23(1975)24

Psathyrotopsis Rydb. (~ *Psathyrotes*). Compositae (Hele.-Cha.). 2 Mex.

Psednotrichia Hiern. Compositae (Ast.-Ast.). 1 Angola

Pseudabutilon R. Fries. Malvaceae. 18 warm Am.

Pseudacanthopale Benoist = *Strobilanthopsis*

Pseudacoridium Ames. Orchidaceae (V 12). 1 Philippines

Pseudactis S. Moore = *Emilia*

Pseudaechmanthera Bremek. = *Strobilanthes*

Pseudaechmea L.B. Sm. & Read (~ *Fernseea*). Bromeliaceae (3). 1 Colombia: *P. ambigua* L.B.Sm. & Read. R: Phytol. 52(1982)53

Pseudaegiphila Rusby = *Aegiphila*

Pseudagrostistachys Pax & K. Hoffm. Euphorbiaceae. 2 trop. Afr.
Pseudaidia Tirv. Rubiaceae (II 1). 1 India: *P. speciosa* (Bedd.) Tirv. R: BMNHN 4,8(1986)286
Pseudais Decne = *Phaleria*
Pseudalthenia Nakai (*Vleisia*). Zannichelliaceae. 1 SW Cape: *P. aschersoniana* (Graebner) den Hartog in vleis
Pseudammi H. Wolff = *Seseli*
Pseudanamomis Kausel. Myrtaceae (Myrt.). 1 Puerto Rico
Pseudananas (Hassler) Harms. Bromeliaceae (3). 1 S trop. Am.: *P. sagenarius* (Arruda) Camargo, like *Ananas* but with stolons; cult. orn. terr. herb
Pseudannona (Baillon) Saff. = *Annona*
Pseudanthistiria (Hackel) Hook.f. Gramineae (40g). 4 India, S China, Thailand
Pseudanthus Sieber ex A. Sprengel. Euphorbiaceae. 8 Aus.
Pseudarabidella O. Schulz = *Arabidella*
Pseudarrhenatherum Rouy = *Arrhenatherum*
Pseudartabotrys Pellegrin. Annonaceae. 1 W trop. Afr.: *P. letestui* Pellegrin – liane
Pseudarthria Wight & Arn. Leguminosae (III 9). 5 trop. OW
Pseudechinolaena Stapf. Gramineae (34b). 6 Madag. with 1 extending pantrop. R: Adansonia 15(1975)123
Pseudelephantopus Rohr (~ *Elephantopus*). Compositae (Vern.-Ele.). 2 trop. Am.
Pseudellipanthus Schellenb. = *Ellipanthus*
Pseudeminia Verdc. Leguminosae (III 10). 4 trop. Afr.
Pseudephedranthus Aristeg. Annonaceae. 1 Brazil, Venezuela
Pseuderanthemum Radlk. Acanthaceae. 60 trop. Cult. orn. greenhouse shrubs
Pseuderemostachys Popov. Labiatae (VI). 1 C As.
Pseuderia Schltr. Orchidaceae (V 15). 4 Mal., W Pacific
Pseuderucaria (Boiss.) O. Schulz. Cruciferae. 2 Morocco to Israel
Pseudeugenia Legrand & Mattos. Myrtaceae. 1 Brazil
Pseudibatia Malme = *Matelea*
Pseudima Radlk. Sapindaceae. 3 trop. Am.
Pseudiosma DC. Rutaceae. 1 SE As.
Pseudoacanthocereus F. Ritter = *Acanthocereus*
Pseudobaccharis Cabrera = *Baccharis*
Pseudobaeckea Niedenzu. Bruniaceae. 4 S Afr.
Pseudobahia (A. Gray) Rydb. Compositae (Hele.-Bac.). 3 Calif. R: Novon 1(1991)122
Pseudobartlettia Rydb. = *Psathyrotopsis*
Pseudobartsia Hong (~ *Bartsia*). Scrophulariaceae. 1 China: *P. yunnanensis* Hong
Pseudoberlinia Duvign. = *Julbernardia*
Pseudobersama Verdc. Meliaceae (I 4). 1 trop. E Afr.
Pseudobetckea (Hoeck) Lincz. Valerianaceae. 1 Caucasus
Pseudoblepharispermum Lebrun & Stork. Compositae (Pluch.). 1 Ethiopia
Pseudoboivinella Aubrév. & Pellegrin = *Englerophyton*
Pseudobombax Dugand. Bombacaceae. 20 trop. Am. R: BJBB 33(1963)28. Some cult. orn. spineless trees incl. *P. ellipticum* (Kunth) Dugand (*Bombax e.*, '*Pachira insignis*', C Am.) – decoction used in treatment of toothache
Pseudobotrys Moeser. Icacinaceae. 2 New Guinea
Pseudobrachiaria Launert = *Urochloa*
Pseudobrassaiopsis Banerjee = *Brassaiopsis*
Pseudobravoa Rose. Agavaceae. 1 Mexico
Pseudobrickellia R. King & H. Robinson (~ *Eupatorium*). Compositae (Eup.-Alom.). 2 Brazil. R: MSBMBG 22(1987)256
Pseudobromus Schumann = *Festuca*
Pseudobrownanthus Ihlenf. & Bittrich (~ *Brownanthus*). Aizoaceae (IV). 1 SW Afr.: *P. nucifer* Ihlenf. & Bittrich
Pseudocadiscus Lisowski = *Stenops*
Pseudocalymma A. Samp. & Kuhlm. = *Mansoa*
Pseudocalyx Radlk. Acanthaceae. 4 trop. Afr., Madagascar
Pseudocamelina (Boiss.) N. Busch. Cruciferae. 3 Iran
Pseudocarapa Hemsley = *Dysoxylum*
Pseudocarpidium Millsp. Labiatae (I; Verbenaceae). 8 WI
Pseudocarum Norman. Umbelliferae (III 8). 1 trop. E Afr.
Pseudocaryophyllus O. Berg = *Pimenta*
Pseudocatalpa A. Gentry. Bignoniaceae. 1 N C Am.

Pseudocedrela Harms. Meliaceae (IV 2). 1 trop. Afr.: *P. kotschyi* (Schweinf.) Harms – gum, dyes & local medic.
Pseudocentrum Lindley. Orchidaceae (III 3). 6 trop. Am.
Pseudochaetochloa A. Hitchc. Gramineae (34e). 1 W Aus.
Pseudochamaesphacos Parsa. Labiatae. = ?
Pseudochimarrhis Ducke = *Chimarrhis*
Pseudochirita W.T. Wang. Gesneriaceae. 1 Guangxi (China): *P. guangxiensis* (Huang) W.T. Wang
Pseudocinchona A. Chev. = *Corynanthe*
Pseudoclappia Rydb. Compositae (Heli. - Flav.). 2 SW US & N Mex.
Pseudoclausia Popov. Cruciferae. 1 C As.
Pseudoclysis Gómez = *Polypodium*
Pseudocoix A. Camus. Gramineae (1b). 1 Madag., Tanz.: *P. perrieri* A. Camus
Pseudoconnarus Radlk. Connaraceae. 6 trop. S Am.
Pseudoconyza Cuatrec. (~ *Blumea*). Compositae (Pluch.). 1 trop.: *P. lyrata* (Kunth) Cuatrec.
Pseudocopaiva Britton & P. Wilson = *Guibourtia*
Pseudocorchorus Capuron. Tiliaceae. 6 Madagascar
Pseudocranichis Garay. Orchidaceae (III 3). 1 Mexico
Pseudocroton Muell. Arg. Euphorbiaceae. 1 C Am.
Pseudocrupina Velen. = *Leysera*
Pseudoctomeria Kraenzlin = *Pleurothallis*
Pseudocunila Brade = *Hedeoma*
Pseudocyclanthera Mart. Crov. Cucurbitaceae. 1 S Am.
Pseudocyclosorus Ching = *Cyclosorus*
Pseudocydonia (C. Schneider) C. Schneider = *Cydonia*
Pseudocylosorus Airy Shaw = *Cyclosorus*
Pseudocymopterus J. Coulter & Rose. Umbelliferae (III 9). 7 SW N Am.
Pseudodanthonia Bor & Hubb. Gramineae (21a). 2 NW Himal., W China. R: KBAS 13(1986)122
Pseudodichanthium Bor. Gramineae (40c). 1 W India
Pseudodicliptera Benoist. Acanthaceae. 2 Madagascar
Pseudodigera Chiov. = *Digera*
Pseudodiphasium Holub = *Lycopodium*
Pseudodiphryllum Nevski = *Platanthera*
Pseudodissochaeta Nayar. Melastomataceae. 5 N India to Hainan
Pseudodracontium N.E. Br. Araceae (VIII 5). 7 SE As. R: Blumea 40(1995)217
Pseudodrynaria (C. Chr.) C. Chr. = *Aglaomorpha*
Pseudo-Elephantopus Rohr = *Pseudelephantopus*
Pseudoentada Britton & Rose = *Adenopodia*
Pseudoeriosema Hauman. Leguminosae (III 10). 6 trop. Afr. Some locally eaten roots
Pseudoernestia (Cogn.) Krasser. Melastomataceae. 1 Venezuela
Pseudoeurya Yamamoto = *Eurya*
Pseudoeurystyles Hoehne = *Eurystyles*
Pseudoeverardia Gilly = *Everardia*
Pseudofortuynia Hedge. Cruciferae. 1 Iran
Pseudofumaria Medikus (~ *Corydalis*). Fumariaceae (I 2). 2 NW Balkans, S Alps, natur. elsewhere. R: OB 88(1986)32. Stems branched & perennial unlike *Corydalis*. *P. lutea* (L.) Borkh. (*C. lutea*, Eur., widely natur. inc. GB) – cult. orn.
Pseudogaillonia Lincz. Rubiaceae (IV 12). 1 SW As.
Pseudogaltonia (Kuntze) Engl. = *Lindneria*
Pseudogardenia Keay. Rubiaceae (II 1). 1 trop. Afr.
Pseudognaphalium Kirpiczn. (~ *Gnaphalium*). Compositae (Gnap.- Gnap.). c. 80 C & S Am. (most), warm OW (few), *P. luteo-album* (L.) Hilliard & B.L. Burtt weedy. R: OB 104(1991)146
Pseudognidia E. Phillips = *Gnidia*
Pseudogomphrena R. Fries. Amaranthaceae (II 2). 1 Brazil
Pseudogonocalyx Bisse & Berazaín = *Schoepfia*
Pseudogoodyera Schltr. Orchidaceae (III 3). 1 C Am.
Pseudogynoxys (Greenman) Cabrera. Compositae (Sen.-Sen.). 14 trop. S Am. R: Phytol. 36(1977)177
Pseudohamelia Wernham. Rubiaceae (inc. sed.). 1 Andes
Pseudohandelia Tzvelev. Compositae (Anth.-Han.). 1 C As., Iran, Afghanistan, China

Pseudohydrosme Engl. Araceae (V 6). 2 trop. W Afr.
Pseudojacabaea (Hook.f.) Mathur = *Senecio*
Pseudokyrsteniopsis R. King & H. Robinson (~ *Eupatorium*). Compositae (Eup.-Alom.). 1 Mex., Guatemala. R: MSBMBG 22(1987)245
Pseudolabatia Aubrév. & Pellegrin = *Pouteria*
Pseudolachnostylis Pax. Euphorbiaceae. 1 trop. Afr.
Pseudolaelia Porto & Brade (~ *Schomburgkia*). Orchidaceae (V 13). 6 trop. Am.
Pseudolarix Gordon. Pinaceae. 1 C & NE China: *P. amabilis* (Nelson) Rehder (*P. kaempferi*, golden larch or pine). Like *Larix* but male cones clustered (not solit.) & females with decid. scales, etc.
Pseudolasiacis (A. Camus) A. Camus = *Lasiacis*
Pseudoligandra Dillon & Sagást. = *Chionolaena*
Pseudolinosyris Novopokr. = *Linosyris*
Pseudoliparis Finet = *Malaxis*
Pseudolitchi Danguy & Choux = *Stadmania*
Pseudolithos Bally. Asclepiadaceae (III 5). 5 NE trop. Afr. R: Bradleya 8(1990)33
Pseudolmedia Trécul. Moraceae (III). 9 trop. Am. R: FN 7(1972)20. Cyanogenesis reported
Pseudolobivia (Backeb.) Backeb. = *Echinopsis*
Pseudolopezia Rose = *Lopezia*
Pseudolophanthus Lelvin = *Marmoritis*
Pseudoludovia Harling = *Sphaeradenia*
Pseudolycopodiella Holub = *Lycopodiella*
Pseudolycopodium Holub = *Lycopodium*
Pseudolysimachion (Koch) Opiz (~ *Veronica*). Scrophulariaceae. 19 temp. & subtemp. Euras. R: OB 75(1984)57. Cult. orn. incl. *P. spicatum* (L.) Opiz (*V. s.*, N Euras.) – protected in GB
Pseudomachaerium Hassler = *Nissolia*
Pseudomacrolobium Hauman. Leguminosae (I 4). 1 Congo basin
Pseudomantalania J. Leroy. Rubiaceae (II 1). 1–2 Madagascar
Pseudomariscus Rauschert = *Courtoisina*
Pseudomarrubium Popov. Labiatae (? VI). Spp?
Pseudomarsdenia Baillon. Asclepiadaceae (III 4). 5 trop. Am.
Pseudomaxillaria Hoehne = *Maxillaria* (but see BJ 97(1977)551 – R)
Pseudomertensia Riedl (~ *Mertensia*). Boraginaceae. 8 Iran to Himal.
Pseudomuscari Garb. & Greuter = *Muscari*
Pseudomussaenda Wernham (~ *Mussaenda*). Rubiaceae (? I 7). 4–5 trop. Afr.
Pseudomyrcianthes Kausel = *Myrcianthes*
Pseudonemacladus McVaugh. Campanulaceae. 1 Mexico
Pseudonesohedyotis Tenn. Rubiaceae (IV 1). 1 trop. E Afr.
Pseudonoseris H. Robinson & Brettell. Compositae (Liab). 3 Peru. R: SCB 54(1983)47
Pseudopachystela Aubrév. & Pellegrin = *Synsepalum*
Pseudopaegma Urban = *Anemopaegma*
Pseudopanax K. Koch. Araliaceae. 10–12 Tasmania, NZ, Chile. Dioec. or monoec. cult. orn. shrubs & trees with lvs simple to compound at diff. ages of pl. in some spp. e.g. *P. crassifolius* (Cunn.) K. Koch (lancewood, NZ) where 4 distinct forms of foliage; lvs on root-suckers more juvenile in form the further from the trunk of *P. simplex* (Forster f.) K. Koch (NZ)
Pseudopancovia Pellegrin. Sapindaceae. 1 trop. W Afr.
Pseudoparis Perrier. Commelinaceae (II 2). 2 Madagascar
Pseudopavonia Hassler = *Pavonia*
Pseudopectinaria Lavranos = *Echidnopsis*
Pseudopentameris Conert (~ *Danthonia*). Gramineae (25). 2 S Afr. R: MBSM 10(1971)303
Pseudopentatropis Costantin. = *Pentatropis*
Pseudopeponidium Homolle ex Arènes = *Pyrostria*
Pseudophegopteris Ching. Thelypteridaceae. 20 OW trop. R: Blumea 17(1969)112. No indusia
Pseudophleum Doğan = *Phleum*
Pseudophoenix H. Wendl. ex Sarg. Palmae (IV 1). 4 Carib. R: GH 10(1968)169. Tapped for wine but *P. ekmanii* Burret (Hispaniola), poss. 'wine palm' of early explorers, prob. extinct since 1926
Pseudopilocereus F. Buxb. = *Pilosocereus*
Pseudopinanga Burret = *Pinanga*

Pseudopiptadenia Rauschert (*Monoschisma*). Leguminosae (II 3). 2 S Am.
Pseudopipturus Skottsb. Older name for *Nothocnide*
Pseudoplantago Süsseng. Amaranthaceae (II 1). 2 Arg., Venez. Pollen cuboid or prismatic (unique in fam.)
Pseudopodospermum (Lipsch. & H. Kraschen.) Kutateladze = *Scorzonera*
Pseudopogonatherum A. Camus = *Eulalia*
Pseudoprosopis Harms. Leguminosae (II 3). 7 trop. Afr. R: BJBB 53(1983)417
Pseudoprotorhus Perrier = *Filicium*
Pseudopteris Baillon. Sapindaceae. 1 Madagascar
Pseudopyxis Miq. Rubiaceae (IV 12). 2 Japan. R: PSB 4(1989)131
Pseudoraphis Griffith. Gramineae (34f). 7 India to Japan & Aus.
Pseudorchis Séguier (*Polybactrum*, *Leucorchis*). Orchidaceae (IV 2). 3 E N Am. to Eur. (2). *P. albida* (L.) Löve & D. Löve (Eur., W Siberia) hybridizes with *Platanthera chlorantha* in Scotland
Pseudorhipsalis Britton & Rose (~ *Disocactus*). Cactaceae (III 2). 5 trop. Am.
Pseudorlaya (Murb.) Murb. Umbelliferae (III 3). 1–2 Eur., Medit.
Pseudorobanche Rouy = *Alectra*
Pseudoroegneria (Nevski) Löve = *Elymus*
Pseudorontium (A. Gray) Rothm. (~ *Antirrhinum*). Scrophulariaceae. 1 SW N Am.
Pseudoruellia Benoist. Acanthaceae. 1 Madagascar
Pseudosabicea Hallé. Rubiaceae (I 8). 12 trop. Afr.
Pseudosagotia Secco. Euphorbiaceae. 1 Venez. R: BMPEG 2(1985)23
Pseudosalacia Codd. Celastraceae. 1 Natal
Pseudosamanea Harms = *Albizia*
Pseudosaponaria (F.N. Williams) Ikonn. = *Gypsophila*
Pseudosarcolobus Costantin = *Gymnema*
Pseudosasa Makino ex Nakai. Gramineae (1a). 4 Taiwan, Japan. R: KBAS 13(1986)47. Like *Sasa* (A 6) but A 3(4). *P. japonica* (Steudel) Nakai (Japan) – much-cult. bamboo to 5 m for screening & hedges
Pseudosassafras Lecomte = *Sassafras*
Pseudosbeckia Fernandes & R. Fernandes. Melastomataceae. 1 trop. E Afr.
Pseudoscabiosa Devesa (~ *Scabiosa*). 4 C & W Med. R: Lagascalia 12(1984)213
Pseudoschoenus (C.B. Clarke) Oteng-Yeboah (~ *Scirpus*). Cyperaceae. 1 S Afr.
Pseudosciadium Baillon. Araliaceae. 1 New Caledonia
Pseudoscolopia Gilg. Flacourtiaceae (6). 1 S Afr.
Pseudosedum (Boiss.) A. Berger. Crassulaceae. 10 C As.
Pseudoselinum Norman. Umbelliferae (III 10). 1 Angola
Pseudosericocoma Cavaco. Amaranthaceae (I 2). 1 SW & S Afr.
Pseudosicydium Harms. Cucurbitaceae. 1 Peru, Bolivia
Pseudosindora Sym. = *Copaifera*
Pseudosmelia Sleumer. Flacourtiaceae (9). 1 Moluccas
Pseudosmilax Hayata = *Heterosmilax*
Pseudosmodingium Engl. Anacardiaceae. 7 Mexico
Pseudosopubia Engl. Scrophulariaceae. 7 trop. Afr.
Pseudosorghum A. Camus. Gramineae (40c). 2 trop. As.
Pseudospigelia Klett = *Spigelia*
Pseudospondias Engl. Anacardiaceae (II). 2 W & C trop. Afr. Bark medic.
Pseudostachyum Munro = *Schizostachyum*
Pseudostelis Schltr. = *Pleurothallis*
Pseudostellaria Pax. Caryophyllaceae (II 1). 16 Eur. (1) & C As., Afghanistan to Japan & N Am. (1). R: JapJB 9(1937)95
Pseudostenomesson Velarde. Amaryllidaceae (Liliaceae s.l.). 2 Andes
Pseudostenosiphonium Lindau = *Strobilanthes*
Pseudostifftia H. Robinson (~ *Moquinia*). Compositae (Moq.). 1 Bahia: *P. kingii* H. Robinson. R: T 43(1994)42
Pseudostreptogyne A. Camus = *Streblochaete*
Pseudostriga Bonati. Scrophulariaceae. 1 SE As.
Pseudotaenidia Mackenzie (~ *Taenidia*). Umbelliferae (III 10). 1 N Am.
Pseudotaxus W.C. Cheng. Taxaceae. 1 EC China
Pseudotectaria Tard. (? = *Tectaria*). Aspleniaceae (Tectarioideae). 6 Madag. R: KB 45(1990)257
Pseudotrimezia R. Foster. Iridaceae (III 3). c. 14 Brazil

Pseudotsuga Carr. Pinaceae. 4 E As., 2 W N Am. R: Phytol.M 7(1984)70. *P. menziesii* (Mirbel) Franco (*P. douglasii, P. taxifolia*, Douglas fir, Oregon pine, W N Am.) – major timber tree (133 m specimen felled in Br. Columbia in 1895, 126.5 m in 1902) introd. GB 1827, used for telegraph poles, railway sleepers, plywood & pulp, source of Oregon balsam; 67.36 m specimen erected as Christmas tree Seattle 1950. Genus allied to *Picea* but with conspic. bracts between cone-scales

Pseudourceolina Vargas = *Urceolina*

Pseudovanilla Garay (~ *Galeola*). Orchidaceae (V 4). 8 Mal. to Pacific (Aus. 1). *P. foliata* (F. Muell.) Garay (E As.) – liane to 15 m long

Pseudovesicaria (Boiss.) Rupr. Cruciferae. 1 Caucasus

Pseudovigna (Harms) Verdc. Leguminosae (III 10). 2 trop. Afr.

Pseudovossia A. Camus = *Phacelurus*

Pseudovouapa Britton & Killip = *Macrolobium*

Pseudoweinmannia Engl. Cunoniaceae. 2 NE Aus. *P. lachnocarpa* (F. Muell.) Engl. (marara) – local timber

Pseudowillughbeia Markgraf = *Melodinus*

Pseudowintera Dandy. Winteraceae. 3 NZ. R: Blumea 18(1970)227

Pseudowolffia Hartog & van der Plas = *Wolffiella*

Pseudoxandra R. Fries. Annonaceae. 10 trop. S Am. Some fish-disp. in Amazon

Pseudoxytenanthera Söderstrom & R. Ellis = *Schizostachyum*

Pseudoxythece Aubrév. = *Pouteria*

Pseudozoysia Chiov. Gramineae (33d). 1 Somalia: *P. sessilis* Chiov., on coastal dunes

Pseuduvaria Miq. Annonaceae. 35 India to Aus.

Psiadia Jacq. Compositae (Ast.-Ast.). 60 OW trop.

Psiadiella Humbert. Compositae (Ast.-Ast.). 1 Madagascar

Psidiopsis O. Berg = seq.

Psidium L. Myrtaceae (Myrt.). 100 trop. Am. Guavas, evergreen trees & shrubs with ed. berries, esp. *P. guajava* L. (common g.) – cult. & natur. (often a pest) throughout trop. & subtrop., fr. usu. tinned or made into jelly, jam & chutney, used to counter diarrhoea in Burundi; evergrowing shoots with capacity to flower at base. Other spp. with ed. fr. include *P. cattleianum* Sabine (*P. littorale*, purple g., strawberry g., Brazil) – sweet purplish red. fr., serious allelopathic weed in Hawaii; *P. friedrichsthalianum* (O. Berg) Niedenzu (Costa Rican g., C Am.) – fr. smaller, *P. guineense* Sw. (guisaro, trop. Am.) – small tart fr. for jelly

Psiguria Necker ex Arn. Cucurbitaceae. 15 trop. Am. Monoec. but changing sex so as to appear dioec.; tough fls visited by birds but also butterflies that mix pollen with nectar, causing it to pre-germinate & release amino-acids taken up by butterflies, which, thereby, live up to 6 months (much more than allied spp.)

Psila Philippi = *Baccharis*

Psilactis A. Gray (~ *Machaeranthera*). Compositae (Ast.- ? Chry.). 6 SW N Am. R: SB 18(1993)290

Psilantha (K. Koch) Tzvelev = *Eragrostis*

Psilanthele Lindau. Acanthaceae. 1 Ecuador

Psilanthopsis A. Chev. = *Coffea*

Psilanthus Hook.f. Rubiaceae (II 3). c. 20 trop. OW to Aus. (India 7: R: BBAS 33(1992)209)

Psilathera Link = *Sesleria*

Psilocarphus Nutt. Compositae (Gnap.-Gnap.). 8 W US, W temp. S Am. R: OB 104(1991)174

Psilocarya Torrey = *Rhynchospora*

Psilocaulon N.E. Br. Aizoaceae (IV). 12 S Afr. Some cult. orn. with stems carrying out most of photosynthesis

Psilochilus Barb. Rodr. (~ *Pogonia*). Orchidaceae (V 3). 7 trop. Am.

Psilochloa Launert = *Panicum*

Psiloesthes Benoist = *Peristrophe*

Psilolaemus I.M. Johnston. Boraginaceae. 1 Mexico

Psilolemma S. Phillips. Gramineae (31d). 1 E Afr.

Psilopeganum Hemsley. Rutaceae. 1 China: *P. sinense* Hemsley – shrub used in Chinese med. for dropsy

Psilostrophe DC. Compositae (Hele.-Gai.). 7 SW N Am. R: Rhodora 79(1977)169

Psilotaceae Eichler. Psilotopsida. 2/12 trop. to warm. Sporophytes often epiphytic, rootless, with dichot. endophytic mycorrhizal rootless rhizomes & aerial branches with protostele or solenostele. Leaf-like or scale-like lateral appendages spirally arr. or appearing distichous. Sporangia homosporous, on short lateral branches; antherozoids multiflagellate.

Gametophyte subterr. x = c. 50
Genera: *Psilotum*, *Tmesipteris* (sometimes placed in its own fam.)
Structure much resembling some of the earliest land pls

Psilothonna E. Meyer ex DC = *Steirodiscus*

Psilotopsida (Psilotatae) One order of 1 fam. – Psilotaceae, q.v. Considered close to Filicopsida by many authors

Psilotrichopsis C. Towns. Amaranthaceae (I 2). 3 Thailand, Malay Pen.

Psilotrichum Blume. Amaranthaceae (I 2). 18 trop. OW esp. Afr.

Psilotum Sw. Psilotaceae. 2 trop., C Japan & SW Spain. *P. nudum* (L.) Pal. (trop. to SW Spain, Hawaii & NZ) – terr. or epiphytic herb with dichot. branches & scales without leaf-traces, cult. orn. Japan for 400 yrs incl. cvs with term. sporangia & no leaf-like appendages thus much resembling fossil *Rhynia*!; *P. complanatum* Sw. (*P. flaccidum*, Mex., Jamaica, Pacific) – epiphyte with flattened branches & leaf-traces from stele but not reaching scales (cf. *Tmesipteris*)

Psiloxylaceae Croizat = Myrtaceae–Psiloxyloideae

Psiloxylon Thouars ex Tul. Myrtaceae (Psiloxylaceae). 1 Mauritius, Réunion. Allied to *Heteropyxis*

Psilurus Trin. Gramineae (17). 1 S Eur. to Pakistan: *P. incurvus* (Gouan) Schinz & Thell.

Psittacanthus C. Martius. Loranthaceae. 50 trop. Am. Hummingbird-poll.

Psomiocarpa C. Presl. Dryopteridaceae (I 3). 1 Philippines: *P. apiifolia* (Kunze) C. Presl – hairs 'intestiniform'

Psophocarpus Necker ex DC. Leguminosae (III 10). c. 10 trop. OW. R: KB 33(1978)191. *P. tetragonolobus* (L.) DC (winged bean or pea, Goa b., asparagus pea, prince's p., dambala, trop. As.) – cult. for ed. legumes high in protein & ed. roots, poss. derived from *P. grandiflorus* Wilczek in Zaire

Psoralea L. Leguminosae (III 2). (Excl. *Bituminaria*, *Cullen*) 20 S Afr. esp. Cape. Some cult. orn. incl. *P. pinnata* L. (blue pea) – fls blue & white

Psoralidium Rydb. = *Orbexilum*

Psorospermum Spach. Guttiferae (I). 40–45 trop. Afr., Madag.

Psorothamnus Rydb. Leguminosae (III 12). 9 deserts SW N Am.

Psychanthus (Schumann) Ridley = *Pleuranthodium*

Psychilis Raf. (~ *Epidendrum*). Orchidaceae (V 13). 15 WI. R: Phytol. 65(1988)1

Psychine Desf. Cruciferae. 1 N Afr.

Psychopsiella Lückel & Braem (~ *Oncidium*). Orchidaceae (V 10). 1 trop. Am.: *P. limminghei* (Lindley) Lückel & Braem – cult. orn.

Psychopsis Raf. (~ *Oncidium*). 4 trop. Am. Cult. orn.

Psychotria L. Rubiaceae (IV 7). (Incl. *Cephaelis*) 800–1500 trop. (Papuasia 115). Trees & shrubs (architectural branching a useful char. at subgeneric or species-group rank), incl. epiphytes (some with adventitious roots) & rheophytes (some fish-disp. in Amazonia), with alks & some heterostyly. *P. ipecacuanha* (Brot.) Stokes (*Cephaelis i.*, ipecacuanha, Brazil) – form. much coll. in Mato Grosso, Brazil & cult. in Mal. for dried rhiz. used med. (alk. = emetine) esp. as expectorant. Other spp. exploited somewhat similarly & also cult. orn. (often with coloured infl. axes) incl. *P. emetica* L.f. (trop. Am.) – roots source of inferior subs. for ipecacuanha. Others locally medic. & *P. viridis* Ruíz & Pavón (Amazonia) mixed with other hallucinogens. *P. douarrei* (Beauvisage) Däniker (New Caled.) – lvs accum. nickel to 4.7% dry wt., the highest recorded in any pl.

Psychrogeton Boiss. Compositae (Ast.-Ast.). 20 SW & C As. R: NRBGE 27(1967)101

Psychrophila (DC) Bercht. & J. Presl = *Caltha*

Psychrophyton Beauverd (~ *Raoulia*). Compositae (Gnap.-Lor.). 10 NZ. Cushion-formers incl. *P. mammillaris* (Hook.f.) Beauverd (*R. mammillaris*) forming dense rounded cushions (vegetable sheep). R: OB 104(1991)61

Psydrax Gaertner (~ *Canthium*). Rubiaceae (III 2). 100 OW trop. (Afr. 34; R: KB 40(1985)687)

Psygmorchis Dodson & Dressler (~ *Oncidium*). Orchidaceae (V 10). 5 trop. Am. Short-lived epiphytes on coffee & guava & in sec. forest

Psylliostachys (Jaub. & Spach) Nevski. Plumbaginaceae (II). 10 E Med. to C As. Some cult. orn. like *Limonium* esp. *P. suworowii* (Regel) Roshk. (Iran & C As.) – C pink

psyllium *Plantago afra*

Psyllocarpus C. Martius & Zucc. Rubiaceae (IV 15). 8 Brazil, white sands & savanna. R: SCB 41(1979)

Ptaeroxylaceae J. Leroy (~ Rutaceae). Dicots – Rosidae – Sapindales. 3/10 trop. & S Afr., Madag. Trees & shrubs, polygamo-dioec. Lvs pinnate, spirally arr. or opp. with entire, oblique leaflets; stipules 0. Fls small, reg., in small axillary cymes; K 4 or 5, basally connate, lobes ± imbricate, C 4 or 5, imbricate or valvate; A 4 or 5, alt. with C, disk s.t. present around G (2–5) with 2–5 short connate styles, each loc. with 1–3 campylotropous ovules. Fr. a samaroid capsule of 2–5 follicles dehiscing along inner suture; seeds 2–5, winged, embryo curved or folded, endosperm thin & fleshy or 0

Genera: *Bottegoa, Cedrelopsis, Ptaeroxylon*

Lying between Meliaceae & Rutaceae; referred to Sapindaceae by Cronquist

Ptaeroxylon Ecklon & Zeyher. Ptaeroxylaceae. 1 S Afr.: *P. obliquum* (Thunb.) Radlk. (*P. utile*, sneezewood) – termite-resistant timber for external use, pepper-scented sawdust a medic. snuff

Ptelea L. Rutaceae. c. 11 N Am. Polygamous trees & shrubs with cordate reticulate-winged samaras & alks. Cult. orn. esp. *P. trifoliata* L. (hop tree, C & E US) – some antifungal alks, frs a hop subs. in brewing beer

Pteleocarpa Oliver. Boraginaceae (Ehretiaceae). 1 W Mal.: *P. lamponga* (Miq.) Heyne – 2 ovules per loc. with upper ascending very odd in B.

Pteleopsis Engl. Combretaceae. 9 trop. Afr.

Ptelidium Thouars. Celastraceae. 2 Madagascar

Pteracanthus (Nees) Bremek. = *Strobilanthes*

Pterachaenia (Benth.) Lipsch. Compositae (Lact.-Scor.). 1 Pakistan & Afghanistan

Pteralyxia Schumann. Apocynaceae. 2 Hawaii. R: W.L. Wagner et al, *Man. Fl. Pls Hawaii* (1990)219

Pterandra A. Juss. Malpighiaceae. 6 trop. Am.

Pteranthus Forssk. Caryophyllaceae (I 2). 1 N Afr. & Cyprus to Iran: *P. dichotomus* Forssk.

Pterichis Lindley. Orchidaceae (III 3). 20 Andes, Costa Rica (1), Jamaica (1)

Pteridaceae Reichb. (Adiantaceae, incl. Parkeriaceae & Platyzomataceae). Filicopsida. 35/825 cosmop. esp. trop. Terr. or aquatic (*Ceratopteris*) ferns, usu. small. Stem with medullated protostele to dictyostele with trichomes &/or scales. Frond usu. pinnate. Sporangia in soral lines along veins (indusium 0) or marginal when covered with marginal indusium

Classification & chief genera:

I. **Platyzomatoideae** (Platyzomataceae; sterile fronds dimorphic): *Platyzoma* (only)

II. **Ceratopteridoideae** (Parkeriaceae; aquatic): *Ceratopteris*

III. **Taenitidoideae** (stem with trichomes only &/or spores with equatorial flange; frond s.t. farinose abaxially, sporangia abaxial): *Eriosorus, Pityrogramma, Taenitis*

IV. **Cheilanthoideae** (petiole with 1 vasc. bundle or if 2 then frond not farinose abaxially): *Cheilanthes, Notholaena, Pellaea*

V. **Adiantoideae** (sporangia on a circular to elongate strongly recurved margin): *Adiantum, Doryopteris,*

VI. **Pteridoideae** (trichomes among sporangia diff. from others): *Acrostichum, Pteris*

App. close to Vittariaceae

Pteridiaceae Ching = Dennstaedtiaceae

Pteridium Gled. ex Scop. Dennstaedtiaceae. 1 cosmop.: *P. aquilinum* (L.) Kuhn (bracken, eagle fern, (hog) brake ('Brache' = Old German for wasted land)) diploid to octoploid, with 2 subspp. – *aquilinum* in N hemisph. & Afr., subsp. *caudatum* (L.) Bonap. in S hemisph., overlapping in WI & Mal., though certain hybrid pops in Aus. sterile suggesting greater speciation. R: Rhodora 43(1941)1, 37, BJLS 73, 1–3(1976), R.T. Smith & J.A. Taylor (eds, 1986) *Bracken*. Creeping rhizome (the vasc. strands in section alleged to resemble G O D, J C, or I H S) with usu. annual fronds (the croziers poss. inspiration for the capitals of Ionic columns) toxic to stock through thiaminase causing acute symptoms of vitamin B_1 deficiency, also producing ecdysones & bearing nectaries (in N Yorks populations with attracted wood-ants had reduced insect predators esp. suckers); rhizomes with tracheid end-walls dissolving to give vessel-like elements, forming clones to 1400 yrs old; allelopathic, the toxins leaching from green fronds all year round in Costa Rica to affect surrounding pls, from litter in W N Am. preventing germination of seeds & from standing fronds in Calif. countering germination in wet season. Often a serious weed; rhizome & young fronds tinned esp. in Japan ('sawarabi', the ed. starch known as 'warabi-ko'), form. much eaten by Maoris, N Am. Indians etc. (ground with oats in Canary Is. – 'goflo') but shikimic acid in them known to promote stomach cancers being carcinogenic & mutagenic; form. locally medic., astringent rhiz. also used in preparation of kid & chamois leathers & form. in glass- & soap-making (potash), the fronds used for thatch & paper; pl. a source of a green dye but also dark yellow in some Scottish tartans & black when rhiz. boiled; boiled rhiz. also used in fibrework (brown when unboiled) of N Am. Indians

Pteridoblechnum Hennipman. Blechnaceae (Blechnoideae). 1 N Queensland: *P. neglectum* (Bailey) Hennipman

Pteridocalyx Wernham. Rubiaceae (I 5). 2 Guyana

Pteridophyllaceae (Murb.) Reveal & Hoogl. (~ Fumariaceae). 1/1 Japan. Herb with oblanceolate pinnate lvs. Infl. a thyrse. Fls small; K 2, early caducous, petaloid; C 4 imbricate; A 4 alt. with C, anthers with slits. G̲ (2) with long style & 2(4) anatropous to subcampylotropous ovules. Fr. a capsule with 2 seeds
Only genus: *Pteridophyllum*

Pteridophyllum Siebold & Zucc. Pteridophyllaceae (Fumariaceae s.l.). 1 Japan: *P. racemosum* Siebold & Zucc. – rare in coniferous woods, cult. orn. with alks

Pteridophyta 232/9800 here arr. in 38 fams. Vasc. pls with spores. Probably a group of unrelated pls having achieved a terr. level of organization without seeds; from this level the seed-pls have been derived, diff groups prob. independently. Many fossil groups & the pls which are the basis of coal deposits (Carboniferous spore-forests). The spores have 'pre-adapted' the group (esp. ferns & clubmosses) to living in the rising canopy of rain forests where they are v. successful (cf. Orchidaceae). Extant groups: Psilotopsida, Lycopsida (clubmosses), Equisetopsida (horsetails) & Filicopsida (ferns)

Pteridrys C. Chr. & Ching. Dryopteridaceae (I 3). 8 trop. As. & Mal.

Pterigeron (DC) Benth. = *Streptoglossa*

Pteris L. Pteridaceae (VI). 250 cosmop. (Eur. 3). Cult. orn. (R: BFG 10(1970)143) terr. ferns esp. *P. cretica* L. (trop. & warm OW) – common pot-pl. with many cvs incl. crested & varieg. ones; *P. ensiformis* Burm.f. (Indomal. to Japan & W Pacific) – young fronds ed., locally medic.; *P. multifida* Poiret (E As.) – locally medic. (vermifuge), many cvs; *P. vittata* L. (trop. & warm OW, introd. Am.) – natur. hot slag-heaps of Forest of Dean, S England

Pterisanthes Blume. Vitaceae. 18 Indomal. (to Philippines)

Pternandra Jack. Melastomataceae. 15 Mal. R: GBS 34(1981)1. Berry

Pternopetalum Franchet. Umbelliferae (III 8). 27 China. R: APS 16, 3(1978)65

Pterobesleria Morton = *Besleria*

Pterocactus Schumann. Cactaceae (II). 9 S & W Arg. R: CSJ 44(1982)51. Aril forming papery wing round seed – wind-disp.

Pterocarpus Jacq. Leguminosae (III 4). 21 trop. R: PM 5(1972). Some water-disp. in Amazonia. Sources of timber (padouk, vermilion wood, bloodwood) & kino. *P. angolensis* DC (ambila, bleedwood tree, kiaat, muninga, trop. & S Afr.) – timber for furniture, drums, canoes etc.; *P. dalbergioides* Roxb. ex DC (Andaman padauk, E I mahogany, Indomal.) – cabinet-work; *P. erinaceus* Poiret ((W) African rosewood, barwood, African kino, W Afr.) – cabinet-work; *P. indicus* Willd. (angsana, Burmese or Papua New Guinea rosewood, Andaman redwood, Amboyna wood, E & SE As. to Mal.) – fl. buds open 3 days after sudden drop in temp., rose-scented timber for furniture, cult. orn. shade-tree rapidly growing from v. long cuttings as in Singapore City; *P. macrocarpus* Kurz (Burma padauk, maidu, SE As.) – inlay etc.; *P. marsupium* Roxb. (gamalu, E I or Malabar kino, India) – timber for outdoor use esp. window- & door-frames, also wood-carving in Sri Lanka; *P. santalinus* L.f. (red sandalwood or saunderswood, poss. algum or almug wood of Bible, India) – cabinet-work, red dye used to mark castes in Hindu religion; *P. soyauxii* Taubert (W Afr. padauk, barwood, camwood, W & C Afr.) – wood for canoes, dye for body-paint

Pterocarya Kunth. Juglandaceae. 6 Cauc. to E & SE As. R: AMBG 65(1978)1073. Wingnut. Cult. orn. trees with winged fr. esp. *P. fraxinifolia* (Poiret) Spach (Cauc. to Iran), *P. stenoptera* C. DC (China) & their hybrid, *P.* × *rehderiana* C. Schneider, hardier than either & v. fast-growing

Pterocaulon Elliott. Compositae (Pluch.). 18 warm Am., SE As. to Aus. & New Caledonia (3). R: Darwiniana 21(1978)185. *P. undulatum* (Walter) Mohr (US) – medic.

Pterocelastrus Meissner. Celastraceae. 4–5 S Afr.

Pteroceltis Maxim. Ulmaceae (II). 1 N & C China: *P. tartarinowii* Maxim.

Pterocephalidium G. López (~ *Pterocephalus*). Dipsacaceae. 1 C Spain, Portugal: *P. diandrum* (Lag.) G. López

Pterocephalus Adans. Dipsacaceae. 25 Medit. (Eur. 5) to C As., Himal. & W China, trop. Afr. Some cult. orn. rock-pls. K with 10 or more pappus-like awns aiding wind-disp.

Pteroceras Hasselt ex Hassk. (~ *Sarcochilus*). Orchidaceae (V 16). 28 Indomal.

Pterocereus MacDoug. & Miranda = ? *Pachycereus*

Pterochaeta Steetz (~ *Waitzia*). Compositae (Gnap.-Ang.). 1 SW Aus.: *P. paniculata* Steetz. R: Nuytsia 8(1992)422

Pterochloris (A. Camus) A. Camus = *Chloris*

Pterocissus Urban & Ekman. Vitaceae. 1 Hispaniola

Pterocladon Hook.f. = *Miconia*
Pterococcus Hassk. (~ *Plukenetia*). Euphorbiaceae. 1 Indomal.
Pterocyclus Klotzsch = *Pleurospermum*
Pterocymbium R. Br. Sterculiaceae. 15 SE As. to Fiji. Winged ovaries for wind-disp.
Pterocypsela Shih (~ *Lactuca*). Compositae (Lact.-Lact.). 11 E (esp. China) & SE As. R: APS 26(1988)385
Pterodiscus Hook. Pedaliaceae. 18 trop. & S Afr. Cult. orn. esp. those with swollen caudex
Pterodon Vogel. Leguminosae (III 3). c. 6 Brazil, Bolivia
Pterogaillonia Lincz. Rubiaceae (IV 12). 3 SW As.
Pterogastra Naudin. Melastomataceae. 4 N trop. S Am.
Pteroglossa Schltr. (~ *Stenorrhynchos*). Orchidaceae (III 3). 8 S Am. R: HBML 28(1982)349
Pteroglossaspis Reichb.f. (~ *Eulophia*). Orchidaceae (V 9). 3 Florida, Cuba, Arg., trop. Afr.
Pterogonum Gross = *Eriogonum*
Pterogyne Tul. Leguminosae (I 1). 1 trop. S Am.: *P. nitens* Tul. – valuable timber for construction, furniture, barrels etc.
Pterolepis (DC) Miq. (~ *Tibouchina*). Melastomataceae. 15 trop. Am. esp. Brazilian savanna
Pterolobium R. Br. ex Wight & Arn. Leguminosae (I 1). 11 OW trop. (Afr. 1). R: BMNHN 3[e] Bot. 15(1974)
Pteroloma Desv. ex Benth. = *Tadehagi*
Pteromanes Pichi-Serm. = *Crepidomanes*
Pteronia L. Compositae (Ast.-Sol.). 80 S Afr. R: ASAM 9(1917)277
Pteropepon (Cogn.) Cogn. Cucurbitaceae. 3 Brazil, Argentina
Pteroptychia Bremek. = *Strobilanthes*
Pteropyrum Jaub. & Spach. Polygonaceae (II 4). 6 SW As. Shrubs
Pterorhachis Harms. Meliaceae (I 4). 1–2 W trop. Afr. *P. zenkeri* Harms – bark (nut flavour) a local aphrodisiac
Pteroscleria Nees = *Diplacrum*
Pteroselinum Reichb. (~ *Peucedanum*). Umbelliferae (III 10). 1 Eur.: *P. austriacum* (Jacq.) Reichb.
Pterosicyos Brandegee = *Sechiopsis*
Pterospermum Schreber. Sterculiaceae. 25 trop. As. Fragrant fls opening for 1 night. Timber esp. *P. acerifolium* (L.) Willd. (India to Java) – durable like teak
Pterospora Nutt. Ericaceae (Monotropaceae). 1 N Am.: *P. andromeda* Nutt. – root-parasite to 1 m tall
Pterostegia Fischer & C. Meyer. Polygonaceae (I 2). 1 W & SW N Am.: *P. drymarioides* Fischer & C. Meyer
Pterostemma Kraenzlin. Orchidaceae (V 10). 1 Colombia
Pterostemon Schauer. Grossulariaceae (Pterostemonaceae). 2 Mexico
Pterostemonaceae Small = Grossulariaceae
Pterostylis R. Br. Orchidaceae (IV 1). 120 W Pacific esp. Aus. (c. 100) & NZ. Fls usu. green ('greenhoods') poll. small flies or fungus-gnats
Pterostyrax Siebold & Zucc. Styracaceae. 4 Burma to Japan. Cult. orn.
Pterotaberna Stapf = *Tabernaemontana*
Pterothrix DC. Compositae (Gnap.-Rel.). 6 S Afr. R: OB 104(1991)78
Pteroxygonum Dammer & Diels = *Fagopyrum*
Pterozonium Fée. Pteridaceae (III). 13 Roraima sandstones (Venez.), some ext. to Surinam, Colombia & N Peru R: MNYBG 17(1967)2
Pterygiella Oliver. Scrophulariaceae. 3 China
Pterygiosperma O. Schulz. Cruciferae. 1 Patagonia
Pterygocalyx Maxim. (~ *Gentiana*). Gentianaceae. Spp.?
Pterygodium Sw. Orchidaceae (IV 3). 15 trop. & S Afr. Elaiophores attractive to bees
Pterygopappus Hook.f. Compositae (Gnap.-Lor.). 1 Tasmania: *P. lawrencii* Hook.f. R: OB 104(1991)62
Pterygopleurum Kitagawa. Umbelliferae (III 8). 1 Korea, S Japan
Pterygopodium Harms = *Oxystigma*
Pterygota Schott & Endl. Sterculiaceae. 15 trop. esp. OW. Like *Brachychiton* & *Sterculia* but seeds winged, those of *P. alata* (Roxb.) R. Br., opium subs. in N India, with wings to 6 cm long. *P. bequaertii* De Wild. (awari, W Afr.) – timber
Pteryxia (Torrey & A. Gray) J. Coulter & Rose. Umbelliferae (III 9). 5 W N Am.
Ptilagrostis Griseb. = *Stipa*
Ptilanthelium Steudel = *Ptilothrix*
Ptilanthus Gleason = *Graffenrieda*

Ptilimnium Raf. Umbelliferae (III 8). 5 E N Am.

Ptilochaeta Turcz. Malpighiaceae. 5 warm S Am.

Ptilopteris Hance = *Monachosorum*

Ptilostemon Cass. Compositae (Card.-Card.). 14 Medit. (Eur. 9). R: Boissiera 22(1973)

Ptilothrix K.A. Wilson (*Ptilanthelium*). Cyperaceae. 1 E Aus.: *P. deusta* (R. Br.) K.A. Wilson

Ptilotrichum C. Meyer = *Alyssum*

Ptilotus R. Br. Amaranthaceae (I 2). 90 arid & semi-arid Aus., 1 ext. to Mal. Some cult. orn. (pussy tail) esp. *P. manglesii* (Lindley) F. Muell. – remarkable jointed hairs on P

Ptycanthera Decne. = *Matelea*

Ptychandra R. Scheffer = *Heterospathe*

Ptychococcus Becc. Palmae (V 4k). 5 Papuasia

Ptychogyne Pfitzer = *Coelogyne*

Ptycholobium Harms (~ *Tephrosia*). Leguminosae (III 6). 3 dry trop. Afr. to S Arabia. R: KB 35(1980)461

Ptychomeria Benth. = *Gymnosiphon*

Ptychopetalum Benth. Olacaceae. 4 trop. Am. (2) & Afr. (2). Locally medic.

Ptychopyxis Miq. Euphorbiaceae. 13 Thailand to W Mal., E New Guinea

Ptychosema Benth. Leguminosae (III 25). 2 C & W Aus.

Ptychosperma Labill. Palmae (V 4k). 28 E Mal., Papuasia & trop. Aus. R: Allertonia 1(1978)415. Cult. orn. unarmed monoec. palms

Ptychotis Koch. Umbelliferae (III 8). 1–2 C & S Eur.

Ptyssiglottis T. Anderson. Acanthaceae. (Incl. *Hallieracantha*) c. 60 Indomal. R: JB 60(1922)355

Pubistylus Thoth. (~ *Diplospora*). Rubiaceae (?II 5). 1 Andamans

Pucara Ravenna. Amaryllidaceae (Liliaceae s.l.). 1 N Peru

Puccinellia Parl. (~ *Phippsia*). Gramineae (17). 25 N temp. (Eur. 13) esp. As., S Afr., Aus. (1). Alkali grass (N Am.)

Puccionia Chiov. Capparidaceae. 1 N Somalia: *P. macradenia* Chiov., on gypsum

puccoon, red *Sanguinaria canadensis*; **yellow p.** *Lithospermum canescens*

puchurin nut *Licaria pucheri*

pudding pipe tree *Cassia fistula*

Puelia Franchet. Gramineae (1b). 5 trop. Afr. R: HIP 37 t. 3642(1967)3. Ligules external. *P. coriacea* Pilger (Cameroun) with root tubers, *P. schumanniana* W. Clayton (Zaire) with only 1 leaf at top of culm

Pueraria DC. Leguminosae (III 10). 17 trop. & E As. Twiners with extrafl. nectaries. R: AUWP 85-1(1985)9. *P. mirifica* Airy Shaw & Suvatabandhu (Thailand) – miroestrol like oestrone but stronger & used in abortion in Thailand; *P. montana* (Lour.) Merr. var. *lobata* (Willd.) Maesen & S. Almeida (*P. lobata, P. thunbergiana*, kudzu vine, Japanese arrowroot, E India, China & Japan) – fodder & cover crop, erosion control, ed. roots, stems to 20 m long with fibres for textiles (anciently used in China) & cordage, invasive pest in parts of US

Pugionium Gaertner. Cruciferae. 5 Mongolia. *P. cornutum* (L.) Gaertner – veg.

pukatea *Laurelia novae-zelandiae*

pukeweed *Lobelia inflata*

pulasan *Nephelium ramboutan-ake*

Pulchranthus Baum, Reveal & Nowicke. Acanthaceae. 4 trop. S Am. R: SB 8(1983)211

Pulicaria Gaertner. Compositae (Inul.). c. 80 temp. (Eur. 5) & warm Euras. Fleabane

Pullea Schltr. = *Codia*

Pulmonaria L. Boraginaceae. 18 Eur. R: BB 131(1975). Cult. orn bristly herbs incl. *P. angustifolia* L. (lvs green) & *P. officinalis* L. (lungwort; lvs white-spotted) with heterostyly & fls changing from red to blue as they mature; *P. officinalis* form. medic. in bronchial complaints (dried lvs, the speckled lvs suggesting diseased lungs in Doctrine of Signatures)

pulque See *Agave*

Pulsatilla Miller (~ *Anemone*). Ranunculaceae (II 2). 38 temp. Euras. (Eur. 9), E N Am. R: FR 60(1957)1. Differing from *A.* in having achenes with persistent long plumose styles. *P. vulgaris* Miller (*A. p.*, pasque-flower, Eur.) – indicative of grassland at least a century old in GB, cult. orn., form. medic. esp. in C Eur.

Pultenaea Sm. Leguminosae (III 24). 100 Aus.

pulu *Cibotium glaucum, Sadleria cyatheoides*

Pulvinaria Fourn. = *Matelea*

pulza *Jatropha curcas*

pumilio pine oil *Pinus mugo*

pummelo *Citrus maxima*

pumpkin *Cucurbita moschata*, *C. pepo*, *C. argyrosperma*
punah *Tetramerista glabra*
Punica L. Punicaceae. 2 SE Eur. to Himal. (*P. granatum*), Socotra (*P. protopunica*). *P. granatum* L. (pomegranate) – anc. cultigen (? wild in S Caspian & NE Turkey) known from Bronze Age Jericho, grown for refreshing fr. (ed. pulp around seeds, which contain 17 mg per kg oestrone), easily fermented to grenadine, a cordial, extract effective against HIV, bark medic. (alks) & it & fr. pericarp used in tanning leather in Egypt, the fls medic. (balustine fls), persistent calyx the inspiration for King Solomon's crown (& hence other crowns in Eur.), allegedly the fr. given to Venus by Paris; each leaf with single apical nectary exuding fructose, glucose & sucrose; cult. orn. esp. 'Nana' to 1.5 m. *P. protopunica* Balf.f. – poss. ancestor (cf. *Ceratonia*), now reduced to a few trees & endangered
Punicaceae Horan. (~ Lythraceae). Dicots – Rosidae – Myrtales. 1/2 Eur. to India. Trees & shrubs, often spiny, with alks. Lvs simple, ± opp., stipules rudimentary. Fls reg., bisexual, solit. or in fascicles at apices, with coloured hypanthium; K 5–8 lobes on hypanthium, C alt. with K, imbricate & crumpled in bud, A ∞ on hypanthial tube, developing centrifugally, anthers with longit. slits, $\overline{G}$ (7–9(–15)) with axile placentation (*P. protopunica*) or becoming superposed in 2 or 3 layers (*P. granatum*), the upper with app. parietal placentation, each placenta with ± ∞ anatropous, bitegmic ovules. Fr. with leathery pericarp & persistent K; seeds in pulpy enlarged sarcotestas, embryo oily, straight, with spirally rolled cotyledons, endosperm 0. x = 8, 9
Genus: *Punica*
Seed & wood anatomy studies support close relationship with Lythraceae though A development like Myrtaceae
Punjuba Britton & Rose (~ *Pithecellobium*). Leguminosae (II 5). 3 trop. Am.
Puntia Hedge. Labiatae (VIII 4). 1 C Somalia
Pupalia Juss. Amaranthaceae (I 2). 4 trop. OW. R: KB 34(1979)131
Pupilla Rizz. = *Justicia*
Purdiaea Planchon. Cyrillaceae. 12 trop. Am. esp. Cuba. R: CGH 186(1960)47
Purdieanthus Gilg = *Lehmanniella*
purging buckthorn *Rhamnus cathartica*; **p. cassia** *Cassia fistula*; **p. flax** *Linum catharticum*; **p. nut** *Jatropha curcas*
Puria Nair = *Cissus*
puriri *Vitex lucens*
purple apple berry *Billardiera longiflora*; **p. heart** *Peltogyne paniculata*
Purpureostemon Gugerli. Myrtaceae (Lept.). 1 New Caledonia
Purpusia Brandegee. Rosaceae. 1 SW US
Purshia DC ex Poiret. Rosaceae. (Incl. *Cowania*) 7 W N Am. Decid. shrubs (antelope bush) with nitrogen-fixing actinomycete root-symbionts, inner layer of outer integument with stomata. Cult. orn. esp. *P. tridentata* (Pursh) DC
purslane, common *Portulaca oleracea* subsp. *sativa*; **sea p.** *Honckenya peploides*; **water p.** *Lythrum portula*; **winter p.** *Claytonia perfoliata*
Puschkinia Adams. Hyacinthaceae (Liliaceae s.l.). 1 Syria, Lebanon, Turkey & Cauc.: *P. scilloides* Adams (striped or Lebanon squill) – oily ant-disp. seeds, cult. orn.
pussy tail *Ptilotus* spp.; **p. willow** *Salix caprea*
pussy's-toes *Antennaria* spp.
Putoria Pers. Rubiaceae (IV 12). 3 Medit. (Eur. 1). Stinking shrubs
Putranjiva Wallich = *Drypetes*
Putranjivaceae Endl. = Euphorbiaceae
Putterlickia Endl. Celastraceae. 3 S Afr.
puttyroot *Aplectrum hyemale*
Puya Molina. Bromeliaceae (1). 168 highland S Am. R: FN 14(1974)66. Terr. often pachycaul pls with spiny-margined lvs, *P. raimondii* Harms (Peru, Bolivia) to 10.7 m, hapaxanthic (cf. *Lobelia*) after 80–150 yrs (world's slowest) in Andes, but after 28 yrs when grown at sea-level, with c. 8000 fls & infl. branch extensions acting as perches for poll. birds (cf. *Babiana*). Some cult. orn. incl. *P. chilensis* Molina (Chile) – lvs yield a fibre used for rot-resistant fishing-nets, but many grubbed up as hazardous to sheep which get entangled in them, as do birds, the nutrients from which, as well as those from their droppings may be absorbed (? foliar)
Pycnandra Benth. Sapotaceae (IV). 12 New Caled. R: FNC 1(1967)36. Mostly pachycaul treelets
Pycnanthemum Michaux. Labiatae (VIII 2). 17 N Am. Some local medic. & flavourings incl. *P. incanum* (L.) Michaux – poss. source of natural rubber

Pycnanthus Warb. Myristicaceae. 7 trop. Afr. *P. angolensis* (Welw.) Warb. (*P. kombo*, akomu, eteng, ilomba, kombo, false nutmeg, W Afr.) – seeds a source of oil, timber useful

Pycnarrhena Miers ex Hook.f. & Thomson. Menispermaceae (I). 9 Indomal. to Aus. R: KB 26(1972)405, 30(1975)97. *P. manillensis* S. Vidal (Philipp.) – powdered root used against snakebite

Pycnobotrya Benth. Apocynaceae. 2 trop. Afr.

Pycnobregma Baillon = *Malatea*

Pycnocephalum (Less.) DC (~ *Eremanthus*). Compositae (Vern.-Vern.). 3 C Brazil. R: SB 10 (1985)461

Pycnocoma Benth. Euphorbiaceae. 15 trop. Afr. to Masc. Litter-collecting rosettes of lvs in W Afr.

Pycnocomon Hoffsgg. & Link (~ *Scabiosa*). Dipsacaceae. 2 W Medit.

Pycnocycla Lindley. Umbelliferae (III 1). 12 trop. W Afr. to NW India

Pycnoloma C. Chr. = *Seliguea*

Pycnoneurum Decne. Asclepiadaceae (III 1). 2 Madagascar

Pycnonia L. Johnson & B. Briggs (~ *Persoonia*). Proteaceae. 6 SW & N Aus.

Pycnophyllopsis Skottsb. Caryophyllaceae (II 1). 2 Andes

Pycnophyllum Remy. Caryophyllaceae (II 2). 17 Andes. R: FR 18(1922)167

Pycnoplinthopsis Jafri (= ? *Pycnoplinthus*). Cruciferae. 2 Bhutan

Pycnoplinthus O. Schulz. Cruciferae. 1 Himal.: *P. uniflorus* (Hook.f. & Thomson) O. Schulz

Pycnorhachis Benth. Asclepiadaceae. 1 Malay Peninsula

Pycnosorus Benth. = *Craspedia*. But see Telopea 5(1992)39

Pycnospatha Thorel ex Gagnepain. Araceae (V 3). 2 SE As.

Pycnosphaera Gilg. Gentianaceae. 5 trop. Afr.

Pycnospora R. Br. ex Wight & Arn. Leguminosae (III 9). 1 OW trop.: *P. lutescens* (Poiret) Schindler

Pycnostachys Hook. Labiatae (VIII 4c). 37 trop. & S Afr., Madag. R: KB 1939:563. Cult orn. herbs esp. *P. urticifolia* Hook. (blue boys, trop. & S Afr.) – fls bright blue

Pycnostelma Bunge ex Decne. Asclepiadaceae (III 1). 2–3 China

Pycnostylis Pierre = *Triclisia*

Pycreus Pal. = *Cyperus*

Pygeum Gaertner = *Prunus*

Pygmaea B.D. Jackson = *Chionohebe*

Pygmaeocereus H. Johnson & Backeb. = *Echinopsis*

Pygmaeopremna Merr. = *Premna*

Pygmaeorchis Brade. Orchidaceae (V 13). 2 Brazil

Pygmaeothamnus F. Robyns. Rubiaceae (III 2). 4 trop. & S Afr. Geoxylic suffrutices

Pygmea Hook.f. = *Chionohebe*

pyinkado *Xylia xylocarpa*

pyinma *Lagerstroemia speciosa*; **Andaman p.** *L. hypoleuca*

pyjama lily *Crinum kirkii*

Pynaertiodendron De Wild. = *Cryptosepalum*

Pyracantha M. Roemer. Rosaceae (III). 9 SE Eur. (1) & As. R: CJB 68(1990)2230. Cult. orn. usu. thorny shrubs esp. *P. coccinea* M. Roemer (firethorn, SE Eur., native in GB in warmer interglacials) – many cvs, often grown espaliered for persistent colourful fr. in winter. Many named cvs hybrids, some unaffected by fireblight &/or with frs not taken by birds

Pyragra Bremek. Rubiaceae (IV 7). 2 Madagascar

Pyramia Cham. = *Cambessedesia*

Pyramidanthe Miq. (~ *Fissistigma*). Annonaceae. 1 W Mal.: *P. prismatica* (Hook.f. & Thomson) Sinclair

Pyramidium Boiss. (*Veselskya*). Cruciferae. 1 Afghanistan

Pyramidoptera Boiss. Umbelliferae (III 7). 1 Afghanistan

Pyrenacantha Wight. Icacinaceae. 20 OW trop.

Pyrenaria Blume. Theaceae. 20 SE As., W Mal.

pyrene oil *Olea europaea*

Pyrenoglyphis Karsten = *Bactris*

pyrethrum *Tanacetum cinerariifolium* (insecticide), *T. coccineum* (cut-fl.)

Pyrethrum Zinn = *Tanacetum*

Pyrgophyllum (Gagnep.) T.L. Wu & Z.Y. Chen. Zingiberaceae. 1 China. R: APS 27(1989)124

Pyriluma Baillon ex Aubrév. = *Pouteria*

Pyrola L. Ericaceae (Pyrolaceae). 35 N hemisph. (Eur. 7) to Sumatra, temp. S Am. Wintergreen. Seeds wind-disp. Some locally used to heal wounds & cult. orn. incl. *P. asar-*

ifolia Michaux (N Am. & As.) – lvs a source of a tea for N Am. Indians in E Canada

Pyrolaceae Dumort. = Ericaceae

Pyrolirion Herbert (~ *Zephyranthes*). Amaryllidaceae (Liliaceae s.l.). 4 Andes. R: BSB 2, 53(1981)1197

Pyrostegia C. Presl. Bignoniaceae. 3–4 trop. S Am. Lianes with 6–8-ribbed branchlets. *P. venusta* (Ker-Gawler) Miers (*P. ignea*, flame vine, golden shower, Brazil & Paraguay) – commonly planted in trop. for orange fls poll. by hummingbirds

Pyrostria Comm. ex Juss. Rubiaceae (III 2). c. 45 Afr. (14), Madag., Mauritius, Rodrigues (? to Mal.). *P. bibracteata* (Baillon) Cavaco (E Afr. to Seychelles) – fr. ed.

Pyrrhanthera Zotov. Gramineae (25). 1 NZ

Pyrrhocactus (A. Berger) Backeb. & Knuth = *Neoporteria*

Pyrrhopappus DC. Compositae (Lact.-Mic.). 3 N Am. R: AJB 77(1990)845. Some roots eaten by N Am. Indians

Pyrrocoma Hook. (~ *Haplopappus*). Compositae (Ast.-Sol.). c. 10 W US & Canada

Pyrrorhiza Maguire & Wurd. Haemodoraceae. 1 Venezuela

Pyrrosia Mirbel. Polypodiaceae (I). 100 trop. OW to NE As. & NZ. R: AFJ 73(1983)73. Scandent epiphytes often with crassulacean acid metabolism & usu. simple leathery/fleshy fronds covered in stellate hairs. Some cult. orn. incl. *P. nummulariifolia* (Sw.) Ching (Indomal.) – pest of coffee in webbing branches together & poss. causing rotting. Some local diuretics etc.

Pyrrothrix Bremek. = *Strobilanthes*

Pyrularia Michaux. Santalaceae. 2–3 SE US (1), China, Himal. *P. pubera* Michaux (buffalo or elk nut, SE US) – ed. fr. but oil acid & poisonous

Pyrus L. Rosaceae (III). c. 25 Euras. (esp. Armenia (>20), Eur. 13), Medit. R: CJB 68(1990)2238. Pears. Form. spp. referred (not on unreasonable grounds) here now placed in *Malus* & *Sorbus*, but *P.* now restricted to trees with sclerenchyma cells in pome (i.e. gritty texture in pear flesh); extrafl. nectaries. Ed. pears (1000 cvs (Theophrastus described 3, Pliny 39, C16 Italy 232, 677 in GB 1831), prop. by grafting (no cuttings), some parthenocarpic) found in Neolithic & Bronze Age Eur. deposits, derive from *P. communis* L. (Eur., W As.) – incl. Bartlett pears (i.e. 'Williams' Bon Chrétien', 1770) which are those most commonly tinned, most requiring cross-poll. (self-incompatible) & some triploids; this has been crossed with *P. pyrifolia* (Burm.f.) Nakai (*P. serotina*, Chinese or Japanese p., nashi (= Jap. for pear), China, natur. Japan & grown in trop. As.) to give *P.* × *lecontei* Rehder esp. 'Kieffer' with enhanced disease resistance; the fr. of *P. pyrifolia* usu. ± apple-shaped & noted for excellent keeping qualities. Fr. prod. c. 7m tons per annum, best-flavoured being 'Doyenne du Comice' (1849) with buds less prone (higher concs of phenolic acids) to bullfinch attack than 'Conference' (1894), the most widely planted of c. 20 in UK commerce; perry (cf. cider from apples) made from fr. with trees 200–300 yrs old surviving in Gloucestershire, UK, though in decline until 1940s when controlled fermentation used at Shepton Mallett, Somerset to produce comm. 'Babycham'. Bark has antibacterial action & timber used for turnery, cutlery handles, inlay & (stained) piano keys – form. much favoured for carving as by the incomparable Grinling Gibbons (1648–1721). Some cult. orn. esp. *P. calleryana* Decne (China) – more salt-tolerant in urban sites than *Platanus* × *hispanica*, many cvs grown as street-trees in US & *P. salicifolia* Pallas (E Turkey, Cauc., NW Iran) – habit of willow with greyish lvs

Pyxidanthera Michaux. Diapensiaceae. 1 E US

Qaisera Omer = *Gentiana*

Qiongzhuea Hsueh & Yi = *Chimonobambusa*

Quadricasaea Woodson = *Bonafousia*

Quadripterygium Tard. = *Euonymus*

quake or **quaking grass** *Briza* spp.

Qualea Aublet. Vochysiaceae. 59 trop. Am. R: ABN 2(1953)150. G

Qualup bell *Pimelea physodes*

quamash *Camassia* spp.

Quamoclidion Choisy. Nyctaginaceae. 1 Mexico

Quamoclit Miller = *Ipomoea*

quandong *Santalum acuminatum*

Quapoya Aublet. Guttiferae (III). 4–5 Guyana, Peru

Quaqua N.E. Br. (~ *Caralluma*). Asclepiadaceae (III 5). 13 SW Afr. R: Bradleya 1(1983)33.

Cult. orn. succ.

Quararibea Aublet. Bombacaceae. 35 trop. Am. *Q. cordata* (Bonpl.) Vischer (S Am.) – poll. by non-flying mammals in SE Peru rain forest; *Q. funebris* (Llave) Vischer (Mexico) & other spp. – fls dried & added as spice to flavour chocolate drinks in Mexico; *Q. pumila* Alverson (Costa Rica) – cauliflorous pachycaul

quaruba *Vochysia hondurensis*

Quassia L. Simaroubaceae. 40 trop. (incl. *Samadera, Simaba*; s.s. 1: *Q. amara* L. (Surinam quassia wood, stave-wood, Brazil) – source of bitters, vermifuge & poison in fly-papers). R: Blumea 11(1962)514. *Q. guianensis* (Aublet) D. Dietr. (*Simaba g.*, trop. S Am.) – fish-disp. in Amazon; *Q. simarouba* L.f. (*Simarouba glauca*, aceituna, C Am., WI) – oilseed; *Q. cedron* (Planchon) D. Dietr. (cedron, Amazonia) – bitter seeds vermifuge; *Q. indica* (Gaertner) Nooteb. (Madag. to Papuasia) – medicinal oil & insecticide

quassia wood *Picrasma excelsa*; **Surinam q. w.** *Quassia amara*

Quaternella Pedersen. Amaranthaceae. 1 S Am.: *Q. confusa* Pedersen. R: BMNHN 4,12(1990)92

quebracha *Aspidosperma tomentosum*

quebracho *Schinopsis* spp.

Queen Anne's lace *Anthriscus sylvestris*

queen's flower *Lagerstroemia speciosa*

Queensland arrowroot *Canna indica*; **Q. bean** *Entada phaseoloides*; **Q. hemp** *Sida rhombifolia*; **Q. maple** *Flindersia brayleyana*; **Q. nut** *Macadamia* spp.; **Q. walnut** *Endiandra palmerstonii*

Queenslandiella Domin (~ *Cyperus*). Cyperaceae. 1 E Afr. to E Aus.: *Q. hyalina* (Vahl) Ballard

queenwood *Daviesia arborea*

Quekettia Lindley. Orchidaceae (V 10). 5 trop. S Am. to Trinidad

Quelchia N.E. Br. Compositae (Mut.-Mut.). 4 Venez., Guyana. Shrubs & trees

Quercifilix Copel. = *Tectaria*

quercitron yellow dye from inner bark of *Quercus* spp. esp. *Q. velutina*

Quercus L. Fagaceae. c. 400 N temp. (Eur. 20) S to Mal. & Colombia at alt. Oaks. Evergreen or decid. trees, few shrubs, some with domatia; trop. spp. with intermittent growth of apical buds developing fls & in next flush lvs photosynthesizing whilst acorns develop, with acorn maturity apical buds produce more fls, lvs, etc.; in spp. where winter stops growth, evergreen oaks fl. in spring & fr. in autumn, decid. ones with wind-poll. fls before new lvs fully grown & shed lvs when acorns have fallen; in subg. *Erythrobalanus* (red oaks, Am.) fert. in spring the year after poll. so fr. matures in 2nd autumn. R: A. Camus (1934–54) *Les chênes*. Cult. orn. & timber trees, acorns used for swine-food, local medic. (preserved at Pompeii) & as tanning agents, bark for dye (quercitron) tannins & cork, galls (oak apples) for tanning & lvs for feeding silkworms; timber for construction but also whisky casks (bourbon in new ones, whisky in old, some 1m per annum in US (*Q. alba*) & 100 000 in Eur.) & inlay (brown o. = timber stained with mycelium of bracket fungus (*Fistulina hepatica*) & green by *Chlorosplenium aeruginascens*); worshipped in S Eur. from earliest times. *Q. alba* L. (white or Quebec oak, E N Am.) – imp. timber & fuel source; *Q. cerris* L. (Turkey or manna o., C & S Eur., W As.) – timber & source of sweetmeats etc.; *Q. coccifera* L. (Kermes o., Medit.) – form. used for feeding cochineal insects, 3 sprigs the crest of the Dyers' Company; *Q. coccinea* Muenchh. (scarlet o., E N Am.) – timber, lvs bright red in autumn; *Q. falcata* Michaux ((American) red o., SE US); *Q. gambelii* Nutt. (shin o., S N Am.) – acorns ed.; *Q. garryana* Douglas ex Hook. (Oregon white o., W N Am.) – imp. timber & fuel; *Q. ilex* L. (incl. *Q. roundifolia* (*Q. i.* ssp. *ballota* (Desf.) Samp., W Medit.), holm o., holly o., ilex, Medit.) – much-planted 'evergreen o.' with ed. acorns, galls (prob. Morea, Greek, Marmora or Italian g.) used in tanning (ilex an old Roman name now transferred to holly, i.e. *Ilex*, while holm, Old English name for holly now transferred to *Q. ilex*!); *Q. ilicifolia* Wangenh. (bear o., E US) – usu. shrubby; *Q. imbricaria* Michaux (shingle o., C & E US) – imp. timber for clapboards, shingles etc.; *Q. lobata* Née (Californian o., Calif.) – acorns form. eaten by Indians; *Q. lusitanica* Lam. (Portugal, Spain) – shrub, often creeping; *Q. macrocarpa* Michaux (burr o., E N Am.) – timber for ship-building, furniture etc.; *Q. macrolepis* Kotschy (*Q. ithaburensis* ssp. *m.*,'*Q. aegilops*', S Eur., W As.) – dried cupules used in tanning (valonea, palamut) esp. for quality heavy leathers; *Q. marilandica* Muenchh. (jack o., C & SE US) – charcoal; *Q. mongolica* Fischer ex Turcz. (Japanese o., NE As.) – imp. local timber; *Q. nigra* L. (possum o., SE US) – acorns form. Indian food; *Q. obtusiloba* Michaux (iron or post o., SE US) – timber, ed. acorns, lvs used as cigarette-papers; *Q. prinus* L. (chestnut o., (American) white o., E US) – imp. tanbark, timber, ed. acorns: *Q. pubescens* Willd. (*Q. infectoria*, gall, E Med.) – galls (Aleppo, Levant, Mecca or Turkish galls) with 36–58% tannin coll. for high tannin content & used in ointments & suppositories esp. in treatment of piles; *Q. robur* L. (English, French, Polish or Slavonian o., Eur., Medit.) – principal hardwood of Britain, selected as 'standards' in coppice & standards silviculture, imp. timber for construction of houses & furniture of past, also warships, flooring, charcoal etc., acorns eaten by pigs (pannage) & ground as coffee subs., replaced in N & W GB by *Q. petraea* (Mattuschka) Liebl. (*Q. esculus* L., durmast or sessile o., Eur., W As.) with similar uses but many trees in GB at least appear to be of hybrid origin (*Q.* × *rosacea* Bechst.), the 'spp.' being imperfectly isolated genetically, growth-rates a good indication of soil quality & depth, sometimes distrib. by jays which select largest acorns they can swallow; *Q. rubra* L. (*Q. borealis*, (American) red o., E N Am.) – imp. timber for general construction; *Q. suber* L. (cork o., S Eur., N Afr.) – thick bark the cork of commerce, stripped off trees older than 20 yrs every 8–10, used for insulation, tiles & bottle-corks, floats, form. cork-tipped cigarettes, when ground a constituent of linoleum, most cult. in Portugal, hybrid with *Q. cerris* = *Q.* × *pseudosuber* Santi ('*Q.* × *hispanica*', Lucombe o.) hardy GB; *Q. velutina* Lam. (*Q. tinctoria*, quercitron, E N Am.) – tanbark, yellow dye used in printing calico; *Q. virginiana* Miller (live o., SE US) – form. imp. shipbuilding timber

Quesnelia Gaudich. Bromeliaceae (3). 15 SE Brazil. R: FN 14(1979)1956. Some cult. orn.

quetsche(n) plum *Prunus* × *domestica*

Quetzalia Lundell. Celastraceae. 9 trop. Am.

Quezelia Scholz = *Quezeliantha*

Quezeliantha Scholz ex Rauschert. Cruciferae. 1 Sahara: *Q. tibestica* (Scholz) Rauschert

Quiabentia Britton & Rose. Cactaceae (II). 2+ Arg. & Paraguay (Chaco)

quickbeam *Sorbus aucuparia*

quickset, quickthorn *Crataegus monogyna*

Quidproquo Greuter & Burdet = *Raphanus*

Quiducia Gagnepain = *Silvianthus*

Quiina Aublet. Quiinaceae. 25 trop. S Am. Some fish-disp. in Amazon

Quiinaceae Choisy ex Engl. Dicots – Dilleniidae – Theales. 4/45 trop. Am. esp. Amazonia. Trees or shrubs, sometimes lianoid. Lvs simple or pinnate (in maturity only in *Froesia* & *Touroulia*), opp. or whorled; stipules interpetiolar, persistent & rigid or leafy. Fls bisexual (or pl. polygamous), ± reg. in panicles or racemes; K 4 or 5, small, unequal, imbricate, C 4 or 5(–8), imbricate, A 15–∞, the filaments sometimes basally connate or adnate to C, anthers with distinct pollen sacs or those back-to-back & latrorse, with longit. slits, $\underline{G}$ (2–13) with free styles or $\underline{G}$ 3 (*Froesia*), each loc. with 2 ascending, anatropous ovules. Fr. fleshy but often dehiscent at maturity when often 1- or 2-loc. by abortion; seeds velutinous (glabrous in *Froesia*) with straight embryo & 0 endosperm

Genera: *Froesia, Lacunaria, Quiina, Touroulia*

quillai *Quillaja saponaria*

Quillaja Molina. Rosaceae. 3 temp. S Am. Evergreen trees with bark used as soap & medic. esp. *Q. saponaria* Molina (quillai, Chile) – source of q. bark with 9% saponin used in fire-extinguishers

quillings See *Cinnamomum verum*

quince *Cydonia oblonga*; **Bengal q.** *Aegle marmelos*; **Japanese q.** *Chaenomeles speciosa*

Quinchamalium Molina. Santalaceae. 25 Andes

Quincula Raf. (~ *Physalis*). Solanaceae (1). 1 N Am.: *Q. lobata* (Torrey) Raf.

Quinetia Cass. Compositae (Gnap.-Ang.). 1 S & SW Aus.: *Q. urvillei* Cass. R: OB 104(1991)124

quinine *Cinchona* spp.; **q. bush** or **tree** *Petalostigma* spp. esp. *P. pubescens*

quinoa, quinua *Chenopodium quinoa*

Quinqueremulus P.G. Wilson. Compositae (Gnap.-Ang.). 1 W Aus.: *Q. linearis* P.G. Wilson. R: Nuytsia 6(1987)1

quinsyberry *Ribes nigrum*

Quintinia A. DC. Grossulariaceae (Escalloniaceae). 20 C Mal. to Aus. (4), New Caled. (6) & NZ. Trees & shrubs incl. *Q. sieberi* A. DC (opossum wood, possumwood, Aus.)

Quiotania Zarucchi. Apocynaceae. 1 Colombia: *Q. colombiana* Zarucchi. R: Novon 1(1991)33

Quisqualis L. (~ *Combretum*). Combretaceae. 16 OW trop. Lianes with eccentric xylem & alks. *Q. indica* L. (Rangoon creeper, Burma to New Guinea) – cult. orn. with fragrant white fls becoming pink with age, anthelmintic (Malay name 'udani' punned via Dutch 'hoedanig' (how, what?) to *Q.*)

Quisqueya Dod (~ *Broughtonia*). Orchidaceae (V 13). 4 WI

Quisumbingia Merr. Asclepiadaceae (III 4). 1 Philippines: *Q. merrillii* (Schltr.) Merr.

Quivisianthe Baillon. Meliaceae (II). 1 or 2 Madagascar

R

rabbit-ears *Thelymitra antennifera*

Rabdosia Hassk. = *Isodon*

Rabdosiella Codd (~ *Isodon*). Labiatae (VIII 4c). 2 S Afr., N India. R: Bothalia 15(1984)9

Rabiea N.E. Br. (~ *Nananthus*). Aizoaceae (V). c. 4 S Afr.

Racemobambos Holttum. Gramineae (1b). 18 Indomal. Montane forests. R: KB 37(1982)661, 44(1989)364

Rachicallis DC = (~ *Arcytophyllum*). Rubiaceae (IV 1). 1 WI

Racinaea M. Spencer & L.B. Sm.(~ *Tillandsia*). Bromeliaceae (2). 51 trop. Am. R: Phytol. 74(1993)151, 429

Racosperma C. Martius = *Acacia*

Radamaea Benth. Scrophulariaceae. 5 Madagascar

Raddia Bertol. Gramineae (4). 5 E Brazil & Guyana. R: KBAS 13(1986)63. Often fern-like

Raddiella Swallen. Gramineae (4). 8 trop. Am. V. small bamboos, like dwarf grasses

Radermachera Zoll. & Moritzi. Bignoniaceae. 15 SE As., Mal. R: Blumea 23(1976)121. Some cult. orn. trees & locally used timber

radicchio, radiccio *Cichorium intybus*

Radinosiphon N.E. Br. Iridaceae (IV 3). 1–2 S trop. & S Afr.

Radiola Hill. Linaceae. 1 Eur., Medit., temp. As., trop. Afr. mts: *R. linoides* Roth (allseed)

radish *Raphanus sativus*; **Chinese** or **Japanese r.** *R. s.* 'Longipinnatus'; **wild r.** *R. raphanistrum*

Radlkofera Gilg. Sapindaceae. 1 trop. Afr.

Radlkoferotoma Kuntze. Compositae (Eup.-Ager.). 3 Uruguay, S Brazil. R: MSBMBG 22(1987)142. Hairy shrubs with opp. lvs

Radyera Bullock (~ *Hibiscus*). Malvaceae. 1 S Afr., 1 Aus. R: BG 132(1971)57

Raffenaldia Godron. Cruciferae. 2 Morocco, Algeria

raffia *Raphia farinifera*

Rafflesia R. Br. Rafflesiaceae (II 1). 15–16 W Mal. (some extinct). R: Blumea 30(1984)209. Parasitic (only through wounds?) on *Tetrastigma* spp. (Vitaceae). *R. arnoldii* R. Br. (*R. titan*, Sumatra) – fl. buds appear 19–21 months before anthesis when fl. to 80 cm diam. (largest fl. known) & 7 kg, smelling of carrion & visited by (?poll.) flies, now an endangered sp.; *R. kerrii* Meijer (Thailand) & *R. pricei* Meijer (N Borneo) poll. bluebottles & other carrion flies

Rafflesiaceae Dumort. Dicots – Magnoliidae – Aristolochiales. (Incl. Mitrastemmataceae) 9/50 trop. to temp. (few). Chlorophyll-less parasites usu. on roots, the veg. body ± filamentous like a fungal mycelium through the host, giving rise to infls endogenously, but s.t. with vestigial vasc. system & scale-like lvs around spike or single fl. Fls solit. & often v. large, or small & embedded in infl. axis (e.g. *Cytinus*), often fleshy & stinking, usu. unisexual & reg.; P 4 or 5 (–10 or more), s.t. basally connate, usu. imbricate, petaloid or not, A 8–∞ with filaments forming a tube around stylar column or usu. adnate to it with anthers in 1–several whorls, opening with slits or pores, stylar column in large fls apically expanded above anthers to form large disk, anthers with longit. slits, apical pore or transverse slits, $\overline{G}$ (sup. in *Mitrastemma*) (4–8) or half-inf. with columnar style with apical disk, 1-loc. with 4–∞ parietal placentas, ovules ∞ (anatropous to) orthotropous, uni- or bitegmic. Fr. often fleshy, irreg. dehiscent or not; seeds ∞ with undifferentiated embryo in endosperm. x = 6?

Genera: *Apodanthes, Bdallophytum, Berlinianche, Cytinus, Mitrastemma, Pilostyles, Rafflesia, Rhizanthes, Sapria*; some with v. disjunct distribs

Re-allied with Aristolochiaceae as proposed by Robert Brown (1821). Parasitic on Leguminosae, Vitaceae, Cistaceae, Fagaceae etc., incl. world's largest flower (*Rafflesia*)

Rafinesquia Nutt. Compositae (Lact.-Step.). 2 SW N Am.

Rafnia Thunb. Leguminosae (III 28). 22 S Afr. esp. SW Cape

Ragala Pierre = *Chrysophyllum*

ragged robin *Silene flos-cuculi*

ragi *Eleusine coracana*

ragweed *Artemisia* spp.

ragwort *Senecio* spp. esp. *S. jacobaea*; **Oxford r.** *S. squalidus*

Rahowardiana D'Arcy (~ *Markea*). Solanaceae (1). 1 Panamá: *R. wardiana* D'Arcy

rai *Brassica juncea*

Raillardella (A. Gray) Benth. Compositae (Hele.-Mad.). 3 W US

Railliardia Gaudich. = *Dubautia*

Raillardiopsis Rydb. (~ *Raillardella*). Compositae (Hele.-Mad.). 2 SW N Am. R: Madrono 37(1990)43

Raimondia Saff. (~ *Annona*). Annonaceae. 2 Colombia to Ecuador. R: AHB 10(1930)81
Raimondianthus Harms = *Chaetocalyx*
Raimundochloa A. Molina = *Koeleria*
rain lily *Zephyranthes carinata*; **r. tree** *Albizia saman*
Rainiera E. Greene (~ *Luina*). Compositae (Sen.-Tuss.). 1 NW US
raisin *Vitis vinifera*; **r. tree** *Hovenia dulcis*
rajah cane *Eugeissona minor*
Rajania L. Dioscoreaceae. 20 WI. Capsule with 1 wing; some ed. tubers
raki *Pimpinella anisum* (also *Foeniculum vulgare*)
rakkyō *Allium chinense*
rakum *Salacca wallichiana*
Ramatuela Kunth = *Terminalia*
rambai *Baccaurea motleyana* & *B. ramiflora*
rambutan *Nephelium lappaceum*
Ramelia Baillon = *Bocquillonia*
Rameya Baillon = *Triclisia*
ramie *Boehmeria nivea*
ramin *Gonystylus bancana*
Ramirezella Rose = *Vigna*
Ramisia Glaz. ex Baillon. Nyctaginaceae (II). 1 SE Brazil: *R. brasilensis* Oliver
ramón *Brosimum alicastrum*
Ramonda Rich. Gesneriaceae. 3 Pyrenees, Balkans. Fls almost reg., A 5. Cult. orn. rock rosette-pls esp. *R. myconi* (L.) Reichb. (Pyrenees) – can survive 2–3 yrs drought & 55 °C
ramontchi *Flacourtia indica*
Ramorinoa Speg. Leguminosae (III 4). 1 W Argentina
Ramosmania Tirv. & Verdc. (~ *Randia*). Rubiaceae (II 5). 2 Rodrigues: *R. heterophylla* (Balf.f.) Tirv. & Verdc., last seen 1938, & *R. rodriguesii* Tirv. R: Comptes Rendus Soc. Biogeogr. 65(1989)13
rampion *Campanula rapunculus*, *Phyteuma orbiculare*
ramsons *Allium ursinum*
ramtil *Guizotia abyssinica*
Ranalisma Stapf (~ *Echinodorus*). Alismataceae. 1 SE As. & Mal., 1 trop. W Afr.
Randia L. Rubiaceae (II 1). Incl. *Basanacantha* c. 100 trop. & warm Am. (Mex. 33, 23 endemic). Some ed. fr. & cult. orn. incl. *R. formosa* (Jacq.) Schumann but most OW spp. form. here referred to other genera (see e.g. *Catunaregam*, *Kailarsenia* & *Rothmannia*)
Randonia Cosson. Resedaceae. 3 N Afr., Arabia, Somalia. Camel fodder
Ranevea L. Bailey = *Ravenea*
Rangaeris (Schltr.) Summerh. Orchidaceae (V 16). 1 trop. & S Afr.
Rangoon creeper *Quisqualis indica*
Ranopisoa J. Leroy (~ *Oftia*). Scrophulariaceae. 1 Madagascar
Ranunculaceae Juss. Dicots – Magnoliidae – Ranunculales. (Incl. Hydrastidaceae & Kingdoniaceae) 62/2450 temp. & boreal. Lianes (*Clematis* spp.), small shrubs (*Xanthorhiza*), & (usu.) herbs, s.t. aquatic, usu. with benzyl-isoquinoline or aporphine alks or various other toxins. Lvs simple to compound, spirally arr. (opp. in *Clematis*); stipules minute or 0. Fls usu. bisexual & insect-poll. (wind-poll. in *Thalictrum* etc.), (solit. or) in basically cymose infls, with ± elongate receptacle & spirally arr. perianth; K (3–)5–8 or more often petaloid & caducous, C (0) few–∞, app. staminodal & often with basal nectaries (nectar sucrose-rich), A usu. ∞ & spirally arr., centripetal (1 or 2 whorls of 5 in *Xanthorhiza*, 6–10 such in *Aquilegia*) with unprolonged connective & anthers with longit. slits, $\underline{G}$ (1–)several, rarely ± connate (e.g. *Nigella*) with axile placentation, each carpel with several – ∞ marginal, or 1 nearly basal pendulous, anatropous or hemitropous, bitegmic or unitegmic ovules. Fr. usu. follicles, achenes or berries (capsule in *Nigella*); seeds with minute undeveloped embryo or linear & with cotyledons often basally connate, s.t. 1 suppressed, in copious oily & proteinaceous (sometimes starchy) endosperm. x = 6–10, 13. R: APG 41(1990)93, (Am.) Phytol. 70(1991)24

Classification & principal genera:

I. **Helleboroideae** (G with many ovules; 4 tribes):
   1. **Helleboreae** (lvs simple to palmate or pedate, fls reg.): *Caltha, Eranthis, Helleborus, Trollius*
   2. **Cimicifugeae** (lvs ternately compound, fls usu. reg.): *Actaea, Cimicifuga*
   3. **Nigelleae** (lvs pinnately compound, fls reg.): *Nigella*
   4. **Delphinieae** (fls zygomorphic, K spurred): *Aconitum, Consolida, Delphinium*

II. **Ranunculoideae** (G with 1(2) ovules; 3 tribes):
 1. **Adonideae** (K petaloid, C longer): *Adonis, Callianthemum*
 2. **Anemoneae** (K petaloid, C 0 or small): *Anemone, Clematis, Hepatica, Kingdonia, Pulsatilla*
 3. **Ranunculeae** (K sepaloid or petaloid, C usu. present): *Ranunculus* (limits debated)

III. **Isopyroideae** (G with 2 to many ovules; 3 tribes):
 1. **Isopyreae** (lvs ternately compound): *Aquilegia, Isopyrum*
 2. **Dichocarpeae** (lvs usu. pedately compound): *Dichocarpum*
 3. **Coptideae** (lvs variously compound, follicles without transverse veins): *Coptis, Xanthorhiza*

IV. **Thalictroideae** (C 0, 1 ovule per carpel): *Thalictrum*

V. **Hydrastidoideae** (C 0, 2–4 ovules per carpel): *Hydrastis*

Glaucidium here referred to Paeoniaceae

Many cult. orn. & some med. (also ed. seeds) etc., though many are v. poisonous. Some bad weeds e.g. *Ranunculus* spp.

Ranunculus L. Ranunculaceae (II 3). c. 600 temp. (Eur. 133, NZ 43) incl. trop. mts & boreal. Buttercups, crowfoot, paigle. Herbs with yellow, white or red fls, all poisonous (ranunculin, see *Anemone*) & usu. avoided by stock though harmless in hay. Cult. orn. & weeds, some aquatics (water crowfoot), esp. *R. aquatilis* L. (Eur. & bipolar) with submerged dissected lvs & lobed floating ones, while *R. fluitans* Lam. (Eur.) with only dissected ones & *R. omiophyllus* Ten. (Eur.) on mud with only lobed ones. *R. aconitifolius* L. (Fair maids of France (or Kent), C Eur. mts) – cult. orn. with tubers & white fls; *R. acris* L. (Euras.) – double-flowered forms ('Flore Pleno') cult. orn. ('bachelor's buttons'); *R. asiaticus* L. (garden ranunculus, SE Eur. & SW As.) – cult. orn. with claw-like tubers & variously coloured, often double fls, form. imp. florists' flower first developed by Turks & 1100 cvs listed for GB by 1777; *R. auricomus* L. (goldilocks, Euras.) – many lines (some sexual diploids but many polyploids, facultative or obligate apomicts) recog. as spp. by some authorities; *R. ficaria* L. (celandine, pilewort, Eur., W As., natur. N Am.) – cotyledon 1, blanched lvs sometimes eaten, cult. orn. with several cvs, tubers resembling haemorrhoids (Doctrine of Signatures), fls with glistening P due to epidermal cells full of pigment & cells beneath packed with white starch grains, tetraploid subsp. *bulbilifer* Lambinon with conspic. bulbils; *R. glacialis* L. (Eur. mts) – calcifuge cult. orn. holding alt. record for Scandinavia (2370 m) & Alps (4275 m); *R. lingua* L. (spearwort, Eur., Siberia) – cult. orn. with ± ovate lvs; *R. lobatus* Jacquem. (Himal.) – with *Christolea himalayensis* the fl. pl. reaching highest alts (7756 m); *R. ophioglossifolius* Villars (Euras., Med., natur. NZ) – annual sp., the subject of the world's smallest nature reserve (0.04 ha) at Badgeworth, Gloucestershire, GB; *R. polyphyllus* Waldst. & Kit. ex Willd. (E Eur., W As.) – aquatic with 'Hippuris syndrome'; *R. repens* L. (creeping b., Euras., natur. N Am.) – bad weed, rain-poll. (rain washes pollen to base of glossy petals & capillary action draws it up to stigmas)

Ranzania Itô. Berberidaceae (II 2). 1 Japan: *R. japonica* (Maxim.) Itô – cult. orn. herb with nodding purple fls

Raoulia Hook.f. ex Raoul. Compositae (Gnap.-Cass.). Excl. *Psychrophyton* (q.v.) 11 New Guinea, NZ (Aus. spp. = *Ewartia*). R: OB 104(1991)83. Tufted or creeping herbs & subshrubs; some cult. orn. incl. *R. australis* Hook.f. ex Raoul with silver-hairy lvs forming mat

Raouliopsis S.F. Blake. Compositae (Gnap.-Lor.). 2 Colombian Andes. R: OB 104(1991)63

Rapanea Aublet = *Myrsine*

Rapatea Aublet. Rapateaceae. 20 trop. S Am.

Rapateaceae Dumort. Monocots – Commelinidae – Commelinales. 16/84 trop. S Am. esp. Guayana Highland, with monotypic *Maschalocephalus* in W Afr. Perennial often large rhizomatous herbs often accum. aluminium & with vessel-elements throughout stems & roots. Lvs distichous, basal, with folded open sheath & narrow parallel-veined lamina. Fls reg., 3-merous, bisexual, insect-poll. in spikelets of single term. fl. with spirally arr. imbricate bracts beneath it, the spikelets forming a head or 1-sided raceme often with involucre of (1) 2–several large bracts; K 3, ± basally connate, imbricate, C 3 usu. basally connate forming tube with imbricate lobes, yellow (red), nectaries & nectar 0, A 3 + 3 with short filaments often connate & adnate to C-tube, anthers with 1, 2 or 4 apical pores or short slits, pollen grains usu. monosulcate, $\underline{G}$ (3), 3-loc. with 1 style & several–∞ (or 1–2 axile – basal) anatropous ovules on axile placentas in each loc. Fr. a loculicidal capsule; seeds with small embryo near hilum alongside copious starchy (mostly simple grains) endosperm & s.t. with caruncle or elaiosome. 2n = 22

Classification & principal genera:

Saxofridericieae (several ovules per loc., C yellow; 5 genera): *Saxofridericia, Stegolepis*

Schoenocephalieae (several ovules per loc., C reddish; 3 genera): *Schoenocephalium*

Rapateae (1 ovule per loc., seeds longit. striate; 4 genera): *Rapatea*

Monotremeae (1 ovule per loc.; seeds not striate; 4 genera): *Maschalocephalus, Monotrema*

Perhaps closest to Xyridaceae

rape *Brassica* spp. esp. *B. napus*; **oilseed r.** *B. napus* (summer races)

Raphanorhyncha Rollins. Cruciferae. 1 Mexico

Raphanus L. Cruciferae. 3 W & C Eur. (1). Medit. to C As. Radish. R: JAA 66(1985)328. *R. sativus* L. (cultigen grown since time of Assyrians & prob. selected from *R. raphanistrum* L. ssp. *landra* (DC) Bonnier & Layens (*R. landra*)) – taproots cherry-sized to 1 m long & 60 cm diam., common garden radish with ed. taproot used in salad; some cvs cooked, lvs used like spinach & young lomenta (fr.) pickled – 'Caudatus' with lomenta to 30 cm long, pickled, 'Longipinnatus' (daikon, mooli, mula, muli, Chinese or Japanese r.) with large, long-lasting roots to 50 kg much grown for human & stock food in E

Raphia Pal. Palmae (II 1g). 28 trop. Am. (1, also in Afr.), Afr. & Madag. Solit. or clustered hapaxanthic stems with fr. covered with prominent shiny scales. Fibre (raffia) & palm wine, those with small fr. a source of beads, petioles a bamboo subs., some cult. orn. (bamboo palm). *R. farinifera* (Gaertner) N. Hylander (*R. ruffia*, raphia palm, trop. Afr., Madag.) – world's largest lvs (lamina to 19.8 m long with 3.96 m petiole), young lvs the source of raffia form. imp. hort. tying material also used in basketry etc., older lvs source of raffia wax, now afforded 'special protection' in Zimbabwe; *R. hookeri* G. Mann & H. Wendl. (W Afr.) – wine palm, source (with *R. palma-pinus* (Gaertner) Hutch., W Afr.) of African piassava

Raphidiocystis Hook.f. Cucurbitaceae. 5 trop. Afr. & Madag. R: BJBB 37(1967)319

Raphiocarpus Chun = *Didissandra*

Raphiolepis Lindley = *Rhaphiolepis*

Raphionacme Harvey. Asclepiadaceae (I). 30 trop. & S Afr., Arabia (1: *R. arabica* A.G. Miller & Bingi – rootstock ed.). *R. utilis* N.E. Br. & Stapf (Bitinga rubber) – rubber source

Raphistemma Wallich. Asclepiadaceae (III 1). 2 Indomal.

Rapicactus F. Buxb. & Oehme = *Turbinicarpus*

Rapistrum Crantz. Cruciferae. 2 C Eur. (2), Medit. & W As. R: JAA 66(1985)335

Rapona Baillon. Convolvulaceae. 1 Madag.: *R. tiliifolia* (Baker) Verdc. – seeds with irritant hairs

Raputia Aublet. Rutaceae. 4 NE trop. Am. R: Brittonia 42(1990)175. Medic. bark

Raritebe Wernham. Rubiaceae (I 8). 1 C & N S Am.

rasamala *Altingia excelsa*

Raspalia Brongn. Bruniaceae. 16 S Afr.

raspberry *Rubus idaeus*; **black r.** *R. occidentalis*; **r. jam (tree)** *Acacia acuminata*; **Mauritius r.** *R. rosifolius*; **Mysore r.** *R. niveus*; **purple r.** *R.* × *neglectus*

Rastrophyllum Wild & Pope. Compositae (Vern.-Vern.). 2 Zambia, Tanzania

rata *Metrosideros robusta*

Rathbunia Britton & Rose = *Stenocereus*

rati *Abrus precatorius*

Ratibida Raf. Compositae (Helia.-Rudb.). 7 N Am. R: Rhodora 70(1968)348. Cult. orn. coarse-hairy herbs (cone flowers) esp. *R. columnifera* (Nutt.) Wooton & Standley (W N Am., natur. E N Am.)

rat's tail (cactus) *Disocactus flagelliformis*

rattan Stripped stems of lianoid palms esp. spp. of *Calamus, Daemonorops, Korthalsia* etc.

rattle, yellow *Rhinanthus minor*

rattlepod *Crotalaria* spp. (Aus.)

Rattraya J. Phipps = *Danthoniopsis*

Ratzeburgia Kunth. Gramineae (40h). 1 Burma

rau ram *Persicaria odorata*

Rauhia Traub. Amaryllidaceae (Liliaceae s.l.). 1 Peru

Rauhiella Pabst & Braga (~ *Ornithocephalus*). Orchidaceae (V 10). 2 Brazil

Rauhocereus Backeb. = *Weberbauerocereus*

Rauia Nees & C. Martius = *Angostura*

rauli *Nothofagus nervosa*

Raulinoa Cowan. Rutaceae. 1 Brazil

Raulinoreitzia R. King & H. Robinson (~ *Eupatorium*). Compositae (Eup.-Dis.) 3 E S Am. R: MSBMBG 22(1987)74

Rautanenia Buchenau = *Burnatia*

Rauvolfia L. Apocynaceae. c. 60 trop. (Afr. 7, Madag. 3; R: BJBB 61(1991)21). Lvs often in whorls of 3–5; v. many alks. Cult. orn.; tranquillizing drugs from roots; some inks & dyes from fr. *R. serpentina* (L.) Kurz (Indomal.) – source of medic. alks esp. reserpine; *R. vomitoria* Afzel. (trop. Afr.) – similar, though this defence chemical overcome by colobine monkeys; reserpine reduces high blood pressure & used in treatment of mental illness

Rauwenhoffia R. Scheffer = *Melodorum*

Ravenala Adans. Strelitziaceae. 1 Madag.: *R. madagascariensis* Sonn. (traveller's palm) – clump-forming tree with palm-like trunks (pith ed., also stockfeed) usu. in sec. forest but widely planted trop. & warm; fl. bracts & leaf-bases hold rainwater useful in emergency; lvs in 2 ranks; fls explosively bird-poll.; aril blue, laciniate (seeds ed.)

Ravenea Bouché. Palmae (IV 2). 17 Madag., Comoro Is. Cult. orn. esp. Aus. *R. rivularis* Jum. & Perrier (Madag.) – exploited for drugs

Ravenia Vell. Conc. Rutaceae. 14 trop. Am. Alks.

Raveniopsis Gleason. Rutaceae. 17 Guayana Highland to Brazil. R: Brittonia 32(1980)47, 39(1987)409

Ravensara Sonn. = *Cryptocarya*

Ravnia Oersted. Rubiaceae (I 3). 3 C Am. Epiphytes, some cult. orn. looking like gesneriads

Rawsonia Harvey & Sonder. Flacourtiaceae (2). 2 trop. Afr.

Raycadenco Dodson. Orchidaceae (V 10). 1 Ecuador: *R. ecuadorensis* Dodson

Rayleya Cristóbal. Sterculiaceae. 1 Bahia

Raynalia Soják = *Alinula*

Razisea Oersted. Acanthaceae. 3 C Am.

Rea Bertero ex Decne. = *Dendroseris*

Readea Gillespie. Rubiaceae (IV 7). 3 Fiji

Reaumuria Hasselq. ex L. Tamaricaceae. 12 E Medit. (Eur. 1) to C As. & Pakistan. R: BZ 51(1966)1057. Halophytes

Reboudia Cosson & Durieu. Cruciferae. 2 N Afr., SE Medit.

Rebutia Schumann. Cactaceae (III 4). c. 35 S Bolivia, NW Arg. Cult. orn. small tubercled cacti

Recchia Moçiño & Sessé ex DC. Simaroubaceae. 3 Mex. R: Brittonia 37(1985)219. Stipules

Rechsteineria Regel = *Sinningia*

Recordia Mold. Verbenaceae. 1 Bolivia

Recordoxylon Ducke. Leguminosae (I 1). 3 Amazonia. R: ABV 21(1984)3. *R. amazonicum* (Ducke) Ducke – timber v. hard used for durable construction work

Rectanthera Degener = *Callisia*

red almond , **r. ash** (Aus.) *Alphitonia* spp.; **r. ash (Am.)** *Fraxinus pennsylvanica*; **r. bartsia** *Odontites vernus*; **r. bean** *Dysoxylum mollissimum*; **r. beech** *Nothofagus fusca*; **r. birch** *Betula nigra*; **r. boppel nut** *Hicksbeachia pinnatifida*; **r. box** *Lophostemon confertus*; **r. buckeye** *Aesculus pavia*; **r. bud** *Cercis canadensis*; **r. bugles** *Conostylis canescens*; **r. cedar** *Toona ciliata*, (US) *Juniperus virginiana*; **r. chokeberry** *Photinia pyrifolia*; **r. clover** *Trifolium pratense*; **r.currant** *Ribes rubrum*; **r. deal** *Pinus sylvestris*; **r. dhup** *Parishia insignis*; **r. fir** *Abies magnifica*; **r. gum, Cape York** *Eucalyptus brassiana*; **r. g., forest** *E. tereticornis*; **r. g., Murray** or **river** *E. camaldulensis*; **r. hot poker** *Kniphofia* spp.; **r. lauan** *Shorea* spp. esp. *S. negrosensis*; **r. louro** *Ocotea rubra*; **r. mahogany** *Eucalyptus resinifera*; **r. maple** *Acer rubrum*; **r. oak** *Quercus rubra*; **r. pine** *Dacrydium cupressinum*, *Pinus* spp.; **r. puccoon** *Sanguinaria canadensis*; **r. rice** *Oryza rufipogon*; **r. sandalwood** *Adenanthera pavonina*, *Pterocarpus santalinus*; **r. siris** *Paraserianthes toona*; **r. squill** *Drimia maritima*; **r. top** *Agrostis gigantea*

Redfieldia Vasey. Gramineae (31d). 1 C US

redleg *Persicaria maculosa*

Redowskia Cham. & Schldl. Cruciferae. 1 NE Siberia

redshank *Persicaria maculosa*

redwood, Andaman *Pterocarpus indicus*; **Baltic r.** *Pinus sylvestris*; **Brazilian r.** *Brosimum rubescens*, *Caesalpinia echinata*; **Californian r.** *Sequoia sempervirens*; **Indian r.** *Soymida febrifuga*, *Chukrasia tabularis*; **WI r.** *Guarea guidonia*; **Zambesi r.** *Baikiaea plurijuga*

reed *Phragmites australis*, (musical instruments) *Arundo donax*; **r. mace** *Typha* spp.

Reederochloa Söderstrom & H. Decker. Gramineae (31c). 1 Mex. Dioec.

Reedia F. Muell. Cyperaceae. 1 SW Aus.

Reediella Pichi-Serm. = *Crepidomanes*

Reedrollinsia J. Walker = *Stenanona*

Reesia Ewart = *Polycarpaea*

Reevesia Lindley. Sterculiaceae. 3–4 Himal. to Taiwan. R: NRBGE 15(1926)121

Regelia Schauer. Myrtaceae (Lept.). 5–6 N & SW Aus. Some cult. orn.

Registaniella Rech.f. Umbelliferae (III 3). 1 Afghanistan

Regnellidium Lindman. Marsileaceae. 1 SE Brazil & adjacent Arg.: *R. diphyllum* Lindman – like *Marsilea* but latex present & pinnae 2, cult. orn. aquatic

Rehdera Mold. Verbenaceae. 3 C Am.

Rehderodendron Hu. Styracaceae. 9 S & W China

Rehderophoenix Burret = *Drymophloeus*

Rehia Fijten (~ *Olyra*). Gramineae (4). 1 trop. S Am.: *R. nervata* (Swallen) Fijten

Rehmannia Libosch. ex Fischer & C. Meyer. Gesneriaceae. 9 E As. R: Taiwania 1(1948)71, Plantsman 8(1987)193. Some cult. orn. esp. *R. elata* N.E. Br. ('*R. angulata*', Chinese foxglove, China) to 2 m & resembling Scrophulariaceae; *R. glutinosa* Libosch. ex Fischer & C. Meyer (China) – medic. in modern Chinese herbalism

Rehsonia Stritch = *Wisteria*

Reichardia Roth. Compositae (Lact.-Son.). 8 Medit. (Eur. 4). R: Lagascalia 9(1980)159

Reichea Kausel = *Myrcianthes*

Reicheella Pax. Caryophyllaceae (II 1). 1 Chile: *R. andicola* (Philippi) Pax

Reicheia Kausel = *Myrcianthes*

Reichenbachanthus Barb. Rodr. Orchidaceae (V 13). 3 trop. Am.

Reichenbachia Sprengel. Nyctaginaceae (II). 2 trop. S Am.

Reimaria Humb. & Bonpl. ex Fluegge = *Paspalum*

Reimarochloa A. Hitchc. (~ *Paspalum*). Gramineae (34b). 4 S US to Arg. R: JAA suppl. 1(1991)290

Reineckea Kunth. Convallariaceae (Liliaceae s.l.). 1 China & Japan: *R. carnea* (Andrews) Kunth – cult. orn. like *Ophiopogon*

Reinhardtia Liebm. Palmae (V 4d). 7 C Am. R: GH 8(1957)541. Small understorey palms suited to house cult. esp. *R. gracilis* (H. Wendl.) Dammer (C Am.)

Reinwardtia Dumort. Linaceae. 1 mts N India & China: *R. indica* Dumort. – glabrous subshrub with tristylous yellow fls., cult. orn.

Reinwardtiodendron Koord. Meliaceae (I 5). 7 Indomal. R: Blumea 31(1985)144

Reissantia Hallé. Celastraceae. 6 OW trop.

Reissekia Endl. Rhamnaceae. 1 Brazil: *R. smilacina* (Sm.) Steudel

Reitzia Swallen (~ *Olyra*). Gramineae (4). 1 S Brazil: *R. smithii* Swallen

Rejoua Gaudich. = *Tabernaemontana*

Relbunium (Endl.) Hook.f. Rubiaceae (IV 16). 30 trop. Am. R: BJ 76(1955)516. Like *Galium* but fls involucrate; roots sources of dyes in Peru

Relchela Steudel (~ *Briza*). Gramineae (21b). 1 Arg., Chile

Reldia Wiehler. Gesneriaceae. 5 Panamá to Peru. R: NJB 8(1989)601

Relhania L'Hérit. Compositae (Gnap.-Rel.). 13 S Afr. R: OB 40(1976)

Remijia DC. Rubiaceae (I 1). 25 trop. S Am. Some with ant-infested hollow stem domatia. Bark (cuprea bark) a source of quinine, *R. pedunculata* Flueckiger & *R. purdieana* Wedd. esp. exploited

Remirea Aublet. Cyperaceae. 1 trop.

Remirema Kerr (? = *Operculina*). Convolvulaceae. 1 SE As.

Remusatia Schott. Araceae (VII 2). 3 trop. Afr., Himal. to Taiwan. R: ABY 13(1991)113. *R. vivipara* (Roxb.) Schott (Afr. to S Arabia, As.) has bulbils in special shoots, the bulbils with hooks which aid animal disp., rarely fls in As. (fruiting not recorded in India or Java, where pollen sterile), never in Afr. or Dhofar, tubers ed.

Remya Hillebrand ex Benth. (~ *Olearia*). Compositae (Ast.-Ast.). 3 Hawaii (Kaui & Maui). *R. kauiensis* Hillebrand (Kaui) – long believed extinct but re-found 1983. R: SB 12(1987)601

Renanthera Lour. Orchidaceae (V 16). 15 Indomal. Cult. orn. epiphytes with many-flowered infls, hybridized with spp. in other genera

Renantherella Ridley (~ *Renanthera*). Orchidaceae (V 16). 2 Thailand, Malay Pen.

Renata Ruschi = *Pseudolaelia*

Rendlia Chiov. = *Microchloa*

Renealmia L.f. Zingiberaceae. 61 trop. Am. (59; R: FN 18(1977), NRBGE 44(1987)237, 46(1990)315), trop. Afr. Some ant-disp.?; some cult. orn. & locally medic. (piles)

rengas *Gluta renghas*, but also used for many Anacardiaceae in Mal.

Renggeria Meissner. Guttiferae. 3 Brazil

Rennellia Korth. (~ *Morinda*). Rubiaceae (IV 10). 4 W Mal. R: Blumea 34(1989)3

Rennera Merxm. Compositae (Anth.-Mat.). 3 Namibia. R: BJLS 96(1988)308

rennet, vegetable *Withania coagulans*

Renschia Vatke. Labiatae (V). 1 N Somalia: *R. heterotypica* (S. Moore) Vatke

Rensonia S.F. Blake. Compositae (Helia.). 1 El Salvador

Reptonia A. DC = *Sideroxylon*

Requienia DC (~ *Tephrosia*). Leguminosae (III 6). 3 drier trop. & S Afr. R: KB 35(1980)469

rescue grass *Ceratochloa cathartica*

Reseda L. Resedaceae. 60 Eur. (20), Medit. to C As. R: MLW 78–14(1978). Herbs with alks & source of yellow dyes (luteolin); cult. orn. esp. *R. lutea* L. (wild mignonette, Eur., Medit., natur. N Am.), *R. luteola* L. (dyer's rocket or weld, Medit., Macaronesia) – yellow dye used since Neolithic esp. for silk & wedding garments by Romans; *R. odorata* L. (mignonette, N Afr.) – cult. orn. & for essential oil used in scent-making, though fls scentless to some

Resedaceae Bercht. & J.Presl. Dicots – Dilleniidae – Capparidales. 6/80 N hemisph. esp. OW. Herbs & (few) shrubs with scattered myrosin cells throughout & producing mustard-oil glucosides. Lvs entire to deeply lobed, spirally arr.; stipules represented by glands. Fls ± strongly zygomorphic, bisexual, with short androgynophore or gynophore; K (4–)6(–8), valvate or slightly imbricate, C (0, 2)4–8 (usu. 6), unequal, valvate, yellow to white, innermost (upper) large & usu. fringed, outer ones progressively smaller & with fewer appendages, androgynophore often with dilated nectary-disk below A 3–50+, anthers with longit. slits, $\underline{G}$ ((2)3–6(7)), 1-loc. but open apically & bearing small stigmas (G free in *Sesamoides* where only 1(2) ovule, united only basally in *Caylusea* where several ovules crowded on short axile placentas) but usu. ovules campylotropous, bitegmic, on parietal placentas. Fr. an open capsule, berry or (*Sesamoides*) distinct radiating carpels; seeds reniform with large curved or folded oily embryo & ± 0 endosperm. x = 6–15. R: Belmontia n.s. 8, 26A(1967) B(1978)

Genera: *Caylusea, Ochradenus, Oligomeris, Randonia, Reseda, Sesamoides* (all exc. *S.* occur in Sahara)

The open ovary of *Reseda* & allies unique. Some dyes & cult. orn. (*Reseda*)

Resia H. Moore. Gesneriaceae. 1 Colombia

Resnova J. Merwe = *Drimiopsis*

Restella Pobed. Thymelaeaceae. 1 C As.

restharrow *Ononis* spp.

Restio Rottb. Restionaceae. 88 S (40 endemics in Cape) & trop. Afr., Madag. R: Bothalia 15(1985)437

Restionaceae R. Br. Monocots – Commelinidae – Restionales. 41/420 S hemisph. esp. Aus. & S Afr. (10 endemic genera & 180 endemic spp. in Cape region) to Vietnam (1 sp.). Perennial xeromorphic herbs with rhizomes & solid or hollow internodes, the stems photosynthetic; vessel elements generally throughout & epidermal cells usu. with silica-bodies. Lvs usu. an open sheath (with terete or flattened blade in *Anarthria*), spirally arr. (all basal in *Anarthria*); ligule usu. 0. Fls small, reg., usu. unisexual & pl. dioec., wind-poll. in axils of chaffy bracts in (1–) several–∞-flowered spikelets or much-branched infls s.t. with leafy bracts & bracteoles, spikelets usu. with sheath-like basal spathe (not in *Ecdeiocolea*), the male & female infls usu. dissimilar, the fls often with rudiments of opp. sex; P (0–)3 + (0–)3 scale-like, A (1–)3(4) opp. inner P, the filaments usu. free with anthers usu. 2-sporangiate (rarely 4-) & monothecal (rarely 2-), usu. introrse, with longit. slits & monoporate ± graminoid pollen, $\underline{G}$ ((1–)3), (1–) 3-loc., the styles s.t. basally connate, with 1 pendulous, apical-axile, orthotropous, bitegmic ovule per loc. Fr. an achene, small nut or loculicidal capsule; seeds 1–3 with copious mealy endosperm. x = 6–13+

Principal genera: *Calopsis, Elegia, Ischyrolepis, Restio, Thamnochortus*

Anarthria & *Ecdeiocolea* with 4-sporangiate, 2-thecal anthers are sometimes split off as segregate families. Taxonomy difficult, the diff. sexes often v. diff. in aspect

Typical of nutrient-poor soils & imp. component of vegetation of SW Cape, where germ. improved by smoke. Wind-disp. aided by persistent P in *Restio* etc., ant-disp. aided by elaiosomes on nuts of *Cannomois, Hypodiscus* & *Willdenowia* (S Afr.) or on persistent P in some *Restio* spp.

Some locally used for thatching, broom-making etc.

Restrepia Kunth. Orchidaceae (V 13). 30 trop. Am. Cult. orn. epiphytes

Restrepiella Garay & Dunsterv. Orchidaceae (V 13). 1 C Am.: *R. ophiocephala* (Lindley) Garay & Dunsterv. – cult. orn. tufted epiphyte. R: MSB 39(1991)83

Restrepiopsis Luer. Orchidaceae (V 13). 15 trop. Am. R: MSB 39(1991)87

resurrection plant *Anastatica hierochuntica, Selaginella lepidophylla*

Retama Raf. Leguminosae (III 31). 4 Canary Is. & Medit. (Eur. 3, '*Lygos*') to Middle E. Alks. *R. raetam* (Forssk.) Webb (Canary Is., Medit.) – retem, the juniper of the Bible, a desert shrub with wood used for charcoal, dangerous med.

retamo wax *Bulnesia retamo*

Retanilla (DC) Brongn. Rhamnaceae. 2 Peru, Chile. Locally medic.

retem *Retama raetam*

Retiniphyllum Bonpl. Rubiaceae (III 1). 20 trop. S Am. Locally medic.

Retinispora Siebold & Zucc. = *Chamaecyparis*. Name form. used for many scale-leaved conifers in the juvenile needle-leaved stage (*'Retinospora'*)

Retispatha Dransf. Palmae (II 1d). 1 Borneo: *R. dumetosa* Dransf. Only Bornean endemic palm genus

Retrophyllum Page (*Decussocarpus*). Podocarpaceae. 2 New Caled., 1 Moluccas to Fiji, 2 S Am. R: NRBGE 45(1988)379

Retzia Thunb. Stilbaceae (Retziaceae, Loganiaceae s.l.). 1 Cape: *R. capensis* Thunb.

Retziaceae Bartling (~ Loganiaceae) = Stilbaceae

Reussia Endl. = *Pontederia*

Reutealis Airy Shaw. Euphorbiaceae. 1 Philippines

Reutera Boiss. = *Pimpinella*

Revealia R. King & H. Robinson = *Carpochaete*

Reverchonia A. Gray. Euphorbiaceae. 1 S N Am.

rewa-rewa *Knightia excelsa*

Reyemia Hilliard. Scrophulariaceae (Manul.). 2 NW Cape. R: EJB 49(1993)291

Reyesia C. Gay (~ *Salpiglossis*). Solanaceae (2). 4 N Chile. R: HBML 27(1980)24

Reynaudia Kunth. Gramineae (34e). 1 Cuba, Hispaniola

Reynoldsia A. Gray. Araliaceae. 2 Samoa, 2 Marquesas, 1 Society Is., 1 Hawaii

Reynosia Griseb. Rhamnaceae. 18 Florida (1), C Am. (1), WI. Endosperm ruminate. Some timber & ed. fr.

Reynoutria Houtt. (~ *Fallopia*). Polygonaceae. 15 temp. As. *R. japonica* Houtt. (*F. japonica*, *Polygonum cuspidatum*, Japanese knotweed, Japan) – cult. orn. with bamboo-like stems to 3 m, often becoming an aggressive weed in orn. plantations, locally medic., roots a source of a yellow dye, introd. GB from von Siebold's Nursery (Holland) & wild by 1886 but in Br. Is. all octoploids & almost all seed is of hybrid origin involving *F. baldschuanica*

Rhabdadenia Muell-Arg. Apocynaceae. 3 trop. Am. Alks

Rhabdocaulon (Benth.) Epling. Labiatae (VI 2). 7 trop. S Am.

Rhabdodendraceae (Huber) Prance. Dicots – Rosidae – Rosales. 1/3 trop. S Am. Shrubs often with unusual sec. growth, parenchyma with resin-filled scattered secretory cavities. Lvs simple, leathery, spirally arr., with peltate hairs & glandular dots; stipules minute or 0. Fls reg., usu. bisexual, in racemes or racemoid cymes; K ± entire or with 5 imbricate lobes, C 5 sepal-like, A 25–50 in ± 3 whorls with short flat filaments & anthers with longit. slits, disk 0, $\underline{G}$ 1 with long almost basal style & 1 (s.t. another aborted) basal hemitropous unitegmic ovule. Fr. a drupe with somewhat woody endocarp; seed with 2 fleshy cotyledons & 0 endosperm. x = 10

Genus: *Rhabdodendron*

Of disputed affinity with some similarities with Caryophyllales though seed & sieve-tube plastid characters against this

Rhabdodendron Gilg & Pilger. Rhabdodendraceae. 3 trop. S Am. R: FN 11(1972)3

Rhabdophyllum Tieghem = *Gomphia*

Rhabdosciadium Boiss. Umbelliferae (III 2). 5 Eur. to W As.

Rhabdothamnopsis Hemsley. Gesneriaceae. 1 China: *R. sinensis* Hemsley

Rhabdothamnus Cunn. Gesneriaceae. 1 NZ

Rhabdotosperma Hartl = *Verbascum*

Rhachicallis Spach = *Arcytophyllum*

Rhacodiscus Lindau = *Justicia*

Rhadamanthus Salisb. Hyacinthaceae (Liliaceae s.l.). 10 S Afr. R: BN 123(1970)155

Rhadinopus S. Moore. Rubiaceae (II 5). 2 New Guinea. R: Blumea 25(1979)297

Rhadinothamnus P.G. Wilson (~ *Nematolepis*). Rutaceae. 1 W Aus.

Rhaesteria Summerh. Orchidaceae (V 16). 1 Uganda: *R. eggelingii* Summerh. – coll. only once

Rhagadiolus Juss. Compositae (Lact.-Hyp.). 2 Medit. (Eur. 1) to Iran. R: T 28(1979)133. Fr. linear encased in involucral bract, pappus 0

Rhagodia R. Br. Chenopodiaceae (I 2). 11 Aus. R: FA 4(1984)164. *R. parabolica* R. Br. – red, yellow or white fr. or combinations of these, foraged at random by *Zosterops* sp., principal avian consumer but seeds from diff. morphs respond differently, red germ. fastest, then yellow but white greatest response to passage through bird-gut, these diff. factors poss.

explaining polymorphism

Rhammatophyllum O. Schulz = *Erysimum*

Rhamnaceae Juss. Dicots – Rosidae – Rhamnales. 49/900 cosmop. esp. trop. & warm. Trees & shrubs, sometimes lianoid (hooks in *Ventilago*, tendrils in *Gouania*), rarely herbs (*Crumenaria*), often with anthraquinone glycosides & alks; some with nitrogen-fixing Actinomycetes in roots (*Ceanothus, Trevoa*). Lvs simple (s.t. much reduced & stems photosynthetic), spirally arr. or opp., pinnately veined or with several main veins from base; stipules small or spiny or 0. Fls usu. small & unisexual, (4 or) 5-merous with hypanthium resembling K-tube to epigyny with disk adnate to G & hypanthium, which is often circumscissile & decid. above middle, the fls in cymes, thyrses or fascicles (–solit.); K lobes valvate, C often hooded & concave holding anthers or 0 (rare), A alt. with K, filaments adnate to C, anthers with longit. slits & pollen-grains often ± triangular, disk within A, often adnate to hypanthium & s.t. G ((2, 3)–5)), ± pluriloc. with term. deeply cleft style, each loc. with 1 (2 in *Karwinskia*) anatropous, bitegmic ovule. Fr. a drupe with separate stones or 1 pluriloc. one, or dry-dehiscent or separating into mericarps; seeds exotestal, s.t. with dorsal groove, embryo usu. straight, large & oily with little or 0 (ruminate in *Reynosia*) endosperm. x = (9–)12(13, 23)

Principal genera: *Ceanothus, Colletia, Colubrina, Cryptandra, Gouania, Phylica, Pomaderris, Rhamnus, Sageretia, Spyridium, Ventilago, Ziziphus*

Medic. (*Colubrina, Hovenia, Karwinskia, Krugiodendron. Rhamnus* (cascara)) etc., soap (*Gouania*), dyes (*Rhamnus, Ventilago*), timber (*Alphitonia, Karwinskia, Maesopsis*), ed. fr. (*Ziziphus*) & cult. orn. esp. *Ceanothus* & *Colletia*

Rhamnella Miq. Rhamnaceae. 10 W Himal. to Japan, Fiji, trop. Aus. & New Caled. (1)

Rhamnidium Reisseck. Rhamnaceae. 12 trop. Am.

Rhamnoneuron Gilg. Thymelaeaceae. 1 SE As.

Rhamnus L. Rhamnaceae. 125 N hemisph. (Eur. 16 incl. *Frangula*) to Brazil & S Afr. Usu. decid. trees (to 50 m) & shrubs with spirally arr. or opp. lvs & 4-merous polygamous & dioec. (usu. subg. *Rhamnus*) or 5-merous bisexual (usu. subg. *Frangula*) fls; cult. orn., dyes (fr. gives blue-green, bark yellow) & medic. (purgatives). Buckthorns; R: TBIANSSSR I, 8(1949)243. *R. cathartica* L. (common, purging or European b., Euras., natur. N Am.) – drupes (Rhine berries) medic. & source of artist's pigment 'sap green'; *R. davurica* Pallas (Japan, Korea) & *R. utilis* Decne. (C & E China) – sources of Chinese green indigo used in dyeing silk; *R. erythroxyloides* Hoffsgg. (*R. pallasii*, Pallas's b., W As.) – cult. orn.; *R. frangula* L. (*Frangula alnus*, alder b., Euras., Medit., natur. N Am.) – excellent charcoal form. used for small-arms gunpowder (esp. slow fuses in World War II) & cult. for this, medic. as in bowel regulators; *R. purshiana* DC (cascara sagrada, W N Am.) – bark the source of comm. purgatives (anthraquinones) with *R. frangula* used since 1870 but not analysed until 1975; *R. saxatilis* Jacq. (*R. infectoria*, Avignon or yellow berry or Persian berry, S & SC Eur.) – form. imp. source of yellow dye. Some timbers for turnery

Rhamphicarpa Benth. Scrophulariaceae. 6 trop. & S Afr., India, trop. Aus. R: BT 70(1975)103

Rhamphogyne S. Moore. Compositae (Ast.-Ast.). 2 Rodrigues & New Guinea

Rhamphorhynchus Garay. Orchidaceae (III 3). 1 Brazil

Rhanteriopsis Rauschert (*Postia*). Compositae (Inul.). 2 Lebanon & Syria, 2 Iran & Iraq. R: BJLS 95(1987)27

Rhanterium Desf. Compositae (Inul.). 3 NW Afr. to Pakistan. R: BJLS 93(1986)231

Rhaphidophora Hassk. Araceae (III 2). 140 Indomal. to New Caled. Lvs sometimes perforate; fls bisexual. Some cult. orn. & locally medic., long roots for cordage. *R. pertusa* (Roxb.) Schott (W Pacific) – veg. indistinguishable from juvenile or small cvs of *Monstera deliciosa*

Rhaphidophyton Iljin. Chenopodiaceae (III 3). 1 C As.: *R. regelii* (Bunge) Iljin

Rhaphidospora Nees = *Justicia*

Rhaphidura Bremek. Rubiaceae (I 9). 1 Borneo

Rhaphiodon Schauer (~ *Hyptis*). Labiatae (VIII 4a). 1 E Brazil

Rhaphiolepis Lindley. Rosaceae (III). 9 E & SE As. R: CJB 68(1990)2240. Evergreen shrubs incl. some rheophytes & cult. orn. with white or pink fls esp. *R. umbellata* (Thunb.) Makino – bark a source of brown dye

Rhaphiostylis Planchon ex Benth. Icacinaceae. 6 trop. Afr.

Rhaphispermum Benth. Scrophulariaceae. 1 Madagascar

Rhaphithamnus Miers. Verbenaceae. 2 Chile, Arg., Juan Fernandez. Cult. orn. trees & shrubs

Rhapidophyllum H. Wendl. & Drude. Palmae (I 1a). 1 SE US: *R. hystrix* (Pursh) H. Wendl. & Drude – low polygamo-dioec. or dioec. (rarely monoec.) palm with needle-like appendages on leaf-sheaths, v. hardy

Rhapis L.f. ex Aiton. Palmae (I 1a). c. 12 S China to W Mal. R: ARBGC 13(1931)242. Cult. orn. tub-pls esp. *R. excelsa* (Thunb.) Rehder (*R. flabelliformis*, Japanese peace palm, S China) – culture a cult in Japan on a par with tulipomania, esp. varieg. forms today, purchased as hedge against inflation (kan-non-chiku)

Rhaponticum Hill = *Stemmacantha*

Rhaptonema Miers. Menispermaceae (V). 4 Madagascar

Rhaptopetalum Oliver. Scytopetalaceae. 10 trop. Afr.

rhatany (root) *Krameria* spp.

Rhazya Decne. Apocynaceae. 2 SE Eur. to Pakistan & Arabia. Alks. *R. stricta* Decne. (SW As. to Pakistan & Arabia – source of lacquer, wax from fr. used in candles, form. most imp. desert medic. in Arabia; *R. orientalis* (Decne.) A. DC (*Amsonia o.*, Greece to Turkey) – cult. orn. perennial with blue to lilac fls

Rheedia L. = *Garcinia*

Rhektophyllum N.E. Br. = *Cercestis*

Rheome Goldbl. Iridaceae (III 2). 3 SW Cape to Namaqualand

Rheopteris Alston. Vittariaceae. 1 New Guinea: *R. cheesemanniae* Alston

Rhetinodendron Meissner = *Robinsonia*

Rhetinolepis Cosson. Compositae (Anth.-Art.). 1 Algeria, Tunisia, Libya

Rhetinosperma Radlk. = *Chisocheton*

Rheum L. Polygonaceae (II 3). 30 Eur. (2), temp. & warm As. Rhubarb; R: TBIANSSSR I, 3(1936)67; C.M. Foust (1992) *Rhubarb: the wondrous drug*. Coarse herbs with large lvs & entomophilous fls (cf. *Rumex* but large stigmas like anemophilous (? ancestors)). Medic., ed. & cult. orn. *R. australe* D. Don (*R. emodi*, Himalayan or Indian rhubarb, Himal.) – drug (roots); *R.* × *hybridum* Murray (*R.* × *cultorum*, garden rhubarb) – prob. hybrid with *R. rhabarbarum* L. (Mongolia) as a parent, petioles stewed as pudding, used in jam & wine; *R. nobile* Hook.f. & Thomson (Nepal to SE Tibet) – infls densely covered with bracts 15 cm across; *R. officinale* Baillon (W China & Tibet) – roots & rhizome medic. (purgative); *R. palmatum* L. (Chinese or Turkish r., NW China) – medic.; *R. ribes* L. (SW As.) – stalks taste like currants

Rhexia L. Melastomataceae. 13 N Am. R: Brittonia 8(1956)201. Cult. orn. herbs incl. *R. virginica* L. – lvs a source of a tea for N Am. Indians in E Canada

Rhigiocarya Miers. Menispermaceae (III). 3 trop. W Afr.

Rhigiophyllum Hochst. Campanulaceae. 1 S Afr.

Rhigospira Miers (~ *Tabernaemontana*). Apocynaceae. 1 N S Am. to Andes. R: AUWP 87-1(1987)

Rhigozum Burchell. Bignoniaceae. 7 trop. & S Afr., Madag. Much-branched spiny shrubs

Rhinacanthus Nees. Acanthaceae. 15 trop. OW. Medic. (skin diseases)

Rhinactinidia Novopokr. = *Aster*

Rhinanthus L. Scrophulariaceae. 45 N hemisph. esp. Eur. (25 incl. many ecotypes). Hemiparasitic annual herbs esp. *R. minor* L. (yellow or hay rattle, Eur., E N Am.)

Rhine berry *Rhamnus cathartica*

Rhinephyllum N.E. Br. Aizoaceae (V). 14 Cape. Cult. orn.

Rhinerrhiza Rupp. Orchidaceae (V 16). 2 Aus., New Guinea, Solomon Is.

rhinoceros bush *Elytropappus rhinocerotis*

Rhinopetalum Fischer ex D. Don = *Fritillaria*

Rhinopterys Niedenzu = *Acridocarpus*

Rhipidantha Bremek. Rubiaceae (I 10). 1 trop. Afr.

Rhipidia Markgraf = *Condylocarpon*

Rhipidocladum McClure. Gramineae (1b). c. 12 trop. Am. R: SCB 9(1973)101. Bamboos

Rhipidoglossum Schltr. = *Diaphananthe*

Rhipogonum Sprengel = *Ripogonum*

Rhipsalidopsis Britton & Rose = *Hatiora*

Rhipsalis Gaertner. Cactaceae (III 5). 50 trop. Am. (esp. Brazil), 1 ext to Afr., Madag. & Sri Lanka. Epiphytic or saxicolous jointed cacti with cylindrical, angled or flattened joints & spherical translucent mucilaginous fr. Cult. orn., specific limits unclear (mistletoe cactus) incl. *R. baccifera* (J.S. Miller) Stearn (*R. cassutha*, *R. cassytha*, trop. Am. to Sri Lanka)

Rhizanthella R. Rogers. Orchidaceae (IV 1). 2 SW & E Aus. *R. gardneri* R. Rogers almost completely subterr., known from 5 localities always in stands of *Melaleuca uncinata* R. Br., infl. growing up to soil surface with bracts pushing away soil to expose fls poss. poll. by tiny flies, infr. emerging & seeds disp. by wind

Rhizanthes Dumort. Rafflesiaceae (II 1). 2 W Mal. (on 2 spp. of *Tetrastigma*). R: Blumea

33(1988)329

Rhizobotrya Tausch. Cruciferae. 1 W Dolomites: *R. alpina* Tausch – cult. orn. rock-pl.

Rhizocephalum Wedd. Campanulaceae. 5 Andes

Rhizocephalus Boiss. Gramineae (21d.). 1 Medit. to Iran

Rhizoglossum C. Presl = *Ophioglossum*

Rhizomonanthes Danser. Loranthaceae. 1–2 New Guinea

Rhizophora L. Rhizophoraceae. 8–9 (? some hybrids) trop. coasts (Aus. 4). Mangrove trees with arching aerial roots from stem & branches, their subaerial parts with conspic. lenticels, app. moving from animal- to wind-poll.; seeds germ. in fr. & producing elongate radicle before being dropped. *R. mangle* L. (SW Pacific & NW) – ? distrib. achieved in Eocene. Bark used in tanning, timber, esp. from *R. mucronata* Lam. (boriti poles, OW) for building but also for pulp & rayon in Mal.

Rhizophoraceae Pers. Dicots – Rosidae – Rhizophorales. 15/120 trop. esp. OW. Trees & shrubs, often mangroves, at least s.t. with alks. Lvs simple, opp.; stipules interpetiolar, caducous, sheathing term. bud. Fls reg., 4- or 5-merous, usu. unisexual, s.t. with elongated hypanthium, in axillary infls or solit.; K (3)4, 5(–16), valvate, thick, C alt. with K, often fleshy & shorter than K, convolute or infolded in bud, A 2(–4) × K or ∞, often paired opp. C in single whorl with filaments often basally connate, attached around or on to disk or disk 0 or anthers sessile, G (2–5(6)) with as many loc. or partitions aborting to give 1-loc., with term. style & 2(4 +) pendulous, anatropous, bitegmic ovules with zig-zag micropyles per loc. Fr. a berry with 1 seed (or 1 per loc.), rarely a capsule; seeds s.t. arillate with straight, linear, often green (viviparous in mangroves) embryo in copious oily endosperm. x = 8, 9. R: AMBG 75(1988)1278

Principal genera: *Bruguiera, Carallia, Cassipourea, Crossostylis, Pellacalyx, Rhizophora*

Embryologically heterogeneous; seed-structure suggests segregation of *Gynotroches* & *Pellacalyx* as Legnotidaceae. Affinities disputed but embryology & many other features against traditional position in Myrtales

Timber, charcoal & tanbarks

Rhodalsine Gay = *Minuartia*

Rhodamnia Jack. Myrtaceae (Myrt.). 28 Indomal. to W Pacific (Aus. 13 endemic). R: KB 33(1979)429. *R. arborea* Jack (Indomal.) – timber & charcoal

Rhodanthe Lindley (*'Helipterum'*). Compositae (Gnap.-Ang.). 46 Aus. R: Nuytsia 8(1992)383. Cult. orn. 'everlastings' esp. *R. manglesii* Lindley (*Helipterum m.*, W Aus.) – heads pink

Rhodanthemum (Vogt) Bremer & Humphries (~ *Chrysanthemum*). Compositae (Anth.-Leuc.). 12 Spain (1), Morocco, Algeria

Rhodax Spach = *Helianthemum*

Rhodiola L. = *Sedum*

Rhodocalyx Muell. Arg. Apocynaceae. 1 Brazil, campos. K coloured

Rhodochiton Zucc. ex Otto & Dietr. Scrophulariaceae. 3 Mex., C Am. R: D.A. Sutton, *Rev. Antirrhineae* (1988)505. *R. atrosanguineum* (Zucc.) Rothm. (Mex.) – twiner with sensitive petioles & peduncles, K persistent

Rhodocodon Baker. Hyacinthaceae (Liliaceae s.l.). 8 Madagascar

Rhodocolea Baillon. Bignoniaceae. 6 Madag. *R. perrieri* Capuron with reg. fls & A5

Rhodocoma Nees (~ *Restio*). Restionaceae. 3 SW & E Cape. R: Bothalia 15(1985)478

Rhododendron L. Ericaceae. Incl. *Ledum* (marsh rosemary), 850 temp. N hemisph. (Eur. 7, China 650, largest genus there) esp. Himal., SE As. & Mal. mts (155 endemics in New Guinea), Aus. (1). R: NRBGE 39(1980)1, (1982)209, 40(1982)225, 44(1986)1, EJB 47(1990)89, 50(1993)249, 52(1995)1. Evergreen to decid. poisonous (diterpenes) trees & shrubs (pachycaul to leptocaul) incl. epiphytes, with lvs in term. rosettes & various coloured fls – the rhododendrons (name from the Greek for similar-leaved oleander) & azaleas of gardens & florists intimately assoc. with an almost unparalleled vainglory & exclusiveness in hort. of the last century through the concentrated efforts to introd. & hybridize particularly the Asiatic spp. by the wealthy & privileged of W Eur., over 500 spp. having been cult. in GB alone. Hybrids (azaleas) sold as pot-pls for a wider market incl. crosses between Am. & As. spp. At least 8 (some monotypic) subg. incl.: *Rhododendron* (lepidote r.) – some 500 evergreen spp. incl. most Mal. spp., with glandular hairs (scales) & lvs adaxially rolled in bud, polyploidy common, incl. all epiphytes, the most northerly & southerly spp. & all 'blue'-flowered spp.; subg. *Hymenanthes* (Blume) K. Koch – evergreen, elepidote with abaxially rolled lvs, unique nodal anatomy & ± winged seeds, i.e. most spp. in cult. incl. *R. ponticum* L. (W & E Medit., introd. from Gibraltar region to GB 1763 & now an aggressive weed in forestry plantations on light acid soils (1988 cost to clear Snowdonia alone was £30m); in Ireland in last interglacial) but up to 120 spp. used in hybridizing with over 1000 named cvs in commerce; subg. *Pentanthera* G. Don f. (subg. *Azalea*) – elepidote & usu. decid., azaleas of commerce; subg. *Azaleastrum* Planchon – like *Azalea* but infls axillary. Nectar in some spp. containing antibiotic acetylandromedol (diterpene) as low sugar content otherwise does not prevent bacterial activity so that excessive eating of rhododendron honey by humans can lower blood pressure, promoting dizziness ('Mad honey poisoning', US). Noted wild & useful spp. incl. *R. anthopogon* D. Don (E Himal.) – lvs & twigs burnt as incense in temples in Sikkim; *R. arboreum* Sm. (Himal., China to Thailand, S India, Sri Lanka (ssp. *zeylanicum* (Booth) Tagg – 1st pl. record for the is. (1343–4!)); *R. falconeri* Hook.f. (Bhutan to Nepal) – lvs to 30 cm long, rusty-tomentose abaxially; *R. ferrugineum* L. (alpenrose, alpine rose, C Eur.) – form. medic. but dangerous; *R. indicum* (L.) Sweet (S Japan) – long cult. Japan where over 200 cvs by 1900; *R. molle* (Blume) G. Don f. (C & E China) – leaf-extracts effective insect-repellents; *R. simsii* Planchon (NE Burma, China, Taiwan) – origin of widely cult. 'Indian azalea' usu. with ± double fls; *R. stenopetalum* (Hogg) Mabb. (*R. macrosepalum* 'Linearifolium', spider azalea, Japan) – mutant with narrow lvs & distinct petals; *R. tomentosum* Harmaja (*Ledum palustre*, circumpolar) – lvs used as tea by Am. Indians ('Labrador tea'); *R. viscosum* (L.) Torrey (swamp honeysuckle, E US)

Rhododon Epling. Labiatae (VIII 2). 1 Texas

Rhodogeron Griseb. = *Sachsia*

Rhodognaphalon (Ulbr.) Roberty = *Bombax*

Rhodognaphalopsis Robyns = *Pachira*

Rhodohypoxis Nel. Hypoxidaceae. 6 SE Afr. R: NRBGE 36(1978)43, Plantsman 6(1984)53. Cult. orn. rock-pls esp. *R. baurii* (Baker) Nel

Rhodolaena Thouars. Sarcolaenaceae. 5 Madagascar

Rhodoleia Champ. ex Hook. Hamamelidaceae (III). 1–7 S China to Sumatra. Stipules on young shoots; bird-poll.

Rhodoleiaceae Nakai = Hamamelidaceae

Rhodomyrtus (DC) Reichb. Myrtaceae (Myrt.). 11 Indomal. to Aus. (7 endemic) & New Caled. R: KB 33(1978)311. Some ed. fr. esp. *R. tomentosa* (Aiton) Hassk. (SE As.) but also *R. macrocarpa* Benth. (finger cherry, Queensland) though alleged to promote blindness

Rhodophiala C. Presl = *Hippeastrum*

Rhodopis Urban. Leguminosae (III 10). 1 Hispaniola

Rhodosciadium S. Watson. Umbelliferae (III 10). 14 Mexico

Rhodosepala Baker = *Dissotis*

Rhodospatha Poeppig. Araceae (III 2). 12 trop. Am. Cult. orn. climbers

Rhodosphaera Engl. Anacardiaceae (IV). 1 NE Aus.: *R. rhodanthema* (F. Muell.) Engl. – allied to *Rhus*, timber beautifully marked & used in cabinet-making

Rhodostachys Philippi = *Ochagavia*

Rhodostegiella Li = *Cynanchum*

Rhodostemonodaphne Rohwer & Kubitzki (~ *Ocotea*). Lauraceae (I). 20 trop. S Am. R: MIABH 20(1986)82

Rhodothamnus Reichb. Ericaceae. 1 E Alps (*R. chamaecistus* (L.) Reichb. – cult. orn.) & 1 NE Turkey (stomata on both sides of lvs)

Rhodotypos Siebold & Zucc. Rosaceae. 1 Korea, China, Japan: *R. scandens* (Thunb.) Makino – cult. orn. decid. shrub with *opp.* lvs (unknown otherwise in Rosaceae exc. seedling *Prunus* spp.), epicalyx & solit. term. white fls with C 4 & (1–)4(–6) shining black achenes

Rhoeo Hance = *Tradescantia*

Rhoiacarpos A. DC. Santalaceae. 1 S Afr.

Rhoicissus Planchon. Vitaceae. 12 trop. & S Afr. Lianes & shrubs with tendrils, like *Cissus* but fls 5-merous, like *Vitis* but disk adnate to G. *R. capensis* Planchon (S Afr.) – cult. orn. house-pl. with fr. ed. when cooked

Rhoiptelea Diels & Hand.-Mazz. Rhoipteleaceae. 1 SW China & N Vietnam

Rhoipteleaceae Hand.-Mazz. Dicots – Hamamelidae – Juglandales. 1/1 SW China, N Vietnam. Aromatic wind-poll. decid. tree. Lvs pinnate, spirally arr. with sunken aromatic resinous peltate glands & stipules. Fls small, in triplets with central bisexual flanked by female, sterile or abortive, in axils of bracts, triplets in catkins arranged in term. nodding panicles; P 2 + 2, small, persistent in fr., disk 0, A 6 with short filaments & anthers with longit. slits, $\underline{G}$ (2), 1-loc. apically, with separate styles & 1 hemitropous to anatropous, bitegmic ovule. Fr. a 2-winged samaroid nut; seed 1 with straight oily embryo & 0 endosperm

Species: *Rhoiptelea chiliantha* Diels & Hand.-Mazz.

Rhombochlamys Lindau. Acanthaceae. 2 Colombia

Rhombolythrum Airy Shaw = *Rhombolytrum*

Rhombolytrum Link. Gramineae (17). 3 S Brazil, Uruguay, Chile. R: KBAS 13(1986)100

Rhombonema Schltr. = *Parapodium*

Rhombophyllum (Schwantes) Schwantes. Aizoaceae (V). 3 E Cape. Cult. orn. succ. ± shrubby pl.

Rhoogeton Leeuwenb. Gesneriaceae. 3 NE S Am.

Rhopalephora Hassk. (~ *Aneilema*). Commelinaceae (II 2). 4 Madag., Indomal. to Fiji. R: Phytologia 37(1977)479

Rhopaloblaste R. Scheffer. Palmae (V 4m). 6 Nicobars & Malay Pen. to Solomon Is. R: Principes 14(1970)75. Some cult. orn. monoec. unarmed palms

Rhopalobrachium Schltr. & K. Krause. Rubiaceae (inc. sed.). 1 New Caled.: *R. fragrans* Schltr. & K. Krause. R: Candollea 46(1991)253

Rhopalocarpus Bojer. Ochnaceae (Diegodendraceae). 14 Madagascar

Rhopalocnemis Junghuhn. Balanophoraceae. 1 Indomal.

Rhopalocyclus Schwantes = *Leipoldtia*

Rhopalopilia Pierre. Opiliaceae. 3 C Afr. R: BJ 108(1987)273

Rhopalopodium Ulbr. = *Ranunculus*

Rhopalosciadium Rech.f. Umbelliferae (III 2). 1 Iran

Rhopalostylis H. Wendl. & Drude. Palmae (V 4h). 3 NZ (1), Norfolk Is. (1), Raoul (Sunday) Is. (1). *R. sapida* H. Wendl. & Drude (nikau palm, NZ) – stemless for 40–50 yrs & mature at 150–250, lvs used for hut-building by Maoris

Rhopalota N.E. Br = *Crassula*

Rhuacophila Blume = *Dianella*

rhubarb *Rheum* spp. esp. *R. × hybridum*; **Chinese r.** *R. palmatum*; **Himalayan** or **Indian r.** *R. australe*; **monk's r.** *Rumex pseudoalpinus*; **Turkish r.** *Rheum palmatum*

Rhus L. Anacardiaceae (IV). (Inc. *Toxicodendron*) c. 200 temp. (Eur. 3) & warm. Usu. dioec. trees, shrubs or lianes with clinging roots, simple or pinnate lvs & drupes. Dyes, lacquers & tanning materials etc., some (poison ivy, p. oak) promoting dermatitis due to 3-*n* pentadecycatechnol in over 350 000 cases per annum in US alone, others cult. orn. (sumac(h)s). *R. copallina* L. (E US) – lvs used for tanning & dyeing; *R. coriaria* L. (sumac, S Eur.) – Middle E spice, ground dried lvs form. v. imp. tanning material for sheepskin etc. in S Italy (tannin content c. 26%), the dye for Cordoba & Morocco leather; *R. glabra* L. (E N Am.) – drink made from fr., dyes from roots; *R. hirta* (L.) Sudw. (*R. typhina*, staghorn or velvet sumac, E N Am.) – high in oils & polyphenols, imp. tannin source, fr. used in drinks, cult. orn. with reg. pseudo-n-chotomous branching & serrate (f. *typhina* (L.) Reveal) to laciniate (f. *hirta*) leaflets with brilliant autumn tints, young shoots used as pipes for tapping sugar maple; *R. javanica* L. (*R. chinensis*, *R. semialata*, Chinese sumac, E & SE As. to Sumatra) – Chinese or Japanese galls known in Eur. since C15, used in tanning & dyeing blue silk; *R. natalensis* Krauss (trop. & S Afr.) – fr. ed. Uganda; *R. radicans* L. (*Toxicodendron r.*, *R. toxicodendron*, poison ivy, p. oak, C China to Japan, S Canada to Guatemala) – liane, shrub or tree causing dermatitis, marks linen; *R. succedanea* L. (*T. s.*, Japanese tallow or wax tree, E As.) – cult. for wax from fr. used as subs. for beeswax in polishes etc., stem exudes a natural lacquer; *R. trilobata* Nutt. (Calif.) – stems used for 3500 yrs in basketry & for figures, bushes burnt to produce the necessary long sucker-shoots; *R. tripartita* (Ucria) Grande ('*R. oxyacantha*', Sahara) – fr. ed.; *R. verniciflua* Stokes (*T. v.*, *R. vernicifera*, (Chinese or Japanese) lacquer tree, temp. E As.) – cult. in SW Japan where major source of lacquer (lactone causing allergy), obtained by cutting bark, also wax source for candles; *R. vernix* L. (*T. v.*, poison elder, E N Am.) – dermatitis

Rhynchanthera DC. Melastomataceae. 15 Mexico to Bolivia & Paraguay

Rhynchanthus Hook.f. Zingiberaceae. 5–6 Burma, SW China, W Mal.

Rhyncharrhena F. Muell. (~ *Pentatropis*). Asclepiadaceae. 1 Aus.: *R. linearis* (Decne) K.L. Wilson

Rhynchelytrum Nees = *Melinis*

Rhynchocalycaceae L. Johnson & Briggs (~ Crypteroniaceae). Dicots – Rosidae – Myrtales. 1/1 S Afr. Lately segregated from polygamo-dioec. Crypteroniaceae (as fls bisexual) & Lythraceae (as A as many as & alt. with K)
Genus: *Rhynchocalyx*

Rhynchocalyx Oliver. Rhynchocalycaceae (Crypteroniaceae s.l.). 1 Natal & Transkei

Rhynchocladium T. Koyama. Cyperaceae. 1 Venezuela

Rhynchocorys Griseb. Scrophulariaceae. 6 S Eur. (1) to Iran. R: NRBGE 30(1970)97. *R. elephas* (L.) Griseb. (Medit., W As.) – seedlings are underground parasites for at least 2 yrs

Rhynchodia Benth. Apocynaceae. 8 India & S China to Java

Rhynchoglossum Blume. Gesneriaceae. 11–13 SE As., S Am. (1–3). Marked anisophylly, fls in 1-sided racemes. Cult. orn. (*Klugia*)

Rhynchogyna Seidenf. & Garay. Orchidaceae (V 16). 2 SE As., Mal. R: OB 95(1988)192

Rhyncholacis Tul. Podostemaceae. 25 N trop. S Am. *R. nobilis* Van Royen – powdered lvs used like salt

Rhyncholaelia Schltr. (~ *Brassavola*). Orchidaceae (V 13). 2 Mex., C Am. Cult. orn.

Rhynchophora Arènes. Malpighiaceae. 1 Madagascar

Rhynchophreatia Schltr. (~ *Phreatia*). Orchidaceae (V 14). 5 New Guinea, Micronesia

Rhynchopsidium DC & A. DC (~ *Relhania*). Compositae (Gnap.-Rel.). 2 S Afr.

Rhynchoryza Baillon (~ *Oryza*). Gramineae (9). 1 Paraguay to Arg.: *R. subulata* (Nees) Baillon – like *Oryza* but lemma with aerenchymatous beak, a flotation device

Rhynchosia Lour. Leguminosae (III 10). c. 300 trop.

Rhynchosida Fryx. Malvaceae. 2 Am.

Rhynchosinapis Hayek = *Coincya*

Rhynchospermum Reinw. Compositae (Ast.-Ast.). 1 E & SE As.

Rhynchospora Vahl. Cyperaceae. (Incl. *Dichromena* = sect. *D.* (R: MNYBG 37(1984)23) with at least 7 spp. insect-poll.) c. 250 subcosmop. (Eur. 3) esp. trop. & warm S Am. ('Mesoamerica' 81, R: Brittonia 44(1992)14). R: BJ 74(1949)375 (beak rush). Some ant-disp. frs

Rhynchostele Reichb.f. = *Leochilus*

Rhynchostigma Benth. = *Toxocarpus*

Rhynchostylis Blume. Orchidaceae (V 16). 3 trop. As. R: Selbyana 9(1986)169. Cult. orn. epiphytes

Rhynchotechum Blume (*Rhynchotoechum*). Gesneriaceae. 12 Indomal.

Rhynchotheca Ruíz & Pavón. Geraniaceae (Ledocarpaceae). 1 Andes. Spiny, C 0

Rhynchothecaceae Endl. = Geraniaceae

Rhynchotoechum Blume = *Rhynchotechum*

Rhynchotropis Harms. Leguminosae (III 8). 2 SC Afr.

Rhynea DC = *Tenrhynea*

Rhysolepis S.F. Blake. Compositae (Helia.-Helia.). 3 Mexico

Rhysopterus J. Coulter & Rose. Umbelliferae (III 9). 3 N Am.

Rhysotoechia Radlk. Sapindaceae. 15 C Mal. to trop. Aus. (4) . R: Blumea 39(1994)41

Rhyssocarpus Endl. = *Melanopsidium*

Rhyssolobium E. Meyer. Asclepiadaceae (III 4). 1 S Afr.: *R. dumosum* E. Meyer

Rhyssopteris Blume ex. A. Juss. = *Ryssopteris*

Rhyssostelma Decne. Asclepiadaceae (III 1). 1 Argentina: *R. nigricans* Decne

Rhytachne Desv. Gramineae (40h). 12 trop. Am. & Afr. R: KB 32(1978)767

Rhyticalymma Bremek. = *Justicia*

Rhyticarpus Sonder = *Anginon*

Rhyticaryum Becc. Icacinaceae. 12 E Mal., W Pacific. Some local potherbs esp. *R. oleraceum* Becc. in Moluccas & W New Guinea & *R. longifolium* Schumann & Lauterb. in Solomon Is.

Rhyticocos Becc. = *Syagrus*

Rhytidanthera (Planchon) Tieghem (~ *Godoya*). Ochnaceae. 5 Colombia. R: BJ 113(1991)173. Lvs pinnate

Rhytidocaulon Bally. Asclepiadaceae (III 5). 8 NE trop. Afr., Arabia. R: KB 36(1981)51. Cult. orn. succ.

Rhytidophyllum C. Martius (~ *Gesneria*). Gesneriaceae. 21 WI, S Am. (1). Pseudostipules; mucilage from lvs scented & (?) attractive to bat-pollinators

ribbon grass *Phalaris arundinacea* 'Picta'

Ribes L. Grossulariaceae. 150 temp. N hemisph. (Eur. 9), Andes. Currants. Low shrubs (some evergreen), ± prickles with hermaphr. fls or dioec. & fr. crowned with persistent K. Many ed. fr. & some cult. orn. *R. americanum* Miller (American blackc., N US) – fr. ed.; *R. aureum* Pursh (golden or buffalo c., W US to Mex.) – fr. ed., form. used with buffalo meat as pemmican; *R. curvatum* Small (granite gooseberry, S & SE US); *R. cynosbati* L. (American or prickly g., E N Am.) – fr. prickly, unarmed cvs cult. for fr. used in pies etc.; *R. divaricatum* Douglas (Worcesterberry, W N Am.) – nat. hybrid?, fr. form. dried; *R. glandulosum* Grauer ex Weber (skunk c., US mts) – foetid; *R. grossularioides* Maxim. (catberry, Japan) – cult. orn.; *R. hirtellum* Michaux (E N Am.) – ed. gooseberry used in hybridizing; *R. laurifolium* Jancz. (W China) – evergreen; *R. nigrum* L. (blackcurrant, Euras.) – fr. stewed or made into jam or a liqueur (cassis) or soft drinks with reputation for high levels of vitamin C but active flavour principles unknown (sometimes adulterated with buchu which has similar 'catty' odour), since Mediaeval times used to colour wine, a trad. treatment for colds (quinsyberry), alt. host for pine blister rust (*Cronartium ribicola* J. Fischer) prob. originally restricted to *Pinus cembra* in W Siberia & poss. Swiss Alps, though modern blackcurrants involving *R. bracteosum* Douglas ex Hook. (NW N Am.) & *R. sanguineum* etc.; *R. odoratum* H.L. Wendl. (buffalo c., E US) – cult. for large yellow fls, fr. ed.; *R. rubrum* L. (redcurrant, garnetberry, Euras.) – cult. for fr. like blackc. in Eur., rarely in Am., esp. for jelly eaten with turkey, whitecurrant a form of this sp. with greenish fr., many redcurrants also involving *R. petraeum* Wulf. (W & C Eur.) & *R. silvestre* (Lam.) Mertens & Koch (*R. sativum*, W Eur.) & other spp.; *R. sanguineum* Pursh (flowering or American c., W N Am.) – in Canada fls in March as first hummingbirds arrive; cult. orn. shrub of suburbia; *R. speciosum* Pursh (Californian fuchsia, Calif.) – bright red drooping fls in winter, visited by Anna hummingbirds which breed before other spp. in Calif.; *R. uva-crispa* L. (*R. grossularia*, gooseberry, (arch.) feaberry, Eur., natur. or poss. native GB) – spiny bushes reaching great age, fr. much used in pies & jam, subject of 'Gooseberry Clubs' in C & N England in late C18 & early C19 so that greatly increased fr. size was selected for through competitions; hybrid berries incl. jostaberry a hybrid between a gooseberry–blackcurrant cross & a blackcurrant–*R. divaricatum* cross

ribwort *Plantago* spp. esp. *P. lanceolata*

rice *Oryza sativa*; **Canadian wild r.** *Zizania aquatica*; **r. flower** *Pimelea* spp.; **r. grass** *Spartina* spp., *Paspalum scrobiculatum*; **hungry r.** *Digitaria exilis*; **Indian r.** *Z. aquatica*; **jungle r.** *Echinochloa colona*, *E. frumentacea*; **Manchurian water r.** *Z. aquatica*; **mountain r.** *Oryzopsis* spp.; **r. paper** *Tetrapanax papyrifer*; **Malayan r. p.** *Scaevola taccada*; **red r.** *Oryza glaberrima*, *O. rufipogon*; **Tuscarora r.** *Z. aquatica*

Richardia Kunth = *Zantedeschia*

Richardia L. Rubiaceae (IV 5). 15 trop. Am. R: Brittonia 26(1974)271. Some cult. orn. incl. *R. scabra* L. – source of false ipecacuanha

Richardsiella Elffers & Kenn.-O'Byrne. Gramineae (31d). 1 Zambia

Richea R. Br. Epacridaceae. 11 SE Aus. R: AP 5(1969)121. Lvs with parallel veins; some pls arborescent incl. *R. pandanifolia* Hook.f. (pandani, pandanni, pandanny, Tasmania) to 15 m

Richella A. Gray. Annonaceae. 1 Borneo, 1 New Caledonia, 1 Fiji

Richeria Vahl. Euphorbiaceae. c. 5–7 trop. S Am. *R. grandis* Vahl (WI) – timber & local medic.

Richeriella Pax & K. Hoffm. Euphorbiaceae. 2 SE China, W & C Mal.

Richteria Karelin & Kir. (~ *Chrysanthemum*). Compositae (Anth.- Chr.). 3 SW to E As.

Ricinocarpodendron Amman ex Boehmer. Prob. oldest name for *Aphanamixis*

Ricinocarpos Desf. Euphorbiaceae. 15 Aus., 1 New Caled.: *R. pinifolia* Desf. (wedding bush)

Ricinodendron Muell. Arg. Euphorbiaceae. 1 trop. & S Afr.: *R. heudelotii* (Baillon) Heckel (erimado) – trade timber, ed. nuts & oilseeds; see also *Schinziophyton*

Ricinus L. Euphorbiaceae. 1 E & NE Afr. to Middle E, now natur. throughout trop.: *R. communis* L. (castor oil, palma-Christi) – fast-growing (Jonah alleged to have sat under one, which then grew rapidly) colonist shrub to 4 m with regular pseudo-*n*-chotomous branching, the term. infls usu. with male fls (open for 1 day) at apex & females (open for 14 days) at base, the males with much-branched A, protogynous, at least partially insect-poll., extrafl. nectaries present, seeds with caruncles. Many alks & toxic principles incl. ricin, a toxalbumin binding to carbohydrates (10^{-7} body wt. fatal; used with a sharpened umbrella-ferrule in a political assassination in London & suggested use in cancer treatment), 2–6 seeds being a fatal dose. Cult. for at least 6000 yrs (one of the few major crops with an Afr. origin), the Ancient Egyptians using the seed-oil as an illuminant, but most familiar as purgative; diff. fractions used in high speed aero-engines (viscosity changes little with temp.), soap, paint, varnish, candles, cosmetics, crayons, carbon paper & in dyeing textiles, preserving leather & waterproofing fabrics, 'Rilson' being a nylon-type polyamide fibre prepared from oil; residue used for fert. & stems for fibre & board; some forms cult. orn. esp. red-leaved 'Gibsonii'; silkworms (Eri) fed on lvs in Afr.

Ricotia L. Cruciferae. 9 E Med. (Eur. 2). R: KB 1951(1952)123

Riddellia Nutt. = *Psilostrophe*

Ridleyella Schltr. Orchidaceae (V 14). 1 New Guinea

Ridolfia Moris (~ *Carum*). Umbelliferae (III 8). 1 Medit. incl. Eur.: *R. segetum* (L.) Moris

Riedelia Oliver. Zingiberaceae. 60 Mal.

Riedeliella Harms. Leguminosae (III 4). 3 Paraguay, Brazil. R: Rodriguesia 58(1984)9

Riencourtia Cass. Compositae (Helia.). 8 N S Am.

Riesenbachia C. Presl = *Lopezia*

Rigidella Lindley. Iridaceae. 4 Mex., Guatemala. R: Brittonia 23(1971)217. Have been

hybridized with *Tigridia* spp.
Rigiolepis Hook.f. = *Vaccinium*
Rigiopappus A. Gray. Compositae (Ast.-Ast.). 1 SW N Am.
Rikliella Raynal = *Lipocarpha*
Rimacola Rupp. Orchidaceae (IV 1). 1 NSW: *R. elliptica* (R. Br.) Rupp
Rimaria N.E. Br. = *Gibbaeum*
rimu *Dacrydium cupressinum*; **mountain r.** *Lepidothamnus laxifolius*
Rindera Pallas. Boraginaceae. c. 25 Medit. (Eur. 3) to C As. Alks
ringworm senna *Cassia alata*
Rinorea Aublet. Violaceae. c. 200 trop. Some potherbs in S Am.
Rinoreocarpus Ducke. Violaceae. 1 Amazonia
Rinzia Schauer (~ *Baeckea*). Myrtaceae. 12 SW Aus. R: Nuytsia 5(1986)415. Most with arils
Riocreuxia Decne. (~ *Ceropegia*). Asclepiadaceae (III 5). 9 trop. & S Afr., India (1)
ripgut *Bromus rigidus*
Ripogonaceae J. Conran & H. Clifford = Smilacaceae
Ripogonum Forster & Forster f. Smilacaceae (Ripogonaceae; Liliaceae s.l.). 8 New Guinea, Aus. (6, 4 endemic) & NZ. *R. scandens* Forster & Forster f. (NZ) – cult. orn. liane locally medic. & used in basketry
Riqueuria Ruíz & Pavón. Rubiaceae (inc. sed.). 1 Peru
Risleya King & Pantl. Orchidaceae (V 7). 1 Himal., W China
Ristantia P.G. Wilson & Waterhouse (~ *Tristania*). Myrtaceae (Lept.). 3 Queensland. R: Telopea 3(1988)265
Ritchiea R. Br. ex G. Don f. Capparidaceae. 30 trop. Afr. A ∞
Ritonia Benoist. Acanthaceae. 3 Madagascar
Rivea Choisy. Convolvulaceae. 4 Indomal. Alks. *R. corymbosa* Hall.f. (S Am.) – Aztecs smoked it (hallucinogens); *R. hypocrateriformis* (Desr.) Choisy (India, Sri Lanka) – fls fragrant
river oak *Casuarina cunninghamiana*
Rivina L. Phytolaccaceae (II). 1 S US to trop. Am.: *R. humilis* L. (bloodberry) – cult. orn. herb with red. fr., source of red dye. R: JAA 66(1985)26
roast beef plant *Iris foetidissima*
Robbairea Boiss. (~ *Polycarpaea*). Caryophyllaceae (I 1). 2 N Afr. to Arabia
Robeschia Hochst. ex O. Schulz. Cruciferae. 1 Sinai to Afghanistan
robin, ragged *Silene flos-cuculi*
Robinia L. Leguminosae (III 7). 4 N Am. R: Castanea 49(1984)187. Locust. Decid. trees & shrubs with extrafl. nectaries & thorny stipules, cult. orn. incl. *R. hispida* L. (rose acacia, SE US) – fls rose or purple & *R. pseudoacacia* L. (false or bastard a., black locust, C & E US, widely planted & ± natur. as in C Eur.) – dominating early succession in Appalachians, shoots growing to 8 m in 3 yrs, timber for construction & fuel, fls good bee-forage, a manganese indicator in Arkansas, many orn. cvs incl. the bilious yellow-leaved 'Frisia' of suburbia & the twisted-stemmed 'Tortuosa', the 350 yr-old tree of the wild form from Robin's garden still at Jardin des Plantes, Paris, in 1980s
Robinsonella Rose & Baker f. Malvaceae. 14 C Am. R: GH 11(1973)1. Some cult. orn. shrubs & trees
Robinsonia DC. Compositae (Sen.-Sen.). 7 Juan Fernandez. Woody, pachycaul, s.t. epiphytic on tree-ferns, dioec., allied to S Am. *Senecio* spp.
Robinsoniodendron Merr. = *Maoutia*
Robiquetia Gaudich. Orchidaceae (V 16). 40 Indomal. to Fiji
roblé (roble) Furniture timber of various spp. incl. *Nothofagus obliqua*, *Platymiscium pinnatum*, *Tabebuia rosea*
Roborowskia Batalin = *Corydalis*
Robynsia Hutch. Rubiaceae. 1 trop. W Afr.
Robynsiella Süsseng. = *Centemopsis*
Robynsiochloa Jacq.-Fél. = *Rottboellia*
Robynsiophyton R. Wilczek. Leguminosae (III 28). 1 SC Afr.
rocambole *Allium sativum* var. *ophioscorodon*
Rochea DC = *Crassula*
Rochefortia Sw. Boraginaceae (Ehretiaceae). 12 WI, 1 C Am. to Colombia. Dioec.
Rochelia Reichb. Boraginaceae. 20 Euras. (Eur. 1)
Rochonia DC. Compositae (Ast.-Ast.). 4 Madagascar
rock cress *Arabis* spp.; **r. elm** *Milicia excelsa*; **r. foil** *Saxifraga* spp.; **r. jasmine** *Androsace* spp.; **r. lily** *Dendrobium speciosum*; **r. maple** *Acer saccharum*; **r. melon** (NSW) *Cucumis melo*

Reticulatus Group; **r. orchid** *D. speciosum*; **r. rose** *Cistus* spp., *Helianthemum* spp.

rocket *Eruca vesicaria* ssp. *sativa*; **r.** or **dame's r.** *Hesperis matronalis*; **dyer's r.** *Reseda luteola*; **garden r.** *E. vesicaria* ssp. *sativa*; **London r.** *Sisymbrium irio*; **salad r.** *E. vesicaria* subsp. *sativa*; **sea r.** *Cakile maritima*; **wall r.** *Diplotaxis muralis*; **white r.** *D. erucoides*; **yellow r.** *Barbarea vulgaris*

Rockia Heimerl = *Pisonia*

Rockinghamia Airy Shaw. Euphorbiaceae. 2 NE Queensland

Rodgersia A. Gray. Saxifragaceae. 6 Nepal & E As. R: NRBGE 34(1975)113. Pls with stout rhizomes & peltate, palmate or pinnate lvs & large term. infls of small white fls; most cult. orn.

Rodrigoa Braas = *Masdevallia*

Rodriguezia Ruíz & Pavón. Orchidaceae (V 10). 40 trop. Am. esp. Brazil. Cult. orn. epiphytes

Rodrigueziella Kuntze (~ *Theodora*). Orchidaceae (V 10). 5 Brazil

Rodrigueziopsis Schltr. Orchidaceae (V 10). 2 Brazil

Roegneria K. Koch = *Elymus*

Roella L. Campanulaceae. 25 S Afr.

Roemeria Medikus (~ *Papaver*). Papaveraceae (IV). 3 Medit. (Eur. 1) to Afghanistan (all in Turkey). R: Flora 179(1987)135. Annuals with solit. red or violet fls; *R. hybrida* (L.) DC (*R. violacea*, Medit. to SW As.) – cult. orn.

Roentgenia Urban (~ *Cydista*). Bignoniaceae. 2 N S Am.

Roepera A. Juss. = *Zygophyllum*

Roeperocharis Reichb.f. Orchidaceae (IV 2). 5 trop. E Afr.

Roezliella Schltr. = *Sigmatostalix*

Roger, stinking *Tagetes minuta*

Rogeria Gay ex Del. Pedaliaceae. 6 Brazil, trop. & S Afr.

Rogersonanthus B. Maguire & Boom (~ *Irlbachia*). Gentianaceae. 3 NE S Am., Trinidad. R: MNYBG 51(1989)3. Small trees

Roggeveldia Goldbl. Iridaceae (III 2). 2 Karoo. R: SAJB 58(1992)211

Rogiera Planchon (~ *Rondeletia*). Rubiaceae (I 5). 18 Mex. to Panamá. R: ABAH 28(1982)66, 33(1987)301, 35(1989)311

Rohdea Roth. Convallariaceae (Liliaceae s.l.). 1 SE China & Japan: *R. japonica* (Thunb.) Roth (omoto) – cult. orn. esp. in Japan for 500 yrs with 1500 named cvs incl. forms with twisted, varieg. or curled lvs, poll. by slugs & snails feeding on fleshy P smelling of bad bread

Roigella Borh. & Fernández (~ *Rondeletia*). Rubiaceae (I 5). 1 W Cuba

Rojasia Malme = *Oxypetalum*

Rojasianthe Standley & Steyerm. Compositae (Helia.-Verb.). 1 C Am.

Rojasimalva Fryxell. Malvaceae. 1 Venez.: *R. tetrahedralis* Fryxell. R: Ernstia 28(1984)11

Rolandra Rottb. Compositae (Vern.-Rol.). 1 trop. Am., introd. SE As., Japan

Roldana Llave & Lex. (~ *Senecio*). Compositae (Sen.-Tuss.). c. 55 Mex. & C Am. R: Phytologia 27(1974)408. Trees to herbs; some cult. orn. esp. *R. petasitis* (Sims) H. Robinson & Brettell (*Senecio p.*)

Rolfeella Schltr. = *Benthamia*

Rollandia Gaudich. = *Cyanea*

Rollinia A. St-Hil. Annonaceae. 60 trop. Am. Incl. *Rolliniopsis*, R: AHB 12(1934)112,190. Some ed. fr. esp. *R. mucosa* (Jacq.) Baillon (*R. deliciosa*, *R. pulchrinervis*, biribá – fr. up to size of football

Rolliniopsis Saff. = *Rollinia*

Rollinsia Al-Shehbaz (~ *Thelypodium*). Cruciferae. 1 warm N Am.

Romanoa Trevis. (*Anabaena*). Euphorbiaceae. 1 E Brazil

Romanschulzia O. Schulz. Cruciferae. 14 C Am.

Romanzoffia Cham. Hydrophyllaceae. 4 W N Am. to Aleutians. Some cult. orn. rock-pls

Romeroa Dugand. Bignoniaceae. 1 Colombia

Romnalda P. Stevens. Xanthorrhoeaceae (Liliaceae s.l.). 2 Papuasia & Queensland

Romneya Harvey. Papaveraceae (IV). 1 S Calif. & N Baja Calif.: *R. coulteri* Harvey (Matilija poppy) – cult. orn. glaucous rabbit-proof perennial with large white fls & alks; poll. bees become 'fuddled' in fls & only slowly recover

Romulea Maratti. Iridaceae (IV 3). c. 90 Eur. (8), Medit., S Afr. (68, R: JSAB suppl. 9(1972)1). Cormous herbs like *Crocus* but without keeled lvs & with long peduncle above ground. Fls close as temperature falls & will open in dark if warmed; some gynodioecy. Cult. orn., several natur. Aus. ('onion-grass') e.g. *R. rosea* (L.) Ecklon (Guildford grass, SW Cape) also natur. Channel Is. & St Helena

Ronabea Aublet = *Psychotria*

Rondeletia L. Rubiaceae (I 5). 130 WI, 1 Panamá, 7 S Am. Cult. orn. trees & shrubs esp. *R. odorata* Jacq. (Cuba). *R. anguillensis* Howard & Kellogg (Anguilla) – long-styled fls with 3- & 4-colpate pollen, short-styled with only 3-colpate; *R. strigosa* Hemsley (Guatemala) – fls with protein- & starch-rich false pollen

Rondonanthus Herzog = *Paepalanthus* (but see AMBG 78(1991)441)

Ronnbergia C.J. Morren & André. Bromeliaceae (3). 8 C Am., NW S Am. R: FN 14(1979)1497. *R. explodens* L.B. Sm. (Panamá & Peru) – fr. explosive

Roodia N.E. Br. = *Argyroderma*

rooibos tea *Aspalathus linearis*

rooigras *Themeda triandra*

Rooseveltia Cook = *Euterpe*

Roridula Burm.f. ex L. Byblidaceae (Roridulaceae). 2 S Afr. Viscid subshrubs trapping, but not absorbing contents of, insects (? defence) & dying insects preyed on by crab spiders which are not caught by pl.; poll. by small Heteroptera showered with pollen when they touch sensitive anthers which suddenly spring upright

Roridulaceae Bercht. & J. Presl. ? Corr. name for Byblidaceae

Rorippa Scop. Cruciferae. 80 subcosmop. (Euras. 25, incl. *Nasturtium*, Afr. 12, N Am. 23). R: JAA 69(1988)144. Watercress esp. *R. nasturtium-aquaticum* (L.) Hayek (*N. officinale*, Eur., natur. N Am.) – salad pl. grown in inundated beds, & the larger-flowered *R. microphylla* (Boenn.) Löve & D. Löve (*R. m.* = *R. nasturtium-aquaticum* × ? *Cardamine* sp., W Eur., natur. US) as well as their sterile hybrid *R.* × *sterilis* Airy Shaw; extracts used in cosmetics. *R. montana* (Wallich) Small (China) – effective in respiratory complaints

Rosa L. Rosaceae. 100–150 N temp. (Eur. c. 45) to trop. mts. Roses. Prickly shrubs s.t. climbing or trailing with fls in panicles or solit., followed by hips – achenes encl. in fleshy receptacle; K quincuncial: 2 'bearded', 2 not, 1 half so ('the five brethren of the rose'). Sources of imp. essential oils used in scent-making, also local medic. but esp. imp. as cult. orn. with so many thousands of cvs that formal classification is now inexact & several hort. schemes proposed. Many of the orig. 'spp.' prob. ancient hybrids of complex & disputed ancestry (see W.J. Bean (1980) *Trees & shrubs hardy in the British Isles* ed. 8, vol. 4: 131). Essentially hybridization has brought together the genomes of eastern diploid (2n = 14) 'tea' roses & western tetraploid 'cabbage' roses. Generally these are the modern garden roses used for display & cutting (incl. Hybrid Teas derived from crosses between older Hybrid Perpetuals & the Teas & Chinas, the first cv. so classified being 'La France' (1867) but many hundreds of new ones raised each year now; Floribundas differing in usu. clustered fls & derived from crosses between Hybrid Teas & Hybrid Polyanthas (these being crosses between Polyanthas, Hybrid Teas etc.); such cvs as the popular 'Queen Elizabeth' (c. 1950) somewhat intermediate while Miniature Roses derived from crosses between Hybrid Teas & *R. chinensis* Jacq. 'Minima' (*R. roulettii*, fairy rose, China); Climbing Roses incl. mutants of the above groups as well as hybrids between wild climbing spp.). Old Garden Roses incl. Hybrid Perpetuals, *the* roses a century ago prob. derived from a number of spp. incl. *R.* × *borboniana* Desp. (Bourbon roses, *R. chinensis* (a Chinese cv., the wild type in W China) × *R.* × *damascena* Miller (damask rose, poss. hybrid between *R. gallica* L. – Eur., W As. (red rose of Lancaster ('Officinalis') in War of the Roses) & *R. moschata* Herrm., a cultigen)); Tea Roses derived largely from *R.* × *odorata* (Andrews) Sweet (*R. chinensis* × *R. gigantea* Crépin (NE Burma to China) raised in China), Polyanthas from *R.* × *rehderana* Blackburn (*R. chinensis* × *R. multiflora* Thunb. ex Murray (Korea, Japan)). More recent developments include Hybrid Musks, 'English Roses' combining old flower-forms with repeat-flowering, Patio Roses bred from Floribunda-miniature crosses & ground-cover roses. Modern roses are basis of enormous cut-fl. & automated rose-bush industry; essences (otto or attar of roses) still obtained by steam distillation of rose petals esp. of *R.* × *damascena* & others esp. *R.* × *alba* L. (*R. gallica* × *R. canina*/*R. arvensis* Hudson (C & SW Eur. to Turkey)) – incl. white rose of York ('Semi Plena') as in War of the Roses & the Jacobite rose of Bonnie Prince Charlie ('Maxima'), and roses cult. by Romans; *R. banksiae* Aiton f. (Banksian r., W & C China) – wild type cult. for 80 yrs without flowering & meantime double cv. introduced & named, fls yellow or white, bark of roots used in tanning in China, at Tombstone Arizona allegedly the biggest of all roses (2.74 m tall & covering 499 m^2 in 1991 from a 1884 cutting); *R. bracteata* Wendl. (Macartney r., SE China); *R. canina* L. (dog rose, Euras., Medit., natur. N Am.) & other Br. spp. – hips coll. by children after World War II to boost supplies of vitamin C, hairs & achenes a vermifuge, bad weed in NSW where controlled by goats; *R. carolina* L. (pasture r., C & E N Am.); *R.* × *centifolia* L. (cabbage r.) – prob. derived from *R. gallica* & Damask roses etc., source of essence, cvs incl. 'Muscosa' (moss r.) – K & pedicels with 'mossy' outgrowths (much-branched scent-glands); *R. chinensis* 'Viridiflora' (green rose) – C green streaked reddish, A & G replaced by toothed leafy organs, cult. orn. curiosity; *R. davurica* Pallas (shi mei, NE As.) – fls used as herbal tea; *R. filipes* Rehder & E. Wilson (W China) – rampant esp. 'Kiftsgate' growing several m per annum; *R. glauca* Pourret (*R. rubrifolia*, C & S Eur. mts) – reddish shoots; *R. luciae* Franchet & Rochebr. (*R. wichuraiana*, E As., natur. N Am.) – scrambler from which common 'ramblers' 'Dorothy Perkins' & 'American Pillar' derived; *R. majalis* Herrm. (*R. cinnamomea*, cinnamon r., Euras.); *R. moschata* Herrm. (musk r., Medit.) – parent with *R. chinensis* of *R.* × *noisettiana* Thory (Noisette r.); *R. moyesii* A. Henry & E. Wilson (W China) – favoured for spectacular hips; *R. rubiginosa* L. (*R. eglanteria*, eglantine, sweetbriar, Euras., Medit., natur. N Am.) – foliage scented, fls medic., pest in NZ; *R. rugosa* Thunb. (Japanese r., Korea, China, Japan, natur. GB, US) – v. hardy r. planted in Russia etc., used to flavour some Chinese wines, hips eaten by Ainu; *R. setigera* Michaux (prairie r., C & E N Am.) – dioec. (unique) but cryptic though more fls per infl. in males & greater petal exapansion in females; *R. spinosissima* L. (*R. pimpinellifolia*, burnet or Scotch r., Euras.) – form. many cvs cult.

Rosaceae Juss. Dicots – Rosidae – Rosales. 95/2825 subcosmop. but esp. temp. & warm N. Trees, shrubs & herbs, rarely with alks. Lvs simple, compound or dissected, usu. spirally arr. (opp. in *Rhodotypos*); stipules usu. present & often adnate to petiole (0 in *Spiraea* etc.). Fls usu. reg. & bisexual, perigynous with a hypanthium to epigynous, solit. or in cymes; K (3–)5(–10) imbricate, often appearing as lobes on hypanthium, C(3–)5(–10), imbricate, often large, rarely 0, A (1, 5)20 – ±∞ usu. in sets of 5 often originating centripetally, filaments usu. free & attached to hypanthium, anthers with longit. slits, rarely term. pores, inner surface of hypanthium often nectariferous, G (usu. inf. in Maloideae) 1 (mainly Prunoideae) to ∞ free or united with 2–5 separate styles & axile placentas, 1 or 2 (several–∞ on marginal placenta in Spiraeoideae), anatropous to hemitropous or campylotropous bi-(uni-)tegmic ovules. Fr. a head of follicles or achenes (s.t. on enlarged fleshy receptacle as in *Fragaria*) or encl. in swollen hypanthium (as in rose hips) or a head of drupelets (as in *Rubus*) or a pome (Maloideae), rarely a capsule (*Lindleya*); seeds mesotestal with embryo straight or bent usu. without endosperm. x = 7–9, 17+

Classification & principal genera:

I. **Spiraeoideae** ($\underline{G}$, pistils 2 or more simple, each with several–∞ ovules, ripening as follicles, x = 7–9): 3 tribes incl. *Spiraeeae* (*Spiraea*) & *Exochordeae* (*Exochorda*) – seeds winged

II. **Rosoideae** ($\underline{G}$, pistils 2 or more each with 1 ovule, ripening as achenes or drupelets, x = 7–9): c. 6 tribes incl. Ulmarieae (receptacle ± flat, filaments ± club-shaped; *Filipendula*), Kerrieae (receptacle flat or convex, filaments broad-based; *Kerria*, *Rhodotypos*), Potentilleae (like K. but G ∞ on convex gynophore; *Fragaria*, *Geum*, *Potentilla*, *Rubus*), Sanguisorbeae (receptacle urn-shaped, usu. hard, encl. 2 or more achenes; *Acaena*, *Agrimonia*, *Alchemilla*, *Lachmilla*, *Sanguisorba*), Roseae (receptacle soft at maturity encl. many free carpels; *Rosa*)

III. **Maloideae** ($\overline{G}$ (2–5) ripening as pome; x = 17 (R: CJB 68(1990)2209, Syst.B 16(1991)376): *Amelanchier*, *Chaenomeles*, *Cotoneaster*, *Crataegus*, *Cydonia*, *Eriobotrya*, *Malus*, *Mespilus*, *Photinia*, *Pyracantha*, *Pyrus*, *Sorbus*

IV. **Prunoideae** ($\underline{G}$, pistil 1 with 1 ovule, ripening as drupe; x = 7–9): *Prunus*

Poss. orig. in W Gondwanaland. Held by some that Maloideae the allopolyploid results of cross between a Spiraeoidea and a Rosoidea but *Kageneckia* & *Quillaja* (Spiraeoideae) have x = 17 & 14 resp., prob. best treated as separate subfam. (Goldblatt)) & *Lindleya* somewhat intermediate between S. & M. has x = 17, it now being held that M. have been derived from S.-like ancestors. The taxonomy of the family is confounded by apomixis (e.g. *Rubus*, *Alchemilla*), non-mixing of parental chromosomes (*Rosa*) & at generic level by excessive splitting of temperate genera of economic imp. e.g. *Pyrus* & *Malus*, a hangover from pre-Linnaean folk taxonomies

Many comm. imp. frs esp. *Malus* (apples), *Prunus* (almonds, apricots, cherries, peaches & plums), *Rubus* (blackberries, loganberries, raspberries etc.), *Pyrus* (pears), *Fragaria* (strawberries), *Cydonia* (quince), *Eriobotrya* (loquat), *Mespilus* (medlar), *Oemleria* etc.; scents (*Rosa*) & many cult. orn. shrubs & herbs esp. *Acaena*, *Alchemilla*, *Amelanchier*, *Aruncus*, *Chaenomeles*, *Cotoneaster*, *Crataegus*, *Dryas*, *Exochorda*, *Geum*, *Kerria*, *Malus*, *Photinia*, *Potentilla*, *Prunus*, *Pyracantha*, *Rosa*, *Rubus*, *Sorbaria*, *Spiraea*. Temperate gardens would be v. different if R. did not exist

Roscheria H. Wendl. ex Balf.f. Palmae (V 4a). 1 Seychelles

Roscoea Sm. Zingiberaceae. 17 Himal. & W China. R: KB 36(1982)747. Herbs with thick fleshy roots, cult. orn.

rose *Rosa* spp. but in compound names often applied to pls with fls superficially like *R.*; **r. acacia** *Robinia hispida*; **alpine r.** *Rhododendron ferrugineum*; **Andes r.** *Bejaria* spp.; **r. apple** *Syzygium jambos, S. malaccense*; **Banksian r.** *Rosa banksiae*; **r. bay** *Nerium oleander*; **r.b. willowherb** *Epilobium angustifolium*; **burnet r.** *R. spinosissima*; **cabbage r.** *R.* × *centifolia*; **r. campion** *Silene* spp.; **Chinese r.** *Hibiscus rosa-sinensis*; **Christmas r.** *Helleborus niger*; **cinnamon r.** *R. majalis*; **Confederate r.** *Hibiscus mutabilis*; **damask r.** *R.* × *damascena*; **desert r.** *Adenium* spp.; **eglantine r.** *R. rubiginosa*; **fairy r.** *R. chinensis* 'Minima'; **green r.** *R. c.* 'Viridiflora'; **guelder r.** *Viburnum opulus*; **r. gum** *Eucalyptus grandis*; **Honolulu r.** *Clerodendrum chinense*; **Japanese r.** *R. rugosa*; **r. of Jericho** *Anastatica hierochuntica, Selaginella lepidophylla*; **kiwi r.** *Telopea speciosissima*; **Lady Nugent's r.** *Clerodendrum chinense*; **Lenten r.** *Helleborus orientalis*; **Macartney r.** *R. bracteata*; **r. mallow** *Hibiscus* spp. esp. *H. rosa-sinensis*; **marsh r.** *Orothamnus zeyheri*; **moss r.** *R. centifolia* 'Muscosa'; **musk r.** *R. moschata*; **noisette r.** *R.* × *noisettiana*; **r. pink** *Sabatia angularis*; **r. plantain** *Plantago major* 'Rosularis'; **prairie r.** *R. setigera*; **rock r.** *Helianthemum* & *Cistus* spp.; **r. root** *Sedum rosea*; **Scotch r.** *R. spinosissima*; **r. of Sharon** *Hypericum calycinum*, (in Bible) *Lilium candidum* or *Tulipa agenensis* subsp. *boissieri*, (in N Am.) *Hibiscus syriacus*; **stock r.** *Sparrmannia africana*; **Sturt('s) desert r.** *Gossypium sturtianum*; **wood(en) r.** *Argyreia nervosa, Dactylanthus taylorii, Merremia tuberosa*

roseau, black *Bactris major*; **white r.** *Gynerium sagittatum*

roselle *Hibiscus sabdariffa*

rosemary *Rosmarinus officinalis*; **bog r.** *Andromeda polifolia*; **marsh r.** *Limonium vulgare*

Rosenbergiodendron Fagerl. = *Randia*

Rosenia Thunb. Compositae (Gnap.-Rel.). 4 S Afr. Shrubs with opp. lvs. R: BN 129(1976)97

Rosenstockia Copel. = *Hymenophyllum*

Roseodendron Miranda = *Tabebuia*

roseroot *Sedum rosea*

rosewood Orig. *Dalbergia nigra* (**Brazilian, Rio** or **Bahia r.**) but used for timbers with similar qualities; **African r.** *Guibourtia demeusei, Pterocarpus erinaceus*; **Australian r.** *Dysoxylum fraserianum, Alectryon oleifolius*; **bastard r.** *Synoum glandulosum*; **Burmese r.** *Pterocarpus indicus*; **Honduras r.** *Dalbergia stevensonii*; **Indian** or **Malabar r.** *D. latifolia*; **Papua New Guinea r.** *P. indicus*; **scentless r.** *S. glandulosum*; **W African r.** *P. erinaceus*

Rosifax C. Townsend. Amaranthaceae (I 2). 1 Somalia

Rosilla Less. = *Dyssodia*

rosin *Pinus* spp.

Rosmarinus L. Labiatae (VIII 2). 2 Medit. incl. Eur. Rosemary. *R. officinalis* L. – cult. orn. & herb with lvs dried for use in stews, sausages etc., the flowering shoots distilled for oil used in scent-making (with bergamot (citrus) & neroli oil the chief constituent of eau-de-Cologne) & medic., excellent honey-pl., form. worn at funerals & weddings

Rossioglossum (Schltr.) Garay & G. Kenn. (~ *Odontoglossum*). 4 C Am. Cult. orn. showy epiphytes

Rostellularia Reichb. = *Justicia*

Rostkovia Desv. Juncaceae. 1 Tristan da Cunha, 1 temp. S Am., NZ

Rostraria Trin. (*Lophochloa*; ~ *Koeleria*). Gramineae (21b). 5 Eur. (5), Medit. & S Afr. Annuals derived from *Koeleria* incl. *R. cristata* (L.) Tzvelev (*Koeleria phleoides*, cat-tail grass, Med.) – natur. GB

Rostrinucula Kudô (~ *Elsholtzia*). Labiatae (VII). 2 E As.

Rosularia (DC) Stapf. Crassulaceae. 27 N Afr. & Eur. (1) to C As. esp. SE Turkey & Himal. R: Bradleya 6 suppl. (1988). Some cult. orn. succ. rosette pls, usu. with taproots

Rotala L. Lythraceae. 44 temp. to trop. Wet places, several with 'Hippuris syndrome' incl. *R. hippuris* Makino (Japan) & *R. mexicana* Cham. & Schldl. (pantrop.). R: Boissiera 29(1979). *R. indica* (Willd.) Koehne (Indomal. to Japan) – s.t. grown in aquaria; *R. repens* (Hochst.) Koehne (E & NE Afr. highlands) – on contact with water, seeds put out hairs which become viscid & may be attached to animal disp. agents

rotan(g) = rattan

Rotheca Raf. (*Cyclonema*; ~ *Clerodendrum*). Labiatae (II). c. 5 trop. Afr., As. Cult. orn. esp. *R. myricoides* (Hochst.) Steane & Mabb. 'Ugandensis' (Oxford & Cambridge bush, trop. E Afr.)

Rothia Pers. Leguminosae (III 28). 1 dry Afr., 1 Baluchistan to Aus.: *R. indica* (L.) Druce

Rothmaleria Font Quer. Compositae (Lact.-Cat.). 1 S Spain

Rothmannia Thunb. Rubiaceae (II 1). 40 trop. & S Afr. to Seychelles & Malay Pen. (3). Cult. orn. gardenia-like small trees, some Afr. spp. with ant-infested stems

Rothrockia A. Gray = *Matelea*

Rottboellia L.f. Gramineae (40h). (Incl. *Manisuris*) 4 trop. OW. *R. cochinchinensis* (Lour.) W. Clayton (Kelly grass) – fodder, serious trop. weed since 1970s

Rotula Lour. Boraginaceae (Ehretiaceae). 1 E Brazil, OW trop.: *R. aquatica* Lour. – rheophyte with tough branches, allied to *Ehretia*

Roubieva Moq. = *Chenopodium*

Roucheria Planchon. Linaceae. 9 trop. S Am.

roucou *Bixa orellana*

Rouen lilac *Syringa × persica*

Rouliniella Vail = *Cynanchum*

roundwood *Sorbus americana*

Roupala Aublet. Proteaceae. 90+ trop. Am. Some locally used timber, cult. orn. & medic.

Roupellina (Baillon) Pichon = *Strophanthus*

Rourea Aublet. Connaraceae. 40–70 trop. (Afr. 12, R: AUWP 89-6(1989)310). *R. mimosoides* (Vahl) Planchon (Mal.) – sleep movements; some spp. with poisonous seeds

Roureopsis Planchon = *Rourea*

Roussea Sm. Grossulariaceae (Brexiaceae). 1 Mauritius

Rousseaceae DC = Grossulariaceae

Rousseauxia DC. Melastomataceae. 13 Madagascar. R: Adansonia 13(1973)180

Rousselia Gaudich. Urticaceae (IV). 3 C Am. to Colombia & WI

Rouya Coincy. Umbelliferae (III 12). 1 N Afr., Corsica, Sardinia: *R. polygama* (Desf.) Coincy

rowan *Sorbus aucuparia*

royal fern *Osmunda* spp. esp. *O. regalis*; **r. palm** *Roystonea* spp. esp. *R. regia*

Roycea C. Gardner. Chenopodiaceae (I 5). 3 temp. & subtrop. W Aus. R: FA 4(1984)216

Royena L. = *Diospyros*

Roylea Wallich ex Benth. Labiatae (VI). 1 Himalaya

Roystonea Cook. Palmae (V 4g). 10–12 Carib. & NE S Am. Unarmed monoec. palms with pinnate lvs. R: GH 8(1949)114. Cult. orn. esp. as avenues in trop., usu. *R. regia* (Kunth) Cook (Cuba, Florida (incl. *R. elata*)) – royal palm, source of palm-hearts as is *R. oleracea* (Jacq.) Cook (palmiste, Barbados & Trinidad); some ed. fr., sago & thatch

Ruagea Karsten. Meliaceae (I 6). 5 trop. Am. montane rain forest. R: FN 28(1981)242

rubber Coagulated latex of various trees, usu. *Hevea brasiliensis*; **abbo r.** (W Afr.) *Ficus lutea*; **African r.** *Funtumia elastica, Landolphia* spp. etc.; **Assam r.** *Ficus elastica*; **Bitinga r.** *Raphionacme utilis*; **Bolivian r.** *Sapium aucuparium*; **Borneo r.** *Willughbeia coriacea*; **ceara r.** *Manihot* spp. esp. *M. glaziovii*; **C American r.** *Castilla elastica*; **chilte r.** *Cnidoscolus elasticus*; **Congo r.** *Ficus lutea*; **couma r.** *Couma* spp.; **Dahomey r.** *F. lutea*; **E African r.** *Landolphia* spp.; **Esmeralda r.** *Sapium jenmanii*; **guayule r.** *Parthenium argentatum*; **India(n) r.** *F. elastica*; **intisy r.** *Euphorbia intisy*; **Iré r.** *F. vogelii*; **kok-saghyz r.** *Taraxacum bicorne*; **krim-saghyz r.** *T. megalorhizon*; **Lagos r.** *Funtumia elastica*; **Madagascar r.** *Cryptostegia, Landolphia, Marsdenia & Mascarenhasia* spp.; **mangabeira r.** *Hancornia speciosa*; **Manicoba r.** *Manihot glaziovii*; **milkweed r.** *Asclepias* spp.; **Panamá r.** *Castilla elastica*; **Pernambuco r.** *Hancornia speciosa*; **r. plant** *Ficus elastica*; **Rangoon r.** *Urceola maingayi*; **silk r.** *Funtumia elastica*; **tau-saghyz r.** *Scorzonera tausaghyz*; **teke-saghyz r.** *S. acanthoclada*; **tirucalli r.** *Euphorbia tirucalli*; **Tongking r.** *Streblus tonkinensis*; **Ulé r.** *Castilla elastica*

Rubia L. Rubiaceae (IV 16). 60 Medit. (Eur. 4), Afr., temp. As., Am. Source of dyes & local medic. etc. esp. *R. cordifolia* L. (Indian madder, munjeet, Afr. to As.) – red dye from roots, medic. in China (menstrual disorders); *R. peregrina* L. (Levant m., wild m., Eur., Medit.) – medic.; *R. tinctorum* L. (madder, S Eur., W As.) – form. cult. for alizarin dye from roots fixed by mordants (alum to give dark red, iron alum for brown-red, chrome alum for red-violet), known to Persians & Anc. Egyptians, but now supplanted by aniline dyes (a derivative of anthracene in coal tar 1868), though used until recently for 'rose madder' & 'm. brown' tints in paint-boxes; *R. yunnanensis* Diels – subpachycaul scrambler with long lvs

Rubiaceae Juss. Dicots – Asteridae – Rubiales. 630/10 200 cosmop. but esp. trop. & warm. Trees, shrubs, lianes & few herbs, s.t. ant-inhabited (as *Cuviera, Duroia, Hydnophytum, Myrmecodia, Myrmeconauclea* etc.), epiphytic or rarely aquatic (1 *Limnosipanea* sp.), s.t. with unusual sec. growth & with wide range of chemical repellents incl. isoquinoline alks. Lvs simple & usu. entire, usu. decussate & with usu. connate stipules or less often app. whorled with lvs in place of interpetiolar stipules (e.g. *Galium*), rarely (e.g. *Didymochlamys*) lvs spirally arr. through suppression of 1 member of each pair; stipules often bearing colleters with slime protecting growing buds, petiole s.t. with medullary bundles. Fls usu. bisexual & epigynous & insect-poll., often heterostylous, in cymes or rarely solit.; K (4 or 5), the lobes open & often small or 0 (enlarged & brightly coloured in *Mussaenda* etc.), C ((3)4 or 5(8–10)) reg. to 2-labiate, lobes valvate, imbricate or convolute, A attached to C-tube & alt. with lobes, anthers with longit. slits, nectary-disk often present at top of $\overline{G}$ (2(3–5+)) (but $\underline{G}$ in *Gaertnera* & *Pagamea*) with axile placentas, rarely 1-loc. with parietal placentas (in *Gardenia* etc.) with term. style (free styles in *Galium* etc.), each loc. with 1–∞ anatropous to hemitropous unitegmic ovules often with a funicular obturator. Fr. a capsule, berry, drupe or schizocarp etc.; seeds with usu. straight embryo embedded in usu. copious oily endosperm sometimes with reserves of starch & hemicellulose, 0 in Guettardoideae etc. x = (6–)11(–17)

Classification & principal genera (OBB 1(1988)178, though some genera (e.g. *Manettia*) not clearly placed):

I. **Cinchonoideae** (usu. woody, often large trees; stipules usu. entire; raphides usu. 0; fr. usu. dry, endosperm present; 10 tribes):
  1. **Cinchoneae** (seeds winged; 2 subtribes): *Cinchona, Hymenodictyon, Ladenbergia, Uncaria*
  2. **Naucleeae** (Naucleaceae; fls in heads; 3 subtribes, R: Blumea 24(1977)315): *Nauclea, Neolamarckia, Neonauclea*
  3. **Hillieae** (seeds plumose): *Hillia*
  4. **Henriquezieae** (Henriqueziaceae): *Platycarpum*
  5. **Rondeletieae** (not clearly distinct from 7.): *Arachnothryx, Rondeletia, Simira, Wendlandia*
  6. **Sipaneeae** (herbaceous forms of 5.): *Sipanea*
  7. **Condamineeae** (capsules with many horizontal seeds): *Portlandia*
  8. **Isertieae** (fr. fleshy with many small angular seeds): *Acranthera, Isertia, Mussaenda, Sabicea*
  9. **Urophylleae** (like 8. but raphides present): *Praravinia, Urophyllum*
  10. **Pauridiantheae** (?distinct from 9): *Pauridiantha*

II. **Ixoroideae** (woody; stipules usu. entire; raphides usu. 0; fr. usu. fleshy; 5 tribes):
  1. **Gardenieae** (C lobes usu. contorted to left; 2 subtribes): *Alibertia, Duroia, Gardenia, Randia, Rothmannia*
  2. **Pavetteae** (like 1. but fr. fleshy): *Ixora, Pavetta, Tarenna*
  3. **Coffeeae** (like 2. but 2 seeds per fr.): *Coffea*
  4. **Aulacocalyceae** (like 3. but seeds without seed-coat): *Heinsenia*
  5. **Hypobathreae** (like 4. but toughened exotesta): *Hypobathrum*

III. **Antirheoideae** (like II. but exotestal cells parenchyma-like; 8 tribes):
  1. **Retiniphylleae** (infls term.): *Retiniphyllum*
  2. **Vanguerieae** (infls axillary): *Canthium, Cuviera, Fadogia, Keetia, Pavetta, Psydrax, Pyrostria*
  3. **Guettardeae** (Guettardiodeae; like 1. & 2. but seeds often elongated): *Antirhea, Bobea, Guettarda, Timonius*
  4. **Chiococceae** (like 3. but endosperm oily): *Chiococca*
  5. **Alberteae** (fls ±irreg.): *Nematostylis*
  6. **Cephalantheae** (like 3. but seeds with apical stony aril): *Cephalanthus*
  7. **Craterispermeae** (trees or shrubs, C valvate): *Craterispermum, Rudgea*
  8. **Knoxieae** (like 7. but herbs): *Knoxia*

IV. **Rubioideae** (raphides usu. present, heterostyly common; 16 tribes):
  1. **Hedyotideae** (usu. herbs; exotestal cells usu. parenchyma-like): *Hedyotis, Kohautia, Oldenlandia, Pentas*
  2. **Ophiorrhizeae** (woody to herbaceous; exotestal cells (?)thick-walled): *Ophiorrhiza*
  3. **Coccocypseleae** (like 2. but creeping herbs): *Coccocypselum*
  4. **Argostemmateae** (like 3. but unbranched plantlets): *Argostemma*
  5. **Hamelieae** (shrubs or small trees, fls monomorphic): *Hoffmania*
  6. **Schradereae** (scramblers with adhesive roots): *Schradera*
  7. **Psychotrieae** (stipules inter- to intra-petiolar, C valvate): *Amaracarpus, Chassalia, Gaertnera, Hydnophytum, Myrmecodia, Palicourea, Psychotria*
  8. **Triainolepideae** (like 7. but 2 or 3 ovules per loc.): *Triainolepis*
  9. **Lathraeocarpeae** (fls solit.): *Lathraeocarpa*
  10 **Morindeae** (fr. often connate in syncarps): *Lasianthus, Morinda*
  11. **Coussareeae** (fr. 1-loc., 1-seeded): *Coussarea, Faramea*
  12. **Paederieae** (fr. dry dehiscent with mericarps or operculum or fleshy indehiscent): *Paederia*

13. **Anthospermeae** (fls unisexual, wind-poll.; 3 subtribes): *Anthospermum, Coprosma*
14. **Theligoneae** (Theligonaceae; like 13. but marked anisophylly): *Theligonum*
15. **Spermacoceae** (herbs, fr. dry with mericarps): *Spermacoce*
16. **Rubieae** (herbs with lvs & leaflike stipules appearing whorled): *Asperula, Crucianella, Galium, Relbunium, Rubia*

Although so widespread and common particularly as trees & shrubs (frequently poll. Lepidoptera) in the lower reaches of trop. forests, & varying from pachycaul treelets (*Captaincookia*) to creeping herbs, this huge fam. contributes little to Man (cf. Compositae) exc. as drugs & stimulants (*Psychotria* – ipecacuanha, *Cinchona* – quinine, *Coffee* – coffee, *Pausinystalia* – yohimbine), some (usu. light) timbers (*Burchellia, Hymenodictyum, Mitragyna, Mussaendopsis, Neolamarckia*), ed. fr. (*Sarcocephalus, Vangueria*), dyes (*Morinda, Rubia*) & tan (*Uncaria*), as well as cult. orn. – *Asperula, Bouvardia, Coprosma, Galium, Gardenia, Hoffmannia, Ixora, Manettia, Mussaenda, Nertera, Pentas, Rothmannia, Wendlandia* etc.

Rubiteucris Kudô (~ *Teucrium*). Labiatae (II). Spp.?

Rubrivena Král = *Persicaria*

Rubus L. Rosaceae. 250 (+ ∞ apomictic lines) cosmop. esp. N. R (N Am.): GH 5(1941–5)1. Shrubs, often stoloniferous with subpachycaul shoots (brambles), thorns, conspic. insect-poll. fls & bird- or other animal-disp. fr. of aggregates of drupelets, some with nitrogen-fixing actinomycete root symbionts, NZ spp. dioec.; x = 7. Divisible into 12 subgg. of which subg. *R.* (brambles) has sexual diploids & apomicts (S(C) Eur., S(W) Am.) & higher ploidies apomicts (C & N Eur., N Am.), subg. *Idaeobatus* Focke (raspberries) has sexual diploids (trop. to temp.), triploid apomicts & sexual tetraploids & subg. *Malachobatus* Focke (trop. & subtrop.) has sexual 4–14-ploids. Many cult. fr. & orn., though many complex hybrids, some of obscure ancestry. *R. allegheniensis* L. Bailey (Alleghany blackberry, E N Am.) – ed. usu. long or oblong fr.; *R. arcticus* L. (N Euras.) – crossed with *R. stellatus* Sm. (*R. a.* ssp. *s.*, Alaska) in 1950s to give comm. fr. crop cvs; *R. australis* Forster f. (bush lawyer, NZ) – dioec. liane; *R. caesius* L. (dewberry, Eur., SW As.) – ed.; *R. chamaemorus* L. (cloudberry, circumboreal) – fr. ed., coll. from wild; *R. flagellaris* Willd. (dewberry (Am.), E N Am.) – black round ed. fr. with many cvs; *R. fruticosus* L. (bramble, blackberry, Eur., Med.) – aggregate of pseudogamous facultative apomicts & other forms incl. tetraploids & hexaploids as well as triploids & pentaploids (precisely applied *R. fruticosus* s.s. is corr. name for *R. plicatus* Weihe & Nees (C & NW Eur.), also sexual *R. ulmifolius* Schott ('*R. inermis*', SW & C Eur.) – most toothsome fr. ('Bellidiflorus' cult. orn. with double fls) but in autumn in Spain fl. buds producing lvs instead of perianth & carpels, the rest in Eur. 'not profitable to treat fully & by conventional means' (*Flora Europaea*, where the whole array of agamospecies grouped about 66 'circle-species' used as 'nodes' of variation), aggressive weeds in Aus. (where controlled by goats) & NZ – the thornless cvs of blackberry derived from crosses between *R. ulmifolius* & *R. hastiformis* W.C.R. Watson (GB), cut-leaved bramble with sweet black fr. (*R. laciniatus* Willd.) origin unknown; *R. glaucus* Benth. (trop. Am.) – raspberry cult. Ecuador & Colombia but retains 'plug' so poss. hybrid with blackberry; *R. idaeus* L. (raspberry, Euras., N Am.) – ed. fr. red or sometimes yellow or white, orig. coll. for flavouring wine, the leaves for making tea, now comm. imp. as fresh fr. (best flavour in 'Lloyd George' found in a wood at Corfe Castle, Dorset, England), in jams, pies, sweets, drinks, liqueurs & vinegar, subsp. distinguished pl. in N Am. also imp. (cross between them prob. ancestry of 'Cuthbert' standard comm. cv. for nearly 100 yrs in US), the purple raspberries being crosses between Am. ssp. & *R. occidentalis* L. (black r., thimbleberry, C & E N Am.) = *R.* × *neglectus* Peck, selection of 2nd generation cross between 'Cuthbert' & *R. vitifolius* being 'phenomenal berry' and this crossed with *R. caesius* 'Austin Mayes' = youngberry much grown in S Afr. as 'S Afr. loganberry', *R. idaeus* 'November Abundance' crossed with *R. rusticanus* E. Merc. *R. ulmifolius* s.l.) = veitchberry, 'Malling Jewel' (first virus-resistant cv) with *R. vitifolius* = sunberry, tetraploid form with an Am. blackberry ('Aurora') = tayberry and this crossed with a sister seedling = tummelberry; *R. laciniatus* Willd. (Eur., natur. N Am.) – leaflets deeply lobed, parent of some garden blackberries; *R. loganobaccus* L. Bailey (*R. ursinus* var. *loganobaccus*, loganberry, app. cross (Calif. 1881) between 8n form of *R. vitifolius* & *R. idaeus* cv. like 'Red Antwerp') – large red ed. fr., some thornless cvs, 'Boysen' = boysenberry (fr. to 2 cm) poss. involving hybridization with blackberry & raspberry, nectarberry being a form (? mutant; marionberry similar but origin unknown), while hildaberry = tayberry × boysenberry; silvaberry is a complex hybrid involving loganberry, youngberry, marionberry & boysenberry, raised in Aus.; *R. niveus* Thunb. (*R. albescens*, Mysore r., India, W China) – fr. red, ed. 'Mysore' comm. cult. Florida; *R. phoenicolasius* Maxim.

((Jap.) wineberry, Korea, China, Japan) – fr. small, ed.; *R. rosifolius* Sm. (Mauritius r., E As.) – ed. fr. cult. trop., roots med.; *R. vitifolius* Cham. & Schldl. (*R. ursinus*, dewberry (Am.), W N Am.) – ed. fr., parent of loganberry etc. Cult. orn. incl. *R. cissoides* A. Cunn. (NZ) – liane to 10 m, juvenile form (poss. *R. australis*) with lvs reduced to prickly midribs forming a bolster-like bush without fls., like other 4 endemic NZ spp. heteroblastic in habit (in raspberries juvenile features may be maintained through infection with crown gall) cult. in cool house, *R.* 'Benenden' ('*R. tridel*', *R. trilobus* Ser. (S Mex.) × *R. deliciosus* Torrey (W US) poss. not specifically distinct) – vogue hardy shrub

Ruckeria DC = *Euryops*

Rudbeckia L. Compositae (Helia.-Rudb.). 15 N Am. Coneflowers, black-eyed Susan, Mexican hat. Cult. orn. herbs esp. 'annual' *R. hirta* L. & perennial *R. laciniata* L. – toxic, medic. E Canadian Indians

Ruddia Yakovlev. See *Ormosia*

ruddles *Calendula officinalis*

Rudgea Salisb. Rubiaceae (IV 7). 150 trop. Am. Some heterostyly, some domatia

Rudolfiella Hoehne (~ *Bifrenaria*). Orchidaceae (V 10). 7 trop. Am.

Rudua F. Maek. = *Phaseolus*

rue *Ruta graveolens*

Ruehssia Karsten ex Schldl. = *Marsdenia*

Ruellia L. (incl. *Aporuellia*, *Dipteracanthus* etc.). Acanthaceae. c. 150 trop., temp. N Am. Seeds mucilaginous-pubescent when wet though appearing glabrous when dry. Cult. orn. shrubs & herbs. *R. prostrata* Poiret (*D. p.*, OW trop.) – C falls before midday; *R. radicans* (Nees) Lindau (NE S Am.) – poll. hummingbirds attracted by nectar but after C falls nectar higher in sugars & attractive to (? protecting) ants

Ruelliola Baillon = *Brillantaisia*

Ruelliopsis C.B. Clarke. Acanthaceae. 2–3 trop. & S Afr.

Rufodorsia Wiehler. Gesneriaceae. 4 C Am. R: Selbyana 1(1975)138

Rugelia Shuttlew. ex Chapman (~ *Senecio*). Compositae (Sen.-Tuss.). 1 SE US

Ruilopezia Cuatrec. See *Espeletia*

Ruizia Cav. Sterculiaceae. 3 Réunion

Ruizodendron R. Fries. Annonaceae. 1 Bolivia, Peru: *R. ovale* (Ruíz & Pavón) R. Fries

Ruizterania Marcano-Berti (~ *Qualea*). Vochysiaceae. 18 trop. S Am.

rukam *Flacourtia rukam*

Rulingia R. Br. Sterculiaceae. 20 Aus. Some fibres

rum spirit distilled from sugar-cane; **r. cherry** *Prunus serotina*

Rumex L. Polygonaceae (II 3). 200 temp. esp. N (Eur. 44; R (N Am.): FMB 17(1937)1–49). Docks, sorrel. Weeds (e.g. natur. Aus. from contents of camel-saddles) esp. subg. *R.* but this subg. parasitized by leaf-fungus (*Ramularia rubella* (Bon.) Nannf.) a poss. mycoherbicide, culinary herbs & tannin, lvs trad. (esp. *R. obtusifolius* L. (Eur.; app. allelopathic to germ. of competitors) applied to nettle-stings. *R. acetosa* L. (sorrel, Euras.) – ed.lvs cooked or in salads, sauces etc. ('*R. rugosus* Campd.', cultigen derived from *R. acetosa*), juice form. used to remove rust stains on linen; *R. acetosella* L. (sheep('s) s., N temp.) – weed of acid soils; *R. crispus* L. (yellow dock, Euras.) – weed with long-lived seeds, these lasting 80 but not 90 yrs in Dr Beal's experiment set up in 1879, contains hydroxyanthraquinone (like *Senna*) used as laxative & in arthritis treatment; *R. hymenosepalus* Torrey (canaigre, ganagra, tanner's dock, SW N Am.) – tubers rich in tannin (30–35%) form. used on leather, also yellow dye, lvs ed., petioles like rhubarb; *R. lunaria* L. (Canary Is.) – dioec. evergreen shrub planted as hedge in S Eur.; *R. patientia* L. (patience (dock), Euras., natur. N Am.) – lvs like sorrel; *R. pseudoalpinus* Hoefft ('*R. alpinus*', monk's rhubarb, Eur. mts) – lvs ed. cooked or salad; *R. pulcher* L. (fiddle d., Eur.); *R. scutatus* L. ((French) s., Euras.) – used like *R. acetosa*

Rumfordia DC. Compositae (Helia.-Mel.). 12 Mex. & C Am. mts. R: SB 2(1977)302

Rumia Hoffm. Umbelliferae (III 8). 1 Crimea

Rumicastrum Ulbr. (~ *Calandrinia*). Portulacaceae. 50 Aus.

Rumicicarpus Chiov. = *Triumfetta*

Rumohra Raddi. Dryopteridaceae (I 1). 7: 1 circumaustral to Bermuda, Zimbabwe, Madag., New Guinea: *R. adiantiformis* (Forster f.) Ching – cult. orn., 5 Madag. (1 obligate epiphyte on *Pandanus*), 1 Juan Fernandez

Rungia Nees. Acanthaceae. 50 trop. OW. *R. klossii* S. Moore (New Guinea) – cult. potherb

running buffalo clover *Trifolium stoloniferum*

Runyonia Rose = *Agave*

Rupicapnos Pomel. Fumariaceae (I 2). 7 S Spain (1), NW Afr., on cliffs. R: OB 88(1986)92

Rupicola Maiden & Betche. Epacridaceae. 4 NSW. R: Telopea 5(1992)234

Rupiphila Pimenov & Lavrova (~ *Selinum*). Umbelliferae (III 8). 1 E As.

Ruppia L. Ruppiaceae. 7+ temp. (Eur. 2) & subtrop. salt, alkaline or brackish water. R: AB 14(1982)325. Pollen in rafts to stigma, cf. *Halodule*, *Halophila*, *Lepilaena*. Duckfood

Ruppiaceae Horan. (~ Potamogetonaceae). Monocots – Alismatidae – Najadales. 1/7+ temp. & subtrop. coastal waters, S Am. & NZ. Submerged, rooted, glabrous herbs without vessel-elements; stem a sympodium formed by development of lateral buds beneath term. infls, usu. with 4 lvs to each segment. Lvs opp. or spirally arr., ± linear with midvein & basal open sheath & 2 intravaginal scales at each axil. Fls small, bisexual, in usu. 2-flowered term. spikes, subtended by small sheathing prophyll s.t. later assoc. with upper fl. only, the peduncle eventually elongated; P 0 (or a tiny appendage near tip of anther connective, A 2 opp. with subsessile anthers with expanded connective & longit. slits, G (2–)4(–8), each stipitate in fr. & with 1 pendulous, campylotropous, bitegmic ovule. Fr. of often asymmetric drupelets; seeds with 0 endosperm. x = 8, 10. R: PCAS IV, 25(1946)469

Genus: *Ruppia*

Some in ± fresh water in NZ & Andes, where found at up to 4000 m. Poll. underwater or pollen floating in some races to floating stigmas. Specific distinctions dubious

Ruprechtia C. Meyer. Polygonaceae (II 1). 20 trop. Am. Dioec. trees & shrubs

Ruptiliocarpon Hammel & Zamora (~ *Lepidobotrys*). Lepidobotryaceae. 1 trop. Am.: *R. caracolito* Hammel & Zamora. R: Novon 3(1993)408

rupturewort *Herniaria glabra*

Rusbya Britton. Ericaceae. 1 Bolivia

Rusbyanthus Gilg = *Macrocarpaea*

Rusbyella Rolfe ex Rusby. Orchidaceae (V 10). 1 Bolivia

Ruscaceae M. Roemer = Asparagaceae

Ruschia Schwantes. Aizoaceae (V). c. 360 S Afr. Succ. shrubs, older stems often bearing dry leaf sheaths of old lvs; some cult. orn. & natur. S Eur.

Ruschianthemum Friedrich. Aizoaceae (V). 2 SW Afr.

Ruschianthus L. Bolus. Aizoaceae (V). 1 S Namibia

Ruscus L. Asparagaceae (Ruscaceae, Liliaceae s.l.). 6 Macaronesia & W Eur. (3), Medit. to Iran. R: NRBGE 28(1968)237. Dioec. (some herm. clones in cult.) evergreen shrubs with shoots from rhizomes behaving rather like perennial infls. of asparagus, the term. leaf-like 'phylloclade' a flattened apex, the laterals similar in axils of scale lvs; each 'phylloclade' bearing a scale-leaf with axillary fl. on midrib. *R. aculeatus* L. (butcher's broom (form. used to decorate meat), pettigree, pettigrue, range of genus) – form. young shoots ed. like asparagus (Greeks & Romans), dead shoots sold with wax fls by gypsies, cult. orn. as is *R. hypoglossum* L. (S Eur. to Turkey) – the 'laurel' of Caesar

rush *Juncus* spp. esp. *J. effusus*; **beak r.** *Rhynchospora* spp.; **chair-maker's r.** *Scirpus americanus*; **corkscrew r.** *J. effusus* 'Spiralis'; **flowering r.** *Butomus umbellatus*; **r. grass** *Sporobolus* spp.; **scouring r.** *Equisetum* spp.; **spike r.** *Eleocharis* spp.; **toad r.** *J. bufonius*; **wood r.** *Luzula* spp.

rusha grass *Cymbopogon martinii*

Ruspolia Lindau. Acanthaceae. 4 trop. Afr. Cult. orn. shrubs; moth-poll. but hybrids formed with bird-poll. & butterfly-poll. spp. of *Ruttya*

Russelia Jacq. Scrophulariaceae. 52 Cuba & Mexico to Colombia. R: FMBot. 29(1957)231. Stems sometimes pendent & bearing scale lvs as in *R. equisetiformis* Schldl. & Cham. (*R. juncea*, firecracker plant, coral pl., fountain pl., Mex., widely natur.) – cult. orn. with red fls, shoots s.t. bearing broad lvs

Russian olive *Elaeagnus angustifolia*; **R. thistle** *Salsola australis*; **R. vine** *Fallopia baldschuanica*; **R. wild rye** *Elymus farctus*

Russowia Winkler. Compositae (Card.-Cent.) 1 C As.

Rustia Klotzsch. Rubiaceae (I 7). 15 trop. Am.

rusty-back fern *Asplenium officinarum*

Ruta L. Rutaceae. 7 Macaronesia, Medit. (Eur. 5) to SW As. Rue. Alks. Cult. orn. & (esp. form.) for flavouring & form. medic. esp. *R. chalepensis* L. (Medit.) – rue of Bible, but usu. *R. graveolens* L. (S Eur.) – rue of gardens ('herb of grace'), strong-smelling (ethereal oils in lvs) with infls where term. fl. 5-merous & laterals 4-merous, stamens bend over stigma & dehisce, fall back while stigma matures & return effecting self-poll., cult. since time of Pliny for strong flavour (form. much used in sausages) & medic. but causing dermatitis in some; *R. montana* (L.) L. (Medit.) – in US causes blisters & ulcers on skin when absorbed & exposed to sun (furcocoumarins) when thiamine & pyrimidine bond covalently causing cross-linking in DNA strands leading to antimitotic effect

rutabaga *Brassica napus* Napobrassica Group

Rutaceae Juss. Dicots – Rosidae – Sapindales. 156/1800 cosmop. esp. trop. Aromatic trees & shrubs, s.t. lianoid, rarely herbs, s.t. thorny & often with bitter terpenoids, alks & lysigenous secretory cavities with aromatic ethereal oils scattered through parenchyma & pericarp. Lvs usu. spirally arr., pinnate or 3-foliolate, rarely simple or pinnatisect, with pellucid gland-dots; stipules 0. Fls usu. bisexual & reg., hypogynous to (rarely) epigynous (*Platyspermation*), in cymes or racemes, rarely solit. or epiphyllous (*Erythrochiton*); K (2–)5, s.t. basally connate, often imbricate, C alt. with & 2 × K, s.t. basally connate, imbricate or valvate, rarely 0, A usu. diplostemonous (1 whorl s.t. staminodes, or those alt. C absent, or 3–4 × C to 60, rarely only 2 or 3 fertile), filaments ± basally connate (s.t. completely (cf. Meliaceae), anthers with longit. slits, nectary-disk annular, s.t. 1-sided or a gynophore, around G (2–)4 or 5(–∞), ± united & pluriloc. or separate except for coherent style, rarely 1-loc. or G 1, each loc. with (1)2(–∞) usu. superposed, anatropous or hemitropous, usu. bitegmic (unitegmic in *Glycosmis*) ovules. Fr. schizocarps, berries, drupes etc.; seed with large straight or curved ± endosperm (oily). x = 7–11+, anciently (?) 9

Classification & principal genera:

I. **Rutoideae** (G (1–)4 or 5(+), often united only by styles; fr. with loculicidal dehiscence, rarely fleshy drupes): 5 tribes
   1. **Zanthoxyleae** (woody pls; fls usu. reg., often unisexual; loc. usu. with 2 ovules): *Choisya, Fagara, Melicope, Zanthoxylum*
   2. **Ruteae** (usu. ± herbs, fls bisexual s.t. ± irreg.; loc. usu. with more than 2 ovules): *Dictamnus, Ruta*
   3. **Boronieae** (shrubs, usu. with reg. bisexual fls., Aus.): *Boronia, Eriostemon, Phebalium*
   4. **Diosmeae** (usu. shrubby; lvs simple; endosperm 0; esp. S Afr. with 10/150 endemics): *Agathosma, Diosma*
   5. **Cusparieae** (trees & shrubs; fls reg. or not.; endosperm little or 0, embryo curved): *Cusparia, Esenbeckia, Galipea*

II. **Dictyolomatoideae** (G united only at base, each carpel with several ovules separating in fr. when 3–4-seeded) – Dictyolomateae (trees with 2-pinnate lvs): *Dictyoloma*

III. **Spathelioideae** (G (3), each loc. with 2 pendulous ovules; drupe winged) – Spathelieae (pachycaul hapaxanthic trees): *Spathelia*

IV. **Toddalioideae** (G (5–2) or 1, each with 2–1 ovules; fr. a drupe or dry & winged) – Toddalieae: *Ptelea, Skimmia*

V. **Aurantioideae** (fr. a berry with pulp derived from hairs within ovary (hesperidium); endosperm 0) – Aurantieae: *Aegle, Citrus, Glycosmis, Limonia*

Close to Ptaeroxylaceae, Meliaceae & Simaroubaceae

Imp. fr. trees (esp. *Citrus*, but also *Aegle, Clausena, Feroniella, Glycosmis, Limonia*) & flavourings (*Agathosma, Casimiroa, Esenbeckia, Galipea* (angostura), *Murraya, Ruta, Zanthoxylum*); some timbers esp. *Amyris, Balfouriodendron, Chloroxylon, Euxylophora* & *Flindersia* (s.t. these 2 put in sep. Flindersiaceae), *Euxylophora* & *Limonia* & medic. incl. *Agathosma, Pilocarpus* (jaborandi) & *Raputia* with some cult. orn. esp. *Boronia, Choisya, Citrus, Dictamnus, Poncirus, Ptelea, Skimmia, Tetradium* ('*Evodia*')

Rutaneblina Steyerm. & Luteyn. Rutaceae. 1 Venezuela

Ruthalicia C. Jeffrey. Cucurbitaceae. 2 trop. W Afr.

Ruthea Bolle = *Rutheopsis*

Rutheopsis Hansen & Kunkel (*Ruthea, Gliopsis*). Umbelliferae (III 8). 1 Canary Is.

Ruthiella Steenis (*Phyllocharis*). Campanulaceae. 5 New Guinea. Like *Lobelia* but fls epiphyllous

Rutidea DC. Rubiaceae (II 2). 22 trop. Afr. & Madag. R: KB 33(1978)243

Rutidosis DC. Compositae (Gnap.-Ang.). 9 Aus. R: OB 104(1991)122

rutin material used in haemorrhage treatment, derived from several pls incl. *Eucalyptus macrorhyncha* & *Sophora japonica*

Ruttya Harvey. Acanthaceae. 3 trop. & S Afr. to Yemen, 3 Madag. Cult. orn. shrubs esp. *R. fruticosa* Lindau (E Afr.) – fls yellow to scarlet with darker markings

Ruyschia Jacq. Marcgraviaceae. 6 trop. Am.

Ryania Vahl. Flacourtiaceae (9). 8 trop. Am. R: FN 22(1980)258. Toxic: *R. angustifolia* (Turcz.) Monach. used to poison alligators; ryanodine also an efficient insecticide

rye *Secale cereale*; **giant r.** *Triticum turgidum* Polonicum Group; **Russia wild r.** *Elymus farctus*

ryegrass *Lolium* spp.; **Italian r.** *L. multiflorum*; **perennial r.** *L. perenne*

Rylstonea R. Baker = *Homoranthus*

Ryncholeucaena Britton & Rose = *Leucaena*

Ryparosa Blume. Flacourtiaceae (4). 18 Mal.

Ryssopterys Blume ex A. Juss. (*Rhyssopteris*). Malpighiaceae. 6 Mal. to trop. Aus. & New Caled.

Ryticaryum Becc. = *Rhyticaryum*

Rytidocarpus Cosson. Cruciferae. 1 Morocco

Rytidosperma Steudel = *Danthonia*

Rytidostylis Hook. & Arn. Cucurbitaceae. 5 trop. Am.

Rytigynia Blume (~ *Vangueria*). Rubiaceae (III 2). 60–70 trop. Afr. Some ed. fr. Uganda

Rytilix Raf. ex A. Hitchc. = *Hackelochloa*

Rzedowskia Gonz. Medr. Celastraceae. 1 Mexico

S

Saba (Pichon) Pichon. Apocynaceae. 3 trop. Afr. to Madag. R: BJBB 59(1989)189. *S. comorensis* (Bojer) Pichon (*S. florida*, trop. Afr.) – handsome liane with strong-smelling fls, source of rubber in World War II, fr. ed. Mafia Is., E Afr., 'sponges' from stem-fibres

sabadilla *Schoenocaulon officinale*

Sabal Adans. Palmae (I 1d). 16 SE US to S Am. (also Eocene of Britain). Dwarf or stout unarmed palms with bisexual fls. R: Aliso 12(1990)583. Cult. orn. with comm. imp. fibres & thatch esp. *S. causiarum* (Cook) Becc. (Hispaniola, Virgin Is., Puerto Rico) – lvs split for matting, basketry & hat-making; *S. palmetto* (Walter) Schultes & Schultes f. (palmetto, cabbage palmetto palm, SE US to Bahamas & Cuba) – palm cabbage, lvs for thatch, mats etc., stems for furniture, wharf-piles

Sabatia Adans. Gentianaceae. 17 N Am., WI. R: Rhodora 73(1971)309. Cult. orn. (rose pinks) esp. *S. angularis* (L.) Pursh (E US) – cut-fl. in Netherlands, bitter principle medic.

Sabaudia Buscal. & Muschler. Labiatae (?VIII). 2 NE & SE trop. Afr.

Sabaudiella Chiov. Convolvulaceae. 1 NE trop. Afr.

Sabazia Cass. Compositae (Helia.-Helia.). 17 C Am. mts. R: PMSUBS 4(1970)321

Sabia Colebr. Sabiaceae. 19 SE As., Mal. R: Blumea 26(1980)1

Sabiaceae Blume. Dicots – Magnoliidae – Ranunculales. 3/80 SE As., Mal., trop. Am. Trees, shrubs & lianes. Lvs simple or pinnate, spirally arr.; stipules 0. Fls small, bisexual (or pl. polygamo-dioec.) in panicles or thyrses; K (3–)5, imbricate, unequal, s.t. basally connate, C (4 or) 5 opp. or alt. with K, the 2 inner often smaller, A opp. C, all fertile (*Sabia*) or only 2 inner so, nectary-disk small, annular, around $\underline{G}$ (2(3)) with ± connate styles & 2(3) loc. each with 1 or 2 axile, hemitropous, unitegmic ovules. Fr. 1-loc. or ± dicoccous, fleshy or not, indehiscent; seeds with large oily embryo with curved radicle & 2 folded or coiled cotyledons in little or 0 endosperm. 2n = 24, 32

Genera: *Meliosma*, *Ophiocaryon*, *Sabia*

Assignment to Ranunculales 'tentative' (Cronquist) & often assoc. with Sapindales. *Ophiocaryon* & *Meliosma* sometimes separated as Meliosmaceae, *Sabia* differing in its 5 fertile stamens

Sabicea Aublet. Rubiaceae (I 8). 120 trop. Afr., Madag. (4), Am. (45). R: H.F. Wernham (1914) *Monog. genus S.* Some spp, with fls on prostrate shoots. *S. amazonensis* Wernham (Brazil) – frs used to symbolize hearts in manhood initiations in Amazon

sabicu *Lysiloma latisiliqua*

Sabina Miller = *Juniperus*

Sabinea DC = *Poitea*

sabre bean *Canavalia ensiformis*

Saccardophytum Speg. = *Benthamiella*

Saccellium Bonpl. Boraginaceae (Ehretiaceae). 3 trop. S Am.

Saccharum L. Gramineae (40a). (Incl. *Erianthus*) 35–40 trop. & warm (Eur. 3), s.s. 5 (3 wild, 2 cultigens; R: Baileya 23(1991)109). Sugar-cane the source of half of world's sugar, in last century almost all cvs grown derived from *S. officinarum* L., a complex agg. of hybrids domesticated in New Guinea ('noble' canes) where used for weaving house-walls, one selection being *S. edule* Hassk. ((lowland) pit pit) – immature infls a veg. in New Guinea, but now hybrids with other spp. or cultigens also grown esp. in WI, Hawaii etc.; hybrids formed with *Sorghum* spp. too. Crusaders took sugar from Israel to Medit. is. & Iberia; later taken by colonists to Madeira, Canary Is. & then Am. & islands of Indian Ocean where plantations have done much to destroy the orig. vegetation; cult. responsible for transmigration of peoples from Afr. to Carib., China to C Am., Polynesia to Aus. etc. In refining sugar, uncrystallizable sugars a byproduct – molasses or treacle from unrefined (muscovado); fermented juice gives betsa-betsa (Madag.) & when distilled is rum, the depth of its colour dependent on level of caramel, & aguadent (cachaça) in Brazil; the crushed fibre after extraction (bagasse) mixed with molasses is used as cattle-feed, alone is used as fuel in mills or for fibreboard & some kinds of paper; wax is yielded by lvs; at times of high oil-prices (1970s–80s), cane much grown in Brazil & used as alcohol source for automobiles; used in treatment of diarrhoea as water & salt absorption improved in presence of sugar; uba cane an old cv. now used only for fodder. Cult. orn. incl. *S. bengalense* Retz. (eker, ekra, Indomal.) & all spp. can be used as hosts for orn. parasite *Aeginetia indica*

Saccifoliaceae Maguire & Pires = Gentianaceae

Saccifolium Maguire & Pires. Gentianaceae (Saccifoliaceae). 1 Guayana Highland: *S. bandeirae* Maguire & Pires

Sacciolepis Nash. Gramineae (34b). 30 trop. & warm esp. Afr. R: JAA suppl. 1(1991)252

Saccocalyx Cosson & Durieu (~ *Satureja*). Labiatae (VIII 2). 1 NW Afr.

Saccoglossum Schltr. Orchidaceae (V 16). 2 New Guinea

Saccoglottis Walp. = *Sacoglottis*

Saccolabiopsis J.J. Sm. Orchidaceae (V 16). 13 Indomal.

Saccolabium Blume. Orchidaceae (V 16). 4 Sumatra, Java. R: KB 41(1986)833. Small to tiny epiphytes even growing on lvs

Saccolena Gleason = *Salpinga*

Saccoloma Kaulf. Dennstaedtiaceae. 12 trop. exc. Afr. (3 Am., 1 Madag.)

Saccopetalum Benn. = *Miliusa*

Saccularia Kellogg = *Gambelia*

Sachsia Griseb. Compositae (Pluch.). 4 Florida, Bahamas, Cuba

Sacleuxia Baillon. Asclepiadaceae (I). 2 trop. E Afr.

Sacoglottis C. Martius. Humiriaceae. 8 trop. Am., 1 W Afr. (*S. gabonensis* (Baillon) Urban (Liberian cherry)). R: CUSNH 35(1961)161. Noisy slash; bat-disp. in Amazon

Sacoila Raf. (~ *Spiranthes*). Orchidaceae (III 3). 10 warm Am. R: HBML 28(1982)351. Hummingbird-poll., but some apomicts

Sacosperma G. Taylor. Rubiaceae (IV 1). 2 trop. Afr.

Sadiria Mez. Myrsinaceae. 6 E Himalaya & Assam

Sadleria Kaulf. (~ *Blechnum*). Blechnaceae (Blechnoideae). 4 Hawaii. Low tree-ferns with 2-pinnate fronds, often on lava; only endemic fern genus in Hawaii. Sometimes cult. esp. *S. cyatheoides* Kaulf. – form. source of red dye; scales of this & other spp. the 'pulu' used as packing material

safflower *Carthamus tinctorius*

Saffordia Maxon = *Trachypteris*

Saffordiella Merr. = *Myrtella*

saffron *Crocus sativus*; **s. thistle** *Carthamus tinctorius*

sagapenum *Ferula* spp.

sage *Salvia officinalis*; **blue s.** *S. azurea*; **s. brush** *Seriphidium tridentatum*; **Jerusalem s.** *Phlomis fruticosa*; **pineapple (-scented) s.** *S. elegans*; **red** or **scarlet s.** *S. coccinea*, *S. splendens*; **wood s.** *Teucrium scorodonia*

Sageraea Dalz. Annonaceae. 9 India & Sri Lanka to Philippines

Sageretia Brongn. Rhamnaceae. 35 Somalia & SW As. to Taiwan, trop. & warm Am. Some ed. fr. & bonsai subjects imported to Eur. *S. thea* (Osbeck) M. Johnston (*S. theezans*, C & E India & China to Japan) – lvs used as tea in Vietnam

Sagina L. Caryophyllaceae (II 1). c. 20 N temp. (Eur. 12), trop. mts. Herbs, usu. tufted, sometimes C 0. Pearlwort esp. *S. procumbens* L. (N temp.) – bad weed of paving etc., but some cult. orn. rock-pls

Sagittaria L. Alismataceae. 20 cosmop. (Eur. 4) esp. Am. R: MNYBG 9(1955)179, AZBR 76(1972)1, 78(1972)1. Monoec. stoloniferous, often tuberiferous herbs of aquatic habitats – some cult. orn. esp. *S. sagittifolia* L. (*S. trifolia*, arrowhead, Euras.) with ed. tubers (kyor, Kashmir) usu. fed to pigs in Mal., where grown in paddy-fields; lvs of various types, submerged being ribbon-shaped, floating ones with ovate lamina, the projecting ones sagittate. Other spp. with ed. tubers incl. *S. cuneata* E. Sheldon & *S. latifolia* Willd. (wapato, N Am.)

Saglorithys Rizz. = *Justicia*

sago palm *Metroxylon* spp. (see also *Caryota*, *Cycas*, *Encephalartos* & *Eugeissona*)

Sagotia Baillon. Euphorbiaceae. 2 NE S Am.

Sagraea DC = *Clidemia*

saguaragy bark *Colubrina glandulosa*

saguaro *Carnegiea gigantea*
Sahagunia Liebm. = *Clarisia*
sailor, blue *Cichorium intybus*
sainfoin *Onobrychis viciifolia*
Saint Augustine grass *Stenotaphrum secundatum;* **S. Barnaby's thistle** *Centaurea solstitialis;* **S. Bernard's lily** *Anthericum liliago;* **S. Bruno's l.** *Paradisea liliastrum;* **S. Dabeoc's heath** *Daboecia cantabrica;* **S. Ignatius's bean** *Strychnos ignatii;* **S. John's wort** *Hypericum* spp.; **S. Lucie cherry** *Prunus mahaleb;* **S. Patrick's cabbage** *Saxifraga spathularis;* **S. Vincent arrowroot** *Maranta arundinacea*
Saintpaulia H. Wendl. Gesneriaceae. 20 trop. E Afr. with high local endemism in crystalline mts of Tanzania. R: BC 14(1978)45. Cult. orn. (African violets) with over 2000 cvs (spreading, trailing or dwarf, 'boy-type' plain, 'girl-type' with basal white spot on leaf; fls single to double, petals entire to ruffled, white or pink to violet or deep blue) derived esp. from *S. ionantha* H. Wendl. (coastal Tanzania) – now rare in wild but many cvs & followers; usu. pls of wet places in forest but *S. rupicola* B.L. Burtt on limestone outcrops in more open country in SE Kenya; some smaller spp. now popular as novelties & involved in hybridization. Enantiostyly like *Exacum* (q.v.). Journal: *African Violet Mag.*
Saintpauliopsis Staner (~ *Staurogyne*). Acanthaceae. 1 trop. Afr.: *S. lebrunii* Staner. R: BJBB 61(1991)154
Saionia Hatusima. Burmanniaceae. 1 Japan: *S. shinzatoi* Hatusima. R: JG 24(1976)2
Sairocarpus D.A. Sutton (~ *Antirrhinum*). Scrophulariaceae (Antirrh.). 13 SW N Am. R: D.A. Sutton, Rev. Antirrh. (1988)461
Sajanella Soják. Umbelliferae (III 8). 1 S Siberia
Sajania Pim. = *Sajanella*
sakaki *Cleyera ochnacea*
sake *Oryza sativa*
Sakoanala R. Viguier. Leguminosae (III 2). 2 NW & E Madagascar
sal *Shorea robusta*
salab-misri = salep
Salacca Reinw. Palmae (II 1d). c. 20 Indomal. (Mal. 17). Subsessile spiny dioec. palms (salak) esp. *S. zalacca* (Gaertner) Voss (*S. edulis*, Mal.) – poll. weevils, cult. for ed. fr., s.t. pickled. Infls root to form new pls in *S. flabellata* Furtado (Malay Pen.) & *S. wallichiana* C. Martius (rakum) – several cvs, sarcotesta ed., petioles used for walls & hen-coops in S Thailand
Salacia L. Celastraceae. c. 200 trop. Often lianes with dimorphic branching; some locally ed. fr. Some known to reduce glucose levels in (rat) blood
Salacicratea Loes. = *Salacia*
Salacighia Loes. Celastraceae. 1 W trop. Afr.
Salaciopsis Baker f. Celastraceae. 5 New Caledonia
salad, millionaire's = palm hearts
salak *Salacca* spp.
salal *Gaultheria shallon*
Salaxis Salisb. Ericaceae. 8 SW & S Cape
Salazaria Torrey = *Scutellaria*
Saldanhaea Bur. = *Cuspidaria*
Saldinia A. Rich. ex DC. Rubiaceae (IV 7). 2 Madagascar
salep Dried starchy tubers of various orchids form. used in cooking & medic., esp. from spp. of *Eulophia* (As.), *Dactylorhiza* (Iran) & *Orchis* (Eur.)
Salicaceae Mirbel. Dicots – Dilleniidae – Violales. 2/435 subcosmop. esp. N, not in Mal. nor Aus. Decid., dioec. trees & shrubs, s.t. creeping, usu. with phenolic heterosides like salicin, rarely with alks; ectomycorrhizae common. Lvs simple, spirally arr. but often forming planar sprays in maturity (*Salix*); stipules usu. present but often caducous (± absent in '*Chosenia*' = *Salix*). Fls wind- or insect-poll., in (often precocious) catkins, with bracts but 0 bracteoles; K 0 to cupulate (females of *Populus*) or 1 or 2(–5) often unequal & often united nectaries (*Salix*), C 0, A (1)2(most *Salix*)∞, s.t. ± connate, anthers with longit. slits. $\underline{G}$ (2(–4)) with distinct stigmas, parietal placentation, each placenta with (2–) ∞ anatropous, unitegmic (± bitegmic in some *Populus* spp.) testal ovules. Fr. a 2–4-valved capsule; seeds minute, wind-disp., not long-lived, with tuft of hairs derived from funicle or placenta, embryo straight in little or 0 oily endosperm. x = 11, 12, 19

Genera: *Populus*, *Salix*

Salicin also found in Flacourtiaceae (*Idesia*), which also are hosts to similar rust-fungi. The simple fls were once considered primitive but now seen as simplified & advanced

Timber, pulp, osiers, medic. & cult. orn. (aspen, poplars, sallows, willows) with many hybrids

Salicornia L. Chenopodiaceae (II 3). (Incl. *Sarcocornia*) 28 cosmop. (Eur. 12, Aus. 3 (*Sarcocornia*), sea-coasts & other salt habitats (glasswort, marsh samphire) – succ. herbs with scale-like lvs & jointed nodes; fls v. small, in 3s in axils, sunken, K 4 fleshy, C 0, A 1 or 2, in *S. europaea* L. (chicken claws) the median fls largest producing 1 large seed (0.78 ± 0.1 mg), the smaller laterals producing 1 small seed (0.24 ± 0.4 mg), the larger ones more salt-tolerant. Some eaten as spinach or pickled

Salicorniaceae J. Agardh = Chenopodiaceae

saligna gum *Eucalyptus saligna*

saligot *Trapa natans*

Salix L. Salicaceae. c. 400 cold (*S. arctica* Pallas (circumpolar) with *Papaver radicatum* most northerly vasc. pl. (83° N)) & temp. N (Eur. 64, China 257) few S (Aus. 0). Willows, osiers, sallows; v. many hybrids (59 combinations (10 involving 3 spp.) in GB). R: C. Newsholme (1992) *Willows*. Trees 30 m high to creeping shrublets of 1–2 cm, some rheophytes; usu. insect-poll., creeping *S. herbacea* L. (circumboreal) poll. by 'biting midges', catkins of Arctic spp. with temp. several degrees higher within than outside; foliage often disfigured by sawfly galls in summer; male seedlings of *S. myrsinifolia* Salisb.–*S. phylicifolia* L. complex in N Sweden consumed by voles 3 times as much as females, latitudinal variation in maleness assoc. with vole herbivory (also more males on vole-less is.); some timbers, coppice for comm. fuel, pliable branches (withies or withes) esp. of *S. fragilis* & *S. viminalis* used in basketry (earliest weave for chairs & still the best for hot-air balloon baskets), where coppice shoots from trees grown in osier-beds often used (in GB usu. *S. viminalis*, *S. triandra* (*S. amygdalina*) & *S. purpurea* with locally favoured clones; bark the origin of aspirin (now synthesized, acetylsalicylic since 1853), salicin having similar effect as quinine, German: *a* (*cetylirte*, acetylated) + *spir* (*säure*, salicylic acid) *in*; roots yield blue-red dyes; many cult. orn. readily prop. by cuttings (root in water); long considered a symbol of chastity ('spontaneous generation' & (males) no seeds. *S. alba* L. (white willow, Euras., Medit.) – twigs for basketry, timber form. for Dutch clogs, lvs form. used as tea, cvs incl. 'Coerulea' (cricket-bat w.) – timber for c.bs (handles of cane), 'Chermesina' ('Britzensis') & 'Vitellina' (golden w.) – grown for coloured bark (orange & yellow resp.) consp. in winter; *S. babylonica* L. (? China) – inspiration for willow-pattern plates, prob. parent of common weeping willows (*S.* × *sepulcralis*) & believed to be the tree under which children of Israel mourned & wept (prob. *Populus euphratica*) though 'weeping' to describe trees with trailing growth may stem from this, while all willows are associated with sadness; *S. caprea* L. (pussy or goat w., Eur.) – charcoal, poss. source of polyphenols, the 'palm' of Palm Sunday in rural GB, the week before known as w. week in Russia, 'Pendula' (Kilmarnock w.) – crooked drooping branches; *S. chilensis* Molina (*S. humboldtiana*, Humboldt w., Mex. & trop. Am.) – fastigiate forms like Lombardy poplars; *S. fragilis* L. (? *S. excelsa* S. Gmelin, crack or brittle w., Euras.) – basketry & charcoal; *S. humilis* Marshall (prairie w., E N Am.); *S. lasiolepis* Benth. (W N Am.) – shrubby growth-form in Arizona maintained by galling sawflies; *S. matsudana* Koidz. (N As.) – 'Tortuosa' (corkscrew w.) – branchlets spirally twisted; *S. mucronata* Thunb. (*S. capensis*, Cape w., S Afr.) – locally medic.; *S. nigra* Marshall (black w., N Am.) – catkins appear with lvs, pulp, timber for artificial limbs etc.; *S. purpurea* L. (basket w., Eur. & Medit. to Japan) – osiers; *S. repens* L. (creeping w., Euras.); *S.* × *sepulcralis* Simonkai (*S. alba* × *S. babylonica*, weeping w.) – hardier than *S. babylonica*, esp. 'Chrysocoma' (*S. alba* 'Tristis', *S.* 'Chrysocoma'), largest weeping tree in cult.; *S. tetrasperma* Roxb. (? subtrop. As.) – cult. Mal. where all male; *S. triandra* L. (Eur.) – osier; *S. viminalis* L. (basket w., Euras.) – sp. most used for basketry

sallow *Salix* spp. esp. *S. caprea* etc.

Salmalia Schott & Endl. = *Bombax*

Salmea DC. Compositae (Helia.- Zinn.). 10 trop. Am. R: SB 16(1991)462

Salmeopsis Benth. = *Salmea*

salmon gum *Eucalyptus salmonophloia*

salmwood *Cordia alliodora*

Salomonia Lour. Polygalaceae. 8 Indomal. to Aus. Some parasites

Salpianthus Bonpl. Nyctaginaceae (I). 1 Mex. & C Am.: *S. arenarius* Bonpl.

Salpichlaena Hook. Blechnaceae (Blechnoideae). 1–2 trop. Am. *S. volubilis* (Kaulf.) Hook. – rachis scandent, twining

Salpichroa Miers. Solanaceae (1). 17 SW US to S Am. *S. origanifolia* (Lam.) Baillon (*S. rhomboidea*, Arg.) – ed. fr. of poor flavour but cult. S US as ground-cover & bee-forage

Salpiglossis Ruíz & Pavón. Solanaceae (2). 2 S Andes. R: HBML 27(1980)4. *S. sinuata* Ruíz & Pavón – cult. orn. with yellow or dark purple to scarlet or almost blue fls with darker markings, many cvs

Salpinctes Woodson. Apocynaceae. 2 Venezuela

Salpinctium T. Edwards = *Asystasia* (but see SAJB 55(1989)6)

Salpinga C. Martius ex DC. Melastomataceae. 8 trop. S Am.

Salpingostylis Small = *Calydorea*

Salpistele Dressler (V 13). Orchidaceae. 6 C Am. to Ecuador. R: MSB 39(1991)123

Salpixantha Hook. Acanthaceae. 1 Jamaica

salsify *Tragopogon porrifolius*; **Spanish s.** *Scorzonera hispanica*

salsilla *Bomarea edulis*

Salsola L. Chenopodiaceae (III 3). 150 cosmop. sea-coasts or other saline habitats (Eur. 26). Alks; phytoecdysones in some Egyptian spp.; ash form. used in soap- & glass-making (barilla); some locally eaten or cattle-fodder. *S. kali* L. (blackbush, N temp., Aus.) can grow away from saline habitats, a tumbleweed, since c. 1900 an agricultural pest in N Am., where introd. 1871 in flaxseed (ssp. *ruthenica* (Iljin) Sóo; var. *targus*, *S. pestifera*, *S. australis*, Russian thistle), poll. by bees & wasps; *S. komarovii* Iljin (E As.) – local veg. (C Japan), in wet conditions long-winged 'frs' with green seeds & short-winged with yellow but only latter in dry (or when abscisic acid applied), yellow seeds more cold-tolerant & with longer dormancy than green; *S. nitraria* Pallas (Russia, SW & C As.) – used as boron indicator in Russia

salt fern *Asplenium acrobryum*

saltbush *Atriplex* spp. esp. *A. vesicaria* (Aus.)

Saltera Bullock. Penaeaceae. 1 S Afr.

Saltia R. Br. ex Moq. Amaranthaceae (I 2). 1 S Arabia

saltwort *Batis maritima*, *Glaux maritima*

Salvadora L. Salvadoraceae. 5 warm Afr. to trop. As. *S. persica* L. (trop. Afr. to As.) – 'toothbrush tree', the branches with bundles of vessels in thick-walled fibres, separated by thin-walled parenchyma with included phloem which two are destroyed when twigs beaten such that fibres & vessels form 'bristles' of chewing-stick much used by Bedouin & others as toothbrushes, antiseptic reducing tooth decay, healing & hardening gums (chlorine, trimethylamine, resin, silica, sulphur & vitamin C; 'peelu extract' used in comm. toothpastes in UK); shoots ed. & camel-fodder; ash yields salt, seeds a wax used on skin & for candles, fr. ed. (sweet but peppery, the mustard seed of Bible), wood useful (poss. fuelwood source on saline soils)

Salvadoraceae Lindley. Dicots – Dilleniidae – Capparidales. 3/11 warm OW esp. dry. Small trees or shrubs, s.t. scrambling, s.t. with mustard oils or alks; interxylary phloem often present. Lvs simple, opp., often leathery; stipules small or 0, axillary thorns s.t. present. Fls small, bisexual or not (pls then dioec. or polygamous), reg., in varied infls; K (2–4(5)) with often imbricate lobes, C 4(5), imbricate, free (basally connate in *Salvadora*), A alt. with C, free (*Azima*), basally united (*Dobera*) or adnate to C base (*Salvadora*), anthers with longit. slits, disk 0 or glands alt. with A, $\underline{G}$ (2) with term. short style, 1-loc. (2- in *Azima*), each loc. with 1 or 2 basal erect anatropous bitegmic ovules. Fr. usu. drupe or 1-seeded berry; embryo with thick cordate oily cotyledons in 0 endosperm. x = 12

Genera: *Azima*, *Dobera*, *Salvadora*

Affinities unclear. Some locally useful pls

Salvadoropsis Perrier. Celastraceae. 1 Madagascar

Salvertia A. St-Hil. Vochysiaceae. 1 S Brazil, campos

Salvia L. Labiatae (VIII 2). 900 trop. (Afr. 59) to temp. (Eur. 36) esp. Am., Sino-Himal. & SW As. (Turkey 86, 50% endemics). Shrubs & herbs with A 2, each with elongate connective & single fertile anther cell brought into contact with back of poll. bees etc. by other end of versatile anther being pushed upwards by pollinator probing for nectar; protandry, position of A being taken up by style & stigma later; *S. brandegei* Munz (Calif.) – heterostylous (only zygomorphic example). Cult. orn. & medic. (Latin: *salveo* = I heal), some culinary (sage) etc. *S. hispanica* L., etc. (chia, C Mex.) – seeds form. used in a drink, oil in painting etc., though *S. reflexa* Hornem. (N to C Am.) & other spp. toxic to stock in Aus.; R. (subg. *Jungia (Calosphace)*): FRB 110(1938)1. *S. apiana* Jepson (? *S. camphorata* Hort. Huber, Calif.) – bee forage; *S. azurea* Michaux ex Lam. (blue sage, SE US) – cult. orn.; *S. coccinea* Etl. (scarlet s., trop. S Am.) – cult. orn., many cvs; *S. columbariae* Benth. (chia, SW US) – seeds form. made into a cake, soups & a drink; *S. divinorum* Epling & Játiva (Mex., cultigen) – hallucinogen used only when no other h. available (active principle ? diterpenes), also contains loliolide a potent ant-repellent; *S. elegans* Vahl (*S. rutilans*, pineapple (-scented) s., Mex., Guatemala) – cult. orn.; *S. microphylla* Kunth (*S. grahamii*, Mex.) – cult. orn. hardy shrub with red fls; *S. miltiorhiza* Bunge (China) – locally used in heart conditions; *S. offici-*

nalis L. (sage, S Eur. & Medit.) – lvs used, esp. dried, in cooking (that in US often adulterated with 50–95% *S. fruticosa* Miller (*S. cypria*, *S. triloba*, Sicily to Syria), form. medic., local drinks (chahomilia, Cyprus), flavour due to terpineol & thujone; *S. sclarea* (clary, S Eur. to C As.) – aromatic oil used in soaps & scent (esp. lavender & bergamot), eau-de-cologne, vermouth & liqueurs, fruits form. placed under eyelids as eyewashes; *S. splendens* Sellow ex Roemer & Schultes (scarlet s., S Brazil) – the common red bedding salvia of gardens, several cvs; *S. viridis* L. (*S. horminum*, bluebeard, S Eur.) – cult. orn.

Salviastrum Scheele (*non* Heister ex Fabr.) = *Salvia*

Salvinia Séguier. Salviniaceae. 10 trop. & warm temp. (Eur. 1 with 2 escapes; not native SE As. to Aus.). R: Hedwigia 74(1935)257. Water ferns with water-repellent hairs on adaxial surface – those of *S. auriculata* Aublet (trop. Am.) with the tips of 4 arms united like a lantern. Some cult. orn. incl. *S. natans* (L.) All. (Euras., N Afr.) – in GB in earlier interglacials; *S. molesta* D. Mitch. ('*S. auriculata*', Kariba weed) – sterile triploid, poss. of hybrid origin in SE Brazil, a pestilential weed of waterways in Afr. (doubling time of 4.5 days in Lake Naivasha in 1980s), where prob. introd. as cult. orn. (now also Sri Lanka, Aus. etc.) but reduced by weevil introd. from Brazil c. 1930. High in tannin, lignin etc. so restricted use as fodder

Salviniaceae Lestib. Filicopsida. 1/10 trop. & warm. Free-floating water ferns. Stems with rudimentary protostele & lvs in whorls of 3, of which 2 floating & 1 submerged & finely divided so as to resemble roots. Sporocarp on submerged lvs representing a single sorus with indusium forming sporocarp wall, the first-formed with up to 25 megasporangia, the later with microsporangia in large numbers; all but 1 megaspore abort & this is surrounded by thick perispore resembling integument of gymnosperm seed; 64(32) microspores lie on surface of a frothy massula, the male prothalli projecting all round after germination & each 1 with 2 antheridia giving 8 sperms. Megaspore remains in sporangium, the prothallus protruding after dispersal of sporocarp & bearing several archegonia

Genus: *Salvinia*

Close to Azollaceae, the combined group known from Cretaceous (5 genera incl. *Salvinia* & *Azolla*)

Salweenia Baker f. Leguminosae (III 2). 1 SE Tibet

Salzmannia DC. Rubiaceae (III 4). 1 E Brazil

sama *Panicum sumatrense*

Samadera Gaertner = *Quassia*

Samaipaticereus Cárdenas. Cactaceae (III 4). 1 Bolivia: *S. corroanus* Cárdenas

saman *Albizia saman*

Samanea (Benth.) Merr. = *Albizia*

Sambirania Tard. = *Lindsaea*

Sambucaceae Batsch ex Borkh. See Caprifoliaceae

Sambucus L. Caprifoliaceae (Sambucaceae, Adoxaceae). 9 temp. (Eur. 3) & subtrop. R: DB 223(1994). Elder. Shrubs & small trees with pithy stems & alks; extrafl. nectaries. Cult. orn., some ed. fr. (some toxic), some locally med. *S. caerulea* Raf. (blue e., W N Am.) – cult. orn., ed. fr.; *S. canadensis* L. (American e., E N Am.) – cult. for fr. (the elderberry of US) used in jellies, pies, sauces & wines, several cvs, also cult. orn. incl. 'Maxima' with fl. heads to 40 cm diam.; *S. ebulus* L. (danewort, dwarf e., Euras., Medit. with distinct ssp. on trop. Afr. mts) – ± herbaceous, drupes source of a blue dye used in colouring leather etc., trop. Afr. ssp. a subpachycaul shrubby pl. with marcescent lvs; *S. nigra* L. (elder, e.berry, Euras., Medit.) – foetid pithy shrub, weed associated with superstition, drupes used for making wine, form. (Romans) used as hair-dye (Pan dyed himself with it), fls form. medic. & ed. as elderflower pancakes, pith used for holding specimens when sectioning botanical material, fly-control since C2 in W Eur.; *S. pubens* Michaux (stinking e., E N Am.) – red fr. inedible; *S. racemosa* L. (Euras.) – cult. orn. with several cvs incl. 'Tenuifolia', a dwarf suckering shrub with lvs like a Jap. maple

Sameraria Desv. (~ *Isatis*). Cruciferae. 11 SW to C As. & Afghanistan

Samolus L. Primulaceae. 15 cosmop. esp. salt-marshes (Eur. 1: *S. valerandi* L.)

Sampantaea Airy Shaw. Euphorbiaceae. 1 SE As.

samphire *Crithmum maritimum*; **golden s.** *Limbarda crithmoides*; **marsh s.** *Salicornia* spp.

Samuela Trel. Agavaceae. 2 SW N Am. *S. carnerosana* Trel. – fls & fr. ed., lvs source of palma fibre used for twine, brushes etc.

Samuelssonia Urban & Ekman. Acanthaceae. 1 Hispaniola

Samyda Jacq. Flacourtiaceae (9). 9 C Am., WI. R: FN 22(1980)225

Samydaceae Vent. = Flacourtiaceae

San Domingo boxwood *Phyllostylon brasiliensis*

Sanango Bunting & Duke. Gesneriaceae (II). 1 Ecuador: *S. racemosum* (Ruíz & Pávon) Barringer – tree to 12 m

Sanblasia L. Andersson. Marantaceae. 1 Panamá

Sanchezia Ruíz & Pavón. Acanthaceae. 20 trop. Am. R: JWAS 16(1926)484. Some cult. orn. *S. thinophyllum* Leonard – decoction used to bathe heads of girls dehaired (by pulling) during initiation ceremonies in NW Amaz.

sanchi *Panax pseudoginseng*

Sanctambrosia Skottsb. Caryophyllaceae (I 1). 1 S Ambrosio Is. (Chile): *S. manicata* (Skottsb.) Skottsb. – small tree or shrub, K 4, C 4, A 4 + 4, G (3)

sand box tree *Hura crepitans*; **s. cherry** *Prunus pumila*; **s. leek** *Allium scorodoprasum*; **s. myrtle** *Leiophyllum buxifolium*; **s. spurrey** *Spergularia* spp.; **s. wort** *Arenaria* spp.; **Teesdale s.w.** *Minuartia stricta*

sandalwood or **Indian s.** *Santalum album*; **Australian s.** *Santalum* spp.; **s. fan** *Vetiveria zizanioides*; **red s.** *Adenanthera pavonina*, *Pterocarpus santalinus*

sandarac *Tetraclinis articulata*; **Australian s.** *Callitris endlicheri*

Sandbergia E. Greene = *Halimolobos*

Sandemania Gleason. Melastomataceae. 1 Amaz. savannas of Brazil, Venezuela, Peru: *S. hoehnei* (Cogn.) Wurdack. R: Brittonia 39(1987)441

Sanderella Kuntze. Orchidaceae (V 10). 2 Brazil

Sandersonia Hook. Colchicaceae (Liliaceae s.l.). 1 Natal: *S. aurantiaca* Hook. (Christmas bells) – cult. orn.

sanderswood *Pterocarpus santalinus*

Sandoricum Cav. Meliaceae (I 7). 5 Mal. R: Blumea 31(1985)146. *S. koetjape* (Burm.f.) Merr. (santol, sentul, Mal.) – cvs with ed. fr. prob. orig. W Mal. now widely planted in Indian Ocean, Florida etc., some timber (katon), antifeedant limonoids in seeds, cytotoxic triterpenes in stems

Sandwithia Lanj. (~ *Sagotia*). Euphorbiaceae. 2 NE S Am.

Sandwithiodoxa Aubrév. & Pellegrin = *Pouteria*

sang *Panax ginseng*

Sanguinaria L. Papaveraceae (I). 1 E N Am.: *S. canadensis* L. (bloodroot, red puccoon) – woodland herb with alks & rhizome producing 1 leaf & 1 term. fl., medic. (emetic etc.); K 2 ephemeral, C 8–16, A ∞, seeds arillate, ant-disp.; cult. orn. esp. 'Multiplex' with A & G replaced by C

Sanguisorba L. Rosaceae. c. 10 (incl. *Dendriopoterium*, *Poterium*) N temp. (Eur. 7). Shrubs (monoec. in Canary Is.) to herbs. Some supposed to have styptic qualities, some cult. orn. & ed. lvs esp. *S. officinalis* L. (burnet, Euras., natur. N Am.); *S. minor* Scop. ssp. *muricata* (Gremli) Briq. (S Eur.) – form. fodder-pl. in GB

Sanhilaria Baillon = *Paragonia*

sanicle, wood *Sanicula europaea*

Sanicula L. Umbelliferae (II 1). 39 subcosmop. (Eur. 2; not Papuasia, Australasia). R: UCPB 25(1951)1. Herbs with fr. covered with hooked bristles or tubercles (animal-disp.) & alks. Some form. medic. esp. *S. europaea* L. (wood sanicle, Euras.)

Saniculaceae (Drude) Löve & D. Löve = Umbelliferae (II)

Saniculophyllum C.Y. Wu & Ku. Saxifragaceae (Saniculophyllaceae). 1 China: *S. guangxiense* C.Y. Wu & Ku. R: APS 30(1992)194

Saniella Hilliard & B.L. Burtt. Amaryllidaceae (Liliaceae s.l.). 1 Lesotho, Cape: *S. verna* Hilliard & B.L. Burtt

Sansevieria Thunb. Dracaenaceae (Agavaceae s.l.). 100+ dry trop. & S Afr., Madag., S As., Arabia. Perennial herbs with short rhizomes & tough thick lvs, often marked; extrafl. nectaries; cult. orn. pot-pls & sources of fibre ((African) bowstring hemp) used for ropes, sails & paper. *S. hyacinthoides* (L.) Druce (*S. guineensis*, *S. thyrsiflora*, S Afr.) – imp. fibre; *S. trifasciata* Prain (mother-in-law's tongue, Nigeria) – cult. pot-pl. esp. 'Laurentii' with golden yellow marginal stripes on lvs; *S. zeylanica* (L.) Willd. (Sri Lanka) – fibre

sansho see *Zanthoxylum*

sant pods *Acacia nilotica*

Santa Maria *Calophyllum brasiliense*

Santalaceae R. Br. Dicots – Rosidae – Santalales. 34/540 subcosmop. esp. trop. & warm dry. Hemiparasitic trees, shrubs & herbs usu. on roots (*Dendrotrophe* on branches), sometimes thorny or xeromorphic. Lvs simple, entire, s.t. scale-like, usu. opp.; stipules 0. Fls small, reg., bisexual or not, hypogynous to epigynous, often greenish, in various types of infl. (often small dichasium in axil of each bract); P (?C) (3)4 or 5(–8) free or valvate lobes of a fleshy tube, A opp. & often adnate to base of C, anthers with longit. slits or 1 apical pore, disk often present on ovary etc., $\underline{G}$–$\overline{G}$ ((2)3(–5)) with term. style, 1-loc. or with basal partitions, the free-central placenta with 1–4 pendulous, anatropous (to hemitropous), unitegmic (st. integument not differentiated) ovule. Fr. a nut or drupe; seed 1 without testa, embryo straight, embedded in copious fleshy, oily or starchy (*Thesium*) endosperm. x = 5–7, 12, 13+

Classification & principal genera:

I. **Santaleae** ($\overline{G}$; receptacle saucer- or cup-shaped, lined with nectary-disk): c. 27 genera incl. *Osyris, Phacellaria, Santalum*

II. **Thesieae** ($\overline{G}$; receptacle ± tubular, disk 0): *Arjona, Thesium* etc.

III. **Anthoboleae** ($\underline{G}$ to $\overline{G}$; ovules not clearly differentiated from placenta; pedicel becoming swollen & fleshy as fr. develops): *Anthobolus, Exocarpos* etc.

Poss. Olacaceae should be incl.

Timber, oil & ed. fr. (*Santalum, Exocarpos, Leptomeria, Osyris*), ed. tubers (*Arjona*), tanning material (*Iodina*)

Santaloides Schellenb. = *Rourea*

Santalum L. Santalaceae. (Incl. *Eucarya*) c. 25 Indomal. to Aus. (6, 5 endemic) & Hawaii (*S. freycinetianum* Gaudich. – basis of Hawaiian sandalwood industry from 1791, peaking 1810s, exhausted 1840), Juan Fernandez (*S. fernandezianum* F. Philippi, rare through overexploitation for timber by 1740, extinct by 1916). Fragrant timber, esp. *S. album* L. ((Indian) sandalwood, India, widely cult.), used for making chests, the distilled oil used in scent-making & medic., ground wood used in cosmetics & one of the pigments used in making caste-marks, the timber burnt at Buddhist funerals; *S. spicatum* (R. Br.) A. DC (*Eucarya s.*, S & E Aus.), now much reduced through exploitation, & other spp. = Aus. s., distilled for incense & medic.; *S. acuminatum* (R. Br.) A. DC (*E. acuminata*, Aus.) has ed. fr. & seeds (dong or quandong nuts), used in C19 jewellery

Santapaua Balakr. & Subram. = *Hygrophila*

Santiria Blume. Burseraceae. 24 OW trop. Some locally used timbers & ed. fr., which also yields oil

Santiriopsis Engl. = *Santiria*

Santisukia Brummitt (*Barnettia* Santisuk). Bignoniaceae. 2 Thailand. R: NHBSS 33(1986)82

santol *Sandoricum koetjape*

Santolina L. Compositae (Anth.-Art.). 18 Medit. (Eur. 5). Holy flax, lavender cotton esp. *S. chamaecyparissus* L. (Medit.) – cult. orn. silvery grey shrub, app. insecticidal, form. used as vermifuge, in scent & brushed over clothes by Victorians; *S. rosmarinifolia* L. (S Eur.) – sold in Algarve markets for flavouring

Santomasia N. Robson (~ *Hypericum*). Guttiferae (I). 1 Guatemala & Mexico

santonica *Artemisia cina*

Santosia R. King & H. Robinson. Compositae (Eup.-Crit.). 1 E Brazil. Liane

Sanvitalia Lam. Compositae (Helia.-Zinn.). 7 C Am., Mex. R: Brittonia 16(1964)417. *S. procumbens* Lam. – cult. orn. annual with dark purple disk-florets & yellow to orange ray-florets

Sanwa millet *Echinochloa utilis*

sapele *Entandrophragma* spp.

sapgreen *Rhamnus cathartica*

Saphesia N.E. Br. Aizoaceae (V). 1 SW Cape

Sapindaceae Juss. Dicots – Rosidae – Sapindales. 131/1450 trop. & warm, few temp. Trees, shrubs, lianes & herbaceous climbers, these often with tendrils in place of infls & with unusual vasc. structure; toxic saponins usu. present. Lvs (bi-)pinnate, 3-foliolate or simple, usu. spirally arr.; petiolules often swollen basally, stipules 0 or small (climbers). Fls small, usu. reg. & unisexual in cymes or thyrses, rarely solit.-axillary; K 4 or 5, s.t. basally connate, usu. imbricate, C (3)4 or 5(+) or 0, often with basal scale-like appendage concealing a nectary, annular or often 1-sided disk around A or bearing A, rarely (*Dodonaea*) intrastaminal, A (4–)8(–10+), filaments often hairy, anthers with longit. slits, pollen-exine v. variable, $\underline{G}$ ((2)3(–6)), usu. pluriloc. with distinct or 1 style(s) & 1(2) ascending (pendulous) or several spreading, bitegmic, anatropous to hemitropous or campylotropous ovules, often without a clear funicle but attached to placental protuberance. Fr. fleshy or dry, dehiscent or not; seeds with arils or sarcotestas, embryo oily & starchy, curved, often with plicate or twisted cotyledons, endosperm 0. x = 10–16

Classification & principal genera:

I. **Sapindoideae** (ovule 1 per loc., erect): (lvs usu. paripinnate, disk annular) *Alectryon, Arytera, Blighia, Cupania, Cupaniopsis, Dimocarpus, Lepisanthes, Litchi,*

Matayba, Nephelium, Pappea; (lvs usu. imparipinnate, climbers with tendrils): *Paullinia, Serjania*

II. **Dodonaeoideae** (ovules usu. 2 or several per loc., erect or pendulous if 2, or spreading): *Dodonaea, Harpullia, Koelreuteria*

Over half of genera mono-(65) or oligo-typic; Ptaeroxylaceae here kept separate though incl. by Cronquist. It is likely that both Aceraceae & Hippocastanaceae should be included here. Many seem to have male phase then female & finally male ((duo)dichogamy)

Many ed. fr. esp. *Blighia* (akee), *Cubilia, Dimocarpus* (longan), *Litchi, Melicoccus, Pancovia, Nephelium* (rambutan), *Pappea;* timber (*Cupania, Harpullia, Hypelate, Matayba, Melicoccus, Pappea, Schleichera*), oilseeds (*Dilodendron, Pappea, Paranephelium* (medic.), *Schleichera*), beads (*Cardiospermum, Sapindus*), soap subs. (*Sapindus*), fish-poisons (*Jagera*), stimulating drinks (*Paullinia*) & cult. orn. esp. *Koelreuteria* & *Xanthoceras*

Sapindopsis How & Ho = *Lepisanthes*

Sapindus L. Sapindaceae. c. 13 trop. & warm. Berries rich in saponins & used as soap subs. (*sapindus* = Indian soap), soap nuts, soapberry, esp. *S. marginatus* Willd. (S N Am.) – split timber used for cotton baskets, *S. mukorossi* Gaertner (India to C Japan), *S. rarak* DC (SE As. & Java) with antifungal & molluscicidal saponins, *S. saponaria* L. (trop. Am.) – fr. also used as beads, *S. trifoliatus* L. (OW trop.), all cult. orn.

Sapium Jacq. Euphorbiaceae. 100 trop. & warm (Afr. 13) to Patagonia. Some latex good as rubber & as bird-lime esp. *S. aucuparium* Jacq. (trop. Am.) – Bolivian rubber, & *S. jenmanii* Hemsley (NE S Am.) – Esmeralda r.; some timber. *S. sebiferum* (L.) Roxb. (*Stillingia s.*, Chinese or vegetable tallow-tree, China, Japan, natur. US) – shade-tree with poplar habit, fatty seed-covering (? caruncle, 54.5% tallow) used for candlewax & soap. Some spp. with 'jumping beans' (see *Sebastiania*)

Sapium P. Browne = *Gymnanthes*

sapodillo (plum) *Manilkara zapota*

Saponaria L. Caryophyllaceae (III 1). 40 temp. Euras. (Eur. 10). R: DAWW 85(1910)433. Soapwort, form. used as soap subs. & still in bodyscrubs esp. *S. officinalis* L. (bouncing bet, fuller's herb, Euras., natur. N Am.) – cult. orn., form. medic. *S. ocymoides* L. (S Eur. mts) – common rock-pl.

Saposhnikovia Schischkin (~ *Ledebouriella*). Umbelliferae (III 10). 1 NE As.: *S. divaricata* (Turcz.) Schischkin

Sapotaceae Juss. Dicots – Dilleniidae – Sapotales. 53/975 trop., few temp. Trees & shrubs with well-developed latex-system & 2-armed hairs (1 arm often reduced or 0). Lvs simple, usu. spirally arr.; stipules rare. Fls usu. reg., bisexual, in cymes (large panicles in *Sarcosperma*); K (4)5(–12) ± free & imbricate or in 2 whorls of 2, 3 or 4, C (4–18) with imbricate lobes, s.t. with paired appendages, A in 1–3 whorls, epipetalous, s.t. incl. petaloid staminodes, anthers with longit. slits, G̲ (1–15(–30)), usu. hairy, with axile or axile-basal placentation (1-loc. in *Diploon*), with 1 style & 1 anatropous to hemitropous, unitegmic ovule per loc. Fr. fleshy, indehiscent; seeds often with large hollow hilum & shiny integument, embryo large with flat cotyledons in oily endosperm or thick ones without endosperm. x = 7, 9–13. R: T.D. Pennington (1991) *S.*

Classification & principal genera:

I. **Mimusopeae** (K usu 2 whorls of 3 or 4 with A same no.; 17 genera in 3 subtribes): *Baillonella, Manilkara, Mimusops, Vitellaria*

II. **Isonandreae** (K usu. 2 whorls of 2 or 3 with A 2–3 times no.; 7 genera): *Madhuca, Palaquium*

III. **Sideroxyleae** (K 5, C rotate or cyathiform with lobes often 3-fid; 6 genera): *Argania, Sideroxylon*

IV. **Chrysophylleae** (K 4 or 5, C tubular to rotate with entire lobes; 19 genera): *Capurodendron, Chrysophyllum, Micropholis, Pouteria, Synsepalum*

V. **Omphalocarpeae** (like 4. but A in groups of 2–6 opp. each C lobe); 4 genera): *Omphalocarpum*

Many ed. fr. (esp. *Chrysophyllum, Manilkara* (sapodilla, chewing-gum), *Pouteria* (sapote)), oils (*Argania, Diploknema, Madhuca, Vitellaria*), timber (*Argania, Baillonella, Manilkara, Micropholis, Sideroxylon, Tieghemella*), gutta-percha (*Palaquium, Payena*) & sweetener (*Synsepalum*)

sapote *Pouteria sapota;* **black s.** *Diospyros digyna;* **white s.** *Casimiroa edulis*

sappan wood *Caesalpinia sappan*

Sapphoa Urban. Acanthaceae. 2 Cuba

Sapranthus Seemann. Annonaceae. 7 C Am.

sapree wood *Widdringtonia nodiflora*
Sapria Griffith. Rafflesiaceae (II 1). 1–3 trop. SE As.
Saprosma Blume. Rubiaceae (IV 7). 30 Indomal.
sapu *Michelia champaca*
sapucaia nut *Lecythis zabucajo*
Sapucaya F. Knuth = *Lecythis*
Saraca L. Leguminosae (I 4). 11 Indomal., char. of particular streams. R: Blumea 15(1967)413, 27(1981)235. Young lvs limp (cf. *Amherstia, Brownea*); fls with brightly coloured bracts & K; C 0 with some A derived from C primordia. Fls used in temple offerings, esp. *S. asoca* (Roxb.) Wilde (asoka, Indomal.) – form. medic. & the tree under which Buddha was born
Saracha Ruíz & Pavón. Solanaceae (1). 3 S Am.
Saranthe (Regel & Koern.) Eichler. Marantaceae. 10 Brazil. R: Brittonia 13(1961)212
Sararanga Hemsley. Pandanaceae. 2 Philippines, Solomon Is. Trees to 20 m
Sarawakodendron Ding Hou. Celastraceae. 1 Borneo. Links the form. recognized fam. Hippocrateaceae to C.; seeds with filamentous aril
Sarcandra Gardner. Chloranthaceae. 2 E As., Indomal. Vessels only in primary xylem of stems & sec. of roots
Sarcanthemum Cass. (~ *Psiadia*). Compositae (Ast.-Ast.). 1 Rodrigues
Sarcanthidion Baillon = *Citronella*
Sarcanthopsis Garay. Orchidaceae (V 16). 7 Indomal. to New Caledonia (1)
Sarcanthus Lindley = *Cleisostoma*
Sarcaulus Radlk. Sapotaceae (IV). 5 trop. S Am.
Sarcinanthus Oersted = *Asplundia*
Sarcobatus Nees. Chenopodiaceae (III 1). 1–2 W N Am. *S. vermiculatus* (Hook.) Torrey (greasewood) – hard yellow fuelwood, poss. shellac source, fodder but toxic in excess
Sarcoca Raf. = *Phytolacca* (but see Preslia 57(1985)372)
Sarcocapnos DC. Fumariaceae (I 2). 4 W Medit. incl. Eur. R: OB 88(1986)33
Sarcocaulon (DC) Sweet. Geraniaceae. 14 S Afr. R: Bothalia 12(1979)581. Fleshy shrublets with spines representing persistent toughened petioles & with lvs in their axils; A (15)
Sarcocephalus Afzel. ex Sabine (~ *Nauclea*). Rubiaceae (I 2). 2 trop. Afr. R: Blumea 22(1975)546. *S. latifolius* (Sm.) E.A. Bruce (*S. esculentus, Nauclea latifolia*, Afr. or Guinea peach, trop. Afr.) – fr. ed. with apple taste, seeds germ. best when passed through baboons, used in treatment of diabetes in Zaire & as chewing-stick in Nigeria
Sarcochilus R. Br. Orchidaceae (V 16). 17 Aus. (16, endemic) to New Caled. Some cult. orn. diminutive epiphytes
Sarcochlamys Gaudich. Urticaceae (III). 1 Indomal.: *S. pulcherrima* Gaudich. Ramie subs.
Sarcococca Lindley. Buxaceae. 11 Afghanistan to China & Philippines. R: BJLS 92(1986)117. Monoec. shrubs, evergreen; like *Buxus* but lvs spirally arr., fr. a drupe etc. Cult. orn. esp. ground-cover; alks
sarcocolla or **gum s.** *Astracantha gummifera*
Sarcocornia A.J. Scott = *Salicornia* (but see BJLS 75(1978)366)
Sarcodes Torrey. Ericaceae (Monotropaceae). 1 W US: *S. sanguinea* Torrey (snow plant)
Sarcodraba Gilg & Muschler. Cruciferae. 4 Andes
Sarcodum Lour. (~ *Clianthus*). Leguminosae (III 6). 2 SE As. to Solomon Is.
Sarcoglottis C. Presl. Orchidaceae (III 3). 40 trop. Am. R: HBML 28(1982)352
Sarcoglyphis Garay. Orchidaceae (V 16). 11 Indomal.
Sarcolaena Thouars. Sarcolaenaceae. 11 Madagascar
Sarcolaenaceae Caruel (Chlaenaceae). Dicots – Dilleniidae – Malvales. 9/51 Madag. Trees & shrubs usu. with stellate indumentum & mucilage-cells in pith, cortex etc. Lvs simple, spirally arr., petioles often with medullary bundles; stipules usu. caducous. Fls bisexual, ± reg., usu. with involucel of bracteoles around 1 or 2 fls, these in thyrsoid infls, rarely solit., the involucel often cupulate (? pedicel tip); K 3 (–5, when outer ones smaller), C 5(6), convolute in bud, s.t. basally connate, A (5–)∞ usu. inside a ± cupulate disk (disk rarely 0), filaments s.t. basally connate in 5–10 bundles, anthers with longit. slits, $\underline{G}$ ((1–)3 or 4(5)) with term. style & (1)2–∞ erect to pendulous anatropous ovules per loc. on (basal, apical or) axile placentas. Fr. a many-seeded loculicidal capsule or indehiscent with 1 or few seeds (only 1 carpel maturing); embryo straight, endosperm starchy, copious to (rarely) 0. n = 11
Principal genera: *Leptolaena, Sarcolaena, Schizolaena*
Beautiful trees of the eastern rain forests of Madag. & form. dominating the western slopes of the high plateaux until destroyed by burning & replaced by *Heteropogon* grassland. Fossils in Tertiary of Afr.

Sarcolobus R. Br. Asclepiadaceae (III 4). 6 Indomal., W Pacific. R: Blumea 26(1980)65. Some pericarps eaten in Malay Pen., also candied; some poisons

Sarcolophium Troupin. Menispermaceae (III). 1 trop. Afr.: *S. tuberosum* (Diels) Troupin

Sarcomelicope Engl. Rutaceae. 9 New Caled., 1 E Aus. to Fiji. R: Aus. JB 30(1982)359, BMNHN 4,8(1986)183

Sarcopetalum F. Muell. Menispermaceae (V). 1 E Aus.: *S. harveyanum* F. Muell.

Sarcophagophilus Dinter = *Quaqua*

Sarcopharyngia (Stapf) Boit. = *Tabernaemontana*

Sarcophrynium Schumann. Marantaceae. 3 trop. Afr.

Sarcophytaceae Kerner = Balanophoraceae

Sarcophyte Sparrmann. Balanophoraceae (Sarcophytaceae). 1 E trop. Afr.: *S. sanguinea* Sparrmann. R: BJ 106(1986)364

Sarcophyton Garay. Orchidaceae (V 16). 2 SE As.

Sarcopilea Urban. Urticaceae (II). 1 Hispaniola: *S. domingensis* Urban – habit like *Aeonium*

Sarcopoterium Spach. Rosaceae. 1 Italy, E Medit.: *S. spinosum* (L.) Spach – often dominant in the phrygana & v. common near Jerusalem & poss. the crown of thorns of Jesus Christ

Sarcopteryx Radlk. Sapindaceae. 11 E Mal., Aus. R: Austrobaileya 2(1984)53, Blumea 36(1991)87

Sarcopygme Setch. & Christoph. Rubiaceae (IV 10). 5 Samoa

Sarcopyramis Wallich. Melastomataceae. 1 Indomal. Disp. by raindrops forcing seeds from capsule

Sarcorhachis Trel. Piperaceae. 4 trop. Am. R: Pittiera 3(1971)31

Sarcorhynchus Schltr. = *Diaphananthe*

Sarcorrhiza Bullock. Asclepiadaceae (I, Periplocaceae). 1 trop. Afr.: *S. epiphytica* Bullock

Sarcosperma Hook.f. Sapotaceae (III; Sarcospermataceae). 8 Indomal. R: T.D. Pennington, *S.* (1991)179

Sarcospermataceae H.J. Lam = Sapotaceae

Sarcostemma R. Br. Asclepiadaceae (III 1). (Incl. *Funastrum*, excl. *Oxystelma*) 30 trop. & warm (Afr. to Aus. 8, Am. 22). Leafless trailing or twining, jointed succ. shrubs, 1 sp. a favourite food of rhinoceros. Some cult. orn. esp. *S. viminale* R. Br. (trop. & S Afr., SW Arabia to Burma) – outer stem-layers ed. cooked; latex used for flykillers in Costa Rica; *S. intermedium* Decne. (India) – ? a major ingredient of 'soma' of old India

Sarcostigma Wight & Arn. Icacinaceae. 2 Indomal. *S. kleinii* Wight & Arn. (Indomal.) – seed-oil used in rheumatism

Sarcostoma Blume. Orchidaceae (V 14). 2 Malay Pen., Java, Sulawesi

Sarcotheca Blume. Oxalidaceae. 11 W Mal. Acid fr. eaten with curry. *S. celebica* Veldk. (Sulawesi) – app. evolving dioecy from distyly with short-styled pls contributing genes largely through pollen, long-styled through ovules

Sarcotoechia Radlk. Sapindaceae. 11 New Guinea (5, R: Blumea 33(1988)198), NE Aus. (R: Austrobaileya 33(1988)198)

Sarcoyucca (Trel.) Lindinger = *Yucca*

Sarcozona J. Black = *Carpobrotus*

Sarcozygium Bunge = *Zygophyllum*

Sargentia S. Watson = *Casimiroa*

Sargentodoxa Rehder & E. Wilson. Lardizabalaceae (Sargentodoxaceae). 1–2 China, Laos, Vietnam. Lvs simple; andromonoec.; cult. orn.

Sargentodoxaceae Stapf ex Hutch. = Lardizabalaceae

Sarinia Cook = *Attalea*

Saritaea Dugand. Bignoniaceae. 1 Colombia & Ecuador: *S. magnifica* (W. Bull) Dugand – cult. orn. liane with large rose to purple fls, scent coll. male euglossine bees

Sarmienta Ruíz & Pavón. Gesneriaceae. 1 S Chile: *S. scandens* (J. Brandis) Pers. – cult. orn. liane with reddish fls; only Am. G. with A2

Sarojusticia Bremek. = *Justicia*

Sarothamnus Wimmer = *Cytisus*

Sarracenella Luer = *Pleurothallis*

Sarracenia L. Sarraceniaceae. 8 E N Am. Pitcher plants of swamps; many natural hybrids. R: Preslia 64(1992)9. Pitchers in place of lvs (cf. *Nepenthes*) & absorbing nutrients from insects, in *S. flava* L. (SE US) at least, imp. for nitrogen & phosphorus but not calcium, magnesium & potassium, trapping flying insects (*S. psittacina* Michaux (SE US) sub-merged in winter & trapping creeping ones; pitcher lips with nectariferous veins inside, pitcher inner surfaces with 5 zones – just inside a zone of downward-pointing hairs, then one of cells with elongated downward-pointing processes, then smooth zone, then a long zone of sinuous hairs & a 2nd smooth zone. Cult. orn. esp. *S. purpurea* L. (huntsman's cup or horn, E N Am., natur. Ireland (introd. late C19 & planted in Roscommon 1906) & Switzerland) – fls purple, obligate commensals incl. midge larvae feeding on mixed detritus in bottom of pitchers of this & other spp., while a maggot living at the surface of the water inside eats up to half of the captured prey

Sarraceniaceae Dumort. Dicots – Dilleniidae – Theales. 3/14 W & E N Am., NE S Am. Perennial, usu. stemless, insectivorous herbs with rhizomes with rather irreg. anatomy; alks s.t. present. Lvs rep. by rosetted (or on scrambling stem in *Heliamphora*) pitcher-like traps with digestive liquid (not proved for *H.*), short petioles, a well-developed wing or ridge adaxially & flattened but rather small hood-like prolongation abaxially, the traps with nectar-glands without & within, where retrorse stiff hairs near top & smooth lower down; stipules 0. Fls large, reg., bisexual, nodding, solit. on scape (few-flowered racemes in *H.*); K (3–)5(6), imbricate, persistent & ± petaloid, C (0 – *H.*) 5 imbricate, decid., A (10–)∞, anthers basifixed (versatile in *Sarracenia*), G ((3–*H.*) 5), basally placentation axile, apically intruded-parietal, with term. style, with 5 short branches with term. stigmas in *Darlingtonia* or branches peltate with stigmas underneath in *Sarracenia*, each loc. with ∞ anatropous unitegmic or bitegmic ovules. Fr. a loculicidal capsule with ∞ small, often winged seeds with linear embryos in copious endosperm rich in oil & protein. x = 13 (*Sarracenia*), 15 (*Darlingtonia*), 21 (*Heliamphora*)

Genera: *Darlingtonia, Heliamphora, Sarracenia*

Heliamphora has many primitive features & poss. most similar to Nepenthaceae (Cronquist); it is tempting to see *H.* as a relic of the stock of Sarraceniaceae which up the W side of N Am. led to *Darlingtonia* & up the E to *Sarracenia*; the species of *H.* are v. close suggesting recent speciation, however. The wood of *H.* is like that in Theales to which the fam. is now referred (Taxon 24(1975)297)

sarsaparilla *Smilax* spp.; **bristly s.** *Aralia hispida*; **false s.** *Hardenbergia violacea*; **wild s.** *S. glauca*, (US) *A. nudicaulis*

sarson *Brassica rapa*

Sartidia de Winter. Gramineae (28). 4 trop. & S Afr., Madag.

Sartoria Boiss. & Heldr. (~ *Onobrychis*). Leguminosae (III 17). 1 S Turkey

Sartorina R. King & H. Robinson (~ *Eupatorium*). Compositae (Eup.-Flei.). 1 Mex. (?). R: MSBMBG 22(1987)289

Sartwellia A. Gray. Compositae (Hele.-Flav.). 4 S N Am. R: Sida 4(1971)265

Saruma Oliver. Aristolochiaceae (I). 1 NW to SW China: *S. henryi* Oliver – apocarpous, pollen sulcate

Sarx H. St. John = *Sicyos*

Sasa Makino & Shib. Gramineae (1a). (Incl. *Sasaella, Sasamorpha*) c. 60 temp. E As. esp. Japan. Small bamboos; some cult. orn. incl. *S. palmata* (Burb.) Camus (Japan, Sakhalin) – rapid spreader to 2.5 m tall, sometimes used in cardboard-making & *S. ramosa* (Makino) Makino & Shib. (*Arundinaria vagans, Sasaella ramosa*, Japan) – first flowered outside Japan 1981, 89 yrs after introd.

Sasaella Makino = *Sasa*

Sasamorpha Nakai = *Sasa*

sassafras *Sassafras albidum*; **Brazilian s.** *Aniba* spp.; **Chinese s.** *S. tzumu*; **Orinoco s.** *Ocotea cymbarum*

Sassafras Nees & Eberm. Lauraceae (II). 3 E As. (2), E N Am. (1). *S. albidum* (Nutt.) Nees (*S. officinale*, sassafras, E N Am.) – the first export of N Am. colonies (2 shiploads saturated the market), antiscorbutic but abortifacient & now linked to liver cancers so banned from pharmacy & cosmetics throughout EU, oil used medic. incl. killing lice & for insect-bites, & in scent-making, bark & twigs made into a tea by N Am. Indians in E Canada, light timber, cult. orn.; *S. tzumu* (Hemsley) Hemsley (Chinese s., C China) – imp. local timber

sassy bark *Erythrophleum suaveolens*

Satakentia H. Moore (~ *Clinostigma*). Palmae (V 4m). 1 Ryukyu Is.: *S. liukiuensis* (Hatusima) H. Moore – cult. orn. monoec. palm with swollen trunk-base

Satanocrater Schweinf. Acanthaceae. 4 trop. Afr.

satin oak *Alloxylon* spp.; **s. walnut** *Liquidambar styraciflua*; **s.wood** *Chloroxylon swietenia*; **African s.w.** *Zanthoxylum macrophyllum*; **Cairns s.w.** *Dysoxylum pettigrewianum*; **Jamaican** or **WI s.w.** *Z. flavum*; **Nigerian s.w.** *Distemonanthus benthamianus*

satiné *Brosimum rubescens*

Satorkis Thouars. Older name for *Coeloglossum*

satsuma *Citrus reticulata*

Sattadia Fourn. = *Tassadia*

Satureja L. Labiatae (VIII 2). 30 temp. (Eur. 12; R (N Am.): Brittonia 18(1966)244) & warm. Gynodioec. herbs cult. for condiments (see also *Micromeria*) esp. *S. hortensis* L. (summer savory, SE Eur.) – flavouring (dried lvs) like sage often in 'mixed herbs', trad. used with legumes esp. broad beans; *S. montana* L. (winter s., S Eur.) – less used, flavour poorer

Satyria Klotzsch. Ericaceae. 23 trop. Am. R: OB 92(1987)121

Satyridium Lindley. Orchidaceae (IV 3). 1 S Afr.

Satyrium Sw. Orchidaceae (IV 3). 100+ trop. OW to Arabia & S Afr., China. *S. bicallosum* Thunb. (S Afr.) – poll. sciarid flies. Some cult. terr. orchids

sau *Paraserianthes falcataria*

sauerkraut *Brassica oleracea*

Saugetia A. Hitchc. & Chase = *Enteropogon*

Saundersia Reichb.f. Orchidaceae (V 10). 2 Brazil

Saurauia Willd. Actinidiaceae (Saurauiaceae). 300 trop. Am. (32, R: FMBot. n.s. 2(1980)), trop. As. Cryptic dioecy. Some locally eaten fr. *S. purgans* B.L. Burtt (Solomon Is.) – fls make a crying noise when opening. Some rheophytes

Saurauiaceae Griseb. = Actinidiaceae

Sauroglossum Lindley. Orchidaceae (III 3). 9 S Am. R: HBML 28(1982)355

Sauromatum Schott. Araceae (VIII 6). 2 trop. OW. Lvs pedate, solit., emerging after infls, which are disgustingly foetid. *S. venosum* (Aiton) Kunth (*Arum cornutum*, incl. *S. guttatum*, voodoo lily, monarch-of-the-east, OW) – sold as dried tubers which produce infl. without water, weedy in wild; *S. nubicum* Schott (trop. Afr.) – roasted tubers eaten locally

Sauropus Blume. Euphorbiaceae. 25 Indomal. to Aus. (19, 18 endemic). Some local fr. & medic.; *S. androgynus* (L.) Merr. (India to C Mal.) – used for fencing & vegetable (soup alleged to encourage lactation in Sumatra)

Saururaceae Voigt. Dicots – Hamamelidae – Piperales. 4/6 E As., N Am. Perennial aromatic herbs with articulated stems, rhizomes & vasc. bundles in 1 or 2 concentric rings. Lvs simple, spirally arr. with stipules adnate to petioles. Fls bisexual, small, in dense bracteate spikes or racemes resembling a single fl. when basal bracts large & petaloid; P 0, A 3 or 3 + 3 or 4 + 4, sometimes adnate to G, anthers with longit. slits, G (3–5), 1-loc. with parietal placentation (or distinct above connate bases in *Saururus*; sunk in infl. axis in *Anemopsis*) with free styles & 6–10 ((1)2–4 in *Saururus*) hemitropous to orthotropous, bitegmic ovules on each placenta. Fr. an apically dehiscent capsule (head of 1-seeded carpels in S.); seeds endotegmic with minute embryo in little endosperm & copious perisperm. x = 11 (?12)

Genera: *Anemopsis, Gymnotheca, Houttuynia, Saururus*

Some cult. orn.

Saururus L. Saururaceae. 1 E N Am., 1 E As. Bog pls, locally medic. incl. *S. cernuus* L. (E N Am.) – cult. orn. with fragrant white fls

sausage tree *Kigelia africana*

Saussurea DC. Compositae (Card.-Card.). c. 300 Euras. (China 300, Eur. 9). R: S. Lipschitz, *Rod Saussurea* (1979). Some cult. orn. esp. *S. alpina* (L.) DC (Euras.) – sweetly-scented purple fls; *S. gossypiphora* D. Don (Himal.) – dwarf pachycaul with solit. hollow hairy stem in alpine belt (cf. *Lobelia*); *S. costus* (Falc.) Lipsch. (*S. lappa*, costus root, kuth, E Himal.) – medic. but esp. used in scents because of its strong lingering smell

Sautiera Decne. Acanthaceae. 1 Timor

Sauvagesia L. Ochnaceae. 39 trop. (Afr. 2 (incl. *S. erecta*), Am. 35, Mal. 2). R: BJ 113(1991)184. Buzz-poll. by euglossine bees in Am. *S. erecta* L. (pantrop. ruderal) – creole tea

Sauvagesiaceae (DC) Dumort. = Ochnaceae

Sauvallea C. Wright (~ *Commelina*). Commelinaceae (II 1). 1 Cuba

Sauvallella Rydb. = *Poitea*

sauwi *Panicum sonorum*

Savannosiphon Goldbl. & Marais. Iridaceae (IV 2). 1 S trop. Afr.

Savia Willd. Euphorbiaceae. 25 warm Am., S & E Afr. (1), Madagascar (9)

Savignya DC. Cruciferae. 2 Morocco to Afghanistan

savin *Juniperus sabina*

savonette *Lonchocarpus sericeus*

savory, summer *Satureja hortensis*; **winter s.** *S. montana*

Savoy cabbage *Brassica oleracea* Capitata Group

saw *Millettia* sp. (*Pongamia pinnata*)

sawai *Eulaliopsis binata*

saw-wort *Serratula tinctoria*

saxaul *Haloxylon persicum, H. aphyllum*

Saxegothaea Lindley. Podocarpaceae. 1 S Chile & Arg.: *S. conspicua* Lindley (Prince Albert's yew) – cult. orn. monoec. evergreen with some features transitional to Araucariaceae; timber used locally

Saxicolella Engl. Podostemaceae. 2 trop. W Afr.

Saxifraga L. Saxifragaceae. c. 440 N temp., esp. Eur. (123, R: D.A. Webb & R.J. Gornall (1989) *Saxifrages of Eur.*), Himal. & E As. (China 200) & W N Am., to arctic, few S to Thailand, Ethiopia & Andes to Tierra del Fuego. Saxifrages, rockfoil. Usu. perennial herbs, some rather pachycaul & hapaxanthic with large rosettes, few annuals; insect-poll. with $\underline{G}$ to $\overline{G}$. Cult. orn., some locally eaten salads, some alpine spp. used in Chartreuse. F. Kohlein (1984) *Saxifrages & related genera.* 15 sects (R: BJLS 95(1987)273) recog. incl. *Irregulares* (*Diptera*, IRR) – fls irreg., *Gymnopera* (*Robertsonia*, GYM) – lvs not pitted nor lime-secreting, obovate or orbicular, *Cymbalaria* (CYM) – annuals, fls yellow or white, *Saxifraga* (SAX) – usu. biennial with bulbils & white fls (incl. *Dactyloides* – cushion-forming without bulbils (mossy s.), *Ligulatae* (*Euaizoonia*, LIG) – lvs lime-secreting, offsets separating, *Porphyrion* (*Kabschia*, POR) incl. *Engleria* – offsets remaining attached. Cult. orn. spp. incl.: *S. burseriana* L. (POR, E Alps) – fls large, white; *S. callosa* Sm. (LIG, *S. lingulata*, S Eur.) – fls in large much-branched panicles; *S. cernua* L. (SAX, Arctic to N Am. & Eur.) – in GB a protected sp. with no seed or fr., reproducing by bulbils; *S. cespitosa* L. (SAX, N N temp.) – protected sp. in GB; *S. cymbalaria* L. (CYM, SE Eur., SW As., N Afr.) – self-sowing annual with yellow fls; *S. granulata* L. (SAX, meadows., Eur.) – bulbils, fls white, gynodioec. & in N England females veget. more vigorous but producing only 57% seeds of hermaphr. though seeds 1.28 times 'fitter'; *S. hypnoides* L. (SAX, NW Eur.) – fls white, one of original mossy s. spp.; *S. longifolia* Lapeyr. (LIG, Pyrenees, E Spain, Morocco) – hapaxanthic unbranched calcicole, some cvs incl. hybrids like 'Tumbling Waters' (*S. l.* × *S. cotyledon* L. (Eur. mts)) with rosettes to 30 cm diam.; *S. paniculata* Miller (LIG, *S. aizoon*, N N temp.) – many cvs of encrusted s.; *S. stolonifera* Meerb. (IRR, *S. sarmentosa*, mother-of-thousands, wandering Jew, China, Japan) – hanging baskets, *S.* × *urbium* D. Webb (GYM, *S. spathularis* Brot. (W Eur.) × *S. umbrosa* L. (Pyrenees)) – the common London pride (after London & Wise, C18 nurserymen) or none-so-pretty confused with *S. umbrosa* but differing in larger more deeply crenated lvs & larger fls etc.; besides the last the most frequently seen are the mossy s. (SAX) – many hybrids & the encrusted s. (LIG, POR) – choice rock-pls (usu. hybrids in cult.) with hydathodes secreting chalky water on to surface where chalk deposited

Saxifragaceae Juss. Dicots – Rosidae – Rosales. 35/660 subcosmop. esp. N temp. & cold. Usu. perennial herbs, s.t. rather succ., rarely suffrutescent; cortical &/or medullary bundles s.t. present. Lvs usu. spirally arr., often basal rosettes, simple to pinnate or palmate, often with hydathodes; stipules usu. 0 (exc. e.g. *Francoa*). Fls usu. reg. & bisexual, in cymes, racemes or solit., ± hypo- to epigynous; K (3–)5(–10) imbricate or valvate, often lobes on hypanthium, C (0) same as & alt. with K, imbricate or convolute, sometimes lobed or early decid. or small, A in 2 whorls with 1 whorl sometimes staminodes or 0, anthers with longit. slits, nectary-disk or annulus often around G (2–4(–7)), rarely ± free, each lobe term. with a stigma & with marginal, axile or parietal placentas etc. with several–∞ (1 in *Eremosyne*) anatropous (uni- or) bi-tegmic ovules sometimes with zig-zag micropyle. Fr. dry dehiscent usu. septicidally; seeds small with straight embryo in copious (rarely 0) oily endosperm. x = 6–15, 17

Principal genera: *Astilbe, Chrysosplenium, Heuchera, Mitella, Saxifraga* (the bulk of the spp.), over half of genera monotypic

The limits of this group are controversial & a number of genera have been segregated as mono- or oligo-typic fams – Eremosynaceae, Francoaceae, Lepuropetalaceae, Parnassiaceae (excl. here), Penthoraceae, Saniculophyllaceae, Vahliaceae, while woody groups like Grossulariaceae, Hydrangeaceae & *Bauera* (Cunoniaceae) have sometimes been incl. Cronquist keeps the woody groups associated with Pittosporaceae & Cunoniaceae, but combines all the herbs because many of the characters used to keep them distinct are found in genera classically incl. in S. Close to Rosaceae, *Astilbe* often confused with *Filipendula*

Many cult. orn. in above genera, *Bergenia, Darmera, Francoa, Lithophragma, Rodgersia, Tellima, Tiarella, Tolmiea* etc.

saxifrage *Saxifraga* spp.; **burnet s.** *Pimpinella* spp.; **golden s.** *Chrysosplenium oppositifolium*; **meadow s.** *S. granulata*; **mossy s.** *S.* sect. *Saxifraga*

Saxifragella Engl. Saxifragaceae. 2 Antarctic S Am.

Saxifragodes D.M. Moore. Saxifragaceae. 1 Tierra del Fuego to 51° S in Chile: *S. albowiana* (Kurtz) D.M. Moore

Saxifragopsis Small (~ *Saxifraga*). Saxifragaceae. 1 W US: *S. fragarioides* (E. Greene) Small

Saxiglossum Ching = *Pyrrosia*

Saxofridericia Schomb. Rapateaceae. 9 trop. S Am.

Scabiosa L. Dipsacaceae. (Incl. *Sixalix*) 80 temp. Euras. (Eur. 43), Medit., E Afr. mts & S Afr. Scabious; herbs (*S. cretica* L. (W Medit.) – shrubby) with epicalyx extension acting as an umbrella in wind-disp. of fr. Cult. orn. esp. annual *S. atropurpurea* L. (*Sixalix a.*, sweet s., S Eur.) – fls dark purple to white (see also *Lomelosia*)

Scabiosella Tieghem = *Scabiosa*

Scabiosiopsis Rech.f. (~ *Scabiosa*). Dipsacaceae. 1 W Iran: *S. enigmatica* Rech.f. R: Willd. 19(1989)153

scabious *Scabiosa* spp.; **devil's bit s.** *Succisa pratensis*; **field s.** *Knautia arvensis*

scabrin *Heliopsis* spp.

Scadoxus Raf. (~ *Haemanthus*). Amaryllidaceae (Liliaceae s.l.). 9 trop. Afr. Like *Haemanthus* but 2n = 18 (*H.* = 16) & lvs with distinct midvein & not distichous; alks toxic to stock; large chromosomes suited to cytological study. Cult. orn. with red (at least some bird-poll.) fls esp. *S. multiflorus* (T. Martyn) Raf. (blood flower, trop. Afr.) & *S. puniceus* (L.) I. Friis & Nordal (Ethiopian highlands, Tanz., S Afr.)

Scaevola L. Goodeniaceae. 96 trop. Indo-Pacific esp. Aus. (71, 70 endemic, R: Telopea 3(1990)489, FA 35(1992)84) with 2 widespread trop. beach spp.: *S. plumieri* (L.) Vahl (Indo-Atlantic) & *S. taccada* (Gaertner) Roxb. (*S. sericea*, Indo-Pacific), their seeds viable for long periods in seawater but germinate only in freshwater, i.e. when washed up on a rainy beach; extrafl. nectaries. Pith of *S. taccada* (taccada) used for making Malayan rice-paper, artificial fls etc.

scag *Papaver somniferum*

Scagea McPherson. Euphorbiaceae. 2 New Caled. R: BMNHN 4,7(1986)247

scald *Cuscuta* spp.

Scalesia Arn. Compositae (Helia.-Helia.). 11 Galápagos. R: OB 36(1974). Trees allied to *Helianthus*, showing adaptive radiation comparable with Darwin's finches

Scaligeria DC. Umbelliferae (III 5). 3 Medit. (Eur. 1) to SW As.

scallion Orig. = shallot, later = spring onion, (US) leek

scallopini Variously used for small forms of courgette

Scambopus O. Schulz. Cruciferae. 1 S & E Aus.: *S. curvipes* (F. Muell.) O. Schulz. R: TRSSA 89(1965)219

scammony *Ipomoea orizabensis*, **(Levant s.)** *Convolvulus scammonia*

Scandia Dawson. Umbelliferae (III 8). 2 NZ. Woody, gynodioec.

Scandicium (K. Koch) Thell. = *Scandix*

Scandivepres Loes. Celastraceae. 1 Mexico

Scandix L. Umbelliferae (III 2). 15–20 Eur. (3), Medit. *S. pecten-veneris* L. (shepherd's needle, Venus's or Lady's comb, Euras.) – v. long mericarps, which separate violently

Scaphiophora Schltr. Burmanniaceae. 1 Philippines, 1 New Guinea

Scaphispatha Brongn. ex Schott. Araceae (VII 1). 1 Bolivia. R: Aroideana 3(1980)4

Scaphium Schott & Endl. Sterculiaceae. 6 W Mal. Ovaries winged (wind-disp.). Monkeys eat jelly formed around germ. seeds; in Malay Pen. seeds soaked overnight swell in producing mucilage which is drunk as febrifuge etc.

Scaphocalyx Ridley. Flacourtiaceae (4). 1 Malay Peninsula, Sumatra

Scaphochlamys Baker. Zingiberaceae. 30 Indomal.

Scaphopetalum Masters. Sterculiaceae. 15 trop. Afr. *S. amoenum* A. Chev. (W Afr.) – axis with vertical growth then leaning over, the tip rooting & thickest of epicormics growing up like original trunk

Scaphosepalum Pfitzer. Orchidaceae (V 13). 30 trop. Am. R: MSB 26(1988)21, 39(1991)147, 44(1992)42. Cult. orn.

Scaphospermum Korovin = *Parasilaus*

Scaphyglottis Poeppig & Endl. Orchidaceae (V 13). (Incl. *Hexisea*) 85 trop. Am. Cult. orn. epiphytes

Scapicephalus Ovcz. & Chukavina. Boraginaceae. 1 C As.

Scarborough lily *Cyrtanthus elatus*

Scariola F.W. Schmidt (~ *Lactuca*). Compositae (Lact.-Lact.). 10 S Eur., N Afr., SW & C As.

scarlet globe mallow *Sphaeralcea coccinea*, *S. angustifolia* ssp. *cuspidata*; **s. runner bean** *Phaseolus coccineus*

Scassellatia Chiov. = *Lannea*

Sceletium N.E. Br. = *Phyllobolus*

Scelochiloides Dodson & M. Chase. Orchidaceae (V 10). 1 Bolivia: *S. vasquezii* Dodson & M. Chase

Scelochilus Klotzsch. Orchidaceae (V 10). 35 trop. Am.

scented orchid *Gymnadenia conopsea*

scentless rosewood *Synoum glandulosum*

Schachtia Karsten (~ *Duroia*). Rubiaceae (II 1). Spp.? trop. Am.

Schaefferia Jacq. Celastraceae. 16 trop. & warm Am. Some box subs. esp. *S. frutescens* Jacq. (Florida boxwood, Florida to WI)

Schaenomorphus Thorel ex Gagnepain = *Tropidia*

Schaetzellia Schultz-Bip. = *Macvaughiella*

Schaffnerella Nash. Gramineae (33c). 1 Mexico

Schaffneria Fée ex T. Moore = *Asplenium*

Schaueria Nees. Acanthaceae. 8 Brazil. *S. flavicoma* (Lindley) N.E. Br. – cult. orn. with yellow fls

Schedonnardus Steudel. Gramineae (33b). 1 S US: *S. paniculatus* (Nutt.) Trel. (tumble grass) – a conspic. feature of deserted towns in Western films, introd. Arg.

Scheelea Karsten. Palmae (V 5c). 28 trop. Am. R: Phytologia 37(1977)219. Large unarmed monoec. palms, cult. orn. Some poss. oilseeds, but slow-growing

Schefferomitra Diels. Annonaceae. 1 New Guinea: *S. subaequalis* (Scheffer) Diels

Schefflera Forster & Forster f. Araliaceae (1). c. 650 trop. & warm. Trees, shrubs, lianes & epiphytes (*S. gemma* Frodin (New Guinea) a herb), prob. congeneric with *Hedera* & incl. *Brassaia, Crepinella, Didymopanax, Dizygotheca, Tupidanthus*. Many cult. orn. trees (trop.) & house-pls esp. *S. actinophylla* (Endl.) Harms (*Brassaia a.*, New Guinea, trop. Aus.) – sometimes epiphytic, street-tree, *S. arboricola* (Hayata) Merr. (Taiwan) – frequent pot-pl., *S. elegantissima* (Masters) Lowry & Frodin (New Caled.) – juvenile forms with deeply lobed lvs, adult ones with more entire; *S. morototoni* (Aublet) Maguire, Steyermark & Frodin (*Didymopanax m.*, jereton, trop. Am.) – timber used for pulp, matches etc.; *S. pueckleri* (K. Koch) Frodin (*Tupidanthus calyptratus*, trop. As.) – cult. orn. liane or tree

Schefflerodendron Harms. Leguminosae (III 6). 3–4 trop. Afr.

Scheffleropsis Ridley = *Schefflera*

Schelhammera R. Br. Convallariaceae (Liliaceae s.l.). 2 E Aus., 1 ext. to New Guinea. R: FA 45(1987)412. Alks

Schellenbergia C.E. Parkinson = *Vismianthus*

Schellolepis J. Sm. = *Goniophlebium*

Schenckia Schumann = *Deppea*

Schenckochloa Ortíz. Gramineae (31d). 1 NE Brazil: *S. barbata* (Hackel) Ortíz. R: Candollea 46(1991)241

Scherya R. King & H. Robinson. Compositae (Eup.-Ager.). 1 E Brazil: *S. bahiensis* R. King & H. Robinson. R: MSBMBG 22(1987)150

Scheuchzeria L. Scheuchzeriaceae. 1 N temp. & Arctic – bogs

Scheuchzeriaceae Rudolphi (~ Juncaginaceae). Monocots – Alismatidae – Najadales. 1/1 N temp. & Arctic. Perennial herb with rhizome; vessels confined to roots. Stem with long intravaginal hairs at nodes. Lvs with long semi-terete lamina, open sheath & ligule at their junction, spirally arr. Fls wind-poll., bisexual, in term. bracteate racemes; P 3 + 3, yellow-green, A 6, with elongate anthers with longit. slits, pollen in diads, $\underline{G}$ 3(–6), basally weakly connate, each with sessile stigma & 2(–several) erect, anatropous, bitegmic ovules. Fr. a head of follicles; seeds with 0 endosperm. n = 11

Only species: *Scheuchzeria palustris* L.

Schickendantzia Pax. Alstroemeriaceae (Liliaceae s.l.). 1 Argentina

Schickendantziella Speg. Alliaceae (Liliaceae s.l.). 1 Argentina

Schiedea Cham. & Schldl. Caryophyllaceae (II 1). (Excl. *Alsinidendron*) 22 Hawaii. R: W. Wagner et al., *Man. Fl. Pl. Hawaii* (1990)508. Shrubs & lianes. *S. salicaria* Hillebrand (Maui) long believed extinct but lately rediscovered, the similar *S. adamantis* St John (Oahu) restricted to 60–65 individuals, cryptically dioec.

Schiedeella Schltr. Orchidaceae (III 3). 10 SN & C Am. R: FFG 36(1991)14

Schiekia Meissner. Haemodoraceae. 1 trop. S. Am.: *S. orinocensis* (Kunth) Meissner – soap-subs., local medic.

Schima Reinw. ex Blume. Theaceae. 1 (v. variable) trop. & warm As.: *S. wallichii* (DC) Korth. – timber for construction, imp. tree for reafforestation in Java; contact with bark (source of a fish-poison) causes intense itching. R: Reinw. 2(1952)133

Schimpera Hochst. & Steudel ex Endl. Cruciferae. 1 E Medit. to S Iran

Schimperella H. Wolff = *Oreoschimperella*

Schindleria H. Walter. Phytolaccaceae (II). 2 Peru, Bolivia

Schinopsis Engl. Anacardiaceae (IV). 7 S Am. R: Lilloa 33(1973)205. *S. quebracho-colorado* (Schldl.) F. Barkley & T. Meyer & other spp. with bark & timber (quebracho) high in tannin used in tanning

Schinus L. Anacardiaceae (IV). 27 trop. Am. R: Brittonia 5(1944)160. Usu. dioec. trees, some planted in trop. & warm as shade etc. *S. molle* L. (pepper tree, California p., Peruvian Andes) – lvs 7–20-jugate, cult. shade-tree in Medit. etc., locally medic., fr. used to adulterate pepper, exudates chewed (American mastic), fert. control in Uruguay, harbours black scale of citrus, molle a rendering of native name 'mulli'; *S. terebinthifolius* Raddi (Brazilian pepper tree, Brazil) – lvs 2- or 3(–6)-jugate with domatia, cult. orn. natur. Florida where a cause of dermatitis & respiratory problems, intoxicating birds but has good nectar for bees, serious weed in Hawaii (first coll. 1911)

Schinziella Gilg. Gentianaceae. 2 trop. Afr.

Schinziophyton Hutch. ex A.R.-Sm. (~ *Ricinodendron*). Euphorbiaceae. 1 trop. & S Afr.: *S. rautanenii* (Schinz) A.R.-Sm. (*R. r.*, mongonogo, mugongo) – seeds a staple of Sou bushmen (!Kung) & source of manketti nut oil used in food, varnishes etc., timber a balsa subs. & poss. use in paper-making. R: KB 45(1990)157

Schippia Burret. Palmae (I 1a). 1 Belize: *S. concolor* Burret – cult. orn.

Schisandra Michaux. Schisandraceae. 25 E As. & E N Am. (1). R: Sargentia 7(1947)86. Cult. orn. dioec. or monoec. (*S. chinensis* (Turcz.) Baillon (China) with labile sexuality) lianes with attractive fr.; some effective tonics & sedatives

Schisandraceae Blume. Dicots – Magnoliidae – Illiciales. 2/47 E As. & E N Am. (*Schisandra glabra* (Brickell) Rehder). Dioec. or monoec. aromatic, glabrous lianes. Lvs simple, usu. toothed & often with pellucid dots, spirally arr.; stipules 0. Fls small, with ± elongate receptacle, solit. to few-flowered infls. in axils; P 5–24 spirally arr. in 2–several series, s.t. outer ± K-like & inner C-like, A 4–80 ± spirally arr., ± connate basally, anthers basifixed with ± expanded connective, $\underline{G}$ 12–100(–300) conduplicate, unsealed, spirally arr. with decurrent stigmas & each with 1–5(–11) marginal, anatropous to campylotropous, bitegmic ovules. Fr. a head (*Kadsura*) or elongate axis of berries with usu. 2 seeds; embryo minute in copious endosperm rich in oil & starch. x = 13, 14

Genera: *Kadsura*, *Schisandra*

Fr. analogous to that in *Sargentodoxa*

Cult. orn.

Schischkinia Iljin. Compositae (Card.-Cent.). 1 SW & C As.

Schischkiniella Steenis = *Silene*

Schismatoclada Baker. Rubiaceae (I 1/IV 24). 20 Madagascar

Schismatoglottis Zoll. & Moritzi. Araceae (VI 1). 100 trop. As. & Am. (1). Herbs, some rheophytic, with spathe-tube persistent around female fls, the blade dropped early exposing males & term. sterile appendix of spadix; some cult. orn.

Schismocarpus S.F. Blake. Loasaceae. 1 Mexico: *S. pachypus* S.F. Blake

Schismus Pal. Gramineae (25). 5 Afr., Medit. (Eur. 2) to NW India

Schistocarpaea F. Muell. Rhamnaceae. 1 Queensland

Schistocarpha Less. Compositae (Helia.-Helia.). 10 trop. Am. R: Phytol. 59 (1986)272

Schistocaryum Franchet = *Microula*

Schistogyne Hook. & Arn. Asclepiadaceae (III 1). 12 S Am.

Schistolobos W.T. Wang. Gesneriaceae. 1 China

Schistonema Schltr. = *Tweedia*

Schistophragma Benth. ex Endl. (~ *Leucocarpus*). Scrophulariaceae. 2 Am.

Schistostemon (Urban) Cuatrec. Humiriaceae. 7 trop. S Am. R: CUSNH 35(1961)146

Schistostephium Less. Compositae (Anth.-Mat.) 12 trop. & S Afr. R: KB 1916:99

Schistotylus Dockr. Orchidaceae (V 16). 1 NSW: *S. purpuratus* (Rupp) Dockr.

Schivereckia Andrz. ex DC. Cruciferae. 2 N Russia & Balkans (Eur. 2) to Turkey. Some cult. orn. herbs like *Alyssum*

Schizachne Hackel. Gramineae (19). 1 Arctic Eur., NE As., temp. N Am., SW US (mts)

Schizachyrium Nees. Gramineae (40f). 60 trop. savannas. *S. scoparium* (Michaux) Nash (blue-stem, bunchgrass, N Am.) – imp. erosion control & grazing in Great Plains

Schizaea Sm. Schizaeaceae. 30 trop., S temp., N Am. Usu. on nutrient-poor soils, sometimes on decaying wood. Fronds simple or dichot. lobed with sporangia in rows along segments. Prothalli merely uniseriate filaments, in *S. dichotoma* (L.) Sm. (Madag. to Pacific) multiseriate & subterr. or cylindrical, subterranean & becoming tuberous with age

Schizaeaceae Kaulf. Filicopsida. 4/180 subcosmop. usu. warm & trop. Upright or creeping ferns with protostelic to dictyostelic stems & simple, dichot. or pinnately divided fronds. Sporangia borne singly on modified segments of whole fronds & not in sori, the annulus merely a group of thick-walled cells, dehiscence longit.

Genera: *Anemia, Lygodium, Mohria, Schizaea*

Schizanthus Ruíz & Pavón. Solanaceae (2). 12 Chile. R: MBSM 20(1984)121. Fls irreg., resupinate, with upper 2 petals forming 3–4 lobed lip, laterals 4-lobed, lower one forming simple or weakly bilobed upper lip, A 4 (2 sterile); fl. mechanism like Leguminosae (III) with explosion (cf. *Genista*). Cult. orn. (poor man's orchid, butterfly flower) esp. *S. pinnatus* Ruíz & Pavón & its hybrids

Schizeilema (Hook.f.) Domin. Umbelliferae (I 2b). 11 NZ, 1 E Aus., 1 temp. S Am. Allied to *Diplaspis* & *Huanaca*

Schizenterospermum Homolle ex Arènes. Rubiaceae (? II 2). 4 Madagascar

Schizobasis Baker. Hyacinthaceae (Liliaceae s.l.). 5 Ethiopia to S Afr. Compound twining infls (cf. *Bowiea*)

Schizoboea (Fritsch) B.L. Burtt. Gesneriaceae. 1 trop. Afr.

Schizocaena J. Sm. ex Hook. = *Cyathea*

Schizocalomyrtus Kausel = *Calycorectes*

Schizocalyx Wedd. Rubiaceae (I 1). 2 Colombia

Schizocapsa Hance = *Tacca*

Schizocarphus J. Merwe = *Scilla*

Schizocarpum Schrader. Cucurbitaceae. 8 C Am.

Schizochilus Sonder. Orchidaceae (IV 2). 26 trop. & S Afr. R: JSAB 46(1980)379

Schizococcus Eastw. = *Arctostaphylos*

Schizocodon Siebold & Zucc. = *Shortia*

Schizocolea Bremek. Rubiaceae (IV 11). 1 trop. W Afr.

Schizodium Lindley. Orchidaceae (IV 3). 6 Cape. R: JSAB 47(1981)339

Schizoglossum E. Meyer. Asclepiadaceae (III 1). 14 S Afr. R: KB 38(1984)599

Schizogyne Cass. (~ *Inula*). Compositae (Inul.). 2 Canary Is.

Schizolaena Thouars. Sarcolaenaceae. 12 Madag. R: Boissiera 24(1975)339

Schizolepton Fée = *Taenitis*

Schizolobium Vogel. Leguminosae (I 1). 1–2 trop. Am. *S. parahyba* (Vell. Conc.) S.F. Blake – spectacular pachycaul flowering tree, unbranched when young & bearing large bipinnate lvs, the fls almost reg., 5-merous, cult. orn.

Schizoloma Gaudich. = *Lindsaea*

Schizomeria D. Don. Cunoniaceae. 18 E Aus., Papuasia. *S. ovata* D. Don (E Aus.) – timber for coffins, veneers etc.

Schizomeryta R. Viguier = *Meryta*

Schizomussaenda Li = *Mussaenda*

Schizonepeta (Benth.) Briq. (*nom. illeg.*). Labiatae (VIII 2). 3 temp. As. *S. tenuifolia* (L.) Briq. (China) – painkiller used with *Stenocoelium divaricatum* in treatment of arthritis & toothache in modern Chinese herbalism

Schizopepon Maxim. Cucurbitaceae. 8 Himal. to E As. R: APS 23(1985)110. Fls sometimes bisexual

Schizopetalon Sims. Cruciferae. 10 NC Chile & nearby Arg. Pinnatsect C unique in fam. *S. walkeri* Sims (Chile) – cult. orn. annual with bifid twisted cotyledons as have 2 other spp.

Schizophragma Siebold & Zucc. Hydrangeaceae. 2 Himal. to Japan & Taiwan. Decid. climbers with aerial clinging rootlets. Cult. orn. esp. *S. hydrangeoides* Siebold & Zucc. (Korea, Japan) – much like *Hydrangea anomala* but sterile fls with K 1 not 4

Schizopsera Turcz. Compositae (Helia.-Verb.). 1 Ecuador

Schizoptera Benth. = *Schizopsera*

Schizoscyphus Schumann ex Taubert = *Maniltoa*

Schizosepala G. Barroso. Scrophulariaceae. 1 Brazil

Schizosiphon Schumann = *Maniltoa*

Schizospatha Furt. = *Calamus*

Schizostachyum Nees. Gramineae (1c). (Incl. *Cephalostachyum, Neohouzeaua, Pseudostachyum, Teinostachyum*) 60 Madag., Indomal. to Hawaii. R: KBAS 13(1986)56. Bamboos; some with 3-yr cycle of flowering (but *S. dullooa* (Gamble) Majumdar (*N. d.*, SE As.) on a 14–17 yr cycle). *S. brachyandrum* Kurz (Mal.) – stems used for flutes in Sumatra; *S. jaculans* Holttum (Malay Pen.) – blowpipes; *S. latifolium* Gamble (Indomal.) – up to 10 lodicules grading into staminodes; *S. polymorphum* (Munro) Majumdar (*P. p.*, Himal. to Borneo) – source of paper pulp

Schizostegopsis Copel. = *Pteris*

Schizostigma Arn. ex Meissner. Rubiaceae (I 8). 1 Sri Lanka. G 5–7-loc.

Schizostylis Backh. & Harvey. Iridaceae (IV 3). 1 S Afr.: *S. coccinea* Backh. & Harvey (kaffir lily) – cult. orn. with red fls, some cvs with washed-out pink

Schizotorenia Yamaz. = *Lindernia*

Schizotrichia Benth. Compositae (Hele.-Pect.). c. 5 Peru

Schizozygia Baillon. Apocynaceae. 1 trop. E Afr.

Schkuhria Roth. Compositae (Hele.-Cha.). 6 warm Am., *S. pinnata* (Lam.) Thell. weedy elsewhere. R: AMBG 32(1945)265

Schlagintweitiella Ulbr. = *Thalictrum*

Schlechtendalia Less. Compositae (Barn.). 1 (variable) Brazil, Uruguay, Arg. Habit *Eryngium*-like with opp. or whorled linear lvs

Schlechteranthus Schwantes. Aizoaceae (V). 2 Cape

Schlechterella Schumann. Asclepiadaceae (I, Periplocaceae). 1 E Afr.: *S. africana* (Schltr.) Schumann

Schlechteria Bolus ex Schltr. Cruciferae. 1 S Afr.

Schlechterina Harms (~ *Crossostemma*). Passifloraceae. 1 trop. E Afr.

Schlechterosciadium H. Wolff = *Chamarea*

Schlegelia Miq. Scrophulariaceaee. 12 trop. Am. Form. incl. in Bignoniaceae

Schleichera Willd. Sapindaceae. 1 Indomal.: *S. oleosa* (Lour.) Oken (*S. trijuga*, kussum, kosumba) – timber (Ceylon oak) hard & used for mortars, etc., bark for tanning, lvs ed. as veg. with rice, unripe fr. pickled, seeds the source of Macassar oil, used for candles, hairdressing, batik-work, soap & illumination; tree a host of lac insects

Schleinitzia Warb. ex Nevling & Niezgoda (~ *Prosopis*). Leguminosae (II 3). 3–4 Pacific. R: Adansonia 18(1978)345

Schliebenia Mildbr. = *Isoglossa*

Schlimmia Planchon & Linden. Orchidaceae (V 10). 7 N Andes

Schlumbergera Lemaire (*Zygocactus*). Cactaceae (III 5). 6 Brazil (all near Rio). Some cult. orn. (millions a year sold in Denmark & Holland – some 200 cvs incl. some with fls yellow under heat but pink when cool) are hummingbird-poll. epiphytic cacti with flat-jointed stems esp. *S.* × *buckleyi* (T. Moore) Tjaden ('*S. bridgesii*', *S. truncata* × *S. russelliana* (Hook.) Britton & Rose, Christmas cactus) – stem-joints crenate, fls almost reg., ovary 4–5-angled, flowering in winter & *S. truncata* (Haw.) Moran (crab cactus) – stem-joints 2–4-serrate, fls irreg., ovary cylindrical, flowering in autumn

Schlumbergera C.J. Morren = *Guzmania*

Schmalhausenia Winkler. Compositae (Card.-Card.). 1 Tienshan

Schmaltzia Desv. ex Small = *Rhus*

Schmardaea Karsten. Meliaceae (IV 2). 1 Andes: *S. microphylla* (Hook.) C. Mueller

Schmidtia Steudel ex J.A. Schmidt. Gramineae (29). 2 Afr., Cape Verde Is., Pakistan. R: BSB II,39(1965)303

Schmidtottia Urban. Rubiaceae (inc. sed.). 15 E Cuba (serpentine)

Schnabelia Hand.-Mazz. Labiatae (II; Verbenaceae). 2 S & SW China. Cleistogamy

Schnella Raddi = *Bauhinia*

Schoenefeldia Kunth. Gramineae (33b). 2 Afr. to India. Awns curiously braided

Schoenia Steetz (~ *Helichrysum*). Compositae (Gnap.-Ang.). 5 temp. Aus. R: Nuytsia 8(1992)371. *S. cassiniana* (Gaudich.) Steetz – cult. orn. annual 'everlasting'

Schoenobiblus C. Martius. Thymelaeaceae. 8 trop. Am. Dioec. *S. peruvianus* Standley (Amaz.) – fish- & arrow-poison

Schoenocaulon A. Gray. Melanthiaceae (Liliaceae s.l.). 10 Florida to Peru. Alks incl. veratrin. *S. officinale* (Schldl. & Cham.) A. Gray (sabadilla, cevadilla, Mex.) – seeds insecticidal, used in veterinary medicine

Schoenocephalium Seub. Rapateaceae. 5 NE S Am.

Schoenocrambe E. Greene (~ *Sisymbrium*). Cruciferae. 4 N Am. R: CGH 212(1982)93

Schoenoides Seberg = *Oreobolus*

Schoenolaena Bunge (~ *Xanthosia*). Umbelliferae (I 1b). 2 W Aus.

Schoenolirion Torrey. Hyacinthaceae (Liliaceae s.l.). 3 S US. R: Madrono 38(1991)132

Schoenomorphus auctt. = *Tropidia*

Schoenoplectus (Reichb.) Palla (~ *Scirpus*). Cyperaceae. 60 cosmop. (Aus. 11). *S. californicus* (C. Meyer) Palla (*Scirpus c.*, Calif. to Chile & Easter Is., ? Hawaii) – used for mats, houses, floats, boats etc. & minor food-source at Lake Titicaca; *S. corymbosus* (Roemer & Schultes) Raynal (OW warm & trop.) – used by Anc. Egyptians in funeral wreaths; *S. lacustris* (L.) Palla (*Scirpus l.*, clubrush, bulrush, N temp.) – shoots regenerate in total absence of oxygen for up to 90 days allowing spread into anaerobic muds denied to other spp., chair-seats, mats, hassocks, baskets, rhizomes eaten by N Am. Indians, used in water purification in Netherlands & Germany; *S. mucronatus* (L.) Kerner (*Scirpus m.*, warm OW) – cult. Sumatra for bag prod.; *S. triqueter* (L.) Palla (*Scirpus t.*, OW) – matting

Schoenorchis Reinw. Orchidaceae (V 16). 24 Indomal. & China to Fiji. R: OB 95(1988)66. Some cult. orn. with narrow lvs

Schoenoxiphium Nees (~ *Kobresia*). Cyperaceae. 12 trop. & S Afr., Madag. R: Bothalia 14(1983)819

Schoenus L. Cyperaceae. 100 subcosmop. (not N Am.; Eur. 2) esp. Mal. & Aus. (90, 85 endemic). *S. nigricans* L. (black bog rush, Eur., S Afr.) – still used for thatch in Ireland; *S. asperocarpus* F. Muell. (Aus.) – one of few Aus. sedges reputed to be toxic

Schoepfia Schreber. Olacaceae (Schoepfiaceae). 23 trop. esp. Am. (20). Often root-parasites

Schoepfiaceae Blume = Olacaceae

Scholleropsis Perrier. Pontederiaceae. 1 Madagascar

Scholtzia Schauer. Myrtaceae (Lept.). 15 SW Aus.

Schomburgkia Lindley (~ *Cattleya*). Orchidaceae (V 13). 22–24 trop. Am. (WI 6). R: C. Withner (1993) *The Cs III*: 169. Cult. orn. epiphytes with extrafloral nectaries incl. *S. thomsoniana* Reichb.f. (*Laelia t.*, WI) – pseudobulbs carved into pipe-bowls locally & *S. tibicinis* (Bateman) Bateman (Mex. to Panamá) – hollow ant-filled pseudobulbs to 55 cm tall absorbing nutrients brought in by ants

Schotia Jacq. Leguminosae (I 4). 4–5 S Afr., open woodland & scrub. Cult. orn. trees. Some ed. seeds (Boer beans) esp. *S. afra* (L.) Thunb. (*S. speciosa*)

Schoutenia Korth. Tiliaceae. 8 Thailand to C Mal. & N Aus. Some local timbers

Schouwia DC. Cruciferae. 2 Sahara to India. *S. purpurea* (Forssk.) Schweinf. – cult. orn. annual, lvs ed. Sahara

Schradera Vahl. Rubiaceae (IV 6). 25 trop. Am. R: MNYBG 10(1963)259. Epiphytes

Schrameckia Danguy = *Tambourissa*

Schranckiastrum Hassler = *Mimosa*

Schrankia Willd. = *Mimosa*

Schrebera Roxb. Oleaceae. 10 trop. (Am. 1)

Schreiteria Carolin (~ *Calandrinia*). Portulacaceae. 1 Arg.: *S. macrocarpa* (Speg.) Carolin. R: Parodiana 3(1985)330

Schrenkia Fischer & C. Meyer. Umbelliferae (III 4). 12 C As.

Schtschurowskia Regel & Schmalh. Umbelliferae (III 4). 2 C As.

Schubertia C. Martius. Asclepiadaceae (III 3). 6 S Am.

Schultesia C. Martius. Gentianaceae. 20 trop. Am.

Schultesianthus Hunz. Solanaceae (1). 5 trop. Am.

Schultesiophytum Harling. Cyclanthaceae. 1 NW trop. S Am.

Schulzia Sprengel. Umbelliferae (III 8). 2 C As., ? 2 NW India

Schumacheria Vahl. Dilleniaceae. 3 Sri Lanka. *S. castaneifolia* Vahl – 'throw-away' branches, stilt-roots in wet sites

Schumannia Kuntze = *Ferula*

Schumannianthus Gagnepain. Marantaceae. 2 Indomal.

Schumanniophyton Harms. Rubiaceae (II 1). 5 trop. Afr. Stimulants & fish-poison esp. *S. magnificum* (Schumann) Harms (W Afr.)

Schumeria Iljin = *Serratula*

Schuurmansia Blume. Ochnaceae. 3 C Mal. to Papuasia

Schuurmansiella Hallier. Ochnaceae. 1 NW Borneo: *S. angustifolia* (Hook.f.) Hallier. R: BJ 113(1991)181

Schwabea Endl. & Fenzl = *Monechma*

Schwackaea Cogn. Melastomataceae. 1 Mexico, C Am.

Schwalbea L. Scrophulariaceae. 1 E US: *S. americana* L. – hemiparasitic perennial of savannas, not host-specific

Schwannia Endl. = *Janusia*

Schwantesia Dinter. Aizoaceae (V). 3–5 SW Afr. Dwarf succ. with marcescent lvs around internodes, cult. orn.

Schwartzia Vell. = *Norantea*

Schwartzkopffia Kraenzlin = *Brachycorythis*

Schweiggeria Sprengel. Violaceae. 2 Mexico, Brazil

Schweinfurthia A. Braun. Scrophulariaceae (Antirr.). 6 NE Afr. to India, desert & semi-d. R: NRBGE 40(1982)23

Schwenckia Vahl = *Schwenkia*

Schwenckiopsis Dammer (~ *Schwenkia*). Solanaceae. 1 Andes

Schwendenera Schumann. Rubiaceae (IV 15). 1 SE Brazil

Schwenkia L. Solanaceae (2). 25 trop. Am., 1 (*S. americana* L.) extending to W Afr.

Schwenkiopsis Dammer = *Schwenckiopsis*

Sciadocephala Mattf. Compositae (Eup.-Aden.). 5 Panamá & N S Am. R: MSBMBG 22(1987)60

Sciadodendron Griseb. Araliaceae. 1 trop. Am.

Sciadopanax Seemann = *Polyscias*

Sciadophyllum P. Browne = *Schefflera*

Sciadopityaceae Luerss. (~ Taxodiaceae). Pinopsida. 1/1 C & S Japan. Resinous monoec. tree with short shoots borne on principal ones. Lvs flattened, spirally arr. in seedlings, adults with scalelike spurs subtending 'phylloclades' forming false whorls. Male cones subglobose in term. raceme-like clusters; females solit. with numerous scales each with 7–9 anatropous ovules. Seeds ovate-elliptic with v. narrow wing & 2 cotyledons

Genus: *Sciadopitys*

Not like other Taxodiaceae (rbcl data) & app. distinct from the Upper Triassic when more widespread.

Sciadopitys Siebold & Zucc. Sciadopityaceae (Taxodiaceae s.l.). 1 C & S Japan: *S. verticillata* (Thunb.) Siebold & Zucc. (parasol pine) – widely planted around temples & perhaps native only in 2 small areas of C Honshu, though in Tertiary common in Eur. & a char. fossil of some brown coal. Photosynthetic organs appear to be united needles which occ. branch – pigeonholed as 'cladodes' or 'phylloclades' they are best considered organs intermediate between lvs & stems. Gametophytes rather like Pinaceae & Podocarpaceae, while n = 10 (11, 33 etc. in rest of Taxodiaceae); cones mature over 2 yrs. Timber used for ship-building etc., oil for varnish; cult. orn. for acid soil

Sciadotenia Miers. Menispermaceae (I). 19 trop. Am. R: MNYBG 22, 2(1971)15. Curare sources

Sciaphila Blume. Triuridaceae. (Incl. *Hyalisma*) 31 trop. & warm. R: PR IV, 18(1938)30, BB 140(1991)9. Endotrophic mycorrhiza

Sciaphyllum Bremek. Acanthaceae. 1 cult.: *S. amoenum* Bremek. (orig. unknown)

Sciaplea Rauschert = *Dialium*

Scilla L. Hyacinthaceae (Liliaceae s.l.). (Excl. *Ledebouria*) 40 Euras. (Eur. 16), temp. Afr. Distinct from *Chionodoxa*, *Hyacinthoides* & *Puschkinia* in P-segments free. In fr. peduncle of many spp. extends & flops on ground when ant-disp. seeds released. Cult. orn. bulbous pls esp. *S. autumnalis* L. (autumn squill, Eur. to N Afr. & Cauc.), *S. mischtschenkoana* Grossh. (*S. tubergeniana*, NW Iran & Transcaucasia) – v. early white fls with blue stripes, the pls in cult. prob. a clone producing v. few seeds, suggesting meiotic irregularities, *S. peruviana* L. (Cuban lily, W Medit.(!)), *S. siberica* Haw. (Iran–Russia) esp. 'Spring Beauty' (sterile so fls last longer), *S. verna* L. (spring squill, W Eur.)

Scindapsus Schott. Araceae (III 2). 40 SE As., Mal., Brazil (1). Some cult. orn. lianes esp. *S. pictus* Hassk. 'Argyraeus' (W Mal.) & local medic.

Sciodaphyllum P. Browne = *Schefflera*

Sciothamnus Endl. (~ *Peucedanum*). Umbelliferae (III 10). 4 S Afr.

Scirpodendron Zipp. ex Kurz. Cyperaceae. 1 Indomal., Pacific: *S. ghaeri* (Gaertner) Merr. (*S. costatum*) – hats & mats locally, seeds ed. (Samoa)

Scirpoides Séguier (*Holoschoenus*; ~ *Scirpus*). Cyperaceae. 1 Eur., SW As.: *S. holoschoenus* (L.) Soják

Scirpus L. Cyperaceae. (Excl. *Bolboschoenus*, *Isolepis*, *Schoenoplectus*, *Scirpoides*, *Trichophorum*) c. 20 subcosmop. (200 cosmop. s.l., R: JFSUTB 3, 7(1958)271). Stems used for thatch, matting etc. esp. *S. americanus* Pers. (sword grass, chairmaker's rush, N Am.) – seating rush; *S. atrovirens* Willd. (Mex.) – hallucinogen; *S. paludosus* Nelson (bayonet grass, N Am.) – rhizomes form. eaten

Sclerachne R. Br. Gramineae (40j). 1 Thailand to W Mal.: *S. punctata* R. Br. – ant-disp. (elaiosome)

Sclerandrium Stapf & C. Hubb. = *Germainia*

Scleranthera Pichon = *Wrightia*

Scleranthopsis Rech.f. Caryophyllaceae (III 1). 1 Afghanistan: *S. aphanantha* (Rech.f.) Rech.f.

Scleranthus L. Caryophyllaceae (II 5). 15 Eur. (3), As., Afr., Aus. C 0, autogamous, weedy esp. *S. annuus* L. (knawel, temp. Euras., Medit., introd. N Am.)

Scleria P. Bergius. Cyperaceae. (Excl. *Diplacrum*) 200 trop. & warm (Aus. 23, S Afr. 23 – R: Bothalia 15(1985)505). Scrambling spp. often a fire-risk & objectionable, e.g. *S. boivinii* Steudel (W Afr.) with cutting leaf-edges

Sclerobassia Ulbr. = *Bassia*

Scleroblitum Ulbr. (~ *Chenopodium*). Chenopodiaceae (I 2). 1 SE Aus.: *S. atriplicinum* (F. Muell.) Ulbr. R: FA 4(1984)175

Sclerocactus Britton & Rose (~ *Pediocactus*). Cactaceae (III 9). (Incl. *Ancistrocactus*) 19 SW N Am. R: CSJ 38(1966)50, 100. Small cult. orn. undulate-ribbed cacti s.t. with areoles with extrafl. nectaries

Sclerocarpus Jacq. Compositae (Helia.-Helia.). 8 trop. & warm Am., Afr. (1)

Sclerocarya Hochst. Anacardiaceae (II). 4 trop. & S Afr. Ed. fr. esp. *S. birrea* (A. Rich.) Hochst. subsp. *caffra* (Sonder) Kokwaro (maroola plum)

Sclerocaryopsis Brand = *Lappula*

Sclerocephalus Boiss. Caryophyllaceae (I 2). 1 Macaronesia to Iran: *S. arabicus* Boiss. R: MBMHRU 285(1968)61

Sclerochiton Harvey. Acanthaceae. 12 trop. & S Afr.

Sclerochlamys F. Muell. = *Sclerolaena*

Sclerochloa Pal. Gramineae (17). 1 S Eur., W As.: *S. dura* (L.) Pal.

Sclerochorton Boiss. Umbelliferae (III 8). 1 Iran: *S. haussknechtii* Boiss.

Sclerodactylon Stapf. Gramineae (31d). 1 E Afr. coast, is. of Ind. Ocean on coral rocks near shore

Sclerodeyeuxia Pilger = *Calamagrostis*

Scleroglossum Alderw. (~ *Grammitis*). Grammitidaceae. 8 Indomal. to Aus. & W Pacific. Superficially resembling small *Vittaria* spp. but spores chlorophyllous etc.

Sclerolaena R. Br. Chenopodiaceae (I 5). (Incl. *Stelligera* A.J. Scott) 64 Aus. (not Tasmania). Copper-burr. R: FA 4(1984)236, Telopea 3(1988)142

Sclerolepis Cass. Compositae (Eup.-Tric.). 1 E US. R: MSBMBG 22(1987)192

Sclerolinon C. Rogers. Linaceae. 1 W US

Sclerolobium Vogel. Leguminosae (I 1). 35 trop. Am. esp. Amazonia. R: Lloydia 20(1957)67, 266. Lvs with continuous growth

Scleronema Benth. Bombacaceae. 5 trop. S Am.

Sclerophylax Miers. Solanaceae (2). 12 Arg. R: Kurtziana 1(1961)9. Succ. halophytes; lvs opp.

Scleropoa Griseb. = *Catapodium*

Scleropogon Philippi. Gramineae (31d). 1 SW US, Mex., Arg., Chile: *S. breviflorus* Philippi (burrograss) – fr. buried by awn of floret. R: Phytol. 62(1987)267

Scleropyrum Arn. Santalaceae. 6 Indomal.

Sclerorhachis (Rech.f.) Rech.f. Compositae (Anth.-Han.). 4 Afghanistan

Sclerosciadium Koch ex DC = *Capnophyllum*

Sclerosperma G. Mann & H. Wendl. Palmae (V 4o). 3 trop. W Afr.

Sclerostachya (Hackel) A. Camus = *Miscanthus*

Sclerostegia P.G. Wilson. Chenopodiaceae (II 2). 5 Aus. R: Nuytsia 3(1980)1

Sclerostephane Chiov. (~ *Pulicaria*). Compositae (Inul.). 5 Somalia. R: BJ 104(1983)91

Sclerotheca A. DC. Campanulaceae. 3 Cook Is., Society Is. Woody

Sclerothrix C. Presl = *Klaprothia*

Sclerotiaria Korovin. Umbelliferae (III 4). 1 C As.

Scobinaria Seib. (~ *Arrabidaea*). Bignoniaceae. 1 trop. Am.

Scoliaxon Payson. Cruciferae. 1 NE Mexico

Scoliopus Torrey. Trilliaceae (Liliaceae s.l.). 2 W N Am. Cult. orn. with foetid fls

Scoliosorus T. Moore = *Antrophyum*

Scoliotheca Baillon = *Monopyle*

Scolochloa Link (~ *Festuca*). Gramineae (17). 1 N temp. inc. Eur.: *S. festucacea* (Willd.) Link (sprangletop) – fodder & hay

Scolophyllum Yamaz. = *Lindernia*

Scolopia Schreber. Flacourtiaceae (6). 37 OW trop. R: Blumea 20(1972)25. Some S Afr. spp. with hard timber for axles etc. esp. *S. ecklonii* (Nees) Harvey (*S. zeyheri*, thorn pear). *S. nitida* C. White (New Guinea) – fls open all at once so trees white for 2–3 days

Scolosanthus Vahl. Rubiaceae (III 4). 21 WI

Scolymus L. Compositae (Lact.-?). 3 Medit. incl. Eur. *S. hispanicus* L. (Spanish oyster, golden thistle, S Eur. to NW France) – roots form. ed. (like salsify)

Scoparia L. Scrophulariaceae. 20 trop. Am., *S. dulcis* L. a pantrop. weed used to sweeten well-water & for snakebite (C Am.)

Scopelogena L. Bolus (~ *Ruschia*). Aizoaceae (V). 2 SW Afr.

Scopolia Jacq. Solanaceae (1). (excl. *Anisodus*) 5 Medit. (Eur. 1) to Himal. R: FR 82(1972)617. Alks, some sources of atropine

Scopulophila M.E. Jones. Caryophyllaceae (I 2). 1 SW US: *S. nitrophiloides* M.E. Jones

Scorodocarpus Becc. Olacaceae. 1 Mal.: *S. borneensis* (Baillon) Becc. – hard onion-scented timber for construction, ed. fr.

Scorodophloeus Harms. Leguminosae (I 4). 1 Guinea coast, 1 E Afr. coast

scorpion grass *Myosotis* spp.; **s. senna** *Hippocrepis emerus*

Scorpiothyrsus Li. Melastomataceae. 6 Hainan, SE As.

Scorpiurus L. Leguminosae (III 19). 2–4 Macaronesia & Medit. (Eur. 2) to Iran. R: Lagascalia 4(1974)259. Legume twisted, indehiscent

Scorzonera L. Compositae (Lact.-Scor.). c. 175 Medit. (Eur. 28) to C As. *S. hispanica* L. (scorzonera, Spanish salsify, viper's grass, Eur.) – ed. roots like salsify, coffee subs.; *S. tausaghyz* Lipsch. & Bosse (former USSR) – poss. rubber source tried in World War II

Scotch, Scots or **Scottish asphodel** *Tofieldia pusilla*; **S. fir** or **pine** *Pinus sylvestris*; **S. mint** *Mentha × gracilis*; **S. thistle** *Cirsium vulgare* or *Carduus nutans* now tending to be applied to *Onopordum acanthium*

Scottellia Oliver. Flacourtiaceae (2). 3 trop. Afr. *S. coriacea* A. Chev. ex Hutch. & Dalz. (W Afr.) – comm. timber (odoko)

screwpines *Pandanus* spp.

Scribneria Hackel. Gramineae (18). 1 W US: *S. bolanderi* (Thurber) Hackel

Scrithacola Alava. Umbelliferae (III 11). 1 Afghanistan, Pakistan

Scrobicaria Cass. (~ *Gynoxys*). Compositae (Sen.-Tuss.). 2 NE S Am.

Scrobicularia Mansf. = *Poikilogyne*

Scrofella Maxim. Scrophulariaceae. 1 C China: *S. chinensis* Maxim.

Scrophularia L. Scrophulariaceae. 200 N temp. (Eur. 30) to trop. Am. Coarse foetid herbs & shrubs locally medic. (figwort), few cult. orn. e.g. *S. auriculata* L. (*S. aquatica*, water f., Eur., Med.). Style & stamens lying along lower lip of C, posterior stamen rep. by staminode; some poll. by wasps

Scrophulariaceae Juss. Dicots – Asteridae – Scrophulariales. 268/5100 cosmop. esp. temp., trop. mts. Trees, shrubs &, usu., herbs, s.t. climbing, aquatic or hemiparasitic on roots of hosts (*Lathraea* & *Harveya* wholly parasitic, as are genera here referred to Orobanchaceae, q.v.). Lvs simple, s.t. dissected, spirally arr. or opp.; stipules 0. Fls usu. irreg. & bisexual, solit. or in spikes, racemes or thyrses; K (2)4- or 5-lobed with valvate or imbricate segments, C ((0,4)5(–8)), often bilabiate, s.t. basally spurred, lobes valvate or imbricate, A 5 or 4 ± staminode (uppermost) or 2(3) with lower pair reduced or 0, anthers with longit. slits, disk often around $\underline{G}$ (2(3)), 2(3)-loc. with term. style, each loc. with (2–)±∞ anatropous or hemitropous (rarely amphitropous or campylotropous) unitegmic ovules on axile placentas. Fr. often a septicidal capsule (s.t. loculicidal or with pores), rarely a berry or schizocarp (*Lagotis*); seeds winged or angled with straight or weakly curved embryo in oily endosperm. x = 6+

Classification & principal genera:

Verbascoideae (2 posterior C-lobes outside laterals in bud; lvs spirally arr., A often 5; 2 tribes): *Verbascum* etc.

Scrophularioideae (C-lobes similar; at least lower lvs opp., A usu. 4 ± staminode; 7 tribes (Selagineae in part referred to Globulariaceae): *Antirrhinum, Asarina, Bacopa, Calceolaria, Diascia, Jamesbrittenia, Linaria, Lindernia, Mimulus, Nemesia, Penstemon, Russelia, Scrophularia, Selago, Sutera, Torenia, Wightia* (strangler)

Rhinanthoideae (2 posterior C-lobes covered by 1 or both laterals in bud; many hemiparasites; 3 tribes): *Bartsia, Buchnera, Castilleja, Digitalis, Euphrasia, Hebe, Melampyrum, Pedicularis, Rhinanthus, Veronica*

Limits still not clear: Orobanchaceae prob. best incl., some elements of Globulariaceae s.l. also; recently added are *Oftia* & *Ranopisoa* form. in Myoporaceae, while *Schlegelia*, often allied with *Gibsoniothamnus* & *Paulownia* here once more in Bignoniaceae, *Rehmannia* in Gesneriaceae

Range of poll. mechanisms from open short-tubed fls suited to bees & flies (*Veronica, Verbascum*) to long-tubed bee-poll. (*Digitalis, Linaria, Antirrhinum*) or to fls which shower visitors with loose pollen (*Euphrasia*), bird-poll. etc.

In some ways the temp. counterpart of Gesneriaceae, which are v. close, but often thought of as largely herbaceous assemblage close to Bignoniaceae. These fams are intimately allied. Despite its size, the fam. of little consequence to mankind exc. as (sometimes serious) weeds (*Kickxia, Striga, Veronica*) save as ornamentals (cf. Compositae, Orchidaceae) exc. for a few drugpls (esp. *Digitalis, Picrorhiza*) & dyes (*Escobedia*); cult. pls in genera other than those above inc. *Chelone, Collinsia, Erinus, Mazus, Paulownia, Phygelius, Zaluzianskya*

Scrotochloa Judz. = *Leptaspis*

scrub bloodwood *Baloghia lucida*
Scurrula L. Loranthaceae. 20 S As. to China & Moluccas (Mal. 8; R: Blumea 36(1991)65)
scurvy grass *Cochlearia officinalis*
Scutachne A. Hitchc. & Chase. Gramineae (34b). 2 Cuba, Hispaniola
Scutellaria L. Labiatae (V). c. 350 cosmop. (Eur. 13) exc. S Afr. R: NRBGE 46(1990)345. Some locally medic. & cult. orn. (skull cap, helmet flower) incl. *S. baicalensis* Georgi (Siberia to Japan) – medic. in modern Chinese herbalism & *S. mexicana* (Torrey) A. Paton (*Salazaria m.*, SW N Am.) – shrub for arid areas
Scutia (DC) Brongn. Rhamnaceae. 3 trop. S Am., 1 OW trop. to S Afr. R: BTBC 101(1974)64. Alks
Scuticaria Lindley. Orchidaceae (V 10). 7 trop. S Am. Cult. orn. epiphytes
Scutinanthe Thwaites. Burseraceae. 2 Sri Lanka to Sulawesi
Scybalium Schott & Endl. Balanophoraceae. 4 trop. Am. R: FN 23(1980)25
Scyphanthus D. Don. Loasaceae. 2 Chile
Scyphellandra Thwaites = *Rinorea*
Scyphiphora Gaertner f. Rubiaceae (III). 1 coasts of Indomal. to Aus. & New Caled.
Scyphocephalium Warb. Myristicaceae. 4 trop. W Afr.
Scyphochlamys Balf.f. Rubiaceae (III 2). 1 Rodrigues: *S. revoluta* Balf.f. R: KM 6(1989)102 – heterostyly
Scyphocoronis A. Gray = *Millotia*
Scyphogyne Decne. Ericaceae. 12 S Afr.
Scyphonychium Radlk. Sapindaceae. 1 NE Brazil: *S. multiflora* (C. Martius) Radlk. – monoec. R: Bonpl. 6(1989)117
Scyphopappus R. Nordenstam = *Argyranthemum*
Scyphostachys Thwaites. Rubiaceae (II 5). 2 Sri Lanka
Scyphostegia Stapf. Scyphostegiaceae. 1 NW Borneo
Scyphostegiaceae Hutch. Dicots – Dilleniidae – Violales. 1/1 Borneo. Dioec. glabrous tree. Lvs simple, toothed, spirally arr.; stipules small, caducous. Fls v. small, subtended by tubular bracts, forming racemes or spikes arranged in racemes; males with P (3) + (3), 3 large nectary glands opp. inner P, A 3 in a column also opp. inner P, anthers with long connective & longit. slits; females with K 3, C 3, $\underline{G}$ (8–13), basally 1-loc. with sessile stigma with as many rays as G & ∞ basal erect, anatropous, bitegmic ovules with consp. funicles. Fr. a fleshy capsule dehiscing apically; seeds arillate with straight embryo in scanty oily endosperm & thin perisperm. n = 9
Only species: *Scyphostegia borneensis* Stapf
Most recent authors ally this with Flacourtiaceae, though seed structure suggests Celastraceae (Corner). Only fam. restricted to Mal.
Scyphostelma Baillon = *Cynanchum*
Scyphostrychnos S. Moore = *Strychnos*
Scyphosyce Baillon. Moraceae (IV). 2 W trop. Afr.
Scyphularia Fée = *Davallia*
Scytopetalaceae Engl. Dicots – Dilleniidae – Theales. 5/20 trop. W Afr. Trees, shrubs & lianes. Lvs simple, spirally arr.; stipules 0. Fls reg., bisexual in term. panicles, axillary racemes or in fascicles on branches; K cupular with truncate or toothed margin, C (0)3–10(–16) often basally connate, sometimes forming a decid. calyptra, lobes linear & valvate, A (10–)∞ in c. 3–6 whorls around or on disk, free or connate only at base, arising centrifugally, anthers with longit. slits or term. pores, $\underline{G}$ (3–8) with curved style & as many loc. as G (partitions sometimes incomplete apically), each loc. with 2–8 apical-axile pendulous, anatropous, bitegmic ovules. Fr. a loculicidal capsule or 1-seeded drupe; embryo embedded in ruminate endosperm. n = 11, 18
Genera: *Brazzeia, Oubanguia, Pierrina, Rhaptopetalum, Scytopetalum*
Scytopetalum Pierre ex Engl. Scytopetalaceae. 3 trop. W Afr. Some locally used timber
sea aster *Tripolium vulgare*; **s. bean** *Entada gigas*; **s. beet** *Beta vulgaris* subsp. *maritima*; **s. blite** *Suaeda vera, S. maritima*; **s. buckthorn** *Hippophae rhamnoides*; **s. campion** *Silene uniflora*; **s. daffodil** *Pancratium maritimum*; **s. heath** *Frankenia* spp. esp. *F. laevis*; **s. holly** *Eryngium maritimum*; **s. island cotton** *Gossypium barbadense*; **s. kale** *Crambe maritima*; **s. lavender** *Limonium* spp.; **s. onion** *Drimia maritima*; **s. pink** *Armeria maritima*; **s. poppy** *Glaucium flavum*; **s. purslane** *Honckenya peploides*; **s. rocket** *Cakile maritima*; **s. samphire** *Crithmum maritimum*; **s. wormwood** *Seriphidium maritimum*
sea-grass *Zostera marina*; **s. matting** (Aus.) *Oryza sativa*
sealing-wax palm *Cyrtostachys renda*
Sebaea Sol. ex R. Br. Gentianaceae. 60 warm Afr. to India, Aus., NZ. Some cult. orn., usu.

annuals e.g. *S. albens* (L.f.) Sm. & *S. aurea* (L.f.) Sm. (S Afr.)

Sebastiania Sprengel. Euphorbiaceae. c. 100 trop. Am., E US, W Afr. to Aus. (1), W Mal. (3). Some, esp. *S. pavoniana* Muell. Arg. (Mex.), with fr. when occupied by larvae of a small moth, *Cydia* (*Carpocapsa*) *saltitans*, give 'jumping beans'; *S. bilocularis* S. Watson (yerba de la feche, SW N Am.) – source of arrow-poison

Sebastiano-schaueria Nees. Acanthaceae. 1 Brazil

sebastião-de-arruda *Dalbergia decipularis*

Sebertia Pierre ex Engl. = *Niemeyera*

sebesten plum *Cordia myxa*

Secale L. Gramineae (23). 3 Euras., S Afr.: Eur. 2 + *S. cereale* L. (rye) derived from *S. montanum* Guss., a weed of wheat & barley, giving rise to non-shattering-eared forms, from which large-fruited forms selected in E Turkey, where rye orig. used as crop in places difficult for other cereals – grain rich in gluten & used in schwarzbröt & crispbreads, in whisky (US), gin (Holland) & beer (Russia); attacked by ergot but good stock fodder & stems used for the best straw-matting & archery targets; root-system in 0.051 m^3 found to be 622.8 km long!

Secamone R. Br. Asclepiadaceae (II). 62 Madag. & W Indian Ocean. R: OB 112(1992)5

Secamonopsis Jum. Asclepiadaceae (II). 1 Madag.: *S. madagascariensis* Jum. – some rubber

Sechiopsis Naudin. Cucurbitaceae. 5 C Am. R: SB 17(1992)395. Fr. winged

Sechium P. Browne. Cucurbitaceae. c. 6 (incl. *Frantzia*) trop. Am. *S. edule* (Jacq.) Sw. (chayote, choco, cho-cho, choko, chow chow, christophine, Madeira marrow, vegetable pear) – fr. & tubers cooked, each fr. with 1 v. large seed; in Java grown as cover for fish-ponds

Secondatia A. DC. Apocynaceae. 7 trop. Am. R: AMBG 22(1935)224

Securidaca L. Polygalaceae. 80 trop. (Afr. 2; R: SAJB 59(1987)5) exc. Aus. Trees & scramblers; fr. a samara with dorsal wing. Some cult. orn. for showy fragrant fls incl. *S. longipedunculata* Fres. (trop. Afr.) – small tree with allegedly up to 100 medical uses, twigs source of a fibre (buaze) used in fishing-nets

Securigera DC (~ *Coronilla*). Leguminosae (III 19). 12 Medit. to Somalia. R: Willd. 19(1989)59. Stems angled, those in *Coronilla* rounded

Securinega Comm. ex Juss. Euphorbiaceae. 20 trop. & warm (Eur. 1). Alks. *S. durissima* J. Gmelin (Otaheite myrtle, Masc.); *S. flexuosa* Muell. Arg. (Mal.) – timber for house-posts in Solomon Is. See also *Flueggea*

Sedastrum Rose = *Sedum*

Seddera Hochst. Convolvulaceae. 15 trop. & warm Afr., Madagascar & Arabia

Sedella Fourr. = *Sedum*

sedge *Carex* spp. & other Cyperaceae

Sedirea Garay & H. Sweet (~ *Aerides*). Orchidaceae (V 16). 2 E As.

Sedopsis (Legrand) Exell & Mendonça = *Portulaca*

sedra *Ziziphus jujuba*

Sedum L. Crassulaceae. c. 280 N temp. (Eur. 55), trop. mts, Madag., Mex. Stonecrops, livelong. R: JFSUTB III, 12(1978)175; R. L. Praeger (repr. 1967) *An account of the genus S. as found in cultivation*; R. Stephenson (1994) *S. Cultivated stonecrops*. Generic limits disputed (*Rhodiola* (but see JFSUTB III, 12(1978)182) & *Hylotelephium, Sedella & Telmissa* incl. here); hybrids formed with spp. of *Echeveria* & *Villadia*. Many cult. orn. succ. herbs & shrublets, some locally eaten in salads or medic. *S. acre* L. (wall pepper, stonecrop, Euras., Medit.) – natur. N Am.; *S. morganianum* Walther (Mex.) – trailing tail-like shoots, hanging-basket pl.; *S. populifolium* Pallas (*Hylotelephium p.*, W Siberia) – decid. subshrub to 30 cm; *S. praealtum* DC (*S. dendroideum* ssp. *p.*, Mexico) – shrub to 2 m or trailing to 6 m with trunk to 10 cm diam. & branches with term. bunches of lvs, natur. Medit. & S GB; *S. rosea* (L.) Scop. (*Rhodiola r.*, '*S. roseum*', roseroot, N hemisph.) – lvs eaten; *S. rupestre* L. (*S. reflexum* (W & C Eur.) – lvs eaten in salads; *S. spectabile* Boreau (*Hylotelephium s.*, E As.) – border pl., a parent with *S. telephium* of pop. 'Autumn Joy' favoured by late butterflies; *S. suaveolens* Kimnach (Mex.) – 2n = 640!; *S. telephium* L. (*H. t.*, orpine, live-forever, Eur. to Siberia)

seedcake *Carum carvi*

Seemannaralia R. Viguier. Araliaceae. 1 S Afr.

Seemannia Regel = *Gloxinia*

seersucker plant *Geogenanthus poeppigii*

Seetzenia R. Br. ex Decne. Zygophyllaceae. 1 N & S Afr. to Afghanistan: *S. lanata* (Willd.) Bullock (*S. africana*)

Segetella Desv. = *Spergularia*

Seguieria Loefl. Phytolaccaceae (II). 6 trop. S Am. R: MBSM 18(1982)244. Stipules thorny, G1, fr. a samara; strong garlic odour

Sehima Forssk. Gramineae (40d). 5 trop. OW
Seidelia Baillon. Euphorbiaceae. 2 S Afr.
Seidenfadenia Garay. Orchidaceae (V 16). 1 Burma & Thailand: *S. mitrata* (Reichb.f.) Garay. R: OB 95(1988)212
Seidlitzia Bunge ex Boiss. Chenopodiaceae (III 3). 7 W As.
seje *Oenocarpus bataua*
Selaginella Pal. Selaginellaceae. c. 700 trop. & warm, few temp. (Eur. 4). Cult. orn. esp. *S. kraussiana* (Kunze) A. Braun (S Afr.) – creeping & rooting sp. much grown in greenhouses, natur. Cornwall, Ireland etc.; *S. lepidophylla* (Hook. & Grev.) Spring (rose of Jericho, resurrection plant, S US to Peru) – tufted pl. with branches curling up into ball when dry but re-opening when wet (also in some other spp.), sold as curiosity, locally medic.; *S. tuberosa* McAlpin & Lellinger (Costa Rica) – annual reprod. from aestivating tubers arising on aerial stems; *S. willdenowii* (Poiret) Baker (OW trop.) – cult. orn. scandent to 6 m or more, lvs appearing blue due to reflections of cuticle & outer epidermal walls differentially reflecting blue, a phenomenon not fully developed in full light
Selaginellaceae Willk. Lycopsida. 1/700 cosmop. Erect, prostrate, tufted, creeping & rooting or climbing moss-like pls with dichot. branching, s.t. frond-like, & dichot. roots. Lvs scale-like, all similar & spirally arr. or dimorphic & 4-ranked. Sporophylls in term. strobili bearing microsporangia with many microspores & megasporangia usu. with 4 megaspores; embryology v. varied. n = usu. 9 (low in Pteridophyta)
Genus: *Selaginella* (& fossil *Selaginellites* (Carboniferous))
Selago L. Scrophulariaceae. 150 trop. (few) & S Afr. Some cult. orn. esp. *S. spuria* L. (blue haze, S Afr.)
selangan *Shorea* spp. esp. *S. guiso*
Selenia Nutt. Cruciferae. 4 S N Am. R: JAA 69(1988)127. *S. aurea* Nutt. (Montana, Kansas to Texas) – cult. orn. annual with sweetly-scented fls
Selenicereus (A. Berger) Britton & Rose. Cactaceae (III 2). c. 20 S US to trop. Am. Climbers with cardiac glycosides, aerial roots & nocturnal fls; cult. orn. esp. *S. grandiflorus* (L.) Britton & Rose (night-flowering cactus, Cuba & Jamaica, natur. trop. Am.) – cult. in Mex. for drug used in rheumatism, used as heart-stimulant in Costa Rica & now cult. comm. for this in Germany etc.
Selenipedium Reichb.f. Orchidaceae (II). 6 trop. Am. R: C. Cash, *The slipper orchids* (1991)131. Some spp. to 5 m tall (some form. vanilla subs.)
Selenodesmium (Prant) Copel. = *Cephalomanes*
Selenothamnus Melville = *Lawrencia*
Selera Ulbr. = *Gossypium*
self-heal *Prunella vulgaris*
Selinocarpus A. Gray. Nyctaginaceae (III). 10 trop. Am. R: Phytologia 37(1977)177, 75(1993)239. On gypsum outcrops
Selinopsis Coss. & Durieu ex Batt. & Trab. (~ *Carum*). Umbelliferae (III 8). 2 Medit.
Selinum L. Umbelliferae (III 8). 2 Eur. (excl. *Cnidium*) to C As.
Selkirkia Hemsley. Boraginaceae. 1 Juan Fernandez
Selleola Urban = *Minuartia*
Selleophytum Urban = *Coreopsis*
Selliera Cav. (~ *Goodenia*). Goodeniaceae. 1 Aus., NZ, temp. S Am.: *S. radicans* Cav. – fr. indehiscent. R: FA 35(1992)281
Selliguea Bory. Polypodiaceae (II 2). c. 50 India to Japan & New Guinea (most), Aus. (1) & Pacific, S Afr. (1), Madag. (1). *S. feei* Bory (Mal.) a consp. terr. sp. near fumaroles on volcanoes in Java
Selloa Kunth. Compositae (Helia.-Helia.). 3 Mex. to C Am. R: PMMSUBS 4(1970)371. Some locally medic.
Sellocharis Taubert. Leguminosae (III 28). 1 SE Brazil
Selysia Cogn. Cucurbitaceae. 4 N S Am. R: Novon 4(1994)37
semaphore plant *Codariocalyx motorius*
Semecarpus L.f. Anacardiaceae (III). 60 Indomal. to New Caled. (c. 5). Trees, some pachycaul treelets with detritus-coll. heads of large lvs often occupied by ants. *S. anacardium* L.f. (trop. As. to Aus.) – unripe fr. (marking nut) has sap drying as a black resin used as an ink or dye for linen when mixed with lime; green fr. used as bird-lime & in tanning etc.; *S. australiensis* Engl. (tar tree, trop. Aus.) – sap dermatological, swollen peduncle ed. but pericarp toxic
Semeiandra Hook. & Arn. = *Lopezia*
Semeiocardium Zoll. = *Impatiens*

Semele Kunth. Asparagaceae (Ruscaceae; Liliaceae s.l.). 2 Macaronesia. *S. androgyna* (L.) Kunth – cult. orn. dioec. liane to 20 m with shoots arising from underground & bearing flattened leaf-like organs in axils of scale lvs; fls borne in small infls around edge of the leaf-like organs

Semenovia Regel & Herder (*Platytaenia*; ~ *Heracleum*). Umbelliferae (III 11). 18 As.

Semialarium Hallé (*'Hemiangium'*). Celastraceae. 2 S Am. R: BMNHN 4,5(1983)24

Semiaquilegia Makino. Ranunculaceae (III 1). 6 E As. R: KB 1920:165. Like *Aquilegia* but fls ± spurless; some cult. orn. esp. *S. adoxoides* (DC) Makino (E China, Korea, Japan) with cream & chocolate coloured fls

Semiarundinaria Makino ex Nakai. Gramineae (1b). c. 5 China & Japan. R: KBAS 13(1986)28. Cult. orn. bamboos

Semibegoniella C. DC = *Begonia*

Semiliquidambar H.T. Chang (~ *Altingia, Liquidambar*). Hamamelidaceae (IV). c. 3 E China

Seminole tea *Asimina reticulata*

Semiramisia Klotzsch. Ericaceae. 4 N Andes. R: SB 9(1984)359

Semnanthe N.E. Br. = *Erepsia*

Semnostachya Bremek. = *Strobilanthes*

Semnothyrsus Bremek. = *Strobilanthes*

semolina *Triticum turgidum* Durum Group

Semonvillea Gay = *Limeum*

Sempervivella Stapf = *Rosularia*

Sempervivum L. Crassulaceae. c. 50 Eur. (excl. *Jovibarba*, 19), Morocco, W As. R: QBAGS 47(1979)75. Houseleeks. Cult. orn. rosetted succ. with offsets (see also *Jovibarba*) grown in rock-gardens etc. esp. *S. tectorum* L. (Eur.) – form. much planted on roofs to keep slates in place & held to ward off thunder, form. crushed in lard for treatment of chilblains & piles & used since time of Pliny in plant-pest control, variable & forms hybrids readily both in cult. & in wild, as in Pyrenees with *S. arachnoideum* L. (S Eur. mts) – commonly cult. its smaller rosettes with cobwebby strands connecting leaf-tips, so that some 'spp.' may represent ancient hybrids. Over 1000 named cvs. Alks incl. nicotine. Journal: S. Society J.

sempilor *Dacrydium elatum*

sen *Kalopanax septemlobus*

Senaea Taubert. Gentianaceae. 2 Brazil

senat seed *Cucumis melo* subsp. *agrestis*

Senecio L. Compositae (Sen.-Sen.). c. 1250 cosmop. (Eur. c. 60, Afr. c. 350, S Am. c. 500) exc. Antarctica. Trees, shrubs, lianes & herbs; although recently shorn (or reshorn) of a number of satellite genera (see *Brachyglottis, Cineraria, Crassocephalum, Delairea, Dendrosenecio, Gynura, Kleinia, Lachanodes, Ligularia, Othonna, Parasenecio, Pericallis, Pladaroxylon, Roldana, Sinacalia, Sinosenecio, Solanecio, Tephroseris* etc.), still one of the largest genera of seed-pls & limits not fully clear. Many alks toxic to stock esp. in *S. jacobaea* L. (ragwort, Eur., Medit.) but when attacked by cinnabar moth producing 'regrowth' shoots with smaller fruits giving less competitive seedlings; some cult. orn. & weeds: *S. cineraria* DC (*S. bicolor*, dusty-miller, W & C Medit., natur. GB) – cult. orn. alleged to be efficacious in eye problems; *S. articulatus* (L.f.) Scultz-Bip. (*Kleinia a.*, candle plant, S Afr.) – familiar window-sill pl. with jointed stems & lobed term. lvs; *S. crassissimus* Humbert (Madag.) – succ. sp. with lvs held vertically; *S. elegans* L. (S Afr.) – cult. orn. ('jacobaea') with purple (or white) ray-florets, natur. Calif.; *S. inaequidens* DC (S Afr.) – natur. along railway lines in Germany etc.; *S. madagascariensis* Poiret (fireweed, S Afr.) – pest in Aus. introd. early C20; *S. rowleyanus* H.J. Jacobsen (SW Afr.) – mat-forming pl. with slender stems & bead-like lvs with no apical or marginal growth, but radial expansion from an adaxial meristem (marked by a narrow 'window', which is thus not the junction of leaf margins); *S. smithii* DC (S Patagonia) – introd. to Orkney & Shetland by whalers; *S. squalidus* L. (Oxford ragwort, S Eur., natur. GB, having escaped from Oxford Botanic Garden along railway lines to London etc., hybridizing with *S. vulgaris* (tetraploid) in N Wales & Edinburgh to form the sexual hexaploid sp., *S. cambrensis* Rosser, a sp. which has thereby arisen polytopically (cf. *Spartina anglica, Tragopogon minus*); *S. vulgaris* L. (groundsel, pantemp. weed) – poss. autotetraploid of *S. vernalis* Waldst. & Kit. (self-incompat. sp.) in E Medit.

Senefeldera C. Martius. Euphorbiaceae. 9–10 trop. S Am. *S. inclinata* Muell.-Arg. – bark for toothache in Amazon

Senefelderopsis Steyerm. Euphorbiaceae. 2 NW trop. S Am.

senega root *Polygala senega*

Senegal tea *Gymnocoronis spilanthoides*

Senegalia Raf. = *Acacia*
Senisetum Honda = *Agrostis*
senna *Senna italica*; **Alexandrian s.** *S. alexandrina*; **American s.** *S. marilandica*; **bastard s.** *Coronilla valentina*; **bladder s.** *Colutea arborescens*; **coffee s.** *S. occidentalis*; **dog** or **Italian s.** *S. italica*; **prairie s.** *Chamaecrista fasciculata*; **ringworm s.** *S. alata*; **scorpion s.** *Hippocrepis emerus*; **Spanish s.** *S. italica*; **Tinnevelly s.** *S. alexandrina*; **wild s.** *S. marilandica*
Senna Miller (~ *Cassia*). Leguminosae (I 2). c. 350 trop. (Afr. 24, R: KB 43(1988)338) & warm temp. Form. *Cassia* subg. *Senna* (Miller) Benth. but differing from *Cassia* s.s. in filaments of all A straight & bracteoles 0. Trees, shrubs & herbs, buzz-poll., many imp. medic. (anthraquinones) & cult. orn. (senna; colonizing spp. in Aus. called 'acacia'). *S. alata* (L.) Roxb. (*C. a.*, ringworm cassia or senna, craw-craw, trop. Am., natur. trop. OW) – for skin disease & vermifuge, cult.; *S. alexandrina* Miller (*C. senna*, Alexandrian or Tinnevelly s., NE Afr. Middle E) – lvs & pods some of the s. of commerce; *S. artemisioides* (DC) Randell (*C. sturtii*, Aus.) – drought-tolerant fodder shrub esp. in Israel; *S. auriculata* (L.) Roxb. (*C. a.*, avaram, Matara tea, turwad bark, tanner's c., Tanz., India, Sri Lanka); *S. corymbosa* (Lam.) Irwin & Barneby (*C. c.*, N S Am., natur. N Am.) – much cult.; *S. didymobotrya* (Fres.) Irwin & Barneby (*C. d.*, trop. Afr., natur. As. & Am.) – cult. as green manure in As., cult. orn. shrub; *S. italica* Miller (*C. i.*, Italian or Spanish or dog s., OW trop.) – lvs & pods some of the s. of commerce; *S. marilandica* (L.) Link (*C. m.*, Am. or wild s., E US); *S. obtusifolia* (L.) Irwin & Barneby (*C. o.*, trop. & warm) – green lvs fermented to give protein source (20% by weight, kawal) by bacteria & *Rhizopus* fungi in Sudan; *S. occidentalis* (L.) Link (*C. o.*, coffeeweed, trop. OW) – medic., seeds a coffee subs.; *S. siamea* (Lam.) Irwin & Barneby (*C. s.*, minjiri, Burma to Mal.) – cult. orn. & firewood crop, timber hard & strong for bridges, mine props, telegraph poles etc.; *S. tora* (L.) Roxb. (*C. t.*, Indomal.) – medic. (cult. India), seeds used as mordant in dyeing blue cloth, seeds a coffee subs.
Sennia Chiov. = *Dialium*
Senniella Aellen = *Atriplex*
Senra Cav. Malvaceae. 1 E Afr. to India. Fruit in persistent K wind-disp.
sensitive plant *Mimosa pudica*
sentul *Sandoricum koetjape*
Sepalosaccus Schltr. = *Maxillaria*
Sepalosiphon Schltr. = *Glossorhyncha*
Separotheca Waterf. = *Tradescantia*
Sepikea Schltr. Gesneriaceae. 1 New Guinea
Septas L. = *Crassula*
Septimia P.V. Heath = *Crassula*
Septogarcinia Kosterm. = *Garcinia*
Septotheca Ulbr. Bombacaceae. 1 Peru
Septulina Tieghem (~ *Taxillus*). Loranthaceae. 2 S Afr.
Sequoia Endl. Taxodiaceae. 1 Oregon to Calif. in foothills within 30 km of coast: *S. sempervirens* (D. Don) Endl. (Californian redwood) – massive tree to 1000 yrs old & 103.6 m with bole to 8.5 m diam. with valuable timber, form. much used for shingles, fences & general building etc., overexploited & now restricted to a few reserves etc.; root-suckers; cotyledons 2; mitochondrial DNA paternally inherited (? unique); cult. orn. incl. dwarf cvs
Sequoiadendron Buchholz. Taxodiaceae. 1 W slopes of Sierra Nevada of California: *S. giganteum* (Lindley) Buchholz (wellingtonia, mammoth or big tree) – discovered 1850, introd. Eur. 1853, reaching 105 m with bole to 12 m diam. (the largest trees on earth) & poss. 3000 yrs ('General Sherman' 84 m tall, 25 m girth with trunk vol. c. 1400 m^3 & wt. c. 2500 t), reduced to 72 groves (some v. small), timber only good for shingles (& fencing); cotyledons usu. 4; cult. orn. (the biggest tree of every county in GB, is never blown down & is indifferent to frost & drought) incl. dwarf cvs & 'Pendulum' with sinuous trunk sometimes 'looping the loop'
serai *Cymbopogon citratus*
Seraphyta Fischer & C. Meyer = *Epidendrum*
Serapias L. Orchidaceae (IV 2). 13 Azores to Medit. (Eur. 5). R: QBAGS 43(1975)188. Terr. *S. vomeracea* (Burm.f.) Briq. (Medit.) – in Israel bees sleep inside fls & are warmed 3 °C more than outside temperature in morning
seraya, red *Shorea* spp.; **white s.** *Parashorea* spp.; **yellow s.** *S.* spp.
Serenoa Hook.f. Palmae (I 1b). 1 SE US: *S. repens* (Bartram) Small (*S. serrulata*, saw palmetto (palm)) – colonial ± stemless palm with bisexual fls & toothed petiole margins often forming pestilential thickets, fossils in Eocene of London Clay, fr. ed. & medic., a tea also made from them, allegedly aphrodisiac (male)

Seretoberlinia Duvign. = *Julbernardia*

Sergia Fed. (~ *Campanula*). Campanulaceae. 2 C As.

Serialbizzia Kosterm. = *Albizia*

Serianthes Benth. Leguminosae (II 5). 17 E Mal. to W Pacific esp. New Caled. R: BMNHN 4,6(1984)84. Some timber

Sericanthe Robbrecht (~ *Tricalysia*). Rubiaceae (II 1). 15 trop. Afr. R: BJBB 48(1978)3, 51(1981)171

Sericocalyx Bremek. = *Strobilanthes*

Sericocarpus Nees (~ *Aster*). Compositae (Ast.-Sol.). 5 N Am. R: Phytol. 75(1993)45

Sericocoma Fenzl. Amaranthaceae (I 2). 2 S & SW Afr.

Sericocomopsis Schinz. Amaranthaceae (I 2). 2 trop. E Afr.

Sericodes A. Gray. Zygophyllaceae. 1 N Mexico

Sericographis Nees = *Justicia*

Sericolea Schltr. Elaeocarpaceae. 16 E Mal. R: Blumea 28(1982)103

Sericorema (Hook.f.) Lopr. Amaranthaceae (I 2). 2 S Afr.

Sericospora Nees. Acanthaceae. 1 WI (= ?)

Sericostachys Gilg. & Lopr. Amaranthaceae (I 2). 1 trop. Afr.

Sericostoma Stocks. Boraginaceae. 1 trop. E Afr. to NW India

Seringia Gay. Sterculiaceae. 1 New Guinea, E Aus.: *S. arborescens* (Aiton) Druce

Seriphidium (Hook.) Fourr. (~ *Artemisia*). Compositae (Anth.-Art.). 130 N temp. R: CN 25(1994)39. Differs from *A.* in all fls bisexual; char. of arid SW US where *S. tridentatum* (Nutt.) W.A. Weber etc. form 'sage-brush'. *S. maritimum* (L.) Soják (*A. m.*, sea wormwood, Euras.) – source of santonin (vermifuge)

Serissa Comm. ex Juss. Rubiaceae (IV 12). 2 China (? & Japan). *S. japonica* (Thunb.) Thunb. (*S. foetida*) – early introd. or native to Japan, cult. orn. shrub with foetid lvs (when bruised), esp. cvs with double fls & varieg. lvs, also sold as bonsai in Eur.

Serjania Miller. Sapindaceae. 215 trop. & warm Am. Polygamous lianes with 'watch-spring' tendrils & 3-winged schizocarps; some cordage & fish-poisons, *S. cuspidata* Cambess. (Brazil) grown on trop. pergolas & fences

serpentaria or **serpentary** = snakeroot

serpolet oil *Thymus polytrichus*

Serpyllopsis Bosch. Hymenophyllaceae (Hym.). 1 S S Am., Juan Fernandez, Falkland Is.: *S. caespitosa* (Gaudich.) C. Chr.

serradella *Ornithopus sativus*

serrated tussock *Nassella trichotoma*

Serratula L. Compositae (Card.-Cent.). c. 70 Eur. (17) to N Afr. & Japan. Pls ± dioec., monoec. or with bisexual fls. Some cult. orn. esp. forms of *S. tinctoria* L. (saw-wort, Euras., N Afr.) – lvs with alum give green or yellow dye for wool, form. medic. (wounds & piles) & *S. seoanei* Willk. (SW Eur.) such as '*S. shawii*', a dwarf rock-pl.

Serruria Burm. ex Salisb. Proteaceae. 65 SW Cape. Some cult. orn. shrubs

Sertifera Lindley & Reichb.f. Orchidaceae (V 13). 6 Andes

service berry *Amelanchier* spp.; **s. tree** *Sorbus torminalis*

sesame or **sesamum** *Sesamum orientale*; **black s.** *Hyptis spicigera*

Sesamoides All. Resedaceae. 4+ W Medit. (inc. Eur.)

Sesamothamnus Welw. Pedaliaceae. 5 trop. Afr. (disjunct: 2 NE Afr., 2 SW & 1 SE). Some with water-storing trunks, cult. orn.

Sesamum L. Pedaliaceae. 15 OW trop. & S Afr. *S. orientale* L. (*S. indicum*, sesame, sesamum, simsim, gingelly, halvah, trop. OW) – widely natur., long cult. (oldest grown oilseed since c. 3500–3050 BC in Indus & Anc. Mesopotamia), the oil used like olive o., the seeds sprinkled on bread & cakes like poppy s. or made into sticky sweetmeats & used in cosmetics, by 1979 the world's 9th most imp. veg. oil; paste of crushed seeds used in Lebanese cooking = tahini. R: JAA suppl. 1(1991)328

sesban *Sesbania sesban*

Sesbania Scop. Leguminosae (III 7). 50 warm & usu. wet. Some cult. orn. herbs, shrubs & shade-trees; some poss. firewood crops incl. *S. bispinosa* (Jacq.) W. Wight (dhaincha, OW trop.) – poss. source of guar, fibres used for sails & fishing-nets, *S. cannabina* (Retz.) Pers. ('*S. aculeata*', '*S. bispinosa*', orig. Aus.) – widely cult., *S. eremurus* (Aublet) Urban (trop. Am.) – annual growing to 6 m; *S. exaltata* (Raf.) A.W. Hill (N Am.) – form. imp. Indian fibre, *S. grandiflora* (L.) Pers. (bakphul, trop. As., natur. Am.) – cult. orn. with large (to 10 cm) red or white fls, ed. as salad (fr. inedible), lvs & bark medic., *S. punicea* (Cav.) Benth. (S Am.) – natur. SE US & pestilential in S Afr. fynbos where controlled by weevils, *S. sesban* (L.) Merr. (sesban, *S. aegyptiaca*, OW trop.) – molluscicidal saponins, imp. fodder & fibre pl., green manure, wood form. used for gunpowder charcoal. *S. rostrata* Bremek. & Oberm. (S Afr.) – nitrogen-fixing nodules on stems

Seseli L. Umbelliferae (III 8). 100–120 Eur. (34) to C As. & N trop. Afr.

Seselopsis Schischkin. Umbelliferae (III 8). 2 C As.

Seshagiria Ansari & Hemadri. Asclepiadaceae (III 1). 1 W India: *S. sahyadrica* Ansari & Hemadri

Sesleria Scop. Gramineae (17). 27 Eur. (26) esp. Balkans, W As. R: OBC 3(1946)1. Spp. closely related & introgressing e.g. *S. caerulea* (L.) Ard. (*S. albicans*, blue moorgrass, W & C Eur.)

Sesleriella Deyl = *Sesleria*

Sessea Ruíz & Pavón. Solanaceae (2). 7 S Am.

Sesseopsis Hassler = *Sessea*

Sessilanthera Molseed & Cruden. Iridaceae (III 4). c. 3 Mexico & C Am.

Sessilistigma Goldbl. = *Homeria*

Sestochilos Breda = *Bulbophyllum*

Sesuvium L. Aizoaceae (II). 12 warm, esp. coastal. R: JAA 51(1970)450. Halophytes – 1 pantrop.: *S. portulacastrum* (L.) L. (mboga (Afr.), veg. sold in markets in As.), 1 Galápagos, 4 Angola area

Setaria Pal. Gramineae (34b). (Incl. *Paspalidium*) c. 150 trop. & warm (Eur. 3 + *S. italica*). R: Sida 15(1993)447. Some are hosts for maize & sugar-cane mosaic viruses. *S. italica* (L.) Pal. (*S. viridis* ssp. *italica*, foxtail, German, Hungarian or Italian millet, Bronze Age cereal cultigen prob. derived from *S. viridis* (L.) Pal. (temp. & warm, like similar *S. verticillata* (L.) Pal. often a noxious weed) in China or poss. several centres in Eur. & As. c. 5000 BC) – hay, pasture & cereal, often seen as birdseed, the millet-spray of petshops; *S. pumila* (Poiret) Roemer & Schultes ('*S. glauca*', yellow foxtail, pigeon grass, cat-tail millet, warm) – cattle fodder; *S. palmifolia* (Koenig) Stapf (Indomal.) – young shoots eaten with rice in Java & New Guinea, where domesticated (highland pitpit), cult. orn. in Eur.; *S. sphacelata* (Schum.) Moss (S Afr.) – diploids to decaploids with hybridization between levels, imp. silage crop

Setariopsis Scribner ex Millsp. Gramineae (34b). 2 Mexico to Colombia

Setchellanthus Brandegee. Capparidaceae. 1 Mexico. Fls blue

Setcreasea Schumann & Sydow = *Tradescantia*

Setiacis S.L. Chen & Y.X. Jin = *Panicum*

Seticleistocactus Backeb. = *Cleistocactus*

Setiechinopsis Backeb. = *Echinopsis*

Setilobus Baillon. Bignoniaceae. 3 Brazil

setterwort *Helleborus foetidus*

Seutera Reichb. = *Cynanchum*

Sevada Moq. (~ *Suaeda*). Chenopodiaceae (III). 1 Sudan, Ethiopia, Somalia, Arabia: *S. schimperi* Moq. R: NJB 11(1991)315

sevadilla *Schoenocaulon officinale*

sevendara (grass) *Vetiveria zizanioides*

Severinia Ten. ex Endl. (~ *Atalantia*). Rutaceae. 5–6 Indomal. R: W.T. Swingle, *Bot. Citrus* (1943) 274. *S. buxifolia* (Poiret) Ten. (S China & Taiwan) – hedge-pl. & citrus stock; *S. linearis* (Blanco) Swingle (*A. l.*, Philippines) – rheophyte

Seychellaria Hemsley. Triuridaceae. 1 Tanzania, 1 Madagascar, 1 Seychelles

Seymeria Pursh. Scrophulariaceae. 25 S N Am. *S. cassioides* (J. Gmelin) S.F. Blake (SE US) – hemiparasite on pine roots

Seymeriopsis Tzvelev = praec.

Seyrigia Keraudren. Cucurbitaceae. 4 Madagascar

shad or **s. bush** *Amelanchier* spp.

shaddock *Citrus maxima*

Shafera Greenman. Compositae (Sen.-Sen.). 1 E Cuba (serpentines). Multiseriate involucral bracts odd in tribe

Shaferocharis Urban. Rubiaceae (III 4). 3 E Cuba (serpentines)

Shaferodendron Gilly = *Manilkara*

shagbark hickory *Carya ovata*

shaggy pea *Oxylobium* spp.

shaking grass *Briza* spp.

shallon bush *Gaultheria shallon*

shallot *Allium cepa* Aggregatum Group

shamrock usu. *Trifolium dubium* but sometimes *T. repens* or *Medicago lupulina* while in US *Oxalis* spp. often used

Shaniodendron M.B. Deng, H.T. Wei & X.K. Wang = *Hamamaelis*
Shantung cabbage *Brassica rapa* Chinensis Group
Sharon fruit early-fruiting seedless persimmons; **rose of S.** poss. orig. *Lilium candidum* or *Tulipa agenensis* subsp. *boissieri*, lately *Hypericum androsaemum* & in Am. *Hibiscus syriacus*
she balsam *Abies fraseri*; **s. oak** *Casuarina* spp.
shea butter *Vitellaria paradoxa*
Sheareria S. Moore. Compositae (Ast.). 2 China
sheep's bit scabious *Jasione montana*; **s. bush** *Geijera parviflora*; **s.'s fescue** *Festuca ovina*; **s. laurel** *Kalmia angustifolia*; **s.('s) sorrel** *Rumex acetosella*; **vegetable s.** *Psychrophyton* spp. esp. *P. mammillaris*
sheesham *Dalbergia sissoo*
Sheilanthera I.J. Williams. Rutaceae. 1 SW Cape mts
shell flower *Moluccella laevis*
shellac or **lac** a resin derived from secretion of lac insect (*Laccifer lacca*) feeding on trees such as *Butea monosperma*, *Schleichera oleosa* & *Ziziphus jujuba* etc. in trop. As., now of little imp. because of synthetics
Shepherdia Nutt. Elaeagnaceae. 3 N Am. Dioec. shrubs & small trees with actinomycete root-symbionts & drupe-like swollen receptacle in fr. *S. argentea* (Pursh) Nutt. (buffalo-berry, beef-suet tree, silverberry, E N Am.) – cult. orn. v. hardy hedge-pl. with ed. fr. used in jelly or dried by Indians & eaten in winter with buffalo meat
shepherd's needle *Scandix pecten-veneris*; **s.'s cress** *Teesdalia nudicaulis*; **s.'s purse** *Capsella bursa-pastoris*; **s.'s rod** *Dipsacus pilosus*; **s.'s weather-glass** *Anagallis arvensis*
Sherardia L. Rubiaceae (4 16). 1 Eur., Medit., W As.: *S. arvensis* L. (field madder) – natur. S Afr., Am.
Sherbournia G. Don f. Rubiaceae (II 1). 10 trop. Afr. K convolute
shi mei *Rosa davurica*
Shibataea Makino ex Nakai. Gramineae (1b). 8 SE China & SW Japan (1). R: APS 26(1988)130. *S. kumasasa* (Steudel) Nakai (Japan, where common garden pl.) – small bamboo for ground-cover
shield fern *Dryopteris* spp.
shingle oak *Quercus imbricaria*
Shinnersia R. King & H. Robinson. Compositae (Eup.-Tric.). 1 S US, Mex.: *S. rivularis* (A. Gray) R. King & H. Robinson – wholly aquatic. R: MSBMBG 22(1987)190
Shinnersoseris Tomb (~ *Lygodesmia*). Compositae (Lact.-Step.). 1 N Am.
shisham wood *Dalbergia sissoo*
shittim wood *Acacia raddiana* or *A. seyal*
Shiuyinghua Paclt. Scrophulariaceae. 1 C China. Like *Paulownia*
shoe-flower *Hibiscus rosa-sinensis*
shola pith *Aeschynomene aspera*, *A. indica*
shoo-fly *Nicandra physalodes*
shooting-star *Dodecatheon* spp.
shore weed *Littorella uniflora*
Shorea Roxb. ex Gaertner f. Dipterocarpaceae. 357 Sri Lanka to S China, Moluccas & Lesser Sunda Is. (Mal. 163). Usu. emergent trees of rain forest – the most imp. timber genus in trop. As.: red meranti, r. seraya, r. lauan (esp. *S. negrosensis* Foxw., chan, Philipp.), being medium or lightweight reddish timbers (sects *Brachypterae*, *Mutica*, *Ovalis*, *Pachycarpae*, *Rubella*) for light construction & veneers, white m. (sect. *Anthoshorea*) imp. veneers; imp. trade timbers incl. *S. albida* Sym. (alan, NW Borneo), *S. almon* Foxw. (almon, C Mal.), *S. contorta* S. Vidal (white lauan, Philippines), *S. glauca* King (balan, W Mal.), *S. guiso* (Blanco) Blume (selangan, red balau, SE As. to Philippines) *S. kunstleri* King (red balau, W Mal.); *S. polysperma* (Blanco) Merr. (Bataan mahogany, bataan, tangile, Philipp.), *S. robusta* Roxb. ex Gaertner f. (sal, N India) – coppice fire-resistant, general construction, bark a source of a black dye, a dammar (resin) used in typewriter ribbons, carbon paper, incense; dammars of other spp. form. imp. in varnish manufacture, e.g. *S. javanica* Koord. & Val. in S Sumatra where retained when forest cleared for cult. & also planted out; fr. boiled as veg. but also exploited (as illipe nuts, when soaked known as black pontianak in trade) for oil (up to 70%) esp. *S. macrophylla* (Vriese) Ashton & other Bornean spp. used as subs. for cocoa butter in chocolate-making & also in cosmetics, form. imp. for soap, candles etc. e.g. Borneo tallow (*S. palembanica* Miq., W Mal.); *S. acuminata* Dyer (W Mal.) – 1 tree can give timber for a house & all its furniture in Sumatra; *S. ovalis* (Korth.) Blume (W Mal.) – tetraploid apomict of hybrid origin & *S. resinosa* Foxw. triploid; *S. trapezifolia* (Thw.) Ashton (Sri Lanka) – flowers annually unlike rest of genus in aseasonal trop. SE

As. 1 month after increase in night temp. (23+ °C), poss. use as plywood sp. in enrichment planting

Shortia Torrey & A. Gray. Diapensiaceae. 1 N Am., 5 E As. (*'Schizocodon'*). R: Taxon 32(1983)420, Plantsman 12(1990)23. Cult. orn. evergreen herbs

Shoshonia Evert & Constance. Umbelliferae (III 8). 1 Wyoming: *S. pulvinata* Evert & Constance. R: SB 7(1982)471

shrimp plant *Justicia brandegeeana*

shungiku *Chrysanthemum coronarium*

Shuteria Wight & Arn. Leguminosae (III 10). 4 Indomal. R: Adansonia 12(1972)291

Siamweed *Chromolaena odorata*

Siamosia Larsen & T.M. Pedersen. Amaranthaceae (I 2). 1 Thailand: *S. thailandica* Larsen & T.M. Pedersen. R: NJB 7(1987)271

Sibangea Oliver (~ *Drypetes*). Euphorbiaceae. 3 trop. Afr.

Sibara E. Greene. Cruciferae. 10 E & S N Am. R: JAA 69(1988)142

Sibbaldia L. Rosaceae. 8 N temp. (Eur. 2). R: TBIANSSR I,2(1936)217

Sibbaldianthe Juz. = *Sibbaldia*

Sibbaldiopsis Rydb. = *Potentilla*

Siberian elm *Ulmus pumila*; **s. saxifrage** *Bergenia* spp.; **s. wallflower** *Erysimum* × *marshallii*

Sibiraea Maxim. (= *Eleiosina*). Rosaceae. 2 SE Eur. (1), C & E As. *S. laevigata* (L.) Maxim. (Balkans, Siberia, China) – cult. orn. decid. shrub

Sibthorpia L. Scrophulariaceae. 5 trop. & S Am., Macaronesia, Eur. (2), Afr. mts. R: BN 108(1955)161. Creeping pls like *Hydrocotyle* spp. esp. *S. europaea* L. (Cornish moneywort, pennywort, W Eur.)

Sicana Naudin. Cucurbitaceae. 1 trop. Am.: *S. odorifera* (Vell. Conc.) Naudin (casabanana, cassabanana) – monoec. perennial liane with ed. fragrant fr. used for preserves & scenting linen

Sichuan pepper *Zanthoxylum piperitum*

Sichuania M. Gilbert & P.T. Li. Asclepiadaceae. 1 China: *S. alterniloba* M. Gilbert & P.T. Li. R: Novon 5(1995)12

Sickingia Willd. = *Simira*

Sicrea (Pierre) Hallier f. Tiliaceae. 1 SE As.

Sicydium Schldl. Cucurbitaceae. 6 trop. Am.

Sicyocarya (A. Gray) H. St John = *Sicyos*

Sicyocaulis Wiggins = *Sicyos*

Sicyos L. Cucurbitaceae. 50 Aus., NZ, Pacific (incl. Hawaii 24), trop. Am. *S. angulata* L. (bur cucumber, E N Am.) – grown as screen

Sicyosperma A. Gray. Cucurbitaceae. 1 S N Am.

Sida L. Malvaceae. c. 200 trop. & warm (Aus. 40) esp. Am. R: CGH 180(1957)1. Shrubs & herbs with schizocarps with 1-seeded mericarps; alks. *S. jatrophoides* L'Hérit. (Peru & Galápagos) & *S. palmata* Cav. (Peru & Ecuador) with glandular trichomes producing secretions toxic to ants & cockroaches. Some fibre-pls esp. *S. rhombifolia* L. (Queensland hemp, Paddy's lucerne, trop.) used e.g. for art basketry as by Margaret Preston (Aus.)

Sidalcea A. Gray. Malvaceae. 20 W N Am. R: AMBG 18(1931)117. Cult. orn. herbs esp. *S. malviflora* (DC) Benth. with pink fls

Sidasodes Fryx. & Fuertes (~ *Sida*). Malvaceae. 2 Colombia to Peru. R: Britt. 44(1992)438

Sidastrum Baker f. = *Sida* (but see Britt. 30(1978)449)

Siderasis Raf. Commelinaceae (II 1e). 2 Brazil. *S. fuscata* (Lodd.) H. Moore (E Brazil) – cult. orn. rosette-herb with violet fls

Sideria Ewart & Petrie = *Melhania*

Sideritis L. Labiatae (VI). 150 N temp. OW, Macaronesia. R (sect. *S.* – 69 spp.): PM 20 & 21 (1992, 1994). Some cult. orn. herbs & shrubs; lvs of spp. in sect. *Empedoclea* used as tea in Turkey

Siderobombyx Bremek. Rubiaceae (IV 1). 1 N Borneo

Sideropogon Pichon = *Arrabidaea*

Sideroxylon L. Sapotaceae (III). (Incl. *Bumelia* & *Mastichodendron*) 75 trop. (Am. 49). Hard wood; some ed. fr., also sources of a kind of chicle e.g. *S. foetidissimum* Jacq. (*Mastichodendron sloaneanum*, Barbados mastic, WI). R: T.D. Pennington, *S.* (1991)169. *S. sessiliflorum* (Poiret) Aubrév. (Mauritius) – v. rarely regenerating & supposed to have been dispersed by dodos extinct for some 300 yrs, so turkeys force-fed with fr. & germination enhanced (but introd. monkeys also eat unripe fr.)

Siebera Gay. Compositae (Card.-Card.). 2 Middle E to Afghanistan

Siegfriedia C. Gardner. Rhamnaceae. 1 SW Aus. (restricted)

Sieglingia Bernh. = *Danthonia*

Siemensia Urban. Rubiaceae (?IV 1). 1 W Cuba

Sierra Leone butter *Pentadesma butyracea*; **S. L. copal** *Guibourtia copallifera*

Sievekingia Reichb.f. Orchidaceae (V 10). 15 trop. Am. Cult. orn. incl. *S. suavis* Reichb.f. (Costa Rica, Panamá) – floral fragrance largely cineole attractive to male euglossine bees

Sieversia Willd. = *Geum*

Sigesbeckia L. Compositae (Helia.-Mel.). c. 3 trop. (orig. OW), Macaronesia (Eur. 2 natur.). Capitula disp. as units, the 5 bracts forming involucre covered with sticky glands. *S. orientalis* L. (OW) – locally medic. in India, poss. oilseed

Sigmatanthus Huber ex Emmerich = *Raputia*

Sigmatochilus Rolfe = *Chelonistele*

Sigmatogyne Pfitzer = *Panisea*

Sigmatostalix Reichb.f. Orchidaceae (V 10). 35 trop. Am. Cult. orn. epiphytes

silage green fodder preserved by pressure in silo or stack

Silaum Miller. Umbelliferae (III 8). 1 temp. Euras.: *S. silaus* (L.) Schinz & Thell. (pepper saxifrage)

Silene L. Caryophyllaceae (III 3). c. 700 N hemisph. (Eur. 194) esp. S Balkan Peninsula (Greece 119) & SW As. R: T 44(1995)543. Incl. *Cucubalus, Lychnis, Melandrium, Viscaria.* Campion, catchfly. Herbs, rarely shrubby (e.g. Hawaii), with extrafl. nectaries, some with ecdysteroids. Cult. orn. annuals & perennials: *S. acaulis* (L.) Jacq. (moss campion, Arctic Euras., W N Am. mts, C Eur.) – rock-pl.; *S. armeria* L. (none-so-pretty, C & S Eur.) – annual or biennial with flat-topped cymes; *S. baccifera* (L.) Roth (*Cucubalus b.*, Euras., N Afr.) – form. medic., fr. a berry (dry when mature); *S. banksia* (Meerb.) Mabb. (*S. fulgens L. f.*, E As.) – ecdysteroids, cult. as '*L.* × *haageana*'; *S. chalcedonica* ((L.) E. Krause (*L.c.*, Maltese cross, c. of Jerusalem, N Russia 49–56 °N) – fls vivid scarlet; *S. coronaria* (L.) Clairv. (*L. c.*, rose campion, mullein pink, S Eur. to N Iran but widely natur.) – woolly herb with purplish fls; *S. dioica* (L.) Clairv. (red campion, Eur., natur. N Am.) – woods & hedges but hybridizing with *S. latifolia* Poiret (*S. alba, S. pratensis*, white campion or cockle, Eur., Medit.) – weed of open ground & fields with dimorphic pollen & sex chromosomes XX, XY with large dominant Y (when half Y removed plant hermaphr. so 1 region involved in female suppression; rust fungus activates male genes in female but pollen grains replaced by rust spores!); *S. flos-cuculi* (L.) Clairv. (*L. f.-c.*, ragged robin, Eur. to C As., natur. N Am.) – fls rose-red; *S. otites* (L.) Wibel (Eur. to Siberia) – fls poll. by nocturnal Lepidoptera & mosquitoes in Dutch sand-dunes; *S. schafta* C. Gmelin ex Hohen. (Cauc.) – commonly cult. rock-pl. with pink fls; *S. suecica* (Lodd.) Greuter & Burdet (*L. alpina*, alpine c., Eur., NE N Am.) – used as copper indicator in Norway & nickel indicator in Finland; *S. uniflora* Roth (*S. maritima*, sea campion, Eur. mts & coasts); *S. vulgaris* (Moench) Garcke (bladder campion, Euras., Medit.) – night-scented fls

Silentvalleya V.J. Nair, Sreekumar, Vajravelu & Bhargavan. Gramineae (31d). 1 India: *S. nairii* V.J. Nair, Sreekumar, Vajravelu & Bhargavan – close to *Lophacme* (Afr.) & *Gouinia* (Am.)

Siler Miller = *Laserpitium*

Silicularia Compton. Cruciferae. 1 S Afr.

Siliquamomum Baillon. Zingiberaceae. 1 SE As.

silk cotton tree *Bombax ceiba, Cochlospermum religiosum*; **Indian s. c. t.** *Bombax ceiba*; **s. tree** *Albizia julibrissin*; **s. vine** *Periploca graeca*; **s. weed** *Asclepias syriaca*

silky bent grass *Apera* spp.; **s. oak** *Grevillea robusta, Cardwellia sublimis*

silkwood *Flindersia pimenteliana*

Siloxerus Labill. (~ *Angianthus*). Compositae (Gnap.-Ang.). 3 SW Aus. R: OD 104(1991)127

Silphium L. Compositae (Helia.-Eng.). 23 E N Am. *S. laciniatum* L. (compass pl., prairies) – young pl. with lvs tipped vertically, avoiding full incidence of midday sun, the surfaces facing N & S (cf. *Lactuca serriola*), locally medic.

silvaberry see *Rubus*

Silvaea Philippi (*Philippiamra*) = *Cistanthe*

silver beech *Nothofagus menziesii*; **s. bell** *Halesia carolina*; **s. birch** *Betula pendula*; **s. fir** *Abies* spp.; **s. wattle** *Acacia dealbata*; **s.weed** *Potentilla anserina*

silverballi *Nectandra* spp.

silverberry *Shepherdia argentea*

silverseed gourd *Cucurbita argyrosperma*

silversword *Argyroxiphium* spp.

Silvianthus Hook.f. Caprifoliaceae (Carlemanniaceae). 2 E India to SE As.

Silviella Pennell. Scrophulariaceae. 2 Mexico

Silvorchis J.J. Sm. Orchidaceae (V 5). 1 Java. Mycotrophic

Silybum Adans. Compositae (Card.-Card.). 2 Medit. inc. Eur. R: JAA 71(1990)426. *S. marianum* (L.) Gaertner (holy or milk thistle) – annual or biennial with white-blotched lvs found in nutrient-rich sites, the cypselas with oil-bodies attractive to harvester ants which deposit them in rich organic material from their nests, cult. orn. now widely natur. N & S Am. incl. pampas & bad weed in NSW, lvs & stems ed. as salad, fr. (kenguel seed) form. a coffee subs. & locally medic. considered to protect liver from hepatitis since time of Dioscorides, now known to contain flavonoids effective as antidotes to *Amanita* poisoning, their being able to displace phalloidin from membrane receptors

Simaba Aublet = *Quassia*

Simarouba Aublet = *Quassia*

Simaroubaceae DC. Dicots – Rosidae – Sapindales. (Incl. Leitneriaceae) 13/110 trop. (*Ailanthus altissima* & *Picrasma* extending to temp. As., *Leitneria* to SE US). Trees & shrubs usu. with bitter bark, wood & seeds (triterpenoid lactones – simaroubilides – present). Lvs pinnate to unifoliolate or rarely simple, usu. spirally arr., without gland dots; stipules 0 . Fls reg., usu. unisexual & often with rudiments of opposite sex (monoec. or dioec.), 3–5(–)8)-merous, small, in racemes, cymes or thyrses; K usu. 5 & basally connate, lobes imbricate or valvate, C (0) usu. 5 free, imbricate or valvate, A (as many as & alt. with) twice C (or more), filaments often with basal appendages, anthers with longit. slits, disk usu. present, s.t. a gynophore or androgynophore, G 1–5(–8) rarely free (*Picrolemma*, *Recchia* etc.), usu. ± united (s.t. only styles connate), s.t. forming pluriloc. ovary with axile placentas, each (rarely only 1) loc. with 1(2) apical or basal, anatropous to hemitropous, bitegmic ovule. Fr. a samara, rarely a drupe; embryo straight or curved, oily, endosperm ± 0. x = 8, 13+. R: T 44(1995)177

Principal genera: *Ailanthus*, *Castela*, *Quassia*, *Soulamea*

Fam. restricted to form. subfam. Simarouboideae; *Harrisonia* here referred to Rutaceae (Thorne), *Irvingia* & allies separated as Irvingiaceae, *Kirkia* & allies as Kirkiaceae; *Suriana* (without simaroubalides) & other genera with simple lvs here referred to Surianaceae, while *Alvaradoa* & *Picramnia* put in own fam. Picramniaceae. In many respects close to Rutaceae, but with similarities to Meliaceae & perhaps more similar to the ancestral group from which these families have been derived

Some medic. (*Brucea*, *Picrasma*, *Quassia*), oilseeds & timber (*Quassia*) & cult. orn. esp. *Ailanthus*

simaruba *Quassia amara*

Simenia Z. Szabó = *Dipsacus*

Simethis Kunth. Asphodelaceae (Liliaceae s.l.). 1 W Eur. & Medit.: *S. planifolia* (L.) Grenier

Simicratea Hallé (~ *Hippocratea*). Celastraceae. 1 Angola: *S. welwitschii* (Oliver) Hallé. R: BMNHN 4,5(1983)20

Similisinocarum Cawet-Marc & Farille = *Pimpinella*

Simira Aublet. Rubiaceae (I 5). 35 trop. Am., 8 trop. Afr. (R: BMNHN 4,6(1984)4). Some local timbers, dyes & febrifuges ('*Sickingia*')

Simirestis Hallé = *Hippocratea*

Simmondsia Nutt. Simmondsiaceae. 1 SW N Am.: *S. chinensis* (Link) C. Schneider (*S. californica*, jojoba, goat nut, pig n.) – wind-poll. shrub or small tree tolerant of arid sites & source of jojoba oil, a subs. for sperm oil used in cosmetics etc., the seeds avoided by desert rodents (cyanoglucosides) but form. ground as coffee subs., the foliage imp. browse

Simmondsiaceae (Muell.Arg.) Reveal & Hoogland (~ Buxaceae). Dicots – Dilleniidae – Buxales. 1/1 SW N Am. Evergreen dioec. shrub with unusual sec. growth. Lvs small, leathery, simple but with jointed base, opp.; stipules 0, Fls small, reg., C 0, the males in axillary clusters, the females usu. solit.; K (4) 5 (6), imbricate, accrescent in females, disk 0, A (8)10(12) free, anthers elongate, with longit. slits, G (3), 3-loc. with 3 long feathery stigmas, each loc. with 1 apical-axile, pendulous, anatropous ovule. Fr. a loculicidal capsule with 2 empty locules & 1 seed; embryo straight, cotyledons fleshy & with much liquid wax, endosperm ± 0

Only species: *Simmondsia chinensis*

Differing from Buxaceae in 5-mery, A 10 & structure, pollen, solit. ovule & unusual sec. thickening (concentric rings of vasc. bundles), sieve-tube plastids etc.

Simocheilus Klotzsch. Ericaceae. 20 SW & S Cape

Simplicia Kirk. Gramineae (21d). 2 NZ. R: NZJB 9(1971)539

Simsia Pers. Compositae (Helia.-Helia.). 18 trop. Am. to S US. R: SBM 30(1990). Shrubs to annuals

simsim *Sesamum orientale*

simul *Bombax ceiba*

Sinacalia H. Robinson & Brettell (~ *Ligularia*). Compositae (Sen.-Tuss.). 4 China. R: KB 39(1984)215. *S. tangutica* (Maxim.) R. Nordenstam (*Senecio t.*) – cult. orn. with panicles of capitula

Sinadoxa C.Y. Wu, Z.L. Wu & R.F. Huang. Adoxaceae. 1 China: *S. corydalifolia* C.Y. Wu, Z.L. Wu & R.F. Huang. R: BBR 7,4(1987)100

Sinapidendron Lowe. Cruciferae. 5–6 Macaronesia

Sinapis L. (~ *Brassica*). Cruciferae. 7 Eur. (4), Medit. R: JAA 66(1985)312. *S. alba* L. (*B. hirta*, black, yellow or white mustard, Medit.) – mustard of mustard & cress, when grown to cotyledon stage only; *S. arvensis* L. (charlock) – farmweed, form. v. common in cornfields, the seeds remaining viable for at least 10 years even after ingestion by animals

Sinarundinaria Nakai. Gramineae (1a). (Incl. *Chimonocalamus, Otatea, Yushania*) 50 trop. As., trop. E Afr. mts (*S. alpina* (Schumann) C.S. Chao & Renvoize (*Arundinaria alpina*) forming thick belts to 19.5 m high on some, e.g. 250 sq. miles in Aberdares), Madag., C Am. (2, (*Otatea*) – R: SCB 44(1980)21). R: KBAS 13(1986)41. S.t. thorns derived from adventitious roots at nodes. Preferred food for pandas; some cult. orn. bamboos like *A.* but with much-branched rhizome etc. esp. *S. nitida* (Stapf) Nakai (*Fargesia n.*, C China) – culms to 6 m flowering after 100 yrs in cult.

Sinclairia Hook. & Arn. Compositae (Liab.). (Incl. *Liabellum*) 23 Mex. to Colombia. R: Phytol. 67(1989)168

Sincoraea Ule = *Orthophytum*

Sindechites Oliver. Apocynaceae. 1 subtrop. China: *S. henryi* Oliver. R:AUWP 88-6(1988)29

Sindora Miq. Leguminosae (I 4). 18–20 SE As., Mal. Some timbers & medic. oils (skin disease)

Sindoropsis Léonard. Leguminosae (I 4). 1 Gabon

Sindroa Jum. = *Orania*

Sineoperculum Jaarsveld = *Dorotheanthus*

Sinephropteris Mickel = *Asplenium*

Singana Aublet = ? Leguminosae

Singapore daisy *Wedelia trilobata*

Singhara nut *Trapa bispinosa*

sing-kwa *Luffa acutangula*

Sinia Diels (~ *Sauvagesia*). Ochnaceae. 1 SE China

Sinningia Nees. Gesneriaceae. 60 trop. Am. Herbs & shrubs, usu. tuberous, s.t. with scented fls (rare in fam.), some bird-poll. ('*Rechsteineria*') others visited by bees, the glossy texture of C due to fine hairs. Many cult. orn. (prop. by leaf-cuttings producing tubers rather than plantlets at cut surface) esp. 'gloxinia', *S. speciosa* (Lodd.) Hiern (Brazil), of which florist's g. differs from wild pl. in having large erect campanulate fls (Fyfiana Group), the wild ones having nodding fls; other cult. spp. incl. *S. canescens* (C. Martius) Wiehler (*S. leucotricha, Rechsteineria l.*, Brazil) – tubers may flower without planting; *S. cardinalis* (Lehm.) H. Moore (*R. c.*, Brazil) – scarlet 2-lipped bird-poll. fls, though many selected lines referred to this sp. may be hybrids

Sinoadina Ridsd. Rubiaceae (I 2). 1 E As.

Sinobacopa Hong. Scrophulariaceae. 1 China: *S. aquatica* Hong. R: APS 25(1987)393

Sinobambusa Makino ex Nakai. Gramineae (1a). 17 China, Taiwan, Vietnam. R: JBR 1(1982)140. Forest bamboos; cult. orn. incl. *S. tootsik* (Makino) Nakai (China)

Sinocalamus McClure = *Dendrocalamus*

Sinocalycanthus (W.C. Cheng & S.Y. Chang) W.C. Cheng & S.Y. Chang (~ *Calycanthus*). Calycanthaceae. 1 China

Sinocarum H. Wolff. Umbelliferae (III 8). 7 China

Sinochasea Keng = *Pseudodanthonia*

Sinocrassula A. Berger (~ *Sedum*). Crassulaceae. 5 Himal. to SW China. Some cult. orn. esp. *S. yunnanensis* (Franchet) A. Berger (Yunnan)

Sinodielsia H. Wolff. Umbelliferae (III 5). 3 Bhutan, Tibet, Yunnan. R: FR 97(1986)753

Sinodolichos Verdc. (= ? *Glycine*). Leguminosae (III 10). 2 Burma, S China

Sinofranchetia (Diels) Hemsley. Lardizabalaceae. 1 C & W China: *S. chinensis* (Franchet) Hemsley – cult. orn. liane with purple berries & black seeds

Sinoga S.T. Blake = *Asteromyrtus*

Sinojackia Hu. Styracaceae. 2 S China. Cult. orn. shrubs

Sinojohnstonia Hu. Boraginaceae. 1 W China

Sinoleontopodium Y.L. Chen. Compositae (Gnap.-Lor.). 1 China: *S. lingianum* Y.L. Chen. R: APS 23(1985)457

Sinolimprichtia H. Wolff. Umbelliferae (III 5). 1 E Tibet: *S. alpina* H. Wolff

Sinomenium Diels. Menispermaceae (V). 1 C China, Japan: *S. acutum* (Thunb.) Rehder & E. Wilson – alks

Sinomerrillia Hu = *Neuropeltis*

Sinopanax Li. Araliaceae. 1 Taiwan

Sinopimelodendron Tsiang = *Cleidiocarpon*

Sinoplagiospermum Rauschert = *Prinsepia*

Sinopodophyllum Ying = *Podophyllum*

Sinopteridaceae Koidz. = Pteridaceae

Sinopteris C. Chr. & Ching = *Cheilanthes*

Sinoradlkofera F. Meyer = *Boniodendron*

Sinorchis S.C. Chen = *Cephalanthera*

Sinosassafras H.W. Li = *Parasassafras*

Sinosenecio R. Nordenstam (~ *Senecio*). Compositae (Sen.-Tuss.). 34 China to SE As. R: KB 39(1984)222

Sinosideroxylon (Engl.) Aubrév. = *Sideroxylon*

Sinowilsonia Hemsley. Hamamelidaceae (I 3). 1 C & W China: *S. henryi* Hemsley – cult. orn. tree

sinqua (melon) *Luffa acutangula*

Sinthroblastes Bremek. = *Strobilanthes*

Siolmatra Baillon. Cucurbitaceae. 3 S Am. Wind-disp. (cf. allied *Fevillea*)

Sipanea Aublet. Rubiaceae (I 6). 17 trop. S Am.

Sipaneopsis Steyerm. Rubiaceae (I 5). 6 NW trop. S Am.

Sipapoa Maguire = *Diacidia*

Siparuna Aublet. Monimiaceae (III; Siparunaceae). 150 trop. S Am. Almost closed hypanthium (cf. *Ficus*) but poss. heterogeneous

Siparunaceae (A. DC) Schodde. See Monimiaceae

Siphanthera Pohl ex DC. Melastomataceae. 16 NE S Am.

Siphantheropsis Brade = *Macairea*

Siphocampylus Pohl. Campanulaceae. 215 trop. Am. Shrubs; some poss. rubber sources

Siphocodon Turcz. Campanulaceae. 2 SW Cape

Siphocranion Kudô = *Hanceola*

Siphokentia Burret. Palmae (V 4l). 1–2 Moluccas

Siphonandra Klotzsch. Ericaceae. 3 Andes

Siphonandrium Schumann. Rubiaceae (?IV 10). 1 New Guinea

Siphonella Small = *Fedia*

Siphoneugena O. Berg (~ *Eugenia*). Myrtaceae (Myrt.). 8 trop. Am. R: EJB 47(1990)239

Siphonochilus J.M. Wood & Franks. Zingiberaceae. 15 trop. Afr. R: NRBGE 40(1982)372. Like *Kaempferia* but infls lateral etc.; dry forests & savannas. Some cult. orn.

Siphonodon Griffith. Celastraceae (Siphonodontaceae). 7 Indomal., Aus. *S. australis* Benth. (ivorywood, Aus.) – timber for rulers, turnery, engraving, etc.

Siphonodontaceae (Croizat) Gagnepain & Tard. = Celastraceae

Siphonoglossa Oersted (~ *Justicia*). Acanthaceae. 7 trop. Am. R: Phytol. 67(1989)227

Siphonosmanthus Stapf = *Osmanthus*

Siphonostegia Benth. Scrophulariaceae. 1 E Med. incl. Eur., 2 E As.

Siphonostelma Schltr. = *Brachystelma*

Siphonostylis W. Schulze = *Iris*

Siphonychia Torrey & A. Gray. Caryophyllaceae. 6 E N Am.

Sipolisia Glaz. ex Oliver. Compositae (Vern.-Vern.). 1 Brazil

Siraitia Merr. Cucurbitaceae. 1 SE As.: *S. grosvenori* (Swingle) C. Jeffrey (*Thladiantha g.*)

Sirhookera Kuntze. Orchidaceae (V 13). 2 S India, Sri Lanka

siris *Albizia lebbeck*; **red s.** *Paraserianthes toona*

sisal *Agave sisalana*

Sison L. Umbelliferae (III 8). 2 Eur. (1), Medit. *S. amomum* L. (Eur., Medit.) – local medic. & food-flavouring

sissoo *Dalbergia sissoo*

Sisymbrella Spach. Cruciferae. 5 W & C Medit. to C France (Eur. 2). Seeds mucilaginous when wet (cf. *Rorippa*)

Sisymbriopsis Botsch. & Tzvelev. Cruciferae. 1 C As.

Sisymbrium L. Cruciferae. 90 Euras. (Eur. 19), Medit., S Afr., N Am., Andes. R: Pfr. IV,105(1924)46. *S. altissimum* L. (Eur.) – tumbleweed in US; *S. irio* L. (London rocket, Medit.) – appeared conspic. in London after Great Fire of 1666 as did *S. orientale* L.

(Medit.) after the blitz of World War II; *S. officinale* (L.) Scop. (hedge mustard, Eur., Medit.) – form. medic. & antiscorbutic

Sisyndite E. Meyer ex Sonder. Zygophyllaceae. 1 S Afr.

Sisyranthus E. Meyer ex Sonder. Asclepiadaceae (III 5). 12 trop. & S Afr.

Sisyrinchium L. Iridaceae (III 1). 80 Am. (esp. C & S) with 1 ext. to Ireland (poss. natur.). Australasian spp. referred to allied *Libertia* (tepals not ± similar). 40 spp. with elaiophores attractive to bees; some with black seeds (unusual in I.). Cult. orn. (blue-eyed grass) incl. *S. bermudiana* L. (E N Am. to Ireland) poss. a pre-glacial relic in Eur. (Godwin) often confused with *S. montanum* E. Greene (E N Am.) with larger fls & erect (not nodding) pedicels in fr., natur. Eur.; *S. striatum* Sm. (Chile & Arg.) – sturdy garden pl. with greenish-yellow fls, natur. Scilly Is.; *S. vaginatum* Sprengel (trop. Am.) – fert. control in Uruguay

Sisyrolepis Radlk. (*Delpya*). Sapindaceae. 1 Thailand, Kampuchea

Sitanion Raf. Gramineae (23). 4 W N Am. R: Britt. 15(1963)303

Sitella L. Bailey = *Waltheria*

Sitka cypress *Chamaecyparis nootkatensis*; **S. spruce** *Picea sitchensis*

Sitopsis (Jaub. & Spach) Löve = *Aegilops*

Sium L. Umbelliferae (III 8). 14 N hemisph. (Eur. 2), Afr. Some cult. orn. (water parsnip), lvs locally eaten, esp. *S. sisarum* L. (crummock, skirret, E As.) – grown for ed. root eaten like salsify or coffee subs.

Sixalix Raf. = *Scabiosa*

Skapanthus C.Y. Wu & Li = *Plectranthus*

skeleton weed *Chondrilla juncea*

Skeptrostachys Garay (~ *Stenorrhynchos*). Orchidaceae (III 3). 13 trop. S Am. R: HBML 28(1982)358

skewerwood *Cornus sanguinea*

Skiatophytum L. Bolus. Aizoaceae (V). 1 SW Cape

Skimmia Thunb. Rutaceae. 4 Himal., China & Japan to Philippines. R: KM 4(1987)168. Evergreen shrubs with alks. Cult. orn. with fragrant fls esp. *S. japonica* Thunb. (Japan, Sakhalin, Philipp.) – dioec., the males with more strongly scented fls, the females with bright red drupes in winter (occ. 'males' produce fr.), many cvs. *S. laureola* (DC) Decne. (Sino-Himal.) – fr. an abortifacient in India

skirret *Sium sisarum*

Skoliopterys Cuatrec. = *Clonodia*

Skottsbergianthus Boelcke (*Skottsbergiella*). Cruciferae. 2 S S Am.

Skottsbergiella Boelcke = praec.

Skottsbergiliana H. St John = *Sicyos*

skull cap *Scutellaria* spp.

skunk *Cannabis sativa*; **s. cabbage** *Lysichiton americanus*, *Symplocarpus foetidus*; **s. currant** *Ribes glandulosum*; **s.weed** *Croton texensis*, *Navarretia squarrosa*

Skytanthus Meyen. Apocynaceae. 3 Brazil, Chile. Alks

Sladenia Kurz. Theaceae (Sladeniaceae). 1 SE As.

Sladeniaceae (Gilg & Werderm.) Airy Shaw. See Theaceae

slash pine *Pinus elliottii*

Sleumerodendron Virot. Proteaceae. 1 New Caledonia

slipper flower *Calceolaria* spp.

Sloanea L. Elaeocarpaceae. 120 Madag. (1), trop. As. to Aus. (52 – R: KB 38(1983)347), New Caled. (9) & Am. (62 – R: CGH 175(1954)1). Some locally used timbers esp. *S. australis* (Benth.) F. Muell. (maiden's blush, E Aus.) – turnery

sloe *Prunus spinosa*

Sloetiopsis Engler = *Streblus*

smack *Papaver somniferum*

Smallanthus Mackenize (~ *Polymnia*). Compositae (Helia.-Mel.). c. 19 trop. & warm Am. R: Phytologia 39(1978)47. Disk-florets with pedicels as in *P. S. sonchifolius* (Poeppig & Endl.) H. Robinson (*P. s.*, *P. edulis*, S Am. esp. Peru) – source of yacou, ed. tubers used as alcohol source etc.; *S. uvedalia* (L.) Mackenzie (E N Am.) – local stimulant & medic.

Smeathmannia Sol. ex R. Br. Passifloraceae. 2 trop. W Afr.

Smelophyllum Radlk. = *Stadmannia*

Smelowskia C. Meyer. Cruciferae. 6 arctic-alpine E As., W N Am. R: Rhodora 40(1938)294. Cult. orn. rock-pls

Smicrostigma N.E. Br. (~ *Ruschia*). Aizoaceae (V). 1 C S Afr.

Smilacaceae Vent. Monocots – Liliidae – Dioscoreales. 3/320 trop. & warm esp. S. Lianes or branching shrubs with starchy rhizomes & usu. vessel-elements throughout veg. body.

Roots mycorrhizal, without root-hairs. Lvs usu. spirally arr. (opp. in I), often with pair of tendrils from base of petiole near sheath, lamina usu. with 3–7 curved-convergent veins, cross-connected. Fls reg., usu. unisexual (dioecy in *Smilax*; bisexual in I.), 3-merous, solit. or in axillary or term. infls, s.t. foetid.; P 3 + 3, usu. petaloid, sometimes basally connate (a 3- or 6-toothed P-tube in *Heterosmilax*), A (3)6(–18) sometimes adnate to P-tube or forming a column, anthers with longit. slits, G usu. (3), 1- or 3-loc. with parietal or axile placentation & 1 or 3 styles, each loc. with 1–∞ anatropous to campylotropous or orthotropous bitegmic ovules. Fr. a berry (rarely dry-dehiscent) with 1–3(–∞) seeds; embryo usu. small in hard endosperm rich in protein, oil & hemicellulose & sometimes starch. x = 10, 13–16
Classification & genera:

I. **Ripogonoideae** (lvs opp.): *Ripogonum*
II. **Smilacoideae** (lvs spirally arr.): *Heterosmilax*, *Smilax*

V. close to Dioscoreaceae & s.t. put in Liliales. Many genera form. here now in Philesiaceae (Asparagales) though *Ripogonum* & *Petermannia* (both sometimes put in monotypic fams) 'combine features' of the two

Smilacina Desf. = *Maianthemum*

Smilax L. Smilacaceae. c. 300 trop. & temp. (Eur. 3). Dioec. lianes (green or cat briar, bamboo vines) or herb. climbers with paired stipular tendrils & often recurved hooks on stems, though *S. biflora* Siebold ex Miq. (S Japan) a bush 10 cm tall. Dried rhizomes of some trop. Am. spp. the sarsaparilla of commerce, used medic. form. as tonic e.g. for gym enthusiasts in US. *S. regelii* Killip & Morton ('*S. officinalis*', N C Am.) – Jamaica s., while rhiz. of other spp. a carbohydrate source e.g. *S. bona-nox* L. (China briar, SE US to Texas) while *S. glyciphylla* Sm. (E Aus.) with 21 mg vitamin C per 100 g fr. equivalent to tomatoes. Some cult. orn. esp. *S. anceps* Willd. (*S. kraussiana*, Afr.), *S. glauca* Walter (wild sarsaparilla, SE US) & *S. rotundifolia* L. (bull briar, U US), though 'smilax' of florists is *Asparagus asparagoides*

Smirnowia Bunge. Leguminosae (III 15). 1 Turkestan: *S. turkestana* Bunge – allied to *Sphaerophysa*; alks

Smithia Aiton. Leguminosae (III 13). 30 OW trop. esp. As. & Madagascar

Smithiantha Kuntze. Gesneriaceae. 4 Mex. Cult. orn. with creeping scaly rhizomes

Smithiella Dunn = *Pilea*

Smithorchis Tang & Wang. Orchidaceae (IV 2). 1 China

Smithsonia J. Saldanha (*Loxoma*). Orchidaceae (V 16). 3 India. R: JBNHS 71(1974)73

Smitinandia Holttum. Orchidaceae (V 16). 3 Indomal.

Smodingium E. Meyer ex Sonder. Anacardiaceae (IV). 1 S Afr.: *S. argutum* E. Meyer ex Sonder

smoke bush *Conospermum* spp.; **s. b.** or **tree** *Cotinus coggygria*

Smyrniopsis Boiss. Umbelliferae (III 5). 1 E Medit. to Iran

Smyrnium L. Umbelliferae (III 5). 7 Eur. (5), Medit. *S. olusatrum* L. (alexanders, W Eur., Medit., natur. GB, Netherlands & Bermuda) – aggressive pl. in SW GB, form. used like celery (fl. buds also ed.) & roots like parsnips

Smythea Seemann ex A. Gray. Rhamnaceae. 7 SE As. to Polynesia. Endosperm 0

snail flower *Vigna caracalla*

snake bark *Colubrina arborescens*; **s. bean** *Vigna unguiculata* ssp. *sesquipedalis*; **s. climber** *Bauhinia scandens* var. *anguina*; **s. gourd** *Trichosanthes cucumerina*; **s.'s head** *Fritillaria* spp. esp. *F. meleagris*, *Hermodactylus tuberosa*; **s. tongue** *Ophioglossum* spp.; **s. vine** *Hibbertia scandens*; **s. wood** *Brosimum rubescens*, *Strychnos minor* etc.

snakeroot, serpentaria or **serpentary** Name used for many pls in US esp. *Cimicifuga racemosa* (black s.) & *Endotheca serpentaria*

snapdragon *Antirrhinum majus*

sneezeweed *Achillea ptarmica*, *Centipeda* spp., *Helenium hoopesii*

sneezewood *Ptaeroxylon obliquum*

sneezewort *Achillea ptarmica*

snodgollion, snot-goblin (!) or **snottygobble** (!!) *Persoonia* spp.

snow, glory of the *Chionodoxa forbesii*; **s. gum** *Eucalyptus pauciflora* ssp. *niphophila*; **s. on the mountain** *Euphorbia marginata*; **s. pea** = mangetout; **s. plant** *Sarcodea sanguinea*; **s. in summer** *Cerastium tomentosum*

snowball tree *Viburnum opulus* 'Roseum'

snowberry *Symphoricarpos alba*

snowbush *Breynia nivosa* 'Roseapicta'

Snowdenia C. Hubb. Gramineae (34f). 4 trop. E Afr.

snowdrop *Galanthus* spp. esp. *G. nivalis*; **Barbados s.** *Habranthus tubispathus*; **s. tree** *Halesia carolina*

snowflake *Leucojum* spp.; **autumn s.** *L. autumnale*; **spring s.** *L. vernum*; **summer s.** *L. aestivum*
snowy mespilus *Amelanchier* spp. esp. *A. rotundifolia*
soap tree *Alphitonia excelsa*
soapbark tree *Quillaja saponaria*
soapberry *Sapindus* spp.
soapwood *Caryocar glabrum*
soapwort *Saponaria officinalis*
Soaresia Schultz-Bip. Compositae (Vern.-Vern.). 1 Brazil
Sobennikoffia Schltr. Orchidaceae (V 16). 4 Mascarenes
Sobolewskia M. Bieb. Cruciferae. 4 E Med. (Eur. 1) to Caucasus. Some cult. rock-pls
Sobralia Ruíz & Pavón. Orchidaceae (V 13). 95 trop. Am. Cult. orn. terr. & epiphytic orchids with large fls esp. *S. macrantha* Lindley (Mex. to Costa Rica) – fls to 25 cm across lasting 1 day
Socotora Balf.f. = *Periploca*
Socotranthus Kuntze = *Cryptolepis*
Socotria Levin = *Punica*
Socratea Karsten. Palmae (V 2a). c. 4 N S Am. R: Britt. 38(1986)55. Beetle-poll. cf. allied *Iriartea*. Wood of *S. exorrhiza* (C Martius) H. Wendl. (paxiuba palm) used for construction; base of trunk with stilt-roots, the trunk eventually free of ground such that plant can 'walk' from under obstacles as it grows
Socratina Balle. Loranthaceae. 2 Madagascar
Soderstromia C. Morton. Gramineae (33c). 1 Mex.: *S. mexicana* (Scribner) C. Morton
Sodiroa André = *Guzmania*
Sodiroella Schltr. = *Stellilabium*
Sodom apple *Solanum linnaeanum* & other spp., *Calotropis procera*
Soehrensia Backeb. = *Echinopsis*
Soemmeringia C. Martius. Leguminosae (III 13). 1 trop. S Am.
softgrass, creeping *Holcus mollis*
soga *Peltophorum pterocarpum*
Sogerianthe Danser. Loranthaceae. 4 E New Guinea to Solomon Is. R: AusJB 22(1974)599
Sohnsia Airy Shaw. Gramineae (13d). 1 Mexico. Dioec.
soh-phlong *Flemingia vestita*
soja *Glycine max*
sola pith *Aeschynomene aspera*
Solanaceae Juss. Dicots – Asteridae – Solanales. 94/2950 subcosmop. esp. S Am. (56 genera of which 25 endemic). Shrubs, trees, lianes & herbs with branched hairs & often prickles & alks (esp. nicotine & steroid groups); internal phloem around pith. Lvs simple, or lobed to pinnate or 3-foliolate, usu. spirally arr.; stipules 0. Fls usu. bisexual, solit. or in app. basically cymose infls; K (5), persistent, C ((4)5(6)), rotate to tubular with lobes usu. plicate (& s.t. convolute) in bud, convolute, imbricate or valvate, s.t. irreg. or even bilabiate, A usu. alt. with C & attached to tube, s.t. 4 or (*Schizanthus*) 2 fertile with staminodes, anthers often connivent, with longit. slits, term. pores or slits, usu. a disk around G̲ (2) orientated diagonally in fl. (sometimes 4-loc. or irreg. 3–5-loc. (*Nicandra*) or pseudomonomerous with 1 ovule (*Henoonia*)) with (1–) ±∞ anatropous to hemitropous or amphitropous unitegmic ovules on axile placentas per loc. Fr. a berry (or drupe) or dehiscent (often a septicidal capsule); seed with usu. linear straight to ± curved embryo in (0 or) usu. oily & proteinaceous (rarely starchy) endosperm. x = 7–12+. R: J.G. Hawkes *et al.* (1979) *The biology & taxonomy of the S.*

Classification & principal genera (after D'Arcy):

1. **Solanoideae** (K accrescent, C reg., fr. baccate, embryo coiled, n = 10: *Brugmansia, Capsicum, Cyphomandra, Datura, Hyoscyamus, Iochroma, Jaborosa, Lycianthes, Lycium, Markea, Nicotiana, Physalis, Solanum, Withania, Witheringia*
2. **Cestroideae** (K not accrescent, C often irreg. capsular fr., embryo straight, n = rarely 12): *Benthamiella, Brunfelsia, Cestrum, Fabiana, Nierembergia, Petunia, Salpiglossis, Schwenkia*

Close to Scrophulariaceae but with internal phloem & oblique G. 1 of only 10 fams with elaiophores attractive to bees (*Nierembergia*)

D'Arcy also makes Nolanaceae (*Nolana*) a subfam. here, while Cronquist includes Goetzeaceae (*Goetzea, Espadaea, Coeloneurum, Henoonia*, Mexico & WI), a course followed here though that group has been allied with many others. Journal: *Solanaceae Newsletter*

Subfam. 1 provides many ed. fr. (*Capsicum* (peppers), *Cyphomandra, Lycopersicon* (tomato), *Physalis, Solanum* (also ed. tubers = potatoes)). Many yield medic. or hallucinogenic alks, sometimes insecticidal (*Atropa, Brugmansia, Brunfelsia, Datura, Duboisia, Mandragora, Nicotiana* (tobacco), *Solandra*, atropine being the standard antidote to some systemic herbicides) while many of these genera, *Browallia, Cestrum, Lycium, Nicandra, Nierembergia, Petunia, Salpiglossis, Schizanthus, Solandra* & *Streptosolen* incl. cult. orn. Some noxious weeds esp. *Physalis, Solanum*

Solandra Sw. Solanaceae. (1). 10 trop. Am. R: NJB 7(1987)639. Alks – source of sacred hallucinogens in Mex. Cult. orn. shrubs & lianes with showy fls fragrant at night esp. *S. maxima* (Sessé & Moçiño) P. Green (*S. hartwegii*, '*S. guttata*', chalice vine, Mex.) – glabrous liane with yellow fls to 24 cm long

Solanecio (Schultz-Bip). Walp. (~ *Senecio*). Compositae (Sen.-Sen.). 16 Yemen, trop. Afr. Madag. R: KB 41(1986)920. Climbers, herbs & pachycaul trees incl. *S. gigas* (Vatke) C. Jeffrey (*Senecio g.*, Ethiopia) & *S. mannii* (Hook.f.) C. Jeffrey (*Senecio m.*, trop. Afr.) – fast-growing candelabriform, used for hedging in Kenya

Solanoa E. Greene = *Asclepias*

Solanopteris Copel. (~ *Microgramma*). Polypodiaceae (II 5). 3–4 Costa Rica to Peru. *S. bifrons* (Hook.) Copel. (potato fern) & other spp. with complex lateral rhizomatous sacs infested by *Azteca* & other ants which bring in débris later exploited by roots entering sacs

Solanum L. Solanaceae (1). 1700 subcosmop. (Eur. 3) esp. warm. Trees, shrubs & herbs, s.t. lianoid, often spiny (incl. *S. sacupanense* Rusby (NE S Am.) – only known thorny rheophyte) with prickliness assoc. with marsupial (esp. wallabies) grazing in Aus. from coasts to upland forest & semi-desert; see also *Lycopersicon*. R: AMBG 59(1972)274, (subg. *Leptostemonum*) GH 12(1984)179. Many alks so often toxic though fr. of e.g. *S. indicum* & *S. torvum* ed. (curried) in Sri Lanka when seeds removed; some with extrafl. nectaries visited by ants; some with 'buzz' fls visited by euglossine bees in Am.; disp. by animals, e.g. *S. lycocarpum* St.-Hil. (Brazil) – principal food of maned wolf in cerrado. Food-pls, cult. orn. & weeds, presence often indicating overgrazing (e.g. *S. incanum* L. (carcinogenic *N*-nitrosodimethylamine in fr.) in S Arabia): *S. aculeatissimum* Jacq. (*S. ciliatum*, Sodom apple (name used for other spp. incl. *S. linnaeanum* Hepper & Jaeger ('*S. ciliatum*', *S. hermannii*, S Afr.), warm Am., natur. OW) – sliced fr. consumed by cockroaches (fatal) in Puerto Rico, medic.; *S. berthaultii* J. Hawkes (Bolivia) – traps small insects & mites in hairs & emits pheromone mimicking scent of aphids in distress; *S.* × *burbankii* Bitter (sunberry, wonderberry, origin unclear) – annual cult. for ed. fr. resembling blackcurrants; *S. capicastrum* Link ex Schauer (Brazil) – cult. orn. ('winter cherry') widely natur. & self-sown on London's pavements, several cvs incl varieg.; *S. centrale* J. Black (arid Aus.) – fr. fresh or dried an Aboriginal staple (also other spp.); *S. cinereum* R. Br. (Narrawa burr, SE Aus.) – noxious weed; *S. dulcamara* L. (bittersweet, woody nightshade, Euras., natur. N Am.) – locally medic., fr. threaded on strips of date-leaf on the collarette of Tutankhamun's third coffin; *S. erianthum* D. Don. ('*S. verbascifolium*', S Am., widely natur.) – lvs used for washing clothes in Amaz. but causes skin rash; *S. gilo* Raddi (Brazil) – ed. fr., medic.; *S. indicum* L. & *S. torvum* (trop.) – ed. curried in S As. once toxic seeds (alks) removed; *S. jasminoides* Paxton (potato vine, Brazil) – cult. orn. liane; *S. laciniatum* Aiton (~ *S. aviculare* Forster f., kangaroo apple, New Guinea, Aus., NZ) – cult. in E Eur. & Hawaii for lvs with 1–2% solasodine, steroidal alk. like diosgenin used in contraceptive pills, fr. poisonous when green; *S. macrocarpon* L. (W Afr.) – cult. Afr. & As. for orange-yellow fr., lvs potherb; *S. mammosum* L. (macaw bush, trop. Am.) – molluscicidal glycoalks, shrub cult. for orn. but poison fr.; *S. melanocerasum* All. (*S. intrusum*, huckleberry, ? cultigen) – cult. trop. W Afr. for fr. used as vegetable etc.; *S. melongena* L. (aubergine, egg plant, Jew's apple, brinjal, badinjan, trop. OW) – fr. an imp. vegetable (round pale forms = 'Thai e.-p.'), essential ingredient of moussaka; *S. muricatum* Aiton (pepino, Andes) – fr. eaten fresh; *S. nigrum* L. (black nightshade, Eur., widely natur.) – poss. trigenomic hexaploid, some forms used as potherbs, some with ed. fr. used in pies; *S. pseudocapsicum* L. (Jerusalem cherry, winter c., Madeira, widely natur.) – cult. pot-pl. for orn. poison fr. in winter; *S. quitoense* Lam. (naranjilla, lulo, Andes) – cult. orn. & for fr. juice; *S. rostratum* Dunal (Kansas thistle, buffalo bur, SW N Am.) – pestilential weed in N Am.; *S. topiro* Humb. & Bonpl. ex Dunal (*S. hyporhodium*, cocona, Upper Amaz.) – fr. ed.; *S. torvum* Sw. (trop.) & *S. viarum* Dunal (Brazil) – comm. sources of solasodine: *S. tuberosum* L. (potato, ancient cultigen derived from pls domesticated c. 8000 BP at c. 3500–4000 m from *S. leptophyes* Bitter (Bolivia) or similar in N Bolivia to S Peru (tuber alks bred out) & hybridizing with at least 3 other spp. (*S. sparsipilum* (Bitter) Juz. & Bukasov (S Peru & N-C Bolivia), *S. acaule* Bitter (N Peru to NW Arg.) & *S. megistacrolobum* Bitter (S Peru to N Arg.)) giving diploid to pentaploid forms incl. autotetraploid *S. tuberosum* (frost resistance from high alt. spp. allowing living at high alts) poss. also with introgression from *S. oplocense* J. Hawkes

(Bolivia & Arg.) introd. from Chile 1840s & crossed with orig. wild weedy forms from NW Arg.) – many cvs (for history see NP 94(1983)479) prop. vegetatively from tubers (UK tuber record 3.2 kg) developing from underground runners from leaf axils (hence value of ridging potatoes, where more axils covered) or from their 'eyes', i.e. buds in axils of rudimentary lvs, first grown in Eur. in C16 (Ireland in 1566), though original introductions now scarce, & providing up to 5% of protein, 7–10% of iron & riboflavin & 25% of vitamin C in Europeans' diet (Britons eat most – over 100 kg each per annum though overconsumption (of solanine) by pregnant women alleged to promote spina bifida in babies), eaten boiled or fried (chips (e.g. 'Russet Burbank' (Idaho Baker) raised by Luther Burbank & used for McDonald's, Kentucky Fried Chicken & Burger King French fries worldwide), crisps (esp. 'Record')), powdered when dried for 'instant' potato, also commercial source of starch & alcohol (e.g. Polish vodka), the world's 4th crop after wheat, rice & maize; *S. uporo* Dunal (*S. anthropophagorum*, Melanesia to Polynesia) – tomato-like ed. fr. used in ceremonies

Solaria Philippi. Alliaceae (Liliaceae s.l.). 2 Chile

Soldanella L. Primulaceae. 10 C & S Eur. mts. R: QBAGS 31(1963)219, 56(1988)205. Rock-pls often growing up through & flowering in snow, cult. orn. incl. hybrids

Soleirolia Gaudich. Urticaceae (IV). 1 W Medit. Is., Italy: *S. soleirolii* (Req.) Dandy (*Helxine s.*, helxine, mind-your-own-business, mother-of-thousands) – cult. orn. creeping pl. with pale green tiny lvs used in greenhouses etc.

Solena Lour. Cucurbitaceae. 1 trop. As. to Mal.: *S. amplexicaulis* (Lam.) Gandhi – lvs, roots & fr. ed, roots & seeds sources of purgatives & stimulants

Solenachne Steudel = *Spartina*

Solenangis Schltr. Orchidaceae (V 16). 5 trop. Afr. to Masc. *S. aphylla* Schltr. leafless, others not & cult. orn.

Solenanthus Ledeb. (~ *Cynoglossum*). Boraginaceae. c. 17 Medit. (6 Eur.) to C As. & Afghanistan

Solenidiopsis Senghas (~ *Odontoglossum*). Orchidaceae (V 10). 2 trop. Am.

Solenidium Lindley. Orchidaceae (V 10). 3 N S Am.

Solenixora Baillon = *Coffea*

Solenocarpus Wight & Arn. = *Spondias*

Solenocentrum Schltr. Orchidaceae (III 3). 3 Costa Rica, Ecuador

Solenogyne Cass. Compositae (Ast.-Ast.). 3 Aus. R: Brunonia 2(1979)43

Solenomelus Miers. Iridaceae (III 1). 2 Chile, Arg. Cult. orn. like *Sisyrinchium* but with P tube; seeds half-winged (cf. *Gladiolus*)

Solenophora Benth. Gesneriaceae. 16 + C Am. Some trees to 12 m tall with boles to 30 cm diam.

Solenopsis C. Presl = *Laurentia*

Solenoruellia Baillon = *Henrya*

Solenospermum Zoll. = *Lophopetalum*

Solenostemma Hayne. Asclepiadaceae (III 1). 1 Egypt, Arabia: *S. argel* (Del.) Hayne

Solenostemon Thonn. = *Plectranthus*

Solidago L. Compositae (Ast.-Sol.). c. 80 N Am. with few in S Am., Macaronesia & Euras. (Eur. 1: *S. virgaurea* L. (Euras.) – form. medic., diuretic action due to flavonol glycosides & saponins). R: Phytol. 75(1993)3. Cult. orn. (goldenrod) & locally medic. esp. *S. canadensis* L. (N Am.) – commonest garden sp. in Eur. & hybrids between it & *S. virgaurea* but often weedy & becoming natur. See also *Euthamia* & *Oligoneuron*

× **Solidaster**' Wehrh. (Actually *Euthamia* × *Oligoneuron* so a new hybrid generic name required if such genera are to be recog. as distinct). Compositae (Ast.-Sol.). × *S. luteus* Green ex Dress (*E. graminifolia* (L.) Nutt. × *O. album* (Nutt.) Nesom) – pop. cutfl. esp. Netherlands, spontaneous hybrid found 1910 in a Lyon nursery

Soliva Ruíz & Pavón. Compositae (Anth.-Mat.). (Incl. *Gymnostyles*) 8 S Am. R: NMLP 14 Bot 70(1949)123. *S. anthemifolia* (A. Juss.) Sweet & others esp. *S. sessilis* Ruíz & Pavón (*S. pterosperma*, jo-jo weed) – troublesome weeds natur. in Aus., California etc. in lawns etc.

Sollya Lindley. Pittosporaceae. 2 SW Aus. R: KB 1948:74. Lianes with bright blue fls (bluebells), cult. orn. esp. *S. heterophylla* Lindley

Solmsia Baillon. Thymelaeaceae. 2 New Caledonia

Solms-Laubachia Muschler. Cruciferae. 13 China & Sikkim (1). R: APS 19(1981)480

Solomon nut oil *Canarium indicum*; **S.'s seal** *Polygonatum* spp.

Solonia Urban. Myrsinaceae. 1 E Cuba

Solori Adans. = *Derris*

soma *Mandragora turcomanica* (?), *Sarcostemma intermedium*

Sommera Schldl. Rubiaceae (inc. sed.). 12 trop. Am.

Sommerfeltia Less. Compositae (Ast.-Ast.). 2 S Am.

Sommieria Becc. Palmae (V 4m). 3 New Guinea

Somphoxylon Eichler = *Odontocarya*

sompong *Tetrameles nudiflora*

Sonchus L. Compositae (Lact.-Son.). 62 Euras. (Eur. 8) to Australasia & trop. Afr. (29 in Macaronesia, incl. subg. *Dendrosonchus*, pachycaul shrubs & treelets). Many weedy spp. now subcosmop. R: BN 125(1972)287, 126(1973)155, 127(1974)7, 407. Milk thistles; some farmweeds esp. *S. oleraceus* L. (sowt., Euras., Medit.). *S. grandifolius* Kirk (NZ) – lvs to 1 m × 20 cm

Sonderina H. Wolff (~ *Stoibrax*). Umbelliferae (III 8). 5 S Afr. R: EJB 48(1991)248

Sonderothamnus R. Dahlgren. Penaeaceae. 2 S Afr.

Sondottia Short. Compositae (Gnap.-Ang.). 2 Aus. R: Muelleria 7(1989)113

Sonerila Roxb. Melastomataceae. c.175 trop. As. Herbs & shrubs with 3-merous fls (only M. thus). Some cult. orn. esp. *S. margaritacea* Lindley (Burma to Java) – herb with lvs bearing rows of puckered pearly spots between veins & hybrids with *Bertolonia* spp.

Sonnea E. Greene = *Plagiobothrys*

Sonneratia L.f. Sonneratiaceae. 6 mangroves of Indian & Pacific Oceans. R: APS 23(1985)313. Mangroves with aerial 'breathing' roots arising from ordinary ones, their aerenchyma used as cork subs. for fishermen's floats etc. when boiled; fls ephemeral, foetid, nocturnal, those of *S. caseolaris* (L.) Engl. (Indomal.) visited by birds, bats & moths, lvs & fr. ed.

Sonneratiaceae Engl. (~ Lythraceae). Dicots – Rosidae – Myrtales. 2/8 trop. OW. Trees with internal phloem next to pith. Lvs simple, entire, opp. or whorled; stipules ± 0. Fls rather large, bat-poll., reg., usu. bisexual, perigynous, solit.–3 (*Sonneratia*) or in cymes, term.; hypanthium thick, K 4–8 valvate, C 4–8 (or 0) crumpled in bud, A (12–)∞ in several whorls on hypanthium or in clusters opp. C, $\underline{G}$ (4–20) with term. style bent in bud, pluriloc. with thick axile placenta bearing ∞ anatropous, bitegmic ovules. Fr. a berry (*Sonneratia*) or capsule with ∞ seeds; embryo straight or curved, endosperm 0. x = 9, 11, 12

Genera: *Duabanga*, *Sonneratia*

Mangrove & rain-forest genera (cf. Meliaceae, Rhizophoraceae etc.). Allied to Lythraceae but with more A & G & diff. chromosome no., though wood anatomy suggests S. be included in L.

Sooia Pócs = *Epiclastopelma*

Sophiopsis O. Schulz. Cruciferae. 4 C As. mts

Sophora L. Leguminosae (III 2). 45 trop. & mostly N temp. (Eur. 2). R: APS 19(1981)1, 143. Alks. *S. japonica* L. (pagoda tree, China & Korea) – cult. orn. decid. tree (several cvs incl. 'Tortuosa' with twisted branches), the fls a source of a yellow dye & the dried fls (sold in Java as sari kuning or s. tijina) medic. (rutin used in haemorrhagic problems); *S. secundiflora* (Ortega) DC (SW N Am.) – cult. orn. evergreen tree, before peyote use principal hallucinogen in SW N Am. but fr. v. toxic (cytisine) & fatal in excess; *S. tetraptera* J.F. Miller (kowhai, NZ (N Is.)) – useful timber; *S. toromiro* (Philippi) Skottsb. (Easter Is.) – only pl. on Easter Is. suitable for building, carving etc., reduced to 1 tree by 1917 & exterminated by grazing by 1962 but seeds grown so that sp. survives in botanic gardens

Sophronanthe Benth. = *Gratiola*

Sophronitella Schltr. = *Isabelia*

Sophronitis Lindley. Orchidaceae (V 13). 9 Paraguay & E Brazil. R: C. Withner (1993) *The Cattleyas* III: 57. Cult. orn. epiphytes; in intergeneric hybrids contributing bright red perianth colour

Sopubia Buch.-Ham. ex D. Don. Scrophulariaceae. 60 trop. OW to S Afr.

Soranthus Ledeb. = *Ferula*

Sorbaria (DC) A. Braun. Rosaceae. 4 E As. R: NJB 8(1989)557. Cult. orn. decid. shrubs, esp. *S. sorbifolia* (L.) A. Braun – stout to 2 m

Sorbus L. Rosaceae (III). 193 N hemisph. Trees & shrubs to shrublets a few cm tall with simple or pinnate lvs; fls foetid (trimethylamine like privet); many apomicts (41 in W Eur. alone incl. *S. latifolia* (Lam.) Pers. (2n = 51) poss. polytopically derived from *S. torminalis* & *S. aria* group (2n = 34)). Can be split into *Aria* Jacq. f. (incl. *Micromeles*) – lvs simple (97 Euras.), *Chamaemespilus* Medikus (1 S Eur.), *Cormus* Spach – lvs pinnate (1 S Eur.), *Torminalis* Medikus – lvs simple-lobed (2 Medit. to Iran) & *Sorbus* s.s. – lvs pinnate (92 N temp.). Cult. orn. (esp. for brightly coloured fruits e.g. 'Joseph Rock' (form of an undescribed sp. from Yunnan) & some ed. fr. etc. *S. americana* Marshall (roundwood, C & E US)

– fr. locally medic.; *S. aria* (L.) Crantz (whitebeam, Eur.) – fr. used in brandy & vinegar etc.; *S. aucuparia* L. (rowan, mountain ash, quickbeam, Eur., SW As.) – fr. used in jellies, brandy etc. though allegedly carcinogenic, wood for turnery etc.; *S. domestica* L. (S Eur., N Afr., SW As.) – timber useful, bark for tanning leather, fr. ed. when bletted like a medlar, the tree grown for this in France & near Genoa, & may be fermented into a kind of cider & distilled as a spirit; *S. intermedia* (Ehrh.) Pers. (Swedish whitebeam, Scandinavia) – cult. orn.; *S. torminalis* (L.) Crantz (service tree, Medit.) – fr. used like that of *S. domestica*, in drink called checkers or chequers (so pubs displaying a chessboard prob. solecism)

Sorghastrum Nash. Gramineae (40c). 17 Afr. (2), trop. & warm Am. (15). *S. nutans* (L.) Nash (*S. avenaceum*, warm N Am.) – imp. forage grass

Sorghum Moench. Gramineae (40c). 24 warm OW (Eur. 1) & Mex. (1). R (sect. S.): AJB 65(1978)477. Guinea corn, kaolang (China), sorghum esp. cvs of *S. bicolor* (L.) Moench (great millet, imphee) domesticated in Sudan c. 1000 BC, the wild progenitor prob. *S. arundinaceum* (Desv.) Stapf & now world's 4th most imp. cereal after wheat, rice & maize, a staple in Afr., India & China, thriving under drier conditions than does maize; 8 major groups of cvs grown for grain (non-saccharine sorghums), sweet juice or forage (saccharine s.) & brush manufacture (broom corns), incl. Caudatum Group (feterita) – large white, yellow or red grains, Durra Group (durra, dari, jowar) – principal grain sorghum of Afr. & India, Saccharatum Group – culms to 4 m with juicy sweet pith, cult. for syrup, forage & silage in US, Subglabrescens Group (milo) – imp. grain in C N Am., Technicum Group (Italian, Florence or Venetian whisk, broom corn) – principal source of domestic brooms & brushes; backcrosses with *S. arundinaceum* gave *S.* × *drummondii* (Steudel) Millsp. & Chase incl. Sudan grass (*S. sudanense* (Piper) Stapf), cult. for forage; *S. halepense* (L.) Pers. (Johnson grass, Medit.) – widely natur. fodder pl., often weedy, often cyanogenic

Soridium Miers. Triuridaceae. 1 N S Am. R: BB 140(1991)10

Soridium Miers ex Henfrey = *Peltophyllum*

Sorindeia Thouars. Anacardiaceae (IV). 50 trop. Afr., Madag. Locally used timber & ed. fr., twigs of *S. warneckei* Engl. (W Afr.) used as (sweet-tasting) chewing-sticks in Nigeria

Sorocea A. St-Hil. Moraceae (II). 16 trop. Am. R: PKNAW, C 88(1985)381

Sorocephalus R. Br. Proteaceae. 11 SW Cape. R: JSAB 7 suppl. (1969)21

Sorolepidium Christ = *Polystichum*

Soromanes Fée = *Polybotrya*

Soroseris Stebb. Compositae (Lact.-Crep.). 8 Himalaya to W China

Sorostachys Steudel = *Cyperus*

sorrel *Rumex acetosa*; **French s.** *R. scutatus*; **red s.** *Hibiscus sabdariffa*; **sheep's s.** *R. acetosella*; **tree s.** *Oxydendrum arboreum*; **wood s.** *Oxalis acetosella*

sorva *Couma macrocarpa*

Sosnovskya Takht. = *Centaurea*

Soterosanthus Lehm. ex Jenny. Orchidaceae (V 10). 1 trop. Am.

sotol *Dasylirion* spp.

souchong *Camellia sinensis*

soufrière plant *Spachea elegans*

Soulamea Lam. Simaroubaceae. 14 Mahé (Seychelles), Mal., New Caled. (7). *S. amara* Lam. (Mal.) – locally medic. (fevers)

Souliea Franchet. Ranunculaceae (I 2). 1 SW China

souphlong *Flemingia vestita*

sour cherry *Prunus cerasus*; **s. orange** *Citrus* × *aurantium*; **s.sop** *Annona muricata*; **s. wood** *Oxydendrum arboreum*

Souroubea Aublet. Marcgraviaceae. 20 trop. Am. G 5-loc. Local medic.

soursob *Oxalis pes-caprae*

southern pine *Pinus echinata*, *P. elliottii*, *P. palustris*, *P. taeda*

southernwood *Artemisia abrotanum*

Southland beech *Nothofagus menziesii*

sow bread *Cyclamen hederifolium*; **s. thistle** *Sonchus oleraceus*

Sowerbaea Sm. Anthericaceae (Liliaceae s.l.). 5 Aus. R: FA 45(1987)264

soy or **soya bean** *Glycine max*

Soyauxia Oliver. Medusandraceae. 7 trop. W Afr.

Soymida A. Juss. Meliaceae (IV 2). 1 India, (? natur. Sri Lanka): *S. febrifuga* (Roxb.) A. Juss. (Indian redwood, bastard cedar) – locally medic., bark a fibre-source & tanning, timber for building

Spachea A. Juss. Malpighiaceae. 6 trop. Am. At least 1 sp. functionally dioec. *S. elegans* (G.

Meyer) A. Juss. (soufrière plant, of St Vincent) – long thought exterminated by volcano but actually native in Guyana

spaghetti *Triticum turgidum* Durum Group; **vegetable s.** *Cucurbita pepo*

Spananthe Jacq. Umbelliferae (I 2b). 1 Andes: *S. paniculata* Jacq. – annual to 2 m tall

Spaniopappus Robinson. Compositae (Eup.-Oxy.). 5 Cuba. R: MSBMBG 22(1987)444

Spanish bluebell *Hyacinthoides hispanica*; **S. broom** *Spartium junceum*; **S. cane** *Arundo donax*; **S. chestnut** *Castanea sativa*; **S. elm** *Cordia gerascanthus*; **S. iris** *Iris xiphium*; **S. jasmine** *Jasminum officinale* f. *grandiflorum*; **S. moss** *Tillandsia usneoides*; **S. reed** *Arundo donax*; **S. senna** *Senna italica*

Sparattanthelium C. Martius. Hernandiaceae (Gyrocarpaceae). 13 trop. S Am. R: BJ 89(1969)193

Sparattosperma C. Martius ex Meissner. Bignoniaceae. 2 trop. S Am. Cult. orn. trees

Sparattosyce Bur. Moraceae (II). 1 New Caledonia

Sparaxis Ker-Gawler. Iridaceae (IV 3). (Incl. *Synnotia*) 12 SW Cape to Karoo. R: JSAB 35(1969)219. Cult. orn. cormous pl. esp. *S. tricolor* (Schneev.) Ker-Gawler but *S. bulbifera* (L.) Ker-Gawler with bulbils in leaf-axils leading to formation of dense stands & resistant weeds in S Aus. poss. serious in native vegetation

Sparganiaceae Hanin. Monocots – Commelinidae – Typhales. 1/14 N temp. to Aus. & NZ. Perennial monoec. emergent hydrophytes with starchy rhizomes & vessels throughout vegetative body. Lvs with sheathing base & parallel-veined linear lamina, distichous. Fls wind-poll. & s.t. by syrphids, sessile, in dense globular unisexual heads with males uppermost & often without bracts, females in axils of leafy bracts; males with P 1–6 (? 1 or more being bracts) scales but in development staminate head becoming a mass of stamens & scales, A 1–8, free or basally connate & opp. P, females usu. in axil of bract but this often not distinguishable from P (2) 3 or 4(5), greenish in 1 or 2 whorls, $\underline{G}$ pseudo-monomerous with 1 reduced sterile carpel or (2 or 3), each loc. with 1 pendulous, anatropous, bitegmic ovule. Fr. a dry drupe usu. with persistent P & style; seed with straight linear embryo in copious mealy oily endosperm with protein & starch & thin perisperm. x = 15

Genus: *Sparganium*

Poss. best united with Typhaceae

Sparganium L. Sparganiaceae. 14 N temp. (Eur. 7, N Am. 10), Mal. to Aus. & NZ. R: BH 96(1986)213, 97(1987)1. Bur reed. Fr. an imp. part of wildfowl diet in autumn & early winter

Sparganophorus Boehmer = *Struchium*

Sparmannia auctt. = seq.

Sparrmannia L.f. Tiliaceae. 3 Afr. & Madag. Cult. orn. esp. *S. africana* L.f. (African hemp, house lime, stock rose, S Afr.) – housepl. tolerant of rough treatment; fls inverted with anthers app. exposed but C reflexed forming a cup behind A – this fills with water during rain & gently overflows a drop at a time, the anthers remaining dry; an insect visiting the fls touches the stamens, which are irritable, springing apart & depositing pollen on insect; bark a fibre-source

sparrow grass = asparagus

Spartidium Pomel (~ *Genista*). Leguminosae (III 28). 1 N Afr.: *S. saharae* (Cosson & Dur.) Pomel – like *Lebeckia*

Spartina Schreber. Gramineae (33b). 17 coastal Am., Eur. (3 + 2 natur.), N Afr. R: ISCJS 30(1956)471. Marsh or cord grass; halophytes with salt-excreting hydathodes in epidermis; some mooted as biofuel crops for poor soils. *S. anglica* C. Hubb. (2n = 120–124) an aggressive pl. spreading up to 5.3 m per annum in NZ, derived c. 1890 from *S.* × *townsendii* Groves & J. Groves, an hybrid (2n = 62) arising c. 1870 at Hythe in Southampton Water, England (& also in Bay of Biscay ('*S.* × *neyrautii*' in 1892) between the introd. *S. alterniflora* Lois. (2n = 62, N Am., introd. c. 1820s) & *S. maritima* (Curtis) Fern. (2n = 60, Eur., Afr.) – a classic case of allopolyploid sp. origin but polytopic (cf. *Tragopogon minus*, *Senecio cambrensis*), its parents almost extinct in GB

Spartium L. Leguminosae (III 31). 1 Medit. (incl. Eur.): *S. junceum* L. (Spanish broom) – cult. orn. with alks, stems form. used in basketry & as fibre-source, fragrant fls a poss. constituent in scent-making & source of yellow dye, introd. S Am. & now occupying large area in Andean highlands

Spartochloa C. Hubb. Gramineae (25). 1 SW Aus.: *S. scirpoidea* (Steudel) C. Hubb. with laminaless sheaths

Spartocytisus Webb. & Berth. = *Cytisus*

Spartothamnella Briq. Labiatae (II; Dicrastylidaceae). 3 Aus. R: JABG 1(1976)1. Switch pls with succ. drupes

Spatalla Salisb. Proteaceae. 20 SW & S Cape

Spatallopsis E. Phillips = *Spatalla*

Spathacanthus Baillon (~ *Justicia*). Acanthaceae. 5 C Am.

Spathandra Guillemin & Perrottet (~ *Memecylon*). Melastomataceae. 6 trop. Afr. R: Adansonia 18(1978)245

Spathantheum Schott. Araceae (VIII 3). 2 Bolivia. Infls epiphyllous; G 6–8-loc.

Spathanthus Desv. Rapateaceae. 2 N S Am. Lvs to 1.5 m long

Spathelia L. Rutaceae. 15 WI to N S Am. *S. sorbifolia* (L.) Fawcett & Rendle (*S. simplex*, mountain pride, Jamaica) – unbranched or sparsely branched hapaxanthic pachycaul with pinnate lvs & term. infls produced after c. 8–10 yrs

Spathia Ewart. Gramineae (40c). 1 trop. Aus.

Spathicalyx J. Gómes. Bignoniaceae. 2 Brazil

Spathicarpa Hook. Araceae (VIII 3). 6 trop. S Am. Infls adnate to spathe. Cult. orn. esp. *S. sagittifolia* Schott (caterpillar pl.)

Spathichlamys R. Parker. Rubiaceae (I 5). 1 Burma

Spathidolepis Schltr. Asclepiadaceae (III 4). 1 New Guinea: *S. torricelliensis* Schltr

Spathionema Taubert. Leguminosae (III 10). 1 trop. Afr.

Spathiostemon Blume. Euphorbiaceae. 3 S Thailand, W Mal., New Guinea

Spathipappus Tzvelev = *Tanacetum*

Spathiphyllum Schott. Araceae (III 4). 36 trop. Am., C Mal. to Solomon Is. (1). R: MNYBG 10,3(1960)1. Peace lilies. Spadix partly adnate to spathe; scent & starchless pollen coll. euglossine bees (cannot digest starch); some with elaiosomes. Cult. orn. with usu. white spathes esp. 'Clevelandii' used in cut-fl. trade (poss. a form of *S. wallisii* Regel, C Am.) & hybrids esp. 'Mauna Loa', also *S. cannifolium* (Dryander) Schott (N S Am., Trinidad) – ash gives a superior alkaline powder for coating pellets of hallucinogenic *Virola elongata*

Spathodea Pal. Bignoniaceae. 1 trop. Afr.: *S. campanulata* Pal. (*S. nilotica*, flame tree, African tulip t., fountain t., Nandi flame) – cult. orn. street tree in trop., aggressive escape in New Guinea; extrafl. nectaries & 3 or 4 buds in each leaf axil, K inflated & full of secreted (?) water, elaiophores attracting bees

Spathodeopsis Dop = *Fernandoa*

Spathoglottis Blume. Orchidaceae (V 11). 30 trop. As. to Aus. Cult. orn. terr. orchids with pseudobulbs; some locally medic.

Spatholirion Ridley. Commelinaceae (II 1b). 3 SE As. 1 erect, 2 climbing

Spatholobus Hassk. Leguminosae (III 10). 29 SE As. to C Mal. R: Reinw. 10(1985)139

speargrass *Aciphylla squarrosa*

spearmint *Mentha spicata*

spearwort *Ranunculus lingua*

Speea Loes. Alliaceae (Liliaceae s.l.). 1–2 Chile

speedwell *Veronica* spp.; **germander s.** *V. chamaedrys*

Speirantha Baker. Convallariaceae (Liliaceae s.l.). 1 E China: *S. convallarioides* Baker – cult. orn.

spelt *Triticum aestivum* Spelta Group

Spenceria Trimen. Rosaceae. 2 W China. Cult. orn. herbs like *Agrimonia*

Speranskia Baillon. Euphorbiaceae. 3 China

Spergella Reichb. = *Sagina*

Spergula L. Caryophyllaceae (I 1). 6 temp. (Eur. 4; 1 endemic N Patagonia). *S. arvensis* L. ((corn) spurrey, Eur., widely natur.) – some forms cult. for forage & green manure as in Holland

Spergularia (Pers.) J. Presl & C. Presl (~ *Spergula*). Caryophyllaceae (I 1). 25 cosmop. esp. halophytes (Eur. 17). Sand spurrey esp. *S. rubra* (L.) J. Presl & C. Presl (Eur., natur. N Am.) – seeds a famine food. *S. marina* (L.) Griseb. (Euras., S S Am.) – disjunct pops hybridize but hybrids almost sterile, form. medic. in Hawaii

Spermacoce L. Rubiaceae (IV 15). 150 warm Am. (incl. *Borreria*). Buttonweeds – some troublesome weeds in S Afr., local medic.

Spermadictyon Roxb. (*Hamiltonia*). Rubiaceae (IV 12). 1 India: *S. suaveolens* Roxb. – cult. orn. shrub with fragrant blue, pink or white fls

Spermatophyta Seed pls, classically divided into Angiospermae & Gymnospermae (q.v.). Extant taxa arranged as 13 200/250 300 in 422 fams

Spermolepis Raf. Umbelliferae (III 8). 5 N Am., Arg., Hawaii (1: *S. hawaiiensis* Wolff – believed extinct but lately re-found)

Sphacanthus Benoist. Acanthaceae. 2 Madagascar

Sphacele Benth. = *Lepechinia*
Sphaenolobium Pim. Umbelliferae (III 8). 3 C As.
Sphaeradenia Harling. Cyclanthaceae. 40 trop. Am. R: AHB 18(1958)341
Sphaeralcea A. St-Hil. Malvaceae. (Incl. *Iliamna*) c. 60 arid Am. R: UCPB 19(1935)1. Cult. orn. herbs & subshrubs incl. *S. coccinea* (Pursh) Rydb. & *S. angustifolia* (Cav.) G. Don f. ssp. *cuspidata* (A. Gray) Kearny (scarlet globe mallow, yerba de la negrita, N Am.) – negrita extract used for hair conditioners in commerce
Sphaeranthus L. Compositae (Pluch). 38 OW trop. to Iran & Egypt (all but 4 in Afr.). R: HIP 36(1955)tt.3501–25
Sphaerantia P.G. Wilson & Hyland. Myrtaceae. 2 N Queensland. R: Telopea 3(1988)260
Sphaereupatorium (O. Hoffm.) Robinson. Compositae (Eup.-Crit.). 1 Bolivia, Brazil. R: MSBMBG 22(1987)321
Sphaerobambos S. Dransf. (~ *Bambusa*). Gramineae (1b). 3 Mal. R: KB 44(1989)428
Sphaerocardamum Schauer. Cruciferae. 8 C Mex. R: CGH 213(1984)12
Sphaerocaryum Nees ex Hook.f. Gramineae (35). 1 India to Taiwan & W Mal.
Sphaerocionium C. Presl (~ *Hymenophyllum*). Hymenophyllaceae (Hym.). c. 80 trop. R: JFSUTB III,13(1982)207
Sphaeroclinium (DC) Schultz-Bip. = *Cotula*
Sphaerocodon Benth. Asclepiadaceae (III 4). 2 Afr.
Sphaerocoma T. Anderson. Caryophyllaceae (I 2). 2 NE Sudan, S Arabia, Iran. R: MBMHRU 285(1968)30
Sphaerocoryne (Boerl.) Ridley = *Melodorum*
Sphaerocyperus Lye (~ *Cyperus*). Cyperaceae. 1 trop. Afr.
Sphaerolobium Sm. Leguminosae (III 24). 12 Aus. esp. SW (11 endemic)
Sphaeromariscus Camus = *Cyperus*
Sphaeromeria Nutt. (~ *Tanacetum*). Compositae (Anth.-Art.). 9 W US, Mex. R: Britt. 28(1976)255
Sphaeromorphaea DC = *Epaltes*
Sphaerophora Schultz-Bip. = *Morinda*
Sphaerophysa DC. Leguminosae (III 15). 2 E Medit. to C As. Halophytes
Sphaeropteris Bernh. = *Cyathea*
Sphaerosacme Wallich ex Royle (~ *Lansium*). Meliaceae (I 5). 1 Himal.: *S. decandra* (Wallich) Penn.
Sphaerosciadium Pim. & Kljuykov. Umbelliferae (III 5). 1 C As.
Sphaerosepalaceae (Warb.) Bullock = Ochnaceae (Diegodendraceae)
Sphaerostephanos J. Sm. = *Cyclosorus*
Sphaerostylis Baillon. Euphorbiaceae. 5 Madagascar, W Mal.
Sphaerothylax Bisch. ex Krauss. Podostemaceae. 10 trop. & S Afr.
Sphaerotylos C.J. Chen = *Sarcochlamys*
Sphagneticola O. Hoffm. Compositae (Helia.-Verb.). 1 SE Brazil
Sphallerocarpus Besser ex DC. Umbelliferae (III 2). 1 S As.
Sphalmanthus N.E. Br. = *Phyllobolus*
Sphalmium B. Briggs, B. Hyland & L. Johnson. Proteaceae. 1 NE Queensland
Sphedamnocarpus Planchon ex Benth. Malpighiaceae. 18 trop. & S Afr., Madagascar
Sphenandra Benth. = *Sutera*
Spheneria Kuhlm. Gramineae (34b). 1 trop. S Am.
Sphenocarpus Korovin = *Seseli*
Sphenocentrum Pierre. Menispermaceae (I). 1 trop. W Afr.: *S. jollyanum* Pierre – roots used as chewing-sticks rendering food eaten afterwards sweet, locally medic. R: NJBB 53(1983)59
Sphenoclea Gaertner. Sphenocleaceae. 1 W Afr.; 1 pantrop. (? orig. Afr.): *S. zeylanica* Gaertner – a weed of rice in 17 countries, shoots locally eaten with rice. R: JAA 67(1986)1
Sphenocleaceae (Lindley) Baskerville (~ Campanulaceae). Dicots – Asteridae – Asterales. 1/2 trop. Annuals of wet places with cortical air-passages. Lvs simple, entire, spirally arr.; stipules 0. Fls reg., bisexual, bibracteolate, in axils of small bracts on term. spikes; K (5) with imbricate lobes, C (5) with urceolate-campanulate tube & imbricate lobes, A alt. with C & attached to tube, anthers with longit. slits, $\overline{G}$ (2) or semi-inf., 2-loc., stigma without pollen-coll. hairs, axile placentas with ∞ anatropous unitegmic ovules. Fr. a circumscissile capsule; seeds ∞ small with straight embryo in ± 0 endosperm. n = 12
Genus: *Sphenoclea* – habit like *Phytolacca*
Sphenodesme Jack. Verbenaceae (Symphoremataceae). 14 SE As., W Mal. R: GBS 21(1966)315

Sphenomeris Maxon = *Odontosoria*
Sphenopholis Scribner. Gramineae (21b). 5 Canada to Mex. R: ISJS 39(1965)289
Sphenopsida = Equisetopsida
Sphenopus Trin. Gramineae (17.). 2 Medit. (Eur. 1). Halophytes
Sphenosciadium A. Gray. Umbelliferae (III 8). 1–2 W N Am.
Sphenostemon Baillon. Aquifoliaceae (Sphenostemonaceae). 7 C Mal. to trop. Aus. & New Caled.
Sphenostemonaceae P. Royen & Airy Shaw = Aquifoliaceae
Sphenostigma Baker = *Gelasine*
Sphenostylis E. Meyer. Leguminosae (III 10). 7 Africa, India. Some ed. seeds & tubers esp. *S. stenocarpa* Harms (girigiri, trop. Afr.) – tubers twice as rich as seeds in protein
Sphenotoma (R. Br.) Sweet. Epacridaceae. 6 SW Aus.
Sphinctacanthus Benth. Acanthaceae. 1 NE India to Burma. R: NJB 5(1985)225
Sphinctanthus Benth. Rubiaceae (II 1). 3 S Am.
Sphinctospermum Rose (~ *Tephrosia*). Leguminosae (III 7). 1 SW N Am.: *S. constrictum* (S. Watson) Rose. R: SB 15(1990)544, 16(1991)162
Sphingiphila A. Gentry. Bignoniaceae. 1 Paraguay: *S. tetramera* A. Gentry. R: SB 15(1990)277
Sphyranthera Hook.f. Euphorbiaceae. 1 Andaman Is.
Sphyrarhynchus Mansf. Orchidaceae (V 16). 1 Tanzania
Sphyrastylis Schltr. Orchidaceae (V 10). 1 Colombia
Sphyrospermum Poeppig & Endl. Ericaceae. 18 trop. Am. mts
spice bush *Lindera benzoin*
Spiculaea Lindley. Orchidaceae (IV 1). 1 W Aus.: *S. ciliata* Lindley
spider azalea *Rhododendron stenopetalum;* **s. flower** *Cleome sesquiorygalis;* **s. lily** *Hemerocallis* spp.; **s. orchid** *Ophrys sphegodes;* **s. plant** *Chlorophytum comosum;* **s.wort** *Tradescantia virginiana*
Spigelia L. Strychnaceae (Loganiaceae s.l.). 50 trop. & warm Am., 1 natur. OW: *S. anthelmia* L. – vermifuge & criminal poison. *S. longiflora* Martens & Galeotti (Mex.) – insect antifeedant. Other spp. medic. esp. *S. marilandica* L. (Maryland pinkroot, SE US) – vermifuge (dried rhizomes etc.), cult. orn.
Spigeliaceae C. Martius = Strychnaceae
spignel *Meum athamanticum*
spiked millet *Pennisetum glaucum*
spikenard *Nardostachys grandiflora;* **American, false** or **wild s.** *Maianthemum racemosum;* **ploughman's s.** *Inula conyzae*
Spilanthes Jacq. Compositae (Helia.-Zinn.). 6 trop. R: SB 6(1981)231. See also *Acmella*
Spiloxene Salisb. (~ *Hypoxis*). Hypoxidaceae. 30 S Afr.
spinach *Spinacia oleracea;* **s. beet** *Beta vulgaris* subsp. *vulgaris;* **Ceylon** or **Indian s.** *Basella alba;* **NZ s.** *Tetragonia tetragonioides;* **Okinawa s.** *Gynura bicolor;* **water s.** *Ipomoea aquatica*
Spinacia L. Chenopodiaceae (I 3). 4 SW As., N Afr. R: BSBG 48(1938)485. *S. oleracea* L. (spinach, orig. SW As.?) – potherb rich in vitamins A, B, C, E & K (also oxalic acid & alks) though its iron cannot be assimilated by body, also used in soup & tinned, introd. Eur. c. AD 1000
spindle tree *Euonymus europaeus*
spinifex *Triodia* spp., *Plectrachne* spp.; **buck s.** *T. longiceps*
Spinifex L. Gramineae (34g). 5 Indomal. to E As. & Pacific. Dioec. grasses, female spikelets 1-flowered with spiny bracts, in heads, which break off at maturity & are blown about like tumbleweeds, dispersing fr.
Spiniluma Baillon ex Aubrév. = *Sideroxylon*
spinks *Cardamine pratensis*
Spiracantha Kunth. Compositae (Vern.-Rol.). 1 C Am. to Venezuela
Spiradiclis Blume. Rubiaceae (IV 2). 16 India & SW China to SE As. & Java. R: ABAS 1(1983)32, Cand. 44(1989)225
Spiraea L. Rosaceae. 80–100 N temp. (Eur. 8) to Mexico & Himal. Cult. orn. decid. shrubs esp. *S.* × *arguta* Zabel (*S.* × *multiflora* Zabel (i.e. *S. crenata* L. (SE Eur. to C As.) × *S. hypericifolia* L. (SE Eur. to Siberia)) × *S. thunbergii* Siebold ex Blume (China, Japan)) – shrub with arching branches of white fls. etc.
Spiraeanthemum A. Gray. Cunoniaceae. (Incl. *Acsmithia*) 20 W Pacific. R: Blumea 25(1978)492, 501. Some locally used timbers
Spiraeanthus (Fischer & C. Meyer) Maxim. Rosaceae. 1 C As.
Spiranthera A. St-Hil. Rutaceae. 4 N S Am. R: BJBB 54(1984)485
Spiranthes Rich. Orchidaceae (III 3). 30 N temp. (Eur. 3–4), few in trop. Am., Mal., Aus. (1) &

Pacific. R: HBML 28(1982)360. Cult. orn. terr. orchids (lady's tresses) incl. *S. romanzoffiana* Cham. & Schldl. (N Am., Ireland & coasts of W Britain but absent from continental Eur.; cf. *Eriocaulon aquaticum*)

Spirella Costantin. Asclepiadaceae (III 4). 2 SE As.

Spiroceratium H. Wolff = *Pimpinella*

Spirodela Schleiden. Lemnaceae. 4 cosmop. (Eur. 1). R: IBM 34(1965)8. Least reduced of Lemnaceae (cf. *Pistia*). In US, grown on dairy farm waste water & subs. for alfalfa in cattle & pig feed

Spirogardnera Stauffer (~ *Choretrum*). Santalaceae. 1 SW Aus.

Spirolobium Baillon. Apocynaceae. 1 SE As.

Spiropetalum Gilg = Rourea

Spirorhynchus Karelin & Kir. Cruciferae. 1 C As. & Iran

Spiroseris Rech.f. Compositae (Lact.-Crep.). 1 Pakistan

Spirospermum Thouars. Menispermaceae (V). 1 Madag.: *S. penduliflorum* Thouars

Spirostachys Sonder. Euphorbiaceae. 2 trop. & S Afr.

Spirostegia Ivanina. Scrophulariaceae. 1 C As.

Spirostigma Nees. Acanthaceae. 1 Brazil

Spirotecoma Baillon ex Dalla Torre & Harms. Bignoniaceae. 5 Cuba, Hispaniola

Spirotheca Ulbr. = *Ceiba*

Spirotropis Tul. Leguminosae (III 2). 1 NE S Am.

spleenwort *Asplenium* spp.; **maidenhair s.** *A. trichomanes*; **mountain s.** *A. montanum*; **sea s.** *A. marinum*; **wall s.** *A. ruta-muraria*

Spodiopogon Trin. Gramineae (40a). 9 As.

Spondianthus Engl. Euphorbiaceae. 1 trop. E Afr.: *S. preussii* Engl. R: BJBB 59(1989)133

Spondias L. Anacardiaceae (II). 10 Indomal. to SE As., trop. Am. Trees with resins coll. euglossine bees for their nests & drupes, some ed. (hog or Jamaica plum, Otaheite apple, imbu (Brazil)) esp. *S. cytherea* Sonn. (*S. dulcis*, ambarella, Jew's plum, golden or Otatheite apple, Pacific) – fr. for jellies, pickles etc.; *S. mombin* L. (*S. lutea*, caja fruit, yellow mombin, jobo, trop. Am.) – bat-disp. fr. eaten fresh, in icecream & liqueurs; *S. purpurea* L. (Spanish plum, red mombin, trop. Am.) – parthenocarpic in SE As., eaten fresh, boiled or dried, fuelwood

Spondogona Raf. = *Sideroxylon*

Spongiocarpella Yakolev & Ulzjkhumag. Leguminosae (III 15). c. 10 Himal. to China

Spongiosperma Zarucchi (~ *Ambalania*). Apocynaceae. 6 N S Am. R: AUWP 87-1(1987)

Spongiosyndesmus Gilli = *Ladyginia*

Sporadanthus F. Muell. Restionaceae. 6 Aus., 1 NZ & Chatham Is. salt-marshes & peatswamps

Sporobolus R. Br. Gramineae (31d). 160 Am., As., Afr. (Eur. 1). Spp. intergrading (cf. *Eragrostis* with closely interrelated but distinct spp.). Some used in revegetation (rush grass) & ed. grains; lvs of *S. festivus* Hochst., *S. lampranthus* Pilger & *S. stapfianus* Gand. (Afr.) revive after reduction of water-content to 5–13%. *S. africanus* (Poiret) Robyns & Tournay (Parramatta grass, trop. & S Afr.) – agric. weed in Aus.; *S. cryptandrus* (Torrey) A. Gray (N Am.) – grains consumed by Indians

Sporoxeia W.W. Sm. Melastomataceae. 6 Burma to SE As.

Spraguea Torrey = *Calyptridium*

Spragueanella Balle. Loranthaceae. 1 trop. E Afr.

sprangletop *Scolochloa festucacea*; **green s.** *Leptochloa dubia*

Sprekelia Heister. Amaryllidaceae (Liliaceae s.l.). 1 Mex., Guatemala: *S. formosissima* (L.) Herbert (Jacobean lily) – cult. orn. with crimson fls & alks

Sprengelia Sm. Epacridaceae. 4 E Aus. *S. incarnata* Sm. – protected sp.

spring beauty *Claytonia virginica*

sprouting, purple or **s. broccoli** *Brassica oleracea* Italica Group

sprouts (**, Brussels**) *B. oleracea* Gemmifera Group

spruce *Picea* spp.; **s. beer** *P. abies*; **black** or **bog s.** *P. mariana*; **hemlock s.** *Tsuga* spp.; **Norway s.** *P. abies*; **Sitka s.** *P. sitchensis*; **white s.** *P. glauca*

spurge *Euphorbia* spp.; **caper s.** *E. lathyris*; **cypress s.** *E. cyparissias*; **ipecacuanha s.** *E. ipecacuanhae*; **s. laurel** *Daphne laureola*; **wood s.** *E. amygdaloides*

Spryginia Popov = *Orychophragmus*

Spuriodaucus Norman. Umbelliferae (inc. sed.). 3 trop. Afr.

Spuriopimpinella Kitag. (~ *Pimpinella*). Umbelliferae (III 8). 4 E As.

spurrey *Spergula arvensis*; **sand s.** *Spergularia* spp.

Spyridium Fenzl. Rhamnaceae. 30 temp. Aus.

Squamellaria Becc. Rubiaceae (IV 7). 3 Fiji. R: Blumea 36(1991)53. ? Ant-pls. 'Male' fls force up & snap off stigma, females with reduced anthers

Squamopappus R. Jansen, Harriman & Urbatsch (~ *Podachaenium*). Compositae (Helia.-Zinn.). 1 SE Mex., Guatemala: *S. skutchii* (S.F. Blake) R. Jansen, Harriman & Urbatsch. R: SB 7(1982)480

squash *Cucurbita* spp.

squaw root *Caulophyllum thalictroides, Conopholis americana*

squill or red s. *Drimia maritima*; **autumn s.** *Scilla autumnalis*; **Lebanon** or **striped s.** *Puschkinia scilloides*; **spring s.** *S. verna*

squinancywort *Asperula cynanchica*

squirting cucumber *Ecballium elaterium*

Sredinskya (Stein) Fed. = *Primula*

Sreemadhavana Rauschert = *Aphelandra*

Staavia Dahl. Bruniaceae. 8 S Afr. R: Bothalia 15(1985)396

Staberoha Kunth. Restionaceae. 9 SW & S Cape

Stachyacanthus Nees = ? *Eranthemum*

Stachyandra Leroy ex Radcl.-Sm. (~ *Androstachys*). Euphorbiaceae. 4 Madag. R: KB 45(1990)562

Stachyanthus DC = *Argyrovernonia*

Stachyanthus Engl. Icacinaceae. 6 trop. Afr.

Stachyarrhena Hook.f. Rubiaceae (II 1). 10 trop. Am. R: RBB 6(1983)109

Stachycephalum Schultz-Bip. ex Benth. Compositae (Helia.). 2 Mex. (1), Arg. (1)

Stachydeoma Small (~ *Hedeoma*). Labiatae (VIII 2). 4 S US

Stachyococcus Standley. Rubiaceae (IV 7). 1 Peru

Stachyophorbe (C. Martius) Liebm. = *Chamaedorea*

Stachyopsis Popov & Vved. Labiatae (VI). 3 C As.

Stachyothyrsus Harms. Leguminosae (I 1). 1 trop. Afr., 2 trop. Am. (Kaoue)

Stachyphrynium Schumann. Marantaceae. 14 Indomal.

Stachys L. Labiatae (VI). 300 temp. (Eur. 58; Turkey 72) & warm exc. Australasia, trop. mts. Cult. orn. herbs & shrubs with alks, some with ed. tubers (hedge nettle or woundwort). *S. affinis* Bunge (*S. sieboldii*, Chinese or Japanese artichoke, crosnes, China) – ed. white tubers salted, pickled or boiled; *S. arvensis* (L.) L. (Eur., Medit.) – natur. N Am. & Aus. where toxic to sheep (stagger-weed); *S. byzantina* K. Koch (*S. lanata*, '*S. olympica*', lamb's ears, SW As.) – much cult. rabbit-proof perennial with silvery densely tomentose lvs; *S. officinalis* (L.) Trev. (betony, Euras.) – locally medic.; *S. palustris* L. (marsh betony, Eur.) – locally medic. & eaten tubers

Stachystemon Planchon. Euphorbiaceae. 4 SW Aus.

Stachytarpheta Vahl. Verbenaceae. 65 trop. & warm Am., trop. OW (few). Fls last 1 day but if spike picked C shed in few mins (traumatochory). Some cult. orn. but usu. widespread weeds, *S. jamaicensis* (L.) Vahl medic. & tea subs. ('devil's coachwhip'), weedy; *S. urticifolia* Sims (trop. As. (natur.?) & Am., natur. Afr.) – hedge-pl. in Afr.

Stachyuraceae J. Agardh. Dicots – Dilleniidae – Violales. 1/5–6 Himal. to Japan. Small trees & shrubs. Lvs simple, toothed, spirally arr.; stipules small, decid. Fls usu. bisexual, small, 4-merous, bibracteolate, in axillary spikes or racemes; K 2 + 2, C 4, imbricate, A 4 + 4 with deeply sagittate anthers with longit. slits, $\underline{G}$ (4), 4-loc. basally but partitions not reaching apex where placentation parietal (Cronquist) with ∞ anatropous, bitegmic ovules. Fr. a 4-loc. dry berry; seeds ∞ arillate, with straight embryo in hard oily proteinaceous endosperm

Genus: *Stachyurus*

Usu. considered near Flacourtiaceae but seed-structure suggests (Corner) alliance with Theales, though separation of the 2 orders 'arbitrary' (Cronquist)

Stachyurus Siebold & Zucc. Stachyuraceae. 5–6 Himal. to Japan. R: BTBC 70(1943)615. Cult. orn. shrubs flowering before lvs expand

Stackhousia Sm. Stackhousiaceae. 14 Aus. (13, 12 endemic 1 ext. to Micronesia, NZ (1). *S. monogyna* Labill. (candles, E Aus.) – cult. orn.; *S. tryonii* Bailey (Queensland) – endemic of serpentine soils of Port Curtis Dist., nickel hyperaccum.

Stackhousiaceae R. Br. Dicots – Rosidae – Celastrales. 3/17 Aus. & NZ, *Stackhousia* ext. to Mal. & Pacific. Usu. perennial rhizomatous ± xeromorphic herbs (annual in *Macgregoria*). Lvs simple, often fleshy, leathery or small; stipules small, decid. or 0. Fls small, bisexual, ± reg., bibracteolate (exc. *Macgregoria*) with cupular hypanthium, in racemes or cymes; K 5 lobes on hypanthium, imbricate, C (5) with free claws, tube & imbricate lobes, nectary-disk within hypanthium, A alt. with C, 3 longer than rest (except *Macgregoria*), anthers with longit. slits, G ((2)3(5)), pluriloc., laterally & usu. apically lobed, each loc. with 1 erect basal-axile, anatropous, bitegmic ovule. Fr. dry, maturing as indehiscent mericarps; seeds with straight embryo in oily endosperm. x = 9, 10, 15

Genera: *Macgregoria* (put in own subfam.), *Stackhousia*, *Tripterococcus*

Stadiochilus R.M. Sm. Zingiberaceae. 1 Burma

Stadmannia Lam. Sapindaceae. 2 trop. E Afr., S Afr., Madagascar, Mascarenes

Staehelina L. Compositae (Card.-Card.). 8 Medit. (Eur. 5)

Staelia Cham. & Schldl. Rubiaceae (IV 15). 12 N S Am.

stagger-weed *Stachys arvensis*

stag(s)horn fern *Platycerium* spp.; **s. sumach** *Rhus hirta*

Stahelia Jonker = *Tapeinostemon*

Stahlia Bello. Leguminosae (I 1). 1 Hispaniola, Puerto Rico: *S. monosperma* (Tul.) Urban. Timber

Stahlianthus Kuntze. Zingiberaceae. 6–7 E Himalaya to SE As. & Hainan

Staintoniella H. Hara. Cruciferae. 2 Himalaya

Stalagmitis Murray = *Garcinia*. Genus based on specimen comprising 2 diff. spp. stuck together with sealing-wax (cf. *Actinotinus* & *Papilionopsis*)

Stalkya Garay. Orchidaceae (III 3). 1 Venezuelan Andes

Standleya Brade. Rubiaceae (I 5). 4 Brazil

Standleyacanthus Leonard. Acanthaceae. 1 Costa Rica

Standleyanthus R. King & H. Robinson (~ *Eupatorium*). Compositae (Eup.-Oxy.). 1 Costa Rica. R: MSBMBG 22(1987)446

Stanfieldiella Brenan. Commelinaceae (II 2). 4 trop. Afr.

Stanfordia S. Watson = *Caulanthus*

Stangea Graebner = *Valeriana*

Stangeria T. Moore. Stangeriaceae. 1 S Afr.: *S. eriopus* (Kunze) Baillon (*S. paradoxa*) first described as a fern (*Lomaria e.*). R: Bothalia 8(1965)429

Stangeriaceae L. Johnson. Cycadopsida. 1/1 S Afr. Dioec. fern-like cycad with underground stem differing from Zamiaceae in convolute (not imbricate) leaf vernation, pinnae with definite midrib & many dichot. branched costae

Only species: *Stangeria eriopus*

Stanhopea Frost. Orchidaceae (V 10). 55 trop. Am. Cult. orn. epiphytes with large v. fragrant fls to 20 cm diam. (*S. tigrina* Bateman ex Lindley, E Mex. to Brazil) incl. *S. candida* Barb. Rodr. (NE S Am.) – poll. euglossine bee, *Eulaema mocsaryi*, which scrapes for scent & brushes off pollinium which is deposited in stigmatic cavity of next fl. visited but *E. ignita* also collects scent but does not brush off pollinia so is a robber

Stanleya Nutt. Cruciferae. 6 W N Am. R: Lloydia 2(1939)109. Some potherbs used by Indians

Stanleyella Rydb. = *Thelypodium*

Stanmarkia Almeda. Melastomataceae. 2 W Guatemala & Mex. R: Britt. 45(1993)187

Stapelia L. Asclepiadaceae (III 5). 44 trop. & S Afr. R: Excelsa Tax. ser. 3(1985)1. Cult. orn. succ. with photosynthetic stems & small or 0 lvs. Fls to 46 cm diam. in *S. gigantea* N.E. Br., often luridly marked & fleshy, foetid, attractive to flies which lay eggs there & carry out poll. (carrion flowers). The anthers contain the pollen united into masses, 'pollinia' (cf. Orchidaceae) & are embedded below the central disk which is the style, bearing the stigmatic surfaces on the underside; poll. effected by legs or proboscides of the flies falling or being pushed through cracks overlying anthers, pollinia being clipped on & finally deposited on stigmas through other cracks. See also *Orbea*, *Stapelianthus*, *Stapeliopsis*, *Tridentia*, *Tromotriche* (somewhat 'one-legged' segregates ('genuslets'))

Stapelianthus Choux ex A. White & B. Sloane. Asclepiadaceae (III 5). 8 Madag. R: Excelsa 7(1977)83

Stapeliopsis Pill. (~ *Orbea*). Asclepiadaceae (III 5). 6 S Afr. R: CSJ 43(1981)73

Stapfiella Gilg. Turneraceae. 5 trop. Afr.

Stapfiola Kuntze = *Desmostachya*

Stapfiophyton Li. Melastomataceae. 4 S China

Staphylea L. Staphyleaceae. 11 N temp. (Eur. 1). R: Arnoldia 40(1980)76. Cult. orn. trees & shrubs (bladder nuts) esp. *S. pinnata* L. (Eur., SW As.) with white fls

Staphyleaceae Martinov. Dicots – Rosidae – Sapindales. 5/27 Am., Euras. to Mal. Trees & shrubs. Lvs pinnate (to unifoliolate), opp. (Staphyleoideae) or spirally arr. (Tapiscioideae), leaflets usu. toothed; stipules decid. or 0. Fls bisexual or not (pls s.t. dioec.), reg., 5-merous, in racemes or panicles; K imbricate, sometimes (Tapiscioideae) connate, often petaloid, C imbricate, A alt. with C on or outside annular nectary-disk (0 in some Tapiscioideae), anthers with longit. slits, G (2 or 3(4)), pluriloc. (free in *Euscaphis*), styles free or ± united, each loc. with (1 or 2–)6–12 anatropous, bitegmic ovules in 2 rows on axile or basal-axile placentas (only 1 or 2 in Tapiscioideae). Fr. a head of follicles (*Euscaphis*), a drupe or berry or an inflated capsule, each loc. often with only 1 or 2 seeds; embryo straight in copious oily endosperm. x = 13

Classification & genera:

1. **Staphyleoideae** (lvs opp., K ± free): *Euscaphis, Staphylea, Turpinia*
2. **Tapiscioideae** (lvs spirally arr., K ± tubular): *Huertea, Tapiscia*

Seed structure (Corner) confirms position in Sapindales though wood like some Cunoniaceae

Cult. orn. (*Euscaphis, Staphylea*) & timbers (*Turpinia*)

star anise *Illicium verum*; **s. apple** *Chrysophyllum cainito*; **s. of Bethlehem** *Ornithogalum angustifolium* & other spp.; **yellow s. of B.** *Gagea lutea*; **s. fruit** *Averrhoa carambola*; **s. grass** *Cynodon* spp. esp. *C. dactylon*; **Texas s.** *Lindheimera texana*; **s.wort** *Aster* spp., *Callitriche* spp., *Stellaria* spp.

starch comm. prepared from potatoes, cassava, maize, wheat & other cereals

starflower oil *Borago officinalis*

Stathmostelma Schumann. Asclepiadaceae (III 1). 12 trop. Afr.

statice *Limonium* spp.

Staudtia Warb. Myristicaceae. 2 W Afr. Some timber (e.g. *S. stipitata* Warb. (niové, trop. Afr.)) & little-known oilseeds, locally medic.

Stauntonia DC. Lardizabalaceae. 24 E As. Evergreen monoec. lianes. *S. hexaphylla* (Thunb.) Decne. (China, Japan) – cult. orn. with fragrant white fls & ed. purple fr.

Stauracanthus Link. Leguminosae (III 31). 2 SW Eur. (2), NW Afr. Gorse-like prickly shrubs with 3-foliolate seedling lvs like *Ulex*

Stauranthera Benth. Gesneriaceae. 8 SE As. to New Guinea. R: JAA 65(1984)129

Stauranthus Liebm. Rutaceae. 1 S Mexico

Staurochilus Ridley ex Pfitzer (~ *Trichoglottis*). Orchidaceae (V 16). 4 Indomal.

Staurochlamys Baker. Compositae (Helia.-Mel.). 1 N Brazil

Staurogyne Wallich. Acanthaceae. 140 trop. esp. Mal. & Am. (Afr. 5; R: BJBB 61(1991)98). Excl. those with epiphyllous infls (= *Saintpauliopsis*)

Staurophragma Fischer & C. Meyer = *Verbascum*

stavesacre *Delphinium staphisagria*

stave-wood *Quassia amara*

Stawellia F. Muell. Anthericaceae (Liliaceae s.l.). 2 SW Aus. R: FA 45(1987)252

Stayneria L. Bolus (~ *Ruschia*). Aizoaceae (V). 1 SW S Afr.

Stebbinsia Lipsch. = *Soroseris*

Stebbinsoseris K. Chambers. Compositae (Lact.-Mic.). 2 SW US

Steenisia Bakh.f. Rubiaceae (I 5). 5 Borneo, Natuna Is. R: NJB 4(1984)333

Steenisioblechnum Hennipman. Blechnaceae (Blechnoideae). 1 Queensland: *S. acuminatum* (C. White & Goy) Hennipman

Stefanoffia H. Wolff. Umbelliferae (III 8). 2 E Medit. (Eur. 1)

Steganotaenia Hochst. Umbelliferae (III 10). 3 Ethiopia to S Afr. *S. araliacea* Hochst. – tree with fls produced before lvs; *S. hockii* (Norman) Norman – perennial herb

Steganthera Perkins. Monimiaceae (V 2). (Inc. *Anthobembix*) 17+ E Mal. to Solomon Is. & trop. Aus. R: Blumea 29(1984)481

Stegia DC = *Lavatera*

Stegnogramma Blume = *Cyclosorus*

Stegnosperma Benth. Stegnospermataceae (Phytolaccaceae s.l.). 4 Baja Calif. to WI & C Am. R: BSBM 46(1984)37. Some roots a soap subs

Stegnospermataceae (A. Rich.) Nakai (~ Phytolaccaceae). Dicots – Caryophyllidae – Caryophyllales. 1/4 SW N & C Am. Small trees & shrubs. Lvs spirally arr., simple; stipules 0. Fls in thyrses, racemes or cymules, reg., bisexual. K 5, C 5 (? staminodes), A 5(–10) united at base, anthers with longit. slits, G3–5, becoming 1-loc. with central column, each with 1 basal amphitropous bitegmic ovule. Fr. a capsule with 3–5 valves; seeds (1–)3–5 covered with red or white aril, perisperm copious, embryo curved

Genus: *Stegnosperma*

Wood anatomy, floral structure etc. suggest affinity with the anthocyanin-containing Caryophyllaceae & Molluginaceae

Stegolepis Klotzsch ex Koern. Rapateaceae. 23 N S Am.

Steinbachiella Harms = *Diphysa*

Steinchisma Raf. (~ *Panicum*). Gramineae (34b). 4 S US to Arg.
Steinheilia Decne. = *Odontanthera*
Steirachne Ekman. Gramineae (31d). 1 NE S Am.
Steiractinia S.F. Blake. Compositae (Helia.-Verb.). 12 Venezuela, Colombia, Ecuador
Steirodiscus Less. (*Gamolepis, Psilothonna*). Compositae (Sen.-Sen.). 5 S Afr. Some cult. orn.
Steiropteris (C. Chr.) Pichi-Serm. = *Cyclosorus*
Steirosanchezia Lindau = *Sanchezia*
Steirotis Raf. = *Struthanthus*
Stelechantha Bremek. Rubiaceae (I 10). 1 Angola
Stelechocarpus Hook.f. & Thomson. Annonaceae. 5 SE As., Mal. Dioec., females cauliflorous. Ed. fr. esp. *S. burahol* (Blume) Hook.f. & Thomson (keppel fr., Mal.) – cult., gives all bodily secretions violet scent temporarily so a favourite of Javanese sultans' harems
Steleocodon Gilli = *Phalacraea*
Steleostemma Schltr. = *Amblystigma*
Stelestylis Drude. Cyclanthaceae. 4 N S Am.
Stelis Sw. Orchidaceae (V 13). 370 trop. Am. R: HBML 26(1980)167. Few cult. orn. epiphytes with small dingy fls
Stellaria L. Caryophyllaceae (II 1). 150–200 cosmop. (Eur. 18). Some cult. orn. herbs & weeds (stitchwort, chickweed, starwort) esp. *S. media* (L.) Vill. (common c., (?) orig. S Eur.) – cosmop. weed usu. autogamous, poss. also s.t. thrips-poll. in Netherlands, s.t. cleistogamous, saponins basis of efficacy in treatment of skin disorders & *S. neglecta* Weihe (greater c., Eur.) & *S. holostea* L. (greater stitchwort, adder's meat, Eur., Medit.)
Stellariopsis (Baillon) Rydb. = *Potentilla*
Stellera L. Thymelaeaceae. 8 Iran to China. *S. chamaejasme* L. (C As., Himal.) – cult. shrub
Stelleropsis Pobed. = *Diarthron*
Stelligera A.J. Scott = *Sclerolaena*
Stellilabium Schltr. Orchidaceae (V 10). 20 Colombia, Peru
Stellularia Benth. = *Buchnera*
Stelmacrypton Baillon. Asclepiadaceae (I). 1 E India, S China: *S. khasianum* (Kurz) Baillon
Stelmagonum Baillon = ? *Matelea*
Stelmanis Raf. Older name for *Oxypappus*
Stelmation Fourn. = *Metastelma*
Stelmatocodon Schltr. = *Philibertia*
Stelmatocrypton Baillon = *Pentanura*
Stemmacantha Cass. (*Rhaponticum* Hill). Compositae (Card.-Cent.). 20 S Eur., As., N Afr., Aus. (1). R: Cand. 39(1984)45. *S. carthamoides* (Willd.) Dittrich (Siberia) – ecdysteroids
Stemmadenia Benth. Apocynaceae. 10 trop. Am. R: A.J.M. Leeuwenberg (1991, 1994) *Rev. Tabernaemontana*: 398. Alks; some local masticatories in Mex. & cult. orn.
Stemmatella Wedd. ex Benth. = *Galinsoga*
Stemmatodaphne Gamble = *Alseodaphne*
Stemodia L. Scrophulariaceae. c. 55 trop. (OW c. 20; Am. 37, R: Phytol. 74(1993)61, 75(1993)281). *S. multifida* (Michaux) Sprengel (*Leucospora* m., E N Am.) – cult. orn. annual
Stemodiopsis Engl. Scrophulariaceae. 3 Afr.
Stemona Lour. Stemonaceae. 25 Indomal. to E As. & trop. Aus. Alks (some effective insecticides). *S. tuberosa* Lour. (Indomal.) – looks like hop pl., fls carrion-scented & visited by flies
Stemonaceae Caruel. Monocots – Liliidae – Liliales. (Excl. Pentastemonaceae) 3/30 Indomal. to E As. & trop. Aus., E N Am. Herbs, climbing or erect, or shrublets with rhizomes or tubers; vasc. bundles in 1 or 2 rings (vessels in at least roots). Lvs with non-sheathing petioles & entire cordate laminas with 5–15 arching, convergent main veins with cross-veins, distichous, opp. or verticillate. Fls usu. bisexual, axillary, solit. or in few-flowered axillary infls, 2-merous; P 2 + 2, s.t. basally united, A 2 + 2, filaments s.t. united or adnate to P, connective s.t. elongate, G̲ (2), semi-inf. in *Stichoneuron*, 1-loc. with usu. few basal (*Stemona*) or apical pendulous, anatropous or orthotropous (*Stemona*) bitegmic ovules. Fr. a 2-valved capsule; seeds 1–few, longit. ribbed bearing arillate structures (elaiosomes, caruncles etc.), embryo small in copious oily proteinaceous endosperm s.t. with hemicellulose. x = 7. R: Blumea 36(1991)239

Genera: *Croomia, Stemona, Stichoneuron*

Pentastemona, excl. to own fam. here is unique in monocots in reg. 5-merous 3-whorled fls

Some insecticides (*Stemona*)

Stemonocoleus Harms. Leguminosae (I 4). 1 Guineo-Congolian forests
Stemonoporus Thwaites. Dipterocarpaceae. 15 Sri Lanka. R: Blumea 20(1972)363

Stemonurus Blume. Icacinaceae. (Incl. *Urandra*) c. 30 Indomal. Some timber
Stemotria Wettst. & Harms ex Engl. Scrophulariaceae. 1 Peru
Stenachaenium Benth. Compositae (Pluch.). 5 S Brazil, Arg., Uruguay, Paraguay
Stenactis Cass. = *Erigeron*
Stenadenium Pax = *Monadenium*
Stenandriopsis S. Moore = *Crossandra*
Stenandrium Nees. Acanthaceae. 25 warm Am. Some cult. orn.
Stenanona Standley. Annonaceae. 10 Costa Rica, Panamá
Stenanthella Rydb. = *Stenanthium*
Stenanthium (A. Gray) Kunth. Melanthiaceae (Liliaceae s.l.). 5 E As., N Am. Some cult. orn.
Stenia Lindley. Orchidaceae (V 10). 8 N S Am. to Trinidad. Some cult. orn.
Stenocactus (Schumann) A.W. Hill (~ *Ferocactus*). Cactaceae (III 9). 10 Mex.
Stenocarpha S.F. Blake (~ *Galinsoga*). Compositae (Helia.-Helia.). 2 Mexico
Stenocarpus R. Br. Proteaceae. c. 25 W Pacific (Aus. 8, endemic). Some timbers esp. *S. salignus* R. Br. (beefwood, E Aus.) – used for furniture; some cult. orn. incl. *S. sinuatus* Endl. (firewheel tree, E Aus.) with red fls
Stenocephalum Schultz-Bip. (~ *Vernonia*). Compositae (Vern.-Vern.) 5 trop. Am.
Stenocereus (A. Berger) Riccob. (*Rathbunia*; ~ *Lemaireocereus*). Cactaceae (III 8). 25 Mex. to WI
Stenochasma Miq. = *Broussonetia*
Stenochilus R. Br. = *Eremophila*
Stenochlaena J. Sm. Blechnaceae (Stenochlaenoideae). 6 trop. OW (Afr. 2 ext. to Madag. & S Afr.). R: AFJ 61(1971)119. Young fronds of some ed. & cult. orn. climbing epiphytes
Stenochlaenaceae Ching = Blechnaceae
Stenochlamys Griffith = ? *Davallia*
Stenocline DC. Compositae (Gnap.-Gnap.). 3 Madag., Mauritius. R: OB 104(1991)138. Some locally medic.
Stenocoelium Ledeb. Umbelliferae (III 8). 3 C As. *S. divaricatum* Turcz. – used with *Schizonepeta tenuifolia* in treatment of toothache & colds in modern Chinese herbalism
Stenocoryne Lindley = *Bifrenaria*
Stenodon Naudin. Melastomataceae. 1 S Brazil
Stenodraba O. Schulz = *Weberbauera*
Stenodrepanum Harms. Leguminosae (I 1). 1 Argentina
Stenofestuca (Honda) Nakai = *Bromus*
Stenoglottis Lindley. Orchidaceae (IV 2). 4 E & S Afr. R: KM 6(1989)9. Cult. orn. terr. orchids
Stenogonum Nutt. (~ *Eriogonum*). Polygonaceae (I 1). 2 W N Am.
Stenogyne Benth. Labiatae (VI). 21 Hawaii. W. Wagner et al., *Man. Fl. Pl. Hawaii* (1990)831, PS 45(1991)30. Lianes & herbs
Stenolepia Alderw. Dryopteridaceae (I 2). 1 C & E Mal.: *S. tristis* (Blume) Alderw.
Stenolirion Baker = *Ammocharis*
Stenoloma Fée = *Odontosoria*
Stenomeria Turcz. Asclepiadaceae (III 1). 3 Colombia
Stenomeris Planchon. Dioscoreaceae. 2 W Mal.
Stenomesson Herbert. Amaryllidaceae (Liliaceae s.l.). c. 35 Andes. R: PL 27(1971)73. Some cult. orn. bulbous herbs. *S. elwesii* (Baker) Macbride – pollen shed in tetrads (only monocot known)
Stenopadus S.F. Blake. Compositae (Mut.-Mut.). 14 Guayana Highland. R: AMBG 76(1989)1002. Pachycaul shrubs & trees
Stenopetalum R. Br. ex DC. Cruciferae. 9 Aus. R: JAA 53(1972)52. *S. velutinum* F. Muell. – Aboriginal veg.
Stenophalium A. Anderb. (~ *Stenocline*). Compositae (Gnap.-Gnap.). 4 Brazil. R: OB 104(1991)141
Stenops R. Nordenstam (~ *Senecio*). Compositae (Sen.-Sen.). 2 trop. Afr. *S. zairensis* (Lisiowski) R. Nordenstam (Zaire) – floating aquatic
Stenoptera C. Presl. Orchidaceae (III 3). 10 Andes. One mycotrophic sp.
Stenorrhynchos Rich. ex Sprengel. Orchidaceae (III 3). 13 trop. Am. R: HBML 28(1982)372
Stenoschista Bremek. = *Ruellia*
Stenosemia C. Presl = *Tectaria*
Stenosemis E. Meyer ex Harvey & Sonder (~ *Annesorhiza*). Umbelliferae (III 8). 2 S Afr.
Stenoseris Shih (~ *Mycelis*). Compositae (Lact.-Lact.). 5 C As. to China. R: APS 29(1991)411
Stenosiphon Spach. Onagraceae. 1 C & S US
Stenosiphonium Nees. (~ *Strobilanthes*) Acanthaceae. c. 3 India, Sri Lanka

Stenosolen (Muell. Arg.) Markgraf = *Tabernaemontana*
Stenosolenium Turcz. Boraginaceae. 1 C As.
Stenospermation Schott. Araceae (III 2). c. 30 trop. Am.
Stenostelma Schltr. (~ *Schizoglossum*). Asclepiadaceae (III 1). 4–5 S Afr.
Stenostemomum Gaertner f. NW spp. of *Antirhea*
Stenostephanus Nees. Acanthaceae. 6 trop. S Am.
Stenotaenia Boiss. (~ *Heracleum*). Umbelliferae (III 11). 5–6 SW As.
Stenotaphrum Trin. Gramineae (34b). 7 trop. & warm. R: Britt. 24(1972)202. Some fodders; *S. secundatum* (Walter) Kuntze (St Augustine grass, buffalo grass (Aus.), warm Am.) – lawn-grass & for binding sand, varieg. 'Variegatum' a hanging-basket pl.
Stenothyrsus C.B. Clarke. Acanthaceae. 1 Malay Peninsula
Stenotus Nutt. (~ *Haplopappus*). Compositae (Ast.-Sol.). 6 W N Am. & Canada
Stephanachne Keng. Gramineae (21a). 2 C As., W China. R: KBAS 13(1986)122
Stephanandra Siebold & Zucc. Rosaceae. 3–4 E As. Cult. orn. decid. shrubs esp. *S. incisa* (Thunb.) Zabel (China, Korea, Japan) with weeping branches of deeply incised lvs
Stephania Lour. Menispermaceae (V). 35 trop. OW (Afr. 5). Many alks. Some locally medic.; *S. japonica* (Thunb.) Miers (*S. hernandiifolia*, Indomal. to E Aus.) – fish-poison
Stephanocaryum Popov. Boraginaceae. 2 C As.
Stephanocereus A. Berger (~ *Cephalocereus*). Cactaceae (III 3). 2 Bahia. Cult. orn. tree or unbranched
Stephanochilus Cosson & Durieu ex Maire (~ *Volutaria*). Compositae (Card.-Cent.). 1 N Afr.
Stephanococcus Bremek. Rubiaceae (IV 1). 1 trop. Afr.
Stephanodaphne Baillon. Thymelaeaceae. 8–9 Madag., Comoro Is.
Stephanodoria E. Greene (~ *Xanthocephalum*). Compositae (Ast.-Sol.). 1 Mexico
Stephanolepis S. Moore = *Erlangea*
Stephanomeria Nutt. Compositae (Lact.-Step.). 17 W N Am. *S. schottii* (A. Gray) A. Gray rediscovered in 1978 after 100 yrs
Stephanopholis S.F. Blake = *Chromolepis*
Stephanophysum Pohl = *Ruellia*
Stephanopodium Poeppig. Dichapetalaceae. 9 trop. S Am. R: FN 10(1972)36. Infls on petioles
Stephanorossia Chiov. = *Oenanthe*
Stephanostegia Baillon. Apocynaceae. 5 Madag. R: FMad. 169(1976)102. Allied to *Aspidosperma* (Am.), not Afr. genera
Stephanostema Schumann. Apocynaceae. 1 Tanz.: *S. stenocarpum* Schumann (?) extinct
Stephanotella Fourn. = *Marsdenia*
Stephanothelys Garay. Orchidaceae (III 3). 4 Andes
Stephanotis Thouars = *Marsdenia*
Steptorhamphus Bunge. Compositae (Lact.-Lact.). 7 E Eur. (1), SW to C As. & Pakistan
Sterculia L. Sterculiaceae. 150 trop. Monoec. or polygamous trees (for New Caledonian remarkable unbranched pachycauls see *Acropogon*); some timbers, gums (used in comm. bowel regulators) & cult. orn. Many spp. with sugar-secreting hairs within calyx, attracting lots of bugs & cicadas which lead to enlarging & sterilizing of some floral parts. *S. africana* (Forssk.) Fiori (S Arabia) – resin coll. to make a lathering agent against lice; *S. balanghas* L. (SE As. to Mal.) – seeds ed.; *S. chicha* A. St-Hil. (NE S Am.) – ed. seeds, oil used in lubrication etc.; *S. fanaiho* Setch. (Tonga) – fibre for textiles & mats; *S. foetida* L. (Java olive, trop. OW) – ed. seeds, oil, fls smell of decaying meat, bole to 2 m girth; *S. monosperma* Vent. (China) – ed. seeds; *S. oblonga* C. Martius (trop. Afr.) – timber (eyong, yellow sterculia) dense, for tables, boats, sleepers etc.; *S. rhinopetala* Schumann (W Afr.) – timber (aye, brown sterculia); *S. urens* Roxb. (Indomal.) – source of karaya gum or Indian tragacanth used as tragacanth subs. in cosmetics & icecream; *S. villosa* Roxb. (India, Burma) – timber for tea-chests, bark-fibre for elephant harness, coarse canvas, bags etc. & form. for paper
Sterculiaceae DC. Dicots – Dilleniidae – Malvales. 67/1500 trop. & warm, few temp. Trees & shrubs, s.t. unbranched pachycaul treelets, rarely herbs or lianes, usu. with stellate indumentum. Lvs simple (pinnately or palmately veined), often palmately lobed, or palmate; petiole often with medullary bundles, stipules s.t. decid. Fls usu. bisexual, usu. reg., often with epicalyx, usu. in complex infls; K 3–5, valvate, usu. briefly connate basally & with tufts of glandular hairs acting as nectaries, C 5 convolute, usu. clawed, s.t. adnate to A-tube at base, or 0, A 5 (often staminodal & petaloid, or 0) + 5 or 5 bundles of 2–3(–10 +) developing centrifugally, filaments connate in tube around ovary, often on androgynophore, anthers with longit. slits (rarely apical pores), $\underline{G}$ ((1–)5(–60)) with as many

loc. (rarely uniloc. but in Sterculieae united only by styles & becoming distinct at maturity or free throughout as in *Cola*), each carpel with (1)2–∞ anatropous or hemitropous bitegmic ovules often with zig-zag micropyle on usu. axile placentas. Fr. dehiscent or not, fleshy to leathery or woody, often separating as mericarps; seeds s.t. arillate with straight or curved embryo in usu. copious oily &/or starchy endosperm (0 in *Cola*). x = (5–)20(–∞)

Principal genera: *Ayenia, Brachychiton, Byttneria, Cola, Dombeya, Helicteres, Heritiera, Hermannia, Lasiopetalum, Leptonychia, Melhania, Melochia, Thomasia, Waltheria*

Rather variable, Sterculieae (*Sterculia, Cola* etc.) s.t. kept separate from the rest (then treated as Byttneriaceae), but this latter group remains next to S. in the system & is still heterogeneous (Cronquist). *Leptonychia* differs in its seed anatomy from the rest (Corner), while some genera incl. here have been placed in allied fams – *Dicarpidium, Maxwellia* (Bombacaceae, to which *Paradombeya* is referred) or considered possible candidates for Malvaceae

Ovarywall in some genera splits & expands into boat-shaped wing so that developing seed in *Firmiana, Pterocymbium* & *Scaphium* is as 'naked' as a gymnosperm (Corner): if such an immature fr. were found fossilized, it could easily be thought 'a primitive angiosperm'

Imp. tree crops incl. cocoa (*Theobroma*), cola nuts (*Cola*), some timbers (*Argyrodendron, Guazuma, Heritiera, Mansonia, Nesogordonia, Pterospermum, Sterculia, Triplochiton* (obeche) etc.), fibres (*Abroma, Helicteres, Sterculia, Waltheria*), medic. (*Scaphium, Sterculia*) & cult. orn. incl. spp. of *Brachychiton, Dombeya, Firmiana, Fremontodendron* (± hardy), *Kleinhovia, Sterculia*

Stereocaryum Burret. Myrtaceae (Myrt.). 2 Malay Pen. (New Caled. spp. = *Arillastrum*)

Stereochilus Lindley (~ *Sarcochilus*). Orchidaceae (V 16). 6 India to Thailand

Stereochlaena Hackel. Gramineae (34d). 5 trop. to S Afr. R: KB 33(1978)295

Stereosandra Blume. Orchidaceae (V 5). 1 SE As., W Mal. Mycotrophic

Stereospermum Cham. Bignoniaceae. 15 trop. OW. Cult. orn. trees with good timber (cf. most B.) esp. *S. colais* (Dillwyn) Mabb. (*S. personatum*, padri, India) – tea-chests, furniture, construction

Sterigmapetalum Kuhlm. Rhizophoraceae. 7 trop. S Am. R: AMBG 70(1983)179

Sterigmostemum M. Bieb. Cruciferae. 7 Eur. (1), SW to C As. R: Boissiera 40(1988)

Steriphoma Sprengel. Capparidaceae. 8 trop. Am.

Steris Adans. = *Silene* (*Viscaria*)

Sternbergia Waldst. & Kit. Amaryllidaceae (Liliaceae s.l.). 7 SE Eur. (2) to SW As. & Kashmir. R: Plantsman 5(1983)1. Cult. orn. with alks, a yellow-flowered 'version' of *Colchicum*, esp. *S. lutea* (L.) Sprengel (SE Eur. to Iran) – contender for 'lilies of the field', also *S. vernalis* (Miller) Gorer & J. Harvey (*S. fischeriana*, Cauc. to Kashmir) with larger fls. Some spp. with ant-disp. seeds with fleshy arils remaining soft & sticky on herbarium specimens to 35 yrs old!

Sterropetalum N.E. Br. = *Nelia*

Stethoma Raf. = *Justicia*

Stetsonia Britton & Rose. Cactaceae (III 1). 1 Bolivia, Paraguay, Arg.: *S. coryne* (Salm-Dyck) Britton & Rose – cult. orn. tree-like

Steudnera K. Koch. Araceae (VII 2). 8 Himalaya to SE As. & Malay Peninsula

Stevenia Adams ex Fischer. Cruciferae. 3 E As.

Steveniella Schltr. Orchidaceae (IV 2). 1 Crimea to W As.

Stevensia Poit. Rubiaceae (I 5). 8 WI. R: Phytol. 70(1991)151

stevia *Piqueria trinervia*

Stevia Cav. Compositae (Eup.-Ager.). c. 235 trop. & warm Am. R: MSBMBG 22(1987) 170. *S. rebaudiana* (Bertoni) Bertoni (caa-ehe, Paraguay) – long used by indigenous people for sweetening drinks, a diterpene glycoside (stevioside) up to 300 times as sweet as sucrose (i.e. up to three fifths as effective as saccharine) now much used in Japan, the shrub being prop. by root-cuttings (seeds difficult)

Steviopsis R. King & H. Robinson (~ *Eupatorium*). Compositae (Eup.-Alom.). 8 SW US, Mex. R: MSBMBG 22(1987)247

Stewartia L. Theaceae. 9+ E As., E N Am. R: JAA 55(1974)182. Cult. orn. decid. trees & shrubs

Stewartiella Nasir. Umbelliferae (III 5). 1 Pakistan: *S. crucifolia* (Gilli) Hedge & Lamond

Steyerbromelia L.B. Sm. Bromeliaceae (1). 3 C Amazonas, Venez. R: AMBG 73(1986)699

Steyermarkia Standley. Rubiaceae (I 6). 1 C Am.

Steyermarkina R. King & H. Robinson (~ *Eupatorium*). Compositae (Eup.-Crit.). 4 Venez. & Brazil. R: MSBMBG 22(1987)368

Steyermarkochloa Davidse & R. Ellis. Gramineae (38). 1 Venez. & Colombia forming monotypic tribe

Stiburus Stapf = *Eragrostis*

Sticherus C. Presl. Gleicheniaceae (Gleich.). c. 80 trop. (Afr. v. few) & S temp. Some cult. orn.

Stichianthus Valeton. Rubiaceae (I 9). 2 Borneo

Stichoneuron Hook.f. Stemonaceae (Croomiaceae). 2 Assam to Malay Peninsula

Stichorkis Thouars = *Liparis*

sticktight *Bidens* spp.

sticky wattle *Acacia howittii*

Stictocardia Hallier f. Convolvulaceae. 9 trop. *S. beraviensis* (Vatke) Hallier f. (Afr.) – liane with crimson fls, cult. orn.

Stictophyllorchis Carnevali & Dodson (*Stictophyllum*). Orchidaceae (V 10). 1 NE S Am., Trinidad: *S. pygmaea* (Cogn.) Carnevali & Dodson. R: Lindleyana 8(1993)101

Stictophyllum Dodson & M. Chase = praec.

Stifftia Mikan. Compositae (Mut.-Mut.). 6 NE Brazil, Fr. Guiana. R: SB 16(1991)685. Trees, shrubs & lianes

Stigmaphyllon A. Juss. Malpighiaceae. 100 trop. Am. Lianes, some cult. orn. esp. *S. ciliatum* (Lam.) A. Juss. (Belize to Uruguay)

Stigmatella Eig = *Eigia*

Stigmatodactylus Maxim. ex Makino (*Pantlingia*). Orchidaceae (IV 1). 4 E As. to Mal. R: FR 94(1983)434

Stigmatopteris C. Chr. Dryopteridaceae (I 2). 20–25 trop. Am.

Stigmatorhynchus Schltr. Asclepiadaceae (III 4). 3 E & S Afr.

Stigmatorthos M. Chase & D. Bennett (~ *Scelochilus*). Orchidaceae (V 10). I Peru: *S. peruviana* M. Chase & D. Bennett. R: Lindleyana 8(1993)3

Stigmatosema Garay. Orchidaceae (III 3). 2 trop. S Am.

Stilaginaceae Agardh = Euphorbiaceae

Stilbaceae Kunth (Incl. Retziaceae, ~ Verbenaceae). Dicots – Asteridae – Lamiales. 7/15 SW Afr. Split from Verbenaceae s.l. & united with *Retzia* excl. from Loganiaceae as lvs usu. opp. acicular

Genera: *Campylostachys, Eurylobium, Eurystachys, Retzia, Stilbe, Thesmophora, Xeroplana*

Stilbanthus Hook.f. Amaranthaceae (I 2). 1 Himalaya

Stilbe P. Bergius. Stilbaceae (Verbenaceae s.l.). 6 SW & S Cape

Stilbocarpa (Hook.f.) Decne. & Planchon. Araliaceae. 1 islands off S NZ: *S. polaris* (Hook.f.) A. Gray – stout *herb*, form. antiscorbutic for sailors

Stillingia Garden ex L. Euphorbiaceae. c. 30 trop. & warm Am. (25), Madag. (1–2), E Mal. & Fiji (1). R: AMBG 38(1951)207, 75(1988)1666. Alks, those of *S. sylvatica* Garden ex L. (SE US) rhizomes medic.

Stilpnogyne DC. Compositae (Sen.-Sen.). 1 S Afr.

Stilpnolepis H. Kraschen. (~ *Artemisia*). Compositae (Anth.-Art.). 2 China, Mongolia. R: APS 23(1985)470

Stilpnopappus C. Martius ex DC. Compositae (Vern.-Vern.). 24 trop. S Am.

Stilpnophleum Nevski = *Calamagrostis*

Stilpnophyllum Hook.f. Rubiaceae (I 1). 1 Peru

Stilpnophyton Less. = *Athanasia*

Stimpsonia C. Wright ex A. Gray. Primulaceae. 1 E As.

stinger *Urtica dioica, Dendrocnide* spp.

stinking cedar or **yew** *Torreya taxifolia*; **s. gladwin** *Iris foetidissima*; **s. Roger** *Tagetes minuta*

stinkweed *Thlaspi arvense*

stinkwood *Coprosma foetidissima, Gustavia augusta, Jacksonia scoparia, Ocotea bullata, Zieria arborescens* etc.; **red s.** *Prunus africana*

stinkwort *Helleborus foetidus*

Stipa L. Gramineae (16). c 300 trop. & temp., often dry (Eur. incl. *Achnatherum* 44; SW & S As. 42, R: NRBGE 42(1985)355; Aus. 61, R: Telopea 3(1986)1), see also *Nassella*; Am. spp. poss. best segrated as *A.*, Aus. spp. to new genus (Jacobs). Feather grass, corkscrew grass (lvs often inrolling in dry conditions covering stomata & green tissues (on adaxial surface only)); awns long (to 50 cm in *S. pulcherrima* K. Koch (C & S Eur.) – cult. orn.) & feathered, hygroscopic, the caryopsis with backward-pointing hairs, such that fr. is driven into ground or into skin, eyes & mouths of stock (cf. *Heteropogon*) & contaminates wool. *S. aphylla* (Rodway) Townrow (SE Tasmania) & *S. muelleri* Tate (S & SE Aus.) – with rudimentary leaf-blades. Some cult. orn., fibres etc. *S. avenacea* L. (black oat grass, N Am.); *S. comata* Trin. & Rupr. (N Am.) – cult. for forage; *S. ichu* (Ruíz & Pavón) Kunth (ichu, Mex. to Arg., highlands) – fodder; *S. pennata* L. (Euras.) – cult. orn.; *S. tenacissima* L. (esparto,

Algerian grass, alfa, halfa, W Medit.) – stems used in paper-making, ropes, sails, mats etc.

Stipagrostis Nees. Gramineae (28). 50 desert & semi-desert OW (Eur. 2). Like *Aristida* but with plumose awns. *S. plumosa* (L.) T. Anders. (warm OW) – much favoured by horses in Middle E

Stipecoma Muell. Arg. Apocynaceae. 1 Brazil

Stiptanthus (Benth.) Briq. = *Anisochilus*

Stipularia Pal. = *Sabicea*

Stipulicida Michaux. Caryophyllaceae (I 1). 1 SE US

Stirlingia Endl. Proteaceae. 6 Aus.

Stironeurum Radlk. = *Synsepalum*

stitchwort *Stellaria* spp.; **greater s.** *S. holostea*

Stixis Lour. Capparidaceae. 7 E Himal. to Hainan & Lesser Sunda Is. R: Blumea 12(1963)5

Stizolobium P. Browne = *Mucuna*

Stizolophus Cass. (~ *Centaurea*). Compositae (Card.-Cent.). 4 SW & C As.

Stizophyllum Miers. Bignoniaceae. 3 trop. Am. Twigs hollow

stock, Brompton, garden or **hoary** *Matthiola incana*; **night-scented s.** *M. longipetala* ssp. *bicornis*; **s. rose** *Sparrmannia africana*; **Virginia s.** *Malcolmia maritima*

Stocksia Benth. Sapindaceae. 1 E Iran, Afghanistan

Stoebe L. Compositae (Gnap.-Rel.). 34 trop. & S Afr. to Masc. R: JSAB 3(1937)1

Stoeberia Dinter & Schwantes (~ *Ruschia*). Aizoaceae (V). 3 SW Afr. R: MBSM 3(1960)560

Stoibrax Raf. (*Brachyapium*). Umbelliferae (III 8). 4 Iberia, 1 S Afr. R: EJB 48(1991)250

Stokesia L'Hérit. Compositae (Vern.-Vern.). 1 SE US: *S. laevis* (Hill) E. Greene (*S. cyanea*) – cult. orn. herb with blue fls, poss. comm. oilseed. R: EB 28(1974)130

Stokoeanthus E. Oliver. Ericaceae. 1 SW Cape

Stolzia Schltr. Orchidaceae (V 14). 4 trop. Afr. R: KB 33(1978)79

Stomandra Standley. Rubiaceae (I 7). 1 C Am.

Stomatanthes R. King & H. Robinson (~ *Eupatorium*). Compositae (Eup.-Eup.). 12 Brazil & Uruguay, 3 trop. Afr. R: MSBMBG 22(1987)69

Stomatium Schwantes. Aizoaceae (V). 40 SW Cape to Namaqualand & Karoo. Cult. orn. tufted succ.

Stomatochaeta (S.F. Blake) Maguire & Wurd. Compositae (Mut.-Mut.). 6 Venez., Guyana, Brazil. R: Brittonia 41(1989)37. Shrubs

Stomatostemma N.E. Br. Asclepiadaceae (I; Periplocaceae). 2 SE Afr.

stone pine *Pinus pinea*; **s. root** *Collinsonia canadensis*

stonecrop *Sedum acre* & other *S.* spp.

Stonesia G. Taylor. Podostemaceae. 4 trop. W Afr. R: Adansonia 13(1973)307

storax *Styrax officinalis*, see also *Liquidambar orientalis*; **American s.** *L. styraciflua*

Storckiella Seemann. Leguminosae (I 2). 5 New Caled. (3), Queensland (1), Fiji (1). Allied to *Koompassia* (Mal.), *Baudouinia* & *Mendoravia* (Madag.), app. an archaic group

storksbill *Erodium* spp.

Storthocalyx Radlk. Sapindaceae. 4 New Caledonia

Stracheya Benth. Leguminosae (III 17). 1 high Himalaya

Strailia T. Durand = *Lecythis*

Stramentopappus H. Robinson & Funk (~ *Vernonia*). Compositae (Vern.-Vern.). 1 Mex.: *S. pooleae* (B. Turner) H. Robinson & Funk. R: BJ 108(1987)227

stramonium *Datura stramonium*

Strangea Meissner. Proteaceae. 3 Aus.

Strangweja Bertol. = *Bellevalia*

Stranvaesia Lindley = *Photinia* (but see CJB 68(1990)2248)

strapwort *Corrigiola litoralis*

Strasburg turpentine *Abies alba*

Strasburgeria Baillon. Strasburgeriaceae (Ochnaceae s.l.). 1 New Caledonia

Strasburgeriaceae Engl. & Gilg (~ Ochnaceae). Dicots – Dilleniidae – Sapotales. 1/1 New Caledonia.

Genus: *Strasburgeria*. Recently re-split from Ochnaceae

Stratiotes L. Hydrocharitaceae. 1 Euras.: *S. aloides* L. (water soldier) – dioec. stoloniferous rosetted perennial aquatic, floating to surface & flowering in summer, sinking in autumn, this rise & fall alleged to be due to levels of lime in lvs; in GB usu. female, occ. bisexual but poss. just 1 clone; cult. orn. & where abundant used as manure

Straussiella Hausskn. Cruciferae. 1 Iran

strawberry *Fragaria ananassa*; **alpine s.** *F. vesca*; **barren s.** *Potentilla sterilis*; **s. guava** *Psidium* *cattleianum*; **hautbois s.** *F. muricata*; **Indian s.** *Potentilla indica*; **s. myrtle** *Ugni molinae*; **s. tomato** *Physalis pruinosa*; **s. tree** *Arbutus unedo*; **wild s.** *F. vesca*

strawflower *Bracteantha bracteata*

Streblacanthus Kuntze (~ *Schaueria*). Acanthaceae. 6 C Am.

Streblochaete Hochst. ex Pilger. Gramineae (19). 1 trop. Afr., Réunion, & Mal. mts: *S. longiarista* (A. Rich.) Pilger – coiling of awn draws out floret exposing a callus which adheres to clothes & pelts so that tangles of florets are disp.

Streblorrhiza Endl. Leguminosae (III 16). 1 Philip Is. (nr. Norfolk Is.): *S. speciosa* Endl. – extinct through grazing by feral animals

Streblosa Korth. (~ *Psychotria*). Rubiaceae (IV 7). 25 W Mal. R: JAA 28(1947)145

Streblosiopsis Valeton. Rubiaceae (I 8?). 2 Borneo

Streblus Lour. Moraceae (I). Excl. *Bleekrodea*, c. 25 Afr., Madag., Indomal. to Norfolk Is. (1, endemic). R: PKNAW C91(1988)356. *S. asper* Lour. (trop. As.) – locally medic., used for paper-making in Thailand, where also a topiary subject in temples; *S. brunonianus* (Endl.) F. Muell. (whalebone tree, NE & E Aus.) – timber; *S. (Bleekrodea) tonkinensis* (Eberh. & Dubard) Corner (SE As.) – poss. rubber source

Strelitzia Banks ex Dryander. Musaceae (Strelitziaceae). 5 S Afr. Woody pls of riverbanks & forest glades with alks, dichotomizing trunks to 10 m tall & 2-ranked banana-like lvs; infl. encl. in spathe from which fls emerge one at a time, inner tepals forming an arrow-like structure traversed by a longit. groove in which the 5 stamens & style lie, the pollen united by filaments into a viscous mass; arils orange, hairy. Cult. orn. greenhouse pls esp. *S. reginae* Banks ex Dryander (bird-of-paradise, crane flower, Cape) – dichotomizing trunk to 1 m, infl. with a conspic. droplet & bright orange & purple fls attractive to a sunbird, *Nectarinia afra*, which alights on perch-like spathe & in getting to nectar, presses against swollen stigma, depositing pollen from another fl.; at same time, pollen lying in the 'arrow' deposited on bird's feathers to be taken to another fl. (poll. poss. also effected by birds' feet); *S. nicolai* Regel & Koern. (S Natal, NE Cape) – trunk to 10 m

Strelitziaceae (Schumann) Hutch. = Musaceae (Strelitzioideae)

Strempelia A. Rich. ex DC. = *Psychotria*

Strempeliopsis Benth. Apocynaceae. 2 Cuba, Jamaica

Strephium Schrader ex Nees = *Raddia*

Strephonema Hook.f. Combretaceae. 6 trop. W Afr.

Streptachne R. Br. = *Aristida*

Streptanthella Rydb. Cruciferae. 1 W US

Streptanthera Sweet = *Sparaxis*

Streptanthus Nutt. Cruciferae. 35 W & S N Am. Some nickel accumulators on serpentine; some cult. orn. annuals

Streptocalyx Beer = *Aechmea* (but see FN 14(1979)1513)

Streptocarpus Lindley. Gesneriaceae. c. 125 trop. & S Afr., Madag. R: O. Hilliard & B.L. Burtt (1971) *S.* Cape primrose. Annual or perennial, s.t. hapaxanthic, herbs & subshrubs. Cotyledons grow unequally after germ. & in *S. grandis* N.E. Br. (SE & S Afr.) & other spp. (as in *Monophyllaea* & *Moultonia*) 1 cotyledon enlarges by basal growth (to 76 cm in some spp.) while plumule development is suppressed, the hypocotyl increasing in thickness to form 'stem' of pl.; fls borne at junction of hypocotyl & midrib with adventitious roots from lower part of hypocotyl. Other spp. develop new 'leaf' & 'petiole' units (phyllomorphs) on the first & these may behave as hapaxanthic units so that a plant, while appearing perennial, may be considered a colony of unifoliates; there are other elaborations of these morphological novelties & the morphological flexibility of these pls is evident in the way that they may be propagated from small fragments of the 'leaf'; phyllomorph development can be suppressed & caulescent growth typical of other spp. induced by supplying gibberellic acid or inhibiting auxin transport. Cult. orn. esp. phyllomorphic *S.* × *hybridus* Voss, the florist's streptocarpus, a complex hybrid group with *S. rexii* (Hook.) Lindley (Natal) predominating, e.g. 'Constant Nymph'; caulescent cult. orn. incl. *S. caulescens* Vatke & *S. stomandrus* B.L. Burtt (E Afr.)

Streptocaulon Wight & Arn. Asclepiadaceae (I, Periplocaceae). 9 Indomal.

Streptochaeta Schrader ex Nees. Gramineae (3). 3 trop. Am. R: SCB 68(1989)29

Streptochaetaceae Nakai = Gramineae

Streptoglossa Steetz ex F. Muell. (~ *Oliganthemum*). Compositae (Pluch.). 8 Aus. R: JABG 3(1981)167

Streptogyna Pal. Gramineae (8). 1 OW trop., 1 Am. R: AMBG 74(1987)871. At maturity disarticulating florets dangle from infl. by long tangled stigmas

Streptolirion Edgew. Commelinaceae (II 1b). 1 E Himal. to Korea & SE As. Climber with infl.

penetrating sheath

Streptoloma Bunge. Cruciferae. 1 C As. to Afghanistan

Streptolophus Hughes. Gramineae (34f). 1 Angola

Streptomanes Schumann. Asclepiadaceae (I, Periplocaceae). 1 New Guinea: *S. nymanii* Schumann

Streptopetalum Hochst. Turneraceae. 6 trop. & S Afr.

Streptopus Michaux. Convallariaceae (Liliaceae s.l.). 7–10 N temp. (exc. W Eur.; Eur. 1) to Himal. & S US. R: Rhodora 37(1935)88. Some cult. orn.

Streptosiphon Mildbr. Acanthaceae. 1 trop. E Afr.

Streptosolen Miers. Solanaceae (2). 1 Colombia, Peru: *S. jamesonii* (Benth.) Miers – cult. orn. with orange fls

Streptostachys Desv. (~ *Panicum*). Gramineae (34b). 3 trop. Am. R: AMBG 78(1991)359

Streptothamnus F. Muell. Flacourtiaceae (1). 1 E Aus.: *S. moorei* F. Muell.

Streptotrachelus Greenman = *Laubertia*

Stricklandia Baker = *Phaedranassa*

Striga Lour. Scrophulariaceae. 40 OW trop. to S Afr. Hemiparasites with minute seeds viable for up to 20 yrs germ. in response to host root exudates, often pestilential (esp. in C_4 cereals where loss can be total) in crops (witchweed) esp. *S. lutea* Lour. (*S. asiatica*, trop. OW) on sugar-cane, maize, sorghum etc., some Am. populations cleistogamous; application of nitrogen to infested crop improves yield because parasites then photosynthesize more & take less from hosts. *S. hermontheca* Benth. (Afr.) – poss. feed in semi-arid regions, germ. improved when seed passed through rumen

Strigina Engl. = *Lindernia*

Strigosella Boiss. = *Malcolmia*

stringybark *Eucalyptus* spp. with fibrous bark; **brown s.** *E. capitellata*; **red. s.** *E. macrorhyncha*; **yellow s.** *E. acmenoides*, *E. muelleriana*

Striolaria Ducke. Rubiaceae (inc. sed.). 1 Amazonia

striped squill *Puschkinia scilloides*

Strobilacanthus Griseb. Acanthaceae. 1 Panamá

Strobilanthes Blume. Acanthaceae. 250 trop. As. Sometimes split into many unsatisfactory segregate genera. Many spp. with gregarious flowering in forest undergrowth: *S. cernua* Blume has a cycle of 5–12 yrs in Java, *S. kunthiana* (Nees) T. Anderson a 12-yr one in India but with sporadic fls in between; when stigma touched, moves downwards becoming pressed against lower lip of fl. Cult. orn. & dye-sources: *S. anisophylla* (Lodd.) T. Anderson (goldfussia, Assam) – shrub with lvs in unequal-sized pairs; *S. cusia* (Nees) Kuntze (*S. flaccidifolia*, India to S China & SE As.) – cult. for blue dye (maigyee, Assam indigo). Some cult. orn.

Strobilanthopsis S. Moore. Acanthaceae. 5 trop. Afr.

Strobilocarpus Klotzsch = *Grubbia*

Strobilopanax R. Viguier = *Meryta*

Strobilopsis Hilliard & B.L. Burtt. Scrophulariaceae. 1 Lesotho & Natal. R: O.M. Hilliard (1994) *Manuleeae*: 532

Strobopetalum N.E. Br. = *Pentatropis*

Stroganowia Karelin & Kir. Cruciferae. 16 C As., 1 Nevada (!)

Stromanthe Sonder. Marantaceae. 13 trop. S Am. Cult. orn. herbs with distichous lvs & racemes or panicles with decid. coloured bracts & zig-zag rachises

Stromatopteridaceae (Nakai) Bierh. = Gleicheniaceae (Stromatopteridoideae)

Stromatopteris Mett. Stromatopteridaceae. 1 New Caled.: *S. moniliformis* Mett. – frond continuous with 'stem' axis, not a lateral appendage, so that there is no clear distinction between stem & leaf. R: Phytomorphology 18(1968)232

Strombocactus Britton & Rose. Cactaceae (III 9). 1 EC Mex.: *S. disciformis* (DC) Britton & Rose – cult. orn. with arillate seeds

Strombocarpa (Benth.) A. Gray = *Prosopis*

Strombosia Blume. Olacaceae. c. 12 trop. OW (Afr. 9). *S. javanica* Blume (W Mal.) – young lvs ed., useful timber

Strombosiopsis Engl. Olacaceae. 1 trop. Afr.

Strongylocaryum Burret = *Ptychosperma*

Strongylodon Vogel. Leguminosae (III 10). 12 Madag. to Polynesia, esp. Philipp. R: AUWP 90-8(1990)1. *S. lucidus* (Forster f.) Seemann (range of genus) – distrib. sea currents; *S. macrobotrys* A. Gray (jade vine, Philipp.) – liane to 18 m with bluish green fls 8 cm long & large indehiscent 3–10-seeded fr., cult. orn.

Strophacanthus Lindau = *Isoglossa*

Strophanthus DC. Apocynaceae. 38 trop. OW. R: MLW 82–4(1982). Usu. lianoid shrubs with long, often twisted C-lobes & spreading follicles; some cult. orn. but seeds imp. sources of arrow-poisons & strophanthin, a cardiac drug, esp. *S. gratus* (Wallich & Hook.) Baillon (W Afr.) & *S. kombe* Oliver (Afr.), *S. sarmentosus* DC (trop. Afr.) being a comm. source of cortisone

Strophioblachia Boerl. Euphorbiaceae. 2 SE As., Hainan to C Mal.

Strophiodiscus Choux = *Plagioscyphus*

Strophocactus Britton & Rose = *Selenicereus*

Stropholirion Torrey = *Dichelostemma*

Strophostyles Elliott (~ *Phaseolus*). Leguminosae (III 10). 3 N Am.

Strotheria B. Turner. Compositae (Hele.-Pect.). 1 N & C Mexico

Struchium P. Browne (*Sparganophorus*). Compositae (Vern.-Vern.). 1 trop. Am., natur. W Afr.: *S. sparganophora* (L.) Kuntze

Strumaria Jacq. ex Willd. Amaryllidaceae (Liliaceae s.l.). 9 S Afr. R: BJ 107(1985)22

Strumpfia Jacq. Rubiaceae (inc. sed.). 1 WI

Struthanthus C. Martius (*Steirotis*). Loranthaceae (V). 50 trop. Am.

Struthiola L. Thymelaeaceae. 30 trop. & S Afr. Cult. orn. evergreen shrubs s.t. with scented fls

Struthiolopsis E. Phillips = *Gnidia*

Strychnaceae DC ex Perleb. (Loganiaceae (III)). Dicots – Asteridae – Gentianales. 8/260 warm. Re-excluded from Loganiaceae s.l. as C valvate

Genera: *Antonia, Bonyunia, Gardneria, Neubergia, Norrisia, Spigelia, Strychnos, Usteria*

Poisons (*Spigelia, Strychnos* – strychnine) & timber (*Strychnos*)

Strychnopsis Baillon. Menispermaceae (V). 1 Madag.: *S. thouarsii* Baillon. Female unknown

Strychnos L. Strychnaceae (Loganiaceae s.l.). 190 trop. & warm (Afr. c. 75). Trees, shrubs & lianes with v. many alks & axillary thorns or (lianes) axillary hooks which twine around support & lignify; fr. berry-like (monkey apples), suggested as the inspiration for the 'apple' of the Garden of Eden, the flesh allegedly harmless but seeds extremely toxic because of strychnine in integuments, causing tetanus-like convulsions & form. used to poison rats etc., also in S Am. arrow-poisons & as ordeal-poisons in Afr. Some timbers & cult. orn., locally medic. *S. ignatii* Berg (St Ignatius's bean, Vietnam, Mal.) – fr. eaten by monkeys & civets, seeds a source of strychnine; *S. minor* Dennst. (*S. colubrina*, snake-wood, India, Sri Lanka) – bark & wood used medic. (malaria etc.); *S. nux-vomica* L. (nux-vomica, S As.) – comm. source of strychnine for rodent control in W Eur. since 1802; *S. potatorum* L.f. (clearing nut, E India, Burma) – rubbed inside water-vessels, it causes precipitation of impurities in cloudy water, locally medic., timber useful; *S. spinosa* Lam. (trop. & S Afr., Madag.) – fr. ed., to 11 cm diam., roots used against chiggers on Mafia Is.; *S. toxifera* R. Schomb. (trop. Am.) – a source of curare obtained by scraping & macerating bark, used medic. (see *Chondrodendron*) but form. imp. arrow-poison

Stryphnodendron C. Martius. Leguminosae (II 3). 25 trop. Am. (Amazonia 12, R: Leandra 10/11(1981)3). Some locally used timbers, tanbarks (esp. *S. adstringens* (C. Martius) Coville (*S. barbatimao*, barbatimao, Brazil) – 20–35% tannin) & medic.

Stuartia L'Hérit. = *Stewartia*

Stuartina Sonder. Compositae (Gnap.-Gnap.). 2 S & E Aus. R: Muelleria 6(1986)255

Stubendorffia Schrenk ex Fischer, C. Meyer & Avé-Lall. Cruciferae. 5 C As. & Afghanistan

Stuckertia Kuntze = *Morrenia*

Stuckertiella Beauverd. Compositae (Gnap.-Gnap.). 2 Arg. R: OB 104(1991)157

Stuessya B. Turner & Davies. Compositae (Helia.-Helia.). 3 Mex. R: Britt. 32(1980)209

Stuhlmannia Taubert (~ *Caesalpinia*). Leguminosae (I 1). 1 coastal forest of Tanz.: *S. moavi* Taubert

Stultitia E. Phillips = *Orbea*

Sturt('s) desert pea *Swainsona formosa*; **S('s). d. rose** *Gossypium sturtianum*

Sturtia R. Br. = *Gossypium*

Stussenia C. Hansen (~ *Neodriessenia*). Melastomataceae. 1 Vietnam. R: Willd. 15(1985)175

Styasasia S. Moore = *Asystasia*

Stylapterus A. Juss. (~ *Penaea*). Penaeaceae. 8 SW Cape

Stylidiaceae R. Br. Dicots – Asteridae – Asterales. 5/154 S & SE As., Australasia, S S Am. Small herbs, rarely shrublets, usu. with basal rosette of linear lvs, storing inulin; laticifers 0. Lvs simple, spirally arr.; stipules 0. Fls usu. bisexual, in term. bracteate cymes or racemes or solit. in upper axils; K ((2–)5(–7)), C (5), irreg. with imbricate lobes & resupinate or half so, or C reg., A 2 free from C but adnate to style forming a column, anthers extrorse with longit. slits, nectary-disk (or pair of glands) atop $\overline{G}$ (2), ± 2-loc. with axile to free-central placentation or posterior loc. reduced or 0 (G pseudomonomerous), the stylar column often irritable moving rapidly from bent position on 1 side of fl. to opp. (trigger pl. due to phloem cells in small groups separating from xylem while other cells prob. change shape & size on bending), ovules ∞ anatropous, unitegmic. Fr. usu. a capsule; seeds (few–)∞ with small (frequently monocot.) embryo in oily endosperm. n = 15, 18. R: R. Erickson (1958) *Triggerpls*

Genera: *Forstera. Levenhookia, Oreostylidium, Phyllachne, Stylidium*

Some cult. orn. (*Forstera, Stylidium*), see Donatiaceae

Stylidium Sw. ex Willd. Stylidiaceae. 136 SE As., Aus. (110), NZ. Trigger pls (see fam.); some with explosive poll. mechanism in that column showers insect with pollen, then after A shrivel again stigma getting pollen from insects. N = 5–16, 26, 28, 30. Few cult. orn.

Stylisma Raf. Convolvulaceae. 6 S & E US. R: Britt. 18(1966)97

Stylites Amstutz = *Isoetes*

stylo *Stylosanthes* spp.; **Townsville s.** *S. humilis*

Stylobasiaceae J. Agardh. See Surianaceae

Stylobasium Desf. Surianaceae (Stylobasiaceae). 2 N & W Aus.

Styloceras Kunth ex A. Juss. Buxaceae (Stylocerataceae). 5 Venez., N Andes. R: Novon 3(1993)142. Some locally eaten fr., timber for joinery

Stylocerataceae (Pax) Rev. & Hoogl. = Buxaceae

Stylochaeton Lepr. Araceae (VIII 1). 15 trop. & warm Afr. esp. Tanz. Roots swollen; infl. below ground with only tip emergent. Some local medic.

Stylochiton Schott = *Stylochaeton*

Stylocline Nutt. Compositae (Gnap.-Gnap.). 7 SW N Am. R: Madrono 39(1992)114. Tiny annuals

Styloconus Baillon = *Blancoa*

Stylodon Raf. (~ *Verbena*). Verbenaceae. 1 SE US

Stylogyne A. DC. Myrsinaceae. 60 trop. Am. Dioec.

Stylolepis Lehm. = *Podolepis*

Stylomecon G. Taylor. Papaveraceae (IV). 1 Calif. & Baja Calif.: *S. heterophylla* (Benth.) G. Taylor – cult. orn. annual with alks, yellow sap & red fls like *Papaver* but with slender style

Stylophorum Nutt. Papaveraceae (I). 2 E As., 1 E US: *S. diphyllum* (Michaux) Nutt. (E US) – cult. orn. with yellow fls. Alks, seeds arillate

Stylophyllum Britton & Rose = *Dudleya*

Stylosanthes Sw. Leguminosae (III 13). 25 trop. & warm. R: AMBG 44(1958)299. Cult. fodder pls (stylo); mycorrhizal, phosphate accumulators, esp. *S. erecta* Pal. (W Afr.) & *S. guianensis* (Aublet) Sw. (trop. Am.); *S. humilis* Kunth (*S. sundaica*, Townsville stylo, Brazil, natur. E Mal. to trop. Aus.) poss. early introd. by Portuguese, form. imp. fodder in N Aus., where cattle ticks immobilized & killed on flowering stems

Stylosiphonia Brandegee. Rubiaceae. 2 C Am. = ?

Stylotrichium Mattf. Compositae (Eup.-Gyp.). 4 E Brazil. R: MSBMBG 22(1987)122

Stylurus Salisb. ex J. Knight = *Grevillea*

Stypandra R. Br. Phormiaceae (Liliaceae s.l.). 1 temp. Aus. (? & New Caled.): *S. glauca* R. Br. (blind grass) – causing blindness in grazing goats & sheep through degeneration of optic nerve etc.

Styphelia Sm. Epacridaceae. 14 Aus. Some ed. fr. (*S. acerosa* Sol. ex Gaertner imp. emu food)

Styphnolobium Schott. ex Endl. = *Sophora*

Styppeiochloa de Winter (~ *Zenkeria*). Gramineae (25). 2 S Afr. & Madag. R: Bothalia 9(1966)134

Styracaceae DC & Sprengel. Dicots – Dilleniidae – Theales. 11/160 warm temp. & trop. Am., Medit., SE As., W Mal. Trees & shrubs with resinous bark & usu. scaly indumentum. Lvs spirally arr., simple; stipules 0. Fls usu. bisexual, reg., ebracteolate in racemose or cymose infls, rarely solit.; K ((2–)4,5(–7)) with open or valvate lobes or 0, C (2–5(–7)) tubular proximally, the lobes imbricate or valvate, shorter than tube in *Halesia*, C free in *Bruinsmia*, A 5 (*Pamphilia*) or 2(–4) × C, filaments usu. adnate to C-tube & basally connate with a tube, anthers with longit. slits & connective s.t. prolonged, $\underline{G}$ to $\overline{G}$ ((2)3–5), basally pluriloc., but apically often 1-loc. with slender style & axile placentation, each placenta with (1–)4–6(–∞) erect or pendulous, anatropous to hemitropous uni- or bitegmic ovules (not more than 2 becoming seeds). Fr. usu. a capsule, s.t. samaroid or a drupe (*Parastyrax*) with persistent K; seeds with large straight to weakly curved embryo in copious oily endosperm. x = 8, 12

Principal genera: *Alniphyllum, Rehderodendron, Styrax*

Resins (*Styrax*) & cult. orn. (esp. *Halesia*, *Sinojackia*, *Styrax*)

Styrax L. Styracaceae. 120 Medit. (incl. Eur.; 1), SE As., Mal., trop. Am. Trees & shrubs with resins used medic. (benzoin, gum Benjamin (a corruption) used in friar's balsam; benzoic acid esters haemolytic) & in incense, obtained by wounding bark of trop. As. spp. esp. *S. benzoin* Dryander (Sumatra), the resin being principally 2 alcohols combined with cinnamic acid & free cinnamonic & benzoic acids, & used in treatment of coughs, as antiseptic, in flavouring cigarettes & in ceremonial. Some with remarkable aphid-induced galls; some cult. orn. *S. officinalis* L. (storax, styrax, Medit. with varietally distinct pl. in Calif.) – orig. balsam in med., seeds used as beads; *S. tessmannii* Perkins (trop. Am.) – crushed lvs used against fungal infections of feet in Colombia

Styrophyton S.Y. Hu = *Allomorphia*

Suaeda Forssk. ex J.F. Gmelin. Chenopodiaceae (III 2). 100 cosmop. coasts & salt steppe (Eur. 15, Aus. 5), in GB *S. vera* Forssk. ex J. Gmelin ('*S. fruticosa*', Euras.) – camel fodder & *S. maritima* (L.) Dumort. (Am., Euras. etc.) – seablite. Ash (barilla) high in sodium carbonate form. used in glass-making; alks. *S. monoica* J.F. Gmelin (Medit. to As. & dry trop. Afr.) – imp. camel fodder in dry season in N Kenya; *S. suffrutescens* S. Watson (SW US) – poll. by bees, butterflies & thrips, source of a black dye used by Indians

suan zau ren *Ziziphus spinosus*

Suarezia Dodson. Orchidaceae (V 10). 1 Ecuador

suari *Caryocar amygdaliferum*

Suberanthus Borh. & Fernández (~ *Rondeletia*). Rubiaceae (I 1). 7 Cuba & Hispaniola. R: ABH 29(1983)29

Subularia L. Cruciferae. 1 N temp. incl. Eur.: *S. aquatica* L. (awlwort) – one of few aquatic annuals; 1 mts trop. E Afr. R: Rhodora 66(1964)127

Succisa Haller. Dipsacaceae. 1 Eur. & W Siberia, N Afr.: *S. pratensis* L. (devil's bit (scabious), blue buttons) – locally medic.; 1 NW Spain; 1 Cameroun Mt. R: AMH n.s. 2(1952)237

Succisella G. Beck. Dipsacaceae. 4 Eur.

succory *Cichorium intybus*; **lamb's** or **swine s.** *Arnoseris minima*

Succowia Medikus. Cruciferae. 1 Canary Is., 1 W Medit.

Suchtelenia Karelin ex Meissner. Boraginaceae. 1–3 Caucasus to C As.

Suckleya A. Gray. Chenopodiaceae (I 3?). 1 Rocky Mts: *S. suckleyana* (Torrey & A. Gray) Rydb.

Sucrea Söderstrom. Gramineae (4). 3 Brazil. R: KBAS 13(1986)63. Intermediate between *Olyra* & *Raddia*. Sterile culms of *S. monophylla* Söderstrom (Brazil) with only 1 leaf per culm; *S. sampaiana* (Hitchc.) Söderstrom (Brazil) – root tubers

sucupira timber from *Bowdichia*, *Diplotropis* & *Sweetia* spp.

Sudan grass *Sorghum × drummondii*

Suddia Renvoize. Gramineae (I ?9). 1 Sudan: *S. sagittifolia* Renvoize – aquatic in sudd, blades sagittate, false petioles to 1 m

Suessenguthia Merxm. Acanthaceae. 1 Bolivia

Suessenguthiella Friedrich (~ *Pharnaceum*). Molluginaceae. 2 NW Cape & Namibia. R: MBSM 2(1955)60

sugar apple *Annona* spp.; **s. beet** *Beta vulgaris* cv.; **s. berry** *Celtis* spp.; **s. bush** *Protea* spp. esp. *P. mellifera*; **s. cane** *Saccharum officinarum*; **s. grass** *Eulalia fulva*; **s. maple** *Acer saccharum*; **s. pine** *Pinus lambertiana*; **s. plum** *Uapaca guineensis*

sugi *Cryptomeria japonica*

Suksdorfia A. Gray. Saxifragaceae. 3 NW N Am. & Andes of N Arg. & S Bolivia. R: BJLS 90(1985)60. Small cult. orn. rock-pls

Sukunia A.C. Sm. Rubiaceae (II 1). 2 Fiji

Sulaimania Hedge & Rech.f. Labiatae (VI). 1 Pakistan

Sulcorebutia Backeb. = *Rebutia*

Sulitia Merr. Rubiaceae (II 1). 1 Philippines

sulla *Hedysarum coronarium*

Sullivantia Torrey & A. Gray. Saxifragaceae. 4 C US. R: Brittonia 43(1991)27. Cult. orn.

sultan, sweet *Amberboa moschata*

sultana *Vitis vinifera* 'Sultana'

sultan's flower *Impatiens walleriana*

sumac(h) *Rhus coriaria*; **Chinese s.** *R. javanica*; **Indian s.** *Cotinus coggygria*; **staghorn** or **velvet s.** *R. hirta*

Sumatra camphor *Dryobalanops aromatica*

Sumatroscirpus Oteng-Yeboah (~ *Scirpus*). Cyperaceae. 1 Sumatra

Sumbaviopsis J.J. Sm. Euphorbiaceae. 1 Assam; 1 SE As. to W Mal.

sumbul *Ferula sumbul*

summer cypress *Bassia scoparia*; **s. hyacinth** *Galtonia candicans*

Summerhayesia Cribb. Orchidaceae (V 16). 2 trop. Afr. Cult. orn. epiphytes

sun berry *Physalis minima, Solanum × burbankii*; see also *Rubus*; **s.dew** *Drosera* spp.; **s.flower** *Helianthus* spp. esp. *H. annuus*; **Mexican s.flower** *Tithonia diversifolia*; **s.(n) hemp** *Crotalaria juncea*

Sundacarpus (Buchholz & N.E. Gray) Page (~ *Podocarpus*). Podocarpaceae. 1 Sumatra & Philippines to N Queensland & New Ireland: *S. amara* (Blume) Page. R: NRBGE 45(1988)378

Sunipia Buch.-Ham. ex Lindley. Orchidaceae (V 15). 18 India & SE As. to Taiwan

Supushpa Suryan. Acanthaceae. 1 W India

Suregada Roxb. ex Rottler. Euphorbiaceae. 40 OW trop. (Afr. 8)

surette *Byrsonima* spp.

Surfacea Moldenke = *Premna*

Suriana L. Surianaceae. 1 pantrop., littoral: *S. maritima* L.

Surianaceae Arn. Dicots – Rosidae – Rosales. 4/5 trop. & warm esp. Aus. Trees & shrubs. Lvs simple, spirally arr.; stipules small & decid. or 0. Fls usu. bisexual, 5-merous, in thyrses or solit. in axils; K ± free, imbricate, C free & imbricate or 0 (*Stylobasium* – wind-poll.), A in 2 whorls but some or all of whorl opp. C in *Suriana* staminodal or 0, anthers with longit. slits, disk 0, $\underline{G}$ 1 (*Stylobasium*), 2 (*Guilfoylia*), 5 (*Cadellia, Suriana*). Fr. a berry, drupe, or nut-like; seeds with curved or folded embryo with starchy or oily cotyledons in little or 0 endosperm

Genera: *Cadellia, Guilfoylia, Stylobasium, Suriana*

Some genera form. in Simaroubaceae but 'nothing ... out of harmony with the rather amorphous order Rosales' (Cronquist)

Susanna E. Phillips = *Amellus*

Sussenia C. Hansen. Melastomataceae. 1 As.: *S. membranifolia* (Li) C. Hansen. R: Willd. 15(1985)175

Sutera Roth. Scrophulariaceae. 49 S Afr. R: O. Hilliard (1994) *Manuleeae*: 220. Some cult. orn. esp. *S. cordata* (Thunb.) Kuntze ('*Bacopa* 'Snowflake''). See also *Camptoloma* & *Jamesbrittenia*

Sutherlandia R. Br. Leguminosae (III 15). 3 S Afr. R: Plantsman 14(1992)65. Fr. large, bladder-like. *S. frutescens* (L.) R. Br. (balloon pea, duck pl.) – cult. orn. shrub with red fls, the fr. floated in water reminiscent of ducks, pl. form. considered a cancer cure

Sutrina Lindley. Orchidaceae (V 10). 1 Peru

Suttonia A. Rich. = *Myrsine*

Suzukia Kudô. Labiatae (VI). 2 Ryukyu Is., Taiwan

Svenkoeltzia Burns-Bal. = *Funkiella*

Svensonia Moldenke = *Chascanum*

Sventenia Font Quer (~ *Sonchus*). Compositae (Lact.-Son.). 1 Canary Is.

Svitramia Cham. Melastomataceae. 1 S Brazil

Swainsona Salisb. Leguminosae (III 15). 50 dry Aus., mts of S Is. NZ (1). R: CNSWNH 1(1948)131. Toxic because of mannose-based oligosaccharides which animals cannot break down. Some cult. orn. esp. *S. formosa* (G. Don f.) J. Thompson (*Clianthus f.*, (Sturt's) desert pea, Aus.) – protected sp.; *S. galegifolia* (Andrews) R. Br. (Darling pea, E Aus.) – perennial herb with red fls

Swallenia Söderstrom & H. Decker. Gramineae (31c). 1 Calif. Sand dunes

Swallenochloa McClure = *Chusquea* (but see Brittonia 30(1978)303)

swallow wort *Asclepias curassavica, Chelidonium majus*

swamp cottonwood *Populus heterophylla*; **s. cypress** *Taxodium distichum*; **s. mahogany** *Eucalyptus botryoides* & *E. robusta*; **s. oak** *Casuarina* spp.

Swan River daisy *Brachycome iberidifolia*

swarri nut *Caryocar amygdaliferum*

Swartzia Schreber. Leguminosae (III 1). 140 trop. Am. (R: FN 1(1968)), Afr. (2). Some cult. orn. trees; many local medic. NW Amaz. *S. madagascariensis* Desv. (Madag.) & *S. simplex* (Sw.) Sprengel (trop. Am.) with molluscicidal saponins (triterpenes); *S. polyphylla* DC (Amazon) – air-filled cavity between cotyledons allowing hydrochory

swede *Brassica napus* Napobrassica Group

Swedish whitebeam *Sorbus intermedia*

sweet alyssum *Lobularia maritima*; **s. balm** *Melissa officinalis*; **s. basil** *Ocimum basilicum*; **s. bay** *Laurus nobilis*; **s.briar** *Rosa rubiginosa*; **s. chestnut** *Castanea sativa*; **s. cicely** *Myrrhis odorata*; **s.corn** *Zea mays*; **s. flag** *Acorus calamus*; **s. gale** *Myrica gale*; **s. galingale** *Cyperus longus*; **s.**

grass *Glyceria* spp. esp. *G. fluitans*; **s. vernal g.** *Anthoxanthum odoratum*; **s. gum** *Liquidambar styraciflua*; **s. hearts** *Galium aparine*; **s. marjoram** *Origanum majorana*; **s. pea** *Lathyrus odoratus*; **s. potato** *Ipomoea batatas*; **s. rocket** *Hesperis matronalis*; **s. scabious** *Scabiosa atropurpurea*; **s.sop** *Annona squamosa*; **s. sultan** *Amberboa moschata*; **s. william** *Dianthus barbatus*

Sweetia Sprengel. Leguminosae (III 2). 1 trop. S Am.: *S. fruticosa* Sprengel – timber (sucupira)

sweetie *Citrus limetta*

Swertia L. Gentianaceae. 50 N temp. (Eur. 1), Afr. & Mal. mts. Herbs s.t. with spirally arr. lvs, fls with 2 nectar-producing pits at base of each petal; distinction from some *Gentiana* spp. unclear; alks. – some medic. (chiretta) as tonics; few cult. orn.

Swida Opiz = *Cornus*

Swietenia Jacq. Meliaceae (IV 2). 3 trop. Am. R: FN 28(1981)389. Mahogany (name poss. derived from 'oganwo' (i.e. *Khaya ivorensis* in Yoruba, SW Nigeria) modified by Portuguese & English in C17 Jamaica where Yoruba were transported as slaves), esp. *S. mahagoni* (L.) Jacq. (Cuba m.) – much exploited for fine cabinet timber, first used extensively by Chippendale & Hepplewhite, & form. much used for ship-building etc., but trade reduced to zero 1908–60 & wild populations now of poor form through selection of best pls for felling (genetic erosion); *S. macrophylla* King (baywood) & *S. humilis* Zucc. (Mexican m.) – similar uses, the first in plantation & described from cult. material in As.

swine('s) cress *Coronopus didymus*; **s. succory** *Arnoseris minima*

Swinglea Merr. Rutaceae. 1 Philippines: *S. glutinosa* (Blanco) Merr. – locally medic. (skin disease), cult. orn.

Swintonia Griffith. Anacardiaceae (I). 12 Indomal. C accrescent & persistent, forming wings on fr. Some timbers, esp. *S. floribunda* Griffith (SE As., W Mal.) – match-boxes & -sticks in Burma

Swiss chard *Beta vulgaris* subsp. *vulgaris*; **S. cheese plant** *Monstera deliciosa*

sword bean *Canavalia ensiformis*; **s. fern** *Nephrolepis* spp.; **s. grass** *Scirpus americanus*; **s. plant** *Echinodorus* spp.; **s. sedge** *Lepidosperma gladiatum*

Swynnertonia S. Moore. Asclepiadaceae (III 5). 1 Zimbabwe: *S. cardinea* S. Moore

Syagrus C. Martius. Palmae (V 5b). 32 trop. S Am. R: IBM 56(1987)16. Monoec. palms with lvs arranged as to appear in 3 ranks; sources of a palm kernel oil esp. *S. coronata* (C. Martius) Becc. (ouricuri, nicuri (palm nut), arid Brazil) – oil (urucury wax) used in soap & subs. for carnauba wax, there being perhaps 5 billion trees (many more than carnauba), lvs for weaving into sleeping mats; *S. romanzoffiana* (Cham.) Glassman (Brazil) – used as sago

sycamore *Acer pseudoplatanus*, (Am.) *Platanus* spp. esp. *P. occidentalis*

sycomore *Ficus sycomorus*

Sycopsis Oliver. Hamamelidaceae (I 4). 2–3 Assam to Taiwan. *S. sinensis* Oliver (C China) – cult. orn.

Sydney blue gum *Eucalyptus saligna*

Symbegonia Warb. (~ *Begonia*). Begoniaceae. 14 New Guinea. R: SCB 60(1986)252

Symbolanthus G. Don f. Gentianaceae. 15 trop. Am. *S. latifolius* Gilg – bat-poll.

Symingtonia Steenis = *Exbucklandia*

Symmeria Benth. Polygonaceae (II 1). 1 N S Am., W Afr.: *S. paniculata* Benth.

Symonanthus Haegi. Solanaceae (2). 2 W Aus.

Sympa Ravenna. Iridaceae. 1 Brazil

Sympagis (Nees) Bremek. = *Strobilanthes*

Sympegma Bunge. Chenopodiaceae (III 3). 1 C As.: *S. regelii* Bunge

Sympetalandra Stapf. Leguminosae (I 1). 5 W Mal. R: Blumea 22(1975)159

Sympetaleia A. Gray = *Eucnide*

Symphionema R. Br. Proteaceae. 2 NSW. R: FNSW 2(1991)19

Symphonia L.f. Guttiferae (III). 17 Madag., trop. Am. (1: *S. globulifera* L.f. (hog plum, Am.) – hummingbird-poll., locally medic.)

Symphorema Roxb. Verbenaceae (Symphoremataceae). 3 Indomal.

Symphoremataceae (Meissner) Rev. & Hoogl. See Verbenaceae

Symphoricarpos Duhamel. Caprifoliaceae. 17 N Am. (16), China (1). R: JAA 21(1940)201. Cult. orn. decid. suckering shrubs esp. *S. albus* (L.) S.F. Blake (snowberry, E N Am.) – spreading by suckers in GB, rarely by seed, fr. a mushy white berry; *S. orbiculatus* Moench (coral berry, N Am.) – lvs locally medic., fr. red

Symphostemon Hiern. Labiatae (VIII 4c). 2 Angola

Symphyandra A. DC. Campanulaceae. 12 E Medit. (Eur. 3) to C As. R: QBAGS 45(1977)246.

Differing from *Campanula* merely in anthers united around style. Cult. orn. herbs

Symphyglossum Schltr. Orchidaceae (V 10). 5 trop. S Am.

Symphyllarion Gagnepain = *Hedyotis*

Symphyllocarpus Maxim. Compositae (? tribe, excl. from Inul.). 1 Manchuria, E Siberia

Symphyllophyton Gilg. Gentianaceae. 2 Brazil

Symphyobasis K. Krause = *Goodenia*

Symphyochaeta (DC) Skottsb. = *Robinsonia*

Symphyochlamys Guerke. Malvaceae. 1 NE trop. Afr.

Symphyoloma C. Meyer. Umbelliferae (III 11). 1 high Caucasus

Symphyonema Sprengel = *Symphionema*

Symphyopappus Turcz. Compositae (Eup.-Dis.). 11 Brazil, campos. R: MSBMBG 22(1987)81

Symphyosepalum Hand.-Mazz. Orchidaceae (IV 2). 1 China

Symphysia C. Presl. Ericaceae. 2 WI

Symphytonema Schltr. = *Camptocarpus*

Symphytum L. Boraginaceae. 35 Eur. (11), Medit. to Caucasus. Comfrey. *S. officinale* L. (Euras., natur. N Am.) – form. used as styptic, young shoots eaten like asparagus; hybrid with *S. asperum* Lepechin (Russia to Iran) – cult. as forage, = *S.* × *uplandicum* Nyman (Russian or blue c.), a common pl. of roadsides in GB, resynthesized through artificial hybridization of parents; also hybridizing with *S. tuberosum* L. (Eur., W As.). Liver failure from tea or tablets prepared from comfrey due to pyrrolizidine alks which break down in an hour giving toxins blocking veins in liver. L.D. Hills (1976) *Comfrey*

Sympieza Lichtenst. ex Roemer & Schultes. Ericaceae. 8 SW & S Cape

Symplectochilus Lindau = *Anisotes*

Symplectrodia Lazarides (~ *Triodia*). Gramineae (31a). 2 N Territory, Aus. R: Nuytsia 5(1984)273. Sandy soils

Symplocaceae Desf. Dicots – Dilleniidae – Cornales. 1/250 trop. & warm Am. & E OW. Usu. evergreen trees often accumu. aluminum. Lvs spirally arr., often sweet-tasting; stipules 0. Fls usu. bisexual, reg., bibracteolate, in axils of bracts in racemes (less often panicles, solit. etc.); K (3–)5, basally connate, the lobes valvate or imbricate, C (3–)5(–11) with short tube & imbricate lobes s.t. in ± 2 rows, A (4–)12–∞ attached to C-tube, usu. in 2 or more whorls or in bundles alt. with C, s.t. forming a tube with anthers inside, anthers with longit. slits, $\overline{G}$, rarely half-inf. (2–5), usu. pluriloc., the style often surrounded by nectary-disk, each loc. with 2–4 pendulous anatropous unitegmic ovules on axile or deeply intruded placentas. Fr. usu. a 1-stoned drupe, or a berry, topped with persistent K, endocarp with 1 apical germ. pore per loc.; seeds with large straight or curved embryo in copious oily & proteinaceous endosperm (s.t. also with starch). x = 11–14
Genus: *Symplocos*

Symplocarpus Salisb. ex W. Barton. Araceae (V 1). 1 NE As., NE N Am.: *S. foetidus* (L.) W. Barton (skunk cabbage) – cult. orn. herb with alks, spadix reaching 35 °C above ambient

Symplococarpon Airy Shaw. Theaceae. 9 trop. Am. $\overline{G}$

Symplocos Jacq. Symplocaceae. 250 trop. & warm Am. & E OW (108 excl. New Caled.). Some polyembryony. All aluminum accumulators have yellow-green lvs & blue fr.; spp. with sweet foliage relished by stock. Chewing-sticks; some sources of dye e.g. bark for dyeing batik red or brown in Java, mordant (aluminium) in turkey red etc.; *S. racemosa* Roxb. (lodh bark, S & SE As.) – lvs & bark for dye, bark medic.; *S. theiformis* (L.f.) Oken (Bogota tea, C Am.) – tea subs. in Colombia; *S. tinctoria* (L.) L'Hérit. (E N Am.) – lvs & fr. sources of yellow dyes, wood for turnery

Synadenium Boiss. Euphorbiaceae. 20 trop. Am. to Mascarenes (E & S trop. Afr. 19)

Synammia C. Presl = *Polypodium*

Synandra Nutt. Labiatae (VI). 1 E US

Synandrina Standley & L.O. Williams = *Casearia*

Synandrodaphne Gilg. Thymelaeaceae. 1 W trop. Afr.

Synandrogyne Buchet = *Arophyton*

Synandropus A.C. Sm. Menispermaceae (III). 1 NE Brazil: *S. membranaceus* A.C. Sm.

Synandrospadix Engl. Araceae (VIII 3). 1 NE Andes (N Arg. & Bolivia): *S. vermitoxica* (Griseb.) Engl. – v. poisonous

Synanthes Burns-Bal., H. Robinson & S. Foster = *Eurystyles*

Synaphea R. Br. Proteaceae. 10 SW Aus. Lignotubers

Synapsis Griseb. Bignoniaceae (? Scrophulariaceae). 1 E Cuba

Synaptantha Hook.f. (~ *Hedyotis*). Rubiaceae (IV 1). 1 warm Aus.: *S. tillaeacea* (F. Muell.) Hook.f.

Synaptolepis Oliver. Thymelaeaceae. 4–5 trop. Afr., 1 Madagascar

Synaptophyllum N.E. Br. Aizoaceae (IV). 1 SW Namibia
Synardisia (Mez) Lundell = *Ardisia*
Synassa Lindley = *Sauroglossum*
Syncalathium Lipsch. Compositae (Lact.-Lact.). 4 Tibet, China. R: APS 10(1965)283
Syncarpha DC ('*Helipterum*' *p.p.*). Compositae (Gnap.-Gnap.). 25 Cape. R: OB 104(1991)150
Syncarpia Ten. Myrtaceae (Lept.). 5 C Mal. to NE Aus. (2). *S. glomulifera* (Sm.) Niedenzu (*S. laurifolia*, turpentine wood, NE Aus.) – timber resistant to white ants & marine borers used for ship-building etc., cult. shade tree in S US
Syncephalantha Bartling = *Dyssodia*
Syncephalum DC. Compositae (Gnap.-Gnap.). 5 Madagascar. R: OB 104(1991)137
Synchaeta Kirpiczn. = *Gnaphalium*
Synchoriste Baillon = *Lasiocladus*
Synclisia Benth. Menispermaceae (I). 1 C Afr.: *S. scabrida* Miers
Syncolostemon E. Meyer. Labiatae (VIII 4b). 10 SE Afr., R: Bothalia 12(1976)21
Syncretocarpus S.F. Blake. Compositae (Helia.-Helia.). 2 Peru. Xeromorphic
Syndesmanthus Klotzsch (~ *Scyphogyne*). Ericaceae. 18 SW & S Cape
Syndiclis Hook.f. = *Potameia*
Syndyophyllum Schumann & Lauterb. Euphorbiaceae. 1 Sumatra, Borneo, New Guinea
Synechanthus H. Wendl. Palmae (IV 3). 2 trop. Am. R: Principes 15(1971)10. Unarmed monoec. palms, cult. orn.
Synedrella Gaertner. Compositae (Helia.-Verb.). 2 trop. Am., natur. OW esp. *S. nodiflora* (L.) Gaertner – fr. disp. on clothes, weedy but young lvs eaten as potherb in Indonesia, local medic.
Synedrellopsis Hieron. & Kuntze. Compositae (Helia.-Verb.). 1 Argentina
Syneilesis Maxim. Compositae (Sen.-Tuss.). 7 E As. R: Phytol. 27(1973)269. = ? *Ligularia*
Synelcosciadium Boiss. = *Tordylium*
Synepilaena Baillon = *Kohleria*
Syngonanthus Ruhl. Eriocaulaceae. 200 trop. Am., Afr. & Madag. *S. elegans* (Bong.) Ruhl. (Brazil) – an 'immortelle' dyed for commerce
Syngonium Schott. Araceae (VII 5). 33 trop. Am. R: AMBG 68(1981)565. Cult. orn. lianes with milky latex (used to counter *Papaponera* ant bites) esp. *S. podophyllum* Schott with several cvs
Syngramma J. Sm. Pteridaceae (III). 15 Mal. (Borneo 7) to Carolines
Syngrammatopsis Alston = *Pterozonium*
Synima Radlk. Sapindaceae. 1 NE Aus.: *S. cordierorum* (F. Muell.) Radlk.
Synisoon Baillon = *Retiniphyllum*
Synnema Benth. = *Hygrophila*
Synnotia Sweet = *Sparaxis*
Synosma Raf. ex Britton & Brown = *Hasteola*
Synostemon F. Muell. = *Sauropus*
Synotis (C.B. Clarke) C. Jeffrey & Y.L. Chen. Compositae (Sen.-Sen.). 50 Sino-Himal. & China (1)
Synotoma (G. Don f.) R. Schulz = *Physoplexis*
Synoum A. Juss. Meliaceae (I 6). 1 NE Aus.: *S. glandulosum* (Sm.) A. Juss. (bastard or scented rosewood) – dark red rose-scented timber for cabinet-work; seeds united by common 'aril'
Synsepalum (A. DC) Daniell. Sapotaceae (IV). 20 trop. Afr. R: T.D. Pennington, *S.* (1991)242. *S. dulcificum* (Schum. & Thonn.) Daniell (miraculous berry, W Afr.) – berries with miraculin causing sour & salt things to taste sweet by affecting taste-buds, a glycoprotein also depressing appetite
Synstemon Botsch. Cruciferae. 1 NW China
Synthlipsis A. Gray. Cruciferae. 3 S N Am.
Synthyris Benth. Scrophulariaceae. 9 W N Am. mts. R: PPANS 85(1933)83. Cult. orn. perennial herbs
Syntriandrium Engl. Menispermaceae (III). 1 trop. W Afr.: *S. preussii* Engl.
Syntrichopappus A. Gray. Compositae (Hele.-Cha.). 2 SW US. R: Novon 1(1991)123
Syntrinema H. Pfeiffer = *Rhynchospora*
Synurus Iljin. Compositae (Card.-Cent.). 1 E As.
Sypharissa Salisb. = *Tenicroa*
Syreitschikovia Pavlov. Compositae (Card.-Cent.). 2 C As.
Syrenia Andrz. ex Besser. (~ *Erysimum*). Cruciferae. 10 E Eur. (4) to W Siberia
Syrenopsis Jaub. & Spach = *Thlaspi*

syringa *Philadelphus* spp.

Syringa L. Oleaceae. 23 SE Eur. (2) to E As. R: F.R. Fiala (1988) *Lilacs, the genus S.* Decid. shrubs & small trees with showy thyrses or panicles of usu. strongly fragrant fls & extrafl. nectaries; cult. orn. (lilac; nilak = bluish in Persian) esp. *S. vulgaris* L. (common l., SE Eur.) – some 1500 cvs incl. double-flowered (since 1843) esp. due to Lemoine & fils (1850–1955) at Nancy & US cvs with C 5 or 6 in 1 whorl, depithed stems form used for pipes (p. tree); other cult. hybrids & spp. incl. *S. emodi* Wallich ex Royle (Himalayan l., Afghanistan, Himal.), *S. laciniata* Miller (*'S. afghanica'*, W China) – juvenile lvs pinnately lobed, introd. to Eur. via Iran in C17, poss. a form of *S.* × *persica; S. meyeri* C. Schneider 'Palibin' (N China, ? cultigen) – commonly seen dwarf l. but poss. a form of *S. pubescens* Turcz. (N China), *S.* × *persica* L. (incl. *S.*× *chinensis*, Rouen lilac, *S. vulgaris* × *S. protolaciniata* P.S. Green & M.C. Chang (*S. buxifolia* Nakai), China) – first synth. c. 1777 at Rouen but other times since & *S. reticulata* (Blume) H. Hara (incl. *S. amurensis*, Japanese l., N Japan)

Syringantha Standley. Rubiaceae (I 1). 1 Mexico

Syringidium Lindau = *Habracanthus*

Syringodea Hook.f. Iridaceae (IV 3). 7 S Afr. R: JSAB 40(1974)201

Syringodium Kütz. Cymodoceaceae. 1 Indian & W Pacific Oceans, 1 Carib. R: JB 77(1939)114

Syrrheonema Miers. Menispermaceae (I). 3 trop. W Afr.

Systeloglossum Schltr. Orchidaceae (V 10). 5 Costa Rica

Systemonodaphne Mez. Lauraceae (I). 1 Guianas, Amazon: *S. geminiflora* (Meissner) Mez. R: BJ 110(1988)161

Systenotheca Rev. & Hardham (~ *Centrostegia*). Polygonaceae (I 1). 1 Calif.: *S. vortriedei* (Brandegee) Rev. & Hardham

Syzygiopsis Ducke = *Pouteria*

Syzygium Gaertner. Myrtaceae (Myrt.) c. 1000 trop. OW. Evergreen trees & shrubs, incl. some rheophytes, diff. from *Eugenia* in axile vasc. supply to ovules (not transseptal) & a set of differential characters. Some with ant-infested hollow stems. Cult. orn. & fr. trees (jambu, esp. *S. jambos*) incl. *S. aqueum* (Burm.f.) Alston (water rose-apple, W Mal.), *S. cumini* (L.) Skeels (jambolan, Java plum, Indomal.) – ed. fr. much used for juice, tanbark, fuelwood, *S. jambos* (L.) Alston (rose apple, SE As.) – fr. best preserved, *S. malaccense* (L.) Merr. & Perry (rose or Malay apple, pomerac, Otaheite a., ohia, Malay Pen.) – fr. ed. fresh or cooked; *S. samarangense* (Blume) Merr. & Perry (wax jambu, Mal.) – ed. fresh. *S. aromaticum* (L.) Merr. & Perry (*Eugenia caryophyllus*, cloves, Moluccas) – sundried fl. buds the cloves of commerce much exported from Zanzibar (Z. red heads) etc. (all in Réunion & Madag. thought to derive from 1 tree) known to Chinese by 200 BC when courtiers had to have them in their mouths when addressing the emperor, now used in flavouring pickles, cakes & apple pies etc., in making pomanders, the oil (eugenol) having analgesic properties much used in relief of toothache, in toothpaste as well as principal source of vanillin; fragments added to kretek cigarettes in Indonesia (some 30 000 t per annum)

Szovitsia Fischer & C. Meyer. Umbelliferae (III 8). 1 Caucasus, Armenia, Iran

T

tabardillo *Piqueria trinervia*

Tabascina Baillon = *Justicia*

tabasco *Capsicum frutescens*

tabasheer or **tabashir** Siliceous concretions found in hollow stems of *Bambusa bambos* & other As. bamboos – alleged medic. value

Tabebuia Gomes ex DC. Bignoniaceae. 100 trop. Am. Trees & shrubs; some excellent timbers, poss. most durable Am. wood (dead specimens of *T. guayacan* (Seemann) Hemsley (Mex. to Colombia) still standing in the Panamá Canal), 400 yr-old beams in Panamá still in excellent condition; some cult. orn. & some barks medic. (allegedly cures for cancers, now patented) incl. malaria. *T. aquatilis* (E. Meyer) Sprague & Sandwith (Amazon) – air-filled seed-coat promoting water-disp.; *T. billbergii* (Bur. & Schumann) Standley (Ecuador) – most sought after timber for carving; *T. chrysantha* (Jacq.) Nicholson (Mexico to Venezuela) – most imp. tree of coastal Ecuador; *T. impetiginosa* (DC) Standley (incl. *T. avellanedae*, N Mex. to Arg.) – cabinet timber (lapacho) form. used as ball bearings; *T. pallida* (Lindley) Miers (WI) – cult. orn. with 4 or 5 buds per axil; *T. rosea* (Bertol.) DC (*'T. pentaphylla'*, roble, Mex. to Venez.) – cult. orn. with large pink fls, timber valued; *T. serratifolia* (Vahl) Nicholson (WI to Bolivia) – valuable timber (washiba), cult. for consp. yellow fls

Taberna Miers = *Tabernaemontana*

Tabernaemontana L. Apocynaceae. 99 trop. (OW 55, incl. *Conopharyngia*, *Ervatamia*). A.J.M. Leeuwenberg, *Rev. of T. I, II* (1991, 1994). Trees & shrubs with alks; some rubber sources & local medic. False dichotomy in branching due in at least *T. crassa* Benth. (W & C Afr.) to slowing of apical meristem & growing out of laterals. *T. divaricata* (L.) Roemer & Schultes (*T. coronaria*, *Ervatamia c.*, crepe jasmine, moon beam, India) – cult. orn.; *T. pachysiphon* Stapf (trop. E Afr.) – bark source of dodo cloth, lvs of a black dye for hair

Tabernanthe Baillon. Apocynaceae. 2 C Afr. R: AUWP 89-4(1989)3. Alks: *T. iboga* Baillon (W Afr.) – hallucinogen (ibogo)

tacamahac *Populus balsamifera*

Tacarcuna Huft. Euphorbiaceae. 3 trop. Am. R: AMBG 76(1989)1080

tacay *Caryodendron orinocense*

Tacazzea Decne. Asclepiadaceae (I, Periplocaceae). 4 trop. & S Afr. R: SAJB 56(1990)93. Riparian lianes & shrubs

Tacca Forster & Forster f. Taccaceae. 10 OW trop. (Mal. to Papuasia 9). Fls poll. by flies & may act as traps. Tubers a source of starch used for bread etc., once bitter taccalin removed, esp. *T. leontopetaloides* (L.) Kuntze (*T. pinnatifida*, E Indian or Tahiti arrowroot, trop. OW) – ? orig. a beach pl. (fr. float for many months), lvs used in hat-making; *T. palmata* Blume (Mal.) – locally medic.

Taccaceae Bercht. & J. Presl. Monocots – Liliidae – Haemodorales. 1/10 trop. Tuberous herbs without secondary growth; vessels restricted to roots. Lvs with long petioles & elliptic to dissected laminas, basal, venation parallel or palmate. Fls bisexual, reg., 3-merous in cymose umbel with involucre at top of scape; P 3 + 3, ± petaloid, s.t. basally connate, brown-purple to greenish, A 3 + 3 attached to base of P with short, flat ± petaloid filaments forming hoods over anthers which have longit. slits, $\overline{G}$ (3), 1-loc., 6-ribbed with ± intruded parietal placentas & ∞ anatropous to campylotropous bitegmic ovules. Fr. usu. a berry (loculicidal capsule) with 10–∞ seeds; embryo small in copious starchless endosperm rich in fat & protein. n = 15. R: Blumea 20(1972)366
Genus: *Tacca*

taccada pith *Scaevola taccada*

Taccarum Brongn. ex Schott. Araceae (VIII 3). 5 N S Am. R: Willd. 19(1989)191

Tachia Aublet. Gentianaceae. 9 trop. S Am. R: JAA 56(1975)103. Some tree-like; some locally medic.

Tachiadenus Griseb. Gentianaceae. 11 Madag. R: BMNHN 4,9(1987)43. Bitters used in beer-making

Tachigali Aublet. Leguminosae (I 1). 24 trop. Am. esp. Amaz. Trees with extrafloral nectaries; some hapaxanthic but much-branched (cf. *Cerberiopsis*). Some medic.

Tachigalia Juss. = *Tachigali*

Tacinga Britton & Rose. Cactaceae (II). 2 E Brazil. *T. funalis* Britton & Rose – cult. orn. clambering to 13 m like *Opuntia* but A long-exserted

Tacitus Moran = *Graptopetalum*

Tacoanthus Baillon. Acanthaceae. 1 Bolivia

Tadehagi Ohashi (*Pteroloma*). Leguminosae (III 9). 3 Indomal. Intergrades with *Droogmansia*

Taeckholmia Boulos = *Sonchus*

Taeniandra Bremek. = *Strobilanthes*

Taenianthera Burret = *Geonoma*

Taeniatherum Nevski. Gramineae (23). 1 Spain to C As. & Pakistan: *T. caput-medusae* (L.) Nevski. R: NJB 6(1986)389

Taenidia (Torrey & Gray) Drude. Umbelliferae (III 5). 2 E US. R: Brittonia 34(1982)365

Taeniopetalum Vis. (~ *Peucedanum*). Umbelliferae (III 10). 2 C to SE Eur.

Taeniophyllum Blume. Orchidaceae (V 16). 170 trop. Afr. (1) to Japan, Aus. (5, 3 endemic) & Tahiti. Photosynthetic roots

Taeniopleurum J. Coulter & Rose = *Perideridia*

Taeniorrhiza Summerh. Orchidaceae (V 16). 1 Gabon. Leafless

Taenitidaceae (C. Presl) Pichi-Serm. = Pteridaceae (III)

Taenitis Willd. ex Schkuhr. Pteridaceae (III; Taenitidaceae). 15 Indomal. to trop. Aus. & Fiji. R: Blumea 16(1968)87

tagasaste *Chamaecytisus proliferus* ssp. *palmensis*

Tagetes L. Compositae (Hele.-Pect.) c. 50 trop. & warm Am., Afr. (1). Foetid herbs, some medic. etc.; alpha-terphenyl considered poss. effective against AIDS & as effective as DDT in controlling mosquito larvae; some cult. orn. *T. erecta* L. (African marigold, Mex. & C Am., widely natur.) – to 1 m, cult. orn. in pre-Conquest Mex., still used in ceremonies, fls source of a poor quality yellow dye, locally medic., many cvs for formal bedding; *T. lucida* Cav. (Mex., Guatemala) – tea a tonic in Mex., hallucinogenic (agent unidentified but not alk.); *T. minuta* L. (stinking Roger, S Am., natur. Afr. etc.) – locally medic. & insecticidal, alleged to protect potatoes from eelworm & to be a weedkiller (allelopathic) & effective at controlling 'damping off' (*Pythium* spp.); *T. patula* L. (~ *T. erecta*, French marigold) – cult. orn. with many cvs c. 30 cm with yellow, orange & brownish fls, also triploid hybrids with *T. erecta* known as 'Afro-French'

taggar *Cinnamosma fragrans*

tahini *Sesamum orientale*

Tahiti arrowroot *Tacca leontopetaloides*; **T. chestnut** *Inocarpus fagifer*

Tahitia Burret = *Berrya*

Tahitian bridal veil *Gibasis pellucida*

Taihangia T.T. Yu & C.L. Li = *Geum*

Tainia Blume. Orchidaceae (V 11). 14 Indomal. to China. R: OM 6(1992)73

Tainionema Schltr. = ? *Orthosia*

Tainiopsis Schltr. = *Eriodes*

Taitonia Yamamoto = *Gomphostemma*

Taiwania Hayata. Taxodiaceae. 2–3 NE Burma, Yunnan, Taiwan. *T. cryptomerioides* Hayata – cult. orn. to 55 m allied to *Sequoiadendron* but rare as much felled

Takeikadzuchia Kitagawa & Kitam. = *Olgaea*

Takhtajania M. Baranova & J. Leroy. Winteraceae (Takhtajaniaceae). 1 Madag.: *T. perrieri* (Capuron) M. Baranova & J. Leroy – coll. once

Takhtajaniaceae (J. Leroy) J. Leroy = Winteraceae

Takhtajaniantha Nazarova = *Scorzonera*

Takhtajanianthus De = *Rhanteriopsis*

tak-out galls tamarisk g.

Talamancalia H. Robinson & Cuatrec. Compositae (Sen.-Sen.). 3 C Am. & Ecuador. R: Novon 4(1994)50, CN 27(1995)34

talas *Colocasia esculenta*

Talauma Juss. = *Magnolia*

Talbotia S. Moore = *Afrofittonia*

Talbotia Balf. (*Talbotiopsis*, ~ *Vellozia*). Velloziaceae. 1 S Afr.: *T. elegans* Balf. – cult. orn.

Talbotiella Baker f. (~ *Hymenostegia*). Leguminosae (I 4). 3 trop. W Afr. *T. gentii* Hutch. & Greenway – forms monospecific stands in drier forests of Ghana (canopy & understorey) but endangered (charcoal)

Talbotiopsis L.B. Sm. = *Talbotia* Balf.

Talguenea Miers ex Endl. Rhamnaceae. 1 Chile

Talinaria Brandegee (~ *Talinum*). Portulacaceae. 3 warm Am.

Talinella Baillon. Portulacaceae. 2 Madagascar

Talinopsis A. Gray. Portulacaceae. 1 arid S US & Mex.

Talinum Adans. Portulacaceae. 39–40 S Afr., C & S Am. Herbs, ± succ.; R: FR 35(1934)1. Some cult. orn. & potherbs esp. *T. fruticosum* (L.) Juss. (*T. triangulare*, trop. Am.) – tastes like purslane. Some with molluscicidal saponins

talipot palm *Corypha umbraculifera*

Talisia Aublet. Sapindaceae. 40 trop. Am. Some cult. for ed. fr. sold in markets esp. *T. oliviformis* (Kunth) Radlk. (Mex. to Colombia, Trinidad)

tallicona *Carapa guianensis*

tallow, Mafura *Trichilia emetica*; **Malabar t.** *Vateria indica*; **t. nut** *Ximenia americana*; **t. shrub** *Myrica cerifera*; **t. tree** *Detarium senegalense*, *Pentadesma butyracea*; **Chinese** or **vegetable t.t.** *Sapium sebiferum*; **Japanese t.t.** *Rhus succedanea*; **t. wood** *Eucalyptus microcorys*

Tamamschjania Pim. & Kljuykov. Umbelliferae (III 8). 2 S Eur. to SW As.

Tamananthus Badillo. Compositae (Helia.-?). 1 Venezuela

Tamania Cuatrec. See *Espeletia*

tamarack larch *Larix laricina*

Tamaricaceae Bercht. & J. Presl. Dicots – Dilleniidae – Violales. 4/78 Euras. & Afr. esp. Medit. to C As. Shrubs & trees, usu. halophytes, xerophytes or rheophytes with slender branches. Lvs small, often scale-like & centric & with salt-excreting glands, spirally arr.; stipules 0. Fls small, ebracteolate, solit. or usu. in bracteate spikes to panicles; K 4 or 5(6) rarely basally connate, imbricate, persistent with C alt. & s.t. persistent but often seated on nectary-disk with A, or disk intrapetiolar or 0, A usu. 2 × C or more & connate in 5 bundles, anthers with longit. slits, $\underline{G}$ ((2)3 or 4(5)), 1-loc. with parietal or (*Tamarix*) basal or parietal basal placentation s.t. almost pluriloc. with 2–∞ anatropous, bitegmic ovules per placenta. Fr. a loculicidal capsule; seeds hairy or the hairs forming a tuft at 1 end, embryo straight in 0 or scanty starchy endosperm often with thin perisperm. N = 12

Genera: *Hololachna, Myricaria, Reaumuria, Tamarix*

Dyes, medic., manna & cult. orn. (*Tamarix*)

Tamaricaria Qaiser & Ali = *Myricaria*

tamarilla, tamarillo *Cyphomandra betacea*

tamarind *Tamarindus indica;* **Manila t.** *Pithecellobium dulce;* **velvet t.** *Dialium guineense, D. ovoideum*

Tamarindus L. Leguminosae (I 4). 1 cultigen (? orig. trop. Afr.): *T. indica* L. (tamarind, Indian date) – nyctitropic movements described by Theophrastus, fr. disp. by ruminants like gazelles (fr. high in vitamin C, protein, tartaric, malic & citric acids) cult. orn. shade tree with useful timber & ed. pulp around seeds (laxative), sharp-tasting & used in drinks, chutney etc. (an essential ingredient of Worcester sauce), when overripe used to clean brass & copper; walking-sticks made from roots

tamarisk *Tamarix* spp. esp. *T. chinensis;* **t. galls** *T. aphylla & T. gallica;* **manna t.** *T. mannifera*

Tamarix L. Tamaricaceae. 54 Euras. (Eur. 14), Afr. R: B.R. Baum (1978) *The genus T.* Tamarisk. Spp. diff. to delimit; many rheophytes, cult. orn. & galls (teggaout or tak-out g.) for tanning; some manna sources esp. *T. mannifera* Ehrenb. (manna t., Iran to Arabia) – manna of the Bedouin produced as result of punctures in stems made by scale-insects. Most commonly cult. orn. is *T. chinensis* Lour. (temp. E As., natur. N Am.) – stems used for making lobster-pots in Cornwall, often confused with *T. gallica* L. (*T. anglica*, Medit.) differing in decid. C etc., with active principle combating liver damage, parasitized fr. behaving like 'jumping beans' (cf. *Sebastiana*). Morocco leather tanned with galls of *T. aphylla* (L.) Karsten (S & E Med. to India) & *T. gallica* (40–45% tannin)

Tamaulipa R. King & H. Robinson (~ *Eupatorium*). Compositae (Eup.-Gyp.). 1 Texas, N Mex. R: MSBMBG 22(1987)132

Tamayoa Badillo = *Lepidesmia*

Tambourissa Sonn. Monimiaceae (V 1). 43 Madag., Masc. R: AMBG 72(1985)90

Tamilnadia Tirv. & Sastre (~ *Gardenia*). Rubiaceae (II 1). 1 S & SE As.: *T. uliginosa* (Retz.) Tirv. & Sastre

Tammsia Karsten. Rubiaceae (inc. sed.). 1 Venez., Colombia: *T. anomala* Karsten

Tamonea Aublet (*Ghinia*). Verbenaceae. 4–5 trop. Am. R: Phytol. 47(1981)404, 448, 48(1981)111

tampala *Amaranthus tricolor*

tampico fibre *Agave* spp.

Tamus L. Dioscoreaceae. 4–5 Macaronesia, Eur. (1), Medit. *T. communis* L. (black bryony, murraim berries, Eur., Medit.) – tuber formed by enlargement of first 2 internodes form. medic. & eaten when boiled, fr. toxic

Tanacetopsis (Tzvelev) Kovalevsk. (~ *Cancrinia*). Compositae (Anth.-Tan.). 21 Iran to C As.

Tanacetum L. (~ *Chrysanthemum*). Compositae (Anth.-Tan.). (Incl. *Balsamita*) 150 N temp. esp. OW (Eur. 14). Cult. orn. aromatic herbs esp. *T. balsamita* L. (*Balsamita major, C. b.*, alecost, camphor pl., costmary, Eur. to C As.) – aromatic lvs form. flavour in ale, *T. coccineum* (Willd.) Grierson (pyrethrum of gardens, Cauc., Iran), *T. parthenium* (L.) Schultz-Bip. (feverfew, SE Eur. to Cauc.) – dried capitula used medic. (febrifuge & efficacious in treatment of migraine) & as tea etc., 'Aureum' (golden feather) – commonly cult. orn. *T. cinerariifolium* (Trev.) Schultz-Bip. (Albania & former Yugoslavia) – self-incompatible diploid clones in cult., fl. heads the source of pyrethrum insecticide (monoterpenes called pyrethrins), cult. esp. highlands of E Afr.; *T. corymbosum* (L.) Schultz-Bip. (SE Eur.) – oil antibacterial; *T. vulgare* L. (tansy, Euras.) – tea form. a tonic & vermifuge, crystallized for treatment of gout, lvs form. placed in winding-sheets as deterrent to worms, rubbed in to meat to keep off flies, ingredient of drisheen, an Irish sausage made of sheep's blood & in modern cosmetics

Tanaecium Sw. Bignoniaceae. 7 trop. Am. *T. nocturnum* (Barb. Rodr.) Bur. & Schumann (Brazil) – hallucinogen; other spp. toxic to stock

Tanakaea Franchet & Sav. Saxifragaceae. 1 E As.: *T. radicans* Franchet & Savi – cult. orn. dioec. evergreen perennial herb

Tanaosolen N.E. Br. = *Tritoniopsis*

tanekaha *Phyllocladus trichomanoides*

tangelo *Citrus* × *aurantium* Tangelo Group

tangerine *Citrus reticulata*

Tanghinia Thouars = *Cerbera*

tangile *Shorea polysperma*

tangor *Citrus × aurantium* Tangor Group
tang-shen *Codonopsis tangshen*
Tangtsinia S.C. Chen = *Cephalanthera*
Tanner grass *Brachiaria arrecta*
tanner's cassia *Senna auriculata*; **t. dock** *Rumex hymenosepalus*
tannia *Xanthosoma sagittifolium*
Tannodia Baillon. Euphorbiaceae. 3 trop. Afr., Comoro Is.
Tanquania H. Hartman & Liede (~ *Pleiospilos*). Aizoaceae (V). 3 Cape. R: BJ 106(1986)479
Tansaniochloa Rauschert = *Setaria*
tansy *Tanacetum vulgare*
Tanulepis Balf.f. = *Camptocarpus*
tapa cloth *Broussonetia papyrifera*
táparos *Orbignya cuatrecasana*
Tapeinanthus Herbert = *Narcissus*
Tapeinia Comm. ex Juss. Iridaceae (III 1). 1 S Chile & Arg.
Tapeinidium (C. Presl) C. Chr. Dennstaedtiaceae (Lindsaeoideae). 17 SE As. to Samoa
Tapeinochilos Miq. Zingiberaceae (Costaceae). c. 12 Mal. to Aus.
Tapeinoglossum Schltr. = *Bulbophyllum*
Tapeinosperma Hook.f. Myrsinaceae. 4 E Mal. & trop. Aus., 39 New Caledonia, 11 Fiji. Pachycaul treelets
Tapeinostemon Benth. Gentianaceae. 7 NE S Am. R: MNYBG 32(1981)356
tapeweed *Posidonia* spp.
Tapheocarpa Conran (~ *Aneilema*). Commelinaceae (II). 1 N Queensland: *T. calandrinioides* (F. Muell.) Conran. R: AusSB 7(1994)585. Fr. hypogynous
Taphrospermum C. Meyer. Cruciferae. 2 C As.
Tapinanthus (Blume) Reichb. Loranthaceae. 250 trop. Afr. Some with explosive fr. like *Globimetula* (q.v.)
Tapinopentas Bremek. = *Otomeria*
Tapinostemma (Benth.) Tieghem = *Plicosepalus*
tapioca *Manihot esculenta*
Tapiphyllum F. Robyns. Rubiaceae (III 2). c. 20 trop. Afr.
Tapirira Aublet. Anacardiaceae (II). 16 trop. Am.
Tapirocarpus Sagot. Burseraceae. 1 Guyana
Tapiscia Oliver. Staphyleaceae (Tapisciaceae). 1 China
Tapisciaceae (Pax) Takht. = Staphyleaceae
Taplinia Lander. Compositae (Gnap.-Ang.). 1 W Aus. R: Nuytsia 7(1989)37
Tapoides Airy Shaw. Euphorbiaceae. 1 Borneo
Taprobanea E. Christ. (~ *Vanda*). Orchidaceae (V 16). 1 S India, Sri Lanka: *T. spathulata* (L.) E. Christ. R: Lindleyana 7(1992)90
Tapura Aublet. Dichapetalaceae. 28 trop. Am. (21), trop. Afr. (7). R: Brittonia 35(1983)49, AUWP 86-3(1986)43. Infls axillary, on petioles or midribs. Some timber
tar tree *Semecarpus australiensis*
tara *Caesalpinia spinosa*
taraire *Beilschmiedia taraira*
Taraktogenos Hassk. = *Hydnocarpus*
Taralea Aublet = *Dipteryx*
Tarasa Philippi. Malvaceae. 25 Andes
Taravalia E. Greene = *Ptelea*
Taraxacum Weber ex Wigg. Compositae (Lact.-Crep.). 60 temp. inc. S Am. (2), some cosmop. weeds. Many hundreds of apomictic lines named ('microspp.'; 30 'groups' in Eur.), *T.* poss. orig. Himal. Some cult. orn., salad & rubber pls. *T. bicorne* Dahlst. (*T. koksaghyz*, kok-saghyz, Russian dandelion, Turkestan) – obligate outbreeding diploid cult. for rubber in Russia & N China; *T. megalorhizon* (Forssk.) Hand.-Mazz. (krim-saghyz, W As.) – similar; *T. officinale* Weber ex Wigg. s.l. (*T. vulgare* (Lam.) Schrank, dandelion, Euras., cosmop. triploid weed where endosperm does not necessarily develop before embryo (all identical!), *T. o.* s.s. restricted to Lapland; the debate over the correct name for this most common pl. should be resolved by conservation with a new type) – cult. (blanched) for salad lvs (several cvs), taproot form. ground as coffee subs., somewhat diuretic (Fr.: pissenlit) though agent unknown, alledgedly a liver & kidney remedy, unopened fl. heads used like capers, open heads dried for dandelion wine-making, good bee fodder (in orchards can distract bees from poll. fr. trees), form with pitcher-shaped lvs found in England 1980s

Tarchonanthus L. Compositae (Mut.-Mut./Tarch.). 2 Afr., Arabia. Dioec. *T. camphoratus* L. (camphor wood, E & S Afr.) – lvs chewed medic., timber for musical instruments etc.

tare *Vicia* spp. esp. *V. sativa*

Tarenna Gaertner. Rubiaceae (II 2). 180 OW trop. Some timber; alks

Tarennoidea Tirv. & Sastre (~ *Randia*). Rubiaceae (II 1). ? Spp.

tares *Lolium temulentum*

tarhui *Lupinus mutabilis*

Tarigidia Stent. Gramineae (34d). 1 S Afr.

taro *Colocasia esculenta*; **giant t.** *Alocasia macrorrhizos*; **swamp t.** *Cyrtosperma merkusii*

Tarphochlamys Bremek. = *Strobilanthes*

tarragon *Artemisia dracunculus*

Tarrietia Blume = *Heritiera*

Tartarian lamb *Cibotium barometz*

tarweed *Madia* spp.

tarwi *Lupinus mutabilis*

Tashiroea Matsum. = *Bredia*

Tasmanian beech *Nothofagus cunninghamii*; **T. cedar** *Athrotaxis* spp.; **T. myrtle** *N. cunninghamii*; **T. oak** *Eucalyptus delegatensis*

Tasmannia R. Br. ex DC = *Drimys*

Tassadia Decne. (~ *Cynanchum*). Asclepiadaceae (III 1). 17 trop. Am. R: AJBRJ 21(1977)235. Mostly riparian

tasua *Aphanamixis polystachya*

Tateanthus Gleason. Melastomataceae. 1 Venezuela

Tatianyx Zuloaga & Söderstrom. Gramineae (34b). 1 Brazil. Savanna

tau foo *Glycine max*

Taubertia Schumann = *Disciphania*

tau-saghyz *Scorzonera tausaghyz*

Tauscheria Fischer ex DC. Cruciferae. 1 E Eur. to India: *T. lasiocarpa* Fischer ex DC

Tauschia Schldl. Umbelliferae (III 5). 31 W US to W trop. S Am.

Tavaresia Welw. (~ *Decabelone*). Asclepiadaceae (III 5). 2 trop. & S Afr. R: Kirkia 9(1974)349. Cult. orn. succ.

Taverniera DC. Leguminosae (III 17). 15 SW As., NE Afr. R: SBU 25,1(1985)45

Taveunia Burret = *Cyphosperma*

tawa *Beilschmiedia tawa*

Taxaceae Gray. Pinopsida. 4/16 N temp. to Malesia & New Caled. Dioec. evergreen trees & shrubs s.t. without resin-canals. Lvs linear, needle-like, spirally arr. but often appearing 2-ranked. Microsporangiophores 6–14 in small cones, scale-like, with 3–6 pollen-sacs; ovules solit., arillate, term. dwarf shoots. Embryo with 2 cotyledons

Genera: *Austrotaxus, Pseudotaxus, Taxus, Torreya*

For other genera form. here, see Cephalotaxaceae & Podocarpaceae

Timber & cult. orn.

Taxillus Tieghem. Loranthaceae. 30 S Afr. & Madag., Pakistan to Borneo (Mal. 1)

Taxodiaceae Saporta (~ Cupressaceae). Pinopsida. 10/16 temp. N Am., E As., Tasmania. Monoec. resinous evergreen or decid. trees. Lvs not clustered, spirally arr. though app. s.t. opp., linear or needle-like. Males cones catkin-like with spirally arr. microsporophylls bearing usu. 2 pollen-sacs, pollen without air-sacs; females (often not prod. until plant older) usu. globose with wedge-shaped or pelate scales, each scale with 2–9 sometimes winged seeds usu. maturing in first season; cotyledons 2–15

Genera: *Athrotaxis, Cryptomeria, Cunninghamia, Glyptostrobus, Metasequoia, Sciadopitys, Sequoia, Sequoiadendron, Taiwania, Taxodium* (5 monotypic)

Form. imp. & widespread fam. in N hemisph., now a classic relict distrib. DNA evidence suggests amalgamation with Cupressaceae

Many imp. timbers & cult. orn.

Taxodium Rich. Taxodiaceae. 2 E N Am., highland Mex. R: Phytol.M 7(1984)72. Cult. orn. timber trees esp. *T. distichum* (L.) Rich. (swamp, southern or bald cypress, E N Am.) – to 45 m with conspic. aerating 'knees', seeds water-disp., timber for sleepers etc.; *T. mucronatum* Ten. (C Plateau, Mex.) – girth to 35.8 m (1982, largest tree-girth known), timber & medic. resins

taxol *Taxus* spp. esp. *T. brevifolia*

Taxopsida = Pinopsida

Taxus L. Taxaceae. 7 N temp. (Eur. 1) to C Mal. & Mex. R: OARDCRB 1086 (1976). Dioec. evergreens with fleshy scarlet aril; foliage & seeds toxic (alks), though bird-disp., the aril being harmless. Yew esp. *T. baccata* L. (English y., Eur., Medit.) – cult. orn. sombre slow-growing tree (allegedly 4000–5000 yr-old tree in Clwyd, Wales) assoc. with churchyards but excellent hedge, timber form. much used for bows in GB & knife-handles, lvs rich in beta-ecdysones, many cvs commonly planted incl. 'Aurea' (golden y.) & 'Stricta' ('Fastigiata', 'Hibernica', Irish y.) – female, all trees from cuttings from one at Florence Court, County Fermanagh (where 2 trees found on moors in 1778), Ireland, columnar habit; other cult. orn. etc. incl. *T. brevifolia* Nutt. (Californian y., W N Am.) – much exploited for taxol (less conc. in other spp.) effective in control of ovarian cancer due to phenylisoserine side-chain but 6 trees needed for 1 dose (10 000 for 1 kg) though baccatin-3 from *T. baccata* lvs now used in taxol synthesis, *T. canadensis* Marshall (American y., E N Am.), *T. mairei* (Lemée & Lév.) Liu (*T. chinensis*, p.p., *T. sumatrana*, Chinese y., S China to C Mal.), *T. cuspidata* Siebold & Zucc. (Japanese y., E As.) – cult. orn., timber for furniture, marquetry, pencils etc., brown dye from heartwood & *T. × media* Rehder (*T. baccata × T. cuspidata*) – many cvs

tayberry see *Rubus*

Tayloriophyton Nayar. Melastomataceae. 2 W Mal.

Tchihatchewia Boiss. = *Neotchihatchewia*

tchirisch *Asphodelus aestivus*

tchitola *Oxystigma oxyphyllum*

tea *Camellia sinensis*; **t. balm** *Melissa officinalis*; **Bogotá t.** *Symplocos theiformis*; **chamomile t.** *Chamaemelum nobile*; **Fukien t.** *Carmona retusa*; **Jesuits's t.** *Otholobium glandulosum*; **Labrador t.** *Rhododendron tomentosum*; **Matara t.** *Senna auriculata*; **Oswego t.** *Monarda didyma*; **Paraguay t.** *Ilex paraguariensis*; **Rooibos t.** *Aspalathus linearis*; **t. seed oil** *Camellia sasanqua*; **Senegal t.** *Gymnocoronis spilanthoides*; **t. tree** *Leptospermum* spp.; **Duke of Argyll's t. t.** *Lycium barbarum*; **yaupon t.** *Ilex vomitoria*; **yerba maté t.** *I. paraguariensis*

Teagueia (Luer) Luer (~ *Platystele*). Orchidaceae (V 13). 6 Andes. R: MSB 39(1991)139

teak *Tectona grandis*; **African t.** *Milicia excelsa*, *Oldfieldia africana*; **Australian t.** *Flindersia australis*; **bastard t.** *Butea monosperma*; **Borneo t.** *Intsia bijuga*; **Brunei t.** *Dryobalanops* spp.; **grey t.** *Gmelina* spp.; **Malacca t.** *Intsia palembanica*; **New Guinea t.** *Vitex cofassus*; **Sudan t.** *Cordia myxa*

teasel (teazel or **teazle)** or **fuller's t.** *Dipsacus sativus*

Teclea Del. Rutaceae. 30 trop. Afr., Madagascar. Alks

Tecleopsis Hoyle & Leakey = *Vepris*

Tecoma Juss. Bignoniaceae. (Incl. Tecomaria) 13 trop. & S Afr. (1), trop. Am. esp. Andes, to Arizona. Cult. orn. trees & shrubs with alks & extrafl. nectaries esp. *T. capensis* (Thunb.) Lindley (*Tecomaria c.*, Cape honeysuckle, E & S Afr.) – fls scarlet, powdered bark medic. in S Afr. & *T. stans* (L.) Kunth (trop. Am.) – fls bright yellow, locally used diuretic

Tecomanthe Baillon. Bignoniaceae. 5 Mal., Aus., NZ (Three Kings Is.: *T. speciosa* W. Oliver now reduced to a single tree from which being propagated)

Tecomaria Spach = *Tecoma*

Tecomella Seemann. Bignoniaceae. 1 Arabia to W India.: *T. undulata* (Sm.) Seemann – cult. orn. with orange fls

Tecophilaea Bertero ex Colla. Tecophilaeaceae (Liliaceae s.l.). 2 Chilean Andes. Cult. orn. esp. *T. cyanocrocus* Leyb. (Chilean crocus)

Tecophilaeaceae Leyb. (~ Liliaceae s.l.). Monocots – Liliidae – Asparagales. Incl. Ixioliriaceae 7/24 Turkey to C As. (*Ixiolirion*), Afr., S Am. N to Calif. Herb. pls with tunicated corms to rhiz. Lvs lanceolate to linear, parallel-veined; vessels in roots. Fls bisexual in racemes or thyrses, sometimes umbel-like, ± reg.; P 3 + 3 shortly connate at base, A 3 + 3, some (esp. upper) sterile, anthers with apical (longit. in *Ixiolirion*) slits or pores. G usu. semi-inf., 3-loc. each with several–∞ ovules in 2 rows. Fr. a loculicidal capsule with small black seeds

Genera: *Conanthera, Cyanella, Ixiolirion, Odontostomum, Tecophilaea, Walleria, Zephyra*
Lanaria (Haemodoraceae) poss. here too

Tectaria Cav. Dryopteridaceae (I 3). (Incl. *Amphiblestra*, *Dictyoxiphium*, *Fadyenia*, *Stenosemia* etc.) c. 150 trop. (Afr. 12, Am. 40). Terrestrial, some acrostichoid, some with simple fronds, some with apical buds on fronds. Some cult. orn.

Tectariaceae Panigrahi = Dryopteridaceae

Tectaridium Copel. = *Tectaria*

Tecticornia Hook.f. Chenopodiaceae (II 2). 3 Mal. (1) to Aus., esp. Nuyts Is. R: Nuytsia 1(1972)277

Tectiphiala H. Moore. Palmae (V 4n). 1 Mauritius

Tectona L.f. Labiatae (III; Verbenaceae). 4 SE As. to Mal. *T. grandis* L.f. (teak, India to Laos, introd. to Indonesia 400–600 yrs ago & natur.) – imp. timber much used for ship-building, bridges, flooring, furniture etc.; sinks in water unless dried so in India tree 'girdled' i.e. a ring of bark & living tissue removed near base, so that it dies & is left standing for 2 yrs before extraction; sawdust in water an antidote to *Gluta* lesions; lvs source of brown-red dye in C Mal. See D.N. Tewari (1992) *A monograph on teak*

Tecunumania Standley & Steyerm. Cucurbitaceae. 1 Guatemala

Tedingea D. & U. Mueller-Doblies (~ *Hessea*). Amaryllidaceae (Liliaceae s.l.). 1 S Afr.: *T. tenella* (L.f.) D. & U. Mueller-Doblies. R: BJ 107(1985)45

Teedia Rudolphi. Scrophulariaceae. 2 S Afr. Cult. orn. foetid shrubs

Teesdalia R. Br. Cruciferae. 2 Eur. (2), Medit. incl. *T. nudicaulis* (L.) R. Br. (shepherd's cress)

Teesdaliopsis (Willk.) Gand. (~ *Iberis*). Cruciferae. 1 Spain: *T. conferta* (Lag.) Rothm.

teff *Eragrostis tef*

Tegicornia P.G. Wilson (~ *Halosarcia*). Chenopodiaceae (II 2). 1 Aus.: *T. uniflora* P.G. Wilson. R: FA 4(1984)300

Teijsmanniodendron Koord. Labiatae (I; Verbenaceae). 14 SE As., Mal. (exc. S). R: Reinwardtia 1(1951)75

Teinosolen Hook.f. = *Heterophyllaea*

Teinostachyum Munro = *Schizostachyum*

Teixeiranthus R. King & H. Robinson (~ *Eupatorium*). Compositae (Eup.-Ager.). 2 Brazil

Telanthophora H. Robinson & Brettell (~ *Senecio*). Compositae (Sen.-Tuss.). 14 C Am. R: Phytologia 27(1974)424. Trees & shrubs. *T. grandifolia* (Less.) H. Robinson & Brettell (Mex.) – cult. orn. shrub

Telectadium Baillon. Asclepiadaceae (I, Periplocaceae). 3 SE As. Rheophytes. *T. edule* Baillon (Laos) – ed.

telegraph plant *Codariocalyx motorius*

Telekia Baumg. Compositae (Inul.). 1 C Eur. to Cauc.: *T. speciosa* (Schreber) Baumg. (SE Eur. to S Russia) – cult. orn. coarse herb

Telemachia Urban = *Elaeodendron*

Telephium L. Caryophyllaceae (I 3). 5 Medit. (Eur. 1), Madag. R: JB 44(1906)289. *T. imperati* L. (W Med.) – cult. orn. rock-pl.

tequila *Agave* spp.

Telesonix Raf. = *Boykinia*

Telfairia Hook. Cucurbitaceae. 3 trop. Ed. oil seeds esp. *T. occidentalis* Hook.f. (krobonko, trop. Afr.) & *T. pedata* (Sm.) Hook. (oyster nut, kweme, trop. Afr.) – oil used in soap- & candle-making

Teline Medikus = *Genista*

Teliostachya Nees. Acanthaceae. 10 trop. S Am.

Telipogon Mutis ex Kunth. Orchidaceae (V 10). 100 C & trop. S Am. Pseudocopulation (?). Some cult. orn.

Telitoxicum Mold. Menispermaceae (II). 6 trop. S Am. R: MNYBG 22,2(1971)76. Form. curare sources

Tellichery bark *Holarrhena pubescens*

Tellima R. Br. Saxifragaceae. 1 W N Am.: *T. grandiflora* (Pursh) Douglas ex Lindley – cult. orn. ground cover

Telmatophila C. Martius ex Baker. Compositae (Vern.-Vern.). 1 Brazil

Telminostelma Fourn. = *Cynanchum*

Telmissa Fenzl = *Sedum*

Telopea R. Br. Proteaceae. 5 E Aus. Cult. orn with red fls in dense heads surrounded by coloured involucres esp. *T. speciosissima* (Sm.) R. Br. (waratah, NSW) – protected (floral emblem NSW but cut-fl. exported from NZ as kiwi rose!)

Telosma Cov. Asclepiadaceae (III 4). 10 trop. OW. *T. cordata* (Burm.f.) Merr. (India to SE As.) – liane with roots, lvs & fls locally eaten

Teloxys Moq. = *Chenopodium*

temiche cap *Manicaria saccifera*

Temmodaphne Kosterm. = *Cinnamomum*

Temnadenia Miers. Apocynaceae. 3 trop. S Am.

Temnocalyx F. Robyns. Rubiaceae (III 2). 1 Tanzania

Temnopteryx Hook.f. Rubiaceae (I 8). 1 trop. W Afr.

temple tree *Plumeria rubra*

Templetonia R. Br. Leguminosae (III 25). 11 Aus. R: Muelleria 5(1982)1. Cult. orn. shrubs with alks esp. *T. retusa* (Vent.) R. Br. (coral bush, S & W Aus.)

Temu O. Berg = *Blepharocalyx*

Tenagocharis Hochst. = *Butomopsis*

Tenaris E. Meyer. Asclepiadaceae (III 5). 7 trop. & S Afr.

Tengia Chun. Gesneriaceae. 1 SW China: *T. scopulorum* Chun

Tenicroa Raf. (~ *Drimia*). Hyacinthaceae (Liliaceae s.l.). 5 S Afr. R: JSAB 47(1981)577

Tennantia Verdc. Rubiaceae (II 2). 1 E Afr.

Tenrhynea Hilliard & B.L. Burtt (*Rhynea*). Compositae (Gnap.-Gnap.). 1 S Afr.: *T. phylicifolia* (DC) Hilliard & B.L. Burtt. R: OB 104(1991)140

teosinte *Zea mays* subsp. *mexicana*

tepa *Laurelia sempervirens*

tepary bean *Phaseolus acutifolius*

tepescohuite *Mimosa tenuiflora*

Tephroseris (Reichb.) Reichb. (~ *Senecio*). Compositae (Sen.-Tuss.). 50 temp. & Arctic Euras., 1 ext. to N Am. Some cult. orn. incl. *T. integrifolia* (L.) Holub (*Senecio i.*, fleawort, Euras.)

Tephrosia Pers. Leguminosae (III 6). 400 seasonal trop. esp. Afr. (incl. *Lupinophyllum* – geocarpic fr.). Cult. cover crops & green manures, fish-poisons & cult. orn. *T. candida* DC (boga medalo, India) – green manure; *T. macropoda* Harvey (lozane, S Afr.) – roots a fish-poison & insecticide; *T. purpurea* (L.) Pers. (OW trop.) – seeds ed. Mafia Is., Tanz.; *T. sinapou* (Buc'hoz) A. Chev. (*T. toxicaria*, yarroconalli, trop. Am.) – roots a fish-poison, green manure; *T. virginiana* (L.) Pers. (goat's rue, cat gut, E N Am.) – roots medic. etc.; *T. vogelii* Hook. (trop. Afr.) – cult. for lvs used as fish-poison in small-scale coastal fishing in Comoro Is.

Tepualia Griseb. Myrtaceae (Lept.). 1 Chile. Only Leptospermoidea in Am. Hard wood

Tepuia Camp. Ericaceae. 8 Guayana Highland. R: MNYBG 29(1978)153

Tepuianthaceae Maguire & Steyerm. (~ Linaceae). Dicots – Rosidae – Linales. 1/5 Guayana Highland. Trees & shrubs with resin cells. Lvs simple, entire, spirally arr. to opp. Infls androdioec; K 5 imbricate, C 5 imbricate, disk of 5–10 suborbicular fleshy glands, A 5,12–16 in 1–3 whorls, anthers versatile with longit. slits, ± appendaged connective, $\underline{G}$ (3) with 3 forked styles, each loc. with 1 pendulous, anatropous ovule. Fr. a 3-loc. loculicidal capsule; seeds with consp. ridged raphe. R: MNYBG 32(1981)8
Only genus: *Tepuianthus*

Tepuianthus Maguire & Steyerm. Tepuianthaceae. 5 Guayana Highland. R: MNYBG 32(1981)9

tequila *Agave* spp.

Teramnus P. Browne. Leguminosae (III 10). 8 trop.

terap *Artocarpus odoratissimus*

Teratophyllum Mett. ex Kuhn. Lomariopsidaceae). 12 S Burma to Queensland & E Polynesia

Terauchia Nakai. Liliaceae s.l. 1 Korea = ?

terebinth *Pistacia terebinthus*

Terebraria Kuntze = *Neolaugeria*

Terminalia L. Combretaceae. 150 trop. Timber, dyes, tannin, gums (Ashanti g.) & cult. orn. with char. planar foliage (pagoda tree shape): *T. alata* Roth (asna, Indian laurel, India) – comm. timber, food for silkworms; *T. arjuna* (DC) Wight & Arn. (arjun, kumbuk, kahua, India, Sri Lanka) – timber for construction, bark with heart stimulants, lime extractable for chewing with betel; *T. arostrata* Ewart & O. Davies (nutwood, NW Aus.) – ed. seeds; *T. australis* Camb. (Brazil, Arg., Paraguay) – rheophyte used for erosion control; *T. bellirica* (Gaertner) Roxb. (Indomal.) – fr. a tannin source & black dyes & inks, timber good; *T. bialata* (Roxb.) Steudel (Andamans) – grey heartwood for fancy work like picture-framing (chuglam) & construction (white chuglam wood); *T. catappa* L. (Indian, Barbados or wild almond, Indomal.) – widely planted & natur., salt-tolerant street-tree, sea- & bat-disp. (rat-disp. on Krakatoa), timber red & good, lvs (with domatia) for silkworms, roots & bark for tanning, seeds ed. like almonds with similar oil; *T. chebula* (Gaertner) Retz. (myrobalan (name also used for other spp.), Indomal.) – dried fr. used in tanning, giving soft mellow leather (tannin c. 32%), medic.; *T. ferdinandiana* Exell (Aus.) – small ed. fr. with 31 mg vitamin C per g (oranges with 0.5 mg); *T. glaucescens* Planchon ex Benth. & *T. laxiflora* Engl. (W Afr.) – roots chewing-sticks in Nigeria; *T. ivorensis* A. Chev. (black afara, emeri, idigbo, W Afr.) – weather-resistant timber for shingles etc. & plywood; *T. kaernbachii* Warb. (*T. okari*, okari, S New Guinea, (?) natur. Polynesia) – ed. fr., the kernel a good flavoured 'nut'; *T. lucida* Hoffsgg. ex C. Martius (trop. Am.) – assoc. with *Azteca* ants in C Am.; *T. oblonga* (Ruíz & Paón) Steudel (trop. Am.) – timber for flooring & panelling (up to 13% calcium by wt.); *T. procera* Roxb. (badam, white bombway, Andamans) – timber; *T. superba* Engl. & Diels (afara, akom, limba, ofram, white afara, W Afr.) – general construction timber

Terminaliopsis Danguy. Combretaceae. 2 Madagascar

Terminthia Bernh. = *Rhus*

Terminthodia Ridley = *Tetractomia*

Terniola Tul. = *Dalzellia*

Terniopsis H.C. Chao = *Dalzellia*

Ternstroemia Mutis ex L.f. Theaceae. 85 trop. (Afr. 2). *T. japonica* Thunb. (Japan) – hedge-pl.

Ternstroemiaceae Mirbel ex DC = Theaceae

Ternstroemiopsis Urban = *Eurya*

Terpsichore A.R. Sm. (~ *Grammitis*). Grammitidaceae. 50 trop. Am. & Afr. (1). R: Novon 3(1993)478

Tersonia Moq. Gyrostemonaceae. 1 SW Aus.: *T. cyathiflora* (Fenzl) A.S. George

Terua Standley & F.J. Herm. = *Lonchocarpus*

tesota *Olneya tesota*

Tessarandra Miers = *Chionanthus*

Tessaria Ruíz & Pavón (~ *Pluchea*). Compositae (Pluch.). 1 warm & trop. Am. (arrow-weed)

Tessmannia Harms. Leguminosae (I 4). 11 trop. Afr. Furniture timbers

Tessmanniacanthus Mildbr. Acanthaceae. 1 E Peru

Tessmannianthus Markgraf. Melastomataceae. 7 trop. Am. Trees to 45 m

Tessmanniodoxa Burret = *Chelyocarpus*

Testudinaria Salisb. = *Dioscorea*

Testudipes Markgraf = *Tabernaemontana*

Testulea Pellegrin. Ochnaceae. 1 Gabon: *T. gabonensis* Pellegrin. R: BJ 113(1991)175

Tetilla DC. Saxifragaceae (Francoaceae). 1 Chile

Tetraberlinia (Harms) Hauman. Leguminosae (I 4). 4 Guineo-Congolian forest

Tetracanthus A. Rich. = *Pectis*

Tetracarpaea Hook. Grossulariaceae (Tetracarpaeaceae). 1 Tasmania: *T. tasmannica* Hook.

Tetracarpaeaceae Nakai = Grossulariaceae

Tetracarpidium Pax. Euphorbiaceae. 1 W trop. Afr.: *T. conophorum* (Muell. Arg.) Hutch. & Dalz. (conophor or awusa nut) – source of a drying oil

Tetracentraceae A.C. Sm. = Trochodendraceae

Tetracentron Oliver. Trochodendraceae (Tetracentraceae). 1 Nepal, SW & C China, N Burma: *T. sinense* Oliver – tracheids to 45 mm long

Tetracera L. Dilleniaceae. 40 trop. Some cryptic dioecy. Rough lvs of some spp. used like glasspaper. *T. scandens* (L.) Merr. (Madag.) – local medic.

Tetrachaete Chiov. Gramineae (33d). 1 Tanzania, Ethiopia, Arabia

Tetrachne Nees (~ *Uniola*). Gramineae (31b). 1 S Afr. & Pakistan

Tetrachondra Petrie ex Oliver. Labiatae (Tetrachondraceae). 2 NZ (1), temp. S Am. (1). R: BSAB 13(1970)2

Tetrachondraceae Wettst. See Labiatae

Tetrachyron Schldl. (~ *Calea*). Compositae (Helia.-Mel.). c. 7 E Mex., Guatemala. R: SB 4(1979)297

Tetraclea A. Gray (~ *Clerodendrum*). Labiatae (II). 2 S N Am. R: BSAB 13(1970)2

Tetraclinis Masters. Cupressaceae. 1 S Spain, Malta, N Afr.: *T. articulata* (Vahl) Masters (arar, alerce, thuya) – timber used for building since antiquity (& burnt on Calypso's fire in the Odyssey), the citrus-wood of antiquity particularly valued for tables (Pliny), resin (sandarac(h)) used in picture varnish, for glazing paper, for pounce & incense

Tetraclis Hiern = *Diospyros*

Tetracme Bunge. Cruciferae. 8 E Medit. (Eur. 1) to C As. & Baluchistan

Tetracoccus Engelm. ex C. Parry. Euphorbiaceae. 5 SW N Am. Dioec. shrubs

Tetractomia Hook.f. (~ *Melicope*). Rutaceae. 6 Mal. R: JAA 60(1979)127

Tetracustelma Baillon = *Matelea*

Tetradema Schltr. = *Agalmyla*

Tetradenia Benth. Labiatae (VIII 4). 9 trop. Afr. (3), Madag. R: Bothalia 14(1983)177, 15(1984)1. Dioec. (sometimes hermaphr.) cult. orn. (incl. *Iboza*)

Tetradiclidaceae (Engl.) Takht. = Zygophyllaceae

Tetradiclis Steven ex M. Bieb. Zygophyllaceae. 1 SE Russia & E Med. to C As.

Tetradium Lour. Rutaceae. 9 Himal. to Japan, Philippines, Sumatra & Java. R: GBS 34(1981)91. *T. daniellii* (Bennett) Hartley (*Euodia d.*, *Evodia d.*, SW China to Korea) – cult. orn. tree; *T. fraxinifolium* (Hook.) Hartley (Nepal to SE As.) – fr. for chutney & medic., cult. orn.; *T. ruticarpum* (A. Juss.) Hartley (China, Taiwan) – drug (wu-chu-yu used for 2000 yrs, stimulant to anthelmintic, alks with uterotonic activity

Tetradoa Pichon = *Hunteria*
Tetradoxa C.Y. Wu = *Adoxa*
Tetradyas Danser = *Cyne*
Tetradymia DC. Compositae (Sen.-Tuss.). 10 N Am. R: Brittonia 26(1974)177. Toxic to sheep – if eaten in sunlight leads to 'bighead' swelling, liver damage & photosensitization
Tetraedrocarpus O. Schwarz = *Echiochilon*
Tetraena Maxim. Zygophyllaceae. 1 Mongolia: *T. mongolica* Maxim.
Tetragamestus Reichb.f. = *Scaphyglossum*
Tetragastris Gaertner. Burseraceae. 9 trop. Am. R: KB 45(1990)179. Some fish-disp. in Amazon. *T. balsamifera* (Sw.) Oken (WI) – timber for cabinet-work, panelling, flooring etc.
Tetraglochidium Bremek. = *Strobilanthes*
Tetraglochin Poeppig. Rosaceae. 8 Andes, temp. S Am.
Tetragoga Bremek. = *Strobilanthes*
Tetragompha Bremek. = *Strobilanthes*
Tetragonia L. Aizoaceae (III; Tetragoniaceae). 85 trop. & warm S esp. S Afr., N Am. Fr. with thorny projections sometimes producing fls. *T. tetragonioides* (Pallas) Kuntze (NZ spinach, warrigal (cabbage), Japan & Pacific to NZ & temp. S Am.) – grown as spinach in sites too hot for true s., fed to his crew by Captain Cook
Tetragoniaceae Link = Aizoaceae
Tetragonocalamus Nakai = *Bambusa*
Tetragonolobus Scop. = *Lotus*
Tetragonotheca L. Compositae (Helia.-Mel.). 4 SE US. R: Sida 8(1980)296
Tetralix Griseb. Tiliaceae. 2 E Cuba (serpentine)
Tetralocularia O'Don. Convolvulaceae. 1 Colombia
Tetralopha Hook.f. = *Gynochthodes*
Tetramelaceae (Warb.) Airy Shaw = Datiscaceae
Tetrameles R. Br. Datiscaceae (Tetramelaceae). 1 Indomal. to Aus.: *T. nudiflora* R. Br. (kapong, sompong) – buttresses large, wood for matches, canoes etc.; seeds winged
Tetrameranthus R. Fries. Annonaceae. 6 trop. S Am. R: PKNAW C88(1985)449. Lvs spirally arr.
Tetramerista Miq. Tetrameristaceae. 3 W Mal. *T. glabra* Miq. (punah) – timber for beams, ed. fr.
Tetrameristaceae Hutch. Dicots – Dilleniidae – Theales. 2/4 W Mal., Guayana Highland. Trees & shrubs. Lvs simple, spirally arr.; stipules 0. Fls bisexual, bibracteolate in axillary racemes; K 4 or 5, imbricate, persistent or not, C 4 or 5 imbricate, scarcely longer than K, A 4 or 5 alt. with C, filaments basally connate, anthers with longit. slits, $\underline{G}$ (4 or 5), 4- or 5-loc. with term. style, each loc. with 1 axile-basal anatropous bitegmic ovule. Fr. a 4- or 5-seeded berry; seeds with straight embryo in copious endosperm
Genera: *Pentamerista*, *Tetramerista*
Pentamerista discovered in 1972 giving the remarkable known distrib. & demonstrating the relic nature of the Guayana Highland flora
Tetramerium Nees. Acanthaceae. 28 C Am. R: SBM 12(1986)
Tetramicra Lindley. Orchidaceae (V 13). 11 WI. Cult. orn. terr. orchids
Tetramolopium Nees. Compositae (Ast.-Ast.). 37 New Guinea, Hawaii (some extinct), Cook. Is. Yellow-flowered spp. sometimes seg. as *Luteidiscus*; rest with purple fls
Tetranema Benth. Scrophulariaceae. 2 Mex. & Guatemala. Cult. orn. esp. *T. roseum* (M. Martens & Galeotti) Standley & Steyerm. (*T. mexicanum*)
Tetraneuris E. Greene = *Hymenoxys*
Tetranthera Jacq. = *Litsea*
Tetranthus Sw. Compositae (Helia.-Pin.). 4 WI
Tetrapanax (K. Koch) K. Koch. Araliaceae (1). 1 S China (? native), Taiwan: *T. papyrifer* (Hook.) K. Koch – cult. orn. unarmed clump-forming shrub, pith an imp. source of fine rice-paper in China
Tetrapathaea (DC) Reichb. = *Passiflora*
Tetraperone Urban. Compositae (Helia.-Pin.). 1 Cuba
Tetrapetalum Miq. Annonaceae. 2 Borneo
Tetraphyllaster Gilg. Melastomataceae. 1 trop. W Afr.
Tetraphyllum Griffith ex C.B. Clarke. Gesneriaceae. 2 NE India, Thailand
Tetraphysa Schltr. = *Pentacyphus*
Tetrapilus Lour. Oleaceae. 10–15 SE As.
Tetraplandra Baillon. Euphorbiaceae. 5 Brazil
Tetraplasandra A. Gray (~ *Gastonia*). Araliaceae. 6 Hawaii. R: W. Wagner et al., *Man. Fl. Pls*

Hawaii (1990) 232. Some cult. orn.

Tetraplasia Rehder = *Damnacanthus*

Tetrapleura Benth. Leguminosae (II 3). 2 trop. Afr. Timber. *T. tetraptera* (Schum.) Taubert – smelly fr. elephant-disp., saponins (triterpenes) antifungal & molluscicidal in field trials

Tetrapodenia Gleason = *Burdachia*

Tetrapogon Desf. Gramineae (33b). 5 warm Afr., Middle E, India

Tetrapollinia Maguire & Boom (~ *Lisianthus*). Gentianaceae. 1 NE trop. Am.: *T. caerulescens* (Aublet) Maguire & Boom. R: MNYBG 51(1989)31

Tetrapteris Cav. = *Tetrapterys*

Tetrapterocarpon Humbert. Leguminosae (I 1). 1 Madagascar

Tetrapterys Cav. (*Tetrapteris*). Malpighiaceae. 90 trop. Am. Hallucinogenic bark (cf. *Banisteriopsis*); some medic.

Tetrardisia Mez. Myrsinaceae. 5 W Mal.

Tetraria Pal. Cyperaceae. 35 S Afr., Aus. (4), Mal.

Tetrariopsis C.B. Clarke = *Tetraria*

Tetrarrhena R. Br. (~ *Ehrharta*). Gramineae (11). 6 Aus.

Tetraselago Jun. Scrophulariaceae. 4 S Afr. R: NRBGE 35(1977)175

Tetrasida Ulbr. Malvaceae. 2 Peru. R: Britt. 44(1992)444

Tetrasiphon Urban. Celastraceae. 1 Jamaica

Tetraspidium Baker. Scrophulariaceae. 1 Madagascar

Tetrastigma (Miq.) Planchon. Vitaceae. 90 Indomal. to trop. Aus. Decid. or evergreen dioec. lianes ± tendrils, hosts for *Rafflesia* spp. Cult. orn. housepls esp. *T. harmandii* Planchon (SE As. to Philippines) – ed. fr. for jellies etc. & *T. voinieranum* (Nichols. & Mottet) Gagnepain (chestnut vine, Laos) – screening pl. in Calif.; *T. lauterbachianum* Gilg (Papuasia) – liane to 40 m with aerial roots 10–40 m long used for weaving in Bismarck Is.

Tetrastylidium Engl. Olacaceae. 3 S Brazil

Tetrastylis Barb. Rodr. = *Passiflora*

Tetrasynandra Perkins. Monimiaceae (V 2). 3 NE Aus.

Tetrataenium (DC) Manden. Umbelliferae (III 11). 7–8 SW to C As.

Tetrataxis Hook.f. Lythraceae. 1 Mauritius: *T. salicifolia* (Tul.) Baker – only 7 individuals left of this tree rediscovered in 1970s

Tetratelia Sonder = *Cleome*

Tetrathalamus Lauterb. = *Zygogynum*

Tetratheca Sm. Tremandraceae. 39 Aus. R: Telopea 1(1976)139. Cult. orn. small shrubs (pink-eyes)

Tetrathylacium Poeppig. Flacourtiaceae (9). 6 trop. Am. *T. costaricense* Standley (Costa Rica) – ant-infested twigs in 68% of pls

Tetrathyrium Benth. (~ *Loropetalum*). Hamamelidaceae (I 1). 1 Hong Kong

Tetraulacium Turcz. Scrophulariaceae. 1 Brazil

Tetrazygia Rich. ex DC. Melastomataceae. 25 Florida & WI. Trees & shrubs, cult. orn. esp. *T. bicolor* (Miller) Cogn. with small fls but conspic. yellow stamens

Tetrazygiopsis Borh. = *Tetrazygia*

Tetroncium Willd. Juncaginaceae. 1 Straits of Magellan to 40° S in Andes: *T. magellanicum* Willd. – dioec., fls 2-merous

Tetrorchidiopsis Rauschert = seq.

Tetrorchidium Poeppig. Euphorbiaceae. 20 trop. Am. & Afr.

Teucridium Hook.f. Labiatae (II). 1 NZ

Teucrium L. Labiatae (II). 100 cosmop. esp. Medit. (Eur. 49). C with 1 5-lobed lip. Germander, esp. *T. chamaedrys* L. (Eur. to Cauc.) – locally medic. tea used since antiquity. Cult. orn. herbs & shrubs incl. *T. canadense* L. (N Am.) – poss. source of natural rubber; *T. marum* L. (cat thyme, W Medit. is.) – attractive to cats; *T. scordium* L. (Euras., Medit.) – source of yellow-green dye for cloth in Danube area; *T. scorodonia* L. (wood g. or w. sage, Eur., natur. N Am.) – form. medic.

Teuscheria Garay. Orchidaceae (V 10). 6 N S Am.

Texas blue grass *Poa arachnifera*; **T. millet** *Panicum texanum*; **T. star** *Lindheimera texana*

Teyleria Backer. Leguminosae (III 10). 3 SE As., Sumatra, Java. R: AUWP 85-1(1985)119

Thai basil *Ocimum tenuiflorum*; **T. eggplant** *Solanum melongena*

Thaia Seidenf. Orchidaceae (? tribe). 1 Thailand. Mycotrophic

Thalassia Banks ex C. Koenig. Hydrocharitaceae (Thalassiaceae). 1 Carib., 1 trop. Indian & Pacific Oceans. Submerged marine aquatics to 30m depth; nitrogen-fixing bacteria in rhizosphere. *T. hemprichii* (Solms) Asch. (OW) – mulch for paddy-fields & coconuts

Thalassiaceae Nakai = Hydrocharitaceae

Thalassodendron Hartog. Cymodoceaceae. 1 Red Sea, W Indian Ocean, E Mal., NE Aus.: *T. ciliatum* (Forssk.) Hartog with submarine poll. & disp. by free-floating detatched fruiting branchlets; 1 SW Aus.

Thaleropia P.G. Wilson (~ *Metrosideros*). Myrtaceae. 3 New Guinea, Queensland. R: AusSB 6(1993)257

Thalestris Rizz. = *Justicia*

Thalia L. Marantaceae. 7 trop. Am., *T. geniculata* L. ext. to trop. Afr. – pollen deposited on style in bud & insect proboscis touching style promotes explosive S-shaped movement, in 0.03 secs style becomes erect, scrapes pollen from proboscis into stigmatic hollow & deposits home pollen on proboscis; cult. orn.

Thalictrum L. Ranunculaceae (IV). 330 N temp. (most; Eur. 15), New Guinea, trop. Am., trop. & S Afr. R: BSRBB 24(1885)78. Meadow rue. Herbs with alks; C 0, A often coloured & conspic., insect-poll., others wind-poll. Some locally medic.; cult. orn. esp. *T. delavayi* Franchet (W China) – K red or lilac, A yellow

Thaminophyllum Harvey. Compositae (Anth.-Tham.). 3 SW Cape. R: JSAB 46(1980)157

Thamnea Sol. ex Brongn. Bruniaceae. 7 SW Cape

Thamnocalamus Munro (~ *Arundinaria*). Gramineae (1a). (Incl. *Fargesia*) 6 China (2), Himal. (3) S Afr. & Madag. (1). R: KBAS 13(1986)43. Forest bamboos. Cult. orn. esp. *T. spathaceus* (Franchet) Söderstrom (*A. murielae, A. s.*, umbrella bamboo, C China) – fls c. every century while *T. spathiflorus* (Trin.) Munro (NW Himal.) every 16 or 17 yrs; *T. tesselatus* (Nees) Söderstrom & Ellis (S Afr., Madag.) – canes used for Zulu shields

Thamnocharis W.T. Wang. Gesneriaceae. 1 SW China: *T. esquirolii* (Léveillé) W.T. Wang

Thamnochortus P. Bergius. Restionaceae. 33 SW Afr. to E Cape & Namaqualand. R: Bothalia 15(1985)471

Thamnojusticia Mildbr. = *Justicia*

Thamnosciadium Hartvig (~ *Seseli*). Umbelliferae (III 8). 1 Greece: *T. junceum* (Sm.) Hartvig. R: Willd. 14(1984)321

Thamnoseris F. Philippi. Compositae (Lact.-Dend.). 1 S Ambrosio Is. (Chile)

Thamnosma Torrey & Frémont. Rutaceae. 6 SW N Am. (2), S W Afr. deserts (2), Socotra(1), S Arabia & Somalia (1)

Thamnus Klotzsch (~ *Erica*). Ericaceae. 1 S Cape

thanaka *Murraya paniculata*

Thapsia L. Umbelliferae (III 12). 3 Medit. (Eur. 3). *T. garganica* L. (Spanish turpeth root, W & S Medit.) – resin used in plasters

Thaspium Nutt. Umbelliferae (III 8). 3 N Am.

Thaumasianthes Danser. Loranthaceae. 2 Philippines

Thaumastochloa C. Hubb. Gramineae (40h). 7 Aus. to New Guinea. R: GBS 36(1983)137

Thaumatocaryon Baillon. Boraginaceae. 4 Brazil

Thaumatococcus Benth. Marantaceae. 1 trop. W Afr.: *T. daniellii* (Bennett) Benth. – aril contains thaumatin a protein with a taste 1600 times as sweet as sucrose; thought a promising sweetener but breaks down on heating food, grown in tissue culture & genetic engineering using bacteria attempted but now dipeptides synthesized from aspartate used in preference to natural product

Thaumatophyllum Schott = *Philodendron*

Thayeria Copel. = *Aglaomorpha*

Theaceae Mirbel ex Ker-Gawler. Dicots – Dilleniidae – Theales. 22/610 trop. with few warm temp. Usu. evergreen trees & shrubs, few lianes, hairs usu. unicellular. Lvs simple, entire to toothed, usu. spirally arr.; stipules 0. Fls usu. large & bisexual, hypogynous to (rarer) epigynous, usu. solit. & axillary but s.t. in term. racemes or panicles (esp. Asteropeioideae), bibracteolate; K (4)5(–7), imbricate & usu. basally connate, s.t. persistent (accrescent & wing-like in fr. in *Asteropeia*), C (4)5(–∞), imbricate, sometimes basally connate, s.t. not clearly distinguishable from K, A usu. ±∞ developing centrifugally, free or basally connate in a ring or in 5 bundles opp. & adnate to C (rarely 10 or even 5 (*Eurya*)), often basally nectariferous, anthers with longit. slits (term. pore-like slits in *Eurya*), G ((2)3–5(–10)) with as many loc. (united only basally in some *Camellia*) & axile placentation, styles s.t. basally united, each loc. with (1)2–±∞ anatropous or weakly campylotropous bitegmic ovules. Fr. a usu. loculicidal capsule or indehiscent & s.t. fleshy; 1–∞ mesotestal seeds per loc., embryo straight or curved in little or 0 endosperm (copious oily & proteinaceous in *Visnea*). x = 15, 18, 21, 25

Classification & principal genera (after Cronquist, but Bonnetioideae (C convolute, capsule septicidal, seed structure diff. etc.) here in Guttiferae):

1. **Asteropeioideae** (fr. dry-indehiscent; K accrescent & wing-like in fr., Madag.):

Asteropeia (only) – sometimes treated as Asteropeiaceae

2. **Theoideae** (fr. a capsule or drupe; K not accrescent; anthers usu. small & versatile; embryo usu. straight): *Camellia, Gordonia, Stewartia* etc.
3. **Ternstroemioideae** (fr. baccate or dry-indehiscent; K not accrescent; anthers usu. long & basifixed; embryo usu. ± curved): *Adinandra, Eurya, Ternstroemia*

Limits of fam. unclear & related Pellicieraceae, Pentaphylacaceae & Tetrameristaceae here kept distinct but fam. usu. considered close to Dilleniaceae which have arils (arils in *Anneslea*, T.) & raphides unlike T.

Tea is *Camellia sinensis*, other spp. of C. oil-sources & cult. orn. as are spp. of *Cleyera, Eurya, Franklinia, Gordonia, Stewartia, Ternstroemia* etc.; some timbers (*Anneslea, Schima*)

Thecacoris A. Juss. Euphorbiaceae. 20 trop. Afr., Madagascar

Thecagonum Babu (*Gonotheca*) = *Oldenlandia*

Thecanthes Wikström. (~ *Pimelea*). Thymelaeaceae. 5 Philippines to N Aus. (3). R: Nuytsia 6(1988)262

Thecocarpus Boiss. Umbelliferae (III 1). 2 Iran. Mericarps not separating

Thecophyllum André = *Guzmania*

Thecopus Seidenf. (~ *Thecostele*). Orchidaceae (V 9). 2 Indomal. R: OB 72(1983)101

Thecorchus Bremek. Rubiaceae (IV 1). 1 trop. Afr.

Thecostele Reichb.f. Orchidaceae (V 9). 2 Burma to W Mal. R: OB 72(1983)98

theetsee *Gluta usitata*

Theileamea Baillon = *Chlamydacanthus*

Theilera E. Phillipps (~ *Wahlenbergia*). Campanulaceae. 1 SW Cape

Thelasis Blume. Orchidaceae (V 14). 20 Indomal. to S China & Solomon Is.

Thelechitonia Cuatrec. (~ *Wedelia*). Compositae (Helia.-Verb.). 4 E As., warm Am. R: Phytologia 72(1992)142

Theleophyton (Hook.f.) Moq. Chenopodiaceae (1). 1 SE Aus., NZ

Thelepaepale Bremek. = *Strobilanthes*

Thelepogon Roth ex Roemer & Schultes. Gramineae (40d). 1 Ethiopia to India: *T. elegans* Roth ex Roemer & Schultes – alks, local veterinary medic.

Thelesperma Less. Compositae (Helia.-Cor.). 15 W N Am., S S Am. Involucral bracts in 2 rows, the inner united in lower ⅓ or more; pappus of 2 recurved barbed awns like *Bidens*. Cult. orn. (*Cosmidium*) esp. *T. burridgeanum* (Regel, Koern. & Rach) S.F. Blake (Texas)

Thelethylax C. Cusset. Podostemaceae. 2 Madagascar

Theligonaceae Dumort. = Rubiaceae

Theligonum L. Rubiaceae (IV 14; Theligonaceae). 3 Macaronesia & Medit. (Eur. 1), SW China & Japan. Form. assoc. with Haloragidaceae, Hippuridaceae, Portulacaceae etc. Seeds disp. by ants which feed on an elaiosome attached to seed & derived from pericarp. *T. cynocrambe* L. (Medit.) – young shoots a veg., form. laxative

Thelionema R. Henderson (~ *Stypandra*). Phormiaceae (Liliaceae s.l.). 3 E Aus.

Thellungia Stapf (~ *Eragrostis*). Gramineae (31d). 1 NE Aus.: *T. advena* Stapf (coolibah grass)

Thellungiella O. Schulz = *Arabidopsis*

Thelocactus (Schumann) Britton & Rose. Cactaceae (III 9). 11 SW US, N & C Mex. R: Bradleya 5(1987)49. Cult. orn. small ovoid cacti with large fls

Thelychiton Endl. = *Dendrobium*

Thelycrania (Dumort.) Fourr. = *Cornus*

Thelymitra Forster & Forster f. Orchidaceae (IV 1). 46 Mal., Aus. (37), New Caled., NZ (12). Some cult. orn. terrestrial orchids. *T. antennifera* Hook.f. (rabbit-ears, S Aus.) – general 'mimic' of yellow- or cream-flowered spp. of *Hibbertia, Goodenia* etc., brown 'ears' on top of column

Thelypodiopsis Rydb. (~ *Thelypodium*). Cruciferae. 17 N Am. to Guatemala. R: CGH 212(1982)74, 214(1984)26

Thelypodium Endl. Cruciferae. (excl. *Thelypodiopsis*) c. 3 W N Am. R: AMBG 9(1922)233

Thelypteridaceae Ching ex Pichi-Serm. Filicopsida. 6/c. 900 subcosmop. (Eur. 5 spp., Mal. c. 440, NW c. 300, Afr. 55). Terr. ferns, sometimes with upright trunks, with simple to pinnate (rarely more divided) fronds, the pinnae almost equal in length; veins simple. less often forked; unicellular acicular hairs on adaxial surfaces. Sori on abaxial surface of veins, ± indusium. x = 27, 29–36

Genera: *Craspedosorus, Cyclosorus* (most form. recog. genera now incl. here), *Macrothelypteris, Phegopteris, Thelypteris*

Poss. closest to Cyatheaceae

Thelypteris Schmidel. Thelypteridaceae. (s.l.) c. 280 trop. & temp., (incl. *Amauropelta, 'Lastrea', Metathelypteris, Parathelypteris*) in 6 subgg. (s.s. 2 subpantemp.)

Thelyschista Garay. Orchidaceae (III 3). 1 Brazil

Themeda Forssk. Gramineae (40g). 18 trop. OW. S.t. forming the principal cover in tropical fire-climax 'steppe' as *T. triandra* Forssk. (rooigras) in trop. & S Afr., *T. australis* (R. Br.) Stapf (~ *T. triandra*). & other spp. in Aus. (kangaroo grass); *T. gigantea* (Cav.) Duthie (ulla, trop. As.) – poss. use in paper-making; *T. villosa* (Poiret) A. Camus (Indomal.) – huge cult. orn.

Themistoclesia Klotzsch. Ericaceae. 25 N Andes

Thenardia Kunth. Apocynaceae. 4 C Am.

Theobroma L. Sterculiaceae. 20 trop. Am. R: CUSNH 35(1964)379. Trees with cauliflorous infl. & usu. large woody fr. of monkey-disp. seeds; some cult. for cocoa esp. *T. cacao* L. (cacao, Andean foothills) – cult. since antiquity in Am. for seeds, the source of (addictive, esp. in women) chocolate cont. stimulating alks (up to 4 'squares' of chocolate improve mind action & strengthen immune responses but more reverses this) incl. theobromine, caffeine & theophylline (muscle stimulant) & over 700 compounds (incl. phenolics (as in red wine) efficacious in slowing fat build-up in arteries but also tyramine & phenylethylamine responsible for migraine in some by causing platelets to clump, releasing serotonin which constricts blood vessels reducing blood to brain) analysed but still unclear what is responsible for flavour, the butter also used medic. & in cosmetics though causing urticaria in some, the seeds currency in Yucatán until 1850 & still valued in 1923; poll. by biting midges bred in the decaying fr., now much cult. in W Afr., Mal., Brazil etc. Chocolate houses were fashionable in GB before coffee h. but the way of separating the fat from the drink had not been perfected before introduction; milk chocolate is C19 invention, deriving from a glut of milk & thus milk-powder used in its manufacture. Consumption in UK equivalent to 9 kg per person per annum: industry based on confectionery utilizing chocolate derived from numerous cvs (s.t. polyembryonous) of *T. cacao* s.t. crossed with, or the product mixed with that of, *T. bicolor* Bonpl. (tiger c.), *T. angustifolium* DC (monkey c.), *T. grandiflorum* (Sprengel) Schumann (cupuaçu) worth £3 billion per annum (1993) in UK alone

Theophrasta L. Theophrastaceae. 2 Hispaniola. R: NJB 7(1987)529. Pachycaul trees with serial buds in axils & thorny scales in upper parts of stems; sapromyophilous fls in axils of scale lvs of buds

Theophrastaceae Link. Dicots – Dilleniidae – Primulales. 4/90 trop. Am. Trees & shrubs, often pachycaul, without resin-ducts. Lvs simple, entire or often spine-toothed, spirally arr., often in dense term. spirals; stipules 0. Fls usu. large & bisexual (polygamodioecy in *Clavija*) in usu. term. infls; K (4)5, basally connate in *Clavija*, C ((4)5) with short tube & imbricate lobes, both K & C marked with glandular dots or dashes, A opp. C attached to C-tube, filaments s.t. basally connate, anthers with longit. slits, the connective often shortly prolonged, whorl of staminodes alt. with A inserted further up C-tube, $\underline{G}$ (5), 1-loc. with term. style & $\pm\infty$ anatropous to s.t. hemitropous or almost campylotropous bitegmic ovules usu. borne on apically sterile free-central placenta in mucilage. Fr. a dry many-seeded berry or rarely a 1-seeded drupe; seeds orangeish, large with straight embryo in copious oily endosperm. n = 18

Genera: *Clavija, Deherainia, Jacquinia, Theophrasta*

Some cult. orn.; fish-poison & timber (*Jacquinia*)

Thereianthus G. Lewis. Iridaceae (IV 2). 7 SW Cape

Theriophonum Blume. Araceae (VIII 6). 5 C & S India, Sri Lanka. R: KB 37(1982)277

Thermopsis R. Br. Leguminosae (III 30). 13 E As., 10 N Am., usu. montane. Cult. orn. with alks

Therocistus Holub = *Tuberaria*

Theropogon Maxim. Convallariaceae (Liliaceae s.l.). 1 Himal.: *T. pallidus* (Kunth) Maxim. – cult. orn. ground-cover

Therorhodion (Maxim.) Small = *Rhododendron*

Thesidium Sonder. Santalaceae. 8 SW to E Cape

Thesium L. Santalaceae. 325 OW (Eur. 25). Herbaceous root-parasites with epiphyllous infls & alks, incl. *T. humifusum* DC (bastard toadflax, W Eur.). *T. humile* Vahl (Medit.) – s.t. a bad weed of barley in Iraq

Thesmophora Rourke. Stilbaceae. 1 SW Cape: *T. scopulosa* Rourke. R: EJB 50(1993)89

Thespesia Sol. ex Corr. Serr. Malvaceae. 17 (incl. *Azanza*) trop. R: P.A. Fryxell (n.d.) *Nat. hist. Cotton tribe*: 84. Some cult. orn. incl. *T. garckeana* F. Hoffm. (E & S Afr.) – fr. ed.; *T. lampas* (Cav.) Dalz. & A. Gibson (E Afr. to Philipp.) – fibre like *Crotalaria juncea*; *T. populnea* (L.) Corr. Serr. (pantrop., littoral) – hard wood for gunstocks, wheel-frames (Sri Lanka where used also for 'hopper' bowls) etc., street-tree, fls turning yellow to purple in 24 hrs

Thespesiopsis Exell & Hillc. = *Thespesia*

Thespidium F. Muell. Compositae (Pluch.). 1 trop. Aus.
Thespis DC. Compositae (Ast.-Ast.). 3 SE As.
Thevenotia DC. Compositae (Card.-Carl.). 2 SW As.
Thevetia L. Apocynaceae. 8 trop. Am. Lvs spirally arr. Cult. orn. trees (all but type (*T. ahouai* (L.) DC) sometimes considered a separate genus, *Cascabela*) esp. *T. peruviana* (Pers.) Schumann (*T. neriifolia*, yellow oleander, lucky bean or nut, trop. Am.) – rheophyte with thevetin (glucoside), heart-depressant, taken by suicides in S India
Thibaudia Ruíz & Pavón. Ericaceae. 60 trop. Am.
thickhead *Crassocephalum crepidioides*
Thieleodoxa Cham. = *Alibertia*
Thilachium Lour. Capparidaceae. 10 E Afr., Madagascar
Thiloa Eichler. Combretaceae. 3 N S Am.
thimbleberry *Rubus occidentalis*
thingadu *Parashorea stellata*
thingan *Hopea odorata*
Thinopyrum Löve (~ *Elymus*). Gramineae (23). 20 Euras. *T. bessarabicum* (Săvul. & Rayss.) Löve (W As.) – poss. salt-tolerant cereal
Thinouia Triana & Planchon. Sapindaceae. 12 warm S Am.
thinwin *Millettia* sp. (*Pongamia pinnata*)
Thiseltonia Hemsley (~ *Hyalosperma*). Compositae (Gnap.-Ang.). 2 W Aus. R: OD 104(1991)125
Thismia Griffith. Burmanniaceae. c. 29 trop. As., Aus. & NZ (1), trop. Am. & nr. Chicago (*T. americana* N. Pfeiffer – extinct). Mycotrophs
Thismiaceae J. Agardh = Burmanniaceae
thistle usu. *Cirsium* or *Carlina* spp., (Bible) *Centaurea iberica*; **blessed t.** *Centaurea benedicta*; **bull t.** *Cirsium vulgare*; **Canadian t.** *C. arvense*; **carline t.** *Carlina vulgaris*; **cotton t.** *Onopordum acanthium*; **creeping t.** *Cirsium arvense*; **globe t.** *Echinops* spp.; **golden t.** *Scolymus hispanicus*; **holy t.** *Silybum marianum*; **marsh t.** *C. palustre*; **milk t.** *Sonchus* spp., *Silybum marianum*; **musk t.** *Carduus nutans*; **Russian t.** *Salsola kali*; **saffron t.** *Carthamus lanatus*, *C. tinctorius*; **St Barnaby's t.** *Centaurea solstitialis*; **Scotch t.** *Cirsium vulgare*, *Carduus nutans* but now tending to be applied to *Onopordum acanthium*; **sow t.** *Sonchus* spp. esp. *S. oleraceus*; **spear t.** *Cirsium vulgare*; **star t.** *Centaurea calcitrapa*
thitka *Pentace burmannica*
thitsi *Gluta usitata*
Thladiantha Bunge. Cucurbitaceae. 22 (excl. *Microlagenaria*) E As. to Mal. R: BBR 1(1981)61. Dioec. tendril-climbers with root-tubers; elaiophores attractive to bees. Some cult. orn.
Thlaspeocarpa C.A. Sm. Cruciferae. 2 S Afr.
Thlaspi L. Cruciferae. 60 N temp. (Eur. 27) & mts of N hemisph. Some Eur. & As. spp. acccum. nickel to over 1000 μg/g dry matter, some Eur. spp. zinc also. *T. arvense* L. (penny cress, Mithridate mustard, stinkweed, Eur., natur. N Am.) – form medic., seeds with 30–40% oil suitable for illumination
Thodaya Compton = *Euryops*
Thogsennia Aiello. Rubiaceae (I 7). 1 E Cuba, Hispaniola
Thomandersia Baillon. Acanthaceae. 6 trop. Afr.
Thomasia Gay. Sterculiaceae. 32 Aus. (all but 1 endemic SW)
Thompsonella Britton & Rose (~ *Echeveria*). Crassulaceae. 3 Mex. R: CSJAm 64(1992)37. Cult. orn.
Thomsonia Wallich = *Amorphophallus*
Thonningia Vahl. Balanophoraceae. 1 trop. Afr.: *T. sanguinea* Vahl – tubers sold in W Afr. markets for medic., fr. with alks. R: BJ 106(1986)367
Thoracocarpus Harling. Cyclanthaceae. 1 trop. S Am.
Thoracosperma Klotzsch (~ *Eremia*). Ericaceae. 10 S Cape
Thoracostachyum Kurz = *Mapania*
Thoreldora Pierre = *Glycosmis*
Thorella Briq. = *Caropsis*
thorn *Crataegus* spp.; **t. apple** *Datura stramonium*; **blackt.** *Prunus spinosa*; **buffalo** or **Cape t.** *Ziziphus mucronata*; **Christ's t.** *Z. spina-christi*, *Paliurus s.-c.*; **cockspur t.** *Crataegus crusgalli*; **crown of t.s** *Euphorbia milii*; **devil t.** *Tribulus terrestris*; **kangaroo t.** or **t. tree** *Acacia paradoxa*; **t. pear** *Scolopia ecklonii*
Thorncroftia N.E. Br. Labiatae (VIII 4b). 3 Transvaal
Thornea Breedlove & McClint. Guttiferae (I). 2 Mex., Guatemala
Thorntonia Reichb. = *Pavonia*

thorow-wax *Bupleurum rotundifolium*

Thottea Rottb. Aristolochiaceae (II 1). 26 (incl. *Apama*) Indomal. Some medic. (alpam root)

Thouarsiora Homolle ex Arènes = *Ixora*

Thouinia Poit. Sapindaceae. 28 Mexico to WI. Lianes

Thouinidium Radlk. Sapindaceae. 7 Mexico to WI

Thozetia F. Muell. ex Benth. Asclepiadaceae (III 4). 1 E Queensland: *T. racemosa* F. Muell. ex Benth.

Thrasya Kunth. Gramineae (34b). 19 trop. Am. R: ABV 14,4(1987)7

Thrasyopsis L. Parodi. Gramineae (34b). 2 S Brazil. R: Phyton 23(1983)101

Thraulococcus Radlk. = *Lepisanthes*

Threlkeldia R. Br. Chenopodiaceae (I 5). 2 Aus., coasts. R: FA 4(1984)230

thrift *Armeria maritima*

Thrinax Sw. Palmae (I 1a). 7 Carib. R: SCB 19(1975), ABH 3(1985)225. Some wind-poll. (rare in palms). Lvs for thatch; fibres. Cult. orn. esp. *T. parviflora* Sw. (thatch palm, Jamaica) – dried lvs used for decoration in temp. regions; *T. radiata* Lodd. ex Schultes & Schultes f. (*T. wendlandiana*, Florida to N S Am.) – fr. ed., fibre for mattress-stuffing

Thrincia Roth = *Leontodon*

Thrixspermum Lour. Orchidaceae (V 16). 140 Indomal. to Taiwan & W Pacific. Some cult. orn.

thrumwort *Damasonium alisma*

Thryallis C. Martius. Malpighiaceae. 3 warm Am.

Thryothamnus Philippi. Older name for *Junellia*

Thryptomene Endl. Myrtaceae (Lept.). 32 S, C & NE Aus.

Thuarea Pers. Gramineae (34b). 1 Indomal. to New Caledonia, coastal; 1 Madagascar

Thuja L. Cupressaceae. 5 E As., N Am. R: Phytol.M 7(1984)73. Arbor-vitae; monoec. cult. orn. & timber trees esp. *T. occidentalis* L. (white cedar, American a.-v., E N Am.) – soft fragrant wood used for fencing, railway-sleepers, etc., oil medic., planted as screens & windbreaks, many orn. cvs incl. varieg. & dwarf ones; *T. plicata* Donn ex D. Don (western red cedar, w. or giant a.-v., W N Am.) – timber weather-resistant & used for shingles, 'cedar' greenhouses & frames, bee-hives, boats, bungalows, ladders, totem poles though sawdust can cause asthma (plicatic acid), foliage used in floristry, introd. GB 1853 & now much planted for hedging, 'Gracilis' haploid (n = 11); *T. standishii* (Gordon) Carrière (Japanese a.-v., C Japan). See also *Platycladus*

Thujopsis Siebold & Zucc. ex Endl. Cupressaceae. 1 N Japan: *T. dolabrata* (L.f.) Siebold & Zucc. (hiba) – cult. orn., durable timber for construction, cabinet-work, railway sleepers etc.

Thulinia Cribb. Orchidaceae (IV 2). 1 Nguru Mts, Tanzania

Thunbergia Retz. Acanthaceae (Thunbergiaceae). 90 OW trop. Climbing or erect herbs & shrubs with bracteoles encl. K. Cult. orn. incl. *T. alata* Bojer ex Sims (black-eyed Susan, trop. Afr.) – precise origin unclear as long cult. in Afr., common greenhouse climber with cream, white or orange fls with darker middle, natur. As. & Mal.; *T. grandiflora* Roxb. (Bengal clock vine, N India) – liane with blue 2-lipped fls grown as arbour vine; *T. mysorensis* T. Anderson ex Beddome (Nilgiris, India) – similar but with red & yellow fls

Thunbergiaceae Bremek. = Acanthaceae (II)

Thunbergianthus Engl. Scrophulariaceae. 1 São Tomé (W Afr.), 1 trop. E Afr.

Thunbergiella H. Wolff = *Itasina*

thunderbolt See Gramineae

Thunia Reichb.f. Orchidaceae. c. 5 Himal. to Burma. Cult. orn. terr. esp. *T. alba* (Lindley) Reichb.f. (incl. *T. marshalliae* B.S. Williams)

Thuranthos C.H. Wright. Hyacinthaceae (Liliaceae s.l.). 3 S Afr.

Thurberia A. Gray = *Gossypium*

Thurnia Hook.f. Thurniaceae. 2 Amazon basin & Guyana

Thurniaceae Engl. Monocots – Commelinidae – Juncales. 1/2 N S Am. Tough perennial herbs with upright rhizomes & no sec. growth; vessels in all veg. organs. Lvs with sheathing base & long leathery flat or canaliculate lamina s.t. with marginal prickles, basal, spirally arr. Fls bisexual, small, in dense term. racemose heads subtended by leafy bracts, wind-poll.; P 3 + 3, chaffy, A 3 + 3, filament ± adnate to P-base, anthers basifixed with longit. slits, pollen (monoporate) in tetrads, G̲ (3), 3-loc., each loc. with 1 or more erect anatropous ovules. Fr. a loculicidal capsule with 1 seed per loc.; seeds hispid with processes at both ends, embryo small, ± cylindric, embedded in copious mealy starchy endosperm

Genus: *Thurnia*; kept separate from Juncaceae because leaf vasc. bundles are paired 1 above another with phloem strands adjacent (unique in monocots), presence of silica-bodies in lvs as well as seed-characters above

Thurovia Rose (~ *Gutierrezia*). Compositae (Ast.-Sol.). 1 S US

Thurya Boiss. & Bal. Caryophyllaceae (II 1). 1 SW As.: *T. capitata* Boiss. & Bal.

Thuspeinanta T. Durand. Labiatae (VI). 2 C As. to Afghanistan & Iran

thuya *Tetraclinis articulata*; see also *Thuja*

Thylacanthus Tul. Leguminosae (I 4). 1 Amazonian Brazil

Thylacodraba (Náb.) O. Schulz = *Draba*

Thylacophora Ridley = *Riedelia*

Thylacopteris Kunze ex J. Sm. Polypodiaceae (II 5). 2 Mal. R: Blumea 39(1994)351

Thylacospermum Fenzl. Caryophyllaceae (II 1). 1 C As. to Himal. & W China: *T. rupifragum* Schrenk

Thymbra L. Labiatae (VIII 2). 4 Medit. (Eur. 3) to SW As. R: AJBM 44(1987)348. Cult. orn. shrubs

thyme *Thymus vulgaris*; **basil t.** *Clinopodium acinos*; **caraway t.** *T. herba-barona*; **cat t.** *Teucrium marum*; **lemon t.** *T.* × *citriodorus*; **wild t.** *T. polytrichus* subsp. *britannicus*

Thymelaea Miller. Thymelaeaceae. 30 N temp. OW (Eur. 17) esp. Medit. R: NRBGE 38(1980)89. *T. hirsuta* (L.) Endl. (*Passerina h.*, Medit.) – dioec. or monoec. with male fls first, bark fibre for rope & (since 1979) paper in Israel; *T. tatonraira* (L.) All. (Medit.) – ship cordage in Anc. Greece

Thymelaeaceae Juss. Dicots – Dilleniidae – Euphorbiales. 53/750 cosmop. esp. Aus. & trop. Afr. Toxic trees & shrubs, rarely lianes or herbs, producing glycosides & accum. daphnin (a coumarin) or related substances; usu. internal phloem next to pith. Lvs simple, entire, s.t. ericoid or even sheath-like (*Struthiola*), spirally arr., opp. or ± whorled; stipules ± 0. Fls usu. bisexual, reg. or not, often ± perigynous with hypanthium s.t. coloured, in racemes or heads or solit., (3)4 or 5(6)-merous; K imbricate or valvate lobes on hypanthium or ± free (fl. hypogynous), C small & often scale-like in throat of hypanthium, alt. with K or pairs opp. K (∞ in Gonystyloideae) or often 0, A opp. K or in 2 whorls or ∞ (Gonystyloideae) or even 2 (*Pimelea*), filaments short or 0, anthers with longit. slits, disk often present around $\underline{G}$ (2–5(–12)) with as many loc. or (G 2) pseudomonomerous & 1-loc., style often eccentric, each loc. with 1 pendulous anatropous to hemitropous bitegmic ovule. Fr. dry-indehiscent, baccate or drupaceous, less often a loculicidal capsule; seeds often carunculate, embryo oily, straight in (0 or copious) endosperm. x = 9 (often)

Classification & principal genera (after Cronquist):

1. **Gonystyloideae** (capsule; C usu. ∞, A 8–∞; disk usu. 0; hypanthium ± 0; lvs often dotted; internal phloem 0): *Gonystylus*
2. **Aquilarioideae** (capsule; C small or 0; A 5 or 10; disk usu. 0; lvs undotted; internal phloem 0): *Aquilaria*
3. **Synandrodaphnoideae** (fr. indehiscent; hypanthium ± 0): *Synandrodaphne* (only)
4. **Thymelaeoideae** (fr. indehiscent, disk usu. present; hypanthium ± well-developed): *Daphne, Daphnopsis, Gnidia, Passerina, Pimelea, Struthiola, Thymelaea, Wikstroemia*

Affinities unclear, former position in Myrtales 'intuitive & arbitrary' (Cronquist) & not supported by embryological studies, greater similarity with Malvales but most allied (?) to Euphorbiales; Gonystyloideae with distinctive exine structure once thought perhaps close to Sphaerosepalaceae, i.e. Ochnaceae (Diegodendraceae) & sometimes, like Aquilarioideae, seg. as separate fam.

Timber (*Gonystylus* – ramin, *Gyrinops*), incense (*Aquilaria, Wikstroemia*), bark fibre for paper (*Daphne, Edgeworthia, Thymelaea, Wikstroemia*), cordage (*Dais, Daphne, Daphnopsis, Dirca, Gyrinops, Thymelaea*) or ornament (*Lagetta*) & cult. orn. esp. *Daphne, Dirca, Pimelea* etc.

Thymocarpus Nicolson, Steyerm. & Sivadasan = *Calathea*

Thymophylla Lagasca (~ *Dyssodia*). Compositae (Hele.-Pect.). 17 SW US & Mex. *T. tenuiloba* (DC) Small (*D.t.*, Dahlberg daisy) – strongly scented cult. orn. bedding pl.

Thymopsis Benth. Compositae (Hele.-Cha.). 2 WI

Thymus L. Labiatae (VIII 2). c. 350 temp. Euras. (Eur. 66). Herbs & shrubs, gynodioec., protandrous. Cult. orn. & flavourings esp. *T. vulgaris* L. (thyme, W Medit. to SE Italy, gynodioec., with females prod. more seeds than hermaphr. & on burnt sites percentage of females high at first but then declining) – dried lvs used to flavour meat dishes, sausages etc., oil medic. (suppressing lung infection, easing bronchial spasms & helping sufferers to expel phlegm from lungs); also cult.: *T.* × *citriodorus* Pers. (lemon thyme = *T. pulegioides* L. (Eur.) × *T. vulgaris*), *T. herbabarona* Lois. (caraway t., Sardinia & Corsica), *T. polytrichus*

A. Kerner ex Borbas (Eur.) esp. subsp. *britannicus* (Ronn.) Kerguélen (subsp. *arcticus*, *T. drucei*, '*T. praecox*', wild t., W & N Eur.) – oil medic. (serpolet oil)

Thyrasperma N.E. Br. = *Hymenogyne*

Thyridachne C. Hubb. Gramineae (34b). 1 trop. Afr.

Thyridocalyx Bremek. Rubiaceae (IV 8). 1 Madagascar

Thyridolepis S.T. Blake. Gramineae (34a). 3 arid Aus.

Thyrocarpus Hance. Boraginaceae. 3 China

Thyrsanthella Pichon. Apocynaceae. 1 SE US

Thyrsanthemum Pichon. Commelinaceae (II 1f). 3 Mexico

Thyrsanthera Pierre ex Gagnepain. Euphorbiaceae. 1 SE As.

Thyrsia Stapf = *Phacelurus*

Thyrsodium Salzm. ex Benth. Anacardiaceae (IV). 6 trop. Am. R: Britt. 45(1993)115

Thyrsopteridaceae C. Presl. See Dicksoniaceae

Thyrsopteris Kunze. Dicksoniaceae (Thyrsopteridoideae). 1 Juan Fernandez (400–700 m): *T. elegans* Kunze – long-lived in cult. Fossils referred to the genus widespread, *T. elegans* a relic

Thyrsosalacia Loes. Celastraceae. 1–2 W trop. Afr.

Thyrsostachys Gamble. Gramineae (1b). 2 Burma & Thailand. *T. siamensis* Gamble – stems used for umbrella handles; *T. oliveri* Gamble – flowering cycle of 48 yrs

Thysanella A. Gray = *Polygonella*

Thysanocarpus Hook. Cruciferae. 4–5 W US

Thysanoglossa Porto & Brade. Orchidaceae (V 10). 2 Brazil

Thysanolaena Nees. Gramineae (26). 1 SE As.: *T. latifolia* (Hornem.) Honda (*T. maxima*, tiger grass) – panicles used as brooms

Thysanosoria Gepp (~ *Lomariopsis*). Lomariopsidaceae. 1 NW New Guinea: *T. pteridiformis* (Cesati) C. Chr. (? atavistic form of *Lomariopsis*)

Thysanostemon Maguire. Guttiferae (III). 2 Guyana

Thysanostigma Imlay. Acanthaceae. 2 S Thailand, Malay Pen. R: NJB 8(1988)227

Thysanotus R. Br. Anthericaceae (Liliaceae s.l.). 49 Aus., esp. SW, 2 ext. to New Guinea, 1 to China. R: FA 45(1987)308. Cult. orn. (fringed lilies)

Thysanurus O. Hoffm. = *Geigeria*

Thyselium Raf. (~ *Peucedanum*). Umbelliferae (III 10). 1 Eur., N As.

ti *Cordyline fruticosa*

Tianschaniella B. Fedtsch. ex Popov. Boraginaceae. 1 C As.

Tiarella L. Saxifragaceae. 3–7 mts & coasts of N Am. (R: AJB 24(1937)344), 1 E As. Cult. orn. rhizomatous herbs esp. *T. cordifolia* L. (coolwort, foamflower, E N Am.) – alleged diuretic

Tiarocarpus Rech.f. (~ *Cousinia*). Compositae (Card.-Card.). 3 Afghanistan

Tibestina Maire = *Dicoma*

Tibetia (Ali) Tsui = *Gueldenstaedtia* (but see BBLNEFI 5(1979)48)

Tibouchina Aublet. Melastomataceae. 243 trop. Am. Cult. orn. shrubs esp. *T. urvilleana* (DC) Cogn. ('*T. semidecandra*', glory bush, Brazil) – fls purple to violet with purple anthers

Tibouchinopsis Markgraf. Melastomataceae. 2 NE Brazil

tickseed *Bidens* spp.

Ticodendraceae Gómez-Laurito & Gómez P. Dicots – Hamamelidae – Fagales. 1/1 C Am. Dioec. (polygamodioec.) evergreen . Lvs serrate; stipules encircling stem, decid. Male fls in spike-like thyrses of 1–3-flowered cymules in trimerous whorls subtended by a bract, P 0, A 8–10 with longit. dehiscence. Female fls solit. subtended by pair of bracteoles, P an inconsp. rim, $\overline{G}$ (2), 4-loc., each loc. with 1 hemitropous ovule, styles 2(3) stigmatic throughout. Fr. a drupe; seed with thin endosperm & massive oily embryo

Genus: *Ticodendron*

Most primitive wood in Fagales

Ticodendron Gómez-Laurito & Gómez P. Ticodendraceae. 1 S Mex. to C Panamá: *T. incognitum* Gómez-Laurito & Gómez P. – overlooked until 1980s

Ticoglossum Lucas Rodr. ex Halb. (~ *Odontoglossum*). Orchidaceae (V 10). 2 Costa Rica

Ticorea Aublet. Rutaceae. 6 Costa Rica to NE S Am.

Tidestromia Standley. Amaranthaceae (II 2). 3 SW N Am.

tidy-tips *Layia platyglossa*

Tieghemella Pierre. Sapotaceae (I). 2 trop. W Afr. *T. heckelii* Pierre ex A. Chev. (makoré, cherry mahogany, bacu, baku) – fr. eaten by elephant, timber a mahogany subs., oilseed for soap ('Dumori butter')

Tieghemia Balle = *Oncocalyx*

Tieghemopanax R. Viguier = *Polyscias*

Tienmuia Hu. Orobanchaceae. 1 E China

Tietkensia Short. Compositae (Gnap.-Ang.). 1 C & WC Aus.: *T. corrickiae* Short. R: Muelleria 7(1990)248

tigasco oil *Campnosperma* spp.

tiger cocoa *Theobroma bicolor*; **t. flower** *Tigridia pavonia*; **t. grass** *Thysanolaena latifolia*; **t.'s jaws** *Faucaria tigrina*; **t. lily** *Lilium lancifolium*; **t. nut** *Cyperus esculentus*; **t. pear** *Opuntia aurantiaca*; **t. wood** *Astronium fraxinifolium*, *Lovoa trichilioides*

Tigridia Juss. Iridaceae (III 4). c. 35 C & S Am. R: UCPB 54(1970)1. Elaiophores attractive to bees in some. Cult. orn. bulbous herbs esp. *T. pavonia* (L.f.) Ker-Gawler ((peacock) tiger fl., Mex., natur. Guatemala etc.) – lvs plicate, wild pl. with red fls to 15 cm across (yellow & purple spots in tube), starchy bulbs eaten since Aztec times, selected cvs with white, yellow, pink & orange fls, each lasting only 8–12 hrs

Tigridiopalma C. Chen. Melastomataceae. 1 Kwangtung

Tikalia Lundell = *Blomia*

Tilia L. Tiliaceae. 45 N temp. (Eur. 5). Lime, linden. R: Plantsman 5(1984)206. Decid. trees with 2-ranked lvs (some with domatia) & fragrant fls on infl. emergent from a large bract; imp. timbers, cult. orn., bee-fodder (though some sugars (mannose) toxic in excess, so dead bees (often bumblebees) often found below flowering trees), phloem bast used for cordage, ed. oil extractable from fr. (endosperm) used in comm. cough mixtures. Timber pale & used for piano-keys & decorative carving as by the incomparable Grinling Gibbons (1648–1721). *T. americana* L. (American lime or basswood, whitewood, C & E N Am.) – wood for cheap furniture & excelsior, inner bark for mats etc.; *T. cordata* Miller (small-leaved l., pry, Eur.) – form. imp. tree in lowland England, characterizing the original forest as the last major tree to enter the island after the last Ice Age & not reaching Ireland & Scotland (seeds sterile in NW England where too cold for pollen-tube growth for successful fert.), timber for tables, plates, spoons, musical instruments, excelsior, charcoal, fls used in a medic. tea also used as mouthwash, bark for 'Archangel' mats; *T.* × *europaea* L. (*T.* × *vulgaris*, common l., *T. cordata* × *T. platyphyllos* Scop. (Eur. to SW As.), natural hybrid) – much cult. street-tree but disagreeable because aphids secrete large amounts of honeydew on pavements & parked motors at a rate of 1 kg of sugars per m^2 per annum, poss. stimulating growth of nitrogen-fixing bacteria in ground around tree & thus enhancing its nitrogen availability at the cost of some carbohydrate taken from the phloem by the aphids, 'Pallida' with larger lvs the linden of Unter den Linden, Berlin; *T. japonica* (Miq.) Simonkai (Japanese l., Japan) – much like Eur. but smaller; *T. tomentosa* Moench (SE Eur., As.) esp. 'Petiolaris' (silver or weeping l.) with pendent branches & fls opening later than *T. cordata*

Tiliaceae Juss. Dicots – Dilleniidae – Malvales. 46/680 subcosmop. Trees & shrubs, rarely herbs often with stellate hairs or peltate scales; mucilage cells & cavities present. Lvs usu. spirally arr. (s.t. lying distichously at maturity), simple to lobed & usu. palmately-veined; stipules often decid. Fls usu. bisexual, reg., s.t. with epicalyx, in cymes, paired or solit.; K (3–)5 valvate, s.t. basally connate, C alt. with K, imbricate, valvate, convolute or 0, tufts of glandular hairs serving as nectaries, A (10–) ± ∞ developing centrifugally, filaments s.t. joined in 5 or 10 groups, s.t. 5 or more staminodes, anthers with longit. slits or apical pores, $\underline{G}$ ($\overline{G}$ in *Neotessmannia*), (2 – ∞) with as many locs, rarely 1-loc. with incomplete partitions, style 1, each loc. with (1)2–several anatropous to almost orthotropous bitegmic ovules with zig-zag micropyles on usu. axile placentas. Fr. dry or not, dehiscent or not, seeds sometimes arillate (*Mortoniodendron*, *Westphalina*), embryo straight or with folded cotyledons in little to copious oily endosperm. x = 7–41

Principal genera: *Corchorus*, *Grewia*, *Microcos*, *Tilia*, *Triumfetta*

Although placed in Malvales by Cronquist this fam. has much in common with Dipterocarpaceae (esp. Pakaraimaeoideae; q.v.) form. in Theales, but some genera placed here or in Sterculiaceae (*Corchoropsis*) or Flacourtiaceae in Violales (*Hasseltia*, *Hasseltiopsis*, *Macrohasseltia*, *Pleuranthodendron*, *Prockea*) from which fam. *Muntingia* reacquired

Timbers (*Apeiba*, *Berrya*, *Entelea*, *Luehea*, *Pentace*, *Schoutenia*, *Tilia*), fibres (*Clappertonia*, *Corchorus* (jute), *Erinocarpus*, *Grewia*, *Heliocarpus*, *Tilia*, *Trichospermum*, *Triumfetta*) & cult. orn. esp. *Muntingia*, *Sparrmannia* & *Tilia*

Tiliacora Colebr. Menispermaceae (II). 22 OW trop. (Afr. 19, SE As., 2, Aus. 1). Alks. *T. triandra* (Colebr.) Diels (SE As.) – cordage, local medic., flavouring (Thailand)

Tilingia Regel & Tiling (~ *Ligusticum*). Umbelliferae (III 8). 3 N & E As., Alaska

Tillaea L. = *Crassula*

Tillandsia L. Bromeliaceae (2). (Excl. *Racinaea*) c. 380 trop. Am. R: FN 14(1977)665, 1392. *T. capillaris* Ruíz & Pavón – cleistogamy. Cult. orn. epiphytes with rosetted lvs & poor root-systems (nutrients being absorbed all over pls) but some with pendent stems esp. *T. usneoides* L. (Spanish or Florida moss, old man's beard, S Virginia to Arg., over 8000 km latitude, an almost unique distrib.) – hanging in festoons from trees & overhead wires like a lichen such as *Usnea*, the base attached to support but dying as apex grows downwards leaving an axile strand of sclerenchyma, the whole pl. absorbing water & nutrients (esp. calcium) dripping over it; rarely flowering but distrib. by wind as fragments & by birds using it as nesting material; dried pl. used as packing material & like horsehair in upholstery (e.g. car seats in Model T Ford), some 5000 t used per annum in US alone; *T. latifolia* Meyen (Peru & N Chile desert) – rootless & blown about

Tillospermum Salisb. = *Kunzea*

timbo *Lonchocarpus* spp. esp. *L. nicou*

Timonius DC. Rubiaceae (III 3). 150 Seychelles & Mauritius (2), Sri Lanka (1), Andamans (1), Mal. (esp. New Guinea) to Pacific. *T. timon* (Sprengel) Merr. (*T. rumphii*, Mal. to Aus.) – ed. fr.

timothy *Phleum pratense*

Timouria Rosch. = *Stipa*

Tina Schultes. Sapindaceae. 16 Madagascar

Tinantia Scheidw. Commelinaceae (II 1f). c. 10 Texas to trop. Am. Cult. orn. esp. *T. erecta* (Jacq.) Fenzl – fls pink to blue

tingo fibre *Pouzolzia mixta*

Tinguarra Parl. Umbelliferae (III 8). 2 Canary Is.

tinker's weed *Triosteum perfoliatum*

Tinnea Kotschy ex Hook.f. Labiatae (V). 19 trop. Afr. R: BT 70(1975)1. Some fish-poisons

Tinnevelly senna *Senna alexandrina*

Tinomiscium Miers ex Hook.f. & Thomson. Menispermaceae (IV). 1 Indomal. to SE As.: *T. petiolare* Miers ex Hook.f. & Thomson – alks (fr. a fish-poison but seeds ed.), medic. e.g. dental caries. R: KB 40(1985)542

Tinospora Miers. Menispermaceae (III). 32 trop. OW. R (not Afr.): KB 36(1981)379. Locally medic., *T. crispa* (L.) Hook.f. & Thomson (*T. rumphii*, Indomal. – used to flavour cocktails & cordials), *T. sinensis* (Lour.) Merr. (India to SE As.) & other spp. act as host-pls for noctuid moths attacking *Dimocarpus longan* & citrus & are typical of sec. forest so clearance of primary forest leads to their increase & more damage to fr. crops

Tintinnabularia Woodson. Apocynaceae. 1 C Am.

tipu *Tipuana tipu*

Tipuana (Benth.) Benth. Leguminosae (III 4). 1 Brazil, Bolivia & Arg.: *T. tipu* (Benth.) Kuntze (tipu) – widely planted street-tree in trop. with samara-like 1–3-seeded winged legumes & useful timber

Tipularia Nutt. Orchidaceae (V 8). 3 N Am. & As. Cult. orn. terr. orchids with corm & 1 leaf

Tiquilia Pers. (~ *Coldenia*). Boraginaceae (Ehretiaceae). 27 Am. deserts. R: Rhodora 79(1977)467

Tirania Pierre. Capparidaceae. 1 SE As.

tirite *Ischnosiphon arouma*

Tirpitzia Hallier f. Linaceae. 2 SW China, SE As.

tirucalli *Euphorbia tirucalli*

Tischleria Schwantes = *Carruanthus*

Tisonia Baillon. Flacourtiaceae (8). 14 Madagascar

Tisserantia Humbert = *Sphaeranthus*

Tisserantiodoxa Aubrév. & Pellegrin = *Englerophytum*

Tisserantodendron Sillans = *Fernandoa*

tisso flowers *Butea monosperma*

Titanopsis Schwantes (~ *Aloinopsis*). Aizoaceae. 6 S Afr. Cult. orn. stemless succ. with coloured lvs

Titanotrichum Soler. Gesneriaceae. 1 China & Taiwan: *T. oldhamii* (Hemsley) Soler. – cult. orn. terr. herb with fls sometimes replaced by green scale-like veg. propagules

Tithonia Desf. ex Juss. Compositae (Helia.-Helia.). 11 SW US to C Am. R: Rhodora 84(1982)453. Cult. orn. herbs & shrubs esp. *T. diversifolia* (Hemsley) A. Gray (Mexican sunflower) – natur. OW where much cult. as hedge-pl. esp. in Kenya, green manure in Sri Lanka

Tithymalus Gaertner = *Pedilanthus*

titoki *Alectryon excelsus*

Tittmannia Brongn. Bruniaceae. 4 SW Cape

tlanochtle *Lycianthes moziniana*

Tmesipteridaceae Nakai. See Psilotaceae

Tmesipteris Bernh. Psilotaceae (Tmesipteridaceae). c. 10 SE As. to (esp.) Aus. (tetraploids & octoploids) & NZ (tetraploids) to Tahiti. Terr. or often epiphytic & then often on tree-ferns with dichotomizing rhizome with rhizoids & mycorrhiza; aerial stem usu. unbranched, basally with scale lvs like *Psilotum* but elsewhere with bilaterally symmetrical (not dorsiventral) broadly lanceolate lvs to 2 cm long; prothalli like sporophytic rhizomes; n = 102–105, 204–210. All Aus. spp. protected

toad lily *Tricyrtis* spp.; **t. rush** *Juncus bufonius*

toadflax *Linaria* spp.; **bastard t.** *Thesium humifusum*; **common t.** *L. vulgaris*; **ivy-leaved t.** *Cymbalaria muralis*

tobacco *Nicotiana* spp. esp. *N. tabacum*; **Aztec t.** *N. rustica*; **Indian t.** *Lobelia inflata*; **mountain t.** *Arnica montana*; **t. plant** or **flowering t.** *N.* × *sanderae*; **wild t.** *N. rustica*

Tobagoa Urban. Rubiaceae (IV 15). 1 Panamá, Tobago, Venezuela

tobira *Pittosporum tobira*

tobosa grass *Hilaria mutica*

Tococa Aublet. Melastomataceae. 54 trop. Am. Some fish-disp. in Amazon

Tocoyena Aublet. Rubiaceae (II 1). 20 trop. Am. Alks – root an ipecacuanha subs.

Todaroa Parl. Umbelliferae (III 8). 1 Canary Is.

Toddalia Juss. Rutaceae. 1 OW trop.: *T. asiatica* (L.) Lam. – prickly scrambler, source of Lopez root, medic. (alks) & yellow dye

Toddaliopsis Engl. (~ *Vepris*). Rutaceae. 2–4 trop. Afr.

toddy palm wine esp. from *Arenga, Borassus, Caryota, Cocos, Elaeis, Hyphaene, Nypa, Phoenix* & *Raphia* spp. etc.; **t. palm** *Caryota urens*

Todea Willd. ex Bernh. Osmundaceae. 6 New Guinea (1), Aus., NZ, S. Afr., S temp. OW. *T. barbara* (L.) T. Moore (crape fern, Aus., NZ, S Afr.) – cult. orn. with massive erect rhizome & fronds over 1 m long, pls weighing up to 1.5 t, protected sp. in Aus.

Toechima Radlk. Sapindaceae. 8 Aus. (6, 5 endemic), New Guinea, Flores

Toelkenia P.V. Heath = *Crassula*

Tofieldia Hudson. Melanthiaceae (Liliaceae s.l.). Spp.? N temp. (Eur. 2). Tufted rhizomatous pls with colchicine; some cult. orn. incl. *T. pusilla* (Michaux) Pers. (Scotch asphodel, N cool OW) – 3-lobed involucre beneath K

tofu *Glycine max*

Tolbonia Kuntze (~ *Calotis*). Compositae (Ast.-Ast.). 1 SE As.

tollon *Heteromeles arbutifolia*

Tolmiea Torrey & A. Gray. Saxifragaceae. 1 W N Am.: *T. menziesii* (Hook.) Torrey & A. Gray (pickaback or piggyback pl.) – cult. orn. herb with plantlets developing at junction of petiole & lamina

Tolpis Adans. Compositae (Lact.-Hier.). 20 Macaronesia (most), Medit. (Eur. 5) to Ethiopia & Somalia with *T. barbata* (L.) Gaertner (Medit.) – widely natur.

tolu balsam *Myroxylon balsamum*

Tolumnia Raf. (~ *Oncidium*). Orchidaceae (V 10). 35 Am.

Tolypanthus (Blume) Reichb. Loranthaceae. 4 India to SE China

Tomanthea DC = *Centaurea*

Tomanthera Raf. = *Agalinis*

tomatillo *Physalis philadelphica*

tomato *Lycopersicon esculentum*; **cherry t.** *L. e.* var. *cerasiforme*; **currant t.** *L. pimpinellifolium*; **husk** or **strawberry t.** *Physalis pruinosa*

Tomentaurum Nesom (~ *Heterotheca*). Compositae (Ast.-Sol.). 1 Mex.: *T. vandevenderorum* Nesom. R: Phytol. 71(1991)128

Tommasinia Bertol. (~ *Peucedanum*) Umbelliferae (III 10). 1 Eur.: *T. verticillaris* (L.) Bertol.

Tonalanthus Brandegee = *Calea*

Tonduzia Pittier = *Alstonia*

Tonella Nutt. ex A. Gray. Scrophulariaceae. 2 W US. Cult. orn. annuals like *Collinsia* but C-lobes not 2-lipped

Tonestus A. Nelson (~ *Haplopappus*). Compositae (Ast.-Sol.). 7 W N Am. R: Phytol. 68(1990)177. some cult. orn.

Tongan oil *Calophyllum inophyllum*

Tongking cane *Arundinaria amabilis*

Tongoloa H. Wolff. Umbelliferae (III 8). 8–10 C As. to India

Tonina Aublet. Eriocaulaceae. 1 trop. Am.

Tonka bean *Dipteryx odorata*

Tonningia Necker ex A. Juss. = *Cyanotis*

Tontelea Aublet = *Elachyptera*

Tontelea Miers. Celastraceae. 30 trop. Am.

toon *Toona ciliata*

Toona (Endl.) M. Roemer. Meliaceae (IV 1). 4–5 Indomal. to N Aus. R: FM I,12(1995)358. Timber trees esp. *T. ciliata* M. Roemer (*Cedrela toona*, toon, Australian (or) red cedar, Indomal. to Aus.) – furniture & building; *T. sinensis* (A. Juss.) M. Roemer (*Cedrela s.*, China to Mal. mts) – similar, coffee-shade, street-tree in Eur.; *T. sureni* (Blume) Merr. (Mal.) – similar timber, bark medic.

toothache grass *Ctenium aromaticum*

toothbrush tree *Salvadora persica*

toothwort *Lathraea squamaria*

Toowoomba canary grass *Phalaris aquatica*

topee-tampo, topinambour, topi-tamboo *Calathea allouia*

Topobea Aublet. Melastomataceae. 62 trop. Am. Allied to *Blakea*

toquilla *Carludovica palmata*

torch ginger *Etlingera elatior*; **t. lily** *Kniphofia* spp.

Tordyliopsis DC. Umbelliferae (III 11). 1 Himal.: *T. brunonis* (Wallich) DC

Tordylium L. Umbelliferae (III 11). 18 Eur. (5), Medit., SW As. R: BJLS 97(1988)357. Lvs of *T. apulum* L. (Medit.) eaten as vegetable in Greece

Torenia L. Scrophulariaceae. 40 trop. OW (Afr. 4), S Am. (1). Cult. orn. esp. *T. fournieri* Linden ex Fourn. (wishbone fl., Vietnam) – pot-pl. with pale blue, purple & yellow fls, anthers shed pollen by lever action of flange-like outgrowths of lateral pollen-sac wall - when pressed causes buckling of wall nearby & pollen forced out (1–1.5 g force against 4 levers of anther pair releases <3000 grains

Toricellia DC = *Torricellia*

Torilis Adans. Umbelliferae (III 3). 15 Canary Is., Medit. (Eur. 6) to E As.. trop. & S Afr. *T. japonica* (Houtt.) DC (hedge parsley, Euras.)

tormentil *Potentilla erecta*

Torminalis Medikus = *Sorbus*

Tornabenea Parl. Umbelliferae (III 12). 3 Cape Verde Is.

Toronia L. Johnson & B. Briggs. Proteaceae. 1 NZ

Torralbasia Krug & Urban. Celastraceae. 2 WI

Torrenticola Domin. Podostemaceae. 1 New Guinea, Queensland

Torresea Allemão = *Amburana*

Torreya Arn. Taxaceae. 7 E As., S US (2). R: Phytol.M 7(1984)74. Dioec. & monoec. trees with drupe-like seeds covered in fleshy arils; fossils from mid-Jurassic referred here. *T. californica* Torrey (Californian nutmeg, Calif.) – cult. orn.; *T. nucifera* (L.) Siebold & Zucc. (kaya nut, Japan) – seeds ed., oil used for cooking in Japan; *T. taxifolia* Arn. (stinking cedar or yew, 3 counties of Florida, 1 site in Georgia) – timber for fencing, bruised foliage foetid, endangered sp.

Torreyochloa Church (~ *Puccinellia*). Gramineae (17). 4 N Am., NE As. R: KBAS 13(1986)99

Torricellia DC. Cornaceae (Torricelliaceae). 3 E Himalaya to W China

Torricelliaceae (Wangerin) Hu. See Cornaceae

Tortuella Urban. Rubiaceae (IV 15). 1 Isle Tortue (nr. Hispaniola)

Torularia O. Schulz = *Neotorularia*

Torulinium Desv. = *Cyperus*

totara *Podocarpus totara*

totora *Scirpus californicus*

Toubaouate Aubrév. & Pellegrin = *Didelotia*

Toubasuate Airy Shaw = *Didelotia*

Touchardia Gaudich. Urticaceae (III). 1 Hawaii: *T. latifolia* Gaudich. (olona) – fibre used for fishing-lines & nets

touch-me-not *Impatiens* spp.

tou-fou *Glycine max*

Toulicia Aublet. Sapindaceae. 14 N S Am.

Tournefortia L. Boraginaceae. 100 trop. & warm. R: JAA suppl. 1 (1991)65. Mostly lianes, fr. fleshy (some ed.); local purgatives & other med. See also *Argusia*

Tournefortiopsis Rusby = *Guettarda*

tournesol *Chrozophora tinctoria*

Tourneuxia Cosson. Compositae (Lact.-Scar.). 1 Algeria

Tournonia Moq. Basellaceae. 1 Colombia: *T. hookeriana* Moq.

Touroulia Aublet. Quiinaceae. 4 trop. S Am.

Tourrettia Foug. Bignoniaceae. 1 Andes to Mex.: *T. lappacea* (L'Hérit.) Willd. – annual climber with some characters like Pedaliaceae

tous-les-mois *Canna indica*

Toussaintia Boutique. Annonaceae. 3 trop. Afr.

Tovaria Ruíz & Pavón. Tovariaceae. 2 trop. Am.

Tovariaceae Pax. Dicots – Dilleniidae – Capparidales. 1/2 trop. Am. Foetid herbs & shrubs with mustard-oil glucosides. Lvs 3-foliolate, spirally arr.; stipules 0. Fls bisexual, reg., (6–)8(9)-merous in long term. racemes; K imbricate, decid., C sessile, A within lobed nectary-disk, anthers sagittate with longit. slits, $\underline{G}$ ((5)6(–8)), pluriloc., ± gynophore, style short, with ∞ ± campylotropous bitegmic ovules with zig-zag micropyles on expanded placenta. Fr. a berry with central placental mass separating from pericarp & ∞ small seeds with embryo curved around periphery in thin layer of oily endosperm. n = 14
Genus: *Tovaria*
Close to Capparidaceae but endosperm, pluriloc. ovary & high-mery different

Tovarochloa T. Macfarl. & But. Gramineae (21b). 1 Peruvian Andes

Tovomita Aublet. Guttiferae (III). 12–25 trop. Am. Fl. tea for treatment of diarrhoea

Tovomitidium Ducke. Guttiferae (III). 2 Brazil

Tovomitopsis Planchon & Triana (III). Guttiferae. 50 trop. Am.

tow waste fibre after preparation of flax, hemp, etc.

towai bark *Weinmannia racemosa*

tower mustard *Arabis glabra*

Townsendia Hook. Compositae (Ast.-Ast.). 25 W N Am. R: CGH 183(1957)1. Some cult. orn. herbs

Townsonia Cheeseman (~ *Acianthus*). Orchidaceae (IV 1). 2 Tasmania, NZ

Townsville stylo *Stylosanthes humilis*

Toxanthes Turcz. = *Millotia*

Toxicodendron Miller = *Rhus*

Toxicoscordion Rydb. = *Zigadenus*

Toxocarpus Wight & Arn. Asclepiadaceae (II). 40 trop. OW

Toxosiphon Baillon (~ *Erythrochiton*). Rutaceae. 4 trop. Am. R: Britt. 44(1992)117

toyon *Heteromeles arbutifolia*

Tozzia L. Scrophulariaceae. 1 Alps, Carpathians: *T. alpina* L. – hemiparasite, fly-poll.

trac *Dalbergia cochinchinensis*

Trachelanthus Kunze (~ *Lindelofia*). Boraginaceae. 3 Iran to C As.

Tracheliopsis Buser = *Campanula*

Trachelium L. Campanulaceae. 7 Medit. (Eur. 3). Cult. orn.

Trachelospermum Lemaire. Apocynaceae. 20 India to Japan, SE US. Cult. orn. with fragrant fls

Trachoma Garay = *Tuberolabium*

Trachomitum Woodson = *Apocynum* (but see AMBG 17(1930)156)

Trachyandra Kunth (~ *Anthericum*). Asphodelaceae (Liliaceae s.l.). 65 trop. & S Afr., Madag. (1). R: Bothalia 7(1962)669. Some tumbleweeds. *T. adamsonii* (Compton) Oberm. (S Afr.) – woody stem to 180 cm

Trachycalymma (Schumann) Bullock (~ *Asclepias*). Asclepiadaceae. 4 trop. Adr.

Trachycarpus H. Wendl. Palmae (I 1a). 4 (? +) Himal. region. R: Principes 21(1977)155. Usu. dioec. palms, the trunks often covered with marcescent lvs. Cult. orn. esp. *T. fortunei* (Hook.) H. Wendl. (*'T. excelsus'*, chusan, windmill palm, N Burma, C & E China) – hardy in GB, fibre (Chinese coir) of leaf-bases used for cordage, brooms & capes (NW Yunnan), lvs used for hats, fls for food, drugs from seeds, wax from fr.

Trachydium Lindley. Umbelliferae (III 5). 1 Himalaya: *T. roylei* Lindley

Trachylobium Hayne = *Hymenaea*

Trachymene Rudge (*Didiscus*). Umbelliferae (I 1a). 45 SE As. to Aus. (35), New Caled. & Fiji. Cult. orn. esp. *T. coerulea* Graham (blue lace flower, W Aus.)

Trachynia Link = *Brachypodium*

Trachyphrynium Benth. = *Hypselodelphys*

Trachypogon Nees. Gramineae (40b). 3 trop. Am., Afr.

Trachypteris André ex Christ. Pteridaceae (IV). 3 S Am., Galápagos, Madag. (*T. drakeana* Christ – sterile lvs with apical buds (veg. reproduction))

Trachyrhizum (Schltr.) Brieger = *Dendrobium*

Trachys Pers. Gramineae (34f). 1 S India & Sri Lanka to Burma, coastal

Trachyspermum Link. Umbelliferae (III 8). 15 trop. & NE Afr. to C As., India & W China. *T. ammi* (L.) Sprague (*T. copticum*, ajowan, ?As.) – fr. medic. & spice containing thymol, the principal flavour of 'Bombay Mix' of nuts, pulses & crisp sticks

Trachystemon D. Don. Boraginaceae. 2 Medit. (Eur. 1). *T. orientalis* (L.) G. Don f. (E Bulgaria & As. Minor) – cult. orn. perennial with bold foliage

Trachystigma C.B. Clarke. Gesneriaceae. 1 trop. Afr. Infls epiphyllous

Trachystoma O. Schulz. Cruciferae. 3 Morocco

Trachystylis S.T. Blake. Cyperaceae. 1 E Aus.

Tractocopevodia Raiz. & Naray. Rutaceae. 1 Burma

Tracyina S.F. Blake. Compositae (Ast.-Ast.). 1 California

Tradescantia L. Commelinaceae (II 1g). 70 Am. R: KB 35(1980)437 (incl. *Cymbispatha*, *Rhoeo*, *Setcreasea* & *Zebrina*); erect or trailing, rather pachycaul to v. slender cult. orn. esp. hardy *T. virginiana* L. (spiderwort, E N Am.) – erect with violet purple fls though most garden material is referable to *T.* × *andersoniana* W. Ludwig & Rohw. (invalid name, *T. v.* × *T. ohiensis* Raf. (blue jacket, E N Am.) × *T. subaspera* Ker-Gawler (E N Am.)) – many cvs; long staminal hairs, in which Robert Brown first observed & described protoplasmic streaming (1828), eaten by poll. insects. Many greenhouse or house pls ('wandering Jews'; see also *Gibasis*) of trailing habit incl. *T. albiflora* Kunth (S Am.) – fls white, lvs 3–4 times as long as wide, glabrous adaxially, *T. fluminensis* Vell. Conc. (SE Brazil) – similar but lvs c. twice as long as wide, aggressive weed in NZ & *T. zebrina* Loudon (*Zebrina pendula*, *T. p.*, Mex.) – fls red-purple, lvs with 2 silvery bands separated by green, glabrous adaxially, purple abaxially, these some of the commonest of all housepls. Other cult. spp. incl. *T. cerinthoides* Kunth (*T. blossfeldiana*, SE Brazil) – trailing & ascending purplish stems, densely white-villous; *T. pallida* (Rose) D. Hunt (E Mex.) – erect or sprawling esp. 'Purple Heart' ('Purpurea', *Setcreasea purpurea* Boom) with intense violet-purple lvs for bedding; *T. discolor* L'Hérit. (*Rhoeo d.*, *R. spathacea.*, boat lily, C Am., WI) – pachycaul usu. unbranched with lvs to 30 cm. See also *Callisia*

tragacanth *Astracantha* spp. esp. *A. gummifera*

Traganopsis Maire & Wilczek. Chenopodiaceae (III 3). 1 Morocco: *T. glomerata* Maire & Wilczek

Traganum Del. Chenopodiaceae (III 3). 2 N Afr., E Medit.

tragasol *Ceratonia siliqua*

Tragia L. Euphorbiaceae. 100 trop. & warm. Some sting

Tragiella Pax & K. Hoffm. (~ *Sphaerostylis*). Euphorbiaceae. 5 NE trop. to S Afr.

Tragiola Small & Pennell = *Gratiola*

Tragiopsis Pomel = *Stoibrax*

Tragoceros Kunth = *Zinnia*

Tragopogon L. Compositae (Lact.-Scar.). c. 110 temp. Euras. (Eur. 20), Medit. Taprooted herbs with monocot-like lvs & solit. capitula opening only in morning, so weedy *T. pratensis* L. (goat's-beard, Eur., natur. N Am.) known as Johnny-go-to-bed (-at-noon); *T. porrifolius* L. (salsify, vegetable oyster, S Eur.) – cult. for ed. root & for young flowering shoots ('chards'), parent with *T. dubius* Scop. (Eur.) of *T. mirus* Ownbey, an allopolyploid with multiple origins (cf. *Senecio cambrensis*, *Spartina anglica*)

Tragus Haller. Gramineae (33d). 6 OW trop. R: KB 36(1981)55; 1 ext. to Eur.

Trailliaedoxa W.W. Sm. & Forrest. Rubiaceae (III). 1 SW China

Transcaucasia Hiroe = *Astrantia*

Transvaal daisy *Gerbera jamesonii*

Trapa L. Trapaceae. c. 15 (or 1 polymorphic) C & SE Eur., As., Afr. Water chestnut; the fr. rich in starch & fat esp. *T. natans* L. (water caltrops, saligot, horn or Jesuit's nut, Euras., Afr., natur. N Am.) – used in GB for food in Neolithic, weedy as in Caspian where a threat to sturgeon feeding-grounds; *T. bicornis* Osbeck (ling nut, E As.) – staple in As.; *T. bispinosa* Roxb. (~ *T. bicornis*, singhara n., trop. As.)

Trapaceae Dumort. Dicots – Rosidae – Myrtales. 1/1–15 OW exc. Pacific. Annual aquatics sometimes free-floating; vasc. bundles with internal phloem. Submerged stem with ± opp. elongate dissected green organs ('lvs', 'stipules', 'photosynthetic roots') & adventitious roots. Lvs with elongate petiole with aerenchymatous float & rhombic blade on short aerial stem; stipules decid., cleft. Fls bisexual, reg., 4-merous, solit. in axils, aerial; K valvate, basally forming hypanthial tube & 2 or 4 persistent & accrescent as horns or spines on fr., C imbricate, A with short filaments & anthers with longit. slits, G (2), 2-loc., half-inf., sunk in receptacle & lying within 8-lobed cupular disk, each loc. with 1 pendulous anatropous bitegmic ovule. Fr. indehiscent with 1 loc. suppressed persistent stony endocarp; embryo with uneven cotyledons, 1 large & starchy kept in fr., other small scale-like growing out with plumule through pore left by dehiscence of style; endosperm 0. n = c. 18, 24

Genus: *Trapa*

Trapella Oliver. Pedaliaceae (Trapellaceae). 1–2 E As. Aquatics; $\overline{G}$

Trapellaceae Honda & Satisake = Pedaliaceae (Trapelloideae)

Trattinnickia Willd. Burseraceae. 11 N S Am. Resins coll. euglossine bees for their nests

Traubia Mold. Amaryllidaceae (Liliaceae s.l.). 1 Chile

Traunia Schumann = *Toxocarpus*

Traunsteinera Reichb. Orchidaceae (IV 2). 1 Eur., Medit.

Trautvetteria Fischer & C. Meyer. Ranunculaceae (II 3). 1 E As., W & S N Am.

traveller's joy *Clematis vitalba*; **t. palm** or **tree** *Ravenala madagascariensis*

Traversia Hook.f. Compositae (Sen.-Tuss.). 1 NZ

treacle mustard *Erysimum cheiranthoides*

Trechonaetes Miers = *Jaborosa*

Treculia Decne. ex Trécul. Moraceae (II). 3 trop. Afr. & Madag. R: BJBB 47(1977)378. *T. africana* Decne. ex Trécul (African breadfruit) – seeds ed., ground into flour

tree cotton *Gossypium arboreum*; **t. dahlia** *Dahlia excelsa*; **t. mallow** *Lavatera arborea*; **t. medick** *Medicago arborea*; **t. of heaven** *Ailanthus altissima*; **t. peony** *Paeonia suffruticosa*; **t. tomato** *Cyphomandra betacea*

trefoil, bird's-foot *Lotus corniculatus*; **hop t.** *Trifolium campestre*; **yellow t.** *Medicago lupulina*

Treichelia Vatke. Campanulaceae. 1 SW Cape

Trema Lour. Ulmaceae (II). 10–15 trop. & warm. Fast-growing pioneer trees with alks, incl. *T. orientalis* (L.) Blume (*T. guineensis*, OW) – lvs (mpesi) 8% dry wt. tannin so avoided by colobus monkeys but used for tanning fish-nets in W Afr. & elsewhere, wood for charcoal & fireworks, & *T. micrantha* (L.) Blume (guacimilla, Florida trema, trop. Am.) – growing in forest gaps at least 376 m^2 & reaching 13.5 m in 2 yrs, pre-Hispanic barkcloth now for tourism & export (San Pablito, Mex.) with soft timbers for tea-chests & matches, tea & coffee shade & poss. in soil conservation (as in S Afr.)

Tremacanthus S. Moore. Acanthaceae. 1 Brazil

Tremacron Craib. Gesneriaceae. 6 China

Tremandra R. Br. ex DC. Tremandraceae. 2 SW Aus.

Tremandraceae R. Br. ex DC. Dicots – Rosidae – Polygalales. 3/43 temp. Aus., esp. SW. Usu. ericoid shrubs often with winged stems. Lvs simple, sometimes v. small, spirally arr., opp. or whorled; stipules 0. Fls bisexual, reg., (3)4- or 5-merous, solit. in axils; K valvate, usu. free, C induplicate-valvate, disk with lobes opp. C in *Tremandra*, A twice C with basifixed anthers with single apical pore or small slit, $\underline{G}$ (2), 2-loc. with (apical-) axile placentas, each loc. with 1 or 2(3) pendulous, anatropous, bitegmic ovules. Fr. a compressed loculicidal (s.t. also septicidal) capsule; seeds s.t. hairy, arillate (exc. *Platytheca*), embryo straight, small, embedded in copious endosperm

Genera: *Platytheca, Tetratheca, Tremandra*

Tremastelma Raf. = *Lomelosia*

Trematocarpus A. Zahlbr. = *Trematolobelia*

Trematolobelia A. Zahlbr. ex Rock (~ *Lobelia*). Campanulaceae. 4 Hawaii. R: W. Wagner et al., *Man. Fl. Pls Hawaii* (1990)485

Trembleya DC. Melastomataceae. 11 S Brazil

Trepocarpus Nutt. ex DC. Umbelliferae (III 8). 1 S US

Tresanthera Karsten. Rubiaceae (I 7). 2 Venezuela, Tobago

Treutlera Hook.f. Asclepiadaceae (III 4). 1 E Himalaya: *T. insignis* Hook.f.

Trevesia Vis. Araliaceae. 12 Indomal. Cult. orn. shrubs with palmate leaflets attached to a lamina

Trevoa Miers ex Hook. Rhamnaceae. 6 Andes. Spiny trees with actinomycete root-symbionts incl. *T. trinervia* Miers ex Hook. (Chile) – bark used medic. on burns

Trevoria F. Lehm. Orchidaceae (V 10). 6 Colombia, Ecuador

Trewia L. (*Trevia*). Euphorbiaceae. 2 Himal. to Sri Lanka & Hainan. Drupes. *T. nudiflora* L. (India, Nepal) – fr. eaten by rhinoceros in rainy season in S Nepal & disp., soft timber for tea-chests, packing-cases etc., seed-oil like tung

Triadenum Raf. Guttiferae. 6–10 E As., E N Am.

Triadodaphne Kosterm. = *Endiandra*

Triaenanthus Nees = *Strobilanthes*

Triaenophora (Hook.f.) Soler. Scrophulariaceae. 2 China

Triainolepis Hook.f. Rubiaceae (IV 8). 2 trop. E Afr., Madagascar

Trianaea Planchon & Linden. Solanaceae (1). 4 trop. Am. *T. speciosa* (Drake) Soler. (Ecuador) – bat-poll.

Trianaeopiper Trel. Piperaceae. 18 N Andes

Trianoptiles Fenzl ex Endl. Cyperaceae. 3 S Afr.

Triantha (Nutt.) Baker (~ *Tofieldia*). Melanthiaceae (Liliaceae s.l.). 2 N Am. R: Novon 3(1993)278

Trianthema L. Aizoaceae (II). c. 17 warm esp. Aus. (12, 10 endemic). R: JAA 51(1970)453. Alks – some medic., soap, lvs used like spinach

Triaristella Brieger = *Trisetella*

Triaristellina Rauschert = *Trisetella*

Trias Lindley. Orchidaceae (V 15). 10 SE As (7 endemic in Thailand). R: BT 71(1976)19

Triaspis Burchell. Malpighiaceae. 18 trop. & S (2) Afr.

Tribelaceae (Engl.) Airy Shaw = Grossulariaceae

Tribeles Philippi. Grossulariaceae (Tribelaceae). 1 temp. S Am. Prostrate shrublet with contorted C

Triblemma (J. Sm.) Ching = *Diplazium*

Tribolium Desv. (~ *Lasiochloa*). Gramineae (25). 11 S Afr. bushland. R: KB 40(1985)795

Tribonanthes Endl. Haemodoraceae. 4 SW Aus. R: FA 45(1987)132

Tribroma Cook = *Theobroma*

Tribulocarpus S. Moore. Aizoaceae (III; Tetragoniaceae). 1 SW & NE Afr. R: KB 12(1957)348

Tribulopis R. Br. (~ *Tribulus*). Zygophyllaceae. 5 trop. Aus. Geocarpy (unique in fam.)

Tribulus L. Zygophyllaceae. 25 trop. & warm (Eur. 1), esp. dry Afr. R: Taeckholmia 9(1978)59. Fr. with G 3–5 with sharp spines separating when mature & dispersed on animals & feet (burnut). Weeds esp. *T. terrestris* L. (caltrops, devil's thorn, OW, widely natur.) – bad weed in Calif., where, after 50 yrs of herbicides, Californian Dept. of Agriculture introduced 2 weevil spp. from India (1961) & these fed selectively on *T. terrestris*, controlling it; ingestion in stock leads to hepatogenic photosensitization perhaps involving nitrate & selenium poisoning manifest as 'bighead'

Tricalistra Ridley. Convallariaceae (Liliaceae s.l.). 1 Malay Peninsula

Tricalysia A. Rich. ex DC. Rubiaceae (II 1). 95 trop. Afr., Madag. R: BJBB 49(1979)239, 52(1982)311, 53(1983)299, 57(1987)39. (As. spp. = *Diplospora*, *Discospermum*)

Tricardia Torrey. Hydrophyllaceae. 1 SW US

Tricarpelema J.K. Morton. Commelinaceae (II 2). 7 As., Afr.

Tricarpha Longpre = *Sabazia*

Triceratella Brenan. Commelinaceae (I 2). 1 Zimbabwe

Triceratorhynchus Summerh. Orchidaceae (V 16). 1 trop. E Afr.

Tricerma Liebm. = *Maytenus* (but see Wrightia 4(1971)158)

Trichacanthus Zoll. = *Blepharis*

Trichachne Nees = *Digitaria*

Trichadenia Thwaites. Flacourtiaceae (4). 1 Sri Lanka, 1 E Mal.

Trichantha Hook. (~ *Columnea*). Gesneriaceae. 70 trop. Am. Cult. orn.

Trichanthemis Regel & Schmalh. Compositae (Anth.-Can.). 9 C As.

Trichanthera Kunth. Acanthaceae. 2 N S Am. *T. gigantea* (Humb. & Bonpl.) Nees – fodder-tree in Colombia

Trichanthodium Sonder & F. Muell. (~ *Gnephosis*). Compositae (Gnap.-Ang.). 4 Aus. R: Muelleria 7(1990)213

Trichapium Gilli = *Clibadium*

Trichilia P. Browne. Meliaceae (I 4). 84 trop. (Am. 70 (R: FN 28(1981)25)., Afr. 14). For As. spp. see *Heynea*. Timber & oilseeds esp. *T. emetica* Vahl (Afr.) – timber (Cape mahogany) & seed-oil (mafurra or mafoureira tallow) used for candle- & soap-making; *T. moschata* Sw. (trop. Am.) – source of pameroon bark

Trichipteris C. Presl = *Cyathea*

Trichlora Baker. Alliaceae (Liliaceae s.l.). 1 Peru

Trichloris Fourn. ex Benth. (~ *Chloris*). Gramineae (33b). 2 S Am. *T. crinita* (Lagasca) L. Parodi – cult. orn.

Trichocalyx Balf.f. Acanthaceae. 2 Socotra

Trichocaulon N.E. Br. Asclepiadaceae (III 5). c. 15 S Afr., Madag. Cult. orn. succ.

Trichocentrum Poepping & Endl. Orchidaceae (V 10). 30 trop. Am. Cult. orn. epiphytes

Trichocereus (A. Berger) Riccob. = *Echinopsis*

Trichoceros Kunth. Orchidaceae (V 10). 5 N S Am. *T. antennifer* Kunth (Ecuador) – pseudocopulation by *Paragymnomma* flies

Trichochiton Komarov = *Cryptospora*

Trichocladus Pers. Hamamelidaceae (I 1). 4 trop. & S Afr.

Trichocline Cass. Compositae (Mut.-Mut.). 22 S Am., SW Aus. (1). R: Darwiniana 19(1975)618. Herbs; some used like tobacco, esp. *T. reptans* (Wedd.) Robinson (coro)

Trichocoronis A. Gray. Compositae (Eup.-Tric.). 2 SW N Am. R: MSBMBG 22(1987)188. Wholly aquatic

Trichocoryne S.F. Blake. Compositae (Hele.-Hym.). 1 Mex.: *T. connata* S.F. Blake

Trichocyamos Yakovlev. See *Ormosia*

Trichodesma R. Br. Boraginaceae. 45 trop. & warm OW. Alks (*T. africanum* (L.) Sm. (trop. Afr.) – local medic.). *T. zeylanicum* (Burm.f.) R. Br. (camel bush, trop. As. to Aus.) – reputedly favoured by introd. grazing camels in Aus., poss. comm. oilseed

Trichodiadema Schwantes. Aizoaceae (V). 30 Ethiopia, S Afr. Some cult. orn. often with turnip-like roots

Trichodypsis Baillon = *Dypsis*

Trichoglottis Blume. Orchidaceae (V 16). c. 60 Indomal. to Taiwan & Polynesia. Cult. orn. epiphytes

Trichogonia (DC) Gardner (~ *Eupatorium*). Compositae (Eup.-Gyp.). 30 N S Am. R: MSBMBG 22(1987)102

Trichogoniopsis R. King & H. Robinson (~ *Eupatorium*). Compositae (Eup.-Gyp.). 2 Brazil. R: MSBMBG 22(1987)104

Trichogyne Less. (~ *Ifloga*). Compositae (Gnap.-Gnap.). 8 S Afr. R: OB 104(1991)155

Tricholaena Schrader. Gramineae (34c). 4 Medit. (Eur. 1) & Macaronesia to Afr. R: BB 138(1988)36. *T. vestita* (Balf.f.) Stapf & C.E. Hubb. (Socotra) – only known from type. See also *Melinis*

Tricholaser Gilli. Umbelliferae (III 11). 2 S & SW As.

Tricholepidium Ching = *Neocheiropteris* (but see APG 29(1978)41)

Tricholepis DC. Compositae (Card.-Cent.). 18 C As. to Burma. Some medic. (skin disease, 'seminal debility')

Trichomanaceae Burmeister = Hymenophyllaceae

Trichomanes L. Hymenophyllaceae (Hym.). (Incl. *Didymoglossum*, *Microgonium*) c. 80 trop. & warm Am, few OW; (s.s.) 25 trop. & warm Am. Bristle or kidney ferns. See also *Crepidomanes*; R: PJSB 51(1933)119. Some with ± peltate fronds (*Microgonium*). Some cult. orn.

Trichomeriopsis auctt. = *Trochomeriopsis*

Trichoneura Andersson. Gramineae (31d). 7 Arabia, trop. Afr., S US (1), Peru (1), Galápagos (1). R: AB 11,9(1912)8

Trichoneuron Ching. Aspleniaceae (Tectarioideae). 1 SW China

Trichopetalum Lindley (*Bottionea*). Anthericaceae (Liliaceae s.l.). 1 Chile: *T. plumosum* (Ruíz & Pavón) Macbr. – cult. orn.

Trichophorum Pers. (~ *Scirpus*.). Cyperaceae. Spp.?

Trichopilia Lindley. Orchidaceae (V 10). 30 trop. Am. Cult. orn. epiphytes

Trichopodaceae Hutch. = Dioscoreaceae

Trichopteryx Nees. Gramineae (39). 5 trop. & S Afr., Madagascar

Trichoptilium A. Gray. Compositae (Hele.-Gai.). 1 SW US

Trichopus Gaertner. Dioscoreaceae (Trichopodaceae). 1 Indomal.: *T. zeylanicus* Gaertner – stem with 1 app. term. leaf. R: KB 45(1990)353

Trichosacme Zucc. Asclepiadaceae (III 3). 1 Mexico: *T. lanata* Zucc.

Trichosalpinx Luer (~ *Pleurothallis*). Orchidaceae (V 13). 90 trop. Am. R: Phytol. 54(1983)393. Some cult. orn.

Trichosanchezia Mildbr. Acanthaceae. 1 E Peru

Trichosandra Decne. Asclepiadaceae (III 4). 1 Réunion: *T. borbonica* Decne. R: BMNHN 4,12(1990)131

Trichosanthes L. Cucurbitaceae. 15 Indomal. to Pacific. *T. cucumerina* L. (*T. anguina*, snake gourd, India) – monoec. annual climber cult. for ed. slender fr. 30–200 cm long, often coiled, peptides used as abortifacient in China; *T. ovigera* Blume (*T. cucumeroides*, Indomal.) – dried fr. a soap subs.; *T. kirilowii* Maxim. (*T. japonica*, E As.) – roots a starch source; others locally medic.

Trichoschoenus Raynal. Cyperaceae. 1 Madagascar

Trichoscypha Hook.f. Anacardiaceae (IV). 50 trop. Afr.

Trichospermum Blume. Tiliaceae. (Incl. *Belotia*) 36 Mal., W Pacific, 3 trop. Am. R: TBSE 41(1972)401. Bark for cord & rope

Trichospira Kunth. Compositae (Vern.-Vern.). 1 trop. Am.

Trichostachys Hook.f. Rubiaceae (IV 10). 10 trop. Afr.

Trichostelma Baillon = *Gonolobus*

Trichostema L. Labiatae (II). 16 N Am. R: Brittonia 5(1945)276. Cult. orn. herbs & shrubs esp. *T. lanatum* Benth. (blue-curls, Calif.) – char. of chaparral, toxic volatiles inhibiting other plant growth, fls in woolly infls.

Trichostephania Tard. = *Ellipanthus*

Trichostephanus Gilg. Flacourtiaceae (?). 2 W trop. Afr. R: BJBB 60(1990)143. Corona!

Trichostigma A. Rich. (~ *Rivina*). Phytolaccaceae (II). 3 trop. Am. Glabrous shrubs, some locally medic., cult. orn.

Trichostomanthemum Domin = *Melodinus*

Trichotaenia Yamaz. = *Lindernia*

Trichothalamus Sprengel = *Potentilla*

Trichotolinum O. Schulz. Cruciferae. 1 Patagonia

Trichotosia Blume (~ *Eria*). Orchidaceae (V 14). 45 India to SE As. Cult. orn. epiphytes

Trichovaselia Tieghem = *Elvasia*

Trichuriella Bennet (*Trichurus* C. Towns., ~ *Aerva*). Amaranthaceae (I 2). 1 S & SE As.: *T. monsoniae* (L.f.) Bennet

Trichurus C. Towns. = praec.

Tricliceras Thonn. ex DC (*Wormskioldia*). Turneraceae. 12 warm Afr.

Triclisia Benth. Menispermaceae (I). 10 trop. Afr. & Madag. Alks – some arrow-poisons & medic. (allegedly antimalarial)

Tricomaria Gillies ex Hook. & Arn. Malpighiaceae. 1 Argentina

Tricoryne R. Br. Anthericaceae (Liliaceae s.l.). 7 Aus., 1 ext. to New Guinea. Aerial parts app. all inflorescence, P 6 subequal, persistent & spirally twisted after flowering

Tricostularia Nees ex Lehm. Cyperaceae. 5 Aus., 1 ext. to S As.

Tricuspidaria Ruíz & Pavón = *Crinodendron*

Tricycla Cav. = *Bougainvillea*

Tricyclandra Keraudren. Cucurbitaceae. 1 Madagascar

Tricyrtis Wallich. Liliaceae. 20 Himal. to Taiwan & Japan. R: Plantsman 6(1985)193. Rather links L. to 'Uvulariaceae'. Cult. orn. (toad lilies) esp. *T. hirta* (Thunb.) Hook. (Japanese t.l., Japan) with purple-spotted P, several cvs

Tridactyle Schltr. Orchidaceae (V 16). 38 trop. & S Afr. R: KB 3(1948)282

Tridactylina (DC) Schultz-Bip. (~ *Chrysanthemum*). Compositae (Anth.-Art.). 1 E Siberia

Tridax L. Compositae (Helia.-Helia.). 30 Am. esp. Mex. R: Brittonia 17(1967)47. *T. procumbens* L. (Mex.) – now pantrop. weed

Tridens Roemer & Schultes. Gramineae (31d). 18 E US to Arg., Angola (1). R: AJB 48(1961)565

Tridentea Haw. (~ *Stapelia*). Asclepiadaceae (III 5). 17 S Afr. R: Excelsa Tax. ser. 2(1981?)1. Cult. orn. succ.

Tridesmostemon Engl. Sapotaceae (V). 2–3 C Afr. R: T.D. Pennington, *S.* (1991)261

Tridianisia Baillon = *Cassinopsis*

Tridimeris Baillon. Annonaceae. 3 Mexico

Tridynamia Gagnepain (~ *Porana*). Convolvulaceae. 4 India to Hainan & W Mal. R: Novon 3(1993)200

Trieenia Hilliard (~ *Phyllopodium*). Scrophulariaceae. 9 S Afr. R: NRBGE 45(1988)489. Commemorates E.E. Esterhuysen!

Trientalis L. Primulaceae. 4 N temp. (Eur. 1). Perennial rhizomatous herbs with term. tufts of 4–7 lvs, K (5), A 5; cult. orn. esp. *T. europaea* L. (chickweed wintergreen, Euras.)

triffid weed *Chromolaena odorata*

Trifidacanthus Merr. (~ *Desmodium*). Leguminosae (III 9). 1–2 Vietnam, Hainan & C Mal.

Trifolium L. Leguminosae (III 22). 238 temp. (Eur. 99) & subtrop. exc. Aus. R: M. Zohary & D. Heller (1984) *The genus T.*, Acta Univ. Carol. 33(1989)257. Clover. Bee-poll. herbs with A & style emerging when keel depressed by insect, returning when insect leaves; imp. fodder pls, some cyanogenic esp. *T. nigrescens* Viv. (W Medit. to Russia), rich in isoflavones affecting mammal reproductive capacity if taken in excess (oestrogenic). *T. amabile* Kunth (Aztec c., Mex. to S Am.) – eaten with maize etc.; *T. burchellianum* Ser. subsp. *johnstonii* (Oliver) J.B. Gillett (Uganda c., E Afr.); *T. campestre* Schreber (hop trefoil, Eur., Medit., natur. N Am.) – forage; *T. dubium* Sibth. (Eur. to Cauc.) – the shamrock (though many mimics & pretenders); *T. fragiferum* L. (strawberry c., Medit.) – salt-tolerant fodder, bladdery wing formed by persistent K (cf. persistent C in *T. badium* Schreber, C & S Eur. mts); *T. hybridum* L. (alsike c., Eur., natur. US) – fodder; *T. incarnatum* L. (Italian c., S & W Eur.) – fodder; *T. medium* L. (zig-zag c., Eur. to Iran) – fodder; *T. pannonicum* Jacq. (Hungarian c., E Eur.) – drought resistant forage; *T. pratense* L. (red or purple c., Eur. to Afghanistan, natur. N Am.) – bumblebee-poll. fodder, nectar sucked out by country children, 'sprouts' sold in Aus., antispasmodic cigarettes; *T. reflexum* L. (buffalo c., US) – fodder; *T. repens* L. (white or Dutch c., Eur., Medit., natur. N Am.) – most imp. fodder &

rotational (10% cover gives 50 kg nitrogen per ha per annum, enough to make a sward self-sustaining) allopolyploid pl. widely introd. (early in Peru) with a number of strains adapted to co-existence with particular other spp. such that a field will hold a number of discrete 'clones', resembling a polymorphism with different leaf-markings, cyanogenesis capacity, fl. colour etc., imp. bee-forage (clover honey), lvs s.t. used as shamrock (cf. *T. dubium*), 'Purpurascens Quadrifolium' – cult. orn. with lvs with 4 bronzy-red green-margined leaflets (up to 14 leaflets recorded in wild pops; also in *T. pratense*), f. *lodigense* Gams (Ladino c.) a 'giant' form used in pastures, some forms with plantlets in axils of petals; *T. resupinatum* L. (Persian c., Medit. to Iran) – fodder; *T. stoloniferum* Muhl. ex Eaton (running buffalo clover, N Am.) – almost extinct with 2 pops in W Virginia & 1 in Kentucky poss. due to disappearance of bison; *T. subterraneum* L. (subterranean c., N Eur. to Med.) – aerial infls & geocarpic with few fls the rest with hooked K acting as grapnels, fodder e.g. worth up to £200m per annum in 1930s though orig. introduced as packing around wine-bottles; *T. wormskioldii* Lehm. (NW Am.) – rhizome eaten

Trifurcia Herbert = *Herbertia*

trigger plant *Stylidium* spp.

Triglochin L. Juncaginaceae. 17 cosmop. (Eur. 4) esp. S temp. (Aus. 14 with most endemic), freshwater & salt-marshes. Arrowgrass. Plantain-like herbs, wind-poll., spiny prob. animal-disp. fr. Lvs allegedly toxic (hydrogen cyanide) in W US, though young lvs of *T. maritima* L., N temp.) cooked & fr. sold as birdseed in Paris markets

Trigonachras Radlk. Sapindaceae. 8 Mal. R: Blumea 33(1988)204

Trigonanthe (Schltr.) Brieger = *Dryadella*

Trigonella L. Leguminosae (III 22). c. 50 Medit. (spp. with explosive fls referred to *Medicago*), Macaronesia, S Afr., Aus. (1). R: TBIANSSR I, 10(1953)124. *T. foenum-graecum* L. (fenugreek, S Eur., W As.) – cult. since time of Assyrians for seeds used like lentils & medic. (allegedly contraceptive – steroids (diosgenin)), dyes etc., preserved in Tutankhamun's tomb (1325 BC)

Trigonia Aublet. Trigoniaceae. 24 trop. Am.

Trigoniaceae Endl. Dicots – Rosidae – Polygalales. 3/26 trop. Trees, shrubs & lianes without internal phloem. Lvs simple, opp. or spirally arr. (*Trigoniastrum*); stipules decid., often united. Fls bisexual, irreg., 2- or 3-bracteolate, in cymes; K (5) with unequal imbricate lobes, C papilionoid convolute in bud with 2 lower ones forming a keel, the upper a spurred or saccate standard, the laterals spathulate, A 5–12 on anterior (lower) side of fl. (up to 4 staminodal) ± united as a tube, anthers with longit. slits, (1)2 nectary glands in front of standard, $\underline{G}$ (3(4)), pluriloc. with axile placentas or 1-loc. with ± deeply intruded parietal placentas, style simple, each loc. with 1 or 2 to ∞ anatropous bitegmic ovules. Fr. a 3-winged samara or (*Trigonia*) a septicidal capsule; seeds with straight embryo & 0 endosperm. n = c. 10

Genera: *Humbertiodendron* (Madag.), *Trigonia* (trop. Am.), *Trigoniastrum* (W Mal.) – a distribution diff. to explain

Ovule & seed structure suggest Linales

Trigoniastrum Miq. Trigoniaceae. 1 W Mal.: *T. hypoleucum* Miq. – timber for furniture-making

Trigonidium Lindley. Orchidaceae (V 10). 14 trop. Am. Cult. orn. epiphytes

Trigonobalanus Forman. Fagaceae. 3 W Mal. (1), Thailand, Laos & China (1), Colombia (1), fossils in Eocene of Eur. Fr. like *Fagus* & eaten by rhinoceros

Trigonocapnos Schltr. Fumariaceae (I 2). 1 S Afr.: *T. lichtensteinii* (Cham. & Schldl.) Lidén

Trigonocaryum Trautv. Boraginaceae. 1 Caucasus

Trigonophyllum (Prantl) Pichi-Serm. (~ *Trichomanes*). Hymenophyllaceae. 2 trop. Am.

Trigonopleura Hook.f. Euphorbiaceae. 1 W Mal.

Trigonopyren Bremek. = *Psychotria*

Trigonosciadium Boiss. Umbelliferae (III 11). 4 Turkey, Iraq, Iran

Trigonospermum Less. Compositae (Helia.-Mel.). 5 Mex. R: CUMH 9, 6(1972)495

Trigonospora Holttum = *Cyclosorus*

Trigonostemon Blume. Euphorbiaceae. 45 Indomal. Some unbranched pachycauls

Trigonotis Steven. Boraginaceae. 50 E Eur. (1), C As. to New Guinea

Triguera Cav. Solanaceae (1). 1 W Medit. (incl. Eur.): *T. osbeckii* (L.) Willk.

Trigynaea Schldl. Annonaceae. 5 N S Am.

Trigynia Jacq.-Fél. = *Leandra*

Trihesperus Herbert = *Echeandia*

Trikeraia Bor. Gramineae (16). 1 Pakistan to Tibet

Trilepidea Tieghem. Loranthaceae. 1 NZ (N Is.): *T. adamsii* (Cheeseman) Tieghem – extinct.

R: Conserv. Biol. 5(1991)52

Trilepis Nees. Cyperaceae. 5 NE S Am. Erect pl. with woody stems & 3-ranked persistent ligulate lvs

Trilepisium Thouars (*Bosquiea*). Moraceae (IV). 1 trop. & S Afr. to Masc.: *T. madagascariense* DC.

Trilisa (Cass.) Cass. (~ *Carpephorus*). Compositae (Eup.-Gyp.). 2 SE US. R: MSBMBG 22(1987)275

Trilliaceae Chevall. (~ Liliaceae). Monocots – Liliidae – Asparagales. 3/65 N temp. Rhiz. herbs, aerial stems with 1 verticil with lvs usu. same no. as floral -mery (*Scoliopus* with 2 basal ones); vessels in roots. Fls solit., bisexual, 3–8(–10)-merous P (usu. 2 whorls, outer sepaloid, though inner sometimes rudimentary) & A marcescent in fr., anthers extrorse (introrse in *Scoliopus*). Ovules anatropous. Fr. a berry or capsule; seeds sometimes with elaiosomes, sometimes sarcotestal, endosperm copious oily & starchy, embryo undifferentiated

Genera: *Paris, Scoliopus, Trillium*

Paris & *Trillium* said to approach Convallariaceae to which fam. is referred *Medeola* form. here & held, with *Scoliopus*, to approach Liliales: another example of an infelicitous consequence of splitting Liliaceae s.l.

Cult. orn.

Trillidium Kunth = seq.

Trillium L. Trilliaceae (Liliaceae s.l.). 42 N Am. (esp. S Appalachians), Himalaya, E As. R: Plantsman 10(1989)221, 11(1989)67, 133, 12(1990)44, 13(1992)219. Some ant-disp. (elaiosomes). Cult. orn., some locally medic. (wake robin, esp. *T. grandiflorum* (Michaux) Salisb. (E N Am.) with frequent wild abnormal variants with bizarre fls etc.), seeds ant-disp. *T. chloropetalum* (Torrey) Howell (Calif.) – much used in chromosome studies where, after cold treatment, segments of heterochromatin stick together at anaphase, the adhesion persisting to telophase forming a bridge between daughter nuclei, a phenomenon prob. caused by subchromatic breakage followed by partial inverted reunion; *T. erectum* L. (birthroot, stinking Benjamin, E N Am.) – locally medic.

Trilobachne Schenk ex Henrard. Gramineae (40j). 1 W India

Trimenia Seemann. Trimeniaceae. (Incl. *Piptocalyx*) 5 C Mal. to Marquesas & Samoa

Trimeniaceae (Perkins & Gilg) Gibbs. Dicots – Magnoliidae – Magnoliales. 1/5 C Mal. to SE Aus., Marquesas & Samoa. Polygamous trees, shrubs & lianes often accum. aluminium. Lvs simple, ± opp., with translucent dots; stipules 0. Fls small, in cymes or panicles, with ± flat receptacle continuous with pedicel bearing 2–38 opp. bracteoles (?P) passing imperceptibly to spirally arr. P, all decid. at anthesis, A 7–25 in (1) 2 or 3 rows inserted spirally, anthers elongate, basifixed, with longit. slits, weakly extended connective, G̲ 1(2) with 1 pendulous, anatropous ovule. Fr. a berry; embryo small in copious endosperm. 2n = 16. R: JAA 64(1983)447

Genus: *Trimenia*

Trimeria Harvey. Flacourtiaceae (5). 2 trop. & S Afr. R: BJ 94(1974)302

Trimeris C. Presl (~ *Lobelia*). Campanulaceae. 1 St Helena: *T. scaevolifolia* (Roxb.) Mabb. – shrub, affinities with Pacific & S Am. taxa

Trimerocalyx (Murb.) Murb. = *Linaria*

Trimezia Salisb. ex Herbert. Iridaceae (III 3). c. 20 trop. Am. Elaiophores attractive to bees. Local laxatives. *T. martinicensis* (Jacq.) Herbert – cult. orn. & widely natur. trop.

Trimorpha Cass. (~ *Erigeron*). Compositae (Ast.-Ast.). c. 45 N temp. & Arctic. R: Phytol. 67(1989)63

Trimorphopetalum Baker = *Impatiens*

Trinacte Gaertner = *Jungia*

Trincomali wood *Berrya cordifolia*

Trinia Hoffm. Umbelliferae (III 8). c. 10 Eur. (9), Medit. to C As. *T. glauca* (L.) Dumort. (honewort, C & S Eur.)

Triniochloa A. Hitchc. Gramineae (19). 5 trop. Am. R: Novon 5(1995)36

Triocles Salisb. = *Kniphofia*

Triodanis Raf. Campanulaceae. 7 N Am., 1 Medit. R: JAA 67(1986)33

Triodia R. Br. Gramineae (31a). 45 Aus. R: PLNSW 96(1971)175. The pungent lvs dominate much of Aus. vegetation – 'spinifex' (name also applied to *Plectrachne* spp.) e.g. *T. longiceps* J. Black (black s., S Aus.)

Triodoglossum Bullock = *Raphionacme*

Triolena Naudin. Melastomataceae. 22 trop. Am. Some cult. orn. like *Bertolonia* but with 3

thread-like appendages on connectives of larger A & 3-winged capsule

Triomma Hook.f. Burseraceae. 1 W Mal.

Trioncinia (F. Muell.) Veldk. (~ *Glossocardia*). Compositae (Helia.-Cor.). 1 Aus. R: Blumea 35(1991)481

Triopteris L. = seq.

Triopterys L. Malpighiaceae. 3 trop. Am.

Triosteum L. Caprifoliaceae. 5–6 E As., E N Am. Perennial woody herbs incl. *T. perfoliatum* L. (fever root, tinker's weed, E N Am.) – form. a coffee subs.

Tripetaleia Siebold & Zucc. (~ *Elliottia*). Ericaceae. 2 Japan. Cult. orn. decid. shrubs

Tripetalum Schumann = *Garcinia*

Triphasia Lour. Rutaceae. 3 SE As., Philippines. R: W.T. Swingle (1943) *Bot. Citrus*: 236. *T. trifolia* (Burm.f.) P. Wilson (limeberry, myrtle lime, orig. unclear (? Malay Pen.)) – widely cult. hedge-pl. in trop., fr. ed. esp. preserved (China)

Triphora Nutt. Orchidaceae (V 3). 19 Am.

Triphylleion Süsseng. = *Niphogeton*

Triphyophyllum Airy Shaw. Dioncophyllaceae. 1 W Afr.: *T. peltatum* (Hutch. & Dalz.) Airy Shaw – carnivorous liane to 40 m long on soil of pH 4.2, lvs on sterile branches ephemeral, elongate, circinate & with stalked & sessile glands; other lvs with 2 small hooks (long shoots) or larger & hookless (short shoots) hence name; glands with hydrolytic enzymes the most elaborate of all known angiosperm glands, mucilage blocks insects' spiracles

Triphysaria Fischer & C. Meyer (~ *Orthocarpus*). Scrophulariaceae. 5 Calif., with 1 to Br. Columbia: *T. pusilla* (Benth.) Chuang & Heckard – ant-poll.?

Triplachne Link. Gramineae (21d). 1 Medit.

Tripladenia D. Don (*Schelhammera*). Convallariaceae (Liliaceae s.l.). 1 E Aus.: *T. cunninghamii* D. Don. R: FA 45(1987)416

Triplaris Loefl. ex L. Polygonaceae (II 1). 18 trop. Am. R: NJB 6(1986)545. Dioec. trees with hollow stems inhabited by ants; outer P accrescent as wings on fr. – wind-disp. Cult. orn. (long jack) with fls in v. long panicles incl. *T. weigeltiana* (Reichb.f.) Kuntze (*T. surinamensis*, NE S Am.) – timber for interior work

Triplasis Pal. Gramineae (31d). 3 SE US to Costa Rica

Tripleurospermum Schultz-Bip. (~ *Matricaria*). Compositae (Anth.-Mat.). 38 N temp. (Eur. 8 incl. *T. inodorum* (L.) Schultz-Bip. (*T. maritimum* subsp. *inodorum*, *T. perforatum*))

Triplisomeris Aubrév. & Pellegrin = *Anthonotha*

Triplocephalum O. Hoffm. Compositae (Pluch.). 1 trop. E Afr.

Triplochiton Schumann. Sterculiaceae. 3 trop. Afr. *T. scleroxylon* Schumann (obeche, arere, ayous, wawa, whitewood, W & C Afr.) – plantation timber tree, wood used for plywood & veneers (when stained, superficially like mahogany); induced to flower when a seedling by hormonal treatment in breeding programmes

Triplochlamys Ulbr. (~ *Pavonia*). Malvaceae. 5 N S Am.

Triplophyllum Holttum (~ *Ctenitis*). Dryopteridaceae (I 3). 15 trop. Afr., Madag., trop. S Am.

Triplopogon Bor. Gramineae (40d). 1 W India

Triplostegia Wallich ex DC. Valerianaceae (Triplostegiaceae). 2 E As., Mal.

Triplostegiaceae (Hoeck) Airy Shaw. See Valerianaceae

Triplotaxis Hutch. = *Vernonia*

Tripodandra Baillon = *Rhaptonema*

Tripodanthus (Eichler) Tieghem (~ *Loranthus*). Loranthaceae. c. 6 S Am. *T. acutifolius* (Ruíz & Pavón) Tieghem – lvs yield a blackish dye

Tripodium Medikus = *Anthyllis*

Tripogandra Raf. Commelinaceae (II 1g). 22 trop. Am. R: Rhodora 77(1975)213. Cult. orn. incl. *T. grandiflora* (J.D. Sm.) Woodson (C Am.) – climber to 3 m

Tripogon Roemer & Schultes. Gramineae (31d). 20 trop. OW with 1 in trop. Am. R (Afr.): KB 25(1971)301. Wet flushes. *T. minimus* (A. Rich.) Steudel (Afr.) – lvs revive after reduction of water-content to 7%

Tripolium Nees (~ *Aster*). Compositae (Ast.-Ast.). 1 Euras., N Afr., N Am.: *T. vulgare* Besler ex Nees (*A. t.*, sea aster) – fleshy halophyte

Tripsacum L. Gramineae (40i). 13 S US to Paraguay esp. C Am. R: Phytologia 33(1976)203. *T. fasciculatum* Trin. ex Asch. (*T. laxum*, Guatemala grass, C Am.) – fodder. *T. dactyloides* (L.) L. (gama grass, C Am.) – form. considered to have contributed to evolution of maize, though only *T. andersonii* J.R. Gray (cult. in Guatemala) may have some *Zea* chromosomes, some hybrids having them only in endosperm

Tripteris Less. (~ *Osteospermum*). Compositae (Cal.). c. 20 S Afr. to Middle E

Tripterocalyx Hook. ex Standley = *Abronia*

Tripterococcus Endl. (~ *Stackhousia*). Stackhousiaceae. 2 SW Aus.

Tripterodendron Radlk. (~ *Dilodendron*). Sapindaceae. 1 Brazil

Tripterospermum Blume (~ *Gentiana*). Gentianaceae. 20 E As. Climbers

Tripterygium Hook.f. Celastraceae. 2 E As. Cult. orn. scandent shrubs incl. *T. wilfordii* Hook.f. (E China to Taiwan) – alks, roots insecticidal & antitumour uses, salaspermic acid with significant anti-HIV activity

Triptilion Ruíz & Pavón. Compositae (Mut.-Nass.). 12 C Chile (1 in Patagonia). Herbs

Triptilodiscus Turcz. (~ *'Helipterum'*). Compositae (Gnap.-Ang.). 1 Aus.: *T. pygmaeus* Turcz. R: Nuytsia 8(1993)420

Triraphis R. Br. Gramineae (31d). 7 trop. Afr., Arabia & Aus. (1)

Trirostellum C.P. Wang & Xie = *Gynostemma*

Triscenia Griseb. (~ *Panicum*). Gramineae (34b). 1 Cuba

Triscyphus Taubert ex Warm. = *Thismia*

Trisepalum C.B. Clarke. Gesneriaceae. 14 China, Burma, Thailand & Malay Pen. R: NRBGE 41(1984)441

Trisetaria Forssk. Gramineae (21b). (Incl. *Avellinia, Parvotrisetum*) 15 Medit. (Eur. 3) to W Himal.

Trisetella Luer. Orchidaceae (V 13). 11 trop. Am. R: MSB 31(1989)69. Cult. orn. small epiphytes

Trisetobromus Nevski = *Bromus*

Trisetum Pers. Gramineae (21b). c. 70 temp. (Eur. 24) excl. Afr. R: BN 118(1965)210. Like *Helictotrichon* but G glabrous, lemma thin & palea free. R: Webbia 17(1963)569. Hairgrass esp. *T. flavescens* (L.) Pal. (Eur., Med.) – forage grass, producing vitamin D in ultraviolet as animals do; *T. pensylvanicum* (L.) Pal. (swamp oat, E N Am.)

Trismeria Fée = *Pityogramma*

Tristachya Nees. Gramineae (30). 22 trop. & S Afr., Madag., trop. Am. Some fodders esp. *T. leiostachya* Nees (trop. Am.)

Tristagma Poeppig. Alliaceae (Liliaceae s.l.). (Incl. *Ipheion*) c. 20 temp. S Am. Cult. orn. esp. *T. uniflorum* (Lindley) Traub (*Brodiaea u., I. u., Milla u., Triteleia u.*, Arg., Uruguay) – spring bulb

Tristania R. Br. Myrtaceae (Lept.). (S.s.) 1 NSW: *T. neriifolia* (Sims) R.Br. (water gum); see *Lophostemon, Ristantia, Tristaniopsis, Welchiodendron* for other spp.

Tristaniopsis Brongn. & Gris (~ *Tristania*). Myrtaceae (Lept.). 40 SE As., Mal. to E Aus. & New Caled. (13). Some rheophytes

Tristellateia Thouars. Malpighiaceae. c. 20 Madag., Afr. (1), Indomal. to Aus. & New Caled. (1)

Tristemma Juss. Melastomataceae. 15 trop. Afr. to Masc. R: BMNHN 3° Bot 28(1977)137

Tristemonanthus Loes. (~ *Campylostemon*). Celastraceae. 2 trop. W Afr.

Tristerix C. Martius = *Macrosolen*

Tristicha Thouars. Podostemaceae (Tristichaceae). 2 trop. Am., Afr. to India, N Aus. (*T. trifaria* (Willd.) Sprengel pantrop. W of Wallace's Line). Some described *Frullania* spp. (mosses) & *Crassula* truly *T.*

Tristichaceae Willis = Podostemaceae

Tristira Radlk. Sapindaceae. 4 C Mal.

Tristiropsis Radlk. Sapindaceae. 2 C & E Mal., W Pacific. R: Blumea 13(1966)395

Triteleia Douglas ex Lindley (~ *Brodiaea*). Alliaceae (Liliaceae s.l.). 18 W N Am. R: Four Seasons 6,1(1980)17. Cult. orn. cormous pls esp. *T. laxa* Benth. with blue fls. See also *Tristagma*

Triteleiopsis Hoover (~ *Brodiaea*). Alliaceae (Liliaceae s.l.). 1 W US

Trithecanthera Tieghem. Loranthaceae. 5 W Mal.

Trithrinax C. Martius. Palmae (I 1a). 5 Brazil, Bolivia, Paraguay, Uruguay, Arg. Many allegedly primitive features, fls bisexual. Ed. fr. & oilseeds; some cult. orn.

Trithuria Hook.f. Hydatellaceae. 3–4 Aus. (1), NZ

triticale × *Triticosecale*

× **Triticosecale** Wittm. ex A. Camus. Hybrids between spp. of *Triticum* & *Secale* (triticale). Hexaploid forms approaching wheat yields with consistently higher contents of proteins & essential amino-acids

Triticum L. Gramineae (23). 4 Medit. (Eur. 1) to Iran. Wheat. Most imp. temp. cereals (17 000 cvs) of complex (& still not completely understood) ancestry involving closely allied *Aegilops* spp., the most. cult. of all pls (20% of all food calories consumed by humans; 8–14% protein with gluten allowing 'rising' of dough). *T. monococcum* L. (einkorn, 2n = 14, E Medit.) – diploid fodder wheat of Palaeolithic derived from forms (ssp. *boeoticum*, Near

E esp. Turkey) with brittle ear, hulled grain, prob. coll. before grown, little cult. now; *T. turgidum* L. (2n = 28, a genome app. from *A. speltoides* Tausch W As.) – the allotetraploid wheats selected from forms (ssp. *dicoccoides*) with brittle ears & hulled grains (esp. Upper Jordan Valley) arr. in 8 groups incl. Dicoccon Group (*T. dicoccon*, emmer) – cult. by Babylonians & Swiss lake-dwellers, now usu. for livestock or breakfast food, Durum Group (*T. durum*, durum w., flint, hard or macaroni w.) – gluten-rich. cult. for cracked wheat (boughal, bulgur, burghul, esp. N Afr.), macaroni, semolina, spaghetti, vermicelli & other pasta (27 kg per head eaten annually in Italy, 2 kg in GB), Polonicum Group (*T. polonicum*, Polish w., giant or Jerusalem rye), Turgidum Group (Poulard w.) – tall winter or spring wheats, little cult.; *T. aestivum* L. (common bread wheat, 2n = 42, as *T. turgidum* with addition of genome from *A. tauschii* Cosson ('*A. squarrosa*', *T. aegilops*, W As.) once tetraploids taken into its range & probably arose several times (cf. *Spartina anglica*)) – allo-hexaploid wheats arr. in 6 groups incl. Aestivum Group – spring & winter grain wheats for bread and fermenting to 'white' beer, distilling to vodka (Russia), Compactum Group – protein-poor cvs used for pastry-flour, Spelta Group (*T. spelta*, spelt) – ancient cvs with hulled grains tolerant of poor soils; wheatgerm a health food & used in cosmetics. The optimistic certainty surrounding the cytological results of the 1950s & 1960s has gone (cf. *Tripsacum* & *Zea*) & the definitive schemes in many textbooks need revision as the wild spp. from which the various domesticated pls were thought to have been derived do not have the exact genomes predicted for them (NB. If hexaploid breadwheat is truly allopolyploid as above this genus should be considered to comprise intergeneric hybrids and written with a '×', the diploid unigeneric wheats referred to a separate genus!)

Tritonia Ker-Gawler. Iridaceae (IV 3). 28 S trop. Afr. to Cape. R: JSAB 48(1982)105, 49(1983)347. Some cult. orn. but see *Crocosmia*

Tritoniopsis L. Bolus. Iridaceae (IV 3). 22 S Afr. R: SAJB 56(1990)580. Cult. orn.; some ('*Anapalina*') bird-poll. with long tube & exserted A

Triumfetta L. Tiliaceae. 70 trop. Some locally used fibres

Triumfettoides Rauschert = *Triumfetta*

Triunia L. Johnson & B. Briggs (~ *Helicia*). Proteaceae. 4 warm E Aus. R: Muelleria 6(1986)195, (1987)302

Triuranthera Backer = *Driessenia*

Triuridaceae Gardner. Monocots – Alismatidae – Triuridales. (Incl. Lacandoniaceae) 5/40 trop. & warm. Glabrous usu. monoec. or dioec. achlorophyllous mycotrophic herbs to 1 m tall (usu. much smaller), white, yellow or purplish; slender rhizomes with scales; vessel-elements only in roots. Fls reg., usu. in term. bracteate racemes; P (3–)6(–10) valvate, s.t. basally connate & with apical appendages, A (2)3 or 6, some staminodes, anthers ± sessile & connate with longit. or tranverse slits, the connective often exended into long appendage, $\underline{G}$ 6–50, each with 1 basal erect anatropous bitegmic ovule. Fr. a head of follicles; seeds with small undifferentiated embryo in copious oily & proteinaceous endosperm. x = 11, 12, 14. R: BB 140(1991)

Genera: *Peltophyllum, Sciaphila, Seychellaria, Soridium, Triuris*

So reduced in structure as to obscure their relationships, though some are bold enough to isolate them from other 'superorders', the high number of free carpels suggesting they retain primitive features typical of early monocotyledons. Lacandoniaceae with G outside A now seen as homoeotic mutation in *Triuris*

Triuris L. Triuridaceae. 3 Guatemala, Guyana, Brazil. R: BB 140(1991)10

Triurocodon Schltr. = *Thismia*

Trivalvaria (Miq.) Miq. Annonaceae. 5 Assam to W Mal.

Trixis P. Browne. Compositae (Mut.-Nass.). c. 50 SW N Am. (17) to Chile. Some fert. control in Uruguay

Trizeuxis Lindley. Orchidaceae (V 10). 1 trop. Am.: *T. falcata* Lindley – small cult. orn. epiphyte

Trochetia DC. Sterculiaceae. 6 Mauritius, Réunion. Some cult. orn. small trees

Trochetiopsis Marais (~ *Trochetia*). Sterculiaceae. 3 St Helena: *T. erythroxylon* (Forster f.) Marais – extinct in wild & *T. melanoxylon* (R. Br.) Marais extinct since end C18, pls re-discovered in 1970 (2 trees left 1993) referred to *T. ebenus* Cronk now hybridized with *T. erythroxylon* in cult.; dead trunks used to burn lime for mortar for fortifications in Napoleonic era & wood from old roots still used in inlay work. Affinities with Madag. & Masc. pls

Trochiscanthes Koch. Umbelliferae (III 8). 1 S Eur.

Trochiscus O. Schulz. Cruciferae. 1 NE India

Trochocarpa R. Br. Epacridaceae. 12 Mal., E Aus. (7–8)

Trochocodon Candargy = ? *Campanula*

Trochodendraceae Prantl. Dicots – Hamamelidae – Trochodendrales. Incl. Tetracentraceae 2/2 Himal. to Taiwan & Japan. Evergreen andro-dioec. glabrous trees without vessels. Lvs simple in term. spirals; stipules 0 or small & adnate to petiole. Fls in term. racemoid cymes, bisexual, with small scales around swollen pedicel; P 0 (rudimentary in *Trochodendron*), A 4 or 40–70 developing centripetally in a spiral, anthers basifixed with longit. slits, connective not prolonged, G 4–11(–17) laterally connate but adaxial surfaces nectariferous, each with 5–30 anatropous, bitegmic ovules in 2 lateral series. Fr. a head of laterally cohering follicles; seeds dust-like, testa thin, embryo straight & minute in copious oily & proteinaceous endosperm. 2n = 40

Genera: *Tetracenton, Trochodendron*

Order interm. between Magnoliidae & Hamamelidae

Trochodendron Siebold & Zucc. Trochodendraceae. 1 Korea & Japan to Taiwan: *T. aralioides* Siebold & Zucc. – cult. orn. tree with trunk to 20 m, s.t. an epiphyte on *Cryptomeria japonica* at first & growing at alts to 3000 m, cryptic dioecy; bark locally used as bird-lime. R: JAA 26(1945)130

Trochomeria Hook.f. Cucurbitaceae. 7 Afr. Some ed. roots

Trochomeriopsis Cogn. Cucurbitaceae. 1 Madagascar

Troglophyton Hilliard & B.L. Burtt (~ *Gnaphalium*). Compositae (Gnap.-Gnap.). 6 S Afr. R: OB 104(1991)168

Trogostolon Copel. = *Davallia*

Trollius L. Ranunculaceae (I 1). 31 N temp. (Eur. 2). R: MB 41(1974)1. Herbs with alks & ranunculin (see *Anemone*); cult. orn. (globe flowers) with term. usu. solit. fls with petaloid K 5–15 & C 5+ rep. by nectaries, esp. *T. ranunculinus* (Sm.) Stearn (*T. caucasicus*, Turkey, Cauc. & NW Iran) – K 5–10 yellow & *T. asiaticus* L. (Siberia to NW China) – K 10–15 orange but usu. its hybrids (*T. × cultorum* Bergmans) with *T. europaeus* L. (N temp.) & *T. chinensis* Bunge (NE As.) to 1 m tall

Tromotriche Haw. (~ *Stapelia*). Asclepiadaceae (III 5). 3 S Afr. R: JSAB 50(1984)549

Tropaeastrum auctt. = *Trophaeastrum*

Tropaeolaceae Bercht. & J. Presl. Dicots – Rosidae – Capparidales. 3/89 C & S Am. Herbs, ± succ. & often climbing (petioles twining), s.t. with tuberous roots; mustard-oils. Lvs peltate or palmately lobed or divided, spirally arr., ± stipules esp. in seedlings. Fls large, bisexual, ± irreg., solit. & axillary; K 5 imbricate, the adaxial 1 (or 3) extended into nectariferous spur (almost 0 in *Trophaeastrum*), C 5 imbricate, clawed, the 3 abaxial usu. diff. (or 0) often with hairy claw, A 4 + 4 with small anthers with longit. slits, G (3), 3-loc. with 3-fid style, each loc. with 1 pendulous, apical-axile, anatropous, bitegmic ovule. Fr. separating into 1-seeded mericarps, fleshy or dry (samaroid in *Magallana* with only 1 maturing); seed with straight embryo in 0 endosperm. x = 12–14. R: OB 108(1991)1

Genera: *Magallana, Trophaeastrum, Tropaeolum*

Tropaeolum L. Tropaeolaceae. 87 S Mex. to Brazil & Patagonia. R: OB 108(1991)19. *T. majus* L. (nasturtium (because of mustard-oil like Cruciferae), Indian cress, cultigen (poss. spontaneous hybrid between Peruvian spp., *T. minus* L. & *T. ferreyae* Sparre, in Lima area) introd. from Peru to Eur. in 1684 (prob. the origin of all modern Eur. stock), though after 1845 crossed with *T. peltophorum* & backcrossed with *T. minus* (non-climbing), fast-growing cult. orn. climbing annual with fibre capable of being made into delicate lace & signifying patriotism in 'Language of Fls'; *T. peltophorum* Benth. (Andes) – cult. orn., antibiotic action, fl.-buds & young fr. used as caper subs.; *T. peregrinum* L. (*T. canariense*, canary creeper, C & S Peru) – cult. orn. with adaxial C fimbriate, yellow; *T. speciosum* Poeppig & Endl. (Chile) – cult. orn. perennial with red fls & blue fr.; *T. tuberosum* Ruíz & Pavón (añu, osañu, Andes) – tubers ed. when boiled

Trophaeastrum Sparre. Tropaeolaceae. 1 S Patagonia: *T. patagonicum* (Speg.) Sparre. R: OB 108(1991)18

Trophis P. Browne. Moraceae (I). 9 trop. Am., Madag., Mal. to New Caled. R: PKNAW C91(1988)352

Tropidia Lindley. Orchidaceae (III 2). 35 Indomal. to Taiwan & W Pacific, C Am. to WI

Tropidocarpum Hook. Cruciferae. 2 California

Tropilis Raf. = *Dendrobium* (but see Willdenowia 12(1982)249)

Trouettia Pierre ex Baillon = *Niemeyera*

Trudelia Garay (~ *Vanda*). Orchidaceae (V 16). 5 India, Thailand. Cult. orn.

Trukia Kanehira (~ *Randia*). Rubiaceae (II 1). 5+ Indomal. to W Pacific

trumpet climber, creeper or **vine** *Campsis radicans*; **Chinese t. flower** *C. grandiflora*; **t. tree** *Cecropia peltata*

Trungboa Rauschert (*Cyphocalyx*). Scrophulariaceae. 1 SE As.
Trybliocalyx Lindau = *Chileranthemum*
Trychinolepis Robinson = *Ophryosporus*
Tryginia Jacq.-Fél. = *Leandra*
Trymalium Fenzl. Rhamnaceae. 11 Aus., esp. SW (7)
Trymatococcus Poeppig & Endl. Moraceae (IV). 3 trop. Am. R: FN 7(1972)208
Tryonella Pichi-Serm. = *Doryopteris*
Tryphostemma Harvey = *Basananthe*
Tryssophyton Wurd. Melastomataceae. 1 Guyana
Tsaiorchis Tang & Wang. Orchidaceae (IV 2). 2 China
Tsavo Jarmol. = *Populus*
Tsebona Capuron. Sapotaceae (V). 1 Madag.: *T. macrantha* Capuron. R: T.D. Pennington, S. (1991)257
Tsiangia But, Hsue & P.T. Li (~ *Gaertnera*). Rubiaceae (?IV 7). 1 Hong Kong: *T. hongkongensis* (Seemann) But, Hsue & P.T. Li. R: Blumea 31(1986)311
Tsimatimia Jum. & Perrier = *Garcinia*
Tsingya Capuron. Sapindaceae. 1 Madagascar
Tsoongia Merr. Labiatae (I; Verbenaceae). 1 S China to SE As.
Tsoongiodendron Chun = *Michelia*
tsubaki oil *Camellia japonica*
Tsuga (Antoine) Carrière. Pinaceae. c. 14 temp. N Am. & E As. S to Vietnam. R: Phytol.M 7(1984)75. Hemlock (spruce). Evergreen monoec. timber trees, esp. *T. canadensis* (L.) Carrière (Canada or eastern or white h., E N Am.) – to 350 yrs old in Connecticut, coarse timber, tanbark (form. most imp. in N Am. – 'hemlock bark'), bark & twigs made into refreshing medic. tea by N Am. Indians in E Canada, many orn. cvs; *T. caroliniana* Engelm. (Carolina h., SE US mts) – cult. orn.; *T. heterophylla* (Raf.) Sarg. (western h., Alaskan pine, W N Am.) – imp. timber & pulp tree to 65 m or more introd. GB 1851; *T. sieboldii* Carrière (Japanese h., S Japan) – cult. orn. See also *Hesperopeuce*
tsukemono *Pteridium aquilinum*
Tsusiophyllum Maxim. Ericaceae. 1 Japan: *T. tanakae* Maxim. – cult. orn. rock garden shrub
tualang *Koompassia excelsa*
tuart *Eucalypus gomphocephala*
tuba root *Derris elliptica*
Tuberaria (Dunal) Spach (*Xolantha*). Cistaceae. 12 W & C Eur. (10), Medit. Herbs like *Helianthemum* but lvs in basal rosette & stigma sessile. A sensitive. Some cult. orn.
Tuberolabium Yamamoto. Orchidaceae (V 16). 12 Indomal. R: NJBV 10(1990)481
tuberose *Polianthes tuberosa* (corruption of specific name, cf. japonica)
Tuberostylis Steetz. Compositae (Eup.-Crit.). 2 Panamá, Colombia. R: MSBMBG 22(1987)376. Epiphytes with fleshy lvs in mangrove
Tubilabium J.J. Sm. Orchidaceae (III 3). 2 Indonesia
Tubocapsicum Mak. = *Capsicum*
tuckeroo *Cupaniopsis anarcardioides*
Tucma Ravenna = *Ennealophus*
Tuctoria Reeder. Gramineae (30). 3 Calif. in spring-pools. R: AJB 69(1982)1090
tucuma palm *Astrocaryum tucuma*
Tuerckheimocharis Urban. Scrophulariaceae. 1 Hispaniola
Tugarinovia Iljin. Compositae (Card.-Carl.). 1 Mongolia
Tulasnea Naudin = *Siphanthera*
Tulasneantha P. Royen. Podostemaceae. 1 W Brazil
Tulbaghia L. Alliaceae (Liliaceae s.l.). 22 trop. & S Afr. R: NRBGE 36(1978)77. Some cult. orn. with delicate, sometimes sweetly-scented, fls
Tulestea Aubrév. & Pellegrin = *Synsepalum*
tulip *Tulipa* spp., (New Guinea) *Gnetum* spp.; **t. oak** *Argyrodendron* spp.; **t. poplar** or **t. tree** *Liriodendron tulipifera*; **African t. t.** *Spathodea campanulata*; **Chinese t. t.** *L. chinense*; **water-lily t.** *Tulipa kaufmanniana*; **t. wood** *Dalbergia cearensis*; **Australian t.w.** *Harpullia pendula*; **Brazilian t.w.** *D. decipularis*
Tulipa L. Liliaceae. c. 100 Euras. E to C As., N Afr. Tulips. Cult. since C13 in Iran when there were multicoloured forms; nothing of them in Eur. art before mid C16 suggesting that all Eur. tulips (exc. *T. sylvestris* L. (tetraploid; incl. ssp. *australis* (Link) Pamp. (*T. australis*, diploid)), Medit.) are escapes from cult. & weeds of crops etc. A.D. Hall (1940) *The genus T.*; W. Blunt (1950) *Tulipomania*; Z.P. Botschantzeva (1982) *Tulips*. Introd. via Turkey (not grown there before 1500 but prominent in design etc.) to Eur. 1554 (Vienna), in Holland

(1571) leading to 'tulipomania' (1634) with great speculation in bulbs (later also in Turkey & as late as 1836 a bulb of 'Citadel of Antwerp' cost £650) esp. those (Rembrandt ts) with 'broken' fls (i.e. infested with aphid-transmitted virus), these were already complex garden hybrids & have given rise to the common tulip of modern gardens but by 1950s Dutch growers had selected forms of wild spp. collected in C As. since World War I & hybridized them with one another & old cvs to give e.g. 'Darwin Hybrids', esp. *T. fosteriana* Hoog ex W. Irv. (scarlet black-blotched fls) & its crosses with garden tulips, *T. kaufmanniana* Regel (waterlily tulip, fls opening flat), *T. greigii* Regel (lvs streaked purple-brown) etc. Besides these groups, garden cvs arranged according to flowering-time, the 'late or May-flowering' ts incl. 'Single Late' (incl. Darwin ts, single fls usu. rectangular in outline at base, bred at Lille), Rembrandt ts (fls striped or marked), parrot ts (P laciniate) & 'Double Late' or 'Peony-flowered' with many P used in pots. With pinks the first pls to have cv. names – now some 5000 named cvs & c. 300 grown on comm. scale; imp. cut-fl. industry esp. in Netherlands, the bulbs eaten during the Nazi occupation (disturbing menstrual cycles), the most widely grown cvs today being 'Monte Carlo' (Double Early, yellow) & 'Apeldoorn' (Darwin Hybrid (Mid-season) scarlet), a total of over 2 billion bulbs exported from the country per annum by 1990s; also imp. in Spalding area of GB & in W N Am. Many 'wild' spp. cult. in rock gardens etc. incl. *T. agenensis* DC subsp. *boissieri* (Regel) Feinbrun (W & S Turkey, NW Iran etc., natur. W Medit., perhaps the 'Rose of Sharon'), *T. persica* (Lindley) Sweet (*T. eichleri*, *T. undulatifolia*, E Medit. to C As.), *T. praestans* Hoog (C As.), etc.

tulipwood *Dalbergia decipularis*

Tuloclinia Raf. = *Metalasia*

Tulotis Raf. = *Platanthera*

Tumamoca Rose. Cucurbitaceae. 1 Arizona & Mex. (Sonora): *T. macdougalii* Rose – once considered poss. extinct

tumble grass *Schedonnardus paniculatus*; **t.weed** *Amaranthus caudatus*, *Salsola kali*

Tumidinodus Li = *Anna*

tummelberry See *Rubus*

tuna *Opuntia* spp.

Tunaria Kuntze = *Cantua*

tung oil *Aleurites* spp.

Tupeia Cham. & Schldl. Loranthaceae. 1 NZ

tupelo *Nyssa sylvatica*

Tupidanthus Hook.f. & Thomson = *Schefflera*

Tupistra Ker-Gawler. Convallariaceae (Liliaceae s.l.). Incl. *Campylandra* c. 35 E Himal. & China to W Mal. *T. nutans* Wallich (Himal.) – fl. spikes (cf. *Ornithogalum*) a veg. in Sikkim, petioles smoked in hookahs in Bhutan

Turanga (Bunge) Kimura = *Populus*

Turaniphytum Polj. Compositae (Anth.-Art.). 2 C As.

Turbina Raf. Convolvulaceae. 15 trop. Am. (5), trop. & S Afr. (9), New Caled. (1). R: BTBC 118(1991)265. *T. corymbosa* (L.) Raf. (trop. Am.) – cult. orn. climber, seeds used by Indians (source not identified until 1940s) as hallucinogen, active principles being ergoline alks & lysergic acid derivatives previously known in ergot (leading to 'St Anthony's fire' in rye-flour thus tainted)

Turbinicarpus (Backeb.) F. Buxb. & Backeb. = *Neolloydia*

Turczaninovia DC (~ *Aster*). Compositae (Ast.-Ast.). 1 E As.

Turczaninowiella Kozo-Polj. Umbelliferae. Spp.?

Turgenia Hoffm. (~ *Caucalis*). Umbelliferae (III 3). 2 C Eur., Medit. to C As.

Turgeniopsis Boiss. = *Glochidotheca*

Turkey oak *Quercus cerris*; **T. red** *Peganum harmala*, *Morinda citrifolia*

Turkish beech *Fagus sylvatica* ssp. *orientalis*; **T. delight** Form. salep (*Ophrys* & *Orchis* spp.); **T. galls** *Quercus pubescens*; **T. hazel** *Corylus colurna*

Turk's cap (cactus) *Melocactus intortus*; **T. c. gourd** *Cucurbita maxima* 'Turbaniformis'

turmeric *Curcuma longa*; **t. root** *Hydrastis canadensis*

Turnera L. Turneraceae. 50 trop. & warm Am., E & SW Afr. (1). Herbs & shrubs with alks. *T. diffusa* Willd. (damiana, trop. Am.) – dried lvs laxative & stimulant, tea, liqueur flavour; *T. ulmifolia* L. (trop. Am.) – variable weed (several spp. recog. by some authors) s.t. cult. orn., in Puerto Rico with distyly & incompatibility mechanisms but in Jamaica with variable anther heights though no 'thrum' pl.

Turneraceae Kunth ex DC. Dicots – Dilleniidae – Violales. 10/100 trop. & warm Am. & Afr. to Rodrigues. Trees but usu. shrubs & herbs. Lvs simple, spirally arr., often with 2 glands or extrafl. nectaries at base of lamina; stipules 0 or small. Fls bisexual, reg., bibracteolate, with often tubular hypanthium, solit. or in infls: K 5 imbricate at top of hypanthium, C 5 s.t. convolute & ephemeral, s.t. with fringed corona (*Erblichia*, *Piriqueta*) or 5 glands or lobes between C & A, A 5 attached in throat of hypanthium, anthers with longit. slits, G (3), s.t. half-inf., 1-loc. with parietal placentas & distinct styles, each placenta with (3–)∞ anatropous, bitegmic ovules. Fr. a (usu. loculicidal) capsule; seeds (1–)3–∞, arillate, embryo straight or weakly curved in copious oily endosperm. x = 7, 10

Genera: *Adenoa, Erblichia, Hyalocalyx, Loewia, Mathurina, Piriqueta, Stapfiella, Streptopetalum, Tricliceras, Turnera*

turnip *Brassica rapa* Rapifera Group; **t.wood** *Akania bidwillii*

turpentine *Pinus* spp.; **t. bush** *Beyeria* spp.; **Chian t.** *Pistacia terebinthus*; **Jura t.** *Picea abies*; **Strasburg t.** *Abies alba*; **Venice t.** *Larix decidua*; **t. wood** *Syncarpia glomulifera*

turpeth root, turpethum *Operculina turpethum*; **Spanish t. r.** *Thapsia garganica*

Turpinia Vent. Staphyleaceae. 10 Indomal. to Japan, trop. Am. Some timbers

Turraea L. Meliaceae (I 1). 60 trop. & S Afr. (24), Madag., Masc., trop. As. to Aus. (1). Arillate seeds disp. by birds, those of grassland spp. (*T. pulchella* (Harms) Penn. etc.) in S Afr. small & ant-disp.

Turraeanthus Baillon. Meliaceae (I 6). c. 2 trop. W Afr. *T. africanus* (C. DC) Pellegrin (avodiré, C & W Afr.) – furniture timber

Turricula Macbr. Hydrophyllaceae. 1 SW N Am.: *T. parryi* (A. Gray) Macbr. – cult. orn.

Turrigera Decne = *Tweedia*

Turrillia A.C. Sm. = *Bleasdalea*

Turritis L. = *Arabis*

turtlehead *Chelone* spp.

turwad bark *Senna auriculata*

Tussilago L. Compositae (Sen.-Tuss.). (Excl. *Nardosmia*) 1 Euras., N Afr.: *T. farfara* L. (coltsfoot, Euras., Medit., natur. E N Am.) – infls prod. in spring before lvs, each head with 40 nectariferous male fls with pollen-presenting styles but no stigmas surrounded by c. 300 nectarless females, highly protogynous; lvs smoked, form. in treatment of asthma; tincture an ingredient of comm. hairsprays

tussock grass *Deschampsia cespitosa*; **serrated t.** *Nassella trichotoma*

Tutcheria Dunn = *Pyrenaria*

tutsan *Hypericum androsaemum*

Tuxtla Villaseñor & Strother. Compositae (Helia.-Verb.). 1 Mex.: *T. pittieri* (Greenm.) Villaseñor & Strother. R: SB 14(1989)529

Tuyamaea Yamaz. = *Lindernia*

twayblade *Listera ovata*

Tweedia Hook. & Arn. Asclepiadaceae (III 1). 6 temp. S Am. R: Parodiana 5(1989)375. *T. caerulea* D. Don – cult. orn. herbaceous climber

twinflower *Linnaea borealis*

twinleaf *Zygophyllum* spp. (Aus.)

twistwood *Viburnum lantana*

twitch *Elytrigia repens*

Tylanthera C. Hanson. Melastomataceae. 2 Thailand. R: NJB 9(1990)631. Habit of *Cincinnobotrys*

Tylecodon Toelken (~ *Cotyledon*). Crassulaceae. 31 S Afr. R: Bothalia 12(1978)378. Some cult. orn. succ. esp. *T. papillaris* (L.) Rowley (*T. cacalioides*, *C. c.*, Cape) – toxic

Tyleria Gleason. Ochnaceae. 13 Guayana Higland, Venezuela. R: BJ 113(1991)182

Tyleropappus Greenman. Compositae (Helia.-Mel.). 1 Venezuela

Tylocarya Nelmes = *Fimbristylis*

Tylodontia Griseb. = *Cynanchum*

Tylopetalum Barneby & Krukoff = *Sciadotenia*

Tylophora R. Br. Asclepiadaceae (III 1). 50 OW trop. (Mal. c. 20), S Afr. Seeds with wing allowing water-disp.; alks. *T. indica* (Burm.f.) Merr. (India) – medic. roots with effect of ipecacuanha

Tylophoropsis N.E. Br. = *Tylophora*

Tylopsacas Leeuwenb. Gesneriaceae. 1 trop. Am.

Tylosema (Schweinf.) Torre & Hillc. = *Bauhinia*

Tylosperma Botsch. = *Potentilla*

Tylostigma Schltr. Orchidaceae (IV 2). 3 Madagascar

Tynnanthus Miers. Bignoniaceae. 14 trop. Am.

Typha L. Typhaceae. 10–12 cosmop. (Eur. 6). Reed mace, cat-tail, bulrush (q.v.), cumbungi (reeds); starch-rich rhizomes emergency food, plush of female fls form. used as kapok subs., pollen (unusual carbohydrates) eaten locally, lvs for mats & chair-seating esp. *T. angustifolia* L. (N temp.) – bulk of natural 'rush' available in US, *T. elephantina* Roxb. (India) etc., some used for paper-making & dried infl. orn.; *T. latifolia* L. (N temp.) – pollen (golden) in honey sold as a sweetmeat in Anc. China

Typhaceae Juss. Monocots – Commelinidae – Typhales. 1/10–12 cosmop. Perennial marsh-herbs, monoec., wind-poll., glabrous, with starchy rhizomes; stems erect, term. by infl., interior pithy; vessel-elements throughout veg. body. Lvs with sheathing base & narrow parallel-veined spongy lamina, distichous, mostly basal. Infl. of ∞ fls forming dense cylindrical spike, the upper part distinct & male, the lower female; males with P 0–3(–8) slender bristles. A (1–)3(–8) with short largely connate filaments, the anthers basifixed with broad connective prolonged beyond pollen-sacs, pollen in monads or tetrads; females with P ±∞ slender bristles or scales in 1–4 irreg. whorls, s.t. ± connate in groups & adnate to gynophore bearing G̲ 1 (pseudomonomerous) with 1 pendulous, anatropous, bitegmic ovule (some with abortive ovaries). Fr. with slender stipe formed by accrescent gynophore & accrescent style, wind-disp., 1-seeded, eventually dehiscent; embryo straight in copious starchy endosperm (with protein & oil) & thin perisperm. x = 15

Genus: *Typha*

The disp. unit is the whole fl., the P acting as 'pappus'. Sparganiaceae perhaps referable here

Typhoides Moench = *Phalaris*

Typhonium Schott. Araceae (VIII 6). 30 SE As., Indomal., NE Aus. (Australasia 14, R: Blumea 37(1993)345)

Typhonodorum Schott. Araceae (VI 6). 1 E Afr. coast, Madag., Masc.: *T. lindleyanum* Schott – large viviparous water-pl. to 4 m tall, with 'trunk' (pseudostem) to 30 cm diam. & huge lvs, of banana-like aspect allied to *Peltandra*; seeds & tuber ed.

Tyrimnus (Cass.) Cass. Compositae (Cyan.-Carn.). 1 Medit. incl. S Eur.

Tysonia Bolus = *Afrotysonia*

Tytonia G. Don f. Older name for *Hydrocera*

Tytthostemma Nevski = *Stellaria*

Tzellemtinia Chiov. = *Bridelia*

Tzvelevia Alexeev = *Festuca*

U

Uapaca Baillon. Euphorbiaceae (Uapacaceae). 61 trop. Afr. & Madag. *U. guineensis* Muell. Arg. (sugar plum, trop. Afr.) – medlar-flavoured fr., timber valuable; *U. kirkiana* Muell. Arg. (wild loquat, C Afr.) – ed. fr., timber good, charcoal

Uapacaceae (Muell. Arg.) Airy Shaw = Euphorbiaceae

uba cane *Saccharum officinarum*

Ubochea Baillon. Verbenaceae. 1 Cape Verde Is.

ucahuba or **ucuhuba** *Virola surinamensis*

udo *Aralia cordata*

Uebelinia Hochst. Caryophyllaceae (III 3). 6–7 trop. Afr. mts. R: BJBB 55(1985)421. Some ± aquatic

Uebelmannia Buin. Cactaceae (III 5). 5 E Brazil mts. Cult. orn.

Uechtritzia Freyn. Compositae (Mut.-Mut.). 3 Armenia & C As. to W Himal. & China. R: NJB 8(1988)72. Herbs

Ugamia Pavlov. Compositae (Anth.-Can.). 1 C As.

Uganda grass *Cynodon transvaalensis*

Ugli proprietary name of a hybrid citrus fruit, widely misapplied to *Citrus* × *aurantium* Tangelo Group

Ugni Turcz. Myrtaceae (Myrt.). 5–15 trop. & warm Am. *U. molinae* Turcz. (*Myrtus ugni*, Chilean guava, strawberry myrtle) – cult. shrub with blue-black berries c. 6 mm diam. used for jam

Uittienia Steenis = *Dialium*

Uladendron Marc.-Berti. Malvaceae. 1 Venezuela

Ulbrichia Urban = *Thespesia*

Uldinia J. Black. Umbelliferae (I 1a). 1 C Aus.: *U. certatocarpa* (W. Fitzg.) N. Burb. R: KB 9(1954)451

Uleanthus Harms. Leguminosae (III 2). 1 Amazonia

Ulearum Engl. Araceae (VIII 4). 1 Upper Amazonia

Uleiorchis Hoehne. Orchidaceae (V 5). 1 Brazil: *U. ulaei* (Cogn.) Handro – mycotrophic
Uleodendron Rauschert = *Naucleopsis*
Uleophytum Hieron. Compositae (Eup.-Crit.). 1 Peru. R: MSBMBG 22(1987)373. Liane
Ulex L. Leguminosae (III 31). 20 W Eur. (7), N Afr. R: BJ 72(1941)69. Seedlings with 3-foliolate lvs, later stages with them v. reduced with branches forming spines. Poll. mechanism explosive – in newly opened fls keel C adhere by upper edges & keel held straight by A tube & style, no nectar but foraging bees cause keel to break apart releasing A & G; fr. dehisces explosively with valves curving back & becoming twisted, expelling seeds bearing elaiosomes carried off by ants. Shrubs with alks, dominating much of the fire-climax heathland of Atlantic Eur., a formation replacing the original decid. forest long-cleared, esp. *U. europaeus* L. (gorse, whin), now weedy elsewhere (e.g. pest in NZ), but also smaller *U. minor* Roth & *U. gallii* Planchon – dwarf gorse (W Eur.)
uli *Castilla ulei*
ulla grass *Themeda gigantea*
Ulleria Bremek. = *Ruellia*
ullucu *Ullucus tuberosus*
Ullucus Caldas. Basellaceae. 1 Andes: *U. tuberosus* Caldas (ullucu, oca quina) – prostrate pl. with ed. potato-like tubers from rhizome, anthers with apical pores, fr. a berry
Ulmaceae Mirbel. Dicots – Dilleniidae – Urticales. 16/175 trop. to temp., esp. N. Trees & shrubs without laticifers but often with mucilage cells or canals. Lvs simple, often basally oblique, usu. spirally arr. or distichous; decid. stipules lateral or interpetiolar, protecting bud. Fls small, wind-poll., unisexual (monoec.) or bisexual (e.g. *Ulmus*), ± reg., solit. (females) & axillary or in cymes to panicles; P (2–)5(–9), subcampanulate, A opp. P, or 2 × P (–16) anthers dorsifixed & somewhat versatile, G̲ (2(3)), a pistillode in males, 1- (or 2-) loc. with separate styles, each loc. with 1 pendulous, anatropous to amphitropous bitegmic ovule. Fr. a samara, nut or drupe; seeds with straight or curved embryo in little or 0 endosperm. x = 10, 11, 14

Classification & principal genera (Cronquist). Two subfams:

I. **Ulmoideae** (fr. dry, often a samara, x = 14): *Ulmus, Zelkova*

II. **Celtidoideae** (fr. a drupe, x = 10, 11): *Celtis, Trema*

Imp. timber trees, esp. *Ulmus*, but also *Aphananthe, Celtis, Chaetachme, Chaetoptelea, Gironniera, Holoptelea, Phyllostylon, Trema, Zelkova* spp. etc., ed. fr. (*Celtis* – hackberry), fibres & cult. orn.

ulmer pipes *Acer campestre*
ulmo *Eucryphia cordifolia*
Ulmus L. Ulmaceae (I). 25–30 N temp. to N Mex. (Eur. 6). Elms. Form. imp. timber & street-trees but many susceptible to beetle-borne Dutch Elm Disease (*Ophiostoma* (*Ceratocystis*) *ulmi*); rapid disappearance of Eur. elm in pollen record c. 3000 yrs BC poss. partly attributable to this, but also selective pollarding & other management by Neolithic man, *U. wallichiana* Planchon (NW Himal.) still imp. cattle fodder tree, cf. *Ficus semicordata*). Since c. 1965 most in N Am. & Eur. killed (11m in GB 1970–8) alone). Timber rot-resistant under-water, that of Eur. spp. for beams where oak too dear, furniture, bellows, coffins, chair-seats, wheel-hubs, mallet-heads, village pumps & orig. water-pipes as in London before metal & now plastic ones; bark form. used for Indian houses in Kansas. Most fl. before lvs expand, though *U. parvifolia* Jacq. (Japan, China) etc. autumn-flowering while some spp. evergreen (cf. *Quercus*). *U. alata* Michaux (red elm, E N Am.) – timber, bast used to tie cotton-bales; *U. americana* L. (American or white elm, E N Am.) – timber, many orn. cvs; *U. crassifolia* Nutt. (cedar elm, S US) – allegedly resistant to Dutch Elm Disease; *U. davidiana* Planchon (N China, Japan) incl. Japanese e. – timber; *U. glabra* Hudson (wych (i.e. pliant) or Scotch e., N Eur. to SW As.) – timber form. used for desks ('wyches'), form. common woodland elm in GB, not suckering, many orn. cvs incl. 'Camperdown' with drooping branches usu. grafted on an erect stock; *U.* × *hollandica* Miller (*U. minor* Miller (S & C Eur.) × *U. glabra*, Dutch e.) – many clones form. in cult.; *U. procera* Salisb. (*U. minor* var. *vulgaris*, English e., ? origin, allegedly introd. N Spain by Celts from S France but otherwise known only from England & S Wales) – form. char. of English landscape as depicted by Constable, etc. but almost all lost to Dutch Elm Disease in 1970s–80s, clonal & largely reprod. by suckers (fr. usu. sterile); *U. pumila* L. (Siberian e., E As.) – resistant to Dutch Elm Disease; *U. thomasii* Sarg. (cork or hickory or rock e., NE N Am.) – timber for chairs, tool-handles, etc.; many hybrids incl. *U.* × *vegeta* (Loudon) Ley (*U.* 'Vegetata', ? = *U.* × *hollandica*, Huntingdon e.) – cult. orn. with ascending branches, a magnificent avenue tree susceptible to Dutch Elm Disease & 'Sapporo Autumn Gold' (? derived from *U. pumila*) app. resistant but a shrub. R.H. Richens (1983) *Elm*

Ultragossypium Roberty = *Gossypium*
Ulugbekia Zak. = *Arnebia*
ulva marina *Zostera marina*
Umbelliferae Juss. (Apiaceae). Dicots – Rosidae – Apiales. 446/3540 cosmop. esp. N temp. & trop. mts. Herbs, often with pithy scapes, or less often shrubs or even trees, often with unusual sec. thickening, aromatic & s.t. poisonous (17-carbon-skeletoned polyacetylenes, rarely alks, e.g. *Conium*). Lvs pinnately or ternately compound or dissected (rarely palmately so or simple & even phyllodic), s.t. spiny, usu. spirally arr., often with broad sheathing base s.t. with stipular flanges. Fls usu. small, bisexual (*Acronema* dioec.), 5-merous exc. G, reg. in compound umbels (Apioideae) when often subtended by involucre of free or united bracts & the marginal fls s.t. sterile with C expanded marginally, heads or simple umbels s.t. reduced to single fls, or dichasia; K often small teeth around ovary apex to 0, C often white, yellow, purple etc., valvate (rarely 0), A alt. with C on nectary-disk, anthers with longit. slits, $\overline{G}$ (2), 2-loc. (rarely 1-loc & pseudomonomerous) with distinct styles, each loc. with 1 (+ 1 abortive) apical-axile, pendulous, anatropous unitegmic ovule. Fr. usu. a schizocarp of 2 mericarps facially united; integument often adherent to pericarp, embryo usu. small, in copious oily endosperm. x = (4)8(–)11(12). R: M.G. Pimenov & M.V. Leonov (1993) *The genera of the U.*
Classification & principal genera (after Drude – in need of modern review; some genera not placed in this scheme):

I. **Hydrocotyloideae** (stipules present; fr. with woody endocarp; mainly S temp., 60% in Am., 90% of those in S) – 2 tribes:
   1. **Hydocotyleae** (2 subtribes): *Centella, Hydrocotyle, Platysace, Trachymene* (a, Hydocotylinae); *Actinotus, Xanthosia* (b, Xanthosiinae)
   2. **Mulineae** (3 subtribes): *Bowlesia* (a, Bowlesiinae); *Azorella, Bolax* (b, Azorellinae); *Hermas, Mulinum* (c, Asterisciinae)

II. **Saniculoideae** (stipules 0; fr. with soft endocarp; style surrounded by ring-like disk) – 2 tribes:
   1. **Saniculeae (Eryngieae)**: *Alepidea, Astrantia, Eryngium, Sanicula*
   2. **Lagoecieae**: *Lagoecia*

III. **Apioideae** (stipules 0; fr. with soft endocarp; style on apex of disk; 80% in OW – 12 tribes with details of principal ones:
   1. **Echinophoreae**: *Thecocarpus*
   2. **Scandiceae** (parenchyma around carpophore with crystal layer): *Anthriscus, Chaerophyllum, Myrrhis, Osmorhiza, Scandix*
   3. **Caucalideae** (Dauceae; mericarps with spines on ridges): *Caucalis, Cuminum, Daucus, Torilis*
   4. **Coriandreae** (parenchyma without crystal layer): *Coriandrum*
   5. **Smyrnieae** (mericarps rounded outwards): *Arracacia, Cachrys, Conium, Oreomyrrhis, Prangos, Scaligeria, Smyrnium, Tauschia*
   6. **Hohenackerieae**: *Hohenackeria* (only)
   7. **Pyramidoptereae**: *Pyramidoptera* (only)
   8. **Apieae** (primary ridges of mericarps all similar; seeds semi-circular in cross-section; poss. splittable into subtribes): *Aciphylla, Acronema, Aegopodium, Anisotome, Apium, Bunium, Bupleurum, Elaeosticta, Foeniculum, Heteromorpha, Ligusticum, Oenanthe, Pimpinella, Seseli*
   9. **Angeliceae**: *Angelica, Cymopterus*
   10. **Peucedaneae**: *Anethum, Ferula, Ferulago, Lomatium, Peucedanum, Steganotaenia*
   11. **Tordylieae**: *Heracleum, Pastinaca*
   12. **Laserpitieae (Thapsieae)**: *Laserpitium, Thapsia*

Only c. 20% of genera have more than 9 spp. (many tribes with 1 large genus & many 'satellites') & many are monotypic & some of these with odd affinities, their closest allies being geographically v. distant now. There is no real justification for separating U. from Araliaceae in which U. are 'nested' & debate over whether particular woody genera should be squeezed into either is sterile. Future investigation of the phylogeny of the group could well reduce the number of genera recog. in temp. Floras. U. have been long recognized as an assemblage based on morphological (& chemical) characters & poss. first designated a fam. in C16; Morison (1672) wrote the first (western) systematic study (of any group) on them, utilizing fr. characters which are v. imp. in their classification. Poll. by insects, the enlarged marginal fls act as attractants (e.g. *Daucus*; cf. *Hydrangea, Viburnum*) or involucral bracts resembling a C such that infl. looks like single fl. (e.g. *Astrantia, Bupleurum, Eryngium, Hacquetia*), in *Mathiasella* the true fls unisexual with

females without C. Hybrids v. rare (but see *Aciphylla*). R: V.H. Heywood (ed. 1971) *The biology & chemistry of the U.*
Many imp. foods, herbs, spices & flavourings incl. *Anethum* (dill), *Angelica*, *Anthriscus* (chervil), *Apium* (celery), *Arracacia*, *Carum* (caraway), *Chaerophyllum*, *Conopodium*, *Coriandrum* (coriander), *Crithmum*, *Cryptotaenia* (mitsuba), *Cuminum* (cumin), *Daucus* (carrot), *Foeniculum* (fennel), *Laserpitium*, *Levisticum* (lovage), *Lomatium*, *Myrrhis*, *Oenanthe*, *Opopanax*, *Pastinaca* (parsnip), *Petroselinum* (parsley), *Pimpinella* (aniseed) etc., though many are toxic (e.g. *Aethusa*, *Cicuta*, *Conium* (hemlock), *Oenanthe*) while *Deverra* & *Heracleum* can cause serious dermatitis; medic. & other gums are derived from *Ferula*, toothpicks from *Ammi*, scent from *Dorema*; few cult. orn. incl. *Aciphylla*, *Ammi*, *Astrantia*, *Eryngium*, *Hacquetia*, *Trachymene* etc. *Aegopodium*, *Aethusa* & *Cyclospermum* weedy

Umbellularia (Nees) Nutt. Lauraceae (II). 1 W N Am.: *U. californica* (Hook. & Arn.) Nutt. (California laurel or olive, Calif.) – fine timber (pepperwood), sniffing crushed lvs can lead to sneezing & headaches

Umbilicus DC. Crassulaceae. 18 Eur. (6) & Medit. to Iran & Afr. mts. Some cult. orn. esp. *U. rupestris* (Salisb.) Dandy (navelwort, penny pies, (wall) pennywort, GB to Macaronesia & SW As.)

umbrella bamboo *Thamnocalamus spathaceus*; **u. fir** *Sciadopitys verticillata*; **u. pine** *Pinus pinea*; **u. palm** *Hedyscepe canterburyana*; **u. plant** *Darmera peltata*, *Cyperus involucratus*; **u. tree** *Cussonia* spp., *Musanga cecropioides*, *Polyscias* spp., *Schefflera actinophylla*

umburana *Amburana cearensis*

umiry balsam *Humiria* spp.

Umtiza Sim. Leguminosae (I 4?). 1 E Cape

umzimbeet *Millettia grandis*

Unanuea Ruíz & Pavón ex Pennell = *Stemodia*

Uncaria Schreber. Rubiaceae (I 1; Naucleaceae). 34 trop. R: Blumea 24(1978)68. Lianes with alks & accrescent clasping hooks in place of infl. axes, 1 Afr. sp. with hollow ant-infested stems. Some locally medic. & dye-pls; *U. gambir* (Hunter) Roxb. (cultigen in trop. As.) – imp. tan-source, in C17 lvs & young shoots used to adulterate tea ('catechu')

Uncarina (Baillon) Stapf. Pedaliaceae. 9 Madag. R: F Mad. 179(1979)7. Frs make effective mouse-traps

Uncifera Lindley. Orchidaceae (V 16). 7 India to W Mal.

Uncinia Pers. Cyperaceae. 54 Mal., Pacific (Aus. 11), C & S Am., S Indian & Atlantic Oceans. Dispersal facilitated by hook from infl. axis projecting beyond utricle

Ungeria Schott & Endl. Sterculiaceae. 1 Norfolk Is.

Ungernia Bunge. Amaryllidaceae (Liliaceae s.l.). 8 C As. to Japan. Alks

Ungnadia Endl. Sapindaceae. 1 S N Am.: *U. speciosa* Endl. (Mexican buck-eye) – cult. orn. decid. tree

Ungula Barlow = *Amyema*

Ungulipetalum Mold. Menispermaceae (I). 1 Brazil: *U. filipendulum* (C. Martius) Mold.

unicorn root *Aletris farinosa*, *Chamaelirium luteum*

Unigenes F. Wimmer. Campanulaceae. 1 S Afr.

Uniola L. Gramineae (31b). 4 S US to Ecuador. Sand-binders (esp. *U. paniculata* L.), poss. material for paper

Unona L.f. = *Xylopia*

Unonopsis R. Fries. Annonaceae. 43 trop. Am. (Amazon 15, R: BMPEG 5(1989)207). *U. veneficiorum* (C. Martius) R. Fries (Amazonia) – arrow-poison & contraceptive

Unxia L.f. (~ *Villanova*). Compositae (Helia.-Mel.). 3 Panamá & N S Am. R: Brittonia 21(1969)314

upas tree *Antiaris toxicaria*

Upudalia Raf. = *Eranthemum*

Upuna Sym. Dipterocarpaceae. 1 Borneo: *U. borneensis* Sym. – many app. primitive features incl. arillate seeds; heavy construction timber

Uralepis Nutt. = *Triplasis*

Urandra Thwaites = *Stemonurus*

Uranodactylus Gilli = *Winklera*

Uranthoecium Stapf. Gramineae (34b). 1 trop. Aus.: *U. truncatum* (Maiden & Betche) Stapf

Uraria Desv. Leguminosae (III 9). 20 trop. OW. Some locally medic.

Urariopsis Schindler = *Uraria*

Urbananthus R. King & H. Robinson (~ *Eupatorium*). Compositae (Eup.-Crit.). 2 Cuba, Jamaica. R: MSBMBG 22(1987)300

Urbania Philippi. Verbenaceae. 1 Chile: *U. pappigera* Philippi. R: BSAB 25(1988)478

Urbanodendron Mez. Lauraceae (I). 1 E trop. S Am. R: BJ 110(1988)165
Urbanodoxa Muschler = *Cremolobus*
Urbanoguarea Harms = *Guarea*
Urbanolophium Melchior = *Haplolophium*
Urbanosciadium H. Wolff = *Niphogeton*
Urbinella Greenman. Compositae (Helia.-Pect.). 1 Mexico
Urceola Roxb. Apocynaceae. 16 Burma, W Mal. Rubber sources esp. *U. esculenta* Benth. ex Hook.f. (kyetpaung, Burma) – fr. ed., source of blue dye
Urceolina Reichb. Amaryllidaceae (Liliaceae s.l.). 2 Andes. Cult. orn. bulbous pls. See also *Stenomesson*
urd *Vigna mungo*
Urechites Muell. Arg. = *Pentalinon*
Urelytrum Hackel. Gramineae (40h). 7 trop. Afr.
Urena L. Malvaceae. 6 trop. & warm. Schizocarp with each carpel hooked; some fibres esp. *U. lobata* L. (Congo jute, cousin mahoe, guaxima, aramina, trop.) – jute subs.
Urera Gaudich. Urticaceae (I). 35 trop. & S Afr., Madag., Hawaii, trop. Am. Trees & shrubs with powerful stinging hairs with fr. encl. in persistent fleshy P; alks; often used for hedges, local medic. *U. baccifera* (L.) Wedd. (S Am.) – bark used for amate paintings sold to tourists in Mex. & *U. laciniata* (Goudot) Wedd. (Panamá) also with large spines
Urginea Steinh. = *Drimia*
Urgineopsis Compton = *Drimia*
Uribea Dugand & Romero. Leguminosae (III 2). 1 C Am.
Urmenetea Philippi. Compositae (Mut.-Mut.). 1 N Chile & NW Arg. – Andean annual herb
urn plant *Platyaechmea fasciata*
Urnularia Stapf = *Willughbeia*
Urobotrya Stapf. Opiliaceae. 7 trop. Afr. (2), SE As. to Flores (5). R: BJ 107(1985)137
Urocarpidium Ulbr. Malvaceae. 12 Mex., C Andes, Galápagos Is.
Urocarpus J.L. Drumm. ex Harvey = *Asterolasia*
Urochlaena Nees (~ *Tribolium*). Gramineae (25). 1 W Cape, Karoo
Urochloa Pal. Gramineae (34b). 110 OW trop. (Aus. 18). Most spp. form. referred to *Brachiaria*. Some with pedicel tips acting as elaiosomes. Some fodder esp. *U. maxima* (Jacq.) R. Webster (Afr.)
Urochondra C. Hubb. Gramineae (31e). 1 Sudan & Somalia to Sind. Grain with beak (accrescent connate style-bases)
Urogentias Gilg & C. Benedict. Gentianaceae. 1 Tanz.: *U. ulugurensis* Gilg & C. Benedict – C with v. long fimbriate processes
Urolepis (DC) R. King & H. Robinson (~ *Eupatorium*). Compositae (Eup.-Gyp.) 1 S Am.
Uromyrtus Burret. Myrtaceae (Myrt.). 12–15 Borneo to New Guinea, New Caled., trop. Aus.
Uropappus Nutt. (~ *Microseris*). Compositae (Lact.-Mic.). 1 SW US
Urophyllum Jack ex Wallich. Rubiaceae (I 9). 200 OW trop. to Japan
Urophysa Ulbr. Ranunculaceae (III 1). 2 China
Uroskinnera Lindley. Scrophulariaceae. 4 Mex. & C Am. R: Madrono 39(1992)131
Urospatha Schott. Araceae (V 3). 20 trop. Am. Seeds corky, water-disp. *U. antisylleptica* R. Schultes (Colombia) – ground spadix an oral contraceptive; *U. caudata* (Poeppig & Endl.) Schott (Brazil) – spongy rhizome ed.
Urospathella Bunting = praec.
Urospermum Scop. Compositae (Lact.-Hypo.). 2 Medit. incl. Eur. to Pakistan
Urostachya (Lindley) Brieger = *Eria*
Urostemon R. Nordenstam = *Brachyglottis*
Urostephanus Robinson & Greenman = *Matelea*
Urotheca Gilg. = *Gravesia*
Ursia Vassilcz. = *Trifolium*
Ursinia Gaertner. Compositae (Anth.-Urs.). 38 S Afr. & Ethiopia(1). R: MBSM 6(1967)363, 531. Shrubs to herbs; some ray-florets absent exposing iridescent involucral bracts poss. mimicking beetles (cf. *Gorteria*). Some annual spp. cult. for orange or yellow fls
Ursiniopsis E. Phillips = *Ursinia*
Urtica L. Urticaceae (I). 80 subcosmop. esp. N temp. (Eur. 11). Nettles. Usu. herbs, incl. annuals, with opp. lvs, monoec. or dioec., usu. with alks & stinging hairs (bulb-like projections which break off along a weak line leaving sharp bevelled edge which pierces skin & injects (in *U. urens* n., N temp.) at least), histamine causing itching, a neurotoxin (prob. a sodium channel toxin), 5-hydroxytryptamine & acetylcholine a burning sensation

effective even in dried pl. In *U. urens*, panicle of male & female fls at each node, in *U. pilulifera* L. (Roman n., S Eur.) male catkin-like infl. & female pseudo-head infl. at each node, in *U. dioica* L. (stinging n., stinger, Euras., widely natur., tetraploid weed poss. derived from stingless diploid ('var. *inermis*') of forests) usu. dioec.; A bent down in bud & when ripe spring upwards violently, the anthers turning inside out & ejecting a cloud of smooth poll. borne away by wind. Some young shoots eaten like spinach (nettle pudding), some fibre-sources esp. for fishermen's nets: *U. dioica* used for cloth until C18 & in Silesia until C20, an infusion used as a massage to cure dandruff; *U. parviflora* Roxb. (Himal.) – shoots sold as vegetable in Sikkim; *U. pilulifera* 'var. *dodartii*' passed off as 'Spanish marjoram' in C18 as a 'joke'

Urticaceae Juss. Dicots – Dilleniidae – Urticales. 48/1050 trop. to temp. (few). Usu. dioec. or monoec. wind-poll. herbs & shrubs, lianes or trees (few), often with stinging hairs but usu. without milky latex. Lvs simple, spirally arr. or opp., usu. with 3 subequal veins from base & with stipules. Fls small, usu. reg., in axillary cymes s.t. reduced to 1 fl., s.t. unisexual; males with P(1–)4 or 5(6) s.t. united by tube & A (1 or) as many as & opp. them, filaments violently reflexed at pollen maturity, anthers with longit. slits, G vestigial; females with P 4(5), often unequal, free to ± united or 0, G pseudomonomerous, 1-loc. with 1 style & 1 basal orthotropous (to hemitropous) bitegmic ovule. Fr. an achene, nut or drupe often encl. in accrescent P (sometimes fleshy), rudimentary A in females acting to eject achene; seeds with reduced testa & straight embryo in thin oily or starchy endosperm (or 0). x = 6–14

Classification & principal genera: 5 tribes

1. **Urticeae** (trees to herbs, stinging hairs): *Dendrocnide, Laportea, Obetia, Urera, Urtica*
2. **Lecantheae** (Procrideae; no stinging hairs; P of females 3-lobed, stigma like a paint-brush): *Elatostema, Pilea*
3. **Boehmerieae** (no stinging hairs; males usu. with A 4 or 5; involucre 0): *Boehmeria, Leucosyke, Maoutia* (females with P 0)
4. **Parietarieae** (no stinging hairs; P present; bracts often forming involucre): *Parietaria* (fls bisexual, stipules 0)
5. **Forsskaoleeae** (no stinging hairs; A 1; R: NJB 8(1988)34): *Forsskaolea*

Close to Moraceae & Cecropiaceae, which with Cannabaceae should probably be reunited with U. Position in Dilleniidae or Hamamelidae controversial

Fibres (*Boehmeria* – ramie, *Debregeasia, Forsskaolea, Girardinia, Laportea, Oreocnide, Sarcochlamys, Touchardia, Urtica* etc.), some ed. as spinach (*Laportea, Urtica* etc.) & some cult. orn. esp. spp. of *Pilea* & *Soleirolia*

urucu *Bixa orellana*

urunday *Astronium urundeuva*

Urvillea Kunth. Sapindaceae. 13 trop. Am. Lianes

Usteria Willd. Strychnaceae (Loganiaceae s.l.). 1 W & C Afr.

utile *Entandrophragma utile*

Utleria Beddome ex Benth. Asclepiadaceae (I; Periplocaceae). 1 S India: *U. salicifolia* Bedd. ex Hook. f.

Utleya Wilbur & Luteyn. Ericaceae. 1 Costa Rica

Utricularia L. Lentibulariaceae. 180 cosmop. (Eur. 6) esp. trop. Aquatics, epiphytes or even twiners. R: KB 18(1964)1. No clear distinction between roots, stems & lvs; *U. vulgaris* L. (bladderwort, N temp.), a submerged aquatic with no roots (not even in seedling), the stem bearing photosynthetic appendages carrying bladders with trap-doors; when these are triggered by small animals, the animal is sucked into the bladder & the trapdoor recloses (recovery time less than 1/500 sec.), the pitchers secreting digestive enzymes & absorbing the nutrients from the corpse; other spp. (trop.) have runners bearing bladders or (epiphytes) tuberous branches holding water; in aquatics the fls are aerial on axes with small lvs. All lvs, bladders, runners, tubers, erect shoots etc.'homologous', developing from similar primordia under diff. conditions, though not all primordia are totipotent; the stolons of *U. longifolia* Gardner (Brazil) & terr. spp. strongly resemble adventitious roots like some rootcapless ones in *Pinguicula*. Many spp. with cleistogamous fls in spring & resting buds (turions) tolerant of dry conditions. Cult. orn. for enthusiasts. *U. humboldtii* Schomb. (trop. S Am.) – grows in *Brocchinia* pitchers at high alts

Utsetela Pellegrin. Moraceae (IV). 1 W trop. Afr.

uva grass *Gynerium sagittatum*

uvalha *Eugenia uvalha*

Uvaria L. Annonaceae. 110 OW trop. (Afr. 69). Usu. lianes with recurved infl.-axis hooks. Acetogenins with antitumour action. Some local medic. esp. *U. chamae* Pal. (finger root,

W Afr.) – eyewash etc.; some ed. fr.

Uvariastrum Engl. Annonaceae. 7 trop. Afr.

Uvariodendron (Engl. & Diels) R. Fries. Annonaceae. 12 trop. Afr.

Uvariopsis Engl. Annonaceae. 11 trop. Afr.

uva-ursi *Arctostaphylos uva-ursi*

uvilla *Pourouma cecropiifolia*

Uvularia L. Colchicaceae (Uvulariaceae; Liliaceae s.l.). 5 E N Am. Bellwort. Cult. orn. rhizomatous herbs incl. *U. sessilifolia* L. – young shoots ed. like asparagus (cf. *Ornithogalum*)

Uvulariaceae A. Gray ex Kunth = Colchicaceae; see also Convallariaceae

V

Vaccaria Wolf. Caryophyllaceae (III 1). 1 Euras., Medit.: *V. hispanica* (Miller) Rauschert (*V. pyramidata*, *V. segetalis*, cow cockle) – cult. orn., appearing from sown birdseed

Vacciniaceae DC ex Gray = Ericaceae

Vacciniopsis Rusby = *Disterigma*

Vaccinium L. Ericaceae. c. 450 circumpolar (4), Eur. (8), N Am. (65), trop. Am. (30), C & SE Afr. mts & Madag. (6), Japan (22), trop. As. (76) to Mal. (240). Decid. or evergreen shrubs, small trees & lianes, many with ed. fr. (blueberries, buckberries, huckleberries, bluets, N Am.), some cult. orn. *V. angustifolium* Aiton (NE N Am.) – fr. much-coll. from wild pls managed for production; *V. arboreum* Marshall (farkleberry, S & SE US) – fr. inedible; *V. arctostaphylos* L. (Broussa tea, Cauc.); *V. corymbosum* L. (E US) – cult. in N Am. & the blueberry grown comm. in W & C Eur. for pies, syrups, tinning etc.; *V. macrocarpon* Aiton (cranberry, N As., E N Am.) – fr. eaten with poultry, grown comm. by Joseph Banks in GB 1808, before US (1840s); *V. mortinia* Benth. (mortiña, Ecuador) – fr. in local commerce; *V. myrtillus* L. (bilberry, whortleberry, blaeberry, Euras.) – ed. fr. for pies etc., wine-making, fr. skin alleged to improve night vision; *V. ovalifolium* Sm. (mathers, N N Am.) – cult. orn., fls opening before lvs; *V. oxycoccos* L. (cranberry, N temp.) – lvs revolute, fr. an inferior subs. for *V. macrocarpon*; *V. reticulatum* Sm. ('ohelo berry, Hawaii) – disp. Hawaiian geese (nénés), fr. in comm. preserves; *V. stramineum* L. (deerberry) – buzz-poll., cult. orn., fr. ined.; *V. vitis-idaea* L. (cowberry, lingberry, foxberry, Euras., N Am.) – fr. a cranberry subs.

Vagaria Herbert. Amaryllidaceae (Liliaceae s.l.). 4 Syria & Israel (1: *V. parviflora* Herbert – cult. orn.), N Am. (3)

Vaginularia Fée = *Monogramma*

Vahadenia Stapf. Apocynaceae. 2 trop. W Afr.

Vahlia Thunb. Saxifragaceae (Vahliaceae). 5 trop. & S Afr. R: KB 30(1975)163

Vahliaceae Dandy. See Saxifragaceae

Vahlodea Fries = *Deschampsia*

vahy *Landolphia madagascariensis*

Vailia Rusby. Asclepiadaceae (III 4). 1 Bolivia: *V. mucronata* Rusby

Valantia L. Rubiaceae (IV 16). 3–4 Macaronesia to Iran (Eur. 3)

Valdivia C. Gay ex Remy. Grossulariaceae (Escalloniaceae). 1 Chile

Valentiana Raf. = *Thunbergia*

Valentiniella Speg. = *Heliotropium*

Valenzuelia Bertero ex Cambess. = *Guindilia*

Valeria Minod = *Stemodia*

valerian *Valeriana officinalis*; **Greek v.** *Polemonium caeruleum*; **red v.** *Centranthus ruber*

Valeriana L. Valerianaceae. c. 200 N temp. (Eur. 20), S Afr., Andes (incl. *Phyllactis*, R: NJB 6(1986)435). Herbs incl. pachycaul rosette-pls (*Phyllactis*) & shrubs with alks; K forms pappus on fr. Valerian esp. *V. officinalis* L. (Euras.) – rhizome medic. form. used in scent etc., attractive to cats, Canidae & rats (? Pied Piper of Hamelin), poss. due to actinidine; other spp. locally medic. or cult. orn. e.g. *V. pratensis* (Benth.) Steudel in Mex.

Valerianaceae Batsch. Dicots – Asteridae – Dipsacales. Incl. Triplostegiaceae, 10/300 ± cosmop. esp. N temp. & Andes. Herbs & rarely shrubs with char. foetid odour (monoterpenoid & sesquiterpenoid ethereal oil-cells, detectable after 100 yrs in herbarium material). Lvs simple (entire to pinnatifid) or pinnate, opp.; stipules 0. Fls usu. bisexual with bracts & bracteoles but no epicalyx (exc. *Triplostegia*), in cymes but not involucrate heads; K 0, inconsp. teeth (accrescent to form pappus in fr. of *Valeriana*), rarely 5 & distinct (*Nardostachys*), C ((3–)5), ± reg. to irreg. & s.t. bilabiate with tube ± spurred & nectariferous basally, lobes imbricate, A (1–)3(4), attached to tube, anthers versatile with longit. slits. $\overline{G}$ (3) with term. style & 2 locules reduced to 0, fertile 1 with 1 pendulous anatropous, unitegmic ovule. Fr. like an achene, s.t. with wing or plumose K, with 1 seed, embryo straight, oily in 0 endosperm. x = (7–)9(–12)

Genera: *Aligera, Centranthus, Fedia, Nardostachys, Patrinia, Plectritis, Pseudobetckea, Triplostegia* (G more like Dipsacaceae), *Valeriana, Valerianella*

Prob. best incl. in Caprifoliaceae (shorn of *Sambucus* & *Viburnum*)

Some ed. (*Valerianella*), medic. & scents (*Nardostachys* – spikenard, *Valeriana*), few cult. orn. (*Centranthus, Patrinia* etc.)

Valerianella Miller. Valerianaceae. 50 N temp. to N Afr. (Eur. 22). Fr. disp. mech. various – inflated K, or sterile loc., K parachute or hooks (diff. spp.). Some potherbs etc. esp. *V. locusta* (L.) Laterr. (cornsalad, lamb's-lettuce, Eur. & Medit.) – salad

Valerioa Standley & Steyerm. = *Peltanthera*

Valerioanthus Lundell (~ *Ardisia*). Myrsinaceae. 2 C Am. R: Wrightia 7(1982)50

Vallariopsis Woodson. Apocynaceae. 1 Malay Pen., Sumatra: *V. lancifolia* (Hook.f.) Woodson. R: AUWP 86-5(1986)89

Vallaris Burm.f. Apocynaceae. 3 trop. As. R: MLW 82-11(1982)

Vallea Mutis ex L.f. Elaeocarpaceae. 2 Colombia to Bolivia. R: NJB 8(1988)19

Vallesia Ruíz & Pavón. Apocynaceae. 8 trop. Am. Shrubs & small trees with alks

vallis See *Vallisneria*

Vallisneria L. Hydrocharitaceae. 2–10 trop. & warm temp. (Eur. 1). Submerged dioec. grass-like herbs; male fls in head subtended by spathe, breaking off & floating up to open on surface; female fls solit. & sessile in tubular spathe on long peduncle, tightly spiralling in fr. & maturing fr. at bottom; good oxygenators for aquaria ('vallis')

Vallota Salisb. ex Herbert = *Cyrtanthus*

valonea *Quercus macrolepis*

Valvanthera C. White = *Hernandia*

Vanasushava P.K. Mukherjee & Constance. Umbelliferae (III 11). 1 S India

Vanclevea E. Greene. Compositae (Ast.-Sol.). 1 US. R: GBN 34(1974)151

Vancouveria Morren & Decne. (~ *Epimedium*). Berberidaceae (II 2). 3 W N Am. R: JLSBot. 51(1938)445. Cult. orn. herbs with ant-disp. seeds incl. *V. hexandra* (Hook.) Morren & Decne (inside-out flower, W US) – seeds also distrib. *Vespula vulgaris* (L.), common yellow jacket

Vanda Jones ex R. Br. Orchidaceae (V 16). 45 Himal. to Mal. Cult. orn. epiphytes with extrafl. nectaries; see also *Papilionanthe* & *Taprobanea*

Vandasia Domin = *Vandasina*

Vandasina Rauschert (*Vandasia*). Leguminosae (III 10). 1 New Guinea, Queensland. Allied to *Kennedia*

Vandellia P. Browne = *Lindernia*

Vandenboschia Copel. = *Crepidomanes*

Vandopsis Pfitzer. Orchidaceae (V 16). 5 SE As. to Mal. Some cult. orn.

Vangueria Comm. ex Juss. Rubiaceae (III 2). c. 15 trop. Afr. & Madag. Trees & shrubs, some with ed. fr. esp. *V. madagascariensis* J. Gmelin (*V. edulis*, Madag.)

Vangueriella Verdc. (~ *Vangueriopsis*). Rubiaceae (III 2). 21 W trop. Afr. R: KB 42(1987)189

Vangueriopsis F. Robyns. Rubiaceae (III 2). 4 trop. Afr. R: KB 42(1987)187

Vanheerdia L. Bolus ex Hartmann. Aizoaceae (V). 2 Cape

Vanhouttea Lemaire. Gesneriaceae. 3 Brazil. R: Selbyana 6(1982)180

vanilla *Vanilla planifolia* & other spp.

Vanilla Miller. Orchidaceae (V 4). 100 trop. & warm. Epiphytic scandent & s.t. leafless lianes to 30+ m cult. for orn. & for long fr. (pods), which are cured & source of v. extract (esp. in Madag. where anther & stigma have to be pressed together as poll. bee absent) – usu. endozoochorous. *V. planifolia* Andr. (trop. Am.) used by Aztecs to flavour cocoa; although vanillin synthesized in 1874 & from eugenol from cloves in 1891, pods still used in cooking by the discriminating though most flavouring today from wood pulp as a byproduct of paper-making & from coal-tar (toluene)

Vanillosmopsis Schultz-Bip. = *Eremanthus*

Vanoverberghia Merr. Zingiberaceae. 1 Philippines: *V. sepulchrei* Merr. – ? extinct, fr. & seeds said to be ed.

Van-royena Aubrév. = *Pouteria*

Vantanea Aublet. Humiriaceae. 16 trop. Am. R: CUSNH 35(1961)49. Bat-disp. in Amazon. *V. barbourii* Standley (Costa Rica) – to 65 m tall & 2 m diam.

Vanwykia Wiens. Loranthaceae. 1 SE Afr.

Vanzijlia L. Bolus. Aizoaceae (V). 1 SE Cape to Namaqualand

Vargasiella C. Schweinf. Orchidaceae (V 10). 2 Venez., Peru

Varilla A. Gray. Compositae (Hele.-Flav.). 2 S N Am. R: Phytol. 69(1990)4

Varronia P. Browne = *Cordia*

Varthemia DC. Compositae (Inul.). 1 Iran, Afghanistan, Pakistan

Vaseyanthus Cogn. Cucurbitaceae. 2 SW N Am.

Vaseyochloa A. Hitchc. Gramineae (31d). 1 Texas

Vasivaea Baillon. Tiliaceae. 2 Brazil, Peru

Vasquezia Philippi = *Villanova*

Vasqueziella Dodson. Orchidaceae (V 10). 1 Bolivia

Vassilczenkoa Lincz. Plumbaginaceae (II). 1 C As., Afghanistan: *V. sogdiana* (Lincz.) Lincz.

Vassobia Rusby (~ *Witheringia*). Solanaceae. 4 S Am. R: Kurtziana 17(1984)91

Vatairea Aublet. Leguminosae (III 4). 7 trop. Am.

Vataireopsis Ducke = *Vatairea*

Vateria L. Dipterocarpaceae. 2 S India (1), Sri Lanka (1). *V. indica* L. (S India) – resin (white dammar, piney varnish) form. imp., seed-fat (Malabar tallow, dhupa fat) used for candles etc.

Vateriopsis Heim. Dipterocarpaceae. 1 Seychelles: *V. seychellarum* (Dyer) Heim reduced to 50 trees through felling for timber

Vatica L. Dipterocarpaceae. 65 Indomal. Some timbers & dammars

Vatovaea Chiov. Leguminosae (III 10). 1 E trop. Afr. to Oman: *V. pseudolablab* (Harms) Gillett – pods & lvs ed.

Vauanthes Haw. = *Crassula*

Vaughania S. Moore = *Indigofera*

Vaupelia Brand = *Cystostemon*

Vaupesia R. Schultes. Euphorbiaceae. 1 Colombia, W Brazil: *V. cataractarum* R.E. Schultes – seeds ed. once boiled

Vauquelinia Corr. Serr. ex Humbl. & Bonpl. Rosaceae. 3 SW N Am. R: Sida 12(1987)101

Vausagesia Baillon = *Sauvagesia*

Vavaea Benth. Meliaceae (I 3). 4 Philippines to W Pacific. R: Blumea 17(1969)351. Pagoda trees, some bird-poll.?

Vavara Benoist. Acanthaceae. 1 Madagascar

Vavilovia Fed. (~ *Pisum*). Leguminosae (III 20). 1 Turkey to Cauc.: *V. formosa* (Steven) Fed.

Veconcibea (Muell. Arg.) Pax & K. Hoffm. = *Conceveiba*

Veeresia Monach. & Mold. = *Reevesia*

Vegaea Urban. Myrsinaceae. 1 Hispaniola. Diageotropic-rooted epiphyte

vegetable gold *Coptis trifolia*; **v. hair** *Chamaerops humilis, Carex brizoides*; **v. ivory** *Phytelephas macrocarpa, Palandra aequatorialis*; **v. i. substitute** *Hyphaene thebaica*; **v. lamb of Tartary** *Cibotium barometz*; **v. marrow** *Cucurbita pepo*; **v. oyster** *Tragopogon porrifolius*; **v. pear** *Sechium edule*; **v. sheep** *Psychrophyton* spp. esp. *P. mammillaris*; **v. spaghetti** *C. pepo*; **v. sponge** *Luffa aegyptiaca*; **v. sulphur** *Lycopodium clavatum*

Veillonia H. Moore (~ *Burretiokentia*). Palmae (V 4m). 1 New Caled.: *V. alba* H. Moore. R: Allertonia 3(1984)390

veitchberry see *Rubus*

Veitchia H. Wendl. Palmae (V 4k). 18 Palawan (1), Vanuatu, Fiji. Some cult. orn.

Velezia L. Caryophyllaceae (III 1). 6 Medit. (Eur. 2) to Afghanistan. R: FT 2(1967)135

Vella L. Cruciferae. 4 W Medit. (Eur. 2). Some spiny. R: BJLS 82(1981)165

Velleia Sm. Goodeniaceae. 21 Aus. (20 endemic), New Guinea. R: PLSNSW 92(1967)27. G ± superior

Vellereophyton Hilliard & B.L. Burtt (~ *Helichrysum*). Compositae (Gnap.-Gnap.). 7 S Afr., *V. dealbatum* (Thunb.) Hilliard & B.L. Burtt natur. Aus. & NZ. R: OB 104(1991)168

Vellosiella Baillon. Scrophulariaceae. 3 Brazil

Vellozia Vand. Velloziaceae. 124 trop. Am., esp. campos. Some hummingbird poll. in SE Brazil

Velloziaceae D. Don ex Endl. Monocots – Liliidae – Liliales. 8/288 S Am., Afr., Madag. & S Arabia. R: SCB 30(1976)3. Often ± pachycaul xeromorphic shrubs usu. with marcescent leaf-sheaths & adventitious roots; vessel-elements in roots & s.t. shoots; sec. growth 0. Lvs narrow, parallel-veined, clustered at branch-tips, blade eventually falling from sheath. Fls usu. bisexual, reg., 3-merous, usu. consp., solit. axillary; P 3 + 3 petaloid, ± basally united with tube often bearing 6 united appendages forming a corona outside & s.t. adnate to A 3 + 3 (6 bundles of 2–∞ in *Vellozia*) free of or adnate to P, anthers with longit. slits with pollen often in tetrads, $\overline{G}$ (3), 3-loc. with axile placentation & slender style, the placentae lamellar bearing ∞ ovules. Fr. a loculicidal capsule with compressed hard bitegmic seeds; embryo small in copious hard endosperm with hemicellulose, protein & oil or s.t. starchy. x = 9

Genera: *Barbacenia, Barbaceniopsis, Burlemarxia, Nanuza, Pleurostima, Talbotia, Vellozia, Xerophyta*

Both affinities of fam. & its internal classification controversial

Veltheimia Gled. Hyacinthaceae (Liliaceae s.l.). 2 S Afr. R: JRHS 97(1952)483. Cult. orn. greenhouse-pls

velvet bean *Mucuna pruriens* var. *utilis*; **v.leaf** *Abutilon theophrasti*; **v. sumac(h)** *Rhus hirta*; **v. tamarind** *Dialium guineense, D. ovoideum*

Velvitsia Hiern = *Melasma*

Venegasia DC. Compositae (Hele.-Cha.). 1 SW N Am.: *V. carpesioides* DC. R: Sida 15(1992)223

Venice turpentine *Larix decidua*

Venidium Less. = *Arctotis*

Ventenata Koeler. Gramineae (21b). 5 S Eur. (2), Medit. to Caspian. Linked to *Trisetaria*

Ventilago Gaertner. Rhamnaceae. 35 OW trop. (Afr. 1, Madag. 1). Some hook-climbers; fr. winged (accrescent style). *V. calyculata* Tul. (India) – cooking oil from seeds; *V. madraspatana* Gaertner (trop. As.) – bark a source of a red dye for textiles

Ventricularia Garay. Orchidaceae (V 16). 1 India: *V. tenuicaulis* (Hook.f.) Garay. R: OB 95(1988)56

Venus's comb *Scandix pecten-veneris*; **V. flytrap** *Dionaea muscipula*; **V. looking-glass** *Legousia hybrida*

Veprecella Naudin = *Gravesia*

Vepris Comm. ex A. Juss. Rutaceae. 15 trop. & S Afr. to Masc. Alks

Veratrilla Baillon ex Franchet. Gentianaceae. 2 E Himal. to W China

veratrin *Schoenocaulon officinale*

Veratrum L. Melanthiaceae (Liliaceae s.l.). 15–20 N temp. (Eur. 2). R: Plantsman 11(1989)35. Coarse rhizomatous herbs of wet places. Cult. orn. with plicate lvs & term. panicles of white, green, brown or purplish fls, the lowermost bisexual, the upper male, occ. with all-male pls. V. many alks: locally medic. esp. *V. album* L. (white hellebore, Euras.) – lambs born of ewes which have fed on it have a single central eye, suggesting an inspiration for the Polyphemus of Homer, used in control of rodents & pl. pests in Eur. since C1

verawood *Bulnesia arborea*

Verbascum L. Scrophulariaceae. c. 360 Euras. (Eur. 87, Turkey 228), Ethiopian & E Afr. highlands. R: LUA n.s. 29,2(1933–4), 32,1(1936). Mulleins; usu. biennial herbs, rarely annual or shrubby, with large rosettes of lvs & pachycaul infls typical of much dry Medit. veg. as in Greece & Turkey. Incl. *Celsia* (often with long pedicels & deeply lobed lvs, e.g. *V. daenzeri* (Fauché & Chaub.) Fenzl (*C. d.*, SE Eur.)). Many hybrids though usu. sterile. Cult. orn.; some fr. & seeds used to kill fish in Greece & form. in Spain (verbasco, barbasco). *V. blattaria* L. (moth m., Euras., natur. N Am.) – cult. orn., seeds long-lived (germ. after 90 yrs in Dr W.J. Beal's experiment set up in 1879 with 20% success & some still viable after 100 yrs); *V. lychnitis* L. (white m., W & C Eur., W As.) – cult. orn.; *V. nigrum* L. (black or dark m., Euras.); *V. thapsus* L. (Aaron's rod, common m., flannel plant, hag-taper, Euras., natur. N Am.) – seeds germ. after 100 yrs in Dr Beal's experiment, lvs used medic. incl. in cigarettes for the asthmatic

verbena, garden v. *Glandularia × hybrida*; **lemon v.** *Aloysia citrodora*

Verbena L. Verbenaceae. 200 trop. & temp. Am., 2–3 OW (Eur. 2) incl. *V. officinalis* L. (vervain, Juno's tears, Euras., Am.) – form. medic., bright-eyed fls suggesting cure for eye disease ('Doctrine of Signatures'). Cult. orn. herbs esp. *V. canadensis* (L.) Britton (rose vervain, N Am.) – some highly fragrant forms; *V. litoralis* Kunth used in fert. control in Uruguay. See also *Aloysia, Glandularia*

Verbenaceae J. St-Hil. Dicots – Asteridae – Lamiales. (Incl. Symphoremataceae) 41/950 trop. esp. S Am. with few temp. Trees lianes, shrubs & herbs, s.t. dioec. (*Citharexylum, Lippia*) sometimes thorny, young twigs often 2-angled. Lvs usu. simple, usu. opp.; stipules 0. Fls usu. bisexual in racemes, cymes or heads, often with involucre of coloured bracts; K ((4)5(–8)) with teeth or lobes s.t. irreg., C ((4)5(–8)) with imbricate lobes, ± irreg. to 2-lipped, often with slender tube, A (2)4 sometimes with staminodes, filaments arising from C-tube alt. with lobes, anthers with longit. slits, $\underline{G}$ usu. (2), initially 2-loc. but later subdivided by intrusive partitions into 4 uni-ovulate, style 1; ovules usu. anatropous, unitegmic. Fr. a head of 1-seeded separating nutlets, a drupe with 2 or 4 stones or a 2- or 4-valved capsule; seeds with straight oily embryo usu. in 0 endosperm (occ. oily). x = 5–12

Principal genera: *Aloysia, Duranta, Glandularia, Junellia, Lantana, Lippia, Petrea, Stachytarpheta, Verbena*

The limits of the fam. have been much debated but recent evaluations of Lamiales remove many genera to Labiatae s.l. & split off smaller groups as fams – Avicenniaceae,

Cyclocheilaceae, Nesogenaceae, Phrymaceae & Stilbaceae (q.v.), though a 'greater Labiatae' would probably include them all

Timber (*Citharexylum*), flavourings & medic. teas etc. (*Aloysia, Nashia, Stachytarpheta* etc.), cult. orn. (*Aloysia, Duranta, Glandularia* (garden verbena) *Lantana, Petrea, Verbena*) but some widespread weeds of warm countries esp. *Lantana, Lippia, Stachytarpheta* & *Verbena* spp.

Verbenoxylum Tronc. Verbenaceae. 1 Brazil

Verbesina L. Compositae (Helia.-Verb.). c. 300 warm Am. Trees, shrubs & herbs. Some cult. orn. (crown-beard) incl. *V. helianthoides* Michaux (*Actinomeris h.*, S & SC US) – caruncles

Verdcourtia R. Wilczek = *Dipogon*

Verdickia De Wild. = *Chlorophytum*

Verena Minod = *Stemodia*

Verhuellia Miq. = *Peperomia*

Verlotia Fourn. = *Marsdenia*

Vermeulenia Löve & D. Löve = *Orchis*

vermicelli *Triticum turgidum* Durum Group

Vermifrux J.B. Gillett = *Lotus*

vermilion wood *Pterocarpus* spp.

vernal grass, sweet *Anthoxanthum odoratum*

Vernicia Lour. = *Aleurites*

Vernonanthera H. Robinson (~ *Vernonia*). Compositae (Vern.-Vern.). 65 trop. Am.

Vernonia Schreber. Compositae (Vern.-Vern.). c. 500 OW trop. (E Afr. 126, R: KB 43(1988)199) & warm to N Am. (Am. trop. spp. = *Lepidoploa, Lessingianthus, Vernonanthera* etc. eventually to incl. N am. spp. too). See also *Baccharoides*. Timber trees (e.g. *V. arborea* Buch.-Ham., Indomal.), sometimes pachycaul (cabbage trees) with v. large lvs, shrubby or herbs (iron weed, US); some locally medic. (antitumour sesquiterpene lactones – vernoniosides) & chewing-sticks esp. *V. amygdalina* Del. (trop. Afr.) – stem pith taken by sick chimpanzees (vernoniosides effective against drug-resistant malarial parasites rich there) which avoid app. toxic lvs though these used in soup by humans, & insecticidal. *V. galamensis* (Cass.) Less. (*V. pauciflora*, trop. Afr.) – seed-oil low viscosity, rich in epoxy acid for plastics, coatings better than solvent-based paints causing smog etc. etc.

Vernoniopsis Humbert. Compositae (Ast.-Ast.). 1 Madag.: *V. caudata* (Drake) Humbert – tree

Veronica L. Scrophulariaceae. c. 180 N temp. (Eur. 62, Turkey 79, N Am. 12), few trop. mts (Afr. 25) & S (Aus. 7) with 9 subcosmop. weeds. R: BMOIPB 82,1(1977)151, OB 75(1984)57. Speedwell (i.e. 'goodbye' because C falls as soon as gathered), bird's-eye. Herbs (cf. *Hebe*) with rotate C & A 2 held laterally & grasped by visiting insects such that they come together dusting poll. agents with pollen. Some cult. orn. incl. *V. beccabunga* L. (brooklime, Euras., Medit.) – helophyte, locally medic., shoots ed.; *V. catenata* Pennell (water speedwell, N Temp. to E Afr.); *V. chamaedrys* L. (germander speedwell, Euras., natur. N Am.); *V. filiformis* Sm. (As. Minor & Cauc.) – escaped in GB in 1927 (first record 1838), a persistent weed with rooting stems in lawns in GB & US, esp. common in churchyards; *V. longifolia* L. (N to C Eur., natur. N Am.) – common garden perennial; *V. officinalis* L. (fluellen (arch.), N temp.) – form. medic. tea etc.; *V. persica* Poiret (subcosmop.) – allotetraploid weed (2n = 28) derived from *V. polita* E. Fries (2n = 14, N Iran, a Neolithic weed spreading to Europe) × *V. ceratocarpa* C. Meyer (2n = 14, Cauc. & Iran); *V. prostrata* L. (Eur.) – common rock pl. with deep blue fls. Some spp. turn black after c. 100 yrs as herbarium specimens. See also *Pseudolysimachion*

Veronicastrum Heister ex Fabr. Scrophulariaceae. 1 temp. NE As.; 1 temp. NE Am.: *V. virginicum* (L.) Farw. (Culver's root) – cult. orn., violent purgative, emetic & cathartic (!). R: OB 75(1984)56

Verreauxia Benth. Goodeniaceae. 3 SW Aus. R: FA 35(1992)298. Nut

Verrucifera N.E. Br. = *Titanopsis*

Verrucularia A. Juss. Malpighiaceae. 2 Brazil. R: MNYBG 32(1981)45

Verrucularina Rauschert = *Verrucularia*

Verschaffeltia H. Wendl. Palmae (V 4n). 1 Seychelles: *V. splendida* H. Wendl. – monoec. palm with spiny trunk & stilt-roots

Versteegia Valeton. Rubiaceae (II 2). 2 Papuasia

Verticordia DC. Myrtaceae (Lept.). 97 Aus. esp. SW. R: Nuytsia 7(1991)291. Cult. orn. heath-like shrubs (feather flowers, morrison)

vervain *Verbena officinalis*; **rose v.** *V. canadensis*

Vescisepalum (J.J. Sm.) Garay, Hamer & Siegerist = *Bulbophyllum*

Veselskya Opiz = *Pyramidium*
Veseyochloa J. Phipps = *Tristachya*
Vesicarex Steyerm. = *Carex*
Vesicaria Adans. = *Alyssoides*
Vesselowskya Pampan. Cunoniaceae. 1 NSW: *V. rubifolia* (F. Muell.) Pampan.
Vestia Willd. Solanaceae (2). 1 Chile: *V. foetida* (Ruíz & Pavón) Hoffsgg. (*V. lycioides*) – cult. orn. shrub
vetch *Vicia* spp. esp. *V. sativa*; **bitter v.** *V. ervilea*; **bush v.** *V. sepium*; **chickling v.** *Lathyrus sativus*; **horseshoe v.** *Hippocrepis comosa*; **kidney v.** *Anthyllis vulneraria*; **milk v.** *Astragalus glycophyllos*; **point-v.** *Oxytropis* spp.; **Russian** or **Siberian v.** *V. villosa*; **tufted v.** *V. cracca*
vetchling, common or **meadow** *Lathyrus pratensis*; **yellow v.** *L. aphaca*
vetiver *Vetiveria zizanioides*
Vetiveria Bory ex Lem. Gramineae (40c). 10 OW trop. *V. zizanioides* (L.) Nash (vetiver, khuskhus, cuscus, sevendara (grass), India) – planted for erosion control in Sri Lanka tea estates, fragrant 'roots' used in scent-making (allegedly a constituent of Chanel No. 5) & woven into mats, baskets, screens, ('sandalwood') fans (used by Moghul emperors on wetted punkas) etc. which give off scent when sprinkled with water, containing insect-repellent zizanol & epizizanol (terpenes)
Vexatorella Rourke (~ *Leucospermum*). Proteaceae. 4 SW Cape to Namaqualand. R: SAJB 50(1984)373
Vexillabium Maekawa. Orchidaceae. 2 Korea, Japan
Viburnaceae Raf. = Caprifoliaceae
Viburnum L. Caprifoliaceae (or Adoxaceae). 150 temp. (Eur. 3) & warm esp. As. & N Am. Small trees & shrubs, s.t. with domatia, with panicles or umbel-like cymes, these s.t. with pollinator-attractant marginal sterile fls with enlarged C (cf. K in *Hydrangea* with which s.t. confused), in some garden forms all fls thus, extrafl. nectaries, ed. or poisonous drupes; v. imp. cult. orn. garden shrubs with consp. often scented fls & orn. fr., locally medic. etc. *V. acerifolium* L. (arrowwood, E N Am.) – bark medic., wood for arrows; *V.* × *bodnantense* Aberc. ex Stearn (*V. farreri* × *V. grandiflorum* Wallich, Himal.) – fragrant winter fls.; *V. carlesii* Hemsley (Korea, Japan) – white v. fragrant fls; *V. costaricense* Hemsley (C Am.) – pH of nectar c. 10 (most nectar acidic); *V. dentatum* L. (arrowwood, E N Am.) – wood for arrows; *V. ellipticum* Hook. (W N Am.) – with stipules! (cf. × *Fatshedera*); *V. farreri* Stearn (*V. fragrans*, N China) – fragrant winter fls; *V. lantana* L. (wayfaring tree, twist-wood, Euras., Medit.) – arrowshafts found with Neolithic 'Iceman' preserved in Alps ice; *V. lantanoides* Michaux (*V. alnifolium*, hobblebush, mooseberry, moosewood, E N Am.) – marginal sterile fls; *V. lentago* L. (E N Am.) – fr. cooked by N Am. Indians; *V. opulus* L. (guelder rose, crampbark, Euras., Medit.) – wood form. for skewers, fr. ed. & subs. for cranberries, bark medic., 'Roseum' (Whitsuntide boss, snowball tree) – all fls (white) sterile making a globose infl. the original Guelder rose cult. near Guelders, inner fls of wild form with nectar containing indole & a sweet-fishy scent attractive to Diptera; *V. plicatum* Thunb. (China & Japan) – long cult. in sterile form & first introduced thus, resembling *V. opulus* 'Roseum' but lvs unlobed, the wild form (f. *tomentosum* (Thunb.) Rehder) often taken for a hydrangea; *V. prunifolium* L. ((black) haw, E & EC N Am.) – fr. ed. esp. after frost; *V. setigerum* Hance (*V. theiferum*, C & W China) – local 'tea'; *V. tinus* L. (laurustinus, Medit.) – imp. constituent of orig. Medit. vegetation (the 'plane' of Isaiah) – winter-flowering evergreen with poisonous fr.
Vicatia DC. Umbelliferae (III 8). 4 Himalaya to W China
Vicia L. Leguminosae (III 20). 140 N temp. (Eur. 54) with extensions to S Am., Hawaii (*V. menziesii* Sprengel – endangered) & trop. E Afr. but paraphyletic. Usu. scrambling herbs (vetch, tare) with alks. Forage, green-manure, ed. seeds etc. *V. cracca* L. (tufted v., N temp.) – cult. orn. scrambler, forage; *V. ervilia* (L.) Willd. (bitter v., Macaronesia & Medit. to Afghanistan, ? orig. SW As.) – fodder; *V. faba* L. (broad bean, field b., horse b. (2n = 12), usu. considered cultigen derived from *V. narbonensis* L. (Medit. to C As., 2n = 14) but poss. domesticated in C As.) – primitively black-seeded, the bean of antiquity & that of 'peas & beans' in mediaeval crop-rotation in GB, ed. seeds but when uncooked can lead to hepatitis in Italians & some Jewish people due to biochemical deficiency in red blood cells, var. *equina* Pers. (pigeon bean) – small-seeded form fed to racing pigeons; *V. sativa* L. (vetch, tare, Euras., natur. N Am.) – green manure & fodder; *V. sepium* L. (bush v., Eur. to Himal.); *V. villosa* Roth (Russian or Siberian v., Euras.) – manure, hay & silage
Vicoa Cass. = *Pentanema*
Victoria Lindley. Nymphaeaceae (Euryalaceae). 2 trop. S Am. Rooted aquatics prickly except on adaxial leaf surfaces, lvs peltate, floating with upturned margin & remarkably

architectural leaf-venation, the fragrant fls opening white late in afternoon & staying open but turning pink on 2nd day. In temp. regions cult. as annuals under glass, a patriotic vogue ('*V. regia*') in last century leading to construction of Victoria Houses as at Chatsworth (Derbyshire) & Oxford Botanic Garden, their designs being forerunners of Paxton's Crystal Palace of the Great Exhibition (1851). *V. amazonica* (Poeppig) Sowerby (*V. regia*, giant waterlily, Guyana & Amazonia) – germ. at 30 °C, lvs to 2 m diam., supporting weight of a child, 4 spp. of dynastid beetle attracted by scent & temp. (11 °C above ambient) are trapped for 24 hrs & feed on starchy carpellary appendages before leaving with pollen, seeds ed. if roasted; *V. cruziana* Orb. (N Arg., Paraguay, Bolivia) – germ. at 20°C, often cult. in place of *V. amazonica* (upturned leaf margin to 20 cm, outer P prickly only at base) as is their hybrid, *V.* 'Longwood Hybrid'

Victorinia Léon = *Cnidoscolus*

Vieraea Schultz-Bip. (*Vieria*). Compositae (Inul.). 1 Canary Is.

Viereckia R. King & H. Robinson (~ *Eupatorium*). Compositae (Eup.-Crit.). 1 Mex. (hybrid?, coll. once). R: MSBMBG 22(1987)312

Vieria Webb & Berth. = *Vieraea*

Vietnam mint *Persicaria odorata*

Vietsenia C. Hansen. Melastomataceae. 4 Vietnam. R: BMNHN 4,9(1987)259

Vigethia W.A. Weber. Compositae (Helia.-Verb.). 1 Mexico

Vigia Vell. = *Fragariopsis*

Vigna Savi. Leguminosae (III 10). 150 trop. esp. OW. R: ARES 20(1989)199. Erect or twining herbs with extrafl. nectaries, like *Phaseolus* but stipules often with appendages, thickened part of style less strongly twisted, etc.; incl. *Voandzeia* (spp. with subterr. poll. by ants & geocarpy). Imp. pulses, green manure etc. *V. aconitifolia* (Jacq.) Maréchal (*Phaseolus a.*, moth bean, S As.) – beans ed., forage; *V. angularis* (Willd.) Ohwi & Ohashi (*P. a.*, adzuki b., As.) – long cult. for beans boiled or made into curd, constituent of comm. 'washing grains'; *V. caracalla* (L.) Verdc. (*P.c.*, snail flower, trop. Am.) – cult. orn. perennial with fragrant fls & keel coiled like snail-shell; *V. lanceolata* Benth. (maloga bean, Aus.) – taproot to 40 cm ed.; *V. mungo* (L.) Hepper (*P. m.*, urd, black gram, trop. As.) – imp. pulse in As.; *V. radiata* (L.) R. Wilczek (*P. r.*, *P. aureus*, mung bean, green or golden gram, ? Indonesia) – ed. seeds & pods, seedlings the 'bean sprouts' of commerce, var. *sublobata* (Roxb.) Verdc. the poss. ancestor; *V. subterranea* (L.) Verdc. (*Voandzeia s.*, Bambara groundnut, W Afr.) – geocarpic fr. with imp. ed. seeds; *V. umbellata* (Thunb.) Ohwi & Ohashi (*P. calcaratus*, rice bean, S As.) – beans ed.; *V. unguiculata* (L.) Walp. (cowpea, yawa, OW) – imp. pulse (chowlee – India, gubgub – WI), forage & fibre poss. domesticated Ethiopia, subsp. *sesquipedalis* (L.) Verdc. (asparagus b., snake bean, yard-long b.) – legume to 90 cm eaten; *V. vexillata* (L.) A. Rich. (trop. OW) – roots ed.

Viguiera Kunth. Compositae (Helia.-Helia.). 180 warm & trop. Am. R: CGH 54(1918)1. Few cult. orn. with yellow fls

Viguierella A. Camus. Gramineae (31d). 1 Madagascar

Villadia Rose. Crassulaceae. 25–30 Texas to Peru. Cult. orn. succ.

Villanova Lagasca (excl. *Unxia*). Compositae (Hele.-Hym.). 10 Mex. to Chile

Villaresiopsis Sleumer = *Citronella*

Villaria Rolfe. Rubiaceae (II 5). 5 Philippines

Villarsia Vent. Menyanthaceae. 16 SE As. to Aus. (12, R: Muelleria 2(1969)3), 1 Afr. Ant-disp. in Aus.

Villebrunea Gaudich. ex Wedd. = *Oreocnide*

Villocuspis (A.DC) Aubrév. & Pellegrin = *Chrysophyllum*

Vilobia Strother. Compositae (Hele.-Pect.). 1 Bolivia

Viminaria Sm. Leguminosae (III 24). 1 Aus.: *V. juncea* (Schrader) Hoffsgg. (*V. denudata*) – cult. orn. shrub with broom-like branches of lvs mostly reduced to thread-like petioles

Vinca L. Apocynaceae. 7 Eur. (5) to N Afr. & C As. R: BMNHN II, 23(1951)439, W.I. Taylor & N. Farnsworth (1973) *The V. alkaloids*. Periwinkles, cult. orn. esp. *V. major* L. (greater p.) – lvs ciliate & *V. minor* L. (lesser p.) – lvs glabrous of Eur., natur. GB, with v. many alks., locally medic., vincamine used in cerebral vascular disorders though now comm. synthesized from tabersonine from *Voacanga* spp. See also *Catharanthus*

Vincentella Pierre = *Synsepalum*

Vincentia Gaudich. = *Machaerina*

Vincetoxicopsis Costantin. Asclepiadaceae (III 4). 1 SE As.: *V. harmandii* Costantin

Vincetoxicum Wolf (~ *Cynanchum*). Asclepiadaceae (III 1). 15 temp. Euras. (Eur. 11)

Vindasia Benoist. Acanthaceae. 1 Madagascar

vine *Vitis vinifera*, (Am.) any liane; **balloon v.** *Cardiospermum halicacabum*; **kangaroo v.** *Cissus* *antarctica;* **kudzu v.** *Pueraria lobata;* **Madeira** or **mignonette v.***Anredera cordifolia;* **potato v.** *Ipomoea pandurata;* **Russian v.** *Fallopia aubertii;* **silk v.** *Periploca graeca;* **wonga-wonga v.** *Pandorea pandorana*

vinhatico *Plathymenia reticulata*

Vinkia Meijden. Haloragidaceae. 1 N Aus.

Vinkiella R. Johns. Gnetaceae. 1 W New Guinea

Vinticena Steudel = *Grewia*

Viola L. Violaceae. c. 400 temp. (Eur. 91) esp. N & Andes. Violets. R: JRHS 55(1930)223, 57(1932)212. Herbs, rarely subshrubs, often with cleistogamous fls at end of season. Consp. fls with insect landing-place formed by anterior petal often spurred & containing nectar; in some spp. pollen shed on to that petal & lower edge of stigma protected from it by a flap which closes as insect leaves fl. thus preventing self-poll. (in weedy spp. like *V. arvensis* Murray (field pansy, Eur.) there is no flap & self-poll. frequent); some with ant-disp. seeds with caruncles (arils). Cult. orn. (violets, violas (rayless = violettas), pansies) comm. since 400 BC in Attica & local medic. *V. arborescens* L. (W Medit.) – stems woody; *V. canina* L. (dog violet, Euras.) – cult.; *V. cornuta* L. (viola, Spain & Pyrenees) – tufted or bedding pansies & violas largely derived from this sp.; *V. oahuensis* C. Forbes (Oahu, Hawaii) – unbranched shrub to 40 cm in cloud forest; *V. odorata* L. (sweet violet, Euras. to Afr.) – essential oil used in flavouring & in scent-making, 100 kg of flowers giving 31 g of oil, fls crystallized as food decorations etc., incl. Parma (prob. involving *V. suavis* M. Bieb. (E Eur.) & poss. other spp.) & Devon v. & the common v. of florists (prob. hybrids), signifying modesty in 'Language of Fls'; *V. palustris* L. (marsh v., N temp.); *V. tricolor* L. (heart's-ease, love-in-idleness, Johnny-jump-up, Eur. & natur. in N Am.) – locally medic. (risky) & a parent of the garden pansies (*V.* × *wittrockiana* Gams ex Kappert) raised c. 1830 & involving *V. lutea* Hudson subsp. *sudetica* (Willd.) W. Becker (C Eur. mts) & *V. altaica* Ker-Gawler (As. Minor), backcrossing with *V. tricolor* in wild

Violaceae Batsch. Dicots – Dilleniidae – Violales. 20/800 cosmop. but only *Viola* temp. Herbs, shrubs & even lianes or trees often with alks. Lvs simple, entire to dissected, usu. spirally arr. with stipules. Fls reg. to irreg., usu. bisexual (s.t. cleistogamous), bibracteolate, solit. & axillary or in racemes, panicles or heads; K 5, imbricate, ± free, often persistent, C 5 imbricate or convolute, the lowermost often with spur in irreg. fls, A (3)5 with filaments free or ± connate & anthers often connivent around G & with nectaries on back, $\underline{G}$ ((2)3(–5)), 1-loc. with parietal placentation & 1 style, each placenta with 1–∞ anatropous, bitegmic ovules. Fr. a berry or loculicidal capsule, rarely a nut (*Leonia*); seeds often arillate with straight embryo embedded in oily endosperm. x = 6–13, 17, 21, 23

Principal genera: *Hybanthus, Rinorea, Viola*

Mostly woody; prob. related to Flacourtiaceae

Some medic. esp. *Corynostylis* & *Hybanthus*; cult. orn. esp. *Hymenanthera* & *Viola*, which also yields essential oils

violet *Viola* spp.; **African v.** *Saintpaulia ionantha;* **Arabian v.** *Exacum affine;* **bog v.** *V. palustris, Pinguicula vulgaris;* **v. cress** *Ionopsidium acaule;* **dame's v.** *Hesperis matronalis;* **Devon v.** see *V. odorata;* **dog v.** *V. canina;* **d. tooth v.** *Erythronium dens-canis;* **essence of v.** *Iris* × *germanica* 'Florentina'; **Parma v.** see *V. odorata;* **sweet v.** *V. odorata;* **Usambara v.** *Saintpaulia ionantha;* **water v.** *Hottonia palustris*

violetta See *Viola*

viper's bugloss *Echium vulgare;* **v. grass** *Scorzonera hispanica*

Viposia Lundell = *Plenckia*

Virecta Sm. = *Virectaria*

Virectaria Bremek. Rubiaceae (IV 1). 7 trop. Afr.

Virga Hill = *Dipsacus*

Virgilia Poiret. Leguminosae (III 26). 2 S & SW Cape. R: SAJB 52(1986)347. *V. oroboides* (Bergius) Salter (*V. capensis*) – cult. orn., timber for rafters etc.; alks

virgin oil *Olea europaea*

Virginia(n) bluebell or **cowslip** *Mertensia virginica;* **V. creeper** *Parthenocissus quinquefolia;* **V. snakeroot** *Endotheca serpentaria;* **V. stock** *Malcolmia maritima*

Virgulaster Semple = *Aster*

Virgulus Raf. = *Aster*

Viridivia J.H. Hemsley & Verdc. Passifloraceae. 1 Zambia, SW Tanz.

Virola Aublet. Myristicaceae. 45 trop. Am. Timber, hallucinogenic snuff (active principles in cambium) blown up nostrils through *Ischnosiphon* tubes (though often taken as pills covered in Lecythidaceous (e.g. *Eschweilera*), Cyclanthaceous or (best) *Spathiphyllum cannifolium* ash) & comm. fats (see also *Bicuiba*). Alks. *V. elongata* (Benth.) Warb. (*V. theiodora,*

NE S Am.) – resin boiled, dried & powdered to give hallucinogenic snuff rich in tryptamines, also arrow-poison; *V. koschnyi* Warb. (banak, C & N S Am.) – timber, promising tree for reafforestation; *V. sebifera* Aublet (N S Am.) – seed-oil used for candles & soap; *V. surinamensis* (Rottb.) Warb. (NE S Am.) – bird- & water-disp., timber (baboen, dalli) for plywood etc., seed-oil (caihuba, ucahuba, ucuhuba) like cocoa butter

Virotia L. Johnson & B. Briggs. Proteaceae. 6 New Caledonia

Viscaceae Batsch. Dicots – Rosidae – Santalales. 7/385 cosmop. esp. trop. & warm. Photosynthesizing monoec. or dioec. hemiparasites on trees; haustoria penetrating & ramifying in host tissues, aerial branching often pseudo-dichotomous & jointed. Lvs simple, entire, s.t. scale-like, opp.; stipules 0. Fls reg., small, usu. yellow or green, wind- or insect-poll., ± sessile in (1–)3-flowered dichasia in axil of 1 bract in spikes or branched infls; P 4 (males) or 3 (females), valvate but often merely teeth or bumps on ovary rim, A opp. & often adnate to P, anthers with longit. slits or reduced to 1 or 2 sporangia & opening with term. pores or slits or partitioned & with transverse slits or confluent; $\overline{G}$ (3 or 4), 1-loc. with term style & massive placenta bearing 2 8-nucleate embryo-sacs. Fr. a berry (s.t. explosive) with 1(2) testa-less seeds with viscid tissue at 1 end & large embryo embedded in starchy chlorophyllous endosperm. x = 10–14 (polyploidy rare)

Genera: *Arceutholobium, Dendrophthora, Ginalloa, Korthalsella, Notothixos, Phoradendron, Viscum*

Form. incl. in Loranthaceae but embryology v. different though *Lepidoceras* (incl. here) with some features of L. (Cronquist)

Some troublesome parasites in plantations, *Phoradendron* & *Viscum* providing Am. & OW decorative mistletoe respectively, the first dried also sold as 'wood fls' in C Am.

Viscainoa E. Greene. Zygophyllaceae. 1 Baja California

Viscaria Bernh. = *Silene*

Viscum L. Viscaceae. 65 OW trop. to temp. (few; Eur. 2, Aus. 4). *V. album* L. (mistletoe, Eur., W As.) – parasitic (3 sspp., 2 restricted to conifers in C & S Eur. & N Turkey) esp. on lime, hawthorn, poplars, apple & other aliens, v. rarely on oak (etc. (even hyperparasitizing *Loranthus europaeus* app. reducing it in oakwoods when artificially increased using blackcaps)), its distrib. poss. increased by Man, dioec. with fly-poll. fls, fr. taken by birds but viscid tissue on seed prevents seed being swallowed & bird scrapes it off into bark-crevice etc. where it germinates, berries used as bird-lime & locally medic. (cytotoxic & immunostimulatory) but with cardiotoxic polypeptide (viscotoxin) affecting mammals but not birds, v. large chromosomes, since mid-C19 (Herefordshire) with evergreen holly & ivy ancient wonder at evergreen-ness in N winter accommodated in Christianized winter festival of Christmas, poss. 'Golden Bough' of Virgil's Aeneid also imp. in Viking sagas but modern significance from assoc. with Druids since time of Pliny (the kissing habit app. English & poss. assoc. with general views on fertility symbol presented by paired frs); *V. articulatum* Burm.f. (Indomal.) – on other Viscaceae, Loranthaceae & *Exocarpos* in NSW, lvs sold as paste for fractures & bruises in Sikkim

Vismia Vand. Guttiferae (I). 35 trop. Am. & Afr. (few). Some resins locally medic.

Vismianthus Mildbr. Connaraceae. 1 SE Tanz., 1 SW Burma. R: AUWP 89-6(1989)369

visnaga *Ammi visnaga*

Visnaga Miller = *Ammi*

Visnea L.f. Theaceae. 1 Canary Is., Madeira: *V. mocanera* L.f. – cult. orn.

Vitaceae Juss. (Vitidaceae). Dicots – Rosidae – Rhamnales. 14/850 trop. & warm. Lianes usu. with tendrils opp. lvs, rarely succ. treelets or herbs; alks 0. Lvs simple, less often palmate or pinnate but often palmately lobed or veined; stipules often decid. Fls reg., small, bisexual or not, (3)4 or 5(–7)-merous, in cymes or panicles, term., opp. lvs or rarely axillary; K small & ± reduced to lobes or a collar, C valvate, usu. free (apically a decid. calyptra in *Vitis* etc.), A opp. C, nectary-disk of 5 glands or annular to cupulate around $\underline{G}$ (2), 2-loc. with simple style & s.t. ± sunk in disk, each loc. with 2 anatropous bitegmic ovules ascending from carpellary margins. Fr. a 1- or 2-loc. berry, each loc. with 1 or 2 seeds; seeds endotestal with a deep groove either side of raphe, the embryo small, straight embedded in ruminate oily & proteinaceous endosperm. x = 11–20

Principal genera: *Ampelocissus, Ampelopsis, Cayratia, Cissus, Cyphostemma, Tetrastigma, Vitis*

The seed structure suggests affinities with Sapindales rather than Rhamnales (Corner)

Grapes & currants (*Vitis*) & cult. orn. esp. *Ampelopsis, Cissus, Parthenocissus* & *Rhoicissus*

Vitaliana Sesler = *Douglasia*

Vitellaria Gaertner f. (*Butyrospermum*). Sapotaceae (I). 1(–2) W trop. Afr. & Cameroun: *V.*

paradoxa Gaertner f. (*B. parkii*) – pulp ed. (sweet), kernel source of shea butter rich in vitamin E, used in food (form. in chocolate as has high melting-point) & illumination, cosmetics & aromatherapy, root a chewing-stick in Nigeria

Vitellariopsis Baillon ex Dubard. Sapotaceae (I). 6 E Afr.

Vitex L. Labiatae (I; Verbenaceae s.l.). 250 trop. to temp. (few; Eur. 1). Trees & shrubs incl. monocaul pachycauls (Madag., but pollen distinctive), some with domatia, some with ecdysteroids (e.g. *V. glabrata* R. Br. (Indomal.) – medic. in Thailand). Some timber (fiddle-wood) & cult. orn. *V. agnus-castus* L. (chaste tree, S Eur., widely natur.) – rheophyte, twigs s.t. used in basketwork, fr. a pepper subs., 'Alba' with white fls long considered a symbol of chastity; *V. altissima* L.f. (Pakistan to Sri Lanka) – timber for construction esp. window-frames & cabinet-work; *V. cofassus* Reinw. ex Blume (New Guinea teak, vitex, Mal. to W Pacific) – wood for carving, drums, pestles; *V. cymosa* Benth. (trop. Am.) – water-disp. in Amazon (fr. with aerenchyma); *V. divaricata* Sw. (Venez., WI) – timber for shingles, lvs for tanning; *V. doniana* Sweet (trop. Afr.) – common savanna tree, local timber, fr. ed.; *V. lucens* Kirk (puriri, NZ) – cult. orn. tree; *V. negundo* L. (Indomal.) – shrub, largely sterile, soil-stabilizer in Nepal; *V. pinnata* L.f. (*V. pubescens*, milla, trop. As.) – timber for construction, living tree carrying many epiphytic orchids in Malay Pen. & so much searched by collectors; *V. rotundifolia* L.f. (~ *V. trifolia* L., Indomal.) – used for 'flu, colds & sore eyes in modern Chinese herbalism; *V. triflora* Vahl (trop. S Am.) – sweet-tasting fr. poss. for conserves

Viticipremna H.J. Lam (~ *Vitex*). Labiatae (I; Verbenaceae s.l.). 5 Mal. to W Pacific. R: JABG 7(1985)181

Vitidaceae Juss. = Vitaceae

Vitiphoenix Becc. = *Veitchia*

Vitis L. Vitaceae. c. 65 N hemisph. Lianes cult. for fr. eaten fresh, dried or drunk fermented etc. (distrib. birds, foxes, bears, box turtles etc. incl. primates attracted by alcohol – now known to be beneficial in reasonable quantities) & orn. with domatia (often) & tendrils in place of infls esp. *V. vinifera* L. (grape vine, poss. derived from 'subsp. *sylvestris*' (Medit., though that name strictly refers to escaped pls in Rhine Valley)) & now cult. to 52° N in Poland) – cult. clones prob. arose in SW As. (all with bisexual fls & not dioec.) grown from 4th millenium BC in Syria & Egypt, from 2500 BC in Aegean, but after outbreak of attack by *Phylloxera* rootlouse (1867) Am. spp. introduced & used as resistant stocks; these esp. *V. labrusca* also involved in hybrids giving American cultivar grapes much grown in US & crossed with *V. vinifera* to give French hybrids, largely wine grapes, but also v. hardy dessert grape 'Brant' (= *V.* 'Clinton' (*V. labrusca* × *V. riparia* Michaux (C N Am.)) × *V. vinifera* 'Black St Peters'); for early ripening necessary in e.g. UK hybrids with *V. amurensis* Rupr. (NE As.) grown. Classical cvs of *V. vinifera* grown for wine in France etc. incl. 'Pinot Noir', 'Cabernet Sauvignon' (red) & 'Chardonnay' & 'Riesling' (white) etc. fermented for wines, claret (or red Bordeaux) being made from 'Cabernet Sauvignon', 'Merlot' etc., white Bordeaux & sancerre from 'Sauvignon Blanc', sauternes from 'Sémillon' etc., burgundy from 'Pinot Noir', beaujolais from 'Gamay', chablis from 'Chardonnay', champagne from 'Chardonnay' & 'Pinot Noir' (white & pink), fitou & minervois etc. from 'Carignan', Alsace from 'Gewürztraminer', 'Pinot Blanc', 'Sylvaner' etc.; elsewhere 'Riesling' for Mosel & 'hock' (Eiswein from grapes frozen on vines, Germany), 'Sangiovese' (& increasingly with 'Cabernet Sauvignon' for chianti & 'Corvina' (& many others)) for valpolicella, 'Gorganega' for soave, 'Malvasia' & 'Trebbiano' for frascati (Italy); world's most planted red cv. is 'Garnacha' (esp. Spain, being a component of rioja) with 'Pinotage' widely grown in S Afr., NZ etc. (monoterpenes responsible for aroma & flavour of Muscat grapes, also 'Riesling', 'Traminer' & 'Müller-Thurgau'; 2-methoxy-3-isobutylpyrazine in 'Cabernet Sauvignon', 'Sauvignon Blanc' & 'Shiraz' gives 'vegetative' flavour (also in broad beans, spinach & parsnips), poss. connected with anti-herbivory & disp. attraction etc.), the concentration of organic acids esp. malic acid, anthocyanin & phenolics etc. affected by rain régime so that latter 2 higher in water deficit; with added brandy or other alcohol such wines are 'fortified' giving sherry, madeira, port etc.; with added flavourings they give vermouths & martinis; with quinine, iron etc., they are medicinal. Grape wine has been associated with religion & other ceremonial since antiquity (though drunk diluted by Anc. Greeks) & despite its dangers (in excess) is an intrinsic feature of Western society. Fresh dessert grapes imp. export of S Afr., Israel etc.; dried fr. = raisins used in cooking, sultanas being seedless forms ('Sultana' (Kishmish, Thompson Seedless)) & the most widely grown, currants being smaller ones much used in cakes etc. (currant a corruption of Corinth(ian grape); muscatel grapes are those left to dry on vine in Med.; seed-oil used in cosmetics & cooking (grapeseed oil). Other cult. spp. incl. *V. acerifolia* Raf. (bush g., SC US) – fr. sweet; *V. arizonica* Engelm. (canyon g., SW N Am.) – fr. sweet; *V. coignetiae* Pull. ex Planchon (Japan, Korea) – rapid-growing screening liane with red autumn colours; *V. labrusca* L. (fox or skunk g., NE US) – fr. with musky taste; *V. monticola* Buckley (mountain g., SW Texas); *V. palmata* Vahl (cat g., C & S US); *V. rotundifolia* Michaux (bullace or fox g., SE US) – fr. thick-skinned, ed.; *V. rupestris* Scheele (sand g., SC US) – fr. sweet; *V. vulpina* L. (chicken or frost g., C & E US) – fr. v. acid but becoming sweet & ed. after frost

Vittadinia A. Rich. Compositae (Ast.-Ast.). 29 New Guinea (2), Aus., New Caled. (1), NZ (1). R: Brunonia 5(1982)1. Some cult. orn.

Vittaria Sm. Vittariaceae. 50–80 trop. & warm. Epiphytes with linear pendent fronds with linear sori in parallel rows; some cult. orn.

Vittariaceae (C. Presl) Ching (~ Adiantaceae). 6/130 trop. (few in Afr. & Aus., NZ 0) to Japan, Arg. & SE US. Epiphytic ferns. Rhiz. protostelic to dictyostelic with clathrate scales (often with metallic sheen). Fronds simple with usu. simple or forked veins. Sporangia usu. in simple or branched soral lines s.t. in grooves but without true indusium

Genera: *Anetium, Antrophyum, Hecistopteris, Monogramma, Rheopteris* (? poss. to be excl.), *Vittaria*

Vittetia R. King & H. Robinson (~ *Eupatorium*). Compositae (Eup.-Gyp.). 2 Brazil. R: MSBMBG 22(1987)110

Viviania Cav. Geraniaceae (Vivianiaceae). 6 S Brazil, Chile. R: UCOP 2(1975)231

Vivianiaceae Klotzsch. See Geraniaceae

Vladimiria Iljin = *Dolomiaea*

Vleisia Toml. & Posluszny = *Pseudalthenia*

Voacanga Thouars. Apocynaceae. 12 trop. OW (Afr. 7, As. 5). R: AUWP 85–3(1985)5. Seeds of *V. africana* Stapf & *V. thouarsii* Roemer & Schultes (trop. Afr.). coll. for industrial production of tabersonine (alk.) used as depressor of central nervous system activity in geriatric patients & source of precursor for vincamine (see *Vinca*) for cerebral vascular disorders

Voandzeia Thouars = *Vigna*

Voaniola Dransf. Palmae (V 5b). 1 Madag.: *V. gerardii* Dransf. R: KB 44(1989)191. 2n = 596+ (twice that in any other monocot; polyploidy rare in palms)

Voatamalo Capuron ex Bosser. Euphorbiaceae. 2 Madag. R: Adansonia 15(1976)333

Vochysia Aublet. Vochysiaceae. 100 trop. Am. R: RTBN 41(1948)423. Some timbers esp. *V. hondurensis* Sprague (quaruba, C Am.); many local medic.

Vochysiaceae A. St-Hil. Dicots – Rosidae – Polygalales. (Incl. Euphroniaceae) 8/210 trop. Am. & W Afr. (1/3). Trees, shrubs, rarely herbs, with resin, often with internal phloem & accum. aluminium. Lvs simple usu. opp. or whorled; stipules small to 0. Fls bisexual, obliquely irreg., bibracteolate, in racemes or thyrses; K 5 basally connate, imbricate with 1 often large & spurred basally, C (0) 1–3(5), ± unequal, convolute or imbricate, A 1–5(–7) usu. only 1 fertile with filaments free or connate in 2 groups, anthers with longit. slits, $\underline{G}$ to $\overline{G}$ (3), 3-loc. with axile placentas or pseudomonomerous with 1 loc., style 1, each loc. with (1)2–∞ anatropous or hemitropous bitegmic ovules. Fr. a loculicidal capsule or samara winged with accrescent K; seeds often winged or hairy with straight embryo ± endosperm. n = 11

Genera: *Callisthene, Erisma, Erismadelphus, Euphronia, Qualea, Ruizterania, Salvertia, Vochysia*

Ovule structure suggests alignment with Linales

Some timbers

vodka *Solanum tuberosum* (Poland), *Triticum aestivum* (Russia)

Voharanga Costantin & Bois = *Cynanchum*

Vohemaria Buchenau = *Cynanchum*

Voladeria Benoist = *Oreobolus*

Volkensia O. Hoffm. = *Bothriocline*

Volkensiella H. Wolff = *Oenanthe*

Volkensinia Schinz. Amaranthaceae (I 2). 1 E Afr. R: KB 33(1979)417

Volkiella Merxm. & Czech. Cyperaceae. 1 SW Afr., Zambia

Volutaria Cass. (~ *Centaurea*). Compositae (Card.-Cent.). 16 Medit. (Eur. 1), SW As., trop. E Afr.

Vonitra Becc. Palmae (V 4e). 4 Madag. Stems often forked. *V. fibrosa* (C. Wright) Becc. – source of Madagascar piassava; *V. utilis* Jum. prob. reduced to c. 6 trees in rain forest through overexploitation for palm cabbage etc.

Vossia Wallich & Griffith. Gramineae (40h). 1 trop. & SW Afr., E India to Burma: *V. cuspidata* (Roxb.) Griffith – floating grass, part of the Nile sudd & poss. the orig. bulrush (q.v.) of the

Bible

Votomita Aublet. Melastomataceae. 9 trop. Am. R: FN 16(1976)256. Elaiophores attractive to bees

Votschia Ståhl (~ *Jacquinia*). Theophrastaceae. 1 NE Panamá: *V. nemophila* (Pittier) Ståhl. R: Brittonia 45(1993)204

Vouacapoua Aublet (~ *Andira*). Leguminosae (I 1). 3 trop. S Am. *V. americana* Aublet (*A. excelsa*) – good timber

Vouarana Aublet. Sapindaceae. 1 NE S Am.

Voyria Aublet. Gentianaceae. 20 trop. Am., W Afr. (1: *V. primuloides* Baker). Seeds unitegmic & anatropous to ategmic & orthotropous

Voyriella Miq. Gentianaceae. 2 NE S Am. R: MNYBG 51(1989)47

Vriesea Lindley. Bromeliaceae (2). 280 trop. Am. but few in Amazonia. R: FN 14(1977)1068, 1395. Usu. large epiphytes with conspic. bracts much cult. incl. many hybrids, some of unknown parentage. *V. imperialis* Carrière (Rio state, Brazil) – 3–5 m in montane grassland taking 20 yrs to flower; *V. incurva* (Griseb.) Read (trop. Am.) – leaf-bases secrete a mucilage said to be proteolytic; *V. ranifera* L.B. Sm. (Costa Rica cloud forest) – spp. of frog restricted to the pl. where they breed

Vrydagzynea Blume. Orchidaceae (III 3). 40 Indomal. to Taiwan & W Pacific

Vulpia C. Gmelin (~ *Festuca*). Gramineae (17). 22 temp. esp. Med. (Eur. 12) & W Am. R: KBAS 13(1986)97

Vulpiella (Battand. & Trabut) Burollet. Gramineae (17). 1 W Medit. incl. Eur.

Vvedenskya Korovin. Umbelliferae (III 8). 1 C As.

Vvedenskyella Botsch. Cruciferae. 2 C As. to Himalaya

W

Wachendorfia Burm. Haemodoraceae. 5 SW & S Cape. Tubers; lvs plicate

Wagatea Dalz. = *Moullava*

Wagenitzia Dostál = *Centaurea*

Wahlenbergia Schrader ex Roth. Campanulaceae. c. 200 subcosmop. esp. S temp. (Eur. 2, trop. Afr. & Madag. 51, St Helena 4 (2 extinct), Aus. 26 (21 endemic – R: Telopea 5(1992)91). Like *Campanula* & *Edraianthus* but capsule with apical valves; on St Helena pachycaul at high alts. Some cult. orn. ('bluebell', harebell, NZ, S Afr.) incl. *W. hederacea* (L.) Reichb. (ivy-leaved bell flower, W Eur.)

Waitzia Wendl. Compositae (Gnap.-Ang.). 5 temp. S & W Aus. R: Nuytsia 8(1992)461. Annuals with 'everlasting' capitula & ectomycorrhizae

Wajira Thulin. Leguminosae (III 10). 1 E Kenya

wake-robin *Trillium grandiflorum, Arum maculatum*

Wakilia Gilli = *Phaeonychium*

Walafrida E. Meyer. Globulariaceae. 40 trop. & S Afr.

Waldheimia Karelin & Kir. = *Allardia*

Waldsteinia Willd. Rosaceae. 6 N temp. (Eur. 2). Cult. orn. herbs incl. *W. ternata* (Stephan) Fritsch (Carpathians & 5000 km away in Siberia & Japan)

Walidda (A. DC) Pichon = *Wrightia*

walking fern *Asplenium rhizophyllum*

wall cress *Arabis* spp.; **w. lettuce** *Mycelis muralis*; **w. pennywort** *Umbilicus rupestris*; **w. pepper** *Sedum acre*; **w. rocket** *Diplotaxis muralis*; **w. rue** or **spleenwort** *Adiantum ruta-muraria*

wallaba *Eperua* spp.

wallaby grass *Danthonia* spp.

Wallacea Spruce ex Hook.f. Ochnaceae. 2 N S Am. R: BJ 113(1991)179

Wallaceodendron Koord. Leguminosae (II 5). 1 Philipp., Sulawesi: *W. celebicum*. R: BMNHN 4,5(1983)347

wallapata *Gyrinops walla*

Wallenia Sw. = *Cybianthus*

Walleniella P. Wilson = *Solonia*

Walleria J. Kirk. Tecophilaeaceae (Liliaceae s.l.). 3 trop. & S Afr. R: KB 16(1962)185

wallflower *Erysimum cheiri*; **alpine** or **fairy w.** *E.* spp.; **Siberian w.** *E.* × *marshallii*; **western w.** *E. asperum*

Wallichia Roxb. Palmae (V 1). 7 E Himal. to S China. Monoec., hapaxanthic

wallum *Banksia aemula*

walnut *Juglans regia*; **African or Benin w.** *Lovoa trichilioides*; **American** or **black w.** *J. nigra*;

w. **bean** *Endiandra palmerstonii*; **Black Sea, Carpathian, European** or **Persian w.** *J. regia*; **E Indian w.** *Albizia lebbeck*; **Japanese w.** *J. ailantifolia*; **New Guinea w.** *Dracontomelon* spp.; **Nigerian golden w.** *L. trichilioides*; **Otaheite w.** *Aleurites moluccana*; **Pacific** or **Papuan w.** *D.* spp.; **poison w.** *Cryptocarya glabella*; **Queensland** or **Australian w.** *Endiandra palmerstonii*; **satin w.** *Liquidambar styraciflua*; **yellow w.** *Beilschmiedia bancroftii*

Walsura Roxb. Meliaceae (I 4). 16 Indomal. (to Sulawesi). R: Blumea 38(1994)257. *W. monophylla* Elmer ex Merr. (Palawan) – on ultramafics accum. nickel to 7000 μg/g & tested for use in AIDS & cancer treatment

Walteranthus Keighery. Gyrostemonaceae. 1 W Aus.: *W. erectus* Keighery. R: BJ 106(1985)108

Waltheria L. Sterculiaceae. 30–50 trop. Am., Afr. (1), Madag. (1), Malay Pen. (1), Taiwan (1). Cult. orn. with alks esp. *W. indica* L. (*W. americana*, pantrop.) – form. medic. (Hawaii), fibre like jute

Wamalchitamia Strother. Compositae (Helia.-Verb.) 5 C Am. R: SBM 33(1991)30

wampee *Pontederia cordata*

wampi *Clausena lansium*

wand flower *Dierama* spp.

wandering Jew *Tradescantia* spp., *Saxifraga stolonifera*; **w. sailor** *Cymbalaria muralis*

wandoo *Eucalyptus wandoo*

wangee *Phyllostachys nigra*

Wangenheimia Moench (~ *Vulpia*). Gramineae (17). 2 Spain, N Afr. R: KBAS 13(1986)97

Wangerinia Franz = *Microphyes*

wapato *Sagittaria cuneata, S. latifolia*

wara *Calotropis gigantea*

warabi *Pteridium aquilinum*

waratah *Telopea speciosissima*

Warburgia Engl. Canellaceae. 3 E Afr. *W. salutaris* (Bertol.f.) Chiov. (*W. ugandensis*) – resin for fixing tool-handles, bark a purgative & with antifeedant sesquiterpenes (e.g. warburganal) active against army worm, lvs eaten in curries

Warburgina Eig. Rubiaceae (IV 16). 1 Syria, Israel

Wardaster Small = *Aster*

Wardenia King = *Brassaiopsis*

Warea Nutt. Cruciferae. 4 SE US. R: JAA 66(1985)99

Warionia Benth. & Cosson. Compositae (Cich.). 1 NW Sahara. Herb

Warmingia Reichb.f. Orchidaceae (V 10). 3 trop. Am. R: Lindleyana 7(1992)196. Cult. orn.

Warneckea Gilg (~ *Memecylon*). Melastomataceae. 31 trop. Afr. R: Adansonia 18(1978)228. Elaiophores attractive to bees

Warpuria Stapf = *Podorungia*

Warrea Lindley. Orchidaceae (V 10). 4 trop. Am. Cult. orn. terr. orchids with large fls

Warreella Schltr. Orchidaceae (V 10). 2 Colombia. *W. cyanea* (Lindley) Schltr. – cult. orn.

Warreopsis Garay. Orchidaceae (V 10). 3 trop. Am.

warrigal (cabbage) *Tetragonia tetragonioides*

Warszewiczia Klotzsch. Rubiaceae (I 5). 4 trop. Am. *W. coccinea* (Vahl) Klotzsch – cult. orn. greenhouse shrub with 1 sepal in a group of fls enlarged to 6 cm long, red, a mutant known with all fls with such a lobe

wartcress *Coronopus* spp. esp. *C. squamatus*

Wasabia Matsum. (~ *Eutrema*). Cruciferae. 4 E As. *W. wasabi* (Siebold) Makino (*W. japonica*, Japan, Sakhalin) – cult. condiment (wasabi) for eating with raw fish (sashimi), used like horseradish in Japan

washiba wood *Tabebuia serratifolia*

Washington thorn *Crataegus phaenopyrum*

Washingtonia H. Wendl. Palmae (I 1b). 2 arid SW N Am. R: GH 4(1936)53. Massive palms with bisexual fls & palmate lvs with spiny petioles (spineless above 14m, over height of present (or past?) herbivores). Used as street trees. *W. filifera* (André) H. Wendl. (SW US) – the palm of Palm Springs, fr. & seeds ed., leaf-fibre for basketry; *W. robusta* H. Wendl. (*W. sonorae*, Mex.) – fr. ed.

water arum *Calla palustris*; **w. ash** *Fraxinus caroliniana*; **w. avens** *Geum rivale*; **w. beech** *Carpinus caroliniana*; **w. blinks** *Montia fontana*; **w. caltrop** *Trapa natans*; **w. chestnut** *T.* spp.; **Chinese w. c.** *Eleocharis dulcis*; **chinquapin w. c.** *Nelumbo lutea*; **w.cress** *Rorippa* spp. esp. *R. nasturtium-aquaticum*; **w. crowfoot** *Ranunculus aquatilis*; **w. dropwort** *Oenanthe* spp.; **w. elm** *Planera aquatica*; **w. feather** *Myriophyllum brasiliense*; **w. fern** *Azolla* spp., *Salvinia* spp.; **w. figwort** *Scrophularia auriculata*; **w. forget-me-not** *Myosotis scorpioides*; **w. fringe** *Nymphoides peltata*; **w. grass** *Panicum molle*; **w. gum** *Tristania neriifolia*; **w. hair** *Catabrosa* *aquatica*; **w. hawthorn** *Aponogeton distachyos*; **w. hemlock** *Oenanthe crocata*; **w. hyacinth** *Eichhornia crassipes*; **w. lettuce** *Pistia stratiotes*; **w.lily** *Nymphaea* spp.; **giant w.l.** *Victoria amazonica*; **w. lobelia** *Lobelia dortmanna*; **w. meal** *Wolffia* spp.; **w. melon** *Citrullus lanatus*; **w. milfoil** *Myriophyllum* spp.; **w. mint** *Mentha aquatica*; **w. parsnip** *Sium* spp.; **w. pepper** *Persicaria hydropiper*; **w. plantain** *Alisma* spp.; **w. rice** *Zizania aquatica*; **w. rose apple** *Syzygium aqueum*; **w. soldier** *Stratiotes aloides*; **w. speedwell** *Veronica catenata*; **w. spinach** *Ipomoea aquatica*; **w. starwort** *Callitriche stagnalis*; **w. tupelo** *Nyssa aquatica*; **w. violet** *Hottonia palustris*; **w.weed, Canadian** or **American** *Elodea canadensis*; **w. wisteria** *Hygrophila difformis*; **w.wort** *Elatine* spp.; **w. yam** *Dioscorea alata*

Waterhousea B. Hyland (~ *Syzygium*). Myrtaceae (Myrt.). 4 trop. Aus. R: AusJB supp. ser. 9(1983)138

Watsonia Miller. Iridaceae (IV 2). 52 S Afr. Cult. orn. like *Gladiolus* but style-branches 2-fid, etc. esp. *W. borbonica* (Poiret) Goldb. (*W. ardernei*, *W. pyramidata*, SW Cape). Some noxious weeds in temp. Aus. esp. *W. meriana* (L.) Miller 'Bulbillifera', large-flowered sterile triploid with axillary cormlets

Wattakaka Hassk. = *Dregea*

wattle *Acacia* spp.; **black w.** *A. mearnsii*; **blue w.** *A. dealbata*; **Cootamundra w.** *A. baileyana*; **golden w.** *A. pycnantha*; **golden rain w.** *A. prominens*; **hairy w.** *A. pubescens*; **Mudgee w.** *A. spectabilis*; **silver w.** *A. dealbata*; **sticky w.** *A. howittii*; **Sydney golden w.** *A. longifolia*

wawa *Triplochiton scleroxylon*

wax bean *Phaseolus vulgaris*; **Carnauba w.** *Copernicia prunifera*; **w. flower** *Hoya carnosa*, *Chamelaucium* spp.; **Geraldton w. (f.)** *C. uncinata*; **w. gourd** *Benincasa hispida*; **w. jambu** *Syzygium samarangense*; **Japanese w.** *Rhus succedanea*; **w. myrtle** *Myrica* spp., **w. palm** *Ceroxylon alpinum*; **w. plant** *Hoya carnosa*, (Aus.) *Eriostemon* spp.; **w. tree, Japanese** *R. succedanea*; **urucury w.** *Syagrus coronata*

wax-lip (orchid) *Glossodia* spp.

wayfaring tree *Viburnum lantana*

weasel's snout *Misopates orontium*

Weberbauera Gilg & Muschler. Cruciferae. (Incl. *Stenodraba*) 17 W & S S Am. R: JAA 71(1990)221. Some cult, orn. alpines

Weberbauerella Ulbr. Leguminosae (III 13). 2 coastal Peru

Weberbauerocereus Backeb. = *Haageocereus*

Weberocereus Britton & Rose. Cactaceae (III 2). 9 C Am. esp. Costa Rica. Cult. orn. climbers with aerial roots, night-flowering intergrading with *Selenicereus*

Websteria S. Wright. Cyperaceae. 1 trop.

Weddellina Tul. Podostemaceae (Tristichaceae). 1 N S Am.: *W. squamulosa* Tul. – shoots to 75 cm & branched with short unbranched flowering ones. R: BMNHN 4,10(1988)169

wedding bush *Ricinocarpos* spp. esp. *R. pinifolia*; **w. flower** *Francoa sonchifolia*

Wedelia Jacq. Compositae (Helia.-Verb.). (incl. *Aspilia*) 100 trop. & warm. R (N Am.): SBM 33(1991)38. Some spp. with elaiosomes, *W. hispida* Kunth (Mex.) at least, myrmechorous. Many weeds, some locally medic., stopping bleeding, while some ed. chimpanzees (e.g. some forms of *W. mossambicensis* Oliver (*A. m.*, trop. Afr.)) contain thiarubine A effective against bacteria & intestinal parasites, some with molluscicidal saponins; extracts of *W. mossambicensis* in vitro effective against AIDS & herpes viruses, kaurenoic & grandiflorenic acids effective uterostimulators. Cult. orn. incl. *W. trilobata* (L.) A. Hitchc. (Singapore daisy, Florida & trop. Am.) – widely natur., pest in Queensland

wedge pea *Gompholobium* spp.

weed *Cannabis sativa* ssp. *indica*

Wehlia F. Muell. = *Homalocalyx*

Weigela Thunb. Caprifoliaceae. 10 E As. Cult. orn. shrubs with some hybrids, esp. *W. florida* (Bunge) A. DC (*W. rosea*, N China, Korea, Japan) with pink fls, many cvs

Weigeltia A. DC = *Cybianthus*

Weihea Sprengel = *Cassipourea*

Weingartia Werderm. = *Rebutia*

Weinmannia L. Cunoniaceae. 190 Andes, Madag., Masc., Mal., Pacific, NZ. Timbers, local medic. & tanbarks esp. *W. racemosa* L.f. (towai, NZ)

Welchiodendron P.G. Wilson & Waterhouse (~ *Tristania*). Myrtaceae (Lept.). 1 New Guinea & Queensland

weld *Reseda luteola*

Weldenia Schultes f. Commelinaceae (II 1f). 1 Mex., Guatemala: *W. candida* Schultes f. – cult. orn. with tuberous roots, underground stem & rosette of linear-lanceolate lvs

Welfia H. Wendl. Palmae (V 6). 2 C Am. R: VKNAW II,18(1968)76. *W. georgii* H. Wendl. ex

Burret with fronds 6–8 m long

wellingtonia *Sequoiadendron giganteum*

Wellstedia Balf.f. Boraginaceae (Wellstediaceae). 3 Somalia, Socotra, Ethiopia

Wellstediaceae (Pilger) Novák = Boraginaceae (Wellstedioideae)

Welsh onion *Allium fistulosum*; **W. poppy** *Meconopsis cambrica*

Welwitschia Hook.f. Welwitschiaceae. 1 S Angola & SW Afr. deserts, extending into mopane woodland: *W. mirabilis* Hook.f. (*W. bainesii*). Long-lived (at least 1500 yrs with stem to 1 m across), deriving moisture from sea-fog dew. Winged seeds germ. in wet years, the cotyledons photosynthesizing for 1.5 yrs; foliage lvs decussate with respect to cotyledons, of indefinite growth (up to 13.8 cm a year), wearing away at tips, the apical meristem being lost; 2 scale-like lvs develop & engulf shoot apex which then becomes meristematically inactive. Fertile buds produced on the 'crowns' between the leaf-bases; pollen-wall development like that in angiosperms. C.H. Bornman (1978) *W.*

Welwitschiaceae Caruel. Gnetopsida. 1/1 SW Afr. Dioec. perennial with short stem & taproot. Lvs 2 opp., parallel-veined. Infls dichasial; insect-poll. Females subtended by cone-scales making up a red cone & consisting of single nucellus encl. in an integument & another layer derived from 2 confluent primordia ('perianth') with 2 'bracts'; megaspore mother-cell grows into prothallus without archegonia. Male fls subtended by cone-scales making up a red cone & consisting of 2 lateral 'bracts' & a 'perianth' formed by union of 2 bract-like organs, 6 microsporangiophores & a sterile unitegmic ovule; male gametophyte merely a tube-nucleus, an abortive sterile cell & a cell giving rise to 2 sperm nuclei. As pollen tubes penetrate prothallus, prothallial tubes grow up to meet them in the nucellus, fusion of females & males allegedly occuring in the tubes; many zygotes but only 1 matures & is disp. with 'perianth' as wing

Only species: *Welwitschia mirabilis*

Welwitschiella O. Hoffm. Compositae (Hele.-Cha.). 1 Angola, Zambia

Wenchengia C.Y. Wu & S. Chow. Labiatae (IV/V). 1 Hainan (SE China)

Wendelboa Soest = *Taraxacum*

Wendlandia Bartling ex DC. Rubiaceae (I 5). 70 warm As. (Iraq 1), Indomal. R: NRBGE 16(1932)233. Lvs opp. or whorled (stipules leafy); some cult. orn. shrubs & trees

Wendlandiella Dammer. Palmae (IV 3). 3 Peru, Brazil

Wendtia Meyen. Geraniaceae (Ledocarpaceae). 3 Chile, Argentina

wenge *Millettia laurentii*

Wenzelia Merr. Rutaceae. 9 Philippines to Solomons, Fiji. R: W.T. Swingle, *Bot. Citrus* (1943)214

Wercklea Pittier & Standley. Malvaceae. 12 trop. Am. R: JAA 62(1981)457. Shrubs & trees, *W. insignis* Pittier & Standley (Costa Rica) – cult. orn.

Werckleocereus Britton & Rose = *Weberocereus*

Werdermannia O. Schulz. Cruciferae. 3–4 N Chile

Werneria Kunth. Compositae (Sen.-Sen.). 40 Andes

Wernhamia S. Moore. Rubiaceae (? I 1). 1 Bolivia

West Indian arrowroot *Maranta arundinacea*; **WI birch** *Bursera gummifera*; **WI boxwood** *Casearia praecox*; **WI cherry** *Malpighia glabra*; **WI ebony** *Brya ebenus*; **WI elemi** *Bursera gummifera*; **WI gherkin** *Cucumis anguria*; **WI gooseberry** *Pereskia aculeata*; **WI locust-tree** *Hymenaea courbaril*; **WI mahogany** *Swietenia mahagoni*; **WI redwood** *Guarea guidonia*; **WI sandalwood** *Amyris balsamifera*

western arbor-vitae or **red cedar** *Thuja plicata*; **w. hemlock** *Tsuga heterophylla*

Westoniella Cuatrec. Compositae (Ast.-Ast.). 6 Panamá, Costa Rica, paramos

Westphalina Robyns & Bamps. Tiliaceae. 1 Guatemala. Arils

Westringia Sm. Labiatae (III). 25 Aus., Lord Howe Is. R: PRSQ 60(1949)99. Shrubs

Wetria Baillon. Euphorbiaceae. 1 S Burma & Thailand to W Mal., New Guinea

Wettinia Poeppig. Palmae (V 2b). 9 trop. S Am. esp. Colombia. R: NRBGE 36(1978)259

Wettiniicarpus Burret = *Wettinia*

Wettsteiniola Süsseng. Podostemaceae. 3 S Brazil, Argentina

Weymouth pine *Pinus strobus*

whalebone tree *Streblus brunonianus*

whangee *Phyllostachys nigra*

wheat (bread w.) *Triticum aestivum*; **crack or durum w.** *T. turgidum* Durum Group; **emmer w.** *T. t.* Dicoccon Group; **flint, hard** or **macaroni w.** *T. t.* Durum Group; **Polish w.** *T. t.* Polonicum Group; **Poulard w.** *T. t.* Turgidum Group

whin *Ulex europaeus*

Whipplea Torrey. Hydrangeaceae (Philadelphaceae). 1 W N Am.: *W. modesta* Torrey – decid.

trailing shrub for rock garden

whisk, French *Chrysopogon gryllus*

whisk(e)y *Hordeum vulgare*

whistling jacks *Gladiolus* spp.; **w. pine** *Casuarina equisetifolia*; **w. thorn** *Acacia drepanolobium*

white ash *Fraxinus americana*; **w.beam** *Sorbus aria* & other spp.; **w.beam, Swedish** *S. intermedia*; **w. bombway** *Terminalia procera*; **w. bryony** *Bryonia dioica*; **w. cedar** *Calocedrus decurrens, Chamaecyparis thyoides, Melia azedarach*; **w. chuglam (wood)** *Terminalia bialata*; **w. clover** *Trifolium repens*; **w.currant** *Ribes rubrum* cv.; **w. dammar** *Vatica indica*; **w. dead nettle** *Lamium album*; **w. deal** *Picea abies*; **w. elm** *Ulmus americana*; **w. fir** *Abies amabilis*; **w. gum** *Eucalyptus viminalis*; **w. hellebore** *Veratrum album*; **w. horehound** *Marrubium vulgare*; **w. mahogany** *E. acmenoides, E. robusta*; **w. maple** *Acer saccharinum*; **w. melilot** *Melilotus alba*; **w. mulberry** *Morus alba*; **w. mullein** *Verbascum lychnitis*; **w. mustard** *Sinapis alba*; **w. oak** *Quercus alba*; **w. pear** *Apodytes dimidiata*; **w. pepper** *Piper nigrum*; **w. pine** *Pinus strobus, Podocarpus* spp.; **w. poplar** *Populus alba*; **w. Sally** *Eucalyptus pauciflora*; **w. spruce** *Picea glauca*; **w. thorn** *Crataegus monogyna*; **w. willow** *Salix alba*; **w.wood** *Abies alba, Atalaya hemiglauca, Liriodendron tulipifera, Triplochiton scleroxylon, Tilia americana*; **w.w., African** *Annickia chlorantha*; **w.w., American** *L. tulipifera*; **w.w., Baltic** *Picea abies*; **w. yam** *Dioscorea alata*

Whiteheadia Harvey. Hyacinthaceae (Liliaceae s.l.). 1 S Afr.: *W. bifolia* (Jacq.) Baker

Whiteochloa C. Hubb. Gramineae (34b). 5 trop. Aus. R: Brunonia 1(1978)69

Whiteodendron Steenis. Myrtaceae (Lept.). 1 Borneo

Whitesloanea Chiov. Asclepiadaceae (III 5). 1 Somalia: *W. crassa* (N.E. Br.) Chiov.

Whitfieldia Hook. Acanthaceae. 10 trop. Afr. Some cult. orn. shrubs & local dyes

Whitfordiodendron Elmer = *Callerya*

whitlow-grass *Erophila* spp., esp. *E. verna*

Whitmorea Sleumer. Icacinaceae. 1 Solomon Is.

Whitneya A. Gray. Compositae (Hele.-Cha.). 1 California

Whittonia Sandw. Peridiscaceae. 1 NE S Am.

Whitsuntide boss *Viburnum opulus* 'Roseum'

whortleberry *Vaccinium myrtillus*

Whyanbeelia Airy Shaw & B. Hyland. Euphorbiaceae. 1 Queensland

Whytockia W.W. Sm. Gesneriaceae. 3 China & Taiwan. R: NRBGE 40(1982)359

Wiasemskya Klotzsch = *Tammsia*

Wiborgia Thunb. Leguminosae (III 28). 10 Cape. R: OB 28(1975). Close to *Lebeckia*

wickup *Epilobium angustifolium*

Widdringtonia Endl. Cupressaceae. 3 trop. & S Afr. R: Phytol.M 7(1984)76. African cypress; some timber & cult. orn. esp. *W. cedarbergensis* J. Marsh (*W. juniperoides*, Clanwilliam cedar, SW Cape) & *W. nodiflora* (L.) Powrie (*W. whytei*, Mlanje c., sapree wood, S & S trop. Afr.). *W. cupressoides* (L.) Endl. (Cape) – germ. enhanced by smoke

widdy *Potentilla fruticosa*

Widgrenia Malme. Asclepiadaceae (III 1). 1 Brazil: *W. corymbosa* Malme. R: Darw. 30(1990)279

widow, black or **mourning** *Geranium phaeum*

Wiedemannia Fischer & C. Meyer. Labiatae (VI). 3 SW As.

Wielandia Baillon. Euphorbiaceae. 1 Seychelles

Wiesneria M. Micheli. Alismataceae. 3 trop. Afr. to India. Like *Aponogeton*

wig tree *Cotinus coggygria*

Wigandia Kunth. Hydrophyllaceae. 2–3 trop. Am. Shrubs & trees grown for large lvs (covered in stinging hairs) in sapling stages & used in 'subtrop.' gardening

Wigginsia D. Porter = *Parodia*

Wightia Wallich. Scrophulariaceae. 2–3 Indomal. Epiphytic shrubs, eventually becoming independent trees, like strangling figs

Wikstroemia Endl. Thymelaeaceae. 50 SE As. to Pacific. Close to *Daphne*. Some cryptic dioec. (dioecy evolved twice in Hawaii), *W. indica* (L.) C. Meyer (Indomal. to Aus.) apomictic. Bark used for rope, bank-notes & strong paper, wood a source of incense. *W. ovata* C. Meyer (C Mal.) – strong purge; *W. sikokiana* Franchet & Savigny (*Diplomorpha s.*, gampi, Korea) – bark used for paper in Japan

Wilbrandia J. Silva Manso. Cucurbitaceae. 2 N S Am.

Wilcoxia Britton & Rose = *Echinocereus*

wild oat *Avena fatua*; **w. rice** *Zizania aquatica*; **w. thyme** *Thymus polytrichus* spp. *britannicus*

Wildemaniodoxa Aubrév. & Pellegrin = *Englerophytum*

wilga *Geijera parviflora*

Wilhelminia Hochr. = *Hibiscus*

Wilhelmsia Reichb. Caryophyllaceae (II 1). 1 NE As., NW Am.: *W. physodes* (Sér.) McNeill

Wilkesia A. Gray. Compositae (Hele.-Mad.). 1–2 Kauai (Hawaii). R: Allertonia 4,1(1985)59. *W. gymnoxiphium* A. Gray (iliau), not or weakly branched pachycaul tree, s.t. hapaxanthic, hybridizes with *Argyroxiphium* & *Dubautia* spp.

Wilkiea F. Muell. Monimiaceae (V 2). 6 E Aus.

Willardia Rose (~ *Lonchocarpus*). Leguminosae (III 6). 6 C Am. Cult. orn. small trees with lilac fls

Willbleibia Herter = *Willkommia*

Willdenowia Thunb. Restionaceae. 11 S & SW Cape, Namaqualand. R: Bothalia 15(1985)493

Willemetia Necker (*Calycocorsus*). Compositae (Lact. – Crep.). 2 C Eur. to Iran

Williamodendron Kubitzki & H. Richter (~ *Mezilaurus*). Lauraceae (I). 3 trop. Am. R: BJ 109(1987)49

Williamsia Merr. = *Praravinia*

Willisia Warm. Podostemaceae. 1 S India

Willkommia Hackel. Gramineae (33b). 3 trop. & S Afr., 1 Texas, introd. Arg. R: KBAS 13(1986)239

willow *Salix* spp.; **basket w.** *S. viminalis, S. purpurea*; **black w.** *S. nigra*; **brittle w.** *S. fragilis*; **Cape w.** *S. mucronata*; **corkscrew w.** *S. matsudana* 'Tortuosa'; **crack w.** *S. fragilis*; **creeping w.** *S. repens*; **cricket-bat w.** *S. alba* 'Coerulea'; **flowering w.** *Chilopsis linearis*; **goat w.** *S. caprea*; **golden w.** *S. alba* 'Chermesina'; **w. herb** *Epilobium* spp.; **Humboldt w.** *S. chilensis*; **Indian w.** *Polyalthia longifolia* cv.; **Kilmarnock w.** *S. caprea* 'Pendula'; **w. myrtle** *Agonis* spp.; **Port Jackson w.** *Acacia saligna*; **prairie w.** *S. humilis*; **pussy w.** *S. caprea*; **weeping w.** *S. × sepulcralis*, (willow-pattern) *S. babylonica*, (Bible) *Populus euphratica*; **white w.** *S. alba*

Willughbeia Roxb. Apocynaceae. 15 Indomal. R: Blumea 38(1993)1. Some rubber sources esp. *W. coriacea* Wallich (*W. firma*, Borneo rubber, W Mal.)

Willughbeiopsis Rauschert = *Willughbeia*

Wilsonia R. Br. Convolvulaceae. 4 Aus.

Wimmeria Schldl. & Cham. Celastraceae. 14 C Am. Lvs torn across hang together because of pulled-out xylem wall-thickenings (cf. *Cornus*)

Winchia A. DC. = *Alstonia*

wind flower *Anemone* spp.

windmill palm *Trachycarpus fortunei*

Windsorina Gleason. Rapateaceae. 1 Guyana

wine see *Vitis vinifera*

wineberry *Rubus phoenicolasius*

wingnut *Pterocarya* spp.

Winifredia L. Johnson & B. Briggs. Restionaceae. 1 Tasmania: *W. sola* L. Johnson & B. Briggs. R: Telopea 2(1986)737

Winklera Regel (*Uranodactylus*). Cruciferae. 2 C As. to Himalaya

Winklerella Engl. Podostemaceae. 1 W Afr.

winter aconite *Eranthis hyemalis*; **W.'s bark** *Drimys winteri*; **w. cherry** *Physalis alkekengi, Solanum pseudocapsicum, S. capicastrum*; **w. cress** *Barbarea* spp.; **w. green** *Gaultheria* spp. but now usu. *Betula lenta, Pyrola* spp.; **w. heliotrope** *Petasites fragrans*; **w. purslane** *Claytonia perfoliata*; **w. sweet** *Chimonanthus praecox*

Winteraceae R. Br. ex Lindley. Dicots – Magnoliidae – Illiciales. 4/60 S Am., Madagascar (1/1), Aus., New Guinea, SW Pacific. Evergreen trees & shrubs, s.t. with alks; hairs (except on G) & vessels 0. Lvs simple, entire, spirally arr.; stipules 0. Fls usu. bisexual, reg. to irreg., poll. by insects or wind, solit. & term. or in cymes, receptacle short; K 2–4(–6) valvate, sometimes basally connate or completely so & forming calyptra, C (2–)5–∞, often in 2 or more whorls, A 3–∞, initiated centripetally but maturing centrifugally, often ± laminar though s.t. with distinct filament, often with term. bisporangiate pollen-sacs, pollen grains usu. in tetrads, $\underline{G}$ (1–) usually ∞ in 1 whorl, s.t. weakly connate, conduplicate & often not completely closed with stigma along margins to closed with term. stigma, 1–several anatropous, bitegmic ovules laminar or near margin. Fr. berry-like or follicles in heads, s.t. ± connate into multiloc. capsule or syncarp; seeds exotestal with small embryo in copious oily endosperm. n = 13, 43

Genera: *Drimys, Pseudowintera, Takhtajania, Zygogonum*

Often recognizable in the field by abaxial dotting of lvs due to plugging of stomata with wax. Hydrolysis of polysaccharide granules in cells of petals of several *Zygogynum* spp. leads to rapid uptake of water & subsequent opening or closing of fls; may also act as pollen rewards for beetles. Poss. with more archaic features than any other fam. Fossils in

Tertiary of Afr.
Some cult. orn. & medic. (*Drimys*)

wirilda *Acacia retinodes*

wirra *Acacia salina*

wishbone flower *Torenia fournieri*

Wislizenia Engelm. Capparidaceae. 1 SW N Am. (polymorphic). R: Brittonia 31(1979)333

Wissadula Medikus. Malvaceae. 40 trop. esp. Am. R: KSVAH n.s. 43,4(1908)1. Some locally used bark-fibres esp. *W. spicata* C. Presl (paco-paco, trop. Am.) & *W. amplissima* (L.) R. Fries (*W. rostrata*, maholtine, trop.) – fibre like jute

Wissmannia Burret = *Livistona*

wisteria *Wisteria* spp.; **Chinese w.** *W. sinensis*; **Japanese w.** *W. floribunda*; **water w.** *Hygrophila difformis*

Wisteria Nutt. Leguminosae (III 6). c. 6 E As., N Am. R: P. Valder (1995) *Ws.* Lianes with pendent racemes of showy often scented fls, much planted against houses esp. *W. floribunda* (Willd.) DC (Japanese wisteria, Japan) climbing clockwise with scented racemes to 45 (–150 in 'Macrobotrys') cm long & lvs with 13–19 leaflets, many cvs, & *W. sinensis* (Sims) Sweet (Chinese w., China) climbing anticlockwise with scentless racemes to 30 cm & lvs with usu. 11 leaflets, a specimen at Sierra Madré, Calif. the largest blossoming pl. in cult. (planted 1892, 152 m long with 1.5 m infls (30 000 people a year pay to see it)). All have poisonous lvs, fr. & seeds

witch grass *Elytrigia repens*; **w. hazel** *Hamamelis virginiana*; **w. weed** *Striga lutea*

Withania Pauquy. Solanaceae (1). 10 OW (Eur. 2). Alks. *W. coagulans* (Stocks) Dunal (India, Afghanistan) – fr. (panirband, vegetable rennet) used to coagulate milk in cheese-making; *W. somnifera* (L.) Dunal (Afr., Medit. to India) – source of ashwagandha, a drug used in Ayurvedic med., narcotic & diuretic

Witheringia L'Hérit. Solanaceae (1). 15 trop. Am. R: Kurtziana 5(1969)110

witloof *Cichorium intybus*

Witsenia Thunb. Iridaceae (II). 1 Cape: *W. maura* Thunb. – rare, shrubby to 2 m, seed water-disp., cult. orn.

Wittia Schumann = *Disocactus*

Wittiocactus Rauschert = *Disocactus*

Wittmackanthus Kuntze (*Pallasia*). Rubiaceae (I 1). 1 trop. Am.

Wittmackia Mez = *Aechmea*

Wittrockia Lindman. Bromeliaceae (3). 10 E Brazil. R: FN 14(1979)1725. Some cult. orn. like *Canistrum* but C united

Wittsteinia F. Muell. Alseuosmiaceae. 3 New Caled., New Guinea, SE Aus.

woad *Isatis tinctoria*

Wodyetia Irvine. Palmae (V 4k). 1 Queensland (Melville Range): *W. bifurcata* Irvine

Woehleria Griseb. Amaranthaceae (II 2). 1 Cuba

Woikoia Baehni = *Pouteria*

Wokoia Baehni = *Pouteria*

Wolffia Horkel ex Schleiden. Lemnaceae. c. 7 trop. & warm to temp. Water meal. Minute thalloid aquatics rarely flowering, *W. arhiza* (L.) Wimmer (Euras., Afr., Aus.) c. 1.5 mm long & the smallest angiosperm by repute but *W. brasiliensis* Wedd. (Brazil), growing with *Victoria amazonica* (!), & *W. microscopica* (Griffith) Kurz (India) even smaller & poss. *W. angusta* Landolt (SE Aus.) described 1980 the minutest (0.6 mm × 0.33 mm). R: IBM 34(1965)41. *W. arhiza* has stomata in parallel rows, a deep reproductive pocket & a meristem for new thalloid growths which are 'budded' off, the pl. entirely vegetative in Eur. & SE As. where (?) introduced, used as a quantitative test for herbicide pollution in Poland; *W. brasiliensis* distrib. on birds' feathers

Wolffiella (Hegelm.) Hegelm. Lemnaceae. (Incl. *Wolffiopsis*) c. 7 trop. & S Afr. (1), trop. & warm Am. R: IBM 34(1965)34

Wolffiopsis Hartog & van der Plas = praec.

wolfsbane *Aconitum lycoctonum*

Wollastonia DC ex Decne. (~ *Wedelia*). Compositae (Helia.-Verb.). 1 Indopacific strand: *W. biflora* (L.) DC (*Wedelia b.*). R: SCB 45(1980)32

Wollemi pine *Wollemia nobilis*

Wollemia W.G. Jones, K. Hill & J.M. Allen (~ *Agathis*). Araucariaceae. 1 NSW (v. rare): *W. nobilis* W.G. Jones, K. Hill & J.M. Allen (Wollemi pine) – undetected until 1990s. R: Telopea 6(1995)173. Char. spongy nodular bark, lvs trimorphic (cf. *Glyptostrobos*)

wombat berry *Eustrephus latifolius*

wonderberry *Solanum × burbankii*

wonga-wonga vine *Pandorea pandorana*

wood apple *Limonia acidissima;* **w. avens** *Geum urbanum;* **w.bine** *Lonicera periclymenum;* **w. flowers** Loranthaceae esp. *Phoradendron* spp.; **w. melick** *Melica uniflora;* **w. oil** *Aleurites fordii;* **w. rose** *Argyreia nervosa, Merremia tuberosa* (see also *Dactylanthus*); **w. rush** *Luzula* spp.; **w. sorrel** *Oxalis acetosella;* **w. spurge** *Euphorbia amygdaloides*

Woodburnia Prain. Araliaceae. 1 Burma

Woodfordia Salisb. Lythraceae. 1 NE Afr., S Arabia: *W. uniflora* (A. Rich.) Koehne – used in tanning, 1 Madag. to China & Timor: *W. fruticosa* (L.) Kurz (*W. floribunda*) – source of tragacanth-like gum & dye (from fls). K tubular, red (C v. small)

Woodia R. Br. = *Woodsia*

Woodia Schltr. Asclepiadaceae (III 1). 3 S Afr.

Woodiella Merr. = *Woodiellantha*

Woodiellantha Rauschert (*Woodiella*). Annonaceae. 1 Borneo: *W. sympetala* (Merr.) Rauschert. R: MNJ 42(1989)267

Woodrowia Stapf = *Dimeria*

woodruff *Galium odoratum*

Woodsia R. Br. Dryopteridaceae (II 1). 25 temp. (not Aus.) & cool-temp. (Eur. 3). R: BNH 16(1964). Tufted saxicolous ferns, some cult. orn. incl. *W. alpina* (Bolton) Gray (Euras., N Am.) & *W. ilvensis* (L.) R. Br. (Euras., N Am.), both protected spp. in GB

Woodsiaceae (Diels) Herter. Older name for Dryopteridaceae

Woodsonia L. Bailey = *Neonicholsonia*

Woodwardia Sm. Blechnaceae (Blechnoideae). (Incl. *Lorinseria*) 13 N hemisph. (S Eur. 1) esp. E As. (some gaps) S to Costa Rica & Mal. Cult. orn. incl. *W. fimbriata* Sm. (W N Am.) stems dyed red & those of *W. spinulosa* M. Martens & Galeotti (Mex., Guatemala) white before use in N Am. Indian weaving

woody pear *Xylomelum* spp.

Wooleya L. Bolus. Aizoaceae (V). 1 Namaqualand

Woollsia F. Muell. (~ *Lysinema*). Epacridaceae. 1 NE Aus.: *W. pungens* (Cav.) F. Muell.

woolly bush *Adenanthos* spp.

Wootonella Standley = *Verbesina*

Wootonia E. Greene = *Dicranocarpus*

Worcesterberry *Ribes divaricatum*

Wormia Rottb. = *Dillenia*

wormseed *Chenopodium ambrosioides;* **Levant w.** *Artemisia cina*

Wormskioldia Thonn. = *Tricliceras*

wormwood *Artemisia absinthium,* (Bible) *A. herba-alba;* **Roman w.** *A. pontica;* **sea w.** *Seriphidium maritimum*

Woronowia Juz. = *Geum*

Worsleya (Traub) Traub (~ *Hippeastrum*). Amaryllidaceae (Liliaceae s.l.). 1 Brazil (Organ Mts): *W. procera* (Lem.) Traub (*W. rayneri*, blue amaryllis) – cult. orn. with lilac fls & bulb-neck to 1.5 m

woundwort *Stachys* spp.

Woytkowskia Woodson. Apocynaceae. 2 trop. S Am. R: MMNHN 30(1985)120

Wrightia R. Br. Apocynaceae. 24 trop. OW (Afr. (S & E) 2, R: AUWP 87-5(1987)35). R: AMBG 52(1965)114. Cult. orn. trees & shrubs, some timber. *W. arborea* (Dennst.) Mabb. (*W. tomentosa*, India) – fls smell of decaying fr. (? poll. flies), white wood for carving etc., bark medic., yellow dye used as styptic by Nepalese; *W. tinctoria* (Rottler) R. Br. (C India to Timor) – building timber, indigo-like dye from lvs

Wrixonia F. Muell. Labiatae (III). 2 Aus. R: JABG 1(1976)27

Wulfenia Jacq. Scrophulariaceae. 3 SE Eur., Turkey. Cult. orn. rock-pls

Wulfeniopsis Hong (~ *Wulfenia*). Scrophulariaceae. 2 E Afghanistan, Pakistan, N India, Nepal. R: OB 75(1984)56

Wulffia Necker ex Cass. Compositae (Helia.-Verb.). 5 trop. Am. *W. baccata* (L.f.) Kuntze (*W. stenoglossa*, trop. Am.) – infr. a head of fleshy cypselas looking like a blackberry, bird-disp.

Wullschlaegelia Reichb.f. Orchidaceae (V 5). 2 trop. Am. Some mycotrophs

Wunderlichia Riedel ex Benth. Compositae (Mut.-Mut.). 5 Brazil. R: RBB 33 (1973)379. Pachycaul shrubs & trees

Wunschmannia Urban = *Distictis*

Wurdackanthus Maguire (~ *Irlbachia*). Gentianaceae. 2 NE S Am., WI. R: MNYBG 51(1989)8

Wurdackia Mold. = *Rondonanthus*

Wurmbea Thunb. (incl. *Anguillaria*). Colchicaceae (Liliaceae s.l.). c. 40 Afr. (19, Cape 13 – R: OB 87(1985)), Aus. (R: FA 45(1987)387). Alks. Cult. orn. incl. *W. dioica* (R. Br.) F. Muell. (*A.*

d., early Nancy, Aus.)

wych elm *Ulmus glabra*

Wyethia Nutt. Compositae (Helia.-Verb.). 14 W US. R: AMN 35(1946)400. Coarse herbs; cypselas & taproots form. eaten by N Am. Indians. Some cult. orn.

Xantheranthemum Lindau. Acanthaceae. 1 Peruvian Andes: *X. igneum* (Regel) Lindau – cult. orn. foliage pl.

Xanthisma DC. Compositae (Ast.-Sol.). 1 Texas: *X. texana* DC – cult. orn. with capitula closing at night

Xanthium L. Compositae (Helia.). c. 3 now cosmop. Monoec., the female fls in pairs in a prickly involucre with only styles projecting; fr. enclosed in accrescent involucre covered with hooks aiding disp. on animal fur (burweed) but lowering value of fleeces etc. e.g. *X. occidentale* Bertol. (*X. strumarium* s.l., Noogoora burr, orig. WI?) in Aus. Some locally medic. – *X. spinosum* L. (clotbur, Bathurst burr) & *X. strumarium* L. (cocklebur) – much used in genetic analysis (cf. *Datura*). *X. canadense* Miller (Am.) – lactone acting as growth inhibitor in *Drosophila melanogaster* larvae

Xanthocephalum Willd. = *Gutierrezia*

Xanthoceras Bunge. Sapindaceae. 1 N China: *X. sorbifolium* Bunge – small cult. orn. decid. tree with white fls & ed. seeds

Xanthocercis Baillon. Leguminosae (III 2). 2 S Afr., N Madag. *X. zambesiaca* (Baker) Dumaz (nyala tree, S Afr.) – typical tree of low veld

Xanthogalum Avé-Lall. = *Angelica*

Xanthomyrtus Diels. Myrtaceae (Myrt.). 23 Borneo to New Caled. R: KB 33(1979)461

Xanthopappus Winkler. Compositae (Card.-Card.). 1 NW China, Mongolia

Xanthophyllaceae (Chodat) Rev. & Hoogland (~ Polygalaceae). Dicots – Rosidae – Polygalales. 1/93 Indomal. Small trees often accum. aluminium. Lvs simple, spirally arr.; stipules 0 or crateriform glands. Fls bisexual, v. irreg., in racemes of dichasial fascicles or thyrses; K 5 imbricate, the inner 2 longest, C 4 or 5 imbricate with the lowermost folded to form a keel encl. by rest, A 8 s.t. adnate to clawed bases of C, anthers with longit. slits, nectary-disk around $\underline{G}$ (2), stipitate, 1-loc. with 2 parietal placentas & 2–16 anatropous bitegmic ovules, style term. Fr. indehiscent, dry or fleshy-fibrous, globose, with 1 seed, embryo oily, endosperm 0

Genus: *Xanthophyllum*

Xanthophyllum Roxb. Xanthophyllaceae. 93 Indomal. R: LBS 7(1982). Genus originating in Aus. (van der Meijden). Some dyes & oilseeds

Xanthophytopsis Pitard = *Xanthophytum*

Xanthophytum Reinw. ex Blume. Rubiaceae (IV 1). 30 Java & Borneo to Fiji. R: Blumea 34(1990)425. Some monocaul treelets

Xanthorhiza Marshall. Ranunculaceae (III 3). 1 E N Am.: *X. simplicissima* Marshall. Decid. *shrub* with 1–2-pinnate lvs, brownish-purple fls & bitter yellow roots, source of a yellow dye; alks

Xanthorrhoea Sm. Xanthorrhoeaceae. 28 Aus. Yacca, (Australian) grass tree, black boy (blackened by fire). Slow-growing, long-lived fire-tolerant pachycaul pls with thick stems producing acaroid resins at bases of old lvs – resin form. used to fix spear-heads to shafts, now used to varnish or lacquer metals & leather; wood used for making bowls etc. *X. australis* R. Br. (SE Aus.) – resistant to fire when young as bud 12 cm below ground, later bud protected by old lvs. *X. preissii* Endl. (SW Aus.) lives to at least 350 yrs & flowering may be delayed until 200 yrs old, though stimulated by fire, other spp. at least flowering most frequently near habitation as fires are more frequent there; *X. pumilio* R. Br. (Queensland) lvs only 3 cm long; *X. resinosa* Pers. (SE Aus.) – yellow resin ('yellow gum', also from other spp.) medic. & in glue

Xanthorrhoeaceae Dumort. (~ Liliaceae). Monocots – Liliidae – Liliales. (Incl. Calectasiaceae, Dasypogonaceae, Lomandraceae) 10/90 Aus. to New Guinea & New Caled. Pachycaul treelets to stemless herbs with thick underground stocks; vessels usu. only in roots, occ. lvs; some sec. growth. Lvs narrowly oblong to linear, xeromorphic often spine-tipped &/or sheathing, spirally arr., leaf-bases often marcescent. Fls bisexual or not (dioec. in *Lomandra*), 3-merous, usu. small & in spikes, heads or thyrses (solit. & large in *Baxteria*); P 3 + 3, dry & chaffy, the inner rarely coloured, A 3 + 3, the inner often adnate to C, anthers basifixed or versatile, with longit. slits, $\underline{G}$ (3), 3-loc. with axile or axile-basal placentation or 1-loc. with 3 basal ovules, styles 3 or 1 with 3 stigmas, each loc.

with 1–∞ hemitropous to campylotropous (? anatropous) bitegmic ovules. Fr. a loculicidal capsule or 1-seeded nut; embryo usu. straight with term. cotyledon embedded in copious oily & proteinaceous endosperm often also rich in hemicellulose. x = 7, 8, 11, 17, 24

Genera: *Acanthocarpus, Baxteria, Calectasia, Chamaexeros, Dasypogon, Kingia, Lomandra, Romnalda, Xanthorrhoea, Xerolirion*

Poss. 'that several genera do not all properly belong together' (Cronquist) & *Xanthorrhoea* & *Calectasia* have been removed to monotypic fams leaving the rest in Dasypogonaceae 'with reservations' & now the latter restricted to *Dasypogon* & *Kingia*, the remainder in Lomandraceae. The splitting of these fams makes necessary much of the fragmentation in Liliidae generally & points up the problem of the difficulty of applying criteria used to separate 'more recently evolved' groups to 'older' ones, hence the lability of fams thought to be 'basal' in subclasses of angiosperms

Some varnishes (*Xanthorrhoea*)

Xanthosia Rudge. Umbelliferae (I 1b). 25 Aus. esp. SW. Umbels in some spp. 1-flowered

Xanthosoma Schott. Araceae (VII 1). 57 trop. Am. Stemless or caulescent herbs with thick rhizomes or tubers & milky sap; lvs sagittate or hastate (cf. *Colocasia*); cult. orn. & food-pls (yautia) esp. *X. violaceum* Schott ('*X. nigrum*', ~ *X. sagittifolium*) & *X. sagittifolium* (L.) Schott (tannia, tania, original range unclear) – grown for ed. tubers & young lvs ed. like spinach

Xanthostachya Bremek. = *Strobilanthes*

Xanthostemon F. Muell. Myrtaceae (Lept.). c. 45 C Mal. to NE Aus. (13, R: Telopea 3(1990)451) & New Caled. *X. oppositifolius* Bailey (penda, SE Queensland) – timber tree

Xanthoxylum Miller = *Zanthoxylum*

Xantolis Raf. Sapotaceae (IV). 14 S India, SE As., Philippines. R: Blumea 8(1957)207

Xantonnea Pierre ex Pitard. Rubiaceae (II 5). 3 SE As.

Xantonneopsis Pitard. Rubiaceae (II 1). 1 SE As.

Xatardia Meissner & Zeyher. Umbelliferae (III 9). 1 Pyrenees

Xenacanthus Bremek. = *Strobilanthes*

Xenia Gerbaulet (~ *Anacampseros*). Portulacaceae . 1 Arg. R: BJ 113(1992)552

Xenikophyton Garay. Orchidaceae (V 16). 1 W Pacific

Xenophya Schott = *Alocasia*

Xenopoma Willd. (~ *Micromeria*). Labiatae (VIII 2). Spp.?

Xenostegia D. Austin & Staples. Convolvulaceae. 2 trop. OW

Xeranthemum L. Compositae (Card.-Carl.). 5–6 Medit. (Eur. 3) to SW As. Non-spiny annual herbs, cult. orn. 'everlastings' esp. *X. annuum* L. (SE Eur. to Iran)

Xeroaloysia Tronc. Verbenaceae. 1 Argentina

Xerocarpa H.J. Lam = *Teijsmanniodendron*

Xerochlamys Baker = *Leptolaena*

Xerochloa R. Br. Gramineae (34g). 3 Aus. R: BJ 35(1904)64

Xerocladia Harvey. Leguminosae (II 3). 1 SW Afr.

Xerococcus Oersted = *Hoffmannia*

Xerodanthia J. Phipps = *Danthoniopsis*

Xeroderris Roberty = *Ostryocarpus*

Xerodraba Skottsb. Cruciferae. 6 S Arg. Some like veg. sheep

Xerolekia A. Anderb. (~ *Telekia*). Compositae (Inul.). 1 C Eur.: *X. speciosissima* (L.) A. Anderb. (*Telekia* s.)

Xerolirion A.S. George. Xanthorrhoeaceae (Lomandraceae). 1 SW Aus.

Xeromphis Raf. = *Catunaregam*

Xeronema Brongn. & Gris. Phormiaceae (Liliaceae s.l.). 2 New Caled. & Poor Knights Is. (N NZ – *X. callistemon* W. Oliver, P.K. lily)

Xerophyllum Michaux. Melanthiaceae (Liliaceae s.l.). 2 N Am. Cult. orn. tall herbs incl. *X. tenax* (Pursh) Nutt. (bear grass, W N Am.) – lvs form. used by Indians to make water-tight baskets

Xerophyta Juss. Velloziaceae. 28 trop. Afr. & Madag. R: KB 29(1974)184

Xeroplana Briq. Stilbaceae (Verbenaceae s.l.). 2 SW & S Cape. R: JSAB 43(1977)1

Xerorchis Schltr. Orchidaceae (? tribe). 2 N S Am.

Xerosicyos Humbert. Cucurbitaceae. 4 Madag. R: Ashingtonia 2(1977)177

Xerosiphon Turcz. (~ *Gomphrena*). Amaranthaceae (II 2). 2 S Am. R: BMNHN 4,12(1990)94

Xerospermum Blume. Sapindaceae. 2 Indomal. R: Blumea 28(1983)389. Some cryptic dioecy. Some tough timbers

Xerospiraea Henrickson (~ *Spiraea*). Rosaceae. 1 Mexico

Xerotecoma J. Gómes = *Godmania*
Xerothamnella C. White. Acanthaceae. 2 E Aus. R: JABG 9(1986)166
Xerotia Oliver. Caryophyllaceae (I 1). 1 Arabia: *X. arabica* Oliver – *Ephedra*-like subshrub
xianmu *Burretiodendron hsienmu*
Ximenia L. Olacaceae. 8 trop. Root-parasites. *X. americana* L. (hog or monkey plum, tallow nut, trop.) – timber a sandalwood subs., fr. ed. though kernel contains a strong purgative, oilseed a subs. for ghee
Ximeniopsis Alain = praec.
Xiphidium Aublet. Haemodoraceae. 1 trop. Am.: *X. coeruleum* Aublet – buzz-fls visited by male euglossine bees collecting oils
Xiphion Miller = *Iris*
Xiphochaeta Poeppig (~ *Stilpnopappus*). Compositae (Vern.-Vern.). 1 NE S Am. ± Aquatic
Xiphopteris Kaulf. = *Grammitis*
Xiphotheca Ecklon & Zeyher (~ *Priestleya*). Leguminosae (III 27). 9 Cape fynbos. R: T 42(1993)45
Xixangia Hong. Scrophulariaceae. 1 China: *X. serrata* Hong. R: APS 24(1986)139
Xolantha Raf. Older name for *Tuberaria*
Xolocotzia Miranda. Verbenaceae. 1 Mexico
Xylanche G. Beck = *Boschniakia*
Xylanthemum Tzvelev (~ *Chrysanthemum*). Compositae (Anth.-Tan.). 9 S & C As.
Xylia Benth. (*Esclerona*). Leguminosae (II 3). 13 trop. OW esp. Afr. *X. xylocarpa* (Roxb.) Theob. (*X. dolabriformis*, pyinkado, ironwood, Burma to Mal.) – hard reddish timber second only to teak in Burma, used for ship-building, bridges etc.
Xylinabaria Pierre = *Urceola*
Xylinabariopsis Pitard = *Urceola*
Xylobium Lindley. Orchidaceae (V 10). 29 trop. Am. Cult. orn. epiphytic orchids
Xylocalyx Balf.f. Scrophulariaceae. 5 Somalia (3), Socotra (2). R: KB 16(1962)147, NJB 7(1987)267
Xylocarpus Koenig. Meliaceae (IV 3). 3 coasts trop. OW, E Afr. eastwards to Pacific. R: MF 45(1982)448, FM I,12(1995)371. *X. granatum* Koenig (range of genus) sympatric in mangroves with *X. moluccensis* (Lam.) M. Roemer (Indomal. to trop. Aus.) the first with snake-like surface roots, the second with pneumatophores; *X. rumphii* (Kostel.) Mabb. restricted to rocky coasts through range of genus. Fr. with corky seeds fitted together like a Chinese puzzle; these irreg. shaped & sized seeds float just beneath surface in seawater; timber hard; tannin for nets etc.
Xylococcus Nutt. = *Arctostaphylos*
Xylomelum Sm. Proteaceae. 2 SW, 2 SE Aus. Woody pears, protected spp.; fr. woody, pear-shaped, splitting to release winged seeds
Xylonagra J.D. Sm. & Rose. Onagraceae. 1 Baja California
Xylonymus Kalkman ex Ding Hou. Celastraceae. 1 W New Guinea
Xyloolaena Baillon. Sarcolaenaceae. 1 Madagascar
Xylophragma Sprague. Bignoniaceae. 5 trop. Am.
Xylopia L. Annonaceae. c. 160 trop. (Afr. 61, Am. 51). Only pantrop. genus in fam.; perigyny unique. Timbers, fr. used as condiments, local med.; alks, soporifics in Amazonia. *X. aethiopica* (Dunal) A. Rich. (Afr., Guinea or negro pepper, trop. W Afr.) – form. much used in Eur., now shown to have antimalarial activity ; *X. brasiliensis* Sprengel (trop. S Am.) – pepper source (piperine)
Xylorhiza Nutt. (~ *Machaeranthera*). Compositae (Ast.-Sol.). 9 W US, Mex. R: Brittonia 29(1977)199
Xylosma Forster f. Flacourtiaceae (8). 85 trop. (Am. 49, R: FN 22(1980)128; SE As. 5, Mal. 4, Aus. 4, New Caled. 15, Vanuatu 1, Polynesia 7, Guam 1). Usu. dioec. trees; nickel accumulators in New Caled.
Xylosterculia Kosterm. = *Sterculia*
Xylothamia Nesom, Suh, Morgan & B. Simpson (~ *Euthamia*). Compositae (Ast.-Sol.). 9 Chihuahua Desert. R: Sida 14(1990)101
Xylotheca Hochst. Flacourtiaceae (3). 3 E & SE Afr. *X. tettensis* (Klotzsch) Gilg (E Afr.) – supposed aphrodisiac on Mafia Is.
Xymalos Baillon. Monimiaceae (V 1). 1 trop. & S Afr.: *X. monospora* (Harvey) Baillon (lemonwood) – timber useful
Xyridaceae Agardh. Monocots – Commelinidae – Commelinales. 5/260 trop. & warm, few temp. Herbs, usu. in wet places, s.t. rhizomatous; vessel-elements throughout vegetative body. Lvs with open sheath & narrow flat to terete, usu. distichous & basal. Fls bisexual,

3-merous, poll. by pollen-coll. bees (no nectar or scent), usu. sessile in axils of tough imbricate bracts in cylindrical heads or dense spikes; K 3, outer 1 thin, membranous & reflexed at anthesis, to 0, other 2 chaffy, boat-shaped, C 3 clawed or basally connate, usu. yellow & ephemeral, A 3 (6, or + 3 staminodes), anthers with longit. slits, G (3), 1(3)-loc. with parietal (to axile) placentation, simple or 3-fid style, each placenta with (1–)∞ orthotropous to anatropous or weakly campylotropous bitegmic ovules. Fr. a loculicidal to irreg. dehiscent capsule, s.t. encl. within persistent C-tube; seeds many, small with scarcely differentiated lenticular to scutelliform embryo lying alongside starchy, proteinaceous endosperm s.t. with oil. x = 8, 9, 13, 17

Genera: *Abolboda, Achlyphila, Aratitiyopea, Orectanthe, Xyris*. R: AMBG 71(1984)300

Abolboda & *Orectanthe* with usu. spirally arr. lvs etc. have been segregated in own fam.

Xyridopsis Welw. ex R. Nordenstam (~ *Emilia*). Compositae (Sen.- Sen.). 2 Angola

Xyris L. Xyridaceae. 200–400 trop. & warm (Aus. 20, 17 endemic; Afr. 25). Few cult. orn. (yellow-eyed grass) & locally medic. *X. capensis* Thunb. (trop.) – culms used to make figures for Hindu temples in Java

Xyropteris Kramer (~ *Tapeinidium*). Dennstaedtiaceae (Lindsaeoideae). 1 Sumatra, Borneo

Xyropterix Airy Shaw = *Xyropteris*

Xysmalobium R. Br. (~ *Asclepias*). Asclepiadaceae (III 1). 10 trop. & S Afr. Some ed. roots

Y

Yabea Kozo-Polj. (~ *Caucalis*). Umbelliferae (III 3). 1 W N Am.

yacca *Xanthorrhoea* spp.

yachan *Ceiba insignis*

yacou *Smallanthus sonchifolius*

Yadakea Makino = *Pseudosasa*

yajé *Banisteriopsis* spp. esp. *B. caapi*

Yakirra Lazarides & R. Webster (~ *Panicum*). Gramineae (34b). 6 Burma, Aus. R: Brunonia 7(1984)289. Auricles poss. elaiosomes attracting ant-dispersers

yam *Dioscorea* spp., (US) *Ipomoea batatas*; **acom** or **aerial y.** *D. bulbifera*; **y. bean** *Pachyrhizus* spp.; **Chinese y.** *D. batatas*; **cush-cush y.** *D. trifida*; **Guinea y.** *D.* × *cayenensis*; **Otaheite y.** *D. bulbifera*; **potato-y.** *D. esculenta*; **water** or **white y.** *D. alata*; **yampi y.** *D. trifida*; **yellow y.** *D.* × *cayenensis*

yampee *Dioscorea trifida*

yanagi *Debregeasia edulis*

yang *Dipterocarpus* spp.

yang(g)ona *Piper methystichum*

yangtao *Actinidia deliciosa*

yar *Casuarina equisetifolia*

yard grass *Eleusine indica*; **y.-long bean** *Vigna unguiculata* subsp. *sesquipedalis*

yareta *Azorella compacta*

Yarina Cook = *Phytelephas*

yarran *Acacia homalophylla*

yarroconalli *Tephrosia sinapou*

yarrow *Achillea millefolium*

yate *Eucalyptus cornuta*

yaupon *Ilex vomitoria*

yautia *Xanthosoma* spp., *Caladium lindenii*

yawa *Vigna unguiculata*

Yeatesia Small. Acanthaceae. 3 SE US to NE Mex. R: SB 14(1989)427

ye'eb nut *Cordeauxia edulis*

yeenga *Geodorum densiflorum*

yegoma oil *Perilla frutescens*

yellow archangel *Lamium galeobdolon*; **y. ash** *Cladrastis lutea, Eucalyptus luehmanniana*; **y. avens** *Geum urbanum*; **y. bark** *Cinchona calisaya*; **y. berries** *Rhamnus infectoria*; **y. birch** *Betula alleghaniensis*; **y.bird's-nest** *Monotropa hypopitys*; **y. box** *Eucalyptus moluccana*; **y. buckeye** *Aesculus flava*; **y. cedar** *Chamaecyparis* spp.; **y. cypress** *C. nootkatensis*; **y. deal** or **Baltic y. d.** *Pinus sylvestris*; **y.-eyed grass** *Xyris* spp.; **y. flag** *Iris pseudacorus*; **y. foxtail** *Setaria pumila*; **y. gentian** *Gentiana lutea*; **y. gum** *Xanthorrhoea resinosa*; **y. loosestrife** *Lysimachia vulgaris*; **y. monkey flower** *Mimulus luteus*; **y. pereira** *Howardia glaucescens*; **y. pimpernel** *L. nemorum*; **y. pine** *Pinus echinata*; **y. poplar** *Liriodendron tulipifera*; **y. rocket** *Barbarea verna*; **y. stringybark** see **stringybark**; **y. walnut** *Beilschmiedia bancroftii*; **y. water-**

lily *Nuphar luteum*; **y.wood** spp. of *Cladrastis, Flindersia, Milicia, Ochrosia, Podocarpus, Zanthoxylum* etc.; **y.w., Kentucky** *Cladrastis lutea*; **y.wort** *Blackstonia perfoliata*; **y. yam** *Dioscorea × cayenensis*

yemtani *Gmelina arborea*

yerba buena *Micromeria chamissonis*; **y. de la feche** *Sebastiania bilocularis*; **y. maté** *Ilex paraguariensis*; **y. de la negrita** *Sphaeralcea coccinea, S. angustifolia* ssp. *cuspidata*; **y. reuma** *Frankenia salina*; **y. de la sangre** *Cordia globosa*; **y. sancta** *Eriodictyon californicum*; **y. del zorillo** *Chenopodium graveolens*

yercum *Calotropis gigantea*

Yermo Dorn. Compositae (Sen.-Tuss.). 1 Wyoming: *Y. xanthocephalus* Dorn. R: Madrono 38(1991)198

yesterday, today and tomorrow *Brunfelsia australis, B. pauciflora*

yew *Taxus baccata*; **American y.** *T. canadensis*; **Californian y.** *T. brevifolia*; **Chinese y.** *T. mairei*; **English y.** *T. baccata*; **golden y.** *T. b.* 'Aurea'; **Irish y.** *T. b.* 'Stricta'; **Japanese y.** *T. cuspidata*; **plum y.** *Cephalotaxus* spp.; **Prince Albert's y.** *Saxegothaea conspicua*; **stinking y.** *Torreya californica*

yggdrasill Scandinavian mythological 'world-tree' with branches spread over the world and reaching above heaven

yinma *Chukrasia tabularis*

Yinquania Z.Y. Zhu = *Cornus*

Yinshania Ma & Zhao (~ *Sophiopsis*). Cruciferae. 8 China. R: APS 25(1987)204

ylang-ylang *Cananga odorata*

Ynesa Cook = *Attalea*

Yoania Maxim. Orchidaceae (V 8). 2 Himal., Japan, NZ (1). Mycotrophic

yoco *Paullinia yoco*

yohimbe bark *Pausinystalia johimbe*

Yolanda Hoehne = *Brachionidium*

yomhin *Chukrasia tabularis*

yon *Anogeissus acuminata*

York road poison *Gastrolobium calycinum*

Yorkshire fog *Holcus lanatus*

young fustic *Cotinus coggygria*

youngberry See *Rubus*

Youngia Cass. Compositae (Lact.-Crep.). c. 40 As. R: CIWP 484(1939). *Y. japonica* (L.) DC – lawnweed S US

Ypsilandra Franchet = *Helonias*

Ypsilopus Summerh. (~ *Tridactyle*). Orchidaceae (V 16). 4 E Afr. Cult. orn.

Ystia Compère = *Schizachyrium*

Yua C.L. Li = *Parthenocissus*

yuca *Manihot esculenta*

Yucaratonia Burkart = *Gliricidia*

Yucca L. Agavaceae. 30 warm N Am. (incl. *Hesperoyucca*). R (anat. etc.): UCPB 35(1962)1. Woody pachycauls ± trunks with stiff sword-like lvs & panicles of usu. white fls (bear-grass, Adam's needle; *Yucca* being taken from yuca which Gerard thought was these pls). Most spp. app. dependent on yucca moths for poll. (though seed set in cult. in other countries), visited by sibling spp. of & incl. *Tegeticula* (*Pronuba*) *yuccasella*, which is active by night, resting in fls (similar colour) by day; fls most strongly scented at night & nectar s.t. secreted at base of ovary though moths do not take it (poss. serves to attract other insects away from stigma); female y. moth climbs up a stamen & bends her head over the anther steadied by her uncoiling tongue, the pollen then being scraped together in a ball under the head by the maxillary palps; as many as 4 stamens may be processed thus before moth flies to another flower, where the ovary is inspected & if it is of the right age & does not already have eggs in it, the moth usu. lays 1 egg in each loc. & after laying each 1 deposits some pollen into the tube formed by the stigmas. As unpoll. fls are soon dropped, the deposition of the pollen ensures that there will be a continuing food supply for larvae provided by abnormal growth of ovules near larvae; other ovules develop normally & larvae emerge to pupate in ground when the seeds are ripe. The adults emerge over a period up to 3 yrs with the effect that even if the *Y.* does not flower every year, some moths survive to reproduce. Bogus y. moths (*Prodoxus*) also breed in ovaries but do not pollinate, relying on the true y. moth for survival. Some fibres, consituents of comm. shampoos, ed. parts etc. (see Excelsa 7(1977)45) & cult. orn. incl. hybrids esp. *Y. aloifolia* L. (Spanish bayonet, SE US, WI) – rope-fibre from lvs, molluscicidal glycosides; *Y. baccata*

Torrey (SW US, N Mex.) – tough fibre, fr. ed., fl. buds roasted; *Y. brevifolia* Engelm. (Joshua tree, SW US) – grows c. 10 cm per annum, lvs function for c. 12–20 yrs, fibre can be used as newsprint; *Y. elata* Engelm. (SW N Am.) – fibre, 'root' extract mooted as foaming agent for drinks; *Y. elephantipes* Regel (C Am., Mex.) – fuelwood; *Y. filamentosa* L. (lvs to 2.5 cm wide) & *Y. gloriosa* L. (lvs to 7 cm wide) of E N Am. – commonly seen cult. spp. in Eur.; *Y. glauca* Nutt. ex J. Fraser (WC US) – fibre used in kraft paper; *Y. schidigera* Roezl ex Ortgies (palmilla, SW US) – juice a source of preservative, deodorant etc.; *Y. whipplei* Torrey (*Hesperoyucca w.*, SW US, Mex.) – infl. to 3.65 m in 14 days (world record pl. growth), strong fibre, fls ed.

yulan *Magnolia heptapeta*

Yunckeria Lundell = *Ctenardisia*

Yunquea Skottsb. = *Centaurodendron*

Yushania Keng f. = *Sinarundinaria*

Yutajea Steyerm. Rubiaceae (I 8). 1 Guayana Highland: *Y. liesneri* Steyerm.

Yuyba (Barb.-Rodr.) L. Bailey = *Bactris*

Yvesia A. Camus (~ *Brachiaria*). Gramineae (34b). 1 Madagascar

Z

Zabelia (Rehder) Makino = *Abelia*

zacate *Nolina longifolia*

Zacateza Bullock. Asclepiadaceae (I: Periplocaceae). 1 trop. Afr.: *Z. pedicellata* (Schumann) Bullock

Zaczatea Baillon = *Raphionacme*

Zahlbrucknera Reichb. = *Saxifraga*

zakaton *Muhlenbergia macroura*

Zalacca Blume = *Salacca*

Zalaccella Becc. = *Calamus*

Zaleya Burm.f. (~ *Trianthema*). Aizoaceae (II). 3 NE & E Afr., India, Aus.

zalil *Delphinium semibarbatum*

Zaluzania Pers. Compositae (Helia.-Verb.). 10 SW US, Mex. R: Rhodora 81(1979)449. See also *Kingianthus*

Zaluzianskya F.W. Schmidt. Scrophulariaceae. 55 W S Afr., R: O. Hilliard (1994) *Manuleeae*: 460. Some cult. orn. herbs with evening-fragrant fls superfic. resembling Caryophyllaceae

zambac *Jasminum sambac*

Zambesi redwood *Baikiaea plurijuga*

Zamia L. Zamiaceae (III). 40 trop. & warm Am. R: JAA suppl. 1(1991)371. Cult. orn. & starch sources (often toxic until boiled) – Florida arrowroot (esp. *Z. integrifolia* Aiton, Florida, WI). *Z. furfuracea* L.f. (E Mex.) – poll. host-specific *Rhopalotria mollis* Sharp (snout weevil) mating, feeding & ovipositing in male cones rich in starch, larvae pupating & adults chewing through pollen-sacs before merely visiting starch-poor females & thereby effecting poll.; *Z. pseudoparasitica* Yates (Panamá) – often epiphytic; *Z. pumila* L. (Carib.) – survives repeated scrub-fires in Florida, beetle-poll. as in *Z. f.*; *Z. roezlii* Linden (*Z. chigma*, Mex.) – egg to 3 mm long, spermatazoids to 400 µm diam. & with 40 000 flagella

Zamiaceae Horan. Cycadopsida. 8/125 trop. & warm Afr., Aus. & Am. Dioec. pachycaul trees to 18 m or with underground stems. Lvs pinnate (pinnae rarely dichotomously divided), leaflets midrib-less (cf. Cycadaceae) but with ± parallel longit. veins. Sporophylls in determinate cones, scales on females ± peltate & with 2(3) ovules on adaxial margins

Classification & genera:

1. **Encephalarteae** (sporophylls imbricate, pinnae narrowing to attachment point): *Encephalartos, Lepidozamia, Macrozamia*
2. **Dioeae** ((like 1. but pinnae not narrowing): *Dioon*
3. **Zamieae** (sporophylls app. valvate): *Ceratozamia, Chigua, Microcycas, Zamia*

See also Boweniaceae

Cult. orn., often poisonous but starch & seeds in many ed. after treatment

Zamioculcas Schott. Araceae (V 4). 1 SE trop. Afr.: *Z. zamiifolia* (Lodd.) Engl. Lvs pinnate

Zandera D.L. Schulz. Compositae (Helia.-Mel.). 3 Mexico

Zanha Hiern. Sapindaceae. 1–2 trop. Afr. *Z. africana* (Radlk.) Exell – local medic. in E Afr. for all ailments

Zannichellia L. Zannichelliaceae. 1–5 cosmop. (Eur. 1–5). R: Lagascalia 14(1986)241, JAA

68(1987)264. *Z. palustris* L. (horned pondweed) – monoec. herb of fresh or brackish water with male & female fls in same axil; poll. underwater, when selfed pollen dropped in gelatinous mass directly on stigma

Zannichelliaceae Chevall. (~ Potamogetonaceae). Monocots – Alismatidae – Alismatales. 4/c. 12 cosmop. Monoec. (to dioec.?) submerged glabrous herbs in fresh, brackish or alkaline water; stems much-branched, arising from rhizomes with unbranched roots; single vasc. bundle with vessel-less xylem surrounded by phloem. Lvs linear, 1-(or ± 3-) veined, with open basal sheath encl. stem & apically ligulate (free or 0 in *Zannichellia*), spirally arr. in app. whorls of 3 or 4 or app. distichous to subopp.; usu. a pair of filiform scales at each node. Fls small, water-poll. in usu. complex axillary sympodial infls, these separated by 1–3 lvs; P 0 or 3 scale-like basally connate, A 1 (or ?2 or 3 with 4–6 thecae united), anthers with 1 or more longit. slits, pollen spherical, G (1–)3 or 4(–9) each with ± short style & 1 apical, pendulous anatropous bitegmic ovule. Fr. a head of achenes or drupelets; embryo with circinate cotyledon in 0 endosperm. X = 6–8. R: Taxon 25(1976)273

Genera: *Althenia, Lepilaena, Pseudalthenia, Zannichellia*

Allied to (& best united with) Cymodoceaceae (saline water)

Zanonia L. Cucurbitaceae. 1 Indomal.: *Z. indica* L.

Zantedeschia Sprengel (*Richardia*). Araceae (VI 5). 6 trop. & S Afr. R: Bothalia 11(1973)5. Cult. orn. rhizomatous pls with showy spathes (calla of florists) esp. *Z. aethiopica* (L.) Sprengel (arum lily, S Afr.) – spathes white, much used at & assoc. with funerals in GB; also *Z. elliottiana* (Will. Watson) Engl. (yellow a. l., cultigen poss. hybrid) – spathe yellow, lvs spotted white; *Z. odorata* P. Perry (S Afr.) – freesia-scented

Zanthorhiza L'Hérit. = *Xanthorhiza*

Zanthoxylum L. (*Xanthoxylum*). Rutaceae. (Incl. *Fagara*) c. 250 Am., Afr., As. (incl. Mal., R: JAA 47(1986)221), Aus. Decid. or evergreen prickly shrubs & trees with aromatic bark, s.t. with knobs tipped with spines (knobthorn); alks incl some with antitumour activity. Some timbers (yellowwood), spices etc. *Z. acanthopodium* DC (Himal.) – fr. sold as spice in Sikkim; *Z. americanum* Miller (E N Am.) – hardy shrub with locally medic. bark; *Z. capense* (Thunb.) Harvey (S Afr.) – locally medic. ('fever tree'); *Z. clava-herculis* L. (Hercules' club, C & S US) – toothache cure & other local medic.; *Z. flavum* Vahl (trop. Am.) – timber (Jamaican or WI satinwood) used for cabinet-work etc.; *Z. gillettii* (De Wild.) Waterm. (*Z. macrophyllum* Oliver, trop. Afr., African satinwood) – fine timber; *Z. piperitum* (L.) DC (Sichuan or Japan pepper, E As.) – fr. used as condiment in Japan, one of the few in Jap. cuisine; *Z. simulans* Hance (*Z. bungei*, China) – cult. as condiment; *Z. xanthoxyloides* (Lam.) Zepernick & Timler (W Afr.) – chewing-stick from roots with marked antimicrobial activity against oral flora

Zanzibar copal *Hymenaea verrucosa*; **Z. redheads** *Syzygium aromaticum*

zapatero *Casearia praecox*

Zapoteca H. Hernández (~ *Calliandra*). Leguminosae (II 5). 17 SW N Am. to N Arg. R: AMBG 76(1989)806. Moth-poll. but moths preyed on by crab spiders

Zataria Boiss. Labiatae (VIII 2). 1 Iran, Afghanistan

Zauschneria C. Presl = *Epilobium*

zawa *Lophira* spp.

Zdravetz oil *Geranium macrorrhizum*

Zea L. Gramineae (40i). 4 C Am. R: AJB 67(1980)1000. *Z. mays* L. (maize) – cultigen with controversial history in that geneticists attempted to explain its origin through the then fashionable phenomena of introgression etc. involving *Tripsacum* grasses, though app. derived merely by selection from wild forms of teosinte (subsp. *mexicana* (Schrader) Iltis) called ssp. *parviglumis* Iltis & Doebley with distichous lateral ears, once considered to belong in a separate genus (*Euchlaena*) so that origin of the cultigen form. thought to have given rise to a new genus; cult. pl. with male fls in term. plumes (tassels), the females in lower axils & bearing silks (styles) which pollen-grains penetrate, fert. leading to development of cob from which seeds cannot escape except through human activities. Earliest known forms (c. 5600 yrs old) with small cobs but each grain subtended by long glumes as in modern popcorn. Sculptures of C12 & C13 in India diff. to explain in view of post-Columbian introd. OW. Next to wheat & rice, the most imp. cereal cult. in trop. & warm ('corn' (US), mealies, i.e. Port. 'milho [i.e. millet] (maiz)' (Afr.), Indian corn) with many cvs used as human (incl. cornflakes, polenta, cornflour, corn on the cob, babycorn (immature cobs for stir-fry usu. from Taiwan)) & animal food, cooking oil, beer & spirits, (esp. bourbon, national spirit of USA first made in B. County, Kentucky), industrial alcohol etc.; locally the staple carbohydrate esp. in Afr. where eaten like dough made from m. flour, crushed grains (US) being hominy (grits). Cvs with hard endosperm exploding when heated = popcorn ('var. *praecox*'), those with ± translucent horny endosperm = sweetcorn ('var. *rugosa*'), usu. harvested when immature & tinned; gall prod. *Ustilago maydis* (rust) a vegetable; some cult. orn. with grains of diff. colours in same cob, all red ('Strawberry c.') etc., blue-black ones ('blue c.') held sacred by Pueblos & now extracts used in US cosmetics

zebrano *Microberlinia* spp.

zebrawood *Astronium fraxinifolium*, *Centrolobium robustum*, *Diospyros* spp., *Microberlinia* spp.

Zebrina Schnizl. = *Tradescantia*

Zederbauera H.P. Fuchs = *Erysimum*

zedoary *Curcuma zedoaria*

Zehnderia Cusset. Podostemaceae. 1 Cameroun

Zehneria Endl. Cucurbitaceae. 30 OW trop. Allied to *Melothria* (Am.). *Z. anomala* C. Jeffrey (NE Afr. to S Arabia) – fr. sweet

Zehntnerella Britton & Rose = *Facheiroa*

Zelkova Spach. Ulmaceae (I). 4 Crete (1), W & E As. R: Plantsman 11(1989)80. Cult. orn. like elms (some now attacked by Dutch Elm disease) with some timber esp. *Z. serrata* (Thunb.) Makino (keaki, keyaki, E China, Taiwan, Japan) – imp. timber tree for building, also a bonsai subject

Zemisne Degener & Sherff = *Scalesia*

Zenia Chun. Leguminosae (I 2). 1 S China, Thailand

Zenkerella Taubert. Leguminosae (I 4). 5 trop. Afr. with 1 on Guinea coast, rest in E Afr. coastal mts

Zenkeria Trin. Gramineae (25). 4 India, Sri Lanka. R: HIP t. 3597(1962)

Zenobia D. Don. Ericaceae. 1 SE US: *Z. pulverulenta* (Willd.) Pollard (*Z. speciosa*) – cult. orn. shrub

zephyr lily *Zephyranthes atamasco*

Zephyra D. Don. Tecophilaeaceae (Liliaceae s.l.). 1 Chile

Zephyranthella (Pax) Pax = *Habranthus*

Zephyranthes Herbert. Amaryllidaceae (Liliaceae s.l.). 40 trop. & warm Am. R: JRHS 62(1937)195, Plantsman 2(1980)8. Cult. orn. bulbous herbs with *Colchicum*-like white, yellow, pink or red fls esp. *Z. atamasco* (L.) Herbert (atamasco or zephyr lily, SE US) – fls white, bulbs eaten in emergency, *Z. carinata* Herbert (*Z. grandiflora*, '*Z. rosea*', rain lily, S Mex. to Guatemala) – most commonly cult. sp., fls pink after stimulus of rain & associated cooling. See also *Habranthus*

Zeravschania Korovin. Umbelliferae (III 10). 6 SW to C As. R: BMOIPB 93,4(1988)80

Zerdana Boiss. Cruciferae. 1 mts of Iran: *Z. anchonioides* Boiss. R: Cand. 40(1985)363

Zerna Panzer = *Vulpia* but often applied to *Bromus* (*Bromopsis*)

Zetagyne Ridley (~ *Panisea*). Orchidaceae (V 12). 1 SE As.

Zeugandra P. Davis. Campanulaceae. 2 Iran

Zeugites P. Browne. Gramineae (24). 12 trop. Am. Ovate blades & false petioles much resembling dicot lvs

Zeuktophyllum N.E. Br. Aizoaceae (V). 1 Little Karoo: *Z. suppositum* (L. Bolus) N.E. Br. – coll. once

Zeuxanthe Ridley = *Prismatomeris*

Zeuxine Lindley. Orchidaceae (III 3). 26 trop. & warm OW (Aus. 1). Some cult. orn. incl. *Z. strateumatica* (L.) Schltr. (Indomal.) – one of few annual orchids

Zexmenia Llave & Lex. (~ *Lasianthaea*, *Wedelia*). Compositae (Helia.-Verb.). 2 C Am. R: SBM 33(1991)86

Zeyheria C. Martius. Bignoniaceae. 2 Brazil. *Z. tuberculosa* Bur. – construction timber

Zeylanidium (Tul.) Engl. = *Hydrobryum*

Zhumeria Rech.f. & Wendelbo. Labiatae (VIII 2). 1 Iran

Zieria Sm. Rutaceae. 44 Aus. (43 endemic), New Caled. (1). *Z. arborescens* Sims (stinkwood, E Aus.) – lvs unpleasantly scented when crushed

Zieridium Baillon = *Euodia*

Zigadenus Michaux. Melanthiaceae (Liliaceae s.l.). 18 N Am., Urals (1) & E As. (1). Alkali grass. Many poison (alks) to stock (esp. W US), few cult. orn., incl. *Z. densus* (Desr.) Fern. (crow poison, N Am.), *Z. glaucus* Nutt. (white camash, N Am.) & *Z. nuttallii* A. Gray (death c., N Am.)

Zilla Forssk. Cruciferae. 3 N Afr. to Arabia

Zimmermannia Pax. Euphorbiaceae. 6 trop. E Afr. R: KB 9(1954)38, 36(1981)129. *Z. capillipes* Pax (Tanz.) – roots anthelmintic

Zimmermanniopsis Radcl.-Sm. Euphorbiaceae. 1 Tanz.: *Z. uzungwaensis* Radcl.-Sm. R: KB 45(1990)192

zingana *Microberlinia brazzavillensis*

Zingeria Smirnov. Gramineae (21d). 4 Eur. (2) to Iran. x = 2

Zingeriopsis Probat. = *Zingeria*

Zingiber Boehmer. Zingiberaceae. 60 Indomal. (Borneo 19; R: NRBGE 45(1988)409) to E As. & trop. Aus. Herbs with aromatic rhizomes esp. *Z. officinale* Roscoe (ginger, cultigen of Indian origin?) – sterile, the rhiz. used fresh (green g.) or preserved in syrup or crystallized (esp. in Hong Kong) or dried. & powdered, used in biscuits, cakes, sweets, g. beer, g. wine & brandy, also shampoos & cosmetics & much prescribed for travel sickness & much cult. in Jamaica (first exported 1547); other spp. used incl. *Z. mioga* (Thunb.) Roscoe (Japanese or mioga g., Japan) – bergamot flavour, anc. medic. in China (vermifuge (earliest recorded), antimalarial, insect-bites etc.), *Z. purpureum* Roscoe (*Z. cassumar*, cassumar g., Indomal.) & *Z. zerumbet* (L.) Sm. (Indomal.) ethereal oils etc. shown to have antineoplastic principles & to inhibit prostaglandin synthesis; *Z. squarrosum* Roxb. (Indomal.) – used to tranquillize *Apis dorsata* bees in Andaman Is. so honey can be coll.

Zingiberaceae Martinov (incl. Costaceae). Monocots – Zingiberidae – Zingiberales. 52/1100 trop. esp. Indomal. Usu. caulescent aromatic herbs with thickened rhizomes & secretory cells with ethereal oils; vessel-elements in roots & s.t. also in stems etc.; aerial stem short. Lvs spirally arr. or distichous with usu. open sheath, the sheaths arr. so as to form a pseudostem, ± petiole & with simple lamina rolled from 1 side to another, usu. with prominent midrib & costae pinnate-parallel; adaxial ligule usu. at junction of sheath & lamina. Fls bisexual, zygomorphic, insect-poll., usu. ephemeral, solit. or in cymes in axils of bracts, s.t. on a short separate sheath-covered stem from rhiz.; K 3 not petaloid, basally united with a tube or spathe-like & split to base on 1 side only, C (3) with short lobes the median adaxial often larger than rest, A 1 (median adaxial inner of a presumed 3 + 3 now rep. by consp. labellum (other 2 inner; cf. Orchidaceae where lip is P, or all other 5 (Costoideae)) & staminodes), ± petaloid with elongate thecae & longit. slits, $\overline{G}$ (3), (1–) 3-loc. with (parietal or) axile placentation & usu. apical nectaries & ± ∞ anatropous, bitegmic ovules. Fr. a loculicidal capsule or indehiscent (fleshy or dry); seeds with arils, embryo straight in copious starchy endosperm & perisperm. x = (7–)9–26. R: NRBGE 31(1972)171

Classification & principal genera:

Costoideae (aerial parts not aromatic; lvs spirally arr. or 4-ranked with closed sheaths; labellum of 5-stamen origin(?)): 4 genera incl. *Costus*, *Tapeinochilus*; often treated as separate fam.

Zingiberoideae (aerial parts aromatic; lvs distichous; labellum of 3-stamen origin) – 4 tribes:

Hedychieae (G 3-loc., lvs parallel to rhiz., style not far exserted beyond anther): 16 genera incl. *Boesenbergia*, *Curcuma*, *Hedychium*, *Kaempferia*, *Roscoea* & *Scaphochlamys*

Zingibereae (G 3-loc., lvs parallel to rhiz., style exserted beyond anther & enveloped by elongate process (anther crest): *Zingiber* (only)

Alpinieae (G usu. 3-loc., lvs at right angles to rhiz.): *Aframomum*, *Alpinia*, *Amomum*, *Elettaria*, *Hornstedtia*, *Renealmia*, *Riedelia*

Globbeae (G 1-loc.; anther usu. long-exserted on arched ascending filament): 4 genera incl. *Globba*

Many imp. spice-pls incl. spp. of *Aframomum*, *Amomum*, *Curcuma* (turmeric), *Elettaria* (cardamom), *Kaempferia*, *Zingiber* (ginger); scent (*Hedychium*, *Kaempferia*), diosgenin (*Costus*), dyes (*Curcuma*) & cult. orn. in these genera & *Cautleya*, *Roscoea* etc.

Zinnia L. Compositae (Helia.-Zinn.). 11 US to Arg. esp. Mex. R: Brittonia 15(1963)1. Cult. orn. herbs (& shrubs) with opp. or whorled lvs & alks, esp. *Z. violacea* Cav. (*Z. elegans*, Mex.) – allopolyploid with *Z. angustifolia* Kunth bringing disease resistance, cult. by Aztecs & now with v. many cvs with fls of all colours save blue. C of female fls of some spp. (*Tragoceros*, R: Brittonia 15 (1963)290) persistent & hooked for disp.

Zinowiewia Turcz. Celastraceae. 17 Mexico to Venezuela

Zippelia Blume (~ *Piper*). Piperaceae. 1 Java. X = 19 (in *Piper* 13)

ziricote *Cordia dodecandra*

Zizania L. Gramineae (9). 2–3 E India to E As. (1), N Am. R: JAA 69(1988)265. Aquatic grasses. *Z. aquatica* L. (*Z. palustris*, (Canadian) wild rice, (Manchurian) water rice, wild oat, Tuscarora r., N Am.) – trad. N Am. Indian food, still grown comm. in Minnesota & Far E, sometimes 'popped' (cf. popcorn); *Z. latifolia* (Griseb.) Turcz. (As.) – young shoots eaten in Chinese food to give texture, esp. forms with culms swollen by infection by a smut ('gaausun'), spores from infecting smut used as eyebrow & hair black, lvs made into mats

Zizaniopsis Doell & Asch. Gramineae (9). 5 warm & trop. Am. R: Hickenia 1(1976)39

Zizia Koch. Umbelliferae (III 8). 4 N Am. *Z. aurea* (L.) Koch cult. orn.

Ziziphora L. Labiatae (VIII 2). 25 Medit. (Eur. 6, dry places) to C As. & Afghanistan. Some local medic. (e.g. *Z. tenuior* L. in Turkey – oil mainly pulegone) & flavouring for yoghurt etc.

Ziziphus Miller (*Zizyphus*). Rhamnaceae. 86 trop. & warm (Eur. 1). Decid. & evergreen shrubs & trees, with alks & usu. with stipular thorns, 1 straight, the other recurved. Some timber & ed. fr. (jujubes), often dried like dates. *Z. chloroxylon* (L.) Oliver (cogwood, Jamaica) – tough wood; *Z. joazeiro* C. Martius (Brazil) – fodder; *Z. jujuba* Miller (*Z. zizyphus*, sedra, French jujube, Chinese date, SE Eur. to China) – fr. eaten fresh or cooked, form. imported to GB for cough-cures & extracts for comm. hair-dyes ('neutral henna'), source of shellac, many pomological cvs esp. in E; *Z. lotus* (L.) Lam. (Medit.) – lotus fr. of the Ancients; *Z. mauritiana* Lam. (ber, India) – ed. fr., source of lac, lvs for tanning; *Z. mistol* Griseb. (Andes) – ed. fr. (mistol) used in alcoholic drink; *Z. mucronata* Willd. (buffalo or Cape thorn, trop. Afr.) – medic., seeds used in rosaries; *Z. spina-christi* (L.) Willd. (ilb, Christ's thorn, crown of thorns, Medit. to Arabia) – poss. Christ's crown of thorns; *Z. spinosa* Hu (China) – seeds (suan zau ren) imp. trad. Chinese med. for insomnia etc.

Zizyphus Adans. = *Ziziphus*

Zoegea L. Compositae (Card.-Cent.). 3 SW to C As., Egypt

Zoellnerallium Crosa. Hyacinthaceae (Liliaceae s.l.). 1 Chile

Zollernia Wied-Neuw. & Nees. Leguminosae (III 1). 14 trop. Am. R: BZ 61(1976)1304. Some timber for cutlery-handles, brush-backs etc.

Zollingeria Kurz. Sapindaceae. 3 SE As., 1 Borneo

Zombia L. Bailey. Palmae (I 1a). 1 Hispaniola: *Z. antillarum* (B.D. Jackson) L. Bailey. Suckering palm with persistent fibrous sheaths with spine-like marginal fibres & bisexual fls

Zombitsia Keraudren. Cucurbitaceae. 1 Madagascar

Zomicarpa Schott. Araceae (VIII 4). 3 S Brazil

Zomicarpella N.E. Br. Araceae (VIII 4). 1 Colombia

Zonanthus Griseb. Gentianaceae. 1 Cuba

Zonotriche (C. Hubb.) J. Phipps. Gramineae (39). 3 trop. Afr.

Zootrophion Luer = *Pleurothallis*. But see Selbyana 7(1982)80

Zornia J. Gmelin. Leguminosae (III 13). 86 warm (Aus. 19 endemic). R: Webbia 16(1961)1

Zosima Hoffm. Umbelliferae (III 11). 4 W As.

Zostera L. Zosteraceae. 12 warm to cool (Eur. 3, Aus. 3) to New Guinea. R: C. den Hartog (1970) *Seagrasses of the world*: 42. Nitrogen-fixing bacteria in rhizosphere. *Z. marina* L. (eel-grass, grasswrack, alva or ulva marina, Eur.) – dried lvs used for matting etc. (sea-grass m.), packing (as for glass at Venice), pillows etc., promising grain (imp. bird food in winter)

Zosteraceae Dumort. (~ Potamogetonaceae). Monocots – Alismatidae – Alismatales. 3/18 warm to cool coasts. Monoec. or dioec. (*Phyllospadix*) perennial submerged halophytes to depths of 50 m or exposed at high tide, glabrous & with creeping rhizomes; vessel-elements, stomata & lignin 0. Lvs linear or filiform with open or closed sheath often with stipuloid flanges, ligule at junction with parallel-veined lamina (s.t. with midrib); intravaginal scales at nodes. Fls small, water-poll., sessile in 2 rows (in each alt. male & female in *Zostera* & *Heterozostera*) on 1 side of spadix encl. in spathe, the axis with bract-like lobes (retinacula) that fold over fl(s); P 0, A 1 (= ?(2)), 1-loc. with basally united styles & 1 pendulous orthotropous bitegmic ovule. Fr. a small drupe or irreg. dehiscing; endosperm 0. x = 6, 10

Genera: *Heterozostera*, *Phyllospadix*, *Zostera*

Poll. with grains same density as seawater, captured by feathery stigmas. Marine derivatives of Potamogetonaceae with which Z. should be united

Zosterella Small = *Heteranthera*

Zoutpansbergia Hutch. = *Callilepis*

Zoysia Willd. Gramineae (33d). 10 SE As. to NZ. Some lawn-grasses esp. *Z. matrella* (L.) Merr. (trop. As.) in S US

Zschokkea Muell. Arg. = *Lacmellea*

Zuccagnia Cav. Leguminosae (I 1). 1 Chile, Arg.: *Z. punctata* Cav.

Zuccarinia Blume. Rubiaceae (II 5). 1 W Mal.

zucchini *Cucurbita pepo* 'Zucchini'
Zuckia Standley. Chenopodiaceae (I 3). 1 Arizona: *Z. arizonica* Standley
Zuelania A. Rich. Flacourtiaceae (9). 1 trop. Am. Allied to *Casearia*
Zunilia Lundell = *Ardisia*
Zycona Kuntze = *Schistocarpha*
Zygia P. Browne. Leguminosae (II 5). 20 trop. Am.
Zygocactus Schumann = *Schlumbergera*
Zygochloa S.T. Blake. Gramineae (34g). 1 C Aus.: *Z. paradoxa* (R. Br.) S.T. Blake on desert sand-dunes
Zygodia Benth. = *Baissea*
Zygogynum Baillon. Winteraceae. (Incl. *Belliolum, Bubbia* & *Exospermum*) c. 50 C Mal. to New Caled. & Aus. R: Blumea 31(1985)39. Some rheophytes; poll. by thrips, Coleoptera & primitive moths etc.
Zygonerion Baillon = *Strophanthus*
Zygoon Hiern = *Tarenna*
Zygopetalon auctt. = seq.
Zygopetalum Hook. Orchidaceae (V 10). 15 trop. Am. Cult. orn. epiphytes
Zygophlebia L.E. Bishop. Grammitidaceae. 7 trop. Am. (6), Afr. R: AFJ 79(1989)103
Zygophyllaceae R. Br. Dicots – Rosidae – Sapindales. Incl. Balanitaceae but excl. Nitrariaceae, 27/285 trop. & warm esp. arid, s.t. saline. Small trees, shrubs (usu.) to herbs, s.t. with alks or mustard-oils; stem often swollen at nodes & s.t. with unusual sec. thickening. Lvs usu. paripinnate, often 2-foliolate, less often simple, opp. or spirally arr., often resinous & usu. with stipules, often persistent & s.t. spiny. Fls usu. bisexual & reg., (4)5(6)-merous; K s.t. basally connate, imbricate or valvate, C usu. free, imbricate or convolute, rarely valvate or 0, A in (1)2(3) whorls, filaments with basal glands, anthers with longit. slits, disk usu. intrastaminal & sometimes a gynophore, $\underline{G}$ ((2,4)5(6)), pluriloc. with axile placentas & 1 style, each loc. with 1–several usu. pendulous, anatropous to s.t. hemitropous, campylotropous or orthotropous, bitegmic ovules. Fr. often a capsule or schizocarp, rarely a drupe or berry; embryo straight to weakly curved, in hard oily (or 0) endosperm. x = 6, 8–13+

Principal genera: *Balanites, Fagonia, Kallstroemia, Tribulus, Zygophyllum*

'Loosely knit' (Cronquist) with several genera assigned to diff. fams or made into monotypic ones, e.g. *Balanites* (stipules 0, spines axillary, drupes – Balanitaceae). The seed structure is like that in Malpighiaceae not Sapindales

Most genera in OW, many monotypic; 3 with major disjunctions esp. in N & S Am. – *Fagonia, Larrea, Porlieria*

Timber (*Guaiacum* – lignum-vitae), wax (*Bulnesia*), ed. fr. (*Balanites*), medic. & dyes (*Peganum*), weeds (*Tribulus*)

Zygophyllum L. Zygophyllaceae. 120 Medit. (Eur. 4) to C As., NE & S Afr., Aus., often in deserts or other arid areas. Twinleaf (Aus.). C_3 & C_4 pl. with alks & succ. shoots. Some ed. fl. buds a caper subs. but some in Aus. high in nitrate & suggested to cause sudden death of cattle
Zygoruellia Baillon. Acanthaceae. 1 Madagascar
Zygosepalum Reichb.f. Orchidaceae (V 10). 7 trop. S Am. Cult. orn. epiphytes
Zygosicyos Humbert. Cucurbitaceae. 2 Madagascar
Zygostates Lindley. Orchidaceae (V 10). 7 Brazil. Cult. orn. dwarf epiphytes
Zygostelma Benth. Asclepiadaceae (I; Periplocaceae). 1 Thailand: *Z. benthamii* Baillon
Zygostigma Griseb. Gentianaceae. 2 Brazil, Argentina
Zygotritonia Mildbr. Iridaceae (IV 3). 4 trop. Afr. R: BMNHN 4,11(1989)199. Some corms ± ed.
Zyzyxia Strother. Compositae (Helia.-Verb.). 1 Guatemala, Belize

System for arrangement of vascular plants

This book follows, in general, K. Kubitzki's *The Families and Genera of Vascular Plants* (1990–). The arrangement for Pteridophyta and Gymnosperms as well as part of Angiosperms follows the published works, the remainder being Kubitzki's proposed modifications of Cronquist's 1981 system as published in issues of FGVP Dilleniid, Rosid & Monocot Newsletters, which Prof. dr. Kubitzki has kindly sent me. For a key to the classic literature, see: M.T. Davis (1957) A guide [to] & an analysis of Engler's "Das Pflanzenreich" in *Taxon* 6: 161–182

[PTERIDOPHYTA]

PSILOTOPSIDA (Psilotatae)

Family 1. Psilotaceae

LYCOPSIDA (Lycopodiatae)

Family 1. Isoetaceae
2. Lycopodiaceae
3. Selaginellaceae

EQUISETOPSIDA (Equisetatae)

Family 1. Equisetaceae

FILICOPSIDA (Filicatae)

Family 1. Aspleniaceae
2. Azollaceae
3. Blechnaceae
4. Cheiropleuriaceae
5. Cyatheaceae
6. Davalliaceae
7. Dennstaedtiaceae
8. Dicksoniaceae
9. Dipteridaceae
10. Dryopteridaceae (Woodsiaceae)
11. Gleicheniaceae
12. Grammitidacaceae
13. Hymenophyllaceae
14. Hymenophyllopsidaceae
15. Lomariopsidaceae
16. Lophosoriaceae
17. Loxsomataceae
18. Marattiaceae
19. Marsileaceae
20. Matoniaceae
21. Metaxyaceae
22. Monachosoraceae
23. Nephrolepidaceae
24. Oleandraceae
25. Ophioglossaceae
26. Osmundaceae
27. Plagiogyriaceae
28. Polypodiaceae
29. Pteridaceae

30. Salviniaceae
31. Schizaeaceae
32. Thelypteridaceae
33. Vittariaceae

[GYMNOSPERMAE]

GINKGOOPSIDA (Ginkgoatae)

Family 1. Ginkgoaceae

PINOPSIDA (Pinatae, incl. Taxopsida)

Family 1. Araucariaceae
2. Cephalotaxacae
3. Cupressaceae
4. Phyllocladaceae
5. Pinaceae
6. Podocarpaceae
7. Sciadopityaceae
8. Taxaceae
9. Taxodiaceae

CYCADOPSIDA (Cycadatae)

Family 1. Boweniaceae
2. Cycadaceae
3. Stangeriaceae
4. Zamiaceae

GNETOPSIDA (Gnetatae)

Family 1. Ephedraceae
2. Gnetaceae
3. Welwitschiaceae

[ANGIOSPERMAE]

NB. Besides obvious amalgamations noted in the text below, compared with the first edition of *The Plant-book*, Magnoliales include Laurales; Winteraceae are moved to Illiciales; Chloranthaceae to Magnoliales; Aristolochiales include Hydnoraceae & Rafflesiaceae (Rafflesiales); Nelumbonaceae & Ceratophyllaceae are afforded Orders of their own; Papaverales are included in Ranunculales; Coriariaceae are moved to Sapindales; Cercidiphyllaceae & Eupteleaceae are now in Trochodendrales; Daphniphyllales & Didymelales are combined with Balanopaceae (form. Fagales) & Buxaceae (from Euphorbiales) in Buxales (Dilleniidae), to which Subclass Urticales are referred; Leitneriales are combined with Sapindales (Simaroubaceae s.s.); Juglandales include Myricales; Sarcolaenaceae & Dipterocarpaceae are now in Malvales, Actinidiaceae (with Byblidaceae from Rosales) in Ericales; Theales include Diapensiales & Ebenales (save Sapotaceae put in its own order Sapotales & Symplocaceae now in Cornales), Aquifoliaceae from Celastrales, Parnassiaceae from Rosales (Saxifragaceae) & Sarraceniaceae (from Nepenthales); Guttiferae with Elatinaceae are excluded thence & comprise Guttiferales; Malvales also include Peridiscaceae, Bixaceae & Huaceae (from Violales, to which Elaeocarpaceae are moved), Neuradaceae (Rosales) & Geraniaceae from Rosidae; Nepenthales include Dioncophyllaceae & Ancistrocladaceae from Violales; Violales include Salicales & Dipentodontaceae from Santalales (Rosidae); Euphorbiales are moved to Dilleniidae from Rosidae & include Thymelaeaceae from Myrtales; Capparidales include Batales as well as Akaniaceae & Bretschneideraceae from Sapindales & both Limnanthaceae & Tropaeolaceae from `Geraniales' & Salvadoraceae from Celastrales (all Rosidae); Tepuianthaceae are moved from Sapindales to Linales; in Asteridae Calycerales & Campanulales are merged in Asterales which include Menyanthaceae from Solanales. Moreover, the order of subclasses, orders & families now follows Kubitzki and therefore seems very different from Cronquist's in many respects.

In Monocotyledones, Hydrocharitales and Najadales are merged with Alismatales; Dioscoreales, Haemodorales, Velloziales & Asparagales are separated from Liliales to which Burmanniaceae (Orchidales) are added.

This arrangement is far from stable and many families have controversial positions. For general concerns about these re-arrangements, see Paeoniaceae & Liliaceae (and segregates) in main text.

Goldberg (SCB 58(1986)) has produced a new system for the angiosperms, little followed, while Thorne (*Aliso* 13(1992)365) has updated his. The latter has much to commend it but as Kubitzki's book (which often follows Thorne in amending Cronquist) has full descriptions and is a guide to the literature it is being followed in this and future editions of *The Plant-book.*

DICOTYLEDONAE (Magnoliopsida)

Subclass I. Magnoliidae

Order 1. Magnoliales (incl. Laurales)
Family
1. Himantandraceae
2. Eupomatiaceae
3. Austrobaileyaceae
4. Degeneriaceae
5. Magnoliaceae
6. Annonaceae
7. Myristicaceae
8. Canellaceae
9. Lactoridaceae
10. Amborellaceae
11. Trimeniaceae
12. Chloranthaceae
13. Monimiaceae
14. Gomortegaceae
15. Lauraceae
16. Hernandiaceae
17. Calycanthaceae (incl. Idiospermaceae)

Order 2. Illiciales
Family
1. Winteraceae
2. Illiciaceae
3. Schisandraceae

Order 3. Aristolochiales (incl. Rafflesiales)
Family
1. Aristolochiaceae
2. Hydnoraceae
3. Rafflesiaceae (incl. Mitrastemmataceae)

Order 4. Piperales
Family
1. Saururaceae
2. Piperaceae

Order 5. Nelumbonales
Family
1. Nelumbonaceae

Order 6. Ranunculales (incl. Papaverales)
Family
1. Lardizabalaceae (incl. Sargentodoxaceae)
2. Berberidaceae
3. Menispermaceae
4. Ranunculaceae
5. Circaeasteraceae
6. Pteridophyllaceae
7. Papaveraceae
8. Fumariaceae
9. Sabiaceae (to be moved from here)

Order 7. Nymphaeales (poss. incl. 8. & raised to higher rank)
Family
1. Nymphaeaceae (incl. Barclayaceae)
2. Cabombaceae

Order 8. Ceratophyllales
Family
1. Ceratophyllaceae

Subclass II. Caryophyllidae

Order 1. Caryophyllales (Centrospermae)
Family
1. Caryophyllaceae
2. Molluginaceae
3. Aizoaceae
4. Amaranthaceae

	5. Chenopodiaceae
	6. Halophytaceae
	7. Stegnospermataceae
	8. Achatocarpaceae
	9. Phytolaccaceae
	10. Nyctaginaceae
	11. Cactaceae
	12. Portulacaceae
	13. Didiereaceae
	14. Basellaceae
	15. Hectorellaceae
Order 2.	Polygonales (poss. to be given higher rank)
Family	1. Polygonaceae
Order 3.	Plumbaginales (poss. to be given higher rank)
Family	1. Plumbaginaceae
Subclass III.	Hamamelidae
Order 1.	Trochodendrales
Family	1. Trochodendraceae (incl. Tetracentaceae)
	2. Eupteleaceae
	3. Cercidiphyllaceae
Order 2.	Hamamelidales
Family	1. Platanaceae
	2. Hamamelidaceae
	3. Myrothamnaceae (poss. in Order 1.)
Order 3.	Fagales
Family	1. Fagaceae
	2. Betulaceae
	3. Ticodendraceae
Order 4.	Juglandales (incl. Myricales)
Family	1. Rhoipteleaceae
	2. Juglandaceae
	3. Myricaceae
Order 5.	Casuarinales (poss. not here)
Family	1. Casuarinaceae
Order 6.	Eucommiales (to be moved from here)
Family	1. Eucommiaceae
Subclass IV.	Dilleniidae
Order 1.	Buxales (incl. Daphniphyllales & Didymelales)
Family	1. Buxaceae
	2. Simmondsiaceae
	3. Daphniphyllaceae
	4. Balanopaceae
	5. Didymelaceae
Order 2.	Dilleniales
Family	1. Dilleniaceae
	2. Paeoniaceae
Order 3.	Violales (incl. Salicales)
Family	1. Flacourtiaceae (incl. Lacistemataceae)
	2. Salicaceae
	3. Violaceae
	4. Elaeocarpaceae
	5. Stachyuraceae
	6. Plagiopteraceae
	7. Dipentodontaceae
	8. Scyphostegiaceae
	9. Passifloraceae
	10. Turneraceae

11. Malesherbiaceae
12. Achariaceae
13. Caricaceae
14. Tamaricaceae
15. Frankeniaceae
16. Cucurbitaceae
17. Datiscaceae
18. Begoniaceae
19. Cistaceae (poss. better in Malvales)
20. Fouquieriaceae (? to be moved)
21. Hoplestigmataceae (? to be moved)
22. Loasaceae

Order 4. Nepenthales
Family
1. Droseraceae
2. Dioncophyllaceae
3. Ancistrocladaceae
4. Nepenthaceae

Order 5. Malvales
Family
1. Peridiscaceae (poss. better in Violales)
2. Bixaceae
3. Dipterocarpaceae
4. Sarcolaenaceae
5. Tiliaceae
6. Sterculiaceae
7. Huaceae
8. Bombacaceae
9. Malvaceae
10. Geraniaceae (incl. Dirachmaceae)
11. Neuradaceae

Order 6. Urticales
Family
1. Ulmaceae
2. Moraceae
3. Cecropiaceae
4. Urticaceae
5. Barbeyaceae
6. Cannabaceae

Order 7. Euphorbiales
Family
1. Pandaceae
2. Euphorbiaceae
3. Thymelaeaceae

Order 8. Theales (incl. Diapensiales & Ebenales)
Family
1. Paracryphiaceae
2. Theaceae
3. Pellicieraceae
4. Marcgraviaceae
5. Pentaphylacaceae
6. Tetrameristaceae
7. Parnassiaceae
8. Caryocaraceae
9. Oncothecaceae
10. Ebenaceae
11. Aquifoliaceae
12. Diapensiaceae
13. Styracaceae
14. Lissocarpaceae
15. Sarraceniaceae
16. Ochnaceae (incl. Diegodendraceae (Sphaerosepalaceae, poss. better in Malvales) & Sauvagesiaceae (poss. better in Violales))

	17. Quiinaceae
	18. Scytopetalaceae
	19. Medusagynaceae
Order 9.	Lecythidales (poss. incl. in Theales)
Family	1. Lecythidaceae
Order 10.	Sapotales
Family	1. Strasburgeriaceae
	2. Sapotaceae
Order 11.	Guttiferales
Family	1. Guttiferae (Clusiaceae, incl. Bonnetiaceae)
	2. Elatinaceae
Order 12.	Primulales
Family	1. Theophrastaceae
	2. Myrsinaceae
	3. Primulaceae
Order 13.	Ericales
Family	1. Actinidiaceae
	2. Cyrillaceae (incl. Clethraceae)
	3. Grubbiaceae
	4. Ericaceae (incl. Pyrolaceae & Monotropaceae)
	5. Empetraceae
	6. Epacridaceae
	7. Byblidaceae (Roridulaceae)
Order 14.	Capparidales (incl. Batales)
Family	1. Resedaceae
	2. Capparidaceae
	3. Tovariaceae
	4. Pentadiplandraceae
	5. Cruciferae (Brassicaceae)
	6. Salvadoraceae
	7. Gyrostemonaceae
	8. Bataceae
	9. Bretschneideraceae
	10. Moringaceae
	11. Akaniaceae
	12. Tropaeolaceae
	13. Limnanthaceae
Subclass V.	Rosidae
Order 1.	Rosales
Family	1. Connaraceae
	2. Eucryphiaceae
	3. Cunoniaceae
	4. Brunelliaceae
	5. Davidsoniaceae
	6. Dialypetalanthaceae
	7. Pittosporaceae
	8. Hydrangeaceae
	9. Columelliaceae
	10. Grossulariaceae
	11. Greyiaceae
	12. Bruniaceae
	13. Anisophylleaceae
	14. Crassulaceae
	15. Cephalotaceae
	16. Saxifragaceae
	17. Alseuosmiaceae
	18. Rosaceae
	19. Crossosomataceae
	20. Chrysobalanaceae (? better in Theales)

	21. Surianaceae
	22. Rhabdodendraceae
Order 2.	Fabales
Family	1. Leguminosae
Order 3.	Proteales
Family	1. Elaeagnaceae
	2. Proteaceae
Order 4.	Podostemales
Family	1. Podostemaceae
Order 5.	Haloragidales
Family	1. Haloragidaceae
	2. Gunneraceae
Order 6.	Myrtales
Family	1. Sonneratiaceae
	2. Lythraceae
	3. Punicaceae
	4. Penaeaceae
	5. Crypteroniaceae
	6. Rhynchocalycaceae
	7. Trapaceae
	8. Myrtaceae
	9. Onagraceae
	10. Oliniaceae
	11. Melastomataceae
	12. Combretaceae
Order 7.	Rhizophorales
Family	1. Rhizophoraceae
Order 8.	Cornales
Family	1. Symplocaceae
	2. Alangiaceae
	3. Cornaceae (incl. Nyssaceae)
	4. Garryaceae
Order 9.	Santalales
Family	1. Medusandraceae
	2. Olacaceae
	3. Loranthaceae
	4. Opiliaceae
	5. Santalaceae
	6. Misodendraceae
	7. Viscaceae
	8. Eremolepidaceae
	9. Balanophoraceae
Order 10.	Celastrales
Family	1. Geissolomataceae
	2. Celastraceae (incl. Hippocrateaceae)
	3. Stackhousiaceae
	4. Icacinaceae
	5. Aextoxicaceae
	6. Cardiopteridaceae
	7. Corynocarpaceae
	8. Dichapetalaceae (poss. better in Euphorbiales)
Order 11.	Rhamnales
Family	1. Rhamnaceae
	2. Leeaceae
	3. Vitaceae
Order 12.	Linales
Family	1. Erythroxylaceae
	2. Humiriaceae

3. Ixonanthaceae
4. Linaceae (incl. Hugoniaceae)
5. Tepuianthaceae

Order 13. Polygalales
Family
1. Malpighiaceae
2. Vochysiaceae
3. Trigoniaceae
4. Tremandraceae
5. Polygalaceae
6. Xanthophyllaceae
7. Krameriaceae

Order 14. Sapindales (incl. Leitneriales)
Family
1. Staphyleaceae
2. Melianthaceae
3. Sapindaceae
4. Hippocastanaceae
5. Aceraceae
6. Burseraceae
7. Anacardiaceae (incl. Julianiaceae)
8. Simaroubaceae (incl. Leitneriaceae)
9. Picramniaceae
10. Irvingiaceae
11. Cneoraceae
12. Meliaceae
13. Ptaeroxylaceae
14. Rutaceae
15. Zygophyllaceae
16. Nitrariaceae
17. Coriariaceae
18. Lepidobotryaceae
19. Oxalidaceae
20. Balsaminaceae

Order 15. Apiales
Family
1. Araliaceae
2. Umbelliferae (Apiaceae)

Subclass VI. Asteridae

Order 1. Gentianales
Family
1. Gentianaceae (incl. Saccifoliaceae)
2. Strychnaceae
3. Loganiaceae
4. Geniostomaceae
5. Apocynaceae (incl. Duckeodendraceae)
6. Asclepiadaceae

Order 2. Solanales
Family
1. Nolanaceae
2. Solanaceae
3. Convolvulaceae (incl. Cuscutaceae)
5. Polemoniaceae
6. Hydrophyllaceae

Order 3. Lamiales
Family
1. Lennoaceae
2. Boraginaceae
3. Verbenaceae
4. Avicenniaceae
5. Cylocheilaceae
6. Stilbaceae
7. Phrymaceae
8. Labiatae (Lamiaceae)

Order 4. Callitrichales

Family	1. Hippuridaceae
	2. Callitrichaceae
	3. Hydrostachyaceae
Order 5.	Plantaginales
Family	1. Plantaginaceae
Order 6.	Scrophulariales
Family	1. Oleaceae
	2. Buddlejaceae
	3. Plocospermataceae
	4. Scrophulariaceae
	5. Orobanchaceae
	6. Globulariaceae
	7. Myoporaceae
	8. Gesneriaceae
	9. Acanthaceae
	10. Mendonciaceae
	11. Pedaliaceae
	12. Bignoniaceae
	13. Lentibulariaceae
Order 7.	Asterales (incl. Calycerales & Campanulales)
Family	1. Menyanthaceae
	2. Pentaphragmataceae
	3. Sphenocleaceae
	4. Campanulaceae
	5. Stylidiaceae
	6. Donatiaceae
	7. Goodeniaceae (incl. Brunoniaceae)
	8. Compositae (Asteraceae)
	9. Calyceraceae
Order 8.	Rubiales
Family	1. Gelsemiaceae
	2. Desfontainiaceae
	3. Rubiaceae (incl. Theligonaceae)
Order 9.	Dipsacales
Family	1. Caprifoliaceae
	2. Adoxaceae
	3. Valerianaceae
	4. Dipsacaceae
	5. Morinaceae

MONOCOTYLEDONAE (Liliopsida)

Subclass I.	Alismatidae
Order 1.	Alismatales (incl. Hydocharitales & Najadales)
Family	1. Butomaceae
	2. Alismataceae (incl. Limnocharitaceae)
	3. Hydrocharitaceae
	4. Aponogetonaceae
	5. Scheuchzeriaceae
	6. Juncaginaceae
	7. Potamogetonaceae
	8. Ruppiaceae
	9. Najadaceae
	10. Zannichelliaceae
	11. Posidoniaceae
	12. Cymodoceaceae
	13. Zosteraceae
Order 2.	Triuridales
Family	1. Triuridaceae

Subclass II.	Arecidae
Order 1.	Arecales
Family	1. Palmae (Arecaceae)
Order 2.	Cyclanthales
Family	1. Cyclanthaceae
Order 3.	Pandanales
Family	1. Pandanaceae
Order 4.	Arales
Family	1. Acoraceae
	2. Araceae
	3. Lemnaceae
Subclass III.	Commelinidae
Order 1.	Commelinales
Family	1. Rapateaceae
	2. Xyridaceae
	3. Mayacaceae
	4. Commelinaceae
Order 2.	Eriocaulales
Family	1. Eriocaulaceae
Order 3.	Restionales
Family	1. Flagellariaceae (incl. Hanguanaceae)
	2. Joinvilleaceae
	3. Restionaceae
	4. Centrolepidaceae
Order 4.	Juncales
Family	1. Juncaceae
	2. Thurniaceae
Order 5.	Cyperales
Family	1. Cyperaceae
	2. Gramineae (Poaceae)
Order 6.	Hydatellales
Family	1. Hydatellaceae
Order 7.	Typhales
Family	1. Sparganiaceae
	2. Typhaceae
Subclass IV.	Zingiberidae
Order 1.	Bromeliales
Family	1. Bromeliaceae
Order 2.	Zingiberales
Family	1. Musaceae (incl. Strelitziaceae & Heliconiaceae)
	2. Lowiaceae
	3. Zingiberaceae (incl. Costaceae)
	4. Cannaceae
	5. Marantaceae
Subclass V.	Liliidae
Order 1.	Haemodorales
Family	1. Philydraceae
	2. Pontederiaceae
	3. Haemodoraceae
	4. Taccaceae
	5. Pentastemonaceae
	6. Stemonaceae
Order 2.	Asparagales
Family	1. Agavaceae
	2. Alliaceae

3. Amaryllidaceae
4. Anthericaceae
5. Aphyllanthaceae
6. Asparagaceae (incl. Ruscaceae)
7. Asphodelaceae (incl. Aloaceae)
8. Asteliaceae
9. Blandfordiaceae
10. Convallariaceae
11. Cyanastraceae
12. Doryanthaceae
13. Dracaenaceae
14. Eriospermaceae
15. Hemerocallidaceae
16. Hyacinthaceae
17. Hypoxidaceae
18. Luzuriagaceae
19. Philesiaceae
20. Phormiaceae
21. Tecophilaeaceae
22. Trilliaceae
23. Xanthorrhoeaceae

Order 3. Dioscoreales
Family 1. Smilacaceae
2. Dioscoreaceae

Order 4. Velloziales
Family 1. Velloziaceae

Order 5. Liliales (incl. Burmanniales & Melanthiales)
Family 1. Alstroemeriaceae
2. Burmanniaceae
3. Calochortaceae
4. Campynemataceae
5. Colchicaceae
6. Iridaceae (incl. Geosiridaceae)
7. Liliaceae
8. Melanthiaceae (incl. Petrosaviaceae)

Order 6. Orchidales
Family 1. Corsiaceae
2. Orchidaceae

Acknowledgement of sources

1. Floras and handbooks

Adams, C.D. (1972). *Flowering plants of Jamaica*. University of the West Indies, Mona
Agnew, A.D.Q. (1974). *Upland Kenya wild flowers*. Oxford University Press, London
Allan, H.H. (1961). *Flora of New Zealand*. Volume I. Government Printer, Wellington
Amherst, A. (1895). *A history of gardening in England*. Quaritch, London
Aubréville, A. *et al.* (eds, 1961–). *Flore du Gabon*. Muséum national d'Histoire naturelle, Paris
Aubréville, A. *et al.* (eds, 1963–). *Flore du Cameroun*. Muséum national d'Histoire naturelle, Paris
Aubréville, A. *et al.* (eds, 1967–). *Flore de la Nouvelle-Calédonie et Dépendances*. Muséum national d'Histoire naturelle, Paris
Backer, C. & R.C. Bakhuizen van den Brink (1963–1968). *Flora of Java*. 3 vols. Noordhoff, Groningen, etc.
Baker, J.G. (1887). *Handbook of the fern-allies*. Bell, London
Baumann, H. (1982). *Die griechische Pflanzenwelt in Mythos, Kunst und Literatur*. Hirmer, Munich
Beadle, N.C.W. *et al.* (1982). *Flora of the Sydney Region*. Reed, Balgowlah, NSW
Bean, W.J. (1970–1988). *Trees and shrubs hardy in the British Isles*, ed. 8. 4 vols + suppl. Murray, London
Benedix, E.H. *et al.* (1986). *Rudolf Mansfeld's Verzeichniss Landwirtschaftlicher und Gärtnerischer Kulturpflanzen (ohne Zierpflanzen)*. 4 vols. Akademie-Verlag Berlin
Bentham, G. (1863–1878). *Flora australiensis*. 7 vols. Reeve, London
Bews, J.W. (1925). *Plant forms and their evolution in South Africa*. Longman, London, etc.
Blamey, M. & C. Grey-Wilson (1989). *The illustrated flora of Britain and northern Europe*. Hodder, London
Borg, J. (1959). *Cacti*, ed. 3. Blandford, London
Bosser, J. *et al.* (eds, 1976–). *Flore des Mascareignes*. Royal Botanic Gardens, Kew; ORSTOM, Paris, etc.
Bramwell, D. & Z.I. Bramwell (1974). *Wild flowers of the Canary Islands*. Thornes, London & Burford
Briggs, D. & S.M. Walters (1984). *Plant variation and evolution*, ed. 2. Cambridge University Press
Bureau of Flora and Fauna (1981–). *Flora of Australia*. Canberra
Burbidge, N. (1963). *Dictionary of Australian plant genera (gymnosperms and angiosperms)*. Angus & Robertson, Sydney
Burbidge, N.T. & M. Gray (1979). *Flora of the Australian Capital Territory*. Australian National University, Canberra
Burkill, I.H. (1935). *A dictionary of the economic products of the Malay Peninsula*. 2 vols. Crown Agents, London
Castrovejo, S. *et al.* (1986–). *Flora iberica*. CSIC, Madrid
Chittenden, F.J. (ed., 1951). *Dictionary of gardening*. 4 vols. Clarendon Press, Oxford
Christensen, C. *et al.* (1906–1965). *Index filicum*, with four supplements. Hagerup, Copenhagen, etc.
Church, A.H. (1908). *Types of floral mechanism*. Part I. Clarendon Press, Oxford
Clapham, A.R., T.G. Tutin & E.F. Warburg (1962). *Flora of the British Isles*. Cambridge University Press
Clapham, A.R., T.G. Tutin & E.F. Warburg (1981). *Excursion flora of the British Isles*, ed. 3. Cambridge University Press
Clay, S. (1937). *The present-day rock garden*. Jack, London

Cook, C.D.C *et al.* (1974). *Water plants of the world*. Junk, The Hague
Cooke, M.C. (1882). *Freaks and marvels of plant life; or, curiosities of vegetation*. SPCK, London
Corner, E.J.H. (1952). *Wayside trees of Malaya*, ed. 2. 2 vols. Government Printer, Singapore; (1988) *idem*, ed. 3. 2 vols. Malayan Nature Society, Kuala Lumpur
Corner, E.J.H. (1964). *The life of plants*. Weidenfeld & Nicholson, London
Corner, E.J.H. (1966). *The natural history of palms*. Weidenfeld & Nicolson, London
Corner, E.J.H. (1976). *The seeds of dicotyledons*. 2 vols. Cambridge University Press
Cranbrook, Earl of (1988). *Malaysia*. Pergamon, Oxford
Crawford, R.M.M. (1989). *Studies in plant survival*. Blackwell, Oxford
Cronquist, A. (1981). *An integrated system of classification of flowering plants*. Columbia University Press
Dale, I.R. & P.J. Greenway (1961). *Kenya trees and shrubs*. Buchanan's Kenya Estates Ltd
Dallimore, W. & A.B. Jackson (1966). *A handbook of Coniferae and Ginkgoaceae*, ed. 4. Arnold, London
Dassanayake, M.D. (ed., 1980–). *A revised handbook to the flora of Ceylon*. Amerind Publishing, New Delhi
Davies, R. & S. Ollier (1989). *Allergy. The facts*. Oxford University Press
Davis, P.H. (1965–). *Flora of Turkey and the East Aegean Islands*. Edinburgh University Press
Drummond, J.C. & A. Wilbraham (1958). *The Englishman's food*. Revised by D. Hollingsworth. London
Duthie, R. (1988). *Florists' flowers and societies*. Shire, Princes Risborough
Dyer, R.A. (1975–1976). *The genera of southern African flowering plants*. 2 vols. Department of Agricultural Technical Services, Pretoria
Dyer, R.A. *et al.* (1963–). *Flora of southern Africa*. Government Printer, Pretoria
Engler, A. *et al.* (eds, 1900–). *Das Pflanzenreich*. Engelmann, Leipzig. See M.T. Davis (1957) A guide and an analysis of Engler's `Das Pflanzenreich'. *Taxon* 6: 161–182
Engler, A. *et al.* (eds, 1924). *Die Natürlichen Pflanzenfamilien*, ed. 2. Engelmann, Leipzig, etc.
Exell, A.W. *et al.* (eds, 1960–). *Flora Zambesiaca*. Crown Agents, London
Fahn, A. (1974). *Plant anatomy*, ed. 2. Pergamon Press, Oxford, etc.
Farr, E., J.A. Leussink & F.A. Stafleu (eds, 1979). *Index nominum genericorum (plantarum)*. 3 vols. Bohn, Scheltema & Holkema, Utrecht, etc.
Farr, E., J.A. Leussink & G. Zijlstra (eds, 1986). *Idem*, Supplementum I
Farrer, R. (1928). *The English rock garden*, 4th impr. 2 vols. Jack, London
Fernald, M.L. (1950). *Gray's manual of botany*. ed. 8 (centennial). American Book Co., New York, etc.
Flora of North America Editorial Committee (1993–). *Flora of North America north of Mexico*. Oxford University Press, New York and Oxford
Folkard, R. (1892). *Plant lore, legends, and lyric*, ed. 2. Sampson Low, London
Forey, P.L. (1981). *The evolving biosphere*. British Museum (Natural History), London
Frodin, D.G. (1984). *Guide to the standard floras of the world*. Cambridge University Press
Godwin, H. (1975). *The history of the British flora*, ed. 2. Cambridge University Press
Good, R. (1974). *The geography of the flowering plants*, ed. 4. Longman, London
Grierson, A.J.C. *et al.* (1983–). *Flora of Bhutan*. Royal Botanic Garden Edinburgh
Grigson, G. (1958). *The Englishman's flora*. Phoenix House, London
Hallé, F., R.A.A. Oldeman & P.B. Tomlinson (1978). *Tropical trees and forests: an architectural analysis*. Springer, Berlin, etc.
Hara, H., W.T. Stearn, L.H.J. Williams & A.O. Chater (1978–1982). *Enumeration of the flowering plants of Nepal*. 3 vols. British Museum (Natural History), London
Harborne, J.B. (1988). *Introduction to ecological biochemistry*, ed. 3. Academic Press, London
Harborne, J.B. & F.A. Tomas-Barberan (1991). *Ecological chemistry and biochemistry of plant terpenoids*. Clarendon Press, Oxford
Harden, G.J. (ed., 1990–1993). *Flora of New South Wales*. 4 vols. NSW University Press
Harling, G. & B. Sparre (eds, 1973–). *Flora of Ecuador. Opera botanica* B. Lund.
Harrison, S.G., G.B. Masefield & M. Wallis (1969). *The Oxford book of food plants*. Oxford University Press
Haslam, S.M., P.D. Sell & P.A. Wolseley (1977). *A flora of the Maltese islands*. Malta University Press
Hawthorne, W. (1990). *Field guide to the forest trees of Ghana*. ODA, London
Heywood, V.H. (ed., 1978). *Flowering plants of the world*. Oxford University Press
Heywood, V.H. & S.R. Chant (1982). *Popular encyclopedia of plants*. Cambridge University Press

Holm, L.G. *et al.* (1979). *The world's worst weeds*. Honolulu

Hooker, J.D. (ed., 1872–1897). *The flora of British India*. 7 vols. Reeve, London

Hubbard, C.E. (1968). *Grasses*, ed. 2. Penguin, Harmondsworth

Humbert, H. *et al.* (eds, 1936–). *Flore de Madagascar et des Comores (Plantes vasculaires)*. Muséum national d'Histoire naturelle, Paris

Hunt, T. (1989). *Plant names of mediaeval England*. Brewer, Cambridge

Hutchinson, J. (1964–1967). *The genera of flowering plants*. 2 vols. Clarendon Press, Oxford

Hutchinson, J. & J.M. Dalziel (1954–1972). *Flora of west tropical Africa*, ed. 2, revised by R.W.J. Keay & F.N. Hepper. 3 vols with *The ferns and fern allies* (1959, by A.H.G. Alston) as supplement. Crown Agents, London

Huxley, A. (1967). *Mountain flowers*. Blandford, London

Jackson, B.D. *et al.* (1895–). *Index kewensis plantarum phanerogamarum*. Clarendon Press, Oxford

Jacobsen, H. (1960). *A handbook of succulent plants*. 3 vols. Blandford, London

Janzen, D.H. (ed., 1983). *Costa Rican natural history*. University of Chicago Press

Jex-Blake, A.J. (1950). *Gardening in East Africa*, ed. 3. Longman, London

Juniper, B.E., R.J. Robins & D.M. Joel (1989). *The carnivorous plants*. Academic Press, London

Kerner von Marilaun, A. (1904). *The natural history of plants*, trans. F. W. Oliver. 2 vols. Gresham, London

Komarov, V.L. (ed., 1968–). *Flora of the U.S.S.R.*, translated from the Russian. Israel Program for Scientific Translations, Jerusalem

Liberty Hyde Bailey Hortorium (1976). *Hortus third. A concise dictionary of plants cultivated in the United States and Canada*. Macmillan, New York, etc.

Lucas, G. & H. Synge (1978). *The IUCN Red Data Book*. IUCN, Morges

Mabberley, D.J. (1991). *Tropical rain forest ecology*, ed. 2. Blackie, Glasgow & Chapman & Hall, New York

McFarlane, D. (ed., 1988). *The Guiness Book of Records*. Guiness, Enfield

Macmillan, H.F. (1935). *Tropical planting and gardening*, ed. 4. Macmillan, London

Macmillan, H.F. (1991). *Idem*, ed. 6 (revised by H.S. Barlow, I. Enoch & R.A. Russell). Malayan Nature Society, Kuala Lumpur

Matthews, J.R. (1955). *Origin and distribution of the British flora*. Hutchinson, London

Meiggs, R. (1982). *Trees and timbers in the ancient Mediterranean world*. Clarendon Press, Oxford

Meikle, R.D. (1977–1985). *Flora of Cyprus*. 2 vols. Bentham-Moxon Trust, Kew

Menninger, E.A. (1967). *Fantastic trees*. Viking, New York

Menninger, E.A. (1977). *Edible nuts of the world*. Horticultural Books, Stuart, Florida

Metcalfe, C.R. & L. Chalk (1950). *Anatomy of the Dicotyledons*. 2 vols. Clarendon Press, Oxford; ed. 2 (1979–), Clarendon Press, Oxford

Miller, A.G. & M. Morris (1988). *Plants of Dhofar*. Office of the Advisor for Conservation of the Environment, Oman

Mitchell, A. (1974). *A field guide to the trees of Britain and northern Europe*. Collins, London

Moore, D.M. (1983). *Flora of Tierra del Fuego*. Nelson, Oswestry

Moore, L.B. & E. Edgar (1970). *Flora of New Zealand*. Volume II. Government Printer, Wellington

Morley, B.D. & H.R. Toelken (eds, 1983). *Flowering plants in Australia*. Rigby, Adelaide, etc.

Munz, P.A. (1959). *A California flora*. University of California Press

Needham, J. (1986). *Science and civilization in China* 6(1) Botany. Cambridge University Press

Ohwi, J. (1965). *Flora of Japan*, ed. by F.G. Meyer & E.H. Walker. Smithsonian Institution, Washington

Organisation for Flora Neotropica (1968–). *Flora neotropica*. Hafner, New York, etc.

Ozenda, P. (1977). *Flore du Sahara*, ed. 2. CNRS, Paris

Page, M. & W.T. Stearn (1974). *Culinary herbs*. Royal Horticultural Society, London

Peekel, P.G. (1984). *Flora of the Bismarck Archipelago for naturalists*. Trans. E.E. Henty. Office of Forests, Lae

Perry, F. (1972). *Collins' guide to border plants*, ed. 3. Collins, London

Plant Resources of South-East Asia [PROSEA] (1989–) . Pudoc, Wageningen

Polunin, N. (1959). *Circumpolar Arctic flora*. Clarendon Press, Oxford

Polunin, O. & A. Stainton (1984). *Flowers of the Himalaya*. Oxford University Press

Prance, G.T. & T.E. Lovejoy (eds, 1984). *Amazonia*. Pergamon, Oxford

Pratt, A. (1899–1905). *Flowering plants, grasses, sedges and ferns of Great Britain*. New ed. by

E. Step. 4 vols. Warne, London
Proctor, M. & P. Yeo (1973). *The pollination of flowers*. Collins, London
Purseglove, J. (1968–1972). *Tropical crops*. 4 vols. Longman, London
Putz, F.E. & H.A. Mooney (1991). *The biology of vines*. Cambridge University Press
Quezel, P. & S. Santa (1962–1963). *Nouvelle flore de l'Algérie et des régions désertiques meridionales*. CNRS, Paris
Rackham, O. (1986). *History of the British countryside*. Dent, London
Ramson, W.S. (ed., 1988). *The Australian national dictionary*. Oxford University Press, Melbourne
Rattauf, R.F. (1970). *A handbook of alkaloids and alkaloid-containing plants*. Wiley, New York, etc.
Rechinger, K.H. (ed., 1963–). *Flora Iranica*. Akademische Druck-and Verlagsanstalt, Graz
Ridley, H.N. (1922–1925). *The flora of the Malay Peninsula*. 5 vols. Reeve, London
Ridley, H.N. (1930). *The dispersal of plants throughout the world*. Reeve, Ashford
Roach, F.A. (1985). *Cultivated fruits of Britain*. Blackwell, Oxford
Ruskin, F.R. (ed., 1975). *Underexploited tropical plants with promising economic value*. National Academy of Sciences, Washington
Schultes, R.E. & R.F. Rattauf (1990). *The healing forest*. Dioscorides, Portland
Sculthorpe, C.D. (1967). *The biology of aquatic vascular plants*. Arnold, London
Sharma, B.D. et al. (eds, 1993–), *Flora of India*. Botanical Survey of India, Calcutta
Simmonds, N.W. (ed., 1976). *Evolution of crop plants*. Longman, Harlow
Simpson, J.A. & E.S.C. Weiner (preps, 1989). *The Oxford English Dictionary*, ed. 2. Clarendon Press, Oxford
Smith, D.C. & A.E. Douglas (1987). *The biology of symbiosis*. Arnold, London
Sporne, K.R. (1974). *The morphology of gymnosperms*, ed. 2. Hutchinson, London
Sporne, K.R. (1975). *The morphology of pteridophytes*, ed. 4. Hutchinson, London
Stace, C. (1991). *New flora of the British Isles*. Cambridge University Press
Stainton, A. (1988). *Flowers of the Himalaya. A supplement*. Oxford University Press
van Steenis, C.G.G.J. (ed., 1948–). *Flora malesiana*. Noordhoff-Kolff, Jakarta, etc.
van Steenis, C.G.G.J. (1972). *The mountain flora of Java*. Brill, Leiden
van Steenis, C.G.G.J. (1981). *Rheophytes of the world*. Sijthoff & Noordhoff, Alphen aan den Rijn
Strid, A. (ed., 1986–1991). *Mountain Flora of Greece*. Vol. 1, Cambridge University Press; vol. 2 (ed. with Kit Tan), Edinburgh University Press
Sudworth, G.B. (1908). *Forest trees of the Pacific slope*. US Department of Agriculture, Washington
Synge, P.M. (1969). *Supplement to the dictionary of gardening*. Clarendon Press, Oxford
Tang, S. & M. Palmer (1986). *Chinese herbal prescriptions*. Rider, London
Tansley, A.G. (1939). *The British Islands and their vegetation*. Cambridge University Press
Tomlinson, P.B. (1986). *The botany of mangroves*. Cambridge University Press
Tomlinson, P.B. & M.H. Zimmermann (eds, 1978). *Tropical trees as living systems*. Cambridge University Press
Turrill, W.B. *et al*. (eds, 1952–). *Flora of tropical East Africa*. Crown Agents, London
Tutin, T.G., V.H. Heywood *et al*. (1964–1980). *Flora europaea*. 5 vols. Ed. 2 (1993–). Cambridge University Press
Uphof, J.C.T. (1968). *Dictionary of economic plants*, ed. 2. Cramer, Lehre
Valdés, B. *et al*. (1987). *Flora vascular de Andalucia Occidental*. 3 vols. Barcelona
Verdcourt, B. & E.C. Trump (1969). *Common poisonous plants of East Africa*. Collins, London
Wagner, W.L. et al. (1990). *Manual of the flowering plants of Hawaii*. 2 vols. University of Hawaii Press
Walters, S.M. *et al*. (eds, 1984–). *The European garden flora*. Cambridge University Press
Watt, J.M. & M.G. Breyer-Brandwijk (1962). *The medicinal and poisonous plants of Southern and Eastern Africa*, ed. 2. Livingstone, Edinburgh and London
Webb, C.J. *et al*. (1988). *Flora of New Zealand*. Vol. IV. DSIR, Christchurch
White, F. (1983). *The vegetation of Africa*. UNESCO, Paris
Whitmore, T.C. & F.S.P. Ng (eds, 1972–1989). *Tree flora of Malaya*. 4 vols. Longman Malaysia, Petaling Jaya
Wiggins, I.R. & D.L. Porter (1971). *Flora of the Galápagos Islands*. Stanford University Press
Williams, R.O. (1949). *Useful and ornamental plants of Zanzibar and Pemba*. Zanzibar Protectorate
Willis, J.C. (1931). *A dictionary of the flowering plants and ferns*, ed. 6. Cambridge University Press

Willis, J.C. (1973). *A dictionary of the flowering plants and ferns*, ed. 8, revised by H.K. Airy Shaw. Cambridge University Press

Zohary, D. & M. Hopf (1988, 1993). *Domestication of plants in the Old World*, eds 1 & 2. Clarendon Press, Oxford

Zohary, M. (1982). *Plants of the Bible*. Cambridge University Press

2. Periodicals

Abhandlungen der Akademie der Wissenschaften und der Literatur Mainz: Mathematisch-Naturwissenschaftliche Klasse: tropische und Subtropische Pflanzenwelt
Acanthus
Acta agrobotanica
Acta amazonica
Acta biologica Paranaense, formerly Boletim Universidade do Paraná Botânica
Acta biologica Venezuelica
Acta botanica Academiae Scientiarum Hungaricae, later Acta botanica Hungarica
Acta botanica Austro Sinica
Acta botanica Barcinonensia, formerly Acta phytotaxonomica Barcinonensia
Acta botanica Brasilica
Acta botanica Cubana
Acta botanica Fennica
Acta botanica Gallica, formerly Bulletin de la Société botanique de France
Acta botanica Hungarica, formerly Acta botanica Academiae Scientarum Hungaricae
Acta botanica Islandica
Acta botanica Malacitana
Acta botanica Mexicana
Acta botanica Neerlandica
Acta botanica Sinica
Acta botanica Slovaca Academiae Scientiarum Slovacae, Ser A. Taxonomica Geobotanica
Acta botanica Venezuelica
Acta botanica Yunnanica
Acta cientifica Potosina
Acta Facultatis rerum naturalium Universitatis Comenianae (Bratislava) Series Botanica; Series Physiologia Plantarum
Acta geobotanica Barcinonensia
Acta Horti Bergiani
Acta Musei Reginaehradecensis, A
Acta phytogeographica Suecica
Acta phytotaxonomica Barcinonensia, continued as Acta botanica Barcinonensia
Acta phytotaxonomica et geobotanica
Acta phytotaxonomica Sinica
Acta Societatis Botanicorum Poloniae
Acta Universitatis Carolinae Biologica Prague
Adansonia, formerly Notulae systematicae, later Bulletin du Muséum national d'Histoire naturelle. Section B, Adansonia
Advances in botanical Research
Advances in ecological Research
Advances in economic Botany
Advances in Genetics
Agricultural University, Wageningen. Papers, formerly Mededelingen van de Landbouwhogeschool te Wageningen
Agrobotanika
Agronomia lusitana
Albertoa
Aliso
Allertonia
Allionia
Amazoniana
Ambio
American Fern Journal

Acknowledgement of sources

American Journal of Botany
American Midland Naturalist
American Naturalist
Anales de la Escuela Nacional de Ciencias biológicas
Anales del Instituto de Biologia. Universidad de México
Anales del Instituto botánico A.J. Cavanilles
Anales del Jardin botánico de Madrid
Anales del Museo de Historia natural de Valparaiso
Anales de la Sociedad cientifica Argentina
Anales de la Sociedad Mexicana de Historia de la Ciencia de la Teconologia
Annales Bogorienses
Annales botanici Fennici, formerly Annales botanici Societatis zoologicae-botanicae fennicae Vanamo
Annales botanici Societatis zoologicae-botanicae fennicae Vanamo, continued as Annales botanici Fennici
Annales de la Faculté des Sciences, Université de Dakar
Annales Musei Goulandris
Annales de Physiologie Végétale, Bruxelles
Annales des Sciences naturelles (Paris)
Annales scientifiques de L'Université de Besançon
Annali di Botanica (Roma)
Annals of applied Biology
Annals of Botany
Annals Kirstenbosch Botanical Garden, formerly Journal of South African Botany, supplement
Annals of the Missouri botanical Garden
Annonaceae Newsletter
Annual Report Huntingdonshire Fauna and Flora Society
Annual Review of Biochemistry
Annual Review of Ecology and Systematics
Annual Review of Genetics
Annual Review of Plant Physiology
Aqua Planta
Aquarium
Aquatic Botany
Aquilo, Seria botanica
Arab Gulf Journal of Scientific Research
Arboretum Kórnickie
Arboretum Leaves
Arboricultural Journal
Archives of natural History, formerly Journal of the Society for the Bibliography of natural History
Archivio botânico e biogeograpfico Italiano
Arena
Arkiv för Botanik
Arnoldia
Aroideana
Arquivos do Jardim botânico do Rio de Janeiro
Asclepiadaceae, continued as Asklepios
Ashingtonia
Asklepios, formerly Asclepiadaceae
Astarte
Atas de Sociedade Botânica do Brasil, Secção Rio de Janeiro
Atoll Research Bulletin
Atti dell'Istituto botânico della Università e Laboratorio crittogamico di Pavia
Australian Acacias
Australian Journal of agricultural Research
Australian Journal of Botany
Australian Journal of Ecology
Australian systematic Botany
Australian systematic Botany Society, Newsletter
Austrobaileya, formerly Contributions from the Queensland Herbarium

Baileya
Bangladesh Journal of Botany
Bartonia
Bauhinia
Bean Bag
Beiträge zur Biologie der Pflanzen
Belgian Journal of Botany, formerly Bulletin de la Société Royale de Botanique de Belgique
Belmontia
Berichte der Deutschen botanischen Gesellschaft, continued as Botanica Acta
Berichte der Schweizerischen botanischen Gesellschaft, continued as Botanica Helvetica
Bibliotheca botanica
Biologia
Biologia Plantarum
Biological Conservation
Biological Journal of the Linnean Society of London, formerly Proceedings of the Linnean Society
Biological Reviews
Biological Sciences Review
Biota
Biotica
Biotropica
Bishop Museum Bulletins in Botany
Blancoana
Blumea
Blyttia
Bocconea
Boissiera
Boletim de Botânica
Boletim Museu botânico Municipal. Curitiba
Boletim do Museu municipal do Funchal
Boletim do Museu nacional de Rio de Janerio. Botânica
Boletim do Museu Paraense `Emílio Goedi' New Ser. Botânica
Boletim da Sociedade Broteriana
Boletim técnico Instituto agronomico do Norte
Boletim Universidade do Paraná Botânica, continued as Acta biológica Paranaense
Boletín de los Jardines Botánicos de América Latina
Boletín Museo nacional de Historia natural, Chile
Boletín de la Sociedad Argentina de Botánica
Boletín de la Sociedad botánica de México
Bollettino, Museo Regionale di Scienze naturali
Bonplandia
Botanica Acta, formerly Berichte der Deutschen botanischen Gesellschaft
Botanica Helvetica, formerly Berichte der Schweizerischen botanischen Gesellschaft
Botanica Macaronesica
Botanical Bulletin of Academia Sinica
Botanical Gazette, later International Journal of Plant Sciences
Botanical Journal of the Linnean Society, formerly Journal of the Linnean Society. Botany
Botanical Journal of Scotland, formerly Transactions of the botanical Society of Edinburgh
Botanical Magazine
Botanical Magazine (Tokyo), continued as Journal of Plant Research
Botanical Museum Leaflets. Harvard University
Botanical Review
Botanical Society of the British Isles News
Botaničeskie Materialy̆ Gerbarija Instituta, botaniki Akademii nauk Uzbekskoj SSR
Botanika Chronika
Botanische Jahrbücher für Systematik, Pflanzengeschichte und Pflanzengeographie
Botaniska Notiser
Botanisk Tidsskrift
Botany Bulletin, Department of Forests (Papua New Guinea)
Bothalia
Bradea
Bradleya

Acknowledgement of sources

Brenesia
British Cactus and Succulent Journal, formerly National Cactus and Succulent Journal
British Fern Gazette, continued as Fern Gazette
Brittonia
Bromélia
Brunonia, formerly Contributions from Herbarium Australiense
Buletin Kebun Raya
Bulletin Auckland Institute and Museum
Bulletin of botanical Laboratory [Research] of North-Eastern Forestry Institute (China), continued as Bulletin of botanical Research
Bulletin of botanical Research, formerly Bulletin of botanical Laboratory [Research] of North-Eastern Forestry Institute (China)
Bulletin of the botanical Society of Bengal, later Journal of the national botanical Society
Bulletin of the botanical Survey of India
Bulletin of the British Museum (Natural History) (Series E). Botany
Bulletin. Fairchild Tropical Garden
Bulletin of the Forest Research Institute Chittagong, Plant Taxonomy Series
Bulletin de l'Institut fondamental d'Afrique noire. Sér. A
Bulletin of the international Group for the Study of Mimosoideae
Bulletin du Jardin botanique national de Belgique
Bulletin mensuel de la Société Linnéenne de Lyon
Bulletin du Muséum national d'Histoire naturelle
Bulletin du Muséum national d'Histoire naturelle. Section B, Adansonia. Formerly Adansonia
Bulletin National Tropical Botanical Garden, formerly Bulletin Pacific Tropical Garden
Bulletin. New Zealand Department of scientific and industrial Research
Bulletin de la Société botanique de France, later Acta botanica Gallica, incl. Actualités botaniques & Lettres botaniques
Bulletin de la Société Royale de Botanique de Belgique, later Belgian Journal of Botany
Bulletin of the Sugadaira biological Laboratory
Bulletin of the Torrey botanical Club
Bulletin of the Wellington botanical Society
Byulleten Gosudarstvennogo Nikitskogo Opytnogo Botanicheskogo Sada
Byulleten Moskovskogo Obshchestva Ispȳtateleī Prirodȳ. Biol.
Byulleten Vsesoyuznogo Ordena Lenina Instituta Rastenievodstva Imeni N.I. Vavilova, continued as Nauchno-tekhnicheskii Byulleten Vsesoyuznogo Ordena Lenina i Ordena Drozhby Narodov Instituta Rastenievodstva Imeni N.I. Vavilova

Cactaceas y Suculentas Mexicanas
Cactus and Succulent Journal of Great Britain
Cactus and Succulent Journal (US)
Caldasia
Calyx
Canadian Field-Naturalist
Canadian Journal of Botany
Canadian Journal of Plant Science
Candollea
Carnegie Institution of Washington Publications
Castanea
Cathaya
Ceiba
Ceylon Forester
Chemical Plant Taxonomy Newsletter
Chinese Bulletin of Botany
Chinese Journal of Botany
Ciencias. Ser. 4. Ciencias biológicas, Universidad de la Habana
Ciencias. Ser. 10. Botánica, Universidad de la Habana
Ciencias biológicas, Academia de Ciencias de Cuba
Cladistics
Commentationes biologicae
Commonwealth Forestry Review
Communicaciones Botanicas del Museo de Historia natural de Montevideo

Communicaciones del Museo Argentino de Ciencias Naturales e Instituto nacional de Investigación de las Ciencias naturales
Compositae Newsletter
Compte rendu des Séances mensuelles. Société des Sciences naturelles du Maroc
Compte rendu des Séances. Société de Biogéographie
Conservation Biology
Contributions from the Bolus Herbarium
Contributions from the Gray Herbarium of Harvard University
Contributions from Herbarium Australiense, continued as Brunonia
Contributions from the National Botanic Garden Glasnevin, continued as Glasra
Contributions from the New South Wales National Herbarium, continued as Telopea
Contributions from the New York Botanical Garden
Contributions from the Queensland Herbarium, continued as Austrobaileya
Contributions in Science
Contributions from the United States National Herbarium
Contributions from the University of Michigan Herbarium
Cuadernos de Botanica Canaria
Cunninghamia
Cuscatlania
Cyperaceae Newsletter

Dansk Botanisk Arkiv
Darwiniana
Davidsonia
Desert Plants
Deserta
Development
Dinteria
Dominguezia
Dumortiera

Ecological Monographs
Ecological Review
Ecology
Economic Botany
Edinburgh Journal of Botany, formerly Notes from the Royal Botanic Garden Edinburgh
Egyptian Journal of Botany, formerly United Arab Republic Journal of Botany
Elliottia
Endeavour
Englera
Environmental Conservation
Ernstia
Essex Naturalist
Evolution
Evolutionary Biology
Evolutionary Ecology
Evolutionary Trends in Plants
Excelsa
Excelsa, Taxonomic Series

Families and Genera of vascular Plants Dilleniid Newsletter
Fauna och Flora
Feddes Repertorium Zeitschrift für botanische Taxonomie und Geobotanik
Fern Gazette, formerly British Fern Gazette
Field Studies
Fieldiana (Botany)
Fitoterapia
Flavour and Fragrance Journal
Flora
Flora of China Newsletter
Flora og Fauna

Acknowledgement of sources

Flora Malesiana Bulletin
Flora Mediterranea
Floribunda
Folia botanica et geobotanica Correntesiana
Folia botanica miscellanea
Folia geobotanica et phyto-taxonomica Bohemoslovaca
Folia Musei rerum Naturalium Bohemiae Occidentalis
Fontqueria
Forestry
Four Seasons
Fragmenta floristica et geobotanica
Functional Ecology

Garcia de Orta. Serie de Botanica
Garden, The
Gardens' Bulletin Singapore
Gaussenia, formerly Travaux de Laboratoire forestier de Toulouse
Gayana. Botánica
Genetica
Gentes Herbarum
Ginkgoana
Giornale botanico Italiano
Glasgow Naturalist
Glasra, formerly Contributions from the National Botanic Garden Glasnevin
Gleditschia
Global Ecology and Biogeography Letters
Göttinger floristische Rundbriefe
Gorteria
Grana
Gymnocalycium

Harvard Papers in Botany
Hebe News
Herbertia
Herbs, Spices and medicinal Plants
Hercynia
Hickenia. Boletin del Darwinion
Hikobia
Hoehnea
Hooker's Icones Plantarum
Human Ecology
Huntia

IAWA Bulletin, later IAWA Journal
Iheringia, série Botânica
Independent, The
Indian Forest Records, (new Series) Botany
Indian Forester
Indian Journal of Botany
Indian Journal of Forestry
Insula
International Journal of Plant Sciences, formerly Botanical Gazette
International Tree Crops Journal
Iowa State Journal of Research
Iranian Journal of Botany
Irish Naturalists' Journal
Iselya
Israel Journal of Botany, later Israel Journal of Plant Sciences
Istanbul Üniversitesi Eczacilik Fakültesi Mecmuasi
Istanbul Üniversitesi Fen Fakültesi Mecmuasi, Seri B: Tabii Ilimber, later Biyoloji Dergisi
Itinera geobotanica
Izvestiya Akademii Nauk Kazakhskoi SSR, Seriya biologicheskaya

Izvestiya Akademii Nauk Turkmenskoi SSR, Seriya biologicheskikh Nauk
Izvestiya na Botanicheskiya Insitut (Sofiya), continued as Phytology

Japanese Journal of Botany
Japanese Journal of historical Botany
JARE, continued as Memoirs of National Institute of Polar Research, Series E
Journal of the Adelaide botanic Gardens
Journal d'Agriculture tropicale et de Botanique appliquée
Journal of Animal Ecology
Journal of applied Ecology
Journal of arid Environments
Journal of the Arnold Arboretum
Journal of Biogeography
Journal of biological Sciences Research (Baghdad)
Journal of Biosciences
Journal of the Bombay natural History Society
Journal of the botanical Society of South Africa
Journal of the East Africa natural History Society and national Museum
Journal of Ecology
Journal of economic and taxonomic Botany
Journal of the Elisha Mitchell scientific Society
Journal of evolutionary Biology
Journal of experimental Botany
Journal of the Faculty of Science, University of Tokyo. Botany
Journal of Forestry
Journal of Garden History
Journal of Geobotany, continued as Journal of Phytogeography and Taxonomy
Journal of the Indian botanical Society
Journal of Japanese Botany
Journal of the Kew Guild
Journal of the Korean Forestry Society
Journal of Korean Plant Taxonomy, later Korean Journal of Plant Taxonomy
Journal of Life Sciences, Royal Dublin Society
Journal of the Linnean Society. Botany, continued as Botanical Journal of the Linnean Society
Journal of Nanjing Technological College of Forest Products
Journal of the national botanical Society, formerly Bulletin of the botanical Society of Bengal
Journal of natural Products, formerly Lloydia
Journal of the Orissa botanical Society
Journal of Phytogeography and Taxonomy, formerly Journal of Geobotany
Journal of Plant Anatomy and Morphology
Journal of Plant Research, formerly Botanical Magazine (Tokyo)
Journal and Proceedings of the Royal Society of New South Wales
Journal of the Royal Society of Western Australia
Journal of Science of the Hiroshima University, Series B, Division 2 Botany
Journal of the Society for the Bibliography of natural History, continued as Archives of natural History
Journal of the South African biological Society
Journal of South African Botany
Journal of the Taiwan Museum, formerly Quarterly Journal of the Taiwan Museum
Journal of tropical Ecology
Journal of Zhejiang Forestry College

Kakteen und andere Sukkulenten
Kalikasan
Kalmia
Kew Bulletin
Kew Bulletin, Additional Series
Kew Magazine, later (Curtis's) Botanical Magazine
Kew Magazine, Monograph Series
Kew Record

Kew Scientist
Kingia
Kings Park Research Notes
Kirkia
Kongelige Danske Videnskabernes Selskabs Skrifter, Biological Series
Korean Journal of Botany
Korean Journal of Plant Taxonomy, formerly Journal of Korean Plant Taxonomy
Kulturpflanze
Kurtziana
Kurzmitteilungen der Deutschen dendrologischen Gesellschaft

Lagascalia
Laitsch
Lamiales Newsletter
Landbouwhogeschool te Wageningen, Miscellaneous Papers
Lasca Leaves
Lavori dell' Istituto botanico dell' Università di Milano
Lavori dell' Istituto e Orto botanico dell' Università di Palermo
La-Ya'aran
Lazaroa
Leandra
Leiden botanical Series
Lejeunia
Lesovedenie
Lidia
Lilloa
Lindleyana
Linnean, The
Linnean Society Symposium Series
Lloydia, continued as Journal of natural Products
Loefgrenia
London Naturalist
Lorentzia
Ludoviciana
Lutukka
Lyonia

Madrono
Malayan Nature Journal
Malaysian Forester
Mededelingen van de Landbouwhogeschool te Wageningen continued as Agricultural University, Wageningen. Papers
Mediterranéa
Mémoires de l'Academie Royale des Sciences d'Outre-Mer, Classe des Sciences naturelles et médicales
Mémoires du Muséum nationale d'Histoire naturelle
Mémoires de la Société botanique de France
Mémoires de la Société botanique de Genève
Mémoires de la Société royale botanique de Belgique
Memoirs of the botanical Survey of South Africa
Memoirs of the Ehime University Section II, Natural Science, Series B (Biology)
Memoirs of the Faculty of Agriculture, Kagoshima University
Memoirs of National Institute of Polar Research, Series E, formerly JARE
Memoirs of the National Science Museum, Tokyo
Memoirs of the New York botanical Garden
Memoirs and Proceedings of the Manchester literary and philosophical Society
Memoirs of the Torrey botanical Club
Memórias da Sociedad Broteriana
Memórias de la Sociedad de Ciencias Naturales la Salle
Mentzelia
Michigan Botanist
Micronesica

Mitteilungen aus der botanischen Staatssammlung München
Mitteilungen der Deutschen dendrologischen Gesellschaft
Mitteilungen aus dem Staatsinstitut für allgemeine Botanik in Hamburg
Miyabea
Monographiae biologicae Canarienses
Monographiae botanicae
Monographs in systematic Botany from the Missouri Botanical Garden
Morris Arboretum Bulletin
Morton Arboretum Quarterly
Moscosoa
Muelleria
Musées de Genève
Mutisia

Napaea
National Cactus and Succulent Journal, continued as British Cactus and Succulent Journal
Natural History Bulletin of the Siam Society
Natural History Research
Naturalia Monspeliensia, Série botanique
Naturalist
Naturaliste Canadien
Nature in Wales
Nauchno-tekhnicheskii Byulleten Vsesoyuznogo Ordena Lenina i Ordena Drozhby Narodov Instituta Rastenievodstva Imeni N.I. Vavilova, formerly Byulleten Vsesoyuznogo Ordena Lenina Instituta Rastenievodstva Imeni N.I. Vavilova
New Botanist
New Phytologist
New Plantsman, formerly Plantsman
New Scientist
New Zealand Journal of Botany
New Zealand Journal of Science
Newsletter. Botanical Society of Otago
Newsletter. Friends of the Royal Botanic Gardens Sydney
Newsletter. Hawaiian Botanical Society
Nigerian Field
Nordic Journal of Botany
Norrlinia
Northern Territory botanical Bulletin
Norwegian Journal of Botany
Notas del Museo de la Plata. Botánica
Notes from the Royal Botanic Garden, Edinburgh, continued as Edinburgh Journal of Botany
Notulae botanicae Horti agrobotanici Cluj-Napoca
Notulae Naturae of the Academy of Natural Sciences of Philadelphia
Notulae systematicae, continued as Adansonia
Nova Acta Regiae Societatis Scientiarum Upsaliensis
Novitates botanicae Universitatis Carolinae
Novon
Novosti Sistematiki Vȳsshikh Rastenii
Nuytsia

Occasional Papers of the Bernice P. Bishop Museum
Occasional Papers of the Californian Academy of Sciences
Oecologia Plantarum
Österreichische botanische Zeitschrift, continued as Plant Systematics and Evolution
Ohio Journal of Science
Oikos
Onira
Opera botanica
Opera botanica Belgica
OPTIMA Leaflets
OPTIMA Newsletter

Opuscula botanica Pharmaciae complutensis, later Rivasgodaya
Orchid Monographs
Oréades
Orquidea
Orquideologia

Pabstia
Pacific Science
Pakistan Journal of Botany
Pakistan Systematics
Paleobiology
Palmengarten
Parodiana
Pesquisas
Phanerogamarum Monographiae
Philippine Agriculturist
Philippine Flora Newsletter
Philosophical Transactions of the Royal Society
Phyta
Phytochemical Society of Europe Symposium Series
Phytologia
Phytologia Memoirs
Phytology, formerly Izvestiya na Botanischeskiya Institut (Sofiya)
Phytomorphology
Phyton
Pittieria
Plant, Cell and Environment
Plant Press
Plant Species Biology
Plant Systematics and Evolution, formerly Österreiche botanische Zeitschrift
Planta
Planta medica
Plants and Gardens
Plantsman, continued as New Plantsman
Polish botanical Studies
Portugaliae Acta biologia
Prace Botaniczne
Preslia
Principes
Proceedings of the Academy of Natural Sciences of Philadelphia
Proceedings of the biological Society of Washington
Proceedings of the California Academy of Sciences
Proceedings of the ecological Society of Australia
Proceedings of the Indian Academy of Sciences, Plant Sciences
Proceedings of the Iowa Academy of Science
Proceedings Koninklijke Nederlandse Akademie van Wetenschappen Series C, formerly Proceedings of the Section of Sciences, Koninklijke Nederlandse Akademie van Wetenschappen
Proceedings of the Linnean Society, continued as Biological Journal of the Linnean Society at London
Proceedings of the Linnean Society of New South Wales
Proceedings of the Nova Scotian Institute of Science
Proceedings of the Royal Microscopical Society
Proceedings of the Royal Society, Series B
Proceedings of the Royal Society of New Zealand
Proceedings of the Royal Society of Queensland
Proceedings of the Royal Society of Victoria
Proceedings of the Section of Sciences, Koninklijke Nederlandse Akademie van Wetenschappen, continued as Proceedings Koninklijke Nederlandse Akademie van Wetenschappen Series C
Protecção de Natureza
Publications in Botany, National Museum of Natural Sciences (Canada)

Publications of the Cairo University Herbarium, continued as Taeckholmia
Publications from the Department of Botany, University of Helsinki

Quarterly Bulletin of the Alpine Garden Society
Quarterly Journal of Forestry
Quarterly Journal of the Taiwan Museum, continued as Journal of the Taiwan Museum
Queensland Botany Bulletin
Queensland Naturalist

Raymondiana
Rea
Regnum vegetabile
Reinwardtia
Revista Brasileira de Botânica
Revista del Jardin Botanique nacional, Universidad de la Habana
Revista del Museo Argentino de Ciencias Naturalis `Bernardino Rivadavia'
Revista del Museo de la Plata (Botanica)
Revista de la Sociedad Mexicana de Historia Natural
Revue de Cytologie et de Biologie Végétales, continued as Revue de Cytologie et de Biologie Végétales, le Botaniste
Revue de Cytologie et de Biologie Végétales, le Botaniste, formerly Revue de Cytologie et de Biologie Végétales and Botaniste
Revue général de Botanique
Revue Roumaine de Biologie, Serie de Botanique
Rheedea
Rhodora
Rivasgodaya, formerly Opuscula botanica Pharmaciae complutensis
Rodriguésia
Ruizia

Sandakania
Sarsia
Schlechteriana
Science in New Guinea
Science and public Affairs
Science Reports of the Tôhoku University, Fourth Series, Biology
Scientific Proceedings of the Royal Dublin Society, Series A
Selbyana
Sellowia
Sida
Sida botanical Miscellany
Silva Fennica
Silvae genetica
Sind University Research Journal (Science Series)
Smithsonian Contributions to Botany
Solanaceae Newsletter
Sommerfeltia
South African Journal of Botany
South Australian Naturalist
Southwestern Naturalist
Stapfia
Studia Botanica (Salamanca)
Studies from the Herbarium, California State University Chico
Sultania
Surinaamse Landbouw
Svensk botanisk Tidskrift
Syesis
Symbolae botanicae Upsalienses
Systematic Botany
Systematic Botany Monographs

Taeckholmia, formerly Publications of the Cairo University Herbarium
Taiwania

Taxon
Telopea, formerly Contributions from the New South Wales National Herbarium
Thaiszia
Threatened Plants Committee Newsletter
Trabajos del Departamento de Botanica y Fisiologia vegetal, Universidad de Madrid
Trabajos del Departamento de Botanica (Salamanca)
Transactions (and Proceedings) of the Botanical Society of Edinburgh, continued as Botanical Journal of Scotland
Transactions of the Royal Society of Canada
Transactions of the Royal Society of South Africa
Transactions of the Royal Society of South Australia
Transactions of the Wisconsin Academy of Sciences, Arts and Letters
Travaux de l'Institut scientifique, Série botanique (Rabat)
Travaux du Laboratoire forestier de Toulouse, continued as Gaussenia
Travaux de la Section scientifique et Technique, Institut Français de Pondichéry
Treballs de l'Institut Botànic de Barcelona
Trees in South Africa
Tropical Agriculture
Tropical Ecology
Tuatara
Tulane Studies in Zoology and Botany
Turrialba

Ukraïnskii Botanichniĭ Zhurnal
Unasylva
United Arab Republic Journal of Botany, continued as Egyptian Journal of Botany
United States Department of Agriculture Publications
University of California Publications in Botany
Utafiti

Växtodling
Vegetatio
Verslagen en Mededelingen van de Koninklijke Nederlandse botanische Vereniging, Jaarboek
Verslagen en Mededelingen Plantenziektekundige Dienst, Wageningen

Wahlenbergia
Watsonia
Webbia
Wentia
Western Australian Herbarium Research Notes
Western Australian Naturalist
Willdenowia
Wood and Fibre
Wrightia

Abbreviations

1. General

NB A number of abbreviations refer to journals not published since 1970 and therefore not appearing under `Acknowledgement of sources', a list of materials thoroughly scanned. Acknowledgement to the editors and proprietors of the additional journals listed is made here.

A	androecium, stamens
AA	Acta amazonica
AB	Aquatic Botany
ABAS	Acta botanica Austro Sinica
ABA(S)H	Acta botanica Academiae Scientarum Hungaricae
ABF	Acta botanica Fennica
ABH	Acta botanica Hungarica
ABiV	Acta biologica Venezuelica
ABM	Acta botanica Mexicana
ABN	Acta botanica Neerlandica
ABY	Acta botanica Yunnanica
accum.	accumulating
AD	Anno Domini
AFJ	American Fern Journal
Afr.	Africa(n)
AHB	Acta Horti Bergiani
AHP	Acta Horti Petropolitani
AIBC	Anales del Instituto botánico A.J. Cavanilles
AJB	American Journal of Botany
AJBM	Anales del Jardin botanico de Madrid
AJBRJ	Arquivos do Jardim botânico do Rio de Janeiro
AK	Arboretum Kórnickie
AKBG	Annals of Kirstenbosch Botanic Gardens
AL	Annonaceae Newsletter
alks	alkaloids
alt.	alternative, alternate, altitude
Am(er).	America(n)
Amaz.	Amazon(ia(n))
AMBG	Annals of the Missouri botanical Garden
AMH	Annales historico naturales-musei nationalis Hungarici
AMHNV	Anales del Museo de Historia natural de Valparaiso
AMN	American Midland Naturalist
anatr.	anatropous
anc.	ancient
ANMW	Annalen des K. K. naturhistorischen (Museums) Hofmuseums. Wien
ann.	annual, anniversary
AOSB	American Orchid Society Bulletin
AP	Australian Plants
APG	Acta phytotaxonomica et geobotanica
app.	apparent(ly)

Abbreviations

APS	Acta phytotaxonomica Sinica
ARBGC	Annals of the royal botanic Garden Calcutta
arch.	archaic, archipelago
ARES	Annual Review of Ecology and Systematics
Arg(ent).	Argentina
Ark.B.	Arkiv für Botanik
Aroid.	Aroideana
arr.	arranged, arrangement
As.	Asia(tic)
ASAM	Annals of the South African Museum
ASBSN	Australian systematic Botany Society Newsletter
ASNB	Annales des Sciences naturelles, Botanique
assoc.	associated
attrib.	attributed, attribution
auctt.	of authors
Aus.	Australia(n)
AusJB	Australian Journal of Botany
AusSB	Australian Systematic Botany
Austr.Pl.	Australian Plants
AUWP	Agricultural University, Wageningen. Papers
AZBB	Annotationes zoologicae et botanicae. Bratislava
BAIM	Bulletin Auckland Institute and Museum
BANCC	Boletín de Academia nacional de Ciencias en Córdoba
BB	Bibliotheca botanica
BBAS	Botanical Bulletin of Academia Sinica
BBC	Beihefte zum Botanischen Centralblatt
BBGB	Bulletin of the botanic Gardens Buitenzorg
BBLNEFI	Bulletin of the botanical Laboratory of the north-eastern Forestry Institute
BBMNH(B)	Bulletin of the British Museum (Natural History), series E – Botany
BBPBM	Bulletin of the Bernice P. Bishop Museum
BBR	Bulletin of botanical Research
BBSI	Bulletin of the botanical Survey of India
BC	Biological Conservation, before Christ
BdeB	Boletim de Botânica
BFG	British Fern Gazette
BG	Botanical Gazette
BH	Botanica Helvetica
BHB	Bulletin de l'Herbier Boissier
BI	British Isles
bienn.	biennial
bisex.	bisexual
BJ	Botanische Jahrbücher
BJB	Bangladesh Journal of Botany
BJBB	Bulletin du Jardin botanique à Bruxelles (*later* national de Belgique)
BJBBuit.	Bulletin du Jardin botanique de Buitenzorg
B(ot.)JLS	Botanical Journal of the Linnean Society
BMNHNAdans.	Bulletin du Muséum national d'Histoire naturelle, Adansonia
BMNR(J)B	Boletim do Museu nacional de Rio de Janeiro. Botânica
BMOIPB	Byulleten Moskovskogo Obshchestva Ispȳtateleĭ Priorodȳ. Biol.
BMPEG	Boletim do Museu Paraense 'Emílio Goedi', Ser. Botânica
BM(T)	Botanical Magazine (Tokyo)
BMac.	Botanica Macaronesica
BN	Botaniska Notiser
BNH	Beihefte zur Nova Hedwigia
Br.	Britain, British
BR	Botanical Review

Braz.	Brazil(ian)
BRCI	Bulletin of the Reseach Council of Israel
BSAB	Boletin de la Sociedad Argentina de Botánica
BSB	Boletim da Sociedade Broteriana
BSBF	Bulletin de la Société botanique de France
BSBG	Berichte der Schweizerischen botanischen Gesellschaft
BSBM	Boletín de la Sociedad botánica de México
BSRBB	Bulletin de la Société Royale de Botanique de Belgique
BT	Botanisk Tidsskrift
BTBC	Bulletin of the Torrey botanical Club
BYUSBB	Brigham Young University Science Bulletin, Biology
BZ	Botanicheskii Zhurnal
c.	about
C	corolla (members), central, century
Caled.	Caledonia(n)
Calif.	California(n)
campylot.	campylotropous
Carib.	Caribbean
Cauc.	Caucasus, Caucasian
CB	Chronica botanica
CBH	Contributions from the Bolus Herbarium
CDH	Contributions from the Dudley Herbarium (of Stanford University)
cent.	century
cerem.	ceremony, -ies, -ial
cf.	compare
CGH	Contributions from the Gray Herbarium of Harvard University
CHA	Contributions from Herbarium Australiense
char(ac).	character(s), characteristic(ally)
chem.	chemical(ly), chemistry
CIWP	Carnegie Institution of Washington Publications
CJB	Canadian Journal of Botany
CJPS	Canadian Journal of Plant Science
Clad.	Cladistics
class.	classified, classification
cli.	climber, climbing
CNSWNH	Contributions from the New South Wales national Herbarium
coll.	collected, collecting
comm.	commercial(ly)
Cons. Biol.	Conservation Biology
consp(ic).	conspicuous(ly)
constr.	construction
cont.	contain(ing)
Cont.	The Continent of Europe, excluding British Isles
cosmop.	cosmopolitan
CQH	Contributions from the Queensland Herbarium
CRSS(S)B	Compte rendu Sommaire des Séances. Société de Biogéographie
CSJ	Cactus and Succulent Journal of Great Britain
CSJAm.	Cactus and Succulent Journal of the Cactus and Succulent Society of America
CSM	Cactaceas y Succulentas Mexicanas
CTRFBS	Contributions from the Texas Research Foundation, Botanical Studies
cult.	cultivated, cultivation
CUMH	Contributions from the University of Michigan Herbarium
CUSNH	Contributions from the United States National Herbarium
cv.	cultivar

Darw.	Darwiniana
DAWW	Denkschriften der (kaiserlichen) Akademie der Wissenschaften. Wien
DB	Dissertationes botanicae
DBA	Dansk botanisk Arkiv
decid.	deciduous
dev.	developed
diam.	diameter
dichot.	dichotomous(ly)
dicots	dicotyledons (Dicotyledonae)
diff.	difference(s), different, differing, difficult
dioec.	dioecious
disp.	dispersal, dispersed (by)
distich.	distichous
distr(ib).	distribution(s), distributed
do	as above
DP	Desert Plants
DTYB	Daffodil and Tulip Yearbook
E	eastern, East
EB	Economic Botany
ed.	editor, edited, edible (or eaten by)
EI	East Indies, Indian
EJB	Edinburgh Journal of Botany
encl.	enclosed, enclosing
epipet.	epipetalous
esp.	especially
ess.	essential(ly)
EU	European Union
Eur.	Europe(an)
Euras.	Eurasia(n)
everg.	evergreen
ex (after an authority)	name validly published by (following name)
exc.	except
excl.	excluding
exstip.	exstipulate
ext.	extending
extrafl.	extrafloral
f.	forma
F.	French
FA	Flora of Australia
fam(s).	family(-ies)
fav.	favourite
FBA	Folia biologica Andina
FE	Flora Europaea
fert.	fertilized (by), fertilization, fertile
FFG	Fragmenta floristica et geobotanica
FG	Fern Gazette
FGP	Folia geobotanica phytotaxonomica
fl(s).	flower(s), flowering, active
-fld	-flowered
Fl. Masc.	Flore des Mascareignes
FM	Flora Malesiana
FMB	Flora Malesiana Bulletin
FMBot.	Field Museum of Natural History. Botanical Series.
F Mad(ag).	Flore de Madagascar
FMNHB	Field Museum of natural History. Botanical Series.
FN	Flora neotropica
FNC	Flore de la Nouvelle-Calédonie et Dépendances
FNSW	Flora of New South Wales
form.	formerly

fr.	fruit
Fr.	French
FR	Feddes Repertorium. Zeitschrift für botanische Taxonomie und Geobotanik
FRB(eih.)	Feddes Repertorium. Zeitschrift für botanische Taxonomie und Geobotanik. Beiheft
freq.	frequent(ly)
frs	fruits
FT	Flora of Turkey
G	gynoecium ($\underline{G}$ superior, $\overline{G}$ inferior)
GB	Great Britain
GBIUS	Godišnjak biološkog Institute u Sarajevu
GBN	Great Basin Naturalist
GBS	Gardens' Bulletin Singapore
Ger(m).	German(y)
germ.	germination, germinating
GH	Gentes Herbarum
Gk.	Greek
GOB	Garcia de Orta. Serie de Botanica
half-inf.	half-inferior
hapax.	hapaxanthic
HBML	Harvard Botanical Museum Leaflets
hemisph.	hemisphere
herm(aphr).	hermaphrodite
Himal.	Himalaya(n)
HIP	Hooker's Icones Plantarum
horiz.	horizontal(ly)
hort.	horticulture, horticulturally
hypog.	hypogynous
IBM	Illinois biological Monographs
IF	Indian Forester
illus.	illustrated
imbr.	imbricate
imp.	important, importance
inc(l).	including, included
inconsp.	inconspicuous
inc. sed.	of uncertain affinity (incerta sedis)
Ind.	India(n), Indies
indehisc.	indehiscent
Indom(al).	Indomalesia(n), i.e. India to New Guinea
ined.	unpublished, inedible
infl(s).	inflorescence(s)
infr(s).	infructescence(s)
INHSB	Illinois natural History Survey Bulletin
intr(od).	introduced
Ir.JB	Iranian Journal of Botany
irreg.	irregular(ly)
is.	island(s)
ISCJS	Iowa State College Journal of Science
IS(U)JS	Iowa State (University) Journal of Science
ITCJ	International Tree Crops Journal
JAA	Journal of the Arnold Arboretum
JABG	Journal of the Adelaide botanic Gardens
Jap.	Japan(ese)
JapJB	Japanese Journal of Botany
JB	Journal of Botany
JBNHS	Journal of the Bombay natural History Society
JBR	Journal of Bamboo Research

JEMSS	Journal of the Elisha Mitchell scientific Society
JETB	Journal of economic and taxonomic Botany
JFSUTB	Journal of the Faculty of Science, University of Tokyo. Botany
JG	Journal of Geobotany
JJB	Journal of Japanese Botany
JLSBot.	Journal of the Linnean Society. Botany
JNTCFP	Journal of Nanjing Technological College of Forest Products
JRHS	Journal of the royal horticultural Society
JRSWA	Journal of the royal Society of Western Australia
JSAB	Journal of South African Botany
JWAS	Journal of the Washington Academy of Sciences
K	calyx (members)
KB	Kew Bulletin
KBAS	Kew Bulletin, Additional Series
KM	Kew Magazine
KMMS	Kew Magazine, Monograph Series
KNAWC	Koninklijke Nederlandse Akademie van Wetenschappen, Series C
KSV(A)H	Kungliga Svenska Vetenskapsacademiens Handlinga
KUSB	Kansas University Science Bulletin
l	litre
lat.	lateral
Lat.	Latin
LBS	Leiden botanical Series
loc.	locule, loculate
longit.	longitudinal
LUA	Lund Universitets Årsskrift
lvs	leaves
LWB	Leaflets of Western Botany
Madag.	Madagascar
MAISP	Mémoires de l'Académie impériale des Sciences de St.-Pétersbourg
Mal.	Malesia(n)
MAM	Morris Arboretum Monographs
Masc.	Mascarenes
MB	Monographiae botanicae
MBMHR	Mededeelingen van het botanisch Laboratorium der Rijksuniversiteit te Utrecht
MBMUZ	Mitteilungen aus dem botanischen Museum der Universität Zürich
MBSM	Mitteilungen aus der botanischen Staatssammlung München
MCSUK	Memoirs of the College of Science. University of Kyoto
med(ic).	medicine, medicinal(ly), medical(ly)
Med(it).	Mediterranean
Mex.	Mexico, Mexican
MF	Malaysian Forester
MHB	Mémoires de l'Herbier Boissier
MIABH	Mitteilungen aus dem Institut für allgemeine Botanik in Hamburg
microscop.	microscopic
MLW	Mededelingen van de Landbouwhogeschool te Wageningen
MMNHN(P)	Mémoires du Muséum nationale d'Histoire naturelle (Paris)
MNJ	Malayan Nature Journal
MNYBG	Memoirs of the New York botanical Garden

monocot.	monocotyledonous
monocots	monocotyledons (Monocotyledonae)
monoec.	monoecious
morph.	morphological(ly), morphology
MRL	Mededelingen van 's Rijksherbarium Leiden
MSABH	Mitteilungen aus dem Staatsinstitut für allgemeine Botanik in Hamburg
MSB	Memórias da Sociedade Broteriana
MSBMBG	Monographs in systematic Botany, Missouri Botanical Garden
MSPS	Minnesota Studies in Plant Science
mt(s).	mountain(s)
MTBC	Memoirs of the Torrey botanical Club
n	haploid chromosome number
N	northern, North
natur.	naturalized
NB	note that
NBGB	Notizblatt des botanischen Gartens und Museums zu Berlin
NBP	Novitationes botanicae. Prague
NCSJ	National Cactus and Succulent Journal
n.d.	no date
NDSNG	Neue Denkschriften der Schweizerischen Naturforschenden Gesellschaft
NHBSS	Natural History Bulletin of the Siam Society
NJB	Nordic Journal of Botany
nm.	nothomorph
NM	Naturalia monspeliensia
NMBA	Notas del Museo (de la Plata) Buenos Aires
no.	number
nom. illeg.	name not in accordance with the rules of nomenclature (nomen illegitimum)
NP	New Phytologist
nr.	near
NRBGE	Notes from the Royal Botanic Garden, Edinburgh
n.s.	new series
NS, NS Paris	Notulae systematicae (Paris)
NSL, NSPV	Notulae systematicae (Leningrad)
NSW	New South Wales
NW	New World
NZ	New Zealand
NZJB	New Zealand Journal of Botany
OB	Opera botanica
OBB	Opera botanica Belgica
OBC	Opera botanica Cechica
OBPG	Opuscula botanica Pharmaciae complutensis
obs.	obscure(ly), obsolete
OBZ	Österreichische botanische Zeitschrift
occ.	occasional(ly)
OD	Orchid Digest
oft.	often
OJS	Ohio Journal of Science
O(p).L(ill).	Opera Lilloana
OM	Orchid Monographs
opp.	opposite
orig.	origin(s), original(ly)
orn.	ornamental(ly)
orthotr.	orthotropous
OW	Old World

p(p).	page(s)
P	perianth (members)
PAAAS	Proceedings of the American Academy of Arts and Sciences
Pac(if).	Pacific
Pac. Sci.	Pacific Science
PANSP	Proceedings of the Academy of Natural Sciences of Philadelphia
palaeotrop.	palaeotropics, palaeotropical
pantrop.	pantropical
PAPS	Proceedings of the American Philosophical Society
Papuas.	Papuasia(n)
Par.	Parodiana
PBS	Polish botanical Studies
PCAS	Proceedings of the California Academy of Sciences
PCIW	Publications. Carnegie Institute, Washington
PCUH	Publications of the Cairo University Herbarium
pen(in).	peninsula
pend.	pendulous
perenn.	perennial
Pfl.	Die natürlichen Pflanzenfamilien
PFMB	Publications of the Field Museum of Natural History, Botanical Series
PFSUM	Publications de la Faculté des Sciences de l'Université Masaryk
Philipp.	Philippines
Phytol.	Phytologia
Phytol.M	Phytologia Memoirs
PIAS	Proceedings of the Indiana Academy of Science
PJB(J(S))	Palestine Journal of Botany. Jerusalem Series
PJS(ci).	Philippine Journal of Science
PJSB	Philippine Journal of Science, Section C. Botany
PKNAW	Proceedings Koninklijke Nederlandse Akademie van Wetenschappen, Series C
pl(s).	plant(s)
PL	Plant Life
PLPH	Pengumuman istimewa. Lembaga Pusat Penjelidikan Kehutanan
PLSNSW	Proceedings of the Linnean Society of New South Wales
PM	Phanerogamarum Monographiae
PMSUB(S)	Publications. Michigan State University Museum (Biological Series)
poll.	pollination, pollinating, pollinated (by)
pop.	population
poss.	possible, possibly
p.p.	in part
p.p.m.	parts per million
pr(s)	pair(s)
PR	Das Pflanzenreich
praec.	preceding
prep.	preparing, preparation, prepared
prob.	probable, probably
prod(s).	product(s), producing, production
prop.	propagated, propagation
prot.	protected, protection
PRSQ	Proceedings of the Royal Society of Queensland
PSB	Plant Species Biology
PSE	Plant Systematics and Evolution
publ.	publication(s), published
QBAGS	Quarterly Bulletin of the Alpine Garden Society
QJTM	Quarterly Journal of the Taiwan Museum
q.v.	see

r.	root
R	review, revision, synopsis, key, monograph
RAA	Revista Argentina de Agronomía
RBB	Revista Brasiliera de Botânica
recog.	recognize(d)
reg.	regular(ly)
rep.	represented, representing
repr.	reprinted
reprod.	reproduced, reproducing, reproduction
resp.	respective(ly), respiratory
rhiz.	rhizome, rhizomatous
RMBG	Report Missouri Botanical Garden
RM(L)P	Revista del Museo de la Plata (Botanica)
RTBN	Recueil des Travaux botaniques Néerlandais
S	southern, South
SAJB	South African Journal of Botany
saxic.	saxicole, saxicolous
SB	Systematic Botany
SBi	Senckenbergiana biologica
SBNat.	Stuttgarter Beiträge zur Naturkunde aus dem Staatlichen Museum für Naturkunde in Stuttgart
SBU	Symbolae botanicae Upsalienses
SCB	Smithsonian Contributions to Botany
SE As.	mainland Asia centred on Indochina & Thailand
sec.	secondary
sect.	section
seg.	segregated
Selb.	Selbyana
semi-inf.	semi-inferior
semi-sup.	semi-superior
sep.	separate
seq.	following
S.G.	specific gravity
s.l.	in the broad sense
solit.	solitary
sp.	species
spp.	species (plural)
s.s.	in the narrow sense
ssp.	subspecies
sspp.	subspecies (plural)
s.t.	sometimes
stip.	stipule(s), stipular
subg.	subgenus, subgenera
subs.	substitute(s)
subsp.	subspecies
subseq.	subsequent(ly)
subtemp.	subtemperate
subterr.	subterranean
subtrop.	subtropical
succ.	succulent(s)
sup.	superior
superf.	superficial(ly)
suppl.	supplement(ary)
SWNat.	Southwestern Naturalist
symp.	sympodial(ly)
Syst.B	Systematic Botany
T	Taxon
Tasm.	Tasmania(n)
TBIANSSR	Trudȳ Botanicheskogo Instituta. Akademiy nauk SSR
TBSE	Transactions of the botanical Society of Edinburgh

temp.	temperate, temperature
term.	terminal, terminating
TFB	Thai Forest Bulletin
trad.	traditional(ly)
trop.	tropical, tropics
TRSNZ	Transactions of the Royal Society of New Zealand
TRSSA	Transactions of the Royal Society of South Australia
t.s.	transverse section
TSDSNH	Transactions of the San Diego Society for Natural History
TSP	Abhandlungen der Akademie...Mainz...tropische und subtropische Pflanzenwelt
UCOP	University of Connecticut Occasional Papers, biological Science Series
UCPB	University of California Publications in Botany
UKSB	University of Kansas Science Bulletin
Urug.	Uruguay
US	United States
USDA Agric. Handbk	United States Department of Agriculture, Agriculture Handbooks
USDA Agric. Mon.	United States Department of Agriculture, Agriculture Monographs
USDATB	United States Department of Agriculture, Technical Bulletin
USK	University Studies, University of Karachi
usu.	usually
UWPSB	University of Wyoming Publications in Science. Botany
v.	very
var.	variety, various
varieg.	variegated
vasc.	vascular
veg.	vegetable(s), vegetative, vegetation
Venez.	Venezuela(n)
vern.	vernacular
VGIETHZ	Veröffentlichungen des Geobotanischen Instituts, Eidgenössiche technische Hochschule Rübel in Zürich
VKNAW	Verhandelingen der Koninklijke Nederlandsche Akademie van Wetenschappen
VKZGW	Verhandlungen der kaiserlich-königlichen zoologisch-botanischen Gesellschaft in Wien
vol.	volume
W	western, West
WI	West Indies
WJB	Wasmann Journal of Biology
x	basic chromosome number
xeroph.	xerophyte, xerophytic
yr(s).	year(s)
zygom.	zygomorphic
2n	diploid chromosome number
±	more-or-less, approximately, with or without
×	hybrid
⚥	bisexual
∞	numerous
+	graft hybrid

Abbreviations

2. Authors' names

(As far as could be ascertained with respect to full names and dates) largely following *Draft index of author abbreviations compiled at the Herbarium, Royal Botanic Gardens Kew* (second imprint 1984 with further corrections found in R.K. Brummitt & C.E. Powell (eds, 1992) *Authors of plant names*, Royal Botanic Gardens, Kew)

Abel	Abel, Gottlieb Friedrich (1763–?)
Abel, C.	Abel, Clarke (1780–1826)
Aberc.	McLaren, Henry Duncan, Lord Aberconway (1879–1953)
Abrams	Abrams, Le Roy (1874–1956)
Ackerman	Ackerman, James D. (1950–)
Acuña	Acuña Galé, Julián Baldomero (1900–1973)
Adams	Adams, Michael Friedrich (1780–1833)
Adams, C.	Adams, Charles Dennis (1920–)
Adamson	Adamson, Robert Stephen (1885–1965)
Adans.	Adanson, Michel (1727–1806)
Adelb.	Adelbert, Aalbert George Ludwig (1914–1972)
Adema	Adema, Frits A.C.B. (1938–)
Adlam	Adlam, Richard Wills (1853–1903)
Aellen	Aellen, Paul (1896–1973)
Afzel.	Afzelius, Adam (1750–1837)
Afzel., K.	Afzelius, Karl Rudolf (1887–1971)
Agardh	Agardh, Carl Adolf (1785–1859)
Agardh, J.	Agardh, Jakob Georg (1813–1901)
Aguiar	Aguiar, Joaquim Macedo do (1854–1882)
Aiello	Aiello, Annette (1941–)
Airy Shaw	Airy Shaw, Herbert Kenneth (1902–1985)
Aitch.	Aitchison, James Edward Tierney (1836–1898)
Aiton	Aiton, William (1731–1793)
Aiton, f.	Aiton, William Townsend (1766–1849)
Akers	Akers, John F. (1906–)
Alain	Alain, Brother (né Liogier, Enrique) (1916–)
Alava	Alava, Reino Olavi (1915–)
Albers	Albers, F. (fl. 1979)
Albert, V.	Albert, Victor A. (fl. 1990s)
Albov	Albov (Alboff), Nikolai Michaïlovich (1866–1897)
Alderw.	Alderwerelt van Rosenburgh, Cornelis Rogier Willem Karel van (1863–1936)
Alef.	Alefeld, Friedrich Georg Christoph (1820–1872)
Alexander	Alexander, Edward Johnston (1901–1985)
Alexeev	Alexeev, E.B. (1946–1976)
Ali	Ali, Syed Irtifaq (1930–)
All.	Allioni, Carlo (1728–1804)
Allam.	Allamand, Jean Nicholas Sébastien (1731–1793)
Allan	Allan, Harry Howard Barton (1882–1957)
Allemão	Allemão e Cysneiro, Francisco Friera (1797–1874)
Allemão, M.	Allemão, Manoel (?–1863)

Allen, C.	Allen, Caroline Kathryn (1904–1975)
Allen, J.	Allen, James (c. 1830–1906)
Allen, P.	Allen, Paul Hamilton (1911–1963)
Alm	Alm, Carl Gustav (1888–)
Almeda	Almeda, Frank (1946–)
Al-Shehbaz	Al-Shehbaz, lhsan Ali (1939–)
Alston	Alston, Arthur Hugh Garfit (1902–1958)
Alverson	Alverson, William S. (fl. 1981)
Ames	Ames, Oakes (1874–1950)
Amman	Amman, Johann (1707–1741)
Amstutz	Amstutz, Erika (fl. 1957)
Anderb.	Anderberg, Arne A. (1954–)
Anderson, L.	Anderson, Loran Crittendon (1936–)
Anderson, R.	Anderson, Robert Henry (1899–1969)
Anderson, T.	Anderson, Thomas (1832–1870)
Anderson, W.	Anderson, William (1750–1778)
Anderson, W.R.	Anderson, William Russel (1942–)
Andersson	Andersson, Nils Johan (1821–1880)
Andersson, L.	Andersson, Lennart (1948–)
André	André, Edouard-François (1840–1911)
Andréanszky	Andréanszky, Gábor (1895–1967)
Andres	Andres, Heinrich (1883–1970)
Andrews	Andrews, Henry Charles (fl. 1794–1830)
Andrz.	Andrzejowski, Antoni Lukianowicz (1785–1868)
Añon	Añon, Delia C. Suarez de Cullen (1917–)
Ansari	Ansari, M.Y. (1929–)
Antoine	Antoine, Franz (1815–1886)
Appan	Appan, Subramanian G. (1937–)
Arbo	Arbo, Maria Mercedes (1945–)
Ard.	Arduino, Pietro (1728–1805)
Arech(av).	Arechavaleta, José (1838–1912)
Arènes	Arènes, Jean (1898–1960)
Aristeg.	Aristeguieta, Leandro (1923–)
Armstr.	Armstrong, Joseph Beattie (1850–1926)
Armstr., J.A.	Armstrong, James Andrew (1950–)
Arn.	Arnott, George Arnott Walker (1799–1868)
Arn., S.	Arnott, Samuel (1852–1930)
Arnold, J.F.	Arnold, Johann Franz Xaver (fl. 1785)
Arrill.	Arri(l)laga de Maffei, Blanca Renée (1917–)
Arruda	Arruda de Cámara Manoel (1752–1810)
Asch.	Ascherson, Paul Friedrich August (1834–1913)
Ashton	Ashton, Peter Shaw (1934–)
Askerova	Askerova, Rosa K. (1929–)
Aspl.	Asplund, Erik (1888–1974)
Asso	Asso y del Rio, Ignacio Jordán de (1742–1814)
Aublet	Aublet, Jean Baptiste Christophore Fusée (1720–1778)
Aubrév.	Aubréville, André (1897–1982)
Audubon	Audubon, John James (1785–1851)
Auq.	Auquier, Paul Henri (1939–1980)
Austin, D.	Austin, Daniel Frank (1943–)
Autran	Autran, Eugène John Benjamin (1855–1912)
Avé-Lall.	Avé-Lallemant, Julius Léopold Edouard (1803–1867)
Ayensu	Ayensu, Edward Solomon (1935–)
Baas	Baas, Pieter (1944–)
Bab.	Babington, Charles Cardale (1808–1895)
Babu	Babu, Cherukuri Raghavendra (1940–)
Bacig.	Bacigalupo, Nélida María (1924–)
Backeb.	Backeberg, Curt (1894–1966)
Backer	Backer, Cornelis Andries (1874–1963)
Backh.	Backhouse, James (1794–1869)
Baden	Baden, Claus (1952–)

Badillo	Badillo, Victor Manuel (1920–)
Baehni	Baehni, Charles (1906–1964)
Baijnath	Baijnath, Himansu (1943–)
Bailey	Bailey, Frederick Manson (1827–1915)
Bailey, D.	Bailey, Dana K. (1916–)
Bailey, I.	Bailey, Irving Widmer (1884–1967)
Bailey, J.	Bailey, John Frederick (1866–1938)
Bailey, L.	Bailey, Liberty Hyde (1858–1954)
Baillon	Baillon, Henri Ernest (1827–1895)
Baker	Baker, John Gilbert (1834–1920)
Baker f.	Baker, Edmund Gilbert (1864–1949)
Baker, R.	Baker, Richard Thomas (1854–1941)
Bakh.f.	Bakhuizen van den Brink, Reinier Cornelis (1911–1987)
Bal.	Balansa, Benedict (1825–1892)
Balakr.	Balakrishnan, Nambiyath Puthansurayil (1935–)
Balf.	Balfour, John Hutton (1808–1884)
Balf.f.	Balfour, Isaac Bayley (1853–1922)
Ball, P.	Ball, Peter William (1932–)
Ballard	Ballard, Francis (1896–1975)
Balle	Balle, Simone (1906–)
Bally	Bally, Peter René Oscar (1895–1980)
Balslev	Balslev, Henrik (1951–)
Baltet	Baltet, Charles (1830–1908)
Bamps	Bamps, Paul Rodolphe Joseph (1932–)
Bancr.	Bancroft, Edward Nathaniel (1772–1842)
Banerjee	Banerjee, Rabindra Nath (1935–)
Bange	Bange, A.J. (1896–1950)
Banks	Banks, Joseph (1743–1820)
Baranova, M.	Baranova, M.V. (1932–)
Barbey	Barbey, William (1842–1914)
Barbier	Barbier, E. (fl. 1973)
Barb. Rodr.	Barbosa Rodrigues, João (1842–1909)
Barfod	Barfod, Anders S. (1957–)
Barker, W.	Barker, Winsome Fanny (1907–)
Barkley, F.	Barkley, Fred Alexander (1908–1989)
Barkworth	Barkworth, Mary Elizabeth (1941–)
Barlow	Barlow, Bryan Alwyn (1933–)
Barnades	Barnades (Barnardez), Miguel (1717–1771)
Barneby	Barneby, Rupert Charles (1911–)
Barney	Barney, E.E. (fl. 1877–9)
Barnhart	Barnhart, John Hendley (1871–1949)
Barratte	Barratte, Jean François Gustave (1857–1920)
Barringer	Barringer, Kerry A. (1954–)
Barroso	Barroso, Liberato Joaquim (1900–1949)
Barroso, G.	Barroso, Graziela Maciel (1912–)
Barthlott	Barthlott, Wilhelm A. (1946–)
Bartlett	Bartlett, Harley Harris (1886–1960)
Bartling	Bartling, Friedrich, Gottlieb (1798–1875)
Barton	Barton, Benjamin Smith (1766–1815)
Barton, W.	Barton, William Paul Crillon (1786–1856)
Bartram	Bartram, William (1739–1823)
Bary	Bary, Heinrich Anton de (1831–1888)
Baskerville	Baskerville, Thomas (1812–c. 1840)
Bassi	Bassi, Ferdinando (1710–1774)
Batalin	Batalin, Alexander Theodorowicz (1847–1896)
Bateman	Bateman, James (1811–1897)
Bates (, D.)	Bates, David Martin (1935–)
Batista	Batista, Augusto Chaves (1916–1967)
Batsch	Batsch, August Johann Georg Karl (1761–1802)
Battand.	Battandier, Jules Aimé (1848–1922)
Baudet	Baudet, Jean C. (fl. 1970)
Bauer	Bauer, Franz Andreas (1758–1840)

Baum	Baum, Vicki M. (fl. 1982)
Baumg.	Baumgarten, Johann Christian Gottlob (1765–1843)
Bausch	Bausch, Jan (1917–)
Baxter	Baxter, William Hart (1816?–1890)
Baylis	Baylis, Geoffrey Thomas Sandford (1913–)
Beadle	Beadle, Chauncey Delos (1866–1950)
Beaman	Beaman, John Homer (1929–)
Bean	Bean, William Jackson (1863–1947)
Bean, P.	Bean, Patricia Anne (1930–)
Beauv., P. or Pal.	Palisot de Beauvois, Ambroise Marie François Joseph (1752–1820)
Beauverd	Beauverd, Gustave (1867–1942)
Beauvis.	Beauvisage, Georges Eugène Charles (1852–1925)
Becc.	Beccari, Odoardo (1843–1920)
Beck, G.	Beck, Günther Ritter (Beck) von Mannagetta und Lerchenau (1856–1931)
Beck, H.	Beck, Hans T.
Becker, W.	Becker, Wilhelm (1874–1928)
Beddome	Beddome, Richard Henry (1830–1911)
Beer	Beer, Johann Georg (1803–1873)
Beer, E.	Beer, Eva
Bég.	Béguinot, Augusto (1875–1940)
Beille	Beille, Lucien (1862–1946)
Bél.	Bélanger, Charles Paulus (1805–1881)
Bellair	Bellair, Georges Adolphe (1860–1939)
Bellardi	Bellardi, Carlo Antonio Lodovico (1741–1826)
Bello	Bello y Espinosa, Domingo (1817–1884)
Benedict, C.	Benedict, Charlotte
Benj.	Benjamin, Ludwig (1825–1848)
Benl	Benl, Gerhard (1910–)
Bennet	Bennet, S.S.R. (1940–)
Bennett	Bennett, John Joseph (1801–1876)
Bennett, D.	Bennett, David E. (fl. 1989)
Bennett, E.	Bennett, Eleanor Marion (1942–)
Benn(ett), G.	Bennett, George (1804–1893)
Benoist	Benoist, Raymond (1881–1970)
Benson, L.	Benson, Lyman David (1909–)
Benth.	Bentham, George (1800–1884)
Berazaín	Berazaín, Rosalina (1947–)
Bercht.	Berchtold, Friedrich von (1781–1876)
Berg	Berg, Ernst von (1782–1855)
Berg, C.	Berg, Cornelis C. (1934–)
Berg, O.	Berg, Otto Carl (1815–1866)
Berger, A.	Berger, Alwin (1871–1931)
Berggren, S.	Berggren, Sven (1837–1917)
Bergius	Bergius, Benedictus (1723–1784)
Bergius, P.	Bergius, Peter Jonas (1730–1790)
Bergmans	Bergmans, John (Johannes Baptista?) (1892–1980)
Berland.	Berlandier, Jean Louis (1805–1851)
Bernh.	Bernhardi, Johann Jakob (1774–1850)
Berry	Berry, Andrew (fl. 1780s–1810s)
Bertero	Bertero, Carlo Luigi Giuseppe (1789–1831)
Berth.	Berthelot, Sabin (1794–1880)
Bertol.	Bertoloni, Antonio (1775–1869)
Bertol. f.	Bertoloni, Giuseppe (1804–1879)
Bertoni	Bertoni, Moisés de Santiago (1857–1929)
Bertrand	Bertrand, Marcel C.
Besser	Besser, Gilibald Swibert Joseph Gottlieb von (1784–1842)
Betche	Betche, Ernst (1851–1913)
Betcke	Betcke, Ernst Friedrich (1815–1865)
Beyerle	Beyerle (fl. 1938)
Bezerra	Bezerra, José Luiz (fl. 1970)

Bhand.	Bhandari, Madan Mal (1929–)
Bhargavan	Bhargavan, P. (1939–)
Bieb., M.	Marschall von Bieberstein, Friedrich August (1768–1826)
Bien.	Bienert (Binert), Theophil (?–1873)
Bierh.	Bierhorst, David William (1924–1997)
Bigelow	Bigelow, Jacob (1787–1879)
Binnend.	Binnendijk, Simon (1821–1883)
Birdsey	Birdsey, Monroe Roberts (1922–)
Birdw.	Birdwood, George Christopher Molesworth (1832–1917)
Bisch.	Bischoff, Gottlieb T.G. (1797–1854)
Bishop, L.E.	Bishop, Luther Earl (1943–1991)
Bisse	Bisse, Johannes (1935–1984)
Bitter	Bitter, Friedrich August Georg (1873–1927)
Bittrich	Bittrich, Volker (fl. 1984–94)
Black, G.A.	Black, George Alexander (1916–1957)
Black, J.	Black, John McConnell (1855–1951)
Blackburn	Blackburn, Benjamin Coleman (1908–)
Blake, S.F.	Blake, Sidney Fay (1892–1959)
Blake, S.T.	Blake, Stanley Thatcher (1910–1973)
Blakeley	Blakeley, William Faris (1875–1941)
Blanca	Blanca, Gabriel (1954–)
Blanche	Blanche, Emanuel (1824–1908)
Blanco	Blanco, Francisco Manuel (1778–1845)
Blatter	Blatter, Ethelbert (1877–1934)
Blaxell	Blaxell, Donald Frederick (1934–)
Bleck	Bleck, M.B. (fl. 1984–)
Bluff	Bluff, Matthias Joseph (1805–1837)
Blume	Blume, Carl Ludwig von (1796–1862)
Bobrov	Bobrov, Evgenij Grigorievićz (1920–1983)
Bocquillon	Bocquillon, Henri Théophile (1834–1883)
Bod., M.	Bodard, Marcel (1927–1988)
Bodin	Bodin, Nicolaus Gustavus (fl. 1798)
Boeck	Boeck, Christian Peter Bianco (1798–1877)
Bödecker	Bödecker, Friedrich (1867–1937)
Boehmer	Boehmer, Georg Rudolf (1723–1803)
Boelcke	Boelcke, Osvaldo (1920–)
Boenn.	Boenninghausen (Bönninghausen) Clemens Maria Franz von (1785–1864)
Boerl.	Boerlage, Jacob Gijsbert (1849–1900)
Boerner	Boerner (Börner), Karl Julius Bernhard (1880–1953)
Bogner	Bogner, Josef (1939–)
Bois	Bois, Désiré Georges Jean Marie (1856–1946)
Boiss.	Boissier, Pierre Edmond (1810–1885)
Boissev.	Boissevain, Charles Hercules (1893–1946)
Boissieu	Boissieu (de la Martinière), Claude Victor (1784–1868)
Boissieu, H.	Boissieu, Henri de (1871–1912)
Boit.	Boiteau, Pierre L. (1911–)
Boivin	Boivin, Louis Hyacinthe (1808–1852)
Boivin, B.	Boivin, Joseph Robert Bernard (1916–1985)
Bojer	Bojer, Wenceslas or Wenzel (1797–1856)
Bolle	Bolle, Karl August (1821–1909)
Bolle, F.	Bolle, Friedrich Franz August Albrecht (1905–)
Bolton	Bolton, James (c.1758–1799)
Bolus	Bolus, Harry (1834–1911)
Bolus, L.	Bolus, Harriet Margaret Louisa (née Kensit) (1877–1970)
Bommer	Bommer, Joseph Edouard (1829–1895)
Bonap.	Bonaparte, Roland Napoléon (1858–1924)
Bonati	Bonati, Gustave Henri (1873–1927)
Bondar	Bondar, Gregório Gregorievich (1881–1959)
Bong.	Bongard, August (Gustav) Heinrich von (1786–1839)
Bonnet	Bonnet, Edmond (1848–1922)
Bonnier	Bonnier, Gaston Eugène Marie (1851–1922)

Bonpl.	Bonpland, Aimé Jacques Alexandre (née Goujaud) (1773–1858)
Boom	Boom, Boudewijn Karel (1903–1980)
Boom, B.	Boom, Brian Morley (1954–)
Booth	Booth, William Beattie (c. 1804–1874)
Bor	Bor, Norman Loftus (1893–1972)
Borbás	Borbás, Vinczé von (1844–1905)
Boreau	Boreau, Alexandre (1803–1875)
Borh.	Borhidi, Attila L. (1932–)
Boriss.	Borissova-Bekrjasheva, Antonina Georgievna (1903–1970)
Borkh.	Borkhausen, Mortiz (Moriz) Balthasar (1760–1806)
Bornm.	Bornmüller, Joseph Friedrich Nicolaus (1862–1948)
Bort	Bort, Katherine Stephens (1870–?)
Bory	Bory de Saint-Vincent, Jean Baptiste Georges (Geneviève) Marcellin (1778–1846)
Borzi	Borzi, Antonino (1852–1921)
Bosc	Bosc, Louis Augustin Guillaume (1759–1828)
Bosch	Bosch, Roelof Benjamin van den (1810–1862)
Bosse	Bosse, Julius Friedrich Wilhelm (1788–1864)
Bosser	Bosser, Jean M. (1922–)
Botsch.	Botschantzev, Victor Petrovič (1910–1990)
Botsch., V.V.	Botschantzeva, Vera Viktorovna (1946–)
Bouché	Bouché, Peter Carl (1783–1856)
Boulos	Boulos, Loutfy (1932–)
Boutelje	Boutelje, Julius B. (fl. 1954)
Boutique	Boutique, Raymond (1906–1985)
Bowd., S.	Bowdich, Sarah (1791–1856)
Bowdich	Bowdich, Thomas Edward (1791–1824)
Bower	Bower, Frederick Orpen (1855–1948)
Bowles	Bowles, Edward Augustus (1865–1954)
Box	Box, Harold Edmund (1898–)
Boynton	Boynton, Kenneth Rowland (1891–)
Br., E.	Brown, Elizabeth Dorothy W. (1880–1972)
Br., F.	Brown, Forest Buffen Harkness (1873–1954)
Br., N.E.	Brown, Nicholas Edward (1849–1934)
Br., R.	Brown, Robert (1773–1858)
Br., R.W.	Brown, Roland Wilbur (1893–1961)
Braas	Braas, Lothar A. (fl. 1977)
Brackenr.	Brackenridge, William Dunlop (1810–1893)
Brade	Brade, Alexander Curt (1881–1971)
Braem	Braem, Guido J. (fl. 1980)
Braga	Braga, Ruby (fl. 1964)
Braggins	Braggins, John E. (1944–)
Bramw.	Bramwell, David (1942–)
Brand	Brand, August (1863–1930)
Brandegee, (M.)	Brandegee, Mary Katharine (formerly Curran: née Layne) (1844–1920)
Brandis	Brandis, Dietrich (1824–1907)
Braun, A.	Braun, Alexander Carl Heinrich (1805–1877)
Braun-Blanquet	Braun-Blanquet, Josias (1884–1980)
Breda	Breda, Jacob Gijsbert Samuel van (1788–1867)
Breedlove	Breedlove, Dennis E. (1939–)
Bremek.	Bremekamp, Cornelis Elisa Bertus (1888–1984)
Bremer	Bremer, Kåre (1948–)
Brenan	Brenan, John Patrick Micklethwait (1917–1985)
Brettell	Brettell, R.D.
Brickell	Brickell, John (1748–1809)
Brickell, C.	Brickell, Christopher David (1932–)
Bridson	Bridson, Diane Mary (1942–)
Brieger	Brieger, Friedrich Gustav (1900–1985)
Briggs, B.	Briggs, Barbara Gillian (1934–)
Brign.	Brignoli di Brunnhoff, Giovanni de (1774–1857)

Briq.	Briquet, John Isaac (1870–1931)
Brittan	Brittan, Norman Henry (1920–)
Britton	Britton, Nathaniel Lord (1859–1934)
Briz.	Brizicky, George Konstantine (1901–1968)
Bromf.	Bromfield, William Arnold (1801–1851)
Brongn.	Brongniart, Adolphe Théodore (1801–1851)
Brot.	Brotero, Felix de (Silva) Avellar (1744–1828)
Brouss.	Broussonet, Pierre Auguste Marie (1761–1807)
Browicz	Browicz, Kasimierz (1925–)
Brown, A.	Brown, Addison (1830–1913)
Browne, P.	Browne, Patrick (1720–1790)
Bruce	Bruce, James (1730–1794)
Bruce, E.A.	Bruce, Eileen Adelaide (1905–1955)
Bruegger	Bruegger (von Churwalden) Christian Georg (1833–1899)
Brühl	Brühl, Paul Johannes (1855–?)
Brullo	Brullo, Salvatore (1947–)
Brummitt	Brummitt, Richard Kenneth (1937–)
Bruyns	Bruyns, Peter Vincent (1957–)
Bubani	Bubani, Pietro (1806–1888)
Buch., J.	Buchanan, John (1855–1896)
Buchenau	Buchenau, Franz Georg Philipp (1831–1906)
Buchet	Buchet, Samuel (1875–1956)
Buch.-Ham.	Buchanan-Hamilton, Francis (1762–1829)
Buchheim	Buchheim, Arno Fritz Günther (1924–)
Buchholz	Buchholz, Fedor Vladimirovic (1872–1924)
Buc'hoz	Buc'hoz, Pierre Joseph (1731–1807)
Buckley	Buckley, Samuel Botsford (1809–1884)
Buen	Buen y del Cos, O. de (1863–1945)
Buese	Büse, Lodewijk Hendrik (1819–1888)
Buhse	Buhse, Fedor Aleksandrovich (1821–1898)
Buijsen	Buijsen, J.R.M. (fl. 1988)
Buin.	Buining, Albert Frederick Hendrik (1901–1980)
Buist	Buist, George
Bull, W.	Bull, William (1828–1902)
Bullock	Bullock, Arthur Allman (1906–1980)
Bunge	Bunge, Alexander Andrejewitsch von (Andreevic, Aleksandrovic) (1803–1890)
Bunting	Bunting, George Sydney (1927–)
Burb.	Burbidge, Frederick William Thomas (1847–1905)
Burb., N.	Burbidge, Nancy Tyson (1912–1977)
Burchell	Burchell, William John (1781–1863)
Burck	Burck, William (1848–1910)
Burdet	Burdet, Hervé Maurice (1939–)
Bureau (or Bur.)	Bureau, Louis Edouard (1830–1918)
Burger	Burger, William Carl (1932–)
Burkart	Burkart, Arturo Erhado (1906–1975)
Burkill	Burkill, Isaac Henry (1870–1965)
Burm(an)	Burman, Johannes (1707–1779)
Burm.f.	Burman, Nicolaas Laurens (1734–1793)
Burnett	Burnett, Gilbert Thomas (1800–1835)
Burns-Balogh	Burns-Balogh, Pamela (fl. 1983)
Burollet	Burollet, Pierre Andre (1889–)
Burret	Burret, (Maximilian) Karl Ewald (1883–1964)
Burtt, B.L.	Burtt, Brian Lawrence (1913–)
Buscal.	Buscalioni, Luigi (1863–1954)
Busch	Busch, Anton (1823–1895)
Busch, N.	Busch, Nicolai Adolfowitsch (1869–1941)
Buser	Buser, Robert (1857–1931)
Bush	Bush, Benjamin Franklin (1858–1937)
But	But, Paul Pui-Hay (fl. 1982)
Butzin	Butzin, Friedhelm Reinhold (1936–)
Buxb., F.	Buxbaum, Franz (1900–1979)

Buxton	Buxton, Bertram Henry (1852–1934)
Byles	Byles, Ronald Stewart (fl. 1957)
Byrnes	Byrnes, Norman Bryce (1922–)
Caball.	Caballero, Arturo (1877–1950)
Cabezudo	Cabezudo, Baltasar (1946–)
Cabrera	Cabrera, Angel Lulio (1908–1998)
Caldas	Caldas y Tenorio, Francisco José de (1771–1816)
Calderón	Calderón, Graciela Calderón de Rzedowski (1931–)
Calderón, C.	Calderón, Cléofe E.
Calest.	Calestani, Vittorio (1882–)
Camargo	Camargo, Felisberto Cardoso de (1896–1943)
Cambess.	Cambessèdes, Jacques (1799–1863)
Camp	Camp, Wendell Holmes (1904–1963)
Campderá	Campderá, Francisco (1793–1862)
Camus	Camus, Edmond Gustav(e) (1852–1915)
Camus, A.	Camus, Aimée Antoinette (1879–1965)
Candargy	Candargy, Paléologos C. (1870–?)
Cannon	Cannon, John Francis Michael (1930–)
Cannon, M.	Cannon, Margaret Joy (1928–)
Cao, T.R.	Cao, Te-Ru (1940–)
Capanema	Capanema (fl. 1862)
Capuron	Capuron, René (1921–1971)
Cárdenas	Cárdenas, Hermosa Martin (1899–1973)
Carlq.	Carlquist, Sherwin (1930–)
Carnevali	Carnevali, Germán (fl. 1987)
Caro	Caro, José Aristida (1919–1985)
Carolin	Carolin, Roger Charles (1929–)
Carr	Carr, Cedric Errol (1892–1936)
Carrick	Carrick, John (1914–1978)
Carrière	Carrière, Élie Abel (1818–1896)
Carse	Carse, Harry (1857–1930)
Carter, A.	Carter, Annetta Mary (1907–1991)
Caruel	Caruel, Théodore (1830–1898)
Casar.	Casaretto, Giovanni (1812–1879)
Caspary	Caspary, Johann Xaver (Robert) (1818–1887)
Cass.	Cassini, Alexandre Henri Gabriel de (1781–1832)
Cauwet-Marc	Cauwet-Marc, Ann Marie
Cav.	Cavanilles, Antonio José (1745–1804)
Cavaco	Cavaco, Alberto Judice Leote (1916–)
Cavalc.	Cavalcanti, Wlandemir de Albuquerque (fl. 1967)
Cavara	Cavara, Fridiano (1857–1929)
Čelak.	Čelakovský, Ladislav Josef (1834–1902)
Cerv.	Cervantes, Vicente (Vincente) de (1755/59–1829)
Chaix	Chaix, Dominique (1730–1799)
Chakrab.	Chakrabarty, T. (fl. 1984)
Chakrav.	Chakravarty, Hira Lal (1907–)
Cham.	Chamisso, Ludolf Karl Adelbert von (1781–1838)
Chambers, K.	Chambers, Kenton Lee (1929–)
Champ.	Champion, John George (1815–1854)
Chandra	Chandra, Vinod (1953–)
Chang, C.C.	Chang, Chao-Chien (1900–)
Chang, H.T.	Chang, Hung-Ta (1914–)
Chang, M.C.	Chang, Mei-Chen (1933–)
Chang, S.Y.	Chang, Shao-Yao (fl. 1983)
Chao, A.C.	Chao, Ai-Cheng (fl. 1958)
Chao, C.S.	Chao Chi-Son (1936–)
Chao, H.C.	Chao, Hsin-Chien (1918–)
Chapman	Chapman, Alvin Wentworth (1809–1899)
Charadze	Charadze, Anna Lukianovna (1905–1977)
Charif	Charif (fl. 1952)
Chase	Chase, Mary Agnes (née Merrill) (1869–1963)

Chase, M.	Chase, Mark W. (fl. 1981)
Châtel.	Châtelain, Jean Jacques (1736–1822)
Chatin	Chatin, Gaspard Adolphe (1813–1901)
Chatterjee	Chatterjee, Debabarta (1911–1960)
Chaub.	Chaubard, Louis Athanase (1785–1854)
Chav.	Chavannes, Edouard Louis (1805–1861)
Cheek	Cheek, Martin Roy (1960–)
Cheeseman	Cheeseman, Thomas Frederick (1846–1923)
Chen, C.	Chen, Cheih (1928–)
Chen, C.J.	Chen, Chia-Jui (1935–)
Chen, S.C.	Chen, Sing-Chi (1931–)
Chen, S.H.	Chen, Su-Hwa (1948–)
Chen, S.L.	Chen, Shou-Liang (1921–)
Chen, Y.L.	Chen, Yi-Ling (1930–)
Chen, Z.Y.	Chen, Zhong-Yi (fl. 1989)
Cheng, W.C.	Cheng, Wan-Chun (1904–1983)
Cheng f.	Cheng, Sze-Hsu (fl. 1959)
Chermezon	Chermezon, Henri (1885–1939)
Chev., A.	Chevalier, Auguste Jean Baptiste (1873–1956)
Chevall.	Chevallier, François Fulgis (1796–1840)
Chew	Chew, Wee-Lek (1932–)
Chia	Chia, Liang-Chi (1921–)
Chien	Chien, Jian-Ju (fl. 1957–1984)
Ching	Ching, Ren-Chang (1898–1986)
Chinnock	Chinnock, Robert James (1943–)
Chiov.	Chiovenda, Emilio (1871–1941)
Chippindale	Chippindale, George McCartney (1921–)
Chitt.	Chittenden, Frederick James (1873–1950)
Chodat	Chodat, Robert Hippolyte (1865–1934)
Choisy	Choisy, Jacques Denys (Denis) (1799–1859)
Chouard	Chouard, Pierre (1903–1983)
Choux	Choux, Pierre (1890–1983)
Chow, S.	Chow, S. (fl. 1962)
Chr., C.	Christensen, Carl Frederick Albert (1872–1942)
Christ	Christ, Konrad Hermann Heinrich (1833–1933)
Christenson	Christenson, Eric A. (fl. 1985)
Christm.	Christmann, Gottlieb Friedrich (1752–1836)
Christoph.	Christophersen, Erling (1898–)
Chrshan.	Chrshanovski (Khrzhanowski, Chrzanovskij), Vladimir Gennadievich (1912–1985)
Chrtek	Chrtek, Jindřich (1930–)
Chu, G.L.	Chu, Ge-Lin (1934–)
Chuang	Chuang, Tsan-Lang
Chukavina	Chukavina, Anna Prokofevna
Chun	Chun, Woon-Young (1889–)
Church	Church, George Lyle (1903–)
Cif.	Ciferri, Raffaele (1897–1964)
Cirillo	Cirillo, Domenico Maria Leone (1739–1799)
Clairv.	Clairville, Joseph Philippe de (1742–1830)
Clapham	Clapham, Arthur Roy (1904–1990)
Clarion	Clarion, Jacques (1776–1844)
Clarke, C.B.	Clarke, Charles Baron (1832–1906)
Clarkson	Clarkson, John Richard (1950–)
Clausen, R.T.	Clausen, Robert Theodore (1911–1981)
Clayton	Clayton, John (1686–1733)
Clayton, W.	Clayton, William Derek (1926–)
Clements	Clements, Frederic Edward (1874–1945)
Clifford	Clifford, Harold Trevor (1927–)
Clos	Clos, Dominique (1821–1908)
Coaz	Coaz, Johann Wilhelm Fortunat (1822–1918)
Cochet	Cochet, Pierre Charles Marie (known as `Cochet Cochet') (1866–1936)

Cockayne	Cockayne, Leonard C. (1855–1934)
Cockerell	Cockerell, Theodore Dru Alison (1866–1948)
Codd	Codd, Leslie Edward Wastell (1908–)
Coëm.	Coëmans, Henri Eugène Lucien Gaëtan (1825–1871)
Cogn.	Cogniaux, Célestin Alfred (1841–1916)
Coincy	Coincy, Auguste Henri Cornut de la Fontaine de (1837–1903)
Colebr.	Colebrooke, Henry Thomas (1765–1837)
Colenso	Colenso, John William (1811–1899)
Colla	Colla, Luigi Aloysius (1766–1848)
Collett	Collett, Henry (1836–1901)
Comm.	Commerson, Philibert (1727–1773)
Compère	Compère, Pierre (1934–)
Compton	Compton, Robert Harold (1886–1979)
Conert	Conert, Hans Joachim (1929–)
Conran	Conran, John Godfrey (1960–)
Console	Console, Michelangelo (1812–1897)
Constance	Constance, Lincoln (1909–)
Cook	Cook, Orator Fuller (1867–1949)
Cook, C.	Cook, Christopher David Kentish (1933–)
Cooke	Cooke, David Alan (1949–)
Coombes	Coombes, Allen J.
Copel.	Copeland, Edwin Bingham (1873–1964)
Corb.	Corbière, François Marie Louis (1850–1941)
Cordemoy	Cordemoy, Philippe Eugène Jacob de (1837–1911)
Core	Core, Earl Lemley (1902–1984)
Corner	Corner, Edred John Henry (1906–1996)
Correa	Correa, Maevia Noemi (1914–)
Corr. Serr.	Corrêa de Serra, José Francisco (1751–1823)
Cortés	Cortés, Santiago (1854–1924)
Cory	Cory, Victor Louis (1880–1964)
Cosson	Cosson, Ernest Saint-Charles (1819–1889)
Costa	Costa y Cuxart, Antonio Cipriano (1817–1886)
Costantin	Costantin, Julien Noël (1857–1936)
Coulter	Coulter, Thomas (1793–1843)
Coulter, J.	Coulter, John Merle (1851–1928)
Courchet	Courchet, Lucien Désiré Joseph (1851–1924)
Courtois	Courtois, Richard Joseph (1806–1835)
Cout.	Coutinho, António Xavier Pereira (1851–1939)
Coutts	Coutts, John (1872–1952)
Cov.	Coville, Frederick Vernon (1867–1937)
Covas	Covas, Guillermo (1915–)
Cowan	Cowan, John Macqueen (1891–1960)
Cowan, R.	Cowan, Richard Sumner (1921–1997)
Craib	Craib, William Grant (1882–1933)
Crantz	Crantz, Heinrich Johann Nepomuk von (1722–1797)
Craven	Craven, Lyndley Alan (1945–)
Crepet	Crepet, W.L. (fl. 1990)
Crépin	Crépin, François (1830–1903)
Cretz.	Cretzoiu, Paul (1909–1946)
Cribb	Cribb, Phillip James (1946–)
Crisci	Crisci, Jorge Victor (1945–)
Crisp	Crisp, Michael Douglas (1950–)
Cristóbal	Cristóbal, Carmen Lelia (1932–)
Critchf.	Critchfield, William Burke (1923–1989)
Croat	Croat, Thomas Bernard (1938–)
Croizat	Croizat, Léon Camille Marius (1894–1982)
Cronq.	Cronquist, Arthur John (1919–1992)
Crosa	Crosa, Orfeo (fl. 1975)
Crosswh.	Crosswhite, Frank Samuel (1940–)
Cruden	Cruden, Robert William (1936–)
Crüger	Crüger, Hermann (1818–1864)

Cuatrec.	Cuatrecasas, José (1903–1996)
Cuf.	Cufodontis, Georg (1896–1974)
Cullen	Cullen, James (1936–)
Cunn. (, A.)	Cunningham, Allan (1791–1839)
Cunn., R.	Cunningham, Richard (1793–1835)
Curran	Curran, Mary Katherine (later Brandegee) (1844–1920)
Curtis	Curtis, William (1746–1799)
Curtis, C.	Curtis, Charles Henry (1869–1958)
Curtis, W.	Curtis, Winifred Mary (1905–)
Cusset, C.	Cusset, Colette (1944–)
Cusset, G.	Cusset, Gérard Henri Jean (1936–)
Cusson	Cusson, Pierre (1727–1783)
Cutler	Cutler, Hugh Carson (1912–)
Cutler, D.	Cutler, David Frederick (1939–)
Czech	Czech, Gerald (1930–)
Czerep.	Czerepanov, Sergei Kirillovich (1921–)
Czerniak.	Czerniakowska, Ekaterina Georgiewna (née Reineke) (1892–1942)
Czuk.	Czukavina, Anna Prokofevna (1929–1985)
Dahl	Dahl, Andreas or Anders (1751–1789)
Dahlgren, R.	Dahlgren, Rolf Martin Theodor (1932–1987)
Dahlst.	Dahlstedt, Hugo Gustav Adolf (or Gustav Adolf Hugo) (1856–1934)
Dale	Dale, Ivan Robert (1904–1963)
Dalla Torre	Dalla Torre, Karl Wilhelm von (1850–1928)
Dallimore	Dallimore, William (1871–1959)
Dalz.	Dalzell, Nicolas Alexander (1817–1878)
Dalziel	Dalziel, John McEwan (1872–1948)
Dammer	Dammer, Karl Lebrecht Udo (1860–1920)
Dandy	Dandy, James Edgar (1903–1976)
Danert	Danert, Siegfried (1926–1973)
Danguy	Danguy, Paul Auguste (1862–1942)
Daniel, T.	Daniel, Thomas Franklin (1954–)
Daniell	Daniell, William Freeman (1818–1865)
Danser	Danser, Benedictus Hubertus (1891–1943)
D'Arcy	D'Arcy, William Gerald (1931–)
Darl. C.	Darlington, Cyril Dean (1903–1981)
Darwin, S.	Darwin, Steven P. (1949–)
Dasuki	Dasuki, Undang A. (1943–)
Daubeny	Daubeny, Charles Giles Bridle (1795–1867)
Davidse	Davidse, Gerrit (1942–)
Davidson	Davidson, Anstruther (1860–1932)
Davidson, C.	Davidson, Carol (1944–)
Davies	Davies, Frances G. (1944–)
Davies, O.	Davies, Olive B. (fl. 1917)
Davis, P.	Davis, Peter Hadland (1918–1992)
Davy	Davy, Joseph Burtt (1870–1940)
Dawe	Dawe, Morley Thomas (1880–1943)
Dawson	Dawson, John Wyndham (1928–)
Day, A.	Day, Alva George (later Grant, Alva Day) (1920–)
Dayton	Dayton, William Adams (1885–1958)
DC	de Candolle, Augustin-Pyramus (1778–1841)
DC, A.	de Candolle, Alphonse Louis Pierre Pyramus (1806–1893)
DC, C.	de Candolle, Anne Casimir Pyramus (1836–1918)
De	De, A.B. (fl. 1983)
Dean	Dean, Richard (1830–1905)
Deb	Deb, Debandra Bijoy (1924–)
Decker	Decker, Paul (1867–?)
Decker, D.	Decker, Deena S. (fl. 1988)
Decker, H.	Decker, Henry Fleming (1930–)
Decne.	Decaisne, Joseph (1807–1882)

Deflers	Deflers, Albert (1841–1921)
Degener	Degener, Otto (1899–1988)
Degener, I.	Degener, Isa (1924–)
Dehnh.	Dehnhardt, Friedrich (1787–1870)
Del.	Delile, Alire Raffeneau (1778–1850)
Delarbre	Delarbre, Antoine (1724–1813)
Delaroche	Delaroche, Daniel (1743–1813)
Delavay	Delavay, Pierre Jean Marie (1834–1895)
Deng, M.B.	Deng, Mao-Bin (fl. 1987)
Dennst.	Dennstaedt, August Wilhelm (1776–1826)
Des Moul.	Des Moulins, Charles Robert Alexandre (1798–1875)
Descoings	Descoings, Bernard M. (1931–)
Desf.	Desfontaines, Réné Louiche (1750–1833)
Desp.	Desportes, Jean Baptiste Réné Pouppé (1704–1748)
Desr.	Desrousseaux, Louis Auguste Joseph (1753–1838)
Desv.	Desvaux, Auguste Nicaise (Augustin) (1784–1856)
Desv., E.	Desvaux, Etienne-Emile (1830–1854)
Determan	Determan, R.O. (fl. 1981)
Devesa	Devesa, Juan Antonio (1955–)
Dewar	Dewar, Daniel (c. 1860–1905)
Dewey, L.	Dewey, Lyster Hoxie (1865–1944)
Deyl	Deyl, Miloš (1906–1985)
Díaz de la Guardia	Díaz de la Guardia Guerrero, Consuelo (1952–)
Díaz-Miranda	Díaz-Miranda, David (1946–)
Didr.	Didrichsen, Didrik Ferdinand (1814–1887)
Diels	Diels, Friedrich Ludwig Emil (1874–1945)
Dieterle	Dieterle, Jennie van Akkern (1909–)
Dietr.	Dietrich, Friedrich Gottlieb (1768–1850)
Dietr., A.	Dietrich, Albert Gottfried (1795–1856)
Dietr., D.	Dietrich, David Nathaniel Friedrich (1799–1888)
Dillon	Dillon, Michael O. (1947–)
Dillwyn	Dillwyn, Lewis Weston (1778–1855)
Ding Hou	Ding Hou (1921–)
Diniz, A.	Diniz, Manuel (de) Assunção (fl. 1957)
Dinter	Dinter, Moritz Kurt (1868–1945)
Dippel	Dippel, Leopold (1827–1914)
Dittr.	Dittrich, Manfred (1934–)
Dixit	Dixit, Ram Das (1942–)
Dockr.	Dockrill, Alick William (1915–)
Dod	Dod, Donald D. (1912–)
Dode	Dode, Louis-Albert (1875–1943)
Dodson	Dodson, Calaway H. (1928–)
Doebley	Doebley, John F. (fl. 1980)
Doell	Döll, Johann(es) Christoph (Christian) (also Doell) (1808–1885)
Doerfler	Doerfler, Ignaz (also Dörfler) (1866–1950)
Dogan	Dogan, Musa (fl. 1982)
Dombey	Dombey, Joseph (1742–1796)
Domin	Domin, Karel (1882–1953)
Domke	Domke, Friedrich Walter (1899–1988)
Don, D.	Don, David (1799/1800–1841)
Don f., G.	Don, George (1798–1856)
Donn	Donn, James (1758–1813)
Dop	Dop, Paul Louis Amans (1876–1954)
Dorn	Dorn, Robert D. (fl. 1970)
Dostál	Dostál, Josef (1903–1999)
Doty	Doty, Maxwell Stanford (1911–1996)
Douglas	Douglas, David (1798–1834)
Downs	Downs, Robert Jack (1923–)
Doyle	Doyle, Conrad Bartling (1884–1973)
Drake	Drake del Castillo, Emmanuel (1855–1904)
Dransf.	Dransfield, John (1945–)

Dransf., S.	Dransfield, Soejatmi (1939–)
Dressler	Dressler, Robert Louis (1927–)
Drobov	Drobov, Vasilii Petrovich (1885–1956)
Druce	Druce, George Claridge (1850–1932)
Drude	Drude, Carl Georg Oscar (1852–1933)
Drumm.	Drummond, Thomas (c. 1780–1835)
Drumm. J.L.	Drummond, James Lawson (1783–1853)
Drumm. J.R.	Drummond, James Ramsey (1851–1921)
Dryander	Dryander, Jonas Carlsson (1748–1810)
Dubard	Dubard, Marcel Marie Maurice (1873–1914)
Duby	Duby, Jean Etienne (1798–1885)
Duchartre	Duchartre, Pierre Etienne Simon (1811–1894)
Duchesne	Duchesne, Antoine Nicolas (1747–1827)
Ducke	Ducke, Walter Adolpho (1876–1959)
Dudley, T.	Dudley, Theodore Robert (1936–1994)
Dufr.	Dufresne, Pierre (1786–1836)
Dugand	Dugand, Armando (1906–1971)
Duhamel	Duhamel du Monceau, Henri Louis (1700–1782)
Duke	Duke, James A. (1929–)
Dumaz	Dumaz-le-Grand, Noëlle (fl. 1953)
Dum.-Cours.	Dumont de Courset, George(s) Louis Marie (1746–1824)
Dummer	Dummer, Richard Arnold (1887–1922)
Dumort.	Dumortier, Barthélemy Charles Joseph (1797–1878)
Dunal	Dunal, Michel Félix (1789–1856)
Dunkley	Dunkley, Harvey Lawrence (1910–)
Dunlop	Dunlop, Clyde Robert (1946–)
Dunn	Dunn, Stephen Troyte (1868–1938)
Dunsterv.	Dunsterville, Galfrid Clemens Keyworth (1905–1988)
Durand, T.	Durand, Théophile Alexis (1855–1912)
Durande	Durande, Jean François (1732–1794)
Durazz.	Durazzini, Antonio (fl. 1772)
Durieu	Durieu de Maisonneuve, Michel Charles (1796/97–1878)
Duroi	Duroi, Johann Philipp (1741–1785)
Dusén	Dusén, Per Karl Hjalmar (1855–1926)
Duthie	Duthie, John Firminger (1845–1922)
Dutta	Dutta, S. (fl. 1951)
Duval	Duval, Charles Jeunet (1751–1828)
Duvign.	Duvigneaud, Paul Auguste (1913–)
Dvořák	Dvořák, Frantisek (1921–)
Dwyer	Dwyer, John Duncan (1915–)
Dyer	Dyer, William Turner Thiselton (or Thistleton-) (1843–1928)
Dyer, R.A.	Dyer, Robert Allen (1900–1987)
Eastw.	Eastwood, Alice (1859–1953)
Eaton	Eaton, Amos (1776–1842)
Eaton, D.	Eaton, Daniel Cady (1834–1895)
Eberh.	Eberhardt, Philippe Albert (1874–1942)
Eberm.	Ebermaier, Johann Erdwin Christopher (1769–1825)
Ecklon	Ecklon, Christian Frederich (1795–1868)
Edgew.	Edgeworth, Michael Pakenham (1812–1881)
Edmondson	Edmondson, John Richard (1948–)
Edwards	Edwards, Trevor J. (fl. 1989)
Eggers	Eggers, Henrik Franz Alexander von (1844–1903)
Egli	Egli, Bernhard (fl. 1990)
Ehrenb.	Ehrenberg, Christian Gottfried (1795–1876)
Ehrend.	Ehrendorfer, Friedrich (1927–)
Ehrh.	Ehrhart, (Jacob) Friedrich (1742–1795)
Eichler	Eichler, August Wilhelm (1839–1887)
Eig	Eig, Alexander (1894–1938)
Eiten	Eiten, George (1923–)

Ekman	Ekman, Hedda Maria Emerence Elisabeth (née Akerhielm) (1862–1936)
Ekman, E.	Ekman, Erik Leonard (1883–1931)
Elffers	Elffers, Joan (1928–)
El Gazzar	El Gazzar, Adel Ibrahim Hamed (1942–)
Elliott	Elliott, Stephen (1771–1830)
Ellis	Ellis, John (1711–1776)
Ellis, R.	Ellis, Roger P. (fl. 1984)
Elmer	Elmer, Adolph Daniel Edward (1870–1942)
Emeric	Emeric (fl. c. 1828)
Emmerich	Emmerich, Margarete (1933–)
Endl.	Endlicher, Stephan Friedrich Ladislaus (1804–1849)
Endress	Endress, Peter Karl (1942–)
Engelhorn	Engelhorn, Tamra (1945–)
Engelm.	Engelmann, Georg (1809–1884)
Engl.	Engler, Heinrich Gustav Adolf (1844–1930)
Epling	Epling, Carl Clawson (1894–1968)
Erdtman	Erdtman, Otto Gunnar Elias (1897–1973)
Eriksson	Eriksson, Roger (1958–)
Ernst	Ernst, Alfons (1875–1968)
Ertter	Ertter, Barbara Jean (1953–)
Escal.	Escalante, Manuel G.
Eschsch.	Eschscholtz, Johann Friedrich Gustav von (1793–1831)
Eselt.	Eseltine, Glen Parker van (1888–1938)
Eskuche	Eskuche, Ulrich G. (1926–)
Esser	Esser, Hans-Joachim
Evans, W.E.	Evans, William Edgar (1882–1963)
Everett	Everett, Joy (1953–)
Evert	Evert, Erwin F. (1940–)
Ewan	Ewan, Joseph Andorfer (1909–)
Ewart	Ewart, Alfred James (1872–1937)
Exell	Exell, Arthur Wallis (1901–1993)
Eyma	Eyma, Pierre Joseph (1903–1945)
Fabr.	Fabricius, Philipp Conrad (1714–1774)
Faden	Faden, Robert B. (1942–)
Fagerl.	Fagerlind, Folke (1907–)
Falc.	Falconer, Hugh (1808–1865)
Farille	Farille, Michel A. (1945–)
Farron	Farron, Claude (1935–)
Farw.	Farwell, Oliver Atkins (1867–1944)
Fasano	Fasano, Angelo (fl. 1787)
Fassett	Fassett, Norman Carter (1900–1954)
Fauché	Fauché, M. (fl. 1832)
Fawcett	Fawcett, William (1851–1926)
Fayed	Fayed, A.-A. (fl. 1979)
Fed.	Fedorov, Andrej Aleksandrovich (1909–1987)
Fedde	Fedde, Friedrich Karl Georg (1873–1942)
Fedtsch., B.	Fedtschenko, Boris Alexeevich (or Alexjewitsch) (1872–1947)
Fée	Fée, Antoine Laurent Apollinaire (1789–1874)
Feer	Feer, Heinrich (1857–1892)
Feinbrun	Feinbrun, Naomi (1900–1995)
Felger	Felger, Richard S. (fl. 1968)
Fenzi	Fenzi, Emanuele Orazio (1843–1924)
Fenzl	Fenzl, Eduard (1808–1879)
Fern.	Fernald, Merritt Lyndon (1873–1950)
Fernandes	Fernandes, Abílio (1906–)
Fernandes, R.	Fernandes, Rosette Mercedes Saraiva Batarda (1916–)
Fernández	Fernández, Z. Mayra (1948–)
Fernández-Villar	Fernández-Villar, Celestino (1838–1907)
Fern. Casas.	Fernández Casas, Francisco Javier (1945–)

Fernando	Fernando, Edwino S. (1953–)
Ficalho	Ficalho, Francisco Manoel Carlos de Mello de (1837–1903)
Field	Field, Barron (1786–1846)
Field, D.V.	Field, David Vincent (1937–)
Fijten	Fijten, F. (fl. 1975)
Filgueiras	Filgueiras, Tarciso S. (1950–)
Finet	Finet, Achille Eugène (1863–1913)
Fischer	Fischer, Friedrich Ernst Ludwig von (1782–1854)
Fischer, C.	Fischer, Cecil Ernest Claude (1874–1950)
Fischer, E.	Fischer, Eberhard (fl. 1989)
Fischer, J.	Fischer, Johann Carl (1804–1885)
Fitzg.	Fitzgerald, Robert Desmond (David) (1830–1892)
Fitzg., W.	Fitzgerald, William Vincent (1867–1929)
Fletcher	Fletcher, Harold Roy (1907–1978)
Florence	Florence, E. Jacques M. (1951–)
Floret	Floret, J.J. (1939–)
Florin	Florin, Carl Rudolf (1894–1965)
Flueckiger	Flueckiger, Friedrich August (1828–1894)
Fluegge	Fluegge, Johannes (also Flügge) (1775–1816)
Focke	Focke, Wilhelm Olbers (1834–1922)
Focke, H.	Focke, Hendrik Charles (1802–1856)
Fomin	Fomin, Aleksander Vasiljevich (1869–1935)
Fonnegra	Fonnegra G., Ramiro (fl. 1985)
Font Quer	Font Quer, Pio (1888–1964)
Fontana	Fontana, Felice (1730–1805)
Forbes	Forbes, James (1773–1861)
Forbes, F.B.	Forbes, Francis Blackwell (1839–1908)
Forbes, H.M.	Forbes, Helena M.L. (1900–1959)
Ford	Ford, Neridah Clifton (1926–)
Ford-Lloyd	Ford-Lloyd, Brian V. (fl. 1976–7)
Forman	Forman, Lewis Leonard (1929–1998)
Forrest	Forrest, George (1873–1932)
Forssk.	Forsskaol, Pehr (1732–1763)
Forster	Forster, Johann Reinhold (1729–1798)
Forster f. (or G. Forster)	Forster, Johann Georg Adam (1754–1794)
Fosb.	Fosberg, Francis Raymond (1908–1993)
Foster, M.	Foster, Mercedes S.
Foster, R.	Foster, Robert Crichton (1904–1986)
Foug.	Fougeroux de Bondaroy, Auguste Denis (1732–1789)
Fourn.	Fournier, Eugène Pierre Nicolas (1834–1884)
Fourn., P.	Fournier, Paul-Victor (1877–1964)
Fourr.	Fourreau, Jules Pierre (1844–1871)
Foxw.	Foxworthy, Frederick William (1877–1950)
Franchet	Franchet, Adrien René (1834–1900)
Francis	Francis, William Douglas (1889–1959)
Franco	Franco, João Manuel Antonio Paes do Amaral (1921–)
Frank	Frank, Joseph C. (1782–1835)
Franks	Franks, M. (1886–1961)
Franquet	Franquet, Robert Fernand (1897–1930)
Franz	Franz, C.
Frappier	Frappier, Charles (fl. 1853–95)
Fraser	Fraser, John (1750–1811)
Fraser, J.	Fraser, James (1854–1935)
Frederiksen	Frederiksen, Signe (1942–)
Freeman, G.	Freeman, George Fouché (1876–1930)
Freire	Freire, Susana E. (fl. 1986)
Freitag	Freitag, Helmut E. (1932–)
Frémont	Frémont, John Charles (1813–1890)
Fres.	Fresenius, Johann Baptist Georg Wolfgang (1808–1866)
Freyn	Freyn, Josef Franz (1845–1903)
Frič	Frič, Alberto Vojtech (1882–1944)
Friedmann	Friedmann, F. (1941–)

Friedrich	Friedrich, Hans-Christian (1925–)
Friedrichsthal	Friedrichsthal, Emanuel von (1809–1842)
Fries, R.(E.)	Fries, Klas Robert Elias (1876–1966)
Fries, T.C.E.	Fries, Thore Christian Elias (1886–1930)
Friis, I.	Friis, Ib (1945–)
Fritsch	Fritsch, Karl (1864–1934)
Friv.	Frivaldsky von Frivald, Emerich (1799–1870)
Frodin	Frodin, David Gamman (1940–)
Froebel	Froebel Karl Otto (1844–1906)
Fröhner	Fröhner, Eugene (1858–1940)
Frost	Frost, John (1803–1840)
Fryx.	Fryxell, Paul Arnold (1927–)
Fuchs, H.P.	Fuchs, Hans-Peter (1928–1999)
Fuertes	Fuertes, Javier
Fung	Fung Hok-Lam. (fl. 1988)
Funk	Funk, Victoria Ann (1947–)
Furnari	Furnari, Francesco (1933–)
Furt.	Furtado, Caetano Xavier, Dos Remedios (1897–1980)
Gaertner	Gaertner, Joseph (1732–1791)
Gaertner, P.	Gaertner, Philipp Gottfried (1754–1825)
Gaertner f.	Gaertner, Carl Friedrich von (1772–1850)
Gagnebin	Gagnebin, Abraham (1707–1800)
Gagnepain	Gagnepain, François (1866–1952)
Galeotti	Galeotti, Henri Guillaume (1814–1858)
Gallaud	Gallaud, Ernest-Isodore (fl. 1907)
Galushko	Galushko, Anatol I. (1926–)
Gamble	Gamble, James Sykes (1847–1925)
Gams	Gams, Helmut (1893–1976)
Gand.	Gandoger, Michel (1850–1926)
Gao	Gao, Cheng-Zhi (fl. 1981)
Garay	Garay, Leslie Andrew (1924–)
Garb.	Garbari, Fabio (1937–)
García-Martín	García-Martín, Felipe (1954–)
Garcke	Garcke, Christian August Friedrich (1819–1904)
Garden	Garden, Alexander (1730–1792)
Garden, J.	Garden, Joy, later J. Thompson (1923–)
Gardner	Gardner, George (1812–1849)
Gardner, C.	Gardner, Charles Austin (1896–1970)
Garnock-Jones	Garnock-Jones, Philip John (1950–)
Garsault	Garsault, François Alexandre Pierre de (1691–1778)
Gates	Gates, David M. (1921–)
Gauba	Gauba, Erwin (1891–1964)
Gaudich.	Gaudichaud-Beaupré, Charles (1789–1854)
Gaudin	Gaudin, Jean François Aimé Gottlieb Philippe (1766–1833)
Gay	Gay, Jacques Etienne (1786–1864)
Gay, C.	Gay, Claude (1800–1873)
Geel	Geel, Petrus Cornelius van (or Pierre Corneille) (1796–1836)
Geer.	Geerinck, Daniel (1945–)
Geesink	Geesink, Robert (1945–1992)
Geld.	Geldikhanov, A.M. (1953–)
Gentry, A.	Gentry, Alwyn Howard (1945–1993)
George, A.S.	George, Alexander Segger (1939–)
Georgi	Georgi, Johann Gottlieb (1729–1802)
Gepp	Gepp, Anthony (1862–1955)
Gérardin	Gérardin, Sébastien (1751–1816)
Gerbaulet	Gerbaulet, Maike
Gereau	Gereau, Roy E. (1947–)
Gerrard	Gerrard, William Tyrer (?–1866)
Gerstb.	Gerstberger, Pedro (1951–)
Gibbs	Gibbs, Lilian Suzette (1870–1925)

Gibbs, P.	Gibbs, Peter Edward (1938–)
Gibson, A.	Gibson, Alexander (1800–1867)
Giesenh.	Giesenhagen, Karl Friedrich Georg (1860–1928)
Gilbert, M.	Gilbert, Michael George (1943–)
Gilg	Gilg, Ernst Friedrich (1867–1933)
Gilib.	Gilibert, Jean-Emmanuel (1741–1814)
Gillespie	Gillespie, John Wynn (?–1932)
Gillett, G.	Gillett, George Wilson (1917–1976)
Gillett, J.B.	Gillett, Jan Bevington (1911–1995)
Gilli	Gilli, Alexander (1904–)
Gillies	Gillies, John (1792–1834)
Gillis	Gillis, William Thomas (1933–1979)
Gilly	Gilly, Charles Louis (1911–1970)
Girard	Girard, Frédéric de (1810–1851)
Giri	Giri, G.S. (1950–)
Giroux	Giroux, Mathilde (fl. 1933)
Giseke	Giseke, Paul Dietrich (1741–1796)
Given	Given, David Roger (1943–)
Glass	Glass, Charles (fl. 1975)
Glassman	Glassman, Sidney Frederick (1919–)
Glaz.	Glaziou, Auguste François Marie (1828–1906)
Gleason	Gleason, Henry Allan (1882–1975)
Gled.	Gleditsch, Johann Gottlieb (1714–1786)
Glen	Glen, Hugh Francis (1950–)
Gmelin	Gmelin, Johann Georg (1709–1755)
Gmelin, C.	Gmelin, Carl Christian (1762–1837)
Gmelin, J.	Gmelin, Johann Friedrich (1748–1804)
Gmelin, S.	Gmelin, Samuel Gottlieb (1743/5–1774)
God.-Leb.	Godefroy-Lebeuf, Alexandre (1852–1903)
Godron	Godron, Dominique Alexandre (1807–1880)
Goebel	Goebel, Karl Immanuel Eberhard von (1855–1932)
Goeppert	Göppert, Johann Heinrich Robert (1800–1884)
Goetgh.	Goetghebeur, Paul (1952–)
Goldbl(att)	Goldblatt, Peter (1943–)
Goldie	Goldie, John (1793–1886)
Golosk.	Goloskokov, Vitaliy Petrovich (1913–)
Gomes	Gomes, Bernardino António (1769–1823)
Gomes, B.A.	Gomes, Bernardino António (1806–1877)
Gómes, J.	Gómes, José Corrêa (1919–1965)
Gómez (P.), L.D.	Gómez (Pignataro), Luis Diego (1944–)
Gómez-Campo	Gómez-Campo, César (1933–)
González	González, Francisco (fl. 1877)
Gonz. Medr.	González Medrano, Francisco (1939–)
Good, R.	Good, Ronald d'Oyley (1896–1992)
Goodman	Goodman, George Jones (1904–1999)
Goodspeed	Goodspeed, Thomas Harper (1887–1966)
Goossens	Goossens, Antonie Petrus Gerhardy (1896–1972)
Gordon	Gordon, George (1806–1879)
Gorer	Gorer, Richard
Gottl.-Tann.	Gottlieb-Tannenhain, Paul von (1879–?)
Gouan	Gouan, Antoine (1733–1821)
Goudot	Goudot, Justin (1822–1845)
Gould	Gould, Frank Walton (1913–1981)
Goy	Goy, Doris Alma (1912–)
Goyder	Goyder, David John (1959–)
Goyena	Goyena, Miguel Ramírez (1857–1927)
Graebner	Graebner, Karl Otto Robert Peter Paul (1871–1933)
Graham	Graham, Robert C. (1786–1845)
Graham, J.	Graham, John (1805–1839)
Graham, S.	Graham, Shirley Ann Tousch (1935–)
Grande	Grande, Loreto (1878–1965)
Grant, A.D.	Grant, Alva Day (formerly Day, Alva George) (1920–)

Grant, V.	Grant, Verne Edwin (1917–)
Grashoff	Grashoff, J.L.
Grau	Grau, Hans Rudolph Jürke (1937–)
Grauer	Grauer, Sebastian (1758–1820)
Gray	Gray, Samuel Frederick (1766–1828)
Gray, A.	Gray, Asa (1810–1888)
Gray, N.	Gray, Netta Elizabeth (1913–1970)
Grayum	Grayum, Michael Howard (1949–)
Greb.	Grebenshchikov, Igor Sergeevich (1912–1986)
Green	Green, Mary Letitia (1886–1978)
Green, J.	Green, John William (1930–)
Green, P.	Green, Peter Shaw (1920–)
Greene, E.	Greene, Edward Lee (1843–1915)
Greenman	Greenman, Jesse More (1867–1951)
Gremli	Gremli, August(e) (1833–1899)
Gren.	Grenier, Jean Charles Marie (1808–1875)
Greuter	Greuter, Werner Rodolfo (1938–)
Grev.	Greville, Robert Kaye (1794–1866)
Greves	Greves, S. (fl. 1923–7)
Grey-Wilson	Grey-Wilson, Christopher (1944–)
Grierson	Grierson, Andrew John Charles (1929–1990)
Griess.	Griesselich, Ludwig (1804–1848)
Griffith	Griffith, William (1810–1845)
Griffiths	Griffiths, David (1867–1935)
Grimes	Grimes, James Walter (1953–)
Gris	Gris, Jean Antoine Arthur (1829–1872)
Griseb.	Grisebach, August Heinrich Rudolf (1814–1879)
Groenl.	Groenland, Johannes (1824–1891)
Grondona	Grondona, Eduardo M. (1911–)
Gronov.	Gronovius, Jan Fredrik (1686–1762)
Grosourdy	Grosourdy, Renato de (fl. 1864)
Gross	Gross, Hugo (1888–)
Grossh.	Grossheim, Alexander Alfonsovich (1888–1948)
Groves	Groves, Henry (1855–1891)
Groves, J.	Groves, James (1858–1933)
Grulich	Grulich, Vìt (1956–)
Guag.	Guaglianone, Encarnación Rosa (1932–)
Guédès	Guédès, Michel (1942–1985)
Guého	Guého, E.L. Joseph (1937–)
Gueldenst.	Gueldenstaedt, Anton Johann (von) (also Güldenstädt) (1745–1781)
Guerke	Gürke, Robert Louis August Maximilian (1854–1911)
Guerra	Guerra, P. (fl. 1938)
Guett.	Guettard, Jean Etienne (1715–1786)
Gugerli	Gugerli, Karl (fl. 1939)
Guill.	Guillarmod, Amy Frances May Gordon Jacot
Guillaumin	Guillaumin, André (1885–1974)
Guillemin	Guillemin, Jean Baptiste Antoine (1796–1842)
Guinea	Guinea, Emilio (1907–1985)
Guinet	Guinet, Philippe (fl. 1951)
Guinier	Guinier, Philibert (1876–1962)
Gumbleton	Gumbleton, William Edward (1840–1911)
Guss.	Gussone, Giovanni (1787–1866)
Guşul.	Guşuleac, Michail (1887–1960)
Ha	Ha, Thi Dung (fl. 1970)
Hackel	Hackel, Eduard (1850–1926)
Hadač	Hadač, Emil (1914–)
Haegi	Haegi, Laurence Arnold Robert (1952–)
Hágsater	Hágsater, Eric (1945–)
Haines	Haines, Henry Haselfoot I. (1867–1945)
Haines, R.	Haines, Richard Wheeler (1906–1982)

Halb.	Halbinger, Federico (1925–)
Hall, H.M.	Hall, Harvey Monroe (1874–1932)
Hall, J.B.	Hall, John Bartholomew (1932–1984)
Hallberg	Hallberg, Mildred (later Jones)
Hallé	Hallé, Nicolas (1927–)
Haller	Haller, Victor Albrecht von (1708–1777)
Hallier, H. or Hallier f.	Hallier, Johannes Gottfried (`Hans') (1868–1932)
Hamann	Hamann, Ole Jorgen (1944–)
Hamer	Hamer, Fritz (fl. 1970)
Hamet	Raymond-Hamet (1890–1972)
Hammer	Hammer, K. (1944–)
Hance	Hance, Henry Fletcher (1827–1886)
Hancock	Hancock, Thomas (1783–1849)
Hand.-Mazz.	Handel-Mazzetti, Heinrich (1882–1940)
Handro	Handro, Osvaldo (1908–1986)
Hanin	Hanin, L. (fl. 1800)
Hansen, C.	Hansen, Carlo (1932–1991)
Hanst.	Hanstein, Johannes Ludwig Emil Robert von (1822–1880)
Hara	Hara, Kanesuke (1885–1962)
Hara, H.	Hara, Hiroshi (1911–1986)
Hardham	Hardham, Clare B. (fl. 1964)
Hardw.	Hardwicke, Thomas (1757–1835)
Harley	Harley, Raymond Mervyn (1936–)
Harling	Harling, Gunnar Wilhelm (1920–)
Harmaja	Harmaja, Harri (1944–)
Harms	Harms, Hermann August Theodor (1870–1942)
Harper	Harper, Roland McMillan (1878–1966)
Harriman	Harriman, Neil A. (1938–)
Hartl	Hartl, Dimitri (1926–)
Hartley	Hartley, Thomas Gordon (1931–)
Hartman	Hartman, Carl Johann (1790–1849)
Hartmann, H.	Hartmann, Heidrun, Elsbeth Klara Osterwald (1942–)
Hartog	Hartog, Cornelis den (1931–)
Hartvig	Hartvig, Per (1941–)
Hartweg	Hartweg, Karl Theodor (1812–1871)
Harvey	Harvey, William Henry (1811–1866)
Harvey, J.	Harvey, John H.
Harz	Harz, Carl Otto (1842–1906)
Hasselq.	Hasselquist, Fredric (1722–1752)
Hasselt	Hasselt, Johan Coenraad van (1797–1823)
Hassk.	Hasskarl, Justus Carl (1811–1894)
Hassler	Hassler, Emile (1861–1937)
Hatch	Hatch, Edwin Daniel (fl. 1946)
Hatch, S.	Hatch, Stephan LaVor (1945–)
Hatusima	Hatusima, Sumihiko (1906–)
Hauman	Hauman, Lucien Leon (Hauman-Merck) (1880–1965)
Hausskn.	Haussknecht, Heinrich Carl (1838–1903)
Havil.	Haviland, George Darby (1857–1901)
Haw.	Haworth, Adrian Hardy (1768–1833)
Hawkes, A.	Hawkes, Alex Drum (1927–1977)
Hawkes, J.	Hawkes, John Gregory (1915–)
Hay, A.	Hay, Alistair James Montagu (1955–)
Hayata	Hayata, Bunzô (1874–1934)
Hayek	Hayek, August Edler von (1871–1928)
Hayne	Hayne, Friedrich Gottlob (1763–1832)
Heads	Heads, Michael J. (1957–)
Heath	Heath, Paul V. (fl. 1983–94)
Hebenstr.	Hebenstreit, Johann Christian (1720–1795)
Heckard	Heckard, Lawrence Ray (1923–1991)
Heckel	Heckel, Edouard Marie (1843–1916)
Hedb.	Hedberg, Karl Olov (1923–)
Hedge	Hedge, Ian Charleson (1928–)

Hedr.	Hedrick, Ulysses Prentiss (1870–1951)
Hedwig f. (or R. Hedwig)	Hedwig, Romanus Adolf (1772–1806)
Heer	Heer, Oswald von (1809–1883)
Heering	Heering, Wilhelm Christian August (1876–1916)
Heese	Heese, Emil (1862–1914)
Hegelm.	Hegelmaier, Christoph Friedrich (1833–1906)
Heil	Heil, Hans Albrecht (1899–)
Heilborn	Heilborn, Otto (1892–1943)
Heim	Heim, Frédéric Louis (1869–?)
Heimerl	Heimerl, Anton (1857–1942)
Heine	Heine, Heino Hermann (1923–1996)
Heiser	Heiser, Charles Bixler (1920–)
Heister	Heister, Lorenz (1683–1758)
Heldr.	Heldreich, Theodor Heinrich Hermann von (1822–1902)
Heller	Heller, Franz Xaver (1775–1840)
Heller, A.A.	Heller, Amos Arthur (1867–1944)
Hellwig	Hellwig, Frank
Hemadri	Hemadri, Koppula (fl. 1970)
Hemsley, J.H.	Hemsley, James Hatton (1923–)
Hemsley	Hemsley, William Botting (1843–1924)
Henckel	Henckel von Donnersmarck, Leo Victor Felix (1785–1861)
Henderson, A.	Henderson, Archibald (1921–)
Henderson, L.	Henderson, Louis Forniquet (1853–1942)
Henderson, M.D.	Henderson, Mayda Doris (1928–)
Hendrych	Hendrych, Radovan (1926–)
Henfrey	Hentrey, Arthur (1819–1859)
Henkel	Henkel, Heinrich (fl. 1897–1914)
Hennipman	Hennipman, Elbert (1937–)
Henrard	Henrard, Jan Theodoor (1881–1974)
Henrickson	Henrickson, James Solberg (1940–)
Henry, A.	Henry, Augustine (1857–1930)
Henry, A.N.	Henry, Ambrose Nathaniel (1936–)
Henry, L.	Henry, Louis (1853–1903)
Henschel	Henschel, August Wilhelm Eduard Theodor (1790–1856)
Hepper	Hepper, Frank Nigel (1929–)
Herbert	Herbert, William (1778–1847)
Herbich	Herbich, Franz (1791–1865)
Herder	Herder, Ferdinand Gottfried Maximilian Theobold von (1828–1896)
Herm., F.J.	Hermann, Frederick Joseph (1906–1987)
Hernández, H.	Hernández, Héctor Manuel (1954–)
Herre	Herre, Adolar Gottlieb Julius (Hans) (1895–1979)
Herrera	Herrera, Alarcón de Loja, Berta (1930–)
Herrm.	Herrmann, Johann (or Jean) (1738–1800)
Herter	Herter, Wilhelm Gustav Franz (Guillermo Gustavo Francisco) (1884–1958)
Herzog	Herzog, Theodor Karl Julius (1880–1961)
Heward	Heward, Robert (1791–1877)
Hewson	Hewson, Helen Joan (1938–)
Heyne	Heyne, Benjamin (1770–1819)
Heyne, K.	Heyne, Karel (1877–1947)
Heynhold	Heynhold, Gustav (1800–1860)
Heyw.	Heywood, Vernon Hilton (1927–)
Hicken	Hicken, Cristóbal Mariá (1875–1933)
Hiern	Hiern, William Philip (1839–1925)
Hieron.	Hieronymus, Georg Hans Emmo Wolfgang (1846–1921)
Hildebr.	Hildebrand, Friedrich Hermann Gustav (1835–1915)
Hildm.	Hildmann, H. (?–1895)
Hilger	Hilger, Hartmut H. (1948–)
Hill	Hill, John (1716–1775)
Hill, A.W.	Hill, Arthur William (1875–1941)
Hill, K.	Hill, Kenneth D. (1948–)

Hill, W.	Hill, Walter (1820–1904)
Hillc.	Hillcoat, Jean Olive Dorothy (1904–1990)
Hillebrand	Hillebrand, Wilhelm B. (1821–1886)
Hilliard	Hilliard, Olive Mary (1926–)
Himmelb.	Himmelbauer, Wolfgang (1886–1937)
Hiriart	Hiriart, Patricia (fl. 1981)
Hiroe	Hiroe, Minosuke (1914–)
Hitchc., A.	Hitchcock, Albert Spear (né Jennings) (1865–1935)
Hitchc., C.	Hitchcock, Charles Leo (1902–1986)
Hjelmq.	Hjelmquist, Karl Jesper Hakon (1905–)
Ho	Ho, Chun-Nien (fl. 1955)
Ho, T.N.	Ho, Ting-Nung (1938–)
Hochr.	Hochreutiner, Bénédict Pierre Georges (1873–1959)
Hochst.	Hochstetter, Christian Ferdinand Friedrich (1787–1860)
Hoehne	Hoehne, Frederico Carlos (1882–1959)
Hoeck	Hoeck, Fernando (1858–1915)
Hoefft	Höfft, Franz M.S.V. (fl. 1826)
Hoffm.	Hoffmann, George Franz (1761–1826)
Hoffm., F.	Hoffmann, Ferdinand (1860–1914)
Hoffm., K.	Hoffmann, Käthe (1883–c. 1931)
Hoffm., O.	Hoffmann, Karl August Otto (1853–1909)
Hoffm., P.	Hoffmann, Philipp (fl. 1868)
Hoffsgg.	Hoffmannsegg, Johann Centurius von (1766–1849)
Hogg	Hogg, Thomas (1777–1855)
Hohen.	Hohenacker, Rudolph Friedrich (1798–1874)
Holl	Holl, Friedrich (fl. 1820–50)
Holland	Holland, John Henry (1869–1950)
Holmboe	Holmboe, Jens (1880–1943)
Holttum	Holttum, Richard Eric (1895–1990)
Holub	Holub, Josef Ludvík (1930–1999)
Homolle	Homolle, Anne Marie (fl. 1937–50)
Honck.	Honckeny, Gerhard August (1724–1805)
Honda	Honda, Masaji (1897–1984)
Hong	Hong, De-Yuang (1936–)
Hong, S.P.	Hong, Suk-Pyo (fl. 1989)
Hoog	Hoog, Johannes Marius Cornelis (1865–1950)
Hoogl.	Hoogland, Ruurd Dirk (1922–1994)
Hook.	Hooker, William Jackson (1785–1865)
Hook.f.	Hooker, Joseph Dalton (1817–1911)
Hoover	Hoover, Robert Francis (1913–1970)
Hoppe	Hoppe, David Heinrich (1760–1846)
Horan.	Horaninow, Paul Fedorowitsch (1796–1865)
Horkel	Horkel, Johann (1769–1846)
Hornem.	Hornemann, Jens Wilken (1770–1841)
Hort. Allw.	Allwood Bros, Haywards Heath, England
Hort. Burkw. & Skipw.	Burkwood & Skipwith Ltd, Kingston-on-Thames, England (mid-C20)
Hort. Huber	C. Huber & Cie, Hyères, France (C19)
Hort. Lemoine	V. Lemoine & fils, Nancy, France (1850–C20)
Hort. Sander	F. Sander & Co., St Albans, England
Hort. Tubergen	C.G. van Tubergen Ltd, Haarlem, Netherlands (1869–)
Hort. Veitch	James Veitch & Sons, Chelsea, England (early C19–1914)
Hort. Vilm.	Vilmorin-Andrieux & Cie, Paris, France (c. 1745–)
Horvat	Horvat, Ivo (1897–1963)
Hosok.	Hosokawa, Takahide (1909–)
Hossain	Hossain, Mosharraf (1928–)
Host	Host, Nicolaus Thomas (1761–1834)
Hotta	Hotta, Teikichi (1899–)
Houllet	Houllet, R.J.B. (1811/18–1890)
Houlston	Houlston, John (fl. 1848–52)
Houtt.	Houttuyn, Martin (1720–1798)
van Houtte	van Houtte, Louis Benoît (1810–1876)

Houz.	Houzeau de Lehaie, Jean Charles (1820–1888)
How	How, Foon-Chew (fl. 1956)
Howard, R.	Howard, Richard Alden (1917–)
Howell	Howell, Thomas Jefferson (1842–1912)
Howell, J.	Howell, John Thomas (1903–1999)
Howitt	Howitt, Beatrice F. (1891–1981)
Hoyle	Hoyle, Arthur Clagne (1905–1986)
Hsiao	Hsiao, Pei-Ken (fl. 1964)
Hu	Hu, Hsen-Hsu (1894–1968)
Hu, S.Y.	Hu, Shiu-Ying (1910–)
Hua	Hua, Henri (1861–1919)
Huang	Huang, Rong-Fu (fl. 1981)
Huang, S.H.	Huang, Shu-Hua (1941–)
Hubb.	Hubbard, Frederic Tracy (1875–1962)
Hubb., C.(E.)	Hubbard, Charles Edward (1900–1980)
Huber	Huber, Jakob (Jacques) E. (1867–1914)
Huber, H.	Huber, Heribert Franz Josef. (1931–)
Hudson	Hudson, William (1730–1793)
Huegel	Huegel, Carl Alexander Anselm von (Hügel) (1794–1870)
Hürl.	Hürlimann, Hans (1921–)
Hughes	Hughes, Dorothy Kate (1899–1932)
Hull	Hull, John G. (1761–1843)
Hultén	Hultén, Oskar Eric Gunnar (1894–1981)
Humb.	Humboldt, Friedrich Wilhelm Heinrich Alexander von (1769–1859)
Humbert	Humbert, Jean-Henri (1887–1967)
Humbl.	Humblot, Léon (fl. 1848–?)
Humphries	Humphries, Christopher John (1947–)
Hunt	Hunt, Trevor Edgar (1913–)
Hunt, D.	Hunt, David Richard (1938–)
Hunt, P.	Hunt, Peter Francis (1936–)
Hunter	Hunter, William (1755–1812)
Hunz.	Hunziker, Armando Theodoro (1919–)
Hurst	Hurst, Charles Chamberlain (1870–1947)
Hutch.	Hutchinson, John (1884–1972)
Hutch., J.B.	Hutchinson, Joseph Burtt (1902–1988)
Huth	Huth, Ernst (1845–1897)
Huxley	Huxley, Camilla Rose (1952–)
Hyland, B.	Hyland, Bernard Patrick Matthew (1937–)
Hylander	Hylander, Hjalmer (1877–1965)
Hylander, N.	Hylander, Nils (1904–1970)
Hylmö	Hylmö, Bertil (1915–)
Ignatov	Ignatov, Mikhail S. (1956–)
Ihlenf.	Ihlenfeldt, Hans-Dieter (1932–)
Iinuma	Iinuma, Yokusai (1782–1865)
Ikonn.	Ikonnikov, Sergei Sergeevich (1931–)
Iljin	Iljin, Modest Mikhailovich (Ilyin) (1889–1967)
Iljinsk.	Iljinskaja, J.A. (1921–)
Iltis	Iltis, Hugh Hellmut (1925–)
Imlay	Imlay, Joan B. (fl. 1939)
Ingram, J.	Ingram, John William (1924–)
Irmscher	Irmscher, Edgar (1887–1968)
Irv., W.	Irving, Walter (1867–1934)
Irvine	Irvine, Alexander (1793–1873)
Irvine, A.	Irvine, Anthony Kyle (1937–)
Ising	Ising, Ernest Horace (1884–1973)
Itô	Itô, Tokutarô (1868–1941)
Itô, H.	Itô, Hirosi (1909–)
Itô, Y.	Itô, Yoshi (1907–)
Ivanina	Ivanina, L.I. (1917–)
Iwarsson	Iwarsson, M. (1948–)

Iwatsuki	Iwatsuki, Kunio (1934–)
Jaarsveld	Jaarsveld, Ernst Jacobus van (1953–)
Jabl.	Jablonszky, Eugene, also Jablonski, E. (1892–1975)
Jack	Jack, William (1795–1822)
Jackson	Jackson, George (c. 1780–1811)
Jackson, A.B.	Jackson, Albert Bruce (1876–1947)
Jackson, B.D.	Jackson, Benjamin Daydon (1846–1927)
Jackson, R.	Jackson, Raymond Carl (1928–)
Jacobs	Jacobs, Marius (1929–1983)
Jacobs, S.	Jacobs, Surrey Wilfred Laurance (1946–)
Jacobsen, H.J.	Jacobsen, Hermann Johannes Heinrich (1898–1978)
Jacq. (Jacquin)	Jacquin, Nicolaus (or Nicolaas) von (1727–1817)
Jacques	Jacques, Henri Antoine (1782–1866)
Jacq.f. (Jacquin f.)	Jacquin, Joseph Franz von (1766–1839)
Jacq.-Fél.	Jacques-Félix, Henri (1907–)
Jafri	Jafri, Saiyad Masudal Hasan (1927–1986)
James	James, Lois Elsie (1914–)
Janchen	Janchen, Erwin Emil Alfred (1882–1970)
Jansen	Jansen, Pieter (1882–1955)
Jansen, M.	Jansen, M.E. (fl. 1979)
Jansen, R.	Jansen, Robert K. (1954–)
Járai-Komlódi	Járai-Komlódi, Magda (1931–)
Jaramillo	Jaramillo, Víctor (fl. 1984)
Jarmol.	Jarmolenko, A.V. (1905–1944)
Jarvis	Jarvis, Charles Edward (1954–)
Játiva	Játiva, Carlos D. (fl. 1963)
Jaub.	Jaubert, Hippolyte François (1798–1874)
Jayaw.	Jayaweera, Don Martin Arthur (1912–1982)
Jebb	Jebb, Matthew (1958–)
Jeffrey, C.	Jeffrey, Charles (1934–)
Jenny	Jenny, R. (fl. 1985)
Jepson	Jepson, Willis Linn (1867–1946)
Jérémie	Jérémie, J. (1944–)
Jessen	Jessen, Karl Friedrich Wilhelm (1821–1889)
Jessop	Jessop, John Peter (1939–)
Jin	Jin, Yue-Xing (1934–)
Jir., V.	Jirásek, Václav (1906–1991)
Johansson	Johansson, Jan Thomas (fl. 1988)
Johnson, A.M.	Johnson, Arthur Monrad (1878–1943)
Johnson, J.(H.)	Johnson, Joseph Harry (1894–)
Johnson, L.	Johnson, Lawrence Alexander Sidney (1925–1997)
Johnson, P.	Johnson, Peter Neville (1946–)
Johnston, I.M.	Johnston, Ivan Murray (1898–1960)
Johnston, M.	Johnston, Marshall Conring (1930–)
Johow	Johow, Friedrich (or Federico) Richard Adelbert (1859–1933)
Jonch.	Joncheere, Gerardus J. de (1909–1989)
Jones	Jones, William (1746–1794)
Jones, H.	Jones, Henry G. (fl. 1961)
Jones, M.E.	Jones, Marcus Eugene (1852–1934)
Jonker	Jonker, Fredrik Pieter (1912–1995)
Jonker, A.	Jonker, (Verhoef) Anni Margriet Emma (1920–)
Jordan	Jordan, Claude Thomas Alexis (1814–1897)
Joseph	Joseph, J. (fl. 1964–1979)
Jovet	Jovet, Paul Albert (1896–1991)
Jowitt	Jowitt, John F.
Judz.	Judziewicz, Emmet J. (1953–)
Juel	Juel, Hans Oscar (1863–1931)
Jum.	Jumelle, Henri Lucien (1866–1935)
Jun.	Junell, Sven Albert Brynolt (1901–)
Junghuhn	Junghuhn, Franz Wilhelm (1809–1864)

Jung-Mendaçolli	Jung-Mendaçolli, Sigrid Luiza (1952–)
Jusl.	Juslenius, Abrahamus Danielis (1732–1803)
Juss.	Jussieu, Antoine Laurent de (1748–1836)
Juss., A.	Jussieu, Adrien Henri Laurent de (1797–1853)
Juz.	Juzepczuk, Sergei Vasilievich (1893–1959)
Källersjö	Källersjö, Mari (1954–)
Kalkman	Kalkman, Cornelis (1928–1998)
Kallunki	Kallunki, Jacquelyn A. (1948–)
Kam	Kam, Yee-Kiew (?–1981)
Kamelin	Kamelin, R.V. (1938–)
Kamm.	Kammathy, R.V. (1932–)
Kaneh.	Kanehira, Ryözö (1882–1948)
Kanis	Kanis, Andrew (1934–1986)
Kanitz	Kanitz, August (1843–1896)
Kao, P.C.	Kao, Pao-Chun
Kappert	Kappert, Hans (1890–)
Karav.	Karavaev, Mikhail Nikolaevich (1903–)
Karelin	Karelin, Grigorij Silyč (1801–1872)
Karis	Karis, Per Ola (1955–)
Karsten	Karsten, Gustav Karl Wilhelm Hermann (1817–1908)
Karw.	Karwinsky von Karwin, Wilhelm Friedrich von (1780–1855)
Kasapl.	Kasapligil, Baki (1918–)
Kato	Kato, Masahiro (1946–)
Kaulf.	Kaulfuss, Georg Friedrich (1786–1830)
Kausel	Kausel, Eberhard Max Leopold (1910–1972)
Kazmi	Kazmi, Syed Muhammad Anwar (1926–)
Kearney	Kearney, Thomas Henry (1874–1956)
Keay	Keay, Ronald William John (1920–1998)
Keck	Keck, David Daniels (1903–1995)
Keigh.	Keighery, Gregory John (1950–)
Kellerman	Kellerman, Maude (1888–)
Kellogg	Kellogg, Albert (1813–1887)
Kem.-Nat	Kemularia-Natadze, Liubov Manucharovna (1891–1985)
Keng	Keng, Yi-Li (1897–1975)
Keng, H.	Keng, Hsuan (1923–)
Keng f.	Keng, Pai-Chieh (previously Keng, Kwan-Hou) (1917–)
Kenn., G.	Kennedy, George C. (fl. 1976)
Kenn., H.	Kennedy, Helen Alberta (1944–)
Kenn.-O'Byrne	Kennedy-O'Byrne, John Kevin Patrick (1927–)
Kensit	Kensit, Harriet Margaret Louisa (later Bolus) (1877–1970)
Kent	Kent, Douglas Henry (1920–1998)
Keraudren	Keraudren, Monique (1928–1981)
Ker(-Gawler)	Ker-Gawler, John Bellenden (also J. Gawler) (1764–1842)
Kern	Kern, Johannes Hendrikus (1903–1974)
Kerner	Kerner, Johann Simon von (1755–1830)
Kerr	Kerr, Arthur Francis George (1877–1942)
Khokhr.	Khokhrjakov, Michael Kuzmich (1905–)
Kiaerskov	Kiaerskov, Hjalmar Frederik Christian (1835–1900)
Kiew	Kiew, Ruth (1946–)
Killeen	Killeen, Timothy
Killip	Killip, Ellsworth Paine (1890–1968)
Kimnach	Kimnach, Myron William (1922–)
Kimura	Kimura, Arika (1900–)
King	King, George (1840–1909)
King, R.	King, Robert Merrill (1930–)
Kingdon-Ward	Kingdon-Ward, Francis (or Ward, Frank Kingdon) (1885–1958)
Kinzik.	Kinzikaëva, G.K. (1931–)
Kipp.	Kippist, Richard (1812–1882)
Kir.	Kirilov, Ivan Petrovich (1821–1842)

Kirchner	Kirchner, Georg (1837–1885)
Kirk	Kirk, Thomas (1828–1898)
Kirk, J.	Kirk, John (1832–1922)
Kirpiczn.	Kirpicznikov, Moisey Elevich (1913–)
Kit.	Kitaibel, Paul (1757–1817)
Kit Tan	Kit Tan (1953–)
Kitagawa	Kitagawa, Masao (1909–)
Kitam.	Kitamura, Siro (1906–)
Kjellm.	Kjellman, Frans Reinhold (1846–1907)
Klack.	Klackenberg, Jens (1951–)
Klad.	Kladiwa, Leo (1920–)
Klatt	Klatt, Friedrich Wilhelm (1825–1897)
Klein, E.	Klein, Erich (fl. 1989)
Klett	Klett, Gustav Theodor (–1827)
Kljuykov	Kljuykov, E.V. (1950–)
Klotzsch	Klotzsch, Johann Friedrich (1805–1860)
Knight, J.	Knight, Joseph (1777?–1855)
Knobloch	Knobloch, Irving William (1907–)
Knoche	Knoche, Edward Louis Herman (1870–1945)
Knorr.	Knorring, O.E. (later Knorring-Neustrvjeva) (1896–1979)
Knowles	Knowles, George Beauchamp (fl. 1829–52)
Knuth, (F.)	Knuth, Frederik Marcus (1904–1970)
Kobuski	Kobuski, Clarence Emmeren (1900–1963)
Koch	Koch, Wilhelm Daniel Joseph (1771–1849)
Koch, K.	Koch, Karl Heinrich Emil Ludwig (1809–1879)
Koehne	Koehne, Bernhard Adalbert Emil (1848–1918)
Koeler	Koeler, Georg Ludwig (1765–1807)
Koelle	Koelle, Johann Ludwig Christian (1763–1797)
Koenig, (J.G.)	König (Koenig), Johann Gerhard (1728–1785)
Koenig, C.	Koenig, Carl Dietrich Eberhard (1774–1851)
Koern.	Koernicke (Körnicke), Friedrich August (1828–1908)
Koerte	Koerte, Franz Friedrich Ernst (1782–1845)
Kofoid	Kofoid, Charles Atwood (1865–1947)
Koidz.	Koidzumi, Gen'ichi (1883–1953)
Kokwaro	Kokwaro, John Ongayo (1940–)
Komarov	Komarov, Nikolai Fedrovič (1901–1942)
Koord.	Koorders, Sijfert Hendrik (1863–1919)
Korovin	Korovin, Eugenii (or Yevgeni) Petrovich (1891–1963)
Korsh.	Korshinsky (Korzinskij), Sergei Ivanovitsch (1861–1900)
Korth.	Korthals, Pieter Willem (1807–1892)
Kostel.	Kosteletzky, Vincenz Franz (1801–1887)
Koster, J.	Koster, Joséphine Thérèse (1902–1986)
Kosterm.	Kostermans, André Joseph Guillaume Henri (1907–1994)
Kotschy	Kotschy, Karl Georg Theodor (1813–1866)
Kovalevsk.	Kovalevskaja, S.S. (1929–)
Koyama, T.	Koyama, Tetsuo Michael (1935–)
Kozo-Polj.	Kozo-Poljansky, Boris Mikhailovic (1890–1957)
Kraenzlin	Kränzlin, Friedrich Wilhelm Ludwig (1847–1934)
Král	Král, Miloš (1932–)
Kramer	Kramer, Karl Ulrich (1928–1994)
Krapov.	Krapovickas, Antonio (1921–)
Kraschen. H.	Krascheninnikov, Ippolit Mikhailovich (1884–1947)
Krasser	Krasser, Fridolin (1863–1923)
Krause, E.	Krause, Ernst (?–1858)
Krause, K.	Krause, Kurt (1883–1963)
Krauss	Krauss, Otto (?–1935)
Krecz.	Kreczetovicz, Lev Melkhisedekovich (1878–?)
Kress	Kress, Walter John (1951–)
Krug	Krug, Carl Wilhelm Leopold (1833–1898)
Krukoff	Krukoff, Boris Alexander (1898–1983)
Ku	Ku, Tsue-Chih (1931–)
Kuang	Kuang, Ko-Rjên (fl. 1941)

Kuber	Kuber, G. (fl. 1964)
Kudô	Kudô, Yûshun (1887–1932)
Kudr.	Kudrjaschev, S.N. (1907–1943)
Kükenthal	Kükenthal, Georg (1864–1955)
Kündig	Küundig, Jakob (1863–1933)
Kütz.	Kützing, Friedrich Traugott (1807–1893)
Kuhl	Kuhl, Heinrich (1796–1821)
Kuhlm.	Kuhlmann, Joâo Geraldo (1882–1958)
Kuhn	Kuhn, Maximilian Friedrich Adalbert (1842–1894)
Kuijt	Kuijt, Job (1930–)
Kung, H.W.	Kung, Hsien-Wu (1897–)
Kunkel	Kunkel, Günther W.H. (1928–)
Kunth	Kunth, Karl Sigismund (1788–1850)
Kuntze	Kuntze, Carl Ernst (sometimes Eduard) Otto (1843–1907)
Kunze	Kunze, Gustav (1793–1851)
Kupicha	Kupicha, Frances Kristina (1947–)
Kupper	Kupper, Walter (1874–1953)
Kuprian.	Kuprianova, Lyndmila Andreyevna (1914–1987)
Kurz	Kurz, Wilhelm Sulpiz (1834–1878)
Kuschel	Kuschel, G. (fl. 1963)
Kuzn.	Kuznetsov, Nikolai Ivanovich (1864–1932)
L.	Linnaeus (von Linné), Carl (1707–1778)
L.f.	Linnaeus (von Linné), Carl (1741–1783)
Laan, van der	Laan, F.M. van der (fl. 1986)
Lab(ill).	Labillardière, Jacques Julien Houttou de (1755–1834)
Labouret	Labouret, J. (fl. 1853–8)
Lack	Lack, Hans Walter (1949–)
Lackey	Lackey, James A. (fl. 1978)
Ladiz.	Ladizinsky, Gideon (1936–)
LaFrankie	LaFrankie, James V. (fl. 1986)
Lagasca	Lagasca y Segura, Mariano (1776–1839)
Lagerh.	Lagerheim, Nils Gustaf (1860–1926)
Lagr.-Fossat	Lagrèze-Fossat, Adrian Rose Arnaud (1818–1874)
Lam.	Lamarck, Jean Baptiste Antoine Pierre de Monnet de (1744–1829)
Lam, H.J.	Lam, Herman Johannes (1892–1977)
Lamb.	Lambert, Aylmer Bourke (1761–1842)
Lambinon	Lambinon, Jacques (1936–)
Lamond	Lamond, Jenifer M. (1936–)
Lander	Lander, Nicholas Sean (1948–)
Lang, K.Y.	Lang, Kai-Yung (1936–)
Lange	Lange, Johan Martin Christian (1818–1898)
Lanj.	Lanjouw, Joseph (1902–1984)
Lapeyr.	Lapeyrouse, Philippe Picot de (1744–1818)
Large	Large, Mark Frederick (1959–)
Larréat.	Larréategui, José Dionisio (fl. 1795–c. 1805)
Larsen	Larsen, Kai (1926–)
Lassen	Lassen, Per (1942–)
Laubenf.	Laubenfels, David John de (1925–)
Launert	Launert, Georg Oskar Edmund (1926–)
Lauterb.	Lauterbach, Carl Adolf Georg (1864–1937)
Lavallée	Lavallée, Pierre Alphonse Martin (1836–1884)
Lavin	Lavin, Matt (1956–)
Lavranos	Lavranos, John Jacob (1926–)
Law	Law, Yuh-Wu (1917–)
Lawalrée	Lawalrée, André Gilles Célestin (1921–)
Lawson	Lawson, Charles (1794–1873)
Lawson, C.	Lawson, Cheryl A. (1947–)
Lawson, P.	Lawson, Peter (fl. 1770s–1820)
Laxm.	Laxmann, Erik G. (1737–1796)
Layens	Layens, Georges (1834–1897)

Lazarides	Lazarides, Michael (1928–)
Leach	Leach, Leslie Charles (1909–1996)
Leakey	Leakey, D.G.B. (fl. 1932)
Leal	Leal, Carlos G. (fl. 1951)
Leandri	Leandri, Jacques Désiré (1903–1982)
Leandro	Leandro do Sacromento, P. (1778–1829)
Lebrun	Lebrun, Jean-Paul Antoine (1906–1985)
Lecomte	Lecomte, Paul Henri (1856–1934)
Ledeb.	Ledebour, Carl Friedrich von (1785–1851)
Ledoux	Ledoux, E.P. (1898–)
Lee, A.T.	Lee, Alma Theodora (1912–1990)
Lee, S.K.	Lee, Shu-Kang (1915–)
Leeke	Leeke, Georg Gustav Paul (1883–1933)
Leenh.	Leenhouts, Pieter Willem (1926–)
Leeuwenb.	Leeuwenberg, Anthonius Josephus Maria (1930–)
Lefor	Lefor, Michael William (fl. 1975)
Legrand	Legrand, Carlos Maria Diego Enrique (1901–)
Lehm.	Lehmann, Johann Georg Christian (1792–1860)
Lehm. F.	Lehmann, Friedrich Carl (1850–1903)
Leichtlin	Leichtlin, Maximilian (1831–1910)
Leighton, F.M.	Leighton, Frances Margaret (later Mrs William Edwin Isaac) (1909–)
Leistner	Leistner, Otto Albrecht (1931–)
Lejoly	Lejoly, Jean (1945–)
Lellinger	Lellinger, David Bruce (1937–)
Lemaire	Lemaire, Antoine Charles (1801–1871)
Lemmon	Lemmon, John Gill (1832–1908)
Lemoine	Lemoine, Pierre Louis Victor (1823–1911)
Lenné	Lenné, Peter Joseph (1789–1866)
Lenz, L.	Lenz, Wayne Lee (1915–)
Léon	Léon, Jules (fl. 1949)
Leonard	Leonard, Emery Clarence (1892–1968)
Léonard	Léonard, Jean Joseph Gustave (1920–)
Lepechin	Lepechin, Ivan Ivanovich (1737–1802)
Lepr.	Leprieur, M.F.R. (1799–1869)
Leroy, J.	Leroy, Jean-François (1915–1999)
Leschen.	Leschenault de la Tour, Jean Baptiste Louis Claude Théodore (1773–1826)
Less.	Lessing, Christian Friedrich (1809–1862)
Lestib.(oudois)	Lestiboudois, Thémistocle Gaspard (1797–1876)
Letouzey	Letouzey, Réné (1918–1989)
Leute	Leute, Gerfried Horand (1941–)
Léveillé	Léveillé, Joseph-Henri (1796–1870)
Léveillé, A.	Léveillé, Augustin(e) Abel Hector (1863–1918)
Levier	Levier, Emilio (1839–1911)
Levin	Levin, Ernst Ivar (1868–)
Levyns	Levyns, Margaret Rutherford Bryan (née Mitchell) (1890–1975)
Lewis (, G.)	Lewis, Gwendoline Joyce (1909–1967)
Lewis, J.	Lewis, John (1921–)
Lewis, W.	Lewis, Walter Hepworth (1930–)
Lewton	Lewton, Frederick Lewis (1874–1959)
Lex.	Lexarza, Juan Joseé Martinez de (1785–1824)
Leyb.	Leybold, Friedrich (1827–1879)
Leysser	Leysser (or Leyser), Friedrich Wilhelm von (1731–1815)
L'Hér(it).	L'Héritier (de Brutelle), Charles Louis (1746–1800)
Li	Li, Hsi-Wen (1902–)
Li, A.J.	Li, An-Jen (fl. 1981)
Li, C.L.	Li, Chao-Luan(g) (1938–)
Li, H.L.	Li, Hui-Lin (1911–)
Li, P.T.	Li, Ping-Tao (1936–)
Li, Z.Y.	Li, Zheng-Yu (fl. 1987)

Liais	Liais, E.
Liben	Liben, Louis (1926–)
Libosch.	Liboschitz, Joseph (1783–1824)
Lichtenst.	Lichtenstein, Martin Heinrich Karl von (1780–1857)
Liebl.	Lieblein, Franz Caspar (1744–1810)
Liebm.	Liebmann, Frederick Michael (1813–1856)
Liede	Liede, Sigrid (1957–)
Lilja	Lilja, Nils (1808–1870)
Lillo	Lillo, Miguel (1862–1931)
Lima	Lima, Haroldo Cavalcante de (1955–)
Lincz.	Linczevski, Igor Alexandrovich (1908–1997)
Lindau	Lindau, Gustav (1866–1923)
Lindb.	Lindberg, Sextus Otto (1835–1889)
Linden	Linden, Jean Jules (1817–1898)
Linden, L.	Linden, Lucien (1851–1940)
Linden, van der	van der Linden, B.L. (fl. 1959)
Linder	Linder, Hans Peter (1954–)
Lindinger	Lindinger, Karl Hermann Leonhard (1879–?)
Lindley	Lindley, John (1799–1865)
Lindman	Lindman, Carl Axel Magnus (1856–1928)
Lindsay, G.	Lindsay, George Edmund (1916–)
Ling	Ling, Yong-Yuan (1903–1981)
Ling, Y.R.	Ling, Yeou-Ruenn (1937–)
Link	Link, Johann Heinrich Friedrich (1767–1851)
Lipsch.	Lipschitz, Sergej Julievitsch (1905–1983)
Lipsky	Lipsky, Vladimir Ippolitovich (also Lipskij) (1863–1937)
Lisowski	Lisowski, Stanislaw (1924–)
Little	Little, Elbert Luther (Jr.) (1907–)
Litv.	Litvinov, Dmitrij Ivanovitsch (1854–1929)
Liu	Liu, Tung-Shui (1911–)
Liu, H.	Liu, Hou (fl. 1932)
Liu, S.W.	Liu, Sang-Wu (1934–)
Liv.	Livera, E.J.
Llave	Llave, Pablo de la (1773–1833)
Lodd.	Loddiges, Conrad (1738–1826)
Loefgren	Loefgren (Löfgren), Johan Alberto Constantin (1854–1918)
Loefl.	Loefling, Pehr (1729–1756)
Loes.	Loesener, Ludwig Eduard Theodor (1865–1941)
Löve	Löve, Askell (1916–1994)
Löve, D.	Löve, Doris Benta Maria (née Wahlen) (1918–)
Loher	Loher, August (1874–1930)
Lois.	Loiseleur–Deslongchamps, Jean Louis August(e) (1774–1849)
Lojac.	Lojacono-Pojero, Michele (1853–1919)
Londoño	Londoño, Ximena (fl. 1987)
Long	Long, Bayard Henry (1885–1969?)
Long, G.R.	Long, Guang-Ri
Longpre	Longpre, Edwin Keith (fl. 1970)
Looser	Looser, Gualterio (1898–1982)
López	López González, Ginés Alejandro (1950–)
Lopr.	Lopriore, Guiseppe (1865–1928)
Lorence	Lorence, David H. (1946–)
Lorentz	Lorentz, Paul Günther (1835–1881)
Loret	Loret, Henri (1811–1888)
Loscos	Loscos y Bernál, Francisco (1823–1886)
Losink.	Losinkaja, A.S. Losina (1903–1958)
Lotsy	Lotsy, Johannes Paulus (1867–1931)
Lott	Lott, Emily J. (fl. 1982)
Loud.(on)	Loudon, John Claudius (1783–1843)
Loudon, J.W.	Loudon, Jane Webb (?Wells) (1807–1858)
Louis	Louis, Jean Laurent Prosper (1903–1947)
Lour.	Loureiro, Joâo de (1717–1791)

Lourteig	Lourteig, Alicia (1913–)
Lowe	Lowe, Richard Thomas (1802–1874)
Lowry	Lowry, Porter Peter (1956–)
Lucas Rodr.	Lucas Rodríguez, Rafael (1915–1981)
Lückh.	Lückhoff, Carl August (1914–1960)
Ludwig	Ludwig, Christian Gottlieb (1709–1773)
Ludwig, W.	Ludwig, Wolfgang (1923–)
Lückel	Lückel, Emil (fl. 1978)
Luer	Luer, Carlyle A. (1922–)
Luerssen	Luerssen, Christian (1843–1916)
Lundell	Lundell, Cyrus Longworth (1903–1994)
Luteyn	Luteyn, James Leonard (1948–)
Lye	Lye, Kåre Arnstein (1940–)
Ma	Ma, Yu-Chuan (1916–)
Maack	Maack, Richard Karlovich (1825–1886)
Maas	Maas, Paulus Johannes Maria (1939–)
Mabb.	Mabberley, David John (1948–)
McAlpin	McAlpin, Bruce (fl. 1986)
Macarthur	Macarthur, William (1800–1882)
Macbr.	Macbride, James (1784–1817)
Macbr., J.F.	Macbride, James Francis (1892–1976)
McClint.	McClintock, Elizabeth May (1912–)
McClint., D.	McClintock, David Charles (1913–)
McClure	McClure, Floyd Alonzo (1897–1970)
McDaniel	McDaniel, Sidney T. (1940–)
McDonald	McDonald, J. Andrew (fl. 1989)
MacDoug.	MacDougal, John (1954–)
Macfad.	Macfadyen, James (1798–1850)
Macfarl., T.	Macfarlane, Terry D. (1953–)
McGillivray	McGillivray, Donald John (1935–)
McKen	McKen, Mark Johnston (1823–1872)
Mackenzie	Mackenzie, Kenneth Kent (1877–1934)
MacLeish	MacLeish, Nanda F. Fleming (1953–)
McNeill	McNeill, John (1933–)
McPherson	McPherson, Gordon (1947–)
McVaugh	McVaugh, Rogers (1909–)
Madison	Madison, Michael (fl. 1977)
Maek., F.	Maekawa, Fumio (1908–1984)
Maekawa	Maekawa, Tokujirô (1886–)
Maguire	Maguire, Bassett (1904–1991)
Maiden	Maiden, Joseph Henry (1859–1925)
Maingay	Maingay, Alexander Carroll (1836–1869)
Maire	Maire, René Charles Joseph Ernest (1878–1949)
Majumdar	Majumdar, R.B. (fl. 1971)
Makino	Makino, Tomitarô (1862–1957)
Maleki	Maleki, Zeynol-Abedin (1913–)
Malinv.	Malinvaud, Louis Jules Ernst (1836–1913)
Malme	Malme, Gustaf Oskar Andersson (1864–1937)
Manden.	Mandenova, Ida P. (1907–1995)
Mani	Mani, K.J. (fl. 1985)
Mann, G.	Mann, Gustav (1836–1916)
Mann, H.	Mann, Horace (1844–1868)
Mansf.	Mansfeld, Rudolf (1901–1960)
Marais	Marais, Wessel (1929–)
Maratti	Maratti, Giovanni Francesco (1723–1777)
Marcano-Berti	Marcano-Berti, Luis (fl. 1967)
Marchal	Marchal, Elie (1839–1923)
Marchand	Marchand, Nestor Léon (1833–1911)
Marchant	Marchant, William James (1886–1952)
Maréchal	Maréchal, Robert Joseph Jean-Marie (1926–)
Margot	Margot, Henri (1807–1894)

Mariz	Mariz, Joaquim de (1847–1916)
Markgraf	Markgraf, Friedrich (1897–1987)
Marloth	Marloth, Hermann Wilhelm Rudolf (1855–1931)
Marquand	Marquand, Cecil Victor Boley (1897–1943)
Marsh, J.	Marsh, Judith Anne (1951–)
Marshall	Marshall, Humphry (1722–1801)
Marsili	Marsili, Giovanni M. (1727–1794)
Mart. Crov.	Martinez Crovetto, Raul (1921–1988)
Martelli	Martelli, Ugolino (1860–1934)
Martens	Martens, Georg Matthias von (1788–1872)
Martens, M.	Martens, Martin (1797–1863)
Martic.	Marticorena, Clodomiro (1929–)
Martínez	Martínez, Maximino (1888–1964)
Martínez G.	Martínez Garcia, Julieta (fl. 1989)
Martinov	Martinov, Ivan Ivanovič (fl. 1826)
Martius, C. or Mart.	Martius, Carl Friedrich Phillip von (1794–1868)
Martyn, T.	Martyn, Thomas M. (1736–1825)
Masam.	Masamune, Genkei (1899–)
Masf.	Masferrer y Arquimbau, Ramón (1850–1884)
Mason, C.	Mason, Charles Thomas (1918–)
Masson	Masson, Francis (1741–1805)
Masters	Masters, Maxwell Tylden (1833–1907)
Mathew, B.	Mathew, Brian Frederick (1936–)
Mathez	Mathez, Joël (fl. 1969)
Mathias	Mathias, Mildred Esther (1906–1995)
Mathur	Mathur, R. (1948–)
Maton	Maton, William George (1774–1835)
Matsum.	Matsumura, Jinzô (1856–1928)
Mattei	Mattei, Giovanni Ettore (1865–1943)
Mattf.	Mattfeld, Johannes (1895–1951)
Mattos	Mattos, Joâo Rodrigues de (1926–)
Mattos, A.	Mattos, Filho, Armando de
Mattuschka	Mattuschka, Heinrich Gottfried von (1734–1779)
Matuda	Matuda, Eizi (1894–1978)
Maxim.	Maximowicz, Carl Johann (1827–1891)
Maxon	Maxon, William Ralph (1877–1948)
Mazel	Mazel, Eugène (fl. 1981)
Mazzucc.	Mazzuccato, Giovanni (1787–1814)
Medikus	Medikus, Friedrich Casimir (1736–1808)
Medrono	Medrono, Francisco González (1939–)
Medw.	Medwedew, Jakob Sergejevitsch (1847–1923)
Meerb.	Meerburgh, Nicolaas (1734–1814)
Meeuse	Meeuse, Adrianus Dirk Jacob (1914–)
Meeuwen van	Meeuwen, M.S. Knaap van (1936–)
Meijden	Meijden, R. van der (1945–)
Meijer	Meijer, Willem (1923–)
Meikle	Meikle, Robert Desmond (1923–)
Meissner	Meissner, Carl Daniel Friedrich (né Meisner) (1800–1874)
Melchior	Melchior, Hans (1894–1984)
Méllo	Méllo, Joaquim Correia de (1816–1877)
Melville	Melville, Ronald (1903–1985)
Mendonça	Mendonça, Francisco de Ascençâo (1889–1982)
Menezes, N.	Menezes, Nanuza Luiza de (1934–)
Merc.	Mercier, Philippe (1781–1831)
Merr.	Merrill, Elmer Drew (1876–1956)
Merwe, J.	Merwe, Jacoba Johanna Maria van der (1946–)
Merxm.	Merxmüller, Hermann (1920–1988)
Mett.	Mettenius, Georg Heinrich (1823–1866)
Meyen	Meyen, Franz Julius Ferdinand (1804–1840)
Meyer	Meyer, Bernhard (1767–1836)
Meyer, C.	Meyer, Carol Anton Andreevič von (1795–1855)
Meyer, E.	Meyer, Ernst Heinrich Friedrich (1791–1858)

Meyer, F.	Meyer, Frederick Gustav (1917–)
Meyer, G.	Meyer, Georg Friedrich Wilhelm (1782–1856)
Meyer, G.L.	Meyer, G.L. (fl. 1881)
Meyer, T.	Meyer, Teodore (1910–)
Mez	Mez, Carl Christian (1866–1944)
Miau	Miau, Ru-Huai (1943–)
Michaux	Michaux, André (1746–1803)
Michaux f.	Michaux, François André (1770–1855)
Micheli, M.	Micheli, Marc (1844–1902)
Mickel	Mickel, John Thomas (1934–)
Middleton	Middleton, David John (1963–)
Miégev.	Miégeville, Joseph (1819–1901)
Mielcarek	Mielcarek, R. (fl. 1982)
Miers	Miers, John (1789–1879)
Mikan	Mikan, Josef Gottfried (1743–1814)
Miki	Miki, Shigeru (1901–1974)
Milano	Milano, V.A. (1921–)
Mildbr.	Mildbraed, Gottfried Wilhelm Johannes (1879–1954)
Milde	Milde, Carl August Julius (1824–1871)
Millán	Millán, Aníbal Roberto (1892–)
Miller	Miller, Philip (1691–1771)
Miller, A.	Miller, A.G. (1951–)
Miller, J.F.	Miller, John Frederick (1715–1794)
Miller, J.S.	Miller, J.S. (fl. 1760–1790)
Millsp.	Millspaugh, Charles Frederick (1854–1923)
Milne	Milne, Colin (c. 1743–1815)
Milne-Redh.	Milne-Redhead, Edgar Wolston Bertram Handsley (1906–1996)
Mimeur	Mimeur, Geneviève
Minkw.	Minkwitz, Zenaida Alexandrovna (1878–1918/19)
Minod	Minod, Marcel Maurice (1887–1939)
Miq.	Miquel, Friedrich Anton Wilhelm (1811–1871)
Miranda	Miranda González, Faustino (1905–1964)
Mirbel	Mirbel, Charles François Brisseau de (1776–1854)
Mitch.	Mitchell, John (1711–1768)
Mitch., D.	Mitchell, David Searle (1935–)
Mitchell, T.	Mitchell, T.L. (fl. 1927)
Mitford	Mitford, Algernon Bertram Freeman (1837–1916)
Miyabe	Miyabe, Kingo (1860–1951)
Mizg.	Mizgireva, O.F. (1908–)
Mob.	Mobayen, Sadegh (1919–)
Moçiño	Moçiño, José Mariano (1757–1820)
Möller, H.	Möller, Hjalmar August (1866–1941)
Moench	Moench, Conrad (1744–1805)
Moeser	Moeser, Walter
Mohr	Mohr, Daniel Matthias Heinrich (1780–1808)
Molau	Molau, Ulf (1949–)
Mold.	Moldenke, Harold Norman (1909–)
Molina	Molina, Giovanni Ignazio (1737–1829)
Molina, A.	Molina, Ana María (1947–)
Molseed	Molseed, Elwood Wendell (1938–1967)
Momose	Momose, Sizuo (1906–1968)
Monach.	Monachino, Joseph Vincent (1911–1962)
Moncada	Moncada Ferrera, Milagros (1937–)
Montr.	Montrouzier, Xavier (1820–1897)
Monv.	Monville, de (fl. 1838–40)
Moon	Moon, Alexander (?–1825)
Moore, C.	Moore, Charles (1820–1905)
Moore, D.M.	Moore, David Moresby (1933–)
Moore, H.	Moore, Harold Emery (1917–1980)
Moore, J.	Moore, John William (1901–)
Moore, L.	Moore, Lucy Beatrice (1906–1987)
Moore, S.	Moore, Spencer le Marchant (1850–1931)

Moore, T.	Moore, Thomas (1821–1887)
Moq.	Moquin-Tandon, Christian Horace Bénédict Alfred (1804–1863)
Morales	Morales, Sebastiàn Alfredo de (1823–1900)
Morales Torres	Morales Torres, Concepción (1944–)
Moran	Moran, Reid Venable (1916–)
Moran, R.C.	Moran, Robbin C. (fl. 1986)
Morandi	Morandi, Giambattista (fl. 1740s)
Morat	Morat, Philippe (1937–)
Morden	Morden, Clifford W. (fl. 1986)
Morelet	Morelet, Pierre Marie Arthur (1809–1892)
Morgan	Morgan, David R. (fl. 1990)
Moris	Moris, Giuseppe Giacinto (1796–1869)
Moritz	Moritz, Otto (1904–)
Moritzi	Moritzi, Alexander (1807–1850)
Morong	Morong, Thomas (1827–1894)
Morren	Morren, Charles François Antoine (1807–1858)
Morren, C.J.	Morren, Charles Jacques Edouard (1833–1886)
Morrone	Morrone, Osvaldo (1957–)
Morton	Morton, Julius Sterling (1832–1902)
Morton, C.	Morton, Conrad Vernon (1905–1972)
Morton, J.K.	Morton, John Kenneth (1928–)
Moss	Moss, Charles Edward (1870–1930)
Mottet	Mottet, Séraphin Joseph (1861–1930)
Moyaben	Moyaben, Sadegh (1919–)
Muell., C.	Müller, Carl Alfred (1855–)
Muell., F.	Mueller, Ferdinand Jacob Heinrich von (1825–1896)
Muell. Arg. (or J. Mueller)	Mueller, Jean (Müller) of Aargau (1828–1896)
Münchh.	Münchhausen, Otto von (1716–1744)
Muhlenb.	Muhlenberg, Gotthilf Heinrich Ernest (1753–1815)
Mukherjee, P.K.	Mukherjee, Pronob Kumar (1934–)
Muldashev	Muldashev, A.A. (1954–)
Munro	Munro, William (1818–1889)
Munster	Munster, R. (fl. 1990)
Munz	Munz, Philip Alexander (1892–1974)
Murb.	Murbeck, Svante Samuel (1859–1946)
Murillo	Murillo, Adolfo (1840–1899)
Muroi	Muroi, Hiroshi (1914–)
Murr., A.	Murray, Andrew (c. 1805–1850)
Murray	Murray, Johan Andreas (1740–1791)
Muschler	Muschler, Reinhold (Reno) (1883–1957)
Mutis	Mutis, José Celestino (1732–1808)
Mutis, S.	Mutis, Sinforoso (also known as Sinforoso Mutis y Conswegra) (1773–1822)
Náb.	Nábělek, František (1884–1965)
Nabiev	Nabiev, M.M. (1926–)
Nad.	Nadeaud, Jean (1834–1898)
Nair	Nair, V.J. (1940–)
Nakai	Nakai, Takenoshin (1882–1952)
Nakaike	Nakaike, Toshiyuki (1943–)
Nannenga-Bremek.	Nannenga-Bremekamp, Neeltje Elizabeth (1916–1996)
Napper	Napper, Diana Margaret (1930–1972)
Naray.	Narayanaswami, V. (fl. 1949)
Nard.	Nardina, N.S. (fl. 1965)
Nash	Nash, George Valentine (1864–1921)
Naudin	Naudin, Charles Victor (1815–1899)
Nayar	Nayar, Madhavan Parameswaran (1932–)
Naz.(arova)	Nazarova, Estella A. (1936–)
Necker	Necker, Noël Martin Joseph de (1730–1793)
Née	Née, Louis (fl. 1789–94)
Nees	Nees von Esenbeck, Christian Gottfried Daniel (1776–1858)

Neger	Neger, Franz Wilhelm (1868–1923)
Neilr.	Neilreich, August (1803–1871)
Nel	Nel, Gert Cornelius (1885–1950)
Nelmes	Nelmes, Ernest (1895–1959)
Nelson	Nelson, Aven (1859–1952)
Nelson, C.	Nelson, Cirilo H. (1938–)
Nestler	Nestler, Christian Gottfried (1778–1832)
Neto	Guarim Neto, Germano (1950–)
Nevski	Nevski, Sergei Arsenjevic (1908–1938)
Newman	Newman, Edward (1801–1876)
Nicholas	Nicholas, Ashley (1954–)
Nicholls	Nicholls, William Henry (1885–1951)
Nicholson	Nicholson, George (1847–1908)
Nicolson	Nicolson, Dan Henry (1933–)
Nicora	Nicora de Panza, Elisa G. (1912–)
Niedenzu	Niedenzu, Franz Josef (1857–1937)
Nielsen	Nielsen, Etlar Lester (1905–)
Nielsen, I.	Nielsen, Ivan Christian (1946–)
Nieuw.	Nieuwland, Julius Arthur (1878–1936)
Niezgoda	Niezgoda, Christine J. (1950–)
Nilsson, O.	Nilsson, Orjan Eric Gustaf (1933–)
Nimmo	Nimmo, Joseph (fl. 1830s–1854)
Nixon	Nixon, Kevin C. (1953–)
Noot(eb).	Nooteboom, Hans Peter (1934–)
Nordal	Nordal, Inger
Nordenstam, R.	Nordenstam, Rune Bertil (1936–)
Norlindh	Norlindh, Nils Tycho (1906–)
Norman	Norman, Cecil (1872–1947)
Normand	Normand, Didier (1908–)
Noronha	Noronha, Francisco (c. 1748–1787)
Norton	Norton, John Bitting Smith (1872–1966)
Not(aris), de	Notaris, Giuseppe de (1805–1877)
Novák	Novák, František Antonín (1892–1964)
Novopokr.	Novopokrovsky, Ivan Vassiljevich (1880–1951)
Nowicke	Nowicke, Joan W. (1938–)
Nutt.	Nuttall, Thomas (1786–1859)
Nyár.	Nyárády, Erasmus Julius (1881–1966)
Nyman	Nyman, Carl Fredrik (1820–1893)
Oberm.	Obermejer, Anna Amelia (Obermeyer) (1907–)
Oberprieler	Oberprieler, Christoph
O'Brien	O'Brien, James (1842–1930)
Occh.	Occhioni, Paul (1915–)
O'Don.	O'Donnell, Carlos Alberto (1912–1954)
Oehme	Oehme, Hans (fl. 1940)
Oersted	Ørsted, Anders Sandøe (1816–1872)
Ohashi	Ohashi, Hiroyoshi (1936–)
Ohba, H.	Ohba, Hideaki (1943–)
Ohwi	Ohwi, Jisaburo (1905–1977)
Okamoto	Okamoto, Motoharu (1947–)
Oken	Oken, Lorenz (1779–1851)
Oliver	Oliver, Daniel (1830–1916)
Oliver, E.	Oliver, Edward George Hudson (1938–)
Oliver, F.	Oliver, Francis Wall (1864–1951)
Oliver, W.	Oliver, Walter Reginald Brook (1883–1957)
Olivier	Olivier, Guillaume Antoine (1756–1814)
Omer	Omer, Saood (1957–)
Oostr.	Oostroom, Simon Jan van (1906–1982)
Opiz	Opiz, Philipp Maximilian (1787–1858)
Orb.	d'Orbigny, Alcide Dessalines (1802–1857)
Orch.	Orchard, Anthony Edward (1946–)
Orph.	Orphanides, Theodoros Georgios (1817–1886)

Ortega	Ortega, Casimiro Gómez de (1740–1818)
Ortgies	Ortgies, Karl Eduard (1829–1916)
Ortíz	Ortíz, J. Javier (1957–)
Osbeck	Osbeck, Pehr (1723–1805)
Osp.	Ospina, H. Mariano (fl. 1973)
Ossa	Ossa, José Antonio de la (?–1829)
Ostenf.	Ostenfeld, Carl Emil Hansen (1873–1931)
Oteng-Yeboah	Oteng-Yeboah, A.A. (fl. 1970)
Otto	Otto, Christoph Friedrich (1783–1856)
Ovcz.	Ovczinnikov, Pavel Nikolaevich (1903–1979)
Overk.	Overkott, Ortrud (1914–)
Ownbey	Ownbey, Francis Marion (1910–1974)
Pabst	Pabst, Guido Frederico João (1914–1980)
Paclt	Paclt, Jiří (1925–)
Page	Page, Christopher Nigel (1942–)
Painter	Painter, Joseph Hannum (1879–1908)
Pakhomova	Pakhomova, M.G. (1925–)
Pal.	See Beauv., P.
Palau	Palau, Pedro (1881–1956)
Palib.	Palibin, Ivan Vladimirovich (1872–1949)
Palla	Palla, Eduard (1864–1922)
Pallas	Pallas, Peter Simon von (1741–1811)
Pampan.	Pampanini, Renato (1875–1949)
Pan, K.Y.	Pan, Kai-Yu (1937–)
Pancher	Pancher, Jean Armand Isidore (1814–1877)
Pang.	Pangalo, Konstantin Ivanovič (1883–1965)
Panigr.	Panigrahi, Gopinath (1924–)
Pantl.	Pantling, Robert (1856–1910)
Panzer	Panzer, Georg Wolfgang Franz (1755–1829)
Pappe	Pappe, Carl Wilhelm Ludwig (1803–1862)
Pardo	Pardo de Tavera, Trinidad Herménégilde José (1857–1925)
Parish	Parish, Samuel Bonsall (1838–1928)
Parker, R.	Parker, Richard Neville (1884–1958)
Parkinson	Parkinson, Sydney C. (1745–1771)
Parkinson, C.E.	Parkinson, Charles Edward (1890–1945)
Parl.	Parlatore, Filippo (1816–1877)
Parnell, J.	Parnell, John Adrian Naicker (1954–)
Parodi	Parodi, Domingo (1823–1890)
Parodi, L.	Parodi, Lorenzo Raimundo (1895–1966)
Parry	Parry, William Edward (1790–1855)
Parry, C.	Parry, Charles Christopher (1823–1890)
Parsa	Parsa, Ahmed (Ahmad) (1907–)
Pascher	Pascher, Adolf A. (1881–1945)
Paton, A.	Paton, Alan James (1963–)
Pau	Pau y Español, Carlos (1857–1937)
Paul	Paul, T.K. (fl. 1983)
Pauquy	Pauquy, Charles Louis Constant (1800–1854)
Pavlov	Pavlov, Nikolai Vasilievich (1893–1971)
Pavón	Pavón, José Antonio (1754–1844)
Pavone	Pavone, Petro (1948–)
Pax	Pax, Ferdinand Albin (1858–1942)
Paxton	Paxton, Joseph (1803–1865)
Payson	Payson, Edwin Blake (1893–1927)
Pearson	Pearson, William Henry (1849–1923)
Peck	Peck, Charles Horton (1833–1917)
Pedersen	Pedersen, Throels Myndel (1916–)
Pedro	Pedro, José Gomes (1915–)
Pellegrin	Pellegrin, François (1881–1965)
Pellet., J.	Pelletier, Pierre Joseph (1788–1842)
Pelt., M.	Peltier, M. (fl. 1965)
Penn.	Pennington, Terence Dale (1938–)

Pennell	Pennell, Francis Whittier (1886–1952)
Pépin	Pépin, Pierre Denis (c. 1802–1876)
Perkins	Perkins, John Russell (1868–)
Perleb	Perleb, Karl Julius (1794–1845)
Perrier	Perrier de la Bâthie, Eugène Pierre (1825–1916)
Perrottet	Perrottet, Georges Samuel (1793–1870)
Perry	Perry, Lily May (1895–1992)
Perry, P.	Perry, P.L. (fl. 1984)
Pers.	Persoon, Christiaan Hendrik (1761–1836)
Persson	Persson, Nathan Petter Herman (1893–1978)
Peter	Peter, Gustav Albert (1853–1937)
Petersen	Petersen, Otto George (1847–1937)
Petit	Petit, Felix (fl. 1824)
Petit, E.	Petit, Ernest M.A. (1927–)
Petrie	Petrie, Donald (1846–1925)
Petrovsky	Petrovsky, V.V. (1930–)
Peyr.	Peyritsch, Johann Joseph (1835–1889)
Pfeiffer	Pfeiffer, Louis (Ludwig) Karl Georg (1805–1877)
Pfeiffer, H.	Pfeiffer, Hans Heinrich (1890–)
Pfeiffer, N.	Pfeiffer, Norman Etta (1889–)
Pfitzer	Pfitzer, Ernst Hugo Heinrich (1846–1906)
Philcox	Philcox, David (1926–)
Philippi	Philippi, Rulolf Amandus (Rodolfo Amando) (1808–1904)
Philippi, F.	Philippi, Federico (Friedrich Heinrich Eunom) (1838–1910)
Philipson	Philipson, William Raymond (1911–)
Phillips, E.	Phillips, Edwin Percy (1884–1967)
Phillips, S.	Phillips, Sylvia Mabel (1945–)
Phipps, J.	Phipps, James Bird (1934–)
Pichi-Serm.	Pichi-Sermolli, Rodolfo Emilio Giuseppe (1912–)
Pichon	Pichon, Marcel (1921–1954)
Pickersgill	Pickersgill, Barbara (1940–)
Pieper	Pieper, Gustav Robert (fl. 1908)
Pierce	Pierce, John Hwett (1912–)
Pierre	Pierre, Jean Baptiste Louis (1833–1905)
Pilger	Pilger, Robert Knud Friedrich (1876–1953)
Pill.	Pillans, Neville Stuart (1884–1964)
Pim.	Pimenov, Michael Georgievich (1937–)
Piper	Piper, Charles Vancouver (1867–1926)
Pippen	Pippen, Richard W. (1935–)
Pires	Pires, Joâo Murça (1917–1994)
Pitard	Pitard, Charles-Joseph Marie (1873–1937)
Pittier	Pittier (de Fábrega), Henri François (1857–1950)
Planchon	Planchon, Jules Emile (1823–1888)
Plancke	Plancke, Jacqueline (1937–)
van der Plas	Plas, F. van der (fl. 1970)
Plitm.	Plitmann, Uzi (1936–)
Plowes	Plowes, Darrel C.H. (fl. 1986)
Pobed.	Pobedimova, Eugenia Georgievna (1898–1973)
Pócs	Pócs, Tamas (1933–)
Podlech	Podlech, Dieter (1931–)
Poelln.	Poellnitz, Karl von (1896–1945)
Poeppig	Poeppig, Eduard Friedrich (1798–1868)
Pogg.	Poggenburg, Justus Ferdinand (1840–1893)
Pohl	Pohl, Johann Baptist Emanuel (1782–1834)
Poiret	Poiret, Jean Louis Marie (1755–1834)
Poisson	Poisson, Jules (1833–1919)
Poit.	Poiteau, Pierre Antoine (1766–1854)
Poitr.	Poitrasson (fl. 1873–8)
Pojark.	Pojarkova, Antonina Ivanovna (1897–1980)
Polatschek	Polatschek, Adolf (1932–)
Polj.	Poljakov, Petr Petrovich (1902–1974)

Pollard	Pollard, Charles Louis (1872–1945)
Pollard, G.	Pollard, Glenn E. (1901–1976)
Pomel	Pomel, Auguste Nicolas (1821–1898)
Pope	Pope, Willis Thomas (1873–1961)
Popov	Popov, Mikhail Grigorévich (1893–1955)
Porta	Porta, Pietro (1832–1923)
Porter	Porter, Thomas Conrad (1822–1901)
Porter, C.L.	Porter, Charles Lyman (1889–)
Porter, D.	Porter, Duncan MacNair (1937–)
Porto	Porto (Porte), Paulo Campos (1889–)
Poselger	Poselger, Heinrich (1818–1883)
Posluszny	Posluszny, Usher (fl. 1976)
Post	Post, George Edward (1838–1909)
Potztal	Potztal, Eva Hedwig Ingeborg (1924–)
Pourret	Pourret de Figeac, Pierre André (1754–1818)
Powell, A.M.	Powell, Albert Michael (1937–)
Powell, J.	Powell, Jocelyn Marie (1939–)
Powrie	Powrie, Elizabeth (1925–1977)
Pradhan	Pradhan, Udai (fl. 1974)
Praeger	Praeger, Robert Lloyd (1865–1953)
Prain	Prain, David (1857–1944)
Praminik	Praminik, B.B. (1933–)
Prance	Prance, Ghillean ('Iain') Tolmie (1937–)
Prantl	Prantl, Karl Anton Eugen (1849–1893)
Presl, C.	Presl, Carel Bořivoj (1794–1852)
Presl, J.	Presl, Jan Swatopluk (1791–1849)
Preuss	Preuss, Paul Rudolf (1861–)
Price, M.	Price, Michael Greene (1941–)
Pritzel, E.	Pritzel, Ernst Georg (1875–1946)
Probat.	Probatova, N.S. (1939–)
Progel	Progel, August (1829–1889)
Pruski	Pruski, John Francis (1955–)
Puff	Puff, Christian (1949–)
Pugsley	Pugsley, Herbert William (1868–1947)
Pull.	Pulliatt, Victor (1827–1866)
Purdy	Purdy, Carlton Elmer (1861–1945)
Pursh	Pursh, Frederick Traugott (Pursch, Friedrich Traugott) (1774–1820)
Putterl.	Putterlick, Alois (1810–1845)
Puttock	Puttock, Christopher F.
Putzeys	Putzeys, Jules (or Julius) Antoine Adolphe Henri (1809–1882)
Qaiser	Qaiser, Mohammad (1946–)
Qi	Qi, Cheng-Jing (1932–)
du Quesnay	du Quesnay, M.C. (fl. 1971)
Quezada	Quezada, Max (1936–)
Quinn	Quinn, Christopher John (1936–)
Quis.	Quisumbing y Argüelles, Eduardo (1895–1986)
Rach	Rach, Louis Theodor (1821–1859)
Raddi	Raddi, Giuseppe (1770–1829)
Radlk.	Radlkofer, Ludwig Adolph Timotheus (1829–1927)
Räusch.	Räuschel, Ernst Adolf (fl. 1772–1797)
Raf.	Rafinesque-Schmaltz, Constantine Samuel (1783–1840)
Raiz.	Raizada, Mukat Behari (1907–)
Ralph	Ralph, Thomas Shearman (1813–1891)
Ramach.	Ramachandran, V.S. (fl. 1982)
Ramam.	Ramamurthy, Kandasamy (1933–)
Ramamoorthy	Ramamoorthy, T.P. (1945–)
Ramat.	Ramatuelle, Thomas Albin Joseph d'Audibert de (1750–1794)

Ramírez	Ramírez, José (1852–1904)
Ramírez, I.	Ramírez de Carnavali, Ivón (fl. 1987)
Ramond	Ramond de Caronbonnière, Louis François Elisabeth (1753–1827)
Rao, A.	Rao, Aragula Sathyanarayana (1924–1983)
Rao, A.N.	Rao, A. Nageswara (fl. 1985)
Rao, R.	Rao, Rolla Seshagiri (1921–)
Raoul	Raoul, Edouard Fiacre Louis (1815–1852)
Rapaics	Rapaics, Raymund (1885–1953)
Rattray	Rattray, James McFarlane (1907–1974)
Raup	Raup, Hugh Miller (1901–)
Rauschert	Rauschert, Stephan (1931–1986)
Raven	Raven, Peter Hamilton (1936–)
Ravenna	Ravenna, Pedro Felix (1938–)
Raynal	Raynal, Jean (1933–1979)
Raynal, A.	Raynal, Aline Marie Roques (1937–)
Raynaud	Raynaud, Christian (1939–)
Rayss	Rayss, Tscharna (1890–1965)
Read	Read, Robert William (1931–)
Rec.	Record, Samuel James (1881–1945)
Rech.f.	Rechinger, Karl Heinz (1906–1998)
Reed, C.	Reed, Clyde Franklin (1918–1999)
Reeder	Reeder, John Raymond (1914–)
Reeder, C.	Reeder, Charlotte Olive (née Goodding) (1916–)
Rees, B.	Rees, Bertha
Regel	Regel, Eduard August von (1815–1892)
Rehder	Rehder, Alfred (1863–1949)
R(ei)chb.	Reichenbach, Heinrich Gottlieb Ludwig (1793–1879)
R(ei)chb.f.	Reichenbach, Heinrich Gustav (1824–1889)
Reiche	Reiche, Carlos Frederico (1860–1929)
Reichst.	Reichstein, Tadeus (1897–1996)
Reim.	Reimers, Hermann Johann O. (1893–1961)
Reinw.	Reinwardt, Caspar George Carl (1773–1854)
Reisseck	Reisseck, Siegfried (1819–1871)
Remy	Remy, Esprit Alexandre (1826–1893)
Rendle	Rendle, Alfred Barton (1865–1938)
Renvoize	Renvoize, Stephen Andrew (1944–)
Req.	Requien, Esprit (1788–1851)
Retz.	Retzius, Anders Johan (1742–1821)
Reuter	Reuter, Georges François (1805–1872)
Rev.	Reveal, James Lauritz (1941–)
Reynaud	Reynaud, A.A. (1804–?)
Rheede	van Reede tot Drakenstein, Hendrik Adriaan (1637–1691)
Riccob.	Riccobono, Vincenzo (1861–1943)
Rich.	Richard, Louis Claude Marie (1754–1821)
Rich., A.	Richard, Achille (1794–1852)
Richardson	Richardson, Ian Bertram Kay (1940–)
Richter	Richter, August Gottlieb (1742–1812)
Richter, H.	Richter, H.G. (fl. 1987)
Rickett	Rickett, Harold William (1896–1989)
Riddell	Riddell, John Leonard (1807–1865)
Ridley	Ridley, Henry Nicholas (1855–1956)
Ridsd.	Ridsdale, Colin Ernest (1944–)
Riedel	Riedel, Ludwig (1790–1861)
Riedl	Riedl, Harald Udo von (1936–)
Riley	Riley, John (c. 1796–1846)
Risso	Risso, Joseph Antoine (1777–1845)
Ritter, F.	Ritter, Friedrich (1898–1989)
Riv.	Rivière, Charles Marie (1845–)
Riv., M	Rivière, Marie Auguste (1821–1877)
Rivas-Martínez	Rivas-Martínez, Salvador (1935–)
Rizz.	Rizzini, Carlos Toledo (1921–)

Robbrecht	Robbrecht, Elmar (1946–)
Robertson, K.	Robertson, Kenneth R. (1941–)
Roberty	Roberty, Guy Edouard (1907–1971)
Robinson	Robinson, Benjamin Lincoln (1864–1935)
Robinson, C.	Robinson, Charles Budd (1871–1913)
Robinson, H.	Robinson, Harold Ernest (1932–)
Robson, N.	Robson, Norman Keith Bonner (1928–)
Robyns	Robyns, André Georges Marie Walter Albert (1935–)
Robyns, F.	Robyns, Frans Herbert Edouard Arthur Walter (1901–1986)
Rochel	Rochel, Anton (1770–1847)
Rock	Rock, Joseph Francis Charles (1884–1962)
Rodigas	Rodigas, Emile (1831–1902)
Rodionenko	Rodionenko, Georgi Ivanovich (1913–)
Rodr.	Rodrigues, William Antonio (1928–)
Rodrigues, L.	Rodrigues, L.
Rodriguez	Rodriguez, José Demetrio (1780–1846)
Rodway	Rodway, Leonard (1853–1936)
Roehl.	Röhling, Johann Christoph (1757–1813)
Roemer	Roemer, Johann Jacob (1763–1819)
Roemer, M.	Roemer, Max Joseph (1791–1849)
Roessler	Roessler, Helmut (1926–)
Roezl	Roezl, Benedikt (Benito) (1824–1885)
Rogers, C.	Rogers, Claude Marvin (1919–)
Rogers, D.	Rogers, David James (1918–)
Rogers, R.	Rogers, Richard Sanders (1862–1942)
Rohr	Rohr, Julius Bernard von (1686–1742)
Rohr, J.P.	Rohr, Julius Philip Benjamin von (1737–1793)
Rohw.	Rohweder, Otto (1919–)
Rohwer	Rohwer, Jens G. (fl. 1985)
Roiv.	Roivainen, Heikki (1900–1983)
Rojas	Rojas, Teodoro (1877–1954)
Rojas, N.	Rojas Acosta, Nicolás (1873–1947)
Rolander	Rolander, Daniel (1725–1793)
Rolfe	Rolfe, Robert Allen (1855–1921)
Rollins	Rollins, Reed Clark (1911–1998)
Romero	Romero, Rafael Castaneda (1910–1973)
Romero, G.	Romero, Gustavo A. (1955–)
Romero García	Romero García, Ana Teresa (1957–)
Romero-Zarco	Romero-Zarco, Carlos (1954–)
Ronn.	Ronniger, Karl (1871–1954)
Roos	Roos, Marco C. (1955–)
Rosch.	Roschevicz, Roman Julievich (or Roshevitz, Rozevic) (1882–1949)
Roscoe	Roscoe, William (1753–1831)
Rose	Rose, Joseph Nelson (1862–1928)
Roseng.	Rosengurtt, Bernardo (1916–1985)
Roshk.	Roshkova, Olga Ivanovna (1909–1989)
Ross,J.	Ross, James Henderson (1941–)
Ross, R.	Ross, Robert (1912–)
Rosser	Rosser, Effie Moira (1923–)
Rossow	Rossow, Ricardo A. (1956–)
Roth	Roth, Albrecht Wilhelm (1757–1834)
Rothm.	Rothmaler, Werner Hugo Paul (1908–1962)
Rothr.	Rothrock, Joseph Trimble (1839–1922)
Rottb.	Rottbøll (Rottboell), Christen Friis (1727–1797)
Rottler	Rottler, Johan Peter (1749–1836)
Roul.	Rouleau, Joseph Albert Ernest (1916–1991)
Rourke	Rourke, John Patrick (1942–)
Rouy	Rouy, Georges C. Chr. (1851–1924)
Rowley, G.	Rowley, Gordon Douglas (1921–)
Roxb.	Roxburgh, William (1751–1815)
Roy	Roy, G.P. (1939–)

Royen	Royen, David van (1727–1799)
Royen, P.	Royen, Pieter van (1923–)
Royle	Royle, John Forbes (1798–1858)
R.-Sm.	Radcliffe-Smith, Alan (1938–)
Rudge	Rudge, Edward (1763–1846)
Rudolphi	Rudolphi, Karl Asmund (1771–1832)
Rudolphi, F.	Rudolphi, Friedrich Karl Ludwig (1801–1849)
Ruempler	Ruempler (or Rümpler, Karl Theodor (1817–1891)
Rüssmann	Rüssmann, M. (fl. 1984)
Rúgolo	Rúgolo de Agrasar, Sulma E. (1940–)
Ruhl.	Ruhland, Wilhelm Otto Eugen (1878–1960)
Ruíz	Ruíz López, Hipólito (1754–1815)
Rupp	Rupp, Herman Montague Rucker (1872–1956)
Rupr.	Ruprecht, Franz Josef Ivanovich (1814–1870)
Rusby	Rusby, Henry Hurd (1855–1940)
Ruschi	Ruschi, Augusto (1915–)
Rydb.	Rydberg, Per Axel (1860–1931)
Rzazade	Rzazade, Rza Jakhja Ogly (1909–)
Rzed.	Rzedowski, Jerzy (1926–)
Sabine	Sabine, Joseph (1770–1837)
Sachet	Sachet, Marie-Hélène (1922–1986)
Saff.	Safford, William Edwin (1859–1926)
Sagást.	Sagástegui Alva, Abundio (1932–)
Sagot	Sagot, Paul Antoine (1821–1888)
Sakis.	Sakisaka, Michiji (1895–)
Salazar	Salazar, Gerardo A. (1961–)
Saldanha, J.	Saldanha da Gama, José de (1839–1905)
Salisb.	Salisbury, Richard Anthony (né Markham) (1761–1829)
Salm-Dyck	Salm-Reifferscheid-Dyck, Joseph Franz Maria Anton Hubert Ignatz Fürst zu (1773–1861)
Salzm.	Salzmann, Philipp (1781–1851)
Samp.	Sampaio, Gonçalo António da Silva Ferreira (1865–1937)
Samp., A.	Sampaio, Alberto José de (1881–1946)
Samuelsson	Samuelsson, Gunnar (1885–1944)
Sander	Sander, Henry Frederick Conrad (1847–1920)
Sandw.	Sandwith, Noel Yvri (1901–1965)
Sands, M.	Sands, Martin Jonathan Southgate (1938–)
Santi	Santi, Giorgio (1746–1822)
Santisuk	Santisuk, Thawatchai (1944–)
Sarg.	Sargent, Charles Sprague (1841–1927)
Sastry	Sastry, A.R.K. Ramakrishna (1938–)
Sauer, G.	Sauer, G. (fl. 1980)
Sauer, W.	Sauer, Wilhelm (1935–)
Savat.	Savatier, Paul Alexandre (1824–1886)
Savat., P.A.L.	Savatier, Paul Amedée Ludovic (1830–1891)
Savi	Savi, C. Gaëtano (1769–1844)
Savigny	Savigny, Marie Jules César Lélorgne de (1777–1851)
Săvul.	Săvulescu, Traian (1889–1963)
Schaeffer	Schäffer (Schaeffer), Jacob Christian (H. von) (1718–1790)
Schauer	Schauer, Johannes Conrad (1813–1848)
Scheele	Scheele, George Heinrich Adolf (1808–1864)
Scheffer, R.	Scheffer, Rudolph Herman Christiaan Carel (1844–1880)
Scheidw.	Scheidweiler, Michael Joseph François (1799–1861)
Schellenb.	Schellenberg, Gustav August Ludwig David (1882–1963)
Scheng.	Schengelia, E.M. (fl. 1953)
Schenk	Schenk, Joseph August (1815–1891)
Scherb.	Scherbius, Johannes (1769–1813)
Schery	Schery, Robert Walker (1917–1987)
Schiede	Schiede, Christian Julius Wilhelm (1798–1836)
Schindler	Schindler, Anton Karl (1879–1964)
Schinz	Schinz, Hans (1858–1941)

Schipcz.	Schipczinski, Nikolaj Valerianovich (1886–1955)
Schischkin	Schischkin, Boris Konstantinovich (1886–1963)
Schkuhr	Schkuhr, Christian (1741–1811)
Schldl.	Schlechtendal, Diederich Franz Leonhard von (1794–1866)
Schltr.	Schlechter, Friedrich Richard Rudolf (1872–1925)
Schleiden	Schleiden, Matthias Jacob (1804–1881)
Schmalh.	Schmalhausen, Johannes Theodor (1849–1894)
Schmarse	Schmarse, Helmut (fl. 1933)
Schmeiss	Schmeiss, Oskar (fl. 1906)
Schmidel	Schmidel, Casimir Christoph (1718–1792)
Schmidt	Schmidt, Franz (1751–1834)
Schmidt, F.W.	Schmidt, Franz Wilibald (1764–1796)
Schmidt, J.A.	Schmidt, Johann Anton (1823–1905)
Schnack	Schnack, Benno Julio Christian (1910–1981)
Schneev.	Schneevoogt, George Voorhelm (Schneevoight) (1775–1850)
Schneider, C.	Schneider, Camillo Karl (1876–1951)
Schnitzl.	Schnitzlein, Adalbert Carl Friedrich Hellwig Conrad (1814–1868)
Schnizlein	Schnizlein, (Karl Friedrich Christoph) Wilhelm (1780–1856)
Schodde	Schodde, Richard (1936–)
Schönl.	Schönland, Selmar (1860–1940)
Scholz	Scholz, Joseph B. (1858–1915)
Scholz, H.	Scholz, Hildemar Wolfgang (1928–)
Schomb.	Schomburgk, Robert Hermann (1804–1865)
Schot, A.M.	Schot, Anne M.
Schott	Schott, Heinrich Wilhelm (1794–1865)
Schottky	Schottky, Ernst Max (1888–1915)
Schousboe	Schousboe, Peder Kofod Anker (1766–1832)
Schrader	Schrader, Heinrich Adolph (1767–1836)
Schrank	Schrank, Franz von Paula von (1747–1835)
Schreber	Schreber, Johann Christian Daniel von (1739–1810)
Schrenk	Schrenk, Alexander Gustav von (1816–1876)
Schröder	Schröder, Richard Iwanowitch (1822–1903)
Schroedinger	Schrödinger, Rudolf (1857–1919)
Schubert, B.G.	Schubert, Bernice Giduz (1913–)
Schuebler	Schübler (Schuebler), Gustav (1787–1834)
Schultes	Schultes, Josef August (1773–1831)
Schultes, R.	Schultes, Richard Evans (1915–)
Schultes f.	Schultes, Julius Hermann (1804–1840)
Schultz, C.H.	Schultz, Carl Heinrich `Schultzenstein' (1798–1871)
Schultz, F.W.	Schultz, Friedrich Wilhelm (1804–1876)
Schultz-Bip.	Schultz, Carl Heinrich `Bipontinus' (1805–1867)
Schulz, D.L.	Schulz, Dorothea L. (1931–)
Schulz, O.	Schulz, Otto Eugen (1874–1936)
Schulz, R.	Schulz, Roman (1873–1926)
Schulze, G.	Schulze, Georg Martin (1906–1985)
Schulze, W.	Schulze, Werner (1930–)
Schum.	Schumacher, Heinrich Christian Friedrich (1757–1830)
Schumann	Schumann, Karl Moritz (1851–1904)
Schur	Schur, Philip Johann Ferdinand (1799–1878)
Schwacke	Schwacke, Carl August Wilhelm (1848–1904)
Schwantes	Schwantes, Martin Heinrich Gustav Georg (1891–1960)
Schwartz	Schwartz, Oskar (1901–1945)
Schwarz	Schwarz, August Friedrich (1852–1915)
Schwarz, O.	Schwarz, Otto Karl Anton (1900–1983)
Schweick.	Schweickerdt, Herold Georg Wilhelm Johannes (1903–1977)
Schweigger	Schweigger, August Friedrich (1783–1821)
Schwein.	Schweinitz, Ludwig David von (1780–1834)
Schweinf.	Schweinfurth, Georg August (1836–1925)
Schweinf., C.	Schweinfurth, Charles (1890–1970)

Scop.	Scopoli, Giovanni Antonio (1723–1788)
Scott, A.J.	Scott, Andrew John (1950–)
Scott-Elliot	Scott-Elliot, George Francis (1862–1934)
Scribner	Scribner, Frank Lamson (1851–1938)
Sealy	Sealy, Joseph Robert (1907–)
Seberg	Seberg, Ole (1952–)
Sébert	Sébert, Hippolyte (1839–1930)
Secco	de Secco, Ricardo (fl. 1985)
Seemann	Seemann, Berthold Carl (1825–1871)
Séguier	Séguier, Jean François (1703–1784)
Seib.	Seibert, Russell Jacob (1914–)
Seidel	Seidel, Johann Heinrich (fl. 1779)
Seidenf.	Seidenfaden, Gunnar (1908–)
Seidenschnur	Seidenschnur, Christiane Eva (1944–)
Seidl	Seidl, Wenzel Benno (1773–1842)
Sell	Sell, Peter Derek (1929–)
Sello	Sello, Hermann Ludwig (1800–1876)
Sellow	Sellow, Friedrich (1789–1831)
Semir	Semir, S. João (1937–)
Semple	Semple, John Cameron (1947–)
Sendtner	Sendtner, Otto (1813–1859)
Senghas	Senghas, Karlheinz (1928–)
Ser.	Seringe, Nicolas Charles (1776–1858)
Sesler	Sesler, Leonard (?–1785)
Sessé	Sessé y Lacasta, Martín de (1751–1808)
Setch.	Setchell, William Albert (1864–1943)
Setten	van Setten, A.K. (fl. 1985)
Seub.	Seubert, Moritz August (1818–1878)
Seward	Seward, Albert Charles (1863–1941)
Shan	Shan, Ren Hwa (1909–)
Shang	Shang, Chih-Bei (1935–)
Sharp	Sharp, Aaron John (1904–1997)
Sharp, W.	Sharp, Ward McClintic (1904–)
Shaw, E.	Shaw, Elizabeth Anne (1938–)
Sheh	Sheh, Men(g)-Lan (fl. 1986)
Sheldon, E.	Sheldon, Edmund Perry (1869–1947)
Shen	Shen, Lian-Dai (Tai) (fl. 1970)
Sherff	Sherff, Earl Edward (1886–1966)
Shevock	Shevock, James R. (1950–)
Shib.	Shibata, Keita (1877–1949)
Shih (C.)	Shih, Chu (1932–)
Shim, P.S.	Shim, Phyau Soon (fl. 1982)
Shinn.	Shinners, Lloyd Herbert (1918–1971)
Short	Short, Philip Sydney (1955–)
Shrestha	Shrestha, T.B.
Shuttlew.	Shuttleworth, Robert James (1810–1874)
Sibth.	Sibthorp, John (1758–1796)
Siebenl.	Siebenlist
Sieber	Sieber, Franz Wilhelm (1789–1844)
Siebert	Siebert, August (1854–1923)
Siebold	Siebold, Philipp Franz von (1796–1866)
Siegerist	Siegerist, E.S. (1925–)
Sillans	Sillans, Roger (fl. 1952)
Silva Manso, A.	Silva Manso, António-Luiz Patricio da (1788–1818)
Silva Manso, J.	Silva Manso, José da
Silverside	Silverside, Alan James (1947–)
Silvestre	Silvestre Domingo, Santiago (1944–)
Sim	Sim, Robert (1791–1878)
Simon	Simon, Bryan Keith (1943–)
Simonkai	Simonkai, Lájos von (1851–1910)
Simon-Louis	Simon-Louis, Léon L. (1834–1913)
Simpson, B.	Simpson, Beryl Britnall (1942–)

Sims	Sims, John (1749–1831)
Sincl., James	Sinclair, James (1913–1968)
Singh	Singh, D.N.
Sivadasan	Sivadasan, M. (1948–)
Skeels	Skeels, Homer Collar (1873–1934)
Skip.	Skipworth, John Peyton (1934–)
Skog	Skog, Laurence Edgar (1943–)
Skottsb.	Skottsberg, Carl Johan Fredrik (1880–1963)
Skovsted	Skovsted, Åge Thorsen (1903–1983)
Skvortzov	Skvortzov, Boris Vassilievich (1890–1980)
Sleumer	Sleumer, Hermann Otto (1906–1993)
Sloane, B.	Sloane, Boyd Lincoln (1885–)
Slooten	Slooten, Dirk Fox van (1891–1953)
Sm.	Smith, James Edward (1759–1828)
Sm., A.C.	Smith, Albert Charles (1906–1999)
Sm., A.R.	Smith, Alan Reid (1943–)
Sm., C.	Smith, Christen (1785–1816)
Sm., C.A.	Smith, Christo Albertyn (1898–1956)
Sm., E.W.	Smith, Elmer William (1920–1981)
Sm., F.D.	Donnell Smith, F.
Sm., H.	Smith, Karl August Harald ('Harry') (1889–1971)
Sm., J.	Smith, John (1798–1888)
Sm., J.D.	Donnell Smith, John (1829–1928)
Sm., J.J.	Smith, Johannes Jacobus (1867–1947)
Sm., L.B.	Smith, Lyman Bradford (1904–)
Sm., L.S.	Smith, Lindsay Stewart (1917–1970)
Sm., R.M.	Smith, Rosemary Margaret (1933–)
Sm., W.(W.)	Smith, William Wright (1875–1956)
Sm., W.G.	Smith, Worthington George (1835–1917)
Small	Small, John Kunkel (1869–1938)
Small, E.	Small, Ernest (1940–)
Smirnov	Smirnov, Pavel Aleksandrovich (1896–1980)
Smoljan.	Smoljaninova, L.A. (1904–)
Snijman	Snijman, Deidré A. (1949–)
Söderstrom	Söderstrom, Thomas Robert (1936–1987)
Soeg.-Reks.	Soegeng-Reksodiharjo, Wertit (1935–)
Soest	Soest, Johannes Leendert van (1898–1983)
Sohmer	Sohmer, Seymour Hans (1941–)
Soják	Soják, Jiří (1936–)
Sol.	Solander, Daniel Carl (1733–1782)
Sole	Sole, William (1741–1802)
Soler.	Solereder, Hans (1860–1920)
Solms-Laub.	Solms-Laubach, Hermann Maximilian Carl Ludwig Friedrich zu (1842–1915)
Somers	Somers, Carl (1963–)
Sommier	Sommier, Carlo Pietro Stefano (1848–1922)
Sonder	Sonder, Otto Wilhelm (1812–1881)
Sonn.	Sonnerat, Pierre (1748–1814)
Soriano	Soriano, Alberto (1920–)
Sosn.	Sosnowsky, Dimitrii Ivanovich (1885–1952)
Sota	Sota, Elías Ramón de la (1932–)
Soto	Soto Arenas, Miguel Angel (1963–)
Soul.-Bod.	Soulange-Bodin, Etienne (1774–1846)
Souster	Souster, John Eustace Sirett (1912–)
Southworth	Southworth, Effie Almira (1860–1947)
Sowerby	Sowerby, James De Carle (1787–1871)
Spach	Spach, Edouard (1801–1879)
Span.	Spanoghe, Johan Baptist (1798–1838)
Sparre	Sparre, Benkt (1918–1986)
Sparrman	Sparrman, Anders (1748–1820)
Speg.	Spegazzini, Carlo Luigi (1858–1926)
Spencer, M.	Spencer, Michael A.

Speta	Speta, Franz (1941–)
Spire, A.	Spire, André (fl. 1903)
Spire, C.	Spire, Camille Joseph (fl. 1903)
Splitg.	Splitgerber, Frederik Louis (1801–1845)
Sprague	Sprague, Thomas Archibald (1877–1958)
Sprengel	Sprengel, Curt Polycarp Joachim (1766–1833)
Sprengel, A.	Sprengel, Anton (1803–1851)
Spring	Spring, Anton Friedrich (1814–1872)
Spruce	Spruce, Richard (1817–1893)
Sreek.	Sreekumar, P.V. (fl. 1982)
Sreemad.	Sreemadhaven, C.P.
St John, (H.)	St John, Harold (1892–1991)
Stace	Stace, Clive Anthony (1938–)
Stadman	Stadman (fl. 1810)
Stadtm(ann)	Stadtmann, Jean Frédéric (1762–1807)
Stafleu	Stafleu, Frans Antonie (1921–1997)
Ståhl	Ståhl, Bertil (1957–)
St-Amans	de Saint-Amans, Jean Florimond Boudon (1748–1831)
Standley	Standley, Paul Carpenter (1884–1963)
Staner	Staner, Pierre (1901–1984)
Stapf	Stapf, Otto (1857–1933)
Staples	Staples, George William (1953–)
Stapleton	Stapleton, C.M.A.
Stauffer	Stauffer, Hans Ulrich (1929–1965)
Steane	Steane, Dorothy L.
Stearn	Stearn, William Thomas (1911–)
Stebb.	Stebbins, George Ledyard (1906–2000)
Steck	Steck, Abraham (fl. 1757)
Steenis	Steenis, Cornelis Gijsbert Gerrit Jan van (1901–1986)
Steetz	Steetz, Joachim (1804–1862)
Stefanoff	Stefanoff, Boris (1894–1979)
Stehlé	Stehlé, Henri (1909–1983)
Stein	Stein, Berthold (1847–1899)
Steinh.	Steinheil, Adolphe (1810–1839)
Stent	Stent, Sydney Margaret (1875–1942)
Stephan	Stephan, Christian Friedrich (1757–1814)
Stephens	Stephens, Edith Layard (1884–1966)
Stern	Stern, William Louis (1926–)
Stern, F.	Stern, Frederick Claude (1884–1967)
Sternb.	Sternberg, Caspar Maria von (1761–1838)
Sterns	Sterns, Emerson Ellick (1846–1926)
Steudel	Steudel, Ernst Gottlieb von (1783–1856)
Steven	Steven, Christian von (1781–1863)
Stevens, P.	Stevens, Peter F. (1944–)
Stevenson	Stevenson, Dennis William (1942–)
Steyerm.	Steyermark, Julian Alfred (1909–1988)
St-Hil., A.	Saint-Hilaire, Auguste François César Prouvençal de (1779–1853)
St-Hil., J.	Saint-Hilaire, Jean Henri Jaume (1772–1845)
Stirton	Stirton, Charles Howard (1946–)
St-Lager	Saint-Lager, Jean Baptiste (1825–1912)
Stocks	Stocks, John Ellerton (1822–1854)
Stokes	Stokes, Jonathan S. (1755–1831)
Stone	Stone, Benjamin Clemens Masterman (1933–1994)
Stopp	Stopp, Klaus Dieter (1926–)
Stork	Stork, Adélaïde Louise (1937–)
Stoy.	Stoyanoff, Nikolai Andreev (1883–1968)
Strack	Strack, Dieter (fl. 1987)
Strasb.	Strasburger, Eduard Adolf (1844–1912)
Straw	Straw, Richard Myron (1926)
Stritch	Stritch, Lawrence R. (fl. 1982)
Strother	Strother, John Lance (1941–)

Struwe	Struwe, Lena
Stschegl.	Stschegleew, Serge S. (fl. 1851)
Stuntz	Stuntz, Stephen Conrad (1875–1918)
Subils	Subils, Rosa (1929–)
Subram.	Subramanyam, Krishnaier (1915–1980)
Suckow	Suckow, Georg Adolph (1751–1813)
Sudw.	Sudworth, George Bishop (1864–1927)
Süsseng.	Süssenguth (Suessenguth), Karl (1893–1955)
Suh	Suh, Young-Bae (fl. 1990)
Suksd.	Suksdorf, Wilhelm Nikolaus (1850–1932)
Summerh.	Summerhayes, Victor Samuel (1897–1974)
Sun, Y.Z.	Sun, Yon Zai
Suresh	Suresh, C.R. (fl. 1988)
Suryan.	Suryanarayana, M.C.
Susanna	Susanna, Alfonso (1956–)
Sutô	Sutô, Tiharu (1910–)
Suvatabandhu	Suvatabandhu, Kasin (1916–)
Svent.	Sventenius, Eric R.Svensson (1910–1973)
Sw.	Swartz, Olof Peter (1760–1818)
Swallen	Swallen, Jason Richard (1903–1991)
Swart	Swart, Jan Johannes (1901–1974)
Sweet	Sweet, Robert (1783–1835)
Sweet, H.	Sweet, Herman Royden (1911–1991)
Swingle	Swingle, Walter Tennyson (1871–1952)
Sydow	Sydow, Hans (1879–1946)
Sym.	Symington, Colin Fraser (1905–1943)
Syme	Syme, John Thomas Irvine Boswell (1822–1888)
Szabó, Z.	Szabó, Zoltán von (1882–1944)
Szlach.	Szlachetko, Dariusz L. (1961–)
Szyszyl.	Szyszylowicz, Ignaz von (1857–1910)
Tag.	Tagawa, Motozi (1908–1977)
Tagg	Tagg, Harry Frank (1874–1933)
Takah.	Takahashi, Yoshinao (?–1914)
Takeda	Takeda, Hisayoshi (1883–1972)
Takht.	Takhtadjan, Armen Leonovich (1910–)
Tamamschjan	Tamamschjan, Sophia G. (1901–1981)
Tamura	Tamura, Michio (1927–)
Tanaka (, T.)	Tanaka, Tyôzaburô (1885–)
Tanf.	Tanfani, Enrico (1848–1892)
Tang	Tang, T. (1897–1984)
Tard.	Tardieu, Marie Laure (also Tardieu-Blot) (1902–)
Targ.-Tozz.	Targioni-Tozzetti, Ottaviano (1755–1829)
Tate	Tate, Ralph (1840–1901)
Tateoka	Tateoka, Tuguo (1931–)
Tatew.	Tatewaki, Misao (1899–)
Taubert	Taubert, Paul Hermann Wilhelm (1862–1897)
Tausch	Tausch, Ignaz Friedrich (1793–1848)
Taylor	Taylor, Richard (1781–1858)
Taylor, D.	Taylor, Dean W.
Taylor, G.	Taylor, George (1904–1993)
Taylor, N.	Taylor, Norman (1883–1967)
Taylor, P.	Taylor, Peter Geoffrey (1926–)
Teijsm.	Teijsmann, Johannes Elias (1809–1882)
Ten.	Tenore, Michele (1780–1861)
Tenn.	Tennant, James Robert (1928–)
Terrell	Terrell, Edward E. (1923–)
Tharp	Tharp, Benjamin Carroll (1885–1964)
Thell.	Thellung, Albert (1881–1928)
Thénint	Thénint, André
Theob.	Theobald, William (1829–1908)
Theob., W.L.	Theobold, William Louis (1936–)

Thieret	Thieret, John William (1926–)
Thomas, H.	Thomas, Hugh Hamshaw (1885–1962)
Thomas, Le	Le Thomas-Hommay, Annick (1936–)
Thomas, W.	Thomas, William Wayt (1951–)
Thompson, C.	Thompson, Charles Henry (1870–1931)
Thompson, J.	Thompson, Joy (née Garden) (1923–)
Thomson	Thomson, Thomas (1817–1878)
Thonn.	Thonning, Peter (1775–1848)
Thore	Thore, Jean (1762–1823)
Thorel	Thorel, Clovis (1833–1911)
Thory	Thory, Claude Antoine (1759–1827)
Thoth.	Thothathri, K. (1929–)
Thouars	du Petit Thouars, Louis Marie Aubert (1758–1831)
Thuill.	Thuiller, Jean Louis (1757–1822)
Thulin	Thulin, Mats (1948–)
Thunb.	Thunberg, Carl Peter (1743–1828)
Thurb.(er)	Thurber, George (1821–1890)
Thwaites	Thwaites, George Henry Kendrick (1812–1882)
Tichom.	Tichomirov, Vadim Nikolaevich (1932–)
Tieghem	Tieghem, Phillippe Edouard Léon van (1839–1914)
Tiên Bân	Bân, Nguyên Tiên (fl. 1973)
Tilloch	Tilloch, Alexander (1759–1825)
Timbrook	Timbrook, Steven (fl. 1986)
Timler	Timler, Friedrich Karl (1914–)
Tindale	Tindale, Mary Douglas (1920–)
Tineo	Tineo, Vincenzo (1791–1856)
Tirv.	Tirvengadum, D.D. (fl. 1986)
Tobe	Tobe, Hiroshi (fl. 1987)
Tobler	Tobler, Friedrich (1879–1957)
Tod.	Todaro, Agostino (1818–1892)
Toelken	Toelken, Helmut R. (1939–)
Tol.	Toledo, Joaquim Franco de (1905–1952)
Tomb	Tomb, Andrew Spencer (1943–)
Toml.	Tomlinson, Philip Barry (1932–)
Tong	Tong, Shao-Quan (1935–)
Torre	Torre, Antonio Rocha da (1904–)
Torrey	Torrey, John (1796–1873)
Townrow	Townrow, John A. (1927–)
Towns., C.	Townsend, Clifford Charles (1926–)
Toyok.	Toyokuni, Hideo (1932–)
Trabut	Trabut, Louis (Charles) (1853–1929)
Tracey	Tracey, John Geoffrey (1930–)
Trail	Trail, James William Helenus (1851–1919)
Tralau	Tralau, Hans (1932–1977)
Tratt.	Trattinick, Leopold (1764–1849)
Traub	Traub, Hamilton Paul (1890–1983)
Trautv.	Trautvetter, Ernst Rudolf von (1809–1889)
Trécul	Trécul, Auguste Adolphe Lucien (1818–1896)
Trel.	Trelease, William (1857–1945)
Trev.	Treviranus, Ludolf Christian (1799–1864)
Trevis.	Trevisan de Saint-Léon, Vittore Benedetto Antonio (1818–1897)
Trew	Trew, Christoph Jakob (1695–1769)
Triana	Triana, José Jéronimo (1834–1890)
Triest	Triest, Ludwig J. (1957–)
Trimen	Trimen, Henry (1843–1896)
Trin.	Trinius, Carl Bernhard von (1778–1844)
Tronc.	Troncoso, Nélida Sara (1914–)
Trotter	Trotter, Alessandro (1874–1967)
Trotzky	Trotzky, Petrus Kornuch (1803–1877)
Troupin	Troupin, Georges M.D.J. (1923–1997)
Trudgen	Trudgen, Malcolm Eric (1951–)

Tryon	Tryon, Rolla Milton (1916–)
Tscherneva	Tscherneva, O.V. (1929–)
Tsi	Tsi, Zhan-Huo (1937–)
Tsiang	Tsiang, Ying (1898–1982)
Tsui	Tsui Yon-Wen (1907–)
Tul.	Tulasne, Louis René (1815–1885)
Türpe	Türpe, Anna Maria (1946–)
Turcz.	Turczaninow, Porphir Kiril Nicolai Stepanowitsch (1796–1863)
Turner, B.	Turner, Billie Lee (1925–)
Turner, H.	Turner, Hubert (1955–)
Turner, M.	Turner, Melvin D. (fl. 1988)
Turpin	Turpin, Pierre Jean François (1775–1840)
Turra	Turra, Antonio (1730–1796)
Turrill	Turrill, William Bertram (1890–1961)
Tussac	Tussac, François Richard de (1751–1837)
Tutcher	Tutcher, William James (1867–1920)
Tutin	Tutin, Thomas Gaskell (1908–1987)
Tuyama	Tuyama, Takasi (1910–)
Tzvelev	Tzvelev, Nikolai Nikolaievich (1925–)
Ucria	da Ucria, Bernardino (1739–1796)
Uitew.	Uitewaal, Antonius Josephus Adrianus (1899–1963)
Uittien	Uittien, Hendrik (1898–1944)
Ulbr.	Ulbrich, Oskar Eberhard (1879–1952)
Ule	Ule, Ernst Heinrich Georg (1854–1915)
Underw.	Underwood, John (?–1834)
Underw., L.	Underwood, Lucien Marcus (1853–1907)
Unger	Unger, Franz (Joseph Andreas Nicolaus) (1800–1870)
Ung.-Sternb.	Ungern-Sternberg, Franz (1808–1885)
Urban	Urban, Ignatz (1848–1931)
Urbatsch	Urbatsch, Lowell Edward (1942–)
Urmi-König	Urmi-König, Katherina (fl. 1975)
Ursch	Ursch, Eugène (1882–1962)
Urv.	Urville, Jules Sébastian César Dumont d' (1790–1842)
Vahl	Vahl, Martin (1749–1804)
Vail	Vail, Anna Murray (1863–1955)
Vajravelu	Vajravelu, E. (1936–)
Valcken.	Valckenier, Suringar, Jan (1865–1932)
Valdebenito	Valdebenito, Hugo A. (fl. 1986)
Valdés	Valdés, Benito (1942–)
Valdés R.	Valdés-Reyna, Jésus (1948–)
Valeton	Valeton, Theodoric (1855–1929)
Vand.	Vandelli, Domingos (1735–1816)
Vargas	Vargas Calderón, Julio César (1907–1960)
Vasc.	Vasconcellos, João de Carvalho e
Vasey	Vasey, George (1822–1893)
Vassal	Vassal J. (1932–)
Vassilcz.	Vassilczenko, I.T. (1903–)
Vassiljeva, A.	Vassiljeva, A.N. (fl. 1969)
Vasudeva	Vasudeva Rao, M.K. (fl. 1979)
Vatke	Vatke, Georg Carl Wilhelm (1849–1889)
Vatt.	Vattimo-Gil, Ida de (1928–)
Veitch	Veitch, John Gould (1839–1870)
Veken	Veken, Paul A.J.B. van der (1928–)
Velarde	Velarde, Octavio (Octavio Velarde Nuñez) (fl. 1945–1959)
Veldk.	Veldkamp, Jan Frederik (1941–)
Velen.	Velenovský, Josef (1858–1949)
Vell.	Velloso de Miranda, Joaquim (1733–1815)
Vell. Conc.	Vellozo, José Mariano da Conceição (1742–1811)
Vent, W.	Vent, Walter (1920–)

Vent.	Ventenat, Etienne Pierre (1757–1808)
Verdc.	Verdcourt, Bernard (1925–)
Verlot	Verlot, Jean-Baptiste (1825–1891)
Verma	Verma, D.M. (1937–)
Vermeulen	Vermeulen, Pieter (1899–1981)
Vermoesen	Vermoesen, François Marie Camille (1882–1922)
Vernet	Vernet (fl. 1904)
Vest	von Vest, Lorenz Chrysanth (1776–1840)
Vick.	Vickery, Joyce Winifred (1908–1979)
Vidal, J.E.	Vidal, Jules Eugène (1914–)
Vidal, S.	Vidal y Soler, Sebastian (1842–1889)
Vieill.	Vieillard, Eugène (1819–1896)
Viguier	Viguier, L.G. Alexandre (1790–1867)
Viguier, R.	Viguier, René (1880–1931)
Villar	del Villar y Serratacó, Emile Huguet
Villars	Villars, Domínique (1745–1814)
Villaseñor	Villaseñor, José Luis (1954–)
Villiers	Villiers, J.-F. (1943–)
Vilm.	Vilmorin, Pierre Philippe André Lévêque de (1776–1862)
Vilm., P.L.	Vilmorin, Pierre Louis François Lévêque de (1816–1860)
Vines	Vines, Sydney Howard (1849–1934)
Vink	Vink, Willem (1931–)
Virot	Virot, Robert (1915–)
Vis.	Visiani, Roberto de (1800–1878)
Vischer	Vischer, Wilhelm (1890–1960)
Viv.	Viviani, Domenico (1772–1840)
Vogel	Vogel, Benedict Christian (1745–1825)
Vogel, de	Vogel, Eduard Ferdinand de (1942–)
Vogel, J.	Vogel, Julius Rudolph Theodor (1812–1841)
Vogt	Vogt, Robert M. (1957–)
Voigt	Voigt, Friedrich Sigismund (1781–1850)
Voigt, J.	Voigt, Joachim Otto (1798–1843)
Volkens	Volkens, Georg Ludwig August (1855–1917)
Vos, de	de Vos, Miriam Phoebe (1912–)
Voss	Voss, Andreas (1857–1924)
Votsch	Votsch (fl. 1904)
Vural	Vural, M. (fl. 1983)
Vriese	Vriese, Willem Hendrik de (1806–1862)
Vved.	Vvedensky, Aleksej Ivanovič (1898–1972)
Wagenitz	Wagenitz, Gerhard Werner Friedrich (1927–)
Wager	Wager, Vincent Athelstan (1904–)
Wagner	Wagner, Warren Herbert (1920–1999)
Wahlenb.	Wahlenberg, Georg (Göran) (1780–1851)
Waldst.	Waldstein (-Wartemburg), Franz de la Paula Adam von (1759–1823)
Walker, J.	Walker, James Willard (1943–)
Walker, T.	Walter, Trevor George (1927–)
Wall	Wall, Arnold (1869–1966)
Wallace	Wallace, Alfred Russel (1823–1913)
Wallace, A.	Wallace, Alexander (1829–1899)
Wallich	Wallich, Nathaniel (or Nathan Wolf) (1786–1854)
Wallr(oth)	Wallroth, Carl Friedrich Wilhelm (1792–1857)
Walp.	Walpers, Wilhelm Gerhard (1816–1853)
Walter	Walter, Thomas (1740–1789)
Walter, H.	Walter, Hans Paul Heinrich (1882–)
Walther	Walther, Edward Eric (1892–1959)
Walton	Walton, Frederick Arthur (1853–1922)
Wang	Wang, Chen Hwa (1908–)
Wang, C.P.	Wang, Cheng-Ping (fl. 1982)
Wang, F.T.	Wang, Fa-Tsuan (1929–)
Wang, W.T.	Wang, Wen-Tsai (1926–)

Wang, X.Q.	Wang, Xi-Qu
Wang, Z.R.	Wang, Zhong-Ren (1939–)
Wangenh.	Wangenheim, Friedrich Adam Julius von (1749–1800)
Wangerin	Wangerin, Walther Leonhard (1884–1938)
Warb.	Warburg, Otto (1859–1938)
Warb., E.	Warburg, Edmund Frederic (1908–1966)
Warder	Warder, John Aston (1812–1883)
Warm.	Warming, Johannes Eugen Bülow (1841–1924)
Warsc.	Warscewicz, Josef von (1812–1866)
Wassh.	Wasshausen, Dieter Carl (1938–)
Waterf.	Waterfall, Umaldy Theodore (1910–1971)
Waterhouse	Waterhouse, John Teast (1924–1983)
Watson, H.	Watson, Hewett Cottrell (1804–1881)
Watson, J.	Watson, John Forbes (1827–1892)
Watson, L.	Watson, Leslie (1938–)
Watson, S.	Watson, Sereno (1826–1892)
Watson, W.	Watson, William (1858–1925)
Watson, W.C.R.	Watson, William Charles Richard (1885–1954)
Watson, Will.	Watson, William (1858–1925)
Watt	Watt, David Allan Poe (1830–1917)
Wawra	Wawra von Fernsee, Heinrich (1831–1887)
Webb	Webb, Philip Barker (1793–1854)
Webb, D.	Webb, David Allardice (1912–1994)
Weber, A.	Weber, Anton (1947–)
Weber, C.	Weber, Jean-Germaine Claude (1922–)
Weber, F.	Weber, Frédéric Albert Constantin (1830–1903)
Weber, G.	Weber, Georg Heinrich (1752–1828)
Weber, W.A.	Weber, William Alfred (1918–)
Weberb.	Weberbauer, Otto (1846–1881)
Webster	Webster, Grady Linder (1927–)
Webster, R.	Webster, Robert D. (1950–)
Wedd.	Weddell, Hugh Algernon (1819–1877)
Wehrh.	Wehrhahn, Heinrich Rudolf (1887–1940)
Wei, F.N.	Wei, Fa-Nan (1941–)
Wei, H.T.	Weii, Hong Tu (fl. 1982)
Weihe	Weihe, Carl Ernst August (1779–1834)
Weiller	Weiller, Marc (1880–1945)
Welw.	Welwitsch, Friedrich Martin Josef (1806–1872)
Wen, T.H.	Wen, Tai-Hui (1924–)
Wendelbo	Wendelbo, Per Erland Berg (1927–1981)
Wender.	Wenderoth, George Wilhelm Franz (1774–1861)
Wendl.	Wendland, Johann Christoph (1755–1828)
Wendl., H.(A.)	Wendland, Hermann A. (1825–1903)
Wendl., H.L.	Wendland, Heinrich Ludolph (1792–1869)
Wendt	Wendt, Thomas Leighton (1950–)
Werderm.	Werdermann, Erich (1892–1959)
van der Werff	van der Werff, Henk (fl. 1980)
Wernham	Wernham, Herbert Fuller (1879–1941)
Wernisch.	Wernischeck, Johann Jacob (1743–1804)
Wesm.	Wesmael, Alfred (1832–1905)
Westc.	Westcott, Frederic (?–1861)
Weston	Weston, Richard (1733–1806)
Weston, P.	Weston, Peter Henry (1956–)
Westra	Westra, Lübbert Ybele Theodoor (1932–)
de Wet	de Wet, Johannes Martenis Jacob (1927–)
Wettst.	Wettstein von Westersheim, Richard von (1863–1931)
Whiffin	Whiffin, Trevor Paul (1947–)
White, A.	White, Alain Campbell (1880–?)
White, C.	White, Cyril Tenison (1890–1950)
White, F.	White, Frank (1927–1994)
White, R.	White, Richard Alan (1935–)
Whitm.	Whitmore, Timothy Charles (1935–)

Wibel	Wibel, August Wilhelm Eberhard Christoph (1775–1814)
Wied-Neuw.	Wied-Neuwied, Maximilian Alexander Philipp zu (1782–1867)
Wiehler	Wiehler, Hans Joachim (1930–)
Wiens	Wiens, Delbert (1932–)
Wierzb.	Wierzbicki, Piotr Pawlus (1794–1847)
Wigg.	Wiggers, Friedrich Heinrich (1746–1811)
Wiggins	Wiggins, Ira Loren (1899–1987)
Wight	Wight, Robert (1796–1872)
Wight, W.	Wight, William Franklin (1874–1954)
Wijnands	Wijnands, D. Onno (1945–1993)
Wiklund	Wiklund, Annette (1953–)
Wikström	Wikström, Johan Emanuel (1789–1856)
Wilbur	Wilbur, Robert Lynch (1925–)
Wilczek	Wilczek, Ernst (1867–1948)
Wilczek, R.	Wilczek, Rudolf (1903–1984)
Wild	Wild, Hiram (1917–1982)
De Wild.	De Wildeman, Emile August(e) Joseph (1866–1947)
Wilde	Wilde, Earle Irving (1888–1949)
Wilhelm	Wilhelm, Karl Adolf (1848–1933)
Willd.	Willdenow, Carl Ludwig von (1765–1812)
Willemet	Willemet, Pierre Remi (1735–1807)
Williams, F.N.	Williams, Frederic Newton (1862–1923)
Williams, I.J.	Williams, Ion James Muirhead (1912–)
Williams, J.	Williams, John Beaumont (1932–)
Williams, L.O.	Williams, Louis Otho (1908–1991)
Williams, N.	Williams, Norris H. (1943–)
Williamson	Williamson, Phyllis Alison (1925–)
Willis	Willis, John Christopher (1868–1958)
Willis, J.H.	Willis, James Hamlyn (1910–1995)
Willk.	Willkomm, Heinrich Moritz (1821–1895)
Wilson, E.	Wilson, Ernest Henry (1876–1930)
Wilson, G.	Wilson, George Fox (1896–1951)
Wilson, K.	Wilson, Karen Louise (1950–)
Wilson, P.	Wilson, Percy (1879–1944)
Wilson, P.G.	Wilson, Paul Graham (1928–)
Wimmer	Wimmer, Christian Friedrich Heinrich (1803–1868)
Wimmer, F.	Wimmer, Franz Elfried (1881–1961)
Windham	Windham, Michael D. (1954–)
Winkler	Winkler, Constantin Georg Alexander (1848–1900)
Winkler, H.	Winkler, H. (fl. 1868)
de Winter	de Winter, Bernard (1924–)
de Wit	de Wit, Hendrik Cornelis Dirk (1909–1999)
With.	Withering, William (1741–1799)
Wittm.	Wittmack, Marx Carl Ludwig (1839–1929)
Wodehouse	Wodehouse, Roger Philip (1889–1978)
Wolf	Wolf, Nathanael Matthaeus von (1724–1784)
Wolff, H.	Wolff, Karl Friedrich August Hermann (1866–1929)
Wol.	Woloszczak, Eustach (1835–1918)
Wollenw.	Wollenweber, Hans Wilhelm (1879–1949)
Wong	Wong Khoon Meng (1954–)
Wood, Alph.	Wood, Alphonso W. (1810–1881)
Wood, J.J.	Wood, Jeffrey James (1952–)
Wood, J.M.	Wood, John Medley (1827–1915)
Wood, J.R.I.	Wood, John Richard Ironside (1944–)
Wood, K.	Wood, Kenneth R.
Woodson	Woodson, Robert Everard (1904–1963)
Woolls	Woolls, William (1814–1893)
Wooton	Wooton, Elmer Ottis (1865–1945)
Wormsk.	Wormkskjöld, Martin (1783–1845)
Woronow	Woronow, Georg Jurij Nikolaewitch (1874–1931)
Worsley	Worsley, Arthrington (1861–1944)

Wright	Wright, William (1735–1819)
Wright, C.	Wright, Charles (Carlos) (1811–1885)
Wright, C.H.	Wright, Charles Henry (1864–1941)
Wright, S.	Wright, Samuel Hart (1825–1905)
Wu, C.Y.	Wu, Cheng Yih (1916–)
Wu, S.K.	Wu, Su-Kung (1935–)
Wu, T.L.	Wu, Te-Lin(g) (1934–)
Wu, Z.L.	Wu, Zhen-Lan (1939–)
Wuerttemb.	Württemberg, Friedrich Paul Wilhelm von (1797–1860)
Wulfen	Wulfen, Franz Xavier von (1728–1805)
Wulff	Wulff, Eugen Vladimirowitsch (1885–1941)
Wurd.	Wurdack, John Julius (1921–1998)
Wurmb	Wurmb, Friedrich von (?–1781)
Wydler	Wydler, Heinrich (1800–1883)
Wyk	van Wyk, Abraham Erasmus (1952–)
Yakovlev	Yakovlev, G.P. (1938–)
Yamamoto	Yamamoto, Yoshimatsu (1893–1947)
Yamaz.	Yamazaki, Takasi (1921–)
Yang, J.L.	Yang, Jun-Liang (fl. 1988)
Yang, Y.C.	Yang, Yen Chin (1913–1984)
Yang, Y.L.	Yang, Ya-Ling (1933–)
Yatabe	Yatabe, Ryôkichi (1851–1899)
Yates	Yates, Harris Oliver (1934–)
Yatsk., Yatskievych	Yatskievych, George A. (1957–)
Yen	Yen, Chi (fl. 1983)
Yi	Yi, Tong-Pei (fl. 1980)
Yu, T.T.	Yu, Tse-Tsun (1908–1986)
Yuncker	Yuncker, Truman George (1891–1964)
Zabel	Zabel, Hermann (1832–1912)
Zahlbr., A.	Zahlbruckner, Alexander (1860–1938)
Zak.	Zakirov, K.Z. (1906–)
Zamora	Zamora, Nelso A. (fl. 1988)
Zardini	Zardini, Else Matilde (1949–)
Zareh	Zareh, M. (fl. 1989)
Zauschner	Zauschner, Johann Baptista Josef (1737–1799)
Zepernick	Zepernick, Bernhard (1926–)
Zeyher	Zeyher, K. Johann Michael (1770–1843)
Zhang	Zhang, Yuffna (fl. 1986)
Zhao	Zhao, Zheng Yu (1928–)
Zhu, Z.Y.	Zhu, Zheng-Yin (fl. 1982)
Zinn	Zinn, Johann Gottfried (1727–1759)
Zipp.	Zippelius, Alexander (1797–1828)
Zizka	Zizka, Georg (fl. 1987)
Zoellner	Zoellner, Otto
Zoll.	Zollinger, Heinrich (1818–1859)
Zotov	Zotov, Victor Dmitrievich (1908–1977)
Zou, S.Q.	Zou, Shou-Qing (fl. 1984)
Zoz	Zoz, I.G. (1903–)
Zucc.	Zuccarini, Joseph Gerhard (1797–1848)
Zuloaga	Zuloaga, Ferando Omar (1951–)